DIGITAL COMMUNICATIONS
WITH EMPHASIS ON DATA MODEMS

DIGITAL COMMUNICATIONS WITH EMPHASIS ON DATA MODEMS

Theory, Analysis, Design, Simulation, Testing, and Applications

RICHARD W. MIDDLESTEAD

Published by John Wiley & Sons, Inc., Hoboken, New Jersey
Published simultaneously in Canada

Library of Congress Cataloging-in-Publication Data

Middlestead, Richard W.
Digital communications with emphasis on data modems : theory, analysis, design, simulation, testing, and applications / Richard W. Middlestead.
pages cm
Includes bibliographical references and index.
ISBN 978-0-470-40852-0 (cloth)
1. Digital modulation. 2. Digital communications. 3. Data transmission systems. I. Title.
TK5103.7.M544 2017
621.39'8–dc23

2014050193

Cover Design: Wiley
Cover Image: Richard W. Middlestead

Set in 10/12pt Times by SPi Global, Pondicherry, India

10 9 8 7 6 5 4 3 2 1

Thanks to:
My Lord and Savior Jesus Christ.
My Father in heaven without whose creative genius
science would be rudderless.
The Holy Spirit who inspires men and women beyond their own understanding.

Dedicated to:
My mother and father
and loving wife Donna
My children Lynn, Mark, and Leighann
and my grandchildren
Kasey, Erik, Jacob, Kaylee, Zachary, Mackenzie, Tyler, Hannah,
and great grandchildren Kaleigh and Isla

My prayer for my children is that wisdom, hard work, perseverance, and compassion will be the hallmark of their character.

CONTENTS

PREFACE

The material in this book contains sufficient mathematical background to challenge the graduate student and to allow the experienced communication systems engineer to analyze, specify, and select optimum solutions for a variety of applications. Furthermore, it is hoped that the many performance plots, tables, and design concepts will prove beneficial to system design engineers who are less inclined to delve into the mathematical rigors. The importance of communication systems performance simulations cannot be overstated as a developmental tool prior to the software and hardware designs. Computer simulations allow the designer to predict the system performance beyond the assumptions leading to mathematical models. In nearly every chapter to follow, the results presented have been verified using computer simulations and found to be in agreement with the theoretical results.

Although the material presented in this book focuses on the modulator and demodulator (modem) subsystems, the transmitter and receiver (transceiver) subsystems and the corresponding antenna subsystems must be considered in order to evaluate the overall communication link performance. Furthermore, the design of these subsystems is influenced by the communication channels that are broadly characterized as wireline and wireless or radio frequency (RF) channels.

The modulator, transmitter, and transmitter antenna subsystems function together with inputs from an information source. Similarly, the receiver antenna, receiver, and demodulator subsystems perform the complementary functions providing optimum data estimates to the information destination or sink. In these roles, the modulator performs the important functions of preparing the source information data for subsequent transmission. This involves source coding, channel coding, and optimal mapping of the coded data to the selected symbol modulation format for transmission. Design considerations regarding the symbol modulation format selection involve bandwidth conservation and power efficiency that are largely influenced by the channel.

Communication systems are broadly defined in terms of point-to-point and network communications. Within these systems, there are three basic forms of transmissions: *simplex* refers to information transmission in one direction only as in commercial broadcast radio; *half-duplex* refers to transmission is one direction at a time as in push-to-talk radios; *full-duplex* refers to simultaneous transmissions in both directions. These forms of transmission require increasing degrees of complexity and performance capabilities. The data link performance analysis presented in the book typically applies to the physical layer corresponding to single carrier per channel (SCPC) point-to-point communication links. With the inclusion of the network controls [1] and the associated overhead functions, the physical layer performance can be applied directly to time division multiple access (TDMA) networks. The performance of frequency division multiple access (FDMA) and code division multiple access [2] (CDMA) networks can also be evaluated with the respective inclusion of adjacent channel interference (ACI) and co-channel interference (CCI) losses. In general, network centric protocols are specialized for specific applications and as such are beyond the scope of the book. However, because of its relative simplicity and utility in providing virtually error-free data transmission, the performance of automatic repeat request (ARQ) protocols requiring half or full-duplex networking capabilities is discussed.

A major performance measure of a digital communication system is the bit-error probability characterized as a function of the ratio of the received signal energy per source bit to the noise density. Many of these performance plots are obtained using Monte Carlo simulations with the goal of providing a smooth eye-appealing curve approximating the theoretical performance.[1] To this end, the number of Monte Carlo trials over the entire range of signal-to-noise ratios is selected to achieve a reasonable confidence level at the lowest bit-error probability of interest. The resulting performance curve is very accurate for the lower signal-to-noise ratios; however, the performance at the highest signal-to-noise ratio may appear as an outlier not conforming to the expected theoretical result. In these cases, the outlier data is brought into compliance with the smooth performance curve in one of two ways; the simulation is re-run using a minimum of 10 times the number of Monte Carl trials or the data point is brought into compliance manually by adjusting the bit-error probability; this approach is comparable to fitting a *French curve* to the lower signal-to-noise ratio data points and appropriately adjusting the outlier. Flaring of the performance curve due to nonlinearities or intersymbol interference is also taken into account.

The following is a brief description of the subjects in each chapter that often includes case studies to illustrate the methodology of the evaluation and the characteristics of the underlying performance measure.

Chapter 1 focuses on the description of techniques and analytical methods used throughout the book including real and analytic characterization of waveform modulations, Fourier transforms, an introduction to statistical analysis, optimum waveform detection and parameter estimation, a brief look at ARQ protocols, spectral control using windows, vector and matrix operations, and lists of commonly mathematical relationships. The chapter can be considered as reference material associated with the remaining chapters; however, the notion of the complex envelope or analytic representation of baseband signals should be thoroughly understood.

Chapter 2 discusses many fundamental relationships required for analysis involving digital signal processing. The notion of Nyquist sampling and the requirement for capturing the information contained in the received signal are described using baseband and bandpass sampling techniques. Multirate signal processing and rate conversion is examined using various rate conversion filters including the cascade integrate and comb (CIC) filter. The chapter includes a discussion of polyphase filters and Lagrange interpolation leading to the Farrow filter. The chapter concludes with the derivation of a parabolic interpolator for improvement of time and frequency estimation errors used during waveform acquisition and information detection and tracking. The material in the chapter provides essential insights into fundamental digital signal processing requirements for the design of the modem subsystem and, as such, each topic should be examined in sufficient detail to result in an optimum design.

[1]To aid in the simulation of a smooth eye-appealing performance curve, all of the noise generator seeds are reset at each signal-to-noise ratio.

Chapter 3 introduces the fundamental concepts of digital communication systems involving source and channel coding and optimum techniques for information recovery. The concepts are discussed using discrete memoryless and binary symmetric channels. Shannon's channel capacity limit forms the basis of the analysis and bounds on the bit error are examined using the computational cutoff rate. The chapter concludes with a discussion of the probability integral and the error function. The chapter, like Chapter 1, may be considered reference material; however, Shannon's error-free performance limit must be understood along with the many applications and forms of the error function.

Chapters 4 through 7 provide a comprehensive analysis of various waveform modulations and the corresponding power spectral density (PSD) and bit-error performance. Constant envelope modulations ranging from multiphase shift keying (MPSK) to Gaussian minimum shift keying (GMSK) are examined. Spectral efficiency achieved through phase shaping while maintaining a constant envelope is a major topic. The simplicity and robustness of differentially coherent waveform modulation and demodulation is also examined. The spectral root-raised-cosine (SRRC) waveform, although not a constant envelope modulation, results in superior spectral efficiency. The spectrums of these modulated waveforms are compared to industry-standard transmit spectral masks. Chapter 5 examines the bit-error performance and PSD of frequency shift keying with coherent and noncoherent detection. Chapter 6 examines the performance of amplitude shift keying (ASK) including binary on–off keying (OOK), pulse amplitude modulation (PAM), and quadrature amplitude modulation (QAM). Chapter 6 concludes with a discussion of partial response modulation focusing on the modified and multilevel duobinary modulations. Chapter 7 discusses *M*-ary coded waveforms focusing on the coherent and noncoherent detection of *M*-ary orthogonal and biorthogonal waveforms.

Chapter 8 focuses on coding for improved communications, beginning with the description of commonly used pulse code modulation (PCM) waveforms. Coding techniques that are generally applied to a variety of waveform modulations include gray and differential coding, binary cyclic coding, cyclic redundancy check (CRC) coding, data randomizers, and block and convolutional interleaving. The chapter also includes descriptions and performance results for Wagner coding, convolutional coding with Viterbi decoding, turbo and turbo-like parallel concatenated convolutional coding (PCCC) or turbo coding, serial concatenated convolutional coding (SCCC), low-density parity-check (LDPC) coding, and turbo product codes (TPCs). The chapter concludes with Bose–Chaudhuri–Hocquenghem (BCH), Reed–Solomon (RS), and Reed–Solomon Viterbi (RSV) coding.

Chapter 9 focuses on forward error correction (FEC) codes without bandwidth expansion. In these cases, additional modulation states are included as redundant states for maximum likelihood error correction in a trellis decoder. The implementation and performance of multi-*h* *M*-ary continuous phase modulation (CPM) is compared to that of MPSK trellis-coded modulation (TCM).

Chapter 10 provides a detailed analysis and performance evaluation of carrier acquisition and tracking using first through third-order phaselock loops. The tracking performance using classical control theory is compared to that using the maximum a posteriori probability (MAP) estimation procedure. Detailed implementations diagrams and performance results are provided for MPSK and MSK waveform modulations. A procedure is discussed to optimize the phaselock loop gains in simulation and hardware designs to conform to the theoretical response under prescribed input conditions. Case studies are provided that demonstrate the dynamic and steady-state tracking conditions.

Chapter 11 discusses the signal presence detection and acquisition of a received waveform using a data preamble; the topics including automatic gain control (AGC), coarse and fine carrier frequency estimation and acquisition, symbol synchronization, and start-of-message (SOM) detection. The chapter concludes with various methods of estimating signal and noise powers and the received signal-to-noise ratio.

Chapter 12 analyzes various adaptive estimation algorithms beginning with a discussion of the orthogonality principle and Wiener's solution to the optimum filtering problem. Various estimation techniques are examined with algorithms developed for the finite impulse response (FIR) and least mean-square (LMS) adaptive filters. Adaptive equalization algorithms include the zero-forcing, linear feedforward, nonlinear decision feedback, fractionally spaced and the recursive least-squares (RLS) equalizers. Interference and noise cancellation algorithms are also discussed.

Chapter 13 provides a detailed study of the spread spectrum communications focusing on direct-sequence spread-spectrum (DSSS), frequency-hopping spread-spectrum (FHSS), and time-hopping spread-spectrum (THSS). The link geometry and link margins are characterized for anti-jam (AJ) and low probability of intercept (LPI) communications. The probability of detecting and intercepting spread-spectrum waveforms is examined for various interceptor detection algorithms. The communicator's performance with intentional jamming is evaluated under the following conditions: BPSK and QPSK with a continuous wave (CW) jammer; *M*-ary FSK and DC-MPSK with partial band noise jammers; and FHSS with multitone jammers. A simplified analytical computation of the upper bound on the bit-error performance with various modulations and jammers is outlined. The chapter concludes with a case study of a terrestrial jammer encounter using the Longley–Rice irregular terrain model (ITM).

Chapter 14 describes various acceptance and rejection procedures suitable for modem pre- and postproduction testing. Modem subsystem modeling and Monte Carlo simulation techniques are also described leading to the bit-error performance evaluation through various types of channels. The chapter concludes with the description of the bit-error performance evaluation using quadrature integration. Several case studies are provided to demonstrate the methodology and utility of the evaluation procedures.

Chapter 15 describes the link budget analysis using the communication range equation. This is an essential chapter that outlines fundamental system requirements and related analysis that must be established prior to a detailed subsystem design. The link budget essentially establishes the cost-effective subsystem conditions to ensure that the signal-to-noise ratio at the receiver input is sufficient for reliable communications under the specified channel condition. The chapter highlights the important topics involving high-power amplifier (HPA) nonlinearities, transmitter and antenna effective isotropic radiated power (EIRP), receiver antenna gain-temperature ratio (G/T), receiver noise figure, antenna polarization, system phase noise, and channel rain losses.

Chapter 16 analyzes various satellite orbits focusing on the link range and signal dynamics. The results correspond to terrestrial or airborne satellite links and satellite cross-links and are used in establishing a satellite link budget as discussed in Chapter 15.

Chapter 17 discusses the transmission information through a bandlimited time-invariant channel. The channel response to an input data symbol is examined in terms of the channel amplitude and phase functions. This analysis forms the bases for evaluating the performance of baseband PCM modulated waveforms using wireline medium.

Chapter 18 discusses communicating through a Ricean fading channel characterized in terms of the specular-to-random power ratio with limits corresponding to the Rayleigh and Gaussian channel models. The bit-error performance of fast and slowing nonselective fading channels is examined for coherent BPSK, differentially coherent BPSK, and noncoherent BFSK waveform modulations. The relationship between the channel coherence time and bandwidth is, respectively, associated with Doppler spread and time dispersion. The channel concludes with a discussion to diversity combining techniques to mitigate the loss associated with fading channels. This material is considered to be a prerequisite for the material in Chapters 19 and 20.

Chapter 19 discusses various aspects of atmospheric propagation using the spherical 4/3 effective Earth radius model. The topics discussed include line of sight (LOS) propagation, reflection from the Earth's surface, tropospheric and ionospheric refraction, and diffraction. Several propagation loss models are characterized for urban, suburban, rural environments with applications to land mobile and satellite communications. The chapter concludes with the characterization of communication links involving impulsive noise induced by lightning strikes, ocean wind-waves, and dispersion of optical pulses through clouds.

Chapter 20 discusses various aspects of ionospheric propagation beginning with the characterization of electron densities in the natural and nuclear-disturbed environments. The refractive index is characterized in terms of the magnetic field strength and electron collisions from which the signal absorption and phase functions are established. Signal fading is characterized in terms of the scintillation index and the Rytov parameter. The dependence of the signal-decorrelation time and the frequency-selective bandwidth on the carrier frequency is also identified. A methodology is described for seamlessly concatenating an unlimited number of fading channel temporal FFT generated records to facilitate accurate Monte Carlo performance simulations in a stressed environment. Based on the electron density profiles, the Rayleigh channel fading model corresponding to severe signal scintillation is used to evaluate the link performance using several robust waveform modulations. Monte Carlo simulations are used to compare the performance of DEPSK, DCBPSK, and 8-ary FSK modulations with and without FEC coding in the slow Rayleigh fading channel. The simulated performance of interleaved RSV coded DCBPSK and DCQPSK is also examined over the range of fast to slow Rayleigh fading regimes.

The book concludes with three appendices:

Appendix A discusses the following classical analog filters: Butterworth, Chebyshev, Bessel, and Elliptic. The filter functions are implemented as digital filters and used in several chapters and case studies.

Appendix B is a brief discussion of the design and implementation of digital filters. The filter functions are used in several chapters and case studies.

Appendix C discusses the theoretical detection and false-alarm probabilities of signals in noise under a variety of conditions are examined, including coherent and noncoherent detection using single and multiple pulse integration with and without fading. The results are based on the classical radar related work of J.I. Marcum and P. Swerling and are selectively applied principally to the detection and acquisition of communication waveforms.

REFERENCES

1. F. Halsall, *Data Communications, Computer Networks and Open Systems*, Fourth Edition, Addison-Wesley Publishing Company, Harlow, UK, 1996.
2. A.J. Viterbi, *CDMA Principles of Spread Spectrum Communications*, Addison-Wesley Publishing Company, Reading, MA, 1995, Seventh Printing 1998.

ACKNOWLEDGMENTS

The writing of the book would not have been possible without the many distinguished mathematicians, scientists, and engineers whose work forms the basis of the depth and breadth of much the material presented. The names of these distinguished men and women, upon whose shoulders I have stood throughout my careers in radar and communication systems, appear in the many references provided in each chapter. The preparation of the manuscript was encouraged by my colleagues James J. Merkel, Mark A. Cusano, and Kevin J. Lynaugh, who also proofread several chapters and provided helpful suggestions. Finally, I want to acknowledge my wife Donna and Jacob R. Pearson who proofread the manuscript and provided editorial suggestions and the editorial and production staff at Wiley in the arduous tasks leading to the production of the book.

SUMMARY OF NOTATIONS

Notation	Description
T	Modulation symbol duration[†] ($= 1/R_s$)
T_b	Information bit duration ($= 1/R_b$)
Tx	Transmitter subsystem
Rx	Receiver subsystem
N_o	One-sided white noise spectral density
$=$	Equal
$\cong$	Approximately equal
$\neq$	Not equal
$\sim$	Asymptotically equal
$<, >; \leq, \geq$	Inequality; inclusion
o.w.	Otherwise all values not indicated
$\rightarrow$	Approaches in the limit
$\forall$	For all
$\triangleq$	Definition
$\Rightarrow$	Implies
$\Leftrightarrow$	Transform pair
$1/ab$	Inline division: $1 \div ab$
$\lfloor x \rfloor$	Floor: greatest integer $\leq x$
$\lceil x \rceil$	Ceiling: smallest integer $\geq x$
$(x_{t1}, \ldots, x_{tm})$	Finite time sequence of elements x_t
$\{x_1, \ldots, x_n\}$	Finite set of elements x
$<x(t)>$	Time average
$\Sigma'_i x_i$	Summation over all i except $i = 0$
$[x]$, $int(x)$	Integer value of x
$[x]_n$, $nint(x)$	Nearest integer to x computed as: $[x+.5]$: $x > 0$ and $[x-.5]$; $x \leq 0$
$\binom{m}{n}$, ${}_mC_n$	Binomial coefficient
$[A]^T$	Matrix transpose
$[A]^*$	Matrix conjugation
$[A]^+$	Matrix complex conjugate transpose
$\lvert A \rvert$, $\lvert A \rvert_{det}$	Determinate of matrix
δ_{ij}	Kronecker delta function: $\delta_{ij} = 1$: $i = j$; *o.w.* $\delta_{ij} = 0$
$\delta(t)$	Delta function: $\int_{-\infty}^{\infty} \delta(t)dt = 1$
$sinc(fT)$	$\sin(\pi fT)/\pi fT$
$rect(t/T)$	$= 1 : \lvert t \rvert \leq T/2$; $= 0$ o.w.
$rect_T(t/T,n)$	$rect((t - nT)/T)$[‡]
$Sign(a,x)$	$= a$: $x \geq 0$; $= -a$: *o.w.*
$N(m,\sigma)$	Gaussian (normal) distributed random variable
$E[x]$	Statistical average (expectation)
pdf	Probability density function
pmf	Probability mass function
cdf	Cumulative distribution function
iid	Independently and identically distributed
id	Identically distributed
iff	If and only if

[†]Almost exclusively used as the symbol duration.

[‡]Repetition notation for P.M. Woodward's *rect(t/T)* function.

ABOUT THE COVER

The three-dimensional plot on the cover depicts the bit-error performance as a function of the number of rows and columns of a deterministic block interleaver and deinterleaver operating in a Rayleigh fading channel. The plot corresponds to a signal-to-noise ratio of $E_b/N_o = 6.5$ dB and is generated using a Monte Carlo simulation with 10 M bits at each row–column combination. The source data bits are coded using a rate 1/2, constraint length 7, convolutional encoder and are decoded using the Viterbi algorithm with a trellis depth or memory of 31 coded bits. The coded bits are sequentially entered into the interleaver row by row and removed column by column and applied to a binary phase shift keying (BPSK) symbol modulator. The channel is modeled as a correlated Gaussian random process with the correlation coefficient selected to result in a decorrelation time corresponding to six coded bits. In the region of the plot corresponding to the lowest bit-error performance, the loss in E_b/N_o at $P_{be} = 10^{-5}$ compared to the additive white Gaussian noise channel is 2.7 dB. Based on these conditions and results, a reasonably sized (row,column) block interleaver is (11,51).

ABOUT THE COMPANION WEBSITE

This book is accompanied by a companion website:

www.wiley.com/go/digitalcommunications

The website includes:

- Solution Manual

1

MATHEMATICAL BACKGROUND AND ANALYSIS TECHNIQUES

1.1 INTRODUCTION

This introductory chapter focuses on various mathematical techniques and solutions to practical problems encountered in many of the following chapters. The discussions are divided into three distinct topics: deterministic signal analysis involving linear systems and channels; statistical analysis involving probabilities, random variables, and random processes; miscellaneous topics involving windowing functions, mathematical solutions to commonly encountered problems, and tables of commonly used mathematical functions. It is desired that this introductory material will provide the foundation for modeling and finding practical design solutions to communication system performance specifications. Although this chapter contains a wealth of information regarding a variety of topics, the contents may be viewed as reference material for specific topics as they are encountered in the subsequent chapters.

This introductory section describes the commonly used waveform modulations characterized as amplitude modulation (AM), phase modulation (PM), and frequency modulation (FM) waveforms. These modulations result in the transmission of the carrier- and data-modulated subcarriers that are accompanied by negative frequency images. These techniques are compared to the more efficient suppressed carrier modulation that possesses attributes of the AM, PM, and FM modulations. This introduction concludes with a discussion of *real* and *analytic* signals, the Hilbert transform, and demodulator heterodyning, or frequency mixing, to baseband.

Sections 1.2–1.4, *deterministic signal analysis*, transform in the context of a uniformly weighted pulse $f(t)$ and its spectrum $F(f)$ and the duality between ideal time and frequency sampling that forms the basis of Shannon's sampling theorem [1]. This section also discusses the discrete Fourier transform (DFT), the fast Fourier transform (FFT), the pipeline implementation of the FFT, and applications involving waveform detection, interpolation, and power spectrum estimation. The concept of paired echoes is discussed and used to analyze the signal distortion resulting from a deterministic band-limited channel with amplitude and phase distortion. These sections conclude on the subject of autocorrelation and cross-correlation of real and complex deterministic functions; the corresponding covariance functions are also examined.

Sections 1.5–1.10, *statistical analysis*, introduce the concept of *random variables* and various probability density functions (*pdf*) and cumulative distribution functions (*cdf*) for continuous and discrete random variables. *Stochastic* processes are then defined and the properties of *ergodic* and *stationary* random processes are examined. The characteristic function is defined and examples, based on the summation of several underlying random variables, exhibit the trend in the limiting behavior of the *pdf* and *cdf* functions toward the normal distribution; thereby demonstrating the central limit theorem. Statistical analysis using distribution-free or nonparametric techniques is introduced with a focus on *order statistics*. The random process involving narrowband white Gaussian noise is characterized in terms of the noise spectral density at the input and output of an optimum detection filter. This is followed by the derivation of the *matched filter* and the

Digital Communications with Emphasis on Data Modems: Theory, Analysis, Design, Simulation, Testing, and Applications, First Edition. Richard W. Middlestead.

Companion website: www.wiley.com/go/digitalcommunications

equivalence between the matched filter and a correlation detector is also established. The next subject discussed involves the *likelihood ratio* and *log-likelihood ratio* as they pertain to optimum signal detection. These topics are generalized and expanded in Chapter 3 and form the basis for the optimum detection of the modulated waveforms discussed in Chapters 4–9. Section 1.9 introduces the subject of parameter estimation which is revisited in Chapters 11 and 12 in the context of waveform acquisition and adaptive systems. The final topic in this section involves a discussion of modem configurations and the important topic of automatic repeat request (ARQ) to improve the reliability of message reception.

Sections 1.11–1.14, *miscellaneous topics*, include a characterization of several *window functions* that are used to improve the performance the FFT, decimation filtering, and signal parameter estimation. Section 1.12 provides an introductory discussion of matrix and vector operations. In Section 1.13 several mathematical procedures and formulas are discussed that are useful in system analysis and simulation programming. These formulas involve *prime factorization* of an integer and determination of the *greatest common factor* (GCF) and *least common multiple* (LCM) of two integers, *Newton's approximation* method for finding the roots of a transcendental function, and the definition of the *standard deviation* of a sampled population. This chapter concludes with a list of frequently used mathematical formulas involving *infinite* and *finite summations*, *the binomial expansion theorem*, *trigonometric identities*, *differentiation and integration rules*, *inequalities*, and other *miscellaneous relationships*.

Many of the examples and case studies in the following chapters involve systems operating in a specific frequency band that is dictated by a number of factors, including, the system objectives and requirements, the communication range equation, the channel characteristics, and the resulting link budget. The system objectives and requirements often dictate the frequency band that, in turn, identifies the channel characteristics. Table 1.1 identifies the frequency band designations with the corresponding range of frequencies. The designations low frequency (LF), medium frequency (MF), and high frequency (HF) refer to low, medium, and high frequencies and the prefixes E, V, U, and S correspond to *e*xtremely, *v*ary, *u*ltra, and *s*uper.

TABLE 1.1 Frequency Band Designations

Designation	Frequency	Letter Designation	Frequency (GHz)
ELF	3–30 Hz	*L*	1–2
SLF	30–300 Hz	*S*	2–4
ULF	0.3–3 kHz	*C*	4–8
VLF	3–30 kHz	*X*	8–12
LF	30–300 kHz	*Ku*	12–18
MF	0.3–3 MHz	*K*	18–27
HF	3–30 MHz	*Ka*	27–40
VHF	30–300 MHz	*V*	40–75
UHF	0.3–3 GHz	*W*	75–110
SHF	3–30 GHz	*mm (millimeter)*	110–300
EHF	30–300 GHz		

1.1.1 Waveform Modulation Descriptions

This section characterizes signal waveforms comprised of baseband information modulated on an arbitrary carrier frequency, denoted as f_c Hz. The baseband information is characterized as having a lowpass bandwidth of B Hz and, in typical applications, $f_c >> B$. In many communication system applications, the carrier frequency facilitates the transmission between the transmitter and receiver terminals and can be removed without effecting the information. When the carrier frequency is removed from the received signal the signal processing sampling requirements are dependent only on the bandwidth B.

The signal modulations described in Sections 1.1.1.1 through 1.1.1.4 are amplitude, phase, frequency, and suppressed carrier modulations. The amplitude, phase, and frequency modulations are often applied to the transmission of analog information; however, they are also used in various applications involving digital data transmission. For example, these modulations, to varying degrees, are the underlying waveforms used in the U.S. Air Force Satellite Control Network (AFSCN) involving satellite uplink and downlink control, status, and ranging.

In describing the demodulator processing of the received waveforms, the information, following removal of the carrier frequency, is associated with in-phase and quadphase (I/Q) baseband channels or rails. Although these I/Q channels are described as containing quadrature real signals, they are characterized as *complex* signals with *real* and *imaginary* parts. This complex signal description is referred to as complex envelope or analytic signal representations and is discussed in Section 1.1.1.5. Suppressed carrier modulation and the analytic signal representation emphasize quadrature data modulation that leads to a discussion of the Hilbert transform in Section 1.1.1.6. Section 1.1.1.7 discusses conventional heterodyning of the received signal to baseband followed by data demodulation.

1.1.1.1 Amplitude Modulation Conventional amplitude modulation (AM) is characterized as

$$s(t) = A(1 + m_I m(t)\sin(\omega_m t))\sin(\omega_c t) \qquad (1.1)$$

where A is the peak carrier voltage, $m_I > 0$ is the modulation index, $m(t)$ is the information modulation function, ω_m is the modulation angular frequency, and ω_c is the AM carrier angular frequency. Upon multiplying (1.1) through by sin

($\omega_c t$) and applying elementary trigonometric identities, the AM-modulated signal is expressed as

$$s(t) = A\sin(\omega_c t) + \frac{Am_I}{2}m(t)\cos((\omega_c - \omega_m)t) - \frac{Am_I}{2}m(t)\cos((\omega_c + \omega_m)t) \quad (1.2)$$

Therefore, $s(t)$ represents the conventional double sideband (DSB) AM waveform with the upper and lower sidebands at $\omega_c \pm \omega_m$ equally spaced about the carrier at ω_c. With the information modulation function $m(t)$ normalized to unit power, the power in each sideband is $m_I P_S/4$ where P_S is the power in the carrier frequency f_c.

1.1.1.2 *Phase Modulation*

Conventional phase modulation (PM) is characterized as

$$s(t) = A\sin(\omega_c t + \varphi(t)) \quad (1.3)$$

where A is the peak carrier voltage, ω_c is the carrier angular frequency, and $\varphi(t)$ is an arbitrary phase modulation function containing the information. The commonly used phase function is expressed as

$$\varphi(t) = \phi\sin(\omega_m t) \quad (1.4)$$

where ϕ is the peak phase deviation. Substituting (1.4) into (1.3), the phase-modulated signal is expressed as

$$s(t) = A(\sin(\omega_c t + \phi\sin(\omega_m t)) \quad (1.5)$$

and, upon applying elementary trigonometric identities, (1.5) yields

$$s(t) = A\sin(\omega_c t)\cos(\phi\sin(\omega_m t)) + A\cos(\omega_c t)\sin(\phi\sin(\omega_m t)) \quad (1.6)$$

The trigonometric functions involving sinusoidal arguments can be expanded in terms of Bessel functions [2] and (1.6) simplifies to

$$s(t) = AJ_0(\phi)\sin(\omega_c t) + A\sum_{n=1}^{\infty}[J_n(\phi)\sin((\omega_c + n\omega_m)t) + (-1)^n J_n(\phi)\sin((\omega_c - n\omega_m)t)] \quad (1.7)$$

Equation (1.7) is characterized by the carrier frequency with peak amplitude $AJ_0(\phi)$ and upper and lower sideband pairs at $\omega_c \pm n\omega_m$ with peak amplitudes $AJ_n(\phi)$. For small arguments the Bessel functions reduce to the approximations $J_0(\phi) \cong 1, J_1(\phi) \cong \phi/2$ with $J_n(\phi) \cong 0 : n > 1$ and (1.7) reduces to

$$s(t) \cong A\sin(\omega_c t) + \frac{A\phi}{2}\sin((\omega_c + \omega_m)t) - \frac{A\phi}{2}\sin((\omega_c - \omega_m)t) \quad : \text{small } \phi \quad (1.8)$$

Under these small argument approximations, the similarities between (1.8) and (1.2) are apparent.

1.1.1.3 *Frequency Modulation*

The frequency-modulated (FM) waveform is described as

$$s(t) = A\sin\left(\omega_c t + \frac{\Delta f}{f_m}\sin(\omega_m t)\right) \quad (1.9)$$

where A is the peak carrier voltage, ω_c is the carrier angular frequency, Δf is the peak frequency deviation of the modulation frequency f_m, and ω_m is the modulation angular frequency. The ratio $\Delta f/f_m$ is the frequency modulation index. Noting the similarities between (1.9) and (1.5), the expression for the frequency-modulated waveform is expressed, in terms of the Bessel functions, as

$$s(t) = AJ_0\left(\frac{\Delta f}{f_m}\right)\sin(\omega_c t) + A\sum_{n=1}^{\infty}\left[J_n\left(\frac{\Delta f}{f_m}\right)\sin((\omega_c + n\omega_m)t) + (-1)^n J_n\left(\frac{\Delta f}{f_m}\right)\sin((\omega_c - n\omega_m)t)\right] \quad (1.10)$$

with the corresponding small argument approximation for the Bessel function expressed as

$$s(t) \cong A\sin(\omega_c t) + \frac{A\Delta f}{2f_m}\sin((\omega_c + \omega_m)t) - \frac{A\Delta f}{2f_m}\sin((\omega_c - \omega_m)t) \quad : \text{small } \frac{\Delta f}{f_m} \quad (1.11)$$

The similarities between (1.11), (1.8), and (1.2) are apparent.

1.1.1.4 *Suppressed Carrier Modulation*

A commonly used form of modulation is suppressed carrier modulation expressed as

$$s(t) = Am(t)\sin(\omega_c t + \varphi(t)) \quad (1.12)$$

In this case, when the carrier is mixed to baseband, information modulation function $m(t)$ does not have a direct current (DC) spectral component involving $\delta(\omega)$. So, upon multiplication by the carrier, there is no residual carrier component ω_c in the received baseband signal. Because the carrier is suppressed it is not available at the receiver/demodulator to provide a coherent reference, so special considerations must be given to the carrier recovery and subsequent data demodulation. Suppressed carrier-modulated waveforms are efficient, in that, all of the transmitted power is devoted to the information. Suppressed carrier modulation and the various methods of carrier recovery are the central focus of the digital communication waveforms discussed in the following chapters.

1.1.1.5 *Real and Analytic Signals* The earlier modulation waveforms are described mathematically as *real* waveforms that can be transmitted over real or physical channels. The general description of the suppressed carrier waveform, described in (1.12), can be expressed in terms of in-phase and quadrature modulation functions $m_c(t)$ and $m_s(t)$ as

$$s(t) = m_c(t)\cos(\omega_c t) - m_s(t)\sin(\omega_c t) \tag{1.13}$$

The quadrature modulation functions are expressed as

$$m_c(t) = Ad_c m(t)\cos(\varphi(t)) \tag{1.14}$$

and

$$m_s(t) = Ad_s m(t)\sin(\varphi(t)) \tag{1.15}$$

With PM the data $\{d_c, d_s\}$ may be contained in a phase function $\varphi_d(t)$, $m(t)$ is a unit energy symbol shaping function that provides for spectral control relative to the commonly used $rect(t/T)$ function, and A represents the peak carrier voltage on each rail. With quadrature modulations, unique symbol shaping functions, $m_c(t)$ and $m_s(t)$, may be applied to each rail; for example, unbalanced quadrature modulations involve different data rates on each quadrature rail. With quadrature amplitude modulation (QAM) the data is described in terms of the multilevel quadrature amplitudes $\{\alpha_c, \alpha_s\}$ that are used in place of $\{d_c, d_s\}$ in (1.14) and (1.15).

Equation (1.13) can also be expressed in terms of the *real* part of a *complex* function as

$$s(t) = \mathrm{Re}\{\tilde{s}(t)e^{j\omega_c t}\} \tag{1.16}$$

where

$$\tilde{s}(t) = m_c(t) + jm_s(t) \tag{1.17}$$

The function $\tilde{s}(t)$ is referred to as the complex envelope or analytic representation of the baseband signal and plays a fundamental role in the data demodulation, in that, it contains all of the information necessary to optimally recover the transmitted information. Equation (1.17) applies to receivers that use linear frequency translation to baseband. Linear frequency translation is typical of heterodyne receivers using intermediate frequency (IF) stages. This is a significant result because the system performance can be evaluated using the analytic signal without regard to the carrier frequency [3]; this is particularly important in computer performance simulations.

Evaluation of the *real* part of the signal expressed in (1.16) is performed using the complex identity No. 4 in Section 1.14.6 with the result

$$s(t) = \frac{1}{2}\{\tilde{s}(t)e^{j\omega_c t} + \tilde{s}(t)*e^{-j\omega_c t}\} \tag{1.18}$$

A note of caution is in order, in that, the received signal power based on the analytic signal is twice that of the power in the carrier. This results because the analytic signal does not account for the factor of 1/2 when mixing or heterodyning with a locally generated carrier frequency and is directly related the factor of 1/2 in (1.18). The signal descriptions expressed in (1.12) through (1.18) are used to describe the narrowband signal characteristics used throughout much of this book.

1.1.1.6 *Hilbert Transform and Analytic Signals* The Hilbert transform of the real $s(t)$ is defined as

$$\breve{s}(t) \triangleq \frac{1}{\pi}\int_{-\infty}^{\infty}\frac{s(\tau)}{t-\tau}d\tau = s(t)*\frac{1}{\pi t} \tag{1.19}$$

The second expression in (1.19) represents the convolution of $s(t)$ with a filter with impulse response $h(t) = 1/\pi t$ where $h(t)$ represents the response to a Hilbert filter with frequency response $H(\omega)$ characterized as

$$h(t) \Leftrightarrow H(\omega) = \begin{cases} -jsign(\omega) & : |\omega| > 0 \\ 0 & : o.w. \end{cases} \tag{1.20}$$

The Hilbert transform of $s(t)$ results in a spectrum that is zero for all negative frequencies with positive frequencies representing a complex spectrum associated with the real and imaginary parts of an analytic function. Applying (1.20) to the signal spectrum $S(\omega) \Leftrightarrow s(t)$ results in the spectrum of the Hilbert transformed signal

$$\breve{S}(\omega) = \begin{cases} jS(\omega) & : \omega < 0 \\ 0 & : \omega = 0 \\ -jS(\omega) & : \omega > 0 \end{cases} \tag{1.21}$$

Applying (1.21) to the spectrum $S(\omega)$ of (1.12) or (1.13), the bandwidth B of $m(t)$ must satisfy the condition $B << f_c$. In this case, the inverse Fourier transform of the spectrum $\breve{S}(\omega)$ yields the Hilbert filter output $\breve{s}(t)$ given by

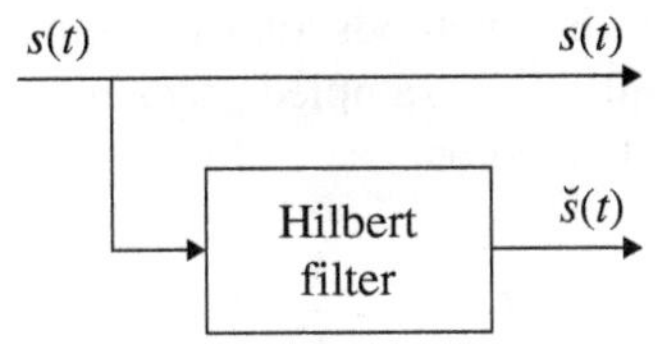

FIGURE 1.1 Hilbert transform of carrier-modulated signal $s(t)$ ($B/f_c \ll 1$).

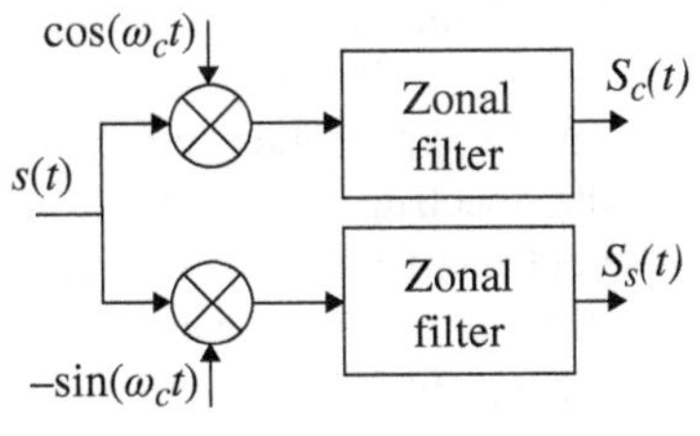

FIGURE 1.2 Heterodyning of carrier-modulated signal $s(t)$ ($B/f_c \ll 1$).

$$\breve{s}(t) = T_H[Am(t)\sin(\omega_c t + \phi(t))] = Am(t)\sin(\omega_c t + \phi(t) - \pi/2)$$
$$= -Am(t)\cos(\omega_c t + \phi(t)) \tag{1.22}$$

where $T_H[s(t)]$ represents the Hilbert transform of $s(t)$.

The function $\breve{s}(t)$ expressed by (1.22) is orthogonal to $s(t)$ and, if the carrier frequency were removed following the Hilbert transform, the result would be identical to the imaginary part of the analytic signal expressed by (1.17). The processing is depicted in Figure 1.1.

1.1.1.7 Conventional and Complex Heterodyning Conventional heterodyning is depicted in Figure 1.2. The zonal filters are ideal low-pass filters with frequency response given by

$$H(f) = rect\left(f - \frac{f_c}{2}\right) \quad \text{: zonal lowpass filter} \tag{1.23}$$

These filters remove the $2\omega_c$ term that results from the mixing operation and, for $s(t)$ as expressed by (1.13), the quadrature outputs are given by

$$s_c(t) = \frac{A}{2}[m_c(t)\cos(\phi(t)) - m_s(t)\sin(\phi(t))] \tag{1.24}$$

and

$$s_s(t) = \frac{A}{2}[m_c(t)\sin(\phi(t)) + m_s(t)\cos(\phi(t))] \tag{1.25}$$

With ideal phase tracking the phase term $\phi(t)$ is zero resulting in the quadrature modulation functions $m_c(t)$ and $m_s(t)$ in the respective low-pass channels.

1.2 THE FOURIER TRANSFORM AND FOURIER SERIES

The Fourier transform is so ubiquitous in the technical literature [4–6], and its application are so widely used that it seems unnecessary to dwell at any length on the subject. However, a brief description is in order to aid in the understanding of the parameters used in the applications discussed in the following chapters.

The Fourier transform $F(f)$ of $f(t)$ is defined over the interval $|t| \le \infty$ and, if $f(t)$ is absolutely integrable, that is, if

$$\int_{-\infty}^{\infty} |f(t)|dt < \infty \tag{1.26}$$

then $F(f)$ exists, furthermore, the inverse Fourier transform of $F(f)$ results in $f(t)$. In most applications* of practical interest, $f(t)$ satisfies (1.26) leading to the Fourier transform pair $f(t) \Leftrightarrow F(f)$ defined as

$$F(f) = \int_{-\infty}^{\infty} f(t)e^{-j2\pi ft}dt \Leftrightarrow f(t) = \int_{-\infty}^{\infty} F(f)e^{j2\pi ft}df \tag{1.27}$$

In general, $f(t)$ is real and the Fourier transform $F(f)$ is complex and Parseval's theorem relates the signal energy in the time and frequency domains as

$$\int_{-\infty}^{\infty} f(t)^2 dt = \int_{-\infty}^{\infty} |F(f)|^2 df \tag{1.28}$$

The Fourier series representation of a periodic function is closely related to the Fourier transform; however, it is based on orthogonal expansions of sinusoidal functions at discrete frequencies. For example, if the function of interest is periodic, such that, $f(t) = f(t - iT_o)$ with period T_o and is finite and single valued over the period, then $f(t)$ can be represented by the Fourier series

$$f(t) = \sum_{n=-\infty}^{\infty} C_n e^{jn\omega_o t} \tag{1.29}$$

where $\omega_o = 2\pi/T_o$ and C_n is the n-th Fourier coefficient given by

*For special cases refer to Papoulis (Reference 7, Chapter 2).

$$C_n = \frac{1}{T_o} \int_{-T_o/2}^{T_o/2} f(t) e^{-jn\omega_o t} dt \tag{1.30}$$

Equation (1.29) is an interesting relationship, in that, $f(t)$ can be described over the time interval T_o by an infinite set of frequency-domain coefficient C_n; however, because $f(t)$ is contiguously replicated over all time, that is, it is periodic, the spectrum of $f(t)$ is completely defined by the coefficients C_n. Unlike the Fourier transform, the spectrum of (1.29) is not continuous in frequency but is zero except at discrete frequencies occurring at multiples of $n\omega_o$. This is seen by taking the Fourier transform of (1.29) and, using (1.27), the result is expressed as

$$\begin{aligned} F(f) &= \sum_{n=-\infty}^{\infty} C_n \int_{-\infty}^{\infty} e^{-j2\pi(f-nf_o)t} dt \\ &= \sum_{n=-\infty}^{\infty} C_n \delta(f-nf_o) \end{aligned} \tag{1.31}$$

where $\delta(f-nf_o)$ is the Fourier Transform* of $e^{jn\omega_o T}$. Equation (1.31) is applied in Chapter 2 in the discussion of sampling theory and in Chapter 11 in the context of signal acquisition.

Alternate forms of (1.29) that emphasize the series expansion in terms of harmonics of trigonometric functions are given in (1.32) and (1.33) when $f(t)$ is a real-valued function. This is important because when $f(t)$ is *real* the complex coefficients C_n and C_{-n} form a complex conjugate pair such that $C_{-n} = C_n^*$ which simplifies the evaluation of $f(t)$. For example, using the complex notations $C_n = \alpha_n + j\beta_n$ and $C_n^* = \alpha_n - j\beta_n$, the function $f(t)$ is evaluated as

$$f(t) = C_o + 2\sum_{n=1}^{\infty} [\alpha_n \cos(n\omega_o t) - \beta_n \sin(n\omega_o t)] \tag{1.32}$$

this simplifies to

$$f(t) = C_o + 2\sum_{n=1}^{\infty} |C_n| \cos(n\omega_o t + \phi_n) \tag{1.33}$$

where $|C_n| = \sqrt{\alpha^2 + \beta^2}$ and $\phi_n = \arctan(\beta/\alpha)$.

An important consideration in spectrum analysis is the determination of signal spectrums involving random data sequences, referred to as stochastic processes [8]. A stochastic process does not have a unique spectrum; however, the power spectral density (PSD) is defined as the Fourier transform of the autocorrelation response. Oppenheim and Schafer [9] discuss methods of estimating the PSD of a real finite-length (N) sampled sequence by averaging periodograms, defined as

$$I_N(\omega) = \frac{1}{N} |F(\omega)|^2 \tag{1.34}$$

where $F(\omega)$ is the Fourier transform of the sampled sequence. This method is accredited to Bartlett [10] and is used in the evaluation of the PSD in the following chapters. For a fixed length (L) of random data, the number of periodograms (K) that can be averaged is $K = L/N$. As K increases the variance of the spectral estimate approaches zero and as N increases the resolution of the spectrum increases, so there is a tradeoff between the selection of K and N. To resolve narrowband spectral features that occur, for example, with nonlinear frequency shift keying (FSK)-modulated waveforms, it is important to use large values of N. Fortunately, many of the spectrum analyses presented in the following chapters are not constrained by L so K and N are chosen to provide a low estimation bias, that is, low variance, and high spectral resolution. Windowing† the periodograms will also reduce the estimation bias at the expense of decreasing the spectral resolution.

1.2.1 The Transform Pair *rect(t/T)* ⇔ *Tsinc(fT)*

The transform relationship *rect(t/T)* ⇔ *Tsinc(fT)* occurs so often that it deserves special consideration. For example, consider the following function:

$$s(t) = A\cos(\omega_c(t-\tau) - \phi) rect\left(\frac{(t-\tau)}{T}\right) \tag{1.35}$$

where ω_c, τ, and ϕ represent arbitrary angular frequency, delay, and phase parameters. The signal $s(t)$ is depicted in Figure 1.3.

The Fourier transform of $s(t)$ is evaluated as

$$S(f) = A \int_{-T/2+\tau}^{T/2+\tau} \cos(\omega_c(t-\tau) - \phi) e^{-j2\pi ft} dt \tag{1.36}$$

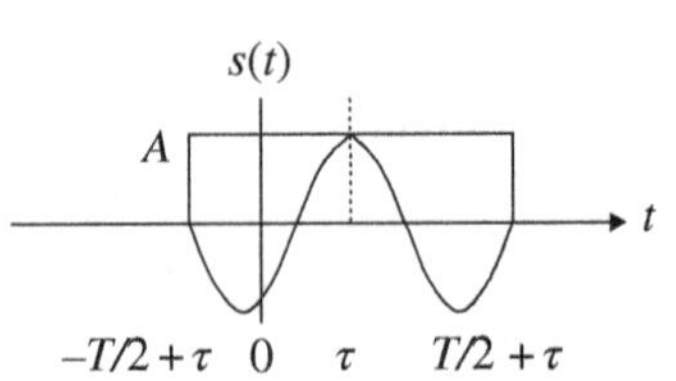

FIGURE 1.3 Pulse-modulated carrier.

*A summary of Fourier transforms pairs is given in Section 1.2.11.

†Windows are discussed in Section 1.11.

Expressing the cosine function in terms of complex exponential functions and performing some simplifications results in the expression

$$S(f) = \frac{A}{2}\left\{ e^{-j(2\pi f\tau+\phi)} \left[\int_{\tau-T/2}^{\tau+T/2} e^{-j2\pi(f-f_c)t} dt \right] + e^{-j(2\pi f\tau-\phi)} \left[\int_{\tau-T/2}^{\tau+T/2} e^{-j2\pi(f+f_c)t} dt \right] \right\} \tag{1.37}$$

Evaluation of the integrals in (1.37) appears so often that it is useful to generalize the solutions as follows:

Consider the integral

$$I(y) = \int_{x_1}^{x_2} e^{-j(y\pm\bar{y})x} dx = \frac{e^{-j(y\pm\bar{y})x_2} - e^{-j(y\pm\bar{y})x_1}}{-j(y\pm\bar{y})} \tag{1.38}$$

The general solution involves multiplying the last equality in (1.38) by the factors $e^{-j(y\pm\bar{y})(x_2+x_1)/2}$ and $e^{j(y\pm\bar{y})(x_2+x_1)/2}$, having a product of one, where $(x_2+x_1)/2$ is the average of the integration limits. Distributing the second factor over the numerator of (1.38) and then simplifying yields the result

$$I(y) = (x_2-x_1)e^{-j(y\pm\bar{y})(x_2+x_1)/2} \frac{\sin((y\pm\bar{y})(x_2-x_1)/2)}{(y\pm\bar{y})(x_2-x_1)/2} \tag{1.39}$$

Applying (1.39) to (1.37) and simplifying gives the desired result

$$S(f) = \frac{AT}{2}\left\{ e^{-j(2\pi f\tau+\phi)} sinc((f-f_c)T) + e^{-j(2\pi f\tau-\phi)} sinc((f+f_c)T) \right\} \tag{1.40}$$

When $f_c \gg 1/T$, the positive and negative frequency spectrums do not influence one another and, in this case, the positive frequency spectrum is defined as

$$S^+(f) = \frac{AT}{2} sinc((f-f_c)T) \quad : f_c \gg \frac{1}{T} \tag{1.41}$$

On the other hand, when the carrier frequency and phase are zero, (1.40) simplifies to the baseband spectrum, evaluated as

$$S_{bb}(f) = ATe^{-j2\pi f\tau} sinc(fT) \quad : f_c = 0, \ \phi = 0 \tag{1.42}$$

Using (1.42), the baseband Fourier transform pair, corresponding to of (1.35) with $f_c = 0$, is established as

$$rect\left(\frac{(t-\tau)}{T}\right) \Leftrightarrow Te^{-j2\pi f\tau} sinc(fT) \quad : |t-\tau| \le \frac{T}{2} \tag{1.43}$$

and, with $\tau = 0$,

$$rect\left(\frac{t}{T}\right) \Leftrightarrow Tsinc(fT) \quad : |t| \le \frac{T}{2} \tag{1.44}$$

1.2.2 The *sinc*(*x*) Function

The *sinc*(*x*) function is defined as

$$sinc(x) = \frac{\sin(\pi x)}{\pi x} \tag{1.45}$$

and is depicted in Figure 1.4. When x is expressed as the normalized frequency variable $x = fT$ then (1.45), when scaled by T, is the frequency spectrum of the unit amplitude pulse *rect* (t/T) of duration T seconds such that $t \le |T/2|$. This function is symmetrical in x and the maximum value of the first sidelobe occurs at $x = 1.431$ with a level of $10\log(sinc^2(x)) = -13.26$ dB; the peak sidelobe levels decrease in proportion to $1/|x|$. The noise bandwidth of a filter function $H(f)$ is defined as

$$B_n \triangleq \frac{\int_{-\infty}^{\infty} |H(f)|^2 df}{|H(f_o)|^2} \tag{1.46}$$

where f_o is the filter frequency corresponding to the maximum response. When a receiver filter is described as $H(f) = sinc(fT)$ the receiver low-pass noise bandwidth is evaluated as $B_n = 1/T$ where T is the duration of the filter impulse response.

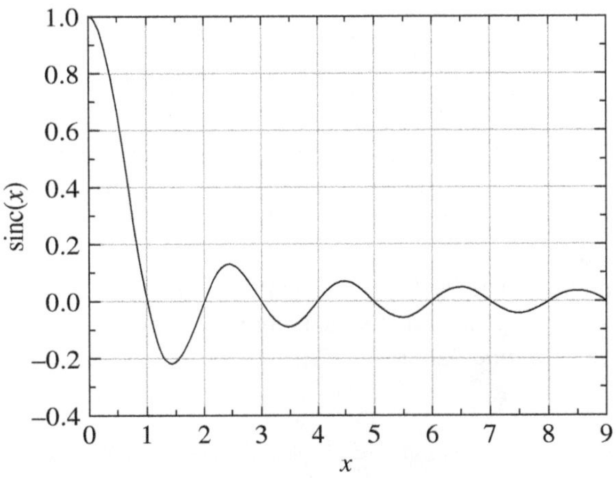

FIGURE 1.4 The *sinc*(*x*) function.

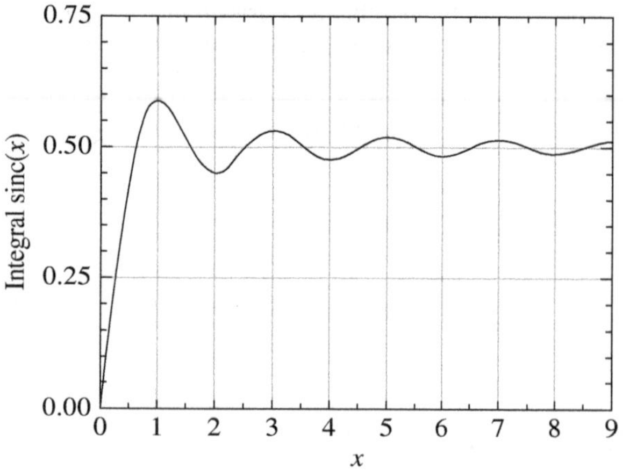

FIGURE 1.5 Integral of *sinc*(*x*).

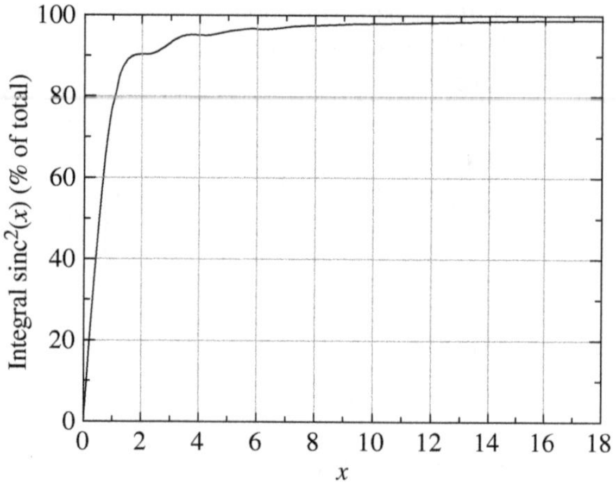

FIGURE 1.6 Integral of $sinc^2(x)$ function.

It is sometimes useful to evaluate the area of the *sinc*(*x*) function and, while there is no closed form solutions, the solution can be evaluated in terms of the *sine-integral* $S_i(x)$* as

$$\int_0^z sinc(x)dx = \int_0^z \frac{\sin(\pi x)}{\pi x}dx = \frac{1}{\pi}\int_0^{\pi z}\frac{\sin(\lambda)}{\lambda}d\lambda = \frac{1}{\pi}S_i(\pi z) \quad (1.47)$$

where the *sine-integral* is defined as the integral of $\sin(\lambda)/\lambda$. Equation (1.47) is shown in Figure 1.5. The limit of $S_i(\pi z)$ as $|z| \to \infty$ is† $\pi\, sign(1,z)/2$ so the corresponding limit of (1.47) is $0.5sgn(z)$.

A useful parameter, often used as a benchmark for comparing spectral efficiencies, is the area under $sinc^2(x)$ as a function of x. The area is evaluated in terms of the *sine-integral* as

$$\int_0^z sinc^2(x)dx = \int_0^z \frac{\sin^2(\pi x)}{(\pi x)^2}dx = \frac{1}{\pi}\left[S_i(2\pi z) - \frac{\sin^2(2\pi z)}{(2\pi z)}\right] \quad (1.48)$$

Equation (1.48) is plotted in Figure 1.6 as a percent of the total area and it is seen that the spectral containment of 99% is in excess of 18 spectral sidelobes, that is, $x = fT = 18$. In the following chapters, spectral efficient waveforms are examined with 99% containment within 2 or 3 sidelobes, so the *sinc*(*x*) function does not represent a spectrally efficient waveform modulation.

*The arguments x and z may be complex; however, the following analysis uses only real arguments.

†The *sign*(*a*, *x*) function is defined in Section 1.14.7.

1.2.3 The Fourier Transform Pair $\sum_n \delta(t-nT) \Leftrightarrow \omega_o \sum_n \delta(\omega - n\omega_o)$

The evaluation of this Fourier transform pair is fundamental to Nyquist sampling theory and is demonstrated in Section 2.3 in the evaluation of discrete-time sampling. In this case, the function $f(t)$ is an infinite repetition of equally spaced delta functions $\delta(t)$ with intervals T seconds as expressed by

$$f(t) = \sum_{n=-\infty}^{\infty} \delta(t-nT) \quad (1.49)$$

The challenge is to show that the Fourier transform of (1.49) is equal to an infinite repetition of equally spaced and weighted frequency domain delta functions expressed as

$$F(\omega) = \omega_o \sum_{n=-\infty}^{\infty} \delta(\omega - n\omega_o) \quad (1.50)$$

with weighting ω_o and frequency intervals $\omega_o = 2\pi/T$. Direct application of the Fourier transform to (1.49) leads to the spectrum $\sum_n e^{-jn\omega T}$ but this does not demonstrate the equality in (1.50). Similarly, evaluation of the inverse Fourier transform of (1.50) results in the time-domain expression

$$g(t) = \frac{1}{T}\sum_{n=-\infty}^{\infty} e^{jn\omega_o t} \quad (1.51)$$

So, by showing that $g(t) = f(t)$, the transform pair between (1.49) and (1.50) will be established. Consider $g_N(t)$ to be a finite summation of terms in (1.51) given by

$$g_N(t) = \frac{1}{T}\sum_{n=-N}^{N} e^{jn\omega_o t} = \frac{\sin((2N+1)\omega_o t/2)}{T\sin(\omega_o t/2)} \quad (1.52)$$

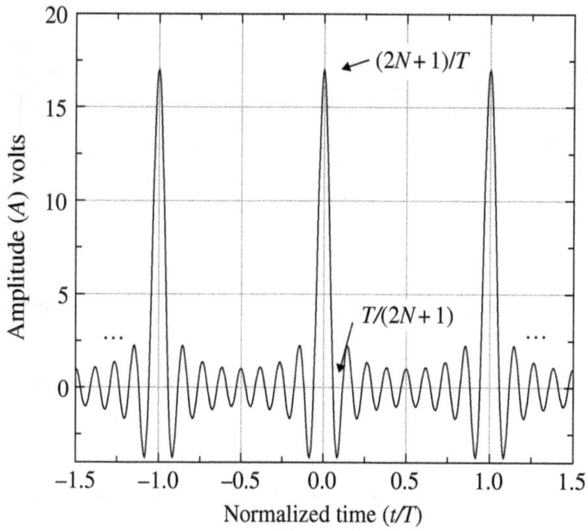

FIGURE 1.7 The Fourier-series kernel $g_N(t)$ ($N = 8$).

The second equality in (1.52) can be shown using the finite series identity No. 12, Section 1.14.1. Equation (1.52) is referred to by Papoulis [7] as the *Fourier-series kernel* and appears in a number of applications involving the Fourier transform.

The function $g_N(t)$ is plotted in Figure 1.7 for $N = 8$. The abscissa is time normalized by the pulse repetition interval $T = 1/f_o$ such that, $g_N(t) = g_N(t - nT)$, and there are a total of $2N + 1$ peaks of which three are shown in the figure. Furthermore, there are eight time sidelobes between $t/T = 0$ and 0.5 with the first nulls from the peak value at $t/T = 0$ occurring at $\pm T/(2N+1)$; the peak values are $(2N + 1)/T = 17/T$ in this example.

The maximum values of $2N + 1$, occurring at $t/T = n$, are determined by applying L'Hospital's rule to (1.52), which is rewritten as

$$g_N(t) = (2N+1)\frac{\sin((2N+1)\omega_o t/2)}{T(2N+1)\sin(\omega_o t/2)}$$
$$\cong (2N+1)\frac{\sin((2N+1)\omega_o t/2)}{T(2N+1)\omega_o t/2} \quad (1.53)$$

The approximation in (1.53) is obtained by noting that as N increases the rate of the sinusoidal variations in the numerator term increases with a frequency of $(2N+1)f_o$ Hz while the rate of sinusoidal variation in the denominator remains unchanged. Therefore, in the vicinity of $t/T = n$, $\sin(\omega_o t/2) \cong \omega_o t/2$ and (1.53) reduces to a $\sin(x)/x$ function with $x = (2N+1)\omega_o t/2$ and a peak amplitude $(2N + 1)$. The proof of the transform pair is completed by showing that $f(t) = g(t)$. Referring to (1.51) $g(t)$ is expressed as

$$g(t) = \lim_{N\to\infty} g_N(t) \quad (1.54)$$

From (1.53) as N approaches infinity the $\sin(x)/x$ sidelobe nulls converge to $t/T = n$, the peak values become infinite, and the corresponding area over the interval $|t/T| = n \pm 1/2$ approaches unity. Therefore, $g(t)$ resembles a periodic series of delta functions resulting in the equality

$$g(t) = \lim_{N\to\infty} g_N(t) = \sum_{n=-\infty}^{\infty} \delta(t - nT)$$
$$= f(t) \quad (1.55)$$

thus completing the proof that (1.49) and (1.50) correspond to a Fourier transform pair. Papoulis (Reference 7, pp. 50–52) provides a more eloquent proof that the limiting form of $g_N(t)$ is indeed an infinite sequence of delta functions.

1.2.4 The Discrete Fourier Transform

The DFT pair relating the discrete-time function $f(m\Delta t) \equiv f(m)$ and discrete-frequency function $F(n\Delta f) \equiv F(n)$ is denoted as $f(m) \Leftrightarrow F(n)$ where

$$F(n) = \Delta t \sum_{m=0}^{M-1} f(m) e^{-j2\pi n\Delta f m\Delta t} \Leftrightarrow f(m) = \Delta f \sum_{n=0}^{N-1} F(n) e^{j2\pi n\Delta f m\Delta t}$$

: DFT

(1.56)

With the DFT the number of time and frequency samples can be chosen independently. This is advantageous when preparing presentation material or examining fine spectral or temporal details, as might be useful when debugging simulation programs, by the independent selection of the integers m and n.

1.2.5 The Fast Fourier Transform

As discussed in the preceding section, the DFT pair, relating the discrete-time function $f(m\Delta t) \equiv f(m)$ and the discrete-frequency function $F(n\Delta f) \equiv F(n)$, is denoted as $f(m) \Leftrightarrow F(n)$ where $f(m)$ and $F(n)$ are characterized by the expressions for the DFT. The FFT [11–17], is a special case corresponding to m and n being equal to N as described in the remainder of this section. In these relationships N is the number of time samples and is defined as the power of a fixed radix-r FFT or as the powers of a mixed radix-r_j FFT.* The fixed radix-2 FFT, with $r = 2$ and $N = 2^i$, results in the most processing efficient implementation.

*Mixed radix FFTs provide an efficient method of computing the Fourier transform when the number of samples is not a power of r. In general, $N = r_1^{i_1} r_2^{i_2} \ldots$ and the radices of the FFT are determined as the prime factors of N. For example, $N = 31 = 1{*}31$ requires a single radix = 31 FFT, $N = 32 = 2{*}2{*}2{*}2{*}2$ requires a radix-2 FFT, and $N = 33 = 3{*}11$ requires a radix-3 and a radix-11 FFT.

Defining the time window of the FFT as T_w results in an implicit periodicity of $f(t)$ such that $f(t) = f(t \pm kT_w)$ and $\Delta t = T_w/N$. The sampling frequency is defined as $f_s = 1/\Delta t = N/T_w$ and, based on Shannon's sampling theorem, the periodicity does not pose a practical problem as long as the signal bandwidth is completely contained in the interval $|B| \le f_s/2 = N/(2T_w)$. Since the FFT results in an equal number of time and frequency domain samples, that is, $\Delta f = f_s/N$ and $\Delta t = T_w/N$, it follows that $\Delta f \Delta t = f_s T_w/N^2 = 1/N$. Normalizing the expression of the time function, $f(m)$, in (1.56), that is, multiplying the inverse DFT (IDFT) by Δt requires dividing the expression for $F(n)$ by Δt. Upon substituting these results into (1.56), the FFT transform pairs become

$$F(n) = \sum_{m=0}^{N-1} f(m)e^{-j2\pi nm/N} \Leftrightarrow f(m) = \frac{1}{N}\sum_{n=0}^{N-1} F(n)e^{j2\pi nm/N} \quad : \text{FFT} \tag{1.57}$$

The time and frequency domain sampling characteristics of the FFT are shown in Figure 1.8. This depiction focuses on a communication system example, in that, the time samples over the FFT window interval T_w are subdivided into N_{sym} symbol intervals of duration T seconds with N_s samples/symbol.

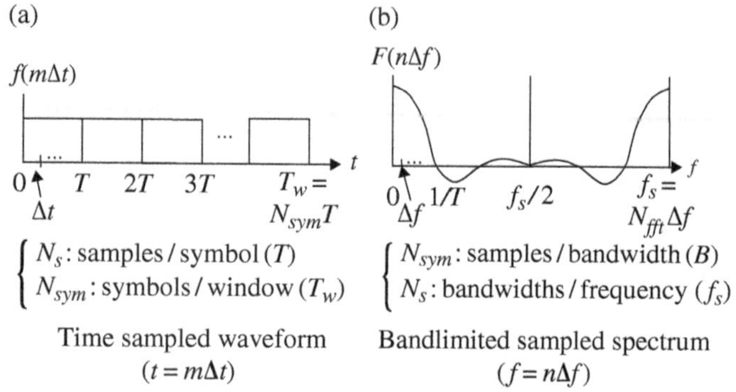

FIGURE 1.8 FFT time and frequency domain sampling.

Typically the bandwidth of the modulated waveform is taken to be the reciprocal of the symbol duration, that is, 1/T Hz; however, the receiver bandwidth required for low symbol distortion is typically several times greater than $1/T$ depending upon the type of modulation. Referring to Figure 1.8 the sampling frequency is $f_s = 1/\Delta t$, the sampling interval is $\Delta t = T/N_s$, the size of the FFT is $N_{fft} = N_s N_{sym}$, and the frequency sampling increment is $\Delta f = f_s/N_{fft}$. Upon using these relationships, the frequency resolution, or frequency samples per symbol bandwidth $B = 1/T$, is found to be

$$\frac{B}{\Delta f} = N_{sym} \quad : \text{determines frequency resolution} \tag{1.58}$$

and the number of spectral sidelobes* or symbol bandwidths over the sampling frequency range is

$$\frac{f_s}{B} = N_s \quad : \text{determines spectral sidelobes} \tag{1.59}$$

Therefore, to increase the resolution of the sampled signal spectrum, the number of symbols must be increased and this is comparable to increasing T_w. On the other hand, to increase the number of signal sidelobes contained in the frequency spectrum the number of samples per symbol must be increased and this is comparable to decreasing Δt. Both of these conditions require increasing the size (N) of the FFT. However, for a given size, the FFT does not allow independent selection of the frequency and time resolution as determined, respectively, by (1.58) and (1.59). This can be accomplished by using the DFT as discussed in Section 1.2.4. Since the spectrum samples in the range $0 \le f < f_s/2$ represent the positive frequency signal spectrum and those over the range $f_s/2 \le f < f_s$ represent the negative frequency signal spectrum, the range of signal sidelobes of interest is $\pm f_s/(2B) = \pm N_s/2$. As a practical matter, if the signal carrier frequency is not zero then the sampling frequency must be increased to maintain the signal sidelobes aliasing criterion. The sampling frequency selection is discussed in Chapter 11 in the context of signal acquisition when the received signal frequency is estimated based on locally known conditions.

The following implementation of the FFT is based on the Cooley and Tukey [18] decimation-in-time algorithm as described by Brigham and Morrow [19] and Brigham [20]. Although (1.57) characterizes the FFT transform pairs, the real innovation leading to the *fast* transformation is realized by the efficient algorithms used to execute the transformation. Considering the radix-2 FFT with $N = 2^n$, this involves defining the constant

$$W \triangleq e^{-j2\pi/N} \tag{1.60}$$

and recognizing that

$$F(n) = \Delta t \sum_{m=0}^{N-1} f(m)W^{nm} \tag{1.61}$$

Equation (1.61) can be expressed in matrix form, using $N = 4$ for simplicity, as

*These results are based on the underlying $rect(t/T)$ window and the $sinc(fT)$ frequency function that includes the principal lobe and the positive and negative spectral side lobes.

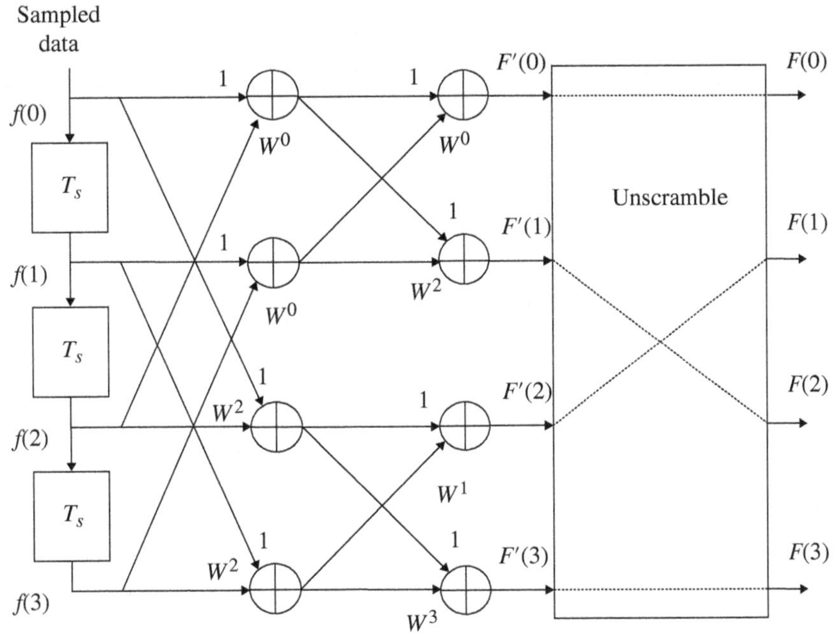

FIGURE 1.9 Radix-2, $N = 4$-point FFT implementation tree diagram.

$$\begin{bmatrix} F(0) \\ F(1) \\ F(2) \\ F(3) \end{bmatrix} = \Delta t \begin{bmatrix} W^0 & W^0 & W^0 & W^0 \\ W^0 & W^1 & W^2 & W^3 \\ W^0 & W^2 & W^4 & W^6 \\ W^0 & W^3 & W^6 & W^9 \end{bmatrix} \begin{bmatrix} f(0) \\ f(1) \\ f(2) \\ f(3) \end{bmatrix} \tag{1.62}$$

Recognizing that $W^0 = 1$ and the exponent nm is modulo (N), upon factoring the matrix in (1.62) into the product of two submatrices (in general the product of $\log_2 N$ submatrices) leads to the implementation involving the minimum number of computations expressed as

$$\begin{bmatrix} F(0) \\ F(2) \\ F(1) \\ F(3) \end{bmatrix} = \Delta t \begin{bmatrix} 1 & W^0 & 0 & 0 \\ 1 & W^2 & 0 & 0 \\ 0 & 0 & 1 & W^1 \\ 0 & 0 & 1 & W^3 \end{bmatrix} \begin{bmatrix} 1 & 0 & W^0 & 0 \\ 0 & 1 & 0 & W^2 \\ 1 & 0 & W^2 & 0 \\ 0 & 1 & 0 & W^2 \end{bmatrix} \begin{bmatrix} f(0) \\ f(1) \\ f(2) \\ f(3) \end{bmatrix} \tag{1.63}$$

The simplifications result in the outputs $F(2)$ and $F(1)$ being scrambled and the unscrambling to the natural-number ordering simply involves reversing the binary number equivalents, that is, with $F'(1) = F(2)$ and $F'(2) = F(1)$; therefore, the unscrambling is accomplished as $F(1) = F(01) = F'(2) = F'(10)$ and $F(2) = F(10) = F'(1) = F'(01)$. The radix-2 with $N = 4$ FFT, described by (1.63), is implemented as shown in the diagram of Figure 1.9.

The inverse FFT (IFFT) is implemented by changing the sign of the exponent of W in (1.60), interchanging the roles of $F(n)$ and $f(m)$, as described earlier, and replacing Δt by Δf. Recognizing that $\Delta t \Delta f = 1/N$, it is a common practice not to weight the FFT but to weight the IFFT by $1/N$ as indicated in (1.57). The number of complex multiplication is determined from (1.63) by recognizing that $W^2 = -W^0$ and not counting previous products like $W^0 f(2)$ from row 1 and $W^2 f(2) = -W^0 f(0)$ from row 3 in the first matrix multiplication on the *rhs* of (1.63). For the commonly used radix-2 FFT, the number of complex multiplications is $(N/2)\log_2(N)$ and the number of complex additions is $N\log_2(N)$. By comparison, the number of complex multiplications and additions in the direct Fourier transform are N^2 and $N(N-1)$, respectively. These computational advantages are enormous for even modest transform sizes.

1.2.5.1 The Pipeline FFT The FFT algorithm discussed in the preceding section involves decimation-in-time processing and requires collecting an entire block of time-sampled data prior to performing the Fourier transform. In contrast, the pipeline FFT [21] processes the sampled data sequentially and outputs a complete Fourier transform of the stored data at each sample. The implementation of a radix-2, $N = 8$-point pipeline FFT is shown in Figure 1.10. The pipeline FFT inherently scrambles the outputs $F'(n)$ and the unscrambled outputs are not shown in the figure; the unscrambling is accomplished by simply reversing the order of the binary representation of the output locations, n, as described in the preceding section.

In general, the number of complex multiplications for a complete transform is $(N/2)(N-1)$. In Chapter 11 the pipeline FFT is applied in the acquisition of a waveform where a

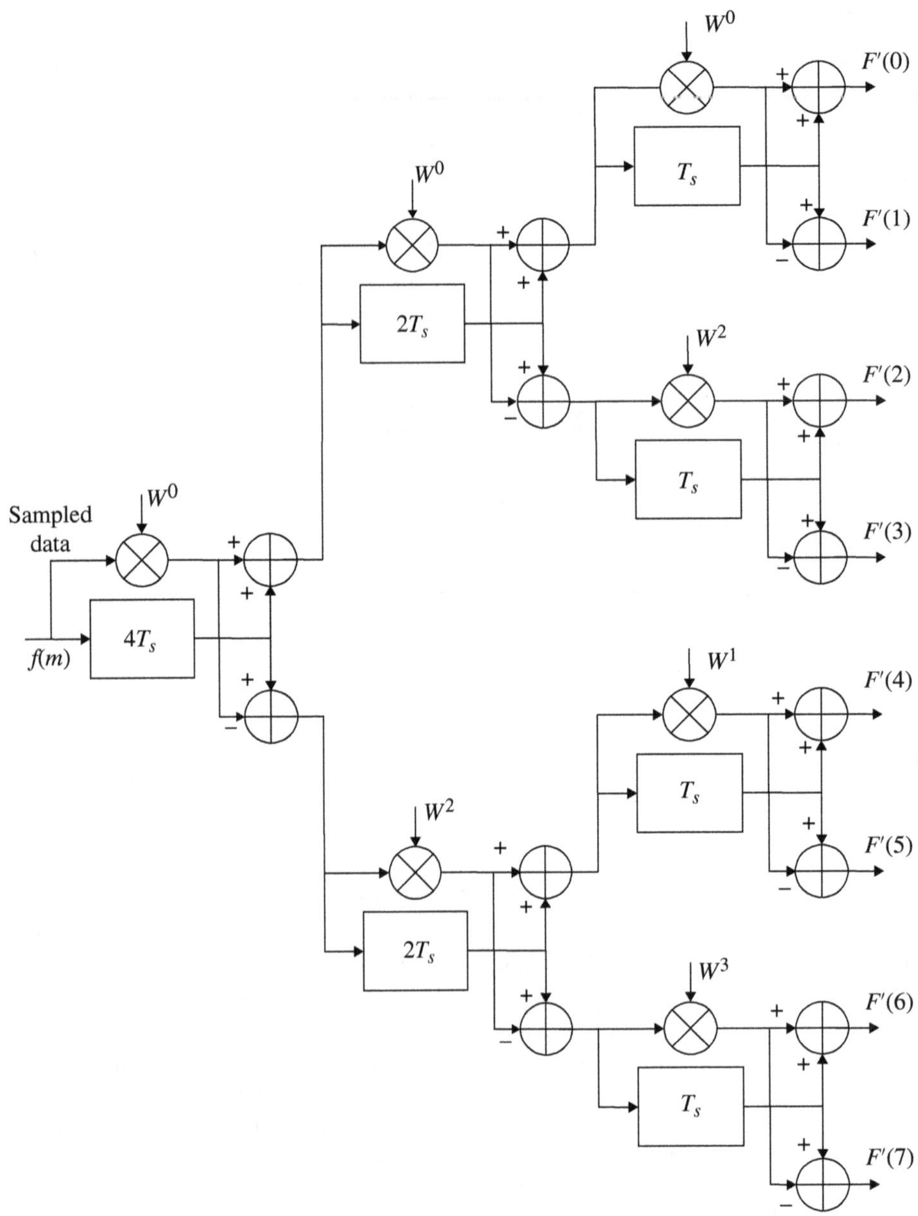

FIGURE 1.10 Radix-2, N = 8-point pipeline FFT implementation tree diagram.

complete N-point FFT output is not required at every sample. For example, if the complete N-point FFT is only required at sample intervals of N_sT_s, the number of complex multiplications can be significantly reduced (see Problem 10). The pipeline FFT can be used to interpolate between the fundamental frequency cells by appending zeros to the data samples and appropriately increasing the size of the FFT; it can also be used with data samples requiring mixed radix processing. The pipeline FFT is applicable to radar and sonar signal detection processing [21] using a variety of spectral shaping windows; however, the intrinsic *rect*(t/T) FFT window is nearly matched for the detection of orthogonally spaced M-ary FSK modulated frequency tones.

1.2.6 The FFT as a Detection Filter

The pipeline Fourier transform is made up of a cascade of transversal filter building blocks shown in Figure 1.10. The transfer function of this building block is

$$T_i(s) = \frac{e_o(s)}{e_i(s)} = W_i + e^{-sk_iT_s} \tag{1.64}$$

The overall transfer function from the input to a particular output is evaluated as

$$T_\ell(s) = \frac{e_\ell(s)}{e_i(s)} = \prod_{i=1}^{k}\left(W_{\ell,i} + e^{-sk_iT_s}\right) \tag{1.65}$$

where $k = \log_2(N)$ and $k_i = 2i - 1$, $i = 1, \ldots, k$. The complex weights are given by

$$W_{\ell,i} = e^{-j\phi_{\ell,i}} \tag{1.66}$$

where

$$\phi_{\ell,i} = \frac{2\pi\ell i}{N} \tag{1.67}$$

Substitution of $W_{\ell,i}$ into (1.65) results in

$$T_\ell(s) = e^{-j\Phi_\ell}\sum_{k=0}^{N-1} e^{-k(sT_s - j\phi_\ell)} \tag{1.68}$$

where

$$\Phi_\ell = \sum_{i=1}^{k}\phi_{\ell,i} \tag{1.69}$$

This transfer function is expressed in terms of a magnitude and phase functions in ω by substituting $s = j\omega$ with the result

$$T_\ell(\omega) = \frac{\sin\left(N(\omega T_s - \phi_{\ell,k})/2\right)}{\sin\left((\omega T_s - \phi_{\ell,k})/2\right)} e^{-j\Phi'_\ell}\, e^{-j(N-1)\omega T_s/2} \tag{1.70}$$

where

$$\Phi'_\ell = \Phi_\ell - (N-1)\phi_{\ell,k}/2 \tag{1.71}$$

Therefore, the FFT forms N filters, $\ell = 0, \ldots, N-1$ each having a maximum response $|T_\ell(\omega)|_{max} = N$ that occurs at the frequencies $\omega = \phi_{\ell,k}/T_s$. As N increases these transfer functions result in the response

$$T_\ell(\omega) = N\frac{\sin\left(N(\omega T_s - \phi_{\ell,k})/2\right)}{N(\omega T_s - \phi_{\ell,k})/2} e^{-j\Phi'_\ell}\, e^{-j(N-1)\omega T_s/2} \quad : N \to \infty \tag{1.72}$$

The magnitude of (1.72) is the *sinc*(x) function associated with the uniformly weighted envelope modulation function and, therefore, the FFT filter functions as a matched detection filter for these forms of modulations. Examples of these modulated waveforms are binary phase shift keying (BPSK), quadrature phase shift keying (QPSK), offset quadrature phase shift keying (OQPSK), and M-ary FSK.

The FFT detection filter loss relative to the ideal matched filter is examined as N increases. The input signal is expressed as

$$s(t) = \sqrt{2P}\cos(\omega_c t - \phi)rect\left(\frac{t}{T}\right) \tag{1.73}$$

and the corresponding signal spectrum for positive frequencies with $\omega_c \gg 2\pi/T$ is

$$S(\omega) = \frac{T}{2}\frac{\sin((\omega-\omega_c)T/2)}{(\omega-\omega_c)T/2}e^{-j\phi} \tag{1.74}$$

The matched filter for the optimum detection of $s(t)$ in additive white noise with spectral density N_o is defined as

$$H(\omega) = KS^*(\omega)e^{-j\omega T_o} \tag{1.75}$$

where K is an arbitrary scale factor and T_o is an arbitrary delay influencing the causality of the filter. By letting $K = 2N\sqrt{2P}/T$, $\phi = -\Phi'_\ell$, $T_o = (N-1)T_s/2$, and $\omega_c = \phi_{\ell,n}/T_s$ it is seen that the FFT approaches a matched filter as N increases.

The question of how closely the FFT approximates a matched filter detector is examined in terms of the loss in signal-to-noise ratio. The filter loss is expressed in dB as

$$\rho = 10\log_{10}\left(\frac{\mathrm{SNR}_o}{(\mathrm{SNR}_o)_{opt}}\right) \tag{1.76}$$

where $(\mathrm{SNR}_o)_{opt} = 2E/N_o$ is the signal-to-noise ratio out of the matched filter and E is the signal energy. The signal-to-noise ratio out of the FFT filter is expressed in terms of the peak signal output of the detection filter and the output noise power as

$$\mathrm{SNR}_o = \frac{|g_\ell(t)\max|^2}{N_o B_n} \tag{1.77}$$

where B_n is the detection filter noise bandwidth. For convenience the zero-frequency FFT filter output is considered, that is, for $\ell = 0$, and letting the signal phase $\phi = 0$, the response of interest is

$$T_o(\omega) = \frac{\sin(N\omega T_s/2)}{\sin(\omega T_s/2} \tag{1.78}$$

and, from (1.74),

$$S(\omega) = \frac{T}{2}\frac{\sin(\omega T/2)}{\omega T/2} \tag{1.79}$$

TABLE 1.2 *N*-ary FSK Waveform Detection Loss Using an *N*-Point FFT Detection Filter

N	ρ (dB)
2	0.452
3	0.236
8	0.116
16	0.053

To evaluate SNR_o at the output of the FFT filter, $g_o(t)_{max}$ and B_n are computed as

$$g_o(t)_{max} = \frac{1}{2\pi N}\int_{-\infty}^{\infty} T_o(\omega)S(\omega)d\omega$$
$$= \frac{1}{2\pi N}\int_{0}^{\pi/T_s} \frac{\sin(\omega T/2)\sin(N\omega T_s/2)}{(\omega T/2)\sin(\omega T_s/2)}d\omega \tag{1.80}$$

and

$$B_n = \frac{1}{2\pi}\frac{\int_{-\infty}^{\infty}|T_o(\omega)|^2 d\omega}{|T_o(0)|^2}$$
$$= \frac{1}{\pi N}\int_0^{\pi/T_s}\frac{\sin^2(N\omega T_o/2)}{\sin^2(\omega T_o/2)}d\omega \tag{1.81}$$
$$= \frac{1}{2NT_s}$$

Substituting these results into (1.77) and using (1.76), the parameter ρ is evaluated as

$$\rho = 20\log\left(\frac{2}{N\pi}\int_0^{\pi/2}\frac{\sin^2(Nx)}{x\sin(x)}dx\right) \quad : \text{dB} \tag{1.82}$$

Equation (1.82) is evaluated numerically for several values of N and the results are tabulated in Table 1.2. These results indicate, for example, that detecting an 8-ary FSK-modulated waveform with orthogonal tone spacing using an N = 8-point FFT results in a performance loss of 0.116 dB relative to an ideal matched filter.

1.2.7 Interpolation Using the FFT

When an FFT is performed on a uniformly weighted set of N data samples a set of N $sinc(fT_w)$ orthogonal filters is generated where $T_w = NT_s$ is the sampled data window and T_s is the sampling interval. The N filters span the frequency range $f_s = 1/T_s$ and provide N frequency estimates that are separated by $\Delta f = f_s/N$ Hz. Frequency interpolation is achieved if the FFT window is padded by adding nN zero-samples, thereby increasing the window by nNT_s seconds. In this case, a set of $(n + 1)N$ $sinc(fT_w)$ filters spanning the frequency f_s is generated that provides n-point interpolation between each of the original N filters.

The FFT can also be used to interpolate between time samples. For example, consider a sampled time function characterized by N samples over the interval $T_w = NT_s$ where T_s is the sampling interval. The corresponding N-point FFT has N filters separated by $\Delta f = f_s/N$ where $f_s = 1/T_s$. If nN zero-frequency samples are inserted between frequency samples $N/2$ and $N/2 + 1$ and the IFFT is taken on the resulting $(n + 1)N$ samples, the resulting time function contains n interpolation samples between each of the original N time samples. These interpolations methods increase the size of the FFT or IFFT and thereby the computational complexity.

1.2.8 Spectral Estimation Using the FFT

Many applications involve the characterization of the PSD of a finite sequence of random data. A random data sequence represents a stochastic process, for which, the PSD is defined as the Fourier transform of the autocorrelation function of the sequence. If the random process is such that the statistical averages formed among independent stochastic process are equal to the time averages of the sequences, then the Fourier transform will converge in some sense* to the true PSD, $S^2(\omega)$; however, this typically requires very long sequences that are seldom available. Furthermore, the classical approach, using the Fourier transform of the autocorrelation function, is processing intense and time consuming, requiring long data sequences to yield an accurate representation to the PSD. A much simpler approach, analyzed by Oppenheim and Schafer [22], is to recognize that the Fourier transform of a relatively short data sequence $x(n)$ of N samples is

$$X(e^{j\omega}) = \sum_{n=0}^{N-1} x(n)e^{-j\omega n} \tag{1.83}$$

and, defining the Fourier transform of the autocorrelation function $C_{xx}(m)$ of $x(n)$ as the periodogram

$$I_N(\omega) = \sum_{m=-(N-1)}^{N-1} C_{xx}(m)e^{-j\omega m}$$
$$= \frac{1}{N}|X(e^{j\omega})|^2 \tag{1.84}$$

*These are referred to as ergodic process and, under some circumstances, converge to the mean of the stochastic process.

However, the periodogram is not a consistent estimate* of the true PSD, having a large variance about the true values resulting in *wild* fluctuations. Oppenheim and Schafer then show that Bartlett's procedure [10, 23] of averaging periodograms of independent data sequences results in a consistent estimate and, if K periodograms are averaged, the resulting variance is decreased by K. In this case, the PSD estimate is evaluated as

$$\hat{S}_{xx}^2 = \frac{1}{K}\sum_{i=1}^{K} I_N^{(i)}(\omega) \tag{1.85}$$

Oppenheim and Schafer also discuss the application of windows to the periodograms and Welch [17] describes a procedure involving the averaging of modified periodograms.

1.2.9 Fourier Transform Properties

The following Fourier transform properties are based on the transform pairs $x(t) \Leftrightarrow S(f)$ and $y(t) \Leftrightarrow Y(f)$ where $x(t)$ and $y(t)$ may be real or complex.

1.2.9.1 Linearity

$$ax(t) + by(t) \Leftrightarrow aX(f) + bY(f) \tag{1.86}$$

1.2.9.2 Translation

$$x(t-\tau) \Leftrightarrow X(f)e^{-j2\pi f\tau} \tag{1.87}$$

and

$$X(f-f_o) \Leftrightarrow x(t)e^{-j2\pi f_o t} \tag{1.88}$$

1.2.9.3 Conjugation

$$x^*(t) \Leftrightarrow X^*(-f) \tag{1.89}$$

and

$$X^*(f) \Leftrightarrow x^*(-t) \tag{1.90}$$

1.2.9.4 Differentiation With $z(t) \triangleq d^n x(t)/dt^n$ and $Z(f) \triangleq d^n X(f)/df^n$ then

$$z(t) \Leftrightarrow (j2\pi f)^n X(f) \tag{1.91}$$

*A consistent estimate is one in which the variance about the true value and the bias approaches zero as N increases.

and

$$Z(f) \Leftrightarrow (-j2\pi t)^n x(t) \tag{1.92}$$

1.2.9.5 Integration Defining $z(t) \triangleq \int_{-\infty}^{t} \cdots \int_{-\infty}^{\tau_1} x(\tau)\, d\tau \ldots d\tau_n$ and $Z(f) \triangleq \int_{-\infty}^{f} \cdots \int_{-\infty}^{f_1} Z(f)df \ldots df_n$ then

$$z(t) \Leftrightarrow X(f)\left(\frac{\delta(t)}{2} + \frac{1}{(j2\pi f)^n}\right) \tag{1.93}$$

and

$$Z(f) \Leftrightarrow x(t)\left(\frac{\delta(t)}{2} + \frac{1}{(-j2\pi t)^n}\right) \tag{1.94}$$

1.2.10 Fourier Transform Relationships

The following Fourier transform relationships are based on the transform pairs $x(t) \Leftrightarrow X(f)$ and $y(t) \Leftrightarrow Y(f)$ where $x(t)$ and $y(t)$ may be real or complex.

1.2.10.1 Convolution Defining the Fourier transforms $z(t) \Leftrightarrow X(f)Y^*(f)$ and $Z(f) \Leftrightarrow x(t)y^*(t)$ then

$$z(t) = \int_{-\infty}^{\infty} x(t-\tau)y^*(\tau)d\tau = \int_{-\infty}^{\infty} x(\tau)y^*(t-\tau)d\tau \tag{1.95}$$

and

$$Z(f) = \int_{-\infty}^{\infty} X(f-\bar{f})y^*(\bar{f})d\bar{f} = \int_{-\infty}^{\infty} X(\bar{f})Y^*(f-\bar{f})d\bar{f} \tag{1.96}$$

1.2.10.2 Integral of Product (Parseval's Theorem)

$$\int_{-\infty}^{\infty} x(t)y^*(t)dt = \int_{-\infty}^{\infty} X(f)Y^*(f)df \tag{1.97}$$

Letting $y(t) = x(t)$ results in Parseval's Theorem that equates the signal energy in the time and frequency domains as

$$\int_{-\infty}^{\infty} |x(t)|^2 dt = \int_{-\infty}^{\infty} |X(f)|^2 df \quad \text{: Parseval's theorem} \tag{1.98}$$

TABLE 1.3 Fourier Transforms for $f(t) \Leftrightarrow F(f)$

Waveform $f(t)$	Spectrum $F(f)$
1	$\delta(f)$
$f(t-\tau)$	$F(f)\exp(-j2\pi f\tau)$
$f(t)e^{j2\pi f_o t}$	$f(f-f_o)$
$\delta(t)$	$\frac{1}{2\pi}\int e^{j\omega t}dt = 1$
$\delta(t-\tau)$	$\exp(-j2\pi f\tau)$
$f(at)$	$(1/a)F(f/a)$
$e^{j2\pi f_o t}$	$\delta(f-f_o)$
$\cos(2\pi f_o t)$	$\left(\frac{1}{2}\right)(\delta(f+f_o)+\delta(f-f_o))$
$\sin(2\pi f_o t)$	$\left(\frac{j}{2}\right)(\delta(f+f_o)-\delta(f-f_o))$
$\sum_{n=-\infty}^{\infty}\delta(t-n\tau)$	$\frac{1}{\tau}\sum_{n=-\infty}^{\infty}e^{-j2\pi nf\tau} = \frac{1}{\tau}\sum_{n=-\infty}^{\infty}\delta\left(f-\frac{n}{\tau}\right)$
$\frac{d^n f(t)}{dt^n}$	$(j2\pi f)^n F(f)$
$(-j2\pi t)^n f(t)$	$\frac{d^n F(f)}{df^n}$
$\int_{-\infty}^{t} f(\xi)d\xi$	$(1/2)(\delta(f)+1/(j\pi f))F(f)$
$(1/2)(\delta(t)-1/(j\pi t))s(t)$	$\int_{-\infty}^{f} F(\lambda)d\lambda$
$e^{-(t/a)^2/2}$	$\sqrt{2\pi}ae^{-(2\pi fa)^2/2}$
$f(t) = x(t)\, y(t)$	$X(f)*Y(f) = \int X(f-\lambda)Y(\lambda)d\lambda$
$f(t) = x(t)*y(t)$	$F(f) = X(f)\, Y(f)$
$u(t) = \begin{cases} 1 & : t \ge 0 \\ 0 & : o.w. \end{cases}$	$U(f) = \frac{1}{2}\left(\delta(f)+\frac{1}{j\pi f}\right)$
$u(t-\tau) = \begin{cases} 1 & : t-\tau \ge 0 \\ 0 & : o.w. \end{cases}$	$U(f)\exp(-j2\pi f\tau)$
$sgn(t) = 2u(t)-1$ $= \begin{cases} 1 & : t>0^a \\ -1 & : t<0 \end{cases}$	$sgn(f) = \frac{1}{j\pi f}$
$sgn(t) = \frac{1}{j\pi t}$	$sgn(f) = 2u(f)-1$
$rect\left(\frac{t}{T}\right) = 1 \;:\; \lvert t\rvert < \frac{T}{2}$ $= 0 \;:\; o.w.$	$Tsinc(fT)^b$
$sinc(2t/T)$	$\left(\frac{T}{2}\right)rect\left(\frac{fT}{2}\right) = 1 \;:\; \lvert f\rvert < \frac{1}{T}$ $= 0 \;:\; o.w.$

*Denotes convolution.
[a]The *signum* function $sgn(x)$ is also denoted as $signum(x)$.
[b]Woodward [24].

1.2.11 Summary of Some Fourier Transform Pairs

Some often used transform relationships are listed in Table 1.3.

1.3 PULSE DISTORTION WITH IDEAL FILTER MODELS

In this section the distortion is examined for an isolated baseband pulse after passing through an ideal filter with uniquely prescribed amplitude and phase responses. In radar applications isolated pulse response leads to a loss in range resolution; however, in communication application, where the pulse is representative of a contiguous sequence of information-modulated symbols, the pulse distortion leads to intersymbol interference (ISI) that degrades the information exchange. The following two examples use the baseband pulse, or symbol, as characterized in the time and frequency domains by the familiar functions

$$s(t) = rect\left(\frac{t}{T}\right) \Leftrightarrow S(f) = Tsinc(fT) \tag{1.99}$$

1.3.1 Ideal Amplitude and Zero Phase Filter

In this example, the filter is characterized in the frequency domain as having a constant unit amplitude over the bandwidth $f \le |B|$ with zero amplitude otherwise and a zero phase function. Using the previous notation the filter is characterized in the frequency and time domains as

$$H(f) = rect\left(\frac{f}{2B}\right) \Leftrightarrow h(t) = 2Bsinc(2Bt) \tag{1.100}$$

The frequency characteristics of the signal and filter are shown in Figure 1.11.

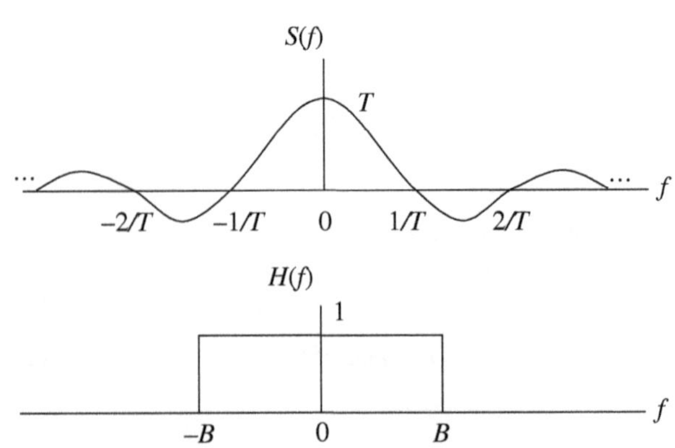

FIGURE 1.11 Ideal signal and filter spectrums.

The easiest way to evaluate the filter response to a pulse input signal is by convolving the functions as

$$g(t) = \int_{-\infty}^{\infty} h(\tau)s(\tau - t)d\tau$$
$$= 2B\int_{-\infty}^{\infty} sinc(2B\tau)rect\left(\frac{\tau - t}{T}\right)d\tau \qquad (1.101)$$

The $rect(\bullet)$ function determines the integration limits with the upper and lower limits evaluated for τ when the argument equals ±½, respectively. This evaluation leads to the integration

$$g(t) = 2B\int_{t-T/2}^{t+T/2} \frac{\sin(2\pi B\tau)}{2\pi B\tau}d\tau \qquad (1.102)$$

Equation (1.102) is evaluated in terms of the *sine integral* [25]

$$S_i(y) = \int_0^y \frac{\sin(x)}{x}dx \qquad (1.103)$$

resulting in the filter output $g(t)$ expressed as

$$g(t) = \frac{1}{\pi}S_i(2\pi B(t + T/2)) - \frac{1}{\pi}S_i(2\pi B(t - T/2)) \qquad (1.104)$$

Defining the normalized variable $y = t/T$ and the parameter $\rho = BT$, Equation (1.104) is expressed as

$$g(y) = \frac{1}{\pi}S_i(2\pi\rho(y + 1/2)) - \frac{1}{\pi}S_i(2\pi\rho(y - 1/2)) \qquad (1.105)$$

Equation (1.105) is plotted in Figure 1.12 for several values of the time-bandwidth (*BT*) parameter. Range resolution is proportional to bandwidth and the increased rise time or *smearing* of the pulse edges with decreasing bandwidth is evident. The ISI that degrades the performance of a communication system results from the symbol energy that occurs in adjacent symbols due to the filtering.

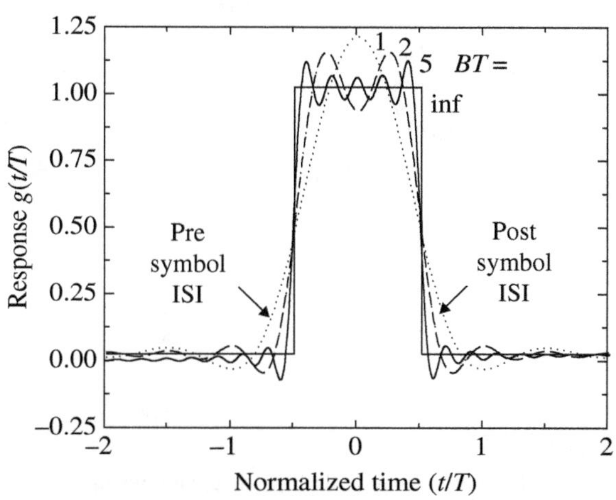

FIGURE 1.12 Ideal band-limited pulse response (constant-amplitude, zero-phase filter).

This analysis considers only the pulse distortion caused by constant amplitude filter response and, as will be seen in the following section, filter amplitude ripple and nonlinear phase functions also result in additional signal distortion. If the filter were to exhibit a linear phase function $\phi(f) = -2\pi fT_o$ where T_o represents a constant time delay, then, referring to Table 1.3, the output is simply delayed by T_o without any additional distortion. If T_o is sufficiently large, the filter can be viewed as a causal filter, that is, no output is produced before the input signal is applied.

1.3.2 Nonideal Amplitude and Phase Filters: Paired Echo Analysis

In this section the pulse distortion caused by a filter with prescribed amplitude and phase functions is examined using the analysis technique of paired echoes [26]. A practical application of paired echo analysis occurred when a modem production line was stopped at considerable expense due to noncompliance of the bit-error test involving a few tenths of a decibel. The required confidence level of the bit-error performance under various IF filter conditions precluded the use of Monte Carlo simulations; however, much to the pleasure of management, the paired echo analysis was successfully applied to identify the cause of the subtle filter distortion losses.

Consider a filter with amplitude and phase functions expressed as

$$H(f) = A(f)e^{-j\phi(f)}rect\left(\frac{f}{2B}\right) \qquad (1.106)$$

where the amplitude and phase fluctuations with frequency are expressed as

$$A(f) = 1 + a\sin(2\pi f\tau_a) \qquad (1.107)$$

and

$$\phi(f) = 2\pi fT_o + b\sin(2\pi f\tau_b) \qquad (1.108)$$

The parameters a and τ_a represent the amplitude and period of the amplitude ripple and b and τ_b represent the amplitude and period of the phase ripple. Using these functions in (1.106) and separating the constant delay term involving T_o, results in the filter function

$$H(f) = (1 + a\cos(2\pi f\tau_a))e^{-jb\sin(2\pi f\tau_b)}rect\left(\frac{f}{2B}\right)e^{-j2\pi fT_o} \qquad (1.109)$$

Equation (1.109) is simplified by using the trigonometric identity

$$\cos(2\pi f\tau_a)=\frac{1}{2}\left(e^{-j2\pi f\tau_a}+e^{j2\pi f\tau_a}\right) \tag{1.110}$$

and the Bessel function identity [27]

$$\begin{aligned} e^{-jb\sin(2\pi f\tau_b)} &= \sum_{n=-\infty}^{\infty} J_n(-b)e^{j2\pi nf\tau_b} \\ &= J_o(b)+J_1(b)\left(e^{-j2\pi f\tau_b}-e^{j2\pi f\tau_b}\right)+O_2(\pm n) \end{aligned} \tag{1.111*}$$

In arriving at the last expression in (1.111), the following identities were used

$$J_n(-b)=J_{-n}(-b)=(-1)^n J_n(b) \tag{1.112}$$

Upon substituting (1.110) and (1.111) into (1.109), and performing the multiplications to obtain additive terms representing unique delays results in the filter frequency response

$$\begin{aligned} H(f)=\Big\{&J_0(b)e^{-j2\pi fT_o}+\frac{a}{2}J_0(b)e^{-j2\pi f(T_o+\tau_a)} \\ &+\frac{a}{2}J_0(b)e^{-j2\pi f(T_o-\tau_a)}+J_1(b)e^{-j2\pi f(T_o+\tau_b)} \\ &+\frac{a}{2}J_1(b)e^{-j2\pi f(T_o+\tau_a+\tau_b)}+\frac{a}{2}J_1(b)e^{-j2\pi f(T_o-\tau_a+\tau_b)} \\ &-J_1(b)e^{-j2\pi f(T_o-\tau_b)}-\frac{a}{2}J_1(b)e^{-j2\pi f(T_o+\tau_a-\tau_b)} \\ &-\frac{a}{2}J_1(b)e^{-j2\pi f(T_o-\tau_a-\tau_b)}\Big\} \\ &rect\left(\frac{f}{2B}\right)+\text{higher order filter terms involving } J_n(b) \end{aligned} \tag{1.113}$$

Upon performing the inverse Fourier transform of each term in (1.113), the filter impulse response, $h(t)$, becomes a summation of weighted and delayed $sinc(x)$ functions of the form $2BKsinc(2B(t-T_d))$ where K and T_d are the amplitude and delay associated with each of the terms. Performing the convolution indicated by the first equality in (1.101), that is, for an arbitrary signal $s(t)$, the ideally filtered response $g(t)$ is expressed as

$$g(t)=2B\int_{-\infty}^{\infty} sinc(2B\tau)s(\tau-t)d\tau \tag{1.114}$$

*The notation $O_2(\pm n)$ refers to higher order terms involving $|n| \ge 2$. These terms can be neglected for small values of b, that is, $b < 0.2$.

When $g(t)$ is passed through the filter $H(f)$ with amplitude and phase described, respectively, by (1.107) and (1.108), the distorted output $g_o(t)$ is evaluated as

$$\begin{aligned} g_o(t)=&J_0(b)g(t-T_o)+\frac{a}{2}J_0(b)g(t-T_o-\tau_a) \\ &+\frac{a}{2}J_0(b)g(t-T_o+\tau_a)+J_1(b)g(t-T_o-\tau_b) \\ &+\frac{a}{2}J_1(b)g(t-T_o-\tau_a-\tau_b)+\frac{a}{2}J_1(b)g(t-T_o+\tau_a-\tau_b) \\ &-J_1(b)g(t-T_o+\tau_b)-\frac{a}{2}J_1(b)g(t-T_o-\tau_a+\tau_b) \\ &-\frac{a}{2}J_1(b)g(t-T_o+\tau_a+\tau_b) \end{aligned} \tag{1.115}$$

If the input signal is described by the $rect(t/T)$ function, then $g(t)$ is the response expressed by (1.104) and depicted in Figure 1.12. The distortion terms appear as paired echoes of the filtered input signal and Figure 1.13 shows the relative delay and amplitude of each *echo* of the filtered output $g(t)$. For $b \ll 1$ the approximations $J_0(b) = 1.0$ and $J_1(b) = b/2$ apply and when $a = b = 0$ the filter response is simply the

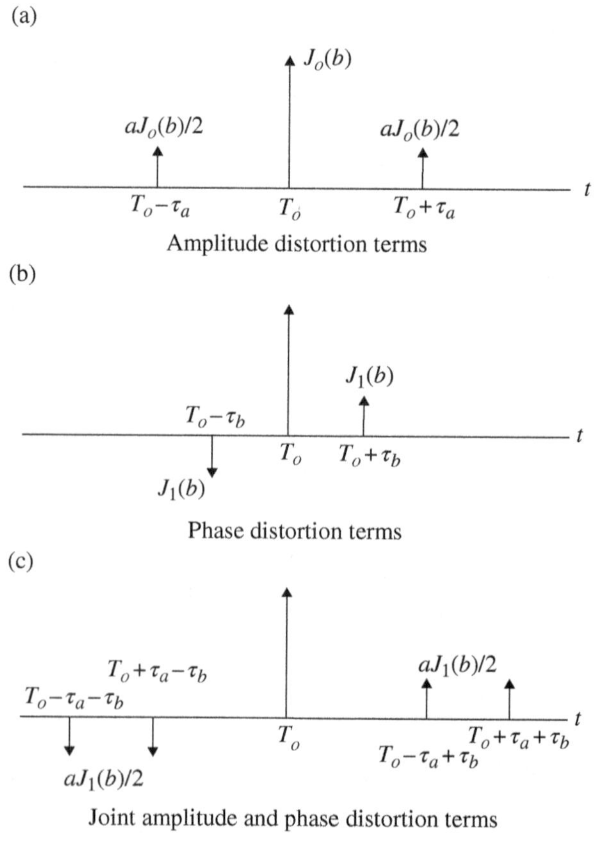

FIGURE 1.13 Location of amplitude and phase distortion paired echoes relative to delay T_o.

delayed but undistorted replica of the input signal, that is, $g_o(t) = g(t - T_o)$. More complex filter amplitude and phase distortion functions can be synthesized by applying Fourier series expansions that yield paired echoes that can be viewed as noisy interference terms that degrade the system performance; however, the analysis soon becomes unwieldy so computer simulation of the echo amplitudes and delays must be undertaken.

1.3.3 Example of Delay Distortion Loss Using Paired Echoes

The evaluation of the signal-to-interference ratio resulting from the delay distortion of a filter is examined using paired echo analysis. The objective is to examine the distortion resulting from a specification of the filters peak phase error and group delay within the filter bandwidth. The filter phase response is characterized as

$$\phi(f) = 2\pi f T_o - \phi_o \sin(2\pi f \tau) \tag{1.116}$$

where T_o is the filter delay resulting from the linear phase term, ϕ_o is the peak phase deviation from linearity over the filter bandwidth, and τ is the period of the sinusoidal phase distortion function. The linear phase term introduces the filter delay T_o that does not result in signal distortion; however, the sinusoidal phase term does cause signal distortion. In this example, the phase deviation over the filter bandwidth is specified parametrically as $\phi_o(\text{deg}) = 3$ and $7°$. The parameter τ is chosen to satisfy the peak delay distortion defined as

$$\begin{aligned} T_d(f) &= -\frac{\partial\phi(f)}{2\pi\partial f} \\ &= \tau\phi_o\cos(2\pi f\tau) \end{aligned} \tag{1.117}$$

where ϕ_o is in radians. The peak delay, evaluated for $f\tau = 0$, is specified as $T_d = 34$ and 100 *ns* and, using (1.117), the period of the sinusoidal phase function, $\tau = T_d/\phi_o$, is tabulated in Table 1.4 for the corresponding peak phase errors and peak delay specification. Practical maximum limits of the group delay normalized by the symbol rate, R_s, are also specified.

Considering an ideal unit gain filter with amplitude response of $A(\omega) = 1$, the filter transfer function is expressed as

$$\begin{aligned} H(f) &= e^{-j\phi(f)} \\ &= e^{-j2\pi f T_o} e^{j\phi_o \sin(2\pi f\tau)} \\ &= e^{-j2\pi f T_o} \sum_{n=-\infty}^{\infty} J_n(\phi_o) e^{jn\pi f\tau} \end{aligned} \tag{1.118}$$

TABLE 1.4 Values of τ for the Phase and Delay Specifications

ϕ_o(deg)	$T_d(ns)$	$\tau(ns)$	T_g/R_s [a]
3	34	649	±0.15
7	100	818	

[a] Normalized group delay over filter bandwidth.

Upon taking the inverse Fourier transform of (1.118), the filter impulse response is evaluated as

$$\begin{aligned} g(t) &= J_o(\phi_o)\delta(t-T_o) \\ &+ \sum_{n=1}^{\infty} J_n(\phi_o)[\delta(t-T_o+n\tau/2) - \delta(t-T_o-n\tau/2)] \end{aligned} \tag{1.119}$$

The parameter τ determines the delay spread of all the interfering terms; however, for small arguments the interference is dominated by the $J_1(\phi_o)$ term and the signal-to-interference ratio is defined as

$$\gamma_i = 20\log\left(\frac{J_1(\phi_o)}{J_o(\phi_o)}\right) \tag{1.120}$$

For $\phi_o(\text{deg}) = 3$ and $7°$, the respective signal-to-interference ratios are 32 and 24.3 dB and under these conditions, a 10 dB filter input signal-to-noise ratio results in the output signal-to-noise ratio degraded by 0.02 and 0.17 dB, respectively.

1.4 CORRELATION PROCESSING

Signal correlation is an important aspect of signal processing that is used to characterize various channel temporal and spectral properties, for example, multipath delay and frequency dispersion profiles. The correlation can be performed as a time-averaged autocorrelation or a time-averaged cross-correlation between two different signals. Frequency domain, autocorrelation, and cross-correlation are performed using frequency offsets rather than time delays. The Doppler and multipath profiles are characteristics of the channel that are typically based on correlations involving statistical expectations as opposed to time-averaged correlations that are applied to deterministic signal waveforms and linear time-invariant channels. The following discussion focuses on the correlation of deterministic waveforms and linear time-invariant channels.

The autocorrelation of the complex signal $\tilde{x}(t)$ is defined as*

$$R_{\tilde{x}\tilde{x}}(\tau) \triangleq \int_{-\infty}^{\infty} \tilde{x}(t)\tilde{x}^*(t-\tau)dt = \int_{-\infty}^{\infty} \tilde{x}(t+\tau)\tilde{x}^*(t)dt \tag{1.121}$$

*The asterisk denotes complex conjugation. The double subscripts on $R_{xx}(\tau)$ are not always included for the autocorrelation notation; however, they are important when describing the cross-correlation response.

TABLE 1.5 Properties of Correlation Functions

Property	Comments
$R_{\tilde{x}\tilde{x}}(-\tau)=R^*_{\tilde{x}\tilde{x}}(\tau)$	
$C_{\tilde{x}\tilde{x}}(-\tau)=R_{\tilde{x}\tilde{x}}(\tau)-\|\tilde{m}_x\|^2$	
$R_{xx}(-\tau)=R_{xx}(\tau)$	x : real
$C_{xx}(-\tau)=R_{xx}(\tau)-m_x^2$	x : real
$R_{\tilde{x}\tilde{y}}(-\tau)=R^*_{\tilde{y}\tilde{x}}(\tau)$	
$C_{\tilde{x}\tilde{y}}(\tau)=R_{\tilde{x}\tilde{y}}(\tau)-\tilde{m}_x\tilde{m}_y^*$	
$C_{xy}(\tau)=R_{xy}(\tau)-m_xm_y$	x,y : real
$R_{\tilde{z}\tilde{z}}(\tau)=R_{\tilde{x}\tilde{x}}(\tau)+R_{\tilde{y}\tilde{y}}(\tau)+R_{\tilde{x}\tilde{y}}(\tau)+R_{\tilde{y}\tilde{x}}(\tau)$	$\tilde{z}=\tilde{x}+\tilde{y}$

The autocorrelation function implicitly contains the mean value of the signal and the autocovariance is evaluated, by removing the mean value, as

$$\begin{aligned} C_{\tilde{x}\tilde{x}}(\tau) &= \int_{-\infty}^{\infty} [\tilde{x}(t)-\tilde{m}_x][\tilde{x}(t-\tau)-\tilde{m}_x]^* dt \\ &= R_{\tilde{x}\tilde{x}}(\tau)-|\tilde{m}_x|^2 \end{aligned} \tag{1.122}$$

where $\tilde{m}_x = m_{xc} + jm_{xs}$ is the complex mean of the signal $\tilde{x}(t)$. The cross-correlation of the complex signals $\tilde{x}(t)$ and $\tilde{y}(t)$ is defined as

$$R_{\tilde{x}\tilde{y}}(\tau) \triangleq \int_{-\infty}^{\infty} \tilde{x}(t)\tilde{y}^*(t-\tau)dt = \int_{-\infty}^{\infty} \tilde{x}(t+\tau)\tilde{y}^*(t)dt \tag{1.123}$$

Similarly, the corresponding cross-covariance is evaluated as

$$\begin{aligned} C_{\tilde{x}\tilde{y}}(\tau) &= \int_{-\infty}^{\infty} [\tilde{x}(t)-\tilde{m}_x][\tilde{y}(t-\tau)-\tilde{m}_y]^* dt \\ &= R_{\tilde{x}\tilde{y}}(\tau)-\tilde{m}_x\tilde{m}_y^* \end{aligned} \tag{1.124}$$

The properties of various correlation functions applied to complex and real valued functions are summarized in Table 1.5. The properties of correlation functions are also discussed in Section 1.5.9 in the context of stochastic processes.

1.5 RANDOM VARIABLES AND PROBABILITY

This section contains a brief introduction to random variables and probability [6, 8, 28–30]. A random variable is described in the context of Figure 1.14 in which an event χ in the space S is mapped to the real number x characterized as $X(\chi) = x$ or $f(x)$: $x_a \le x \le x_b$. The function $X(\chi)$ is defined as a random variable which assigns the real number x or $f(x)$ to each event $\chi \in S$.* The limits $[x_a, x_b]$ of the mapping are dependent upon the physical nature or definition of the event space. The second depiction shown in Figure 1.14 comprises disjoint, or nonintersecting, subspaces, such that, for $i \neq j$ the intersection $S_i \cap S_j = \emptyset$ is the null space. Each subspace possesses a unique mapping $x|S_j$ conditioned on the subspace $S_j : j = 1, \ldots, J$. The union of subspaces is denoted as $S_i \cup S_j$. This is an important distinction since each subspace can be analyzed in a manner similar to the mapping of $\chi \in S$. The three basic forms of the random variable X are continuous, discrete, and a mixture of continuous and discrete random variables as distinguished in the following sections.

1.5.1 Probability and Cumulative Distribution and Probability Density Functions

The mathematical description [6, 8, 24, 28, 30–32] of the random variable X resulting from the mapping $X(\chi)$ given the random event $\chi \in S$ is based on the statistical properties of the random event characterized by the probability $P(\{X \le x\})$ where $\{X \le x\}$ denotes all of the events $X(\chi)$ in S. For continuous random variables $P(X = x) = 0$. The probability function $P(X_i \in S_i)$ satisfies the following axioms:

A1. $P(X(\chi) \in S) \ge 0$
A2. $P(\{X(\chi) \in S\}) = 1$
A3. If $P(S_i \cap S_j) = \emptyset \ \forall i \neq j$ then $P(S_1 \cup, \ldots, S_J) = \sum_{j=1}^{J} P(Sj)$

Axiom A3 applies for infinite event spaces by letting $J = \infty$. Several corollaries resulting from these axioms are as follows:

C1. $P(\chi^c) = 1 - P(\chi)$ where χ^c is the complement of χ such that $\chi^c \cap \chi = \emptyset$
C2. $P(\chi) \le 1$
C3. $P(\chi_i \cup \chi_j) = P(\chi_i) + P(\chi_j) - P(\chi_i \cap \chi_j)$
C4. If $P(\emptyset) = 0$

The *cumulative distribution function* (*cdf*) of the variable X is defined in terms of the value of x on the real line as

$$F_X(x) \triangleq P(X \le x) \quad : -\infty < x < \infty, \text{ cumulative distribution function} \tag{1.125}$$

*A particular outcome $x' = f(x)$ is a random variable resulting from the mapping $X(\chi)$ onto the real line; however, $X(\chi)$ is also referred to as a random variable. Wozencraft and Jacobs (Reference 30, p. 39) point out that this nomenclature stems from practical applications and is somewhat misleading.

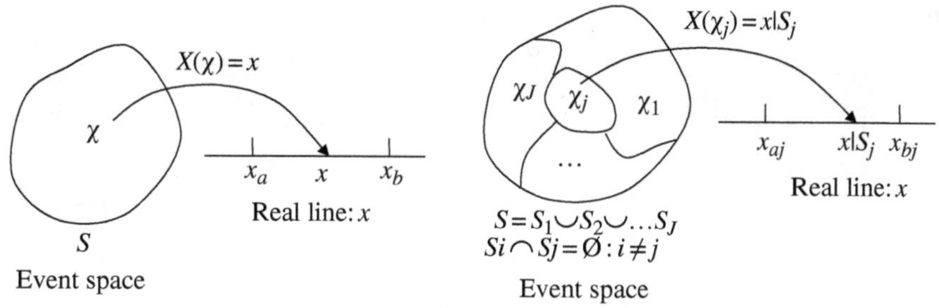

FIGURE 1.14 Mapping of random variable $X(\chi)$ on the real line x.

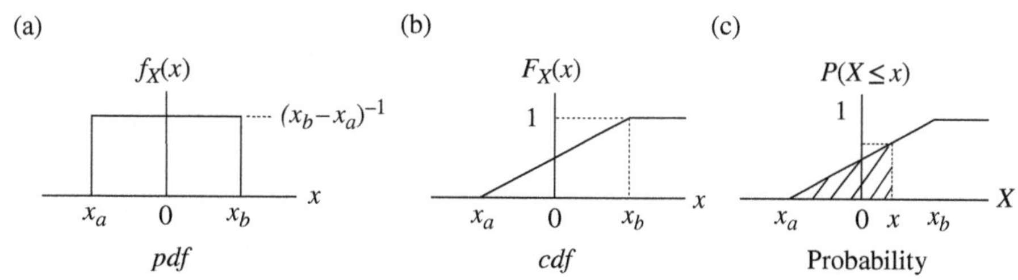

FIGURE 1.15 Uniformly distributed continuous random variable.

where $F_X(x)$ has the following properties:

P1. $0 \leq F_X(x) \leq 1$
P2. In the limit as x approaches ∞, $F_X(x) = 1$
P3. In the limit as x approaches $-\infty$, $F_X(x) = 0$
P4. $F_X(x)$ is a nondecreasing function of x
P5. In the limit as ε approaches 0, $F_X(x_i) = F_X(x_i + \varepsilon)$
P6. The probability in the interval $x_i < x \leq x_j$ is: $P(x_i < x \leq x_j) = F_X(x_j) - F_X(x_i)$
P7. In the limit as ε approaches 0, the probability of the event x_i is $P(x_i - \varepsilon < x \leq x_i) = F_X(x_i) - F_X(x_i - \varepsilon)$.

Property P5 is referred to as being *continuous from the right* and is particularly important with discrete random variables, in that, $F_X(x_i)$ includes a discrete random variable at x_i. Property P7, for a continuous random variable, states that $P(x_i) = 0$; however, for a discrete random variable, $P(x_i) = p_X(x_i)$ where $p_X(x_i)$ is the *probability mass function* (*pmf*) defined in Section 1.5.1.2.

The *probability density function** (*pdf*) of X is defined as

$$f_X(x) \triangleq \frac{dF_X(x)}{dx} \quad \text{: probability density function} \tag{1.126}$$

The *pdf* is frequency used to characterize a random variable because, compared to the *cdf*, it is easier to describe and visualize the characteristics of the random variable.

*The *pdf* is formally denoted $f_X(x)$ and in the notation $f(x)$ the random variable X is understood by the usage; the notation $p(x)$ is also used to denote the *pdf*; however, these notations are sometimes justified by notational simplicity.

1.5.1.1 Continuous Random Variables A random variable is continuous if the *cdf* is continuous so that $F_X(x)$ can be expressed by the integral of the *pdf*. The mapping in Figure 1.14 results in the continuous real variable x. From (1.125) and (1.126) it follows that

$$P(X \leq x) = F_X(x) = \int_{-\infty}^{x} f_X(x')dx' \tag{1.127}$$

A frequently encountered and simple example of a continuous random variable is characterized by the uniformly distributed *pdf* shown in Figure 1.15 with the corresponding *cdf* and probability function.

From property P7, the probability of $X = x_i$ is evaluated as

$$P(X=x_i) = \lim_{\varepsilon\to 0}(F_X(x_i-\varepsilon)) - F_X(x_i) \tag{1.128}$$

However, for continuous random variables, the limit in (1.128) is equal to $F_X(x_i)$ so $P(X = x_i) = 0$; this event is handled as described in Section 1.5.2.

1.5.1.2 Discrete Random Variables The *probability mass function* [8, 28, 29] (*pmf*) of the discrete random variable X is defined in terms of the discrete probabilities on the real line as

$$p_X(x_i) \triangleq P(X=x_i) \tag{1.129}$$

The corresponding *cdf* is expressed as

$$F_X(x) = \sum_i p_X(x_i)u(x-x_i) \tag{1.130}$$

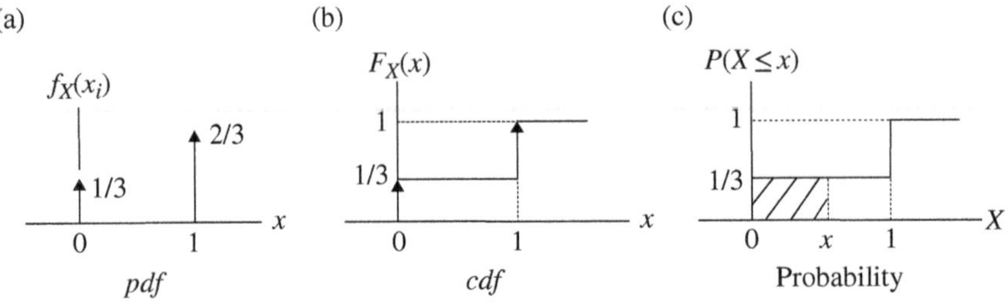

FIGURE 1.16 Discrete binary random variables.

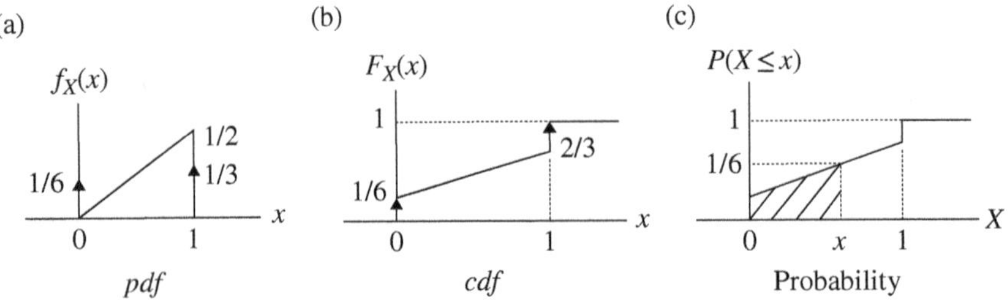

FIGURE 1.17 Mixed random variables.

where $u(x - x_i)$ is the *unit-step function* occurring at $x = x_i$ and is defined as

$$u(x-x_i) \triangleq \begin{cases} 0 & : u < x_i \\ 1 & : u \ge x_i \end{cases} \tag{1.131}$$

Using (1.126), and recognizing that the derivative of $u(x - x_i)$ is the delta function $\delta(x - x_i)$, the *pdf* of the discrete random variable is expressed as

$$f_X(x) = \sum_i p_X(x_i)\delta(x-x_i) \tag{1.132}$$

The *pmf* $p_X(x_i)$ results in a weighted delta function and, from (1.130), (1.131), and property P2, the summation must satisfy the condition $\sum_{\forall i} p_X(x_i) = 1$.

The *pdf, cdf,* and the corresponding probability for the discrete random variable corresponding to binary data {0,1} with *pmf* functions $p_X(0) = 1/3$ and $p_X(1) = 2/3$ are shown in Figure 1.16. The importance of property P5 is evident in Figure 1.16, in that, the delta function at $x = 1$ is included in the *cdf* resulting in $P(X \le 1) = 1$. Regarding property P7, the limit in (1.128) approaches $X = x_i$ from the left, corresponding to the *base* of the discontinuity, so that $P(X = x_i) = p_X(x_i)$.

1.5.1.3 Mixed Random Variables Mixed random variables are composed of continuous and discrete random variables and the following example is a combination of the continuous and discrete random variables in the examples of Sections 1.5.1.1 and 1.5.1.2. The major consideration in this case is the determination of the event *pmf* for the continuous (*C*) and discrete (*D*) random variables to satisfy property P2. Considering equal *pmfs*, such that, $p_X(S = C) = p_X(S = D) = 1/2$, the *pdf, cdf,* and probability are depicted in Figure 1.17.

1.5.2 Definitions and Fundamental Relationships for Continuous Random Variables

For the continuous random variables *X*, such that the events $X(\chi_j) \in S_i$, the joint *cdf* is determined by integrating the joint *pdf* expressed as

$$F_{X_1,\ldots,X_N}(x_1,\ldots,x_N) = \int_{-\infty}^{x_1} \cdots \int_{-\infty}^{x_N} f_{X_1,\ldots,X_N}(x_1'),\ldots,x_N')dx_1',\ldots,dx_N' \tag{1.133}$$

and, provided that $F_{X_1,\ldots,X_N}(x_1,\ldots,x_N)$ is continuous and exists, it follows that

$$f_{X_1,\ldots,X_N}(x_1,\ldots,x_N) = \frac{\partial^N F_{X_1,\ldots,X_N}(x_1,\ldots,x_N)}{\partial x_1 \ldots \partial x_N} \tag{1.134}$$

The probability function is then evaluated by integrating x_i over the appropriate regions $x_{i1} < r_i \le x_{i2}$: $i = 1, \ldots, N$ with the result

$$P(X_{r1},\ldots,X_{rN}) = \int_{r_1} \cdots \int_{r_N} f_{X_1,\ldots,X_N}(x_1',\ldots,x_N')dx_1' \ldots dx_N' \tag{1.135}$$

1.5.2.1 *Marginal pdf* of Continuous Random Variables

The marginal *pdf* is determined by integrating over the entire region of all the random variables except for the desired marginal *pdf*. For example, the marginal *pdf* for x_1 is evaluated as (see Problem 17)

$$\begin{aligned} M_{X_1}(x_1) &= \frac{d}{dx_1}\int_{-\infty}^{x_1}\int_{-\infty}^{\infty}\cdots\int_{-\infty}^{\infty} f_{X_1\cdots X_N}(x_1',\ldots,x_N')dx_1',\ldots,dx_N' \\ &= f_{X_1}(x_1)\int_{-\infty}^{\infty}\cdots\int_{-\infty}^{\infty} f_{X_2\cdots X_N}(x_2',\ldots,x_N')dx_2',\ldots,dx_N' \\ &= f_{X_1}(x_1) \end{aligned} \tag{1.136}$$

The random variables X_i are independent *iff* the joint *cdf* can be expressed as product of the each *cdf*, that is

$$F_{X_1,\ldots,X_N}(x_1,\ldots,x_N) = F_{X_1}(x_1)F_{X_2}(x_2)\cdots F_{X_N}(x_N) \tag{1.137}$$

In addition, if $X_i \ \forall\ i$ are jointly continuous, the random variables are independent if the joint *pdf* can be expressed as the product of each *pdf* as

$$f_{X_1,\ldots,X_N}(x_1,\ldots,x_N) = f_{X_1}(x_1)f_{X_2}(x_2)\cdots f_{X_N}(x_N) \tag{1.138}$$

Therefore, the joint *pdf* of independent random variables is the same as the product of each marginal *pdf* computed sequentially as in (1.136).

The joint *cdf* of two continuous random variables is defined as

$$F_{X,Y}(x,y) \triangleq P(X \le x, Y \le y) \tag{1.139}$$

with the following properties,

$$\left.\begin{aligned} &F_{X,Y}(x,\infty) = F_X(x),\ F_{X,Y}(\infty,y) = F_Y(y),\ F_{X,Y}(\infty,\infty) = 1 \\ &F_{X,Y}(x,-\infty) = 0,\ F_{X,Y}(-\infty,y) = 0 \end{aligned}\right\} \tag{1.140}$$

and the joint *pdf* is defined as

$$f(X,Y) = \frac{\partial^2}{\partial x \partial y}F_{X,Y}(x,y) \tag{1.141}$$

with the following properties,

$$\left.\begin{aligned} &\int_{-\infty}^{\infty}\int_{-\infty}^{\infty} f_{X,Y}(x,y)dxdy = 1,\ \frac{\partial F_{X,Y}(x,y)}{\partial x} = \int_{-\infty}^{y} f_{X,Y}(x,y')dy', \\ &\frac{\partial F_{X,Y}(x,y)}{\partial y} = \int_{-\infty}^{x} f_{X,Y}(x',y)dx' \end{aligned}\right\} \tag{1.142}$$

1.5.2.2 *Conditional pdf* and *cdf* of Continuous Random Variables

The conditional *pdf* is expressed as

$$f_{X_1\ldots X_i}(x_1,\ldots,x_i|x_{i+1},\ldots,x_n) = \frac{f_{X_1\ldots X_n}(x_1,\ldots,x_n)}{f_{X_{i+1}\ldots X_n}(x_{i+1},\ldots,x_n)} \tag{1.143}$$

and the conditional *cdf* is evaluated as

$$F_{X_1\ldots X_i}(x_1,\ldots,x_i|x_{i+1},\ldots,x_n) = \frac{\int_{-\infty}^{x_1}\cdots\int_{-\infty}^{x_i} f_{X_1\ldots X_n}(x_1',\ldots,x_i',x_{i+1},\ldots,x_n)dx_1',\ldots,dx_i'}{f_{X_{i+1}\ldots X_n}(x_{i+1},\ldots,x_n)} \tag{1.144}$$

A *basic rule* for removing random variables from the *left and right* side of the conditional symbol (|) is given by Papoulis [33]. To remove random variables from the *left side* simply integrate each variable x_j from $-\infty$ to ∞: $j \le i$. To remove random variables from the *right side*, for example, x_j and x_k: $i + 1 \le j,k \le n$, multiply by the conditional *pdfs* of x_j and x_k with respect to the remaining variables and integrate x_j and x_k from $-\infty$ to ∞. For example, referring to (1.143) and considering $f_{X1}(x_1|x_2,x_3,x_4)$, eliminating the random variables x_3 and x_4 from the *right side* is evaluated as

$$f_{X_1}(x_1|x_2) = \int_{-\infty}^{\infty}\int_{-\infty}^{\infty} f_{X_1}(x_1|x_2,x_3,x_4)f(x_3,x_4|x_2)dx_3dx_4 \tag{1.145}$$

The conditional probability of $Y \in S_1$ given $X(\chi) = x$ is expressed as

$$P(Y \in S_1|X=x) = \frac{P(Y \in S_1, X=x)}{P(X=x)} \tag{1.146}$$

Since $P(X = x) = 0$ for the continuous random variable X, (1.146) is undefined; however, if X and Y are jointly continuous with continuous joint *cdfs*, as defined in (1.139), then the conditional *cdf* of Y, given X, is defined as

$$F_Y(y|X \le x) \triangleq \frac{P(X \le x, Y \le y)}{P(X \le x)} = \frac{F_{X,Y}(x,y)}{F_X(x)} \tag{1.147}$$

and differentiating (1.147) with respect to y results in

$$\begin{aligned} f_Y(y|X \le x) &= \frac{\partial y F_{X,Y}(x,y)/dy}{F_X(x)} = \frac{\int_{-\infty}^{x} f_{X,Y}(x',y)dx'}{\int_{-\infty}^{\infty}\int_{-\infty}^{x} f_X(x',y)dx'dy} \\ &= \frac{f_{X,Y}(x,y)}{f_X(x)} \end{aligned} \tag{1.148}$$

If $f_X(x) \neq 0$, the conditional *cdf* of y, given $X = x$, is expressed as [34]

$$F_Y(y|X=x) = \frac{\int_{-\infty}^{y} f_{X,Y}(x,y')dy'}{f_X(x)} \quad (1.149)$$

and the corresponding conditional *pdf* is evaluated by differentiating (1.149) with respect to y and is expressed as

$$f_Y(y|X=x) = \frac{f_{X,Y}(x,y)}{f_X(x)} = \frac{f_{X,Y}(x,y)}{\int_{-\infty}^{\infty} f_{X,Y}(x,y)dy} \quad (1.150)$$

If X and Y are independent random variables then $f_{Y,X}(y,x) = f_Y(y)f_X(x)$ and (1.147) and (1.150) become $F_Y(y|x) = F_Y(y)$ and $f_Y(y|x) = f_Y(y)$.

Upon rearranging (1.150), the joint *pdf* of X and Y is expressed as

$$f_{Y,X}(y,x) = f_Y(y|x)f_X(x) \quad (1.151)$$

Considering the probability space $S_1 = S_{Y|X} \cap S_X$, such that $S_X^c \neq \emptyset$, the probability $P(Y \in S_X)$ is determined by the *total probability law* defined as

$$\begin{aligned} P(Y \in S_X) &= \int_{S_X} \left[\int_{S_{Y|X}} f_Y(y'|x)dy' \right] f_X(x)dx \\ &= \int_{S_X} P(Y \in S_Y|x) f_X(x)dx \end{aligned} \quad (1.152)$$

In this case, the subspace S_X can be examined as if it were a total probability space obeying the axioms, corollaries, and properties stated earlier.

1.5.2.3 Expectations of Continuous Random Variables

In general, the k-th moment of the random variable X is defined as the expectation

$$E[X^k] \triangleq \int_{-\infty}^{\infty} x^k f_X(x)dx \quad (1.153)$$

and the k-th central moments are defined as the expectation

$$E\left[(X-m_x)^k\right] \triangleq \int_{-\infty}^{\infty} (x-m_x)^k f_X(x)dx \quad (1.154)$$

The mean value m_x of X is defined as the expectation

$$m_x \triangleq E[X] = \int_{-\infty}^{\infty} x f_X(x)dx \quad (1.155)$$

The second central moment of X is evaluated as

$$\begin{aligned} E\left[(X-m_x)^2\right] &\triangleq \int_{-\infty}^{\infty} (x-m_x)^2 f_X(x)dx \\ &= E\left[x^2\right] - m_x^2 \\ &= Var[x] \end{aligned} \quad (1.156)$$

where $Var[x]$ is the variance of x. An efficient approach in evaluating the k-th moments of a random variable, without performing the integration in (1.153) or (1.155), is based on the *moment theorem* as expressed by the *moment generation function* (1.241) in Section 1.5.6.

The expectation of the function $g(x)$ is evaluated as

$$E[g(X)] = \int_{-\infty}^{\infty} g(x)f_X(x)dx \quad (1.157)$$

and the expectation of the function $g(X,Y)$ of two continuous random variables is

$$E[g(X,Y)] = \int_{-\infty}^{\infty}\int_{-\infty}^{\infty} g(x,y)f_{X,Y}(x,y)dxdy \quad (1.158)$$

The expectation is distributive over summation so that

$$E[X+Y] = E[X] + E[Y] \quad (1.159)$$

and

$$E\left[(X+Y)^2\right] = E\left[X^2\right] + 2E[XY] + E\left[Y^2\right] \quad (1.160)$$

The following relationships between X and Y apply under the indicated conditions:

$$E[XY] = \begin{cases} E[X]E[Y] & : X \text{ and } Y \text{ are uncorrelated} \\ 0 & : X \text{ and } Y \text{ are orthogonal} \end{cases} \quad (1.161)$$

From (1.160) and (1.161) it is seen that if X and Y are uncorrelated random variables they are also orthogonal random variables if the mean of either X or Y is zero. The following example demonstrates that if two jointly Gaussian distributed random variables are orthogonal they are also independent.

The conditional expectation of X given Y is defined as

$$E[X|Y] = \int_{-\infty}^{\infty} x f_X(x|y)dx \quad (1.162)$$

However, if Y is a random variable the function $g_2(Y) = E(X|Y)$ is also a random variable and, using (1.157), the expectation (1.162) becomes

$$E[g_2(Y)] = E[E[X|Y]] = \int_{-\infty}^{\infty} E[x|y]f_Y(y)dy \quad (1.163)$$

Papoulis [35] establishes the basic theorem for the conditional expectation of the function $g(X,Y)$ conditioned on $X = x$, expressed as the random variable $E[g(X,Y)|X = x]$. The theorem is:

$$E[E[g(X,Y)|X]] = E[g(X,Y)] \tag{1.164}$$

with the corollary relationship

$$E[g_1(X)g_2(Y)] = E[E[g_1(X)g_2(Y)|X]] = E[g_1(X)E[g_2(Y)|X]] \tag{1.165}$$

Papoulis refers to (1.165) as a *powerful formula.*

The Bivariate Distribution—An Example of Conditional Distributions Consider that x_1 and x_2 are Gaussian random variables with means m_1, m_2 and variances σ_1, σ_2, respectively, with the joint *pdf* is expressed as [36]

$$f_{x_1,x_2}(x_1,x_2) = \frac{\exp\left[-\dfrac{\sigma_2^2(x_1-m_1)^2 - 2\sigma_1\sigma_2\rho(x_1-m_1)(x_2-m_2) + \sigma_1^2(x_2-m_2)^2}{2\,\sigma_1^2\,\sigma_2^2(1-\rho^2)}\right]}{2\pi\sigma_1\sigma_2\sqrt{1-\rho^2}} \tag{1.166}$$

where ρ is the correlation coefficient, such that, $|\rho| \le 1$, expressed as

$$\rho = \frac{E[(x_1-m_1)(x_2-m_2)]}{\sigma_1\sigma_2} = \frac{E[x_1x_2] - m_1m_2}{\sigma_1\sigma_2} \tag{1.167}$$

Using (1.150), the distribution of x_1 conditioned on x_2 is expressed as

$$\begin{aligned} f_{x_1|x_2}(x_1|x_2) &= \frac{f_{x_1,x_2}(x_1,x_2)}{f_{x_2}(x_2)} \\ &= \frac{1}{\sqrt{2\pi(1-\rho^2)}\sigma_1}\exp\left[-\frac{((x_1-m_1)-\rho(\sigma_1/\sigma_2)(x_2-m_2))^2}{2\,\sigma_1^2(1-\rho^2)}\right] \\ &= \frac{1}{\sqrt{2\pi(1-\rho^2)}\sigma_1}\exp\left[-\frac{((x_1-m_1)-\rho(x_2-m_2))^2}{2\,\sigma_1^2(1-\rho^2)}\right] \quad :\sigma_2=\sigma_1 \end{aligned} \tag{1.168}$$

If x_1 and x_2 are uncorrelated random variables then $E[x_1x_2] = E[x_1]E[x_2]$ and, from (1.167), the correlation coefficient is zero and (1.168) reduces to the Gaussian distribution of x_1 with $f_{x_1|x_2}(x_1|x_2) = f_{x_1}(x_1)\ :\rho=0$. Therefore, two jointly Gaussian distributed random variables are orthogonal and independent if they are uncorrelated.

Referring to (1.165), the first and second conditional moments of the second equality in (1.168) are evaluated using as $E[g_1(X_1)g_2(X_2)]$ and $E\left[g_1^2(X_1)g_2^2(X_2)\right]$, respectively, with $g_1(X_1) = X_1$ and $g_2(X_2) = X_2$ In the evaluation, the conditional mean of the Gaussian distribution is established from (1.168) by observation as

$$E[X_1|X_2=x_2] = \rho\frac{\sigma_1}{\sigma_2}x_2 - \rho\frac{\sigma_1}{\sigma_2}m_2 + m_1 \tag{1.169}$$

and the desired result is evaluated as

$$\begin{aligned} E[X_2E[X_1|X_2=x_2]] &= E\left[x_2\left\{\rho\left(\frac{\sigma_1}{\sigma_2}\right)x_2 - \rho\left(\frac{\sigma_1}{\sigma_2}\right)m_2 + m_1\right\}\right] \\ &= \rho\left(\frac{\sigma_1}{\sigma_2}\right)E\left[x_2^2\right] - \left(\rho\left(\frac{\sigma_1}{\sigma_2}\right)m_2 - m_1\right)E[x_2] \\ &= \rho\sigma_1\sigma_2 + m_1m_2 \end{aligned} \tag{1.170}$$

where $E\left[x_2^2\right] = \sigma_2^2 + m_2^2$ and $E[x_2] = m_2$. The evaluation of $E\left[g_1^2(X_1)g_2^2(X_2)\right]$ is left as an exercise in Problem 12. The evaluation of (1.169) could have been performed using the integration in (1.155); however, it is significantly easier and less prone to error to simply associate the required parameters with the known form of the conditional Gaussian distribution as indicated in (1.168).

With zero-mean random variables X_1 and X_2, that is, when $m_1 = m_2 = 0$, the second equality in (1.168) results in (see Papoulis [37])

$$E[X_1X_2] = \rho\sigma_1\sigma_2 \quad : \text{zero-mean Gaussain conditional pdf} \tag{1.171}$$

and

$$E\left[X_1^2X_2^2\right] = E\left[X_1^2\right]E\left[X_2^2\right] + 2E^2[X_1X_2] \tag{1.172}$$

The time correlated zero-mean, equal-variance Gaussian random variables denoted as x_i and x_{i-1} taken at $t_i = t_{i-1} + \Delta t$ are characterized, using the last equality in (1.168), as

$$f_{X_i|X_{i-1}}(x_i|x_{i-1}) = \frac{1}{\sqrt{2\pi}\sigma_1(1-\rho^2)}\exp\left[-\frac{(x_i-\rho x_{i-1})^2}{2\,\sigma_1^2(1-\rho^2)}\right] \tag{1.173}$$

Equation (1.173) is used to model Gaussian fading channels with the fade duration dependent on Δt and ρ and the fade depth dependent on σ_1.

1.5.3 Definitions and Fundamental Relationships for Discrete Random Variables

In the following relationships, x_i, y_i, x, and y are considered to be discrete random variables corresponding to the event probabilities $P_X(x_i)$, $P_Y(y_i)$, $P_X(x)$, and $P_Y(y)$ with the corresponding *pmfs* $p_Z(z) = P_Z(Z = z) : Z = \{X,Y\}, z = \{x_i, y_i, z, y\}$ corresponding to the amplitude of the discrete delta functions. In general, the characterization of discrete random variables is similar to that of continuous random variables with the integrations replaced by summations and the *pdf* replaced with the *pmf*.

1.5.3.1 *Statistical Independence*

If $X(\chi_i) = x$ with $\chi_i \in S$ and the events χ_i are independent $\forall\, i$, then the joint probabilities are expressed as the product

$$P(X_1 = x_1, \ldots, X_N = x_N) = \prod_{i=1}^{N} P_i(X_i = x_i) \tag{1.174}$$

or, in terms of the *pmf*, $p_X(x_i) = P(X = x_i)$

$$p_{X_1}, \ldots, p_{X_N}(x_1, \ldots, x_N) = \prod_{i=1}^{N} p_{Xi(x_i)} \tag{1.175}$$

If $S = S_1 \cap S_2$ such that $X(\chi_i) = x_i$ with $\chi_i \in S_1$, $Y(\chi_j) = y_j$ with $\chi_j \in S_2$, and the individual *mdfs* satisfy (1.175), then

$$\begin{aligned} P[S] &= \sum_i \sum_j p_{X,Y}(x_i, y_j) = \sum_i \sum_j p_X(x_i) p_Y(y_j) \\ &= \sum_i p_X(x_i) \sum_j p_Y(y_j) \\ &= P(S_1)P(S_2) \end{aligned} \tag{1.176}$$

Therefore, if the joint *pmfs* are independent, X and Y are also independent and, from the last equality in (1.176), S_1 and S_2 are also independent. Consequently, $\{X,Y\}$ are independent *iff* the *pmfs* of X and Y can be expressed in the product form as in (1.175).

The expectation of x is evaluated as

$$E[X] = \int_{-\infty}^{\infty} x \sum_i p_X(x)\delta(x - x_i) = \sum_i x_i p_X(x_i) = \sum_i x_i P(X = x_i) \tag{1.177}$$

For the discrete sampled function $g(X,Y)$, the expectation value is evaluated as

$$E[g(X,Y)] = \sum_{i,j} g(x_i, y_j) p_{XY}(X = x_i, Y = y_j) \tag{1.178}$$

where the *pmf* is expressed as $p_{XY}(x_i, y_j) = P(X = x_i, Y = y_j)$.

1.5.3.2 *Conditional Probability*

The conditional probability of X given $Y = y_j$ is expressed as

$$P(X = x_i | Y = y_j) = \frac{P(X = x_i, Y = y_j)}{P(Y = y_j)} \tag{1.179}$$

and, in terms of the conditional *pmfs*, (1.179) becomes

$$p_X(x_i | y_j) = \frac{p_{X,Y}(x_i, y_j)}{p_Y(y_j)} \tag{1.180}$$

The *pmf* behaves like the *pdf* of continuous random variables, in that, if the event $X(\chi_i) = x_i$ with $\chi_i \in S_1$, the probability of $X \in S_1$ given $Y = y_j$ is evaluated as

$$P(X \in S_1 | Y = y_j) = \sum_i p_X(x_i | y_j) \tag{1.181}$$

If X and Y are independent (1.180) becomes

$$p_X(x_i | y_j) = \frac{p_X(x_i) p_Y(y_j)}{p_Y(y_j)} = p_X(x_i) \tag{1.182}$$

1.5.3.3 *Bayes Rule*

Bayes rule is expressed, in terms of the condition probability, as

$$P(X = x_i | Y = y_j) = \frac{P(Y = y_j | x_i) P(X = x_i)}{P(Y = y_j)} \tag{1.183}$$

and, in terms of probabilities and *pmfs*, Bayes rule is expressed as

$$p_X(x_i | y_j) = \frac{p_Y(y_j | x_i) P(X = x_i)}{P(Y = y_j)} \tag{1.184}$$

The probability state transition diagram is shown in Figure 1.18 for N-dimensional input and output states x_i and y_i, respectively. The outputs are completely defined by the *conditional*, or *transition, probabilities* $P(y_j|x_i)$ and the input *a priori probabilities* $P(x_i)$. Upon choosing the state y_j, that is, given y_j, the *a posteriori probability* $P(x_i|y_j)$ is

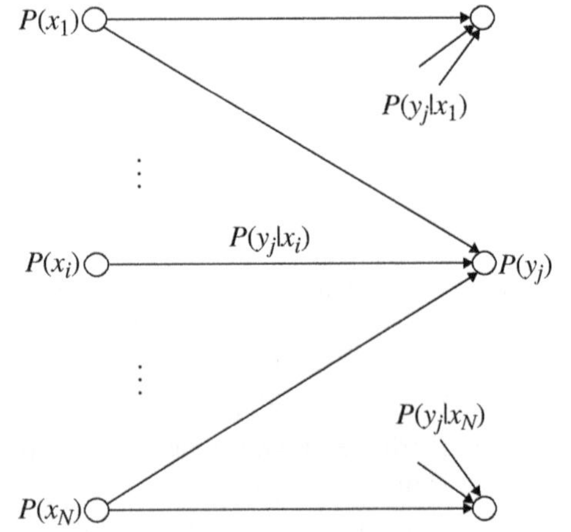

FIGURE 1.18 Probability state transition diagram.

the conditional probability that the input state was x_i. Wozencraft and Jacobs (Reference 30, p. 34) point out that, "The effect of the transmission [decision] is to alter the probability of each possible input from its a priori to its a posteriori value."

The conditional expectation of X given $Y = y$ is

$$E(X|Y) = \sum_{x_i} x_i p_X(x_i|y) \tag{1.185}$$

where the *pmf* $p_X(x_i|y) = P(X = x_i|y)$.

1.5.4 Functions of Random Variables

Applications involving random variables that are functions of random variables, that is, $z = g(x_1, \ldots, x_M)$, require that the density function $f_Z(z)$ be determined given $f_{X_m}(x_m)$: $n = 1, \ldots, M$. In the following subsections, the transformation from $f_{X_m}(x_m)$ to $f_Z(z)$ is discussed for the relatively easy case involving functions of one random variables, that is, $M = 1$. More complicated cases are also discussed involving functions of two random variables and M random variables of the form $Z = \sum_{m=1}^{M} X_m$. The following descriptions involve continuous random variables and cases involving discrete and mixed random variables are discussed in References 6, 8, 29.

1.5.4.1 Functions of One Random Variable In the following description, the mapping of the random variable $X = x$ is continuous and $F_X(x)$ is differentiable at x as in (1.126), with finite values of $f_X(x)$. The transformation from X to Z can be based on the functional relationships $z = g(x)$ or $x = h(z)$ with the requirements that $P(X = \infty) = P(Z = \infty) = 1$ corresponding to unit areas under each transformation. These transformations correspond, respectively, to

$$f_Z(z) = \frac{f_X(x = h(z))}{|dg(x)/dx|_{x=h(z)}} \tag{1.186}$$

and

$$f_Z(z) = f_X(x = h(z)) \left| \frac{dh(z)}{dz} \right| \tag{1.187}$$

Equations (1.186) and (1.187) require the inverse relationship

$$\left| \frac{dh(z)}{dz} \right| = \left| \frac{dg(x)}{dx} \right|^{-1}_{x=h(z)} \tag{1.188}$$

The function $z = h(x)$ typically has a finite number of solutions x_n, corresponding to the roots $z = h(x_1), h(x_2), \ldots, h(x_N)$ of the transformation and, under these conditions, the solution to $f_Z(z)$ given $f_X(x_n)$ is determined using the fundamental theorem [38, 39],

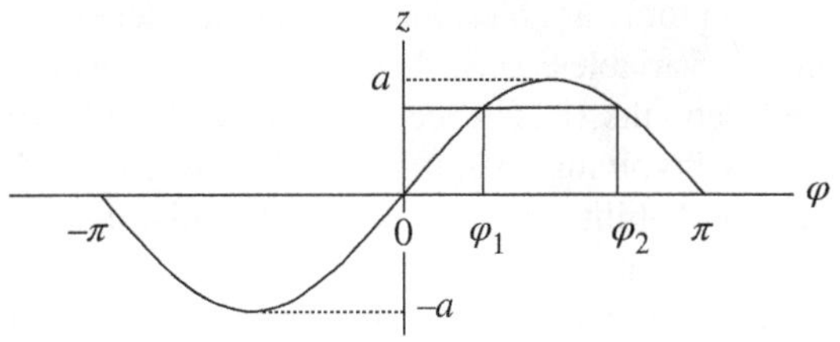

FIGURE 1.19 Random variable $x = a\sin(\varphi)$ ($f_\Phi(\varphi) = 1/(2\pi)$).

$$f_Z(z) = \sum_{n=1}^{N} f_X(x_n = h(z_n)) \left| \frac{dh(z_n)}{dz_n} \right| = \sum_{n=1}^{N} f_X(x_n = h(z_n)) |h'(z_n)| \tag{1.189}$$

where $h(z_n)$ corresponds to the transformation of x_n expressed in terms of z_n and $h'(z_n) = dh(z_n)/dz_n$.

As an example, consider a sinusoidal signal z, with constant amplitude a and random phase φ uniformly distributed between $\pm\pi$, expressed as

$$z = a\sin(\varphi) \tag{1.190}$$

Referring to Figure 1.19, and noting that $z = a\sin(\varphi_1) = a\sin(\varphi_2)$, the problem is to determine the *pdf* $f_Z(z)$ using the two roots of $\varphi_1 = h(z_1)$ and $\varphi_2 = h(z_2)$. Using (1.190), $\varphi = h(z)$ is evaluated as

$$\varphi = h(z) = \sin^{-1}\left(\frac{z}{a}\right) \quad : \varphi = \{\varphi_1, \varphi_2\} \tag{1.191}$$

and

$$h'(z) = \frac{dh(z)}{dz} = \frac{1}{\sqrt{a^2 - z^2}} \quad : \varphi = \{\varphi_1, \varphi_2\} \text{ and } |z| < a \tag{1.192}$$

Therefore, evaluating (1.189) with $f_\Phi(\varphi) = 1/(2\pi)$ results in

$$\begin{aligned} f_Z(z) &= f_\Phi(\varphi_1)|h'(z_1)| + f_\Phi(\varphi_2)|h'(z_2)| \\ &= \frac{1}{2\pi}\left(\frac{1}{\sqrt{a^2 - z^2}} + \frac{1}{\sqrt{a^2 - z^2}}\right) \\ &= \frac{1}{\pi}\frac{1}{\sqrt{a^2 - z^2}} \quad : |z| < a \end{aligned} \tag{1.193}$$

1.5.4.2 Functions of Two or More Random Variables The concepts involving a function of one random variable can also be applied when the random variable Z is a function of several random variables; for example, the dependence on two random variables, such that, $z = g(x,y)$ is discussed at length by Papoulis (Reference 8, Chapters 6 and 7) where the subjects involving marginal distributions, joint density functions, probability masses, conditional distributions and densities, and independence are introduced. According to (1.126), the probability density function $f_Z(z)$ is determined from the distribution function $F_Z(z)$ as

$$f_Z(z) = \frac{\partial F_Z(z)}{\partial z} \tag{1.194}$$

and the joint *pfd* of X and Y is characterized for continuous distributions as

$$f_{XY}(x,y) = \frac{\partial^2 F_{XY}(x,y)}{\partial x \partial y} \tag{1.195}$$

where the joint *cdf* is given by

$$F_{XY}(x,y) = \int_{-\infty}^{y} \int_{-\infty}^{x} f_{XY}(x',y')dx'dy' \tag{1.196}$$

Based on the conditions for the equality of the probabilities, that is,

$$P_Z(Z \le z) = P_{XY}(x, y \in g(x,y) \le z)$$

the *pdfs* are equated as

$$F_Z(z) = F_{XY}(g(x,y) \le z) = \int_{g(x,y) \le z} \int f_{X,Y}(x,y)dxdy \tag{1.197}$$

Upon differentiating (1.197) with respect to z yields the desired result expressed as

$$f_Z(z) = \frac{\partial F_Z(z)}{\partial z} = \frac{\partial}{\partial z} \int_{g(x,y) \le z} \int f(x,y)dxdy \tag{1.198}$$

As an example application consider the random variable $Z = X + Y$; Papoulis states that, "*This is the most important example of a function involving two random variables.*" Upon letting $y = z - x$ and using (1.198) the density function of Z is evaluated as

$$f_Z(z) = \frac{\partial}{\partial z} \int_{-\infty}^{\infty} \int_{-\infty}^{z-x} f(x,y)dxdy = \int_{-\infty}^{\infty} f_{XY}(x, z-x)dx \tag{1.199}$$

and, when X and Y are independent, (1.199) is simply the convolution of $f_X(x)$ with $f_Y(y)$. Several examples involving the use of (1.199) are given in Section 1.5.6.1.

Using the joint probability density function of two continuous random variables x and y, as expressed in (1.195), the marginal *pdfs* $f_X(x)$ and $f_Y(y)$ are obtained by integrating over y and x, respectively, resulting in

$$f_X(x) = \int_{-\infty}^{\infty} f_{XY}(x,y)dy \tag{1.200}$$

and

$$f_Y(y) = \int_{-\infty}^{\infty} f_{XY}(x,y)dx \tag{1.201}$$

These results can also be generalized to apply to the joint density function of any number of continuous random variables by integrating over each of the undesired variables.

1.5.5 Probability Density Functions

The following two subsections examine the probability density function [40] of the magnitude and phase of a sinusoidal signal with additive noise and the probability density function of the product of two zero-mean equal-variance Gaussian distributions. In these cases, the random variables of interest involve functions of two random variables. In Section 1.5.6, the characteristic function is defined and examined for several probability distribution functions demonstrating the central limit theorem with increasing summation of random variables. In Section 1.5.7, many of the probability distributions used in the following chapters are summarized and compared.

1.5.5.1 Distributions of Sinusoidal Signal Magnitude and Phase in Narrowband Additive White Gaussian Noise

This example involves the evaluation of the *pdf* of the magnitude and phase at the output of a narrowband filter when the input is a sinusoidal signal with uniformly distributed phase and zero-mean additive white Gaussian noise [41] (AWGN). In this case, the output of the narrowband filter is a narrowband random process. The evaluation involves three random variables: the input signal phase φ and the two independent-identically distributed (*iid*) zero-mean quadrature noise random variables with variance σ_n^2. The signal plus noise out of the filter is expressed as

$$\begin{aligned} s(t) &= A\cos(\omega_c t + \varphi) + n(t)\cos(\omega_c t + \phi(t)) \\ &= A\cos(\varphi)\cos(\omega_c t) - A\sin(\varphi)\sin(\omega_c t) \\ &\quad + n(t)\cos(\phi(t))\cos(\omega_c t) - n(t)\sin(\phi(t))\sin(\omega_c t) \\ &= (A\cos(\varphi) + n_c(t))\cos(\omega_c t) - (A\sin(\varphi) + n_s(t))\sin(\omega_c t) \end{aligned} \tag{1.202}$$

where the third equality in (1.202) emphasizes the in-phase and quadrature functions of the signal and noise terms and, when sampled at $t = iT_s$, represent the random variables

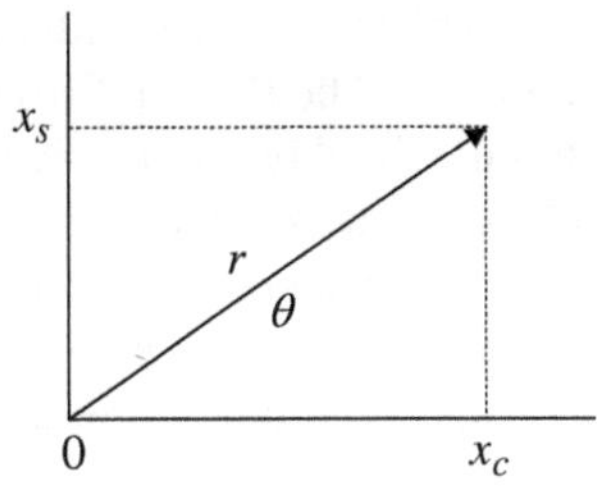

FIGURE 1.20 Relationship between transformation variables.

x_c, n_c, x_s, and n_s. The functional relationships are $x_c = A\cos(\varphi) + n_c$ and $x_s = A\sin(\varphi) + n_s$ with n_c and n_s representing zero-mean quadrature Gaussian random variables. The signal phase, φ, is uniformly distributed between 0 and 2π. Under these conditions, the quadrature signal and noise components x_c and x_s are independent random variables* and the *pdfs* of x_c and x_s are expressed as

$$f_{X_c}(x_c) = \frac{1}{\sqrt{2\pi}\sigma_n} e^{-(x_c - A\cos(\varphi))^2/2\sigma_n^2} \tag{1.203}$$

and

$$f_{X_s}(x_s) = \frac{1}{\sqrt{2\pi}\sigma_n} e^{-(x_s - A\sin(\varphi))^2/2\sigma_n^2} \tag{1.204}$$

The *pdf* of the phase is

$$f_\Phi(\varphi) = \frac{1}{2\pi} \quad : 0 \le \varphi \le 2\pi \tag{1.205}$$

Using (1.203), (1.204), and (1.205) the joint *pdf* is expressed as

$$\begin{aligned} f_{X_c, X_s, \Phi}(x_c, x_s, \varphi) &= f_{X_c}(x_c) f_{X_s}(x_s) f_\Phi(\varphi) \\ &= \frac{1}{4\pi^2\sigma^2} e^{-\frac{(x_c - A\cos(\varphi))^2 + (x_s - A\sin(\varphi))^2}{2\sigma^2}} \end{aligned} \tag{1.206}$$

The evaluation of the joint *pdf* of the magnitude and phase of the sampled sine-wave plus noise involves the transformation of variables from (x_c, x_s) to (r, θ) as depicted in Figure 1.20. The magnitude is described as

$$r = |s(iT_s)| = \sqrt{x_c^2 + x_s^2} \tag{1.207}$$

and the in-phase and quadrature components, x_c and x_s, are described in terms of the angle θ as

$$x_c = r\cos(\theta) \quad \text{and} \quad x_s = r\sin(\theta) \tag{1.208}$$

*Orthogonal Gaussian random variables are also independent.

Expressing the phase angle in (1.208) as a function of x_c and x_s leads to the expressions

$$\theta = \cos^{-1}\left(\frac{x_c}{\sqrt{x_c^2 + x_s^2}}\right) \tag{1.209}$$

and

$$\theta = \sin^{-1}\left(\frac{x_s}{\sqrt{x_c^2 + x_s^2}}\right) \tag{1.210}$$

The Jacobian of the transformation is defined as [6, 8, 28, 29]

$$J(x_c, x_s) \triangleq \begin{vmatrix} \frac{\partial g_{11}(x_c, x_s)}{\partial x_c} & \frac{\partial g_{12}(x_c, x_s)}{\partial x_s} \\ \frac{\partial g_{21}(x_c, x_s)}{\partial x_c} & \frac{\partial g_{22}(x_c, x_s)}{\partial x_s} \end{vmatrix}_{\text{det}} \tag{1.211}$$

and, using the Jacobian, the transformation from (x_c, x_s) to (r, θ) is expressed as

$$f_{R, \Theta, \Phi}(r, \theta, \varphi) = \frac{f_{X_c, X_s, \Phi}(h_1(r, \theta), h_2(r, \theta), \varphi)}{|J(x_c, x_s)|} \tag{1.212}$$

To evaluate the Jacobian for this transformation, the functions $g_{ij}(x_c, x_s)$ are defined in terms of (1.207), (1.209), and (1.210) as follows:

$$g_{11}(x_c, x_s) = g_{12}(x_c, x_s) = \sqrt{x_c^2 + x_s^2} \tag{1.213}$$

$$g_{21}(x_c, x_s) = \cos^{-1}\left(\frac{x_c}{\sqrt{x_c^2 + x_s^2}}\right) \tag{1.214}$$

and

$$g_{22}(x_c, x_s) = \sin^{-1}\left(\frac{x_s}{\sqrt{x_c^2 + x_s^2}}\right) \tag{1.215}$$

Upon evaluating the partial derivatives in (1.211), the Jacobian is found to be†

$$J(x_c, x_s) = \frac{1}{\sqrt{x_c^2 + x_s^2}} = \frac{1}{r} \tag{1.216}$$

†The phase angle can also be expressed as $\theta = \tan^{-1}(x_s/x_c)$ with the Jacobian evaluated as earlier using, $g_{21}(x_s, x_c) = g_{22}(x_s, x_c) = \tan^{-1}(x_s/x_c)$ (see Problem 16).

and, using (1.208), the functions $h_1(r,\theta)$ and $h_2(r,\theta)$ are expressed as

$$h_1(r,\theta) = x_c|_{r\cos(\theta)} = r\cos(\theta) \tag{1.217}$$

and

$$h_2(r,\theta) = x_s|_{r\sin(\theta)} = r\sin(\theta) \tag{1.218}$$

Substituting (1.216), (1.217), and (1.218) into (1.212) and applying the independence of x_c, x_s, and φ, as in (1.206), the *pdf* of the transformed variables r and θ is expressed as

$$\begin{aligned} f_{R,\Theta,\Phi}(r,\theta,\varphi) &= f_R(r) f_\Theta(\theta) f_\Phi(\varphi) \\ &= \frac{r}{4\pi^2\sigma^2} e^{-\frac{(r\cos(\theta)-A\cos(\varphi))^2+(r\sin(\theta)-A\sin(\varphi))^2}{2\sigma^2}} \\ &= \frac{r}{4\pi^2\sigma^2} e^{-\frac{r^2+A^2-2Ar\cos(\theta-\varphi)}{2\sigma^2}} \end{aligned} \tag{1.219}$$

where $r \ge 0$, otherwise the *pfd* is zero, and θ and φ are uniformly distributed over the range $0 \le \theta$, $\varphi \le 2\pi$. The *pdf* for the magnitude r is determined by computing the marginal distribution $M_R(r)$ by integrating over the ranges of θ and φ. Defining $\psi = \theta - \varphi$, the marginal is evaluated as

$$f_R(r) = \frac{r}{\sigma^2} e^{-\frac{r^2+A^2}{2\sigma^2}} \frac{1}{2\pi} \int_0^{2\pi} \left(\frac{1}{2\pi} \int_{\varphi}^{2\pi-\varphi} e^{Ar\cos(\psi)/\sigma^2} d\psi \right) d\varphi \tag{1.220}$$

Davenport and Root [42] point out that the integrand of the bracketed integral is periodic in the uniformly distributed phase ψ and can be integrated over the interval 0 to 2π. With this integration range, the bracketed integral is identified as the zero-order modified Bessel function expressed as [43]

$$I_o\left(\frac{Ar}{\sigma^2}\right) = \frac{1}{2\pi} \int_0^{2\pi} e^{Ar\cos(\psi)/\sigma^2} d\psi \tag{1.221}$$

Therefore, upon using (1.221) and performing the integration over φ, the marginal distribution function $M_R(r)$ simplifies, at least in notation, to

$$f_R(r) = \frac{r}{\sigma^2} e^{-\frac{r^2+A^2}{2\sigma^2}} I_o\left(\frac{Ar}{\sigma^2}\right) \tag{1.222}$$

Equation (1.222) is the Rice distribution or, as referred to throughout this book, the Ricean distribution that, as developed in the forgoing analysis, characterizes the baseband magnitude distribution of a CW signal with narrowband additive white Gaussian noise. The Ricean distribution also characterizes the magnitude distribution of a received signal from a channel with multipath interference; this channel is referred to as a Ricean fading channel. The Ricean distribution becomes the Rayleigh distribution as $A \to 0$ and the Gaussian distribution as $A \to \infty$; the proof of these two limits is the subject of Problems 19 and 20. The Rayleigh distribution characterizes the amplitude distribution of narrowband noise or, in the case of multipath interference, the composite signal magnitude of many random scatter returns without a dominant specular return or signal component. The multipath interference is the subject of Chapter 18. Defining the signal-to-noise ratio as $\gamma = A^2/(2\sigma^2)$, (1.222) is expressed as

$$f_R(r) = \frac{r}{\sigma^2} e^{-(r^2/2\sigma^2+\gamma)} I_o\left(\frac{r\sqrt{2\gamma}}{\sigma}\right) \tag{1.223}$$

The *pdf* of the phase function is evaluated by computing the marginal distribution $M_{\Theta\Phi}(\theta,\varphi)$ by integrating over the range of the magnitude r. By forming or completing the square of the exponent in the last equality in (1.219) the integration is performed as

$$f_{\Theta\Phi}(\theta,\varphi) = \frac{1}{4\pi^2\sigma^2} e^{-A\sin^2(\theta-\varphi)/2\sigma^2} \int_0^\infty r e^{-(r-A\cos(\theta-\varphi))^2/2\sigma^2} dr \tag{1.224}$$

Davenport and Root [44] provide an approximate solution to (1.224), under the condition $A\cos(\theta-\varphi) >> \sigma$. The approximation is expressed as

$$f_{\Theta\Phi}(\theta,\varphi) \cong \frac{A\cos(\theta-\varphi)}{(2\pi)^{3/2}\sigma} e^{-\gamma\sin^2(\theta-\varphi)} \quad : 0 \le \theta, \varphi \le 2\pi \tag{1.225}$$

where γ is the signal-to-noise ratio defined earlier. An alternate solution, without the earlier restriction, is expressed by Hancock [45], with $\psi = \theta - \varphi$, as

$$f_\Psi(\psi) = \frac{e^{-\gamma}}{2\pi}\left(1 + \sqrt{4\pi\gamma}\cos(\psi) e^{\gamma\cos^2(\psi)} P\left(\sqrt{2\gamma}\cos(\psi)\right)\right) \quad : -\pi \le \psi \le \pi \tag{1.226}$$

where $P(z)$ is the probability integral defined in Section 3.5. Hancock's phase function is used in Section 4.2.1 to characterize the performance of phase-modulated waveforms.

As $\gamma \to 0$ in (1.226) the function $f_\Psi(\psi) \to 1/2\pi$ resulting in the uniform phase *pdf*. However, for γ greater than about 3, the probability integral is approximated as [26]

$$P(z) \cong 1 - \frac{e^{-z^2/2}}{\sqrt{2\pi}z} \quad : z > 3 \tag{1.227}$$

Using (1.227), the phase *pdf* is approximated as

$$f_\Psi(\psi) \cong \begin{cases} \sqrt{\gamma/\pi}\cos(\psi)e^{-\gamma\sin^2(\psi)} & : 1 > \cos(\psi) > 2.5/\sqrt{\gamma} \\ 0 & : -2.5/\sqrt{\gamma} > \cos(\psi) > -1 \end{cases} \quad (1.228)$$

With $|\psi| \cong 0$ such that $\sin^2(\psi) \cong \psi^2$ and defining $\gamma = 1/2\sigma_\psi^2$ (1.228) is approximated as

$$f_\Psi(\psi) \cong \frac{1}{\sqrt{2\pi}\sigma_\psi} e^{-\psi^2/2\sigma_\psi^2} \quad : |\psi| \cong 0 \text{ and } \gamma = 1/2\sigma_\psi^2 \quad (1.229)$$

Equation (1.229) describes a zero-mean Gaussian phase *pdf* with the phase variance $\sigma_\psi^2 = 1/2\gamma$. Hancock's phase function, expressed in (1.226), is plotted in Figure 4.3 for various signal-to-noise ratios.

1.5.5.2 Distribution of the Product of Two Independent Gaussian Random Variables In this section the *pdf* of the product $z = xy$ of two zero-mean equal-variance *iid* Gaussian random variables X and Y is determined. The solution involves defining an auxiliary random variable $w = h(x) = x$ with $z = g(x,y) = xy$ and evaluating $f_{Z,W}(w,z)$ characterized as

$$f_{W,Z}(w,z) = \frac{f_{X,Y}(w,z/w)}{|J_{X,Y}(x,y)|} \quad (1.230)$$

where $J_{X,Y}(x,y)$ is the Jacobian of the transformation evaluated as

$$J_{X,Y}(x,y) = \begin{vmatrix} \dfrac{\partial g(x,y)}{\partial x} & \dfrac{\partial g(x,y)}{\partial y} \\ \dfrac{\partial h(x,y)}{\partial x} & \dfrac{\partial h(x,y)}{\partial y} \end{vmatrix}_{\text{det}} = -x \quad (1.231)$$

Using (1.231) and the joint Gaussian *pfd* of X and Y, expressed by (1.230), with $x = w$ and $y = z/w$, the marginal *pdf* of z is evaluated as

$$f_Z(z) = \int_{-\infty}^{\infty} \frac{1}{|w|} f_{X,Y}(w,z/w)dw \quad (1.232)$$

However, since X and Y are independent

$$f_{X,Y}(x,y) = f_X(x)f_Y(y) = \frac{1}{2\pi\sigma^2} e^{-\frac{(x^2+y^2)}{2\sigma^2}} \quad (1.233)$$

and, upon substituting $x = w$ and $y = z/w$ into (1.233), (1.232) is expressed as

$$\begin{aligned} f_Z(z) &= \frac{1}{2\pi\sigma^2}\int_{-\infty}^{\infty} \frac{1}{|w|} e^{-\frac{w^2+(z/w)^2}{2\sigma^2}} dw \\ &= \frac{1}{\pi\sigma^2}\int_0^{\infty} \frac{1}{w} e^{-\frac{w^2+(z/w)^2}{2\sigma^2}} dw \end{aligned} \quad (1.234)$$

where the second equality recognizes that the first equality is symmetrical in w. Letting $\lambda = w^2/2\sigma^2$ (1.234) is expressed as

$$f_Z(z) = \frac{1}{2\pi\sigma^2}\int_0^{\infty} \frac{1}{\lambda} e^{-\left(\lambda + \frac{z^2}{4\sigma^4\lambda}\right)} d\lambda \quad (1.235)$$

The solution to the integral in (1.235) appears in the table of integrals by Gradshteyn and Ryzhik (Reference 46, p. 340, pair No. 12) as

$$\int_0^{\infty} \lambda^{\nu-1} e^{-\left(\lambda + \frac{u^2}{4\lambda}\right)} d\lambda = 2\left(\frac{u}{2}\right)^{\nu} K_{-\nu}(u) \quad (1.236)$$

where $K_\nu(u)$ is the modified Bessel function of the second kind of order ν. With $\nu = 0$ and $u = z/\sigma^2$, (1.235) is evaluated as

$$f_Z(z) = \frac{1}{\pi\sigma^2} K_o\left(\frac{|z|}{\sigma^2}\right) \quad (1.237)$$

The magnitude of z in (1.237) is used because of the even symmetry of $f_Z(z)$ with respect to z. The symmetry of $f_Z(z)$ results in a zero-mean value so the variance is evaluated as

$$\begin{aligned} Var[z] &= E\left[z^2\right] \quad : \text{zero-mean} \\ &= \frac{2}{\pi\sigma^2}\int_0^{\infty} z^2 K_o\left(\frac{z}{\sigma^2}\right)dz \end{aligned} \quad (1.238)$$

The solution to the integral in (1.238) is found in Gradshteyn and Ryzhik (Reference 46, p. 684, Integral No. 16) and the variance $f_Z(z)$ is evaluated as

$$Var[z] = \frac{4}{\pi}\Gamma^2\left(\frac{3}{2}\right)\sigma^4 = \sigma^4 \quad (1.239)$$

where the second equality in (1.239) results from the value of the Gamma function $\Gamma(3/2) = \sqrt{\pi}/2$. In Example 4 of Section 1.5.6.1, the *pdf* of the summation of N *iid* random variables with *pdfs* expressed by (1.237) is examined.

1.5.6 The Characteristic Function

The characteristic function of the random variable X is defined as the average value of $e^{j\nu x}$ and is expressed as

$$C_X(\nu) \triangleq E\left[e^{j\nu x}\right] = \int_{-\infty}^{\infty} f_X(x)e^{j\nu x}dx \quad (1.240)$$

With $v = -\omega$ and $x = t$ (1.240) is similar to the Fourier transform of a time-domain function. The characteristic function is also referred to as the moment-generating function, in that, the nth moment of the random variable X, defined as the expected value $E[x^n]$, is evaluated (see Problem 26) as

$$E[x^n] = (-j)^n \frac{d^n C_X(v)}{dv^n}\bigg|_{v=0} \tag{1.241}$$

The Fourier transform relationship between time domain convolution and frequency domain multiplication also applies to the convolution of random variables and the multiplication of the corresponding characteristic functions. Therefore, based on the discussion in Section 1.5.6.1, the summation of N identically distributed (*id*) random variables corresponds to the product of their individual characteristic functions, that is,

$$C_Z(v) = \prod_{i=1}^{N} C_{X_i}(v) \quad : Z = \sum X_i, \; i = 1, \ldots, N; \; X_i \in id \tag{1.242}$$

This is a very useful result, in that, the distribution of the summation of N independent random variables is obtained as the inverse transform [47] of (1.242) expressed as

$$f_Z(z) = \int_{-\infty}^{\infty} C_Z(v) e^{-jvz} dz \tag{1.243}$$

Campbell and Foster [47] provide an extensive listing of Fourier transform pairs defined as

$$G(g) = \int_{-\infty}^{\infty} F(f) e^{j2\pi fg} df \quad \text{and} \quad F(f) = \int_{-\infty}^{\infty} G(g) e^{-j2\pi fg} dg \tag{1.244}$$

and, by defining $v = -2\pi f$, the Fourier transform pairs apply to the transform pairs between $f_X(x)$ and $C_X(v)$ as expressed in (1.240).

1.5.6.1 Summation of Independently Distributed Random Variables If two random variables X and Y are independent then the probability density $f_Z(z)$ of their sum $Z = X + Y$ is determined from the convolution of $f_X(x)$ with $f_Y(y)$ so that*

$$f_Z(z) = \int_{-\infty}^{\infty} f_X(z-y) f_Y(y) dy = \int_{-\infty}^{\infty} f_X(x) f_Y(z-x) dx \tag{1.245}$$

For multiple summations of a random variable, the convolution is repeated for each random variable in the summation.

*For proof see Reference 8, p. 189.

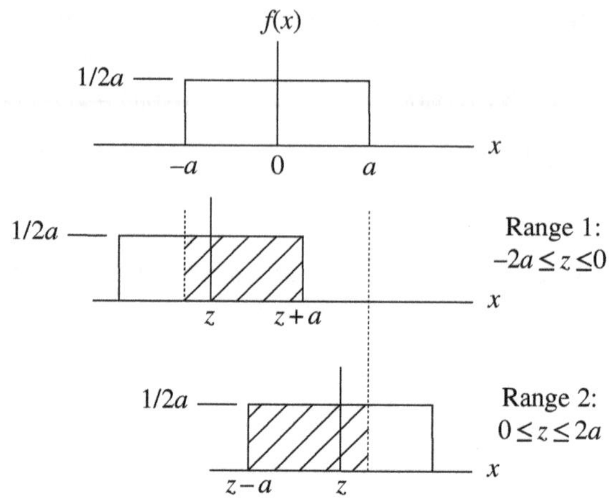

FIGURE 1.21 Convolution of two zero-mean uniform distributions.

Example 1 Consider the summation of N zero-mean uniformly distributed random variables X_i expressed as

$$Z = \sum_{i=1}^{N} X_i \tag{1.246}$$

with

$$f_X(x) = \frac{1}{2a} \quad : -a \le x \le a \tag{1.247}$$

For $N = 2$ the convolution involves two ranges of the variable z as shown in Figure 1.21 and the integrations are evaluated as

$$f_Z(z) = \frac{1}{4a^2} \int_{-a}^{z+a} dx \quad : -2a \le z \le 0 \tag{1.248}$$

and

$$f_Z(z) = \frac{1}{4a^2} \int_{z-a}^{a} dx \quad : 0 \le z \le 2a \tag{1.249}$$

Upon evaluation of (1.248) and (1.249) and recognizing the symmetry about z the density function is expressed as

$$f_Z(z) = \frac{1}{2a}\left(-\frac{|z|}{2a} + 1\right) \quad : |z| \le 2a \tag{1.250}$$

Repeating the application of the convolution for $N = 3$ and 4 (see Problem 24) results in the probability density functions shown in Figure 1.22 with the corresponding *cdf* results shown in Figure 1.23. As $N \rightarrow \infty$ the probability density and characteristic functions will approach those of the Gaussian distributed random variable (see Problem 23).

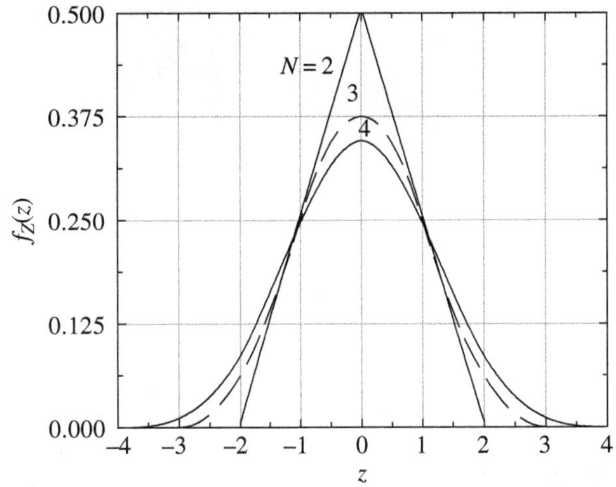

FIGURE 1.22 *pdf* for sum of N = 2, 3 and 4 independent zero-mean uniform distributions (a = 1).

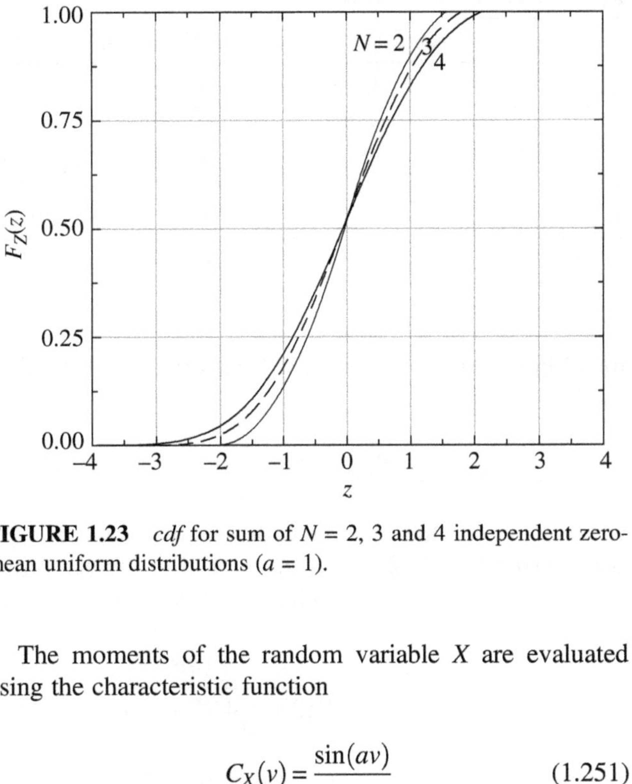

FIGURE 1.23 *cdf* for sum of N = 2, 3 and 4 independent zero-mean uniform distributions (a = 1).

The moments of the random variable X are evaluated using the characteristic function

$$C_X(v) = \frac{\sin(av)}{av} \tag{1.251}$$

In regions where the characteristic function converges, the moments $E[x^n]$ completely define the characteristic function and the *pdf* of the random variable X, so, upon expanding (1.251) as the power series

$$C_X(v) = 1 - \frac{(av)^2}{3!} + \frac{(av)^4}{5!} - \cdots \tag{1.252}$$

TABLE 1.6 Moments of $f_Z(z_N)$ for $Z = \sum X_i$ $: i = 1, \ldots, 4$, X_i *iid* Zero-Mean Uniform Distributions

N	$E[z]$	$E[z^2]$
1	0	$a^2/3$
2	0	$2a^2/3$
3	0	a^2
4	0	$4a^2/3$

the moments are easily evaluated using (1.241). The moments for the random variable Z, formed as in (1.246), are determined using (1.242) and, with $X_i : i = 1, \ldots, N$ *iid* random variables, the characteristic function for Z is approximated as

$$C_Z(v) = C_X^N(v) \cong 1 - \frac{N(av)^2}{3!} \tag{1.253}$$

The first and second moments for $N = 1, \ldots, 4$ are listed in Table 1.6. These results are also obtained by evaluating $f_Z(z)$ using (1.250) and then evaluating the moments (see Problem 25) as

$$E[z^n] = \int_{-Na}^{Na} z^n f_Z(z)dz \tag{1.254}$$

However, it is much easier to use the characteristic function.

Example 2 As another example, consider the summation of N random variables X_i characterized as the sinusoidal function

$$X_i = A_i \sin(\Phi_i) \tag{1.255}$$

with constant amplitudes A_i and zero-mean uniformly distributed phase, expressed as

$$f_\Phi(\varphi) = \frac{1}{2\phi} \quad : |\varphi| \le \phi \tag{1.256}$$

The resulting *pdf* of the random variable X_i for $\phi = \pi$, is evaluated in (1.193) as

$$f_X(x_i) = \frac{1}{\pi\sqrt{A_i^2 - x_i^2}} \tag{1.257}$$

and is plotted in Figure 1.24.

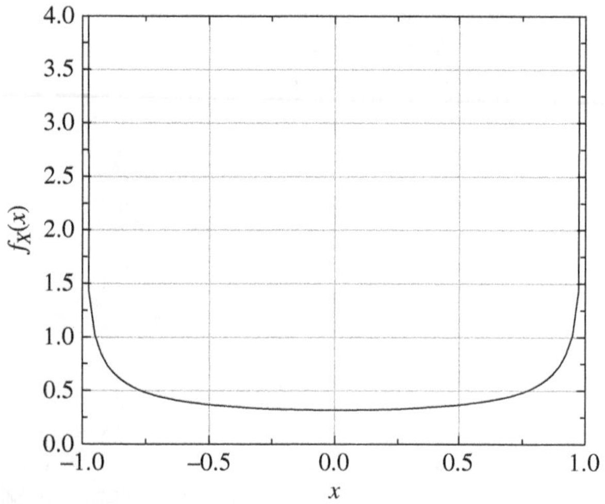

FIGURE 1.24 *pdf* of $x = A\sin(\varphi)$ with zero-mean uniformly distributed phase, $A = 1$ and $\phi = \pi$.

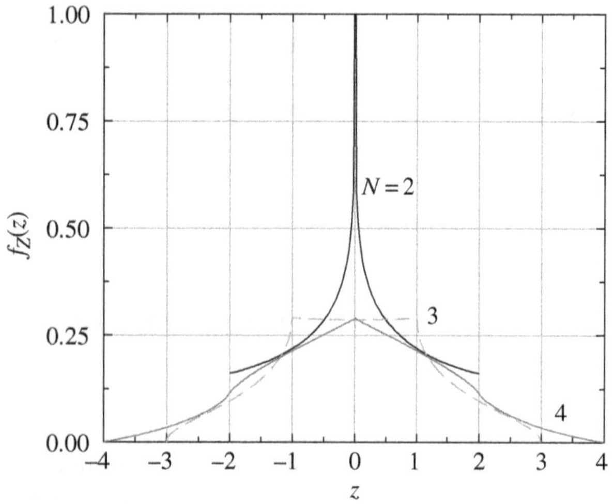

FIGURE 1.25 *pdf* of N = 2, 3, and 4 successive convolutions of $f_X(x)$.

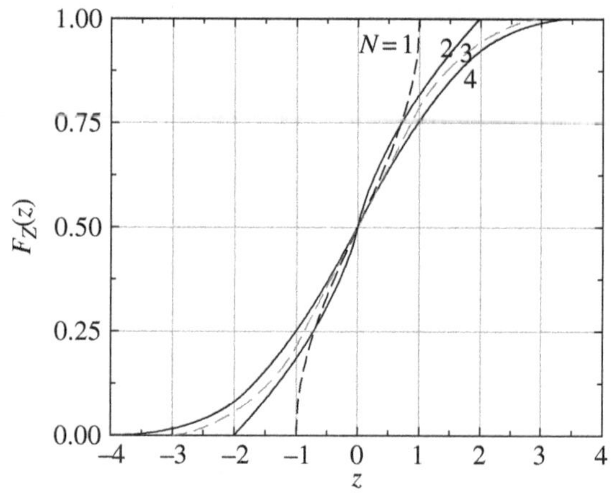

FIGURE 1.26 *cdf* of N = 2, 3, and 4 successive convolutions of $f_X(x)$.

TABLE 1.7 Moments of $f_Z(z_N)$ for $Z = \sum X_i \; : i = 1, \ldots, N$, X_i *iid* Random Variables Expressed by (1.255)

	Theoretical		Numerical[a] (A = 1)	
N	$E[z]$	$E[z^2]$	$E[z]$	$E[z^2]$
1	0	$A^2/2$	0	0.4999
2	0	A^2	0	1.0044
3	0	$3A^2/2$	0	1.5055
4	0	$2A^2$	0	2.0146

[a]Numerical values are sampled with $\Delta z = 2.5 \times 10^{-5}$.

The *pdf* of the random variable Z, expressed as in (1.246),* is evaluated by successive convolutions as in (1.245) and the results for N = 2, 3, and 4 are plotted in Figure 1.25 with the corresponding *cdf* functions shown in Figure 1.26. The results in Figures 1.25 and 1.26 for $N > 1$ are obtained by numerical evaluations of the convolutions using incremental values of $\Delta z = 2.5 \times 10^{-5}$; this is a reasonable compromise between simulation time and fidelity in dealing with the infinite value at $|x| = 1.0$.

In this case, the mean and variance of the random variable X are evaluated using the characteristic function of (1.257) found in (Reference 47, p. 123, Transform Pair 914.5); the result is

$$C_X(v) = I_o(-jAv) \tag{1.258}$$

where $I_o(-)$ is the modified Bessel function of order zero. Expanding (1.258) for $Av < 1$ as a power series, (Reference 46, p. 375, Ascending Series 9.6.10), results in†

$$C_X(v) = 1 - \frac{(Av)^2}{4(1!)^2} + \frac{(Av)^4}{4^2(2!)^2} - \frac{(Av)^6}{4^3(3!)^2} \cdots \tag{1.259}$$

and the moments are easily evaluated using (1.241). The first and second moments are listed as the theoretical values in Table 1.7. The moments for the random variable Z, formed as in (1.246) with X_i *iid* random variables for all i as expressed by the *pdf* in (1.255), are determined using the characteristic function expressed as

*By forming the average summation $Z = (1/N)\sum_{i=1}^{N} x_i$, the range of the *pdf* is limited to $\pm a$ with an associated decrease in the standard deviation.

†By comparing the ascending series expansion of $I_o(z)$ with that of the Bessel function $J_o(z)$ it is found that $I_o(-jz) = J_o(z)$.

$$C_Z(v) = C_X^N(v) \cong 1 - \frac{N(Av)^2}{4(1!)^2} \quad (1.260)$$

The corresponding first two moments of the random variable Z for N = 2, 3, and 4 are also listed in Table 1.7. The numerical results listed in Table 1.7 are based on computer evaluations of the various convolutions resulting in the *pdfs* shown in Figures 1.24 and 1.25.

A major observation in these two examples is that the probability distribution of the random variable Z approaches a Gaussian distribution as N increases (see Problem 27). This is evidence of the central limit theorem which states that (see Davenport and Root, Reference 6, p. 81) *the sample mean of the sum of N arbitrarily distributed statistically independent samples becomes normally distributed as N increases*. This is referred to the equal-components case of the central limit theorem. However, as pointed out by Papoulis (Reference 8, p. 266), a consequence of the central limit theorem is that *the distribution $f_Z(z)$ of the sum of N statistically independent distributions having arbitrary pdf's tends to a normal distribution as N increases*. This is a stronger statement and suggests that the probability $P(z) = f_Z(Z < z)$ can be considered a Gaussian distribution for all z as is frequency assumed to be the case in practice. Davenport and Root also point out that, even though N is seemingly large, the tails of the resulting distribution may result in a *poor* approximation to the Gaussian distribution.

Upon computing the mean and variance using the power series expansion of $C_Z(v)$ expressed by (1.252) with $av << 1$, the approximate expression for the corresponding Gaussian distribution is easily obtained. After summing N uniformly distributed amplitudes the expression for the *pdf* is

$$f_Z(z) = \frac{1}{\sqrt{2\pi Na^2/3}} e^{-z^2/(2Na^2/3)} \quad : z = \sum x_i, \; i = 1, \ldots, N,$$
$$p(x_i) = \frac{1}{2a}, \quad |x_i| \le a \quad (1.261)$$

Similarly, for the summation of N sinusoids with $Av << 1$, the *pdf* in Example 2 is expressed as

$$f_Z(z) = \frac{1}{\sqrt{2\pi NA^2/2}} e^{-z^2/(2NA^2/2)}$$
$$: z = \sum x_i, \; i = 1, \ldots, N, \; x_i = A\sin(\varphi_i) \quad (1.262)$$
$$p(\varphi_i) = \frac{1}{2\pi}, \quad |\varphi_i| \le \pi$$

It is interesting to note that the second moments are $N\lambda^2$ for all values of N including those for which the *pdf* does not have the slightest resemblance to the Gaussian *pdf*. In these cases, the important difference is that the corresponding probabilities $P(x) = F_X(X < x)$ are entirely different from those of the Gaussian distribution with the possible exception of the median value. Finally, it is noted that the limiting behavior for $\lambda v << 1$ and $N \to \infty$ applies to the summation of *independently distributed* distributions that may, or may not, be *identically distributed* distributions.

Example 3 This example involves the summation of random chips {±1} in a direct-sequence spread-spectrum (DSSS) waveform. In this case, the chips occur with equal probabilities according to the *pdf* expressed as

$$f_X(x) = \frac{1}{2}\delta(x-1) + \frac{1}{2}\delta(x+1) \quad (1.263)$$

Using (1.240), the characteristic function is evaluated as

$$\begin{aligned} C_X(v) &= \int_{-\infty}^{\infty} f_X(x) e^{jvx} dx \\ &= \frac{1}{2}\left(\int_{-\infty}^{\infty} e^{jvx}\delta(x-1)dx + \int_{-\infty}^{\infty} e^{jvx}\delta(x+1)dx\right) \\ &= \cos(v) \end{aligned} \quad (1.264)$$

The DSSS waveform uses N chips per bit and the demodulator correlation sums the N chips to form the correlation output $y = \sum_{n=0}^{N-1} x_n$ with the corresponding characteristic function given by

$$\begin{aligned} C_Y(v) &= C_X^N(x) = \cos^N(v) \\ &= \left(1 - \frac{v^2}{2!} + \frac{v^4}{4!} - \frac{v^6}{6!} + \cdots\right)^N \\ &= 1 - \frac{Nv^2}{2!} + \cdots \end{aligned} \quad (1.265)$$

To evaluate the first and second moments of y only the first two terms in the expansion of $\cos^N(v)$ are required and, upon using (1.241), these moments are evaluated as

$$E[x] = j\frac{2Nv}{2!}\bigg|_{v=0} = 0 \quad : \text{first moment (mean value, } \bar{x}) \quad (1.266)$$

and

$$E[x^2] = (-1)\frac{-2N}{2!}\bigg|_{v=0} = N \quad : \text{second moment} \quad (1.267)$$

The variance of y is defined as the second central moment $E\left[(x-\bar{x})^2\right] = E[x^2]-\bar{x}^2$ and with zero-mean the variance is $\sigma_y^2 = E[x^2] = N$.

Example 4 The *pdf* $f_Z(z)$ of the product of two, zero-mean, equal-variance, *iid* Gaussian random variables, $z = xy$, is expressed in (1.237) as a function of the zero-order modified Bessel function $K_o(|z|/\sigma^2)$ where the magnitude of z provides for the range: $-\infty \le z \le \infty$. In this example, the *pdf* $f_{\bar{Z}}(\bar{z})$ is evaluated where $\bar{Z} = \sum_i Z_i \, : \, i = 1, \ldots, N$. The evaluation is based on the N-th power of the characteristic function $C_Z(v)$ and, from the work of Campbell and Foster (Reference 47, p. 60, pair No. 558), the characteristic function is evaluated as*

$$C_Z(v) = \frac{1/\sigma^2}{\sqrt{1/\sigma^4 - p^2}} \tag{1.268}$$

The characteristic function of $\bar{Z}$ is the N-th power of (1.268) expressed as

$$C_{\bar{Z}}^N(v) = \frac{1/\sigma^{2N}}{(1/\sigma^4 - p^2)^{N/2}} \tag{1.269}$$

and, using the transform pair of Campbell and Foster (Reference 47, p. 61, pair No. 569.0), the *pdf* of $\bar{Z}$ is evaluated as

$$f_{\bar{Z}}(\bar{z}) = \frac{1/\sigma^{2N}}{\sqrt{\pi}\Gamma(N/2)(2/\sigma^2)^{(N-1)/2}} \left\{ |\bar{z}|^{(N-1)/2} K_{(N-1)/2}\left(\frac{|\bar{z}|}{\sigma^2}\right) \right\} \tag{1.270}$$

As in the case for $f_Z(z)$, the *pdf* $f_{\bar{Z}}(\bar{z})$ applies for $-\infty \le \bar{z} \le \infty$ and is symmetrical with respect to $\bar{z}$ resulting in a zero-mean value with the variance expressed as

$$Var[\bar{z}] = E\left[\bar{z}^2\right] \quad \text{: with zero-mean}$$

$$= \frac{2/\sigma^{2N}}{\sqrt{\pi}\Gamma(N/2)(2/\sigma^2)^{(N-1)/2}} \int_0^\infty \bar{z}^2 \bar{z}^{(N-1)/2} K_{(N-1)/2}\left(\frac{\bar{z}}{\sigma^2}\right) dz \tag{1.271}$$

The solution to the integral in (1.271) is found in Gradshteyn and Ryzhik (Reference 46, p. 684, Integral No. 16) and the variance $f_Z(z)$ is evaluated using

$$\int_0^\infty \bar{z}^u K_v(a\bar{z}) d\bar{z} = \frac{2^{u-1}}{a^{u+1}} \Gamma\left(\frac{1+u+v}{2}\right) \Gamma\left(\frac{1+u-v}{2}\right) \tag{1.272}$$

*The tables of Campbell and Foster are Fourier transform pairs that correspond to characteristic function pairs with $p = -jv$.

Substituting the solution to the integral in (1.272) into (1.271), with $u = (N + 3)/2$, $v = (N - 1)/2$, and $a = 1/\sigma^2$, the solution to variance simplifies to

$$Var[\bar{z}] = N\sigma^4 \tag{1.273}$$

In the earlier evaluation, the integer argument Gamma function is related to the factorial as $\Gamma(N) = (N-1)!$ and $\Gamma(3/2) = \sqrt{\pi}/2$. This result could also be evaluated using the movement generating function of (1.241), however, using the integral solution as in (1.272) it is sometimes easier to evaluate the moments. With a sufficiently large value of N the *pdf* $f_{\bar{Z}}(\bar{z})$ is approximated as the Gaussian *pfd* expressed as

$$f_{\bar{Z}}(\bar{z}) \underset{N\to\infty}{\sim} \frac{1}{\sqrt{2\pi N\sigma^4}} e^{-\bar{z}^2/(2N\sigma^4)} \quad \text{: zero-mean Gaussian pdf} \tag{1.274}$$

The probability density functions discussed earlier and others encountered in the following chapters are summarized in Table 1.8 with the corresponding mean values, variances, and characteristic functions.

1.5.7 Relationships between Distributions

In the following two subsections, the relationship between various probability density functions is examined by straightforward parameter transformations, allowing parameters to approach limits, or simply altering various parameter values. The most notable relationship is based on the central limit theorem in which a distribution approaches the Gaussian distribution by increasingly summing the operative random variable.

1.5.7.1 Relationship between Chi-Square, Gaussian, Rayleigh, and Ricean Distributions A random variable has a chi-square (χ^2) distribution with N degrees of freedom if it has the same distribution as the sum of the squares of N-independent, normally distributed random variables, each with zero-mean and unit variance.†

Consider the zero-mean Gaussian or normal distributed random variable x with variance σ_x^2 and *pdf* expressed as

$$p_X(x) = \frac{1}{\sqrt{2\pi}\sigma_x} e^{-x^2/2\sigma_x^2} \tag{1.275}$$

The *pdf* of a new random variable $y = x^2$, obtained by simply squaring x, is determined by considering the positive and negative regions of $x = \pm\sqrt{y}$ as shown in Figure 1.27.

†A normally distributed random variable x with mean value m_x and variance σ_x^2 can be transformed into a zero-mean, unit-variance normally distributed random variable y by substituting $y = (x - m_x)/\sigma_x$.

TABLE 1.8 Probability Distributions and Characteristic Functions

Name	$f_X(x)$	$E[x]$	$Var[x]$	$C_X(v)$	Conditions
Uniform	$\frac{1}{b-a}$	$\frac{a+b}{2}$	$\frac{(b-a)^2}{12}$	$\frac{\left(e^{jvb}-e^{jva}\right)}{jv(b-a)}$	$a \ge x \le b$
Bernoulli	$\begin{cases} p & : i=1 \\ 1-p & : i=2 \end{cases}$	p	$p(1-p)$	$(1-p)+pv$	Discrete binary variable $x = k_i : i = 1, 2$
Binomial	$\binom{n}{k} p^k(1-p)^{n-k}$	np	$np(1-p)$	$((1-p)+pv)^n$	Discrete variable $x=k=1,2,\ldots,n$
Poisson	$\alpha^k e^{-\alpha}/k!$	α	α	$e^{\alpha(v-1)}$	Discrete variable $x=k=0,1,\ldots;\alpha>0$
Exponential	$\alpha e^{-\alpha x}$	$1/\alpha$	$1/\alpha^2$	$\alpha/(\alpha-jv)$	$x \ge 0;\ \alpha>0$
Gaussian (normal)	$\frac{e^{-\frac{(x-m)^2}{2\sigma^2}}}{\sqrt{2\pi}\sigma}$	m	σ^2	$e^{-jmv-\frac{\sigma^2 v^2}{2}}$	$-\infty \le x \le \infty$
Chi-square (N = 2) Exponential (α = 1/2)	$\frac{e^{-x/2}}{2}$	2	4	$(1-j2v)^{-1}$	$x \ge 0$
Chi-squared (N-degrees)	$\frac{x^{N/2-1}e^{-x/2}}{2^{N/2}(N/2-1)!}$	N	$2N$	$(1-j2v)^{-N/2}$	N-degrees of freedom $x \ge 0$
Rayleigh	$\frac{xe^{-x^2/2\sigma^2}}{\sigma^2}$	$\sqrt{\pi/2}\sigma$ as $\gamma\to\infty$	$\frac{(4-\pi)\sigma^2}{2}$ as $\gamma\to\infty$	$(1-jv/\beta)^{\alpha}$	$x > 0$
Ricean	$\frac{xe^{-\frac{(x^2-A^2)}{2\sigma^2}}}{\sigma^2} I_o\left(\frac{xA}{\sigma^2}\right)$	[a]	[a]	[a]	$x > 0$ $\begin{cases} A^2/2\sigma^2 \to \infty \\ A^2/2\sigma^2 = 0 \end{cases}$
Gamma	$\frac{\beta(\beta x)^{\alpha-1}}{\Gamma(\alpha)}e^{-\beta x}$	α/β	α/β^2	$\frac{1}{\left(1-j\frac{v}{\beta}\right)^{\alpha}}$	$x > 0$ $\beta > 0;\ \lambda > 0$
Lognormal	$\frac{e^{-\frac{(\ln(y)-m)^2}{2\sigma^2}}}{\sqrt{2\pi}y\sigma}$	$e^{m+\sigma^2/2}$	$e^{m+2\sigma^2}$ $-e^{m+\sigma^2}$	[b]	y is lognormal $y=e^x \ge 0$ $x = N(m,\sigma)$
Nakagami-m	$\frac{2m^m x^{2m-1}e^{-mx^2/\Omega}}{\Gamma(m)\Omega^m}$	[c]	[c]	[c]	$x \ge 0$ $m=\frac{E[x^2]^2}{E[x^2-E(x^2)]^2}$ $\ge 1/2$

Notes: $\gamma = A^2/(2\sigma^2)$ is the signal-to-noise ratio. $\gamma\to 0$ $f_X(x)$ = Rayleigh with $E[x] = \sqrt{\pi/2}\sigma$, $Var[x] = (4-\pi)\sigma^2/2$.

[a] $\gamma\to\infty$ $f_X(x)$ = Gaussian with $E[x] = A$, $Var[x] = \sigma^2$.

[b] Approximated using a series expansion of e^{jvy}.

[c] Refer to special cases in Section 1.5.7.2.

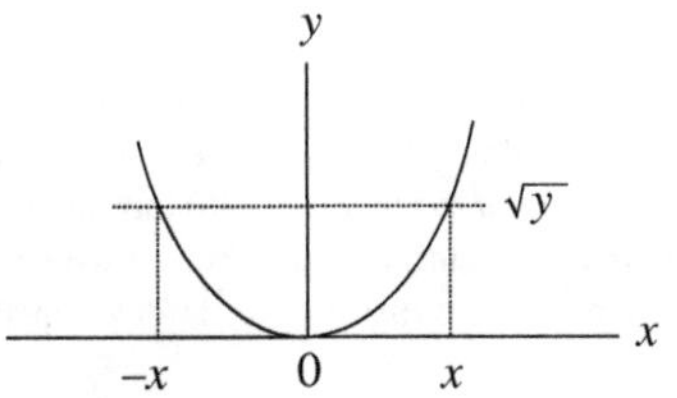

FIGURE 1.27 Transformation of the random x to $y = x^2$.

The *pdf* of y is determined using the incremental intervals $dy = 2xdx$ at $x = \pm\sqrt{y}$ such

$$\begin{aligned} p_Y(y) &= p_X\left(x=\sqrt{y}\right)\left|\frac{dx}{dy}\right| + p_X\left(x=-\sqrt{y}\right)\left|\frac{dx}{dy}\right| \quad : U(y) \\ &= \frac{1}{\sqrt{2\pi}\sigma_x}\left[e^{-y/2\sigma_x^2}+e^{-y/2\sigma_x^2}\right]\left|\frac{1}{2\sqrt{y}}\right| \quad : U(y) \\ &= \frac{e^{-y/2\,\sigma_x^2}}{\sqrt{2\pi y}\sigma_x} \quad : U(y) \end{aligned} \tag{1.276}$$

The characteristic function of (1.276) is evaluated as

$$C_Y^1(v) = E\left[p_Y(y)e^{jvy}\right] = \frac{1}{\left(1-j2\,\sigma_x^2 v\right)^{1/2}} \tag{1.277}$$

Consider now the random variable z resulting from the summation of N independent random variables y_i such that

$$z = \sum_{i=1}^{N} y_i = \sum_{i=1}^{N} x_i^2 \tag{1.278}$$

The characteristic function of z is simply the N-th power of $C_Y(v)$ so that

$$\begin{aligned} C_Z^N(v) &= \left[C_Y^1(v)\right]^N \\ &= \frac{1}{\left(1-j2\,\sigma_x^2 v\right)^{N/2}} \end{aligned} \tag{1.279}$$

Equation (1.279) transforms to the *pdf* of z, resulting in

$$p_Z(z) = \frac{z^{N/2-1}e^{-z/2\,\sigma_x^2}}{\left(2\,\sigma_x^2\right)^{N/2}\left(\frac{N}{2}-1\right)!} \quad : U(y) \tag{1.280}$$

In conforming to the earlier definition, the chi-square distribution is expressed by letting $\sigma_x^2 = 1$ in (1.280) or, more formally, using the transformation $\chi = z/\sigma_x^2$; therefore, the *pdf* of the chi-square random variable χ with N degrees of freedom is

$$p_X(\chi) = \frac{\chi^{N/2-1}e^{-\chi/2}}{2^{N/2}\left(\frac{N}{2}-1\right)!} \quad : U(y) \ \text{Chi-square distribution} \tag{1.281}$$

and the corresponding characteristic, or moment generating, function is

$$C_X^N(v) = \frac{1}{(1-j2v)^{N/2}} \quad : \text{Chi-square characteristic function} \tag{1.282}$$

Equation (1.281) is occasionally referred to as the *central* χ^2 distribution because it is based on noise only, that is, the underlying zero-mean Gaussian random variables x_i with distribution given by (1.275) do not contain a signal component.*

Special Case for N = 2 Under this special case $z = x_1^2 + x_2^2$ (1.280) reduces to the exponential distribution

$$p_Z(z) = \frac{e^{-z/2\,\sigma_x^2}}{2\,\sigma_x^2} \quad : U(y) \ \text{exponential distribution} \tag{1.283}$$

So the resulting chi-square χ^2 distribution is obtained from (1.281) with $N = 2$. This is an important case because x_1 and x_2 can be thought of as orthogonal components in the complex description of a baseband data sample. Urkowitz [48] shows that the energy of a wide-sense stationary narrowband white noise Gaussian random process with bandwidth $-W$ to W Hz and measured over a finite interval of T seconds is approximated by $N = 2WT$ terms or degrees of freedom. The frequency W is the noise bandwidth of the narrowband baseband filter and the approximation error in the energy measurement decreases with increasing $2WT$. The factor of two can be thought of as the computation of complex orthogonal baseband functions $\tilde{z} = x_c^2 + jx_s^2$ so $N = 2$ degrees of freedom correspond to $WT = 1$. For example, the *rect*(t/T) function observed over the interval T seconds has a noise bandwidth of $W = 1/T$ Hz corresponding to $WT = 1$ resulting in 2 degrees of freedom.

Upon letting $w = |\tilde{z}| = \sqrt{x_c^2 + x_s^2}$, the random variable w is described in terms of the Rayleigh distribution

$$p_W(w) = \frac{w}{\sigma_x^2}e^{-w^2/2\sigma_x^2} \quad : U(y) \ \text{Rayleigh distribution} \tag{1.284}$$

So the Rayleigh distribution is derived from the magnitude of the quadrature zero-mean Gaussian distributed random variables, $x = N(0,\sigma)$.†

1.5.7.2 *Relationship between Nakagami-*m*, Gaussian, Rayleigh, and Ricean Distributions* The Nakagami-*m* distribution [49] was initially derived from experimental data to characterize HF fading; however, subsequent experimental observations demonstrate its application to rapid fading at carrier frequencies from 200 MHz to 4 GHz. It is considered to be a generalized distribution from which other distributions can be derived, for example, $m = 1$ results in the Rayleigh power distribution, $m = ½$ results in the one-sided

*When the underlying Gaussian distributed random variable is composed of signal plus noise, the signal amplitude represents the mean value of the distribution. An analysis, similar to that of (1.276) through (1.280), containing a mean value, results in a noncentral χ^2 distribution with 1 through N degrees of freedom. The noncentral χ^2 distribution is developed and discussed in Appendix C.

†The notation $N(m,\sigma)$ is used to denote the distribution of a Gaussian, or normal, random variable with mean value m and standard deviation σ.

zero-mean Gaussian distribution, and as $m \to \infty$ the m-distribution approaches the Gaussian distribution with a unit mean value. In the region $1 \le m \le \infty$, the Nakagami-m distribution behaves much like the Ricean distribution; however, the normalized distributions are subtly different when plotted for various signal-to-noise ratios less than about 10 dB. The Ricean distribution, referred to as the n-distribution by Nakagami, is derived from concepts involving narrowband filtering of a continuous wave (CW) signal with additive Gaussian noise, whereas the Nakagami-m distribution is derived from experimental data involving multipath communication links.

1.5.8 Order Statistics

Communication systems analysis and performance evaluations often involve a large number of random samples taken from an underlying continuous or discrete probability distribution function. The various parameters, used to characterize the system performance, result in limiting distributions with associated means, variances, and confidence levels as dictated, for example, by an underlying distribution. Order statistics [31, 50, 51], on the other hand, involves a *distribution-free* or *nonparametric* analysis that requires only that the probability distribution functions be continuous and not necessary related to the underlying distribution from which the samples are taken. However, the randomly drawn samples are considered to be statistically independent.

Consider that the n random samples $\{X_1, X_2, \ldots, X_n\}$ are taken from the continuous *pdf* $f_X(x)$ over the range $a \le x \le b$. Now consider reordering the random variables $X_i : i = 1, \ldots, n$ to form the random variables $\{Y_1, Y_2, \ldots, Y_n\}$ arranged in ascending order of magnitude, such that, $a \le Y_1 < Y_2 < \cdots < Y_n \le b$ where $f_{Y_i}(y_i) = 1/(b-a)$ is uniformly distributed over the interval $b - a$. The joint *pdf* of the ordered samples [52] is expressed as

$$g_{Y_1,Y_2,\ldots,Y_n}(y_1,y_2,\ldots,y_n) = n!\prod_{i=1}^{n} f_{Y_i}(y_i) \tag{1.285}$$

for $a \le y_1 < y_2 < \cdots < y_n \le b$ and $n!$ is the number of mutually disjoint sets of $x_1, x_2, \ldots, x_n$. For example, for $n = 4$ the set x_1, x_2, x_3, x_4 results in $n! = 24$ mutually disjoint sets determined as shown in Table 1.9. The first six mutually disjoint sets are determined by cyclically left shifting the indicated subsets of original set x_1, x_2, x_3, x_4; a cyclic left shift of a subset is obtained by shifting each element of the subset to the left and replacing the leftmost element in the former position of the rightmost element. Following the first six sets shown in the figure, the original set is cyclically shifted three more times each leading to six mutually disjoint sets by shifting subsets resulting in a total of 24 mutually disjoint sets.

TABLE 1.9 Example of Mutually Disjoint Sets (n = 4, 24 Mutually Disjoint Sets)

No.	Mutually Disjoint Sets	Shifting[a]
1	x_1, x_2, x_3, x_4	Original set
2	x_1, x_2, x_4, x_3	Shift subset x_3, x_4
3	x_1, x_3, x_4, x_2	Shift subset x_2, x_3, x_4
4	x_1, x_3, x_2, x_4	Shift subset x_4, x_2
5	x_1, x_4, x_2, x_3	Shift subset x_3, x_4, x_2
6	x_1, x_4, x_3, x_2	Shift subset x_2, x_3
7	x_2, x_3, x_4, x_1	Shift original set
8	x_2, x_3, x_1, x_4	Shift subset x_4, x_1
9	x_2, x_4, x_1, x_3	Shift subset x_3, x_4, x_1
...	...	...

[a] *Shift* denotes a cyclic left shift of a previous set or subset.

The ordered sample Y_i is referred to as the i-th order statistic of the sample set. The marginal *pfd* of the n-th order statistic Y_n, that is, the maximum of $\{X_1, X_2, \ldots, X_n\}$, is evaluated using (1.285) by performing the integrations in the ascending order $i = 1, 2, \ldots, n - 1$ as follows*:

$$\begin{aligned} g_{Y_n}(y_n) &= n!\int_a^{y_n}\cdots\int_a^{y_4}\int_a^{y_3}\left[\int_a^{y_2} f(y_1)dy_1\right]f(y_2)\cdots f(y_n)dy_2\cdots dy_{n-1} \\ &= n!\int_a^{y_n}\cdots\int_a^{y_4}\left[\int_a^{y_3} F(y_2)f(y_2)dy_2\right]f(y_3)\cdots f(y_n)dy_3\cdots dy_{n-1} \\ &= n!\int_a^{y_n}\cdots\left[\frac{1}{2}\int_a^{y_4} F^2(y_3)f(y_3)dy_3\right]f(y_4)\cdots f(y_n)dy_4\cdots dy_{n-1} \\ &\cdots \\ &= n!f_{Y_n}(y_n)\left[\frac{1}{(n-2)!}\int_a^{y_n} F^{n-2}(y_{n-1})f(y_{n-1})dy_{n-1}\right] \end{aligned} \tag{1.286}$$

The solution (see Problem 15) to (1.286) is

$$g_{Y_n}(y_n) = n[F(y_n)]^{n-1}f_{Y_n}(y_n) \quad : a < y_n \le b \tag{1.287}$$

where $F^{n-1}(y_n)$ is the *cdf* evaluated as

$$F^{n-1}(y_n) = (n-1)\int_a^{y_n} F^{n-2}(y_{n-1})f_Y(y_{n-1})dy_{n-1} = \frac{(y_n-a)^{n-1}}{(b-a)^{n-1}} \tag{1.288}$$

Using the marginal *pdf* of Y_n given by (1.287), the probability of selecting the maximum of value Y_n is determined as

$$P(y') = n\int_a^{y'} F^{n-1}(y_n)f_Y(y_n)dy_n\bigg|_{y'=b} = 1 \tag{1.289}$$

*For notational simplicity $f(y)$ is used to denote $f_Y(y)$.

TABLE 1.10 Order Statistics False-Detection Probability for Gaussian Distributed Random Variables

Ordered $S+N$ and N Statistics (j,k)	False-Detection Probability (P_{fd})		
	$\gamma_{dB}=10$	$\gamma_{dB}=15$	$\gamma_{dB}=20$
1,1	2.440e^{-2}	6.645e^{-5}	1.408e^{-12}
1,2	4.226e^{-2}	1.308e^{-4}	2.815e^{-12}
1,4	6.980e^{-2}	2.547e^{-4}	5.629e^{-12}
1,8	1.089e^{-1}	4.874e^{-4}	1.126e^{-11}

These results are distribution free, in that, the *pdf* has not been defined; however, from a practical point of view (1.289) can be evaluated for any continuous *pdf*.

The distributions from which the x_i are taken need not be identical*; for example, the samples x_1 through x_j can be taken from a distribution involving signal plus noise (or clutter) and those from x_{j+1} through x_n corresponding to noise (or clutter) only. Using this example the distribution in (1.287) is expressed as

$$f_{Y_n}(y_n)=\frac{n!}{(j-1)!(n-j)!}\left[F_{sn}(y_j)\right]^{j-1}\left[1-F_n(y_j)\right]^{n-j}f_{Y_n}(y_n) \tag{1.290}$$

where $F_{sn}(y)$ is the distribution corresponding to signal plus noise and $F_n(y)$ is the noise-only distribution.

Example distributions used to evaluate the performance of communication and radar systems are Gaussian, Ricean, lognormal, and Weibull distributions. Table 1.10 lists the false-detection probabilities, for the indicated signal-to-noise ratios γ_{dB}, associated with the detection of $j=1$ signal-plus-noise event and $k=n-j=(1,2,4,$ and $8)$ noise-only events.

1.5.9 Properties of Correlation Functions

Correlation processing is used in nearly every aspect of demodulator signal detection from energy detection, waveform acquisition, waveform tracking, parameter estimation, and information recovery processing. With this wide range of applications, the theoretical analyst, algorithm developer, software coder, and hardware developer must be thoroughly familiar with the properties and implementation of waveform correlators. An equally important processing function is that of convolution or linear filtering. The equivalence between matched filtering and correlation is established in Section 1.7.2 and involves a time delay in the correlation response; with this understanding, the properties of correlation can be applied to convolution or filtering. The correlation response can be exploited to determine the signal signature regarding the location of a signal in time and frequency, the duration and bandwidth of the signal, the shape of the modulated signal waveform, and the estimate of the information contained in the modulated waveform.

The correlation function† is evaluated for the complex functions $\tilde{x}(t)$ and $\tilde{y}(t)$ as the integral

$$R_{\tilde{x}\tilde{x}}(\tau)=\int_{-\infty}^{\infty}\tilde{x}(t)\tilde{x}^*(t-\tau)dt \quad \text{: autocorrelation} \tag{1.291}$$

and

$$R_{\tilde{x}\tilde{y}}(\tau)=\int_{-\infty}^{\infty}\tilde{x}(t)\tilde{y}^*(t-\tau)dt \quad \text{: cross-correlation} \tag{1.292}$$

where the asterisk denotes complex conjugation.

Autocorrelation processing examines the correlation characteristics of a single random process with the maximum magnitude corresponding to the zero-lag condition $R_{\tilde{x}\tilde{x}}(\tau=0)$ that is equal to the maximum energy over the correlation interval. The correlation response $R_{\tilde{x}\tilde{x}}(\tau)$ is indicative of the shape of $\tilde{x}(t)$ and the duration, τ_d, of the principal correlation response is indicative of the correlation time. For deterministic signals, the correlation time (τ_o) is usually characterized in terms of the one-sided width of the principal correlation lobe; however, for stochastic processes the correlation interval is defined when $|R_{\tilde{x}\tilde{x}}(\tau)|$ decreases monotonically from $R_{\tilde{x}\tilde{x}}(0)$ to a defined level; for example, when the *normalized correlation response* first reaches the level $R_{\tilde{x}\tilde{x}}(\tau_o)/R_{\tilde{x}\tilde{x}}(0)=e^{-1}$. The normalized correlation response is referred to as the correlation coefficient as defined in (1.295) or (1.296). The parameters related to the correlation of the function $\tilde{x}(t)$ have equivalent Fourier transform frequency-domain definitions. In the case of stochastic processes, the Fourier transform of $R_{\tilde{x}\tilde{x}}(\tau)$ is defined as the PSD of the process.

Expanding (1.292) in terms of the real and imaginary with $\tilde{x}(t)=x_r(t)+jx_i(t)$ and $\tilde{y}(t)=y_r(t)+jy_i(t)$ results in

$$\begin{aligned}R_{\tilde{x}\tilde{y}}(\tau)=&\int_{-\infty}^{\infty}x_r(t)y_r(t-\tau)dt+\int_{-\infty}^{\infty}y_i(t)y_i(t-\tau)dt\\&+j\left(\int_{-\infty}^{\infty}x_i(t)y_r(t-\tau)dt-\int_{-\infty}^{\infty}x_r(t)y_i(t-\tau)dt\right)\end{aligned} \tag{1.293}$$

This evaluation requires four real multiplies and integrations for each lag, whereas, if $\tilde{x}(t)$ and $\tilde{y}(t)$ were real functions only one multiplication and integration is required for each

*Equation (1.286) allows for different distributions; however, (1.287) through (1.289) are based on independent identically distributed (*iid*) random variables.

†A stationary stochastic processes is characterized by the first- and second-order moments corresponding to the mean $E[\tilde{x}(t)]=m_{xr}+jm_{xi}$ and autocorrelation response $R_{\tilde{x}\tilde{x}}(\tau)=E[\tilde{x}(t)\tilde{x}^*(t-\tau)]$.

lag. With discrete-time sampling, the integrations are replaced by summations over the finite sample values $\tilde{x}_n$ and $\tilde{y}_n$ where $t = nT_s$: $n = 0, \ldots, N-1$ and T_s is the sampling interval; in this case, the computational complexity is proportional to N^2. The computation complexity can be significantly reduced by performing the correlation in the frequency domain using FFT [53], in which case, for a radix-2 FFT with $N = 2^k$, the computation complexity is proportional to $N\log_2(N)$. Brigham [54] provides detailed descriptions of the implementation and advantages of FFT correlation and convolution processing. The correlation results throughout the following chapters use the direct and FFT approaches without distinction.

Referring to (1.291) the zero-lag correlation is expressed as

$$R_{\tilde{x}\tilde{x}}(0) = \int_{-\infty}^{\infty} \tilde{x}(t)\tilde{x}^*(t)dt + \int_{-\infty}^{\infty} |\tilde{x}(t)|^2 dt \\ = E_x \tag{1.294}$$

where E_x is the total energy in the received signal. Using (1.294), the normalized correlation is defined in terms of the *normalized autocorrelation coefficient* as

$$\rho_x(\tau) \triangleq \frac{R_{\tilde{x}\tilde{x}}(\tau)}{E_x} \quad \text{: normalized autocorrelation coefficient} \tag{1.295}$$

with $|\rho_x(\tau)| \le 1$. From (1.292), the normalized cross-correlation coefficient is defined as

$$\rho_{xy}(\tau) \triangleq \frac{R_{\tilde{x}\tilde{y}}(\tau)}{\sqrt{E_x E_y}} \quad \text{: normalized cross-correlation coefficient} \tag{1.296}$$

with $|\rho_{xy}(\tau)| \le 1$.

The correlation may also be defined in terms of the long-term average over the interval T as

$$R_{\tilde{x}\tilde{x}}(\tau) = \lim_{T\to\infty} \frac{1}{T}\int_{-T/2}^{T/2} \tilde{x}(t)\tilde{x}^*(t-\tau)dt \quad \text{: autocorrelation} \tag{1.297}$$

However, most practical waveforms are limited to a finite duration $T_c = NT_s$ and, in these cases, $\tilde{x}(t)$ is zero outside of the range T_c. Therefore, dividing the zero-lag correlation by T_c results in the second-order moment $E(|\tilde{x}|^2) = \sigma_x^2 + m_x^2$ where m_x^2 is the DC or mean signal power. Removing the mean signal level prior to performing the correlation results in the autocovariance with $E(|\tilde{x}|^2) = \sigma_x^2$. Table 1.11 summarized several properties of correlation functions.

TABLE 1.11 Correlation Function Properties of Deterministic and Stochastic Processes

Property	Comments
$R_{\tilde{x}\tilde{x}}(-\tau) = R^*_{\tilde{x}\tilde{x}}(\tau)$	Autocorrelation
$R_{xx}(-\tau) = R_{xx}(\tau)$	$x(t)$ is real
$C_{\tilde{x}\tilde{x}}(-\tau) = R_{\tilde{x}\tilde{x}}(\tau) - \|m_x\|^2$	Autocovariance
$C_{xx}(-\tau) = R_{xx}(\tau) - m_x^2$	$x(t) \in$ real
$R_{xx}(\tau) = R_{yy}(\tau)$	$x(t) = m(t)\cos(\omega_c t)$ $y(t) = m(t)\sin(\omega_c t)$
$R_{xy}(\tau) = -R_{yx}(\tau)$	$x(t) = m(t)\cos(\omega_c t)$ $y(t) = m(t)\sin(\omega_c t)$
$R_{\tilde{x}\tilde{y}}(-\tau) = R^*_{\tilde{y}\tilde{x}}(\tau)$	$x(t), y(t) \in$ complex
$C_{\tilde{x}\tilde{y}}(\tau) = R_{\tilde{x}\tilde{y}}(\tau) - \tilde{m}_x\tilde{m}_y^*$	Cross-covariance
$C_{xy}(\tau) = R_{xy}(\tau) - m_x m_y$	$x(t), y(t) \in$ real
$R_{\tilde{z}\tilde{z}}(\tau) = R_{\tilde{x}\tilde{x}}(\tau) + R_{\tilde{y}\tilde{y}}(\tau) + R_{\tilde{x}\tilde{y}}(\tau) + R_{\tilde{y}\tilde{x}}(\tau)$	$\tilde{z}(t) = \tilde{x}(t) + \tilde{y}(t)$

Consider, for example, that $\tilde{y}(t) = \tilde{x}(t) + \tilde{n}(t)$ is a received signal plus AWGN, the correlation $R_{\tilde{x}\tilde{y}}(\tau)$ is performed in the demodulator using the known reference signal $\tilde{x}(t)$. The dynamic range of the demodulator detection processing is minimized by the normalization in (1.296) and the optimum signal detection corresponds to $\rho_{xy}(0)$. On the other hand, if the optimum timing is not known, near optimum detection can be achieved by choosing the maximum correlation output over the uncertainty range of the correlation lag about $\tau = 0$. During initial signal acquisition, the constant false-alarm rate (CFAR) threshold, described in Section 11.2.2.1, is an effective algorithm for signal presence detection and coarse synchronization.

1.6 RANDOM PROCESSES

Many of the signal descriptions and processing algorithms in the following chapters deal exclusively with the signal and neglect the additive noise under the reasoning that the noise detracts from the fundamental signal processing requirements and complicates the notation which has the same effect. On the other hand, understanding the impact of the noise on the system performance is paramount to the waveform selection and adherence to the system performance specifications. To this end, the performance evaluation is characterized by detailed analysis of the signal-plus-noise conditions and confirmed by computer simulations.

The following descriptions of noise and signal plus noise are provided to illustrate the assumptions and analysis associated with the inclusion of the most basic noise source—AWGN. The reference to narrowband Gaussian noise simply means that the carrier frequency f_c is much greater than the signal modulation Nyquist bandwidth B_N so that the $2f_c$ heterodyning or homodyne mixing terms are completely eliminated through filtering. In such cases, the white noise in the baseband demodulator bandwidth is denoted by the single-sided noise density N_o watts/Hz, where *single-sided* refers to positive frequencies.

1.6.1 Stochastic Processes

The subject of stochastic processes is discussed in considerable detail by Papoulis [55] and Davenport and Root [56] and the following definitions are often stated or implied in the applications discussed in throughout the following chapters. A stochastic process is defined as a random variable that is a function of time and the random events χ in S as depicted in Figure 1.14. In this context the random variable is characterized as $x(t,\chi)$. For a fixed value of $t = t_i$, $x(t_i,\chi)$ is a random variable and $\chi = \chi_i$, $x(t,\chi_i)$ denotes as the real random process $x(t)$ such that $x(t_i)$ is a random variable with *pdf* $f_X(x{:}t_i)$; in general, the *pdf* of $x(t)$ is defined as $f_X(x{:}t)$.

1.6.1.1 Stationarity There are several ways to define the stationarity of a stochastic process, for example, stationarity of finite order, asymptotic stationary, and periodic stationarity; however, the following two are the most frequently encountered.

Strict-Sense Stationary Process The stochastic process $x(t)$ is strict-sense stationary, or simply stationary, if the statistics are unaltered by a shift in the time axis. Furthermore, two random variables are jointly stationary if the joint statistics are unaltered by an equal time shift of each random variable, that is, the probability density function $f(x\ ;\ t)$ is the same for all time shifts τ. This is characterized as

$$f(x_1,x_2,\ldots;\ t_1,t_2,\ldots) = f(x_1,x_2,\ldots;\ t_1+\tau,\ t_2+\tau,\ldots) \tag{1.298}$$

Wide-Sense Stationary Process The stochastic process $x(t)$ is wide-sense stationary (WSS) if its expected value is constant and autocorrelation function is a function of the time shift $\tau = t_2 - t_1\ \forall\ t_1$ and t_2. WSS stationarity is characterized as

$$E[x(t)] = m_x = \text{constant} \tag{1.299}$$

and

$$E[x(t)x(t-\tau)] = R_x(\tau) \tag{1.300}$$

Because wide-sense stationarity depends on only the first and second moments it is also referred to as *weak* stationarity. A function of two random processes is wide-sense stationary if each process is wide-sense stationary and their cross-correlation function is dependent only the time shift, that is,

$$E[x(t_1)y(t_2)] = R_{xy}(t_1-t_2) = R_{xy}(\tau) \tag{1.301}$$

1.6.1.2 Ergodic Random Process The random process $x(t)$, defined earlier, is an ergodic random process if the statistics of $x(t)$ are completely defined by the statistics of $x(t,\chi)$. Denoting the random process $x(t_i,\chi)$ as an ensemble of $x(t,\chi)$, then ergodicity ensures that the statistics $x(t_i)$ are identical to those of the ensemble; in short, the time statistics are identical to the ensemble statistics.* Ergodicity of the mean, of the stochastic process $x(t,\chi)$, exists under the condition

$$\langle x(t,\chi_i)\rangle = E[x(t_i,\chi)] \quad : \forall i \tag{1.302}$$

where the time average is defined as

$$\langle x(t,\chi_i)\rangle \triangleq \lim_{T\to\infty}\frac{1}{2T}\int_{-T}^{T} x(t,\chi_i)dt \tag{1.303}$$

and the ensemble average is defined as

$$E[x(t_i,\chi)] \triangleq \int_{-\infty}^{\infty} \chi f_\chi(t_i,\chi)d\chi \tag{1.304}$$

Since the mean value of a random process must be a constant, the *ergodic of the mean* theorem states that the equality condition in (1.302) is satisfied when $E[x(t_i,\chi)] = \eta : \forall i$ where η is a constant. This is a nontrivial task to prove, however, following the discussion by Papoulis [57], the *ergodic of the mean theorem* states that

$$\lim_{T\to\infty}\frac{1}{2T}\int_{-T}^{T} x(t)dt = E[x(t)] = \eta \qquad : \textit{iff} \lim_{T\to\infty}\frac{1}{2T}\int_{-T}^{T} R_x(\tau)d\tau = \eta^2 \tag{1.305}$$

The *iff* condition in (1.305) is formally expressed in terms of the autocovariance function for which the limit $T \to \infty$ is expressed as the variance $\sigma_x^2 = \eta^2 - E[x(t)]^2$. However, from (1.305), the expectation $E[x(t)] = \eta$ resulting in $\sigma_x^2 = 0$. Therefore, the limit $T \to \infty$ of the autocovariance function *converges in probability* with the conclusion that $E[x(t,\chi_i)] = E[x(t_i,\chi)]$ proving *ergodicity of the mean*.† Demonstration of *ergodicity of the autocorrelation function*

*Papoulis (Reference 8, Chapter 9) discusses the ergodicity of a stochastic process with respect to certain parameters.

†Convergence in probability is also discussed by Davenport and Root (Reference 6, pp. 66–71).

is considerably more involved, requiring the fourth-order moments.

1.6.2 Narrowband Gaussian Noise

Consider the noise described by the narrowband process [58] with bandwidth $B << f_c$ expressed as

$$n(t)=N(t)\cos(\omega_c t+\theta(t))-N(t)\sin(\omega_c t+\theta(t)) \qquad (1.306)$$

where $N(t)$ and $\theta(t)$ represent, respectively, the envelop and phase of the noise and $\omega_c = 2\pi f_c$ is the angular carrier frequency. Upon expanding the trigonometric functions, (1.306) can also be expressed as

$$n(t)=n_c(t)\cos(\omega_c t)-n_s(t)\sin(\omega_c t) \qquad (1.307)$$

where

$$n_c(t)=N(t)\cos(\theta(t)) \qquad (1.308)$$

and

$$n_s(t)=N(t)\sin(\theta(t)) \qquad (1.309)$$

The noise terms $n_c(t)$ and $n_s(t)$ are uncorrelated with spectrum $S(f)$ and bandwidth B, such that $S(f) = 0$ for $|f-f_c| > B/2$. This is the general characterization of a narrowband noise process; however, in the following analysis, $n_c(t)$ and $n_s(t)$ are also considered to be statistically independent, stationary zero-mean white noise Gaussian processes with one-sided spectral density N_o watts/Hz.

Because of the stationarity, the noise autocorrelation is dependent only on the correlation lag τ and is evaluated as

$$\begin{aligned}R_{nn}(\tau)&=E[n(t)n(t-\tau)]\\&=E[\{n_c(t)\cos(\omega_c t)-n_s(t)\sin(\omega_c t)\}\\&\quad\{n_c(t-\tau)\cos(\omega_c(t-\tau))-n_s(t-\tau)\sin(\omega_c(t-\tau))\}]\end{aligned} \qquad (1.310)$$

Upon evaluating the product in (1.310) and distributing the expectation, it is found that the conditions for stationarity require* $R_{ss}(\tau) = R_{cc}(\tau)$ and $R_{cs}(\tau) = -R_{sc}(\tau)$ so that (1.310) reduces to

$$R_{nn}(\tau)=R_{cc}(\tau)\cos(\omega_c\tau)-R_{ss}(\tau)\sin(\omega_c\tau) \qquad (1.311)$$

The noise power is evaluated using (1.311) with $\tau=0$ with the result $R_{nn}(0) = R_{cc}(0) = \sigma_n^2$. This evaluation can be carried further using the Wiener–Khinchin theorem† which states that the power spectral density of a WSS random process is the Fourier transform of the autocorrelation function, that is,

$$S_n(f)=\int_{-\infty}^{\infty} R_{nn}(\tau)e^{-j2\pi f\tau}d\tau \qquad (1.312)$$

From (1.312) the inverse Fourier transform is

$$R_{nn}(\tau)=\int_{-\infty}^{\infty} S_n(f)e^{j2\pi f\tau}df \qquad (1.313)$$

and, substituting the condition that the single-sided noise spectral density is defined as N_o watts/Hz, (1.313) becomes

$$R_n(\tau)=\frac{N_o}{2}\int_{-\infty}^{\infty} e^{j2\pi f\tau}df=\frac{N_o}{2}\delta(\tau) \qquad (1.314)$$

In (1.314) the single-sided noise density is divided by two because of the two-sided integration, that is, the integration includes negative frequencies. In this case, the noise power, defined for $\tau = 0$, is infinite, however, when the ideal band-limited filter, with bandwidth B, is considered the noise power in the filter centered at f_c is computed as

$$R_n(0)=N_o\int_{f_c-B/2}^{f_c+B/2} df=N_oB \qquad (1.315)$$

In this case the one-sided noise density N_o is used instead of $N_o/2$ because the one-sided integration is over positive frequencies.

If a linear filter with impulse response $h(t)$ is used, the frequency response is given by

$$H(f)=\int_{-\infty}^{\infty} h(t)e^{-j2\pi ft}dt \qquad (1.316)$$

The corresponding unit gain normalizing factor is $|H(0)|$. With the stationary noise process $n(t)$ applied to the input of the filter, the output is determined using the convolution integral and the result is as follows:

$$n_o(t)=\int_{-\infty}^{\infty} n(t-\lambda)h(\lambda)d\lambda \qquad (1.317)$$

*See Problem 29.

†Leon-Garcia (Reference 29, p. 404) refers to this theorem as the Einstein–Wiener–Khinchin theorem based on the discovery of an earlier paper by Albert Einstein.

Using (1.317) it can be shown (see Problem 33) that the normalized spectrum of the output noise is expressed, in terms of the input noise PSD $S_n(f)$, as

$$S_{no}(f) = S_n(f)\left|\frac{H(f)}{H(0)}\right|^2 \tag{1.318}$$

where $|H(0)|$ is the normalizing gain of the filter. Using (1.318), with $S_n(f) = N_o/2$ corresponding to white noise, the output noise power is evaluated as

$$\begin{aligned} R_{no}(0) &= \int_{f_c-B/2}^{f_c+B/2} S_n(f)\frac{|H(f)|^2}{|H(f_c)|^2}df \\ &= \frac{N_o}{2}\int_{f_c-B/2}^{f_c+B/2}\frac{|H(f)|^2}{|H(f_c)|^2}df \\ &= N_oB_n \end{aligned} \tag{1.319}$$

where the second integral in (1.319) is recognized as the definition of the noise bandwidth of the bandpass filter with lowpass bandwidth B_n.

1.7 THE MATCHED FILTER

The problem in the detection of weak signals in noise is one of deciding whether the detection filter output is due to the signal and noise or simply noise only. The matched filter [59, 60], provides for the optimum signal detection in AWGN noise based on the maximum instantaneous signal-to-noise ratio when sampled at the optimum time.* The matched filter, for an AWGN channel, is characterized as having an impulse response equal to the delayed time-reverse replica of the received signal. To maximize the signal detection probability the matched filter output must be sampled at T_o as defined in the following analysis. The matched filter can be implemented at a convenient receiver IF or in the demodulator using quadrature baseband-matched filters.

Considering the received signal, $s_r(t)$, the matched filter impulse response depicted in Figure 1.28 is expressed as

$$h(t) = Gs_r(T_o - t) \tag{1.320}$$

The gain G is selected for convenience; however, it must be a constant value. The delay T_o is required to result in a causal impulse response, that is, the response of $h(t \le 0) = 0$ for $h(t)$ to be realizable; consequently, $s_r(t \ge T_o)$ must be zero. Usually the selection of T_o is not an issue since many symbol modulation functions are time limited or can be truncated without a significant impact on the transmitted signal

*The matched filter was first derived by D.O. North [59] and is also referred to as the North filter.

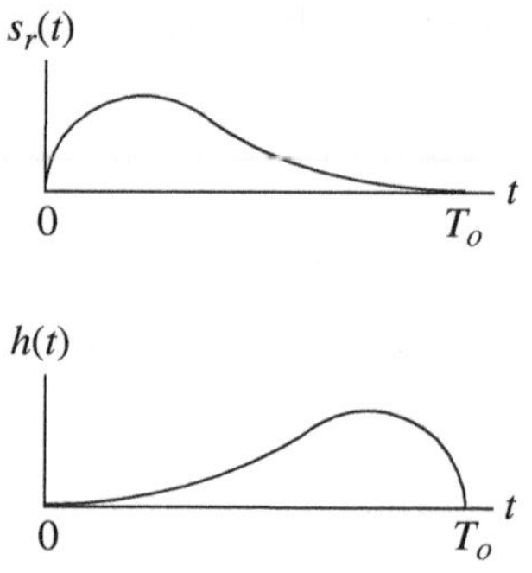

FIGURE 1.28 Example received signal and corresponding matched filter.

spectrum; however, the matched filter delay results in a throughput delay. To the extent that the impulse approximates (1.320) a detection loss will be encountered.

The criterion of the matched filter is to provide the maximum signal-to-noise ratio in the AWGN channel when sampled at the optimum time T_o. The following matched filter analysis follows that of Skolnik [61]. The signal-to-noise ratio of interest is

$$\gamma_f = \frac{|s_r(t)|^2_{max}}{N} \tag{1.321}$$

where $|s_r(t)|^2_{max} = |s_r(T_o)|^2$ is evaluated as

$$|s_o(T_o)|^2 = \left|\int_{-\infty}^{\infty} S(f)H(f)e^{j2\pi fT_o}df\right|^2 \tag{1.322}$$

and N is the noise power evaluated as

$$N = \int_{-\infty}^{\infty} N(f)|H(f)|^2 df = \frac{N_o}{2}\int_{-\infty}^{\infty}|H(f)|^2 df \tag{1.323}$$

In these expressions, the filter spectrum $H(f)$ is normalized, such that $H(0) = 1$, and the last equality in (1.323) results because the channel noise is white with one-sided constant power density of N_o watts/Hz. Substituting (1.322) and (1.323) into (1.321) results in the expression for the signal-to-noise ratio

$$\gamma_f = \frac{\left|\int_{-\infty}^{\infty} S(f)H(f)e^{j2\pi fT_o}df\right|^2}{(N_o/2)\int_{-\infty}^{\infty}|H(f)|^2 df} \tag{1.324}$$

The maximum signal-to-noise ratio is evaluated by applying Schwarz's inequality (see Section 1.14.5, Equation 5) to the numerator of (1.324). Upon substituting $f^*(f) = S(f)e^{j2\pi fT_o}$ and $g(f) = H(f)$ into the Schwarz inequality, (1.324) is expressed as

$$\gamma_f \leq \frac{\int_{-\infty}^{\infty} |S(f)|^2 df \int_{-\infty}^{\infty} |H(f)|^2 df}{(N_o/2)\int_{-\infty}^{\infty} |H(f)|^2 df} = \frac{\int_{-\infty}^{\infty} |S(f)|^2 df}{N_o/2} \tag{1.325}$$

The equality condition of the signal-to-noise ratio in (1.325) applies when $f(f) = cg(f)$, where $c > 0$ is a conveniently selected constant resulting in the matched filter frequency response expressed as

$$H(f) = GS^*(f)e^{-j2\pi fT_o} \quad \text{: matched filter frequency response} \tag{1.326}$$

where $G = 1/c$ is an arbitrarily selected constant gain greater that zero. Upon applying Parseval's theorem and recognizing that the numerator of the second equality in (1.325) is the signal energy, E, the optimally sampled matched filter output signal-to-noise ratio is simply expressed as

$$\gamma_f(\max) = \frac{2E}{N_o} \quad \text{: baseband modulated signal} \tag{1.327}$$

Therefore, for the AWGN channel, the optimally sampled matched filter output results in a signal-to-noise ratio that is a function the signal energy and noise density and is independent of the shape of the signal waveform. The factor of two in (1.327) results from the analytic or baseband signal description in the derivation of the matched filter. Typically, the received signal spectrum is modulated onto a carrier frequency with an average power equal to one-half the peak carrier power. In this case, the signal-to-noise ratio at the output of the matched filter is one-half of that in (1.327) resulting in

$$\gamma_f(\max) = \frac{E}{N_o} \quad \text{: carrier moduated signal} \tag{1.328}$$

Referring to (1.326), the inverse Fourier transform of the complex conjugate of the signal spectrum results in the filter impulse response corresponding to the time reverse of the signal. In addition, the inverse Fourier transform of the exponential function in (1.326) results in a signal time delay of T_o seconds, so the resulting filter impulse response, $h(t)$, corresponds to the example depicted in Figure 1.28. Consequently, the matched filter impulse response can be expressed in the time domain by (1.320) or in the frequency domain by (1.326).

The detection loss associated with a filter that is not matched to the received signal is evaluated as

$$\rho_f \triangleq \frac{\gamma_f}{\gamma_f(\max)} = \frac{|s_o'(t)|^2_{max}/N_o'}{2E/N_o} \tag{1.329}$$

where $s_o'(t)$ and N_o' are the output signal and mean noise power density at the output of the unmatched filter. Typically the matched filter is based on the transmitted waveform; however, the received signal into the matched filter may be distorted by the channel or receiver filtering* resulting in a detection loss. The matched filter implementation may also result in design compromises that result in a detection loss.

1.7.1 Example Application of Matched Filtering

In this example, a BPSK-modulated received signal is considered with binary source data bits $b_i = \{0,1\}$ expressed as the unipolar data $d_i = (1 - 2b_i) = \{1,-1\}$ over the data intervals $iT \leq t \leq (i+1)T$ of the bit duration T. The received signal plus noise is expressed as

$$r(t) = d_i s(t) + n(t) \tag{1.330}$$

The signal is described as

$$s(t) = A\cos(\omega_c t) \tag{1.331}$$

The noise is zero-mean additive white Gaussian noise with one-sided spectral density N_o described as

$$n(t) = n_c(t)\cos(\omega_c t) - n_s(t)\sin(\omega_c t) \tag{1.332}$$

The receiver-matched filter impulse response and Fourier transform are given by

$$h(t) = As(T_o - t) \underset{\Im}{\Leftrightarrow} H(f) = GS^*(f)e^{-j2\pi fT_o} \tag{1.333}$$

In (1.333) the signal spectrum defined as $S(f)$ and the squared magnitude of the matched filter output at the optimum sampling point is

$$\begin{aligned} |s_o(T_o)|^2 &= \int_{-\infty}^{\infty} S(f)H(f)e^{j2\pi fT_o} df \\ &= \int_{-\infty}^{\infty} |S(f)|^2 df = E \end{aligned} \tag{1.334}$$

where the gain $G = |H(0)|$ is normalized to one resulting in a unit gain-matched filter response $H(f)$.

Referring to the additive noise described by (1.332) and Section 1.6.2, the noise power at the output of the matched filter is expressed as

$$R_{no}(0) = \frac{N_o}{2}\int_{-B/2}^{B/2} |H(f)|^2 df \tag{1.335}$$

where $B/2$ is the baseband bandwidth of the matched filter.

*Adaptive channel equalizers are often used to compensate for the channel distortion.

The received signal, as expressed in (1.330), can be rewritten in terms of the optimally sampled matched filter output as

$$l(r(iT_o)) = \sqrt{E}d_i + \sqrt{\frac{N_o}{2}}\, n_i \tag{1.336}$$

where n_i are *iid* zero-mean, unit variance, white Gaussian noise samples. Upon dividing (1.336) by $\sqrt{N_o/2}$ the sampled matched filter output is expressed as

$$l'(r_i) = \sqrt{\frac{2E}{N_o}}d_i + n_i \tag{1.337}$$

The sampled values $l(r(iT_o))$ and $l'(r_i)$ are referred to as *sufficient statistics*, in that, they contain all of the information in $r(t)$, expressed in (1.330), to make a maximum-likelihood estimate $\hat{d}_i$ of the source-bit d_i. The normalized form in (1.337) is used as the turbo decoder input discussed in Section 8.12. In Section 1.8 the sufficient statistic is seen to be a direct consequence of the log-likelihood ratio.

1.7.2 Equivalence between Matched Filtering and Correlation

Consider the receiver input as the sum of the transmitted signal plus noise expressed as

$$r(t) = s(t) + n(t) \tag{1.338}$$

The cross-correlation of $r(t)$ with a replica of the received signal is computed as

$$R(\tau) = \int_{-\infty}^{\infty} r(t)s(t-\tau)dt \quad \text{: cross-correlation} \tag{1.339}$$

Defining the matched filter impulse response as $h(t)$, the matched filter output response to the input $r(t)$ is

$$y_o(t) = \int_{-\infty}^{\infty} r(\lambda)h(t-\lambda)d\lambda \quad \text{: convolution} \tag{1.340}$$

However, referring to the preceding matched filter discussion, the matched filter response is equal to the delayed image of the signal, such that,

$$h(t) = s(T_o - t) \tag{1.341}$$

As mentioned previously, the delay T_o ensures that the filter response is causal and, therefore, realizable. To substitute (1.341) into (1.340) first let $t' = t - \lambda$ so that $h(t-\lambda) = s(T_o - (t-\lambda)) = s(\lambda - (t - T_o))$ and substitute this result in (1.340) to get

$$\begin{aligned} y_o(t) &= \int_{-\infty}^{\infty} r(\lambda)h(t-\lambda)d\lambda \\ &= \int_{-\infty}^{\infty} r(\lambda)s(\lambda-(t-T_o))d\lambda \\ &= R(t-T_o) \end{aligned} \tag{1.342}$$

So that the convolution response is equal to the cross-correlation response delayed by T_o. If the input noise is zero, so that $r(t) = s(t)$, the same conclusion can be drawn regarding the autocorrelation response.

1.8 THE LIKELIHOOD AND LOG-LIKELIHOOD RATIOS

Bayes criterion is based on two events, referred to as hypothesis H_1 and H_0, that are dependent upon a priori probabilities P_1 and P_0 and the, respective, associated costs (C_{01},C_{11}) and (C_{10},C_{00}). Letting m correspond to the decision and n correspond to the hypothesis, the range of the cost is $0 \le C_{mn} \le 1$ with $C_{mn} + C_{nn}|_{m\neq n} = 1$. The cost of a correct decision is C_{nn} and an incorrect decision is $C_{mn}|_{m\neq n}$. For communication links the cost of incorrect decision is typically higher than a correct decision so that $C_{mn}|_{m\neq n} > C_{nn}$. For example, when $C_{mn}|_{m\neq n} = 1$ and $C_{nn} = 0$ the decision threshold minimizes the probability of error which is the goal of communication demodulators. In summary,

$$C_{mn} = \begin{cases} m = n & \text{: cost of correct decision} \\ m \neq n & \text{: cost of decision error} \end{cases} \tag{1.343}$$

and the a priori probabilities are typically known and equal.

In the following example, the hypotheses correspond to selecting $d_i = \{1,-1\}$, such that, under the two hypotheses

$$\begin{aligned} H_1 &\ : d_i = +1 \text{ with a prioi probability } P_1 \\ H_0 &\ : d_i = -1 \text{ with a prioi probability } P_0 \end{aligned} \tag{1.344}$$

with the observations

$$r_i = d_i + n_i \quad : t = iT_o \tag{1.345}$$

corresponding to the optimally sampled outputs of the matched filter. In terms of the a priori, the transition probabilities, and the cost functions, the hypothesis H_1: with $d_i = 1$ is chosen if the following inequality holds,

$$P_{r|H_1}(r|H_1)\ P_1(C_{01} - C_{11}) \ge P_{r|H_0}(r|H_0)\ P_0(C_{10} - C_{00}) \tag{1.346}$$

otherwise, chose H_0 with $d_i = -1$. The decisions are made explicit under the following rearrangement of (1.346)

$$\frac{P_{r|H_1}(r|H_1)}{P_{r|H_0}(r|H_0)} \mathop{\gtrless}_{H_0}^{H_1} \frac{P_0(C_{10}-C_{00})}{P_1(C_{01}-C_{11})} \tag{1.347}$$

Left and right sides of (1.347) are defined as the likelihood ratio (LR) $\Lambda(r)$ and decision threshold η or, alternately, as the log-likelihood ratio (LLR) $\ln\Lambda(\boldsymbol{r})$ with the threshold $\ln\eta$, so (1.347) is also expressed as

$$\Lambda(\boldsymbol{r}) \mathop{\gtrless}_{H_0}^{H_1} \eta \quad \text{or} \quad \ln\Lambda(\boldsymbol{r}) \mathop{\gtrless}_{H_0}^{H_1} \ln\eta \quad : \text{alternate decisions thresholds} \tag{1.348}$$

1.8.1 Example of Likelihood and Log-Likelihood Ratio Detection

Consider the two hypotheses H_1 and H_0 mentioned earlier with $d_i = \{1,-1\}$ and the observation r_i in (1.345) with the additive noise n_i characterized as *iid* zero-mean white Gaussian noise, denoted as $N(0,\sigma_n)$. The transition probabilities are expressed in terms of the Gaussian noise *pdf* as

$$p(n) = \frac{1}{\sqrt{2\pi}\sigma_n} e^{-(r-\bar{m})^2/2\sigma_n^2} \tag{1.349}$$

Upon forming the likelihood ratio and recognizing that $\bar{m} = \pm d_i$, the likelihood ratio decision simplifies to

$$\exp\left(\frac{2r_i}{\sigma_n^2}\right) \mathop{\gtrless}_{H_0}^{H_1} \frac{P_0(C_{10}-C_{00})}{P_1(C_{01}-C_{11})} \tag{1.350}$$

and the log-likelihood ratio decision simplifies to

$$\frac{2}{\sigma_n^2} r_i \mathop{\gtrless}_{H_0}^{H_1} \ln\left(\frac{P_0(C_{10}-C_{00})}{P_1(C_{01}-C_{11})}\right) \tag{1.351}$$

Recognizing that $l(r_i)$ is a sufficient statistic, (1.351) is rewritten as

$$l(r_i) \mathop{\gtrless}_{H_0}^{H_1} \frac{\sigma_n^2}{2} \ln\left(\frac{P_0(C_{10}-C_{00})}{P_1(C_{01}-C_{11})}\right) \tag{1.352}$$

When $C_{10} = C_{01} = 1$, $C_{00} = C_{11} = 0$, and $P_0 = P_1$ the LLR simplifies to

$$l(r_i) \mathop{\gtrless}_{H_0}^{H_1} 0 \quad : \ln(\eta = 1) = 0 \tag{1.353}$$

Therefore, the data estimate is $\hat{d}_i = 1$ when $l(r_i) > 0$, otherwise, $\hat{d}_i = -1$. Recall that observations $r_i : t = iT_o$ are made at the output of the matched filter. These concepts involving the LR and LLR surface again in Section 3.2 and the notion of the natural logarithm of the transitions probabilities is discussed in the following section involving parameter estimation.

1.9 PARAMETER ESTIMATION

The subject of optimum signal detection in noise was examined in the preceding section in terms of a pulsed-modulated carrier and it resurfaces throughout the following chapters in the context of a number of different waveform modulations. However, signal detection is principally based on the signal energy without regard to specific signal parameters, although frequency and range delay must be estimated to some degree to declare signal presence and subsequently detection. Signal detection uses concepts involving *direct probabilities*, whereas the subject of parameter estimation uses concepts involving *inverse probabilities* as discussed by Feller [32], Slepian [62], Woodward and Davies [63], and others. The distinction between these concepts is that *direct probability* is based on the probability of an event happening, whereas *inverse probability* formulates the best estimate of an event that has already occurred. With this distinction, it is evident that parameter estimation involves inverse probabilities. The major characteristic of inverse probability is the use of a priori information associated with the available knowledge of each source event. At the receiver the a posteriori probability is expressed in terms of the inverse probability using Bayes rule that associates the transition probability and a priori knowledge of the source events.

The subject of this section is signal parameter estimation and, although the major parameter of interest in communications is the estimation of the source information, the estimation of parameters like, frequency, delay, and signal and noise powers are important parameters that aid in the estimation of the source information. For example, estimation of the received signal and noise powers form the basis for estimating the receiver signal-to-noise ratio that is used in the network management to improve and maintain communication reliability. Furthermore, characterizing the theoretical limits in the parameter estimates provides a bench mark or target for the accuracy of the parameter estimation during the system design.

The following discussion of statistical parameter estimation is largely based on the work of Cramér [64], Rao [65], Van Trees [66], and Cook and Bernfeld [67]. The received signal is expressed in terms of the transmitted signal with M unknown parameters $a_1, a_2, \ldots, a_M$ and additive noise, as

$$r(t) = s(t; a_1, a_2, \ldots, a_M) + n(t) \tag{1.354}$$

Considering that N discrete samples of the received signal and additive noise are used to estimate the parameters, the joint probability density function (*pdf*) of the samples is

$$p_{\mathbf{r}}(r_1,r_2,\ldots,r_N|a_1,a_2,\ldots,a_M)=p_{\mathbf{n}}(r_1-s_1,r_2-s_2,\ldots,r_N-s_N) \tag{1.355}$$

where the noise samples $n_i = r_i - s_i$ are substituted into the joint *pdf* of the noise. The noise samples are statistically independent and the statistical characteristics of the noise are assumed to be known. Therefore, based on the received signal-plus-noise samples r_i, the receiver must determine the estimates $\hat{a}_1, \hat{a}_2, \ldots, \hat{a}_M$ of the M unknown parameters. The probability density function $p_r(r_1,\ldots|a_1,\ldots)$ in (1.355) is called the *likelihood function*.

Van Trees discusses several estimation criteria* and the following focuses on the optimum estimates for the *mean-square* (MS) error† and *maximum a posteriori probability* (MAP) criterion that are defined, respectively, for a single parameter a as

$$\hat{a}_{ms}(\boldsymbol{r})=\int_{-\infty}^{\infty} a p_{a|r}(a|\boldsymbol{r})da \quad \text{: MS estimate} \tag{1.356}$$

and

$$\left.\frac{\partial}{\partial a}\ln p_{a|r}(a|\boldsymbol{r})\right|_{a=\hat{a}_{map}(\boldsymbol{r})}=0 \quad \text{: MAP equation} \tag{1.357}$$

The estimate $\hat{a}_{ms}(\boldsymbol{r})$ is optimum in the sense that it results in the minimum MS error over all s_i and a. The MAP estimate $\hat{a}_{map}(\boldsymbol{r})$ is the solution to (1.357) and is optimum in the sense that it locates the maximum of the *a posteriori* probability density function; however, the solution must be checked to determine if it corresponds to the global maximum in the event of a multimodal distribution.

By applying Bayes rule to (1.357), the MAP estimate is expressed in terms of the a priori *pdf*, $p_{\mathbf{a}}(a)$, and the likelihood function, $p_{r|a}(\boldsymbol{r}|a)$, as

$$\left.\frac{\partial}{\partial a}\ln p_{r|a}(\boldsymbol{r}|a)+\frac{\partial}{\partial a}\ln p_{a}(a)\right|_{a=\hat{a}_{map}(\boldsymbol{r})}=0 \tag{1.358}$$

When the a priori probabilities are unknown, that is, as the a priori knowledge approaches zero, (1.358) becomes the *maximum-likelihood equation* and $\hat{a}_{ml}(\boldsymbol{r})$ is the *maximum-likelihood* estimate, evaluated as the solution to

$$\left.\frac{\partial}{\partial a}\ln p_{r|a}(\boldsymbol{r}|a)\right|_{a=\hat{a}_{ml}(\boldsymbol{r})}=0 \quad \text{: ML estimate} \tag{1.359}$$

To make use of these estimates it is necessary to determine the bias and the variance of the estimate. The mean value of the estimate is computed as

$$E[\hat{a}(\boldsymbol{r})]=\int_{-\infty}^{\infty}\hat{a}(\boldsymbol{r})p_{r|a}(\boldsymbol{r}|a)d\boldsymbol{r} \tag{1.360}$$

The bias of the estimate is defined as $B(a)=E[\hat{a}(\boldsymbol{r})]-a$. If, as indicated, the bias is a function of a, the estimate has an *unknown bias*, however, if the bias is B, independent of a, the estimate has a *known bias* that can be removed from the observation measurements r. In general, for any known biased estimate $\hat{a}(\boldsymbol{r})$ of the real random variable a, the variance is defined as

$$\sigma_{\hat{a}}^2=Var[\hat{a}(\boldsymbol{r})-a]\triangleq E\left[(\hat{a}(\boldsymbol{r})-a)^2\right]-B^2 \tag{1.361}$$

Although the bias and variance are often difficult to determine, the Cramér–Rao inequality provides a lower bound on the variance of the estimate. For a biased estimate of the random parameter a with a priori *pdf* $p_a(a)$, the variance is lower bounded by the Cramér–Rao inequality [64, 66]

$$\sigma_{\hat{a}}^2\geq\frac{(\partial E[\hat{a}(\boldsymbol{r})]/\partial a)^2}{E\left[\left\{\frac{\partial\ln(p_{r|a}(\boldsymbol{r}|a))}{\partial a}+\frac{\partial\ln(p_a(a))}{\partial a}\right\}^2\right]} \quad \text{: biased; variable} \tag{1.362}$$

or, the equivalent result,

$$\sigma_{\hat{a}}^2\geq\frac{(\partial E[\hat{a}(\boldsymbol{r})]/\partial a)^2}{-E\left[\frac{\partial^2\ln(p_{r|a}(\boldsymbol{r}|a))}{\partial a^2}+\frac{\partial^2\ln(p_a(a))}{\partial a^2}\right]} \tag{1.363}$$
$$\text{: biased; variable}$$

When the estimate is unbiased, that is, when $E[\hat{a}(\boldsymbol{r})]=a$, the estimation variance of the random variable a simplifies to

$$\sigma_{\hat{a}}^2\geq\frac{1}{E\left[\left\{\frac{\partial\ln(p_{r|a}(\boldsymbol{r}|a))}{\partial a}+\frac{\partial\ln(p_a(a))}{\partial a}\right\}^2\right]} \tag{1.364}$$
$$\text{: unbiased; variable}$$

or, the equivalent result,

$$\sigma_{\hat{a}}^2\geq\frac{1}{-E\left[\frac{\partial^2\ln(p_{r|a}(\boldsymbol{r}|a))}{\partial a^2}+\frac{\partial^2\ln(p_a(a))}{\partial a^2}\right]} \tag{1.365}$$
$$\text{: unbiased; variable}$$

*The criteria are based on Bayes estimation procedure that minimizes the risk associated with the cost or weight assigned to various kinds of statistical decisions. In communication systems, the decision costs are assigned to minimize the demodulator bit-error probability.

†The MS error is also referred to as the *minimum mean-square error* (MMSE).

The Cramér–Rao bound in these relationships is formulated in terms of the Schwarz inequality and the equality condition applies when

$$\frac{\partial \ln(p_{r|a}(\boldsymbol{r}|a))}{\partial a} + \frac{\partial \ln(p_a(a))}{\partial a} = k(\hat{a}(\boldsymbol{r}) - a) \quad (1.366)$$

$$: \text{MAP efficient estimate condition}$$

where k is a constant. Therefore, (1.366) guarantees that the equality condition for the variance applies in (1.362) through (1.365); in this case, the MAP estimate is defined as an efficient estimate. Furthermore, an unbiased estimate, excluding the trivial case $k = 0$, requires that $\hat{a}(\boldsymbol{r}) = a$ leading to (1.358).

When the a priori knowledge $p_a(a)$ is constant, that is, the parameter a is nonrandom, or unknown, then (1.359) also requires that $\hat{a} = a$ or $\hat{a}_{ml}(\boldsymbol{r}) = a$. Under the maximum-likelihood (ML) criteria Schwarz's equality condition applies when

$$\frac{\partial \ln(p_{r|a}(\boldsymbol{r}|a))}{\partial a} = k(a)(\hat{a}(\boldsymbol{r}) - a) \quad (1.367)$$

$$: \text{ML efficient estimate condition}$$

In this case, the constant $k(a)$ may be a function of a; this condition only applies when parameter a is a constant which corresponds to the ML estimate.

Van Trees lists three principles based on the forgoing results:

1 The mean-square (MS) error estimate is always the mean of the a posteriori density, that is, the conditional mean.
2 The MAP estimate corresponds to the value of a for which the a posteriori density is maximum.
3 For a large class of cost functions, the optimum estimate is the conditional mean whenever the a posteriori density is a unimodal function which is symmetric about the conditional mean. The Gaussian *pdf* is a commonly encountered example.

By way of review, the estimates are evaluated using the a posteriori *pdf*; however, if the parameter is a random variable, the a posteriori *pdf* must be expressed in terms of the transition distribution and the a priori *pdf* of the random parameter using Bayes rule. If the estimate is unbiased, that is, if $B = E[\hat{a}(\boldsymbol{r})] - a = 0$, evaluation of the Cramér–Rao bound simplifies to (1.364); it is sometimes necessary to use the equivalent expression in (1.365). The Cramér–Rao equality condition is established if the left-hand side of (1.366) can be expressed in terms of the right-hand side where k is a constant parameter resulting from Schwarz's condition for equality.

If the a priori knowledge is unknown then the maximum-likelihood equation given in (1.359) is used to determine maximum-likelihood estimate. In this case, the Cramér–Rao bound is established by omitting the dependence of $p_a(a)$ in, (1.362) through (1.365) and the equality condition is established if the left-hand side of (1.367) can be expressed in terms of the right-hand side where, in this case, the constant $k(a)$ is a function of the parameter a. With either the MAP or ML estimates, if the bias is zero and the equality condition applies, the estimate is referred to as an efficient estimate.

Van Trees shows that for the MS estimate to be an efficient estimate, the a posteriori probability density $p_{a|r}(a|r)$ must be Gaussian for all r and, for efficient MAP estimates, $\hat{a}_{ms}(r) = \hat{a}_{map}(r)$. However, it may be easier to solve the MAP equation than to determine the conditional mean as required by the MS estimation procedure.

1.9.1 Example of MS and MAP Parameter Estimation

As an example application of the parameter estimation procedures discussed earlier, consider the Poisson distribution that is used to predict population growth, telephone call originations, gamma ray emissions from radioactive materials, and is central in the development of queueing theory [68]. For this example, the Poisson distribution is characterized as

$$p_a(a|n) = \frac{a^n}{n!} e^{-a} \quad : a \geq 0; \; n = 0, 1, 2, \ldots \quad (1.368)$$

In the application of (1.368) to queueing theory, $a = \lambda t$ is the average number of people entering a queueing line in the time interval 0 to t and λ is the arrival rate. The a posteriori distribution $p_{a|n}(a|n)$ is the probability of a conditioned on exactly n arrivals occurring in the time interval. A fundamental relationship in the Poisson distribution is that the time interval *between* people entering the queueing line is exponentially distributed and is characterized by the a priori distribution

$$p_a(a) = e^{-a} \quad : a \geq 0 \quad (1.369)$$

The a posteriori *pdf* in (1.368) is expressed in terms of the a priori and transition *pdfs* as

$$p_{a|n}(a|n) = \frac{p_{n|a}(n|a)p_a(a)}{p_n(n)} = \frac{k}{n!} a^n e^{-2a} \quad (1.370)$$

where the constant k is a normalizing constant that includes $1/p_n(n)$. Integrating of the second equality in (1.370) with respect to da over the range 0 to ∞ and setting the result equal to one, the value of k is found to be $k = 2^{n+1}$ and (1.370) becomes

$$p_{a|n}(a|n) = \frac{2^{n+1}}{n!} a^n e^{-2a} \tag{1.371}$$

Using (1.356) and (1.371) the MS estimate is evaluated as

$$\hat{a}_{ms}(n) = \frac{2^{n+1}}{n!} \int_0^\infty a^{n+1} e^{-2a} da = \frac{n}{2} \tag{1.372}$$

Also, using (1.357) and (1.371) the MAP estimate is evaluated as

$$\frac{\partial}{\partial a}\left(\ln\left\{\frac{2^{n+1}}{n!}a^n e^{-2a}\right\}\right) = \frac{\partial}{\partial a}\left(\ln\left\{\frac{2^{n+1}}{n!}\right\} + n\ln(a) - 2a\right)$$
$$= \frac{n}{a} - 2\bigg|_{a=\hat{a}_{map}(n)} = 0 \tag{1.373}$$

and solving the second equality in (1.373) for a results in $\hat{a}_{map}(n) = n/2$. As is typical in many cases, the MS and MAP estimation procedures result in the same estimate. It is left as an exercise (see Problem 38) to determine the bias of the estimates, compute the Cramér–Rao bound, and using (1.366), determine if the estimates are efficient, that is, if the Cramér–Rao equality condition applies.

1.9.2 Constant-Parameter Estimation in Gaussian Noise

To simplify the description of the estimation processing, the analysis in this section considers the single constant-parameter case with zero-mean narrowband additive Gaussian noise. Under these conditions, the estimation is based on the solution to the maximum-likelihood equation with the joint *pdf* of the received signal and noise written as

$$p_{\mathbf{r}}(r_1, r_2, \ldots, r_N; a) = p_{\mathbf{n}}(r_1 - s_1, r_2 - s_2, \ldots, r_N - s_N)$$
$$= \prod_{i=1}^{N} p_{\mathbf{n}}(r_i - s_i) \tag{1.374}$$

where a is the constant parameter to be estimated and $r_i = s_i + n_i$ represents the received signal samples. The sampling rate satisfies the Nyquist sampling frequency and the second equality in (1.374) recognizes that the noise samples are independent. The following analysis is based on the work of the Woodard [24] and Skolnik [61] and uses the maximum-likelihood estimate of (1.359) with the Cramér–Rao bound expressed by (1.365).

Using (1.374) with zero-mean AWGN, the minimum Cramér–Rao bound on the variance of the estimate is expressed as

$$\sigma\hat{a}^2(\min) = -E\left[\frac{\partial^2}{\partial a^2}\ln\left\{\prod_i^N \frac{1}{\sqrt{2\pi}\sigma_n}\exp\left(-\frac{(r_i - s_i)^2}{2\sigma_n^2}\right)\right\}\right]^{-1}$$
$$= -E\left[\frac{\partial^2}{\partial a^2}\ln\left\{k\exp\left(\frac{-1}{2N_o}\sum_{i=1}^N (r_i - s_i)^2 \Delta t\right)\right\}\right]^{-1}$$
$$= -E\left[\frac{\partial^2}{\partial a^2}\ln\left\{k\exp\left(\frac{-1}{2N_o}\int_0^{T_e}(r(t) - s(t))^2 dt\right)\right\}\right]^{-1} \tag{1.375}$$

In arriving at the third equality in (1.375) the factor k is independent of the parameter a and, it is recognized that, $\sigma_n^2 = N_o B$ where $B = 1/T_e$ is the bandwidth corresponding to the estimation time. The integral is formed by letting $\Delta t \to 0$ as the number of samples $N \to \infty$ over the estimation interval T_e. Upon taking the logarithm of the product $k\exp(-)$ and performing the partial derivatives on the integrand, (1.375) simplifies to

$$\sigma\hat{a}^2(\min) = E\left[\frac{-1}{N_o}\int_0^{T_e}\left\{(r(t) - s(t))\frac{\partial^2 s(t)}{\partial a^2} - \left(\frac{\partial s(t)}{\partial a}\right)^2\right\}dt\right]^{-1}$$
$$= \left[\frac{1}{N_o}\int_0^{T_e}\left(\frac{\partial s(t)}{\partial a}\right)^2 dt\right]^{-1} \tag{1.376}$$

The last equality in (1.376) is the basis for determining the variance and is obtained by moving the expectation inside of the integral and recognizing that $E[r(t) - s(t)] = E[n(t)] = 0$. The following example outlines the procedures for estimating the variance of the estimate using the ML procedures.

1.9.2.1 Example of ML Estimate Variance Evaluation

Consider the signal $s(t)$ expressed as

$$s(t) = A\cos(\omega_o t + \dot{\omega}t^2/2 + \phi) \tag{1.377}$$

where A is the peak carrier voltage, ω_o is the IF angular frequency, $\dot{\omega}$ is the angular frequency rate, and ϕ is a constant phase angle; the signal power is defined $P_s = A^2/2$.

The variance of the frequency estimate is determined by squaring the partial derivative of $s(t)$ respect to ω_o and integrating over the estimation interval T_e as indicated in (1.376). Under these conditions the analysis of the Cramér–Rao lower bound is performed as follows.

$$\sigma^2_{\hat{\omega}_o}(\min) = \left[\frac{A^2}{N_o}\int_0^{T_e} t^2 \sin^2(\omega_o t + \dot{\omega}t^2/2 + \phi)\,dt\right]^{-1}$$
$$= \left[\frac{P_s}{N_o}\int_0^{T_e} t^2\left(1 - \cos(2\omega_o t + \dot{\omega}t^2 + 2\phi)\right)dt\right]^{-1} \tag{1.378}$$

Upon neglecting the term involving $2\omega_o$ and performing the integration, (1.378) becomes

$$\sigma^2_{\hat{\omega}_o}(\min) = \frac{3}{T_e^2 \gamma_e} \quad : \text{rad}^2/\text{s}^2 \tag{1.379}$$

where $\gamma_e = P_s T_e / N_o$ is the signal-to-noise ratio in the estimation bandwidth of $1/T_e$. In terms of the carrier frequency f_o in hertz, the standard deviation of the estimate is

$$\sigma_{\hat{f}_o}(\min) = \frac{\sqrt{3}}{2\pi T_e \sqrt{\gamma_e}} \quad : \text{Hz} \tag{1.380}$$

In a similar manner, the standard deviation of the frequency rate and phase are evaluated as

$$\sigma_{\hat{\dot{f}}}(\min) = \frac{\sqrt{5}}{\pi T_e^2 \sqrt{\gamma_e}} \quad : \text{Hz/s} \tag{1.381}$$

and

$$\sigma_{\hat{\phi}}(\min) = \frac{1}{\sqrt{\gamma_e}} \quad : \text{rad} \tag{1.382}$$

The evaluation of the standard deviation of the signal amplitude (A) estimation error is left as an exercise (see Problem 39).

1.9.3 Received Signal Delay and Frequency Estimation Errors

Accurate estimation of the signal delay and frequency are essential in aiding the signal acquisition processing by minimizing the overall time and frequency search ranges. The delay estimation accuracy (σ_{Td}) is inversely related to the signal bandwidth (B) and the signal frequency estimation accuracy (σ_{fd}) is inversely related to the time duration (T) of the signal. Neglecting the signal-to-noise dependence of each measurement, these inverse dependencies are evident, in that, the product $\sigma_{Td}\sigma_{fd} \propto 1/TB$ where TB is the time-bandwidth product of the waveform. For typical modulated waveforms T and B are inversely related so that simultaneous accurate time and frequency estimates are not attainable. However, the use of spread-spectrum (SS) waveform modulation provides for arbitrarily large BT products with simultaneous accurate estimates of T_d and f_d. The analysis of delay and frequency estimation errors in the following sections is based on the work of Skolnik [61] and can be applied to conventional or SS-modulated waveforms. In Section 1.9.3.3 delay and frequency estimation is examined using a DSSS-modulated waveform.

1.9.3.1 Delay Estimation Error Based on Effective Bandwidth The signal delay measurement accuracy using the effective signal bandwidth was introduced by Gabor [69] and is discussed by Woodward [24] and defined as the standard deviation of the delay measurement expressed as*

$$\sigma_{T_d} = \frac{1}{\beta\sqrt{2\gamma_e}} \quad : \text{second} \tag{1.383}$$

where $\gamma_e = P_s/N_oW = E/N_o$ is the signal-to-noise ratio† measured in the two-sided bandwidth W, N_o is the one-sided noise density, P_s is the signal power, and β is the effective bandwidth of the signal. β^2 is the normalized second moment of the waveform spectrum $|S(f)|^2$, defined as

$$\beta^2 \triangleq \frac{\int_{-\infty}^{\infty} (2\pi f)^2 |S(f)|^2 df}{\int_{-\infty}^{\infty} |S(f)|^2 df} \tag{1.384}$$

The denominator in (1.384) is the signal energy and the integration limits extend over the frequency range $f \leq |W/2|$ corresponding to the nonzero signal spectrum. The one-way range error corresponding to (1.383) is $\sigma_{rng} = c\sigma_{T_d}$ meters where c is the free-space velocity of light in meter/second.

For the rectangular symbol modulation function *rect* (t/T), band limited to W Hz with $\beta^2 \cong W/T$ and large time-bandwidth products $WT/2$, (1.383) is evaluated as

$$\sigma_{T_d} = \frac{T}{\sqrt{WT\gamma_e}} \quad : \text{band-limited } rect(t/T) \text{ symbol; } WT/2 \gg 1 \tag{1.385}$$

1.9.3.2 Frequency Estimation Error Based on Effective Signal Duration In a manner similar to the analysis of the delay estimation error in the preceding section, Manasse [70] shows that the, minimum root-mean-square (*rms*) error in the frequency estimate is given by‡

$$\sigma_{f_d} = \frac{1}{\alpha\sqrt{2\gamma_e}} \quad : \text{Hz} \tag{1.386}$$

where $\gamma_e = E/N_o$ is the signal-to-noise ratio measured in the two-sided bandwidth W, N_o is the one-sided noise density, P_s is the signal power, and α is the effective time duration of the

*Woodward refers to the delay error as the standard deviation of the error. Slepian's analysis [62] is based on the likelihood function and also arrives at (1.383).

†The factor of two in γ_e results because N_o is the one-sided noise density and W is the two-sided bandwidth.

‡The notation σ_{fd} is used to denote the standard deviation of the frequency estimation error.

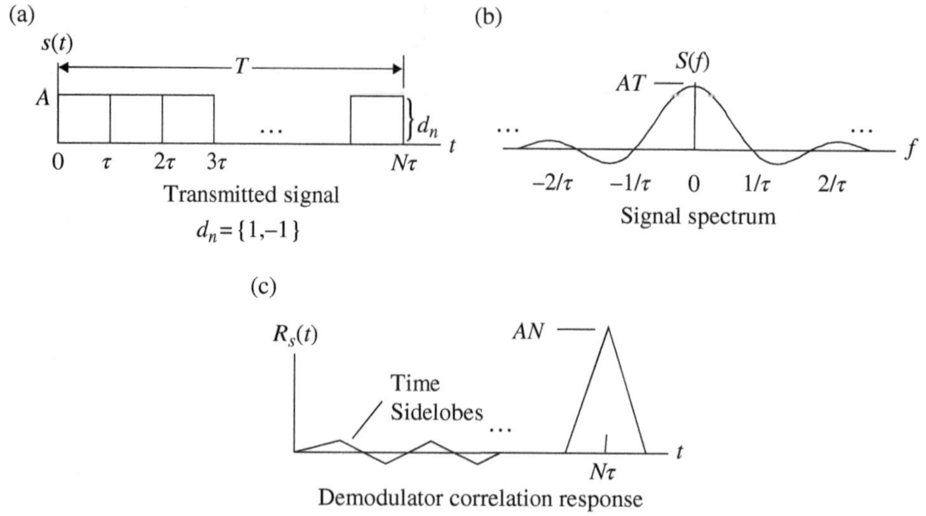

FIGURE 1.29 Time–frequency estimation using DSSS waveform.

received signal. The parameter α^2 is the normalized second moment of the waveform $s(t)$ and is defined as

$$\alpha^2 \triangleq \frac{\int_{-\infty}^{\infty} (2\pi t)^2 s^2(t)dt}{\int_{-\infty}^{\infty} s^2(t)dt} \tag{1.387}$$

The Doppler frequency results from the velocity (v) and the carrier frequency (f_c) and is expressed as $f_d = (v/c)f_c$. Frequency errors resulting from hardware oscillators are usually treated separately and combined as the root-sum-square (RSS) of the respective standard deviations.

For the band-limited $rect(t/T)$ symbol modulation used in the preceding section, the normalized second moment is evaluated as $\alpha^2 \cong (\pi T)^2/3$ and (1.386) is expressed as

$$\sigma_{fd} = \frac{\sqrt{3}}{\pi T\sqrt{2\gamma_e}} \quad : \text{band-limited } rect(t/T) \text{ symbol; } WT/2 \gg 1 \tag{1.388}$$

Comparison of (1.385) and (1.388) demonstrates the inverse relationship between the estimation accuracy of the range-delay and frequency errors for conventional (unspread) modulations. For example, for a given time bandwidth (WT) product and signal-to-noise ratio (γ_e), the delay estimate error decreases with decreasing symbol duration; however, the frequency estimate error increases. The issue resolves about the signal-to-noise ratio in the estimation bandwidth. For example, with conventional waveform modulations, $WT = 2BT = 2$ so $BT = 1$ and the bandwidth changes inversely with the symbol duration. Consequently, by decreasing symbol duration, the bandwidth increases resulting in a signal-to-noise γ_s, measured in the symbol bandwidth, degradation of B/B' where $B' > B$. Therefore, in the previous example, to maintain a constant signal-to-noise ratio γ_e the estimation interval must be appropriately adjusted. As mentioned previously, the solution to simultaneously obtaining accurate estimates of range delay and frequency while maintaining a constant γ_s, involves the use of a SS-modulated signals with an inherently large WT product as discussed in the following section.

1.9.3.3 Improved Frequency and Time Estimation Errors Using the DSSS Waveform The DSSS waveform uses a pseudo-noise (PN) sequence of chips with and instantaneous bandwidth (W) over the estimation interval (T) as shown in Figure 1.29.* The resulting large WT product signal provides for arbitrarily low time and frequency estimation errors. This is accomplished by the respective selection of a high bandwidth (short duration) chip interval (τ) and the low bandwidth (long duration) estimation interval T. The estimation interval can be increased to improve the frequency estimate; conversely, the chip interval can be decreased to improve the range-delay estimate; however, to maintain the accuracy of the other, the number of chips per estimation interval (N) must be increased. These relationships are described in terms of the pulse compression ratio, defined as $\rho = T/\tau = N$. In Figure 1.29 the chips are depicted as appropriately delayed $Ad_n rect((t-n)/\tau - 0.5)$: $n = 0, \ldots, N-1$ functions and, because of the equivalence between the correlator and matched filter, the peak correlator output is a triangular function with a peak value† of AN. When sampled at $t = N\tau$, the correlator output

*Although not shown in the figure, the chips are ± binary pulses which suggest that the received signal carrier frequency is BPSK modulated.

†This suggests that the local reference is an identical unit amplitude PN sequence that is synchronized with the received PN sequence.

results in the maximum signal-to-noise ratio measured in the bandwidth of $1/T$ Hz.

Based on the fundamental principles for jointly achieving accurate time and frequency estimates as stated earlier, the triangular shape of the wide bandwidth correlator output is related to the accuracy of the time estimate and the low bandwidth sampled output at interval of $T = N\tau$ determines the accuracy of the frequency estimate. Therefore, evaluation of the time and frequency estimation accuracies of the DSSS waveform involves evaluating, respectively, the effective bandwidth (β) of the triangular function and the effective time duration (α) of the $rect(t/T - 0.5)$ function.

Delay Estimation Error of the DSSS Waveform The delay estimation error is based on detecting the changes in the leading and trailing edge of wideband signals. This does not require that the signal has a short duration but that the bandwidth is sufficiently wide to preserve the rapid rise and fall times of the correlator response. On the other hand, received signals with additive noise must be detected and the parameters estimated under the optimum signal-to-noise conditions as provided by matched filtering or correlation. In this regard, the correlator output in Figure 1.29 is examined in the context of the signal delay estimate error.

The triangular function, corresponding to the correlator output, is an isosceles triangle with base and height equal to 2τ and AN, respectively, and is described as

$$R_s(\xi) = AN(1 - |\xi|/\tau) \quad : |\xi| \le \tau \tag{1.389}$$

where, for convenience, $\xi = t - N\tau$ such that the time axis is shifted so that the isosceles triangle is symmetrical about $\xi = 0$. The effective bandwidth of $R_s(\xi)$ is evaluated (see Problem 41) as

$$\beta = \frac{\sqrt{3}}{\tau} \quad : \text{triangular function} \tag{1.390}$$

and the corresponding standard deviation of the delay estimate is

$$\sigma_{T_d} = \frac{\tau}{\sqrt{3}\sqrt{2\gamma_e}} \quad : \text{triangular function} \tag{1.391}$$

Frequency Estimation Error of the DSSS Waveform The frequency estimation error is based on the interval T of the PN sequence under the conditions corresponding to the local PN reference being exactly synchronized with and multiplied by the received signal; in other words, with zero frequency and phase errors, the integrand of the correlation integral is constant over the interval T. However, with a frequency error of f_e Hz the correlator response is computed as

$$\begin{aligned} R_s(T;f_e) &= A\int_0^T e^{j2\pi f_e \xi} d\xi \\ &= ATe^{j\pi f_e T}\frac{\sin(\pi f_e T)}{\pi f_e T} \quad : \text{correlator response with } f_e \end{aligned} \tag{1.392}$$

The principal frequency error in the main lobe of the *sinc* $(f_e T)$ function corresponds to $|f_e| \le 1/T$ which defines the fundamental resolution accuracy of the frequency estimate. However, the effective duration of the correlator of length $T = N\tau$ is evaluated (see Problem 42) as

$$\alpha = \frac{2\pi T}{3} \quad : \text{rectangular function} \tag{1.393}$$

and the corresponding standard deviation of the frequency estimate is

$$\sigma_{fd} = \frac{\sqrt{3}}{2\pi T\sqrt{2\gamma_e}} \quad : \text{rectangular function} \tag{1.394}$$

Considering the SS pulse compression ratio, or processing gain, $\rho = T/\tau$, the correlator output signal-to-noise ratio (γ_e) in (1.391) and (1.394) is measured in the bandwidth of $1/T$. The product of the estimation accuracies of the SS waveform is

$$\sigma_{T_d}\sigma_{fd} = \frac{\tau}{2\pi T\gamma_e} = \frac{1}{2\pi\rho\gamma_e} \tag{1.395}$$

Therefore, the time and frequency estimates accuracies can be made arbitrarily low, even in low signal-to-noise ratio environments, by designing a SS waveform with a sufficiently high pulse compression ratio.

1.9.3.4 Effective Bandwidth of SRRC and SRC Waveforms In view of the increasing demands on bandwidth, the spectral containment of the spectral raised-cosine (SRC) waveform meets the corresponding need for spectrum conservation. Although the spectral root-raised-cosine (SRRC) waveform has a slightly wider bandwidth than the SRC waveform, it is preferred because of the improved matched filter detection* and, in the context of range delay estimation, provides for a somewhat better range delay estimate. The spectrum of the SRC waveform is characterized, in the context of a spectral windowing function, in Section 1.11.4.1 and the spectrum of the SRRC is characterized in Section 4.3.2 in the context of the optimum transmitted waveform for root-raised-cosine (RRC) waveform modulation. The following analysis compares the effective

*The optimally sampled SRRC matched filter output results in the maximum signal energy and orthogonal symbol samples resulting in zero intersymbol interference.

bandwidth of the SRC and SRRC waveforms with the understanding that the SRC delay estimate is based on the matched filter out samples taken symmetrically about the optimum matched filter sample at $t = T_o$.

The dependence of the effective bandwidth of the SRRC and SRC waveforms on the excess bandwidth parameter α is expressed as

$$\beta T = \pi\sqrt{\frac{1}{3} + \left(\frac{\pi^2 - 8}{\pi^2}\right)\alpha^2} \quad : \text{SRRC waveform} \qquad (1.396)$$

and

$$\beta T = \pi\sqrt{\frac{4 - 3\alpha + 12\alpha^2 - \alpha^3}{3(4-\alpha)} - \frac{8\alpha^2(16-\alpha)}{(2\pi)^2(4-\alpha)}} \quad : \text{SRC waveform} \qquad (1.397)$$

Equations (1.396) and (1.397) are plotted in Figure 1.30 that demonstrates the advantages of the wider bandwidth SRRC waveform in providing short rise-times symbols with the associated improvement in range-delay detection. In this regard, the *rect(t/T)* modulated received symbol, as characterized by the BPSK-modulated waveform, has zero rise-time and results in perfect range-delay detection in a noise-free channel and receiver; however, infinite bandwidth is required to achieve this performance. The *dashed* curve in Figure 1.30 shows the normalized effective bandwidth of the *rect(t/T)* modulated symbol after passing through an ideal (1/*B*)*rect* (*f/B*) filter with one-sided bandwidth *B*/2 Hz; this is referred to a *filtered* BPSK and is discussed in the following section.

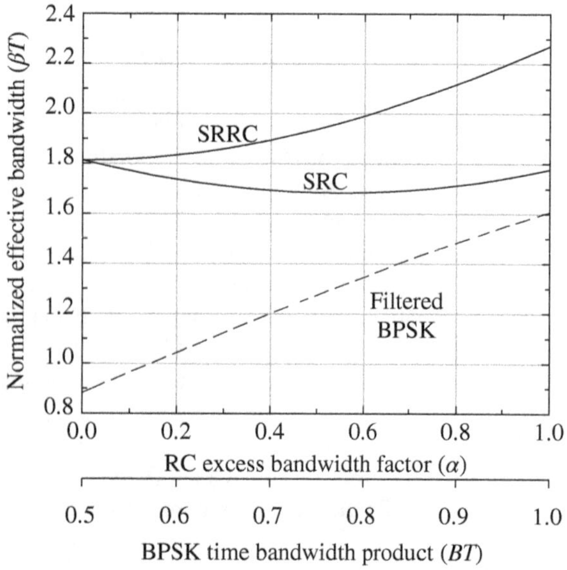

FIGURE 1.30 Normalized effective bandwidths for SRRC and SRC waveforms.

The noise bandwidth of the SRRC frequency function is significant, in that, it corresponds to the demodulator-matched filter response used in the detection of the SRRC-modulated waveform. On the other hand, the interest in the noise bandwidth of the SRC is more academic in nature because of its application as a windowing function. In either event, the noise bandwidth of the SRRC and SRC frequency functions is examined in Problem 44.

1.9.3.5 Effective Bandwidth of the Ideally Filtered rect (t/T) Waveform

In this case the effective bandwidth of the ideal symbol modulation, characterized by *rect(t/T)*, is evaluated after passing through an ideal filter with frequency response (1/*B*)*rect*(*f/B*) where *B*/2 is the one-sided or low-pass bandwidth of the filter. The filter response is examined in Section 1.3 and the normalized effective bandwidth of the filtered symbol is characterized by Skolnik [71] as

$$\beta T = \sqrt{\frac{\pi BT - \sin(\pi BT)}{Si(\pi BT) + (\cos(\pi BT) - 1)/\pi BT}} \qquad (1.398)$$

$$: \text{ideally filtered } rect(t/T) \text{ waveform}$$

This result is plotted in Figure 1.31 as a function of *BT* where *T* is the symbol duration. From this plot it is evident that as *BT* approaches infinity the standard deviation of the range-delay estimate approaches zero resulting in an exact estimate of the true range delay. A practical application is to define a finite bandwidth which is sufficiently wide so as not to degrade the symbol detection through intersymbol interference.

Defining the excess bandwidth factor for the ideal filter as $\alpha' = B/R_s = BT$, where $R_s = 1/T$ is the *rect(t/T)* symbol rate, in

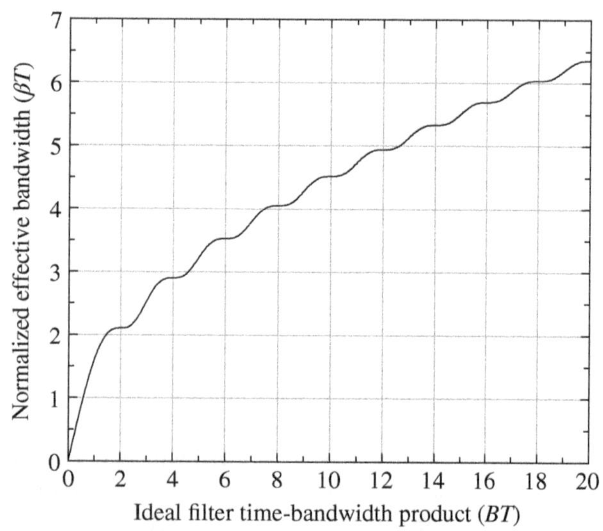

FIGURE 1.31 Normalized effective bandwidth for filtered *rect(t/T)* waveform.

terms of the excess bandwidth factor α of the raised-cosine (RC) waveform, $\alpha' = (1+\alpha)/2$. The corresponding range of BT is $1/2 \le BT \le 1$; this range of the filtered *rect(t/T)* effective bandwidth from Figure 1.31 is plotted as the *dashed* curve in Figure 1.30. The range of BT results in significant intersymbol interference and received symbol energy loss even under the ideal conditions of symbol time and frequency correction. However, under the same conditions, if the SRRC waveform and matched filter responses are sufficiently long, the intersymbol interference and symbol energy loss will be negligible. At the maximum SRRC normalized effective bandwidth of $\beta T = 2.27$, the filter time bandwidth product corresponds to $BT = 2.64$ or a 32% increase with the filter bandwidth spanning the main signal spectral lobe and 16% of the adjacent sidelobes. In other words, the one-sided filter bandwidth spans 1.32 lobes and, referring to Appendix A, this results in a performance loss of about 1.25 dB for BPSK waveform modulation; for a loss of less than 0.3 dB the BT product should be greater than 5 with a resulting effective bandwidth of $\beta T = 3.23$ corresponding to a $(3.23/2.27 - 1)100 = 42\%$ improvement relative to the best SRRC range-delay estimation error; however, the required bandwidth is 150% wider. The bandwidth and range-delay estimation accuracy are design trade-off in the waveform selection.

1.10 MODEM CONFIGURATIONS AND AUTOMATIC REPEAT REQUEST

The three basic modulator and demodulator configurations are simplex, half-duplex, and full-duplex. The definition of simplex communications involves communication in one direction between a modulator/transmitter and a remote receiver/demodulator. Examples of simplex communications include broadcasting from radio and television stations or from various types of monitoring devices. Half-duplex communications is a broader definition including two-way communications but only in one direction at a time. In these cases, transceivers and modems are required at each location. A common application of half-duplex operation is the *push-to-talk* handheld radios. Full-duplex communications provide the capability to communicate in both directions simultaneously. In these cases the bidirectional communications may use identical transceivers and modems operating at the same symbol rate; however, as is often the case, the communication link in one direction may be designated as the reverse channel and operated at a lower symbol rate. In either event, the forward and reverse channels must operate at different, noninterfering, frequencies.

The transfer of data is often performed using information frames or packets, each containing a cyclic redundancy check (CRC) code for error checking. If an error is detected the receiving terminal requests that the frame be retransmitted, otherwise an acknowledgment may be returned indicating that the frame was received without error. These protocols are referred to as automatic repeat request [72] (ARQ). The ARQ protocol requires either a half-duplex or full-duplex communication capability. The two commonly used variations of the ARQ protocol are generally referred to as idle-repeat request (RQ) and continuous-RQ.[*] However, more complex variations involving point-to-point and multipoint protocols are also defined.[†]

The remainder of this section analyzes the idle-RQ protocol which is the simplest ARQ system to implement and evaluate, in that, when a data frame is transmitted a timer is initiated and a new frame is transmitted only after acknowledgment (ACK) that the current frame was received without errors and/or the timer has not exceeded a maximum timeout T_{max}. However, the current frame is retransmitted if the timer exceeds T_{max}, a negative acknowledgment (NAK) is received, indicating the receipt of an incorrect frame, or the ACK or NAK code is received in error. The timer limit is based on the information bits per frame, the bits in the ACK and NAK codes, the date rates, and the expected two-way link propagation delay through the media. The idle-RQ implementation also has the advantage of requiring less data storage compared to the continuous-RQ protocol and the Go-back-N protocol [73]. However, the performance cost of these advantages is that the end-to-end transmission efficiency is lower and more sensitive to the link propagation delay. The end-to-end transmission efficiency is defined as

$$\eta_{trans} \triangleq \frac{\bar{R}_{bf}}{R_{bf}} \tag{1.399}$$

where $\bar{R}_{bf}$ is the average bit rate over the forward channel and R_{bf} is the uninterrupted forward channel bit rate.

The idle-RQ is modeled as shown in Figure 1.32 with the delays and other parameters defined in Table 1.12.

Using the parameters described earlier, the average time associated with the transmission and acknowledgment is described as

$$\bar{T} = \bar{N}\left(\left(T_{df} + T_{sf}\right) + 2T_p + T_{cs} + \left(T_{dr} + T_{sr}\right) + T_{cp}\right) \tag{1.400}$$

where $\bar{N}$ is the average number of frame repetition based on the specified bit-error probability and the number of frame information and CRC bits N_B. The computation of $\bar{N}$ is based on *iid* additive white Gaussian channel noise over all of the N_B bits. This provides for the probability of a correct message to be expressed in terms of the discrete binomial distribution[‡] given the bit-error probability P_{bef}; the result is expressed as

[*]Reference 72, Chapter 4, Protocol Basics.

[†]Reference 72, Chapter 5, Data Link Control Protocols.

[‡]See Table 1.42.

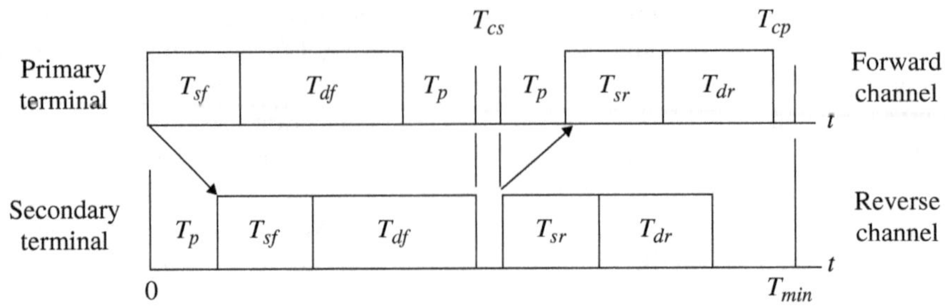

FIGURE 1.32 Model of idle-RQ implementation.

TABLE 1.12 Idle-ARQ Parameter Definitions

Delay	Value	Description
T_{df}	$(N_b + N_{crc})/R_{bf}$	Forward message duration
T_{sf}	N_{sf}/R_{bf}	Forward synchronization duration
T_p	*range/c*	Propagation delay between terminals
T_{dr}	$(N_{br}+N_{chk})/R_{br}$	Reverse message duration
T_{sr}	N_{sr}/R_{br}	Reverse synchronization duration
T_{cs}	0	Secondary terminal computational delay (ms)[a]
T_{cp}	0	Primary terminal computational delay (ms)[a]
T_{max}	$>T_{min}$	Idle-RQ maximum idle time (ms)
N_B	Variable	Message bits: (N_b = info) + (N_{crc} = CRC)
N_{sf}	30	Forward synchronization bits
R_{bf}	100	Forward bit rate (kbps)
range	Parameter	One-way: 18.5, 200, 600, 35,800 (km)
c	3×10^8	Free-space velocity (m/s)
N_{brt}	30	ARQ bits:(N_{br}=ACK) + (N_{chk} = parity)
N_{sr}	10	Reverse synchronization bits
R_{br}	$= R_{bf}$	Reverse bit rate (kbps)
P_{bef}	Parameter	Forward channel bit-error probability: 10^{-4}, 10^{-5}
P_{ber}	$= P_{bef}$	Reverse channel bit-error probability

[a]The CRC and parity check codes provide instantaneous error decisions.

$$P_{cm} = \left(1 - P_{bef}\right)^{N_B} \tag{1.401}$$

Therefore, using (1.401), the average number of transmissions required to obtain an error-free frame with N_B bits is evaluated as

$$\bar{N} = P_{cm}^{-1} = \left(1 - P_{bef}\right)^{-N_B} \tag{1.402}$$

The idle-RQ transmission efficiency, as defined in (1.399), is expressed as

$$\eta_{trans} = \frac{T_{df} + T_{sf}}{\bar{T}} \tag{1.403}$$

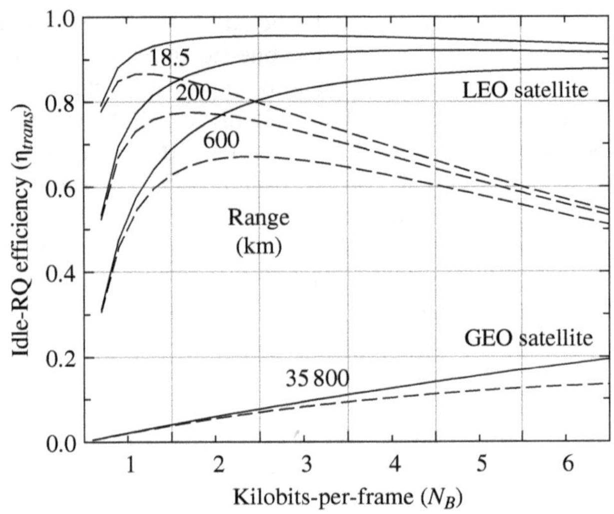

FIGURE 1.33 Idle-RQ efficiency as function of bits/frame (R_{bf} = 100 kbps, $P_{bef} = 10^{-5}$ *solid*, 10^{-4} *dashed* curve).

The number of forward channel bits per frame is defined as*

$$N_B \triangleq \left(T_{df} + T_{sf}\right) R_{bf} \tag{1.404}$$

With this definition, the transmission efficiency is expressed explicitly in terms of N_B, by substituting (1.400) and (1.402) into (1.403) and, after some simplifications, the efficiency is expressed as

$$\eta_{trans} = \frac{N_B \left(1 - P_{bef}\right)^{N_B}}{N_B + K_B} \tag{1.405}$$

where K_B is defined as

$$K_B \triangleq \left(2Tp + Tcr + (Tdr + Tsc) + Tcf\right) R_{bf} \tag{1.406}$$

The idle-RQ efficiency expressed in (1.405) is plotted in Figure 1.33 for several one-way communication link ranges;

*Inclusion of the synchronization bits $N_{sf} = T_{sf} R_{bf}$ in the definition of N_B is optional.

TABLE 1.13 Optimum N_B[a] Corresponding to the Maximum Efficiency Conditions in Figure 1.33

	Range (km)			
P_{bef}	18.5	200	600	35,800
1e–3	204	338	478	960
1e–4	697	1,232	1,889	7,589
1e–5	2,261	4,077	6,416	38,380
1e–6	7,208	13,079	20,757	143,125
1e–7	22,850	41,546	66,112	477,137

[a]N_B in bits.

the *solid* curves correspond to $P_{bef} = 10^{-5}$ and the *dashed* curves correspond to $P_{bef}=10^{-4}$. The impact of the link propagation delay (T_p) is significant and results in long idle times for the ACK/NAK response. The performance is also dependent on the link bit-error probabilities, the bit rate, and the number of bits per frame.

The optimum number of forward channel bits per frame, N_B, corresponding to the maximum efficiency, η_{trans}(max), is evaluated by differentiating (1.405) with respect to N_B and setting the result equal to zero and then solving for $N_B(opt)$. Following this procedure, the optimum N_B, corresponding to the maximum efficiency, is evaluated as the solution to the quadratic equation

$$N_B^2+N_BK_B-K_B/\lambda=0 \qquad (1.407)$$

where $\lambda=\ln(1/(1-P_{be}))$. The solution to (1.407) is

$$N_B(opt)=-\frac{K_B}{2}+\sqrt{\left(\frac{K_B}{2}\right)^2+\frac{K_B}{\lambda}} \qquad (1.408)$$

Using the example parameters, $N_B(opt)$ is listed in Table 1.13 for the conditions shown in Figure 1.33.

Optimizing the message frame size using $N_B(opt)$ has limited practical appeal because the resulting maximum efficiency may be unacceptable or because of the broad range of N_B over which the slope around η_{trans}(max) is virtually unchanged; this latter point is seen in Figure 1.33 corresponding to $P_{bef}= 10^{-5}$. Selecting N_B from a range that satisfies an acceptable minimum transmission efficiency is a preferable criterion and operating at low bit-error probabilities offers a wider range of selections.

1.11 WINDOWS

Windows have been characterized and documented by a number of researchers [74, 75], and this section focuses on the windows that are used in the simulation codes in the following chapters to enhance various performance objectives. Windows are applied in radar and communication systems to achieve a variety of objectives including antenna sidelobe reduction, improvements in range resolution and target discrimination, spectral control for adjacent channel interference (ACI) reduction [76], ISI control, design of linear phase filters, and improvements in parameter estimation algorithms. Windows can be applied as time or frequency functions to achieve complementary results depending on the application.

Windows are described in terms of the discrete-time samples* $w(n)$ where n is indexed over the finite window length of N samples. When the window is applied as a discrete-frequency sampled window the notation $W(n)$ is used. The discrete-time sampled rectangular window, defined, for example, as $w(n) = 1$ for $|n| \le N/2$ and zero otherwise, is typically used as the reference window by which the performance measures of other windows are compared. The spectrum of the time-domain rectangular window is described in terms of the *sinc*(x) function as $S(f) = sinc(nf/N)$. The rectangular window is also referred to as a uniformly weighted or simply as an unweighted window.

Several window parameters [75] that are useful in selecting a window for a particular application are the gain, the noise bandwidth, and the scalloping loss. The window voltage gain is defined as

$$G_v \triangleq \sum_n w(n) \qquad (1.409)$$

The noise bandwidth of the window follows directly from the definition of the noise bandwidth defined by Equation (1.46). In terms of the discrete-time sampling and application of Parseval's theorem, the normalized noise bandwidth is expressed as

$$B_n' = \frac{\sum_n w^2(n)}{\left(\sum_n w(n)\right)^2} \qquad (1.410)$$

In terms of Hertz, the bandwidth is given by

$$B_n(\text{Hz}) = \frac{B_n'}{T_w} \qquad (1.411)$$

where $1/T_w$ Hz is the fundamental frequency resolution of the window with duration T_w seconds. The theoretical normalized noise bandwidth of the rectangular window is $B_n' = 1$, so the noise bandwidth is $B_n = 1/T_w$ Hz. Harris [75] defines the scalloping loss (SL) of a time-domain window, as the frequency domain loss, relative to the maximum level, mid-way between two maximum adjacent DFT outputs. The scalloping loss is expressed as

*The notation $w(n)$ is a time-normalized sample corresponding to $w(nT_s)$ where T_s is the sampling interval. Similarly, the frequency-domain samples $W(n)$ are normalized by $f_s/N = 1/NT_s$.

TABLE 1.14 Summary of Window Performance Results

Window	Max. Sidelobe (dB)	At $\pm fT_w$	Scalloping Loss (dB) with T_w Zero-Padding 0	1	2	3
Rectangular	−13.26	1.5	3.92	0.91	0.4	0.18
Bartlett	−26.4	3.0	1.81	0.44	0.20	0.09
Blackman	−58.2	3.6	1.09	0.27	0.12	0.05
Blackman–Harris	−92.0	4.52	0.82	0.20	0.09	0.04
Hamming	−42.6	4.5	1.74	0.43	0.19	0.09
Cosine $k=1$	−23.0	1.89	2.08	0.51	0.22	0.10
$k=2$	−31.5	2.4	1.41	0.35	0.15	0.07
$k=3$	−39.3	2.83	1.06	0.26	0.12	0.05
$k=4$	−46.7	3.33	0.85	0.21	0.09	0.04

Using $N = 200$ samples/window.

$$SL \triangleq \frac{|W(\pi/N)|}{W(0)} = \frac{\left|\sum_n w(n)e^{-j\pi n/N}\right|}{\sum_n w(n)} \tag{1.412}$$

The characteristics of the various windows considered in the following sections are summarized in Tables 1.14 and 1.15. The maximum sidelobes correspond to those adjacent to the central, or main, lobe, and apply to the time (or frequency) domains depending on whether the window is applied, respectively, in the frequency (or time) domains. The scalloping loss results from the frequency or time domain ripple resulting from contiguous repetitions of the window functions. Harris has compiled an extensive table of windows and their performance characteristics.

The duration of the window and the manner in which it is sampled is dependent upon the application. In the following descriptions, the sampling is applied to windows that are symmetrical and asymmetrical about $t = 0$ as described by (1.413) using the *rect*(*x*) window. The *rect*(*x*) window is a uniformly weighted window* that is used to describe the window delay† and duration T_w for all arbitrarily weighted windows.

$$w(t) = \begin{cases} rect\left(\dfrac{t-T_d}{T_w} - m - \dfrac{1}{2}\right) & : \text{asymmetrcial about } t=0 \\ rect\left(\dfrac{t-T_d}{T_w} - m\right) & : \text{symmetrical about } t=0 \end{cases} \tag{1.413}$$

The parameter T_d introduces a delay and m is an integer corresponding to a contiguous sequence of windows; in the following analysis T_d and m are zero. Considering N to be an integer number of samples over the window duration with the sampling interval of $T_s = T_w/N$ seconds, the windows are characterized, with a maximum value of unity, in terms of the sample index n. In the following examples, the Bartlett or triangular window‡ is used and the respective odd and even values of N are 9 and 8. For odd integers N, the asymmetrical triangle window is expressed as

$$w(n)|_{N\in odd} = 1 - \frac{|n-(N-1)/2|}{(N-1)/2} : 0 \le n \le N-1, \text{ asymmetrical} \tag{1.414}$$

and the symmetrical triangle window is expressed as

$$w(n)|_{N\in odd} = 1 - \frac{|n|}{(N-1)/2} \quad : |n| \le (N-1)/2, \text{ symmetrical} \tag{1.415}$$

For the even integers, the asymmetrical triangle window is expressed as

$$w(n)|_{\substack{N\in even \\ n=n'+.5}} = \begin{cases} \dfrac{2n'+1}{N} & : 0 \le n' < N/2 \\ 2 - \dfrac{2n'+1}{N} & : N/2 \le n' \le N-1 \end{cases}, \text{ asymmetrical} \tag{1.416}$$

and the symmetrical triangle window is expressed as

$$w(n)|_{\substack{N\in even \\ n=n'+.5}} = \begin{cases} 1 + \dfrac{1-2|n'|}{N} & : -N/2 \le n' < 0 \\ 1 - \dfrac{1+2|n'|}{N} & : 0 \le n' \le N/2-1 \end{cases}, \text{ symmetrical} \tag{1.417}$$

Equations (1.414) through (1.417) are plotted in Figure 1.34 with the *circles* indicating the window sampling instances. The distinction between the symmetrical and asymmetrical sampling is evident and must be applied commensurately to the sampled data. In this regard, the windows can be applied, for example, to the transmitted data-modulated symbols for spectrum control, to the received data symbol for detection, or to the FFT window for spectrum evaluation. With minimum shift keying (MSK) modulation, discussed in Section 4.2.3.4, a cosine window is applied to each quadrature rail that are delayed, or offset, by one-half symbol period, so the notion of symmetrical and asymmetrical windows applies.§ The examples using an odd number of window samples, shown in Figure 1.34a and b, include the first and last window samples that are zero and increase

*Since the weighting is constant over T_w, the uniformly weighed window is also referred to as an unweighted window.

†The application of a window to a sequence of received symbols prior to spectral analysis requires the window to be appropriately delayed.

‡The triangular window is used to emphasize the sampling with respect to a nonuniformly weighted window.

§Windows applied to transmitter symbols for spectral control are generally referred to as symbol shaping functions and denoted as *p*(*t*).

TABLE 1.15 Summary of Window Performance Results

	Bartlett		Blackman		Blackman–Harris		Hamming	
Samples (N)	B'_n	G_v	B'_n	G_v	B'_n	G_v	B'_n	G_v
1000	1.3357	0.4995	1.728	0.4196	2.006	0.3584	1.364	0.5395
500	1.335	0.499	1.730	0.419	2.008	0.358	1.365	0.539
200	1.340	0.498	1.735	0.418	2.014	0.357	1.368	0.538
100	1.347	0.495	1.744	0.416	2.025	0.355	1.373	0.535
50	1.361	0.490	1.762	0.412	2.045	0.352	1.383	0.531
25	1.394	0.480	1.799	0.403	2.088	0.344	1.403	0.522
12	1.467	0.455	1.884	0.385	2.186	0.329	1.450	0.502
	Cosine k = 1		Cosine k = 2		Cosine k = 3		Cosine k = 4	
1000	1.235	0.636	1.502	0.4995	1.737	0.424	1.946	0.3746
500	1.236	0.635	1.503	0.499	1.738	0.4236	1.948	0.3742
200	1.240	0.633	1.507	0.498	1.744	0.422	1.954	0.373
100	1.246	0.630	1.515	0.495	1.752	0.420	1.964	0.371
50	1.260	0.624	1.531	0.490	1.770	0.416	1.984	0.368
25	1.290	0.610	1.562	0.480	1.807	0.407	2.026	0.360
12	1.364	0.580	1.636	0.458	1.892	0.389	2.121	0.344

Noise bandwidth and voltage gain.

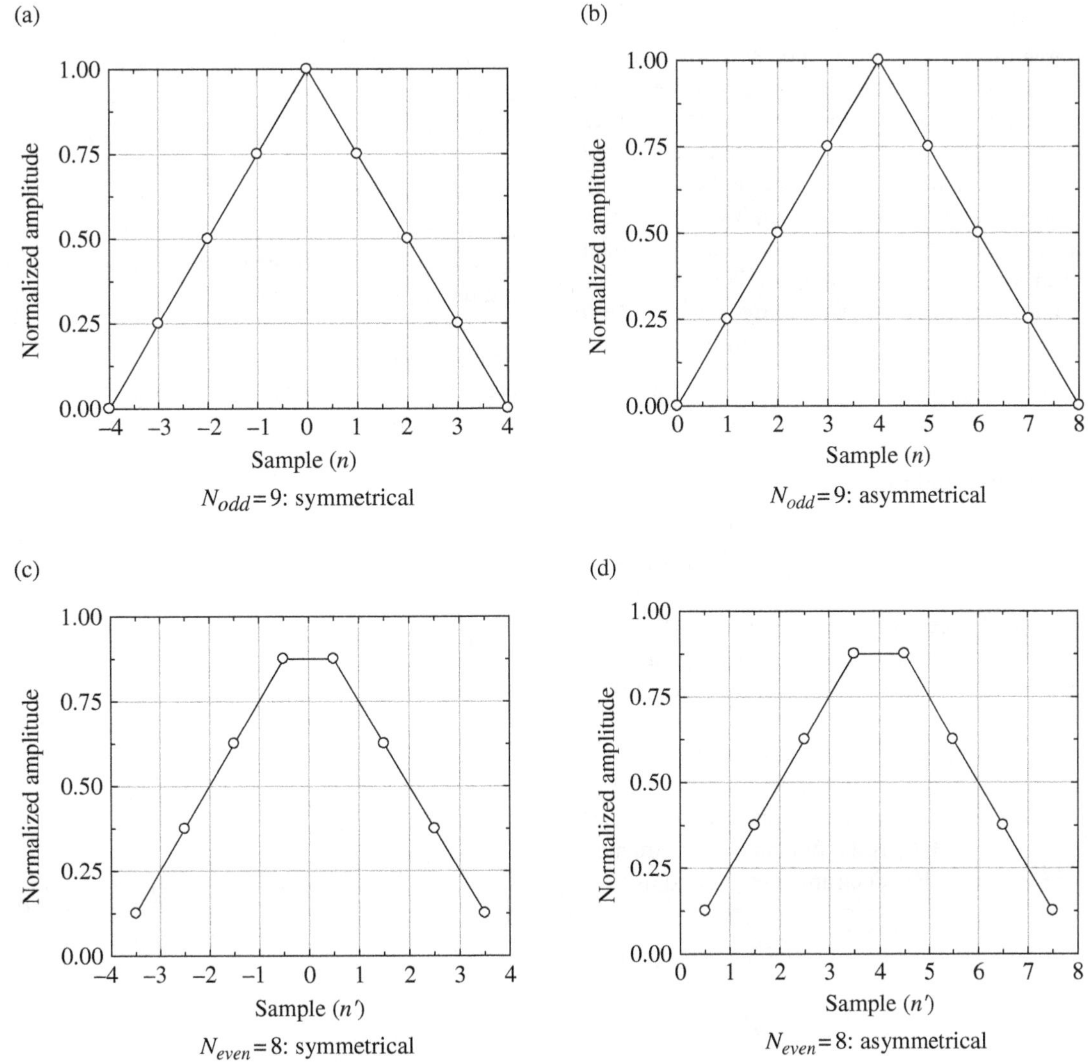

FIGURE 1.34 Examples of triangular window sampling.

symmetrically to the peak value. Although these cases are visually appealing, the sampled windows in Figure 1.34c and d use an even number of samples per window and result in same performance commensurate with the Nyquist sampling rate. Furthermore, using an even number of samples is suitable for analysis using the efficient FFT.

The case for using an even number of samples can be made based on the down-sampled output of a high sample rate analog-to-digital converter. For example, relative to a received data-modulated symbol, the down-sampled, or rate-reduced, output is the average symbol amplitude of the down-sampled interval from nT_s to $(n + 1)T_s$, so the window weighting should be applied to the point midway between the two samples as is often done using rectangular integration. This is accommodated by the even number of samples over the window interval shown in Figure 1.34c and d, with the understanding that the data sample at $n = n' + 0.5$ results from the linear interpolation of the data between sample n' and $n' + 1$. In addition to applying the correctly aligned window with interpolated data samples, this sampling arrangement also removes the delay estimation bias thereby improving the symbol tracking performance. Of course, the roles of the odd and even sampling can be reversed; however, it is convenient to use an even number of samples per symbol into the detection-matched filter for symbol tracking and the detection of quadrature symbol offset modulations. However, as long the Nyquist sampling criterion is satisfied the symbol information can be extracted in either case. In the following sections, the spectrums of the various windows are evaluated using both symmetrically and asymmetrically sampled windows with an odd number of samples. The reason for this choice is simply based on esthetics or eye appeal which is particularly noticeable when only a few number of samples is used.

1.11.1 Rectangular Window

In the time domain, the rectangular window is a uniformly weighted function described by the $rect(t/T_w)$ function with amplitude equal to unity over the range $|t/T_w| \le 1/2$ and zero otherwise. Expressing the time in terms of the discrete samples $t = nT_s$, where T_s is the sampling interval, results in the sample range $|n| \le (N - 1)/2$ for the symmetrical window and $0 \le n \le N - 1$ for the asymmetrical window, where $N = T_w/T_s$ is the total number of samples per window.

The spectrum of the rectangular window is described in terms of the $sinc(fT_w)$ function and, upon letting $f = n\delta f$ and $\delta f = \Delta f/L$, where $\Delta f = 1/T_w$ is the fundamental frequency resolution of the window, the spectrum is expressed as

$$W(n) = \frac{\sin(\pi n/L)}{\pi n/L} = sinc\left(\frac{n}{L}\right) \tag{1.418}$$

Defining the frequency increment in this way allows for L samples per spectral sidelobe. The Fourier transforms relationship between the time and frequency domains is discussed in Section 1.2. Special applications involving the *rect* (x) window in the time and frequency domains are discussed in Sections 1.11.4, 4.4.1, 4.4.4, and 4.4.5.

1.11.2 Bartlett (Triangular) Window

The sampling of the Bartlett or triangular window is discussed earlier under a variety of conditions that depend largely on the application and signal processing capabilities. The Bartlett window is shown in Figure 1.35 for $N = 21$ samples per symbol. Considering the frequency dependence of the spectral attenuation, expressed as $sinc^2(fT_w/2)$, the spectral folding about the Nyquist band $f_N = f_s/2 = N/2T_w$ is negligible for $N = 201$ and the peak spectral sidelobes are virtually identical to the theoretical Bartlett spectrum over the range of frequencies shown in Figure 1.36b, although the sidelobe level at $fT_w = 29$ is −65.8 dB and the theoretical value is −66.5 dB. Furthermore, upon close examination, the peak values of the sidelobe are progressively shifted to the right with increasing fT_w. These observations are more evident for the case involving $N = 21$ where the first sidelobe level is −26.8 dB (1 dB higher that theory) and the sidelobe at $fT_w = 9$ is significantly skewed to the right with a level 6 dB higher than the theoretical value. The sidelobe skewing is a direct result of the odd symmetry of the folded spectrum about $2kfT_w : k = 1, 2, \ldots$ which does not occur when N is even. However, with even values of N, the sidelobes are still altered by the folded spectral sidelobes. This phenomenon is a direct result of the sampling and the implicit periodicity of the window when using the discrete Fourier transform. These observations do not alter the utility of windows for spectral control; however, they may influence the spectral detail in applications involving spectral analysis.

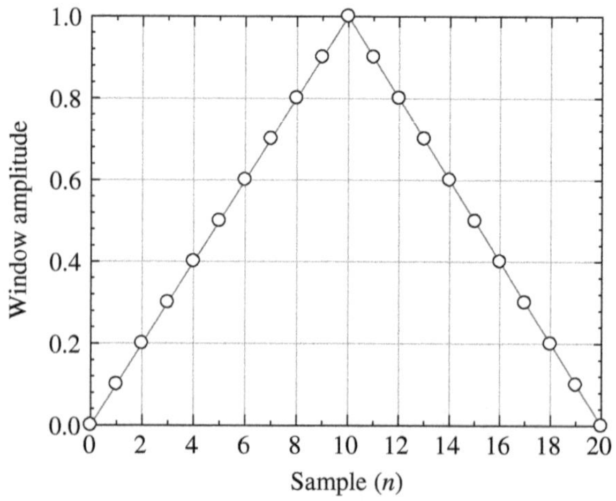

FIGURE 1.35 Bartlett window with $N = 21$ samples.

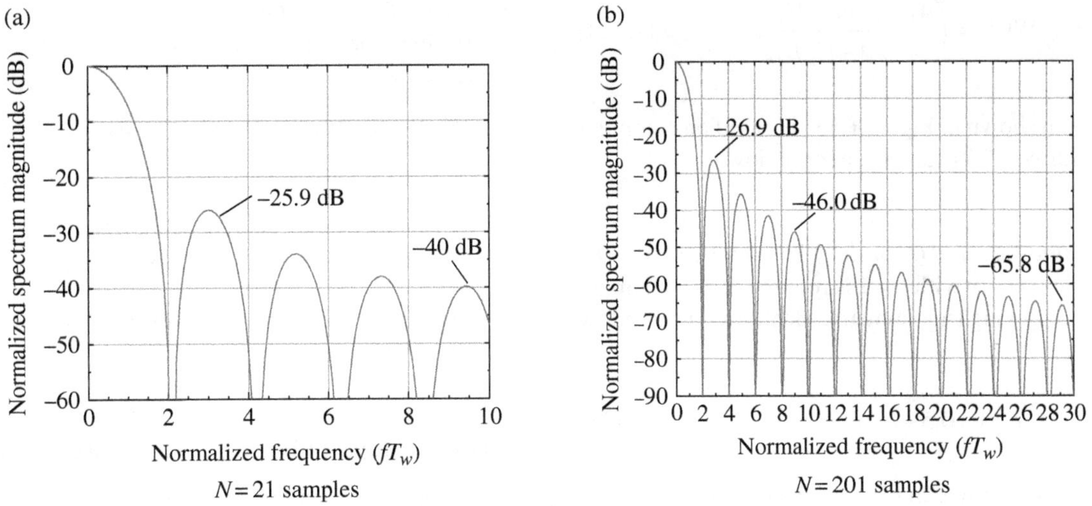

FIGURE 1.36 Bartlett (triangular) window spectrums.

1.11.3 Cosine Window

The sidelobes of the window spectrum can be reduced by providing additional shaping or tapering of the window function at the expense of a wider main spectral lobe that results in a higher noise bandwidth. For example, the rectangular window has an abrupt change in amplitude leading to a first sidelobe level of −13 dB, a spectral roll-off proportional to $1/f^2$ (6-dB/octave), and a noise bandwidth of $1/T_w$ Hz. On the other hand, the Bartlett window has an abrupt change in the slope of the amplitude leading to a first sidelobe level of −26.9 dB, a spectral roll-off of $1/f^4$ (12-dB/octave), and a noise bandwidth of $1.336/T_w$. The cosine window results in the k-th derivative of the amplitude, that is, the slope of the amplitude at the edges of the window is zero resulting in even greater spectral sidelobe roll-off.

The cosine window is characterized as

$$w(n) = \cos^k(\phi_n + \delta_1) \tag{1.419}$$

where $\delta_1 = 0$ when $|n| \le \lfloor (N-1)/2 \rfloor$ and $\delta_1 = -\pi/2$ when $n = 0, \ldots, N-1$; this corresponds to a $\sin^k(\phi_n)$ window function. The phase function in (1.419) is expressed as

$$\phi_n = \frac{\pi}{N-1}(n+\delta_0) \quad : |n| \le \lfloor (N-1)/2 \rfloor \tag{1.420}$$

where $\delta_0 = 0$ when $n \in$ odd and $\delta_0 = 0.5$ when $n \in$ even. The cosine window is shown in Figure 1.37 for $n > 0$ and $k = 1, \ldots, 4$. In the following subsections, the cosine windows for various values of k are described for the conditions $\delta_0 = \delta_1 = 0$.

The gain for the finite sampled cosine window with $k = 1$ and N samples per window is given by

$$G_v = \frac{2}{\pi}\left(\frac{N-1}{N}\right) \tag{1.421}$$

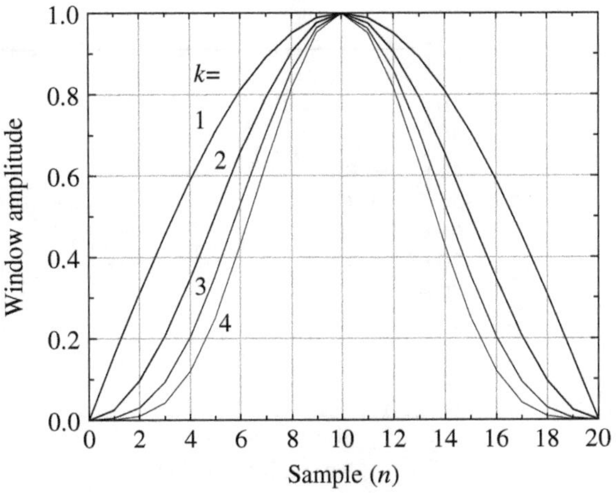

FIGURE 1.37 Cosine windows ($\delta_0 = 0$, $\delta_1 = -\pi/2$).

and the normalized noise bandwidth is given by

$$B'_n = \frac{\pi^2}{8}\left(\frac{N}{N-1}\right) \tag{1.422}$$

G_v and B'_n are recorded in Table 1.15 for various values of N. Equations (1.421) and (1.422) approach their theoretical limits as $N \to \infty$.

1.11.3.1 *Cosine Window* (k = *1*)

With $k = 1$ the cosine window is expressed as

$$w(n) = \cos(\phi_n) \tag{1.423}$$

This window is used as the symbol weighting function for MSK modulation and, in terms of the window duration T_w, the theoretical spectrum is given by

$$W(f) = \left(\frac{4T_w}{\pi}\right)\frac{\cos(\pi f T_w)}{1-(2fT_w)^2} \tag{1.424}$$

The spectrum shown in Figure 1.38a is based on computer simulation with $N = 200$ samples per window.

1.11.3.2 Cosine-Squared (Hanning) Window (k = 2)

The cosine-squared window with $k = 2$ is referred to as a Hanning window [74]. Applying trigonometric identities, this window is expressed as

$$w(n) = 0.5 + 0.5\cos(2\phi_n) \tag{1.425}$$

and the spectrum is shown in Figure 1.38b for $N = 200$ samples per window.

1.11.3.3 Cosine Window (k = 3 and 4)

Applying trigonometric identities, the cosine windows for $k = 3$ and 4 are expressed as

$$w(n) = 0.75\cos(\varphi_n) + 0.25\cos(3\varphi_n) \quad : k = 3 \tag{1.426}$$

and

$$w(n) = 0.375 + 0.5\cos(2\phi_n) + 0.125\cos(4\phi_n) \quad : k = 4 \tag{1.427}$$

The spectrums for these two cases are shown in Figure 1.38c.

1.11.4 Temporal Raised-Cosine (TRC) Window

The temporal raised-cosine (TRC) window applies a cosine shaping function symmetrically about each end of the

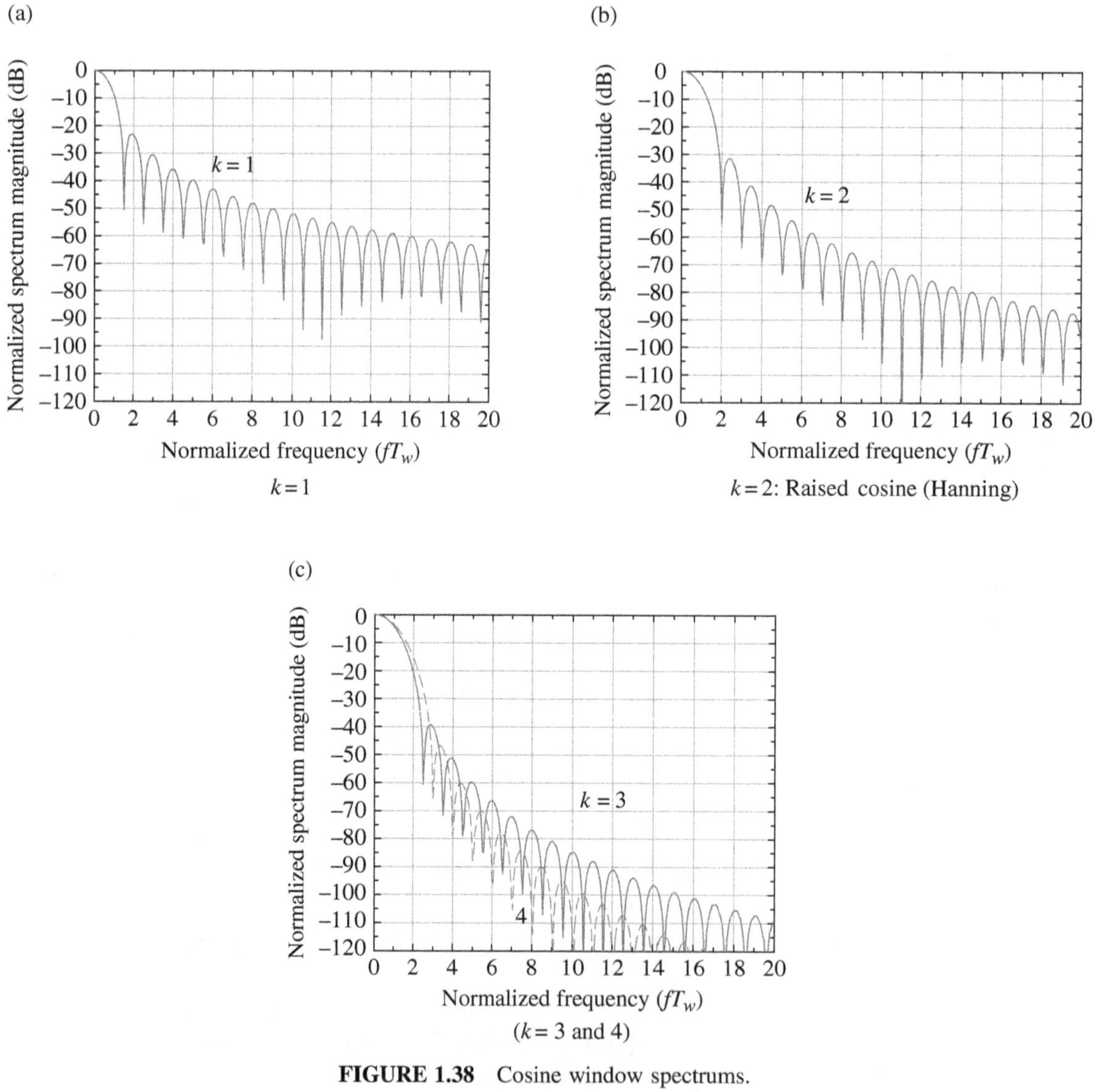

FIGURE 1.38 Cosine window spectrums.

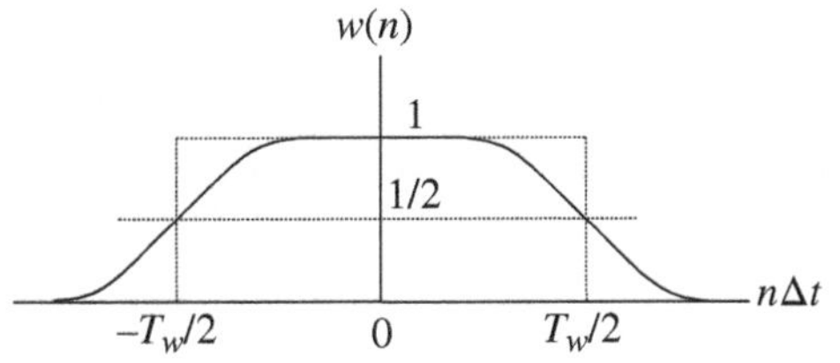

FIGURE 1.39 Temporal raised-cosine window.

rectangular window function, $rect(n\Delta t/T_w)$, as shown in Figure 1.39. In this case, the integer n corresponds to the samples N over the interval T_w such that $N = T_w/\Delta t$.

The TRC window is expressed as

$$w(n') = \begin{cases} 1 & : |n'| < \dfrac{1-\alpha}{2} \\ \dfrac{1}{2}\left\{1 - \sin\left(\dfrac{\pi}{\alpha}\left(|n'| - \dfrac{1}{2}\right)\right)\right\} & : \dfrac{1-\alpha}{2} \le |n'| \le \dfrac{1+\alpha}{2} \end{cases} \tag{1.428}$$

where $n' = t/T_w = n\Delta t/T_w$ is the normalized sample index and α is a design parameter limited to $0 \le \alpha \le 1$. The TRC window is applied to the phase function of PSK-modulated waveforms in Sections 4.2.8 and 4.4.3.9 to improve the waveform spectral containment while maintaining a constant signal amplitude; in this application α is referred to as the excess phase factor.

Letting $m = f/T_w$, the spectrum of the TRC amplitude window is evaluated as

$$W(m) = \frac{\sin(\pi m)}{\pi m}\left[\frac{\cos(\alpha\pi m)}{1-(2\alpha m)^2}\right] \tag{1.429}$$

1.11.4.1 Spectral Raised-Cosine Window The raised-cosine window when applied in the frequency domain is referred to as the spectral RC (SRC) window and the parameter α is referred to as the excess bandwidth. With $m = fT_w$, the SRC frequency response is expressed as

$$W(m) = \begin{cases} T_w & : |m| < \dfrac{1-\alpha}{2} \\ \dfrac{T_w}{2}\left\{1 - \sin\left(\dfrac{\pi}{\alpha}\left(|m| - \dfrac{1}{2}\right)\right)\right\} & : \dfrac{1-\alpha}{2} \le |m| \le \dfrac{1+\alpha}{2} \end{cases} \tag{1.430}$$

and upon letting $n = t/T_w$, the SRC impulse response is evaluated as

$$w(n) = \frac{\sin(\pi n)}{\pi n}\left[\frac{\cos(\alpha\pi n)}{1-(2\alpha n)^2}\right] \tag{1.431}$$

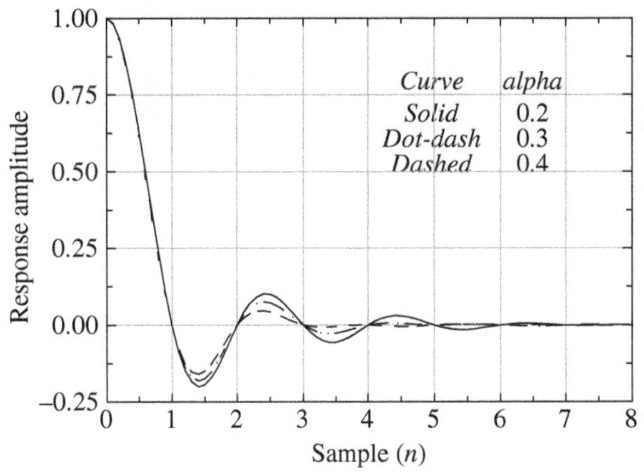

FIGURE 1.40 Impulse response of SRC window.

This response has indeterminate solutions of the form 0/0 at $n = 0$ and $1/(2\alpha)$ that are evaluated as $w(0) = 1$ and

$$w\left(\frac{1}{2\alpha}\right) = \frac{\alpha}{2}\sin\left(\frac{\pi}{2\alpha}\right) \tag{1.432}$$

Using these results, (1.431) is plotted in Figure 1.40 for several values of the excess bandwidth factor. The SRRC window, associated with spectral shaping of a modulated symbol, is discussed in detail in Section 4.3.2. In this application the symbol duration is $T = T_w$ and the transmitted symbol and demodulator-matched filter spectrums have a square-root RC, or simply root RC, frequency response of $\sqrt{W(n)}$. The matched filter output results in the SRC impulse response shown in Figure 1.40 with symbol spacing corresponding to integer values of n that results in zero ISI $\forall n \neq 0$. With an AWGN channel and optimum matched filter sampling the demodulator performance corresponds to maximum-likelihood detection.

1.11.5 Blackman Window

The Blackman window uses three terms to provide more tapering of the window at the expense of a narrower time response and a wider noise bandwidth. The Blackman window is expressed as

$$w(n) = 0.42 - 0.5\cos(\phi_n) + 0.08\cos(2\phi_n) \tag{1.433}$$

where the phase function in (1.433) is expressed as

$$\begin{aligned} \phi_n &= \frac{2\pi}{N-1}n & &: 0 \le n \le N-1 \\ &= \frac{2\pi}{N-1}(n + \lfloor N/2 \rfloor) & &: -\lfloor N/2 \rfloor \le n \le \lfloor N/2 \rfloor \end{aligned} \tag{1.434}$$

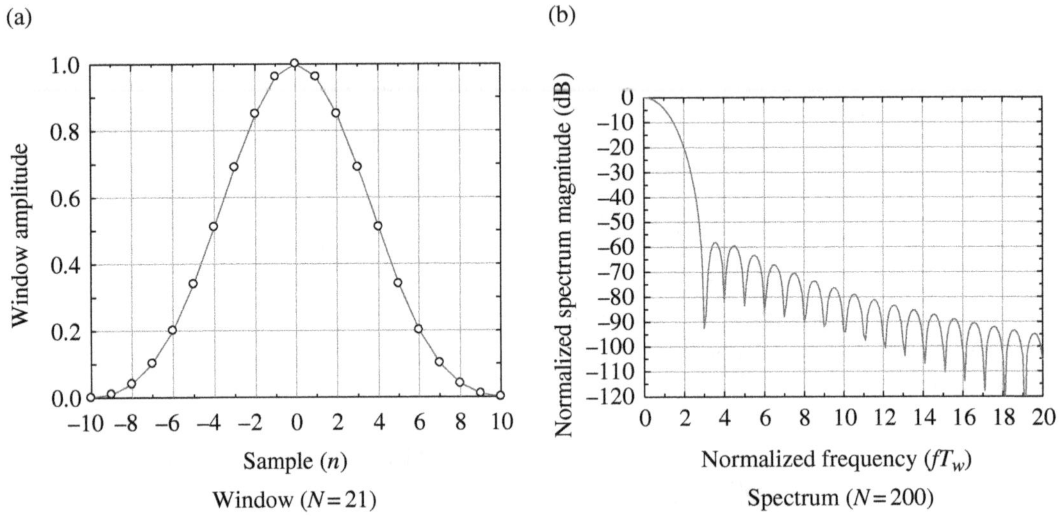

FIGURE 1.41 Blackman window and spectrum.

TABLE 1.16 Blackman–Harris Window Coefficients[a]

Coefficient	3-Term 67 dB	4-Term 92 dB
c_0	0.42323	0.35875
c_1	0.49755	0.48829
c_2	0.07922	0.14128
c_3	0.0	0.01168

[a]Harris [75]. Reproduced by permission of the IEEE.

This definition of the phase function results in a symmetrical window for all values of N with indexing over $\pm \lfloor N/2 \rfloor$ as shown in Figure 1.41a for $N = 21$. The spectrum of the Blackman window corresponding to $N = 200$ samples per window is shown in Figure 1.41b.

1.11.6 Blackman–Harris Window

The Blackman–Harris window is expressed as

$$w(n) = c_o - c_1\cos(\phi_n) + c_2\cos(2\phi_n) - c_3\cos(3\phi_n) \quad (1.435)$$

Harris [75] performed a gradient search on the coefficients c_i to minimize spectral sidelobe level and the results are given in Table 1.16 for a three- and four-term window.

The Blackman–Harris window, as formulated by Harris, divides the phase function by N that results in an asymmetrical window. To provide a symmetrical window for all N, the phase function ϕ_n is divided by $N - 1$ and expressed as

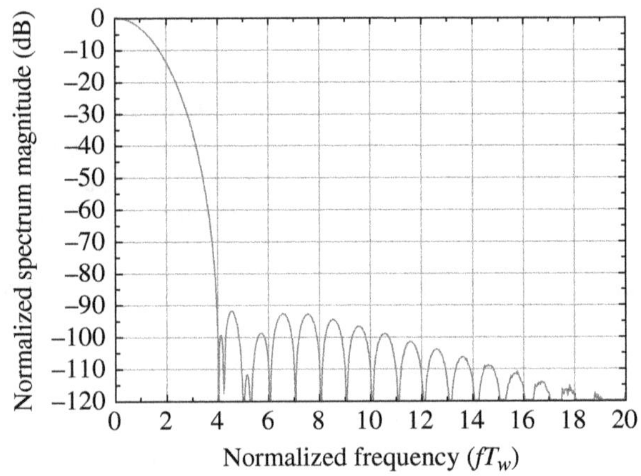

FIGURE 1.42 Blackman–Harris window spectrum (N = 200, 92 dB, 4-Term).

$$\begin{aligned} \phi_n &= \frac{2\pi}{N-1}n & &: 0 \le n \le N-1 \\ &= \frac{2\pi}{N-1}(n + \lfloor N/2 \rfloor) & &: -\lfloor N/2 \rfloor \le n \le \lfloor N/2 \rfloor \end{aligned} \quad (1.436)$$

The spectrum of the 4-Term 92 dB Blackman–Harris window is shown in Figure 1.42 for $N = 200$ samples per symbol and $L = 40$ samples in the bandwidth $1/T_w$ Hz.

The coefficients for the Blackman–Harris function, as given in Table 1.16, do not result in window values of zero for the first and last sample resulting in a window *pedestal* of 6×10^{-5}; this is a direct result of the coefficients not summing to zero at the window edges. The pedestal plays a critical role in the control of the sidelobes and noise bandwidth and windows with more dramatic pedestals are examined in the following sections.

1.11.7 Hamming Window

The Hamming window is expressed as

$$w(n) = 0.54 - 0.46\cos(\phi_n) \quad (1.437)$$

where the phase function ϕ_n is expressed as

$$\phi_n = \frac{2\pi}{N-1}n \quad : 0 \le n \le N-1 \quad (1.438)$$

The window is shown in Figure 1.43a with indexing over the range $0 \le n \le N-1$ and the spectrum is shown in Figure 1.43b.

1.11.8 Kaiser (Kaiser–Bessel) Window

The Kaiser window, also referred to as the Kaiser–Bessel window, is expressed as

$$w(n) = \frac{I_o\left(\pi\beta\sqrt{1-\left(\frac{n}{N/2}\right)^2}\right)}{I_o(\pi\beta)} \quad : |n| \le N/2 \quad (1.439)$$

where $I_o(x)$ is the modified Bessel function of order zero and β is the time-bandwidth product equal to T_wB where T_w is the window duration and B is the corresponding baseband bandwidth. The Kaiser window is shown in Figure 1.44a, using $N = 51$ samples per window, as a

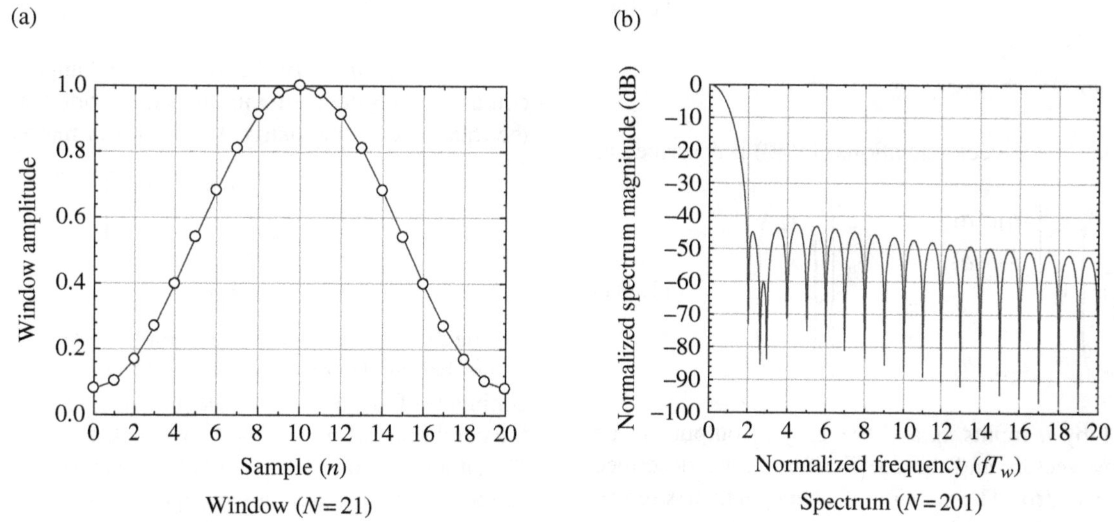

FIGURE 1.43 Hamming window and spectrum.

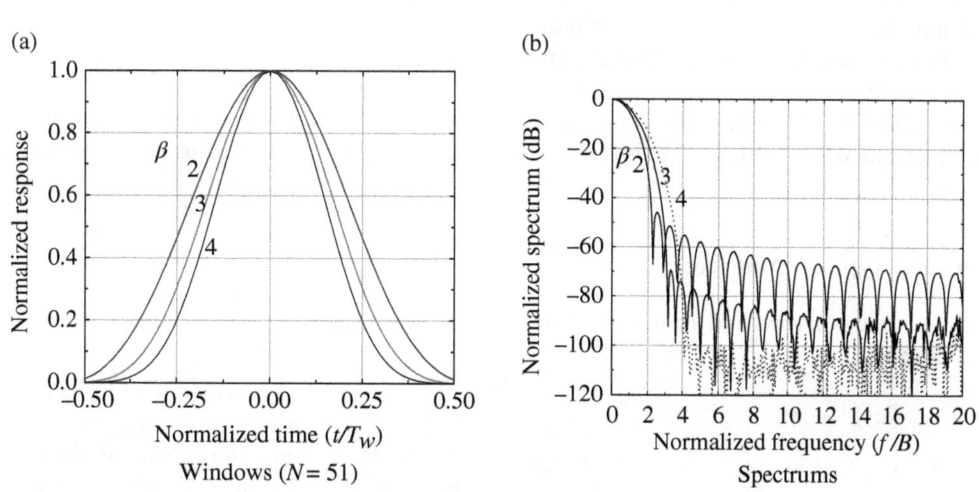

FIGURE 1.44 Kaiser windows and spectrums (β = 2, 3, and 4).

symmetrical window for $\beta = 2$, 3, and 4; the range $-N/2 \le n \le N/2$ corresponds to $-T_w/2 \le t \le T_w/2$. The corresponding spectrum of the Kaiser window is shown in Figure 1.44b.

1.12 MATRICES, VECTORS, AND RELATED OPERATIONS

A matrix is a convenient way to describe complicated linear systems and is used extensively in the analysis of control systems. For example, a large linear system is generally described in terms of the inputs x_j, outputs y_i, and the system coefficients a_{ij} as

$$\begin{array}{lllllllll} y_1 & = & a_{11}x_1 & + & a_{12}x_2 & + & \cdots & + & a_{1n}x_n \\ y_2 & = & a_{21}x_1 & + & a_{22}x_2 & + & \cdots & + & a_{2n}x_n \\ \vdots & \vdots & \vdots & \vdots & \vdots & \vdots & \vdots & \vdots & \vdots \\ y_m & = & a_{m1}x_1 & + & a_{m2}x_2 & + & \cdots & + & a_{mn}x_n \end{array} \tag{1.440}$$

and using matrix and vector notations (1.440) is described as

$$\begin{pmatrix} y_1 \\ y_2 \\ \vdots \\ y_m \end{pmatrix} = \begin{bmatrix} a_{11} & a_{12} & \cdots & a_{1n} \\ a_{21} & a_{22} & \cdots & a_{2n} \\ \vdots & \vdots & \vdots & \vdots \\ a_{m1} & a_{m2} & \cdots & a_{mn} \end{bmatrix} \begin{pmatrix} x_1 \\ x_2 \\ \vdots \\ x_n \end{pmatrix} \tag{1.441}$$

where the 1 by n inputs x_j, and the 1 by m outputs y_i are described by vectors and the coefficients a_{ij} are described by the m by n matrix. The notation of an m by n matrix refers, respectively, to the number of rows and columns of the matrix. The linear system may include time-varying inputs, outputs, and coefficients as found in control systems and time-varying parameter estimation and tracking applications. The following descriptions of matrices and vectors are introductory and targeted to the description of adaptive systems in Chapter 12. More in-depth discussions on the subject and applications are given by Derusso, Roy, and Close [77], Haykin [78], Sage and White [79], and Gelb [80].

1.12.1 Definitions and Types of Matrices

- A $m \times n$ matrix A contains m rows and n columns of elements a_{ij} and is denoted as

$$A = \begin{bmatrix} a_{11} & \cdots & a_{1n} \\ \vdots & \cdots & \vdots \\ a_{m1} & \cdots & a_{mn} \end{bmatrix} \tag{1.442}$$

- A *diagonal matrix* (Λ) is a $m \times m$ *square matrix* with $a_{ij} = 0$: $i \neq j$, otherwise, the elements on the principal diagonal are identified as a_{ii} with $\Lambda = \text{diag}(a_{ii})\ \forall\ i$.
- The *unit matrix* (I) is similar to the diagonal matrix with $a_{ij} = 0$: $i \neq j$, and $a_{ii\cdot} = 1$.
- A *null matrix* or *zero matrix* (O) is a matrix for which all of the elements are zero.
- A *symmetric matrix* is a square matrix of real elements if $A = A^T$, that is, $a_{ji} = a_{ij}\ \forall\ i,j$.
- A *skew matrix* is a square matrix of real elements if $A = -A^T$, that is, $a_{ij} = -a_{ji}\ \forall\ i,j$.
- A *nonsingular matrix* A has an inverse A^{-1}.
- A *complex matrix* is denoted as $\tilde{A}$ and has complex elements $\tilde{a}_{ij} = a_{cij} + ja_{sij}$ where a_{cij} and a_{sij} denote the respective real and imaginary values of the complex elements. Conjugation the complex matrix is denoted as $\tilde{A}^*$ with elements $\tilde{a}_{ij} = a_{cij} - ja_{sij}$.

The *transposition* of the matrix interchanges the row and columns and is denoted with the superscript T. For example, the transpose of the matrix A is the $n \times m$ matrix denoted as

$$A^T = \begin{bmatrix} a_{11} & \cdots & a_{m1} \\ \vdots & \cdots & \vdots \\ a_{1n} & \cdots & a_{mn} \end{bmatrix} \tag{1.443}$$

The *Hermitian matrix* is a complex matrix with the elements below the principal diagonal equal to the complex conjugate of the those about the principal diagonal; satisfying the conditions $(A^*)^T = A^+ = A^H$ where the superscripts + and H denote the complex conjugate transposition. The superscript + generally denotes complex transposition and H is used to emphasize the Hermitian complex transposition.

The order, or rank, of an $(m \times n)$ matrix is denoted as m-by-n and the order, or rank, of the square $(m \times m)$ matrix is denoted as m.

1.12.2 The Determinant and Matrix Inverse

The determinant [81] of an $m \times m$ square matrix A is denoted as $|A|$ and defined in terms of the *cofactors*, A_{ij}, of A as

$$|A| = \sum_{i=1}^{m} a_{ij}A_{ij} = \sum_{j=1}^{m} a_{ij}A_{ij} \tag{1.444}$$

The cofactors are defined as

$$A_{ij} = (-1)^{i+j}M_{ij} \tag{1.445}$$

where M_{ij} is the determinant of the *minor matrix* of the element a_{ij}. The $m' \times m' : m' = m-1$ minor matrix of a_{ij} is the matrix formed by the m' rows and columns of A

excluding the row and column containing a_{ij}. For example, considering the 4×4 square matrix A, the 3×3 minor matrix A' of the element a_{32}, and the determinant M_{32} are expressed in (1.446) as

$$A = \begin{bmatrix} a_{11} & a_{12} & a_{13} & a_{14} \\ a_{21} & a_{22} & a_{23} & a_{24} \\ a_{31} & a_{32} & a_{33} & a_{34} \\ a_{41} & a_{42} & a_{43} & a_{44} \end{bmatrix}, \quad A' = \begin{bmatrix} a_{11} & a_{13} & a_{14} \\ a_{21} & a_{23} & a_{24} \\ a_{41} & a_{43} & a_{44} \end{bmatrix}, \quad M_{32} = |A'| \tag{1.446}$$

Therefore, using (1.446), the cofactor A_{32} is evaluated as $A_{32} = (-1)^5 M_{32}$. Following the evaluation of the remaining $m-1$ cofactors $A_{i,2}$: $i = 1,2,4$ or the cofactors A_{3j}: $j = 1,3,4$, the determinant $|A|$ is evaluated using, respectively, the first or second equality in (1.444).

For square matrices of order ≤ 3, the determinant can be computed in terms of the elements a_{ij} as expressed, for example, in the 3×3 matrix described in (1.447). In this example, the determinant is formed by summing the $n = 3$ products of the elements parallel to the principal diagonal elements (a_{11},a_{22},a_{33}) and subtracting the $n = 3$ products of the elements parallel to the diagonal elements (a_{13},a_{22},a_{31}). The three positive products are indicated by *solid arrows* pointing to the diagonal elements parallel to the principal diagonal and the three negative products are indicated by *dashed arrows* pointing to the diagonal elements parallel to the complement of the principal diagonal. In both cases, the elements below the diagonals are *wrapped around* to form the three element products. In general, this procedure does not give the correct results for orders > 3.

$$|A| = \begin{vmatrix} a_{11} & a_{12} & a_{13} \\ a_{21} & a_{22} & a_{23} \\ a_{31} & a_{32} & a_{33} \end{vmatrix} = \begin{cases} a_{11}a_{22}a_{33} + a_{12}a_{23}a_{31} + a_{13}a_{21}a_{32} \\ -a_{13}a_{22}a_{31} - a_{12}a_{21}a_{33} - a_{11}a_{23}a_{32} \end{cases} \tag{1.447}$$

Determinants are defined only for square matrices and if $|A| = 0$ the matrix A is referred to as a singular matrix, otherwise, A is nonsingular. The inverse of a *nonsingular* $n \times n$ matrix A, with elements a_{ij}, is expressed as

$$A^{-1} = \frac{[A_{ji}]}{|A|} \tag{1.448}$$

where the matrix $[A_{ji}]$ is defined as the adjoint of the matrix A. Considering the $n \times n$ matrix of cofactors $[A_{ij}]$ of the matrix A, with the cofactor of each element a_{ij} evaluated using (1.445), the adjoint matrix is the transpose of the matrix $[A_{ij}]$ so that

$$[A_{ji}] = [A_{ij}]^T \tag{1.449}$$

Premultiplication of A^{-1} by A results in the unit matrix, that is, $AA^{-1} = I$

1.12.3 Definition and Types of Vectors

A *vector* is an $m \times 1$ column matrix with elements a_i and is denoted as

$$a = \begin{bmatrix} a_1 \\ \vdots \\ a_m \end{bmatrix} \tag{1.450}$$

A *complex vector* is denoted as $\tilde{a}$ with complex elements $\tilde{a}_k = a_{ck} \pm ja_{sk}$ where a_c and a_s denote the real and imaginary values of the complex element. The transposition of the matrix a is the $1 \times m$ complex row vector a^T. *Conjugation* and *conjugate transposition* are denoted by the respective superscripts $*$ and $+$.

1.12.4 Matrix and Vector Operations

Addition of two matrices of the same order $(m \times n)$ is

$C = A + B$: The elements of C are $c_{ij} = a_{ij} + b_{ij}$

Multiplication of the two matrices must conform to the following rule: $[m \times n][n \times m'] = [m \times m']$ where the inner dimension n must be identical.

$y = Ax$: Matrix postmultiplication by a vector results in a $[m \times n][n \times 1] = [m \times 1]$ vector

The element multiplication is $y_i = \sum_k^n a_{ik}x_k \ : i = 1,\ldots,m$

$y = x^T A$: Matrix premultiplication by a vector results in a $[1 \times n][n \times m] = [1 \times m]$ vector.

The element multiplication is $y_i = \sum_k^m a_{ki}x_k \ : i = 1,\ldots,n$

$C = AB$: Matrix multiplication results in a $[m \times \ell][\ell \times n] = [m \times n]$ matrix

The element multiplication is $c_{ij} = \sum_k^\ell a_{ik}b_{kj} : i = 1,m;$ $j = 1,\ldots,n$

In general, multiplication is not communicative $AB \neq BA$

$A(BC) = (AB)C$: Associative

$A(B + C) = AB + AC$: Distributive

Multiplication by diagonal matrix D

$C = AD$: Postmultiplication of a real $(m \times m)$ matrix A by D results in a diagonal matrix C with the diagonal element of A scaled by those of D; the elements of C are $c_{ij} = a_{ij}d_{jj} : i,j = 1, \ldots, m$

$C = DA$: Premultiplication of a real $(m \times m)$ matrix A by D results in the same diagonal matrix C; the elements of C

are $c_{ij} = d_{ii}a_{ij} : i,j = 1, \ldots, m$. In this case multiplication is communicative.

Multiplication by unit matrix I

$IA = A$: Multiplication of a $(m \times m)$ matrix A by I does not alter the matrix A.

$IA = AI$: Multiplication by I is communicative.

Transposition of AB

$C = (AB)^T = A^T B^T$

Scalar, or inner, product of two $(m \times 1)$ complex vectors $\tilde{x}$ and $\tilde{y}$

$\tilde{y}^T\tilde{x}^* = \tilde{x}^+\tilde{y} = \sum_i^m \tilde{y}_i\tilde{x}_i^*$, $\tilde{x}^T\tilde{y}^* = \tilde{y}^+\tilde{x} = \sum_i^m \tilde{x}_i\tilde{y}_i^*$ are complex values and $\tilde{y}^T\tilde{x}^* \neq \tilde{x}^T\tilde{y}^*$

Scalar product of two $(m \times 1)$ real vectors x and y

$y^T x = \sum_i^m y_i x_i$ is a real scalar value and $y^T x = x^T y = \sum_i^m x_i y_i$

Orthogonality of two vectors is defined if their scalar product is zero.

Length of a complex vector $\tilde{x}$ is defined as magnitude, denoted as $\|\tilde{x}\|$, of the inner product

$$\|\tilde{x}\| = \sqrt{\sum_i^m \tilde{x}_i\tilde{x}_i^*} = \sqrt{\sum_i^m |\tilde{x}_i|^2} \tag{1.451}$$

Outer, or dyadic, product of two $(m \times 1)$ and $(1 \times n)$ complex vectors $\tilde{x}$ and $\tilde{y}$ forms an $(m \times n)$ complex matrix $\tilde{C} = xy^+$ with elements $c_{ij} = \sum_j^n x_i y_j^* : i = 1, \ldots, m$. The autocorrelation matrix of a complex vector is computed as the outer product.

Differentiation of Complex Matrices: Differentiation of a complex matrix $\tilde{A}(t)$ results in the complex matrix $\tilde{C} = (d/dt)\tilde{A}(t)$ with elements evaluated as $\tilde{c}_{ij} = (da_{cij}/dt) \pm j(da_{sij}/dt)$. The differentiation of sums and products of matrices is evaluated as

$$\frac{d}{dt}\left(\tilde{A}(t) + \tilde{B}(t)\right) = \frac{d}{dt}\tilde{A}(t) + \frac{d}{dt}\tilde{B}(t) \tag{1.452}$$

$$\frac{d}{dt}\left(\tilde{A}(t)\tilde{B}(t)\right) = \left(\frac{d}{dt}\tilde{A}(t)\right)\tilde{B}(t) + \tilde{A}(t)\left(\frac{d}{dt}\tilde{B}(t)\right) \tag{1.453}$$

Differentiation of a complex vector follows by considering the vector as a single column matrix. Therefore, differentiation of real matrices and vectors is identical with $a_{ij} = 0$.

Differentiation of quadratic transformation with the matrix $\tilde{Q}(t)$ expressed as

$$\tilde{Q}(t) = \tilde{x}^H(t)\tilde{A}(t)\tilde{x}(t) \tag{1.454}$$

where the elements of the $(m \times m)$ complex matrix $\tilde{A}(t)$ and the complex $(m \times 1)$ vector $\tilde{x}(t)$ are functions of t. The derivative of $\tilde{Q}(t)$ is evaluated as

$$\frac{d\tilde{Q}(t)}{dt} = \left(\frac{d}{dt}\tilde{x}^H(t)\right)\tilde{A}(t)\tilde{x}(t) + \tilde{x}^H(t)\left(\frac{d\tilde{A}(t)}{dt}\right)\tilde{x}(t) + \tilde{x}^H(t)\tilde{A}(t)\left(\frac{d}{dt}\tilde{x}(t)\right) \tag{1.455}$$

For a real matrix $A(t)$ and vector $x(t)$, the matrix $Q(t)$ is symmetric and the derivative of the quadratic function $Q(t)$ simplifies to

$$\frac{dQ(t)}{dt} = 2\left(\frac{d}{dt}x^T(t)\right)A(t)x(t) + x^T(t)\left(\frac{dA(t)}{dt}\right)x(t) \tag{1.456}$$

Bilinear transformation with the real $(m \times m)$ matrix $B(t)$, expressed in terms of the $(m \times 1)$ real vectors $x(t)$ and $y(t)$, is evaluated as

$$B(t) = x^T(t)A(t)y(t) \tag{1.457}$$

Differentiation of $B(t)$ with respect to t is evaluated as

$$\frac{dB(t)}{dt} = \left(\frac{d}{dt}x^T(t)\right)B(t)y(t) + x^T(t)\left(\frac{dB(t)}{dt}\right)y(t) + x^T(t)B(t)\left(\frac{d}{dt}y(t)\right) \tag{1.458}$$

1.12.5 The Quadratic Transformation

The quadratic matrix Q is similar to the bilinear matrix, expressed in (1.457), with the vector $y = x$. The following description is based on the $(n \times n)$ correlation matrix*

$$R = E\left[\tilde{u}\tilde{u}^H\right] \tag{1.459}$$

where $\tilde{u}$ is an $(n,1)$ complex vector of arbitrary data and $\tilde{u}^H$ is the complex conjugate transpose† of $\tilde{u}$. The n columns of the matrix Q represent the $(n,1)$ characteristic vectors of

*In Chapter 12 the data vector is denoted as $\tilde{x}$ and represents the discrete-time samples of a stationary stochastic process and $\tilde{w}$ is used to denote an arbitrary nonzero vector with $\tilde{y} = \tilde{w}^H\tilde{x}$.

†The Hermitian transpose $\tilde{u}^H$ is an alternate notation to $\tilde{u}^+$ denoting the complex conjugate transpose.

the correlation matrix R that is transformed to the diagonal matrix Λ as

$$Q^H R Q = \Lambda \quad (1.460)$$

The diagonal elements of Λ are the characteristic values* of the matrix R.

The derivation of (1.460) follows the work of Haykin [82] and is based on the linear transformation of the $(n,1)$ complex vector $\tilde{x}$ to the vector $\lambda\tilde{x}$ using the transformation

$$R\tilde{x} = \lambda\tilde{x} \quad (1.461)$$

where λ is a constant. Equation (1.461) can be expressed as $(R-\lambda I)\tilde{x} = O$ which has nontrivial solutions with $\tilde{x} \neq 0$, *iff* the determinant of $R - \lambda I$ is zero, that is,

$$|R-\lambda I|_{\det} = 0. \quad (1.462)$$

Equation (1.462) is the characteristic equation of the matrix R and has n solutions corresponding to the λ_i characteristic values or roots of (1.462). In general, the characteristic values are distinct; however, for the correlation matrix R, with the complex vector $\tilde{u}$ based on samples of a discrete-time weakly stationary stochastic process, the mean-square value of the scalar $\tilde{y}' = \tilde{x}^H \tilde{u}$ is evaluated as

$$E\left[|\tilde{y}'|^2\right] = \tilde{x}^H R \tilde{x} \geq 0 \quad (1.463)$$

Equation (1.463) corresponds to a nonnegative definite quadratic form and Haykin points out that the equality condition rarely applies in practice so that (1.463) is almost always positive definite.† Consequently, in these cases, characteristic values are real with and $\lambda_i > 0 \; \forall \, i$ and, for the weakly stationary stochastic process, the characteristic values are equal with $\lambda_i = \sigma^2 \; \forall \, i$.

Referring to (1.461), the n solutions to the characteristic equation are determined using

$$R\tilde{x}_i = \lambda_i \tilde{x}_i \quad (1.464)$$

where $\tilde{x}_i$ are the characteristic vectors corresponding to the characteristic values λ_i. Based on the correlation matrix R, it can be shown [83] that the characteristic values are all real and nonnegative and the characteristic vectors are linearly independent and orthogonal to each other. A corollary to the proof of independence is that when the characteristic value is multiplied by a scalar constant the characteristic vectors remain independent and orthogonal. This allows for the independent scaling of all characteristic values so that

$$\tilde{x}_i^H \tilde{x}_j = \begin{cases} 1 & : i = j \\ 0 & : i \neq j \end{cases} \quad (1.465)$$

resulting in an orthonormal set of characteristic vectors satisfying (1.464). Recognizing that the matrix Q is a column of characteristic vectors with $Q = [\tilde{x}_1, \tilde{x}_2, \ldots, \tilde{x}_n]$ and Λ is a diagonal matrix of characteristic values $\Lambda = diag(\lambda_1, \lambda_2, \ldots, \lambda_n)$, (1.464) is expressed as

$$RQ = Q\Lambda \quad (1.466)$$

and the orthonormal matrix Q satisfies the relationship $Q^H Q = I$, therefore, $Q^H = Q^{-1}$ and premultiplying both sides of (1.466) by Q^H results in (1.460) as stated. Also, by post-multiplying both sides of (1.466) the correlation matrix is expressed as

$$R = Q\Lambda Q^H = \sum_i^n \lambda_i \tilde{x}_i \tilde{x}_i^H \quad (1.467)$$

Haykin states that there is no best way to compute the characteristic values and suggest that the use of the characteristic equation should be avoided except for the simple case involving the 2 × 2 matrix. However, a number of authors [77, 78, 84–89] describe computationally efficient methods for computing the characteristic values.

Consider the quadratic form $\tilde{Q} = \tilde{w}^H R \tilde{w}$ discussed in Section 12.3, where $\tilde{w}$ is a $(m \times 1)$ complex vector with elements $\tilde{w}_i = w_{ci} + jw_{si}$ and $R = E[\tilde{x}\tilde{x}^*]$ is a $(m \times m)$ Hermitian matrix with elements $\tilde{r}_{ji} = \tilde{r}_{ij}^*: i \neq j$ corresponding to the autocorrelation matrix of the $(m \times 1)$ complex vector $\tilde{x}$. The following derivatives with respect to the complex vector $\tilde{w}$ are evaluated as:

$$\frac{d}{d\tilde{w}}\tilde{Q} = \frac{d}{d\tilde{w}}\tilde{x}^H R \tilde{w} = 2R\tilde{w} \quad (1.468)$$

$$\frac{d}{d\tilde{w}}\tilde{c}^H \tilde{w} = 0 \quad (1.469)$$

and

$$\frac{d}{d\tilde{w}}\tilde{w}^H \tilde{c} = 2\tilde{c} \quad (1.470)$$

where $\tilde{c}$ is a $(m \times 1)$ complex vector. The solutions to the derivative of the quadratic form with respect to the vector w are used in Chapter 12 in the analysis of adaptive systems. The proofs of (1.468), (1.469), and (1.470) are given by Haykin [78].

*The characteristic vectors and values are also referred to as the eigenvectors and eigenvalues.

†A quadratic form is positive definite *iff* the characteristic values of R are all positive.

1.13 OFTEN USED MATHEMATICAL PROCEDURES

This section outlines the processing and formulas encountered in several of the chapters throughout this book.

1.13.1 Prime Factorization and Determination of Greatest Common Factor and Least Common Multiple

Prime factors are used in various coding applications where polynomials are used as code generators. The greatest common factor (GCF) and least common multiple (LCM) are often used where signal processing sample-rate changes are required to improve performance or reduce processing complexity. They are also used in the implementation of mixed radix fast Fourier transforms. An integer p is prime if $p \neq \pm 1$ and the only divisors are $\pm p$ and ± 1. The procedures involved in determining the prime factors of a number and the GCF and LCM between two or more numbers are simply given by way of the following examples. The algorithms are easy to generalize and implement in a computer program.

1.13.1.1 Prime Factors of Two Numbers The following examples, presented in Table 1.17, demonstrate the procedure for determining the prime factors of the numbers 120 and 200. The results are used in the following two examples to determine the GCF and LCM. The prime factors of a number are determined, starting with repeated division of the number by 2.

1.13.1.2 Determination of the Greatest Common Factor This example outlines the procedures in determining the GCF of the two numbers 120 and 200 using the prime factors listed in Table 1.17; inclusion of more than more than two numbers is straightforward. The GCF is also referred to as the greatest common divisor (GCD). The GCF (or GCD) between 120 and 200 is determined as the product of the prime factors taken the *minimum* number of times that they occurred in any one of the prime factorizations, that is,

$$\mathrm{GCF}(120,200) = 2^3 \times 3^0 \times 5^1 = 8 \times 1 \times 5 = 40.$$

TABLE 1.17 Example of Prime Factorization

Using 120	Using 200
$120 \div 2 = 60$	$200 \div 2 = 100$
$60 \div 2 = 30$	$100 \div 2 = 50$
$30 \div 2 = 15$	$50 \div 2 = 25$
$15 \div 3 = 5$	$25 \div 5 = 5$
$5 \div 5 = 1$	$5 \div 5 = 1$
The prime factors are:	
$2 \times 2 \times 2 \times 3 \times 5$	$2 \times 2 \times 2 \times 5 \times 5$

1.13.1.3 Determination of the Least Common Multiple This example outlines the procedures in determining the LCM between 120 and 200. In this case, the LCM is determined as the product of the prime factors taken the *maximum* number of times that they occurred in any one of the prime factorizations, that is,

$$\mathrm{LCM}(120,200) = 2^3 \times 3^1 \times 5^2 = 8 \times 3 \times 5 = 600.$$

1.13.2 Newton's Method

A transcendental equation involves trigonometric, exponential, logarithmic, and other functions that do not lend themselves to solutions by algebraic means. Newton's method [90] of solving transcendental equations is used extensively in arriving at solutions to problems characterized by nonalgebraic equations. The method provides a rapid and accurate solution to determine the value of functions having the form $f(x) = h(x)$, by finding the solution of the auxiliary function $g(x) = f(x) - h(x) = 0$. The solution begins by starting with an estimate $\hat{x}_o$ of the solution and performing iterative updates to the estimate, described as

$$x_{i+1} = x_i - \frac{g(x_i)}{g'(x_i)} \quad : i = 0, \ldots \tag{1.471}$$

where $g'(x_i) = \partial g(x_i)/\partial x_i$. The evaluation is terminated when $|f(x_{i+1}) - h(x_{i+1})| \le \varepsilon$ where ε is an acceptable error in the solution and the corresponding x_{i+1} is the desired value of x satisfying $f(x) \cong h(x)$.

1.13.3 Standard Deviation of Sampled Population

When a finite number of samples n comprises the entire population, the standard deviation is computed as

$$\sigma = \sqrt{\frac{1}{n}\sum x^2 - \left(\frac{1}{n}\sum x\right)^2} \tag{1.472}$$

where the summation is over the entire population. However, when the samples n are a subset of the entire population, the standard deviation is computed as

$$s = \sqrt{\frac{1}{n-1}\sum x^2 - \frac{n}{n-1}\left(\frac{1}{n}\sum x\right)^2} \tag{1.473}$$

where the summations are over the sample size n of the subset.

1.13.4 Solution to the Indeterminate Form 0/0

The frequently encountered functional form

$$h(x) = \frac{f(x)}{g(x)} \tag{1.474}$$

evaluated at $x = a$ often results in the indeterminate form 0/0 resulting from $f(a) = g(a) = 0$. Application of L'Hospital's rule of repeated differentiation of $f(x)$ and $g(x)$ and evaluating the result as

$$h(x) = \lim_{x \to a} \frac{f^{(n)}(x)}{g^{(n)}(x)} \tag{1.475}$$

often leads to a solution for $n = 1$ or 2. Solutions to other indeterminate forms involving ∞/∞, $0*\infty$, $\infty-\infty$, 0^0, ∞^0, 0^∞, and 1^∞ may also be found using similar techniques [90, 91].

1.14 OFTEN USED MATHEMATICAL RELATIONSHIPS

In this section a number of mathematical relationships are listed as found in various mathematical handbooks. The references frequently referred to are as follows: Burington [92], Korn and Korn [93], Milton and Stegun [94], and Gradshteyn and Ryzhik [46].

1.14.1 Finite and Infinite Sums

1 $\sum_{m=1}^{M} m = \frac{M(M+1)}{2}$: summation of integers

2 $\sum_{m=1}^{M} (2m-1) = M^2$: summation of odd integers

3 $\sum_{m=1}^{M} 2m = M(M+1)$: summation of even integers

4 $\sum_{m=1}^{M} m^2 = \frac{M(M+1)(2M+1)}{6}$

5 $\sum_{m=0}^{M-1} x^m = \frac{1-x^M}{1-x}$: $|x| < 1$ finite geometric series

6 $\sum_{m=0}^{\infty} x^m = \frac{1}{1-x}$: $|x| < 1$ infinite geometric series

7 $\sum_{m=0}^{\infty} m x^m = \frac{x}{(1-x)^2}$: $|x| < 1$

8 $\sum_{m=0}^{M} \sin(m\alpha + \beta) = \frac{\sin\left(\frac{M+1}{2}\alpha\right)}{\sin(\alpha/2)} \sin\left(\frac{M}{2}\alpha + \beta\right)$
: M = positive integer

9 $\sum_{m=0}^{M} \cos(m\alpha + \beta) = \frac{\sin\left(\frac{M+1}{2}\alpha\right)}{\sin(\alpha/2)} \cos\left(\frac{M}{2}\alpha + \beta\right)$
: M = positive integer

10 $1 + 2\sum_{m=1}^{M} \cos(m\alpha) = \frac{\sin((M+1/2)\alpha)}{\sin(\alpha/2)}$

11 $\sum_{m=0}^{M} e^{j(m\alpha+\beta)} = \frac{\sin\left(\frac{M+1}{2}\alpha\right)}{\sin(\alpha/2)} e^{j\left(\frac{M}{2}\alpha+\beta\right)}$
: M = positive integer

12 $\sum_{m=-M}^{M} e^{\pm jm\alpha} = \frac{\sin((M+1/2)\alpha)}{\sin(\alpha/2)}$

13 $(1 \pm x)^m = 1 \pm mx + \frac{m(m-1)}{2!} x^2 + \cdots + (\pm 1)^n \frac{m(m-1)\ldots(m-n+1)}{n!} x^n + \cdots \quad : m > 0$

14 $(1 \pm x)^{-m} = 1 \mp mx + \frac{m(m+1)}{2!} x^2 + \cdots + (\pm 1)^n \frac{m(m+1)\ldots(m+n-1)}{n!} x^n + \cdots \quad : m > 0$

1.14.2 Binomial Theorem and Coefficients

1 $(x \pm y)^m = x^m \pm \binom{m}{1} x^{m-1} y + \binom{m}{2} x^{m-2} y^2 \pm \cdots (\pm 1)^{m-2} \binom{m}{m-1} x y^{m-1} (\pm 1)^{m-1} \binom{m}{m} y^m = \sum_{i=0}^{m} (\pm 1)^i \binom{m}{i} x^{m-i} y^i$

2 $(1 \pm y)^m = 1 - \sum_{i=1}^{m} (\pm 1)^{i+1} \binom{m}{i} y^i$

3 $\binom{m}{n} = {}_mC_n = \frac{m!}{(m-n)!n!}$: coefficient evaluation

4 $\binom{m}{n} = \binom{m}{m-k}$: coefficient symmetry

5 $\sum_{n=0}^{m} \binom{m}{n} = 2^m$: coefficient summation

6 $\sum_{n=0}^{m} \binom{m}{n} (-1)^n = 0$: alternating sign coefficient summation

TABLE 1.18 Brief List of Binomial Coefficients

m \ n	0	1	2	3	4	5	6	7	8	9	10
1	1	1									
2	1	2	1								
3	1	3	3	1							
4	1	4	6	4	1						
5	1	5	10	10	5	1					
6	1	6	15	20	15	6	1				
7	1	7	21	35	35	21	7	1			
8	1	8	28	56	70	56	28	8	1		
9	1	9	36	84	126	126	84	36	9	1	
10	1	10	45	120	210	252	210	120	45	10	1

7 $\sum_{n=1}^{m-1}\binom{m-1}{n}\frac{1}{n+1}=\frac{1}{m}\sum_{n=2}^{m}\binom{m}{n}$

: coefficient summation identity

1.14.3 Trigonometric Identities

1 $\sin(2\alpha)=2\sin(\alpha)\cos(\alpha)$

2 $\cos(2\alpha)=2\cos^2(\alpha)-1=1-2\sin^2(\alpha)=\cos^2(\alpha)-\sin^2(\alpha)$

3 $\sin^2(\alpha)=\frac{1}{2}(1-\cos(2\alpha))$

4 $\sin^3(\alpha)=\frac{1}{4}(3\sin(\alpha)-\sin(3\alpha))$

5 $\sin^4(\alpha)=\frac{1}{8}(3-4\cos(2\alpha)+\cos(4\alpha))$

6 $\cos^2(\alpha)=\frac{1}{2}(1+\cos(2\alpha))$

7 $\cos^3(\alpha)=\frac{1}{4}(3\cos(a)+\cos(3\alpha))$

8 $\cos^4(\alpha)=\frac{1}{8}(3+4\cos(2a)+\cos(4\alpha))$

9 $\sin(\alpha)+\sin(\beta)=2\sin((\alpha+\beta)/2)\cos((\alpha-\beta)/2)$

10 $\sin(\alpha)-\sin(\beta)=2\cos((\alpha+\beta)/2)\sin((\alpha-\beta)/2)$

11 $\cos(\alpha)+\cos(\beta)=2\cos((\alpha+\beta)/2)\cos((\alpha-\beta)/2)$

12 $\cos(\alpha)-\cos(\beta)=-2\sin((\alpha+\beta)/2)\sin((\alpha-\beta)/2)$

13 $\sin(\alpha)\sin(\beta)=\frac{1}{2}\cos(\alpha-\beta)-\frac{1}{2}\cos(\alpha+\beta)$

14 $\cos(\alpha)\cos(\beta)=\frac{1}{2}\cos(\alpha-\beta)+\frac{1}{2}\cos(\alpha+\beta)$

15 $\sin(\alpha)\cos(\beta)=\frac{1}{2}\sin(\alpha+\beta)+\frac{1}{2}\sin(\alpha-\beta)$

16 $\cos(\alpha)\sin(\beta)=\frac{1}{2}\sin(\alpha+\beta)-\frac{1}{2}\sin(\alpha-\beta)$

17 $\tan^{-1}(z_1)\pm\tan^{-1}(z_2)=\tan^{-1}\left(\frac{z_1\pm z_2}{1\mp z_1z_2}\right)$

18 $\sinh(z)=\frac{1}{2}(e^z-e^{-z})=1/\text{csch}(z)$

19 $\cosh(z)=\frac{1}{2}(e^z+e^{-z})=1/\text{sech}(z)$

20 $\tanh(z)=\left(\frac{e^z-e^{-z}}{e^z+e^{-z}}\right)=\frac{\sinh(z)}{\cosh(z)}=1/\text{ctnh}(z)$

1.14.4 Differentiation and Integration Rules

The notations u and v are functions of x

1 $\frac{d}{dx}(uv)=u\frac{dv}{dx}+v\frac{du}{dx}$

2 $\frac{d}{dx}(u/v)=\frac{v\frac{du}{dx}-u\frac{dv}{dx}}{v^2}$

3 $\frac{d}{dx}(u^v)=u^v\left(\frac{v\,du}{u\,dx}+\ln(u)\frac{dv}{dx}\right)$

4 $\frac{d}{dx}\ln(u)=\frac{1}{u}\left(\frac{du}{dx}\right);\ \frac{d}{dx}\log_a(u)=\frac{\log_a(e)}{u}\left(\frac{du}{dx}\right)$

5 $\frac{d}{dx}a^u=a^u\log_e(a)\left(\frac{du}{dx}\right);\ \frac{d}{dx}e^u=e^u\left(\frac{du}{dx}\right)$

6 $\frac{d}{dx}\sin(u)=\cos(u)\left(\frac{du}{dx}\right);\ \frac{d}{dx}\sin^{-1}(u)=\frac{1}{\sqrt{1-u^2}}\left(\frac{du}{dx}\right)$

7 $\frac{d}{dx}\cos(u)=-\sin(u)\left(\frac{du}{dx}\right);\ \frac{d}{dx}\cos^{-1}(u)=\frac{-1}{\sqrt{1-u^2}}\left(\frac{du}{dx}\right)$

8 $\frac{d}{dx}\tan(u)=\sec^2(u)\left(\frac{du}{dx}\right);\ \frac{d}{dx}\tan^{-1}(u)=\frac{1}{1+u^2}\left(\frac{du}{dx}\right)$

9 $\frac{d}{dx}\sinh(u)=\cosh(u)\left(\frac{du}{dx}\right);\ \frac{d}{dx}\sinh^{-1}(u)=\frac{1}{\sqrt{1+u^2}}\left(\frac{du}{dx}\right)$

10 $\frac{d}{dx}\cosh(u)=\sinh(u)\left(\frac{du}{dx}\right);$

$\frac{d}{dx}\cosh^{-1}(u)=\frac{1}{\sqrt{-1+u^2}}\left(\frac{du}{dx}\right)\quad :u>1$

11 $\frac{d}{dx}\tanh(u)=\ \mathrm{sech}^2(u)\left(\frac{du}{dx}\right);$

$\frac{d}{dx}\tanh^{-1}(u)=\frac{1}{1-u^2}\left(\frac{du}{dx}\right)$

12 $\int u\,dv=uv-\int v\,du$: integration by parts with $u = f(x)$ and $dv = g(x)dx$

13 $\int_0^{\infty} x^b e^{-ax}dx=\frac{\Gamma(b+1)}{a^{b+1}}\quad :a>0,\ b>-1$

$=\frac{b!}{a^{b+1}}\quad :a>0,\ b=0,1,2,\ldots$

14 $\int_0^{\infty} e^{-ax}dx=\frac{1}{a}$

15 $\int_0^{\infty} \sqrt{x}e^{-ax}dx=\frac{\sqrt{\pi/a}}{2a}$

16 $\int_0^{\infty} \frac{1}{\sqrt{x}}e^{-ax}dx=\sqrt{\frac{\pi}{a}}$

17 $\int_0^{\infty} x^b e^{-ax^2}dx=\frac{\Gamma((b+1)/2)}{2a^{(b+1)/2}}\quad :a>0,\ b>-1$

$=\frac{1(3)\ldots(b-1)\sqrt{\pi}}{2^{b/2+1}a^{(b+1)/2}}\quad :a>0,\ b=0,2,4,\ldots$

$=\frac{((b-1)/2)!}{2a^{(b+1)/2}}\quad :a>0,\ b=1,3,5,\ldots$

18 $\int_0^{\infty} e^{-a^2x^2}dx=\frac{\sqrt{\pi}}{2a}$

19 $\frac{d}{dy}\int_{a(y)}^{b(y)} f(x,y)dx=\int_{a(y)}^{b(y)}\frac{\partial}{\partial y}f(x,y)dx$

$+f(b,y)\frac{db}{dy}-f(a,y)\frac{da}{dy}\quad$: Leibniz's theorem

20 $\int_{-\infty}^{\infty} |s(t)|^2dt=\frac{1}{2\pi}\int_{-\infty}^{\infty} |F(\omega)|^2d\omega$

: Parseval's Theorem

21 $g(t)=\int_{-\infty}^{\infty} x(\lambda)h(t-\lambda)d\lambda$

$=\int_{-\infty}^{\infty} x(t-\lambda)h(\lambda)d\lambda\quad$: convolution

22 $R(\tau)=\int_{-\infty}^{\infty} x(t)y(t-\tau)dt$

$=\int_{-\infty}^{\infty} x(t+\tau)y(t)dt\quad$: correlation

1.14.5 Inequalities

1 $|a-b|\le|a\pm b|\le|a|+|b|$: triangle inequality

2 $\left|\sum_{n=1}^{N} a_n\right|\le\sum_{n=1}^{N}|a_n|$: triangle inequality

3 $\left(\prod_{n=1}^{N} b_n\right)^{1/N}\le\frac{1}{N}\sum_{n=1}^{N} a_n$: geometric mean ≤ arithmetic mean

$b_n > 0$; equality holds *iff* $b_n = a_n$

4 $\left[\sum_{n=1}^{N} a_nb_n\right]^2\le\sum_{n=1}^{N}a_n^2\sum_{n=1}^{N}b_n^2$: Cauchy's inequality

Equality holds for c = constant > 0 *iff* $a_n = c\ b_n$

5 $\left[\int_a^b f^*(x)g(x)dx\right]^2\le\left[\int_a^b |f(x)|^2dx\right]\left[\int_a^b |g(x)|^2dx\right]$

: Schwarz's inequality

Equality holds for c = constant > 0 *iff* $f(x) = c\ g(x)$.

1.14.6 Relationships between Complex Numbers

For $A=M_ae^{j\alpha}$ and $B=M_be^{j\beta}$

1 $||A|-|B||\le|A\pm B|\le|A|+|B|$

2 $|AB|=|A||B|;\ |A/B|=|A|/|B|$

3 $A/B=AB/|B|^2$

4 $\mathrm{Re}[A]=\frac{1}{2}[A+A^*]$

5 $\mathrm{Re}[A]\mathrm{Re}[B]=\frac{1}{2}\mathrm{Re}[AB]+\frac{1}{2}\mathrm{Re}[AB^*]$

6 $\text{Re}[A]\text{Im}[B] = \frac{1}{2}\text{Im}[AB] - \frac{1}{2}\text{Im}[AB^*]$
$= \frac{1}{2}\text{Im}[AB] + \frac{1}{2}\text{Im}[A^*B]$

7 $\text{Im}[A]\text{Re}[B] = \frac{1}{2}\text{Im}[AB] + \frac{1}{2}\text{Im}[AB^*]$
$= \frac{1}{2}\text{Im}[AB] - \frac{1}{2}\text{Im}[A^*B]$

8 $\text{Im}[A]\text{Im}[B] = \frac{-1}{2}\text{Re}[AB] + \frac{1}{2}\text{Re}[AB^*]$

1.14.7 Miscellaneous Relationships [94]

1 $\Gamma(z) = \int_0^\infty t^{z-1}e^{-t}dt$:Euler's integral, $\text{Re}(z) > 0$

2 $\Gamma\left(\frac{1}{2}\right) = \left(\frac{-1}{2}\right)! = \sqrt{\pi}$

3 $\Gamma\left(\frac{3}{2}\right) = \left(\frac{1}{2}\right)! = \frac{\sqrt{\pi}}{2}$

4 $\Gamma(n+1) = n! = n(n-1)(n-2)\ldots 1$:integer factorial; $0! = 1$

5 $(2n+1)!! = 1{\cdot}3{\cdot}5\cdots(2n+1)$:odd integer factorial [95]

6 $(2n)!! = 2{\cdot}4{\cdot}6\cdots(2n)$:even integer factorial [95]

7 $\log_{10}(x) = \log(x)$: common logarithm of x

8 $\log_e(x) = \ln(x)$:natural logarithm of x

9 $\log_a(b) = \frac{1}{\log_b(a)}$

10 $\log_a(x) = \frac{\log_b(x)}{\log_b(a)}$

11 $a^{\log_a(x)} = x$

12 $r = \text{mod}(a,b) = a - \text{int}\left(\frac{a}{b}\right)b$:$r,a,b \neq 0$
are same type $\in$ real or integer*
Example: for $b = 3$: $a = \ldots -4\ -3\ -2\ -1\ 0\ 1\ 2\ 3\ 4 \ldots$
$r = \ldots -1\ 0\ -2\ -1\ 0\ 1\ 2\ 0\ 1 \ldots$

13 $r = \text{modulo}(a,b)$

$$= \begin{cases} a - \lfloor a/b \rfloor b & :r,a,b \in \text{real} \\ a - bq & :r,a,b,q \in \text{integer with } q \text{ selected* so that } r \text{ is nearer to zero than } b \end{cases}$$

Example: for $b = 3$: $a = \ldots -4\ -3\ -2\ -1\ 0\ 1\ 2\ 3\ 4 \ldots$
$r = \ldots\ 2\ 0\ 1\ 2\ 0\ 1\ 2\ 0\ 1 \ldots$

14 $sign(a,b) = \begin{cases} |a| & : b \geq 0 \\ -|a| & : b < 0 \end{cases}$: a, b are same type $\in$ real or integer*

15 The solutions to the quadratic equation $x^2 + bx + c = 0$ is

$$x_1, x_2 = -b/2 \pm \sqrt{(b/2)^2 - c}$$

16 Completing the square of: $x^2 + bx = (x + b/2)^2 - (b/2)^2$

17 Completing the square of: $ax^2 + bx = (\sqrt{a}x + b/(2\sqrt{a}))^2 - (b/2)^2/a$

*Fortran 95 Language Reference, Revision D, Lahey Computer Systems, Inc., Incline Village, NV, 1994. The notations a modulo(b) and a modulo-b are also used.

ACRONYMS

ACI	Adjacent channel interference
ACK	Acknowledgment (protocol)
AFSCN	U.S. Air Force Satellite Control Network
AM	Amplitude modulation
ARQ	Automatic repeat request
AWGN	Additive white Gaussian noise
BPSK	Binary phase shift keying
BT	Time bandwidth (product, low pass)
CFAR	Constant false-alarm rate
CRC	Cyclic redundancy check (code)
DC	Direct current
DFT	Discrete Fourier transform
DSB	Double sideband
DSSS	Direct-sequence spread-spectrum (waveform)
EHF	Extremely high frequency
ELF	Extremely low frequency
FFT	Fast Fourier transform
FM	Frequency modulation
FSK	Frequency shift keying
GCD	Greatest common divisor
GCF	Greatest common factor
HF	High frequency
I/Q	In-phase and quadrature (channels or rails)
IDFT	Inverse discrete Fourier transform
IF	Intermediate frequency
IFFT	Inverse Fast Fourier transform
ISI	Intersymbol interference
LCM	Least common multiple
LF	Low frequency
LLR	Log-likelihood ratio
LR	Likelihood ratio
MAP	Maximum a posteriori
MF	Medium frequency
ML	Maximum likelihood
MMSE	Minimum mean-square error
MS	Mean square
MSK	Minimum shift keying
NAK	Negative acknowledgment (protocol)
OQPSK	Offset quadrature phase shift keying
PM	Phase modulation
PN	Pseudo-noise (sequence)
PSD	Power spectral density
PSK	Phase shift keying
QAM	Quadrature amplitude modulation

QPSK	Quadrature phase shift keying
RC	Raised-cosine
RQ	Repeat request
RRC	Root-raised-cosine (temporal)
RSS	Root-sum-square
SHF	Super high frequency
SLF	Super low frequency
SRC	Spectral raised-cosine
SRRC	Spectral root-raised-cosine
SS	Spread-spectrum
TRC	Temporal raised-cosine
UHF	Ultra-high frequency
ULF	Ultra-low frequency
VLF	Very low frequency
WSS	Wide-sense stationary
WT	Time bandwidth (product, bandpass)

PROBLEMS

1. Show that the amplitude-modulated waveform given by (1.2), when heterodyned by a receiver local oscillator that is phase locked to the received carrier angular frequency ω_c, recovers the modulation function $s(t)=A(1+m_I\sin(\omega_m t))$, except for a factor of 1/2.

   Hint: Mix (1.2) with $\sin(\omega_c t+\phi)$ and show that ϕ must be zero.

2. Show that the real signal given by (1.13) is a form of suppressed carrier modulation. Under what conditions of $M(t)$ and $\phi(t)+\psi(t)$ does (1.13) reduce to the form of the suppressed carrier modulation given by (1.12)? What can be said about the information capacity between the suppressed carrier modulations given by (1.12) and (1.13)?

3. Compute the Hilbert transform of $s_c(t)=\cos(\omega_c t)$ and $s_s(t)=\sin(\omega_c t)$.

4. Given that the bandwidth of the modulation function $A(t)$ satisfies the condition $B << f_c$, compute the Hilbert transform of $s(t)=A(t)\cos(\omega_c t+\phi(t))$.

5. Show that the Fourier coefficients C_n and C_{-n}, expressed in (1.30), form complex conjugate pairs when $f(t)$ is real.

6. Show that the real-valued function $f(t)$ can be expressed in terms of the Fourier series real coefficient $C_o=\alpha_o$ and the complex coefficients $C_n=\alpha_n+j\beta_n$: $1\le n\le\infty$ as

$$f(t)=C_o+2\sum_{n=1}^{\infty}|C_n|\cos(n\omega_o t+\phi_n)$$

   where $|C_n|=\sqrt{\alpha_n^2+\beta_n^2}$ and $\phi_n=\arctan(\beta_n/\alpha_n)$. Note: This solution is based on Problem 5.

7. Show that the finite summation $(1/T)\sum_{n=-N}^{N}e^{jn\omega_o t}$ is equal to the second equality in (1.52).

   Hint: Expand the summation and combine the exponential terms to yield a series involving $\cos(n\omega_o T)$ terms and then evaluate the closed form of the corresponding trigonometric series as identified in Section 1.14.1 Identity No. 12.

8. With $\omega_o = 2\pi/T$ show that the integral $(2N+1)\int_{-T/2}^{T/2}\frac{\sin((2N+1)\omega_o t/2)}{(2N+1)\omega_o t/2}dt$ is equal to unity as $N\to\infty$.

9. Referring to Figure 1.7 and using $\omega_o=2\pi/T$, show that the maximum value of (1.52) is $(2N+1)/T$ and that the closest zero or null removed from a maximum occurs at $t=nT\pm T/(2N+1)$: $|n|=0,1,\ldots$.

10. Consider a radix-2, N-point, pipeline FFT with the output sampled at intervals of $T=N_sT_s$ seconds, where N_s is the number of samples per symbol. If the sequential input samples are simply passed through the FFT delay elements with the complex multiplications and additions only performed at the output sampling instants: (A) determine the percentage of complex multiplies relative to the pipeline FFT sampled every T_s seconds. Examine the result as a function of increasing N_s with $1\le N_s\le 32$ and $N\ge N_s$; (B) determine the minimum number of complex multiplications when 100 % zero padding is used for frequency estimation and tracking and discuss the pipeline FFT sampling requirements.

11. Compute the second moment, $E[X^2]$, of the Gaussian random variable X with mean m and variance σ^2.

12. Referring to (1.165) compute $E\left[g_2^2(X_2)E\left[g_1^2(X_1)|X_2=x_2\right]\right]$ for the conditional Gaussian *pdf*, expressed by the second equality of (1.168), with $g_2^2(X_2)=X_2^2$ and $g_1^2(X_1)=X_1^2$. Express the result in terms of the expectations as $E\left[X_2^2X_1^2\right]=C_1E\left[X_2^2\right]+C_2E\left[X_2^3\right]+C_3E\left[X_2^4\right]$ and express C_1, C_2, and C_3 in terms of the parameters ρ, σ_1^2, σ_2^2, m_1, and m_2.

    Using $E\left[X_2^2\right]=\sigma_2^2+m_2^2$, $E\left[X_2^3\right]=3\sigma_2^2m_2+m_2^3$, and $E\left[X_2^4\right]=3\sigma_2^4+6\sigma_2^2m_2^2+m_2^4$ evaluate $E\left[X_2^2X_1^2\right]$ with $m_1=m_2=0$.

13. Repeat Problem 12 using (1.172) and show that

$$E\left[x_1^2x_2^2\right]=E\left[x_1^2\right]E\left[x_2^2\right]+2E^2[x_1x_2]$$

    when x_1 and x_2 are zero-mean Gaussian random variables.

14. In the transformation from $f_X(x)$ to $f_Z(z)$, discussed in Section 1.5.4.1, show that the inverse relationship in (1.188) applies for the function $z=ax^2$.

    Part B: Express $f_Z(z)$ using (1.186) or (1.187).

Part C: Express $f_Z(z)$ when the *pdf* of $f_X(x)$ is Gaussian with mean value m and variance σ^2. Plot or sketch your expression $f_Z(z)$ as a function of z.

Note: $e^{\lambda} + e^{-\lambda} = 2\cosh(\lambda)$.

Part D: Express $f_Z(z)$ when $m = 0$ in Part C and plot or sketch as a function of z.

15. Given the statistically independent ordered random variables $\{X_1, X_2, \ldots, X_n\}$ such that $a \le X_1 < X_2 < \cdots < X_n \le b$ and characterized by the uniformly distributed *pdf* $f_X(x_i) = 1/(b-a)$: $\forall i$, $i = 1,2,\ldots,i',\ldots,n$ with the corresponding *cdf* expressed as $F_X(x) = \int_a^x f_X(x_i)dx_i$. Show that

$$\frac{1}{(i'-2)!}\int_a^{x_{i'}} F_X^{i'-2}(x_{i'-1})f_X(x_{i'-1})dx_{i'-1} = \frac{1}{(i'-1)!}F_X^{i'-1}(x_{i'})$$

with $x_n \le b$.

Hint: Start with $F_X^1(x_2) = \int_a^{x_2} F_X^0(x_1)f_{X_1}(x_1)dx_1$ with $F_X^0(x_1) \triangleq 1$ and note that $F_X^1(x_2) = (x_2-a)/(b-a)$.

16. For the transformation in Section 1.5.5, evaluate the Jacobian in (1.211) using the phase angle expressed as $\theta = \tan^{-1}(x_s/x_c)$.

Hint: use $g_{11}(x_c,x_s) = g_{12}(x_c,x_s) = \sqrt{x_c^2 + x_s^2}$ and $g_{21}(x_c,x_s) = g_{22}(x_c,x_s) = \tan^{-1}(x_s/x_c)$.

17. Given the joint *pdf* $f_{X,Y}(x,y)$, expressed in (1.166), compute the marginal *pdf* $M_X(x)$.

Hint: Complete the square using: $a^2 + ba + c^2 = (a + b/2)^2 - b^2/4 + c^2$.

18. Given the *pdf* $f_{X,}(x)$, perform the following:

A. Compute the *pdf* $f_{Y,}(y)$ under the condition $y = |x|$. Note that $f_{Y,}(y) = 0$ for $y < 0$.

B. Determine and sketch $f_{Y,}(y)$ when $f_{X,}(x)$ is described by the normal distribution $N(m_x,\sigma_x)$

C. Repeat Part B with $m_x = 0$

19. Show that the limiting form of the Ricean distribution, expressed by (1.222), corresponds to the Rayleigh distribution as $A \to 0$. Refer to Table 1.8.

Hint: Use the ascending series expression

$$I_o(z) = 1 + (z^2/4)/(1!)^2 + (z^2/4)^2/(2!)^2 + (z^2/4)^3/(3!)^2 + \cdots \text{ with } z = Ar/\sigma^2.$$

20. Show that the limiting form of the Ricean distribution, expressed by (1.222), corresponds to the Gaussian distribution as $A \to \infty$. Refer to Table 1.8.

Hint: Use the asymptotic expansion of $I_o(z)$ for large arguments expressed as

$$I_o(z) \sim \frac{e^z}{\sqrt{2\pi z}}\left(1 + 1/8z + 9/2!(8z)^2 + \cdots\right) \quad : z \to \infty$$

with $z = Ar/\sigma^2$.

Recognize that as $r \to A$ the condition $r = A$ results in the Gaussian distribution.

21. Determine the marginal *pdf* of $Y_1 = \min\{X_1, X_2, \ldots, X_n\}$ given the joint *pdf* $g_Y(y_1, y_2, \ldots, y_n)$ of the uniformly distributed ordered samples $a \le y_1 < y_2 < \cdots < y_n \le b$ corresponding to $f_{Y_j}(y_j) = 1/(b-a)$: $\forall j$.

Hint: Show that the *cdfs* in the descending order $y_n, y_{n-1}, \ldots, y_{j'}, \ldots y_2$ are expressed as

$$\frac{1}{(n-j'+1)!}\bar{F}_Y^{n-j'+1}(y_{j'-1}) = \frac{1}{(n-j')!}\int_{y_{j'-1}}^{b} \bar{F}_Y^{n-j'}(y_{j'})f(y_{j'})dy_{j'}$$

$$\text{with } \bar{F}_Y^{n-j'}(y_{j'}) = \left[\frac{(b-y_{j'})}{(b-a)}\right]^{n-j'} \text{ and } \bar{F}_Y^0(y_n) \triangleq 1.$$

Also show that $\bar{F}_Y^{n-j'}(y_{j'}) = [1 - F_Y(y_{j'})]^{n-j'}$ where $F_Y(y_{j'}) = \frac{(y_{j'}-a)}{(b-a)}$.

22. Show that the Nakagami-m distribution is the same as the Rayleigh power distribution.

Hint: Use the transformation $y = x/\sqrt{2}$ in the Rayleigh distribution.

23. Derive the expression for the characteristic function $C_X(v)$ for the Gaussian distribution $f_X(x)$ with mean value x_o and variance σ^2.

24. Set up the integrations identifying the integration limits and ranges of the variable z for the evaluation of $f_Z(z)$ where the random variable Z is the summation of three (3) zero-mean uniformly distributed random variables X between $-a$ and a.

Hint: There are three unique ranges on z. The evaluation of the integrations is optional; however, the application of Mathsoft's Mathcad® symbolic formula evaluation is an error-free time saver.

25. Using $f_Z(z)$ evaluated in Problem 24 for $N = 3$, compute the first and second moments of the random variable Z using (1.254) and compare the results with those in Table 1.6.

Hint: It is much easier and less prone to mistakes to use Mathsoft's Mathcad symbolic formula evaluation.

26. Show that the moments of the random variable X are determined from the characteristic function as expressed in (1.240).

Hint: Take the first derivative of $C_X(v)$ with respect to v and evaluate the result for $v = 0$; and observe that the resulting integral is $E[x]$. Repeat this procedure for additional derivatives of $C_X(v)$ and show that (1.240) follows.

27. Plot the *cdf* of a zero-mean Gaussian distribution with variances corresponding to the second moments in

Table 1.6 for $N = 3$ and 4 and compare the results with the corresponding *cdf's* in Figure 1.23; comment on the quality of the match in light of the central limit theorem. Repeat this exercise using the theoretical second moments from Table 1.7 for $N = 3$ and 4 and compare with the corresponding *cdf's* in Figure 1.26.

28. Show that equations (1.261) and (1.262) apply for $\lambda v \ll 1$ as N increases in the respective summation of N *iid* distributions in Examples 1 and 2 of Section 1.5.6.1.

29. The narrowband noise process $n(t)$, given by (1.307), is expressed in terms of the baseband analytic noise $\tilde{n}(t)$ as

$$n(t) = \frac{1}{2}\left(\tilde{n}(t)e^{j\omega_c t} + \tilde{n}^*(t)e^{-j\omega_c t}\right)$$

Using this relationship, express the correlation function $R_{nn}(\tau) = E[n(t)n(t-\tau)]$ in terms of the individual correlation functions $R_{\tilde{n}\tilde{n}^*}(\tau)$, $R_{\tilde{n}\tilde{n}}(\tau)$, $R_{\tilde{n}^*\tilde{n}^*}(\tau)$ and $R_{\tilde{n}^*\tilde{n}}(\tau)$. What are the required conditions on these correlation functions to satisfy the stationarity property of the narrowband process $n(t)$?

30. Express the individual correlation functions in Problem 29 in terms of the correlation functions $R_{cc}(\tau)$, $R_{cs}(\tau)$, $R_{sc}(\tau)$, and $R_{ss}(\tau)$, where the baseband analytic noise is given by $\tilde{n}(t) = n_c(t) + jn_s(t)$. Use these results and the conditions for stationarity found in Problem 29 to express $R_{nn}(\tau)$ in terms of the $R_{cc}(\tau)$ and $R_{sc}(\tau)$.

31. Referring to (1.315), that applies to the noise power out of a bandpass filter centered at the positive frequency f_c. When the bandpass filter output is mixed to baseband, express the noise power out the baseband filter in terms of the bandwidth B and the one-sided noise spectral density N_o.

32. Given the noise input, expressed by (1.307), to a linear filter with impulse response $h(t)$, show that the respective input and output of the correlation responses $R_{nn}(\tau)$ and $R_{n'n'}(\tau)$ are related by the convolutions $R_{n'n'}(\tau) = h^*(-\tau)*h(\tau)*R_{nn}(\tau)$. Using this result with Fourier transform pairs $h(\tau) \leftrightarrow H(f)$ and $h^*(-\tau) \leftrightarrow H^*(f)$, show the relationship between the input and output noise spectrums.

Hint: Using the convolution integral $n'(t) = \int_{-\infty}^{\infty} n(t-\lambda)h(\lambda)d\lambda$ show that $n'(t)$ has zero-mean. Then from the correlation

$$R_{n'n}(\tau) = E[n'(t)n^*(t-\tau)] = E\left[\int n(t-\lambda)h(\lambda)n^*(t-\tau)d\lambda\right]$$

and show that $R_{n'n}(\tau) = R_{nn}(\tau)*h(\tau)$ and, as the final step, form the correlation

$$R_{n'n'}(\tau) = E\left[n'(t+\lambda)n'^*(t-\tau)\right] = E\left[\int n'(t+\tau)n^*(t-\lambda)h^*(\lambda)d\lambda\right]$$

and show that $R_{n'n'}(\tau) = R_{n'n}(\tau)*h^*(-\tau)$.

33. Derive the expression for the matched filter output signal-to-noise ratio when the additive noise is not white noise, that is, the noise power spectral density into the matched filter is $N_i(f) \neq N_o$.

34. Under the condition stated in Section 1.7.1 show that (1.332) is a wide-sense stationary random process.

35. Given the random process $x(t_i) = a$ where t_i is a discrete-time sample and a is a discrete random variable such that $a = 1$ with probability p and $= -1$ with probability $q = 1 - p$. Using (1.303) and (1.304) determine if $x(t_i)$ is ergodic.

36. Show that the random process $y(t) = x(t)\cos(\omega_c t + \varphi)$ is wss if f_c is constant and $x(t)$ is a wss random process independent of the random variable φ uniformly distributed over the interval 0 to 2π. Also, express the PSD $S_y(\omega)$ in terms of the autocorrelation $R_x(\tau)$ and the PSD $S_x(\omega)$.

37. The risk for the mean-square estimate is defined as

$$\Re_{ms} \triangleq \int_{-\infty}^{\infty}\left\{\int_{-\infty}^{\infty} [a-\hat{a}(s_r)]^2 p_{\mathrm{a}|\mathrm{s}_r}(\mathrm{a}|\mathrm{s}_r)\mathrm{da}\right\} p_{\mathrm{s}_r}(s_r)ds_r$$

Show that $\partial\Re_{ms}/\partial\hat{a}(r_s) = 0$ and results in the optimum estimate given by (1.356).

38. Determine if the MS and MAP estimates in the example of Section 1.9.1 are unbiased estimates. If not, what is the bias of the estimate? Also, evaluate the Cramér–Rao bound for the estimates and, using (1.366), determine if the estimates are efficient.

39. Given that the received baseband signal amplitude is A volts, using the ML estimate, determine the following: Part 1, the variance σ_{a}^2 of the estimation error of A given the baseband samples $r_i = A + n_i$: $i = 1, \ldots, N$ where n_i are *iid* Gaussian random variables characterized as $N(0,\sigma_n)$, Part 2, show that the estimate $\hat{\mathrm{a}}_{ml}(r)$ is efficient, Part 3 show the condition for which the estimate $\hat{\mathrm{a}}_{ml}(r)$ is unbiased.

40. Repeat Problem 39 under the following condition: the baseband signal amplitude is Gaussian distributed with a priori *pdf* $p_a(A)$ characterized by $N(A,\sigma_a)$.

41. Using (1.384) determine the effective bandwidth (β) for the isosceles triangle shaped pulse with base equal to 2τ and peak amplitude of AN volts.

Hints: The solution to the integral $\int_0^{\infty} \frac{\sin^{2m}(ax)}{x^2}dx = \frac{(2m-3)!!}{(2m-2)!!}\left(\frac{a\pi}{2}\right)$ is encountered with $m = 2$ and the double factorial [96] is defined as $(2m+1)!! = 1 \times 3 \times 5 \ldots (2m+1)$. The denominator in the expression for α^2 is the signal energy E.

42. Determine the normalized effective bandwidth (βT) and the corresponding standard deviation (σ_{Td}) for the SRC and SRRC waveforms with 100% excess bandwidth, that is, $\alpha = 1$.

43. Determine the normalized effective bandwidth (βT) and the corresponding standard deviation (σ_{Td}) of the delay estimate for the SRC and SRRC waveforms with zero excess bandwidth, that is, $\alpha = 0$.

44. Determine the noise bandwidth for the SRRC and SRC frequency functions.

Note: the noise bandwidth is defined by (1.46).

45. Using (1.387) determine the normalized effective time duration ($\bar{\alpha}T$) for the rectangular pulse $Arect(t/T - 0.5)$.

REFERENCES

1. C.E. Shannon, "Communication in the Presence of Noise," Proceedings of the IEEE, Vol. 86, Issue 2, pp. 447–457, February, 1998.
2. M. Abramowitz, I.A. Stegun, *Handbook of Mathematical Functions with Formula, Graphs and Mathematical Tables*, National Bureau of Standards, Applied Mathematical Series 55, Washington, DC, U.S. Government Printing Office, p. 361, June 1964.
3. R.A. Manske, "Computer Simulation of Narrowband Systems," IEEE Transactions on Computers, Vol. C-17, No. 4, pp. 301–308, April 1968.
4. E.C. Titchmarsh, *The Theory of Functions*, 2 ed., Oxford University Press, New York, 1939.
5. E.C. Titchmarsh, *Introduction to the Theory of Fourier Integrals*, 2 ed., Oxford University Press, New York, 1948.
6. W.B. Davenport, Jr., W.L. Root, *An Introduction to the Theory of Random Signals and Noise*, McGraw Hill Book Company, New York, 1958.
7. A. Papoulis, *The Fourier Integral and Its Applications*, pp. 9, 42–47, McGraw-Hill Book Company, New York, 1987.
8. A. Papoulis, *Probability, Random Variables and Stochastic Processes*, McGraw-Hill Book Company, New York, 1965.
9. A.V. Oppenheim, R.W. Schafer, *Digital Signal Processing*, Chapter 11, "Power Spectrum Estimation", Prentice-Hall, Inc., Englewood Cliffs, NJ, 1975.
10. M.S. Bartlett, *An Introduction to Stochastic Processes with Special Reference to Methods and Applications*, Cambridge University Press, New York, 1953.
11. B.P. Bogert, Guest Editor, "The Fast Fourier Transform and Its Application to Digital Filtering and Spectral Analysis," Special Issue of the IEEE Transactions on Audio and Electroacoustics, Vol. AU-15, No. 2, June 1967.
12. J.W. Cooley, P.A.W. Lewis, P.D. Welch, "The Finite Fourier Transform," IEEE Transactions on Audio and Electroacoustics, Vol. AU-17, No. 2, pp. 77–85, June 1969.
13. R.C. Singleton, "A Short Bibliography on the Fast Fourier Transform," IEEE Transactions on Audio and Electroacoustics, Vol. AU-17, No. 2, pp. 166–169, June 1969.
14. G.D. Bergland, "A Guided Tour of the Fast Fourier Transform," IEEE Spectrum, Vol. 6, pp. 41–52, July 1969.
15. T.H. Glisson, C.I. Black, A.P. Sage, "The Digital Computation of Discrete Spectra Using the Fast Fourier Transform," IEEE Transactions on Audio and Electroacoustics, Vol. AU-18, No. 3, pp. 271–287, September 1970.
16. J.D. Merkel, "FFT Pruning," IEEE Transactions on Audio and Electroacoustics, Vol. AU-19, No. 4, pp. 305–311, September 1971.
17. P.D. Welch, "The Use of the Fast Fourier Transform for the Estimation of Power Spectra: A Method Based on Time Averaging over Short, Modified Periodograms," IEEE Transactions on Audio and Electroacoustics, Vol. AU-15, pp. 70–73, June 1967.
18. J.W. Cooley, J.W. Tukey, "An Algorithm for Machine Calculation of Complex Fourier Series," Mathematical Computation, Vol. 9, No. 5, pp. 297–301, April 1965.
19. E.O. Brigham, R.E. Morrow, "The Fast Fourier Transform," IEEE Spectrum, Vol. 4, No. 2, pp. 63–70, December 1967.
20. E.O. Brigham, *The Fast Fourier Transform and Its Applications*, Prentice-Hall, Inc., Englewood Cliffs, NJ, 1988.
21. H.L. Groginsky, G.A. Works, "A Pipeline Fast Fourier Transform," IEEE Transactions on Computers, Vol. C-19, No. 11, pp. 1015–1019, November 1970.
22. A.V. Oppenheim, R.W. Schafer, *Digital Signal Processing*, pp. 542–554, Prentice-Hall, Inc., Englewood Cliffs, NJ, 1975.
23. A.V. Oppenheim, R.W. Schafer, *Digital Signal Processing*, pp. 548–549, Prentice-Hall, Inc., Englewood Cliffs, NJ, 1975.
24. P.M. Woodward, *Probability and Information Theory, with Applications to Radar*, Pergamon Press, London, 1960.
25. M. Abramowitz, I.A. Stegun, *Handbook of Mathematical Functions with Formula, Graphs and Mathematical Tables*, National Bureau of Standards, Applied Mathematical Series 55, Washington, DC, U.S. Government Printing Office, p. 231, June 1964.
26. P.F. Panter, *Modulation, Noise, and Spectral Analysis Applied to Information Transmission*, McGraw-Hill Book Company, New York, 1965.
27. S. Goldman, *Transformation calculus and Electrical Transients*, Chapter 9, "Bessel Functions," Prentice-Hall, Inc., Englewood Cliffs, NJ, 1950.
28. H. Cramér, *Mathematical Methods of Statistics*, Princeton University Press, Princeton, NJ, 1974.
29. A. Leon-Garcia, *Probability and Random Processes for Electrical Engineering*, Second Addition, Addison-Wesley Publishing Company, Inc., New York, May 1994.
30. J.M. Wozencraft, I.M. Jacobs, *Principles of Communication Engineering*, John Wiley & Sons, Inc., New York, 1967.
31. H.D. Brunk, *An Introduction to Mathematical Statistics*, Blaisdell Publishing Company, Waltham, MA, 1965.
32. W. Feller, *An Introduction to Probability Theory and Its Applications*, John Wiley & Sons, Inc., New York, 1957.

33. A. Papoulis, *Probability, Random Variables, and Stochastic Processes*, p. 236, McGraw-Hill Book Co., New York, 1965.
34. A. Papoulis, *Probability, Random Variables, and Stochastic Processes*, p. 176, McGraw-Hill Book Co., New York, 1965.
35. A. Papoulis, *Probability, Random Variables, and Stochastic Processes*, pp. 207–209, McGraw-Hill Book Co., New York, 1965.
36. W.B. Davenport, Jr., W.L. Root, *An Introduction to the Theory of Random Signals and Noise*, p. 149, McGraw Hill Book Co., New York, 1958.
37. A. Papoulis, *Probability, Random Variables, and Stochastic Processes*, p. 221, McGraw-Hill Book Co., New York, 1965.
38. W.B. Davenport, Jr., W.L. Root, *An Introduction to the Theory of Random Signals and Noise*, pp. 33–35, McGraw Hill Book Co., New York, 1958.
39. A. Papoulis, *Probability, Random Variables, and Stochastic Processes*, pp. 126–127, McGraw-Hill Book Co., New York, 1965.
40. N.A.J. Hastings, J.B. Peacock, *Statistical Distributions: A Handbook for Students and Practitioners*, A Halsted Press Book, John Wiley & Sons, Inc., New York, 1975.
41. W.B. Davenport, Jr., W.L. Root, *An Introduction to the Theory of Random Signals and Noise*, p. 165, McGraw Hill, New York, 1958.
42. W.B. Davenport, Jr., W.L. Root, *An Introduction to the Theory of Random Signals and Noise*, p. 166, McGraw Hill, New York, 1958.
43. M. Abramowitz, I.A. Stegun, *Handbook of Mathematical Functions with Formulas, Graphs, and Mathematical Tables*, National Bureau of Standards, Applied Mathematics Series No. 55, U.S. Government Printing Office, Washington, DC, p. 376, Integral 9.6.16, June 1964.
44. W.B. Davenport, Jr., W.L. Root, *An Introduction to the Theory of Random Signals and Noise*, pp. 165–167, McGraw Hill Book Co., New York, 1958.
45. J.C. Hancock, *An Introduction to the Principles of Communication Theory*, McGraw-Hill Book Co., New York, 1961.
46. I.S. Gradshteyn, I.M. Ryzhik, *Table of Integrals, Series, and Products*, Corrected and Enlarged Edition, Academic Press, Inc., New York, 1980.
47. G.A. Campbell, R.M. Foster, *Fourier Integrals for Practical Applications*, Fourth Printing, D. Van Norstrand Company, Inc., New York, 1948.
48. H. Urkowitz, "Energy Detection of Unknown Deterministic Signals," Proceeding of the IEEE, Vol. 55, No. 4, pp. 523–531, April 1967.
49. M. Nakagami, "The *m*-distribution—A general formula for Intensity Distribution of Rapid Fading," W.C. Hoffman, Editor, *Statistical Methods in Radio Wave Propagation*, pp. 3–36, Pergamon Press, New York, 1960.
50. M.G. Kendall, A. Stuart, *The Advanced Theory of Statistics*, Vol. I, II, Hafner Publishing Company, New York, 1958, 1961 and *Advanced Theory of Statistics*, Vol. III, Hafner Press, New York, 1982.
51. H.A. David, H.N. Nagaraja, *Order Statistics*, 3rd Addition, John Wiley & Sons, Hoboken, NJ, 2003.
52. B.W. Lindgren, *Statistical Theory*, The Macmillan Company, New York, 1962.
53. R.C. Borgioli, "Fast Fourier Transform Correlation versus Direct Discrete Time Correlation," Proceeding of the IEEE, Vol. 56, No. 9, pp. 1602–1604, September, 1968.
54. E.O. Brigham, Editor, *The Fast Fourier Transform and its Applications*, Chapter 10, "FFT Convolution and Correlation," Prentice-Hall, Englewood Cliffs, NJ, 1988.
55. A. Papoulis, Editor, *Probability, Random Variables, and Stochastic Processes*, Chapter 9, "Stochastic Processes: General Concepts," McGraw-Hill Book Co., New York, 1965.
56. W.B. Davenport, Jr., W.L. Root, *An Introduction to the Theory of Random Signals and Noise*, pp. 38–42, 66–71, McGraw Hill Book Co., New York, 1958.
57. A. Papoulis, *Probability, Random Variables, and Stochastic Processes*, pp. 323–332, McGraw-Hill Book Co., New York, 1965.
58. A. Papoulis, Editor, *Probability, Random Variables, and Stochastic Processes*, Chapter 10, "Stochastic Processes: Correlation and Power Spectrum of Stationary Processes," McGraw-Hill Book Co., New York, 1965.
59. D.O. North, "Analysis of the Factors Which Determine Signal/Noise Discrimination in Radar," Radio Corporation of America (RCA), Technical Report PTR-6-C, June, 1943; reprinted in Proceeding of the IRE, Vol. 51, pp. 1016–1028, July 1963.
60. G.L. Turin, "An Introduction to the Matched Filter," IRE Transactions on Information Theory, Vol. 6, No. 3, pp. 311–329, June 1960.
61. M.L. Skolnik, *Introduction to Radar Systems*, McGraw-Hill Book Company, Inc., New York, 1962.
62. D. Slepian, "Estimation of Signal Parameters in the Presence of Noise," IRE Transactions on Information Theory, Vol. PGIT-3, No. 4, pp. 68–89, March 1954.
63. P.M. Woodward, I.L. Davies, "Information Theory and Inverse Probability in Telecommunication," Proceedings of the IEE, Vol. 99, Part III, pp. 37–44, March 1952.
64. H. Cramér, *Mathematical Methods of Statistics*, Chapter 32, "Classification of Estimates," Princeton University Press, Princeton, NJ, 1974.
65. C.R. Rao, "Information and Accuracy Attainable in the Estimation of Statistical Parameters," Bulletin Calcutta Mathematical Society, Vol. 37, pp. 81–91, 1945.
66. H.L. Van Trees, *Detection, Estimation, and Modulation Theory: Part I*, John Wiley & Sons, New York, 1968.
67. C.E. Cook, M. Bernfeld, *Radar Signals: An Introduction to Theory and Application*, Academic Press, New York, 1967.
68. L. Kleinrock, *Queueing Systems, Volume I: Theory*, John Wiley & Sons, New York, 1975.
69. D. Gabor, "The Theory of Communication," Journal of the IEE, Vol. 93, Part III, pp. 429–441, 1946.
70. R. Manasse, "Range and Velocity Accuracy from Radar Measurements," MIT Lincoln Laboratory, Lexington, MA, February, 1955. (This unpublished internal report is not generally available.)
71. M.L. Skolnik, *Introduction to Radar Systems*, pp. 467–469, McGraw-Hill Book Co., Inc., New York, 1962.

72. F. Halsall, *Data Communications, Computer, Networks and Open Systems*, Fourth Edition, Addison-Wesley Publishing Company, Harlow, UK, 1996.
73. C. Fujiwara, K. Yamashita, M. Kasahara, T. Namekawa, "General Analyses Go-Back-N ARQ System," Electronics and Communications in Japan, Vol. J59-A, No. 4, pp. 24–31, 1975.
74. A.V. Oppenheim, R.W. Schafer, *Digital Signal Processing*, Chapter 5, "Digital Filter Design Techniques," Prentice-Hall, Inc., Englewood Cliffs, NJ, 1975.
75. F.J. Harris, "On the Use of Windows for Harmonic Analysis with the Discrete Fourier Transform," Proceeding of the IEEE, Vol. 66, No. 1, pp. 51–83, January 1978.
76. L.S. Metzger, D.M. Boroson, J.J. Uhran, Jr., I. Kalet, "Receiver Windows for FDM MFSK Signals," IEEE Transactions on Communications, Vol. 27, No. 10, pp. 1519–1527, October 1979.
77. P.M. Derusso, R.J. Roy, C.M. Close, *State Variables for Engineers*, John Wiley & Sons, Inc., New York, 1965.
78. S. Haykin, *Adaptive Filter Theory*, Prentice-Hall, Englewood Cliffs, NJ, 1986.
79. A.P. Sage, C.C. White, III, *Optimum System Control*, 2nd Edition, Prentice-Hall, Inc., Englewood Cliffs, NJ, 1977.
80. A. Gelb, Editor, *Applied Optimum Estimation*, The M.I.T. Press, Massachusetts Institute of Technology, Cambridge, MA, 1974.
81. C.R. Wylie, Jr., *Advanced Engineering Mathematics*, McGraw-Hill Book Company, Inc., New York, 1960.
82. S. Haykin, *Adaptive Filter Theory*, Chapter 2, "Stationary Discrete-Time Stochastic Processes," Prentice-Hall, Englewood Cliffs, NJ, 1986.
83. S. Haykin, *Adaptive Filter Theory*, pp. 54–56, Prentice-Hall, Englewood Cliffs, NJ, 1986.
84. G.W. Stewart, *Introduction to Matrix Computations*, Academic Press, New York, 1973.
85. J.S. Frame, "Matrix Functions and Applications, Part I – Matrix Operations and Generalized Inverses," IEEE Spectrum, Vol. 1, No. 3, pp. 209–220, March 1964.
86. J.S. Frame, "Matrix Functions and Applications, Part II – Functions of a Matrix," IEEE Spectrum, Vol. 1, No. 4, pp. 102–110, April 1964.
87. J.S. Frame, H.E. Koenig, "Matrix Functions and Applications, Part III – Applications of Matrices to Systems Analysis, a Matrix," IEEE Spectrum, Vol. 1, No. 5, pp. 100–109, May 1964.
88. J.S. Frame, "Matrix Functions and Applications, Part IV – Matrix Functions and Constituent Matrices," IEEE Spectrum, Vol. 1, No. 6, pp. 123–131, June 1964.
89. J.S. Frame, "Matrix Functions and Applications, Part V – Similarity Reductions by Rational or Orthogonal Matrices," IEEE Spectrum, Vol. 1, No. 7, pp. 103–109, June 1964.
90. L.L. Smail, *Calculus*, Appleton-Century-Crofts, Inc., New York, 1949.
91. O.W. Eshbach, *Handbook of Engineering Fundamentals*, John Wiley & Sons, Inc., New York, 1952.
92. R.S. Burington, *Handbook of Mathematical Tables and Formulas*, 3rd Addition, Handbook Publishers, Inc., Sandusky, OH, 1957.
93. G.A. Korn, T.M. Korn, *Mathematical Handbook for Scientists and Engineers*, 2nd Addition, McGraw-Hill Book Co., New York, 1961.
94. M. Milton, I.A. Stegun, Editors, *Handbook of Mathematical Functions with Formulas, Graphs, and Mathematical Tables*, National Bureau of Standard Applied Mathematical Series 55, U.S. Government Printing Office, Washington, DC, 1964.
95. I.S. Gradshteyn, I.M. Ryzhik, *Table of Integrals, Series, and Products*, Corrected and Enlarged Edition, p. xliii, Academic Press, Inc., New York, 1980.
96. I.S. Gradshteyn, I.M. Ryzhik, *Table of Integrals, Series, and Products*, Corrected and Enlarged Edition, p. 446, Integral No. 10, Academic Press, Inc., New York, 1980.

2

DIGITAL SIGNAL PROCESSING AND MODEM DESIGN CONSIDERATIONS

2.1 INTRODUCTION

This chapter describes the conversion of the analog signal at the input of the demodulator to a digitally sampled version of the signal required for the application of digital signal processing techniques in the recovery of the desired information. This signal conversion involves the discrete-time and amplitude sampling of the analog signal and represents the window into the world of digital signal processing with unprecedented capabilities in terms of algorithms for reliable and efficient waveform processing and information detection. The information is typically in the form of binary data formatted to convey text messages using, for example, the American Standard Code for Information Interchange (ASCII)* that uses 7- or 8-bit characters to describe 128 or 256 characters. However, advances in computer technology using 32 and 64 bits provide virtually unlimited character sets in the foreseeable future.

A general description of discrete amplitude and time sampling is given in Sections 2.2 and 2.3, with a case study of discrete amplitude sampling in Section 2.2.1. Specialized applications of discrete-time sampling involving baseband and bandpass sampling are described, respectively, in Sections 2.5 and 2.6. Subcarrier sampling is described in Section 2.6.1 and Hilbert transform sampling is described in Section 2.6.2. In Section 2.7 nonideal modulator and demodulator implementations are discussed and various methods of correction are examined that reduce the loss in received information.

With these concepts *in-hand* the remainder of this chapter describes special applications of digital signal processing (DSP) that are referred to in various chapters and used extensively in computer simulations leading to demodulator performance evaluation results. The DSP topics are described in Section 2.8 under the general heading of multirate signal processing and interpolation. The topics include signal up and down sample rate conversion and the associated filtering requirements, including the cascaded integrate and comb (CIC) filter, polyphase-matched filtering, and Lagrange interpolation. The signal processing functions are referred to and used throughout this book in the context of efficient implementations for the demodulator recovery of the transmitted data.

2.2 DISCRETE AMPLITUDE SAMPLING

Each discrete-time sample of an analog signal is characterized by a discrete-amplitude representation of the signals amplitude. The discrete amplitude is typically represented by a binary sequence of N_b-bits. In the following description, the sign-magnitude binary representation of the sampled signal is used with M_b bits representing the range of the sampled voltage magnitude and an additional bit that is used to signify the sign of the sampled input. Therefore, with the sign-magnitude representation, an N_b-bit analog-to-digital

*Also the International Organization for Standardization (ISO) designation ISO-646.

Digital Communications with Emphasis on Data Modems: Theory, Analysis, Design, Simulation, Testing, and Applications,
First Edition. Richard W. Middlestead.

Companion website: www.wiley.com/go/digitalcommunications

converter (ADC) uses $M_b = N_b - 1$ magnitude bits and 1 sign-bit. With this characterization the most significant bit (MSB) is used to represent the maximum ADC voltage magnitude V_m and the least significant bit (LSB) is used to represent the smallest voltage magnitude. When the peak voltage of the input signal and noise is less than V_m the ADC is operating in the linear range; however, if the analog input exceeds V_m the ADC will clip the signal and noise resulting in nonlinear operation and some degree of performance degradation.

Using the binary number representation of the sampled amplitude with uniform quantization, the number of quantization levels over the ADC voltage range $\pm V_m$ is $N_\ell = 2^{N_b}$ and the corresponding amplitude resolution or minimum quantization level is

$$\ell = \frac{2V_m}{N_\ell} = \frac{2V_m}{2^{N_b}} \tag{2.1}$$

The dynamic range of the ADC is defined and evaluated as

$$\begin{aligned} \rho(\text{dB}) &\triangleq 20\log_{10}\left(\frac{2V_m}{\ell}\right) \\ &= 6.03N_b \end{aligned} \tag{2.2}$$

Therefore, uniform quantization results in 6.02 dB/bit or a dynamic range of 48.16 and 96.32 dB for an 8- and 16-bit ADC, respectively.

When the ADC is used to convert a received carrier modulation signal to a digital format, the signal dynamic range at the input to the ADC is considerably more limited than that at the input to the receiver. For example, because the signal power is based on the magnitude of the received carrier-modulated signal voltage, the dynamic range, corresponding to the linear range of the ADC, is evaluated as

$$\begin{aligned} \rho_s(\text{dB}) &\triangleq 20\log_{10}(2^{M_b}) - \Delta\text{dB} \\ &= 6.03M_b - \Delta\text{dB} \end{aligned} \tag{2.3}$$

where ΔdB is the signal backoff required to avoid clipping of the input signal and noise.

For example, if the receiver gain is adjusted so that the peak voltage, V_p, of a noise-free carrier-modulated binary phase shift keying (BPSK) waveform is equal to V_m, then the root-mean-square (*rms*) signal voltage is $V_{rms} = V_p/\sqrt{2} = V_m/\sqrt{2}$ corresponding to $\Delta\text{dB} = 3$ dB, or 1/2-bit below saturation. If the received signal includes channel or receiver noise then an additional 6 or 9 dB may be required; further limiting the signal dynamic range by 1–1.5 bits. A more extreme case occurs when the received signal includes multipath fading. In this case, ΔdB must be increased to include the average fade depth relative to the maximum power of the fading

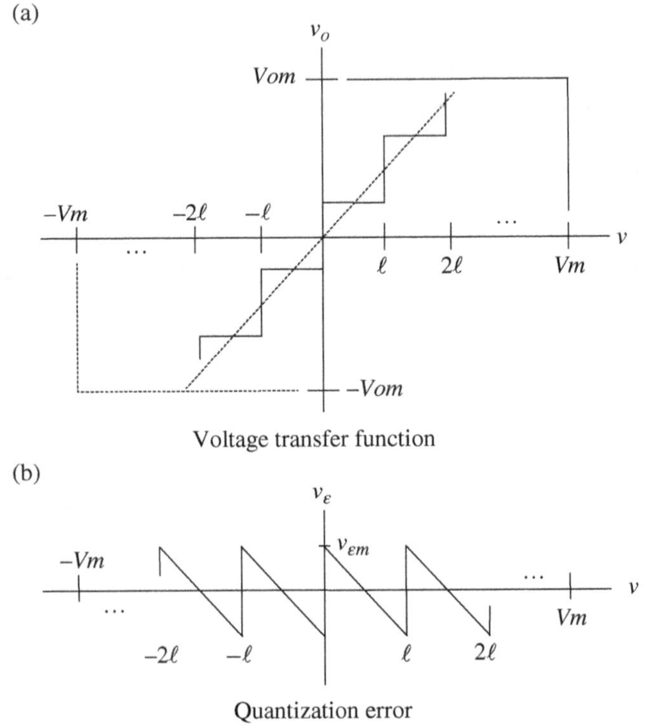

FIGURE 2.1 Unit gain uniform quantization zero-level transition ADC.

signal. This example emphasizes the task of the automatic gain control in maintaining a precisely controlled average signal-plus-noise power at the input to the ADC.

Two unit gain ADC voltage transfer functions with uniform quantization are shown in Figures 2.1 and 2.2 distinguished, respectively, as having a zero level transition and a distinct zero rest level. The effective dynamic range* of the uniformly sampled ADC can be increased somewhat by applying a low-level pseudo-random amplitude jitter with a peak voltage level of $\ell/2$ to the input signal. This intentionally added noise ensures that a quantized signal does not lie within a fixed quantization interval that would otherwise lead to a phase or power estimation error. For example, referring to voltage transfer function of Figure 2.1, a quantized signal within the quantization interval 0–ℓ will result in a positive fixed output level, whereas the additive noise-like amplitude jitter will force a quantization level change resulting in an improved estimate of the true input signal level. When the ADC is used to provide quadrature baseband signal samples the amplitude jitter will improve the estimation accuracy of the received signal including the phase and power estimates. The random chip modulation of a direct-sequence spread-spectrum received waveform will have a similar impact on

*The effective dynamic range is the linear ADC dynamic range as expressed in (2.2) plus the effective range into the underflow region provided by the jitter signal. Because the jitter signal alters the signal level with the truncation limit unchanged, the net effect is that the jitter improves the ADC resolution.

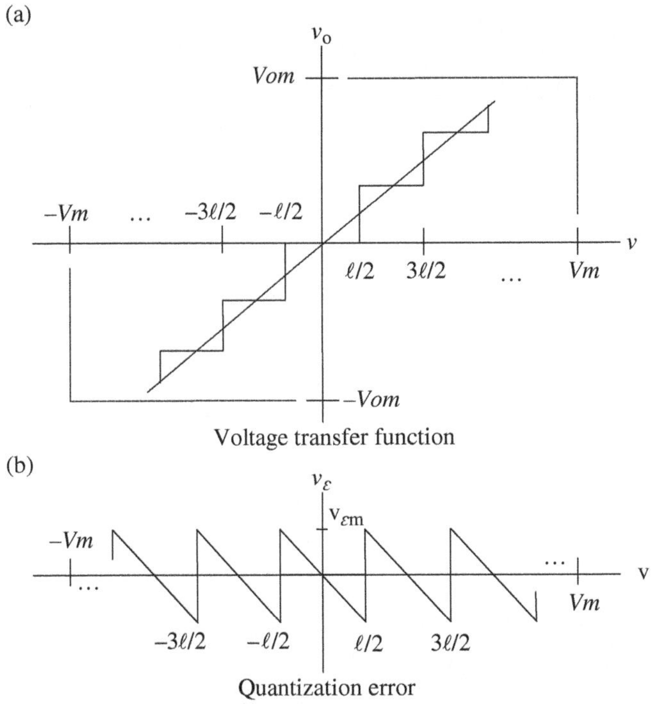

FIGURE 2.2 Unit gain uniform quantization zero-level ADC.

the detection of the correlation response, in that, the randomly modulated chips that are buried in noise provide inherent jitter to the ADC quantization enhancing the dynamic range of the ADC; this is examined in the following case study.

In the linear range of the uniformly quantized ADC, the sampled amplitude quantization error, $v_{\varepsilon j}$, is related to the quantization level ℓ, such that, for an input voltage $j\ell < v_j < (j+1)\ell$, $v_{\varepsilon j} = v_\varepsilon = \pm\ell/2$ independent of j with the result that $v_{\varepsilon m} = |\ell/2|$.* The quantization noise is often evaluated assuming a uniform amplitude distribution over the sampling interval ℓ for which the quantization noise power or variance is evaluated as

$$\sigma_q^2 = \frac{\ell^2}{12} : \text{uniform sampling quantization noise power}$$
$$= V_m^2 \frac{2^{-2N_b}}{3} \tag{2.4}$$

The quantization noise power, expressed by (2.4), represents an additive noise source that is independent of the additive receiver thermal noise and in most applications a 12 or more bit ADC will result in negligible quantization noise. Subsequent digital signal processing should use word lengths in excess of the ADC to ensure that rounding, overflow, and truncation effects do not degrade the quality of the signal detection; this is a particular concern with fixed word size signal processors [1]. Floating point digital signal processors virtually eliminate the concerns regarding signal quality following the ADC; however, because the ADC establishes a limit on the ultimate quality of the signal processing, special applications involving high quality signal processing frequently use more than 12-bit digital-to-analog converters (DACs) requiring special amplitude conversion techniques [2–6].

In the following chapters, when the impact of quantization on the demodulator performance is examined, the zero level transition ADC is used with the sign-magnitude representation of the quantized time sample. A fast amplitude quantization algorithm is described in Appendix 2A.

*Nonuniform sampling results in $v_{\varepsilon j} = \pm\ell_j/2$ and is used to increase the resolution at the lower voltage levels of the sampled signal.

2.2.1 Case Study: ADC Quantization of Direct-Sequence Spread-Spectrum Waveform

This case study examines the detection performance of a direct-sequence spread-spectrum (DSSS) waveform after sampling and ADC conversion. The objective is to determine the correlation performance of the DSSS waveform as the level is attenuated to and below the LSB of the ADC. The first evaluation is based on a noise-free DSSS waveform at the input to the ADC and LSB rounding and the addition of a noise-like jitter signal is examined and shown to extend the dynamic range of the ADC by 1–1.5 bits or 6–9 dB. The second evaluation is based on the DSSS waveform *buried* in additive white Gaussian noise (AWGN) and the dynamic range of the ADC is examined as the composite signal plus noise is attenuated to and below the LSB of the ADC. In this case, the dynamic range of the ADC is extended by 2-bits or 12 dB for the example considered.

For these evaluations, the DSSS waveform and the ADC are specialized as follows. The DSSS waveform is a pseudorandom noise (PRN) sequence of 256 bipolar (±1 V) chips and, to center the zero-lag correlation response, the received chip sequence is shifted relative to the demodulator stored replica by 128 chips. The receiver replica of the PRN sequence is stored as bipolar unit amplitude chips. The ADC uses a total of $M_b = 8$ bits with $M = 2$ integer bits. Although the voltage level of the received chip sequence is gain adjusted to examine the performance with ADC quantization, the maximum chip voltage of ±1 V corresponds to $M_b - M = 6$ bits or 36.12 dB of fractional dynamic range of the ADC. The amplitude range of the DSSS received waveform relative to the ADC bits is depicted in Figure 2.3.

The first evaluation considers a noise-free received DSSS waveform characterized by the unit amplitude bipolar chips that coincide with the 0 dB attenuation scale of Figure 2.3 and, upon passing through the ADC with underflow truncation, the resulting correlation response is shown in Figure 2.4a. The standard deviation of the correlation

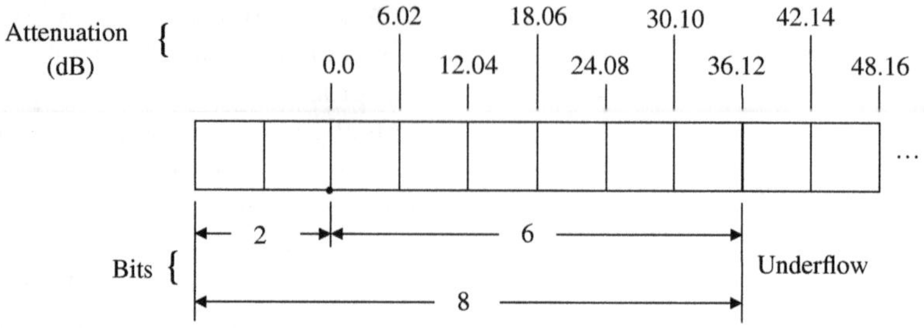

FIGURE 2.3 ADC magnitude bit allocation and dynamic range ($M = 2$, $M_b = 8$).

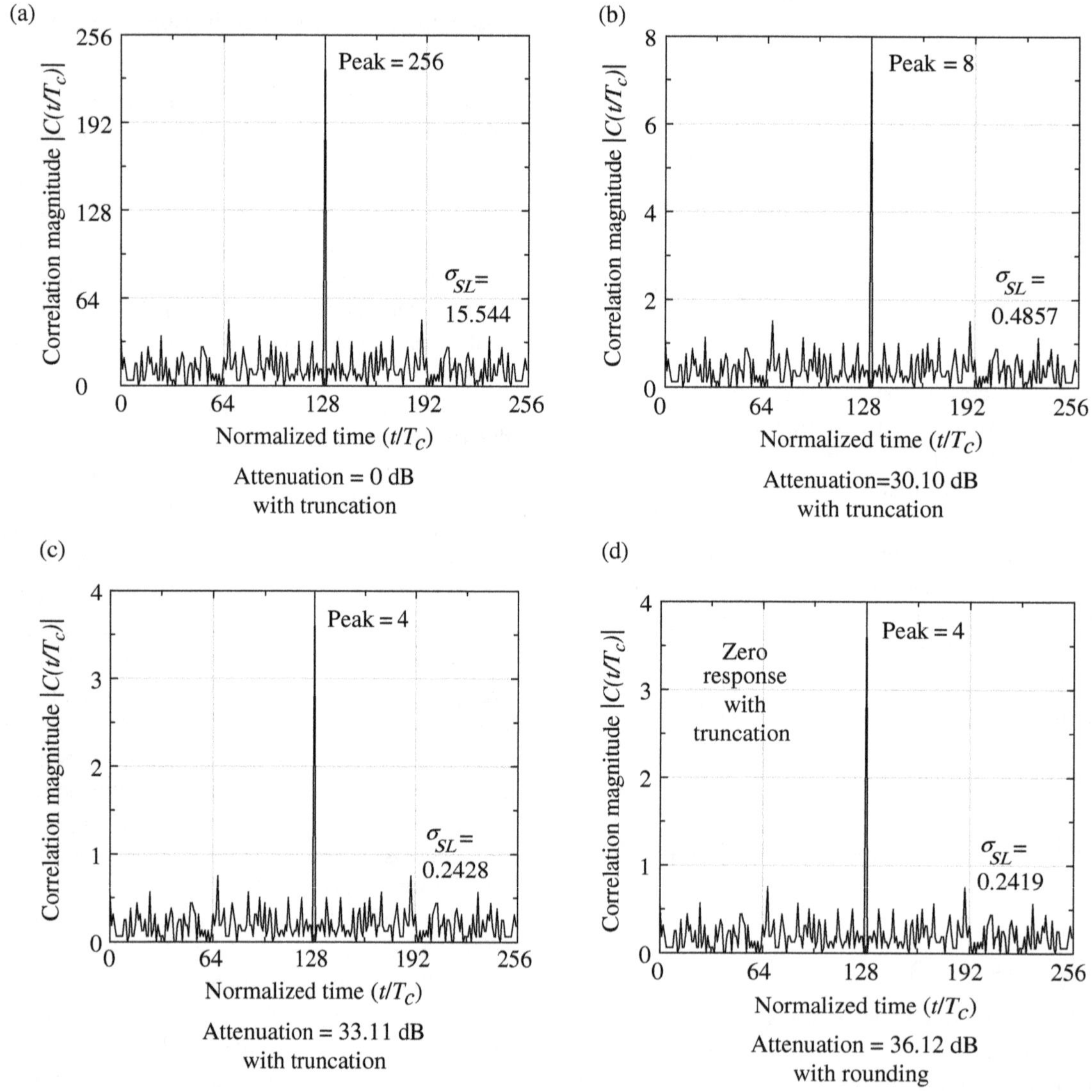

FIGURE 2.4 Correlation dynamic range with truncation and rounding.

sidelobes is evaluated as* $\sigma_{SL} = 15.44$ and the resulting correlation peak-to-sidelobe ratio is 24.39 dB. In keeping with the objective of this case study, the noise-free DSSS waveform is attenuated by 30.1 and 33.11 dB and then amplitude quantized by the ADC with the respective correlation responses shown in Figure 2.4b and c; the peak-to-sidelobe ratios are 24.33 and 24.36, respectively. Although the case with 33.11 dB of attenuation is only 3.01 dB above the lower range of the ADC, these results clearly demonstrate that the DSSS waveform is operating within the linear range of the ADC. When the DSSS waveform is attenuated by 36.12 dB and passed through the truncating ADC the correlation response is zero, that is, the received signal chips are in the underflow region so the output of the ADC is zero. On the other hand, when rounding is applied to the ADC the received chips do appear in the LSB of the ADC and the resulting correlation response is shown in Figure 2.4d; in this case the peak-to-sidelobe ratio is 24.37 dB again operating in the linear range of the ADC. When the received signal is attenuated by 42.14 dB the correlation response is again zero suggesting that rounding will extend the dynamic range of the ADC by 3 dB or one-half bit.

Another evaluation regarding extending the dynamic range of the ADC involves the addition of a noisy amplitude jitter signal to the received DSSS waveform prior to the ADC processing. In this case, a noisy jitter signal consisting of a random sequence of bipolar low-amplitude chips, independent of the DSSS chips, is added to the received signal. The chip amplitude of the noisy jitter signal is 11.05 *mv* corresponding to an attenuation 39.13 dB in Figure 2.3, that is, one-half the level of the first underflow bit. Upon repeating the ADC and correlation processing for the DSSS received waveform with the additive jitter noise the correlation result for a signal attenuation of 45.15 dB is shown in Figure 2.5. The peak-to-sidelobe ratio in this case is 21.35 dB, that is, 3 dB less than in the cases operating in the linear range of the ADC. The reason for this loss is that one-half of the DSSS chips fall into the underflow region of the ADC and result in zero output contributing nothing to the correlation peak. As the attenuation is increased beyond 45.15 dB the ADC output is zero, therefore, with 45.15 dB of signal attenuation, the selected noisy chip amplitude extends the dynamic range of the ADC by 9 dB or 1.5 bits.†

The utility of adding a low-level jitter signal to increase the sensitivity of the LSB impacts all of the quantization levels of the ADC and enhances the estimation of various signal parameters by ensuring that a signal-dependent parameter does not rest on a single quantization level resulting in an averaging that results in an improved estimate of the true parameter value. This low-level jitter signal effectively results in a signal gain variation that presumably has a negligible impact insofar as the dynamic range of the ADC is concerned. Sklar [7] discusses additional applications of amplitude dithering (jitter) including improvements in estimating the true signal level that might appear, for example, at the output of a baseband ADC.

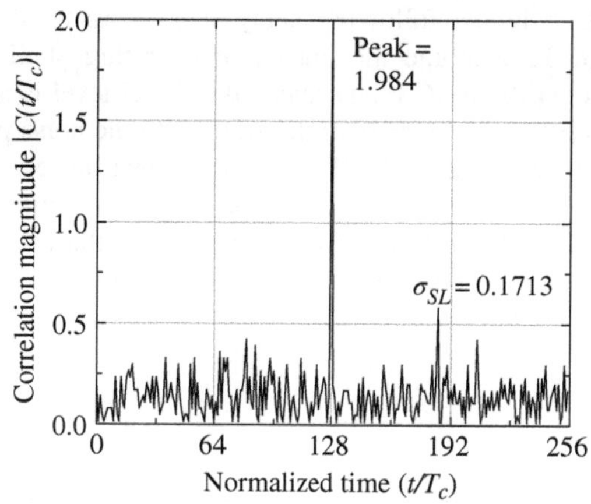

FIGURE 2.5 Correlation dynamic range with additive jitter noise (attenuation = 45.15 dB).

Although the preceding evaluation using the DSSS waveform to demonstrate the capability of rounding and the addition of a noise-like jitter signal to increasing the dynamic range of the ADC, the conditions of the evaluation are somewhat unrealistic. For example, received signals typically involve additive channel or receiver Gaussian noise and most practical receivers use some form of automatic gain control (AGC) to establish and maintain a nearly constant received power level at the input to the ADC. In the case of the DSSS-modulated waveform, the additive noise generally overwhelms the desired signal and the process of correlation despreading results in a net positive signal-to-noise ratio sufficient for signal detection. In this regard, the final evaluation in this case study involves the DSSS received waveform buried in additive white Gaussian noise and examines the despreading correlator output for various AGC level setting relative to the ADC input range as depicted in Figure 2.3. In keeping with the previous examples, the signal correlation response is examined as the composite signal-plus-noise level is attenuated over the linear range of the ADC. The variance of the additive noise is defined as

$$\sigma_n^2 = \frac{V_p^2}{2\gamma} \tag{2.5}$$

where V_p is the peak carrier-modulated signal voltage and γ is the signal-to-noise ratio measured in the input bandwidth of

*The sidelobe standard deviation is based on the, nearly zero mean, bipolar correlation sidelobe response.

†The dynamic range can be increased further by increasing the amplitude of the noisy chips; however, there will still be a 3 dB correlation processing loss. Further increasing of the jitter signal level will eventually result in an undesirable mapping of the input signal onto the ADC bits that will result in signal clipping.

the ADC; in the following analysis, γ is assumed to be -10 dB. To examine the impact of operating at different levels over the ADC linear range, the signal level is attenuated over the linear range of the ADC with the noise power, as expressed in (2.5), 10 dB above the signal power.

In Figure 2.6a the signal power is attenuated by 30.10 dB and, while the noise appearing in the underflow region will be truncated, the composite input signal is essentially operating in the linear range of the ADC with a peak-to-sidelobe ratio of the correlation response equal to 17.56 dB. Figure 2.6b shows the correlation response when the signal level is attenuated by 42.14 dB corresponding to 6 dB (1-bit) into the underflow region. In this case, the correlation peak-to-sidelobe ratio is virtually unchanged at 17.4 dB. Figure 2.6c and d corresponds to further reductions in the signal level by 3 dB corresponding

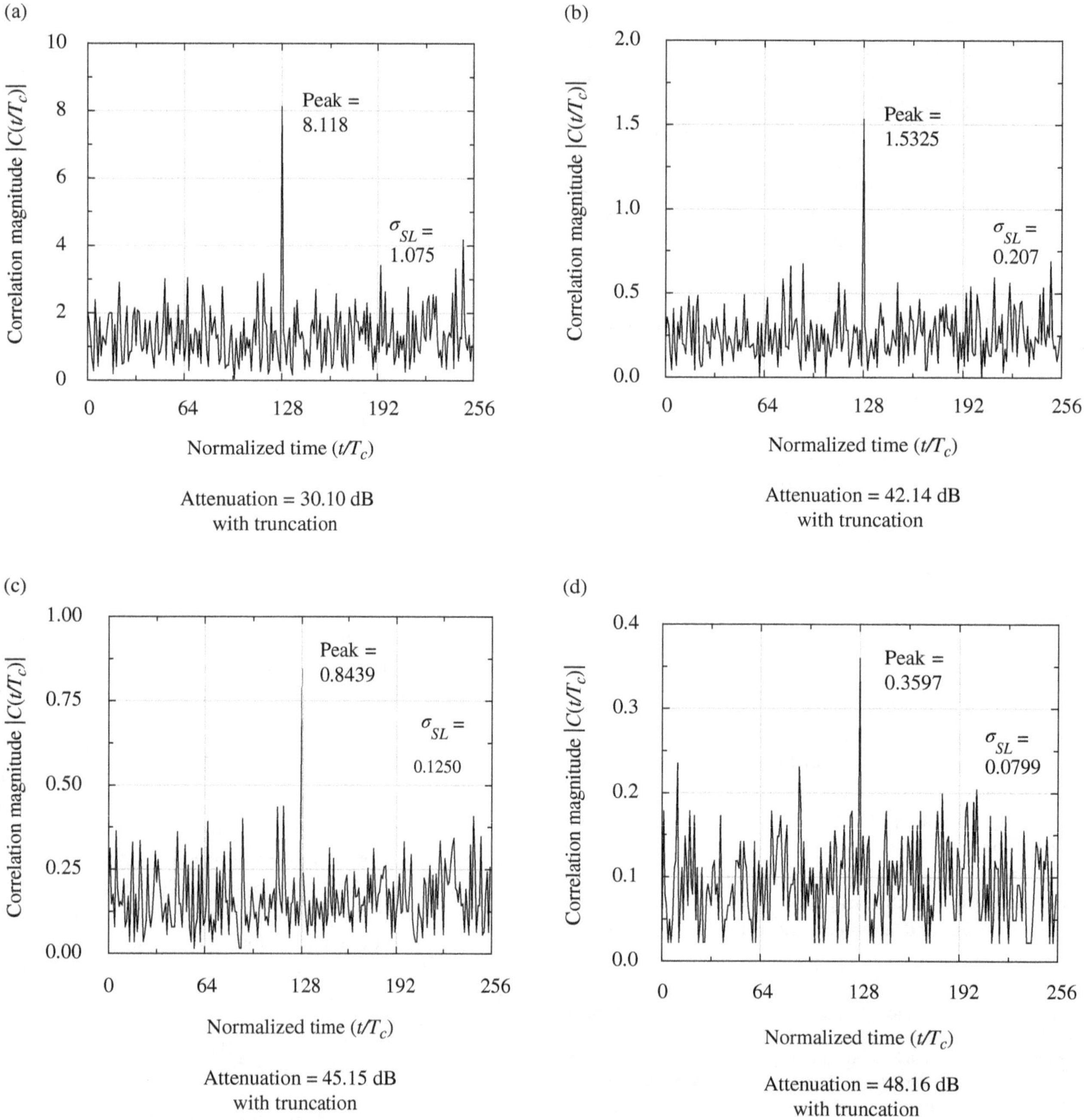

FIGURE 2.6 Correlation dynamic range of DSSS waveform with noise.

to 1.5 and 2-bits into the underflow region with corresponding correlation peak-to-sidelobe ratios of 16.58 and 13.07 dB. Another 3 dB of signal attenuation results in zero ADC output so the composite signal and noise is entirely in the underflow region. The phenomenon resulting in the preservation of the correlation response is understood if the additive noise is viewed as a jitter signal that *carries* the low-level DSSS signal into the higher order bits of the ADC effectively providing gain to the DSSS signal. Because the noise is 10 dB higher than the signal this phenomenon persists until the noise is essentially in the underflow region resulting in the rapid loss in the peak-to-sidelobe ratio from 13.07 dB to zero output when the attenuation is increased by an additional 3.01 dB from 48.16 dB. This is an interesting and distinctly different view of the low-amplitude jitter noise discussed earlier and can only be taken advantage of if subsequent signal processing can *lift* the desired signal above the noise for subsequent signal detection. It follows from this explanation that a lower signal-to-noise ratio will result in a larger useful linear region of operation over the ADC linear range.

2.3 DISCRETE-TIME SAMPLING

The sampling of an analog signal is fundamental to the success of digital signal data processing and various discrete-time sampling criteria are established in this section based on the work and conclusions of Nyquist [8] and Shannon [9]. Shannon's sampling theory concludes that the original analog signal can be perfectly reconstructed from the sampled signal if Nyquist sampling criteria are satisfied. This is an important conclusion because it implies that all of the information contained in the analog signal is also contained in the sampled version so the desired received information can be recovered through digital signal processing techniques. Ideal sampling occurs when the sampling pulse is described in terms of the delta function $\delta(t)$ and, when applied using uniform sampling intervals T_s over all time, the sampling is expressed as

$$f_s(t) = \sum_{n=-\infty}^{\infty} \delta(t - nT_s) \tag{2.6}$$

In Chapter 1 it is shown that Fourier transform of (2.6) is

$$F_s(f) = f_s \sum_{n=-\infty}^{\infty} \delta(f - nf_s) \tag{2.7}$$

where $f_s = 1/T_s$ is the sampling frequency so that (2.6) and (2.7) form a Fourier transform pair as shown in Figure 2.7.

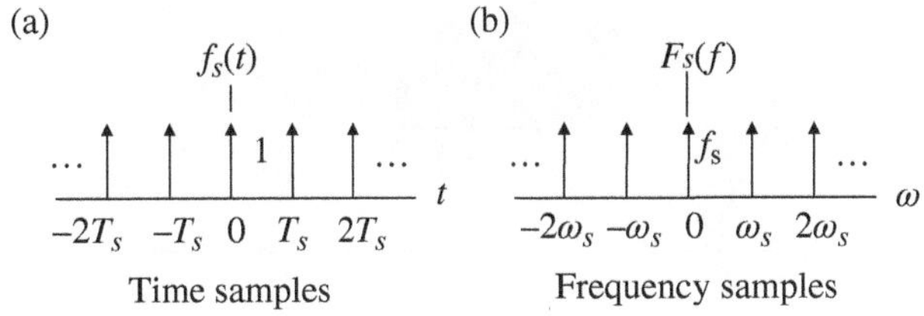

FIGURE 2.7 Ideal sampling Fourier transform Pairs.

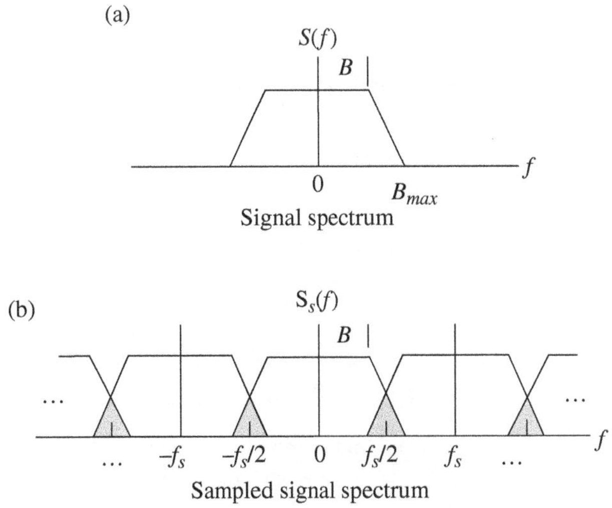

FIGURE 2.8 Spectrum of signal and under-sampled signal (*shaded* regions are aliased spectrums).

Consider that a baseband signal $s(t)$ has the spectrum $S(f)$, shown in Figure 2.8a, with a 3 dB baseband bandwidth of B and a bandwidth B_{max} such that

$$S(f) = 0 \quad : |f| \geq B_{max} \tag{2.8}$$

where f, B, and B_{max} have units of Hz. In practice, it is required that $S(f)$ have negligible spectral power in the range $|f| \geq B_{max}$ as determined by the distortion loss resulting from the spectral aliasing as described in the following discussion.

The sampled signal is the product of $s(t)f_s(t)$ with the corresponding frequency response given by the convolution $S(f)*F_s(f)$ and evaluated as

$$\begin{aligned} S_s(f) &= \int_{-\infty}^{\infty} S(\lambda)F_s(f-\lambda)d\lambda \\ &= \frac{1}{T_s}\sum_{n=-\infty}^{\infty} S(f-nf_s) \end{aligned} \tag{2.9}$$

The spectrum of the sampled signal is shown in Figure 2.8b for a sampling frequency $f_s < 2B_{max}$. Sampling

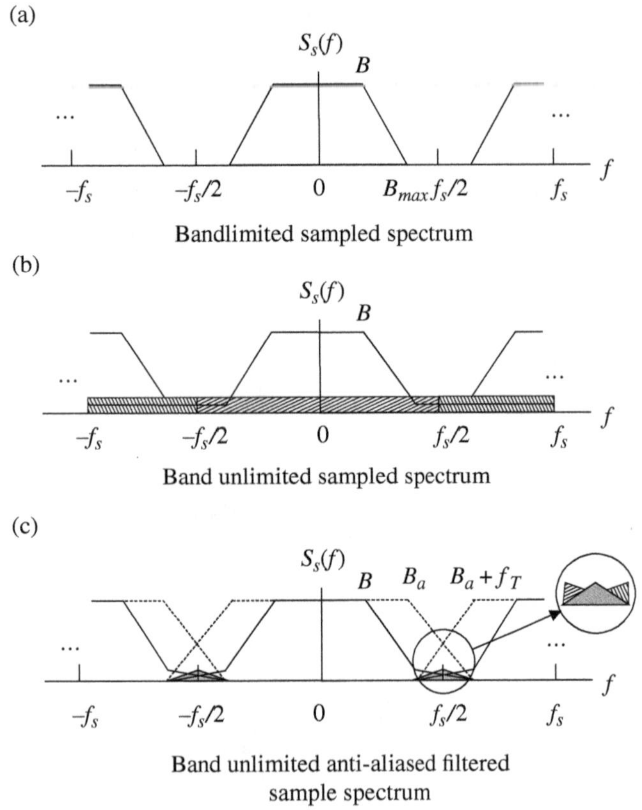

FIGURE 2.9 Signal spectrums with and without antialiasing filter (*shaded* regions are aliased spectrums).

of the signal $s(t)$ occurs at the sampling rate f_s and the signal spectrum is repeated without limit and centered about nf_s. Because the sampling frequency depicted in Figure 2.7b corresponds to $f_s < B_{max}$, spectral interference from the duplicated spectrums at $\pm f_s$ appears in the baseband signal spectrum. This phenomenon is referred to as spectrum aliasing and results in distortion of the baseband signal resulting in degraded signal processing performance.* It is also evident from the figure that with $f_s \geq 2B_{max}$ spectrum aliasing will not occur. This is the genius of Shannon's sampling theorem which states, in effect, that $s(t)$ can be completely reconstructed from samples $s(nT_s)$ $\forall n$ if $T_s = 1/2B_{max}$ or less, that is, $f_s \geq 2B_{max}$; this is depicted in Figure 2.9a in which there is no spectrum aliasing. From a practical point of view, because of the antialiasing filter transition bandwidth and received signal frequency uncertainties, including Doppler, the sampling frequency is selected to satisfy the condition $f_s > 2B_{max}$. With $f_s/2 \geq B_{max}$, the baseband spectrum corresponds to negative frequencies in the range $-f_s/2 \leq f < 0$ and the sampled spectrum corresponds to negative frequencies in the ranges

$$(n-1)f_s + f_s/2 \leq f \leq nf_s \quad :\forall n \text{ sampled spectrum negative frequencies} \tag{2.10}$$

Shannon's sampling theorem applies for an infinite number of samples so $s(t)$ must exist for all time. Functions of infinite time duration are the only functions that results in band-limited spectrums. This is evident in the time–frequency duality between the functions $sinc(t/2T) \Leftrightarrow (1/2T)\,rect(fT/2)$ and $rect(t/T) \Leftrightarrow T\,sinc(fT)$. Therefore, functions with an infinite frequency response are generally band limited by using an antialiasing filter prior to the sampling; however, a digitally implemented antialiasing filter can be used after the ADC conversion [10]. The design of the antialiasing filter is critical, in that, the filter must not result in significant signal distortion relative to the aliased spectrum. Consequently, there is a system performance trade-off between the antialiasing filter bandwidth, transition band, sampling frequency selection, and the resulting aliasing distortion. These characteristics are depicted in Figure 2.9b and c. Figure 2.9b shows the aliasing of large bandwidth sampled signal without an antialiasing filter, in which, the spectrum aliasing is over the entire baseband bandwidth. Figure 2.9c shows the performance improvement when an antialiasing filter (*dashed lines*) is used with a bandwidth $B_a > B$ and transition bandwidth f_T.†

2.3.1 Nonideal Sampling

The ideal sampling described in the previous section involved delta function sampling in both the time and frequency domains as characterized by the Fourier transform pair (2.6) and (2.7). In this section, nonideal time-domain sampling [11, 12], is examined based on a finite sampling width of τ seconds with $f(t)$ described as the unit amplitude sampling pulse

$$f(t) = rect\left(\frac{t - nT_s}{\tau}\right) \quad : -\infty \leq n \leq \infty \tag{2.11}$$

Figure 2.10 depicts the nonideal finite width sampling function and two different techniques involving *average* and *exact sampling*‡ of the signal $s(t)$. In both cases, the sampling pulse is defined of the $|t - nT_s| \leq \tau/2$. For sufficiently narrow sample pulses, that is, $\tau \ll 1/B$, average sampling

*Spectrum aliasing also impacts all of the repeated signal spectrums in the same way; however, the baseband signal is of interest in this discussion.

†Refer to Section 2.8.2 for more details on the design of antialiasing filters.

‡*Average* sampling is also referred to as *flat-top* and *square-top* sampling and *exact* sampling as *natural* and *top* sampling.

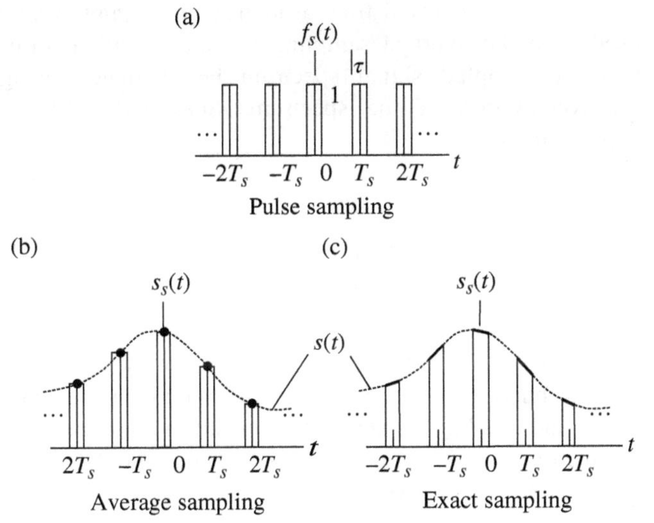

FIGURE 2.10 Nonideal sampling.

can be thought of as providing a single sample at nT_s that corresponds to the average signal level over the interval $nT_s - \tau/2 \le t \le nT_s + \tau/2$. The following discussions focus on the evaluation of the sampled signal spectrum $S_s(f)$ for each of the sampling techniques and, in both cases, the sampling frequency satisfies the Nyquist criterion $f_s \ge 2B_{max}$.

2.3.1.1 ***Signal Spectrum Using Average Sampling*** With average sampling, the sampled signal is the product of the signal sample at the center of each sample pulse, $s(nT_s)$, and the sampled function is expressed as

$$s_s(t) = \sum_{n=-\infty}^{\infty} s(nT_s) f_s(t-nT_s) \tag{2.12}$$

Normally the spectrum of $S_s(f)$ is evaluated as the convolution of the signal spectrum, $S(f)$, with the sampling function spectrum, $F_s(f)$, however, because $s(nT_s)$ represents a constant weight associated with each time sample, the evaluation of $S_s(f)$ is simply the product of $S(f)$ and $F_s(f)$ evaluated as

$$\begin{aligned} S_s(f) &= \int_{-\infty}^{\infty} s_s(t) e^{j2\pi ft} dt \\ &= \sum_{n=-\infty}^{\infty} s(nT_s) \int_{nT_s-\tau/2}^{nT_s+\tau/2} f_s(t-nT_s) e^{j2\pi ft} dt \end{aligned} \tag{2.13}$$

The integral in the second equality of (2.13) is evaluated using the substitution of variables $\xi = t - nT_s$, $d\xi = dt$, with the integration range $|\xi| \le \tau/2$ yielding

$$\begin{aligned} \int_{nT_s-\tau/2}^{nT_s+\tau/2} f_s(t-nT_s) e^{-j2\pi ft} dt &= \int_{-\tau/2}^{\tau/2} f_s(\xi) e^{-j2\pi f(\xi+nT_s)} d\xi \\ &= F(f) e^{-j2\pi f nT_s} \end{aligned} \tag{2.14}$$

where

$$F(f) = \tau\, sinc(f\tau) \tag{2.15}$$

Using $s(nT_s)$ in (2.13), the sampled signal is evaluated in terms of the signal spectrum, $S(f)$, as

$$s(nT_s) = \int_{-\infty}^{\infty} S(\lambda) e^{j2\pi\lambda nT_s} d\lambda \tag{2.16}$$

Substituting (2.16), (2.15), and (2.14) into (2.13) results in the equalities

$$\begin{aligned} S_s(f) &= \tau\, sinc(f\tau) \sum_{n=-\infty}^{\infty} \int_{-\infty}^{\infty} S(\lambda) e^{j2\pi(\lambda-f)nT_s} d\lambda \\ &= \tau\, sinc(f\tau) \int_{-\infty}^{\infty} S(\lambda) \left[\sum_{n=-\infty}^{\infty} e^{j2\pi(\lambda-f)nT_s} \right] d\lambda \\ &= (\tau f_s)\, sinc(f\tau) \int_{-\infty}^{\infty} S(\lambda) \delta(f-\lambda-nf_s) d\lambda \\ &= (\tau f_s)\, sinc(f\tau) S(f-nf_s) \int_{-\infty}^{\infty} \delta(f-\lambda-nf_s) d\lambda \end{aligned} \tag{2.17}$$

The second equality in (2.17) simply moves the summation inside the integral, from which, the delta function is established using the identity in Section 1.2.3 resulting in the third equality using

$$\begin{aligned} \sum_{n=-\infty}^{\infty} e^{j2\pi(\lambda-f)nT_s} &= \sum_{n=-\infty}^{\infty} e^{-j2\pi(f-\lambda)nT_s} \\ &= f_s \delta(f-\lambda-nf_s) \end{aligned} \tag{2.18}$$

The fourth equality uses the sifting property of the delta function corresponding to $\lambda = f - nT_s$. The final result, using average sampling, is expressed in (2.19) with $f_s = 1/T_s$ and the property that the delta function integrated over the infinite range is unity.

$$S_s(f) = \left(\frac{\tau}{T_s}\right) sinc(f\tau)S(f - nf_s)$$

: spectrum using average sampling (2.19)

Equation (2.19) indicates that the each replication of the signal spectrum is weighted by $(\tau/T_s)sinc(f\tau)$ as depicted in Figure 2.11. Since this weighting is a function of the frequency variable f, the repeated spectrums are increasingly diminished and distorted over the frequency bands $2B_{max}$ as $|f|$ increases. The first spectral null of the weighting function occurs at $f_\tau = 1/\tau$ and, in the figure, corresponds to $f_\tau/f_s = T_s/\tau \cong 3.5$; however, in practice $T_s/\tau \gg 10$, so the attenuation and distortion of the baseband signal spectrum and nearby spectral repetitions is negligible.

2.3.1.2 *Signal Spectrum Using Exact Sampling* With exact sampling, the sampled signal is the product of signal $s(t)$ and the sampling function and is expressed as

$$s_s(t) = s(t) \sum_{n=-\infty}^{\infty} f_s(t - nT_s) \tag{2.20}$$

Equation (2.20) is distinct from the signal sample $s(nT_s)$ used with the average sampling technique and requires that the sampled signal spectrum be evaluated using the convolution of the spectrums $S(f)$ and $F_s(f)$ as expressed by

$$S_s(f) = \int_{-\infty}^{\infty} S(\lambda)F_s(f - \lambda)d\lambda \tag{2.21}$$

The evaluation of (2.21) is left as exercise by noting that $s(t)$ is a function of t and is not dependent on the summation over n and that $F_s(f)$ is the Fourier transform involved in the summation in (2.20). Upon working through the details the solution to (2.21) is expressed as

$$S_s(f) = \left(\frac{\tau}{T_s}\right) \sum_{n=-\infty}^{\infty} sinc(nf_s\tau)S(f - nf_s) \tag{2.22}$$

Equation (2.22) is plotted in Figure 2.12 and it is noted that each spectral repetition at $f = nf_s$ is weighted by a constant given by $(\tau/T_s)\,sinc(nf_s\tau)$. In this case, the spectral repetitions are attenuated as $|n|$ increases; however, they are not distorted.

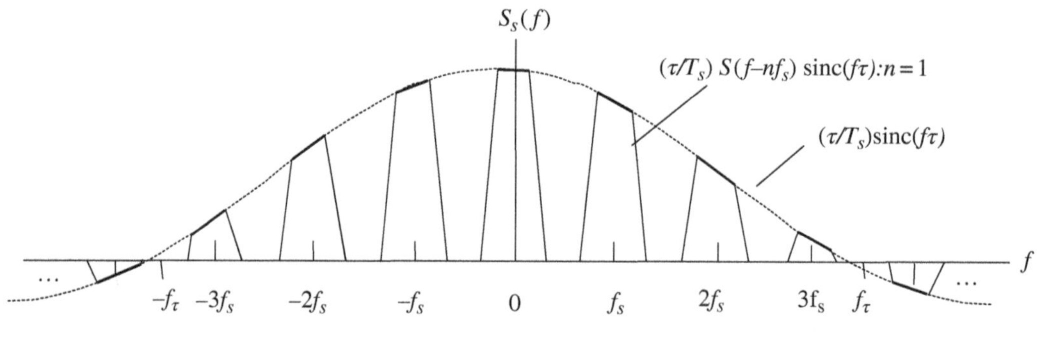

FIGURE 2.11 Signal spectrum with average sampling.

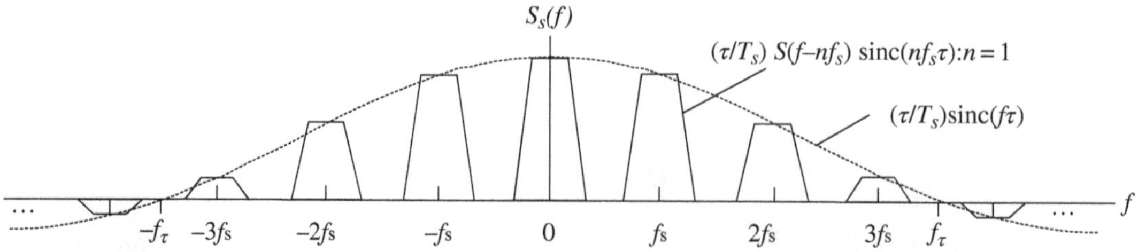

FIGURE 2.12 Signal spectrum with exact sampling.

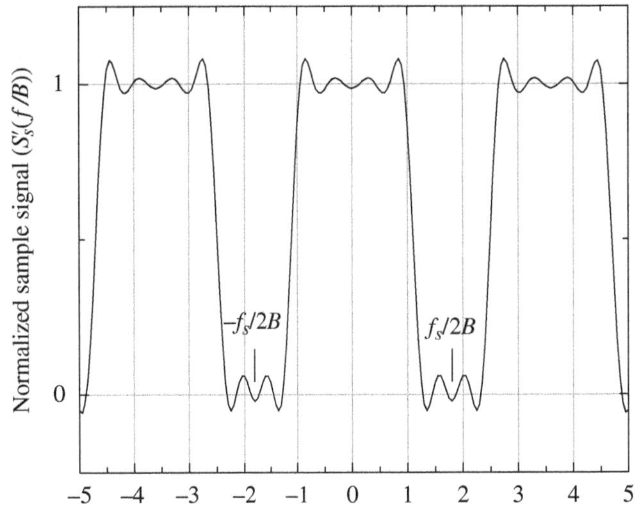

FIGURE 2.13 Windowed sampled signal spectrum (repeated spectrums not shown).

2.3.1.3 *Exact Sampling with Finite Sample Window* In this section, the sampling is considered to exist over a finite time interval or window resulting in the nonideal sampled signal expressed as

$$\begin{aligned} s'_s(t) &= w(t)s(t)f_s(t) \\ &= w(t)s_s(t) \quad : t = nT_s \end{aligned} \tag{2.23}$$

where the $f_s(t)$ is the ideal sampling function expressed in (2.6) and $s(t)$ and $s_s(t)$ are the signal of interest and the sampled signal with respective spectrums shown in Figure 2.9a.

The sampling window $w(t)$ is considered to be a uniformly weighted unit amplitude function of duration T_w seconds, expressed by the Fourier transform pair

$$w(t) = rect(t/T_w) \Leftrightarrow W(f) = T_w sinc(fT_w) \tag{2.24}$$

The spectrum of the windowed sampled signal, $s'_s(t)$, is evaluated as the convolution of the frequency functions of $s_s(t)$ and $w(t)$ and is expressed as

$$S'_s(f) = \int_{-\infty}^{\infty} S_s(\lambda)W(f-\lambda)d\lambda \tag{2.25}$$

The numerical evaluation of (2.25) is shown in Figure 2.13 for three repeated sampled spectrums about the zero frequency. The signal spectrum corresponds to that shown in Figure 2.9a with $B = 1$ kHz, $B_{max} = 1.2$ kHz, and oversampling corresponding to $f_s/2 = 1.8$ kHz; however, the normalized abscissa in Figure 2.13 applies to any frequency scale. The frequency resolution of the window is $f_{res} = 1/T_w$ and this numerical example corresponds to $f_{res} = B/4$, so, if B were to represent the symbol bandwidth R_s, there are only four symbols per window. Typically the window contains many symbols and the impact of the $sinc(fT_w)$ sidelobes is significantly diminished by increasing T_w. In addition to increasing the window duration, the impact of the spectral distortion can be further reduced by using a weighted window as discussed in Chapter 1. In any event, the signal distortion resulting from a finite sampling window must be examined in terms of the demodulator acquisition and detection processing.

Referring to the ranges of negative frequency samples as expressed by (2.10), consider, for example, an N_{fft}-sample fast Fourier transform (FFT) generated spectrum and the negative frequency range corresponding to $n = 1$ in (2.10). The FFT samples are taken over the positive sample window interval $0 \le t \le T_w$ and a right cyclic shift of the resulting FFT samples by $N_{fft}/2$ corresponds to the normalized frequency range given by

$$\frac{f_{n'}}{B} = -\frac{N_s}{N_{fft}}\left(\frac{N_{fft}}{2} - (n'-1)\right) \quad : n' = 1, \ldots, N_{fft} \tag{2.26}$$

where B is signal bandwidth and $N_s = 1/(BT_s)$ is the number of samples in the signal bandwidth. For example, if B is equal to the modulated symbol bandwidth, that is, $B = R_s = 1/T$, then $N_s = T/T_s$ is the number of samples per symbol. For a radix-2 FFT, $N_{fft} = 2^k$ and N_s is selected to be integrally related* to N_{fft}. Equation (2.26) indicates that, after the cyclic shift, the frequencies limits correspond to $f_0/B = -N_s/2$, $f_{Nfft/2+1}/B = 0$, and $f_{Nfft}/B = N_s/2 - N_s/N_{fft}$.

2.4 SIGNAL RECONSTRUCTION FOLLOWING DISCRETE-TIME SAMPLING

Shannon's sampling theorem states that if the Fourier transform of a signal $s(t)$ is zero at or above a certain frequency, for example, B_{max} with $S(f) = 0$ for $|f| \ge B_{max}$, then $s(t)$ can be uniquely determined from a function $g(nT_s)$ of the uniformly

*In general, the ratio N_{fft}/N_s should be a rational number to simplify the sample rate reduction processing. However, if N_s is integrally related to N_{fft} then an integrate-and-dump (I&D) matched filter simply sums N_s symbol samples; furthermore, if N_s is an even integer then E/L symbol tracking processing is also simplified.

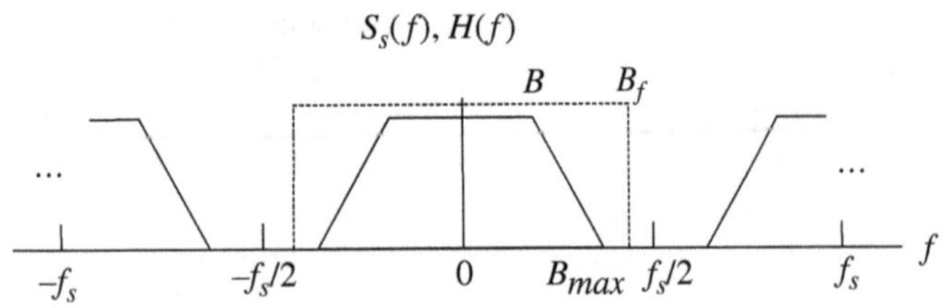

FIGURE 2.14 Ideally filtered sampled signal spectrum.

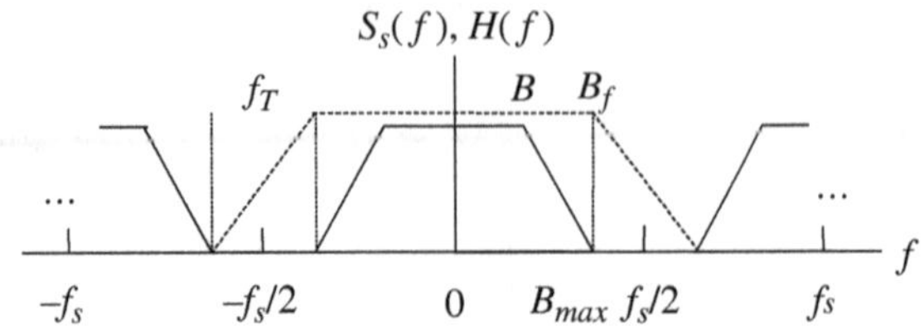

FIGURE 2.15 Filtered sampled signal spectrum.

spaced samples nT_s with $T_s = 1/2B_{max}$. Under these conditions, the reconstructed signal is expressed as [9, 13]

$$s(t) = \sum_{n=-\infty}^{\infty} g(nT_s)\frac{\sin((\pi/T_s)(t-nT_s))}{(\pi/T_s)(t-nT_s)} \tag{2.27}$$

To demonstrate (2.27), an ideal, or zonal low-pass filter, $H(f)$, is used to recover the baseband spectrum of the sampled signal as shown in Figure 2.14, where the *dashed* lines represents the low-pass filter with bandwidth confined to the range $B_{max} \le B_f \le f_s - (B + B_{max})$.

The filter output is determined by the convolution of the sampled signal $s_s(t)$ with the filter impulse response $h(t)$ expressed as

$$s(t) = \int_{-\infty}^{\infty} s_s(\xi)h(t-\xi)d\xi \tag{2.28}$$

Therefore, using (2.6), the ideally sampled signal is expressed as

$$\begin{aligned} s_s(t) &= s(t)f_s(t) \\ &= s(nT_s)\sum_{n=-\infty}^{\infty}\delta(t-nT_s) \end{aligned} \tag{2.29}$$

and the filter impulse response is evaluated as

$$h(t) = 2B_f T_s \frac{\sin(2\pi B_f t)}{2\pi B_f t} \tag{2.30}$$

Upon substituting (2.29) and (2.30) into (2.28) and using the properties of the delta function the reconstructed signal is evaluated as

$$s(t) = 2B_f T_s \sum_{n=\infty}^{\infty} s_s(nT_s)\frac{\sin(2\pi B_f(t-nT_s))}{2\pi B_f(t-nT_s)} \tag{2.31}$$

with $g(nT_S) = 2B_f T_S s_S(nT_S)$. Comparing (2.31) with (2.27) it is seen that $s(t)$ is exactly reconstructed when $B_f = B_{max}$ corresponding to $g(nT_S) = s_S(nT_S)$ and, for $t = mT_s$, this occurs when

$$\frac{\sin(2\pi B_{max}(m-n)T_s)}{2\pi B_{max}(m-n)T_s} = 1 \tag{2.32}$$

Equation (2.32) corresponds to $m = n$ and is equal to zero : $\forall(m-n) \neq 0$; these conditions are only satisfied when $B_f = B_{max}$.

The zonal or low-pass filter in Figure 2.14 is not realizable; however, the practical filter shown in Figure 2.15 can be constructed based on a specified passband attenuation (δ_{pb}), passband cutoff frequency (f_c), transition bandwidth (f_T), and stopband attenuation (δ_{sb}) as discussed in Section 2.8.2. The selection of these parameters is subject to the system performance requirements; however, the low-pass filter, shown as the *dashed* lines in Figure 2.15, will recover the signal spectrum from which the original signal can be reconstructed.* The three ideal conditions are as follows: $\delta_{pb} = 0$ for $|f| \le B_{max}$, $\delta_{sb} = \infty$ for $|f| \ge f_s - B_{max}$, and the signal spectrum is zero for $|f| \ge B_{max}$. The selection of the sampling frequency is a major consideration in the design, in that, the transition frequency can be chosen to simplify the filter.

2.5 BASEBAND SAMPLING

Baseband sampling involves mixing the demodulator intermediate frequency (IF) input frequency to baseband using conventional procedures discussed in Section 1.1.1.7. However, the resulting analog in-phase and quadrature baseband signal must be orthogonal, gain balanced, and direct current (DC) offset balanced as discussed in Section 2.7. Typically, high quality mixers and low-pass filters satisfy these conditions; however, temperature sensitivities are issues to be considered in the design.

2.6 BANDPASS SAMPLING

The minimum software loading on a demodulator results when an IF data-modulated waveform is mixed to baseband using analog quadrature mixers; however, analog signal

*The raised-cosine attenuation function can be used as an alternate to using a linear transition band attenuation. Refer to the discussion of Nyquist filtering in Section 4.4.4.1.

conversion to baseband is subject to a number of sensitivities as mentioned in the preceding section. To mitigate these effects DSP techniques are used to generate the baseband signal as an integral part of the ADC processing. ADC involves discrete-time and discrete-amplitude sampling; however, the following sections focus on the discrete-time sampling of the carrier-modulated signal referred to as bandpass sampling. The bandpass sampling techniques discussed are subcarrier sampling and Hilbert transform sampling. Each of these techniques use a carrier-modulated input signal at a suitably selected IF.*

The subcarrier sampling involves sampling the IF-modulated waveform with bandpass bandwidth W resulting in a lower frequency as a consequence of spectral aliasing. To achieve the aliased signal spectrum without spectral interference requires judicious selection of the sampling frequency. The resulting aliased sampled waveform is then processed to recover the information in the signal bandwidth.

The Hilbert transform processing is also similar, in that, discrete-time sampling of the IF-modulated waveform is performed and passed through a Hilbert filter that produces a quadrature set of I/Q data-modulated carrier frequencies corresponding to the input signal within the bandwidth of the Hilbert filter. The I/Q-modulated waveforms are then mixed to baseband using a complex mixer and the information is recovered.

2.6.1 Subcarrier Sampling

In this section the bandpass sampling of a carrier-modulated signal using subcarrier sampling is described. As in the case of bandpass sampling, the sampling frequency must be chosen such that aliasing distortion does not occur; however, the selection of the sampling frequency is much more rigorous and restrictive.

The analysis of the bandpass sampling criteria involves the IF frequency (f_{if}) and the corresponding bandwidth W, defined as the A dB bandwidth of the antialiasing filter shown in Figure 2.16. The bandwidth W is influenced by the system information bandwidth and may involve a single carrier or multiple carriers as with frequency division multiple access (FDMA) implementations and the value of A dB is chosen so as not to degrade the system performance in consideration of the aliasing distortion. In the context of Figure 2.16, the aliasing distortion is represented by the signal levels below A dB and outside of the bandwidth W. Typically, the bandwidth W is somewhat greater than the useable or 3-dB bandwidth, W_3, of the IF filter, for example, specifying A dB = 30 dB, with a 4-pole Butterworth antialiasing filter $W = 2.3W_3$ and with a 4-pole, 0.1 dB ripple Chebyshev filter $W = 1.8W_3$. As shown in Figure 2.16, W includes the filter transition bandwidth; however, W_3 must also include the frequency uncertainty range of the received waveform.

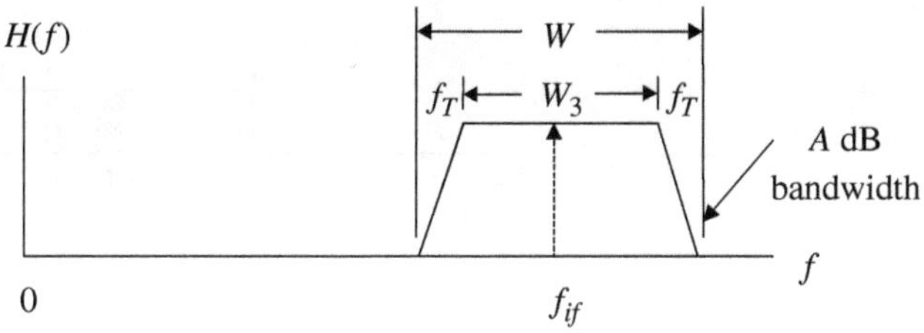

FIGURE 2.16 Bandpass antialiasing filter characteristics.

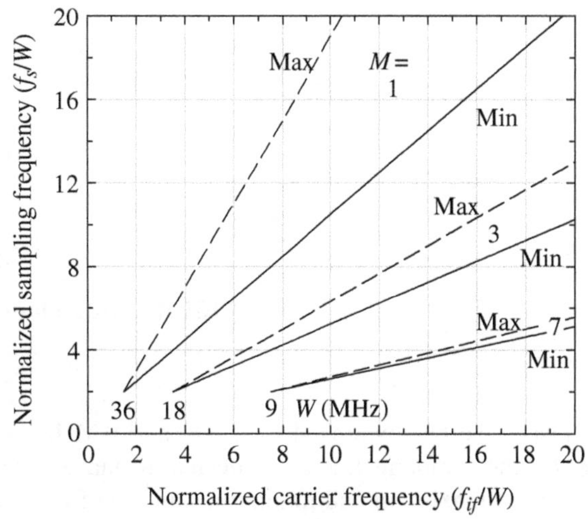

FIGURE 2.17 Bandpass sampling characteristics ($f_{if} = 70$ MHz, $W = 36$, 18, and 9 MHz).

In the following analysis, the IF and sampling frequencies are normalized by the IF filter bandwidth. The sampling frequency is established in consideration of the antialiasing requirement. For a filter centered at the f_{if} with bandwidth W, the criteria for the bandpass sampling frequency, f_s, so as not to result in aliasing, are given by [14, 15],

$$\frac{2f_{if}/W+1}{m+1} \le \frac{f_s}{W} \le \frac{2f_{if}/W-1}{m} \tag{2.33}$$

where $m = 1, 2, \ldots, M$, and M is the maximum value of m. Referring to Problems 4 and 5, M is evaluated as

$$M = \left\lfloor \frac{f_{if}}{W} - 0.5 \right\rfloor \quad : m \le M \tag{2.34}$$

The maximum value of M results in the lowest range of sampling frequencies that does not result in spectral aliasing, so, in (2.33), m is chosen as $m = M$. Figure 2.17 is a plot of (2.33) showing the normalized maximum and minimum sampling frequencies as a function of the normalized IF frequency for $M = 1$, 3, and 7 corresponding to $f_{if} = 70$ MHz

*The IF frequency is usually the carrier-modulated frequency into the demodulator subsystem.

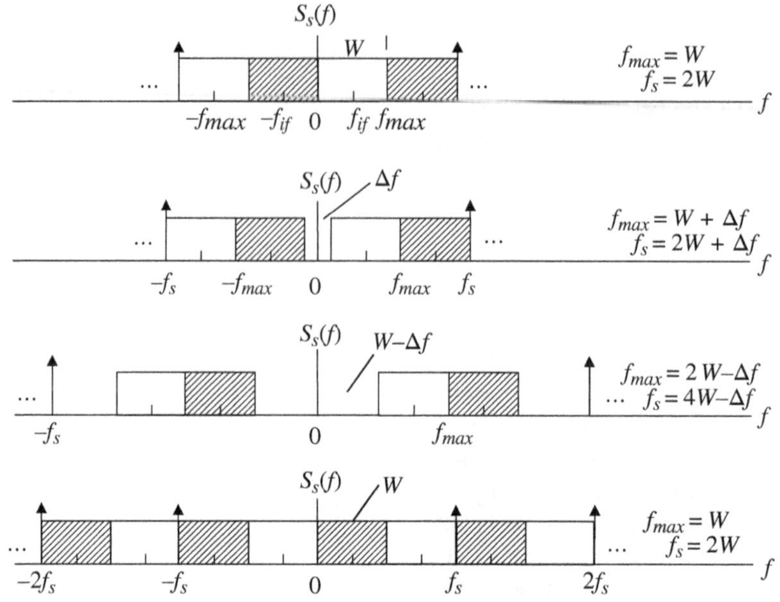

FIGURE 2.18 Illustration of minimum bandpass sampling.

with respective bandwidths $W = 36$, 18, and 9 MHz. As M increases the sampling frequency decreases and the spread between the maximum and minimum sampling frequencies decreases; the starting points, that is, the intersection of the maximum and minimum curves, in these plots correspond to the upper and lower limits in (2.33) being equal.

Based on (2.33) and (2.34) the minimum sampling frequency is expressed, in the normalized form, as

$$\frac{f_{smin}}{W} = \frac{2f_{if}/W + 1}{M + 1} \tag{2.35}$$

When M is exactly equal to argument of the floor(−) function, then (2.35) is evaluated as (see Problem 6) $f_{smin}/W = 2$ so, the minimum sampling frequency is expressed in terms of the lower limit as

$$f_{smin} \geq 2W \tag{2.36}$$

with the equality condition corresponding to $M = f_{if}/W - 1/2$.

To illustrate the relationship between the minimum sampling frequency (f_{smin}) and the maximum signal frequency* (f_{max}), Figure 2.18 shows the sampling requirements as f_{max} increases from W to $2W - \Delta f$ where Δf is a small frequency increment. As f_{max} increases linearly over this range the sampling frequency increases linearly over the range $2W \leq f_{samp} \leq 4W - \Delta f$; this is a direct result of M being constant as seen from (2.34). As Δf approaches zero, the upper limit of f_{max} approaches $2W$ and the sampling frequency approaches $4W$. However, when $\Delta f = 0$ the integer M is indexed by one and the minimum sampling frequency abruptly changes to $f_{smin} = 2W$ as expressed by (2.36). This abrupt change is seen to occur in Figure 2.18 corresponding to $f_{max} = 2W$. This linear increase in f_{smin} followed by an abrupt change to $f_{smin} = 2W$ continues with increasing f_{max} and defines the minimum sampling requirements to avoid spectral aliasing or interference between any repeated spectrums of bandwidth W Hz; these repeated spectrums are packed contiguously with no separation when $f_{max} = kW$: $k \in$ integer. For this reason, as a practical matter, the bandwidth W must be chosen to include oversampling to provide for the antialiasing filter transition bandwidth and signal frequency errors.

The range of the minimum sampling frequencies is evaluated numerically using (2.35) with $f_{if} = f_{max} - W/2$ and is plotted in Figure 2.19 as a function of f_{max}. Based on these results, the minimum sampling frequency is limited to the range $2W \leq f_{smin} \leq 4W$ with the normalized maximum values evaluated as

$$\frac{f_{smin}(max)}{W} = \frac{2(f_{max}/W)}{M} \quad : \frac{f_{max}}{W} = 2, 3, \ldots; M = f_{max}/W - 1 \tag{2.37}$$

*The maximum signal frequency f_{max} is used in the following analysis and corresponds to a translation of the IF frequency defined as $f_{max} = f_{if} + W/2$.

2.6.1.1 ***Example of Subcarrier Sampling*** This example of subcarrier sampling is considered to demonstrate the dependence of the maximum and minimum normalized sampling f_s/W on the normalized IF frequency f_{if}/W as depicted in Figure 2.17. The example considers the case corresponding

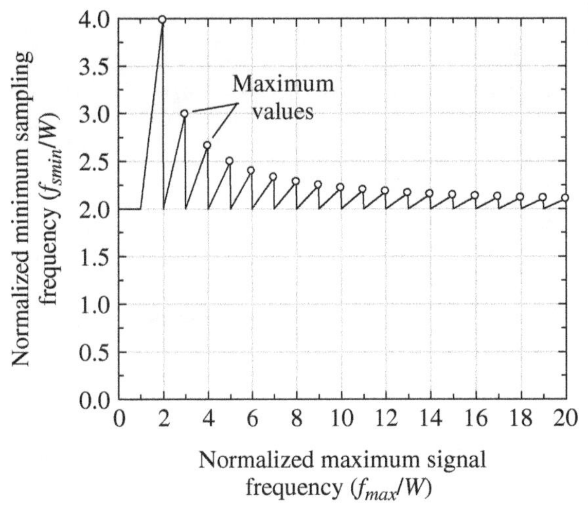

FIGURE 2.19 Minimum bandpass sampling characteristics.

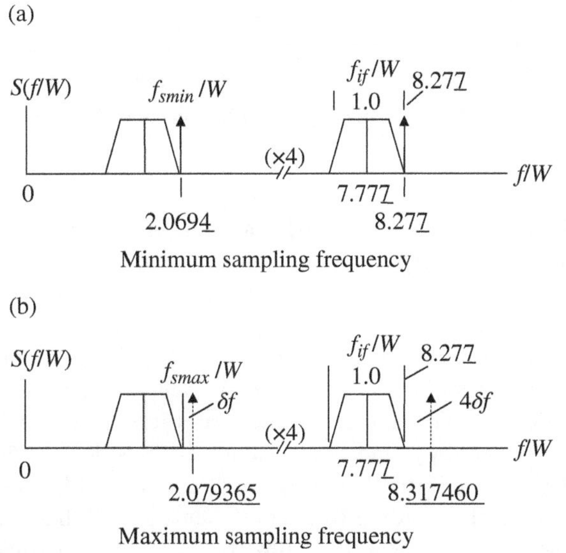

FIGURE 2.20 Minimum and maximum sampling frequencies (f_{if} = 70 MHz, W = 9 MHz, M = 7).

to f_{if} = 70 MHz, W = 9 MHz, and M = 7 computed using (2.34). The normalized IF frequency parameter is $f_{if}/W = 7.77\underline{7}$ and the range of normalized sampling frequencies f_s/W is determined using (2.33), with the result

$$2.069444444 \leq \frac{f_s}{W} \leq 2.079365079 \tag{2.38}$$

The spectrums and sampling frequencies are shown in Figure 2.20; the repeated spectrums at multiples of two and three times f_{smin}/W are not shown. In this example, sampling frequency $f_s'/W = 4f_{smin} = 8.27\underline{7}$ corresponds exactly to the upper frequency of the IF bandwidth and the repeated spectrum of interest lies in the band f_{smin}/W resulting in a 4 : 1 reduction in the signal IF carrier frequency. The sampling frequency corresponding to f_{smax}/W is $f_s''/W = 4f_{smax} = 8.317460317$. The sampling frequency difference is defined as $\delta f = f_{smax} - f_{smin} = 0.009920635$ and $4\delta f = 0.03968254$. Therefore, the maximum IF signal spectrum, relative to $4f_{smax}$ corresponds to $4f_{smax} - 4\delta f = 8.317460317 - 0.03968254 = 8.27\underline{7}$. This is exactly the same results as for the $f_s'/W = 4f_{smin}$ sampling frequency so the desired repeated spectrum again results in a 4 : 1 reduction in the signal IF carrier frequency. The frequency of the desired signal spectrum at $f_{if}/(4W)$ is independent of f_s/W and is not distorted by spectral aliasing as long as f_s/W satisfies (2.38). The accuracy of sampling frequency, to avoid spectrum aliasing, expressed as a percent of $(f_{smax} - f_{smin})/f_{smax}$ is 0.48%.

2.6.2 Hilbert Transform Sampling

Consider the signal at the radio or receiver IF output described as

$$s(t) = a(t)\cos\left(\omega_{if}t + \phi(t)\right) \tag{2.39}$$

where ω_{if} is the IF carrier frequency in radians per second and $a(t)$ and $\phi(t)$ are time-dependent amplitude and phase functions that describe the waveform modulation and channel filtering affects upon the received signal. The received signal is considered to be contained entirely within in the bandpass bandwidth W and the antialiasing filter serves to reject out-of-band interfering signals and limit the receiver noise power into the ADC to $(N_o/2)W_n$ where W_n is the two-sided noise bandwidth of the filter and N_o is the one-sided noise density. When the filtered signal is sampled at the rate $f_s \geq 2W$ and passed through the Hilbert transformer, the resulting quadrature output signals are expressed as

$$s_c(t) = a(t)\cos(\omega_c t + \phi(t)) \tag{2.40}$$

$$s_s(t) = a(t)\sin(\omega_c t + \phi(t)) \tag{2.41}$$

where $\omega_c = \omega_{if} - \omega_{lo}$ is the intermediate frequency prior to the ADC and ω_{lo} is the local oscillator used as a reference to translate the radio IF. The demodulator functional implementation is shown in Figures 2.21 and 2.22 shows the spectral characteristics for the frequencies and bandwidths of interest. This example uses a demodulator input IF frequency of 70 MHz that is mixed to 18 MHz using the local oscillator frequency of 52 MHz. The information bandwidth ($2B < W$) of the received signal may correspond to a single user-modulated signal or multiple frequency division multiplex (FDM) users with independent data.

2.6.2.1 *Hilbert Filter Characteristics* Based on the efforts of many researchers [16–18], the design of Hilbert filters has matured to the point where sets of finite impulse

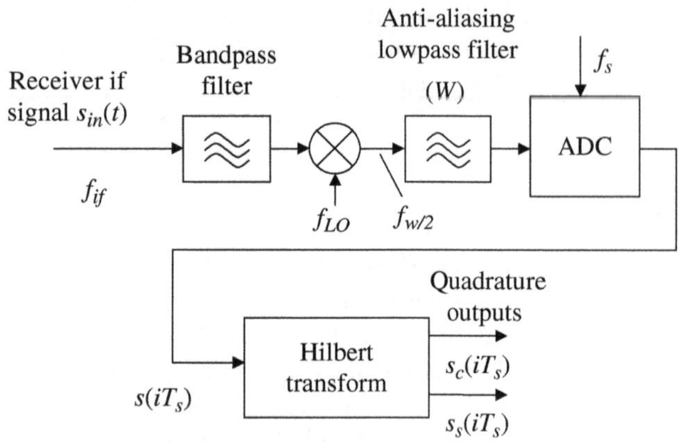

FIGURE 2.21 Functional implementation bandpass sampling with Hilbert transformer.

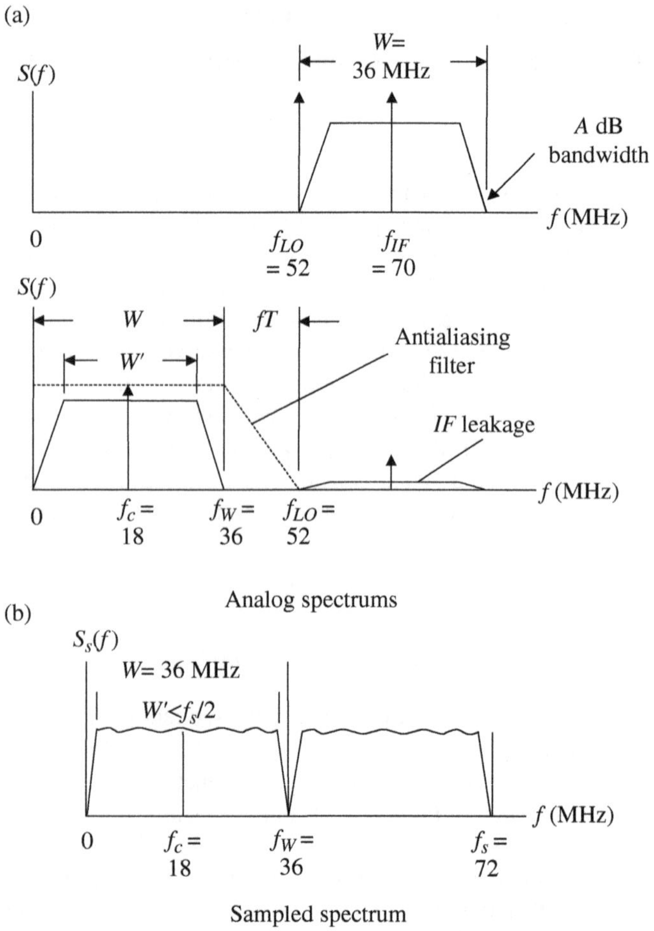

FIGURE 2.22 Filtering and Hilbert transform frequency characteristics.

response (FIR) filter coefficients are tabulated based on various design requirements and signal processing capabilities. Rabiner and Schafer [18] provide extensive tables of the filter coefficients and Tables 2B.1, 2B.2, 2B.3,

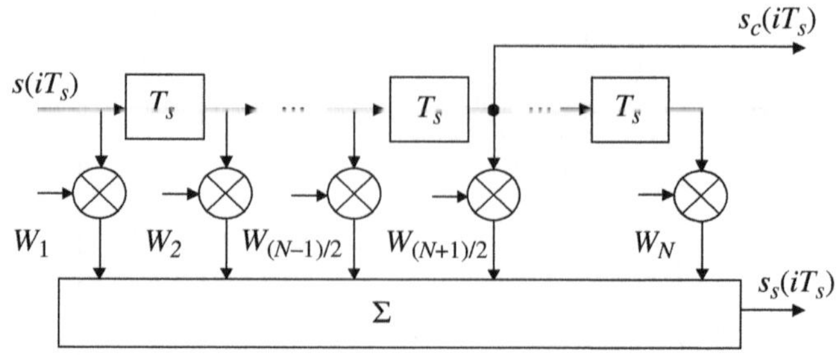

FIGURE 2.23 Functional implementation of Hilbert transform (N = odd integer).

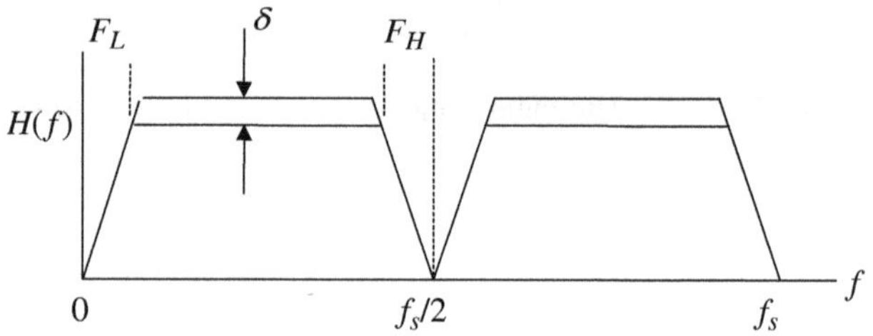

FIGURE 2.24 Typical Hilbert filter response for N = odd.

and 2B.4 in lists the coefficients for filters with an odd number of coefficients for various values corresponding to the lower cutoff frequency F_L. A typical Hilbert filter implementation is shown in Figure 2.23 and the frequency response is shown in Figure 2.24 identifying the filter design parameters when using an odd number of filter coefficients. The number of filter coefficients N, the sampling frequency f_s, the upper and lower transition bands F_H and F_L, the gain ripple δ over the filter bandwidth $F_H - F_L$, and the corresponding filter coefficients are all that is necessary to implement the Hilbert filter. The coefficients listed in the tables apply for $F_H = f_s/2 - F_L$ so there are only four design parameters: N, f_s, F_L, and δ. When N = odd integer the filter cutoff frequencies are equal corresponding to $F_H = f_s/2 - F_L$. Also, when N is an odd integer the filter response at $f_s/2$ is zero as shown in Figure 2.24. The Hilbert filter response for even N is shown in Figure 2B.1, in which case, the filter response at $f_s/2$ is not zero and the upper and lower transition bands are not equal so $F_H \neq f_s/2 - F_L$. Rabiner and Schafer present compelling arguments for using an odd number of coefficients; the most notable is that more efficient realizations are possible because every other coefficient is zero.

For N = odd integer, the filter impulse response, $h(n)$, has odd symmetry about the sample $n = (N-1)/2$ and the frequency response is given by

$$H\left(e^{j\omega}\right) = e^{-j(\omega(N-1)/2+\pi/2)} H^*\left(e^{j\omega}\right) \tag{2.42}$$

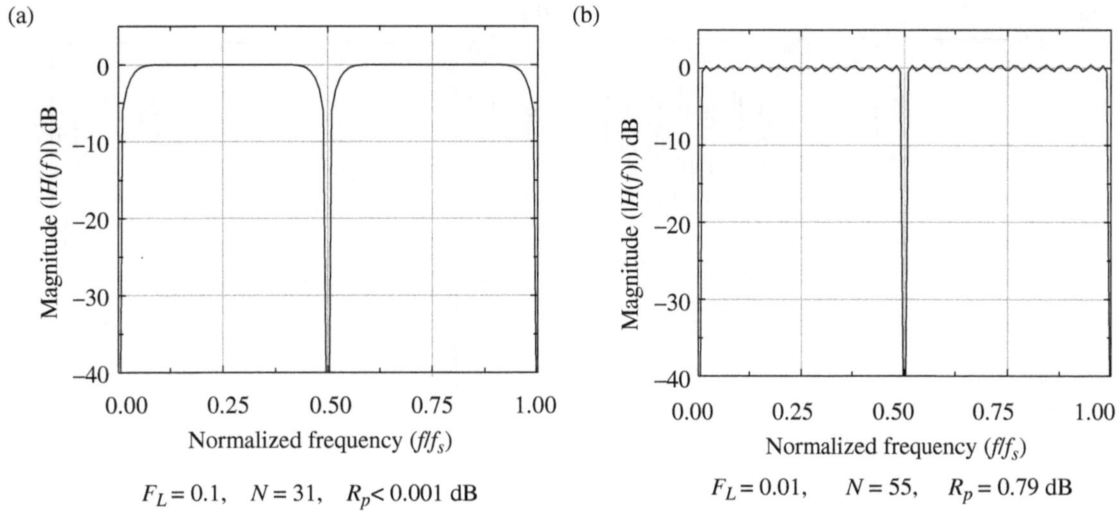

FIGURE 2.25 Hilbert filter frequency responses.

where

$$H^*(e^{j\omega}) = 2\sum_{i=1}^{(N-1)/2} a(n)\sin(n\omega) \qquad (2.43)$$

is a real function of ω and $a(n) = 2h((N-1)/2-n)$. Substituting these results into (2.42) the magnitude of the frequency response is evaluated as

$$|H(e^{j\omega})| = 2\sum_{n=1}^{(N-1)/2} h((N-1)/2-n)\sin(n\omega) \qquad (2.44)$$

The magnitude of the frequency response expressed by (2.44) is plotted in Figure 2.25 for the coefficients listed in Tables 2B.1 and 2B.4 corresponding to a unit sampling frequency $f_s = 1$ with $F_L = 0.1$, $N = 31$, and $D = 8\times10^{-6}$ and $F_L = 0.01$, $N = 55$, and $D = 0.095498$ as indicated; the peak-to-peak in-band ripple is defined as $R_p = 20\log_{10}(1+D)$.

Figure 2.26 shows the quadrature outputs of Hilbert filter in response to a continuous wave (CW) carrier at the center of the filter response and at the upper and lower transition band edges. The in-phase carriers are shown as the *solid* curves and the quadrature carriers as the *dashed* curves. This quadrature relationship is exhibited over the entire band from f_L to f_H. With data-modulated carriers, significant demodulator performance loss occurs from the quadrature unbalance when the waveform bandwidth extends into the transitions regions. Quadrature rail unbalance is shown in Figure 2.26d when the CW input signal is in the lower transition region.

2.6.3 Case Study: Hilbert Filtering

In this case study, the Hilbert transform is applied at the output of a wideband radio receiver for the subsequent detection of up to four data-modulated subchannels. The radio covers the entire ultra-high frequency (UHF) band of interest with an instantaneous bandwidth of 36 MHz and outputs an intermediate frequency of 70 MHz. The concepts discussed here are also applicable to tri-band radios operating in the C, X, and Ku bands that also provide a 70 MHz IF output. In this case study, the 70 MHz radio IF is translated to 18 MHz where analog-to-digital conversion and digital Hilbert filtering is performed using a sampling frequency of $f_s = 72$ MHz to produce I/Q bandpass outputs for subsequent subchannel filtering and data detection.

The Hilbert filter has $N = 31$ taps, equal lower and upper transition bands of $f_T = 0.05\,f_s = 3.6$ MHz, and a maximum in-band peak ripple of 0.023 dB. The resulting Hilbert bandwidth that satisfies this ripple condition is $W' = [0.5-2(0.05)]\,f_s = 28.8$ MHz and the 3 dB bandwidth is about 32.4 MHz.* The Hilbert filter bandwidth W' over the equal ripple specification can be increased by using a 55 tap Hilbert filter with transition bands $\Delta = F_L = 0.01$ with a corresponding in-band ripple to 0.8 dB. In this case, $f_T = 0.01\,f_s = 0.72$ MHz and the resulting ripple bandwidth is $W' = 34.56$ MHz with a 3 dB bandwidth of approximately 35 MHz. The multi subchannel demodulator is shown in Figure 2.27 with the M complex mixer referenced inputs (not shown) equal to the assigned carrier frequencies within the Hilbert bandwidth W'.

*The amplitude of the I and Q responses may differ by the ripple at the operating frequency of the subchannel carrier and the impact of *rail* unbalance must be considered in the waveform demodulation.

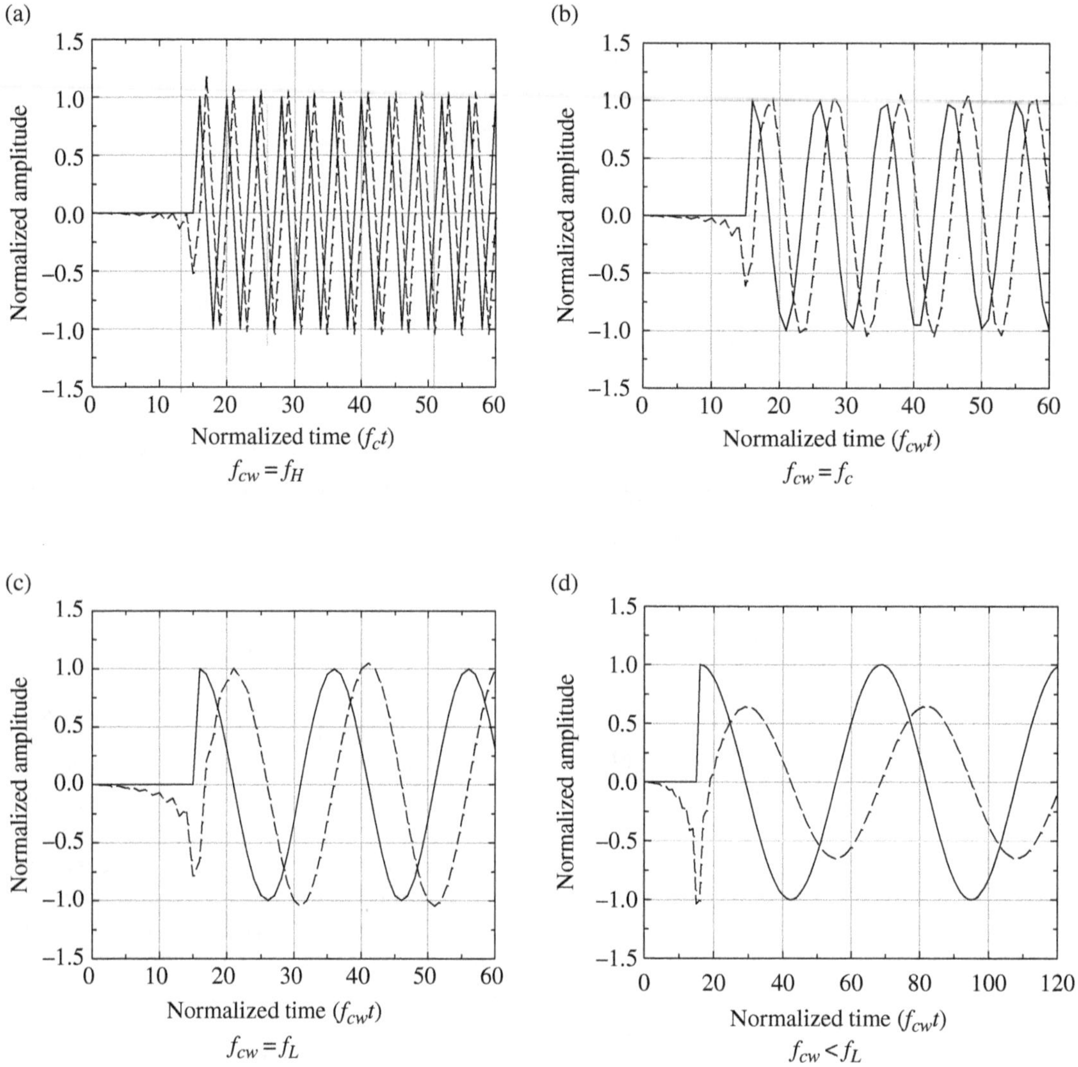

FIGURE 2.26 Hilbert filter subchannel carrier responses.

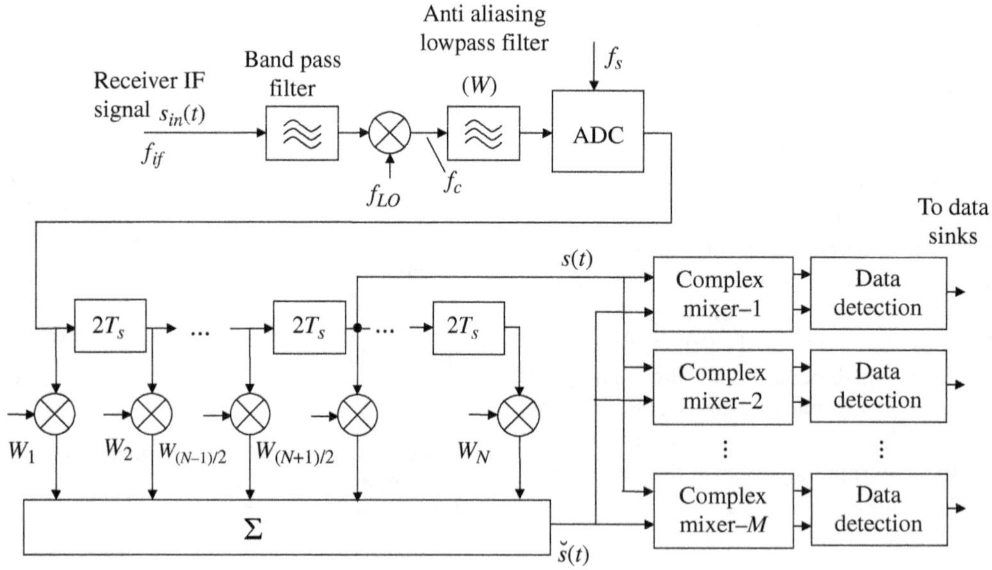

FIGURE 2.27 Multichannel demodulator implementation using Hilbert transform (N = odd integer).

2.7 CORRECTIONS FOR NONIDEAL MODULATORS AND DEMODULATORS

In the preceding sections, the carrier-modulated waveforms are ideal, in that, the quadrature rails have equal gains and are separated by exactly 90°. Furthermore, there is no DC offset on either rail. In the modulator this corresponds to no carrier frequency leakage and, in the demodulator, zero DC offset corresponds to symmetrical data levels on each rail. Carrier leakage results in signal interference and lower power efficiency relative to suppressed carrier modulation and, in the demodulator, quadrature rail offsets result in suboptimal-matched filter sampling and carrier phase and symbol tracking [20–23], Analog quadrature mixers use the nonlinear properties of a balanced set of diodes to generate the difference frequency between the input and a local oscillator (LO) reference frequency and any imbalance in the diodes results in nonideal waveform modulation. The diode imbalances are associated with quadrature rail gain and phase imbalance and DC offset imbalance as depicted in Figure 2.28.* High quality mixers typically have a phase imbalance of less than 3° and generally do not require phase correction; with quadrature phase shift keying (QPSK) modulation a three degree phase imbalance results in about 0.06 dB degradation in E_b/N_o at $P_{be} = 10^{-6}$.

The principal interest in this section is on methods for correcting analog mixer imbalances and the resulting impact on the demodulator bit-error performance. The mixer imbalances are also encountered in the demodulator when analog down-conversion is used to generate quadrature baseband signals following the final IF frequency stage. However, demodulator imbalances may also occur when using digital signal conversion as, for example, when using the Hilbert transform discussed in Section 2.6.2 where, referring to Figures 2.33 and 2.34, the amplitude of the Hilbert transform output $s_s(t)$ will differ from $s_c(t)$ based on the received signal carrier frequency and the quantization of the Hilbert filter coefficients.

In the following sections, nonideal modulators and demodulators are examined in the context of the imbalances shown in Figure 2.28 and the effectiveness of the compensation is examined and characterized in terms of the resulting demodulator bit-error performance. In the modulator, the compensation algorithms are implemented in the sampled data, or digital signal processing domain, with DAC before the nonideal analog up-conversion mixers. In the demodulator, the compensation algorithms are implemented following the analog-to-digital conversion to baseband. The DAC and ACD sampling is assumed to have infinite amplitude resolution with discrete-time sampling $T_s = 1/f_s$ where f_s is the sampling frequency with $t = iT_s$. Furthermore, without loss of generality, the analysis is based on analytic or complex signal representations as discussed in Section 1.1.1.

*This analysis refers to the inphase and quadrature rails as the cosine and sine baseband channels, respectively.

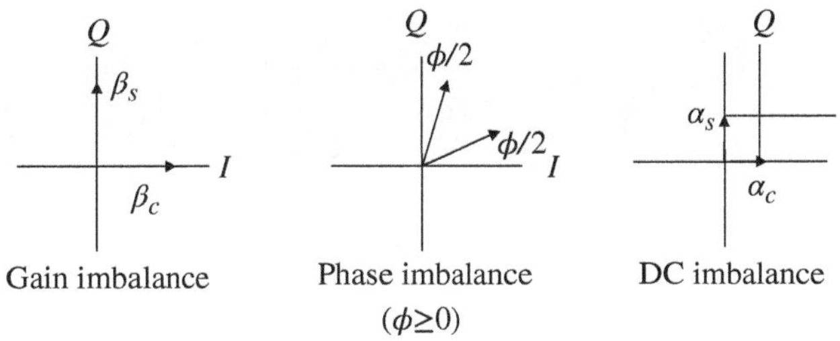

FIGURE 2.28 Analog frequency conversion error sources.

2.7.1 Nonideal Waveform Modulator

The transmitted signal for an imbalanced modulator operating at a carrier frequency of ω_c radian per second is described as

$$s_T(t) = \mathrm{Re}\{\tilde{s}_T(t)e^{j\omega_c t}\} \tag{2.45}$$

where the baseband analytic signal $\tilde{s}_T(t)$ is expressed as

$$\tilde{s}_T(t) = [(m_c(t) + \alpha_c)\beta_c \cos(\phi/2) + (m_s(t) + \alpha_s)\beta_s sin(\phi/2)] \\ + j[(m_c(t) + \alpha_c)\beta_c sin(\phi/2) + (m_s(t) + \alpha_s)\beta_s \cos(\phi/2)] \tag{2.46}$$

The functions $m_c(t)$ and $m_s(t)$ represent the in-phase and quadrature source data modulations. With an ideal modulator the DC offsets (α) and the phase imbalance (ϕ) are zero and the rail imbalances (β) are unity, the ideal modulator is characterized by the analytic signal

$$\tilde{s}_T(t)|_{ideal} = m_c(t) + jm_s(t) \tag{2.47}$$

and the carrier-modulated signal, expressed by (2.45), becomes

$$s_T(t)|_{ideal} = m_c(t)\cos(\omega_c t) - m_s(t)\sin(\omega_c t) \tag{2.48}$$

The nonideal up-converter, shown in Figure 2.29, depicts the functional implementation of (2.46) with the gain and phase imbalance terms contained in the matrix $\boldsymbol{D}$ given by

$$\boldsymbol{D} = \begin{pmatrix} \beta_c \cos(\phi/2) & \beta_s \sin(\phi/2) \\ \beta_c \sin(\phi/2) & \beta_s \cos(\phi/2) \end{pmatrix} \tag{2.49}$$

The first step in the distortion correction processing is to predistort the I/Q data modulation functions by the inverse of (2.49) using the measured gain and phase imbalances

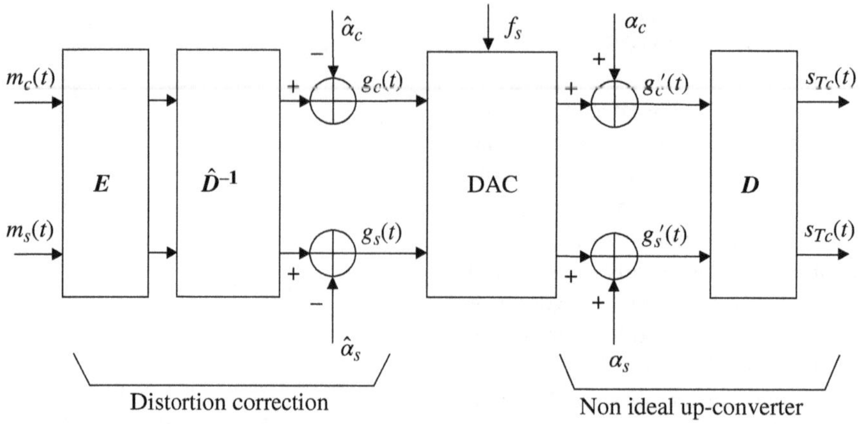

FIGURE 2.29 Analytic signal implementation of nonideal up-converter with distortion correction.

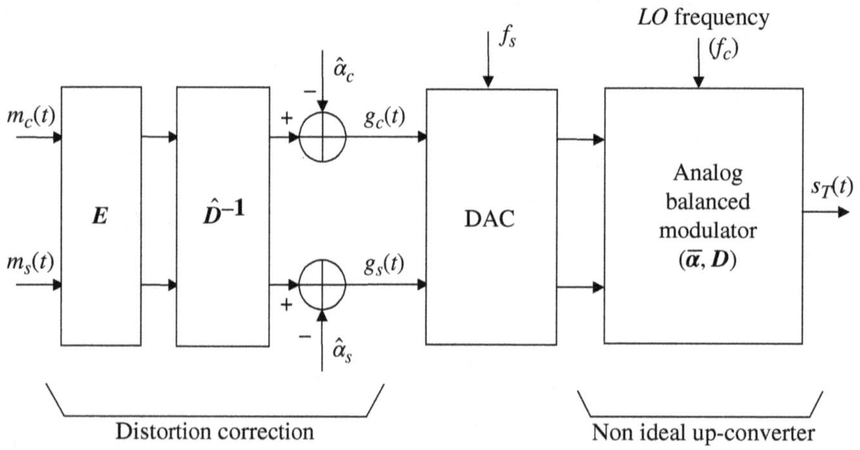

FIGURE 2.30 Modulator operational configuration.

denoted as $\bar{\hat{\boldsymbol{\beta}}}$ and $\bar{\hat{\phi}}$, respectively.* Using these measured values, the inverse distortion matrix is defined as

$$\hat{\boldsymbol{D}}^{-1} \triangleq \begin{pmatrix} \dfrac{\cos(\hat{\phi}/2)}{\hat{\beta}_c} & -\dfrac{\sin(\hat{\phi}/2)}{\hat{\beta}_c} \\ -\dfrac{\sin(\hat{\phi}/2)}{\hat{\beta}_s} & \dfrac{\cos(\hat{\phi}/2)}{\hat{\beta}_s} \end{pmatrix} \tag{2.50}$$

and the correction matrix $\boldsymbol{E}$, used to reduce the cross-channel distortion, is given by

$$\boldsymbol{E} \triangleq \begin{pmatrix} 0 & \dfrac{\beta_c}{\hat{\beta}_c}\sin(\hat{\phi}) \\ \dfrac{\beta_c}{\hat{\beta}_c}\sin(\hat{\phi}) & 0 \end{pmatrix} \tag{2.51}$$

*The **boldface** notation with an over-bar signifies a column vector with two rows corresponding to the *I*/*Q* rails, respectively.

Removal of the DC offsets is accomplished using the estimates $\bar{\hat{\boldsymbol{\alpha}}}$ as indicated in Figure 2.29 and, using the earlier notations, the output is expressed as

$$\bar{\bar{s}}_T = \left(\boldsymbol{D}\hat{\boldsymbol{D}}^{-1} - \boldsymbol{E}\right)\bar{\boldsymbol{m}} + \boldsymbol{D}\left(\bar{\alpha} - \bar{\hat{\alpha}}\right) \tag{2.52}$$

Determination of the estimates of the up-converter imbalances requires observing the level of the carrier frequency f_c as in (2.45) under the prescribed conditions of the source data modulation inputs $m(t)$. The order of the parameter estimation is important, in that, the DC offset must first be eliminated, followed by the gain imbalance, and ending with the phase imbalance correction. These procedures are depicted in terms of the baseband functions in Figure 2.29, however, because the carrier frequency output of the analog balanced modulator is observed, the test and operational configuration is as shown in Figure 2.30. There is virtually no noise involved in the measurement of the distortion parameters so the measurement accuracies are limited only by the instrumentation.

In Section 2.7.2 the distortion correction is examined for the nonideal demodulator down-converter and evaluation of the demodulator correction is similar and is suggested as an exercise. The modulator involves evaluation of the matrix product $\boldsymbol{D}\hat{\boldsymbol{D}}^{-1}$ which is somewhat more involved than evaluating $\hat{\boldsymbol{C}}^{-1}\boldsymbol{C}$ as required in the demodulator down-converter.

2.7.1.1 *Modulator DC Offset Measurement* To measure the DC offset correction parameter $\hat{\boldsymbol{\alpha}}$, the source data I/Q inputs $m(t)$ are each grounded and the carrier power (P_T) is minimized by adjusting $\hat{\alpha}_c$ and $\hat{\alpha}_s$; the values corresponding the minimum power are the correct estimates for use in Figure 2.30.

2.7.1.2 *Modulator Gain Imbalance Measurement* The gain imbalance parameter $\hat{\boldsymbol{\beta}}$ is determined by holding the $m_c(t)$ input at a constant level, grounding the $m_s(t)$ input rail, and selecting $\hat{\beta}_c$ to achieve a prescribed power level. The role of $m_c(t)$ and $m_s(t)$ is then reversed and $\hat{\beta}_s$ is selected to achieve the prescribed carrier power level.

2.7.1.3 *Modulator Phase Imbalance Measurement* The evaluation of the phase imbalance ($\hat{\phi}$) must be determined after the DC offset and gain estimation parameters have been established. Referring to Figure 2.31, the angle between the vector V_c and V_s is less than 90°* so the I/Q source data will not be orthogonal resulting in degraded demodulator performance. Furthermore, the vectors V_0 and V_1 bisect two identical equilateral parallelograms† and the ratio of the vector magnitudes is defined as

$$\rho \triangleq \frac{|V_0|}{|V_1|} \tag{2.53}$$

Using the geometry of the equilateral parallelogram, the angle ϕ is related to the parameter ρ as

$$\tan\left(\frac{\phi}{2}\right) = \frac{\rho - 1}{\rho + 1} \tag{2.54}$$

The desired estimate of the phase imbalance is the solution to (2.54) with the result

$$\hat{\phi} = 2\tan^{-1}\left(\frac{\rho - 1}{\rho + 1}\right) \tag{2.55}$$

Therefore, to determine $\hat{\phi}$ it is necessary to determine the parameter ρ using the measured magnitudes of the vectors V_0 and V_1 as expressed in (2.53). The vector magnitude $|V_0|$ is measured using constant and equal source data inputs, that is, $m_c(t) = m_s(t) = A$ volts, and recording the balanced modulator output carrier voltage level as $|V_0|$, using peak or rms voltage units. The vector magnitude $|V_1|$ is then measured using constant but antipodal source data inputs, with $m_c(t) = -A$ and $m_s(t) = A$ volts, and, using the same units as before, recording the balanced modulator output carrier level as $|V_1|$. The phase estimate is then computed using (2.55).

*When the phase imbalance is negative, the vectors V_s and V_c are in the second and fourth quadrants, respectively; however, the same estimation procedures apply.

†An equilateral parallelogram is a parallelogram with equal length sides.

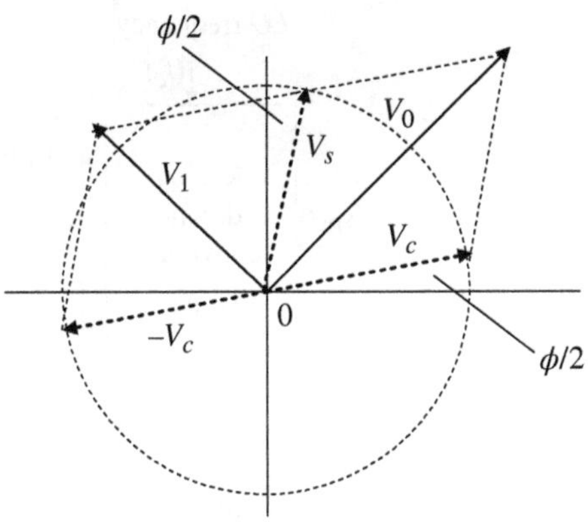

FIGURE 2.31 Geometrical description of phase imbalance.

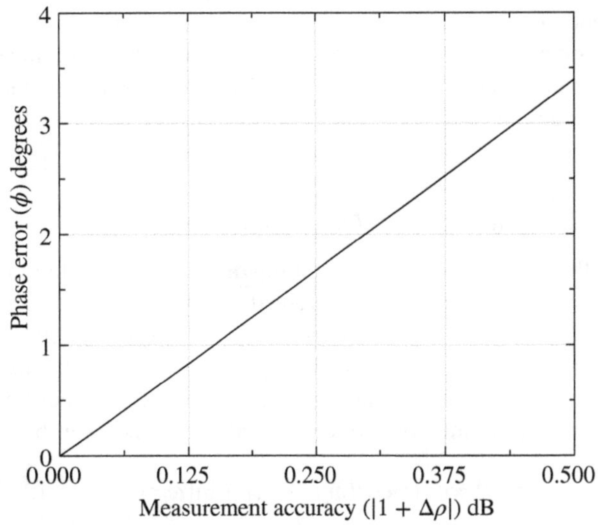

FIGURE 2.32 Sensitivity of phase imbalance to amplitude measurement.

The sensitivity of the phase estimate to the measurement of the parameter ρ is evaluated using

$$\Delta\hat{\phi} = \frac{\partial\hat{\phi}}{\partial\rho}\Delta\rho \tag{2.56}$$

The solution to (2.56) is plotted in Figure 2.32 as a function of $(\rho + \Delta\rho)/\rho$ expressed in decibels under the ideal

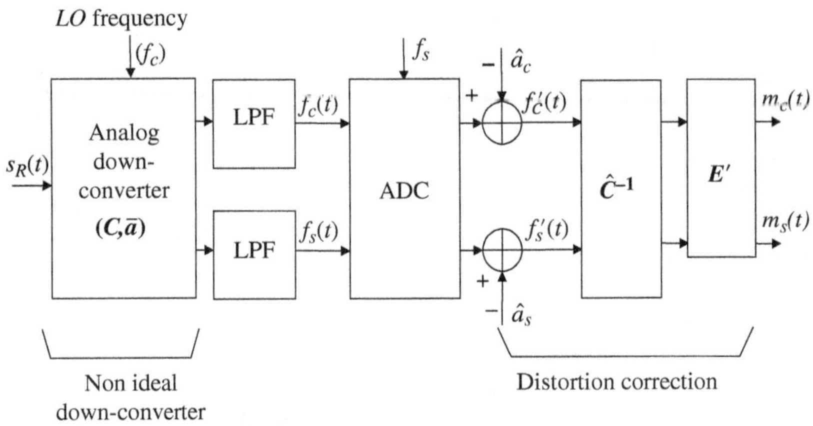

FIGURE 2.33 Demodulator operational configuration.

condition $\rho = 1$. Because there is essentially no noise involved in this measurement, it is reasonable to expect measurement accuracies less than 0.25 dB so a phase estimate accuracy of less than 2° is a realistic expectation.

2.7.2 Nonideal Waveform Demodulator

In this section the demodulator down-conversion of the final IF output to baseband is examined in terms of the corrected gain, the phase imbalances, and the DC offsets that often exist in analog mixers. The received signal is assumed to be ideal, that is, all of the modulator up-conversion distortion has been removed,* so the noise-free received signal† is given by

$$s_R(t) = m_c(t)\cos(\omega_c t) - m_s(t)\sin(\omega_c t) \tag{2.57}$$

Following the measurement and application of the down-conversion distortion parameters, the channel phase function must be removed using a phaselock loop (PLL) prior to the optimum I/Q data decision processing; this is implicit in (2.57) by the absence of the phase function of the form $\theta(t) + \theta_o$ in the sinusoidal arguments.

The demodulator analog baseband down-converter encounters similar distortion parameters as those in the modulator up-converter. In this case, however, the test input used to measure the distortion parameters is a CW signal. The baseband outputs, in response to the test signal, are used to determine the amplitude and phase distortion parameters that are used to compensate for the baseband rail imbalances. In addition, the CW test signal is used to determine the DC offset distortion parameters that are used to compensate for the DC bias on each rail. To distinguish the demodulator distortion parameters from those used in the modulator up-converter, the gain, phase, and DC offset parameters in the demodulator down-converter are denoted, respectively, as b_c, b_s, φ and a_c, a_s. These parameters are implicit in Figure 2.33, in that, **C** describes the down-converter gain and phase imbalance and the vector $\bar{\boldsymbol{a}}$ describes the DC offset.

Except for the parameter notation and a factor of one-half arising from the analog down-conversion of the carrier frequency, the matrix **C** is otherwise identical to (2.49); the matrix **C**, the inverse matrix $\hat{C}^{-1}$, and the matrix E' are expressed as‡

$$C = \frac{1}{2}\begin{pmatrix} b_c\cos(\varphi/2) & b_s\sin(\varphi/2) \\ b_c\sin(\varphi/2) & b_s\cos(\varphi/2) \end{pmatrix} \tag{2.58}$$

$$\hat{C}^{-1} \triangleq \begin{pmatrix} \dfrac{\cos(\hat{\varphi}/2)}{\hat{b}_c} & -\dfrac{\sin(\hat{\varphi}/2)}{\hat{b}_c} \\ -\dfrac{\sin(\hat{\varphi}/2)}{\hat{b}_s} & \dfrac{\cos(\hat{\varphi}/2)}{\hat{b}_s} \end{pmatrix} \tag{2.59}$$

and

$$E' \triangleq \begin{pmatrix} 0 & \dfrac{b_s}{\hat{b}_c}\sin(\hat{\varphi}) \\ \dfrac{b_c}{\hat{b}_s}\sin(\hat{\varphi}) & 0 \end{pmatrix} \tag{2.60}$$

The DC-offset vectors $\bar{\boldsymbol{a}}$ and $\hat{\bar{\boldsymbol{a}}}$ are expressed as $(a_c, a_s)^T$ and $(\hat{a}_c, \hat{a}_s)^T$, respectively.

*Inclusion of the modulator compensation distortion in the up-converter needlessly complicates the analysis; however, the formulation of the demodulator output using (2.52) as the received signal is suggested as an exercise in Problem 15.

†This analysis is concerned with compensation of the desired signal distortion in a linear mixer and the inclusion of additive noise simply complicates the notation with no impact on the conclusion.

‡The matrix inverse $\hat{C}^{-1}$ is normalized by the factor $2/\cos(\varphi)$ that is associated with the matrix inverse $\mathbf{C}^{-1}$. This factor simply represents a constant gain and plays no role in the parameter compensation.

Under normal operation, the input signal $s_R(t)$ is the received carrier-modulated waveform; however, during the measurement of the distortion parameters $s_R(t)$ is a CW test signal described as

$$s_R(t) = A\cos((\omega_c + \omega_d)t + \theta) \quad \text{: input test signal} \tag{2.61}$$

The carrier offset frequency f_d is used for the parameter measurements and the low-pass filters are used to remove the tone at $2f_c + f_d$ and minimize aliasing resulting from the sampling. In general, the filter outputs are expressed as

$$\begin{aligned} f_c(t) &= s_R(t)[b_c\cos(\omega_d t - \varphi/2)] + a_c \\ &= \frac{Ab_c}{2}\cos(\omega_d t + \theta - \varphi/2) + a_c \end{aligned} \tag{2.62}$$

and

$$\begin{aligned} f_s(t) &= s_R(t)[-b_s\sin(\omega_d t + \varphi/2)] + a_s \\ &= \frac{Ab_s}{2}\sin(\omega_d t + \theta + \varphi/2) + a_s \end{aligned} \tag{2.63}$$

The following measurements are based on observations of the baseband signals given by (2.62) and (2.63) and, because the observations are virtually noiseless, averaging of the measurements is not necessary so, theoretically, only one cycle of $f_d = \omega_d/2\pi$ is necessary to perform each of the three measurements.* The selection of f_d is influenced by the low-pass filter (LPF) response and the ADC sampling frequency. For example, because the measurements involve determining maximum and minimum values of the sinusoidal test signal, a measurement accuracy of 0.03 dB requires a sampling resolution of less than 5° or $f_d < 0.014f_s$; however, f_d must also be high enough for the LPFs to reject the $2f_d$ term.† The time for all three measurements will be on the order $2/f_d$ seconds, so, under these conditions, for f_s = 500 kHz the test measurement time will be on the order of 0.3 *ms* plus any additional test overhead time.

2.7.2.1 *Demodulator DC Offset Measurement* Using the test signal described by (2.61) with $\hat{a}_c = \hat{a}_s = 1$, the DC offsets a_c and a_s are measured by recording the maximum and minimum values of $f_c'(t)$ and $f_s'(t)$ at the output of the ADC as shown in Figure 2.33. These functions are sampled representations of (2.62) and (2.63) and the unknown parameters b_c, b_s, and φ will not influence the DC offsets measurements. The correction parameters $\hat{a}_c$ and $\hat{a}_s$ are computed as

$$\hat{a}_c = \frac{A_c(\max) + A_c(\min)}{2} \tag{2.64}$$

and

$$\hat{a}_s = \frac{A_s(\max) + A_s(\min)}{2} \tag{2.65}$$

where the maximum and minimum amplitudes are the recorded values indicated earlier.

2.7.2.2 *Demodulator Gain Imbalance Measurement*

The gain distortion parameters $\hat{b}_c$ and $\hat{b}_s$ are measured in a similar way and the maximum and minimum values from the DC offset test can be used to compute the gain imbalance as

$$\hat{b}_c = \frac{A_c(\max) - A_c(\min)}{2} \tag{2.66}$$

and

$$\hat{b}_s = \frac{A_s(\max) - A_s(\min)}{2} \tag{2.67}$$

These computations are independent of $\hat{a}_c$ and $\hat{a}_s$ so the DC offset corrections do not have to be applied before the gain imbalance measurement. The correction factors are adjusted to result in equal output levels on each rail. For example, by letting $b_c/\hat{b}_c = g$, where g is an arbitrary gain, the gain corrections are computed as $\hat{b}_c = b_c/g$ and $\hat{b}_s = b_s/g$. A simplification occurs if $\hat{b}_c$ is always considered to be unity so that $\hat{b}_s = b_s/b_c$.

2.7.2.3 *Demodulator Phase Imbalance Measurement*

Before measuring the phase imbalance, the DC offset and gain imbalance corrections must be applied. Under these conditions, the measurement involves recording the maximum and minimum values of the magnitude of the baseband outputs $f_c'(t)$ and $f_s'(t)$ as they rotate in response to the test signal phase. With the gain imbalance adjusted for unit gain ($g = 1$) and with ideal DC offset compensation, these baseband functions are obtained from (2.62) and (2.63) with the result

$$f_c'(t) = \frac{A}{2}\cos(\omega_d t + \theta - \varphi/2) \quad \text{: offset and gain corrected} \tag{2.68}$$

and

$$f_s'(t) = \frac{A}{2}\sin(\omega_d t + \theta + \varphi/2) \quad \text{: offset and gain corrected} \tag{2.69}$$

*Upon application of the test tone several cycles will be necessary to allow for settling of the transients.

†These may be conflicting requirements that require a digitally implemented notch filter at $2f_d$ and/or sample rate up-conversion as discussed in Section 2.8.1.1.

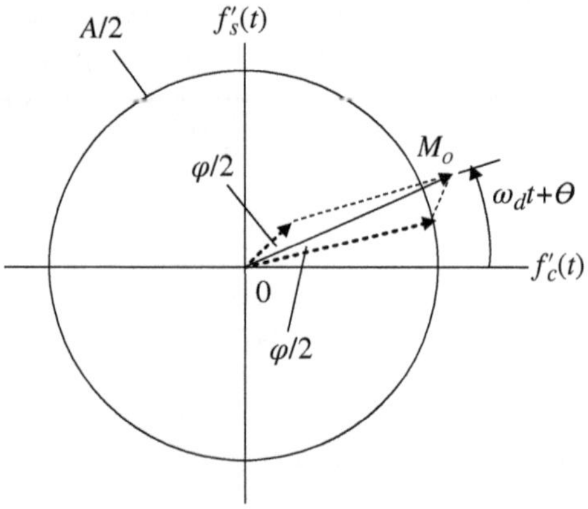

FIGURE 2.34 Demodulator phase imbalance ($\omega t = 0$).

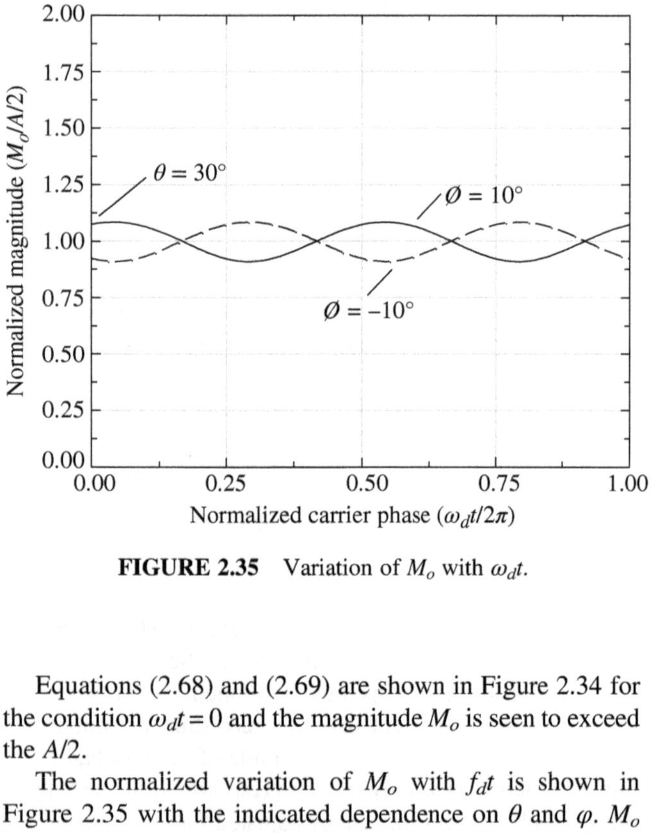

FIGURE 2.35 Variation of M_o with $\omega_d t$.

Equations (2.68) and (2.69) are shown in Figure 2.34 for the condition $\omega_d t = 0$ and the magnitude M_o is seen to exceed the $A/2$.

The normalized variation of M_o with $f_d t$ is shown in Figure 2.35 with the indicated dependence on θ and φ. M_o is seen to vary in amplitude about the normalized magnitude $M_o/(A/2)$ at the frequency $2f_d$ with a sign dependence on the sign of the phase imbalance.

The magnitude M_o is evaluated as

$$M_o(t) = \sqrt{f_c'^2(t) + f_s'^2(t)} \tag{2.70}$$

Based on these observations, the phase imbalance is measured by recording the maximum and minimum values of $M_o(t)$ and forming the ratio

$$\rho = \frac{M_o(\max)}{M_o(\min)} \tag{2.71}$$

and recording the sign of the product of (2.68) and (2.69) as $\kappa = \text{sign}(f_c'(t_m)f_s'(t_m))$ where t_m is the time corresponding to the recording of $M_o(\max)$. Using these recorded values the phase imbalance is computed as

$$\hat{\varphi} = 2\kappa \tan^{-1}\left(\frac{\rho - 1}{\rho + 1}\right) \tag{2.72}$$

Problems 11, 12, and 13 provide further insights into the phase measurement in the context of the frequency dependent test inputs.

2.7.3 Demodulator Down-Conversion Imbalance Error Analysis

The imbalance in the demodulator down-converter after the measured correction parameters have been applied is a result of the measurement inaccuracies associated with each of the distortion parameters. The overall transfer function is expressed as

$$\bar{\hat{m}} = \left(\hat{C}^{-1}C - E'\right)\bar{s}_R + \hat{C}^{-1}\left(\bar{a} - \bar{\hat{a}}\right) \tag{2.73}$$

and is evaluated in terms of the measured distortion parameters and the measurement errors. The imbalances after application of the corrections are embodied in (2.73) and, upon using small argument approximations for the sine and cosine functions, the approximate expression is

$$\bar{\hat{m}} \cong \frac{1}{2}\begin{pmatrix} \frac{b_c}{\hat{b}_c}\left(1 - \frac{\Delta\varphi^2}{8}\right) & -\frac{b_s}{\hat{b}_c}\left(\frac{\Delta\varphi}{2}\right) \\ -\frac{b_c}{\hat{b}_s}\left(\frac{\Delta\varphi}{2}\right) & \frac{b_s}{\hat{b}_s}\left(1 - \frac{\Delta\varphi^2}{8}\right) \end{pmatrix}\bar{s}_R + \begin{pmatrix} \frac{1}{\hat{b}_c}\left(1 - \frac{(\varphi - \Delta\varphi)^2}{8}\right) & \frac{1}{\hat{b}_c}\left(\frac{\varphi - \Delta\varphi}{2}\right) \\ \frac{1}{\hat{b}_s}\left(\frac{\varphi - \Delta\varphi}{2}\right) & \frac{1}{\hat{b}_s}\left(1 - \frac{(\varphi - \Delta\varphi)^2}{8}\right) \end{pmatrix}\Delta\bar{a} \tag{2.74}$$

where the phase and DC offset measurement errors are defined as $\Delta\varphi = \varphi - \overleftarrow{\varphi}$ and $\Delta\bar{a} = \bar{a} - \bar{\hat{a}}$ and the baseband functions are:

$$\bar{\hat{m}} = (\hat{m}_c, \hat{m}_s)^T \tag{2.75}$$

and

$$\bar{s}_R = \left(m_c(t)^c, m_s(t)\right)^T \tag{2.76}$$

In the following case study, these parameters are examined in terms of the measurement accuracy expressed in decibels and the resulting bit-error probability is characterized for the QPSK-modulated waveform.

2.7.4 Case Study: Bit-Error Performance with Baseband Mixer Imbalance

In this case study, the bit-error performance of a QPSK-modulated waveform is examined in terms of the imbalance in a demodulator analog down-converter to baseband. The performance is based on (2.74) with the quadrature modulation functions expressed in the simplest forms as

$$m_c(t) = AdI_i \text{ and } m_s(t) = AdQ_i \tag{2.77}$$

where A is the received carrier frequency peak amplitude and dI_i and dQ_i represent the in-phase and quadrature bipolar binary source data bits equal to ±1 in the symbol interval $iT \le t \le iT + T$; the demodulated quadrature data estimates are $\hat{m}_c = \hat{dI}_i$ and $\hat{m}_s = \hat{dQ}_i$. In the following performance characterization, the phase measurement error is defined as the ratio $\eta_\varphi = \Delta\varphi/\varphi$ and, expanding (2.74) in terms of η_φ, the I/Q rails are expressed as

$$\begin{aligned}\hat{m}_c = \frac{A}{2}\left(\frac{b_c}{\hat{b}_c}\left(1 - \frac{(\varphi\eta_\varphi)^2}{8}\right)dI_i + \frac{b_s}{\hat{b}_c}\left(\frac{\varphi\eta_\varphi}{2}\right)dQ_i\right) \\ + \frac{\Delta a_c}{\hat{b}_c}\left(1 - \frac{\varphi^2(1-\eta_\varphi)^2}{8}\right) + \frac{\Delta a_s}{\hat{b}_c}\left(\frac{\varphi}{2}(1-\eta_\varphi)\right)\end{aligned} \tag{2.78}$$

and

$$\begin{aligned}\hat{m}_s = \frac{A}{2}\left(\frac{b_c}{\hat{b}_s}\left(\frac{\varphi\eta_\varphi}{2}\right)dI_i + \frac{b_s}{\hat{b}_s}\left(1 - \frac{(\varphi\eta_\varphi)^2}{8}\right)dQ_i\right) \\ + \frac{\Delta a_c}{\hat{b}_s}\left(\frac{\varphi}{2}(1-\eta_\varphi)\right) + \frac{\Delta a_s}{\hat{b}_s}\left(1 - \frac{\varphi^2(1-\eta_\varphi)^2}{8}\right)\end{aligned} \tag{2.79}$$

The parameters b and $\hat{b}$ represent voltage gains and are unitless, whereas, the terms A and Δa represent voltage levels. The gain and DC offset distortion terms in (2.78) and (2.79) are formulated in terms of normalized ratios as defined in Table 2.1.

TABLE 2.1 Gain and DC Offset Parameter Definitions

Parameter	Definition	Example Values	Description
		Phase imbalance	
φ		10	Degrees
η_φ	$\frac{\Delta\varphi}{\varphi}$	10	Measurement accuracy (%)
		Gain imbalance	
b_c		−0.25	I-ch gain (dB)
b_s		0.25	Q-ch gain (dB)
ρ_{cs}	$\frac{b_c}{b_s}$	0.5	I/Q gain imbalance (computed) (dB)
k_c	$\frac{\hat{b}_c}{b_c}$	−0.2	I-ch gain measurement accuracy (dB)
k_s	$\frac{\hat{b}_s}{b_s}$	0.2	Q-ch gain measurement accuracy (dB)
		DC—offset imbalance	
ρ_{ac}	$\frac{a_c}{A/2}$	−25	I-ch DC offset imbalance (dB)
ρ_{as}	$\frac{a_s}{A/2}$	−25	Q-ch DC offset imbalance (dB)
K_c	$\frac{\hat{a}_c}{a_c}$	0.2	I-ch DC offset measurement accuracy (dB)
K_s	$\frac{\hat{a}_s}{a_s}$	−0.2	Q-ch DC offset measurement accuracy (dB)

Using these definitions (2.78) and (2.79) are rewritten as

$$\begin{aligned}\hat{m}_c = \frac{A}{2}\left\{\frac{1}{k_c}\left(1 - \frac{(\varphi\eta_\varphi)^2}{8}\right)dI_i + \frac{1}{k_c\rho_{cs}}\left(\frac{\varphi\eta_\varphi}{2}\right)dQ_i \right. \\ + \frac{\rho_{ac}(1-K_c)}{b_c k_c}\left(1 - \frac{\varphi^2(1-\eta_\varphi)^2}{8}\right) \\ \left. + \frac{\rho_{as}(1-K_s)}{b_c k_c}\left(\frac{\varphi}{2}(1-\eta_\varphi)\right)\right\}\end{aligned} \tag{2.80}$$

and

$$\begin{aligned}\hat{m}_s = \frac{A}{2}\left\{\frac{\rho_{cs}}{k_s}\left(\frac{\varphi\eta_\varphi}{2}\right)dI_i + \frac{1}{k_s}\left(1 - \frac{(\varphi\eta_\varphi)^2}{8}\right)dQ_i \right. \\ + \frac{\rho_{ac}(1-K_c)}{b_s k_s}\left(\frac{\varphi}{2}(1-\eta_\varphi)\right) \\ \left. + \frac{\rho_{as}(1-K_s)}{b_s k_s}\left(1 - \frac{\varphi^2(1-\eta_\varphi)^2}{8}\right)\right\}\end{aligned} \tag{2.81}$$

The bit-error performance of QPSK with demodulator imbalances is expressed in terms of the I and Q channel error probabilities as

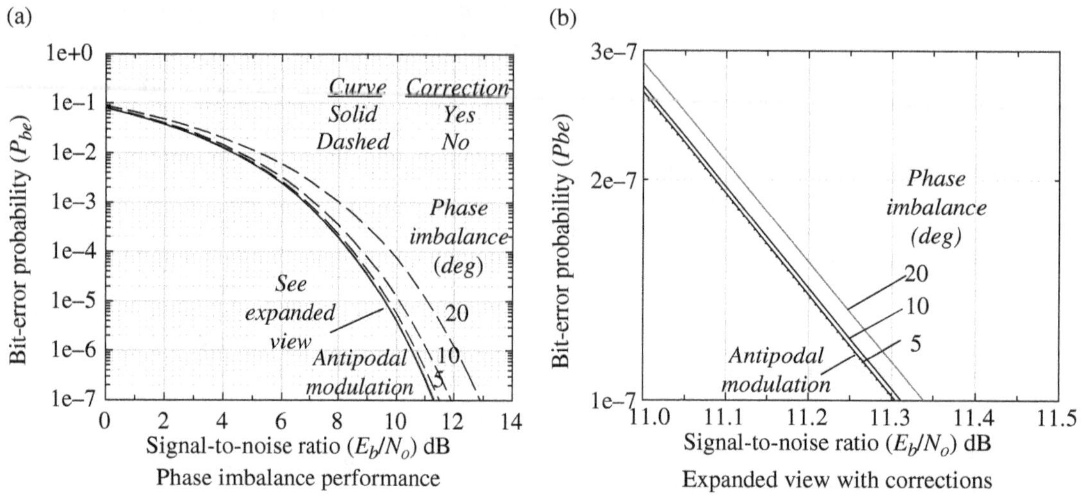

FIGURE 2.36 Phase imbalance performance of QPSK with and without corrections (10% phase measurement accuracy, ideal amplitude, and DC offsets).

$$P_{be} = \frac{1}{2}P_{be}(I) + \frac{1}{2}P_{be}(Q) \quad (2.82)$$

where the I and Q channel error probabilities are evaluated as

$$P_{be}(I) = \frac{1}{2}erfc\left(\sqrt{\gamma_b}\left|\hat{m}_c\left(\hat{d}I,\hat{d}Q\right|\right)\right. \quad (2.83)$$

and

$$P_{be}(Q) = \frac{1}{2}erfc\left(\sqrt{\gamma_b}\left|\hat{m}_s\left(\hat{d}I,\hat{d}Q\right|\right)\right. \quad (2.84)$$

where $erfc$ is the error function discussed in Section 3.5, γ_b is the signal-to-noise ratio measured in the information-bit bandwidth, $\hat{m}_c$ and $\hat{m}_s$ are given by (2.80) and (2.81) and the I/Q data estimates consider all four combinations of the estimates $(dI_i, dQ_i) = \{(1, 1), (1, -1), (-1, 1), (-1, 1)\}$: $i = 1, 4$ so that

$$erfc\left(\sqrt{\gamma}\left|\hat{m}\left(\hat{d}I,\hat{d}Q\right|\right)\right. = \frac{1}{4}\sum_{i=1}^{4} erfc\left(\sqrt{\gamma}\left|\hat{m}\left(\hat{d}I_i,\hat{d}Q_i\right|\right)\right. \quad (2.85)$$

Equation (2.82) is evaluated using these relationships and plotted in Figures 2.36 and 2.37 under the indicated conditions. These results indicate that performance losses in excess of 1 dB occurs if the imbalance corrections are not applied and with the corrections the loss can be reduced to less than 0.05 dB. The imbalance conditions and measurement accuracies listed in Table 2.1 are considered to be typical values and particular mixer devices and measurement conditions must be considered for each unique application.

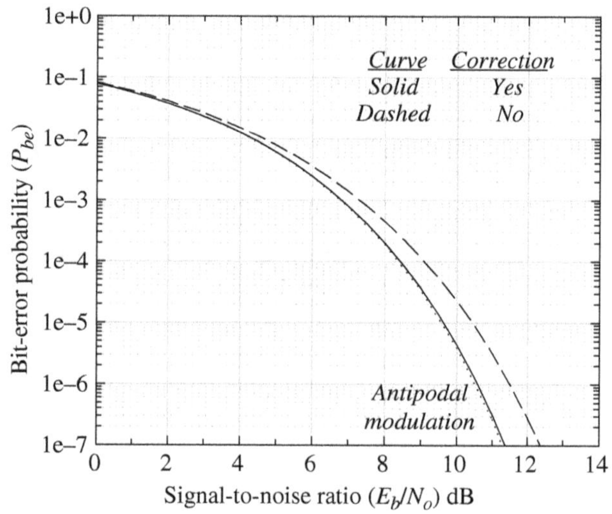

FIGURE 2.37 Performance with imbalance parameters in Table 2.1 (E_b/N_o loss with corrections <0.05 dB).

2.8 MULTIRATE SIGNAL PROCESSING AND INTERPOLATION

Waveform sampling is derived from a system oscillator or clock that provides the timing reference for many receiver and demodulator functions including LO frequency generation and baseband input and output (I/O) interface for data and control. The focus of multirate signal processing [1, 24, 25] in this section is on sample rate conversion from the system oscillator frequency to various sampling frequencies required for waveform acquisition and subsequently to the sampling frequency required for the baseband signal-matched filtering. As discussed in the following section, the sampling rate during acquisition must be high enough to avoid aliasing distortion in view of the high carrier frequency uncertainty resulting from various clocks and the received signal Doppler frequency. The clock frequency uncertainties include those of the remote transmitter and, with relays or repeaters, all intermediate clock uncertainties throughout the link. Following acquisition, the received waveform is mixed to the baseband where it is maintained with the aid of a frequency tracking phaselock loop. In this case, the sampling frequency is typically two to four times the symbol rate; however, to accommodate low-loss symbol tracking, the sample time resolution is typically one-sixteenth of the symbol interval or less depending on the symbol waveform modulation. For example, with the $rect(t/T)$ symbol weighting function a symbol timing error of one-sixteenth of a symbol corresponds to a symbol energy loss of 0.28 dB. Sample time interpolation is used to bring the matched filter symbol samples inline with the symbol edge transitions so that all of symbol energy captured with no intersymbol interference (ISI). These subjects, involving multirate signal processing and interpolation, are discussed in the following sections in the context of specific applications and demodulator capabilities. For example, applications requiring large rate reductions, that is, rate reduction in excess of several hundred, are efficiently achieved using a CIC filter, whereas, applications requiring lower rate reductions and more stringent filtering performance are efficiently implemented using techniques involving FIR filters.

2.8.1 Sample Rate Conversion

A fundamental consideration in rate conversion [26, 27] between two input clock frequencies is the evaluation and application of the greatest common divisor (GCD) discussed in Section 1.13.1. This is best explained by the following example, wherein, a received signal with a symbol modulation rate of $R_s = 4.8$ ksps is sampled at $f_{s1} = 70$ MHz. The example uses two sample rate conversions, one requiring a sample rate of $f_{sa} = 48$ kHz during acquisition to accommodate a large frequency uncertainty and the second at $f_{sd} = 19.2$ kHz for matched filter detection at four samples per symbol.

During signal acquisition, the GCD between 70 MHz and 48 kHz is $f_{gcd} = 16$ kHz and this is used to determine the ratios $k_1 = f_{sa}/f_{gcd} = 3$ and $k_2 = f_{s1}/f_{gcd} = 4375$; these are the required up and down sampling rate conversions of the input 70 MHz sampling frequency to provide the desired 48 kHz sampling frequency during acquisition. The rate conversion processing is depicted in Figure 2.38 where the *up* and *down* arrows indicate rate increase and reduction, respectively. The image reject and antialiasing filters are shown as separate filters to facilitate the following descriptions of the individual rate conversion processes; however, when an overall rate

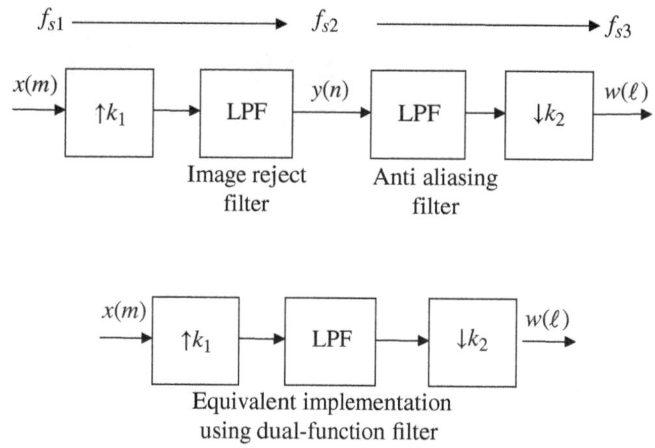

FIGURE 2.38 Rate conversion processing description.

reduction by the rational number $k_1/k_2 < 1$ is required, the two filters can be combined using a single dual-function filter having the most restrictive requirements of the two individual filters as shown by second implementation in Figure 2.38. As mentioned in Section 2.8.1.1 the process of sample rate up-conversion is referred to as interpolation and the image reject filter functions as an interpolation filter; on the other hand, in Section 2.8.1.2, the process of sample rate reduction is referred to as decimation and the antialiasing filter is referred to as the decimation filter. The terms interpolation and decimation are applied to these two processes in the following descriptions.

When the signal acquisition is completed and the frequency uncertainty has been removed, the sample rate can be altered once again to simplify the symbol-matched filter detection processing. In this example, the symbol-matched filter sample rate is $f_{sd} = 19.2$ kHz so an additional rate change is required; the details of the rate change from f_{sa} to f_{sd} is left as an exercise in Problem 16.

The decimation described in Section 2.8.1.2 applies to one stage of decimation as shown in Figure 2.38. However, for applications requiring large decimations, as in the earlier example, it is more efficient in terms of processor loading and memory to cascade several stages of decimations. For example, considering the earlier example, with an input frequency of 210 MHz and $k_2 = 4375$, a multistage decimator using factors $k_{21} = 125$, $k_{22} = 7$, $k_{23} = 5$ is implemented as shown in Figure 2.39a. The implementation efficiency in Figure 2.39a can be improved by replacing the first decimation of 125 with the CIC filter shown in Figure 2.39b. Ifeachor and Jervis [24] and Fliege [25] are recommended reading for the use of other filter types and rate-conversion implementations.

The *natural* rate-conversion analysis begins with the input signal sampled at the highest rate to be converted. For example, referring to Figure 2.38, for up-conversion or down-conversion the highest sampling rate is f_{s2}. Of course,

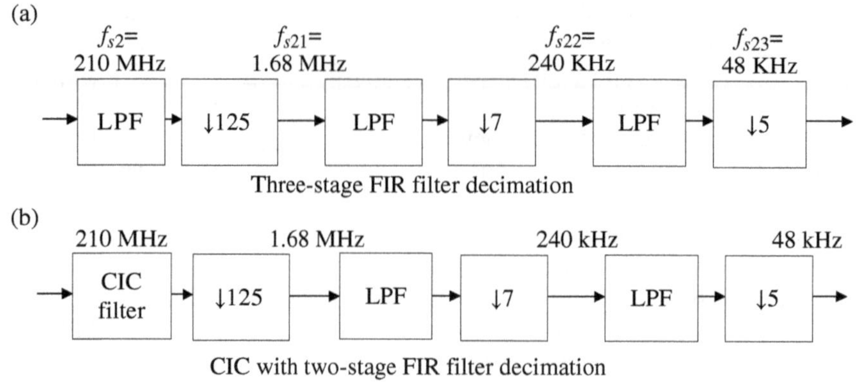

FIGURE 2.39 Rate-conversion processing examples.

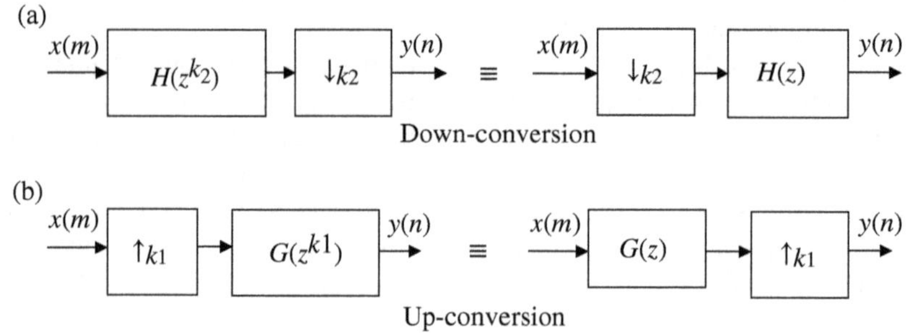

FIGURE 2.40 Noble identities for sample rate conversion.

if the down-conversion were not preceded by up-conversion, as shown in the figure, then the down-conversion sampling rate would be f_{s1}, which may limit the final sampling rate options. The rate-conversion analysis proceeds with the design of a bank of polyphase filters with z-transforms: $G_i(z^{k1})$: $i = 0, 1, \ldots, k_1$ for the up-conversion filter and $H_i(z^{k2})$: $i = 0, 1, \ldots, k_2$ for the down-conversion filter. In other words, the filtering is performed at the highest sampling rate corresponding to the z-transform delay z^k. However, by applying the *noble identities* [27, 28] depicted in Figure 2.40,* the filtering can be performed at the output sampling rate corresponding to the z-transform delay z resulting in considerable computational savings.

The z-transform rate-conversion filters are expressed as

$$G(z) = \sum_{i=0}^{k_1-1} z^{-i} G_i\left(z^{k_1}\right) \quad \text{: up-conversion} \tag{2.86}$$

and

$$H(z) = \sum_{i=0}^{k_2-1} z^{-i} H_i\left(z^{M}\right) \quad \text{: down-conversion} \tag{2.87}$$

*There are six noble identities, three each related to up- and down-conversion; Fliege refers to those depicted in Figure 2.40 as identities 2 and 5.

Equations (2.86) and (2.87) are the polyphase representations of $G(z)$ and $H(z)$ expanded in terms of the polynomials $G_i\left(z^{k_1}\right)$ and $H_i\left(z^{k_2}\right)$, respectively. $G(z)$ and $H(z)$ are similarly partitioned in terms of their discrete filter coefficients $c(n)$; this partitioning is denoted as $F(z)$ and evaluated as

$$\begin{aligned}
F(z) = \quad & c(0) + \quad c(k)z^{-k} + \quad c(2k)z^{-2k} + \quad \cdots \\
& c(1)z^{-1} + \quad c(k+1)z^{-(k+1)} + \quad c(2k+1)z^{-(2k+1)} \quad \cdots \\
& c(k-1)z^{-(k-1)} + c(2k-1)z^{-(2k-1)} + c(3k-1)z^{-(3k-1)} + \cdots \\
= \quad & z^0[c(0) + \quad c(k)z^{-k} + \quad c(2k)z^{-2k} + \quad \cdots] \\
& z^{-1}[c(1) + \quad c(k+1)z^{-k} + \quad c(2k+1)z^{-2k} + \cdots] \\
& z^{-(k-1)}[c(k-1) + c(2k-1)z^{-k} + c(3k-1)z^{-2k} + \cdots]
\end{aligned} \tag{2.88}$$

Equation (2.88) follows directly from (2.86) and (2.87) where the coefficients are associated with $F(z) = G(z)$ and $k = k_1$ for up-conversion and $F(z) = H(z)$ with $k = k_2$ for down-conversion. The *ellipsis* in (2.88) extends over the length or number of taps, N_c, required by the digital filter to satisfy the design requirements as discussed in Section 2.8.2. Typically, the number of taps required to

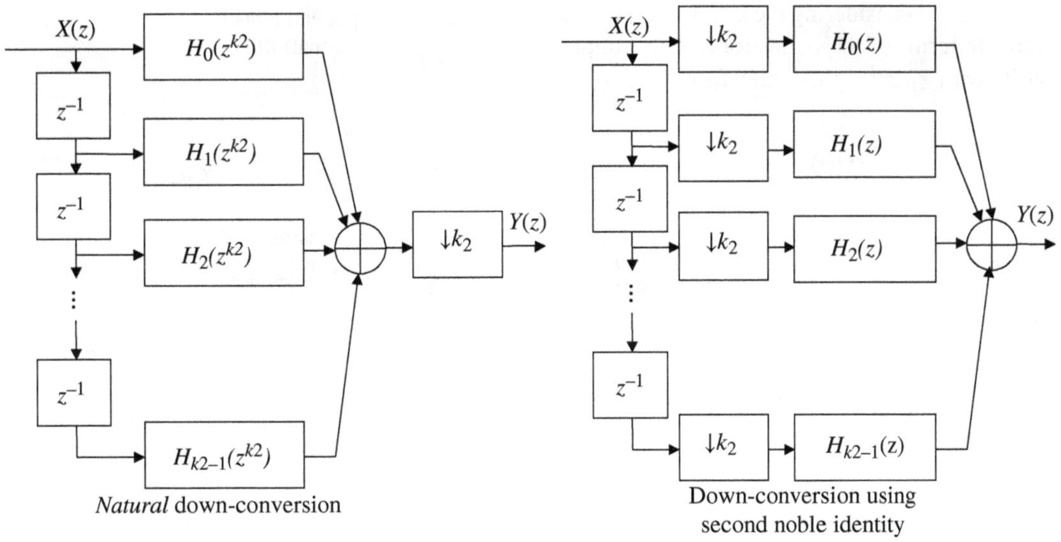

FIGURE 2.41 Down-conversion for natural and efficient processing using second noble identity.

satisfy the design requirements is adjusted to satisfy the condition $N_c = Nk$ where N is an integer. Figure 2.41 compares the sample rate down-conversion implementations using the natural and more efficient processing using the second noble identity where the filtering is performed at the rate-reduced sample rate.

2.8.1.1 *Interpolation (Up-Conversion)* Sample rate up-conversion by k_1 is accomplished by including $k_1 - 1$ uniformly spaced zero samples between each of the input samples separated by T_{s1} seconds resulting in a sampling interval of $T_{s2} = T_{s1}/k_1$. Using the impulse-invariant z-transform $z = e^{sT_{s1}}$, the sample delay is given by* z^{-1} and the z-transforms of $x(m)$ and $y(n)$ are evaluated as

$$X(z) = \sum_{m=-\infty}^{\infty} x(m)z^{-m} \tag{2.89}$$

and

$$Y(z) = \sum_{m=-\infty}^{\infty} x(m)z^{-mk_1} \tag{2.90}$$

The effect of the added zeros is most easily seen by examining the spectrums of $X(f)$ and $Y(f)$ shown in Figure 2.42.† Essentially the addition of the zero samples in $x(n)$ corresponds to selecting the sampled spectrum of $X(f)$ located at $f = k_1 f_{s1}$ and removing the spectral images at f_{s1} through $(k_1 - 1)f_{s1}$ shown in the figure as the *dotted* spectrums. The

*In terms of the Laplace transform variable s, the transform for a sample delay is $x(t - T_s) \Leftrightarrow X(s)e^{-sT_s}$ where $x(t) \Leftrightarrow X(s)$.

†The spectrums are evaluated using the Laplace variable $s = j2\pi f$.

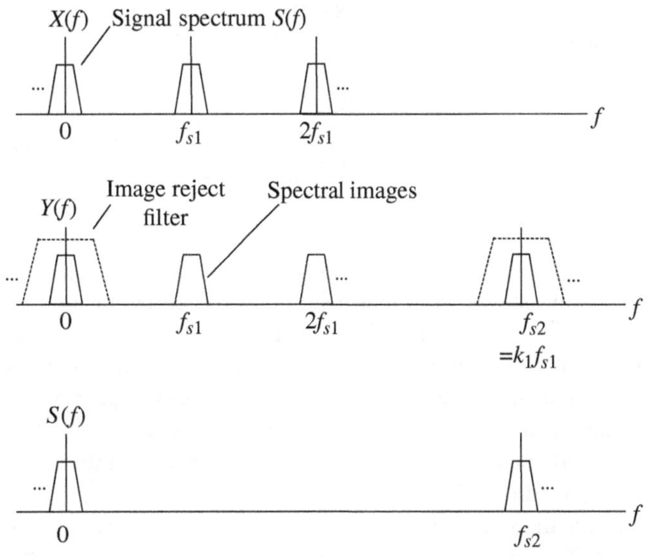

FIGURE 2.42 Spectrum of the sampled functions $x(m)$ and $y(n)$ and image reject filter.

filtering of the spectral images functions as an interpolator for the $k_1 - 1$ samples points in the response $y(n)$ corresponding to the zero added in $x(m)$, therefore, the process of up-conversion by k_1 involves the addition of $k_1 - 1$ zeros to $x(m)$ and image reject low-pass filtering as depicted in Figure 2.42.

2.8.1.2 *Decimation (Down-Conversion)* The sample rate decimation, or rate reduction by k_2, is described in terms of the down-sampler shown in Figure 2.38 preceded by an anti-aliasing low-pass filter. In the time domain, decimation by an integer k_2 is accomplished by choosing every k_2-th sample of the sampled signal $y(n)$ and considering the $k_2 - 1$ intervening samples to be zero. In terms of the z-transform with sample interval T_{s2}, that is, with $z = e^{sT_{s2}}$, the z-transform of $y(n)$ is

$$Y(z) = \sum_{n=-\infty}^{\infty} y(n)z^{-n} \tag{2.91}$$

However, from the sample set $y(n)$ every k_2-th sample is selected so that the z-transform of the down-sampled sequence of $y(n)$ samples is

$$Y_k\left(z^{k_2}\right) = \sum_{\ell=-\infty}^{\infty} y_k(\ell k_2)z^{-\ell k_2} \tag{2.92}$$

where the subscript $k = 0, 1, \ldots, k_2 - 1$ represents the starting phase or sample of the down-sampled sequence of $y(n)$. The parameter k identifies the polyphase filter functions discussed in Section 2.8.5 and, for the present discussion, it is assumed the polyphase filter of interest corresponds to $k = 0$. With this understanding, the z-transform of the down-sampled sequence $w(\ell)$ is given by

$$W(z') = Y_0\left(z^{k_2}\right) = \sum_{\ell=-\infty}^{\infty} w(\ell)z^{-\ell k_2} \tag{2.93}$$

where $z' = e^{sT_{s3}}$ corresponds to the sample interval T_{s3}.

As in the case with interpolation, the decimation filtering requirements are best explained in terms of the sampled spectrums as shown in Figure 2.43. The first or top axis depicts the input signal spectrum $Y(f)$ as it would appear when sampled at f_{s2} without any preceding up-conversion, that is, if only down-conversion is being performed. In this case, the function of the input antialiasing, or zonal,* filter is to confine the sampled spectrum within the low-pass bandwidth $f_{s2}/2$. The second axis in Figure 2.43 depicts the role of the decimation or antialiasing filter which serves to eliminate all of the image spectral content between the desired baseband signal spectrum and that centered around the Nyquist sampling frequency f_{s2} in preparation for the decimated signal spectrum as shown on the third axis. When preceded by up-sampling the image rejection filter may also serve as the antialiasing filter which results in a more efficient design. Therefore, the down-sampled signal results in equally spaced replicas of the input signal spectrum centered at nf_{s3}; however, an important difference is that the magnitude of the spectra is reduced by a factor of k_2; this is an important consideration when unit gain filtering or digital signal processing dynamic range is a concern.

*The antialiasing filter is described in Section 2.3 as an ideal filter with $H(f) = rect((f - B/2)/B)$ and zero phase response.

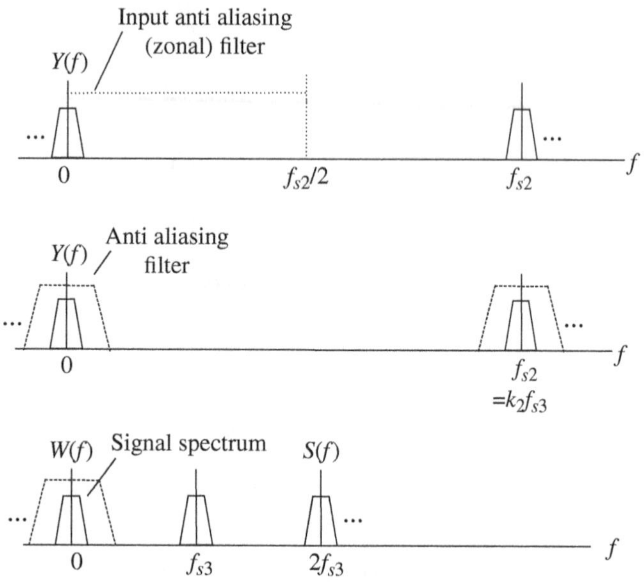

FIGURE 2.43 Spectrum of the sampled functions $y(m)$ and $w(n)$ and antialiasing filter.

Referring to Figure 2.42, the up-sampled signal spectrum $S(f)$ is seen to be ideally replicated at the new sampling frequency $f_{s2} = k_1 f_{s1}$. This ideal spectrum replication is evidence of the ideal interpolation provided by the image reject filter. Similarly, from Figure 2.43 the down-sampled perfect replication of the signal spectrum $S(f)$ is evidence of the complete elimination of all distortion signals by the antialiasing or zonal filter. With nonideal filtering, the signal spectrums will be corrupted and analysis, performance simulations, and testing must be performed to determine the impact on the system performance.

2.8.2 Sample Rate Conversion Filter Specifications

The interpretation and decimation filters, with frequency response $H(f)$, are typically specified in terms of various combinations of the following parameters: the passband and stopband frequencies f_{pb} and f_{sb}; the transition bandwidth, $f_T = f_{sb} - f_{pb}$; the passband and stopband† ripple, δ_{pb} and δ_{sb}; the maximum loss over the passband, δ_{pl} and the minimum stopband attenuation, A_{sb}. These parameters are depicted in Figure 2.44. The ripple specifications δ_{pb} and δ_{sb} are one-sided or symmetrical deviations about the filter amplitudes of 1 and 0, respectively.

These filters are typically implemented using transversal or FIR filters with $H(f)$ consisting only of *zeros*, giving rise to a linear phase response, with the filter complexity determined by the number of filter coefficients. The number of

†The stopband is also referred to as the rejection band.

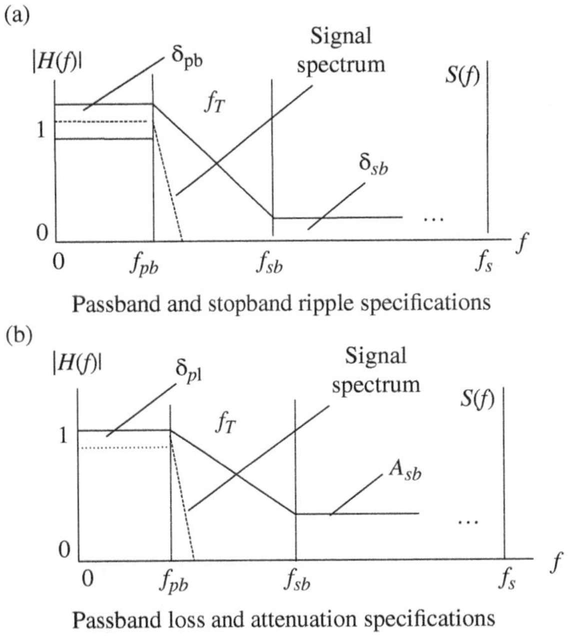

FIGURE 2.44 Lowpass interpolation filter characteristics.

coefficients, N_c, has been empirically characterized by various authors, for example,

$$N_c = \frac{2}{3(f_T/f_s)}\log_{10}\left[\frac{1}{10\delta_{pb}\delta_{sb}}\right] \quad (2.94)[29]$$

$$N_c = 1 + \frac{10\log_{10}(1/(\delta_{pb}\delta_{sb})) - 13}{14.6(f_T/f_s)} \quad (2.95)[30]$$

and

$$N_c = \frac{f_s}{f_T}\left(\frac{A_{sb}(\text{dB})}{22}\right) \quad (2.96)[31]$$

Considerable research has been devoted to designing filters with a flat passband, high stopband attenuation, low transitional bandwidth, and linear phase response. The FIR filters are generally used in these applications* and the design reduces to identifying the filter taps weights or coefficients. The filter design focuses on synthesizing a zonal or prototype filter by curve fitting passband and stopband samples with prescribed ripple specifications while using a minimum number of transition band samples. For example, the procedure of Herrmann and Schuessler [32] involves fixing N_c, δ_{pb}, and δ_{sb} and adjusting f_{pb} and f_{sb} while Parks and McClellan [33] fix N_c, f_{pb}, and f_{sb} and adjust δ_{pb} and δ_{sb}. Parks and McClellan use the iterative Remez algorithm to minimize the maximum absolute difference between the approximation samples and the underlying target or prototype function. These design procedures are reviewed and summarized by Rabiner, Gold, and McGonegal [34], Ifeachor and Jervis [35], Oppenheim and Schafer [36], and Fliege [24]. Another effective way to establish the transversal filter coefficients is through the use of windows to realize a desired filter response. The Kaiser window is used in the following case study to generate an interpolation filter with 60 dB of stopband attenuation.

Half-band FIR decimation filters [24, 25] are efficient implementations because they require fewer multiplies, for example, with N_c odd, aside from the center tap coefficient, the even coefficients are zero so the multiplications are reduced by nearly 2 : 1. Half-band filters exhibit symmetry about one-half the Nyquist frequency requiring: $\delta_{pb} = \delta_{sb}$; $f_{pb} = f_s/2 - f_{sb}$. The last condition also requires that the transition bandwidth is symmetrical about $f_s/2$ and these conditions taken together result in $H(f)$ being symmetrical about $f_s/4$.

2.8.3 Case Study: Sample Rate Conversion Filter Design

In this case study, the Kaiser window described in Section 1.11.8 is used to generate FIR filter coefficients that are used for an antialiasing interpolation filter. The transmitted signal modulation uses Nyquist root-raised-cosine (RRC) frequency shaping with 40% excess bandwidth.† The prototype filter spectrum is characterized as a Nyquist filter with zero excess bandwidth and frequency response given by

$$H(f) = \begin{cases} rect\left(\frac{f}{2B}\right) & : |f| \le B \\ 0 & : o.w. \end{cases} \quad (2.97)$$

Although this ideal prototype filter response is unrealizable, application of a window function results in a realizable and practical filter design.‡ The desired result is obtained by choosing a window with a suitable frequency response $W(f)$ and convolving it with the prototype filter response $H(f)$. The window is selected based on the design criteria given in Figure 2.44b and identified in Table 2.2. The transition bandwidth is not explicitly specified but is computed as $f_T = f_{sb} - f_{pb} = 72$ kHz.

The Kaiser window meets the requirements with a time-bandwidth product of $\beta = 2$. In this example, the desired transversal filter coefficients are determined by multiplying the frequency response of the prototype filter with the Kaiser window and then taking the inverse discrete Fourier

*Infinite impulse response (IIR) filters inherently result in a nonideal phase response that gives rise to signal distortion when the filter bandwidth is not sufficiently greater than the signal bandwidth.

†Spectral RRC-shaped-modulated waveforms are discussed in Section 4.3.2.

‡Causality issues are overcome by including a finite delay.

TABLE 2.2 Example Filter Requirements

Parameter	Specification	Units
Sampling frequency (f_s)	192.0	kHz
Symbol bandwidth (R_s)	48.0	kHz
Prototype filter bandwidth (B)	48.0	kHz
Passband frequency (f_{pb})	24.0	kHz
Stopband frequency (f_{sb})	96.0	kHz
Passband attenuation (δ_{pl})	≤1.0	dB
Stopband attenuation (A_{sb})	≥60.0	dB

transform to obtain the desired filter impulse response.* The impulse response of the prototype filter is the $sinc(2tB)$ function shown in Figure 2.45a over the duration of the window $|t| \leq T_w/2$ and the Kaiser window is shown in Figure 2.45b. Based on the definition of the time-bandwidth product $\beta = T_w B$, where B is the baseband bandwidth, the impulse response of the prototype filter has zero-crossing intervals of $1/(2B)$ seconds and there are β such intervals over one-half the window duration, that is, exactly two intervals in this example.

The frequency response of the interpolation filter is shown in Figure 2.46 using $N_c = 51$ coefficients. For this plot the frequency response is oversampled by 36 samples to improve the sidelobe resolution for viewing; the time responses in Figure 2.45 are also oversampled for the same reason. The sidelobe roll-off with frequency is virtually unchanged by using fewer samples. When N_c is reduced to 13 coefficients, the 60 dB transition bandwidth is increased by 8% and the passband loss is decreased by about 0.1 dB.

The prototype filter baseband bandwidth was selected to be two times the Nyquist bandwidth of $R_s/2$ so that $B = R_s$. The interpolation filter meets the following design requirements: minimum stopband attenuation of 60 dB at $f_{sb} = 96$ kHz; $\delta_{pl} = 0.88$ dB loss over Nyquist band of $R_s/2$. Figure 2.46 also shows the aliased signal folding into the signal spectrum with a maximum level of about −72 dB. The RRC detection-matched filter will eliminate the aliased signal in the transitions band. By choosing a Kaiser window with lower β, the aliased signal level at $R_s/2$ can be specified as −60 dB; this lower value of β will also result in a narrower transition bandwidth; however, the passband attenuation will increase requiring a larger prototype filter bandwidth. The interpolation filter design requirements are based on the system performance requirements that are verified by simulations. The filter design in this example will tolerate a high level of adjacent channel interfering signals relative to the desired channel signal [37].

*This process is the same as convolving the window spectrum with the prototype filter impulse response.

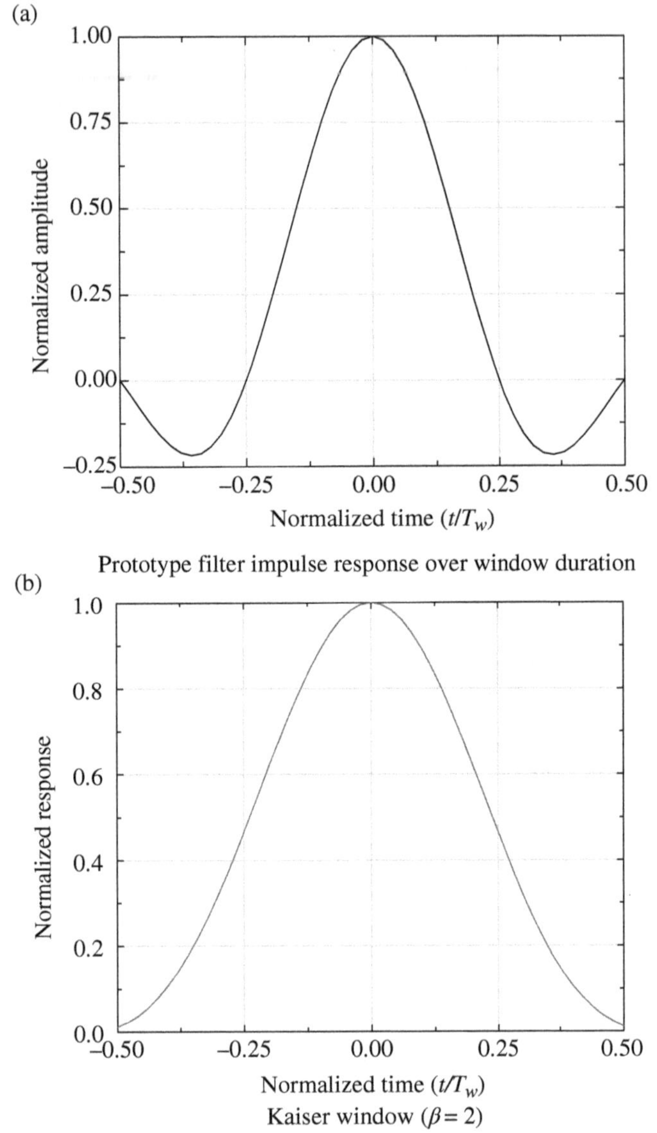

FIGURE 2.45 Prototype filter impulse response and Kaiser window.

2.8.4 Cascaded Integrate and Comb Filter

Cascaded integrate and comb (CIC) filters [38] are a class of FIR filters that provides large rate conversions in applications involving interpolation or decimation. The CIC filter requires very little memory and no coefficient multipliers are required. However, the limited control over the passband bandwidth and the stopband attenuation relegate CIC filters to large initial rate conversions with the final rate change being left to FIR implemented interpolation or decimation filters as discussed in the preceding sections. The filters are based on a fundamental building block, or stage, comprising a single integrator and a comb filter. Control over the stopband attenuation is provided by cascading M-stages denoted as

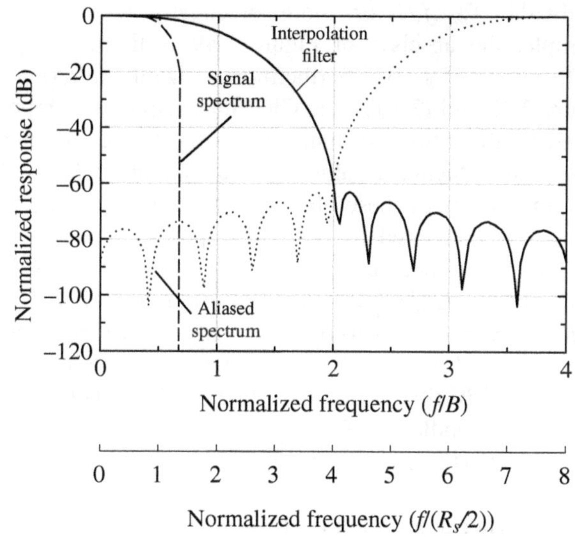

FIGURE 2.46 Interpolation filter and signal spectrums (Kaiser window with $\beta = 2$).

an M-th order CIC filter. In the following section, the analysis of CIC filter decimation is examined and its application as an interpolator is left as an exercise in Problem 19.

2.8.4.1 CIC Filter Decimation The basic building block stage of a CIC filter is characterized by considering the sampled sequence $x(n)$ of equally spaced sampled of $x(t)$ for $t = nT'_s$: $n = 0, \ldots, RN - 1$ with the filter output $y(m)$ formed by the summation

$$y(m) = \sum_{n=0}^{RN-1} x(n) \tag{2.98}$$

The product RN refers to the number of samples of the input $x(n)$ before rate conversion by R and the parameter N is the number of samples that are spanned by the comb filter; N is also referred to as the sample duration or differential delay of the comb filter. Typical values of N are 1 and 2. With, $z' = e^{sT'_s}$ the z-transform of (2.98) is a finite geometric series* in z'^{-1} and the CIC filter z-plane transfer function is evaluated as

$$H(z') = \frac{Y(z')}{X(z')} = \frac{1 - z'^{-RN}}{1 - z'^{-1}} \tag{2.99}$$

However, the interest is in the frequency response of the decimated output, so, using $z = e^{sTs}$ with $T_s = RT'_s$, the desired result is

*Reference: Section 1.14.1.

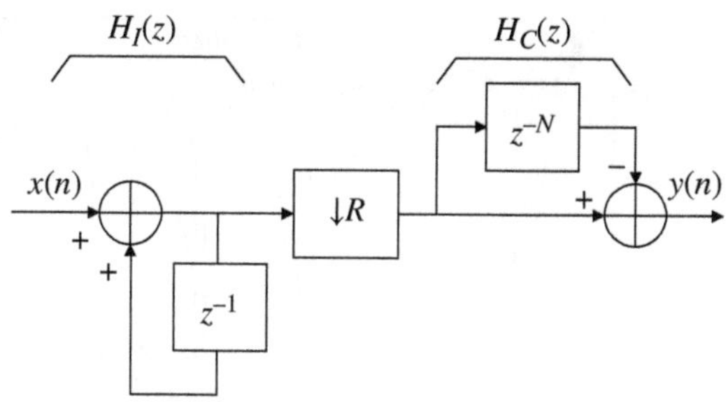

FIGURE 2.47 Single-stage decimation cascade integrate and comb filter.

$$H(z) = \frac{Y(z)}{X(z)} = \frac{1 - z^{-N}}{1 - z^{-1}} \tag{2.100}$$

Equation (2.100) is expressed as an ideal integrator $H_I(z)$ cascaded with an ideal comb filter $H_C(z)$ as shown in Figure 2.47 so the single-stage or first-order CIC filter is expressed as

$$H(z) = H_I(z) H_C(z) \tag{2.101}$$

In terms of the frequency response $H(f)$, (2.100) is evaluated by substituting $s = j2\pi f$ and using $z = e^{j2\pi f Ts}$ with the result

$$H(f) = \frac{\sin(\pi N f T_s)}{\sin(\pi f T_s / R)} e^{j\pi(N-1)fT_s} \tag{2.102}$$

For large rate conversions R such that $R \gg 1$ and $fT_s \ll R/2$, the magnitude of the frequency response is approximated as

$$\begin{aligned} |H(f)| &\cong RN \frac{\sin(\pi N f T_s)}{\pi N f T_s} \quad : R \gg 1 \text{ and } fT_s \ll \frac{R}{2} \\ &= RN\, sinc(NfT_s) \end{aligned} \tag{2.103}$$

Significant improvement in the aliasing rejection is achieved when the CIC filter response is raised to an integer power of M giving rise to an M-th order CIC filter expressed as

$$H(f)^M = \left[\frac{\sin(\pi N f T_s)}{\sin(\pi f T_s / R)}\right]^M e^{j\pi M(N-1/R)fT_s} \tag{2.104}$$

The magnitude of the M-th order CIC filter is shown in Figure 2.48 for $R = 64$, $N = 1$, and $M = 1, \ldots, 5$. The squared magnitude of the frequency spectrum is given by

$$|H(f)|^{2M} = \left[\frac{\sin(\pi N f T_s)}{\sin(\pi f T_s / R)}\right]^{2M} \tag{2.105}$$

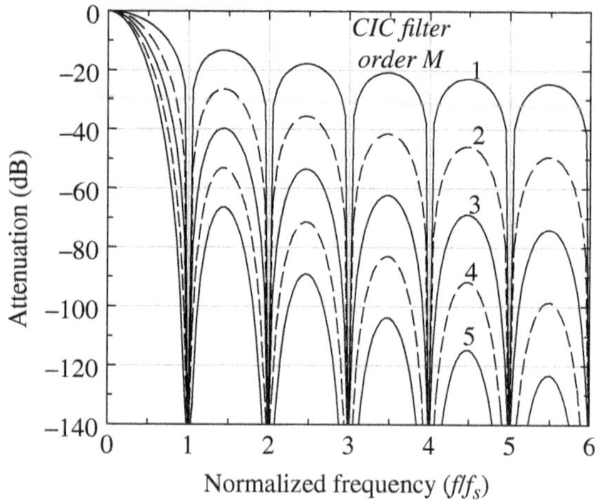

FIGURE 2.48 Frequency response of M-th order cascaded integrate and comb filter ($R = 64$, $N = 1$).

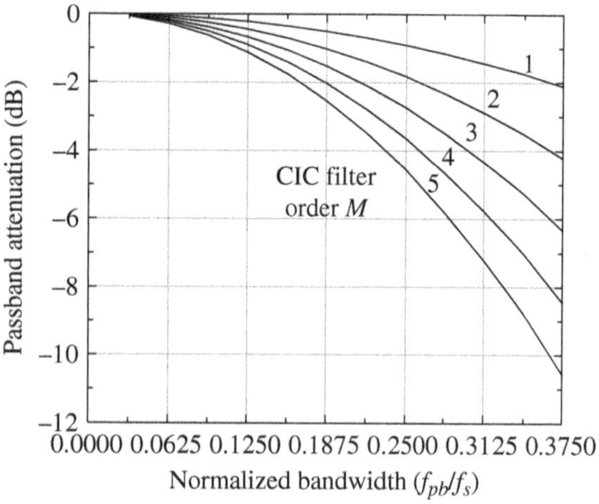

FIGURE 2.49 Passband attenuation of M-th order cascaded integrate and comb filter ($R = 64$, $N = 1$).

Referring to the approximation (2.103), the M-th order CIC filter attenuation rolls off with frequency as $1/f^{2M}$, so the first spectral sidelobe level of −13 dB for $M = 1$ corresponds −65 dB for $M = 5$. The negative aspect of the M-th order CIC filter is that the passband or band edge attenuation at f_{pb} increases significantly with M as shown in Figure 2.49. Because the passband roll off severely limits the signal bandwidth, a gain compensation filter is frequently used to compensate for the loss as discussed in Section 2.8.4.2.

An important observation in Figures 2.48 and 2.49 is that the sampling frequency is normalized by the decimated or down-sampled output frequency $f_s = f_s'/R$ where f_s' is the sampling frequency of the input signal. Therefore, $f/f_s = 1$ is the decimated normalized sampling frequency and, based on Nyquist's criterion, the signal must be confined to the bandwidth $f_{pb} \leq f_s/2$ or more practically $f_{pb} < f_s/2$.* For example, the abscissa of Figure 2.49 is limited to $f_{pb} = 0.375\ f_s = 0.75(f_s/2)$. Furthermore, again referring to Figure 2.48 and (2.103), the CIC filter response is approximated by the frequency function $sinc(Nf/f_s)$ for sufficiently large rate reductions,† albeit scaled in amplitude by RN. The importance of these observations is that the CIC filter aliasing is computed as the signal power about each of the $sinc(Nf/f_s)$ spectral nulls at $f_{null} = nf_s/N$: $1 \leq n \leq RN - 1$, such that, $|f - f_{null}| \leq f_{pb}$. In this description, $f \leq RNf_s/2$ and $f > RNf_s/2$ corresponds to positive and negative frequencies, respectively. The scale factor or power gain $(RN)^{2M}$ of the M-th order CIC filter is discussed in depth by Hogenauer [38] and can be handled by floating-point or two's-complement processing; in this analysis the filter responses are normalized for unit gain.

The earlier description of the alias signal folding is used to evaluate the alias signal level relative to the passband signal defined as

$$\rho_a \triangleq 10M\log_{10}\left(\frac{P_a}{P_s}\right) \tag{2.106}$$

where P_a is the alias power evaluated as

$$P_a = \sum_{n=1}^{RN-1}\left(\int_{-f_{pb}}^{f_{pb}} sinc^2(Nf/f_s - n)df\right) \tag{2.107}$$

and P_s is the signal power over the passband bandwidth evaluated as

$$P_s = \int_{-f_{pb}}^{f_{pb}} sinc^2(\pi Nf/f_s)df \tag{2.108}$$

The continuous integrals in (2.107) and (2.108) are evaluated as discrete summations with 20 samples per spectral sidelobe to provide visual fidelity for simulations and hardware evaluations. Equation (2.106) is plotted in Figure 2.50 for the indicated conditions and improvement in the aliased signal level is evident with increasing CIC filter order. On the other hand, as the passband bandwidth, f_{pb}, increases the alias level increases due to the increased power alias around each spectral null and because of the lower signal level due to the CIC filter passband attenuation.

*This condition applies for $N = 1$ and for $N > 1$ the Nyquist criterion is $f_{pb} < f_s/2N$.

†For rate reductions $R \geq 8$ the $sinc(NfT_s)$ approximation results in negligible error for $fT_s < 1$, that is, within the principal spectral lobe. All of the plots in this section are based on the exact response given by (2.105).

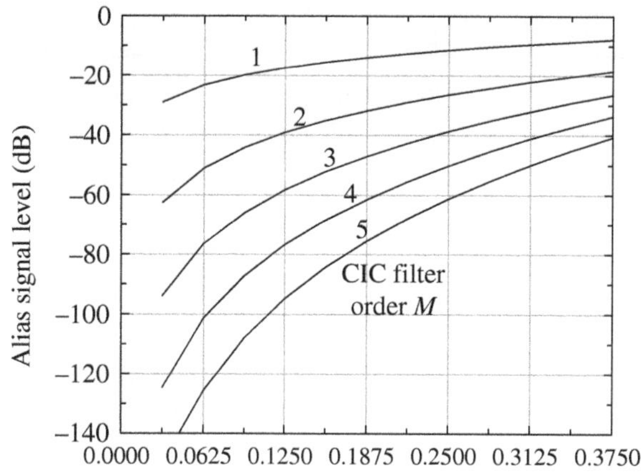

FIGURE 2.50 Alias signal level of M-th order cascaded integrate and comb filter ($R = 64$, $N = 1$).

2.8.4.2 *CIC Filter Passband Gain Compensation*

The CIC filter loss across the passband results in a demodulator performance loss due to signal distortion. The signal loss within the passband can be reduced by choosing the sampling frequency of the rate-reduced CIC filter output to be two to four times that required to satisfy the Nyquist criterion and then use FIR filter decimation to achieve the final sampling frequency; however, it is necessary to provide gain compensation for the $sinc(NfT_s)$* gain roll-off within the Nyquist band $f_N = f_s/2N = f_s'/2$ where f_s and f_s' are the input and rate-reduced sampling frequencies, respectively.

The CIC filter gain compensation function is designed to have nearly unit gain over the band $f < f_s'/2$. To this end, the inverse of the frequency *roll-off* expressed in (2.103) is approximated by the function

$$\begin{aligned}|H'(f)| &\triangleq \frac{\pi NfT_s}{\sin(\pi NfT_s)}\\ &\cong 1 + \frac{1}{6}(\pi NfT_s)^2 + \frac{7}{360}(\pi NfT_s)^4 + \frac{31}{15120}(\pi NfT_s)^6\\ &\quad + \frac{127}{604800}(\pi NfT_s)^8\end{aligned} \tag{2.109}$$

The approximation in (2.109) corresponds to a ninth-order polynomial approximation to the *sine* function and provides for a gain compensated loss of less than 0.1 dB over the Nyquist band $f_N = f_s'/2$. By way of comparison, a third-order polynomial approximation results in a comparable loss over the band $f_N/2$. To prevent the response in (2.109) from increasing without bound with increasing frequency a denominator term is included so the entire gain compensation function becomes

$$H''(fT_s') = \frac{\bar{H}'(fT_s')}{G(fT_s')} \tag{2.110}$$

where $\bar{H}'(fT_s')$ is the approximation in (2.109) normalized by the down-sampled sampling interval $T_s' = NT_s$ and the function $G(fT_s')$ is defined as

$$G(fT_s') \triangleq 1 + b(\pi fT_s')^k \tag{2.111}$$

The parameter $k > k_o$, where k_o is the order of the *sine* approximation in (2.109), ensures that the gain of the compensated response decreases with increasing frequency. Selecting $k = k_o + 1$ minimizes the impact of $G(fT_s')$ on the compensated gain in the Nyquist band. The parameter b is chosen so that (2.111) results in a specified maximum compensated gain loss at f_N; in the following analysis a maximum compensated gain loss of 0.1 dB is specified at f_N. The overall compensated response of the M-th order CIC decimation filter is expressed as

$$|H(f)|^M_{comp} = |H(f)|^M |H''(f)|^M \tag{2.112}$$

The compensated gain losses at f_N and the parameters k and b are listed in Table 2.3 for CIC orders $M \le 5$ and the normalized frequency responses given by (2.103) and (2.112) for $M = 3$ and 5 are shown in Figures 2.51 and 2.52, respectively.

TABLE 2.3 CIC Filter Gain Compensation Parameters and Maximum Nyquist Band Loss

Order (M)	k	b	Loss at f_N (dB)[a]
1	9	1.5e−4	−0.090
2	9	6.0e−3	−0.089
3	9	3.6e−5	−0.098
4	9	2.0e−5	−0.098
5	10	6.0e−4	−0.095

[a]Maximum specified loss = 0.1 dB.

2.8.5 Polyphase-Matched Filter Interpolation for Symbol Timing Control

The notion of a polyphase filter response was introduced in Section 2.8.1.2 in terms of the down-sampled sequence $y_k(k + \ell k_2)$ expressed in (2.92). In this section, however, the samples $y(n)$ are considered to be the finite-length unit-pulse response samples $h(n)$ of a filter characterized by $H(z)$. As an example of the application, the filter impulse response is considered to be matched to the received signal and the optimally sampled output is used to determine the received

*The $sinc(NfT_s)$ function is an approximation for large rate reductions as expressed in (2.103).

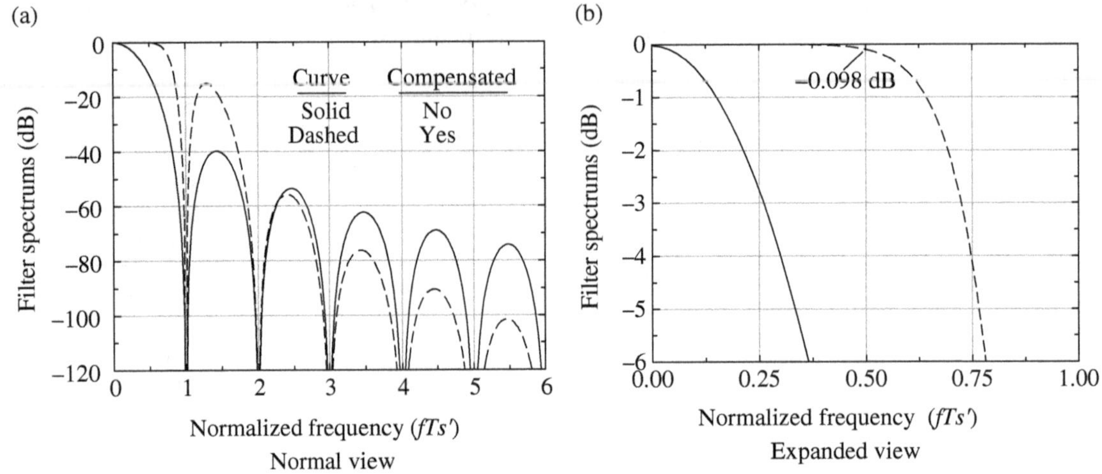

FIGURE 2.51 Gain compensated third-order CIC filter.

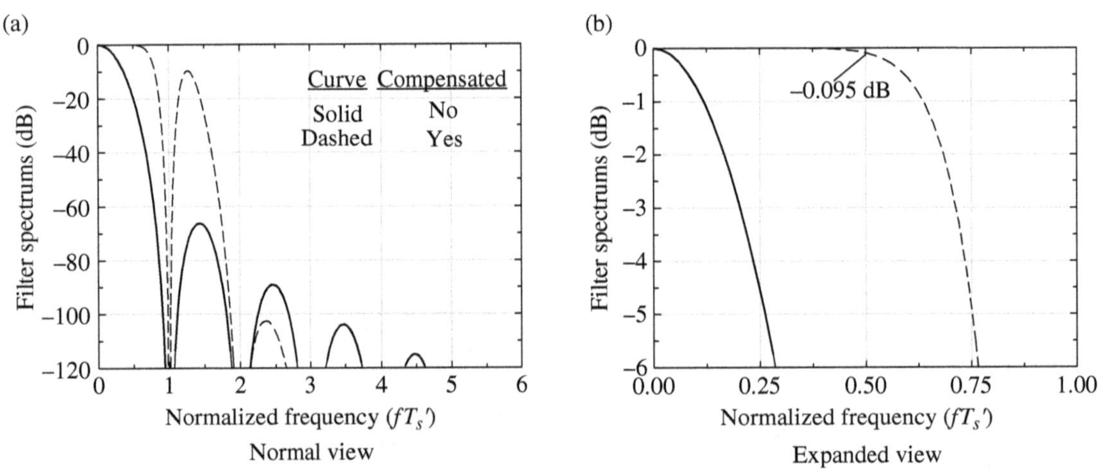

FIGURE 2.52 Gain compensated fifth-order CIC filter.

data estimate. This application applies for band-limited waveforms so spectral root-raised-cosine (SRRC) modulation is used with 40% excess bandwidth. The number of matched filter samples is considered to be N_s and the filter impulse response is oversampled by N_s' samples per symbol. The objective of this example is to correct a symbol timing error *not* by adjusting the underlying sampling time but rather by altering the matched filter coefficients to match those of the received symbol given the timing error. The altering of the matched filter coefficients corresponds to selecting a polyphase filter response as discussed in the remainder of this section.

In the context of the earlier description, the sequence $h(n)$ is down-sampled by $k_2 = N_s'$ so that the samples 0, k_2, $2k_2$, …, ℓk_2, … form the $k = 0$-th polyphase down-sampled sequence $h_0(\ell k_2)$: $\ell = 0, 1, \ldots, \ell_m - 1$ and, in general, the samples k, $k + k_2$, $k + 2k_2$, …, $k + \ell k_2$: $0 \le k \le k_2 - 1$, form the k-th polyphase down-sampled sequence $h_k(k + \ell k_2)$: $\ell = 0, 1, \ldots, \ell_m - 1$. The parameter ℓ_m is related to the maximum number of matched filter samples spanned by the impulse response $h(n)$. The k_2 uniquely sampled sequences represent the polyphase filter impulse responses $h_k(k + \ell k_2)$ as characterized in Figure 2.53.

The following example is specialized for $N_s = 4$ matched filter samples per symbol with an underlying waveform oversampling of $N_s' = 36$ samples per symbol; this corresponds to a symbol timing resolution of 1/36 symbol. Although not obvious, these are the conditions of the 40% SRRC-matched filter response shown in Figure 2.54. When this response is correlated with the identical SRRC received waveform the matched filter output response is that of the *sinc*(t/T) response with zero correlation between adjacent symbols when optimally sampled at the symbol intervals of $t = iT$. Appling these conditions to the earlier polyphase filter responses, shown in

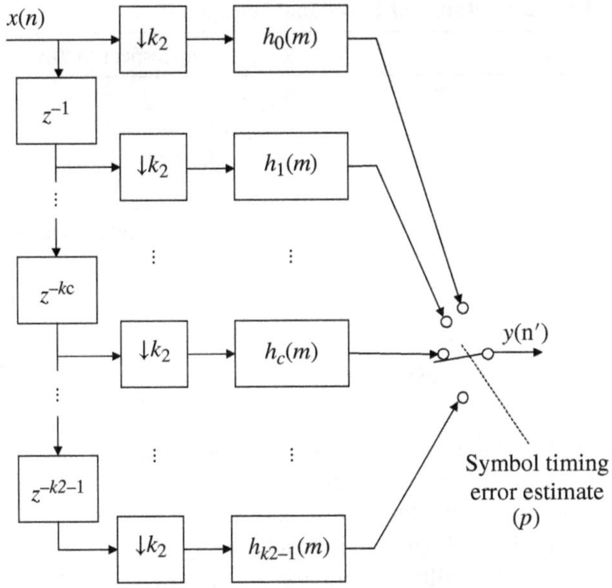

FIGURE 2.53 Polyphase decimation filters.

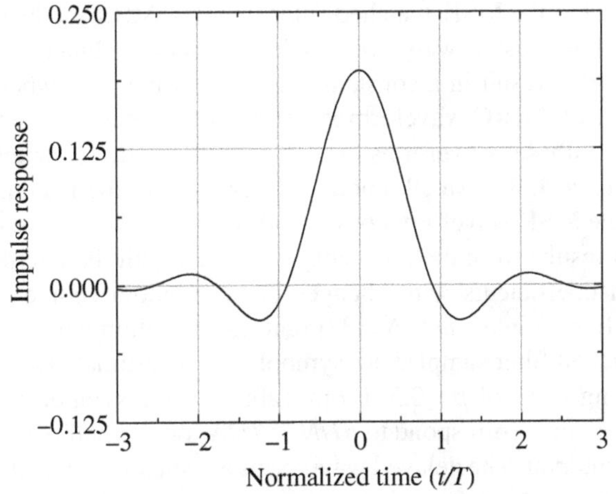

FIGURE 2.54 40% SRRC oversampled-matched filter response ($N_s' = 36$ samples/symbol spanning 6 symbols).

Figure 2.54, there are a total of $N_s'/N_s = 9$ polyphase filters each having $N_s = 4$ matched filter samples per symbol. Choosing N_s'/N_s to be a odd integer is convenient because the center polyphase filter with delay z^{-kc}: $k_c = \lfloor N_s'/2N_s \rfloor + 1 = 5$ can be selected to correspond to the polyphase filter response with zero symbol timing error. Responses for delays $<k_c$ correspond to a timing advance and responses for delays $>k_c$ correspond to a timing delay. Actually, the range of the polyphase timing adjustment is $\pm T/2N_s$ about each matched filter sample so, if k_c does not result in zero symbol timing error, adjustments by T/N_s must be made to bring the timing error into alignment with the received symbols. However,

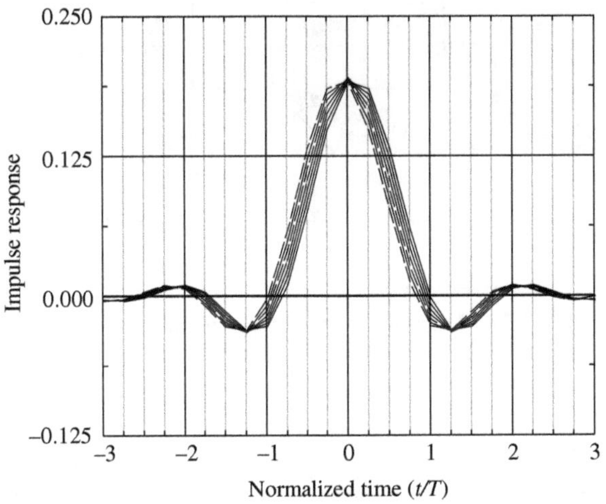

FIGURE 2.55 Nine 40% SRRC polyphase-matched filter responses ($N_s = 4$ samples/symbol spanning 6 symbols).

during symbol acquisition the timing is frequently estimated to be within $\pm T/2N_s$ of the optimum alignment. The sign of the timing error estimate p, indicated in Figure 2.54 and discussed in Section 2.8.7, controls the direction of the symbol timing correction and the magnitude of p can be used for course timing adjustments of T/N_s.

The nine polyphase filter responses depicted in Figure 2.55 represent the responses about the optimum filter response corresponding to zero timing error. The application of the polyphase filter responses to adjust symbol timing requires considerable memory for the filter coefficients. The following section provides an alternative solution requiring less memory.

2.8.6 Lagrange Interpolation

In this section the Lagrange interpolation [39, 40] formulas are characterized in terms of a specified, or estimated, symbol delay error and the resulting interpolation coefficients are modified to achieve a symbol timing adjustment of the matched filter samples to compensate for the error. This approach is distinctly different from the polyphase approach to symbol timing adjustment discussed in the preceding section. Although there are considerably fewer coefficients using Lagrange interpolation the coefficients must be recomputed for each timing adjustment.

The Lagrange interpolation formulas and coefficients are based on those given by Abramowitz and Stegun [41]; however, the natural numbering notation … x_{k-1}, x_k, x_{k+1} … given in the reference is reversed to reflect the sequential time sampling given by t_{k+1}, t_k, x_{k-1}. The coefficient formulas for N-point Lagrange interpolation with, equally spaced sampling increments T_s, are expressed as

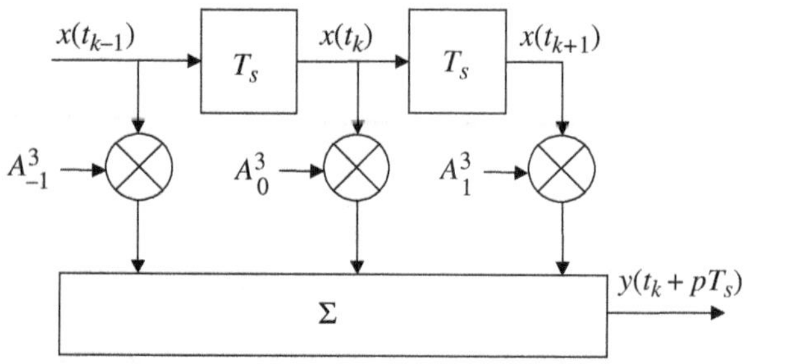

FIGURE 2.56 Lagrange interpolator implementation of fractional sample delay (example for $N = 3$).

TABLE 2.4 ***N*-Point Lagrange Interpolation Coefficients**[a]

N	2	3	4	5
A^N_{-2}				$\frac{(p^2-1)p(p-2)}{24}$
A^N_{-1}		$\frac{p(p-1)}{2}$	$\frac{-p(p-1)(p-2)}{6}$	$\frac{-p(p-1)(p^2-4)}{6}$
A^N_0	$1-p$	$1-p^2$	$\frac{(p^2-1)(p-2)}{2}$	$\frac{(p^2-1)(p^2-4)}{4}$
A^N_1	p	$\frac{p(p+1)}{2}$	$\frac{-p(p+1)(p-2)}{2}$	$\frac{-p(p+1)(p^2-4)}{6}$
A^N_2			$\frac{p(p^2-1)}{6}$	$\frac{(p^2-1)p(p+2)}{24}$

[a]Abramowitz and Stegun [41]. Courtesy of U.S. Department of Commerce.

$$y(t_k + pT_s) = \sum_i A^N_i(p)x(t_{k-i}) + R_{N-1} \tag{2.113}$$

where the index i is over the range

$$-(N-2)/2 \le i \le N/2 \quad :N\text{even} \tag{2.114}$$

and

$$-(N-2)/2 \le i \le (N-1)/2 \quad :N\text{odd} \tag{2.115}$$

The parameter p determines the fractional delay and the coefficient $A^N_i(p)$ and the remainder, or error, R_{N-1} is given in the reference and are functions involving i, p, and even or odd values of N. Factional timing adjustments based on the samples $x(t_{k-i})$ are implemented for a three-point interpolation example as shown in Figure 2.56.

Abramowitz and Stegun provide interpolation coefficients for two- through eight-point Lagrange interpolation and the coefficients for two- through five-point interpolations are listed in Table 2.4. The range of the fractional delay and the corresponding interpolation error are also dependent on N as tabulated in Table 2.5. The interpolation remainder or error is given by

TABLE 2.5 Range of Fractional Interpolation[a]

N	Range with Respect to T/N_s
2	$0 \le p \le 1$
3	$-1 \le p \le 1$
4	$-1 < p < 0,\ 0 < p < 1,\ 1 < p < 2$
5	$-2 < p < -1,\ -1 < p < 1,\ 1 < p < 2$

[a]Abramowitz and Stegun [41]. Courtesy of U.S. Department of Commerce.

$$R_{N-1}(p) = C_N(p)T_s^N x^{(N)} \tag{2.116}$$

where $C_N(p)$ is a constant over the various ranges of p and $x^{(N)}$ is the N-th derivative of $x(t)$ in the range of t over the interpolation interval; additional details are given in Ref. [41].

Examples of Lagrange interpolation symbol timing delay at the matched filter output for an minimum shift keying (MSK) modulated waveform are shown in Figure 2.57 and the 40% SRRC BPSK-modulated waveform is shown in Figure 2.58. In both examples the underlying binary source data is random. The *circled* data points in Figure 2.57 correspond to the MSK-matched filter samples. MSK modulation involves cosine weighted quadrature data-modulated symbols that result in a constant amplitude waveform, whereas, the 40% SRRC waveform is not a constant envelope waveform; these waveforms are discussed in more detail in Chapter 4. The small amount of variation in the magnitude of the MSK waveform corresponds to the delayed waveform and results from the transients associated with the interpolation coefficients with changes in the random source data. Both examples use $N = 3$ point interpolation and $N_s = 4$ matched filter samples per symbol with an estimated symbol timing error of $p = 0.5$. From Table 2.5 the symbol timing corrections correspond to $pT/N_s = T/2N_s$ or one-half of a sample interval. The delayed waveforms are shown as the *dashed* curves and $p > 0$ corresponds to a symbol timing delay. Application of the Lagrange interpolator to provide symbol timing adjustments is more efficient than the polyphase filter approach because of the lower data storage requirement and the capability to provide continuous timing adjustments over the range $-1 \le k \le 1$.

2.8.6.1 The Farrow Filter Farrow [42] refers to the Lagrange interpolator shown in Figure 2.56 as a continuously variable digital delay (CVDD) that eliminates the explicit computations of the FIR filter coefficients, A^N_i, for each value of p. This is accomplished by expressing the coefficients in Table 2.4 as an $N - 1$ degree polynomial in p with coefficients C_i being functions of the sampled inputs $x(t_{k+j})$. The Farrow implementation is shown in Figure 2.59 for the $N = 3$-point interpolator. The Farrow implementation requires several additions and fixed multiplications with a net computational

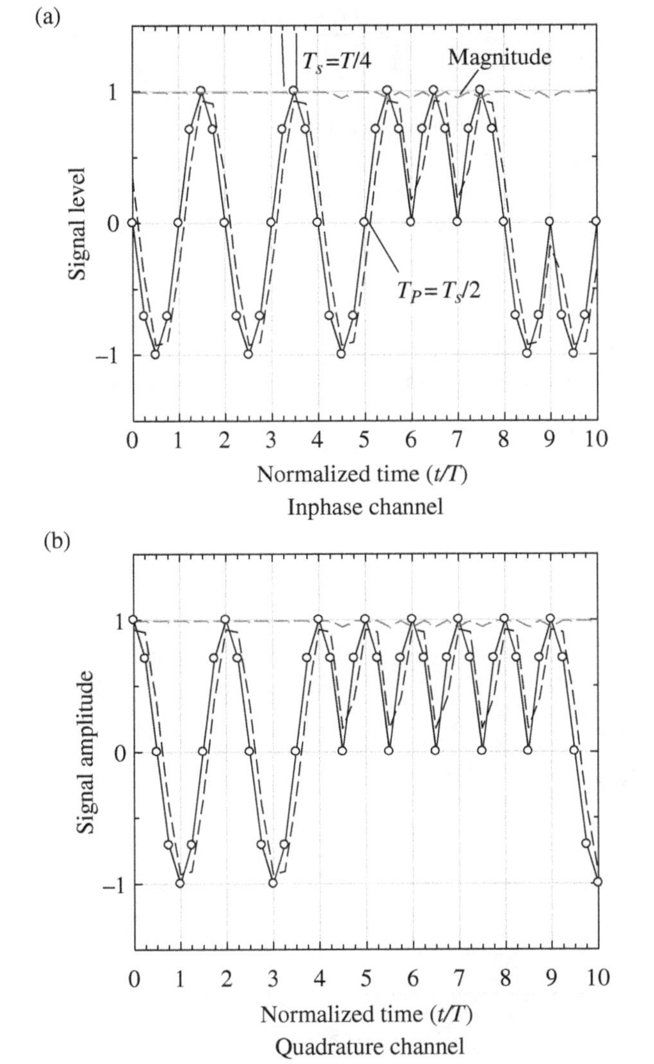

FIGURE 2.57 Lagrange three-point MSK symbol sample interpolator ($N_s = 4$, $p = 0.5$, curves: *solid*—input, *dashed*—output).

savings. For example, in the 3-point interpolator, shown in Figure 2.59, the multiplications, except for those involving the generation of the error *p*, simply involve binary shifts and sign changes.

2.8.7 Symbol Time and Frequency Error Estimation

Estimation of the symbol timing error parameter is necessary to align the demodulator time base with that of the received symbol timing. During acquisition of the received signal, coarse symbol time estimates are made and, during data detection, fine symbol time corrections are made for symbol time tracking. The estimation of the correct symbol timing during acquisition typically is within ±T/4 and subsequent symbol tracking will bring the receiver time base into the correct alignment with the received symbol.

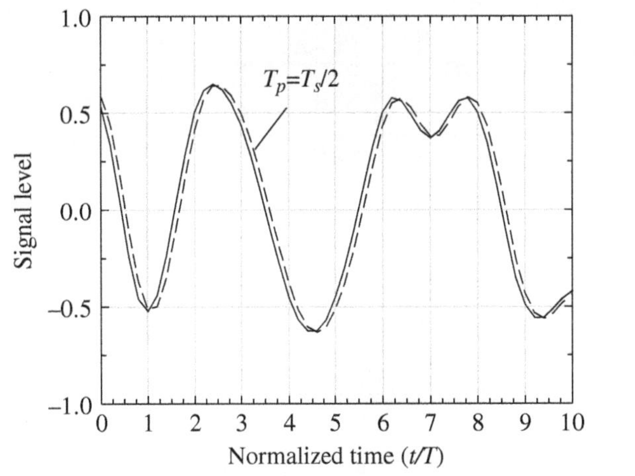

FIGURE 2.58 Lagrange three-point BPSK 40% SRRC symbol sample interpolator ($N_s = 4$, $p = 0.5$, curves: *solid*—input, *dashed*—output).

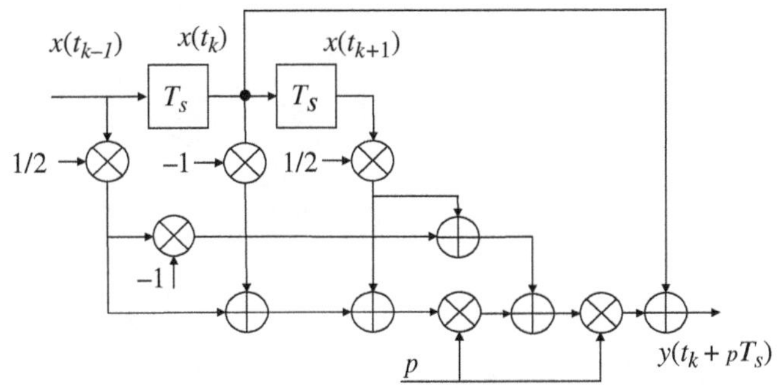

FIGURE 2.59 Farrow implementation of three-point Lagrange interpolator.

Generally, estimation of the parameter *p* involves comparing the signal energy of an *early* integrator output with that of a *late* integrator output relative to the demodulator time base; this process is referred to as early–late (E/L) gate processing.* The simplest application of the E/L gate processing for symbol time estimation involves processing the received signal $s_r(t)$ with a *rect*(*t*/*T*) weighting as shown in Figure 2.60. In this example, the early and late integrator outputs are:

$$I_e = A\frac{T}{2} \quad \text{and} \quad I_\ell = A\left(\frac{T}{2} - \tau\right) \tag{2.117}$$

Defining $I_m = AT$ as the maximum integrator output with ideal symbol timing, that is, when $\tau = 0$, the timing error is defined and evaluated as

*Early/late gate processing is also used for frequency and antenna acquisition and tracking.

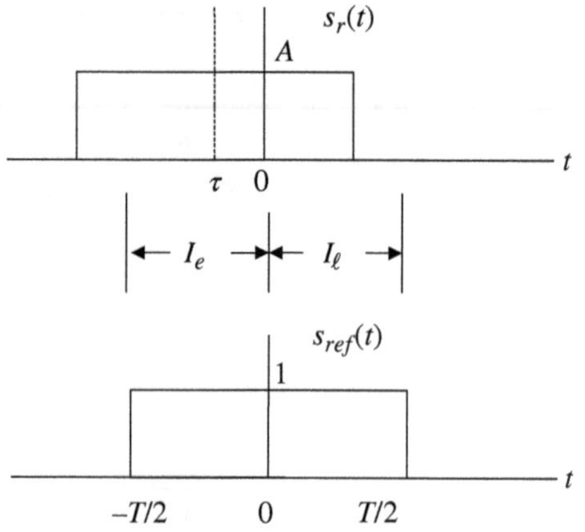

FIGURE 2.60 Early–late gate processing for *rect(t/T)* symbol weighting.

$$p \triangleq \left\langle \frac{I_e - I_\ell}{I_m} \right\rangle = \frac{\tau}{T} \tag{2.118}$$

The notation < > denotes a time average that is required to eliminate the influence of the adjacent random data symbols and the additive noise that are not included in (2.117) or shown in Figure 2.60. Equation (2.118) forms a discriminator response centered about zero when the early and late gate responses are equal. In this regard, the timing is adjusted in the correct direction to result in zero error; this is fundamental to early–late gate processing. The timing error is used in a feedback timing loop to provide adaptive timing adjustments.

The following example uses a parabolic curve to estimate the frequency using three frequency domain samples as shown in Figure 2.61a. In this example, a 100% zero

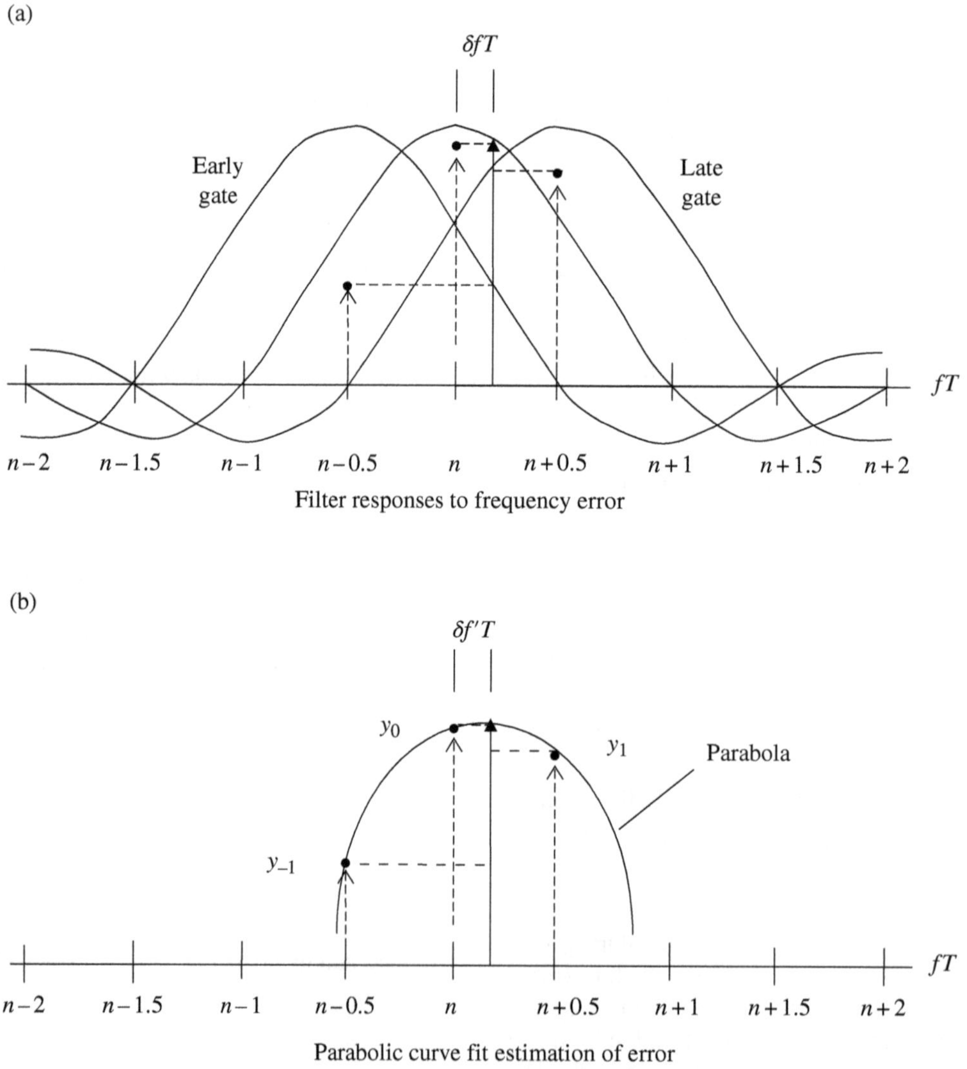

FIGURE 2.61 Early–late gate processing for 100% zero padded FFT with uniform data weighting.

padded FFT* is used to determine the frequency of a received CW signal. The FFT window is uniformly weighted so the frequency response of each cell is expressed as *sinc(fT)*. The received frequency is shown to be at $f_r = n/T + \delta f$: $|\delta f| \le 1/2T$ and the maximum output occurs in cell n as indicated by the *filled-circle* data point.† The corresponding responses of the adjacent E/L gate cells are also indicated by *filled-circle* data points.

Figure 2.61b shows the application of a parabola that is curve fit to the three data points with the peak of the curve-fit parabola corresponding to the estimate $\delta f'$ of the frequency error.

Using the filter responses (y_{-1}, y_0, y_1), indicated in Figure 2.61b, it is shown in Appendix 2A that the frequency estimate is given by the discriminator function

$$|\delta f'| \le \frac{\Delta f}{2}\left(\frac{y_1 - y_{-1}}{2y_o - (y_1 + y_{-1})}\right) \tag{2.119}$$

where $\Delta f = 1/2T$ corresponds to the 100% zero padded FFT filter frequency spacing.‡ In this case, the normalized delay estimation parameter is $p = \delta f' T$. Because the parabolic interpolator estimates the frequency error including the sign, when the frequency is adjusted using a feedback tracking loop the sign of the frequency error must be reversed to result in a zero steady-state tracking error.

APPENDIX 2A AMPLITUDE QUANTIZATION FUNCTION SUBPROGRAM

The function subprogram described in this appendix is a fast amplitude quantization function of a time-sampled analog function $f(t)$. The amplitude quantization is represented by the number of magnitude bits (n) and excludes the sign. The assignment of the integer part of the magnitude suggests that a decimal point (.) appears after the integer bits (m) with the magnitude having the form *m.n-m*. Therefore, the selection of m determines the placement of the decimal point. This function subprogram provides a mechanism for mapping the dynamic range throughout the digital signal processor simulation to minimize the loss of information contained in $x = f(t_s)$. The parameters of the fast amplitude quantization subprogram are identified and described in Table 2A.1.

TABLE 2A.1 Fast Amplitude Quantization Function Subprogram Parameters

Parameter	Input/Output	Description
x	Input	Analog sample
m	Input	Number of bits representing the integer portion of the input magnitude sample
n	Input	Number of bits representing the input sample magnitude
type	Input	= 0 No quantization = 1 Truncated under flow = 2 Round under flow
fold	Input	= 0 Saturates on overflow = 1 Folds magnitude: modulo(max magnitude)
quant	Output	Quantized output representation of *x*

The function subprogram *quant* is coded as follows.

```
function quant(x, m, n, type, fold)
! x     is the input to be quantized
! n     is the number of magnitude bits
        not including sign bit.
! m     is the number of bits representing
        the integer portion of
!       of the magnitude saturates at xm=
        2^m.
! n-m   is the fractional portion to the
        magnitude bits.
! type  =0 no quantization  : quant=x
!       =1 truncates underflow
!       =2 rounds underflow

! fold =0 saturates
!       =1 folds modulo(max magnitude)
   implicit none
   real,      intent(in)  :: x
   integer, intent(in)  :: m, n, type, fold
   real                 :: quant
   real                 :: y, xm, xnp, xp
   if(type = = 0) then
       quant=x              ! no quantization
                              is applied
   else
       xm=2.**m
       snp=2.**(n- m)
       y=abs(x)
       if(y>=xm) then
```

*Expanding the FFT window using 100% zero padding results in interpolated FFT cells between each of the unpadded cells. Although the frequency resolution is increased from *n/T* to *n/2T*, the bandwidth of each FFT cell is unchanged.

†Additive noise may result in the maximum occurring in an adjacent cell; however, the parabolic estimation procedure is still applicable.

‡If the FFT does not include zero padding the filter spacing is $\Delta f = 1/T$.

```
!         overflow processing
            if(fold = = 1) then
              y=mod(y, xm)
            else
              y=xm
            endif
          endif
!       underflow processing
          xpp=y*snp
          xp=int(xpp)
          if(type = = 2) then ! round if
                                fractional
                                part>= 0.5
            if(xpp - xp>= 0.5) xp=xp+1
          endif
          xp=xp/xm              ! quantized
                                magnitude |x|
          quant=sign(xp, x)!  include sign
                                and return
      endif
    end function quant
```

APPENDIX 2B HILBERT TRANSFORM PARAMETERS

The following tables of Hilbert transform coefficients are selected from the extensive lists of Hilbert transform coefficients tabulated by Rabiner and Schafer [19]. To approximate the ideal Hilbert transform without phase distortion it in necessary to use a finite impulse response (FIR) filter that is characterized by the filter coefficients over the normalized frequency band $0 \le f/f_s \le 1$, where f_s is the selected sampling frequency of the system under consideration. However, the FIR filter impulse response coefficient, $h(n)$, optimization that results in the minimum peak distortion error, denoted as D, is applied over a restricted normalized frequency band: $F_L \le f/f_s \le F_H$ with $F_H \le f_N/f_s$ where $f_N = f_s/2$ is the Nyquist band resulting in $F_H \le 0.5$. So, F_L and F_H are the normalized lower and upper cutoff frequencies over the optimized bandwidth of the filter. With an odd number of coefficients (N) the upper and lower transition bands are defined as F_L and $0.5 - F_H$.

Example frequency response magnitudes for even and odd values of FIR filter coefficients are shown in Figure 2B.1. With even values of N the response at $F = 0.5$ is not zero so, for the same values of F_H, the aliased distortion is somewhat more severe than with odd values of N. Furthermore, odd values of N result in the FIR filter delay corresponding to an integer number of samples, whereas, the delay with even values of N result in the output the delay corresponding to an additional one-half sample interval. The responses in the figures are shown to be flat with equal peak distortion over the usable frequency range. This condition occurs with equal lower and

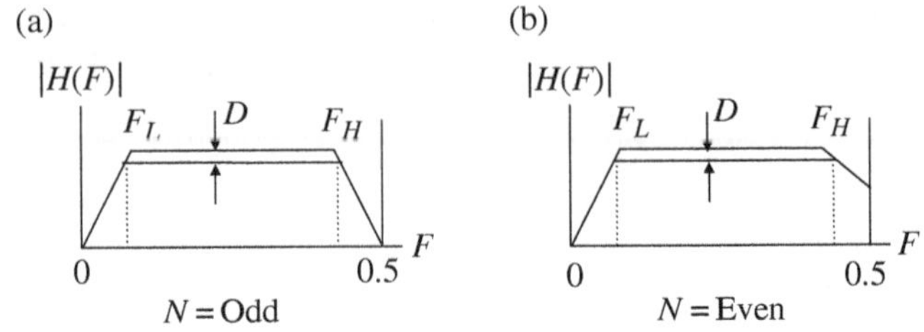

FIGURE 2B.1 Example of Hilbert transformer magnitude responses.

upper transition bands, that is, when $F_L = 0.5 - F_H$. When the transition bands are unequal, the responses will exhibit an undesirable peak at the lower or upper frequency ranges. For these reasons, the following tables identify Hilbert filters corresponding to an odd number of coefficients, a lower frequency range of F_L, and equal transition bands.

Rabiner and Schafer [43] conclude that there are substantial processing advantages, in terms of the multiplications per sample, in using a Hilbert transformer with the largest possible transition bandwidth, F_L, the largest peak distortion error, D, compatible with the system application, and the shortest odd length, N, impulse response. The best approximation to D is based on computer simulations [16] that result in equal ripple over the band F_L to F_H with the smallest number of coefficients. Since only the even numbered coefficients are required for an odd length impulse response, the required number of coefficient multiplications is one-half that required for a comparable even length impulse response.

The following tables of filter coefficients cover a portion of the Hilbert transformers given by Rabiner and Schafer and correspond exclusively to odd values of N. However, the tables include the coefficients for the maximum value of N corresponding to a specified F_L and generally include those for the minimum value of N with several designs for intermediate values of N. The maximum in-band distortion error, expressed in decibels, is given by $20\log_{10}(D)$ and the impact of the designs that are not included in the following tables can be assessed by the peak distortion error D between those designs that are listed. For example, the distortion error for $F_L = 0.05$ and $N = 31$ is $D = 2.68(-3)$ or -51.4 dB. The next higher value of N given in the tables is $N = 39$ with $D = 6.79(-4)$ or -63.4 dB but Rabiner and Schafer also list the coefficients for $N = 35$ with $D = 1.349(-3)$ or -57.4 dB. The parameter D is the ratio of the in-band gain ripple relative to the unit amplitude response and the corresponding peak ripple, expressed in decibels, is defined as

$$R_p \triangleq 20\log_{10}(1 + D) \tag{2B.1}$$

Kaiser [17] expresses the approximate number of filter coefficients, in terms of $\delta = D$ and the transition parameter $\Delta f = F_L = 0.5 - F_H$, as

$$N \cong \frac{-0.61\log_{10}(\delta)}{\Delta f} \quad (2B.2)$$

Figure 2B.2 is a plot of the in-band peak ripple R_p in decibels as a function of the number of filter coefficients for the values of the transition parameter F_L. The dependence on N of these results for the peak ripple is based on Kaiser's approximation as expressed in (2B.2).

The tables of Rabiner and Schafer include a portion of the coefficients of the Hilbert filter starting at $n=0$ from which the entire set of coefficient values is computed as

$$h(N-1-n) = -h(n) \quad : \begin{cases} n=0,\ 2,\ 4,\ \ldots,\ (N-3)/2 & : N = \text{odd} \\ n=0,\ 1,\ 2,\ \ldots,\ (N-2)/2 & : N = \text{even} \end{cases} \quad (2B.3)$$

The following tables correspond to odd values of N and the filter impulse response has odd symmetry about the sample at $(N-1)/2$ that corresponds to a zero sample value.

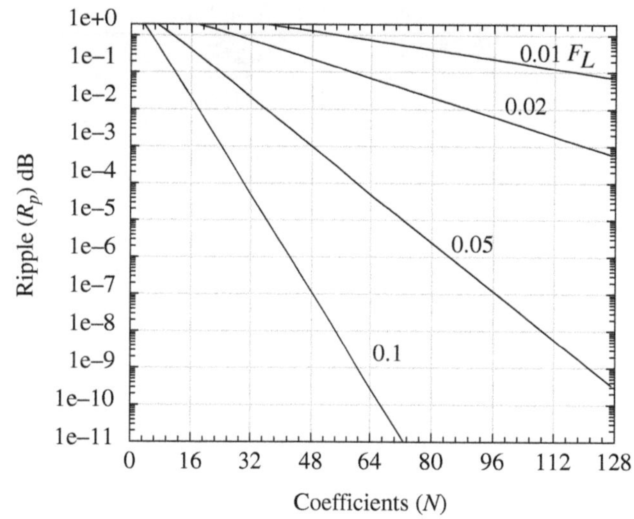

FIGURE 2B.2 Approximate in-band peak ripple dependence on the number of Hilbert filter coefficients ($\delta = D$, $\Delta f = F_L = 0.5 - F_H$).

TABLE 2B.1 Wideband Hilbert Transform Coefficients for $F_L = 0.1$[a]

	$N = 15$	$N = 23$	$N = 31$	$N = 39$
n	$D = 2.546(-3)$	$D = 1.420(-4)$	$D = 8.0(-6)$	$D = 1.0(-6)$
0	−1.25869(−2)	−1.5643(−3)	−2.098(−4)	−2.92(−5)
2	−5.17464(−2)	−8.2383(−3)	−1.3764(−3)	−2.304(−4)
4	−1.563345(−1)	−2.69557(−2)	−5.2972(−3)	−1.0348(−3)
6	−6.159002(−1)	−7.02312(−2)	−1.53339(−2)	−3.4149(−3)
8		−1.722057(−1)	−3.71398(−2)	−9.1741(−3)
10		−6.221851(−1)	−8.14346(−2)	−2.12775(−2)
12			−1.810706(−1)	−4.45889(−2)
14			−6.255683(−1)	−8.89475(−2)
16				−1.867128(−1)
18				−6.276691(−1)

[a]Rabiner and Schafer [19]. Reproduced by permission of Alcatel-Lucent USA Inc.

TABLE 2B.2 Wideband Hilbert Transformers for $F_L = 0.05$[a]

	$N = 23$	$N = 31$	$N = 39$	$N = 47$
n	$D = 1.1071(-2)$	$D = 2.68(-3)$	$D = 6.79(-4)$	$D = 1.73(-4)$
0	−1.44218(−2)	−4.1956(−3)	−1.2787(−3)	−3.957(−4)
2	−2.72241(−2)	−9.2821(−3)	−3.2636(−3)	−1.1534(−3)
4	−5.25858(−2)	−1.88358(−2)	−7.1031(−3)	−2.7042(−3)
6	−9.71984(−2)	−3.44010(−2)	−1.35513(−2)	−5.4585(−3)

(*continued overleaf*)

TABLE 2B.2 *(continued)*

	$N=23$	$N=31$	$N=39$	$N=47$
n	$D=1.1071(-2)$	$D=2.68(-3)$	$D=6.79(-4)$	$D=1.73(-4)$
8	−1.929460(−1)	−5.95516(−2)	−2.37704(−2)	−9.9769(−3)
10	−6.299931(−1)	−1.030376(−1)	−3.95684(−2)	−1.70000(−2)
12		−1.968315(−1)	−6.44154(−2)	−2.75785(−2)
14		−6.313536(−1)	−1.070280(−1)	−4.34280(−2)
16			−1.994533(−1)	−6.79619(−2)
18			−6.322687(−1)	−1.098913(−1)
20				−2.013159(−1)
22				−6.329151(−1)

	$N=55$	$N=63$	$N=71$	$N=79$
n	$D=4.5(-5)$	$D=1.2(-5)$	$D=3.0(-6)$	$D=1.0(-6)$
0	−1.243(−4)	−3.96(−5)	−1.27(−5)	−4.1(−6)
2	−4.080(−4)	−1.447(−4)	−5.12(−5)	−1.79(−5)
4	−1.0288(−3)	−3.910(−4)	−1.476(−4)	−5.50(−5)
6	−2.2014(−3)	−8.853(−4)	−3.531(−4)	−1.389(−4)
8	−4.2215(−3)	−1.7821(−3)	−7.456(−4)	−3.074(−4)
10	−7.4746(−3)	−3.2906(−3)	−1.4366(−3)	−6.182(−4)
12	−1.24599(−2)	−5.6838(−3)	−2.5777(−3)	−1.1532(−3)
14	−1.98467(−2)	−9.3111(−3)	−4.3663(−3)	−2.0239(−3)
16	−3.06162(−2)	−1.46260(−2)	−7.0545(−3)	−3.3761(−3)
18	−4.64284(−2)	−2.22481(−2)	−1.09630(−2)	−5.3956(−3)
20	−7.06666(−2)	−3.31126(−2)	−1.65113(−2)	−8.3167(−3)
22	−1.120458(−1)	−4.88448(−2)	−2.42838(−2)	−1.24372(−2)
24	−2.027050(−1)	−7.28120(−2)	−3.51847(−2)	−1.81511(−2)
26	−6.333949(−1)	−1.137361(−1)	−5.08176(−2)	−2.60178(−2)
28		−2.037871(−1)	−7.45415(−2)	−3.69200(−2)
30		−6.337675(−1)	−1.150864(−1)	−5.24475(−2)
32			−2.046464(−1)	−7.59556(−2)
34			−6.340625(−1)	−1.161821(−1)
36				−2.053402(−1)
38				−6.343000(−1)

[a]Rabiner and Schafer [19]. Reproduced by permission of Alcatel-Lucent USA Inc.

TABLE 2B.3 Wideband Hilbert Transformers for $F_L = 0.02$[a]

	$N=31$	$N=39$	$N=47$	$N=55$
n	$D=7.50950(-2)$	$D=4.13090(-2)$	$D=2.31120(-2)$	$D=1.30540(-2)$
0	−5.1061(−2)	−2.86598(−2)	−1.64094 (−2)	−9.5117(−3)
2	−3.15637(−2)	−1.87406(−2)	−1.14516(−2)	−7.1168(−3)
4	−4.25110(−2)	−2.50998(−2)	−1.54645(−2)	−9.7423(−3)
6	−5.77251(−2)	−3.32813(−2)	−2.04740(−2)	−1.29935(−2)
8	−8.05394(−2)	−4.41044(−2)	−2.67802(−2)	−1.70107(−2)
10	−1.196632(−1)	−5.91324(−2)	−3.48570(−2)	−2.19890(−2)
12	−2.075778(−1)	−8.16868(−2)	−4.55267(−2)	−2.82298(−2)
14	−6.350674(−1)	−1.206017(−1)	−6.03428(−2)	−3.61989(−2)
16		−2.081174(−1)	−8.26818(−2)	−4.67260(−2)

TABLE 2B.3 *(continued)*

n	$N=31$ $D=7.50950(-2)$	$N=39$ $D=4.13090(-2)$	$N=47$ $D=2.31120(-2)$	$N=55$ $D=1.30540(-2)$
18		−6.352463(−1)	−1.213191(−1)	−6.13745(−2)
20			−2.085553(−1)	−8.35102(−2)
22			−6.353986(−1)	−1.219297(−1)
24				−2.089363(−1)
26				−6.355268(−1)

n	$N=63$ $D=7.4430(-3)$	$N=75$ $D=3.2330(-3)$	$N=87$ $D=1.4170(-3)$	$N=95$ $D=8.240(-4)$
0	−5.5706(−3)	−2.5265(−3)	−1.1577(−3)	−6.935(−4)
2	−4.4618(−3)	−2.2373(−3)	−1.1264(−3)	−7.156(−4)
4	−6.2078(−3)	−3.1981(−3)	−1.6565(−3)	−1.0724(−3)
6	−8.3775(−3)	−4.4073(−3)	−2.3363(−3)	−1.5362(−3)
8	−1.10475(−2)	−5.9065(−3)	−3.1921(−3)	−2.1265(−3)
10	−1.43195(−2)	−7.7446(−3)	−4.2528(−3)	−2.8655(−3)
12	−1.83266(−2)	−9.9820(−3)	−5.5519(−3)	−3.7775(−3)
14	−2.32716(−2)	−1.26922(−2)	−7.1273(−3)	−4.8903(−3)
16	−2.94397(−2)	−1.59751(−2)	−9.0270(−3)	−6.2348(−3)
18	−3.73177(−2)	−1.99615(−2)	−1.13048(−2)	−7.8492(−3)
20	−4.77262(−2)	−2.48494(−2)	−1.40366(−2)	−9.7764(−3)
22	−6.22273(−2)	−3.09326(−2)	−1.73154(−2)	−1.20734(−2)
24	−8.41979(−2)	−3.86871(−2)	−2.12753(−2)	−1.48104(−2)
26	−1.224340(−1)	−4.89418(−2)	−2.61107(−2)	−1.80826(−2)
28	−2.092445(−1)	−6.32654(−2)	−3.21163(−2)	−2.20234(−2)
30	−6.356280(−1)	−8.50334(−2)	−3.97698(−2)	−2.68256(−2)
32		−1.230441(−1)	−4.99010(−2)	−3.27847(−2)
34		−2.096158(−1)	−6.40805(−2)	−4.03791(−2)
36		−6.357524(−1)	−8.56865(−2)	−5.04390(−2)
38			−1.235215(−1)	−6.45363(−2)
40			−2.099064(−1)	−8.60506(−2)
42			−6.358499(−1)	−1.237874(−1)
44				−2.100684(−1)
46				−6.359043(−1)

[a]Rabiner and Schafer [19]. Reproduced by permission of Alcatel-Lucent USA Inc.

TABLE 2B.4 Wideband Hilbert Transformers for $F_L = 0.01$[a]

n	$N=55$ $D=9.54980(-2)$	$N=63$ $D=7.04360(-2)$	$N=71$ $D=5.22480(-2)$	$N=79$ $D=3.88830(-2)$
0	−5.55675(−2)	−4.11397(−2)	−3.06629(−2)	−2.29388(−2)
2	−1.67258(−2)	−1.26851(−2)	−9.7354(−3)	−7.5151(−3)
4	−1.94725(−2)	−1.47109(−2)	−1.13179(−2)	−8.7784(−3)
6	−2.26148(−2)	−1.71191(−2)	−1.31328(−2)	−1.01565(−2)
8	−2.64560(−2)	−1.97933(−2)	−1.51642(−2)	−1.17808(−2)
10	−3.10395(−2)	−2.29817(−2)	−1.75162(−2)	−1.35612(−2)
12	−3.66783(−2)	−2.67924(−2)	−2.02240(−2)	−1.55902(−2)
14	−4.39134(−2)	−3.13469(−2)	−2.33958(−2)	−1.79182(−2)

(continued overleaf)

TABLE 2B.4 *(continued)*

	$N=55$	$N=63$	$N=71$	$N=79$
n	$D=9.54980(-2)$	$D=7.04360(-2)$	$D=5.22480(-2)$	$D=3.88830(-2)$
16	−5.35532(−2)	−3.69689(−2)	−2.71460(−2)	−2.06260(−2)
18	−6.71782(−2)	−4.41620(−2)	−3.16681(−2)	−2.37742(−2)
20	−8.81528(−2)	−5.37578(−2)	−3.72851(−2)	−2.74953(−2)
22	−1.253143(−1)	−6.73427(−2)	−4.44587(−2)	−3.19865(−2)
24	−2.109951(−1)	−8.82996(−2)	−5.40183(−2)	−3.75627(−2)
26	−6.362148(−1)	−1.254324(−1)	−6.75540(−2)	−4.47012(−2)
28		−2.110732(−1)	−8.84541(−2)	−5.42333(−2)
30		−6.362430(−1)	−1.255376(−1)	−6.77331(−2)
32			−2.111303(−1)	−8.85965(−2)
34			−6.362585(−1)	−1.256401(−1)
36				−2.111964(−1)
38				−6.362830(−1)

	$N=87$	$N=95$
n	$D=2.90440(-2)$	$D=2.17910(-2)$
0	−1.72371(−2)	−1.30099(−2)
2	−5.8530(−3)	−4.5718(−3)
4	−6.8525(−3)	−5.3689(−3)
6	−7.9733(−3)	−6.2800(−3)
8	−9.2241(−3)	−7.2616(−3)
10	−1.06323(−2)	−8.3873(−3)
12	−1.22073(−2)	−9.6455(−3)
14	−1.39949(−2)	−1.10350(−2)
16	−1.60152(−2)	−1.26022(−2)
18	−1.83265(−2)	−1.43770(−2)
20	−2.10006(−2)	−1.63895(−2)
22	−2.41199(−2)	−1.86922(−2)
24	−2.78194(−2)	−2.13465(−2)
26	−3.22920(−2)	−2.44424(−2)
28	−3.78402(−2)	−2.81175(−2)
30	−4.49441(−2)	−3.25665(−2)
32	−5.44359(−2)	−3.80864(−2)
34	−6.79017(−2)	−4.51608(−2)
36	−8.87289(−2)	−5.46233(−2)
38	−1.257349(−1)	−6.80547(−2)
40	−2.112513(−1)	−8.88468(−2)
42	−6.363005(−1)	−1.258168(−1)
44		−2.112989(−1)
46		−6.363167(−1)

[a]Rabiner and Schafer [19]. Reproduced by permission of Alcatel-Lucent USA Inc.

APPENDIX 2C DERIVATION OF PARABOLIC INTERPOLATION ERROR

The parabolic interpolation algorithm is based on the equation for a parabola given by

$$(x-\bar{x})^2 = 2a(y-\bar{y}) \qquad (2C.1)$$

where a is the distance from the vertex at $(\bar{x},\bar{y})$ to the focus of the parabola. Increasing the magnitude of a increases the latus rectum or spread of the parabola and $a>0$ results in a concave upward parabola and $a<0$ results in a concave downward parabola; the case for $a<0$ is used for parabolic interpolation. For the purpose of interpolation, the data points (y_{-1}, y_0, y_1) are associated with the ordinate values of the received signal $sinc(x)$ spectral function* with the corresponding abscissa values (x_{-1}, x_0, x_1), that is,

$$y_i = \frac{\sin(\pi x_i)}{\pi x_i} \quad : i = -1,\ 0,\ 1 \tag{2C.2}$$

With frequency interpolation of the received modulated symbols, the normalized frequency variable is $x_i = f_i T$ where T is the received symbol duration and $1/T$ corresponds to the frequency interval between the zeros of the $sinc(x)$ function.

Considering the three sample pairs (x_{-1}, y_{-1}), (x_0, y_0), and (x_1, y_1), a set of three simultaneous equations are solved for the three unknowns parameters $(\bar{x}, \bar{y}, a)$. To simplify the evaluation, without any loss in meaning, the parameter x_0 is set to zero and the desired normalized frequency interpolation is computed as $\bar{x} = p = \delta f T$.

The three simultaneous equations corresponding to (2C.1) are

$$\begin{aligned} (x_{-1}-\bar{x})^2 &= 2a(y_{-1}-\bar{y}) \\ (x_0-\bar{x})^2 &= 2a(y_0-\bar{y}) \\ (x_1-\bar{x})^2 &= 2a(y_1-\bar{y}) \end{aligned} \tag{2C.3}$$

Upon expanding the squared terms, with the simplifying assumption $x_0 = 0$, (2C.3) becomes

$$\begin{aligned} x_{-1}^2 - 2x_{-1}\bar{x} + \bar{x}^2 &= 2a(y_{-1}-\bar{y}) \\ \bar{x}^2 &= 2a(y_0-\bar{y}) \\ x_1^2 - 2x_1\bar{x} + \bar{x}^2 &= 2a(y_1-\bar{y}) \end{aligned} \tag{2C.4}$$

Substituting $\bar{x}^2$ from the second equation into the first and third equations and collecting the terms involving $2a$ gives

$$\begin{aligned} x_{-1}^2 - 2x_{-1}\bar{x} &= 2a(y_{-1}-y_0) \\ x_1^2 - 2x_1\bar{x} &= 2a(y_1-y_0) \end{aligned} \tag{2C.5}$$

This step has eliminated the unknown parameter $\bar{y}$ and solving for $2a$ in the first equation in (2C.5) and substituting the result into the second equation and solving for $\bar{x}$ yields

$$\bar{x} = \frac{1}{2}\frac{x_{-1}^2(y_1-y_0) - x_1^2(y_{-1}-y_0)}{x_{-1}(y_1-y_0) - x_1(y_{-1}-y_0)} \tag{2C.6}$$

*This description of the parabolic interpolator is specialized to the example involving the $sinc(x)$ frequency spectrum. The concept can be applied to a variety of spectral functions or, in the case of parabolic symbol timing interpolation, to symbol weighting functions. However, to be effective, the functions must generate a useful interpolation gain or slope at the zero-crossing of the interpolation S-curve.

With the exception of the simplifying assumption that $x_0 = 0$, (2C.6) is a general solution to the parabolic interpolator that can be specialized for the 100% zero padded FFT by recognizing that $x_{-1} = x_0 - 1/2$ and $x_1 = x_0 + 1/2$ or, with the simplifying condition, $x_{-1} = -1/2$ and $x_1 = 1/2$. Therefore, using these conditions, (2C.6) becomes

$$\bar{x} = \frac{1}{4}\frac{(y_1 - y_{-1})}{(2y_0 - (y_1 + y_{-1}))} \quad : 100\% \text{ zero padded FFT} \tag{2C.7}$$

In terms for the un-normalized frequency estimate $|\delta f'| \le \bar{x}/T$, (2C.7) becomes

$$|\delta f'| \le \frac{1}{2}\frac{(y_1 - y_{-1})}{(2y_0 - (y_1 + y_{-1}))}\Delta f \quad : 100\% \text{ zero padded FFT} \tag{2C.8}$$

where $\Delta f = 1/2T$ is the frequency resolution of the 100% zero padded FFT.

ACRONYMS

ADC	Analog-to-digital converter
AGC	Automatic gain control
ASCII	American Standard Code for Information Interchange
AWGN	Additive white Gaussian noise
BPSK	Binary phase shift keying
CIC	Cascaded integrate and comb (filter)
CVDD	Continuously variable digital delay
CW	Continuous wave (frequency)
DAC	Digital-to-analog converter (or conversion)
DC	Direct current
DSP	Digital signal processing
DSSS	Direct-sequence spread-spectrum (waveform)
E/L	Early–late (gate sampling)
FDM	Frequency division multiplex
FDMA	Frequency division multiple access
FFT	Fast Fourier transform
FIR	Finite impulse response (filter)
GCD	Greatest common divisor
I&D	Integrate-and-dump (filter)
I/O	Input and output (interface)
I/Q	Inphase and quadrature (channels or rails)
IF	Intermediate frequency
ISI	Intersymbol interference
LO	Local oscillator
LPF	Low-pass filter
LSB	Least significant bit
MSB	Most significant bit
MSK	Minimum shift keying
PLL	Phaselock loop
PRN	Pseudo-random number (sequence)

QPSK	Quadrature phase shift keying
rms	Root-mean-square
RRC	Root raised-cosine (temporal)
SRRC	Spectral root-raised-cosine
UHF	Ultra-high frequency

PROBLEMS

1. Consider that a noise-free received CW carrier signal with a phase error of ϕ degrees relative the receiver local oscillator frequency is mixed to baseband and that two uniform quantizing 8-bit ADCs, that is, $N_b = 8$ bits, are used to sample the in-phase and quadrature phase rails. Considering that the rails are within the linear range of the ADC, determine the maximum phase error at the ADC output when no jitter signal or rounding is used. Repeat this problem using $N_b = 10$ and 12 bit ADCs.

2. Starting with the convolution integral expressed in (2.21), derive the expression for the sampled signal spectrum given in (2.22). Hint: Note that $s(t)$ is a function of t and is not dependent on the summation over n and that $F_s(f)$ is the Fourier transform involving the summation in (2.20). Show and explain all of the details in the derivation.

3. Sketch the sampled spectrums using subcarrier bandpass sampling for $f_{if} = 70$ MHz and $W = 36$ MHz. Identify the upper and lower sampling frequencies, in MHz, relative to the IF filter centered at 70 MHz. Include in the sketch the first sampled spectrum and the sampled spectrum just below the 70 MHz filter for the upper and lower sampling frequencies.

4. Substituting the real parameter ρ into (2.33) for the integer m, show that the maximum value of $\rho = \rho_{max}$ corresponds to $\rho_{max} = f_{if}/W - 0.5$. State why it follows that the integer $M = \text{floor}(\rho_{max})$, where M is defined as $max(m)$.

5. Using ρ_{max} determined in Problem 4, it follows that the sampling frequency is lower bounded by $f_s \geq f_{if}/W - 0.5$. Show that this condition also corresponds to $f_s \geq f_{max}/W$ where f_{max} is the maximum signal frequency.

6. Show that $f_s = f_{smin} = 2W$ when $M = f_{if}/W - 1/2$.

7. Repeat the subcarrier sampling example in Section 2.6.1.1 for the following conditions $f_{if} = 70$ MHz and $W = 18$ and 36 MHz.

8. Derive the expression for the carrier leakage out of an imbalanced modulator with ideal rail and phase balance corresponding to $\beta_c = \beta_s = 1$, and $\phi = 0$, with DC offsets of α_c and α_s. Express the result in the form $M\cos(\omega_c t + \psi)$ and expresses the magnitude, M, and phase, ψ, as functions of the DC offsets α_c and α_s.

9. Using Figure 2.29 derive the output of the imbalanced modulator with precorrection as expressed by (2.52).

10. Derive equation (2.54) for the phase imbalance in terms of the parameters ρ using the geometry of the two equilateral parallelograms in Figure 2.31.

11. Derive the expression for the modulator phase measurement sensitivity given by (2.56) in terms of the parameters ρ and $\Delta\rho$.

12. Using (2.68) and (2.69) express the magnitude M_o for the demodulator phase imbalance measurement given by (2.70) in terms of $\omega_d t$, θ, and φ. Evaluate the following:

    Determine the general conditions of the argument $\omega_d t + \theta$ corresponding to the maximum and minimum magnitudes and enter the angles of the argument corresponding to quadrants I, II, III, and IV in the following table.

    Using $\varphi = 10°$, evaluate the parameter ρ given by (2.71) using $\omega_d t + \theta$ found earlier and enter the results in the following table.

    Determine the parameter κ as discussed in the text and enter the value in the table corresponding to the quadrants.

    Compute the measured phase imbalance using (2.72) and enter the results in the following table.

    Discuss the consequences of changing the sign of φ.
    Measurement of Phase Imbalance $\hat{\varphi}$ for $\varphi = 10°$.

Quadrant	$\omega_d t + \theta$	ρ	κ	$\hat{\varphi}$
I				
II				
III				
IV				

13. Using algebra and trigonometry show that $\hat{\varphi} = 2\tan^{-1}\left(\frac{\rho-1}{\rho+1}\right)$ when $\rho = \frac{\sqrt{1+\sin\hat{\varphi}}}{\sqrt{1-\sin\hat{\varphi}}}$.

14. Figure 2.34 is shown for a positive phase imbalance, $\varphi > 0$, and corresponds to the *solid* curve in Figure 2.35; redraw Figure 2.34 corresponding for $\varphi < 0$.

15. Using (2.52) as the noise-free receiver input, reformulate the demodulated quadrature data estimates $\hat{m}_c(t)$ and $\hat{m}_s(t)$ in terms of the up-converter imbalance compensated parameters. Hint: Substitute the distortion corrected baseband transmitter output expressed in (2.52) into (2.74) using $\bar{S}_R = \bar{\bar{S}}_T$; this corresponds to an ideal noise-free channel.

16. Referring to the example in Section 2.8.1, analyze the sample rate change from the 48 kHz rate during

acquisition to the 19.2 kHz rate for the matched filter detection. Specify the requirements for the image rejection and antialiasing filters by specifying their 3 dB and transition bandwidths.

17. Referring to the case study in Section 2.8.3, derive the expression of the impulse response of the prototype filter given by (2.97) in terms of the low-pass bandwidth B. Then show that the Kaiser window time-bandwidth product parameter $\beta = T_w B$ corresponds to the number impulse response zero-crossing intervals in one-half of the window duration T_w.

18. Using the half-band antialiasing filter response shown in the following figure, select f_{pb} and f_T and determine the number of FIR coefficients using (2.96) with $A_{sb} = 30$ dB for the following implementation.

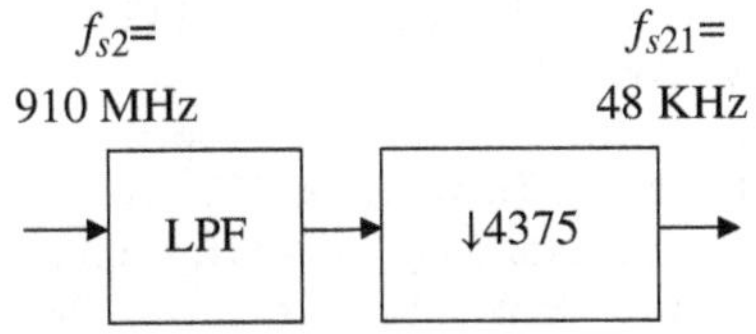

Repeat the earlier analysis using the three-stage FIR filter decimation shown in Figure 2.39a and using the CIC filter with the two-stage FIR filter decimation shown in Figure 2.39b. For these three implementations complete the following table.

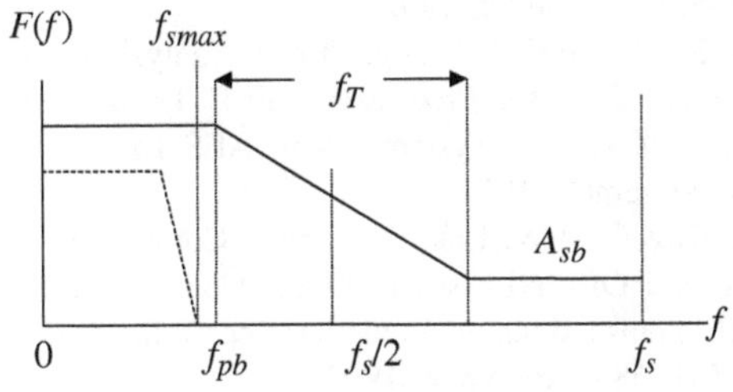

Half-band anti aliasing filter.

19. Referring to the CIC decimation diagram in Figure 2.47, sketch a similar implementation diagram for a rate R single-stage CIC *interpolation* filter with N samples per filter. Derive the z-plane transfer function $H(z) = Y(z)/X(z)$ for $z = e^{sT_s}$ where $T_s = 1/f_s$ and f_s is the output sampling frequency. Also, derive the magnitude of the frequency response $|H(f)|$.

20. Without using the approximations leading to (2.103), determine the maximum magnitude of (2.102) and identify the frequencies at which they occur.

21. Starting with $H(z)$ as given in (2.100) derive $H(f)$ given by (2.102) and the approximation to $|H(f)|$ given in (2.103).

22. Using (2.104) show that the maximum value of $|H(f)|^M$ is $(RN)^M$ and identify the location(s) of the maximum values. Do not use any approximations on R and fT_s as used in arriving at (2.103). Hint: Use L'Hospital's rule for indeterminate forms.

23. Show that the Nyquist criterion for $N > 1$ requires that the signal spectrum be confined to $f_{pb} < f_s/2N$.

24. Sketch the unit pulse response of the single-stage decimation filter shown in Figure 2.47.

25. Redraw the single-stage or first-order decimation filter shown in Figure 2.47 to represent an M-th order CIC decimation filter. Show the first two states with *ellipses* (…) to the M-th stage.

26. Derive the approximation to the gain compensation function $|H'(f)|M$ given in (2.109). Hint: Start with the approximation $\sin(\phi) \cong 1 - \phi^3/3!$.

27. Derive the equation for the $N = 3$ point Lagrange interpolator using the Farrow implementation shown in Figure 2.59.

28. Derive the equation for the $N = 4$ point Lagrange interpolator using the Farrow implementation and sketch the multiplier and adder connections similar to that shown in Figure 2.59.

Table Design and Performance Parameters for Various Decimation Configurations

	Stage 1			Stage 2			Stage 3			Total	
Configuration	f_{pb}	f_T	N_c	f_{pb}	f_T	N_c	f_{pb}	f_T	N_c	Multiplies	Memory
1				—	—	—	—	—	—		
2											
3											
	CIC Filter										

29. Referring to Appendix 2C, evaluate the expression for the parabolic interpolated frequency estimate for an FFT that does not use zero padding interpolation.

30. Derive the linear interpolation for a 100% zero padded FFT. Hint: Referring to Figure 2.61, use the sampled values $y_{-1} = sinc(\pi(x_o - x_{-1} + \bar{x}))$ and $y_1 = sinc(\pi(x_o + x_1 - \bar{x}))$ and, letting $x_o = 0$, solve for $\bar{x}$. The solution involves expressing the numerator of the $sinc(-)$ function in terms of sine and cosine functions and combining y_{-1} and y_1 using the surviving cosine function. Then convert to the frequency $|\delta f'| \leq 2\Delta f \bar{x}$ and express the result in terms of y_1, y_{-1}, and Δf where Δf is the frequency resolution of the 100% zero padded FFT.

31. Repeat Problem 30 for an FFT that does not use zero padding.

REFERENCES

1. E.C. Ifeachor, E.W. Jervis, *Digital Signal Processing: A Practical Approach*, Prentice Hall, Upper Saddle River, NJ, 2002.
2. T.A.C.M. Claasen, W.F.G. Mecklenbrauker, J.B.H. Peek, N. Van Hurck, "Signal Processing Method for Improving the Dynamic Range of A/D and D/A Converters," IEEE Transactions on Acoustics, Speech and Signal Processing, Vol. 28, No. 5, pp. 529–538, May 1980.
3. R.W. Adams, "Design and Implementation of an Audio 18-bit Analog-to-Digital Converter using Overlapping Techniques," Journal of the Audio Engineering Society, Vol. 34, No. 3, pp. 153–166, March 1986.
4. Y. Matsuya, K. Uchimura, A. Iwata, T. Kobayashi, M. Ishikawa, T. Yoshitome, "A 16-bit Over Sampling A-to-D Conversion Technology using Triple Integration Noise Shaping," IEEE Journal on Solid State Circuits, Vol. 22, No. 6, pp. 921–928, June 1987.
5. D.R. Welland, B.P. Del Signore, E.J. Swanson, T. Tanake, K. Hamashita, S. Hara, K. Takasuka, "A Stereo 16-bit Delta Sigma A/D Converter for Digital Audio," Journal of the Audio Engineering Society, Vol. 37, No. 6, pp. 476–486, June 1989.
6. B.P. Agrawal, K. Shenoi, "Digital Methodology for $\Sigma\Delta M$," IEEE Transactions on Communications, Vol. 31, No. 3, pp. 360–370, March 1983.
7. B. Sklar, *Digital Communications Fundamentals and Applications*, Chapter 11, F.J. Harris, Editor "Source Coding," Prentice-Hall, Inc., Englewood Cliffs, NJ, 1988.
8. H. Nyquist, "Certain Topics in Telegraph Transmission Theory," Transactions of the AIEE, Vol. 47, pp. 617–644, April 1928. Reprinted as a classic paper: Proceedings of the IEEE, Vol. 90, No. 2, pp. 280–305, February 2002.
9. C.E. Shannon, "Communication in the Presence of Noise," Proceedings of the IRE, Vol. 37, No. 1, pp. 10–21, January 1948. Reprinted as a classic paper: Proceedings of the IEEE, Vol. 86, No. 2, pp. 442–446, February 1998.
10. A.V. Oppenheim, *Applications of Digital Signal Processing*, Prentice-Hall, Inc., Englewood Cliffs, NJ, 1978.
11. P.F. Panter, *Modulation, Noise, and Spectral Analysis*, McGraw-Hill, New York, 1965.
12. B. Sklar, *Digital Communications: Fundamentals and Applications*, Prentice-Hall, Inc., Englewood Cliffs, NJ, 1988.
13. A. Papoulis, *The Fourier Integral and Its Applications*, McGraw-Hill, New York, 1962, A McGraw-Hill Classic Textbook Reissue, 1987.
14. A. Kohlenberg, "Exact Interpolation of Band-Limited Functions," Journal of Applied Physics, Vol. 24, No. 12, pp. 1432–1436, July 1949.
15. C.B. Feldman, W.R. Bennett, "Bandwidth and Transmission Performance," BSTJ, Vol. 28, pp. 490–595, July 1949.
16. J.H. McClellan, T.W. Parks, L.R. Rabiner, "A Computer Program for Designing Optimum FIR Linear Phase Digital Filters," IEEE Transactions on Audio and Electroacoustics, Vol. AU-21, No. 6, pp. 506–526, December 1973.
17. F.F. Kuo, J.F. Kaiser, Editors, *System Analysis by Digital Computers*, Chapter 7, J.F. Kaiser, "Digital Filters," John Wiley & Sons, New York, 1966.
18. L.R. Rabiner, R.W. Schafer, "On the Behavior of Minimax FIR Digital Hilbert Transformers," The Bell System Technical Journal, Vol. 53, No. 2, pp. 363–391, February 1974.
19. L.R. Rabiner, R.W. Schafer, "On the Behavior of Minimax FIR Digital Hilbert Transformers," The Bell System Technical Journal, Vol. 53, No. 2, pp. 363–390, February 1974.
20. M. Faulkner, T. Mattsson, W. Yates, "Automatic Adjustment of Quadrature Modulators," Electronic Letters, Vol. 27, No. 3, pp. 214–216, January 31, 1991.
21. M. Faulkner, T. Mattsson, "Spectral Sensitivity of Power Amplifiers to Quadrature Modulator Misalignment," IEEE Transactions on Vehicular Technology, Vol. 41, No. 4, pp. 516–525, November 1992.
22. A.I. Sinsky, P.C.P. Wang, "Error Analysis of a Quadrature Coherent Detector Processor," IEEE Transactions on Aerospace Electronics Systems, Vol. AES-10, No. 6, pp. 880–883, November 1974.
23. J.K. Caves, M.W. Liao, "Adaptive Compensation for Imbalance and Offset Losses in Direct Conversion Transceivers," IEEE Transactions on Vehicular Technology, Vol. 42, No. 4, pp. 581–588, November 1993.
24. L.R. Rabiner, G. Gold, *Theory and Applications of Digital Signal Processing*, Prentice-Hall, Inc., Englewood Cliffs, NJ, 1975.
25. N.J. Fliege, *Multirate Digital Signal Processing: Multirate Systems, Filter Banks, Wavelets*, John Wiley & Sons, New York, June 1995.
26. R.E. Crochiere, L.R. Rabiner, "Interpolation and Decimation of Digital Signals—A Tutorial Review," Proceeding of the IEEE, Vol. 69, No. 5, pp. 300–331, March 1981.
27. P.P. Vaidyanathan, "Multirate Digital Filters, Filter Banks, Polyphase Networks and Applications: A Tutorial," IEEE Proceedings, Vol. 78, No. 1, pp. 56–93, January 1990.
28. N.J. Fliege, *Multirate Digital Signal Processing: Multirate Systems, Filter Banks, Wavelets*, Chapter 1, "Sampling Rate Conversion," John Wiley & Sons, New York, June 1995.

29. M. Bellanger, *Digital Processing of Signals*, John Wiley & Sons, New York, 1984.
30. L.R. Rabiner, J.H. McClellan, T.W. Parks, "FIR Digital Filter Design Techniques Using Weighted Chebyshev Approximation," IEEE Proceedings, Vol. 63, No. 4, pp. 595–610, April 1975.
31. F. Harris, "Digital Filters," Lecture Notes, 1996.
32. O. Herrmann, H.W. Schuessler, "Design of Nonrecursive Digital Filters with Minimum Phase," Electronic Letters, Vol. 6, No. 11, pp. 329–330, November 1970.
33. T.W. Parks, J.H. McClellan, "A Program for the Design of Linear Phase Finite Impulse Response Filters," IEEE Transactions on Audio Electroacoustics, Vol. AU-20, No. 3, pp. 195–199, August 1972.
34. L.R. Rabiner, B. Gold, C.A. McGonegal, "An Approach to the Approximation Problem for Nonrecursive Digital Filters," IEEE Transactions on Audio Electroacoustics, Vol. 18, No. 1, pp. 83–106, 1970.
35. L.R. Rabiner, G. Gold, *Theory and Applications of Digital Signal Processing*, Chapter 6, "Time-Domain Methods for Speech Processing," Prentice-Hall, Inc., Englewood Cliffs, NJ, 1975.
36. A.V. Oppenheim, R.W. Schafer, *Digital Signal Processing*, Prentice-Hall, Inc., Englewood Cliffs, NJ, 1975.
37. L.S. Metzger, D.M. Boroson, J.J. Uhran, Jr., I Kalet, "Receiver Windowing for FDM MFSK Signals," IEEE Transactions on Communications, Vol. 27, No. 10, pp. 1519–1527, October 1979.
38. E.B. Hogenauer, "An Economical Class of Digital Filters for Decimation and Interpolation," IEEE Transactions on Acoustics, Speech and Signal Processing, Vol. ASSP-29, No. 2, pp. 155–162, April 1981.
39. F.M. Gardner, "Interpolation in Digital Modems—Part I: Fundamentals," IEEE Transactions on Communications, Vol. 41, No. 3, pp. 501–501, March 1993.
40. L. Erup, F.M. Gardner, R.A. Harris, "Interpolation in Digital Modems—Part II: Implementation and Performance," IEEE Transactions on Communications, Vol. 41, No. 6, pp. 998–1008, June 1993.
41. M. Abramowitz, I.A. Stegun, *Handbook of Mathematical Functions with Formulas, Graphs, and Mathematical Tables*, National Bureau of Standards Applied Mathematics Series, No. 55, pp. 878–879, U.S. Government Printing Office, Washington, DC, June 1964.
42. C.W. Farrow, "A Continuously Variable Digital Delay Element," Proceedings of the IEEE, International Symposium on Circuits and Systems (ISCAS), Vol. 3, pp. 2641–2645, Espoo, Finland, June 6–9, 1988.
43. L.R. Rabiner, R.W. Schafer, "On the Behavior of Minimax Relative Error FIR Digital Differentiators," The Bell System Technical Journal, Vol. 53, No. 2, pp. 333–361, February 1974.

3

DIGITAL COMMUNICATIONS

3.1 INTRODUCTION

A functional description of a communication system is shown in Figure 3.1. Generally the information source data is represented by discrete-time and discrete-amplitude digital data; however, if the information source is represented by continuous-time and continuous-amplitude analog data, then the source encoder includes time-domain sampling and amplitude quantization. The resulting data is referred to as the source or user *bit* of duration T_b. Depending upon the nature of the data and the system performance requirements, the resulting digital information sequence may then undergo data compression. For example, if the information source is a facsimile system, then data compression is applied to eliminate redundant or unnecessary data or, in the case of voice data, the sampled voice signal is typically processed using a voice-coding algorithm to minimize the required amount of data to be transmitted. The resulting source coded data sequence $\{v_o, v_1, \ldots, v_{Q-1}\}$ undergoes channel coding to mitigate impact of distortion on the received signal. Examples of channel coding are forward error correction (FEC) coding, data interleaving, and pseudo-noise (PN) spread-spectrum encoding. The FEC-coded output consists of one digit per T_{cb} seconds where T_{cb} is the *code-bit* duration. When direct-sequence spread-spectrum (DSSS) is applied each interval of duration T_c seconds is referred to as a *chip*. The channel coded data is then applied to the data modulator that assigns the resulting bits or chips to an amplitude and/or phase constellation that characterizes the transmitted *symbol* of duration T seconds.* The modulator may interface at baseband with the transmitter; however, the interface commonly follows digital-to-analog conversion (DAC) and up-conversion to an intermediate frequency (IF) on the order of 70 MHz. The functions of the transmitter provide frequency up-conversion commensurate with the communication channel, power amplification, and spectral control; however, the more advanced waveform modulators include spectrally efficient waveforms that inherently satisfy specific transmit spectral masks. The transmit antenna plays a central role in establishing the directive power to establish the communication link. The functions following the channel perform the inverse of those just described, starting with the receiver antenna and ending with user data estimates provided to the information sink. However, these inverse functions are generally more complex and, in the case of the demodulator, more processing intense. The receive antenna interfaces with the receiver input low-noise power amplifier (LNPA) and the subsequent frequency down-conversion to the demodulator IF includes filtering for interference rejection and automatic gain control (AGC).† Typically, the first tasks of the demodulator are to convert the IF input to baseband and provide analog-to-digital conversion (ADC) in preparation for the various functions of digital signal

*Unless otherwise noted, the symbol T is used to denote the transmitted symbol interval with the corresponding symbol rate denoted as $R_s = 1/T$.
†The AGC control processing is often performed in the demodulator following analog-to-digital conversion (ADC).

Digital Communications with Emphasis on Data Modems: Theory, Analysis, Design, Simulation, Testing, and Applications,
First Edition. Richard W. Middlestead.

Companion website: www.wiley.com/go/digitalcommunications

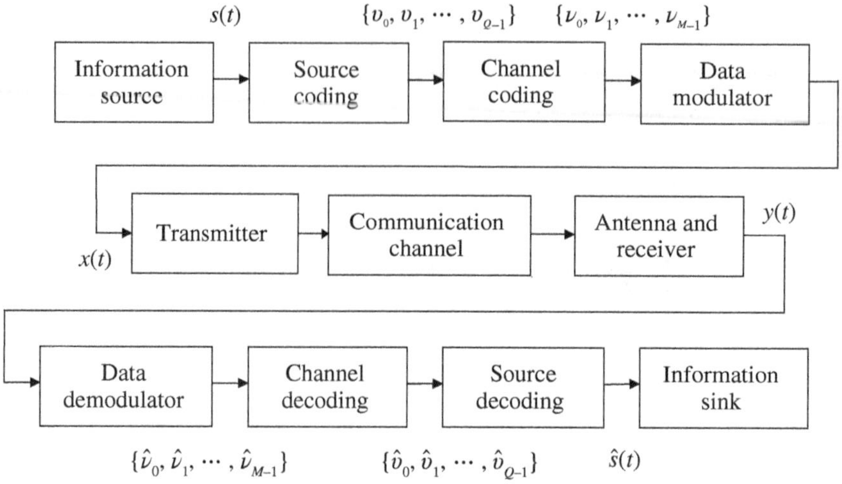

FIGURE 3.1 General functional diagram of a communication system.

processing to accomplish signal acquisition, frequency, phase and symbol tracking, message synchronization, optimum symbol detection, and channel and source decoding to provide the best estimates of the source information.

Figure 3.1 serves as an outline of the subjects covered in this book. For example, a number of techniques for channel coding and decoding are discussed including the block and convolutional coding and their utility in different applications. Code concatenation, involving the application of several codes, has culminated in the powerful turbo codes (TCs) that perform very close to the theoretical coding limit introduced by Shannon [1] in 1948. The choice of waveform modulation is extremely important and establishes the transmission efficiency in terms of the number of bits/hertz and the spectral containment of the transmitter emission. A variety of digitally modulated waveforms are considered with applications to radio frequency (RF) transmission as in satellite communications and noncarrier or baseband transmission as used between many computer systems. The waveform demodulator is driven by the modulator selection and must perform a variety of functions for optimally detecting the received signal. In this regard, demodulator optimum filtering, carrier frequency and phase acquisition, and symbol synchronization and tracking are discussed in considerable detail. The ubiquitous channel is extremely important to understand and characterize to accurately design and predict the performance of communication systems. The channel characteristics and behavior are determined largely by the carrier frequency selection and channels ranging from very low frequency (VLF) to extremely high frequency (EHF) and optical frequencies are discussed and characterized. The waveform selection is also driven by the expected channel conditions. The most benign channel, and the easiest one to characterize, is the additive white Gaussian noise (AWGN) channel. In addition to its frequent occurrence in many applications, the AWGN channel forms the basis for comparing the performance of systems under more severe channel conditions. For example, the performance with selective and nonselective frequency fading is discussed, along with mitigation techniques, and compared to that of the AWGN channel.

In addition to the functional considerations of the system design as outlined in Figure 3.1, a heavy emphasis is placed on computer simulation results to characterize the communication systems performance. Computer simulations serve several essential functions: they provide a basis for the system performance evaluation in applications which defy detailed analysis, for example, evaluation of an entire communication link including complex channels and nonlinear acquisition and tracking functions; they provide a direct method for generating code for firmware and digital signal processor (DSP) implementations; they provide design and evaluation flexibility by means of relatively simple code changes; and, if properly written, they provide a form of self-documentation throughout the code development.

Figure 3.2 shows a simplified application involving binary data from the information source occurring at a rate $R_b = 1/T_b$ where T_b is the bit duration of the source information. In this example, the source data undergoes rate $r_c = 1/2$ FEC encoding resulting in a code-bit rate of $R_{cb} = R_b/r_c = 2R_b$. Assuming that a quadrature phase shift keying (QPSK) waveform is used to transmit the data over the channel, the function of the channel encoder is to map two consecutive code bits into one of four carrier phases resulting in a transmitted symbol rate of $R_s = R_{cb}/2 = R_b$. The symbol rate is important in that it must be compatible with the channel bandwidth requirements so as not to significantly distort the received signal and thereby degrade the detection performance.

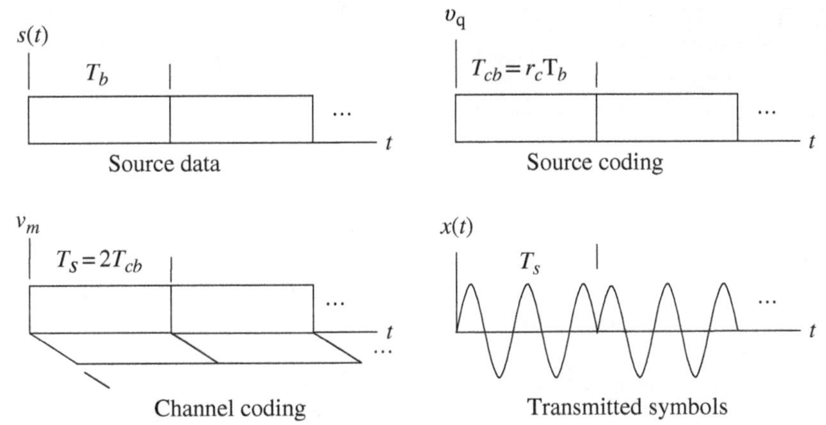

FIGURE 3.2 A simplified example of source and channel encoding.

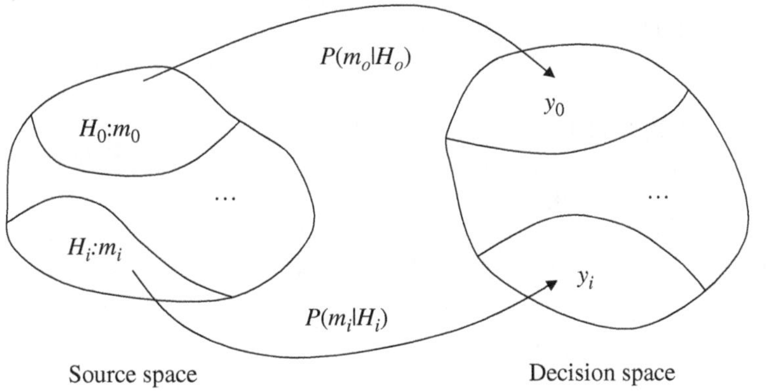

FIGURE 3.3 *M*-ary hypothesis testing.

3.2 DIGITAL DATA MODULATION AND OPTIMUM DEMODULATION CRITERIA

Establishing an optimum decision rule involves multiple hypotheses testing of a statistical event given various statistical characteristics of the system model. The statistics generally involves a priori probabilities, transition probabilities, and a posteriori probabilities and the decision test involves minimizing the risk in the decision that involves choosing the event m given that the event m actually occurred. For example, if the decision is based on some measurement y, and the estimate of the actual event is characterized as $\hat{m}(y)$, then it is desired to minimize the risk in saying that $\hat{m}(y) = m$. This concept is shown in Figure 3.3 that depicts a space containing all possible hypotheses H_m of the events m: $m = 0, \ldots, M-1$ in the source. When the source events undergo a transformation to the decision space, the outcomes of the decisions represent estimates of the source events with associated probabilities of error. The transformation from the source to decision spaces can be viewed as source messages undergoing transmission over a noisy channel and, after some manipulation, interpreted by the receiver as to what message was sent. The receiver must have a priori knowledge, that is, know in advance that the messages will occupy unique decision regions in a noise-free or deterministic transformation. The probabilities $P(m_i|H_i)$ are referred to as transition probabilities and represent the transformation from the source to the decision space as shown in Figure 3.3. Associated with each hypothesis there is also a probability $P(H_i)$ representing the probability that the source event H_i will occur among all of the possible events; this probability is called the *a priori* probability of the source event.

For the *M*-ary hypotheses the risk of a decision is defined as the weighted summation of probabilities, expressed as

$$\Re \triangleq \sum_{i,j} C_{ij} P(i|j) P(j) \tag{3.1}$$

where $P(i|j)$ are the transition probabilities and represent the probability of event i when event j actually occurred. The transition probabilities correspond to an error event when $i \neq j$ and a correct decision is made when $i = j$.

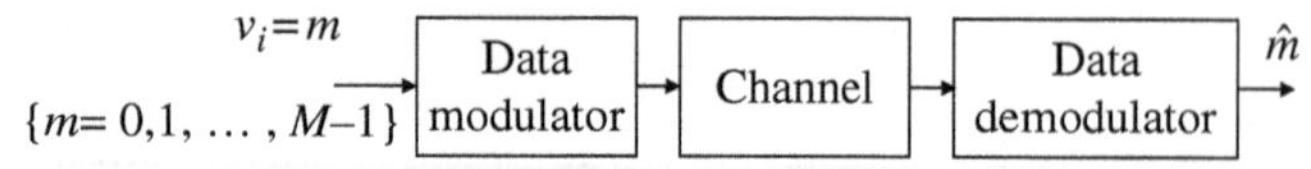

FIGURE 3.4 Waveform transmission with AWGN channel.

The probabilities $P(j)$ represent the a priori probabilities of the individual events. The weights C_{ij} represent costs associated with the corresponding decisions. In communication systems the emphasis is placed on minimizing errors so an error event is given the highest cost corresponding to $C_{ij} = 1$ when $i \neq j$ and $C_{ij} = 0$ when $i = j$. Under these conditions, the risk represents the overall system error probability and is expressed by the total probability law,

$$P_e = \sum_{i \neq j} P(i|j)P(j) \tag{3.2}$$

In the following analysis, the multiple hypothesis error criteria are applied to a communication system and, in the process, the requirements for making a decision that minimizes the overall error, as expressed by (3.2), are established. Consider, for example, that the source information has undergone source and channel coding resulting in contiguous M-ary symbol messages $\nu_i = m$ as shown in Figure 3.4. Based on the earlier discussion, the a priori probability of a message is defined as

$$P_r(\text{message } m \text{ transmitted}) = P(m) = P_m \tag{3.3}$$

and the system error probability is characterized as

$$P_e = P_r(\text{message error}) = P_r(\hat{m} \neq m) \tag{3.4}$$

Applying the total probability law, the error is expressed as

$$\begin{aligned} P_e &= \sum_{m=0}^{M-1} P(\hat{m} \neq m|m)P_m \\ &= 1 - \sum_{m=0}^{M-1} P(\hat{m} = m|m)P_m \end{aligned} \tag{3.5}$$

The probability that $\hat{m} = m$ is simply the integral of the probability density function (*pdf*) of the received signal taken over the decision region for the message event m and is expressed as

$$P(\hat{m} = m|m) = \int_{y_m} p(y|m)dy \tag{3.6}$$

so the error becomes

$$P_e = 1 - \int_{y_m} \sum_{m=0}^{M-1} p(y|m)P_m dy \tag{3.7}$$

The error is minimized by maximizing the integrand, or the summation under the integral, and is expressed as

$$P_e(\min) = 1 - \int_{y_m} \max \left[\sum_{m=0}^{M-1} p(y|m)P_m \right] dy \tag{3.8}$$

Therefore, the optimum decision rule is simply

$$\text{Choose } \hat{m}(y) = m \textit{ iff } p(y|m)P_m = \max\ [p(y|m')P_{m'}] \quad \forall m' \tag{3.9}$$

This rule states that given the message m was transmitted the correct decision is made by choosing the message estimate $\hat{m}(y)$ that minimizes the error. However, at the demodulator the test statistic y is available, so the optimum decision must be based on y alone. To facilitate the use y, the decision rule can be reformulated into the desired result by applying a mixed form of Bayes rule to obtain the a posteriori probability of message m given by

$$P(m|y) = \frac{p(y|m)P_m}{p(y)} \tag{3.10}$$

This result states that the probability of declaring message m, given the test statistic y, can be accomplished through the demodulator processing by substituting (3.10) into the decision rule (3.9) resulting in the reformulated decision rule

$$\begin{aligned} \text{Choose } \hat{m}(y) = m \textit{ iff } P(m|y) &= \max\ [P(m'|y)] &&: \forall m' \\ &= \max \left[\frac{p(y|m')P_{m'}}{p(y)} \right] &&: \forall m' \end{aligned} \tag{3.11}$$

The second equality in (3.11) is the formal characterization of the maximum a posteriori (MAP) decision rule; however, because $p(y)$ is simply a constant, independent of the message, it does not influence the selection of the maximum value so the second equality corresponds to the decision rule given by (3.9), thus confirming that y is a sufficient statistic for making an optimal decision.

The MAP decision rule explicitly involves the a priori probabilities P_m; however, if the messages are equally likely, so that $P_m = 1/M$ for all m, the decision rule is referred to as the maximum-likelihood (ML) rule. The ML decision rule is expressed as

$$\begin{aligned} \text{Choose } \hat{m}(y) = m \textit{ iff } P(m|y) &= \max[P(m'|y)] &&: \forall m' \\ &= \max \left[\frac{p(y|m')/M}{p(y)} \right] &&: \forall m' \\ &= \max[p(y|m')] &&: \forall m' \end{aligned} \tag{3.12}$$

The decision rules can be formulated in terms of the likelihood ratio test (LRT) defined as

$$\Lambda(y;m,m') = \frac{p(y|m)}{p(y|m')} \tag{3.13}$$

for which the MAP decision rule is

$$\text{Choose } \hat{m}(y) = m \text{ iff } \Lambda(y;m,m') > \frac{P_{m'}}{P_m} \quad : \forall m' \tag{3.14}$$

The logarithm is a monotonically increasing function of its argument and because the logarithm of the likelihood ratio often leads to a more intuitive result, the decision rule is also conveniently characterized in terms of the log-likelihood ratio test (LLRT) as

$$\text{Choose } \hat{m}(y) = m \text{ iff } \ln\,[\Lambda(y;m,m')] > \ln\left[\frac{P_{m'}}{P_m}\right] \quad : \forall m' \tag{3.15}$$

3.2.1 Example Using Binary Data Messages

As an example of the application of the decision rules, consider the binary hypothesis test defined as

H_0: message $m = 0$, with a priori probability $P(H_0) = P_0$

H_1: message $m = 1$, with a priori probability $P(H_1) = P_1$

The message is transmitted over an AWGN channel as indicated in Figure 3.5.

In this description, the channel is formulated as a simple vector channel and does not refer to a waveform type. For this characterization of the channel, the noise power is specified in terms of the one-sided noise spectral density N_o (watts/hertz) and is consistent with the vector characterization of the signal in terms of the signal energy-per-symbol E (watt-second). Because of the zero-mean additive Gaussian noise channel, the *pdf* of the received vector is*

$$\begin{aligned} p_Y(y|m) &= p_N(n = y - x) \\ &= \frac{1}{\sqrt{\pi N_o}} e^{-(y-x_m)^2/N_o} \end{aligned} \tag{3.16}$$

Based on the MAP decision rule, $\hat{m}(y) = 0$ is chosen *iff* $p(y|0)P_o > p(y|1)P_1$ and, using the likelihood ratio $\Lambda(y) = p(y|0)/p(y|1)$, the optimum decision is formulated as

$$\begin{aligned} &\text{If } \Lambda(y) > \frac{P_1}{P_o} \text{ or } \ln[\Lambda(y)] > \ln\left[\frac{P_1}{P_o}\right] \\ &: \text{choose } H_o \ o.w. \text{ choose } H_1 \end{aligned} \tag{3.17}$$

Therefore, using the *pdf* for Y, the LRT becomes

$$\begin{aligned} \Lambda(y) &= \exp\left(\frac{-(y-x_o)^2 + (y-x_1)^2}{N_o}\right) \\ &= \exp\left(\frac{2y(x_o - x_1) - (x_o^2 - x_1^2)}{N_o}\right) \end{aligned} \tag{3.18}$$

The utility of the log-likelihood ratio (LLR) is evident in forming the LLRT test for the AWGN channel that results in the optimum decision rule

$$\begin{aligned} \ln(\Lambda(y)) &= \frac{2y(x_o - x_1) - (x_o^2 - x_1^2)}{N_o} > \ln\left(\frac{P_1}{P_o}\right) \\ &: \text{choose } H_o \ o.w. \text{ choose } H_1 \end{aligned} \tag{3.19}$$

Solving (3.19) in terms of the test statistic y gives the useful result for the MAP decision rule, with respect to the threshold y_t, expressed as

$$\begin{aligned} y > y_t &= \frac{x_o + x_1}{2} + \frac{N_o}{2(x_o - x_1)} \ln\left(\frac{P_1}{P_o}\right) \\ &: \text{choose } H_o \ o.w. \text{ choose } H_1 \end{aligned} \tag{3.20}$$

With equally probable messages and antipodal signaling, that is, with $P_0 = P_1$ and $X_0 = -X_1$, the decision threshold with respect to the origin of the x,y plane is shown in Figure 3.6a. The effect of different a priori probabilities tends to bias the threshold away from the message with the higher probability of occurrence.

For this example involving binary messages, the error probability is expressed as

$$P_e = P_e(y \in y_1|0)P_o + P_e(y \in y_o|1)P_1 \tag{3.21}$$

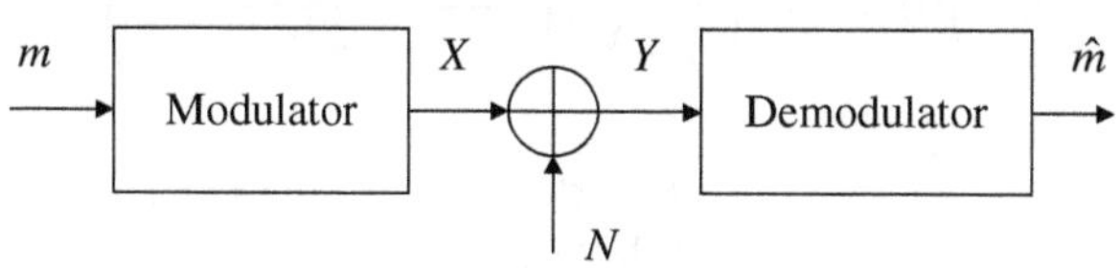

$$m = \{0,1\};\ X = \begin{cases} \sqrt{E} & : m = 0 \\ -\sqrt{E} & : m = 1 \end{cases}$$

$$N = N\left(0, \sqrt{N_o/2}\right);\ Y = X + N$$

FIGURE 3.5 Binary message with AWGN channel.

*Gaussian *pdf* with zero mean and unit variance is often referred to as a normal distribution.

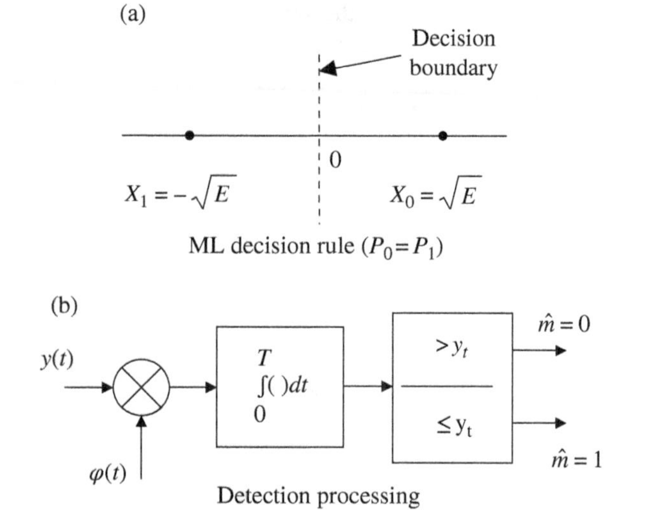

FIGURE 3.6 Optimum decision boundary for binary detection processing.

Using the decision threshold y_t, the individual error probabilities are evaluated as

$$P_e(y \in y_1 | 0) = \frac{1}{\sqrt{\pi N_o}} \int_{-\infty}^{y_t} e^{-(y-x_o)^2/N_o} dy \tag{3.22}$$

and

$$P_e(y \in y_o | 1) = \frac{1}{\sqrt{\pi N_o}} \int_{y_t}^{\infty} e^{-(y-x_1)^2/N_o} dy \tag{3.23}$$

Changing variables in these integrals by letting $\xi = \sqrt{2/N_o}(y-x)$ and recognizing that the integral is simply the probability integral* for the Gaussian *pdf* results in the expressions

$$\begin{aligned} P_e(y \in y_1 | 0) &= \frac{1}{\sqrt{2\pi}} \int_{-\infty}^{\sqrt{2/N_o}(y_t - x_o)} e^{-\xi^2/2} d\xi \\ &= erf_*\left(\sqrt{\frac{2}{N_o}}(y_t - x_o)\right) \end{aligned} \tag{3.24}$$

$$\begin{aligned} P_e(y \in y_o | 1) &= \frac{1}{\sqrt{2\pi}} \int_{\sqrt{2/N_o}(y_t - x_1)}^{\infty} e^{-\xi^2/2} d\xi \\ &= \frac{1}{\sqrt{2\pi}} \int_{-\infty}^{-\sqrt{2/N_o}(y_t - x_1)} e^{-\xi^2/2} d\xi \\ &= erf_*\left(-\sqrt{\frac{2}{N_o}}(y_t - x_1)\right) \end{aligned} \tag{3.25}$$

Upon evaluating $\sqrt{2/N_o}(y_t - x)$ for each message state, it is found that

$$\sqrt{\frac{2}{N_o}}(y_t - x_o) = -\sqrt{\frac{2}{N_o}}\left[\frac{x_o - x_1}{2} - \frac{N_o \ln(P_1/P_o)}{2(x_o - x_1)}\right] \tag{3.26}$$

$$\sqrt{\frac{2}{N_o}}(y_t - x_1) = \sqrt{\frac{2}{N_o}}\left[\frac{x_o - x_1}{2} + \frac{N_o \ln(P_1/P_o)}{2(x_o - x_1)}\right] \tag{3.27}$$

and substituting these results into (3.24) and (3.25) gives

$$\begin{aligned} P_e(y \in y_1 | 0) &= erf_*\left(-\sqrt{\frac{2}{N_o}}\left[\frac{x_o - x_1}{2} - \frac{N_o \ln(P_1/P_o)}{2(x_o - x_1)}\right]\right) \\ &= erfc_*\left(\sqrt{\frac{2}{N_o}}\left[\frac{x_o - x_1}{2} - \frac{N_o \ln(P_1/P_o)}{2(x_o - x_1)}\right]\right) \end{aligned} \tag{3.28}$$

$$\begin{aligned} P_e(y \in y_o | 1) &= erf_*\left(-\sqrt{\frac{2}{N_o}}\left[\frac{x_o - x_1}{2} + \frac{N_o \ln(P_1/P_o)}{2(x_o - x_1)}\right]\right) \\ &= erfc_*\left(\sqrt{\frac{2}{N_o}}\left[\frac{x_o - x_1}{2} + \frac{N_o \ln(P_1/P_o)}{2(x_o - x_1)}\right]\right) \end{aligned} \tag{3.29}$$

Defining the distance between the two modulated signals as $d = |x_o - x_1|$ these results simplify to

$$P_e(y \in y_1 | 0) = erfc_*\left(\sqrt{\frac{d^2}{2N_o}\left(1 - \frac{N_o \ln(P_1/P_o)}{d^2}\right)^2}\right) \tag{3.30}$$

$$P_e(y \in y_o | 1) = erfc_*\left(\sqrt{\frac{d^2}{2N_o}\left(1 + \frac{N_o \ln(P_1/P_o)}{d^2}\right)^2}\right) \tag{3.31}$$

A particularly simple expression results for the ML detection case with $P_1 = P_0 = 1/2$. In this case, the conditional-error probabilities given by (3.30) and (3.31) are equal and the total-error probability becomes

*The notation $erf_*(x)$ for the probability integral is used by Van Trees, and Section 3.5 examines the relationship to several other conventional definitions of the error function.

$$P_{be} = P_e = erfc_*\left(\sqrt{\frac{d^2}{2N_o}}\right) \quad : \text{ML decisions} \tag{3.32}$$

where P_{be} denotes the bit-error probability resulting from the binary message source, that is, each message represents a sequence of binary information bits. Recall that the modulator being considered in this example corresponds to $x_o = \sqrt{E}$ and $x_1 = -\sqrt{E}$ so the squared distance is $d^2 = 4E$, where the energy represents the energy-per-bit (E_b), and (3.32) is expressed as

$$P_{be} = erfc_*\left(\sqrt{\frac{2E_b}{N_o}}\right) = \frac{1}{2}erfc\left(\sqrt{\frac{E_b}{N_o}}\right) \tag{3.33}$$

The relationship to the complementary error function *erfc* (•) is discussed in Section 3.5. The argument E_b/N_o is the signal-to-noise ratio measured in a bandwidth equal to the data rate $R_b = 1/T_b$.

An interesting variation of the modulator can be examined by representing the squared distance in terms of the normalized correlation coefficient (ρ) between the two modulated messages as [2]

$$d^2 = x_o^2 + x_1^2 - 2\rho|x_o||x_1| \tag{3.34}$$

where $|\rho| \le 1$ and ρ is defined as

$$\rho = \frac{x_o x_1}{|x_o||x_1|} \tag{3.35}$$

Upon using these results with $|x_o| = |x_1| = \sqrt{E_b}$ the bit-error probability becomes

$$P_{be} = erfc_*\left(\sqrt{\frac{E_b(1-\rho)}{N_o}}\right) \tag{3.36}$$

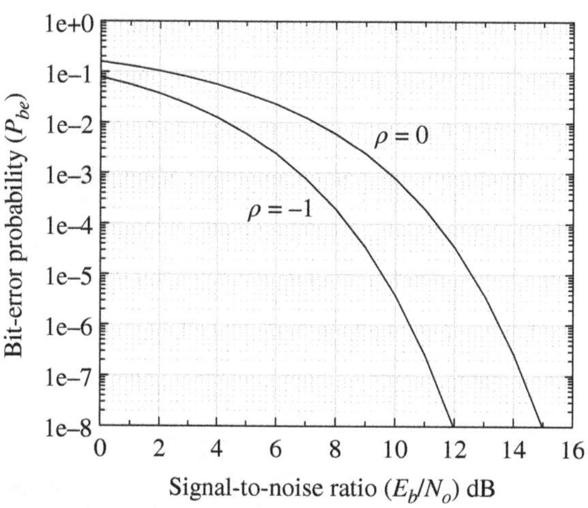

FIGURE 3.7 Bit-error performance for antipodal and orthogonal binary signaling.

When $\rho = -1$ the modulated signals are antipodal as in the preceding example; however, for $\rho = 0$ the signals are orthogonal and the bit-error probability is degraded by 3 dB as seen from

$$P_{be} = erfc_*\left(\sqrt{\frac{E_b}{N_o}}\right) = \frac{1}{2}erfc\left(\sqrt{\frac{E_b}{2N_o}}\right) \quad : \rho = 0 \tag{3.37}$$

The bit-error probability is plotted in Figure 3.7 as a function of the signal-to-noise ratio E_b/N_o, expressed in dB, for antipodal and orthogonal signaling.

3.3 INFORMATION AND CHANNEL CAPACITY

In this section a quantitative definition of information is introduced [1] and applied to a data source and, after passing through a discrete memoryless channel (DMC), the information content at a data sink is evaluated. The DMC channel is shown in Figure 3.8 and is characterized as having a fixed number of input and output letters X_i and Y_i: $i = 0, \ldots, I-1; j = 0, \ldots, J-1$ with source probabilities $P_X(x_i)$ and transitions from x_i to y_i defined in terms of the conditional probabilities $P_{Y|X}(y_j|x_i)$. The conditional probabilities of the source data given the sink data are the a posteriori probabilities, denoted as $P_{X|Y}(x_i|y_j)$; these a posteriori probabilities are the performance measures to be optimized for $i = j$. The sink probabilities $P_Y(y_i)$ are of ancillary interest in the application of Bayes rule.

With the information content of the source data characterized in terms of the self-information and average self-information or source entropy, the remaining task is to define and characterize similar measures of information at the communication receiver or data sink when the source data is passed through the channel. The DMC channel is memoryless, in that, each channel use of a source letter x_i is statistically independent of all the source letter uses. An example application of a DMC channel is the time and amplitude sampling of an analog signal $s(t)$ such that $t = i\Delta T$ where ΔT represents the discrete channel-use interval and the finite number of discrete amplitudes samples, ΔA_i, are represented by source letters $x_i = f(\Delta A_i)$.

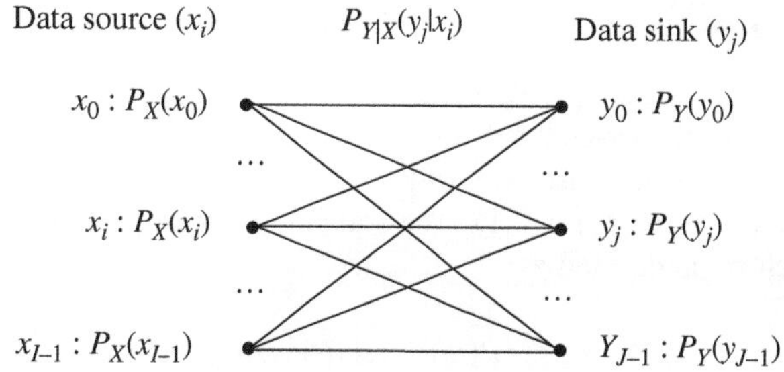

FIGURE 3.8 Discrete memoryless channel.

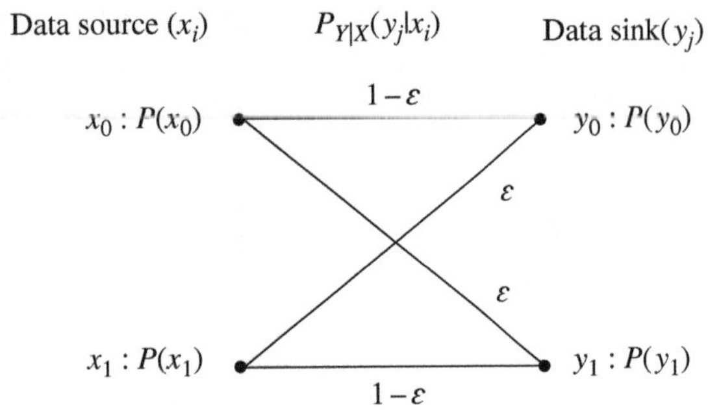

FIGURE 3.9 Binary symmetric channel.

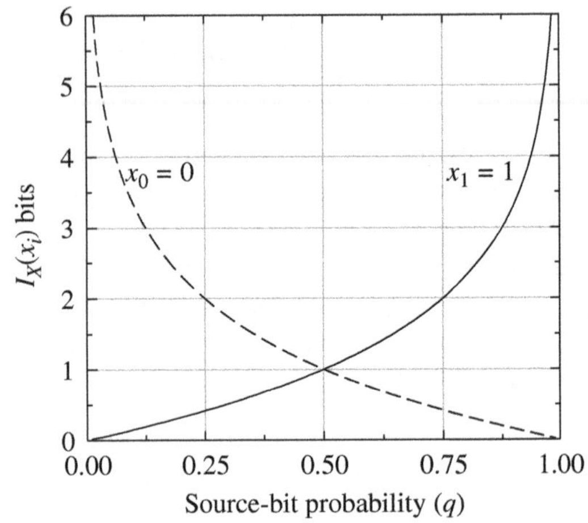

FIGURE 3.10 Self-information associated with one source bit.

The binary symmetric channel (BSC) channel, shown in Figure 3.9, is a special case of the DMC channel that uses binary input and output data, that is, the discrete amplitudes during each channel use have the form $x_i = \{b_i\}$: $i = 0,1$ with the natural bit mapping $(b_0, b_1) = (0,1)$. In this case, the conditional probabilities $P_{Y|X}(y_i|x_i)$ represent the channel transition probability defined as $P_{Y|X}(y_i|x_i) = \varepsilon$ when $j \neq i$ and $1 - \varepsilon$ when $j = i$.

The discrete-time and amplitude analog signal samples, described earlier, can be applied to the BSC channel by quantizing the source samples into $M = 2^K$ discrete binary levels every ΔT seconds and then mapping each of the binary digits into a serial binary data sequence representing all of the discrete-time source data samples. This technique is referred to a pulse code modulation (PCM) and the resulting bits are applied to the BSC channel as though originating from a binary data source.

The BSC channel capacity is defined as the maximum average mutual information between the source and sink data maximized over the a priori probabilities, $P_X(x_i)$. In Section 3.3.1 the capacity of the BSC channel is evaluated using a binary source, and in Section 3.3.2 the results are extended to include an M-ary source data. This section concludes by examining Shannon's capacity limit for coded and uncoded waveforms using the Gaussian channel model.

3.3.1 Binary Symmetric Channel with Binary Data Source

Consider a binary data source $X(x_i)$ consisting of one information bit-per-channel use corresponding to $K = 1$ and $M = 2$. In the case $x_i = \{b_i\}$: $i = 0,1$ with the natural bit mapping $(b_0, b_1) = (0,1)$. The self-information contained in each bit is defined as

$$I_X(x_i) \triangleq \log_2(1/P(x_i)) \quad \text{: self-information} \\ = -\log_2 P(x_i) \tag{3.38}$$

where the base-2 logarithm is used specifically to denote binary data in *bits* of information,* that is, $K = 1$ binary bit corresponds to $M = 2^K = 2$ bit combinations giving rise to $\log_2(2) = 1$ bit of information. Note that (3.38) expresses the uncertainty of the source data, in that, for $P(x_i)$ approaching 1 the uncertainty, $I_X(x_i)$, is nearly 1 and as $P(x_i)$ approaches 0 $I_X(x_i)$ increases indicating more uncertainty. Because the source will generate either a binary 0 or 1, it follows from the total probability law that $P(x_0 = 0) + P(x_1 = 1) = 1$ so $P(x_0 = 1) = 1 - P(x_1 = 0)$. Figure 3.10 shows the self-information associated with each bit for $0 \le P(x_1 = 1) \le 1$, and it is seen that each source-bit conveys one bit of information when each source-bit is equally probable, that is, when $P(x_0 = 1) = P(x_1 = 0) = 1/2$.

The average self-information is defined as the entropy of the source and is given by

$$H_X \triangleq E[I_X(x_i)] \\ = -\sum_{i=0}^{1} P(x_i)\log_2 P(x_i) \quad \text{: average self-information} \tag{3.39}$$

As an example, suppose that a binary source generates the binary sequence $x_i = \{0,1\}$ and that it is required to evaluate the entropy of the source data for $P(x_1) = q$: $0 \le q \le 1$. In this case, $M = 2$ so (3.39) is evaluated as

$$H_X = -(1-q)\log_2(1-q) - q\log_2(q) \tag{3.40}$$

*Any logarithmic base can be used; however, base 2 results in the information expressed in terms of bits. If the natural logarithm is used, then the information is expressed in *nats* where there are ln2 = 0.693 nats/bit. To change between logarithmic bases use: $\log_a(x) = \log_b(x)/\log_b(a)$.

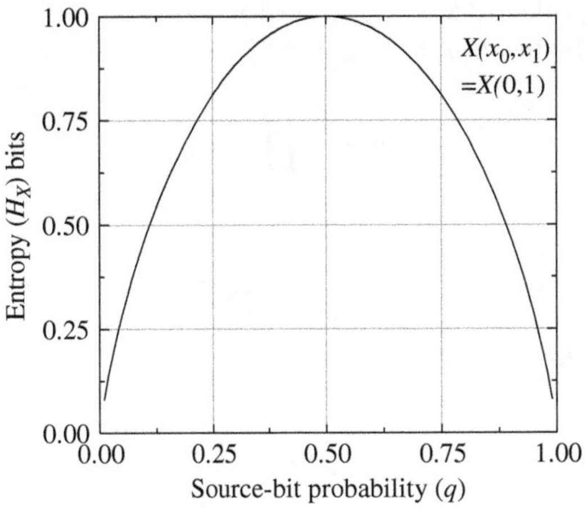

FIGURE 3.11 Source entropy or average self-information of Figure 3.10.

This result is shown in Figure 3.11 and verifies that the maximum average self-information or maximum entropy is one bit and occurs when the source bits are equally probable, that is, when $q = 1/2$.

The mutual information corresponding to the source data x_i and channel output or sink data y_j is defined as

$$I_{X,Y}(x_i,y_j) \triangleq \log_2\left(\frac{P_{X,Y}(x_i,y_j)}{P_X(x_i)P_Y(y_j)}\right) \quad : \text{mutual information} \tag{3.41}$$

The mutual information can also be expressed in terms of the channel conditional transition probabilities between the sink and source data with the result

$$I_{Y,X}(y_j,x_i) = \log_2\left(\frac{P_{Y|X}(y_j|x_i)}{P_Y(y_j)}\right) \tag{3.42}$$

Because $I_{Y,X}(y_j,x_i) = I_{X,Y}(x_i,y_j)$ the probability space notation in (3.42) is changed to be consistent with the conditioning. The average mutual information is defined as

$$I_{Y,X} \triangleq \sum_{i,j} P_{Y,X}(y_j,x_i) I_{Y,X}(y_j,x_i) \quad : \text{average mutual information} \tag{3.43}$$

where the subscript ranges are $i, j = 0,1$. Substituting (3.42) into (3.43) with the change in notation gives

$$I_{Y,X} = \sum_{i,j} P_{Y|X}(y_j|x_i)P_X(x_i)\log_2\left(\frac{P_{Y|X}(y_j|x_i)}{\sum_\ell P_{Y|X}(y_j|x_\ell)P_X(x_\ell)}\right) \tag{3.44}$$

where

$$P_Y(y_j) = \sum_\ell P_{Y|X}(y_j|x_\ell)P_X(x_\ell) \tag{3.45}$$

An alternate solution, and an easier solution to evaluate, is obtained by conditioning the mutual information, expressed by (3.41), on the sink data with the result

$$\begin{aligned} I_{X,Y}(x_i,y_j) &= \log_2\left(\frac{P_{X|Y}(x_i|y_j)}{P_X(x_i)}\right) \\ &= I_X(x_i) - I_{X|Y}(x_i|y_j) \end{aligned} \tag{3.46}$$

Using this result the average mutual information becomes

$$I_{X,Y} = H_X - H_{X|Y} \tag{3.47}$$

where the average self-information of the source or the source entropy is given by (3.39) and the conditional entropy of the source given the sink data is evaluated as

$$\begin{aligned} H_{X|Y} &\triangleq -\sum_{i,j=0}^{1} P_{X,Y}(x_i,y_j)\log_2(P_{X|Y}(x_i|y_j)) \\ &= -\sum_{i,j=0}^{1} P_{Y|X}(y_j|x_i)P_X(x_i)\log_2(P_{X|Y}(x_i|y_j)) \end{aligned} \tag{3.48}$$

Evaluation of (3.48) in terms of the source probability q and the BSC channel transition error probability ε results in

$$H_{X|Y} = -(1-\varepsilon)\log_2(1-\varepsilon) - \varepsilon\log_2(\varepsilon) \tag{3.49}$$

This result shows that, for binary source data, the conditional entropy of the BSC is independent of the a priori probabilities $P(x_i)$.* The channel capacity is defined as the maximum average mutual information between the source and sink data maximized with respect to the a priori source data probabilities and is expressed as

$$\begin{aligned} C_1 &\triangleq \max_{P_X(x_i)} I_{Y,X} \\ &= \max_{P_X(x_i)} (H_X - H_{X/Y}) \quad : \text{channel capacity} \end{aligned} \tag{3.50}$$

Referring to Figure 3.11, the maximum source entropy, H_X, corresponds to $P_X(x_i) = 1/2 \ \forall \ i$ and, using (3.49) for the conditional entropy, the capacity of the BSC channel is evaluated as

$$C_1 = 1 - (1-\varepsilon)\log_2(1-\varepsilon) - \varepsilon\log_2(\varepsilon) \tag{3.51}$$

*In the next section it is shown that this is also true for M-ary source data.

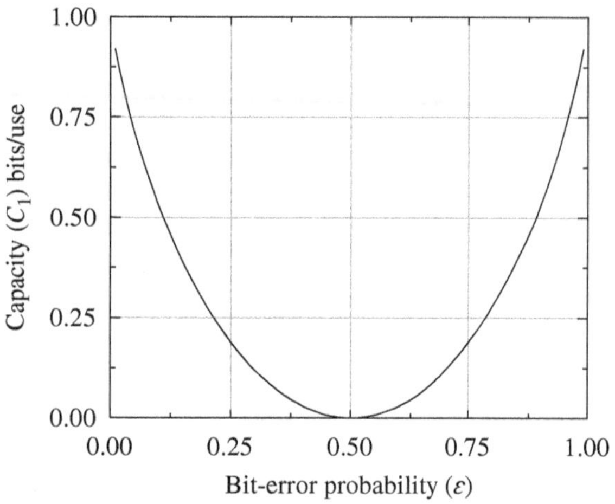

FIGURE 3.12 Channel capacity of BSC ($K = 1$, $P_X(x_i) = 1/2$).

This result is plotted in Figure 3.12 as a function of the channel error. Referring to Figure 3.11 the maximum conditional entropy occurs under the condition $\varepsilon = 1/2$ corresponding to the most uncertainty in the received data. Not surprisingly, this also corresponds to condition for the least amount of information through the channel giving rise to the minimum channel capacity. The conditions corresponding to the minimum conditional entropy, that is, $\varepsilon = 0$ and 1, also correspond to the maximum channel capacity. While the definition of capacity does not distinguish between correct and incorrect data, if the data is known to be completely incorrect, $\varepsilon = 1$, then a correct decision can be made with certainty by simply reversing the decision. In practice, the bit-error probability of an uncoded waveform will range between 1/2 and 1.0 as the signal-to-noise ratio increases from $-\infty$ to ∞, thus limiting the practical range of the abscissas in Figure 3.12.

3.3.2 Binary Symmetric Channel with *M*-ary Data Source

In this section the channel capacity is evaluated when K binary digits of source data are transmitted during each channel use. For an arbitrary integer K this corresponds to $x_i = \{b^i{}_{K-1}, \ldots, b^i{}_1, b^i{}_0\}$: $b^i{}_k \in (1, 0)$, $k = 0, \ldots, K-1$ where $i = 0, \ldots, M-1$ and $M = 2^K$. By convention the least significant bit, b_0, is on right-hand side with the index i corresponding to the natural order of the binary bits, that is, x_0 corresponds to $= \{0,\ldots,0, 0)$, x_1 corresponds to $(0,\ldots,0,1)$, and so on. The average self-information of the M-ary source data involves averaging over all of the M source data sequences and is computed as

$$H_X = -\sum_{i=0}^{M-1} P(x_i)\log_2 P(x_i) \tag{3.52}$$

Equation (3.52) is evaluated in terms of the source-bit probabilities $P(b^i_k = 1) = q$ and $P(b^i_k = 0) = 1 - q$ with

$$P(x_i) = \prod_{k=0}^{K-1} P(b^i_k) \tag{3.53}$$

It is easy to show (see Problem 4a and 4b) that the a priori probabilities of the M source sequences $P(x_i)$ are not all equal and that the summation of $P(x_i)$: $i = 0, \ldots < M-1$ is unity, that is,

$$\sum_{k=0}^{M-1} P(x_i) = 1 \tag{3.54}$$

so that each channel use necessarily corresponds to one of the source sequences. The thrust of this analysis and Problem 4 is to demonstrate that when the binary (1,0) source data probabilities are equal, that is, when $q = 1 - q = 1/2$ the source probabilities are identical and equal to $P(x_i) = 1/M \,\forall\, i$ and that this condition results in the maximum source entropy.

Substituting (3.53) into (3.52) and summing over the M source sequences using the source bit probabilities q and $1 - q$ results in the following expression for the source entropy:

$$H_X = -\sum_{\ell=0}^{K} \binom{K}{\ell} q^\ell (1-q)^{K-\ell} \log_2\left(q^\ell (1-q)^{K-\ell}\right) \tag{3.55}$$

where the first term in the brackets denotes the binomial coefficients. By differentiating (3.55) with respect to q and setting the result equal to zero, the maximum source entropy is found to occur when $q = 1/2$ and the result is

$$H_X = \frac{K}{2^K}\sum_{\ell=0}^{K} \binom{K}{\ell} = K \quad : \text{maximum when } q = \frac{1}{2} \tag{3.56}$$

From (3.55), the unique source bit a priori probabilities are

$$P(x_\ell) = q^\ell (1-q)^{K-\ell} \quad : \ell = 0, \ldots, K-1 \tag{3.57}$$

and the number of sources with probability $P(x_\ell)$ corresponds to the value of the binomial coefficient given in (3.56). With $q = 1/2$, (3.57) is $P(x_\ell) = 1/2^K = 1/M : \forall \ell$ and, considering the duplicated occurrences, this corresponds to $P(x_i) = 1/M : \forall i$.

The conditional entropy of the source data, given the sink data, is evaluated using (3.48) by summing over the M source and sink sequences so that

$$H_{X|Y} = -\sum_{i,j=0}^{M-1} P_{Y|X}(y_j|x_i) P_X(x_i) \log_2\left(P_{X|Y}(x_i|y_j)\right) \tag{3.58}$$

The conditional entropy is evaluated using the same rationale as in the evaluation of the source entropy with the source-bit probabilities $P(b_k^i = 1) = q$ and $P(b_k^i = 0) = 1 - q$ and the transition probabilities ε and $(1 - \varepsilon)$ corresponding the BSC channel. Performing the summations in (3.58) with the appropriate source and sink sequences x_i and y_j the conditional entropy is evaluated as

$$H_{X|Y} = -\sum_{i=0}^{M-1} P_X(x_i) \sum_{\ell=0}^{K} \binom{K}{\ell} \varepsilon^{\ell}(1-\varepsilon)^{K-\ell} \log_2\left(\varepsilon^{\ell}(1-\varepsilon)^{K-\ell}\right)$$
$$= -\sum_{\ell=0}^{K} \binom{K}{\ell} \varepsilon^{\ell}(1-\varepsilon)^{K-\ell} \log_2\left(\varepsilon^{\ell}(1-\varepsilon)^{K-\ell}\right) \quad (3.59)$$

The final expression in (3.59) uses (3.54) so the condition entropy is independent of the a priori probabilities; this is explicit in (3.49) for the binary source case. Using (3.56) and (3.59) the average mutual information becomes

$$I_{Y,X} = K - \sum_{\ell=0}^{K} \binom{K}{\ell} \varepsilon^{\ell}(1-\varepsilon)^{K-\ell} \log_2\left(\varepsilon^{\ell}(1-\varepsilon)^{K-\ell}\right) \quad (3.60)$$

The capacity over the BSC with an M-ary data source is evaluated as

$$C_K \triangleq \max_{P(x_i)} (I_{Y,X})$$
$$= K - \sum_{\ell=0}^{K} \binom{K}{\ell} \varepsilon^{\ell}(1-\varepsilon)^{K-\ell} \log_2\left(\varepsilon^{\ell}(1-\varepsilon)^{K-\ell}\right) \quad (3.61)$$

Evaluation of (3.61) for $\varepsilon = 1$ and 0 using the limiting value $x\log_2(x) = 0$: lim $x \to 0$ results in the channel capacity $C = K$ bits/use and the capacity at $\varepsilon = 1/2$ is found to be zero.

As an example application, consider the BSC channel with $P_X(x_i) = 1/M$: $i = 0,\ldots,M-1$. In this case the conditional entropy $H_{X|Y}$ and the capacity are plotted in Figures 3.13 and 3.14, respectively. The maximum conditional entropy occurs under the condition $\varepsilon = 1/2$ corresponding to the most uncertainty in the received data. Not surprisingly, this also corresponds to the condition for the least amount of information through the channel giving rise to the minimum channel capacity. The conditions corresponding to the minimum conditional entropy are $\varepsilon = 0$ and 1 that also correspond to the maximum channel capacity. Although the definition of capacity does not distinguish between correct and incorrect data, if the data is known to be completely incorrect, $\varepsilon = 1$, then a correct decision can be made with certainty by simply reversing the results. In practice, the bit-error probability will range between 1/2 and 1.0 as the signal-to-noise ratio

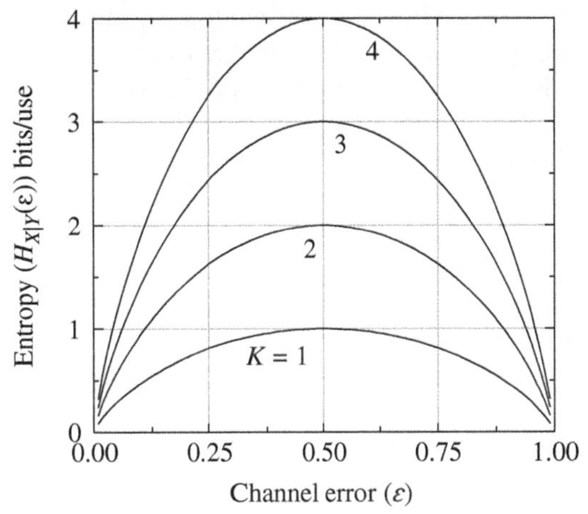

FIGURE 3.13 Conditional entropy of BSC for $K = 1, \ldots, 4$ ($P_X(x_{ki}) = 1/M$).

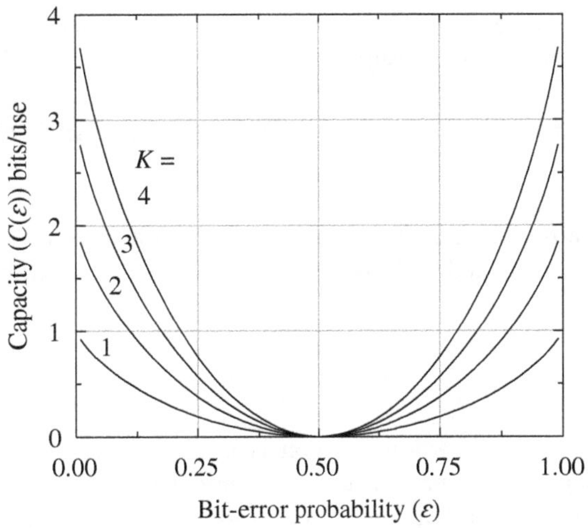

FIGURE 3.14 Channel capacity of BSC for $K = 1, \ldots, 4$ ($P_X(x_i) = 1/M$).

increases from $-\infty$ to ∞, thus limiting the practical range of the abscissas in Figures 3.13 and 3.14.

3.3.3 Converse to the Noisy-Channel Coding Theorem

The converse to the noisy-channel coding theorem has a very practical application that will be developed in Section 3.3.5. The application of interest is the characterization of the channel capacity for constrained data sources. That is, unlike Shannon's unconstrained capacity limit, in this section, the channel capacity for a specified waveform modulation and coding is developed for the BSC channel. The converse to the coding theorem is related to concepts involving rate-distortion theory and this introductory description focuses exclusively on the BSC channel and follows the more general and thorough description by Gallager [3].

The distortion measure for the BSC channel with source $i=0, \ldots, M-1$ and sink $j=0, \ldots, M-1$ is denoted as $d(i,j)$ and is assigned a numerical value depending on the cost of the association between source x_i and sink y_j. For example, on the BSC channel with transition probability $P_{Y|X}(y_i|x_i)$, the distortion measure assignment $d(i,j)=0$ if $i=j$ and $=1$ if $i \neq j$. This notation indicates that the cost is zero for a correct decision and 1 for each incorrect decision. Assigning costs to various statistical decisions is a common practice in the performance optimization of both communications and radar system designs [4, 5]. The average distortion is defined as

$$\bar{d} \triangleq \sum_{i,j=0}^{M-1} P_X(x_i) P_{Y|X}(y_j|x_i) d(i,j) \tag{3.62}$$

and the rate-distortion function of the source relative to a given distortion measure d' is defined as

$$R(d') \triangleq \min_{P(y_j|x_i):\bar{d} \le d'} (I_{Y,X}) \quad : \text{bits/source symbol} \tag{3.63}$$

where $I_{Y,X}$ is the average mutual information given by (3.60) for the BSC and $R(d')$ is the source rate relative to d' with

$$d'_{max} = \min_j \sum_i P(x_i) d(i,j) \tag{3.64}$$

The rate-distortion function is a measure of the source capacity and, after considerable manipulations, is evaluated by Gallager (Reference 3, pp. 457–469) for a discrete memoryless source as

$$\begin{aligned} R(d') &\ge K - d'\log_2(d) - (1-d')\log_2(1-d') - d'\log_2(M-1) \\ &= C' \end{aligned} \tag{3.65}$$

where d' is the lower bound on the M-ary source bit-error to achieve the capacity C'. Equation (3.65) represents the converse to the channel coding theorem for a discrete memoryless source and the equality condition applies when

$$d' \le (M-1) \min_i P_X(x_i) \tag{3.66}$$

As stated previously, (3.65) and (3.66) will be applied in Section 3.3.5 to evaluate the theoretical capacity limits for multilevel pulse amplitude modulation (MPAM) and multilevel quadrature amplitude modulation (MQAM) modulated waveforms.

3.3.4 Shannon's Channel Capacity Limit

The communications capacity through an AWGN channel is established by Shannon's classical formula [6]

$$C = W\log_2\left(1 + \frac{P_s}{N}\right) \quad : \text{bits/second} \tag{3.67}$$

where W is the one-sided bandwidth of the channel, P_s is the signal power, $N = N_o W$ is the total noise power, and N_o is the one-sided noise power spectral density. Shannon's capacity theorem states that, through proper coding and decoding design, error-free transmission can be achieved through the AWGN channel for a communication source data rate $R_b < C$ where C is the capacity of the channel in bits/second and that error-free transmission cannot be achieved for $R_b > C$.

Shannon's capacity limit applies to an unconstrained channel use, that is, it is not constrained by a particular type of waveform modulation. Although this channel capacity is unattainable in practice, except under the most ideal of circumstances,* there are a number of coding techniques that are used together with waveform modulations that come very close to Shannon's limit. For example, M-ary coded waveforms and waveforms employing convolutional and block coding demonstrate the trend toward Shannon's capacity limit. However, the most notable advances toward Shannon's capacity limit have been in the area of TCs, turbo-like codes, and low-density parity-check (LDPC) coding techniques; TC and LDPC codes are discussed in Sections 8.12 and 8.13. Shannon's formula is used in Section 3.3.5, together with the converse to the coding theorem, to establish the capacity limit for various types of waveform modulations with coding.

Shannon's capacity limit is interpreted in two distinct ways: the bandwidth limited interpretation and the power limited interpretation. In the *bandwidth limited* regime, the available channel bandwidth W is fixed and the capacity increases logarithmically with an increase in the signal power P_s, that is, the signal power must increase exponentially with the channel capacity. In this regime, the channel bandwidth is on the order of the information rate so improvements in the bit-error probability is limited to increasing the signal power, providing waveform coding, within the confines of the channel bandwidth, and demodulator equalization; the wire-line telephone channels are classical examples of fixed bandwidth channels. In the *power limited* regime, the signal power is a premium commodity and bandwidth abounds, that is, $W \gg R_b$. In this case, the available bandwidth is used to provide M-ary waveform coding and FEC coding to improve the

*For example, when using spectral root-raised-cosine shaping, the ideal Nyquist bandwidth of $R_s/2$ is obtained only if the corresponding impulse response is infinite in duration. Placing practical constraints on the impulse response will result in ISI thus degrading the performance.

bit-error probability. The satellite communication channel is a classic example of operating in the power limited regime where power is very costly. Although Shannon's capacity theorem provides performance limits in both regimes, in the power limited regime error-free communications is approached with increasing bandwidth.

Shannon's capacity in the bandwidth-limited region is evaluated by normalizing (3.67), by the bandwidth W to obtain a measure of the channel efficiency expressed as

$$\begin{aligned}\eta &= \frac{C}{W}\\ &= \log_2\left(1+\frac{P_s}{N_o W}\right) \quad : \text{bits/second/hertz} \quad \text{or} \quad \text{bits/use}\end{aligned} \tag{3.68}$$

where $\gamma_W = P_s/N_oW$ is the signal-to-noise ratio measured in a bandwidth of W hertz. In this regime, the bandwidth W is fixed and the signal-to-noise ratio is increased by increasing the signal power P_s. Figure 3.15 shows the channel efficiency as defined by C/W as a function of γ_W. In this case, the number of degrees of freedom, based on the Nyquist bandwidth, is $N_d = 2WT$ where T is the duration of the channel use. Shannon's theoretical capacity, expressed by (3.67), applies for $N_d = 2$ degrees of freedom so $W = 1/T$ is the bandwidth occupied by each channel use. For example, to transmit 4 bits/s/Hz through an AWGN channel requires a minimum signal-to-noise ratio of 12 dB using 2 degrees of freedom.

The BSC channel capacity can be applied when the channel errors result from AWGN simply by substituting $\varepsilon = P_{be}$ where

$$P_{be} = erfc_*\left(\sqrt{\frac{d^2}{2N_o}}\right) \tag{3.69}$$

The *solid line* plots in Figure 3.16 show the results for K = 1, 2, and 3 bits-per-use (or bits-per-symbol) when plotted as a function of the signal-to-noise ratio E_b/N_o and interpreted as representing binary phase shift keying (BPSK), QPSK, and 8PSK waveform modulations, respectively. BPSK modulation uses only one of the two orthogonal dimensions in phase space, that is, only one channel of the in-phase and quadrature (I/Q) baseband channels. Therefore, the fundamental capacity is limited to one half that of higher order phase modulated waveforms like QPSK and 8PSK that use both quadrature channels. The *dashed* curves in Figure 3.16 represent Shannon's modulation independent or unconstrained maximum capacity for one- and two-dimensional (1D and 2D) systems; Shannon's capacity limit for coded modulation waveforms is discussed in Section 3.3.5. The detection loss of the M-ary channel capacity relative to Shannon's capacity measured at $E_b/N_o = 2$ dB is about 2.2 dB for each dimension.

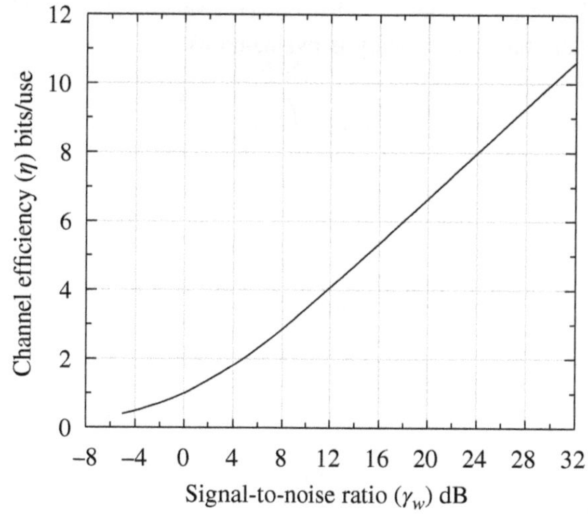

FIGURE 3.15 Shannon's theoretical channel efficiency limit.

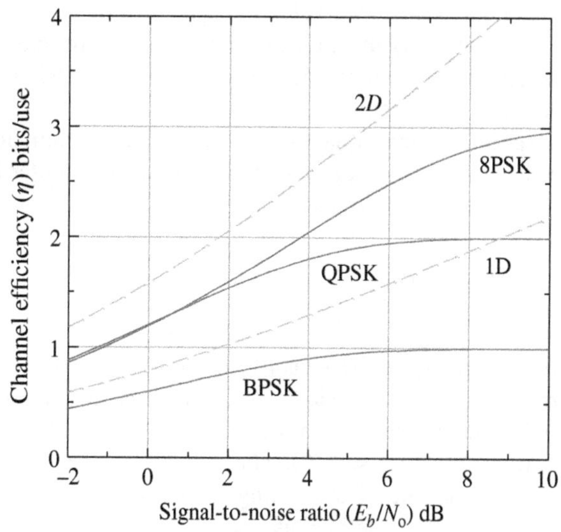

FIGURE 3.16 MPSK capacity of AWGN channel for K = 1, 2, 3 ($P_X(x_i) = 1/M$).

With multiphase shift keying (MPSK) modulation, prior to saturation at the capacity limit, the parameter K corresponds to a relatively high bit-error probability condition so the gray and random decoding formulas* for converting symbol errors to bit errors are not accurate. Consequently, the bit-error performance results are obtained using Monte Carlo simulations to determine the 8PSK capacity performance in Figure 3.16.

In the power-limited regime, the signal power is held constant and the bandwidth W is allowed to increase without limit.

*The formulas are discussed in Section 4.2.1 in connection with MPSK detection.

In this case, carrier-to-noise power density ratio, P_s/N_o, is constant and the capacity is evaluated as

$$\begin{aligned} C &= \lim_{W\to\infty} W\log_2\left(1+\frac{P_s}{N_oW}\right) \\ &= \lim_{W\to\infty} W\log_2(e)\ln\left(1+\frac{P_s}{N_oW}\right) \\ &\approx \lim_{W\to\infty} W\log_2(e)\left(\frac{P_s}{N_oW}\right) \\ &= 1.443\frac{P_s}{N_o} \end{aligned} \quad (3.70)$$

This approximation requires that $P_s/N_oW \le 1^*$ so the communication is in a very low signal-to-noise environment. In this case, as the modulation coding uses more and more bandwidth the capacity increases linearly with the carrier-to-noise density ratio so bandwidth expansion through the use of coding is an effective way to approach the channel capacity limit. This result is based on a value of signal-to-noise ratio less than unity which is the case in a jamming environment so spread-spectrum is an efficient use of bandwidth when the communication is being jammed.

In a period of time equal to T seconds, the maximum number of information bits that can be communicated under low signal-to-noise conditions is given by

$$CT = 1.443\frac{E}{N_o} = 1.443\gamma_b\left(\frac{T}{T_b}\right) \quad : \text{bits} \quad (3.71)$$

where $\gamma_b = E_b/N_o$ is the signal-to-noise ratio in the source information or bit bandwidth $1/T_b$. Equation (3.71) can be viewed as having dimensions of bits/second/hertz in a bandwidth of $W \cong 1/T$ Hz so CT is also a measure of bandwidth efficiency. The rate of the coding, including the waveform symbol mapping, is defined as $r = T/T_b \cong R_b/W \ll 1$.

A commonly used performance measure is the ratio of the energy per information bit to the noise density, E_b/N_o, required to achieve a specified bit-error probability P_{be}. Upon letting $C = R_b$, the channel efficiency is expressed as

$$\frac{R_b}{W} = \log_2\left(1+\frac{E_b}{N_o}\frac{R_b}{W}\right) \quad : \text{bits/second/hertz} \quad (3.72)$$

Defining the channel coding rate as† $r = R_b/W$ and solving (3.72) for E_b/N_o results in

*In general, the approximation requires that $|P_s/N_oW| \le 1$ with $P_s/N_oW \ne -1$; however, in this case, $P_s/N_oW \ge 0$.

†This code rate can be viewed as the rate $r_c = k/m$ of a block code with k information bits and m code bits with $m-k$ parity bits.

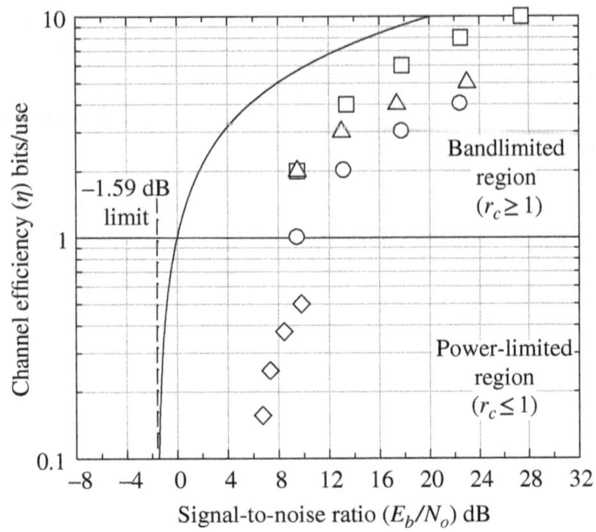

FIGURE 3.17 Theoretical channel efficiency limit at $P_{be} = 10^{-5}$ (MQAM (*squares*), MPAM (*circles*), MPSK (*triangles*), *M-ary* coded BPSK (*diamonds*)).

$$\frac{E_b}{N_o} = \frac{2^r - 1}{r} \quad (3.73)$$

This result is plotted in Figure 3.17 with E_b/N_o as the independent variable. The region $r \ge 1$ represents the bandwidth-limited region corresponding to modulation waveforms without coding like MPSK, MPAM, and MQAM. Typically $M = 2^K$ where K is the number of source bits mapped into an M-ary signal amplitude and phase constellations. The channel efficiency of these modulated waveforms is included in Figure 3.17 corresponding a bit-error probability of $P_{be} = 10^{-5}$; these performance data points are established in Chapters 4 and 6. On the other hand, the region corresponding to $r < 1$ represents the power-limited region. Modulated waveforms having high coding gains and bandwidth requirements apply to this region.

The E_b/N_o asymptote in the power-limited region is determined by letting $r \to 0$ in (3.73). Because the limit is indeterminate, the application of L'Hospital's rule results in

$$\begin{aligned} \left.\frac{E_b}{N_o}\right|_{\text{limit}} &= \ln(2)2^r\big|_{r\to 0} \\ &= 0.693 \quad (= -1.59\ \text{dB}) \end{aligned} \quad (3.74)$$

This is the theoretical limit in E_b/N_o for error-free communications and is the target performance limit for coded waveforms. For example, for $E_b/N_o > -1.59$ dB there exists a coded waveform that will result in an arbitrarily small error probability. Conversely, for $E_b/N_o < -1.59$ dB there is no coding structure that will provide for an arbitrarily small error probability. Until the discovery of TCs [7] in 1993 the practical limit was on the order of 2–3 dB; however, with turbo coding E_b/N_o ratios below 0 dB have been achieved. Shannon's capacity limit does not address the complexity associated with the decoding of low-rate codes; however, prior to 1993 it was thought to be prohibitive; this notion was also overturned through iterative decoding, a hallmark of turbo decoding, resulting in reasonable computational complexity. As mentioned previously, TC, TC-like codes, and LDPC codes are discussed in Chapter 8.

3.3.5 Capacity of Coded Modulated Waveforms

The channel capacity of an M-ary data source with constrained waveform modulation, for example, MPAM, and MQAM, with coding over the BSC channel is evaluated using the rate-distortion function characterized by (3.65) with source bit-error probability corresponding the channel capacity given by (3.66). For the BSC channel, the maximum capacity occurs when $P_X(x_i) = 1/M \ \forall\ i$ so that the equality condition of (3.65) applies when $d' \le (M-1)/M$. Therefore, using $d' = P_{be}$ where P_{be} is the bit-error probability of the waveform modulation of interest corresponding to a prescribed signal-to-noise ratio E_b/N_o. With these conditions C' is expressed as

$$C' = K - P_{be}\log_2(P_{be}) - (1-P_{be})\log_2(1-P_{be}) - P_{be}\log_2(M-1) \quad (3.75)$$

The overall source rate r is defined as the ratio of the symbol rate (R_s) to the source bit-rate (R_b) to the transmitted, that is,

$$r \triangleq \frac{R_s}{R_b} = r_c K \quad (3.76)$$

The second equality includes the FEC coding rate $r_c = R_{cb}/R_b$ where R_{cb} is the rate of the FEC code bits and $K = R_s/R_{cb}$ is number of code bits mapped into a transmitted symbol.* Figure 3.18 shows the relationship between the source coding functions.

For this evaluation Shannon's capacity, expressed in terms of E_b/N_o, r, N_d and normalized by the bandwidth $W = R_b/r$, becomes

$$C = \frac{N_d}{2}\log_2\left(1 + \frac{2r}{N_d}\left(\frac{E_b}{N_o}\right)\right) \quad (3.77)$$

where $N_d = 1$ and 2 corresponding to the degrees of freedom. For example, in phase space, $N_d = 1$ corresponds using one of the quadrature phase channels as in BPSK modulation and

*Sergio Benedetto and Guido Montorsi refer to the rate r as the *codulator* rate.

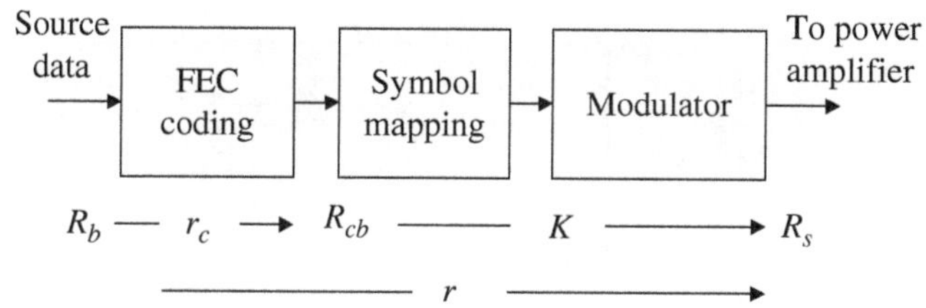

FIGURE 3.18 Source data coding rates.

$N_d = 2$ corresponds to using both the I and Q channels as in QPSK modulation.

For reliable communication, the code rate r_c must satisfy the condition

$$r_c \le \frac{C}{C'} = \frac{\log_2\left(1 + \dfrac{2r}{N_d}\left(\dfrac{E_b}{N_o}\right)\right)}{2C'/N_d} \quad (3.78)$$

where C' is given by (3.65). Solving (3.78) for E_b/N_o under the equality condition results in

$$\frac{E_b}{N_o} = \frac{N_d}{2r}\left(2^{2r_c C'/N_d} - 1\right) \quad (3.79)$$

This result is plotted in Figure 3.19a for coded binary PAM (2PAM) with $K = 1$, $M = 2$ and, because pulse amplitude modulation (PAM) only uses one dimension of the carrier phase, $N_d = 1$. In this figure, and the following figures, the performance is plotted using the FEC code rate $r_c = r/K$. The uncoded performance, shown as the *dashed* curves in the following figures, is included for comparison and is evaluated in Chapter 6. The results show the minimum E_b/N_o that can be obtained for a given code rate r and when $r = 0$ the FEC coding bandwidth is infinite and corresponds to Shannon's limit of −1.59 dB. For this special case, the performance is identical to BPSK. Figure 3.19b shows the performance results using 4PAM with $K = 2$, $M = 4$, and $N_d = 1$. In this case, the theoretical limiting E_b/N_o values are considerably higher than with 2PAM coded modulation.

Figure 3.20 shows the performance of coded 4QAM and 8QAM with $K = 2$ and 3 and the limiting performance with $r = 0$ corresponds to the Shannon limit. The 4QAM case is unique, in that, it also corresponds to the performance of coded QPSK modulation. The uncoded 8QAM performance curve, shown *dashed*, is based on a rectangular rest-point constellation that is not an optimum configuration.

Figure 3.21 is a summary of the minimum E_b/N_o as a function of the codulator rate $r = Kr_c$ with modulation waveforms using $N_d = 1$ and 2 degrees of freedom. These maximum values correspond to $P_{be} = 0$ so that $C' = K$ in (3.79). The limiting performance for $r = 0$ is Shannon's limit of −1.59 dB. Because $r_c \le 1$, values of $r > 1$ correspond to M-ary modulated waveforms with $M > 2$.

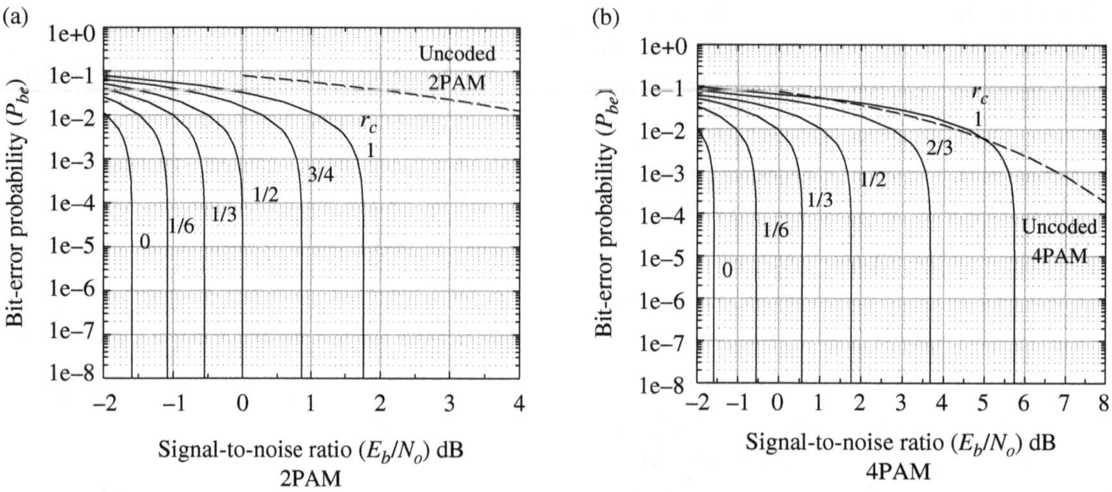

FIGURE 3.19 Performance limits of PAM with FEC coding.

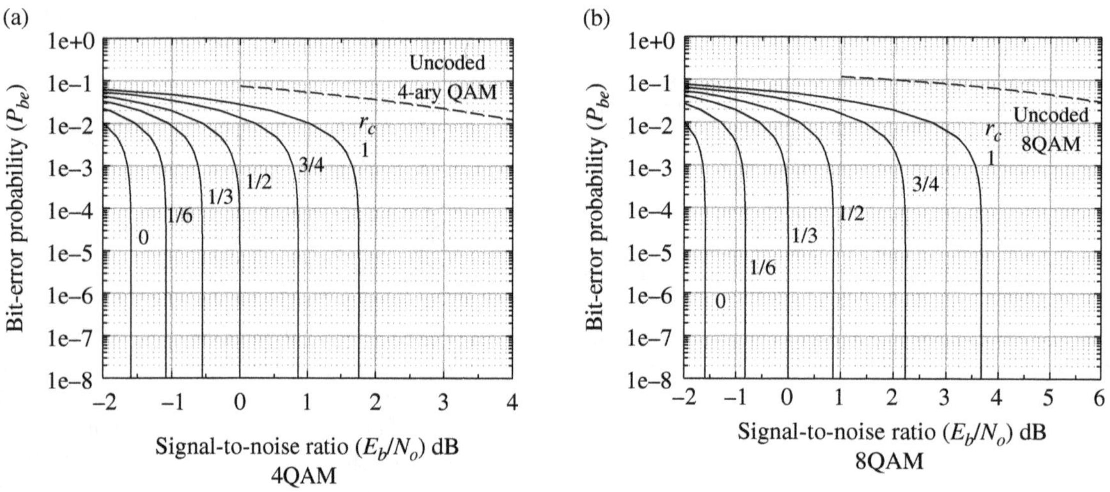

FIGURE 3.20 Performance limits of QAM with FEC coding.

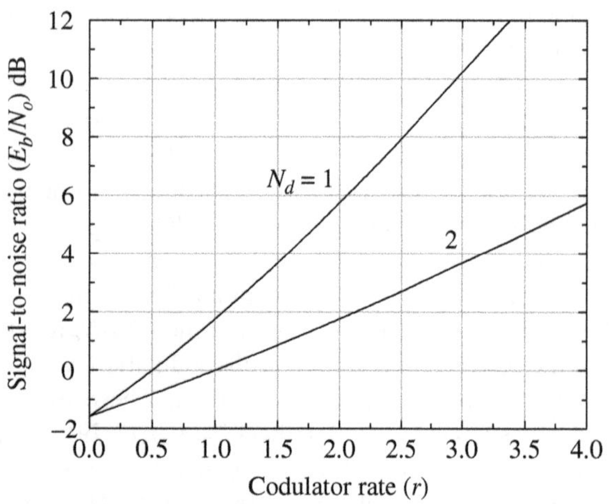

FIGURE 3.21 Performance summary for $P_{be} = 0$ $(r = Kr_c)$.

3.4 BIT-ERROR PROBABILITY BOUND ON MEMORYLESS CHANNEL

The analysis in Sections 3.3.1 and 3.3.2 characterizes the theoretical channel capacity of memoryless BSC channels in terms of the average mutual information as expressed in (3.61) for M-ary modulated waveforms. In Section 3.3.5 the theoretical limit on E_b/N_o as a function of the code rate (r_c) is examined for coded PAM and quadrature amplitude modulation (QAM) modulated waveforms. The results indicate that as r_c approaches zero the limiting E_b/N_o approaches Shannon's limit of −1.59 dB. However, these results do not establish the practical issue regarding the achievable capacity of a particular modulated waveform. An approach involving the *computational cutoff rate* (R_o) [8–11], defined in terms of the memoryless channel transition probabilities $p_{Y|X}(y_j|x_i)$

TABLE 3.1 Computation Cutoff Rate for Selected Waveforms.

Waveform	R_o (Bits/Channel Use)	Comments
Coherent BPSK	$1-\log_2\left(1+\sqrt{4\varepsilon(1-\varepsilon)}\right)$	Hard-decision detection[a]
Coherent BPSK	$1-\log_2\left(1+\exp\left(\frac{-E_b}{N_o}\right)\right)$	Soft-decision detection[b]

[a] $\varepsilon = Q\left(\sqrt{2E_b/N_o}\right)$ for BSC channel with AWGN.

[b] In general, the exponent is E_s/N_o where $E_s = kE_b$ with k bits/symbol or channel use.

introduced in Section 3.3, is used to examine this question.* For example, using a n-bit block coded waveform with n' information bits with code rate $R_c = r_c = n'/n$, the upper bound on the bit-error probability for maximum-likelihood decoding is expressed as

$$P_{be} \le 2^{-n(R_o - R_c)} \tag{3.80}$$

Equation (3.80) indicates that if $R_c < R_o$ the bit-error probability can be made arbitrarily small by increasing the code block size, otherwise an irreducible error event occurs.

The value of R_o for the BSC channel is listed in Table 3.1 [12] for hard- and soft-decision detection of coherent BPSK and for binary, constraint length K, convolutional coding; this convolutional coded case requires that $R_o/R_c > 1$ to achieve an arbitrarily small bit-error probability with increasing constraint length.

The computational cutoff rate is also applied in the design and evaluation of antijam (AJ) waveforms to counter jamming strategies [13–16]. In these descriptions the computational cutoff rate is conveniently characterized, in terms of the decoding metric D, $M = 2^k$ the modulation symbol states, and the code sequences of length n bits. The average bit-error bound for pairs of code sequences of length n with randomly selected symbols is expressed as [17]

$$P_{be} \le \left(\frac{1+(M-1)D}{M}\right)^n \tag{3.81}$$

For (3.81) to conform to (3.80), such that $P_{be} \le 2^{-nR_o}$, the computational code rate is defined as

$$R_o \triangleq \log_2\left(\frac{M}{1+(M-1)D}\right) \tag{3.82}$$

Upon solving (3.82) for D results in

$$D = \frac{M2^{-R_o} - 1}{M-1} \tag{3.83}$$

*This subject is discussed in Chapter 13 in the context of signal jamming.

The upper bound on the bit-error probability is defined in terms of parameter D as

$$\begin{aligned} P_{be} &\le G(D) \\ &= D/2 \quad : \text{with ML detection} \end{aligned} \tag{3.84}$$

For coherent BPSK modulation, corresponding to $M = 2$, with the AWGN channel and hard-decision detection, the decoding metric is computed as [17]

$$\begin{aligned} D &= \min_{\lambda \ge 0}\left(\varepsilon e^{\lambda} + (1-\varepsilon)e^{-\lambda}\right) = \sqrt{4\varepsilon(1-\varepsilon)} \\ &: \text{BPSK, hard-decision detection} \end{aligned} \tag{3.85}$$

Similarly, with soft-decision detection,

$$\begin{aligned} D &= \min_{\lambda \ge 0}\left(e^{-2\lambda E_b + \lambda^2 E_b N_o}\right) = e^{-E_b/N_o} \\ &: \text{BPSK, soft-decision detection} \end{aligned} \tag{3.86}$$

For these binary cases, R_o is computed as

$$R_o \triangleq \log_2\left(\frac{2}{1+D}\right) \quad : \text{bits/symbol} \tag{3.87}$$

and (3.85) and (3.86) correspond to the respective conditions in Table 3.1.

The upper bound on the decoded bit-error probability of the constraint length K, convolutional codes is expressed in terms of R_o as [11, 16]

$$P_{be} < \frac{2^{-KR_o/R_c}}{\left\{1-2^{-((R_o/R_c)-1)}\right\}^2} \tag{3.88}$$

Equation (3.88) applies for $R_o > R_c$. A more general bound, based on the channel coding and demodulator decision metric parameter D, is expressed as

$$P_{be} \le \frac{1}{2}F'(D) \tag{3.89}$$

where

$$F'(D) = \partial T(N, L, D)/\partial N|_{N,L=1} \tag{3.90}$$

The factor of 1/2 in (3.89) applies to all ML demodulation detection processing; typically this applies to antipodal and orthogonal modulated waveforms. The convolutional decoding function $F'(D)$ is expressed as a polynomial in D, with $D^n = Q\left(\sqrt{nE_b/N_o}\right)$. The polynomials in D are listed for constraint lengths 3 through 7 in Tables 8.29 and 8.30 for binary convolutional codes with rates 1/2 and 1/3 and in Tables 8.31 and 8.32 for rate 1/2 convolutional codes with 4-ary and 8-ary coding, respectively.

Wozencraft and Jacobs [18] apply the computational cutoff rate as a measure to determine the bit-error performance degradation with receiver quantization. In this case, a memoryless transition probability diagram is used to characterize L matched filter output levels and Q quantizer levels $(Q > L)$ defining the transition probabilities $q_{\ell,q} : \ell = 1,\ldots,L$; $q = 1, \ldots, Q$ from which R'_o is computed.

3.5 PROBABILITY INTEGRAL AND THE ERROR FUNCTION

Evaluation of the bit-error probability for a communication system in an AWGN channel is essential in determining the performance in the most basic of environments. One of the most important measures of the system performance is the signal-to-noise ratio E_b/N_o required to achieve a specified bit-error probability. This performance measure is applied to most systems early in the design process even though the systems are ultimately to be used in vastly different environments like band-limited channels and fading or impulsive noise channels. One reason for the characterization of the system performance in the AWGN channel is the simplicity associated with mathematical evaluations and the fidelity with which the channel noise can be implemented in the laboratory or modeled in computer simulations. Another reason is that AWGN is an underlying source of noise in almost all systems and must, therefore, be addressed at some point in the evaluation process. In this section various mathematical techniques are examined for describing the integral of the Gaussian *pdf* and clarifying various relationships between them and their application to system performance evaluations.

The Gaussian *pdf* is completely characterized by two parameters: the mean value (m) and standard deviation (σ) and is expressed as

$$p_X(x) = \frac{1}{\sqrt{2\pi}\sigma} e^{-(x-m)^2/2\sigma^2} \tag{3.91}$$

A normalized form with zero mean and unit variance is obtained by letting $y = (x-m)/\sigma$ for which $p(y)$ is evaluated as

$$\begin{aligned} p_Y(y) &= p_X\left(\frac{y^2}{2}\right)\frac{dx}{dy} \\ &= \frac{1}{\sqrt{2\pi}} e^{-y^2/2} \end{aligned} \tag{3.92}$$

The probability integral is defined as

$$\begin{aligned} P(Y) &\triangleq P_r(y \le Y) \\ &= \int_{-\infty}^{Y} p_Y(y)dy = \frac{1}{\sqrt{2\pi}} \int_{-\infty}^{Y} e^{-y^2/2} dy \end{aligned} \tag{3.93}$$

The complement $P_r(Y < y \le \infty)$ is defined as

$$\begin{aligned} Q(Y) &\triangleq P_r(Y < y \le \infty) \\ &= \frac{1}{\sqrt{2\pi}} \int_{Y}^{\infty} e^{-y^2/2} dy \end{aligned} \tag{3.94}$$

The use of $Q(Y)$ to denote the complement of the probability integral* is used by Wozencraft and Jacobs [19] to characterize the bit-error probability of a communication demodulator operating in an AWGN environment.

The error function [20, 21] and its complement are often used in place of $Q(Y)$ to describe the bit-error probability performance. These functions are defined as

$$erf(Y) \triangleq \frac{2}{\sqrt{\pi}} \int_0^Y e^{-y^2} dy \quad \text{and} \quad erfc(Y) \triangleq 1 - erf(Y) \tag{3.95}$$

and are depicted in Figure 3.22 in the context of the Gaussian *pdf* $p_Y(y)$.

Unfortunately, there are several other definitions used for $Q(Y)$, $erf(Y)$, and $erfc(Y)$. For example, Van Trees [22] uses the notation $erf_*(\bullet)$ for $P(Y)$ and $erfc_*(\bullet)$ for $Q(Y)$ and the error function and its complement are alternately defined as

$$erf'(Y) = \frac{2}{\sqrt{2\pi}} \int_0^Y e^{-y^2/2} dy \quad \text{and} \quad erfc'(Y) = 1 - erf'(Y) \tag{3.96}$$

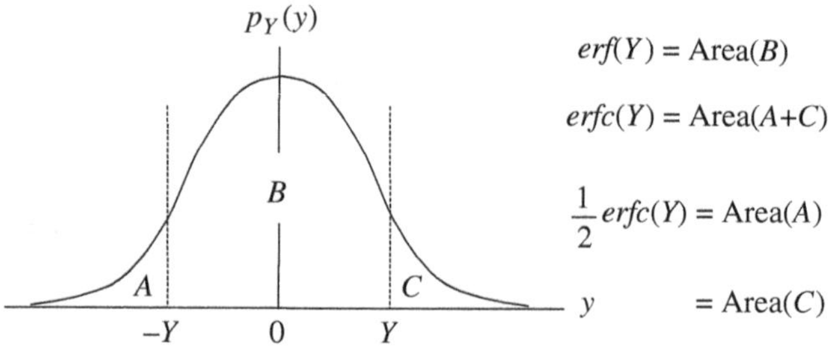

FIGURE 3.22 Depiction of error notations.

*The notation $Q(Y)$ must not be confused with the Q-function which is a two-parameter function used to define the error probability of a noncoherent detector.

In the literature these alternate definitions use identical notations; however, to avoid confusion they are distinguished in this text using the *primed* notation as in (3.96).

Using a simple substitution of variables these definitions are found to be related as

$$erf'(Y) = erf\left(\frac{Y}{\sqrt{2}}\right) \tag{3.97}$$

The error functions *erf*(*Y*) and *erf*′(*Y*) represent the area under the Gaussian *pdf* over the interval $(-Y < y \le Y)$ about the mean value so that one half of the complementary error function corresponds to the complement of the probability integral, that is,

$$Q(Y) = \frac{1}{2}\left(1 - erf\left(\frac{Y}{\sqrt{2}}\right)\right) = \frac{1}{2}(1 - erf'(Y)) = 1 - erf_*(Y)$$
$$= \frac{1}{2}erfc\left(\frac{Y}{\sqrt{2}}\right) = \frac{1}{2}erfc'(Y) = erfc_*(Y) \tag{3.98}$$

and it follows that

$$P(Y) = 1 - \frac{1}{2}erfc\left(\frac{Y}{\sqrt{2}}\right) = 1 - \frac{1}{2}erfc'(Y) = 1 - erfc_*(Y)$$
$$= \frac{1}{2}\left(1 + erf\left(\frac{Y}{\sqrt{2}}\right)\right) = \frac{1}{2}(1 + erf'(Y)) = erf_*(Y) \tag{3.99}$$

A polynomial approximation to the error function *erf*(Y), defined by (3.95), is attributed to Hastings [23], and expressed (Reference 20, Equation 7.1.26, p. 299) as

$$erf(Y) \cong 1 - e^{-Y^2}\sum_{i=1}^{5} a_i\left(\frac{1}{1+pY}\right)^i \quad : 0 \le Y \le \infty \tag{3.100}$$

where $p = 0.3275911$, $a_1 = 0.254829592$, $a_2 = -0.284496736$, $a_3 = 1.421413741$, $a_4 = -1.453152027$, and $a_5 = 1.061405429$. Over the indicated range of Y the approximation error is $|\varepsilon(Y)| \le 1.5 \times 10^{-7}$. Although (3.100) is used extensively in this book to compute the error function, more current applications use the VAX/IBM© intrinsic functions *erf*(*x*) and *erfc*(*x*) available with 32-, 64-, and 128-bit precision [24].

ACRONYMS

ADC	Analog-to-digital conversion
AGC	Automatic gain control
AJ	Antijam (waveform)
AWGN	Additive white Gaussian noise
BPSK	Binary phase shift keying
BSC	Binary symmetric channel
DAC	Digital-to-analog conversion (or converter)
DMC	Discrete memoryless channel
DSP	Digital signal processor
DSSS	Direct-sequence spread-spectrum (waveform)
EHF	Extremely high frequency
FEC	Forward error correction
I/Q	Inphase and quadrature (channels or rails)
IF	Intermediate frequency
LDPC	Low-density parity-check (code)
LLRT	Log-likelihood ratio test
LNPA	Low-noise power amplifier
LRT	Likelihood ratio test
MAP	Maximum a posteriori (decision rule)
ML	Maximum likelihood (decision rule)
MPAM	Multilevel pulse amplitude modulation
MPSK	Multiphase shift keying
MQAM	Multilevel quadrature amplitude modulation
PAM	Pulse amplitude modulation
PCM	Pulse code modulation (baseband)
PN	Pseudo-noise
PSK	Phase shift keying
QAM	Quadrature amplitude modulation
QPSK	Quadrature phase shift keying
RF	Radio frequency
TC	Turbo code
VLF	Very low frequency

PROBLEMS

1. Show that the average conditional information or conditional entropy for the BSC with binary source data given by of (3.48) is independent of the source a priori probabilities. Using this result verify mathematically that the maximum capacity, shown in Figure 3.12, occurs when the source a priori probabilities are equal.

2. Determine the self-information and entropy for the source data sequences consisting of all possible combinations of $K = 2$ bits of binary data $x_i = \{b_1, b_2\}$: $b_\ell \in \{1,0\}$, $\ell = 1,2$ when each bit b_ℓ occurs with equal probability. Express the results in both bits and nats. When each bit b_ℓ occurs with equal probability, and for any positive integer K, show that the source entropy is equal to K bits.

3. Consider that 8-bit ASCII characters are randomly generated with equal probabilities for 1 and 0 bits. Furthermore, each 8-bit character is preceded with a *mark* ($b_1 = 1$) start bit and is terminated with a *space* ($b_{10} = 0$) stop bit. Determine the self-information associated with each bit and the average self-information of the 10-bit sequence.

4. Suppose that a data source associates two bits-per-symbol, that is, $K=2$, $M=4$, and $x_i=\{b_1,b_2\}$: $b_\ell \in \{1,0\}$, $\ell = 1,2$ with the probability of each *mark* bit equal to $P(b_\ell = 1) = q$: $0 \le q \le 1$.

a. Determine the probability of $P(x_i)$: $i=1, \ldots, M$ in terms of q for all 2^K combinations of x_i.

b. Show that $\sum_i P(x_i) = 1$.

c. Express the average self-information in terms of q.

d. Plot a graph of $H_X(q)$ for $0 \le q \le 1$.

e. What is the maximum entropy of the source data bits and at what value of q does the maximum occur?

5. Show that $I_{Y,X}(y_j,x_i) = I_{X,Y}(x_i,y_j)$.

6. For the BSC shown in Figure 3.9, express the sink state probabilities $P(y_j)$: $j=1,2$ in terms of the source bit or a priori probabilities $P(x_i)$: $i=1,2$ and the channel error ε. Sketch $P(y_1)$ as a function of $P(x_1)$ for $\varepsilon = 0$, 1/2, and 1. Under what condition is $P(y_j) = P(x_i)$? What is the relationship between the channel transition probability and the a posteriori probability $P(x_i|y_j)$ under these conditions?

7. Referring to (3.73) derive Shannon's limit for the power-limited channel, that is, shown that $E_b/N_o = -1.59$ dB as $r \to 0$.

8. Derive the relationship between $erf'(Y)$ and $erf(Y)$ given by (3.97).

9. Derive the relationship between $erf(Y)$ and $erf(-Y)$ and plot or sketch $erf(Y)$ for all Y.

10. Repeat Problem 9 for $erfc(Y)$ and $erfc(-Y)$.

REFERENCES

1. C.E. Shannon, "A Mathematical Theory of Communication," Bell System Technical Journal, Part I & II, Vol. 27, No. 7, pp. 379–423, July 1948 and Part III, Vol. 27, No. 10, pp. 623–656, October 1948.
2. J.G. Lawton, "Comparison of Binary Data Transmission Systems," Cornell Aeronautical Laboratory, Inc., Sponsored by the Rome Air Development Center, Contract AF 30(602)-1702, date unknown.
3. R.G. Gallager, *Information Theory and Reliable Communication*, Chapter 9, John Wiley & Sons, Inc., New York, 1968.
4. H.L. Van Trees, *Detection, Estimation, and Modulation Theory*, John Wiley & Sons, New York, 1968.
5. W.B. Davenport, Jr., W.L. Root, *An Introduction to the Theory of Random Signals and Noise*, McGraw-Hill Book Company, Inc., New York, 1958.
6. C.E. Shannon, "Communication in the Presence of Noise," Proceedings of the I.R.E., Vol. 37, No. 1, pp. 10–21, January 1949.
7. C. Berrou, A. Glavieux, P. Thitimajshima, "Near Shannon Limit Error-Correcting Coding and Decoding: Turbo Codes," Proceeding of ICC '93, pp. 1064–1070, Geneva, Switzerland, May 1993.
8. C.E. Shannon, "Probability of Error for Optimal Codes in a Gaussian Channel," Bell System Technical Journal, Vol. 38, pp. 611–656, May 1959.
9. R.G. Gallager, "A Simple Derivation of the Coding Theorem and Some Applications," IEEE Transaction on Information Theory, Vol. IT-11, pp. 3–18, January 1965.
10. J.M. Wozencraft, I.M. Jacobs, *Principles of Communication Engineering*, Chapter 6, John Wiley & Sons, New York, 1967.
11. A.J. Viterbi, J.K. Omura, *Principles of Digital Communication and Coding*, Chapter 2, McGraw-Hill Book Company, New York, 1979.
12. J.M. Wozencraft, I.M. Jacobs, *Principles of Communication Engineering*, pp. 398, 399, John Wiley & Sons, New York, 1967.
13. J.K. Omura, B.K. Levitt, "A General Analysis of Anti-jam Communication Systems," Conference Record, National Telecommunications Conference, November 1981.
14. J.K. Omura, B.K. Levitt, "Coded Error Probability Evaluation for Anti-jam Communication Systems," IEEE Transactions on Communications, Special Issue on Spread-Spectrum Communications, Vol. COM-30, No. 5, pp. 896–903, Part 1 of Two Parts, May 1982.
15. M.K. Simon, J.K. Omura, R.A. Scholtz, B.K. Levitt, *Spread Spectrum Communications*, Volumes I–III, Computer Science Press, Inc., Rockville, MD, 1985.
16. R.L. Peterson, R.E. Ziemer, D.E. Borth, *Introduction of Spread Spectrum Communications*, Prentice Hall, Upper Saddle River, NJ, 1995.
17. M.K. Simon, J.K. Omura, R.A. Scholtz, B.K. Levitt, *Spread Spectrum Communications*, Volume I, Chapter 4, pp. 192–202, Computer Science Press, Inc., Rockville, MD, 1985.
18. J.M. Wozencraft, I.M. Jacobs, *Principles of Communication Engineering*, Chapter 6, pp. 386–405, John Wiley & Sons, New York, 1967.
19. J.M. Wozencraft, I.M. Jacobs, *Principles of Communication Engineering*, Chapter 4, pp. 245–266, John Wiley & Sons, New York, 1967.
20. M. Abramowitz, I.A. Stegun, *Handbook of Mathematical Functions with Formula, Graphs, and Mathematical Tables*, Applied Mathematics Series 55, U.S. Department of Commerce, National Bureau of Standards, Washington, DC, June 1964.
21. H. Urkowitz, "A Note on the Error Function," Research Memorandum RM-1, Philco Scientific Laboratory, Philco Corporation, January 31, 1961.
22. H.L. Van Trees, *Detection, Estimation, and Modulation Theory*, Chapter 2, p. 37, John Wiley & Sons, New York, 1968.
23. C. Hastings, Jr., *Approximations for Digital Computers*, Princeton University Press, Princeton, NJ, 1955.
24. Lahey Software Solutions, "Lahey/Fujitsu Fortran 95 Language Reference, Revision D," Lahey Computer Systems, Inc., Incline Village, NV, 1998.

4

PHASE SHIFT KEYING (PSK) MODULATION, DEMODULATION, AND PERFORMANCE

4.1 INTRODUCTION

The various forms of phase shift keying (PSK) waveform modulations are discussed in this chapter with an emphasis on binary PSK (BPSK) and quadrature PSK (QPSK) corresponding to $M = 2$ and 4 respectively. A significant feature of multiphase PSK (MPSK) waveforms is that they result in a constant envelope waveform that provides for the efficient use of transmitter power amplifiers. Furthermore, they are relatively simple to implement and provide performance robustness with severe channel conditions like: hard limiting, fading, impulse noise, and Doppler rate. The simplest implementations use *rect(t/T)* time-domain symbol shaping resulting in a *sinc(fT)* spectrum. A major drawback with the *sinc(fT)* spectrum is the poor spectral containment leading to more severe intersymbol interference (ISI) with channel filtering and a greater sensitivity to adjacent channel interference (ACI). The QPSK waveform and its variant offset QPSK (OQPSK)* are unique, in that, the bit-error performance, in terms of the energy-per-bit, is identical to that of BPSK. The OQPSK modulation is characterized by the quadrature-phase channel, or rail, being shifted in time by the bit duration $T_b = T/2$ relative to the in-phase channel. This offsetting of the quadrature channels provides: robustness to spectral re-growth in a hard-limited channel; stable phase tracking at 0 and π radians, compared to $n\pi/2$ (n = 0, …, 3) for QPSK, and less sensitivity to demodulator phase tacking. The poor spectral containment associated with the *rect(t/T)* symbol shaping is improved with minimum shift keying (MSK) modulation which uses cosine symbol shaping. MSK modulation, also referred to as fast FSK (FFSK) [1, 2], uses offset quadrature symbols with the advantages listed above for OQPSK. MSK was invented in 1961 [3] and it was not until the early 1980s that further improvements in the modulated signal spectrum were achieved with the introduction of phase shaping over adjacent symbols using phase-shaped BPSK (S-BPSK) [4], phase-shaped MPSK (S-MPSK), and Gaussian MSK (GMSK). During this same time period continuous phase modulation (CPM) [5], using multiple modulation indices (multi-*h* CPM), was finding widespread applications because of the excellent spectral properties and inherent coding gain. MSK embodies binary continuous phase FSK (CPFSK) modulation and can be detected using a much simpler FSK demodulator, albeit with a loss in performance.

In the following sections, the various forms of MPSK are examined focusing primarily on the modulator and demodulator implementations and the spectral characteristics. The implementation of BPSK, QPSK, OQPSK, and MSK is examined using conventional baseband I/Q channel descriptions. These results lead to the generalized implementations involving CPM as discussed in Section 4.2.5 and Chapter 9. Based on more generalized implementations of MSK, GMSK and sinusoidal MSK (SMSK) are examined with a focus on CPM. The notion of continues-phase frequency modulation (CPFM) is introduced and compared to CPM modulation.

The modulations mentioned above all provide a constant amplitude-modulated signal with improvement in the basic *sinc(fT)* spectral response as more exotic phase shaping is

*Offset QPSK is sometimes referred to as staggered QPSK (SQPSK).

Digital Communications with Emphasis on Data Modems: Theory, Analysis, Design, Simulation, Testing, and Applications,
First Edition. Richard W. Middlestead.

Companion website: www.wiley.com/go/digitalcommunications

applied to the symbol phase transitions. These approaches to spectral containment are successful in reducing spectral sidelobes at frequencies $|f| > R_s$ Hz but have little impact for $|f| \leq R_s$. However, an improvement in the spectral characteristics is obtained by using root-raise-cosine (RRC) shaping* in the frequency domain. With spectral RRC [6] (SRRC) shaping,† the signal spectrum is confined, under ideal conditions, to the bandwidth $R_s/(1+\alpha)/2$ where α is the spectral shaping, or excess bandwidth, factor. The range of α is between 0 and 1; when $\alpha = 0$ the waveform is, in theory, completely confined to the Nyquist bandwidth $R_s/2$. The practical aspects of the frequency containment of SRRC modulation are discussed in Section 4.3.2. The penalty in using RRC waveforms is that they result in amplitude modulation (AM) and require a linear power amplifier (PA). When operating with a nonlinear PA, the power must be backed off to operate in the linear range with an associated power backoff loss. The application of SRRC shaping to MPSK-modulated waveforms is examined in Section 4.4.5.

4.2 CONSTANT ENVELOPE PHASE-MODULATED WAVEFORMS

Constant amplitude waveforms allow for operating the transmitter power amplifier in saturation thus providing the maximum available power to be delivered to the antenna. The improvement in the transmitted power is particularly important with disadvantaged terminals that typically operate with low prime power and small antennas. The main focus of this chapter is on PSK-modulated waveforms; however, the description of constant amplitude modulations involving CPFM and FSK is unavoidable.

4.2.1 Multiphase PSK (MPSK) Modulation

The performance of BPSK, corresponding to $M = 2$, was evaluated in Section 3.2.1 and in this section the performance is examined for the general case involving $M = 2^k$. Rather than using the signaling distance as in the previous evaluations, the *pdf* of the received signal phase is established and used to evaluate the symbol and bit-error probability. In this evaluation, the bit-error performance for $M = 2$ and 4 is shown to be identical. Furthermore, the performance is ideal, in that, the demodulator is assumed to have perfect symbol timing and carrier recovery and the memoryless additive white Gaussian noise (AWGN) channel introduces no distortion.

*Formally this is the square-root raised-cosine abbreviated as *root* RC.

†The prefix *spectral* is an important distinction, in that, the root-raised-cosine shaping is applied in the frequency domain resulting in a theoretically bandlimited spectrum.

The waveform modulator applies the channel-coded data to a phase-modulated carrier as shown in Figures 4.1 and 4.2 showing the carrier phase states and the demodulator decision regions, respectively. The transmitted signal for symbol m is

$$s_m(t) = \sqrt{2P}\cos(\omega_c t + \phi_m) \tag{4.1}$$

and, neglecting the $2\omega_c$ term, the energy-per-symbol is simply

$$E_s = \int_{T_i}^{T_i+T} |s_m(t)|^2 dt = PT \tag{4.2}$$

Expressing $s_m(t)$ in terms of the energy-per-symbol and defining the unit energy modulation waveform functions $u_c(t)$ and $u_s(t)$ gives

$$s_m(t) = \sqrt{E_s}\cos(\phi_m)u_c(t) - \sqrt{E_s}\sin(\phi_m)u_s(t) \tag{4.3}$$

where

$$u_c(t) = \sqrt{\frac{2}{T}}\cos(\omega_c t) \tag{4.4}$$

and

$$u_s(t) = \sqrt{\frac{2}{T}}\sin(\omega_c t) \tag{4.5}$$

It is easily shown that these functions have unit energy and are orthogonal to one another.

For the special case of BPSK with a binary source coded alphabet ($M = 2$), $m = \{0, 1\}$, the transmitted signal becomes

$$s_m(t) = (-1)^m \sqrt{E_s} u_s(t) \tag{4.6}$$

The *pdf* of the received carrier phase is evaluated by considering a sinusoidal carrier signal-plus-narrowband Gaussian noise. This has become a conventional problem and the solutions involving the magnitude and phase distributions of the signal-plus-noise are well documented [7]. In general, the *pdf* of the envelope is a Ricean-distributed random variable and a convenient expression for the distribution of the received signal phase, ϕ, is expressed by Hancock [8] as

$$p(\phi) = \frac{e^{-\gamma_s}}{2\pi}\left[1 + \sqrt{4\pi\gamma_s}\cos(\phi)e^{\gamma_s\cos^2(\phi)}erf_*\left(\sqrt{2\gamma_s}\cos(\phi)\right)\right] \tag{4.7}$$

where γ_s is the signal-to-noise ratio in the symbol bandwidth. This function is symmetrical about $\phi = 0$ and is plotted in Figure 4.3. With zero noise ($\gamma_s \rightarrow 0$), the phase *pdf* is uniformly distributed between $-\pi$ and π.

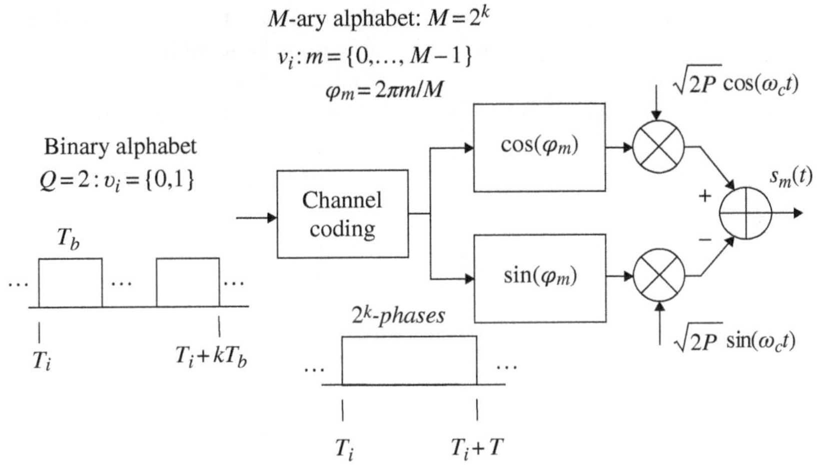

FIGURE 4.1 MPSK channel coding and waveform modulation.

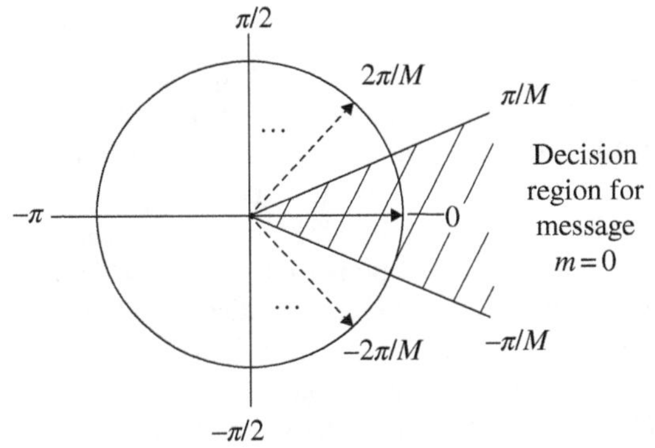

FIGURE 4.2 MPSK phase constellation with demodulator decision boundaries.

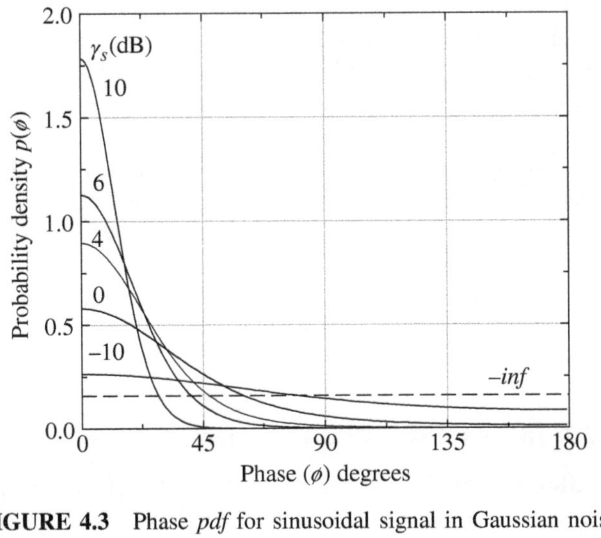

FIGURE 4.3 Phase *pdf* for sinusoidal signal in Gaussian noise (symmetry about $\phi_m = 0$).

For the message $\hat{m}(\phi) = m$, the symbol-error probability is evaluated as

$$P_{se}(m) = 1 - P_{cs}(m)$$
$$= 1 - \int_{\phi_m - \pi/M}^{\phi_m + \pi/M} p(\phi - \phi_m)d\phi \tag{4.8}$$

where $\phi_m = 2\pi m/M$ is the transmitted phase for message m. Based on the symmetry of the decision regions, the error results are identical for all messages m so the average error probability is the same as the error probability for a given message. Based on this observation, the average error probability is most conveniently evaluated for the case $\phi_m = 0$. Therefore, the above integral is numerically evaluated with $\phi_m = 0$ and the results are shown in Figure 4.4 for M = 2, 4, …, 32. The signal-to-noise ratio is in the bit bandwidth where $\gamma_b = E_b/N_o = \gamma_s/k$. The results for M = 2 and 4 modulations are exactly the same as previously mentioned and it is seen that a considerable penalty in signal-to-noise performance is incurred for $M > 4$. For example, M = 8 and 16 require respective increases in the signal-to-noise ratio of about 3.5 and 4.5 dB for reasonably low-error probabilities.

Converting the symbol errors to bit errors is straightforward for M = 2; however, the conversion is not as simple for $M > 2$. The *solid* curves in Figure 4.5 assume that random bit errors occur with each symbol error and any of the $M - 1$ possible symbol errors are equally likely. Therefore, because of the assumed random distribution of the k-bits in each symbol, the upper bound on the bit-error probability is expressed as

$$P_{be} \le \frac{2^{k-1}}{2^k - 1} P_{se} \quad \text{: random error decoding} \tag{4.9}$$

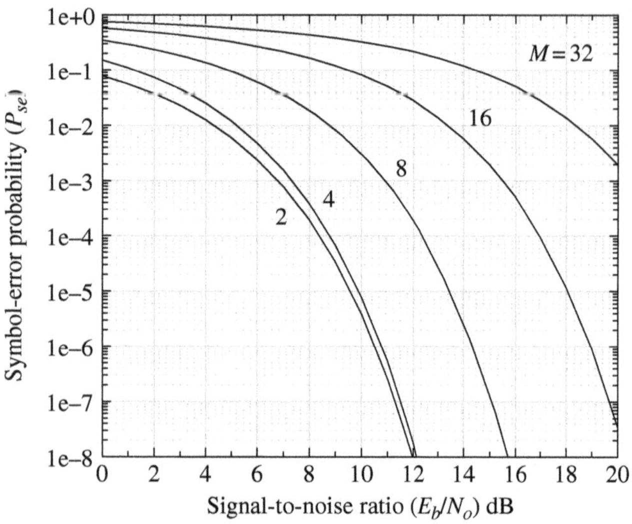

FIGURE 4.4 Coherent MPSK symbol-error performance.

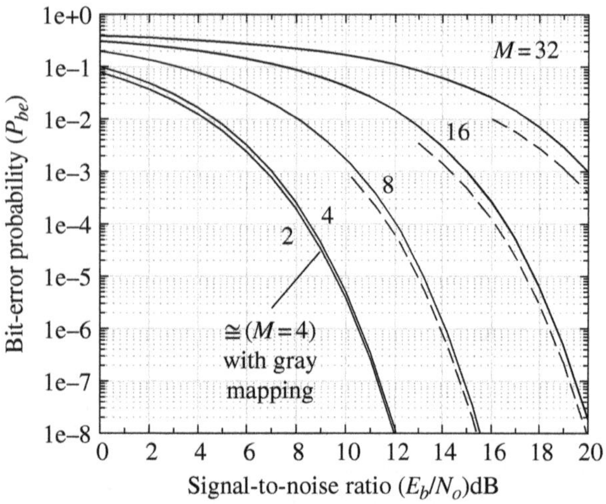

FIGURE 4.5 Coherent MPSK bit-error performance (*solid*: random; *dashed*: gray coding).

In the case of BPSK ($k = 1$), the equality condition applies in (4.9). When the bits are gray coded and mapped to the modulated waveform signal phase and when the decision errors are dominated by errors in adjacent decision regions, as is the case with high signal-to-noise ratios, then a wrong decision results in essentially only one bit-error. Since there are k bits-per-symbol, the approximate bit-error probability with gray coding is evaluated as

$$P_{be} \cong \frac{P_{se}}{k}: \text{ gray coding} \tag{4.10}$$

This result is shown as the *dashed* curves in Figure 4.5 for $M > 4$ and applies only for low-error probabilities. Although not shown in Figure 4.5 as a dashed curve, the performance of QPSK ($M = 4$) with gray coding approaches that of BPSK for low-error probabilities. Because BPSK assigns one bit to each symbol, gray coding is not distinguished, so that $P_{be} = P_{se}$ in (4.10) and the performance is identical to the antipodal signaling results ($\rho = -1$) shown in Figure 3.7.

Numerical evaluation of the integral for $P_{se}(m)$ using Hancock's distribution can be simplified for large signal-to-noise ratios by using the asymptotic expansion for $erf_*(x)$ leading to the result

$$p(\phi) \cong \sqrt{\frac{\gamma_s}{\pi}}\cos(\phi)e^{-\gamma_s \sin^2(\phi)} \quad : \gamma_s \gg 1 \text{ and } |\phi| \le \frac{\pi}{2} \tag{4.11}$$

This approximation is reasonably good for signal-to-noise ratios greater than 10 : 1; the restriction on ϕ simply avoids a negative *pdf*.

The differences between the BPSK and QPSK performances shown in Figure 4.5 result because the QPSK is based on the symbol error computed using (4.8) and there is no convenient way to convert the symbol-error probability to bit-error probability other than by using the approximations (4.9) or (4.10). The fact that the error performance for BPSK and QPSK is identical can be reasoned in two ways; for a fair comparison, the carrier power is considered to be identical for each modulation. First, the QPSK modulation can be viewed as two orthogonal equal power channels of BPSK modulation. Because orthogonal Gaussian distributed noise sources are also independent, the two equivalent BPSK channels are independent with identical signal-to-noise ratios so each exhibits the same bit-error performance, that is, the carrier signal and noise powers divide equally between each quadrature channel so that $P_{beI} = P_{beQ} = P_{be}$ (BPSK). Based on this reasoning, $P_{be}(\text{QPSK}) = (P_{beI} + P_{beQ})/2 = P_{be}(\text{QPSK})$. Another view is to consider the mapping of two BPSK bits, each of duration T_b, to one QPSK symbols of duration $T = 2T_b$. Although the symbol signal-to-noise ratio $E_s/N_o = (E_s/N_o)(T/T_b)$ is 3 dB higher, each of the equivalent BPSK channels has 1/2 of the carrier power. Therefore, it would appear that the bit-error performance of the equivalent BPSK channels would be degraded by 3 dB; however, each of the equivalent QPSK channels is $2T_b$ in duration so the symbol matched filtering reduces the noise by an additional 3 dB resulting in identical performance for BPSK and QPSK. Therefore, referring to Figure 4.5, the QPSK bit-error probability for $M = 4$ is identical to the BPSK bit-error probability corresponding to $M = 2$; this result is also substantiated through Monte Carlo simulations.

4.2.2 Binary PSK (BPSK) Modulation

BPSK is the simplest form of phase shift keying to implement and demodulate and, although not the most efficient in terms of channel efficiency, BPSK provides the most robust

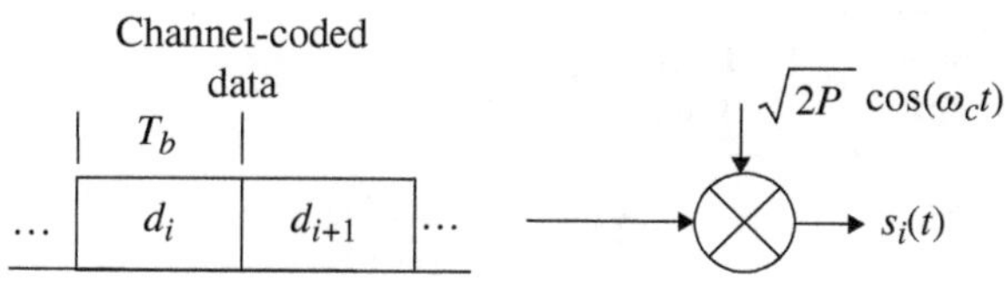

FIGURE 4.6 BPSK implementation.

performance to channel impairments compared to the other forms of MPSK. The poor channel efficiency arises because only one of the two dimensions in phase-space is used. However, these same characteristics provide the tolerance to channel-induced phase variations and nonlinearities that result in the robust performance.

The implementation of BPSK is shown in Figure 4.6. The channel coding converts the binary bits $b_i = \{0,1\}$ to bipolar data $d_i = \{1,-1\}$ as: $d_i = 1-2b_i$, $i = 0, 1, \ldots$ so the transmitted signal is simply

$$\begin{aligned} s_i(t) &= d_i\sqrt{2P}\cos(\omega_c t) \\ &= \sqrt{2P}\cos\left(\omega_c t + \left(\frac{1-d_i}{2}\right)\pi\right) \end{aligned} \tag{4.12}$$

where ω_c is the carrier frequency in radians-per-second and P is the transmitted power level that is related to the peak carrier voltage as $V = \sqrt{2P}$. The second expression in (4.12) maps a binary 0 to a phase of zero radians and binary 1 to π radians.

Figure 4.7 shows the implementation of a BPSK demodulator with symbol time and carrier phase tracking loops. As discussed in Chapter 10, the hard-limiting nonlinearity provides a binary data estimate that results in near optimum performance with phase tracking for E_b/N_o ratios greater than about 6 dB. In applications involving fading channels or forward error correction (FEC) decoding, soft-decision estimates of the data are available before the hard limiter. Under ideal noise-free channel conditions,* the received signal at the baseband input to the demodulator, that is, after passing through the noise-free channel, receiver heterodyning, and demodulator mixing to baseband, is expressed as

$$\tilde{s}(t) = \sqrt{\frac{P_r}{2}}\, e^{j(\omega_\varepsilon t + \phi_i + \varphi)} \tag{4.13†}$$

where P_r is the received signal power, ω_ε is the frequency error after removal of the received carrier frequency ω_c, ϕ_i is the data-related phase corresponding to the i-th symbol and is given by

*It is easier and more instructive to examine the underlying waveform demodulator algorithms without the complications involving the channel noise; the impact of AWGN and other channel impairments is addressed in subsequent sections and chapters.

†The dependence of ϕi on time such that $(i - 1/2)T \le t \le (i + 1/2)T$ is implicit.

$$\phi_i = \left(\frac{1-d_i}{2}\right)\pi \tag{4.14}$$

and φ is constant phase error associated with the channel, receiver, and demodulator heterodyning and filtering. The phaselock loop tracking provides the estimate $\hat{\varphi}(t)$ of the signal time dependent phase $\omega_\varepsilon t + \varphi$ that is removed by the complex phase rotation. With ideal symbol timing, the symbol matched filters perform the symbol integration over the fundamental interval $|t| = \le T/2$ and the corresponding weighting function $w(t) = sign(-1, t)$ provides an early–late (E/L) integrator output for estimating and tracking the symbol timing error. For example, with an initial symbol timing error of τ seconds, the estimate $\hat{\tau}$ is computed and the symbol timing is adjusted to align the demodulator symbol timing to the received symbol timing, that is, τ is driven to zero. Considering the sequence of contiguous received symbols, the integration intervals are identified by the integration limits in Figure 4.7 for i = 0,1,... and the corresponding weighting function is defined as

$$w(t) = \sum_i rect\left(\frac{t-iT-\tau}{T}\right) sign(-1, t-iT-\tau) \tag{4.15}$$

Figure 4.8 shows the required processing for the data detection, symbol timing, and phase-error computations; for simplicity the frequency error is assumed to be zero and $\hat{d}_i = d_i$. The data estimate is the hard limiter ±1 output based on the sign of the input B. The symbol timing and phase errors are generated using the detection filter outputs A and C as indicated in Figures 4.7 and 4.8. With random data the average loop filter outputs are‡ $\hat{\tau} \cong K_\tau E[\varepsilon_\tau] = \tau K_\tau V\cos(\varphi_\varepsilon)$ and $\hat{\varphi} \cong K_\varphi E[\varepsilon_\varphi] = (T-\tau)K_\varphi V \sin(\varphi_\varepsilon)$ respectively where $\varphi_\varepsilon = \varphi - \hat{\varphi}$. The sign of the symbol timing error provides the discriminator response and, under ideal conditions, the symbol timing is adjusted through the dynamics of the closed-loop tracking to result in $\tau = 0$. Similarly, the $\sin(\varphi)$ function provides the phase discriminator response with the phase error ε_φ being driven to zero.

4.2.2.1 *π/2-BPSK Modulation* The *rest* phase-state for BPSK modulation are typically 0 and π radians and, if the input data consist of long strings of mark or space data, the demodulator symbol timing will drift resulting in degraded performance. Data randomizers ensure to some degree that random data are applied to the modulator; however, when the phase modulation between consecutive symbols is shifted by $\pm\pi/2$ radians the demodulator is guaranteed to have a phase transition at every symbol. Based on this description, the rest phase-states for $\pi/2$-BPSK modulation are alternately

‡The loop filter gains K_τ and K_φ are adjusted for the optimal acquisition time and steady-state error.

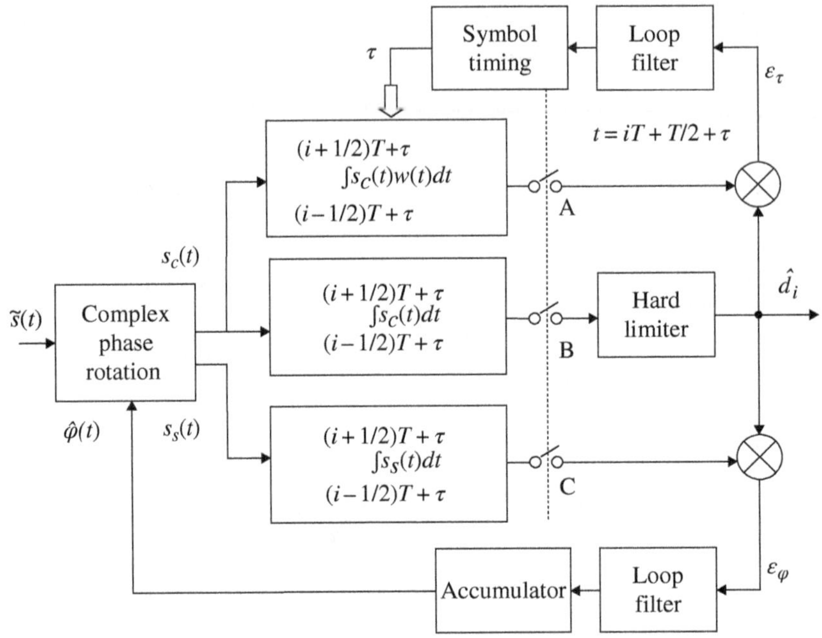

FIGURE 4.7 BPSK demodulator implementation using hard limiter.

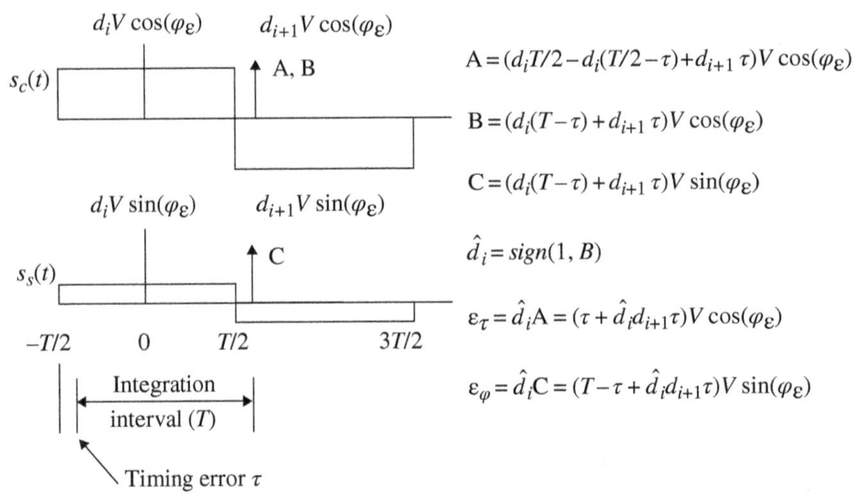

FIGURE 4.8 Example processing for BPSK demodulation.

$(0, \pi)$ and $(\pm\pi/2)$. In addition to the symbol tracking issue, the sensitivity to spectral re-growth with filtering and hard limiting is reduced since the maximum phase change is $|\pi/2|$ instead of π radians. Abrupt modulation phase transition that pass through, or near, the center of the phase constellation produces large amplitude variations when filtered to limit the out-of-band signal energy. However, when a nonlinearity is encountered, as with a hard-limiting repeater, the resulting AM is removed and the spectrum tends to revert to the original unfiltered spectrum. This is particularly apparent with the *sinc*(*fT*) modulation spectrum as characterized in Section 4.4.3.2.

The coherent detection of $\pi/2$-BPSK modulation is accomplished as follows. Given the phase coordinates (x_k, y_k) corresponding to φ_k, the $\pi/2$-BPSK modulator advances or retards the phase to $\varphi_{k+1} = \varphi_k \pm \Delta j$ corresponding to the coordinates (x_{k+1}, y_{k+1}) where

$$\Delta\varphi = \frac{d_k\pi}{2} \tag{4.16}$$

The demodulator uses the estimate $(\hat{x}_{k+1}, \hat{y}_{k+1})$ and the previous data estimate $(\hat{x}'_k, \hat{y}'_k)$ and computes the estimated phase change as

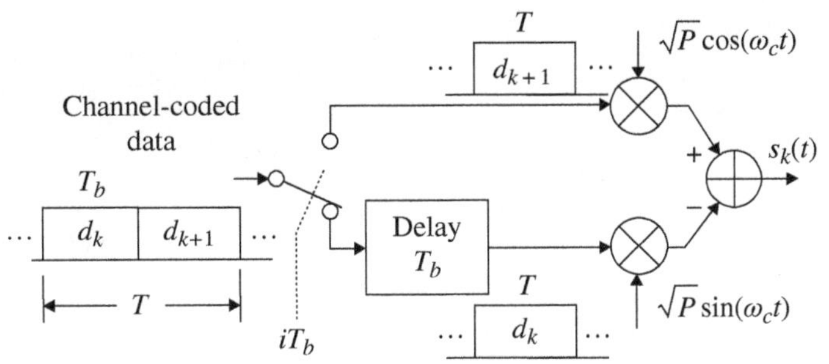

FIGURE 4.9 QPSK modulator implementation.

$$\hat{\Delta\varphi} = \hat{\varphi}_{k+1} - \hat{\varphi}'_k = \tan^{-1}\left(\frac{\hat{y}_{k+1}\hat{x}'_k - \hat{x}_{k+1}\hat{y}'_k}{\hat{y}_{k+1}\hat{y}'_k + \hat{x}_{k+1}\hat{x}'_k}\right) \tag{4.17}$$

Using (4.16) with ρ equal to the argument of the $\tan^{-1}(\bullet)$ function, the demodulated data estimate is computed as

$$\hat{d}_{k+1} = sign(1,\rho) \tag{4.18}$$

The estimates $(\hat{x}'_k, \hat{y}'_k)$ correspond to the rest phase-states of the modulated waveform and simplify the evaluation of ρ. With zero-mean additive noise the rest phase-states correspond to $n\pi/2$ ($n = 0, \ldots, 3$) and the phaselock loop tracking errors are computed based on the average estimate of $(\hat{x}_{k+1}, \hat{y}_{k+1})$ relative to the rest phase-states resulting in coherent carrier tracking; however, the resulting bit-error performance corresponds to that of differentially encoded BPSK (DEBPSK). Differentially coherent detection of BPSK modulation is discussed in Section 4.2.4. The bit-error performance of DEBPSK and differentially coherent BPSK (DCBPSK) is shown in Figure 4.21.

4.2.3 Quadrature Phase-Modulated Waveforms

In this section, the implementation of three forms of QPSK modulation is considered: QPSK, OQPSK, and MSK. Each of these modulations alternately distributes the source data between the in-phase (I) and quadrature-phase (Q) channels. The I and Q channels are associated with the real or *cosine* channel and the imaginary or *sine* channel respectively; these channels are also referred to as I/Q rails. With QPSK the symbol data associated with each channel is time-aligned so that the modulated symbol phase changes abruptly to one of four phase states, typically $\{0, \pm\pi/2, \pi\}$ or $\{\pm\pi/2, \pm 3\pi/2\}$; the selection and associated input data pair is defined by the modulator specification. OQPSK modulation is similar; however, the I/Q channel symbols are offset by one bit interval and associated with the respective sinusoidal quadrature channel. MSK modulation is similar to OQPSK, however, a cosine window or weighting function is applied to each channel symbol that results in a significantly improved transmitted spectrum, that is, the frequency is attenuated as f^{-2} compared to f^{-1} for QPSK and OQPSK. The additional performance differences mentioned in the introduction are examined in various case studies. The QPSK implementations discussed in the following sections represent alternates to the implementation shown in Figure 4.1.

4.2.3.1 Quadrature PSK (QPSK) Modulation An implementation of QPSK in which the channel-coded data are applied directly to the modulator is shown in Figure 4.9. The channel coded binary bits $b_i = \{0,1\}$ are converted to bipolar data $d_i = \{1,-1\}$ as: $d_i = 1 - 2b_i$, $i = 0,1,\ldots$. In this depiction, the parameter k is an even integer and is related to i as: $k = 2\lfloor i/2 \rfloor$ for $i \geq 0$. This simply states that for the natural numbering of the input bits, the even (odd) numbered bits are assigned to the quadrature (in-phase) channel.*

The subscript k corresponds to the d_k symbol being transmitted and the *switch* changes at the data rate $(1/T_b)$ in such a way that the first symbol data bit ($d_k = \pm 1$), $k = \ldots 0,2,\ldots$ modulates the quadrature phase channel as shown. In this case the transmitted waveform is expressed, for $k = \{\text{even}\}$, as

$$\begin{aligned} s_k(t) &= d_{k+1}\sqrt{P}\cos(\omega_c t) - d_k\sqrt{P}\sin(\omega_c t) \\ &= \sqrt{2P}\cos(\omega_c t + \phi_k) \end{aligned} \tag{4.19}$$

where the data-dependent phase is expressed as

$$\phi_k = \tan^{-1}\left(\frac{d_k}{d_{k+1}}\right) = \frac{\pi}{4}(2m+1) \tag{4.20}$$

for $m = \{0,1,2,3\}$. At each bit transition, the change in carrier phase is evaluated as

$$\begin{aligned} \Delta\phi_k &= \phi_{k+1} - \phi_k \\ &= \{0, \pm\pi/2, \pi\} \end{aligned} \tag{4.21}$$

*The use of the index k to denote the data indexing should not be confused with the parameter $k = \log_2(M)$ used previously to denote the bits-per-symbol; the context should make the distinction clear.

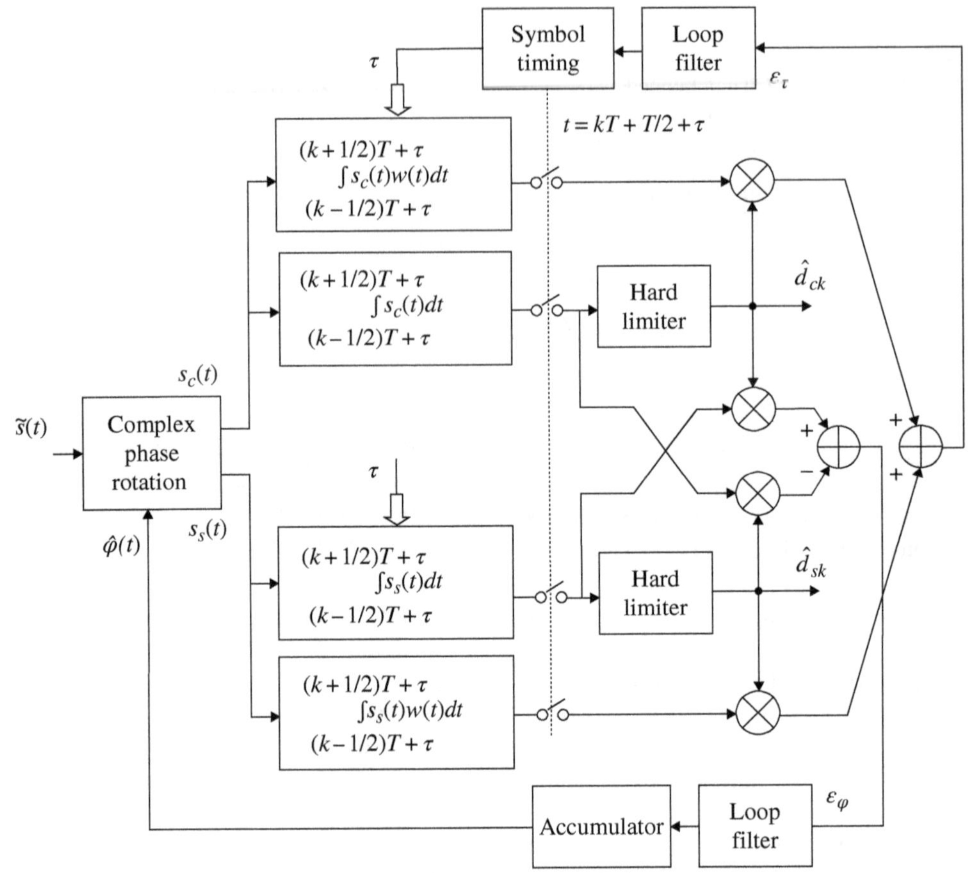

FIGURE 4.10 QPSK demodulator implementation using hard limiter.

As will be seen, the instantaneous changes in the carrier phase play a significant role in determining the spectral characteristics of the waveform and the resulting performance with filtering and hard limiting. With the exception of the constant phase shift of $\pi/4$, (4.20) is identical to the MPSK detection shown in Figure 4.2 with $M = 4$. With the $\pi/4$ phase shift, the decision regions correspond to the first through fourth quadrants of the phase space. On the other hand, if the $\pi/4$ phase shift is removed the decision regions are identical to those defined in Figure 4.2. Since the error performance is invariant with a translation or rotation of the decision space, the performance will be identical in both cases.

Figure 4.10 shows the implementation of a QPSK demodulator with symbol time and carrier phase tracking loops. As in the BPSK demodulator implementation, the hard-limiting nonlinearity provides near optimum performance with phase tracking for high input signal-to-noise ratios. The input signal is described as the baseband function

$$\tilde{s}(t) = \sqrt{\frac{P_r}{2}} e^{j(\omega_e t + \phi_k + \varphi)} \tag{4.22}$$

where P_r is the received signal power, ω_e is the frequency error after removal of the received carrier frequency ω_c, ϕ_k is the data-related phase given by (4.20), and φ is a constant phase error associated with the channel and receiver filtering. The phase and time tracking loops use the phase estimate $\hat{\varphi}(t)$ and symbol time estimate $\hat{\tau}$ to remove the phase and timing errors in the same manner as in the BPSK demodulator. Considering the sequence of contiguous received symbols, the integration intervals are identified by the integration limits in Figure 4.10 and the corresponding weighting function is defined by (4.15) with i replaced by the index k.

4.2.3.2 *π/4-QPSK Modulation* The benefits of $\pi/4$-QPSK modulation are similar to those described in Section 4.2.2.1 for $\pi/2$-BPSK modulation. The rest phase-states for QPSK modulation are expressed in (4.20); however, the instantaneous phase shifts include $\pm\pi$ radians as described by (4.21). These large and abrupt phase changes result in nearly 100% spectral re-growth from hard limiting following spectral control filtering. This sensitivity results from the phase transition passing through the origin of the phase constellation as described in Section 4.2.2.1 for BPSK modulation. With QPSK, the sensitivity to hard limiting can be reduced by including an additional phase shift of $\Delta j = p/4$ at each phase transition in the modulator. The demodulator can remove the phase shift by computing $\hat{\varphi}_k = \hat{\varphi}_{k-2} - \Delta\phi$ and using phase tracking to zero the phase error given by $\varphi_\varepsilon = \hat{\varphi}_k - \hat{\varphi}'_k$, where $\hat{\varphi}'_k$ is the quantized phase corresponding to the rest phase-state. In this case, the bit-error performance corresponds to differentially encoded QPSK (DEQPSK). Alternately, the demodulator can recover the received $\pi/4$-QPSK modulation data using differentially coherent QPSK (DCQPSK) [9] detection [10]. The bit-error performance of DEQPSK and DCQPSK is shown in Figure 4.23.

4.2.3.3 Offset QPSK (OQPSK) Modulation When the channel-coded data are mapped into offset or staggered symbol intervals as shown in Figure 4.11, the resulting modulated waveform is referred to as OQPSK. Under ideal symbol and carrier recovery conditions in the demodulator, the performance of OQPSK modulation is identical to that of QPSK. However, OQPSK is more robust in the presence of filtering followed by a nonlinearity, for example, when using a solid-state power amplifier (SSPA) or a traveling wave tube amplifier (TWTA). In addition, the offset symbols result in less crosstalk between the demodulator channels, which results in more immunity to data transitions during phase and symbol time tracking.

The OQPSK-modulated waveform is expressed, for $k = \{\text{even}\}$, as

$$s_k(t) = \begin{cases} d_{k-1}\sqrt{P}\cos(\omega_c t) - d_k\sqrt{P}\sin(\omega_c t) & : kT_b \le t \le (k+1)T_b \\ d_{k+1}\sqrt{P}\cos(\omega_c t) - d_k\sqrt{P}\sin(\omega_c t) & : (k+1)T_b \le t \le (k+2)T_b \end{cases} \tag{4.23}$$

$$s_k(t) = \begin{cases} \sqrt{2P}\cos(\omega_c t + \phi_{k,o}) & : kT_b \le t \le (k+1)T_b \\ \sqrt{2P}\cos(\omega_c t + \phi_{k,1}) & : (k+1)T_b \le t \le (k+2)T_b \end{cases} \tag{4.24}$$

where

$$\phi_{k,o} = \tan^{-1}\left(\frac{d_k}{d_{k-1}}\right) \tag{4.25}$$

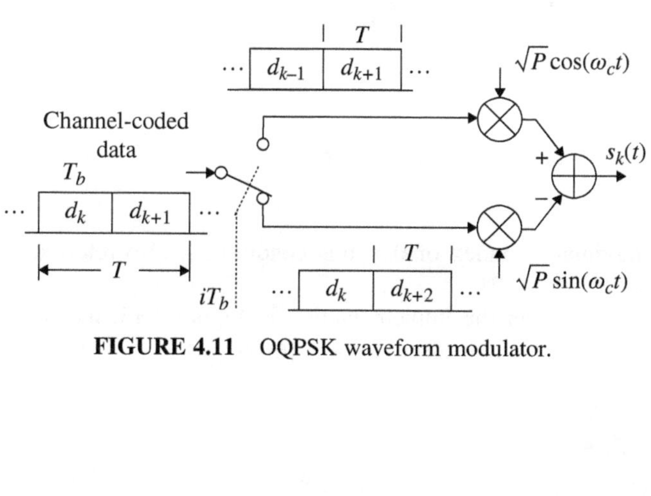

FIGURE 4.11 OQPSK waveform modulator.

and

$$\phi_{k,1} = \tan^{-1}\left(\frac{d_k}{d_{k+1}}\right) \tag{4.26}$$

At each bit transition, the carrier phase change is evaluated as

$$\begin{aligned} \Delta\phi_k &= \phi_{k,1} - \phi_{k,o} \\ &= d_k \tan^{-1}\left(\frac{d_{k-1} - d_{k+1}}{1 + d_{k-1}d_{k+1}}\right) \\ &= \{0,\ \pm\pi/2\} \end{aligned} \tag{4.27}$$

These abrupt phase changes are similar to those occurring in QPSK modulation and result in identical spectral characteristics. However, the improved performance of OQPSK over QPSK with narrowband filtering followed by a nonlinearity is attributed to the fact that an abrupt phase change of π radians cannot occur.

Figure 4.12 shows the implementation of a OQPSK demodulator with symbol time and carrier phase tracking loops. The general description of the QPSK demodulator applies to the OQPSK demodulator with the in-phase symbol matched filter integration over the fundamental interval, that is, with $\tau = 0$, $(k - 1/2)T \le t \le (k + 1/2)T$ and with the symbol time weighting function $w'_s(t) = w(t)$ given by (4.15) and $w_c(t) = w_s(t) = 1$. These in-phase filters are sampled at $t = (k + 1/2)T$. However, because the quadrature channel is offset in time by $T/2$ seconds, the quadrature symbol matched filter integration is over the fundamental interval $kT \le t \le (k + 1)T$ with the symbol time weighting function given by $w'_c(t) = w(t - T/2)$. These quadrature phase filters are sampled at $t = (k + 1)T$. From this description and Figure 4.12, the symbol time and phase loops are updated by the respective errors ε_τ and ε_φ every $T/2$ seconds corresponding to every source bit interval T_b.

4.2.3.4 Minimum Shift Keying (MSK) Modulation The MSK-modulated waveform, shown in Figure 4.13, is also an offset QPSK waveform; however, the symbols in each quadrature channel are cosine weighted resulting in a significant improvement in the out-of-band spectral energy. To obtain the offset, the quadrature or sine channel is delayed by one-half symbol (one-bit) interval relative to the in-phase channel. This delay is implicit in Figure 4.13 by the switching of the input data bits between the two channel, that is, the *switch* operates at the bit rate $R_b = 2/T$ where T is the symbol duration. The *minimum shift* property results because the phase rotation between bit durations is exactly $\pm\pi/2$ radians which is the minimum phase shift resulting in antipodal signaling (π radians) over the symbol duration in each of the quadrature channels. Therefore, with $\omega_m = \pi R_b/2$ each channel can be thought of as cosine weighted BPSK-modulated

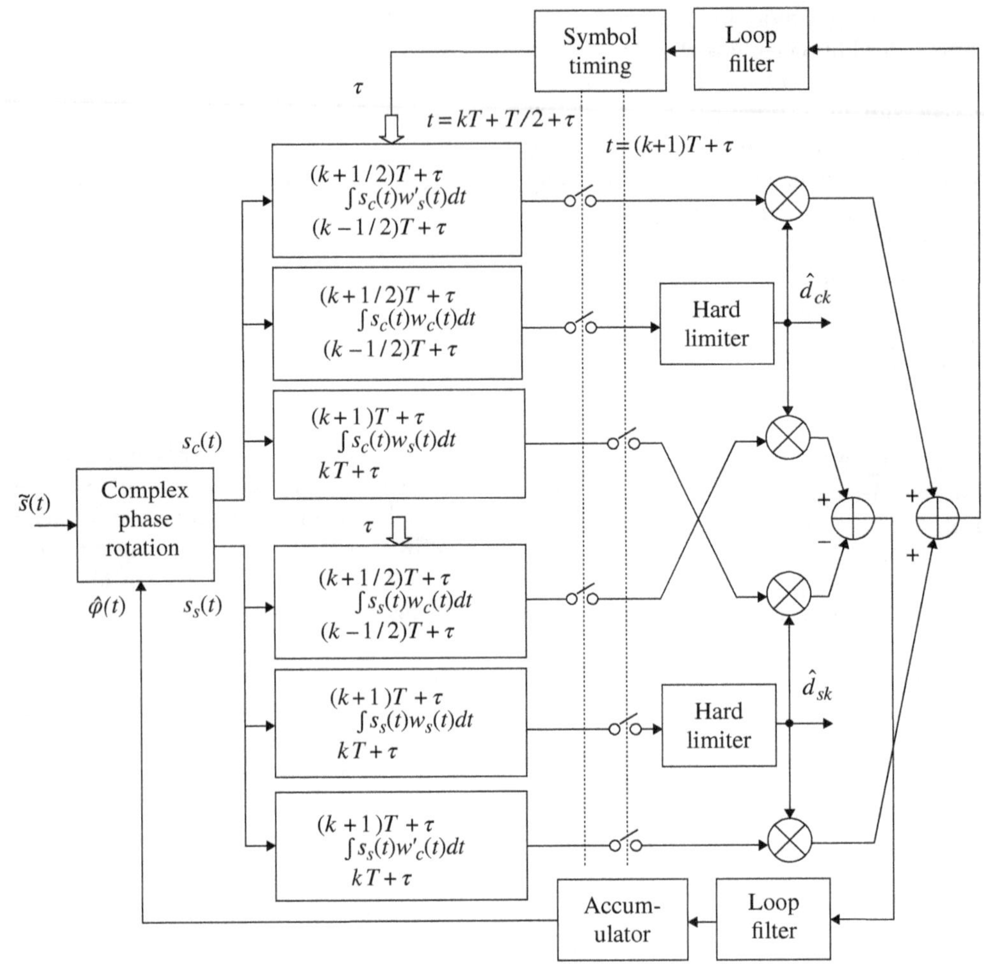

FIGURE 4.12 OQPSK demodulator implementation using hard limiter.

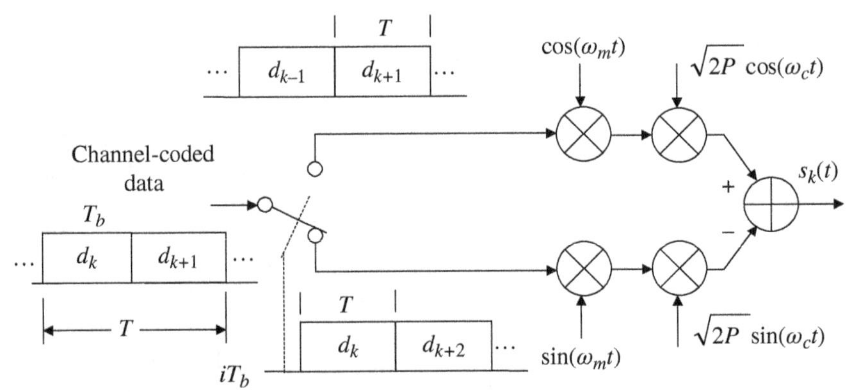

FIGURE 4.13 Classical implementation of MSK waveform modulator.

channels. The implementation shown in Figure 4.13 is the-classical implementation of MSK [3] and provides a mathematical description that conveniently highlights many of the features of the waveform including the unique modulation index of 0.5; this condition is also referred to as FFSK [1, 2].

Based on the implementation in Figure 4.13, the MSK waveform is expressed, for k = {even}, as

$$s_k(t)=\begin{cases} d_{k-1}\sqrt{2P}\cos(\omega_m t)\cos(\omega_c t)-d_k\sqrt{2P}\sin(\omega_m t)\sin(\omega_c t) \\ \qquad : kT_b \le t \le (k+1)T_b \\ d_{k+1}\sqrt{2P}\cos(\omega_m t)\cos(\omega_c t)-d_k\sqrt{2P}\sin(\omega_m t)\sin(\omega_c t) \\ \qquad : (k+1)T_b \le t \le (k+2)T_b \end{cases} \quad (4.28)$$

Upon applying some simple trigonometric identities, this result can be expressed as

$$s_k(t)=\begin{cases} \sqrt{2P}\cos[(\omega_c+(d_{k-1}d_k)\omega_m)t+(d_{k-1}-1)\pi/2] \\ \qquad : kT_b \le t \le (k+1)T_b \\ \sqrt{2P}\cos[(\omega_c+(d_k d_{k+1})\omega_m)t+(d_{k+1}-1)\pi/2] \\ \qquad : (k+1)T_b \le t \le (k+2)T_b \end{cases} \quad (4.29)$$

Defining the phase terms as

$$\phi_{k,0}(t)=d_{k-1}d_k\omega_m t+(d_{k-1}-1)\pi/2 \quad (4.30)$$

and

$$\phi_{k,1}(t)=d_k d_{k+1}\omega_m t+(d_{k+1}-1)\pi/2 \quad (4.31)$$

the time-dependent change in the carrier phase over a bit interval T_b is $\pi k'/2$, where k' is an integer and the data-dependent phase terms ensure a continuous phase function. Referring to (4.29), the phase change over the symbol interval $T = 2T_b$ is evaluated for even integers k as

$$\Delta\phi_k=\phi_{k,0}+\phi_{k,1}=\pm\pi k\mp\pi/2 \quad (4.32)$$

Although (4.32) is developed for even integers k, a similar result is obtained for odd integers, so the data in each channel are antipodal and the additive fixed phase terms ensure a continuous phase transitions. Therefore, the MSK waveform is characterized as having continuous phase transitions giving rise to a significant reduction in the out-of-band spectral energy compared to the QPSK or OQPSK waveforms. It is also apparent that the MSK waveform results in a constant amplitude modulation giving rise to robustness in nonlinear channels similar to that of BPSK, QPSK, and OQPSK. A segment of an MSK-transmitted waveform with $f_cT_b = 1$ is shown in Figure 4.14 for the data sequence $(d_{k-1}, d_k, d_{k+1}, \ldots) = (1,1,-1,1,1,1,-1,\ldots)$. The abscissa is normalized to the bit durations so the transitions occur at integer values along the abscissa with continuous phase transitions.

Further examination of the MSK waveform indicates that the phase transitions are not abrupt but change linearly over the bit interval. This is seen by examining the phase functions $\phi_{k,o}$ and $\phi_{k,1}$ over the appropriate intervals of t and recognizing that $\omega_m T_b=\pi/2$. The modulation index is defined as $m_I = 2\Delta f/R_b$ and, for MSK $\Delta f = f_m$ and $f_m = 1/2T = 1/4T_b$, so $m_I = 1/2$. The allowable phase transitions of the MSK waveform are shown as the *dashed* lines in Figure 4.15 and the *solid* curve corresponds to the data sequence associated with Figure 4.14. The $\pi/2$ phase shift is evident in Figure 4.14 by noting that there is one carrier cycle-per-bit, that is, $f_cT_b = 1$, and that an additional 1/4 cycle occurs in the first, fourth, and fifth bits while the second and third bits each result in 1/4 cycle less of the carrier phase. The carrier frequency shown in Figure 4.14 is conveniently chosen to be equal to the bit duration and with arbitrary carriers the phase change at the bit transitions will not necessarily occur as the waveform goes through zero or a maximum; however, the carrier phase will always be continuous.

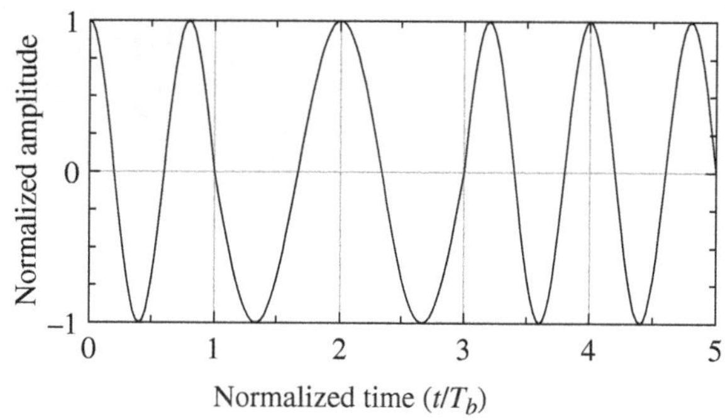

FIGURE 4.14 Example of MSK-modulated waveform.

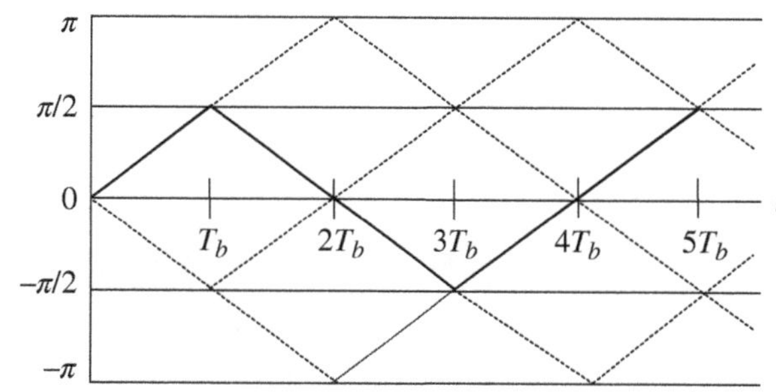

FIGURE 4.15 Example of MSK waveform phase progression.

Figure 4.16 shows an embodiment of the demodulation of the MSK waveform [11] as implemented in Figure 4.13. The received signal is squared to remove the random data and reveal the modulation tones at $2f_c \pm 2f_m$ Hz that are used to phase lock the two voltage controlled oscillators (VCOs); one at $f_c + f_m$ and other $f_c - f_m$ Hz. The phaselocked oscillator outputs are used to generate the matched filter weights for each of the input signal quadrature channels; they are also used to generate the optimum matched filter sampling times for each channel. Because the cosine weighting is applied before the symbol filtering, a simple integrate-and-dump (I&D) detection filter is used; immediately following the sampling of the detection filters the integrators are zeroed,

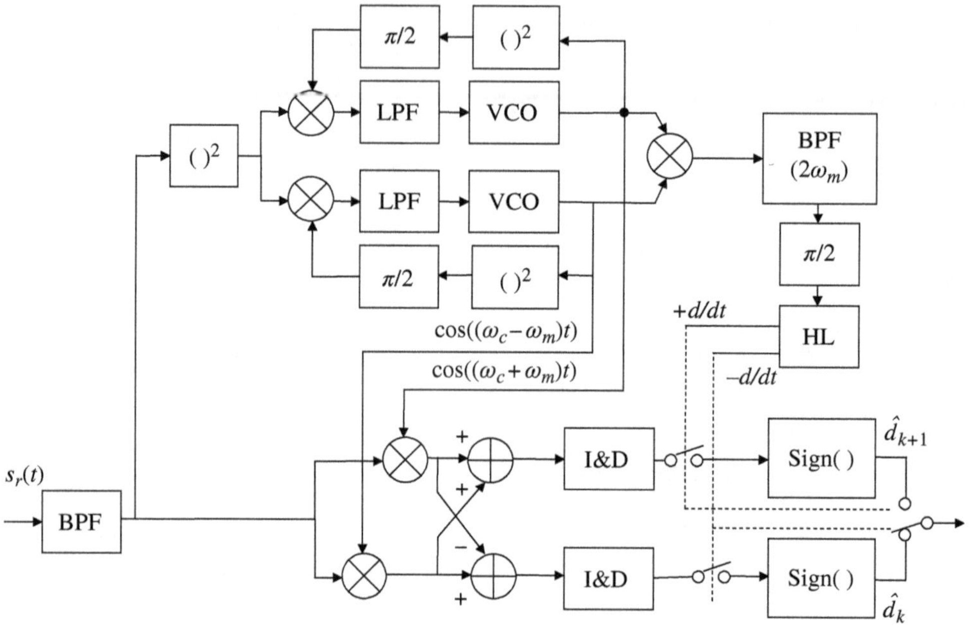

FIGURE 4.16 MSK demodulator implementation for modulator in Figure 4.13.

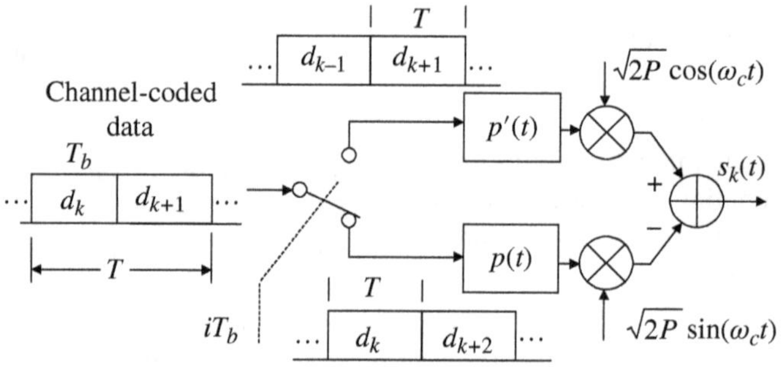

FIGURE 4.17 Conventional implementation of MSK waveform modulator.

or dumped, for the detection of the next symbol. It is implied that the sampling switches are sample-and-hold devices that decouple the filter samples from the next integration interval. The optimum sampling times correspond to the sample impulses $d()/dt$ of the hard-limited (HL) output shown in Figure 4.16. The negative impulses correspond to the optimum sampling instants for the quadrature channel corresponding to d_k in Figure 4.13 with k = even integer and the positive impulses correspond to the optimum samples for the detection of d_{k+1}. The received serial data sequence is $\ldots\hat{d}_k, \hat{d}_{k+1}\ldots$ and the output switch is toggled by the sampling impulses as shown. Although the implementation in Figure 4.16 looks formidable, it includes the carrier frequency phaselock loop and symbol timing acquisition and tracking functions that are equally daunting in other implementations. It is interesting to observe the quadrature matched filter cosine weighting functions adjust as the carrier frequency and channel delay change over time.

Figure 4.17 shows a more conventional implementation of MSK where the cosine weighted symbol shaping is applied to each channel with

$$p'(t) = rect\left(\frac{t-j'T}{T}\right)\cos(\omega_m(t-j'T)) \tag{4.33}$$

and

$$p(t) = rect\left(\frac{t-jT-T/2}{T}\right)\cos(\omega_m(t-jT-T/2)) \tag{4.34}$$

where $j' = (k'+1)/2$ and $j = k/2$ and k' corresponds to the odd numbered source bits. The parameters j' and j simply represent the natural numbering of the bits in the I and Q channels respectively; the relationship between k and k' is characterized in Section 4.2.3.4. Also, the *switch*, operating at the bit rate, directing the source data estimates between the

I and Q channels introduces a one-half symbol delay because of the offset modulation.*

Data detection of the MSK-modulated waveform, as implemented in Figure 4.17, is identical to the OQPSK detection shown in Figure 4.12 with the filter weighting functions changed as follows:

$$w'_s(t) = w_s(t) = \sin(\omega_m t) \tag{4.35}$$

and

$$w'_c(t) = w_c(t) = \cos(\omega_m t) \tag{4.36}$$

where $w_s(t)$ and $w_c(t)$ are equivalent to $p'(t)$ and $p(t)$ respectively as expressed in (4.33) and (4.34) over the offset quadrature symbol intervals.

4.2.3.5 MSK Detection Using Frequency Discriminator

The MSK waveform expressed in (4.29) indicates that MSK can be viewed as a narrow band frequency shift keying (FSK) modulation with a modulation index of $m_I = 2f_mT_b = 1/2$. This is evident from the phase functions which indicate that the waveform frequency during a bit interval is given by

$$\begin{aligned}\omega_i &= \frac{d\phi_i}{dt} \\ &= \omega_c + (d_{i-1}d_i)\omega_m\end{aligned} \tag{4.37}$$

where $i = \{0,1,\ldots\}$. From this result it is apparent that the MSK modulator, as implemented in Figure 4.13, performs the differential encoding operation given by the product $d_{i-1}d_i$ corresponding to $+f_m$ when $d_{i-1} = d_i$ and $-f_m$ when $d_{i-1} \neq d_i$. The MSK demodulator inherently unscrambles the differentially encoded data preserving the antipodal bit-error performance. However, an FSK demodulator must differentially decode the data to recover the source data bits. For example, letting $D_i = d_{i-1}d_i$, the FSK demodulator recovers the data estimate $\hat{D}_i$ and then determines the source data estimate as $\hat{d}_i = \hat{D}_i\hat{d}_{i-1}$; however, this feedback operation involving the previous data estimate $\hat{d}_{i-1}$ results in catastrophic error propagation when a data error occurs. This can be avoided if the source data are differentially encoded as $D_i = D_{i-1}d_i$ and the data sequence D_i is applied to the MSK modulator as shown in Figure 4.18. In this case, the MSK-demodulated tones are given by

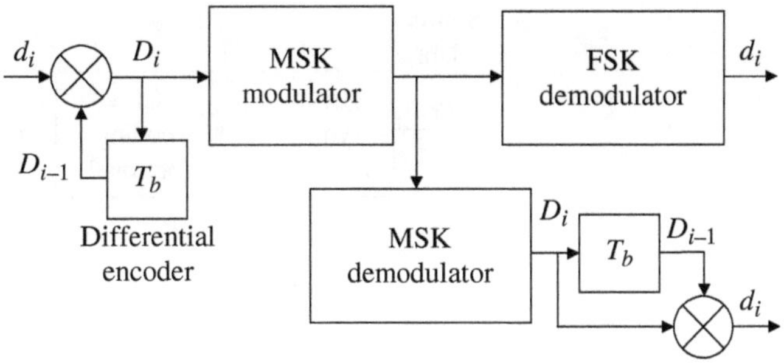

FIGURE 4.18 Differential encoding of MSK modulator (compatible shift keying).

$$\begin{aligned}\omega_i &= \omega_c + (D_{i-1}D_i)\omega_m \\ &= \omega_c + d_i\omega_m\end{aligned} \tag{4.38}$$

where the modulation tone f_m is directly related to the source data d_i. Because this form of modulated signal is compatible with MSK and FSK demodulators, it is referred to as compatible shift keying (CSK). Unfortunately, however, the performance of the MSK demodulator will be degraded because of the necessity to differentially decode the received data estimates $\hat{D}_i$ such that $\hat{d}_i = \hat{D}_i\hat{D}_{i-1}$. This decoding will not result in catastrophic error propagation but results in error multiplication that decreases to 2 : 1 as the signal-to-noise ratio increases. The error performance using an FSK frequency discriminator is examined later; however, it is degraded from that using coherent phase detection of the received MSK waveform.

4.2.4 Differentially Coherent PSK Modulation

Technically, differentially coded PSK refers to the type of phase modulation where the information is contained in the phase difference between the current and the preceding symbol, whereas, differentially coherent PSK refers to the method of detection in the demodulator. The term *differentially coherent detection* is also referred to as *differential comparison* and *phase-comparison detection* so the notation differentially coherent PSK (DCPSK) is appropriate for all of these descriptions; the designation differential PSK (DPSK) is widely used in the literature.†

DCPSK modulation is a robust constant envelope modulation that is simple to demodulate, in that, the information in the current symbol is contained in the phase shift relative to the preceding symbol so there is no need for a phase tracking phaselock loop; although symbol phase decoding must be

*In Figure 4.17, the Q-channel data corresponding to $k = i_{even}$ is depicted as being delayed by $T/2$ from the I-channel data.

†The binary data is often differentially encoded and is denoted by the prefix DE, for example, DEBPSK modulation differentially encodes the binary source bits prior to binary phase modulation. Differential data encoding was introduced in the preceding section and is discussed in more detail in Chapter 8.

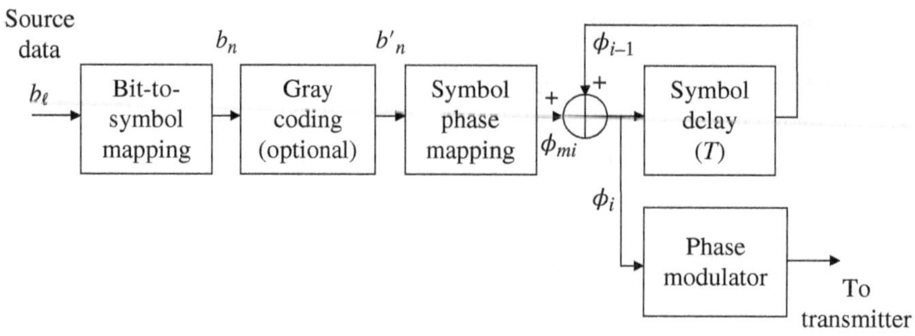

FIGURE 4.19 Differentially coherent MPSK (DCMPSK) modulator.

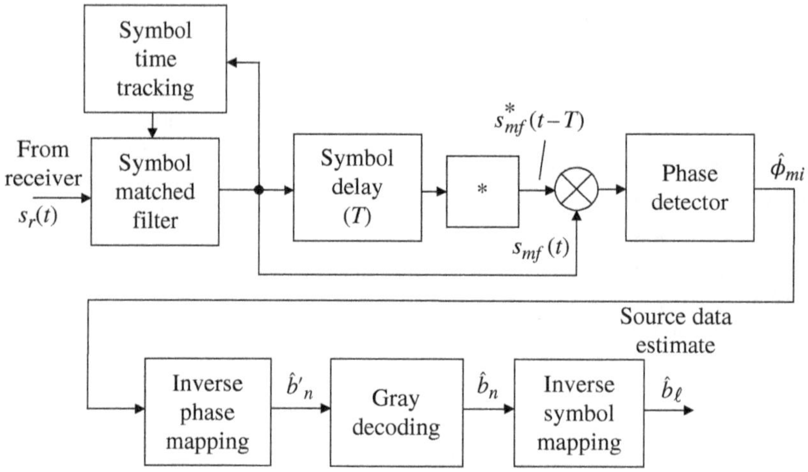

FIGURE 4.20 Differentially coherent MPSK (DCMPSK) demodulator.

established using a reference symbol. The simplicity of detecting a DCPSK-modulated waveform is accompanied by a bit-error performance loss that lies between that of coherent and noncoherent detection. For decreasing signal-to-noise ratios the bit-error performance approaches that of noncoherent detection and as the signal-to-noise ratio increases the bit-error performance approaches that of coherent detection.

Differently coherent MPSK (DCMPSK) modulation is implemented as shown in Figure 4.19 where b_ℓ source bits are mapped into b_n: $n = 0, \ldots, k-1$ symbol bits that are, in turn, gray coded and the resulting b'_n symbol bits are mapped into one of $M = 2^k$ symbol phase states given by

$$\phi_{mi} = \frac{2\pi m}{M} + \frac{\pi}{M} \tag{4.39}$$

The subscript i signifies the i-th symbol and inclusion of the additional π/M phase shift is optional and simply ensures that a phase transition always occurs to aid symbol tracking at the demodulator, although, as mentioned above, the DCMPSK waveform does not require phaselock loop phase tracking. The inherent differential phase encoding is obtained as

$$\phi_i = \phi_{i-1} + \phi_{mi} \tag{4.40}$$

where the phase ϕ_{i-1} corresponds to the reference symbol phase in the demodulator.

The phase modulator produces an intermediate frequency (IF)-modulated waveform that is amplified and frequency translated to the transmitted carrier (f_c) expressed as

$$s(t) = A\cos(\omega_c t + \phi_i + \varphi_o) \tag{4.41}$$

where A is the peak signal level, $\omega_c = 2\pi f_c$, φ_o is an arbitrary phase shift and ϕ_i is the phase of the differentially encoded data.

The demodulator is shown in Figure 4.20 and performs the indicated functions in the differentially coherent demodulation of the received waveform; for simplicity additive noise is not considered and $s_r(t) = s(t)$. The symbol timing recovery can be accomplished in random data; however, a preamble will provide for more rapid synchronization without the loss of data. DCPSK is often used in fast frequency-hopping applications where symbol timing is a byproduct of the hopping sequence acquisition, although acquiring the hopping

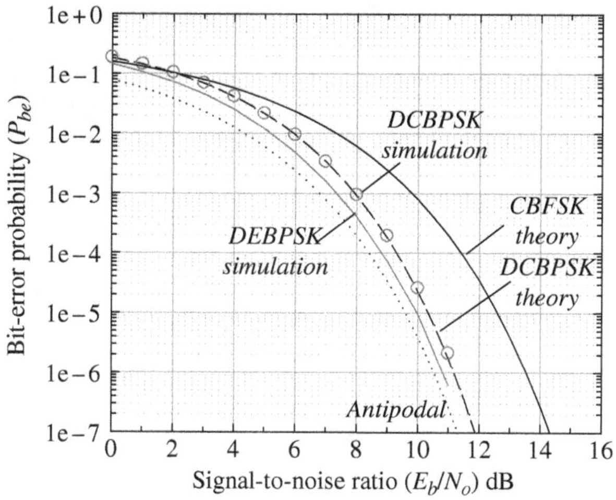

FIGURE 4.21 Comparison of DCBPSK bit-error performance with CBPSK, CBFSK, and DEBPSK.

sequence may take much longer depending on the time and frequency uncertainties.

The theoretical bit-error probability of DCMPSK is difficult to evaluate for all values of M except the binary case corresponding to $M = 2$ or DCBPSK. The theoretical performance of DCBPSK is established by Wozencraft and Jacobs [12], Viterbi [13], and Proakis [14] who present arguments based on the similarity of DCBPSK with the noncoherent detection of orthogonal noncoherent binary-modulated waveforms and conclude that the theoretical bit-error probability for DCBPSK is

$$P_{be} = \frac{1}{2}e^{-\gamma_b} \quad : \text{DCBPSK} \tag{4.42}$$

where $\gamma_b = E_b/N_o$ and E_b is the received signal energy-per-bit.

Equation (4.42) is plotted in Figure 4.21 and compared with the performance of coherently detected BFSK (CBFSK) and the antipodal modulation performance of coherently detected BPSK (CBPSK). The *dashed* theoretical DCBPSK curve is a plot of (4.42) and the *circled* data points are based on Monte Carlo simulations using 10 M bits for each signal-to-noise ratio. The theoretical performance of CBFSK modulation is discussed in Chapter 5 and that of DEBPSK is discussed in Chapter 8. It is interesting to note that if the reference phase of the DCBPSK detected waveform is forced or set to the center of the decision region, that is, the rest phase-state corresponding to the phase of the currently detected bit, then the resulting performance is identical to that of DEBPSK. This has limited utility because the performance is intolerant to frequency errors unless a phaselock loop is used.

Lucky, Salz, and Weldon [15] evaluate the symbol-error probability of DCMPSK using the phase *pdf* given in (4.7). Their evaluation of the symbol-error probability involves the phase difference $\Delta\phi$ between consecutive symbols expressed as

$$\Delta\phi = \tan^{-1}\left(\frac{N_{Ik+1}}{\sqrt{2P}+N_{Qk+1}}\right) - \tan^{-1}\left(\frac{N_{Ik}}{\sqrt{2P}+N_{Qk}}\right) \tag{4.43}$$

where N_I and N_Q are independent quadrature noise samples associated with symbol k and $k+1$. Using (4.43) and (4.7), the symbol-error probability is evaluated as

$$P_{se}(M) = 2\left[1 - \int_{-\pi}^{\pi/M}\left[\int_{-\pi}^{\pi} p(\phi)p(\Delta\phi+\phi)d\phi\right]d\Delta\phi\right] \tag{4.44}$$

Because a closed-form expression for (4.44) does not exist, the integrations must be performed numerically invariably leading to some degree of approximation and the necessity of an additional approximation in converting the symbol errors to bit errors. Because of these approximations it is more convenient to evaluate the symbol and bit-error performance using Monte Carlo simulations as discussed in Chapter 14. It is, however, instructive to evaluate the loss of DCMPSK relative to coherently detected PSK (CMPSK) modulation using the high signal-to-noise asymptotic approximations to $P_{se}(M)$ for DCMPSK and CMPSK expressed by Cahn [16] as

$$P_{se}(M) \approx e^{-2\gamma\sin^2(\pi/2M)} \quad : \text{DCMPSK}, \quad \gamma > 10:1 \tag{4.45}$$

and

$$P_{se}(M) \approx e^{-\gamma\sin^2(\pi/M)} \quad : \text{CMPSK}, \quad \gamma > 10:1 \tag{4.46}$$

Forming the ratio of the exponents in (4.45) and (4.46), the resulting loss is

$$\Delta\gamma = \frac{\sin^2(\pi/M)}{2\sin^2(\pi/2M)} \quad : \gamma > 10 \tag{4.47}$$

This result is plotted in Figure 4.22 as a function of M and indicates that the maximum loss in the symbol-error performance of DCMPSK relative to that of CMPSK approaches 3 dB with increasing M. The performance loss in Figure 4.22 provides a reasonable estimate for signal-to-noise ratios corresponding to $P_{se}(M) \leq 10^{-5}$.

The following evaluations of the performance of DCMPSK are based on Monte Carlo computer simulations using 10 M symbols for each signal-to-noise ratio and gray coding of the bits prior to the symbol phase mapping. Figure 4.23 shows the simulated performance of DCMPSK and differentially encoded MPSK (DEMPSK) for M = 2, 4, and 8. At $P_{be} = 10^{-5}$ the performance relative to antipodal

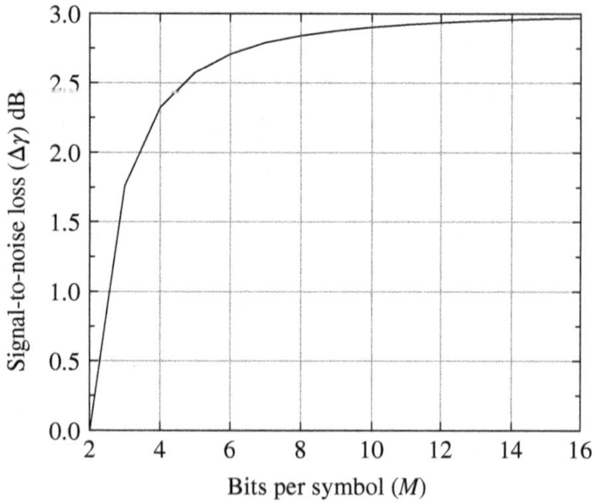

FIGURE 4.22 DCMPSK loss in signal-to-noise ratio relative to CMPSK.

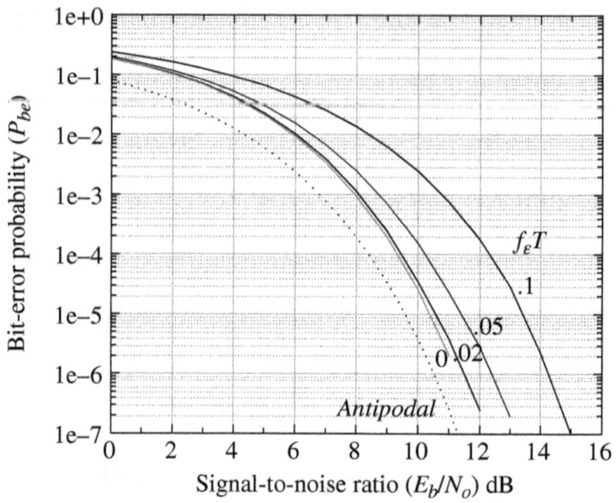

FIGURE 4.24 DCBPSK performance with frequency error f_eT_b.

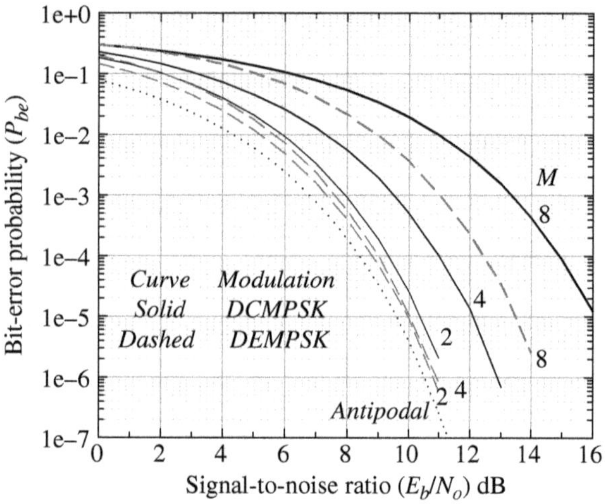

FIGURE 4.23 DCMPSK and DEMPSK bit-error performance (without frequency error).

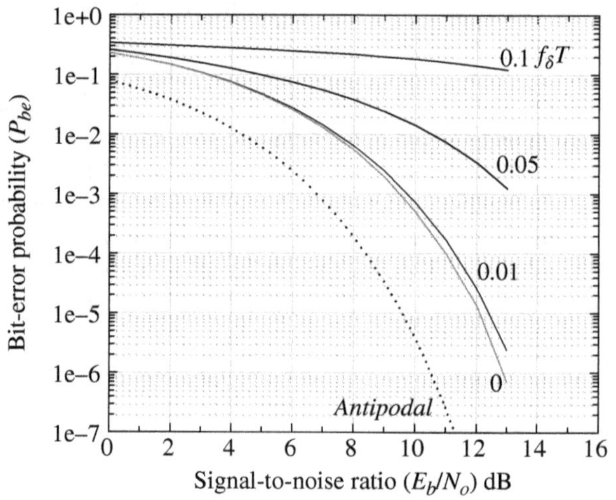

FIGURE 4.25 DCQPSK performance with frequency error f_eT_b.

signaling is degraded by about 1, 2.5, and 7 dB for $M = 2$, 4, and 8 respectively. The equivalent performance of DEMPSK, obtained by resetting the reference phase of the differential coding as described above, degrades by about 0.5 dB for $M = 2$ and 4 and by 4 dB for $M = 8$; however, as mentioned above, the performance advantage is lost with a frequency error, that is, the DEMPSK waveform requires frequency estimation and correction.

The performance degradation of DCMPSK resulting from a frequency error f_e is dependent on the accumulated phase error over the current and reference symbol periods expressed as

$$\phi_e = 4\pi f_e T \tag{4.48}$$

where $T = T_b$ for BPSK. The degradation of DCBPSK and DCQPSK as a function of the normalized frequency error f_eT is shown in Figures 4.24 and 4.25 respectively.

4.2.5 Generalized Modulator Implementations

The constant envelop-modulated waveforms discussed in this section employ phase shaping functions during symbol phase changes that result in significantly lower spectral sidelobes. The improved spectral containment allows for the use of narrower receiver IF bandwidths, reducing the susceptibility to ACI, and providing more efficient use of the radio frequency (RF) bandwidth by allowing closer channel spacing in frequency division multiplex (FDM) applications. In the absence of the inherent waveform modulation spectral control, the modulated waveforms must be passed through a bandpass filter or quadrature rail lowpass filters to achieve a desired spectral response often required to meet a specified spectral mask. However, because of spectral re-growth, filtering before hard-limiting power amplification or a channel with a hard limiter is problematic and power backoff before power amplification defeats the advantage of the constant amplitude modulation. In this regard, the design of symbol phase functions that provide the spectral control and preserve the constant signal amplitude is a major step in bandwidth control and transmitter power efficiency.

The implementations discussed in the proceeding sections focused on specific topologies involving in-phase and quadrature-phase (I/Q) processing of the source coded data. In the discussions, BPSK, QPSK, and OQPSK were shown to result in instantaneous phase changes resulting in discontinuous phase function with a corresponding *sinc(fT)* spectrum. However, in Figure 4.15, the phase transitions of MSK were shown to be continuous with a linear phase change of ±90° over a bit interval or as CPFSK modulation with a constant frequency of $R_b/4$ Hz over a bit interval. The I/Q and CPM implementations are contrasted in Figure 4.26. The modulator implementation shown in Figure 4.26b is used to generate continuous phase-modulated waveforms and is the principal focus of this section. In the following descriptions, the time-dependent functions are understood to correspond to discrete-time samples such that $t = iT_s$ where T_s is the sampling interval.

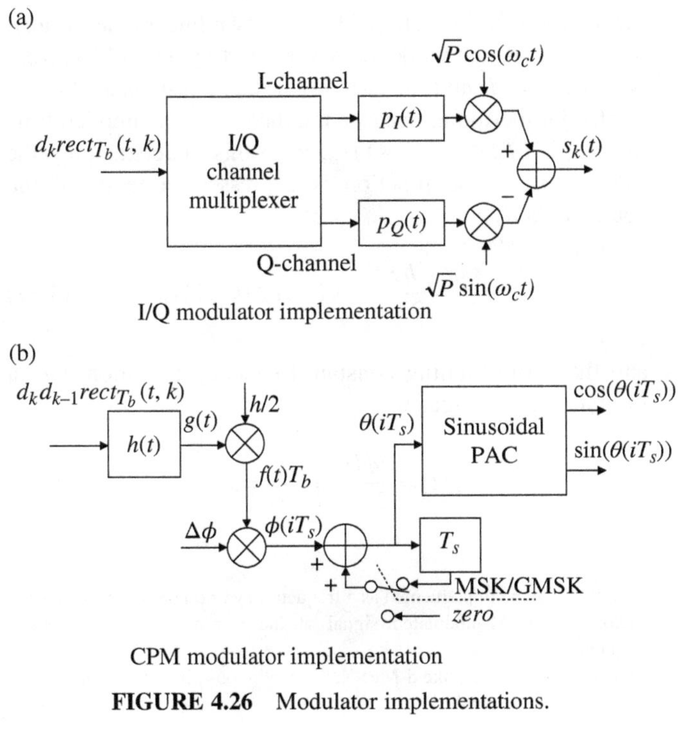

FIGURE 4.26 Modulator implementations.

Referring to Figure 4.26b the function $d_k rect_{T_b}(t,k)$ describes the contiguous sequence of source data bits d_k of duration, T_b and the function *h*(*t*) characterizes the continuous phase modulation shaping to be applied to the waveform. The function *h*(*t*) is specified for a given modulation over the interval T_p and is described as

$$h(t) = m(t)w(t) \quad (4.49)$$

where *m*(*t*) is a stored function that is unique to the specified shaping function *g*(*t*) and *w*(*t*) is a *delta* or *rect* function that influences the duration of *g*(*t*). When the normalizing factor *K* is applied in the computation of *m*(*t*), the stored samples result in unit area over the interval T_p. The function *g*(*t*) is determined as the convolution

$$g(t) = d_k rect_{T_b}(t,k) * h(t) \quad (4.50)$$

and, upon multiplying *g*(*t*) by one-half the modulation index, the normalized frequency function is

$$f(t)T_b = \frac{h}{2} g(t) \quad (4.51)$$

The phase increment $\Delta\phi = 2\pi/N_s$ is multiplied by $f(t)T_b$, and the phase accumulator produces the desired continuous phase function given by

$$\theta(iT_s) = \theta((i-1)T_s) + \Delta\phi f(t) T_b i \quad (4.52)$$

In these relationships, the parameter $N_s = T_b/T_s$ is the number of samples-per-bit and T_s is the sampling interval.

The integration *switch* in Figure 4.26b is shown in the MSK/GMSK position that results in a linear data-dependent phase transition between successive source bits. When GMSK is used, the Gaussian phase function smooths the MSK phase transitions resulting in spectral improvement; the phase function smoothing is similar to that described for SMSK in Section 4.2.6. When the *switch* is in the zero position, the modulator generates various CPFSK waveforms including Gaussian FSK (GFSK) with selectable modulation indices. These functions and parameters and their relationships to the CPM modulator implementation in Figure 4.26b are discussed in the following example applications. The CPM modulator can also be used to generate conventional BPSK, QPSK, OQPSK, and MPSK-modulated waveforms that require instantaneous changes in phase. The details for the implementation of these waveforms are left as an exercise.

4.2.5.1 *MSK I/Q Modulator Characterized as CPM*

This section focuses on the I/Q MSK implementation shown in Figure 4.26a that also depicts the classical implementation shown in Figure 4.13. As discussed in the previous sections, the I/Q implementation multiplexes the source data between the two channels and the symbol weighting functions $p_I(t)$ and $p_Q(t)$ provide for spectral containment; for constant amplitude modulation, this requires that $p_I^2(t)+p_Q^2(t)=1$.

The MSK waveform described by (4.29) contains phase terms $(d_{k-1}-1)\pi/2$ for t in the interval $kT_b, \leq t \leq (k+1)T_b$ and $(d_{k+1}-1)\pi/2$ for t in the interval $((k+1)T_b \leq t \leq (k+2)T_b)$. These terms ensure that the data-dependent phase is continuous as the phase transitions through $n\pi/2$ radians at each bit transition. This continuous phase characteristic can also be described, in terms of the accumulated phase up to $t = (k-1)T_b : k \in$ integer, as follows:

$$\begin{aligned}\theta_{k-1}(\boldsymbol{d}) &= d_0\omega_m T_b + d_1 d_0 \omega_m T_b + d_2 d_1 \omega_m T_b + \cdots + d_{j-1}d_{j-2}\omega_m T_b \\ &= d_0\pi/2 + d_1 d_0 \pi/2 + d_2 d_1 \pi/2 + \cdots + d_{k-1}d_{k-2}\pi/2 \\ &= \frac{\pi}{2}\sum_{n=0}^{k-1} d_n d_{n-1}\end{aligned} \tag{4.53}$$

where $\boldsymbol{d}$ is a vector of past data. For the MSK waveform, $\omega_m T_b = \pi/2$ and at $t = 0$, corresponding to $n = 0$, the data d_{-1} is considered to be $d_{-1} = 1$. In the current kT_b bit interval, the continuous phase function is expressed as

$$\theta(t;\boldsymbol{d}) = \theta_{k-1}(\boldsymbol{d}) + \frac{\pi}{2} d_k d_{k-1}\left(\frac{t-(k-1)T_b}{T_b}\right) \quad : (k-1)T_b \leq t \leq kT_b \tag{4.54}$$

where the time normalization by T_b corresponds to $\omega_m T_b = \pi/2$. During the k-th bit interval, the phase starts at $\theta_{k-1}(\boldsymbol{d})$ and advances linearly by $\pm\pi/2$ over the k-th bit interval.

Based on these results, the CPFSK description of the MSK waveform is expressed as

$$s_k(t) = \sqrt{P}\cos(\omega_c t + d_k d_{k-1}\omega_m t + \theta_{k-1}(\boldsymbol{d})) \tag{4.55}$$

Therefore, as indicated by (4.29), the angular frequency changes instantaneously at the bit transitions and, when differential encoding of the data is used, the radian frequency shift is expressed as $\omega_k = \dot{\phi} = d_k \omega_m$ for $kT_b \leq t \leq (k+1)T_b$ which corresponds to the constant angular frequency

$$\dot{\phi} = d_k\left(\frac{\pi h}{T_b}\right) \tag{4.56}$$

Based on these relationships, the CPM implementation is shown in Figure 4.27.

In Section 9.2, multi-h CPM waveform modulation is examined with the M-ary data into the continuous phase modulator denoted as α_i. This results in M-ary multi-h CPM modulation, however, for the single-h binary data case being considered here $\alpha_i \equiv d_i = \{1, -1\}$.

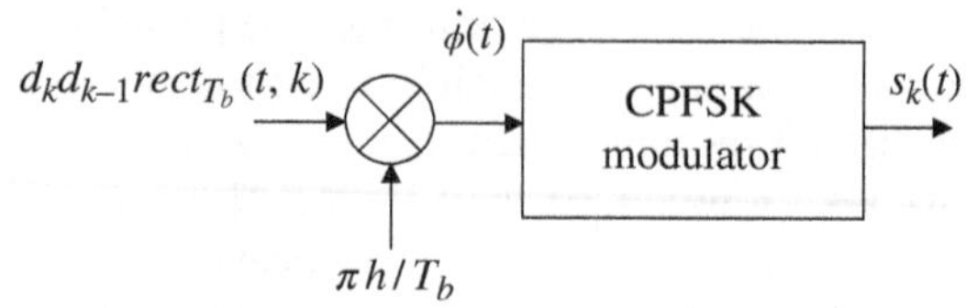

FIGURE 4.27 MSK implementation using CPFSK modulator.

4.2.5.2 MSK Continuous Phase Modulation (CPM) In this section, the implementation of MSK, using the CPM modulator shown in Figure 4.26b, is examined and the general nature of the implementation lends itself to a variety of other CPM-modulated waveforms. Specific examples, examined in the following sections are: sinusoidal FSK (SFSK), alternately referred to as sinusoidal MSK (SMSK), GMSK, and S-MPSK with linear and raised-cosine phase shaping.

Returning to Figure 4.26b, the MSK phase function $\theta(iT_s)$, that is applied to a sinusoidal phase-to-amplitude converter (PAC), results in quadrature MSK-modulated baseband signal $\tilde{s}_k(iT_s) = \cos(\theta(iT_s)) + j\sin(\theta(iT_s))$. After digital-to-analog conversion (DAC) and lowpass filtering, the PAC output signal is applied to a balanced modulator to produce the carrier-modulated output* $s_k(t)$. The sampled phase increment in Figure 4.26b is $\Delta\phi = 2\pi/N_s$, where N_s is the number samples-per-bit† as determined by the sampling frequency and the bit rate such that $N_s = f_s/R_b = T_b/T_s$.

The following description of the CPM modulator implementation of MSK starts with the known linear phase transitions of MSK and works backward starting with (4.51) from which $g(t) = d_k$ and, in turn, $h(t) = \delta(t)$ is established using (4.50). Furthermore, because the data is d_k, the implied time interval is $kT_b \leq t \leq (k+1)T_b$. With this understanding, the MSK linear-phase function is expressed in terms of the source data d_k as

$$\phi(t) = \frac{d_k \pi}{2T_b} t \quad : kT_b \leq t \leq (k+1)T_b \tag{4.57}$$

and the corresponding constant-frequency deviation over a bit interval is evaluated as

$$f(t) = \frac{d\phi(t)}{2\pi dt} = \frac{d_k}{4T_b} = d_k f_m \tag{4.58}$$

*The PAC can be implemented as a frequency synthesizer that produces the analog and CPM-modulated signal at the carrier or an intermediate frequency.

†In many applications, like S-MPSK, the continuous phase function is based on the symbol interval.

TABLE 4.1 CPM Parameters for Different Waveform Modulations ($h(t)=m(t)w(t)$)

Modulation	$m(t)^a$	$w(t)$	T_p
MSK	1.0	$\delta(t)$	T_b
SFSK/ SMSK	$2\sin^2(\pi t/T_b)$	$\delta(t)$	T_b
GMSK	$Ke^{-28.478(f_3 t)^2}$	$rect_{Tp}(t, 0)$	$0.265T_b(t_o/\sigma)/(f_3T_b)^b$

[a] K is a gain constant resulting in unit area over T_p.
[b] Reference, Equation (4.83).

In normalized form, (4.58) is expressed as

$$f(t)T_b = d_k f_m T_b \tag{4.59}$$

The modulation index, h, is defined as $h \triangleq 2f_m/R_b$ so that (4.59) becomes

$$f(t)T_b = \frac{h}{2}d_k \tag{4.60}$$

Referring to (4.51), it is evident that $g(t) = d_k$ and from (4.50) that $h(t) = \delta(t)$, therefore, using (4.52) the continuous phase function is expressed as

$$\theta(iT_s) = \theta((i-1)T_s) + \frac{2\pi}{N_s}(d_k f_m T_b)i \tag{4.61}$$

This form of MSK generation is also referred to as serial MSK [17–20] because there is no explicit quadrature processing. The demodulation can use either conventional quadrature cosine weighted rail matched filter detection or serial detection techniques. Also, the source data can be differentially encoded by substituting $d_{k-1}d_k$ for d_k. Table 4.1 summarizes the required parameters for CPM implementation of MSK, in the context of Figure 4.26b, and includes the SFSK/SMSK and GMSK modulations discussed in the following sections.

4.2.6 Sinusoidal FSK (SFSK) or Sinusoidal MSK (SMSK) Waveform Modulation

With MSK modulation the modulation phase changes linearly by ±90° over successive bit intervals with continuous phase transitions from one bit to the next resulting in abrupt changes in the phase slope. The corollary is that the frequency is constant over a bit interval with abrupt frequency changes of $\pm f_m$ Hz over successive bit intervals. These abrupt changes in phase and frequency influence the spectral shape and improved spectral characteristics can be achieved by eliminating the abrupt changes by applying sinusoidal changes over each bit interval while maintaining the ±90° phase shift and continuous phase properties. The characterization of the sinusoidal phase shaping in the following section is based on the analysis of Amoroso [21] who coined the term *sinusoidal*-FSK (SFSK); however, the U.S. Navy refers to SFSK as *sinusoidal*-MSK (SMSK).* As with MSK, SFSK modulation can be conveniently characterized in terms of the symbol shaping function $p(t)$ and implemented as an I/Q modulator or alternatively by the shaping filter $h(t)$ giving rise to the CPM modulator implementation. The following discussion starts by describing $p(t)$ and then characterizing the I/Q implementation and concludes with a discussion of the CPM modulator implementation of SFSK. Simon [22] and Prabhu [23] elaborate on the work of Amoroso by characterizing the conditions for the selection of $p(t)$ that satisfy the unique conditions of MSK and Simon extends the analysis to include phase-shaped M-ary MSK.

4.2.6.1 *SFSK/SMSK I/Q Modulator Implementation*

The following characterization of the symbol-shaping function is that of Amoroso [20] and the frequency deviation parameter α is used simply to determine the optimum spectrum containment that occurs when $\alpha = 1/4$ referred to as SFSK modulation. In the following, α-FSK is used to refer to the shaping for an arbitrary value of $\alpha \neq 1/4$.

The symbol shaping function for α-FSK is given by

$$p(t) = \cos(\omega_m t - \alpha\sin(4\omega_m t)) \tag{4.62}$$

and, referring to Figure 4.13 and (4.29), the modulated signal is expressed as

$$s_k(t) = \begin{cases} \sqrt{2P}\cos(\omega_c t + d_{k-1}d_k(\omega_m t - \alpha\sin(4\omega_m t)) + (d_{k-1}-1)\pi/2) & : t \in I_0 \\ \sqrt{2P}\cos(\omega_c t + d_k d_{k+1}(\omega_m t - \alpha\sin(4\omega_m t)) + (d_{k+1}-1)\pi/2) & : t \in I_1 \end{cases} \tag{4.63}$$

where the intervals I_0 and I_1 correspond to the first and second halves of the offset symbol interval as defined for MSK in Section 4.2.3.4. When $\alpha = 0$, the symbol shaping function results in MSK modulation. The corresponding phase functions in (4.63) over these intervals are expressed as

$$\phi_{k,o}(t) = d_{k-1}d_k(\omega_m t - \alpha\sin(4\omega_m t)) + (d_{k-1}-1)\pi/2 \tag{4.64}$$

$$\phi_{k,1}(t) = d_k d_{k+1}(\omega_m t - \alpha\sin(4\omega_m t)) + (d_{k+1}-1)\pi/2 \tag{4.65}$$

and, as in the case of the MSK waveform, the change in the carrier phase at the bit transitions is evaluated as

*Some authors refer to MSK as fast FSK or FFSK.

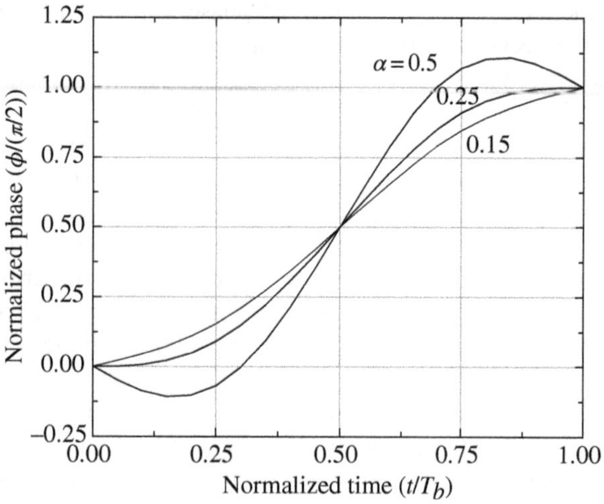

FIGURE 4.28 SFSK phase transitions for various values of α.

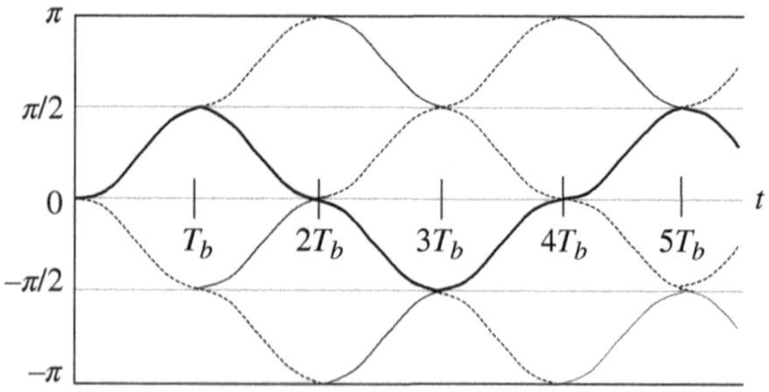

FIGURE 4.29 SFSK phase function for $\alpha = 1/4$ and data sequence $d_k = \{1,1,-1,1,1,1,-1\}$.

$$\begin{aligned}\Delta\phi_k &= \phi_{k,1} - \phi_{k,0} \\ &= \{0, \pm 2k\pi\}\end{aligned} \tag{4.66}$$

resulting in a continuous phase modulated waveform. From (4.62), the phase function is defined as the time-dependent phase

$$\phi(t) \triangleq \omega_m t - \alpha \sin(4\omega_m t) \tag{4.67}$$

Equation (4.67), normalized by $\pi/2$, is shown in Figure 4.28 for several values of α and the optimum response for $\alpha = 0.25$ results in a zero phase slope at the bit transitions. The spectrum of the waveform for the optimum value of α also results in the lowest sidelobe levels for frequencies removed several sidelobes from the peak of the principal spectral lobe and, for this reason, is considered to be the optimum value. The phase transition characteristics for $\alpha = 0.25$ are shown in Figure 4.29 where the *solid* phase path corresponds to the data sequence used in the MSK example of Figure 4.15. These phase transitions have the form of a continuous function over the bit interval with a phase change of ±90° and the corresponding advantage of continuous frequency changes that improve the spectrum containment.

4.2.6.2 SFSK/SMSK CPM Modulator Implementation

In the following analysis, only SFSK is considered for which $\alpha = 0.25$. To examine the implementation of the SMSK waveform using the CPM modulator, consider the phase function

$$\phi_k(t) = d_k[\omega_m t - 0.25\sin(4\omega_m t)] \tag{4.68}$$

and the normalized derivative

$$\begin{aligned} f(t)T_b &= \frac{\dot{\phi}_k(t)T_b}{2\pi} \\ &= d_k f_m T_b\left[1-\cos\left(\frac{2\pi t}{T_b}\right)\right] \\ &= \left(\frac{h}{2}\right)g(t)\end{aligned} \tag{4.69}$$

From (4.69) with $h = 1/2$, the function $g(t)$ is

$$\begin{aligned} g(t) &= d_k\left[1-\cos\left(\frac{2\pi t}{T_b}\right)\right] \\ &= d_k\int_0^t h(\tau)d\tau\end{aligned} \tag{4.70}$$

Using (4.70), the shaping filter impulse response $h(t)$ is evaluated as

$$\begin{aligned} h(t) &= \frac{d}{dt}\left[1-\cos\left(\frac{2\pi t}{T_b}\right)\right] \\ &= \frac{2\pi}{T_b}\sin\left(\frac{2\pi t}{T_b}\right)\end{aligned} \tag{4.71}$$

The shaping filter is seen to have a sinusoidal pulse response and the normalized frequency deviation function, $f(t)T_b = \dot{\phi}(t)T_b/2\pi$, is shown in Figure 4.30 in terms of the normalized variable t/T_b.

4.2.7 Gaussian MSK (GMSK) Waveform Implementation

The GMSK waveform is characterized by a shaping filter having a zero-mean Gaussian frequency response expressed as

$$H(\omega) = \sqrt{2\pi}ae^{-(a\omega)^2/2} \tag{4.72}$$

The parameter a is related to the 3-dB angular-frequency bandwidth ω_3 of the low-pass function $H(\omega)$ and is evaluated in terms of the parameter a as

$$a = \frac{b}{\omega_3} \quad :\text{Hz} \tag{4.73}$$

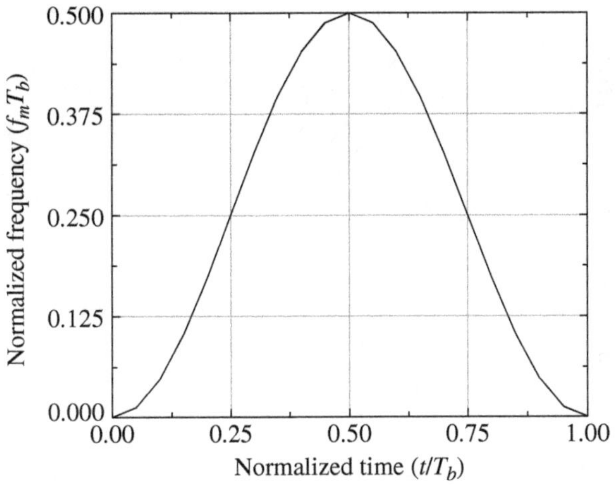

FIGURE 4.30 Frequency function for SMSK waveform modulation ($\alpha = 0.25$).

where $b = \sqrt{2\ln\sqrt{2}}$. The shaping filter response, evaluated as the inverse Fourier transform of $H(\omega)$, is also a zero-mean Gaussian function and, using the Fourier transform pair from Table 1.3, the result is expressed as

$$h(t) = K\left[\frac{1}{2\pi}\int_{-\infty}^{\infty} H(\omega)e^{j\omega t}d\omega\right] rect\left(\frac{t}{T_p}\right)$$
$$= \left[Ke^{-(\omega_3 t/b)^2/2}\right] rect\left(\frac{t}{T_p}\right) \quad (4.74)$$

The constant K is used as a normalizing constant for the frequency shaping function as described below. The range $|t| \leq T_p/2$ is an important consideration and, from a practical point-of-view, T_p must be finite. The paramount requirement is that T_p must be sufficiently large to capture an acceptable percentage of the area under the normalized Gaussian density function defined as

$$p(y) = \frac{1}{\sqrt{2\pi}}e^{-y^2/2} \quad (4.75)$$

where $y = t/\sigma$ and, from (4.74), $\sigma = b/2\pi f_3$. Table 4.2 lists a typical range of t_o/σ, the level of truncation at $p(t_o/\sigma)$, and the corresponding area under $p(t_o/\sigma)$ over the range of $\pm t_o/\sigma$ given in column 1. Truncating at $\pm t_o/\sigma$ results in a discontinuity or pedestal and, in the following applications, the pedestal is removed* prior to normalization by K. The area under the truncated distribution is important in approximating the Gaussian shaping function and, from Table 4.2, truncation in the range $3 \leq |t_o/\sigma| \leq 4$ represent reasonable conditions.

*By removing the pedestal the spectral attenuation for $fT >> 1$ was improved in cases involving $t_o/\sigma < 3$.

TABLE 4.2 Normal Distribution Parameters

t_o/σ	$p(t_o/\sigma)$	Area($\lvert t\rvert \leq T_p/2$)
2	0.053991	0.9545
2.5	0.017528	0.987581
3	0.004432	0.9973
3.5	0.000873	0.999535
4	0.000134	0.999937
4.5	1.60e−5	0.999993
5	1.49e−6	0.999999

Expressing (4.74) in terms of the normalized variables† t/T_p and $\omega_3 T_b = 2\pi f_3 T_b$ and, upon defining $\rho = T_p/T_b$, $x = t/T_p$, and $c = 2\pi/b$, the normalized shaping function becomes

$$h(t) = \left[Ke^{-(c\rho f_3 T_b(t/T_p))^2/2}\right] rect\left(\frac{t}{T_p}\right) \quad (4.76)$$

After some transformation of variables, $g(t)$ is evaluated as

$$g(t) = d_k rect_{T_b}(t,k) * h(t)$$
$$= K\int_{\lambda_L}^{\lambda_U} e^{-\lambda^2/2}d\lambda \quad (4.77)$$

where the integration limits are

$$\lambda_L = \frac{c\rho f_3 T_b}{2} \quad \text{and} \quad \lambda_U = c\rho f_3 T_b\left(\frac{t}{T_p}\right) \quad (4.78)$$

Equation (4.77) is evaluated in terms of the probability integral $P(x)$ with the result

$$g(t) = K\{P(\lambda_U) - P(\lambda_L)\} \quad (4.79)$$

where the scale constant K is chosen such that

$$\int_{-\infty}^{\infty} g(t)dt = 1 \quad \text{:normalization condition} \quad (4.80)$$

Using (4.80), the normalized frequency function applied to the CPM modulator is evaluated as

$$f(t)T_b = \left(\frac{h}{2}\right)g(t) \quad (4.81)$$

where $h = 1/2$ for the MSK modulation. The function $g(t)$ is plotted in Figure 4.31 for several values of the normalized

†Normalizing by t/T_b results in $f(x) = h(t)T_b$.

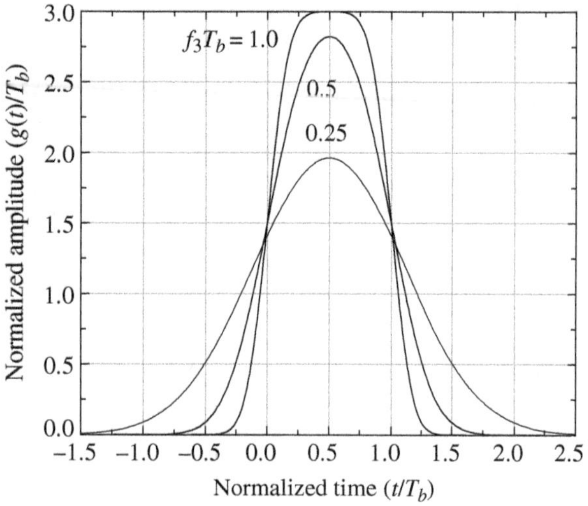

FIGURE 4.31 GMSK shaping filter response function.

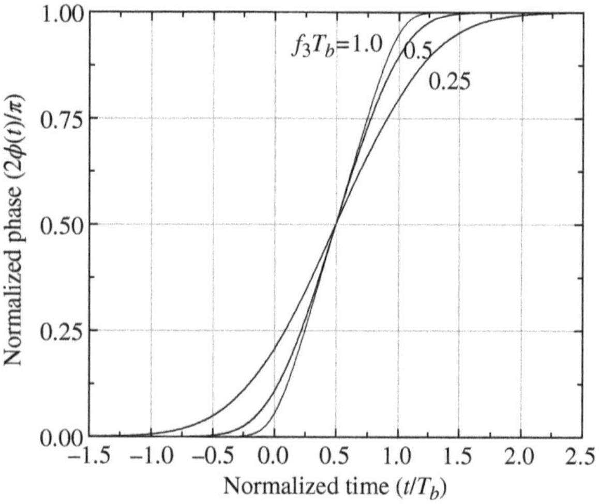

FIGURE 4.32 GMSK shaping filter impulse response function.

filter 3-dB bandwidth f_3T_b for an isolated source data bit $d_k = 1$ and with K chosen to satisfy (4.80).

The corresponding phase function is evaluated as

$$\phi(t) = \int \dot{\phi}(t)dt$$
$$= d_k\left(\frac{\pi h}{T_b}\right)\int_{-\infty}^{t} g(\tau)d\tau \tag{4.82}$$

and, using $h = 1/2$, this result approaches $d_i(\pi/2)$ as $t \rightarrow T_p/2$. The phase function, normalized to $\pi/2$, is shown in Figure 4.32 as a function of the normalized variable t/T_b for several values of the normalized phase-shaping filter bandwidth.

In these plots the isolated input bit corresponds to the interval $0 \le t/T_b \le 1$ so, unlike the previous waveforms, adjacent symbols will interfere with each other giving rise to intersymbol interference. On the other hand, increasing the range of the phase transitions results in a significant reduction in the spectral sidelobes. When using a single-symbol I&D matched filter, the E_b/N_o loss is about 0.8 dB for $f_3T_b = 0.35$; however, using a trellis decoder with metric updates integrated over the interval $-1 \le t/T_b \le 2$ essentially corresponds to maximum-likelihood sequence estimation (MLSE) [24] detection with a reduced loss in the signal-to-noise ratio. These subjects are considered in more detail is Section 4.4.3.7 and in the following.

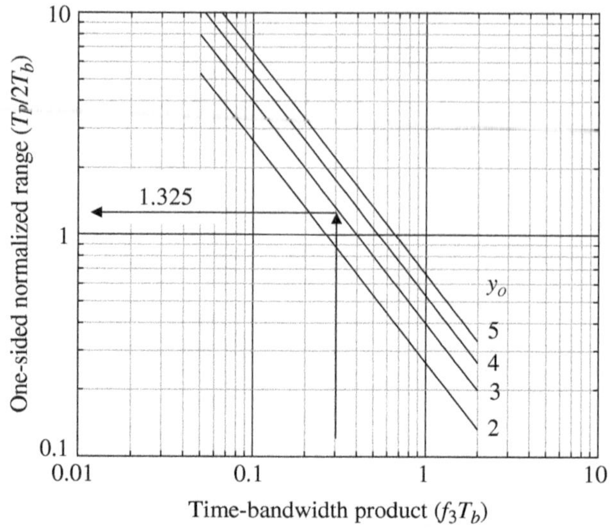

FIGURE 4.33 Dependence of $T_p/2T_b$ on GMSK time-bandwidth product.

For a specified value of $t_o/\sigma = y_o$, the parameter $\rho = T_p/T_b$ is inversely proportional to the selected time bandwidth product f_3T_b. The expression determining the one-sided range $(T_p/2T_b)$, given y_o and f_3T_b, is

$$\frac{T_p}{2T_b} = \frac{y_o}{c(f_3T_b)} \tag{4.83}$$

Equation (4.83) is plotted in Figure 4.33 as a function of the time-bandwidth product f_3T_b for various y_o values. The one-sided range corresponds to the amount of ISI from the preceding and succeeding symbols. The example shown in Figure 4.33 for $f_3T_b = 0.3$, corresponding to $y_o = 3$, indicates that the Gaussian response occupies $1.325T_b$ of the preceding and succeeding symbols. With N_s samples-per-bit, the number of one-sided samples of the phase-shaping function $h(t)$ is defined as

$$N_s' \triangleq \text{floor}\left(\frac{y_oN_s}{c(f_3T_b)}\right) \tag{4.84}$$

In the preceding example $N_s = 16$, so $N_s' = 21$ and the total span of the phase-shaping function corresponds to

$N_{span} = 2N_s' + 1 = 43$ samples. For values of $N_s' > N_s$ the loss in E_b/N_o becomes significant when using a detection filter that integrates over one bit interval; in these cases MLSE detection processing is necessary to maintain an acceptable loss. The power density spectrums for GMSK are examined in Sections 4.4.3.6 through 4.4.3.8.

4.2.8 Phase-Shaped PSK Modulation

The application of phase shaping to PSK-modulated waveforms results in a significant improvement in the spectrum sidelobes relative to the *sinc(fT)* spectrum of conventional PSK modulation. Phase shaping of the BPSK-modulated waveform is referred to as phase-shaped PSK (S-PSK) and is discussed by Dapper and Hill [4], Andren [25], and Cofer, Franke, Johnson, and Erman [26]. As in the previous applications, the phase shaping smooths the abrupt data-dependent symbol phase transitions resulting in an improvement of spectral containment while preserving the constant amplitude of the transmitted signal. In this section, the *linear* and *raised-cosine* phase shaping functions, defined by (4.85) and (4.86), are examined. In Section 4.4.3.9, the waveform spectral re-growth and bit-error characteristics are examined with the phase-shaped modulations: S-BPSK, S-QPSK, S-8PSK, and S-OQPSK modulated waveforms.

The linear and raised-cosine phase shaping functions are characterized as

$$\phi(t) = \begin{cases} \phi_m & : |t| \le \frac{T}{2}(1-\alpha) \\ \frac{\phi_m}{2} - \frac{\phi_m}{\alpha T}(|t| - T/2) & : \frac{T}{2}(1-\alpha) < |t| < \frac{T}{2}(1+\alpha) \quad \text{linear shaping} \\ 0 & : o.w. \end{cases} \tag{4.85}$$

and

$$\varphi(t) = \begin{cases} \varphi_m & : |t| \le \frac{T}{2}(1-\alpha) \\ \frac{\varphi_m}{2}\left\{1 - \sin\left(\frac{\pi}{\alpha T}(|t| - T/2)\right)\right\} & : \frac{T}{2}(1-\alpha) < |t| < \frac{T}{2}(1+\alpha) \\ & \text{RC shaping} \\ 0 & : o.w. \end{cases} \tag{4.86}$$

In these descriptions, the phase $\phi_m = 2\pi/M$ corresponds to the fundamental phase change associated with the underlying MPSK modulation and the symbol phase-shaping parameter α is constrained to the range $0 \le \alpha \le 1$ where $\alpha = 0$ corresponds to no shaping and $\alpha = 1$ corresponds to the maximum shaping. For example, considering a symbol centered at $t = 0$, the span of the symbol phase shaping is constrained to the range $-\alpha T \le t \le \alpha T$. This is depicted in

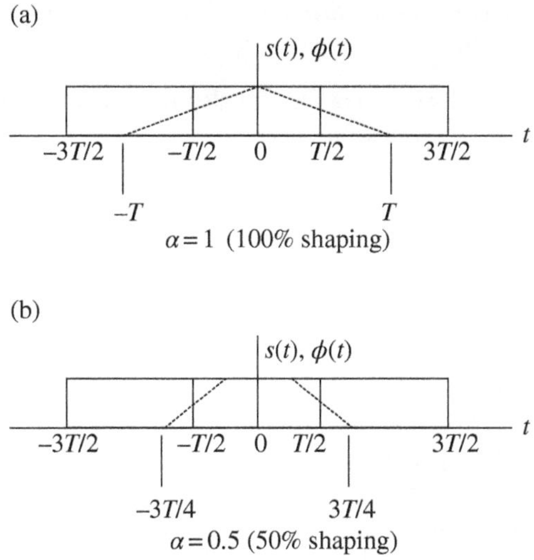

FIGURE 4.34 Symbol linear phase shaping characteristics.

Figure 4.34 for linear phase shaping where it is seen that for $\alpha = 1$ the phase-shaping function $\phi(t)$, corresponding to the *dashed* curve, extents over the entire symbol centered at $t = 0$ with 50% corresponding to the symbol phase change from ϕ_{m-1} to ϕ_m and 50% corresponding to the change from ϕ_m to ϕ_{m+1}. This is referred to as 100% phase shaping and the ISI represents a 50% contribution from each of the preceding and succeeding symbols. Similarly, the linear phase shaping for $\alpha = 0.5$ corresponds to 50% intra-symbol phase shaping with 25% ISI contribution from each of the preceding and succeeding symbols. The phase-shaped raised cosine (S-RC) shaping behaves in a similar manner; however, the *dashed* curve does not have an abrupt phase discontinuity leading to much lower spectral sidelobes as in the case of SMSK modulation. Phase shaping, applied with a monotonically changing phase, results in a shift of the modulated signal spectrum; this spectral shift is avoided by alternately or randomly changing the sign of the phase shift (see Problem 12).

4.3 NON-CONSTANT ENVELOPE PHASE-MODULATED WAVEFORMS

This section examines several spectrally efficient waveform modulations that do not result in constant envelopes and require power amplifiers operating in the linear range with linear channels or repeaters to preserve the spectral characteristics. Typically, this requires power amplifier input signal backoff resulting in less power efficient operation. The SRRC shaping discussed in Section 4.3.2 is a benchmark for spectral conservation and is used in many communication systems.

4.3.1 Waveform Modulation with $sinc^n(t/2T)$ Shaping Filter

Amoroso [27] examines the waveform modulation with the symbol shaping function expressed as*

$$p_n(t) = \begin{cases} \left[\dfrac{\sin(\pi t/T)}{\pi t/T}\right]^n & : |t| \le T \\ 0 & : o.w. \end{cases} \tag{4.87}$$

Equation (4.87) is shown in Figure 4.35 for several values of the parameter n. The motivation for using (4.87) is based on improving the spectral containment of MSK by eliminating the constant amplitude requirement and relaxing the symbol time restriction; MSK is a constant amplitude modulation and the shaping filter response occupies the interval $|t| \le T/2$ as described by (4.33) with $j' = 0$. However, the argument of the MSK shaping function is the same as that of the $sinc(-)$ function in (4.87), that is, for MSK $\omega_m t = \pi t/T$. The cosine-shaping function for MSK is also shown in Figure 4.35 for comparison. Because the modulated waveform involves intersymbol interference the spectrum for a random sequence of data is not simply the spectrum of $p(t)$. Amoroso examines the spectrum and discusses the performance through a channel with filtering and hard limiting. A value of $n = 3$ appears to be a reasonable compromise in the detection loss associated with the amplitude modulation, hard-limiter distortion, the spectral containment, and ISI.

The performance of Amoroso's quasi-bandlimited MSK shaping function is not examined in detail here; however, some salient features [26] are listed in Tables 4.3 and 4.4. Typically, phase shaping functions do a good job in reducing the spectrum level at frequencies far removed from the carrier and one of the objectives using (4.87) is to lower the near-in sidelobes, that is, for the baseband spectrum at frequencies in the range $|f| <= 1/T_b = 2/T$. As a point of reference, the first MSK spectral sidelobe occurs at $|f| \cong 1/T_b$ and is −23 dB relative to the peak spectrum level or main lobe. For the case $n = 3$ and considering the linear channel, the first sidelobe level occurs at a frequency $<1/T_b$ and is −30 dB below the first MSK sidelobe or −53 dB below the main lobe. For the linear channel case with $n = 4$, the first spectral sidelobe is located slightly higher in frequency than the first MSK sidelobe; however, the level is lowered by an additional 44 dB corresponding to −67 dB below the main lobe. The detection loss results, listed in Table 4.4, are equally impressive considering that the main source of loss is the ISI and that adaptive equalization is not used.

*Amoroso uses the notation $T = T_b$ to denote the bit duration, whereas, in this description T is the symbol duration equal to $2T_b$.

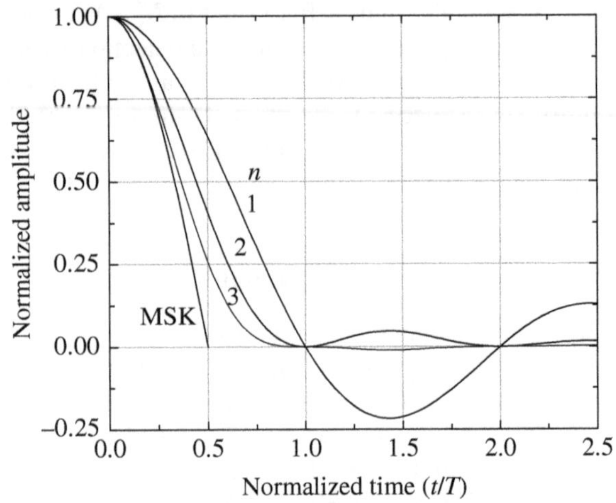

FIGURE 4.35 $[sinc(t/2T)]^n$ shaping filter impulse response.

TABLE 4.3 Spectral Sidelobe Levels Relative to MSK Spectrum

	n = 3		n = 4	
Channel	1st SL (dB)	2nd SL (dB)	1st SL (dB)	2nd SL (dB)
Linear	−30	−33	−44	−54
Hard limiting	−12	−20	−8	−21

TABLE 4.4 Detection Loss at $P_{be} = 10^{-7}$ with Respect to Ideal MSK

	Detection Loss (dB)	
Channel	$n = 3$	$n = 4$
Linear	0.54	0.12
Hard limiting	0.66	0.15

4.3.2 Spectral Root-Raised-Cosine (SRRC) Waveform Modulation

In Section 1.11.4.1 the impulse response of the spectral raised-cosine (SRC) filter $H(\omega)$ with a Nyquist bandwidth of $1/2T$ was shown to be zero at the sampling instants $t = nT$: $n \neq 0$. If this response were to correspond to the demodulator matched filter output, then the optimum sampling instants nT: $n \neq 0$ are orthogonal to the adjacent symbols resulting in zero ISI and optimum detection. Since the matched filter response is the product the received signal and matched filter frequency responses, that is, $H(\omega) = H_s(\omega)H_{mf}(\omega)$, by letting $H_s(\omega) = H_{mf}(\omega) = \sqrt{H(\omega)}$, the received symbols will be optimally detected using the matched filter sampled output. Therefore, the SRRC filter is used to design the modulator waveform and demodulator matched filter as described in this section.

Referring to the discussion in Section 1.11.4.1, the expression for the SRRC frequency response is

$$H_s(\omega)=\begin{cases} AT & :|\omega|\le \pi(1-\alpha)/T \\ \dfrac{AT}{\sqrt{2}}\left\{1-\sin\left[\dfrac{T}{2\alpha}(|\omega|-\pi/T)\right]\right\}^{1/2} & \\ & :\pi(1-\alpha)/T\le|\omega|\le\pi(1+\alpha)/T \\ 0 & :o.w. \end{cases} \tag{4.88}$$

This theoretical response is completely contained within the angular frequency range $\pm\pi(1+\alpha)/T$ where α is defined as the excess bandwidth factor. The impulse response or symbol shaping function of the SRRC filter is evaluated by taking the inverse Fourier transform of (4.88), see Problem 13, and the result is expressed as

$$p(t)=A\frac{\sin((\pi t/T)(1-\alpha))}{\pi t/T} + \left(\frac{4A\alpha}{\pi}\right)\frac{\cos((\pi t/T)(1+\alpha))+(4\alpha t/T)\sin((\pi t/T)(1-\alpha))}{1-(4\alpha t/T)^2} \tag{4.89}$$

There are three conditions that result in the indeterminate form 0/0 and these occur at $t/T = 0$ and $\pm 1/(4\alpha)$. Evaluation of the response under these special conditions results in

$$p(0)=A\left[1+\left(\frac{4-\pi}{\pi}\right)\alpha\right] \tag{4.90}$$

and

$$p\left(\frac{\pm 1}{4\alpha}\right)=\frac{A}{\sqrt{2}\pi}\left[(\pi-\alpha)\cos\left(\frac{\pi}{4\alpha}\right)+(\pi+\alpha)\sin\left(\frac{\pi}{4\alpha}\right)\right] \tag{4.91}$$

The one-sided SRRC symbol shaping function is shown in Figure 4.36 for several values of the excess bandwidth factor. Although the responses are shown as finite one-sided responses, they actually extend over the infinite range $|t| = \infty$, that is, a bandlimited frequency response has an infinite duration impulse response and vice versa. In practice, this is problematic because it requires a matched filter with infinite delay. This issue is overcome by limiting or truncating the two-sided span of the symbol shaping function, for example, the symbol shaping function depicted in Figure 4.36 is truncated to a two-sided span of N_{span} = 12 symbols. The loss in symbol energy, although dependent on α, becomes insignificant for reasonably large values of

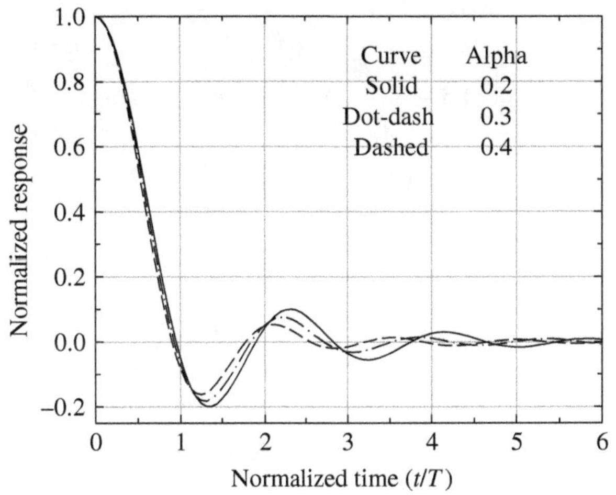

FIGURE 4.36 Theoretical SRRC symbol impulse responses (α = 0.2, 0.3, and 0.4).

N_{span}. The resulting finite time response gives rise to an infinite frequency response which is examined in Section 4.4.4.1; the matched filter detection loss for practical values of N_{span} is also examined.

Upon examination of the responses in Figure 4.36 it is apparent that the values at nT: $n \neq 0$ are not zero and that the amplitude of the temporal sidelobes increase with decreasing α, for example, with $\alpha = 0, H_s(w) = (AT/p)$ $rect(w/(4p/T))$ and $p(t) = Asinc(t/T)$. As reasoned above, the matched filter output satisfies the zero response condition nT: $n \neq 0$, so it follow that this condition is also satisfied when the responses in Figure 4.36 are convolved with themselves; a process performed by the matched filter. The resulting RC responses are shown in Figure 1.40 for the same values of α.

The preceding analysis addresses the characteristics of an isolated symbol and Figures 4.37 and 4.38 show the AM associated with the SRRC-shaped waveform applied to a sequence of BSPK- and QPSK-modulated symbols for excess bandwidth factors α = 0.25 and 1.0. These results are based on the in-phase and quadrature baseband samples and do not include the fluctuations of the carrier. The AM results are quantified in Table 4.5 where the first parameter is the ratio of the second moment to the square of the first moment of the signal magnitude and corresponds to the total-to-average power ratio of the AM. The second parameter corresponds to the ratio of the maximum or peak power to the average power. For constant envelope modulation both of these parameters have a ratio of one or 0 dB. There is not a great deal of difference between these two shaping functions; however, the frequency of the AM is significantly greater for the α = 1.0 case. The SRRC-modulated waveforms are spectrally efficient; however, the power levels into the transmitter PA must be backed off to preserve the AM and the spectral

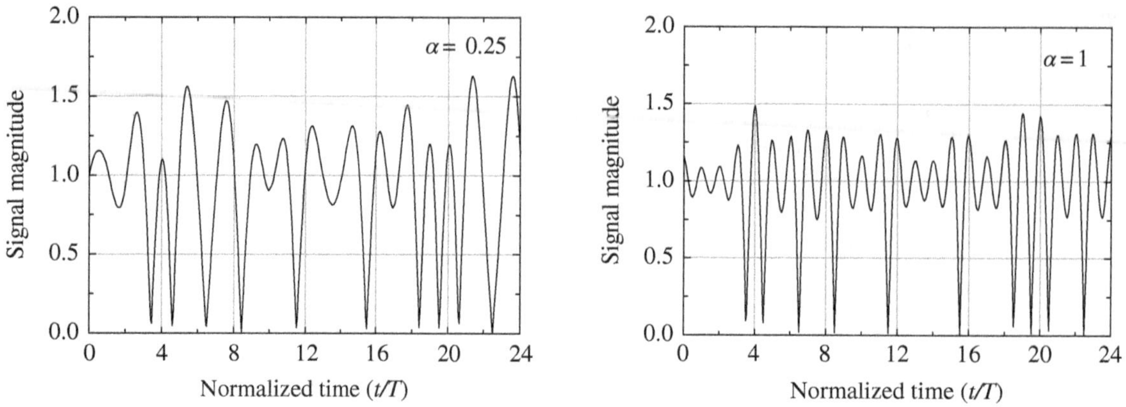

FIGURE 4.37 Envelope of SRRC BPSK-modulated waveform.

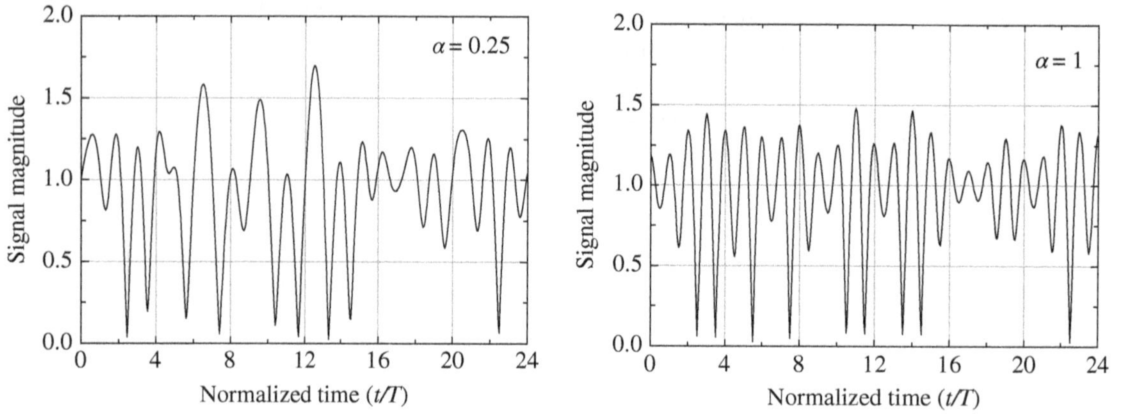

FIGURE 4.38 Envelope of SRRC QPSK-modulated waveform.

TABLE 4.5 Summary of AM Characteristics (dB) with SRRC Shaping

	BPSK α		QPSK α	
Parameter (dB)	0.25	1.0	0.25	1.0
$10\log\left(1-\sigma_m^2/P_a\right)$	0.72	0.61	0.35	0.47
$10\log(P_m/P_a)$	5.63	4.21	4.81	3.97

efficiency. Consequently, SRRC-modulated waveforms are not power efficient compared to constant envelop-modulated waveforms. The spectral efficiency of the SRRC-modulated waveform with PA backoff is discussed in Section 15.14.2.

4.4 PHASE-MODULATED WAVEFORM SPECTRUMS AND PERFORMANCE

In this section the spectral characteristics and bit-error performance of the various forms of PSK modulation, discussed in the previous sections, are examined. The following section identifies various definitions of signal bandwidth and discusses spectral mask specifications that must be met to conform to national and international regulations. Subsequent sections address the spectral characteristics of constant envelope PSK-modulated waveforms and their conformance to the spectral masks. The impact of conventional filtering as a means of spectral control is examined in terms of the spectral re-growth following power amplifier hard limiting and shown to be beneficial with OQPSK and MSK modulations. The application of phase-shaping to MPSK-modulated waveforms is shown to result in significant spectral containment while preserving the constant amplitude characteristic resulting in spectral and power efficient waveforms. The bit-error performance of these modulations is also examined under selected conditions. In Section 4.4.4.1, the SRRC-modulated waveform is examined for spectral containment, and the case study in Section 4.4.5 examines the bit-error performance of SRRC-modulated BPSK, DCBPSK, QPSK, and OQPSK waveforms.

4.4.1 Spectral Masks

A significant performance measure of a modulated waveform is the spectrum containment that specifies the maximum level

TABLE 4.6 Various Definition of Signal Bandwidth

Bandwidth	Definition[a]
Occupied	Two-sided bandwidth containing 99% of the signal power.
Bounded[b]	Two-sided bandwidth outside of which the PSD does not exceed a specified level (dBsd). Often applied to spectral re-growth.
Necessary[c]	Two-sided bandwidth that is just sufficient to ensure the transmission of information at the rate and quality required under specified conditions.
Noise[c]	Two-sided bandwidth of an ideal rectangular spectrum with level equal to the maximum value of the signal PSD and area equal to the total area under the PSD; see Equation (1.46).

[a]PSD is the power spectral density.
[b]dBsd is level relative to the maximum of the modulated waveform PSD.
[c]The notation B_n is used to denote the noise bandwidth so B'_n is used to denote the necessary bandwidth.

of the signal spectrum at frequencies removed from the transmitted carrier frequency. These specifications or spectral masks impact, for example, the tolerance to channel filtering, the allowable ACI, and the channel spacing in frequency division multiple access (FDMA) applications.* The important underlying parameter of the modulated signal is the signal bandwidth as defined in Table 4.6. To standardize the spectral containment for specific applications and waveform modulations, unique spectral masks are specified and several examples are given in the remainder of this section.

The spectral masks specified by the Consultative Committee for Space Data Systems (CCSDS) Space Frequency Coordination Group (SFCG) recommendation† Rec-21-2R3 [28] for RF emissions for space-to-earth communication links are shown in Figure 4.39 and the specified *break-points* are listed in Table 4.7. These specifications apply to the suppressed carrier-modulated waveforms discussed in this chapter; however, for symbol rates in less than 2 Msps the specification also applies to Bi-Phase and non-return to zero (NRZ) modulations. These masks are for space-to-ground satellite links operating in the bands: 2.2–2.29 GHz, 8.025–8.4 GHz, and 8.45–8.5 GHz. The spectral containment requirements apply to bandwidth efficient modulations [29, 30].

The National Telecommunications Industry Association (NTIA) defines spectral masks that are tailored to specific classes of modulations [31]. The following specifications apply to space services, discussed in Section 5.6 of Reference 30, and include associated earth terminals and space stations

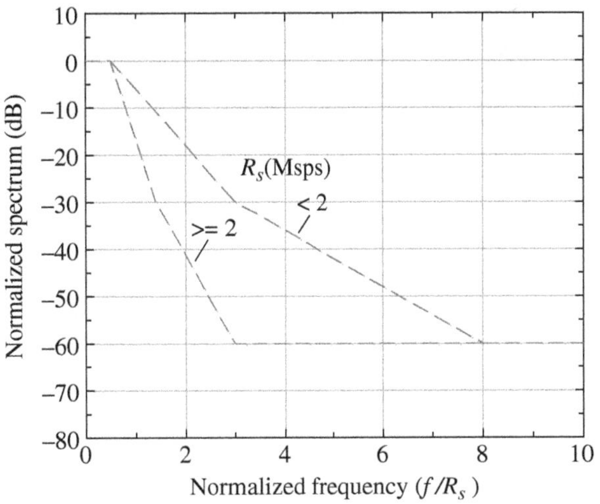

FIGURE 4.39 SFCG Rec-21-2R3 one-sided spectral mask (suppressed carrier modulation, R_s < 2 MHz also includes PCM/PM/Bi-Phase and NRZ).Space Frequency Coordination Group (SFCG) [28]. Reproduced by permission of the Consultative Committee for Space Data Systems (CCSDS).

TABLE 4.7 Specifications for SFCG Rec-21-2R3 Spectral Mask[a]

R_s > 2 Msps		R_s < 2 Msps	
$\lvert f/R_s \rvert$	Mask (dBsd)	$\lvert f/R_s \rvert$	Mask (dBsd)
0.0	0.0	0.0	0.0
0.5	0.0	0.5	0.0
1.4	−30.0	3.0	−30.0
3.0	−60.0	8.0	−60.0
>3.0	−60.0	>8.0	−60.0

[a]Space Frequency Coordination Group (SFCG) 2013 [28]. Reproduced by permission of the Consultative Committee for Space Data Systems (CCSDS).

operating in portions of the spectrum allocated to space services above 470 MHz.‡ The form of the spectral mask attenuation is expressed as

$$A(\text{dBsd}) = \begin{cases} 0 & : |f| < B'_n/2 \\ 8 & : |f| = B'_n/2 \\ 40\log\left(2|f|/B'_n\right) + 8 & : |f| > B'_n/2 \end{cases} \qquad (4.92)$$

where dBsd is the level of the power spectral density§ (PSD) of the modulated waveform relative to the maximum level in the band $|f| < B'_n/2$ and B'_n is the necessary bandwidth of the

*Transmit power control specifications also limit the permissible degradation in the E_b/N_o performance from adjacent channels.
†The SFCG recommendations are subject to change and should be reviewed periodically for updates.

‡The reader should review the NTIA Red Book specifications thoroughly to ensure the compliance of a selected modulation. For example, power output, frequency tolerance, and maximum Doppler shift should be considered as discussed in Reference 30, Chapter 10.
§The power spectral density is also referred to as [7]: power spectrum and spectral density.

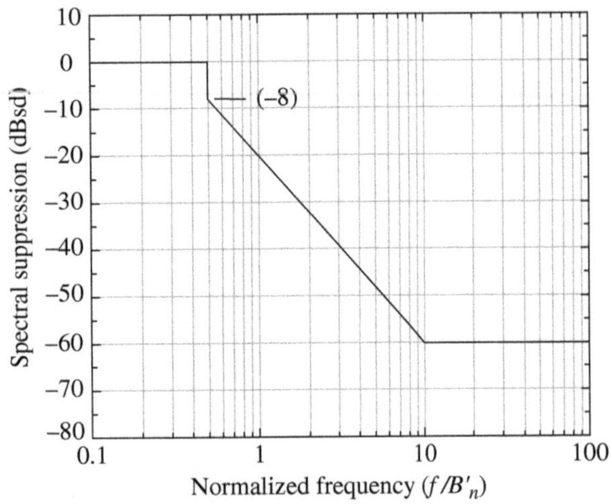

FIGURE 4.40 NTIA maximum emission levels for space services.National Telecommunications and Information Administration (NTIA) [31]. Reproduced by permission of the U.S. Department of Commerce.

modulated signal. The upper bound of the spectral density based on (4.92) is plotted as a function of the normalized frequency f/B'_n as shown in Figure 4.40. In Section 4.4.3.1, this mask is applied to several modulated waveforms based on the computed necessary bandwidths.

The necessary bandwidths for several commonly used modulations are given in Table 4.8 and the following parameter definitions apply:

D = Peak frequency deviation, that is, one-half the difference between the maximum and minimum values of the instantaneous frequency.

R = Total bit rate including data, encoding, and any other overhead bits.

S = Number of equivalent non-redundant signaling states.

K = An overall numerical factor which varies according to the emission and which depends upon the allowable signal distortion.

N_s = Number of sub-carriers.

C_s = Separation in frequency between adjacent sub-carriers or carriers of a multi-carrier modulation.

Nyquist signaling is a bandwidth efficient waveform modulation and the theoretical power spectral density and spectral mask are discussed in the remainder of this section. The theoretical results are unrealizable, in that, the impulse response is infinite in duration; however, practical implementations involving finite duration windowed impulse responses are discussed in Section 4.4.4.1.* Nyquist signaling corresponds to a minimum baseband bandwidth of $B = R_s/2$ Hz without ISI and represents an ISI-free binary rate of two bits-per-channel-use.

The power spectral density of the Nyquist waveform is characterized as a raised-cosine (RC) function; however, for maximum-likelihood (ML) symbol detection, the modulated or transmitted waveform is characterized in terms of the spectral properties of the RRC or SRRC waveform. The spectral mask, defined in this section for the SRRC-modulated waveform, is not an accepted standard; however, it is based on the Intelsat IESS-30 [35] standard spectral mask shown in Figure 4.41 for QPSK modulation and plotted as a function of the normalized frequency deviation from the carrier frequency as fT where the symbol duration is $T = 1/R_s = 2/R_b$ and R_b is the information bit-rate.† The specified *break-points* in Figure 4.41 are listed in Table 4.9. The Intelsat standard spectral mask characterizes the modulator out spectrum at the IF to the radio transmitter. It also applies to waveforms using FEC codes with an appropriate adjustment in the symbol rate R_s. The Intelsat document IESS-310 [36] specifies the same mask, with an appropriate adjustment in R_s, to characterize the spectrum of the rate 2/3 TCM/8PSK-modulated waveform with a Reed–Solomon outer code.

The usual procedure in designing a waveform starts by specifying the symbol shaping function $p(t)$ and then proceeds to determine the spectrum $H(f)$. This procedure has the benefit of controlling the intersymbol interference by limiting the symbol shaping function to the symbol interval or, as in the case of partial response modulation, designing the symbol function with a known intersymbol interference response that can be taken advantage of in the demodulator. In either case, restricting the duration of the symbol shaping function places limits on the waveform spectrum. When designing the SRRC waveform,‡ the procedure is reversed by first specifying the waveform spectrum and then proceeding to characterize the symbol weighting function. With an emphasis on spectrum conservation, this is a logical way to proceed; however, the selection of a frequency spectrum with an impulse, or symbol shaping function, response with zero intersymbol interference is limited to the class of Nyquist filters, or Nyquist spectrums, and the resulting modulation is often referred to as Nyquist signaling.

The theoretical RC spectrum is a bandlimited function such that $H(f) = 0$ for $|f| > B$ where B is the baseband bandwidth given by§

*Nyquist signals are also discussed in Section 6.6 in the context of partial response modulation.

†The abscissa of the standard Intelsat mask is plotted in terms of normalized frequency fT_b with $T_b = T/2$.

‡The goal is to design an RC waveform corresponding to the output of the demodulator matched filter that does not result in ISI. To accomplish this, the RC frequency response is proportioned or distributed between the modulator and demodulator matched filter responses that are characterized as SRRC frequency responses.

§This bandwidth also applies to the RRC spectrum.

TABLE 4.8 Evaluation of Necessary Bandwidth for Commonly Used Modulations[a]

Modulation	$B'^{\,b}_n$	Conditions	Comments
BFSK[c] [32–34]	$3.86D+0.27R$ $2.4D+1.0R$	$0.03<2D/R<1.0$ $1.0<2D/R<20$	
Multilevel FSK	$\frac{R}{\log_2 S}+2DK$	$K\le 0.89$	$K = 0.89$ corresponds to 99% bandwidth
GMSK[d]	$\frac{R}{\log_2 S}+0.5RK$	$K\le 0.28$	$K = 0.28$ corresponds to 99% bandwidth ($S = 2$)
MSK[d]	$\frac{R}{\log_2 S}+0.5RK$	$K\le 0.36$	$K = 0.36$ corresponds to 99% bandwidth ($S = 2$)
PSK	$\frac{2RK}{\log_2 S}$	$0.5\le K\le 1$	Typically: $0.7 \le K \le 0.8$ $K > 0.7$ should be justified for fixed microwave systems
QAM	$\frac{2RK}{\log_2 S}$	$K\le 0.81$	$K = 0.81$ corresponds to 99% bandwidth
OFDM	$(N_s+16.25)C_s$	$N_s > 16$	

[a]NTIA Appendix J. National Telecommunications and Information Administration (NTIA) [31]. Reproduced by permission of the U.S. Department of Commerce.
[b]Example calculations are given in Appendix J.
[c]The following three references contain additional details concerning the calculation of B'_n for microwave radio relay systems using BFSK.
[d]The parameter $S = 2$ in determining B'_n for GMSK and MSK.

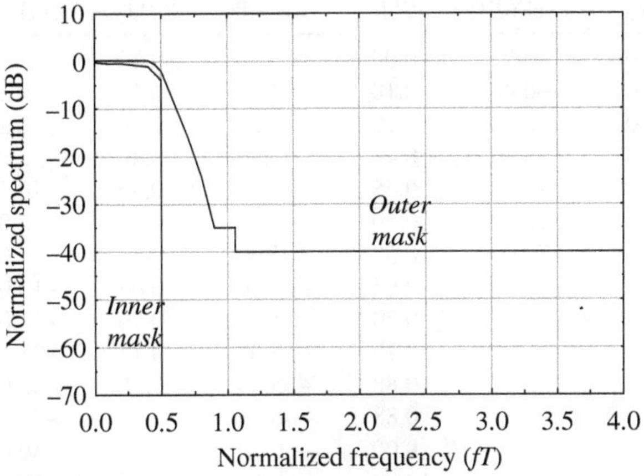

FIGURE 4.41 Intelsat one-sided PSD mask at modulator output. Intelsat Earth Station Standards [35]. Reproduced by permission of Intelsat Global Services LLC).

TABLE 4.9 Specifications for Intelsat Spectral Mask[a]

Inner Mask			Outer Mask		
$\|f/R_b\|$	$\|f/R_s\|$	Level (dB)	$\|f/R_b\|$	$\|f/R_s\|$	Level (dB)
0.00	0.00	−0.25	0.00	0.00	0.25
0.05	0.10	−0.40	0.05	0.10	0.25
0.10	0.20	−0.40	0.10	0.20	0.25
0.20	0.40	−1.00	0.20	0.25	0.25
0.25	0.50	−4.00	0.225	0.45	−0.50
			0.25	0.50	−2.00
			0.35	0.70	−16.00
			0.40	0.80	−24.00
			0.45	0.35	−35.00
			0.53	1.60	−40.00

[a]Intelsat Earth Station Standards [35]. Reproduced by permission of Intelsat Global Services LLC.

$$B=\frac{(1+\alpha)}{2T}=\frac{R_s}{2}(1+\alpha)\quad :\text{Hz} \tag{4.93}$$

with $0\le\alpha\le 1$ and $R_s = 1/T$ is the modulation symbol rate. The factor α defines the excess bandwidth $\alpha R_s/2$ and is referred to as the excess bandwidth factor. In practice, the RRC spectrum is used as the modulator or transmitter filter so that the demodulator matched filter output results in the RC spectrum* that, theoretically, does not exhibit intersymbol interference; this theoretical condition also requires ideal symbol, frequency, and phase tracking. From this description, the product of the transmitted and received SRRC spectrums is equal to the RC spectrum of the demodulator matched filter output which is a bandlimited function with sampled impulse response $x(nT)$ such that $x(nT) = 0\ \forall\ n$ except $n = 0$, where T is the matched filter output symbol sampling interval. These conditions were established by Nyquist and T is referred to as the Nyquist interval and $R_s/2$ as the Nyquist bandwidth. Upon examining the RC spectrum shown in Figure 4.42, it is seen that sampling at the rate $R_s/2$ corresponds to half-band sampling discussed in Section 2.8.2. When the folded spectrum segments A, indicated as *dashed* lines, exactly matches the void spectrum segments B, a $rect(f/R_s)$ spectrum is formed having a $x(t) = sinc(t/T)$ impulse response with zeros occurring at $x(nT)$ as described above. The *sinc* function is the only function that

*The characteristics of the temporal and spectral RC functions are discussed in Sections 1.11.4 and 1.11.4.1 in the context of windows.

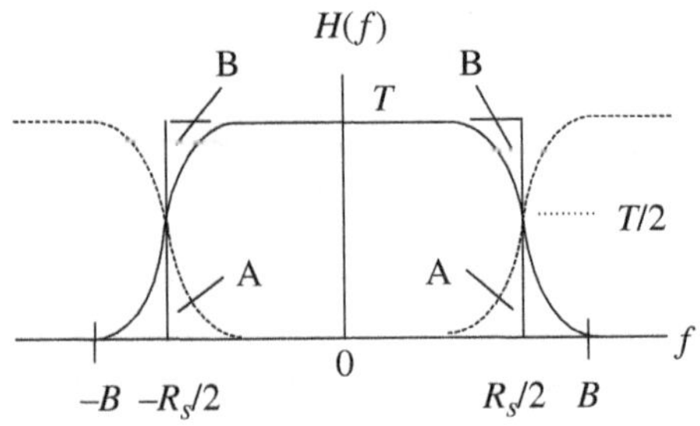

FIGURE 4.42 Nyquist RC waveform spectrum.

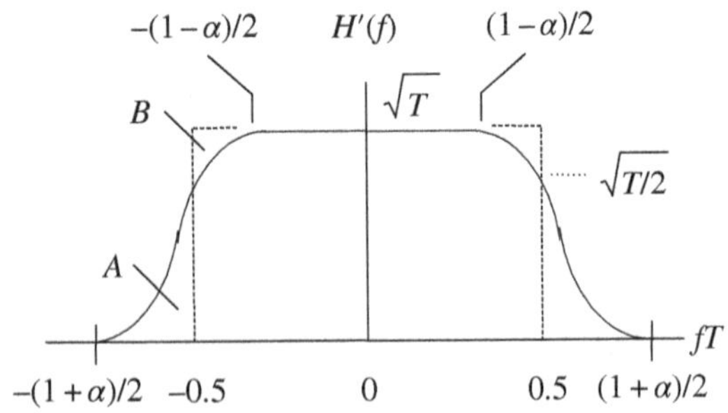

FIGURE 4.43 Theoretical SRRC waveform spectrum.

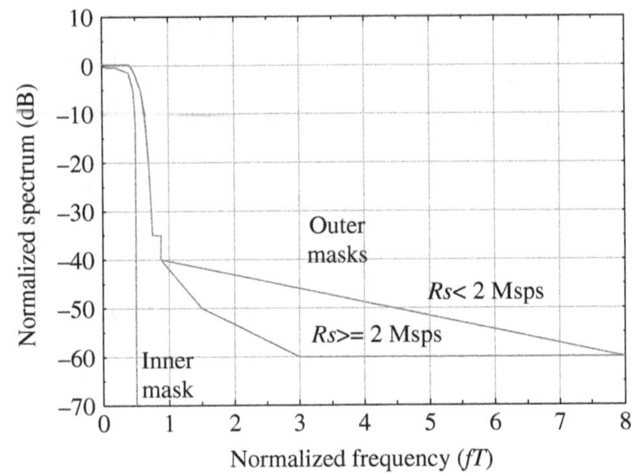

FIGURE 4.44 SRRC one-sided spectral masks ($\alpha \le 0.4$).

possesses these Nyquist properties. Lucy, Salz, and Weldon [14] and Papoulis [37] provide a more in-depth treatment of the Nyquist criterion.

Unfortunately, the SRRC waveform does not have a constant magnitude so the input to the modulator power amplifier must be backed off to avoid clipping that deteriorates the transmitted spectrum and the demodulator detection performance. The spectrum and bit-error performance of SRRC modulation are evaluated in Sections 15.14.2 and 15.14.3 for various backoff levels using a solid-state power amplifier.

TABLE 4.10 Specifications for SRRC Spectral Masks[a] **($\alpha \le 0.4$)**

Inner Mask		Outer Mask $R_s < 2$ Msps		Outer Mask $R_s \ge 2$ Msps	
$\lvert f/R_s \rvert$	Level (dB)	$\lvert f/R_s \rvert$	Level (dB)	$\lvert f/R_s \rvert$	Level (dB)
0.00	−0.25	0.00	0.25	0.00	0.25
0.10	−0.40	0.10	0.25	0.10	0.25
0.20	−0.40	0.20	0.25	0.20	0.25
0.40	−1.50	0.40	0.25	0.40	0.25
0.47	−5.00	0.45	−0.50	0.45	−0.50
0.50	−12.00	0.50	−2.00	0.50	−2.00
0.50	−70.00	0.58	−5.00	0.58	−5.00
		0.64	−10.00	0.64	−10.00
		0.70	−21.00	0.70	−21.00
		0.75	−35.00	0.75	−35.00
		0.88	−35.00	0.88	−35.00
		0.88	−40.00	0.88	−40.00
		8.00	−60.00	1.50	−50.00
		10.00	−60.00	3.00	−60.00
				10.00	−60.00

[a]The last value in the "Level(dB)" columns are for plotting only and represent infinite limits.

4.4.1.1 *Spectral Root-Raised-Cosine (SRRC) Shaping Function*

The theoretical SRRC spectrum $H(\omega)$ is expressed by (4.88) and plotted in Figure 4.43 as a function of the normalized frequency fT for an arbitrary excess bandwidth factor α. In terms of the normalized bandwidth BT in (4.93), the abscissa of Figure 4.43 is identical to the RC spectrum shown in Figure 4.42; however, the amplitude is related by the square-root.

The spectral mask for Nyquist signaling using the SRRC-modulated waveform is based on a modification of the Intelsat mask in Figure 4.41. The modifications reflect the theoretical spectrums that are dependent on the selection of the excess bandwidth factor. The resulting SRRC mask is shown in Figure 4.44 with various *break-points* listed in Table 4.10. This mask represents the spectrum at the output of the waveform modulator. The two symbol rate-dependent variations for $R_s \ge 2$ and <2 Msps are motivated by the requirements of the CCSDS SFCG recommendation shown in Figure 4.39.

Figure 4.45 shows the theoretical baseband SRRC power spectral density for several values of the parameter α. The SRRC spectral mask is designed to satisfy the theoretical spectrum for $\alpha = 0.4$ with a slight margin based on the practical implementations. The final test of the SRRC power spectral density evaluation, and to some extent the spectral mask evaluation, is examined in Section 4.4.4.1 where realizable implementations of the SRRC waveforms are examined using windowed SRRC impulse responses.

4.4.2 Power Spectral Density Characterization

The spectrum of a modulated waveforms is evaluated in terms of the PSD defined as the Fourier transform of the waveform correlation function $R(t)$ [38], that is,

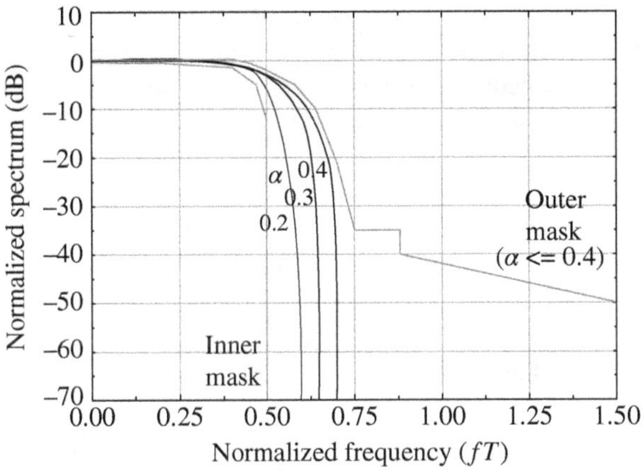

FIGURE 4.45 Theoretical SRRC power spectral density.

$$G(\omega)=\int_{-\infty}^{\infty} R(t)e^{-j\omega t}dt \tag{4.94}$$

The function $R(t)$ has the property of being real and even so $G(\omega)$ is also real and even. In terms of the waveform function $s(t)$ the PSD is given by

$$G(\omega)=\lim_{T\to\infty}\frac{1}{T}|S_T(\omega)|^2 \tag{4.95}$$

where $S_T(\omega)$ is the Fourier transform of a finite or truncated interval of $s(t)$ evaluated as

$$S_T(\omega)=\int_{-T/2}^{T/2} s(t)e^{-j\omega t}dt \tag{4.96}$$

If $s(t)$ is fundamentally limited to the interval $|t| \le T/2$, that is, it is zero outside of this interval, then the power spectrum is simply the squared magnitude of the spectrum of $s(t)$ divided by T. In this case, $G_T(\omega)$ is defined as the average power of $S_T(\omega)$ in the interval T expressed as

$$G_T(\omega)=\frac{1}{T}|S_T(\omega)|^2 \tag{4.97}$$

$S_T(\omega)$ has dimensions of volt-seconds and $G_T(\omega)$ has dimensions of watt-seconds or watts/hertz; therefore, $G_T(\omega)$ is often referred to as the PSD of $s(t)$.

Typically, the baseband spectrums are evaluated because the carrier modulation is linear and simply translates or shifts the spectrum to the carrier frequency, although, the influence of the negative frequency portion of the spectrum must be considered when the bandwidth to carrier frequency ratio is not sufficiently high to ensure that the folding of the spectral sidelobes about the direct current (DC) is negligible.

As long as the symbols in the data sequence are uncorrelated, the power spectral density can be evaluated in terms of an isolated symbol so that $\tilde{s}(t)=p(t)$ where $p(t)$ is the symbol shaping function discussed in the previous sections. However, when intersymbol interference is inherent in the signal waveform, as, for example, with GMSK, S-MPSK, and severely filtered modulated waveforms, the spectrum must be evaluated using the Fourier transform of the autocorrelation function of a data sequence or the method involving Markov processes described by Gronemeyer and McBride [39]. In the following sections, the power spectral density for the commonly used PSK-modulated waveforms is examined.

4.4.3 Constant Envelope PSK Waveforms: Spectrums and Bit-Error Performance

In this section, the spectrums of constant amplitude-modulated waveforms are examined and compared to the appropriate spectral mask. The waveforms examined are: BPSK, QPSK, OQPSK, MSK, SMSK, and GMSK. The spectral re-growth is examined when linear filtering is applied in the modulator and transmitter before power amplifier hard limiting. The spectrum containment of phase-shaped BPSK, QPSK, 8PSK, and OQPSK modulations is examined for linear and raise-cosine phase shaping functions and the resulting bit-error performance examined using a single-symbol detection filter.

4.4.3.1 BPSK, QPSK, and OQPSK Power Spectral Densities In view of the introductory remarks, the power spectral density of the BPSK waveform is determined from the Fourier transform of $p(t)=\sqrt{P}rect(t/T)$ with T equal to the bit duration T_b. The signal spectrum is evaluated as,

$$\begin{aligned} S_{T_b}(\omega) &= \sqrt{P}\int_{-T_b/2}^{T_b/2} e^{-\omega t}dt \\ &= \sqrt{P}T_b\frac{\sin(\omega T_b/2)}{\omega T_b/2} \end{aligned} \tag{4.98}$$

and the power spectral density is

$$G_T(\omega)=PT_b\left(\frac{\sin(\omega T_b/2)}{\omega T_b/2}\right)^2 \quad :\text{BPSK} \tag{4.99}$$

The QPSK and OQPSK spectrums are evaluated in the same way using $T=2T_b$ and recognizing that the symbol on each quadrature rail is characterized as $p(t)=(\sqrt{P}/2)rect(t/T)$. Upon evaluating the signal spectrum, the PSD is evaluated using (4.97) as

$$G_T(\omega)=PT_b\left(\frac{\sin(\omega T_b)}{\omega T_b}\right)^2 \quad :\text{QPSK and OQPSK} \tag{4.100}$$

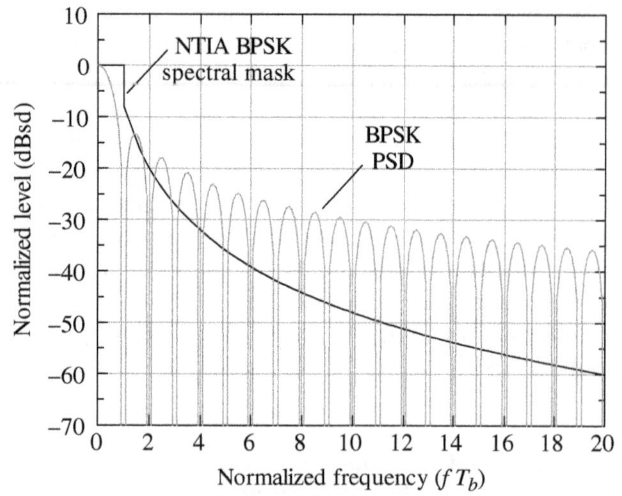

FIGURE 4.46 One-sided BPSK PSD and NTIA spectral mask (K = 1.0).

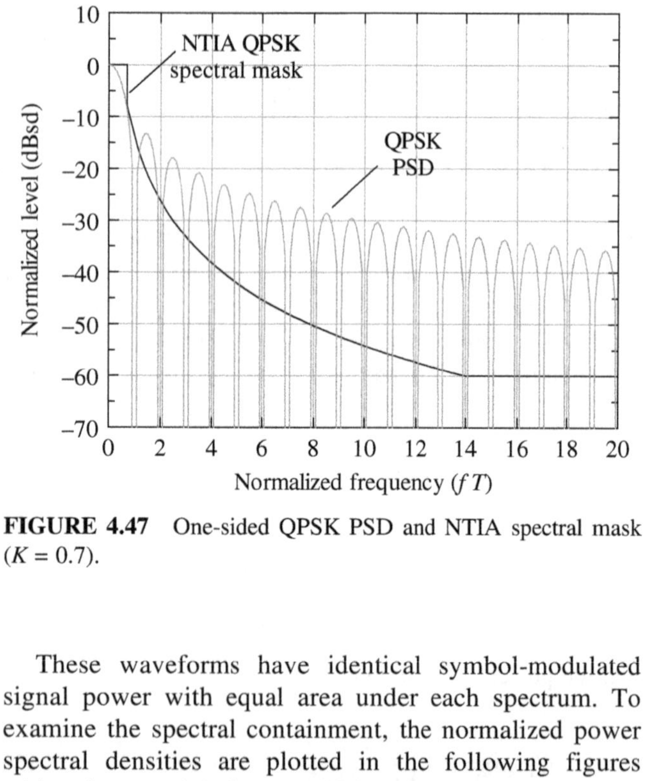

FIGURE 4.47 One-sided QPSK PSD and NTIA spectral mask (K = 0.7).

These waveforms have identical symbol-modulated signal power with equal area under each spectrum. To examine the spectral containment, the normalized power spectral densities are plotted in the following figures with the modulation-specific NTIA spectral mask expressed by (4.92) and the necessary bandwidth corresponding to Table 4.8. Figures 4.46 and 4.47 show the results for BPSK and QPSK. The abscissa of the BPSK PSD and mask are normalized to the bit interval and the QPSK PSD and mask are normalized to symbol interval $T = 2T_b$ so, aside from the different values of the factor K, the two plots appear to be identical. The selection of K demonstrates the differences in the mask when plotted as a function of the fT compared to f/B'_n as shown in Figure 4.40. For example, with $K = 1.0$ the −8 dBsd step occurs at the first spectral null and the −60 point occurs at $fT_b = 20$ compared to $fT = 14$ for QPSK with $K = 0.7$.

The BPSK and QPSK modulations use the *rect(t/T)* symbol weighing and do not meet the masks that are designed for spectrally efficient modulations. In Section 4.4.3.9, the PSD of S-BPSK and S-QPSK-modulated waveforms is examined and found to meet the NTIA spectral masks for properly chosen phase-shaping factors.

4.4.3.2 BPSK, QPSK, and OQPSK Spectral Re-growth and Bit-Error Performance

The increasing demand for bandwidth in commercial radio, including terrestrial and satellite applications, has nurtured new development in waveform designs that increase the bandwidth efficiency. The spectral masks are intended to provide a frequency environment where a variety of users with different bandwidth requirements can coexist without interfering with each other. For example, applications involving FDM and FDMA* use spectrally efficient modulations to allow for closer channel spacing thus conserving bandwidth or increasing user data rates within the same bandwidth.

One way to generate a spectrally efficient modulated signal is simply to pass the signal through a linear filter with an appropriately selected 3-dB frequency and transition bandwidth. However, the filter introduces signal distortion that is characterized by AM and phase distortion resulting in demodulator performance degradation. Although the linear filter distortion can be removed through equalization in the demodulator, if the waveform is amplified by a hard-limiting transmitter power amplifier or encounters a hard-limiting channel, as in a bent-pipe satellite transponder, the AM is removed and the resulting nonlinear waveform spectrum encounters re-growth that tends to restore the unfiltered waveform spectrum. In the case of BPSK modulation the filtered spectrum essentially returns to the original *sinc(fT)* spectrum before the hard limiting.

The spectral re-growth is examined in this section for BPSK, QPSK, and OQPSK modulations following linear filtering with a 3-pole Butterworth filter with 3-dB cutoff frequency equal to three times the symbol rate, that is, $f_c = 3R_s$; for this filter the attenuation at $2f_c$ is about 18 dB. The spectrums and bit-error performance are evaluated using a Monte Carlo simulation with 32 samples-per-symbol. The Butterworth filter is perfectly

*The use of orthogonal frequency spacing in these cases is referred to as OFDM and OFDMA. However, these are distinctly different waveforms that result in high peak-to-rms power variations unlike the constant amplitude waveforms discussed in this section.

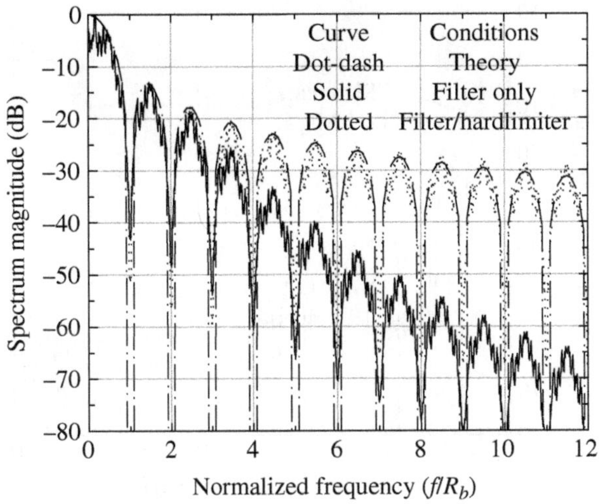

FIGURE 4.48 BPSK power spectral density with filtering followed by hard limiting.

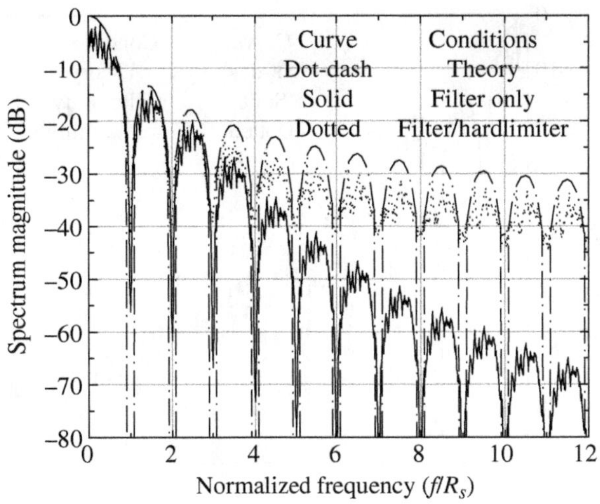

FIGURE 4.50 QPSK power spectral density with filtering followed by hard limiting.

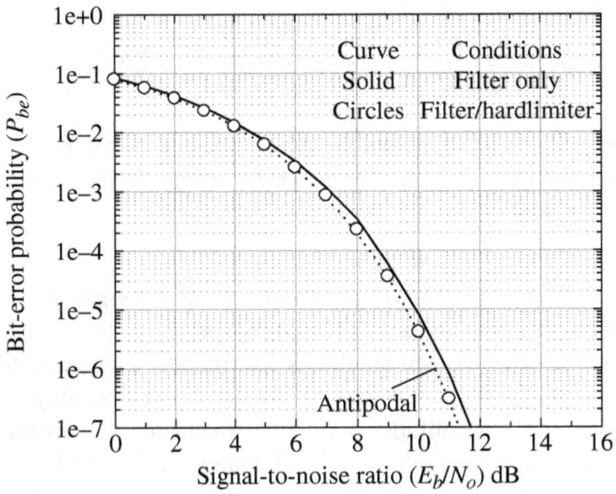

FIGURE 4.49 BPSK bit-error performance with filtering followed by hard limiting.

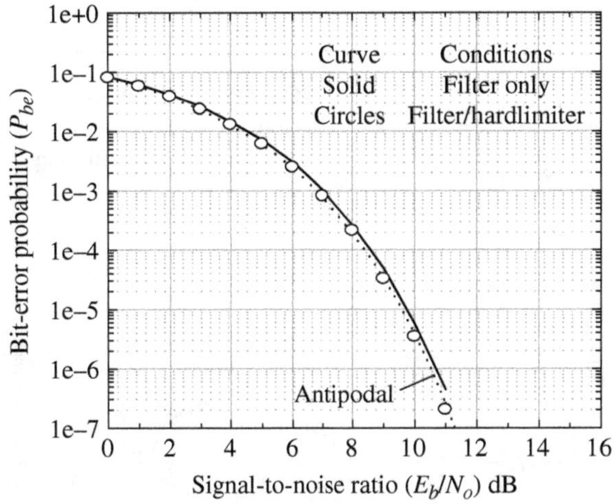

FIGURE 4.51 QPSK bit-error performance with filtering followed by hard limiting.

phase equalized and is implemented as a 64-tap linear-phase finite impulse response (FIR) filter with a delay of one symbol. The number of samples-per-symbol is well in excess of the Nyquist sampling criterion and is used to examine the spectrum over ±16 symbol intervals; however, only the positive frequencies over 11 sidelobes are displayed in the following graphs. The Monet Carlo bit-error performance uses 500K symbols for each signal-to-noise ratio for $P_{be} > 5e^{-4}$ otherwise 10 M symbols are used.

Referring to Figures 4.48, 4.49, 4.50, 4.51, 4.52, and 4.53 the spectral re-growth of BPSK, following the hard limiter, essentially results in the *sinc(fT)* spectrum before the filter and the bit-error performance loss is recovered. The QPSK spectrum behaves in a similar way, however, the re-growth is less than that of BPSK and the bit-error performance loss due to the filtering is essentially recovered. The OQPSK spectrum is considerably less sensitive to spectral re-growth and the bit-error performance following the hard limiter shows some signs of the ISI loss as a result of the spectral sidelobe reduction. The progressively reduced spectral re-growth for these modulations is a direct result of the lower phase transitions encountered during the symbol data changes. In the following section, the spectral re-growth of MSK is seen to be considerably less sensitive even than that of OQPSK.

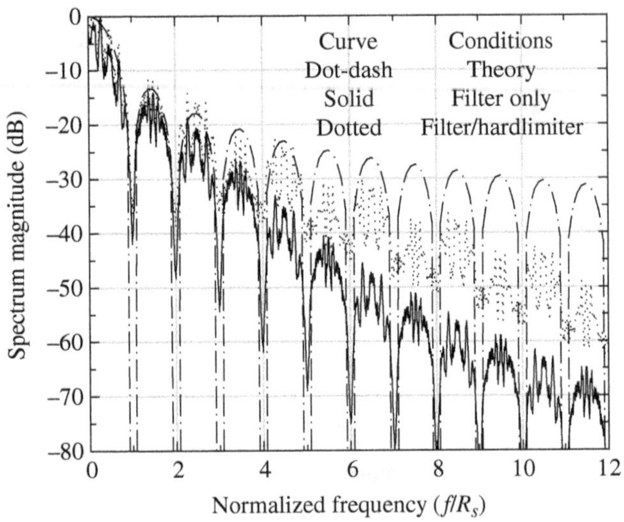

FIGURE 4.52 OQPSK power spectral density with filtering followed by hard limiting.

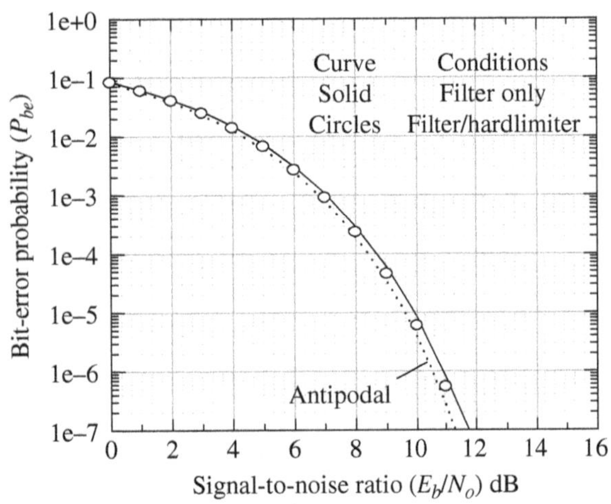

FIGURE 4.53 OQPSK bit-error performance with filtering followed by hard limiting.

4.4.3.3 *MSK Power Spectral Density*

For MSK waveform modulation $p(t)=\sqrt{P}\cos(\omega_m t)rect(t/2T_b)$ and the signal spectrum is evaluated as

$$S_T(\omega)=\sqrt{P}\int_{-T_b}^{T_b}\cos(\omega_m t)e^{-j\omega t}dt$$
$$=\frac{4\sqrt{P}T_b}{\pi}\left(\frac{\cos(\omega T_b)}{1-4(\omega T_b/\pi)^2}\right) \tag{4.101}$$

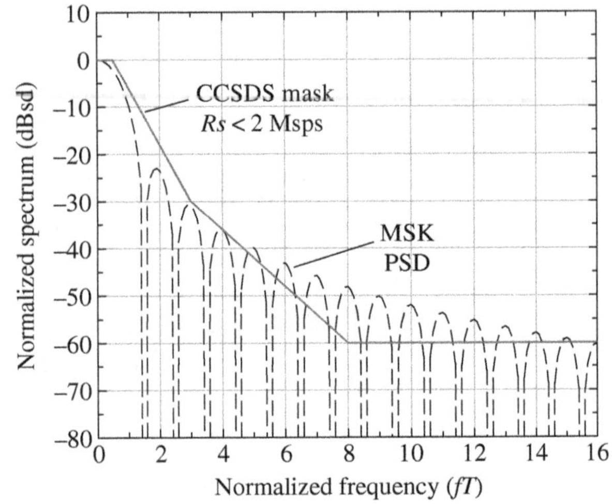

FIGURE 4.54 One-sided MSK PSD and CCSDS spectral mask ($R_s < 2$ Msps).

and the power spectral density is

$$G_T(\omega)=\frac{8PT_b}{\pi^2}\left(\frac{\cos(\omega T_b)}{1-4(\omega T_b/\pi)^2}\right)^2 \quad :\text{MSK} \tag{4.102}$$

The PSD of MSK meets the NTIA spectral mask, however, does not meet the CCSDS mask for $R_s < 2$ Msps as shown Figure 4.54.

4.4.3.4 *Filtered MSK (FMSK) Spectral Re-growth and Bit-Error Performance*

The MSK spectrum can be filtered to reduce the spectral sidelobes for frequencies greater than $3R_s$; however, the filtering introduces AM resulting in a non-constant amplitude-modulated waveform. Nevertheless, because of the excellent spectral characteristic of MSK, the low power levels of the filtered sidelobe will result in negligible bit-error performance degradation. If a hard-limiting PA or repeater is encountered, the issue of spectral re-growth must be considered. For example, if the filter precedes a hard limiter that removes the AM from the filtered signal, nothing is gained if 100% spectral re-growth occurs. As characterized in the previous section, spectral re-growth is sensitive to the waveform modulation and the symbol shaping function and it was demonstrated that BPSK with *rect(t/T)* symbol shaping results in nearly 100% spectral re-growth. Fortunately, MSK is less sensitive to spectral re-growth because the phase change between phase rest-states does not exceed $\pi/2$ radians and occurs when the quadrature offset symbol weight functions $\cos(\omega_m t)$ are zero. To demonstrate the performance of filtered MSK (FMSK), the spectral and the bit-error performance characteristics before and after hard limiting are shown in Figures 4.55 and 4.56 respectively. In both cases, the CCSDS spectral mask for symbol rates less than 2 Msps is met and the bit-error probability degradation is negligible;

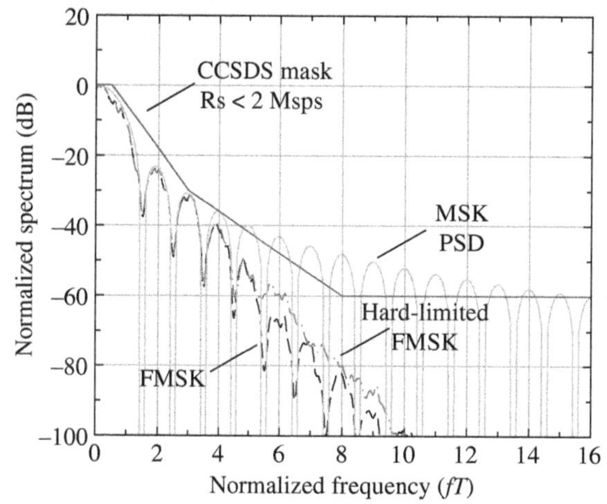

FIGURE 4.55 One-sided FMSK PSD with and without hard limiting (CCSDS spectral mask for $R_s < 2$ Msps).

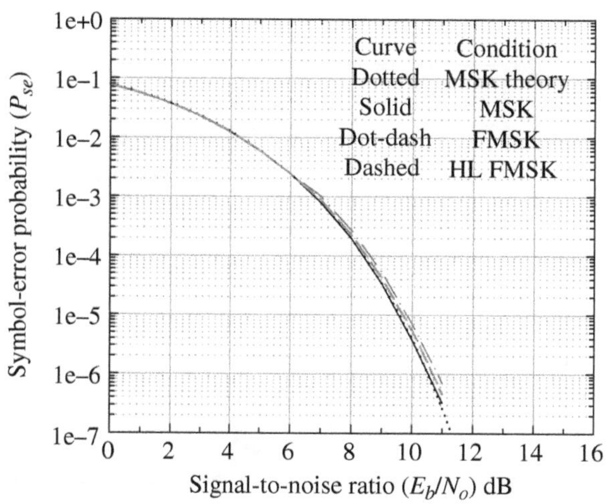

FIGURE 4.56 Bit-error performance of FMSK with and without hard limiting.

less than 0.1 dB. The MSK filter corresponds to a phase-equalized 6-pole Butterworth filter with a 3-dB normalized frequency of $f_3T = 3$.

The simulated bit-error performance of FMSK is shown in Figure 4.56 with and without limiting and compares the theoretical and simulated performance of MSK. The Monte Carlo simulations are based on 500K symbols or trials for $P_{be} > 5 \times 10^{-4}$ otherwise 100 M symbols or trials are used. From the Monte Carlo simulation performance shown in Figure 4.56 the corresponding losses at $P_{be} = 10^{-6}$, with and without limiting, are about 0.3 and 0.12 dB. Upon examination of the noise-free matched filter *eye closure** of the FMSK and hard-limited FMSK-modulated signals, the respective worst-case detection losses are found to be 0.1 and 0.05 dB.

*The eye closure is a common metric used to characterize the demodulator loss at the sampled matched filter output. The eye closure, observed with signal distortion, is also a reliable estimate of the performance loss for low levels of distortion and, in this regard, is similar to estimating the paired echo distortion discussed in Section 1.3.2.

4.4.3.5 SFSK/SMSK Power Spectral Density The spectrum of the SFSK/MFSK-modulated waveform is evaluated using the Fourier transform of $p(t)$ and is expressed by

$$G_T(\omega) = \frac{1}{2T_b}\left|\sqrt{2P}\int_{-T_b}^{T_b} \cos(\omega_m t - \alpha\sin(4\omega_m t))e^{-j\omega t}dt\right|^2 \quad : \text{SMSK} \qquad (4.103)$$

Although Amoroso [20] has derived an expression for this spectrum involving the summation of Bessel functions, the results shown in Figure 4.57a are based on numerical integration of the integral in (4.103). The SFSK spectrum in Figure 4.57b is obtained using the Fourier transform of the modulated waveform generated using either the I/Q or CPM implementations of Figure 4.26. Both techniques (the I/Q and CPM) yield identical results and the spectrum details are in better agreement with Amoroso's results. The spectral sidelobes of the α-FSK spectrums for $\alpha = 0.2$ and 0.3 become increasingly higher than those of SFSK ($\alpha = 0.25$) for $|fT| \geq 6$. As α continues to decrease the spectral sidelobes continue to increase resulting in the MSK spectrum when $\alpha = 0$ and, when α continues to increase, the spectral sidelobes also increase slightly exceeding those of the MSK spectrum when $\alpha = 0.5$. In the range $|fT| < 6$ the spectral sidelobes behave somewhat differently with a significant broadening of the main lobe and adjacent sidelobes with the level of the adjacent sidelobes exceeding those of the first two adjacent MSK sidelobes. Because the SFSK spectrum does not meet the CCSDS or the NTIA spectral masks no further consideration is given to SFSK except to note that it is an optional modulation mode, referred to as SMSK by the U.S. Navy.

4.4.3.6 Gaussian MSK (GMSK) Power Spectral Density GMSK modulation modifies the linear phase function of MSK modulation using a Gaussian phase function that spans several symbol intervals as described in Section 4.2.7. The span of the Gaussian phase function is defined as the total number of symbol, including the current and adjacent symbols, over which the phase function in non-zero. This corresponds to the span of the symmetrical Gaussian shaping filter response $g(t)$ that is centered on the current symbol and extends into the adjacent symbols as determined by the time-bandwidth parameter f_3T_b. The theoretical span of the Gaussian function is infinite and symmetrical about the current symbol; however, for practical considerations, the span is symmetrically truncated and the resulting pedestal is removed prior to the normalization of the phase function to $\pi/2$ radians.

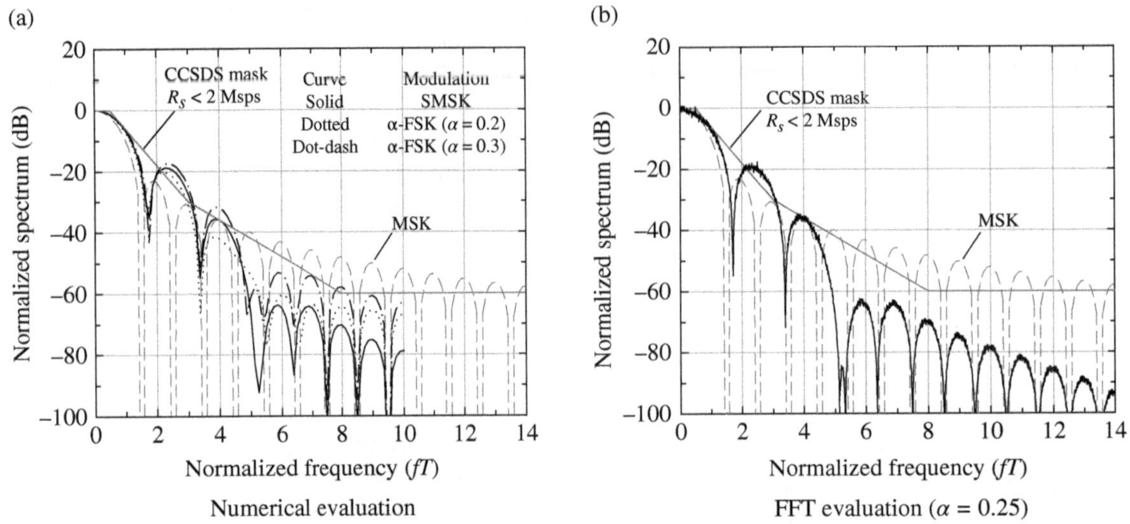

FIGURE 4.57 SFSK/SMSK power spectral density.

For example, referring to Figure 4.31, if the span is limited to three symbols the impulse response $g(t)$ corresponding to $f_3T_b = 0.25$ results in a pedestal or step at $t/T_b = -1$ and 2. Removing the pedestal and performing the phase normalization results in a 10 dB reduction in the power spectral density sidelobes corresponding to $|fT| > 1.5$. The encroachment of the symbol phase function into the adjacent symbols results in ISI and suboptimum symbol detection when a single-symbol detection filter is used. In the following performance evaluations, the span of the phase function is examined for specific values of f_3T_b and reducing the span results in an increasing spectral sidelobe floor. In general, as f_3T_b is decreased the span should be increased to preserve the low spectral sidelobe characteristics; however, increasing the span results in greater ISI loss when using a single-symbol I&D detection filter. The use of MLSE detection as described by Forney can be used to reduce the detection loss at the expense of increased system complexity.

Murota and Hirade [40] have evaluated the power spectral density, the fractional power ratio, the ACI performance, and the single-symbol detection filter loss of GMSK using machine computations and their results are used as a baseline or standard for comparison with the following Monte Carlo simulated performance results. Figure 4.58 shows the PSD of GMSK for $f_3T_b = 0.5$, 0.3, 0.2, and 0.1 based on averaging and smoothing one-hundred, 92 dB Blackman-Harris* windowed, 1024-point fast Fourier transforms (FFTs) of a sequence of 64 randomly generated bits with 16 samples-per-bit. The spectral smoothing is the average of a current sample and five adjacent samples using a uniformly weighted 11-tap sliding window FIR filter. The theoretical PSD for

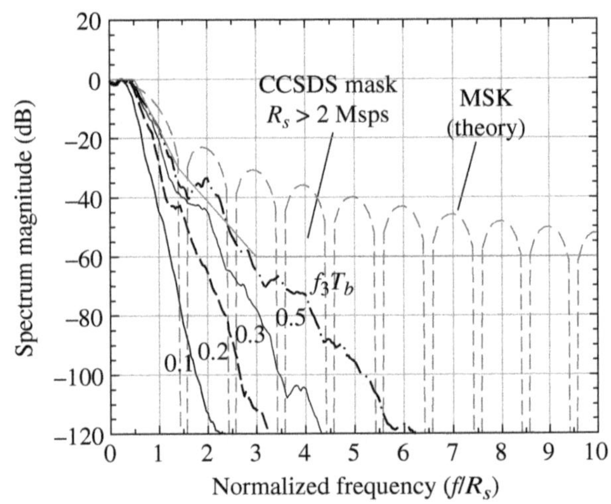

FIGURE 4.58 GMSK power spectral density (total span: 5, 5, 7 and 13 bits).

MSK and the CCSDS spectral mask for symbol rates greater than 2 Msps are also shown. The total spans corresponding to the selected f_3T_b values are: 5, 5, 7, and 13 bits. The GMSK waveform corresponding to $f_3T_b = 0.5$ does not meet the spectral mask; however, the case with $f_3T_b = 0.3$ is within the mask. The GMSK spectrum results of Murota and Hirade are limited to $|f/R_s| \le 5.0$ and over this range the simulated results in Figure 4.58 are in good agreement with their results.

Figure 4.59 compares the PSD when the total spans† are reduced as indicated. With shorter spans the spectral sidelobe levels tend to flare resulting in a sidelobe floor; however, this does not impact the CCSDS spectral mask results as indicated

*Windowing the FFT is important for spectral control and the Hanning window yields similar results.

†The total span in bits is 2(one-sided span) + 1 where the one-sided span is the extent of the phase function over the one-sided adjacent bits.

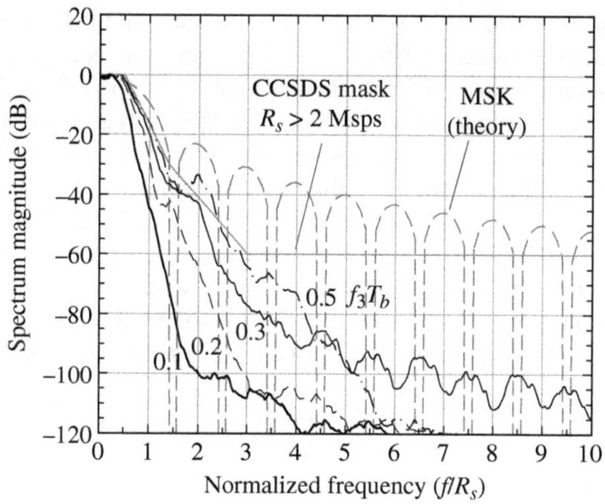

FIGURE 4.59 GMSK power spectral density (total span: 3, 3, 5 and 9 bits).

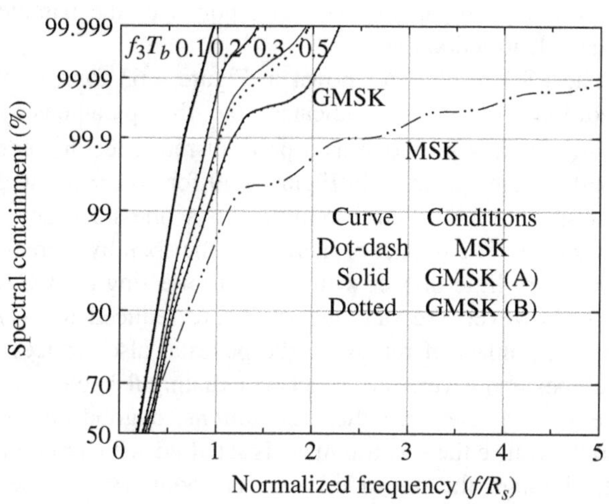

FIGURE 4.60 GMSK spectral containment.

above. Furthermore, in both of these cases, the most spectrally efficient PSD corresponding to $f_3T_b = 0.1$ does not meet the Intelsat IESS-308 SRRC spectral mask defined in Section 4.4.1.1.

4.4.3.7 GMSK Spectral Containment and Bit-Error Performance

Figure 4.60 compares the spectral containment of GMSK with that of MSK shown as a reference. The GMSK cases, denoted as A and B, correspond to the difference spans of the phase function as indicated in Figures 4.58 and 4.59 respectively. The spectral containment corresponding to $f_3T_b = 0.1$, 0.2, and 0.5 are not discernable between the two cases. Under the conditions $f_3T_b = 0.3$, 0.2, and 0.1, the percentage of the total spectral power contained within the lowpass frequency band, equal to the symbol rate, corresponds, respectively, to 99.9, 99.99, and 99.999.* This is a remarkable concentration of power that results in a very bandwidth efficient modulation; however, there is a significant increase in processing complexity when attempting to capture the modulation power in a ML detector.

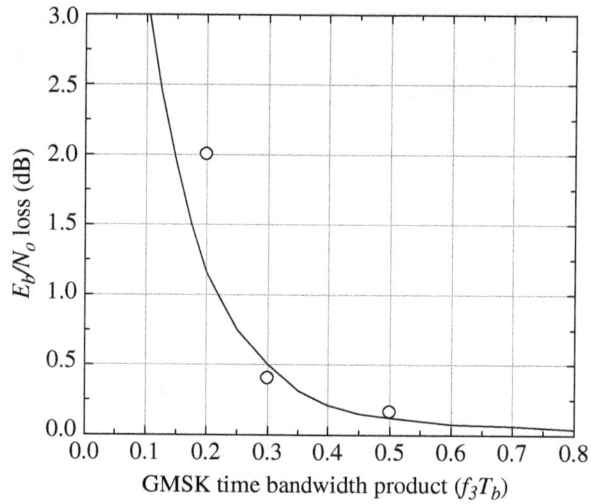

FIGURE 4.61 GMSK theoretical degradation in E_b/N_o at $P_{be} = 10^{-5}$. Murota and Hirade [40]. Reproduced by permission of the IEEE.

Referring to the theoretical bit-error performance expressed by (3.32), Murota and Hirade evaluate the performance bound on the bit-error probability for MSK using the minimum distance $d_{min} = min(d)$ between the complex mark and space received signals, $s_m(t)$ and $s_s(t)$, where d^2 is defined as

$$d^2 \triangleq \frac{1}{2}\int_{-T_1}^{T_2} |s_m(t)-s_s(t)|^2 dt \tag{4.104}$$

and the energy-per-bit, E_b, given by

$$E_b \triangleq \frac{1}{2}\int_0^{T_b} |s_x(t)|^2 dt \quad : x=m,s \tag{4.105}$$

Their results apply to coherent detection of the received MSK waveform and the use of d_{min} applies under high signal-to-noise conditions. Upon machine evaluation of d_{min} as a function of f_3T_b, Murota and Hirade evaluate the bit-error probability and resulting degradation in E_b/N_o relative to antipodal signaling as shown in Figure 4.61. The data points are based on the single-symbol† detection filter using the

*The percent bandwidth is the two-sided or bandpass bandwidth equal to $2R_s$. The occupied bandwidth is defined as the 99% bandwidth.

†The simulation uses quadrature symbol-detection filters.

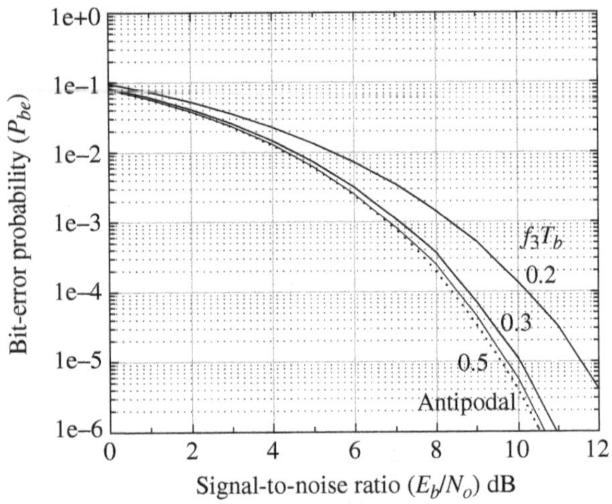

FIGURE 4.62 GMSK bit-error performance (single-symbol detection filter).

Monte Carlo simulation results shown in Figure 4.62. The simulations use 1 M bits for each signal-to-noise ratio for $P_{be} > 5e^{-4}$ and 10 M bits otherwise and the total span of the phase functions correspond to those used in Figure 4.59 (case *B*) as described above. The total span for the case $f_3T_b = 0.5$ and 0.3 is three bits, that is, the phase function extends over the two adjacent bits. As f_3T_b increases the ISI decreased and the detection performance approaches the ML detection of MSK. It appears that these two cases are reasonably close to the performance of ML detection. The case involving $f_3T_b = 0.2$ uses a total span of five bits, that is, the phase function extends over four adjacent bits (two on each side of the current bit) and the ISI significantly degrades the detection performance. The performance using the case *A* phase function spans result in flaring of the bit-error performance resulting in an irreducible bit-error probability. In these cases, the ML detector must be implemented as a MLSE that typically involves a finite-state trellis detector.

4.4.3.8 Revisiting the GMSK Power Spectral Density In Section 4.4.3.6, the PSD of the GMSK-modulated waveform was examined for fixed shaping-function spans equal to an integer number of bits or equivalently an integer number of samples $N(N_s)$ where N_s is the number of samples-per-bit. In this section, the span is examined in terms of the number standard deviations of the Gaussian phase-shaping function, defined in Section 4.2.7 as*

$$\sigma = \frac{b}{2\pi f_3} \tag{4.106}$$

where $b = \sqrt{2\ln\sqrt{2}}$, f_3 is the 3-dB frequency of the Gaussian frequency function. This approach provides for specifying an arbitrary number of one-sided samples over the Gaussian phase-shaping function. Furthermore, the PSDs are evaluated by averaging periodograms which is less cumbersome than averaging and smoothing the raw spectrums.

Using the normalized parameter $y_o = t_o/\sigma$, defined in Section 4.2.7 and characterized in Table 4.2, the focus in this section is on the case $f_3T_b = 0.3$, $N_s = 16$, with $y_o = 2$, 3, and 4. The number of one-sided samples over the Gaussian phase-shaping function, defined in (4.84), is repeated here as

$$N_s' = \text{floor}(\chi_{y_o}) \tag{4.107}$$

with

$$\begin{aligned} \chi_{y_o} &= y_o \sigma N_s \\ &= 7.0669 y_o \end{aligned} \tag{4.108}$$

The last equality in (4.108) corresponds to the parameter values identified above.

The PSD results are shown in Figure 4.63 for the three conditions of y_o as indicated by the parameter χ_{y_o}. Figure 4.63a is provided as a point-of-reference and corresponds to the spectrum in Figure 4.59 for $f_3T_b = 0.3$ with a total span of 3-bits corresponding to a one-sided span of N_s samples. The impact of increasing the span by increasing y_o has very little impact with regard to meeting the CCSDS mask; however, the sidelobe levels are reduced for $t/T_b > 1.5$. The impact of removing the pedestal also reduces the sidelobes; however, the affect is indistinguishable for the case $y_o = 4$. Between these conditions, a good choice is $y_o = 2$ because the spectral mask is satisfied with a minimum ISI, although the one-sided span is about 88% of the bit interval.

4.4.3.9 S-MPSK and S-OQPSK Performance Characteristics In the following sections, the performance of linear and raised-cosine phase-shaped MPSK ($M = 2, 4, 8$) and OQPSK is examined in terms of the spectral containment† and the bit-error probability. All of the performance results are obtained using computer simulations with 32 samples-per-symbol. The time and frequency are normalized to the symbol rate and, because the channel is modeled as a memoryless AWGN channel, the bit-error results can be applied to any bit or symbol rate. The signal is modeled as a baseband-modulated waveform and the number of samples-per-symbol is chosen to provide ±16 spectral sidelobes for the spectrum evaluations. The signal spectrum

*The parameter σ is defined as the parameter a in Section 4.2.7.

†The spectral containment when measured as a percent of the total spectral energy is also referred to as the spectral efficiency.

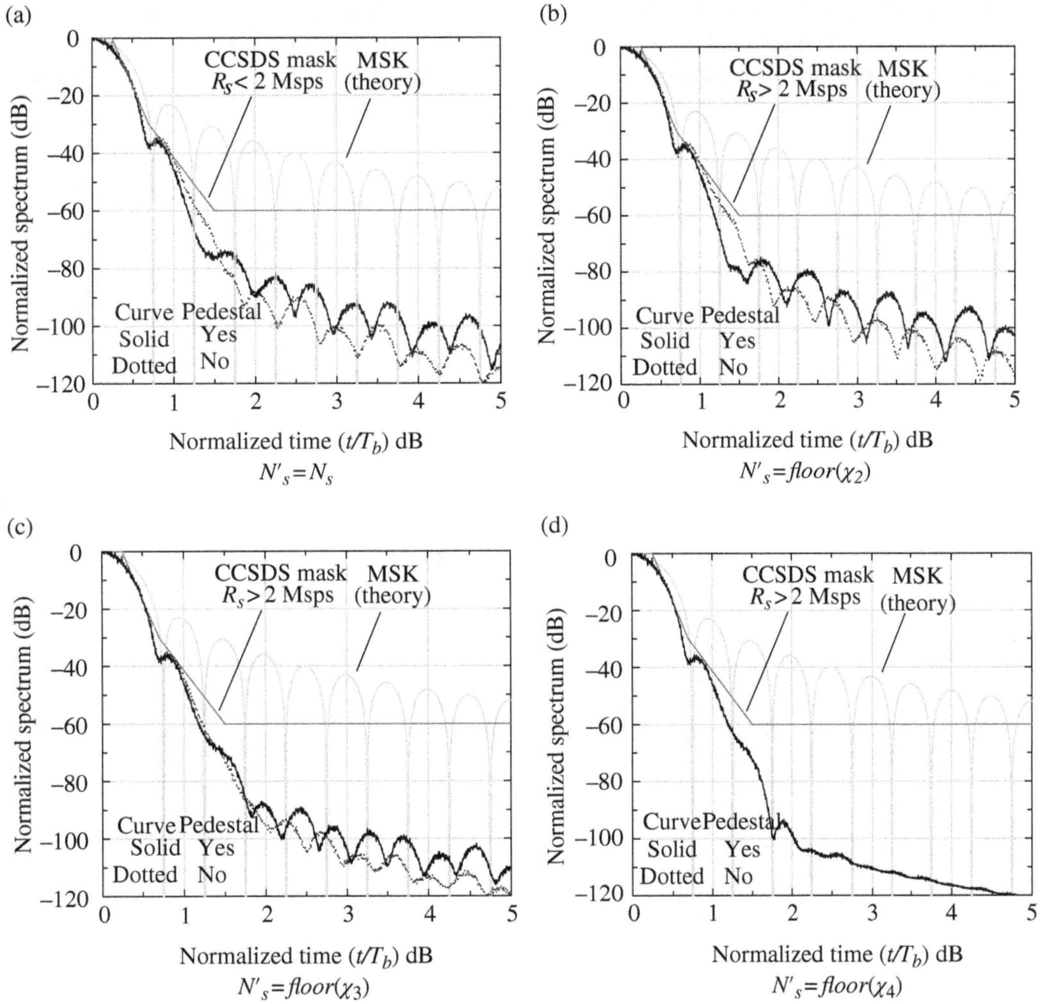

FIGURE 4.63 GMSK power spectral density results ($f_3T_b = 0.3$ and $N_s = 16$).

corresponds to the output of a linear transmitter and is evaluated by averaging and smoothing the FFT outputs. The spectral containment is the integrated spectral power over positive frequencies expressed in percent of the total positive frequency spectral power. The bit-error probability is based on 600K symbols, or Monte Carlo trials, for each signal-to-noise ratio corresponding to bit-error probabilities greater than about $5e^{-4}$, otherwise, 10 M symbols are used. The demodulator uses ideal phase and symbol tracking, that is, symbol timing, frequency, and phase errors are perfectly removed, and the symbol detection uses a single-symbol weighted I&D filter. The filter weighting is expressed as

$$w(t) = \sin^n\left(\frac{\pi p(t)}{2}\right) \quad : \frac{T(1-\alpha)}{2} \le |t| \le \frac{T}{2} \tag{4.109}$$

where a value of $n = 2$ results in the best performance and is used in the performance simulations. The filter weighting improves the signal-to-noise ratio of the sampled detection filter output by minimizing the impact of the adjacent symbols phase function. As mention previously, MLSE detection provides the optimum performance at the expense of processing complexity.

Referring to (4.85) and (4.86) the parameter α is defined in terms of the symbol duration T and, when $\alpha = 1$, the phase shaping extends over an entire symbol as depicted in Figure 4.34a. The performance evaluations of S-MPSK use phase shaping corresponding to $\alpha = 0.25$, 0.5, and 1.0. However, with S-OQPSK the symbols in the quadrature rails change at alternating intervals of the bit duration T_b. Consequently, with $\alpha > 0.5$, the phase changes simultaneously on each rail degrading the spectrum containment, so, in the case of S-OQPSK, the shaping parameter is limited to the bit interval corresponding to $\alpha \le 0.5$. For this reason, the spectral characteristics with phase shaping applied to OQPSK offers little performance improvement.

The following figures show the simulated performance for the three S-MPSK waveforms and the S-OQPSK waveform mentioned above. For each modulation, three performance curves are given: the power spectral density; the spectral containment in percentage of the total spectrum power; and the bit-error performance. In addition, each of the three performance curves is characterized for the linear and raised-cosine phase shaping functions. The results for each of the three MPSK modulations exhibit the following common performance trends:

- The spectral containment improves with increasing α
- The occupied bandwidth, defined as 99% containment, is lower with linear phase shaping
- The bounded bandwidth is lower for RC phase shaping provided that $|fT| > 2$
- The RC phase shaping requires that $\alpha \geq 0.5$ for a significant spectral containment advantage
- The bit-error performance degrades with increasing α
- The bit-error performance is less sensitive with RC shaping

S-BPSK Performance Characteristics

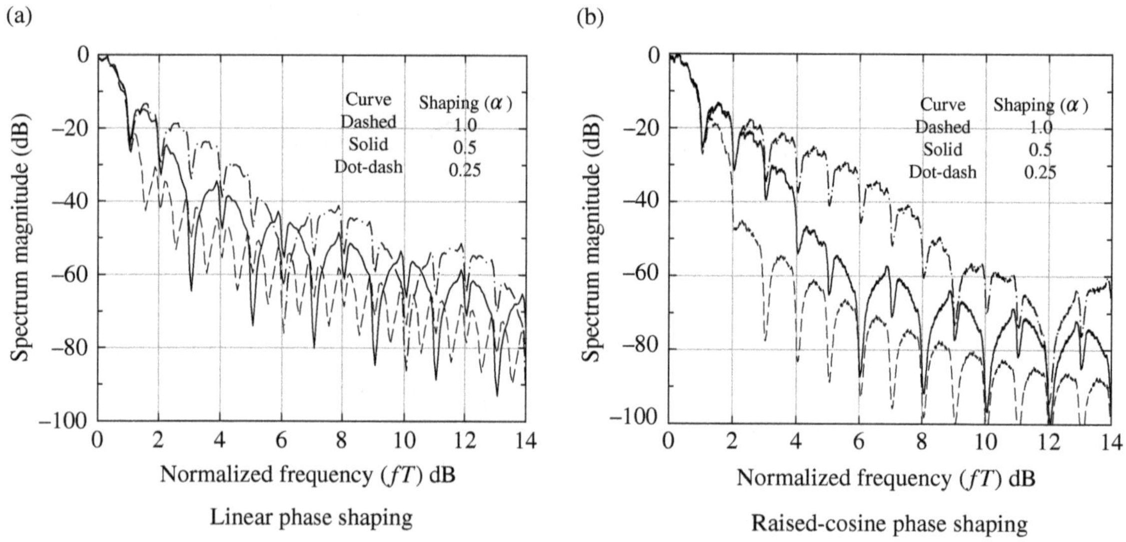

FIGURE 4.64 S-BPSK power density spectrum.

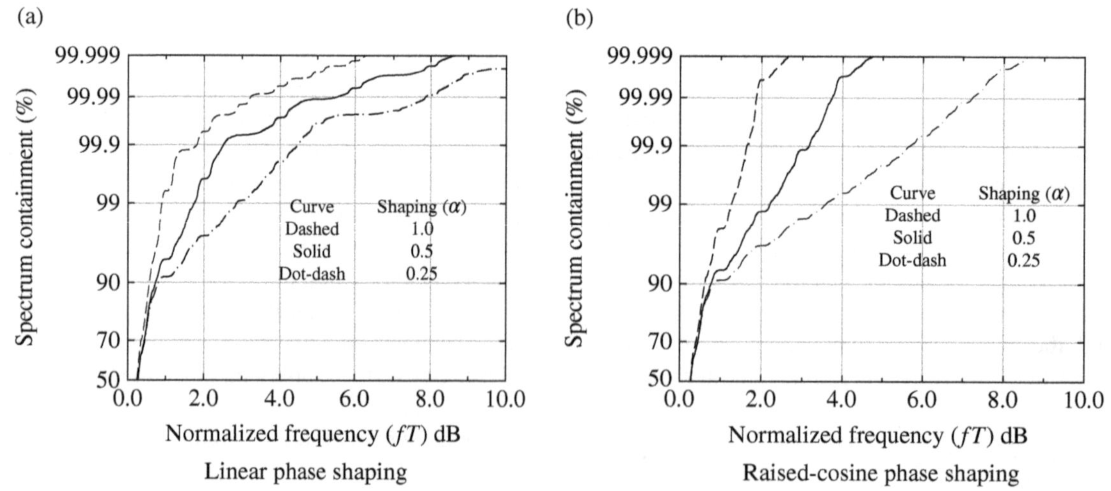

FIGURE 4.65 S-BPSK spectral containment.

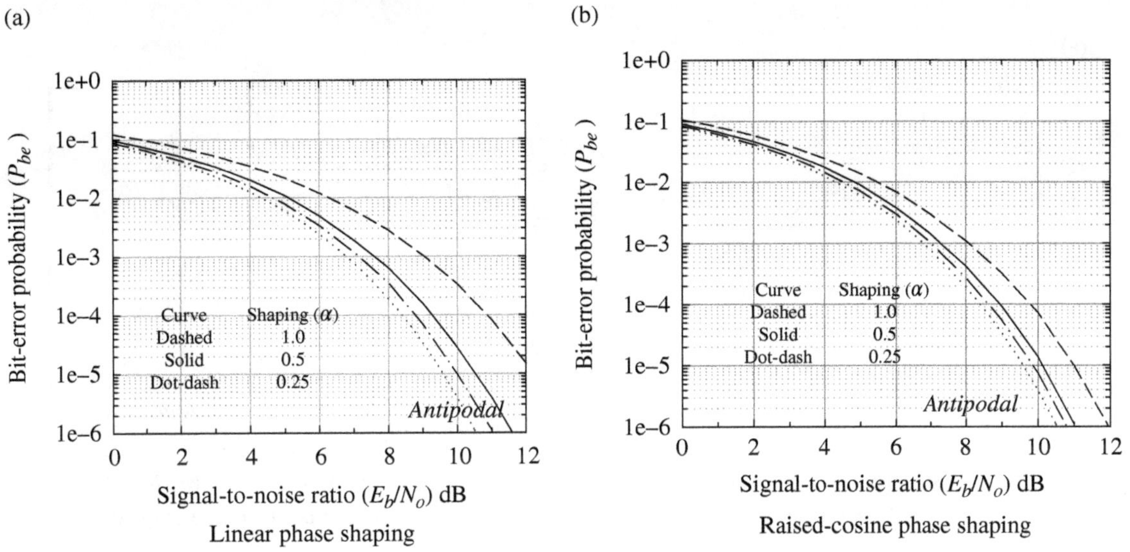

FIGURE 4.66 S-BPSK bit-error probability.

S-QPSK Performance Characteristics

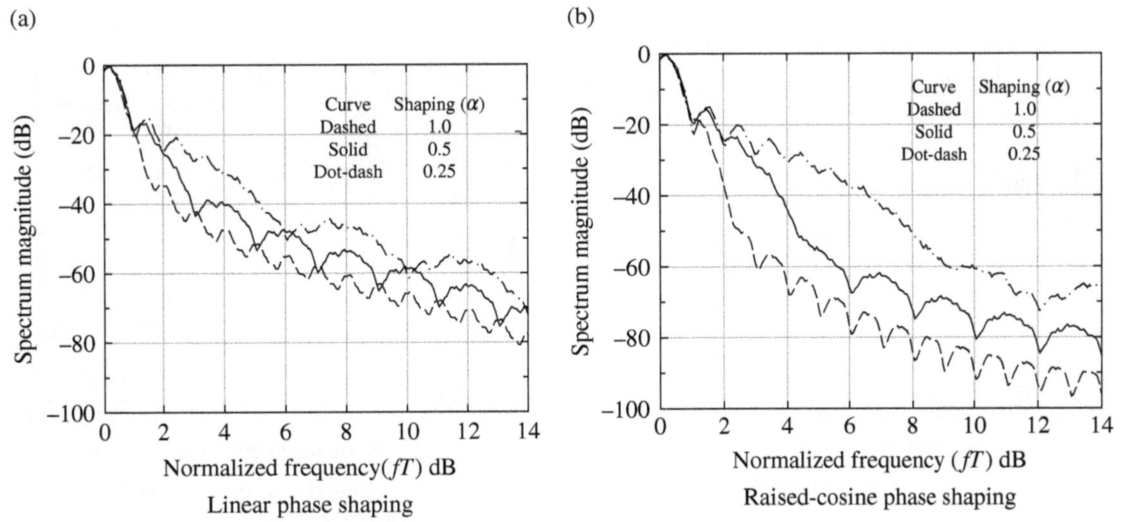

FIGURE 4.67 S-QPSK power density spectrum.

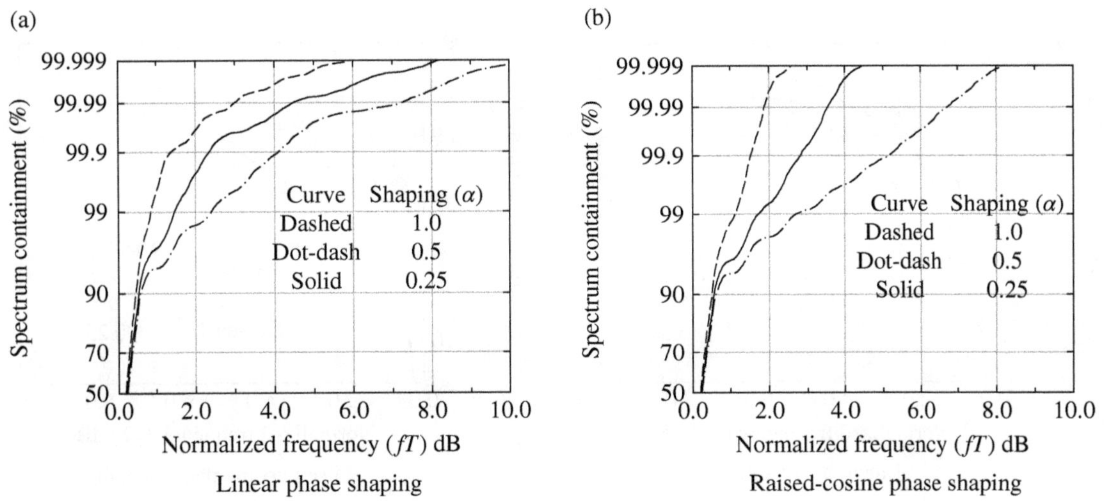

FIGURE 4.68 S-QPSK spectral containment.

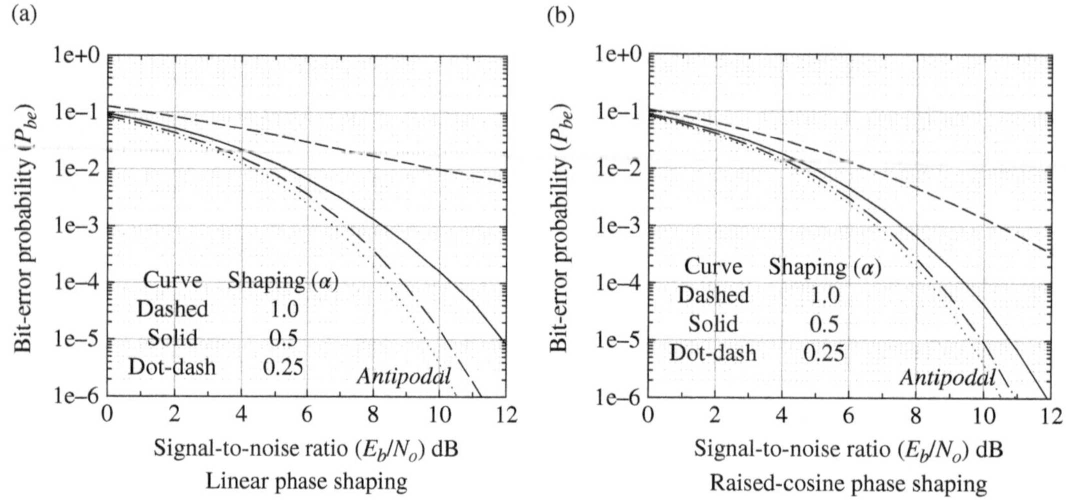

FIGURE 4.69 S-QPSK bit-error probability.

S-8PSK Performance Characteristics

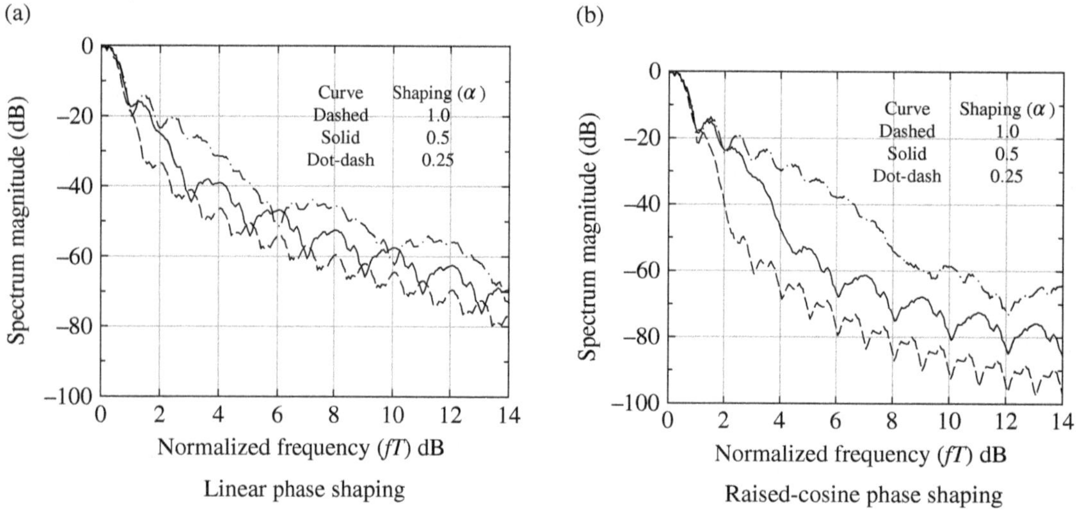

FIGURE 4.70 S-8PSK power density spectrum.

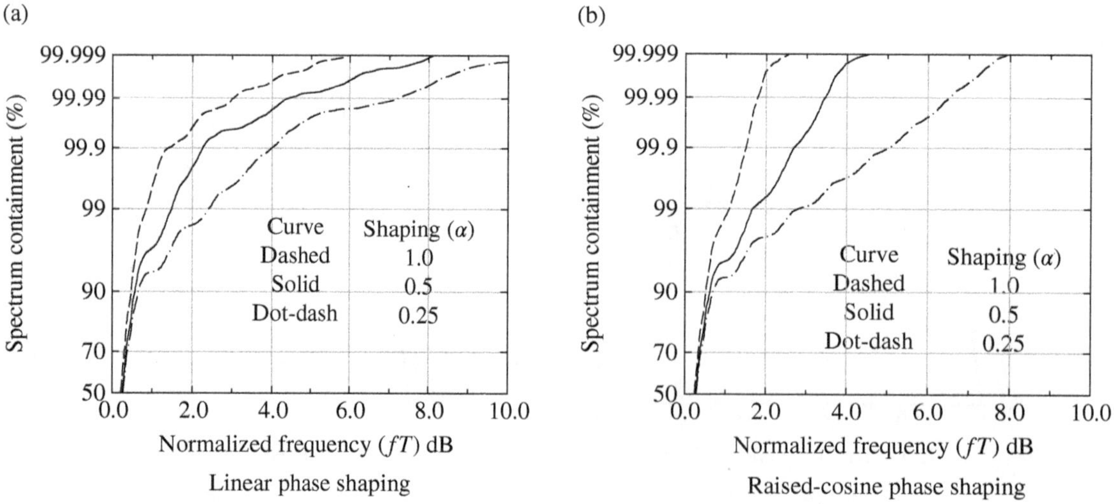

FIGURE 4.71 S-8PSK spectral containment.

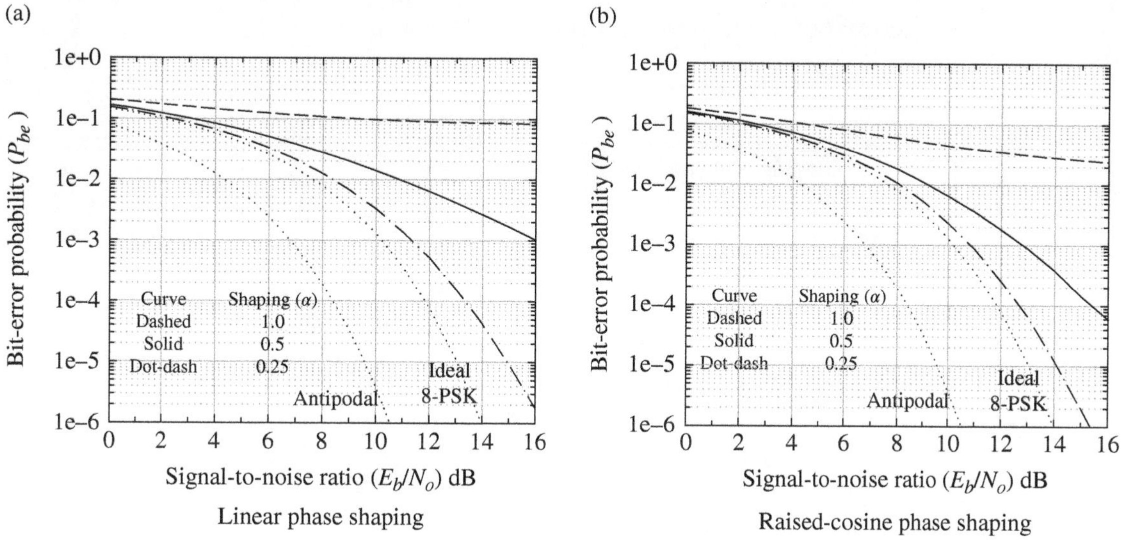

FIGURE 4.72 S-8PSK bit-error probability.

S-OQPSK Performance Characteristics

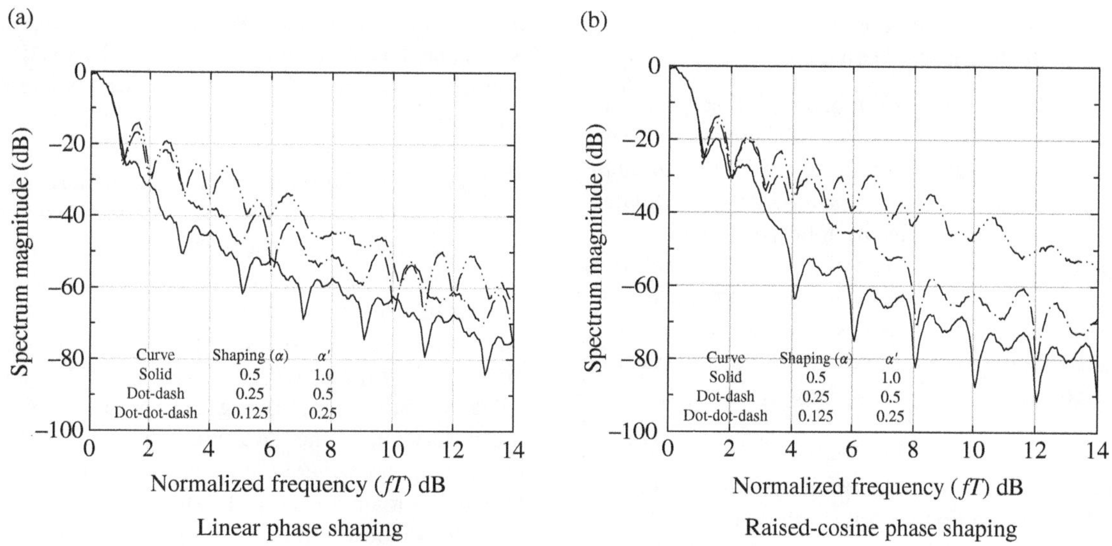

FIGURE 4.73 S-OQPSK power density spectrum.

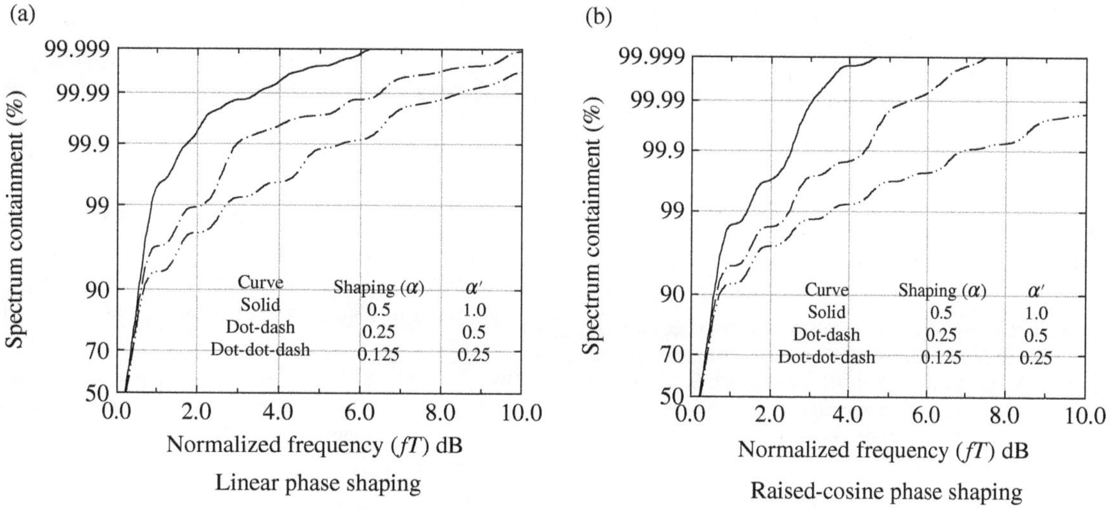

FIGURE 4.74 S-OQPSK spectral containment.

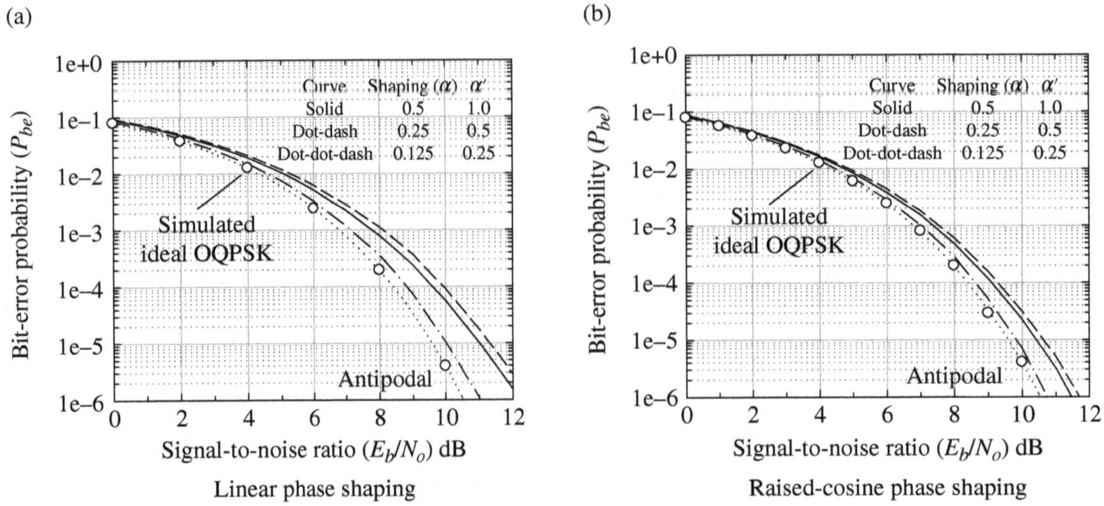

FIGURE 4.75 S-OQPSK bit-error probability.

4.4.4 Non-Constant Envelope PSK Waveforms: Spectrums and Bit-Error Performance

Amoroso's symbol shaping function, discussed in Section 4.3.1, falls in this category of modulations; however, because of the superior spectral efficiency of SRRC modulation, the results in this section focus on the SRRC-modulated waveform defined by (4.88). In the following sections, the spectrum and bit-error performance of the SRRC waveforms are examined with simulated performance results for SRRC BPSK and SRRC QPSK presented as a case study. The modified Intelsat IESS-308 spectral mask is used as the spectral efficiency reference.

4.4.4.1 Spectral Root-Raised-Cosine (SRRC) Shaping Function The inverse Fourier transform $p(t)$ of the SRRC function is expressed in (4.89) with the time response extending over the infinite range $|t| = \infty$. In practice, the theoretical response is truncated by a finite window over a practical range of symbols. The window span is denoted as* N_{span} and defined in terms of symbols with N_s samples-per-symbol. Therefore, the span is computed as,

$$N_{span} \triangleq \frac{N_s(\text{window})}{N_s} \tag{4.110}$$

where N_s(window) is the number of samples-per-window. For example, with N_s(window) = 34 and N_s = 4 samples-per-symbol, the window span is N_{span} = 8.5 symbols. The frequency response $H''(\omega)$ of the windowed impulse response is evaluated as the convolution of $H'(\omega)$ with $W(\omega)$ or as the Fourier transform of the product $p(t)w(t)$ where the window functions $W(\omega)$ and $w(t)$ form a Fourier transform pair.

The following results for a realizable implementation of the SRRC waveform are evaluated using a rectangular modulator window with various spans (N_{tspan}) to demonstrate the compliance with the indicated transmission spectral masks. Although it is demonstrated that a rectangular window span can be selected to meet the requirements of the frequency masks, application of a temporal RC (TRC) window, depicted in Figure 4.79, reduces the out-of-band spectral sidelobes further improving the spectral efficiency. If the mean duration of the TRC window is equal to the span of the rectangular window, then the width of the main spectral lobe will remain unchanged; however, the two-sided span of the TRC window will increase by the factor of $1 + \alpha_t$ where α_t is the excess duration of the TRC window.

Figure 4.76 shows the SRRC PSD using an excess bandwidth factor of α = 0.2 and a rectangular impulse response window spanning 6, 9, and 12 symbols. The effect of the *sinc(fT)* frequency sidelobes of the rectangular window is apparent in these responses and is dependent on the span of the widow. The response for N_{tspan} = 6 does not meet the SRRC spectral mask for symbol rates $R_s \geq 2$ Msps, the case with N_{tspan} = 9 slightly exceed the mask at fT = 3.0 and the N_{tspan} = 12 case satisfies the mask.† The cases for N_{span} = 9 and 12 meet the spectral mask for $R_s < 2$ Msps; however, the N_{span} = 6 case does not.

Figure 4.77 shows the SRRC PSD under similar conditions used in Figure 4.76 for an excess bandwidth factor of

*The span discussed in this section is the modulator or transmitted window span denoted as N_{tspan}. The demodulator or receiver span is the matched filter window N_{rspan} where $N_{rspan} \leq N_{tspan}$. The transmitted span is chosen for transmission bandwidth efficiency and the receiver span is chosen to minimize the detection loss and implementation complexity.

†As in all for these comparisons, the PSD results are bases on ideal window coefficients that must be examined for the practical case involving finite coefficient quantization.

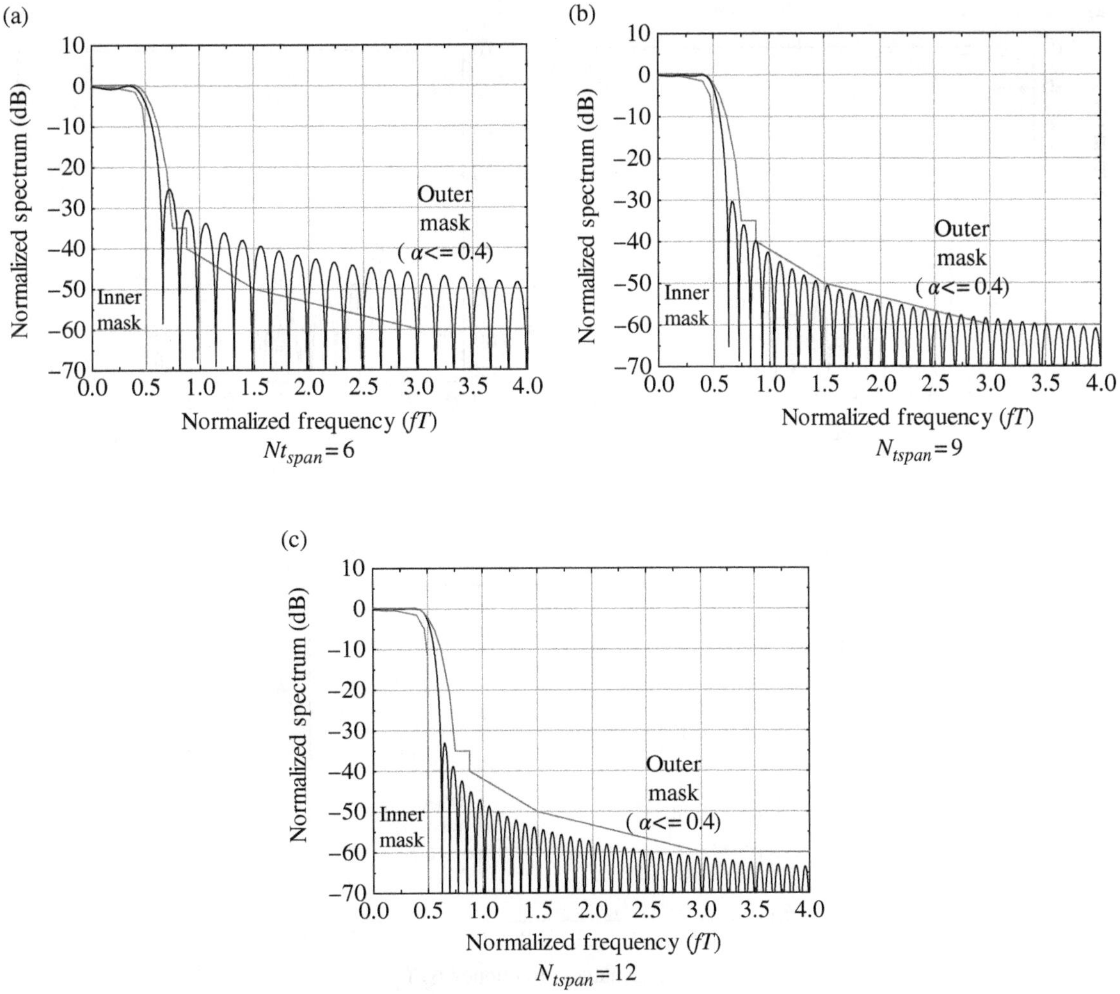

FIGURE 4.76 SRRC PSD with impulse response rectangular windowing ($\alpha = 0.2$).

$\alpha = 0.4$. In these cases, the response for $N_{tspan} = 6$ still does not meet the SRRC spectral mask for symbol rates $R_s \geq 2$ Msps; however, the cases for $N_{tspan} = 9$ and 12 meet the spectral masks.

The improved spectral roll-off with frequency is demonstrated in Figure 4.78 using a raised-cosine modulator window instead of the rectangular window used in Figures 4.76 and 4.77. Part (a) of the figure corresponds to $\alpha = 0.2$ and can be compared to Figure 4.76b that did not meet the spectral mask. Part (b) corresponds to $\alpha = 0.4$ and can be compared to Figure 4.76b. In both cases, there is a significant reduction in the spectral levels for normalized frequencies (fT) greater than about 3.0. Choosing the raise-cosine response to be symmetrical about the ends of the average window, as shown in Figure 4.79, minimizes the impact on the width of main spectral lobe; however, for the low values of α_t being considered, there is very little influence over the near-in spectral sidelobes by varying α_t. The impact of the TRC window on the demodulator matched filter detection is minimal, in that, the orthogonality loss of the detected symbols is similar to that observed with the rectangular window. For example, for cases involving $\alpha_t = 0.2$–0.4, the noise-free maximum level of the matched filter response at intervals of nT was observed to be about 0.03 dB with typical values of 0.004 dB. The span of the transmitted and matched filter reference symbols is an important consideration in the design and detection of the SRRC-modulated waveform.

The raised-cosine window is defined in terms of the average window duration T_w and the overall window duration T'_w as depicted in Figure 4.79. The average duration is defined in terms of an integer number of symbols with N_s samples-per-symbol, so $T_w = N_{symb}T = N_{symb}N_sT_s$ where the span of the window is defined as $N_{span} = N_{symb}N_s$ and T_s is the sample interval. Referring to Figure 4.79, the raised-cosine intervals at each end of the window are defined as $N_{\alpha t}T_s$ and, using these parameters, the overall window duration is expressed as

$$T'_w = \left(N_{symb}N_s + 2N_{\alpha_t}\right)T_s \qquad (4.111)$$

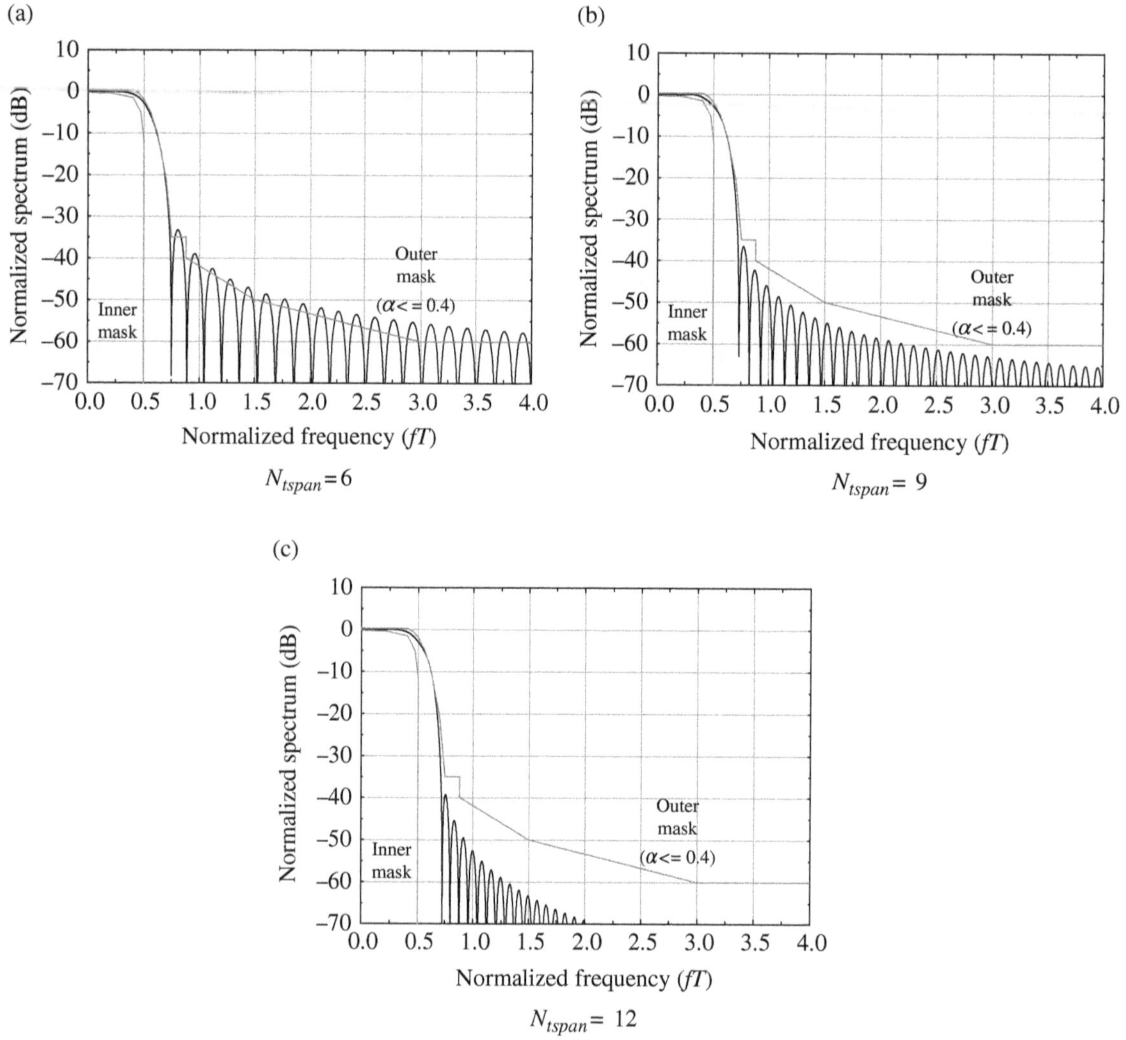

FIGURE 4.77 SRRC PSD with impulse response rectangular windowing ($\alpha = 0.4$).

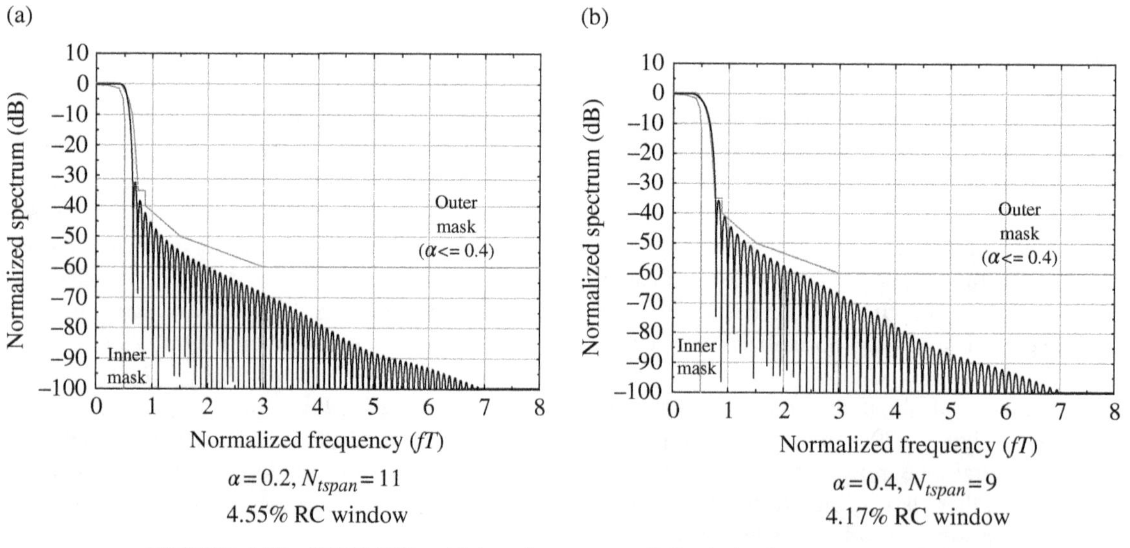

FIGURE 4.78 SRRC PSD with impulse response RC windowing ($\alpha = 0.2$ and 0.4).

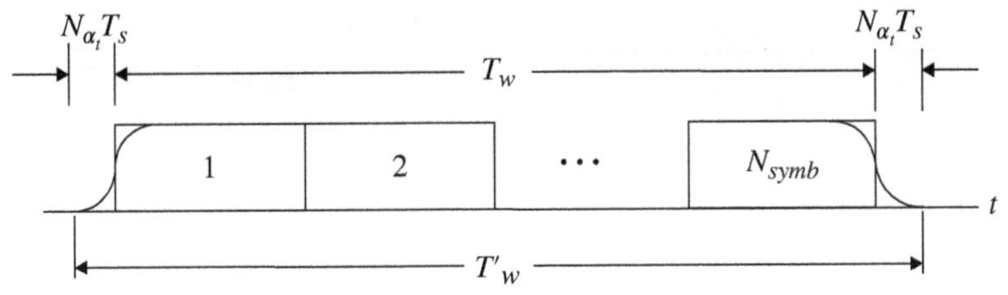

FIGURE 4.79 Raised-cosine window description.

Considering that the window is symmetrical about $t = 0$, such that, $|t| \le T'_w/2$, then, upon defining $t = nT_s$ and dividing by $N_{symb}N_sT_s$, the normalized sampling index is expressed as*

$$n' = \frac{n}{N_{symb}N_s} \tag{4.112}$$

and the range of n' is

$$|n'| \le \frac{1+\alpha_t}{2} \tag{4.113}$$

where $\alpha_t = N_\alpha/(N_{symb}N_s/2)$ is the excess temporal factor; $100\alpha_t$ expresses the percent of the raised-cosine weighting.

Upon comparing these results with the expression for the temporal RC window in Section 1.11.4, the symmetrical window samples are evaluated as

$$w(n') = \begin{cases} 1 & : |n'| < (1-\alpha_t)/2 \\ \dfrac{1}{2}\left\{1 - \sin\left(\dfrac{\pi}{\alpha_t}(|n'| - 1/2)\right)\right\} & : (1-\alpha_t)/2 \le |n'| \le (1+\alpha_t)/2 \end{cases} \tag{4.114}$$

The selection of α_t and N_{tspan} is based on system specification of the ACI. Implementation issues include demodulator processing complexity, matched filter loss, and symbol and frequency tracking losses which are also dependent on the selection of α_t and N_{tspan}. Several of these issues are examined in the following case study.

The SRRC matched filter detection loss is evaluated in terms of the matched filter span (N_{rspan}) relative to a transmitted signal span (N_{tspan}) of 12 symbols. The detection loss is evaluated by varying N_{rspan} while maintaining ideal symbol timing. The loss is shown in Figure 4.80 for the indicated excess bandwidth factors. For example, a matched filter detection loss of 0.1 dB corresponds to an SRRC matched filter spans of approximately: 1, 1.3, and 1.5 symbols for the respective values of $\alpha = 0.45$, 0.3, and 0.2.

*It is convenient for the product $N_{symb}N_s$ to be even, and in the examples in this section $N_s = 16$ and N_{symb} may be even or odd.

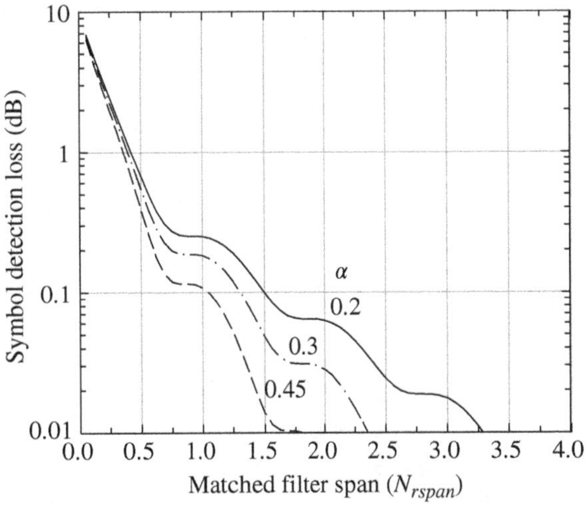

FIGURE 4.80 SRRC symbol detection loss with matched filter span relative to $N_{tspan} = 12$ symbols.

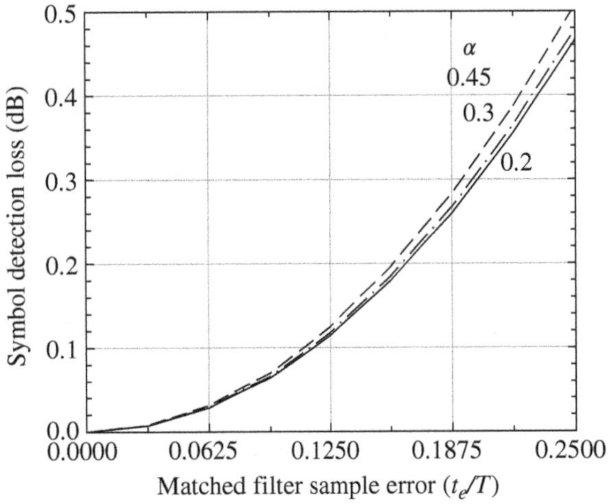

FIGURE 4.81 SRRC isolated symbol detection loss with matched filter sampling error.

Figure 4.81 shows the detection sensitivity to a matched filter symbol timing error. These results correspond to the matched filter loss for an isolated symbol and do not include the intersymbol interference loss associated with a sequence of modulated symbols. In the following case study, the

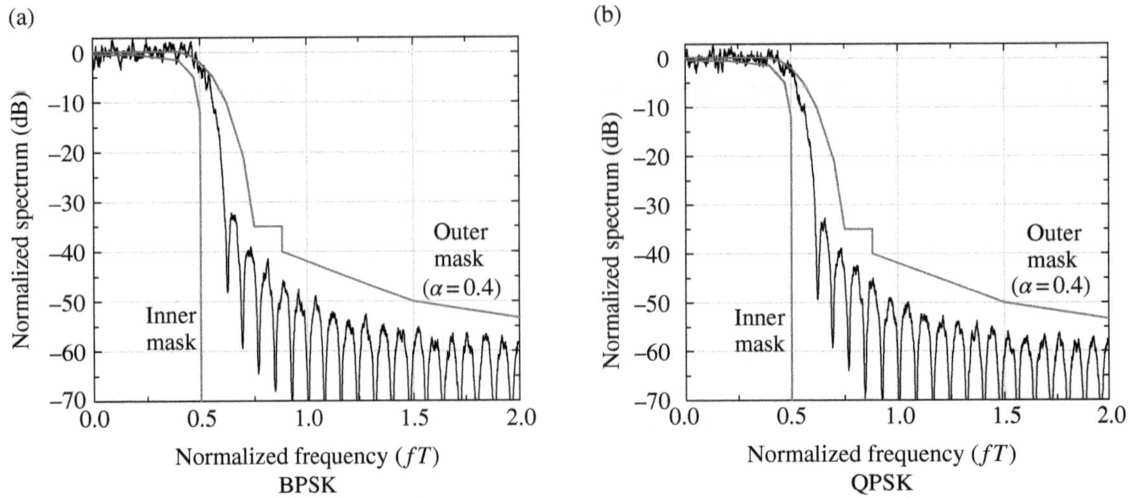

FIGURE 4.82 SRRC simulated power spectral density (α = 0.2, N_{tspan} = 12).

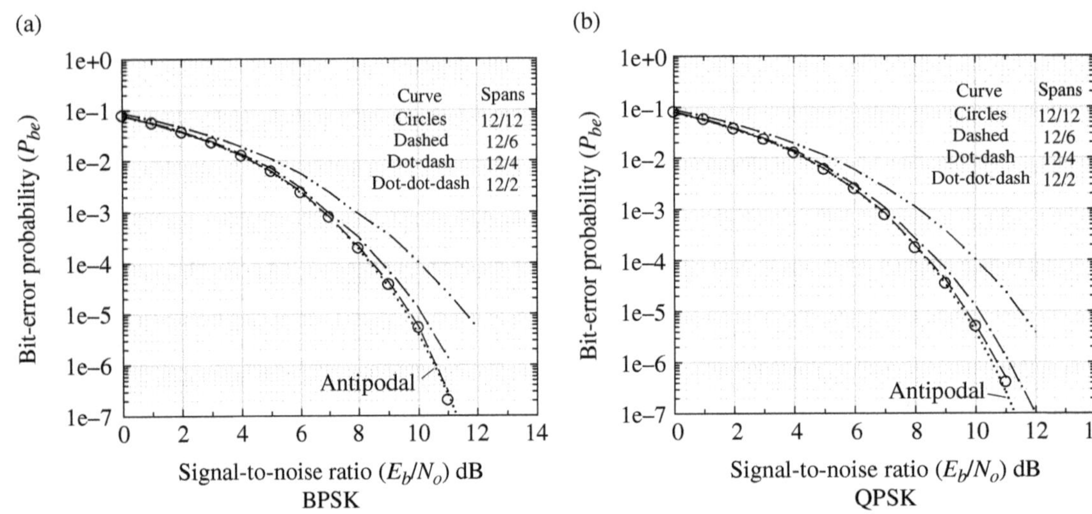

FIGURE 4.83 Uncoded SRRC bit-error performance with matched filter span (α = 0.2, N_{tspan} = 12).

symbol timing sensitivity is examined for SRRC BPSK-modulated data that includes the effect of the ISI.

4.4.5 Case Study: Spectral and Bit-Error Performance of SRRC Applied to BPSK, QPSK, and DCBPSK

Figure 4.82 shows the simulated PSD of SRRC applied to BPSK and QPSK modulations for $\alpha = 0.2$ and $N_{tspan} = N_{rspan} = 12$ symbols. The approximation to the PSD is based on averaging 200 random data spectrums and then smoothing the average spectrum using a 41-tap uniformly weighted sliding window transversal filter. These approximations to the simulated PSD show some residual random variations; however, the average spectrums are similar to that shown in Figure 4.76c and meet the SRRC spectral mask for $R_s \geq$ 2 Msps.*

The simulated bit-error performance of SRRC spectral shaping applied to BPSK and QPSK modulations is shown in Figure 4.83. In each case, the excess bandwidth factor is equal to 0.2, the underlying transmitter symbol span is 12 symbols, and the matched filter span N_{rspan} is varied as denoted by the ratio $Spans = N_{tspan}/N_{rspan}$ in the figures. For these results, and the remaining bit-error performance results in this case study, ideal symbol timing and phase

*The simulated PSD is obtained by taking the Fourier transform of the correlation response of the modulated data sequence; however, additional fidelity can be achieved in these results by averaging periodograms as discussed in Chapter 1.

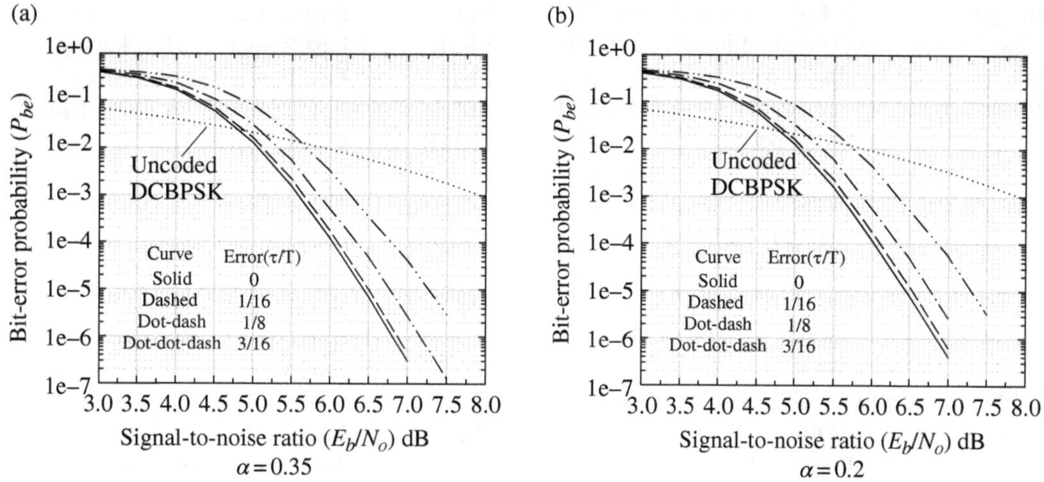

FIGURE 4.84 Coded SRRC DCBPSK bit-error performance with matched filter sample error (N_{tspan}, N_{rspan} = 12, rate 1/2, K = 9, convolutional coding).

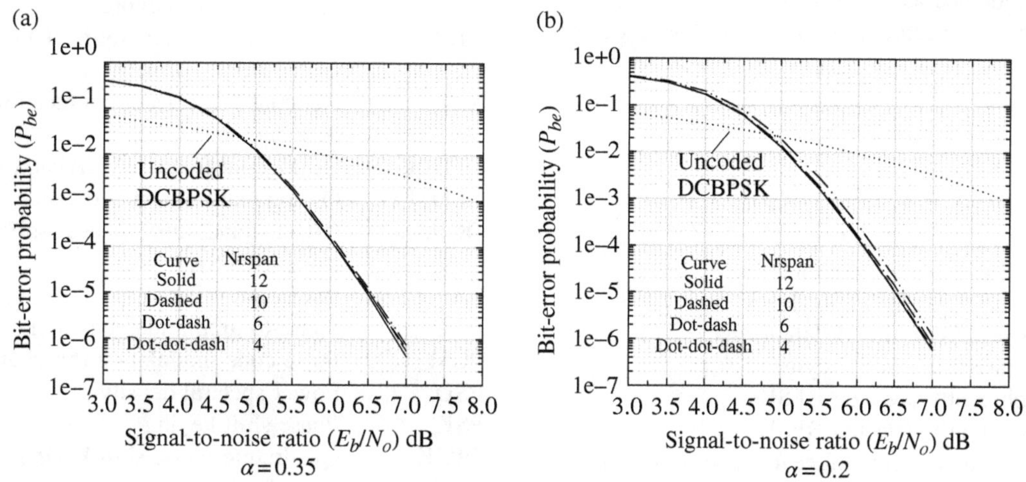

FIGURE 4.85 Coded SRRC DCBPSK bit-error performance with matched filter span (N_{tspan} = 12, ideal timing, rate 1/2, K = 9, convolutional coding).

tracking is used. The bit-error performance characteristics are very similar for all of these modulations and the demodulator matched filter span must be at least six symbols to result in an E_b/N_o loss of less than 0.1 dB. The performance indicated by the *circled* data points in each figure corresponds to a matched filter span of 12 symbols and is virtually the same as the ideal performance of antipodal signaling as indicated by the *dotted* curves. The SRRC waveform is sensitive to filter distortion over the baseband bandwidth $R_s/2$, for example, the cascaded integrate and comb (CIC) rate reduction to four samples-per-symbol with uncompensated gain roll off must be greater than 32 : 1 to result in a performance loss of less than 0.1 dB. These Monte Carlo simulations are based on 500K symbols for each signal-to-noise ratio for $E_b/N_o \leq$ 8 dB, otherwise 5 M symbols are used.

Figures 4.84 and 4.85 compare the performance of coded SRRC DCBPSK modulation for excess bandwidth factors 0.35 and 0.2. The purpose is to examine the performance sensitivities under these two commonly used excess bandwidth factors that satisfy the SRRC spectral mask corresponding to $\alpha = 0.4$. The coding corresponds to a rate ½, constraint length $K = 9$ convolutional code using a Viterbi decoder with a trellis depth of 32 symbols and virtually infinite metric quantization. Figure 4.84 shows the performance sensitivity to the matched filter timing error using transmitter and matched filter spans of 12 symbols. The simulation performs the matched filtering using 16 samples-per-symbol and the performance loss is examined for the normalized timing error $\tau/T = n/16 : n = 0 - 3$. Upon close examination, it is observed that the performance with α = 0.2 is slightly more sensitive and, for fine-tuning the link budget, 0.05 dB additional loss should be included for this case. The more important observation is that, in either case, the absolute timing error should be less than 1/16th of a symbol to result in a loss of less than about 0.1 dB. Figure 4.85 examines the performance sensitivity to the matched filter span for the same two excess bandwidth factors and with a constant transmitter span of 12 symbols. The results also demonstrate the relative performance sensitivity of the $\alpha = 0.2$ case. To keep the detection loss less than 0.1 dB, these results suggest that the minimum matched filter span should be 6 and 10 symbols for the, respective, excess bandwidth factors of 0.35 and 0.2. The differentially coherent waveform modulation is used in a frequency-hopping modem with 32 code-bits/hop (16-information bits/hop) and a matched filter sampling rate of 16 samples-per-symbol. The Monte Carlo simulations of the coded performance use 500K hops for each signal-to-noise ratio for $E_b/N_o \leq 5$ dB, otherwise 4 M hops are used; this corresponds to 8 and 64 M bits for each Monte Carlo trial. The channel is modeled as an AWGN channel and at $P_{be} = 1e^{-7}$ the measurement accuracy is $\Delta P_{be} = 8e^{-8}$ with a confidence of 95%.

ACRONYMS

ACI	Adjacent channel Interference
AM	Amplitude modulation
AWGN	Additive white Gaussian noise
BFSK	Binary frequency shift keying
Bi-Phase	Two-phase (a PCM data format)
BPSK	Binary phase shift keying
CBFSK	Coherently detected BFSK
CBPSK	Coherently detected BPSK
CCSDS	Consultative Committee for Space Data Systems
CIC	Cascaded integrate and comb (filter)
CMPSK	Coherently detected MPSK
CPFM	Continues-phase frequency modulation
CPFSK	Continuous phase frequency shift keying
CPM	Continuous phase modulation
CSK	Compatible shift keying
DAC	Digital-to-analog conversion
DC	Direct current
DCBPSK	Differentially coherent BPSK
DCMPSK	Differentially coherent MPSK
DCPSK	Differentially coherent PSK (also DPSK)
DCQPSK	Differentially coherent QPSK
DEBPSK	Differentially encoded BPSK
DEMPSK	Differentially encoded MPSK
DEQPSK	Differentially encoded QPSK
DPSK	Differential PSK (Same as DCPSK)
E/L	Early–late (sampling)
FDM	Frequency division multiplex (protocol)
FDMA	Frequency division multiple access (protocol)
FEC	Forward error correction (coding)
FFSK	Fast frequency shift keying
FFT	Fast Fourier Transform
FIR	Finite impulse response (filter)
FMSK	Filtered MSK
FSK	Frequency shift keying
GFSK	Gaussian frequency shift keying
GMSK	Gaussian minimum phase shift keying
HL	Hard limiter
I&D	Integrate-and-dump (filter)
I/Q	In-phase and quadrature (channels or rails)
IESS	Intelsat Earth Station Standards
IF	Intermediate frequency
ISI	Intersymbol interference
ML	Maximum-likelihood (estimate or decision rule)
MLSE	Maximum-likelihood sequence estimation (or estimator)
MPSK	Multiphase shift keying
MSK	Minimum shift keying
NRZ	Non-return to zero form of PCM data format (a PCM data format)
NTIA	National Telecommunications Industry Association
OFDM	Orthogonal frequency division multiplexing
OFDMA	Orthogonal frequency division multiple access
OQPSK	Offset quadrature phase shift keying (also called SQPSK)
PA	Power amplifier
PAC	Phase-to-amplitude converter
PCM	Pulse code modulation (baseband)
PSD	Power spectral density
PSK	Phase shift keying
QPSK	Quadrature phase shift keying
RC	Raised cosine
RF	Radio frequency (carrier)
RRC	Root-raise-cosine
S-8PSK	Phase-shaped 8PSK
S-BPSK	Phase-shaped BPSK
SFCG	Space Frequency Coordination Group
SFSK	Sinusoidal FSK
S-MPSK	Phase-shaped MPSK
SMSK	Sinusoidal minimum shift keying
S-OQPSK	Phase-shaped QPSK
S-PSK	Phase-shaped PSK
SQPSK	Staggered QPSK (alternate notation for OQPSK)
S-QPSK	Phase-shaped QPSK
SRC	Spectral RC
S-RC	Phase-shaped RC
SRRC	Spectral root-raised-cosine
SSPA	Solid-state power amplifier
TCM	Trellis-coded modulation

TRC	Temporal RC
TWTA	Traveling wave tube amplifier
VCO	Voltage controlled oscillator

PROBLEMS

1. Show that $u_c(t)$ and $u_s(t)$ given by (4.4) and (4.5) are unit energy functions and are orthogonal to each other.

2. Considering the gray-coded QPSK symbol constellation, the probability of a correct symbol detection is $P_{cs} = (1 - P_{be})^2$, where P_{be} is the bit-error probability. Using this result, express the symbol-error probability, P_{se}, in terms of P_{be} and show the conditions for the approximation $P_{be} \cong P_{se}/2$.

   a. Repeat the above problem for MPSK and indicate the conditions for the approximation $P_{be} \cong P_{se}/\log_2(M)$.

3. Show that Hancock's phase distribution, (4.7), is approximated by (4.11) for $|\phi| \le \pi/2$ and large symbol signal-to-noise ratios, that is, as $\gamma_s \to \infty$.

4. Considering the $\pi/2$-BPSK-modulated waveform, suppose that the matched filter orthogonal output samples (x_k,y_k) and (x_{k+1},y_{k+1}) correspond the phase φ_k and φ_{k+1} with the phase difference $\Delta\varphi = \varphi_{k+1} - \varphi_k$. Express the coordinates (x_o,y_o) of the phase difference in terms of (x_k,y_k) and (x_{k+1},y_{k+1}).

   Hint: Use the trigonometric identity $\tan^{-1}(z_2) - \tan^{-1}(z_1) = \tan^{-1}((z_2 - z_1)/(1 + z_1z_2))$.

5. Referring to the $\pi/4$-QPSK discussion in Section 4.2.3.2, provide answers to the following: (a) Sketch the constellation of the phase rest-states for a noiseless received signal with zero phase error. (b) Determine the phase change from the pervious-to-current phase-state $\Delta\varphi = \varphi_k - \varphi_{k-2}$ defined in terms Cartesian coordinates (x_k,y_k) and (x_{k-2},y_{k-2}). (c) Using the expression for $\Delta\varphi = \tan^{-1}(x_k,y_k;\, x_{k-2},y_{k-2})$ developed in part (b), evaluate $\Delta\varphi$ given the detected points $(x_k,y_k) = (-0.174,0.985)$ and $(x_{k-2},y_{k-2}) = (-0.766,0.643)$. (d) Quantize the previous detected phase x_{k-2},y_{k-2} to the nearest phase rest-point $\left(x'_{k-2},y'_{k-2}\right)$ and compute $\Delta\varphi'$.

6. Equation (4.28) expresses the MSK symbol transmitted in the interval $kT_b \le t \le (k+1)T_b$ corresponding the Q-channel symbol k. Write a similar expression for the I-channel symbol $k-1$ over the interval and reduce the form to correspond to that of (4.29).

7. Referring to (4.29), determine the source bit data patterns $(\ldots\, d_{k+1},\, d_k,\, d_{k-1}\, \ldots)$ required to generate the unique MSK tone sets corresponding to: the upper MSK tone at f_m, the lower MSK tone at $-f_m$, and alternating MSK tones at $\pm f_m$. Note: These data patterns are useful in testing phaselock loop performance and in generating acquisition preambles.

8. Show that the squared MSK noise-free signal $s(t)$ in Figure 4.16 is equal to

$$s^2(t) = P\left[1 + \frac{1}{2}\cos(2(\omega_c + \omega_m)t) + \frac{1}{2}\cos(2(\omega_c - \omega_m)t)\right] + T(d_{k-1}, d_k, d_{k+1})$$

   where $T(-)$ involves independent zero-mean random data terms that are removed by lowpass filtering. Hint: In the solution to this problem it is convenient to use the trigonometric identities:

$$\sin(\alpha)\sin(\beta) = (1/2)(\cos(\alpha-\beta) - \cos(\alpha+\beta)),$$
$$\cos(\alpha)\cos(\beta) = (1/2)(\cos(\alpha-\beta) + \cos(\alpha+\beta)),$$
$$\sin^2(\alpha) = (1/2)(1 - \cos(2\alpha)),\ \text{and}\ \cos^2(\alpha) = (1/2)(1 + \cos(2\alpha)).$$

9. Show the modifications required to Figure 4.16 to differentially decode the data when the source bits are differentially encoded as in Figure 4.18.

10. In the context of the CPM modulator shown in Figure 4.26b, discuss the implementation for conventional MPSK and OQPSK-modulated waveforms that use instantaneous changes in phase. Show all modifications to Figure 4.26b and the necessary parameters including the use of α_m for MPSK, such that, $\alpha_m = m\Delta\phi : m = 0, 1, \ldots, M-1$ and $\Delta\phi = 2\pi/M$. Modify Table 4.1 to include each of these modulated waveforms.

11. Using (4.61) and the definitions of the various parameters with $t = iT_s$ show that $\theta(iT_s) = \theta(t) = \theta(t - T_s) + d_i\omega_m t$.

12. Determine the expression for the frequency shift of a linearly phase-shaped BPSK-modulated waveform when the phase is advanced by π radians for each *mark* and *space* data bit. Express the frequency shift in terms of the shaping factor α under the following conditions: (a) with alternating mark and space data; (b) with random data; (c) determine the frequency shift when the modulation phase of the space data is not advanced but retarded by π radians under the condition of random data.

13. Derive the expression for the SRRC symbol weighting function in (4.89) by performing the inverse Fourier transform of (4.88). Hint: Integrate over the two cosine weighted intervals $-(1+\alpha)/2T \le f \le -(1-\alpha)/2T$ and

$(1-\alpha)/2T \le f \le (1+\alpha)/2T$ to obtain the respective responses $p_1(t)$ and $p_2(t)$. Then integrate over the constant amplitude interval $|f| \le (1-\alpha)/2T$ to obtain $p_0(t)$. Then combine the responses as $p(t) = p_1(t) + p_2(t) + p_0(t)$ and simplify to obtain in the desired response expressed in (4.89).

a. Show all of the steps leading to the responses (4.90) and (4.91) corresponding to $t = 0$ and $\pm 1/4\alpha$.

14. Derive the expression for the noise bandwidth of the SRRC waveform.

15. Derive the Fourier transform $S_T(\omega)$ for QPSK and OQPSK-modulated signals and use the results to verify the power spectral density (PSD) expressed in (4.100).

16. Verify the Fourier transform of the MSK-modulated signal expressed in (4.101) and show all of the steps in the derivation. Hint: Convert $p(t) = \sqrt{P}\cos(\omega_m t)$ to the complex form $p(t) = (\sqrt{P}/2)\left[e^{j\omega_m t} + e^{-j\omega_m t}\right]$ and, at the appropriate point in the simplification, use the trigonometric identities:

$\sin(\alpha) + \sin(\beta) = 2\sin((\alpha+\beta)/2)\cos((\alpha-\beta)/2)$ and
$\sin(\alpha) - \sin(\beta) = 2\cos((\alpha+\beta)/2)\sin((\alpha-\beta)/2)$

When these trigonometric identities are used a major simplification results using the MSK condition $\omega_m T_b = \pi/2$.

17. Upon substituting the symbol interval $T = 2T_b$ and solving the expression for the MSK spectrum in Problem 16 in terms of the normalized frequency fT, the indeterminate form 0/0 is encountered at $fT = 1/2$. Using L'Hospital's rule, determine the value of the MSK spectrum at $fT = 1/2$. Show all of the steps in arriving at your answer.

18. Design a trapezoidal frequency function that satisfies the Nyquist criterion and then calculate and plot the filter impulse response. Comment on the zero-crossing and relate the intervals to the average bandwidth of the spectrum. What is the excess bandwidth for the trapezoidal frequency function that you selected?

REFERENCES

1. R. DeBuda, "Coherent Demodulation of Frequency Shift Keying with Low Deviation Ratio," IEEE Transactions on Communications, Vol. COM-20, pp. 429–435, June 1972.
2. R. DeBuda, "About Optimal Properties of Fast Frequency Shift Keying," IEEE Transactions on Communications, Vol. COM-22, pp. 1726–1728, October 1974.
3. M.L. Doelz, E.H. Heald, "Minimum Shift Data Communication System," U.S. Patent 2977417, Collins Radio Co., March 28, 1961.
4. M.J. Dapper, T.J. Hill, "SBPSK: A Robust Bandwidth Efficient Modulation for Hard-Limited Channels," MILCOM'84, Vol. 3 of 3, Session 31.6, pp. 458–463, Los Angeles, October 21–24, 1984.
5. J.B. Anderson, T. Aulin, C.E. Sundberg, *Digital Phase Modulation*, Plenum Press, New York, 1986.
6. M.C. Austin, M.U. Chang, "Quadrature Overlapped Raised-Cosine Modulation," IEEE Transactions on Communications, Vol. COM-24, No. 3, March 1981, pp. 237–249.
7. W.B. Davenport, Jr., and W.L. Root, *An Introduction to the Theory of Random Signals and Noise*, McGraw-Hill Book Company, New York, 1958.
8. J.C. Hancock, *An Introduction to the Principles of Communication Theory*, McGraw-Hill Book Company, New York, 1961.
9. S. Chennakeshu, G.J. Saulnier, "Differential Detection of π/4-DQPSK for Digital Cellular Radio," IEEE Transactions on Vehicular Technology, Vol. VT-42, pp. 46–57, February 1993.
10. Abu-Dayya, A.A., Beaulieu, N.C., "Diversity π/4-DQPSK on Microcellular Interference Channels," IEEE Transactions on Communications, Vol. 44, No. 10, pp. 1289–1297, October 1996.
11. W.M. Hutchinson, R.W. Middlestead, "Data Demodulator Apparatus," U.S. Patent 3743775, Collins Radio Co., July 3, 1973.
12. J.M. Wozencraft and I.M. Jacobs, *Principles of Communication Engineering*, John Wiley & Sons, Inc., New York, January 1967.
13. A.J. Viterbi, *Principles of Coherent Communication*, McGraw-Hill Book Company, New York, 1966.
14. J.G. Proakis, *Digital Communications*, McGraw-Hill Book Company, New York, 1989.
15. R.W. Lucky, J. Salz, E.J. Weldon, Jr., *Principles of Data Communication*, McGraw-Hill Book Company, New York, 1968.
16. C.R. Cahn, "Performance of Digital Phase-Modulation Communications Systems," Report M1/0-905, Ramo-Wooldridge Corporation, Euclid, OH, April 1959.
17. F. Amoroso, J.A. Kivett, "Simplified MSK Technique," IEEE Transactions on Communications, Vol. COM-25, No. 4, pp. 433–441, April 1977.
18. J.M. Liebetreu, C.R. Ryan, "Performance Simulation of Receiver Filters for Serial Detection of MSK Signals," Allerton Conference Proceedings, University of Illinois, Urbana, IL, pp. 351–358, October 1980.
19. R.E. Ziemer, C.R. Ryan, J.H. Stilwell, "Conversion and Matched Filter Approximations for Serial Minimum-Shift Keyed Modulation," IEEE Transactions on Communications, Vol. COM-30, No. 3, pp. 495–509, March 1982.
20. C.R. Ryan, "Advances in Serial MSK Modems," National Telecommunications Conference, NTC'81, New Orleans, Conference Record, pp. 56.2.1–56.2.5, December 1981.
21. F. Amoroso, "Pulse and Spectrum Manipulation in the Minimum (Frequency) Shift Keying (MSK) Format," IEEE Transactions on Communications, Vol. COM-24, No. 3, pp. 381–384, March 1976.
22. M.K. Simon, "A Generalization of Minimum-Shift-Keying (MSK) Type Signaling Based Upon Input Data Symbol Pulse Shaping," IEEE Transactions on Communications, Vol. COM-24, No. 8, pp. 845–856, August 1976.

23. V.K. Prabhu, "Spectral Occupancy of Digital Angle-Modulated Signals," Bell System Technical Journal, Vol. 55, No. 4, pp. 429–453, April 1976.
24. G.D. Forney, Jr., "Maximum Likelihood Sequence Estimation of Digital Sequences in the Presence of Intersymbol Interference," IEEE Transactions on Information Theory, Vol. IT-18, pp. 363–378, 1972.
25. C. Andren, "PSK Sidebands Reduced by Pre-Modulation Filtering," Microwave Journal, Vol. 6, pp. 69–73, January 1978.
26. R. Cofer, E. Franke, O. Johnson, T. Erman, "Shaped PSK in a Digital Modem with Direct Digital Synthesis," IEEE Military Communications Conference, MILCOM'90, Monterey, CA, September 30 through October 3, 1990.
27. F. Amoroso, "The Use of Quasi-Bandlimited Pulses in MSK Transmission," IEEE Transactions on Communications, Vol. COM-27, No. 10, pp. 1616–1624, October 1979.
28. Consultative Committee for Space Data Systems (CCSDS), Handbook of the Space Frequency Coordination Group (SFCG), Recommendation Rec-21-2R3, Efficient Spectrum Utilization for Space Science Services on Space-to-Earth Links; Category A, Figure 1, p. 3 of 3, June 2012.
29. Consultative Committee for Space Data Systems, "Bandwidth-Efficient Modulations: Summary of Definitions, Implementation, and Performance", CCSDS Informational Report CCSDS 413.0-G-2, Green Book, October 2009.
30. W.L. Martin, T.Y. Yan, A. Gray, D.K. Lee, "CCSDS-SFCG Efficient Modulation Methods Study at NASA/JPL: Phase 4 Interface Susceptibility," SF19-1D, SFCG Meeting, Noordwijk, the Netherlands, September 1999.
31. U.S. Department of Commerce, National Telecommunications and Information Administration, "Manual of Regulations and Procedures for Federal Radio Frequency Management", Red Book, May 2014, Revision of the May 2013 Edition, U.S. Department of Commerce, Washington, DC.
32. L.K. McAdoo, "Speech Volumes on Bell System Message Circuits—1960 Survey," Transactions of the American Institute of Electrical Engineers (AIEE), Part I: Communication and Electronics, Vol. 81, No. 1, pp. 8–13, March 1962.
33. W.C. Ahern, F.P. Duffy, J.A. Maher, "Speech Signal Power in the Switched Message Network," Bell System Technical Journal, Vol. 57, No. 7, Part 2, pp. 2695–2726, September 1978.
34. Federal Communications Commission (FCC), Report and Order, "In the Matter of Amendment of Parts 2 and 21, of the Commissions' Rules Concerning Calculations of Necessary Bandwidth for Frequency Modulation Microwave Radio Relay Systems," General Docket No. 81-743, Rule Making 3625, adopted March 3, 1983.
35. Intelsat Earth Station Standards (IESS), Performance Characteristics for Intermediate Data Rate Digital Carriers Using Convolutional Encoding/Viterbi Encoding and QPSK Modulation (QPSK/IDR), Document IESS-308 (Rev. 10), Intel Corporation, Santa Clara, CA, February 10, 2000.
36. Intelsat Earth Station Standards (IESS), Performance Characteristics for Intermediate Data Rate Digital Carriers Using Rate 2/3 TCM/8PSK and Reed-Solomon Outer Coding (TCM/IDR), Document IESS-310 (Rev. 3), March 10, 2005.
37. A. Papoulis, *Probability, Random Variables, and Stochastic Processes*, McGraw-Hill Book Company, New York, 1965.
38. A. Papoulis, *The Fourier Integral and Its Applications*, McGraw-Hill Book Company, New York, 1987.
39. S.A. Gronemeyer, A.L. McBride, "MSK and Offset QPSK Modulation," IEEE Transactions on Communications, Vol. COM-24, No. 8, pp. 809–820, August 1976.
40. K. Murota, K Hirade, "GMSK Modulation for Digital Mobile Radio Telephone," IEEE Transactions on Communications, Vol. COM-29 No. 7, pp. 1044–1050, July 1981.

5

FREQUENCY SHIFT KEYING (FSK) MODULATION, DEMODULATION, AND PERFORMANCE

5.1 INTRODUCTION

Frequency shift keying (FSK) modulation is a robust modulation, in that, phase coherence is not required to achieve near optimum performance. In this chapter, the focus is on binary FSK (BFSK) and several implementations are discussed in the context of known or estimated frequency and phase information regarding the received signal. In particular, the performance of coherent and noncoherent detection of BFSK is characterized in terms of known frequency and phase; known frequency and unknown phase; unknown frequency and phase. These knowledge-based characteristics of the received signal dictate the use of coherent detection or noncoherent detection and when the frequency and/or phase are unknown noncoherent detection must be used. The theoretical loss of coherently detected BFSK relative to antipodal signaling is 3 dB for all bit-error conditions and, with known frequency and unknown phase, the performance loss of noncoherent BFSK detection is degraded by about 0.9 dB at $P_{be} = 10^{-5}$ relative to coherent BFSK detection. On the other hand, with unknown frequency and phase, the loss is a function of the range of the frequency uncertainty with losses as high as 17 dB at $P_{be} = 10^{-5}$ relative to coherent detection for a frequency uncertainty range of 10^4 times the bit rate. The necessity to estimate and track bit timing for optimum sampling of the detection filtering is required with all implementations. Typically, the bit timing is estimated during a data preamble and tracking is achieved using conventional early–late (E/L) gate processing or the zero crossings of the lowpass filter (LPF) detection filter output. Alternately, bit-timing estimation and tracking can be achieved by selecting a low modulation index amplitude modulation (AM) on the BFSK waveform that is synchronized to the data. In this case, the demodulator provides an AM detector followed by a narrowband filter for recovering bit timing.

The following sections discuss the detection algorithms for each of the knowledge-based conditions. In each case, expressions for the theoretical bit-error probability are developed that provide a baseline for simulated and hardware testing performance evaluations. Several case studies are included that examine the performance sensitivities of specific implementations. The chapter concludes with the characterization of the spectrum of BFSK for various modulation indices. The important subject of M-ary FSK modulation and detection is deferred to Chapter 7.

5.2 COHERENT DETECTION OF BFSK—KNOWN FREQUENCY AND PHASE

Coherent detection of BFSK is easily accomplished when the modulator generates the mark and space tones using continuous wave (CW) frequencies as shown in Figure 5.1. In this case, the phase transitions between the mark and space tones are not continuous resulting in higher out-of-band spectral power; however, receiver phase tracking for coherent detection is relatively easy to achieve. Phase continuity between the mark and space tones can be achieved by applying the binary source data (b_i) to a frequency synthesizer. However, coherent detection can only be achieved for integer values of

Digital Communications with Emphasis on Data Modems: Theory, Analysis, Design, Simulation, Testing, and Applications, First Edition. Richard W. Middlestead.

Companion website: www.wiley.com/go/digitalcommunications

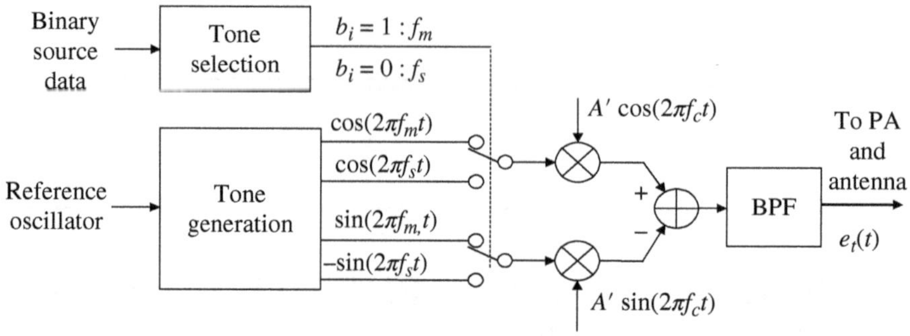

FIGURE 5.1 Continuous frequency FSK tone generation.

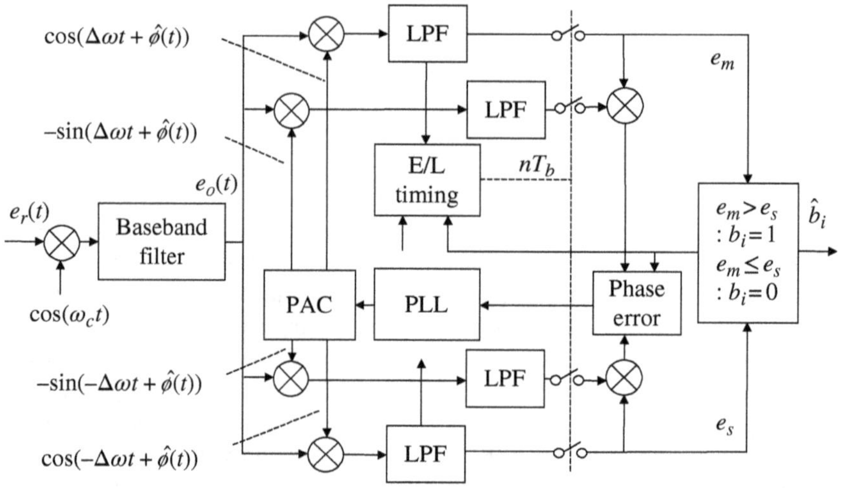

FIGURE 5.2 Functional implementation of coherent BFSK demodulator.

the modulation index. Furthermore, if a first-order phaselock loop is used, the received frequency will be corrected but the resulting unknown phase will require using an enveloped detector resulting in noncoherent detection performance.

The bandpass filter (BPF) shown in Figure 5.1 is not essential and is intended to provide a spectral mask to ensure rejection of spurious interfering signal with negligible impact on the desired FSK modulated waveform. In Sections 4.4.3.6–4.4.3.8, spectral containment is discussed using frequency shaping between the FSK tone transitions which, like phase-shaped PSK (S-PSK), function as a transmitter BPF. The spectrum of the FSK modulate waveform is controlled by shaping the frequency transitions while maintaining the constant amplitude characteristic. Since FSK is ideally a constant amplitude waveform, the transmitter power amplifier (PA) efficiency can be increased by driving it into saturation.

Consider the BFSK transmitted waveform expressed as

$$e_t(t) = A\cos((\omega_c + d_i\Delta\omega)t) \tag{5.1}$$

with the received waveform expressed as

$$e_r(t) = A\cos((\omega_c + d_i\Delta\omega)t + \phi(t)) + n(t) \tag{5.2}$$

where $d_i = 2b_i - 1 = \{1, -1\}$, that is, a binary $b_i = 1$ corresponds to a positive frequency offset $\Delta\omega$ and binary $b_i = 0$ corresponds to a negative frequency offset $-\Delta\omega$. These tones are referred to as *mark* and *space* tones corresponding to the mark data $b_i = 1$ and space data $b_i = 0$. The phase function $\phi(t)$ corresponds to a frequency error (ω_ε) and a constant phase error of ϕ_ε. The additive noise term in (5.2) is typically zero-mean white Gaussian noise characterized as $N(0, \sigma_n)$. A coherent FSK (CFSK) demodulator implementation is shown in Figure 5.2.

After amplification by an input low-noise power amplifier (LNPA) and intermediate heterodyning stages (not shown), the received signal is mixed to baseband and the output of the baseband filter is expressed as

$$e_o(t) = \frac{A}{2}\cos(d_i\Delta\omega t + \phi(t)) + n_o(t) \tag{5.3}$$

where the double frequency term from the final mixing stage is removed by the lowpass detection filters.* The bandwidth of the baseband filter is selected to pass the desired FSK signal and establish the Nyquist band for subsequent digital signal processing; it also removes nearby interfering signals that will degrade the detection performance. If bandpass sampling is used, as discussed in Chapter 2, the baseband filter can be replaced by a bandpass filter to serve the same functions.

The baseband filter shown in Figure 5.2 results in the baseband output corresponding to the FSK mark and space tones with $\omega_m = +\Delta\omega$ and $\omega_s = -\Delta\omega$. The estimate $\hat{\phi}(t)$ of the phase term is provided by a phaselock loop (PLL) that removes the phase error in the baseband outputs corresponding to each tone frequency. The LPF in the quadrature rails eliminate the terms involving $2\Delta\omega$ and correspond to the detection-matched filters. The inphase-matched filter output samples, corresponding to the mark and space data estimates, are expressed in terms of the analytic signals as

$$\tilde{e}_m(t) = \begin{cases} \frac{A}{4}e^{j\phi_e(t)} + N_m(t) & : \text{when } b_i = 1 \\ N_m(t) & : \text{when } b_i = 0 \end{cases} \tag{5.4}$$

and

$$\tilde{e}_s(t) = \begin{cases} \frac{A}{4}e^{j\phi_e(t)} + N_s(t) & : \text{when } b_i = 0 \\ N_s(t) & : \text{when } b_i = 1 \end{cases} \tag{5.5}$$

When the phase error $\phi_e(t) = \phi(t) - \hat{\phi}(t)$ is negligible, the amplitudes of the inphase and quadrature rails corresponding to the mark and space tones are approximately $A/4$ and $(A/4)$ $\sin(\phi_e(t))$ respectively. The inphase-matched filter outputs are sampled at the optimum times $nT_b : n = 0, 1, \ldots$ and the noise-free samples are given by $e_m = AT/4$ when $b_i = 1$ and $e_s = AT/4$ when $b_i = 0$. The E/L gate symbol timing adjustment is shown in Figure 5.2. In this regard, the E/L timing function includes filtering and incremental timing adjustments are a fraction of the bit duration T_b and are derived from the demodulator clocks.

For clarity in describing the signal processing, the channel and receiver noises are often not included in the above descriptions; however, the initial phase and bit-timing estimates in the presence of noise are typically established during a preamble with PLL and bit-time tracking insuring that the estimation errors remain negligible over the message duration. Eliminating the phase and frequency errors using a second-order PLL permits coherent detection of the FSK waveform and the theoretical bit-error performance is evaluated as follows.

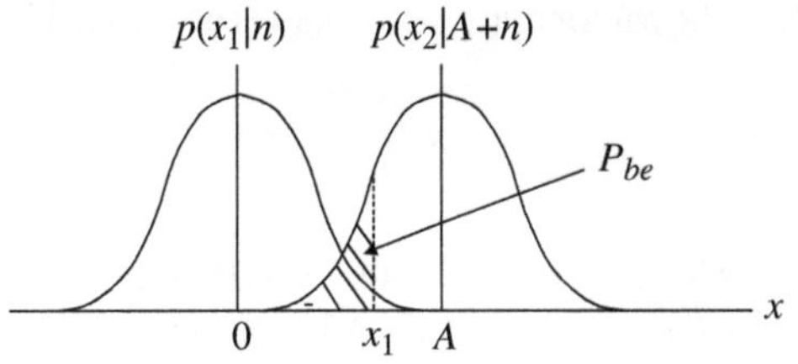

FIGURE 5.3 Bit-error evaluation for coherent BFSK (CBFSK) detection.

The bit-error detection performance of BFSK is evaluated with additive white Gaussian noise (AWGN) in terms of the *pdf* of the mark and space-matched filter outputs. Recognizing that one filter contains noise only (n) and the other contains signal plus noise ($A + n$); the *pfds* are shown in Figure 5.3 and the bit-error probability is evaluated as

$$P_{be} = P_r(x_2 \le x_1 | A + n) \tag{5.6}$$

This result strictly applies under the condition that the upper or positive FSK tone is present; however, the conditions for the lower tone are similar and, assuming equal a priori tone probabilities, (5.6) also applies to the overall bit-error probability.

The random variables x_1 and x_2 in (5.6) are independent with variances σ_n^2 and, upon defining the decision variable $y = x_2 - x_1$, the error condition occurs when $y \le 0$ and upon evaluating *pdf* of the random variable† y, the bit-error probability is computed as

$$P_{be} = \frac{1}{\sqrt{2\pi}\sigma_y}\int_{-\infty}^{0} e^{-(y-A)^2/(2\sigma_y^2)}dy \tag{5.7}$$

Referring to Section 3.5, (5.7) is expressed in terms of the complementary error function as

$$P_{be} = \frac{1}{2}erfc\left(\frac{\gamma_b}{2}\right) \tag{5.8}$$

where the noise power, σ_n^2, is measured in the bandwidth R_b of the data bit and $\sigma_y^2 = 2\sigma_n^2$. This result is shown in Figure 5.4 as the *solid* curve labeled *coherent* and requires a 3 dB higher signal-to-noise ratio to achieve the same performance using antipodal signaling, that is, compared to coherently detected BPSK modulation. The performance of noncoherent detection

*The filtering associated with the sample-rate reduction is used. Typically, the sample-rate reduction involves the processing from the input sampling frequency of about $2.4h + R_b$ Hz to that required by the upper and lower detection filters of about $4R_b$ Hz. Rate reduction reduces the computational intensity and dynamic range requirement of the demodulator detection processing shown in Figure 5.2.

†The *pdf* of y is determined by convolving the *pdfs* of $x_1 = y - x_2$ and x_2.

of BFSK is shown for comparison and is developed in the next section.

5.3 NONCOHERENT DETECTION OF BFSK—KNOWN FREQUENCY AND UNKNOWN PHASE

The optimum noncoherent detection of BFSK when the frequency is known or estimated and the phase is unknown is shown in Figure 5.5. Because of the unknown phase, the data detection decisions must be based on the magnitude of the mark and space tone rails that is not influenced by the phase. However, because of the unknown phase, the conventional phase-error derived from the quadrature rail cannot be used to form the phase discriminator error required by the phase-lock loop. One solution is to form a frequency discriminator error to be used with a frequency-lock loop. A frequency discriminator error is formed by using a frequency discriminator that estimates the frequency error ω_e. The frequency and symbol time tracking are not shown in Figure 5.5; however, with the caveat that a frequency discriminator is used, the processing is similar to that described for the coherent detector. The received frequency error and bit timing must be estimated and tracked as in the coherent demodulator and a preamble is typically used to determine the initial estimates.

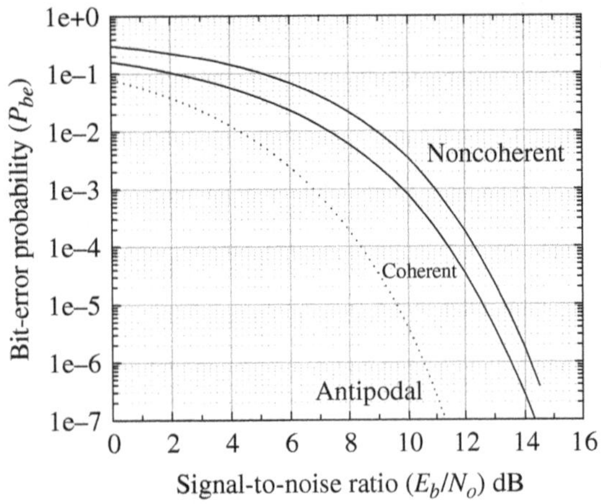

FIGURE 5.4 Theoretical coherent and noncoherent BFSK performance.

The bit-error performance of noncoherently demodulated BFSK is evaluated in the same manner as in the coherent demodulator, in that, using the *pdf* of the mark and space-matched filter outputs recognizing that one filter contains noise only and the other contains signal plus noise. However, with noncoherent detection, the *pfds* are more complicated with the *pdf* of the noise only filter output described by the Rayleigh distribution and the *pdf* of the filter containing signal plus noise described by the Ricean distribution as shown in Figure 5.6.

Considering that the random variable x_1 represents the noise only at the sampled output of the matched filter and the random variable x_2 represents the sample-matched filter output containing the signal and noise; the respective Rayleigh and Ricean distributions, as depicted in Figure 5.6, are expressed as

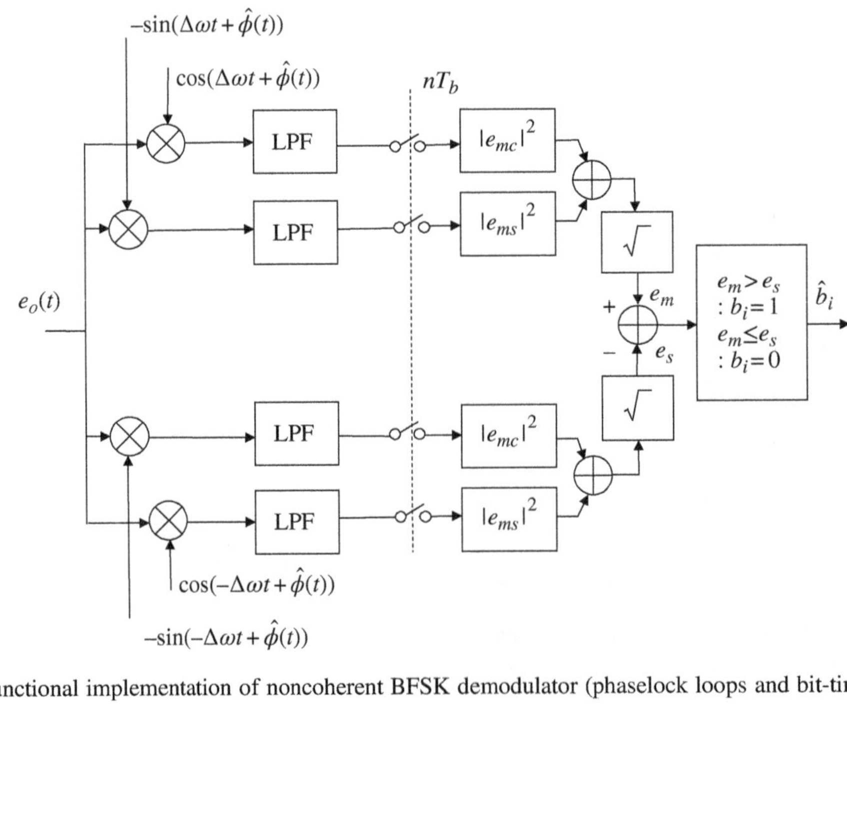

FIGURE 5.5 Functional implementation of noncoherent BFSK demodulator (phaselock loops and bit-timing are not shown).

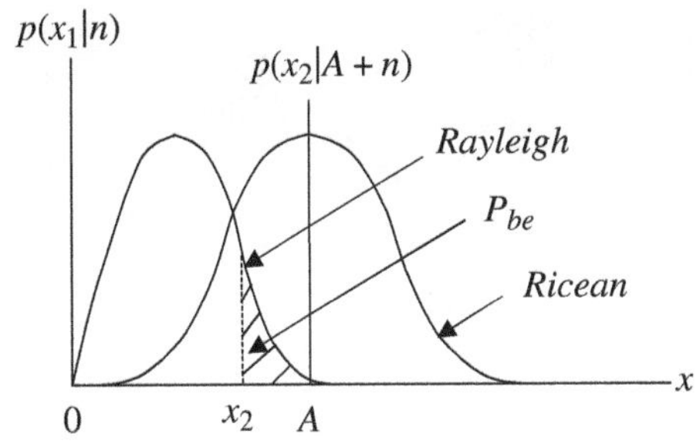

FIGURE 5.6 Bit-error evaluation for noncoherent BFSK (NCBFSK) detection.

$$p(x_1) = \frac{x_1}{\sigma_n^2} e^{-x_1^2/(2\sigma_n^2)} \quad : x_1 \geq 0 \tag{5.9}$$

and

$$p(x_2) = \frac{x_2}{\sigma_n^2} e^{-(x_2^2 + A^2)/(2\sigma_n^2)} I_o\left(\frac{Ax_2}{\sigma_n^2}\right) \quad : x_2 \geq 0 \tag{5.10}$$

In this case, the bit-error is evaluated as

$$\begin{aligned} P_{be} &= P_r(x_1 \geq x_2) \\ &= \int_0^\infty p(x_2)\left[\int_{x_2}^\infty p(x_1)dx_1\right]dx_2 \end{aligned} \tag{5.11}$$

and substitution of (5.9) and (5.10) into (5.11) results in the bit-error expression

$$P_{be} = \int_0^\infty \frac{x_2}{\sigma_n^2} e^{-(2x_2^2 + A^2)/(2\sigma_n^2)} I_o\left(\frac{Ax_2}{\sigma_n^2}\right) dx_2 \tag{5.12}$$

where the second integral in (5.11) is evaluated as $e^{-x_2^2/(2\sigma_n^2)}$. Upon defining the signal-to-noise ratio as $\gamma = A^2/2\sigma_n^2$ and expressing (5.12) in terms of the random variable $y = (\sqrt{2}/\sigma_n)x_2$, (5.12) is evaluated as

$$P_{be} = \frac{1}{2} e^{-\gamma/2} \int_0^\infty y e^{-(y^2 + \gamma)/2} I_o(\sqrt{\gamma} y) dy \tag{5.13}$$

The integral in (5.13) is recognized as the Marcum Q-function denoted as $Q(\sqrt{\gamma}, 0)$. The Marcum Q-function is defined as

$$Q(a,b) = \int_b^\infty y e^{-(y^2 + a^2)/2} I_o(ay) dy \tag{5.14}$$

with the properties $Q(a,0) = 1$, $Q(0,b) = e^{-b^2/2}$ and $Q(0,0) = 1$. Using the first of these properties, (5.13) becomes simply

$$P_{be} = \frac{1}{2} e^{-\gamma/2} \tag{5.15}$$

The bit-error performance of noncoherently detected FSK, expressed by (5.15), is plotted in Figure 5.4. Compared to the performance of coherently detected FSK at $P_{be} = 10^{-5}$, the noncoherent detection loss is about 0.9 dB relative to coherent detection.

5.4 CASE STUDIES: COHERENT AND NONCOHERENT BFSK PERFORMANCE SIMULATION

The simulated performance of coherently and noncoherently detected BFSK discussed in this case study is based on Monte Carlo simulations using 100K bits for each signal-to-noise ratio for $P_{be} > 10^{-4}$ and 10 M bits otherwise. The simulations model the demodulator implementations shown in Figures 5.2 and 5.5. The received FSK waveform is characterized as the sampled baseband analytic signal*

$$\tilde{e}_b(nT_s) = \frac{A}{2} e^{j((2\pi f_e T_b - d_i 2\pi \Delta f T_b)n/(N_s + \phi_e))} \tag{5.16}$$

where $T_s = 1/f_s$ is the sampling interval, f_s is the sampling frequency, f_ε and ϕ_ε represent frequency and phase errors based on the estimation error $\phi_\varepsilon(t) = \omega_\varepsilon t + \phi_\varepsilon$. Also $T_b = 1/R_b$ is the bit interval, R_b is the bit rate, and N_s is the number of samples-per-bit. To characterize the ideal or theoretical performance shown in Figure 5.4, the simulation uses $f_\varepsilon = 0$ and $\phi_\varepsilon = 0$. The parameter Δf is the magnitude of the FSK tone deviation and $\delta f = \pm\Delta f$ is the frequency deviation from the center frequency. The modulation index of BFSK is defined as $h = 2\Delta f/R_b$ and orthogonal tone spacing corresponds to h being an integer and the minimum orthogonal spacing occurs when $h = 1$. Orthogonal frequency spacing for the BFSK waveform with matched filter detection occurs when the modulation tones fall in the spectral nulls at $mR_b : |m| > 0$ of the adjacent tones of the *sinc*(fT_b) spectrum.

5.4.1 Orthogonal Tone Spacing ($h = 19$)

The performance simulation in this case study uses the normalized FSK tone shift $\Delta f T_b = 9.5$ and a bit rate of $R_b = 1$ kbps, so the baseband mark and space tones are: $f_{mk} = 9.5R_b$ and $f_{sp} = -9.5R_b$ corresponding to ±9.5 kHz respectively. In the simulation, the IF BPF shown in Figure 5.2 is modeled as a 6-pole Butterworth lowpass equivalent filter with ideal phase equalization and a normalized 3 dB bandwidth of $f_3T_b = 12.5$. The function of the IF BPF is to reject interfering signals outside of the detection bandwidth and to establish the Nyquist bandwidth and the related sampling frequency.

*For a description of the simulated signal sampling, refer to Sections 14.4–14.6.

The minimum sampling frequency, based on the Nyquist criterion, is $f_s(\min) = 2f_N$ where f_N is Nyquist bandwidth. The 6-pole Butterworth filter provides an attenuation of about 40 dB at $2f_3T_b = 25$ and the 40 dB rejection criterion is taken to be Nyquist bandwidth, that is, $f_N = 25R_b$, so the sampling frequency must be $f_s \geq 50R_b$ corresponding to $N_s \geq 50$ samples-per-bit. Although $f_s = 50R_b$ is sufficient to make the aliasing loss negligible, the number of samples-per-tone frequency is only $f_s/f_{mk} = (f_s/R_b)(R_b/f_{mk}) = 50/9.5 = 5.26$ so, to assist in data viewing and debugging, it is convenient to use $f_s = 128R_b$ that results in $f_{mk}/f_s = 13.47$ samples-per-tone; therefore, the simulation uses $N_s = 128$ samples-per-symbol. These parameter values are depicted in Figure 5.7 in terms of the normalized frequency-domain specification and the simulated received BFSK signal spectrum is shown in Figure 5.8 with and without the lowpass equivalent IF BPF.

Figure 5.9 shows the simulated performance of the coherent and noncoherent demodulator as the *circled* data points without the IF BPF and the *triangular* data points show the performance when IF BPF is included. The noncoherent detector exhibits negligible performance degradation with the filter and the coherent detector is more sensitive to the filter distortion; however, because the filter design has ideal phase equalization, the filter loss is attributed strictly to the filter amplitude response. The *solid* and *dotted* curves represent the theoretical performance from Figure 5.4.

Under the ideal conditions without the input filter, the mark and space tones in this example are orthogonal, in that, the number of $sinc(fT_b)$ spectral side lobes between the FSK tones is $2\Delta fT_b = 19$, so each tone falls in the spectral null of the adjacent tone. This orthogonal spacing results in independent quadrature channel noise and zero adjacent tone interference leading to the optimum performance. Although the input filter will degrade the optimum performance, the degradation caused by the selected filter is considered to be acceptable. The tone spacing of $19R_b$ requires considerably more bandwidth than with the minimum orthogonal spacing; however, the degradation due to the frequency errors is reduced. The impact of frequency errors with minimum orthogonal tone spacing is examined in the following case study.

The final evaluation of this case study involves the simulation of the degradation of the noncoherent BFSK detector with normalized frequency errors $f_eT_b = 0.1$, 0.3, and 0.4. In this evaluation the input filter is not included so the degradation results solely from the frequency errors relative to the demodulator-matched filter. The performance in this case is shown in Figure 5.10. The performance degradation may be tolerable for frequency errors of up to about 20% of the data rate, however, for larger frequency errors, the performance degradation increases dramatically. The coherent demodulator requires a PLL to maintain phase lock and keep the frequency centered on the mark and space-matched filters,

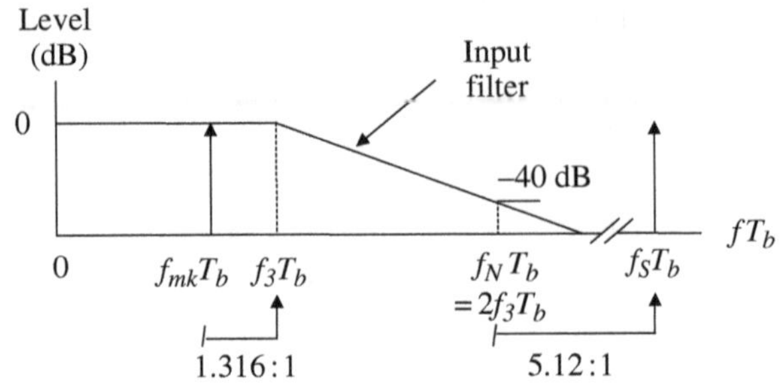

FIGURE 5.7 Demodulator filter characteristics (positive frequencies).

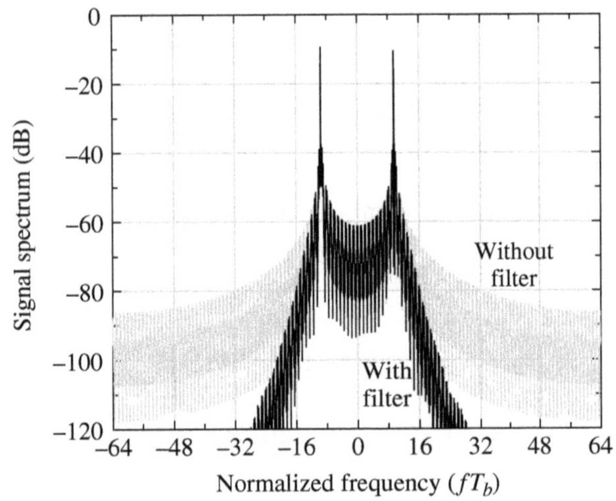

FIGURE 5.8 BFSK spectrum with/without input filter.

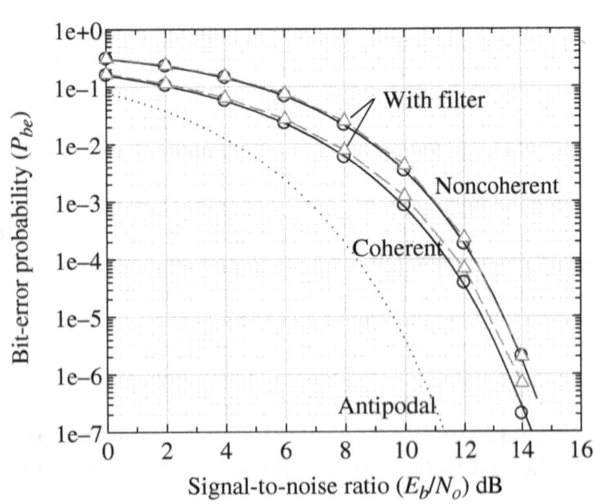

FIGURE 5.9 Simulated BFSK coherent and noncoherent performance with and without input filter.

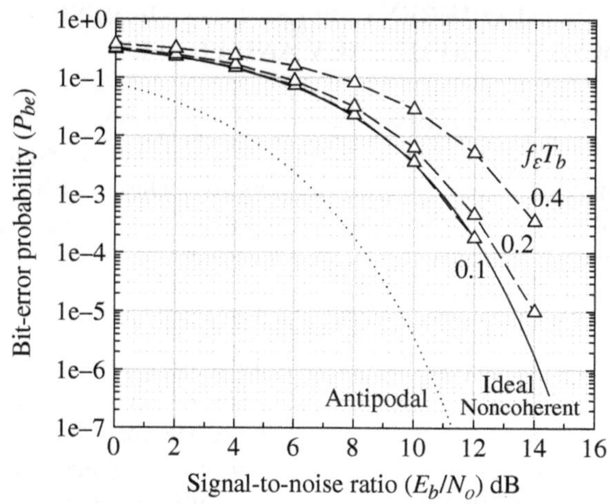

FIGURE 5.10 Simulated noncoherent BFSK performance with frequency error ($h = 19$, no input filter).

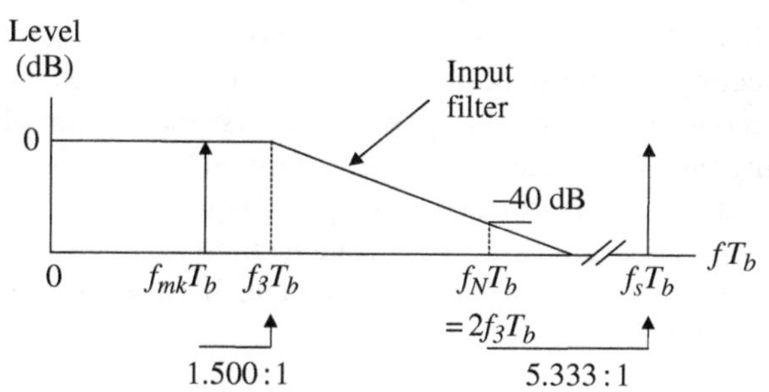

FIGURE 5.11 Demodulator input filter characteristics (positive frequencies).

5.4.2 Minimum Orthogonal Tone Spacing ($h = 1$)

This case study is similar to the previous case, however, minimum orthogonal tone spacing corresponding to $h = 1$ is used. Using a data rate of $R_b = 1$ kbps, the tone shift corresponding to the minimum orthogonal spacing is $|\Delta f| = 0.5$ kHz. The input filter is similar to the filter used in the previous case study with the following exceptions: the normalized 3 dB frequency is $f_3T_b = 1.5$, and the sampling frequency is 5.333 times the Nyquist bandwidth, f_N, and results in 16 samples-per-bit. The filter and sampling characteristics are depicted in Figure 5.11 and the spectrum is shown in Figure 5.12 superimposed on the spectrum of the BFSK waveform. The filter 3 dB bandwidth is chosen based on the theoretical spectrum results shown in Figure 5.25. Referring to the spectrum corresponding to a unit modulation index, the mark tone impulse occurs at the normalized frequency $\beta = 0.5$ and the filter 3 dB bandwidth is located at the first spectral null occurring at $\beta = 1.5$ or $1.5R_b$ Hz.

Figure 5.13 shows the theoretical performance of the coherent and noncoherent demodulator as the *solid* and *dashed* curves respectively. The performance with the filter

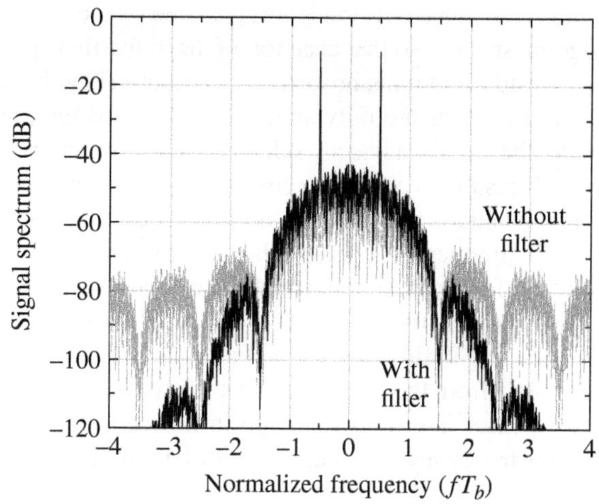

FIGURE 5.12 BFSK spectrum with/without input filter.

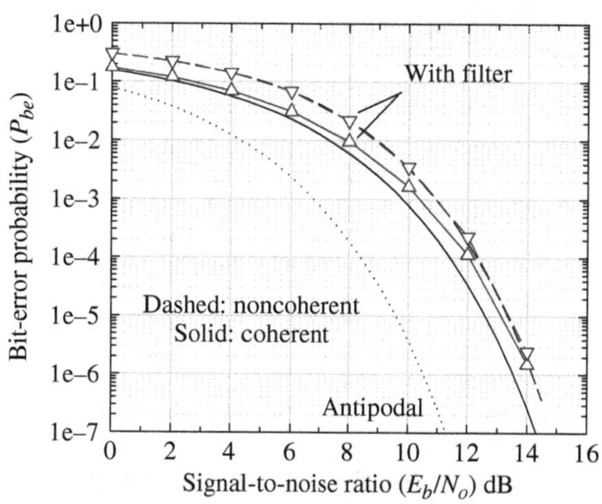

FIGURE 5.13 Simulated coherent and noncoherent performance with and without input filter.

is indicated by the corresponding *solid* and *dashed* curves with the *triangular* data points. As in the previous case, the simulation uses 100K Monte Carlo samples at each signal-to-noise ratio for $P_{be} > 10^{-4}$ and 10 M samples otherwise. The performance of the noncoherent detector exhibits negligible degradation with the filter* and the degradation of the coherent detector is about the same as that for the previous case using $h = 19$. To reduce the impact of the filter on the coherent detector, the filter bandwidth should be increased to include the second spectral sidelobe; for example, by placing the filter 3 dB point at the second spectral null corresponding to $\beta = 2.5$ or $2.5R_b$ Hz. As with the previous case, the filter design has ideal phase equalization so the filter loss is attributed strictly to the filter amplitude response.

*The noncoherent performance curves with and without the filter are nearly identical and are hard to distinguish.

In this case, the BFSK tones are separated by the minimum orthogonal spacing so that each tone falls in the first spectral null of the adjacent tone; however, the spectral sidelobes surrounding the tones are only attenuated by 13 dB and, compared to the 35 dB adjacent sidelobe of the previous case study, this results in a greater sensitivity to frequency errors. The final evaluation of this case involves the signal-to-noise loss with frequency errors. The results are shown in Figure 5.14 for the noncoherent BFSK detector with normalized frequency errors of $f_eT_b = 0.1$, 0.3, and 0.4. In this evaluation, the input filter is not included so the degradation results solely from the frequency error relative to the matched filter response. The maximum performance loss is about 0.9 dB for frequency errors up to about 10% of the data rate; this is nearly the same loss that occurred in the previous case study for a frequency error of 20%.

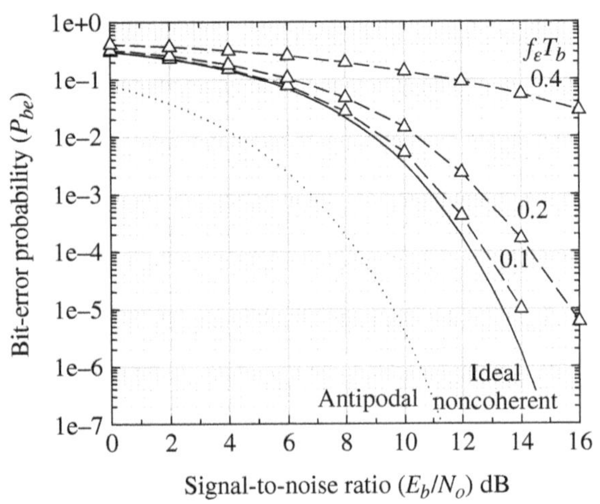

FIGURE 5.14 Simulated noncoherent BFSK performance with frequency error ($h = 1$, no input filter).

5.5 NONCOHERENT DETECTION OF BFSK—UNKNOWN FREQUENCY AND PHASE

The detection of BFSK-modulated waveforms is accomplished in several ways without using complicated frequency acquisition and tracking algorithms. These techniques involve noncoherent detection methods wherein the implementation simplicity is traded for degraded performance. For example, in Section 5.3 the performance of noncoherent detection is examined with known frequency and unknown phase using a *sinc*(fT_b)-matched filter for each tone with a loss of about 0.9 dB at $P_{be} = 10^{-5}$ relative to coherent detection with known frequency and phase. The mark and space-matched filters provide some tolerance to frequency errors; however, in this section, the frequency is unknown over a very wide band $W \gg R_b$ so the matched filters are replaced with envelope detectors and, for a given bit-error probability, the noncoherent detection may require in excess of 10 dB more signal-to-noise ratio.

In the following sections, two methods for noncoherently detecting BFSK-modulated data are examined with unknown frequency and phase. The first method involves using envelope detectors for the mark and space tones; in this case, the performance of linear and square-law envelope detectors is compared. The second method involves using a frequency discriminator to distinguish between the mark and space tones.

5.5.1 Linear Envelope Detection

The functional diagram of the detection circuitry is shown in Figure 5.15 where the detectors are characterized as linear, or envelope, detectors. The implementation does not provide

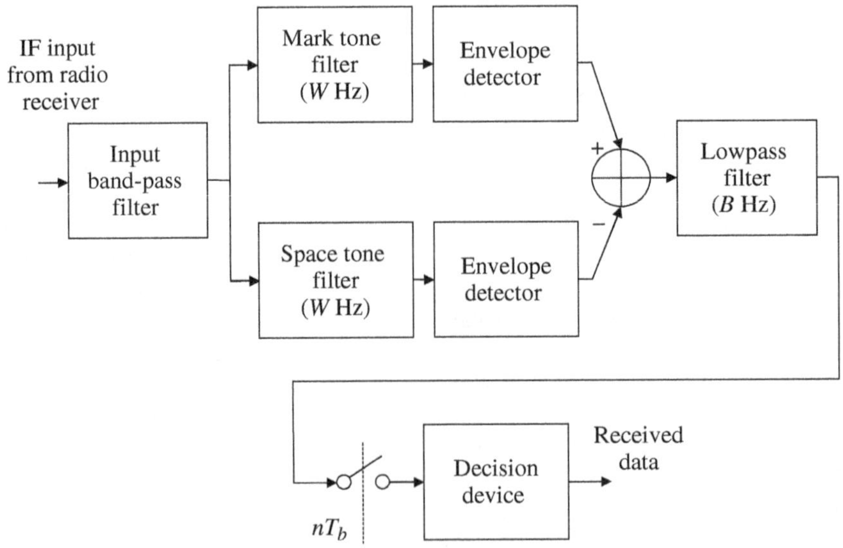

FIGURE 5.15 Noncoherent BFSK demodulator using envelope detection.

optimum noncoherent detection of BFSK-modulated data because the noise bandwidth, W, of the mark and space detection filters is much wider than required for optimum detection, that is, $W \gg B_b$, where $B_b = 1/T_b$ is the bandwidth of the detection filter and T_b is the mark/space bit duration. The wide bandwidth mark and space filters provide a simple implementation when shifts in the received carrier frequency occur and are not corrected.

The performance analysis of this noncoherent detector has been analyzed by Glenn [1] under the condition $WT_b \gg 1$. Glenn's analysis begins with the characterization of the envelope detector output *pdf* as being Ricean when signal and noise is present and Rayleigh when only noise is present. However, the *pdf* of the lowpass post-detection filter output is approximately Gaussian due to the central limit theorem. The lowpass filter with bandwidth $1/T_b$ essentially integrates $n = WT_b$ samples and for $WT_b > 10$ the Gaussian approximation is accurate over a 3-sigma range with rapid improvement in the tails of the distribution with increasing WT_b.

Based on these results, Glenn characterizes the bit-error probability as

$$P_{be} \cong \frac{1}{2}\left[1 - erf\left(\frac{|m_z|}{\sqrt{2}\sigma_z}\right)\right] \quad : WT_b \gg 10 \tag{5.17}$$

where m_z and σ_z are the mean and standard deviation of the approximately Gaussian distributed output of the lowpass filter. Actually, $m_z = m_{z1}$ when a mark or upper tone is present and $m_z = -m_{z0}$ when a space or lower tone is present. With symmetrical filtering and balanced channel gains, the mean values are antipodal with $m_{z1} = -m_{z0}$. Glenn has shown that the argument of the error function is expressed as

$$\frac{|m_z|}{\sqrt{2}\sigma_z} = \left(\frac{\pi}{8}\right)^{\frac{1}{2}} \frac{\sqrt{WT_b}\left[{}_1F_1\left(-\frac{1}{2};1;-(E_b/N_oWT_b)\right) - 1\right]}{\left\{(E_b/N_oWT_b) + 2 - \frac{\pi}{4}\left[1 + {}_1F_1^2\left(-\frac{1}{2};1;(-E_b/N_oWT_b)\right)\right]\right\}^{1/2}} \tag{5.18}$$

where ${}_1F_1()$ is the confluent hypergeometric function [2, 3], E_b is the energy-per-bit, and N_o is the noise power spectral density (PSD). The confluent hypergeometric function ${}_1F_1(-1/2; 1; -z)$ is evaluated using more familiar Bessel functions [4] as

$${}_1F_1\left(-\frac{1}{2};1;-z\right) = e^{-z/2}\left[(1+z)I_o\left(\frac{z}{2}\right) + zI_1\left(\frac{z}{2}\right)\right] \tag{5.19}$$

Using (5.19) and (5.18), the bit-error results given by (5.17) are plotted in Figure 5.16 as a function of the signal-to-noise ratio E_b/N_o for WT_b products of 10, 10^2, 10^3, and 10^4. Glenn

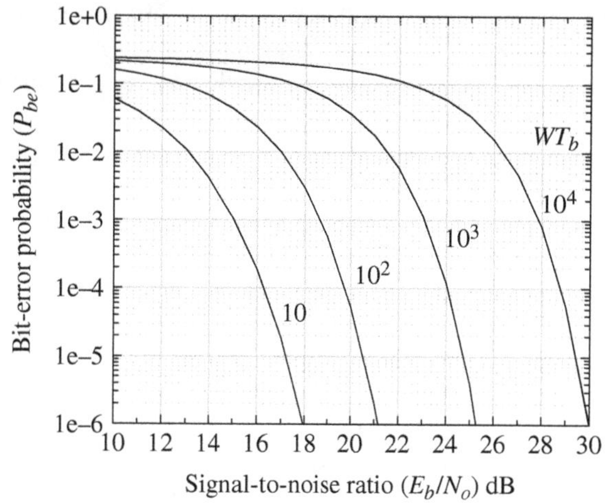

FIGURE 5.16 Noncoherent BFSK performance using linear detector.

has also evaluated approximations for the bit-error probability for large and small values of the signal-to-noise ratio. These approximations are given by

$$P_{be} \cong \frac{1}{2}\left[1 - erf\left(\sqrt{\frac{E_b}{2N_o}} - \sqrt{\frac{\pi WT_b}{8}}\right)\right] \quad : WT_b \gg 10 \text{ and } \frac{E_b}{N_oWT_b} \gg 1 \tag{5.20}$$

and

$$P_{be} \cong \frac{1}{2}\left[1 - erf\left\{\frac{\pi WT_b}{16(4-\pi)}\left(\frac{E_b}{N_oWT_b}\right)\right\}\right] \quad : WT_b \gg 10 \text{ and } \frac{E_b}{N_oWT_b} \ll 1 \tag{5.21}$$

As a practical application of these results, some measured performance data are compared to the theoretical bit-error performance. The measured data are based on a radio receiver followed by a data demodulator used to detected BFSK-modulated data. The function of the radio receiver is to mix the radio frequency (RF) input to a 1 or 2.55 kHz intermediate frequency (IF) which is applied to the demodulator. The demodulator uses a linear detector and the measured data are based on the 1 kHz IF output which applies for modulation frequency shifts ranging from 5 to 100 Hz on each side of the carrier frequency. Under ideal conditions, the received carrier frequency is unimportant as long as the mark and space tones fall with the bandwidth of the respective predetection filters. However, as a practical matter, gain ripple in the input bandpass filter and gain unbalance in the mark and space filters will degrade the performance and result in some dependence on the modulation frequency shifts as well as the carrier frequency.

The measured test results are based on a fixed data rate of $R_b = 50$ bps with modulation frequency shifts of $\Delta f = \pm 25$ and ± 85 Hz. The corresponding modulation indices, defined as $2\Delta f/R_b$, are 1.0 and 3.4. The specified bandwidth of the demodulator input filter is 520 Hz. The measured responses of the mark and space filters and the input bandpass filter are shown in Figure 5.17 and the composite responses of the mark and space channels are shown in Figure 5.18. Based on these measured results, the noise bandwidths of the mark and space filters are computed to be 293 and 246 Hz respectively. In the following comparison, the average of these measured bandwidths is used corresponding to $W = 270$ Hz so that the WT_b product is 5.4.

Based on these conditions and recognizing that the WT_b product is less than required, the measured performance results are compared to the theoretical result in Figure 5.19. In contrast to the theoretical result, the dependence of the performance on the modulation index is clearly evident. This is due to the filter amplitude ripple in each channel. Also, a portion of the approximately 0.5–1.0 dB of degradation from the theoretical prediction is due to the amplitude asymmetry between the mark and space filters which results in a tone unbalance into the post-detection filter. Although performance measurements are not made with various IF frequency offsets, the asymmetries in the mark and space filter frequency responses suggest that the bit-error performance is dependent on the IF carrier frequency offset.

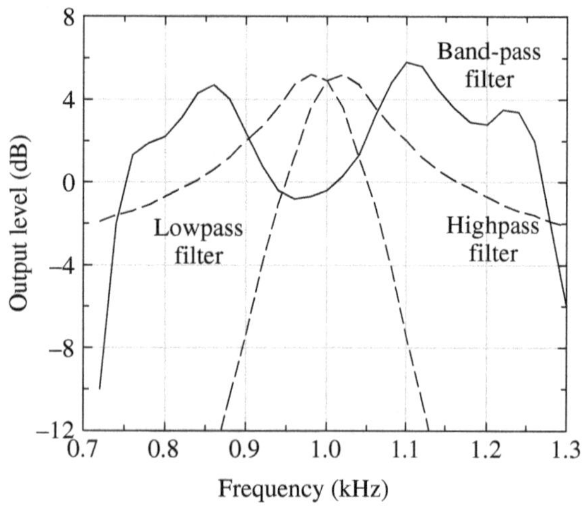

FIGURE 5.17 Individual filter responses.

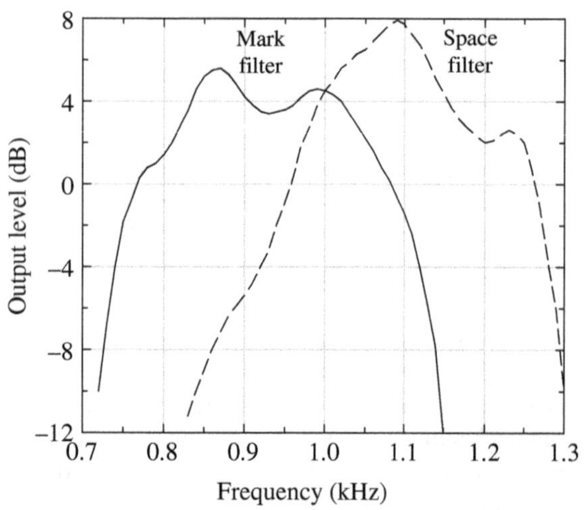

FIGURE 5.18 Composite filter responses.

5.5.2 Square-Law Envelope Detection

The functional diagram for the square-law detection is identical to that shown in Figure 5.15 when the envelope detectors are replaced by square-law detectors. Also, as in the previous section, the pre-detection filter bandwidth is much greater than the bandwidth of the post-detection filter, that is, $W \gg B_b$ The analysis of the square-law detector also applies for $WT_b > 10$ with the understanding that the lowpass filter output *pdf* is approximated by a Gaussian distribution. Using the results of Glenn's analysis of the square-law detector implementation, the bit-error probability is given by

$$P_{be} \cong \frac{1}{2}\left[1 - erf\left(\frac{|m_v|}{\sqrt{2}\sigma_v}\right)\right] \quad : WT_b \gg 10 \tag{5.22}$$

where

$$\frac{|m_v|}{\sqrt{2}\sigma_v} = \frac{E_b}{2N_oWT_b}\sqrt{\frac{WT_b}{1 + E_b/(N_oWT_b)}} \tag{5.23}$$

The performance results for the square-law detector are shown in Figure 5.20 as a function of E_b/N_o for WT_b products of 10, 10^2, 10^3, and 10^4. To achieve $P_{be} = 10^{-5}$ with $WT_b = 10$

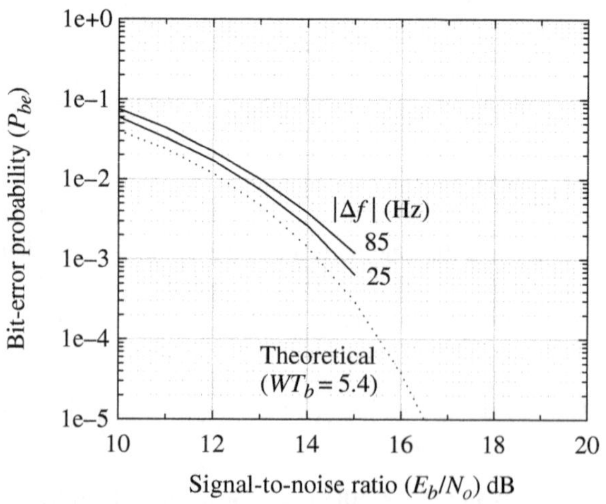

FIGURE 5.19 Noncoherent BFSK detection performance comparison based on measured results ($R_b = 50$ bps, $W = 270$ Hz).

the square-law detector requires a signal-to-noise ratio about 1.5 dB higher than the linear detector; however, this difference is about 0.5 dB for $WT_b = 100$ and is negligible as the time-bandwidth continues to increase.

5.5.3 Noncoherent BFSK Detection Using Frequency Discriminator

The most straightforward implementation of a BFSK detector uses a frequency discriminator to distinguish between the received mark and space tones. The implementation of a conventional BFSK discriminator detector is shown in Figure 5.21. The simplified implementation over that of the noncoherent detectors discussed in the previous sections results in significantly different performance characteristics, the most notable of which is the strong dependence upon the frequency shift of the BFSK-modulated waveform.

Referring to Figure 5.21, the limiter-filter function is not necessary in implementing the frequency discriminator; however, it will result in a signal-to-noise ratio improvement for input signal-to-noise ratios greater than −5.5 dB. The implementation of the limiter-filter is discussed in more detail in Section 10.2. The frequency discriminator can be implemented in a number of ways (see Problems 1 and 2); however, the implementation discussed in Section 11.2.2.4 is preferred. The lowpass filter following the discriminator is essential for accurately estimating the frequency in the presence of noise and, because of the similarities to a correlator response, at a given correlator delay or lag, the combination of the discriminator and filter is referred to as a lag-correlator. The frequency discriminator output function is characterized as an unbiased estimate of the frequency having odd symmetry about zero. Typically the output response corresponds to a *sine* function, often referred to as an *S-curve*. The important feature is that the *positive* or *negative* sense of the estimate is available for the purpose of frequency estimation and control. In addition, the quadrature output corresponds to a *cosine* function that is used for signal presence detection.

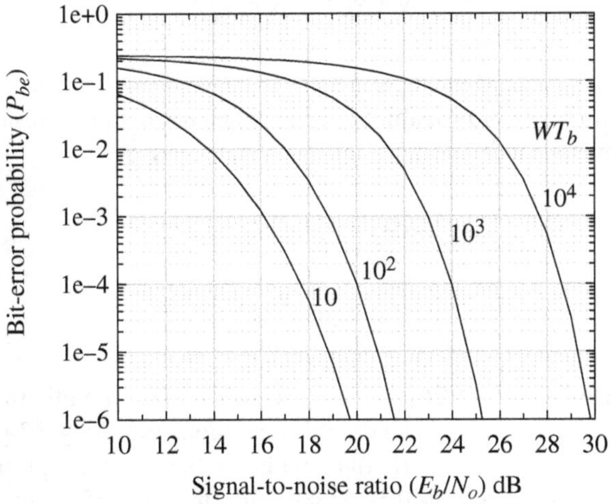

FIGURE 5.20 Noncoherent BFSK performance using square-law detection.

Analyzing the performance of the conventional frequency discriminator is a difficult undertaking. Fortunately, however, Stein [5, 6] has characterized the *pdf* at the output of the limiter/discriminator detector with AWGN. Stein expresses the bit-error probability of the discriminator detector in terms of the Marcum Q-function [7, 8] as

$$P_{be}(a,b) \cong \frac{1}{2}\left[1 - Q\left(\sqrt{b},\sqrt{a}\right) + Q\left(\sqrt{a},\sqrt{b}\right)\right] \tag{5.24}$$

where the Marcum Q-function is defined as

$$Q(a,b) \triangleq \int_b^{\infty} x e^{-(x^2+a^2)/2} I_o(ax) \tag{5.25}$$

The parameters a and b are expressed in terms of the signal-to-noise ratio γ, the modulation frequency shift Δf, and the *rms* frequency f_2, of the normalized IF filter frequency function $G(f)$. The parameters a, b, and f_2 are given by Stein as

$$a = \frac{\gamma}{2}\left[1 - \frac{\Delta f}{f_2}\right]^2 \tag{5.26}$$

and

$$b = \frac{\gamma}{2}\left[1 + \frac{\Delta f}{f_2}\right]^2 \tag{5.27}$$

where

$$f_2 = \left[\int_{-\infty}^{\infty} f^2 G(f) df\right]^{1/2} \tag{5.28}$$

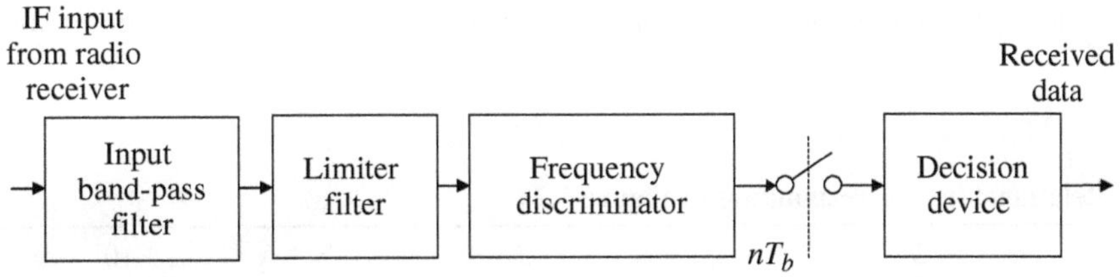

FIGURE 5.21 Conventional BFSK frequency discriminator detector.

with $G(f)$ defined as

$$G(f) \triangleq \frac{|H(f)|^2}{\int_{-\infty}^{\infty} |H(f)|^2 df} \tag{5.29}$$

The bit-error probability, expressed by (5.24), is plotted in Figure 5.22 as a function of the parameter b for various values of $f_2/\Delta f$ where, using (5.26) and (5.27), the parameter a is expressed in terms of b as

$$a = \left[\frac{(f_2/\Delta f) - 1}{(f_2/\Delta f) + 1}\right]^2 b \tag{5.30}$$

For a given modulation frequency, as the IF filter bandwidth is increased the detector noise increases degrading the performance. In the following discussion, it is shown that $f_2/\Delta f = 1$ ($a = 0$) results in the minimum or optimum bit-error probability. Table 5.1 lists the ratios of a/b corresponding to $f_2/\Delta f$ used in Figure 5.22.

These results clearly demonstrate the dependence of the performance on the modulation frequency shift and the IF bandpass filter characteristics. Assuming an N-pole Butterworth IF filter, the 3 dB bandwidth f_B and the *rms* frequency are related to f_B by the expression

$$f_2 = \frac{f_B}{2}\left[\frac{\sin(\pi/2N)}{\sin(3\pi/2N)}\right]^{1/2} \tag{5.31}$$

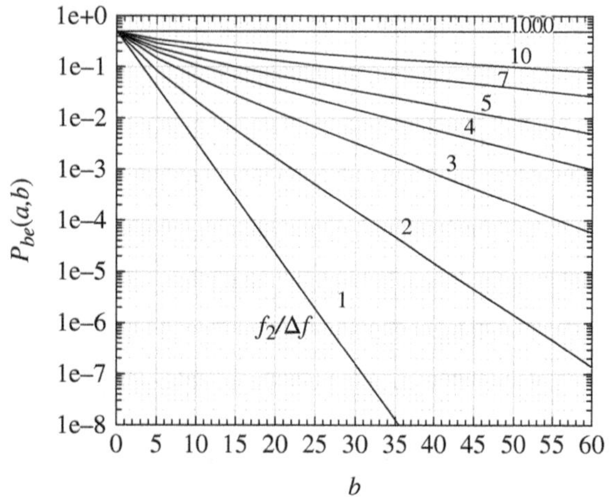

FIGURE 5.22 Bit-error probability as a function of the parameter b.

The signal-to-noise ratio γ is defined at the input to the limiter discriminator. However, it is desirable to characterize the performance as a function of $\gamma_b = E_b/N_o$ as defined in the previous sections so γ must be reflected to the receiver input and expressed in terms of the bandwidth of the modulated waveform, R_b. The necessary relationship for the N-pole Butterworth filter is

$$\gamma_b = \gamma L_f\left(\frac{B_n}{R_b}\right) \tag{5.32}$$

where

$$B_n = f_B \frac{\pi/2N}{\sin(\pi/2N)} \tag{5.33}$$

is the noise bandwidth of the input bandpass Butterworth filter with a filter loss given by

$$L_f = 1 + \left(\frac{\Delta f}{f_B}\right)^{2N} \tag{5.34}$$

Substituting the noise bandwidth and the filter loss in the expression for γ_b gives

$$\gamma_b = \frac{\gamma\pi}{2N}\left(\frac{f_B}{R_b}\right)\left(\frac{1 + (\Delta f/f_B)^{2N}}{\sin(\pi/2N)}\right) \tag{5.35}$$

Schwartz, Bennett, and Stein [6] have shown that the minimum bit-error probability occurs when $a = 0$ and the result is expressed as

$$P_{be}(\min) = P_{be}(0,b) = \frac{1}{2}e^{-b^2/2} = \frac{1}{2}e^{-\gamma^2(1+\Delta f/f_2)^4/8} \tag{5.36}$$

Referring to (5.26) or (5.30), it is seen that the condition $a = 0$ requires that $\Delta f = f_2$. Under this condition, and using a 3-pole ($N = 3$) Butterworth IF filter with $f_2 = 0.35 f_B$, the optimum BFSK frequency shift is $\Delta f = 0.35 f_B$. In terms of the modulation index these conditions yield $m_l = 2\Delta f/R_b = 0.7 f_B/R_b$ and, upon equating the filter bandwidth with the modulation data rate, as is frequently done in practice, the optimum modulation index for BFSK modulation becomes $m_l = 0.7$. This is a frequently assumed condition for optimum performance;

TABLE 5.1 Parameter Relationships Corresponding to Figure 5.22

$f_2/\Delta f$	1	2	3	4	5	7	10	100	1000
a/b	0	0.111	0.250	0.360	0.444	0.562	0.669	0.961	0.996

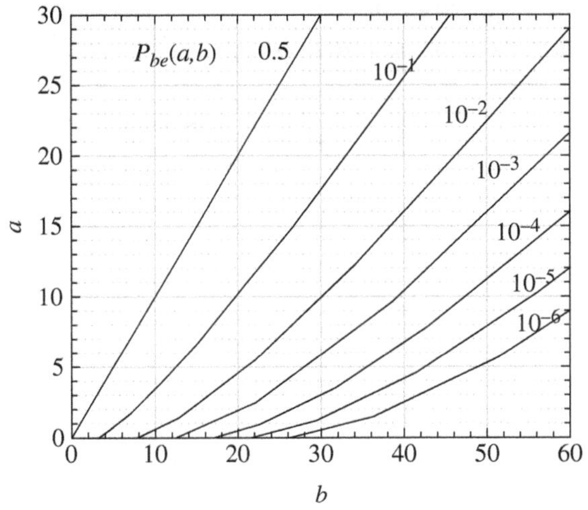

FIGURE 5.23 Universal design curves for BFSK discriminator detector.

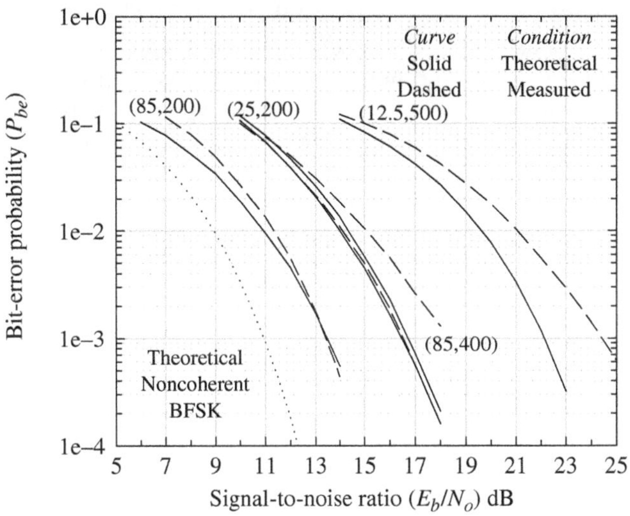

FIGURE 5.24 Error probability performance of VLF radio receiver/demodulator (50 baud FSK with various frequency shifts and IF bandwidths).

however, as seen from the preceding discussion, it applies to a specific receiver/detector implementation.

As an example application, these theoretical results are compared to measured bit-error performance data using a very low frequency (VLF) radio receiver. For this comparison, the input bandpass filter is a 3-pole Butterworth design, so $N=3$. Stein has characterized a set of universal design curves describing the bit-error probability in terms of the parameters a and b. The universal design curves shown in Figure 5.23 are established using Johansen's algorithms [9] for evaluating the Q-functions expressed respectively in (5.24) and (5.25). The resulting contours of constant $P_{be}(a,b)$, expressed in (5.24), are used to evaluate the bit-error probability of noncoherent BFSK detection using the frequency discriminator as described in the following paragraphs.

Figure 5.23 is used by first identifying the receiver filter bandwidth. The radio receiver provides for several bandwidth selections by way of a front panel switch, and bandwidths of $f_B = 200$ and 400 Hz are considered for this comparison. Next, the frequency shift of the received modulated waveform is identified and values of $|\Delta f| = 12.5$, 25, and 85 Hz are used. Using these conditions, a design *load-line* is defined expressing a as a function of b by eliminating γ between (5.26) and (5.27). Therefore, the *load-line* is expressed by (5.30). The *load-line* eliminates the parameter a from the design procedure and upon applying (5.30) to Figure 5.23, a value of $P_{be}(a,b)$ is selected and the corresponding value of b is determined from the intersection with the *load-line*. The selected value of b is used to compute signal-to-noise ratio γ based on the relationship

$$\gamma = \frac{2b}{(1+\Delta f/f_2)^2} \tag{5.37}$$

where $f_2 = 0.35 f_B$ is used for the 3-pole Butterworth IF bandpass filter. Finally, the energy-per-bit to noise-density ratio γ_b corresponding to the selected bit-error probability is determined using (5.35). For the 3-pole Butterworth filter this conversion is

$$\gamma_b = \gamma\left(\frac{\pi f_B}{3R_b}\right)\left(1+\left(\frac{\Delta f}{f_B}\right)^6\right) \tag{5.38}$$

This procedure determines one point on the curve of P_{be} versus γ_b and is repeated as necessary to obtain a performance curve. The results of the theoretical performance for the VLF radio receiver under the conditions discussed above are shown in Figure 5.24 as the *solid* curves and the corresponding measured results are shown as the *dashed* curves. The nomenclature used in Figure 5.24 denotes the frequency shift and the IF filter bandwidths as $(\Delta f, f_B)$. The measured performance results are reasonably close to the theoretical considering that the radio receiver used in these tests was operational in the *field* for a number of years and that no special consideration was given to tuning or adjustments prior to the testing. The theoretical noncoherent performance curve, shown as the *dotted* curve in Figure 5.24, corresponds to noncoherent BFSK detection using a linear receiver with orthogonal frequency shifts, that is, $m_I = 1$, and matched filter detection as discussed in Section 5.3.

5.6 BFSK SPECTRAL DENSITY WITH ARBITRARY MODULATION INDEX

In this section, the spectral density of BFSK is evaluated for arbitrary modulation indices based on the work of Anderson and Salz [10]. This work is also published by Salz [11] and summarized in Lucky, Salz, and Weldon [12]. The notation h is used to denote the modulation index and the symbol duration $T=T_b$ for binary modulated waveforms. Based on these references, the normalized spectral density of BFSK modulation with arbitrary modulation index is expressed as*

$$\frac{S(\beta)}{A^2T}=\begin{cases}\frac{1}{2}\left\langle|F(a_n)|^2\right\rangle+\mathrm{Re}\left\{\frac{\langle F(a_n)\rangle^2}{1-C_a(\omega_dT)e^{-j2\pi\beta}}\right\} & :|C_a(\omega_dT)|<1\\ \frac{1}{2}\left\langle|F(a_n)|^2\right\rangle-\frac{1}{2}|\langle F(a_n)\rangle|^2 \\ \quad+\frac{1}{2}\cos(\lambda)|\langle F(a_n)\rangle|^2\sum_{n=1}^{2}\delta(\lambda-2\pi n) & :|C_a(\omega_dT)|=1\end{cases} \tag{5.39}$$

where

$$F(a_n)=\frac{\sin((\omega-\omega_c)T/2-a_n\omega_dT/2)}{(\omega-\omega_c)T/2-a_n\omega_dT/2}e^{-j[(\omega-\omega_c)T/2-a_n\omega_dT/2]} \tag{5.40}$$

and

$$\lambda=-(\omega-\omega_c)T-b\omega_dT \tag{5.41}$$

In these expressions, b is an arbitrary constant, ω_d is the frequency deviation from the carrier frequency ω_c, $\beta=(f-f_c)T$ is the frequency relative to the carrier normalized by the bit duration $T=T_b$, $a_n=\{1,-1\}:n=0,1$ for binary modulation, and $C_a(\omega_dT)$ is the characteristic function of the random data a_n. In this analysis, f_d is the modulation tone deviation frequency from the carrier frequency; in the preceding sections, the notation $\Delta f=f_d$ is used for the frequency deviation. Assuming equally likely data, the characteristic function is evaluated as

$$\begin{aligned}C_a(v)&=\int_{-T/2}^{T/2}p(a_n)e^{jva_n}da_n\\&=\frac{1}{2}\int_{-T/2}^{T/2}(\delta(a_n-1)+\delta(a_n+1))e^{jva_n}da_n\\&=\frac{1}{2}e^{jv}+\frac{1}{2}e^{-jv}\\&=\cos(v)\end{aligned} \tag{5.42}$$

*<x> denotes the ensemble average.

so that $C_a(\omega_dT)=\cos(\omega_dT)$. Using the modulation index $h=2f_dT$, the function $F(a_n)$ is expressed as

$$F(a_n)=\frac{\sin(\pi(\beta-a_nh/2))}{\pi(\beta-a_nh/2)}e^{-j\pi(\beta-a_nh/2)} \tag{5.43}$$

The statistical average of the squared magnitude of $F(a_n)$ is evaluated as

$$\begin{aligned}\left\langle|F(a_n)|^2\right\rangle&=\left\langle\frac{\sin^2(\pi(\beta-a_nh/2))}{\pi^2(\beta-a_nh/2)^2}\right\rangle\\&=\frac{1}{2}\left[\frac{\sin^2(\pi(\beta-h/2))}{\pi^2(\beta-h/2)^2}+\frac{\sin^2(\pi(\beta+h/2))}{\pi^2(\beta+h/2)^2}\right]\end{aligned} \tag{5.44}$$

This general result is evaluated in a spectrum evaluation program; however, a simpler form results with minimum shift keying (MSK) (h = 1/2) and orthogonal BFSK (h = 1.0). These unique conditions result in

$$\left\langle|F(a_n)|^2\right\rangle=\frac{1}{2\pi^2}\left[\frac{\beta^2+(1/4)^2-(\beta/2)\sin(2\pi\beta)}{\left(\beta^2-(1/4)^2\right)^2}\right] \quad :\text{MSK } h=\frac{1}{2} \tag{5.45}$$

$$\left\langle|F(a_n)|^2\right\rangle=\frac{\cos^2(\pi\beta)}{\pi^2}\left[\frac{\beta^2+(1/2)^2}{\left(\beta^2-(1/2)^2\right)^2}\right] \quad :\text{BFSK } h=1.0 \tag{5.46}$$

Similarly, the squared average of $F(a_n)$ is evaluated as

$$\begin{aligned}\langle F(a_n)\rangle^2=\frac{1}{4}\left\{\frac{\sin(\pi(\beta-h/2))}{\pi(\beta-h/2)}e^{-j\pi(\beta-h/2)}\right.\\\left.+\frac{\sin(\pi(\beta+h/2))}{\pi(\beta+h/2)}e^{-j\pi(\beta+h/2)}\right\}^2\end{aligned} \tag{5.47}$$

The expression in (5.47) does not apply when delta functions appear in the spectrum as in the case of BFSK with h = 1.0; however, for MSK modulation (5.47) is expressed as

$$\langle F(a_n)\rangle^2=\frac{1}{64\pi^2}\left[\frac{\{\cos(2\pi\beta)+j(4\beta-\sin(2\pi\beta))\}^2}{\left(\beta^2-(1/4)^2\right)^2}\right] \quad :\text{MSK } h=1/2 \tag{5.48}$$

Finely, the squared magnitude of the average of $F(a_n)$ is evaluated as

$$|\langle F(a_n)\rangle|^2 = \left|\left\langle \frac{\sin(\pi(\beta-a_nh/2))}{\pi(\beta-a_nh/2)} e^{-j\pi(\beta-a_nh/2)} \right\rangle\right|^2$$

$$= \frac{1}{4}\left|\frac{\sin(\pi(\beta-h/2))}{\pi(\beta-h/2)} e^{-j\pi(\beta-h/2)} + \frac{\sin(\pi(\beta+h/2))}{\pi(\beta+h/2)} e^{-j\pi(\beta+h/2)}\right|^2 \quad (5.49)$$

Equation (5.49) only applies for BFSK with $h = 1$, that is, when impulse functions explicitly appear in the spectral density, and (5.49) is specialized for this case with the result

$$|\langle F(a_n)\rangle|^2 = \frac{\beta^2\cos^2(\pi\beta)}{\pi^2\left(\beta^2-1/4\right)^2} \quad : \text{BFSK } h=1 \quad (5.50)$$

Before evaluating these results for arbitrary modulation indices, it is informative to substitute the various expressions into the normalized spectral density function, $S(\beta)/A^2T$, for the special cases of MSK and orthogonal BFSK modulations.

5.6.1 MSK Power Spectral Density ($h = 1/2$)

For MSK, $C_a(\omega_d T) = C_a(\pi h) = \cos(\pi h) = 0$ and the normalized spectral density is expressed as

$$\frac{S(\beta)}{A^2T} = \frac{1}{2}\left\langle |F(a_n)|^2 \right\rangle + \text{Re}\left\{ \langle F(a_n)\rangle^2 \right\} \quad (5.51)$$

Substituting (5.45) and (5.48) for MSK into the right-hand-side (*rhs*) of (5.51) gives

$$\frac{S(\beta)}{A^2T} = \frac{1}{4\pi^2}\left[\frac{\beta^2+(1/4)^2-(\beta/2)\sin(2\pi\beta)}{\left[\beta^2-(1/4)^2\right]^2}\right] + \frac{1}{64\pi^2}\left[\frac{\cos^2(2\pi\beta)-(4\beta-\sin(2\pi\beta))^2}{\left[\beta^2-(1/4)^2\right]^2}\right] \quad (5.52)$$

and, after some manipulation, (5.52) can be expressed in the form (see Equation 4.101),

$$\frac{S(\beta)}{A^2T} = \frac{8}{\pi^2}\frac{\cos^2(2\pi\beta)}{\left(1-16\beta^2\right)^2} \quad : \text{MSK } h=\frac{1}{2} \quad (5.53)$$

The peak value of the MSK spectral density function is evaluated at $\beta=0$ with $S(0)=0.81057A^2T$ and the value at the MSK tone locations $\beta=\pm 1/4$ is $S(\pm 0.25)=0.5A^2T$.

5.6.2 Orthogonal Binary FSK Power Spectral Density ($h = 1$)

For orthogonal BFSK, $C_a(\omega_d T) = C_a(\pi) = \cos(\pi) = 1$ and the normalized spectral density is expressed as

$$\frac{S(\beta)}{A^2T} = \frac{1}{2}\left\langle |F(a_n)|^2 \right\rangle - \frac{1}{2}|\langle F(a_n)\rangle|^2 + \frac{1}{2}\cos(\lambda)|\langle F(a_n)\rangle|^2 \sum_{n=1}^{2}\delta(\lambda-2\pi n) \quad (5.54)$$

Substituting (5.46) and (5.50) for BFSK into the *rhs* of (5.54) and combining terms gives

$$\frac{S(\beta)}{A^2T} = \frac{\cos^2(\pi\beta)}{8\pi^2\left(\beta^2-(1/2)^2\right)^2} + \frac{\beta^2\cos^2(\pi\beta)\cos(\lambda)}{2\pi^2\left(\beta^2-(1/2)^2\right)^2}[\delta(\lambda)+\delta(\lambda-2\pi)] \quad (5.55)$$

From (5.41) $\lambda = -2\pi(\beta - bh/2)$ and choosing unity for the arbitrary constant b with $h=1$ results in

$$\cos(\lambda)[\delta(\lambda)+\delta(\lambda-2\pi)] = \cos(2\pi(\beta-1/2))[\delta(\beta-1/2)+\delta(\beta+1/2)] = \delta(\beta-1/2)+\delta(\beta+1/2) \quad (5.56)$$

and substituting this result into (5.55) gives

$$\frac{S(\beta)}{A^2T} = \frac{\cos^2(\pi\beta)}{8\pi^2\left(\beta^2-(1/2)^2\right)^2} + \frac{\beta^2\cos^2(\pi\beta)}{2\pi^2\left(\beta^2-(1/2)^2\right)^2}[\delta(\beta+1/2)+\delta(\beta-1/2)] \quad (5.57)$$

The weight on the delta functions at $\beta = \pm 1/2$ is evaluated as 1/8 so the spectral density for orthogonal BFSK becomes

$$\frac{S(\beta)}{A^2T} = \frac{\cos^2(\pi\beta)}{8\pi^2\left(\beta^2-(1/2)^2\right)^2} + \frac{1}{8}[\delta(\beta+1/2)+\delta(\beta-1/2)] \quad : \text{BFSK } h=1 \quad (5.58)$$

The area under the one-sided normalized spectrum is 0.25 so the energy at each of the BFSK tone frequencies represents one-half of the total signal energy.

Returning to the generalized case for arbitrary modulation indices, plots of the normalized spectral densities are shown in Figure 5.25 for $h = 0.5$ (MSK), 1.0 (orthogonal BFSK), and 1.2; the negative frequency portion of the spectrum is mirrored about $\beta = 0$. The energy in the MSK spectrum is much more concentrated about the central lobe with distinct nulls at $\pm(0.75 + 0.5m)T_b$, $m = 0, 1, \ldots$. For modulation indices close to and exceeding unity, the continuous portion of the central lobe extends to $\beta = 1.5$ and the energy tends to concentrate about $\beta = 0.5$. Figure 5.26 shows similar results for $h = 0.5$, 0.7, 0.8, 0.9, 1.0, 1.1, and 1.2 over a wider range of frequencies. Table 5.2 summarizes the computed one-sided area and the zero-frequency value for the various modulation indices considered in Figure 5.26.

The PSD of the BFSK-modulated waveform is shown in Figure 5.27 for a tone spacing of eight times the minimum orthogonal spacing. Large BFSK frequency tone separations, that is, greater than the minimum orthogonal spacing, are typically used in applications involving unknown frequencies with simplified demodulators that do not use frequency tracking loops or even a preamble in some cases; the demodulator performance in these situations is discussed in Section 5.5. When the modulation index is integrally related to the data rate, the FSK tones appear as delta functions about the carrier frequency with an underlying continuous spectrum. However, as indicated in Figure 5.28, these delta-function tones are very sensitive to the modulation index with a significant loss in energy for changes in the modulation index $|\Delta h| > 0.01$. Furthermore, the peak of the spectral energy is mirrored about h for $|\Delta h| \leq 0.5$ approximating a folded $sinc(fT)$ function at $\Delta h = \pm 0.5$ as indicated by the *dash-dot-dot* curves.

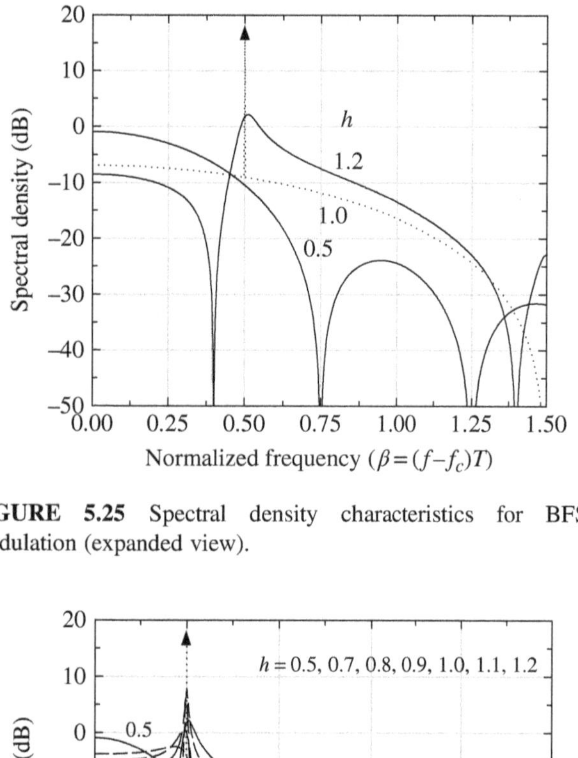

FIGURE 5.25 Spectral density characteristics for BFSK modulation (expanded view).

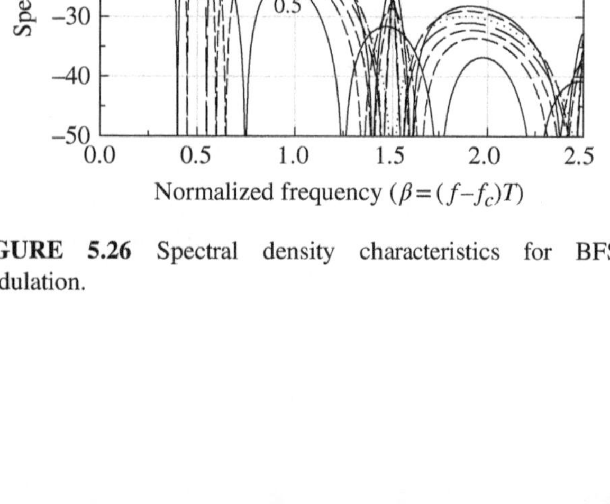

FIGURE 5.26 Spectral density characteristics for BFSK modulation.

5.6.3 Discontinuous FSK Phase Modulation

The preceding section examined the spectrum for continuous phase modulation (CPM) FSK, that is, the phase transitions between the FSK tones that occur at or during* a change in the source data are continuous. In this case, the appearance of delta functions in the spectrum occurs only with integer-valued modulation indices, that is, when $2\Delta f = kR_b$ where k is an integer. Under this condition the mark and space data tones are continuous resulting in the delta functions in the modulated spectrum. With this understanding, if the FSK modulator generates continuous mark and space tone and the modulator simply chooses one or the other tone depending on the source data, then the delta functions will always appear in the spectrum regardless of the modulation index. This situation results in abrupt phase changes or discontinuous phase transitions when the data are changed. For example, Figure 5.29a shows the continuous *mark* and *space* frequency tones for a 2.5-bit segment of data for the normalized tones given by $\Delta fT = 2.105$ corresponding to the non-integer modulation index $h = 4.21$. These tones are generated in the modulator and characterized as baseband tones at $\pm\Delta f$ about zero frequency with mark data associated with the positive frequency tone. Figure 5.29b shows the BFSK modulator output for a 1, 0, 1 or *mark*, *space*, *mark* segment of binary data. The abrupt phase changes corresponding to the data transitions are evident resulting in discontinuous phase modulation.

The PSD for discontinuous phase modulation as described above is shown in Figure 5.30 for modulation indices of 8 and 8.5 corresponding to $\Delta fT = 4$ and 4.25 respectively. The appearance of the delta function in Figure 5.30b is a direct result of the constant *mark* and *space* tones used

*Typically, the phase/frequency change is instantaneous; however, the frequency change is often controlled over a finite interval to improve the spectral efficiency.

TABLE 5.2 Computed Simulation Values

	Modulation Index						
Parameter	0.5	0.7	0.8	0.9	1.0	1.1	1.2
One-sided area	0.250	0.250	0.250	0.250	0.250	0.250	0.250
$S(0)/A^2T$	0.811	0.414	0.317	0.250	0.203	0.167	0.141

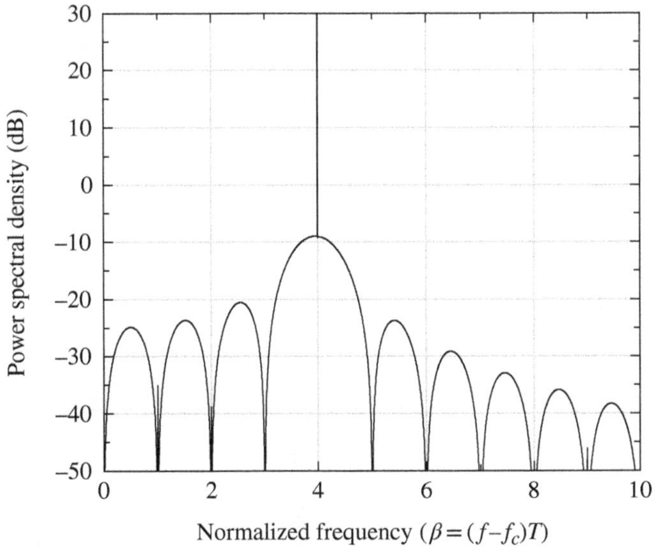

FIGURE 5.27 BFSK power spectral density for modulation index $h = 8.0$.

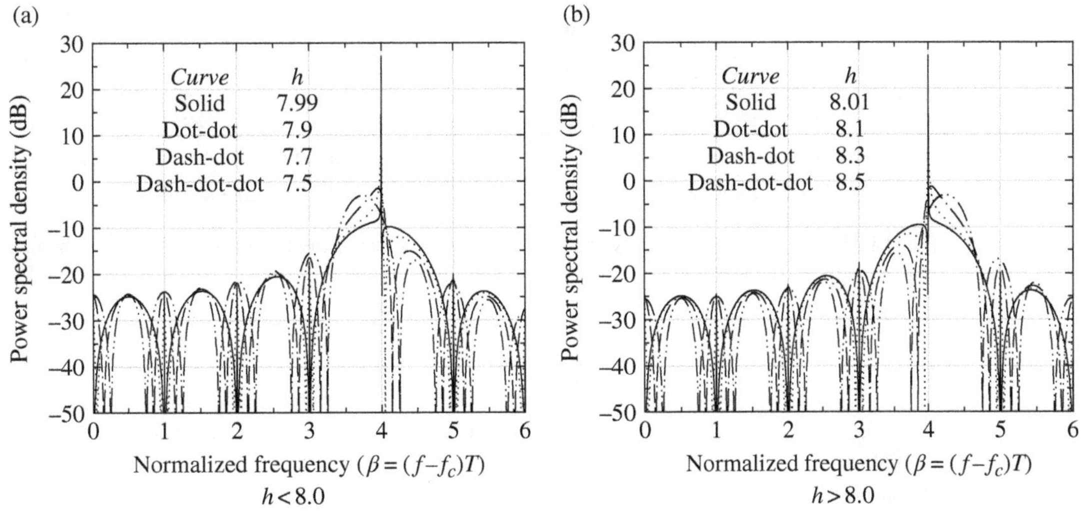

FIGURE 5.28 BFSK power spectral density for modulation indices around $h = 8.0$.

in the modulator and the resulting discontinuous modulation phase with changes in the data.

When the modulation index is an integer, as in Figure 5.30a, there are no modulation phase discontinuities and the spectral efficiency is improved somewhat over the non-integer case shown in Figure 5.30b. In either case, the demodulator can use first- or second-order phaselock loops to achieve phase lock and track the received modulation tones to realize, respectively, either noncoherent or coherent FSK bit-error detection performance. For BFSK modulation, the bandwidth is defined in terms of the *necessary bandwidth* as defined in Table 4.6.

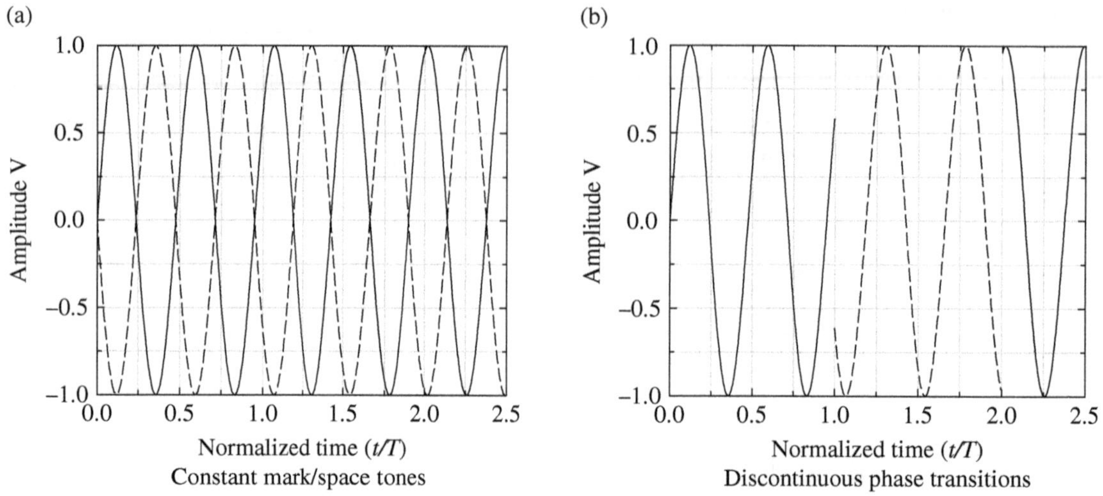

FIGURE 5.29 BFSK discontinuous phase modulation ($h = 4.21$, $R_b = 1$ kbps).

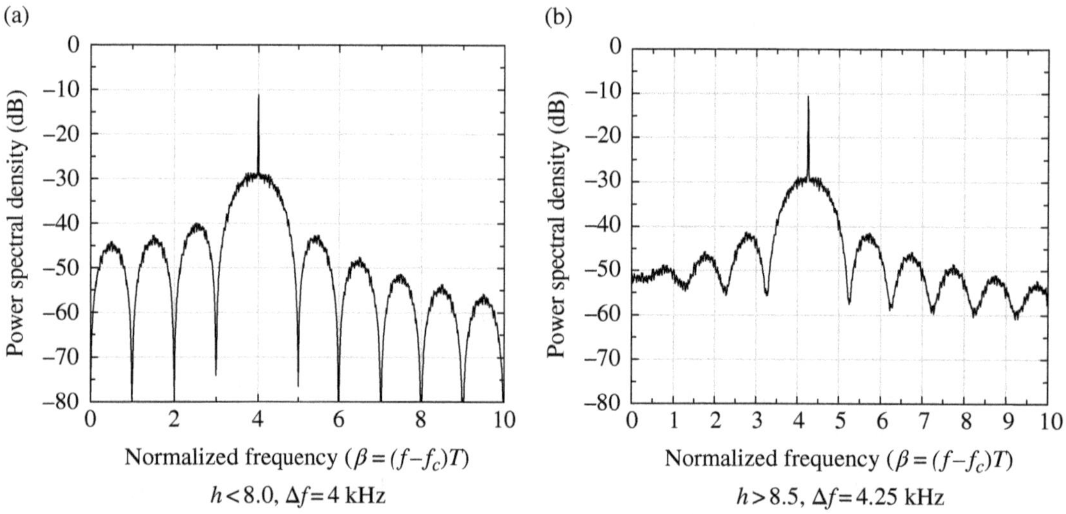

FIGURE 5.30 BFSK power spectral density with discontinuous phase at the bit transitions ($R_b = 1$ kbps).

ACRONYMS

AM	Amplitude modulation
AWGN	Additive white Gaussian noise
BFSK	Binary frequency shift keying
BPF	Bandpass filter
BPSK	Binary phase shift keying
CBFSK	Coherent BFSK
CFSK	Coherent FSK
CPM	Continuous phase modulation
CW	Continuous wave
E/L	Early–late (sampling)
FSK	Frequency shift keying
IF	Intermediate frequency
LNPA	Low-noise power amplifier
LPF	Lowpass filter
MSK	Minimum shift keying
NCBFSK	Noncoherent BFSK (detection)
PA	Power amplifier
PLL	Phaselock loop
PSD	Power spectral density
PSK	Phase shift keying
RF	Radio frequency (carrier)
RHS	right-hand side
S-PSK	Phase-shaped PSK
VLF	Very low frequency

PROBLEMS

1. Referring to the decision variable y in (5.7), show that the mean and variance of the random variable y are A and $2\sigma_n^2$ respectively. With the FSK signal-to-noise ratio $\gamma_b = E_b/N_o$ show that the bit-error probability of coherently detected BFSK is expressed in terms of the error function as given in (5.8).

2. Show that the integral of the Rayleigh *pdf* expressed in (5.9) integrated over the range x_2 to ∞ is equal to $e^{-x_2^2/2\sigma_n^2}$. Using this result determine the solution of the Q-function $Q(0, b)$. Show that the Q-functions $Q(a, 0) = Q(0, 0) = 1$.

3. Compute the loss in signal-to-noise ratio for the signal $s(t) = A rect(t/T_b)\cos(2\pi f_e t)$ in a matched filter detector corresponding to $f_e = 0$. A is the signal amplitude, T_b is the bit duration, and f_e is the frequency error relative to the matched filter. Express the loss in decibels and compare the result with the loss in Figure 5.10.

4. By deriving the output response $s_o(t)$ using the quadrature inputs $s_c(t) = \cos(\omega t + \varphi)$ and $s_s(t) = \sin(\omega t + \varphi)$ show that the processing in the following figure corresponds to a frequency discriminator.

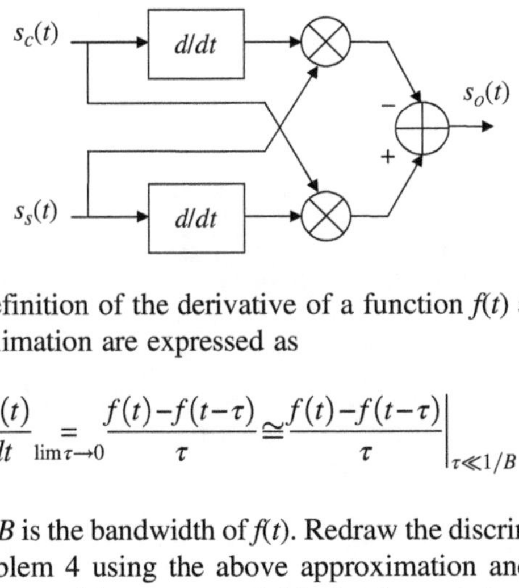

5. The definition of the derivative of a function $f(t)$ and the approximation are expressed as

$$\frac{df(t)}{dt} \underset{\lim \tau\to 0}{=} \frac{f(t)-f(t-\tau)}{\tau} \cong \left.\frac{f(t)-f(t-\tau)}{\tau}\right|_{\tau \ll 1/B}$$

where B is the bandwidth of $f(t)$. Redraw the discriminator in Problem 4 using the above approximation and show that the resulting output of this implementation forms the frequency discriminator function.

REFERENCES

1. A.B. Glenn, "Analysis of Noncoherent FSK System with Large Ratios of Frequency Uncertainties to Information Rates, Part I—Binary Systems," RCA Review, Vol. XXVII, No. 2, pp. 272–303, June 1966.
2. J.I. Marcum, "A Statistical Theory of Target Detection by Pulsed Radar," RAND Research Memo. RM-754, pp. 59–267 (p. 174), The RAND Corporation, Santa Monica, CA, December 1, 1947. Reprinted as a Special Monograph Issue, J.I. Marcum, "Studies of Target Detection by Pulsed Radar," IRE Transactions on Information Theory, Vol. IT-6, No. 2, pp. 59–267, April 1960.
3. M. Abramowitz, I.A. Stegun, Editors, *Handbook of Mathematical Functions With Formulas, Graphs, and Mathematical Tables*, National Bureau of Standards, Applied Mathematics Series 55, Chapter 15, F. Oberhettinger, Editor, "Hypergeometric Functions," U.S. Government Printing Office, Washington, DC, June 1964.
4. S.O. Rice, *Mathematical Analysis of Random Noise*, Selected Papers on Noise and Stochastic Processes by N. Wax, Dover Publication, Mineola, NY, 1945.
5. S. Stein, "Unified Analysis of Certain Coherent and Noncoherent Binary Communications Systems," IEEE Transactions on Information Theory, Vol. II, pp. 43–51, January 1964.
6. M. Schwartz, W.R. Bennett, S. Stein, *Communicator Systems and Techniques*, McGraw Hill, New York, 1966.
7. J.I. Marcum, P. Swerling, "Studies of Target Detection by Pulsed Radar," IRE Transactions on Information Theory, Vol. IT-6, No. 2, pp. 59–308, April 1960.
8. J.I. Marcum, "Table of Q Functions," RAND Research Memorandum, RM-339, Appendix A, Rand Corp., Santa Monica, CA, January 1, 1950.
9. D.E. Johansen, "New Techniques for Machine Computation of the Q-Function, Truncated Normal Deviates and Matrix Eigenvalues," Applied Research Laboratory, Sylvania Electronic Systems, Scientific Report No. 2, United States Air Force Contract AF19(604)-7237, July 1961.
10. R.R. Anderson, J. Salz, "Spectra of Digital FM," The Bell System Technical Journal, Vol. 44, No. 6, pp. 1165–1189, July–August 1965.
11. J. Salz, "Spectral Density Function of Multilevel Continuous-Phase FM," IEEE Transactions on Information Theory, Vol. 11, No. 3, pp. 429–433, July 1965.
12. R.W. Lucky, J. Salz, E.J. Weldon Jr., *Principles of Data Communication*, Chapter 8, McGraw Hill, New York, 1968.

6

AMPLITUDE SHIFT KEYING MODULATION, DEMODULATION, AND PERFORMANCE

6.1 INTRODUCTION

This chapter discusses communication waveforms involving various forms of amplitude shift keying (ASK). As used here ASK is a general term that applies to the modulation of a carrier signal with discrete amplitudes that uniquely identify a symbol of binary data or bits. When the amplitude keying is expressed in terms of $\alpha_m = 2m$: $m = 0, 1$, and applied to one quadrature component of the carrier frequency, the modulation is referred to as on–off keying (OOK) and when $\alpha_m = -1 + 2m$: $m = 0, 1$, it is referred to as antipodal ASK or, when applied to a radio frequency (RF) carrier, OOK is the same as binary phase shift keying (BPSK). The more general case of ASK applied to one quadrature component occurs when $\alpha_m = -(M-1) + 2m$: $m = 0, 1, \ldots, M-1$ and this is referred to as pulse amplitude modulation (PAM) or, more specifically, M-ary PAM. When PAM is applied to each quadrature component, the modulation is referred to as quadrature amplitude modulation (QAM) or M-ary QAM. Based on this description, it is evident that BPSK is a subset of PAM modulation and quadrature phase shift keying (QPSK) is a subset of QAM modulation. QAM necessarily involves a phase-modulated carrier with the resulting signal rest-points at the intersection of the quadrature amplitude terms α_{Im} and α_{Qm}, where I and Q are the inphase and quadrature carrier components. The term *rest-points* refer to the ideal noise-free optimum signal samples at the output of the demodulator inphase and quadrature (I/Q) matched filters. Other variations of ASK are derived by not restricting the rest-points to form a rectangle as in QAM. For example, modulated signals with rest-points that form a regular polygon, or that lie on concentric circles, are sometimes referred to as amplitude PSK (APSK).

The selection of the rest-points impacts the average power required to achieve a given error probability, and minimizing the transmitter power to achieve the performance is a major consideration in the waveform design and selection. Another major consideration, that is dependent on the rest-point selection, is the peak-to-average power ratio. The peak-to-average power ratio influences the amount of power backoff required at the input to the transmitter power amplifier (PA) to avoid waveform distortion and adjacent channel interference. Although ASK-modulated waveforms have a relatively large peak-to-average power ratio, it can be minimized by judicious selection of the rest-points. The spectral efficiency of the transmitted ASK waveform is also a consideration and the M-ary QAM waveform has a distinct advantage as M increases; however, the spectral efficiency and bit-error performance can be lost if the PA is not operated in the linear range. These waveforms and related topics are discussed in the remainder of this chapter that ends with a case study of 16-ary QAM.

6.2 AMPLITUDE SHIFT KEYING (ASK)

In this section, a simple form of ASK modulation is considered that amplitude modulates a carrier based on a direct mapping of the source data bits to the waveform symbol. In this case, the transmitted signal is expressed as

Digital Communications with Emphasis on Data Modems: Theory, Analysis, Design, Simulation, Testing, and Applications, First Edition. Richard W. Middlestead.

Companion website: www.wiley.com/go/digitalcommunications

$$
\begin{aligned}
s_m(t) &= A\alpha_m p(t)\cos(\omega_c t) \\
&= \mathrm{Re}\{\tilde{s}_m(t)e^{j\omega_c t}\} \qquad : |t| \le T/2
\end{aligned} \tag{6.1}
$$

where α_m is the data-dependent scaling factor applied to the signal amplitude A, T is the symbol duration, and $p(t)$ is defined as a unit-power amplitude shaping function satisfying the relationship

$$
\frac{1}{T}\int_{-T/2}^{T/2} p^2(t)dt = 1 \tag{6.2}
$$

This form of ASK modulation is referred as PAM. The shaping function is used to control the transmitted signal spectrum, however, in most applications it is characterized as the unit amplitude rectangular function given by $rect(t/T)$: $|t| \le T/2$ that results in the $sinc(fT)$ frequency spectrum; this is assumed throughout the remainder of this section. The analytic or baseband-modulated waveform is given by

$$
\tilde{s}_m(t) = A\alpha_m p(t) \tag{6.3}
$$

In general, the ASK-modulated waveform is not very efficient in terms of the E_b/N_o required to obtain a specified bit-error result. One notable exception, however, is the situation when $\alpha_m = \{1,-1\}$ that results in antipodal signaling and, with coherent detection, results in the performance of coherently detected BPSK. An understanding of the analysis and performance of PAM has direct application to the more efficient QAM waveform. The relative inefficiency of ASK arises because only one dimension of the orthogonal I/Q signal space is used, whereas QAM uses both dimensions resulting in a theoretical 2 : 1 improvement in the spectrum utilization.

The following analysis considers two special cases of PAM: binary PAM involving 1 bit/symbol and M-ary PAM that uses $k = \log_2(M)$ bits/symbol. To the extent that the modulated waveform is operating in the band-limited region of the Shannon capacity curve, the channel and transmitter amplifiers are considered to be linear and no penalty is incurred for the peak-to-average power ratio. The average power is computed as

$$
P_{avg} = \frac{A^2}{2}\sum_{m=0}^{M-1}\alpha_m^2 P(m) \tag{6.4}
$$

where $P(m)$ is the a priori probability of the signal state α_m. Normally the states α_m are equally likely so that $P(m) = 1/M$, a condition that is assumed throughout this section. The peak signal power is computed as

$$
P_p = A^2 \max_m \left(\alpha_m^2\right) \tag{6.5}
$$

The square of the minimum distance is defined as

$$
d^2 = \min_{m \ne n} \int_{-T/2}^{T/2} |s_m(t) - s_n(t)|^2 dt \tag{6.6}
$$

and, upon substituting the equivalent baseband signal expression into (6.6), d^2 is expressed as

$$
\tilde{d}^2 = \frac{1}{2}\min_{m \ne n}\left\{\int_{-T/2}^{T/2} |\tilde{s}_m(t) - \tilde{s}_n(t)|^2 (1 + \cos(2\omega_c t))dt\right\} \tag{6.7}
$$

When $f_c T \gg 1$, and in consideration of low-pass filtering, the term involving $2\omega_c$ can be neglected with little effect on the result. For example, even without lowpass filtering, when $p(t) = rect(t/T - 1/2)$ the additive term resulting from the integration is given by $Tsinc(2f_c T)$ which decreases in proportion to $1/f_c T$. Therefore, upon eliminating the $2\omega_c$ term (6.7) becomes

$$
\begin{aligned}
\tilde{d}^2 &= \frac{1}{2}\min_{m \ne n}\left\{\int_{-T/2}^{T/2} |\tilde{s}_m(t) - \tilde{s}_n(t)|^2 dt\right\} \\
&= \frac{1}{2}d^2
\end{aligned} \tag{6.8}
$$

where $\tilde{d}^2$ is the minimum squared decision distance of the baseband signal. Based on the previous definitions, d^2 is evaluated as

$$
\begin{aligned}
d^2 &= A^2 T \min_{m \ne n}\left\{|\alpha_m - \alpha_n|^2\right\} \\
&= 4E_b
\end{aligned} \tag{6.9}
$$

In the following analysis of OOK modulation, the normalized minimum distance is $d/\sqrt{E_b} = 1$; however, for antipodal binary PAM, and multilevel PAM and QAM, the normalized minimum distance is equal to 2.

6.2.1 On–Off Keying (OOK) Modulation

The most rudimentary form of ASK is given the special name OOK and is characterized by (6.1) with $\alpha_m = b = \{0,1\}$ where b is the source data and $p(t) = rect(t/T - mT)$ corresponding to $p(t) = 1$ for all t. In this evaluation, the received signal and noise is considered to be

$$
s_r(t) = s_{mr}(t) + n(t) \tag{6.10}
$$

The noise is characterized as white Gaussian noise with spectral density N_o and the received signal is expressed as

$$\begin{aligned} s_{mr}(t) &= A\alpha_m \cos((\omega_c + \Delta\omega)t + \phi) \\ &= \mathrm{Re}\{\tilde{s}_{mr}(t)e^{j\omega_c}\} \end{aligned} \quad (6.11)$$

where $\Delta\omega$ is the received carrier angular frequency error and ϕ is an arbitrary phase shift. The baseband signal in (6.11) is expressed as

$$\tilde{s}_{mr}(t) = A\alpha_m e^{j(\Delta\omega + \phi)} \quad (6.12)$$

6.2.1.1 *Coherent Detection of OOK Modulation* A coherent demodulator for the received OOK-modulated signal with received phase error $\phi(t) = \Delta\omega(t) + \phi$ is shown in Figure 6.1. The processing details, including the sampling requirements before the matched filter, are not shown; however, the details for the phaselock loop (PLL) processing are discussed in Chapter 10. With coherent detection, the received carrier frequency error and phase are removed by the PLL with $\hat{\phi}(t) = \phi(t)$, so the baseband signal into the matched filter simplifies to

$$\tilde{s}_{mr}(t) = A\alpha_m \quad (6.13)$$

With random data, the average and peak powers of OOK modulation are $A^2/4$ and A^2, respectively, and the square of the minimum distance is computed as

$$\begin{aligned} d^2 &= \int_{-T/2}^{T/2} |s_{or}(t) - s_{1r}(t)|^2 dt \\ &= A^2T/2 = E_b \end{aligned} \quad (6.14)$$

This result implicitly assumes a zonal filter has removed the double frequency term. The energy-per-symbol is defined as the average power times the symbol or bit duration so, in this case, $E_b = A^2T/2$.

For equal a priori probabilities of the mark and space data, the optimum decision threshold, based on the signal energy, is $T_{he} = d/2 = \sqrt{E_b}/2$. The relationship between signal energy and amplitude is $A = \sqrt{E_b(2/T)}$, so, in terms of the signal amplitude, the optimum decision threshold is $T_h = A/2$. In practice, the decision threshold is $T_h = \hat{A}/2$, where $\hat{A}$ is the estimate of the received signal amplitude A as determined using an automatic gain control (AGC) or signal level estimation processing; the estimation processing can be aided by using a CW preamble.

Based on the minimum distance, given by (6.14), and the optimum decision threshold $T_h = A/2$, the bit-error probability for coherent detection of OOK is evaluated as

$$P_{be} = Q\left(\sqrt{\frac{E_b}{2N_o}}\right) = \left(\frac{1}{2}\right) erfc\left(\sqrt{\frac{E_b}{4N_o}}\right) \quad (6.15)$$

This bit-error performance is 6 dB worse than that of BPSK and any variations in the estimated signal level from A will further degrade the performance. The bit-error performance of coherently detected OOK, using the optimum threshold $A/2$, is shown as the *solid* curve in Figure 6.3 and the *circled* data point represents the Monte Carlo simulated performance using 1 million bits for each signal-to-noise ratio.

The performance analysis using the minimum distance is simple and direct; however, it is instructive to examine the performance using the known Gaussian probability density functions for noise only and for signal plus noise as depicted in Figure 6.2. This analysis is fairly straightforward and serves as an introduction to the more involved analysis involving noncoherent detection of OOK.

The bit-error probabilities indicated in Figure 6.2 are evaluated as,

$$P_b(\text{error}|\alpha_1) = \frac{1}{\sqrt{2\pi}\sigma_n} \int_0^{T_h} e^{-(x-A)^2/2\sigma_n^2} dx \quad (6.16)$$

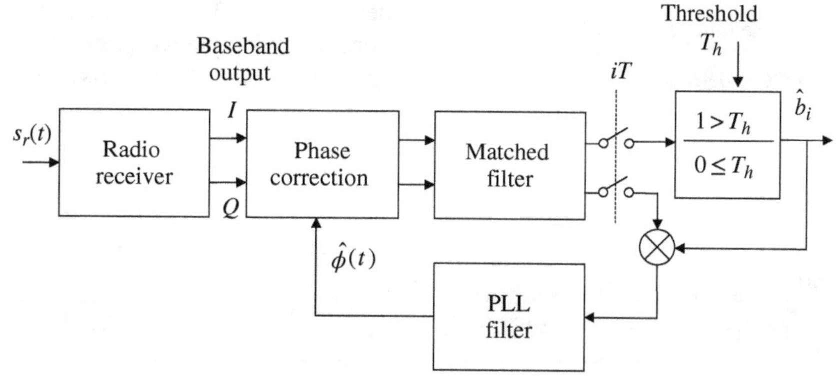

FIGURE 6.1 Implementation of coherent OOK demodulator.

and

$$P_b(\text{error}|\alpha_0) = \frac{1}{\sqrt{2\pi}\sigma_n}\int_{T_h}^{\infty} e^{-x^2/2\sigma_n^2}dx \tag{6.17}$$

Considering equal a priori mark and space source data, such that, $P(\alpha_1) = P(\alpha_0) = 1/2$ and, using the transformation $\lambda = (x-A)/\sqrt{2}\sigma_n$ in (6.16) and $\lambda = x/\sqrt{2}\sigma_n$ in (6.17), the overall bit-error probability is evaluated as

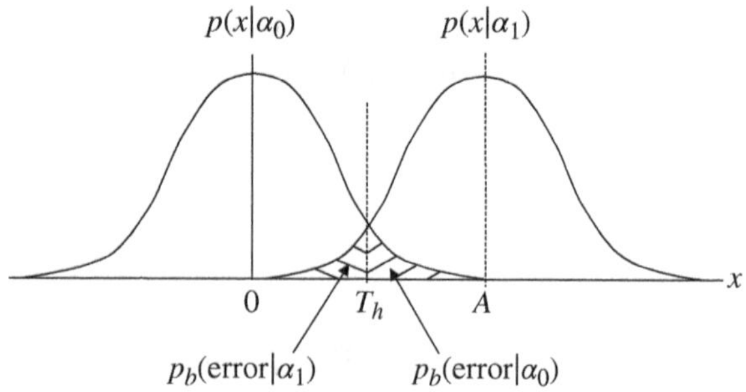

FIGURE 6.2 Mark/space Gaussian *pdf* for coherent detection of OOK.

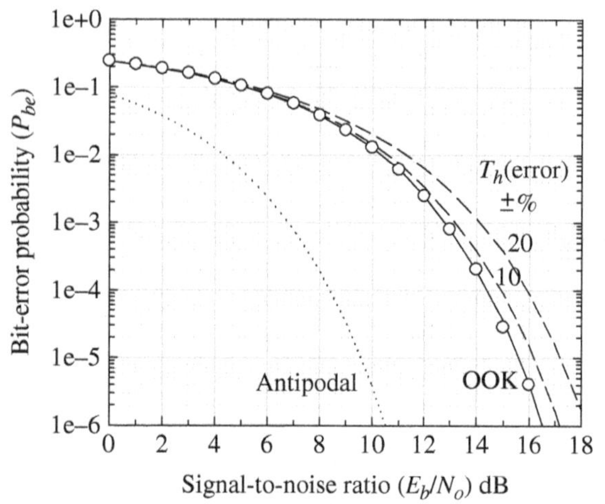

FIGURE 6.3 Coherent OOK detection performance.

$$\begin{aligned} P_{be}(T_h) &= P_b(\text{error}|\alpha_1)P(\alpha_1) + P_b(\text{error}|\alpha_0)P(\alpha_0) \\ &= \frac{1}{2}\left[1 - \frac{1}{\sqrt{\pi}}\int_{\frac{(T_h-A)}{\sqrt{2}\sigma_n}}^{\infty} e^{-\lambda^2}d\lambda + \frac{1}{\sqrt{\pi}}\int_{\frac{T_h}{\sqrt{2}\sigma_n}}^{\infty} e^{-\lambda^2}d\lambda\right] \\ &= \frac{1}{2} - \frac{1}{4}erfc\left(\frac{(T_h-A)}{\sqrt{2}\sigma_n}\right) + \frac{1}{4}erfc\left(\frac{T_h}{\sqrt{2}\sigma_n}\right) \end{aligned} \tag{6.18}$$

By differentiating $P_{be}(T_h)$ with respect to T_h and setting the result equal to zero, the optimum threshold, T_{ho}, corresponding to the minimum bit-error probability is found be $T_{ho} = A/2$ and the corresponding minimum bit-error probability is

$$\begin{aligned} P_{be} &= P_{be}(T_{ho}) \\ &= \frac{1}{2}erfc\left(\frac{\sqrt{\gamma_b}}{2}\right) = Q\left(\sqrt{\frac{\gamma_b}{2}}\right) \end{aligned} \tag{6.19}$$

This result is identical to (6.15) with $\gamma_b = E_b/N_o$ and, as stated above, requires a signal-to-noise ratio that is 6 dB higher to achieve the same bit-error performance as antipodal BPSK signaling. The *solid* curve in Figure 6.3 represents the optimum OOK performance using the threshold $A/2$ and the *circled* data point represents the Monte Carlo simulated performance using 1 million bits for each signal-to-noise ratio. The *dotted* curve is the performance of antipodal signaling and indicates that the performance of OOK requires a 6 dB higher E_b/N_o to achieve the same bit-error performance. The *dashed* curves show the performance sensitivity of OOK with a non-optimum threshold expressed as a percent of the $A/2$. For example, if the received signal amplitude estimation error is 10% of the true amplitude, the performance will be degraded by about 0.6 dB at $P_{be} = 10^{-6}$.

6.2.1.2 *Noncoherent Detection of OOK Modulation*

Noncoherent detection of OOK is depicted in Figure 6.4.

The evaluation of the performance for the noncoherent detection of OOK focuses on the probability density functions involving the reception of mark and space data corresponding to signal plus noise and noise only, respectively.

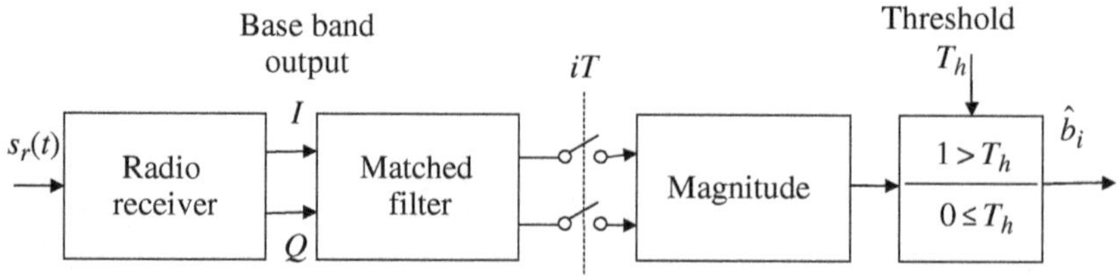

FIGURE 6.4 Noncoherent OOK detection.

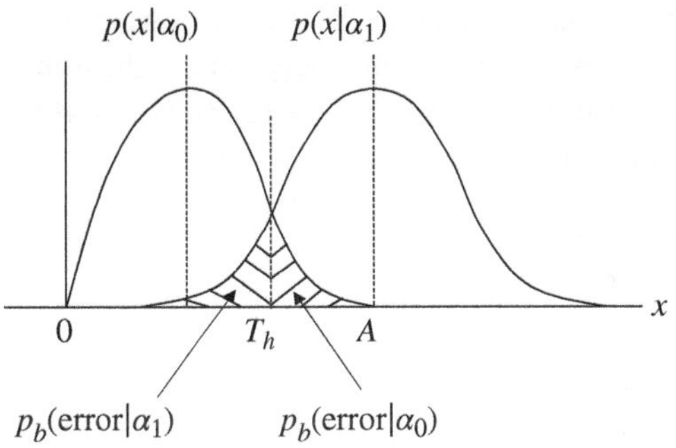

FIGURE 6.5 Mark/space distribution functions for noncoherent detection of OOK.

The respective *pdfs* are characterized by the Ricean *pdf* expressed as

$$p(x|\alpha_1) = \frac{x}{\sigma_n^2} e^{-\frac{x^2+A^2}{2\sigma_n^2}} I_o\left(\frac{xA}{\sigma_n^2}\right) \quad : x \ge 0, \text{ Ricean } pdf \tag{6.20}$$

and the Rayleigh *pdf* expressed as

$$p(x|\alpha_0) = \frac{x}{\sigma_n^2} e^{-\frac{x^2}{2\sigma_n^2}} \quad : x \ge 0, \text{ Rayleigh } pdf \tag{6.21}$$

These distribution functions are shown in Figure 6.5 with the corresponding regions of the conditional-error events.

The bit-error probability is evaluated using equal a priori mark and space data probabilities in a manner similar to that in (6.18) for the coherent OOK and, upon using (6.20) and (6.21), the expression for the threshold-dependent bit-error probability is expressed as

$$\begin{aligned} P_{be}(T_h) &= P_b(\text{error}|\alpha_1)P(\alpha_1) + P_b(\text{error}|\alpha_0)P(\alpha_0) \\ &= \frac{1}{2}\left[1 - \int_{T_h}^{\infty} \frac{x}{\sigma_n^2} e^{-\frac{x^2+A^2}{2\sigma_n^2}} I_o\left(\frac{xA}{\sigma_n^2}\right) dx + \int_{T_h}^{\infty} \frac{x}{\sigma_n^2} e^{-\frac{x^2}{2\sigma_n^2}} dx\right] \end{aligned} \tag{6.22}$$

The optimum threshold (see Problem 4) is found to be the solution to the transcendental equation

$$I_o\left(\frac{AT_{ho}}{\sigma_n^2}\right) = e^{A^2/2\sigma_n^2} \tag{6.23}$$

By defining the normalized threshold as $\rho = T_{ho}/A$ and the signal-to-noise ratio as $\gamma = A^2/2\sigma_n^2$, (6.23) is expressed as

$$I_o(2\rho\gamma) = e^{\gamma} \tag{6.24}$$

The solution to (6.24) for the optimum normalized threshold is evaluated using Newton's method and the result is plotted in Figure 6.6 as a function of the signal-to-noise ratio as

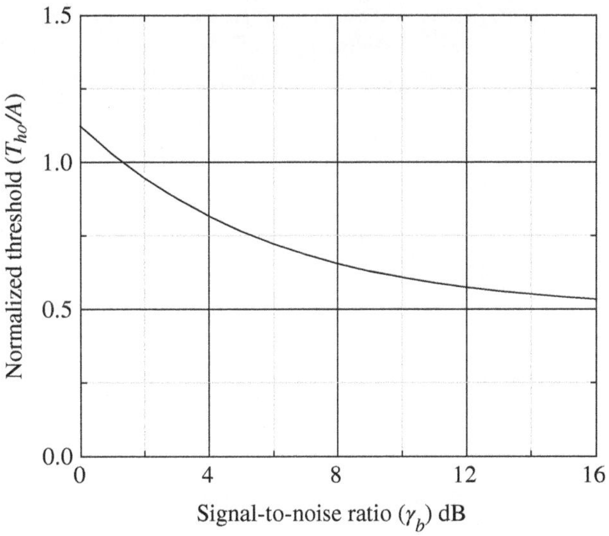

FIGURE 6.6 Optimum threshold for noncoherent OOK detection.

measured in the bandwidth of the data rate. As the noise level becomes smaller the optimum threshold approaches *A*/2 which is the optimum threshold for the coherent detection case. The optimum threshold is approximated by the third-order polynomial

$$\frac{T_{ho}}{A} \cong -0.0000585\gamma_{db}^3 + 0.0038\gamma_{db}^2 - 0.082\gamma_{db} + 1.1 \tag{6.25}$$

with less than 2% error for $0 \le \gamma_{db} \le 28$ dB and less than 4.4% error for $\gamma_{db} \le 30$ dB. By setting $T_{ho}/A = 0.5$ for $\gamma_{db} \ge 23$ dB, the error will be less than 2% for $\gamma_{db} \ge 0$ dB. The estimation accuracy of the signal-to-noise ratio must be factored into the performance evaluation.

Upon changing the integration variable in (6.22) to $\lambda = x/\sigma_n$ and defining the normalized threshold as $\beta = T_h/\sigma_n$ and $\alpha = A/\sigma_n$, the bit-error probability is expressed as

$$\begin{aligned} P_{be}(\beta) &= \frac{1}{2}\left[1 - \int_{\beta}^{\infty} \lambda e^{-\frac{\lambda^2+\alpha^2}{2}} I_o(\alpha\lambda)d\lambda + \int_{\beta}^{\infty} \lambda e^{-\lambda^2/2} d\lambda\right] \\ &= \frac{1}{2}[1 - Q(\alpha,\beta) + Q(0,\beta)] \end{aligned} \tag{6.26}$$

The function $Q(\alpha, \beta)$ [1, 2] is the Marcum *Q*-function or simply as the *Q*-function.* Marcum (Reference [1],

*The notation $Q(a, b)$ is also used to denote the incomplete Gamma function and the notation $Q(a)$ is used to denote the complement of the Probability Integral so the designation *Q-function* is misleading.

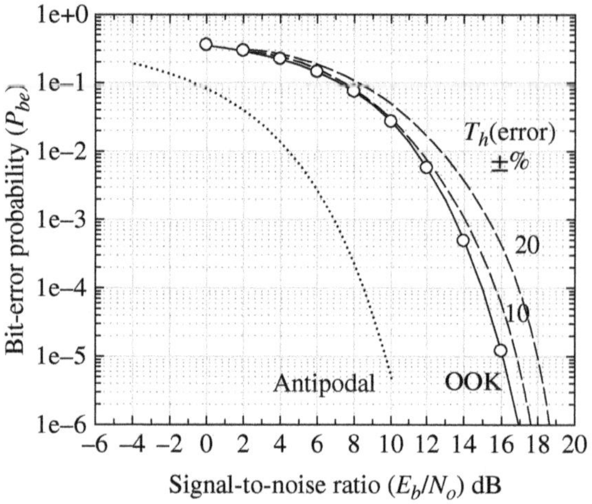

FIGURE 6.7 Summary of noncoherent OOK detection.

pp. 227, 228) provides plots of $Q(\alpha, \beta)$ in terms of the incomplete Toronto function $T_{\beta/\sqrt{2}}(1,\ 0,\ \alpha/\sqrt{2})^*$ as a function of the threshold $\beta/\sqrt{2}$ for various parameter values $\alpha/\sqrt{2}$; Marcum also provides extensive tables of the Marcum Q-function [3] with intervals of $\Delta\beta = 0.1$ and $\Delta\alpha = 0.05$. The Marcum Q-function is difficult to evaluate for the entire range of thresholds and signal levels ≥0; however, Johansen [4] provides a method for computer evaluation with a relative accuracy of 1×10^{-5}. Johansen's method is used to evaluate (6.26) and the results are shown in Figure 6.7 using the optimum threshold and for threshold variations of ±10% and ±20% around the optimum threshold. The performance degradation with threshold error is comparable to that of coherent detected OOK. The *circled* data points are based on Monte Carlo simulations using 500K bits for each signal-to-noise ratio. In this case, the optimum threshold is selected from the minimum bit-error probability corresponding to a range of thresholds with increments of $\Delta\rho = 0.01$.

6.2.2 Binary Antipodal ASK Modulation

The special case of ASK results when $\alpha_m = \{1,-1\}$. Under this condition α_m is related to the input binary data as $\alpha_m = 1 - 2b$, where $b = \{0,1\}$, so the modulated waveform is expressed as

$$\begin{aligned} s_m(t) &= A\alpha_m \cos(\omega_c t) \\ &= A\cos(\omega_c t + \pi(1-\alpha_m)/2) \\ &= A\cos(\omega_c t + \pi b) \end{aligned} \tag{6.27}$$

$^*Q(\alpha,\beta) = T_{\sqrt{v}}(2N-1, N-1, \sqrt{N\gamma})|_{N=1}$, where N is the number of noncoherently integrated pulses, $\sqrt{v} = \beta/\sqrt{2}$ is the threshold, and $\gamma = \alpha^2/2$ is the single-pulse signal-to-noise ratio.

In this case, the average and peak powers are given by $A^2/2$ and A^2, respectively, and the square of the minimum distance is computed as $d^2 = 2A^2T = 4E_b$ so the bit-error performance for coherent detection becomes

$$P_{be} = Q\left(\sqrt{\frac{2E_b}{N_o}}\right) = \left(\frac{1}{2}\right) erfc\left(\sqrt{\frac{E_b}{N_o}}\right) \tag{6.28}$$

This is identical to the coherent detection of BPSK modulation.

6.2.3 Pulse Amplitude Modulation (PAM)

The general case of M-ary ASK is referred to as $M = 2^k$, where k is the number of bits-per-symbol. In this analysis, M is considered to be a positive integer such that $M \ge 2$. The normalized rest-points defining the decision regions are related to M by the relationship

$$\alpha_m = -(M-1) + 2m \quad : m = 0, 1, \ldots, M-1 \tag{6.29}$$

Using (6.4) with $P(m) = 1/M$, the average power and peak powers are given by

$$\begin{aligned} P_{avg} &= \frac{M^2-1}{6}A^2 \\ &= \frac{M^2-1}{3}(A^2/2) \quad : \forall M \end{aligned} \tag{6.30}$$

and

$$P_p = (M-1)^2 A^2 \tag{6.31}$$

The power $A^2/2$ is the *rms* signal power associated with the carrier-modulated signal. Figure 6.8 shows how the peak-to-rms signal power ratio changes with M. The maximum ratio is 6 and is nearly reached for $M = 64$, that is, with 6 bits/symbol.

Figure 6.9 shows the decision regions around each rest-point in the signal space containing points on the *real* line x. The conditional distributions about each point represent the normalized amplitude† of the received signal and in the following analysis the performance is evaluated for the additive white Gaussian noise (AWGN) channel. Figure 6.9a and b represent the unique cases with $M \in$ even and odd, respectively.

The shaded regions, corresponding to the conditional distributions around each decision point, represent a symbol-error event and the symbol-error probability is computed, using the total probability law, as

†Defining the received signal decision statistic as y, the normalized distribution $p(x)$ is obtained by letting $x = y/A$.

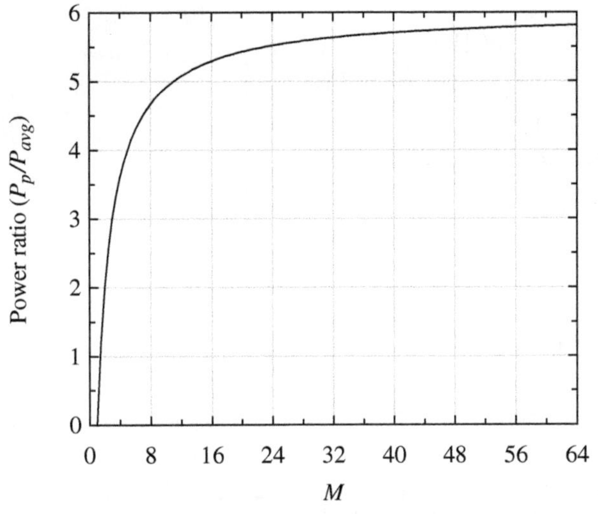

FIGURE 6.8 Peak-to-average power ratio for PAM modulation.

$$P_{se} = \sum_{m=0}^{M-1} P(\text{error}|m)P(m) \tag{6.32}$$

The conditional-error probability in (6.32) is the error associated with each decision region and, from Figure 6.9,* for each of the inner regions ($m = 1, 2, \ldots, M - 2$) this involves two distribution tails, whereas, only one distribution tail is involved for each for the two outer decision regions. Consequently, for $m = 1, 2, \ldots, M - 2$, the conditional-error probability is evaluated as,

$$P(\text{error}|m) = \int_{-\infty}^{A(\alpha_m - 1)} p(y - A\alpha_m)dy + \int_{A(\alpha_m + 1)}^{\infty} p(y - A\alpha_m)dy \quad : m = 1, 2, \ldots, M-2 \tag{6.33}$$

Letting $\xi = (y - A\alpha_m)/\sigma_n$ and recognizing that, for the carrier-modulated waveform, $A = \sqrt{2P}$ where P is the *rms* signal power during a symbol interval T, (6.33) is evaluated as

$$P(\text{error}|m) = \frac{1}{\sqrt{2\pi}} \int_{-\infty}^{-\sqrt{2P/\sigma^2}} e^{-\xi^2/2} d\xi + \frac{1}{\sqrt{2\pi}} \int_{\sqrt{2P/\sigma^2}}^{\infty} e^{-\xi^2/2} d\xi \quad : m = 1, 2, \ldots, M-2 \tag{6.34}$$

*The constellation in Figure 6.9b for M = odd integer does not correspond to an integer number of bits/symbol and does not apply to the conventional M-ary PAM waveform.

Expressing the noise power, in terms of the detection bandwidth B and the one-sided noise power density N_o, as $\sigma_n^2 = N_o B$ and defining the bandwidth in terms of the symbol duration T, such that,† $B = 1/T$, then the signal-to-noise ratio can be expressed in terms of the symbol energy and noise density as $P/\sigma_n^2 = E_s/N_o$. With these caveats, (6.34) is evaluated as

$$P(\text{error}|m) = 2Q\left(\sqrt{\frac{2E_s}{N_o}}\right) \quad : m = 1, 2, \ldots, M-2 \tag{6.35}$$

Following similar substitutions for the cases $m = 0$ and $M - 1$ results in

$$P(\text{error}|0) = \int_{A(\alpha_0+1)}^{\infty} p(y - A\alpha_0)dy = Q\left(\sqrt{\frac{2E_s}{N_o}}\right) \tag{6.36}$$

and

$$P(\text{error}|M-1) = \int_{-\infty}^{A(\alpha_{M-1}-1)} p(y - A\alpha_{M-1})dy = Q\left(\sqrt{\frac{2E_s}{N_o}}\right) \tag{6.37}$$

Substituting these results into the expression for the symbol-error probability, expressed by (6.32), with $P(m) = 1/M$, gives

$$\begin{aligned} P_{se} &= \frac{2(M-2)}{M} Q\left(\sqrt{\frac{2E_s}{N_o}}\right) + \frac{2}{M} Q\left(\sqrt{\frac{2E_s}{N_o}}\right) \\ &= \frac{2(M-1)}{M} Q\left(\sqrt{\frac{2E_s}{N_o}}\right) \\ &= \frac{M-1}{M} erfc\left(\sqrt{\frac{E_s}{N_o}}\right) \end{aligned} \tag{6.38}$$

Equation (6.38) is defined in terms of the signal-to-noise ratio based on the *rms* carrier power $A^2/2$, however, for PAM the symbol-error probability must be based on the average signal power defined over the entire PAM symbol constellation. To accommodate this requirement, (6.30) is used to relate the corresponding symbol energy levels as

$$\bar{E}_s = \frac{M^2 - 1}{3} E_s \tag{6.39}$$

†This assumes that $p(t)$ results in the $sinc(fT)$ frequency response.

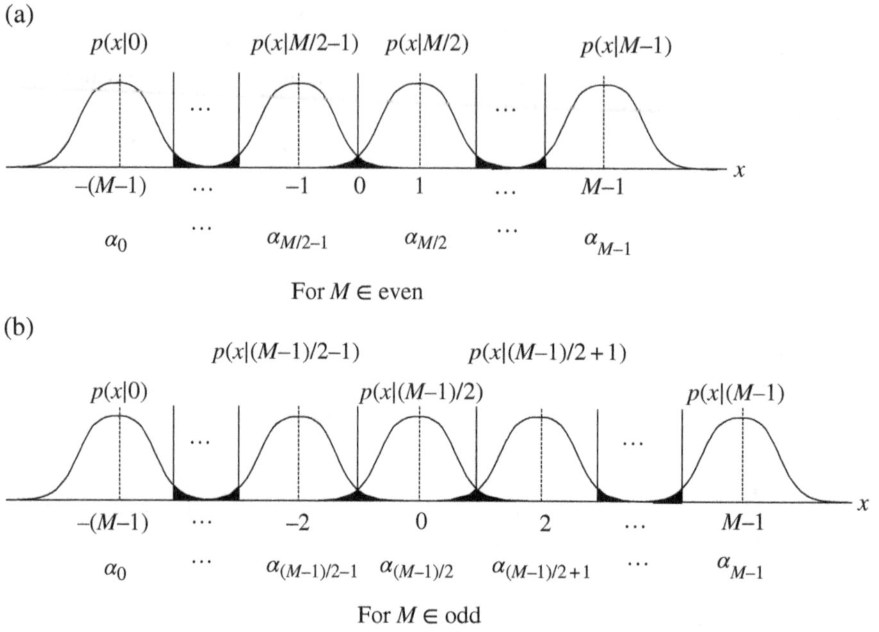

FIGURE 6.9 Signal space and decision regions for PAM modulation.

Furthermore, the energy-per-bit is given by $E_b = \bar{E}_s/k$ and, upon applying these results, the expression for the symbol-error probability becomes*

$$P_{se} = \frac{2(M-1)}{M} Q\left(\sqrt{\frac{6k}{(M^2-1)}\left(\frac{E_b}{N_o}\right)}\right) \quad : M = \text{even integer} \tag{6.40}$$

This expression for the symbol-error probability of M-ary PAM is shown in Figure 6.10 for $k = 1$, 2, 3, 4, and 5 corresponding to $M = 2$, 4, 8, 16, and 32. The performance for $M = 2$ is the same as for coherent detection of binary antipodal ASK PAM which is identical to the coherent detection of BPSK as discussed in Section 6.2.2. These results indicate that E_b/N_o must be increased by about 4 dB for each additional bit per symbol.

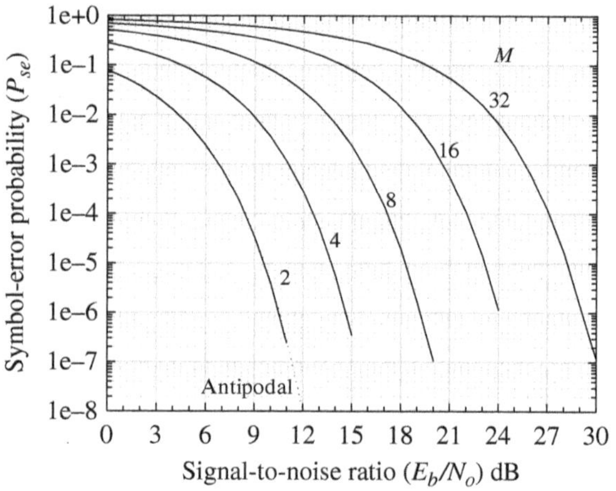

FIGURE 6.10 Symbol-error performance of M-ary PAM modulation.

6.3 QUADRATURE AMPLITUDE MODULATION (QAM)

QAM is composed of two independent PAM baseband modulations placed on the quadrature rails. If the PAM on each rail has the same number of symbols, the signal rest-points form a square constellation decision matrix and the performance of QAM is readily evaluated by extending the analysis of the PAM waveform discussed in the previous section. However, if a different number of symbols is assigned to each quadrature rail, the rest-points form a rectangular decision matrix that is less efficient in terms of the average signal power, required to achieve a specified symbol-error probability, and the peak-to-average power ratio. In the following sections, the QAM-modulated waveform is examined under these two conditions with examples of 8-ary QAM, with 3-bits/symbol and 16-ary QAM with 4-bits/symbol. Following this analysis, other signal constellations are examined that improve the efficiency of the QAM-modulated waveform. Improving the waveform efficiency, defined in terms of a specified minimum distance, involves minimizing the average power and the peak-to-average power ratio while

*This result applies to the conventional M-ary PAM waveform with M = even integer.

maintaining the specified symbol-error probability. A case study of 16-ary QAM is included with some performance simulation results.

6.3.1 QAM as Orthogonal Pulse Amplitude Modulation (PAM) Waveform

In this section, a simple form of QAM is considered that maps two PAM baseband-modulated waveforms onto quadrature rails resulting in a rectangular constellation of rest-points. With this mapping, the transmitted signal is expressed as

$$\begin{aligned} s_m(t) &= A\alpha_{Im}p(t)\cos(\omega_c t) - A\alpha_{Qm}p(t)\sin(\omega_c t) \\ &= \mathrm{Re}\{\tilde{s}_m(t)e^{j\omega_c t}\} \quad : |t| \le T/2 \end{aligned} \tag{6.41}$$

where α_{Im} and α_{Qm} are the quadrature data-dependent scaling factors applied to the signal amplitude A and $p(t)$ is defined as a unit-power amplitude-shaping function satisfying (6.2); as in the preceding sections, $p(t)$ is characterized as $rect(t/T - 1/2)$. In the following discussion, the primed and unprimed designations for M and k refer to rectangular QAM waveform and the underlying quadrature PAM waveforms, respectively. The decision space of a rectangular M'-ary QAM waveform is based on $M' = 2^{k'}$ rest-points α_{Im} and α_{Qm} distributed on the quadrature I and Q rails. When $k' \ge 1$ is even there exists an underlying M-ary PAM constellation with $M = \sqrt{M'}$ rest-points expressed by (6.29), such that, $M_I = M_Q = M$ resulting in a square constellation of rest-points. On the other hand, when k' is an odd integer there is no such underlying PAM constellation and the $M' = 2^{k'}$ rest-points α_{Im} and α_{Qm} result in a rectangular or non-square constellation. In this case, the constellation is selected to minimize the average power and peak-to-average power ratio. Example rest-point constellations for these two cases are shown in Figures 6.11 and 6.12 for $k' = 3$ and 4 corresponding to 8-ary QAM and 16-ary QAM, respectively; these constellations are shown with gray coded symbol assignment.

The symbol-error probability for the cases with k' = even is readily obtained using (6.38) and (6.39) with $\bar{E}'_s = 2E_s$ and recognizing that the quadrature channel is independent and identically distributed with equal symbol-error probability

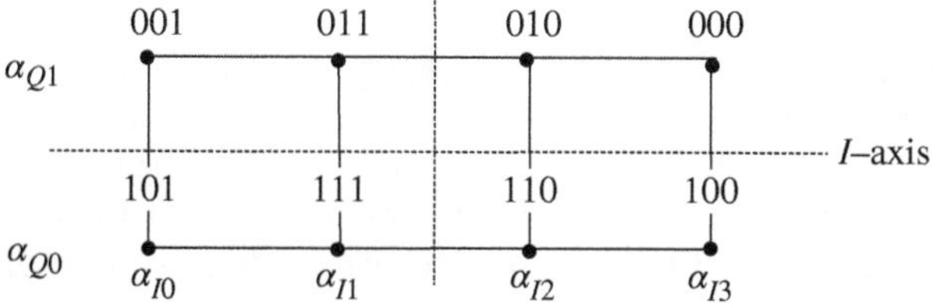

FIGURE 6.11 Rectangular constellation: 8-ary QAM modulation (gray-coded).

and average power so, considering both channels, the overall symbol-error probability is evaluated as

$$\begin{aligned} P'_{se} &= 1-(1-P_{se})^2 \\ &= \frac{4(M-1)}{M}Q\left(\sqrt{\frac{3k'}{M^2-1}\left(\frac{E_b}{N_o}\right)}\right) \\ &\quad \times \left(1-\frac{(M-1)}{M}Q\left(\sqrt{\frac{3k'}{M^2-1}\left(\frac{E_b}{N_o}\right)}\right)\right) \end{aligned} \tag{6.42}$$

where $M = \sqrt{M'}$. Equation (6.42) is plotted as the *solid* curves in Figure 6.13 for M' corresponding to k' = even integer.

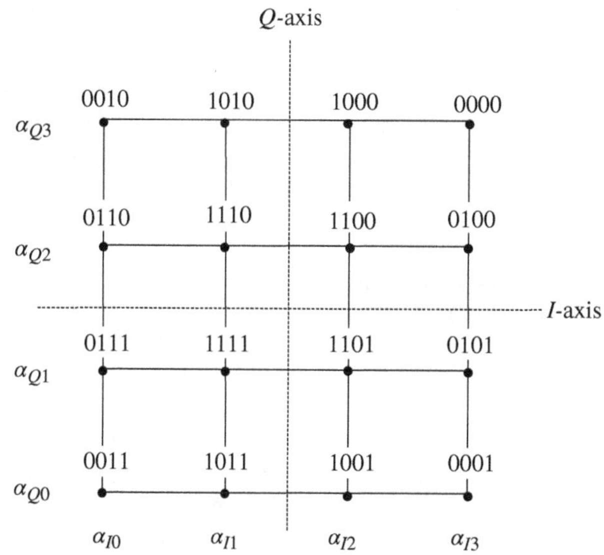

FIGURE 6.12 Square constellation: 16-ary rectangular QAM modulation (gray-coded).

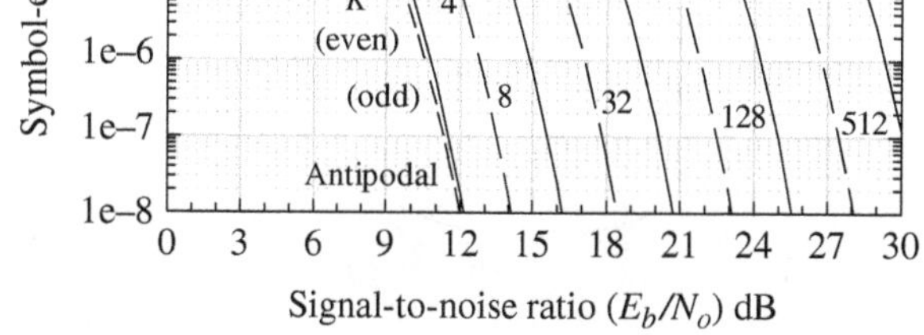

FIGURE 6.13 Symbol-error performance of M-ary QAM modulation.

With gray coding the approximate expression for the bit-error probability, for sufficiently high values of E_b/N_o, is simply $P'_{be} \cong P'_{se}/k'$.

When k' = odd integer the signal rest-point constellation, shown in Figure 6.11 for $k' = 3$, is not square, however, it can be formulated as a rectangular constellation with row-column dimensions M_I and M_Q such that $M_I = 2M_Q$ and $M_I M_Q = M'$. The symbol-error probability is expresses as

$$P'_{se} = 1 - P_{Isc}P_{Qsc} \tag{6.43}$$

where P_{Isc} and P_{Qsc} are the probabilities of a correct symbol in the respective dimensions. Letting $x = \{I,Q\}$, these expressions for the correct symbol detection probability are evaluated as

$$P_{xsc} = 1 - 2\frac{M_x - 1}{M_x} Q\left(\sqrt{\frac{3k'}{M'-1}\left(\frac{E_b}{N_o}\right)}\right) \tag{6.44}$$

Equation (6.43) is evaluated using (6.44), corresponding to the I and Q rails, and plotted as the *dashed* curves in Figure 6.13.

6.4 ALTERNATE QAM WAVEFORM CONSTELLATIONS

The performance for k' = odd integers in Figure 6.13, that is, for $M = 8$, 32, 128, and 512, corresponds to the rectangular rest-point constellations as shown in Figure 6.11 for 8-ary QAM with $k' = 3$. These constellations require a relatively high average power and the performance can be improved through a more judicious selection of the rest-point locations. For example, the constellation in Figure 6.11 is re-drawn in Figure 6.14 as two concentric circles with equal distances between the nearest neighbors. Because this constellation contains rest-points on concentric circles, it is referred to as APSK modulation. Four sets of nearest neighbors result in two bit-errors so gray coding is not fully achievable. On the other hand, if the outer circle constellation were rotated counter-clockwise by 45°, gray coding is achievable; however, the outer radius circle must be decreased slightly to achieve the same distance between all of the rest-points. It is left as an exercise (see Problems 8 and 9) to analyze the performance of this constellation and compare it with the performance of the phase-shifted outer constellation with the adjusted radius to provide equal Euclidian distances between all adjacent rest-points.

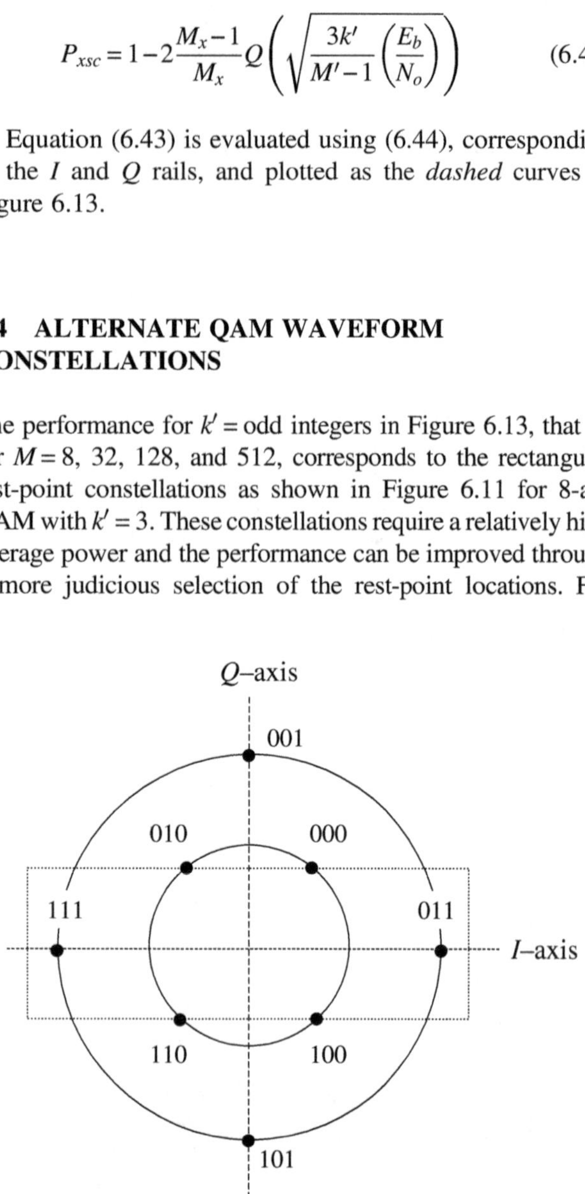

FIGURE 6.14 Concentric circular constellations: 8-ary APSK modulation (partial gray coding).

6.5 CASE STUDY: 16-ary QAM PERFORMANCE EVALUATION

This case study examines the performance of 16-ary QAM using the rest-point configuration shown in Figure 6.12. Referring to the functional demodulator implementation shown in Figure 6.15, the filtering and sample rate reduction provides four complex samples-per-symbol with a bandwidth of $2R_s$ Hz into the phase rotation and matched filtering functions. The fast AGC derives the signal level error in a bandwidth comparable to the input BPF bandwidth and is used during acquisition. The slow narrowband AGC is used to maintain the signal level throughout the data detection processing. The slow AGC, phaselock loop, and symbol timing errors are derived from the decision logic depicted in Figure 6.16; otherwise, these tracking functions operate in a conventional manner. The slow AGC error is computed as the magnitude difference between the rest-point and matched filter vectors; the PLL phase error is the angle between the two vectors. The symbol timing error derivation is not explicitly shown, however, is derived from the difference between early and late samples of the matched filter output.

Figure 6.17 compares the simulated bit-error performance as the *circled* data points and the corresponding theoretical gray-code approximation from Figure 6.13. The *dashed* curve is the simulated symbol-error performance that approaches $4 \times P_{be}$ as the signal-to-noise ratio increases. The Monte Carlo simulations are based on 1 M symbols for each signal-to-noise ratio and use ideal AGC, phase and symbol timing tacking.* The transmitted signal uses the *rect(t/T)* symbol shaping function and the spectrum is shown in Figure 6.18. This traditional *sinc(fT)* frequency response, with first spectral sidelobe levels of −13 dB, can

*The intent of this simulation is to verify agreement between the simulated and theoretical bit-error performances.

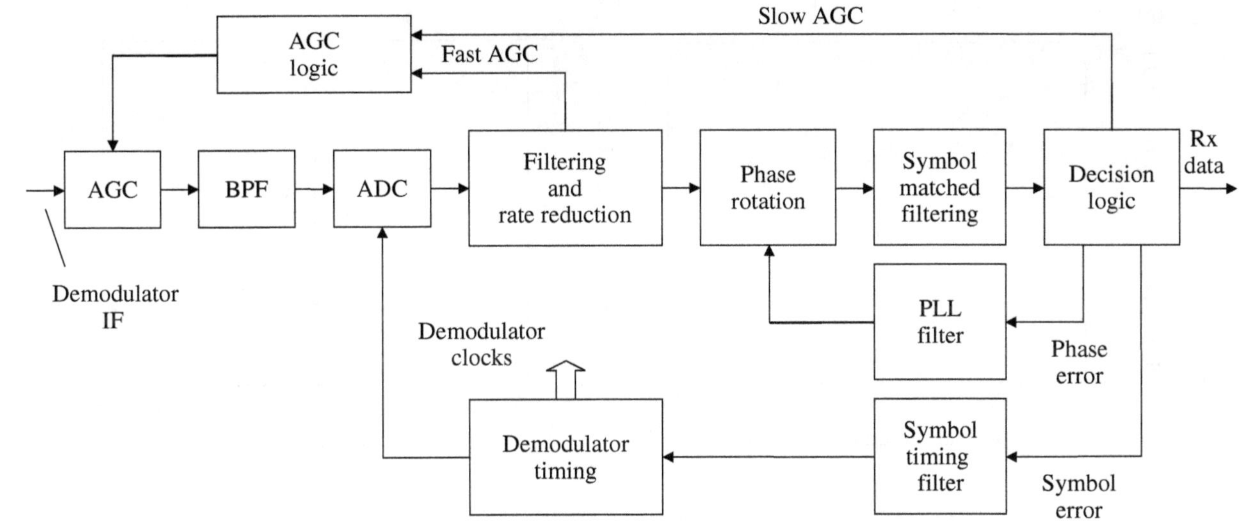

FIGURE 6.15 Functional implementation of an M-ary QAM demodulator implementation.

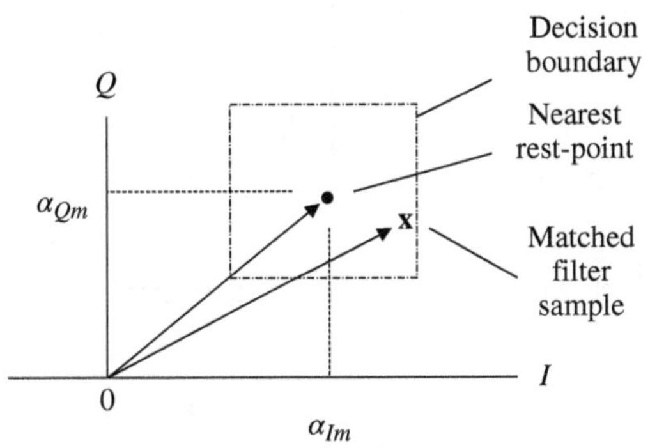

FIGURE 6.16 QAM demodulator decision logic.

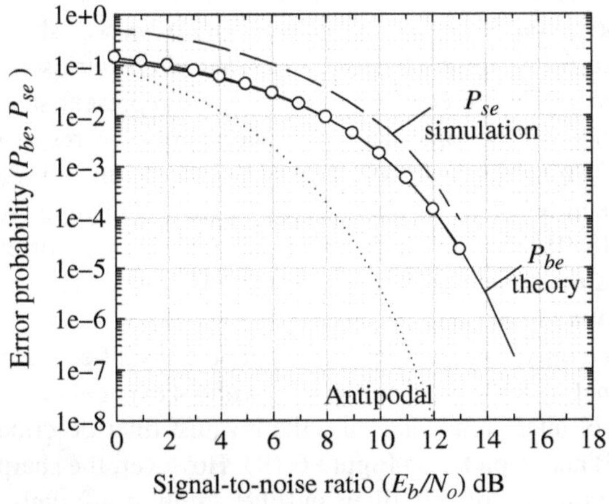

FIGURE 6.17 Simulated performance of 16-ary QAM modulation (Figure 6.12, gray-coded).

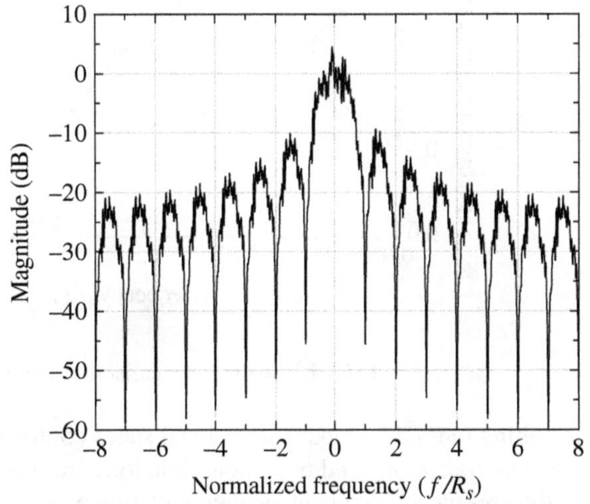

FIGURE 6.18 QAM signal spectrum ($p(t) = rect(t/T)$).

be improved with the use of spectral root-raised-cosine shaping as discussed in Sections 4.3.2, 4.4.1.1, and 4.4.4.1.

6.6 PARTIAL RESPONSE MODULATION

The following concepts dealing with partial response modulations are largely based on the ground breaking work of Nyquist [5, 6]. Partial response modulation is a form of multilevel pulse code modulation (PCM)* or PAM in which intentional intersymbol interference is permitted to confine the power spectral density (PSD) of the transmitted signal.

*PCM modulation is discussed in more detail in Chapter 8.

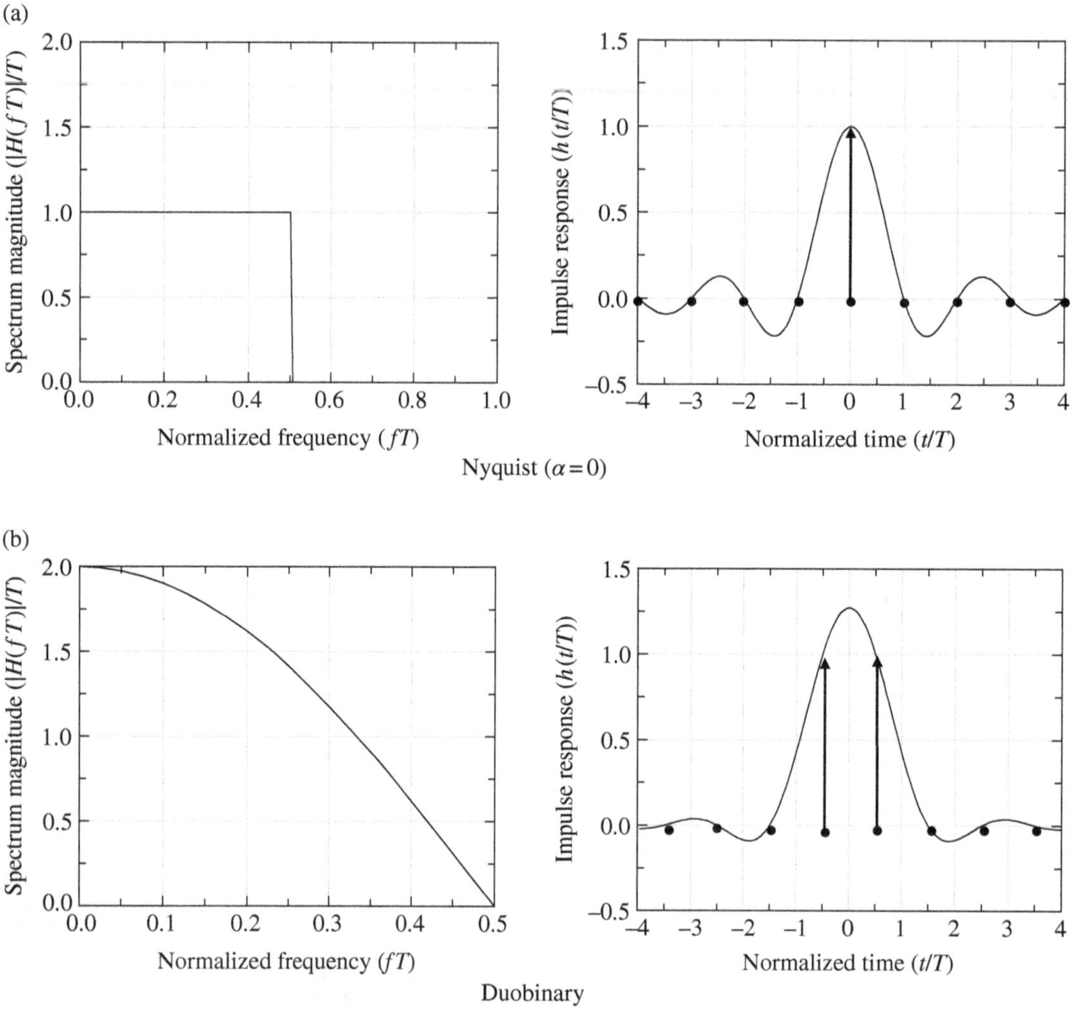

FIGURE 6.19 Comparison of Nyquist and duobinary spectrums and impulse responses.

The resulting redundant modulation signal states conform to unique state transitions and transition violations are used to detect the presence of errors in the demodulation processing. Partial response is also referred to as correlative coding and was introduced by Lender [7] as duobinary modulation in which a data rate of twice that of conventional binary modulation is achieved with a narrower PSD bandwidth. The duobinary modulation is contrasted with Nyquist binary signaling as characterized in (6.45) and (6.46) using the Fourier transform relationship $H(f) \Leftrightarrow h(t)$. These relationships and the following analysis ignore causality:

$$H(f) \qquad\qquad h(t)$$

$$Trect(fT) \;:|fT| \le 1/2 \;\Leftrightarrow\; sinc(t/T) : \text{Nyquist}\,(\alpha=0) \tag{6.45}$$

$$2T\cos(\pi fT) \;:|fT| \le 1/2 \;\Leftrightarrow\; \left(\frac{4}{\pi}\right)\frac{\cos(\pi t/T)}{1-4(t/T)^2} \;: \text{duobinary} \tag{6.46}$$

The frequency and time functions are plotted in Figure 6.19. Figure 6.19a depicts the uncorrelated spacing of the Nyquist impulse response at intervals of $t=\pm nT$: $n>0$. These intervals are contrasted with the duobinary uncorrelated intervals of $t=\pm(n+1/2)T$: $n>0$ shown in Figure 6.19b. In both cases, the *filled circles* are separated by the symbol interval of T seconds and represent sampling instances of the input data. The duobinary filter response, at the sampling instances $t=\pm T/2$, results in correlation between adjacent symbols in the sequence of input source data [8, 9]. The duobinary, or 11* partial response, correlative encoding depicted in Figure 6.19b is also referred to as polybinary and biternary coding.

The only way that $2W$ bits/s can be transmitted through a channel with bandwidth $W=R_b/2$ without experiencing intersymbol interference is to use the Nyquist filter described by (6.45) and depicted in Figure 6.19a. However, the sharp cutoff frequency of this filter and the resulting infinitely long

*Pronounced "one-one."

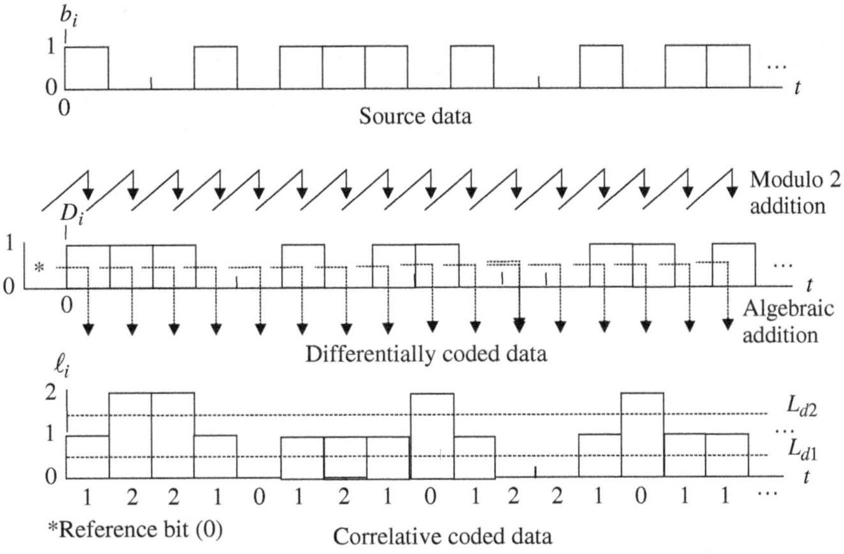

FIGURE 6.20 Example of duobinary baseband modulation and demodulation.

impulse response results in an unrealizable filter that requires time and frequency design compromises as discussed in Section 4.4.4.1. The characteristic of the ideal Nyquist filter, that results in no intersymbol interference, is the orthogonal spacing of the impulse response at instances of $\pm nT$: $|n| > 0$ as depicted by the *filled circles* in Figure 6.19a. As shown in Figure 6.19b, orthogonal spacing also exists for duobinary modulation, however, the symbol responses occurring at $t = \pm 0.5T$ both have unit amplitudes and result in intentional or known intersymbol interference between adjacent symbols. The cosine-shaped frequency response of the duobinary-modulated signal in confined to the same bandwidth as the Nyquist filter; however, the noise bandwidth is $B_n = 1/4T$ Hz. The cosine frequency roll-off also results in lower impulse response sidelobes and less sensitivity to channel impairments and demodulator frequency and timing errors compared to the Nyquist filter.

The duobinary baseband waveform modulation and demodulation using a random sequence of source data bits, denoted as $b_i = \{0,1\}$, is shown in Figure 6.20. The bandwidth of the duobinary spectrum supports an information capacity of 2 bits/Hz by incorporating three distinct amplitude levels obtained by the correlative processing resulting from the known intersymbol interference as described above. For example, referring to Figure 6.20, the source data bits (b_i) are modulo-two added ($\oplus$) to the previously encoded data D_{i-1} resulting in the differentially coded data $D_i = D_{i-1} \oplus b_i$ for $i = 1, 2, \ldots$. The reference data D_0 is set to 0 corresponding to *space* data. This coding prevents catastrophic error propagation when an error occurs in the channel. The three correlative encoded signal levels, constituting the duobinary waveform, are obtained by algebraically adding the differentially encoded data resulting in the modulated signal levels $\ell_i = D_i + D_{i-1}$ indicated in Figure 6.20. The result is a three-level modulated signal corresponding to $\ell_i = \{0,1,2\}$ as depicted in the third sequence in the figure.

The decoding of the received duobinary waveform is based on the detection threshold levels L_{d1} and L_{d2} that are established to correspond to the minimum distance $d_m = A/2(L-1)$ between the $L = 3$ levels corresponding to $\ell_i = \{0,1,2\}$.* Based on this description, the demodulated data estimates $\hat{d}_i = 1$ correspond to the decisions $\ell_i = 1$ and $\hat{d}_i = 0$ correspond to the decisions $\ell_i = 0$ or 2. In general, for the correlative-encoded data decisions as described above, mark and space data correspond, respectively, to odd and even values of ℓ_i; this is useful in decoding the multilevel correlative coding as described by Lender [10] and discussed in Section 6.6.1. This description of the duobinary modulator results in unipolar modulation levels corresponding to $\ell_i \geq 0$ is shown in Figure 6.21a. An alternate coding configuration, resulting in the bipolar modulation levels $\ell_i = \{-2,0,2\}$ is shown in Figure 6.21b.

The correlative coded data in Figure 6.20 must be filtered to restrict the modulated signal spectrum. In keeping with the optimum detected filter in an AWGN channel, the duobinary filter is *split* between the modulator and demodulator filters expressed, respectively, as

$$H_m(fT_b) = H_d(fT_b) = \sqrt{H(fT_b)} \tag{6.47}$$

where $H_d(fT_b)$ represents the demodulator matched filter. The correlative coded data are filtered as shown in Figure 6.22 that also includes the channel, demodulator matched filter,

*In this description the signal amplitude A is 1 V and an ideal noise-free channel is considered.

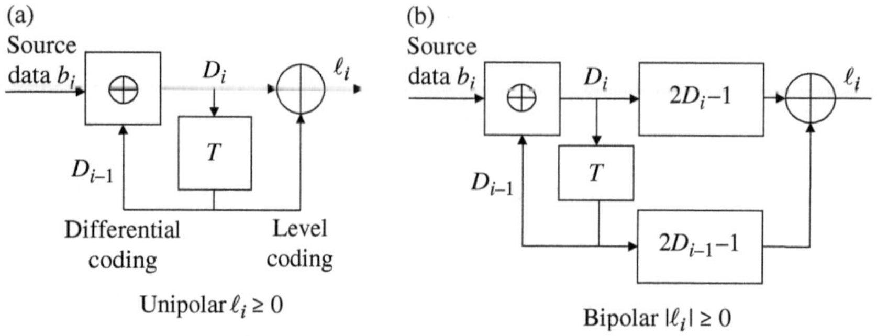

FIGURE 6.21 Encoding of duobinary baseband signal.

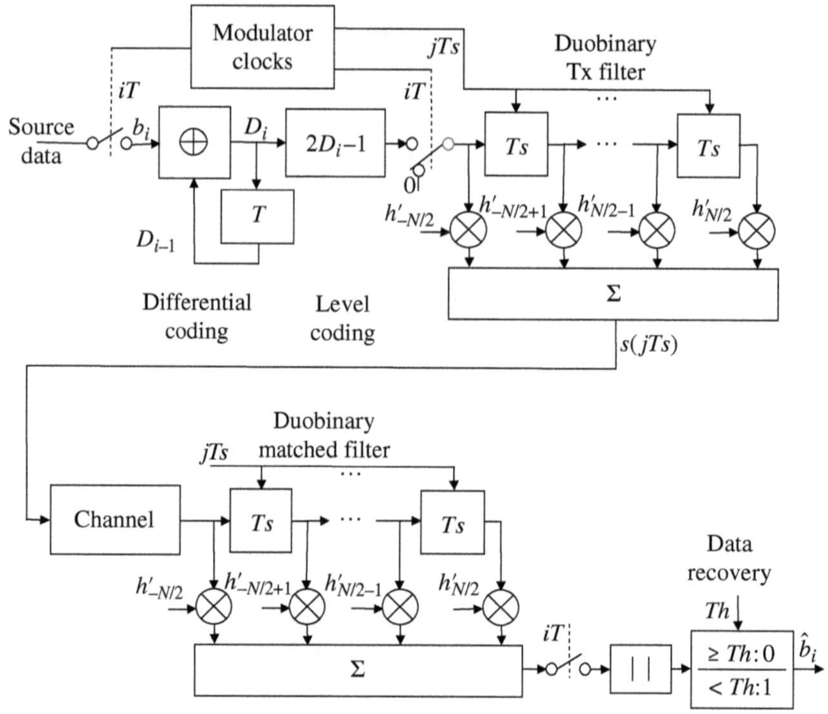

FIGURE 6.22 Simplified implementation of duobinary baseband system.

and data recovery processing. In this case, the unipolar differential encoder output is level coded corresponding to the unipolar-to-bipolar mapping $\{1,0\} \geq \{1,-1\}$. The resulting bipolar sampled data jT_s is then passed through the channel and the duobinary matched filter. For example, with N_s samples-per-symbol, the symbol counter i is derived from the sample counter j according to the condition[*]: if mod $(j,N_s) = 0$ then i is indexed by one, so, at the beginning of each symbol, a unit pulse $2D_i - 1$ of duration T_s is applied to the filter. As indicated in the figure, the remaining $N_s - 1$ samples for each symbol are input as zero. The filter weights h'_n: $-N/2 \leq n \leq N/2$ correspond to $N+1$ symmetrical samples of the duobinary impulse response. The discrete samples are evaluated as $h'_n \Leftrightarrow H_m(m)$ where $H_m(m)$ is the discrete-sample characterization of the duobinary spectrum as expressed in (6.47).[†] After passing the jT_s samples through the channel and demodulator matched filter, the iT received data samples are processed as shown in Figure 6.22[‡] to determine the received bit estimates $\hat{b}_i$. Figure 6.23 shows the duobinary matched filter response, using the 30-bit source data sequence: (111011001110000111010011010011) and a noise-free channel. Upon examining the data recovery processing, it is seen that the recovered data estimates are identical to the source data, that is, $\hat{d}_i = d_i$. Using the magnitude of

[*]The mod(j,i) and modulo(j,i) functions are defined in Section 1.14.7.

[†]The discrete Fourier transform is discussed in Section 1.2.4.

[‡]This simplified implementation does not show the details of time synchronization and gain control for determination of the detection threshold.

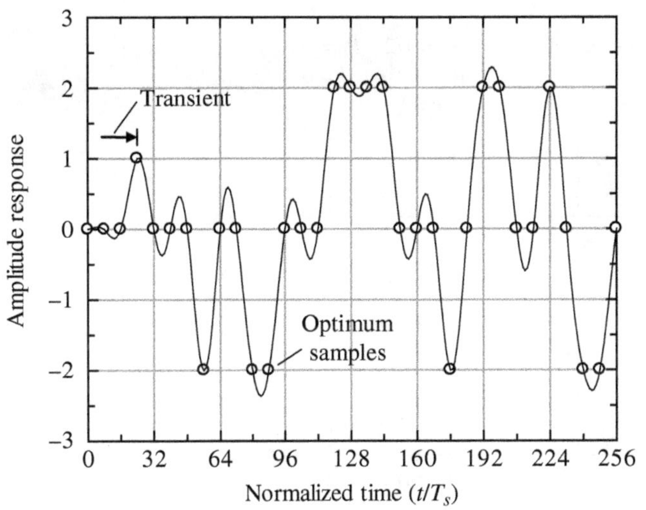

FIGURE 6.23 Duobinary matched filter response and optimum detection samples.

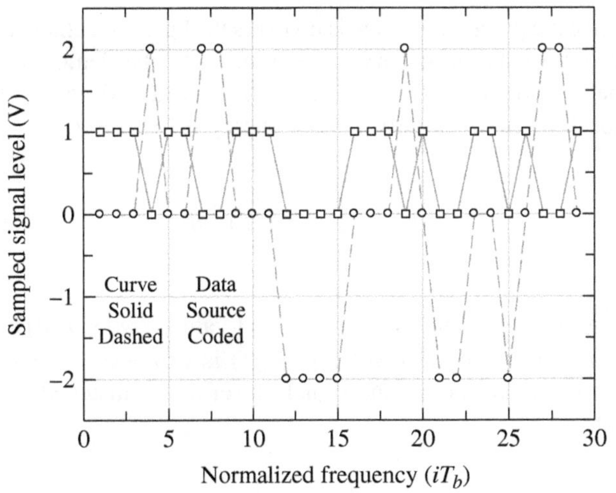

FIGURE 6.24 Duobinary sampled data and three-level demodulator matched filter samples.

the matched filter samples $e_o(iT)$ simplifies the decision processing.

Figure 6.24 shows the one-to-one correspondence between the sampled source data and the $L = 3$ levels of the demodulator matched filter sampled outputs. There is a one-to-one correspondence between the matched filter samples and the data samples such that the magnitude of the matched filter output maps into the optimally detected data estimates expressed as,

$$\hat{d}_i = \begin{cases} 0 & : |\ell_i| \geq 1 \\ 1 & : o.w. \end{cases} \tag{6.48}$$

The bit-error performance of the duobinary-modulated signal is evaluated using the filtered bipolar coded detection levels as characterized in Figure 6.21b. Considering the AWGN channel, the decision regions are depicted in Figure 6.25 for the bipolar and bipolar-magnitude detection processing.

The probability distribution functions for the conditions indicated in Figure 6.25 are expressed as follows:

$$\begin{aligned} p_{N1}(n) &= \frac{1}{\sqrt{2\pi}\sigma_n} e^{-(n+A)^2/2\sigma_n^2} \\ p_{N2}(n) &= \frac{1}{\sqrt{2\pi}\sigma_n} e^{-(n-A)^2/2\sigma_n^2} \\ p_{N3}(n) &= \frac{1}{\sqrt{2\pi}\sigma_n} e^{-n^2/2\sigma_n^2} \\ p_{N4}(n) &= \frac{2}{\sqrt{2\pi}\sigma_n} e^{-n^2/2\sigma_n^2} \end{aligned} \tag{6.49}$$

Using the *a priori* probabilities P_0 and P_1 and integrating the appropriate *pdfs* over the indicated ranges, the conditional probabilities, using the bipolar levels characterized in Figure 6.25a, are evaluated as

$$P_r(\text{error}|b=0) = P_0 \int_{-A/2}^{A/2} p_{N1}(n)dn + P_0 \int_{-A/2}^{A/2} p_{N2}(n)dn \tag{6.50}$$

and

$$P_r(\text{error}|b=1) = P_1 \int_{-\infty}^{-A/2} p_{N3}(n)dn + P_1 \int_{A/2}^{\infty} p_{N3}(n)dn \tag{6.51}$$

Performing the integrations in (6.50) and (6.51) and expressing the result in terms of the complement of the probability integral, the conditional probabilities simplify to

$$P_r(\text{error}|b=0) = \frac{1}{2}Q\left(\sqrt{\frac{\gamma}{4}}\right) - \frac{1}{2}Q\left(\sqrt{\frac{9\gamma}{4}}\right) \tag{6.52}$$

and

$$P_r(\text{error}|b=1) = Q\left(\sqrt{\frac{\gamma}{4}}\right) \tag{6.53}$$

Based on (6.52) and (6.53), the total-error probability is evaluated as

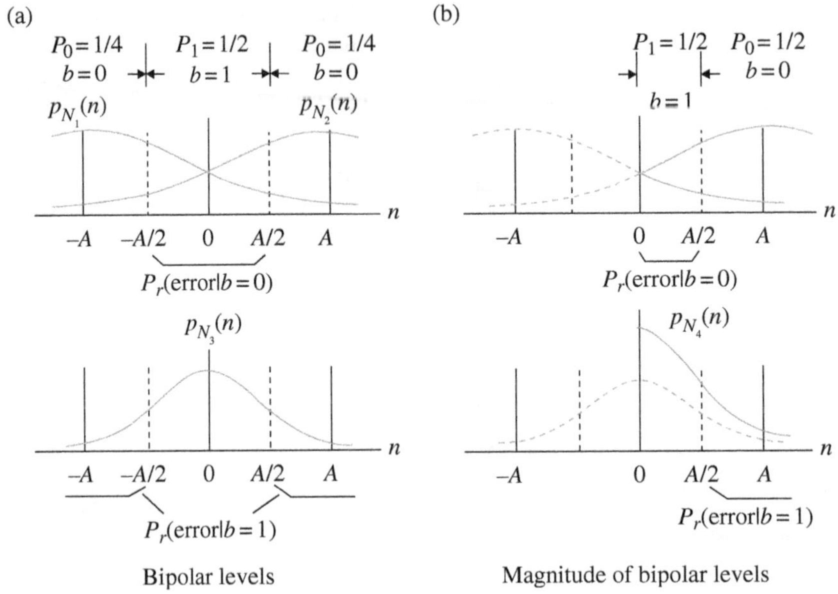

FIGURE 6.25 Demodulator decision regions for duobinary waveform detection.

$$P_e = P_r(\text{error}|b=0) + P_r(\text{error}|b=1)$$
$$= \frac{1}{2}Q\left(\sqrt{\frac{\gamma}{4}}\right) - \frac{1}{2}Q\left(\sqrt{\frac{9\gamma}{4}}\right) + Q\left(\sqrt{\frac{\gamma}{4}}\right)$$
$$= \frac{3}{2}Q\left(\sqrt{\frac{\gamma}{4}}\right) - \frac{1}{2}Q\left(\sqrt{\frac{9\gamma}{4}}\right) \tag{6.54}$$

where $\gamma = A^2/\sigma_n^2$ is the baseband peak signal-to-noise ratio. Equation (6.54) is also obtained when the error probability is evaluated using the magnitude of the bipolar levels as characterized in Figure 6.25b. Forming the magnitude of the received bipolar waveform simplifies the detection processing since a matched filter detection below the single threshold of $L_d = 1$ is declared a mark bit ($\hat{b}_i = 1$), otherwise, a space bit ($\hat{b}_i = 0$) is declared.

The term involving the argument $9\gamma/4$ has a negligible effect of the error probability. For example, when P_e is expressed in scientific notation the *significand* (*mantissa*) is altered in the third and fourth decimal place for $0 \le \gamma_b \le 20$ dB where γ_b is the signal-to-noise ratio measured in the bandwidth equal to R_b. Consequently, indifference to current usage, the error probability is expressed as

$$P_e = \frac{3}{2}Q\left(\sqrt{\frac{\gamma}{4}}\right) \quad : \text{for all practical purposes} \tag{6.55}$$

The argument of the square-root in (6.55) must be expressed in terms of the maximum signal-to-noise ratio, E/N_o, at the output of the demodulator matched filter, where E is the signal energy and N_o is the AWGN noise density. Since the preceding analysis involves the baseband characterization of the duobinary waveform, using the demodulator matched filter $H_d(fT)$, defined by (6.47), and Parseval's relationship, the duobinary symbol energy is expressed as

$$E_s = \int_{-1/2T}^{1/2T} |H_d(fT)|^2 df \tag{6.56}$$

However, since $H_d(fT)$ is real-valued with $H_d^2(fT) = H(fT)$, the integral involving $H(fT)$ is expressed, in terms of the square-root of the signal power at the matched filter output, as

$$\sqrt{2P_s} = \int_{-1/2T}^{1/2T} H(fT)df = \frac{4}{\pi}A \tag{6.57}$$

Therefore, the average received signal power is $A^2/2 = (\pi/4)^2 P_s$ and the corresponding symbol energy is $E_s = (\pi/4)^2 P_s T$ resulting in the error probability expressed as

$$P_e = \frac{3}{2}Q\left(\sqrt{\frac{(\pi/4)^2 P_s T}{N_o}}\right) \tag{6.58}$$

This result is reconciled with (6.55) by considering the definition of the argument given by

$$\frac{\gamma}{4} \triangleq \frac{A^2}{4\sigma_n^2} \tag{6.59}$$

Substituting $A^2/2$ given above and $\sigma_n^2 = N_o/T_b$, where T_b is the bit interval corresponding to 1 bit/symbol, (6.59) is expressed as

$$\frac{\gamma}{4} = \frac{1}{2}\left(\frac{\pi}{4}\right)^2 \frac{P_s T_b}{N_o}\bigg|_{T_b=2T} = \left(\frac{\pi}{4}\right)^2 \frac{P_s T}{N_o}\bigg| \quad (6.60)$$

The second equality in (6.60) is based on the duobinary waveform corresponding to 2 bits/symbol or $T_b = 2T$. Consequently, the argument ($\gamma/4$) of the square-root in (6.55) corresponds to that of the duobinary filtered result in (6.58). However, the duobinary spectrum in (6.46) has a lowpass bandwidth of $W = 1/2T = R_b'/2$ corresponding to the Nyquist rate of $R_b' = 2W$ bits/channel-use. Therefore, the interval T is the bit interval for the duobinary waveform so that $T = T_b'$ and the equality in (6.60) applies to the duobinary bit-error probability expressed as

$$P_{be} = \frac{3}{2} Q\left(\sqrt{\left(\frac{\pi}{4}\right)^2 \gamma_b}\right) \quad (6.61)$$

where $\gamma_b = E_b/N_o = P_s T_b'/N_o$. Lucky, Salz, and Weldon [11] arrive at the same result using the signal distance d and the bit-error probability expressed by (3.32).

Equation (6.61) is plotted as the *solid* curve in Figure 6.26 as a function of γ_b in decibel. The theoretical bit-error probability of unipolar non-return to zero (NRZ) PCM coding is 3 dB worse than antipodal signaling and the duobinary performance is degraded by an additional $-20\log_{10}(\pi/4) = 2.1$ dB from that of unipolar NRZ PCM. The *circled* data points correspond to Monte Carlo simulation of the duobinary performance using 1M bits/signal-to-noise ratio for $\gamma_b <$ 12 dB, otherwise 10M bits are used. The 2.1 dB loss in the duobinary performance can be reduced with maximum-likelihood sequence estimation [12] (MLSE) detection over several bits using a trellis state decoder. Because of the correlation between adjacent bits, the demodulation processing of the 11 duobinary coded waveform provides for error detection by observing violations of the following rule: two consecutive space bits with an intervening even number of mark bits have opposite polarity, otherwise, if the intervening mark bits is odd the polarity of the space bits is the same.

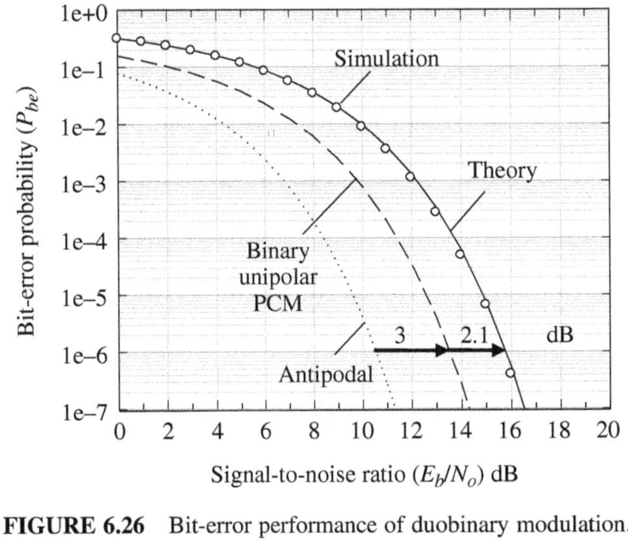

FIGURE 6.26 Bit-error performance of duobinary modulation.

6.6.1 Modified Duobinary Modulation

The modified duobinary response results in a −1 0 1* filter impulse response when sampled at the instances $t = nT$; $n = -1,0,1$; otherwise, the samples are all zero. The obvious difference between the modified duobinary and the 11 duobinary waveform is that the input signal samples are correlated with symbol samples separated by $2T$ instead of the adjacent symbol samples. However, as will be seen, the data corresponding to the modulated levels $\ell_i = 0$ is $d_i = 0$ (space) and $\ell_i = \pm 2$ is $d_i = 1$ (Mark). Another difference is that the transitions from a level can terminate on any of the levels so, unlike the 11 duobinary modulation, a transition from $\ell_i = -2$ to $\ell_{i+1} = +2$ may occur. Lender [10] points out that this large shift between levels reduces the *horizontal eye opening*, thus increasing the demodulator sensitivity to symbol timing errors and channel impairments. The modified duobinary waveform also uses three modulation levels with, as discussed in the preceding section, $T = T_b'$ and provides 2 bits/channel-use. At the optimum symbol sampling instances the *vertical eye opening* has zero intersymbol interference and the degradation in E_b/N_o with an AWGN channel is 2.1 dB. The −1 0 1 duobinary coded waveform provides for error detection by observing violations of the following rule: two consecutive mark bits with an intervening odd number of space bits have opposite polarity; otherwise, if the intervening space bits is even the polarity of the mark bits is the same. These claims are confirmed in the following analysis of the modified duobinary modulation.

The modified duobinary filter and the corresponding impulse response are described by (6.62). These respective functions are plotted in Figure 6.27 in terms of the normalizing parameter T.

$$\begin{array}{ccc} H(fT) & & h(t) \\ \begin{cases} j2T\sin\left(\dfrac{\pi fT}{2}\right) & : |fT| \le 1/2 \\ 0 & : o.w. \end{cases} & \Leftrightarrow & \dfrac{\sin\left(\frac{\pi}{T}(t+T)\right)}{(\pi/T)(t+T)} - \dfrac{\sin\left(\frac{\pi}{T}(t-T)\right)}{(\pi/T)(t-T)} \end{array}$$

: modified duobinary (6.62)

*Pronounced "minus-one-zero-one."

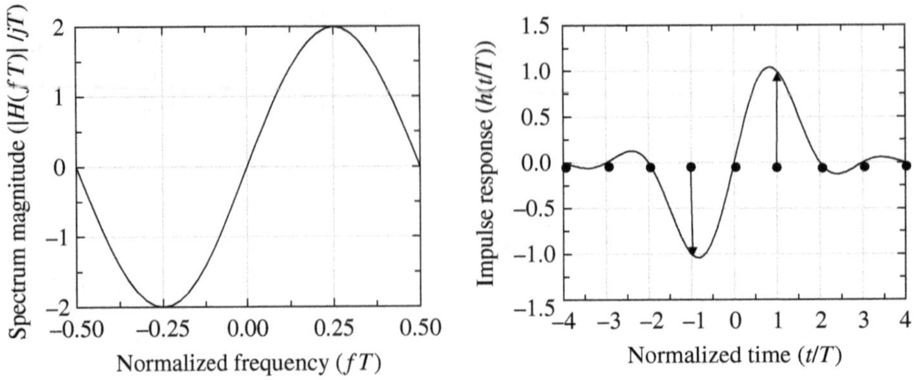

FIGURE 6.27 Modified duobinary spectrum and impulse response.

The frequency response is a sine function contained within the Nyquist band $W = R_b/2$ providing $2W = R_b$ bits/s or 2 bits/symbol or channel use. As with the duobinary cosine filter, the sine filter results in controlled intersymbol interference; however, in this case the non-zero filter impulse responses occur at $t/T = \pm 1$ with a correlation interval of $2T$ resulting in the notation −1 0 1 partial response.* Differential coding of the source data (D_i), the unipolar-to-bipolar translation (B_i), and level coding (ℓ_i) are defined for a source sequence of unipolar binary unipolar non-return to zero level (NRZ-L) data d_i: $i = 1, 2, \ldots$ as follows:

$$\begin{aligned} D_i &= D_{i-2} \oplus d_i \quad &&: \text{modulo 2 addition} \\ B_i &= 2D_i - 1 \quad &&: \text{unipolar-to-bipolar} \\ \ell_i &= -B_{i-2} + B_i \quad &&: \text{algebraic addition} \end{aligned} \tag{6.63}$$

where the two differentially coded bits are defined as $D_{-1} = D_0 = 0$. With $B_i = \{\pm 1\}$, the number of coded levels ℓ is $L = 3$. The correspondence of the coded levels with the source bits and the criteria for detecting demodulator bit-error conditions is verified by examining the coding using the defined sequence of data (see Problem 15). The bit-error probability performance loss of the modified duobinary modulation, under ideal channel and demodulator timing conditions, is 2.1 dB. This loss is established by examining the symbol energy of the modified duobinary modulation (see Problem 20).

The property of the modified duobinary filter corresponding to a zero direct current (DC) component, that is, $H(0) = 0$, is significant, in that, this is a necessary requirement for the generation of a single-sideband (SSB) modulated waveform. The SSB waveform is obtained by passing $h(t)$ through a Hilbert filter [13, 14], forming $\breve{h}(t)$, with the SSB baseband waveform informally expressed as

$$\tilde{s}(t) = \sum_i d_i \left(h(t - iT) + j\breve{h}(t - iT) \right) \tag{6.64}$$

where d_i is the information data.

6.6.2 Multilevel Duobinary Modulation

The multiple levels, exceeding $L = 3$ discussed in the preceding section, can be applied to the duobinary, modified duobinary, and other partial response modulations involving higher order filter functions mentioned in the following section. The inclusion of additional modulation levels results in the information capacity expressed as

$$\frac{R_b}{W} = 2\log_2(n) \quad : \text{bits/symbol or bits/channel-use} \tag{6.65}$$

where $n = L - 1$ is the span of the filter correlation interval.† The PSD is concentrated in an increasingly narrow bandwidth by virtue of a filter correlation length spanning n symbols. The multilevel duobinary frequency function is expressed as

$$H(f) = \frac{T\sin(n\pi fT)}{\sin(\pi fT)} \quad : |fT| \le 1/2 \tag{6.66}$$

The filter frequency function is plotted in Figure 6.28 for values of $L = 3$ through 7 with the *dashed* curve corresponding to the cosine filter in Figure 6.19b. The corresponding impulse responses are left as an exercise (see Problem 20); however, the filter impulse response exhibits n unit-amplitude levels occurring at t/T, such that, for $i = 0, \ldots, n - 1$, t/T is expressed as

$$\frac{t}{T} = \begin{cases} -[n/2] + i & : n \in \text{odd} \\ -[n/2] + 1/2 + i & : n \in \text{even} \end{cases} \tag{6.67}$$

*Some authors characterize the modified duobinary as a 10−1 partial response modulation; although the coding and decoding details are somewhat different, the coding and decoding can be interchanged without altering the correspondence between ℓ_i and d_i.

†All of the filters satisfy the Nyquist bandwidth criterion, $W = 1/2T$.

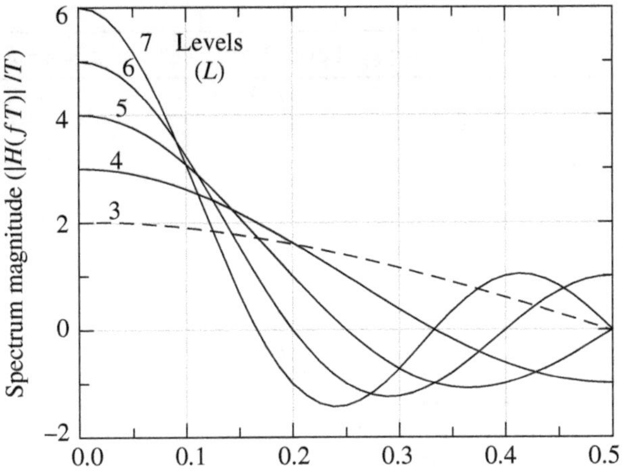

FIGURE 6.28 Multilevel duobinary spectrum.

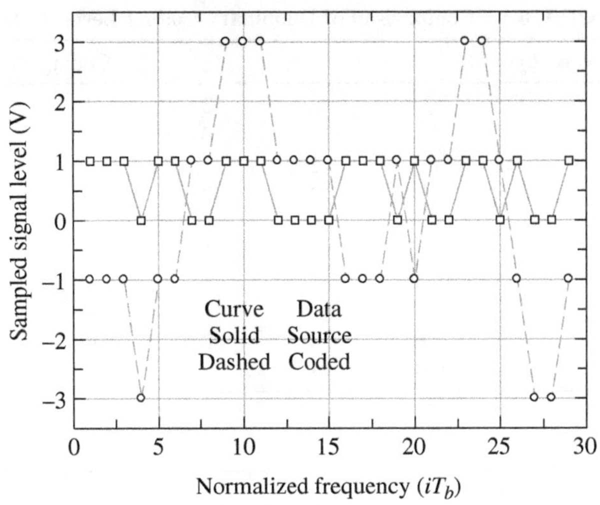

FIGURE 6.29 Duobinary sampled data and four-level demodulator matched filter samples.

The unit-amplitude samples are symmetrical about $t/T = 0$ with zero sample values at all other values of i.

The modulator encoding of the multilevel duobinary waveform is similar to that of the three-level waveform; however, the differential encoding spans n bits including the current source bit d_i: $i = 1, 2, \ldots$ and the $n - 1$ previously differentially encoded bits D_{i-m}: $m = 1, \ldots, n - 1$. The parameters $D_{-m} = 0$: $m = n - 1, \ldots, 0$ are initialized as space bits and the corresponding parameters B_{-m} are initialized as $B_{-m} = 2D_{-m} - 1$. With these normalizations, the encoding in (6.63) is generalized as

$$
\begin{aligned}
\bar{D} &= \sum_{m=1}^{n-1} D_{i-m} && \text{: algebraic addition} \\
D_i &= \bar{D} \oplus d_i && \text{: modulo 2 addition} \\
B_i &= 2D_i - 1 && \text{: unipolar-to-bipolar} \\
\ell_i &= \sum_{m=0}^{n-1} B_{i-m} && \text{: algebraic addition}
\end{aligned}
\qquad (6.68)
$$

The coding described in (6.68) is depicted in Figure 6.29, under noise-free conditions, for $L = 4$ levels ($n = 3$ correlated symbols) corresponding to 111* duobinary modulation. The source data are represented by the *square* data samples with the *circles* representing the demodulator matched filter samples. As in the preceding duobinary examples, there is a one-to-one correspondence between the matched filter sample and the received data; however, the mapping is not as straightforward as in the three-level duobinary case expressed in (6.48). This PAM waveform can be implemented as a QAM waveform by applying differential encoding between the inphase and quadrature rails as discussed in Section 8.4.

The detection levels for $L = 4$ shown in Figure 6.29 are symmetrical about zero; however, there is no optimum sample at $\ell = 0$ as in the $L = 3$ level duobinary case. The optimally sampled duobinary coding levels ℓ are shown in Table 6.1 for the indicated range of L.

The conditions for the decoding of the data d_j corresponding to the levels ℓ_j are given in Table 6.1. The distance between the levels is two and when $\text{mod}(\ell_{max}/2,2) = 1$ the optimally sampled matched filter levels are odd integer values symmetrical about zero. The binary data corresponding to negative levels are inverted with $d_j(\ell_j > 0) = \text{mod}(d_j(\ell_j < 0),2)$. For the case $\text{mod}(\ell_{max}/2,2) = 0$, $d_j(\ell_j > 0) = d_j(\ell_j < 0)$; however, when L is odd the level $\ell_0 = 0$ corresponds to $d_0 = 0$, otherwise $d_0 = 1$. Recognizing that the data $d_j(-\ell_{max}) = 0 \ \forall \ L$, the following procedure outlines the decoding decisions.

```
! ℓ_r is real-valued matched filter
optimum sample.
! ℓ_err is real-valued error used for
matched filter gain control.
! ℓ, ℓ_max, L , and icnt are integer-values
        ℓ_max = L - 1;      icnt = 0
 lp     do ℓ   = -ℓ_max, ℓ_max, 2
            if (ℓ_r >= real(ℓ_max -1)) then
              d̂_i = icnt;  ℓ_err = ℓ_r - real(ℓ)
            else   if(ℓ_r <real(ℓ- 1)) then
              d̂_i = icnt;  ℓ_err = ℓ_r - real(ℓ)
              exit  lp
            endif
          icnt = mod(icnt + 1,2)
        enddo lp
```

*Pronounced "one-one-one."

TABLE 6.1 Comparison of Duobinary Coding Levels ℓ and Data d for $3 \le L \le 7$

Levels (L)			Coding					Levels/Data	$\ell_{max} = n = L - 1$
3			2	0	−2			ℓ	2
			0	1	0			d	
4		3	1	—	−1	−3		ℓ	3
		1	0	—	1	0		d	
5		4	2	0	−2	−4		ℓ	4
		0	1	0	1	0		d	
6	5	3	1	—	−1	−3	−5	ℓ	5
	1	0	1	—	0	1	0	d	
7	6	4	2	0	−2	−4	−6	ℓ	6
	0	1	0	1	0	1	0	d	

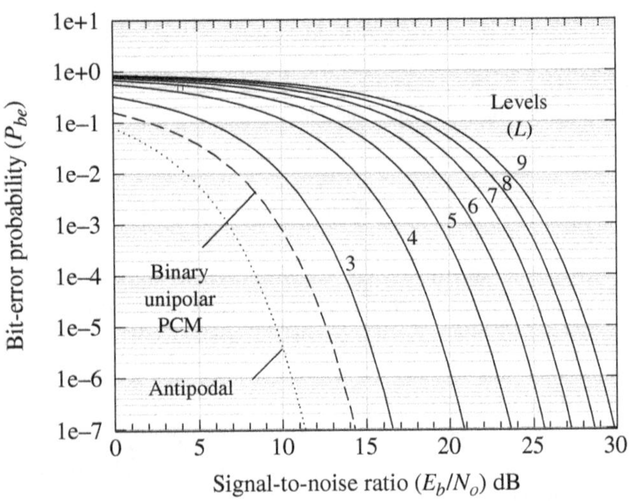

FIGURE 6.30 Multilevel duobinary bit-error performance.

The decision level errors ℓ_{err} are passed through a lowpass filter and applied to the matched filter gain control to ensure that the optimum decision levels are maintained for the duobinary demodulation. In the above examples, the optimum levels are defined in Table 6.1. In a simulation program or hardware test setup, the bit-error probability is computed as the number of bit-errors, $\hat{d}_i \neq d_i$, divided by the total number of received bits.

6.6.3 Multilevel Duobinary Bit-Error Performance

The bit-error performance of multilevel duobinary modulation is expressed as [15]

$$P_{be} = 2\left(1 - \frac{1}{n^2}\right) Q\left(\frac{\pi}{4}\sqrt{\frac{3\gamma_b}{n^2-1}}\right) \tag{6.69}$$

Equation (6.69) is plotted in Figure 6.30 for the indicated values of L and the performance is summarized in Table 6.2.

TABLE 6.2 Summary of Multilevel Duobinary Performance

Levels (L)	n	Loss* γ_b (dB)	γ_b (dB)* ($P_{be} = 10^{-5}$)	Capacity R_b/W bits/Use
3	2	2.1	14.9	2.0
4	3	6.4	19.2	3.2
5	4	9.1	22.0	4.0
6	5	11.1	24.0	4.6
7	6	12.8	25.7	5.2
8	7	14.1	27.0	5.6
9	8	15.3	28.2	6.0
17	16	21.4	34.3	8.0

*Values from Figure 6.30.

6.6.4 Multilevel Duobinary Modulation using k-bit Symbols

In this section, the multilevel duobinary-modulated waveform is described by assigning k gray-coded source bits to an M-ary symbol and performing the duobinary coding described above to the source symbols. In this case, d_i represents the gray-coded source bits and the duobinary coding is expressed in (6.70). The number of demodulator coded levels is $L = 2M - 1$ where $M = 2^k$. This description of the duobinary-modulated symbols results in a PAM waveform; however, the results can also be applied to the inphase and quadrature rails resulting in a QAM waveform. These concepts involving correlative coding have also been applied to frequency shift keying (FSK) and M-ary phase shift keying (PSK) waveform modulations resulting in improved spectral containment [16–18].

$$\begin{aligned}
s_i &= \sum_{m=0}^{k-1} 2^m d_i && \text{: bits-to-symbol conversion} \\
D_i &= \text{modulo}(s_i - D_{i-1}, M) && \text{: differential coding} \\
B_i &= 2D_i - (M-1) && \text{: unipolar-to-bipolar} \\
\ell_i &= B_i + B_{i-1} && \text{: symbol level (algebraic addition)} \\
\hat{s}_i &= \text{modulo}(\ell_i/2 + (M-1), M) && \text{: differential decoding}
\end{aligned} \tag{6.70}$$

TABLE 6.3 Example of Duobinary Symbol Coding ($k = 2$ bits/symbol)

Function	Ref. Data		Duobinary Coded Data														
Gray-coded bits (d_i)	00	00	11	10	11	00	11	10	00	01	11	01	00	11	01	00	11
Source symbols (s_i)	0	0	3	2	3	0	3	2	0	1	3	1	0	3	1	0	3
Differential coding (D_i)	0	0	3	3	0	0	3	3	1	0	3	2	2	1	0	0	3
Bipolar conversion (B_i)	−3	−3	3	3	−3	−3	3	3	−1	−3	3	1	1	−1	−3	−3	3
Level coding (ℓ_i)			0	6	0	−6	0	6	2	−4	0	4	2	0	−4	−6	0
Differential decode ($\hat{s}_i$)			3	2	3	0	3	2	0	1	3	1	0	3	1	0	3

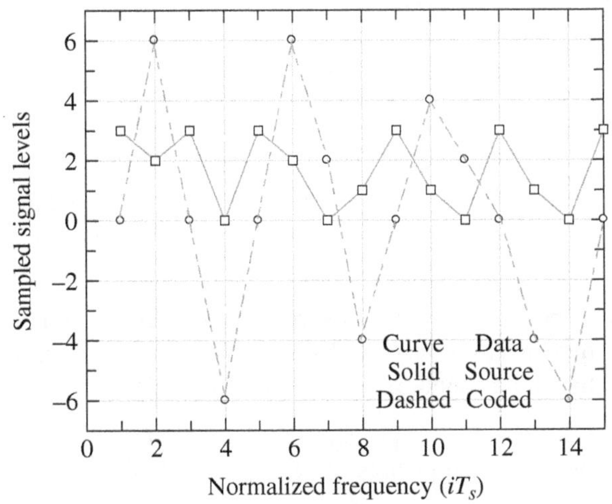

FIGURE 6.31 Example of multilevel duobinary coded symbols ($k = 2$ bits/symbol).

An example of the duobinary symbol coding using $k = 2$ bits/symbol is given in Table 6.3. The previously defined 30-bit binary example data sequence is used with the bits representing the gray-coded source bits. The transmit data and the resulting demodulator matched filter sampled levels ℓ_i are shown, respectively, in Figure 6.31 as the *square* and *circled* sampled data points connected by the respective *solid* and *dashed* lines.

6.6.5 Classification of Partial Response Filters

Kretzmer [19, 20] expands the duobinary modulation of Lender by generalizing the implementation of partial response-modulated waveforms. For example, the duobinary and the multilevel duobinary waveforms are categorized as Class 1 partial response waveforms, characterized by equal-amplitude filter impulses. Class 2 partial response implementations result in a symmetrical triangular-weighted set of filter impulses. The simplest case of the Class 2 partial response is the raised-cosine Nyquist filter with $\alpha = 1$ corresponding to the frequency and impulse responses expressed in (6.71) and shown in Figure 6.32. Although the frequency response extends over the range $|f| \le W$, the Nyquist criterion is satisfied resulting in impulses separated by T with zero responses otherwise as depicted by the *filled circles* in Figure 6.32. The triangular weighting results in a smoother, or more continuous, frequency response with much lower impulse response time sidelobes compared to those in Figure 6.19. This characteristic results in less sensitivity to demodulator symbol timing errors.

$$\overset{H(f)}{T\cos^2(\pi f T/2)\ :|fT| \le 1} \Leftrightarrow \overset{f(t)}{sinc\left(\frac{t}{2T}\right)\frac{\cos(\pi t/2T)}{1-(t/T)^2}}\ :\text{Nyquist}\ (\alpha = 1) \tag{6.71}$$

The Class 3 partial response, defined by Kretzmer, contains positive and negative impulse response components resulting from the ringing of the transient responses. The filters are characterized as causal lowpass filters, with a finite DC component, and result in nonsymmetrical decaying impulse components for $t/T \ge 0$. The number of impulses is chosen by truncating the impulse response. Classes 4 and 5 are quasi bandpass filters with a zero DC component. The modified duobinary waveform discussed in Section 6.6.1 is a Class 4 partial response waveform; the Class 5 is similar, however, the frequency response corresponds to a sine-squared function. For the classes of partial response considered, the best tolerance to signaling rate and channel impairments corresponds to the minimum value of n in each class.

ACRONYMS

AGC	Automatic gain control
APSK	Amplitude PSK
ASK	Amplitude shift keying
AWGN	Additive white Gaussian noise (channel)
BPSK	Binary phase shift keying
DC	Direct current
FSK	Frequency shift keying
I/Q	Inphase and quadrature (channels or rails)
MLSE	Maximum likelihood sequence estimation
NRZ	Non-return to zero (PCM code format)

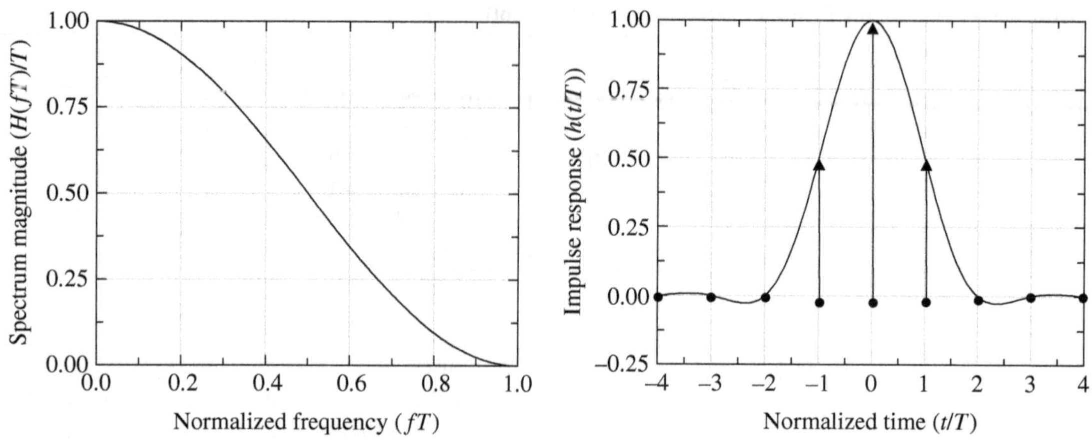

FIGURE 6.32 Class 2, $n = 3$ (1/2, 1, 1/2) partial response.

NRZ-L	Non-return to zero level (PCM code format)
OOK	On–off keying
PA	Power amplifier
PAM	Pulse amplitude modulation
PCM	Pulse code modulation
PLL	Phaselock loop
PSD	Power spectral density
PSK	Phase shift keying
QAM	Quadrature amplitude modulation
QPSK	Quadrature phase shift keying
RF	Radio frequency
SSB	Single-sideband (modulation)

PROBLEMS

1. Using (6.18) for the expressions of $P_{be}(T_h)$ for coherent detection of OOK, derive the optimum detection threshold T_{ho} that results in the minimum bit-error probability.

   Hint: Differentiate $P_{be}(T_h)$ with respect to T_h using the second equality in (6.18) and Leibniz's Theorem and set the result equal to zero and solve for T_{ho}.

2. Derive the optimum bit-error probability (P_{be}) for the coherent detection of OOK by substituting the optimum detection threshold T_{ho} into the expression (6.18) for $P_{be}(T_h)$. That is, show all of the steps used to arrive at $Q\left(\sqrt{E_b/2N_o}\right)$ in (6.19).

3. Consider the performance of a special case of OOK in which the space level is not zero but a small value given by $\alpha_0 = \rho\alpha_1$ where $0 \le \rho \le 1$. This situation was encountered in practice and resulted from a modulator implementation issue involving an analog amplifier bias. Derive the expression for the signal distance and the bit-error probability and comment on the detection loss relative to the ideal OOK modulator.

4. Repeat Problem 1 and find the optimum threshold for the noncoherent detection of OOK using the second equality in (6.22); the same hint applies.

5. Repeat Problem 2 for the bit-error probability of noncoherent detection of OOK by substituting the optimum threshold found in Problem 4 into (6.22).

6. Derive general expressions in terms of M' for the following QAM parameters for k' = even and odd.

   a. The average signal power.

   b. The peak-to-average signal power ratio.

7. Rewrite (6.42) in terms of the complementary error function $erfc(x)$ and E_b/N_o.

8. Derive the expression for the symbol-error probability for the APSK modulation shown in Figure 6.14. Also, determine the average signal power and the peak-to-average power ratio.

9. Determine the outer radius required to provide equal Euclidian distances between all nearest neighbors when the inner constellation in Figure 6.14 is rotated 45° clockwise as shown. Derive the expression for the symbol-error probability for the modified APSK constellation. Also, determine the resulting average signal power and the peak-to-average power ratio. Do the 3-bit eight symbol assignments result in gray coding?

10. The rectangular constellation for the 32-ary QAM waveform with $k' = 5$ and symbol-error probability shown in Figure 6.13, uses eight (8) I-axis and four (4) Q-axis rest-points located symmetrically about the respective axes with a normalized minimum Euclidian distance of two between neighboring rest-points. Reconfigure the 32-ary constellation as shown below using the same Euclidian distances and compute the average power and the peak-to-average power ratio. Compute the same

parameters for the 8×4 constellation and compare the results.

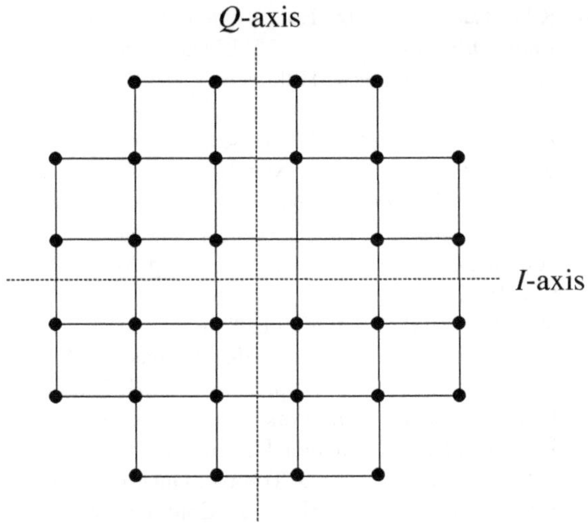

11. Derive the expressions for the symbol-error probabilities for the following 8-ary QAM constellations when all neighboring rest-points are separated by a normalized minimum distance of two. Compute the corresponding average power and the peak-to-average power ratio. Associate the eight 3-bit symbols with the rest-points to minimize the bit-errors when a symbol error occurs. Is gray-coding possible?

(a)

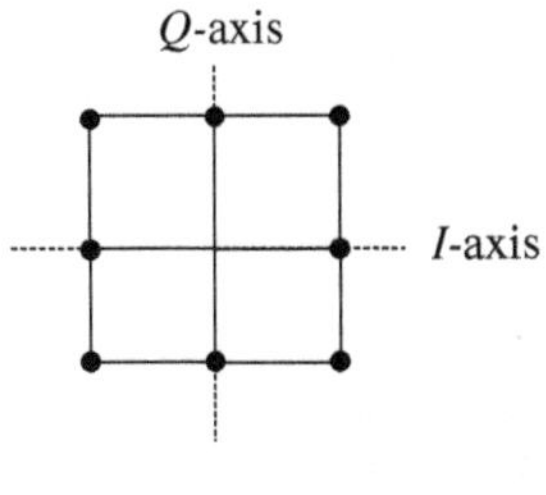

Square constellation

(b)

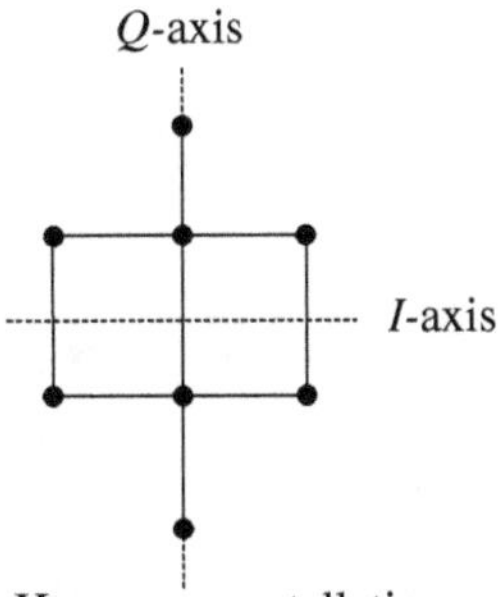

Hexagon constellation

12. Show all of the steps in the determination of the noise bandwidth of the duobinary spectrum expressed in (6.46).

13. Show all of the steps in the evaluation of the inverse Fourier transform of the frequency function $H(f)$ in (6.46) and confirm the corresponding impulse response.

14. Using the probability density distribution $p(y)$ and the random variable transformation $y = |x|$, where the random variable x is characterized as the normal distribution $N(A, \sigma_x)$ and A is the level of bipolar NRZ-L data sequence, show that the total-error probability (P_e) is the same as expressed in (6.54).

Hint: Use the decision regions in Figure 6.25b and modifications to the corresponding *pdfs* in (6.49) based on the magnitude transformation. The final result is obtained by evaluating the conditional probabilities $P_r(\text{error}|b=0)$ and $P_r(\text{error}|b=1)$, performing the integrations over the range $0 \le y \le A/2$, and computing P_e as the total-error probability.

15. With the same source data sequence used in Figure 6.23, show the corresponding coded sequences D_i, B_i, and ℓ_i using the modified duobinary-modulated waveform. Compare the results with each of the related introductory comments in Section 6.6.1. The source sequence is: $d_i =$ (11101100111000011101001101001).

16. Using the same B_i sequence in Problem 15, determine the sequence ℓ_i' based on the $10-1$ modified duobinary filter impulse response. As an alternative to determining the bit-by-bit sequence ℓ_i', simply show the relationship between ℓ_i' and ℓ_i.

17. Evaluate the optimum samples of the received duobinary ($L=3$) response to the binary source data sequence: 1010101010101010. Repeat the evaluation for the multilevel duobinary response using $L=4$, and 5. Discuss how the received modulated sequence might be used and sketch a functional implementation of your idea(s).

18. Determine the noise bandwidth of the modified duobinary-modulated waveform.

19. Show all of the steps in the evaluation of the inverse Fourier transform of the frequency function in (6.62) and confirm the corresponding impulse response.

20. Using the multilevel filter function expressed in (6.66) evaluate and plot (or sketch) the impulse response for $n = 4-7$.

REFERENCES

1. J.I. Marcum, "A Statistical Theory of Target Detection by Pulsed Radars," RAND Research Memorandum, RM-754, RAND Corporation, Santa Monica, CA, December 1, 1947.
2. J.I. Marcum, P. Swerling, "Studies of Target Detection by Pulsed Radar," IRE Transactions on Information Theory, Vol. IT-6, No. 2, Special Monograph Issue, April 1960.
3. J.I. Marcum, "Table of Q Functions," RAND Research Memorandum, RM-339, RAND Corporation, Santa Monica, CA, January 1, 1950.
4. D.E. Johansen, "New Techniques for Machine Computation of the Q-Function, Truncated Normal Deviates and Matrix Eigenvalues," Applied Research Laboratory, Sylvania Electronic Systems, Scientific Report No. 2, Waltham, MA, July 1961.
5. H. Nyquist, "Certain Factors Affecting Telegraph Speed," Bell System Technical Journal, Vol. 3, No. 2, pp. 324–346, April 1924.

6. H. Nyquist, "Certain Topics of Telegraph Transmission Theory," Proceedings of the IEEE, Vol. 90, No. 2, pp. 280–305, February 2002. This paper was first published in the Transaction of the AIEE, Vol. 47, No. 2, pp. 617–644, April 1928.
7. A. Lender, "The Duobinary Technique for High-Speed Data Transmission," AIEE Transactions of the Communication and Electronics, Vol. 82, No. 2, pp. 214–218, May 1963.
8. A. Lender, "Correlative Digital Communications Techniques," IEEE Transactions Communication Technology, Vol. 12, No. 4, pp. 128–135, December 1964.
9. A. Lender, *Digital Communications: Microwave Applications*, Chapter 7, K. Feher, Editor, "Correlative (Partial Response) Techniques and Applications to Digital Radio Systems," Prentice-Hall, Inc., Englewood Cliffs, NJ, 1981.
10. A. Lender, "Correlative Level Coding for Binary-Data Transmission," IEEE Spectrum, Vol. 3, No. 2, pp. 104–115, February 1966.
11. R.W. Lucky, J. Salz, E.J. Weldon, Jr., *Principles of Data Communication*, Chapter 4, "Baseband Pulse Transmission," pp. x83–92, McGraw-Hill, New York, 1968.
12. J.G. Proakis, "Maximum-Likelihood Sequence Estimation of Digital Signals Transmitted over Time-Dispersive Channels," Department of Electrical Engineering, Northeastern University, Boston, MA, August 1976.
13. F.K. Becker, E.R. Kretzmer, J.R. Sheehan, "A New Signal Format for Efficient Data Transmission," Bell System Technical Journal, Vol. 45, No. 5, pp. 755–758, May–June 1966.
14. R.W. Lucky, J. Salz, E.J. Weldon, Jr., *Principles of Data Communication*, Chapter 7, "Linear Modulation Systems," pp. 197–199, McGraw-Hill, New York, 1968.
15. R.W. Lucky, J. Salz, E.J. Weldon, Jr., *Principles of Data Communication*, Chapter 4, "Baseband Pulse Transmission," pp. 88–90, McGraw-Hill, New York, 1968.
16. D. Roth, "Mean Power Density Spectrum of a Continuous FSK Signal Modulated by the Duobinary Sequence," National Telecommunications Conference, NTC74, pp. 572–589, San Diego, CA, 1974.
17. W.J. Melvin, R.W. Middlestead, "Power Density Spectrum of M-ary Correlative Encoded MSK," International Conference on Communications, ICC77, pp. 3.6.60–3.7.63, Chicago, IL, 1977.
18. G.S. Deshpande, P.H. Wittke, "The Spectrum of Correlative Encoded FSK," International Conference on Communications, ICC78, pp. 25.3.1–25.3.5, Toronto, Ontario, 1978.
19. E.R. Kretzmer, "Binary Data Communication by Partial Response Transmission," IEEE Annual Communication Convention, Conference Record, pp. 451–455, Boulder, CO, 1965.
20. E.R. Kretzmer, "Generalization of a Technique for Binary Data Communication," IEEE Transactions on Communication Technology, Vol. 14, No. 1, pp. 67–68, February 1966.

7

M-ary CODED MODULATION

7.1 INTRODUCTION

M-ary coded waveforms operate in the power-limited region of Shannon's capacity curve shown in Figure 3.17 and obtain coding gain through bandwidth expansion. These codes do not provide forward error correction (FEC) as in convolutional and block codes, that achieve bandwidth expansion with the inclusion of parity-check bits, instead, the *M*-ary code waveforms achieve their performance using brute-force bandwidth expansion with code symbols of length $M=2^k$ corresponding to k information bits.* For $k \le 10$ bits, the decoding complexity is manageable; however, for larger values, the decoding complexity becomes unwieldy. The complexity issue arises because a decision is made based on the code symbol having the highest correlation among all of the M possible hypotheses. For example, with $k=14$ information bits per *M*-ary code symbol the decoding must perform over 16,000 correlations before making an optimal decision.

The binary coded *M*-ary sequences [1, 2] can be derived from: maximal length sequences (*M*-sequences), pseudo-random noise (PRN) generated sequences, and Hadamard or Walsh sequences. If the code symbols are sufficiently long, the correlation properties of the first two techniques will result in nearly orthogonal performance; however, the Hadamard [3] and Walsh [4, 5] sequences result in orthogonal codes with the following properties. Two continuous-time unit-energy waveforms $s_i(t)$ and $s_j(t)$ are orthogonal† if their normalized cross-correlation is characterized as

$$C_{ij} = \int_{-\infty}^{\infty} s_i(t)s_j^*(t)dt = \begin{cases} 1 & : i=j \\ 0 & : o.w. \end{cases} \quad \text{orthogonal codes} \tag{7.1}$$

Similarly, in the case of the binary sequences S_i and S_j each with an even number of *M*-bits and having identical bits in M' locations are orthogonal if the cross-correlation satisfies the condition

$$C_{ij} = \frac{2M'-M}{M} = \begin{cases} 1 & : i=j \\ 0 & : o.w. \end{cases} \quad \text{orthogonal codes} \tag{7.2}$$

The orthogonal condition in (7.2) requires that $M' = M/2$. The Gram-Schmidt [6] orthogonalization procedure generally results in non-binary sequences.

The Hadamard orthogonal codes are generated recursively as

$$H_{2n} = \begin{bmatrix} H_n & H_n \\ H_n & \bar{H}_n \end{bmatrix} \tag{7.3}$$

*Referring to Figure 3.1, the source-coded symbols υ_i: $i=0, \ldots, Q-1$, with $Q=k$, are typically binary elements $\upsilon_i = \{0,1\}$. Channel coding provides the *M*-ary code symbols v_j: $j=0, \ldots, M-1$ of length $M=2^k$ with binary elements $v_j=\{0,1\}$.

†Orthogonal signals with equal unit energy are referred to as being *orthonormal*.

Digital Communications with Emphasis on Data Modems: Theory, Analysis, Design, Simulation, Testing, and Applications,
First Edition. Richard W. Middlestead.

Companion website: www.wiley.com/go/digitalcommunications

where the subscript $n = 2^k$: $k > 1$ is the order of the Hadamard matrix. For unipolar binary data $b = \{0,1\}$ the scalar H_1 is initialized to $H_1 = b$ and selecting $H_1 = 0$, the second-order Hadamard matrix H_2 is

$$H_2 = \begin{bmatrix} 0 & 0 \\ 0 & 1 \end{bmatrix} \tag{7.4}$$

where, in general, $\bar{H}_n = 1 - H_n$. Using the unipolar bit to bipolar data translation* $d = 1 - 2b$ results in $d = \{1,-1\}$ with $\bar{H}_n = -H_n$. The data sequence in each row of H_n is referred to as a Walsh sequence or code and the Hadamard matrices form an orthogonal code set. For example, consider the order $n = 4$ Hadamard matrix expressed in terms of the bipolar data as

$$H_4 = \begin{bmatrix} 1 & 1 & 1 & 1 \\ 1 & -1 & 1 & -1 \\ 1 & 1 & -1 & -1 \\ 1 & -1 & -1 & 1 \end{bmatrix} \tag{7.5}$$

The rows of H_4 represent 4-ary orthogonal sequences that are uniquely assigned to $k = 2$ source bits represented by $2^k =$ 4-ary source symbols. In general, when k source bits are assigned to a source symbol, the order of the Hadamard matrix is $n = 2^k$. Therefore, the n-th order Hadamard matrix results in orthogonal binary codes (Walsh codes) of length $L = 2^k$. Randomly generated codes exhibit nearly orthogonal performance with high probability as $L \rightarrow \infty$; whereas, Walsh codes are strictly orthogonal for codes of all lengths of 2^k-bit code symbols.

The M-ary coded waveform modulation is characterized in terms of a vector with bipolar binary elements given by

$$X_i^m = d_i \sqrt{E} \delta_{im} \tag{7.6}$$

where $i = 0, \ldots, M-1$ and δ_{im} is the Kronecker delta function. The coding is shown in Figure 7.1 for the commonly used types of waveforms associated with M-ary coded signaling. The underlying requirement of an optimal orthogonal data decision in a noisy environment is that the noise-free correlation function, or the related normalized correlation coefficient ρ, satisfies the conditions

$$\rho = \frac{E\left(X_i^m X_j^m\right)}{\left|X_i^m\right|\left|X_j^m\right|} = \begin{cases} 1 & : i = j \\ 0 & : o.w. \end{cases} \tag{7.7}$$

*The transformation between unipolar data $b_i = (0,1)$ and polar data $d_i =$ (1,–1) is expressed as $d_i = 1 - 2b_i$; however, in some applications, this bit-mapping convention may be specified as $d_i' = 2b_i - 1$.

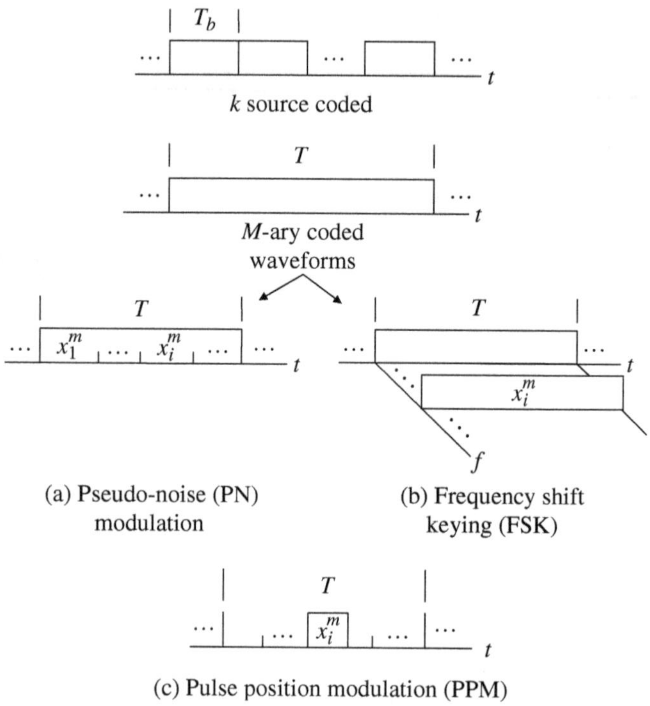

FIGURE 7.1 M-ary coded modulation waveforms.

7.2 COHERENT DETECTION OF ORTHOGONAL CODED WAVEFORMS

In this section, the symbol and bit-error performance is evaluated for the coherent detection of the M-ary coded waveforms operating in the additive white Gaussian noise (AWGN) channel. The performance of the optimum decision processing is evaluated in terms of the symbol-error probability with the corresponding bit-error probability computed based on the occurrence of random bit-errors in each block of M code-bits. Using the procedures and notations developed in Section 3.2 and Appendix C, the symbol-error probability is expressed as

$$\begin{aligned} P_{se}(m) &= 1 - P_{cs}(m) \\ &= 1 - \int_{y_m} p(y_1, \ldots, y_{M-1} | y_m) p(y_m) dy_m \quad : i \neq m \\ &= 1 - \int_{y_m} \prod_i^{M-1} p(y_i < y_m | y_m) p(y_m) dy_m \quad : i \neq m \text{ and independent} \\ &= 1 - \int_{\forall y_m} p(y_m) \left[\int_{y \in y_m} p(y) dy \right]^{M-1} dy_m \quad : i \neq m \text{ and } iid \end{aligned} \tag{7.8}$$

In the last equality, the random variables y_i are statistically independent-identically-distributed (*iid*) random variables.

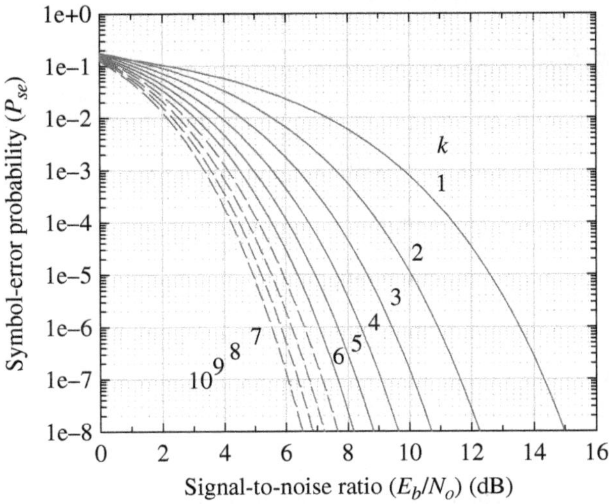

FIGURE 7.2 *M*-ary orthogonal symbol-error performance (coherent detection).

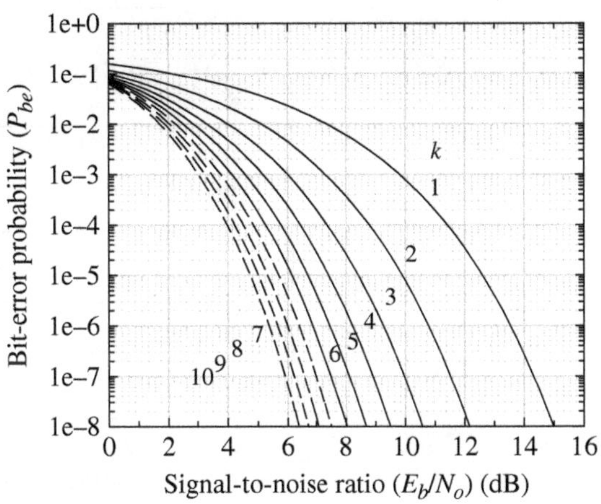

FIGURE 7.3 *M*-ary orthogonal bit-error performance with random bit mapping (coherent detection).

Under these conditions, the joint probabilities for y_i: $i \neq m$ are expressed by the $M-1$ products in the last equality of (7.8). With orthogonal coding and zero-mean AWGN channel noise, the random variables y_i are expressed by the zero-mean Gaussian *pdf*, however, the Gaussian random variable x, defined as $x = y_m$ in the following expressions, with mean value x_m and noise variance σ_n^2 is characterized as

$$p(x) = \frac{1}{\sqrt{2\pi}\sigma_n} e^{-(x-x_m)^2/(2\sigma_n^2)} \tag{7.9}$$

The mean value represents the noise-free peak level $x_m = M$ corresponding to the m-th correlator output associated with the M-ary symbol identifying the estimate of the k source data bits. Substituting these results into (7.8), the expression for the symbol-error is evaluated as

$$P_{se}(m) = 1 - \frac{1}{\sqrt{2\pi}\sigma} \int_{-\infty}^{\infty} e^{-(x-x_m)^2/(2\sigma^2)} \times \left[\frac{1}{\sqrt{2\pi}\sigma} \int_{-x/\sqrt{2}\sigma}^{\infty} e^{-y^2/(2\sigma^2)} dy \right]^{M-1} dx \tag{7.10}$$

The bit-errors associated with each symbol are randomly distributed and equally likely so that the bit-error probability is given by the random-bit mapping conversion

$$P_{be} = \frac{2^{k-1}}{2^k - 1} P_{se} \tag{7.11}$$

The symbol and bit-error performance results are plotted in Figures 7.2 and 7.3 respectively as a function of the signal-to-noise ratio measured in the bit bandwidth.

The coherent demodulation and detection of the *M*-ary orthogonal waveform is shown in Figure 7.4. Code synchronization and carrier frequency and phase acquisition are established using a preamble and symbol time and carrier phase tracking take place throughout the message reception. The detection is based on a *real* signal at an intermediate frequency (IF) filter output. The correlator or symbol integrator interval is $T = kT_b$ and the correlator output samples represent the matched filter outputs for the corresponding code symbol estimates.

Figure 7.5 expands the symbol-error performance for negative signal-to-noise ratios and it is apparent that the performance tends to be bounded by Shannon's coding limit of −1.59 dB; however, the diminishing improvement with increasing k suggests that unrealistically large code lengths are required. There is no known way to determine an optimum code as M approaches infinity. However, Gallager [7] discusses the *coding theorem*, which states that *a randomly selected code will, with high probability, be near optimum and the bound on the error performance is proportional to* $1/M$, so, as M approaches infinity, the error probability becomes zero. However, more practical solutions to approaching Shannon's coding limit are discussed in Section 8.12.

7.3 NONCOHERENT DETECTION OF *M*-ary ORTHOGONAL WAVEFORMS

The symbol and bit-error performance analysis of noncoherently detected *M*-ary orthogonal coded waveforms is similar to that used for the coherent detection analysis. The chief difference is that the decision variables are described by the Ricean *pdf*

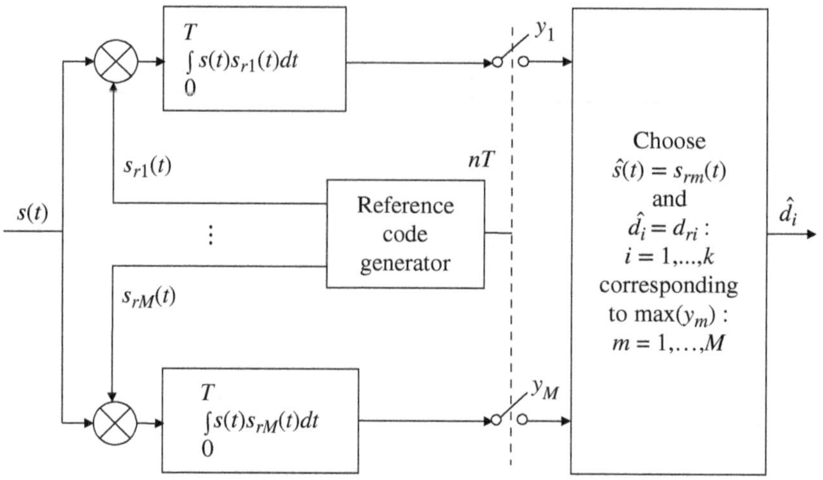

FIGURE 7.4 *M*-ary orthogonal waveform coherent detection with ideal code synchronization.

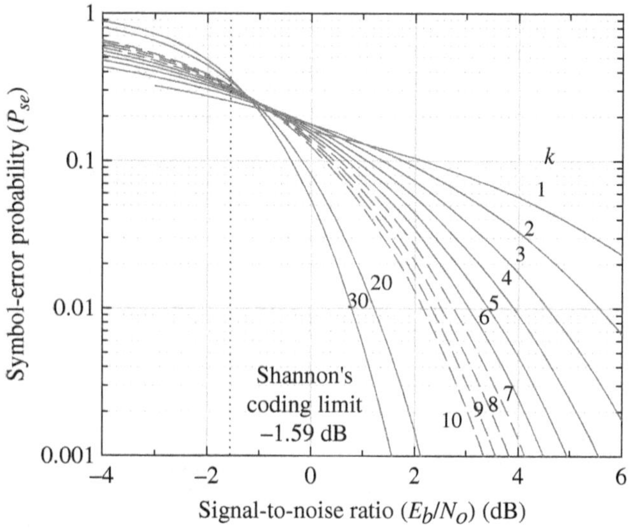

FIGURE 7.5 *M*-ary orthogonal symbol-error performance (coherent detection—expanded view).

$$p_R(r|A) = \frac{r}{\sigma_n^2} e^{-(r^2+A^2)/2\sigma_n^2} I_o\left(\frac{rA}{\sigma_n^2}\right) \tag{7.12}$$

In the following analysis, the bit-error performance is examined using a linear detector characterized by the transformation $y = r/\sigma_n$. In this case, a symbol decision is made based on the largest of the M decision variables y_i and with increasing M there is a negligible performance difference when using a square-law detector characterized by the transformation $y = r^2/\sigma_n^2$. That is, for sufficiently large values of M, selecting the hypothesis corresponding to the largest of M magnitudes is nearly identical to a decision based on selecting the largest of M squared magnitudes.

For the linear detector analysis, the decision variable is normalized as $y = r/\sigma_n$ and the signal amplitude is normalized as $a = A/\sigma_n$. Based on this transformation, the *pdf* for the linear detector is given by

$$p_Y(y|a) = ye^{-(y^2+a^2)/2} I_o(ya) \tag{7.13}$$

The probability of a correct symbol decision is computed as

$$P_{cs} = \int_0^\infty [P(y<y_T|0)]^{M-1} p_Y(y_T|a) dy_T \tag{7.14}$$

where

$$P(y<y_T|0) = \int_0^{y_T} ye^{-y^2/2} dy = 1 - e^{-y_T^2/2} \tag{7.15}$$

Substituting this result in the expression for P_{cs} and replacing y_T with y gives

$$P_{cs} = \int_0^\infty \left[1 - e^{-y^2/2}\right]^{M-1} ye^{-(y^2+a^2)/2} I_o(ya) dy \tag{7.16}$$

Applying the binomial theorem (see Section 1.14.2, No. 2) to the term in the square brackets results in

$$\begin{aligned} P_{cs} &= \int_0^\infty \left[1 - \sum_{m=1}^{M-1} \binom{M-1}{m} (-1)^{m+1} e^{-my^2/2}\right] ye^{-(y^2+a^2)/2} I_o(ya) dy \\ &= 1 - e^{-a^2/2} \sum_{m=1}^{M-1} \binom{M-1}{m} (-1)^{m+1} \int_0^\infty ye^{-(m+1)y^2/2} I_o(ya) dy \end{aligned} \tag{7.17}$$

The integral in (7.17) can be expressed in the form of the Ricean distribution by substituting $y=\lambda/\sqrt{m+1}$, yielding

$$\int_0^\infty y e^{-(m+1)y^2/2} I_o(ya)dy = \frac{1}{m+1}\int_0^\infty \lambda e^{-\lambda^2/2} I_o\left(\frac{\lambda a}{\sqrt{m+1}}\right)d\lambda \tag{7.18}$$

Letting $b=a/\sqrt{m+1}$ and multiplying by $e^{-b^2/2}e^{b^2/2}$ results in the integral

$$\int_0^\infty y e^{-(m+1)y^2/2} I_o(ya)dy = \frac{e^{a^2/2(m+1)}}{m+1}\int_0^\infty \lambda e^{-(\lambda^2+b^2)/2} I_o(\lambda b)d\lambda = \frac{e^{a^2/2(m+1)}}{m+1} \tag{7.19}$$

The integral on the *rhs* of (7.19) is the Ricean distribution integrated over all $\lambda \geq 0$ and is equal to 1. Therefore, upon substituting (7.19) into (7.17) and recognizing that the signal-to-noise ratio in the symbol bandwidth is $\gamma=a^2/2=A^2/2\sigma_n^2$, the symbol-error is evaluated as $P_{se}=1-P_{cs}$ with the result

$$P_{se}=\sum_{m=0}^{M-1}\binom{M-1}{m}\frac{(-1)^m}{m+1}e^{-\gamma m/(m+1)} \tag{7.20}$$

With random bit mapping, the bit-error performance is evaluated as

$$P_{be}=\frac{2^{k-1}}{2^k-1}\left[\sum_{m=1}^{M-1}\binom{M-1}{m}\frac{(-1)^{m+1}}{m+1}e^{-\gamma m/(m+1)}\right] \tag{7.21}$$

where $k=\log_2(M)$ is the number of source bits-per-symbol. The signal-to-noise ratio in the bandwidth of the source bits is related to that in the symbol bandwidth as $\gamma_b=E_b/N_o=\gamma/k$. For binary modulation $k=1$, $M=2$, and (7.21) reduces to

$$P_{be}=\frac{1}{2}e^{-\gamma/2} \quad :\text{binary case} \tag{7.22}$$

where, in this case, $\gamma=\gamma_b=E_b/N_o$.

Evaluation of the symbol-error probability using (7.20) is impractical for values of $M>64$ because of numerical round-off errors associated with the large binomial coefficients. Under this condition numerical integration of the correct-symbol probability given by (7.16) is used and the symbol error is computed as $P_{se}=1-P_{cs}$. The numerical integration is performed using an integration increment of $\Delta y=\sigma/100$ where $\sigma=1$ is the standard deviation of the normalized Ricean distribution. The integration limits are extended over the range $\{0,\ldots,\max(100\sigma,10a)\}$ where $a=\sqrt{2\gamma}$ is the mean value of the normalized distribution. Based on these evaluation conditions, the performance results for the symbol and bit-error probabilities are plotted in Figures 7.6 and 7.7 as a function of $\gamma_b=\gamma/k$ for various values of $k=\log_2(M)$.

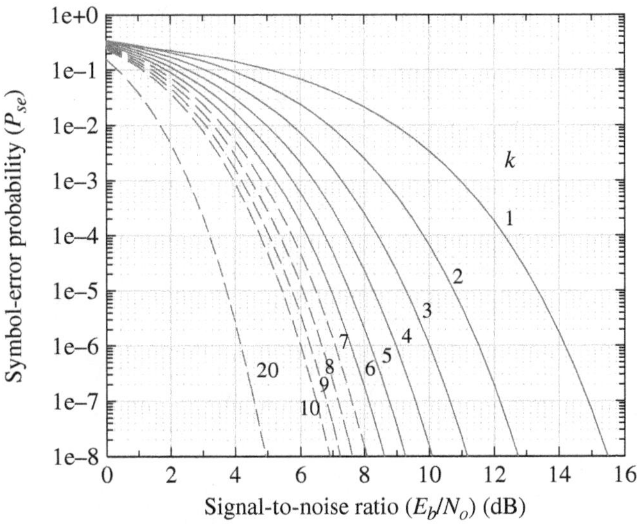

FIGURE 7.6 *M*-ary orthogonal symbol-error performance (noncoherent detection).

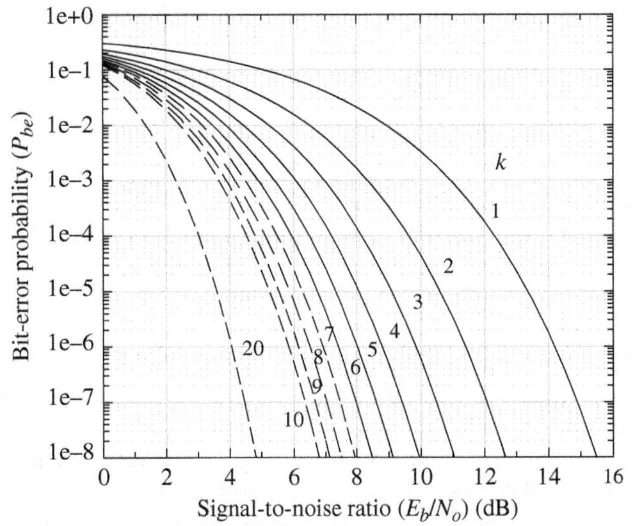

FIGURE 7.7 *M*-ary orthogonal bit-error performance with random bit mapping (noncoherent detection).

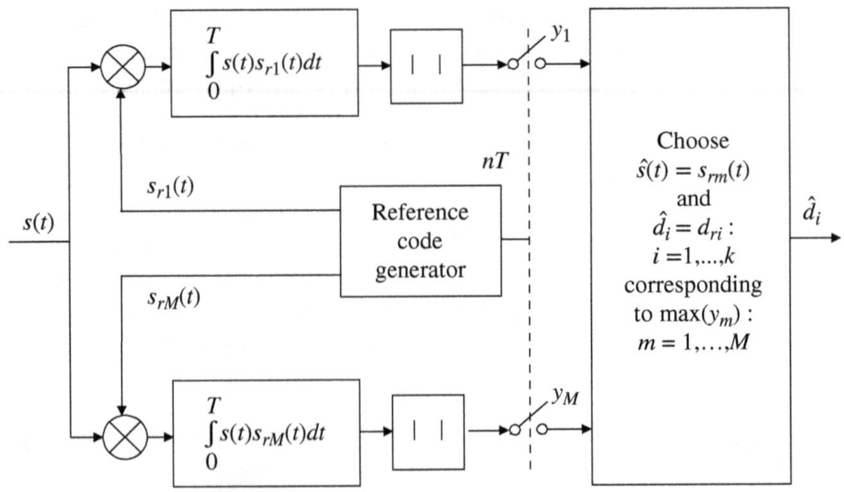

FIGURE 7.8 *M*-ary orthogonal waveform noncoherent detection with ideal code synchronization.

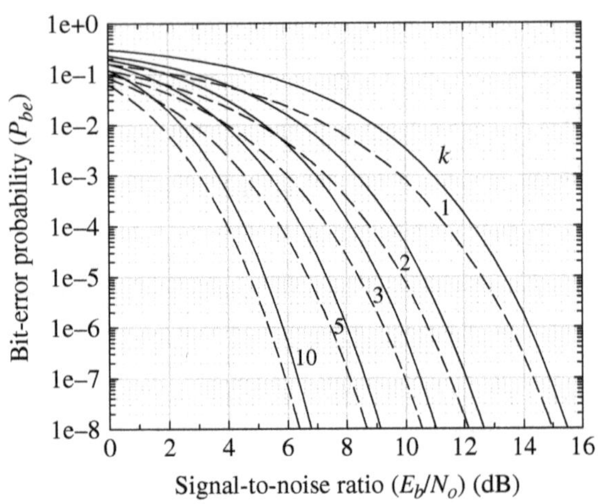

FIGURE 7.9 Comparison of coherent and noncoherent detection of *M*-ary orthogonal coded waveform. Solid—noncoherent, dashed—coherent.

The noncoherent detection of the *M*-ary orthogonal coded waveform is shown in Figure 7.8. Code synchronization and carrier frequency acquisition are established using a preamble and symbol time and frequency tracking take place throughout the message reception. Because of the noncoherent detection processing it is not necessary to provide phase tracking. The correlator or symbol integrator interval is $T = kT_b$ and the output samples y_m: $m = 1, \ldots, M$ represent the magnitude of the optimum matched filter outputs for the corresponding code symbol.

Figure 7.9 compares the performance of coherent and noncoherent detection for selected values of k. These results indicate that the performance gains using coherent detection are less significant as k is increased; however, the difference has a weak dependence on k.

7.4 COHERENT DETECTION OF *M*-ary BIORTHOGONAL WAVEFORMS

Biorthogonal *M*-ary coded waveforms are also referred to as complementary orthogonal (CO) coded waveforms in which $M/2$ orthogonal codes are associated with the transmission and detection of a unique set of *M*-ary orthogonal codes. The major distinction between biorthogonal and orthogonal codes is that the demodulator correlator output decision for biorthogonal codes is based on distinguishing between $\pm M/2$ hypotheses and for orthogonal codes the decision is based on distinguishing between M hypotheses.

Drawing upon the preceding Hadamard matrix discussion with polar source data $d_i = \{1,-1\}$, the square matrix H_n is partitioned as

$$H_n = \begin{pmatrix} G_n \\ \bar{G}_n \end{pmatrix} \tag{7.23}$$

where G_n is a $M/2 \times M$ rectangular matrix formed as $G_n = \left(H_{n/2}\ H_{n/2}\right)$ and for bipolar data $\bar{G}_n = -G_n$. Using the Hadamard matrix H_4 expressed in (7.5) as an example, the submatrices G_4 and $\bar{G}_4$ are constructed as follows:

$$H_4 = \begin{bmatrix} 1 & 1 & 1 & 1 \\ 1 & -1 & 1 & -1 \\ -1 & -1 & -1 & -1 \\ -1 & 1 & -1 & 1 \end{bmatrix};\quad G_4 = \begin{bmatrix} 1 & 1 & 1 & 1 \\ 1 & -1 & 1 & -1 \end{bmatrix};$$

$$\bar{G}_4 = \begin{bmatrix} -1 & -1 & -1 & -1 \\ -1 & 1 & -1 & 1 \end{bmatrix} \tag{7.24}$$

With biorthogonal coding, the 2^k-ary source data symbols are uniquely associated with the $M=2^k$-ary orthogonal codes; however, only $M/2$ of the orthogonal codes corresponds to the rows of G_n with the remaining $M/2$ corresponding to row of $-G_n$. The optimum demodulator decision chooses the largest of the $M/2$ magnitudes of the optimally sampled demodulator correlator outputs and then chooses the correct received symbol based on the binary decision or sign of the correlator output. A significant advantage in using biorthogonal coding is that the number of demodulator correlations is reduced by 2 compared to orthogonal coding.

The symbol-error probability is evaluated in a manner similar to that leading to (7.9) with M replaced by $M/2$ and the additional caveats that: a correct decision among the $M/2$ correlation outputs is based on the decision variable $|x_m|=x_{max}$ and the correct polarity decision is based on the sign of x_m. These conditions lead to the expression for the symbol-error probability evaluated as follows:

$$P_{se}(m)=1-\frac{1}{\sqrt{2\pi}\sigma_n}\int_0^{\infty}e^{-(x_i-x_m)^2/(2\sigma_n^2)}\times\left[\frac{1}{\sqrt{2\pi}\sigma_n}\int_{-x_i/\sqrt{2}\sigma}^{x_i/\sqrt{2}\sigma}e^{-y_i^2/(2\sigma_n^2)}dy_i\right]^{\frac{M}{2}-1}dx_i \quad (7.25)$$

The expression for the symbol-error probability in (7.25) can be simplified for the biorthogonal coding by applying the following transformations. Define the first integral in (7.25) in terms of the new random variable $x=x_i/\sqrt{2}\sigma_n-x_m/\sqrt{2}\sigma_n$ with $dx_i=\sqrt{2}\sigma_n dx$ and the integration limits $-x_m/\sqrt{2}\sigma_n=-\sqrt{\gamma_s}$ to ∞, where γ_s is the signal-to-noise ratio in the symbol bandwidth. The second integral in (7.25) involves the $M/2-1$ zero-mean *iid* random variables y_i and is defined in terms of the new random variable $y=y_i/\sqrt{2}\sigma_n$ with $dy_i=\sqrt{2}\sigma_n dy$ and the integration limits $-x_i/\sqrt{2}\sigma_n=-x-\sqrt{\gamma_s}$ to $x_i/\sqrt{2}\sigma_n=x+\sqrt{\gamma_s}$. Upon applying these transformations, (7.25) simplifies to

$$P_{se}(m)=1-\frac{1}{\sqrt{\pi}}\int_{-\sqrt{\gamma_s}}^{\infty}e^{-x^2}\left[\frac{1}{\sqrt{\pi}}\int_{-x-\sqrt{\gamma_s}}^{x+\sqrt{\gamma_s}}e^{-y^2}dy\right]^{\frac{M}{2}-1}dx \quad (7.26)$$

Equation (7.26) is in the form expressed by Lindsey and Simon [8] who provide an upper bound on the symbol-error probability expressed in terms of the complementary error function as

$$P_{se}\le\frac{M-2}{2}erfc\left(\sqrt{\frac{\gamma_b}{2}}\right)+\frac{1}{2}erfc\left(\sqrt{\gamma_b}\right) \quad (7.27)$$

Referring to Figure 7.12 the probability of making a bit-error is the weighted sum of the probability of a sign error and the symbol-error probability, that is,

$$P_{be}=\frac{1}{2}P_r(\text{sign error})+\frac{1}{2}P_{se} \quad (7.28)$$

Equation (7.28) is evaluated in terms of the complementary error function as [8]

$$P_{be}=\frac{1}{2\sqrt{\pi}}\int_{-\infty}^{-\sqrt{\gamma_b}}e^{-x^2}\left[erfc\left(x+\sqrt{\gamma_b}\right)-1\right]^{\frac{M}{2}-1}dx+\frac{M-2}{4}erfc\left(\sqrt{\frac{\gamma_b}{2}}\right)+\frac{1}{4}erfc\left(\sqrt{\gamma_b}\right) \quad (7.29)$$

The following evaluation of the symbol-error performance is based on the solution of (7.26) using numerical integration and the results are plotted in Figure 7.10 for $k=1$ through 6. Figure 7.11 compares the performance of the biorthogonal and orthogonal coded waveforms. For $k=1$ and 2 bits per symbol, the biorthogonal coded symbol-error performance is identical to that of binary phase shift keying (BPSK) and quadrature phase shift keying (QPSK) modulated waveforms that are also based on the polarity of the matched filter or symbol correlator outputs. This is contrasted with orthogonal coding where, for example, with $k=2$ one of four orthogonal codes is transmitted and the correct decision is made based on the greatest magnitude of the four correlator outputs. The performance advantage of biorthogonal coding is 3 dB for $k=1$ and only a few tenths of a dB for $k=2$. The plot of orthogonal coding with $k=3$ is shown in Figure 7.11; however, the performance is difficult to distinguish from that of the biorthogonal code performance and for $k>3$ the difference is negligible.

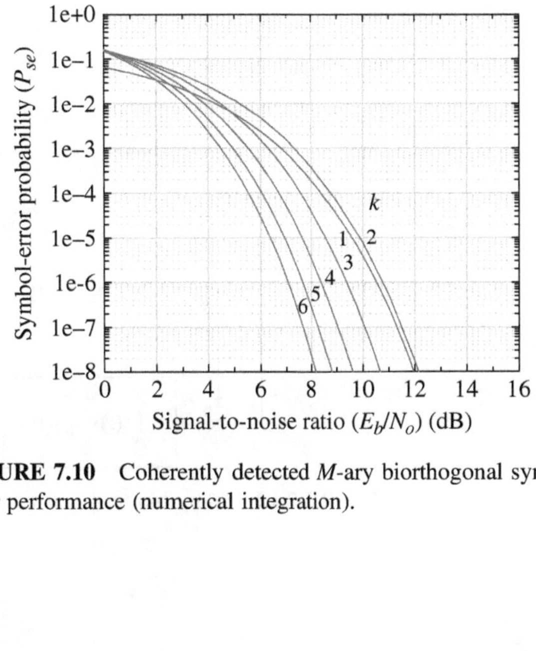

FIGURE 7.10 Coherently detected M-ary biorthogonal symbol-error performance (numerical integration).

The demodulation and detection of the M-ary biorthogonal waveform is shown in Figure 7.12. As indicated, ideal code synchronization including carrier frequency and phase tracking has been established, perhaps through the use of a preamble. The implementation shown is based on a *real* signal, $s(t)$, at the output of the IF filter; however, baseband demodulation involving inphase and quadrature (I/Q) processing will yield identical results. The correlator or symbol integrator interval is $T = kT_b$ and the output samples x_m : $m = 1, \ldots, M/2$ represent the optimum matched filter outputs corresponding to the respective reference coded signals $s_{rm}(t)$.

To verify the theoretical symbol-error performance of the M-ary biorthogonal coded waveform, a Monte Carlo simulation is used and the results are shown in Figure 7.13 as the *circled* data points with the *solid* curves corresponding to the theoretical results from Figure 7.10. The simulation results are based on 1 M symbols for each signal-to-noise ratio for $P_{be} \geq 10^{-4}$ and 10 M symbols for $P_{be} < 10^{-4}$. The simulated results are in good agreement with the theoretical results. The simulations represent $k = 1$ through 5 with 2^k code-bits per symbol. The curve for $k = 1$ corresponds to binary modulation and the performance clearly stands out among the Hadamard codes corresponding to $k = 2$ through 5. The Hadamard matrices corresponding to the values of $k = 2$ through 4 are listed in Table 7.1.

The bit-error performance of coherently detected M-ary orthogonal modulation based on the Monte Carlo simulations provides a good check of the theory and/or the simulation program over the entire range of signal-to-noise ratios of interest. Figure 7.14 shows the simulated bit-error performance using the simulation processing described in Figure 7.12 with the G_n and $-G_n$ submatrices constructed from Table 7.1. The *triangular* data points correspond to the Monte Carlo simulated performance and the *solid* lines connecting the *triangles* are simply linear interpolated plot lines. A detailed comparison with the symbol-error performance in Figure 7.13 reveals that the binary performance

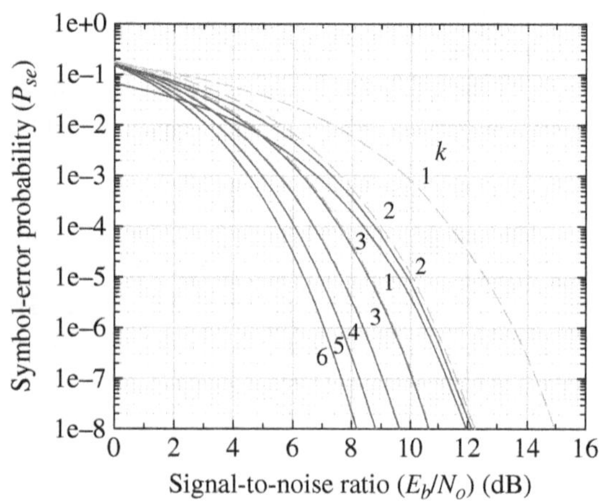

FIGURE 7.11 Comparison of coherently detected M-ary biorthogonal (solid) and orthogonal (dashed) symbol-error performance.

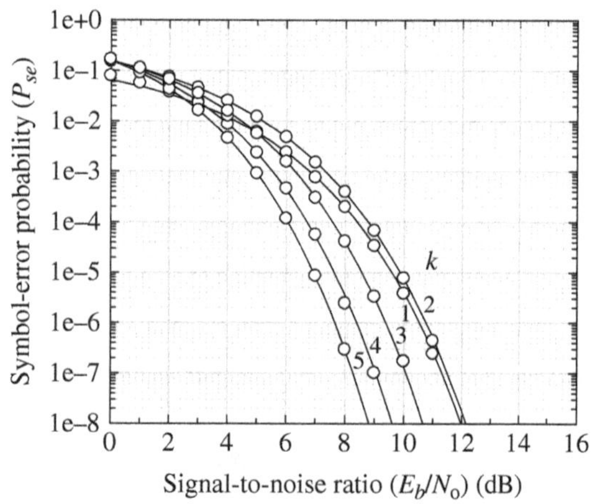

FIGURE 7.13 Coherently detected M-ary biorthogonal symbol-error performance comparison with Monte Carlo simulation. Circled—Monte Carlo, solid—theory.

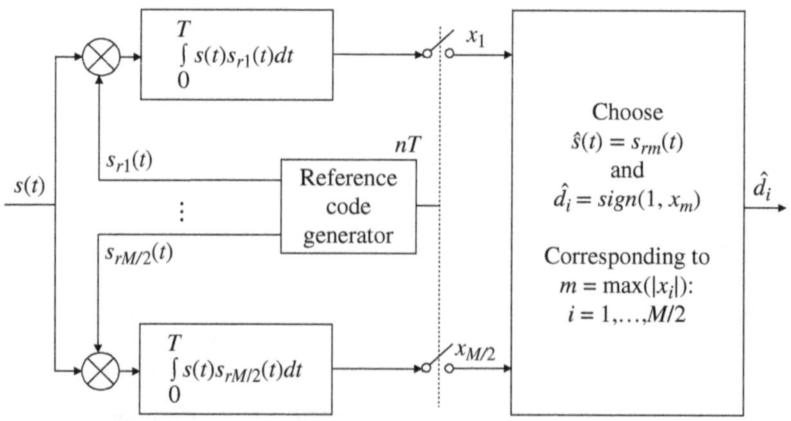

FIGURE 7.12 M-ary biorthogonal waveform coherent detection with ideal code synchronization.

TABLE 7.1 Hadamard matrices for $M/2 \leq 16$.

k	1	2	3						4							
Walsh code	1	2	3	4	5	6	7	8	9	10	11	12	13	14	15	16
1	1	1	1	1	1	1	1	1	1	1	1	1	1	1	1	1
2	1	−1	1	−1	1	−1	1	−1	1	−1	1	−1	1	−1	1	−1
3	1	1	−1	−1	1	1	−1	−1	1	1	−1	−1	1	1	−1	−1
4	1	−1	−1	1	1	−1	−1	1	1	−1	−1	1	1	−1	−1	1
5	1	1	1	1	−1	−1	−1	−1	1	1	1	1	−1	−1	−1	−1
6	1	−1	1	−1	−1	1	−1	1	−1	1	1	−1	−1	1	−1	1
7	1	1	−1	−1	−1	−1	1	1	1	1	−1	−1	−1	−1	1	1
8	1	−1	−1	1	−1	1	1	−1	1	−1	−1	1	−1	1	1	−1
9	1	1	1	1	1	1	1	1	−1	−1	−1	−1	−1	−1	−1	−1
10	1	−1	1	−1	1	−1	1	−1	−1	1	−1	1	−1	1	−1	1
11	1	1	−1	−1	1	1	−1	−1	−1	−1	1	1	−1	−1	1	1
12	1	−1	−1	1	1	−1	−1	1	−1	1	1	−1	−1	1	1	−1
13	1	1	1	1	−1	−1	−1	−1	−1	−1	−1	−1	1	1	1	1
14	1	−1	1	−1	−1	1	−1	1	−1	1	−1	1	1	−1	1	−1
15	1	1	−1	−1	−1	−1	1	1	−1	−1	1	1	1	1	−1	−1
16	1	−1	−1	1	−1	1	1	−1	−1	1	1	−1	1	−1	−1	1

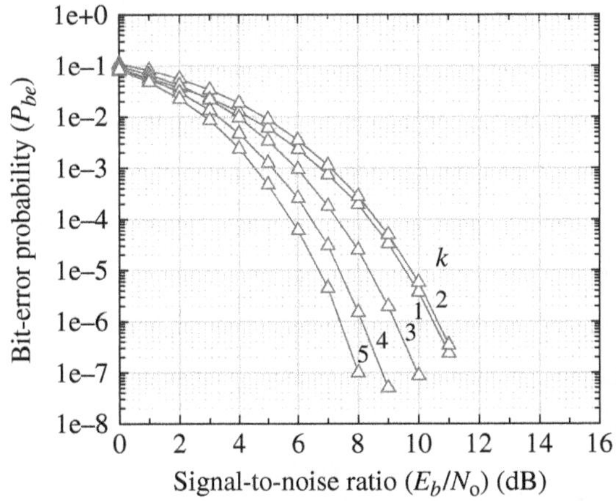

FIGURE 7.14 Coherently detected M-ary biorthogonal bit-error performance using Monte Carlo simulation.

is identical. For $k > 1$, a careful comparison of the results supports the use of the theoretical symbol-to-bit mapping as expressed in (7.11) when gray coding is not used.

ACRONYMS

AWGN	Additive white Gaussian noise
BPSK	Binary phase shift keying
CO	Complementary orthogonal (code)
FEC	Forward error correction (code)
IF	Intermediated frequency
I/Q	Inphase and quadrature (channels or rails)
PRN	Pseudo-random noise (sequences)
QPSK	Quadrature phase shift keying

PROBLEMS

1. For the Hadamard matrix H_3 using polar (±) binary data and, defining the sequences S_i as the row of H_3, evaluate the correlations $C_{i,j}$ for the sequences S_i and S_j for all combinations of $i,j = 1, \ldots, 4$. Note that these sequences are perfectly aligned so there are no correlation lags.
2. Show that the integral in (7.15) is evaluated as $1 - e^{-y_T/2}$.
3. Show that applying the binomial expansion to $\left[1 - e^{-y/2}\right]^{M-1}$ in (7.16) results in the expression given in (7.17).
4. By performing the variable substitutions following (7.17) show that the integral in (7.17) is evaluated as in (7.19). Show all of the details and explain the second equality in (7.19).
5. Show that the symbol-error probability given by (7.26) reduces to the bit-error probability of BPSK when $M = 2$ corresponding to binary modulation.

REFERENCES

1. R. Gold, *"Properties of Linear Binary Encoding Sequences,"* Robert Gold Associates, Los Angeles, CA, July 1975.
2. N. Ahmed, K.R. Rao, *Orthogonal Transforms for Digital Signal Processing*, Springer-Verlag, New York, 1975.
3. Y. Taki, M. Hatori, "PCM Communication System Using Hadamard Transformation," Electronics and Communications in Japan, Vol. 49, No. 11, pp. 347–267, November 1966.

4. H.F. Harmuth, "Applications of Walsh Functions in Communications," IEEE Spectrum, Vol. 6, pp. 82–91, November 1969.
5. R.D. Larsen, W.R. Madych, "Walsh-Like Expansions and Hadamard Matrices," IEEE Transactions on Acoustics, Speech, and Signal Processing, Vol. ASSP-24, No. 1, pp. 71–75, February 1976.
6. P.M. Derusso, R.J. Roy, C.M. Close, *State Variables for Engineers*, John Wiley & Sons, New York, 1965.
7. R.A. Gallager, "A Simple Derivation of the Coding Theorem and Some Application," IEEE Transactions on Information Theory, Vol. II, p. 318, 1965.
8. W.C. Lindsey, M.K. Simon, *Telecommunication Systems Engineering*, pp. 198, 210–212, Dover Publications, Inc., New York, 1991.

8

CODING FOR IMPROVED COMMUNICATIONS

8.1 INTRODUCTION

Before delving into the details of error detection and forward error correction (FEC) techniques, it is useful to describe several baseband coding techniques. In the following sections, various forms of baseband pulse code modulation (PCM) are discussed. PCM codes and data compression codes are source coding techniques. Following the description of PCM waveforms, a variety of important coding-related topics are introduced: gray coding* in Section 8.3 and differential coding in Section 8.4; pseudo-random noise (PRN) sequences in Section 8.5; binary cyclic codes in Section 8.6; cyclic redundancy check (CRC) codes in Section 8.7; data randomizing, or scrambling in Section 8.8; data interleaving in Section 8.9. Following these topics, several forms of FEC channel coding are discussed; Wagner coding and decoding in Section 8.10; convolutional coding and Viterbi decoding in Section 8.11; turbo codes (TCs), parallel concatenated convolutional codes (CCCs) (PCCCs), serially CCCs (SCCCs), double parallel CCCs (DPCCCs), and double serially CCCs (DSCCCs) with reference to the related hybrid CCCs (HCCCs), and self-concatenated codes (SCCs) in Section 8.12† (the nonturbo codes are collectively referred to as *turbo-like codes*); low-density parity-check (LDPC) codes, product codes (PCs), and turbo product codes (TPCs) are introduced in Section 8.13; and Bose–Chaudhuri–Hocquenghem (BCH) codes, including *M*-ary Reed–Solomon (RS) codes and RS Viterbi (RSV) codes, are discussed in Section 8.14.

The subject of coding to improve the performance of communication systems is broad and, in certain instances, requires knowledge in specialized areas like matrix and Boolean algebra and the theory of groups, rings, and fields. For example, an understanding of polynomials in the Galois field (GF) is crucial in the coding and decoding of both block and cyclic codes. These disciplines will be engaged only to the extent that they can be applied in a practical way to the coding and decoding of the various techniques considered. Several coding techniques are listed in Figure 8.1.

8.2 PULSE CODE MODULATION

Baseband signals are characterized as those signals with the signal energy concentrated around zero frequency, although the zero-frequency power spectral density, $S(f = 0)$, may be zero. This is contrasted to bandpass signals that are associated with a carrier frequency around which the signal energy is concentrated. The data from an information source is typically formatted as binary data with a *mark bit* representing logic 1 and *space bit* representing logic 0 data. In cases involving analog information sources, the information is time-sampled and amplitude-quantized and represented as binary data. In either event, the resulting binary data can be represented as a form of PCM [2] that contains all of the source information. The binary formatted data is typically denoted

*Gray and differential coding are introduced in Chapter 4.

†The term turbo code is used in recognition of the pioneering work of Berrou et al. [1] and is constructed as a PCCC with two CCs.

Digital Communications with Emphasis on Data Modems: Theory, Analysis, Design, Simulation, Testing, and Applications,
First Edition. Richard W. Middlestead.

Companion website: www.wiley.com/go/digitalcommunications

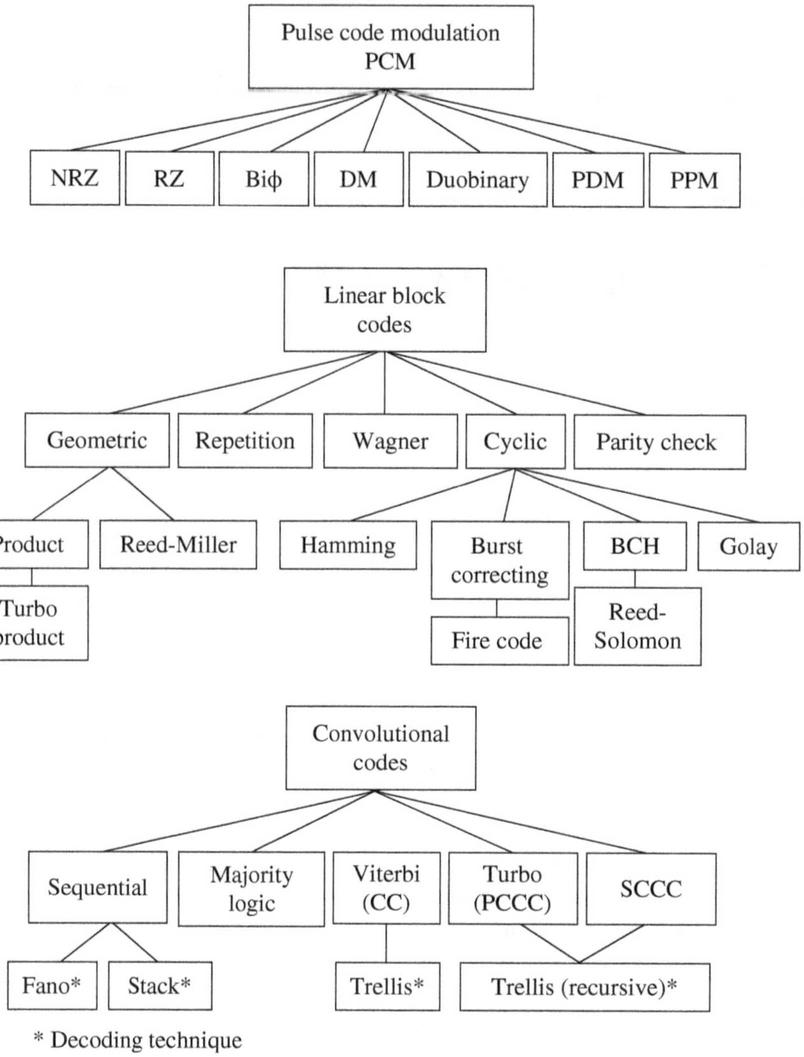

FIGURE 8.1 Channel coding techniques.

as a series of pulses, or bits, of duration T_b with amplitude levels of 1 and 0. A series of data defined in this manner is referred to as unipolar non-return-to-zero (NRZ) formatted data [3], and with the binary amplitude levels of 1 and −1 the data format is referred to as polar NRZ. These and other baseband data formats are the subject of this section. There are a number of PCM codes that have been devised to address issues unique to certain applications; however, the most significant characteristics of the PCM codes are the power spectral density (PSD), bit synchronization properties, and the bit-error performance. Other important characteristics involve the error detection and correction (EDAC) capabilities; however, the more powerful techniques in this regard use redundancy through increased bandwidth or decision space. A review of 25 baseband PCM codes and their properties is presented by Deffebach and Frost [4] and much of this section is based on their work. Another treatment of this subject is given by Stallings [5].

The PSD, with units of watt-seconds/Hz, of binary PCM modulated random data sequences, with mark and space probabilities given by p and $1-p$, respectively, is evaluated by Bennett [6], Titsworth and Welch [7], and Lindsey and Simon [8] as

$$S(f)=\frac{p(1-p)}{T_b}|H_m(f)+H_s(f)|^2 + \frac{1}{T_b^2}\sum_{n=-\infty}^{\infty}|pH_m(f)-(1-p)H_s(f)|^2\,\delta\!\left(f-\frac{n}{T_b}\right) \quad (8.1)$$

where T_b is the bit interval and $H_m(f)$ and $H_s(f)$ are the underlying spectrums of the mark and space bits. With unipolar formatted data, the mark data spectrum is defined as $H_m(f)=H(f)$ and the space data spectrum is zero, that is,

$H_s(f) = 0$, and upon substituting these results into (8.1) yields the unipolar PSD

$$S(f) = \frac{p(1-p)}{T_b}|H(f)|^2 + \frac{p^2}{T_b^2}\sum_{n=-\infty}^{\infty}|H(f)|^2\delta\left(f-\frac{n}{T_b}\right) \quad : \text{unipolar format} \tag{8.2}$$

With polar formatted data the space data spectrum is defined as $H_s(f) = -H_m(f)$ so the polar PSD is given by

$$S(f) = \frac{4p(1-p)}{T_b}|H(f)|^2 + \frac{(1-2p)^2}{T_b^2}\sum_{n=-\infty}^{\infty}|H(f)|^2\delta\left(f-\frac{n}{T_b}\right) \quad : \text{polar format} \tag{8.3}$$

These results are applied in the followings sections to evaluate the spectrums of various forms of unipolar and polar formatted PCM waveforms. Houts and Green [9] compare the spectral bandwidth utilization in terms of the percent of total power in bandwidths of $\{0.5,1,2,3\}R_b$ for a variety of binary baseband coded waveform. In Section 8.2.5, the PCM demodulator bit-error performance is examined.

8.2.1 NRZ Coded PCM

The designation NRZ denotes the binary format that characterizes a *mark* bit as being one amplitude level and *space* bit as another. When the space bit is represented by a zero level, the format is called *unipolar* NRZ. In contrast, when the level of the space bit is the negative of the mark bit level, the designation *polar* NRZ is used. The designation NRZ-L indicates that the level of the binary format is changed whenever the level of the source data changes. These two PCM data formats are shown in Figure 8.2. The *bipolar* NRZ-L format is a special tri-level code for which a space source bit corresponds to a zero output level, while the level corresponding to the mark source data is bipolar, alternating between plus and minus levels.

The designation NRZ-M (NRZ-S) indicates that changes in the formatted data occur when the source data is a mark (space); otherwise, the coded level remains the same. The unipolar NRZ-M and NRZ-S coded waveforms are depicted in Figure 8.3. This baseband coding results in differentially encoded data and is discussed in more detail in Section 8.4.

The advantage of differentially encoded data is that correct bit detection is maintained in the demodulator with a 180° error in the carrier phase as might occur in the channel or the demodulator carrier tracking loops. The penalty for using differential data coding is a degradation of the bit-error performance with a 2 : 1 increase in the bit errors at high signal-to-noise ratios. Because the source data is encoded in the transitions of the PCM coded data, the absolute level is not as important; however, polar NRZ data with binary phase shift keying (BPSK) modulation results in the optimum detection performance through an additive white Gaussian noise (AWGN) channel.

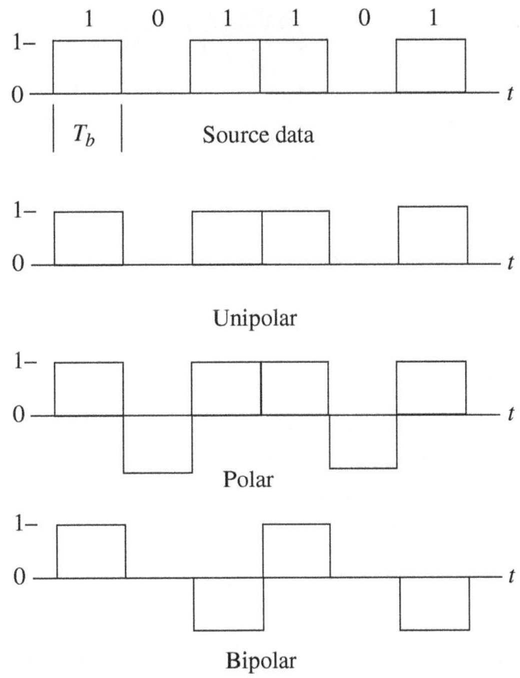

FIGURE 8.2 Unipolar, polar, and bipolar NRZ-L formatted data.

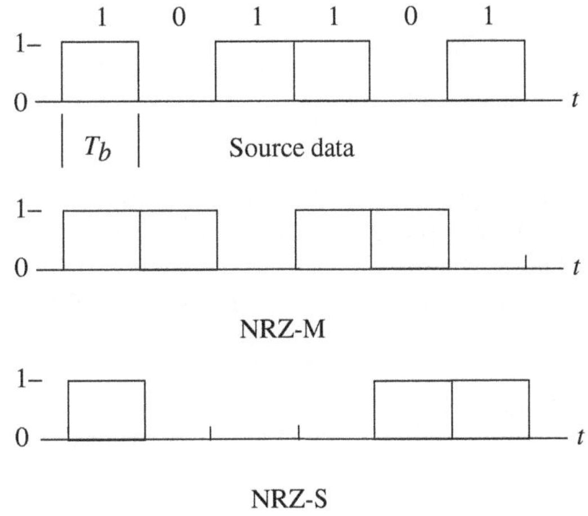

FIGURE 8.3 Unipolar NRZ-M,-S formatted data.

The PSD of the unipolar NRZ level, mark, space NRZ-L,-M,-S formatted data is evaluated using (8.2) with the underlying spectrum characterized by the unit amplitude pulse $rect(t/T_b - 1/2)$ with the square of the spectrum magnitude expressed as

$$|H(f)|^2 = T_b^2 \frac{\sin^2(\pi f T_b)}{(\pi f T_b)^2} \quad : \text{NRZ} \tag{8.4}$$

The sampled spectrum expressed by (8.4) results in zero magnitude at $f = n/T_b$ for all n except $n = 0$, in which case, the magnitude is T_b^2. With random source data, corresponding to $p = 1/2$, (8.2) becomes

$$S(f) = \frac{T_b \sin^2(\pi f T_b)}{4 \ (\pi f T_b)^2} + \frac{1}{4}\delta(f) \quad : \text{unipolar NRZ-L,-M,-S} \tag{8.5}$$

The PSD for the polar NRZ coded waveform is evaluated in a similar way using (8.3) and (8.4) and, assuming random data with $p = 1/2$, the result is

$$S(f) = T_b \frac{\sin^2(\pi f T_b)}{(\pi f T_b)^2} \quad : \text{polar NRZ-L,-M,-S} \tag{8.6}$$

Therefore, the polar NRZ coded waveform has 6 dB more energy in the underlying spectrum, and there is no zero-frequency impulse.

The spectrum for the bipolar NRZ-L,-M,-S is evaluated by Sunde [10] as

$$S(f) = \left(\frac{2p(1-p)}{T_b}\right)(1-\cos(\pi f T_b))|H(f)|^2 \tag{8.7}$$

The PSD in (8.4) is based on the PCM coded data having a constant amplitude over the entire bit interval, that is, the there is no pulse shaping, and, assuming random source data with $p = 1/2$, upon substituting $|H(f)|^2$ using (8.4) into (8.7) yields the PSD for bipolar NRZ coded PCM expressed as

$$S(f) = \frac{T_b}{2} \frac{\sin^2(\pi f T_b)}{(\pi f T_b)^2}(1-\cos(\pi f T_b)) \quad : \text{bipolar NRZ-L, -M, -S} \tag{8.8}$$

Equations (8.5), (8.6), and (8.8) are normalized by T_b and plotted in Figure 8.4 as the PSD for the NRZ coded PCM waveforms. The spectrum of the bipolar NRZ code does not have a direct current (DC) component so transmission over lines that are not DC coupled is possible as, for example, over long-distance lines that use repeater amplifiers.

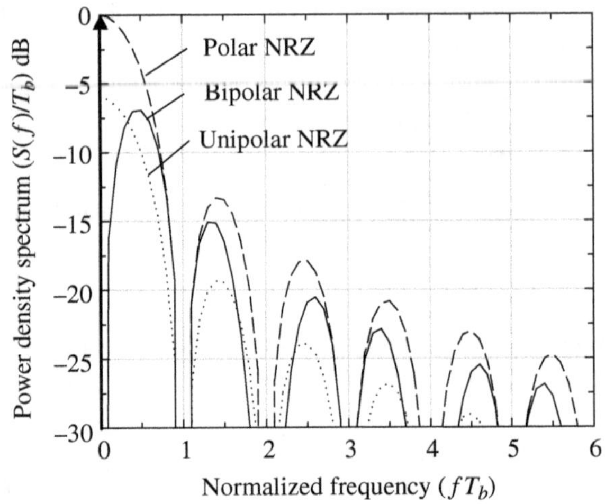

FIGURE 8.4 Power spectral density for NRZ coded PCM waveforms.

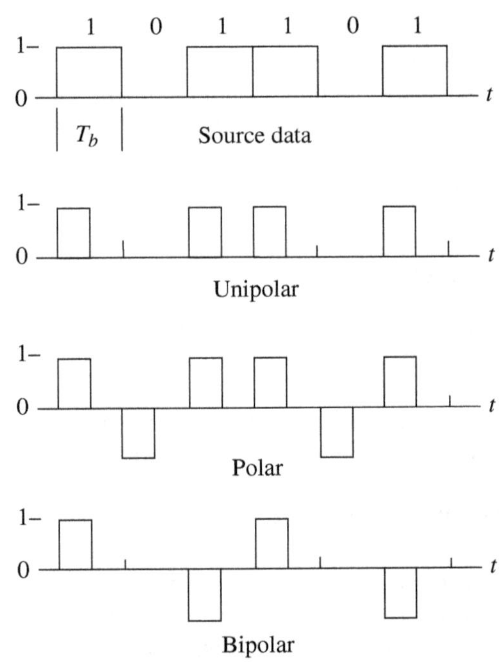

FIGURE 8.5 RZ formatted data.

8.2.2 Return-to-Zero Coded PCM

Another form of baseband coding is return-to-zero (RZ) coding that is characterized by the NRZ-L coding in Figure 8.2 with the exception that the bit interval is *split*, that is, the coded pulse level is nonzero over the first $T_b/2$ data interval and *returns to zero* over the remaining interval.* The RZ coded PCM waveforms are depicted in Figure 8.5. In the unipolar case, the space bit is always zero and the mark bit is split. The polar case is similar; however, the space bit is split with the opposite polarity of the mark bit. The polar RZ code allows the bit timing to be established simply by full-wave rectifying and narrowband filtering the received bit stream.

*The term *split-bit* generally refers to one-half of the bit interval being nonzero with the other half being zero or the negative of the nonzero level.

In the bipolar case, the mark bit is split with alternating polarities and the space bit is always zero. By alternating the polarity of the mark bits, the bipolar RZ code provides for one bit-error detection [11]. The bipolar RZ coded PCM is used in the Bell Telephone T1-carrier system [12].

The PSD of the unipolar and polar RZ coded PCM waveforms are evaluated using (8.2) and (8.3) in a manner similar to that used for the unipolar and polar NRZ code; however, in these cases, the mark bit is split and characterized by the pulse $rect(2t/T_b - 1/2)$ with the underlying square of the spectrum magnitude given by

$$|H(f)|^2 = \frac{T_b^2}{4}\frac{\sin^2(\pi f T_b/2)}{(\pi f T_b/2)^2} \quad : \text{RZ} \tag{8.9}$$

Substituting (8.9) into (8.2) and recognizing that the spectrum is zero for all $n \neq 0$, the PSD of the unipolar RZ coded waveform with random data is evaluated as

$$S(f) = \frac{T_b}{16}\frac{\sin^2(\pi f T_b/2)}{(\pi f T_b/2)^2} + \frac{1}{16}\delta(f) \quad : \text{unipolar RZ-L,-M,-S} \tag{8.10}$$

In a similar manner, the polar RZ PSD is evaluated using (8.3) with the result

$$S(f) = \frac{T_b}{4}\frac{\sin^2(\pi f T_b/2)}{(\pi f T_b/2)^2} \quad : \text{polar RZ-L,-M,-S} \tag{8.11}$$

The PSD of the bipolar RZ coded PCM waveform is characterized by Sunde [10] as in (8.7), with $|N(f)|^2$ given by (8.9) so that, with random data,

$$S(f) = \frac{T_b}{8}\frac{\sin^2(\pi f T_b/2)}{(\pi f T_b/2)^2}(1-\cos(\pi f T_b)) \quad : \text{bipolar RZ} \tag{8.12}$$

The PSD of the bipolar RZ coded PCM waveform is plotted in Figure 8.6.

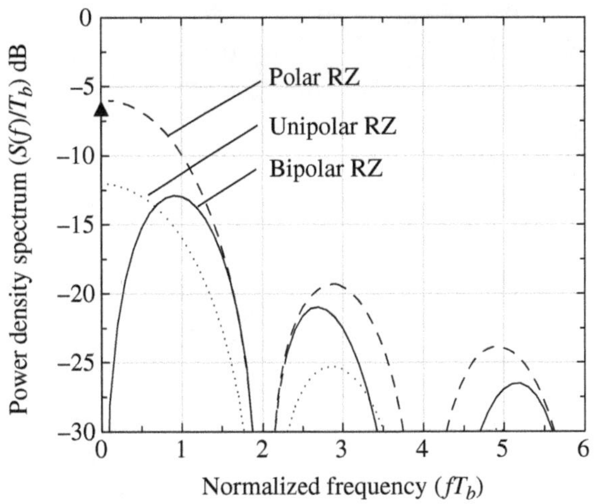

FIGURE 8.6 Power spectral density for RZ coded PCM waveforms.

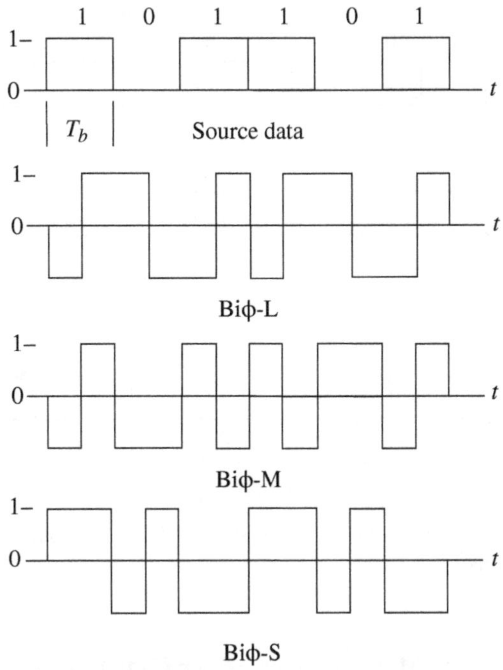

FIGURE 8.7 Biphase Biϕ-L,-M,-S polar formatted data.

8.2.3 Biphase (Biϕ) or Manchester Coded PCM

Biphase (Biϕ), or Manchester, coded PCM is a commonly used waveform because of the robust bit timing recovery even without random data. For example, in an idle mode that uses either mark or space hold data, the demodulator timing can be maintained. The biphase-level, -mark, -space (Biϕ-L,-M,-S) coded waveforms are depicted in Figure 8.7. For the Biϕ-L code, the mark bit is split and the space bit is the inverse of the mark bit. With the Biϕ-M code, a transition occurs at the beginning of each source data bit and the mark bit is split; however there is no change in the code-bit level when a space source bit occurs. The Biϕ-S code is similar to the Biϕ-M with the role of the mark and space bits reversed.

The PSD of the biphase coded PCM waveform, using random data with $p = 1/2$, is evaluated by Batson [13] as

$$S(f) = \frac{T_b}{2\pi}\frac{\sin^4(\pi f T_b/2)}{(\pi f T_b/2)^2} \quad : \text{Bi}\phi\text{-L,-M,-S} \tag{8.13}$$

Equation (8.13) is normalized by T_b and plotted in Figure 8.8.

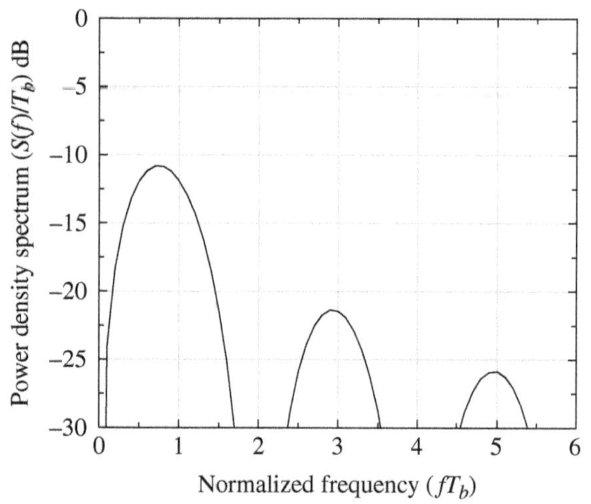

FIGURE 8.8 Power spectral density for Biϕ-L,-M,-S coded PCM waveforms.

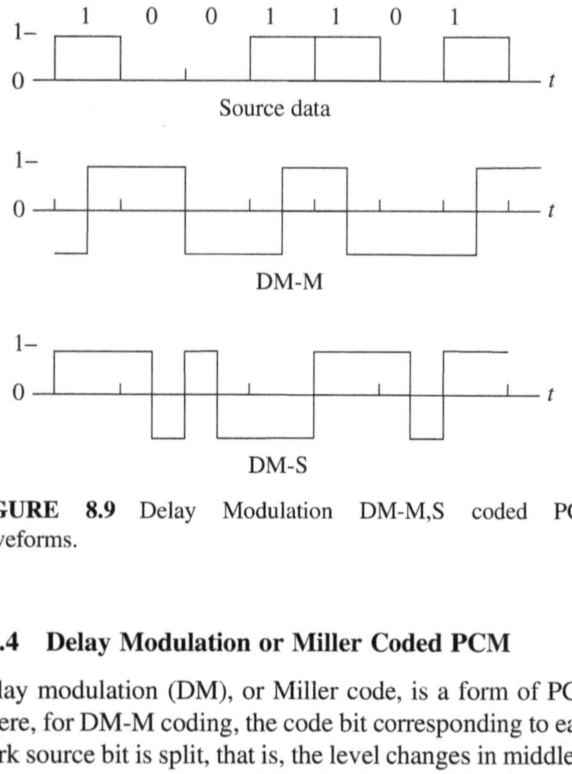

FIGURE 8.9 Delay Modulation DM-M,S coded PCM waveforms.

8.2.4 Delay Modulation or Miller Coded PCM

Delay modulation (DM), or Miller code, is a form of PCM where, for DM-M coding, the code bit corresponding to each mark source bit is split, that is, the level changes in middle of each mark bit. The level of the code bit corresponding to a space source bit is unchanged unless it is followed by another space source bit in which case it is changed at the beginning of the space source bit. The DM-S coding reverses the roles of the mark and space source bits, that is, the code bit corresponding to each space source bit is split and the level of the code bit corresponding to a mark source bit is unchanged unless it is followed by another mark source bit in which case it is changed at the beginning of the mark source bit. DM-M,-S coding is shown in Figure 8.9 for the source data sequence

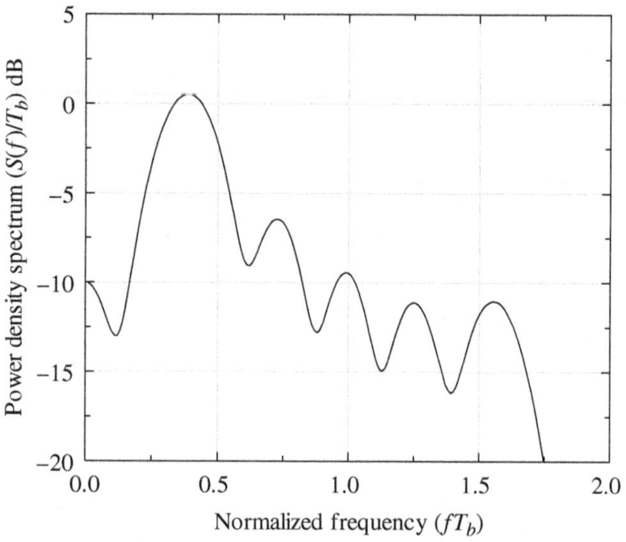

FIGURE 8.10 Power spectral density for DM coded PCM waveforms.

(1,0,0,1,1,0,1). To establish the correct phase of the bit timing in the demodulator an alternating mark-space preamble must be transmitted.

The PSD of the DM coded PCM waveform is evaluated for random data by Hecht and Guida [14] as

$$S(f) = \frac{T_b}{2(\pi f T_b)^2 \cos(\pi f T_b)}[23 - 2\cos(\pi f T_b) - 22\cos(2\pi f T_b) - 12\cos(3\pi f T_b) + 5\cos(4\pi f T_b) + 12\cos(5\pi f T_b) + 2\cos(6\pi f T_b) - 8\cos(7\pi f T_b) + 2\cos(8\pi f T_b)] \tag{8.14}$$

Equation (8.14) is normalized by T_b and plotted Figure 8.10.

8.2.5 Bit-Error Performance of PCM

The bit-error performance of phase modulated* (PM), PCM, and (PCM/PM) coded waveforms [15] are expressed in terms of $Q(x)$, the complement of the probability integral, where x is a function of the signal-to-noise ratio $\gamma_b = E_b/N_o$ measured in the bandwidth corresponding to the information bit rate R_b. The approximate functional dependencies [16] of the indicated PCM/PM coded waveforms are expressed in (8.15) and are based on ideal demodulator bit timing. The

*The term *phase modulation* as used here and in *biphase modulation* refers to the polarity of the signal amplitude and not to carrier phase modulation as used, for example, in BPSK modulation.

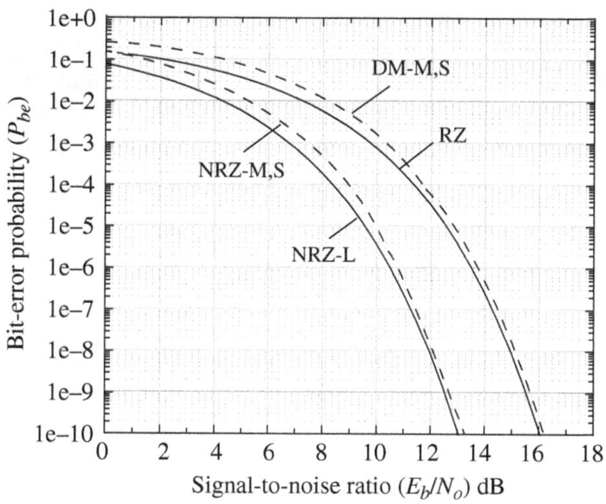

FIGURE 8.11 Bit-error performance of PCM/PM coded waveforms.

approximations improve with increasing signal-to-noise ratio and the bit-error results are shown in Figure 8.11.

$$\begin{aligned} P_{be} &\cong Q(\sqrt{2\gamma_b}) && : \text{NRZ-L} \\ P_{be} &\cong 2Q(\sqrt{2\gamma_b}) && : \text{NRZ-M, -S; Bi}\phi\text{-M, -S} \\ P_{be} &\cong Q(\sqrt{\gamma_b}) && : \text{RZ} \\ P_{be} &\cong 2Q(\sqrt{\gamma_b})[1-Q(\sqrt{\gamma_b})] && : \text{DM-M, -S} \end{aligned} \quad (8.15)$$

The performance of frequency modulated (FM) PCM (PCM/FM) coded waveforms is dependent on the normalized frequency deviation $\Delta f/R_b$, pre-detection filter bandwidth B_{if}/R_b, and post-detection or video bandwidth B_v/R_b, where R_b is the information bit rate. Based these parameters and the pre-detection signal-to-noise ratio γ_b and constants k and k_1, the approximate bit-error probability is expressed [16] as

$$P_{be} \cong 2Q\left(k\left(\frac{R_b}{B_v}\right)^{3/2}\left(\frac{\Delta f}{R_b}\right)^2\sqrt{2\gamma_b}\right) + \frac{(\Delta f/R_b)}{2\pi}e^{-k_1(\Delta f/R_b)\gamma_b} \quad (8.16)$$

The parameters in (8.16) are dependent on the format of the PCM/FM modulation, for example, NRZ-L,-M and Biϕ-L, as indicated in Table 8.1. The signal-to-noise ratio, $\gamma_b = E_b/N_o$, is measured in a bandwidth corresponding to the bit rate R_b.

The peak deviation, Δf, is the modulation frequency deviation from the carrier frequency f_c with $+\Delta f$ usually assigned to binary 1 (Mark) data and $-\Delta f$ assigned to binary 0 (Space) data; the peak-to-peak deviation is $2\Delta f$. Using a five-pole, phase-equalized, Butterworth intermediate frequency (IF) pre-detection filter, the optimum peak deviation

TABLE 8.1 Typical Parameter Sets for PCM/FM Coded Waveforms

PCM/FM	k	k_1	$\Delta f/R_b$	B_v/R_b	B_{if}/R_b
NRZ-L	2.31	1	0.35	1/2	1
NRZ-M,-S	2.22	1	0.35	1/2	1
Biϕ-L[a]	1.89	1/4	0.65	1	2

[a]Manchester code.

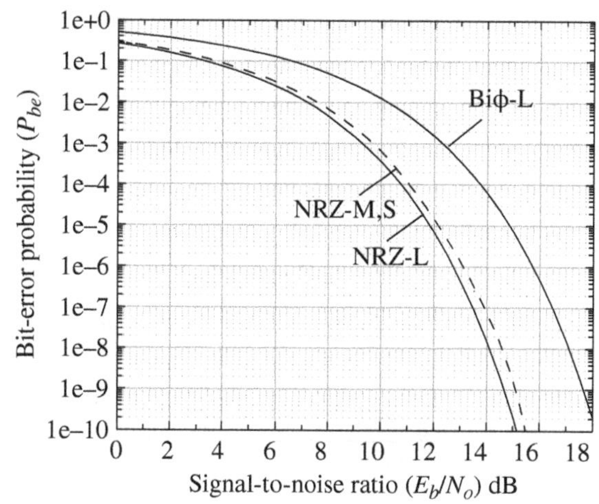

FIGURE 8.12 Approximate bit-error performance of PCM/FM coded waveforms.

[17] for NRZ PCM/FM is $\Delta f = 0.35R_b$ and for Biϕ PCM/FM $\Delta f = 0.7R_b$. The pre-detection filter 3-dB bandwidth [18, 19] is denoted as B_{if} with the condition $B_{if}/R_b \geq 1$. The range of the normalized video bandwidth is $0.5 \leq B_v/R_b \leq 1$; however, the recommended range is $0.7 \leq B_v/R_b \leq 1$. The best values of these parameters are selected by examining the bit-error performance of hardware tests or Monte Carlo computer simulations. Typical parameters sets are shown in Table 8.1 [20] for the indicated PCM/FM coded waveforms and the approximate bit-error results are shown in Figure 8.12 [21, 22]. The bit-error performance results apply to bit-by-bit detection corresponding to a filter matched to the bit duration. Performance improvements can be achieved using a maximum-likelihood detector spanning multiple bits in the form of a decoding trellis [23, 24].

A supportive test, used to establish best parameter values, is the evaluation of the PSD of the modulated waveform. An analytical formulation of the PSD of the NRZ PCM/FM modulated waveform is expressed as [25]

$$S\left(\frac{u}{R_b}\right) = 4\left(\frac{B_{sa}}{R_b}\right)\left(\frac{D}{\pi(D^2-X^2)}\right)^2\frac{(\cos(\pi D)-\cos(\pi X))^2}{1-2\cos(\pi D)\cos(\pi X)+\cos(\pi D)} \quad (8.17)$$

The parameters in (8.17) are defined in Table 8.2 and the spectrums are plotted in Figure 8.13 for peak deviations

$\Delta f/R_b = 0.25$, 0.35, and 0.45. This analytical expression does not include the influence of the pre-detection filter. The radio frequency (RF) bandwidth is defined as the occupied bandwidth corresponding to 99% spectral containment and the objective is to select a set of parameters that minimize the bit-error performance and provide an acceptably low occupied bandwidth [26–29]. Spectral masks are discussed in Section 4.4.1 and examples of spectral containment for various modulations are given in Section 4.4.3 and following. Section 5.6 discusses binary FSK (BFSK) modulation with an emphasis on the PSD using various modulation indices.

TABLE 8.2 NRZ PCM/FM PSD Parameters Used in Equation (8.17)

Normalized Parameter	Description
$S(u)$	Power spectral density *w/r* carrier power[a]
B_{sa}/R_b[b]	Spectrum analyzer resolution = 0.003 for $Q \cong 0.99$ (used); =0.03 for $Q \cong 0.9$
$D = 2\Delta f/R_b$	Peak-to-peak deviation
$X = 2u/R_b$	Frequency deviation from carrier $u = f - f_c$ Hz

[a]Expressed in decibels, the dimension is dBc.
[b]Q is related to narrowband spectral peaking when $D \cong$ integer value.

8.3 GRAY CODING

The performance benefits of gray coding is demonstrated in Section 4.2 where it is stated that the conversion from symbol-error to bit-error probability is expressed by the approximation

$$P_{be} \approx \frac{P_{se}}{k} \tag{8.18}$$

where k is the number of bits mapped into a single symbol. On the other hand, when the source bits are randomly

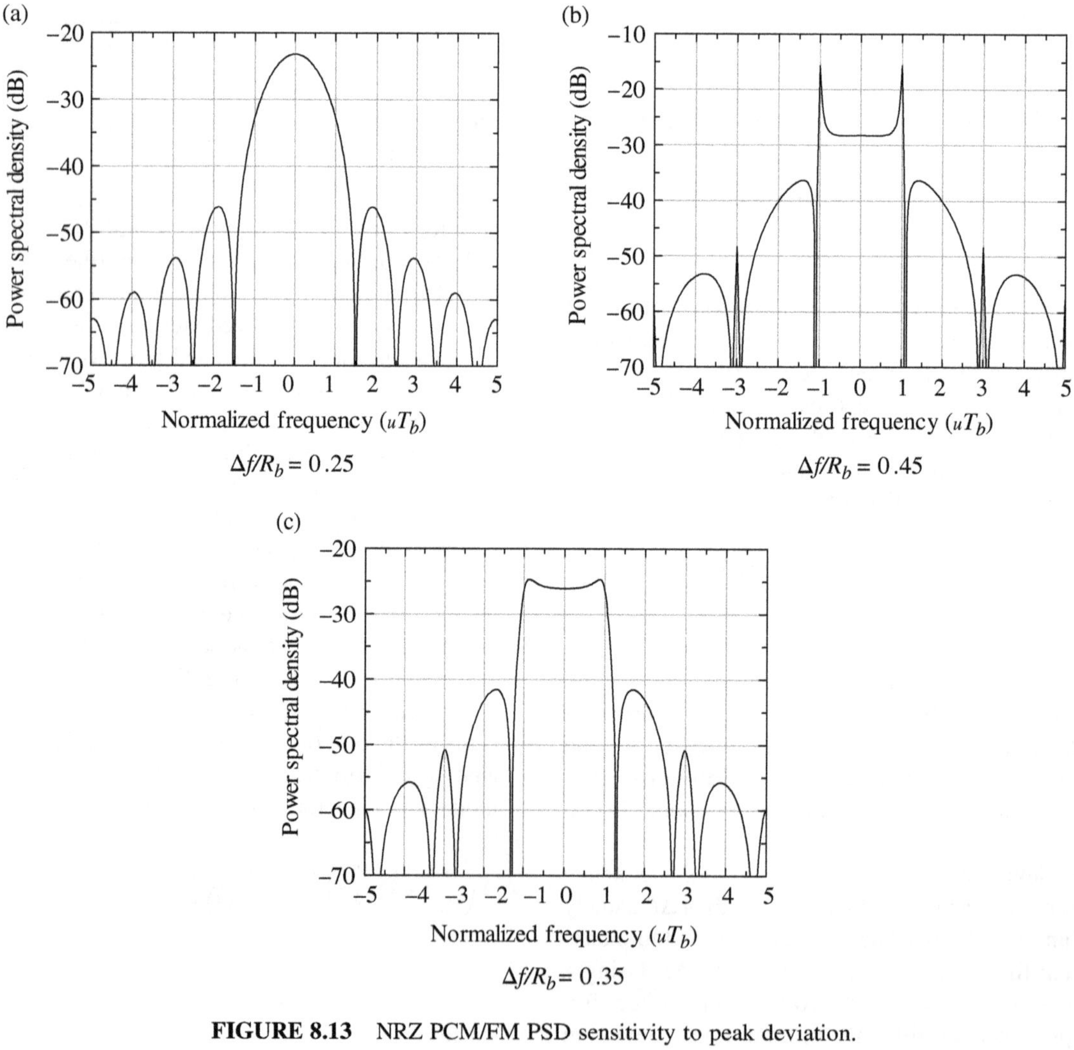

FIGURE 8.13 NRZ PCM/FM PSD sensitivity to peak deviation.

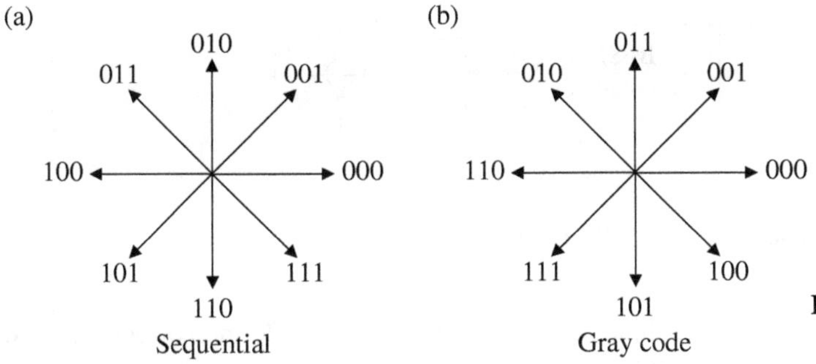

FIGURE 8.14 Mapping of source bits to 8PSK phase constellation.

assigned to a symbol interval, that is, when gray coding is not used, the conversion is bounded by

$$P_{be} \leq \frac{2^{k-1}}{2^k - 1} P_{se} \tag{8.19}$$

The bit-error performance for these two symbol-to-bit-error mapping techniques is shown in Figure 4.5 for multiphase shift keying (MPSK) modulated waveforms.

The function of gray coding is to minimize the number of bit errors in the process of converting from detected symbols to receive bits. For PSK modulation, gray coding ensures that adjacent symbols will differ in only one bit position. For example, consider the 8PSK modulated signal mapped into the two possible phase constellations shown in Figure 8.14.

The most probable error condition occurs when the detected symbol is adjacent to the correctly transmitted symbol. For example, referring to Figure 8.14, if the correct symbol is (000), then the most likely error conditions are the detection of symbols (001) and (111) resulting, on average, in two bit errors. On the other hand, the Gray coded data results in symbol errors (001) and (100) with an average of only one bit error. In general, a k-tuple of source bits $\{b_{k-1}, b_{k-2}, \ldots, b_0\}$ results in the Gray coded k-tuple $\{b'_{k-1}, b'_{k-2}, \ldots, b'_0\}$ established using the encoding rule:

$$b'_n = b_n \oplus b_{n+1} \quad : n = 0, \ldots, k-2 \text{ with } b'_{k-1} = b_{k-1}; \text{ gray coding} \tag{8.20}$$

where $\oplus$ denotes the exclusive-or operation and b_0 is considered to be the least significant bit (LSB).

The modulation of gray coded quadrature PSK (QPSK) unipolar binary data $\boldsymbol{b}'$:$\{1,0\}$ is accomplished in a relatively straightforward manner by translating to bi-polar data $\boldsymbol{d}$:$\{-1,1\}$ as $d_i = 1 - 2b'_i$: $i = 0$, 1 as illustrated by the QPSK phase constellation in Figure 8.15. This mapping results in detected bipolar data estimates corresponding to the sign of the quadrature decisions $\{x,y\}$ such that $\hat{\boldsymbol{b}}$:$\{\hat{d}_1, \hat{d}_0\}$ where $\hat{d}_0 = \text{sign}(y)$ and $\hat{d}_1 = \text{sign}(x)$. The gray decoded received

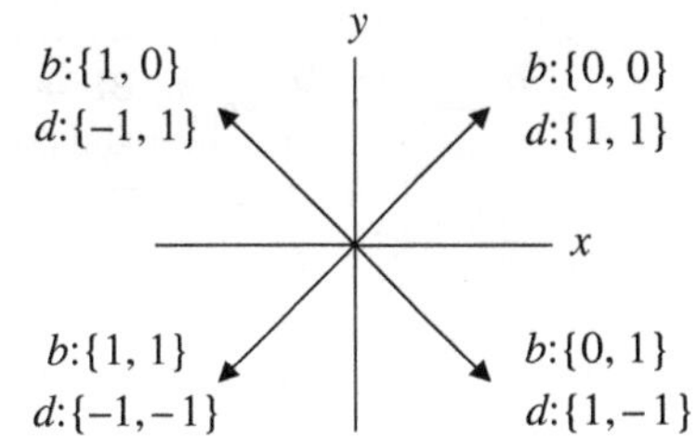

FIGURE 8.15 Quadrature phase constellation using gray coding.

bit estimates are determined as $\hat{b}'_i = (1 - \hat{d}_i)/2$ and the final step is the decoding of the gray coded received bit estimates. The decoding of a unipolar k-tuple of Gray coded bits is accomplished using the following rule:

$$\hat{b}_n = \hat{b}'_n \oplus \hat{b}_{n+1} \quad : n = k-2, \ldots, 0 \text{ with } \hat{b}_{k-1} = \hat{b}'_{k-1}; \text{ gray decoding} \tag{8.21}$$

8.4 DIFFERENTIAL CODING

The need for differential encoding arises to combat catastrophic error propagation resulting from inadvertent initialization or subsequent *false-lock* conditions in the demodulator acquisition or phase tracking process. Although an initial phaselock error can be overcome by using a known preamble, subsequent phase slips, caused by channel noise or phase *hits* are effectively overcome by using differential encoding. Without differential encoding a *phase hit*, causing a false-lock condition to occur, will result in a continual stream of data errors until the phase error is detected and corrected. For example, with MPSK modulation, if the receiver phaselock loop were to settle on a conditionally stable track point π radians out of phase, then all of the received bits will be in error—referred to as a catastrophic error condition. Differential encoding and decoding avoids catastrophic error events and is implemented in the modulator as

$$D_i = b_i \bullet D_{i-1} \quad : \text{differential encoding} \tag{8.22}$$

and, in the demodulator, as

$$\hat{b}_i = \hat{D}_i \bullet' \hat{D}_{i-1} \quad : \text{differential decoding} \tag{8.23}$$

where $\bullet$ represents the differential encoding operator and $\bullet'$ is the inverse operator. For binary data, the $\bullet$ and $\bullet'$ operators are represented by the exclusive-or operation, denoted as $\oplus$.

Consider the simple case of differentially encoded BPSK (DEBPSK)* to prevent catastrophic errors. The encoding,

*The prefix DE is used to denote differentially encoded data modulation and is contrasted to the prefix DC that is used to denote differentially coherent symbol phase modulation.

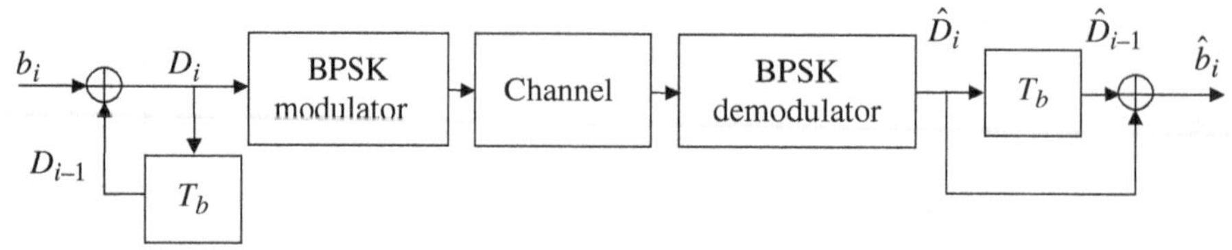

FIGURE 8.16 Differential encoding for DEBPSK modulation.

TABLE 8.3 Transmitted Data Sequence Demonstrating DEBPSK

i	0 1 2 3 4 5 6 7 8 9 10 11 ...
d_i	-- 0 1 0 0 1 1 0 1 0 1 0 ...
D_i	0 0 1 1 1 0 1 1 0 0 1 1 ...
	$\vert\leftarrow \pi$ radian error at $i = 6$
$\hat{D}_i$	0 0 1 1 1 0 0 0 1 1 0 0 ...
$\hat{d}_i$	-- 0 1 0 0 1 0 0 1 0 1 0 ...
	$\uparrow$ Single bit-error at $i = 6$

using the exclusive-or operation, is shown in Figure 8.16 and an example of the received data sequence, with a received error forced at bit $i = 6$, is shown in Table 8.3. The phase error is assumed to result from a 180° phase-step in the carrier frequency caused by the channel or a phase-slip in the demodulator phaselock loop. This example shows that instead of continuing to output inverted data following the phase-step, a single error occurs because of the differential encoding.

In Table 8.3, the encoder and decoder are identically initialized. In Problem 3, the error events are examined under two cases: for a single received bit error in $\hat{D}_i$, that is, the phase-lock loop tracking continues uninterrupted.

As a second example, consider differentially encoded QPSK (DEQPSK) with input data $b_i = \{I_i, Q_i\}$ where I_i and Q_i are the inphase and quadrature bit assignments, respectively. The objective is to encode the data so that receiver phase errors of π and $\pm\pi/2$ will not result in a catastrophic error condition. The encoding and decoding functions shown in Figure 8.17 provide the correct received data for a fixed unknown phase of 0 or π radians provided that unique I and Q symbol synchronization sequences precede the data symbols. Although the differential decoder is self-synchronizing, without the correct initialization,* the first received symbol will be in error and this error is absorbed by the synchronization symbol. In this case, there are two, phase-dependent, correct differential decoder initialization symbols, and because the phase is unknown the correct initialization symbol is unknown.

To accommodate unknown phases of $n\pi/2$: $n = 0, \ldots, 3$, caused by a channel phase hit or a false-lock condition in the demodulator, the differential encoding and decoding algorithms in (8.24) and (8.25) provide the correct received data and are self-synchronizing. As in the preceding example, a synchronization symbol is required to absorb the error in the first received symbol because of the unknown correct initialization state; in this case, there are four possible initialization states.

$$I'_i, Q'_i = \begin{cases} (I_i \oplus I'_{i-1}), (Q_i \oplus Q'_{i-1}) & : I_i \oplus Q_i = 0 \\ (I_i \oplus Q'_{i-1}), (Q_i \oplus I'_{i-1}) & : I_i \oplus Q_i = 1 \end{cases} \quad : \text{encoding} \tag{8.24}$$

$$\hat{I}_i, \hat{Q}_i = \begin{cases} \left(\hat{Q}'_i \oplus \hat{Q}'_{i-1}\right), \left(\hat{I}'_i \oplus \hat{I}'_{i-1}\right) & : \hat{I}'_i \oplus \hat{Q}'_i = 0 \\ \left(\hat{I}'_i \oplus \hat{I}'_{i-1}\right), \left(\hat{Q}'_i \oplus \hat{Q}'_{i-1}\right) & : \hat{I}'_i \oplus \hat{Q}'_i = 1 \end{cases} \quad : \text{decoding} \tag{8.25}$$

The decoding algorithm follows logically from the encoding process and is best seen by examining all of the discrete combinations as shown in Table 8.4. For example, when the data is the same in each channel, that is, $I_i \oplus Q_i = 0$, the phase error affects each channel identically and differential encoding of the I and Q channels individually is required; this is the situation depicted in Figure 8.17. However, when the data is different on each channel, that is, $I_i \oplus Q_i = 1$, the I and Q channels appear to be interchanged so differential encoding across the I and Q channels is required.

The simulated bit-error performance of differentially encoded BPSK and QPSK are shown in Figure 8.18. These results are obtained using Monte Carlo simulations involving 100 Kbits for signal-to-noise ratios ≤6 dB: 1,000 Kbits for 6 dB < signal-to-noise ratios <10 dB, and 10,000 Kbits for signal-to-noise ratios ≥10 dB. The results are essentially the same for both modulations because gray coding is used with the QPSK modulation ensuring that one bit error occurs for each of the most probable symbol-error conditions.

8.5 PSEUDO-RANDOM NOISE SEQUENCES

The subject of PRN† sequences [30, 31] is an important topic in applications involving ranging, FEC coding, information and transmission security, anti-jam, and code division

*The unknown phase can be resolved using an appropriate message preamble and knowledge of the differential encoder initialization state.

†These sequences are sometimes referred to as pseudo-noise (PN) sequences.

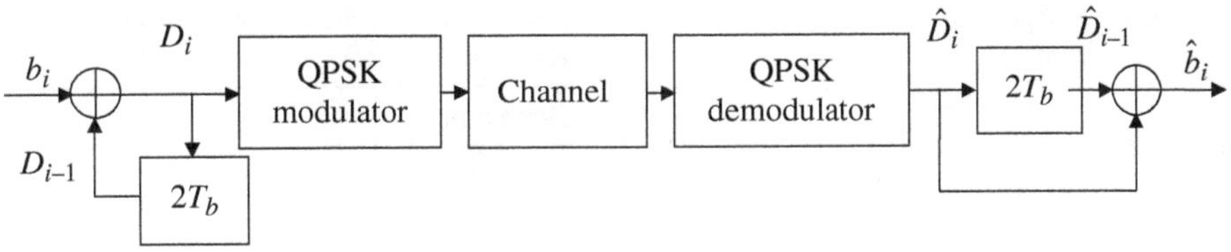

FIGURE 8.17 Differential encoding for DEQPSK modulation.

TABLE 8.4 Discrete Coding and Decoding of DEQPSK for Zero Channel Phase Shift

I_iQ_i	$I'_{i-1}Q'_{i-1}$	$I'_iQ'_i$	$\hat{I'}_i\hat{Q'}_i$	$\hat{I'}_{i-1}\hat{Q'}_{i-1}$	$\hat{I}_i\hat{Q}_i$
00	00	00	00	00	00
00	01	01	01	01	00
00	10	10	10	10	00
00	11	11	11	11	00
01	00	01	01	00	01
01	01	11	11	01	01
01	10	00	00	10	01
01	11	10	10	11	01
10	00	10	10	00	10
10	01	00	00	01	10
10	10	11	11	10	10
10	11	01	01	11	10
11	00	11	11	00	11
11	01	10	10	01	11
11	10	01	01	10	11
11	11	00	00	11	11

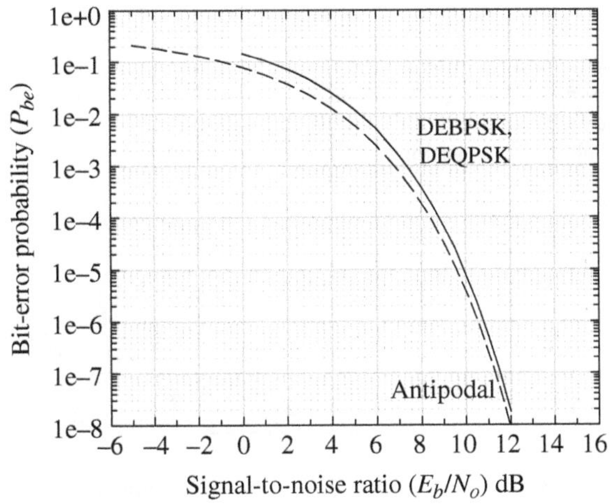

FIGURE 8.18 Performance of differentially encoded BPSK and QPSK.

multiple-access (CDMA) communications. Furthermore, PRN sequences fill an important role in the generation of acquisition preambles for the determination of signal presence, frequency, and symbol timing for data demodulation.

The noise-like characteristics of binary PRN sequences exhibit random properties similar to that generated by the tossing of a coin with the binary outcomes: heads and tails. In this example, the random properties for a long sequence of j events are characterized as having a correlation response c_i characterized by the delta function response $c_i = \delta(j - i)$. A random sequence also exhibits unique run length properties; in that, defining a run of length ℓ as having ℓ contiguous events with identical outcomes, a run of length ℓ occurs with a probability of occurrence approaching $P_r(\ell) = 2^{-\ell}$ as the number of trials increases. This property also indicates that the probability of the number of heads or tails ($\ell = 1$) in the tossing of a coin approaches 1/2 as the number of trials increases. Binary PRN sequences, generated using a linear feedback shift register [32] (LFSR) with module-2 feedback from selected registers, exhibit these noise-like characteristics. A useful property of LFSR generators is that the noise characteristics can be repeated simply by reinitializing the generator.

Sequences can be generated as polynomials over a GF with properties analogous to those of integers. Binary sequences contained in the GF(2)* are particularly useful because of their simplicity and predictable pseudo-random properties. The maximal linear PRN sequences, referred to as m-sequences, generate the longest possible codes for a given number of m shift registers. The code length† is $L_m = 2^m - 1$ and repeats with a repetition time $T_L = L_m/f_{ck}$ where f_{ck} is the shift register clock frequency. The linear characteristic of the LFSR implementation results because of the property: for a given set of initial conditions S_{Ij} with outcomes S_{Oj}: $j = 1$, 2 then, for the initial conditions $S_I = S_{I1} \oplus S_{I2}$, the output sequence is $S_O = S_{O1} \oplus S_{O2}$ where $\oplus$ denotes modulo-2 addition.

The LFSR PRN generators are characterized by m-degree polynomials in x expressed as

$$g(x) = g_m x^m + g_{m-1} x^{m-1} + \cdots + g_1 x + g_0 \tag{8.26}$$

*The elements of the binary field are represented by $\alpha = \{0,1\}$ that conform to the rules of binary multiplication and addition; nonbinary fields GF(2^m) are discussed later in this chapter in the context of Reed-Solomon codes.

†There are 2^m possible binary sequences; however, the all-zero sequence is not included because the LFSR will remain in the zero state indefinitely. In some applications, a test for the zero state is performed and logic is provided to reset the registers.

with $g_m = g_0 = 1$. Some useful properties [33] of $g(x)$ are as follows:

- A generator polynomial $g(x)$ of degree m that is not divisible by any polynomial of degree less than m but if not divisible by a polynomial of any degree greater than 0, then $g(x)$ is an irreducible polynomial.
- A polynomial $g(x)$ of degree m is primitive if it generates all 2^m distinct elements; a field with 2^m distinct elements is called a GF(2^m) field.
- An element, or root, α of $g(x)$ in the GF(2^m) is primitive if all powers $\alpha^j : j \neq 0$ generates all nonzero elements of the field.
- An irreducible polynomial is primitive if $g(\alpha) = 0$ where α is a primitive element.
- Polynomials that are both irreducible and primitive result in m-sequences.

For each m, there exits at least one primitive polynomial of degree m. Irreducible and primitive polynomials are difficult to determine; however, Peterson and Weldon [34] have tabulated extensive list* of irreducible polynomials over GF(2), that is, with binary coefficients, for degrees $m \leq 34$; polynomials that are also primitive are identified.

The m-sequences are generated based on the specific LFSR feedback connections; otherwise, non-maximal sequences will result with periods less than T_L. A feedback connection that does not result in an m-sequence can result in one of several possible non-maximal sequences of varying lengths $<L_m$ depending on the initial LFSR settings. For a sequence of length $L_m = 2^m - 1$ that conforms to an irreducible primitive polynomial, the maximum number m-sequences that can be generated is determined as [35, 36]

$$N_m = \frac{\varphi(L_m)}{m} \tag{8.27}$$

where $\varphi(L_m)$ Euler's phi function evaluated as

$$\varphi(L_m) = \begin{cases} L_m \prod_i \frac{(p_i - 1)}{p_i} & : p_i \text{ is a prime factor of } L_m \\ L_m - 1 & : p_i \text{ is a prime number} \end{cases} \tag{8.28}$$

Table 8.5 lists the number of m-sequences for degrees $m = 2$ through 14.

For a given order, m, the irreducible primitive generator polynomial expressed by (8.26) is established by converting the octal notation to the equivalent binary notation and then associating the taps with the nonzero binary coefficients corresponding ($b_m = 1, b_{m-1}, \ldots, b_1, b_0 = 1$). The LFSR feedback taps are the nonzero taps for the coefficient orders less than m.

*The tables were first published by Peterson, *Error Correction Codes*, MIT Press, 1961, and the updated version of Peterson's tables appear in the publication by Peterson and Weldon.

TABLE 8.5 The Number of m-Sequences and Polynomial Representation[a]

Degree m	L_m	N_m	Prime Factors of L_m
2	3	1	Prime
3	7	2	Prime
4	15	2	3*5
5	31	6	Prime
6	63	6	3*3*7
7	127	18	Prime
8	255	16	3*5*17
9	511	48	7*73
10	1,023	60	3*11*31
11	2,047	176	23*89
12	4,095	144	3*3*5*7*13
13	8,191	630	Prime
14	16,383	756	3*43*127

[a]Ristenbatt [35]. Reproduced by permission of John Wiley & Sons, Inc.

TABLE 8.6 Partial List of Binary Irreducible Primitive Polynomials of Degree ≤ 21 with the Minimum Number of Feedback Connections[a]

m	$g(x)^b$	m	$g(x)^b$
2	7	12	10123
3	13	13	20033
4	23	14	42103
5	45	15	100003
6	103	16	210013
7	211	17	400011
8	435	18	1000201
9	1021	19	2000047
10	2011	20	4000011
11	4005	21	10000005

[a]Peterson and Weldon [34]. Appendix C, Table C.2, © 1961 Massachusetts Institute of Technology. Reproduced by permission of The MIT Press.
[b]Minimum polynomial with root α in octal notation.

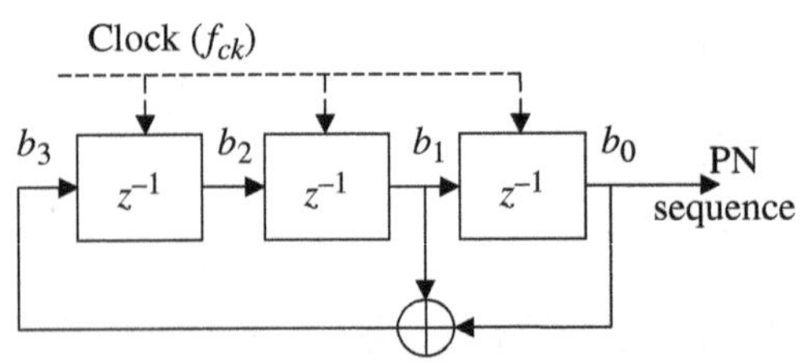

FIGURE 8.19 Third-order LFSR implementation of PN sequence generator $g(x) = x^3 + x + 1$.

For example, from Table 8.6, the octal notation for the irreducible primitive generator polynomial corresponding to $m = 3$ is $(13)_o$ and the binary notation is $(1011)_b$, so the generator polynomial is $g(x) = x^2 + x + 1$. The LFSR is implemented with descending coefficient order from left to right as shown in Figure 8.19. The generator clock frequency f_{ck} is also shown in the figure and the shift register delays correspond to the clock interval $\tau = 1/f_{ck}$, which is denoted as the unit sample delay z^{-1}. The generator must be initialized by the selection of the parameters (b_2, b_1, b_0) recognizing that, for this example, $b_3 = b_1 \oplus b_2$ where $\oplus$ is the exclusive-or operation. The initialization parameters characterize the state of the encoder, generally defined as (b_{n-1}, b_{n-2}, ..., b_0). If $g(x)$ is irreducible and primitive, as in this example, the selection of the initial conditions simply results in one of the $L = 2^n - 1$ cyclic shifts of the output PN sequence.* If, however, $g(x)$ is not primitive then mutually exclusive subsequences with length $<L_m$ are generated and, taken collectively, the subsets contain all of the $2^m - 1$ states of the encoder (see Problem 5).

8.6 BINARY CYCLIC CODES

Cyclic codes are a subset of linear codes that form the basis for a large variety of codes for detecting and correcting isolated single errors, multiple independent errors, and the more general situation involving bursts of errors. Cyclic code encoders are implemented using shift registers with appropriate feedback connections that are easily implemented and operate efficiently at high data rates. The following discussion of cyclic codes focuses on the encoding of systematic cyclic codes that are used for error checking of demodulated received data prior to passing the message data to the user. If errors are detected in the received data, the message may be blocked, sent to the user marked as containing errors, or scheduled for retransmission using an automatic repeat request (ARQ) protocol. In these applications, the cyclic code is referred to as a CRC code.

A systematic code is characterized as containing the uncoded message data followed by the parity-check information as depicted in Figure 8.20. In the following description, the parity and message data, r_i and m_j, are based on the field elements in GF(2) and correspond to the binary bits {1,0}. The notation involving the parameter x denotes the polynomial representation of the cyclic-code block of n transmitted bits, k information bits, and $r = n - k$ parity bits. The information and parity polynomials are expressed as

$$m(x) = m_0 + m_1 x + \cdots + m_{k-1} x^{k-1} \tag{8.29}$$

and

$$r(x) = r_0 + r_1 x + \cdots + r_{k-1} x^{k-1} \tag{8.30}$$

so the cyclic coded message block, or code polynomial, in Figure 8.20 is constructed as

1	x	$\cdots$	x^{n-k-1}	x^{n-k}	x^{n-k+1}	$\cdots$	x^{n-1}
r_0	r_1	$\cdots$	r_{r-1}	m_0	m_1	$\cdots$	m_{k-1}

← Parity check bits → ← Message bits →

FIGURE 8.20 Coded message block.

$$v(x) = r(x) + x^{n-k} m(x) \tag{8.31}$$

The structure of the (n,k) cyclic code is based on the following properties; the proof of the last three properties are given by Lin [37].

- An (n,k) linear code, described by the code polynomial

$$v(x) = v_0 + v_1 x + \cdots + v_{n-1} x^{n-1} \tag{8.32}$$

  is a cyclic code if every polynomial, described by shifting each element by $i = 1, \ldots, n-1$ places, is also a code polynomial of the (n,k) linear code. For example, the polynomial

$$\begin{aligned} v^i(x) &= x^i v(x) \\ &= v_{n-i} + v_{n-i+1}x + \cdots + v_{n-1}x^{i-1} + v_0 x^i + v_1 x^{i+1} + \cdots + v_{n-i-1}x^{n-1} \end{aligned} \tag{8.33}$$

  is also a code polynomial of the (n,k) linear code.
- For an (n,k) cyclic code, there exists a unique polynomial $g(x)$ of degree $n - k$ expressed as

$$g(x) = g_0 + g_1 x + \cdots + g_{n-k-1} x^{n-k-1} + g_{n-k} x^{n-k} \tag{8.34}$$

  with g_0, $g_{n-k} \neq 0$. Furthermore, every code polynomial $v(x)$ is a multiple of $g(x)$ and every polynomial of degree $<= n - 1$, which is a multiple of $g(x)$ is a code polynomial.
- The generator polynomial of a (n,k) cyclic code $v(x)$ is a factor of $x^n + 1$,† that is,

$$x^n + 1 = g(x)v(x) \tag{8.35}$$

- If $g(x)$ is a polynomial of degree $n - k$ and is a factor of $x^n + 1$, then $g(x)$ is the generator polynomial of the (n,k) cyclic code.

The generator polynomials are derived from irreducible polynomials that are difficult to determine; however, as mentioned previously, the tables‡ of Peterson and Weldon [38] provide a list of all irreducible polynomials over GF(2) of degree ≤35.

*This feature is used in the process of synchronizing a locally generated receiver PN sequence to a received PN sequence based on the known system epic time and the modulator PN initialization conditions.

†$x^n + 1$ is the binary equivalent of $x^n - 1$.

‡These tables also identify irreducible polynomials that are primitive. Irreducible primitive polynomials generate maximal length or m-sequences.

The following description of constructing systematic cyclic codes from the message polynomial follows the development by Lin [39]. However, for simplicity and clarity, the construction is described using the (5,3) single-error correcting code with generator polynomial given in octal form as $(7)_o$; the binary equivalent $(111)_b$ corresponds to the polynomial

$$g(x)=x^2+x+1 \tag{8.36}$$

In (8.36), the LSB is the rightmost bit.* The following example evaluates the cyclic code polynomial for the three-bit message $m_4=(001)$ using the (5,3) code. In this case, the message polynomial is $m(x)=0+0x+1x^2=x^2$ and the corresponding code polynomial is obtained by dividing $x^{n-k}m(x)$ by the generator polynomial, that is, in this example, dividing $x^2m(x)=x^4$ by $g(x)=x^2+x+1$ as follows:

$$\begin{array}{r} x^2+x \\ x^2+x+1\overline{)x^4} \\ \underline{x^4+x^3+x^2} \\ x^3+x^2 \\ \underline{x^3+x^2+x} \\ x \end{array} \tag{8.37}$$

The remainder in (8.37) is $r(x)=x$ and, using (8.31), the code polynomial for the message m_4 is evaluated as

$$v_4(x)=x+x^4 \tag{8.38}$$

Therefore, the message bits (001) map into the cyclic code polynomial bits (01001). This result is shown in Table 8.7 with the other seven cyclic coded message blocks. The encoder for the example (5,3) cyclic code corresponding to Table 8.7 is shown in Figure 8.21 and the generalized $r=n-k$ order encoder, generated using (8.34), is shown in Figure 8.22. As the uncoded message bits pass by the encoder, the bits are applied to the taped delay line according to the generator coefficients, 1 or 0 in the binary case. When the last message bit enters the generator, the *switches*, indicated by the *dashed lines*, are changed and the parity information contained in the registers is appended to the information bits forming the coded message block.† These concepts are applied to the generation and performance evaluation of CRC codes discussed in Section 8.7.

*Lin uses the notation in (8.34) or, for the example (5,3) code, $g(x)=1+x+x^2$ with the LSB being the leftmost bit. For compatibility with legacy equipment or a system specification, care must be used in the proper ordering and transmission of the bits.

†The tapped delay line implementations in Figures 8.21 and 8.20 perform polynomial division providing the remainder of $r=n-k$ parity bits.

TABLE 8.7 Code Polynomial Coefficients Corresponding to the (5,3) Cyclic Code

		v_i	
i	m_i	r_i	m_i
0	0 0 0	0 0	0 0 0
1	1 0 0	1 1	1 0 0
2	0 1 0	1 0	0 1 0
3	1 1 0	0 1	1 1 0
4	0 0 1	0 1	0 0 1
5	1 0 1	1 0	1 0 1
6	0 1 1	1 1	0 1 1
7	1 1 1	0 0	1 1 1

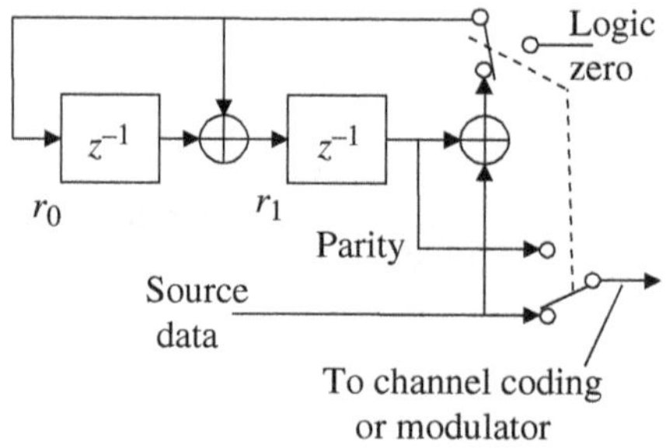

FIGURE 8.21 Binary (5,3) second-order encoder.

8.7 CYCLIC REDUNDANCY CHECK CODES

The CRC code of degree r is generated by multiplying an irreducible primitive polynomial, $p(x)$, by $(x+1)$ resulting in the CRC code generator polynomial

$$g(x)=(x+1)p(x) \quad \text{:CRC code generator} \tag{8.39}$$

The polynomial $p(x)$, with degree $r-1$, corresponds to the maximal length, or m-sequence, of 2^{r-1} bits containing 2^{r-2} ones and $2^{r-2}-1$ zeros. Therefore, the underlying structure of the (n,k) CRC code is the $r-1$ degree m-sequence. CRC code parity bits are appended to the message information bits to form a systematic code; however, the purpose is to detect errors in the received (n,k) cyclic code for a variety of message lengths $<k$, referred to a shortened codes. The performance measure of CRC codes is the undetected error probability (P_{ue}). The theoretical computation of P_{ue} requires the Hamming weight distribution over the n-bits corresponding to the 2^k code words. For large values of n and k, evaluation of the Hamming weight distribution is impractical; however, using MacWilliams' theorem [40, 41], the dual code $(n,n-k)=(n,r)$ requires evaluation of the weight distribution (B_i) over the n-bits corresponding to 2^r code words. In this case, the undetected error probability, over the binary symmetric channel (BSC), with bit-error probability p, is expressed as

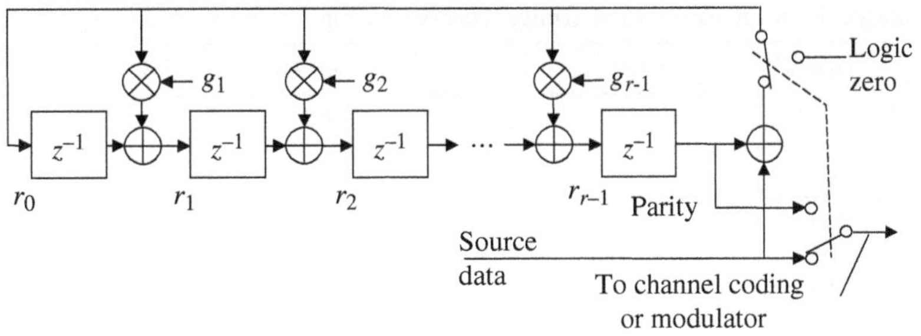

FIGURE 8.22 Generalized r-th-order encoder.

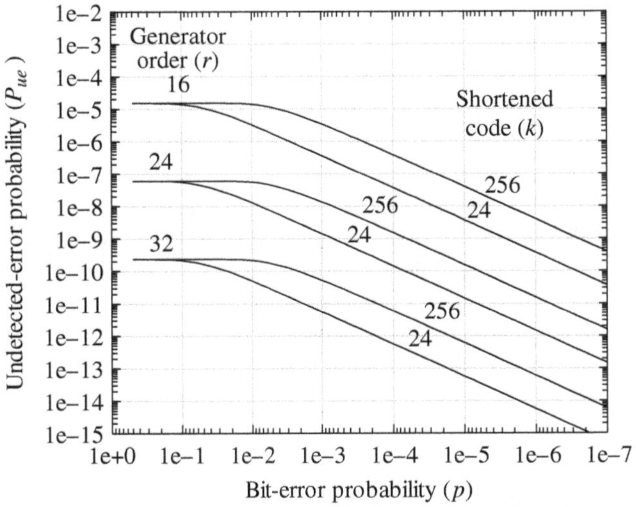

FIGURE 8.23 Ensemble average of undetected error probability.

$$P_{ue} = 2^{-r}\sum_{i=0}^{n} B_i(1-2p)^i - (1-p)^n \quad : p \le 0.5 \qquad (8.40)$$

The theoretical average undetected error probability of a binary (n,k) code, with equally probable messages, is expressed as [42]

$$\bar{P}_{ue} = 2^{-r}\left[1-(1-p)^k\right] \quad : p < 1.0 \qquad (8.41)$$

and is plotted in Figure 8.23 for $r = 16$, 24, and 32 order CRC code generators and shortened messages corresponding to $k = 24$ and 256 bits.

From (8.41), the average value of P_{ue} at $p = 0.5$ is evaluated as

$$\bar{P}_{ue}(p = 0.5) = 2^{-r} - 2^{-k} \cong 2^{-r} \quad : k \gg r \qquad (8.42)$$

Equation (8.40) is used by Fujiwara et al. [43] to evaluate the performance of two International Telecommunication Union Telecommunication Standard

TABLE 8.8 Simulated FCS Bit-Error Probability Corresponding to Undetected Error Probability of 1e–15

		Shortened Length k				
		24	128	512	2048	16,384
FCS Code	$g(x)^a$			p at P_{ue} = 1e–15		
ITU-1	170037	1e–4	4e–5	1e–5	2.7e–6	3.2e–7
ITU-2	140001	1e–4	3e–5	1e–5	2.7e–6	3.2e–7

[a] $g(x)$ designations in octal notation.

(ITU-T X.25)* CRC codes, referred to as the frame check sequence (FCS) codes. Their evaluations provide detailed plots of P_{ue} as a function of p, for the 16th-order generators and various shortened code lengths; however, the results listed in Table 8.8 show the intersection of the bit-error probability p corresponding to P_{ue} = 1e–15. When compared to the r = 16th-order generators in Figure 8.23, it is evident that these CRC codes perform considerably better than the average performance of $\bar{P}_{ue}$. However, the performance trends are similar; in that, the performance for $p = 0.5$ is identical to the value predicted by (8.42) and the trend in the performance degradation as the shortened code length is increased approaching the natural, or underlying, code length of 2^r is consistent with (8.42).

Using a simplified approach in the computation of the Hamming weight distribution (B_i) in (8.40), Wolf and Blakeney [44] verify the simulated performance of the ITU-1 code in Table 8.8 and provide the performance of the additional codes listed in Table 8.9. Upon comparing the performance of the CRC-16 code with that of the ITU-1 code, there is not a significant difference; however, the CRC-16Q code performance crosses the P_{ue} = 1e–15 threshold at a lower bit-error probability by about two and one orders of magnitude, respectively, for the 24- and 32-bit shortened codes; this

*Formerly, it was the International Telegraph and Telephone Consultative Committee (CCITT). At the time of the evaluation, two code recommendations were under consideration; the code denoted as ITU-1 is the current X.25 standard.

TABLE 8.9 Simulated FCS Bit-Error Probability Corresponding to Undetected Error Probability of 1e–15

		Shortened Length k					
		24	32	64	256	512	1024
CRC Codes	$g(x)^a$			p at $P_{ue} = 1\text{e}{-15}$			
CRC-16	100003	1e–4	8e–5	4e–5	1.8e–5	1e–5	—
CRC-16Q	104307	9e–3	2.3e–3	9e–5	2.2e–5	1e–5	—
CRC-24Q	404356 51	—	3e–2	2.3e–3	5.2e–3	2.8e–3	—
CRC-32Q	200601 40231	—	—	2.2e–2	1.3e–3	7e–4	4e–4

[a] $g(x)$ designations in octal notation.

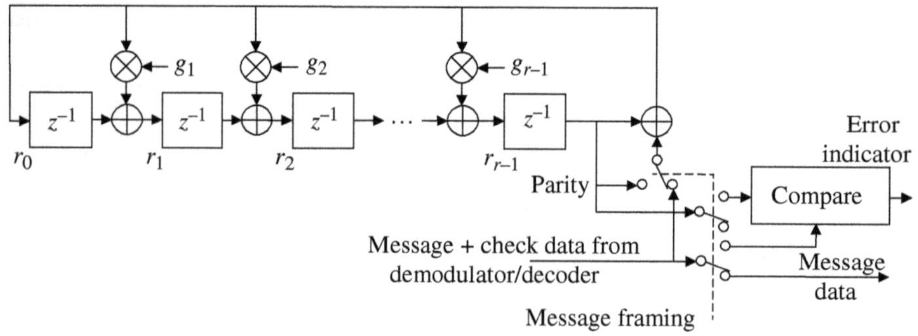

FIGURE 8.24 CRC decoder.

advantage diminishes rapidly as k increases. As seen from Figure 8.23 and demonstrated in Table 8.9 for the CRC-24Q and -32Q codes, the most significant performance improvement over the range of shortened codes is obtained by increasing the degree of the generator polynomial.

The CRC message decoder is shown in Figure 8.24. As the decoded message data passes by the CRC decoder, the processing is identical to that of the CRC encoder creating parity, or check data, corresponding to the received message. Upon completion of the message, the *switches*, indicated by the *dashed lines*, are changed and the newly created message check data is compared to those appended to the received message. If a disagreement between the two sets of check data is detected, the message is declared to be in error.

8.8 DATA RANDOMIZING CODES

Data randomization, or scrambling, is required in many applications to avoid long source data sequences of ones, zeros, and other nonrandom data patterns that may disrupt the demodulator automatic gain control (AGC), symbol and frequency tracking functions, and other adaptive processing algorithms. Nonrandom data patterns may also result in transmitted spectrums containing harmonics that interfere with adjacent frequency channels and randomizers ensure that the transmitted signal spectral energy conforms to the theoretical spectrum of the modulated waveform that is based on random data. Randomizers can be implemented using nonbinary PRN generators initialized with a known starting seed [45]; however, in this section, the focus is on binary PRN generators that use LFSRs that conform to irreducible primitive polynomials [38]. The randomization of the data can be viewed as low-level data security; however, the subject of data encryption provides a high level of communication security and involves specialized and often classified topics regarding implementation and performance evaluation.

Data randomization is implemented using either synchronous or asynchronous configurations. With the binary data randomizer initialized to a known state, the randomizer is synchronized by performing the exclusive-or logical operation of the source bits with the PRN or LFSR generated feedback bits. The derandomization, in the demodulator, is accomplished in the same way; however, the derandomizer must be synchronized to the start-of-message (SOM) or message frame and initialized to the known state used in the modulator. In this case, the loss of demodulator timing during the message results in catastrophic errors and missed messages. The asynchronous randomizer is a self-synchronizing

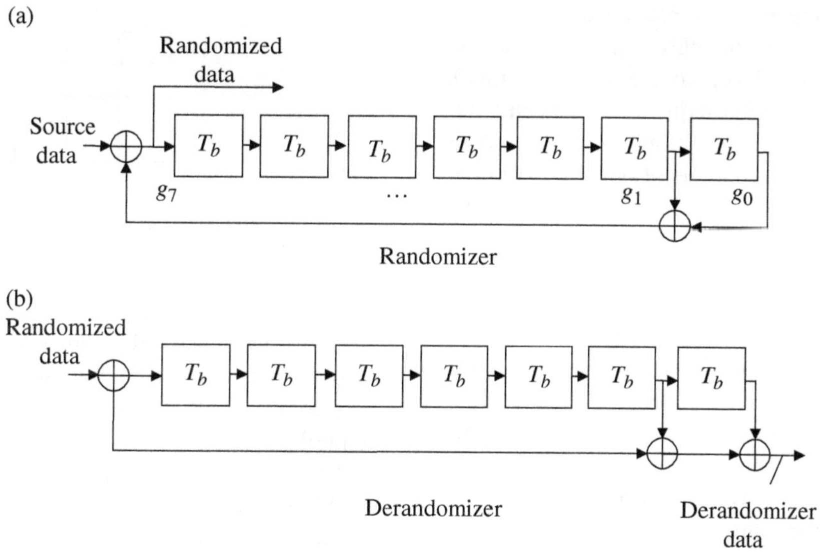

FIGURE 8.25 Binary 7th-order data randomization (T_b = bit interval).

TABLE 8.10 Some Randomizer Generator Polynomials with ≤ 4 LFSR Taps[a]

Order	Generator[b]	Taps	Order	Generator[b]	Taps
6	103	2	13	20033	4
7	203	2		20065	4
8	703	4	14	42103	4
	543	4	15	100003	2
9	1021	2		110013	4
	1131	4		122003	4
10	2011	2	16	210013	4
	3023	4		200071	4
	2431	4		312001	4
11	4005	2	17	400011	2
	5007	4		400017	4
12	10123	4		400431	4
	11015	4	18	1000201	2

[a]Peterson and Weldon [34]. Appendix C, Table C.2, © 1961 Massachusetts Institute of Technology. Reproduced by permission of The MIT Press.
[b]Octal notation.

configuration that avoids the synchronization issues and catastrophic error conditions; however, error multiplication by the number of LFSR taps occurs for each received bit error.

The following example of an asynchronous randomizer generated using the 7th-order irreducible primitive polynomial $g(x) = x^7 + x + 1$ is shown in Figure 8.25a and the derandomizer is shown in Figure 8.25b. Table 8.10 lists some generator polynomials with four or less feedback taps. These generators generate *m*-sequences with periods of $(2^r - 1)T_b$ where r is the order of the generator.

8.9 DATA INTERLEAVING

Controlled correlation between adjacent symbols of a modulated waveform is often used to provide significant system performance advantages for both spectrum control and detection gains. Some examples are partial-response modulation (PRM), continuous phase modulation (CPM), tamed frequency modulation (TFM) [46–49] , and Gaussian minimum shift keying (GMSK). Using these techniques significant system performance improvements are realized through waveform spectral control. On the other hand, communication channels can result in symbol correlation causing significant system performance degradation if mitigation techniques are not included in the system design. The communication channels that result in symbol correlation include narrowband channel (filters) resulting in intersymbol interference (ISI), fading channels resulting in signal multipath and scintillation, and various forms of correlated noise channels including impulse noise cause by lightning and man-made interference including sophisticated jammers that have the potential to result in significant performance losses.

Various types of equalizers provide excellent protection against ISI and multipath. Also, burst-error correction codes, like the RS code, can be used to mitigate these correlated channel errors [37]. However, many FEC codes, including the most commonly used convolutional codes, require randomly distributed errors entering the decoder. Data interleavers and deinterleavers provide an effective way to mitigate the impact of correlated channel errors. An important application of data interleavers, that cannot be understated, is their role in providing coding gain in the construction of turbo and turbo-like codes discussed in Section 8.12.

Interleavers typically operate on source or coded bits; however, deinterleavers may also operate on demodulator soft decisions and require several bits of storage for each symbol. The interleaver accepts an ordered sequence of data and outputs a reordered sequence in which contiguous input symbols are separated by some minimum number of symbols. The minimum interleaved symbol interval between contiguous input symbols is the *span* of the interleaver. At the deinterleaver, the reverse operation is performed restoring the original order of the data. The utility of the interleaver lies in the fact that bursts of contiguous channel errors appearing on the interleaved data will appear as individual or isolated random errors at the output of the deinterleaver. Interleavers are characterized in terms of the block length, delay, and the span of adjacent source bits in the interleaved data sequence. A good interleaver design provides a large span with uniformly distributed symbol errors in the deinterleaved sequence. Interleaver frame synchronization must be established at the demodulator. Commonly used interleavers are the block interleaver and the convolutional interleaver [50, 51], which are discussed in the following sections. The convolutional interleaver is a subset of a more general class of interleavers described by Ramsey [50] and Newman [52].

8.9.1 Block Interleavers

Block interleavers are described in terms of an (M,N) matrix of stored symbols where consecutive input symbols are written to the matrix column by column until the matrix is filled* and then forming the interleaved data sequence by reading the contents of the matrix row by row. A block interleaver is shown in Figure 8.26 and the following examples are specialized for a (4×3) matrix. In these examples, the matrix elements are initialized to zero and the input symbols are represented by a sequence of decimal integers $\{1,2,3,\ldots\}$. Upon interleaving a block of NM symbols, if the process is repeated, the last symbol of the previous block will be followed by the first symbol of the next block resulting in zero span between these two contiguous input symbols. This problem can be circumvented by randomly reordering the addressing of rows and columns from block-to-block. Using two data interleaver matrices in a *ping-pong* fashion is often preferred to the complications associated with clocking the data in and out of the same memory area. The deinterleaving is performed in the reverse order; in that, the received symbols are written to the deinterleaver matrix row by row and read to the output column by column. With block interleavers, bursts of contiguous channel errors longer than N symbols will result in short data bursts in the deinterleaved sequence. For example, if the interleaved sequence contains a burst of errors of length JM bits, then the errors in the deinterleaved

*The output can be initiated as soon as the $N(M-1)+1$ symbol is written to row one of the last column; however, additional memory must be available to support contiguous data over more than one block.

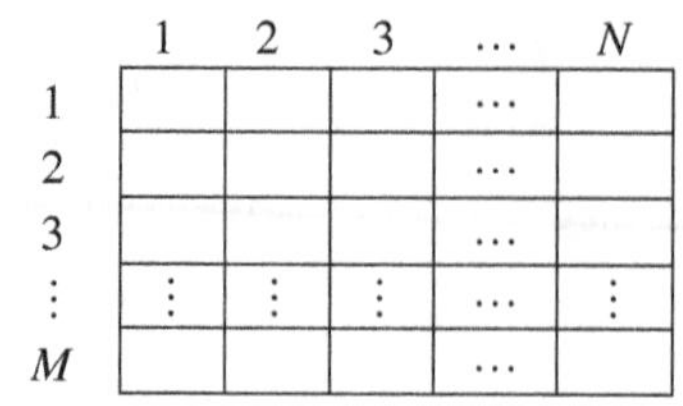

FIGURE 8.26 (M,N) block interleaver.

sequence can be grouped as N bursts of length $\leq J$ bits. However, in this characterization, some of the N deinterleaved output bursts may be contiguous. In general, block interleavers do not result in a uniform distribution of deinterleaved symbol errors.

The following example characterizes the block interleaver for $M=4$ and $N=3$.

The block interleaver characteristics are summarized as follows:

$$\text{Total storage} = NM \tag{8.43}$$

$$\text{Delay} \geq [N(M-1)+1-(M-1)] + [M(N-1)+1-(N-1)], \quad \text{typically} = 2NM \tag{8.44}$$

$$\text{Span} = M-1 \tag{8.45}$$

The expression for the delay simplifies to delay $\geq 2[NM+1]-(N+M)$. As mentioned previously, the *span* is defined as the minimum interleaved symbol interval between contiguous input symbols.

Block interleavers are conveniently applied to block codes of length n by choosing $N=n$ and then selecting the number of interleaver rows M to correspond to the channel correlation time, that is, the channel burst-error length, in view of the error correction capability of the FEC code. For example, consider a t-error correcting M-ary block code denoted as (n,k,t). In this case, choose the interleaver span as $N=n$ with $M \geq 2$ chosen to provide adequate decoder error correction in view of the burst-error length of the channel errors. Large values of M, however, will result in long data throughput delays.

8.9.2 Convolutional Interleavers

The convolutional interleaver is a special case of four types of (N_1, N_2) interleavers discussed by Ramsey and denoted as types I, II, III, and IV. The major difference in the implementation between the different types is the assignment of the input and output taps and the direction of commutation of the input and output data taps. The utility of the Ramsey interleavers is that, with the proper selection of the parameters N_1 and N_2, the output error distribution resulting from a burst of input errors is nearly uniform. The convolutional interleaver is similar to the type II Ramsey interleaver that requires that N_1+1 and N_2 be relatively prime; however, the convolutional interleaver requires that $N_2 = K(N_1 - 1)$ where K is a constant integer. An advantage of convolutional interleavers over Ramsey interleavers is that they are relatively easy to implement.

The characteristics of convolutional interleavers are similar to those of block interleavers; however, the amount of storage required at the modulator and demodulator is $NK(N+1)/2$ or about one-half of that required by a block interleaver with similar characteristics. The interleaver parameters N and K are defined in Figure 8.27. Also, the end-to-end delay, that is, the delay from the interleaver input to the deinterleaver out, is $NK(N-1)$. The operation of the convolutional interleaver is characterized by a commutator that switches the input bits through N positions with the first position passed to the output without delay. With the deinterleaver commutator properly synchronized the first interleaved symbol is switched into the $(N-1)K$ register, while the content of the last storage element is shifted to the deinterleaver output. The convolutional interleaver characteristics are summarized as follows:

$$\text{Total storage} = \frac{NK(N+1)}{2} \tag{8.46}$$

$$\text{Delay} = NK(N-1) \tag{8.47}$$

$$\text{Span} \geq NK + 1 \tag{8.48}$$

Here, the span is defined as the interleaved symbol interval between corresponding contiguous input symbols.

An example of the convolutional interleaver operation is shown in the following data sequences. In this example, the interleaver and deinterleaver are initialized with all zeros and a sequence of symbols consisting of decimal integers {1,2,3,...}, is applied to the input. The resulting out sequences are tabulated below for the case $N = 3$, $K = 1$ and the case $N = 3$, $K = 2$.

$N = 3$, $K = 1$ Interleaver:

Input Sequence:	1, 2, 3, 4, 5, 6, 7, 8, 9, 10, 11, 12, 13, 14, 15, 16, 17, 18, 19, 20, 21, 22, 23, 24, ...
Int'l Out:	1, 0, 0, 4, 2, 0, 7, 5, 3, 10, 8, 6, 13, 11, 9, 16, 14, 12, 19, 17, 15, 22, 20, 18, ...
Deint'l Out:	0, 0, 0, 0, 0, 0, 1, 2, 3, 4, 5, 6, 7, 8, 9, 10, 11, 12, 13, 14, 15, 16, 17, 18, ...

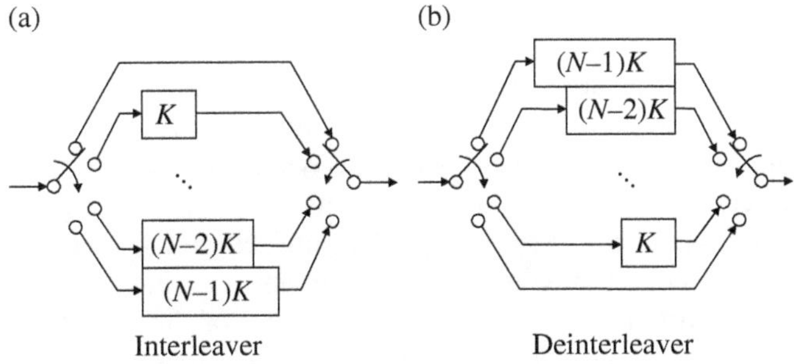

FIGURE 8.27 Convolutional Interleaver Implementation.

$N = 3$, $K = 2$ Interleaver:

Input sequence:	1, 2, 3, 4, 5, 6, 7, 8, 9, 10, 11, 12, 13, 14, 15, 16, 17, 18, 19, 20, 21, 22, 23, 24, ...
Int'l out:	1, 0, 0, 4, 0, 0, 7, 2, 0, 10, 5, 0, 13, 8, 3, 16, 11, 6, 19, 14, 9, 22, 17, 12, ...
Deint'l out:	0, 0, 0, 0, 0, 0, 0, 0, 0, 0, 0, 0, 1, 2, 3, 4, 5, 6, 7, 8, 9, 10, 11, 12, ...

An example of the burst-error characteristics of the convolutional interleaver is shown in the following data sequences for the $N = 3$, $K = 2$ interleaver with a burst of ten (10) contiguous channel errors following the 30th source symbol. In this example, the error bits are denoted as an X and the data record begins at the 30th symbol.

$N = 3$, $K = 2$ interleaver with 10 bit-error burst, denoted as X, beginning at bit 30:

Input sequence:	30, 31, 32, 33, 34, 35, 36, 37, 38, 39, 40, 41, 42, 43, 44, 45, 46, 47, 48, 49,
Deint'l input:	18, X, X, X, X, X, X, X, X, X, X, 35, 30, 43, 38, 33, 46, 41, 36, 49,
Deint'l out:	18, 19, 20, X, 22, 23, X, 25, X, X, 28, X, 30, X, X, 33, X, 35, 36, X, 38, 39, X, 41, 42, 43, ...

Evaluations of the interleaver performance include the uniformity of deinterleaved error events for various channel burst-error lengths and the probabilities associated with the span of the interleaved sequence relative to the contiguous source symbols. However, the ultimate evaluation of the interleaver performance is the bit-error probability for a specified waveform modulation and FEC code under various channel fading conditions.

8.10 WAGNER CODING AND DECODING

The Wagner code is a single-error-correcting block code denoted by $(N, N-1)$ where N is the block length containing $N-1$ information bits. A typical block structure for the $(N, N-1)$ Wagner code is shown in Figure 8.28 where $d_i = \{0,1\}$, $i = 1, 2, \ldots, N-1$ represents the binary information bits and the single parity bit d_N is determined from the modulo-2 sum of the information bits as $d_N = d_1 \oplus d_2 \oplus \cdots \oplus d_{N-1}$. The Wagner decoding process is quite simple. A hard decision is made to obtain an estimate

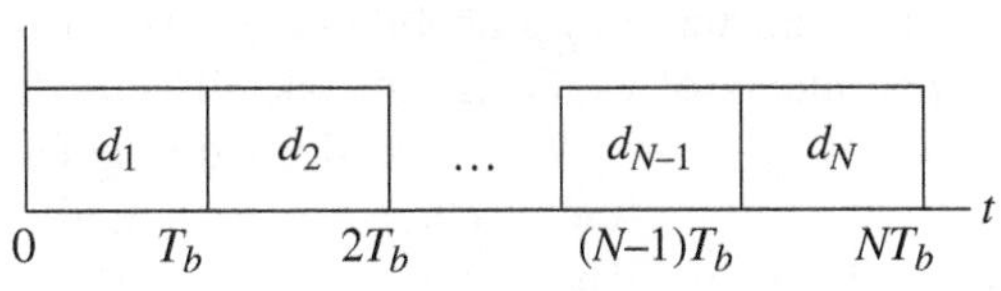

FIGURE 8.28 Wagner code structure.

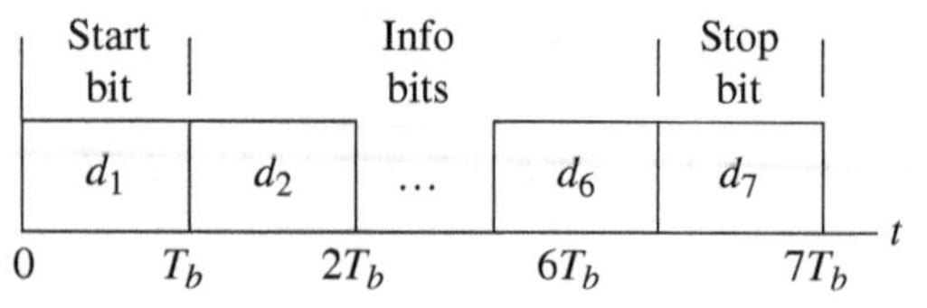

FIGURE 8.29 Seven-bit ASCII character with Wagner coding.

of each received bit and the corresponding soft decisions are saved. A modulo-2 addition is then performed on the five information bits and the parity bit. If the resulting parity bit is zero, the character is assumed to be correct. However, if the parity is one, the bit estimate corresponding to the smallest soft-decision magnitude is inverted and the character is declared as being correct.

The (6,5) Wagner code is sometimes applied to 7-bit American Standard Code for Information Interchange (ASCII) character transmission. The first bit of the ASCII character is the start bit, which is followed by five information bits, and the seventh bit is the stop bit as shown in Figure 8.29. The start and stop bits are *space* or binary zero bits which are normally used for synchronization. In this situation, the Wagner coding is performed by replacing the stop bit with the parity bit formed as $p = d_2 \oplus d_3 \oplus \cdots \oplus d_6$ with $d_7 = p$, so the (6,5) Wagner code includes to the information bits and the stop bit. With this coding structure, the start bit is available for character synchronization.

When differential coding is applied to the source data, as for example with compatible shift keying (CSK), the demodulation processing will result in two contiguous bit errors for every single-error event. Therefore, to ensure detection of single-error events, the parity check must be performed on every other bit. In this case, the start bit is used as the parity bit and is chosen to satisfy odd parity as $p = d_3 \oplus d_5 \oplus d_7$ with $d_1 = \bar{p}$. Note that the parity check will detect all single-error events with differential encoding; however, special consideration must be given to pairs of bit errors that span two adjacent characters because a detection error in the stop bit position will result in an error in the start (parity) bit in the next character. To ensure proper decoding in this case, a total of eight magnitude bits must be compared prior to the error correction. When the parity check fails on the current character block and the smallest magnitude is associated with the stop bit of the preceding character block, then the current parity bit is also assumed to be in error also and no error correction is performed on the current information bits.

In the following sections, the character-error performance of the Wagner code is examined. First, approximate expressions for the character-error probability are developed for the three embodiments of the Wagner code discussed earlier, namely, the raw $(N, N-1)$ Wagner code, and Wagner coding applied to the 7-bit ASCII character with and without differential encoding. The results of these analyses are approximate because the location of the bit in error is assumed to be known. The theoretical character-error performance is then established based on using the soft decisions to identify the most likely bit to be in error. This analysis is exact in the sense that the soft decisions are used to locate the most probable bit in error.

8.10.1 Wagner Code Performance Approximation

The character-error performance of a Wagner coded block can be approximated by assuming that all single errors are corrected and that all other error conditions are uncorrected and result in a character error. In this case, the character-error probability of the raw $(N, N-1)$ Wagner code is simply

$$P_{ce} = 1 - P_{cc} \tag{8.49}$$

where the probability of a correct character, under the assumption of independent error events with probability P_{be}, is given by

$$P_{cc} \cong (1-P_{be})^N + NP_{be}(1-P_{be})^{N-1} \quad \text{: raw Wagner code} \tag{8.50}$$

This result is based on the binomial distribution [53] and P_{be} is the bit-error probability of the underlying waveform modulation.

For the Wagner coded 7-bit ASCII character without differential coding, the probability of a correct character is given by

$$P_{cc} \cong (1-P_{be})^5 + 5P_{be}(1-P_{be})^4 \quad \text{: Wagner coded ASCII character} \tag{8.51}$$

With differential coding, the 7-bit ASCII character is correctly decoded if a single bit-error occurs among the six information bits. An error in the parity bit results in an error in the following information bits so the ASCII character cannot be corrected. The resulting probability of a correct character is then

$$P_{cc} \cong (1-P_{be})^6 + 6P_{be}(1-P_{be})^5 \quad \text{: Wagner differential coded ASCII character} \tag{8.52}$$

The approximate performance results for the raw $(N, N-1)$ Wagner code are shown in Figure 8.30 for antipodal waveform modulation with AWGN. Figure 8.31 shows the approximate performance when the Wagner coded is applied to the 7-bit ASCII character as discussed before. The signal-to-noise ratio (γ_c) for the underlying antipodal signaling of the coded bits is expressed in term to the code-bit bandwidth $1/T_c$, where T_c is the duration of the coded bit. The performance, however, is plotted as a function of the

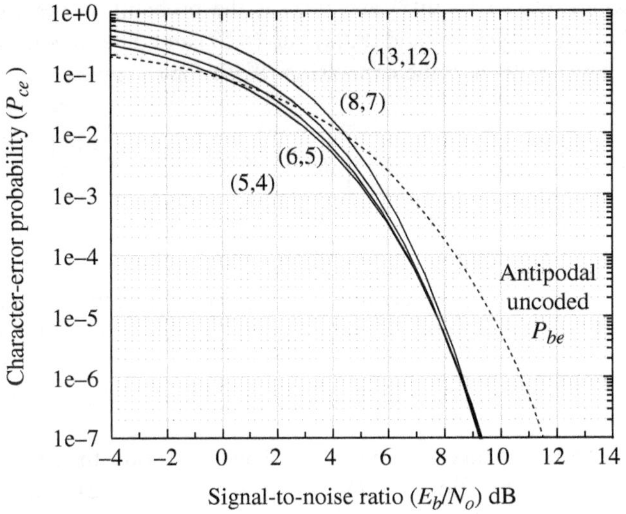

FIGURE 8.30 Approximate performance of Wagner coded data (*N*-bit character).

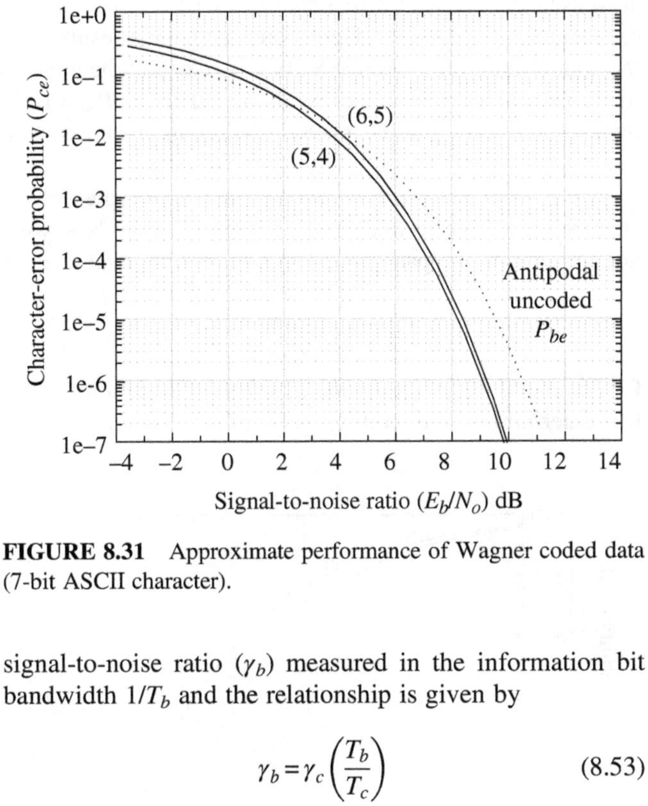

FIGURE 8.31 Approximate performance of Wagner coded data (7-bit ASCII character).

signal-to-noise ratio (γ_b) measured in the information bit bandwidth $1/T_b$ and the relationship is given by

$$\gamma_b = \gamma_c \left(\frac{T_b}{T_c} \right) \tag{8.53}$$

For the raw Wagner code, $T_b/T_c = N/(N-1)$ and for the Wagner coded 7-bit ASCII character $T_b/T_c = 7/5$ (= 1.46 dB).

8.10.2 (*N*, *N* – 1) Wagner Code Performance

In this section, the relationship between the bit-error probability at the input to the Wagner decoder and the resulting character-error probability at the decoder output is established. The theoretical performance in the AWGN channel is examined and the results are compared to Monte Carlo simulation results. The theoretical results apply for antipodal signaling such as BPSK, QPSK, OQPSK, or minimum shift keying (MSK); however, the simulation results are based on MSK waveform modulation with ideal symbol timing and phase tracking in the demodulator.

The general expression for the probability of a correctly received Wagner coded block is given by

$$\begin{aligned} P_{cc} &= P_r(\text{no bit errors}) \\ &\quad + P_r(\text{exactly one bit error}) P_r(\text{correctly locating error}) \end{aligned} \tag{8.54}$$

When the noise is independent and identically distributed (*iid*) from bit to bit, this result can be expressed as

$$\begin{aligned} P_{cc} &= (1-P_{be})^N \\ &\quad + NP_{be}(1-P_{be})^{N-1} P_r(\text{correctly locating error}) \end{aligned} \tag{8.55}$$

In the remainder of this section, the probability of correctly locating a single-error event is analyzed using the stored soft decisions out of the optimally sampled matched filter.

The location of a single-error event in the *j*-th bit is based upon the magnitude z_j of the *j*-th decision variable v_j being less than the magnitudes z_i of all the other bit decision variables v_i, $i = 1, \ldots, N$ such that $i \neq j$. Given the transmitted bit sequence $\{d_i\}$ and the received sequence $\{\hat{d}_i\}$, the probability of correctly locating the error is given by the general expression

$$\begin{aligned} P_r(\text{correct error location}) &= \int_0^\infty dz_j \int_{z_j}^\infty \cdots \int_{z_j}^\infty f\left(z_i, \ldots, z_j, \ldots, z_n \mid \hat{d}_i = d_i, \ldots, \right. \\ &\quad \left. \hat{d}_j \neq d_j, \ldots, \hat{d}_n = d_n\right) dz_i, \ldots, dz_{j-1}, \ldots, dz_{j+1}, \ldots, dz_n \end{aligned} \tag{8.56}$$

where *f*(–) is the conditional joint density function of the random variables $z_i = |v_i|$. With independent source data, a memoryless channel, and ideal receiver timing, the AWGN channel decision variables, v_i, are *iid* with distribution $N(d_i, \sigma_n)$ for which the earlier result specializes to

$$\begin{aligned} P_r(\text{correct error location}) &= \int_0^\infty dz_j f\left(z_j \mid \hat{d}_j \neq d_j\right) \\ &\quad \times \left[\int_{z_j}^\infty dz_i f\left(z_i \mid \hat{d}_i = d_i\right) \right]^{N-1} \end{aligned} \tag{8.57}$$

The two conditioned distributions are found from a form of Bayes rule given by

$$f(v|c) = \frac{P(c|v)f(v)}{\int P(c|v)f(v)dv} \tag{8.58}$$

Conditioning on an error event, that is, $c \equiv \hat{d}_j \neq d_j$, the following relationships apply

$$P(\hat{d}=1, d=-1|v_j) = \begin{cases} 0 & : v_j \leq 0 \\ 1 & : v_j > 0 \end{cases} \tag{8.59}$$

and

$$P(\hat{d}=-1, d=1|v_j) = \begin{cases} 1 & : v_j \leq 0 \\ 0 & : v_j > 0 \end{cases} \tag{8.60}$$

Upon combining these results with the normal distribution $f(v) = N(d, \sigma_n)$, the numerator in (8.58) becomes

$$f(\hat{d}_j \neq d_j|v_j)f(v_j) = \begin{cases} 0 & : v_j < 0 \\ \dfrac{1}{\sqrt{2\pi}\sigma_n} e^{-(vj+|dj|)^2/2\sigma_n^2} & : v_j \geq 0 \end{cases} \tag{8.61}$$

Using this result to evaluate the denominator in (8.58) conditioned on an error, it is noted that the denominator is simply the probability of a bit error, that is,

$$\int_0^\infty P(\hat{d} \neq d|v)dv = P_{be} \tag{8.62}$$

Since these results apply for $v_j > 0$, upon substituting z_j for v_j in (8.61) the solution to (8.58) becomes

$$f(z_j|\hat{d} \neq d) = \frac{1}{\sqrt{2\pi}\sigma_n P_{be}} e^{-(z_j+|d_j|)^2/2\sigma_n^2} \tag{8.63}$$

Evaluation of (8.58) conditioned on a correct decision, that is, $c \equiv \hat{d}_j = d_j$, and proceeding in a similar manner leads to the conditional distribution

$$f(z_i|\hat{d} = d) = \frac{1}{\sqrt{2\pi}\sigma_n(1-P_{be})} e^{-(z_i-|d_i|)^2/2\sigma_n^2} \tag{8.64}$$

Using (8.64) to evaluate the second integral in (8.57) results in

$$\int_{z_j}^\infty dz_i f(z_i|\hat{d}_i = d) = \frac{1}{(1-P_{be})} \int_{z_j}^\infty dz_i \frac{1}{\sqrt{2\pi}\sigma_n} e^{-(z_i-|d_i|)^2/2\sigma_n^2}$$
$$= \frac{Q((z_j-|d_i|)/\sigma_n)}{(1-P_{be})} \tag{8.65}$$

where $Q(x) = 1 - \Phi(x)$ and $\Phi(x)$ is the probability integral. Substituting (8.65) and (8.63) into (8.57) and, in turn, substituting (8.57) into the expression (8.55) for P_{cc} gives

$$P_{cc} = (1-P_{be})^N + \frac{N}{\sqrt{2\pi}\sigma_n} \int_0^\infty dz_j e^{-(z_j+|d_j|)^2/2\sigma_n^2} Q\left(\frac{z_j-|d_i|}{\sigma_n}\right)^{N-1} \tag{8.66}$$

This expression is computed numerically and the result is used to compute the character-error probability expressed as

$$P_{ce} = 1 - P_{cc} \tag{8.67}$$

The *solid* curves in Figures 8.32 and 8.33 show the performance as a function of E_b/N_o for the (6,5) and (13,12) Wagner codes, respectively. The numerical integration is based on a resolution of 20 samples for each standard deviation σ_n over a range of 10 standard deviations above the mean value at $z_j = |d_j|$. The discrete *circled* data points for the (6,5) Wagner code are based on Monte Carlo simulations using 500 Kbits at each signal-to-noise ratio. Upon comparing these results with the approximations in Figure 8.30, it is seen that the approximate results are optimistic by about 1.0 dB at $P_{ce} = 10^{-5}$; owing to the fact that the bit-error locations were assumed to be known.

The *dotted* curve in Figures 8.32 and 8.33 represent the character-error performance if the information bits were not coded with the parity bit. These results are evaluated as

$$P_{ce} = 1 - (1-P_{be})^N \tag{8.68}$$

for $N = 5$ and 12, respectively, and they provide a reference for determining the coding gain of the Wagner coded

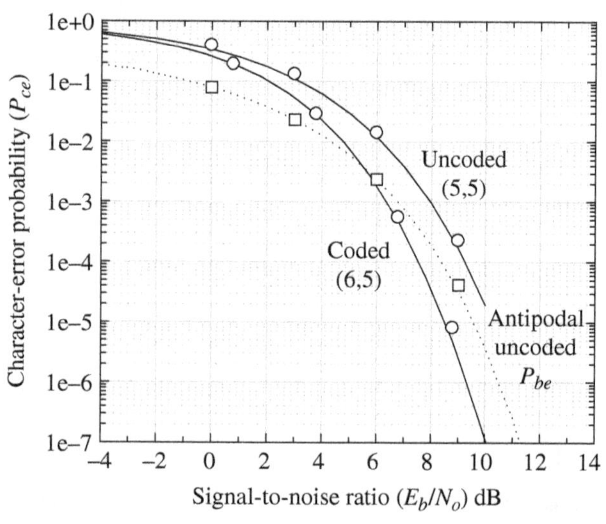

FIGURE 8.32 Performance of MSK demodulator with (6,5) Wagner code.

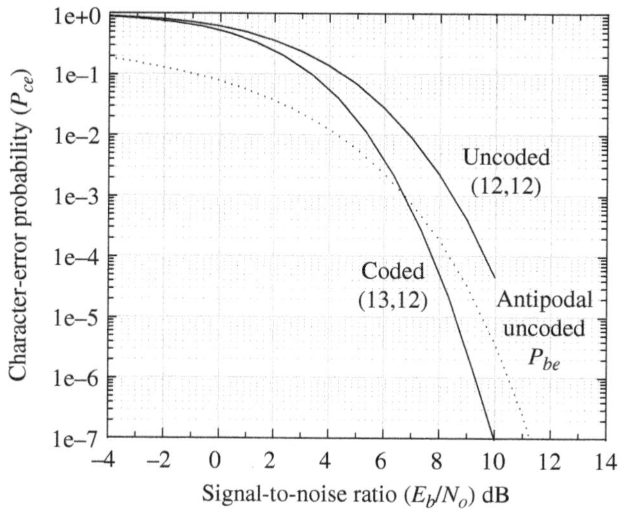

FIGURE 8.33 Performance of MSK demodulator with (13,12) Wagner code.

characters which is about 2.0 dB at $P_{ce} = 10^{-5}$. The *dotted* curve in these figures is simply the uncoded bit-error performance of antipodal signaling and is used to compute uncoded (5,5) and (12,12) P_{ce} as a point of reference.

8.11 CONVOLUTIONAL CODES

The convolutional decoding discussed in this section focuses on the Viterbi decoding algorithm as it applies to a *trellis* decoding structure. Prior to the discovery of the trellis decoding structure, sequential decoding of convolutional codes, described by Wozencraft [54] and Fano [55], was used. Sequential decoding is based on metric computations on various branches through a *code-tree* decoding structure [56]. In this case, the decoded data is associated with the branches or paths through the code tree that yield the largest metric. Heller and Jacobs [57] compare the advantages and disadvantages of the Viterbi and sequential decoding techniques in consideration of error performance, decoder delay, decoder termination sequences, code rates, quantization, and the sensitivity to channel conditions. The processing complexity is a major factor that distinguishes these two decoding procedures and, by this measure, Viterbi decoding is applicable to short constraint length codes ($K \le 9$); otherwise, the sequential decoding is more processing efficient.

The following description of convolutional coding draws upon the vast resource of research and publications on the subject [58–64] and other references cited throughout this section. Convolutional codes, unlike block codes, do not involve algebraic concepts in the decoding process and, therefore, result in more intuitive and, to some extent, straightforward processing algorithms. Throughout the following descriptions, binary data is considered and sequences are represented by polynomials $f(x) = b_o + b_1x^1 + b_2x^3 + \cdots$ with binary coefficients $b_i \in \{1,0\}$. In this case, multiplication and summation of polynomials is performed using GF(2) arithmetic. For example, multiplication of two polynomials $f(x)$ and $f'(x)$ corresponds the coefficient multiplication b_ib_i' and summation is simply the modulo-2 summation of the b_i and b_i' coefficients.

Convolutional encoders accept a continuous stream of source data at the rate R_b bps and generate a continuous stream of code bits, at a rate of $R_{cb} = R_b/r_c$ where r_c is the code rate, defined as

$$r_c \triangleq \frac{k}{n} \tag{8.69}$$

The parameter k represents the number of source data bits, corresponding to a q-ary symbol ($q = 2^k$), that are entered into the encoder each code block and n represents the corresponding number of coded bits.

A single parity bit is influenced by $\upsilon = kK$ source bits and the n encoder parity bits are generated by the convolutional sum, expressed as

$$P_\ell(i) = \sum_{j=1}^{\upsilon} g_\ell(j)S(i-j) \quad : \ell = 1,2,\ldots,n; \;\; i = 1,2,3,\ldots \tag{8.70}$$

where $g_\ell(j)$ represents the ℓ-th subgenerator of the code, $S(j)$ is the stored array of source bits, and $P_\ell(i)$ are the output parity bits corresponding to the ℓ-th subgenerator. The parameter K is the constraint length of the code and corresponds to the number of stored M-ary symbols during each code block. The index i, corresponding to the source data, can be arbitrarily long, distinguishing the convolutional codes from fixed-length block codes. The encoder described by (8.70) can be viewed as a transversal, or finite impulse response (FIR), filter with unit-pulse response described by the code generator.

The following description of the convolutional encoder is characterized in terms of the *coding interval*, or *code block*,* defined as the encoder processing required to generate n parity bits in response to an input symbol of k source bits. This significantly simplifies the convolutional encoding and decoding description and notation. In the demodulator, the convolutional decoding is analogous to characterizing the trellis decoding as a recursion of multistate symbol decoding intervals.† The encoding and decoding recognizes the correlative properties of preceding symbols, for example, the

*In the following descriptions, the *coding interval* is referred to as a *code block*.

†The symbol decoding of the convolutional coded data using a multistate trellis is discussed in Section 8.11.2.

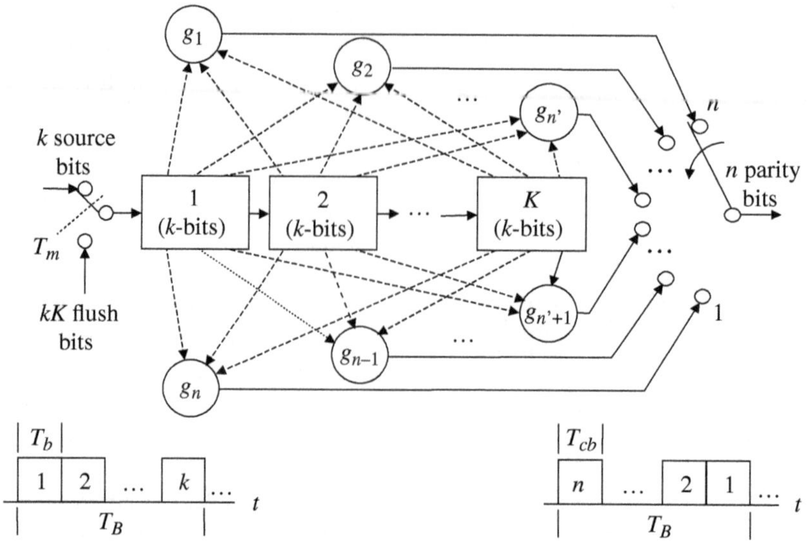

FIGURE 8.34 Generalized GF(2^k) nonsystematic convolutional encoder (rate $r_c = k/n$, constraint length K).

preceding $K-1$ source symbols are processed in the encoder as described in the following.

In general, the convolutional encoder is implemented using binary arithmetic to compute the parity symbols by clocking a block of k source bits into shift registers and computing the corresponding n parity bits over the υ source bits stored in the encoder memory. To maintain real-time operation, the duration of the n-tuple of parity bits must equal the input symbol duration. The implementation of a GF(2^k) convolution encoder is shown in Figure 8.34 where the binary arithmetic involves exclusive-or operations.

The implementation in Figure 8.34 corresponds to one code block with the most recent k-bit source symbol corresponding to the storage location 1 with the previous $K-1$ symbols in the storage locations 2 through K. Following the message of length T_m seconds, the kK flush bits are used to return the encoder to the initial encoder state. Typically, the encoder is initialized to the zero state. The flush bits are also used in the decoder as described in Section 8.11.2. Each source symbol corresponds to a $(k-1)$-dimensional polynomial

$$S_l(x) = b_{l,k-1}x^{k-1} + \cdots + b_{l,2}x^2 + b_{l,1}x + b_{l,0} \quad : l = 1, \ldots, K \tag{8.71}$$

where the bit b_{l0} is the rightmost or LSB of the input symbol. Taken collectively, the entire K symbols are denoted as the $(\upsilon-1)$-dimensional polynomial

$$S(x) = S_1(x)x^{(K-1)k} + \cdots + S_{K-1}(x)x^k + S_K(x) \tag{8.72}$$

In Figure 8.34, the *dotted lines*, connecting the stored symbols to the subgenerators of the code, denote multiple connections from each subgenerator corresponding to the binary-one coefficients of the $(\upsilon-1)$-dimensional generator polynomial, expressed as

$$g_\ell(x) = b'_{\ell,\upsilon-1}x^{\upsilon-1} + \cdots + b'_{\ell,1}x + b'_{\ell,0} \quad : \ell = 1, 2, \ldots, n \tag{8.73}$$

In the context of the code-block processing, the subgenerators and stored data bits can be viewed as vectors with the parity formed by the scalar product $P_\ell = \bar{g}_\ell \cdot \bar{S}^T \; : \ell = 1, \ldots, n$.

An important consideration in the modulator encoding is the mapping of the code bits* to the waveform modulation symbol as depicted in Figure 8.35 [65, 66]. The encoder shown in Figure 8.34 results in a nonsystematic code, in that the coded output contains only parity-check bits; however, the code-bit mapping also applies to systematic codes that include the k source bits plus $n-k$ appended parity-check bits. In either case, the information bits or the most significant bit (MSB) parity-check bits must be mapped to the most protected states of the modulated waveform. If an interleaver is used, the number of interleaver columns is selected to accommodate an integer number of transmitted symbol. For example, for a rate $r_c = 1/n$ code using MPSK symbol modulation with modulation efficiency $r_m = \log_2(M)$ and n = multiples of r_m, a row-column interleaver may be designed with r_m columns and n/r_m rows. In this example, rate matching is not necessary; however, puncturing of low-rate codes has many advantages and requires rate matching to the symbol modulation waveform.

*The code bits are associated with the coding of one q-ary input symbol and are often referred to as a code word.

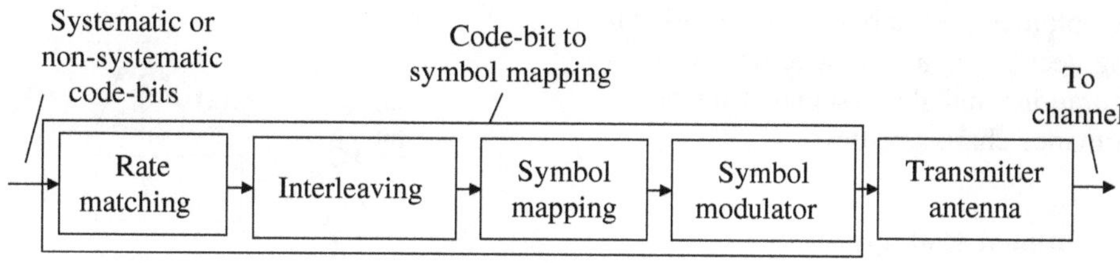

FIGURE 8.35 Mapping of code bits to modulation symbols.

Interleaved parity bits			Gray code
$b_{m2}x^2$ +	$b_{m1}x$ +	b_{m0}	
$P_9=1$	$P_6=1$	$P_3=0$	101
$P_8=0$	$P_5=1$	$P_2=1$	010
$P_7=0$	$P_4=0$	$P_1=1$	001

3 × 3 block interleaver

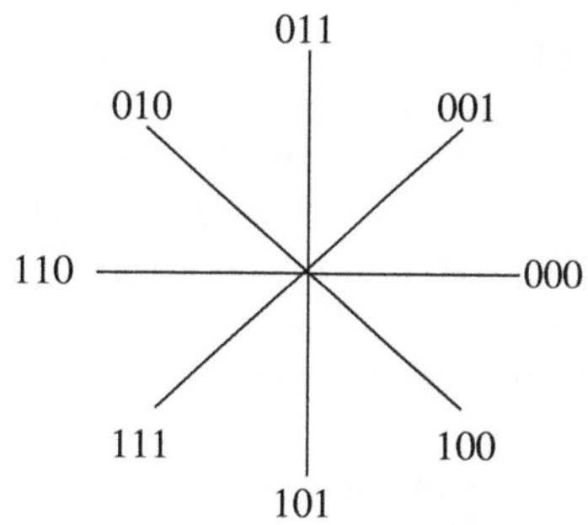

Gray-coded phase mapping

FIGURE 8.36 Example of rate matching for (3,9) convolutional encoder (3 × 3 block interleaver, gray coding, and 8PSK modulation).

Puncturing is a process in which certain parity-check bits are not transmitted to realize a higher code rate than that of the underlying convolutional code. Although the resulting coding gain is reduced, the symbol bandwidth W is also reduced thus increasing the spectral efficiency, defined as, $r_s = R_b/W$ bits/s/Hz. Alternately, the symbol rate may remain the same thus increasing the data throughput time. In either event, puncturing provides for tailoring the message transmission to the channel conditions. When puncturing is used, it is only applied to the parity-check bits, that is, the source bits of a systematic code are not punctured. Optimum puncturing of low-rate convolutional codes is discussed in Section 8.11.7. Appendix 8A provides a list convolutional code puncturing patterns found by Yasuda et al. [67], for punctured code rates $(n-1)/n$ with $n = 3$ through 14 and constraint length $\nu = 2$ through 8. The punctured codes are based on the underlying rate 1/2 codes with the same constraint lengths. The rate matching function in Figure 8.35 identifies and eliminates the punctured parity-check bits and prioritizes and assigns the surviving bits in the interleaver.

Referring again to Figure 8.34, rate-matching parity-check bits are priority ordered and entered column by column into a row–column interleaver with the highest priority bits in the MSB (leftmost) columns. The information bits of a systematic code are always assigned the highest priority. The interleaver is emptied row by row and mapped to the symbol modulator with the highest priority row bits assigned to the MSB of the modulation states, thus providing the most protection. Gray coding prior to the modulation symbol mapping provides the most protection from demodulator symbol errors. For example, consider that three contiguous code blocks of the rate 1/3 convolutional encoder in Figure 8.34 are entered into a 3 × 3 block interleaver. The nine parity-check bits, P_ℓ, are assigned to the block interleaver, gray coded, and then mapped to the 8PSK modulator phase constellation as shown in Figure 8.36. The ordering of the bits (P_9, P_8, P_7) corresponds to the three consecutive MSB bits of the encoder. For example, P_9 corresponds to the parity-check bit b_{m2}, of the first code-block modulator polynomial

$$m(x) = b_{m2}x^2 + b_{m1}x + b_{m0} \tag{8.74}$$

In terms of (8.70), $P_9 = P_3(1)$, $P_8 = P_3(2)$, and $P_7 = P_3(3)$. The mapping of the code bits to the symbol modulation is discussed again in Section 8.12.2 in the context of concatenated convolutional codes where the code block is defined as the interleaver length that ranges from 128 to over 16,384 information bits. This is a significantly different requirement; in that, the number of transmitted code bits corresponding to the interleaver plus flushing bits must be integrally related to the number of transmitted symbols.

Gray coding is used to minimize the error events in a noisy channel and, upon examining the phase constellation, it is seen that the first- and second-order modulator bits are the most protected bits. In this example, the entire interleaved parity set is transmitted in the code-block interval using three consecutive phase modulated symbols. Increasing the number of interleaver columns and filling the interleaver across several code blocks provides for improved mitigation of channel fading at the expense of throughput delay.

Beginning with the commonly used binary convolutional encoder, the following sections examine specific convolutional encoder configurations and the associated decoding complexity and performance characteristics.

8.11.1 Binary Convolutional Coding and Trellis Decoding

The binary convolutional code is characterized by entering one source bit into the encoder memory array each code block. This corresponds to $k = 1$ in Figure 8.34 with the K encoder memory consisting of K bits. The widely used rate 1/2 convolutional code with constraint length $K = 7$ provides excellent coding gain with reasonable complexity; however, the following description of the rate 1/2, $K = 3$ binary convolutional encoder provides all of the essential encoding and decoding concepts necessary to implement the more involved coding structures.

The coding and decoding of the simple, and relatively easy to understand, rate 1/2, $K = 3$ encoder is shown in Figure 8.37. The structure follows directly for the general implementation shown in Figure 8.34 using $k = 1$, $K = 3$ and the two subgenerators $g_1(x)$ and $g_2(x)$.

The $K = 3$ encoder stores $K - 1$ previous bits in locations 2 and 3 and with current bit of the input source data in location 1. The $K - 1$ bits represent the state of the encoder so, for this binary convolution encoder, there are $2^{K-1} = 4$ states. The encoder memory is initialized to zero and the following example uses the input data sequence of 1011010010.... The first bit into the encoder is 1, the leftmost bit, and, in keeping the convolutional sum, the input data is time-reversed yielding,* ...0100101101. Based on this description, the encoder output is developed as shown in Table 8.11. The code bits are then processed as shown in Figure 8.35 and the modulated symbols are transmitted through the channel to the receiver.

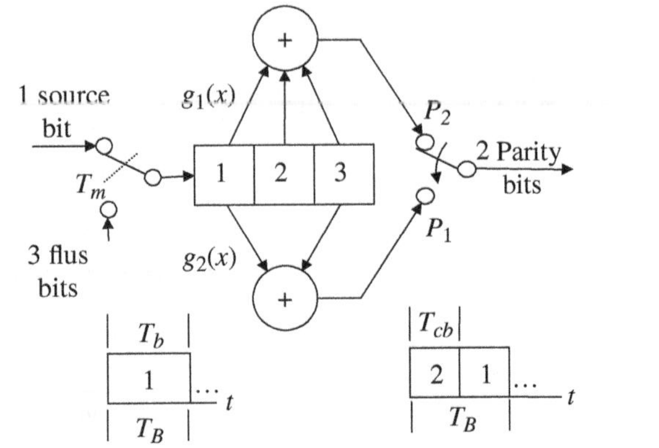

FIGURE 8.37 Binary, rate 1/2, $K = 3$, nonsystematic convolutional encoder ($g_1(x) = x^2 + x + 1$, $g_2(x) = x^2 + 1$).

TABLE 8.11 Example of Binary Rate 1/2, $K = 3$ Convolutional Encoding

Source Bit 1	Encoder State Bits 2,3		Parity Bits (1,2)	
1	0	0	1	1
0	1	0	1	1
1	0	1	0	0
1	1	0	0	1
0	1	1	0	1
1	0	1	0	0
0	1	0	1	0
0	0	1	1	1
1	0	0	1	1
0	1	0	1	0
...	0	1	...	...

8.11.2 Trellis Decoding of Binary Convolutional Codes

The maximum-likelihood decoding of convolutional codes was published by Viterbi [68] in 1967. This paved the way for further research in what has become known as the Viterbi algorithm [69]. Forney [70, 71], rigorously showed that the Viterbi algorithm results in maximum-likelihood decoding, and Heller [72, 73] is credited with the discovery of efficient decoding techniques using the Viterbi algorithm.

In the following, description of the trellis decoding of convolutional codes is based on the relatively straightforward binary, rate 1/2, $K = 3$ code, recognizing that more *powerful* convolutional codes use exactly the same, albeit more involved, decoding algorithms. The decoding starts by identifying the four-state trellis decoder and the corresponding binary representations of the four states as indicated in Figure 8.38. The parameters of the generalized k-ary trellis decoder in Figure 8.34 are identified in Table 8.12 with the numerical values corresponding to the example decoder in Figure 8.38. In general, the trellis diagram of a rate k/n convolutional code has $2^{k(K-1)}$ states with 2^k transitions emanating from each source state and terminating on each termination state.

By way of explanation, the decoding states in Figure 8.38 are represented by the state vectors $\bar{S}_i$ with elements $\alpha_{i,j}: j = 0, \ldots, 3$ denoted as the state metric corresponding to each of the trellis states. It will be seen that the state metrics $\alpha_{i,j}$ are updated with estimates of the received data at each state transition. The *dark gray* state transition lines are shown to indicate that these transitions are *not* required as the trellis is *building up* when starting at the known zero initial state of the encoder. For short messages or short decoding depths, the decoder E_b/N_o performance is improved somewhat by

*The time-reversed data sequence is the *natural ordering* of the data entered into the encoder, that is, the first, second, etc., bits are taken in order.

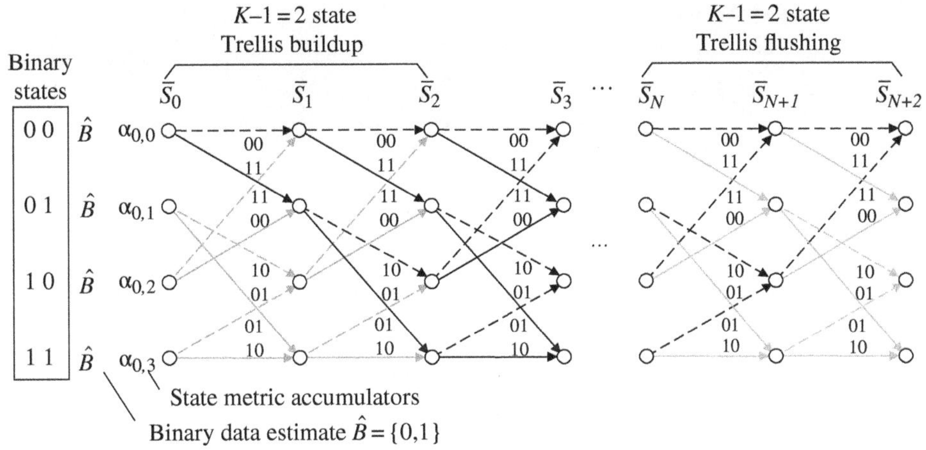

FIGURE 8.38 Description of convolutional code Trellis decoder (binary, rate 1/2, $K = 3$).

TABLE 8.12 Generalized Convolution Decoder Parameters

Parameter	Description	Example Value
k	Input source bits per encoding block	1
n	Output code bits (parity)	2
k/n	Code rate	1/2
K	Constraint length	3
$K-1$	Trellis buildup and flushing transitions	2
kK	Encoder bits and stuff bits	2
$2^{k(K-1)}$	Encoder and trellis decoder states	4
2^k	Trellis branches emanating from and terminating on each state	2
N_{dep}	Trellis depth (symbols) before removing data estimates (4 or 5 constraint lengths)[a]	15

[a]For high-rate codes (3/4 and higher), trellis depths of 8 or 9 constraint lengths are required.

not including these transition; however, the decoder synchronization is more problematic, in that the first code-bit word must be synchronized with the SOM preamble. On the other hand, for long messages, that can tolerate longer trellis depths, the inclusion of these transitions is not as important because, with a trellis depth in excess of four constraint lengths, the trellis processing essentially eliminates the incorrect paths through the maximum-likelihood decisions at each state transition.*

The self-synchronizing characteristic of the trellis decoder allows for message decoding in random data without knowledge of the SOM. For example, after initializing the state metrics, $\alpha_{0,j}$, to zero, if the decoding is not correctly synchronized with the received code-bit word, the surviving metrics at each state, after about four constraint lengths, will appear as random variables. In this event, the state metric accumulators are re-initialized, the code-bit word timing is shifted, and the processing is repeated until a dominant metric is observed indicating code synchronization.

The *dark gray* state transition lines indicate the transitions that are *not* associated with the *trellis flushing* required to drive the trellis to the known all-zero state of the encoder following the message bits. The two (in general kK) zero flushing bits are appended to the message bits as indicate in Figure 8.38.

The details of the trellis decoding are described by the transition processing shown in Figure 8.39. The state transition is depicted from the source state to the termination state denoted by the state vectors $\bar{S}_s = \bar{S}_{i-1}$ and $\bar{S}_t = \bar{S}_i$. As a practical matter, only the two state vectors $\bar{S}_s$ and $\bar{S}_t$, their respective elements $\alpha_{s,j} = \alpha_{i-1,j}$ and $\alpha_{t,j} = \alpha_{i,j}$, and the detected data estimate $\hat{B}$ are required for the entire trellis decoding. This simplicity results from the add–compare–select (ACS) function of the max(a,b) algorithm and, for a sufficiently long trellis depths, leads to maximum-likelihood data decoding [68].

As indicated, the state transitions in Figure 8.39 are only shown for the source data estimate $\hat{B} = 0$ and the remaining processing for $\hat{B} = 1$ is implicit. The source states $j = 0, \ldots, 3$ are represented by the binary equivalents $j = (b_2,b_1)_b$ that are concatenated with the source data estimate $\hat{B} = 0$ denoted† as $b_2,b_1//\hat{B}$. The rightmost bits of the concatenation correspond, or point, to the termination state, and by convention the transitions corresponding to $\hat{B} = 0$ are denoted by the *dashes* state transition lines. When the $\hat{B} = 1$ source data estimate is included, two state transitions will emanate from each source state and terminate on the termination states as indicated in Table 8.12. As mentioned before, the two transitions

*The path elimination procedure applied to a code tree is referred to as the Viterbi decoding algorithm.

†The notation // is used to signify concatenation.

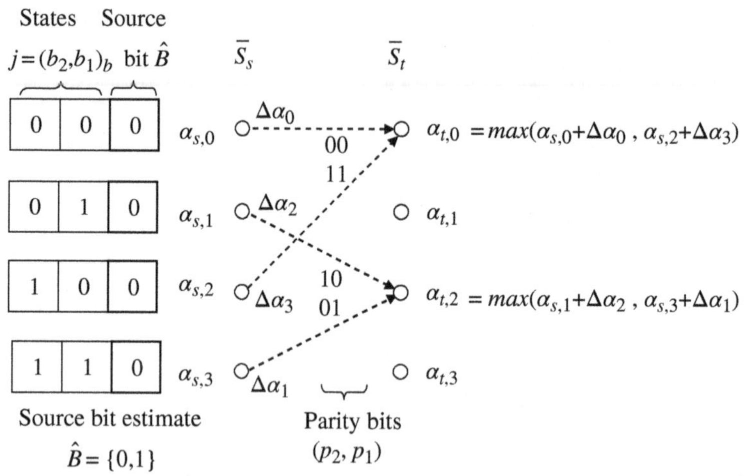

FIGURE 8.39 Trellis decoding state transition processing (binary ($k=1$), rate 1/2, $K=3$, $\hat{B}=0$).

that terminate on state $j=0$ correspond to $(b_1,\hat{B})=(0,0)$ with the source states identified as $(b_2,b_1)=(0,0)$ and $(1,0)$ corresponding to $j=0$ and 2. For each state transition, there is an associated source state metric $\alpha_{s,j}$, metric update $\Delta\alpha_m$, and termination state j'. Therefore, the termination state must select the maximum $\alpha_{s,j}+\Delta\alpha_m$ of all source states terminating on j'. This is shown in Figure 8.39 as the ACS function corresponding to the termination states $j'=0$ and 2.

The parity bits (p_2,p_1) corresponding to each transition are determined from knowledge of the convolutional encoder and are determined in the same manner as in the encoder. Several important observations are that each state transition is characterized as having a Hamming distance* of two, the parity bits associated with transitions terminating on the same state are antipodal, and the metric updates $\Delta\alpha_m$ are associated with the parity bits with $m=(p_2,p_1)_b=0,\ldots,3$. The parity bits form the link to the received data; in that, the received parity bits at the demodulator matched filter output are correlated with each set of transition parity bits (p_2,p_1) with the correlation equal to the state transition update $\Delta\alpha_m$. The details in the computation of $\Delta\alpha_m$ are discussed in the following.

The link between the trellis decoder and the convolutional encoder is found in the received symbols that contain noise corrupted estimates of the modulated waveform symbols. For example, consider the BPSK modulated waveform with phase modulation $\phi_\ell=\pi b_\ell$, where $b_\ell=\{0,1\}:\ell=1,2$† represents the two unipolar consecutive code bits corresponding to a code-bit word; these bits are referred to as the parity bits $p_\ell=\{0,1\}:\ell=1,2$ in the earlier description of the trellis decoding. The corresponding demodulator matched filter samples are denoted as $\{\hat{c}_2,\hat{c}_1\}$, where $\hat{c}_\ell:\ell=1,2$ are the independent noisy estimates of $b_\ell:\ell=1,2$. With ideal AGC and phase tracking, the receiver estimate, $\hat{c}_\ell$, is expressed as

$$\hat{c}_\ell=\sqrt{\frac{2E_s}{N_o}}\cos(\phi_\ell)+n_\ell \tag{8.75}$$

where E_s/N_o is the symbol energy-to-noise density ratio, held constant by the ideal gain control and n_ℓ is a zero-mean, unit-variance, independent Gaussian noise random variable.

The binary code bits corresponding to each state transition in Figure 8.38 are represented by their bipolar equivalents as shown in Table 8.13.

TABLE 8.13 Unipolar to Bipolar State Conversion ($d_\ell = 1-2b_\ell$)

	Trellis State			
	Unipolar		Bipolar	
Index m	b_2	b_1	d_2	d_1
0	0	0	1	1
1	0	1	1	−1
2	1	0	−1	1
3	1	1	−1	−1

The correlation of the received code block estimates $\hat{c}_\ell:\ell=1,2$ with the $2^n|_{n=2}=4$ possible code-bit combinations of the trellis bipolar state code-bits $d_{m,\ell}:\ell=1,2;\ \bar{k}=0,\ldots,3$ results in the correlations‡

$$\Delta\alpha_m=\sum_{\ell=1}^{2} d_{m,\ell}\hat{c}_\ell \quad :m=0,\ldots,3 \tag{8.76}$$

*The Hamming distance is defined, between two vectors with equal number of elements, as the number locations in which the elements differ.
†In the general case, $\ell = 1,\ldots,n_o$

‡The metric correlation or state transition update is denoted as Δm_k in Chapter 9.

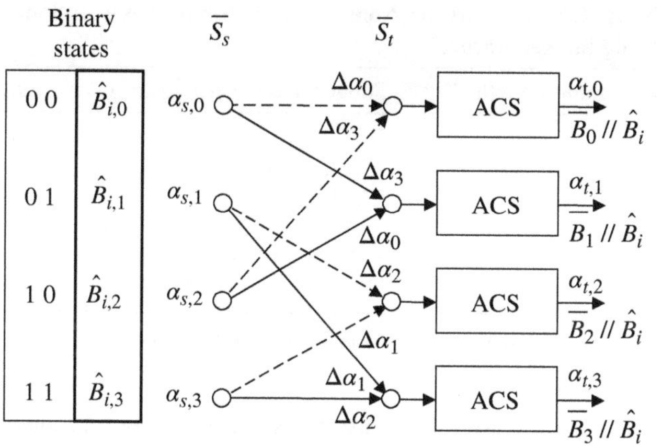

FIGURE 8.40 Summary of convolutional code trellis decoding (binary ($k=1$), rate 1/2, $K=3$).

As the processing continues through the trellis, the surviving metric, determined from the ACS decisions, will eventually integrate to a uniquely large metric providing reliable data decisions. The trellis integration length or depth (N_{dep}) of four to five constraint lengths is usually sufficient for a reliable data decision. Figure 8.40 summarizes the state transition processing described in Figures 8.38 and 8.39 and includes the storage of the data estimate $\hat{B}_i$ that is concatenation with the data vector $\bar{B}_j$. After a trellis depth of N_{dep} transitions, the oldest data is removed from the data vector and output as the maximum-likelihood decoded estimate of the corresponding source data.

As mentioned before, associated with each surviving state metric is a stored sequence of data estimates and, after N_{dep} received bits are process, the oldest bits in the stored sequence are output as the maximum-likelihood estimate of the transmitted bits. This is referred to at the *path history storage* and Rader [74] discusses efficient memory management and *trace-back* to the transmitted data estimate. Through the ACS decisions, made at each trellis state, all of the incorrect paths leading to the estimate of the transmitted data, N_{dep}-bits earlier, tend to be eliminated so, with high probability, any of the stored data vectors associated with the surviving metrics can be used to recover the estimate of the transmitted data. However, for shorter trellis depths or short constraint lengths codes with lower throughput delays, the optimum transmitted data estimate must be taken from the stored data array associated with the maximum metric of the trellis states, that is, $\max(\alpha_{s,j}) : j=0, \ldots, 3$. Upon completion of transition processing, the termination metrics $\alpha_{t,j}$ are assigned to the source metrics $\alpha_{s,j}$ and the decoding continues with a minimum of trellis memory. To avoid metric accumulation overflow, the maximum metric must be examined periodically and each metric reduced by a constant value if overflow is imminent.

8.11.3 Selection of Good Convolutional Codes

In general, the code generators are selected to ensure that the codes result in the maximal free distance* and do not result in catastrophic error propagation [75] following an error event. Exhaustive computer search algorithms are run to determine the best code generator for specific code configurations [76, 77]. The search algorithms examine details of the encoding structure including those involving catastrophic error propagation and equivalent codes. Equivalent codes have properties involving the reciprocal, shifting, and ordering of subgenerators. For example, reversing the order of the subgenerators results in an equivalent code. The following convolutional codes, referred to as *good codes*, are based on the maximal free distance, d_{free}, criterion when used with Viterbi decoding of convolutional codes.

A useful characteristic of convolutional codes is the insensitivity to signal phase reversals when used with differential data coding. Convolutional codes exhibiting this feature are referred to as *transparent* convolutional codes in which complements of code words are also code words. A convolutional code is transparent if each subgenerator has an odd number of binary coefficients [78]. By examining the variations in the surviving metrics through the trellis decoder, an estimate of the channel quality and the received signal-to-noise ratio are established. These estimates take advantage of the surviving state metric that increases at a constant rate in proportion to the received signal-to-noise ratio through the trellis decoder [78].

Good nonsystematic rate $1/n$ binary convolutional code generators for use with the Viterbi decoding algorithm were investigated by Odenwalder [76], and the results for rate 1/2 convolutional codes with constraint lengths 3 through 9 are listed in Table 8.14.† Odenwalder's results for rate 1/3 binary convolutional codes with constraint lengths 3 through 8 are listed in Table 8.15. Larsen [79] has extended Odenwalder's results to include the respective constraint lengths 10 through 14 and 9 through 14, in these tables. The constraint length 9 codes listed in these tables are specified in the North American direct sequence CDMA (DS-CDMA) Digital Cellular System Interim Standard (IS-95) [80]. The rate 1/4 convolutional codes in Table 8.16 were found by Larson.‡

*The maximal free distance (d_{free}) of a convolution code is defined as the unrestricted minimum distance between code words; whereas, a restricted minimum distance is defined as the distance between a predetermined number of code words. For example, upon defining d_{min} as the minimum distance between code words defined over ℓ consecutive code words, or ℓn code-bits, then $d_{free} \geq d_{min}$.

†The code subgenerators are expressed in octal notation with the LSB of the binary equivalent corresponding to the rightmost bit. For example, the octal notation 35 corresponds to 11101 with $g(x)=x^4+x^3+x^2+1$.

‡The ordering listed by Larsen is reversed to be consistent with the ordering of the convolutional code subgenerators used in this chapter and by Odenwalder. For example, the ordering used in this chapter is $g_1(x)$, $g_2(x)$ …; whereas, Larsen lists the order as … $g_2(x)$, $g_1(x)$.

TABLE 8.14 Rate 1/2 Nonsystematic Binary Convolutional Code Subgenerators[a]

Constraint Length, K	Subgenerators (g_1,g_2)	d_{free}
3	7,5	5
4	17,15	6
5	35,23	7
6	75,53	8
7	171,133	10
8	371,247	10
9	753,561[b]	12
10	1545,1167	12
11	3661,2335	14
12	5723,4335	15
13	17661,10533	16
14	27123,21675	16

[a]Larsen [79]. Reproduced by permission of the IEEE.
[b]IS-95 DS-CDMA forward link speech encoder.

TABLE 8.15 Rate 1/3 Nonsystematic Binary Convolutional Code Subgenerators[a]

Constraint Length, K	Subgenerators (g_1,g_2,g_3)	d_{free}
3	7,7,5	8
4	17,15,13	10
5	37,33,25	12
6	75,53,47	13
7	175,145,133[b]	15
8	367,331,225	16
9	711,663,557[c]	18
10	1633,1365,1117	20
11	3175,2671,2353	22
12	6265,5723,4767	24
13	17661,10675,10533	24
14	37133,35661,21645	26

[a]Larsen [79]. Reproduced by permission of the IEEE.
[b]This code found by Larsen and Odenwalder. Odenwalder's published code is (171,145,133) with $d_{free}=14$
[c]IS-95 DS-CDMA reverse link speech encoder.

TABLE 8.16 Rate 1/4 Nonsystematic Binary Convolutional Code Subgenerators[a]

Constraint Length, K	Subgenerators (g_1,g_2,g_3,g_4)	d_{free}
3	7,7,7,5	10
4	17,15,15,13	13
5	37,33,27,25	16
6	75,71,67,53	18
7	163,147,135,135	20
8	357,313,275,235	22
9	745,733,535,463	24
10	1653,1633,1365,1117	27
11	3175,2671,2353,2387	29
12	7455,6265,5723,4767	32
13	16727,15573,12477,11145	33
14	35537,35527,23175,21113	36

[a]Larsen [79]. Reproduced by permission of the IEEE.

TABLE 8.17 Rate 1/5 Nonsystematic Binary Convolutional Code Subgenerators[a]

Constraint Length, K	Subgenerators (g_1,g_2,g_3,g_4,g_5)	d_{free}
3	7,7,7,5,5	13
4	17,17,13,15,15	16
5	37,27,33,25,35	20
6	75,71,73,65,57	22
7	175,131,135,135,147	25
8	257,233,323,271,357	28

[a]Daut et al. [81]. Reproduced by permission of the IEEE.

TABLE 8.18 Rate 1/6 Nonsystematic Binary Convolutional Code Subgenerators[a]

Constraint Length, K	Subgenerators $(g_1,g_2,g_3,g_4,g_5,g_6)$	d_{free}
3	7,7,7,7,5,5	16
4	17,17,13,13,15,15	20
5	37,35,27,33,25,35	24
6	73,75,55,65,47,57	27
7	173,151,135,135,163,137	30
8	253,375,331,235,313,357	34

[a]Daut et al. [81]. Reproduced by permission of the IEEE.

TABLE 8.19 Rate 1/7 Nonsystematic Binary Convolutional Code Subgenerators[a]

Constraint Length, K	Subgenerators $(g_1,g_2,g_3,g_4,g_5,g_6,g_7)$	d_{free}
3	7,7,7,7,5,5,5	18
4	17,17,13,13,13,15,15	23
5	35,27,25,27,33,35,37	28
6	53,75,65,75,47,67,57	32
7	165,145,173,135,135,147,137	36
8	275,253,375,331,235,313,357	40

[a]Daut et al. [81]. Reproduced by permission of the IEEE.

The bit-error performance dependence on E_b/N_o for the rate 1/2 code is examined in Section 8.11.8 for constraint lengths $K=4$ through 9. Duet, Modestino, and Wismer [81] have extended the binary rate $1/n$ convolutional codes with maximum free distances to include code generators for $n=5$, 6, 7, and 8; their results are listed in Tables 8.17, 8.18, 8.19, and 8.20 for constraint lengths 3 through 8.

The code generators in Tables 8.14, 8.15, 8.16, 8.17, 8.18, 8.19, and 8.20 are based on the binary Hamming distance and, as such, represent good codes for binary signaling in which each of the coded bits is mapped into one symbol using, for example, BPSK or BFSK waveform modulation. Alternately, each pair of code bits generated by the rate 1/2 binary convolutional code can be mapped into a 4-ary modulated symbol and consecutive groups of three code bits generated by a rate 1/3 binary convolutional code can be mapped

TABLE 8.20 Rate 1/8 Nonsystematic Binary Convolutional Code Subgenerators[a]

Constraint Length, K	Subgenerators ($g_1,g_2,g_3,g_4,g_5,g_6,g_7,g_8$)	d_{free}
3	7,7,5,5,5,7,7,7	21
4	17,17,13,13,13,15,15,17	26
5	37,33,25,25,35,33,27,37	32
6	57,73,51,65,75,47,67,57	36
7	153,111,165,173,135,135,147,137	40
8	275,275,253,371,331,235,313,357	45

[a]Daut et al. [81]. Reproduced by permission of the IEEE.

TABLE 8.21 Rate 1/2 Nonsystematic Binary Convolutional Code Subgenerators for 4-ary Signaling[a]

Constraint Length, K	Subgenerators (g_1,g_2)
3	7,5
4	15,12
5	32,25
6	75,57
7	133,176

[a]Michelson and Levesque [63]. Reproduced by permission of John Wiley & Sons, Inc.

TABLE 8.22 Rate 1/3 Nonsystematic Binary Convolutional Code Subgenerators for 8-ary Signaling[a]

Constraint Length, K	Subgenerators (g_1,g_2,g_3)
3	4,6,5
4	11,15,13
5	22,33,26
6	46,67,55
7	176,155,127

[a]Michelson and Levesque [63]. Reproduced by permission of John Wiley & Sons, Inc.

to an 8-ary modulated symbol. However, a nonbinary Hamming distance measure should be used when mapping the code bits to an M-ary modulated symbol. In this regard, Trumpis [82] investigated good rate 1/2 and 1/3 binary convolutional codes based on nonbinary Hamming distances with corresponding 4-ary and 8-ary symbol mapping; the resulting code generators are listed in Tables 8.21 and 8.22.

The preceding codes are all binary rate $1/n$ convolutional codes, and Paaske [77] has evaluated nonbinary rate $(n-1)/n$ convolutional codes and the subgenerators are listed in Tables 8.23 and 8.24 for the codes rates 2/3 and 3/4. Paaske's work is based on Forney's formulation [83, 84] of high-rate convolutional codes where the parameter ν is defined as the constraint length. The topology of the rate $(n-1)/n$ encoder is similar to the general encoder shown in Figure 8.34. For

TABLE 8.23 Rate 2/3 Nonsystematic Nonbinary Convolutional Code Subgenerators[a]

Constraint Length, ν	Subgenerators (g_1,g_2,g_3)	d_{free}
2	13,06,16	3
3	41,30,75	4
4	56,23,65	5
5	245,150,375	6
6	266,171,367	7
7	1225,0655,1574	8
9	4231,2550,7345	9
10	5516,2663,6451	10

[a]Paaske [77] by permission of the IEEE.

TABLE 8.24 Rate 3/4 Nonsystematic Nonbinary Convolutional Code Subgenerators[a]

Constraint Length, ν	Subgenerators (g_1,g_2,g_3,g_4)	d_{free}
3	400,630,521,701	4
5	442,270,141,763	5
6	472,215,113,764	6
8	4463,2470,1511,7022	7
9	4113,2246,1744,7721	8

[a]Paaske [77] by permission of the IEEE.

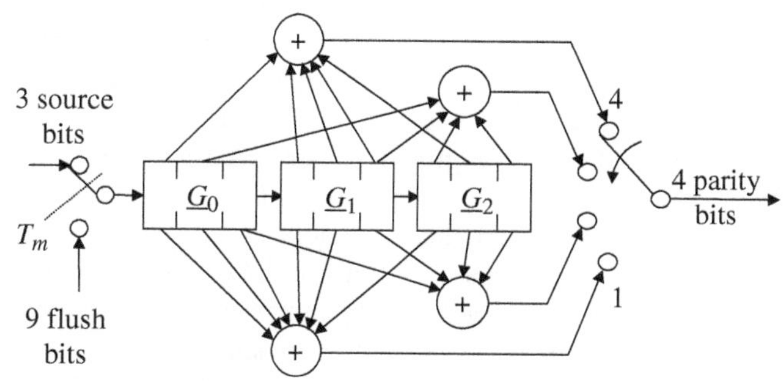

FIGURE 8.41 Rate 3/4, constraint length $\nu=6$, $K=3$, convolutional encoder.

these convolutional codes, the number of source-bits input to the encoder memory each code block is $k=n-1$. The example code, used to clarify the following description, is shown in Figure 8.41 and corresponds to the rate 3/4, constraint length 6, convolutional encoder listed in Table 8.24.

In terms of the code-block parity generator matrix G, Paaske defines as set of ($n-1$ by n) generator matrices $\underline{G_i}$: $i=0, \ldots, K-1$ expressed as

$$G_i = \left(g_1^i(x) \;\; g_2^i(x) \;\; \cdots \;\; g_n^i(x)\right) \tag{8.77}$$

where $g_\ell^i(x): \ell=1,\ldots,n$ are $n-1$-dimensional column vectors with binary elements {1,0}. The subgenerators $g_\ell(x)$, defined by (8.73) and shown in Figure 8.34, are evaluated,

TABLE 8.25 Paaske's Generator Matrices for the Rate 3/4, Constraint Length 6, Convolutional Code

$\underline{G}_0$				$\underline{G}_1$				$\underline{G}_2$			
$g_1^0(x)$	$g_2^0(x)$	$g_3^0(x)$	$g_4^0(x)$	$g_1^1(x)$	$g_2^1(x)$	$g_3^1(x)$	$g_4^1(x)$	$g_1^2(x)$	$g_2^2(x)$	$g_3^2(x)$	$g_4^2(x)$
1	0	0	1	1	0	0	1	0	1	0	1
0	1	0	1	1	0	0	1	1	0	1	0
0	0	1	1	1	1	1	0	0	1	1	0

TABLE 8.26 Subgenerators for the Rate 3/4, Constraint Length 6, Convolutional Code

Subgenerators	Binary	Octal
$g_1(x)$	100111010	472
$g_2(x)$	010001101	215
$g_3(x)$	001001011	113
$g_4(x)$	111110100	764

in terms of the transpose of the $k=n-1$ column vectors $g'^i_\ell(x) = (g^i_\ell(x))^T$, as

$$g_\ell(x) = g'^0_\ell(x)x^{k(K-1)} + g'^1_\ell(x)x^{k(K-2)} + \cdots + g'^{K-2}_\ell(x)x^k + g'^{K-1}_\ell(x) \tag{8.78}$$

In words, $g_\ell(x)$ is the sum of the transpose of the ℓ-th column vector in $\underline{G}_i$ with the significance of the bits in descending order from right to left. For example, Paaske's generator matrices for the rate 3/4, constraint length $v=6$, convolutional code, shown in Figure 8.41, are listed in Table 8.25; the corresponding subgenerators $g_\ell(x)$ are listed in Table 8.26 using binary and octal notation.

Referring to the state transitions in the trellis diagram as emanating from a *source* state and converging on a *termination* state, the code bits associated with each of the transitions to a given termination state are important; in that, they influence the selection of the survivor at the termination states and ultimately the trellis decoding performance. For the rate 1/2, $K=3$ binary trellis decoder, shown in Figure 8.38, there are only two transitions converging on each termination state and the corresponding Hamming distances between any two code words is: $h_d=\{0,1,2\}$. However, with the unipolar-to-bipolar mapping $d_i=1-2b_i$, such that $b_i=\{0,1\}$ and $d_i=\{1,-1\}$, the, noise-free, correlation response between any two code words is, from (8.76), $\Delta m=\{2,0,-2\}$. The maximum cross-correlation, $\Delta m(\max)=2$, corresponds to the received code-word matching the transition code word leading to a correct survivor decision in the noise-free case. Therefore, the matched filter or correlation detector responses dissimilar to the receive code word are either orthogonal or antipodal.

For a trellis decoder with a large number of trellis states with multiple source states converging on each termination state, it is processing efficient to store *pointers* to each source state and the corresponding transition code words. Using the rate 3/4, constraint length 6, $K=3$ convolutional code in Figure 8.41 as an example, the $2^{k=3}=8$ source states (S_i), converging on the selected termination state (S_{i+1}), are depicted in Figure 8.42. The selected termination state identifies the binary source data in the $k=3$-bit encoder memory $B_{-1,n}$ and B and the eight different source states is determined by indexing the 3-bit encoder memory $B_{-2,m}$ from 0 to 7. For each source state, the unique 4-bit code word (CW_m) is determine by applying the subgenerators $g_\ell(x)$ as expressed in (8.73) and identified in Table 8.24.

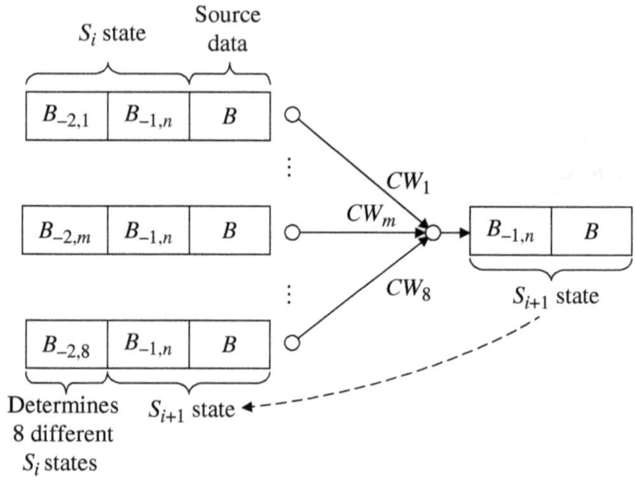

FIGURE 8.42 Convolutional code trellis decoding (8-ary, rate 3/4, $K=3$).

It is relatively easy to write a computer program to evaluate the encoder binary memory that identifies the source states, the input source data B, and the corresponding transition code words given the termination state as described earlier. In this evaluation, it is useful to write a procedure to convert from binary to integer and vice versa using the convention that the LSB is on the right. Computer simulations indicate that, for all $2^6=64$ possible states, $h_d=\{0,1,4\}$ and the noise-free cross-correlation response between the received and transition code-words is $\Delta m=\{4,0,-4\}$. Using the binomial coefficient, the number of combinations of the eight code words taken two at a time is ${}_8C_2=28$ and of these 28 combinations, 4 have cross-correlations of +4 and 24 have cross-correlations of zero.

8.11.4 Dual-k Nonbinary Convolutional Codes

The dual-k code is a nonbinary class of convolutional codes that are designed to be used with M-ary symbol modulation. The dual-k encoder shifts k input bits through dual (two) k-bit registers and outputs dual (two) k-bit symbols. Proakis [85] expresses the subgenerators $g_i(x)\ :\ i = 1,\ \ldots,\ 2k$ for the dual-k codes as follows:

$$\begin{pmatrix} g_1(x) \\ g_2(x) \\ \vdots \\ g_k(x) \end{pmatrix} = \begin{pmatrix} \boldsymbol{I}_k \,\vdots\, \boldsymbol{I}_k \end{pmatrix} \tag{8.79}$$

$$\begin{pmatrix} g_{k+1}(x) \\ g_{k+2}(x) \\ \vdots \\ g_{2k}(x) \end{pmatrix} = \begin{pmatrix} 1 & \boldsymbol{I}_{k-1} & \\ 0 & & \\ \vdots & & \boldsymbol{I}_k \\ 1 & 0 \cdots & 0 \end{pmatrix} \tag{8.80}$$

Here, $\boldsymbol{I}_k$ is the $k \times k$ identity matrix. The dual-3 code provides robust communications in a Rayleigh fading stressed environment using 8-ary FSK symbol modulation and noncoherent detection [86, 87]. The dual-3, -4, and -5 encoder subgenerators are tabulated in Table 8.27 and the dual-3 encoder is shown in Figure 8.43.

TABLE 8.27 Dual-3, -4, and -5 Convolutional Code Subgenerators

Code		Subgenerators[a]
Dual-3	g_1, g_2, g_3	(4,4), (2,2), (1,1)
	g_4, g_5, g_6	(6,4), (1,2), (4,1)
Dual-4	g_1, g_2, g_3, g_4	(10,10), (4,4), (2,2), (1,1)
	g_5, g_6, g_7, g_8	(14,10), (2,4), (1,2), (10,1)
Dual-5	g_1, g_2, g_3, g_4, g_5	(20,20), (10,10), (4,4), (2,2), (1,1)
	$g_6, g_7, g_8, g_9, g_{10}$	(30,20), (4,10), (2,4), (1,2), (20,1)

[a]Groups of k-bits are in octal notation with the LSB on the right.

8.11.5 Convolutional Code Transfer Function and Upper Bound on Performance

An important technique in the performance analysis of a convolutional code is the transfer function described by Odenwalder [76] and in a tutorial paper by Viterbi [88]. The convolutional code transfer function is derived from a state diagram [89] that highlights the Hamming distance properties of the code for various states through the trellis decoder. The transfer function properties are characterized as the ratio of the output to input states over a prescribed code sequence; however, since the convolution code is linear, the Hamming distance is independent of the selected output to input states so the all-zero sequence is chosen because of its relative simplicity in interpreting the transfer function of the code.

The convolutional code transfer function is described using, as an example, the binary rate 1/2, constraint length $K = 3$ code shown in Figure 8.37 and the corresponding trellis diagram shown in Figure 8.38. The state diagram for this code is shown in Figure 8.44 and is based on an all-zero data sequence with the data error branches converging on successive states and reemerging with the correct zero state at different points through the trellis. The distance between the correct code word 00 and the code word corresponding to the transition path is labeled D^d where d denotes the Hamming distance, so the distance from the state 00 to 01 labeled D^2. In a like manner, the number of bit errors and the path length corresponding to the transition path are labeled N and L, respectively. These labels are easily determined from the trellis diagram of the code, as shown for this example code in Figure 8.38 where, for the assumed zero source data, a single bit error is indicated by the *solid* transition lines. In the state diagram description, the path length L is included with each transition path; however, when expressed in terms of the code transfer function, L will take on an integer exponent indicating the path length for a divergent path reemerging with the all-zero path.

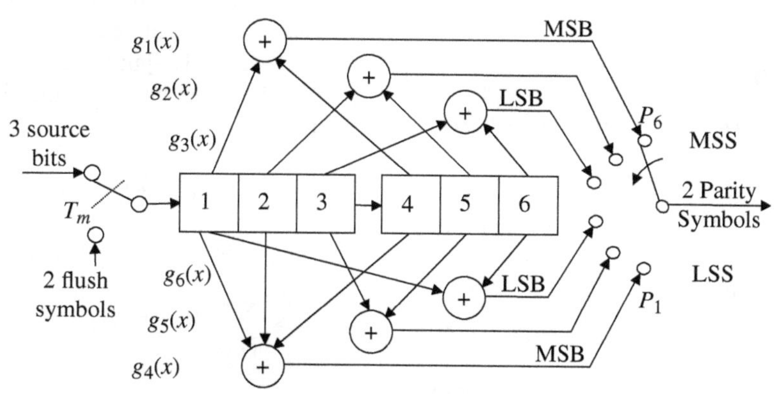

FIGURE 8.43 Dual-3 convolutional encoder.

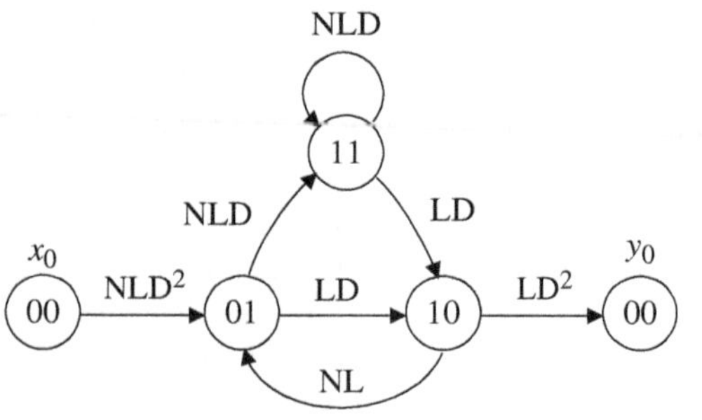

FIGURE 8.44 State diagram for binary, rate 1/2, $K=3$, convolutional code.

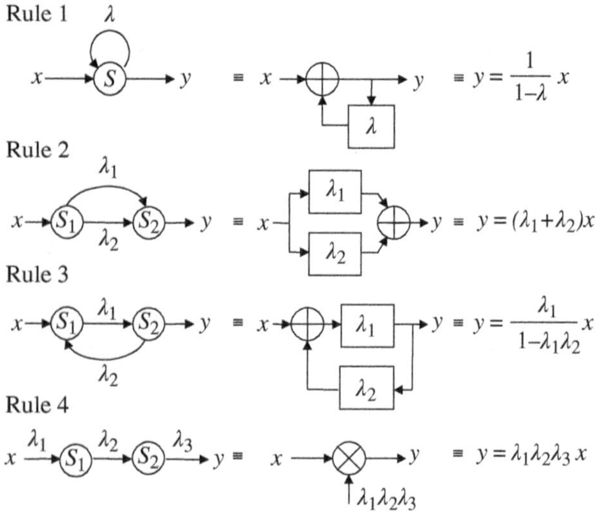

FIGURE 8.45 State diagram reduction rules.

The state diagram is used to compute the convolutional code transfer function by following the few simple rules shown in Figure 8.45.

Defining the transfer function as $T(N,L,D) = y_0/x_0$ and applying the appropriate rules in Figure 8.45 results in the expression

$$\begin{aligned}T(N,L,D) &= \frac{NL^3D^5}{1-NLD-NL^2D}\\ &= NL^3D^5\left(1+N\left(L+L^2\right)D+N^2\left(L^2+2L^3+L^4\right)D^2\right.\\ &\quad \left.+N^3\left(L^3+3L^4+3L^5+L^6\right)D^3+\cdots\right)\\ &= NL^3D^5+N^2\left(L^4+L^5\right)D^6+N^3\left(L^5+2L^6+L^7\right)D^7\\ &\quad +N^4\left(L^6+3L^7+3L^8+L^9\right)D^8+\cdots\end{aligned} \tag{8.81}$$

Referring to the last equality in (8.81), the exponents of the parameters N, L, and D indicate, respectively, the number of decoded bit errors that occurred before reemerging with the all-zero path, the number of path lengths prior to reemerging with the all-zero path, and the Hamming distance of the code words over the path before reemerging with the all-zero path. Therefore, the minimum distance for this example convolutional code is $d_{min} = 5$ and occurred over one path of 3 trellis transitions resulting in one bit error. The next smallest Hamming distance of 6 and occurred over two path lengths of 4 and 5 resulting in two bit errors per path for a total 4 bit-errors, and so on. The number of bit errors corresponding to each Hamming distance is determined by differentiating (8.81) with respect to N and then setting N and L equal to one; the result is expressed as

$$\left.\frac{\partial T(N,L,D)}{\partial N}\right|_{N,L=1} = D^5+4D^6+12D^7+32D^8+\cdots \tag{8.82}$$

For ideal BPSK waveform modulation and the AWGN channel, the probability of error for each incorrect path prior to reemerging with the all zero path is given by

$$P_{be}(\text{path}) = n_p Q\left(\sqrt{2r_c h_d \gamma_b}\right) \tag{8.83}$$

where r_c is the code rate, γ_b is the energy-to-noise-density ratio measured in the data rate bandwidth, h_d and n_p are the Hamming distance and number of bit errors for all paths associated with the Hamming distance. The overall bit-error probability can be upper bounded by applying the union bound and, for the binary, rate $r_c = 1/2$, $K=3$ example code being considered, the bit-error probability is upper bounded by

$$\begin{aligned}P_{be} &\le \sum_{\text{all paths}} P_{be}(\text{path})\\ &= Q\left(\sqrt{5\gamma_b}\right)+4Q\left(\sqrt{6\gamma_b}\right)+12Q\left(\sqrt{7\gamma_b}\right)+32Q\left(\sqrt{8\gamma_b}\right)+\cdots\end{aligned} \tag{8.84}$$

As stated previously, the decoding path length is important with short messages, and (8.84) and the last equality in (8.81) provide some insight on the impact of the trellis decoding length on the decoder performance. However, for long messages the trellis depth of four or five constraint lengths is not a performance limiting issue, aside from the message throughput delay.

Odenwalder [76] provides a convenient closed-form solution for the upper bound on the bit-error probability by using the inequality

$$Q\left(\sqrt{x+y}\right) \le Q\left(\sqrt{x}\right)e^{-y/2} \quad : x,y \ge 0 \tag{8.85}$$

Using this result, (8.84) is expressed as

$$\begin{aligned}P_{be} &\le \sum_{\text{all paths}} P_{be}(\text{path})\\ &\le Q\left(\sqrt{5\gamma_b}\right)\left(1+4e^{-\gamma_b/2}+12e^{-\gamma_b}+32e^{-3\gamma_b/2}+\cdots\right)\end{aligned} \tag{8.86}$$

TABLE 8.28 Derivative of the Transfer Function of the Binary Rate 1/2 Convolutional Code[a]

Constraint Length, K	$\partial T(N,L,D)/\partial N\|_{N,L=1}$
3	$D^5 + 4D^6 + 12D^7 + 32D^8 + 80D^9 + 192D^{10} + 448D^{11} + 1024D^{12} + 2304D^{13} + 5120D^{14} + \cdots$
4	$2D^6 + 7D^7 + 18D^8 + 49D^9 + 130D^{10} + 333D^{11} + 836D^{12} + 2069D^{13} + 5060D^{14} + 12255D^{15} + \cdots$
5	$4D^7 + 12D^8 + 20D^9 + 72D^{10} + 225D^{11} + 500D^{12} + 1324D^{13} + 3680D^{14} + 8967D^{15} + 22270D^{16} + \cdots$
6	$2D^8 + 36D^9 + 32D^{10} + 62D^{11} + 332D^{12} + 701D^{13} + 2342D^{14} + 5503D^{15} + 12506D^{16} + 36234D^{17} + \cdots$
7	$36D^{10} + 211D^{12} + 1404D^{14} + 11633D^{16} + 77433D^{18} + 502690D^{20} + \cdots$

[a]Michelson and Levesque [63]. Reproduced by permission of John Wiley & Sons.

TABLE 8.29 Derivative of the Transfer Function of the Binary Rate 1/3 Convolutional Code[a]

Constraint Length, K	$\partial T(N,L,D)/\partial N\|_{N,L=1}$
3	$3D^8 + 15D^{10} + 58D^{12} + 201D^{14} + 655D^{16} + 2052D^{18} + \cdots$
4	$6D^{10} + 6D^{12} + 58D^{14} + 118D^{16} + 507D^{18} + 1284D^{20} + 4323D^{22} + \cdots$
5	$12D^{12} + 12D^{14} + 56D^{16} + 320D^{18} + 693D^{20} + 2324D^{22} + 8380D^{24} + \cdots$
6	$D^{13} + 8D^{14} + 26D^{15} + 20D^{16} + 19D^{17} + 62D^{18} + 86D^{19} + 204D^{20} + 420D^{21} + 710D^{22} + 1345D^{23} + \cdots$
7	$7D^{15} + 8D^{16} + 22D^{17} + 44D^{18} + 22D^{19} + 94D^{20} + 219D^{21} + 282D^{22} + 531D^{23} + 1104D^{24} + 1939D^{25} + \cdots$

[a]Michelson and Levesque [63]. Reproduced by permission of John Wiley & Sons.

TABLE 8.30 Derivative of the Transfer Function of the Binary Rate 1/2 Convolutional Code for 4-ary Symbol Modulation[a]

Constraint Length, K	$\partial T(N,L,D)/\partial N\|_{N,L=1}$
3	$D^3 + 4D^4 + 12D^5 + 32D^6 + 80D^7 + 192D^8 + 448D^9 + 1024D^{10} + 2304D^{11} + 5120D^{12} + \cdots$
4	$D^4 + 14D^5 + 21D^6 + 94D^7 + 261D^8 + 818D^9 + 2173D^{10} + 6335D^{11} + 17220D^{12} + 47518D^{13} + \cdots$
5	$3D^5 + 15D^6 + 22D^7 + 196D^8 + 398D^9 + 1737D^{10} + 4728D^{11} + 15832D^{12} + 47491D^{13} + 144170D^{14} + \cdots$
6	$9D^6 + 14D^7 + 62D^8 + 212D^9 + 874D^{10} + 2612D^{11} + 9032D^{12} + 28234D^{13} + 93511D^{14} + 288974D^{15} + \cdots$
7	$7D^7 + 39D^8 + 104D^9 + 352D^{10} + 1348D^{11} + 4540D^{12} + 14862D^{13} + 48120D^{14} + 156480D^{15} + 505016D^{16} + \cdots$

[a]Michelson and Levesque [63]. Reproduced by permission of John Wiley & Sons.

TABLE 8.31 Derivative of the Transfer Function of the Binary Rate 1/2 Convolutional Code for 8-ary Symbol Modulation[a]

Constraint Length, K	$\partial T(N,L,D)/\partial N\|_{N,L=1}$
3	$D^3 + 2D^4 + 5D^5 + 10D^6 + 20D^7 + 38D^8 + 71D^9 + 130D^{10} + 235D^{11} + 420D^{12} + \cdots$
4	$D^4 + 2D^5 + 7D^6 + 16D^7 + 41D^8 + 94D^9 + 219D^{10} + 492D^{11} + 1101D^{12} + 2426D^{13} + \cdots$
5	$D^5 + 5D^6 + 8D^7 + 25D^8 + 64D^9 + 170D^{10} + 392D^{11} + 958D^{12} + 2270D^{13} + 5406D^{14} + \cdots$
6	$D^6 + 5D^7 + 7D^8 + 34D^9 + 76D^{10} + 200D^{11} + 557D^{12} + 1280D^{13} + 3399D^{14} + 8202D^{15} + \cdots$
7	$D^7 + 4D^8 + 8D^9 + 49D^{10} + 92D^{11} + 186D^{12} + 764D^{13} + 1507D^{14} + 4198D^{15} + 10744D^{16} + \cdots$

[a]Michelson and Levesque [63]. Reproduced by permission of John Wiley & Sons.

The derivative of the convolutional code transfer functions for the binary rate 1/2 and 1/3 convolutional codes, listed in Tables 8.14 and 8.15, are shown in Tables 8.28 and 8.29 where

$$D^n = Q\left(\sqrt{n\gamma_b}\right) \tag{8.87}$$

Similarly, the derivative of the convolutional code transfer functions for the rate 1/2 and 1/3 convolutional codes with 4-ary and 8-ary symbol modulations are listed in Tables 8.30 and 8.31. The subgenerators for these codes correspond to those listed in Tables 8.21 and 8.22 for 4-ary and 8-ary symbol modulations, respectively. The performance of these codes can be compared, without the need for detailed computer simulations, using (8.84), (8.85), and (8.87) with the appropriate polynomials in D given in the following tables.

8.11.6 The Dual-*k* Convolutional Code Transfer Function

The transfer function of a rate $r_c = 1/n$ dual-k convolutional code, with ℓ transmissions of the code, is expressed as [64, 90],

TABLE 8.32 Derivative of the Transfer Function of the Nonbinary Dual-3 Code[a]

ℓ	$\partial T_\ell(N,L,D)/\partial N\vert_{N,L=1}$
1	$D^4F(D)$
2	$D^8F(D^2)$
3	$D^{12}F(D^3)$
4	$D^{16}F(D^4)$

[a]With $F(D) = C_0 + C_1D + C_2D^2 + \cdots$, the notation $F(D^m) = C_0 + C_1D^m + C_2D^{2m} + \cdots$.

$$T_\ell(N,L,D) = \frac{(2^k-1)NL^2D^{2n\ell}}{1-NL\left(nD^{(n-1)\ell}+(2^k-1-n)D^{n\ell}\right)} \quad :\ell \text{ code diversity} \tag{8.88}$$

Upon differentiating (8.88) with respect to N and evaluating the result for N, $L = 1$ results in the expression

$$\left.\frac{\partial T_\ell(N,L,D)}{\partial N}\right|_{N,L=1} = \frac{(2^k-1)D^{2n\ell}}{\left[1-nD^{(n-1)\ell}-(2^k-1-n)D^{n\ell}\right]^2} \tag{8.89}$$

Equation (8.89) is evaluated for $n = 2$, corresponding to the code rate $r_c = 1/2$, and the resulting polynomial expressions in terms of D are summarized in Table 8.32 for $\ell = 1$ through 4 with

$$\begin{aligned} F(D) = 7 + 28D + 154D^2 + 644D^3 + 2765D^4 \\ + 11144D^5 + 44436D^6 + \cdots \end{aligned} \tag{8.90}$$

In this case, the exponent of D is the Hamming distance of the q-ary symbol with $q = 8$ corresponding to the transmissions of two 8-ary FSK symbols-per-code. The underlying code, corresponding to $\ell = 1$, has a minimum free distance of $d_{free} = 4$.

8.11.7 Code Puncturing and Concatenation

Code puncturing is a technique that involves deleting specified parity bits from an underlying convolutional code, with rate r_c, to provide codes rates greater than r_c, with a commensurate decrease in the coding gain. The puncturing is accomplished by the rate and symbol mapping functions shown in Figure 8.35 and, if the puncturing is performed judiciously, the decoding can be accomplished with very little change in decoding complexity. Cain et al. [91] identify subgenerators for the best codes, derived by periodically puncturing an underlying rate 1/2 convolutional code, to obtain code rates 2/3 and 3/4 with constraint lengths* $\nu = 2, \ldots, 8$. Yasuda et al. [67], identify subgenerators (see Appendix 8A) derived from the underlying rate 1/2 convolutional code, to obtain punctured codes rates $(n-1)/n$ for $n = 3$ through 14 with constraint length $\nu = 2, \ldots, 8$. High-rate codes that are generated in this manner and decoded using a trellis decoder are referred to as *pragmatic trellis codes*. Wolf and Zehavi [92] discuss punctured convolutional codes used to generate high-rate codes in the generation of pragmatic punctured [93] (P^2) trellis codes [94], for PSK and quadrature amplitude modulation (QAM) waveforms.

When the punctured codes are derived from an underlying lower rate code, the trellis decoding is essentially the same as that of the underlying code; however, there is an increase in complexity associated with synchronization and tracking of the puncturing location in the decoding trellis. The ability to select the code rate by simply selecting the puncturing configuration has many advantages. For example, by altering the puncturing configuration, a single convolutional *decoding chip* [95] can be configured to accommodate more efficient modulation waveforms or to adjust the code rate to maintain the message reliability under varying channel conditions.

Code concatenation is a technique in which two codes are operated in a serial configuration to increase the overall coding gain relative to that of a single code. This configuration consists of an outer and inner code that may be identical codes; however, a common configuration is to use an RS outer code with a convolutional inner code. Typically, the codes are separated by an interleaver to ensure that error bursts are randomly distributed between the codes. Forney [96] has proposed the concatenation of a 2^K-ary RS outer code with a constraint length K convolutional inner code separated by an L-by-M row-column block interleaver. The RS symbol contains K bits and the block interleaver is filled row by row with L RS symbols and emptied column by column with $M = 2^K - 1$ symbols. Odenwalder [97] examines the performance this concatenated code configuration. Proakis and Rahman [87] examine the performance of concatenated dual-k codes.

Other applications of code concatenation involve the design of turbo and *turbo-like* codes. These configurations may involve two or more convolutional codes with each code separated by block interleavers that are instrumental in the overall decoding performance. Concatenated convolutional coding is discussed in more detail in Section 8.12 with references to original research and additional reading.

8.11.8 Convolutional Code Performance Using the Viterbi Algorithm

The performance results using the Viterbi algorithm are shown in Figure 8.46 for rate 1/2 binary convolutional coding with constraint lengths of 4 through 9. The decoding trellis depth is four constraint lengths. The results are shown using hard limiting and virtually infinite quantization of

*The constraint length ν corresponds to the number of memory elements as defined by Forney [84].

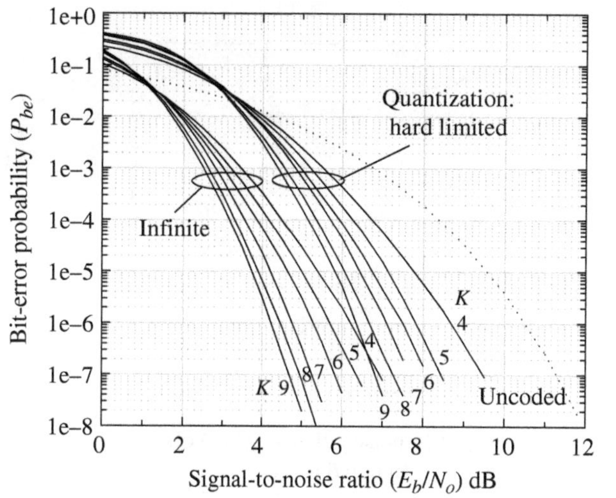

FIGURE 8.46 Rate 1/2, convolutional decoding performance with constraint length and infinite and hard limiting quantization (trellis depth = 5K bits).

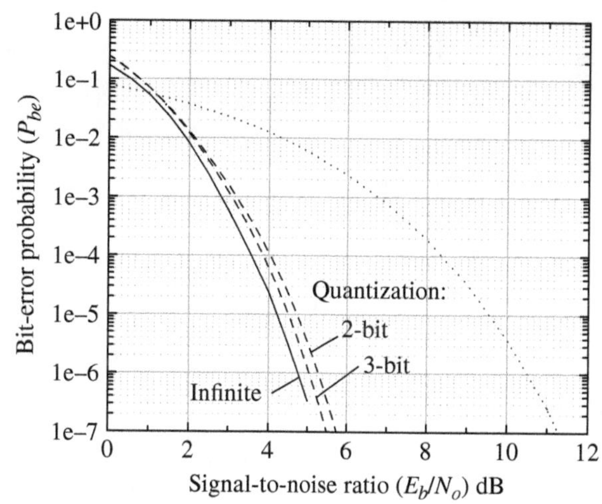

FIGURE 8.47 Rate 1/2, $K = 7$, convolutional decoding performance with input quantization (trellis length = 30 bits).

the matched filtered symbol samples. The hard-limiting performance is equivalent to 1-bit quantization. For comparison, the ideal or theoretical performance of antipodal signaling with AWGN is shown as the *dashed* curve. The Monte Carlo simulation results are based on 5M bits for each signal-to-noise ratio less than 4 dB. To preserve the simulated performance measurement accuracies for larger signal-to-noise ratios, the number is increased 10 fold. Under these conditions, the 90% confidence limit for bit-error probabilities on the order of 5×10^{-7} is $\pm 1.6 \times 10^{-7}$ or within ±32%. For the hard-limited performance evaluation, the 4.5M bit Monte Carlo simulation is increased 10 fold for signal-to-noise ratios ≥6 dB.

An important consideration, regarding the demodulator processing, is the quantization of the matched filter output sample that directly impacts the performance of the convolutional decoder [72]. For example, with the received signal level held constant by the AGC, the two-level hard-limiting simulation performance in Figure 8.46, results in about 2.1 dB loss in E_b/N_o performance at $P_{be} = 10^{-5}$ compared to the infinitely quantized performance. By contrast, the simulation performance results shown in Figure 8.47 indicates that the performance loss with 4- and 8-level (2- and 3-bit) quantization results in a, respective, performance loss of less than 0.2 and 0.1 dB compared to the performance with essentially infinite quantization.

The results indicate that the improvement in signal-to-noise ratio at $P_{be} = 10^{-7}$, relative to the uncoded performance, is about 4.2 dB for $K = 4$ with 8-level quantization and about 2.0 dB with hard limiting. By way of contrast, with $P_{be} = 10^{-7}$ and $K = 9$, the performance with infinite quantization and hard limiting is 7.3 and 4.4 dB respectively. Furthermore, for large signal-to-noise ratios, there is about 0.5 dB of improvement in the signal-to-noise ratio with each unit increase in the constraint length.

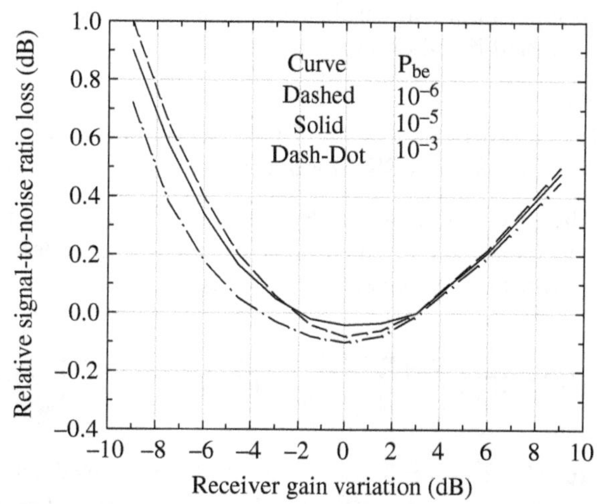

FIGURE 8.48 Rate 1/2, $K = 7$, convolutional decoding sensitivity to demodulator gain changes (input quantization = 3-bits).

Figure 8.48 shows the sensitivity of the performance to receiver AGC gain variations. The 3-bit quantizer includes a sign bit and two magnitude bits with saturation occurring at 1.0 V so the magnitude bits are assigned as $b_1 2^{-1} + b_0 2^{-2}$. It is important to note that under-flow rounding is used in all of the simulation results. In the simulation program, the nominal baseband signal voltage at the quantizer input of is set to 0.707 V or 3 dB below the 1.0 V saturation level.

The results of the simulated performance show that the loss in E_b/N_o performance is within ±0.2 dB over a receiver gain variation of about ±12 dB. Furthermore, the results indicate that symmetrical performance with respect to the receiver gain variation is obtained by setting the nominal

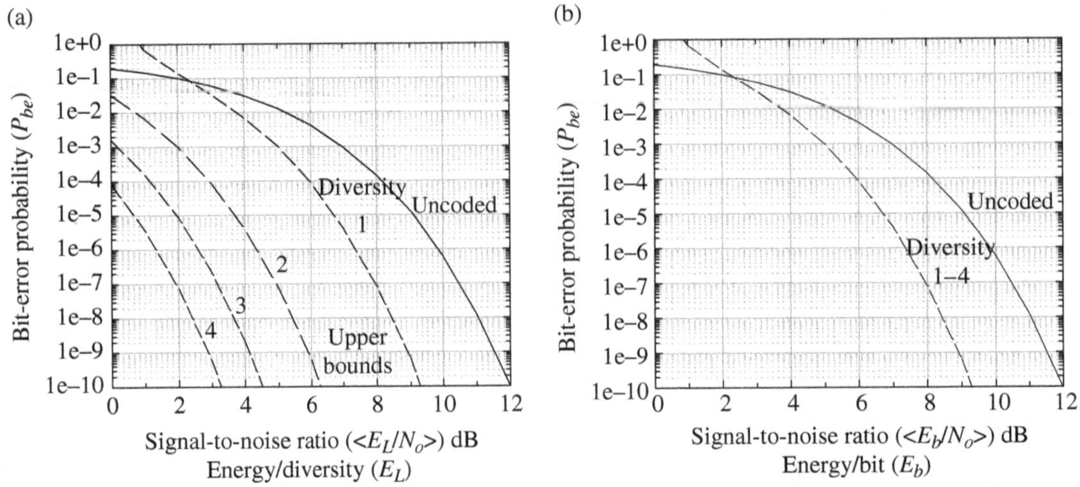

FIGURE 8.49 Dual-3 coded 8-ary noncoherent FSK performance with AWGN.

baseband signal to 0.707 V below the quantizer saturation; this is not a surprising result since the nominal receiver level was adjusted based on this analysis.

8.11.9 Performance of the Dual-3 Convolutional Code Using 8-ary FSK Modulation

The dual-3 convolutional coded 8-ary noncoherent FSK performance with AWGN is evaluated using the uncoded bit-error probability expressed in (7.21) as

$$P_{be}(\gamma) = \frac{2^{k-1}}{2^k - 1}\left[e^{-\gamma}\sum_{i=1}^{M-1}\binom{M-1}{i}\frac{(-1)^{i+1}}{i+1}e^{\gamma/(i+1)}\right] \quad (8.91)$$

where $M = 2^k|_{k=3} = 8$. Equation (8.91) is labeled as the *uncoded* curve in Figure 8.49 and is used as the underlying bit-error probability in the evaluation of the dual-3 convolutional coded performance with AWGN. The signal-to-noise ratio γ is measured in the symbol rate bandwidth and is related to the bandwidth of the bit rate as $\gamma = k\gamma_b$. The performance of the dual-3 coded noncoherent 8-ary FSK waveform is evaluated using (8.89) with $n = 2$, corresponding to the code rate $r_c = 1/2$, and diversity values of $\ell = 1$ through 4.* In the evaluation of (8.89), the terms D^x are replaced by $P_{be}(x\gamma)$ and the results are shown as the *dashed* curves in Figure 8.49 for the indicated diversities; these curves represent upper bounds on the performance as indicate by the first inequality in (8.86).

The *dashed* curves in Figure 8.50 show the performance of the dual-3 coded noncoherent 8-ary FSK in a Rayleigh fading channel for the indicated diversities; the *solid* curve represents

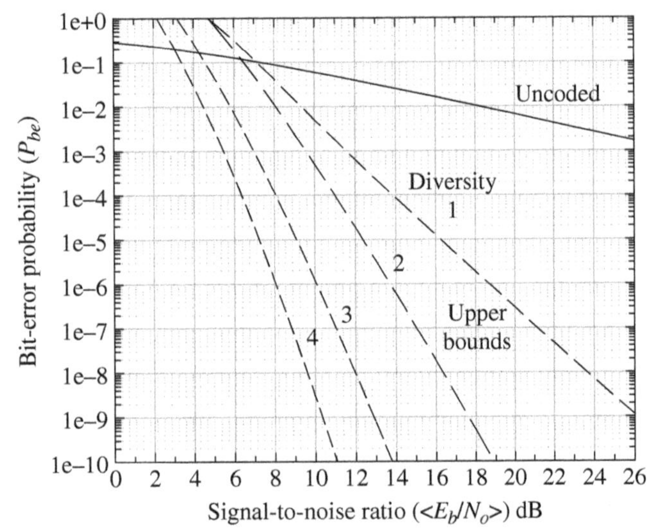

FIGURE 8.50 Dual-3 coded 8-ary noncoherent FSK performance with slow Rayleigh fading.

the uncoded performance with Rayleigh fading. The theoretical expressions for these results are based on the work of Odenwalder [98]. In this case, the bit-error probability is evaluated by summing the coefficients (C_i) of the polynomial

$$P(D) = D^{2n\ell}F(D)\big|_{n=2} = D^{4\ell}\sum_{i=0}^{\infty} C_i D^i \quad (8.92)$$

where $n = 2$ corresponds to the rate 1/2 dual-3 encoder. The coefficients C_i are those of $F(D)$ expressed in (8.90) and the bit-error probability is expressed as

$$P_{be} < \frac{2^{k-1}}{2^k - 1}\sum_{i=4\ell}^{\infty} C_{i-4\ell}P_i \quad (8.93)$$

where the factor involving $k = 3$ is used to convert from symbol errors to bit errors. Typically, only a few coefficient terms

*The equivalent polynomial expression of (8.89) using (8.90) and Table 8.32 yield approximate results when the summation is performed over a finite number of terms.

are required to obtain a good bound on P_{be} and four are used to obtain the results in Figure 8.50. In (8.93), the probability P_i is the probability of an error in comparing two sequences that differ in i symbols and is evaluated as

$$P_i \le p^i \sum_{j=0}^{i-1} \binom{i-1+j}{j} (1-p)^j \tag{8.94}$$

where p is given by [99]

$$p = \left(2 + \frac{k}{n}\left\langle \frac{E_b}{N_o} \right\rangle \right)^{-1} \tag{8.95}$$

An easier evaluation of P_i, which applies for $i \ge 6$, is [64, 90]

$$P_i < \frac{(4p(1-p))^i}{\sqrt{5\pi}2(1-2p)} \quad : i \ge 6 \tag{8.96}$$

Equation (8.95) is plotted as the *solid* curve in Figure 8.50 and, using (8.94), (8.95), and (8.96), Equation (8.93) is plotted as the *dashed* curves. The notation <*s*> denotes the average value of *s* corresponding to the fading channel. The performance of the dual-3 convolutional code in combating Raleigh fading is impressive and, referring to the performance with diversity applied to uncoded 8-ary FSK in Section 20.9.5, the dual-3 code without diversity, that is, with $\ell = 1$, requires about 3 dB less E_b/N_o than the uncoded 8-ary FSK modulation with $\ell = 4$.

8.12 TURBO AND TURBO-LIKE CODES

The discovery by Berrou, Glavieux, and Thitimajshima [1] of near-optimum error correction coding and decoding, referred to as TCs, represents a major step toward achieving Shannon's error-free performance limit of $E_b/N_o = -1.56$ dB. Their revolutionary results describe an implementation involving the parallel concatenation of two convolutional codes, with an intervening random interleaver, that are decoded with multiple forward and backward iterations through a trellis decoder resulting in performance remarkably close to Shannon's limit. This discovery *sparked* a new focus among analysts and practitioners that resulted in a unified theory* of multiple PCCCs [100] with iterative decoding that performed within a few tenths of a decibel of Shannon's limit. Divsalar and Pollara [101], Benedetto and Montorsi, and other researchers have substantiated the near Shannon-limit performance of the PCCC and present generalized descriptions of an encoder/decoder suitable for personal communications service (PCS) and many other applications. This research has also led to a variety of alternative concatenated codes and a unified approach to their implementation [102]. For example, Benedetto, Montorsi, Divsalar, and Pollara have made major contributions to the understanding of the fundamentals of turbo coding and have extended the notion of iterative decoding to SCCCs [103]. Furthermore, their efforts have resulted in the description of a universal soft-in soft-out (SISO) decoding module [104] that is suitable for implementation using a variety of technologies: digital signal processors (DSPs), field programmable gate arrays (FPGAs), and application-specific integrated circuits (ASICs). Divsalar [105] provides detailed descriptions of the coding and decoding of turbo and *turbo-like* codes.

Compared to high-constraint-length convolutional codes, required to achieve moderately high coding gains, turbo-like codes involve relatively short-constraint-length concatenated convolutional codes.† An important distinction is that convolutional codes traditionally use nonrecursive convolutional (NRC) codes requiring, for example, constraint length $K = 9$ bits or 256 states, to achieve modest coding gains at relatively high signal-to-noise ratios. For example, a 256-state convolutional code with a Viterbi decoder achieves $E_b/N_o \cong 3.5$ dB at $P_{be} = 10^{-5}$ and the RSV concatenated code achieves $E_b/N_o \cong$ 3 dB. On the other hand, TCs or PCCCs use recursive systematic convolutional (RSC) constituent codes (CCs) with 8 or 16 states and achieve $E_b/N_o \cong -0.6$ dB at $P_{be} = 10^{-5}$. The SCCCs use two- or four-state NRC and RSC CCs to achieve comparable performance. The selection of the code rates, interleaver lengths, and the number of demodulator decoder iterations is a major factor that influences the resulting performance. There is also an accompanying increase in decoder processing complexity that is inversely proportional to the code rate and proportional to the interleaver length and the number of iterations.

The use of RSC codes is critical in that they result in infinite impulse response (IIR) convolutional codes that yield the largest *effective free distance* [106] for the turbo encoder. This large effective free distance provides the coding gain necessary to overcome the operation in extremely low signal-to-noise conditions when approaching the Shannon limit of $E_b/N_o = -1.59$ dB. Divsalar and McEliece use the notation (r, k, m) for the convolutional CC generators defined as the $k \times r$ matrix $G(D)$ expressed as

$$G(D) = \left(\frac{P_1(D)}{Q(D)} \cdots \frac{P_r(D)}{Q(D)} \right) \tag{8.97}$$

*The term *turbo coding* was originally applied to the parallel concatenation of two convolutional codes; however, it has become more generally used to describe all PCCCs.

†Decoding of high-constraint-length convolutional codes use *sequential decoding* to avoid the computational complexity associated with the high number of states in the trellis decoders.

where $Q(D)$ is a primitive polynomial of degree m and $P_i(D)$ of m-degree polynomials of the form

$$P_i(D) = 1 + \cdots + D^m \neq Q(D) \quad (8.98)$$

The notation D is a unit delay element much like the z-domain notation z^{-1}. In terms of the conventional systematic convolutional code notation (n, k, m) with $n = k + r$, the TC generator $G'(D)$ is defined as

$$G'(D) \triangleq (I_k G(D)) \quad (8.99)$$

where I_k is the $k \times k$ identity matrix and $G'(D)$ the $k \times n$ matrix. These notations are used in the following descriptions of the TC generators.

8.12.1 Interleavers

Interleavers play such a prominent role in the coding gain of turbo-like codes that considerable attention has been focused on their design [107]. Typically, random block interleavers are employed and a number of algorithms have been examined for implementation. In addition to the interleavers impact of the coding gain, other key interleaver characteristics are delay, memory, spreading factor—a measure of the interleavers ability to spread, or distribute, channel burst errors to appear as independent errors over a specified interval at the deinterleaver output, and dispersion—a measure of the "randomness" of the interleaver. Interleavers are discussed in Section 8.12.1 and the selection of an appropriate turbo-like code interleaver is based on the communication channel considerations and the trade-off between interleaver coding gain and delay [108]. Unfortunately, high interleaver gains are associated with long interleavers resulting in long delays. The decoding memory for very long interleavers can be reduced by using overlapping sliding windows [109].

Several interleaver configurations that have been used with turbo-like codes are described in the following sections. In the description of the interleavers and in the general description of the turbo-like code processing to follow, the interleavers are denoted by the symbol π, and the following algorithms describe how the interleavers are filled and read. The turbo-like code performance simulations, discussed in Section 8.12.7.1, use the Jet Propulsion Laboratory (JPL) spread interleaver with various interleaver lengths to demonstrate the dependence on the coding gain.

8.12.1.1 Turbo Interleaver The turbo interleaver is used with the original description of the TC described by Berrou et al. [1]. It is characterized as a square $L \times L$ block interleaver and is filled, or written, row by row and emptied, or read, in a random manner described as follows. The respective row and column indices are i and j, indexed as 0, 1, …, $L - 1$. When the matrix is full, the data is interleaved by randomly reading from location (i_r, j_r) computed as permutations of the row–column indices i and j according to the following rules. The row index is computed as

$$i_r = \left(\frac{L}{2} + 1\right)(i + j) \text{ modulo } L \quad (8.100)$$

where

$$k = (i + j) \text{ modulo } L \quad (8.101)$$

and the column index computed as

$$j_r = (P(k)(i + 1) - 1) \text{ modulo } L \quad (8.102)$$

where $P(k)$, $k = 0, 1, \ldots, M - 1$ is a set of prime numbers. The value of M is dependent on L; typically, $M = 8$ for $L <= 256$. The original set of prime numbers used by Berrou, Glavieux, and Thitimajshima are (7, 17, 11, 23, 29, 13, 21, 19); however, other sets have also been used, for example, (17, 37, 19, 29, 41, 23, 13, 7) [110].

The factor $(L/2 + 1)$ in (8.100) ensures that two neighboring input bits written on two consecutive rows will not remain neighbors in the interleaved output. The factor $(i + j)$ performs a diagonal reading that tends to eliminate regular input low-weight patterns at the input from appearing in the interleaved output.

8.12.1.2 Random Interleaver The random interleaver is a column vector interleaver with elements chosen randomly according to the following algorithm. Define a temporary array of integers $r(i) = i$ and fill the interleaver using the following procedure. Generate a uniformly distributed random number $j : 1 \le j \le L$ and assign to $\pi(1) = r(j)$. Then remove the integer $r(j)$ from the temporary array $r(i)$ by left shifting the elements in $r(i)$ by one position starting at position $\ell = r(j)$, that is, $r(i) = r(i + 1)$ for $i = \ell, \ldots, L - 1$. The size of the temporary integer array $r(i)$ that is of interest is now $L - 1$ with the randomly generated integer $r(j)$ removed. For the next iteration of this procedure, a uniformly distributed random number $j : 1 \le j \le L - 1$ is generated and assigned to $\pi(2) = r(j)$. The integer $r(j)$ is now removed from the temporary array $r(i)$ by left shifting the elements in $r(i)$ by one position starting at position $\ell = r(j)$, that is, $r(i) = r(i + 1)$ for $i = \ell, \ldots, L - 2$. After $L - 1$ iterations, there remains one unassigned element, $r(1)$, in the temporary array $r(i)$. This integer is assigned as $\pi(L) = r(1)$ and completes the algorithms for filling the interleaver.

8.12.1.3 JPL Spread Interleaver The JPL spread interleaver proposed by Divsalar and Pollara [111] for deep-space communications is a variation of the random interleaver and is often referred to as a semi-random interleaver or S-random interleaver. This interleaver is a column vector of L

interleaver locations that are filled in much the same way as the random interleaver; however, if the a random integer is selected that is within a distance S_2 of the past S_1 selections, it is discarded and an additional random integers are tried until this distance condition is satisfied. The parameters S_1 and S_2 are chosen to be larger than the memory, or constraint lengths, of the two related constituent convolutional codes. The time required to determine the interleaver addresses is somewhat longer than that for the random interleaver, and there is no guarantee that all locations will be filled according to this criteria. Divsalar and Pollara suggest that $S_1 = S_2 \le \text{int}\left[\sqrt{L/2}\right]$ is a good choice to complete the addressing within a reasonable time. If the processing fails to complete the array addressing, it is suggested that additional attempts be made with different random number generator seeds. When $S_1 = S_2 = 1$, the spread interleaver reduces to the random interleaver discussed earlier.

8.12.1.4 JPL Interleaver The JPL interleaver [112] is a block interleaver characterized as having a low dispersion $a = 0.0686$ and spreading factors (3,1021), (6,39), (11,38), and (19,37). With M an even integer, the vector of eight prime numbers $p = (31, 37, 43, 47, 53, 59, 61, 67)$ are used to fill the interleaver using the following algorithm. For each $0 \le i < L$,

$$\pi(i) = 2r(i) + Mc(i) - m(i) + 1 \qquad (8.103)$$

where the parameters $m(i)$, $r(i)$, and $c(i)$ are computed as

$$m(i) = i\,\text{mod}(2) \qquad (8.104)$$

$$r(i) = (19r_o + 1)\text{mod}\left(\frac{M}{2}\right) \qquad (8.105)$$

and

$$c(i) = (p(\ell+1)c_o + 21m)\text{mod}(L) \qquad (8.106)$$

The related parameters c_o, r_o, and ℓ are computed as

$$c_o(i) = \left(\frac{(i-m)}{2}\right)\text{mod}(L) \qquad (8.107)$$

$$r_o(i) = \left(\frac{(i-m)}{2} - c_o\right)\text{mod}(L) \qquad (8.108)$$

and

$$\ell = r(i)\text{mod}(8) \qquad (8.109)$$

8.12.1.5 Welch–Costas Interleavers The Welch–Costas interleaver [113] is a column vector interleaver characterized as having unit dispersion. The interleaver is filled using the following algorithm:

$$\pi(i) = \alpha^i \text{mod}(L+1) - 1 \quad : 0 \le i < L \qquad (8.110)$$

Here, $L = p - 1$, where p is a prime number and α is a primitive element [114] of the field $\{1, \alpha, \alpha^2, \ldots, \alpha^{p-1}\}$.

8.12.2 Code Rate Matching

The rate matching of PCCCs is different from that described in Section 8.11 for the convolutional code, in that, the PCCC rate matching must ensure that the number of bits-per-block (N_o) after puncturing is commensurate with an integer number of transmitted symbols. This rate matching criteria is based on the number of information bits-per-block N, the desired modulation efficiency r_s with units of bits/second/Hz, and the modulation efficiency $r_m = \log_2(M)^2$ with units of bits-per-symbol. The parameter N is also equal to the interleaver length. To satisfy the requirement of an integer number of transmitted symbols-per-information block, the parameter N_o is evaluated as

$$N_o = \left\lceil \frac{N}{r_s} \right\rceil r_m \qquad (8.111)$$

The puncturing ratio is defined as

$$p \triangleq \begin{cases} 1 - \dfrac{N_o}{L} & : \text{nonsystematic code} \\ 1 - \dfrac{N_o - N}{L - N} & : \text{systematic code} \end{cases} \qquad (8.112)$$

where L is the total number of parity bits prior to the rate matching. The puncturing ratio for the nonsystematic code uses all of the party bits; whereas, for the systematic code only the party-check bits are used.

An example of the rate matching processing, involving the evaluation of (8.111) and (8.112), is considered using the following design parameters: code rate $r_c = 1/2$, constraint length $K = 4$, PCCC with interleaver length $N = 2048$, and 8PSK symbol modulation. The modulation efficiency is $r_m = \log_2(8) = 3$ bits-per-symbol and the desired modulation or spectral efficiency is $r_s = 2$ bits/s/Hz. In this case, there are $2(K-1) = 6$ flush bits, so there are $L = 2N + 2(K-1) = 4102$ total bits before the rate matching. Under these conditions, the total bits out of the rate matching is $N_o = 3072$ and using N and L the puncturing ratio for the systematic code is evaluated as $p = 0.50146$. Therefore, of the original $L - N = 2054$ parity-check bits only $N_o - N = 1024$ remain after the puncturing and the number of 8PSK symbols required to transmit the punctured block of data is $N_o/r_m = 1024$ symbols.

The block interleaver columns and rows are (r_m, N_o/r_m) and the N_o bits are entered into the interleaver column by

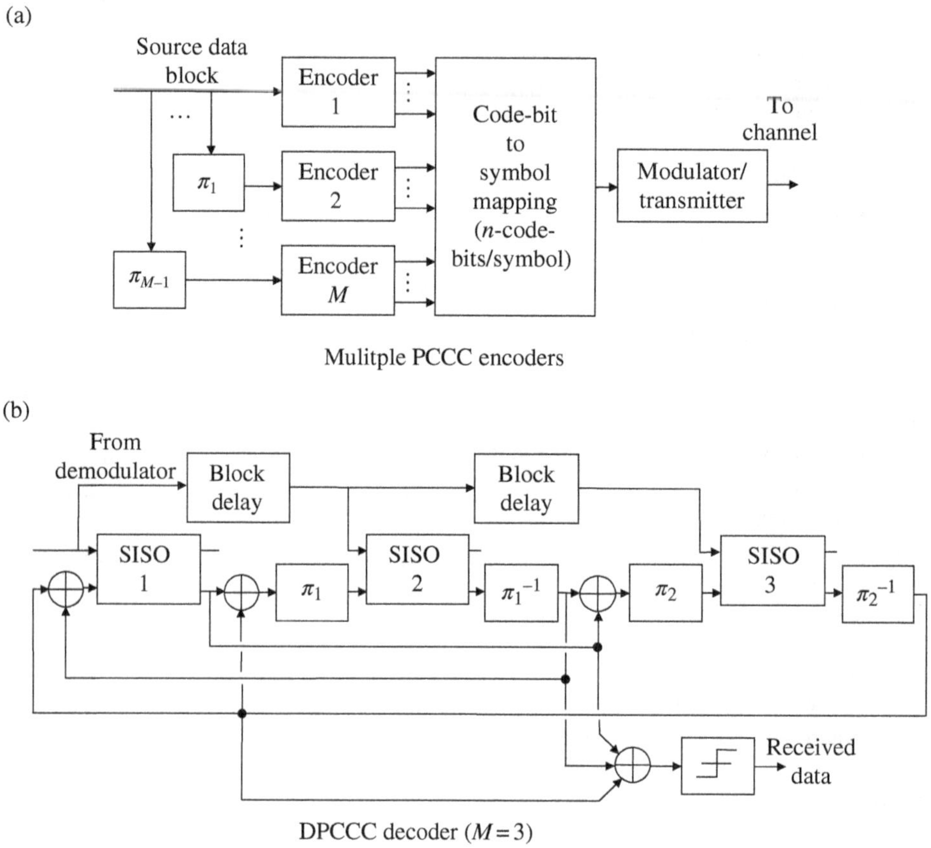

FIGURE 8.51 Multiple parallel concatenated convolutional codes. Benedetto et al. [119]. Reproduced by permission of the IEEE.

column with the highest priority systematic bits assigned to the leftmost columns. As the interleaver is filled, the punctured bits are identified as those bits for which $(i-1)p \geq \lfloor ip \rfloor$: $i = 1, \ldots, L-N$ with the surviving parity-check bits entered into the rightmost columns of the interleaver.

The remaining functions, involving gray coding the interleaver rows and bit mapping of the interleaver row by row to the modulation symbols, are identical to those described in Section 8.11. As discussed in Section 8.12.3, rate matching is provided between each of the SCCC concatenated CCs.

8.12.3 PCCC and SCCC Configurations

As mentioned previously, convolutional code concatenation falls into two generic configurations [65]: turbo or *turbo-like* codes. The HCCC and the SCC configurations are variants of the PCCC and SCCC configurations. These code configurations have been analyzed and the bit-error performance dependence on E_b/N_o (dB) evaluated using computer simulations [103, 105]. This section focuses on the parallel [1, 100, 115–117] and serial [103, 115, 117, 118] configurations, implemented as shown in Figures 8.51 and 8.52, respectively. The codes are implemented as linear block codes composed of a block of input data bits plus *tail-bits* that are used to ensure that the decoding trellis terminates on the all-zero state at the end of each block. For nonrecursive convolutional encoders,* the *tail-bits* are input as a series of zero bits; whereas, for recursive convolutional encoders, the tail-bits are dependent on the encoder feedback taps.† The length of the data interleavers is equal to the number of input data bits and, based on coding theory, long random data sequences or interleaved data sequences are required to approach the channel capacity. Therefore, the interleavers provide the necessary random properties and the concatenated convolutional codes provide the coding structure necessary to approach Shannon's theoretical performance limit. The interleavers are filled with the block of source code-bits as described in Section 8.12.1. The code-bits to symbol mapping is similar to that of the convolutional code shown in Figure 8.35. The PCCC encoder configuration is shown in Figure 8.51a for an arbitrary number (M) of code concatenations and, for simplicity, the PCCC decoder is shown in Figure 8.51b

*Nonrecursive encoders do not use feedback taps; whereas, recursive encoders use feedback taps; these implementations are synonymous, respectively, with FIR and IIR transversal filters.

†The *switches* associated with the recursive encoders shown in Figure 8.57 are changed at the end of each code block, resulting in successive binary zeros in each delay element thus forcing the all-zero encoder state.

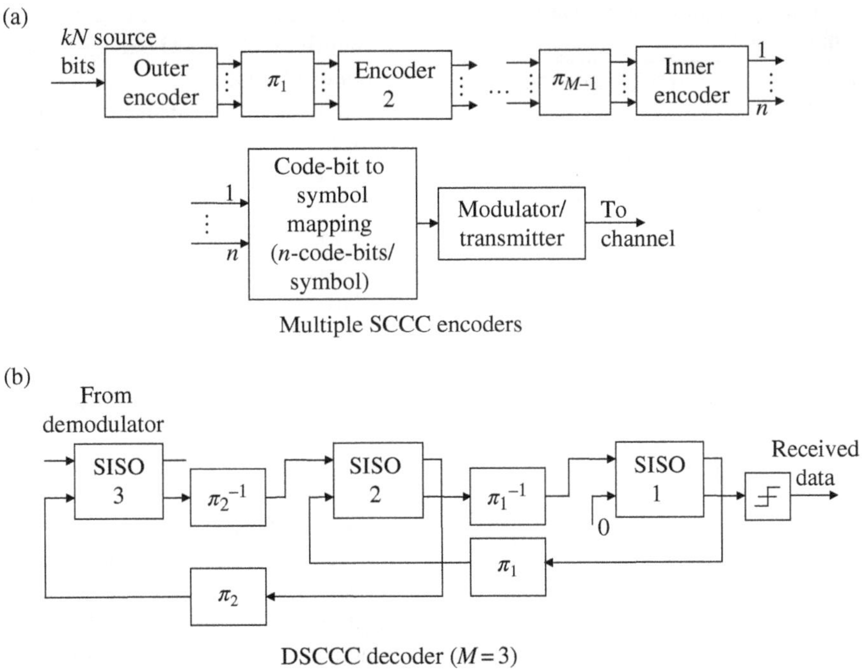

FIGURE 8.52 Multiple serially concatenated convolutional codes. Benedetto et al. [119]. Reproduced by permission of the IEEE.

for $M = 3$; this configuration is referred to as a DPCCC decoder [105].

For an arbitrary number of code concatenations, the basic functions of the PCCC decoder involve the *input-adder*, *interleaver* (π), *SISO*, and *inverse interleaver* (π^{-1}). The SISO newly formed extrinsic information outputs are fed forward through the interleavers and inverse interleavers with feedback connections forming the SISO input reliabilities. These functions are executed for each decoding iteration and repeated in the process of performing multiple iteration data decoding. Upon the last iteration through the decoder, the received data estimates are formed by summing the newly formed extrinsic information from each of the inverse interleavers. The received data block from the demodulator is interleaved data and is applied directly to SISO1 and to subsequent SISOs with the indicated block delays. The four port SISO is described in Section 8.12.4; however, the upper output, $\lambda(\mathbf{c};\mathrm{O})$, is not used in the parallel configurations.

The generalized SCCC encoder is shown in Figure 8.52a for M concatenated convolutional encoders and $M - 1$ interleavers. The number of source data bits entered into the outer encoder for each sample is k and the number of code bits at the output of the inner encoder is n, so the overall code rate is $r_c = k/n$. The individual encoder rates are defined as* k/p_1, p_1/p_2, …, p_{M-1}/p_M, and p_M/n. Furthermore, requiring the interleaver length (L_i) to be integerly related to p_i, such that, $k_i = L_i/p_i : k_i \in \text{integer} > 0$, then, with equal values $k_i = N$: $i = 1, \ldots, M - 1$, the input source-bit block size is kN and the interleaver lengths are determined as

$$L_i = Np_i \quad : k_i = N \forall i \tag{8.113}$$

The n inner code code-bits are mapped onto the modulation symbol using rate matching and interleaving to associate the MSB code-bit mapping with the least vulnerable channel error condition. The SCCC decoder is depicted in Figure 8.52b for $M = 3$ corresponding to the DSCCC [105] configuration. Table 8.33 provides some CC generators used by Benedetto, Divsalar, Montorsi, and Pollara to evaluate rate 1/3, 1/4, and 1/6 SCCC implementations. Their theoretical analysis shows that the DPCCC performs somewhat better than the rate 1/4 SCCC implementation.

8.12.4 SISO Module

The SISO module† decoding structures have been characterized in considerable detail by Benedetto et al. [115, 120]. These modules perform maximum a posteriori (MAP) detection processing using a finite-state trellis decoder and are used in either configuration by appropriately connecting the input and output ports. Figure 8.53 shows an isolated SISO module as a four-port device with two inputs and outputs denoted, respectively, as $\lambda(\boldsymbol{c} : I)$, $\lambda(\boldsymbol{u} : I)$ and $\lambda(\boldsymbol{c} : O)$, $\lambda(\boldsymbol{u} : O)$. In the following descriptions of the decoding, the

*Puncturing the encoder outputs may be performed to achieve the desired overall code rate.

†The description of the SISO processing in this section is based on the work of Divsalar [105].

TABLE 8.33 Constituent Code Generators for Serial Concatenated Convolutional Codes[a]

Code Rate	Code Type	Generator $G(D)$
1/2	Recursive[b]	$\left[1 \quad \frac{1+D^2}{1+D+D^2}\right]$
	Nonrecursive	$\left[1+D+D^2 \quad 1+D^2\right]$
2/3	Recursive	$\begin{bmatrix} 1 & 0 & \frac{1+D^2}{1+D+D^2} \\ 0 & 1 & \frac{1+D}{1+D+D^2} \end{bmatrix}$
	Nonrecursive	$\begin{bmatrix} 1+D & D & 1 \\ 1+D & 1 & 1+D \end{bmatrix}$
3/4	Recursive	$\begin{bmatrix} 1 & 0 & 0 & \frac{D}{1+D} \\ 0 & 1 & 0 & \frac{1+D+D^2}{1+D^2} \\ 0 & 0 & 1 & \frac{1}{1+D^2} \end{bmatrix}$
	Nonrecursive	$\begin{bmatrix} 1+D & 1+D & D & 0 \\ 1+D & D & 1 & 0 \\ 1 & 1 & 1 & 1 \end{bmatrix}$

[a]Benedetto et al. [119]. Reproduced by permission of the IEEE.
[b]Not in referenced table; included here for completeness.

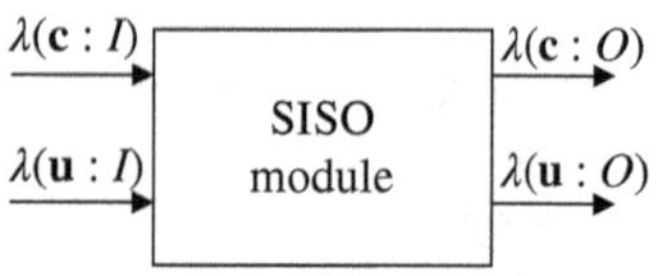

FIGURE 8.53 Soft-in soft-out processing module. Benedetto et al. [121]. Reproduced by permission of Elsevier Books.

number of iterations is established based on the diminishing improvement in the bit-error or code block-error performance and typically 8–20 iterations are used.

Depending upon the requirements of the channel coding, the processing complexity can be reduced if $M = 2$ constituent encoders and SISO decoding modules are used. These simplified implantations follow directly from the $M = 2$ PCCC and SCCC configurations that are shown in Figures 8.54 and 8.55, respectively. Similarly, the encoding with an arbitrary number (M) of CCs is shown in Figures 8.51 and 8.52, and the corresponding decoding configurations can be inferred from the $M = 3$ decoding implementations.

The TC configuration involves two constituent convolutional encoders and one interleaver and the decoder consists of two SISO modules and one interleaver and deinterleaver as shown in Figure 8.54. In this configuration, there is only one source of received code words available for use by the SISO external observations or inputs $\lambda(\boldsymbol{c} : I)$. These inputs are obtained directly from the demodulator matched filter; consequently, with PCCC decoding the decoded outputs $\lambda(\boldsymbol{c} : O)$ is not used. In this case, the SISO1 output $\lambda(\boldsymbol{u} : O)$ represents the newly formed extrinsic information that is interleaved to form the reliability input $\lambda(\boldsymbol{u} : I)$ to SISO2. With subsequent iterations, the deinterleaved output from SISO2 is used as the reliability input to SISO1 and the decoding proceeds as described before. Upon completion of the last iteration, the received data is output as shown in Figure 8.54.

The decoding of the SCCC is described in terms of Figure 8.55 using two serially concatenated encoders. In this case, the decoding involves two SISO modules with the input $\lambda(\boldsymbol{c} : I)$ to the inner SISO corresponding to the demodulator matched filter samples and represents the interleaved symbols received from the channel. The input $\lambda(\boldsymbol{c} : I)$ to the outer SISO corresponds to the deinterleaved output of the newly formed extrinsic information $\lambda(\boldsymbol{u} : O)$ from the inner SISO. The newly formed decoded output $\lambda(\boldsymbol{c} : O)$ of the outer SISO is then interleaved forming the new observations $\lambda(\boldsymbol{u} : I)$ to the inner SISO. In this case, there is no successive SISO from which to obtain extrinsic information, so the outer SISO input $\lambda(\boldsymbol{u} : I)$ is set to zero and is not used. At this point, the first iteration is completed and following the last iteration the received data corresponds to the extrinsic information $\lambda(\boldsymbol{u} : O)$ of outer SISO.

Although the configurations of the PCCC and SCCC implementations are quite different, the processing within the SISO module is very structured with many common algorithms for both codes. The internal processing is based on optimal decoding algorithms for concatenated convolutional codes using the trellis decoding structure. The processing is described in broad terms in the remainder of this section and in detail in Sections 8.12.5 and 8.12.6.

The Bahl–Cocke–Jelinek–Raviv (BCJR) algorithm [122] provides for the MAP decoding of binary convolutional codes using the trellis decoding structure and the SISO processing is a variant of the BCJR algorithm. The MAP algorithm makes symbol-by-symbol decisions and, with equal a priori source symbol probabilities, corresponds to the a posteriori probability (APP) algorithm. However, because the SISO module processes a block of information and parity symbols before outputting the symbol decisions, the SISO module uses an additive APP algorithm rather than the MAP algorithm. Divsalar [105] has developed step-by-step derivations of the generalized additive APP SISO algorithms for the decoding of turbo, serial, hybrid, and self-concatenated codes. These derivations compare the multiplicative and additive APP processing within the SISO module for symbol and bit-level processing. The multiplicative APP forms the product of APPs at each trellis transition and is more computationally complex than the additive APP algorithm, that is based the monotonic property of the logarithm and involves the natural logarithm of the summation of APPs.

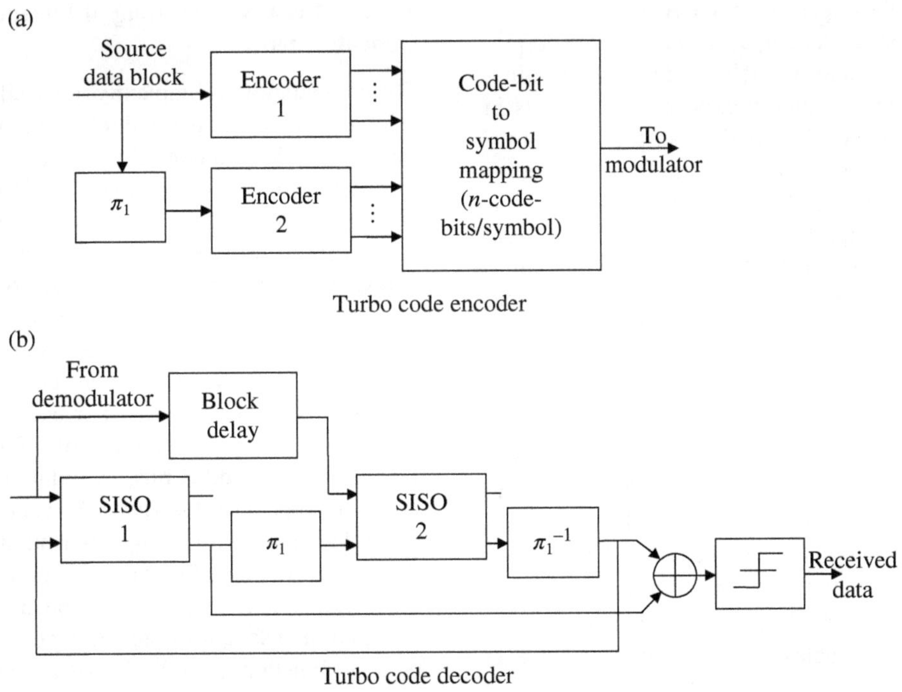

FIGURE 8.54 Turbo code implementation ($M = 2$).

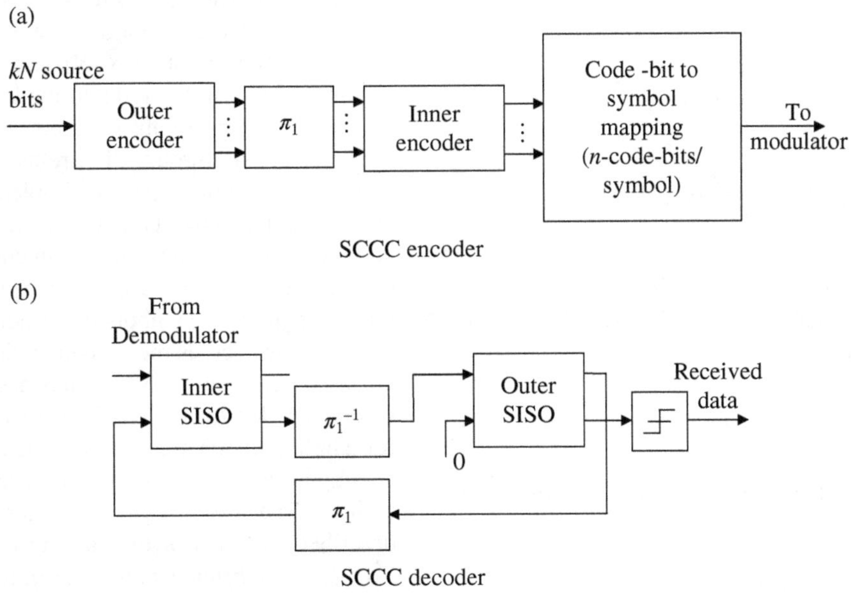

FIGURE 8.55 SCCC code implementation ($M = 2$).

The relative computational simplicity in forming the logarithm of the sum of the APPs, however, involves some degree of complexity. The issue is related to the ACS algorithm used to choose the surviving metric at each of the termination states when using the Viterbi algorithm. With the additive APP algorithm, the comparable processing is referred to as an *ACS plus correction*, or simply the max∗ (*max-star*) operation, defined as

$$\begin{aligned}\max_i{}^*(a_i) &\triangleq \log\left[\sum_{i=1}^{L} e^{a_i}\right]\\ &= \max_i(a_{1,i},a_{2,i}) + \log[1+\exp(-|a_{1i}-a_{2i}|)]\\ &\quad : i = 1,\ldots,L\\ &\cong \max_i(a_{1,i},a_{2,i}) + \delta(\Delta_i)\\ &\cong \max_i(a_{1,i},a_{2,i})\end{aligned} \quad (8.114)$$

where $a_{ji} : j = 1, 2$ are the exponents of the APPs at state i, L is the number of trellis states, and $\delta(\Delta_i)$ is the additive correction term that is stored in a lookup table. The last approximation simply chooses the maximum exponent and results in near-optimum performance for medium-to-high signal-to-noise ratios; however, for very low signal-to-noise ratios an E_b/N_o performance loss [123] of 0.5–0.7 dB is encountered; this loss is not insignificant given the overall coding complexity required to approach the channel coding limit. The first approximation in (8.114) involves a table look, based on the additive term in the second equality, and results in nearly the optimum performance when only eight values are stored.

The demodulator matched filter samples ($y_{k,i}$) represent sufficient statistics that contain all the information required to optimally detect the received information. However, the SISO processing requires that the receiver AGC establish a constant signal level satisfying one of the two following conditions. If the AGC normalizes the signal-to-noise ratio so that the additive noise samples ($n'_{k,i}$) represents a zero-mean, unit-variance random variables, the sufficient statistics is evaluated as

$$y_{k,i} = \sqrt{\frac{2E_s}{N_o}} d_{k,i} + n'_{k,i} \qquad (8.115)$$

where $d_{k,i} = 2c_{k,i} - 1$ with $c_{k,i} = \{0,1\}$ and $d_{k,i} = \{-1,1\}$. The external observations from the channel are denoted as

$$\lambda_k(c_{k,i};I) = 2\sqrt{\frac{2E_s}{N_o}} y_{k,i} \qquad (8.116)$$

On the other hand, if the AGC holds the signal level constant at A, such that the signal-to-noise ratio is $A^2/(2\sigma_n^2)$, then the sufficient statistic is

$$y_{k,i} = Ad_{k,i} + n_{k,i} \qquad (8.117)$$

where $n_{k,i} = N(0,\sigma_n)$ and the external observations from the channel are denoted as

$$\lambda_k(c_{k,i};I) = \left(\frac{2A}{\sigma_n}\right) y_{k,i} \qquad (8.118)$$

These results apply for BPSK; however, QPSK modulation can be used by considering two independent channels of PSK, each with unit noise variance and this requires dividing (8.116) and (8.118) by two.

Sections 8.12.5 and 8.12.6 describe the iterative decoding details for the parallel and serially concatenated codes involving two SISO modules based on the work of Divsalar [105] and Benedetto and Divsalar [65].

8.12.5 Iterative Decoding of Parallel Concatenated Codes

The computational complexity of the additive APP algorithm is lower than that of the multiplicative APP algorithm; so, for this reason, the additive APP algorithm [117] is described in this section. The description of the PCCC processing is based on Figure 8.54 and the focus is on the bit-level processing within the SISO module that is commensurate with BPSK and QPSK waveform modulations. To simplify the description of the SISO algorithms, the state metric computation examples are based on a rate 1/2, four-state binary convolutional encoder; however, the decoding algorithms are expressed in general terms. For example, the code rate* is $r_c = p/q$ where p is the number of information bits; q is the total number of coded bits, including the parity bits; and N is the number of trellis states. For clarity, these parameters are included in the following description. Two major assumptions are made regarding the SISO processing: the channel is memoryless and maps the encoded symbols c_k into the matched filter samples y_k according to the channel probability density function $p(y|c) = P(Y_k|C_k)$; the encoder is time-invariant and corresponds to a fixed block code of length K symbols. In this case, with $k = 1, \ldots, K$, the trellis-state decoding is completely characterized by each trellis section from state $S_{k-1}(\ell)$ to $S_k(\ell) : \ell = (0,\ldots,N-1)$. Based on the encoder convention, the decoding starts at state $S_0(0)$ and ends at state $S_0(K)$ and includes the flush bits required to return the trellis to the known termination state.

The SISO processing is based on the forward and backward recursions through the trellis structure corresponding to the convolutional encoder implementation. The encoder can be implemented as a systematic or nonsystematic code and use recursive encoders, nonrecursive encoders, or both configurations. The recursions take place over the entire code block length of K-symbols corresponding to the interleaver length;† however, as mentioned earlier, the SISO processing is based on each trellis section associated with the q-bit received code symbol. The forward and backward recursions for a trellis section are shown in Figure 8.56. The trellis processing is described using the nomenclature of the referenced authors. The trellis segment for the k-th input code bit is described in terms of the starting edge $S^S(e)$, ending edge $S^E(e)$, and the *branch metric*, $\tilde{\gamma}_k(e)$, for edge "e." These three terms correspond, respectively, to the: *source metric*, *termination metric*, and *metric update* used in Section 8.11.1. To begin the description of the block decoding, the trellis states are initialized as $\alpha_0(\ell = 0)$, $\beta_0(\ell = 0) = 0$ and $\alpha_0(\ell \neq 0)$, $\beta_0(\ell \neq 0) = -\infty$ with h_{α_k} and h_{β_k} initialized to zero.

*In Section 8.11.1, the code rate is defined as k/n and q corresponds to the q-ary source information symbol.

†In Section 8.12.5.1, the column interleaver length is denoted as K.

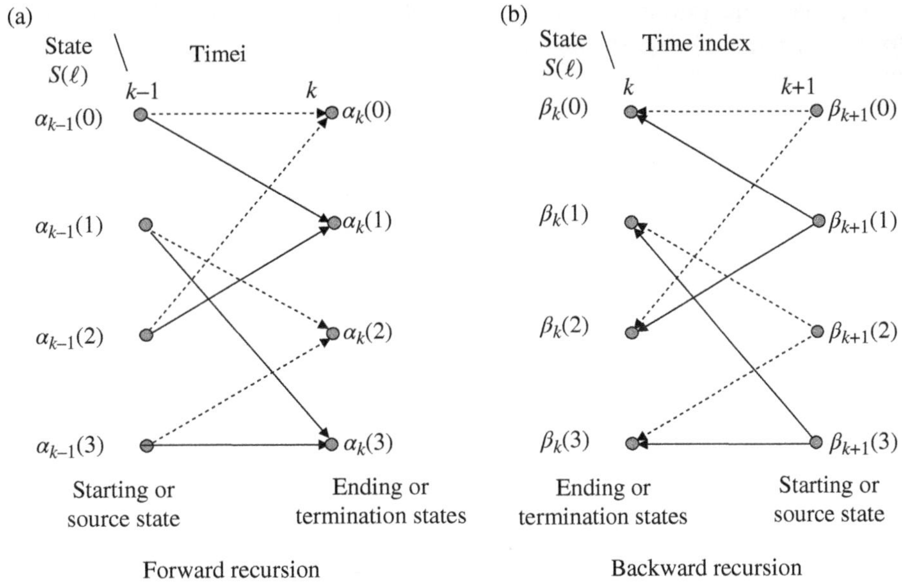

FIGURE 8.56 Example four-state trellis section showing forward and backward recursions.

8.12.5.1 Forward and Backward Recursions The forward recursion processing is very similar to that described in Section 8.11.2, with the inclusion of the reliability input $\lambda(\boldsymbol{u}:I)$. For example, the metric update $\Delta\alpha_m$ in (8.76) is expressed as

$$\gamma_k(e)=\sum_{i=1}^{q} c_{k,i}(e)\lambda_k[C_{k,i}(e);I] \tag{8.119}$$

and, with the inclusion of the reliability input $\lambda(\boldsymbol{u}:I)$, the forward recursion *branch metric* for edge "e" is expressed as

$$\begin{aligned}\tilde{\gamma}_k(e)&=u_k(e)\lambda_k[U_k;I]+\gamma_k(e)\\&=u_k(e)\lambda_k[U_k;I]+\sum_{i=1}^{q} c_{k,i}(e)\lambda_k[C_{k,i}(e);I] \quad \text{:branch metric}\end{aligned} \tag{8.120}$$

For the forward recursion, the SISO processing computes $\alpha_k(s)$ from $\alpha_{k-1}[S^S(e)]$ where $S^S(e)$ *points to* the starting states that terminate on the ending state $S^E(e)=s$. In Figure 8.56a, for each ending state there are two starting states, for example, the ending state $\alpha_k(s=1)$ *points to* the starting states $\alpha_{k-1}(s=0)$ and $\alpha_{k-1}(s=2)$. As in the Viterbi decoding algorithm, a selection algorithm, similar to the ACS algorithm, must be used to determine the surviving metric at each ending state. For the additive APP algorithm, the $\max^*(a_i)$ selection algorithm is used and the surviving metric is computed as

$$\alpha_k(s)=\max_{e:s^E(e)=s}{}^{*}\left\{\begin{array}{l}\alpha_{k-1}[s^S(e)]+\sum_{i=1}^{p}u_{k,i}(e)\lambda_k[U_{k,i};I]\\ \qquad +\sum_{i=1}^{q}c_{k,i}(e)\lambda_k[C_{k,i};I]+h_{\alpha_k}\end{array}\right\} \quad \text{: forward recursion} \tag{8.121}$$

The backward recursion computes $\beta_k(s)$ from $\beta_{k+1}[S^S(e)]$ where $S^S(e)$ *points to* the starting states that terminate on the ending state $S^E(e)=s$. In Figure 8.56b for each ending state, there are two starting states, for example, the ending state $\beta_k(s=1)$ *points to* the starting states $\beta_{k+1}(s=2)$ and $\beta_{k+1}(s=3)$. In this case, the surviving metric is computed as

$$\beta_k(s)=\max_{e:s^S(e)=s}{}^{*}\left\{\begin{array}{l}\beta_{k+1}[s^S(e)]+\sum_{i=1}^{p}u_{k+1,i}(e)\lambda_{k+1}[U_{k+1,i};I]\\ \qquad +\sum_{i=1}^{q}c_{k+1,i}(e)\lambda_{k+1}[C_{k+1,i};I]+h_{\beta_k}\end{array}\right\} \quad \text{: backward recursion} \tag{8.122}$$

The demodulator matched filter observations $\lambda(\boldsymbol{c},I)$ are given by (8.116) and the constants h_{α_k} and h_{β_k} are adjusted as required upon completion of each iteration to prevent accumulation overflow. Overflow in path metric computations using fixed-point processing internal to the SISO must also be considered [124]. The PCCC performance evaluations in this chapter use 32-bit floating point SISO processing and, to avoid computational overflow, the metric adjustments are made at the completion of each iteration.

For the PCCC implementation, the extrinsic bit information output $\lambda(\boldsymbol{u}:O)$ for $U_{k,j}$; $j=1, \ldots, p$ is computed as in (8.123) and the output $\lambda(\boldsymbol{c}:O)$ is not used.

$$\lambda_k(U_{k,j};O) = \max_{e:u_{k,j}(e)=1}{}^{*} \left\{ \begin{aligned} &\alpha_{k-1}[s^S(e)] + \sum_{\substack{i=1\\ i\neq j}}^{p} u_{k,i}(e)\lambda_k[U_{k,i};I] \\ &+ \sum_{i=1}^{q} c_{k,i}(e)\lambda_k[C_{k,i};I] + \beta_k[s^E(e)] \end{aligned} \right\} - \max_{e:u_{k,j}(e)=0}{}^{*} \left\{ \begin{aligned} &\alpha_{k-1}[s^S(e)] + \sum_{\substack{i=1\\ i\neq j}}^{p} u_{k,i}(e)\lambda_k[U_{k,i};I] \\ &+ \sum_{i=1}^{q} c_{k,i}(e)\lambda_k[C_{k,i};I] + \beta_k[s^E(e)] \end{aligned} \right\} \tag{8.123}$$

The data storage requirements involve the forward and backward branch metrics α_k, β_k, the SISO input data, the interleaver and deinterleaver data, the transition code bits $u_{k,i}$, and matched filter code-bit estimates $c_{k,i}$ associated with each branch transition.

8.12.6 Iterative Decoding of Serially Concatenated Convolutional Codes

The serially concatenated code configuration is depicted in Figure 8.55 and the details of the SISO processing and iterative decoding between the two concatenated codes is the subject of this section [118]. Unlike the PCCC decoding algorithm, which only accepts the input bit reliabilities $\lambda(\boldsymbol{c}:I)$ from the demodulator matched filter, the SCCC decoding algorithm uses the matched filter samples on the first iteration; however, on subsequent iterations the input data reliabilities $\lambda(\boldsymbol{c}:I)$ are obtained from the deinterleaved extrinsic bit information $\lambda(\boldsymbol{u}:O)$ from the inner SISO. The outer SISO then computes the extrinsic bit information $\lambda(\boldsymbol{c}:O)$. In describing the decoding algorithms, the respective inner and outer code rates are p_1/q_1 and p_2/q_2, $\boldsymbol{c}_k(e)$ represents $c_{k,i}(e)$: $i=1, 2, \ldots, q_m$ for the inner code ($m=1$), and $\boldsymbol{u}_k(e)$ represents $u_{k,i}(e)$: $i=1,2,\ldots,p_m$ for the outer code ($m=2$). Considering that the p and q input and output bits are binary, represented by $\{0,1\}$, the bit-by-bit decoding for the $M=2$ SCCC code is described as follows.

The max∗ operation is identical to that defined by (8.114) and the intrinsic input $\lambda_k(c_{k,i};I)$ from the demodulator matched filter samples are identical those described by (8.116). The inner and outer trellises are initialized as described for the PCCC decoding, that is, with $\alpha_0(\ell=0)$, $\beta_0(\ell=0)=0$ and $\alpha_0(\ell\neq 0)$, $\beta_0(\ell\neq 0)=-\infty$ with the constants $h_{\alpha_k,m}$ and $h_{\beta_k,m}$ initialized to zero. Under these conditions, the inner and outer trellis recursions are computed as follows:

8.12.6.1 *Inner Code Forward and Backward Recursions*

$$\alpha_k(s) = \max_{e:s^E(e)=s}{}^{*} \left\{ \begin{aligned} &\alpha_{k-1}[s^S(e)] + \sum_{i=1}^{p_1} u_{k,i}(e)\lambda_k[U_{k,i};I] \\ &+ \sum_{i=1}^{q_1} c_{k,i}(e)\lambda_k[C_{k,i};I] + h_{\alpha_k,1} \end{aligned} \right\} \quad \text{: forward recursion} \tag{8.124}$$

and

$$\beta_k(s) = \max_{e:s^S(e)=s}{}^{*} \left\{ \begin{aligned} &\beta_{k+1}[s^E(e)] + \sum_{i=1}^{p_1} u_{k+1,i}(e)\lambda_{k+1}[U_{k+1,i};I] \\ &+ \sum_{i=1}^{q_1} c_{k+1,i}(e)\lambda_{k+1}[C_{k+1,i};I] + h_{\beta_k,1} \end{aligned} \right\} \quad \text{: backward recursion} \tag{8.125}$$

with the extrinsic bit information for $U_{k,j}$: $j=1, 2, \ldots p_1$ is computed as

$$\lambda_k(U_{k,j};O) = \max_{e:u_{k,j}(e)=1}{}^{*} \left\{ \begin{aligned} &\alpha_{k-1}[s^S(e)] + \sum_{\substack{i=1\\ i\neq j}}^{p_1} u_{k,i}(e)\lambda_k[U_{k,i};I] \\ &+ \sum_{i=1}^{q_1} c_{k,i}(e)\lambda_k[C_{k,i};I] + \beta_k[s^E(e)] \end{aligned} \right\} - \max_{e:u_{k,j}(e)=0}{}^{*} \left\{ \begin{aligned} &\alpha_{k-1}[s^S(e)] + \sum_{\substack{i=1\\ i\neq j}}^{p_1} u_{k,i}(e)\lambda_k[U_{k,i};I] \\ &+ \sum_{i=1}^{q_1} c_{k,i}(e)\lambda_k[C_{k,i};I] + \beta_k[s^E(e)] \end{aligned} \right\} \tag{8.126}$$

8.12.6.2 *Outer Code Forward and Backward Recursions*

$$\alpha_k(s) = \max_{e:s^E(e)=s}{}^{*} \left\{ \alpha_{k-1}[s^S(e)] + \sum_{i=1}^{q_2} c_{k,i}(e)\lambda_k[C_{k,i};I] + h_{\alpha_k,2} \right\} \quad \text{: forward recursion} \tag{8.127}$$

and

$$\beta_k(s) = \max_{e:s^S(e)=s}{}^{*} \left\{ \beta_{k+1}\left[s^E(e)\right] + \sum_{i=1}^{q_2} c_{k,i}(e)\lambda_k[C_{k,i};I] + h_{\beta_k,2} \right\} \quad \text{: backward recursion} \tag{8.128}$$

with the extrinsic bit information for $C_{k,j}$: $j = 1, 2, \ldots q_2$ is computed as

$$\lambda_k(C_{k,j};O) = \max_{e:c_{k,j}(e)=1}{}^{*} \left\{ \alpha_{k-1}\left[s^S(e)\right] + \sum_{\substack{i=1 \\ i \neq j}}^{q_2} c_{k,i}(e)\lambda_k[C_{k,i};I] + \beta_k\left[s^E(e)\right] \right\} - \max_{e:c_{k,j}(e)=0}{}^{*} \left\{ \alpha_{k-1}\left[s^S(e)\right] + \sum_{\substack{i=1 \\ i \neq j}}^{q_2} c_{k,i}(e)\lambda_k[C_{k,i};I] + \beta_k\left[s^E(e)\right] \right\} \tag{8.129}$$

where $\lambda_k(C_{k,j};\, O)$ is interleaved and provided to the inner SISO as the input reliabilities $\lambda(\mathbf{c_k};\, I)$. Upon the final iteration, the output bit reliability $U_{k,j}$: $j = 1, 2, \ldots, p_2$ is computed as

$$\lambda_k(U_{k,j};O) = \max_{e:u_{k,j}(e)=1}{}^{*} \left\{ \alpha_{k-1}\left[s^S(e)\right] + \sum_{i=1}^{q_2} c_{k,i}(e)\lambda_k[C_{k,i};I] + \beta_k\left[s^E(e)\right] \right\} - \max_{e:u_{k,j}(e)=0}{}^{*} \left\{ \alpha_{k-1}\left[s^S(e)\right] + \sum_{i=1}^{q_2} c_{k,i}(e)\lambda_k[C_{k,i};I] + \beta_k\left[s^E(e)\right] \right\} \tag{8.130}$$

8.12.7 PCCC and SCCC Performance

In this section, the bit-error performance of the PCCCs and SCCCs are examined and compared.* In the following performance evaluation, the rate 1/3 TC, included in the Consultative Committee for Space Data Systems (CCSDS) multi-rate TC standard for space telemetry [125], is examined using Monte Carlo simulations. The bit-error performance is evaluated for various interleaver lengths with four iterations and then with various iterations using an interleaver lengths of 1024 bits. The performance is also evaluated for short messages with interleaver lengths of 296 and 128 bits using 10 iterations. These results highlight the importance of the interleaver length as discussed in Section 8.12.1. In Section 8.12.7.2, the bit-error performance comparisons between the PCCCs and SCCCs are based on the work of Benedetto, Divsalar, Montorsi, and Pollara. All of the following Monte Carlo performance evaluations correspond to an AWGN channel.

*In response to Berrou et al.'s discovery of the TC, research was supported by the European Space Agency (ESA) and the multi-rate TC (two constituent code PCCC configurations) was recommended by the CCSDS as the channel coding standard for space telemetry.

8.12.7.1 Rate 1/3 Turbo Code Performance The CCSDS [126, 127], recommended TC has selectable code rates of 1/2, 1/3, 1/4, and 1/6 and uses two 16-state convolutional encoders. The interleaver lengths are also selectable among the following options: 1,784, 3,568, 7,136, 8,920, and 16,384. The encoder including the full complement of code rates is shown in Figure 8.57. The source bits are randomly assigned to the interleaver according to the selected interleaver algorithm. This code is a systematic convolutional encoder with one information-bit for each output symbol interval, so the delay registers D correspond to the bit interval T_b. The registers are initialized to zero and, following the last of L information bits, the code input *switches* are changed and four additional clock cycles are executed to terminate the trellis in the zero state. The four additional bits are transmitted and used in the decoder to terminate the decoding trellis in the zero state; consequently, $(L + 4)/r_c$ code bits are transmitted to recover the L information bits.

The generator for the turbo code is $G = (g_0, g_1, g_2, g_3)^T$ and subgenerators (g_i) are listed in Table 8.34 using octal and binary notations. The code consists of recursive generators. The rate 1/2 code is generated by alternately puncturing the connections indicated by the *open circles*, starting with the encoder b connection, that is, the output of generator g_{1b} is punctured first. The *filled circles* are fixed connections so there is no puncturing for the code rates 1/3, 1/4, and 1/6. The subgenerator g_{2b} (indicated by the x) is not used in encoder b. The bits in a code symbol are transmitted in the order of the connections (top to bottom) indicated by the *circles* in Figure 8.57. For example, the information bit is always transmitted first; so, for the rate 1/3 code, the bits associated with the generators (g_{1a}, g_{1b}) are transmitted following the information bit. For the punctured rate 1/2 code, the bit following the information bit in the first symbol is g_{1a}, and in the second symbol is g_{1b} with successive symbols alternating between g_{1a} and g_{1b}. The information and parity bits are sequentially applied to the BPSK waveform modulator and transmitted over the AWGN channel.

The Monte Carlo simulated performance shown in Figure 8.59 is based on the JPL spread interleaver; the length 2048 JPL spread interleaver data was provided by Divsalar† as a reference to validate the simulations. The performance of the JPL spread interleaver was also compared to the turbo and random interleavers, discussed in Section 8.12.1, and found to result in slightly improved bit-error performance. The Monte Carlo bit-error performance evaluations of the PCCC require considerably longer simulation runs compared to the performance evaluations of convolutional codes using the

†Private communications, October 1998.

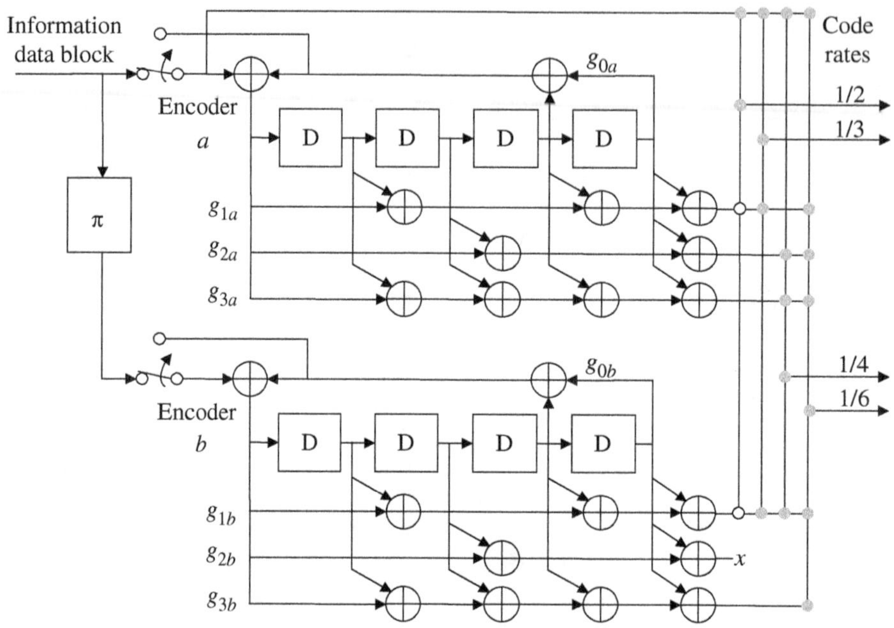

FIGURE 8.57 CCSDS selectable rate turbo code encoder. CCSDS Blue Book [126]. Reproduced by permission of the Consultative Committee for Space Data Systems (CCSDS).

TABLE 8.34 Turbo Code Subgenerators for Encoder in Figure 8.57

Subgenerators	Octal[a]	Binary	Connection
g_0	13	100110	Recursive
g_1	33	110110	Nonrecursive
g_2	52	101010	Nonrecursive
g_3	73	111110	Nonrecursive

[a]Octal notation with LSB on the left.

Viterbi decoding algorithm; this is not surprising in view of the much lower input signal-to-noise ratios and the steepness of the bit-error performance curve. For example, a change in the signal-to-noise ratio of one-tenth of a decibel can result in several orders of magnitude change in the bit-error performance. In this regard, the signal-to-noise ratio (γ_b) in the simulation is established using (8.115), such that, the peak signal level is

$$A = \sqrt{2\gamma_b \sigma_n^2} \tag{8.131}$$

where γ_b is measure in the bandwidth equal to the bit rate. Using (8.131) results in a zero-mean, unit-variance additive noise random variable. As a rule of thumb, the number of frames (N_{sim}) in the Monte Carlo simulations is given by

$$N_{sim} \geq \frac{1}{\rho^2 P_{fe}} \quad : \text{number of frames} \tag{8.132}$$

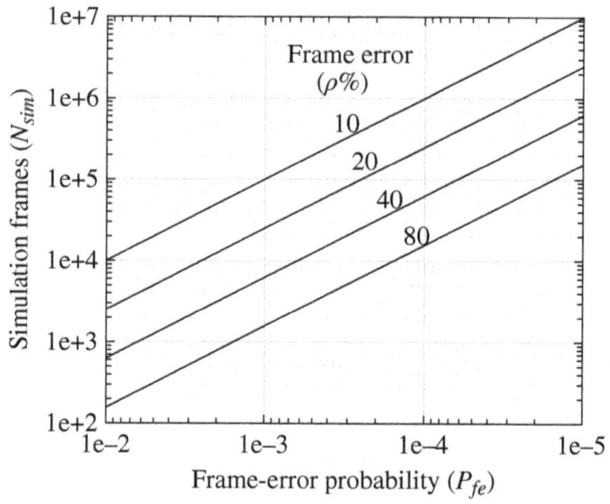

FIGURE 8.58 Rule of thumb for the number of turbo-like code simulation frames for each signal-to-noise ratio.

where ρ is the error in P_{fe}; Equation (8.132) is plotted in Figure 8.58 as a function of P_{fe} for various values of ρ in percent. Based on this rule, a frame-error probability of $P_{fe} = 10^{-4}$ with an 80% accuracy requires a Monte Carlo simulation of 20K frames or code blocks. This rule of thumb is somewhat pessimistic; however, it is useful in establishing a minimum upper limit while monitoring the frame-error performance to terminate the simulation when the performance improvement appears to saturate. Unless save guards are available to restart the code, it is time-consuming to underestimate the simulation run times. The following simulations

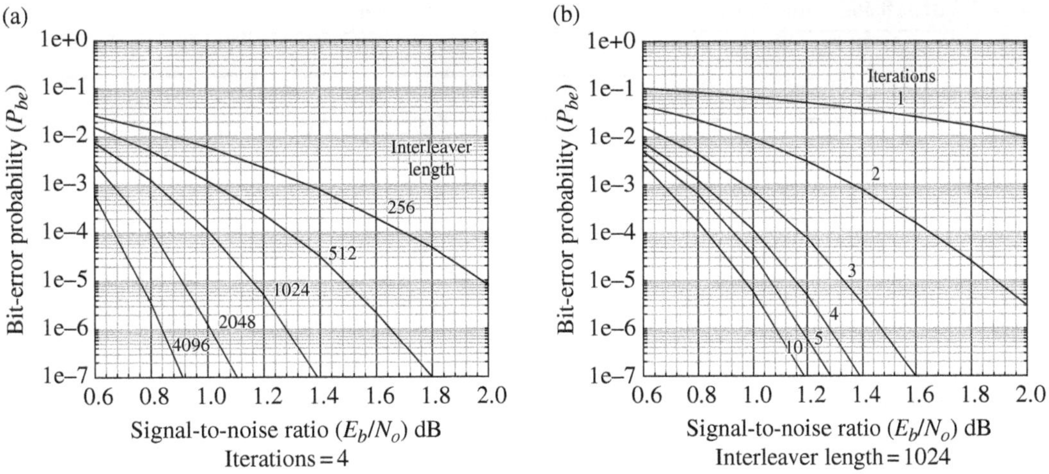

FIGURE 8.59 Rate 1/3 turbo code performance.

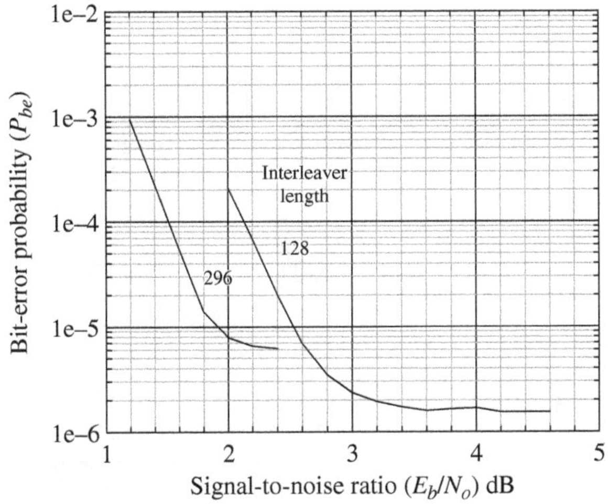

FIGURE 8.60 Rate 1/3 turbo code performance (short messages, iterations = 10).

for the 4096-bit interleaver used about 40M bits for each signal-to-noise ratio.

Figure 8.59a shows the performance using four iterations with interleaver lengths [128] ranging from 256 to 4096 bits. The performance gain, between $P_{be} = 10^{-5}$ and 10^{-6}, in doubling the interleaver length is about 0.1 dB. Figure 8.59b shows the performance for a fixed interleaver length of 1024 bits with iterations ranging from 1 to 10. The diminishing performance gain with additional iterations above 10 is evident in these plots. For example, at $P_{be} = 10^{-6}$ the performance gain between 5 and 10 iteration is about 0.09 dB.

An application involving short-message durations must use commensurately short interleavers, and Figure 8.60 shows the simulated performance using short-message blocks of 296 and 128 bits. The message bits included a

TABLE 8.35 Rate 2/4 TCM Constraint Length $\nu = 4$, 16-State Systematic Recursive Constituent Code Generators

Code	CC	h^2	h^1	h^0	Comments
A	CC1	33	35	31	16-state TCM encoders
	CC2	35	33	31	

CRC code to reduce the false message acceptance probability. The error floor between $P_{be} = 10^{-5}$ and 10^{-6} is evident in both codes and, although the 296-bit interleaver results in a significant coding gain improvement of about 0.6 dB, the error floor is considerably higher than that of the 128-bit code block. These results highlight the importance of selecting good interleavers that are sufficiently random to provide independent estimates of the a posteriori probabilities for an information symbol given the encoder constraint length and channel memory.

8.12.7.2 Comparison of PCCC and SCCC Performance The PCCC bit-error performance examined in this section is based on the work of Benedetto et al. [100], Benedetto and Montorsi [116], and Benedetto et al. [121]. The performance evaluation involves the rate 1/2, 1/3, 1/4, and 1/15 PCCC implementations corresponding to iterations of 18, 11, 13, and 12 and a common a data block length of 16K bits with 16-state encoders. The 2/4 rate code is unique, in that it represents the CCs used in the generation of a trellis-coded modulation (TCM) waveform. For a discussion of the design, implementation, and performance of TCM, refer to Sections 9.4 and 9.5. With TCM the parity-check redundancy is achieved through additional modulation states so there is no bandwidth expansion. The generator polynomials for the rate 2/4 TCM CCs (denoted as Code A) are listed in Table 8.35.

TABLE 8.36 Constituent Code Generators for SCCCs B–E

Code	Rate	Generators[a] CC1	CC2	CC3
B	1/2	$\frac{2}{3}$	$\frac{32}{13}$	
C	1/3	$1\ \frac{33}{13}$	$0\ \frac{33}{13}$	
D	1/4	$1\ \frac{52}{13}\ \frac{73}{13}$	$0\ \frac{33}{13}$	
E[b]	1/15	$1\ \frac{33}{13}\ \frac{52}{13}\ \frac{73}{13}\ \frac{72}{13}$	$1\ \frac{33}{13}\ \frac{52}{13}\ \frac{73}{13}\ \frac{72}{13}$	$1\ \frac{33}{13}\ \frac{52}{13}\ \frac{73}{13}\ \frac{72}{13}$

[a]Octal notation with the LSB on the left.
[b]These generators must be verified (an alternate CC3 parity generator polynomial to 72/13 is 12/13).

Referring to Section 9.4, the two input bits d_2 and d_1 of the rate 2/4 code are sequentially selected from the input source data block of N bits. These bits correspond to the systematic output bits $c_2 = d_2$ and $c_1 = d_1$ of the constituent code CC1. The two CCs are evaluated using

$$\begin{aligned} &CC1: \boldsymbol{c} = (c_2, c_1, c_0), \quad \boldsymbol{H} = (h^2, h^1, h^0) \\ &CC2: \boldsymbol{c} = (d_2', d_1', c_0'), \quad \boldsymbol{H} = (h^2, h^1, h^0) \end{aligned} \qquad (8.133)$$

where d_2' and d_1' are sequentially taken from the independently interleaved block of source data bits. The resulting rate 2/4 systematic encoder results in the four output bits d_2, d_1, c, and c_0' that are mapped to the selected waveform symbol(s) resulting in the rate 1/2 encoder. The generators for the SCCCs denoted as B through E are listed in Table 8.36 using octal notations.

The SCCC implementations are based on the encoder and decoder configurations in Figure 8.61 with the CCs obtained from Table 8.33. The output connections of the inner SCCC CC in Figure 8.61b, indicated by the *open circles*, require puncturing in the order (a,b), starting with connection a and alternating between a and b at the coded symbol rate. The following comparison of the theory, implementation, and bit-error performance of these codes* is based on the work of Benedetto et al. [103, 119, 121, 127] and Dolinar et al. [129].

The performance of the PCCC implementations is shown in Figure 8.62. The performance comparison between the codes is based on comparable complexity in that the codes correspond to 16-state trellis decoders and the interleaver lengths are identical. Under the indicated conditions, the selection of the code rate is a major decision in achieving a specified signal-to-noise condition. However, the performance with longer interleaver block lengths will also result in lower E_b/N_o operating conditions. These simulation results do not show any P_{be} flaring that is customarily associated with a reduction of the slope in the P_{be} vs. E_b/N_o curve. With 16-state encoders, the flaring usually becomes evident for P_{be} below about 10^{-5} or 10^{-6} and results in $P_{be} = 10^{-15}$ at E_b/N_o values of 6 or 7 dB. These determinations are made using theoretical performance bounds based on the number of error events and the weight information sequence [116, 130]. It is noteworthy that SCCCs are less prone to severe flaring.

(a)

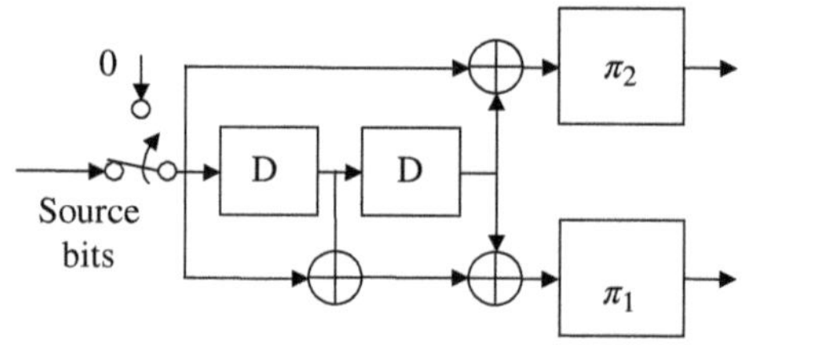

Rate 1/2, 4-state, nonrecursive nonsystematic outer code
(Table 8.33 rate 1/2 constituent code)

(b)

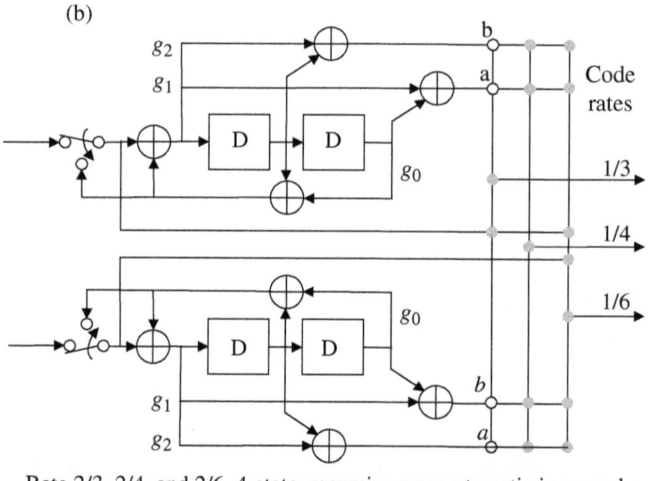

Rate 2/3, 2/4, and 2/6, 4-state, recursive nonsystematic inner code
(Table 8.33 rate 2/3 constituent codes)

FIGURE 8.61 Selectable rate SCCC encoder.

*The details of the (*a*,*b*) rate matching in Figure 8.61b are based on the author's interpretation and should be checked for compatibility.

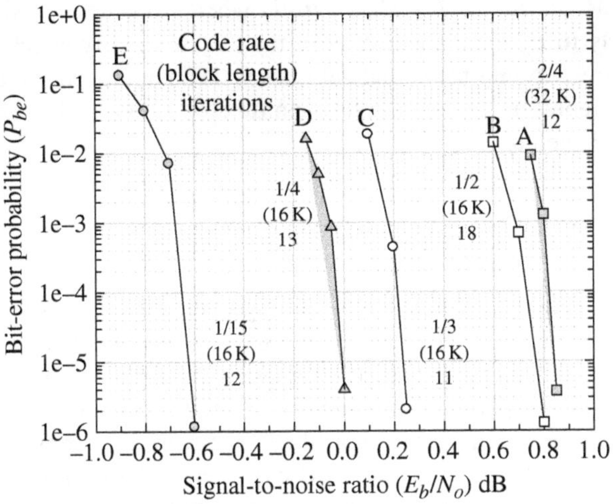

FIGURE 8.62 Simulated PCCC code performance. Benedetto et al. [121]. Reproduced by permission of Elsevier Books.

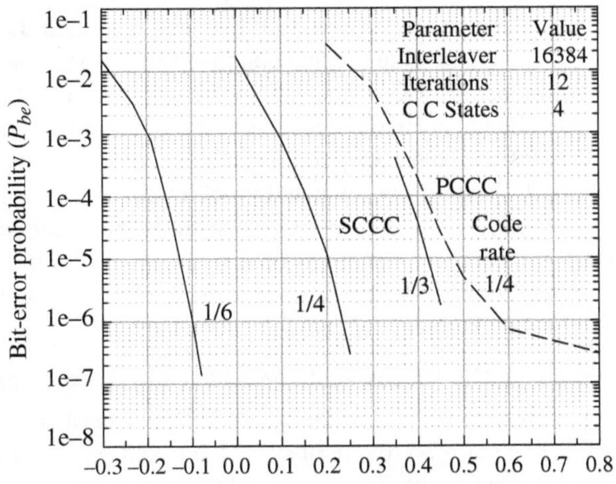

FIGURE 8.63 Theoretical SCCC performance comparison to rate 1/4 PCCC under identical conditions. Benedetto et al. [121]. Reproduced by permission of Elsevier Books.

The performance of the rate 2/4 (or 1/2) TCM waveform is somewhat limited to that shown in Figure 8.62 since bandwidth expansion is not an option and the inclusion of additional redundant modulation states has a rapidly diminishing advantage.

The performance of the SCCC implementations is shown in Figure 8.63. The performance assessments are based on the identical complexity of the codes in that they each have four-state trellis decoders with identical the interleaver lengths and iterations. An advantage of the SCCCs is that the SISO processing can run considerably faster than the PCCCs due to the 4 : 1 reduction in the number of states. The SCCC implementations do not show evidence of any P_{be} flaring, as seen with four-state PCCC performance. In comparing the rate 1/4, four-state PCCC with the rate 1/4, 16-state PCCC in Figure 8.62, there is a significant advantage in the coding gain using the 16-state encoder. Furthermore, even though there is no evidence of P_{be} flaring in the 16-state encoder, it is predicted (see Reference 130, figure 15) that flaring is on the verge of occurring at a signal-to-noise ratios $> \cong 0.5$ dB. In the performance critique of Figure 8.62, it was mentioned that the SCCCs are less prone to severe flaring; however, flaring occurs (see Reference 103, figure 5) with a rate 1/3, four-state, and 2K interleaver SCCC for P_{be} below about 10^{-8} at $E_b/N_o \cong 0$ dB and results in $P_{be} = 10^{-15}$ at $E_b/N_o \cong 8$ dB. For the PCCC and SCCC implementations the flaring conditions improve with increasing interleaver lengths.

8.13 LDPC CODE AND TPC

The LDPC codes were discovered in 1962 by Gallager [131] and gained in recognition and popularity [132] following the discovery of TCs by Berrou, Glavieux, and Thitimajshima in 1993. Costello and Forney [133] have an excellent paper describing the development of coding theory over this period and through 2007. LDPC codes are block codes, formed from a class of parity-check codes [134], and specified by a sparse matrix containing binary information-digits and parity-check digits denoted as *ones*; otherwise, the matrix elements are *zeros*. Using Gallager's designation (n, j, k), the coded matrix is partitioned into j submatrices, each with $k + I$ rows and $n = k(k + I)$ columns, where n is the code block length. Designating the submatrix rows as i, each row of the first submatrix are filled with k consecutive *ones*, such that, $(i-1)k, \ldots, ik$: $i = 1, \ldots, k + I$. Upon completion, each row of the first submatrix contains k consecutive *ones* and each column contains a single *one*. The remaining $j - 1$ submatrices are filled with random permutations of the *ones* in the first submatrix columns, while preserving the condition that the code matrix contains k *ones* in each row and j *ones* in each column. LDPC codes constructed in this manner are referred to as *regular* LDPC codes [135]; however, *irregular* LDPC codes [136, 137] are not restricted to having an equal number of *ones* in the code matrix rows and columns as described earlier and, thereby, result in superior performance by increasing the number of check nodes.

The k consecutive digits in the first submatrix correspond to a coded symbol interval consisting of $k - j$ information-bits and j parity-bits. The k-bit symbol is characterized by starting and ending edges much like the symbol edges defined in the trellis decoding of convolutional codes. With this understanding, the code rate of the LDPC code is defined as

TABLE 8.37 Comparison of δ_{jk} for LDPC Code (n, j, k) for Large n[a]

j	k	Rate r_c	δ_{jk}	δ^b
5	6	1/6	0.225	0.263
4	5	1/5	0.210	0.241
3	4	1/4	0.122	0.214
4	6	1/3	0.129	0.173
3	5	2/5	0.044	0.145
3	6	1/2	0.023	0.110

[a]Gallager [139]. Reproduced by permission of the IEEE.
[b]The same ratio of an ordinary parity-check code of the same rate.

$$r_c \triangleq \frac{k-j}{k} \tag{8.134}$$

The $j-1$ submatrices contain the known parity-bit information needed to determine and correct errors in the received LDPC coded data block. The low-density construction of the code matrix, for large n, results in a minimum distance between nearly all codes of [138]

$$d_{min} = n\delta_{jk} \tag{8.135}$$

where δ_{jk} is a constant listed in Table 8.37 for several LDPC codes. The minimum distance increases linearly with the code length.

Gallager [131] has also shown that maximum-likelihood decoding of LDPC codes over a BSC, corresponding to bit-by-bit decoding, results in a decoding error probability that decreases exponentially with the block length. However, to take advantage of the performance improvement using iterative decoding before making a *hard-bit* decision requires processing of the APP as discussed in Section 8.12.4. The decoding algorithm is formulated in terms of the log-likelihood ratio, and Gallager's solution is expressed as

$$\begin{aligned}\mathrm{LLR} &= \ln\left(\frac{P(x_d=0|\{y\},S)}{P(x_d=1|\{y\},S)}\right)\\ &= \ln\left(\frac{1-P_d}{P_d}\right) + \sum_{i=1}^{j}\left\{\left(\prod_{\ell=1}^{k-1}\alpha_{i\ell}\right) f\left(\sum_{\ell=1}^{k-1} f(\beta_{i\ell})\right)\right\}\end{aligned} \tag{8.136}$$

where x_d is the received bit in position d of the received sequence $\{y\}$, S is the event that x_d satisfies the j parity-check equations associated with position d, P_d is the probability that the transmitted digit x_d is 1 conditioned on the received bit in position d with the log-likelihood ratio $\mathrm{LLR}_d = \ln[(1-P_d)/P_d]$.* Defining the log-likelihood ratio $\mathrm{LLR}_{i\ell} = \ln[(1-P_{i\ell})/P_{i\ell}] = \alpha_{i\ell}\beta_{i\ell}$ where $P_{i\ell}$ is defined similarly to P_d for the ℓ'th digit in the i'th first parity-check set corresponding to second submatrix. The parameters $\alpha_{i\ell}$ and $\beta_{i\ell}$ represent the sign and magnitude of $\mathrm{LLR}_{i\ell}$ and $f(\beta_{i\ell})$ is evaluated as

$$f(\beta_{i\ell}) = \ln\frac{e^{\beta_{i\ell}}+1}{e^{\beta_{i\ell}}-1} \tag{8.137}$$

Using (8.137), LLR_d, and the sign of $\alpha_{i\ell}$, the evaluation of (8.136) is fairly straightforward and can be computed either serially or in parallel for each input bit. Following the evaluation, if a parity error is detected at x_n, the decoder corrects the error based on the sign of (8.136) with the correspondence: $\mathrm{LLR} \geq 0 : x_n = 0$ and $\mathrm{LLR} < 0 : x_n = 1$. These corrections are based on hard decisions and apply to the received information and parity bits.† Using the corrected bits, the process is repeated until a valid code word is detected or after a fixed number of iterations. As an example of the performance, the bit-error probability of a high code rate IEEE 802.16e-compliant LDPC code [140, 141] is shown in Figure 8.64a.

Prior to the acceptance of LDPC codes, PCs and TPCs [143–146] were in wide use and continue to provide competitive solutions for FEC coding. PCs were introduced by Elias in 1954. The PC is characterized as a serial concatenation of block codes with data permutations and row–column parity-check bits. For two concatenated block codes, the designations (n_i, k_i, d_{mi}): $i = 1, 2$, correspond to n_i the length of the code word, k_i the number of information bits, and d_{mi} the minimum Hamming distance. With equal parameters, the PC is denoted as‡ $(n, k, d_m)^2$ where, in general, $n = \prod_i n_i, k = \prod_i k_i$, and $d_m = \prod_i d_{mi}$. The PC row–column matrix (n_1, n_2) is formed using submatrix (k_1, k_2) in the upper-left portion of (n_1, n_2) and forming $(n_2 - k_2)$ row parity-check bits and $(n_1 - k_1)$ column parity-check bits. The resulting PC rate is the product of the individual codes rates. The optimum decoder uses soft decision LLR processing of the rows and columns with iterative decoding to improve the performance. TPC bit-error rates on the order of 10^{-11} can be achieved without experiencing the error floor associated with some PCCC designs. The performance of a high code rate IEEE 802.16-compliant TPC [147] is shown in Figure 8.64b.

LDPC code and TPC are processing efficient high-rate codes with performance approaching within a few tenths of a decibel of the channel capacity. On the other hand, Chung et al. [148] demonstrated that a rate 1/2 irregular LDPC code with block length of 10^7 approaches the channel capacity for binary codes to within 0.04 dB achieving $P_{be} = 10^{-6}$ at

*With equal a priori binary probabilities, as is usually the case, $P_d = 1/2$ and $\mathrm{LLR}_d = 0$.

†The APP processing is based on the noisy soft-inputs $y_n = f(A)d_n + n_n$ from the demodulator matched filter, where A is the peak amplitude of the received signal. Referring to (8.115) or (8.117), the bipolar data $d_n = \{-1,1\}$ is related to x_n as $d_n = 2x_n - 1$.

‡The notation for the PC often omits the d_m designation.

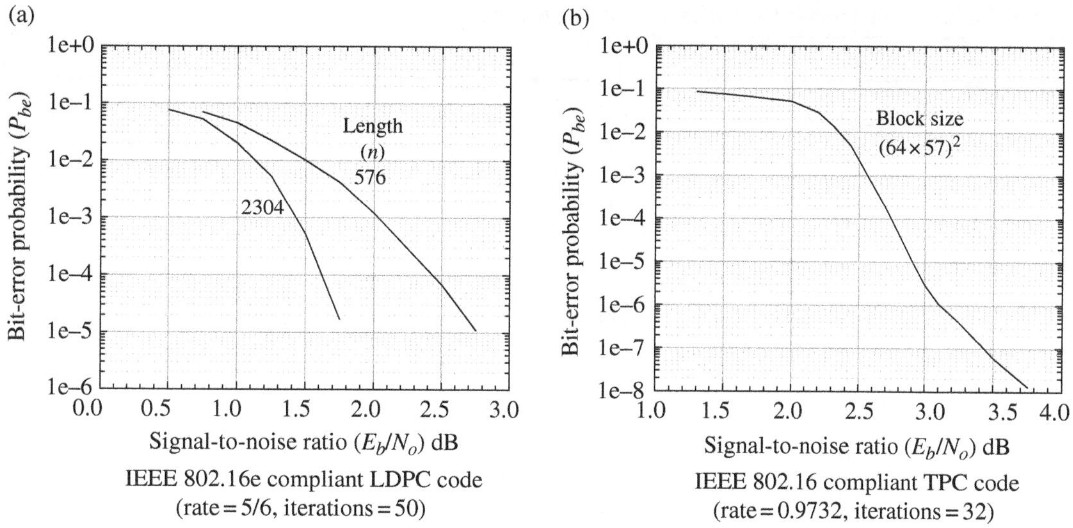

FIGURE 8.64 Performance of LDPC and TPC codes. Gracie et al. [142]. Reproduced by permission of the IEEE.

$E_b/N_o = 0.19$ dB. The long interleaver lengths required for low-rate LDPC code and TPC are restrictive for applications requiring low message latency. However, with parallel processing the decoding speeds can accommodate tens of Mbps data rates. These codes are also better suited for short messages, on the order of 256 bits or less, than concatenated convolution codes with performance limited by the noise floor.

The preceding description of LDPC coding is based on Gallager's original publication on low-density parity-check encoding and decoding and significant advances have been made since the mid-1990s when his work was recognized as providing competitive solutions to the need for low E_b/N_o, nearly error-free, communication systems. The advances in high-speed, high-memory, and low-power signal processing hardware were also a major motivation. In this regard, the vast amount of current literature on FEC coding, including the LDPC code and TPC should be consulted for the latest innovations, for example, the special issue of the IEEE Proceedings [149] contains a wealth of information, including many references, on a variety of FEC coding techniques, designs, and algorithms. TC synchronization [150, 151] is a major consideration in the system design and synchronization requirements for various standard FEC codes are detailed in the CCSDS specifications [152]. The impact on the performance through channels with memory [153] is another important consideration.

8.14 BOSE-CHAUDHURI-HOCQUENGHEM CODES

The binary BCH codes [34, 37] are block codes of length N containing K information bits with an error correction capability of t bit or symbol errors; in the following discussions, the binary BCH codes are denoted as (N, K, t); on occasions, the notation $(N, K, t)_m$ is used where m determines the number of bits or symbols in the code block given by $N = 2^m - 1$. Binary BCH codes are generated over the GF(2^m) with the binary polynomial coefficients in GF(2). The code polynomials are related to the parameters N, K, t, and m as and the number of parity bits satisfies the condition $N - K \le mt$.* The BCH codes will correct all combinations of t errors for $t \le (d-1)/2$ where d is the minimum distance of the code. These parameters are listed in Table 8.38 [154] for binary *BCH* codes generated over GF(2^m) for $3 \le m \le 10$ corresponding to codes of length 7–1023. The parameter $t_{bin} = t$ is used to emphasize the *binary* BCH code error correction capability. However, the parameters N and K also apply to the nonbinary *BCH* RS codes with 2^m-ary symbols generated over GF(2^m). The nonbinary RS code will correct any combination of $t_{rs} = t = \lfloor (N-K)/2 \rfloor$ m-bit symbol-errors or less; the fourth column in Table 8.38 tabulates t_{rs} for the RS code.

A more extensive list of binary cyclic codes of odd length is given by Peterson and Weldon [155] in terms of the parameters N, K, d, d_{BCH} and the roots of the generating polynomials. In this case, d is the true minimum distance of the code and d_{BCH} the minimum distance guaranteed by the BCH bound. The extensive table of results is accredited to Chen [156] and applies to all binary cyclic codes of odd length or degrees ≤65. The results in Table 8.38 represent the codes of a given length for which $d$xmax(d_{BCH}).

Section 8.14.1 focuses on binary BCH codes with the intent of introducing some basic concepts for examining the more powerful error-correction capabilities of nonbinary, or q-ary BCH, codes where the q-ary symbols are generated

*In the following sections, the parameters N and K are also denoted by lowercase n and k.

TABLE 8.38 *BCH* Code Parameters Generated by Primitive Elements[a,b]

N	K	t_{bin}	t_{rs}
7	4	1	1
15	11	1	2
	7	2	4
	5	3	5
31	26	1	2
	21	2	5
	16	3	7
	11	5	10
	6	7	12
63	57	1	3
	51	2	6
	45	3	9
	39	4	12
	36	5	13
	30	6	16
	24	7	19
	18	10	22
	16	11	23
	10	13	26
	7	15	28
127	120	1	3
	113	2	7
	106	3	10
	99	4	14
	92	5	17
	85	6	21
	78	7	24
	71	9	28
	64	10	31
	57	11	35
	50	13	38
	43	14	42
	36	15	45
	29	21	49
	22	23	52
	15	27	56
	8	31	59
255	247	1	4
	239	2	8
	231	3	12
	233	4	11
	215	5	20
	207	6	24
	199	7	28
	191	8	32
	187	9	34
	179	10	38
	171	11	42
255	163	12	46
	155	13	50
	147	14	54
	139	15	58
	131	18	62
	123	19	66
	115	21	70
	107	22	74
	99	23	78
	91	25	82
	87	26	84
	79	27	88
	71	29	92
	63	30	96
	55	31	100
	47	42	104
	45	43	105
	37	45	109
	29	47	113
	21	55	117
	13	59	121
	9	63	123
511	502	1	4
	493	2	9
	484	3	13
	475	4	18
	466	5	22
	457	6	27
	448	7	31
	439	8	36
	430	9	40
	421	10	45
	412	11	48
	403	12	54
	394	13	58
	385	14	63
	376	15	67
	367	16	72
	358	18	76
	349	19	81
	340	20	85
	331	21	90
	322	22	94
	313	23	99
	304	25	103
	295	26	108
	286	27	112
511	277	28	117
	268	29	121
	259	30	126
	250	31	130
	241	36	135
	238	37	136
	229	38	141
	220	39	145
	211	41	150
	202	42	154
	193	43	159
	184	45	163
	175	46	168
	166	47	172
	157	51	177
	148	53	181
	139	54	186
	130	55	190
	121	58	195
	112	59	199
	103	61	204
	94	62	208
	85	63	213
	76	85	217
	67	87	222
	58	91	226
	49	93	231
	40	95	235
	31	109	240
	28	111	241
	19	119	246
	10	127	250
1023	1013	1	5
	1003	2	10
	993	3	15
	983	4	20
	973	5	25
	963	6	30
	953	7	35
	943	8	40
	933	9	45
	923	10	50
	913	11	55
	903	12	60
	893	13	65
	883	14	70
	873	15	75

TABLE 8.38 *(continued)*

N	K	t_{bin}	t_{rs}	K	t_{bin}	t_{rs}	K	t_{bin}	t_{rs}
1023	863	16	80	573	50	225	278	102	372
	858	17	82	563	51	230	268	103	377
	848	18	87	553	52	235	258	106	382
	838	19	92	543	53	240	248	107	387
	828	20	97	533	54	245	238	109	392
	818	21	102	523	55	250	228	110	397
	808	22	107	513	57	255	218	111	402
	798	23	112	503	58	260	208	115	407
	788	24	117	493	59	265	203	117	410
	778	25	122	483	60	270	193	118	415
	768	26	127	473	61	275	183	119	420
	758	27	132	463	62	280	173	122	425
	748	28	137	453	63	285	163	123	430
	738	29	142	443	73	290	153	125	435
	728	30	147	433	74	295	143	126	440
	718	31	152	423	75	300	133	127	445
	708	34	157	413	77	305	123	170	450
	698	35	162	403	78	310	121	171	451
	688	36	167	393	79	315	111	173	456
	678	37	172	383	82	320	101	175	461
	668	38	177	378	83	322	91	181	466
	658	39	182	368	85	327	86	183	468
	648	41	187	358	86	332	76	187	473
	638	42	192	348	87	337	66	189	478
	628	43	197	338	89	342	56	191	483
	618	44	202	328	90	347	46	219	488
	608	45	207	318	91	352	36	223	493
	598	46	212	308	93	357	26	239	498
	588	47	217	298	94	362	16	147	503
	578	49	222	288	95	367	11	255	506

[a]The shaded parameters result in approximate code rates of 1/2 for the given N.
[b]Peterson and Weldon [154]. Reproduced by permission of The MIT Press.

over the GF(q). In particular, the nonbinary case is specialized to the RS code for which $q = 2^m$.

8.14.1 Binary *BCH* Codes

This section briefly reviews binary BCH codes as an introduction to RS codes discussed in Section 8.14.2. To this end, the following introduces the GF(2) involving binary data $d_i \in \{0,1\}$. The set of elements α forms a field [157] with respect to addition and multiplication, if the set is closed with respect to addition and multiplication, the distributive law applies, and it contains a zero and unit element. The key to the generation of the field elements over GF(2) is that the generator, or primitive polynomial $p(x)$, has binary coefficients* $\{0,1\}$. The field element for which $p(\alpha) = 0$ is referred to as the primitive element. Furthermore, the primitive polynomials for which $p(\alpha) = 0$ are irreducible polynomials, that is, $p(x)$ of degree m are not divisible by any other polynomial of degree m': $0 < m' < m$; this ensures that the field is closed.

Construction of the elements over GF(2) is best described by the following example using $m = 3$ and the known primitive polynomial $p(x) = x^3 + x + 1$. Using the primitive polynomial, all of the distinct 2^3 symbols of the code are generated in terms of the primitive element α as shown in Table 8.39.

In the second column of Table 8.39, the primitive polynomial provides the key equality $p(\alpha) = \alpha^3 + \alpha + 1 = 0$ from which $\alpha^3 = \alpha + 1$ is determined. Furthermore, addition and subtraction are indistinguishable, multiplication is performed by adding the exponents, and the field is closed, that is, $\alpha^7 = 1$. The primitive element of the field is the unit element so the set of elements $0, 1, \alpha, \alpha^2, \ldots, \alpha^{2^m-2}$ form the 2^m unique elements of the field. As mentioned earlier, the operations of addition and subtraction are the same in GF(2) with the special case $\alpha^n \pm \alpha^n = 0$: $0 \le n \le m - 1$; these properties are

*The primitive polynomials for nonbinary BCH codes are identical; however, the coefficients are q-ary symbols or words. For RS codes, the coefficients are $q = 2^m$-ary symbols.

TABLE 8.39 Construction of the GF(2^3) Using $p(x) = x^3 + x + 1$ [$f(x) = b_2x^2 + b_1x + b_0$]

Elements	Reduction over Coefficient Field GF(2)	$b_2\ b_1\ b_0$	Integer
0		0 0 0	0
1		0 0 1	1
α		0 1 0	2
α^2		1 0 0	4
α^3	$\alpha + 1$	0 1 1	3
α^4	$\alpha\alpha^3 = \alpha(\alpha+1) = \alpha^2 + \alpha$	1 1 0	6
α^5	$\alpha\alpha^4 = \alpha(\alpha^2+\alpha) = \alpha^2+\alpha+1$	1 1 1	7
α^6	$\alpha\alpha^5 = \alpha(\alpha^2+\alpha+1) = \alpha^2+1$	1 0 1	5
α^7	$\alpha\alpha^6 = \alpha(\alpha^2+1) = 1$	0 0 1	1

TABLE 8.40 Primitive Polynomials for GF(2^m) for m = 2–10[a]

m	$p(x)$
2	x^2+x+1
3	x^3+x+1
4	x^4+x+1
5	x^5+x^2+1
6	x^6+x+1
7	x^7+x^3+1
8	$x^8+x^4+x^3+x^2+1$
9	x^9+x^4+1
10	$x^{10}+x^3+1$

[a]Partial list from Reference 155.

a direct result of the modulo-2 addition of the binary coefficients b_n. Multiplication is performed simply by adding the exponents as $\alpha^n\alpha^{n'} = \alpha^{n+n'}$.

Primitive polynomials are difficult to determine; however, Table 8.40 provides a list, for $m = 3$ through 10; these polynomials are also applicable to 8-ary through 1024-ary RS codes discussed in Section 8.14.2. More extensive tables of primitive polynomials are provided by Peterson and Weldon [155], from which Table 8.40 is compiled, and by Lin [37], Peterson [58], and Marsh [158].

Binary BCH codes are defined for any positive integers m and t with the following properties:

- Block length $n = 2^m - 1$
- Parity-check bits $n - k \le mt$
- Minimum distance $d \ge 2t + 1$

The binary BCH code is denoted as a (n, k, t) t-error correcting code, where t or less errors can be corrected in any block of n bits, where $(n-k)/m \le t \le (d-1)/2$. There is no easy way to evaluate the value $n - k$ that satisfies the equality $mt = (n - k)$; however, the values of t are tabulated in the third column of Table 8.38 as t_{bin}.

The generator polynomial of a, t-error-correcting, binary BCH code is defined as the least common multiple (LCM) of the minimum polynomials* $m_i(x)$: $i = 1, 3, 5, \ldots, 2t-1$, expressed as

$$g(x) = \text{LCM}(m_1(x), m_3(x), \ldots, m_{2t-1}(x)) \tag{8.138}$$

Comparing (8.151), the expression for computing the generator polynomials for nonbinary RS codes, with (8.138), it appears evident that determining the generator polynomials for binary BCH codes is considerably more difficult. The details for determining the minimum polynomials for the primitive element in GF(2) and the generator polynomial $g(x)$ are discussed by Lin [159], Lin and Costello [160], and Peterson and Walden [34]. Peterson and Walden (Appendix C) provide tables of irreducible polynomials over GF(2) in which the minimum polynomials are identified. Michelson and Levesque [63] provide succinct descriptions of these topics with an explanation of the method for determination of the minimum polynomials from Peterson's table. The generator polynomials for primitive binary BCH codes, tabulated by Stenbit [161], are listed in Appendix 8B.

Referring to Appendix 8B, the generator for the length n = 15, t = 2 error-correcting (15,7,2) binary BCH code, in octal notation, is (721) and, considering the binary coefficients $g_i = \{1,0\}$, the generator is computed as

$$\begin{aligned} g(x) &= g_8x^8 + g_7x^7 + g_6x^6 + g_4x^4 + 1 \\ &= x^8 + x^7 + x^6 + x^4 + 1 \end{aligned} \tag{8.139}$$

The generator is implemented as shown in Figure 8.66 for the RS code; however, the generator weights g_i corresponding to the coefficients of (8.139) are equal to binary one.

The decoding involves the following three principal decoding computations:

1. Error syndrome calculation
2. Error location polynomial calculations
3. Error location calculation

The computation of the error values, required for the RS code, is not required because the error correction simply involves changing the binary coefficient at the calculated error location. There are several techniques, including the Berlekamp algorithm, that are used in the decoding computations for the binary BCH codes; these techniques are discussed by several authors [37, 58, 160, 162–165]. The following sections focus on the decoding of the binary BCH codes with the major emphasis on the decoding of M-ary RS codes using the Berlekamp algorithm. The procedures for finding and correcting the errors in the binary BCH code are identified by

*For an element $\alpha \in$ GF(2^m), the minimum polynomial $m(x)$, with binary coefficients, has the smallest degree such that $m(\alpha) = 0$.

TABLE 8.41 Binary BCH Iterative Algorithm for Determining the Error Location Polynomial[a]

υ	$\sigma^{\upsilon}(x)$	d_{υ}	ℓ_{υ}	$2\upsilon-\ell_{\upsilon}$
−1/2	1	1	0	−1
0	1	S_1	0	0
1				
…				
t				

[a]Lin [166]. Reproduced by permission of Pearson Education, Inc., New York, NY.

evaluating the t rows of Table 8.41 as described by Lin [166]. The table is initiated as shown and the rows are completed iteratively, as described by the following algorithm, in the process of determining the error location polynomial.

Table 8.41 is filled out using the following algorithm. Given the previous row υ, the $\upsilon+1$ row is determined as

```
if (d_υ = 0) then
  σ^{υ+1}(x) = σ^υ(x)
else
  Find another row υ' < υ, such that the
  last column 2υ' − ℓ_υ'
  is the largest value with d_υ' ≠ 0, and
  then compute:
  σ^{υ+1}(x) = σ^υ(x) + d_υ d_υ'^{-1} x^{2(υ−υ')} σ^{υ'}(x)
end if
  ℓ_{υ+1} = 2(υ−υ') + υ': the degree of σ^{υ+1}(x) and
  d_{υ+1} = S_{2υ+3} + σ_1^{υ+1} S_{2υ+2} + σ_2^{υ+1} S_{2υ+1} + ··· + σ_{ℓ_{υ+1}}^{υ+1}
  S_{2υ+3−ℓ_{υ+1}}
  where the σ_i^{υ+1} are the coefficients of
  the polynomial
  σ^{υ+1}(x) = 1 + σ_1^{υ+1} x + σ_2^{υ+1} x^2 + ··· + σ_{ℓ_{υ+1}}^{υ+1} x^{ℓ_{υ+1}}
```

The polynomial $\sigma^t(x)$ in the last row of Table 8.41 is the error location polynomial; however, if it has degree greater than t, the errors cannot be located and corrected.

8.14.2 RS Codes

The RS code is a subclass of nonbinary *BCH* codes that have excellent burst- and random error correcting properties. In general, the M-ary RS code has a code or block length of $N = M-1$ symbols where each symbol is composed of an M-ary symbol alphabet. Referring to the number of information symbols as $K < N$, the M-RS code is designated as a (N, K) code. The code rate is defined as $r_c = K/N$. The error-free correction capability of the RS code is $t = (N-K)/2$ symbols, that is, any combination of symbol errors equal to or less than one-half the number of parity symbols can be corrected. In the following discussions, the notation (N, K, t) is used to explicitly identify the error-correction capability being considered. The minimum distance of the RS code is $d_{min} = 2t+1$. In the event that the decoder fails to correct the received symbol errors, additional processing can be performed by using symbol erasures to correct up to $t' = (2t-e)/2$ symbol errors where e is the number of symbols that are erased. The decision as to which symbols to erase must be made on the basis of a maximum-likelihood rule at the symbol detection processor output; noisy threshold margins will influence the erasure selections. The code symbols most commonly used in the implementation of the RS code are taken from the GF(2^m) where $M = 2^m$. The notation $(N, K, t)_m$ is used to identifying the underlying code structure. This is a particularly useful notation when dealing with shortened codes with $N < M-1$ that may lead to confusion as to the applicable GF. Several references are given in the following sections in which the details of GF arithmetic can be found. The RS code characteristics are summarized in Table 8.42.

TABLE 8.42 Summary of M-ary Reed–Solomon Code Characteristics

Parameter	Symbols	Bits
Block length (N)	$M-1$	$m(M-1)$
Bits/symbol (m)	—	$\log_2(M)$
Information	K	mK
Parity	$N-K$	$m(N-K)$
Error correction		
Without erasures (t)	$(N-K)/2$	$m(N-K)/2$
With erasures (t')	$(2t-e)/2$	$m(2t-e)/2$
Min. distance (d_{min})	$2t+1$	—

In the following sections, the message format and encoding requirements are reviewed for a general M-ary RS encoder. Of particular interest is the development of the encoder vector notation from the transmitted sequence. The vector notation represents the encoded message as a polynomial upon which the decoding algorithms are performed and is not an implicit function of time. Also the concept of shortened codes is introduced for which a specific example of a (24,12) code is given. The following discussion focuses on the processing requirements without the use of erasures; the processing requirement with erasures is discussed in Section 8.14.2.3. In either case, with or without erasures, the four principal decoding computations involve:

1. Error syndrome calculation
2. Error location polynomial calculations
3. Error location calculation
4. Error value calculations

The concluding sections outline a computer simulation code for all of the required processing for both the encoder and decoder. This chapter concludes with a discussion of several theoretical and simulated performance results. The performance is characterized in terms of the bit-error probability dependence on the signal-to-noise ratio as measured in the symbol and information-bit bandwidths.

8.14.2.1 ***Encoder Processing*** This section relates the source data information or message sequence to the sequence of encoded symbols including the parity symbols. The parity symbols are appended to the message to form a systematic M-ary RS coded sequence. The relationship between the time-dependent transmitted sequence and the notion of a code vector generated from elements of the GF(2^m) is discussed. Also, the generation of shortened forms of the underlying or parent (N, K) code is discussed.

Binary data comprising the source message is collected in a block of mK bits and taken m at a time to form K message symbols. Each symbol forms a 2^m-ary vector that is represented as an element in the GF having a primitive polynomial $p(\alpha)$ where α is called the "primitive element." Therefore, each nonzero message symbol can be represented as a power of α such that

$$\alpha^n = \text{element of GF}(2^m) \tag{8.140}$$

where $n = i$ modulo-($2^m - 1$) and i is any integer. The complete set consists of the 2^m elements $0, 1, \ldots, \alpha, \alpha^{2^m-2}$.

Most of the simulation examples provided in the following sections are based on the 64-ary RS code generated over GF(2^6) and the corresponding set of elements are listed in Table 8.43. These elements form the basis upon which the encoding and decoding operations are performed and serve to illustrate the algebraic properties of GFs. The nonzero elements of GF(2^6) are generated from the primitive polynomial

$$p(x) = x^6 + x + 1 \tag{8.141}$$

with $p(\alpha) = 0$. The binary values are assigned, from left to right, as $(b_0, b_1, b_2, b_3, b_4, b_5)$ where b_0 is the LSB. The LSB is taken first-in-time from the data source, or parity generator, and is transmitted first-in-time. The relationship between the binary bits and the primitive element α for a typical symbol is

$$\text{Encoded symbol} = \sum_{i=0}^{m-1} b_i \alpha^i \tag{8.142}$$

Because of this assignment of the LSB, it is necessary in the simulation to reverse the binary symbol to accommodate the vector processing as defined throughout the literature and used in the following sections.

In completing Table 8.43, it is noted that the first $m-1$ powers of α are simply evaluated as $\alpha^i = 2^i$, $i = 0, 1, \ldots, m-1$ and α^m is evaluated directly from the solution of the primitive polynomial as $\alpha^m = p(\alpha) - \alpha^m$. Subsequent powers are determined by repeated multiplication by α and subsequent reduction using the first m powers previously evaluated.

TABLE 8.43 Binary and Integer Equivalents of Field Elements in GF(2^6) [$f(x) = b_5x^5 + b_4x^4 + b_3x^3 + b_2x^2 + b_1x + b_0$]

Symbol	Binary	Integer	Symbol	Binary	Integer
0	000000	0	α^{31}	100101	37
α^0	000001	1	α^{32}	001001	9
α^1	000010	2	α^{33}	010010	18
α^2	000100	4	α^{34}	100100	36
α^3	001000	8	α^{35}	001011	11
α^4	010000	16	α^{36}	010110	22
α^5	100000	32	α^{37}	101100	44
α^6	000011	3	α^{38}	011011	27
α^7	000110	6	α^{39}	110110	54
α^8	001100	12	α^{40}	101111	47
α^9	011000	24	α^{41}	011101	29
α^{10}	110000	48	α^{42}	111010	58
α^{11}	100011	35	α^{43}	110111	55
α^{12}	000101	5	α^{44}	101101	45
α^{13}	001010	10	α^{45}	011001	25
α^{14}	010100	20	α^{46}	110010	50
α^{15}	101000	40	α^{47}	100111	39
α^{16}	010011	19	α^{48}	001101	13
α^{17}	100110	38	α^{49}	011010	26
α^{18}	001111	15	α^{50}	110100	52
α^{19}	011110	30	α^{51}	101011	43
α^{20}	111100	60	α^{52}	010101	21
α^{21}	111011	59	α^{53}	101010	42
α^{22}	110101	53	α^{54}	010111	23
α^{23}	101001	41	α^{55}	101110	46
α^{24}	010001	17	α^{56}	011111	31
α^{25}	100010	34	α^{57}	111110	62
α^{26}	000111	7	α^{58}	111111	63
α^{27}	001110	14	α^{59}	111101	61
α^{28}	011100	28	α^{60}	111001	57
α^{29}	111000	56	α^{61}	110001	49
α^{30}	110011	51	α^{62}	100001	33

Multiplication (division) over GF(2^m) is performed simply by adding (subtracting), modulo-($2^m - 1$), the exponents of α, that is, $\alpha^5\alpha^{59} = \alpha^1$ and $\alpha^5\alpha^{58} = \alpha^0 = 1$. Addition over GF($2^m$) is performed by forming the binary representations of the terms being added and summing the corresponding coefficients modulo-2; this is the same as performing an *exclusive-or* on the corresponding binary elements. For example, referring to Table 8.43 and (8.140), $\alpha^5 + \alpha^{59} = (\alpha^5) + (1 + \alpha^2 + \alpha^3 + \alpha^4 + \alpha^5) = 1 + \alpha^2 + \alpha^3 + \alpha^4 = \alpha^{41}$, noting that, $\alpha^5 + \alpha^5 = 0$.

Figure 8.65 depicts an arbitrary message symbol from a data source with each symbol composed of m source data bits. The collection of K source symbols forms the RS code block duration.

The message sequence is described as

$$m(t) = \sum_{n=0}^{K-1} m_{K-1-n} d(t - nT) \tag{8.143}$$

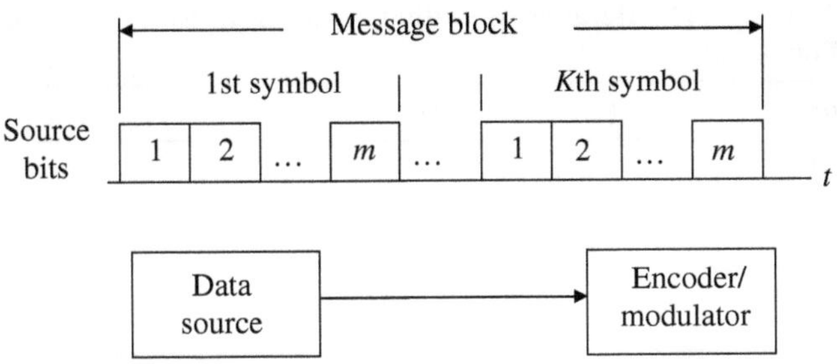

FIGURE 8.65 Relationship between source message bits, symbols, and code block.

and the parity sequence is described as

$$p(t)=\sum_{n=0}^{N-K-1} p_{N-K-1-n}d(t-nT) \tag{8.144}$$

where

$$d(t-nT)=rect_T(t,n-0.5) \tag{8.145}$$

and

$$rect_T(t,n-0.5)=\begin{cases}1 & :nT\le t\le (n+1)T\\ 0 & :o.w.\end{cases} \tag{8.146}$$

For the systematic-code format, the parity is simply appended to the message sequence and results in the transmitted code sequence expressed as

$$r'(t)=m(t)+p(t-KT) \tag{8.147}$$

It is convenient to represent the transmitted sequence as an N-dimensional code vector by making the substitution* $x^{-n}=d(t-nT)$ which gives

$$r'(X)=\sum_{n=0}^{K-1} m_{K-1-n}X^{-n}+\sum_{n=0}^{N-K-1} p_{N-K-1-n}X^{-n} \tag{8.148}$$

Applying the common coefficient notation

$$\alpha_{N-n}=\begin{cases}m_{K-n} & :n=1,\ldots,K\\ p_{N-n} & :n=K+1,\ldots,N\end{cases} \tag{8.149}$$

to each summation in (8.148) and multiplying both sides by X^{N-1} result in the simplified expression

*This substitution is similar to the z-transform of the unit delay defined as $u(t-nT)\leftrightarrow z^{-n}$.

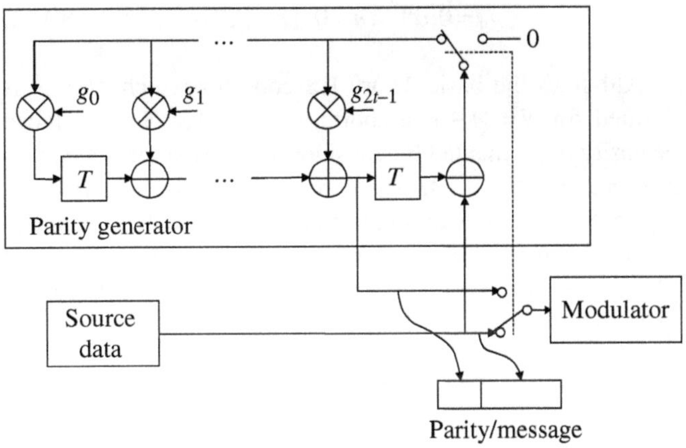

FIGURE 8.66 Reed–Solomon encoder.

$$\begin{aligned}r(X)&=X^{N-1}r'(X)\\ &=\sum_{n=0}^{N-1}\alpha_n X^n\end{aligned} \tag{8.150}$$

where the first $N-K$ terms represent the parity information and the remaining K terms represent the unaltered message information.

It is important to keep in mind the apparent reversal of the message and parity information symbols between the sequence notation of (8.147) and the vector notation of (8.150). In both cases, the message symbols are transmitted first and the first message symbol into the receiver corresponds to the coefficient of the highest order $(N-1)$ term of $r(X)$. The expression for $r(X)$ given by (8.150) presupposes an encoder implementation that stores the sequence prior to the encoding and transmission. This is not required, however, and $r(X)$ can be generated without delay between the data source and modulator as shown in Figure 8.66.

The tap weights, g_n, of the parity generator are determined from the coefficients of the generator polynomial $g(X)$. For the t-error correcting RS code, $g(X)$ is expressed as

$$\begin{aligned}g(X)&=\prod_{n=1}^{2t}(X-\alpha^n)\\ &=\sum_{n=0}^{2t}g_nX^n\end{aligned} \tag{8.151}$$

where $g_{2t}=1$ and $g_0=\prod_{n=1}^{2t}\alpha^n$.

This description of the generator polynomial is not unique and more in-depth discussions can be found in the literature [37, 157, 158, 164, 167]. All of these operations must be performed in accordance with the algebraic rules of the GF(2^m). As in the previous discussion of m_n and p_n, g_n is also an element of the GF for which

$$g_n = 0, \alpha^n \quad : n = 0, 1, \ldots, 2^m - 2 \qquad (8.152)$$

Although the basic M-ary RS code for which $M = 2^m$ is defined for $N = M - 1$, a shortened code for $N < M - 1$ can be easily implemented by considering the first message symbols m_n: $n = N, \ldots, M - 1$ to be zero. As is evident from Figure 8.66, these leading message symbols will not contribute to the parity information. For example, the shortened (32,20) 64-ary RS code assumes that the code vector is expressed as

$$r(X) = \underbrace{\alpha_0 + \cdots + \alpha_{11}X^{11}}_{\text{Parity}} \; \underbrace{+\alpha_{12}X^{12} + \cdots + \alpha_{31}X^{31}}_{\text{Message}} \; \underbrace{+\alpha_{32}X^{32} + \cdots + \alpha_{62}X^{62}}_{\text{Forced zeros}} \qquad (8.153)$$

The shortened RS codes are denoted as $(N', K', t')_m$ codes that are derived from the parent $(N, K, t)_m$ code by reducing the N and K symbols by a constant integer $I < K$. The resulting shorted code corresponds to $N' = N - I$ and $K' = K - I$. Although the symbol-error correction capability remains the same, that is, $t' = t$, the code rate of the shortened code is

$$r_c' = r_c\left(\frac{1 - I/K}{1 - I/N}\right) \qquad (8.154)$$

so that, for $I > 0$, $r_c' < r_c$.

8.14.2.2 *Decoder Processing without Erasures* When errors, introduced by the transmission media, are added to the transmitted code vector $r(X)$ the resulting received vector is expressed as

$$\hat{r}(X) = r(X) + e(X) \qquad (8.155)$$

The first operation of the receiver decoder is to compute the error syndrome to determine the existence of errors. For the RS codes, the syndrome is a $2t$-dimensional vector given by

$$S(X) = \sum_{i=0}^{2t-1} S_i X^i \qquad (8.156)$$

When $S(X) \neq 0$, it is concluded that errors have occurred in the received data and that additional processing must be performed to determine their locations and magnitudes. The components S_i are examined individually to determine the existence of an error condition. These components are given by

$$S_i = \hat{r}(\alpha^i) = \sum_{n=0}^{N-1} \alpha_n (\alpha^i)^n \qquad (8.157)$$

TABLE 8.44 Iterative Algorithm Table for the Error Location Polynomial[a]

υ	$\sigma^\upsilon(x)$	d_υ	ℓ_υ	$\upsilon - \ell_\upsilon$
−1	1	1	0	−1
0	1	S_1	0	0
1				
…				
$2t$				

[a]Lin [166]. Reproduced by permission of Pearson Education, Inc., New York, NY.

where $\alpha_n = 0, a^j : \; j = 0, 1, \ldots, N-1$ are the coefficients of the received vector. Whenever $S_i = 0 \forall i$, the message block is assumed to be received without errors; otherwise, an error condition is declared. The operations involved in computing S_i are referred to as the *parity-check calculations*.

In the event of $\upsilon \leq t$ symbol errors, there exists a set of unique nonzero error location numbers (X_ℓ) and error value numbers (Y_ℓ) that are related to the syndrome components by

$$S_i = \sum_{\ell=1}^{\upsilon} Y_\ell X_\ell^i \quad : 1 \leq i \leq 2t \qquad (8.158)$$

The problem in solving the nonlinear equations (8.158) for X_ℓ and Y_ℓ was the major task facing algebraic coding theorists and the major task facing hardware or software engineers is to efficiently implement the resulting algorithms. The following development follows closely that given by Berlekamp.*

When $\upsilon \leq t$ errors occur a set of *elementary symmetric functions*, σ_ℓ, are defined by the equations

$$\prod_{\ell=1}^{\upsilon}(1 - X_\ell X) = \sum_{\ell=0}^{\upsilon} \sigma_\ell X^\ell \qquad (8.159)$$

where the *elementary symmetric functions* are related to the error syndrome components by the linear equations†

$$\sum_{j=0}^{\upsilon} S_{i+j}\sigma_{\upsilon-j} = 0 \quad : 1 \leq i \leq 2t - \upsilon \qquad (8.160)$$

Berlekamp developed an iterative procedure for solving this set of linear equations for the elementary symmetric functions and Lin presents a simplified outline of the algorithms. The procedure involves completing Table 8.44 after initialization of the first two rows as shown. In Table 8.44, S_i is the error syndrome computed previously and completion of the table, given the υ-th row, is accomplished as follows:

*Refer to Lin [37] for a description of the Berlekamp algorithm.

†Equation (8.160) is developed by Peterson and Weldon (see Reference 34, Equation (9.16), p. 284).

```
if (d_υ = 0), then
  σ^(υ+1)(x) = σ^υ(x)
  ℓ_(υ+1) = ℓ_υ
else
  Find row υ' < υ, such that the last column
  υ' - ℓ_υ'
  is the largest value with d_υ' ≠ 0, and
  then compute:
  σ^(υ+1)(x) = σ^υ(x) - d_υ d_υ'^(-1) x^(υ-υ') σ^υ'(x)
  ℓ_(υ+1) = max(ℓ_υ, ℓ_υ' + υ - υ')
endif
d_(υ+1) = S_(υ+2) + σ_1^(υ+1) S_(υ+1) + ··· + σ_(ℓ_(υ+1))^(υ+1) S_(υ+2-ℓ_(υ+1))
where the σ_i^(υ+1) are the coefficients of the
polynomial
```

$$\sigma^{\upsilon+1}(x) = 1 + \sigma_1^{\upsilon+1}x + \sigma_2^{\upsilon+1}x^2 + \cdots + \sigma_{\ell_{\upsilon+1}}^{\upsilon+1}x^{\ell_{\upsilon+1}} \quad (8.161)$$

Finding the row such that $\upsilon' < \upsilon$ under the indicated conditions requires completing $2t$ rows in Table 8.44 and the polynomial $\sigma^{2t}(x)$ in the last row is the error location polynomial. If the degree of this polynomial is greater than $2t$, the errors cannot be located and corrected.

Upon successful completion of Table 8.44, the resulting polynomial $\sigma^{2t}(x)$ is the required *error location polynomial*. Therefore, having solved for the elementary symmetric functions (σ_ℓ), the *error locations numbers* are determined as follows. The right-hand side of (8.159) is referred to as the *error location polynomial*, and it follows from the *left-hand side* of (8.159) that any valid error location number $X_\ell (1 \le \ell \le \upsilon \le 2t)$ is related to the roots of this polynomial. The roots form a subset of the GF(2^m) field elements given by

$$X_{root} = \alpha^i \quad : i = 0, 1, \ldots, 2^m - 2 \quad (8.162)$$

from which the error location numbers are given by

$$X_\ell = (X_{root})^{-1} = \alpha^{N-i} \quad (8.163)$$

This is equivalent to identifying the error location in terms of the power n of X^n given in (8.150). Therefore, the location of the received symbol α_n that is in error is given by the coefficients of X^n in the received vector where $n = N - i$.

The terminology used here is consistent with that used by Berlekamp. Peterson [58], however, defines the error locations as being the reciprocal of the error location numbers. The difference arises because of the form of (8.159), which Peterson expresses as

$$\prod_{\ell=1}^{\upsilon}(X_\ell - X) \quad (8.164)$$

The location numbers are determined simply by substituting each element of GF(2^m), except zero, into (8.159) and taking those values that are roots.

Having computed the error locations, the error values (Y_ℓ) are determined from (8.158), which is linear in Y_ℓ. For the binary *BCH* codes, the received bits found to be in error are simply inverted to give the corrected information. Determining the error values for the nonbinary *BCH* codes, however, is considerably more involved. The error values are expressed as

$$Y_\ell = \frac{W(X_\ell^{-1})}{\prod_{\substack{j=1 \\ i \ne \ell}}^{\upsilon}(1 - X_j X_\ell^{-1})} \quad (8.165)$$

where $W(X)$ is referred to as the *error evaluation polynomial* and can be expressed in the form

$$W(X) = 1 + \sum_{i=1}^{\upsilon} C_i X^i \quad (8.166)$$

where the coefficients are given by

$$C_i = \sum_{j=0}^{i} S_{i-j}\sigma_j \quad (8.167)$$

where $S_o = 1$, $S_i(1 < i < 2t)$ are the previously computed error syndromes, and σ_j are the *elementary symmetric functions*. An alternate procedure for computing the error values when with erasures is discussed in Section 8.14.2.3.

8.14.2.3 Decoder Processing with Erasures As outlined in this section, the processing requirements with erasures follow the development given by Peterson. In the event that the decoding processing as outlined previously fails, additional attempts can be made with symbol erasures. The advantage of this procedure lies in the fact that the erasure locations are known that eliminates the need to evaluate of the error location polynomial. The computed error locations are then combined with the known erasure locations to compute the error values. The four basic steps listed in the introduction are still required with modifications as outlined in the following discussion.

The first requirement is to compute a new or modified error syndrome taking into consideration the known erasure locations. If symbols are erased in e locations, then, from Table 8.42, there are $\upsilon \le t'$ additional errors where

$$t' = \frac{(2t - e)}{2} \quad (8.168)$$

and (8.158) can be expressed as

$$S_j = \sum_{\ell=1}^{\upsilon} Y_\ell X_\ell^j + \sum_{\ell=1}^{e} V_\ell U_\ell^j \quad : 1 \le j \le 2t \quad (8.169)$$

where U_ℓ are the known erasure locations and V_ℓ are the erasure error values that may be zero. Proceeding as before, a set of elementary symmetric functions is defined in terms of the known erasure locations as

$$\prod_{\ell=1}^{e}(U-U_\ell)=\sum_{\ell=0}^{e}\sigma_\ell U^{e-\ell} \tag{8.170}$$

The elementary symmetric functions (σ_ℓ) are readily determined by expanding both sides of (8.170) and equating the coefficients of the parameter U. The modified syndrome components are then evaluated as

$$T_i=\sum_{\ell=0}^{e}\sigma_\ell S_{i-\ell} \quad : i=e+1, e+2, \ldots, 2t \tag{8.171}$$

where the syndrome components S_j are determined from the received code vector as expressed by (8.157). The remaining unknown error location numbers are related to a modified set of elementary symmetric functions and the modified error syndrome components as given by (8.159) and (8.160) where T_{e+i} is substituted for S_i in (8.160) and the index i is taken over the range $1 \le i \le 2t\text{-}e\text{-}\upsilon$. The new set of elementary symmetric functions is computed using Berlekamp's iterative algorithm. The unknown error location numbers X_ℓ are computed from the roots on the right-hand side of (8.159) as expressed in (8.163).

The error values can now be computed based on the error location numbers X_ℓ and U_ℓ. However, in view of the erasures, only υ nontrivial elementary symmetric functions have been computed; whereas, using (8.165) through (8.167), $\upsilon + e$ such functions are required. For this reason, the error values will be computed using Forney's procedure [165], which is also discussed by Peterson [58]. The result of Forney's evaluation is

$$Y_\ell=\frac{\sum_{j=0}^{\upsilon'-1}\sigma_{\ell j}S_{\upsilon'-j}}{\sum_{j=0}^{\upsilon'-1}\sigma_{\ell j}X_\ell^{\upsilon'-j}} \tag{8.172}$$

where $\upsilon'=\upsilon+e$. The elementary symmetric functions ($\sigma_{\ell j}$) can be determined from the recursive relationship

$$\sigma_{\ell j}=\bar{\sigma}_j+X_\ell\sigma_{\ell(j-1)} \quad : 1\le j\le \upsilon' \tag{8.173}$$

where $\sigma_{\ell 0}=1$ and $\bar{\sigma}_j : 1\le j\le \upsilon'$ are the elementary symmetric functions of the error location numbers $X_\ell : 1\le \ell \le \upsilon'$, namely,

$$\prod_{\ell=0}^{\upsilon'}(X_\ell-X)=\sum_{j=0}^{\upsilon'}\bar{\sigma}_jX^j \tag{8.174}$$

It is not necessary to compute each value of $\bar{\sigma}_j$ based on (8.174) since the coefficients related to υ of the products are also related to the error polynomial coefficients computed using Berlekamp's algorithm. Therefore, defining the coefficients for the partial expansion as γ_j, the relationship of interest is

$$\gamma_j=\sigma_{\upsilon-j} \quad : 0\le j\le \upsilon \tag{8.175}$$

Consequently, the product expansion must only be continued over the erasure locations and (8.174) becomes

$$\sum_{j=0}^{\upsilon}\gamma_jX^j\prod_{\ell=1}^{e}(U_\ell-U)=\sum_{j=0}^{\upsilon}\bar{\sigma}_jX^j \tag{8.176}$$

Actually, the product expansion in (8.176) was also evaluated using (8.170) and could be used to further reduce the computational time. Of particular interest is the case with no erasures ($e=0$), in which, all the coefficients $\bar{\sigma}_j$ are computed using (8.175) and, in the case where only erasures are used ($\upsilon=0$), then (8.170) provides the required coefficients directly. These procedures for determining the error values could also be used without erasures, in which case $\upsilon'=\upsilon$.

8.14.2.4 RS Simulation Program Outline The program used to evaluate the RS code and the required processing functions are listed in Table 8.45. The simulation program focuses of the 64-ary RS code and various shortened forms defined by the parent (N, K) code. However, the code can also be used to evaluate the performance of 16-ary through 1024-ary RS codes by selecting $m=4$ through 10, respectively. The required inputs to the simulation are defined in Table 8.46.

A block of source data consisting of mK message bits is obtained using a uniform number generator. The bits are taken m at a time to form the message coefficients m_n, $n=0, 1, \ldots, K-1$. From this message information, the parity coefficients p_n, $n=0, 1, \ldots, N-K-1$ are generated and appended to the message block to form a systematic block code. The unipolar binary block data (b_n) is converted to

TABLE 8.45 Reed–Solomon Simulation Code Processing

Main program
Error syndrome computations.
Modified error syndrome computations
Berlekamp algorithm, error locations, and error values
Erasure coefficient computations
Error value computations
Block- and bit-error statistics
GF(2^m) field element and encoder coefficients
GF(2^m) addition
GF(2^m) multiplication
Symbol bit reverse
Binary-to-integer conversion
Uniform number generator
Gaussian number generator

TABLE 8.46 Simulation Code Input Parameters

Parameter	Description
SNR	Signal-to-noise ratio (E_b/N_o) dB (SNR ≥ 301 no noise is added)
m	$\text{Log}_2(M)$ for *M*-ary RS code
N,K	Defines block code
mpass	Number of data blocks in simulation (bits per simulation = *mpass*N*m*)
iwrite	=1 Detailed output for evaluation =0 Suppresses detailed output (defaults to zero for *mpass* > 2)
iforce	=1 Forces defined error patterns for evaluation =0 Suppresses forced error patterns (defaults to zero for *mpass* > 2)
nstart	Seed for random source data
mstart	Seed for channel noise

bipolar data as $d_n = 1 - 2b_n$. This translation can be viewed as the baseband equivalent modulation for BPSK for which the signal-to-noise ratio is given by

$$\gamma_c = \frac{(A^2/2)T}{N_o} \tag{8.177}$$

where γ_c is expressed in terms of the transmitted code-bit or symbol* duration, T, and is defined as E_c/N_o. This definition is used for a normalized comparison of the performance; however, an absolute comparison is based on the signal-to-noise ratio defined in terms of the message bit duration, T_b, defined as E_b/N_o. These signal-to-noise ratios are related as

$$\left.\frac{E_b}{N_o}\right|_{\text{dB}} = \left.\frac{E_c}{N_o}\right|_{\text{dB}} + 10\log\left(\frac{N}{K}\right) \quad \text{: BPSK modulation} \tag{8.178}$$

The Gaussian noise is generated as a zero-mean random variable with standard deviation given by

$$\sigma_n = \sqrt{\frac{N_o}{T}} = \sqrt{\frac{A^2}{2\gamma_c}} \tag{8.179}$$

The received signal is then the sequence composed of noisy bipolar data $d_i' = d_i + n_i$. The receiver threshold is set as zero and the received binary data is determined as $\hat{b}_i = 1 : d_i' < 0$ and $\hat{b}_i = 0 : d_i' \geq 0$ where $\hat{b}_i$ is the estimate of the transmitted block data b_i. Therefore, based on the received binary

*The use of the term *symbol* as used here to define the modulated waveform interval should not be confused with the same term used to describe the RS *M*-ary symbol. For the *M*-ary RS code, the latter consists of *m* modulated waveform intervals or *symbols* with BPSK modulation.

TABLE 8.47 Computed Error Locations for Various Erasure Conditions Using Six Error-Correcting 64-ary Reed–Solomon Code

	Theoretical and Simulated	
Number of Erasures (e)	Computed (t')	Total
0	6	6
3	4	7
6	3	9
8	2	10

estimates, the RS decoder outputs estimates the original source message bits.

The RS decoding functions are outlined in Table 8.45. First, the syndromes are computed and a flag is set to one if an error is detected from the syndrome results. The number of decoding attempts to correct a block of data is recorded. The first attempt at correcting errors is made without erasures and subsequent attempts are for various combinations of erasures. Three erasure conditions are considered in the simulation using 3, 6, and 8 symbol erasures, respectively. Table 8.47 outlines the number of computed error locations based on the number of erasures (e). The computed error locations listed in Table 8.47 are based on $t' = (2t - e)/2$ and the total number of correctable symbols is simply the sum of the erasures and the computed error locations.

It may be desirable to compute t' error locations, one less than the maximum values shown in Table 8.47, by using t' = 5, 3, 2, 1. This set of computed error locations reflects a conservative decoding approach based on the minimum code distance and the related false-decoding probabilities. The simulation results discussed in the following section are based on the uncorrectable-error condition corresponding to the degree of the error locator polynomial; however, in the event of a declaration of a successful error correction, the error syndromes are re-computed as a final check of the error-free message.

8.14.2.5 RS Code Performance In this section, the analytical performance of the RS code is examined and the results are compared to the performance based on the Monte Carlo simulation program. Because the RS code is based on symbol alphabets consisting of *m* binary bits, the analytical performance over the BSC is relatively easy to evaluate. In addition to providing a simple method of evaluating the performance over a very basic channel, the analytical performance results serve the important function of validating the implementation of the software code. Once validated, the simulation code can be targeted to an FPGA, ASIC, or DSP device. In addition, the simulation code will provide the user with an important capability to evaluate the performance of the RS code over a variety of channels without having to make simplifying assumptions. For example, the analytical results can be obtained assuming a slowly or rapidly varying channel; however, the simulation program can evaluate the performance over any channel fading condition. The burst-error statistics are integrally related to the channel fading conditions and impact the RS code performance in way that are difficult to model.

The performance of various shortened forms of the parent (63,51) code structure is compared to the performance of antipodal and 64-ary coherent orthogonal signaling. The (24,12) shortened code is used to show the improvement utilizing forced erasure processing. In all cases, the performance is evaluated in terms of the probability of a bit error in the received message as related to the signal-to-noise ratio. The baseline bit-error probability for the underlying waveform modulations have been developed previously and are shown in the following figures as the *dotted* curves. These results are based on uncoded BPSK waveform modulation and the signal-to-noise ratio is expressed in terms of the source-bit energy (E_b).

Under the conditions of a BSC with bit-error probability P_{be}, the analytical bit-error probability, P'_{be}, for a t-error correcting M-ary RS code is evaluated as follows. First, the probability of a symbol error is determined as

$$P_{se} = 1 - (1 - P_{be})^m \tag{8.180}$$

Since t or fewer of the N symbols are corrected by the RS decoding, the resulting corrected symbol-error probability is evaluated as

$$\begin{aligned} P'_{se} &= \sum_{i=t+1}^{N} \binom{N}{i} P_{se}^i (1 - P_{se})^{N-i} \\ &= 1 - \sum_{i=0}^{t} \binom{N}{i} P_{se}^i (1 - P_{se})^{N-i} \end{aligned} \tag{8.181}$$

Using (8.181) the corrected bit-error probability is determined by solving (8.180) for P'_{be} in terms of P'_{se} with the result

$$P'_{be} = 1 - \sqrt[m]{1 - P'_{se}} \tag{8.182}$$

Figure 8.67 shows the results for the first four high-rate 64-ary RS codes denoted as (63,57,3), (63,51,6), (63,45,9), and (63,39,12). Figure 8.68 shows similar results for the first five highest rate 256-ary RS codes denoted as (255,247,4), (255,239,8), (255,231,12), (255,223,16), (255,215,20) and a lower rate code (255,123,66). The uncoded bit-error probability is shown as the *dotted* curve. The abscissa is expressed in terms the energy-per-bit so the correction for the code rate has been applied.

The higher order codes exhibit two advantages: compared to lower order codes having the same code rate, they result in greater coding gain, for example, comparing the coding gain of the rate 0.906 code in Figure 8.68 with the rate 0.905 code in Figure 8.67, and they result in error probability curves with steeper slopes resulting in improved performance with an incremental change in E_b/N_o. These results also show the effect of the diminishing returns as the code rate is reduced and, for a given RS code order, there exists a code rate, below which, the relative coding gain decreases. Figure 8.69 compares the performance of the 64-ary RS codes: (63,57,3), (63,45,9), (63,30,16), (63,16,23), and (63,7,28). The code rate loss for the rate 0.111 code is about 1.8 dB at $P_{be} = 10^{-5}$, which nullifies the coding gain resulting from the error correction. The performance of the rate 0.714 and 0.476 codes is nearly the same. These observations of the coding

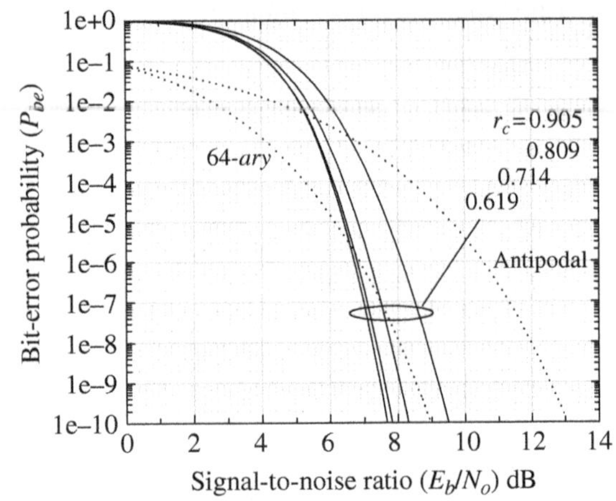

FIGURE 8.67 Analytical performance for several high-rate 64-ary Reed–Solomon codes.

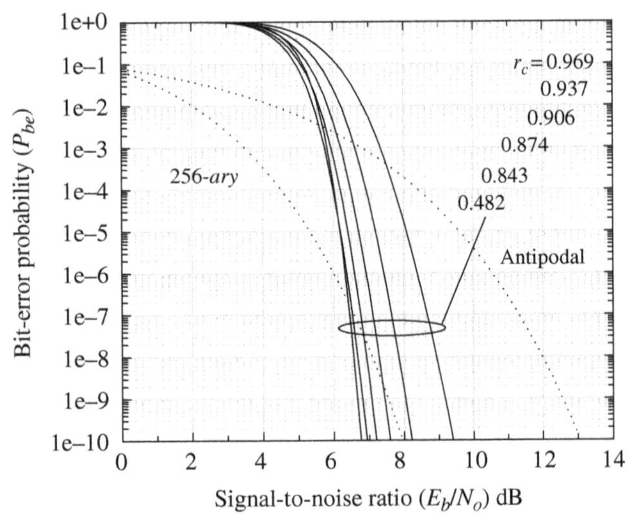

FIGURE 8.68 Analytical performance for several high-rate 256-ary Reed–Solomon codes.

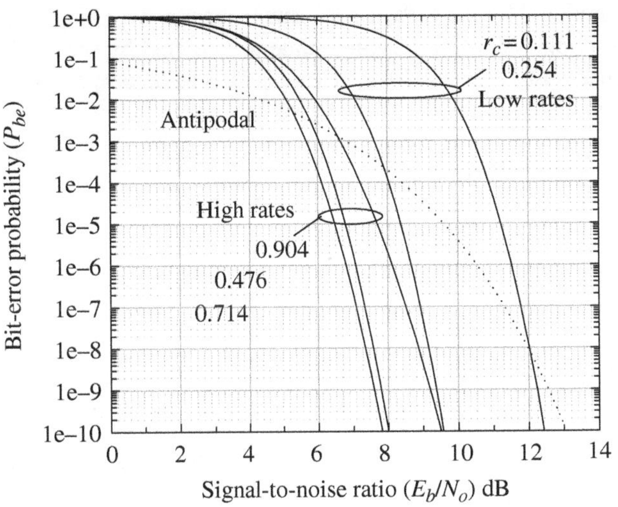

FIGURE 8.69 Analytical performance for high- and low-rate 64-ary Reed–Solomon codes.

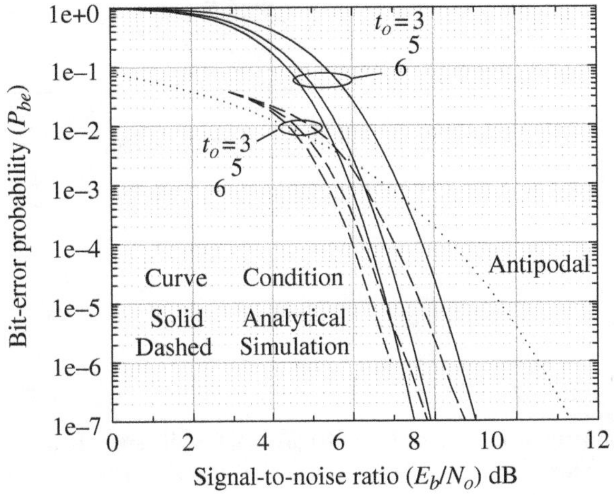

FIGURE 8.70 Comparison of analytical and Monte Carlo simulated performance for (63, 51, t_o)$_6$ 64-ary Reed–Solomon codes.

gain result because of performance dependence on E_b/N_o and, when plotted as function of the signal-to-noise ratio measured in the symbol bandwidth, the plots would continually move to the left with decreasing code rate. These results suggest that the RS code parameters should be selected to achieve the minimum bit-error performance when plotted as a function of E_b/N_o.

The *dashed* curves in Figures 8.70 and 8.71 show, respectively, the Monte Carlo simulation results for the $(63,51,t_o)_6$ code and a shortened $(24,12,t_o)_6$ version of the 64-ary RS code with various symbol-error correction capabilities. These results are compared to the corresponding analytical performance, shown as the *solid* curves. The *dotted* curve is the

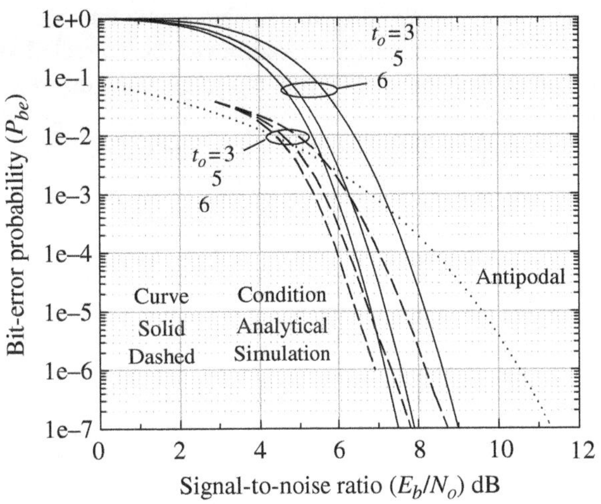

FIGURE 8.71 Comparison of analytical and Monte Carlo simulated performance for shortened (24, 12, t_o)$_6$ 64-ary Reed–Solomon codes.

bit-error performance of the underlying uncoded antipodal or BPSK modulated waveform. The Monte Carlo results are based on 100K RS blocks for each signal-to-noise ratio corresponding to 63(6)100K = 37.8M and 24(6)100K = 14.4M bits, respectively. Whenever an RS message block fails the decoder processing, the bit errors in the received message are counted. This results in the bit-error performance, for low signal-to-noise ratios, bending to the left away from the analytical predictions. If the antipodal waveform reference is plotted in terms of the signal-to-noise ratio in the symbol bandwidth, that is, by moving the *dotted* curve to the right by $10\log(N/K)$ dB, then the coded results would be seen to approach the adjusted antipodal performance curve for low values of E_b/N_o; this is the correct comparison since the uncoded and coded results would both be operating under the same of the additive noise bandwidth.

Figure 8.72 compares the simulated performance of the $(255,247,t_o)_8$ 256-ary RS code for $t_o = 2$ and 4. In this case, the difference between the coded and uncoded signal-to-noise ratio is only $10\log(255/247) = 0.14$ dB, so the coded asymptotic performance for low signal-to-noise ratios is shifted from the uncoded (*dotted*) curve by this amount.

As a final example, Figure 8.73 shows the performance of a concatenated code using an inner convolutional code and RS outer code; the convolutional code is decoded using the Viterbi algorithm so this concatenated configuration is referred to as an RSV concatenated code. The performance is shown for two rate 1/2 convolutional codes with constraint lengths 7 and 9. The RS code performance is based on the theoretical relationship given the hard-decision outputs of the convolutional code. The theoretical bit-error probability out of the RSV code is evaluated using (8.180) to compute the symbol-error probability into the RS decoder given the bit-error probability output of the convolutional code decoder, then using (8.181) to compute the corrected symbol-error probability and (8.182) to compute the desired corrected bit-error probability with the signal-to-noise ratio adjusted to reflect the measurement in the bandwidth of the source-bit interval. Although not evaluated, the Monte Carlo simulated RSV bit-error performance results will bend to the left approaching a limiting value as depicted by the *dashed* curves in Figures 8.70, 8.71, and 8.72. The theoretical results are ideal, in the sense that they assume that the bit-errors are independent; this is the same as assuming the use of an ideal interleaver between the codes. In practice, the RSV code typically uses an interleaver between the inner and outer codes to improve the decoding reliability.

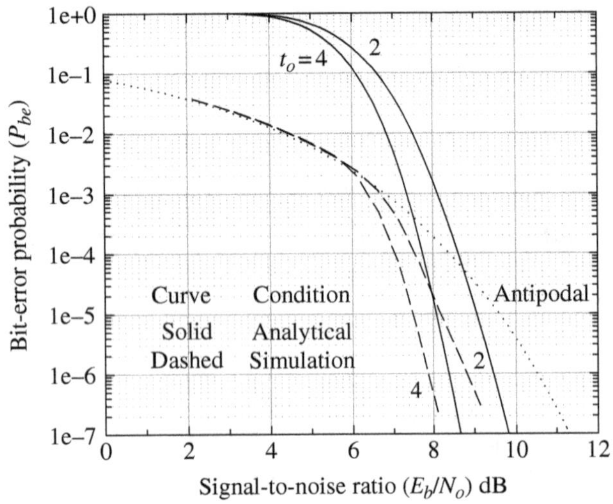

FIGURE 8.72 Comparison of analytical and Monte Carlo simulated performance for (255, 247, t_o)$_8$ 256-ary Reed–Solomon codes.

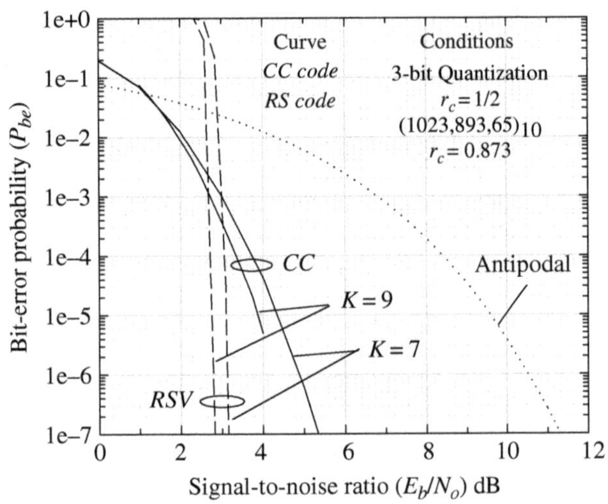

FIGURE 8.73 Performance of RSV concatenated codes.

APPENDIX 8A

TABLE 8A.1 Map of Deleting Bits for Punctured Codes Derived from Rate 1/2 Codes with Constraint Lengths 2 through 8[a]

The constraint length ν is defined by Forney [84]. The code generators for the underlying rate 1/2 codes with maximal free distance are indicated in octal notation by the brackets. The parity-check bits signified by a "0" are punctured and a "1" indicates the bits that are transmitted as the code word.

	Constraint Length ν						
Code Rate	2	3	4	5	6	7	8
1/2	1 (5)	1 (15)	1 (23)	1 (53)	1 (133)	1 (247)	1 (561)
	1 (7)	1 (17)	1 (35)	1 (75)	1 (171)	1 (371)	1 (753)
2/3	10	11	11	10	11	10	11
	11	10	10	11	10	11	10
3/4	101	110	101	100	110	110	111
	110	101	110	111	101	101	100
4/5	1011	1011	1010	1000	1111	1010	1101
	1100	1100	1101	1111	1000	1101	1010
5/6	10111	10100	10111	10000	11010	11100	10110
	11000	11011	11000	11111	10101	10011	11001
6/7	101111	100011	101010	110110	111010	101001	110110
	110000	111100	110101	101001	100101	110110	101001
7/8	101111 1	100001 0	101001 1	101110 1	111101 0	101010 0	110101 1
	110000 0	111110 1	110110 0	110001 0	100010 1	110101 1	101010 0
8/9	101111 11	100000 11	101000 11	111000 10	111101 00	101101 10	111000 00
	110000 00	111111 00	110111 00	100111 01	100010 11	110010 01	100111 11

TABLE 8A.1 (*continued*)

	Constraint Length ν						
Code Rate	2	3	4	5	6	7	8
9/10	101111	101000	111110	100001	111101	101100	111000
	111	000	011	111	110	110	101
	110000	110111	100001	111110	100010	110011 001	100111
	000	111	100	000	001		010
10/11	101111	100000	100000	100111	111011	100100	100010
	1111	0011	0101	0100	0111	0011	1100
	110000	111111	111111	111000	100100	111011	111101
	0000	1100	1010	1011	1000	1100	0011
11/12	101111	100000	101011	100011	111101	101100	110000
	11111	00010	01101	10100	11110	00110	10001
	110000	111111	110100	111100	100010	110011	101111
	00000	11101	10010	01011	00001	11001	01110
12/13	101111	100000	101101	110100	111111	100100	110000
	111111	000011	111011	110110	110101	001100	011010
	110000	111111	110010	101011	100000	111011	101111
	000000	111100	000100	001001	001010	110011	100101
13/14	101111	101000	111011	110001	110100	101010	110000
	111111 1	000000 0	011011 1	100010 0	000111 1	010000 0	010000 1
	110000	110111	100100	101110	101011	110101	101111
	000000 0	111111 1	100100 0	011101 1	111000 0	101111 1	101111 0

[a]Yasuda et al. [67]. Reproduced by permission of the IEEE.

APPENDIX 8B

TABLE 8B.1 Coefficients for Generator Polynomials of Primitive Binary BCH Codes (Octal Notation)[a]

n	k	t	$g(x)$
7	4	1	13
15	11	1	23
	7	2	721
	5	3	2467
31	26	1	45
	21	2	3551
	16	3	107657
	11	5	542332 5
	6	7	313365 047
63	57	1	103
	51	2	12471
	45	3	170131 7
	39	4	166623 567
	36	5	103350 0423
	30	6	157464 165547
	24	7	173232 604044 41
	18	10	136302 651235 1725
	16	11	633114 136723 5453
	10	13	472622 305527 250155
	7	15	523104 554350 327173 7
127	120	1	211
	113	2	41567
	106	3	115547 43
	99	4	344702 3271

(*continued overleaf*)

TABLE 8B.1 (*continued*)

n	k	t	$g(x)$
	92	5	624730 022327
	85	6	130704 476322 273
	78	7	262300 021661 30115
	71	9	625501 071325 312775 3
	64	10	120653 402557 077310 0045
	57	11	335265 252505 705053 517721
	50	13	544465 125233 140124 215014 21
	43	14	177217 722136 512275 212205 74343
	36	15	314607 466652 207504 476457 472173 5
	29	21	403114 461367 670603 667530 141176 155
	22	23	123376 070404 722522 435445 626637 647043
	15	27	220570 424456 045547 705230 137622 176043 53
	8	31	704726 405275 103065 147622 427156 773313 0217
255	247	1	435
	239	2	267543
	231	3	156720 665
	223	4	756266 41375
	215	5	231575 647264 21
	207	6	161765 605676 36227
	199	7	763303 127042 072234 1
	191	8	266347 017611 533371 4567
	187	9	527553 135400 013222 36351
	179	10	226247 107173 404324 163004 55
	171	11	154162 142123 423560 770616 30637
	163	12	750041 551007 560255 157472 451460 1
	155	13	375751 300540 766501 572250 646467 7633
	147	14	164213 017353 716552 530416 530544 101171 1
	139	15	461401 732060 175561 570722 730247 453567 445
	131	18	215713 331471 510151 261250 277442 142024 165471
	123	19	120614 052242 066003 717210 326516 141226 272506 267
	115	21	605266 655721 002472 636364 046002 763525 563134 72737
	107	22	222057 723220 662563 124173 002353 474201 765747 501544 41
	99	23	106566 672534 731742 227414 162015 7433225 524110 764323 03431
	91	25	675026 503032 744417 272363 172473 251107 555076 272072 434456 1
	87	26	110136 763414 743236 435231 634307 172046 206722 545273 311721 317
	79	27	667000 356376 575000 202703 442073 661746 201153 267117 665413 42355
	71	29	240247 105206 443215 155541 721123 311632 054442 503625 576432 217060 35
	63	30	107544 750551 635443 253152 173577 070036 661117 264552 676136 567025 43301
	55	31	731542 520350 110013 301527 530603 205432 541432 675501 055704 442603 547361 7
	47	42	253354 201706 264656 303304 137740 623317 512333 414544 604500 506602 455254 3173
	45	43	152020 560552 341611 311013 463764 237015 636700 244707 623730 332021 570250 51541
	37	45	513633 025506 700741 417744 724543 753042 073570 617432 343234 764435 473740 304400 3
	29	47	302571 553667 307146 552706 401236 137711 534224 232420 117411 406025 475741 040356 5037
	21	55	125621 525706 033265 600177 315360 761210 322734 140565 307745 252115 312161 446651 347372 5
	13	59	464173 200505 256454 442657 371425 006600 433067 744547 656140 317467 721357 026134 460500 547
	9	63	157260 252174 724632 010310 432553 551346 141623 672120 440745 451127 661155 477055 616775 16057

[a]Stenbit [161]. Reproduced by permission of the IEEE.

ACRONYMS

ACS	Add–compare–select
AGC	Automatic gain control
APP	A posteriori probability
ARQ	Automatic repeat request
ASCII	American Standard Code for Information Interchange
ASIC	Application-specific integrated circuit
AWGN	Additive white Gaussian noise
BCH	Bose–Chaudhuri–Hocquenghem (code)
BCJR	Bahl–Cocke–Jelinek–Raviv (algorithm)
BFSK	Binary frequency shift keying
Biϕ	Biphase or Manchester code (PCM format)
Biϕ-L, -M,-S	Biphase-level, -mark, -space (PCM formats)
CC	Constituent code (code concatenation)
CCC	Concatenated convolution code
CCITT	International Telegraph and Telephone Consultative Committee (currently referred to as ITU)
CCSDS	Consultative Committee for Space Data Systems
CDMA	Code division multiple access
CPM	Continuous phase modulation
CRC	Cyclic redundancy check
CSK	Compatible shift keying
DC	Direct current
DEBPSK	Differentially encoded BPSK
DEQPSK	Differentially encoded QPSK
DM	Delay modulation or Miller code (PCM format)
DM-M,-S	Delay modulation-mark, -space (PCM format)
DPCCC	Double parallel concatenated convolutional code
DSCCC	Double serially concatenated convolutional code
DS-CDMA	Direct sequence CDMA
DSP	Digital signal processor
EDAC	Error detection and correction
FCS	Frame check sequence
FEC	Forward error correction
FIR	Finite impulse response
FM	Frequency modulation (PCM format)
FPGA	Field programmable gate array
FSK	Frequency shift keying
GF	Galois field
GMSK	Gaussian minimum shift keying
HCCC	Hybrid concatenated convolutional code
IF	Intermediate frequency
IIR	Infinite impulse response (filter)
IS-95	Interim Standard 95
ISI	Intersymbol interference
ITU-T	International Telecommunication Union (Sector T: Telecommunications and Computer Protocols)
JPL	Jet Propulsion Laboratory (California Institute of Technology)
LCM	Least common multiple
LDPC	Low-density parity-check (code)
LFSR	Linear feedback shift register
LSB	Least significant bit
LSS	Least significant symbol
MAP	Maximum a posteriori (probability)
MPSK	Multiphase shift keying
MSB	Most significant bit
MSK	Minimum shift keying
MSS	Most significant symbol
NRC	Nonrecursive convolutional (code)
NRZ	Non-return-to-zero (PCM format)
NRZ-L, -M,-S	Non-return-to-zero-level, -mark, -space (PCM formats)
NSC	Nonsystematic convolutional (code)
OQPSK	Offset QPSK
PC	Product code
PCCC	Parallel concatenated convolutional code
PCM	Pulse code modulation
PCM/FM	Frequency modulated PCM
PCM/PM	Phase modulated PCM
PCS	Personal communications service
PM	Phase modulation (PCM format)
PRM	Partial response modulation
PRN	Pseudo-random noise (sequence)
PSD	Power spectral density
QAM	Quadrature amplitude modulation
QPSK	Quadrature phase shift keying
RF	Radio frequency (carrier)
RS	Reed–Solomon (code)
RSC	Recursive systematic convolutional (code)
RSV	Reed–Solomon Viterbi (concatenated code)
RZ	Return-to-zero (PCM format)
RZ-L, -M,-S	Return-to-zero-level, -mark, space (PCM formats)
SCC	Self-concatenated code
SCCC	Serially concatenated convolutional code
SISO	Soft-in soft-out (processor module)
SOM	Start-of-message
TC	Turbo code
TCM	Trellis-coded modulation
TFM	Tamed frequency modulation
TPC	Turbo product code

PROBLEMS

1. Referring to the unipolar NRZ-M and NRZ-S baseband data formats shown in Figure 8.3, sketch the formats of the corresponding polar NRZ-M and NRZ-S PCM waveforms.

2. In generating the unipolar NRZ and unipolar RZ coded waveforms, the level of the coded bit corresponding to a

zero source bit is always zero. Given the nonzero mark-bit spectrum, $H_m(f)$ with $H_s(f) = 0$, derive the PSD for these PCM waveforms. Hint: Evaluate the space-bit modulation as $e_s(t) = e_m(t - T_b) - e_m(t)$.

3. Evaluate the bit-error performance for the DEBPSK modulated data sequence in Table 8.3 if a single-error event occurs at bit $i = 6$. In this case, the phaselock loop continues to operate uninterrupted. Show the modified rows $\hat{D}_i$ and $\hat{d}_i$ in Table 8.3 and indicate the error events. Then repeat this example if there are no received bit errors, but the encoder and decoder are initialized differently, that is, using $D_i = 0$ and $\hat{D}_{i-1} = 1$. Is the differential decoder self-synchronizing? If the encoder and decoder cannot be identically initialized, how can the data error event be avoided?

4. For the $m = 21$ degree polynomial listed in Table 8.6, determine the number of m-sequences that can be generated. Show all of the details in determining the prime factors p_i. Hint: The prime factors can be determined using the procedures described in Section 1.13.1.

5. For the degree 4 irreducible and non-primitive polynomial described by the octal notation (37), determine the output sequence when the shift registers are initialized with $(b_3,b_2,b_1,b_0) = (0001)$. Repeat this using the initialization $(b_3,b_2,b_1,b_0) = (1001)$ and (1111). Identifying these respective subsequences as A, B, and C, compute the integer values of the binary states for each cyclic shifted subsequence. Compare the results of the subsequences A, B, and C with the $2^4 - 1 = 15$ states of the 4th-degree nonprimitive polynomial.

6. For the m-sequence generated by the polynomial $g(x) = x^3 + x + 1$, compute or determine the following: (i) the length of the sequence, (ii) a table of all initial conditions, (iii) the run length probabilities $P_\ell(0)$, $P_\ell(1)$ for all run lengths ℓ, (iv) that the m-sequence generator is linear, and (v) the correlation response for two contiguous received sequences, including the leading and trailing correlation responses, that is, for a total correlation response over a length of four sequences.

7. Consider the binary data randomizer implemented as the m-sequence generated by the third-degree polynomial $(13)_o$. Using the randomizer initial state (001) and the source data sequence (10011101011), with the leftmost bit input first, evaluate (i) the randomized data sequence, (ii) the output of the derandomizer, and (iii) the output of the derandomizer if the 5th bit of the randomized data is in error. Suggestion: complete the following tables where D_s is the source data, D_r is the randomized data, $b_3 = D_s \oplus b_0$, the primed designations correspond to the estimations in the derandomizer, and the *shaded cells* correspond to the randomizer and derandomizer initializations.

Randomizing						Derandomizing				
D_s	b_3	b_2	b_1	b_0	D_r	D'_r	d'_2	d'_1	d'_0	D'_s
1		0	0	1			0	0	1	
0										
0										
etc.						etc.				

8. Derive equation (8.21) for the gray decoding of an n-tuple of gray coded binary bit. Hint: Use the gray encoding algorithm given in (8.20).

9. With a word (vector or n-tuple) defined as $(b_n, \ldots, b_0)$: $b_i = \{1,0\}$, the following definitions apply:

Hamming distance (h_d) between two equal-length words is the number differing elements b_i.

Minimum distance (d_{min}) of a linear block code of words is the smallest Hamming distance between pairs of code words.

Hamming weight (h_w) is the number of nonzero elements in a word.

Applying these definitions to a linear block code of three words:

$x = (11000101011)$, $y = (01011100010)$ and $z = (10000000001)$ determine $h_d(x,y)$, $h_d(x,z)$, $h_d(y,z)$, $w(x)$, $w(y)$, $w(z)$ and d_{min} and, using modulo-2 addition, show that $h_d(x,y) = h_w(x + y)$, $h_d(x,z) = h_w(x + z)$, and $h_d(y,z) = h_w(y + z)$.

10. Using the state diagram in Figure 8.44 and the state reduction rules in Figure 8.45, show that the convolutional code transfer function for the binary rate 1/2, $K = 3$, convolutional code is given by the first equality in (8.81). Also, show all of the steps in the division resulting in the first four terms of the infinite series given by the second equality in (8.81).

11. Referring to the punctured code example in Section 8.11.7 corresponding to the rate 1/2, $K = 3$ convolutional code in Figure 8.37, draw a sketch of the four-state trellis decoder for the states $\bar{S}_i, \bar{S}_{i+1}$, and $\bar{S}_{i+2}$ and label the parity-check bits (P_2,P_1) on each state transition, noting the punctured parity-check bits by X.

12. Redraw the SCCC encoder shown in Figure 8.52a for the DSCCC decoder shown in Figure 8.52b. Using outer, middle, inner code rates of 1/2, 2/3, and 3/4, respectively, with $k = 1$ source bit for each input sample and a block length of $kN = 16{,}384$ source bits, determine the overall code rate and the interleaver lengths L_1 and L_2. Hint: for the i-th encoder output, with rate p_i/p_{i+1}, the interleaver length is $L_i = k_i p_{i+1}$ where, using (8.113), k_i is constant for all i.

13. Using Table 8.33 and referring to Figure 8.61, draw the rate 1/2 SCCC outer and inner constituent encoders and identify encoder number of states, if the CCs are both recursive and systematic.

14. For the LDPC code (*n, j, k*), described in Section 8.13, derive the expression for the density of ones in terms of the parameters *n*, *j*, and *k*.

15. For the binary GF(2^8), complete the table for the first 16 field elements including the zero and unity elements. Express the primitive elements $\alpha^{n'} : n' > 7$ in terms of the elements $\alpha^n : 1 \le n \le m-1$. Also include in the table the binary coefficients and the corresponding integer value as shown in Table 8.39.

16. Show that addition and subtraction of powers the primitive element α is the same in the binary Galois field GF(2), that is, show that $\alpha^n + \alpha^{n'} = \alpha^n - \alpha^{n'}$. Also show that $\alpha^n \pm \alpha^n = 0$ for $0 \le n \le m-1$.

REFERENCES

1. C. Berrou, A. Glavieux, P. Thitimajshima, "Near Shannon Limit Error-Correcting Coding and Decoding: Turbo Codes," IEEE Proceedings of ICC '93, pp. 1064–1070, Geneva, Switzerland, May 1993. (This paper is reprinted in IEEE Transactions on Communications, October 1996—see Reference 107).
2. M.R. Aaron, "PCM Transmission in the Exchange Plant," The Bell System Technical Journal, Vol. 41, No. 1, pp. 99–141, January 1962.
3. Inter-Range Instrumentation Group (IRIG) Telemetry Standard, Range Commanders Council (RCC) Document, 106–05.
4. H.L. Deffebach, W.O. Frost, "A Survey of Digital Baseband Signaling Techniques," NASA Technical Memorandum X-64615, George C. Marshall Space Flight Center, Alabama, June 30, 1971.
5. W. Stallings, "Digital Signaling Techniques," IEEE Communications Magazine, Vol. 22, No. 12, pp. 21–25, December 1984.
6. W.R. Bennett, "Statistics of Regenerative Digital Transmission," Bell System Technical Journal, Vol. 37, No. 6, pp. 1501–1542, November 1958.
7. R.C. Titsworth, L.R. Welch, "Power Spectra of Signals Modulated by Random and Pseudorandom Sequences," Technical Report 32–140, Jet Propulsion Laboratory, Pasadena, CA, October 1961.
8. W.C. Lindsey, M.K. Simon, *Telecommunication Systems Engineering*, Dover Publications, Inc., New York, 1773.
9. R.C. Houts, T.A. Green, "Comparing Bandwidth Requirements for Binary Baseband Signals," IEEE Transactions on Communications, Vol. COM-12, No. 6, pp. 776–781, June 1973.
10. E.D. Sunde, "Theoretical Fundamentals of Pulse Transmission," Bell System Technical Journal, Part I, Vol. 33, pp. 721–788, May 1954 and Part II, Vol. 33, pp. 25–97, July 1954.
11. V. Johannes, A. Kaim, T. Walzman, "Bipolar Pulse Transmission with Zero Extraction," IEEE Transactions on Communication Technology, Vol. COM-17, No. 2, pp. 303–310, April 1969.
12. K.E. Fultz, D.B. Penick, "The T1 Carrier System," Bell System Technical Journal, Vol. 44, No. 7, pp. 1405–1451, September 1965.
13. B.H. Batson, "An Analysis of the Relative Merits of Various PCM Code Formats," Report MSC-EB-R-68-5, NASA Manned Spacecraft Center, Houston, TX, 1 November 1968.
14. M. Hecht, A. Guida, "Delay Modulation," Proceeding of the IEEE, Vol. 57, No. 7, pp. 1314–1316, July 1969.
15. Range Commanders Council (RCC), Telemetry Applications Handbook, RCC-119-06, May 2006.
16. W.N. Waggener, *Pulse Code Modulation Systems Design*, Artech House, Boston, MA, 1998.
17. R.M. Gagliardi, *Introduction to Communications Engineering*, John Wiley & Sons, Inc., New York 1978.
18. I. Korn, "Error Probability and Bandwidth of Digital Modulation," IEEE Transactions on Communications, Vol. COM-28 No. 2, pp. 287–290, February 1980.
19. E.L. Law, "Optimized Low Bit Rate PCM/FM Telemetry with Wide IF Bandwidths," Proceedings of the International Telemetering Conference, San Diego, CA, October 2002.
20. Range Commanders Council (RCC), Telemetry Applications Handbook, RCC-119-06, May 2006, Chapter 2, Sections 2.1.1, 2.1.2, and 2.2.
21. C.H. Tan, T.T. Tjhung, H. Singh, "Performance of Narrow-Band Manchester Coded FSK with Discriminator Detection," IEEE Transactions on Communications, Vol. COM-31, No. 5, pp. 659–667, May 1983.
22. D.E. Cartier, "Limiter-Discriminator Detection Performance of Manchester and NRZ Coded FSK," IEEE Transactions on Aerospace and Electronic Systems, Vol. AES-13, No. 1, pp. 62–70, January 1977.
23. W.P. Osborne, M.B. Luntz, "Coherent and Noncoherent Detection of CPFSK", IEEE Transactions on Communications, Vol. COM-22, No. 8, pp. 1023–1036, August 1974.
24. M.S. Geoghegan, "Improving the Detection Efficiency of Conventional PCM/FM Telemetry by using a Multi-Symbol Demodulator," Proceedings of the 2000 International Telemetering Conference, Volume XXXVI, pp. 717–724, San Diego, CA, October 2000.
25. Range Commanders Council (RCC), Telemetry Applications Handbook, RCC-119-06, May 2006, Chapter 2, Section 2.2.8.
26. E. L. Law, K. Feher, "FQPSK versus PCM/FM for Aeronautical Telemetry Applications: Spectral Occupancy and Bit-error Probability Comparisons," Proceedings of the International Telemetering Conference, Las Vegas, NV, October 1997.
27. M.G. Pelchat, "The Autocorrelation Function and Power Spectrum of PCM/FM with Random Binary Modulating Waveforms," IEEE Transactions on Space Electronics and Telemetry, Vol. SET-10, No. 1, pp. 39–44, March 1964.
28. E.L. Law, "RF Spectral Characteristics of Random PCM/FM and PSK Signals," Proceedings of the International Telemetering Conference, Los Vegas, NV, November 4–7, 1991.
29. A.D. Watt, V.J. Zurick, R.M. Coon, "Reduction of Adjacent-Channel Interference Components from Frequency-Shift-Keyed Carriers," IRE Transactions on Communication Systems, Vol. CS-6, No. 12, pp. 39–47, December 1958.

30. S. W. Golomb, "Sequence with Randomness Properties," Glen L. Martin Company, Internal Report, Baltimore, MD, June 14, 1955.
31. R. Gold, "Properties of Linear Binary Encoding Sequences," Lecture Notes, Robert Gold Associates, Los Angeles, CA, July 1975.
32. W.W. Peterson, E.J. Weldon, Jr., *Error-Correcting Codes*, Second Edition, The MIT Press, Cambridge, MA, 1972, Chapter 7.
33. W.W. Peterson, E.J. Weldon, Jr., *Error-Correcting Codes*, Second Edition, The MIT Press, Cambridge, MA, 1972, Chapter 6.
34. W.W. Peterson, E.J. Weldon, Jr., *Error-Correcting Codes*, Second Edition, The MIT Press, Cambridge, MA, 1972.
35. M.P. Ristenbatt, "Pseudo Binary Coded Waveforms," R.S. Berkowitz, Editor, *Modern Radar, Analysis, Evaluation and System Design*, p. 296, Part IV, Chapter 4, John Wiley & Sons, Inc., New York, 1966.
36. R. Gold, "Properties of Linear Binary Encoding Sequences," Lecture Notes, Robert Gold Associates, Los Angeles, CA, July 1975, p. 6–2.
37. S. Lin, *An Introduction to Error-Correcting Codes*, Prentice-Hall, Inc., Englewood Cliffs, NJ, 1970. (This reference includes a discussion of burst-error correcting codes.)
38. W.W. Peterson, E.J. Weldon, Jr., *Error-Correcting Codes*, Second Edition, The MIT Press, Cambridge, MA, 1972, Appendix C.
39. S. Lin, *An Introduction to Error-Correcting Codes*, pp. 58–70, Prentice-Hall, Inc., Englewood Cliffs, NJ, 1970.
40. F.J. MacWilliams, "A Theorem on the Distribution of Weights in Systematic Code," Bell System Technical Journal, Vol. 42, pp. 79–94, 1963.
41. W.W. Peterson, E.J. Weldon, Jr., *Error-Correcting Codes*, Second Edition, The MIT Press, Cambridge, MA, 1972, pp. 64–70.
42. J.L. Massey, "Coding Techniques for Digital Data Networks," Proceedings of the International Conference on Information Theory and Systems," NTG-Fachberichte, Vol. 65, pp. 18–20, Berlin, Germany, September 1978.
43. T. Fujiwara, T. Kasami, A. Kitai, S. Lin, "On the Undetected Error Probability for Shortened Hamming Codes," IEEE Transactions on Communications, Vol. COM-33, No. 6, pp. 570–574, June 1985.
44. J.K. Wolf, R.D. Blakeney, II, "An Exact Evaluation of the Probability of Undetected Error for Certain Shortened Binary CRC Codes," IEEE Military Communications Conference, MILCOM '88, Vol. 1, pp. 0287–0292, San Diego, CA, October 23–26, 1988.
45. M. Abramowitz, I.A. Stegun, Editors, Handbook of Mathematical Functions with Formulas, Graphs, and Mathematical Tables, National Bureau of Standards, Applied Mathematics Series No. 55, pp. 949–950, November 1964.
46. F. Jager, C.B. Dekker, "Tamed Frequency Modulation, A Novel Method to Achieve Spectrum Economy in Digital Transmission," IEEE Transactions on Communications, Vol. COM-26, No. 5, pp. 534–542, May 1978.
47. D. Muilwijk, "Tamed Frequency Modulation—A Bandwidth-Saving Digital Modulation Method, Suited for Mobile Radio," Philips Telecommunication Review, Vol. 37, No. 3, pp. 35–49, March 1979.
48. S. Bellini, M. Sonzogni, G. Tartara, "Noncoherent Detection of Tamed Frequency Modulation," IEEE Transactions on Communications, Vol. COM-32, No. 3, pp. 218–224, March 1984.
49. K. Chung, "Generalized Tamed Frequency Modulation and Its Application for Mobile Radio Communications," IEEE Transactions on Selected Areas in Communications, Vol. SAC-2, No. 4, pp. 487–497, July 1984.
50. J.L. Ramsey, "Realization of Optimum Interleavers," IEEE Transactions on Information Theory, Vol. IT-16, No. 3, pp. 338–345, May 1970.
51. G.D. Forney, Jr., "Burst-Correcting Codes for the Classic Bursty Channel," IEEE Transactions on Communication Technology, Vol. COM-19, pp. 772–781, October, 1971.
52. D.D. Newman, "Observations of the Nature and Effects of Interleavers in Communications Links," Final Report AFWL-TR-82-125, Air Force Weapons Laboratory, Kirtland Air Force Base, NM, February 1983.
53. A. Papoulis, *Probability, Random Variables and Stochastic Processes* pp. 57–63, McGraw-Hill, New York, 1965.
54. J.M. Wozencraft, "Sequential Decoding for Reliable Communications," Technical Report 325, Research Laboratories of Electronics, The MIT Press, Cambridge, MA, 1957.
55. R.M. Fano, "A Heuristic Discussion of Probabilistic Decoding," IEEE Transactions on Information Theory, Vol. IT-9, pp. 64–74, 1963.
56. J.L. Massey, D.J. Costello, Jr., "Nonsystematic Convolutional Codes for Sequential Decoding in Space Applications," IEEE Transactions on Communication Technology, Part II, Vol. COM-19, pp. 806–813, October, 1971.
57. J.A. Heller, I. M. Jacobs, "Viterbi Decoding for Satellite and Space Communications," IEEE Transactions on Communication Technology, Vol. COM-19, No. 5, pp. 835–848, October 1971.
58. W.W. Peterson, *Error-Correcting Codes*, The MIT Press, Cambridge, MA, 1961.
59. A.J. Viterbi, *Principles of Coherent Communications*, McGraw-Hill, New York, 1966.
60. S. Lin, *An Introduction to Error-Correcting Codes*, Chapters 10–13, Prentice-Hall, Inc., Englewood Cliffs, NJ, 1970.
61. W. W. Peterson, E. J. Weldon, Jr., *Error-Correcting Codes*, Chapters 13 and 14, The MIT Press, Cambridge, MA, 1972.
62. A.J. Viterbi, J.K. Omura, *Principles of Digital Communication and Coding*, McGraw-Hill, New York, 1966.
63. A.M. Michelson, A.H. Levesque, *Error-Control Techniques for Digital Communication*, John Wiley & Sons, Inc., New York, 1985.
64. J.P. Odenwalder, "Error Control Coding Handbook," United Sates Air Force, Contract No. F44620-76-C-0056, Final Report, Linkabit Corporation, July 15, 1976.
65. S. Benedetto, D. Divsalar, "Turbo Codes: Principles and Applications," Coarse Notes Presented at Titan Linkabit, July 28–30, 1997.
66. S. Benedetto, G. Montorsi, "Bandwidth Efficient, Versatile Concatenated Coded Schemes with Interleavers," Presented at Titan Linkabit, July 29, 1999.
67. Y. Yasuda, K. Kashiki, Y. Hirata, "High-Rate Punctured Convolutional Codes for Soft Decision Viterbi Decoding," IEEE Transactions on Communications, Vol. COM-32, No. 3, pp. 315–319, March 1984.

68. A.J. Viterbi, "Error Bounds for Convolutional Codes and an Asymptotically Optimum Decoding Algorithm," IEEE Transactions on Information Theory, Vol. IT-13, pp. 260–269, 1967.
69. G.D. Forney, Jr., "The Viterbi Algorithm," Proceedings of the IEEE, Vol. 61, No. 3, pp. 268–278, March 1973.
70. G.D. Forney, Jr., "Convolutional Codes II: Maximum Likelihood Decoding," Technical Report 7004–1, Stanford Electronics Laboratories, Stanford, CA, June 1972.
71. G.D. Forney, Jr., "Coding and Its Application in Space Communications," IEEE Spectrum, June 1970.
72. J.A. Heller, "Short Constraint Length Convolutional Codes," Space Programs Summary 37–54, Vol. III, pp. 171–177, Jet Propulsion Laboratory, California Institute of Technology, October/November, 1968.
73. J.A. Heller, "Improved Performance of Short Constraint Length Convolutional Codes," Space Programs Summary 37–56, Vol. III, pp. 83–84, Jet Propulsion Laboratory, California Institute of Technology, February/March 1969.
74. C.M. Rader, "Memory Management in a Viterbi Decoder," IEEE Transactions on Communications, Vol. COM-29, No. 9, pp. 1399–1401, September, 1981.
75. J.L. Massey, "Catastrophic Error-Propagation in Convolutional Codes," Proceedings, Eleventh Midwest Symposium on Circuit Theory, University of Notre Dame, Notre Dame, Indiana, May 13–14, 1968.
76. J.P. Odenwalder, "Optimum Decoding of Convolutional Codes," Ph.D. dissertation, University of California, Los Angeles, CA, 1970.
77. E. Paaske, "Short Binary Convolutional Codes with Maximal Free Distance for Rates 2/3 and 3/4," IEEE Transactions on Information Theory, Vol. IT-20, pp. 683–689, September 1974.
78. A. M. Michelson, A. H. Levesque, *Error-Control Techniques for Digital Communication*, John Wiley & Sons, Inc., New York, 1985, Section 9.5.4.
79. K.J. Larsen, "Short Convolutional Codes with Maximal Free Distance for Rates 1/2, 1/3, and 1/4," IEEE Transactions on Information Theory, IT-19, pp. 371–372, May 1973.
80. R.L. Peterson, R.E. Ziemer, D.E. Borth, *Introduction to Spread Spectrum Communications*, Chapter 9, Section 4.1, Prentice Hall, Englewood Cliffs, NJ, 1995.
81. D.G. Daut, J.W. Modestino, L.D. Wismer, "New Short Constraint Length Convolutional Code Construction for Selected Rational Rates," IEEE Transactions on Information Theory, Vol. IT-28, pp. 793–799, September 1982.
82. B.D. Trumpis, "Convolutional Coding for M-ary Channels," Ph.D. dissertation, University of California, Los Angeles, CA, 1975.
83. G.D. Forney, Jr., "Structural Analysis of Convolutional Codes via Dual Codes," IEEE Transactions on Information Theory, Vol. IT-19, pp. 512–518, July, 1973.
84. G.D. Forney, Jr., "Convolutional Codes I: Algebraic Structure," IEEE Transactions on Information Theory, Vol. IT-16, pp. 720–738, November 1970.
85. J. G. Proakis, *Digital Communications*, Second Edition, pp. 468–473, McGraw-Hill, New York, 1989.
86. A.J. Viterbi, I.M. Jacobs, "Advances in Coding and Modulation for Noncoherent Channels Affected by Fading, Partial Band, and Multiple-Access Interference," A.J. Viterbi, Editor, *Advances in Communication Systems*, Vol. 4, Academic Press, New York, 1975.
87. J.G. Proakis, I. Rahman, "Performance of Concatenated Dual-*k* Codes on a Rayleigh Fading Channel with a Bandwidth Constraint," IEEE Transactions on Communications, Vol. COM-27, No. 5, pp. 801–806, May 1979.
88. A.J. Viterbi, "Convolutional Codes and Their Performance in Communication Systems," IEEE Transactions on Communication Technology, Vol. COM-19, No. 5, pp. 751–772, October 1971.
89. A.J. Viterbi, "The State-Diagram Approach to Optimal Decoding and Performance Analysis for Memoryless Channels," Space Programs Summary, 37–38, Vol. III, pp. 50–55, Jet Propulsion Laboratory, 1969.
90. J.P. Odenwalder, "Dual-*k* Convolutional Codes for Noncoherently Demodulated Channels," Proceeding of the International Telemetering Conference, Vol. 12, pp. 165–174, September 1976.
91. J.B. Cain, G.C. Clark, Jr., J.M. Geist, "Punctured Convolutional Codes of Rate $(n\text{-}1)/n$ and Simplified Maximum Likelihood Decoding," IEEE Transactions on Information Theory, Vol. IT-25, No. 1, pp. 97–100, January 1979.
92. J.K. Wolf, E. Zehavi, "P2 Codes: Pragmatic Trellis Codes Utilizing Punctured Convolutional Codes," IEEE Communications Magazine, Vol. 33, No. 2, pp. 94–99, February 1995.
93. A.J. Viterbi, J.K. Wolf, E. Zehavi, R. Padovani, "A Pragmatic Approach to Trellis-Coded Modulation," IEEE Communications Magazine, Vol. 27, No. 7, pp. 11–19, July, 1989.
94. G. Ungerboeck, "Channel Coding with Multilevel/Phase Signals," IEEE Transactions on Information Theory, Vol. IT-28, pp. 55–67, January 1982.
95. Qualcomm, *Q1900 Viterbi/Trellis Decoder*, Qualcomm Corporation, ASIC Products, San Diego, CA, 1991.
96. G.D. Forney, Jr., *Concatenated Codes*, The MIT Press, Cambridge, MA, 1966.
97. J. P. Odenwalder, "Optimum Decoding of Convolutional Codes," Ph.D. dissertation, University of California, Los Angeles, CA, 1970, Chapter 6.
98. J.P. Odenwalder, "Error Control Coding Handbook," United Sates Air Force, Contract No. F44620-76-C-0056, Final Report, Linkabit Corporation, July 15, 1976, Section 6.0, pp. 183–189.
99. J.M. Wozencraft, I.M. Jacobs, *Principles of Communication Engineering*, Chapter 7, John Wiley & Sons, Inc., New York, 1967.
100. S. Benedetto, D. Divsalar, G. Montorsi, F. Pollara, "Parallel Concatenated Trellis Coded Modulation," IEEE Proceedings of ICC'96, San Diego, CA, October 28–31, 1996.
101. D. Divsalar, F. Pollara, "Turbo Codes for PCS Applications," Proceedings of IEEE Proceedings of ICC'95, Seattle, Washington, June 1995.
102. C. Heegard, S.B. Wicker, *Turbo Coding*, Kluwer Academic Publishers, Boston, MA, 1999.
103. S. Benedetto, D. Divsalar, G. Montorsi, F. Pollara, "Serial Concatenation of Interleaved Codes: Performance Analysis, Design, and Iterative Decoding," The JPL TDA Progress Report 42–126, August 15, 1996.

104. S. Benedetto, D. Divsalar, G. Montorsi, F. Pollara, "Soft-Out Decoding Algorithms in Iterative Decoding of Turbo Codes," TDA Progress Report 42–124, JPL, CalTech, Pasadena, CA, February 15, 1996. Also appears in IEEE Proceedings of ICC'96, Dallas, TX, pp. 112–117, June 1996.
105. D. Divsalar, "Iterative Decoding of Turbo, Serial, Hybrid, and Self Concatenated Codes," JPL-CALTECH, October, 1998. (Provides performance summaries for: PCCC, p. 71; SCCC p.845-86; DPCCC, p. 94, 95; HCCC, p. 114, 115; Self Concatenated Code, p. 123.)
106. D. Divsalar, R.J. McEliece, *The Effective Free Distance of Turbo Codes*, Jet Propulsion Laboratory, California Institute of Technology, Pasadena, CA, December 1995.
107. B. Vucetic, Y. Li, L.C. Perez, F. Jiang, "Recent Advances in Turbo Code Design and Theory," Proceedings of the IEEE, Vol. 95, No. 6, pp. 1323–1344, June 2007.
108. J.D. Anderson, V.V. Zybolv, "Interleaver Design for Turbo Coding," International Symposium on Turbo Codes, Brest, France, 1997.
109. J. Yuan, S.G. Wilson, "Multiple Output Sliding Window Decoding Algorithm for Turbo Codes," Proceeding of CISS'97, Baltimore, MD, March 1997.
110. C. Berrou, A. Glavieux, "Near Optimum Error Correcting Coding and Decoding: Turbo-Codes," IEEE Transactions on Communications, Vol. 44, No. 10, pp. 1261–1271, October 1996. (This paper was presented at ICC'93, Geneva, Switzerland, May 1993).
111. D. Divsalar, F. Pollara, "Multiple Turbo Codes for Deep-Space Communications," JPL TDA Progress Report 42–121, Jet Propulsion Laboratory, Pasadena, CA, May 1995.
112. S. Dolinar, D. Divsalar, F. Pollara, "Code Performance as a Function of Block Size," TMO Progress Report 42–133, JPL, May 1998.
113. S.W. Golomb, H. Taylor, "Constructions and Properties of Costas Arrays," Proceedings of the IEEE, Vol. 72, No. 9, pp. 1143–1163, September 1984.
114. S.B. Wicker, *Error Control Systems for Digital Communications and Storage*, Prentice Hall, Englewood Cliffs, NJ, 1995.
115. S. Benedetto, G. Montorsi, D. Divsalar, F. Pollara, "A Soft-In Soft-Out Maximum A Posteriori (MAP) Module to Decode Parallel and Serial Concatenated Codes," TDA Progress Report 42–127, November 15, 1996.
116. S. Benedetto, G. Montorsi, "Unveiling Turbo Codes: Some Results on Parallel Concatenated Coding Schemes," IEEE Transactions on Information Theory, Vol. 42, No. 2, pp. 409–429, March 1966.
117. D. Divsalar, "Iterative Decoding of Turbo, Serial, Hybrid, and Self Concatenated Codes," JPL-CALTECH, October 1998. (Provides equations for forward and backward recursion processing within the SISO module).
118. D. Divsalar, "Iterative Decoding of Turbo, Serial, Hybrid, and Self Concatenated Codes," JPL-CALTECH, October, 1998. Presents the algorithms for the SCCC two-SISO inner code recursions, pp. 78, 79, outer code recursions, and output bit reliability, pp. 80–82.
119. S. Benedetto, D. Divsalar, G. Montorsi, F Pollara, "Analysis, Design, and Iterative Decoding of Double Serially Concatenated Codes with Interleaves," IEEE Transactions on Selected Areas in Communications, Vol. 16, No. 2, pp. 231–244, February 1998.
120. S. Benedetto, D. Divsalar, G. Montorsi, F. Pollara, "A Soft-Input Soft-Output AAP Module for Iterative Decoding of Concatenated Codes," Interim Report, Partially supported by NATO Grant RG951208, partially carried out at JPL CALTECH under contract with NASA, January 3, 1997.
121. S. Benedetto, D. Divsalar, G. Montorsi, "Turbo-Like Codes Constructions," Chapter 2, D. Declercq, M. Fossorier, E. Biglieri, Editors, *Channel Coding Theory, Algorithms, and Applications*, pp. 53–140, Academic Press Library in Mobile and Wireless Communications, Elsevier Books, Oxford, 2014.
122. L.R. Bahl, J. Cocke, F. Jelinek, J. Raviv, "Optimal Decoding of Linear Codes for Minimizing Symbol Error Rate," IEEE Transactions on Information Theory, Vol. IT-20, pp. 284–287, March 1974.
123. D. Divsalar, "Iterative Decoding of Turbo, Serial, Hybrid, and Self Concatenated Codes," p. 41, JPL-CALTECH, October 1998.
124. Y. Wu, B.D. Woerner, T.K. Blankenship, "Data Width Requirements in SISO Decoding with Modulo Normalization," IEEE Transactions on Communications, Vol. 49, No. 11, pp. 1861–1868, November 2001.
125. Consultative Committee for Space Data Systems (CCSDS), Blue Book, CCSDS 101.0-B-6, Section 4, Turbo Coding, October 2002.
126. Consultative Committee for Space Data Systems (CCSDS), Recommendation for Space Data System Standards, "Telemetry Channel Coding," CCSDS 101.0-B-6, Blue Book, pp. 4–5, October 2002.
127. C. Douillard, M. Jezequel, "Turbo Codes: From First Principles to Recent Standards," Chapter 1, D. Declercq, M. Fossorier, E. Biglieri, Editors, *Channel Coding Theory, Algorithms, and Applications*, pp. 1–52, Academic Press Library in Mobile and Wireless Communications, Elsevier Books, Oxford, 2014.
128. S. Dolinar, D. Divsalar, F. Pollara, "Code Performance as a Function of Block Size," JPL TDA Progress Report 42–130, Jet Propulsion Laboratory, Pasadena, CA, August 1998.
129. S. Dolinar, D. Divsalar, F. Pollara, "Turbo Codes and Space Communications," Communications Systems and Research Section, Jet Propulsion Laboratory, CalTech, Pasadena, CA, 1998.
130. S. Benedetto, G. Montorsi, "Design of Parallel Concatenated Convolutional Codes," IEEE Transactions on Communications, Vol. 44, No. 5, pp. 591–600, May 1996.
131. R.G. Gallager, "Low-Density Parity-Check Codes," Institute of Radio Engineers, Vol. IT-8, pp. 21–28, January 1962. Also, reprinted in the "The Development of Coding Theory," Editor E.R. Berlekamp, IEEE Press, Part II: Constructions for Block Codes, pp. 121–128, New York, 1974.
132. D.J.C. MacKay, "Good Error-Correcting Codes Based on Very Sparse Matrices," IEEE Transactions on Information Theory, Vol. 45, No. 3, pp. 399–431, March 1999.
133. D.J. Costello, Jr., G.D. Forney, Jr., "Channel Coding: The Road to Channel Capacity," Proceedings of the IEEE, Vol. 95, No. 6, pp. 1150–1177, June 2007.
134. D. Slepian, "A Class of Binary Signaling Alphabets," Bell System Technical Journal, Vol. 35, pp. 203–234, January 1956.

135. D. Sridhara, T. Fuja, R.M. Tanner, "Low Density Parity-Check Codes form Permutation Matrices," Conference on Information Sciences and Systems, The John Hopkins University, March 2001.
136. M. Luby, M. Mitzenmacher, M.A. Shokrollahi, D.A. Spielman, "Improved Low-Density Parity-Check Codes using Irregular Graphs," IEEE Transactions on Information Theory, Vol. 47, No. 2, pp. 585–598, February 2001.
137. T.J. Richardson, M.A. Shokrollahi, R.L. Urbanke, "Design of Capacity-Approaching Irregular Low-Density Parity-Check Codes," IEEE Transactions on Information Theory, Vol. 47, No. 2, pp. 619–637, February 2001.
138. R.G. Gallager, "Low-Density Parity-Check Codes," Sc. D. thesis, Massachusetts Institute of Technology, Cambridge, MA, September 1960.
139. R.G. Gallager, "Low-Density Parity-Check Codes," Institute of Radio Engineers, Vol. IT-8, pp. 21–28, January 1962. Also, reprinted in the "The Development of Coding Theory", Editor E.R. Berlekamp, IEEE Press, Part II: Constructions for Block Codes, page 122, Table 3, New York, 1974.
140. K. Gracie, M.H. Hamon, "Turbo and Turbo-Like Codes: Principles and Applications in Telecommunications," Proceedings of the IEEE, Vol. 95, No. 6, pp. 1228–1254, June 2007.
141. C.-H. Liu, S.-W. Yen, C.-L. Chen, H.-C. Chang, C.-Y. Lee, Y.-S. Hsu, S.-J. Jou, "An LDPC Decoder Based on Self-Routing Network for IEEE 802.16e Applications," IEEE Journal on Solid-State Circuits, Vol. 43, No. 3, pp. 684–694, March 2008.
142. K. Gracie, M.H. Hamon, "Turbo and Turbo-Like Codes: Principles and Applications in Telecommunications," Proceedings of the IEEE, Vol. 95, No. 6, p. 1235, Figure 9 and p. 1236, Figure 12, June 2007.
143. D. Williams, "Turbo-Charging Next-Gen Wireless, Optical Systems," Communication Systems Design, Vol. 7, No. 2, pp. 39–47, February 2001.
144. R. Pyndiah, A. Glavieux, A. Picart, S. Jacq, "Near Optimum Decoding of Product Codes," Proceeding of Global Telecommunications Conference (Globecom'94), pp. 339–343, San Francisco, CA, November 28–December 2, 1994.
145. R.M. Pyndiah, "Near-Optimum Decoding of Product Codes: Block Turbo Codes," IEEE Transactions on Communications, Vol. 46, No. 8, pp. 1003–1010, August 1998.
146. K. Gracie, M.H. Hamon, "Turbo and Turbo-Like Codes: Principles and Applications in Telecommunications," Proceedings of the IEEE, Vol. 95, No. 6, pp. 1233–1234, June 2007.
147. J. Fang, F. Buda, E. Lemois, "Turbo Product Codes: A Well Suitable Solution to Wireless Packet Transmissions for Very Low Error Rates," Proceedings of the 2nd International Symposium on Turbo Codes and Related Topics," pp. 101–111, Brest, France, September 2000.
148. S.Y. Chung, G.D. Forney, Jr., T.J. Richardson, R. Urbanke, "On the Design of Low-Density Parity-Check Codes Within 0.0045 dB from the Shannon Limit," IEEE Communication Letters, Vol. 5, No. 2, pp. 58–60, February 2001.
149. C. Berrou, J. Hagenauer, M. Luise, C. Schlegel, and L. Vandendorpe, Editors, "Turbo-Information Processing: Algorithms, Implementations, and Applications," Special Issue, Proceeding of the IEEE," Vol. 95, No. 6, June 2007.
150. W.W. Peterson, E.J. Weldon, Jr., *Error-Correction Codes*, Second Edition, Chapter 12, The MIT Press, Cambridge, MA, 1972.
151. C. Herzet, N. Noels, V. Lottici, H. Wymeersch, M. Luise, M. Moeneclaey, L. Vandendorpe, "Code-Aided Turbo Synchronization," Proceeding of the IEEE, Vol. 95, No. 6, pp. 1255–1271, June 2007.
152. Consultative Committee for Space Data Systems (CCSDS), Blue Book, CCSDS 101.0-B-6, October 2002. Turbo Code Synchronization, Section 4, Specification 10; Frame Synchronization, Section 5.
153. A. Anastasopoulos, K.M. Chugg, G. Colavolpe, G. Ferrari, R. Raheli, "Interactive Detection for Channels with Memory," Proceedings of the IEEE, Vol. 95, No. 6, pp. 1272–1294, June 2007.
154. W.W. Peterson, E.J. Weldon, Jr., *Error-Correcting Codes*, Second Edition, pp. 274 and 275, The MIT Press, Cambridge, MA, 1972, Chapter 9.
155. W.W. Peterson, E.J. Weldon, Jr., *Error-Correcting Codes*, Second Edition, pp. 493–534, The MIT Press, Cambridge, MA, 1972, Appendix D.
156. C.L. Chen, "Some Results on Algebraically Structured Error-Correction Codes," Ph.D. dissertation, University of Hawaii, 1969.
157. M.J. Weiss, R. Dubisch, *Higher Algebra for the Undergraduate*, John Wiley & Sons, Inc., New York, 1962.
158. R.W. Marsh, *Table of Irreducible Polynomials over GF(2) Through Degree 19*, NSA, Washington, DC, 1957.
159. S. Lin, *An Introduction to Error-Correcting Codes*, pp. 21–23; 112–119, Prentice-Hall, Inc., Englewood Cliffs, NJ, 1970
160. S. Lin, D.J. Costello, Jr., *Error Control Coding: Fundamentals and Applications*, Prentice-Hall, Inc., Englewood Cliffs, NJ, 1983.
161. J.P. Stenbit, "Table of Generators for Bose-Chaudhuri Codes," IEEE Transactions on Information Theory, Vol. IT-10, pp. 390–391, January 1964.
162. J.L. Massey, "Shift-Register Synthesis and BCH Decoding," IEEE Transactions on Information Theory, Vol. IT-15, pp. 122–127, January 1969.
163. W.W. Peterson, E.J. Weldon, Jr., *Error-Correcting Codes*, Second Edition, The MIT Press, Cambridge, MA, 1972, pp. 289–296.
164. E.R. Berlekamp, "On Decoding Binary Bose-Chaudhuri-Hocquenghem Codes," IEEE Transactions on Information Theory, Vol. IT-11, pp. 577–580, October 1965.
165. G.D. Forney, Jr., "On Decoding BCH Codes," IEEE Transactions on Information Theory, Vol. IT-11, pp. 549–557, October 1965.
166. S. Lin, *An Introduction to Error-Correcting Codes*, Prentice-Hall, Inc., Englewood Cliffs, NJ, 1970, pp. 122, 125.
167. E.R. Berlekamp, *Algebraic Coding Theory*, McGraw-Hill, New York, 1968.

9

FORWARD ERROR CORRECTION CODING WITHOUT BANDWIDTH EXPANSION

9.1 INTRODUCTION

This chapter focuses on two modulation waveforms that provide coding gain without requiring additional bandwidth beyond that of the transmitted information symbol [1]. The two waveforms are M-ary multi-h continuous phase modulation (CPM) and multiphase shift keying (MPSK) trellis-coded modulation (TCM). The MPSK TCM is also examined in terms of multilevel quadrature pulse amplitude modulation (MQPAM) TCM. Because the information or data is contained in the signal phase function, both of these techniques provide constant envelope waveforms: a necessary condition for robust performance when used with hard-limiting channels. Furthermore, the capability to provide coding gain without bandwidth expansion is paramount when attempting to achieve the highest possible data rate in narrowband channels with low-power small aperture disadvantaged terminals. The coding gain for these waveforms is provided by the inclusion of redundant phase states that behave much like the redundant parity bits in conventional bandwidth expansion with forward error correction (FEC) coding techniques as discussed in Chapter 8. The multi-h CPM waveform provides the redundant phase states by judicious variations of the modulation index h, whereas the MPSK-TCM waveform provides redundant phase states by the using $M' < M$ phase states for the source information leaving $M - M'$ redundant phase states to resolve errors.

In Section 9.2, the theoretical characteristics of the multi-h M-ary CPM waveform are discussed including the selection of the modulation indices (h), the evaluation of the modulated waveform spectrum, and the waveform demodulation. The options involving full response CPM, in which the information bearing phase is constrained to one symbol, and partial response CPM, where the phase response is spread over adjacent symbols, are also discussed.

Section 9.3 provides a case study that focuses on the 2-h 4-ary full response CPM waveform. Following a detailed description of the waveform modulation and spectrum, the simulated bit-error performance results are provided comparing the performance using the additive white Gaussian noise (AWGN) channel with those using a narrowband hard-limiting satellite channel. With full response modulation, denoted as 1REC, the symbol energy is confined to one symbol interval in contrast to partial-response modulation that distributed symbol energy over several adjacent symbols. Demodulation of the full response modulated multi-h CPM waveform is considerably more computationally efficient.

In Section 9.4, a similar characterization is provided for the MPSK-TCM modulated waveform. In this case, the intrinsic *sinc*(fT) frequency spectrum of the MPSK waveform is improved by applying phase shaping that extends into the adjacent symbols. Section 9.5 provides a case study characterizing the performance of the 8PSK TCM waveform comparing the bit-error using the linear AWGN channel with those using a narrowband hard-limiting satellite channel.

Digital Communications with Emphasis on Data Modems: Theory, Analysis, Design, Simulation, Testing, and Applications, First Edition. Richard W. Middlestead.

Companion website: www.wiley.com/go/digitalcommunications

9.2 MULTI-*h* *M*-ary CPM

In this section, the theoretical characteristics and performance of the multi-*h* *M*-ary CPM waveform are discussed. The modulation index is generally selected to achieve the maximum coding gain. However, the modulation index also influences the waveform spectrum, so the available channel bandwidth and phase distortion must also be considered in the selection of the modulation index. In general, higher modulation indices are associated with increased coding gain and more efficient use of the signal spectrum, that is, the spectral energy is more uniformly distributed over the occupied bandwidth.

The multi-*h* *M*-ary CPM waveform is a constant envelope signal with continuous phase variations that depend on the past and present source data, α, and the modulation indices *h*. The CPM waveform modulation is described as

$$s(t) = A\cos(\omega_c t + \theta(t;\boldsymbol{\alpha},\boldsymbol{h}) + \phi_o) \qquad (9.1)$$

where *A* is the peak signal amplitude, ω_c is the carrier radian frequency, and ϕ_o is an arbitrary signal phase. The phase function $\theta(t;\boldsymbol{\alpha},\boldsymbol{h})$ describes the phase history of the signal starting at $t = 0$ and continuing through the present or *j*-th symbol starting at $t = (j-1)T$ where *T* is the symbol duration and $j = 1, 2, \ldots$. The phase function is described as

$$\begin{aligned}\theta(t;\alpha,\mathbf{h}) &= 2\pi\sum_{n=1}^{j}\alpha_n h_n q(t-(n-1)T) \qquad :(n-1)T \le t \le nT \\ &= 2\pi\alpha_j h_j q(t-(j-1)T) + \theta_{j-1}(\alpha,\mathbf{h}) \quad :(j-1)T \le t \le jT\end{aligned} \qquad (9.2)$$

where $\theta_{j-1}(\boldsymbol{\alpha},\mathbf{h})$ is the accumulated phase history up to $t = (j-1)T$ and is expressed as

$$\theta_{j-1}(\boldsymbol{\alpha},\mathbf{h}) = 2\pi\sum_{n=1}^{j-1}\alpha_n h_n q(t-(n-1)T) \quad :(n-1)T \le t \le nT \qquad (9.3)$$

The data-dependent parameter $\boldsymbol{\alpha} = \{\alpha_n\}^*$ represents the *M*-ary data symbol alphabets $\alpha_i = \{\pm1, \pm3, \ldots, \pm(M-1)\}$, $M = 2^k$ corresponding to *k* source bits-per-symbol. The modulation index $\boldsymbol{h} = \{h_n\}$ is typically cycled through the set $h_j = \{h_1, h_2, h_K\}$ over *K* contiguous symbols and repeated thereafter. An *over-bar* is used to explicitly denote the *modulo*(*K*) repetition of the modulation indices as $\bar{h}_j = h_i$ where $i = (j-1)modulo(K) + 1$. The designation *h* for the modulation index is in keeping with the standard usage for binary frequency shift keying (BFSK) waveforms, and it is defined as $h = \Delta f/R_b$ where Δf is the frequency shift and R_b is the data rate; for orthogonal continuous phase BFSK (CPBFSK) $h = 1$ and for MSK $h = ½$. The CPM modulation can also be characterized as a constant envelope modulation with a time-dependent frequency function $g(t)$ with the phase function $q(t)$ described as

$$q(t) = \int_{-\infty}^{t} g(\tau)d\tau \qquad (9.4)$$

Several standard frequency functions have been used to characterize the multi-*h* waveform [2]. The more common ones are denoted as LREC for rectangular; LRC for raised cosine; LSRC for spectral raised cosine. The designation $L \ge 1$ indicates the time extent of $g(t)$ in terms of the number of symbol intervals. The LSRC function results in an infinite impulse response and is typically truncated after *L* symbols. For $0 \le t \le LT$, these functions are described as

$$g(t) = \frac{1}{2LT}rect\left(\frac{t}{LT} - \frac{1}{2}\right) \quad :\text{LREC} \qquad (9.5)$$

$$g(t) = \frac{1}{2LT}\left[1-\cos\left(\frac{2\pi t}{LT}\right)\right]rect\left(\frac{t}{LT} - \frac{1}{2}\right) \quad :\text{LRC} \qquad (9.6)$$

$$\begin{aligned}g(t) = \frac{1}{LT}&\left[\frac{\sin(2\pi t/(LT))}{2\pi t/(LT)}\right]\left[\frac{\cos(2\pi\beta t/(LT))}{1-(4\beta t/(LT))^2}\right] \\ &rect\left(\frac{t}{LT} - \frac{1}{2}\right) \quad :\text{LSRC}\end{aligned} \qquad (9.7)$$

When $L = 1$, the modulation is referred to as *full response*; otherwise, it is a *partial response* modulation. Full response simply means that the current phase trajectory is uncorrelated with the past or future data so the frequency function, $g(t)$, depends on the source data corresponding to the current symbol interval $(j-1)T \le t \le jT$. Typically, full response multi-*h* waveforms are much easier to demodulate, whereas partial response waveforms have better spectral containment at the expense of demodulator processing.[†] The 1REC CPM waveform described by (9.5) with $L = 1$ is also referred to as continuous phase FSK (CPFSK) and single-*h* CPFSK with $h = 1/2$ corresponds to MSK also known as fast FSK (FFSK). Another class of frequency functions $g(t)$ introduced in the late 1970s is referred to as tamed frequency modulation (TFM) [3–5]. The key parameters of the *M*-ary multi-*h* or *K–h* CPM waveforms are summarized in Table 9.1.

The system performance and complexity of the demodulator processing is influenced in several ways by the selection

[*] The use of α without italics denotes the currently selected symbol from the alphabetic set $\{\alpha_j\}$ denoted by italics; the same notation is used for modulation indices *h* and h_j.

[†] The Gaussian MSK (GMSK) waveform, discussed in Chapter 4, is a single-*h* partial response CPM waveform with $h = 1/2$. In practice, the theoretically infinite phase response is truncated to *L* symbols.

TABLE 9.1 K–h M-ary CPM Parameter Summary

Parameter	Description
M	M-ary symbol, $M = 2^k$
k	Information bits/symbol
α_j	M-ary symbol alphabet $\{\alpha_i\}$: $i = 0, \ldots, M-1$
K	Number of modulation indices h_j
h_i	Modulation index: $i = 1, \ldots, K$
L	Extent of $g(t)$, $L > 1$ is partial response

TABLE 9.2 Maximum Upper-Bound Values for Multi-h Linear Phase Responses[a]

M	K	Coding Gain (dB)[b]
4	1	3.25
4	2	5.15
4	3	5.83
8	1	4.87
8	2	5.74
8	3	6.23

[a]Aulin [7]. Reproduced by permission of the IEEE.
[b]With respect to BPSK.

of these parameters. For example, the coding gain and the corresponding signal-to-noise ratio, E_b/N_o, required to achieve a specified bit-error probability, the spectral containment required by the adjacent channel interference (ACI) and channel bandwidth specifications, and the channel efficiency or the required bits/second/Hz [6]. The coding gain is determined by the minimum Euclidean distance d_{min} with larger values resulting in greater coding gains. The minimum Euclidean distance increases with M, K, and h, and some examples of their influence on the relative coding gain are given in Tables 9.2 and 9.3 for full response modulation.

The spectral characteristics are predominantly determined by the frequency shaping function $g(t)$ with the REC function exhibiting the *scallop* sidelobe behavior associated with the *sinc(fT)* function. Functions with higher order continuous derivatives, like the raised-cosine function, exhibit improved asymptotic or *far-out* sidelobe behavior, for example, the spectrum rolls off asymptotically as $f^{-2(m+1)}$ where m is the highest continuous derivative of $g(t)$. On the other hand, the width of the main spectral lobe decreases with increasing L and M and increases with h.

As mentioned earlier, the complexity of the demodulator is also influenced by the waveform parameters. When using a trellis decoder in the demodulator, it is necessary that the modulation indices be rational numbers, that is, $h_i = p_i/q$ where p_i and q are integers. In the context of the trellis decoder, the constraint length of the decoder is defined as the smallest number of iterations that can remain unmerged. For full response modulation, it has been shown [8, 9] that there exits modulation index sets $\{h_1, h_2, \ldots, h_K\}$ yielding constraint lengths of $K + 1$ provided that $q \ge M^K$ and

TABLE 9.3 Dependence of Relative Coding Gain on Waveform Parameters ($L = 1$, $h \le 0.5$)

Parameters		Relative Coding Gain (dB)
$K = 1$	$M = 2-4$	3.0
	$M = 4-8$	2.1
$K = 2$	$M = 2-4$	3.4
	$M = 4-8$	1.7
$M = 2$	$K = 1-2$	3.0

$\sum_i^K \alpha_i h_i \notin I$ for any α_i: $i = 1, \ldots, K$, selected from the M-ary set $\{\pm1, \pm3, \ldots, \pm(M-1)\}$ where I signifies the set of integers. The total number of trellis states required by the decoder is Kq; however, only q states are updated during any transition. The memory length of the decoder prior to making a data decision with negligible degradation is $D \gg K + 1$ and typically $D = 5(K + 1)$ to $10(K + 1)$. These design details are further clarified in Section 9.2.3 and in the case study in Section 9.3.

9.2.1 Selection of Modulation Indices for Multi-h, M-ary CPM

The selection of the optimum multi-h modulation indices is an involved process for all but the simplest of cases. This section focuses on the 1REC or full response multi-h modulated waveforms using either a single-h or, in the case of multi-h waveforms, the average of the modulation indices. The performance analysis of the partial response multi-h waveforms is considerably more involved and is discussed in detail in the literature [10–12]. Using the average modulation index is advantageous for three important reasons: the analysis is simplified, the indices are typically very close to one another, and optimizing the performance for the average is representative of the overall performance. The performance measure used to determine the optimum modulation indices is the squared Euclidean distance [7, 13] expressed as

$$\begin{aligned} d^2 &= \int_0^T |s_i(t) - s_j(t)|^2 dt \quad : i \neq j \\ &= \int_0^T |s_i(t)|^2 dt + \int_0^T |s_j(t)|^2 dt - 2\int_0^T s_i(t) s_j(t) dt \end{aligned} \tag{9.8}$$

In this characterization, the integration is performed over one symbol interval T and this is extended to include observation intervals of duration NT symbols where N is dependent on several factors described as follows:

The signals $s_i(t)$ and $s_j(t)$ have constant envelopes and are distinguished by the phase functions in the interval $0 \le t \le NT$ such that

$$s_i(t) = A\cos\left(\omega_c t + 2\pi\alpha_i h_i q(t) + \phi_o'\right) \quad : 0 \le t \le LT \qquad (9.9)$$

and

$$s_j(t) = A\cos\left(\omega_c t + 2\pi\alpha_j h_j q(t) + \phi_o'\right) \quad : 0 \le t \le LT \qquad (9.10)$$

where ϕ_o' is the accumulated phase function expressed as

$$\phi_o' = \theta_{j-1}(\alpha, \mathbf{h}) + \phi_o \qquad (9.11)$$

These phase expressions are obtained from (9.2) using the time translation $t' = t - (j-1)T$. Upon substituting (9.9) and (9.10) into (9.8), the normalized squared Euclidean distance becomes

$$\begin{aligned} d'^2 &= \frac{d^2}{2E_b} \\ &= k\left\{1 - \frac{1}{T}\int_0^T \left[\cos\left(\omega_c t + 2\pi\alpha_i h_i q(t) + \phi_o'\right)\right.\right. \\ &\qquad \left.\left. - \cos\left(\omega_c t + 2\pi\alpha_j h_j q(t) + \phi_o'\right)\right] dt\right\} \\ &= k\left\{1 - \frac{1}{T}\int_0^T \cos\left(2\pi\left(\alpha_i h_i - \alpha_j h_j\right) q(t)\right) dt\right\} \end{aligned} \qquad (9.12)$$

In this expression, the trigonometric terms involving $2\omega_c$ are neglected under the condition that the carrier frequency is much, much greater than the symbol rate. E_b is the energy-per-bit and is related to the energy-per-symbol as $E = kE_b$ where k is the number of bits-per-symbol; the symbol energy is $E_b = A^2T_b/2$.

Using the average modulation index,

$$h_a = \frac{1}{K}\sum_{i=1}^{K} h_i \qquad (9.13)$$

simplifies the computation of the squared distance, in that, the terms $(\alpha_i h_i - \alpha_j h_j)$ in (9.12) simplify to $h_a(\alpha_i - \alpha_j)$ and the difference in the data $(\alpha_i - \alpha_j)$ takes on the unique values $0, \pm 2, \ldots, \pm 2(M-1)$. However, for the 1REC single-h M-ary waveform $i \neq j$ in the evaluation of d^2 so the zero condition is not included and the data difference sequence becomes

$$\alpha_i - \alpha_j = \pm 2\ell \quad : \ell = 1, \ldots, M-1; \text{ 1REC single-}h\ M\text{-ary} \qquad (9.14)$$

Using these results, (9.12) simplifies to

$$d'^2 = k\left\{1 - \frac{1}{T}\int_0^T \cos(4\pi\ell h_a q(t))\, dt\right\} \quad : \ell = 1, \ldots, M-1 \qquad (9.15)$$

The cosine function is even, so the minus sign resulting from (9.14) is ignored in the argument of the integrand.

The trajectory of the phase function with time is depicted in Figure 9.1 for the single-h, 1REC waveform; the binary case (k = 1) is depicted in part a and the quaternary case (k = 2) in part b of Figure 9.1. Besides the significantly increased complexity between the binary and quaternary cases, there are two key observations concerning the

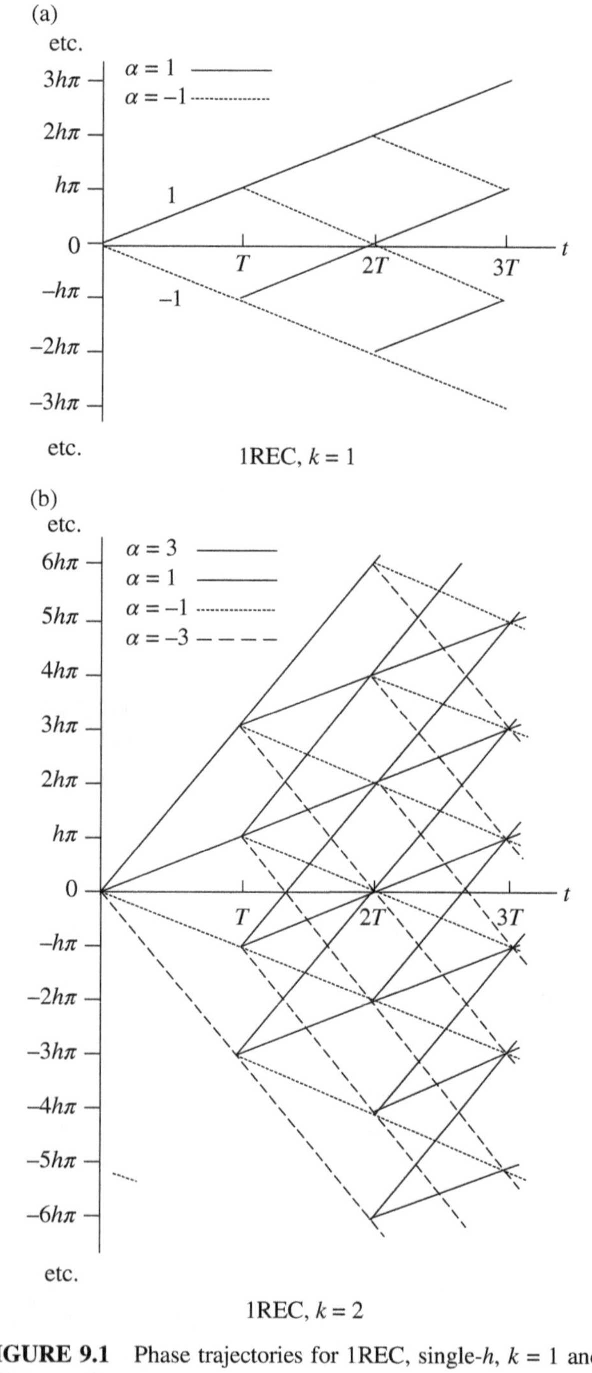

FIGURE 9.1 Phase trajectories for 1REC, single-h, k = 1 and 2 CPM waveform.

selection of the minimum Euclidean distance; first, paths that diverge from points at nT reemerge after two symbols intervals so the computation of the minimum Euclidean distance must be over the interval 2 T corresponding to $N = 2$. The second observation is that there are $\ell_{max} = 2^k - 1$ distances to evaluate and the minimum squared distance is selected as

$$d'^2_{min} = \min\left(d'^2_1, \ldots, d'^2_{\ell_{max}}\right) \tag{9.16}$$

Referring to Figure 9.1a the $M - 1 = 1$ phase difference path corresponds to the unique difference sequences (1, −1; −1, 1) for $\ell = 1$, whereas, in Figure 9.1b the $M - 1 = 3$ phase difference paths correspond to the three difference sequences: (1, −1; −1, 1) for $\ell = 1$, (1, −3; −3, 1) for $\ell = 2$, and (3, −3; −3, 3) for $\ell = 3$; the $\ell = 1, 2$ cases are not unique, in that, other difference sequences exist that yield the same result. As stated earlier, the paths corresponding to $\ell < 0$ yield identical results and are not considered.

The full response CPM waveform being considered allows the computation of the squared distance over the interval NT to be performed as a piecewise integration over N contiguous symbols of duration T seconds. Under this condition, evaluation of $q(t)$ using (9.5) with $L = 1$ simplifies the evaluation of the normalized squared distance as

$$d'^2_{min} = k \min_{\alpha_{ni}, \alpha_{nj}} \left\{ N - \frac{1}{T} \sum_{n=1}^{N} \left[\int_0^T \cos\left(\pi\left(\alpha_{ni} h_{ni} - \alpha_{nj} h_{nj}\right)\left(\frac{t}{T}\right) + \Theta_{n-1}\right) dt \right] \right\} \tag{9.17}$$

where Θ_{n-1} is the phase of the $n - 1$ cosine argument evaluated at $t = T$ and $\Theta_0 = 0$. When using multi-h values the simplifications using the average h_a value leading to (9.15) result in the following single-h expression for the 1REC M-ary modulated waveform

$$d'^2_{min} = k \min_{\ell} \left\{ N - \frac{1}{T} \sum_{n=1}^{N} \left[\int_0^T \cos(2\pi\ell h_a t/T + \Theta_{n-1}) dt \right] \right\} \tag{9.18}$$

Using these simplifying conditions, d'^2_{min}, as expressed in (9.18), is evaluated for the 1REC M-ary waveforms with $N = 2$ as

$$d'^2_{min} = k \min_{\ell} \left\{ 2 - 2\frac{\sin(2\pi\ell h_a)}{2\pi\ell h_a} \right\} \tag{9.19}$$

This result is plotted in Figure 9.2 as a function of a single-h modulation index, that is, $h = h_a$. When $h = ½$, the three M-ary modulations ($M = 2$, 4, and 8) correspond to minimum squared

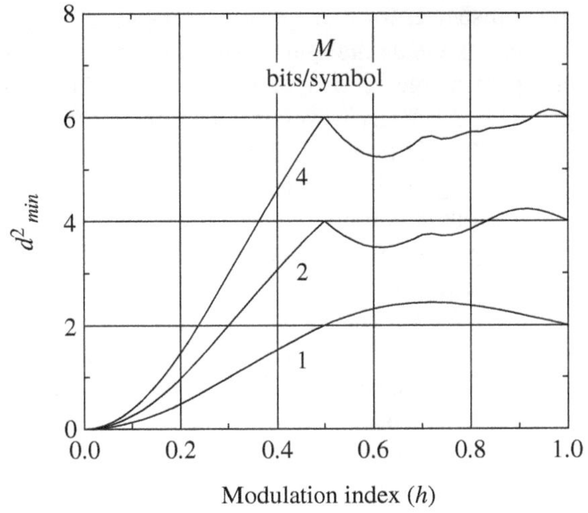

FIGURE 9.2 Minimum squared Euclidean distance for M-ary modulated waveform (1REC, single-h).

distances of 2, 4, and 6, respectively; the quaternary case (M = 4) corresponds to MSK modulation. For the binary case, the minimum Euclidean distance corresponds to $d^2_{min} = 2.4343$ and occurs at $h = 0.72$. This result is comparable to the optimum modulation index of $h = 0.7$ found in Section 5.5.3 for the detection of BFSK modulated signals. Figure 9.2 indicates that the minimum Euclidean distance for the binary case is not sensitive to changes in the modulation index around h = 0.72 so there is considerable design latitude in the selection modulation index.

The bit-error probability is evaluated in terms of the Q-function as

$$P_{be} \le Q\left(\sqrt{\frac{d^2_{min}}{2N_o}}\right) \tag{9.20}$$

where $d^2_{min} = 2d'^2_{min} E_b$ for the binary case; otherwise,* $d^2_{min} = d'^2_{min} E_s$. Increasing the minimum Euclidean distance results in a lower bit-error probability; however, the transmitted bandwidth increases. The trade-off between the bit-error performance and the spectrum occupancy with the selection of the modulation index is the subject of Section 9.3.

In concluding this section, the procedures for selecting multiple modulation indices are examined focusing on the example 1REC 2-h 4-ary waveform used in the case study in Section 9.3. The observation interval of N symbols for the 1REC 1-h M-ary waveform represents the first occurrence of a path merger in the phase trajectory and is unique, in that, it represents the tightest upper bound on the squared

*This results because the binary case uses only one channel, whereas, the quaternary and higher modulated waveforms use both the in-phase and quadrature channels.

Euclidean distance, for example, examination of d'^2_{min} over larger intervals yields the same result. In general, however, the first path merger corresponds to $N = K + 1$ symbols. For the 1REC 2-h M-ary waveform, the first path merger corresponds to $N = 3$; however, this represents a weak upper bound on the signal distance and inclusion of path mergers for $N > 3$ results in a tighter upper bound. Anderson, Aulin and Sundberg [10] indicate that path mergers up to $N = 7$ symbols influence the upper bound for ranges of modulation indices $0 \le h_1, h_2 \le 1$ and identify values of N and the path differences ℓ listed in Table 9.4 as being significant contributors to the Euclidean distance. Exclusion of the path corresponding to $\ell = 0$, that is, the $\alpha_i - \alpha_j : i \neq j$ or the zero phase difference condition, applies only for the initial and final phase differences in the phase difference sequence and is unique to the single-h M-ary case. The intermediate phase differences in the multi-h case may, however, correspond to the condition $i = j$. The distance properties of these path difference sequences are dependent upon the initial modulation index at time zero. Therefore, the bounds corresponding to the difference sequences in Table 9.4 must be evaluated beginning with each h_i: $i = 1, 2$ from which the minimum squared distance d'^2_{min} is selected.

The computation of the 2-h distance bounds is based on (9.18) re-written to include the individual upper bounds on the squared Euclidean distance for each N_m and ℓ_{mn} listed in Table 9.4; the result is

$$d'^2_{B_{m,i}} = N_m - \frac{1}{T}\sum_{n=1}^{N_m}\left[\int_0^T \cos(2\pi\ell_{mn}h_i t/T + \Theta_{n-1})dt\right] \quad (9.21)$$

and the global normalized minimum squared Euclidean distance is determined as

$$d'^2_{min} = k \min_{\forall m,i}\left\{d'^2_{B_{m,i}}\right\} \quad (9.22)$$

TABLE 9.4 Path Mergers and Corresponding Difference Sequences for 1REC 2-*h* *M*-ary CPM Waveform

		Difference Sequences (ℓ_{mn})
M	N_m	$n = 1, \ldots, N_m$
1	3	1, 0, −1
2	4	1, 1, −1, −1
3	4	1, −1, −1, 1
4	5	1, −1, 0, 1, −1
5	7	1, −1, 1, 0, −1, 1, −1

Aulin and Sundberg [7] characterize the normalized minimum Euclidean distance as the z-axis of a three-dimensional surface with the x- and y-axis representing h_1 and h_2 respectively; Figure 9.3 is a similar representation plotted as an interpolated surface image. The modulation indices used in the case study in Section 9.3 and the corresponding minimum squared Euclidean distances are indicated as $(h_1, h_2) = (p_1/q, p_2/q)$ and d'^2_{min}. The selection of the parameters p_1, p_2 with $q = 16$ is shown in Figure 9.3. By selecting larger values of q,

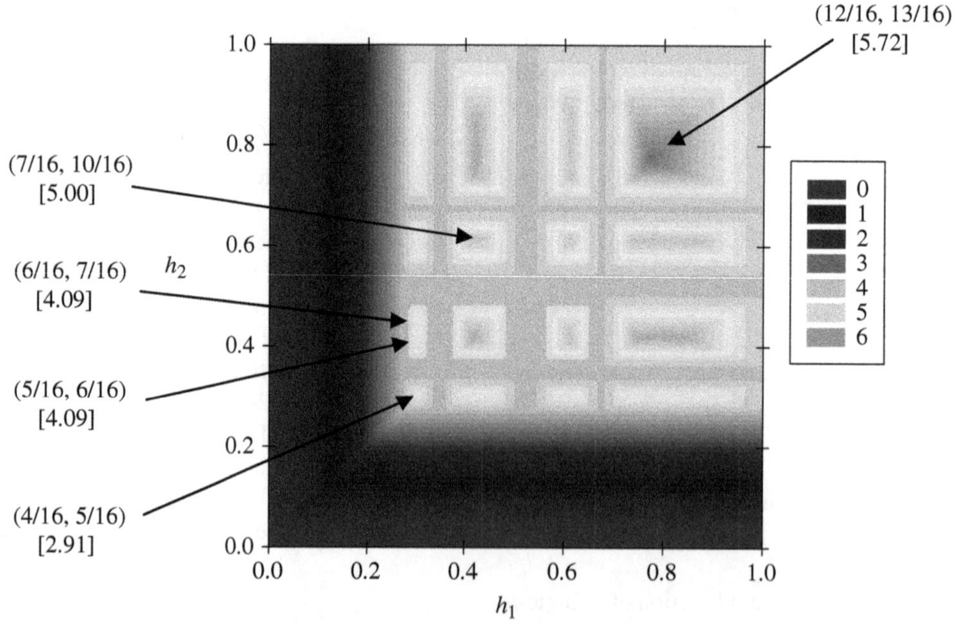

FIGURE 9.3 Interpolated surface image of minimum squared Euclidean distance for 1REC 2-h M-ary CPM waveform $\left(d'^2_{min} \text{ vs } h_1, h_2\right)$.

more resolution is obtained in the placement of h_1 and h_2 closer to the peaks on the image surface plot; however, the required trellis decoding state for 1REC modulation is $2q$ so the trade-off between performance and complexity must be considered.

9.2.2 Multi-*h*, *M*-ary CPM Waveform Spectrum

The power spectral density (PSD) of the 1REC single-h M-ary CPM waveform is characterized by Anderson, Aulin, and Sundberg [14] as

$$S(fT_b)=\frac{2k}{M}\sum_{i=1}^{M}\left(\frac{1}{2}\left(\frac{\sin(\gamma_i)}{\gamma_i}\right)^2+\frac{1}{M}\sum_{j}^{M}A_{ij}\frac{\sin(\gamma_i)}{\gamma_i}\frac{\sin(\gamma_j)}{\gamma_j}\right) \quad (9.23)$$

where

$$\gamma_i=\pi kfT_b-(2i-M-1)\pi h/2 \quad (9.24)$$

$$A_{ij}=\frac{\cos(\gamma_i+\gamma_j)-C_a\cos(2\pi kfT_b-\gamma_i-\gamma_j)}{1-2C_a\cos(2\pi kfT_b)+C_a^2} \quad (9.25)$$

$$C_a=\frac{1}{M}\frac{\sin(M\pi h)}{\sin(\pi h)} \quad (9.26)$$

The PSD is plotted in Figure 9.4 for the 1REC 4-ary waveform using the average of selected modulation indices from Figure 9.3. The average modulation index is computed using (9.13) with $h_i = p_i/16$. These results are based on taking the Fourier transform of the modulated signal autocorrelation function and can be compared to the 1REC 2-h 4-ary CPM spectrums evaluated by windowing and averaging of Fourier transformed signal segments of the simulated CPM waveforms as was done in the case study in Section 9.3.

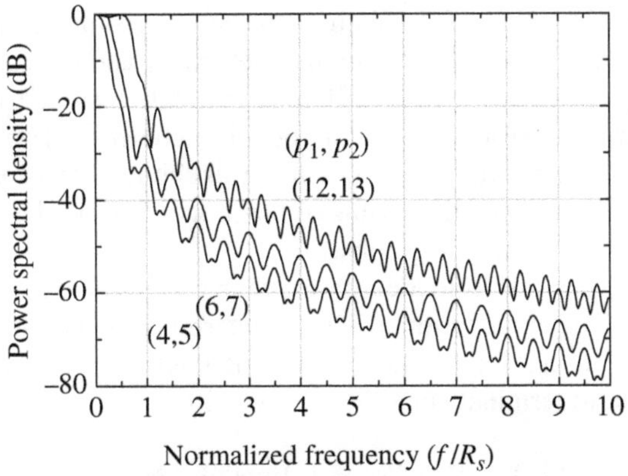

FIGURE 9.4 1REC, single-h, 4-ary CPM power spectral density (h_a = <p_1/16, p_2/16>).

9.2.3 Multi-*h*, *M*-ary CPM Demodulation

Optimum detection of the multi-h CPM waveform involves the maximum-likelihood detection of the current i-th symbol given all past symbols of the message. At face value, this involves the maximum-likelihood sequence estimation (MLSE) spanning the entire message up to the current symbol and choosing the maximum-likelihood estimate from all of the past M^{i-1} possible combinations. With this implementation, at each decision the optimum demodulator forms M likelihood parameters for each $(i - 1)$-tuple $\{\hat{\alpha}_2,\hat{\alpha}_3,\ldots,\hat{\alpha}_i\}$ and chooses $\hat{\alpha}_1$ as the correct symbol estimate corresponding the maximum-likelihood parameter. The recursive property of the Viterbi algorithm greatly simplifies the processing by requiring only M^{L-1} past sequences.

In the following descriptions, it is assumed that the waveform acquisition processing has successfully resolved the carrier frequency and symbol timing to limits within the pull-in range of the appropriate tracking loops. In this regard, the modulation index frame synchronization, also referred to as *superbaud* timing, is known so that the demodulator modulation indices are properly aligned with those of the received waveform.

Consider the received signal as being the transmitted multi-h waveform plus AWGN, expressed as

$$\begin{aligned}y(t)&=s(t)+n(t)\\&=A\cos(\omega_c t+\theta(t;\alpha,\mathbf{h})+\phi_o)+n(t)\end{aligned} \quad (9.27)$$

where the modulated phase term is given by*

$$\theta(t;\alpha,\mathbf{h})=2\pi\alpha_j\bar{h}_j q(t-(j-1)T)+\theta_{j-1}(\alpha,\mathbf{h}) \quad :(j-1)T\le t\le jT \quad (9.28)$$

and $j = 1, 2, \ldots$. It is implicit in this description that the transmitted signal is initiated at $t = 0$ so that $\theta_o(\alpha,\mathbf{h})=0$. The current symbol phase trajectory is given by

$$\theta(t;\alpha_j,\bar{h}_j)=2\pi\alpha_j\bar{h}_j q(t-(j-1)T) \quad :(j-1)T\le t\le jT \quad (9.29)$$

with $\bar{h}_j=\bar{p}_j/q$. For simplicity in explanation, consider a full response modulation ($L = 1$). Under this condition, the possible phase transitions are determined by letting $t = jT$ giving $\theta(jT;\alpha_j,\bar{h}_j)=2\pi\alpha_j\bar{h}_j q(T)=\pi\alpha_j\bar{p}_j/q$. The condition $t = jT$ for the full response waveform results in $q(T) = 1/2$. With $n=\alpha_j\bar{p}_j$, the phase at each state becomes $\theta(jT;a_j,h_j)=n\pi/q$.

*Notation using *normal* type represents the selected h or α for the j-th symbol selected from the set of values denoted by *italic* type h or α, so that, $\mathrm{h_j}=\{h_i\}$: $i = 1, \ldots, K$ and $\mathrm{\alpha_j}=\{\alpha_i\}$: $i = 0, \ldots, M - 1$. Also, the use of the overbar $\bar{\mathrm{h}}_j$ signifies that the modulation indices are repeated modulo (K), that is, for $K = 2$: $\bar{\mathrm{h}}_1,\bar{\mathrm{h}}_2,\bar{\mathrm{h}}_3,\ldots$ represents the sequence $h_1,h_2,h_1,\ldots$ Similarly $\bar{p}_j=\{p_i\}$.

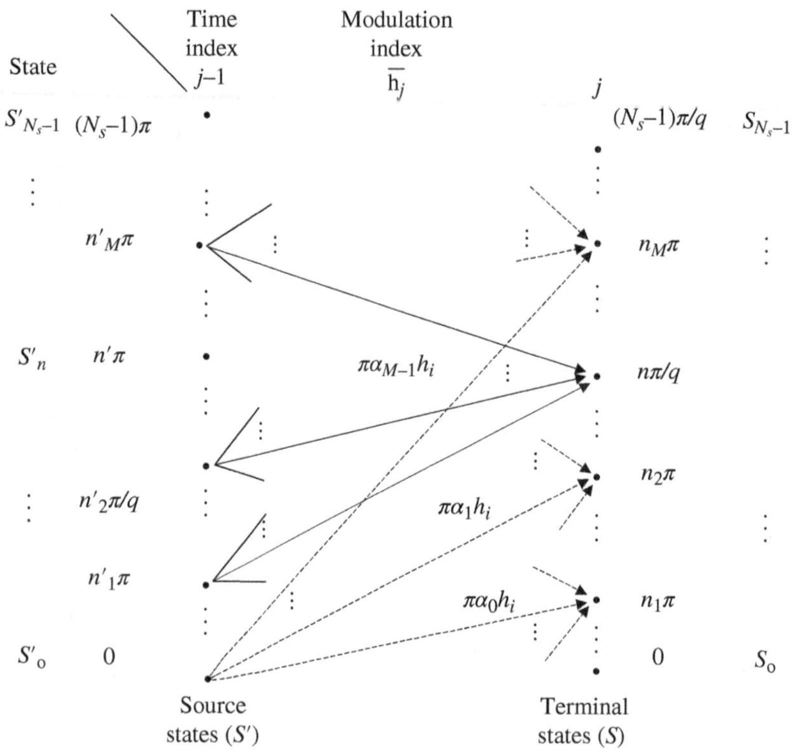

FIGURE 9.5 Typical state transition for trellis decoder.

However, the data symbol alphabets $\alpha_j=\{\pm 1, \pm 3, \ldots, \pm(M-1)\}$ are integer valued so n will also be an integer when $\bar{h}_j=\bar{p}_j/q$ is a rational number. This is an important condition because it results in a finite number of phase states in the trellis decoder processing. When the initial phase error is zero,* that is, when $\phi_o=0$ as required for ideal coherent detection, the possible phase states, *modulo*(2π), at time jT are $n\pi/q$ where $n = 0, 1, \ldots, 2q-1$. This result also applies for the states at time $(j-1)T$ so that $\theta_{j-1}(\alpha,\mathbf{h})=n\pi/q$, that is, the accumulated phase up to $t=(j-1)T$ may fall on any state depending upon the previous random source data. These are the a priori phase states used by the decoder and the presence of noise will result in the signal phase resting at other positions. The number of trellis states for a transition is $2q$ and, considering the L-tuple of symbols over which the frequency response exists for the general case involving partial response modulation, the total states that must be considered before making an optimum estimate is

$$N_s=2qM^{(L-1)} \tag{9.30}$$

The detection processing for the multi-h waveform is most easily described in terms of an N_s-state trellis diagram for which a typical state transition from the states at $(j-1)T$ to those at jT is shown in Figure 9.5. The states at $(j-1)T$ are referred to as the source states denoted as S'_n, and those at jT are referred to as the terminal states denoted as S_n. Each state is associated with a metric m_n and a data storage array D_n, which stores the tentative data estimates $\hat{\alpha}_j$ associated with the state. The length of data storage, D, associated with D_n is typically 5–10 times the constraint length of the multi-h waveform, that is, typically $D = 5(K+1)$ to $10(K+1)$. The processing involves updating the metric and data storage associated with each terminal state. This association is based on the optimally selected metric and data corresponding to the source states having transition branches to the terminal state. There are M such source states and the optimum selection is based on the source state having the minimum metric in consideration of the metric update corresponding to the transition branch. Therefore, the metrics m_n and the contents of D_n are determined from the source state resulting in the minimum metric from the M possible choices. This trellis processing results in a maximum-likelihood decision for the received data; the details of the processing are described in the following paragraphs.

The first task is to identify the source states having branch transitions to the terminal state of interest. Based on the preceding discussion, a source state $S_l(\phi_l)$ will have a transition branch to terminal state $S_n(\phi_n)$ given by

$$\begin{aligned} S_n(\phi_n) &= \left(S'_l(\phi_l)+\pi\alpha_k\bar{h}_j\right)\text{modulo}(2\pi) \\ &= \left(l\pi/q+\pi\alpha_k\bar{p}_j/q\right)\text{modulo}(2\pi) \end{aligned} \tag{9.31}$$

*Any known initial phase condition is acceptable.

where $k = 0, 1, \ldots, M-1$ represents the M-ary source symbols. By removing the π/q dependence in this relationship, the transition from the source to the terminal state indexes is expressed as

$$n = \left(l + \alpha_k \bar{p}_j\right) \text{modulo}(2q) \quad : \text{source state} \tag{9.32}$$

This result can be *inverted* to identify the M source states ℓ_{kn} given a terminal state S_n with the result

$$l_{kn} = \left(n - \alpha_k \bar{p}_j\right) \text{modulo}(2q) \quad : \text{terminal state} \tag{9.33}$$

The subscript k denotes the dependence of ℓ on the M source states that terminate at state n. For example, when evaluating the metric m_n at terminal state S_n, then the source states with branches to S_n are $S'_{l_{kn}}$: $k = 0, 1, \ldots, M-1$ corresponding to the M possible data symbols α_k. This result can be pre-computed for all possible conditions and stored as pointers for subsequent processing.

The state transition processing from $j-1$ to j involves computing the metrics m_n at the terminal states S_n using the metrics $m'_{l_{kn}}$ of the targeted source states $S'_{l_{kn}}$ and the corresponding metric update Δm_k. The metrics are computed as

$$\begin{aligned} m_n &= \min_k \left[m'_{l_{kn}} + \Delta m_k\right] \quad : k = 0, \ldots, M-1 \\ &= m'_{l_{\underline{k}n}} + \Delta m_{\underline{k}} \end{aligned} \tag{9.34}$$

In this case, the index $\underline{k}$ identifies the transition resulting in the minimum metric, that is, $m'_{\ell_{\underline{k}n}} + \Delta m_{\underline{k}} < m'_{\ell_{kn}} + \Delta m_k \forall k \neq \underline{k}$. The source state data estimates at $t = jT$ are expressed in terms of the bipolar data $\hat{\alpha}_k$ as

$$\mathbf{D}'_{l_{kn}} = \left\{\hat{\alpha}_{l_{kn},j-D}, \hat{\alpha}_{l_{kn},j-(D-1)}, \ldots, \hat{\alpha}_{l_{kn},j-1}\right\} \tag{9.35}$$

where D is the depth or delay of the trellis processing before making a data decision. To determine the surviving source data estimates at the terminal states, it is necessary to use the same index $\underline{k}$ corresponding to the minimum metric. For example, the surviving data at state S_n becomes

$$\begin{aligned} \mathbf{D}_n &= \mathbf{D}'_{\ell_{\underline{k}},n} \text{ (left-shifted)} : \hat{\alpha}_{l_{\underline{k}n},j} \\ &= \left\{\hat{\alpha}_{l_{\underline{k}n},j-(D-1)}, \ldots, \hat{\alpha}_{l_{\underline{k}n},j-1}, \hat{\alpha}_{l_{\underline{k}n},j}\right\} \\ &= \left\{\hat{\alpha}_{n,j-(D-1)}, \ldots, \hat{\alpha}_{n,j-1}, \hat{\alpha}_{n,j}\right\} \end{aligned} \tag{9.36}$$

The notation $\mathbf{D}_n(\text{left-shifted}) : \hat{\alpha}_n$ signifies that the elements of the array $\mathbf{D}_n$ are shifted to the left by one element with $\hat{\alpha}_n$ substituted as the rightmost element. When $j < D$, the leftmost $D - j$ elements of $\mathbf{D}_n$ contain no information; however, for $j \geq D$ the leftmost element contains the oldest data estimate $\hat{\alpha}_{n,j-(D-1)}$ and this estimate becomes a candidate for the maximum-likelihood data decision. All of the N_s candidate data estimates are available upon completion of the terminal state processing at which time the maximum-likelihood data decision is determined as

$$\begin{aligned} \hat{\alpha}_{j-(D-1)} &= \min_{m_n} \left[\hat{\alpha}_{n,j-(D-1)}\right] \\ &= \hat{\alpha}_{\underline{n},j-(D-1)} \end{aligned} \tag{9.37}$$

where the terminal state $n = \underline{n}$ results in the minimum metric among the entire N_s terminal states, that is, $m_{\underline{n}} < m_n \forall n \neq \underline{n}$. From this result, it is seen that the maximum-likelihood data estimate is the leftmost element of $\mathbf{D}_{\underline{n}}$. When D is significantly larger than the constraint length $K+1$, then, with very high probability, $\hat{\alpha}_{n,j-(D-1)} = \hat{\alpha}_{\underline{n},j-(D-1)} : \forall n$ so $\hat{\alpha}_{j-(D-1)}$ can be selected from any of the terminal states. The value of D typically ranges between 5 and 10 times the constraint length depending on the number of modulation indices and the values of p_i.

As a final step, the bipolar data estimate is translated to the M unipolar binary source data estimate as

$$\hat{b}_{j,\log_2(M-1)}, \ldots, \hat{b}_{j,1}, \hat{b}_{j,0} = \left[\frac{M - 1 - \hat{\alpha}_{j-(D-1)}}{2}\right]_{binary} \tag{9.38}$$

where the notation $[\text{I}]_{binary}$ denotes the binary equivalent of the integer I. After completion of the terminal state processing, the state metrics and the data estimates contained in the terminal state arrays are transferred to the corresponding source data arrays in preparation of the next iteration through the trellis, that is, $m'_n = m_n$ and $\mathbf{D}'_n = \mathbf{D}_n \forall n$.

The preceding description of the trellis processing has primarily been a bookkeeping effort and the metric updates are the link between the trellis decoder and the received signal. The metric updates Δm_k are defined in terms of the Euclidean distance between the received signal and the M possible reference signals terminating on each trellis state. Initially, the metrics m'_n are all set to zero; however, they must be reset periodically to avoid underflow and overflow. The complex envelope of the received signal is denoted as*

$$\begin{aligned} \tilde{y}(t) &= \tilde{s}(t) + \tilde{n}(t) \\ &= \left(\frac{A}{2}\right)\cos\left(2\pi\alpha_j \bar{h}_j q(t-(j-1)T) + n'\pi/q + \phi_o\right) \\ &\quad + \boldsymbol{j}\left(\frac{A}{2}\right)\sin\left(2\pi\alpha_j \bar{h}_j q(t-(j-1)T) + n'\pi/q + \phi_o\right) + n_c(t) + \boldsymbol{j}n_s(t) \\ &= y_c(t) + \boldsymbol{j}y_s(t) \end{aligned} \tag{9.39}$$

*When confusion may exist in the use of j as an index or to signify the complex root of -1, then boldfaced type is used for the latter.

where $\tilde{n}(t)$ represents zero-mean narrowband Gaussian noise: $N(0, \sigma_n)$. In a similar manner the complex baseband references signal is denoted as

$$\begin{aligned}\tilde{x}_k(t) &= \left(\frac{A}{2}\right)\cos\left(2\pi\alpha_k\bar{h}_j q(t-(j-1)T) + \ell_{kn}\pi/q + \hat{\phi}_o\right) \\ &\quad + j\left(\frac{A}{2}\right)\sin\left(2\pi\alpha_k\bar{h}_j q(t-(j-1)T) + \ell_{kn}\pi/q + \hat{\phi}_o\right) \\ &= x_{ck}(t) + jx_{sk}(t)\end{aligned} \tag{9.40}$$

Using these results, the normalized squared Euclidean distance is computed as

$$\begin{aligned}d_k^2 &= \frac{1}{2E}\int_{(j-1)T}^{jT} |\tilde{y}(t) - \tilde{x}_k(t)|^2 dt \\ &= \frac{1}{2E}\left[\int_{(j-1)T}^{jT} |\tilde{y}(t)|^2 dt - 2\mathrm{Re}\left\{\int_{(j-1)T}^{jT} \tilde{y}(t)\tilde{x}_k^*(t)dt\right\} \right. \\ &\quad \left. + \int_{(j-1)T}^{jT} |\tilde{x}_k(t)|^2 dt\right]\end{aligned} \tag{9.41}$$

The first and third integrals in the second equality of (9.41) are simply positive signal energy terms that are the same for all of the possible received and reference signal conditions and can be neglected in determining the minimum distance. Therefore, the desired metric update* is computed as

$$\begin{aligned}\Delta m_k &= -\frac{1}{E}\mathrm{Re}\left\{\int_{(j-1)T}^{jT} \tilde{y}(t)\tilde{x}_k^*(t)dt\right\} \\ &= -\frac{1}{E}\mathrm{Re}\{\phi_{yx}(0, \varphi_e; n, k)\} \\ &\equiv -d_k^2\end{aligned} \tag{9.42}$$

In general, the correlation function is dependent on the time shift τ and is denoted as $\phi_{yx}(\tau, \varphi_e; n, k)$; however, for the present discussion, $\tau = 0$. The Euclidean distance is seen to be similar to the normalized zero shift correlation function between the received signal and the reference signal and is a maximum value when the reference $\tilde{x}_k(t)$ is matched to $\tilde{s}(t)$ leading to the minimum squared Euclidean distance d_{min}^2. For the coherent detection case $\hat{\phi}_o = \phi_o$ and the maximum correlation corresponds to the conditions $n = \underline{n}$ and $k = \underline{k}$ so that $\alpha_{\underline{k}} = \alpha_j$ and $l_{\underline{kn}} = \underline{n}'$. If the minus sign were omitted in (9.42) the selection of a *minimum* metric must be changed to selecting the *maximum* metric; in the following descriptions the minus sign is used.

*This definition of the negative metric update may cause some confusion since the survivor selection is based on the minimum state metric not the maximum. Note, however, that d_k^2 is always positive and corresponds to the maximum squared distance in the trellis decoding.

The correlation processing involved in the metric update computations can be simplified by identifying the M unique correlations for each transition and applying a phase rotation corresponding to the position of the source state in the trellis. In this evaluation, the time axis is translated by letting $t_1 = t-(j-1)T$ so that $0 \le t_1 \le T$. With this translation, the correlation function is expressed as

$$\begin{aligned}\phi_{yx}(0, \varphi_e; n, k) &= \frac{A}{2}\int_0^T \tilde{y}(t)e^{-j\left(2\pi\alpha_k\bar{h}_j q(t) + \pi l_{k,n}/q + \hat{\phi}_o\right)}dt \\ &= \frac{A^2T}{4}e^{j(\phi_o - \hat{\phi}_o)}\left[\frac{1}{T}\int_0^T \tilde{y}'(t)e^{-j\left(2\pi\alpha_k\bar{h}_j q(t)\right)}dt\right]e^{-j\pi l_{k,n}/q} \\ &= \frac{E}{2}e^{j(\phi_o - \hat{\phi}_o)}\left[\phi'_{yx}(0;k)\right]e^{-j\pi l_{k,n}/q}\end{aligned} \tag{9.43}$$

where $E = A^2T/2$ is the symbol energy and the unknown received signal phase (ϕ_o) is removed from $\tilde{y}(t)$ to explicitly demonstrate its impact on the detection processing. The primed notation $\tilde{y}'(t)$ represents $\tilde{y}(t)$ with the signal phase removed and normalized by $A/2$. With coherent processing, the phase estimate $\hat{\phi}_o$ is approximately equal to ϕ_o with little impact on the detection performance. At this point in the processing, there are M unique correlation values $\phi'_{yx}(0;k)$: $k = 0, \ldots, M-1$ for each received symbol. These complex correlation values are computed and stored once each symbol for the subsequent state transition processing. During the evaluation of each terminal state metric, the stored correlation values must undergo a phase rotation given by $-\pi l_{k,n}/q$ radians as indicated earlier.

9.2.3.1 *Coherent Detection*

For ideal coherent detection [15, 16], the condition $\hat{\phi}_o = \phi_o$ applies and the metric is evaluated as

$$\Delta m_k = -\frac{1}{2}\mathrm{Re}\left\{\phi'_{yx}(0;k)e^{-j\pi l_{kn}/q}\right\} \tag{9.44}$$

where

$$\phi'_{yx}(0;k) = \frac{1}{T}\int_0^T \tilde{y}'(t)e^{-j\left(2\pi\alpha_k\bar{h}_j q(t)\right)}dt \tag{9.45}$$

However, in a practical coherent detector where phase estimation and tracking is employed, there will always be some finite phase error that is defined as $\varphi_e = \phi_o - \hat{\phi}_o$. In this case, the metric update becomes

$$\Delta m_k = -\frac{1}{2}\text{Re}\left\{e^{-j\phi_\varepsilon}\left[\phi'_{yx}(0;k)\right]e^{-j\pi l_{kn}/q}\right\} \tag{9.46}$$

Without additive channel noise, the source state corresponding to $n=\underline{n}$ is $\ell_{k\underline{n}}=\underline{n}'$ and, under these conditions and with $\alpha_{\underline{k}}=\alpha_j$, the corresponding metric update is

$$\Delta m_k = -\frac{\text{Re}\left\{e^{-j\varphi_\varepsilon}\right\}}{2} \quad \text{: minimum metric conditions} \tag{9.47}$$

This suggests that the phase error is available from the correlation function $\phi_{yx}(0, \varphi_\varepsilon; n, k)$ corresponding to the minimum metric over all of the terminal states. To determine the minimum metric, the candidate correlation functions at each terminal state are expressed as

$$\phi_{yx}(0,\varphi_\varepsilon;n,\underline{k}) = \min_k{}^{-1}\left[m'_{l_{kn}} + \Delta m_k : \phi_{yx}(0,\varphi_\varepsilon;n,k)\right] \tag{9.48}$$

where the notation $\min_k{}^{-1}[a_k : b]$ picks the value of b corresponding to the minimum of all a_k values. This processing results in the selection of the phase-error function corresponding to the minimum metric at each terminal state; however, the final, or optimum, selection is based on the minimum metric over all of the terminal states and is expressed as

$$\begin{aligned}\phi_{yx}(0,\varphi_\varepsilon;\underline{n},\underline{k}) &= \min_n{}^{-1}\left[m_n : \phi_{yx}(0,\varphi_\varepsilon;n,\underline{k})\right] \\ &\cong K_d e^{j\varphi_\varepsilon}\end{aligned} \tag{9.49}$$

where K_d is defined as the phase detector gain constant for the multi-h demodulator. Based on this result, the phase detector phase-error response is established from the imaginary part of $\phi_{y,x}(0,\varphi_\varepsilon;\underline{n},\underline{k})$ and the real part is used as the input to the phaselock detector. The phase detector phase error

$$e_j(\varphi_\varepsilon) = \text{Im}\left\{\phi_{yx}(0,\phi_\varepsilon;\underline{n},\underline{k})\right\} \cong K_d\sin(\varphi_\varepsilon) \tag{9.50}$$

is passed through the loop filter $F(s)$ and applied to the input signal $\tilde{y}(t)$. Similarly, the phaselock detector

$$e_{LD_j}(\varphi_\varepsilon) = \text{Re}\left\{\phi_{yx}(0,\varphi_\varepsilon;\underline{n},\underline{k})\right\} \cong K_d\cos(\varphi_\varepsilon) \tag{9.51}$$

is filtered and compared to a fixed threshold to determine a locked condition.

9.2.3.2 *MAP Estimation of Received Signal Phase and Symbol Timing* The received signal phase estimation discussed earlier is based on the application of the phase error in the correlation function $\phi_{yx}(0,\varphi_\varepsilon;\underline{n},\underline{k})$. The maximum a posteriori (MAP) estimation of the received signal phase ϕ_o and the symbol timing error τ are examined in this section [17, 18]. In addition to providing a formal approach to optimally estimating these errors, this analysis yields the sensitivity of the estimates under the joint error conditions.

In keeping with the notion of the MAP estimator for ϕ_o and τ, the corresponding partial derivatives of the log-likelihood functions are used. The equivalent process involves the correlation function $\phi_{yx}(\tau, \varphi_\varepsilon; n, k)$ where the phase and timing errors are explicitly denoted. In the following analysis, the algorithms for the MAP estimation processing are established by considering the noise-free case so that $\tilde{y}(t)=\tilde{s}(t)$. The performance evaluation with noise is then examined through simulations. Figure 9.6 depicts the relationship between the received signal $\tilde{y}(t)$, the local reference $\tilde{x}(t)$, and the receiver integration interval.

As was noted previously, the selection of the time origin at $t = 0$, to correspond to the beginning of the j-th interval of the modulation index, simplifies the notation in the following analysis by eliminating the time dependence on the index j. With this understanding, the local reference signal is expressed as

$$\tilde{x}(t-\tau) = \frac{A}{2}e^{j\left(2\pi\alpha_k\bar{h}_j q(t-\tau) + l_{kn}\pi/q + \hat{\phi}_o\right)} \quad : \tau \le t \le T+\tau \tag{9.52}$$

and the MAP estimation errors associated with the phase and time estimates are given by

$$\varepsilon_j(\varphi_\varepsilon;\tau) = \left.\frac{\partial\phi_{yx}(\tau,\varphi_\varepsilon;n,k)}{\partial\hat{\phi}_o}\right|_{\substack{\phi_o=\hat{\phi}_o \\ \tau=0}} = 0 \tag{9.53}$$

and

$$\varepsilon_j(\tau;\varphi_\varepsilon) = \left.\frac{\partial\phi_{yx}(\tau,\varphi_\varepsilon;n,k)}{\partial\tau}\right|_{\substack{\phi_o=\hat{\phi}_o \\ \tau=0}} = 0 \tag{9.54}$$

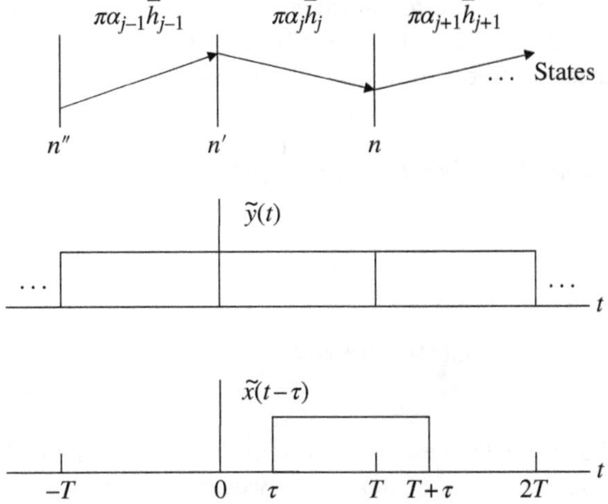

FIGURE 9.6 Receiver processing with timing error τ.

Equations (9.53) and (9.54) are evaluated under the conditions corresponding to the terminal state having the minimum metric, that is, $m_{\underline{n}} = \min[m_n]\ \forall n$. In the ideal or noise-free channel, these conditions are $\ell_{\underline{k}\underline{n}} = \underline{n}'$ and $\alpha_{\underline{k}} = \alpha_j$. These conditions are identical to those required by the selection of k and n resulting in the correlation function $\phi_{yx}(\tau, \phi_\varepsilon; \underline{n}, \underline{k})$ as given by (9.49). Using (9.53) and (9.54), the respective error functions are evaluated as

$$\varepsilon_j(\varphi_\varepsilon;\tau) = -j \int_\tau^{T+\tau} \tilde{y}(t)\tilde{x}^*(t-\tau)dt \Bigg|_{\substack{\alpha_{\underline{k}}=\alpha_j \\ \ell_{\underline{k}\underline{n}}=\underline{n}'}} \tag{9.55}$$

and

$$\varepsilon_j(\tau;\varphi_\varepsilon) = -j2\pi\alpha_j\bar{h}_j \int_\tau^{T+\tau} \tilde{y}(t)\tilde{x}^*(t-\tau)\frac{\partial q(t-\tau)}{\partial \tau}dt \Bigg|_{\substack{\alpha_{\underline{k}}=\alpha_j \\ \ell_{\underline{k}\underline{n}}=\underline{n}'}} + \Delta(\tau,\varphi_\varepsilon) \tag{9.56}$$

The received signal $\tilde{y}(t)$ under the bounds of the integration shown in Figure 9.6 are given by

$$\tilde{y}(t) = \begin{cases} \tilde{y}_{-1}(t+T) = \left(\dfrac{A}{2}\right) e^{j\left(2\pi\alpha_{j-1}\bar{h}_{j-1}q(t+T) + n''\pi/q + \phi_o\right)} & -T \le t \le 0 \\ \tilde{y}_o(t) = \left(\dfrac{A}{2}\right) e^{j\left(2\pi\alpha_j\bar{h}_j q(t) + n'\pi/q + \phi_o\right)} & 0 \le t \le T \\ \tilde{y}_1(t-T) = \left(\dfrac{A}{2}\right) e^{j\left(2\pi\alpha_{j+1}\bar{h}_{j+1}q(t-T) + n\pi/q + \phi_o\right)} & T \le t \le 2T \end{cases} \tag{9.57}$$

For the case $\tau \ge 0$, the phase-error function is evaluated by substituting $\tilde{y}_o(t)$ and $\tilde{y}_1(t-T)$ from (9.57) into (9.55) and integrating over the appropriate limits with the result

$$\varepsilon_j(\varphi_\varepsilon;\tau\ge 0) = -j\left(\frac{A^2}{4}\right)\left\{\int_\tau^T e^{j\left(2\pi\alpha_j\bar{h}_j q(t) + n'\pi/q + \phi_o\right)} e^{-j\left(2\pi\alpha_k\bar{h}_j q(t-\tau) + \ell_{kn}\pi/q + \hat{\phi}_o\right)}dt + \int_T^{T+\tau} e^{j\left(2\pi\alpha_{j+1}\bar{h}_{j+1}q(t-T) + n\pi/q + \phi_o\right)} e^{-j\left(2\pi\alpha_k\bar{h}_j q(t-\tau) + \ell_{kn}\pi/q + \hat{\phi}_o\right)}dt\right\}\Bigg|_{\substack{\alpha_{\underline{k}}=\alpha_j \\ \ell_{\underline{k}\underline{n}}=\underline{n}'}} \tag{9.58}$$

$$= -j\left(\frac{A^2}{4}\right)e^{j\varphi_\varepsilon}\left\{\int_\tau^T e^{j2\pi\alpha_j\bar{h}_j(q(t)-q(t-\tau))}dt + e^{j\pi\alpha_j\bar{h}_j}\int_T^{T+\tau} e^{j2\pi\left(\alpha_{j+1}\bar{h}_{j+1}q(t-T) - \alpha_j\bar{h}_j q(t-\tau)\right)}dt\right\}$$

where the relationships $n' - \ell_{kn} = 0$ and $(n'-n)/q = \alpha_j\bar{p}_j/q = \alpha_j\bar{h}_j$ are used in the evaluation. Similarly, for $\tau \ge 0$, the symbol time-error function is evaluated by substituting $\tilde{y}_{-1}(t+T)$ and $\tilde{y}_o(t)$ from (9.57) into (9.56) and integrating over the appropriate limits with the result

$$\begin{aligned}\varepsilon_j(\tau\ge 0,\varphi_\varepsilon) = -j2\pi\alpha_j\bar{h}_j\left(\frac{A^2}{4}\right)e^{j\varphi_\varepsilon}&\left\{\int_\tau^T e^{j2\pi\alpha_j\bar{h}_j(q(t)-q(t-\tau))}\frac{\partial q(t-\tau)}{\partial\tau}dt\right. \\ &\left.+ e^{j\pi\alpha_j\bar{h}_j}\int_T^{T+\tau} e^{j2\pi\left(\alpha_{j+1}\bar{h}_{j+1}q(t-T) - \alpha_j\bar{h}_j q(t-\tau)\right)}\frac{\partial q(t-\tau)}{\partial\tau}dt\right\} \\ &-\tilde{y}_o(\tau)\tilde{x}^*(0) + \tilde{y}_1(\tau)\tilde{x}^*(T)\end{aligned} \tag{9.59}$$

To evaluate these functions further, the symbol phase function $q(t)$ must be defined; however, when $\tau = 0$, the real part of the phase-error function is identical to the imaginary part of the previous result as expressed in (9.49). This is because of the $-j$ multiplier arising from the differentiation. The phase and symbol time tracking processing is shown in Figure 9.7 together with the lock detector processing. To detect the loss of phase tracking a lower threshold is used to detect the unlock condition as shown in Figure 9.7. In the following case study, these functions are evaluated for the 1REC phase function.

9.3 CASE STUDY: 2-*h* 4-ary 1REC CPM

As mentioned in the introduction the spectrally efficient constant-amplitude multi-*h* *M*-ary CPM waveform is well suited for operation through narrowband hard-limiting relay satellite channels. For this reason, a versatile, low complexity 2-*h* 4-ary full response CPM waveform is used for satellite

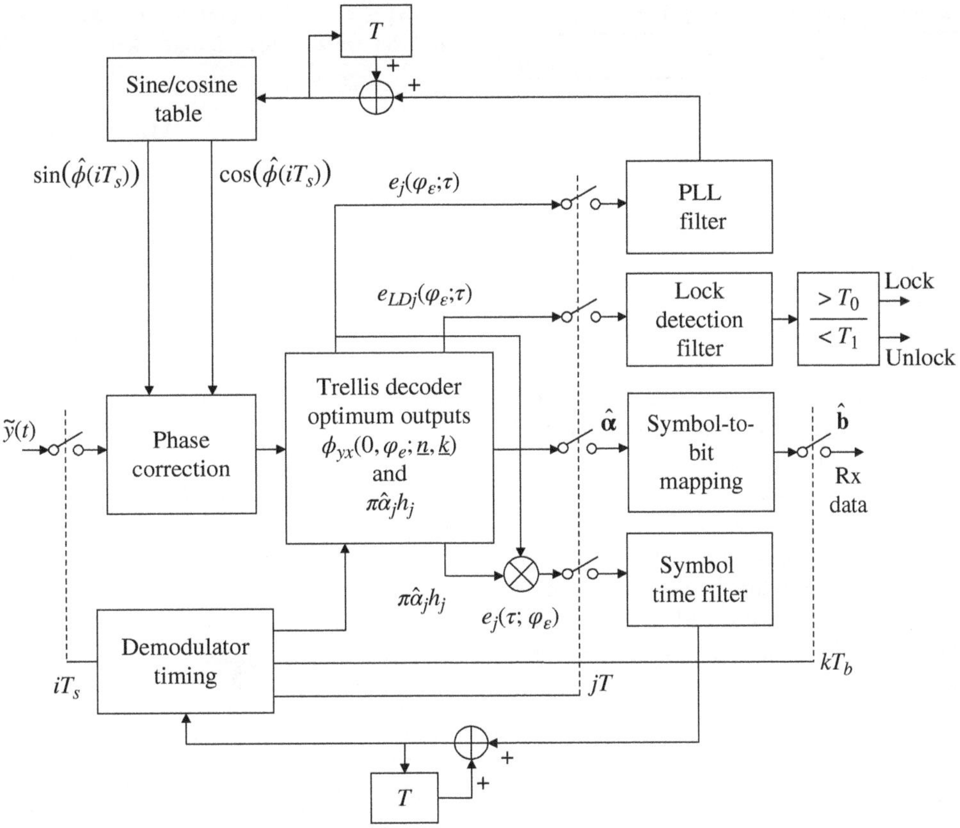

FIGURE 9.7 Symbol tracking and lock indicator processing.

communications through the hard-limiting 5 and 25 kHz channels of the ultra-high-frequency (UHF) relay satellite system. This case study describes the processing requirements for 2-h 4-ary 1REC CPM waveform modulation and demodulation and presents some performance results for the U.S. Navy UHF satellite relay communication system. The focus of this case study is on the 25 kHz channel that is modeled as a 32 kHz, 0.05 dB ripple, 6-pole Chebyshev filter with 6 degrees of phase equalization at ±12.5 kHz and 9.4 degrees at 30 kHz. The hard limiter is followed by a spurious signal rejection filter before routing the uplink data to the assigned downlink. The performance is based on a received signal with AWGN. This study focuses on the communication data rates, modulation indices, and the resulting bit-error and spectral performance through the 25 kHz satellite channel.

The 2-h waveform is characterized in terms of the two modulation indices, $h = \{h_1, h_2\}$ and the performance evaluations are compared for the following sets or pairs of modulation indices: {5,6}, {6,7}, {7,10}, and {12,13}. Furthermore, the waveform is a 4-ary full response modulation with two source bits-per-symbol with the symbol energy contained within one transmitted symbol interval. These pairs of modulation indices result in significantly different spectral characteristics and provide coding gains relative to quadrature PSK (QPSK) waveform modulation. In general, as the data rate is decreased the modulation index is increased so that the waveform spectrum occupies more of the 25 kHz channel bandwidth. The increased modulation index is also accompanied by an increase in the coding gain allowing for operation at a lower E_b/N_o signal-to-noise ratio. On the other hand, for the higher data rates the modulation indices must be decreased to confine the spectrum to the channel bandwidth and this results a lower coding gain. The trade-off between the modulation indices and the coding gain is based on optimizing the bit-error performance in the presence of intersymbol interference (ISI). For the modulation indices {5,6}, {6,7}, {7,10}, and {12,13}, the respective coding gains in E_b/N_o at $P_{be} = 10^{-6}$ are 2.2, 3.1, 3.5, and 4.2 dB. The following analysis does not include the waveform acquisition so the carrier frequency and symbol timing estimates are resolved with sufficient accuracy to allow for phase and symbol tracking during the data demodulation. Acquisition preambles and performance results are discussed in Chapter 11.

In Section 9.3.1, the implementation of the 2-h 4-ary CPM waveform is discussed and Section 9.3.2 shows the resulting spectral characteristics for the selected data rates and the corresponding modulation indices. In Section 9.3.3, the trellis demodulator processing is discussed in considerable detail emphasizing the computation of the symbol timing and phase

TABLE 9.5 Mapping of Source Bits to 4-ary Levels

j	$\{b_1, b_0\}$[a]	α_j
0	{1, 0}	−1
1	{0, 1}	1
2	{1, 1}	−3
3	{0, 0}	3

[a]b_0 is first source bit (LSB).

errors within the trellis for optimally removing and tracking these errors. The case study concludes with Section 9.3.4 presenting simulated bit-error probability performance results with sensitivities to symbol and phase tracking. The performance is also characterized using the relay satellite channel model with the filtering and hard limiting.

Referring to Table 9.1 and Figure 9.3, the corresponding waveform parameters of interest in the design of the demodulator are $M = 4$, $k = 2$, $K = 2$, $L = 1$, $\alpha_j = \{\pm 1, \pm 3\}$, and $h_j = p_j/q$ with $q = 16$ and p_1 and p_2 representing pairs of modulation indices for the 2-h waveform as indicated earlier. In all cases, the modulation indices are rational numbers as required for a practical trellis decoder implementation.

9.3.1 2-h 4-ary 1REC CPM Waveform Modulation

The 1REC waveform with $L = 1$ simplifies the modulator implementation and, more importantly, reduces the complexity of the acquisition and demodulator processing. The modulator signal and the corresponding data-dependent phase function are characterized by (9.1) and (9.2), respectively. The second equality in (9.2) contains the expression of the phase trajectory during the current symbol in the range $(j - 1)T \le t \le jT$ and, upon translating the time by $t = \tau + (j - 1)T$, the current symbol phase trajectory is evaluated as

$$\theta(t;\alpha_j,\bar{h}_j) = 2\pi\alpha_j\bar{h}_j q(t) \tag{9.60}$$

The phase function $q(t)$ is evaluated using (9.4) and (9.5) with $L = 1$ and upon substituting these expressions into (9.60) the phase function is evaluated as

$$\begin{aligned}\theta(t;\alpha_j,\bar{h}_j) &= 2\pi\alpha_j\bar{h}_j\int_0^t g(\tau)d\tau \\ &= \pi\alpha_j\bar{h}_j\left(\frac{t}{T}\right) \quad :0<t\le T\end{aligned} \tag{9.61}$$

The parameter α_j is characterized by the 4-ary symbol alphabet $\{\pm 1, \pm 3\}$ and is based on the two current source data bits $\{b_1, b_0\}$ as defined in Table 9.5; this mapping is described as $\alpha_j = 3 - 2(2b_1 + b_0)$.

The selection of the modulation index set h_j influences the spectrum with larger values resulting in a wider main spectral lobe; therefore, the modulation indices are increased for lower data rates to more fully occupy the available channel bandwidth. This has the additional benefit of providing more coding gain and hence operation at lower E_b/N_o values. Selected sets of modulation indices* corresponding to the source data rates are defined in Table 9.6 where $R_s = 1/T$ is the symbol rate and B_c is the channel bandwidth. For the 25 kHz satellite channel, $B_c = 32$ kHz.†

TABLE 9.6 Correspondence of Selected Modulation Indices to User Data Rates (25 and 5 kHz Satellite Channels)[a]

R_b (kbps)	R_s (ksps)	$\{h_1, h_2\}$	R_b/B_c
	25 kHz channel		
56.0	28	4/16, 5/16	1.75
48.0	24.0	5/16, 6/16	1.50
38.4	19.2	5/16, 6/16	1.20
32.0	16.0	6/16, 7/16	1.00
28.8	14.4	6/16, 7/16	0.90
19.2	9.6	12/16, 13/16	0.60
	5 kHz channel		
9.6	4.8	5/16, 6/16	1.60
6.0	3.0	7/16, 10/16	1.00
4.8	2.4	12/16, 13/16	0.80

[a]Battista et al. [19]. Reproduced by permission of the American Institute of Aeronautics and Astronautics (AIAA).

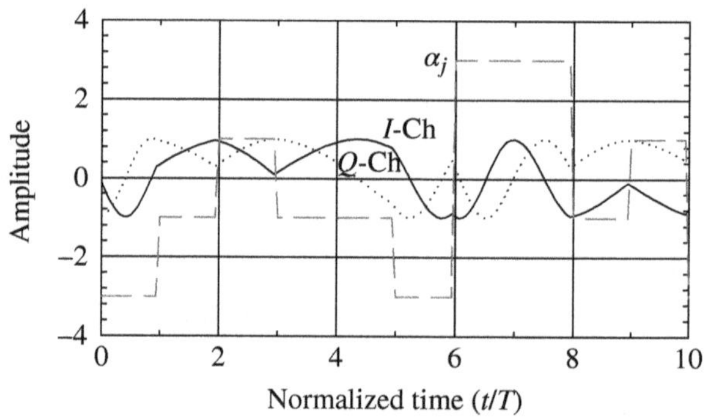

FIGURE 9.8 Typical I/Q waveform with $\{h_1 = 5/16, h_2 = 6/16\}$.

A typical sequence of in-phase and quadrature (I/Q) modulated data is shown in Figure 9.8 where the *solid* and *dashed* curves correspond to the I and Q channels, respectively. The heavy *dashed* curve represents the data-dependent parameter α_j for the 4-ary modulation. The abscissa is normalized by the symbol duration and, although not shown, the magnitude of the I and Q channels is normalized and results in a constant unit value.

*The selected modulation indices do not include those with RS concatenated coding.

†The 3-dB bandwidth of the 5 kHz satellite channel is 6 kHz.

9.3.2 2-*h* 4-ary 1REC CPM Spectral Characteristics

The spectral characteristics for the 2-*h* 4-ary 1REC CPM waveform are examined for several sets of modulation indices and data rates as defined in Table 9.6. To provide a point of comparison, Figure 9.9a shows the spectrum of QPSK modulation that is described by the *sinc(fT)* function. The CPM spectrum results are based on simulations using Welch's method [20] of spectral estimation. In these cases,

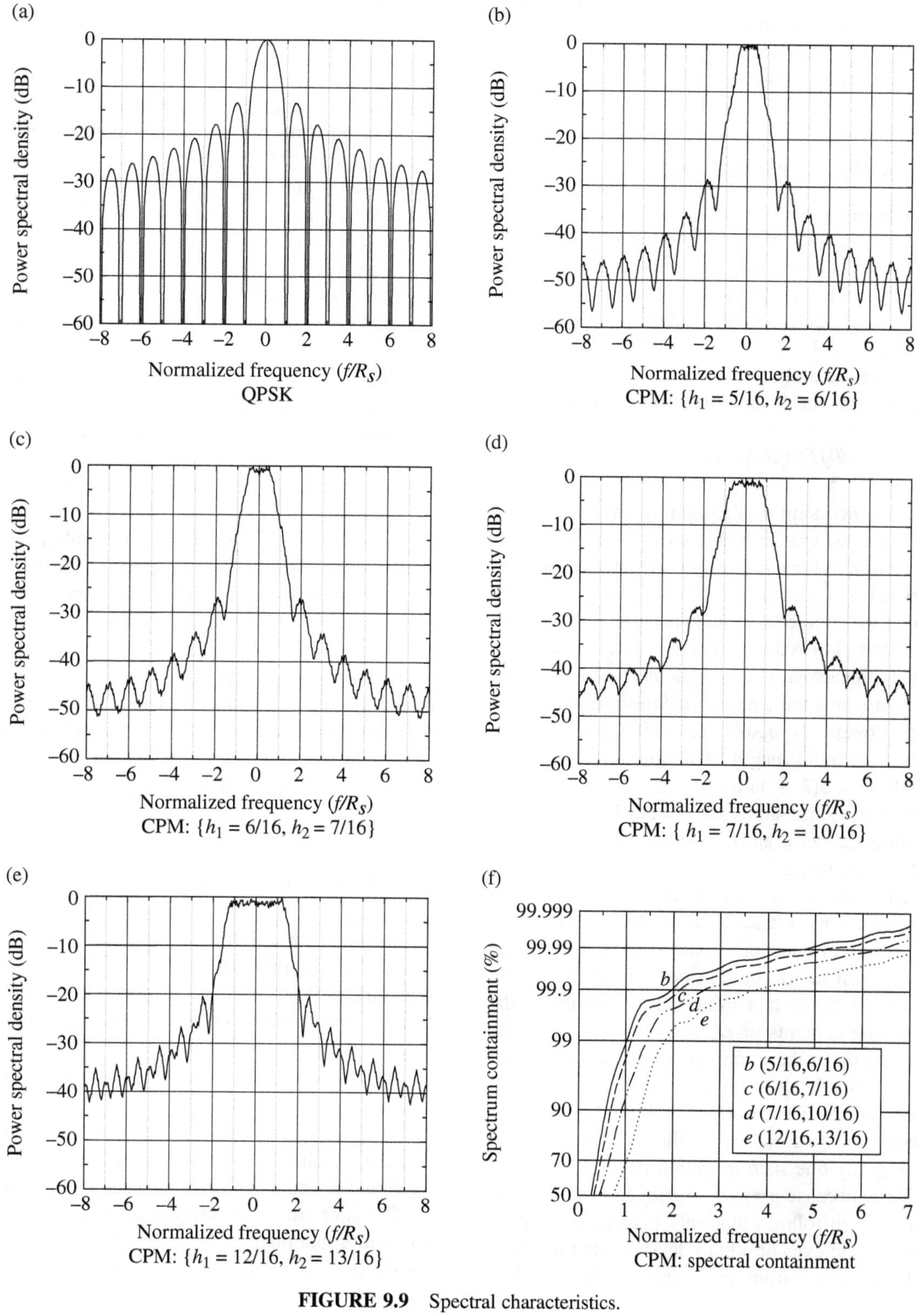

FIGURE 9.9 Spectral characteristics.

one hundred spectrum segments are averaged with each segment consisting of 32 random source bits with 16 samples/bit; cosine windowing is applied to each segment. The CPM spectrums are shown in Figure 9.9b through e. The spectrum containment of the CPM waveforms is relative to the total spectral power over the interval $f = \pm 8R_s$ and is summarized in Figure 9.9f. The spectrum shown in Figure 9.9b results in the best spectral containment and corresponds to the highest data rate of 48 kbps.

9.3.3 2-*h* 4-ary 1REC CPM Demodulator

In this section, the design of the 2-*h* 4-ary CPM demodulator applies to any of the sets of modulation indices listed in Table 9.6; however, the various examples are specialized using the case $\{h_1, h_2\} = \{5/16, 6/16\}$. The received signal, $y(t)$, and the corresponding data-dependent phase function, $\theta(t;\alpha,h)$, are described by (9.27) and (9.28), respectively. With these descriptions and the ensuing discussions, the signal phase at each state, in the absence of noise, is given by

$$\theta(jT;\alpha_j,h_j) = \pi\alpha_j h_j \tag{9.62}$$

The detection processing for the multi-*h* waveform being considered is most easily described in terms of the $N_s = 2q =$ 32-state trellis diagram for which a typical state transition is from the states at $(j-1)T$ to those at jT. The states at $(j-1)T$ are referred to as the source states and are denoted by S_n, and those at jT are referred to as the termination states, denoted by S'_n. Associated with each state is a metric m_n and a data storage array D_n that stores the tentative data estimates $\hat{\alpha}_j$. The length of data storage, D, associated with D_n is typically 5–10 times the constraint length of the multi-*h* waveform, that is, typically $D = 5(K+1)$ to $10(K+1)$; the following bit-error performance simulations use $D = 20$. The processing involves updating the metric and data storage associated with each termination state using the optimally selected metric and data associated with the source states having transition branches to the termination state. There are $M = 4$ such source states and the optimum selection is based on the source state having the minimum metric in consideration of the metric update associated with the transition branch. Therefore, the metrics m'_n and the contents of D'_n are associated with the source state resulting in the minimum metric from the four possible choices. This leads to a maximum-likelihood decision for the received data as described in the following paragraphs. Although the decoding trellis is composed of 32 states, during any one state transition only 16 states are processed. This is because $n = p_1\alpha_j$ or $p_2\alpha_j$ and α_j is an odd integer for all j; so, it follows that only even or odd values of n, corresponding to even or odd trellis states, are involved during a state transition from the source to the termination state.

The trellis diagram is shown in Figure 9.10 for the two modulation indices (5/16, 6/16). In the figure, the state transitions are shown emanating from the source states and branching to the four possible termination states. For source state l the $M = 4$ termination states are determined by the relationship

$$n = \left(l + \alpha_k \bar{p}_j\right)\text{modulo}(32) \quad \text{: terminal state} \tag{9.63}$$

where $k = 0, 1, 2, 3$. The source symbol data $\alpha_i = \{-3, -1, 1, 3\}$ corresponding to each of the branch transitions from a source state are identified in the figure. The odd and even trellis states are identified as *filled* and *unfilled* circles, respectively, and for a given state transition, all of the termination states will be either even or odd so one-half of the N_s states are used during each transition. For the (5/16, 6/16) example code the initial source states are *even* with $p_1 \in$ odd corresponding to the first modulation index; under this condition the termination states involve *even/odd* states according to the sequence: {*odd*, *odd*, *even*, *even*, …}. Depicting the transitions as in Figure 9.10 emphasizes the minimum distance of the waveform as determined by the branch separation and reemergence. For example, considering the source symbol sequence $\boldsymbol{\alpha} = \{1,1,1,\ldots\}$ shown as the *heavy solid* lines it is noted that the sequence $\{-3, -1\}$ diverges and reemerges after two symbols. Several other branches are also seen to diverge and reemerge after two symbols and the minimum Euclidean distance corresponding to these situations represents the minimum distance of the coded waveform.

The trellis decoding actually involves processing each termination state $S'_n : n = 0,\ldots,31$ with $n \in$ even (odd) depending on the initial conditions as discussed earlier. The processing of the termination states $\{S'_{11}, S'_5, S'_{20}, S'_2, \ldots\}$ for $j = 2, 3, 4, 5, \ldots$ is shown in Figure 9.11. The source states corresponding to the termination states are given by

$$l_{k,n} = \left(n - \alpha_k \bar{p}_j\right)\text{modulo}(32) \quad \text{: source state} \tag{9.64}$$

The *heavy* branch lines shown in Figure 9.11 correspond to the source symbol sequence $\boldsymbol{\alpha} = \{1, -1, 3, -3, \ldots\}$. When these transitions result in the maximum metric among the four branches terminating at each state, they identify or point to the source symbol estimates.

For example, when evaluating the metric m'_n at termination state S'_n the source states with branches to S'_n are $S_{l_{k,n}}$: $k = 0, 1, 2, 3$ correspond to the four possible data symbols α_k. This result can be pre-computed for all possible conditions and stored as pointers for subsequent processing.

The state transition processing from $j-1$ to j involves computing the metrics m'_n at the termination states S'_n using the metrics $m_{l_{k,n}}$ of the targeted source states $S_{l_{k,n}}$ and the corresponding metric update Δm_k as

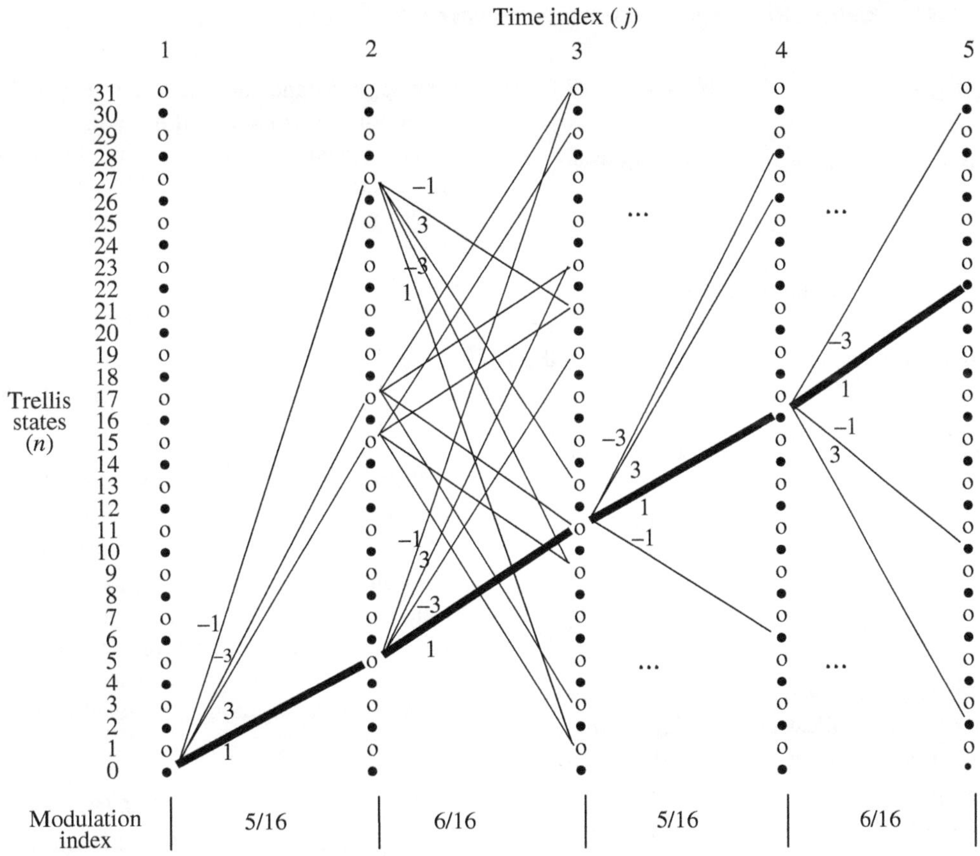

FIGURE 9.10 Trellis coding states for symbol sequence (1, 1, 1, 1) using {5/16,6/16} 2-*h* 4-ary full response CPM waveform.

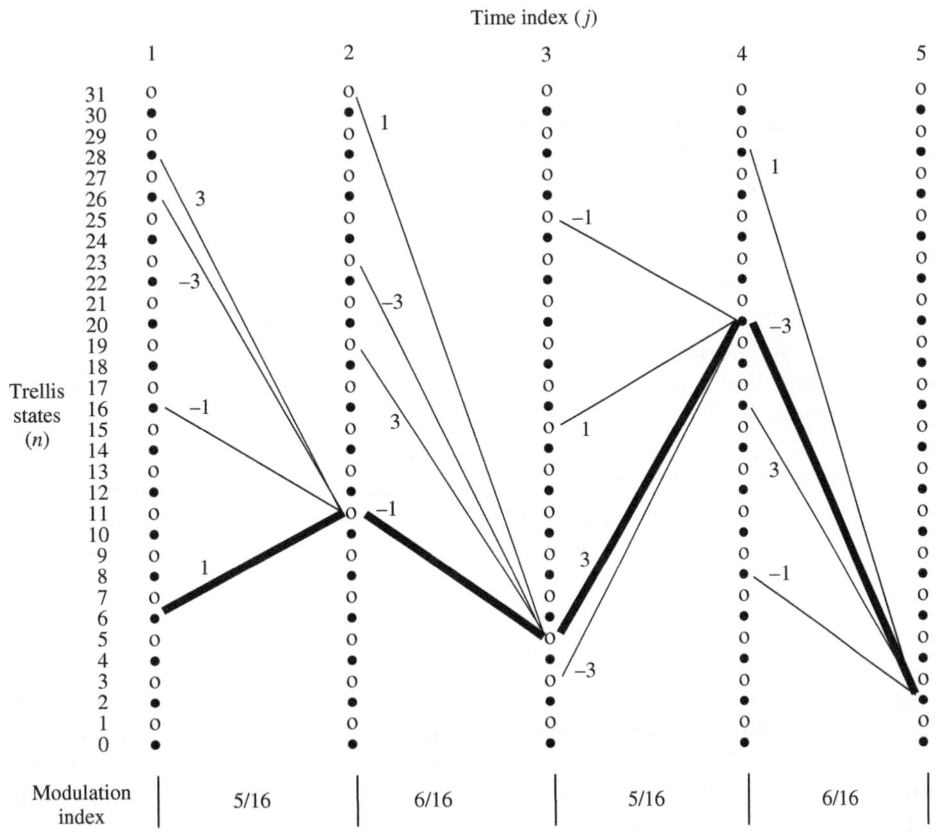

FIGURE 9.11 Trellis decoding states for symbol sequence {1, −1, 3, −3} using {5/16,6/16} 2-*h* 4-ary full response CPM waveform.

$$m'_n = \min_k \left[m_{l_{k,n}} + \Delta m_k\right] \quad : k = 0, \ldots, M-1 \tag{9.65}$$

The decoded source state data estimates are expressed as

$$\mathbf{D}_{l_{k,n}} = \left\{\hat{\alpha}_{l_{k,n},j-D}, \hat{\alpha}_{l_{k,n},j-(D-1)}, \ldots, \hat{\alpha}_{l_{k,n},j-1}\right\} \tag{9.66}$$

where D is the depth or delay of the trellis processing before making a data decision; so, at each termination state, the surviving data sequence estimate is stored corresponding to the metric m'_n.

$$\begin{aligned}\mathbf{D}'_n &= \min_{l_{k,n}} \left\{\mathbf{D}_{l_{k,n}}(\text{left-shifted})\right\} & : k = 0, \ldots, M-1 \\ &= \min_{l_k} \left\{\left(\hat{\alpha}_{l_{k,n},j-(D-1)}, \ldots, \hat{\alpha}_{l_{k,n},j-1}, \hat{\alpha}_{l_{k,n},j}\right)\right\} & : k = 0, \ldots, M-1 \\ &= \left(\hat{\alpha}_{n,j-(D-1)}, \ldots, \hat{\alpha}_{n,j-1}, \hat{\alpha}_{n,j}\right)\end{aligned} \tag{9.67}$$

where $min\{\ \}: \forall\ l_{k,n}$ signifies the left-shifted data sequence $\mathbf{D}_{l_{k,n}}$ corresponding to the smallest or surviving metric m'_n. When $j \le D$, the data is simply left-shifted and $\hat{\alpha}_{l_{k,n},j}$ is set equal to the α_k corresponding to the surviving transition path; however, when $j > D$ the left-shifted data $\hat{\alpha}_{l_{k,n},j-D}$ falling off the left end of the array becomes a candidate for the maximum-likelihood data decision. In general, this data decision is computed as

$$\hat{\alpha}_{j-D} = \min_{m'_n} \left[\hat{\alpha}_{n,j-D}\right] \tag{9.68}$$

However, when D is significantly larger than the constraint length $K + 1$ then, with very high probability, $\hat{\alpha}_{n,j-D} = \hat{\alpha}_{n',j-D} : \forall n'$ so $\hat{\alpha}_{j-D}$ can be selected from any of the termination states.

The metric updates Δm_k are defined in terms of the Euclidean distance between the received signal and the $M = 4$ possible reference signals terminating on each trellis state. Initially, the metrics m_n are all set to zero; however, they must be reset periodically to avoid underflow and overflow. The complex envelope of the received signal $\tilde{y}(t)$ and the baseband reference signal $\hat{x}_k(t)$ are given, respectively, by (9.39) and (9.40) and, using these results, the squared Euclidean distance d_k^2 is computed using (9.41). Upon applying the simplifications leading to (9.42), the metric updates are evaluated as

$$\Delta m_k = -\frac{1}{E}\mathrm{Re}\left\{\phi_{yx}(0,k)\right\} \tag{9.69}$$

where $\phi_{yx}(0, k)$ is the zero lag correlation function between the received and reference signals. The maximum correlation response occurs when the received signal $\tilde{s}(t)$ is matched to the reference $\tilde{x}_k(t)$ and $\Delta m_k = -d_k^2$. For the coherent detection case, $\hat{\phi}_o = \phi_o$ and the maximum correlation corresponds to the conditions: $\alpha_k = \alpha_j$ and $l_{k,n'} = n$.

The correlation processing involved in the metric update computations can be simplified by identifying the four unique correlations for each transition and applying a phase rotation corresponding to the position of the source state in the trellis. With a shift in the time axis so that $0 \le \tau \le T$, the correlation function is evaluated as in (9.43) with the result

$$\phi_{yx}(0;k) = \frac{E}{2} e^{j(\phi_o - \hat{\phi}_o)} \left[\phi'_{yx}(0;k)\right] e^{-j\pi l_{k,n}/q} \tag{9.70}$$

With coherent processing, the phase estimate $\hat{\phi}_o$ is approximately equal to ϕ_o. There are four unique correlation values $\phi'_{yx}(0;k)$: $k = 0, 1, 2, 3$ for each received symbol. The four complex correlation values are computed once for each symbol and stored for the remaining state transition processing. During the evaluation of each termination state metric, the stored correlation values must undergo a phase rotation given by $-\pi l_{k,n}/q$ radians as noted earlier.

9.3.3.1 2-h 4-ary 1REC CPM Coherent Detection

Defining the demodulator phase tracking error as $\varphi_\varepsilon = \phi_o - \hat{\phi}_o$, the metric update becomes

$$\Delta m_k = -\frac{1}{2}\mathrm{Re}\left\{e^{-j\phi_\varepsilon} \left[\phi'_{yx}(0;k)\right] e^{-j\pi l_{k,n}/q}\right\} \tag{9.71}$$

where

$$\phi'_{yx}(0;k) = \frac{1}{T}\int_0^T \tilde{y}'(t) e^{-j(2\pi\alpha_k h_j q(t))} dt \tag{9.72}$$

Under noise-free conditions and when $\alpha_k = \alpha_j$ and $l_{k,n'} = n$, Equation (9.70) results in the metric update $\Delta m_k = -\mathrm{Re}\left\{e^{-j\varphi_\varepsilon}\right\}/2$ and corresponds to the optimum detection condition for termination state n'. In this case, the phase error is available from the correlation function $\phi_{yx}(0, k)$ corresponding to the maximum metric over all of the termination states. The candidate phase error is obtained from the correlation response at each termination state as

$$\phi_{yx}(\phi_\varepsilon, n) = \min_k{}^{-1}\left[m_{l_{k,n}} + \Delta m_k : \phi_{yx}(0,k)\right] \tag{9.73}$$

where the notation $\min_k{}^{-1}[a_k : b]$ picks the value of b corresponding to the minimum of all a_k values. This processing results in the optimum selection of the phase-error function at each termination state and the final selection is based on the optimum selection over all of the termination states, expressed as

$$\phi_{yx}(\varphi_\varepsilon) = \min_n{}^{-1}\left[m'_n : \phi_{yx}(\varphi_\varepsilon, n)\right]$$

$$\cong K_d e^{j\varphi_\varepsilon} \tag{9.74}$$

K_d is defined as the phase detector gain constant. Based on this result, the desired phase error is determined from the imaginary part of $\phi_{y,x}(\varphi_\varepsilon)$ and the real part is used as the input to the phaselock detector. The phase detector error signal

$$e(\varphi_\varepsilon) = \mathrm{Im}\{\phi_{yx}(\varphi_\varepsilon)\} \cong K_d \sin(\varphi_\varepsilon) \tag{9.75}$$

is passed through the loop filter $F(s)$ and applied to the input signal $\tilde{y}(t)$. Similarly, the lock detector signal

$$e_{LD}(\varphi_\varepsilon) = \mathrm{Re}\{\phi_{yx}(\varphi_\varepsilon)\} \cong K_d \cos(\varphi_\varepsilon) \tag{9.76}$$

is filtered using two synchronously tuned low-pass filters and compared to a fixed threshold to determine a locked condition.

9.3.3.2 MAP Estimation of Received Signal Phase and Symbol Timing The received signal phase estimation discussed earlier is based on the intuitive application of the observed phase error in the correlation function $\phi_{yx}(0, \varphi_\varepsilon)$ corresponding to the surviving metric at the termination states. The details of the MAP processing for the signal phase and symbol timing errors [17, 18] are discussed in Section 9.2.3.2 and demonstrate the validity of the preceding estimates. Furthermore, the MAP estimation analysis yields the sensitivity of these estimates under the joint error conditions.

The MAP estimation corresponding to the termination state having the maximum metric results in $m' = \max[m'_n]\ \forall n$ and, under noise-free conditions, the necessary conditions are $\ell_{kn'} = n$ and $\alpha_k = \alpha_j$. These conditions are identical to those required by the selection of k and n resulting in the correlation function $\phi_{yx}(0, \phi_\varepsilon)$ as expressed in (9.74). Using these results, the discriminator function for the phase estimate, given a timing error τ, is evaluated using (9.55) and expressed as

$$e_j(\varphi_\varepsilon;\tau) = \left(\frac{A^2}{4}\right)\left\{(T-|\tau|)\sin\left(\pi\alpha_j\bar{h}_j\left(\frac{\tau}{T}\right)+\varphi_\varepsilon\right)\right.$$

$$+|\tau|\sin\left(\pi\left(\alpha_{j\pm1}\bar{h}_{j\pm1}+\alpha_j\bar{h}_j\right)\left(\frac{\tau}{2T}\right)+\varphi_\varepsilon\right)$$

$$\left.sinc\left(\left(\alpha_{j\pm1}\bar{h}_{j\pm1}-\alpha_j\bar{h}_j\right)\left(\frac{|\tau|}{2T}\right)\right)\right\} \tag{9.77}$$

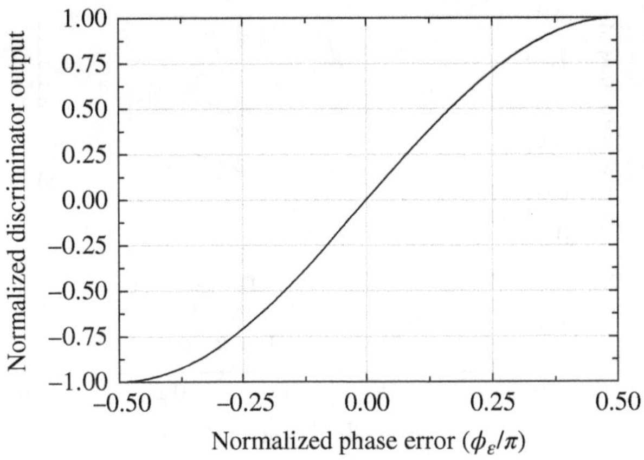

FIGURE 9.12 CPM phase detector discriminator characteristic (all CPM modes, $\tau = 0$).

where the +(−) associated with the index j corresponds to the positive (negative) values of τ. When this result is filtered and applied to an ideal integrator with gain K_o radian/volt-second, the integrator output becomes the MAP phase estimate $\hat{\phi}_o$. Ideally, the filtered result is evaluated as the joint expectation over the *iid* discrete random data $(\alpha_j, \alpha_{j\pm1})$ and the resulting three-dimensional surface characterizes the phase detector discriminator characteristics as a function of phase and symbol time error. The phase discriminator response for $\tau = 0$ is shown in Figure 9.12 and, under this condition, is independent of the CPM modulation indices. The discriminator detector gain is defined as the slope of the response about the origin and is evaluated as $0.25A^2T$ V^2-s/rad.*

Similarly, the discriminator function for the symbol time estimate, given a phase error, is evaluated using (9.56) as

$$e_j(\tau;\varphi_\varepsilon) = \pi\alpha_j\bar{h}_j\left(\frac{A^2}{4}\right)e_j(\varphi_\varepsilon;\tau)$$

$$-\left(\frac{A^2}{2}\right)\sin\left(\pi\left(\alpha_{j\pm1}\bar{h}_{j\pm1}+\alpha_j\bar{h}_j\right)\left(\frac{\tau}{2T}\right)+\varphi_\varepsilon\right)$$

$$sinc\left(\pi\left(\alpha_{j\pm1}\bar{h}_{j\pm1}-\alpha_j\bar{h}_j\right)\left(\frac{|\tau|}{2T}\right)\right) \tag{9.78}$$

where the +(−) associated with the index j corresponds to the positive (negative) values of τ. When this result is filtered and applied to an ideal integrator with gain K_o second/volt-second, then the integrator output becomes the MAP symbol time estimate $\hat{\tau}$. As in the case of the phase error, the filtered result is evaluated as the joint expectation over the *iid* discrete random data $(\alpha_j, \alpha_{j\pm1})$.

*This detector gain is the gain required for a sampled data implementation, that is, K_dT. If the result were normalized by the symbol period and the reference signal were unit amplitude, then the units would be volts/radian as typically associated with analog detectors.

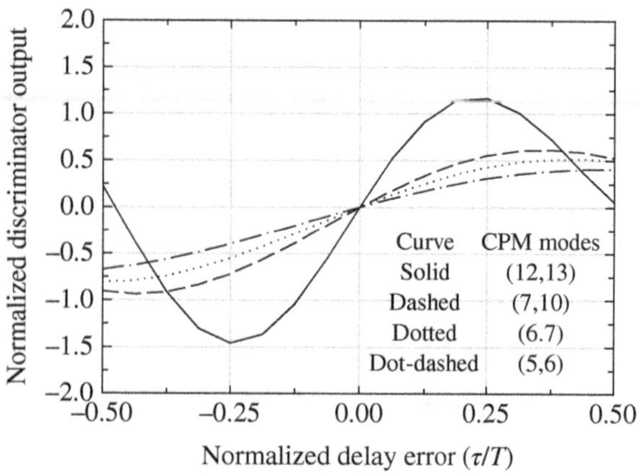

FIGURE 9.13 Symbol time discriminator characteristics for several CPM waveforms with $\varphi_\varepsilon = 0$.

The symbol time discriminator response for $\varphi_\varepsilon = 0$ is not independent of the CPM modulation indices and Figure 9.13 shows the response for the 2-h 1REC CPM waveform used in this case study. In this situation, the loop gain must be adjusted to compensate for the change in detector gain to ensure identical transient responses for the various waveforms. Furthermore, these results indicate that the initial symbol time acquisition must be within 0.5 T, and, for the 2-h modulation indices (12/16,13/16), within 0.25 T, to ensure a correct lock condition.

Based on the 1REC MAP phase and symbol timing estimation, the required simplified processing is shown in Figure 9.7. This implementation recognizes that the expected value of the additive error term in $e_j(\tau;\varphi_\varepsilon)$ is zero resulting in the approximation

$$e_j(\tau;\varphi_\varepsilon) \cong \pi\alpha_j\bar{h}_j e_j(\varphi_\varepsilon;\tau) \tag{9.79}$$

Figure 9.7 also shows the phaselock loop (PLL) detector processing that includes a lower threshold for determining a lost-lock condition.

9.3.4 2-*h* 4-ary 1REC CPM Performance Simulation

The performance of the 2-h 4-ary 1REC CPM waveform is examined for the four sets of modulation indices listed in Table 9.7. To further designate the conditions of the decoding, the notation (h_1,h_2,D) or (p_1,p_2,D) is used to signify the decoding trellis depth. The performance is characterized in terms of the bit-error probability for various signal-to-noise ratios E_b/N_o. For the purpose of comparing various results, the coding gain, relative to ideal QPSK, at a bit-error probability of 10^{-5} is used. The results are based on Monte Carlo simulations using 1 M bits at each signal-to-noise ratio for $P_{be} \geq 10^{-4}$; otherwise, 10 M bits are used. Furthermore, the performance is based on the completion of the waveform acquisition processing so that the carrier frequency and symbol timing are established to an accuracy that ensures carrier frequency and symbol time loop acquisition and tracking. For example, the frequency accuracy is estimated to within the pull-in range of the PLL and symbol timing accuracy is estimated to within one-fourth of the received symbol duration. The CPM start-of-message (SOM) sequence and the start of the multi-h modulation index frame synchronization, referred to as the *superbaud* timing, are also established during the acquisition processing so that the demodulator modulation indices are properly aligned with those of the received waveform. In Section 9.3.4.1, the ideal performance of the CPM modulated waveform is evaluated with the AWGN channel and ideal receiver phase and symbol timing. In Section 9.3.4.2, the performance with phase and symbol tracking are examined using the (5/16, 6/16, 20) CPM waveform with the AWGN channel; the performance is characterized for various time-bandwidth products of the tracking loops. The (5/16, 6/16) modulation index pair is selected for operating with a 48 kbps CPM waveform through a 25 kHz satellite channel, In Section 9.3.4.3, the performance through the satellite transponder is examined with AWGN receiver noise. The transponder is modeled as a bandlimited filter followed by a hard limiter, zonal filter, and traveling wave tube (TWT) downlink amplifier.

TABLE 9.7 CPM Waveform Performance Gain with AWGN Channel (D = 20)

R_b (kbps)	CPM (p_1,p_2,D)	E_b/N_o (dB)	Gain[a] (dB)
56	(4,5,20)	8.4	1.2
48	(5,6,20)	7.5	2.1
32	(6,7,20)	6.5	3.1
19.2	(12,13,20)	5.3	4.3

[a]Relative to 9.6 dB: QPSK @ $P_{be} = 10^{-5}$.

9.3.4.1 Ideal Performance with AWGN Channel The ideal performance of the (5/16, 6/16, D) CPM waveform is shown in Figure 9.14 for various decoding trellis depths D. These results represent ideal tracking of both the received signal phase and symbol timing. A trellis depth of $D = 10$ symbols results in a performance degradation of about 0.3 dB while depths ≥15 result in negligible degradation. These conclusions are based solely on the simulation performance that demonstrates convergence as D is increased. Unless otherwise noted, a trellis depth of 20 symbols is used for all of the following simulation results. Figure 9.15 shows the performance with various modulation indices; the performance gains at $P_{be} = 10^{-5}$ relative to ideal QPSK demodulation are 1.2, 2.1, 3.1, and 4.3 dB. Table 9.7 summarizes the 2-h 4-ary 1REC CPM performance results relative to E_b/N_o = 9.6 dB corresponding to QPSK at $P_{be} = 10^{-5}$.

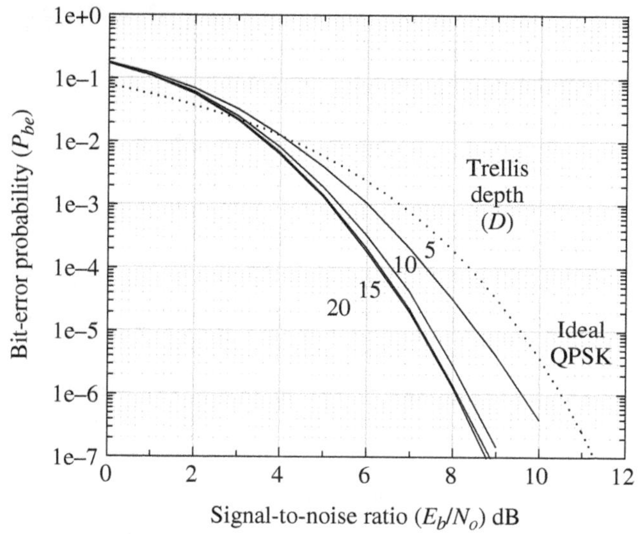

FIGURE 9.14 Performance of (5/16, 6/16,D) CPM waveform with AWGN channel.

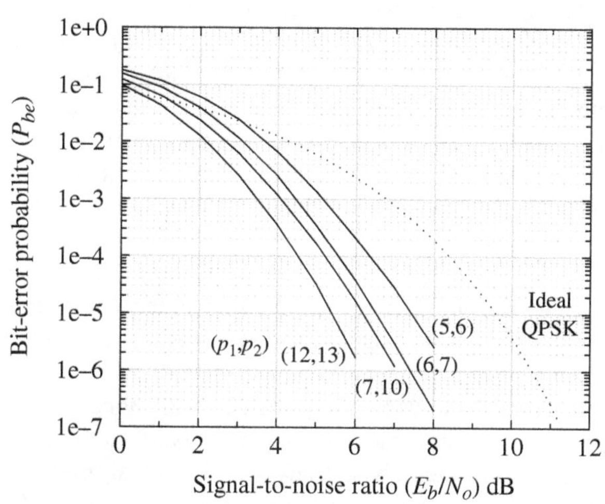

FIGURE 9.15 Performance of CPM waveforms with AWGN channel (ideal carrier and symbol tracking, D = 20).

9.3.4.2 Phase and Symbol Tracking Performance with AWGN Channel Figure 9.16 shows the simulated phase and lock-detector responses for the example 2-h, 4-ary 1REC CPM waveform. A second-order PLL is used and the transient responses are shown for PLL time-bandwidths products of B_LT = 0.025 and 0.005. These responses result from an input phase-step of six degrees, a damping factor of 0.707 and a noise-free channel. The corresponding lock detector responses are also shown for $B_{LD}T = B_LT/2$. Using a lock detector threshold of 0.75, phase-lock is achieved in approximately 20 and 110 symbols respectively. For a continuous wave (CW) input signal, a damping factor of ζ = 0.707 and an initial frequency estimate that is within the loop bandwidth, the approximate normalized lock-in time is given by

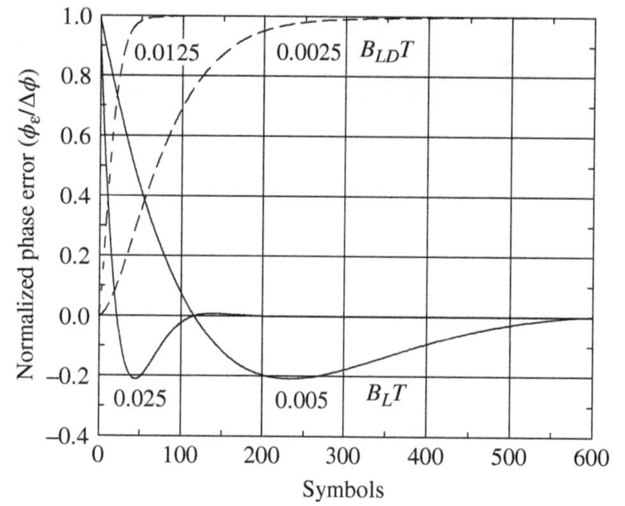

FIGURE 9.16 PLL and lock detector responses for phase step input (ζ = 0.707).

$$T_LR_s \approx \frac{0.5303}{B_LT} \quad \text{: normalized phaselock loop lock time} \tag{9.80}$$

The simulated bit-error performance results discussed in the remainder of this section focus on the (5/16, 6/16, 20) 2-h 4-ary 1REC CPM waveform operating at 48 kbps through the AWGN channel with the second-order PLL as characterized in Figure 9.16. An appropriate delay is included in the simulation before error counting is begun to ensure that steady-state tracking conditions are achieved. The performance degradation is defined relative to the simulated AWGN performance shown in Figure 9.15. First, the performance with AWGN and various PLL bandwidths is considered using ideal symbol timing. Next, the performance with symbol tracking only is examined. Last, the performance with joint phase and symbol time tracking is examined. This section is concluded by the examining the performance degradation with a received carrier frequency error characterized as a linear frequency rate.

Figure 9.17 shows the performance with ideal symbol tracking and with PLL tracking for various loop time-bandwidths products. The results indicate that for $B_LT \leq$ 0.025, the maximum performance loss is about 0.2 dB with a loss of less than 0.1 dB for B_LT = 0.005. This loss increases to 0.6 dB as B_LT is increased to 0.05. For rapid phase convergence, a higher B_LT product is typically used during acquisition that is decreased prior to data detection to achieve a lower detection loss. Typically, a phase tracking loss of 0.2 dB is included in the modem link budget.

The sensitivity to errors in the symbol timing is examined by characterizing the performance degradation with a fixed symbol timing error ΔT. Figure 9.18 shows that the 2-h 4-ary 1REC CPM waveform is relatively sensitive to the

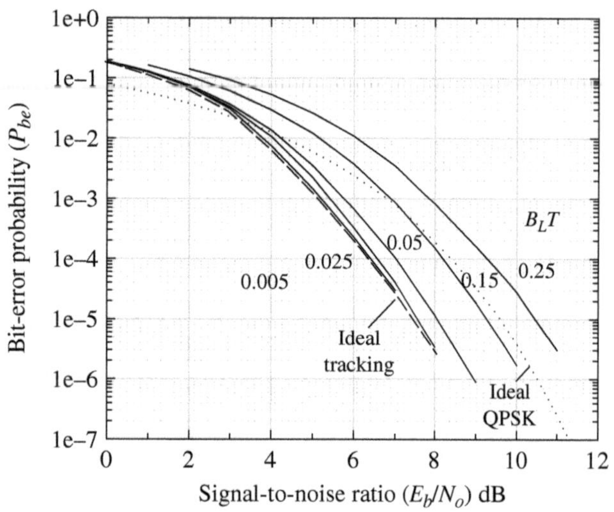

FIGURE 9.17 Performance of (5/16, 6/16, 20) CPM waveform with phase tracking only.

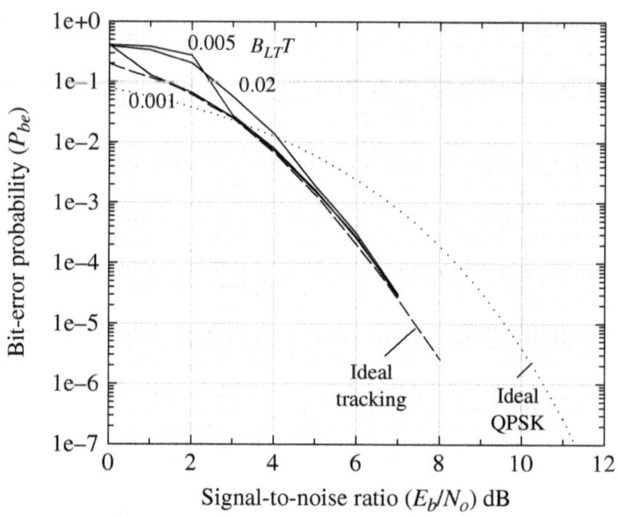

FIGURE 9.19 Performance of (5/16, 6/16, 20) 4-ary CPM with symbol tracking only ($\Delta T = T/16$, $\delta T = T/32$).

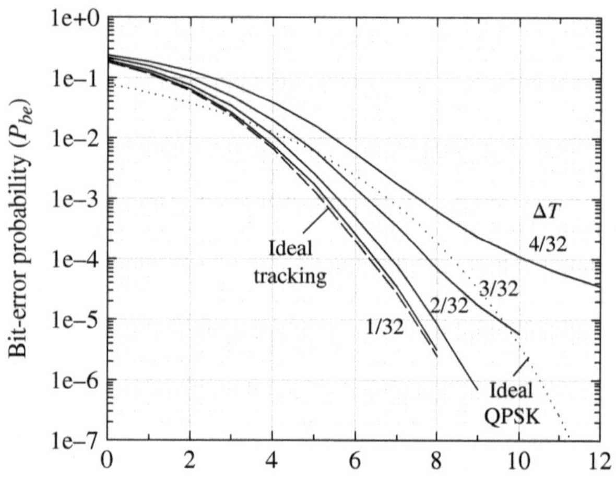

FIGURE 9.18 Performance of (5/16, 6/16, 20) CPM waveform with fixed symbol time error and 32 samples/symbol.

symbol timing errors, for example, a fixed timing error of 3/32 of a symbol results in a degradation of 2 dB. However, with a fixed timing error of 1/32 of a symbol the loss decreased to about 0.1 dB so the maximum timing adjustment with symbol tracking should be 1/32 of a symbol. This suggests that the symbols should be sampled using a minimum of 32 samples-per-symbol and subsequently down sampling to 4 or 8 samples-per-symbol for matched filter detection processing. During the signal acquisition processing, the initial symbol timing should be resolved to within $T/8$ and the acquisition preamble must be long enough to allow tracking to reach the steady-state condition before entering the data detection mode.

The symbol tracking loop time-bandwidth product $(B_{LT}T)$* is used to characterize the bandwidth of the time tracking filter. For all of the following simulations involving symbol time tracking, a single-pole digital filter is used. The filter is sampled at the symbol rate and the filter coefficient is computed as $k = \exp(-2\pi\ B_{LT}T)$. The intrinsic filter gain is adjusted by $(1 - k)$ to ensure a steady-state unit-gain response. At intervals equal to one-half of the filter time constant, given by $\lceil 1/(1-k) \rceil$ symbols, the receiver sampling time is incremented by ±1 sample depending upon the sign of the filter output e_{oj}, that is, the time adjustment is computed as $\text{sign}(1, e_{oj})$.

Figure 9.19 shows the performance with symbol tracking for various symbol tracking loop time-bandwidth products $B_{LT}T$ and with ideal phase tracking. The initial timing error is $\Delta T/T = 1/16$ symbol and the tacking update is $\delta T/T = 1/32$ of a symbol. At a sufficiently high signal-to-noise ratio, greater than 5 dB for the $B_{LT}T$ products considered, the performance degradations are relatively low (<0.1 dB) and nearly identical. However, there is a critical, $B_{LT}T$-dependent, signal-to-noise ratio† below which the performance degrades rapidly. For example, with $B_{LT}T = 0.02$ the performance begins to degrade significantly at a signal-to-noise ratio of 5.0 dB and with $B_{LT}T = 0.001$ the performance degradation remains less than about 0.1 dB for signal-to-noise ratios as low as 1 dB before significant degradation is encountered. The focus of this discussion is on the degradation at $P_{be} = 10^{-5}$, and all three of the $B_{LT}T$ products considered have

*The notation $B_{LT}T$ is used to denote the time-bandwidth product of the symbol tracking loop, whereas B_LT is used to denote the time-bandwidth product of the phase tracking loop.

†The critical signal-to-noise ratio is defined and characterized in Chapter 10 for the PLL.

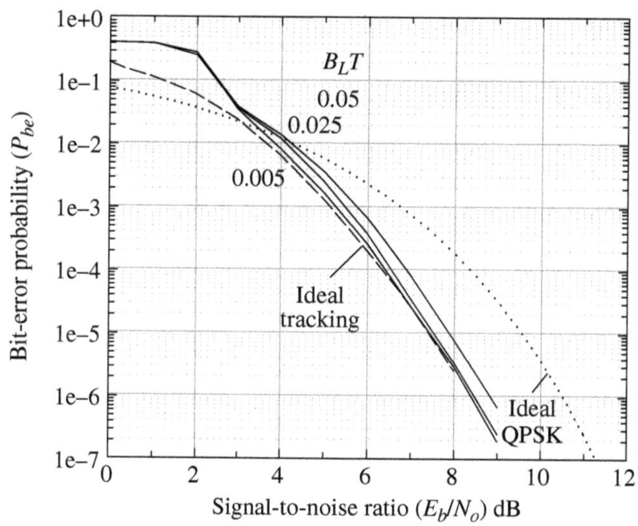

FIGURE 9.20 Performance of (5/16, 6/16, 20) CPM waveform with phase and symbol tracking ($B_{LT}T = 0.005$, $\Delta T = 0$, $\delta T = T/32$).

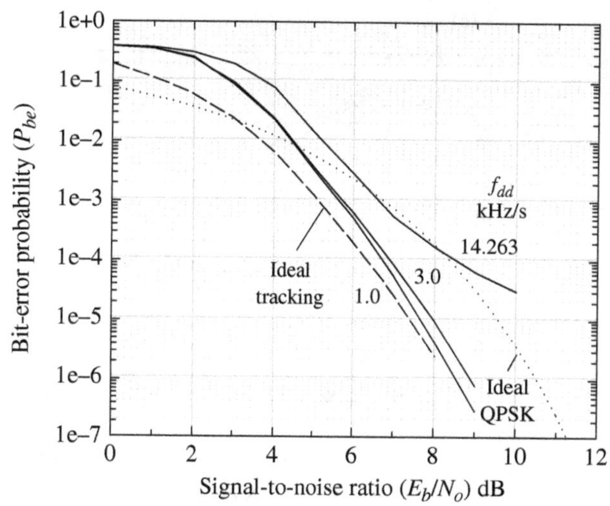

FIGURE 9.21 Performance of (5/16, 6/16, 20) CPM waveform with phase and symbol tracking and frequency-rate error ($B_{LT}T = 0.02$, $\Delta T = 0$, $\delta T = T/32$ and $B_LT = 0.025$).

comparably low degradations.* The selection of an excessively high $B_{LT}T$ product will result in the critical signal-to-noise ratio that results in significant performance degradation even at high operating signal-to-noise ratios. These results show performance degradations of about 0.1 dB that are consistent with those in Figure 9.19 using a fixed timing error of $\Delta T = \pm T/32$; they confirm that the timing adjustment increment should be $\delta T \le T/32$.

Figure 9.20 shows the performance with joint phase and symbol time tracking. For a constant $B_{LT}T$ product equal to 0.02, the B_LT product is varied between 0.005 and 0.05. For $B_LT = 0.05$, the performance degradation is about 0.6 dB and for $B_LT = 0.005$ the degradation is less than 0.1 dB.

The final subject in this section is the performance of the (5/16, 6/16, 20) 2-*h* 4-ary 1REC CPM modulated waveform with a received carrier frequency error characterized as a frequency rate. When an ideal linearly swept carrier frequency with frequency rate f_r Hz/sec is applied to a second-order PLL, the theoretical steady-state loop condition will result in a static phase error of φ_e radians. For a loop with a damping factor of 0.707, the relationship between the frequency rate and phase error is

$$f_r = 0.566\varphi_e(B_LT)^2R_s^2 \tag{9.81}$$

Evaluation of the (5/16, 6/16, 20) CPM waveform performance with a static phase error of $\varphi_e = 0.07$ radians (4 degrees) results in a performance degradation of about 0.2 dB. Therefore, using $\varphi_e = 0.07$ in (9.81) with $B_LT = 0.025$ and $R_s = 24$ ksps results in a frequency rate of $f_r =$ 14.263 kHz/s. However, applying this frequency rate to the (5/16, 6/16, 20) CPM modulated waveform, the degradation at $P_{be} = 10^{-5}$ is in excess of 3 dB; therefore, this theoretical relationship is not applicable to the coded 2-*h* 4-ary 1REC CPM modulated waveform.† Figure 9.21 shows the performance with and received signal frequency rates of 14.263, 3.0 and 1.0 kHz/s when using the phase and symbol tracking parameters $B_LT = 0.025$ and $B_{LT}T = 0.02$, respectively. The performance losses at $P_{be} = 10^{-5}$ for 1.0 and 3.0 kHz/s frequency rates are 0.25 and 0.55 dB, respectively, so the frequency rate should be less than 1.0 kHz/s for a reasonable detection loss.

9.3.4.3 *Performance in Bandlimited Satellite Channel*

The performance of the 2-*h* 4-ary 1REC CPM waveform through a 25 kHz bandwidth UHF satellite transponder is examined in this section. The transponder is modeled‡ as a bandlimited filter followed by a hard limiter with zonal filtering and a TWT downlink amplifier. The bandlimiting filter is modeled as a 6-pole, 0.05 dB ripple Chebyshev filter. The 3 dB bandwidth of the 25 kHz channel is typically 32 kHz. Based on measured filter phase response data, the Chebyshev filter phase is equalized using a cubic phase function with a 6-degree phase error at ±12.5 kHz. The hard limiter is ideal in the sense that the output is passed through a zonal filter to remove all harmonics of the carrier. The effect of this ideal

*For coded systems, that operate at much lower E_b/N_o values, selection of an acceptable $B_{LT}T$ product is equally important.

†The relationship is valid for the uncoded modulated QPSK, in that, (9.81) results in the predicable static phase error for a given frequency rate. Furthermore, the static phase error is a reliable predictor of the detection loss.

‡The channel filtering consists of an input or pre-filter that establishes the channel bandwidth, a noise filter following hard limiter, and a post-filter that attenuates intermodulation interference between the channels prior to the TWT downlink amplifier.

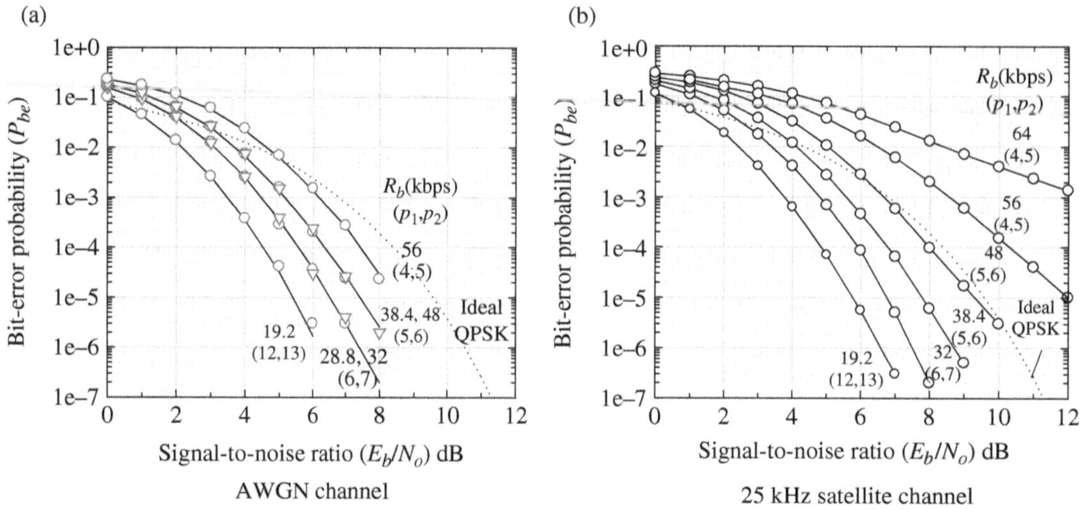

FIGURE 9.22 Bit-error performance of 2-h 4-ary CPM waveforms (ideal carrier and symbol tracking).

limiter is most pronounced when multiple carriers are combined and passed through the TWT amplifier. In this analysis, however, the TWT is operated with a high output backoff and essentially behaves like a linear amplifier.

The performance analysis focuses on selected data rates for use over the 25 kHz bandwidth channel. The simulated performance results of the CPM waveform are shown in Figure 9.22a using the AWGN channel; the *triangle* data points correspond to the higher of the two user data rates. Figure 9.22b shows the performance through the 25 kHz bandwidth channel and clearly demonstrates the degradation resulting from the bandlimited and hard-limiting channel with the 64 and 56 kbps user data rates resulting in a performance loss. A significant advantage of the CPM modulated waveform is the spectral efficiency shown in Figure 9.9. The improved spectral efficiency results in lower adjacent channel interference (ACI) in frequency division multiplex applications. The optional Reed–Solomon (RS) concatenated coding is not considered in this analysis; however, the performance of the 2-h 4-ary 1REC CPM through the 25 kHz satellite channel with a rate 0.987 outer RS code provides an additional coding gain of 0.5 dB at $P_{be} = 10^{-5}$ with $R_b = 38.4$ kbps [19].

9.4 MULTIPHASE SHIFT KEYING TRELLIS-CODED MODULATION

Trellis-coded modulation was first investigated by Ungerboeck [21–23], and achieves coding gain by introducing redundant, or parity check, signaling states in the underlying modulation waveform. Transitions between the various states of the modulated waveform follows a rigid set of coding rules and errors are detected when the state transition rules are violated. The errors are corrected in a trellis decoder similar to that used in Viterbi decoding of convolutional codes. The advantage of this approach is that additional transmission bandwidth, as used in block and convolutional codes, is not required. Ungerboeck recognized that the conventional approach to error correction treats the error-correction coding and waveform modulation as independent functions and that by combining them improvements in the channel capacity can be achieved. Ungerboeck formulated the difference between optimum decoding on discrete channels based on the minimum squared Hamming distance and optimum decoding on continuous or unquantized channels based on the minimum squared Euclidean distance. The squared Hamming and Euclidean distances are equivalent only in the cases of binary and 4-phase modulation; the latter being simply two channels of orthogonal binary modulation. Consequently, binary modulation systems using codes optimized for Euclidean distance and soft-decision decoding have found widespread use in power-efficient transmissions with 2 bits/Hz capacity. This also explains the relatively poor performance encountered when integrating codes designed for maximum Hamming distance with multilevel modulated waveforms. Therefore, as in the case of CPM decoding, the Euclidean distance is used as the metric for the decoding of TCM waveforms.

The squared free-distance of the trellis-coded waveform is the minimum squared Euclidean distance between all code words c_n and is given by

$$d_{free}^2 = \min_{n \neq m} |c_n - c_m|^2 \tag{9.82}$$

For sufficiently high signal-to-noise ratios, such that the probability of error events with distances exceeding d_{free} is negligible, the error probability in an AWGN channel is upper bounded by

$$P_e \le \frac{N_{free}}{2} e^{-d_{free}^2/4N_o} \tag{9.83}$$

where N_{free} is the number of neighboring code words with distance d_{free} and N_o is the one-sided noise power spectral density. The following discussions and examples focus on the application of TCM to PSK modulated waveforms and the potential TCM coding gain is expressed as

$$G = 10\log_{10}\left(\frac{d_{free}^2}{d_{min}^2}\right) \tag{9.84}$$

where d_{min} is the minimum distance of the underlying or uncoded PSK waveform.

A generalized TCM encoder using a convolutional encoder is shown in Figure 9.23. The uncoded input corresponds to m bits and the output corresponds to n bits resulting in a code rate is $r = m/n$. The n bits are mapped onto $M = 2^n$ symbol modulation states based on set partitioning. The modulation states are characterized as MPSK and MQPAM.

The convolutional code rate, as defined in Figure 9.23, is $r_c = \tilde{m}/\tilde{n}$; however, in the following descriptions the encoder is specialized for an overall TCM code rate of $r = m/(m+1)$.

The set partitioning or mapping of the coded bits to the MPSK or MQPAM constellations is critical to ensure optimum distance properties and ultimately good code performance. Calderbank and Mazo [24] provide a generalized theory for the formulation of trellis codes and Biglieri, Divsalar, McLane, and Simon [25] relate the design procedures to Ungerboeck's implementation; they also provide a comprehensive treatment of trellis-coded modulation with optimum codes and performance results under various channel conditions. Tables 9.8 and 9.9 list several 8PSK and 16QPAM modulated codes tabulated by Ungerboeck [23]. In these tables, ν is the convolutional code constraint length, $2^{\nu-1}$ is the number of decoding trellis states, h^i are parity-check coefficients in octal notation, d_{free} is normalized to the unit circle, and G(dB) is the asymptotic coding gain for large signal-to-noise ratios.

As an example of implementing the trellis code using Ungerboeck's parity-check coefficients h^i, consider the systematic eight-state ($\nu = 3$) 8PSK-TCM code given in Table 9.8. The parity-check matrix is used to implement the feedback configuration of the convolutional encoder and is defined in terms of the parity-check coefficients as

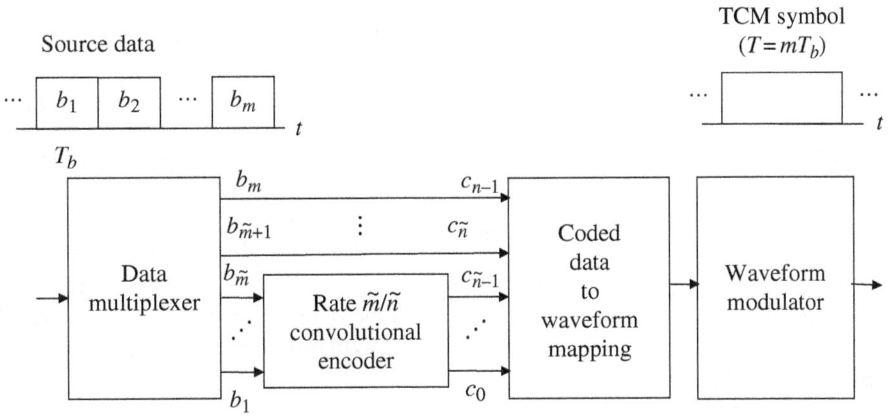

FIGURE 9.23 Generalized systematic rate m/n trellis-coded modulator. Adapted from Ungerboeck [21]. Reproduced by permission of the IEEE.

TABLE 9.8 8PSK-TCM codes (m = 2)[a]

ν	2^ν	$\tilde{m}$	h^2	h^1	h^0	d_{free}^2	G (dB)	N_{free}
2	4	1	—	2	5	4.000	3.01	1
3	8	2	04	02	11	4.586	3.60	2
4	16	2	16	04	23	5.172	4.13	≈2.3
5	32	2	34	16	45	5.758	4.59	4
6	64	2	066	030	103	6.343	5.01	≈5.3
7	128	2	122	054	277	6.586	5.17	≈0.5
8	256	2	130	072	435	7.515	5.75	≈1.5

[a]Adapted from Ungerboeck [26]. Reproduced by permission of the IEEE.

TABLE 9.9 16PSK-TCM Codes (m = 3)[a]

ν	2^ν	$\tilde{m}$	h^2	h^1	h^0	d^2_{free}	G (dB)	N_{free}
2	4	1	—	2	5	1.324	3.54	4
3	8	1	—	04	13	1.476	4.01	2
4	16	1	—	04	23	1.628	4.44	8
5	32	1	—	10	45	1.910	5.13	8
6	64	1	—	024	103	2.000	5.33	2
7	128	1	—	024	203	2.000	5.33	2
8	256	2	374	176	427	2.085	5.51	≈8

[a]Adapted from Ungerboeck [26]. Reproduced by permission of the IEEE.

$$\mathbf{H}=\left[h^2,\ h^1,\ h^0\right] \tag{9.85}$$

The conversion of the parity-check coefficients from octal to binary notation results in h^0 = [1001], h^1 = [0010], h^2 = [0100]. The binary notation is represented by the n-th degrees polynomial $p(D)=b_nD^n+b_{n-1}D^{n-1}+\cdots+b_1D^1+b_0$ with coefficients b_i = (0, 1) and unit delay operator D. Using this notation, the parity-check coefficients are denoted as h^0 = [D^3 + 1], h^1 = [D], h^2 = [D^2] and the parity-check matrix becomes

$$\mathbf{H}=\left[D^2,\ D,\ D^3+1\right] \tag{9.86}$$

The parity-check equation is used to compute the error syndrome vector as* $s=\hat{c}H^T$; however, since $\hat{c}=c$, is a valid code word the error syndrome will be zero, so the code word of interest is the solution to

$$cH^T=0 \tag{9.87}$$

Evaluation of (9.87) using (9.86) with c = (c_2, c_1, c_0) results in

$$c_2D^2\oplus c_1D\oplus c_0D^3\oplus c_0=0 \tag{9.88}$$

Solving for c_0, in consideration of the feedback implementation and using modulo-two arithmetic, yields the desired result

$$c_0=c_2D^{-1}\oplus c_1D^{-2}\oplus c_0D^{-3}=0 \tag{9.89}$$

This is implemented as the feedback configuration of the convolutional encoder shown in Figure 9.24b; the systematic form is characteristic of the feedback configuration. Ungerboeck's implementations for the ν = 2, 3, and 4 TCM waveforms are shown†, respectively, in Figure 9.24a, b, and c.

As a second example, consider implementing the eight-state (ν = 3) 8PSK-TCM encoder as a feed-forward structure.

*The vector $\hat{c}$ is the received estimate of transmitted code vector **c**.

†The addition symbol in these figures signifies modulo-two addition of the binary coefficients and the crossing of drawing lines signifies a connection only when denoted by •.

This involves using the generator matrix G and the binary message vector $\boldsymbol{b}$ = (b_2, b_1) to generate the code vector $\boldsymbol{c}$ = (c_2, c_1, c_0) using

$$\boldsymbol{c}=\boldsymbol{b}\boldsymbol{G} \tag{9.90}$$

where $\boldsymbol{c}=\hat{\boldsymbol{c}}$ and $\boldsymbol{G}$ is determine as the solution to $\boldsymbol{G}\boldsymbol{H}^{\mathrm{T}}$ = 0. Defining the elements of the matrix $\boldsymbol{G}$ as g_{ij}, the solution is formulated as

$$\begin{bmatrix} g_{11} & g_{12} & g_{13} \\ g_{21} & g_{22} & g_{23} \end{bmatrix}\begin{bmatrix} D^2 \\ D^1 \\ D^3+1 \end{bmatrix}=0 \tag{9.91}$$

From (9.91) the set of equations to be solved is

$$\begin{aligned} g_{11}D^2\oplus g_{12}D\oplus g_{13}(D^3+1)&=0 \\ g_{21}D^2\oplus g_{22}D\oplus g_{23}(D^3+1)&=0 \end{aligned} \tag{9.92}$$

The solution to (9.92) is obtained by letting $g_{ij}=p_{ij}(D)=d_nD^n+\cdots+d_1D^1+d_0$ with degree $n \le \nu$, the code constraint length, and allowing various combinations of the binary coefficients d_n. The solution involves the modulo-two sum of the terms involving powers of D, for example, $\sum_{i\in even} D^m=0$. Ungerboeck [23] suggests the following coefficient solutions to (9.92): g_{11} = 1, g_{12} = D, g_{13} = 0, g_{21} = D(D + 1), g_{22} = D^2 + 1, and g_{23} = D. Substitution of these results into (9.92) yields

$$D^2\oplus D^2\oplus 0=0 \tag{9.93}$$

for the first equation and the second equation is

$$\begin{aligned} &D(D\oplus 1)D^2\oplus\left(D^2+1\right)D\oplus D\left(D^3+1\right) \\ &=D^4\oplus D^3\oplus D^3\oplus D\oplus D^4\oplus D=0 \end{aligned} \tag{9.94}$$

Equations (9.93) and (9.94) demonstrate that the computation of the error syndromes with an error-free message (b_2,b_1) is zero. Using these solutions, the nonsystematic convolutional code vector is computed using (9.90) as

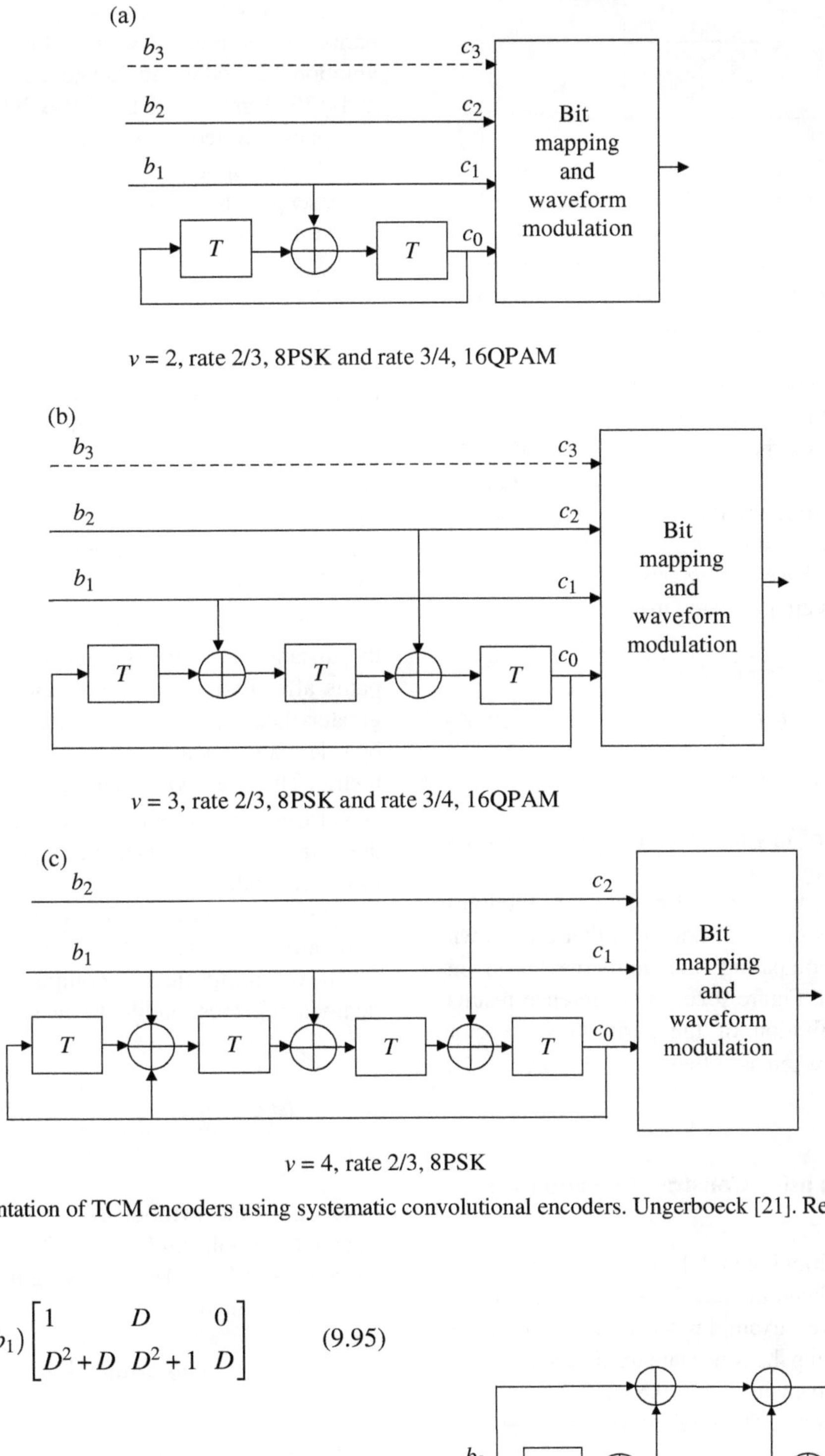

FIGURE 9.24 Implementation of TCM encoders using systematic convolutional encoders. Ungerboeck [21]. Reproduced by permission of the IEEE.

$$(c_2, c_1, c_0) = (b_2, b_1)\begin{bmatrix} 1 & D & 0 \\ D^2 + D & D^2 + 1 & D \end{bmatrix} \tag{9.95}$$

resulting in

$$\begin{aligned} c_2 &= b_2 + b_1 D^2 + b_1 D \\ c_1 &= b_2 D + b_1 D^2 + b_1 \\ c_0 &= D b_1 \end{aligned} \tag{9.96}$$

Equation (9.96) is implemented by the configuration shown in Figure 9.25.

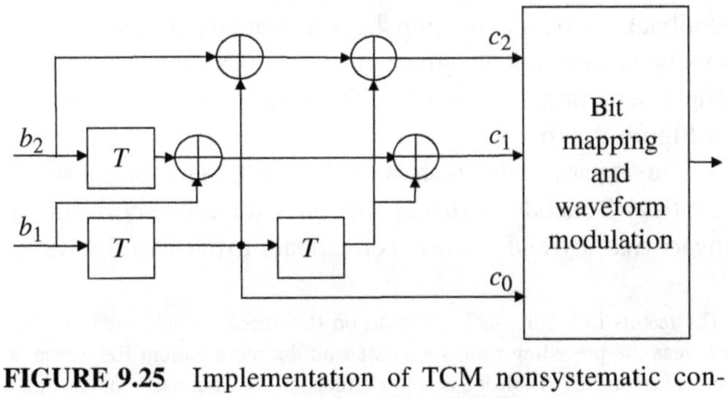

FIGURE 9.25 Implementation of TCM nonsystematic convolutional encoder (v = 3, rate 2/3, 8PSK).

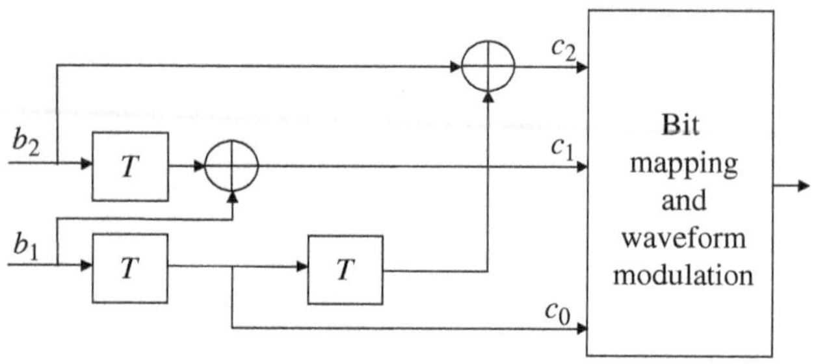

FIGURE 9.26 Implementation of TCM minimal nonsystematic convolutional encoder (v = 3, rate 2/3, 8PSK).

Ungerboeck [23] has also suggested a minimal nonsystematic configuration in which, the number of modulo-two additions are minimized. Using the v = 3, rate 2/3, 8PSK-TCM waveform with the error syndrome expressed in (9.92), the minimal nonsystematic configuration is obtained by choosing $g_{11} = 1$, $g_{12} = D$, $g_{13} = 0$, $g_{21} = D^2$, $g_{22} = 1$, and $g_{23} = D$. It is easily demonstrated that these coefficients satisfy (9.87), and that the parity-check bits are generated using (9.90) as

$$\begin{aligned} c_2 &= b_2 + b_1 D^2 \\ c_1 &= b_2 D + b_1 \\ c_0 &= b_1 D \end{aligned} \tag{9.97}$$

The implementation of this minimal nonsystematic configuration is shown in Figure 9.26.

Ungerboeck has published the configurations of the minimal nonsystematic convolutional structures that are shown in Figure 9.27.* In comparing the implementation in Figure 9.27b with that in Figure 9.26, the difference results from Ungerboeck's definition of the parity-check vector $H = (D \ D^2 \ D^3 + 1)$, whereas that in Figure 9.26 uses $H = (D^2 \ D \ D^3 + 1)$.

9.4.1 Example Design using Constraint Length v = 3, Rate 2/3 8PSK-TCM

This example of set partitioning and data-to-waveform mapping uses the constraint length v, rate 2/3 8PSK-TCM waveform. Following the given examples for implementing the feedback encoder and using the four-state trellis decoder parity-check coefficients given in Table 9.8 and the encoder implementation for the rate 2/3 8PSK waveform is shown in Figure 9.24b.

The set partitioning is shown in Figure 9.28 for the constraint length $v = 3$ encoder with the parity bit designations corresponding to (c_0, c_1, c_2) and where c_0 corresponds to the least significant bits (LSBs) and c_2 is the most significant bits (MSBs). With this parity bit ordering, the two LSB bits are determined by the convolutional encoder and dictate the fine-phase shift between 0 and 135 degrees and the MSB bit determines whether the fine-phase shifted starts at zero or 180 degrees. The bit-to-waveform mapping is shown in Table 9.10 and corresponds to the set partitioned mapping. In term of the natural numbering n the transmitted signal is given by

$$s(t) = Ae^{j(n\pi/4)} \tag{9.98}$$

The decoding corresponds to a four-state trellis with four branches emanating from and terminating at each state as shown in Figure 9.29. The goal of the set partitioning is to provide a phase mapping that result in the greatest distance between the correct and incorrect state transitions terminating on each output state. Referring to Figure 9.29, the four transitions to each termination state are pair-wise antipodal† with minimum free Euclidean distance $d_{free} = 2$; however, the pairs are orthogonal with a distance of $d_{min} = \sqrt{2}$. These orthogonal paths, separated by the distance $\sqrt{2}$, diverge and remerge with the distance 2 paths after two or more transitions resulting in a distance greater than d_{free} and, therefore, are eventually eliminated by selecting the surviving path associated with the minimum metric. The underlying uncoded 4-phase shift keying (4PSK) waveform has a minimum distance of $d_{min} = \sqrt{2}$ so, using these results and (9.84), the expected coding gain relative to 4PSK modulation is 3 dB. In the context of Figure 9.28, the MSBs in the last row are antipodal resulting in the optimal estimation of the two source bits: 11, 10, 01, and 00.

The decoding starts by computing the state transition metric updates corresponding to the squared Euclidean distance‡

$$\Delta m_k = -d_k^2 = \frac{-1}{2E} \int_{(j-1)T}^{jT} |\tilde{y}(t) - \tilde{x}_k(t)|^2 dt \tag{9.99}$$

where E is the symbol energy, j corresponds to the j-th received symbol, and $k = 1, \ldots, 2^m$ corresponds to the k stored reference symbols. The surviving metric at each termination state is computed as

$$m_\ell = \min_k{}^{-1} \left[m_{\ell_{k,n}} + \Delta m_k \right] \tag{9.100}$$

where the subscript $\ell_{k,n}$ denotes the source state of the transitions terminating at state ℓ. After $D \cong 5\nu$ state transitions, the surviving data is selected based on the minimum state metric.

*The results in Figure 9.27 are based on Reference 21 published in 1982, whereas the preceding results are based on the more current Reference 26 published in 1987. Table 9.8 is more current and the parameters for constraint lengths v = 6, 7, and 8 have been updated from those in the 1982 publication.

†The pair-wise antipodal transitions are referred to as parallel transitions, and there are $2^{m-\tilde{m}}$ parallel transitions emanating from and converging on each state; when $m = \tilde{m}$, there are no parallel transitions.

‡The negative metric update requires selecting the optimum state transition survivor as the minimum state transition at each termination state.

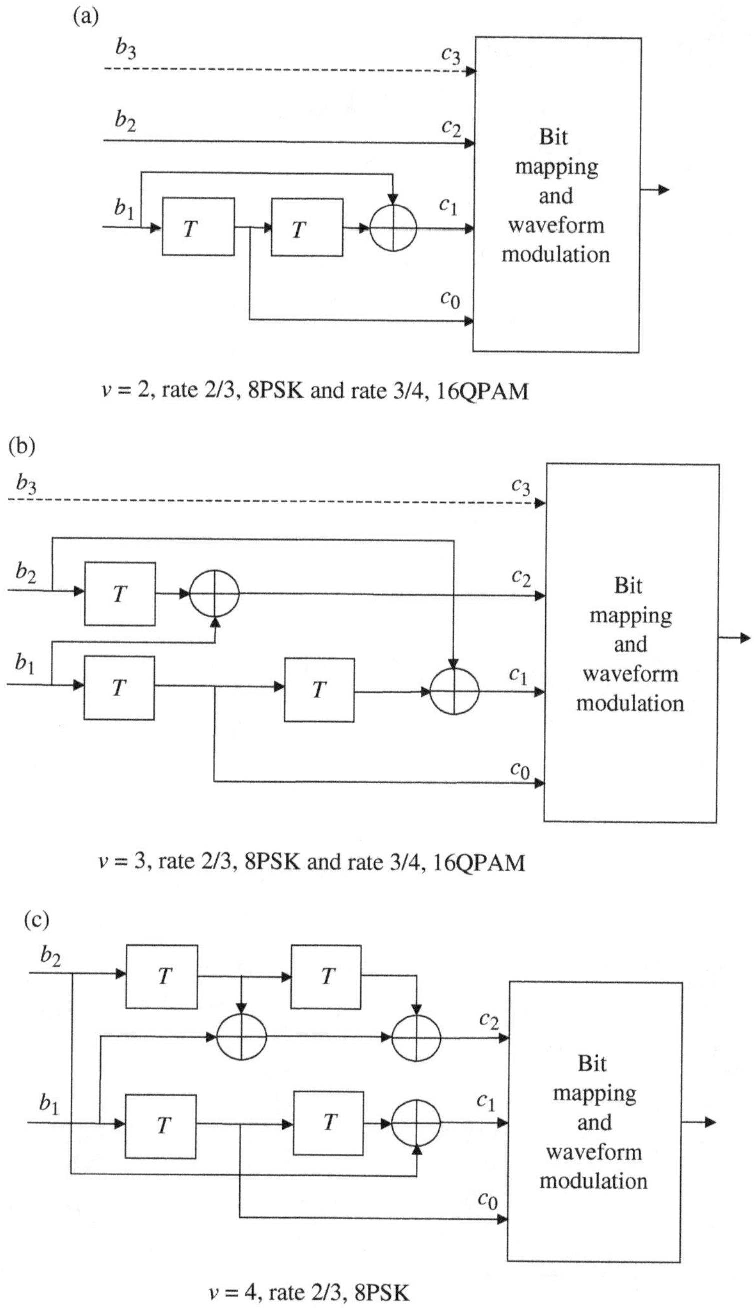

FIGURE 9.27 Implementation of TCM minimal nonsystematic convolutional encoders. Ungerboeck [21]. Reproduced by permission of the IEEE.

9.5 CASE STUDY: FOUR-STATE 8PSK-TCM PERFORMANCE OVER SATELLITE REPEATER

The performance results presented in this section are obtained using a Monte Carlo simulation program, which models the modulator, satellite relay, and demodulator as shown in Figure 9.30. Gaussian channel noise is added at the input to the receiver. The discrete-time sampling generally uses eight samples-per-symbol; however, when phase shaping is used for spectral control 16 samples-per-symbol are used. The Monte Carlo simulations are based on a total of 1 M source bits at each signal-to-noise ratio for $P_{be} \geq 10^{-4}$; otherwise, 10 M bits are used. Various combinations of the functions shown in Figure 9.30 can be selected using an options menu at the beginning of each simulation. This menu also provides for inputting appropriate parameter values for each of the functions.

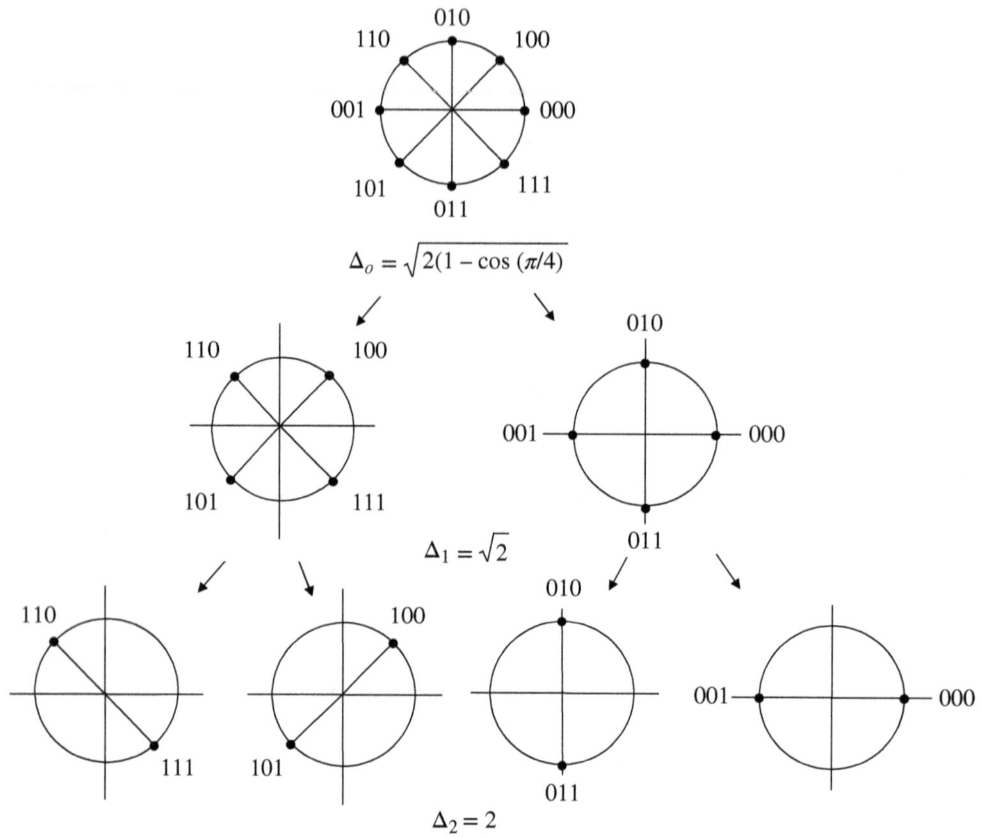

FIGURE 9.28 Set partitioning for rate 2/3 8PSK-TCM. Adapted from Ungerboeck [21]. Reproduced by permission of the IEEE.

TABLE 9.10 Code-Bit Mapping to Signal Phase

Code-bits (c_0,c_1,c_2)	n	Phase (degrees)
000	0	0
100	1	45
010	2	90
110	3	135
001	4	180
101	5	225
011	6	270
111	7	315

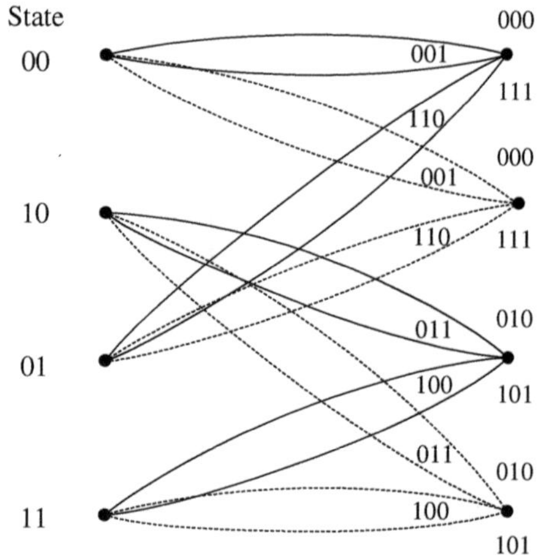

FIGURE 9.29 Four-state 8PSK-TCM trellis decoder.

The RS code is a rate $r_{co} = 0.937$, 256-ary code with 239 source data bits-per-code block. The 8PSK trellis code, shown in Figure 9.31, is a rate 2/3 code with two source bits (or RS code bits) that are mapped to eight modulation phase states corresponding to the three trellis-coded bits for each symbol. A rate 1/2, constraint length seven, binary convolutional encoder is used to generate the trellis code-bits. Linear phase shaping is used to contain the signal spectrum thereby reducing adjacent channel emissions and ISI resulting from channel filtering. The phase-shaped constant envelope transmitted waveform ensures that the spectral containment is maintained with the hard limiting and the traveling wave tube amplifier (TWTA) nonlinearities encountered in the satellite. The 50% linear phase shaping is symmetrical about each symbol transition so the phase shaping extends 25% into the preceding and succeeding symbols. The transmit filter is simply a roofing filter that reduces the effect of the *near/far* interference at co-located transmit and receive ground terminals.

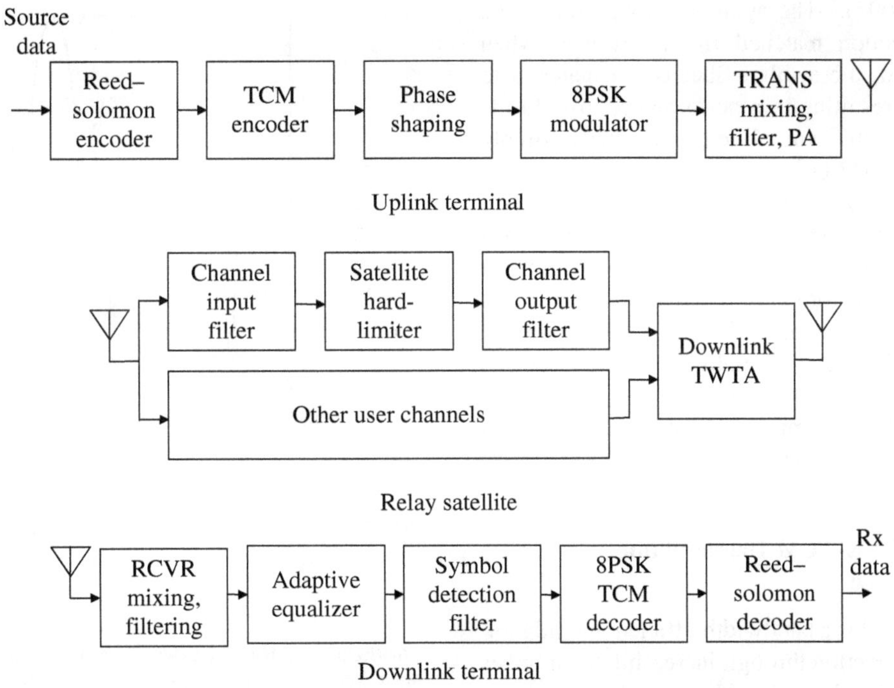

FIGURE 9.30 Satellite link and processing functions.

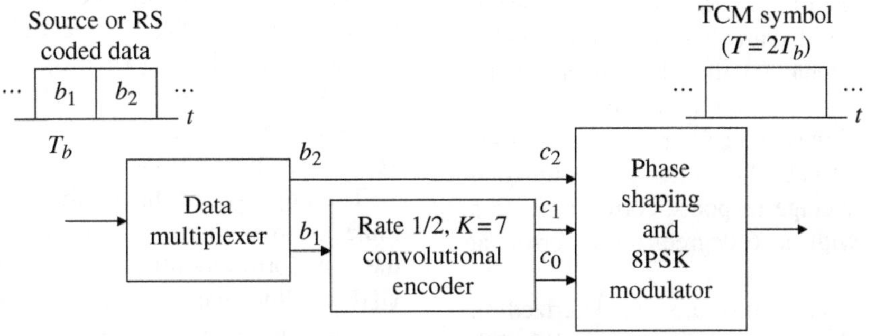

FIGURE 9.31 Rate 2/3 TCM encoder.

The satellite narrowband filtering, with baseband bandwidth B_c Hz, is modeled as a 6-pole Chebyshev filter with 0.05 dB ripple. The hard limiter is ideal in the sense that it is followed by a zonal filter that removes the out-of-band harmonics of the carrier frequency. The output of the ideal hard limiter is described in terms of the input as

$$\tilde{e}_o(t) = A\frac{\tilde{e}_i(t)}{|\tilde{e}_i(t)|} \tag{9.101}$$

This modeling of the hard limiter results in no harmonically related interfering signals so there is no need to implement the output filter following the limiter. The TWTA is characterized by the amplitude modulation to amplitude modulation (AM–AM) and amplitude modulation to phase modulation (AM–PM) characteristics of the amplifier; however, for the simulation results discussed in Section 9.5.1, the TWTA backoff is set sufficiently high to result in linear operation. The receiver input filter is a roofing filter to ensure the rejection of adjacent channel signals and to provide equivalent intermediate frequency (IF) filtering to minimize aliasing resulting from the sampled data processing. The equalizer is used to reduce the ISI introduced by the satellite and receiver filtering; the equalizer is necessary when modulator spectral shaping is not used. The equalization is performed using a five-tap symbol-spaced linear equalizer and the taps are updated using the least mean-square (LMS) algorithm. The equalizer tap gain coefficient is typically μ = 0.001; however, a value of 0.0003 results in improvement with severe ISI when R_b/B_c > 0.78. To account for equalizer acquisition in the performance simulation, the bit-error counting is delayed for 60 symbols (120 symbols for μ = 0.0003).* The symbol filter performs the function of the detection matched filter; however, when phase shaping is used, the symbol filter is not matched to the received symbol resulting in a performance loss. In the following simulations, the symbol detection filter is simply an integrate-and-dump (I&D) filter.

The 8PSK-TCM decoder, shown in Figure 9.29, uses a four-state trellis with a depth of 30 symbols. The RS code is inherently a burst-error-correcting code and will correct various combinations of correlated burst errors out of the 8PSK-TCM decoder. For example, the 256-ary RS code will correct any combination of up to eight code symbol errors. Therefore, because a code symbol contains 8 bits, the decoder will correct up to eight 8-bit random burst-error events or a single burst-error containing up to 64 contiguous bit errors.

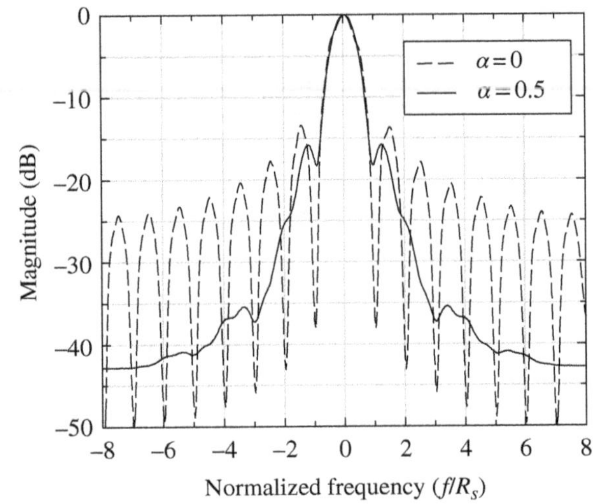

FIGURE 9.32 4-State 8PSK-TCM spectrums with phase shaping parameter α.

9.5.1 Four-State 8PSK-TCM Demodulator Simulated Performance

The rate 2/3 8PSK-TCM is a bandwidth-efficient modulation and achieves error correction through increasing the number of phase states, not through bandwidth expansion. Based on this bandwidth conservation property, the 8PSK-TCM achieves 2 bits/channel-use so a user rate of R_b bps is achieved with a symbol rate of $R_s = R_b/2$ sps. The bit-error probability performance results are based on the 25 kHz UHF satellite transponder channel model using a 6-pole, 0.05 dB ripple, Chebyshev filter with a 3 dB bandwidth of B_c = 32 kHz. A review of measured filter phase responses indicate that the intrinsic Chebyshev filter phase function is modified and that the average response corresponds to a cubic phase function† with a 6-degree phase error at ±12.5 kHz.

The following simulation results are characterized in terms of the normalized data rates: R_b/B_c = (2.0, 1.6, 1.3, 1.0, 0.7, 0.5). The corresponding approximate data rates are 64, 51.2, 42.7, 32, 21.3, and 16.0 kbps and, although these are not standard rates used over the 25 kHz channel, they do span the specified range data rates.

When RS concatenated coding is used, the transmitted data rate in increased by R_b/r_{co} where $r_{co} < 1$ is the RS rate; increasing the transmission rate ensures that the user data rates are unchanged when RS coding is used. This condition also recognizes the ISI performance penalty incurred by the increased symbol bandwidth with the RS code. The performance with TCM and RS code concatenation is not examined in the following sections; however, the performance comparison with and without the RS coding is presented in Reference 27. Typically, the RS coding begins to exhibit a coding gain, relative to the performance of the underlying modulation, for bit-error probabilities in the range of 10^{-2} and 10^{-3} and, because of the relatively steep slope of the P_{be} curve with RS coding, the coding gain begins to increases rapidly with E_b/N_o in this range of bit-error probabilities.

Figure 9.32 shows the spectrum of the 8PSK-TCM modulated waveform with 50% linear phase shaping (α = 0.5) and without phase shaping (α = 0). The spectrum is computed using Bartlett's procedure [28] of averaging periodograms containing 512-point FFTs of randomly modulated data with 16 samples-per-symbol.

The objective of this performance evaluation is to determine the maximum symbol rate that can be supported by the waveform modulation through the 25 kHz hard-limiting UHF satellite channel. The measure of the waveform performance is the signal-to-noise improvement, or coding gain, relative to the underlying QPSK modulated waveform at a bit-error probability of $P_{be} = 10^{-5}$. As the modulation symbol rate increases, more of the available bandwidth is occupied resulting in ISI that decreases the coding gain. Consequently, the coding gain is a qualitative measure of the waveform performance. Figure 9.33a and b show the bit-error performance in an AWGN channel with and without modulation phase shaping. Figure 9.34a and b shows the bit-error performance under the same conditions through the 25 kHz hard-limiting channel. The *dotted* curve in each of the figures corresponds to the AWGN performance of underlying QPSK modulated waveform. The *dashed* curves represent the ideal performance of the 8PSK-TCM modulated waveform in an AWGN channel. These two performance curves represent operational limits for the 8PSK-TCM modulated waveform.

Referring to Figure 9.33a and b, the 8PSK-TCM exhibits no coding gain for R_b/B_c = 2 (64 kbps) without phase shaping and that phase shaping results in a significant performance loss with reasonable coding gains for $R_b/B_c \le 1.3$ (40 kbps).

*In practice, the equalizer convergence is accomplished during the acquisition preamble.

†The cubic phase function is discussed in Sections 17.8.2 and 17.8.3.

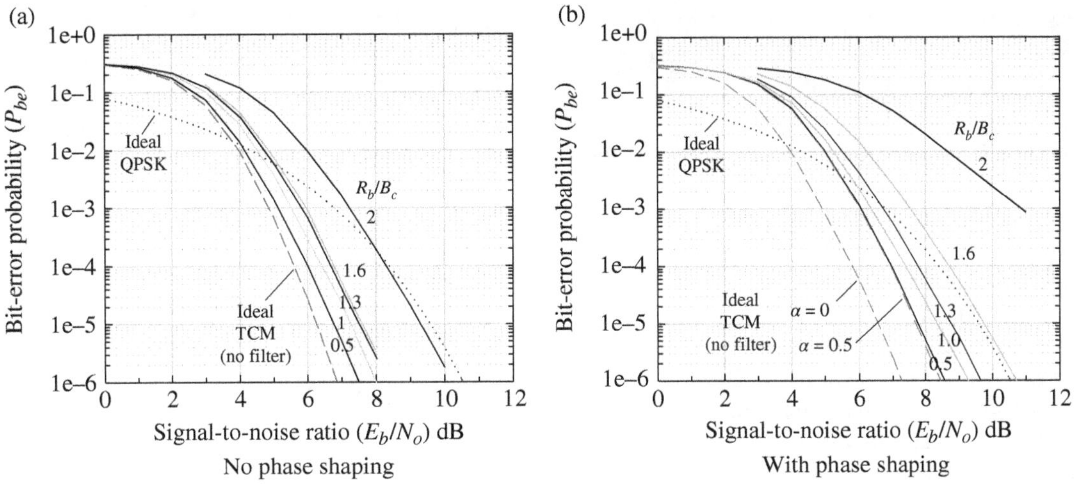

FIGURE 9.33 4-State 8PSK-TCM linear channel performance ($r_{cc} = 1/2$, $K = 7$, trellis depth = 30).

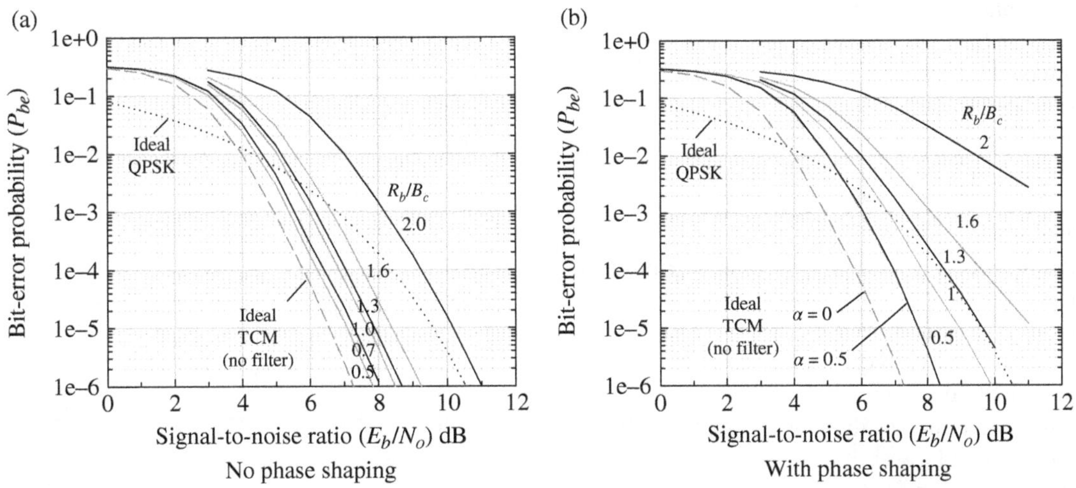

FIGURE 9.34 4-State 8PSK-TCM hard-limited channel performance ($r_{cc} = 1/2$, $K = 7$, trellis depth = 30).

The intrinsic loss associated with 50% linear phase shaping and single-symbol I&D detection filtering is about 0.9 dB.*

The performance with the hard-limiting channel is shown Figure 9.34a and b. With the exception of the case for $R_b/B_c = 2$ (64 kbps), the bit-error performance without phase shaping exhibits coding gain ranging from 2 dB at $R_b/B_c = 1.6$ (51 kbps) to 3 dB at $R_b/B_c = .5$ (16 kbps). However, as shown in Figure 9.33b, with 50% linear phase shaping there is no coding gain for $R_b/B_c \geq 1.3$ (40 kbps) and coding gains of 0.7 dB at $R_b/B_c = 1$ (32 kbps) and 2 dB at $R_b/B_c = 0.5$ (16 kbps). The application of a rate 0.937 RS outer code results in notable improvements [19] with coding gains of 1.6 dB at $R_b/B_c = 1.6$ (51 kbps) and about 2.6 dB for $R_b/B_c \leq 1.28$ (41 kbps).

A major requirement of any advanced modulation waveform is that of spectral containment or efficiency and, unlike the multi-h M-ary CPM waveform, the 8PSK-TCM waveform does not inherently provide spectral control. However, some degree of spectral control can be achieved by altering the phase shaping factor α which can be performed at the transmitter location without impacting the demodulation processing. Also, referring to Section 4.4.3.9, using the raised-cosine phase shaping function is a viable alternative to the linear phase shaping function. The importance of having a constant envelope modulation is also a major requirement in applications involving hard-limiting channels especially for disadvantaged low-power terminals. As a final comment, replacing the single-symbol I&D detection filter with a multiple symbol MLSE detector will result in a significant performance improvement at the expense of demodulator complexity.

*This loss can be reduced to 0.5 dB with an optimally weighted I&D detection filter.

ACRONYMS

1REC	full response modulation
ACI	adjacent channel interference
AM–AM	amplitude modulation to amplitude modulation (conversion)
AM–PM	amplitude modulation to phase modulation (conversion)
AWGN	additive white Gaussian noise
BFSK	binary frequency shift keying
$B_{LD}T$	time-bandwidth product (lock detector)
B_LT	time-bandwidth product (loop filter)
CPBFSK	continuous phase binary frequency shift keying
CPM	continuous phase modulation
CW	continuous wave
FEC	forward error correction
GMSK	Gaussian minimum shift keying
I&D	Integrate-and-dump (filter)
IF	intermediate frequency
ISI	intersymbol interference
LMS	least mean-square
LSB	least significant bit
MAP	maximum a posterior (estimation)
MLSE	maximum-likelihood sequence estimation
MPSK	Multiphase shift keying
MQPAM	multilevel quadrature pulse amplitude modulation
MSB	most significant bit
MSK	minimum shift keying
PLL	phaselock loop
PSD	power spectral density
QPSK	quadrature phase shift keying
RS	Reed–Solomon (code)
SOM	Start-of-message
TCM	trellis-coded modulation
TFM	tamed frequency modulation
TWT	traveling wave tube
TWTA	traveling wave tube amplifier
UHF	ultra-high frequency

PROBLEMS

1. Referring to Figure 9.1a and b, identify all of the phase trajectories, starting at $t = 0$, that contribute to the minimum squared Euclidean distance. Hint: These are 1REC single-h, 2-ary and 2-h 4-ary waveform so examine all the possibilities of $\ell = \alpha_i - \alpha_j$.

2. Show that (9.19) follows from (9.18) for the 1REC, single-h, M-ary modulation with observation interval $N = 2$ symbols.

3. Using (9.18) with $h = 1/2$, show that all of the $2^k - 1$ responses to the parameter ℓ have the same value of d^2_{min} for a given M-ary modulation. What is the implication of this result regarding the bit-error probability expressed by (9.20)?

4. Using the results of Problem 1, determine the exact expression for the bit-error probability for the single-h 4-ary CPM waveform for an arbitrary modulation index.

5. The connections for the convolutional encoder are not unique. For the generator matrices listed in the following as (a) through (d), determine which are valid generators for the 2^ν-state 8PSK-TCM encoder. For those that form a valid generator matrix, show the message bit (b_2,b_1) mapping to generate the parity bits $c = (c_2,\ c_1,\ c_0)$. Hint: Use the parity-check matrix $H = \left(D^2\ \ D\ \ D^3+1\right)$. Also, referring to the 8PSK set partitioning in Figure 9.28, how does the partitioning differ from gray coding?

(a) $G = \begin{pmatrix} 1 & D & 0 \\ D^2+1 & D+1 & D \end{pmatrix}$ (b) $G = \begin{pmatrix} 1 & D & 0 \\ D^2 & 1 & D \end{pmatrix}$

(c) $G = \begin{pmatrix} 1 & D & 0 \\ D^2+D & D^2+1 & D \end{pmatrix}$ (d) $G = \begin{pmatrix} D^2 & 1 & D \\ D^2+1 & D+1 & D \end{pmatrix}$

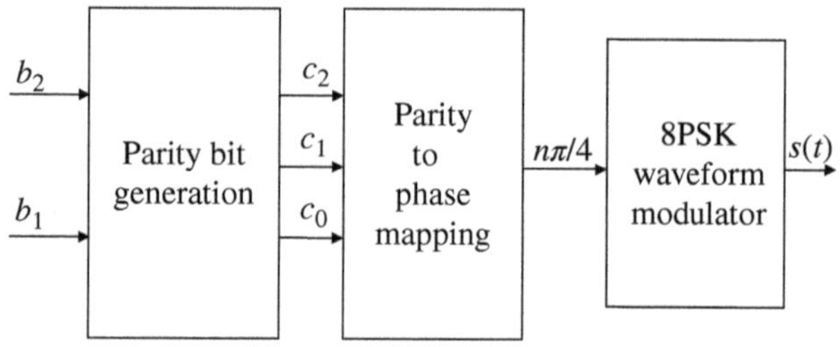

6. Formulate the $\nu = 8$ parity-check matrix H for the 256-state 8PSK-TCM modulated waveform using the parity-check matrix elements given in Table 9.8.

REFERENCES

1. J.B. Anderson, J.R. Lesh, Guest Editors, "Special Section on Combined Modulation and Encoding," IEEE Transactions on Communications, Vol. COM-29, No. 3, pp. 185–285, March 1981.
2. J.B. Anderson, T. Aulin, C.-E. W. Sundberg, *Digital Phase Modulation*, pp. 51–53, Plenum Press, New York, 1986.
3. D. Muilwijk, "Tamed Frequency Modulation—A Bandwidth Saving Digital Modulation Suited for Mobile Radio," Phillips Telecommunication Review, Vol. 37, No. 1, pp. 35–39, March 1979.
4. K.S. Chung, "Generalized Tamed Frequency Modulation and Its Application for Mobile Radio Communications," IEEE Journal on Selected Areas in Communications, Vol. SAC-2, No. 4, pp. 487–497, July 1984.

5. R. Debuda, "Coherent Demodulation of Frequency Shift Keying with Low Deviation Ratio," IEEE Transactions on Communications, Vol. COM-20, pp. 429–435, June 1972.
6. I. Sasase, S. Mori, "Multi-h Phase-Coded Modulation," IEEE Communications Magazine, Vol. 29, No. 12, pp. 46–56, December 1991.
7. T. Aulin, C.-E. W. Sundberg, "On the Minimum Euclidean Distance for a Class of Signal Space Codes," IEEE Transactions on Information Theory, Vol. IT-28, No. 1, pp. 43–55, January 1982.
8. J.B. Anderson, D.P. Taylor, "A Bandwidth-Efficient Class of Signal-Space Codes," IEEE Transactions on Information Theory, Vol. IT-24, No. 6, pp. 76–80, November 1978.
9. J.B. Anderson, C.-E. W. Sundberg, T. Aulin, N. Rydbeck, "Power-Bandwidth Performance of Smoothed Phase Modulation Codes," IEEE Transactions on Communications, Vol. COM-29, No. 3, pp. 187–195, March 1981.
10. J.B. Anderson, T. Aulin, C.-E. W. Sundberg, *Digital Phase Modulation*, Plenum Press, New York, 1986, Chapter 3.
11. M.C. Austin, M.U. Chang, "Quadrature Overlapped Raised-Cosine Modulation," International Conference on Communications, Conference Record, Seattle, Washington, pp. 26.7.1–26.7.5, June 1980.
12. T. Aulin, C.-E. W. Sundberg, "CPM—An Efficient Constant Amplitude Modulation Scheme," International Journal on Satellite Communications, Vol. 2, pp. 161–186, 1984.
13. T. Aulin, C.-E. W. Sundberg, "Minimum Euclidean Distance and Power Spectrum for a Class of Smoother Phase Modulation Codes with Constant Envelope," IEEE Transactions on Communications, Vol. COM-30, No. 7, pp. 1721–1729, July 1982.
14. J.B. Anderson, T. Aulin, C.-E. W. Sundberg, *Digital Phase Modulation*, Chapter 4, p. 167, Plenum Press, New York, 1986.
15. B.A. Mazur, D.P. Taylor, "Demodulation and Carrier Synchronization of Multi-*h* Phase Codes," IEEE Transactions on Communications, Vol. COM-29, No. 3, pp. 257–266, March 1981.
16. A.N. Premji, D.P. Taylor, "A Practical Receiver Structure for Multi-*h* CPM Signals," IEEE Transactions on Communications, Vol. COM-35, No. 9, pp. 901–908, September 1987.
17. J.M. Liebetreu, "Joint Carrier Phase Estimation and Data Detection Algorithms for Multi-*h* CPM Data Transmission," IEEE Transactions on Communications, Vol. COM-34, No. 9, pp. 873–881, September 1986.
18. A.N. Premji, D.P. Taylor, "Receiver Structures for Multi-*h* Signaling Formats," IEEE Transactions on Communications, Vol. COM-35, No. 4, pp. 439–451, September 1987.
19. R.M. Battista, R.R. Jacobson, R.W. Middlestead, "Spectrally Efficient High Data Rate Waveforms for the UFO SATCOM Channel," IEEE Military Communications Conference, MILCOM'98, Boston, MA, pp. 134–143, October 18–21, 1998.
20. A.V. Oppenheim, R.W. Schafer, *Digital Signal Processing*, p. 548, Prentice-Hall Inc., Englewood Cliffs, NJ, 1975.
21. G. Ungerboeck, "Channel Coding with Multilevel/Phase Signals," IEEE Transactions on Information Theory, Vol IT-28, pp. 55–67, January 1982.
22. G. Ungerboeck, "Trellis-Coded Modulation with Redundant Signal Sets Part I: Introduction," IEEE Communications Magazine, Vol. 25, No. 2, pp. 5–11, February 1987.
23. G. Ungerboeck, "Trellis-Coded Modulation with Redundant Signal Sets Part II: State of the Art," IEEE Communications Magazine, Vol. 25, No. 2, pp. 12–21, February 1987.
24. R. Calderband, J.E. Mazo, "A New Description of Trellis Codes," IEEE Transactions on Information Theory, Vol. IT-30, pp. 784–791, November 1984.
25. E. Biglieri, D. Divsalar, P.J. McLane, M.K. Simon, "*Introduction to Trellis-Coded Modulation with Applications*," Macmillan Publishing Company, New York, 1991.
26. G. Ungerboeck, "Trellis-Coded Modulation with Redundant Signal Sets Part II: State of the Art," IEEE Communications Magazine, Vol. 25, No. 2, pp. 12–21, February 1987, page 20, Table II.
27. R.M. Battista, R.R. Jacobson, R.W. Middlestead, "Spectrally Efficient High Data Rate Waveforms for the UFO SATCOM Channel," AIAA Defense and Civil Space Programs Conference and Exhibit, pp. 15–22, Huntsville, Alabama, October 28–30, 1998.
28. A.V. Oppenheim, R.W. Schafer, *Digital Signal Processing*, Prentice-Hall, Englewood Cliffs, NJ, 1975.

10

CARRIER ACQUISITION AND TRACKING

10.1 INTRODUCTION

This chapter focuses on the application of phaselock techniques to acquire and track the carrier frequency of a modulated waveform. Typically, the received carrier frequency error relative to the receiver and demodulator local oscillators far exceed the phaselock loop (PLL) frequency pull-in range requiring frequency estimation and correction. In these cases, the unknown frequency can be resolved by slowly sweeping the PLL over the range of the frequency uncertainty or by estimating the frequency error using a waveform preamble; the choice is one of expediency, however, in most applications a preamble or in some cases a midamble is provided to assist in the received waveform acquisition as discussed in Chapter 11. Phaselock techniques are used in a variety of applications involving the synchronization of various continuous wave (CW) oscillators for coherent heterodyning in a receiver or demodulator; however, although these applications are addressed in this chapter, the main focus is on the acquisition and tracking of data-modulated waveforms involving the baseband or Costas implementation of the PLL. Acquisition and tracking of the carrier phase and frequency of waveform-modulated carrier signals is essential for optimum coherent detection and examples are given for multiphase shift keying (MPSK), quadrature phase shift keying (QPSK), offset QPSK (OQPSK), and minimum shift keying (MSK) modulations. The material in this chapter draws upon Gardner's work [1] regarding the characterization and general applications of phaselock techniques. Emphasis is placed on the Costas implementation of the PLL for which Lindsey and Simon [2] have also laid a firm foundation with respect to data communications.

The configuration of the PLL is shown in Figure 10.1 where $h(t)$ is the loop filter impulse response function, K_d is the gain of the phase detector with dimensions of volts/radian, and K_o is the gain of the voltage controlled oscillator (VCO) with dimensions of radians/volt-seconds; the VCO is characterized as an ideal integrator in Figure 10.1. The product K_dK_o has dimensions of 1/s. Consider the input signal to the phase detector to be a real signal described as a sine wave with amplitude A, angular frequency ω_c, and unknown phase $\phi(t)$. The signal phase is purposely made time dependent which provides for an unknown frequency and frequency rate in addition to an unknown constant phase angle. Similarly, consider the output of the VCO to be a reference cosine function with amplitude normalized to unity and with carrier frequency ω_c, and a phase function, denoted as $\hat{\phi}(t)$, which is an estimate of the input signal phase. The phase estimate is expressed as

$$\hat{\phi}(t) = K_o \int_o^t \varepsilon_o(\tau)d\tau \quad : \text{radians} \tag{10.1}$$

The function of the phase detector is to generate an error signal which will be zero when the estimate $\hat{\phi}(t)$ is an exact replica of the input signal phase function. The error signal is formed by taking the product of the input and reference

Digital Communications with Emphasis on Data Modems: Theory, Analysis, Design, Simulation, Testing, and Applications, First Edition. Richard W. Middlestead.

Companion website: www.wiley.com/go/digitalcommunications

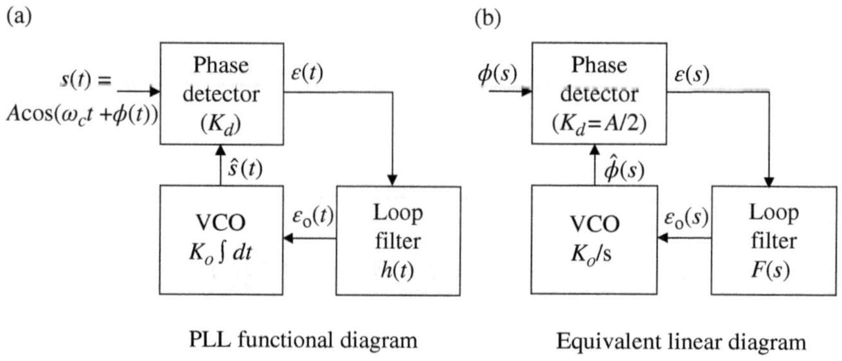

FIGURE 10.1 Phase-lock loop implementation.

signals and, upon neglecting the double frequency term, is given by*

$$\begin{aligned}\varepsilon(t) &= s(t)\hat{s}(t)|_{\text{filtered}} \\ &= \frac{A}{2}\sin(\phi_\varepsilon(t)) \\ &= K_d \sin(\phi_\varepsilon(t))\end{aligned} \tag{10.2}$$

where $\hat{s}(t)$ is a unit amplitude estimate of $s(t)$ and the phase error is $\phi_\varepsilon(t) = \phi(t) - \hat{\phi}(t)$. The phase detector constant is defined as $K_d = A/2$ V/rad. In many digital implementations of the PLL, the input signal is at baseband and is described by the analytic representation $\tilde{s}(t) = Ae^{j\phi(t)}$ so the corresponding baseband reference signal, or estimate of $\tilde{s}(t)$, is computed by taking the *sine* and *cosine* of the phase estimate yielding $\hat{\tilde{s}}(t) = e^{j\hat{\phi}(t)}$. Using these results, the phase detector output is computed by taking the conjugate of the reference signal and computing the phase detector output as $\varepsilon(t) = \text{Im}\left\{\tilde{e}(t)\hat{\tilde{e}}(t)^*\right\}$.

When the phase error is small, that is, less than about 20°, the phase detector output is approximated as

$$\varepsilon(t) \cong K_d \phi_\varepsilon(t) \tag{10.3}$$

Using this small angle linear approximation the equivalent frequency-domain linear model of the PLL is shown in Figure 10.1b with $\varepsilon(s) \cong K_d \phi_\varepsilon(s)$ where s is the Laplace frequency variable $s = j\omega$. In these cases, involving the analytic signal representation, the VCO output is actually the phase estimate given by

$$\hat{\phi}(s) = K_o \varepsilon_o(s)/s \tag{10.4}$$

*In practice, the double frequency term is attenuated by the closed-loop frequency response of the phase-lock loop.

Using the linear model it is easily shown that the closed-loop transfer function is defined as,

$$H(s) \triangleq \frac{\hat{\phi}(s)}{\phi(s)} \tag{10.5}$$

and, in terms of the loop filter transfer function $F(s) = \varepsilon_o(s)/\varepsilon(s)$, is expressed as

$$H(s) = \frac{K_o K_d F(s)}{s + K_o K_d F(s)} \quad \text{: loop filter transfer function} \tag{10.6}$$

The direct current (DC) loop gain, defined as the product $K_v = K_o K_d F(0)$, plays an important role in the dynamics of the PLL. It is apparent that the order of the closed-loop response is one more than the order of the loop filter, so the frequently used second-order loop requires a first-order filter. An equally useful result is the transfer function of the phase error, defined as

$$\begin{aligned}T(s) &\triangleq \frac{\phi_\varepsilon(s)}{\phi(s)} \\ &= \frac{\phi(s) - \hat{\phi}(s)}{\phi(s)} = 1 - H(s) \\ &= \frac{s}{s + K_o K_d F(s)} \quad \text{: phase-error transfer function}\end{aligned} \tag{10.7}$$

Finally, the open-loop transfer function is defined as

$$\begin{aligned}G(s) &\triangleq \frac{\hat{\phi}(s)}{\phi(s)}\Big|_{\text{open-loop}} \\ &= \frac{K_o K_d F(s)}{s} \quad \text{: open loop transfer function}\end{aligned} \tag{10.8}$$

and is used to characterize the root-locus plot of $H(s)$ as a function of the loop gain. The root-locus plot involves a relatively simple procedure to identify the loop stability conditions with varying loop gain and, to some degree, estimating the frequency response and transient behavior of the loop. In establishing the root-locus plot when the loop gain is zero the poles of $H(s)$ correspond to the poles of $G(s)$ and, as the gain increases to infinity, the poles of $H(s)$ terminate on the zeros of $G(s)$ [3]. These three transfer function relationships are used in the following sections to characterize the performance of the first-, second-, and third-order PLLs.

10.2 BANDPASS LIMITER

Preceding the PLL with a bandpass limiter serves as a gain control function that limits the dynamic range requirements following the limiter and mitigates the impact of signal level fluctuation on the PLL dynamics. For example, without some form of gain control, the loop bandwidth and damping factor vary in proportion to the input signal level. In this analysis the received signal-plus-noise, $s_i(t)$, is assumed to be at a convenient IF f_o Hz with bandwidth B_i Hz, such that, $B_i << f_o$, and is expressed as

$$s_i(t) = A\sin(\omega_o t + \phi(t)) + n_{ci}(t)\sin(\omega_o t) + n_{si}(t)\sin(\omega_o t) \tag{10.9}$$

The signal phase function $\phi(t)$ may include a residual intermediate frequency (IF) error, a narrowband phase modulation term, and a constant phase, in any event, the modulation bandwidth B_m of $\phi(t)$ must satisfy the condition $B_m < B_i << f_o$. The additive quadrature noise terms $n_{ci}(t)$ and $n_{si}(t)$ are stationary *iid*, zero-mean, white Gaussian noise processes with single-sided noise density N_o watts/hertz. The bandpass limiter is characterized by passing the input signal $s_i(t)$ through an ideal hard limiter with the output expressed as

$$s'_o(t) = sign(A_L, s_i(t)) \tag{10.10}$$

and then passing $s'_o(t)$ through a zonal bandpass filter centered at f_o with a bandwidth B_o, such that, $B_i < B_o \le f_o$. The input bandwidth is typically determined by an abrupt cutoff filter, so the zonal filter low-pass bandwidth is taken to be $B_o = f_o$ which serves the purpose of eliminating the harmonics of f_o. Under these conditions, the bandpass filtered output $s_o(t)$ consists of the carrier frequency f_o plus the narrowband *iid* additive orthogonal white Gaussian noise processes $N_{co}(t)$ and $N_{so}(t)$. Davenport [4] and Springett and Simon [5] have evaluated the input and output signal-to-noise ratios for the bandpass limiter and Gardner [6] formulates the results in the convenient approximate form

$$\gamma_o \cong \gamma_i \left(\frac{1 + 2\gamma_i}{4/\pi + \gamma_i} \right) \tag{10.11}$$

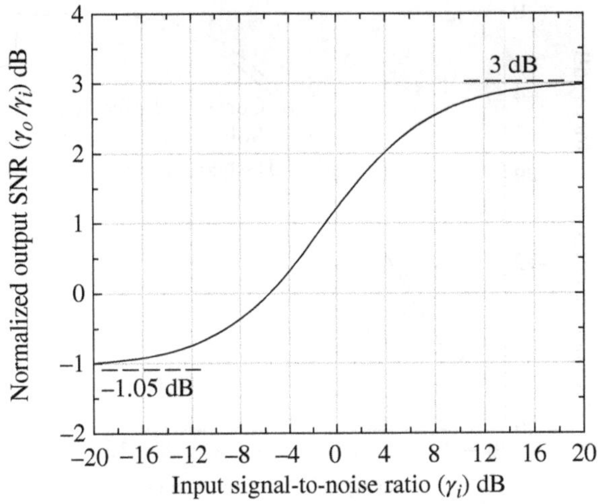

FIGURE 10.2 Baseband limiter input–output signal-to-noise ratio characteristics.

where γ_i and γ_o are the input and output signal-to-noise ratios. Equation (10.11) is plotted in Figure 10.2 as the normalized output signal-to-noise ratio as a function of the input signal-to-noise ratio in decibels. There is a 3 dB signal-to-noise improvement for high input signal-to-noise ratios and for low input signal-to-noise ratios the output signal-to-noise ratio only degraded by 1.05 dB relative to the input signal-to-noise ratio. The crossover point, that is, when $\gamma_o = \gamma_i$, occurs when $\gamma_i = -5.5$ dB.

In the absence of noise, the peak level of the bandpass limiter output is $4A_L/\pi$ volts; however, the noise suppresses the signal level by the limiter signal suppression factor, defined as the ratio of the output signal amplitude at a given input signal-to-noise ratio to the noise-free signal level $4A_L/\pi$.

Gardner expresses the limiter signal suppression factor as

$$\alpha = \sqrt{\frac{\gamma_i}{4/\pi + \gamma_i}} \tag{10.12}$$

and Tausworthe [7] provides the approximation

$$\alpha \cong \sqrt{\frac{0.7854\gamma_i + 0.4768\,\gamma_i^2}{1 + 1.024\gamma_i + 0.4768\,\gamma_i^2}} \tag{10.13}$$

Equations (10.12) and (10.13) are plotted in Figure 10.3 as a function of the input signal-to-noise ratio. When the bandpass limiter is followed by a PLL, the loop gain becomes $\alpha K_o K_d$ which, as mentioned earlier, impacts the loop bandwidth and damping factor. That is, as the input signal-to-noise ratio decreases the loop bandwidth and damping factor also decrease. Although the received signal dynamics influence the PLL performance, decreasing the loop bandwidth in low signal-to-noise environments can help maintain the loop tracking; this is discussed in more detail in Section 10.8

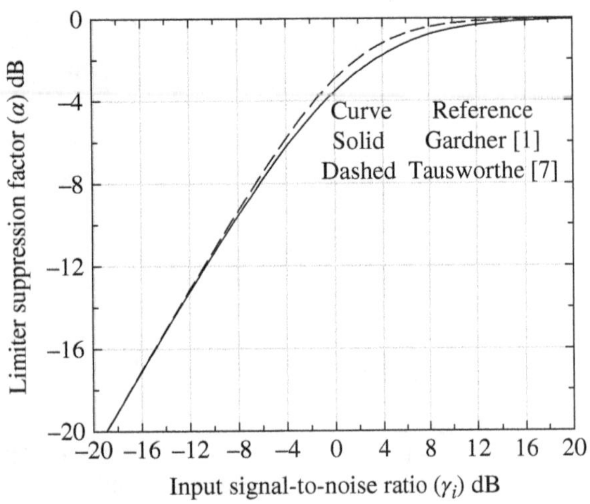

FIGURE 10.3 Limiter signal suppression factor.

where the optimum PLL requires that the signal be weighted by the input signal-to-noise ratio. However, as the input signal-to-noise ratio increases α approaches one and the loop gain is influenced by the product K_oK_d; this condition applies in the following PLL analysis that selects the optimum loop design based on an ideal automatic gain control (AGC) and the constant loop parameter K_oK_d.

10.3 BASEBAND PHASELOCK LOOP IMPLEMENTATION

As mentioned earlier the input to the complex multiplier is typically the baseband or analytic signal representation of the carrier-modulated received waveform. The baseband implementation of the PLL is referred to as a Costas loop in which the in-phase and quadrature phase (or quadphase) rails are phase corrected to facilitate the data demodulation on each rail. In keeping with the previous discussion, consider that the baseband received signal $\tilde{s}(t) = (A/2)\exp(j\phi(t))$ is applied to the input of the complex multiplier shown in Figure 10.4. In addition to a constant phase error, the phase function may include frequency and higher order frequency errors as expressed, for example, by

$$\phi(t) = \frac{\dot{\omega}t^2}{2} + \omega t + \phi \tag{10.14}$$

Referring to Figure 10.4 the inputs to the phase accumulator corresponding to the phase function of Equation (10.14) are defined as

$$\phi_\varepsilon \triangleq \phi - \hat{\phi}, \quad \Delta\phi \triangleq 2\pi\left(f-\hat{f}\right)\Delta T_s \quad \text{and,} \quad \dot{\Delta\phi} \triangleq 2\pi\left(\dot{f}-\hat{\dot{f}}\right)\Delta T_s^2 \tag{10.15}$$

The phase estimate $\hat{\phi}(t)$ is typically derived by squaring and filtering operations discussed in the following section. This is an intuitive implementation of the phase-error detector, in that, for MPSK modulation, the squaring operations remove the received phase modulation yielding a sinusoidal signal reference. In Section 10.9 an optimum implementation is examined based on the minimum mean-square phase-error estimate and several important differences are highlighted that result in significant improvements under various input signal-to-noise conditions.

10.4 PHASE-ERROR GENERATION

When applied to data communications the input signal is not simply a sinusoidal signal but is typically phase modulated in some defined way as in the various forms of MPSK. In these cases, the generation of the phase error must remove the data-dependent phase variations by taking the M-th power of the input signal. For example, with MPSK the signal phase is $\phi'(t) = (2\pi m/M)p(t) + \phi(t), \ m = 0, 1, \ldots, M-1, \ M = 2^k$ where $p(t)$ is a phase weighting function often of the form $rect(t/T)$ where T is the symbol period.* In this case when the input signal is raised to the M-th power the resulting signal will be at a frequency $M\omega$ with the phase equal to $\phi(t) = M\phi'(t) = 2\pi + M\phi(t)$. In terms of the baseband signal, characterized by the normalized in-phase and quadrature components $\cos(\phi'(t))$ and $\sin(\phi'(t))$, the M-th power is obtained by the recursive relationships†

$$\varepsilon_{MI}(t) = \cos(M\phi'(t)) = \cos^2\left(\frac{M}{2}\phi'(t)\right) - \sin^2\left(\frac{M}{2}\phi'(t)\right) \tag{10.16}$$

and

$$\varepsilon_{MQ}(t) = \sin(M\phi'(t)) = 2\sin\left(\frac{M}{2}\phi'(t)\right)\cos\left(\frac{M}{2}\phi'(t)\right) \tag{10.17}$$

Upon repeated application of these relationships the squaring of the signal with the phase function $\phi'(t)$ is obtained as

$$\varepsilon_{2I}(t) = \cos(2\phi'(t)) = \cos^2(\phi'(t)) - \sin^2(\phi'(t)) \tag{10.18}$$

and

*In this section it is assumed that the exact symbol timing is known at the demodulator.

†The subscripts I and Q refer to the in-phase and quadrature phase terms, respectively.

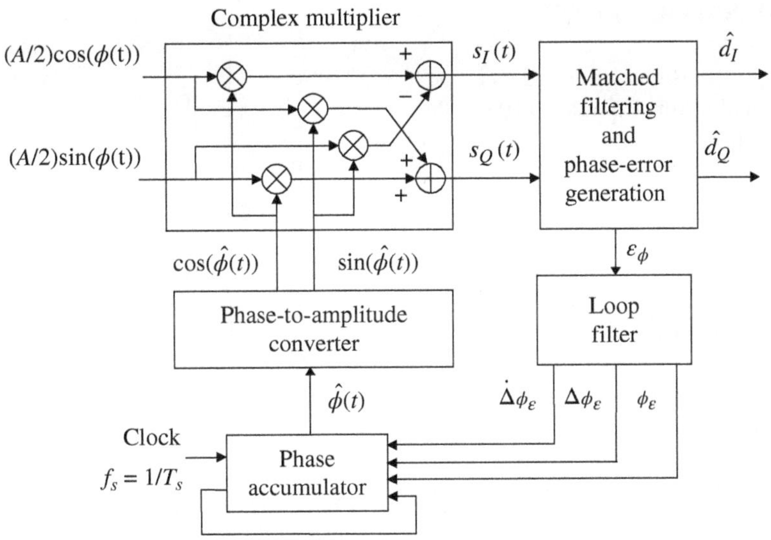

FIGURE 10.4 Baseband implementation of phaselock loop.

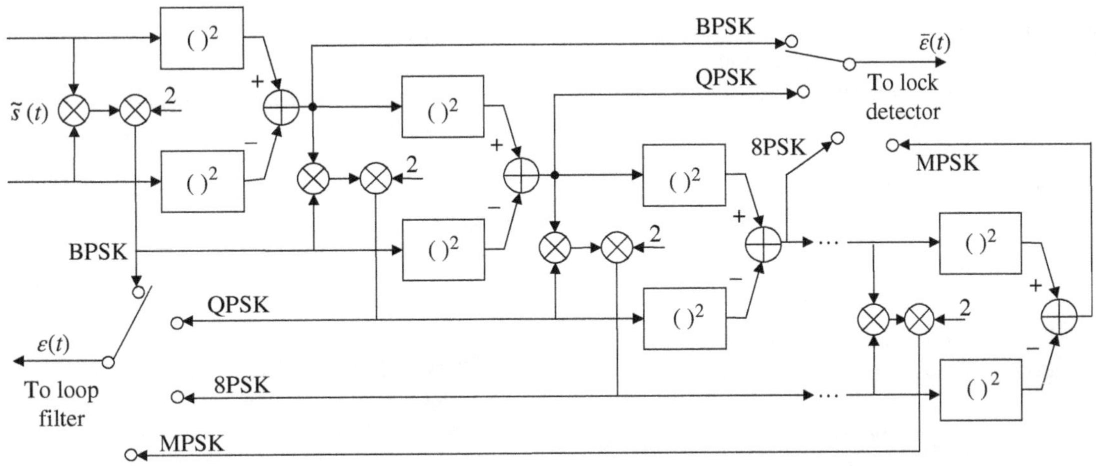

FIGURE 10.5 MPSK phase-error generation.

$$\varepsilon_{2Q}(t) = \sin(2\phi'(t)) = 2\sin(\phi'(t))\cos(\phi'(t)) \qquad (10.19)$$

This last set of quadratic expressions represents the required phase detector output for binary phase shift keying (BPSK) modulation. These relationships are depicted in the MPSK phase-error computations of Figure 10.5. Lindsey and Woo [8] discuss an alternative to the squaring loop for suppressed carrier BPSK modulation. The loss in the signal-to-noise ratio associated with the repeated squaring operations is referred to as the squaring loss and is discussed in Section 10.9.

An alternate phase-error generation algorithm is based on the inverse tangent function which computes the phase detector output phase as*

$$\phi(t) = \tan^{-1}\left(\frac{\text{Im}\{\hat{s}(t)\}}{\text{Re}\{\hat{s}(t)\}}\right) \qquad (10.20)$$

The phase $\phi(t)$ is used to compute the received symbol phase estimate $\hat{\phi}_m = m\Delta\phi$ where $m = 0, \ldots, M-1$ is indexed over the known phase rest states with $\hat{\phi}_m$ corresponding to the minimum magnitude $|\phi(t)-\hat{\phi}_m| < \Delta\phi/2$ where $\Delta\phi = 2\pi/M$. Using this result, the approximate phase-error input to the loop filter is computed as $\varepsilon(t) \cong \phi(t)-\hat{\phi}_m$. The *arctan* operation functions as a hard limiter by removing the amplitude dependence of the input signal so the PLL gain and bandwidth are unchanged. Although this is advantageous, the *arctan* function in low signal-to-noise environments degrades the signal tracking performance relative to the error generation processing shown in Figure 10.5.

*An alternate procedure without the influence of noisy samples and the complexity of performing the arctan function is to determine the angular segment that includes the real and imaginary parts of $\overleftrightarrow{s}(t)$.

10.5 FIRST-ORDER PHASELOCK LOOP

The first-order PLL the loop filter is characterized as a constant gain so that $F(s) = G$ and, from Equation (10.6), the closed-loop transfer function becomes

$$H(s) = \frac{K_o K_d G}{s + K_o K_d G} \quad \text{: first-order loop} \tag{10.21}$$

This response has a real-axis pole at $s = -K_o K_d G$ and, in terms of the root-locus plot,* as the gain increases from zero the pole migrates along the negative real axis from the open-loop pole at $s = 0$ to the open-loop zero at infinity. Therefore, because the pole does not enter the right half of the s-plane, the closed-loop response is unconditionally stable. The closed-loop gain is simply $K_o K_d G$ and the best tracking performance is obtained with a high-gain and low-bandwidth loop to minimize the impact of the noise receiver noise. However, from Equation (10.21), the 3-dB bandwidth of the closed-loop response is also $K_o K_d G$ rad/s so a high gain and low bandwidth are conflicting requirements. For this reason, the first-order PLL has limited applications [9] and the second-order loop, which is not subject to this restriction, is used in most coherent waveform demodulators.

Referring to Problem 3, the one-sided noise bandwidth of the first-order loop is evaluated as

$$B_L = \frac{K_o K_d G}{4} \quad \text{: first-order loop bandwidth (Hz)} \tag{10.22}$$

Referring to Equation (10.7), the phase error, $\phi_e(s)$, for an input phase function $\phi_i(s) = \phi(s)$ is expressed as

$$\phi_e(s) = \frac{s\phi(s)}{s + K_o K_d G} \tag{10.23}$$

Using inverse Laplace transforms [10], the response $\phi_e(t)$ to a phase step, a frequency step, and a frequency ramp are evaluated for the first-order PLL using $K_o K_d G = 4B_L$. The following analysis assumes sufficiently small input levels so that the loop responses correspond to the linear operating range of the PLL.

The response to a phase-step input is evaluated using $\phi_i(s) = \Delta\phi/s$ and results in the normalized phase-error response

$$\frac{\phi_e(t)}{\Delta\phi} = e^{-4B_L t} \quad \text{: phase-step input} \tag{10.24}$$

The response to a frequency-step input is evaluated using $\phi_i(s) = \Delta w/s^2$ and results in the normalized phase-error response

*The generation of root-locus plots is discussed in Section 10.7.1.

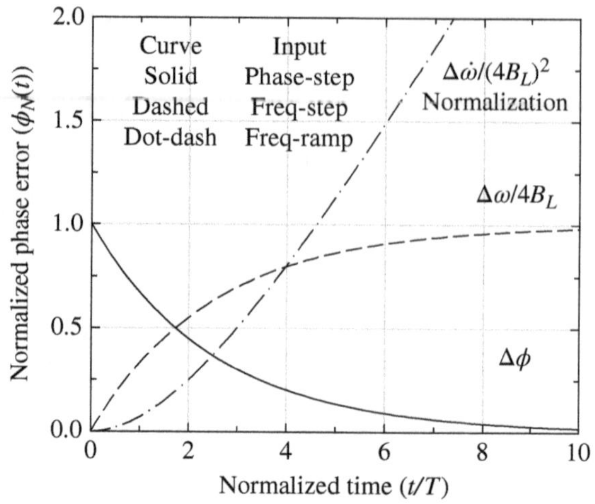

FIGURE 10.6 First-order PLL theoretical responses to phase-step input ($B_L T = 0.1$).

$$\frac{\phi_e(t)}{\Delta\omega/4B_L} = 1 - e^{-4B_L t} \quad \text{: frequency-step input} \tag{10.25}$$

The response to a frequency-ramp input is evaluated using $\phi_i(s) = \Delta\omega/s^3$ and results in the normalized phase-error response

$$\frac{\phi_e(t)}{\Delta\dot{\omega}/(4B_L)^2} = -\left(1 - e^{-4B_L t}\right) + 4B_L t \quad \text{: frequency-ramp input} \tag{10.26}$$

These responses are plotted in Figure 10.6 where the output to a continually applied frequency ramp or Doppler rate increases without bound. In the following section, similar theoretical responses are developed for the second-order PLL that is used to evaluate the transient responses of various modulated waveforms with additive noise.

10.6 SECOND-ORDER PHASELOCK LOOP

The second-order PLL described in this section uses a first-order loop filter having the form

$$F(s) = \frac{\varepsilon_o(s)}{\varepsilon_i(s)} = G_1 \frac{s + \alpha + G_2}{s + \alpha} \tag{10.27}$$

The PLL will acquire and lock to a carrier frequency that is within the *pull-in frequency* of the loop and, once the loop is locked to the carrier frequency, it will remain locked and track the carrier as long as the carrier frequency does not exceed the loop *pull-out frequency*. The loop pull-in and pull-out frequency characteristics are discussed in Section 10.6.9. Analysis of the root-locus plot for the second-order loop shows that the migration of the two poles in the left half of the s-plane as the loop gain is increased. For example, as the gain approaches infinity, one pole terminates on the finite real open-loop zero and the other pole terminates on the real zero at minus infinity. In this case, the loop is unconditionally stable and the gain and bandwidth can be adjusted independently; for these reasons the second-order PLL is preferred in most applications.

10.6.1 Second-Order PLL Loop Filter Implementations

Two implementations of a first-order loop filter are shown in Figure 10.7. Although the transfer functions for these loops are identical, the gains G_1 and G_2 are quite different. The first implementation, shown in Figure 10.7a, provides for a single phase/frequency estimate while the second configuration provides for separate estimates of the phase and frequency. The advantage of having separate estimates is that the corrections can be applied at different points in the demodulator, which is convenient in some cases. The disadvantage in this configuration is that the ratio of the gains, G_1 and G_2, is relatively large and G_2 may be subject to truncation or round-off errors in a digital implementation of the loop. The first configuration requires that the phase estimate be applied at a common point; however, the gain

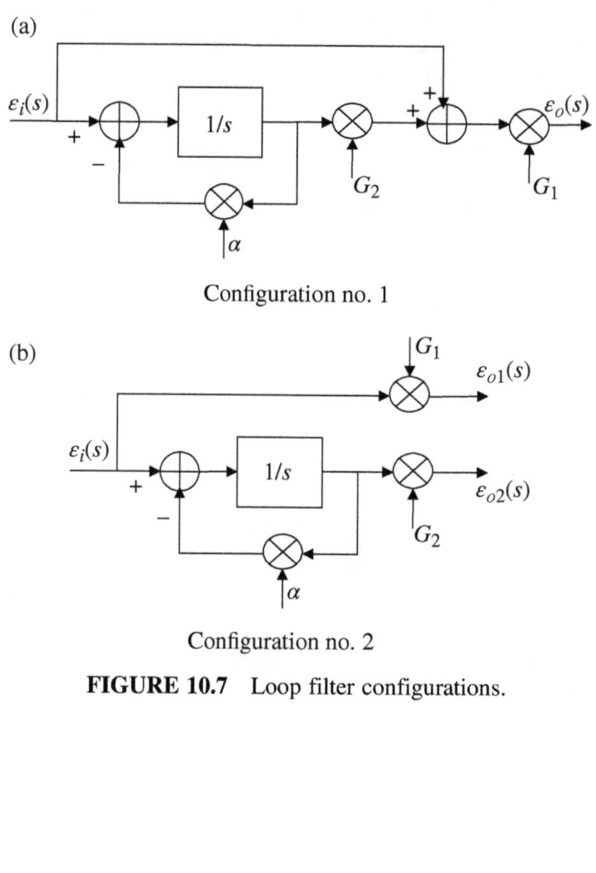

FIGURE 10.7 Loop filter configurations.

ratio is closer to unity thus reducing the possibility of truncation or round-off errors.

These complex frequency domain depictions of the filter functions involving $s=\alpha+j\omega$ are readily characterized in the z-plane for discrete-time implementations amenable to digital signal processing. Considering the discrete-time sampling $t=iT_s$, where T_s is the sampling interval,* and using the impulse-invariant z-transform

$$z=e^{sT_s} \tag{10.28}$$

the filter transfer function $F'(z)$ can be obtained. For example, evaluation of the impulse-invariant z-transform of the loop filters, shown in Figure 10.7, involves characterizing the integrator s-plain transfer function $F'(s)=1/(s+\alpha)$, performing the inverse Laplace transform to determine the impulse response $f(t)$ of the integrator, sampling the impulse response at uniform sampling times iT_s, and then performing the variable transformation in Equation (10.28).

Referring to Figure 10.7, the impulse response of the lossy integrator is evaluated as $f(t)=e^{-\alpha t}$ and the sampled impulse response is expressed as

$$\begin{aligned} f_s(iT_s) &= \sum_i f(t)\delta(t-iT_s) \\ &= \sum_i e^{-\alpha iT_s}\delta(t-iT_s) \end{aligned} \tag{10.29}$$

Taking the Laplace transform of Equation (10.29) results in the discrete sampled s-domain function

$$\begin{aligned} F_s'(s) &= \sum_i e^{-\alpha iT_s}e^{-siT_s} \\ &= \frac{1}{1-e^{-\alpha T_s}e^{-sT_s}} \end{aligned} \tag{10.30}$$

The second expression is obtained from the closed-form result of an infinite power series. Applying the impulse-invariant z-transform to Equation (10.30), the sampled data lossy integrator is expressed as

$$F'(z)=\frac{1}{1-e^{-\alpha T_s}z^{-1}} \tag{10.31}$$

The functional implementation of $F'(s)$ and $F'(z)$ is shown in Figure 10.8 with $k=e^{-\alpha T_s}$. For convenience the parameter α is often normalized by the sampling interval T_s. The PLL transfer functions for the two loop filters in Figure 10.7 are evaluated in the following sections.

*The sample time increment subscript s should not be confused with the Laplace s-plain complex variable $s=\alpha+j\omega$.

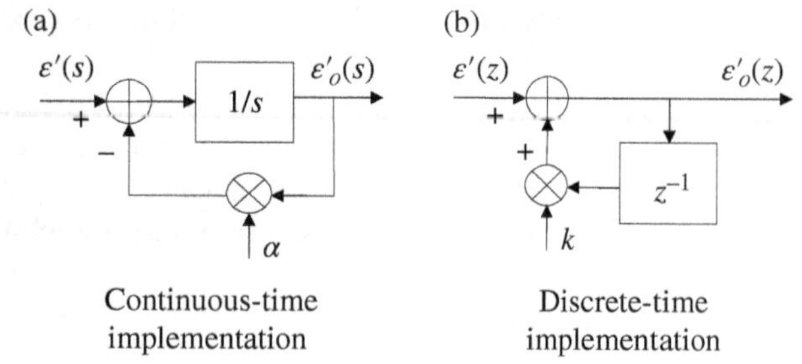

FIGURE 10.8 Discrete-time characterization of lossy integrator.

10.6.2 Loop Filter Configuration No. 1

For the loop filter configuration No. 1, the loop filter transfer function is evaluated as

$$F(s) = \frac{\varepsilon_o(s)}{\varepsilon_i(s)} = G_1 \frac{s+\alpha+G_2}{s+\alpha} \tag{10.32}$$

An important performance consideration in the design is the gain of the loop filter, evaluated as $F(0) = G_1(\alpha + G_2)/\alpha$. It will be seen that a lossy integrator results in a finite steady-state phase error when the input signal has a frequency error, for example, an unknown Doppler frequency. Furthermore, an infinite phase-error results when the input signal has a finite frequency-rate error. On the other hand, with a lossless integrator, corresponding to $\alpha = 0$ and $K_v = \infty$, the steady-state phase error is zero with an input frequency error and finite with an input frequency-rate error. Continuing with this analysis, the closed-loop transfer function for the filter function, given by Equation (10.32), becomes

$$H(s) = \frac{K_o K_d G_1(s + G_2 + \alpha)}{s^2 + (K_o K_d G_1 + \alpha)s + K_o K_d G_1(G_2 + \alpha)} \tag{10.33}$$

The classical, or control theory, expression for the second-order closed-loop transfer function is expressed in terms of the loop damping factor (ζ) and natural resonant frequency (ω_n) as

$$H(s) = \frac{(2\varsigma\omega_n - \alpha)s + \omega_n^2}{s^2 + 2\varsigma\omega_n s + \omega_n^2} \tag{10.34}$$

Upon equating the denominator coefficients in Equations (10.33) and (10.34), the following correspondence between the loop parameters is obtained

$$2\varsigma\omega_n = K_o K_d G_1 + \alpha \tag{10.35}$$

and

$$\omega_n^2 = K_o K_d G_1(G_2 + \alpha) \tag{10.36}$$

Based on the steady-state performance advantages of infinite DC gain loops and their simplicity to implement in digital systems, the following analysis is based on $\alpha = 0$.

It is convenient to specify the PLL design parameters using the loop bandwidth rather than the natural frequency of the loop. In general, the loop bandwidth is defined as*

$$B_L = \frac{1}{2\pi}\int_0^\infty |H(\omega)|^2 d\omega \quad \text{: hertz} \tag{10.37}$$

and, for the infinite-gain ($\alpha = 0$) second-order loop, the bandwidth is expressed as

$$B_L = \frac{\omega_n}{2}\left(\varsigma + \frac{1}{4\varsigma}\right) \quad \text{: hertz} \tag{10.38}$$

Using these results the loop gains, expressed in terms of the design parameters B_L and ζ, are

$$G_1 = \frac{4\varsigma B_L}{K_o K_d(\varsigma + (1/4\varsigma))} \quad \text{and} \quad G_2 = \frac{B_L}{\varsigma(\varsigma + (1/4\varsigma))} \tag{10.39}$$

and normalizing by the symbol interval (T), the gains for the digital implementation become

$$G_1' = \frac{4\varsigma B_L T}{K_o K_d(\varsigma + (1/4\varsigma))} \quad \text{and} \quad G_2' = \frac{B_L T}{\varsigma(\varsigma + (1/4\varsigma))} \tag{10.40}$$

The symbol interval is chosen as the normalizing parameter because the phase tracking is generally performed at the symbol matched filter sampling rate. The ratio of the gains is $G_1/G_2 = 4\varsigma^2/(K_o K_d)$ and, with $K_o = K_d = 1.0$ and the commonly used cases of $\zeta = 1$ and 0.707, the gain ratio is 4 and 2, respectively. In digital implementations gain ratios near unity are less susceptible to performance issues resulting from finite quantization levels.

The sensitivity of the loop bandwidth and damping factor to changes in the signal level through the phase detector gain, that is, $K_d = A/2$, is evaluated by expressing $B_L T$ and ζ as functions of G_1', G_2', K_o, and K_d and solving for the differentials $\Delta B_L T$ and $\Delta\zeta$. The results in the normalized forms are

$$\frac{\Delta B_L T}{B_L T} = \left(\frac{4\varsigma}{\varsigma + 1/4\varsigma}\right)\frac{\Delta A}{A} \tag{10.41}$$

and

$$\frac{\Delta\varsigma}{\varsigma} = \left(\frac{1}{2}\right)\frac{\Delta A}{A} \tag{10.42}$$

*Although ω_n has dimensions of radians/sec, B_L has dimensions of cycles/sec or Hz.

Therefore, for commonly used damping factors, the bandwidth will change by about two to three times the change in the normalized signal level and the loop damping will change by one-half.

10.6.3 Loop Filter Configuration No. 2

For the loop filter configuration No. 2, the loop filter transfer function is found to be

$$F(s) = G_1 \frac{s + \alpha + G_2/G_1}{s + \alpha} \tag{10.43}$$

and the corresponding closed-loop transfer function is

$$H(s) = \frac{K_o K_d G_1 (s + \alpha + G_2/G_1)}{s^2 + (K_o K_d G_1 + \alpha)s + K_o K_d (G_1 \alpha + G_2)} \tag{10.44}$$

Following the same procedures as in the previous section, the sampled data form of the loop filter gains with $\alpha = 0$ is expressed in terms of the damping factor and loop bandwidth as

$$G_1' = \frac{4\varsigma B_L T}{K_o K_d (\varsigma + (1/4\varsigma))} \quad \text{and} \quad G_2' = \frac{4(B_L T)^2}{K_o K_d (\varsigma + (1/4\varsigma))^2} \tag{10.45}$$

In this case, the ratio of the gains is $G_1'/G_2' = \varsigma(\varsigma + 1/4\varsigma)/(B_L T)$. The respective time-bandwidth products ($B_L T$) for BPSK and QPSK are on the order of 0.1–0.01, respectively, so, for the commonly used cases of $\zeta = 1$ and 0.707, gain ratios of 7–125 are encountered.

Upon examining the sensitivity of the loop bandwidth and damping factor to changes in the input signal level the results are found to be

$$\frac{\Delta B_L T}{B_L T} = \left(\frac{4\varsigma - 1}{4(\varsigma + 1/4\varsigma)} \right) \frac{\Delta A}{A} \tag{10.46}$$

and

$$\frac{\Delta \varsigma}{\varsigma} = \varsigma \frac{\Delta A}{A} \tag{10.47}$$

Therefore, for commonly used damping factors, both the bandwidth and damping factor will change in roughly the same proportion to changes in the signal level. Although this loop filter is slightly more sensitive to changes in the damping factor than configuration No. 1, the sensitivity to changes in the loop bandwidth is about 4–6× less sensitive.

10.6.4 Theoretical Closed-Loop Phase-Error Response

In this section, the focus is on the loop characteristics using the loop filter configuration No. 1. Evaluation of the theoretical phase-error response is particularly useful, in that, it is an effective measure for verifying the performance of hardware and software implementations. The phase-error transfer function was expressed in Equation (10.7) as $1 - H(s)$ and the phase-error transfer function for the second-order loop is

$$T(s) = \frac{\phi_\varepsilon(s)}{\phi(s)} = \frac{s(s + \alpha)}{s^2 + 2\varsigma\omega_n s + \omega_n^2} \tag{10.48}$$

The steady-state phase error is evaluated for phase inputs corresponding to a phase step, a linear phase ramp or frequency step, and a quadratic phase function or frequency ramp. The three inputs are characterized in terms of $\phi(s)$ as $\Delta\phi/s$, $\Delta\omega/s^2$, and $\Delta\dot{\omega}/s^3$, respectively. The steady-state value of the phase-error response is determined by applying the final value theorem of Laplace transforms given by

$$\lim_{t \to \infty} \phi_\varepsilon(t) = \lim_{s \to 0} s\phi_\varepsilon(s) \tag{10.49}$$

Evaluation of the steady-state phase error under the three conditions given earlier results in phase errors $\phi_\varepsilon(\infty)$ of: 0, $\alpha\Delta\omega$, and $\alpha\Delta\dot{\omega}t + \Delta\dot{\omega}/\omega_n^2$, respectively. Therefore, an input signal with a frequency step results in a finite frequency error and an input signal with a frequency ramp results in a time-dependent and fixed frequency-rate error. Recalling that the DC gain of the loop filter is evaluated as $F'(0) = 1/\alpha$, by letting $\alpha = 0$ the only steady-state error is the fixed frequency rate resulting from the frequency ramp input. Because of the importance of this result $\alpha = 0$ is used in the following analyses and because digital and active filter implementations readily achieve this condition it is commonly used in practice.

10.6.5 Phase-Error Response to Phase Step

The phase response for an input phase step $\phi_i(s) = \Delta\phi/s$ is given by

$$\phi_\varepsilon(s) = \frac{s\Delta\phi}{s^2 + 2\varsigma\omega_n s + \omega_n^2} \tag{10.50}$$

The time response is determined using inverse Laplace transforms and the solution depends whether the poles are complex or real. For example, the poles of $\phi_\varepsilon(s)$ are evaluated as

$$s_p = -\omega_n \left(\varsigma \pm \sqrt{\varsigma^2 - 1} \right) \tag{10.51}$$

and for $\zeta > 1$ there are two distinct real poles, for $\zeta = 1$ there are two identical real poles, and for $\zeta < 1$ there is a complex pole pair. For the infinite gain loop ($\alpha = 0$), the time response for each of these conditions is evaluated as [11,12],

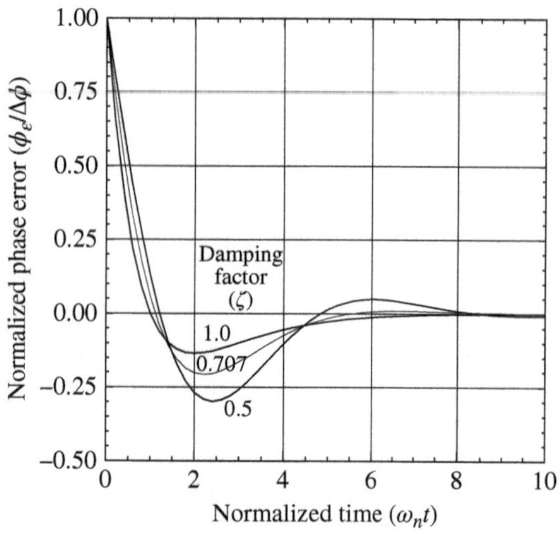

FIGURE 10.9 Theoretical responses for phase-step input (infinite gain: $\alpha = 0$).

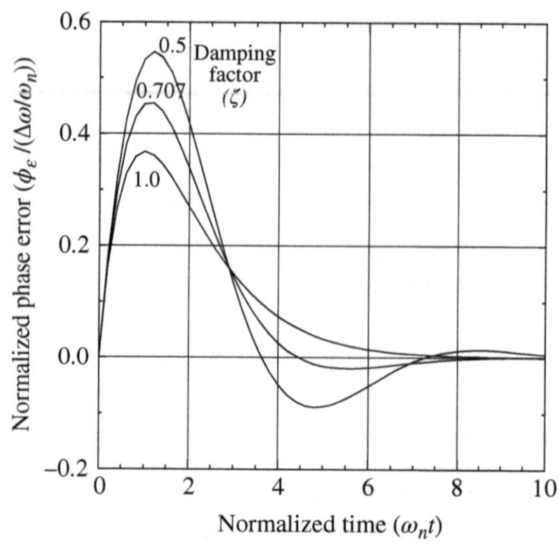

FIGURE 10.10 Second-order PLL theoretical responses for frequency-step input (infinite gain: $\alpha = 0$).

$$\phi_\varepsilon(t) = \begin{cases} \Delta\phi\left[\cosh\left(\omega_n t\sqrt{\varsigma^2-1}\right) - \dfrac{\varsigma}{\sqrt{\varsigma^2-1}}\sinh\left(\omega_n t\sqrt{\varsigma^2-1}\right)\right]e^{-\varsigma\omega_n t} & :\varsigma > 1 \\ \Delta\phi[1-\omega_n t]e^{-\omega_n t} & :\varsigma = 1 \\ \Delta\phi\left[\cos\left(\omega_n t\sqrt{1-\varsigma^2}\right) - \dfrac{\varsigma}{\sqrt{1-\varsigma^2}}\sin\left(\omega_n t\sqrt{1-\varsigma^2}\right)\right]e^{-\varsigma\omega_n t} & :\varsigma < 1 \end{cases} \tag{10.52}$$

The response to a frequency ramp is plotted in Figure 10.9 for damping factors $\zeta = 0.5$, 0.707, and 1.0.

10.6.6 Phase-Error Response to Frequency Step

The phase response, for the infinite gain loop ($\alpha = 0$), with a frequency-step input $\phi_i(s) = \Delta\omega/s^2$ is given by

$$\varphi_\varepsilon(s) = \frac{\Delta\omega(s+\alpha)}{s\left(s^2 + 2\varsigma\omega_n s + \omega_n^2\right)} \tag{10.53}$$

In this case, there is an additional pole at zero and upon evaluating the inverse Laplace transforms the time responses for the three conditions of the damping factor are expressed as [11,12]

$$\phi_\varepsilon(t) = \begin{cases} \dfrac{\Delta\omega}{\omega_n}\left(\dfrac{1}{\sqrt{\varsigma^2-1}}\right)\sinh\left(\omega_n t\sqrt{\varsigma^2-1}\right)e^{-\varsigma\omega_n t} & :\varsigma > 1 \\ \dfrac{\Delta\omega}{\omega_n}(\omega_n t)e^{-\omega_n t} & :\varsigma = 1 \\ \dfrac{\Delta\omega}{\omega_n}\left(\dfrac{1}{\sqrt{1-\varsigma^2}}\right)\sin\left(\omega_n t\sqrt{1-\varsigma^2}\right)e^{-\varsigma\omega_n t} & :\varsigma < 1 \end{cases} \tag{10.54}$$

The response to a frequency ramp is plotted in Figure 10.10 for damping factors $\zeta = 0.5$, 0.707, and 1.0.

10.6.7 Phase-Error Response to Frequency Ramp

The phase response, for the infinite gain loop ($\alpha = 0$), with a frequency-ramp input $\phi_i(s) = \Delta\dot{\omega}/s^3$ is given by

$$\phi_\varepsilon(s) = \frac{\Delta\dot{\omega}(s+\alpha)}{s^2\left(s^2 + 2\varsigma\omega_n s + \omega_n^2\right)} \tag{10.55}$$

In this case, there is an additional pole pair at zero frequency and the time responses for the three conditions of the damping factor are expressed as [11,12]

$$\phi_\varepsilon(t) = \begin{cases} \dfrac{\Delta\dot{\omega}}{\omega_n^2} - \dfrac{\Delta\dot{\omega}}{\omega_n^2}\left[\cosh\left(\omega_n t\sqrt{\varsigma^2-1}\right) + \dfrac{\varsigma}{\sqrt{\varsigma^2-1}}\sinh\left(\omega_n t\sqrt{\varsigma^2-1}\right)\right] e^{-\varsigma\omega_n t} & :\varsigma > 1 \\ \dfrac{\Delta\dot{\omega}}{\omega_n^2} - \dfrac{\Delta\dot{\omega}}{\omega_n^2}(1+\omega_n t)e^{-\omega_n t} & :\varsigma = 1 \\ \dfrac{\Delta\dot{\omega}}{\omega_n^2} - \dfrac{\Delta\dot{\omega}}{\omega_n^2}\left[\cos\left(\omega_n t\sqrt{1-\varsigma^2}\right) + \dfrac{\varsigma}{\sqrt{1-\varsigma^2}}\sin\left(\omega_n t\sqrt{1-\varsigma^2}\right)\right] e^{-\varsigma\omega_n t} & :\varsigma < 1 \end{cases} \tag{10.56}$$

The response to a frequency ramp is plotted in Figure 10.11 for damping factors $\zeta = 0.5$, 0.707, and 1.0.

Referring to Equation (10.56) it is important to note that as $t \to \infty$ there is a constant phase error given by

$$\phi_\varepsilon(t)\big|_{t\to\infty} = \frac{\Delta\dot{\omega}}{\omega_n^2} = \frac{\pi\dot{f}\left(\varsigma + (1/4\varsigma)\right)^2}{2B_L^2} \tag{10.57}$$

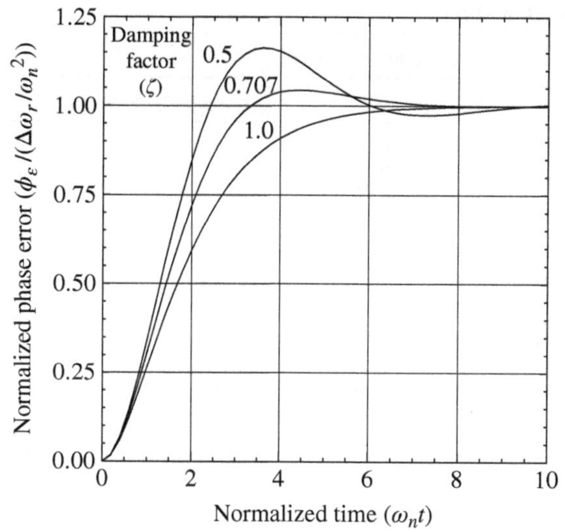

FIGURE 10.11 Second-order PLL theoretical responses for frequency-ramp input (infinite gain: $\alpha = 0$).

where the frequency ramp corresponds to $\dot{f}$ in hertz/second. The second equality in Equation (10.57) is obtained using Equation (10.38) and substituting ω_n in terms of B_L and ζ. The constant phase error in Equation (10.57) is based on the infinite DC loop gain ($K_v = \infty$) that results when $\alpha = 0$ as discussed earlier. For the finite DC loop gain Gardner [1] shows that an additional phase-error term, given by $\Delta\dot{\omega}t/K_v$, is present that increases without bound as $t \to \infty$. The residual phase error with a frequency ramp is eliminated when using a third-order loop; however, the third-order loop results in a constant phase error with an input phase jerk characterized as a $\ddot{f}$ step input with dimensions hertz/second^2.

10.6.8 Simulated MPSK Transient Responses

To demonstrate the application of the theoretical transient responses, the transient response of the PLL is evaluated using a Monte Carlo simulation program for BPSK-, QPSK-, and 8PSK-modulated waveforms and the results are compared to the theoretical responses. However, because the theoretical results correspond to a noise-free CW input signal, to make a direct comparison the simulated MPSK modulators use mark-hold data that results in an unmodulated or CW carrier. A reasonable comparison, either from simulations or hardware tests, is viewed as a validation of the PLL implementation. Generally, the simulation or hardware test results are based on sampled matched filter outputs occurring at the symbol rate $R_s = 1/T$, that is, for $t = nT$, $n = 1, 2, \ldots$, so the response is expressed in terms of the normalized time t/T. For this comparison the theoretical transient responses for a step input in phase and frequency are based on the respective normalized responses for the commonly used case of $\zeta =$ 0.707. The normalized theoretical responses corresponding to Equations (10.52) and (10.54) are:

$$\frac{\phi_\varepsilon(t/T)}{\Delta\varphi} = \left[\cos\left(\omega_n T(t/T)\sqrt{1-\varsigma^2}\right) - \frac{\varsigma}{\sqrt{1-\varsigma^2}}\sin\left(\omega_n T\left(\frac{t}{T}\right)\sqrt{1-\varsigma^2}\right)\right] e^{-\varsigma\omega_n T(t/T)} \quad : \varsigma < 1 \text{ phase step} \tag{10.58}$$

and

$$\frac{\varphi_\varepsilon(t/T)}{\Delta\omega/\omega_n} = \frac{1}{\sqrt{1-\varsigma^2}}\sin\left(\omega_n T\left(\frac{t}{T}\right)\sqrt{1-\varsigma^2}\right) e^{-\varsigma\omega_n T(t/T)} \quad : \varsigma < 1 \text{ frequency step} \tag{10.59}$$

As stated earlier, in Figures 10.9 through 10.11 the abscissas or time axes are normalized by the natural resonant frequency ω_n; however, it is often useful to characterize the normalized time axis in terms of the modulation symbol interval T. This conversion is accomplished by using $\omega_n t = \omega_n T(t/T)$ and solving for t/T with the result

$$\frac{t}{T} = \left(\frac{1}{\omega_n T}\right)\omega_n t = \frac{\varsigma + 1/(4\varsigma)}{2B_L T}\omega_n t \tag{10.60}$$

The second equality is based on the normalized natural resonant frequency $\omega_n T$ using Equation (10.38). Table 10.1 tabulates several key theoretical response conditions in terms of t/T. These conditions correspond to the parameters $B_L T = 0.1$ and 0.01 with and $\zeta = 0.707$ that are frequently used, respectively, with BPSK- and QPSK-modulated waveforms. These conditions provide convenient *benchmarks* for establishing the simulated and hardware loop performance. For example, considering BPSK modulation with a phase-step input the simulated or hardware PLL gain $K_o K_d$ must be adjusted until the test response corresponds to a zero crossing at $t/T = 5.6$ and a maximum normalized undershoot of -0.207 occurring at $t/T = 11.9$. This adjustment corresponds to optimizing G_1' in configuration No. 1 and G_1' and G_2' in configuration No. 2. In many hardware situations, it is more convenient to generate a frequency-step input so the optimum characteristics for a frequency-step input are also listed in the table.

In the following comparisons the phase-error generation algorithm, shown in Figure 10.5, is used with loop filter configuration No. 1. In order for the simulated results to match, or closely match, the theoretical response, it is essential that the PLL gains be appropriately scaled in the simulation, or hardware, implementations as mentioned earlier. Since the phase detectors gain K_d changes in direct proportion to the input signal level, the AGC setting can be used to provide the optimum PLL gain. For these tests the phase and frequency steps must be small, about 20° and $R_s/20$ Hz, respectively, so that the loop operates in the linear region.

TABLE 10.1 Second-Order PLL Selected Theoretical Responses using $\omega_n T$ and t/T

	BPSK with $B_L T = 0.1$ and $\zeta = 0.707$					
	Phase Step			Frequency Step		
Response Conditions	$\omega_n t$	t/T	Level[a]	$\omega_n t$	t/T	Level[b]
Peak	0.0	0.0	1.0	1.05	5.6	0.45[b]
First zero crossing	1.05	5.6	0.0	4.5	23.8	0.0
First undershoot	2.25	11.9	−0.207	5.4	28.6	−0.015
	QPSK with $B_L T = 0.01$ and $\zeta = 0.707$					
Peak	0.0	0.0	1.0	1.05	56	0.45
First zero crossing	1.05	56	0.0	4.5	238	0.0
First undershoot	2.25	119	−0.207	5.4	286	−0.015

[a]Normalized by $\Delta\phi$.
[b]Normalized by $\Delta\omega/\omega_n$.

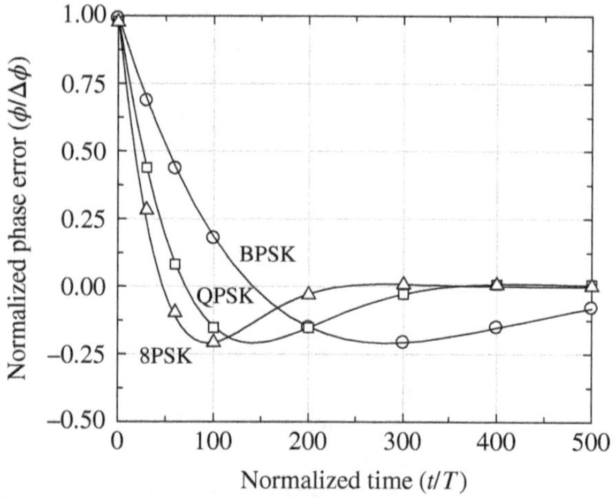

FIGURE 10.12 Comparison of simulated and theoretical responses for phase-step input ($R_b = 12$ kbps, $B_L = 50$ Hz, $\zeta = 0.707$).

The results for a phase-step input are shown in Figure 10.12 using a damping factor of $\zeta = 0.707$ and a loop bandwidth of $B_L = 50$ Hz. The abscissa is normalized to the symbol duration and the source data rate is 12 kbps for all three modulations corresponding to normalized loop bandwidths of $B_L T = 0.00417$, 0.0083, and 0.0125 for BPSK, QPSK, and 8PSK, respectively. The excellent matches between the simulated (*data points*) and theoretical (*solid curves*) results validate the implementation and the loop parameter computations. The phase step used in the simulation is $\Delta\phi = 20°$ and these results clearly indicate that the loop is operating in the linear regime for which the theoretical results apply. The responses Figure 10.12 show that 8PSK has a faster response time than QPSK and BPSK; however, this results solely because to the selected 50 Hz loop bandwidth and the 12 kbps data rate that the results in the highest $B_L T$ product for the 8PSK modulation. It will be seen that, among these modulations, BPSK modulation is the most robust and can tolerate the highest $B_L T$ product; on the order of 0.1 compared to 0.01 for QPSK. The resulting response time for BPSK is recorded in Table 10.1 with the frequency-step response zero crossing occurring at 23.8 symbols which is 10 : 1 faster than that of the QPSK response. Once the optimum PLL response is obtained, the $B_L T$ can be changed without altering the characteristics of the optimum response.

The phaselock detector shown in Figure 10.13 follows directly from the phase detector implementation shown in Figure 10.5. For BPSK modulation, the lock detector computes $\cos(2\phi_\varepsilon)$ and the simulation results showing the lock detector response for BPSK are shown in Figure 10.14. For these results the lock detection filter is implemented as two synchronously tuned single-pole low-pass filters* with a composite bandwidth of B_{LD} (Hz). The normalized form of the lock detector bandwidth is the time bandwidth product $B_{LD}T$. When the lock detector bandwidth is equal to about 1/10 the loop bandwidth and the lock detector threshold is 0.75, the lock indication output is a conservative indication of the phaselocked condition. This condition is used in simulations where the bit-error performance is being evaluated without the influence of the loop transient response. However, with $B_{LD} = B_L/10$, a lock detector threshold of 0.1 is more in keeping with the lock-time criterion of $T_L = 1/\omega_n$. This also corresponds to a lock detector bandwidth of approximately $B_L/4$ and a lock detector threshold of 0.75.

Comparing the implemented PLL response to the theoretical response for a phase step may be inconvenient and, depending upon the instrumentation, it may be easier to observe the response to a step change in frequency. Therefore, the measured response to a frequency step is compared to the corresponding theoretical response in Figure 10.15 for BPSK, QPSK, and 8PSK. The lock detector response is also

*The design of synchronously tuned single-pole low-pass filters is discussed in Appendix B.

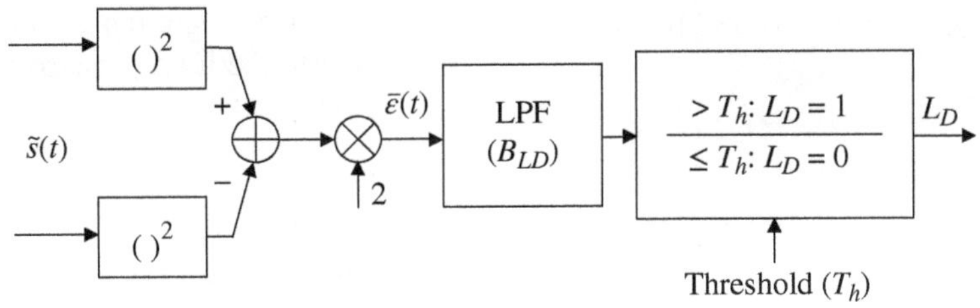

FIGURE 10.13 PLL lock detector implementation for BPSK.

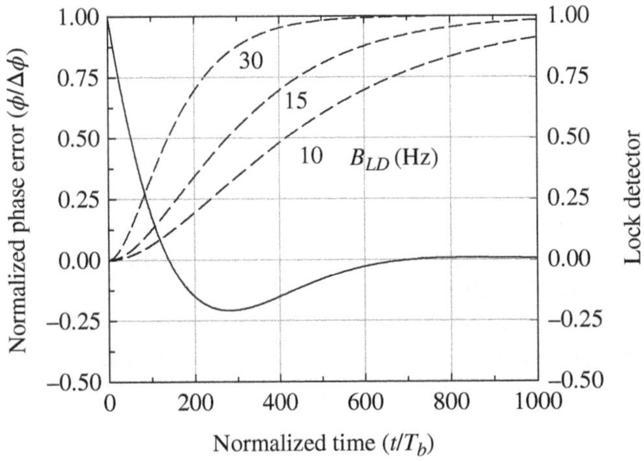

FIGURE 10.14 Simulated BPSK phase and lock detector responses for phase-step input ($R_b = 12$ kbps, $B_L = 150$ Hz, $\zeta = 0.707$, $B_{LD} = 15$, 30, 50 Hz).

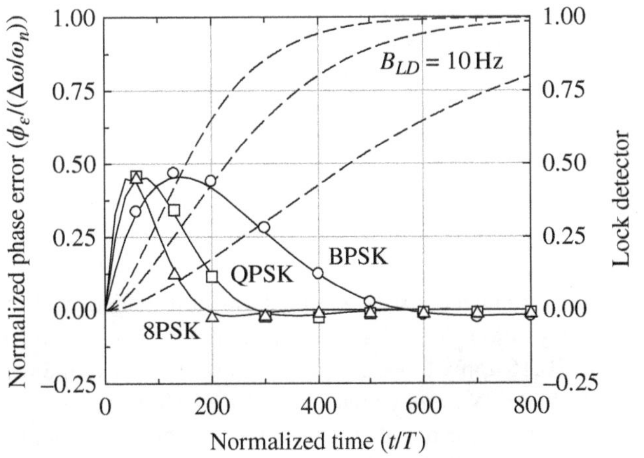

FIGURE 10.15 Comparison of simulated and theoretical responses for frequency-step input ($R_b = 12$ kbps, $B_L = 50$ Hz, $\zeta = 0.707$).

shown for each waveform for a lock detector bandwidth of $B_{LD} = 10$ Hz. As in the preceding examples, although the source data corresponds to *mark-hold* data, the source data rate is 12 kbps so the corresponding normalized loop bandwidths are $B_LT = 0.00417$, 0.0083, and 0.0125.

10.6.9 Initial Frequency Error and PhaseLock Loop Acquisition Time

The maximum input frequency error that can be applied to the PLL and achieve lock without cycle skipping is defined as the *lock-in frequency* F_L and is approximated by Gardner as

$$F_L \cong 2\varsigma\omega_n = \frac{2\varsigma B_L}{\pi(\varsigma + 1/4\varsigma)} \quad \text{: hertz} \tag{10.61}$$

The corresponding *lock-in time* is given by the approximate expression

$$T_L \cong \frac{1}{\omega_n} = \frac{(\varsigma + 1/4\varsigma)}{2B_L} \quad \text{: second} \tag{10.62}$$

The lock-in frequency and time for a damping factor $\varsigma = 0.707$ are plotted in Figure 10.16 for the normalized forms F_LT and T_L/T as a function of the loop time-bandwidth product B_LT.

The pull-in frequency is related to the acquisition performance of the PLL when cycle skipping occurs. In this case the PLL will eventually lock if the input frequency error is within the *pull-in frequency* range F_P of the loop, albeit with cycle skipping. The approximate pull-in frequency for high-gain active filter loops, such that $K_v > 2.5\,\omega_n$, is given by Gardner as

$$F_P \cong \frac{1}{\pi}\sqrt{\varsigma\omega_n K_v} = \frac{1}{\pi}\sqrt{\frac{2\varsigma B_L K_v}{(\varsigma + 1/4\varsigma)}} \quad \text{: hertz} \tag{10.63}$$

and the corresponding *pull-in time* is approximated by

$$T_P \cong 2\sqrt{\varsigma\omega_n K_v} = 15.5\frac{(\varsigma + 1/4\varsigma)^3}{\varsigma B_L^3}\Delta f^2 \quad \text{: second} \tag{10.64}$$

Once the PLL is locked, the loop will remain locked without cycle skipping if the input frequency error is less than the *pull-out frequency* F_{PO} given by

$$F_{PO} = 0.573 B_L \frac{(\varsigma + 1)}{\varsigma + 1/4\varsigma} \tag{10.65}$$

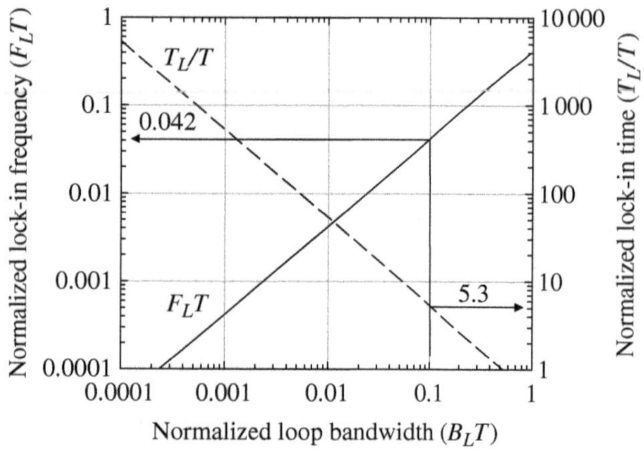

FIGURE 10.16 Lock-in frequency and time for second-order phaselock loop ($\varsigma = 0.707$).

The normalized forms of the loop pull-in and pull-out frequency are F_pT and $F_{po}T$, respectively. The results in Figure 10.16 apply to PLLs operating in the linear range with a noise-free CW input signal and Sections 10.6.11, 10.8, and 10.9 address the acquisition performance when the input is a modulated signal with additive noise.

10.6.10 Optimum Second-Order PLL Configuration

The previous discussions dealt with the second-order closed-loop response in general terms using the damping factor ζ and natural frequency ω_n or the loop bandwidth B_L as arbitrarily selected parameters. For example, the minimum value of the ratio B_L/ω_n occurs at $\zeta = 0.5$; however, this condition results in a transient response overshoot of nearly 20%. Therefore, a damping factor of 0.707 is typically selected as a compromise between overshoot and the critically damped case of $\zeta = 1.0$, which has no overshoot but results in a sluggish response. In this section, the results of Jaffe and Rechtin [13] are used to examine the optimum selection of the damping factor. Their analysis applies the Wiener filter criterion to establish the optimum closed-loop transfer function of the second-order loop when the input phase function is characterized by a frequency step. Based on this criterion, which minimizes the mean-square phase error for a given input condition, Jaffe and Rechtin have shown that, for a frequency-step input and a constant input noise spectral density, the optimum second-order loop transfer function is given by

$$H(s) = \frac{\sqrt{2}Bs + B^2}{s^2 + \sqrt{2}Bs + B^2} \quad : \text{optimum second-order loop} \tag{10.66}$$

where the coefficient B has dimensions of radians/second and is related to the loop noise bandwidth as

$$B_L = \frac{3B}{4\sqrt{2}} \tag{10.67}$$

where B_L is in Hz. The closed-loop transfer function is related to the loop filter as expressed in Equation (10.6) and the corresponding loop filter response is readily evaluated as

$$F(s) = \frac{\sqrt{2}Bs + B^2}{K_dK_os} \quad : \text{optimum second-order loop} \tag{10.68}$$

Upon equating the coefficients of the optimum second-order loop transfer function to the classical parameters of Equation (10.34) with $\alpha = 0$ results in the optimum damping factor $\varsigma = 1/\sqrt{2} = 0.707$, so, in addition to the qualitative advantages previously mentioned, this damping factor also results in the optimum loop performance under the conditions of the Wiener filter criterion with a frequency step input. The subject of the optimum loop response is visited again when the third-order PLL characteristics are analyzed. Gardner [14] discusses other optimization procedures [12,15,16] based on different loop criterion.

10.6.11 Second-Order Phaselock Loop Acquisition Performance with Signal and Noise

The previous analysis of the PLL assumes that the input signal is a CW wave, possibly with a frequency rate and phase jerk, in a noise-free environment. However, as a practical matter, except for special cases of phase locking requirements among various system oscillators, PLLs are required to operate with data-modulated input signals, additive noise, and fading signal environments. The signal modulation is generally characterized as a complex baseband function with independent data and noise on the quadrature rails. The symbol matched filter outputs are used to derive an error signal that is filtered by the loop filter to form a phase-error estimate. In these cases the loop filter must remove the influence of the noise as well as the self-noise introduced by the random data-modulated carrier. The error signal, generated as shown in Figure 10.5, is formed by taking the M-th power of the signal and the resulting loss in the signal-to-noise ratio is referred to as the *squaring loss*. The squaring loss is discussed in detail in Section 10.9; however, in this section the impact of the squaring loss on the PLLs ability to acquire a modulated signal with additive white Gaussian noise (AWGN) is examined. A critical signal-to-noise ratio (γ_c) is identified, below which, the loop will not acquire or remain locked to the modulated signal. The two key parameters in the determination of γ_c are the signal self-noise and additive noise powers in the loop bandwidth B_L; the input

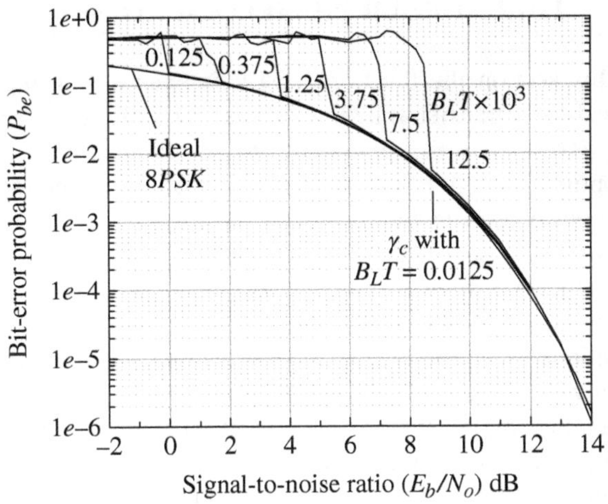

FIGURE 10.17 Phaselock loop acquisition performance with 8PSK modulation.

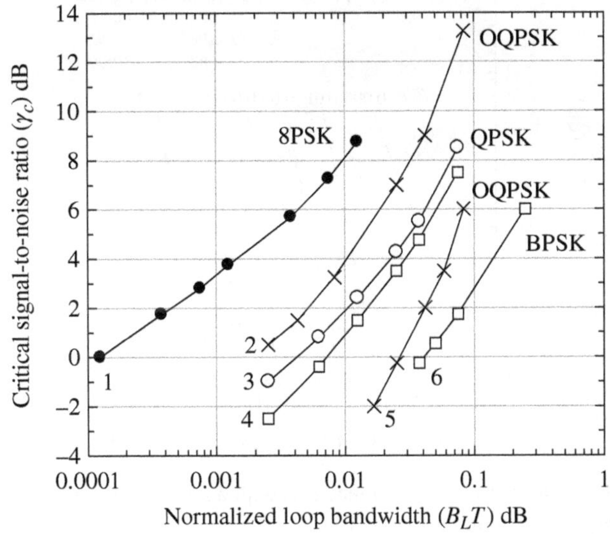

FIGURE 10.18 PLL critical signal-to-noise ratio for acquisition with signal and noise.

signal-to-noise ratio measured in a bandwidth equal to the bit rate, that is, $\gamma_b = E_b/N_o$. The critical signal-to-noise ratio is defined as $\gamma_c = \gamma_b$ (min) where γ_b (min) is the minimum signal-to-noise ratio required to achieve and maintain phase lock with the modulated input signal.

Consider the example shown in Figure 10.17 depicting the PLL acquisition performance of an 8PSK-modulated signal in terms of the bit-error probability as a function of γ_b expressed in decibels with the loop time-bandwidth product (B_LT) as the parameter. The performance results in Figure 10.17 are based on coherent detection with initial conditions of zero phase, frequency, and frequency rate. The critical signal-to-noise ratio is equal to the minimum value of γ_b where the PLL is reliably tracking the carrier as indicated by the bit-error performance being nearly the same as the ideal performance, for example, in Figure 10.17 $\gamma_c = 8.375$ dB for $B_LT = .0125$.

The critical signal-to-noise ratio for this 8PSK example and other waveform modulations is obtained by observing 1000 Monte Carlo acquisition trials for each B_LT product with a signal-to-noise ratio resolution of 0.125 dB and the results are summarized in Figure 10.18. The simulated performance evaluation of the critical signal-to-noise ratio also applies for initial phase errors less than ±20° to ensure that the PLL is operating in the region to avoid cycle skipping during acquisition. In addition to the waveform modulation, the performance is also dependent on the form of the PLL phase detector. In Section 10.8 it is shown that the optimum phase detector for a minimum mean-square phase estimate is the tanh(x) function. Figure 10.18 shows the critical signal-to-noise ratio using the tanh(x) small argument approximation to the phase detector for 8PSK waveform modulation. Table 10.2 summarizes the critical signal-to-noise ratio for various modulations using the tanh(x) phase detector and approximations. These results suggest that the PLL will acquire a modulated signal at an arbitrarily low signal-to-noise ratio if the B_LT product is sufficiently low. The issue is that lowering the B_LT product severely restricts the PLL *pull-in frequency* range requiring very accurate initial frequency estimation during the waveform acquisition processing. Techniques for operating in very low signal-to-noise environments are discussed in Chapter 11.

TABLE 10.2 Phase Detector Functions Associated with Figure 10.18

Note	Phase Detector	Comment
1	$x - x^2/3$	Small argument
2	sign(x)	Large argument
3	$x - x^2/3$	Small argument
4	sign(x)	Large argument
5	tanh(x)	Exact
6	tanh(x)	Exact

10.6.12 Second-Order Phaselock Loop Performance Loss with Frequency-Rate Input

When the input signal to a second-order PLL has a fixed frequency rate the steady-state output of the loop results in a fixed phase error given by Equation (10.57) and expressed as

$$\phi = \frac{\pi \dot{f} T^2 (\varsigma + 1/4\varsigma)^2}{2(B_LT)^2} \quad \text{: radians} \qquad (10.69)$$

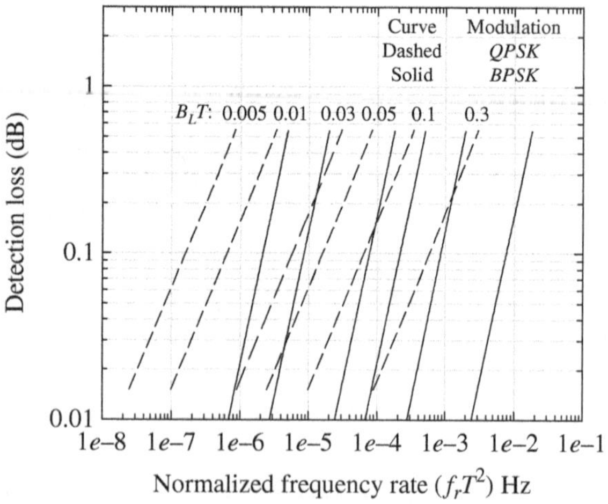

FIGURE 10.19 Approximate loss with a second-order PLL with frequency-rate input ($\zeta = 0.707$).

This fixed phase error will degrade the bit-error performance of a coherently detected BPSK and QPSK for small angles according to the approximations

$$\text{Loss (dB)} \cong \begin{cases} 20\log_{10}(|\cos(\varphi)|) & : \text{BPSK} \\ 20\log_{10}(|\cos(\varphi) - \sin(\varphi)|) & : \text{QPSK} \end{cases} \tag{10.70}$$

Using Equations (10.69) and (10.70) the detection loss is plotted in Figure 10.19* as function of the normalized frequency-rate error $\dot{f}T^2$ for a typical range of PLL time-bandwidth products. The angle error is limited to the range $0 \le \phi \le 0.26$ rad (15°) so Equation (10.70) corresponds to the maximum loss in signal power and the upper limit on ϕ ensures that the approximations apply. The results show that QPSK is considerably less tolerable to a second-order PLL operating with an input signal having a frequency-rate error. For example, consider a source data rate of $R_b = 9.6$ kbps. With BPSK modulation a typical loop time-bandwidth product of $B_LT = 0.1$ will exhibit a 0.5 dB performance loss when $\dot{f}T^2 = 0.002$ corresponding to $\dot{f} = 184$ kHz/s, whereas, with QPSK the typical loop time-bandwidth product is $B_LT = 0.01$ that results in a 0.5 dB performance loss when $\dot{f}T^2 = 4 \times 10^{-6}$ corresponding to $\dot{f} = 92$ Hz/s. In this regard, two factors work against the QPSK modulation: the B_LT product must be about 10 : 1 lower for comparable acquisition and tracking performance and the symbol rate is lower than that of BPSK by a factor of two. Typical Doppler rates for ultra-high frequency (UHF) satellite communication links are on the order of 32 Hz/s and, in some cases, an additional detection loss of 1 dB is provided.

*The B_LT labels in Figure 10.19 correspond to the *dashed* (QPSK) curves but can be shifted to the right and applied to the *solid* (BPSK) curves, also $f_r = \dot{f}$.

10.7 THIRD-ORDER PHASELOCK LOOP

In this section the third-order PLL is examined using the closed-loop transfer function derived by Jaffe and Rechtin [13]. Their analysis applies the Wiener filter criterion [17] to optimize the third-order closed-loop transfer function for an input phase function characterized by a frequency ramp. As in the case of the first- and second-order PLLs, the closed-loop response is dependent upon the input signal level; however, with AGC or a signal amplitude limiting technique, the optimum performance will be maintained for a wide range of input signal levels. Pool [18] examines an analog implementation of the optimum third-order loop using a bandpass limiter for amplitude control and found that the circuit component values were impractical; however, he suggested a practical implementation that closely approximates the optimum transfer function. Pool observed that the acquisition characteristics of the third-order PLL are degraded and suggests using a second-order loop for acquisition and then switching to the third-order loop for tracking.

As stated previously, when using infinite gain integrators, the third-order PLL, unlike the first- and second-order loops, has a zero steady-state phase error when a frequency ramp is applied to the input making its use appealing in more dynamic signal environments. The stability concerns of the third-order loop are mitigated to some extent in digital designs where control over the parameter is more certain; however, the impact of coefficient quantization of the parameter values is always a concern.

10.7.1 Optimum Third-Order Phaselock Loop Configuration

Based on the Wiener filter criterion Jaffe and Rechtin show that, for a frequency ramp input and a constant input phase-noise power spectral density (PSD), the optimum third-order closed-loop transfer function is

$$H_{opt}(s) = \frac{2Bs^2 + 2B^2s + B^3}{s^3 + 2Bs^2 + 2B^2s + B^3} \quad : \text{optimum} \tag{10.71}$$

Although an exact implementation of the corresponding loop filter, $F(s)$, is possible, Pool points out that, for practical realizations, the extreme component values are difficult if not impossible to fabricate. Furthermore, digital implementations of this optimum filter require negative gain components. Pool then proceeds to characterize a suboptimum third-order filter as a cascade of two optimum second-order filters. This approach results in a practical implementation that Gupta [19] and Gupta and Solem [20] have examined and shown to result in nearly optimum performance for a frequency-ramp input. Pool's practical and nearly optimum transfer function is given by

$$H_{near\ opt}(s)=\frac{2Bs^2+2\sqrt{2}B^2s+B^3}{s^3+2Bs^2+2\sqrt{2}B^2s+B^3}\quad\text{: near-optimum}\tag{10.72}$$

Pool defines the threshold filter gain $G_o=(K_oK_d)_{min}$ as the minimum gain required for the loop to maintain phase lock* and expresses the loop filter transfer function as

$$F(s)=\frac{1}{G_o}\left[\frac{2Bs^2+2\sqrt{2}B^2s+B^3}{s^2}\right]\tag{10.73}$$

Using Equation (10.73) to evaluate the closed-loop transfer function and defining $G=K_oK_d$ results in

$$H(s)=\frac{(G/G_o)(2Bs^2+2\sqrt{2}B^2s+B^3)}{s^3+(G/G_o)(2Bs^2+2\sqrt{2}B^2s+B^3)}\tag{10.74}$$

So the optimum closed-loop transfer function, as given by either Equation (10.71) or (10.72), results when $G=G_o$ and is referred to as the threshold closed-loop transfer function.

The parameter B in these expressions has dimensions of rad/s and is used to normalize the Laplace transform variable s by letting $x=s/B$, resulting in the convenient normalized expressions

$$H(x)=\frac{(G/G_o)(2x^2+2\sqrt{2}x+1)}{x^3+(G/G_o)(2x^2+2\sqrt{2}x+1)}\tag{10.75}$$

and

$$F(x)=\frac{B}{G_o}\left[\frac{2x^2+2\sqrt{2}x+1}{x^2}\right]\tag{10.76}$$

The normalized pole-zero locations of $H(x)$ are given by

$$H(x)=\frac{(x-z_0)^2}{(x-p_0)(x-p_1)(x-p_2)}\tag{10.77}$$

where

$$\begin{aligned}z_0&=-\alpha_0'\\p_0&=-\alpha_0\\p_1&=-\alpha_1+j\beta_1\\p_2&=p_1^*=-\alpha_1-j\beta_1\end{aligned}\tag{10.78}$$

With $G/G_o=1$ the repeated zeros corresponding to Equation (10.77) are located at $z_0=-1/\sqrt{2}$ and the pole locations are as follows:

$$\begin{aligned}p_0&=-0.475358\\p_1&=-0.762322+j1.23391\\p_2&=-0.762322-j1.23391\end{aligned}\tag{10.79}$$

and the parameters in Equation (10.78) are:

$$\begin{aligned}\alpha_0'&=0.707107\\\alpha_0&=0.475358\\\alpha_1&=0.762322\\\beta_1&=1.23391\end{aligned}\tag{10.80}$$

The root-locus plot is used to examine the stability and transient characteristics of the closed-loop response by plotting the pole migration of $H(x)$ in the normalized s-plane as the loop gain changes from zero to infinity. Defining the normalized gain as $\rho=G/G_o$ the poles are given as the solution to the denominator polynomial of $H(x)$ as expressed in Equation (10.75) with the result

$$q(x)=x^3+2\rho x^2+2\sqrt{2}\rho x+\rho=0\tag{10.81}$$

The zeros of Equation (10.81) are evaluated using the trigonometric solution given by Eshbach [21] and the results are plotted in Figure 10.20 as the gain parameter ρ varies between zero and infinity. The poles of $H(x)$ start at the three poles of the open-loop response located at the origin and migrate through the normalized s-plane as the gain increases to infinity and end up at the three zeros of the open-loop response as indicated in Figure 10.20. One pole transverses the negative real axis to one of the repeated zeros and the other two depart from the real axis forming complex conjugate poles that bend around to meet the negative real axis at a

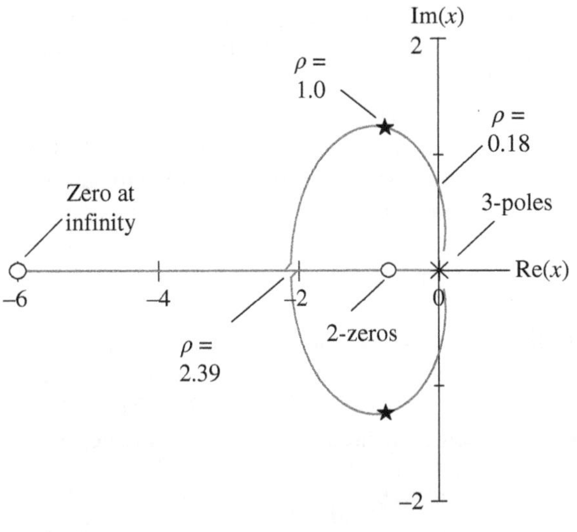

FIGURE 10.20 Root-locus plot for Pool's near-optimum closed-loop transfer function.

*Based on experimental observations, Jaffe and Rechtin note that this occurs when the *rms* phase error out of the VCO is equal to one radian.

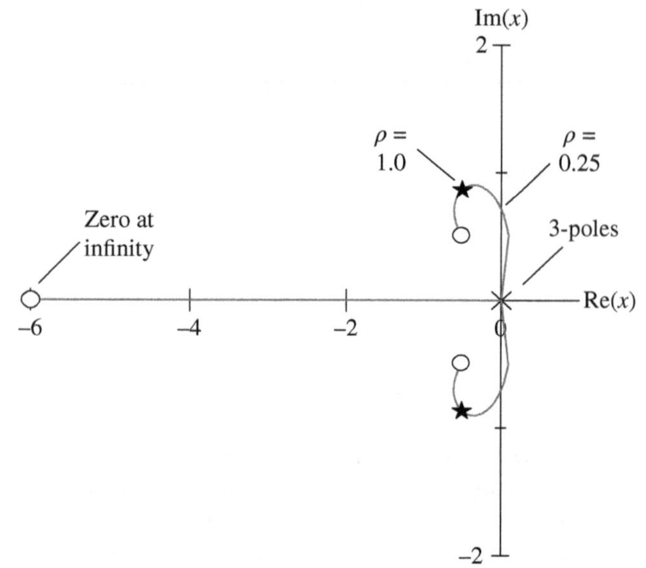

FIGURE 10.21 Root-locus plot for Jaffe and Rechtin optimum closed-loop transfer function.

normalized gain of $\rho = 2.39$ at which point one pole follows the negative real axis to the zero at infinity and the other to the second repeated zero of the open-loop response.* The complex poles that emerge from the origin initially enter the right half of the normalized *s*-plane leading to instability in the closed-loop response; however, a stable system is achieved for normalized gains greater than $\rho = 0.18$ where the complex poles enter into the left half of the normalized *s*-plane. The near optimum or threshold transfer function given by Pool applies for $\rho = 1.0$ and Figure 10.20 indicates that the system is unconditionally stable for $\rho \geq 0.18$. Based on the earlier description, values of $\rho < 1.0$ or $G < G_o$, the system loses phase lock so the question of stability based on the root-locus criteria is of little consequence.

The preceding analysis is repeated for the optimum threshold response given by Equation (10.71) and the root-locus plot is shown in Figure 10.21. In this case, the real part of the poles at $\rho = 1.0$ is somewhat closer to the imaginary axis leading to a more responsive transient response and, as seen from the plot, the response is more sensitive to gain changes. Pool focuses on analog implementations of the third-order PLL and notes that the component values required to implement the optimum loop are difficult, if not impossible, to fabricate leading to his near optimum implementation.

In the following sections, the transient characteristics of the phase response to several input functions are examined using the phase-error transfer function *T*(*x*). These theoretical results are then compared to computer-simulated responses of the third-order PLL as a precursor to the evaluation of the performance with various modulated waveforms.

10.7.2 Evaluation of Loop Bandwidth

The first topic of interest is the relationship of the closed-loop noise bandwidth to the filter coefficients. The loop bandwidth is defined in terms of the one-sided bandwidth of the equivalent low-pass response and is given by†

$$B_L = \frac{1}{2\pi}\int_0^{\infty} |H(\omega)|^2 d\omega \quad \text{: hertz} \tag{10.82}$$

where

$$|H(\omega)|^2 = H(j\omega)H^*(j\omega) = H(j\omega)H(-j\omega) \tag{10.83}$$

Upon letting $s = j\omega$, the bandwidth is expressed in term of the *s*-plane poles as

$$B_L = \frac{1}{2}\left[\frac{1}{2\pi j}\int_{c-j\infty}^{c+j\infty} |H(s)|^2 ds\right] \tag{10.84}$$

The integral inside of the brackets is evaluated using the theory of residues and, in doing so, the magnitude-squared term (10.84) is expressed as $|H(s)|^2 = H(s)H(-s)$. The poles of *H*(*s*) are in the left half of the *s*-plane, whereas, those of *H*(−*s*) are mirrored about the $j\omega$ axis and are in the right half of the *s*-plane. In evaluating the integral along the $j\omega$ axis, the contour of integration is closed by following a semicircular arc either counterclockwise to include the *left half plane* (LHP) poles or clockwise to include the *right half plane* (RHP) poles, in either event only the residues, R_i, of the enclosed poles will influence the integral. Enclosing the LHP poles is implicit in the integral of Equation (10.84) by virtue of the positive offset *c* to account for poles lying on the $j\omega$ axis; however, under the conditions being considered, there are no poles on the $j\omega$ axis so $c = 0$ will include all of the poles of interest in the transfer function. In the following analysis, the residues of the LHP poles are used and the integral is evaluated as

$$\frac{1}{2\pi j}\int_{c-j\infty}^{c+j\infty} H(s)H(-s)ds = \sum_{i=0}^{2} R_i \tag{10.85}$$

In terms of the normalized parameter *x*, the equivalent expressions are

*The migration of individual poles with changing loop gain follows the rules set forth in Reference 22.

†This definition is based on a unit gain closed-loop response, that is, $|H(0)|^2 = 1$.

$$B_L = \frac{B}{2}\left[\frac{1}{2\pi j}\int_{c/B-j\infty}^{c/B+j\infty} |H(x)|^2 dx\right] \quad (10.86)$$

and

$$\frac{1}{2\pi j}\int_{c/B-j\infty}^{c/B+j\infty} H(x)H(-x)dx = \sum_{i=0}^{2} R_i' \quad (10.87)$$

The residue of the i-th LHP pole is evaluated as

$$R_i' = H(x)H(-x)(x-p_i)|_{x=p_i}$$
$$= \frac{4(x-z_0)(-x-z_0)(x-p_i)}{(x-p_0)(x-p_1)(x-p_2)(-x-p_0)(-x-p_1)(-x-p_2)}\Bigg|_{x=p_i} \quad (10.88)$$

Upon substituting the pole locations given above the loop bandwidth for Pool's near optimum threshold transfer function is expressed in terms of the parameter B as*

$$B_L = 0.929475B \quad : \text{near optimum} \quad (10.89)$$

Similarly the expression for the optimum threshold transfer function of Jaffe and Rechtin results in

$$B_L = 0.83333B \quad : \text{optimum} \quad (10.90)$$

These are important relationship and will be used in the simulation program to match the simulated responses of the third-order loop to the equivalent theoretical responses. This matching of the responses provides a method of validating the correct loop parameters in simulation programs, digital signal processors (DSP), or field programmable gate array (FPGA) hardware implementations.

10.7.3 Theoretical Responses of the Third-Order Phase-Lock Loop

As mentioned earlier the theoretical responses of the third-order loop under different input conditions provide a baseline of comparison to validate the correct software or hardware implementations. To this end the response of the phase error, $\phi_e(s)$, to a phase step, frequency step (phase ramp), and frequency ramp input is examined. The analysis procedures are similar to those discussed in the context of the second-order loop; however, in those cases, the resulting expressions are based upon the previously published expressions for the inverse Laplace transforms. For the third-order loop, the procedures for finding the inverse Laplace transform using the normalized Laplace variable x are outlined in this section and the time responses are evaluated in terms of the known normalized pole-zero locations as identified earlier.

The unit-step input is characterized as u_n/s^n where $u_n = \{\Delta\phi, \Delta f, \Delta\dot{f}, \Delta\ddot{f}\}$ for $n = 1, 2, 3, 4$ corresponding to a phase step, frequency step, frequency ramp, and phase jerk input, respectively. Using these inputs the phase error is evaluated as

$$\phi_e(s) = \frac{u_n}{s^n}[1-H(s)]$$
$$= u_n\left[\frac{s^{3-n}}{s^3 + 2Bs^2 + 2\sqrt{2}B^2 s + B^3}\right] \quad (10.91)$$

Using Equation (10.91) the time response is the inverse Laplace transform of $\phi_e(s)$, expressed as

$$\phi_e(t) = \frac{1}{2\pi j}\int_{c-j\infty}^{c+j\infty} \phi_e(s)e^{st}ds$$
$$= \frac{u_n}{2\pi j}\int_{c-j\infty}^{c+j\infty}\left[\frac{s^{3-n}}{s^3 + 2Bs^2 + 2\sqrt{2}B^2 s + B^3}\right]e^{st}ds \quad (10.92)$$

In terms of the normalized variable $x = s/B$ (10.92) becomes†

$$\phi_e(y) = \frac{(u_n/B^{n-1})}{2\pi j}\int_{c-j\infty}^{c+j\infty}\left[\frac{x^{3-n}}{x^3 + 2x^2 + 2\sqrt{2}x + 1}\right]e^{xy}dx \quad (10.93)$$

where the normalized time variable is $y = tB$. Using partial fraction expansion techniques [22] or Heaviside's expansion theorem [23] to evaluate the contributions from individual poles and pole pairs in the integrand of Equation (10.93) results in the following expression for $n \le 3$‡

$$\frac{x^{3-n}}{(x+\alpha_0)\left((x+\alpha_1)^2+\beta_1^{\,2}\right)} = \frac{C_{n0}}{(x+\alpha_0)} + \frac{C_{n1}x + C_{n2}}{\left((x+\alpha_1)^2+\beta_1^{\,2}\right)}$$
$$= \frac{C_{n0}}{(x+\alpha_0)} + \frac{C_{n1}(x+\alpha_1)}{\left((x+\alpha_1)^2+\beta_1^{\,2}\right)}$$
$$+ \frac{-C_{n1}\alpha_1 + C_{n2}}{\left((x+\alpha_1)^2+\beta_1^{\,2}\right)} \quad (10.94)$$

*Equation (10.89) is close to Gupta's result, approximated as $2B_L \approx 1.859B$.

†The normalized responses result from a scale change in the Laplace transformation: $f(ta) \leftrightarrow aF(s/a)$.

‡In general, to apply this expansion n must be $\le$ to the order of the loop. For n greater than the order of the loop, additional poles are included at the origin of the x-plane.

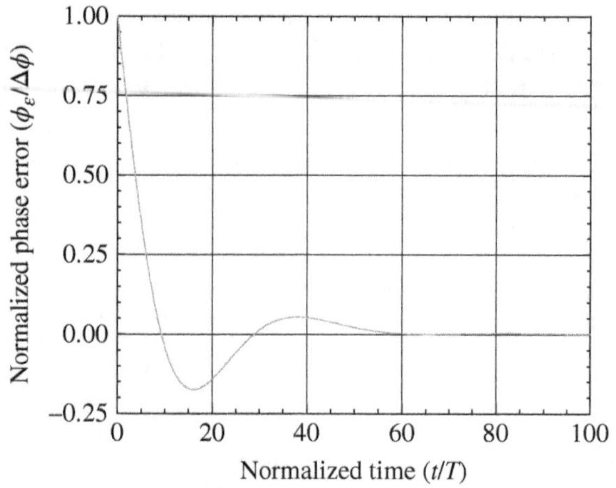

FIGURE 10.22 Third-order PLL response to phase-step input.

The coefficients C_{n0}, C_{n1}, and C_{n2} are determined from the earlier expansion by multiplying both sides by the denominator terms on the left side and collecting common terms in x. The result is

$$x^{3-n} = [C_{n0} + C_{n1}]x^2 + [2\alpha_1 C_{n0} + \alpha_0 C_{n1} + C_{n2}]x + \left[\left(\alpha_1^2 + \beta_1^2\right)C_{n0} + \alpha_0 C_{n2}\right] \tag{10.95}$$

The coefficients are determined by solving the resulting set of simultaneous equations for a specified value of n. Upon applying tabulated inverse Laplace transforms [24] to these elementary terms, resulting from the expansion of the integrand of Equation (10.93), the normalized phase-error response becomes

$$\frac{\phi_\varepsilon(t)}{u_n/B^{n-1}} = C_1 e^{-\alpha_0 Bt} + C_1 e^{-\alpha_1 Bt}\cos(\beta_1 Bt) + \left(\frac{C_2}{\beta_1}\right)e^{-\alpha_1 Bt}\sin(\beta_1 Bt) \tag{10.96}$$

In the following subsections, the time response for the identified inputs is characterized and the corresponding responses are plotted in Figures 10.22 through 10.24. The abscissas in these figures are normalized as t/T using the transformation $t/T = y/BT$.

10.7.4 Phase-Error Response to Phase Step

In this case $n = 1$, $u_1 = \Delta\phi$ and, upon evaluation of the coefficients in Equation (10.96) under these conditions, the normalized time response to a phase-step input is expressed by Equation (10.97) and plotted in Figure 10.22.

$$\frac{\phi_\varepsilon(y)}{\Delta\phi} = \left[\frac{\alpha_0^2}{(\alpha_0 - \alpha_1)^2 + \beta_1^2}\right]e^{-\alpha_0 y} + \left[\frac{(\alpha_0 - \alpha_1)^2 + \beta_1^2 - \alpha_0^2}{(\alpha_0 - \alpha_1)^2 + \beta_1^2}\right]e^{-\alpha_1 y}\cos(\beta_1 y) + \left[\frac{-\alpha_0\left(\alpha_1^2 + \beta_1^2\right)}{\left((\alpha_0 - \alpha_1)^2 + \beta_1^2\right)\beta_1}\right]e^{-\alpha_1 y}\sin(\beta_1 y) \tag{10.97}$$

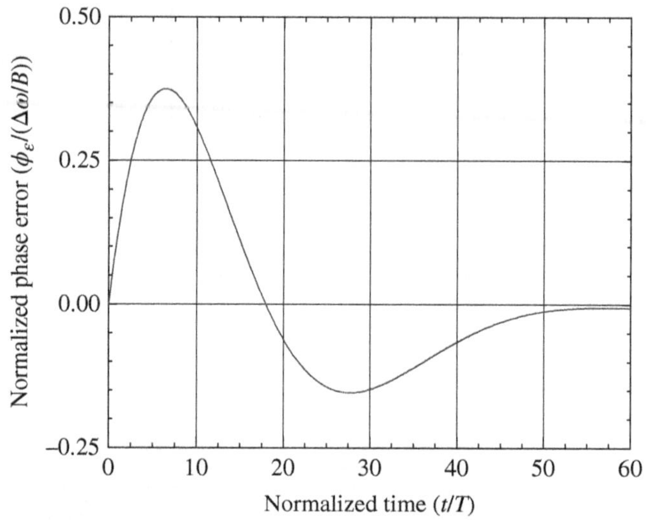

FIGURE 10.23 Third-order PLL response to frequency-step input.

10.7.5 Phase-Error Response to Frequency Step

In this case $n = 2$, $u_2 = \Delta\omega$, and, upon evaluation of the coefficients in Equation (10.96) under these, the normalized time response to a frequency-step input is expressed as

$$\frac{\phi_\varepsilon(y)}{\Delta\omega/B} = \left[\frac{-\alpha_0}{(\alpha_0 - \alpha_1)^2 + \beta_1^2}\right]e^{-\alpha_0 y} + \left[\frac{\alpha_0}{(\alpha_0 - \alpha_1)^2 + \beta_1^2}\right]e^{-\alpha_1 y}\cos(\beta_1 y) + \left[\frac{\left(\alpha_1^2 + \beta_1^2\right)/\beta_1}{(\alpha_0 - \alpha_1)^2 + \beta_1^2}\right]e^{-\alpha_1 y}\sin(\beta_1 y) \tag{10.98}$$

Equation (10.98) is plotted in Figure 10.23.

10.7.6 Phase-Error Response to Frequency Ramp

In this case $n = 3$, $u_3 = \Delta\dot{\omega}$ and, upon evaluation of the coefficients, the normalized time response to a frequency-ramp input is given by

$$\frac{\phi_\varepsilon(y)}{\Delta\dot{\omega}/B^2} = \left[\frac{1}{(\alpha_0 - \alpha_1)^2 + \beta_1^2}\right]e^{-\alpha_0 y} - \left[\frac{1}{(\alpha_0 - \alpha_1)^2 + \beta_1^2}\right]e^{-\alpha_1 y}\cos(\beta_1 y) + \left[\frac{(\alpha_0 - 2\alpha_1)}{\left((\alpha_0 - \alpha_1)^2 + \beta_1^2\right)\beta_1}\right]e^{-\alpha_1 y}\sin(\beta_1 y) \tag{10.99}$$

Equation (10.99) is plotted in Figure 10.24.

10.7.7 Phase-Error Response to Phase Jerk

In this case $n = 4$ and $u_3 = \Delta\ddot{\omega}$ and, because $n > 3$, the partial fraction expansion is evaluated by adding and subtracting

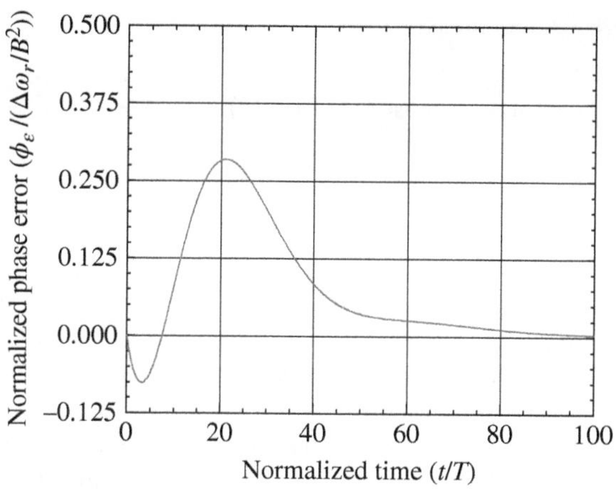

FIGURE 10.24 Third-order PLL response to frequency-ramp input.

$C_{n2}\alpha_1$ and applying partial fraction expansion in terms of the coefficients C_{ni} with the result

$$\frac{1}{x(x+\alpha_0)\left((x+\alpha_1)^2+\beta_1^2\right)}=\frac{C_{n0}}{x}+\frac{C_{n1}}{(x+\alpha_0)}+\frac{C_{n2}(x+\alpha_1)}{\left((x+\alpha_1)^2+\beta_1^2\right)}+\frac{-C_{n2}\alpha_1+C_{n3}}{\left((x+\alpha_1)^2+\beta_1^2\right)} \quad (10.100)$$

Upon evaluation of the coefficients (10.100), the normalized phase-error response to a phase jerk input is given by Equation (10.101) and plotted in Figure 10.25.

$$\frac{\phi_\varepsilon(y)}{\Delta\ddot{\omega}/B^3}=\frac{1}{\alpha_0\left(\alpha_1^2+\beta_1^2\right)}-\left[\frac{1}{\alpha_0\left((\alpha_0-\alpha_1)^2+\beta_1^2\right)}\right]e^{-\alpha_0 y}+\left[\frac{2\alpha_1-\alpha_0}{\left(\alpha_1^2+\beta_1^2\right)\left((\alpha_0-\alpha_1)^2+\beta_1^2\right)}\right]e^{-\alpha_1 y}\cos(\beta_1 y)-\left[\frac{(\alpha_0+\alpha_1)^2+\beta_1^2-4\alpha_1^2-\alpha_0^2}{\left(\alpha_1^2+\beta_1^2\right)\left((\alpha_0-\alpha_1)^2+\beta_1^2\right)}\right]e^{-\alpha_1 y}\sin(\beta_1 y) \quad (10.101)$$

10.7.8 Loop Filter Implementation

The third-order PLL filter is implemented using two cascaded integrators as shown in Figure 10.26 with the transfer functions evaluated as

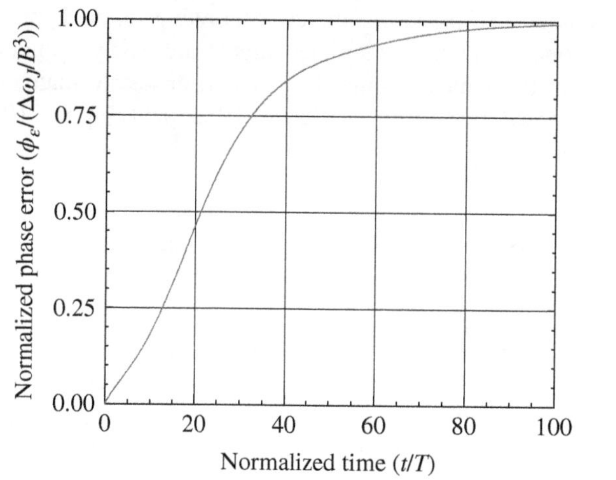

FIGURE 10.25 Third-order PLL response to a phase jerk input.

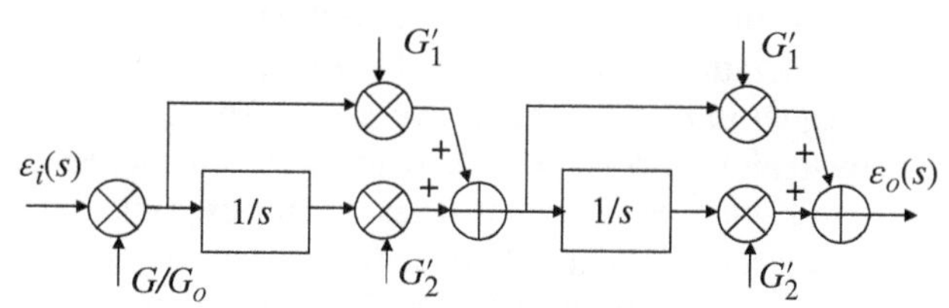

FIGURE 10.26 Loop filter implementation for the third-order phaselock loop.

$$F(s)=\frac{1}{G_o}\left[\frac{G_1^2s^2+2G_1G_2s+G_2^2}{s^2}\right] \quad (10.102)$$

Upon equating the coefficients in Equation (10.102) with those of the near-optimum loop filter, as expressed in Equation (10.73), the filter gains are evaluated as

$$G_1=\sqrt{2B} \text{ and } G_2=\sqrt{B^3} \quad (10.103)$$

Expressed in terms of the loop filter bandwidth B_L these gains become

$$G_1=\sqrt{2.1518B_L} \text{ and } G_2=\sqrt{1.2453B_L^3} \quad (10.104)$$

The normalized gains are denoted as the *primed* values given by

$$G_1'=\sqrt{2.1518B_LT} \text{ and } G_2'=\sqrt{1.2453(B_LT)^3} \quad (10.105)$$

As mentioned earlier, the gain G is adjusted to equal the gain G_o resulting in the threshold response. This gain is adjusted in hardware and software implementations or in computer simulations until the response matches the theoretical response for the corresponding input. Typically, the phase and frequency step inputs are easily implemented so the gain is adjusted to match, or nearly match, the theoretical responses in Figure 10.22 and Figure 10.23, respectively.*

10.8 OPTIMUM PHASE TRACKING ALGORITHMS

In the previous sections, the carrier phase acquisition and tracking is based on classical feedback control theory. In this analysis an estimate of the received carrier frequency is used to mix the received signal to baseband, and the resulting in-phase and quadrature signals are used in a Costas [25] loop feedback implementation. The random data-dependent signal phase is removed by raising the received signal, that is, the transmitted signal plus noise, to an appropriate power and the resulting carrier phase estimate is obtained from the discriminator formed by the quadrature baseband signal $\sin(\phi_e)$; this is approximately equal to ϕ_e for small phase errors. The Costas loop implementation also acquires and tracks a time-dependent baseband signal phase error that results, for example, when the receiver local oscillator frequency is not exactly equal to the received signal frequency. The intuitive solution to the phase detector using the arctan(x) function, where x is the ratio of the quadphase to in-phase baseband components, results in significantly degraded performance at low input signal-to-noise ratios. Because the PLL is essential in most communication demodulators it is important to establish an optimum implementation that applies for all signal-to-noise ratios.

In this section the phase tracking algorithms, based on the maximum a posteriori probability (MAP) estimation procedure, are examined and it is found that the optimum solution involves a signal-to-noise weighted hyperbolic tangent phase detector. The following analysis is based largely on the work of Simon [26–30], and Omura and Simon [31]. For this analysis the received signal from an AWGN channel is considered to be a sequence of K-modulated data symbols expressed as

$$y(t)=\sum_{k=0}^{K-1} y_k(t)rect\left(\frac{t-kT}{T}\right)+n(t)$$
$$=\sum_{k=0}^{K-1} Ap(t-kT)\cos(\omega_o t+\phi_k+\phi)rect\left(\frac{t-kT}{T}\right)+n(t) \quad (10.106)$$

*The gain adjustment is typically required in hardware and software implementations where the phase-lock loop gains K_o and K_d are not known and are difficult to determine. On the other hand, in computer simulations these gains are more easily determined so this adjustment procedure can be used as a check. In either case, the responses should be verified against the theoretical responses.

where $y_k(t)$, expressed as,

$$y_k(t)=Ap(t-kT)\cos(\omega_o t+\phi_k+\phi) \quad : kT-\frac{T}{2}\le t\le kT+\frac{T}{2} \quad (10.107)$$

The function $y_k(t)$ describes the k-th received symbol at the frequency f_o Hz with peak amplitude of A volts and data-dependent phase ϕ_k. The phase ϕ is the unknown phase to be estimated. The frequency $f_o=\omega_o/2\pi$ represents a convenient receiver IF and typically is the input frequency to the demodulator. The function $p(t)$ represents symbol shaping function and is confined to one symbol interval T, such that, $kT-T/2\le t\le kT+T/2$. For BPSK, QPSK, and OQPSK $p(t)=rect(t/T)$; however, with OQPSK modulation, the weighting function in the quadrature channel is offset from the in-phase channel by $T/2$. For MSK modulation $p(t)=\cos(\omega_m t)$ with $\omega_m=\pi/T$ and, like OQPSK, the quadrature channel is offset in time by $T/2$. The AWGN $n(t)$ is denoted as $N(0,\sigma_n)$ with zero-mean and variance σ_n^2. With the understanding that the k-th received symbol interval is characterized by the $rect((t-kT)/T)$ function as expressed in Equation (10.107).

The subject of this section is the characterization of the signal processing required to compute the optimum phase estimate, $\hat{\phi}$, of the unknown received signal phase using the MAP procedure. As in the previous sections, it is assumed that the exact symbol timing is known by the demodulator. Although in the following analysis the unknown phase is treated as a constant, in practice it may be a time-dependent phase expressed, for example, as $\phi(t)=\omega_d t+\phi$. For the PLL to acquire and track the received signal the frequency error must not exceed the pull-in frequency of the loop. The time-dependent phase error may also include frequency rate and phase jerk depending on the conditions of the channel and various oscillators. The results of the following analysis indicate that for low signal-to-noise environments the squaring function of the Costas PLL approaches the performance of the MAP phase detector, whereas, for high signal-to-noise ratios, the decision directed PLL approaches the performance of the MAP phase detector.

Given the a posteriori pdf of the phase conditioned on the received signal, the MAP estimation procedure for the optimal estimate of ϕ is expressed as†

$$\hat{\phi}=\max_{\phi}\,[p(\varphi|y(t))]=\max_{\phi}\,[\ln(p(\varphi|y(t)))] \quad (10.108)$$

where the second expression involves the natural log function as used in the log-likelihood ratio test (LLRT). Applying Bayes rule, the second equality in Equation (10.108) is

†The analysis focuses on the estimation of the unknown fixed phase error ϕ.

evaluated in terms of the *pdf* of the received signal conditioned on the unknown phase and is expressed as

$$\hat{\phi} = \max_{\phi}\left[\ln\left(\frac{p(y(t)\,|\,\phi)p(\phi)}{p(y(t))}\right)\right] = \max_{\phi}[\ln(p(y(t)\,|\,\phi)p(\phi))] \tag{10.109}$$

The second expression follows because $p(y(t))$ is independent of ϕ and, therefore, does not influence the maximization. Assuming that the unknown signal phase is uniformly distributed between $\pm\pi$, that is, the a priori probability is $p(\phi)=1/2\pi$, Equation (10.109) simplifies to

$$\hat{\phi} = \max_{\phi}[\ln p(y(t)\,|\,\phi)] \tag{10.110}$$

At this point the received signal is expressed as a sequence of contiguous symbols and noise given by

$$y(t) = \{y_o(t)+n_o(t),\ldots,y_k(t)+n_k(t),\ldots\} \tag{10.111}$$

where $y_k(t)$ is expressed by Equation (10.107). Because the data in each symbol and the associated noise is independent, the MAP estimation, as expressed in Equation (10.109), becomes

$$\begin{aligned}\hat{\phi} &= \max_{\phi}\left[\prod_{k=0}^{K-1} p(y(t)\,|\,\phi,\phi_k)p(\phi_k)\right]\\ &= \max_{\phi}\left[\prod_{k=0}^{K-1}\ln(p(y(t)\,|\,\phi,\phi_k)p(\phi_k))\right]\end{aligned} \tag{10.112}$$

In Equation (10.112) the dependence of the data-modulated phase angle ϕ_k is explicit. The received signal and noise samples are considered to be vectors spanning one symbol interval T such that

$$y_k = (y_k(t_1), y_k(t_2),\ldots,y_k(t_N)) \tag{10.113}$$

and

$$n_k = (n_k(t_1), n_k(t_2),\ldots,n_k(t_N)) \tag{10.114}$$

where $t_1,t_2,\ldots,t_N \in [kT-T/2,\ kT+T/2]$: $t_{k+1}-t_k=\Delta T\ \forall k$ and $n_k(t_n)$ are *iid* zero-mean Gaussian random variables with variance $\sigma_n^2 = N_o/(N\Delta T)=N_oB$ where $N\Delta T=T$ and $B=1/T$ is the bandwidth of the modulated symbol. The received signal vector is defined as

$$Y_k = y_k + n_k \tag{10.115}$$

and under these conditions, the *pdf* of the received signal vector is evaluated as

$$\begin{aligned}p_Y(Y_k|\phi,\phi_k) &= \prod_{n=1}^{N} p_N(n_k(t_n))\\ &= \frac{1}{(2\pi N_oB)^{N/2}}\prod_{n=1}^{N}\exp\left(-\frac{(Y_k(t_n)-Ap(t_n-kT)\cos(\omega_o t_n+\phi_k+\phi))^2\Delta T}{2N_o}\right)\\ &= \frac{1}{(2\pi N_oB)^{N/2}}\exp\left(-\frac{1}{2N_o}\sum_{n=1}^{N}(Y_k(t_n)-Ap(t_n-kT)\cos(\omega_o t_n+\phi_k+\phi))^2\Delta T\right)\\ &= \frac{1}{(2\pi N_oB)^{N/2}}\exp\left(-\frac{1}{2N_o}\int_{kT-T/2}^{kT+T/2}(Y_k(t)-Ap(t-kT)\cos(\omega_o t+\phi_k+\phi))^2dt\right)\end{aligned} \tag{10.116}$$

The equality in the last expression in Equation (10.116) applies under the limiting conditions $\Delta T\rightarrow 0$ and $N\rightarrow\infty$. The dependence upon the data phase ϕ_k is removed using the *total probability law* with the result

$$p_Y(Y_k|\phi)=\sum_{\phi_k}p_Y(Y_k|\phi,\phi_k)p_\phi(\phi_k)=\frac{1}{(2\pi N_oB)^{N/2}}$$
$$\sum_{\phi_k}\exp\left(\frac{-1}{2N_o}\int_{kT-T/2}^{kT+T/2}(Y_k(t)-Ap(t-kT)\cos(\omega_o t+\phi_k+\phi))^2dt\right)p_\phi(\phi_k) \tag{10.117}$$

At this point, the integrand in Equation (10.117) is simplified by evaluating the squared term as

$$\begin{aligned}&(Y_k(t)-Ap(t-kT)\cos(\omega_o t+\phi_k+\phi))^2\\ &= Y_k^2(t)-2AY_k(t)p(t-kT)\cos(\omega_o t+\phi_k+\varphi)\\ &\quad+(A^2/2)p^2(t-kT)+(A^2/2)p^2(t-kT)\cos(2\omega_o t+2\phi_k+2\phi)\end{aligned} \tag{10.118}$$

Subsequent low-pass or zonal filtering removes the double frequency term and the first and third terms on the right-hand side (*rhs*) of Equation (10.118) are independent of the data phase so only the second term is of interest. Therefore, associating the first and third terms with the data independent constant C, Equation (10.117) simplifies to

$$p_Y(Y_k|\phi)=\frac{C}{(2\pi N_oB)^{N/2}}$$
$$\sum_{\phi_k}\exp\left(-\frac{A}{N_o}\int_{kT-T/2}^{kT+T/2}Y_k(t)p(t-kT)\cos(\omega_o t+\phi_k+\phi)dt\right)p_\phi(\phi_k) \tag{10.119}$$

In the following sections, Equation (10.119) is applied to the BPSK-, QPSK-, OQPSK-, and MSK-modulated waveforms and in each case unique functional implementations of the MAP phase estimator are apparent in the solutions. Under these waveform modulation conditions the square of the symbol weighting function in the matched filter integrand is evaluated as

$$\int_{-T/2}^{T/2} p^2(t)dt = \begin{cases} T & : \text{BPSK, QPSK, and OQPSK} \\ T/2 & : \text{MSK} \end{cases} \tag{10.120}$$

The resulting signal-to-noise ratios for BPSK and MSK are in terms of the bandwidth of the source bit ($1/T_b$) and for QPSK and OQPSK the bandwidth is that of the modulated symbol ($1/T$).

10.8.1 BPSK Modulation

In this section, the MAP phase estimator is applied to the demodulation of a BPSK-modulated waveform with a priori phase probabilities given by

$$p(\phi_k) = \begin{cases} 1/2 & : \phi_k = 0 \\ 1/2 & : \phi_k = \pi \end{cases} \tag{10.121}$$

Appling the a priori phase conditions to Equation (10.119) the conditional probability is expressed as

$$\begin{aligned} p_Y(Y_k|\phi) &= \frac{C}{2(2\pi N_o B)^{N/2}}\left[\exp\left(\frac{A}{N_o}\int_{kT-T/2}^{kT+T/2} Y_k(t)p(t-kT)\cos(\omega_o t+\phi)dt\right)\right. \\ &\quad \left. +\exp\left(-\frac{A}{N_o}\int_{kT-T/2}^{kT+T/2} Y_k(t)p(t-kT)\cos(\omega_o t+\phi)dt\right)\right] \\ &= \frac{C}{(2\pi N_o B)^{N/2}}\cosh\left(\frac{A}{N_o}\int_{kT-T/2}^{kT+T/2} Y_k(t)p(t-kT)\cos(\omega_o t+\phi)dt\right) \end{aligned} \tag{10.122}$$

Using Equation (10.122) and considering the sequence of K contiguous independent data symbols, the *pdf* of Equation (10.111) conditioned on the unknown phase is expressed as

$$p(Y|\phi) = \prod_{k=0}^{K-1} p_Y(Y_k|\phi) \tag{10.123}$$

and, referring to Equation (10.110), the MAP estimate of the unknown signal phase becomes

$$\begin{aligned} \hat{\phi} &= \max_{\phi}\left[\ln\prod_{k=0}^{K-1} p_Y(Y_k|\phi)\right] \\ &= \max_{\phi}\left[\sum_{k=0}^{K-1}\ln\left\{\cosh\left(\frac{A}{N_o}\int_{kT-T/2}^{kT+T/2} Y_k(t)p(t-kT)\cos(\omega_o t+\phi)dt\right)\right\}\right] \end{aligned} \tag{10.124}$$

The last equality in Equation (10.124) is based on Equation (10.122) noting that the constant $C/(2\pi N_o B)^{N/2}$ does not influence the selection of the maximum argument. Evaluation of the signal phase (ϕ) that maximizes this expression involves setting the derivative of the *rhs* of Equation (10.124) to zero and solving for $\phi = \hat{\phi}$. The functional implementation of the PLL is apparent upon noting that $\hat{\phi}$ is the output of the ideal integrator and corresponds to the output of the PLL VCO and that the derivative of the *rhs* of Equation (10.124), when equal to zero, is simply the error $\varepsilon_o(\phi = \hat{\phi}) = 0$ evaluated at the output of the loop filter. This also satisfies the intuitive conclusion since the average of the loop filter output is zero under the phase-locked condition, that is, under the optimum phase estimate condition. With this understanding, the filtered error is evaluated as

$$\begin{aligned} \lim_{\hat{\phi}\to\phi} \varepsilon_o(\phi-\hat{\phi}) &= \frac{d}{d\phi}\sum_{k=0}^{K-1}\left[\ln\left\{\cosh\left(\frac{A}{N_o}\int_{kT-T/2}^{kT+T/2} Y_k(t)p(t-kT)\cos(\omega_o t+\phi)dt\right)\right\}\right]\Bigg|_{\phi=\hat{\phi}} \\ &= \sum_{k=0}^{K-1}\frac{d}{d\varphi}\ln\left\{\cosh\left(\frac{A}{N_o}\int_{kT-T/2}^{kT+T/2} Y_k(t)p(t-kT)\cos(\omega_o t+\phi)dt\right)\right\}\Bigg|_{\phi=\hat{\phi}} \\ &= \lim_{\hat{\phi}\to\phi}\sum_{k=0}^{K-1}\varepsilon_k(\phi-\hat{\phi}) \end{aligned} \tag{10.125}$$

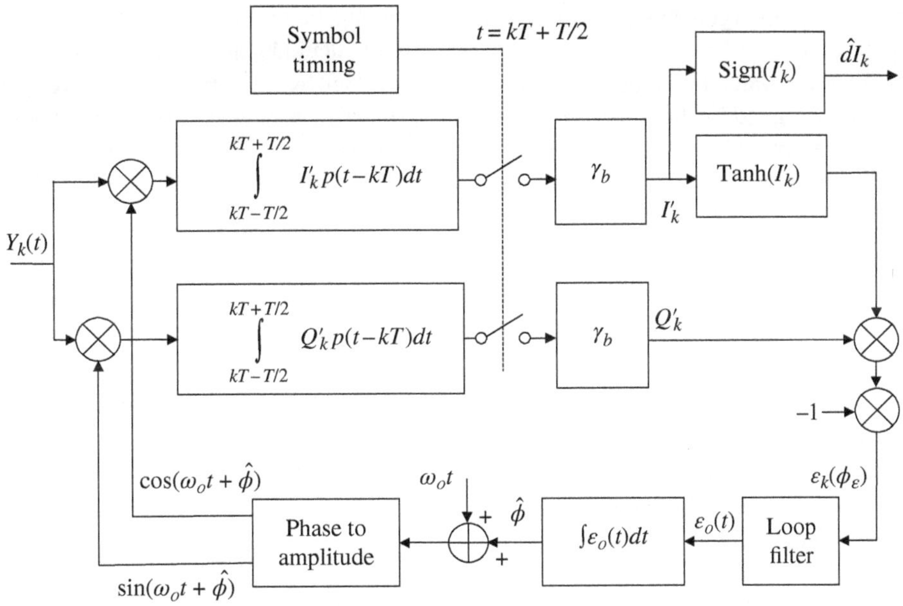

FIGURE 10.27 Implementation of BPSK MAP phase estimator.

The earlier manipulations assume that the summation over the K symbols is equivalent to the filtering performed by the loop filter. The resulting MAP phase estimate is

$$\hat{\phi} = K_o \lim_{\hat{\phi}\to\phi} \int \varepsilon_o(\phi-\hat{\phi})dt \tag{10.126}$$

The input to the PLL filter for the k-th symbol is defined as

$$\begin{aligned}\varepsilon_k(\phi) &= \frac{d}{d\phi}\ln\left\{\cosh\left(\frac{A}{N_o}\int_{kT-T/2}^{kT+T/2} Y_k(t)p(t-kT)\cos(\omega_o t+\phi)dt\right)\right\}\\ &= \tanh\left(\frac{A}{N_o}\int_{kT-T/2}^{kT+T/2} Y_k(t)p(t-kT)\cos(\omega_o t+\phi)dt\right)\\ &\left(-\frac{A}{N_o}\int_{kT-T/2}^{kT+T/2} Y_k(t)p(t-kT)\sin(\omega_o t+\phi)dt\right)\end{aligned} \tag{10.127}$$

Referring to the second equality in Equation (10.127), the integrals involving the cosine and sine functions are the respective in-phase and quadrature matched filter samples of the k-th received symbol. The phase detector error $\varepsilon_k(\phi)$ into the PLL filter, characterized by the last equality in Equation (10.125), is formed by the weighted quadrature matched filter samples as indicated by Equation (10.127). The weighted quadrature matched filter outputs I_k and Q_k are defined as

$$I_k \triangleq \frac{A}{N_o}\int_{kT-T/2}^{kT+T/2} Y_k(t)p(t-kT)\cos\left(\omega_o t+\hat{\phi}\right)dt \tag{10.128}$$

and

$$Q_k \triangleq \frac{A}{N_o}\int_{kT-T/2}^{kT+T/2} Y_k(t)p(t-kT)\sin\left(\omega_o t+\hat{\phi}\right)dt \tag{10.129}$$

The normalization of the I_k and Q_k integrator outputs can be adjusted to provide for normalization by the input signal-to-noise ratio $\gamma_s = E_s/N_o = (A^2T/2)/N_o$. Using Equations (10.128), (10.129), and the phase estimator implementation topology dictated by Equation (10.127), the phase detector error is expressed as

$$\varepsilon_k(\phi) = -\tanh\left(I_k\right)Q_k \tag{10.130}$$

Based on these results, the functional implementation of the MAP phase estimator for BPSK is shown in Figure 10.27.* The local oscillator reference $\omega_o t+\hat{\phi}$ is obtained at the output of the loop filter VCO as shown in the figure.

For small arguments, which corresponds to low input signal-to-noise ratios, $\tanh(I_k) \cong I_k$ and the phase detector error is given by

*The primed values of I_k and Q_k in Figure 10.27 exclude the quadrature sinusoidal signals involving $\omega_o t+\hat{\phi}$.

$$\varepsilon_k(\phi) \cong -I_k Q_k \quad : \text{small tanh arguments} \tag{10.131}$$

With the exception of a constant, Equation (10.131) is identical to the BPSK result obtained by the squaring associated with the Costas loop as discussed previously. Therefore, the BPSK Costas PLL detector results in an optimum MAP phase estimator in low signal-to-noise environments. For large arguments or high signal-to-noise ratios $\tanh(I_k) \cong \text{sign}(I_k) = \hat{d}_k$ and the phase detector error is given by

$$\varepsilon_k(\phi) \cong -\hat{d}_k Q_k \quad : \text{large tanh arguments} \tag{10.132}$$

So the MAP estimator under these conditions is identical to the decision-directed PLL.

It is informative to evaluate the matched filter outputs I_k and Q_k and examine the nature of the phase error resulting from the MAP estimator. To this end, consider $y_k(t)$ as expressed in Equation (10.107) with the additive data-dependent phase ϕ_k replaced by the multiplicative data d_k; the resulting expression is

$$Y_k(t) = Ad_k p(t-kT)\cos(\omega_o t + \phi) + n(t) \quad : kT - \frac{T}{2} \le t \le kT + \frac{T}{2} \tag{10.133}$$

The inputs to the quadrature matched filters corresponding to the integrands in Figure 10.27 are expressed as

$$I_k(t) = Y_k(t)p(t-kT)r_c(t) \tag{10.134}$$

and

$$Q_k(t) = Y_k(t)p(t-kT)r_s(t) \tag{10.135}$$

where $r_c(t) = \cos(\omega_o t + \hat{\phi})$ and $r_s(t) = \sin(\omega_o t + \hat{\phi})$ are the quadrature reference frequencies that are used in the demodulator to mix the received signal to baseband as shown in Figure 10.27. Substituting $r_c(t)$, $r_s(t)$ and Equation (10.133) into Equations (10.134) and (10.135) and performing the integration, the resulting weighted matched filter sampled outputs are evaluated as

$$I_k = \gamma_s d_k \cos(\phi_\varepsilon) + N_{Ik} \tag{10.136}$$

and

$$Q_k = -\gamma_s d_k \sin(\phi_\varepsilon) + N_{Qk} \tag{10.137}$$

The phase error is defined as $\phi_\varepsilon = \phi - \hat{\phi}$ and the signal-to-noise ratio $\gamma_s = (A^2T/2)/N_o = E_s/N_o$ is measured in the symbol bandwidth of $B = R_s = 1/T$ Hz where, for BPSK, $R_s = R_b$. The quadrature additive noise terms are characterized as zero-mean narrow-band white Gaussian noise as discussed in Chapter 1. In arriving at Equations (10.136) and (10.137), it is understood that the terms involving $2\omega_o t$ are removed by the demodulator lowpass matched filters.

Evaluation of the phase detection error $\varepsilon_k(\phi)$ for the small and large argument approximations to $\tanh(x)$ as given by Equations (10.131) and (10.132), the average loop filter output, assuming small angle estimation errors, is approximated as

$$\varepsilon_o(t) = \sum_{k=1}^{N} \varepsilon_k(\phi) \cong 4\gamma_s^2 \phi_e \quad : \text{small argument small error approximation} \tag{10.138}$$

and, for large angle estimation errors, the approximation is

$$\varepsilon_o(t) = \sum_{k=1}^{N} \varepsilon_k(\phi) \cong 2\gamma_s \phi_e \quad : \text{large argument small error approximation} \tag{10.139}$$

These results indicate that the optimum matched filter weighting function is based on the estimate of the received signal-to-noise ratio as discussed in Section 11.5. In the computation of $\varepsilon_k(\phi)$, the additive noise terms in the small argument approximation involve N_{Ik}, N_{Qk}, and the product $N_{Ik}N_{Qk}$. However, because these quadrature noises are zero-mean *iid* random processes, their mean values are zero at the output of the loop filter. With the large argument approximation the only additive noise is N_{Qk} with zero-mean value at the output of the loop filter.

The MAP phase estimate, as expressed by Equation (10.110), is actually an open-loop estimate of the phase and is scaled by a constant gain factor. However, the MAP estimator shown in Figure 10.27 is a closed-loop implementation that Simon substantiates by considering the reference oscillator as a *bumped* oscillator that applies infinitesimal linear corrections over the estimation interval leading to the closed-loop solution. As a practical matter, the constant associated with the phase error influences the closed-loop VCO gain K_o such that the phase estimate out of the VCO is equal to the baseband received signal phase as required to establish the quadrature received channels.

10.8.2 QPSK Modulation

The a priori signal phase distribution for the randomly modulated QPSK waveform is given by the discrete a priori density function

$$p(\phi_k) = \frac{1}{4} \quad : \phi_k = 0, \frac{\pi}{2}, \pi, \frac{3\pi}{2} \tag{10.140}$$

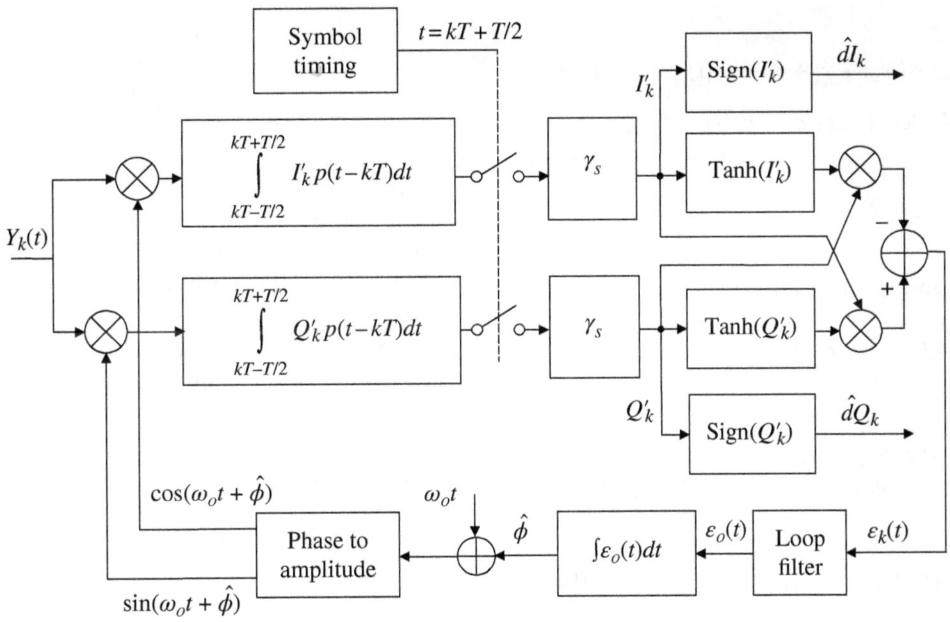

FIGURE 10.28 Implementation of QPSK MAP phase estimator.

and the quadrature symbol weighting function is $p(t) = rect(t/T)$ with unit power as given by Equation (10.120). Following the procedures similar to those in the preceding analysis for BPSK modulation, the phase detector output for QPSK is evaluated as

$$\varepsilon_k(\phi) = \frac{d}{d\phi}\ln\left\{\cosh\left(\frac{A}{N_o}\int_{kT-T/2}^{kT+T/2} Y_k(t)p(t-kT)\cos(\omega_o t+\phi)dt\right) + \cosh\left(\frac{A}{N_o}\int_{kT-T/2}^{kT+T/2} Y_k(t)p(t-kT)\sin(\omega_o t+\phi)dt\right)\right\}$$

$$= \frac{1}{\cosh(I_k)+\cosh(Q_k)}[\sinh(Q_k)I_k - \sinh(I_k)Q_k] \quad (10.141)$$

The weighted quadrature matched filter outputs I_k and Q_k are defined as in the BPSK case by Equations (10.128) and (10.129); however, with QPSK there is data on each quadrature rail resulting in I_k and Q_k being nearly identical except for a sign difference so that $|I_k| \cong |Q_k|$. Using the property that the hyperbolic cosine is an even function and the approximation $\cosh(Q_k) \cong \cosh(I_k)$ then Equation (10.141) becomes

$$\varepsilon_k(\phi) \cong \frac{1}{2}[\tanh(Q_k)I_k - \tanh(I_k)Q_k] \quad (10.142)$$

Therefore, using the phase estimator implementation based on Equation (10.141), the MAP phase estimator for the QPSK-modulated waveform is shown in Figure 10.28.*

With QPSK, the signal quadrupling requires that the small argument approximation to the hyperbolic tangent given by $\tanh(x) \cong x - x^3/3$ be used† and the resulting phase detector error becomes

$$\varepsilon_k(\phi) \cong \frac{1}{6}I_k Q_k\left(I_k^2 - Q_k^2\right) \quad \text{: small tanh argument} \quad (10.143)$$

For large arguments the hyperbolic tangent is approximated as $\tanh(x) \cong \text{sign}(x)$ and the phase detector error simplifies to

$$\varepsilon_k(\phi) \cong \text{sign}(Q_k)I_k - \text{sign}(I_k)Q_k \quad \text{: large tanh argument} \quad (10.144)$$

Following the approach in the evaluation of I_k and Q_k for BPSK modulation, the in-phase and quadrature matched filter outputs for QPSK modulation are evaluated as follows. The input signal for the k-th symbol is represented by quadrature-modulated terms as

$$Y_k(t) = Y_{ck}(t) + Y_{sk}(t) \quad (10.145)$$

*The primed values of I_k and Q_k in Figure 10.28 exclude the quadrature sinusoidal signals involving $\omega_o t + \hat{\phi}$.

†See Problem 6d.

where

$$\begin{aligned} Y_{ck}(t) &= Ap(t-kT)\cos(\omega_o t+\phi_k+\phi)+n_c(t) \quad :\phi_k=\{0,\pi\} \\ &= AdI_k p(t-kT)\cos(\omega_o t+\phi)+n_c(t) \quad :dI_k=\{1,-1\} \end{aligned} \tag{10.146}$$

and

$$\begin{aligned} Y_{sk}(t) &= Ap(t-kT)\sin(\omega_o t+\phi_k+\phi)+n_s(t) \quad :\phi_k=\left\{\frac{\pi}{2},-\frac{\pi}{2}\right\} \\ &= AdQ_k p(t-kT)\sin(\omega_o t+\phi)+n_s(t) \quad :dQ_k=\{1,-1\} \end{aligned} \tag{10.147}$$

The functions $I_k(t)$ and $Q_k(t)$ in the integrands of the quadrature matched filters, shown in Figure 10.28, are evaluated as

$$I_k(t)=[Y_{ck}(t)+Y_{sk}(t)]r_c(t) \tag{10.148}$$

and

$$Q_k(t)=[Y_{ck}(t)+Y_{sk}(t)]\ r_s(t) \tag{10.149}$$

where $r_c(t)=\cos(\omega_o t+\hat{\phi})$ and $r_s(t)=\sin(\omega_o t+\hat{\phi})$ are phaselocked references signals. Performing the integration over the k-th symbol interval results in

$$\begin{aligned} I_k &= \gamma_s(dI_k\cos(\phi_\varepsilon)+dQ_k\sin(\phi_\varepsilon))+N_{Ik} \\ &\cong \gamma_s(dI_k+dQ_k\phi_\varepsilon)+N_{Ik} \end{aligned} \tag{10.150}$$

and

$$\begin{aligned} Q_k &= \gamma_s(-dI_k\sin(\phi_\varepsilon)+dQ_k\cos(\phi_\varepsilon))+N_{Qk} \\ &\cong \gamma_s(-dI_k\phi_\varepsilon+dQ_k)+N_{Qk} \end{aligned} \tag{10.151}$$

where the approximation applies for small angle errors.

For small arguments, or low signal-to-noise ratios the phase detector error is approximated as

$$\begin{aligned} \varepsilon_k(\phi) &\cong \frac{2}{3}\gamma_s^4(\phi_\varepsilon-\phi_\varepsilon^3)+O(\phi_\varepsilon^n)+\text{noise terms} \\ &:\text{small argument, small error} \end{aligned} \tag{10.152}$$

The ordered terms $O(\phi_\varepsilon^n)$: $n=0, 2, 4$ involve various combinations of random data $(dI_k, dQ_k)=\pm1$ that average to zero in the loop filter; however, the $n=1, 3$ terms in Equation (10.152) are constant for all four combinations of (dI_k, dQ_k). Similarly, the noise terms involve various combinations of the zero-mean *iid* orthogonal noise processes N_{Ik} and N_{Qk} that average to zero in the loop filter and terms involving N_{Ik}^2 and N_{Qk}^2 with average values that cancel. With these considerations, the loop filter output is evaluated as

$$\begin{aligned} \varepsilon_o(\phi) &= \frac{2}{3}\gamma_s^4(\phi_\varepsilon-\phi_\varepsilon^3) \\ &\cong \frac{2}{3}\gamma_s^4\phi_\varepsilon \end{aligned} \tag{10.153}$$

This result is functionally equivalent to the squaring loop implementation of the QPSK phase detector so the signal squaring phase detector for QPSK also results in an optimum MAP phase estimator in low signal-to-noise environments.

For large arguments, that is, large signal-to-noise ratios, the phase detector error is evaluated using Equation (10.144) and is approximated for small phase detector errors as

$$\begin{aligned} \varepsilon_k(\phi) &\cong \gamma_s\left(dQ_k dI_k+dQ_k^2\phi_\varepsilon+dI_k^2\phi_\varepsilon-dI_k dQ_k\right) \\ &\quad +dQ_k N_{Ik}-dI_k N_{Qk} \quad :\text{large argument, small error} \end{aligned} \tag{10.154}$$

With zero-mean *iid* noise processes the loop filter output is evaluated as

$$\varepsilon_o(\phi)\cong 2\gamma_s\phi_\varepsilon \tag{10.155}$$

10.8.3 OQPSK and MSK Modulation

The optimum PLL implementation for OQPSK and MSK modulations is derived in a manner similar to that for BPSK modulation; however, the offset or staggering of the in-phase and quadrature phase symbols requires that the quadrature matched filters be sampled at alternating bit intervals corresponding to the detection of one bit per channel. At each bit decision sample the cross-channel contribution contains the phase-error information plus self-noise contributions from the cross-channel data. Several symbols of the offset waveform modulation are depicted in Figure 10.29 showing the timing relationships between the in-phase and quadrature data symbols dI_k: $k \in$ even and dQ_k: $k \in$ odd with dI_0 being the symbol data corresponding to $k=0$. With OQPSK modulation the symbol weighting is simply $p(t)=rect(t/T)$;

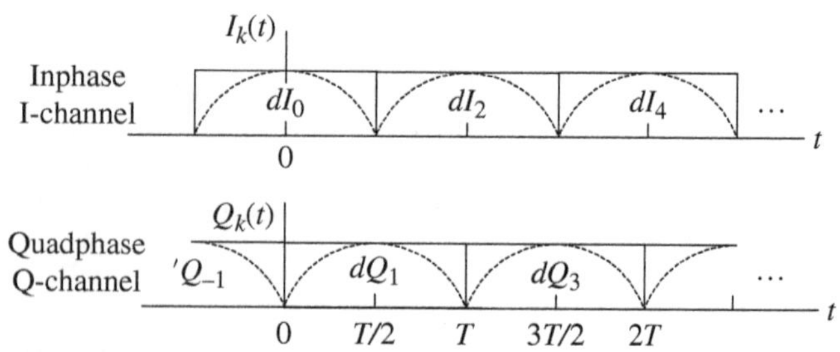

FIGURE 10.29 OQPSK and MSK waveform symbol description.

however, with MSK modulation the symbols are cosine weighted with $p(t)=\cos(\omega_m t)$: $\omega_m=\pi/T$, $-T/2\le t\le T/2$. In both cases, the quadrature symbols are offset or staggered by one bit interval, that is, $T_b=T/2$ s and the symbol weighting is normalized for unit power as expressed in Equation (10.120). The MSK cosine symbol weighting function $p(t)$ is associated with each symbol. This is unlike the description in Section 4.2.3.4 that describes the classical MSK implementation with the symbol weighting functions characterized as quadrature sinusoidal signals with zero crossings aligned with the corresponding data symbol transitions.

To proceed with this analysis the phase-dependent data ϕ_k is replaced by the multiplicative binary symbol data $d_k=\pm 1$ with (10.119) rewritten as

$$p_Y(Y_k|\phi)=\frac{C}{(\pi N_o/\Delta T)^{N/2}}\sum_{d_k}\exp\left(-\frac{A}{N_o}\int_{L_k}Y_k(t)p(t-kT)d_k\cos(\omega_o t+\phi)dt\right)p(d_k) \qquad (10.156)$$

where

$$p(d_k)=\begin{cases}1/2 & :d_k=dI_k=\pm 1;\ k\in\text{even}\\ 1/2 & :d_k=dQ_k=\pm 1;\ k\in\text{odd}\end{cases} \qquad (10.157)$$

The integration limits L_k are defined with the aid of Figure 10.31, $L_{k\in\text{even}}$: $kT-T/2\le t\le kT+T/2$ and $L_{k\in\text{odd}}$: $kT-T\le t\le kT$. Keeping in mind that the two channels are statistically independent, the analysis is similar to that of BPSK with the phase detector error expressed as

$$\varepsilon_k(\phi)=\begin{cases}\dfrac{d}{d\phi}\ln\left\{\cosh\left(\dfrac{A}{N_o}\displaystyle\int_{kT-T/2}^{kT+T/2}Y_k(t)p(t-kT)\cos(\omega_o t+\phi)dt\right)\right\} & :k\in\text{even}\\ \dfrac{d}{d\phi}\ln\left\{\cosh\left(\dfrac{A}{N_o}\displaystyle\int_{kT}^{kT+T}Y_k(t)p(t-kT+T_b)\sin(\omega_o t+\phi)dt\right)\right\} & :k\in\text{odd}\end{cases} \qquad (10.158)$$

The MAP phase estimation occurs when $\phi=\hat{\phi}$ as discussed in the previous sections. Upon evaluation of the derivatives, Equation (10.158) simplifies to

$$\varepsilon_k(\phi_\varepsilon)=\begin{cases}-\tanh(I_k)Q_k & :k\in\text{even}\\ \tanh(Q_k)I_k & :k\in\text{odd}\end{cases} \qquad (10.159)$$

The MAP phase estimator and demodulator for OQPSK and MSK are shown in Figure 10.30.*

The matched filter sample pairs (I_k,Q_k) and (I_{k+1},Q_{k+1}) are evaluated by performing the integrations shown in Figure 10.30 using $I_{k'}(t)$ and $Q_{k'}(t)$ with $k'\in$ even corresponding to the detection of the k-th in-phase symbol data $\hat{d}I_k$ and $k'\in$ odd corresponding the detection of the $(k+1)$ quadrature data symbol data $\hat{d}Q_{k+1}$. Figure 10.31 depicts the quadrature symbol offsets and integration ranges for the detection of the symbols k and $k+1$ used in the following analysis.

The received signal is expressed as

$$Y_{k'}(t)=Y_{ck'}(t)+Y_{sk'}(t) \qquad (10.160)$$

and, referring to Figure 10.31, the in-phase and quadrature terms are expressed for $k\in$ even as

$$Y_{ck}(t)=AdI_kp(t-kT)\cos(\omega_o t+\phi)+n_c(t) \quad :kT-\frac{T}{2}\le t\le kT+\frac{T}{2} \qquad (10.161)$$

and

$$Y_{sk}(t)=AdQ_{k-1}p(t-kT+T/2)\sin(\omega_o t+\phi) \quad :kT-\frac{T}{2}\le t\le kT$$
$$+AdQ_{k+1}p(t-kT-T/2)\sin(\omega_o t+\phi)+n_s(t) \quad :kT\le t\le kT+\frac{T}{2} \qquad (10.162)$$

For $k+1$ the in-phase and quadrature terms are expressed as

$$Y_{ck+1}(t)=AdI_kp(t-kT)\cos(\omega_o t+\phi) \quad :kT\le t\le kT+\frac{T}{2}$$
$$+AdI_{k+2}p(t-kT-T)\cos(\omega_o t+\phi)+n_c(t) \quad :kT+\frac{T}{2}\le t\le kT+T \qquad (10.163)$$

and

$$Y_{sk+1}(t)=AdQ_{k+1}p\left(t-kT-\frac{T}{2}\right)\sin(\omega_o t+\varphi)+n_s(t) \quad :kT\le t\le kT+T \qquad (10.164)$$

Referring again to Figure 10.30 and defining $r_c(t)=\cos(\omega_o t+\hat{\phi})$ and $r_s(t)=\sin(\omega_o t+\hat{\phi})$ the baseband outputs, after the heterodyning by $\omega_o t+\hat{\phi}$, corresponding to $k\in$ even are

$$I_k(t)=(Y_{ck}(t)+Y_{sk}(t))r_c(t) \qquad (10.165)$$

*The primed values of I_k and Q_k in Figure 10.30 exclude the quadrature sinusoidal signals involving $\omega_o t+\hat{\phi}$.

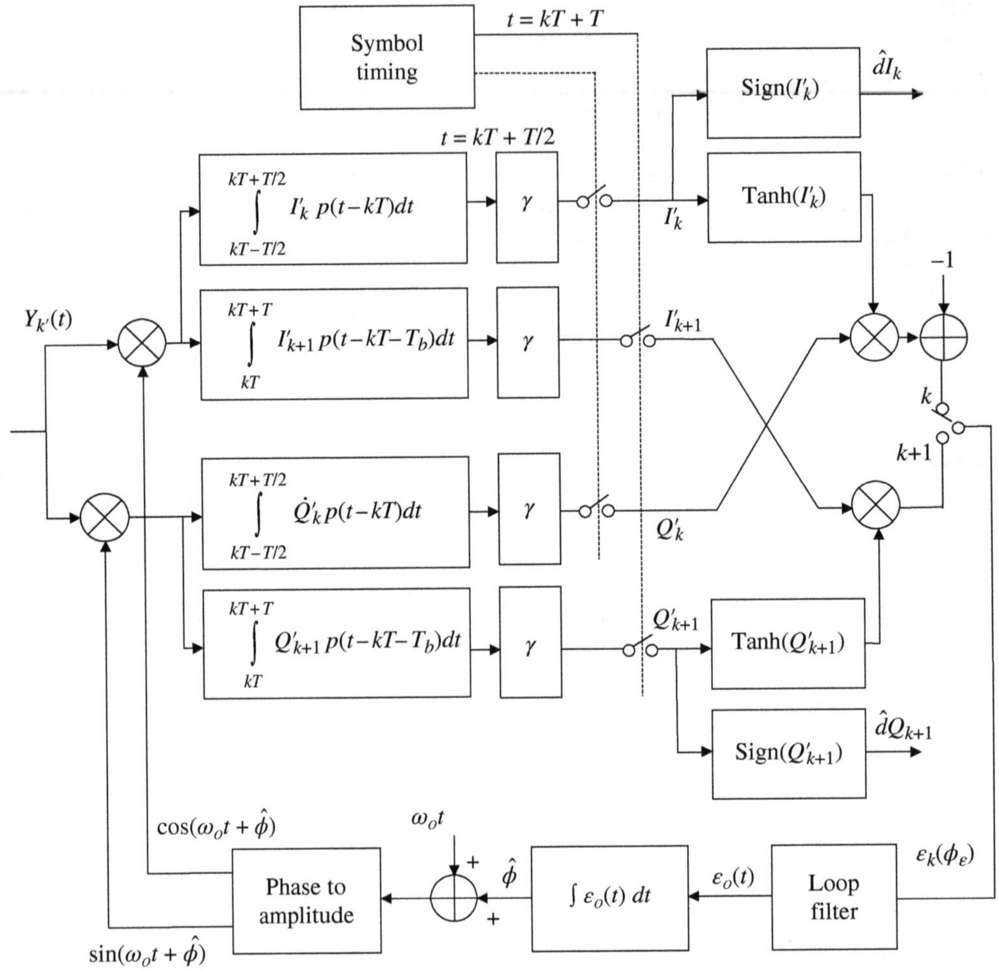

FIGURE 10.30 OQPSK and MSK MAP phase demodulator ($k \in$ even, $\gamma = \{\gamma_s, \gamma_b\}$).

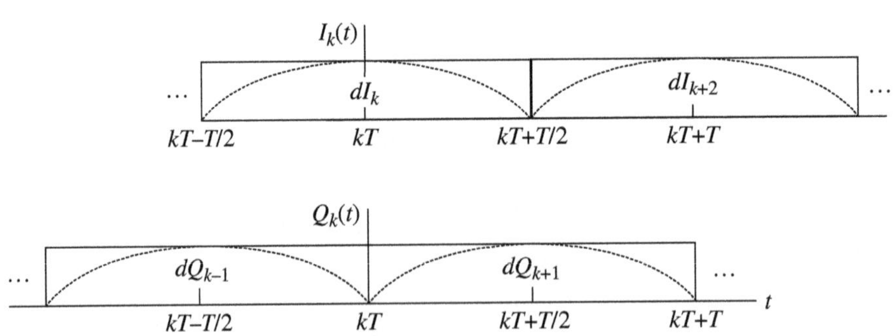

FIGURE 10.31 OQPSK and MSK waveform symbol description ($kT - T/2 \le t \le kT + T$).

and

$$Q_k(t) = (Y_{ck}(t) + Y_{sk}(t))r_s(t) \tag{10.166}$$

For $k + 1$ the corresponding expressions are

$$I_{k+1}(t) = (Y_{ck+1}(t) + Y_{sk+1}(t))r_c(t) \tag{10.167}$$

and

$$Q_{k+1}(t) = (Y_{ck+1}(t) + Y_{sk+1}(t))r_s(t) \tag{10.168}$$

Using Equations (10.165) through (10.168) to form the integrands in Figure 10.30 and performing the integration, recalling from Equation (10.120) that the symbol weighting has unit power, the matched filter outputs for $k \in$ even are expressed as

$$I_k = \gamma\left(dI_k \cos(\phi_\varepsilon) + \frac{dQ_{k-1} + dQ_{k+1}}{2}\sin(\phi_\varepsilon)\right) + N_{Ik} \tag{10.169}$$

and

$$Q_k = \gamma\left(-dI_k \sin(\phi_\varepsilon) + \frac{dQ_{k-1} + dQ_{k+1}}{2}\cos(\phi_\varepsilon)\right) + N_{Qk} \tag{10.170}$$

Similarly, for $k+1$

$$I_{k+1} = \gamma\left(dQ_{k+1} \sin(\phi_\varepsilon) + \frac{dI_k + dI_{k+2}}{2}\cos(\phi_\varepsilon)\right) + N_{Ik+1} \tag{10.171}$$

and

$$Q_{k+1} = \gamma\left(dI_k \cos(\phi_\varepsilon) - \frac{dI_k + dI_{k+2}}{2}\sin(\phi_\varepsilon)\right) + N_{Qk+1} \tag{10.172}$$

These even and odd offset outputs are alternately sampled at the bit rate so the loop filter and the phase estimate corrections are applied at the sampling frequency $f_s = R_b$ Hz. The phase detector estimation error is evaluated using the alternately sampled errors $\varepsilon_k(\phi_e)$ and $\varepsilon_{k+1}(\phi_e)$ expressed by Equation (10.159) where, in keeping with the preceding analysis, k is an even integer. Because these errors are averaged in the loop filter they can be combined by delaying the in-phase phase detector error by one bit interval yielding the combined error

$$\begin{aligned}\varepsilon'_{k+1} &= \varepsilon_{k+1}(\phi_\varepsilon) + \varepsilon_k(\phi_\varepsilon)z^{-1} \\ &= \tanh(Q_{k+1})I_{k+1} - \tanh(I_k)Q_k\end{aligned} \tag{10.173}$$

This implementation is similar to the phase detector error for QPSK modulation given by Equation (10.142) with a loop filter sampling rate of $f_s = R_s$ Hz.

For the small argument case, that is, low signal-to-noise ratios $\tanh(x) \cong x - x^3/3$ and with small estimation errors ϕ_e the phase detector error ε'_{k+1} is expressed in terms of the fourth-order polynomial in ϕ_e as

$$\varepsilon'_{k+1} = \sum_{i=0}^{4} C_i \phi_e^i \quad \text{: small argument, small error} \tag{10.174}$$

Using $dI_{k'}^n = dQ_{k'}^n = 1 : n \in$ even the coefficients are evaluated* as

*These evaluations were made using Mathsoft's Mathcad© symbolic calculation software.

$$\begin{aligned}C_0 &= \left(\frac{-dI_k dQ_{k-1} + dQ_{k+1} dI_{k+2}}{2}\right)\gamma^2 \\ &+ \left(\frac{dI_k dQ_{k-1} - dQ_{k+1} dI_{k+2}}{6}\right)\gamma^4\end{aligned} \tag{10.175}$$

$$\begin{aligned}C_1 &= \left(1 - \frac{dI_k dI_{k+2} + dQ_{k-1} dQ_{k+1}}{2}\right)\gamma^2 \\ &+ \left(\frac{1}{3} + \frac{dI_k dI_{k+2} + dQ_{k-1} dQ_{k+1}}{2}\right)\gamma^4\end{aligned} \tag{10.176}$$

$$C_2 = \left(\frac{dI_k dQ_{k-1} - dQ_{k+1} dI_{k+2}}{2}\right)\gamma^2 \tag{10.177}$$

$$C_3 = \left(-\frac{2}{3} - \frac{dI_k dI_{k+2} + dQ_{k-1} dQ_{k+1}}{3}\right)\gamma^4 \tag{10.178}$$

and

$$C_4 = \left(\frac{dQ_{k+1} dI_{k+2} - dI_k dQ_{k+1}}{6}\right)\gamma^4 \tag{10.179}$$

With random data the average value of the coefficients C_0, C_1, and C_3 is zero and the loop filter output is simply

$$\begin{aligned}\varepsilon_o(t) &\cong \left[\gamma^2 + \frac{1}{3}\gamma^4\right]\phi_\varepsilon - \left[\frac{2}{3}\gamma^4\right]\phi_\varepsilon^3 \\ &\cong \gamma^2 \phi_\varepsilon \quad \text{: small argument, small error}\end{aligned} \tag{10.180}$$

For large arguments $\tanh(x) \cong \text{sign}(x)$ and with small estimation errors, the phase detector error is evaluated as

$$\varepsilon'_{k+1} \cong \left(\frac{dI_k dQ_{k-1} + dI_{k+2} dQ_{k+1}}{2}\right)\gamma + 2\gamma\phi_\varepsilon \quad \text{: large argument, small error} \tag{10.181}$$

and, with random data, the loop filter output is

$$\varepsilon_o(t) \cong 2\gamma\phi_\varepsilon \quad \text{: large argument, small error} \tag{10.182}$$

The signal-to-noise weighting function, revealed by the MAP phase estimation analysis, is an important consideration especially in low signal-to-noise environments. On the other hand, for high signal-to-noise environments the weighing will increase the loop gain and damping factor resulting in a wider carrier acquisition range. However, if it desired to maintain the PLL dynamics over a wide range of signal-to-noise environments a sigmoid weighing function of the form

$$w(\gamma) = \frac{\gamma}{a + \gamma} \tag{10.183}$$

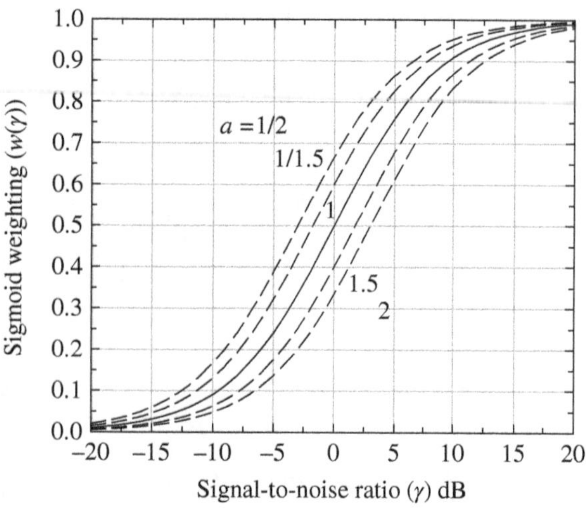

FIGURE 10.32 Phase detector signal-to-noise weighting function.

is a reasonable selection. The signal-to-noise ratio in Equation (10.183) is the linear power ratio, that is, it is not in decibels and the parameter a provides for altering the weighting function as shown in Figure 10.32.

The additive noise terms in these evaluations result from the zero-mean AWGN received noise and consist of independent quadrature noise terms plus various products of quadrature noise terms. The result is that the noise at the loop filter output is represented as narrowband zero-mean white noise that does not alter the average error $\varepsilon_o(t)$ for the various modulated waveforms considered earlier. The noise variance at the loop filter output does, however, have a significant impact on the loop performance and is characterized in terms of the squaring loss as discussed in the following section.

Omura and Simon [31] and Lindsey and Simon [2,32] discuss the implementation of the MAP phase estimation for higher order MPSK with $M > 4$. In general, the approximation to the MAP phase estimator requires that the tanh(x) function be approximated using a polynomial in x of order $M - 1$; lower order polynomial expansions will result in a zero phase detector error. The joint MAP estimation of symbol timing and carrier phase for MSK modulation is analyzed by Booth [33]. The MAP phase estimator for unbalanced QPSK (UQPSK) is examined by Simon [34] and Simon and Alem [35,36]; UQPSK incorporates different data rates on the quadrature channels.

10.9 SQUARING LOSS EVALUATION

In this section the squaring loss is evaluated for BPSK and QPSK modulation. The squaring loss is defined as the loss penalty relative to a linear PLL due to the distortion terms involving the products $S \times S$, $S \times N$ and $N \times N$ where S and N represent the signal and noise terms of the received waveform. When the PLL is operating in the linear range with small phase errors, the squaring loss is expressed as

$$S_L \triangleq \frac{C}{\gamma_L \sigma^2_{\phi_\varepsilon}} \tag{10.184}$$

where C is a phase-dependent constant, $\gamma_L = P_s/(N_o B_L)$ is the signal-to-noise ratio in the PLL bandwidth, and $\sigma^2_{\phi_\varepsilon}$ is the variance of the phase estimation. The received signal is given by

$$y(t) = s(t) + n(t) \tag{10.185}$$

and, referring to Equation (10.106), $s(t)$ is given by

$$s(t) = A\sum_{k=0}^{K-1} p(t-kT)\cos(\omega_o t + \phi_k + \phi)rect\left(\frac{t-kT}{T}\right) \tag{10.186}$$

where the signal phase term ϕ_k corresponds the source data phase. The noise is represented as narrowband quadrature *iid* zero-mean, additive Gaussian noise terms with variance σ_n^2 and is expressed as

$$n(t) = \sqrt{2}\{n_c(t)\cos(\omega_o t + \phi) - n_s(t)\sin(\omega_o t + \phi)\} \tag{10.187}$$

The quadrature noise variances are $\sigma_n^2 = N_o B/2$ watts where N_o is the single-sided noise density in watts/hertz and $B \le f_o/2$ Hz is the single-sided input bandwidth.

In the following sections, the squaring loss is evaluated for the MAP phase estimation as shown in Figures 10.27 and 10.28 for BPSK and QPSK modulation, respectively. The analysis of the squaring loss for the MAP phase estimator is based on the work of Simon [28] using a sampled integrate-and-dump (I&D) symbol filter; however, Lindsey and Simon [2] also evaluated the squaring loss for other filter types including the Butterworth, Gaussian, and ideal bandpass filter. The Butterworth or a single time-constant RC filter results in improved symbol timing acquisition; however, the I&D filter results in improved performance during PLL tracking.

10.9.1 Squaring Loss for BPSK Modulation

Evaluation of the squaring loss for BPSK modulation starts with the evaluation of sampled matched filter outputs I_k and Q_k shown in Figure 10.27. These quadrature responses are then used to form the phase detector error $\varepsilon_k(\phi)$ given by Equation (10.130) that, upon using Equations (10.136) and (10.137), is evaluated as

$$\varepsilon_k(\phi) = (2\gamma_b d_k \sin(\phi) + N_{Qk})\tanh(2\gamma_b d_k \cos(\phi) + N_{Ik}) \tag{10.188}$$

In the following evaluation (10.188) is normalized by the noise standard deviation, σ_n, and the resulting noise terms are denoted by the *iid* zero-mean unit-variance Gaussian random variables X_Q and X_I.

To determine the squaring loss it is necessary to evaluate the variance of the phase error given in Equation (10.184). The phase-error variance is defined as the ratio of the noise power in the PLL bandwidth to the power gain or squared slope of the discriminator curve formed by averaging the doubled signal component of Equation (10.188) expressed, in terms of the normalized noise, as*

$$\eta(\phi) = E_x[(4\gamma_b \sin(\phi))\tanh(2\gamma_b \cos(\phi) + X_I)] \quad (10.189)$$

The slope of the discriminator curve is evaluated at $\phi = 0$ as

$$K_\eta = \left.\frac{d\eta(\phi)}{d\phi}\right|_{\phi=0} = 2\gamma_b E_x\left[\tanh\left(2\gamma_b - \sqrt{2\gamma_b}X_I\right)\right] \quad (10.190)$$

The variance of the noise is evaluated by averaging the noise terms in Equation (10.188) and the result is expressed, in terms of the normalized noise, at $\phi_e = 0$ as

$$\sigma_N^2 = 8\gamma_b E_x\left[\tanh^2\left(2\gamma_b - \sqrt{2\gamma_b}X\right)\right] \quad (10.191)$$

With T_b = bit interval the noise density is expressed as

$$N_o' = 2\sigma_N^2 T = 16\gamma_b T E_x\left[\tanh^2\left(2\gamma_b - \sqrt{2\gamma_b}X\right)\right] \quad (10.192)$$

Therefore, using Equations (10.190) and (10.192), the variance of the phase error is

$$\sigma_{\phi_e}^2 = \frac{N_o' B_L}{K_\eta^2} = \frac{4B_L T E_x[\tanh^2(2\gamma_b - \sqrt{2\gamma_b}X)]}{\gamma_b E_x^2[\tanh(2\gamma_b - \sqrt{2\gamma_b}X)]} \quad (10.193)$$

Substituting Equation (10.193) into (10.184) with $C = 4$ the squaring loss becomes

$$S_L = \frac{E_x^2[\tanh(2\gamma_b - \sqrt{2\gamma_b}X)]}{E_x[\tanh^2(2\gamma_b - \sqrt{2\gamma_b}X)]} = E_x\left[\tanh\left(\sqrt{2\gamma_b}X\right)\sinh\left(\sqrt{2\gamma_b}X\right)\right]e^{-\gamma_b} \quad \text{: BPSK exact } \tanh(x) \quad (10.194)$$

The second equality in Equation (10.194) is a simplification given by Simon.

*The expectation $E_x[\ ]$ is with respect to the normalized Gaussian noise random variable X denoted by $N(0,1)$. The expectation over the random data $d_k = \pm 1$ has been removed using $E(d_k, d_j) = \delta_{kj}$.

Simon [30] has also evaluated the squaring loss for the large argument approximation $\tanh(x) \cong \text{sign}(x)$ with the result

$$S_L = erf^2(\gamma_b) \quad \text{: BPSK large argument approximation} \quad (10.195)$$

The squaring loss for the small argument approximation $\tanh(x) \cong x$ is evaluated by Stiffler [37] as

$$S_L = \frac{2\gamma_b}{1 + 2\gamma_b} \quad \text{: BPSK small argurment approximation} \quad (10.196)$$

The expectation for the exact argument is expressed by Equation (10.194) and is evaluated by integration over the conditional random variable X resulting in

$$S_L = \frac{e^{-\gamma_b}}{\sqrt{2\pi}} \int_{-\infty}^{\infty} \tanh\left(\sqrt{2\gamma_b}X\right)\sinh\left(\sqrt{2\gamma_b}X\right)e^{-X^2/2}dX \quad (10.197)$$

Using Hermite numerical integration [38] Equation (10.197) is evaluated as

$$S_L \cong \frac{e^{-\gamma_b}}{\sqrt{\pi}} \sum_{i=1}^{n} W_i f(\xi_i) \quad \text{: Hermite integration} \quad (10.198)$$

where W_i are the Hermite weight factors and

$$f(\xi) = \tanh\left(\sqrt{4\gamma_b}\xi\right)\sinh\left(\sqrt{4\gamma_b}\xi\right) \quad (10.199)$$

The hyperbolic tangent processing is based on a MAP estimate of the carrier phase and results in the minimum phase variance for all signal-to-noise ratios. Furthermore, the hyperbolic tangent function results in the minimum squaring loss for all signal-to-noise ratios. The squaring loss for each implementation is shown in Figure 10.33 for BPSK modulation. The hard limiter performs better than the linear feedback for higher signal-to-noise ratios and the results suggest that switching between the two approximations based on using an estimate of the signal-to-noise ratio will provide a reasonable compromise to implementing the hyperbolic tangent function over a wide range of signal-to-noise ratios.

10.9.2 Squaring Loss for QPSK Modulation

The squaring loss for QPSK modulation is evaluated for the phase detector error $\varepsilon_k(\phi)$ expressed by Equation (10.142) using Equations (10.150) and (10.151). Also, the signal component (η) of the phase detector error is based on $4\varepsilon_k(\phi)$ and

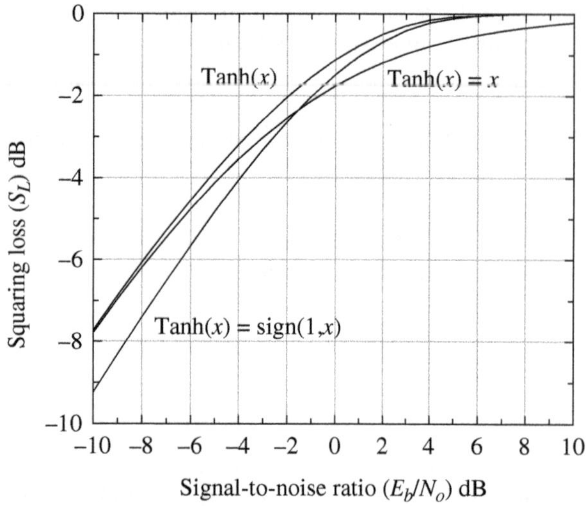

FIGURE 10.33 Squaring loss for BPSK carrier tracking with different phase detectors.Simon [28]. Reproduced by permission of the IEEE.

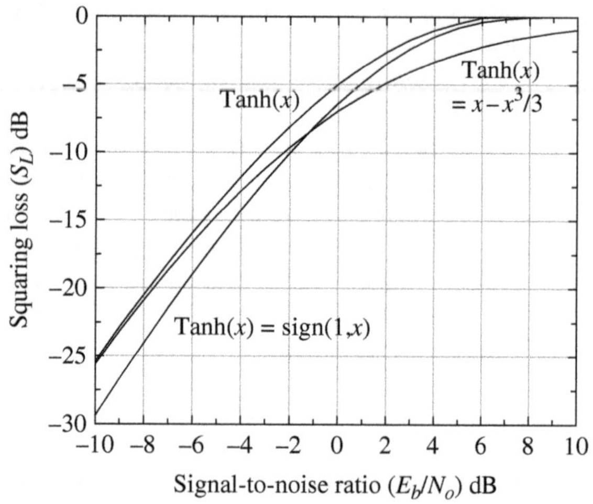

FIGURE 10.34 Squaring loss for QPSK carrier tracking with different phase detectors.Simon [28]. Reproduced by permission of the IEEE.

the constant in Equation (10.184) is $C = 16$. The analysis is similar to that used to characterize the squaring loss of BPSK modulation; however, it is considerably more involved so only the results are given [28].

The squaring loss for the exact tanh(x) function is

$$S_L = E_X\left[\{(1+\gamma_s)\tanh(\sqrt{\gamma_s}X) - \sqrt{\gamma_s}X\}\sinh(\sqrt{\gamma_s}X)\right]e^{-\gamma_s/2} \quad \text{: QPSK exact} \tag{10.200}$$

The squaring loss for the large argument approximation $\tanh(x) \cong \text{sign}(x)$ is

$$S_L = \frac{\left[erf\left(\sqrt{\gamma_s/2}\right) - \sqrt{2\gamma_s/\pi}\,e^{-\gamma_s/2}\right]^2}{1+\gamma_s - \left[\sqrt{\gamma_s}\,erf\left(\sqrt{\gamma_s/2}\right) + \sqrt{2/\pi}\,e^{-\gamma_s/2}\right]^2} \quad \text{: QPSK large argument} \tag{10.201}$$

and the squaring loss for the small argument approximation $\tanh(x) \cong x - x^3/3$ is evaluated by Stiffler [37] as

$$S_L = \frac{1}{1+(9/2\gamma_s)+(6/\gamma_s^2)+(3/2\gamma_s^3)} \quad \text{: QPSK small argument} \tag{10.202}$$

The expectation for the exact argument given by Equation (10.200) is evaluated by integration over the conditional random variable X as

$$S_L = \frac{e^{-\gamma_s/2}}{\sqrt{2\pi}}\int_{-\infty}^{\infty}\{(1+\gamma_s)\tanh(\sqrt{\gamma_s}X) - \sqrt{\gamma_s}X\}\sinh(\sqrt{\gamma_s}X)e^{-X^2/2}dX \tag{10.203}$$

and, as in the case for BPSK modulation, Hermite numerical integration [38] is used to evaluate (10.203) as

$$S_L \cong \frac{e^{-\gamma_s/2}}{\sqrt{\pi}}\sum_{i=1}^{n} W_i f(\xi_i) \quad \text{: Hermite integration} \tag{10.204}$$

where

$$f(\xi) = \left\{(1+\gamma_s)\tanh\left(\sqrt{2\gamma_s}\xi\right) - \sqrt{2\gamma_s}\xi\right\}\sinh\left(\sqrt{2\gamma_s}\xi\right) \tag{10.205}$$

As in the case with BPSK, the hyperbolic tangent function results in the minimum phase variance and squaring loss for all signal-to-noise ratios. The squaring loss for each implementation is shown in Figure 10.34 for the QPSK-modulated waveforms. The hard limiter performs better than the linear feedback for high signal-to-noise ratios and the results suggest that switching between the two approximations as described for BPSK modulation is a reasonable overall performance compromise. Rearranging Equation (10.184) as

$$\sigma_{\phi_e}^2 = \frac{C}{\gamma_L S_L} \tag{10.206}$$

it is evident from Figure 10.34 that the phase variance increases significantly as the squaring loss decreases with decreasing signal-to-noise ratio E_b/N_o. For QPSK modulation the increase is significantly greater than that with BPSK modulation.

10.10 CASE STUDY: BPSK AND QPSK PHASELOCK LOOP PERFORMANCE

The PLL tracking performance is evaluated in this section based on Monte Carlo simulations for BPSK- and

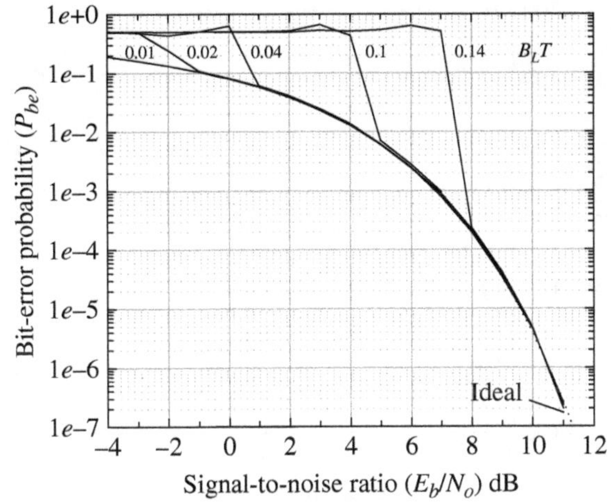

FIGURE 10.35 BPSK carrier tracking performance (phase estimation using Tanh(x) ≅ Sign(x)).

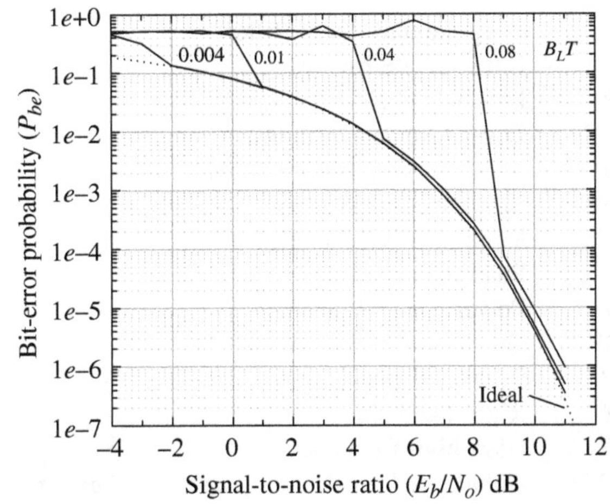

FIGURE 10.36 QPSK carrier tracking performance (phase estimation using Tanh(x) ≅ Sign(x)).

QPSK-modulated waveforms. The simulations involve 100 K Monte Carlo trials (symbols) for each signal-to-noise ratio ≤6 dB and 10 M Monte Carlo trials for each signal-to-noise ratio exceeding 6 dB. The source data rate is 12 kbps; however, when the PLL is characterized in terms of the normalized loop bandwidth, the data rate and modulation are implicit in the normalized parameter B_LT. For each modulation the sampling frequency is based on eight samples per symbol so aliasing occurs around the fourth spectral null. The channel is characterized as an AWGN channel so no channel filter is used.

The bit-error performance with PLL carrier tracking and ideal symbol timing is shown, respectively, in Figures 10.35 and 10.36 for BPSK and QPSK modulations. For these plots the large argument approximation for the tanh(x) function is used, so the performance approaches that of the optimum MAP phase detector for signal-to-noise ratios greater than about 0 dB. However, for signal-to-noise ratios greater than or equal the critical value for the selected B_LT product must achieve an acceptable performance as discussed in Section 10.6.11. For these simulation results the initial carrier phase and frequency errors are zero so the acquisition dependence on the B_LT product is solely dependent on the received waveform random data modulation self-noise, the additive channel noise and, to a lesser extent, the aliasing distortion. To provide near ideal or theoretical performance at lower signal-to-noise ratios, the PLL bandwidth must be decreased; however, the dynamics of the received signal phase function becomes more restrictive as indicated by the decreasing frequency pull-in range of the PLL in Tables 10.3 and 10.4. As expected BPSK modulation is considerably more robust than QPSK and provides for operating with a B_LT product on the order of four to five times higher with less loss in performance once carrier tracking is achieved.

TABLE 10.3 BPSK Loop Bandwidth and Lock-In Frequency

B_LT	B_L (Hz)	F_L (Hz)
0.01	120	50
0.02	240	100
0.04	480	203
0.10	1200	509
0.14	4800	712

TABLE 10.4 QPSK Loop Bandwidth and Lock-In Frequency

B_LT	B_L (Hz)	F_L (Hz)
0.004	24	10
0.01	60	25
0.04	240	100
0.08	480	203

Figure 10.37 shows the performance of BPSK modulation for $B_LT = 0.01$ using the phase-error generation functions corresponding to the large argument approximation tanh(x) ≅ sign(x) (*solid curve*), the small argument approximation tanh (x) ≅ x (*dot-dashed curve*), and exact tanh(x) function (*dashed curve*). These results do not include the signal-to-noise ratio weighting of the phase detector error and the performance differences are nearly indistinguishable except for the small argument case that exhibits a critical signal-to-noise ratio of about –3 dB. Although not shown, when the signal-to-noise ratio weighting is included, the phase detector small argument approximation (the *dot-dash curve* in Figure 10.37) acquires the carrier phase over the entire range of signal-to-noise ratios shown, so the weighting function extends the operating range of the PLL at the lower signal-to-noise ratios. The sigmoid weighting function expressed by Equation (10.183) with $a = 1$ is applied to the phase detector error.

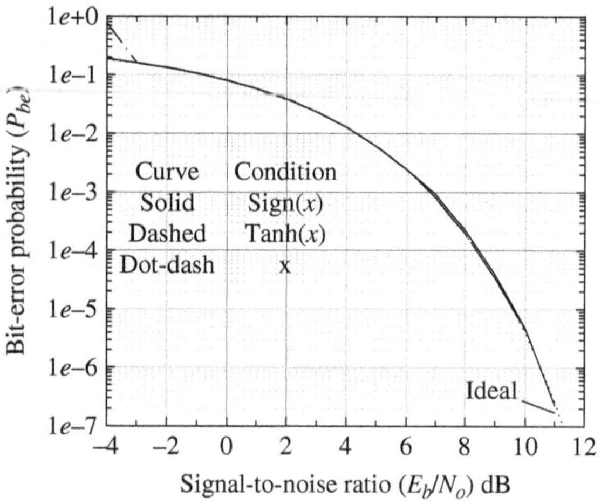

FIGURE 10.37 BPSK carrier tracking performance with various phase-error functions ($B_LT = 0.01$).

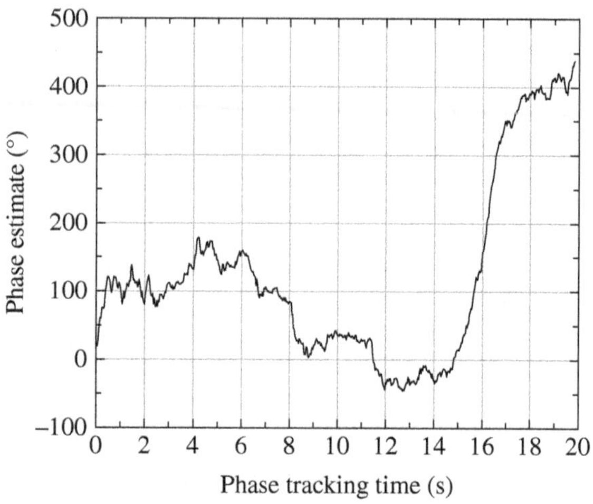

FIGURE 10.38 BPSK carrier tracking of received signal using a commercial grade battery operated transmitter ($B_LT = 0.03$).

10.11 CASE STUDY: BPSK PHASE TRACKING PERFORMANCE OF A DISADVANTAGED TRANSMIT TERMINAL

This case study demonstrates the PLL tracking performance of a received signal from a low-power battery operated transmit terminal. The transmitter uses a low-cost, commercial grade reference oscillator. Following a robust acquisition preamble, the demodulator carrier PLL tracking is initiated and tracks the carrier for coherent demodulation of the received message bits. A rate 1/3 forward error correction (FEC) turbo code using BPSK modulation is applied to the message information bits with an operating signal-to-noise ratio of $E_b/N_o = 3.27$ dB. The demodulator PLL tracks the received signal carrier in a bandwidth equal to the code-bit rate corresponding to a signal-to-noise ratio of $E_{cb}/N_o = -1.5$ dB. The phase tracking performance is shown in Figure 10.38. The results reflect the frequency instability and lack of long-term coherence of the transmitter reference oscillator and the ability of the demodulator second-order PLL to acquire and track the received signal under adverse dynamic conditions. The sigmoid weighting function described in Equation (10.183) was essential in acquiring and maintaining phase-lock throughout the received message.

ACRONYMS

AGC	Automatic gain control
AWGN	Additive white Gaussian noise
BPSK	Binary phase shift keying
CW	Continuous wave
DC	Direct current
DSP	Digital signal processor
FEC	Forward error correction
FPGA	Field programmable gate array
I&D	Integrate-and-dump (filter)
IF	Intermediate frequency
LHP	Left half plane
LLRT	Log-likelihood ratio test
MAP	Maximum a posterior probability
MPSK	Multiphase shift keying
MSK	Minimum shift keying
PLL	Phaselock loop
PSD	Power spectral density
QPSK	Quadrature phase shift keying
RHP	Right half plane
RHS	Right-hand-side
UHF	Ultra-high frequency
UQPSK	Unbalanced quadrature phase shift keying
VCO	Voltage controlled oscillator

PROBLEMS

1. Develop the recursive equations leading to the MPSK phase detector in Figure 10.5 using the normalized baseband analytic signal $\hat{s}(t) = \cos(\phi_m + 2\phi) + \; j\sin(\phi_m + \phi) = s_{cM} + js_{sM}$ where $\phi_m = 2\pi m/M$: $m = 0, \ldots, M-1$, $M = 2^k$, with k bits/symbol, and ϕ is the constant phase error to be estimated, and the time dependence is based on the implicit symbol intervals, such that, $t = iT$. Using the arctan phase detector of Equation (10.20), develop the small angle MPSK phase error $\varepsilon(t)$ into the PLL filter and the MPSK lock detector input error $\bar{\varepsilon}(t)$. Also, express these inputs using the small angle approximations.

   Hint: Use the trigonometric identities $\sin(n\alpha) = 2\sin(n\alpha/2)\cos(n\alpha/2)$ and $\cos(n\alpha) = \cos^2(n\alpha/2) - \sin^2(n\alpha/2)$.

2. Referring to the discussion and procedure leading to the root-locus plot for the third-order PLL shown in Figure 10.20, plot the root loci for the second-order PLL discussed in Section 10.6.

3. Derive the expression for the one-sided noise bandwidth of a first-order PLL as expressed in Equation (10.22).

   Hint: Use the definition of the noise bandwidth given by Equation (1.46).

4. Referring to Equation (10.127) show the steps leading to the second equality.

   Hint: $d\{\ln[\cosh(u(\phi))]\}/d\phi$ is equal to $\tanh(u(\phi))\{du(\phi)/d\phi\}$.

5. Using the normalization $p(t)$ given by Equation (10.120), show that the matched filter outputs I_k and Q_k given by Equations (10.128) and (10.129), when weighted by $2A/N_o$, result in equivalent weighting equal to symbol energy-to-noise density ratio. What are the peak and average input carrier powers?

6. a. Show the steps leading to the second equality in Equation (10.141).
   b. Using the small angle approximations for Equations (10.150) and (10.151) and neglecting the noise terms show the steps in establishing the phase detector error $\varepsilon_k(\phi)$ given by Equation (10.142).
   c. Using the small argument approximation $\tanh(x) \cong x - x^3/3$ show the steps in establishing the phase detector error $\varepsilon_k(\phi)$ given by Equation (10.143).
   d. For BPSK the low signal-to-noise approximation $\tanh(x) \cong x$ is used. What is the approximation to $\varepsilon_k(\phi)$ if this $\tanh(x)$ approximation is used for the QPSK phase detector?

7. Neglecting the noise terms and using the approximations in Equations (10.150) and (10.151) show that the coefficient C_1 of the ϕ_ε term in the polynomial of $\varepsilon_k(\phi) = \sum_i C_i \phi_\varepsilon^i$ is equal to $C_1 = 2\gamma_s^4/3$ and is independent of the data dI_k, $dQ_k = \pm 1$.

REFERENCES

1. F.M. Gardner, *Phaselock Techniques*, John Wiley & Sons, Inc., New York, 1967, p. 56 and the third edition, p. 257, Wiley-Interscience, Hoboken, NJ, 2005.
2. W.C. Lindsey, M.K. Simon, *Telecommunication Systems Engineering*, Dover Publications, Inc., New York, 1991.
3. J.G. Truxal, Automatic Feedback Control System Synthesis, Chapter 4, *Root-Locus Method*, pp. 221–277, McGraw-Hill Book Company, Inc., New York, 1955.
4. W.B. Davenport, Jr., "Signal-to-Noise Ratios in Band-Pass Limiters," Journal of Applied Physics, Vol. 24, No. 6, pp. 720–727, June 1953.
5. J.C. Springett, M.K, Simon, "An Analysis of the Phase Coherent-Incoherent Output of the Bandpass Limiter," IEEE Transactions on Communication Technology, COM-19, No. 1, pp. 42–49, February 1971.
6. F.M. Gardner, *Phaselock Techniques*, John Wiley & Sons, Inc., p. 56, New York, 1967.
7. R.C. Tausworthe, "Theory and Practical Design of Phase-Locked Receivers," JPL Technical Report No. 32-819, Jet Propulsion Laboratory, Pasadena, CA, February 1966.
8. W.C. Lindsey, K.T. Woo, "Analysis of Squaring Circuit Mechanizations in Costas and Squaring Loops," IEEE Transactions on Aerospace and Electronic Systems, Vol. AES-14, No. 5, pp. 756–763, September 1978.
9. W.C. Lindsey, M.K. Simon, "Detection of Digital FSK and PSK Using a First-Order Phase-Locked Loop," IEEE transactions on Communications, Vol. COM-25, No. 2, pp. 200–214, February 1977.
10. G.A. Korn, T.M. Korn, "*Mathematical Handbook for Scientists and Engineers: Definitions, Theorems, and Formulas for Reference and Review*," McGraw-Hill Book Company, New York, 1968.
11. F. M. Gardner, *Phaselock Techniques*, John Wiley & Sons, Inc., p. 33, New York, 1967.
12. L.A. Hoffman, *Receiver Design and the Phase-Locked Loop*, Aerospace Corporation, El Segundo, May 1963.
13. R. Jaffe, R. Rechtin, "Design and Performance of Phaselock Circuits Capable of Near-Optimum Performance Over a Wide Range of Input Signal and Noise Levels," IRE Transactions on Information Theory, Vol. IT-1, pp. 66–76, March 1955.
14. F. M. Gardner, Optimization of Loop Performance, Chapter 6, *Phaselock Techniques*, John Wiley & Sons, Inc., New York, 1967.
15. T. Nishimura, "Design of Phase-Locked Loop Systems with Correlated Noise Input," JPL Space Programs Summary No. 37-26, pp. 234–240, Jet Propulsion Laboratory, Pasadena, CA, April 30, 1964.
16. R. Goldstein, "The Minimization of Oscillator Noise," JPL Research Summary No. 36-14, pp. 61–63, Jet Propulsion Laboratory, Pasadena, CA, May 1, 1962.
17. H. Wiener, *Extrapolation, Interpolation and Smoothing of Stationary Time Series*, The MIT Press, Cambridge, MA, 1949.
18. R.H. Pool, "A Comparison of Second-Order and Third-Order Phase Lock-Loop Receivers," Report CRR-260, Collins Radio, Company, Cedar Rapids, IA, 1962.
19. S.C. Gupta, "Transient Analysis of a Phase-Locked Loop Optimized for a Frequency Ramp Input," IEEE Transactions on Space Electronics and Telemetry, Vol. 10, pp. 79–84, June 1964.
20. S.C. Gupta, R.J. Solem, "Optimum Filters for Second- and Third-Order Phase-Locked Loops by an Error-Function Criterion," IEEE Transactions on Space Electronics and Telemetry, SET-11, pp. 54–62, June 1965.
21. O.W. Eshbach, *Handbook of Engineering Fundamentals*, John Wiley & Sons, New York, 1952.
22. J.G. Truxal, *Automatic Feedback Control System Synthesis*, McGraw-Hill Book Company, Inc., New York, 1955.

23. S. Goldman, *Transform Calculus and Electrical Transients*, Prentice-Hall, Inc., New York, 1949.
24. M.F. Gardner, J.L. Barnes, *Transients in Linear Systems, Volume I: Lumped-Constants Systems*, John Wiley & Sons, New York, 1942. Fourteenth Printing, September 1958. (Contains an extensive table of Laplace transform pairs.)
25. J.P. Costas, "Synchronous Communications," Proceeding of the IEEE, Vol. 44, No. 12, pp. 1713–1718, December 1956.
26. M.K. Simon, "Optimum Receiver Structures for Phase-Multiplexed Modulations," IEEE Transactions on Communications, Vol. COM-26, No. 6, pp. 865–872, June 1978.
27. M.K. Simon, W.C. Lindsey, "Optimum Performance of Suppressed Carrier Receivers with Costas Loop Tracking," IEEE Transactions on Communications, Vol. COM-25, No. 2, pp. 215–227, February 1977.
28. M.K. Simon, "On the Optimality of the MAP Estimation Loop for Carrier Phase Tracking BPSK and QPSK Signals," IEEE Transactions on Communications, Vol. COM-27, No. 1, pp. 158–165, January 1979.
29. M.K. Simon, "The False Lock Performance of Costas Loops with Hard-Limited In-Phase Channel," IEEE Transactions on Communications, Vol. COM-26, No. 1, pp. 23–34, January 1978.
30. M.K. Simon, "Tracking Performance of Costas Loops with Hard-Limited In-Phase Channel," IEEE Transactions on Communications, Vol. COM-26, No. 4, pp. 420–432, June 1978.
31. J.K. Omura, M.K. Simon, "Modern Digital Communication Techniques for Satellite and Radio Channels," Continuing Education Institute, A National Short Course, Los Angeles, CA, pp. 20–23, July, 1982.
32. W.C. Lindsey, M.K. Simon, "Carrier Synchronization and Detection of Polyphase Signals," IEEE Transactions on Communications, Vol. COM-20, No. 3, pp. 441–454, June 1972.
33. R.W.D. Booth, "An Illustration of the MAP Estimation Method for Deriving Closed-Loop Phase Tracking Topologies: The MSK Signal Structure," IEEE Transactions on Communications, Vol. COM-28, No. 8, pp. 1137–1142, August 1980.
34. M.K. Simon, "Two-Channel Costas Loop Tracking Performance for UQPSK Signals with Arbitrary Data Formats," IEEE Transactions on Communications, Vol. COM-29, No. 9, pp. 1275–1289, September 1981.
35. M.K. Simon, W.K. Alem, "Tracking Performance of Unbalanced QPSK Demodulators: Part I – Biphase Costas Loop with Passive Arm Filters," IEEE Transactions on Communications, Vol. COM-26, No. 8, pp. 1147–1156, August 1978.
36. M.K. Simon, "Tracking Performance of Unbalanced QPSK Demodulators: Part II – Biphase Costas Loop with Active Arm Filters," IEEE Transactions on Communications, Vol. COM-26, No. 8, pp. 1157–1166, August 1978.
37. J.J. Stiffler, *Theory of Synchronous Communications*, Prentice-Hall, Inc., Englewood Cliffs, NJ, 1971.
38. M. Abramowitz, I.A. Stegun, *Handbook of Mathematical Functions With Formulas, Graphs, and Mathematical Tables*, National Bureau of Standards, Washington, DC, Applied Mathematics Series 55, June 1964.

11

WAVEFORM ACQUISITION

11.1 INTRODUCTION

Communication link budgets typically focus on the waveform detection requirements; however, an equally important consideration is the message acquisition link budget. The acquisition processing must detect the presence of the received signal and estimate the necessary parameters with sufficient accuracy to provide for message synchronization [1]. The message acquisition is typically designed to operate in a one or two decibel lower signal-to-noise ratio than that required for the message detection. An example link acquisition budget is given in Section 15.15. The more restrictive acquisition performance requirement and the necessity to estimate various received waveform parameters is usually offset by providing a unique preamble* to the message that is tailored to expedite the estimation processing. An important aspect of the preamble is that it provides the necessary integration time for parameter estimation. The fundamental issue in the acquisition processing is the time and frequency error of the received signal relative to the receiver and demodulator clocks and oscillators [2–4]. Time and frequency precorrection requires accurate estimation of the line of sight (LOS) range and range rate and of the receiver/demodulator oscillator and clock accuracies. In ground-to-satellite links, some degree of precorrection is usually required by the ground station to aid in the satellite's uplink acquisition and tracking and thereby to reduce the message overhead and processing complexity in the satellite. The time and frequency precorrection [5, 6] are each dependent on the estimation of two parameters:

- Time precorrection is dependent on the accuracy of the *system clock* and the *propagation delay* estimate.
- Frequency precorrection is dependent on the accuracy of the *system oscillators* and the *Doppler frequency* estimate.

The accuracy with which each parameter can be estimated is based on the transmitter and receiver system capabilities and leads to three fundamental precorrection concepts: *open loop* (OL), *pseudo-open loop* (POL), and *pseudo-closed loop* (PCL). These precorrection concepts can be applied independently, that is, one can be applied to time and another to frequency.

Open loop precorrection generally applies to transmit terminals with very accurate oscillators and clocks, extensive processing capabilities, and knowledge of the receiver terminals location and dynamics, for example, access to satellite orbit and ephemeris data. In this case, the transmit terminal provides autonomous time and frequency precorrections for the receiver to demodulate the message with a minimum amount of uplink acquisition overhead. *Pseudo-open loop* precorrection applies to transmit terminals with fewer capabilities and requires a downlink from the receiving terminal to aid in the uplink precorrection. The transmit terminal then uses estimates of the downlink *propagation delay* and *delay rate* (*Doppler frequency*) and the autonomous estimate of its

*Some applications use mid-ambles that occur at known locations within the message.

Digital Communications with Emphasis on Data Modems: Theory, Analysis, Design, Simulation, Testing, and Applications,
First Edition. Richard W. Middlestead.

Companion website: www.wiley.com/go/digitalcommunications

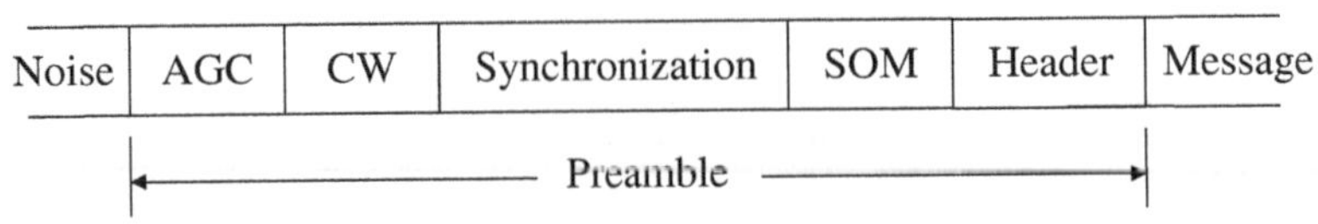

FIGURE 11.1 Message preamble segments.

own *system clocks* (*system oscillators*) to precorrection the uplink time and frequency. *Pseudo-closed-loop* precorrection involves downlink tracking, as in POL, with additional uplink acquisition and tracking by the receiver terminal based on less accurate autonomous estimates. In this case, the transmitting terminal attempts to zero the uplink precorrection error.

Time and frequency precorrection often takes place under the control of the network entry protocol [7] and, upon successful network entry, the time and frequency are maintained throughout the duration of the user's message traffic. Although precorrecting the transmitted waveform time and frequency reduces the communication overhead and message throughput and simplifies the receiver processing, the waveform acquisition discussed in this chapter is general and focuses on the various acquisition algorithms that can be applied under a variety of time and frequency conditions.

The acquisition preamble typically includes several segments to meet the acquisition requirements with a minimum of overhead. For example, the automatic gain control (AGC) and continuous wave (CW) segments provide for receiver gain setting, signal presence detection, and signal power and frequency estimation; the symbol synchronization segment facilitates symbol timing estimation and frequency tracking; the start-of-message (SOM) segment is characterized by uniquely coded pseudo-noise (PN) sequence that is used to determine the first data symbol location for subsequent message or message header detection. The message header is included to identify the message composition and aid in the message detection. In some applications the header bits are used to resolve or verify the correct bit polarity. The AGC segment is typically a short interval of CW transmission and can be thought of as the initial part of the CW segment.

Figure 11.1 depicts the order of the preamble segments and several example specifications are listed in Table 11.1; as indicted in the table, the preamble segments included in the message preamble are application specific. Although the preamble results in an undesirable message overhead, the message acquisition time is considerably less than that required to acquire the message in random data that can take several minutes as various frequency and timing hypotheses are examined. In time division multiple access (TDMA) applications, the users received waveform parameters are determined during network entry and updated or tracked after being assigned a network time slot with an appropriate guard time. Therefore, TDMA can accommodate many user channels with a minimum of overhead following network entry. The overall probability of acquisition is expressed, in terms of the various correct detection probabilities, as

TABLE 11.1 Example Preamble Segment Specifications[a]

AGC[b]	CW	Sync[c]	SOM[c]	Header[d]
10	—	114	74	—
22	—	156	74	—
14	—	111	37	—

[a]Defense Information Systems Agency (DISA) [8]. Courtesy of U.S.A. Department of Defense (DOD).
[b]Maximum time (ms).
[c]Bits.
[d]Header is application dependent.

$$P_{acq} = P_c(agc)P_c(cw)P_c(sync)P_c(som)P_c(header) \quad (11.1)$$

where

$$P_c(sync) = P_c(coarse\,freq)P_c(fine\,freq) \\ P_c(carrier\ track)P_c(symbol\ track) \quad (11.2)$$

With some waveform modulations the CW segment is not necessary because the parameter estimation can be accomplished using specialized symbol synchronization bit patterns; however, it does facilitate reliable AGC, signal presence detection, and estimation of the received signal power and frequency. With coherent data demodulation, the accuracy of the initial, or coarse, frequency estimate is determined by the pull-in frequency of the phaselock loop (PLL). Phase tracking is initiated in the symbol synchronization segment and the PLL must achieve phase-lock before the SOM segment. Referring to Chapter 10, the maximum frequency error to achieve phase-lock without cycle skipping in a second-order PLL is

$$F_L = 0.42(B_LT)R_s \quad : \text{PLL lock-in frequency, } \zeta = 0.707 \quad (11.3)$$

where B_LT is the time-bandwidth product of the PLL. The corresponding lock time is

$$T_L = \frac{0.53}{(B_LT)R_s} \quad : \text{PLL lock-in time, } \zeta = 0.707 \quad (11.4)$$

Using the PLL parameters $\zeta = 0.707$ and $B_LT = 0.1$ and 0.03 for binary phase shift keying (BPSK) and quadrature phase shift keying (QPSK), and phase-shaped offset quadrature

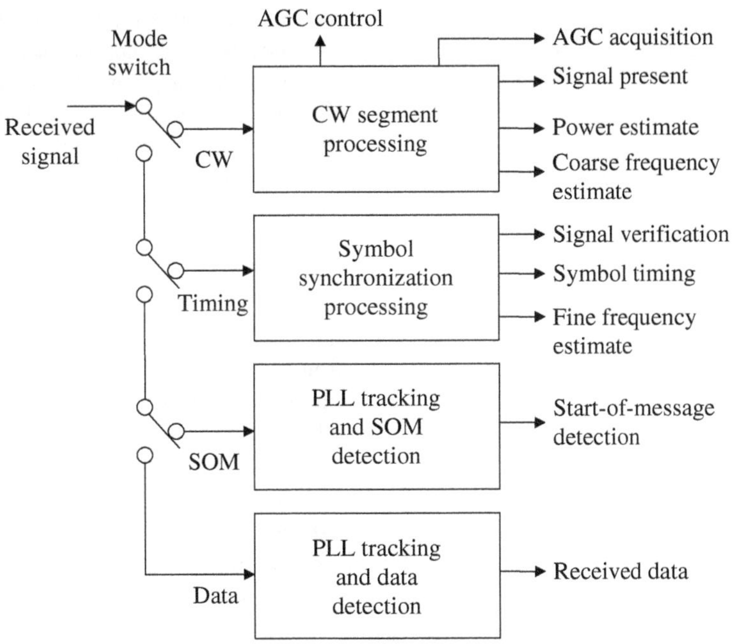

FIGURE 11.2 Functional processing of message preamble.

phase shift keying (S-OQPSK), respectively, requires that the accuracy of the coarse frequency estimate be measured within the lock-in frequency of the PLL given by

$$F_L = \begin{cases} 0.042R_b & : \text{BPSK} \\ 0.006R_b & : \text{QPSK} \end{cases} \tag{11.5}$$

Therefore, for a received carrier frequency of f_c, the accuracy of the coarse frequency estimate $\hat{f}_c$ must satisfy the requirement $\Delta f = |\hat{f}_c - f_c| < F_L$, in order for the PLL to acquire and track the received modulated waveform. The earlier lock-in criteria result in phase-lock without cycle slipping and the received signal-to-noise ratio must exceed the critical value γ_c for the selected B_LT product. The PLL will also acquire phase-lock if the frequency is within the pull-in range, F_P, of the loop; however, cycle skipping will occur resulting in a longer acquisition time and, consequently, a longer symbol synchronization segment.

The symbol synchronization preamble segment serves two important functions in the message acquisition processing: to provide for symbol time acquisition and tracking and, as mentioned earlier, to provide for carrier phase acquisition and tracking. Symbol time and carrier frequency synchronization and tracking are not mutually exclusive, that is, symbol synchronization cannot be achieved without carrier phase-lock and vice versa. Therefore, in the interest of minimizing the preamble overhead, parallel processing of joint timing and phase tracking is used. For example, if the signal processing capability is available, phase acquisition can be attempted in parallel at several symbol timing hypotheses. Or, if the preamble samples are stored in memory, symbol timing hypotheses processing can be performed sequentially by revisiting the stored preamble samples. With a sufficiently high-speed digital signal processor this can be accomplished in real time with an accompanying throughput delay. When a CW segment is not included in the preamble and the symbol synchronization segment is designed with an acceptable time–frequency correlation response, the correlation can be performed at multiple frequency hypotheses over the frequency uncertainty range, whereupon, choosing the time–frequency corresponding to the maximum correlation response will simultaneously provide coarse timing and frequency estimates. In this case, the time and frequency resolution must be adequate when revisiting the corrected stored samples to establish tracking prior to the SOM preamble segment.

The SOM synchronization is established by searching for the peak correlation response of the SOM segment that exceeds the constant false-alarm rate (CFAR) threshold. The SOM sequence is known by the demodulator and selected to provide low autocorrelation sidelobes as provided, for example, by PN codes such as M-sequences, Barker codes, and Neuman–Hofman synchronization codes. Generally, the first information or message bit follows immediately* after the last SOM bit. The SOM sequence correlation processing and the resulting correct SOM detection probability are established using a CFAR detection threshold as described in Section 11.2.2.1.

The functional outputs of the preamble segments shown in Figure 11.1 are depicted in Figure 11.2.

*On occasions the first information bit occurs a know number of bits after the last SOM bit to allow for signal processing software conversion from acquisition to the data detection mode.

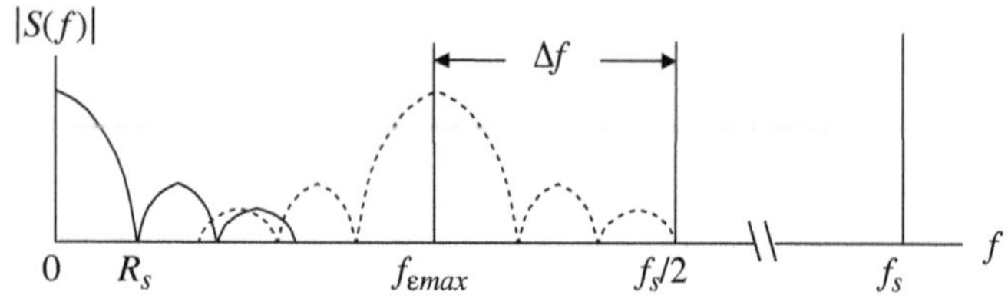

FIGURE 11.3 Received modulated signal spectrum.

The sampling frequency during the acquisition processing is an important design consideration and must account for the received signal modulation bandwidth and the carrier frequency error. For example, with rate r_c denoting the forward error correction (FEC) coding rate and k denoting the modulation bits per symbol, the transmitted symbol rate is given by

$$R_s = \frac{R_b}{r_c k} \tag{11.6}$$

In this case, the received modulated signal spectrum is related to the received symbol rate as shown in Figure 11.3 with the *dashed* spectrum corresponding to the maximum specified frequency error f_{emax}. The *sinc(fT)* spectrum shown in Figure 11.3 suggests a *rect(t/T)* symbol weighting; however, any modulation symbol spectrum can be considered with Δf selected to provide a safeguard against intersymbol interference and the antialiasing filter transition band distortion losses. When the spectrum describes the baseband or analytic signal, as assumed in this section, the frequency band $f_s/2$ to f_s represents the negative frequency band. Based on this depiction the sampling frequency for the modulated signal spectrum is determined from Nyquist's criterion as

$$f_s \geq 2(f_{emax} + \Delta f) \tag{11.7}$$

When the waveform acquisition is completed the coarse frequency error estimate is removed and the carrier frequency is being tracked by the PLL, so that the signal spectrum is the baseband spectrum shown as the *solid* curve in Figure 11.3. Under these circumstances, the sampling frequency is reduced using sample rate conversion to a suitably lower sampling frequency of $f_s = N_s R_s$ where N_s is typically 2 or 4 samples per symbol. Some implementations that achieve symbol time tracking by adding and deleting samples require a suitability higher sampling frequency, for example, because of the sensitivity to symbol timing errors, root-raised-cosine (RRC) frequency shaping requires $N_s = 32$; however, the matched filter integration can be accomplished at a lower sample rate.

The remainder of this chapter discusses and analyzes various processing algorithms for achieving the objectives of each of the preamble sections starting with the AGC processing discussed in the following section. Section 11.2.2 outlines several approaches to estimating the received carrier frequency using the CW preamble segment and Section 11.3 outlines several methods of further resolving the frequency and symbol time estimates using known data patterns including acquisition techniques that do not require the CW preamble segment. This section concludes with a discussion symbol and carrier tracking. Section 11.4 discusses correlation methods for the SOM detection. An important parameter in the correlation detection processing is the two-parameter censored CFAR threshold. Section 11.5 concludes this chapter with a discussion of various methods for estimating signal and noise powers during the CW and synchronization preamble segments as well as in random or unknown data. These power estimates are then used to form estimates of the received signal-to-noise ratio that is used for optimum phase tracking and often required for network centric medium access control layer and TDMA and frequency division multiple access waveform power control management.

11.2 CW PREAMBLE SEGMENT SIGNAL PROCESSING

11.2.1 Automatic Gain Control

The AGC is essential to ensure that the signal level into the analog-to-digital converter (ADC) is maintained to avoid clipping while preserving the dynamic range for signal fluctuations. The dynamic range is related to the number of bits associated with the ADC*; as a rule of thumb each bit of the ADC corresponds to 6 dB so an N_b-bit ADC will have a dynamic range of $6N_b$ dB. If the average signal level at the input to the AGC corresponds to 2 bits below the ADC saturation, then 12 dB is provided for intrinsic signal and noise fluctuations above the average signal level to avoid or minimize clipping. The selection of the average signal and noise power setting depends on the waveform modulation, channel noise, channel fading, and inband interference signal levels. The number of bits below the gain controlled average power level is also an important consideration in maintaining a linear representation of the sampled signal and is especially important when the received signal level is below the average noise level as with applications involving low-rate FEC coding and spread-spectrum waveforms. The AGC is often implemented entirely within the analog receiver; however, when the receiver interfaces with a digital demodulator, it is common to derive the gain control voltage in the digital domain and then use it to control the variable gain analog amplifiers. This section focuses on digitally generated gain control voltages.

*Refer to Chapter 2 for a more in-depth discussion of analog-to-digital conversion.

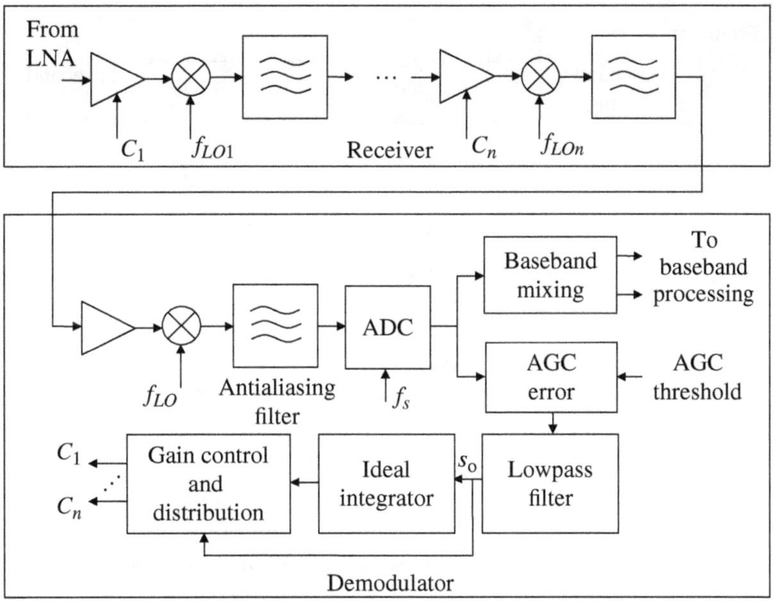

FIGURE 11.4 Bandpass sampled AGC implementation.

The AGC control is generally always operating and when a signal is not present the average receiver noise is maintained at the prescribed level into the ADC. In this case, the voltage controlled amplifiers are typically operating in a high-gain condition and when a high-level signal appears the system gain is reduced to maintain the adjusted signal plus noise at the prescribed level into the ADC. A high-level signal is characterized as having a signal-to-noise ratio greater than 0 dB as measured in the input bandwidth of the ADC; this corresponds to the output bandwidth of the final IF stage. For reasons involving specification control and subsystem testing, the final IF stage and the related local frequency oscillator form part of the demodulator subsystem. The IF frequency at the modem input is often 455 kHz for ultra-high frequency (UHF) modems and 70 MHz for SHF and EHF modems. The AGC time constant is typically characterized in terms of the attack and decay times. The attack time is the time required to adjust the gain to an increase in the signal level and should be as short as possible and the decay time is in response to a drop in the signal level and typically has a much slower response time.

There are a number of ways to generate the AGC control voltage in the demodulator; however, the most responsive control is derived as soon as possible following the ADC. When bandpass sampling or direct IF carrier sampling is used the AGC control can be derived from the digitally sampled carrier as shown in Figure 11.4.

The AGC error is generated by over sampling the carrier of the received modulated waveform and comparing the level of the sampled values to a reference voltage. When the samples are greater than the reference a positive unit-amplitude pulse is output to the low-pass filter (LPF) otherwise a negative unit-amplitude pulse is provided. With an equal number of positive and negative pulses over the period of the carrier frequency, the average LPF output is zero and the power into the ADC corresponds to the *rms* power of the received waveform. For example, consider that the power of a noise-free CW received waveform is to be adjusted by the AGC to be 1 bit below the ADC saturation voltage of $V_m = 1$ V. Assuming a 1-Ω resistive load, the power of the CW signal is given by

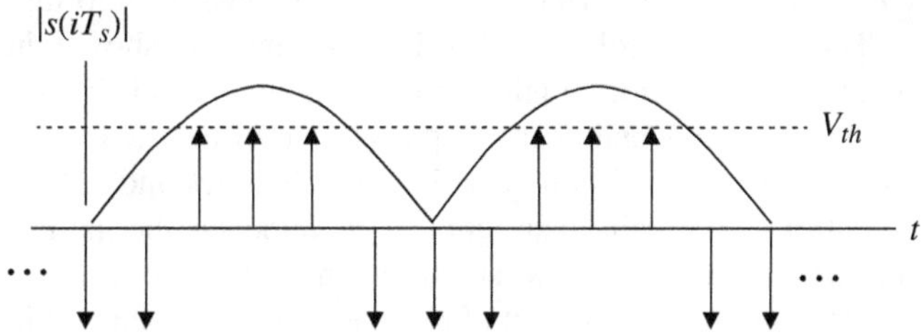

FIGURE 11.5 Sampled CW carrier AGC error discriminator ($V_{th} = V_{rms}$).

$$P_s = V_{rms}^2 \tag{11.8}$$

In this example, the AGC must adjust the signal power level such that $V_{rms}^2 = 0.5$ V. Referring to Figure 11.5, for an AGC threshold of $V_{th} = V_{rms}$ a carrier cycle is divided into equal increments of π radians above and below the threshold and the resulting average discriminator output is zero when the AGC reaches the steady-state condition.* This same

*Although the steady-state output of the AGC low-pass filter is to be zero, the ideal integrator finite output will maintain the gain of the receiver voltage controlled amplifiers corresponding to the prescribed *rms* carrier voltage at the ADC input. The integrator serves the same function as the voltage controlled oscillator (VCO) in the phase-lock loop.

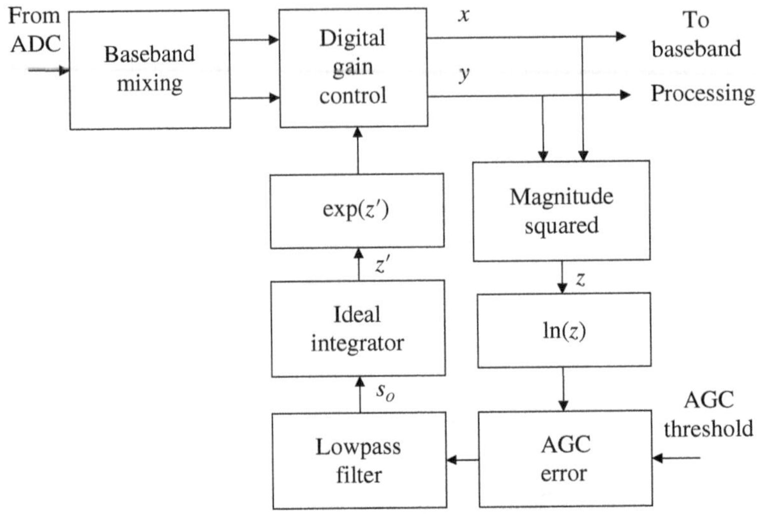

FIGURE 11.6 Baseband sampled AGC implementation.

phenomenon of providing a constant average voltage into the DAC will occur for arbitrary carrier-modulated waveforms.

The gain control and distribution function shown in Figure 11.4 provides logic for controlling the gain increments, the attack and decay response time of the AGC, and the various thresholds for declaring the AGC lock and unlock conditions. The gain distribution logic allocates the gain to the various gain controlled amplifiers in the receiver subsystem to minimize the impact of receiver noise as discussed in Section 15.2.1. The performance of the bandpass sampled AGC is examined in the case study in Section 11.2.1.1.

The gain control voltage can also be generated from the quadrature rails of the baseband received signal obtained by mixing the input carrier frequency directly to baseband. The outputs of the quadrature matched filters provide the optimum, that is, the maximum signal-to-noise samples for estimating the received signal power for the AGC acquisition and tracking. For example, joint power control and PLL tracking can be accomplished with BPSK modulation using the in-phase or $A\cos(\phi_\varepsilon)^*$ rail output and estimating the signal power as $A^2/2$ when phase-lock is achieved. However, this example has limited application because it is often necessary to establish AGC before carrier phase acquisition and tracking. The quadrature rails can be used for AGC acquisition and tracking as shown in Figure 11.6. The functions in the baseband AGC implementation are similar to those of the bandpass sampled AGC shown in Figure 11.4; however, in this case, the logarithmic functions significantly reduce the dynamic range requirements of the LPF and ideal integrator.† The LPF output s_o is input directly into the digital gain control function to provide for versatile gain control as described earlier. When AGC acquisition is declared, the bandwidth of the LPF is reduced to provide a slow decay time for improved tracking performance by providing hysteresis in the response.

*This is the noise-free analytic signal description of the in-phase rail.
†Private conversations with Matthew Davis, Vulcan Wireless, Inc., November, 2009.

11.2.1.1 Case Study: Bandpass Sampled AGC Performance Evaluation This case study examines the performance of a UHF modem AGC with the gain control derived from the demodulator input IF of 455 kHz sampled at a rate of $f_s = 6144$ kHz. The noise bandwidth of the antialiasing filter is 80 kHz. The maximum receiver gain is 135 dB with a minimum detectable input signal level of −135 dBm. Referring to Figure 11.4, a 10-bit DAC is used and the AGC reference input is $V_{ref} = V_{rms}$ where V_{rms} is the root-mean-square voltage of the sampled carrier. The reference voltage is 12 dB below V_m leaving two magnitude bits for additive noise and peak signal fluctuations above V_{ref}. The following simulated performance of the AGC is based on a noise-free CW received signal with the demodulator input signal at 455 kHz under two conditions of the receiver input level: −5 and −60 dBm. In both cases, the receiver gain is set to the maximum gain of 135 dB and, prior to the received signal, the receiver input is zero, that is, receiver noise is not included.

The low-pass AGC filter is a cascade of four synchronously tuned single-pole filters with an overall bandwidth of 200 Hz. In the following description, the LPF output is denoted as s_o. The operation of the AGC control is similar to all adaptive feedback control systems, in that, the error signal s_o forms a discriminator S-curve providing positive and negative gain adjustments resulting in zero average filter output, $\langle s_o \rangle = 0$, under steady-state conditions. Under the

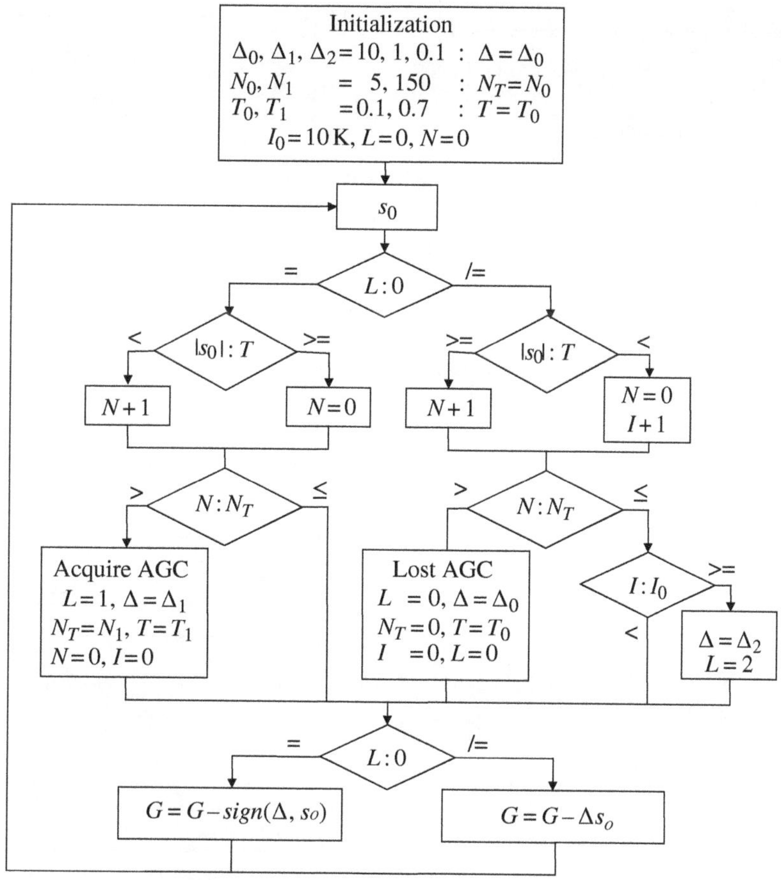

FIGURE 11.7 Gain control processing diagram.

steady-state conditions ideal integrator output corresponds to the optimum receiver gain setting of $V_{rms} = V_{ref}$.

To provide more control over the AGC performance, than by simply letting the ideal integrator output control the receiver gain, the simulated gain control function provides discrete gain adjustment based on the filter output s_o and various thresholds as outlined in Figure 11.7. To this end, the ideal integrator output is replaced by the gain control logic using three fixed gain increments Δ_i that are applied in succession as the filter output falls to zero. The last two gain increments are weighted by the filter output and are proportionally decreased as s_o approaches zero. The thresholds T_0 and N_0 establish the conditions for the declaration of initial AGC acquisition and T_1 and N_1 establish the conditions for declaring the loss of AGC acquisition. The parameter I_0 is the number samples corresponding to one-third of the LPF time constant and invokes the final gain control increment Δ_3 and the declaration of AGC acquisition. Taken together, these parameters establish the AGC attack and decay time. The selection of the AGC control parameters offers considerable design flexibility in the AGC performance and the logic is easily expanded to include additional capabilities. For example, using logic to track signal fade rates will allow for longer and deeper signal fading conditions or temporary loss of signal power before declaring *lost* AGC. The gain control logic can also be used to suspend demodulator symbol time tracking during fading and loss-of-signal conditions. Although not shown in Figure 11.7, the control logic also distributes the gain among the various gain-controlled amplifies so as to preserve the receiver noise figure.*

The simulated performance of the AGC with a noise-free CW received signal is shown in Figure 11.8 for received signal power levels of −5 and −60 dBm. The curves represent the receiver gain and the declaration of AGC acquisition or detection via the parameter $L = 2$. The first 10 ms of the 50 ms AGC simulated response is shown. The AGC reference voltage is set at $V_{ref} = V_{rms}$ where V_{rms} is the voltage of the ADC input signal and is 12 dB below saturation of the 10-bit DAC. The receiver gain is initially set to the maximum gain of 135 dB and, with the noise-free assumption, the receiver input is zero, so under these conditions, application of the CW input signal at $t = 0$ results in an acquisition time 1.9 ms for both the −5 and −60 dBm received signal levels. The acquisition time is essentially determined by the bandwidth of the LPF.

*The function of receiver gain distribution may rightly be the responsibility of the receiver subsystem.

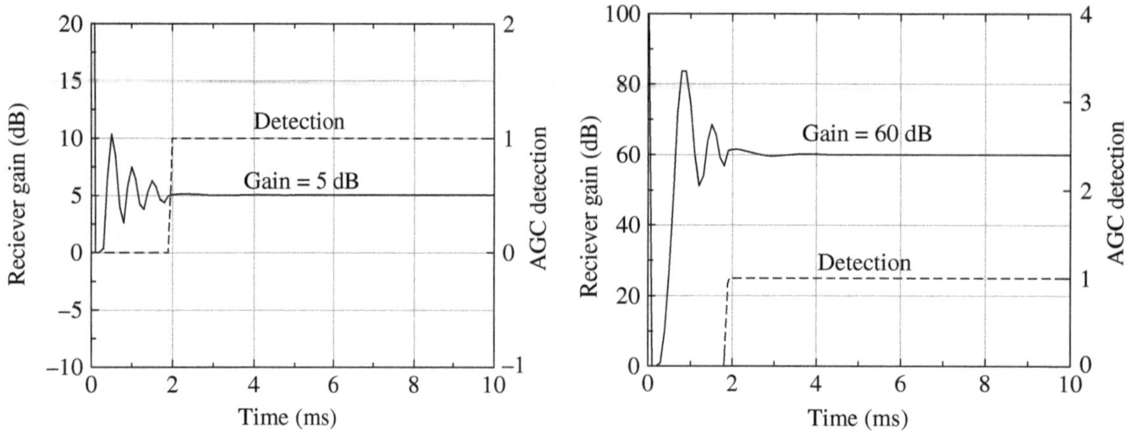

FIGURE 11.8 AGC response for −5 and −60 dBm received signals.

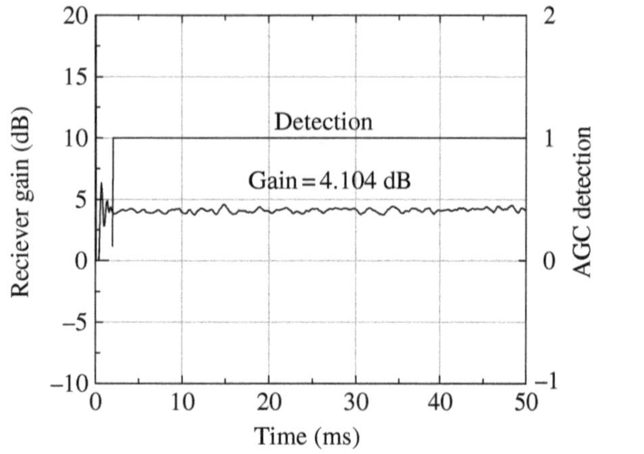

FIGURE 11.9 AGC response for −5 dBm received FEC coded QPSK waveform (19.2 kbps, code rate = 1/2, E_b/N_o = 4.25 dB at $P_{be} = 10^{-5}$, B_n = 80 kHz).

Figure 11.9 shows the AGC response over the full 50 ms of the simulation using a rate 1/2 convolutional coded 19.2 kbps QPSK-modulated waveform. The simulated signal-to-noise performance of this coded waveform, using a constraint length seven Viterbi decoder with infinite quantization, corresponds to an E_b/N_o of 4.25 dB at $P_{be} = 10^{-5}$. Therefore, because the ADC quantization noise is negligible compared to the receiver noise, the signal-to-noise ratio in the 80 kHz noise bandwidth of the antialiasing filter is −1.95 dB. Under these conditions the AGC acquisition time is 2.03 ms.* The samples s_o in Figure 11.9 are obtained from the AGC LPF and occur at a rate of 2 kHz.

*The acquisition times in this case study correspond to the simulation performance for one noise ensemble and several Monte Carlo simulations must be performed to determine the statistical performance.

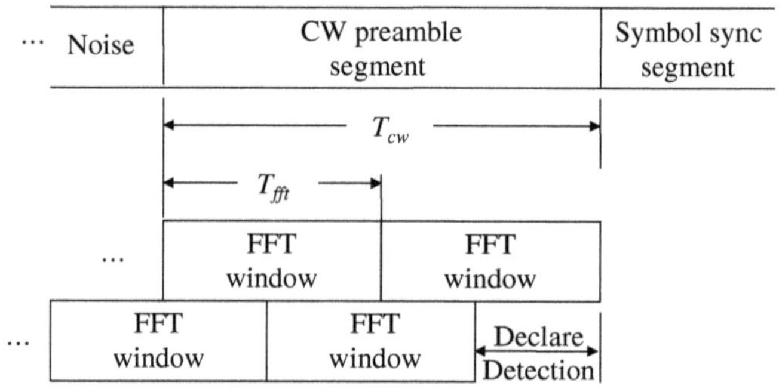

FIGURE 11.10 CW segment FFT processing.

11.2.2 Coarse Frequency Estimation

In this section, several methods of determining the received signal frequency error relative to the demodulator local oscillator frequency are examined. When the message preamble includes the CW segment, as shown in Figure 11.1, the fast Fourier transform (FFT) provides an efficient method for estimating the frequency error as described in Section 11.2.2.1. An alternate method using a frequency discriminator (FD) is discussed in Section 11.2.2.4. When the CW segment is not included, the data pattern in the symbol synchronization segment is often specialized to provide for frequency and symbol time estimation as discussed in Section 11.3.

11.2.2.1 Frequency Estimation Using the FFT In this section, the determination of the received carrier frequency error using the CW preamble segment is accomplished by performing an N_{fft}-point FFT as depicted in Figure 11.10. To account for the uncertainty of not knowing where the preamble starts, FFTs must be performed sequentially until detection is declared. Declaration of signal detection is based on a FFT frequency cell exceeding the CFAR threshold. To maximize the use of the CW preamble duration, overlapping FFTs are used and, to increase the correct frequency detection

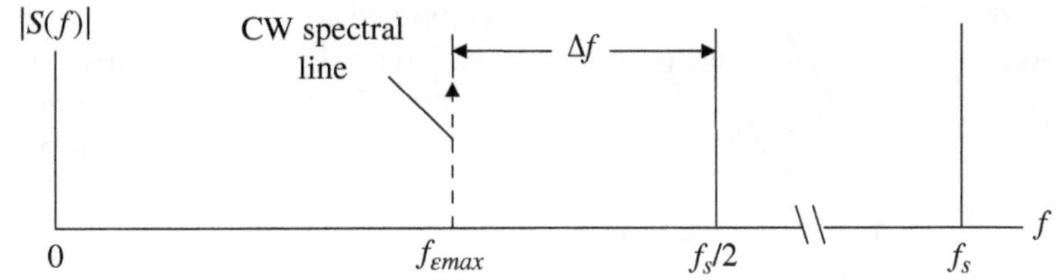

FIGURE 11.11 Received CW segment signal spectrum.

probability, two successive FFT detections are required with the second occurring within ±1 frequency cell of that declared during the previous FFT. Requiring two successive detections under these conditions also reduces the false-detection probability. A CW segment duration of $T_{cw} = 2T_{fft}$ will guarantee that two FFTs with one overlapping FFT can be performed; $T_{cw} = 2.5T_{fft}$ guarantees three FFTs with one overlapping FFT can be performed; $T_{cw} = 3T_{fft}$ guarantees four FFTs with two overlapping FFTs can be performed. In general, for 50% FFT window overlap with $T_{cw} = \rho T_{fft}$ where $\rho = 1, 1.5, 2, 2.5, 3, 3.5, \ldots$ guarantees that N FFTs can be performed with $N = 2\rho - 1 : \rho \in \text{integer}$ and $N = 2\lfloor\rho\rfloor : \rho \notin \text{integer}$. The fundamental frequency resolution of the FFT is defined as the reciprocal of the FFT window, that is,

$$f_{res} \triangleq \frac{1}{T_{fft}} \tag{11.9}$$

For a given CW segment duration, increasing the number of FFTs requires a smaller FFT window resulting in less frequency resolution. Various trade-offs between the CW segment duration, the detection performance, the frequency estimation accuracy, and the FFT parameters are discussed in the remainder of this section.

The received CW signal spectrum shown in Figure 11.11 is similar to that shown in Figure 11.3 with the modulated signal spectrum replaced by the discrete spectral line $S(f) = \delta(f - f_e)$; in the figure the frequency error is shown as the maximum specified error* f_{emax}. In this case, the transition or guard frequency (Δf) is related exclusively to the transition band of the antialiasing filter, that is, there are no spectral sidelobes to contend with as shown in Figure 11.3.

The key performance parameter for the CW signal detection is the carrier power to noise power spectral density ratio (C/N_o) that is related to E_b/N_o as

$$\frac{C}{N_o} = \frac{E_b}{N_o} R_b \quad : \text{hertz} \tag{11.10}$$

For comparable performance, independent of the bit rate, the duration of the CW segment is often specified in terms of the number of baseband data bits (N_B) as

$$T_{cw} = \frac{N_B}{R_b} \quad : \text{seconds} \tag{11.11}$$

Referring to Figure 11.11 the sampling frequency is expressed as

$$f_s = 2(f_{emax} + \Delta f) \tag{11.12}$$

The number of waveform samples in the CW segment is determined as $N_s = T_{cw}/T_s = T_{cw} f_s$ and, upon expressing the bit rate in terms of the symbol rate as $R_b = k r_c R_s$ and using these results and (11.11) with $N_s = \lfloor T_{cw} f_s \rfloor$, the number of CW samples is evaluated as

$$N_s = \left\lfloor \frac{N_B}{k r_c} \left(\frac{f_s}{R_s} \right) \right\rfloor \tag{11.13}$$

Using (11.13) the size of the FFT is determined as

$$N_{fft} = \left\lfloor \frac{N_s}{\rho} \right\rfloor \tag{11.14}$$

where ρ is the number of FFTs to be performed over the CW interval† T_{cw}. Equation (11.14) generally requires using a mixed radix FFT; however, a radix-2 FFT can be used by choosing $N_{fft} = 2^n$: $n \in$ positive integer such that $2^n \le \lfloor N_s/\rho \rfloor$. Unfortunately, the radix-2 FFT often results in $2^n \ll \lfloor N_s/\rho \rfloor$ and corresponds to the inefficient use of the CW interval and less frequency resolution.‡ The CW window utilization efficiency is defined as

$$\eta_{cw} \triangleq \frac{\rho T_{fft}}{T_{cw}} = \frac{\rho N_{fft}}{N_B} \left(\frac{R_b}{f_s} \right) \tag{11.15}$$

where $T_{fft} = N_{fft}/f_s$ is the FFT window duration.

The sampling frequency can be increased to improve the utilization efficiency of the CW interval, thereby improving

*The maximum frequency error is determined by the frequency accuracies of the transmitter and receiver subsystems, the communication relay system if one is used, and the Doppler frequency given the dynamics of the communication channel(s).

†The number ρ is not necessarily an integer and typically $\rho = 2, 2.5, 3$, etc., as described previously.

‡The frequency resolution is equal to the inverse of the FFT window T_{fft}, that is, $f_{res} = 1/T_{fft}$.

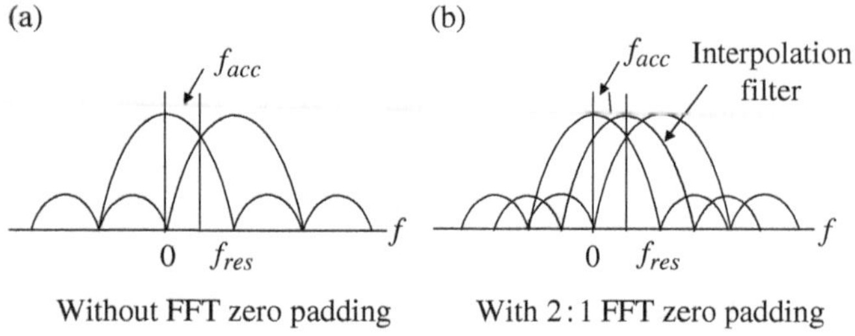

FIGURE 11.12 Frequency resolution and accuracy.

the frequency resolution, using a radix-2 FFT. This is accomplished by choosing a larger number of samples per CW interval. For example, by choosing n' such that $N'_{fft} = 2^{n'} = \lceil N_s/\rho \rceil$, the number of samples per CW interval becomes $N'_s = \rho N'_{fft}$ and, using (11.13), the adjusted normalized sampling frequency becomes

$$\frac{f'_s}{R_s} = kr_c\left(\frac{N'_s}{N_B}\right) \tag{11.16}$$

The FFT provides for a frequency estimation accuracy of $|f_{acc}| = f_{res}/2$; however, this estimation accuracy can be improved by using interpolation between the FFT frequency cells that are separated by f_{res}. Improving the FFT resolution using zero padding is discussed in Section 1.2.7 and 2 : 1 zero padding results in a frequency estimation accuracy of $|f_{acc}| = f_{res}/4$. The frequency resolution and accuracy are depicted in Figure 11.12 for the rectangular weighted FFT with and without zero padding.

An important consideration in the frequency estimation processing is establishing a signal detection algorithm for declaring *signal present* and estimating the signal frequency error. For example, for an N_{fft}-point FFT there are N_{fft} possible frequency locations, however, when the signal is present only one location, or possibly two contiguous locations, correspond to the frequency of the received signal. The signal detection algorithm must then provide for declaring that a signal is present for subsequent acquisition processing, when the signal is not present it must provide for continued searching within the CW signal segment. The importance of the signal detection algorithm cannot be overstated: it must provide for a low probability of false detection and a high probability of correct signal detection with an acceptable coarse frequency estimation.

The signal detection algorithm used in conjunction with the coarse frequency estimation processing is the CFAR algorithm that uses a detection threshold based on the magnitude of the signal plus noise in the frequency cells around a selected frequency cell, referred to as the cell under test. The cell under test is defined as the FFT frequency cell currently being examined under the hypothesis that it corresponds to the correct received signal frequency. The threshold is based on a two-parameter censored CFAR with the two parameters computed as the mean and standard deviation of the cells excluding the cell under test and N_{censor} cells on either side of the cell under test. With this definition N_{censor} represents the number of one-sided censored cells and the threshold is computed as

$$T_{hr} = m_c + \kappa s_c \tag{11.17}$$

where κ is the threshold factor selected to meet the specified system detection and false-alarm probabilities. The mean and standard deviation are based on a finite sample population size as outlined in Section 1.13.3. Denoting the complex spectrum sample in each of FFT frequency cell as c_n: $n = 1, \ldots, N_{fft}$ the mean and standard deviation are computed in consideration of the censoring, as

$$m_c = \frac{1}{N}\sum_{n=1}^{N}{}' |c_n| \tag{11.18}$$

and

$$s_c = \sqrt{\frac{\sum_{n=1}^{\prime N} |c_n|^2 - Nm_c^2}{N-1}} \tag{11.19}$$

The primed summations signify that the summation excludes the cell under test and the $2N_{censor}$ censored cells such that $N = N_{fft} - (2N_{censor} + 1)$. The censoring reduces the influence of the cell under test and the adjacent cells on the censored mean and standard deviation. The influence of the censored cells is related to the spectral sidelobes of the signal. For example, the censoring for a rectangular windowed FFT without interpolation is typically $N_{censor} = 2$ and with a nonuniform weighted FFT window and/or interpolation, censoring values of 4–6 are often used.

The frequency cell identified by the CFAR processing is used to compute the frequency estimate using early–late (E/L) gate interpolation. The frequency estimates from multiple FFTs separated by known intervals of ΔT seconds are used to estimate the received signal Doppler or frequency rate as

$$\dot{f}_{est} = \frac{f_{est}(2) - f_{est}(1)}{\Delta T} \tag{11.20}$$

where $f_{est}(1)$ and $f_{est}(2)$ are successive FFT frequency estimates. These design concepts involving signal present detection and coarse frequency estimation are discussed in more detail using the example in the following case study.

11.2.2.2 Case Study: FFT Signal Detection and Frequency Estimation In this case study, signal present detection and carrier frequency error estimation are examined using an example involving a 19.2 kHz BPSK-modulated waveform without FEC coding. The maximum specified frequency error is $f_{emax} = 10$ kHz and the FFT processing of the CW preamble segment uses $\rho = 2.5$ that guarantees that three FFT can be performed with one overlapping FFT. The signal detection is based on two ($N_{det} = 2$) consecutive FFT detections. The first FFT detection is declared if the maximum cell magnitude exceeds the CFAR threshold. The second FFT detection is declared if the threshold is exceeded in the same or an adjacent cell to that of the first detection; this corresponds to a frequency estimate within f_{acc} of that estimated in the first detection. The CFAR threshold is based on a two-parameter CFAR with $N_{censor} = 2$ cell censoring. The frequency estimation uses parabolic E/L interpolation. In this evaluation, the signal detection and frequency estimation are examined for two FFT windows and interpolation conditions: the rectangular window with zero padding and the Hanning window with and without zero padding.

The sampling frequency is evaluated using (11.12) with a guard frequency $\Delta f = 2R_s = 38.4$ kHz yielding $f_s = 96.8$ kHz. For this uncoded BPSK example $k = r_c = 1$, $R_s = R_b$ and, using $N_B = 88$ information bits per CW segment, the number of samples is found from (11.13) to be $N_s = 443$ and the FFT size is computed using (11.14) with the result $N_{fft} = 177$. Therefore, a 177-point FFT can be used for the signal detection and frequency estimation; however, because 177 is only divisible by 3 and 59 a mixed radix FFT must be used. The more computationally efficient radix-2 FFT with N_{fft} less than 177 uses $N_{fft} = 128$; however, from (11.15) the CW window utilization efficiency is only 72% and the frequency resolution is $f_{res} = 756.25$ Hz.

Because of the poor utilization efficiency and resolution frequency, the FFT size is increased to $N'_{fft} = \lceil N_s/\rho \rceil = 256$. With this modification, the total number of FFT samples becomes $N'_s = \rho N'_{fft} = 640$ and the adjusted sampling frequency is computed using (11.16) and found to be 139.636 kHz; this is rounded up to yield $f'_s = 140$ kHz. In this case, the FFT efficiency is 99.7% with $f_{res} = 546.875$ Hz. The greatest common divisor of the sampling frequency and symbol rate is $gcd(140\text{K}, 19.2\text{K}) = 800$ so these rates are derived from a high-frequency clock of $f_{clk} = 19.2\text{K}(140{,}000/800) = 140\text{K}(19{,}200/800) = 3.36$ MHz. Table 11.2 summarizes the parameters used in this case study for the CW segment acquisition processing.

The CW preamble segment signal detection and frequency estimation performance, operating under the conditions listed in Table 11.2, are evaluated using computer simulations and the results are shown in the following figures. Figures 11.13 and 11.14 show the signal detection performance in terms of the detection and false-alarm probabilities as a function of the CFAR threshold factor k using rectangular and Hanning FFT windows, respectively, with zero padding and equivalent E_b/N_o signal-to-noise ratios of 0 and −3 dB. Referring to Figure 11.12b, the peaks or maximum magnitudes of the FFT outputs occur at $273.4375i$ Hz: $i = 0, \ldots, N_{fft} - 1$ and these conditions correspond to the *best case* performance, whereas the minimum magnitudes correspond to $273.4375i + 136.71875$ Hz and represent the *worst case* performance. The false-alarm probability is conditioned on two

TABLE 11.2 CW Segment Acquisition Processing Parameters

Parameter	Value	Comments
Specified parameters		
Data modulation	—	BPSK
Bit rate	19.2	R_b (kbps)
Symbol rate	19.2	R_s (ksps)
Bits per symbol	1	k
FEC code rate	1	r_c
Maximum frequency error	10	f_{emax} (kHz)
Guard frequency	$2R_s$	Δf (kHz)
Bits per CW segment	88	N_B
FFTs per CW segment	2.5	ρ
Consecutive detections	2	N_{det}
Cell censoring	2	N_{censor}
E/L interpolation	—	Parabolic
FFT window	—	Rectangular and Hanning
Computed parameters		
Sample rate	140	f'_s (kHz)
System clock	3.36	f_{clk} (MHz)
Samples per CW segment	640	N'_s
FFT size[a]	256	N_{fft} without padding
Frequency resolution	546.875	f_{res} (Hz)
Frequency accuracy	273.438	f_{acc} (Hz), without FFT padding
	136.719	f_{acc} (Hz), with FFT padding

[a]FFT size is without zero padding; with 2 : 1 zero padding FFT size is doubled.

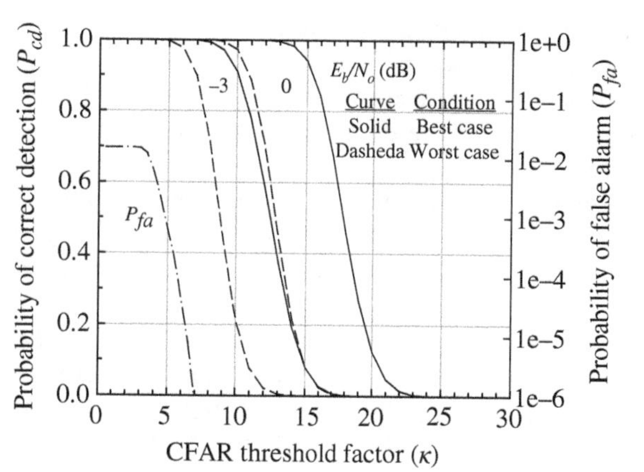

FIGURE 11.13 CW signal detection performance using the FFT (rectangular window with zero padding).

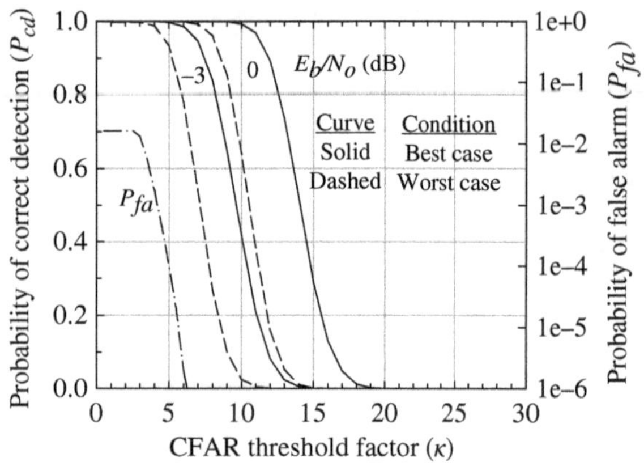

FIGURE 11.14 CW signal detection performance using the FFT (Hanning window with zero padding).

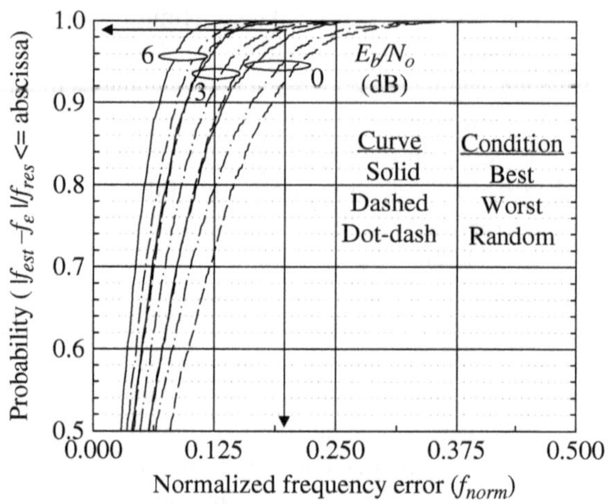

FIGURE 11.15 CW frequency estimation using FFT (rectangular window with zero padding).

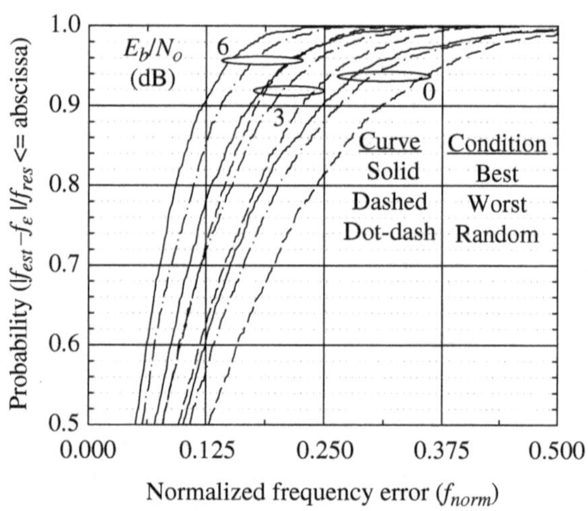

FIGURE 11.16 CW frequency estimation using the FFT (Hanning window with zero padding).

consecutive FFT detections and is obtained with a noise-only input. All of the simulation results are based on Monte Carlo simulations of 5000 trials for each threshold so the false-alarm results are projected below about $P_{fa} = 10^{-4}$.

The simulated frequency estimation performance is characterized in terms of a histogram representing the cumulated distribution function (*cdf*) shown in the following figures. The normalized frequency error is expressed as $|f_{est} - f_\varepsilon|/f_{res}$ where f_{est} is the estimate of the received signal frequency error based on the parabolic E/L estimation algorithm. The histogram consists of 400 bins over the positive frequency range of 0 to f_{res} corresponding to a bin resolution of $\delta f = 1.3671875$ Hz. The abscissa of the *cdf* is the normalized frequency ratio $f_{norm} = i|\delta f|/f_{res}$: $i = 1, \ldots, 400$ with f_{norm} limited to $f_{norm}(\max) = 1/2$ in the *cdf* plots.* The *cdf* is shown for the *best* and *worst* cases as defined earlier and the *random* frequency case where the frequency error is uniformly distributed between $\pm f_{acc}$.

Figure 11.15 shows the three performance conditions for signal-to-noise ratios equivalent to $E_b/N_o = 0$, 3, and 6 dB using a rectangular window with zero padding corresponding to 2 : 1 FFT interpolation. As an example application, the *arrows* in Figure 11.15 correspond to the *worst* case performance at a signal-to-noise ratio of 3 dB and, under this condition, the frequency estimation error is $\delta f \le 0.19 f_{res} = 103.9$ Hz with a probability of 0.99. Referring to (11.3) and (11.4) and using $B_L T = 0.1$ for BPSK modulation, the lock-in frequency and time for a second-order PLL are 806 Hz and 0.28 ms, respectively, and $f_{est} = 103.9$ Hz is well within the lock-in frequency range of the PLL.

These conditions are repeated in Figures 11.16 and 11.17 using a Hanning window with and without zero padding, respectively. The performance of the Hanning window with zero padding is considerably degraded from that of the rectangular window performance shown in Figure 11.15. The performance difference is attributed to the lower discriminator gain that is a consequence of the inherent wider spectral bandwidth of the Hanning window.

11.2.2.3 Frequency Estimation Using the Pipeline FFT The pipeline FFT described in Section 1.2.5.1, although more signal processing intense than the block FFT described earlier, provides an efficient method of simultaneous signal detection and frequency estimation that allows for a shorter CW preamble segment. The benefits are a consequence of the sequential processing that results in a continuous *push-broom* acquisition over the range of the frequency

*Note that f_{norm} represents a frequencies ratio and is dimensionless.

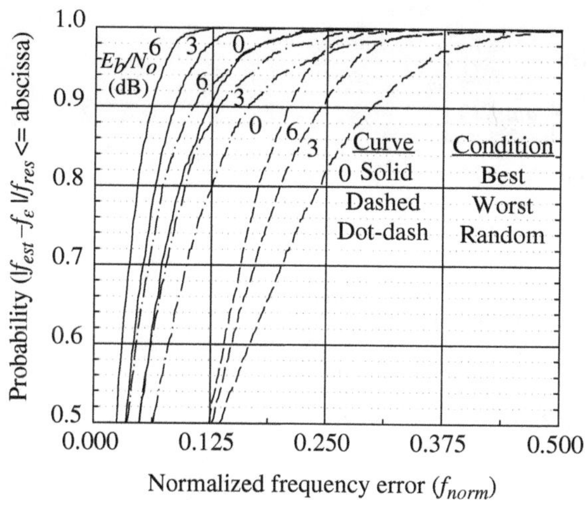

FIGURE 11.17 CW frequency estimation using the FFT (Hanning window without padding).

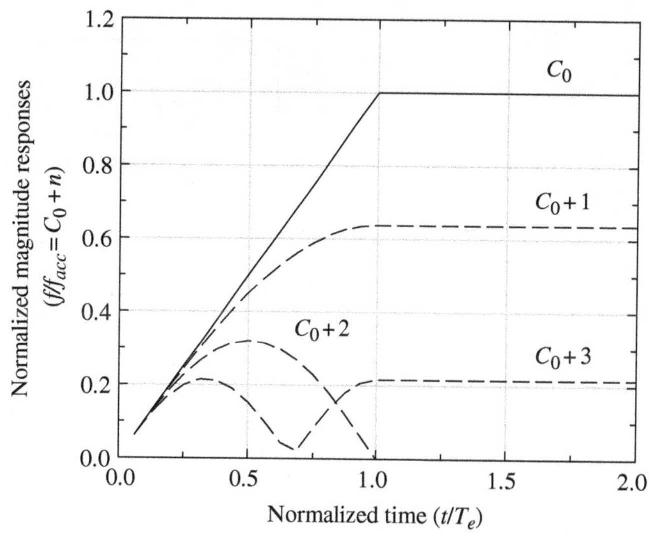

FIGURE 11.19 Pipeline FFT response of optimum and neighboring cells.

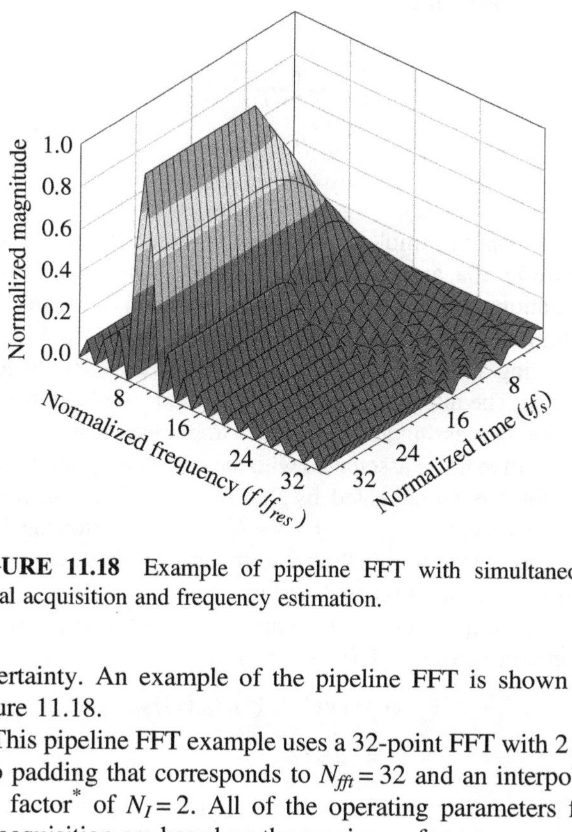

FIGURE 11.18 Example of pipeline FFT with simultaneous signal acquisition and frequency estimation.

uncertainty. An example of the pipeline FFT is shown in Figure 11.18.

This pipeline FFT example uses a 32-point FFT with 2 : 1 zero padding that corresponds to $N_{fft} = 32$ and an interpolation factor* of $N_I = 2$. All of the operating parameters for the acquisition are based on the maximum frequency uncertainty of the received signal f_{emax}, the frequency guard-band Δf, and the sampling frequency f_s as described by (11.7). In this case, the determination of the sampling frequency is somewhat simpler than that discussed in Section 11.2.2.1 because overlapping block FFTs are not involved, also, because of the CW signal, the guard band depends solely on the transition band of the antialiasing filter. Therefore, upon determining the sampling frequency, the estimation interval is determined using the sampling interval $\delta t = 1/f_s$ as†

$$T_e = \frac{N_{fft}/N_I}{f_s} \tag{11.21}$$

and the frequency resolution is $f_{res} = 1/T_e = f_s/N_{fft}$. With zero padding the accuracy of the frequency measurement is $f_{acc} = f_{res}/N_I$. Using these relationships, the frequency axis in Figure 11.18 spans the frequency range $32 f_{acc}$ and the time axis spans the range $32\delta t$ or $2T_e$.

In the noise-free simulation of Figure 11.18, the CW frequency tone is placed in the center of the frequency cell $f/f_{acc} = C_0 = 7$ (cell 0 corresponds to the zero frequency) and Figure 11.19 shows magnitude response of cell C_0 and several neighboring cells. The time is normalized to the estimation interval T_e and the optimum cell output increases linearly, reaching the optimum value at $t = T_e$. The cells $C_0 \pm 1$ are used for E/L gate frequency tracking and are included in the CFAR censoring during acquisition. When searching for an acquisition detection and performing detection verifications the CFAR detection algorithm, discussed in the preceding section, is executed at regular intervals of, for example, $T_e/2$.

The resolution bandwidth, f_{res}, of a uniformly weighted FFT with an underlying size $N_{fft}/N_I \geq 8$ is essentially equal to the noise bandwidth, so increasing the underlying FFT size reduces the noise power in each FFT cell; however, the scalloping, leakage, and aliasing losses must also be dealt with in

*The interpolation factor $N_I = 2$ refers to the estimation improvement resulting from a 100% zero padded FFT.

†The time axis is not influenced by the zero padded interpolation expansion.

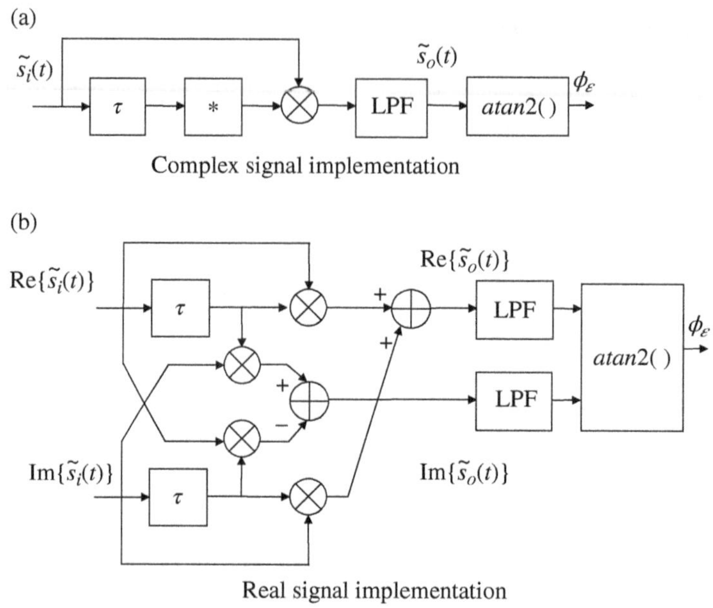

FIGURE 11.20 Phase discriminator implementations.

the acquisition processing. FFT interpolation reduces the scalloping loss and the worst case signal-to-noise loss due to interpolation.

11.2.2.4 Frequency Estimation Using Discriminator In this section, the FD is examined that provides an estimate of a received signal carrier frequency error during acquisition. The basic implementation of the FD is then extended to further resolve the frequency estimate by using smaller delays that are related by powers of two and, in this regard, this implementation is similar to the *butterfly* element of the FFT. A fine frequency estimator is then described that essentially *zooms* in on the initial coarse estimate to provide considerably higher resolution.

To clarify the description of the FD, a noise-free received CW signal is expressed as

$$s_i(t) = A\cos(\omega_o t + \omega_\varepsilon t + \phi_o) \qquad (11.22)$$

where ω_o represents the transmitted carrier angular frequency, ω_ε is an unknown angular frequency error involving the Doppler frequency and various oscillator frequency errors, ϕ_o is an arbitrary phase angle, and A is the peak carrier voltage. The complex envelope of $s_i(t)$ is given by

$$\begin{aligned}\tilde{s}_i(t) &= Ae^{j(\omega_\varepsilon t + \phi_o)}\\ &= s_{ci}(t) + js_{si}(t)\end{aligned} \qquad (11.23)$$

and represents the respective baseband in-phase and quadrature terms are $s_{ci}(t)$ and $s_{si}(t)$. In the simplified noise-free environment, the FD output is computed as the autocorrelation of $\tilde{s}_i(t)$ using a fixed lag-delay τ and is expressed as

$$\begin{aligned}\tilde{s}_o(\tau) &= \frac{1}{T_e}\int_0^{T_e} \tilde{s}(t)\tilde{s}^*(t-\tau)dt\\ &= A^2 e^{j\omega_\varepsilon \tau}\end{aligned} \qquad (11.24)$$

The *complex* implementation of the correlator is shown in Figure 11.20a, and Figure 11.20b shows an equivalent implementation involving *real* functions identified as the in-phase and quadrature components of $\tilde{s}_i(t)$. The LPFs provide time averaging over the estimation interval T_e as indicated in (11.24). The integration or filtering is fundamental to the correlation processing and is intended to reduce the influence of the additive noise associated with the received signal; the output noise is complicated by the multiplication resulting in products involving $S \times N$ and $N \times N$. The correlation lag delay τ is selected to provide the greatest unambiguous range in the frequency estimation as described later.

Referring to (11.24), the angle between the in-phase and quadrature terms of $\tilde{s}_o(t)$ is given by*

$$\begin{aligned}\phi &= \text{atan2}(s_{so}(\tau), s_{co}(\tau))\\ &= 2\pi f_\varepsilon \tau\end{aligned} \qquad (11.25)$$

To avoid ambiguities with ± frequency uncertainties it is necessary that the maximum unknown frequency correspond to a correlator output phase $<\pi$ radians and by solving (11.25) for f_ε with $\phi = \pi$ the condition is

$$|f_\varepsilon| < f_\pi = \frac{1}{2\tau} \quad : \phi = \pi \qquad (11.26)$$

*The atan2(y,x) function computes the angle $\phi = \tan^{-1}(y/x)$ over the range of $\phi = 0$ to $\pm\pi$.

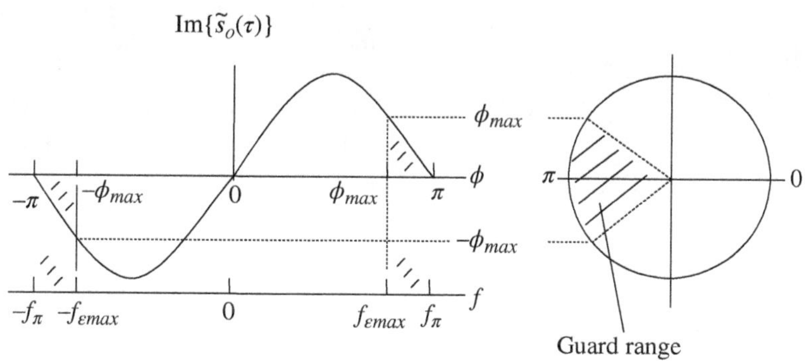

FIGURE 11.21 Phase-frequency response of frequency discriminator.

Allowing for some guard range in a noisy environment, it is prudent to require that

$$|f_\varepsilon| \le f_{\varepsilon max} = \frac{\phi_{max}}{2\pi\tau} \tag{11.27}$$

where $|\phi_{max}| < \pi$ radians. These relationships are depicted by the FD phase diagrams in Figure 11.21 where the frequency error is given by

$$f_\varepsilon = \frac{\phi_\varepsilon}{2\pi\tau} \tag{11.28}$$

Defining ϕ_{max} in terms of the guard range as a fraction, η of π radians, such that,

$$\phi_{max} = \pi(1-\eta) \tag{11.29}$$

then (11.27) is expressed as

$$|f_\varepsilon| \le f_{\varepsilon max} = \frac{(1-\eta)}{2\tau} \tag{11.30}$$

Equation (11.28) is the estimate of the frequency error based on the discriminator phase error. Although evaluation of the phase estimate using the inverse tangent function is computationally complex, there are two major advantages: the estimate is linear with frequency and independent of signal amplitude A.

The imaginary part of $\tilde{s}_o(\tau)$ can also be used to form the discriminator response expressed as

$$\begin{aligned}\alpha_s &= \mathrm{Im}\{\tilde{s}_o(\tau)\}\\ &= A^2\sin(2\pi f_\varepsilon\tau)\end{aligned} \tag{11.31}$$

Approximating (11.31) for small arguments and solving for the frequency error results in

$$f_\varepsilon \cong \frac{\alpha_s}{2\pi A^2\tau} \quad \text{: small argument approximation} \tag{11.32}$$

This form of the discriminator is similar to that used in the Costas implementation of the PLL; however, unlike (11.25), the response given by (11.31) is not linear over the entire unambiguous frequency range as seen by the two responses shown in Figure 11.22. Furthermore, the unambiguous range of (11.31) is limited to $|f\tau| < 0.25$. The following case study uses the linear atan2 discriminator implementation and examines the frequency estimation error under several signal-to-noise conditions.

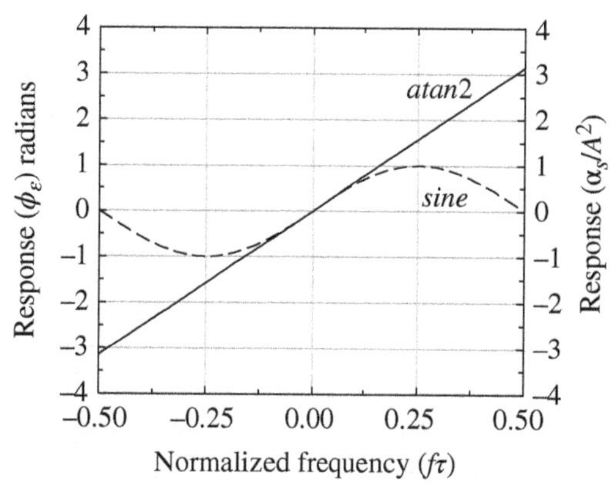

FIGURE 11.22 Frequency discriminator responses.

The FD can also be used to estimate various derivatives of the signal phase function by cascading additional fixed lag-delay correlators. For example, the frequency estimate $(\hat{\omega}_\varepsilon)$ and frequency-rate estimate $\left(\hat{\dot{\omega}}_\varepsilon\right)$ are implemented as shown in Figure 11.23. In this case, the input phase function is expressed as

$$\theta_i(t) = \varphi + \omega_\varepsilon t + \left(\frac{\dot{\omega}_\varepsilon}{2}\right)t^2 \tag{11.33}$$

The details in demonstrating the frequency and frequency rate estimates in Figure 11.23 are left as an exercise (see Problem 9).

11.2.2.5 ***Case Study: Discriminator Frequency Estimation*** The frequency estimation performance and the probability of correctly declaring the frequency using the

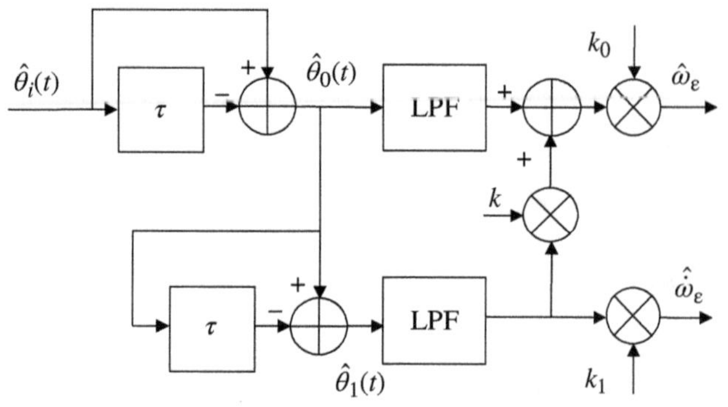

FIGURE 11.23 Frequency and frequency-rate discriminator.

discriminator shown in Figure 11.20 are examined in this case study. The evaluation is based on the normalized form of the key equation (11.26) obtained by dividing by the Nyquist band frequency $f_N = f_\pi = f_s/2$ yielding

$$\left|\frac{f_\varepsilon}{f_N}\right| \le \frac{1}{2f_N\tau} \tag{11.34}$$

Two practical observations are made concerning (11.34). First, to provide a frequency guard range against an unambiguous frequency estimate with additive noise, the normalized form of (11.30) is

$$\frac{f_{emax}}{f_N} = \frac{(1-\eta)}{2f_N\tau} \tag{11.35}$$

The second observation is based on the sampled data processing requiring that the discriminator delay be integrally related to the sampling frequency, that is, $\tau = n/f_s$ where n is an integer. Recognizing that $f_s = 2f_N$, the denominator of the right-hand side of (11.34) and (11.35) is simply $f_s\tau$ and the maximum normalized frequency range satisfying the integer requirement occurs when $f_s\tau = 1$, that is, when the discriminator delay is one sample, and (11.35) becomes

$$\frac{f_{emax}}{f_N} = (1-\eta) \quad \text{: maximum unambiguous frequency} \tag{11.36}$$

The phase and frequency estimates are computed at the output of the LPFs shown in Figure 11.20 and the accuracy of these estimates with signal and noise is a function of the signal-to-noise ratio and the estimation interval $T_e = 1/f_B$ where f_B is the bandwidth of the LPFs as discussed in Section 1.9.2.1.

As an example of this analysis, consider a CW carrier sampled at $f_s = 19.2$ kHz with $f_N = 9.6$ kHz and a guard interval of $\eta = 0.3$ (30%). Based on a simulation of the signal processing in Figure 11.20, with a CW signal source and AWGN channel, the frequency estimation performance of the discriminator is shown in Figures 11.24 and 11.25 as a function for the signal-to-noise ratio measured in a sampling frequency bandwidth of 19.2 kHz. The ordinates are plotted in terms of the normalized frequency estimation performance and the example case using the 19.2 kHz sampling frequency. The performance in Figure 11.24 corresponds to a LPF* bandwidth of $f_B = 38$ Hz with $N = \lfloor f_s/f_B \rfloor = 505$ samples, whereas the performance in Figure 11.25 corresponds to $f_B = 218$ Hz and $N = 88$ samples. This case was selected to correspond to the 88 bit CW preamble length in the FFT case study of Section 11.2.2.2 with a bit rate of 19.2 kbps. In both cases the number of Monte Carlo acquisition trials at each signal-to-noise ratio is 10,000 and the maximum and minimum values correspond to the extremes recorded among the trials for each signal-to-noise ratio.

These results correspond to the worst case frequency error because the input signal frequency error corresponds to $f_{emax} = 6720$ Hz as expressed in the normalized form by (11.36). The impact of the LPF bandwidth is evident in these two figures, for example, the performance using the lower bandwidth results in significantly improved estimations and there is no evidence that an unambiguous phase estimate occurred. To the contrary, the degraded performance for the 218 Hz bandwidth case, shown in Figure 11.25, clearly shows the result of the phase measurement reaching through the guard range and causing ambiguous estimates for signal-to-noise ratios less than 0 dB. The occurrences of a frequency estimate exceeding the unambiguous frequency range of the discriminator are obvious in the simulation program with a fixed frequency error at, for example, a positive value of f_{emax}. That is, referring to the phase diagram in Figure 11.21, when $\phi = \phi_{max}$ a frequency estimate resulting in a phase estimate $\pi + \varphi$ is computed by the atan2(y,x) function as the phase $-\pi + \varphi$ radians that corresponds to a negative frequency most likely in the range $-f_N$ to $-f_{emax}$. Therefore, referring to Figure 11.25, the abrupt negative frequency jump at 0 dB results from the positive frequency exceeding f_N. For the 218 Hz LPF bandwidth case, the probability of a phase over-flow, as described earlier, for signal-to-noise ratios of −4.0 and −0.5 dB is $1.49e^{-2}$ and $1e^{-4}$, respectively.

Although, as described earlier, the ambiguous frequency estimate cannot be discerned in the CW segment, signal verification performed during the synchronization preamble segment will most likely fail because the PLL pull-in frequency is exceeded. However, the initial CW segment ambiguous frequency estimate can also be verified by correcting the frequency and repeating the CW segment estimation processing using a smaller estimation range f_e by increasing the value of τ as in (11.27); this verification processing may have to be

*The low-pass filter is implemented as a cascade of six synchronously tuned single pole filters as described in Appendix B.

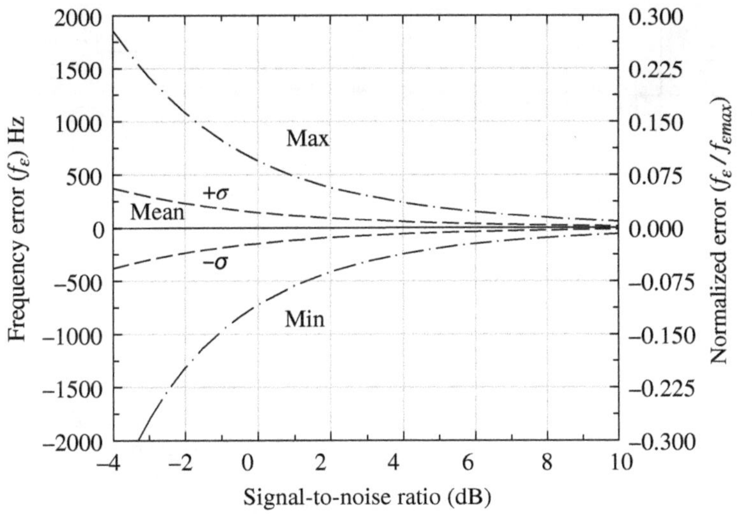

FIGURE 11.24 Frequency discriminator estimation performance ($f_B = 38$ Hz, $N = 505$).

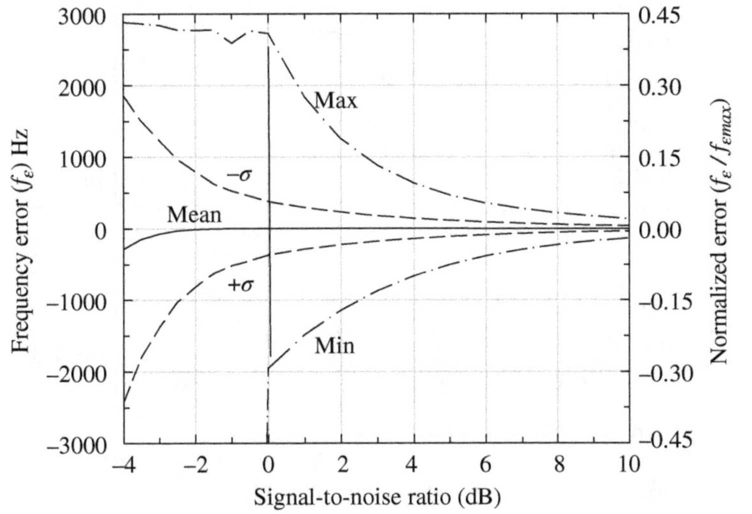

FIGURE 11.25 Frequency discriminator estimation performance ($f_B = 218$ Hz, $N = 88$).

repeated to account for the sign ambiguity of the initial estimate. When τ is varied in multiples of two the processing is similar to that of a radix-2 pipeline FFT thus improving the frequency estimation accuracy.

The frequency detection or CW acquisition probability is evaluated by examining the standard deviation of the phase estimates ϕ_{en} computed by the atan2(y,x) function over the estimation interval of N samples as defined earlier. This criterion is chosen because during the CW preamble, the phase standard deviation is zero without noise and because it increases inversely with the signal-to-noise ratio. The functional processing for the CW acquisition or signal present detection is shown in Figure 11.26; also shown is the frequency correction of the input signal $\tilde{s}_i(t)$ that is used in the symbol synchronization preamble segment. The angular frequency estimate is $\hat{\omega}_\varepsilon = i\hat{\phi}_\varepsilon / \tau f_s$ where the index i is available from the system sample clock generator such that $t = t_i = i\Delta t = i/f_s$.

The frequency detection analysis evaluates the performance for the case of $f_B = 218$ Hz ($N = 88$ samples) corresponding to Figure 11.25. Because the processing of the received signal samples by the FD is sequential, in that, the sampled data is continuously passing through the detection algorithm, much like the pipeline FFT, a detection hypotheses is made at regularly spaced intervals using 75% or $N' = \lceil 0.75N \rceil$ of the most recently collected samples; this

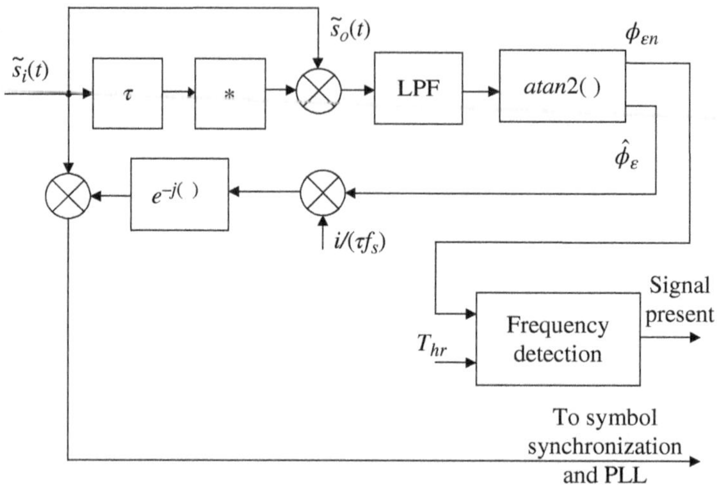

FIGURE 11.26 Frequency discriminator and detection implementation.

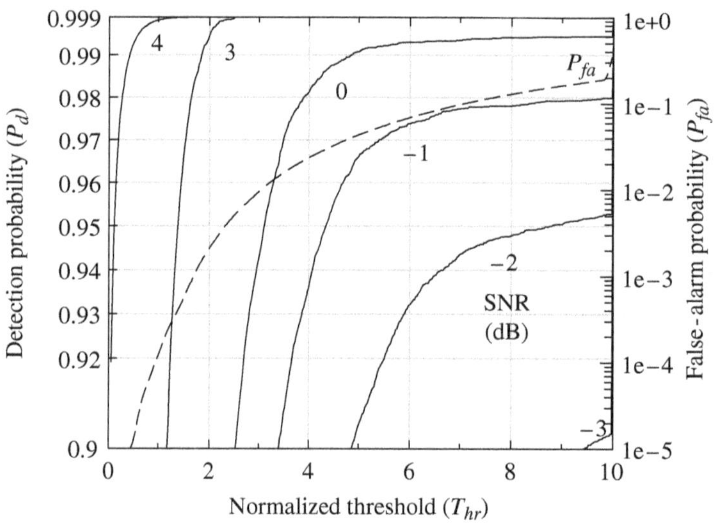

FIGURE 11.27 Frequency detection and false-alarm performance ($f_B = 38$ Hz case).

reduces the possibility of initial transients influencing the detection. With this understanding, the mean and standard deviation are computed over the most recent $N' = 66$ phase samples ϕ_{en} as expressed by (11.18) and (11.19) by substituting $|c_n| = \phi_{en}$ and using the unprimed summations, that is, no censoring of the N' samples is used. Letting $\sigma_f = s_c$ and using the normalizing frequency $f_{norm} = 200$ Hz the threshold, T_{hr}, is selected to meet the detection and false-alarm requirements defined as

$$P_d \triangleq P_r\left(\sigma_f / f_{norm} \le T_{hr} | S+N\right) \tag{11.37}$$

and

$$P_{fa} \triangleq P_r\left(\sigma_f / f_{norm} \le T_{hr} | N\right) \tag{11.38}$$

These probabilities are evaluated using a histogram, with 200 bins spanning $f_{span}/f_{emax} = 0.2976$ or 2 kHz, that is used as a probability distribution function. For each signal-to-noise ratio, the detection probability is based to 10,000 frequency acquisition trials and the false-alarm probability is based on 100,000 acquisitions trials with noise only. The performance with and without noise uses an ideal AGC; however, the atan2(y,x) function also provides immunity to the level of the received signal. Based on this description, the detection and false-alarm performance is shown in Figure 11.27 as the *solid* and *dashed* curves, respectively, as a function of the threshold.

The frequency detection results are summarized in Table 11.3 for the indicated detection probabilities and the

TABLE 11.3 Summary of Worst-Case Detection and False-Alarm Performance (f_B = 38 Hz, N = 88)

	γ = 4 dB		γ = 3 dB		γ = 0 dB		γ = −1 dB	
P_d	P_{fa}	T_{hr}	P_{fa}	T_{hr}	P_{fa}	T_{hr}	P_{fa}	T_{hr}
0.999	2.0e−3	1.92	5e−3	2.52	—	—	—	—
0.99	4.7e−4	1.44	1.5e−3	1.80	4.0e−2	0.48	—	—
0.95	1.7e−4	1.08	4.1e−4	1.44	1.1e−2	3.12	3.0e−2	4.32

corresponding threshold and false-alarm probability. The acceptable operating signal-to-noise ratio depends on the application and system specifications and these results illustrate the relationship between the detection and false-alarm probabilities. Typically the detection probability is specified and the false-alarm probability is chosen as low as possible to meet other system requirements like the message throughput delay and processor loading. Applications involving automatic repeat request (ARQ) and nonreal time message processing may tolerate higher false-alarm probabilities. The detection performance characterized for the 88-bit CW segment, corresponding to f_B = 38 Hz, results in reasonable detection probabilities for signal-to-noise ratios greater than 0 or 3 dB. Recall that these results represent the worst case conditions corresponding to an initial frequency error of $f_e = f_{emax}$ = 6.70 kHz and the performance with uniformly distributed frequency errors will be somewhat better. Furthermore, by decreasing the LPF bandwidth the operating signal-to-noise ratios can be extended into the negative region.

The frequency estimation can be improved by passing the frequency corrected input signal through the FD multiple times, each time improving the frequency of the previous estimate. The estimation improvement is accomplished using the same estimation time T_e by decreasing the sampling frequency after the initial estimate has been removed, thereby, reducing the signal bandwidth uncertainty as shown in Figure 11.26. In other words, instead of passing the initial frequency-corrected signal to the synchronization segment as shown, the corrected signal samples are passed through the frequency decimator a second time using $f_s'/f_s = k, k < 1$, corresponding to $f_N'/f_s' = k/2$ and $f_{emax}/f_N' = 1 - \eta'$. This requires that the signal samples over the *sliding window* of T_e seconds be stored in memory and that the rate of the signal processor is commensurate with real-time processing; the *sliding window* refers to the estimation interval in consideration of the sequential processing. This refinement of the frequency estimate can be repeated until the estimate falls within the PLL bandwidth as given by (11.3). For a given PLL B_LT product, the frequency estimation limit is also dependent on the symbol rate of the underlying received signal modulation and the critical signal-to-noise ratio as discussed in Section 10.6.11.

The frequency estimate resulting from a second pass through the discriminator is shown in Figure 11.28 corresponding to the first pass estimation results shown in Figure 11.25. The parameters for the pass correspond to the example un-normalized conditions: $f_{emax} = 0.7f_N$, f_B = 218 Hz, and N = 88 samples with f_s = 19.2 kHz, f_N = 9.6 kHz, and η = 0.3 (30%). Recalling that the first pass was evaluated for a constant, worse case, frequency offset of $f_\varepsilon = f_{emax}$ = 6,720 Hz with 10,000 Monte Carlo trials for each signal-to-noise ratio, so, for each signal-to-noise ratio there are 10,000 independent randomly distributed frequency estimates. Each of these frequency estimates is applied to the discriminator on the second pass using k = 2 with $f_s' = f_s/2$, $f_N' = f_s'/2$, and $\eta' = \eta$ resulting in the performance in Figure 11.28. The means and standard deviations corresponding, respectively, to the first and second passes are summarized in Table 11.4. These results demonstrate the improvement in

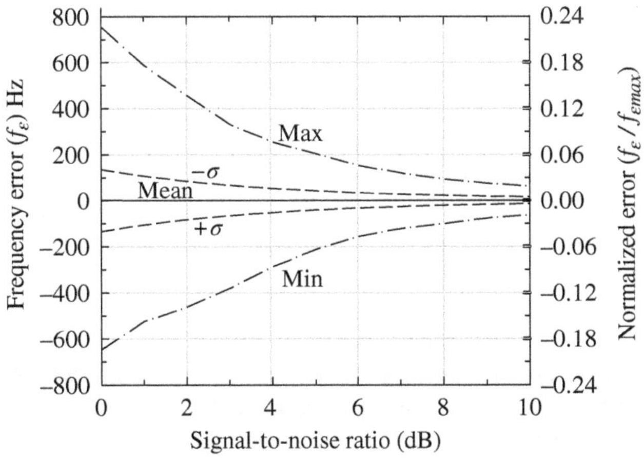

FIGURE 11.28 Second-pass frequency discriminator estimation performance (f_B = 218 Hz, N = 88).

TABLE 11.4 Comparison of First and Second Pass Frequency Discriminator Performance (f_B = 218 Hz, N = 88)

	Mean (Hz)		Standard Deviation (Hz)	
SNR (dB)	Pass 1	Pass 2	Pass 1	Pass 2
0	4.37	0.65	375	135
2	2.54	0.17	228	84
4	1.55	0.11	141	53
6	0.95	0.16	89	33
8	0.59	0.12	56	21
10	0.36	0.12	35	14

the frequency estimate resulting from the second pass through the FD as discussed earlier.

11.3 SYMBOL SYNCHRONIZATION PREAMBLE SEGMENT

11.3.1 Introduction

When a CW preamble segment is available prior to the symbol synchronization segment [9], the AGC provides for a constant signal level into the ADC and an initial indication of the presence of a received signal. Furthermore, estimates of the coarse frequency, power, and signal-to-noise ratio may be established during the CW preamble segment. Although the knowledge of these parameters simplifies the processing and contributes to an overall reduction in the acquisition time, the symbol synchronization and tracking can be established without a CW preamble segment as may be desirable, for example, in applications involving covert communications. Section 11.3.4 examines the acquisition processing without the aid of the CW preamble segment. In these cases, however, the parameter estimation, or integration, times must be increased to provide the estimation accuracies at the expensed of increased signal processing complexity.

For coherent data demodulation, the synchronization preamble segment processing must estimate [10, 11] and correct the fine frequency and symbol timing and provide for carrier phase and symbol tracking prior to entering the SOM segment. To accomplish these functions with the shortest possible preamble, the demodulator often samples and stores the raw preamble data and performs these functions sequentially making the appropriate correction to the stored data. The last pass through the stored data generally involves phaselock and symbol tracking loops requiring that the final frequency and time estimates are within the initial acquisition limits of the loops.* Another important consideration is the required acquisition times for each of the loops to achieve steady-state tracking before the SOM preamble segment; this also influences length of the synchronization segment.

When the CW preamble segment is included, as is often the case, the symbol synchronization segment uses a modulated data sequence that is specifically tailored to aid the demodulator in establishing the symbol timing, verifying the signal presence, and further resolving the frequency estimate. These data sequences typically involve repetitions of short binary data sequence. For example, repeated mark-space or mark-mark-space-space data patterns or pseudo-random synchronization codes. Upon establishing and applying the required parameter estimates, symbol and frequency tracking are initiated in preparation for the SOM processing. The SOM detection is typically based on the correlation response of a unique and known relatively long pseudo-random synchronization sequence with suitable correlation sidelobes so as to minimize false detection of the SOM location. Identifying the time occurrence of the maximum SOM correlation response to within a fraction of a symbol is important because the symbol following the SOM sequence is typically the first information or message header symbol that must be detected correctly or with a sufficiently low probability of error.

Binary sequences with good correlation properties [12–16], principally with low correlation sidelobes, play an important role in the waveform acquisition processing. Commonly used synchronization sequences are as follows: the Barker Codes [17–21], also referred to as perfect, magic, and optimum codes; Williard codes [22, 23]; Neuman–Hofman codes [24, 25]; Gold codes [26]; and Kasami sequences [27]. The number of known Barker codes is limited to those listed in Table 11.5.† Williard codes are listed in Table 11.6 and Neuman–Hofman codes are listed in

TABLE 11.5 Barker Codes and Correlation Sidelobes

Code Length	Binary Level	Lead-In[a] C^+_{max}	Lead-In[a] C^-_{max}	Cyclic C^+_{max}	Cyclic C^-_{max}
2	+ −	—	−1/1	0/1	—
3	+ + −	0/1	−1/1	—	−1/3
4	+ + − +	1/1	−1/1	0/3	—
5	+ + + − +	1/2	—	1/4	—
7	+ + + − − + −	0/3	−1/3	—	−1/6
11	+ + + − − − + − − + −	0/5	−1/5	—	−1/10
13	+ + + + + − − + + − + − +	1/6	—	1/12	—

[a]Repeated analog zeros.

TABLE 11.6 Williard Codes and Correlation Sidelobes[a]

Code Length	Binary Level	Lead-In[b] C^+_{max}	Lead-In[b] C^-_{max}	Cyclic C^+_{max}	Cyclic C^-_{max}
2	+ −	—	−1/1	0/1	—
3	+ + −	0/1	−1/1	—	−1/3
4	+ + − −	1/1	−2/1	0/2	−4/1
5	+ + − + −	1/1	−2/1	1/2	−3/2
7	+ + + − + − −	0/2	−2/1	—	−1/6
11[c]	+ + + − + + − + − − −	2/1	−3/1	—	−1/10
13	+ + + + + − − + − + − − −	3/1	−3/2	1/6	−3/6

[a]Williard [22]. Courtesy of International Society of Automation (ISA).
[b]Repeated analog zeros.
[c]Same as inverted and shifted 11-bit Barker code.

*See for example: Chapter 10; Sections 11.1, 11.3, and 11.3.5.

†The mapping from unipolar bits to bipolar data is $d_i = 2b_i - 1$.

TABLE 11.7 Neuman–Hofman Codes and Correlation Sidelobes[a]

		Lead-In[b]		Cyclic	
Code Length	Binary Level	C^+_{max}	C^-_{max}	C^+_{max}	C^-_{max}
7[c]	− − − + + − +	0/3	−1/3	—	−1/6
8	− − − − + + − +	1/2	−2/1	0/6	−4/1
9	− − + + + + + − +	2/1	−2/1	1/6	−3/2
10	− − − − + + − + − +	2/1	−2/1	2/3	−2/6
11[c]	− − − + + + − + + − +	0/5	−1/5	—	−1/10
12	− − + + − − − − − + − +	2/1	−3/3	4/1	0/10
13	− − − − − − + + − − + − +	2/2	−1/2	1/12	—
14	− − + + − − + + + + + − + −	2/1	−2/2	2/4	−2/9
15	− − + + + + + − − + + − + − +	2/1	−2/3	3/2	−1/12
16	− − − − − + + − − + + − + − + +	2/1	−2/4	0/12	−4/3
17	− − − − + − + + − − + + + − + − +	1/6	−4/2	1/8	−3/8
18	− − + + − − + + + + + − + − − + − +	1/5	−2/4	2/5	−2/12
19	− − − + + + − + + + − + + − + + − + −	2/1	−2/6	3/2	−1/16
20	− − − + − − − + + + + + − − + − + + − +	1/4	−2/1	0/14	−4/5
21	− − − − − − + − + + + − + − − + + + − − +	2/2	−2/2	1/12	−3/8
22	− − − + − − − + + + + + − − + + − + + − + −	1/8	−3/3	2/7	−6/2
23	− − − − − − + − + − + + − − + + − + − − + + +	2/4	−5/1	3/6	−5/4
24	− − − − − + + + − − + + + − + − + − + + − + + −	1/5	−4/2	0/17	−4/6

[a]Neuman and Hofman [24]. Reproduced by permission of the IEEE.
[b]Repeated analog zeros.
[c]Same as Barker code.

Table 11.7 for code lengths up to 24. Walsh codes are discussed in Chapter 7 and correspond to the rows of a Hadamard matrix. Gold codes are generated from linear combinations of *M*-sequences and Kasami sequences are subsets of Gold codes with improved correlation responses; both are widely used in spread-spectrum and code division multiple access (CDMA) applications. *M*-sequences are introduced in Chapter 8 and discussed with Gold and Kasami codes in the context of spread-spectrum waveforms in Chapter 13.

Polyphase codes are nonbinary codes that typically result in nonconstant amplitude waveforms and significantly lower correlation sidelobes [28]. Polyphase codes [29] are as follows: Frank codes [30, 31]; Huffman codes [32–36]. Frank codes are generated from the coefficients of the discrete Fourier transform and, for a code of length 16 sidelobe levels ≤ -33 dB relative to the peak correlation are achieved; −43 dB with windowing [37]. Huffman codes are nonbinary polyphase codes that provide for the detection of signals in high Doppler frequency environments.

Tables 11.5 and 11.6 list all of the known Barker and Williard codes, Table 11.7 lists a partial list of the Neuman–Hofman codes, and Table 11.9 contains two long codes used as SOM codes to identify the start of the message header information. In each of these tables the columns labeled C^+_{max} and C^-_{max} indicate, respectively, the maximum positive and negative correlation sidelobe levels* and the number following the *backslash* is the number sidelobes having these maximum values. The correlation lags represent one code bit. The correlation sidelobes correspond to two noise-free conditions: the first is denoted as the *lead-in correlation* response that results when the received signal is zero preceding the received code; the second condition corresponds to the *cyclic correlation response* that occurs when the correlation interval always involves elements of the input code. For example, the cyclic correlation response is encountered following the correlation of the first of several contiguously repeated codes. Synchronization preambles containing contiguously repeated codes are discussed in Section 11.3.4.

In many applications the acquisition waveform is specified and it is up to the modem designer to implement the acquisition processing to meet a specified correct acquisition probability (P_{ca}) that consists of several successful events as discussed in Section 11.1. However, the correct acquisition probability is also impacted by the correct message delivery (P_{cmd}) specification. For example, for a correct message detection probability of P_{cm}, it is required that $P_{cmd} \geq P_{ca}P_{cm}$ with a specified level of confidence, for example, $P_{cmd} = 0.999$ with a confidence level of 95%, at a specified receiver

*The correlation sidelobes listed in the tables are not normalized; however, the following example correlation plots are normalized so the peak correlation response is unity.

TABLE 11.8 Example of User Data Preambles[a]

			Preamble Segment					
			CW		Sync		SOM	
Modulation	Symbol Rate (ksps)	Ch (*I*/*Q*)	Bits	Pattern	Bits	Pattern	Bits	Pattern[b]
BPSK	9.6	*I*	10	0's	114	110110	74	LPN
BPSK	19.2	*I*	22	0's	156	110110	74	LPN
QPSK	16.0	*I*	14	1's	111	001001	37	LPN
		Q	14	0's	111	110110	37	ILPN
S-OQPSK	3.0 and 3.84	*I*	13	1's	70	101010	37	LPN
		Q	13	0's	70	111111	37	ILPN

[a]Defense Information Systems Agency (DISA) [8]. Courtesy of U.S.A. Department of Defense (DOD).
[b]LPN is Legendre polynomial, ILPN is inverted Legendre polynomial.

TABLE 11.9 SOM LPN Code Bits with Cyclic Correlation Sidelobes[a]

		Lead-In[b]		Cyclic	
Code Length	Code Bits	C^+_{max}	C^-_{max}	C^+_{max}	C^-_{max}
37[c]	111000100001000111101001101110110010**1**	4/1	−3/3	1/18	−3/18
74	1000111010000100111100100001011100011 0100010011010111101111010110010001011	6/2	−7/1	6/8	−6/18

[a]Defense Information Systems Agency (DISA) [38]. Courtesy of U.S.A. Department of Defense (DOD).
[b]Repeated analog zeros.
[c]The last bit, shown in bold type, is not inverted in the 37-bit ILPN pattern.

sensitivity. Typically, the correct message detection is based on the data covered by a cyclic redundancy check code as discussed in Section 8.7.

Implementation techniques and performance specifications are available for many commercial systems. For example, the global system for mobile communications is broadly discussed in Mouly and Pautet [7] with references to specifications* that provide detailed performance and design requirements. The radio interface is the subject of Mouly and Pautet's Chapter 4 that includes acquisition and synchronization, the channel model, source and channel coding, encryption, burst formatting, and the waveform modulation. Another example of commercial communication systems implementation and performance specification is the CDMA2000 system for mobile and personal communications.† The preamble acquisition segment waveforms listed in Table 11.8 are example applications for the indicated modulations and symbol rates and two SOM codes are listed in Table 11.9. SOM codes can also be generated by concatenating shorter fixed length codes with appropriate cyclic shifts of the successive fixed length codes that result in desirable or minimum correlation sidelobes. These waveforms illustrate the complexity of the preamble message structure required prior to the detection of the message information.

The following figures show the lead-in and cyclic normalized correlation response of selected codes. The noise-free correlation responses of the two Barker codes shown in Figure 11.29 demonstrate the lead-in characteristics when attempting to synchronize to a message preceded by at least two Barker codes. For example, by specifying a normalized detection threshold, the low-level lead-in and cyclic correlation sidelobes result in the best synchronization code detection probability among codes of corresponding length. Since the correlation loss with frequency error is a function of the code length, it is typically necessary that a fine-frequency estimate be performed prior to the code correlation. For very low signal-to-noise ratio conditions, as might exist with low-rate FEC coding, it is necessary to combine several contiguously repeated code correlations to increase the signal-to-noise ratio using coherent and/or noncoherent code combining.‡ The lead-in and cyclic correlation sidelobe responses of the Barker codes result in low false-detection performance leading to their widespread use for message synchronization. Unfortunately longer Barker sequences do not exist; however, the shorter codes are more tolerant to channel dynamics.

*For additional specifications visit: www.3gpp.org/specifications.
†For detailed specifications visit: www.tiaonline.org/standards.
‡Coherent and noncoherent detection combining are discussed in Appendix C.

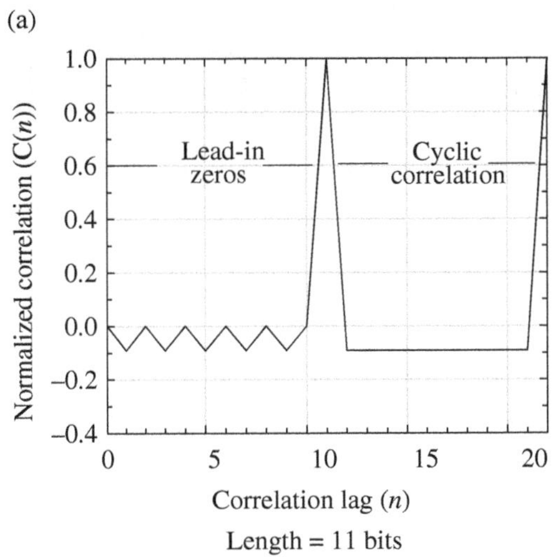

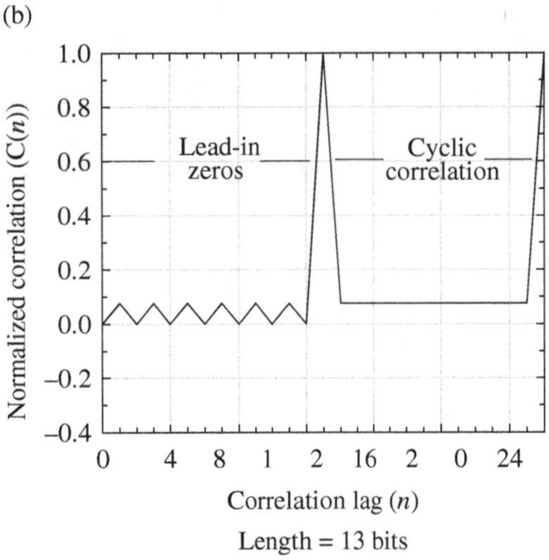

FIGURE 11.29 Barker code correlation responses (selected from Table 11.5).

The noise-free correlation response of four Neuman–Hofman synchronization codes is shown in Figure 11.30 and the lead-in and cyclic correlation sidelobes are contrasted with those of the Barker codes. Unfortunately, however, the only known Barker codes are listed is Table 11.5 so the Neuman–Hofman codes offer a wider selection of code lengths. For example, the 13-bit Neuman–Hofman code has two positive correlation sidelobes with levels that are 8.1 dB below the correlation peak compared to 11 dB for the 13-bit Barker. In contrast, the 24-bit Neuman–Hofman code has a coherent detection advantage with 13.8 dB lead-in correlation sidelobes.

The noise-free correlation responses of the LPN SOM detection codes listed in Table 11.9 are shown in Figure 11.31. The SOM codes are used to locate the first information symbol and, because the SOM codes are not repeated, only the lead-in correlation response needs to be considered. However, coherent correlation can be performed because carrier phase and symbol timing tracking have been established during the symbol synchronization processing. In this case, the lead-in correlations for the length 37 and 74 LPN codes have respective correlation sidelobes of 9.7 and 10.9 dB. However, as indicated in Table 11.8, with BPSK modulation the 74-bit LPN SOM code is preceded by the repeated 110... data pattern that will alter the lead-in correlation sidelobes; this is discussed in more detail in Section 11.4. Furthermore, with QPSK- and S-OQPSK-modulated waveforms, the 37-bit LPN code is used on the in-phase rail with the ILPN code on the quadrature rail. With coherent detection, the lead-in correlation response is improved over the 9.7 dB of the isolated 37-bit LPN.

Although a theoretical correct-detection probability of the synchronization and SOM codes with additive and multiplicative noise (fading) can be analyzed, it is often beneficial to evaluate the performance using a Monte Carlo simulation program in view of the carrier and symbol tracking and the lead-in correlation peaks. However, as is always the case, an accurate theoretical analysis will provide a baseline performance measure for the simulated results. In the following subsections, various details of the symbol synchronization preamble are examined.

11.3.2 Frequency and Symbol Time Estimation

Fine-frequency estimation is performed following the coarse-frequency estimation obtained from a CW preamble segment. In this case, the frequency search range is significantly reduced and the signal processing focuses on the frequency resolution as opposed to the frequency range.* In Sections 11.3.2.1 and 11.3.2.2, the description of the symbol time and frequency estimation is based on the mark-mark-space-space preamble data pattern that is known by the demodulator and repeated over the entire synchronization preamble segment. Section 11.3.2.1 describes the joint symbol time and frequency estimation using a FD with knowledge of the symbol rate. The descriptions in Section 11.3.2.2 first resolves and corrects the fine-frequency estimate over the stored preamble samples and then process the frequency corrected store samples to determine the symbol timing. In this case, the frequency estimation is performed in the frequency domain and, because the symbol rate can also be estimated from the preamble spectrum, it is not necessary to know the symbol rate in advance.

11.3.2.1 Joint Frequency and Symbol Time Estimation Using Discriminator The approach described in this section to estimate the frequency and symbol or bit timing is based on the FD discussed in Sections 11.2.2.4 and

*As discussed in Section 11.3.4 short repeated data sequences can also be used to search over large frequency ranges without the CW preamble segment.

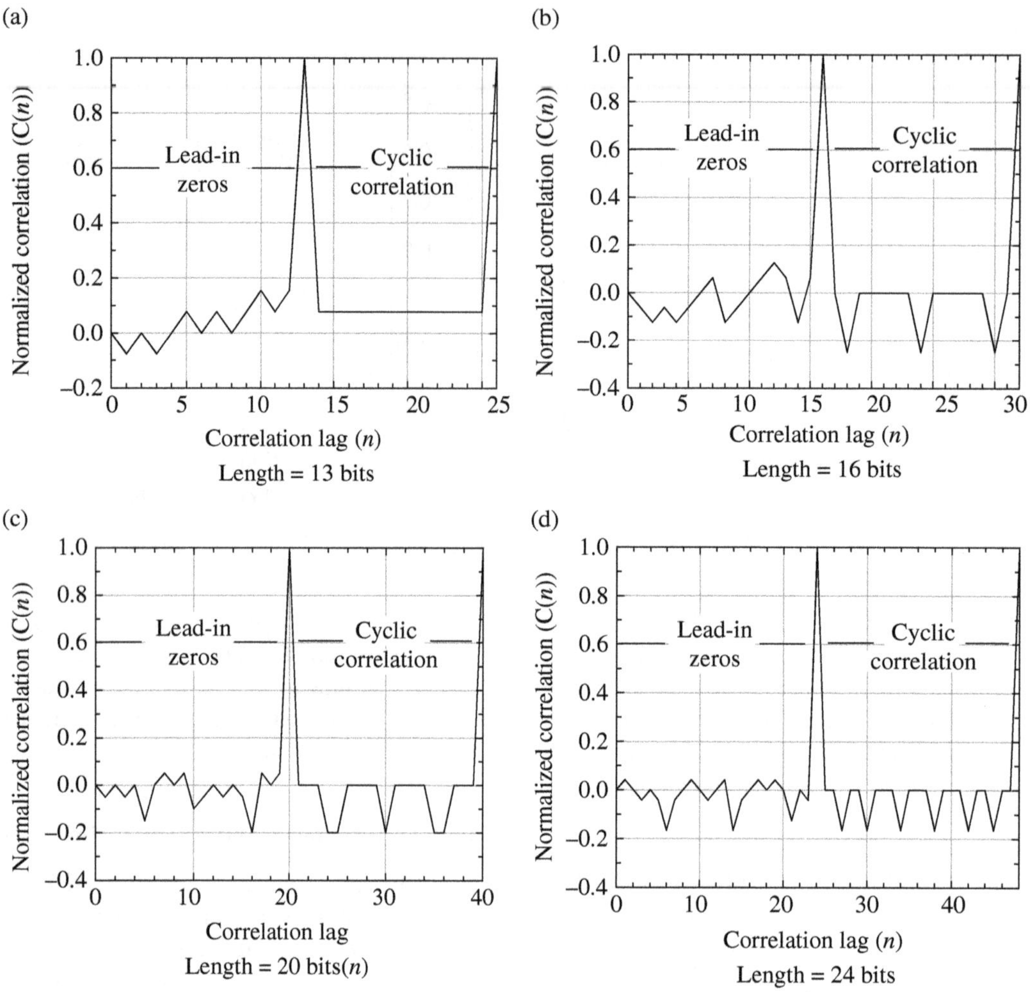

FIGURE 11.30 Neuman–Hofman code correlation responses (selected from Table 11.7).

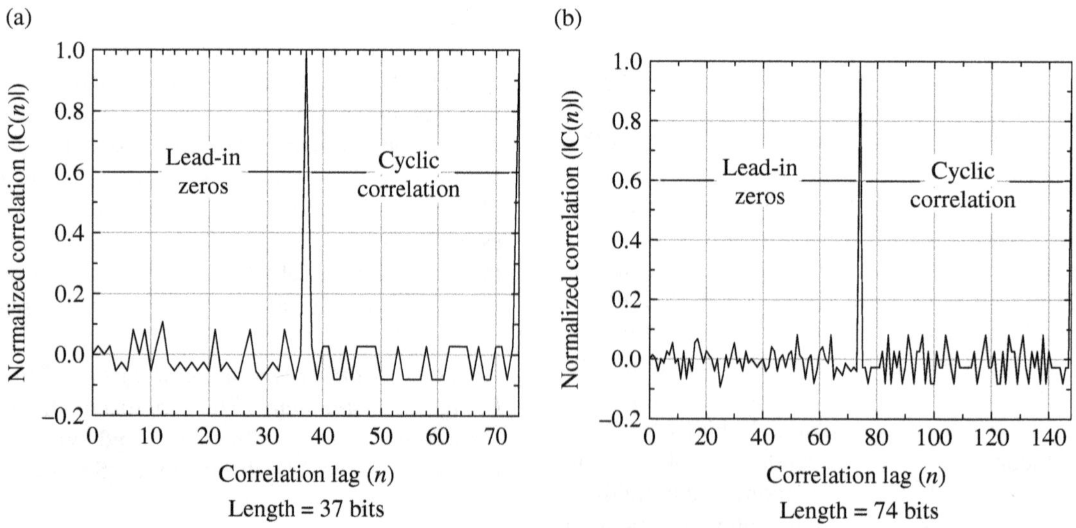

FIGURE 11.31 LPN code correlation response.

11.2.2.5. The preamble synchronization segment in this analysis is composed of the 1100 repeated bit pattern and, because the coarse frequency estimate has been removed, the frequency error is a fraction of the bit rate. The detection processing is based on a parallel repetition of the complex discriminator function shown in Figure 11.20a and depicted in Figure 11.32 as a lag-correlator* (LC).

The concept involves computing LC outputs for $K-1$ hypotheses of the optimum bit timing of the sampled 1100 repeated bit pattern as shown in Figure 11.33. The N_s samples per bit, corresponding to each timing hypothesis, are then used to compute $K-1$ lag-correlator outputs as shown in Figure 11.34. The LC output, corresponding to the optimum bit timing delay relative to the receiver time base, is associated with the imaginary part having the highest mean value and lowest standard deviation or simply having the highest ratio $\rho_k = m_k/\sigma_k$, such that,

$$\rho_{k'} = \max_{\forall k}(\rho_k) \tag{11.39}$$

Referring to Figure 11.33, the bits are sampled at intervals of T_s with N_s samples per bit. Relative to the local demodulator time base, the bipolar data is denoted as $d_n = d_{n+1} = 1$ and $d_{n+2} = d_{n+3} = -1$ corresponds to the repeated binary bit pattern $b_n = (1100)$. The demodulator time base can be defined to start anywhere over the sampled data pattern without affecting the detection of the optimum timing delay.

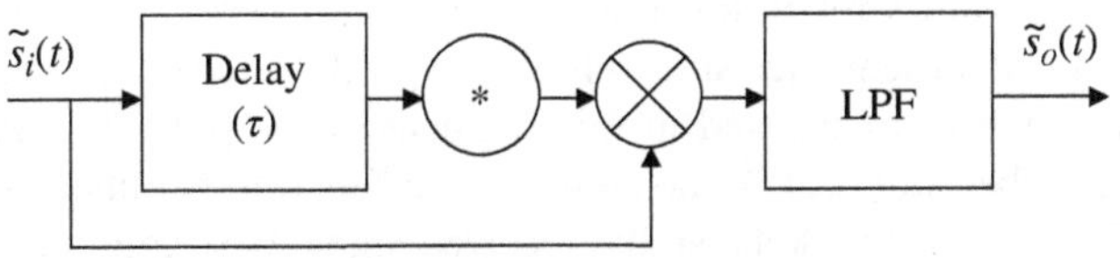

FIGURE 11.32 Lag correlator.

Processing the samples over the entire data pattern as shown results in a discriminator delay $\tau = 2T$ and, from (11.28), the corresponding LC frequency error is $(1-\eta)/4T$ Hz. In this case, the LC spans $4T$ seconds and the sign of the samples corresponding to the negative data must be changed. This is easily accomplished from the knowledge of the hypothesized timing. Defining the parameter N as the number of sample intervals between each bit time hypothesis, the resulting timing error is

$$T_\varepsilon = \pm\frac{NT_s}{2} \tag{11.40}$$

and the number of hypotheses is given by

$$K = \frac{4N_s}{N} \tag{11.41}$$

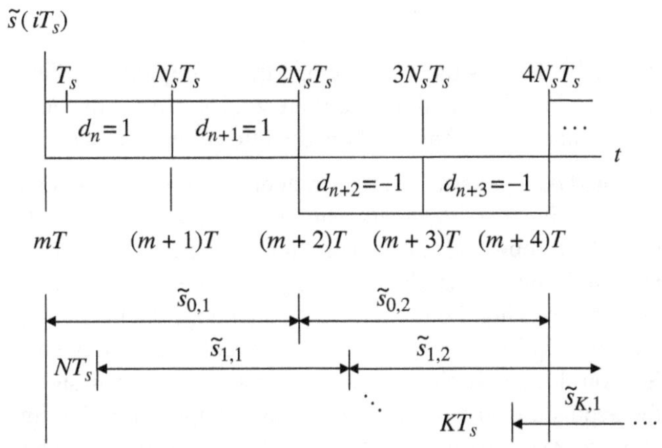

FIGURE 11.33 Signal sampling and processing for joint frequency and bit time estimation.

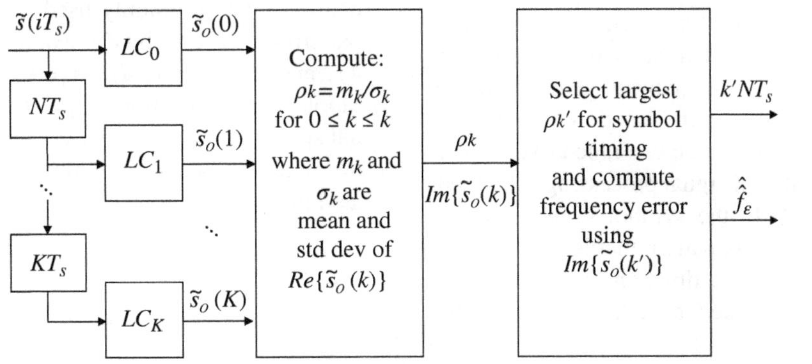

FIGURE 11.34 Lag correlator processing for joint frequency and bit time estimation.

*The lag-correlator implementation is similar to the theoretical definition of the auto-correlation function given in Section 11.3.3 with a correlation *lag* of τ seconds and with the low-pass filter performing the integration.

For example, with $N_s = 8$ and the requirement $T_\varepsilon = \pm T/8$ results in $N = 2$ and $K = 16$ hypotheses. The timing offsets relative to the demodulator time reference are computed as kNT_s

where the parameter k represents the k-th timing hypothesis over the range $0 \le k \le K-1$.

Referring to Figure 11.34 the inputs into the k-th LC, denoted as LC_k, are computed as

$$S_{k,1} = \sum_{i=1}^{2N_s} \tilde{s}(iT_s + kNT_s) \tag{11.42}$$

and

$$S_{k,2} = -\sum_{i=2N_s+1}^{4N_s} \tilde{s}(iT_s + kNT_s) \tag{11.43}$$

The minus sign in (11.43) results from the sign of the data bits d_{n+2} and d_{n+3}. Using (11.42) and (11.43) in the context of Figure 11.32, the output of LC_k is computed as

$$\tilde{S}_o(k) = \left\langle \tilde{S}_{k,2}\tilde{S}^*_{k,1} \right\rangle \tag{11.44}$$

where $\langle\ \rangle$ denotes the time average that is implemented as a LPF. Using the bit timing error estimate $k'NT_s$, the local demodulator time base is brought into alignment with the repeated data pattern. The fine-frequency estimate $\hat{f}_e$ is determined using the imaginary part of $\tilde{s}_o(k')$ as described by (11.23) through (11.30). The fine-frequency estimate is removed from the baseband signal prior to the carrier acquisition and tracking. Following these corrections the bit and frequency acquisition and tracking are executed as described in Section 11.3.5. If the preamble samples have been stored, the acquisition and tracking functions are applied to the time and frequency corrected samples resulting in the most efficient use to the preamble. Reusing the corrected preamble samples results in a shorter preamble and lower message overhead.

11.3.2.2 Signal Detection and Parameter Estimation Using Signal Spectrum In this section the coarse-frequency estimate of the CW preamble segment is refined in the frequency domain for subsequent acquisition and tracking by the PLL. However, as indicated in the introduction, the signal spectrum can also be examined over a wide frequency range with additional signal processing. The focus in this section corresponds to the symbol synchronization segment involving a known data pattern that is used for both fine-frequency and symbol rate estimation. In this case, the data pattern is uniquely designed to enhance the spectral characteristics for determination of the frequency error and the modulation symbol rate. The spectrum also contains information regarding synchronization to the data pattern; however, the unknown signal phase is problematic in making an unambiguous data pattern synchronization decision. In view of these remarks, the entire preamble is often sampled and stored for processing in the frequency domain and then revisited and processed in the time domain for symbol time synchronization and then revisited a third time for frequency, phase, and symbol tracking. Although, this approach requires more memory and more intense processing to maintain real-time throughput, the preambles are typically shorter resulting in lower message overhead. Furthermore, in many network applications, the frequency and timing information is stored during network entry allowing for shorter preambles in subsequent transmissions.

The signal parameters involving unique data patterns are examined in the frequency domain relative to the demodulator local time reference as shown in Figure 11.35. Denoting the received symbols in terms of the binary data $d_m = \pm 1$ suggests that binary PSK modulation is used; however, using the symbol phase notation $\phi_{k,K} = k\pi/K$ to denote higher order modulations with a unique data sequence increases the notational complexity that tends to obscure the synchronization concepts. Therefore, the focus in this description is on BPSK modulation.

The analysis that follows uses the analytic baseband description of the received signal expressed as

$$\tilde{s}_r(t) = rep_{T_o}\left(A\sum_{m=0}^{M-1} d_m rect_T(t-\tau, m) \right) e^{j(\omega_\varepsilon(t-\tau)+\varphi)} \tag{11.45}$$

where τ is the signal delay relative to the demodulator time reference, ω_ε is the frequency error in radians per second, and φ is a constant phase error. The notation $rep_{To}(x)$ is Woodward's repetition function [39] with period $T_o = MT$, so the data sequence shown in Figure 11.35 is repeated every M symbols as defined in this segment of the preamble. As described in (11.45), the signal phase function is outside of the $rep_{To}(x)$ brackets, so the received signal description corresponds to the phase of the carrier frequency.

Equation (11.45) is a general description of the signal and applies to any repeated data pattern; however, to simplify the analysis, the commonly used preamble synchronization bit pattern of 1100... bits is examined for which $M = 4$, $T_o = 4T$, and $d_0 = d_1 = 1$, $d_2 = d_3 = -1$ where the unipolar bit to bipolar data translation is $d_i = 2b_i - 1$. The analysis of the signal spectrum uses the Fourier transform of periodic functions discussed in Section 1.2 for which the spectrum is evaluated as

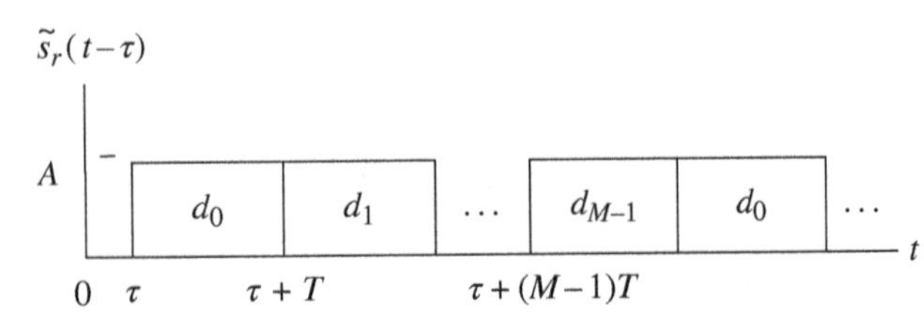

FIGURE 11.35 Received signal timing relative to receiver time-base.

$$F(f) = \sum_{n=-\infty}^{\infty} C_n \delta(f - nf_o) \quad (11.46)$$

where $f_o = 1/T_o$ and the C_n are complex coefficients. In the context of (11.45) and the repeated data pattern 1100..., C_n and $f(t)$ are computed as

$$C_n = \frac{1}{T_o} e^{-jn\omega_o\tau} \int_{-T_o/2}^{T_o/2} f(t) e^{-jn\omega_o t} dt \quad (11.47)$$

and

$$f(t) = A(rect_{2T}(t,0) - rect_{2T}(t,1)) e^{j(\omega_\varepsilon t + \varphi)} \quad (11.48)$$

Substituting (11.48) into (11.47) and performing the appropriate integrations, combining of terms, some simplifications, including $f_o T = T/T_o = 1/4$, and then substituting the expression for C_n into (11.46), the spectrum for the repeated data sequence is expressed as

$$F(f) = A \sum_{n=-\infty}^{\infty} e^{-j\left(2\pi n f_o \tau + 2\lambda_n T - \varphi - \frac{\pi}{2}\right)} \frac{\sin^2(\lambda_n T)}{\lambda_n T} \quad (11.49)$$

where $\lambda_n T = 2\pi(nf_o T - f_\varepsilon T)$. Equation (11.49) is shown in Figure 11.36 for a zero frequency error, that is, $f_\varepsilon T = 0$. The spectrum is simply shifted to the right or left depending on the frequency error. The discrete spectral lines at $f \pm 1/4T$ are of interest and have magnitudes of $2A/\pi$ corresponding to a loss of 3.9 dB; the nearest neighbors at $f \pm 3/4T$ have magnitudes of $2A/3\pi$ resulting in a loss of 13.46 or 9.56 dB below the spectral lines of interest.

The spectral phase function is defined as the phase of $F(f)$ and is expressed as

$$\theta_n(\tau, f_\varepsilon, \varphi) \triangleq 2\pi n f_o \tau + 2\lambda_n T - \varphi - \frac{\pi}{2} \quad (11.50)$$

Upon using $\lambda_n T$, as given earlier, and $f_o T = 1/4$, (11.50) is evaluated as

$$\theta_n(\tau, f_\varepsilon, \varphi) = \pi n \left(1 + \frac{1}{2}\left(\frac{\tau}{T}\right)\right) - 4\pi f_\varepsilon T - \varphi - \frac{\pi}{2} \quad (11.51)$$

The spectral phase function has a considerable amount of structure, being dependent upon each of the three parameters: τ/T, $f_\varepsilon T$, and φ. Noting that the bit rate is $R_s = 1/T$, the frequency tones of interest occur at $f_u = f_\varepsilon + R_s/4$ and $f_l = f_\varepsilon - R_s/4$ corresponding to $n = +1$ and -1, respectively. For example, these tones rotate linearly in opposite directions by $(\pi/2)\tau/T$ radians, so that over the interval of one data pattern (1100) the phase rotates by 2π radians. These delay-dependent phase rotations occur independently of the frequency error; however, an additional frequency-dependent phase of $4\pi f_\varepsilon T$ radians is encountered. The unknown signal frequency error f_ε and phase error φ are problematic in determining the signal delay from the spectral phase information contained in (11.49). However, by determining f_u and f_l corresponding to the two largest spectral lines, the estimates of the frequency error and bit rate are determined as

$$\hat{f}_\varepsilon = \frac{f_u + f_l}{2} \quad (11.52)$$

and

$$\hat{R}_b = 2(f_u - f_l) \quad (11.53)$$

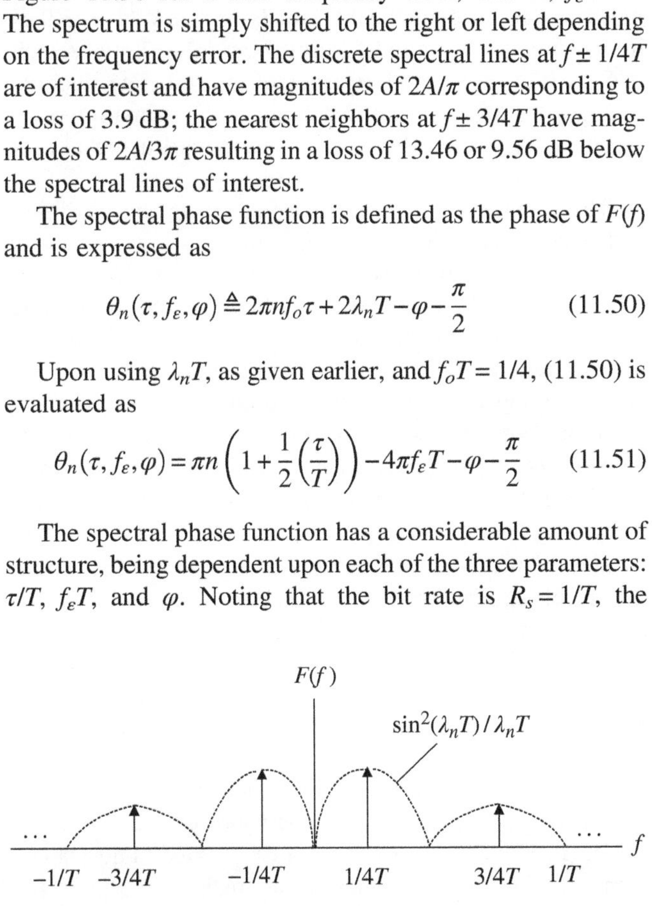

FIGURE 11.36 Magnitude of spectral lines of the repeated preamble data pattern 1100 with $f_\varepsilon = 0$.

In many cases, the symbol rate is known and is used to advantage in locating f_u and f_l when the received signal includes channel and receiver noise. Upon removing the frequency error from (11.51) the signal phase φ remains an impediment to unambiguously determining the signal delay τ/T from the spectral phase function. If, however, the signal phase were successfully removed, the delay estimate can be established to any degree of accuracy with increasing signal-to-noise ratio in the measurement bandwidth. For example, for a symbol timing error $|\tau/T| < 1/8$, the corresponding accuracy of the spectral phase function is $|\theta_n(\tau,0,0)| < \pi/8$ radians. Considering an ℓ-sigma measurement requirement, the required signal-to-noise ratio (γ_{Bm}) in the measurement bandwidth (B_m) is computed as*

$$\gamma_{B_m} = \frac{1}{2}\left(\frac{8\ell}{\pi}\right)^2 \quad (11.54)$$

and using $\ell = 3$ the required signal-to-noise ratio for the 3-sigma delay estimate is 14.6 dB. In the following case study, the FFT is used to determine the spectrum of the repeated 1100 preamble sequence and the probability of correctly estimating the symbol rate† and frequency error are examined.

11.3.2.3 Case Study: Detection, Bit Rate, and Frequency Estimation Using BPSK-Modulated Waveform In this case study, the FFT is used to determine the spectrum of the repeated 1100 preamble data and the resulting probability of correctly estimating the frequency error is examined

*This is based on the relationship $\gamma_{B_m} = 1/2\sigma_\varphi^2$ with $\sigma_\varphi = \pi/8\ell$ radians.

†Baring the distinction between information bits and code bits, the use of the term symbol rate and bit rate are synonymous with BPSK; as are T and T_b in denoting the symbol interval and bit interval.

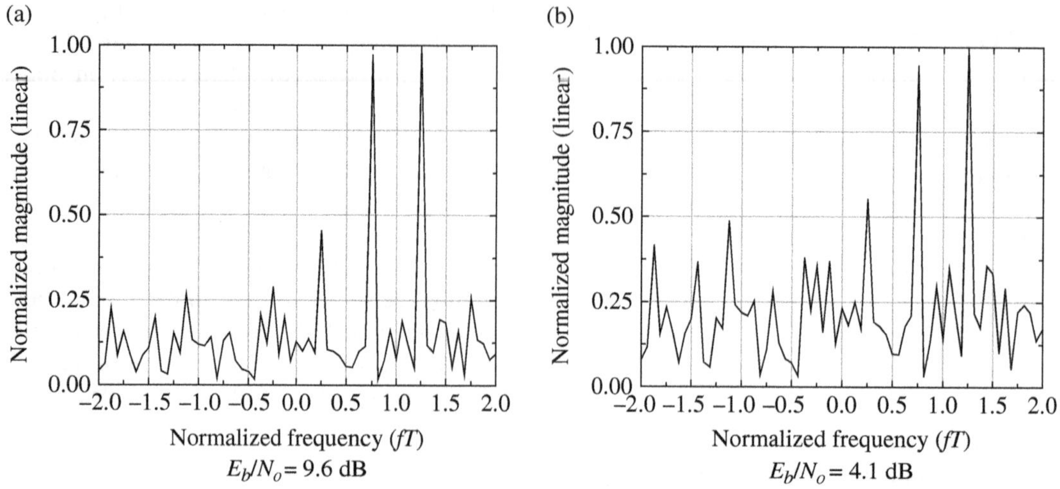

FIGURE 11.37 Simulated spectrum of repeated 1100 repeated ($f_\varepsilon T = 1$).

when the bit rate is unknown.* It is assumed that the coarse frequency error of the received signal has been resolved during the CW preamble segment to an accuracy of $|f_\varepsilon|_{coarse} \le 1.5R_b$. The following description establishes the sampling frequency (f_s), the frequency resolution (Δf), the size (N_{fft}) of the FFT, and the samples per bit (N_s) for the fine-frequency estimation. Given $|f_\varepsilon|_{max} = 1.5R_b$ and referring to Figure 11.36, the Nyquist band is chosen as $B_N = 2R_b$ so the sampling frequency must satisfy the Nyquist condition $f_s > 2R_b$; the value $f_s = 8R_b$ is selected.† The frequency resolution is taken to be one-fourth of the spectral tone at $R_b/4$, that is, $\Delta f = R_b/16$ from which the size of the FFT is determined as $N_{fft} = f_s/Df = 128$. In the time domain, these parameters result in an estimation interval of $T_m = 1/\Delta f = 16/R_b$ or 16 bits corresponding to four 1100 data pattern repetitions with $N_s = 8$ samples per bit. If a wider frequency range is desired while maintaining a fixed resolution bandwidth, the sampling rate and antialiasing filter bandwidth can be increased with a commensurate increase in the size of the FFT; in this case, the estimation interval remains the same. On the other hand, if a finer frequency resolution is necessary, the size of the FFT can simply be increased while maintaining the sampling frequency.

With a uniformly weight FFT, that is, without windowing, the selection of Δf corresponds to the measurement bandwidth, $B_m = \Delta f$, so that a specified received signal-to-noise ratio of $\gamma_b = E_b/N_o$ corresponds to a signal-to-noise ratio in the measurement bandwidth of

$$\gamma_{B_m} = \gamma_b \left(\frac{R_b}{B_m}\right) \tag{11.55}$$

This results in a signal-to-noise ratio improvement of 12 dB. However, as indicated in Figure 11.36, the signal-to-noise ratio in the FFT cells at $f_\varepsilon \pm R_b/4$ is 3.9 dB lower than given by (11.55) resulting in an improvement of 8.1 dB at each tone frequency of interest. The selection of a smaller Δf will allow a more accurate estimation of the frequency error and bit rate as determined using (11.52) and (11.53).

The simulated spectrums under the foregoing conditions and a normalized frequency error of $f_\varepsilon T = 1$ are shown in Figure 11.37 for received signal-to-noise ratios of $E_b/N_o =$ 9.6 and 4.1 dB. These signal-to-noise ratios correspond, respectively, to uncoded and rate 1/2, $K = 7$ convolutional coded antipodal modulation performance at $P_{be} = 10^{-5}$. The simulated signal-to-noise ratio of each spectral tone, corresponding to $n = \pm 1$, is evaluated by establishing the received signal-to-noise ratio, given the signal amplitude A, as expressed in (11.45) and then zeroing the signal into the demodulator and computing the total noise power (σ_n^2) as the sum of the noise variances on the quadrature rails of the complex spectrum. The resulting simulated signal-to-noise ratio, averaged over two independent FFT records, is 17.2 and 12.14 dB for the respective E_b/N_o values in Figure 11.37. These signal-to-noise ratios are simulated in the measurement bandwidth equal to Δf and compare favorably with the theoretical value computed using (11.55) when adjusted by the 8.1 dB loss at the spectral tones at $n = \pm 1$. This simple evaluation, or test, goes a long way in confirming the validity simulation.

The results of this case study are presented in normalized form, that is, the frequencies are normalized by the bit rate as f/R_b or fT. Because of the frequency normalization and the AWGN channel being considered, the following

*The reference to the bit rate can be more broadly applied to mean the symbol rate.

†This sampling frequency is somewhat larger than necessary; however, it is convenient for two reasons: the numerology is compatible with the radix-2 FFT and a larger antialiasing filter transition bandwidth results in reduced filter complexity for a given band-reject attenuation.

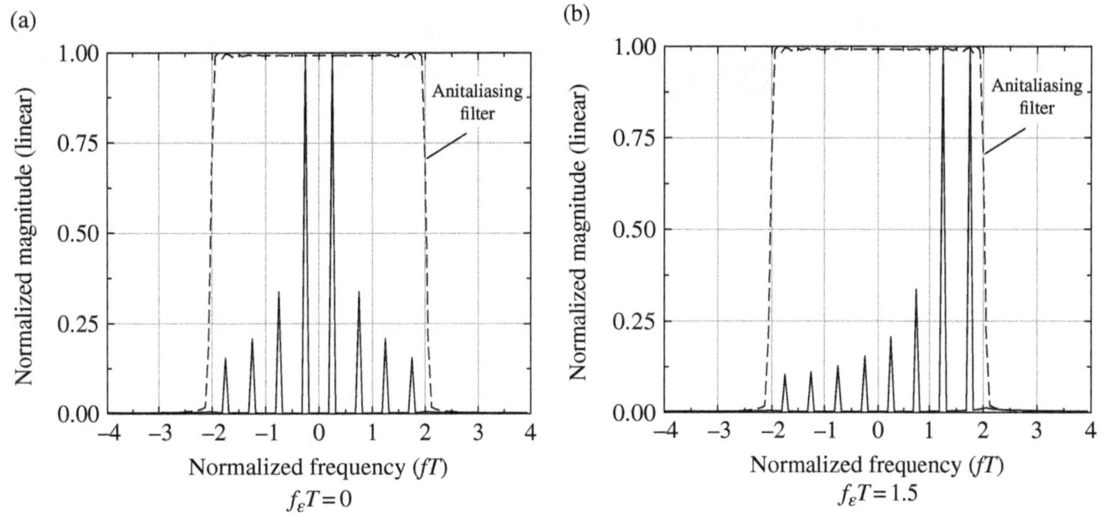

FIGURE 11.38 Simulated spectrum of repeated 1100 repeated ($E_b/N_o = \infty$ dB).

performance results can be applied to any user bit rate of interest; however, the bit rate must be considered when channels with memory are evaluated. The simulation includes a 145 tap transversal antialiasing filter with a normalized sampling frequency of $f_sT = 8$ and normalized 3 dB cutoff and transition frequencies of $f_cT = 2$ and $f_TT = 1.5$ with a band-reject attenuation of 50 dB. Figure 11.38 shows the noise-free spectrums and the antialiasing filter with $f_\varepsilon T = 0$ and 1.5. The relative magnitudes of the spectral tones at $fT \pm n/4 : n \in$ odd integer, when compensated for the 3.9 dB normalizing level used in Figure 11.38, compare favorably with the theoretical values computed using (11.49).

The following simulated synchronization performance results are based on 500 Monte Carlo trials for each signal-to-noise ratio with the random number generators reinitialized for each signal to noise. For each trail the normalized frequency error is uniformly distributed over the range $|f_s/R_b|_{max} = 1.5$. Because the bit rate and frequency error are unknown, the two largest spectral magnitudes found in the bandwidth of the antialiasing filer are used in the estimation process. The frequency estimation for each trial is based on the normalized form of (11.52) where f_uT or f_lT are computed for $x = \{u,l\}$ as

$$\hat{f}_xT = \left[-\left(N_{fft}+1\right)+\ell_x\right]\frac{N_s}{N_{fft}} \qquad (11.56)$$

The FFT array has been *left shifted* by $N_{fft}/2$ so that the zero frequency location corresponds to $\ell = N_{fft}/2 + 1$ and the locations ℓ_l and ℓ_u are determined from the locations of the two largest spectral magnitudes. Using (11.56) the normalized frequency error estimate, expressed in terms of locations ℓ_l and ℓ_u, becomes

$$\hat{f}_\varepsilon T = \left[-\left(\frac{N_{fft}}{2}+1\right)+\frac{\ell_u+\ell_l}{2}\right]\frac{N_s}{N_{fft}} \qquad (11.57)$$

Similarly, the bit rate estimate expressed, in terms of locations ℓ_l and ℓ_u, is evaluated as

$$\hat{f}_xT = 2[\ell_u-\ell_l]\frac{N_s}{N_{fft}} \qquad (11.58)$$

Because the bit rate is unknown, it is essential to estimate the bit rate correctly to establish the two spectral tones for determining the frequency error. The probability that the normalized bit rate is estimated to within 12.5%, that is, $|\hat{R}_b/R_b - 1| \le 12.5\%$, is shown in Figure 11.39a as a function of the signal-to-noise ratio; this resolution corresponds to two frequency resolution cells or $2\Delta f$. Unfortunately, the bit rate estimation is four times more sensitive to errors in the locations ℓ_l and ℓ_u than is the frequency estimation (see Problem 13). This is evident from the normalized frequency and bit rate estimation errors shown in Figure 11.39b and c as a function of the signal-to-noise ratio.* The irregularity in the maximum and minimum errors at low signal-to-noise ratios occurs because they represent the relatively small sample size of outliers in the 500 samples at each signal-to-noise ratio; the random generator seeds are reset for each signal-to-noise ratio. The maximum and minimum values for signal-to-noise ratio ≥8 dB correspond to a variation of one frequency resolution cell for the frequency and bit rate estimation errors.

The performance in this case study represents the use of fundamental algorithms and the performance can be improved by using more processing intense techniques. Some examples are as follows:

- Zero padding the FFT will improve the accuracy of the frequency estimation, although the resolution bandwidth remains unchanged.

*Note that the ordinate values in these plots differ by 2 : 1.

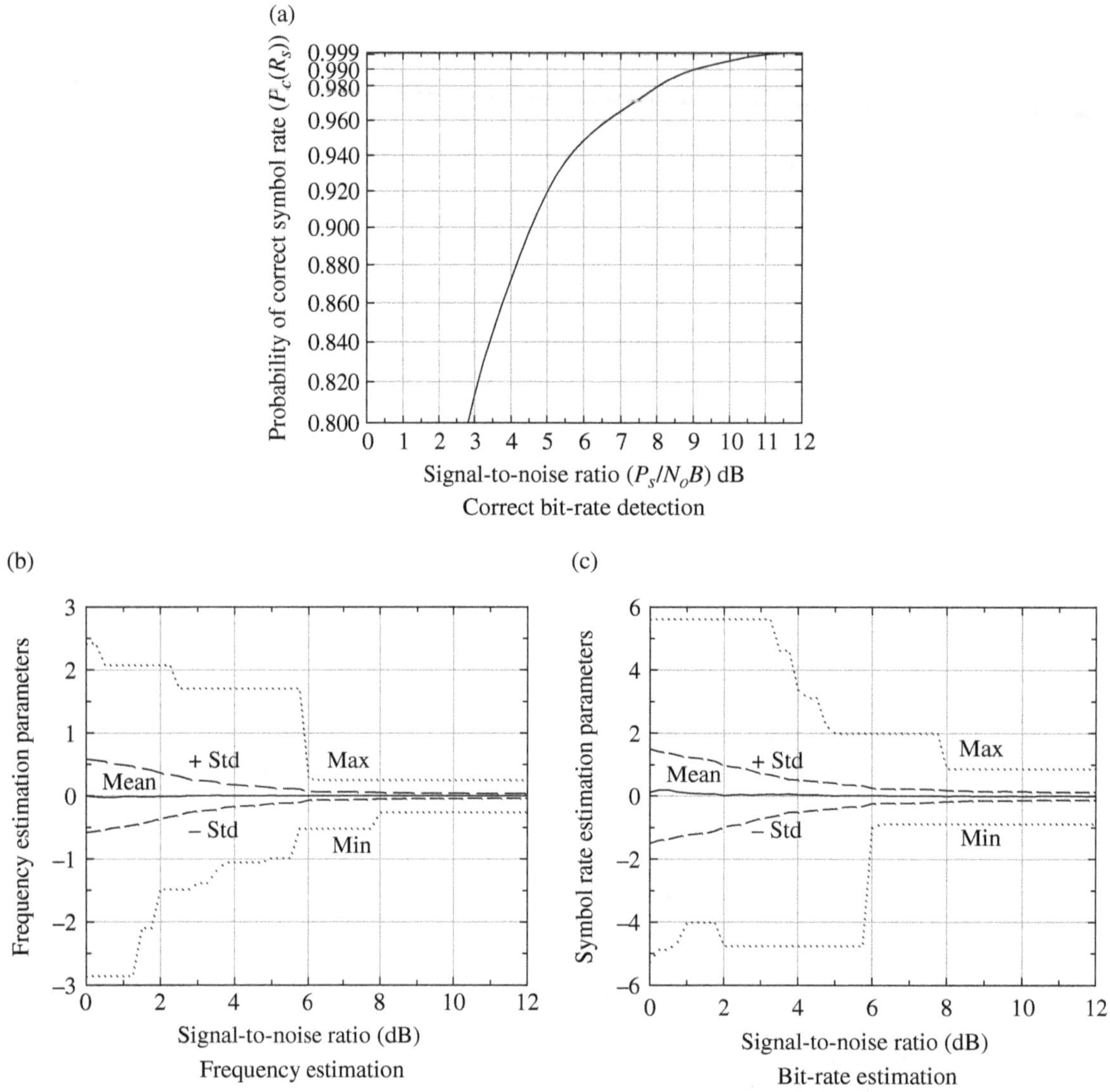

FIGURE 11.39 Detection and estimation results for repeated 1100 data.

- Increasing the measurement window, with a commensurate increase in the FFT size. The resolution bandwidth is improved with an accompanying improvement in the measurement signal-to-noise ratio.
- Using linear or parabolic interpolation to refine the frequency resolution measurement.
- Using an FFT window to reduce the spectrum leakage into adjacent FFT cells when the signal tones at $fT \pm n/4$ do not fall at the center of an FFT cell.
- Using a higher bit rate estimation tolerance for declaring a successful bit rate acquisition, for example, increasing the 12.5–25%. However, increasing the bit rate estimation tolerance is limited by the acquisition and tracking algorithm requirement discussed in Section 11.3.5.
- Using multiple FFTs and an *m*-of-*n* decision criterion for declaring a successful bit rate detection.

In many applications the bit rate or a number of possible bit rates are known and the issue becomes one of finding the ℓ_l and ℓ_u pairs corresponding to the received bit rate. This can be accomplished by searching through pairs of ℓ_l and ℓ_u corresponding to combinations of FFT cells with successively lower magnitudes until the bit rate is found or a limit is reached and a missed bit rate is declared. When the bit rate is declared the frequency estimation is performed using (11.52). The performance improvements listed earlier also apply to the known bit rate case; however, the use of multiple FFTs and the *m*-of-*n* decision criterion is a major consideration for both detection and confirmation before proceeding to the symbol acquisition and tracking processing.

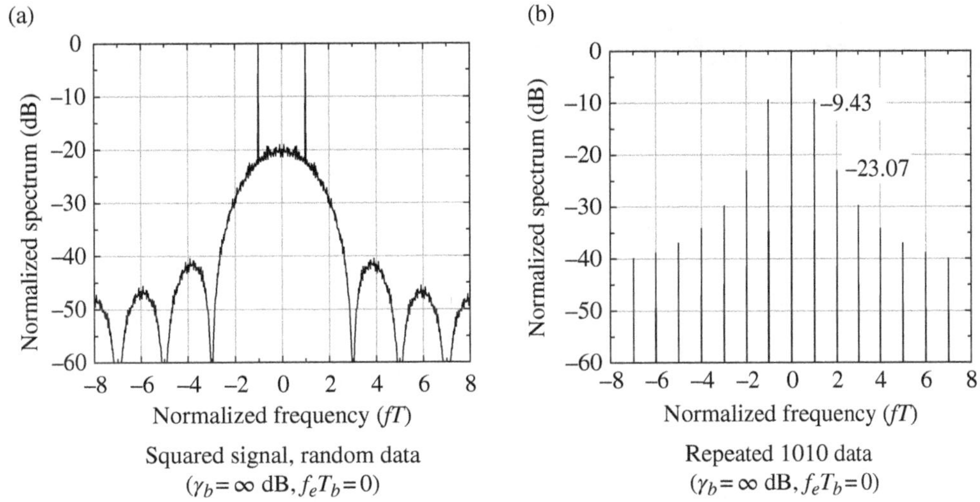

FIGURE 11.40 MSK acquisition preamble spectrums (1024-point FFT).

TABLE 11.10 Summary of FFT Processing Parameters

N_{fft}	N_s	$T_{fft}R_s$	f_{res}/R_b	η (%)	G_{snr} (dB)
1024	16	64	1/32	66.7	15
768	16	48	1/24	50.0	13.8

N_s = sample/symbol, $T_{fft}R_s = N_{fft}/N_s$, $f_{res}/R_b = 2/T_{fft}R_s$, η (%) = 200(T_{fft}/R_s)/192.

11.3.2.4 Detection, Symbol Rate, and Frequency Estimation Using MSK-Modulated Waveform In this section, the characteristics of the MSK-modulated waveform spectrum are examined for detection and estimation of the frequency and symbol rate. In this regard, the objectives are identical to those discussed in Section 11.3.2.2; however, the focus is on the unique spectral characteristics of the MSK waveform. The preamble for the multi-*h* CPM waveform, discussed in Chapter 9, uses MSK with a symbol rate equal to that of the multi-*h*-modulated user data.

The following descriptions of the MSK-modulated signal are based on unique data patterns that result in distinct spectral lines at harmonics related to the symbol rate. These spectral characteristics are exploited to estimate the signal presence, carrier frequency, and symbol rates during the synchronization segment of the preamble. The data patterns examined are the repeated sequences 1100... and 10... where the number of repetitions commensurate with the detection and signal-to-noise specification based on the system application. For example, the repeated 1100... data is specified for the multi-*h* CPM-modulated waveform preamble and is repeated for a total of 192 bits corresponding to the bit rate of the 2-*h* CPM modulation. When mark- or space-hold data is applied to the MSK modulator the transmitted signal is a CW tone corresponding to upper MSK tone $f_c + R_s/2$ and when alternate mark-space data is applied the lower MSK tone at $f_c - R_s/2$ is transmitted; if required, these tones can serve as a CW preamble. However, when a randomly modulated MSK waveform is squared, the resulting spectrum contains discrete spectral tones, at $2f_c \pm R_s$. These tones can be used for signal detection and synchronization without the necessity of having a CW preamble. The downside of the signal squaring is a decrease in the signal-to-noise ratio resulting from the squaring loss and a frequency doubling that requires a higher demodulator sampling frequency. The noise-free spectrums of the squared MSK-modulated signal with random and the repeated 1010... data patterns are shown, respectively, in Figure 11.40a and b. The characteristics of the signal processing and the 1024-point FFT are summarized in Table 11.10. In both of these cases, the spectral lines of interest occur at $f_c \pm R_s$.

Figure 11.41a shows the noise-free spectrum of a randomly modulated MSK waveform and is provided as a point of reference for the spectrum of the MSK-modulated preamble data 11001100... shown in Figure 11.41b; the preamble pattern is repeated for 192 bits. These spectrums are normalized by the maximum level and are based on computer simulations using a 1024-point FFT as detailed in Table 11.10. The efficiency is a measure of the duration of a single FFT relative to the synchronization preamble length of 192 bits. The 1024-point FFT occupies 66.7% of the synchronization interval so, to ensure that at least one FFT captures the interval; the FFTs must be overlapped by 33.3%. On the other hand, the 768-point FFT ensures that at least one FFT

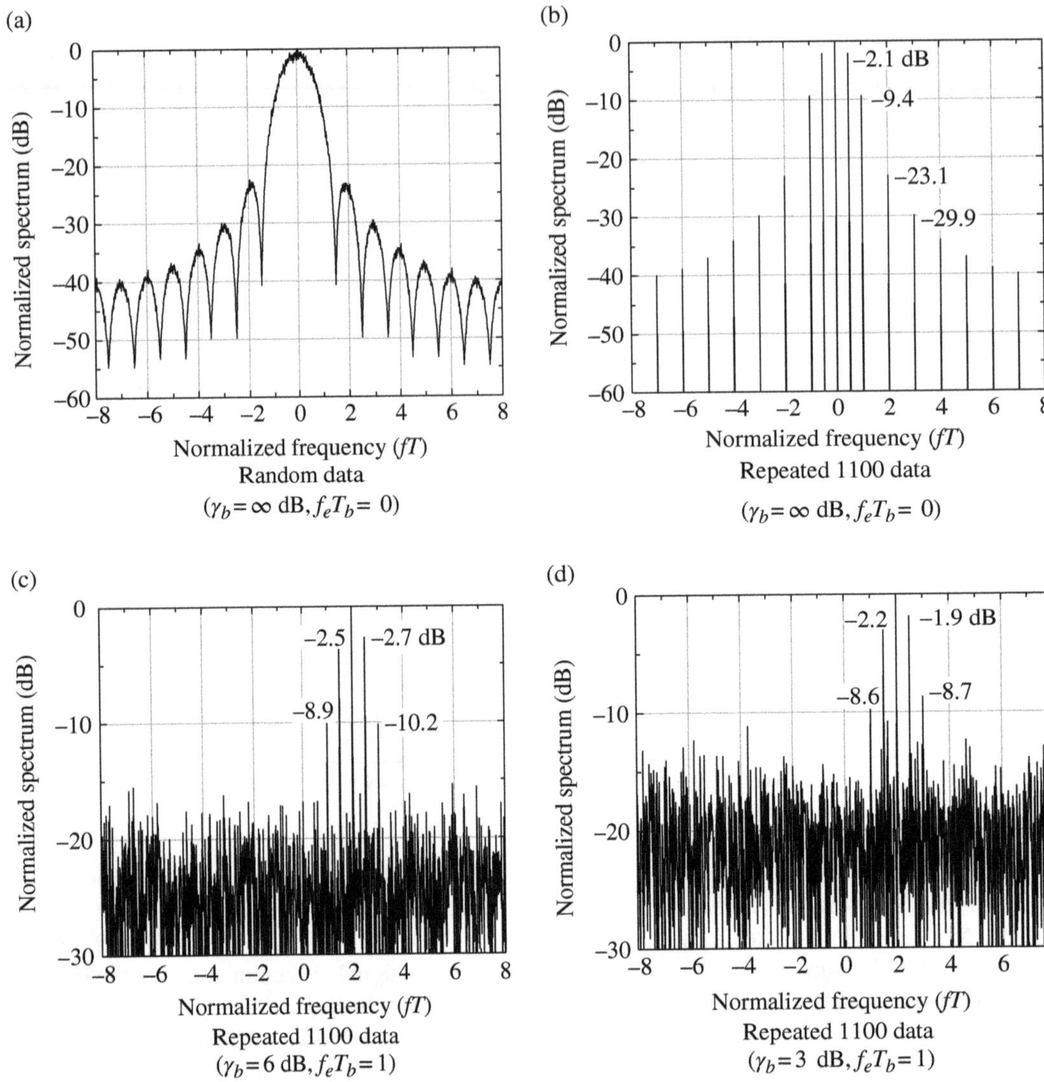

FIGURE 11.41 MSK multi-h CPM acquisition preamble spectrum (1024-point FFT).

captures the interval and, with 50% overlapping, at least two FFTs occupy the synchronization interval. Another important parameter is the signal-to-noise ratio improvement, relative to $\gamma_b = E_b/N_o$, expressed as

$$G_{snr}(\text{dB}) = 10\log_{10}\left(\frac{R_b}{f_{res}}\right) \quad (11.59)$$

Figure 11.41c and d show the spectrum of the MSK synchronization preamble segment using the 1024-point FFT for the signal-to-noise ratios $\gamma_b = 6$ and 3 dB, respectively. In these cases, the normalized frequency error of the received signal is $f_eT_b = 1$. The plots with additive channel noise use different noise seeds so the results are statistically independent. The −23.1 and −29.9 dB spectral lines are *buried* in the noise; however, spectral lines at $f_c \pm R_s/2$ and $f_c \pm R_s$ exhibit a sufficient signal-to-noise ratio, although the variations resulting from the additive noise in the FFT resolution cells are evident.

Figure 11.42a and b are similar to those in Figure 11.41c and d except that the 768-point FFT is used. The increase in the noise floor is evident; however, there is still a *healthy* signal-to-noise ratio in the cells containing the spectral lines of interest. Referring to Section 9.3.4.1, the minimum γ_b for the multi-h CPM modes at $P_{be} = 10^{-5}$ is about 5.3 dB so the synchronization processing for the 3 dB condition using either FFT provides a degree of acquisition robustness. The processing of correctly estimating the frequency error and symbol rate is enhanced by using a weighted sliding window smoothing function across the FFT spectrum cells and selecting the optimum frequency error and symbol rate corresponding to the minimum weighted mean-square error (MSE) between

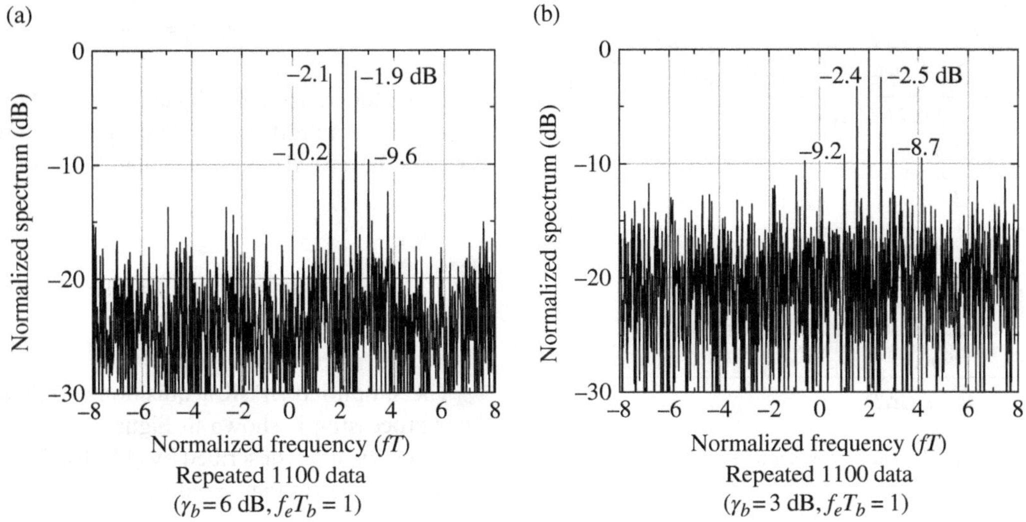

FIGURE 11.42 MSK multi-*h* CPM acquisition preamble spectrum (768-point mixed radix FFT).

the sliding window and the magnitudes of the FFT cells. In this case, the weights of the smoothing function are chosen as the magnitude of the three or five symmetrical cells from Figure 11.41b; this approach is similar to the parabolic interpolation discussed in Appendix 2C.

11.3.3 Signal Detection and Parameter Estimation Using Correlator

Signal correlation is a convenient technique to determine the symbol time and frequency and, as discussed in the following section, can be used without the CW preamble segment. Correlation processing is used extensively in spread-spectrum applications to determine code synchronization as discussed in Chapter 13. However, as mentioned previously, the acquisition time and processing complexity is significantly reduced when all of the preamble segments are available.

In the following analysis, the correlator is examined in terms of the baseband BPSK-modulated received signal $\tilde{s}(t)$ and the demodulator reference signal $\tilde{s}_r(t)$. The correlation function is expressed as

$$R(\tau;\hat{f}_\varepsilon) = \int_{-\infty}^{\infty} \tilde{s}(t;f_\varepsilon,\dot{f}_\varepsilon)\tilde{s}_r^*(t-\tau)dt \qquad (11.60)$$

The delay τ is referred to as the correlation lag and f_ε and $\dot{f}_\varepsilon$ are the input signal frequency and frequency-rate errors following the coarse frequency estimation. In practice, the range of the integration in (11.60) is over the duration of the synchronization data pattern denoted as T_p. The synchronization data pattern must have sufficient energy, either signal power or duration, to reliably detect the synchronization code; however, long codes are subject to unacceptable coherent integration losses requiring coherent integration over shorter intervals with noncoherent combining of multiple intervals to achieve the signal to noise required for reliable synchronization. Frequently, a short synchronization code is simply repeated a number of times to achieve the specified synchronization requirements. For example, if the coarse frequency estimate f_ε is such that the product $f_\varepsilon T_p$ results in an unacceptable signal to noise over the correlation interval $T_p = N_bT_b$ where N_b is the number of code bits of duration T_b, then noncoherent combing of N repeated codes may be performed to achieve the required signal-to-noise ratio for synchronization.*

Consider the *complex* received noise-free baseband BPSK-modulated signal and the corresponding demodulator reference signal expressed, respectively, as

$$\tilde{s}(t) = Ad_n e^{j(\theta_\varepsilon(t)+\varphi)} rect_T(t,n) \quad : \text{received signal} \qquad (11.61)$$

and

$$s_r(t) = d_n rect_T(t-\tau,n) \quad : \text{demodulator reference} \qquad (11.62)$$

where $d_n \in \{\pm 1\}: 0 \le n \le N_b - 1$ represents the received and reference binary data, φ is the unknown constant received signal phase error, and the time-dependent phase function $\theta_\varepsilon(t)$ represents the demodulator angular frequency and frequency-rate errors and is expressed as

$$\theta_\varepsilon(t) = \omega_\varepsilon t + \dot{\omega}_\varepsilon t^2/2 \qquad (11.63)$$

In this analysis the demodulator reference signal has unit voltage amplitude. Considering a repeated code length of N_b bits, the correlation response given by (11.60) is evaluated over the code duration T_p as

*Coherent and noncoherent detection and multiple pulse noncoherent combining loss is the subject of Appendix C.

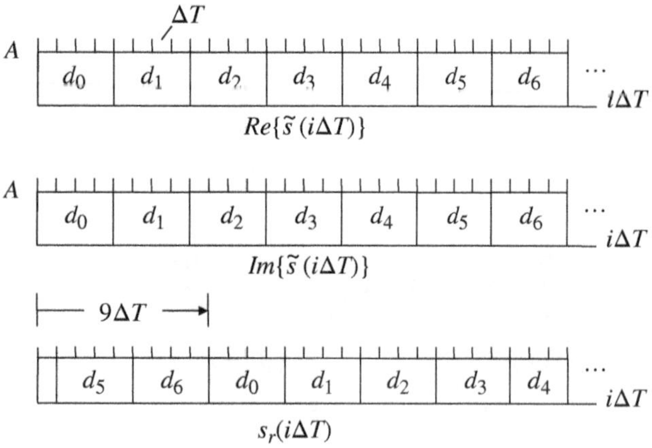

FIGURE 11.43 Repeated 7-bit barker sequences ($f_\varepsilon T_b = 0$, $\varphi = 0$, $N_s = 4$).

$$R(\tau;\hat{f}_\varepsilon) = Ae^{j\varphi}\int_0^{Tp} e^{j\theta_\varepsilon(t)} d_{n:k} d_n rect_T(t,n:k) rect_T(t-\tau,n)dt \tag{11.64}$$

The notation $d_{n:k}$ refers to the k-sample misalignment corresponding to the reference delay τ. This expression is best explained by the example depicted in Figure 11.43 using a 7-bit Barker sequence with $N_s = 4$ samples per bit. The received baseband sampled signal corresponding to one code interval is sequentially stored in the *complex* array $\tilde{s}(i\Delta T)$ and the *real* sampled local reference is stored in the array $s_r(i\Delta T)$. There are $N_s N_b$ contiguous stored samples in each array; however, the data samples are not aligned. For example, the stored reference samples are delayed by $k = 9$ samples from those of the stored received signal samples. The task of the correlator synchronization processing is to determine the optimum correlation lag, τ_{opt}, for which $d_{n:k}|_{k=0} = d_n$ for all samples. The resulting code-bit alignment is used for symbol time and phase tracking in preparation for detecting the SOM location and subsequently the data demodulation. In this example the optimum correlation lag is determined by advancing, or cyclically left shifting, the stored reference bits through $k = 2N_s N_b + 1$ samples and then associating the optimum lag corresponding to the maximum correlation. As shown in Figure 11.43, the optimum correlation lag correction is seen to be $\Delta t = 9\Delta T$.

In the context of Figure 11.43, consider the transmitted BPSK baseband acquisition waveform with Barker code data $dI_n = \pm 1$. The simplified* expression for the transmitted waveform is expressed as

$$\tilde{s}_T(n) = \frac{A}{\sqrt{2}} dI_n \tag{11.65}$$

When the transmitted signal is passed through a noiseless channel that introduces a signal phase shift $\theta_\varepsilon(t) + \varphi$, the received baseband sampled signal is described as

$$\begin{aligned}\tilde{s}(n) &= \tilde{s}_T(i,n)e^{j(\theta_\varepsilon(i\Delta T)+\varphi)} \\ &= \frac{A}{\sqrt{2}} dI_n\{\cos(\theta_\varepsilon(i\Delta T)+\varphi) + j\sin(\theta_\varepsilon(i\Delta T)+\varphi)\}\end{aligned} \tag{11.66}$$

The simplified BPSK acquisition correlator and demodulator processing is shown in Figure 11.44.

Using the input described by (11.66), the correlator output samples $g_c(m)$ are expressed as

$$g_c(m) = \varepsilon_{II}(n-m)\cos(\theta_\varepsilon(i\Delta T)+\varphi) + j\varepsilon_{II}(n-m)\sin(\theta_\varepsilon(i\Delta T)+\varphi) \tag{11.67}$$

where $\varepsilon_{II}(n-m)$ is evaluated as

$$\varepsilon_{II}(m) = \frac{A}{\sqrt{2}}\sum_n dI_n dI_{n-m} = \begin{cases} \dfrac{AN}{\sqrt{2}} & : m = 0 \\ \dfrac{A\varepsilon}{\sqrt{2}} & : m \neq 0 \end{cases} \tag{11.68}$$

When $m = 0$ the ideal correlator output corresponds to the maximum response with $\varepsilon_{II}(0) = AN/\sqrt{2}$. For the cyclically shifted Barker sequence, the correlation error is $\varepsilon = -1$, for the barker sequences.

During the SOM correlation and data detection, the ideal signal phase and timing estimate errors are zero; however, at this stage of the signal acquisition processing the phase is unknown and the effect of the phase-error terms must be removed using noncoherent detection as shown in Figure 11.44. The sensitivity of the normalized correlator output to a frequency error is evaluated at the zero lag condition when the received and reference signals are perfectly aligned. For BPSK modulation, the correlator sensitivity is evaluated† as

$$\bar{R}(0;\hat{f}_\varepsilon,\dot{f}_\varepsilon) = \frac{A}{\sqrt{2}T_p}\int_0^{T_p} e^{j(\varphi + 2\pi f_\varepsilon t + \pi \dot{f}_\varepsilon t^2)}dt \tag{11.69}$$

Substituting the variable $x = t/T_p$ in (11.69) results in the normalized solution

$$\bar{R}(0;\hat{f}_\varepsilon T_p,\dot{f}_\varepsilon T_p^2) = \frac{A}{\sqrt{2}}\int_0^1 e^{j(\varphi + 2\pi f_\varepsilon T_p x + \pi \dot{f}_\varepsilon T_p^2 x^2)}dx \tag{11.70}$$

*The simplifications result from eliminating the *rect*( ) function that describes the BPSK symbols and, in the demodulation processing, the alignment of the receiver reference symbol timing.

†With quadrature modulations the in-phase and quadrature symbol channel *cross-talk* must be considered.

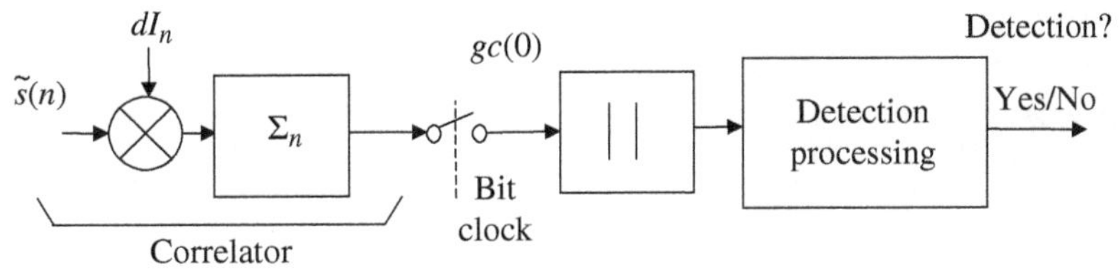

FIGURE 11.44 BPSK acquisition correlator processing.

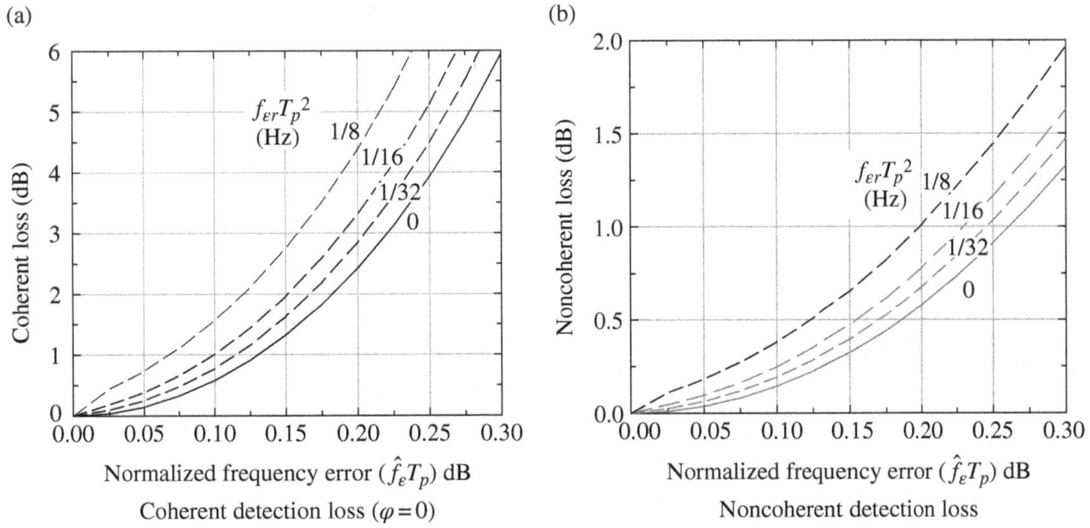

FIGURE 11.45 Acquisition code correlation loss.

The correlation loss resulting from $f_\varepsilon, \dot{f}_\varepsilon$, and φ are evaluated numerically over the ranges of T_p shown in Figure 11.45 where $f_{\varepsilon r} = \dot{f}_\varepsilon$. With coherent integration the phase φ is set to zero, however with noncoherent detection this is not necessary. A theoretical solution to (11.69) and (11.70) is provided by Gradshteyn and Ryzhik [40] and, using their results, the magnitude of the correlation function can be shown to be independent of the signal phase. At this point in the acquisition processing, the constant phase error is the least of the concerns, since the PLL will acquire and track the phase prior to the SOM processing. Furthermore, an estimate of the phase can be obtained from the in-phase and quadrature rails of the peak SOM correlation response to ensure that the PLL does not lock 180 degrees out of phase.

As an example of the correlation processing, the 7-bit Barker code correlation response is evaluated using a simulation program with $\dot{f}_\varepsilon = 0$ and $N_s = 4$ samples per code bit. The simulated coherent detection performance, shown in Figure 11.46a, is ideal sense that f_ε and φ are also zero. The noncoherent correlation response for a normalized frequency error of $f_\varepsilon T_p = 0.2$ is shown in Figure 11.46b. In this case, the simulated peak correlation response is 0.935 corresponding to a loss of 0.58 dB which is confirmed in Figure 11.45b.

The correlator and the CW FFT processing discussed in Section 11.2.2.2 are based on BPSK modulation so, in the following evaluations, the symbol interval corresponds to the bit interval with $T = T_b$. The overall objective of this section is to reduce the estimates of f_ε, $\dot{f}_\varepsilon$, φ, and τ to ensure that the phaselock and symbol timing loops are tracking throughout the SOM correlation processing. This is accomplished using noncoherent estimation with an iterative process of adjusting τ to obtain a peak correlation response and then adjusting the frequency and frequency rate to minimize the loss. The normalized frequency error expressed as the product of the frequency error and the observation or estimation interval is defined as

$$f_\varepsilon T_p = f_{norm} \tag{11.71}$$

where f_{norm} is the normalized CW frequency error that is determined from the specified probability $P_r(\text{error} \le f_\varepsilon|\text{CW})$ for a CW signal and the FFT parameters as depicted in Figures 11.15, 11.16, and 11.17. Equation (11.71) gives rise to three system design issues: the accuracy of the CW frequency estimate (f_{norm}), the duration of the synchronization code (T_p), and the tolerance of the correlator (*Loss*) given the frequency error. Considering the unweighted, zero-padded,

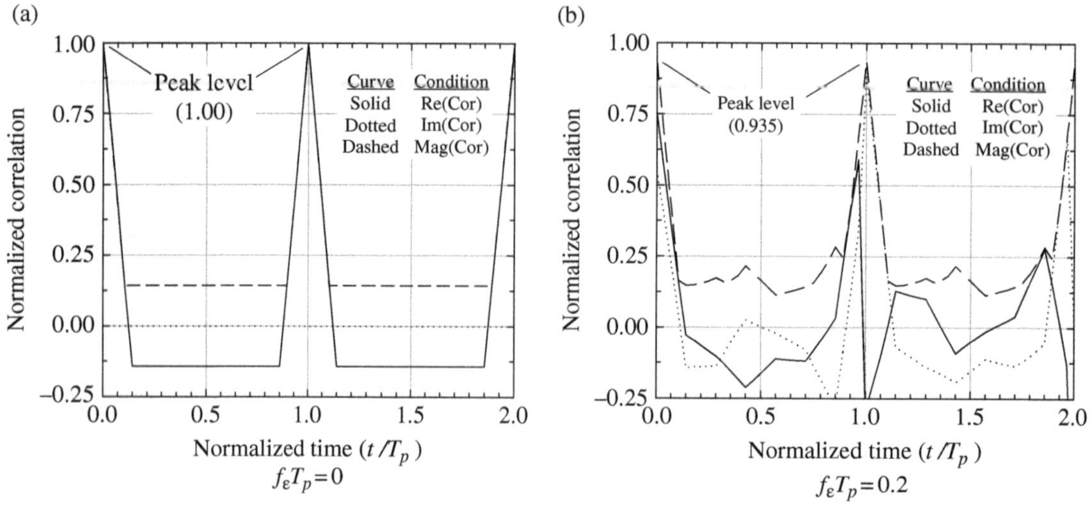

FIGURE 11.46 Repeated 7-bit barker code correlation response ($f_{er}T_p^2 = 0$, $\varphi = 0$, $N_s = 4$).

256-point FFT in Figure 11.15 with a sampling rate of $f_s =$ 140 kHz, the frequency resolution is computed as $f_{res} =$ 546.875 Hz and the estimation interval is $T_p = T_{fft} = 1.828$ ms. Selecting f_{norm} for $P_r(\text{error} \le f_e|\text{CW}) = 0.99$ corresponding to the *best* case error estimate with $E_b/N_o = 3$ dB results in $f_{norm} = 0.15$. Using these parameters and (11.71) results in: $f_\varepsilon T_p = 0.15$, a correlation loss of 0.26 dB, and $f_e = 82$ Hz. The significance of the frequency error depends on the system symbol rate, however, referring to (11.5), for a second order PLL to achieve phase-lock without cycle skipping requires than $f_e < F_L$.

The length of the synchronization code impacts the correlation gain and the correlation loss as indicated earlier. The signal-to-noise ratio at the correlator output is also dependent on the length of the synchronization segment and must satisfy the requirement to declare a correct synchronization probability, $P_c(\text{sync})$. However, the practical length of the synchronization code is dependent on the frequency estimation accuracy and the correlation time of the channel and various system oscillators used for frequency conversion. For example, an off-the-shelf, low-cost, *nonovenized* oscillator with a correlation time of 0.3 s limits the N-bit acquisition correlator length to $N < \lceil 0.3R_b \rceil$ where R_b is the BPSK-modulated bit rate. If the resulting correlator output signal-to-noise ratio is inadequate to provide the specified synchronization probability then repeated synchronization codes must be included in the synchronization segment and noncoherently combined to achieve the performance results.* The noncoherent combining of synchronization codes is discussed in the following section.

*This assumes that the system is operating at the maximum transmitter power and that coherent integration is limited by the phase stability of the system oscillators.

11.3.4 Synchronization Without CW Preamble Segment

The preamble discussed in this section is based on a known synchronization preamble segment of contiguously repeated N-bit Barker codes. The focus is on the detection, synchronization, frequency and symbol timing estimation, and PLL tracking without the aid of the CW preamble. The synchronization objective is identical to those discussed previously, namely, to provide for the coherent detection of the SOM sequence. The demodulator synchronization functional processing is shown in Figure 11.47. The following description applies to very low received signal-to-noise ratios as might be encountered with turbo coded FEC. This is followed by the more conventional description when operating in high signal-to-noise ratio encounters.

The signal acquisition and synchronization processing with *very low input signal-to-noise ratios* is described as follows. Using the known range of the received signal frequency uncertainty ($|f| \le f_{max}$) the demodulator sampling frequency (f_s) is established as in (11.7). The ADC N_{sc} samples spanning the one Barker code interval are mixed to baseband by the coarse reference frequencies ($f_{ref,i}$) and stored as raw data samples over the range of the received signal frequency uncertainty.† The down-sampled and low-pass filtered stored samples result in the equivalent of N_s samples per Barker code bit that are coherently correlated with the stored reference Barker code; cyclic correlation is used to identify the code misalignment. The correlation magnitude is stored in the correlation accumulator forming a

†At this point, the stored raw data samples may consist of noise only or noise only samples followed by Barker code samples. As the synchronization search continues the stored raw data samples are replaced by the N_s newly formed received samples; the N_s samples are not necessarily synchronized with the Barker code bit samples.

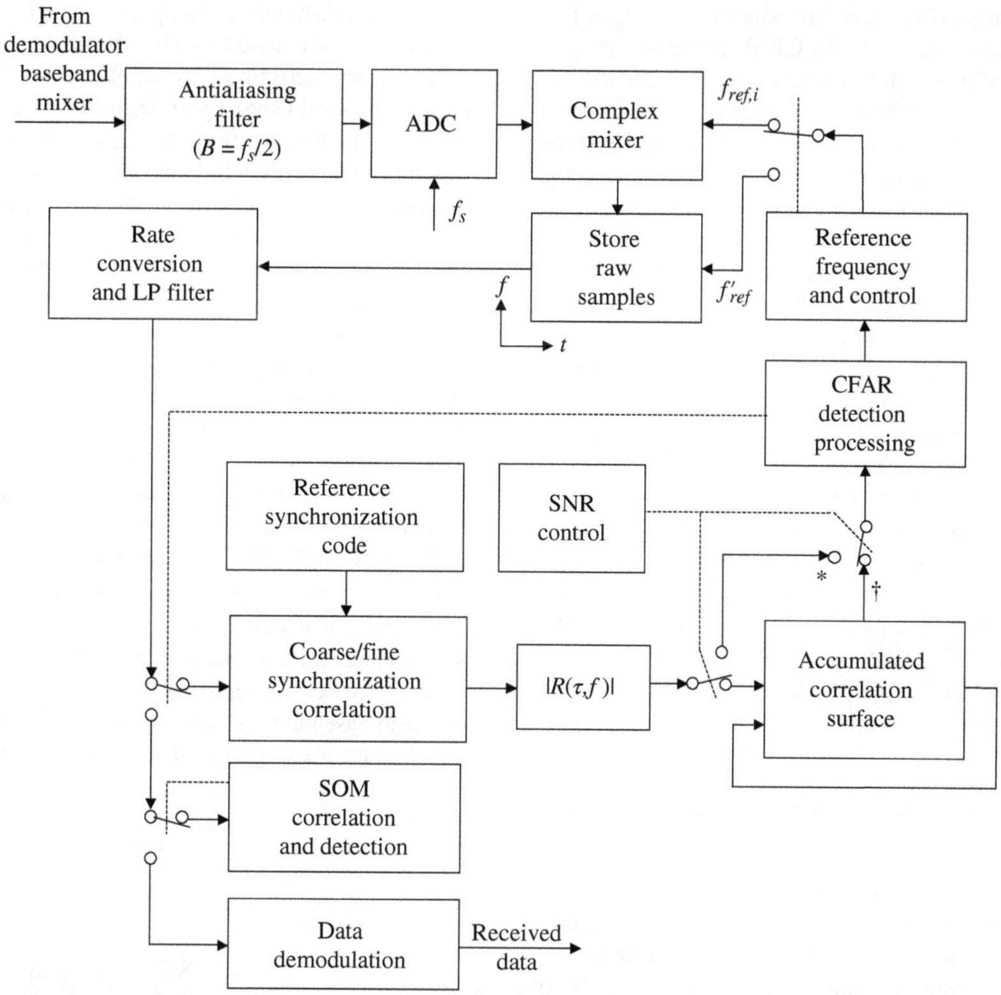

FIGURE 11.47 Correlator synchronization and detection processing (signal-to-noise ratio: *low, †high).

(NN_s, i_{max},) time-frequency noncoherent correlation surface. This provided for a global search over the entire time–frequency uncertainty range. The number of Barker code correlations is determined by the input signal-to-noise ratio and the output signal-to-noise ratio required to achieve the desired synchronization probability as discussed in Appendix C. Upon completing the noncoherent accumulations a two-dimensional time–frequency censored CFAR is performed around the largest correlation magnitude. If the CFAR thresholds are pasted a fine-frequency estimate is established in the same manner around the location of the maximum correlation and phase and symbol time tracking are initiated while searching for the SOM location. If the first CFAR is not passed then the process is repeated for the next four or five largest surface maximums before declaring a missed acquisition.

Examining the entire time–frequency correlation surface prior to performing the CFAF detection processing minimizes the false detections and thereby reduces the average synchronization time when the received signal frequency is uniformly distributed over the entire frequency range. However, if the received signal frequency is, for example, normally distributed about a previously detected mean frequency, the correlation surface search strategy can be improved with a reduction in the average acquisition time. Increasing the length of the synchronization segment by concatenating additional synchronization codes negates the necessity to examine multiple correlation responses over the time–frequency surface; however, the increased preamble duration is unacceptable in many applications. For example, when operating with an input code-bit signal-to-noise ratio of −3.27 dB, choosing the number of synchronization code repetitions so that examining the four highest correlation detections over the time–frequency surface results in a correct synchronization detection of $P_c(\text{sync}) = 0.99$ corresponds to a preamble length that is three to four times shorter than is otherwise required by examining one maximum correlation location.

The signal acquisition and synchronization processing with *high input signal-to-noise ratios* is described as follows.

The time–frequency correlations are established using coherent Barker code combining and the CFAR detections are performed sequentially on each frequency hypothesis without the necessity of creating a time–frequency correlation surface. This allows for fewer Barker coded repetitions and much shorter preambles; however, the correlator output signal-to-noise ratio must be sufficiently high to result in an acceptable false-synchronization probability. The fine-frequency correlation processing is also performed as described earlier to ensure that the frequency estimate is within the lock-in frequency of the PLL.

The correlation processing shown in Figure 11.47 is based on conventional time-domain correlation that is performed at each frequency; however, the processor loading is reduced using an FFT correlator [41] in which the spectrum $H(m)$ of the sampled reference and $S(m)$ of the received synchronization code are appropriately stored in the FFT memory. Under these conditions the correlation response is simply obtained as the IFFT of the product $S(m)H^*(m)$. Furthermore, the correlation corresponding to a correlation lag of $(t - n\Delta\tau)$ is obtained in the frequency domain correlator through a complex phase shift of $\exp(-j\omega n\Delta\tau)$ that is efficiently executed in the frequency domain.

11.3.5 Symbol and Frequency Acquisition and Tracking

Based on the synchronization preamble segment processing discussed in the preceding sections, the estimate of the carrier frequency error has been adjusted to less than the PLL lock-in frequency and the symbol timing error is within about 1/8 of the symbol interval or about 1/32 of the symbol interval with RRC modulation. The final task of the synchronization preamble segment is to use these estimates and initialize frequency and symbol acquisition and tracking and ensure that steady-state tracking occurs prior to the SOM preamble segment. The PLL frequency tracking requirements, functions, and performance are discussed in Chapter 10. The symbol tracking functions are introduced in Sections 2.8.5 through 2.8.7. The E/L gate symbol integrator is used to generate a timing error discriminator response for symbol tracking. In Chapter 4 the symbol timing error generation and tracking functions are shown to be an integral part of the Costas or baseband PLL. Symbol timing adjustments on the order $\Delta T \le T/16$ are typical for most modulations; however, for RRC and phase-shaped modulations adjustments of $\Delta T \le T/64$ are recommended.* The rate of the symbol timing adjust must be controllable with more frequent or larger adjustments during the pull-in range of the symbol tracking processing. To reduce the impact of the channel and receiver noise on the timing adjustment, the output of the E/L gate discriminator is filtered with a controllable bandwidth that is reduced following the detection of symbol tracking. The symbol rate is typically established by an accurate system clock, whereas received signal carrier frequency is dependent on the dynamics of the channel and the link encounter. Therefore, to minimize the mutual interaction between the two tracking functions, the steady-state response of the symbol tracking loop should be 8–10 times slower than the carrier tracking loop.

The general implementation of the frequency and symbol tracking functions is shown in Figure 11.48. A major difference in the PLL implementation, from those previously discussed, is that, the known synchronization preamble data, d_i, is used to aid the carrier frequency acquisition; this is referred to as data-aided synchronization. Although the preamble data is known from the synchronization processing, it is not known when the synchronization code ends and the SOM code starts, so, the PLL must revert to decision-directed synchronization using the demodulator data estimates $\hat{d}_i$ as soon as possible. This transition is made when the loop lock detector indicates the *phaselocked* condition.

The stored samples of the fine-frequency corrected received baseband signal $\hat{s}(nT_s)$ are used to compute the symbol time error using the E/L gate outputs evaluated as

$$I_e = \sum_{k=1}^{N_s/2} \tilde{s}(kT_s' - \tau) \tag{11.72}$$

and

$$I_\ell = \sum_{k=N_s/2+1}^{N_s} \tilde{s}(kT_s' - \tau) \tag{11.73}$$

where $T_s' = T_s/M$, $\tilde{s}(kT_s' - \tau)$ is the down-sampled baseband signal corresponding to a symbol error of τ seconds relative to the demodulator reference, and N_s is the number of down-sampled samples per symbol.† The E/L output is computed as

$$\varepsilon_\tau = \frac{|I_e| - |I_\ell|}{|I_m|} \tag{11.74}$$

where $I_m = I_e + I_\ell$ is the matched filter output sample. Using the magnitudes to determine the timing error corresponds to noncoherent detection and the performance with additive noise is degraded from that of the coherent detector as expressed in (2.33). Referring to (11.74), when $\tau > 0$ the received signal is delayed relative to the demodulator time base and the output error is positive and when $\tau < 0$ the signal leads the demodulator time base and the error is negative. In either case, the respective samples from the succeeding or

*Considering the dynamics of the symbol tracking loop, these discrete adjustment steps result in an average symbol tracking error less than about $T/8$ and $T/32$, respectively.

†Generally N_s is an even number, however, if N_s is odd then the odd sample must be alternately distributed between the early-late outputs to ensure that the timing error is unbiased.

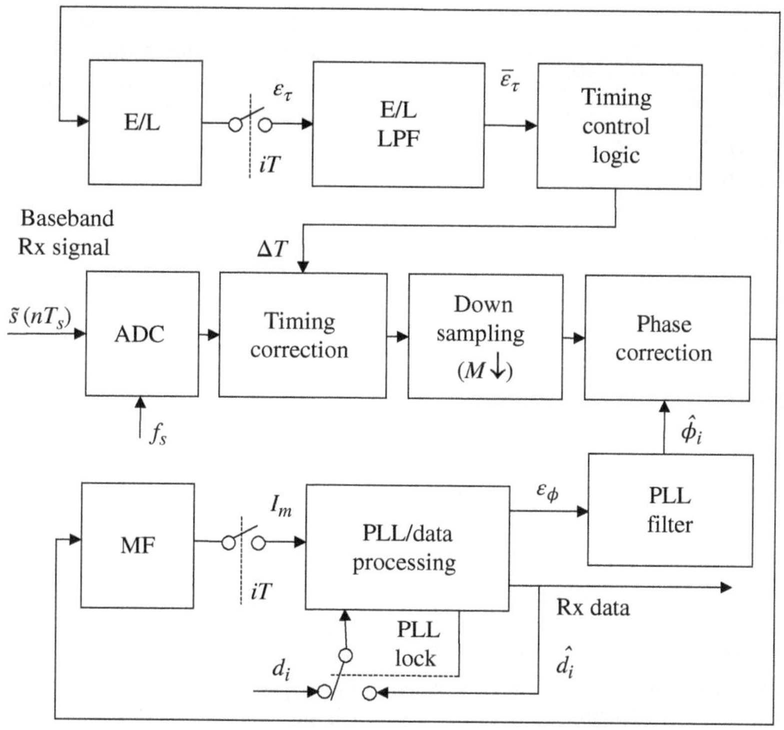

FIGURE 11.48 Signal phase and timing correction.

preceding symbols corrupt the E/L gate error; however, with selected data patterns, including random data, these contributions are averaged to zero in the E/L filter output resulting in an unbiased estimate.

The details of the timing logic in Figure 11.48 are shown in Figure 11.49. The threshold N_{thr} is dependent upon the timing update increment ΔT and is typically selected during the demodulator simulation and confirmed during hardware testing. After each timing update, the symbol counter is reset to zero to provide hysteresis and avoid over-correcting of the symbol timing. Following the detection of symbol-timing lock the threshold is increased to slowdown the timing adjustments during data detection. This can also be accomplished by decreasing the bandwidth of the E/L LPF.* The symbol-timing lock detection is not shown in Figure 11.48, however, is obtained by applying the CFAR algorithm to the E/L filter output with lock detection declared in response to a predetermined low timing error variance.

The single-pole E/L LPF, shown in Figure 11.50, is a recursive filter with the coefficient computed as

$$\alpha = e^{-2\pi BT_s'} \tag{11.75}$$

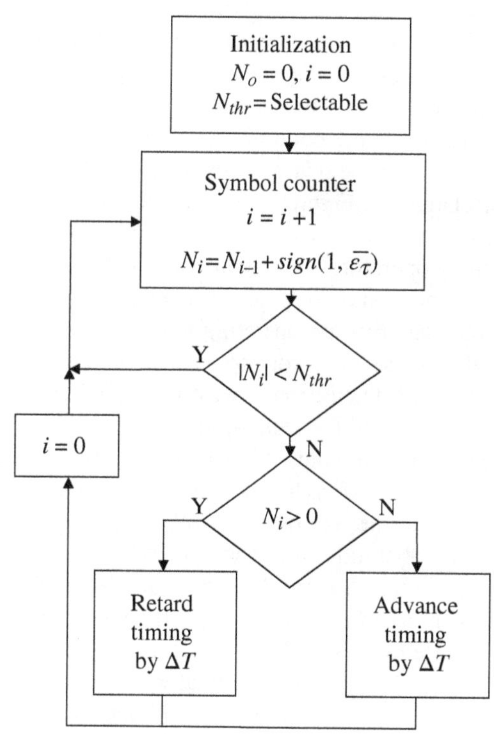

FIGURE 11.49 Timing adjustment logic.

The normalized bandwidth of the filter is BT_s' and the filter gain $G_a = 1 - \alpha$ results in a unit gain response; the gain G and the BT_s' product are available for externally controlled adjustments during system testing.

*When the symbol timing loop is modeled as a first-order loop, the filter is simply a gain followed by an ideal integrator. In this case, the bandwidth is controlled by selecting the filter gain.

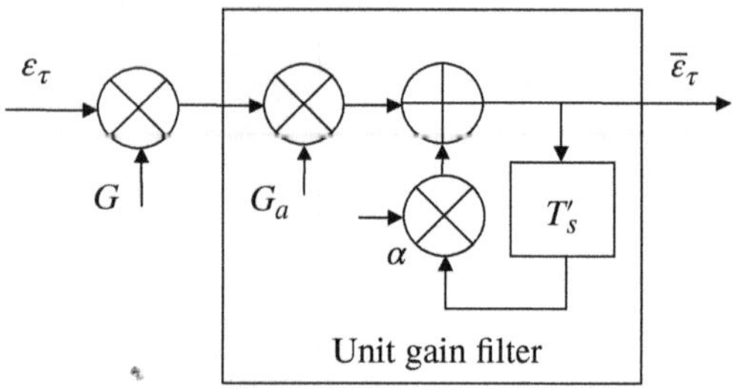

FIGURE 11.50 Symbol E/L low-pass filter.

11.4 START-OF-MESSAGE (SOM) PREAMBLE SEGMENT

The SOM preamble segment is used to locate the first message header bit or the first message information bit if the header is not included in the preamble. This first message bit is identified by the location of the SOM code correlation peak. When the first SOM code bit is encountered, the symbol and frequency tracking loops are locked and operating under their respective steady-state tracking conditions providing for coherent detection of the SOM bits. Upon initiation of the symbol and frequency acquisition during the synchronization preamble segment, the stored synchronization data bits are shifted through the symbol and phase tracking loops eventually encountering the SOM code bits that lead to the detection of the SOM correlation peak. Although the first bits correspond to those of the synchronization code segment, as the SOM code bits are encountered the SOM lead-in correlation begins as illustrated by the LPN SOM code correlations in Figure 11.31.* Throughout the coherent correlation processing a one-dimensional CFAR detection algorithm is operating looking for a threshold crossing to identify the location of the peak correlation. Since the preamble and message bits are operating at the same symbol rate, the signal-to-noise ratio corresponds to the E_b/N_o required for the message bit detection. Therefore, for a SOM sequence of N bits, the gain in the signal-to-noise ratio corresponding to the peak correlation output is $10\log_{10}(N)$ dB. For example, the 74-bit LPN SOM code detection signal-to-noise ratio is 18.7 dB above the operating E_b/N_o. The detection and false-alarm probabilities for coherent detection are characterized in Appendix C. Consider, for example, a SOM false-alarm probability of $P_{fa} \le 10^{-3}$ is specified and the operating signal-to-noise ratio is $E_b/N_o = 6.3$ dB. Using the 74-bit LPN SOM code the SOM detection signal-to-noise ratio is $\gamma_{som} =$ 25 dB and, considering the $N-1=73$ false-alarm opportunities, the false alarm per bit must be $P_{fa}(\text{bit}) \le P_{fa}/N = 1.37 \times 10^{-5}$. Referring to Figures C.2 or C.3 the corresponding correct SOM detection is $P_c(SOM) \ge 0.999$ for a signal-to-noise ratio ≥14.3 dB so the 25 dB signal-to-noise ratio is more than sufficient to achieve a SOM detection probability† of 0.999.

*The lead-in correlation will not be exactly as illustrated in this figure because the bits preceding the SOM sequence are not zero but correspond to the synchronization data.

†This is an optimistic simplification because the false-alarms probabilities computed in Appendix C correspond to zero-mean Gaussian noise in each false-alarm cell, whereas the correlation response in the nonzero lag cells include noise-like uncorrelated signal components.

11.5 SIGNAL-TO-NOISE RATIO ESTIMATION

In the concluding sections of this chapter, various methods of estimating the received signal and noise powers and the resulting signal-to-noise ratio are discussed. Estimation of these received signal parameters is an important aspect of the signal acquisition processing for determination of the link quality. For example, signal-to-noise estimates are often reported to network controllers for the purpose of power control and network configuration management. These estimates are also used in the demodulator for: establishing the optimum PLL operation, diversity combining, as discussed in Chapter 18 and Section 20.9; detection of Reed–Solomon symbol erasures; soft-decision Viterbi decoding; and detection of extended signal fading or loss-of-signal conditions for the purpose of temporally suspending the demodulator tracking functions. System specifications often require that bit count integrity is maintained for a specified time following a signal outage with a specified probability and confidence level of recovering bit synchronization when the signal is recovered. The design challenge is to rapidly detect the signal loss and terminate all tracking functions while maintaining the intrinsic synchronization accuracy based on the parts-per-million (ppm) specification of the local oscillators and system clocks.

Although the focus of this section is on the signal-to-noise ratio estimation, the estimation of the bit error rate (BER) is also an important demodulator signal quality measure. Newcombe and Pasupathy [42] document their survey results of various BER estimation techniques and Rife and Boorstyn [43] discuss multiple tone parameter estimation using discrete-time samples. Pauluzzi and Beaulieu [44] have surveyed several techniques for estimating the signal-to-noise ratio. Their survey includes the analysis and simulation of the various techniques and a performance comparison using a common MSE performance metric; the techniques are listed in Table 11.11. Based on their study the following conclusions are made: if the data is known at the demodulator, as for example, during acquisition and training intervals, the ML and SNV are superior. With unknown data, the best performing estimators are the ML, SNV, and M_2M_4, which, coincidently, are the easiest of the techniques to implement.

TABLE 11.11 Signal-to-Noise Estimation Techniques[a]

Technique	Remarks[b]
Split-symbol moments estimator [45–47] (SSME)	Unknown data, $N_s \geq 1$ BPSK only with *real* AWGN Channel dependent Processing intense
Maximum-likelihood estimator [48, 49] (ML)	Unknown and known data, $N_s \geq 1$ *Real* or *complex*[c] samples Estimation bias compensation
Squared signal-to-noise variance estimator [50] (SNV)	Unknown or known data, $N_s = 1$ (MF samples) *Real* or *complex*[c] samples Estimation bias compensation[c]
Second- and fourth-order moments estimator [51, 52] (M_2M_4)	Independent of data, $N_s = 1$ (MF samples) *Real*[c] or *complex* samples Unaffected by unknown carrier phase
Signal to variation ratio estimator [53] (SVR)	Unknown data, $N_s = 1$ (MF samples) *Real*[c] or *complex* samples (MPSK only) Unaffected by unknown carrier phase

[a]The unknown and independent of data techniques are also referred to as *in-service* estimators. MF refers to matched filter.
[b]*Real* samples imply BPSK with *real* AWGN and *complex* samples imply MPSK, QAM modulations with *complex* AWGN.
[c]Extension by authors Pauluzzi and Beaulieu.

All of the estimators are based on ideal distortion less sampled baseband data-modulated waveforms with AWGN described as

$$\tilde{y}_k = \sqrt{P_s}\tilde{m}_k + \sqrt{P_n}\tilde{n}_k \tag{11.76}$$

where $\sqrt{P_s}$ and $\sqrt{P_n}$ are the signal and noise power scale factors applied, respectively, to the sampled *complex* signal modulation and noise functions $\tilde{m}_k$ and $\tilde{n}_k$. The in-phase and quadrature noise samples are *iid* Gaussian random variables with zero mean and unit variance. Considering *real* and symmetric modulation functions the matched filter output samples are as follows:

$$\begin{aligned}\tilde{y}_n &= S_n + N_n\\ &= \sqrt{P_s}\sum_{\ell=1}^{N_s}\tilde{m}_{k'+\ell}m_{k'+\ell} + \sum_{\ell=1}^{N_s}\tilde{n}_{k'+\ell}m_{k'+\ell}\end{aligned} \tag{11.77}$$

where N_s is the number of samples per symbol, $k' = (n-1)N_s$ is the k-th sample corresponding to the $n-1$ received symbols, and m_ℓ is the matched filter impulse response.*

*The summation over the N_s symbol samples implies a full response modulation; however, the following analysis also applies to Nyquist RRC symbol modulation extending over several symbols, albeit with increased notational complexity.

TABLE 11.12 Reasonable signal-to-noise estimation accuracies ($\gamma = E_b/N_o$ in dB)

Specified	Accuracy
$2.5 \leq \gamma \leq 12$	±0.5
$12 < \gamma \leq 22$	±1.5
$22 < \gamma \leq 32$	±3.0

Referring to the matched filter development in Section 1.7, the matched filer output signal-to-noise ratio for a carrier-modulated waveform is expressed in terms of the signal energy, E_B, and the one-sided noise power spectral density (N_o) as

$$\gamma_B = \frac{E_B}{N_o} \tag{11.78}$$

In general, $E_B = P_s/B$ where B is the bandwidth in which the noise power is measured; in the following evaluation of the signal-to-noise ratio estimation techniques $B = R_b$ so the signal energy corresponds to the energy per bit, denoted as E_b. The following sections outline the estimation processing of several techniques, as formulated by Pauluzzi and Beaulieu and listed in Table 11.11. The concluding sections provide two case studies of the estimation performance. Estimates of the signal-to-noise ratio are often required for network control and, based on Monte Carlo simulations in AWGN channels,† reasonable expectations of the estimation accuracies are indicated in Table 11.12.

11.5.1 Maximum-Likelihood (ML) Estimator

The maximum-likelihood signal-to-noise estimator applies for an arbitrary number of samples per symbol (N_s) so it can be used to estimate the signal-to-noise ratio in an arbitrary bandwidth. The estimators of interest for the *complex* baseband data-modulated waveform are expressed as

$$\hat{P}_s = \left[\frac{(1/K)\sum_{k=0}^{K-1}\left(y_{Ik}m_{Ik}^{(i)} + y_{Qk}m_{Qk}^{(i)}\right)}{(1/K)\sum_{k=0}^{K-1}\left[\left(m_{Ik}^{(i)}\right)^2 + \left(m_{Qk}^{(i)}\right)^2\right]}\right]^2 \tag{11.79}$$

$$\hat{P}_n = \frac{1}{K}\sum_{k=0}^{K-1}\left(y_{Ik}^2 + y_{Qk}^2\right) - \frac{\hat{P}_s}{K}\sum_{k=0}^{K-1}\left[\left(m_{Ik}^{(i)}\right)^2 + \left(m_{Qk}^{(i)}\right)^2\right] \tag{11.80}$$

†The M_2M_4 method exhibited the lowest measurement standard deviation, however, resulted in a bias for $\gamma \leq 12$ dB. A bias correction factor can be applied to remove the bias. The M_2 method, discussed in Section 11.5.5, exhibited a standard deviation within the ±0.5 dB specification for $\gamma \leq 12$ dB without a bias. All of the methods improved with the number of samples per trial and the number of trials; these parameters are dependent on the system application.

and

$$\hat{\gamma}_{ml} = \frac{\hat{P}_s}{\hat{P}_n} = \frac{N_s^2\left[(1/K)\sum_{k=0}^{K-1}\mathrm{Re}\left\{\tilde{y}_k^*\,\tilde{m}_k^{(i)}\right\}\right]^2}{(\rho_c/K)\sum_{k=0}^{K-1}|\tilde{y}_k|^2-(\rho_c N_s/K^2)\left[\sum_{k=0}^{K-1}\mathrm{Re}\left\{\tilde{y}_k^*\,\tilde{m}_k^{(i)}\right\}\right]^2} \tag{11.81}$$

The factor ρ_c in the denominator of (11.81) is included to reduce the estimation bias and, for *complex* signals, is given by

$$\rho_c = \frac{K}{K-3/2} \tag{11.82}$$

In these equations $K = N_{sym}N_s$ where N_{sym} is the number of modulation symbols associated with the estimation interval. The notation $m_k^{(i)}$ corresponds to the k-th symbol with the i-th MPSK phase such that $m_k^{(i)} = p_k^{(i)}e^{j\varphi_i}$, where $p_k^{(i)}$ is the symbol shaping function and $\varphi_i = 2\pi i/M$: $i = 0, \ldots, M-1$ is the symbol phase modulation with $\log_2(M)$ binary bits per symbol. After PLL and symbol tracking, BPSK modulation can be viewed as involving *real* signals with $\tilde{m}_k^{(i)} = m_k^{(i)} = \pm p_k^{(i)}$ and (11.79), (11.80), and (11.81) simply accordingly. For *real* signals the bias reduction factor is $\rho_r = K/(K-3)$.

11.5.2 Squared Signal-to-Noise Variance (SNV) Estimator

As indicated in Table 11.11, the SNV estimator applies for $N_s = 1$ so the estimator data samples correspond to the output of the optimally sampled matched filter. This estimation procedure is based on the first absolute moment and the second moment of the sampled data. In this case, the range of the summation is $K = N_{sym}$. For the *complex* signals, the SNV signal-to-noise ratio estimate is computed as*

$$\hat{\gamma}_{snv} = \frac{\left[(1/K)\sum_{n=0}^{K-1}\mathrm{Re}\left\{\tilde{y}_n^*\,\tilde{a}_n^{(i)}\right\}\right]^2}{(\rho_c/K)\sum_{n=0}^{K-1}|\tilde{y}_n|^2-(\rho_c N_s/K^2)\left[\sum_{n=0}^{K-1}\mathrm{Re}\left\{\tilde{y}_n^*\,\tilde{a}_n^{(i)}\right\}\right]^2}$$

$$: K = N_{sym} \tag{11.83}$$

In this case, the estimation bias reduction factor ρ_c is identical to (11.82) with $K = N_{sym}$. The symbol shaping function has been removed by the matched filter so that $\tilde{m}_n^{(i)} = \tilde{a}_n^{(i)} = e^{j\varphi_i}$. The estimation for *real* signals is similar to that described earlier for the ML estimator with $a_n^{(i)} = \pm 1$ and $\rho_r = K/(K-3)$.

*In contrast to (11.77), the summation in (11.83) over the estimation interval is based on the matched filtered samples n.

11.5.3 Second- and Fourth-Order Moments (M_2M_4) Estimator

The M_2M_4 also applies for $N_s = 1$ sample per symbol so the estimate samples correspond to the optimally sampled matched filter output. Benedict and Soong [51] refer to this method as the square-law moments estimator. In general, the noise power estimate is evaluated in terms of the signal power estimate $\hat{P}_s$ as

$$\hat{P}_n = M_2 - \hat{P}_s \tag{11.84}$$

Using (11.84), the estimates for the *complex* MPSK signal power and signal-to-noise ratio are evaluated as

$$\hat{P}_s = \sqrt{2M_2^2 - M_4} \quad : \text{complex signal} \tag{11.85}$$

and

$$\hat{\gamma}_{M_2M_4} = \frac{\hat{P}_s}{\hat{P}_n} = \frac{\sqrt{2M_2^2-M_4}}{M_2-\sqrt{2M_2^2-M_4}} \quad : \text{complex signal} \tag{11.86}$$

The estimates for the *real* BPSK signal power and signal-to-noise ratio are evaluated as

$$\hat{P}_s = \frac{1}{2}\sqrt{6M_2^2 - 2M_4} \quad : \text{real signal} \tag{11.87}$$

and

$$\hat{\gamma}_{M_2M_4} = \frac{\hat{P}_s}{\hat{P}_n} = \frac{(1/2)\sqrt{6M_2^2-2M_4}}{M_2-(1/2)\sqrt{6M_2^2-2M_4}} \quad : \text{real signal} \tag{11.88}$$

The moments for both the *real* and *complex* signals are computed using the approximate time averages

$$M_2 \cong \frac{1}{N_{sym}}\sum_{n=0}^{N_{sym}-1}|\tilde{y}_n|^2 \tag{11.89}$$

and

$$M_4 \cong \frac{1}{N_{sym}}\sum_{n=0}^{N_{sym}-1}|\tilde{y}_n|^4 \tag{11.90}$$

11.5.4 Case Study Using M_2M_4 Estimator

This case study examines the performance of the M_2M_4 estimator for the *real* signal with BPSK modulation as described by (11.84), (11.87), and (11.88). The results are shown in Figure 11.51 using $N_{sym} = 1000$ and 100 matched filter symbol (or bit) samples, respectively. At each signal-to-noise ratio 100 Monte Carlo trails (or estimates) are evaluated that form the population for establishing the mean, standard deviation, and extreme values indicated in the figures. These results indicate that the signal power estimation results in negligible variation and bias with increasingly high

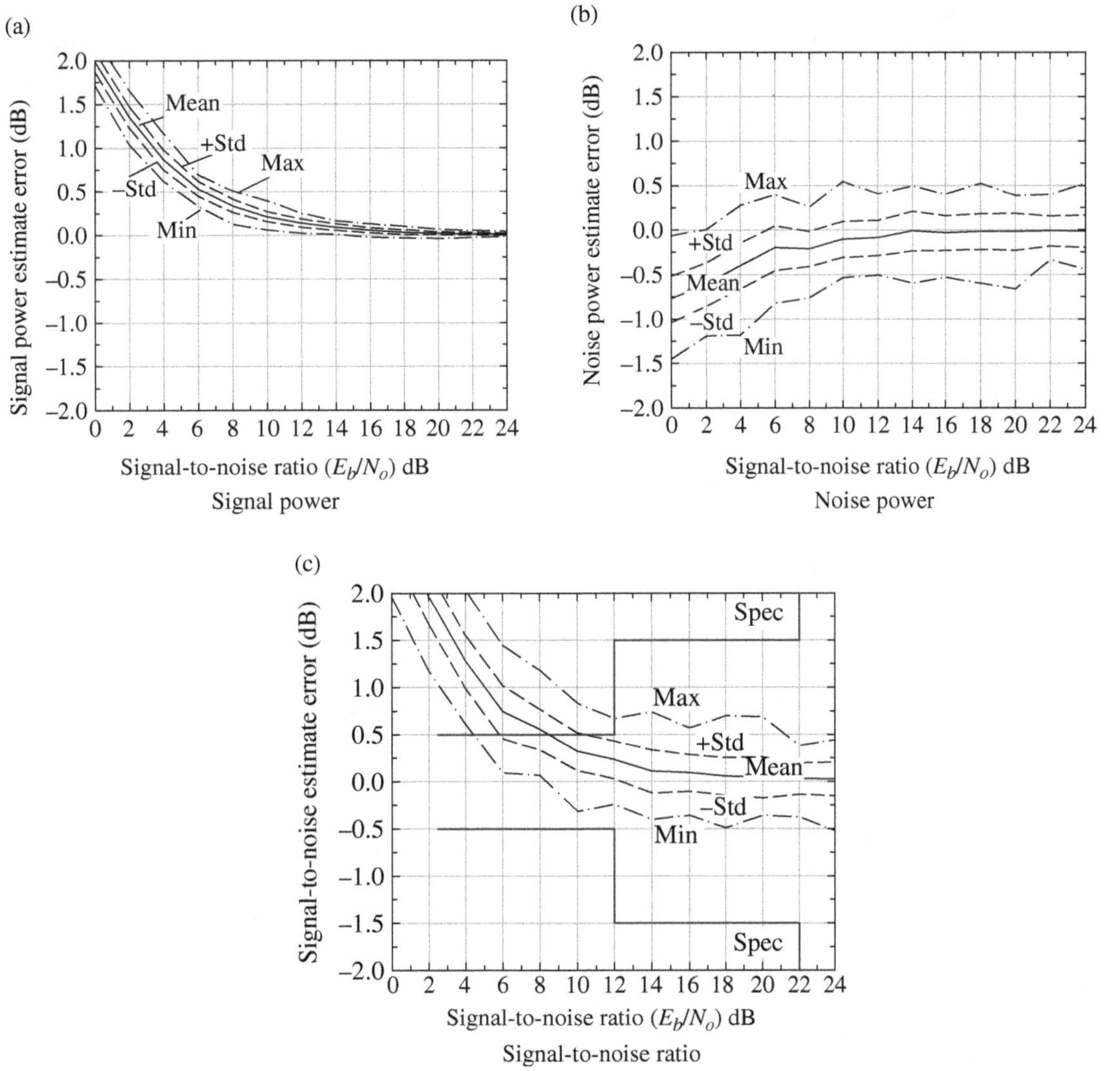

FIGURE 11.51 Baseband M_2M_4 estimation errors (N_{sym} = 1000 bits, N_{trials} = 100).

signal-to-noise ratios; however, the noise power and signal-to-noise ratio estimates, for commonly encountered ranges of signal-to-noise ratio, exhibit large variations with increasing bias at lower signal-to-noise ratios. Referring to Figure 11.51c it is seen that the mean signal-to-noise estimate enters the specification window at $E_b/N_o = 8.5$ dB; the standard deviation at 10.5 dB and the extremes at 12 dB.

Figure 11.52 represents the performance when the estimation is based on $N_{sym} = 100$ symbol samples. The increased variation is a consequence of the shorter estimation interval. The mean signal-to-noise ratio estimation still enters the specification at the same signal-to-noise ratio of about 8 dB. If the specification were to include an associated confidence level then an acceptable estimate standard deviation could be determined, for example, by applying the central limit theorem and associating the standard deviation with the Gaussian distribution. In a network application, in which communication and control channels use repetitive messages, the demodulator can average the estimates over several frames, thereby, reducing the variation. However, in a nonfading AWGN channel, averaging will not improve the mean value of the estimate. By using the complex product $\tilde{y}_n\tilde{y}_n^* = |\tilde{y}_n|^2$ in the M_2M_4 estimator the carrier phase and received data are removed, therefore, the estimation is insensitive to the signal phase and cannot be improved upon using known or demodulator detected data.

The preceding estimated parameter errors are computed based on the accumulated samples at each signal-to-noise ratio. The procedure involves determining the mean, standard deviation, and extremes for each parameter and then performing the following normalization with respect to the true parameter value. For example, using the true or known received power, P_s, the estimated mean signal power is computed as

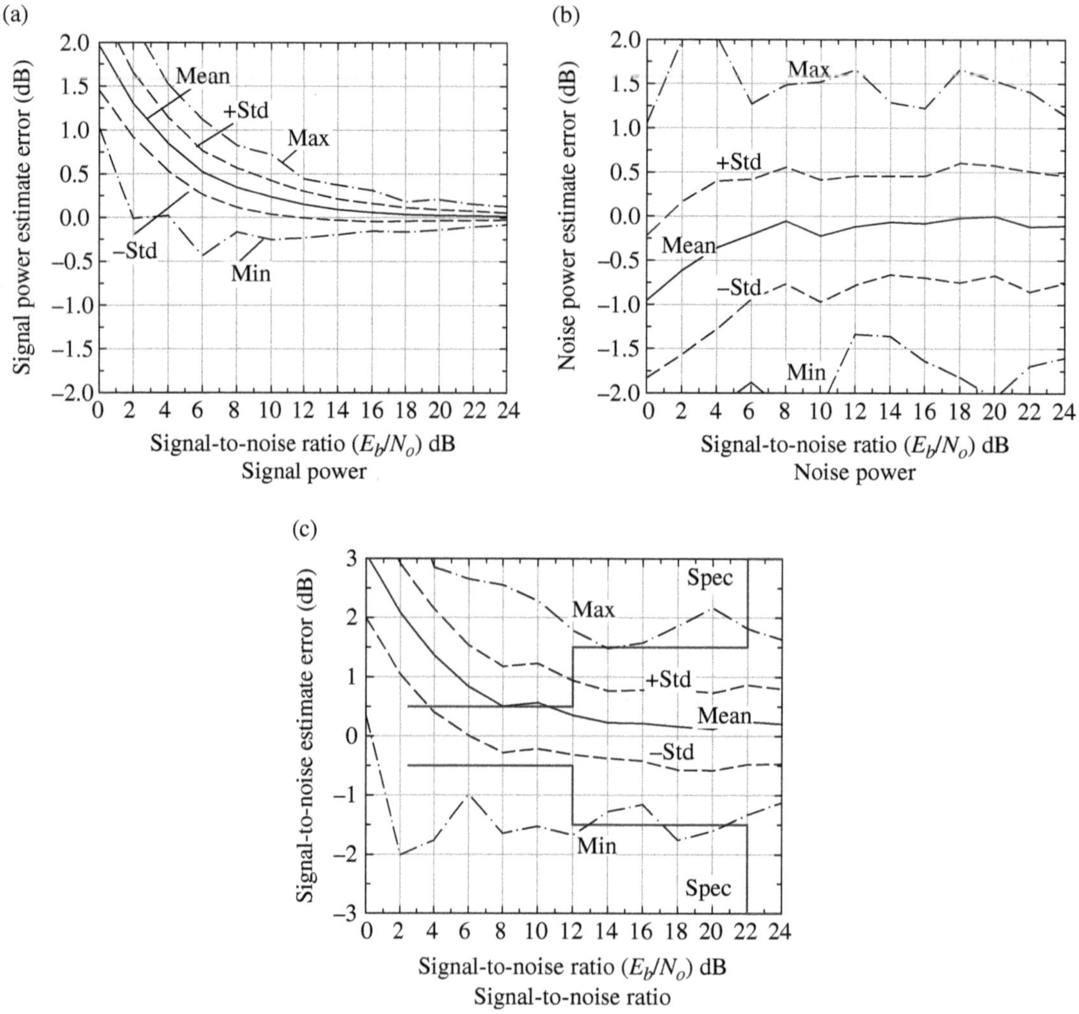

FIGURE 11.52 Baseband M_2M_4 estimation errors (N_{sym} = 100 bits, N_{trials} = 100).

$$\hat{P}_s = \frac{1}{N_{trial}} \sum_{n=1}^{N_{trial}} P_s(n) \tag{11.91}$$

The plus and minus standard deviation and extremes in the power estimates are denoted as $\hat{P}_s \pm \sigma_{\hat{P}_s}$, $\hat{P}_{smax}$, and $\hat{P}_{smin}$. These estimates are then normalized by the true signal power and plotted in decibels as $10\log_{10}(normalized\ estimate)$. The standard deviation of the estimate is computed based on the samples comprising a subset of the entire population as discussed in Section 1.13.3.

11.5.5 Case Study: Estimator Performance Using Independent Signal and Noise Power Estimation

In this case study, the noise power estimate is performed serially at the output of a narrowband filter before the signal is present or in parallel with the signal acquisition during discrete frequency searching over the entire frequency uncertainty range as described in Section 11.3.4. In this case, the noise power estimates are performed at several independent frequency bands to ensure that other signal sources do not influence the estimate. The signal-plus-noise power estimate is then performed using any of the preamble segments or simply on random data when *signal present* is declared. During the independent estimations, the receiver and demodulator AGC levels are monitored and the gain changes are compensated in the estimation processing.

Since there is only one parameter to estimate during an estimation interval a second moment M_2 estimator is used to estimate the noise and signal-plus-noise powers as

$$\hat{P}_n \cong \frac{1}{K}\sum_{k=0}^{K-1} |\tilde{y}_k|^2 \quad : \text{noise-only estimate} \tag{11.92}$$

and

$$\hat{P}_{sn} \cong \frac{1}{K}\sum_{k=0}^{K-1} |\tilde{z}_k|^2 \quad : \text{signal plus noise estimate} \tag{11.93}$$

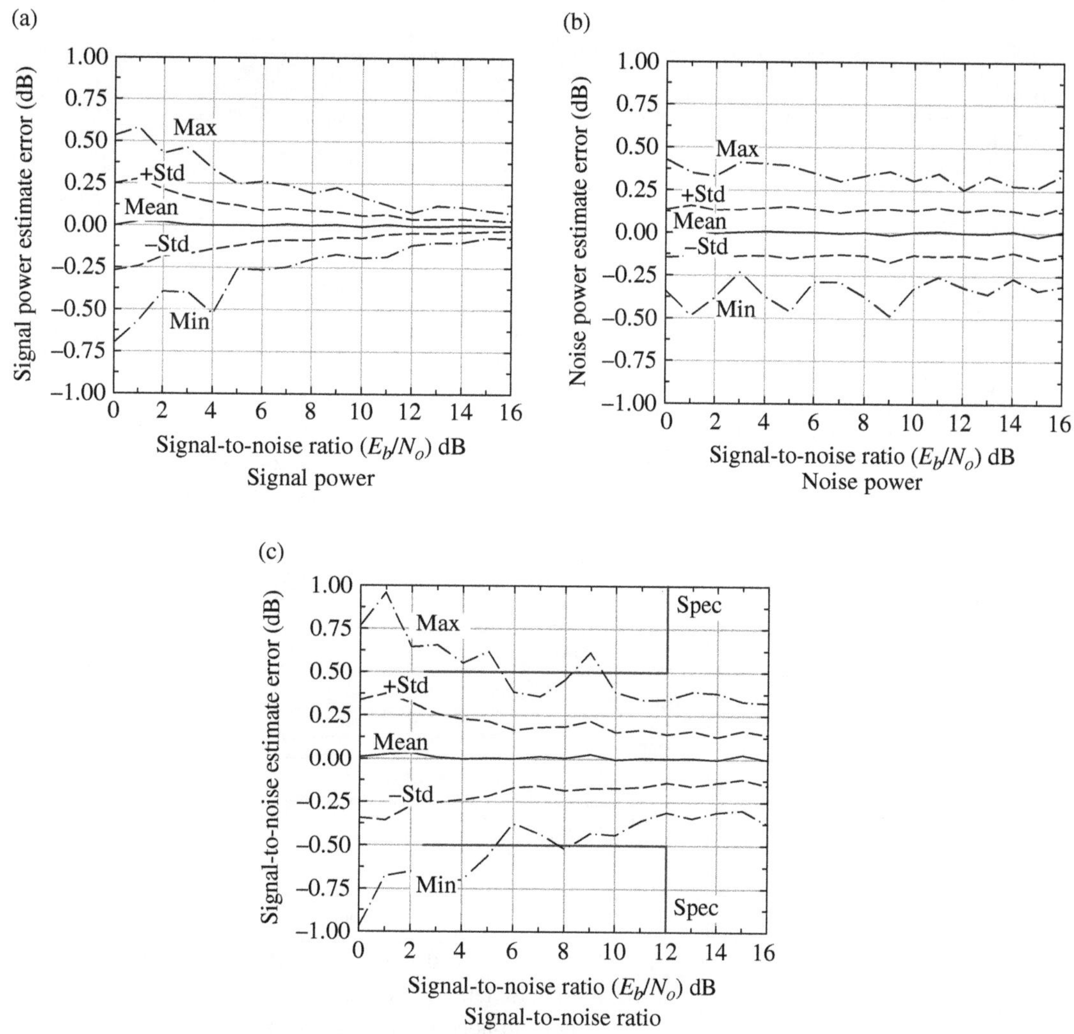

FIGURE 11.53 Baseband M_2 estimation errors ($K = 4000$ samples, $N_{trials} = 100$).

where $\tilde{y}_k$ and $\tilde{z}_k$ are the complex baseband noise and signal-plus-noise samples and $K = N_{sym}N_s$ is the number of samples used in forming the estimates. As with the M_2M_4 estimator, the M_2 estimator removes the MPSK and QAM data and carrier phase dependence so symbol timing and matched filter sampling is of no consequence. The noise power estimate is simply expressed by (11.92) and the signal power and signal-to-noise ratio are computed as

$$\hat{P}_s = \hat{P}_{sn} - \hat{P}_n \quad : \text{noise-only estimate} \tag{11.94}$$

and

$$\hat{\gamma}_B = \frac{\hat{P}_s}{\hat{P}_n} \quad : \text{signal-to-noise ratio estimate} \tag{11.95}$$

where B is the estimation measurement bandwidth and is converted to the user bit rate bandwidth equal to R_b as $\hat{\gamma}_b = \hat{\gamma}_B(B/R_b)$.

Figures 11.53 and 11.54 show the M_2 estimator performance formatted as described for the M_2M_4 estimator in the previous section. The AGC gain compensation is assumed to be ideal. The sample sizes of 4000 and 2000 correspond to 1000 and 500 bits, respectively, with $N_s = 4$ samples per bit BPSK modulation. Therefore, except for the 2 : 1 lower ordinate scale factor, Figure 11.53 is comparable to Figure 11.51 insofar as the sample size is concerned.

With the M_2 estimator the mean signal power estimate is nearly exact over the entire range of the signal-to-noise ratios and the standard deviation and extreme values diminish with increasing signal-to-noise ratios. The mean of the noise power estimate is also nearly exact over the entire range of signal-to-noise ratios; however, the standard deviation and extreme values are relatively constant and demonstrate the fluctuations associated with the narrowband white noise. The signal-to-noise ratio estimate error follows the trends of the signal and inverse noise power estimates. The influence of the estimation interval or sample size is demonstrated

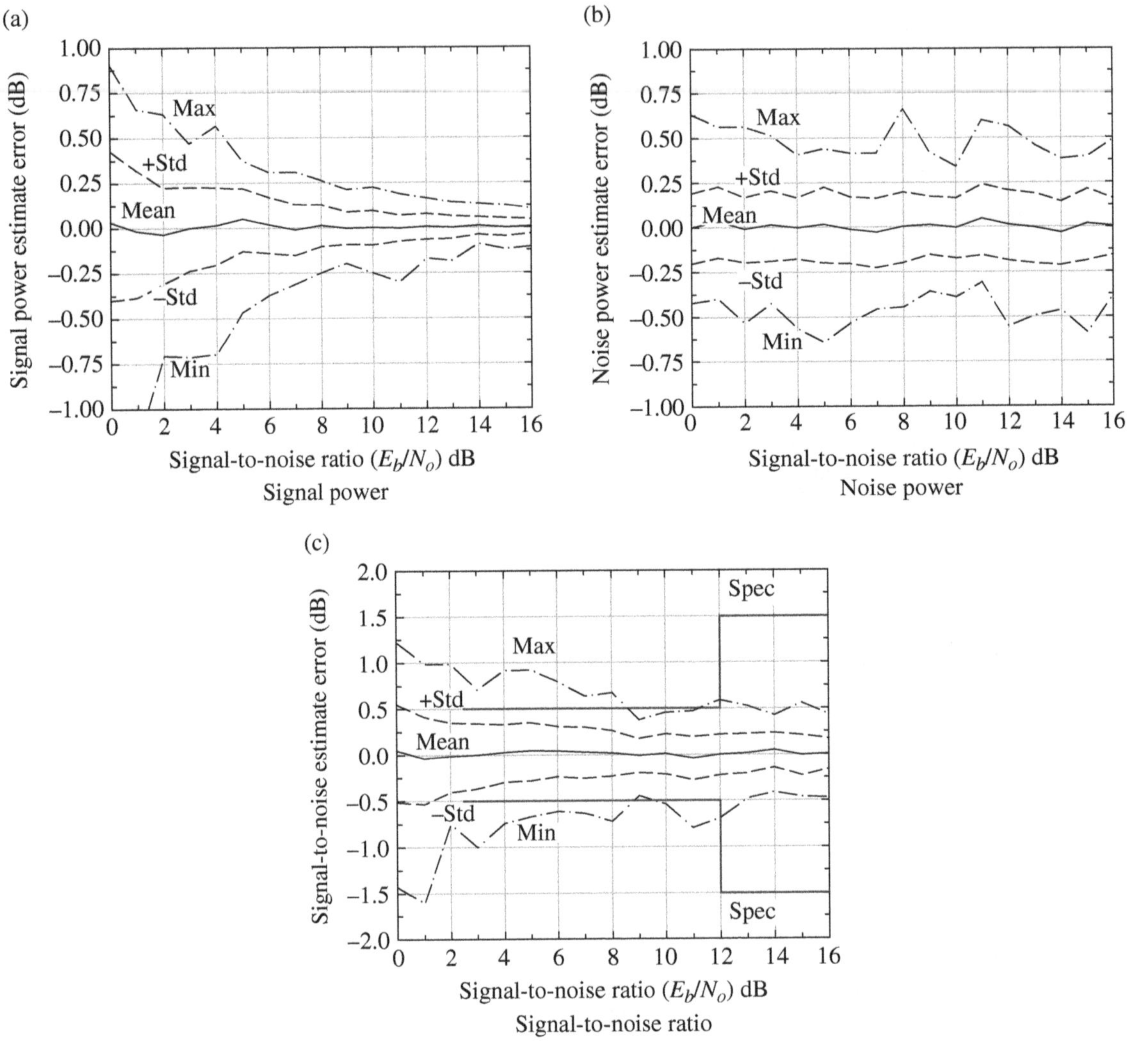

FIGURE 11.54 Baseband M_2 estimation errors ($K = 2000$ samples, $N_{trials} = 100$).

in Figure 11.54. The fluctuations in the estimations are decreased by increasing the sample size either directly or by averaging additional independent measurements of the same sample size.

ACRONYMS

ADC	Analog-to-digital converter
AGC	Automatic gain control
ARQ	Automatic repeat request
AWGN	Additive white Gaussian noise
BCI	Bit count integrity
BER	Bit error rate
BPSK	Binary phase shift keying
CDMA	Code division multiple access
CFAR	Constant false-alarm rate
CRC	Cyclic redundancy check (code)
CW	Continuous wave
DOD	Department of Defense
E/L	Early–late (gate)
FD	Frequency discriminator
FDMA	Frequency division multiple access
FEC	Forward error correction
FFT	Fast Fourier transform
GSM	Global system for mobile communications
ILPN	Inverted Legendre polynomial
ISA	International Society of Automation
LC	Lag correlator
LOS	Line of sight
LPF	Low-pass filter
LPN	Legendre polynomial
M_2M_4	Second- and fourth-order moments estimator
MAC	Medium access control (layer)
MF	Matched filter
ML	Maximum-likelihood estimator
MPSK	Multiphase shift keying

MSE	Mean-square error
OL	open loop
PCL	pseudo-closed loop
PLL	Phaselock loop
PN	Pseudo-noise
POL	pseudo-open loop
PPM	Parts-per-million
QAM	Quadrature amplitude modulation
RRC	Root-raised-cosine
SLM	Square-law moments estimator
SNR	Signal-to-noise ratio
SNV	Squared signal-to-noise variance estimator
SOM	Start-of-message
S-OQPSK	phase-shaped offset quadrature phase shift keying
SSME	Split-symbol moments estimator
SVR	Signal-to-variation ratio estimator
TDMA	Time division multiple access
VCO	voltage controlled oscillator

PROBLEMS

1. When using the Costas PLL to acquire a noise-free QPSK-modulated waveform, express the received baseband power on each quadrature rail and the total baseband power in terms of the peak signal amplitude A and the phase tracking error ϕ_e. Then, considering that joint AGC power control and carrier phaselock tracking are used, express the peak signal amplitude on each rail in terms of the AGC *rms* reference voltage V_{ref} under steady-state AGC conditions and ideal phase tracking, that is, when $\phi_e = 0$.

2. Consider that the input to the sampled ADC with infinite amplitude resolution is the AWGN random variable x described as $N(0,\sigma_n)$. Determine the AGC reference voltage, V_{ref}, which results in zero discriminator output in the steady-state AGC condition.

   Hint: V_{ref} is determined as the median value of $y = |x|$ and is a linear function of σ_n.

3. Repeat Problem 2 when x is described as $N(m,\sigma_n)$.

4. It is often convenient in a simulation program to implement an ideal or theoretical AGC without the signal processing required to implement a particular AGC algorithm using the sampled received signal plus noise. To this end, define the ideal AGC power as P_{agc} and express the signal power P_s and the noise power P_n in terms of P_{agc} and the signal-to-noise ratio $\gamma = P_s/P_n$.

5. Consider the *specified parameters* in Table 11.2, using $\rho = 2$ FFTs for each CW segment instead of 2.5. Evaluate a new set of *computed parameters* in Table 11.2 using a radix-2 FFT that will meet the specifications with $\rho = 2$.

6. Referring to Figure 11.15, the simulated frequency estimates (f_{norm}) and the corresponding probabilities (0.68, 0.9, 0.99) of a correct estimate (P_{ce}) are listed in the following table for the *best* case 6 dB and *worst* case 0 dB conditions. The probabilities correspond to the 1σ, 1.65σ, and 2.58σ sigma values of the normally distributed random variable $N(0,\sigma)$. Show that the simulated error estimates of f_{norm} also conform to the theoretical error based on the normal distribution $N(0,\sigma)$.

Simulated Frequency Estimates

	f_{norm}	
P_{ce}	Best: 6 dB	Worst: 0 dB
0.68	0.042	0.116
0.9	0.068	0.184
0.99	0.108	0.285

7. Evaluate and plot the E/L frequency discriminator responses for the Rectangular, Hanning, and Cosine windows with duration T_w seconds using linear and parabolic interpolation under the following conditions:

   a. The window frequency responses are separated by $\pm 1/T_w$ Hz with E/L samples corresponding to $\pm 1/2\ T_w$.

   b. The window frequency responses are separated by $\pm 1/2T_w$ Hz with E/L samples corresponding to $\pm 1/4T_w$.

   Hint: These windows are characterized Section 1.11.

8. Show that the frequency discriminator guard band is equal to Δf in Equation (11.7).

9. Given the input phase function $\theta_i(t)$ expressed in (11.33), show that the estimation of $\hat{\omega}_\varepsilon$ and $\hat{\dot{\omega}}_\varepsilon$ are established as shown in Figure 11.23 and determine the values of the constants k, k_0, and k_1, in terms of given parameters.

10. Show the detailed steps in deriving the expression for C_n in (11.47).

11. In selecting the FFT parameters for examining the spectrum of the repeated 1100 data pattern show the following:

    a. That commensurate increases in the number of samples/symbol (N_s) and the FFT size (N_{fft}) results in an increased Nyquist bandwidth B_N while maintaining a constant resolution bandwidth with respect to the symbol rate, that is, $\Delta f/R_s$ = constant.

    b. Show that a finer frequency resolution with respect to the symbol rate ($\Delta f/R_s$) is achieved for a constant sampling frequency (f_s) by increasing the measurement window T_m.

12. Following the analysis in Section 11.3.2.2 culminating in (11.49), derive the theoretical spectrum for the repeated 110 repeated data pattern specified for the BPSK-modulated waveforms in Table 11.8. Draw a sketch of the spectrum similar to that shown in Figure 11.37.

13. Referring to the case study in Section 11.3.2.3, show that the sensitivity of the normalized symbol rate estimate $\hat{R}_s/R_s$ is four times more sensitive to a frequency location error than that of the normalized frequency estimate $\hat{f}_s/f_s$.

Hint: Evaluate both estimates using $\ell_u = \ell_c + N_s/2 \pm k$ and $\ell_l = \ell_c - N_s/2$. where k is the frequency location error.

14. The case study in Section 11.3.2.3 examined the frequency estimation for a known symbol rate and the evaluation was based on the FFT samples taken over exactly four (4) repetitions of the 1100 data pattern. However, when the symbol rate is unknown the FFT will most likely not be taken over an integer number of data patterns. Evaluate the spectrum when the spectrum is based on an additional fractional repetition of the 1100 data pattern. For example, evaluate the spectrum for a measurement interval of $T_m = 16T + T'_m$ where $0 < T'_m < 4T$.

15. Referring to (11.65), develop the expression of the correlator output for BPSK modulation using $s_T(n) = dI_n$ and the channel phase function $e^{j\phi}$.

REFERENCES

1. U. Mengali, A.N. D'Andrea, *Synchronization Techniques for Digital Receivers*, Plenum Press, New York, 1977.
2. B.E. Blair, Editor, *Time and Frequency: Theory and Fundamentals*, National Bureau of Standards Monograph, U.S. Department of Commerce, Washington, DC, May 1974.
3. R.E. Cashion, W.J. Klepczynski, K. Putkovich, "The Use of Transit for Time Dissemination," Journal of the Institute of Navigation, Vol. 26, No. 1, pp. 63–69, Spring 1979.
4. D.W. Allan, J. Helwig, "Time Deviation and Time Prediction Error for Clock Specifications, Characterization, and Application," Position Location and Navigation Symposium (PLANS), pp. 22–28, San Diego, CA, November 1978.
5. F.S. Nakamoto, M.P. O'Reilly, C.R. Wolfson, "Time and Frequency Synchronization Issues for Tactical, Strategic, and Mobile Forces EHF SATCOM," Defense Communications Agency, MILSATCOM Systems Office, Contract DCA 100-79-C-008, Fort Meade, MD, 1979.
6. F.S. Nakamoto, R.W. Middlestead, C.R. Wolfson, "Impact of Time and Frequency Errors on Processing Satellites with MFSK Modulation," International Conference on Communications, ICC-81, Denver, CO, June 14–18, 1981.
7. M. Mouly, M.-B. Pautet, *The GSM System for Mobile Communications*, Published by the Authors, Palaiseau, France, 1992.
8. U.S.A. Department of Defense (DOD) Interface Standard, *MIL-STD-188-183A, Interoperability Standard for 25-kHz TDMA/DAMA Terminal Waveform (Including 5- and 25-kHz Slave Channels)*, p. 29, Defense Information Systems Agency (DISA), Fort Meade, MD, March 20, 1998.
9. F.M. Gardner, W.C. Lindsey, "Special Issue on Synchronization, Part I," IEEE Transactions on Communications, Vol. COM-28, No. 8, pp. 1105–1388, August 1980.
10. R. Haeb, H. Meyr, "A Systematic Approach to Carrier Recovery and Detection of Digitally Phase Modulated Signals on Fading Channels," IEEE Transactions on Communications, Vol. 37, No. 7, pp. 748–754, July 1989.
11. A. Vanelli-Coralli, P. Salmi, S. Cioni, G.E. Corazza, A Polydoros, "A Performance Review of PSP for Joint Phase/Frequency and Data Estimation in Future Broadband Satellite Networks," IEEE International Conference on Telecommunications 2001, Bucharest, Romania, June 4–7, 2001.
12. S.W. Golomb, *Sequences with Randomness Properties*, Martin Company, Baltimore, MD, June 1955.
13. S.W. Golomb, *Shift Register Sequences*, Holden Day, Inc., New York, 1967.
14. M.J.E. Golay, "Complementary Series," IRE Transactions on Information Theory, Vol. IT-7, p. 82, April 1961.
15. D.C. Coll, "A Survey of Finite Codes for Pulse Compression and Synchronization," Defense Research Telecommunications Establishment, Communication Laboratory, DRTE Report No. 1152, Ottawa, Canada, October 1963.
16. R.H. Pettit, "Pulse Sequences with Good Autocorrelation Properties," The Microwave Journal, Vol. 10, No. 2, pp. 63–67, February, 1967.
17. R.H. Barker, "Group Synchronization of Binary Digital Systems," W. Jackson, Editor, *Communication Theory*, pp. 273–287, Academic Press, New York, 1953.
18. S.W. Golomb, R.A. Scholtz, "Generalized Barker Sequences," IEEE Transactions on Information Theory, Vol. IT-11, p. 533, October 1965.
19. D.G. Luenberger, "On Barker Codes of Even Length," IEEE Proceeding, Vol. 51, No. 1, p. 230, January 1963.
20. R. Turyn, "On Barker Codes of Even Length," IEEE Proceeding, Vol. 51, No. 9, p. 1256, September 1963.
21. R.S. Berkowitz, *Modern Radar, Analysis, Evaluation and System Design*, John Wiley & Sons, Inc., New York, 1965.
22. M.W. Williard, "Optimum Code Patterns for PCM Synchronization," Proceedings of the 1962 National Telemetering Conference, NTC62, Washington, DC, May 23–25, 1962.
23. M.W. Williard, "Analysis of a System of Mutually Synchronized Oscillators," IEEE Transactions on Communication Technology, Vol. 18, pp. 467–483, October 1970.
24. F. Neuman, L. Hofman, "New Pulse Sequences with Desirable Correlation Properties," Proceedings of the National Telemetering Conference, NTC71, pp. 272–281, Washington, DC, April 12–15, 1971. NTC71 is issued as a publication of record by the IEEE Transactions on Aerospace and Electronic Systems. Abstracts are published in the IEEE Transactions on Aerospace and Electronic Systems, Vol. AES-7, No. 3, pp. 565–571, May 1971.
25. J.J. Spilker, Jr., *Digital Communications by Satellite*, Prentice-Hall, Inc., Englewood Cliffs, NJ, 1977.

26. R. Gold, *Properties of Linear Binary Encoding Sequences*, Robert Gold Associates, Los Angeles, CA, July 1975.
27. E.H. Dinan, B. Jabbari, "Spreading Codes for Direct Sequence CDMA and Wideband CDMA Cellular Networks," IEEE Communications Magazine, pp. 48–54, September 1998.
28. M.R. Schroeder, "Synthesis of Low-Peak-Factor Signals and Binary Sequences with Low Autocorrelation," IEEE Transactions on Information Theory, Vol. IT-16, pp. 85–89, January 1970.
29. U. Somaini, M.H. Ackroyd, "Uniform Complex Codes with Low Autocorrelation Sidelobes," IEEE Transaction on Information Theory, Vol. 20, pp. 689–691, September 1974.
30. R.L. Frank, S.A. Zadoff, "Phase Shift Pulse Codes with Good Periodic Correlation Properties," IEEE Transaction on Information Theory, Vol. IT-8, pp. 381–382, October 1962.
31. R.L. Frank, "Polyphase Codes with Good Nonperiodic Correlation Properties," IEEE Transaction on Information Theory, Vol. IT-9, pp. 43–45, January 1963.
32. D.A. Huffman, "The Generation of Impulse-Equivalent Pulse Trains," IRE Transaction on Information Theory, Vol. IT-8, No. 5, pp. S10–S16, September 1962.
33. D.L. Huffman, "A Modification of Huffman's Impulse-Equivalent Pulse Trains to Increase Signal Energy Utilization," IEEE Transaction on Information Theory, Vol. IT-20, p. 559, July 1974.
34. D.L. Huffman, "Modified Barker Code Approximating Huffman's Impulse-Equivalent Sequence," IEEE Transaction on Aerospace and Electronic Systems, Vol. AES-11, No. 4, pp. 437–442, July 1975.
35. M.H. Ackroyd, "The Design of Huffman Sequences," IEEE Transaction on Aerospace and Electronic Systems, Vol. AES-6, No. 6, pp. 790–796, November 1970.
36. M.H. Ackroyd, "Synthesis of Efficient Huffman Sequences," IEEE Transaction on Aerospace and Electronic Systems, Vol. AES-8, No. 1, pp. 2–8, January 1972.
37. B.H. Cantrell, "Sidelobe Reduction in Polyphase Codes," Naval Research Laboratory, NRL Report 8108, Washington, DC, April 13, 1977.
38. U.S.A. Department of Defense (DOD) Interface Standard, *MIL-STD-188-183A, Interoperability Standard for 25-kHz TDMA/DAMA Terminal Waveform (Including 5- and 25-kHz Slave Channels)*, p. 30, Defense Information Systems Agency (DISA), Fort Meade, MD, March 20, 1998.
39. P.M. Woodward, *Probability and Information Theory, with Applications to Radar*, Pergamon Press Ltd., London, 1953.
40. I.S. Gradshtyen, I.M. Ryzhik, *Tables of Integrals, Series, and Products*, Corrected and Enlarged Edition, p. 147, Integrals 3 and 4, Academic Press, New York, 1980.
41. E.O. Brigham, *The Fast Fourier Transform and Its Applications*, Prentice-Hall, Inc., Englewood Cliffs, NJ, 1988.
42. E.A. Newcombe, S. Pasupathy, "Error Rate Monitoring for Digital Communications," Proceedings of the IEEE, Vol. 70, pp. 805–828, August 1982.
43. D.C. Rife, R.R. Boorstyn, "Multiple Tone Parameter Estimation from Discrete-Time Observations," The Bell System Technical Journal, Vol. 55, No. 9, November 1976.
44. D.R. Pauluzzi, N.C. Beaulieu, "A Comparison of SNR Estimation Techniques for the AWGN Channel," IEEE Transactions on Communications, Vol. 48, No. 10, pp. 1681–1691, October 2000.
45. M.K. Simon, A. Mileant, "SNR Estimation for the Baseband Assembly," Jet Propulsion Laboratory, Telecommunications and Data Acquisition Program Report 42–85, Pasadena, CA, May 15, 1986.
46. B. Shah, S. Hinedi, "The Split Symbol Moments SNR Estimator in Narrow-Band Channels," IEEE Transactions on Aerospace and Electronic Systems, Vol. ASE-26, pp. 737–747, September 1990.
47. B. Shah, J.K. Holmes, "Improving the Split-Symbol Moments SNR Estimator," Jet Propulsion Laboratory, Interoffice Memo. 3338-90-223, Pasadena, CA, December 19, 1990.
48. R.B. Kerr, "On Signal and Noise Level Estimation in a Coherent PCM Channel," IEEE Transactions on Aerospace and Electronic Systems, Vol. AES-2, pp. 450–454, July 1966.
49. R.M. Gagliardi, C.M. Thomas, "PCM Data Reliability Monitoring through Estimation of Signal-to-Noise Ratio," IEEE Transactions on Communications, Vol. COM-16, pp. 479–486, June 1968.
50. C.E. Gilchriest, "Signal-to-Noise Monitoring," JPL Space Programs Summary, Vol. IV, No. 37–27, pp. 169–184, June 1966.
51. T.R. Benedict, T.T. Soong, "The Joint Estimation of Signal and Noise from the Sum Envelope," IEEE Transaction on Information Theory, Vol. IT-13, No. 3, pp. 447–454, July 1967.
52. R. Matzner, F. Engleberger, "An SNR Estimation Algorithm using Fourth-Order Moments," IEEE International Symposium on Information Theory, Trondheim, Norway, June 1994.
53. A.L. Brandao, L.B. Lopes, D.C. McLernon, "In-Service Monitoring of Multipath Delay and Cochannel Interference for Indoor Mobile Communication Systems," IEEE International Conference on Communications, Vol. 3, pp. 1458–1462, May 1994.

12

ADAPTIVE SYSTEMS

12.1 INTRODUCTION

The ever-increasing demand for channel capacity and bandwidth has resulted in the widespread use of adaptive signal processing techniques that compensate for the inevitable signal interference and distortion that results from the competitive needs for capacity and bandwidth. The significant advantages in efficient spectrum utilization, gained by high-order symbol modulation with spectral containment, have been facilitated by the use of adaptive processing algorithms that compensate for the interference and distortion under crowded channel conditions. The development of adaptive processing for improved communications was *jump-started* when wireline telephone networks [1] were under pressure for more capacity and higher data rates, and was well understood and developed when wireless communications entered the marketplace.

In the following sections, the mathematical background and algorithms are developed for adaptive systems as they apply to waveform *equalization* of intersymbol interference (ISI), *cancellation* of interfering signals, and waveform *identification*. These three objectives are obtained using subtle alterations in the adaptive processing configurations. The adaptive processing generally uses finite impulse response* (FIR) filters with weights that are adaptively adjusted to minimize error between the sampled filter output and the received input signal with known or estimated data using the minimum mean-square error [2] (MMSE) algorithm. With known data, a preamble or training sequence is available to aid in the adaptive acquisition or convergence processing. In these cases, the processing following the known data is based on the estimates of the received data using a decision-directed adaptive processing algorithm. After the training interval, the decision error probability is sufficiently low to maintain the MMSE condition. In applications where a training sequence is not available, the adaptive algorithm is referred to as a blind or self-recovering algorithm; however, using the training sequence significantly reduces the adaptation time.

In the next section, the orthogonality principle is introduced and applied in the derivation of Weiner's optimum linear estimation filter [3–5]. In Section 12.2, the Wiener estimation filter is described using the orthogonality principle and the MMSE criteria. In Section 12.3, the optimum FIR filter is examined, with the tap weights adaptively estimated using the lease mean-square (LMS) algorithm. The LMS algorithm results in lower implementation complexity than the MMSE algorithm. Section 12.4 considers various forms of equalizers and Section 12.5 discusses adaptive interference cancellation using the LMS algorithm. The preceding adaptive techniques are based on symbol-by-symbol processing and an entirely different approach, using the recursive least-squares (RLS) algorithm, is discussed in Section 12.6. The RLS algorithm converges to the steady-state condition in considerably less time than that of the MMSE or LMS algorithms. Case studies involving the application of ISI equalization and narrowband interference cancellation are

*The FIR filter is also referred to as a transversal or tapped delay line (TDL) filter.

Digital Communications with Emphasis on Data Modems: Theory, Analysis, Design, Simulation, Testing, and Applications, First Edition. Richard W. Middlestead.

Companion website: www.wiley.com/go/digitalcommunications

given in Sections 12.7 and 12.8; this chapter concludes with Section 12.9 with a case study of the RLS-adaptive equalizer.

12.1.1 The Orthogonality Principle

The orthogonality principle is the fundamental principle used in the evaluation of the MMSE between a set of parameters and their estimates. The principle applies in the analysis of linear or nonlinear estimation, involving constant or time-varying, real or complex parameters. However, to simplify the following description, the linear mean-square estimation (LMSE) of a single, constant, and real parameter is used. The orthogonality principle states that the MMSE results when the estimation error $y-\hat{y}$ is orthogonal to the measurement x, that is, $E[(y-\hat{y})x]=0$ and, under this condition, the MMSE is computed as the expectation $E[(y-\hat{y})y]$.

Consider that a real zero-mean parameter y is observed as* $x=y+n$, where n represents zero-mean additive noise and the estimate of y is formed using linear combination of the observations x_i, expressed as

$$\hat{y}=\sum_i a_i x_i \tag{12.1}$$

The weights a_i form a filter that decreases the noise and, thereby, improves the estimate. In this description, the index i simply associates x with the sampled noisy value of the parameter y. Using (12.1), the mean-square error (MSE) is expressed as

$$\varepsilon=E\left[(y-\hat{y})^2\right]=E\left[\left(y-\sum_i a_i x_i\right)^2\right] \tag{12.2}$$

To establish the MMSE, the derivative of (12.2) with respect to a_i is formed and set to zero resulting in

$$\begin{aligned}\varepsilon_{min}=\frac{d\varepsilon}{da_i}&=E\left[-2(y-\hat{y})\frac{d\hat{y}}{da_i}\right]\\&=-2E\left[(y-\hat{y})\frac{d}{da_i}\sum_i a_i x_i\right]=0\end{aligned} \tag{12.3}$$

Therefore, based on the last equality in (12.3), the requirement for the MMSE is

$$E[(y-\hat{y})x_i]=0 \quad :\forall i \tag{12.4}$$

*The measured data x must be zero-mean to ensure that the correlation corresponds to the covariance. With zero-mean noise, the mean value of x resulting from y can easily be removed.

The first equality in (12.2) is expanded as

$$\varepsilon=E[(y-\hat{y})(y-\hat{y})]=E[(y-\hat{y})y]-E[(y-\hat{y})\hat{y}] \tag{12.5}$$

and, upon substituting (12.1) for the expectation $E[(y-\hat{y})\hat{y}]$, the MMSE error expression is evaluated as

$$\begin{aligned}\varepsilon_{min}&=E[(y-\hat{y})y]-E\left[(y-\hat{y})\sum_i a_i x_i\right]\\&=E[(y-\hat{y})y]-\sum_i a_i E[(y-\hat{y})x_i]\\&=E[(y-\hat{y})y]\end{aligned} \tag{12.6}$$

The last equality in (12.6) follows from (12.4) and confirms the requirement that the MMSE occurs when the expectation of the product of the measurement error and data are orthogonal, that is, when $E[(y-\hat{y})x_i]=0 \quad :\forall i$.

12.2 OPTIMUM FILTERING—WIENER'S SOLUTION

Wiener's approach to determining the optimum filter of the noisy estimate of an event $\tilde{y}(t)$ is based on the filter that provides the MMSE $\hat{\tilde{y}}(t)$ of $\tilde{y}(t)$. The solution is based on the application of the orthogonality principle outline in the introduction. Consider the noisy observation $\tilde{x}(t)$ of the event $\tilde{y}(t)$ expressed as

$$\tilde{x}(t)=\tilde{y}(t)+\tilde{n}(t) \quad :-\infty\le t\le\infty \tag{12.7}$$

with the filtered estimate of $\tilde{y}(t)$ expressed as

$$\hat{\tilde{y}}(t)=\int_{-\infty}^{\infty}h(t,\lambda)\tilde{x}(\lambda)d\lambda \tag{12.8}$$

In the following analysis, the stochastic processes are stationary and characterized as complex baseband functions and the filter is linear and time-invariant. It is desired to determine the optimum linear filter $h_o(t)$ that results in the MMSE of the measurement error.† Referring to Section 12.1.1, the MMSE occurs when the orthogonality principle applies and, under the above conditions, is expressed as

$$E\left[\left(\tilde{y}(t_1)-\int_{-\infty}^{\infty}h_o(t_1-\lambda)\tilde{x}(\lambda)d\lambda\right)\tilde{x}^*(t_2)\right]=0 \tag{12.9}$$

†The conditions of linearity and stationarity result in major analytical simplifications that are applicable to many practical problems.

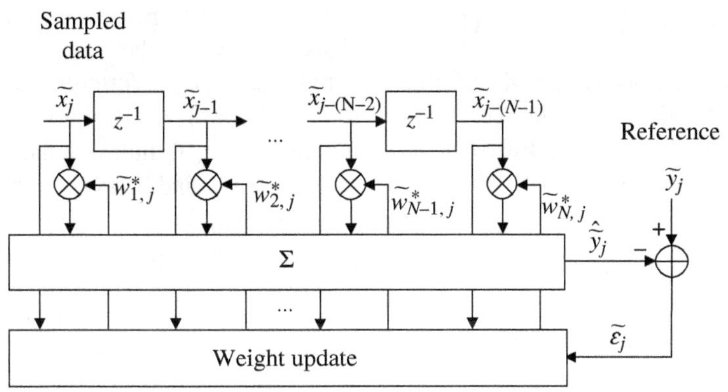

FIGURE 12.1 FIR-adaptive filter.

The solution to (12.9) is expresses in terms of the expectations as

$$E[\tilde{y}(t_1)\tilde{x}^*(t_2)] = \int_{-\infty}^{\infty} h_o(t_1-\lambda)E[\tilde{x}(\lambda)\tilde{x}^*(t_2)]d\lambda \qquad (12.10)$$

Upon using the *wws* property and defining $\tau = t_1 - t_2$, (12.10) simplifies to

$$R_{\tilde{y}\tilde{x}}(\tau) = \int_{-\infty}^{\infty} h_o(t_1-\lambda)R_{\tilde{x}\tilde{x}}(\lambda+\tau-t_1)d\lambda \qquad (12.11)$$

The *rhs* of (12.11) is expressed as the convolution integral by substituting $\zeta = \lambda + \tau - t_1$ with $d\lambda = d\zeta$ resulting in

$$\begin{aligned} R_{\tilde{y}\tilde{x}}(\tau) &= \int_{-\infty}^{\infty} h_o(\tau-\varsigma)R_{\tilde{x}\tilde{x}}(\varsigma)d\varsigma \\ &= h_o(\tau)*R_{\tilde{x}\tilde{x}}(\tau) \end{aligned} \qquad (12.12)$$

Therefore, the optimum filter is expressed as having frequency domain response

$$H_o(\omega) = \frac{S_{\tilde{y}\tilde{x}}(\omega)}{S_{\tilde{x}\tilde{x}}(\omega)} \qquad (12.13)$$

Evaluation of the MMSE corresponding to the optimum filter $h_o(t)$ is left as an exercise. The LMSE algorithm can also be applied to the prediction [6] of $\hat{\tilde{y}}(t'+\lambda)$ from the data $\tilde{x}(t)\ : t \le t'$.

12.3 FINITE IMPULSE RESPONSE-ADAPTIVE FILTER ESTIMATION

In this section, the Wiener filter is applied to the sampled signal $\tilde{x}_i = \tilde{x}(iT_s)$ where the index i represents the uniform sampling of $\tilde{x}(t)$ at increments $t = iT_s$, where T_s is the reciprocal of the sampling frequency f_s. All of the discrete-time functions in this section are considered to be complex baseband functions that are within the Nyquist bandwidth. Furthermore, the analysis involves discrete-time stochastic processes that are ergodic and stationary in the wide-sense, that is, the mean value is constant and the correlation and covariance functions depend only on the sample-time increment T_s. The basic structure of the adaptive filter is shown in Figure 12.1. The filter input is represented by the current sampled complex data $\tilde{x}_j$ and $N-1$ previous samples. If the sampled signal is the output of the demodulator matched filter then T_s is equal to the modulation symbol duration T, otherwise T_s may represent a suitable sub-symbol sampling interval; the unit delay designation z^{-1} corresponds to the normalized sample delay. For each set of N samples, the filter output $\hat{\tilde{y}}_j$ is computed and compared to a known reference signal forming the error $\tilde{\varepsilon}_j$ corresponding to the sample $t = jT_s$. The error is then applied to the weight update function that adjusts all of the weights resulting is a steady-state condition that corresponds to the minimum error. The adaptive processing details are discussed in the remainder of this section.

The sampled complex inputs $\tilde{x}_{j+i} : i = 0,\ldots,N-1$ and the filter weights $\tilde{w}_{i+1,j}$ are defined, respectively, as the complex vectors

$$\tilde{\boldsymbol{x}} = \left(\tilde{x}_j, \tilde{x}_{j-1}, \ldots, \tilde{x}_{j-(N-2)}, \tilde{x}_{j-(N-1)}\right)^T \qquad (12.14)$$

and

$$\tilde{\boldsymbol{w}} = \left(\tilde{w}^*_{1,j}, \tilde{w}^*_{2,j}, \ldots, \tilde{w}^*_{N-1,j}, \tilde{w}^*_{N,j}\right)^T \qquad (12.15)$$

The reference, or desired response, of the adaptive filter is $\tilde{y}_j$ and the difference between $\tilde{y}_j$ and the estimated response* $\hat{\tilde{y}}_j(\tilde{w}) = \tilde{w}^+\tilde{x}$ is the instantaneous estimation error, expressed as

*In this, and the following descriptions, the superscript + denotes complex conjugate transposition, and $*$ and T denote, respectively, complex conjugation and transposition.

$$\tilde{\varepsilon}_j(\tilde{w}) = \tilde{y}_j - \hat{\tilde{y}}_j(\tilde{w}) = \tilde{y}_j - \tilde{w}^+\tilde{x} \quad (12.16)$$

The optimum estimation filter corresponds to the MMSE that results when the filter has settled to the steady-state condition corresponding to the optimum weight vector $\tilde{\boldsymbol{w}}_{\mathbf{o}}$. Based on the orthogonality principle, the MMSE error occurs when the error is orthogonal to the data, that is, when

$$E\left[(\tilde{y}_j - \tilde{\boldsymbol{w}}_{\mathbf{o}}^+\tilde{\boldsymbol{x}})\hat{\tilde{y}}_j^*\right] = E\left[(\tilde{y}_j - \tilde{\boldsymbol{w}}_{\mathbf{o}}^+\tilde{\boldsymbol{x}})\tilde{\boldsymbol{x}}^+\tilde{\boldsymbol{w}}\right] = 0 \quad (12.17)$$

and, referring to (12.6), the corresponding MMSE is

$$\tilde{\varepsilon}_{min} = E\left[(\tilde{y}_j - \tilde{\boldsymbol{w}}_{\mathbf{o}}^+\tilde{\boldsymbol{x}})\tilde{y}_j^*\right] = 0 \quad (12.18)$$

Aside from the trivial condition corresponding to $\tilde{y}_j^* = 0$, the MSE criterion is based on the error defined in (12.16) and is expressed as the expectation

$$J(\tilde{\boldsymbol{w}}) \triangleq E\left[\tilde{\varepsilon}_j(\tilde{w})\tilde{\varepsilon}_j^*(\tilde{w})\right] = E\left[(\tilde{y}_j - \tilde{\boldsymbol{w}}^+\tilde{\boldsymbol{x}})\left(\tilde{y}_j^* - \tilde{\boldsymbol{x}}^+\tilde{\boldsymbol{w}}\right)\right]$$
$$:\text{mean-square error} \quad (12.19)$$

Upon expanding (12.19) and distributing the expectation, the MSE is evaluated as

$$J(\tilde{\boldsymbol{w}}) = E\left[\tilde{y}_j\tilde{y}_j^*\right] - E\left[\tilde{y}_j\tilde{\boldsymbol{x}}^+\tilde{\boldsymbol{w}}\right] - E\left[\tilde{\boldsymbol{w}}^+\tilde{\boldsymbol{x}}\tilde{y}_j^*\right] + E\left[\tilde{\boldsymbol{w}}^+\tilde{\boldsymbol{x}}\tilde{\boldsymbol{x}}^+\tilde{\boldsymbol{w}}\right]$$
$$= \sigma\tilde{y}^2 - \tilde{\boldsymbol{c}}^+\tilde{\boldsymbol{w}} - \tilde{\boldsymbol{w}}^+\tilde{\boldsymbol{c}} + \tilde{\boldsymbol{w}}^+\boldsymbol{R}\tilde{\boldsymbol{w}} \quad (12.20)$$

where $\sigma\tilde{y}^2$ is the variance of the reference, $\tilde{\boldsymbol{c}}$ is the $N \times 1$ cross-correlation vector between the reference and data, and $\boldsymbol{R}$ is an $N \times N$ correlation matrix of the data; $\tilde{\boldsymbol{c}}$ and $\boldsymbol{R}$ are defined, respectively, as

$$\tilde{\boldsymbol{c}} \triangleq E\left[\tilde{\boldsymbol{x}}\tilde{y}_j^*\right] = E\left[\tilde{y}_j^*\tilde{\boldsymbol{x}}\right] \quad (12.21)$$

and

$$\boldsymbol{R} \triangleq E[\tilde{\boldsymbol{x}}\tilde{\boldsymbol{x}}^+] \quad (12.22)$$

The second equality in (12.20) is established using the following relationships:

$$E\left[\tilde{y}_j\tilde{y}_j^*\right] = \sigma_y^2 \quad (12.23)$$

$$E\left[\tilde{y}_j\tilde{\boldsymbol{x}}^+\tilde{\boldsymbol{w}}\right] = E\left[\tilde{y}_j\tilde{\boldsymbol{x}}^+\right]\tilde{\boldsymbol{w}} = \tilde{c}^+\tilde{\boldsymbol{w}} \quad (12.24)$$

$$E\left[\tilde{\boldsymbol{w}}^+\tilde{\boldsymbol{x}}\tilde{y}_j^*\right] = \tilde{\boldsymbol{w}}^+E\left[\tilde{\boldsymbol{x}}\tilde{y}_j^*\right] = \tilde{\boldsymbol{w}}^T\tilde{\boldsymbol{c}} \quad (12.25)$$

and

$$E[\tilde{\boldsymbol{w}}^+\tilde{\boldsymbol{x}}\tilde{\boldsymbol{x}}^+\tilde{\boldsymbol{w}}] = \tilde{\boldsymbol{w}}^+E[\tilde{\boldsymbol{x}}\tilde{\boldsymbol{x}}^+]\tilde{\boldsymbol{w}} = \tilde{\boldsymbol{w}}^+\boldsymbol{R}\tilde{\boldsymbol{w}} \quad (12.26)$$

The matrix $\boldsymbol{R}$ possesses unique properties based on the stationarity of the sampled stochastic processes and is referred to as a *Hermitian* matrix.* The diagonal elements $r_{ii} = E(x_i, x_i^+) = r(0) \; : i = 1, \ldots, N$ are real and the trace of $\boldsymbol{R}$ is $tr[\boldsymbol{R}] = Nr(0)$. Indexing the elements of the row and column vectors $\tilde{\boldsymbol{x}}^+$ and $\tilde{\boldsymbol{x}}$ as x_i and x_j respectively, the elements $\tilde{r}_{ji} = E(\tilde{x}_j, \tilde{x}_i^+) = \tilde{r}_{k=i-j}$. For $k > 0$ ($k < 0$), the matrix elements correspond to positive (negative) correlation lags that lie above (below) the matrix diagonal. Furthermore, the elements of the Hermitian matrix correspond to the condition $\tilde{r}_{-k} = \tilde{r}_k^*$, so the elements above and below the principal diagonal possess complex conjugate symmetry.

The MSE, defined by (12.20), characterizes a concave upward surface with a minimum value corresponding to the derivative $\partial J(\boldsymbol{w})/\partial \boldsymbol{w}$ equal to zero. The derivative is defined as the gradient vector $\nabla_{\tilde{\boldsymbol{w}}}$ and is evaluated as

$$\nabla_{\tilde{\boldsymbol{w}}} \triangleq \frac{\partial J(\tilde{\boldsymbol{w}})}{\partial \tilde{\boldsymbol{w}}} = \frac{\partial \sigma_{\tilde{y}}^2}{\partial \tilde{\boldsymbol{w}}} - \frac{\partial \tilde{\boldsymbol{c}}^+\tilde{\boldsymbol{w}}}{\partial \tilde{\boldsymbol{w}}} - \frac{\partial \tilde{\boldsymbol{w}}^+\tilde{\boldsymbol{c}}}{\partial \tilde{\boldsymbol{w}}} + \frac{\partial \tilde{\boldsymbol{w}}^+\boldsymbol{R}\tilde{\boldsymbol{w}}}{\partial \tilde{\boldsymbol{w}}}$$
$$= 0 - 0 - 2\tilde{\boldsymbol{c}} + 2\boldsymbol{R}\tilde{\boldsymbol{w}} \quad (12.27)$$

The solution to the derivatives in (12.27), indicated by the third equality, is found in Section 1.12.5. Upon evaluation of (12.27) with $\nabla_{\tilde{\boldsymbol{w}}_{\mathbf{o}}} = 0$ the optimum filter weights are expressed as

$$\tilde{\boldsymbol{w}}_{\mathbf{o}} = \boldsymbol{R}^{-1}\tilde{\boldsymbol{c}} \quad (12.28)$$

From the evaluation of (12.20) under the optimum condition $\boldsymbol{R}\tilde{\boldsymbol{w}}_{\mathbf{o}} = \tilde{\boldsymbol{c}}$, the MMSE is expressed as

$$J_{min} = J(\tilde{\boldsymbol{w}}_{\mathbf{o}}) = \sigma_{\tilde{y}}^2 - \tilde{\boldsymbol{c}}^+\tilde{\boldsymbol{w}}_{\mathbf{o}} = \sigma_{\tilde{y}}^2 - \tilde{\boldsymbol{c}}^+\boldsymbol{R}^{-1}\tilde{\boldsymbol{c}} \quad (12.29)$$

Based on the method of steepest-descent [7], the filter tap weights are altered to reduce, or minimize, the slope of the N-dimensional concave upward surface characterized by the weights. At each decision sample jT_s, the weight updated is based on the first-order recursive filter response, expressed as

$$\tilde{\boldsymbol{w}}(j+1) = \tilde{\boldsymbol{w}}(j) - u\nabla_{\tilde{\boldsymbol{w}}} = \tilde{\boldsymbol{w}}(j) + 2u(\tilde{\boldsymbol{c}} - \boldsymbol{R}\tilde{\boldsymbol{w}}(j)) \quad (12.30)$$

The second equality in (12.30) results from (12.27) and the step-size parameter u is used to provide stability and control the rate of convergence. As the expectation of the gradient vector approaches zero, the filter approaches the optimum MSE estimate of the reference $\tilde{y}_j$. The slope of the concave upward surface about the minimum point forms a discriminator S-curve that ensures stable tracking about the optimum tap weights. The updating is identical for each tap weight and the

*The Hermitian transposition $\tilde{x}^H$ of a vector, or R^H of a matrix, is also used to denote complex conjugate transposition.

required processing for the n-th tap weight is determined for the corresponding elements at sample j of the column vectors, $\tilde{c}$ and $R\tilde{w}$, in (12.30); with the subscript n,j representing the n-th row at sample j, the result is expressed as

$$\tilde{w}_{n,j+1} = \tilde{w}_{n,j} + 2u\left((\tilde{c})_{n,j} - (R\tilde{w})_{n,j}\right) \tag{12.31}$$

The complex scalars $(\tilde{c})_{n,j}$ and $(R\tilde{w})_{n,j}$ are evaluated as

$$(\tilde{c})_{n,j} = \tilde{x}_n \tilde{y}_j^* \tag{12.32}$$

and

$$(R\tilde{w})_{n,j} = \sum_{i=1}^{N} r_{n,i}\tilde{w}_{i,j}\delta_{in} = \sigma_x^2 \tilde{w}_{n,j} \tag{12.33}$$

where the Kronecker delta function arises from the evaluation of the gradient vector in (12.27) and the variance in (12.33) is simply $\sigma_x^2 = r_{n.n}(0)$. Substituting these results into (12.31), the weight updating for the n-th tap weight and sample j is expressed as

$$\tilde{w}_{n,j+1} = \left(1 - 2u\sigma_x^2\right)\tilde{w}_{n,j} + 2u\tilde{x}_{j-(n-1)}\tilde{y}_j^* \tag{12.34}$$

The updating of the weight $\tilde{w}_{n,j}^*$ at sample j given the previous weight $\tilde{w}_{n,j-1}^*$ at sample $j-1$ $:j = 1, \ldots$ is shown in Figure 12.2; the asterisk $*$ denotes conjugation. The filter is typically initialized using $\tilde{w}_{n,0}^* = 0 + j0$; however, if feasible, the best guess of the steady-state solution should be used.

The stability and convergence properties of the FIR-adaptive filter have been analyzed by Widrow [8], Ungerboeck [9], Gitlin and Weinstein [10], and Haykin [5]. Their analysis focuses on the eigenvalues of the correlation matrix R and, based on the method of steepest-descent, the adaptive is filter stable and converges to the steady-state condition, regardless of initial conditions, if all of the eigenvalues $\lambda_n : n = 1, \ldots, N$ satisfy the condition

$$|1 - u\lambda_n| < 1, \ \forall n \ \text{: condition for average tap weight convergence} \tag{12.35}$$

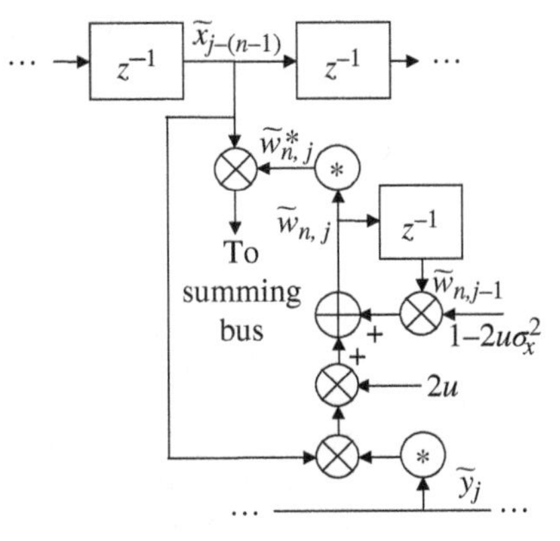

FIGURE 12.2 MSE-adaptive filter weight update algorithm.

Equation (12.35) leads to the conditions $0 < u < 2/\lambda_n \ : \forall n$ and, recognizing that the eigenvalues for the Hermitian matrix R are all positive real values, upon defining the maximum eigenvalue as λ_{max}, the condition on u is

$$0 < u < \frac{2}{\lambda_{max}} \ \text{: condition for average tap weight convergence} \tag{12.36}$$

Haykin [11] examines the transient behavior of the steepest-descent algorithm for a two-tap FIR filter with varying eigenvalue spread,* fixed step-sizes u, and varying step-sizes u with fixed eigenvalue spread. Haykin's salient observations are:

- *The steepest-descent algorithm converges most rapidly when the eigenvalues, λ_1 and λ_2 are equal or when the initial conditions are chosen properly.*
- *For a fixed step-size u, as the eigenvalue spread increases the correlation matrix R becomes more ill conditioned.*
- *For small step-sizes u, the transient behavior of the steepest-descent algorithm becomes overdamped. For large step-sized, approaching $2/\lambda_{max}$, the transient behavior becomes underdamped and exhibits oscillations.*

Haykin concludes that *the transient behavior of the steepest-descent algorithm is highly sensitive to the step-size parameter and the eigenvalue spread.* In the selection of the adaptive processing gain, a performance tradeoff exists between the variance of the tap weights and the responsiveness of adaptive filter to the channel dynamics. For example, decreasing the gain decreases the variance but increases the system performance sensitivity to the received signal dynamics and increasing the gain has the inverse affects. Gitlin and Weinstein have investigated the adaptive equalizer MSE as it relates to the tap weight precision, the number of taps, the channel characteristics, and the digital resolution. Gitlin and Weinstein conclude that *the number of taps should be kept to a minimum, consistent with the steady-state performance objectives and the weight precision should be ten to twelve bits.* Insufficient processing precision will result in tap accumulation round-off errors and dominate the system performance as the number of filter taps is increased. In general, the gain should be no greater than necessary to track the time-varying fading as determined by the signal decorrelation time and the length or span of the equalizer should not exceed the multipath delay spread; these topics are also discussed in Chapters 18 and 20. Adaptive filtering in the frequency

*The eigenvalue spread is defined as the ratio $\lambda_{max}/\lambda_{min}$.

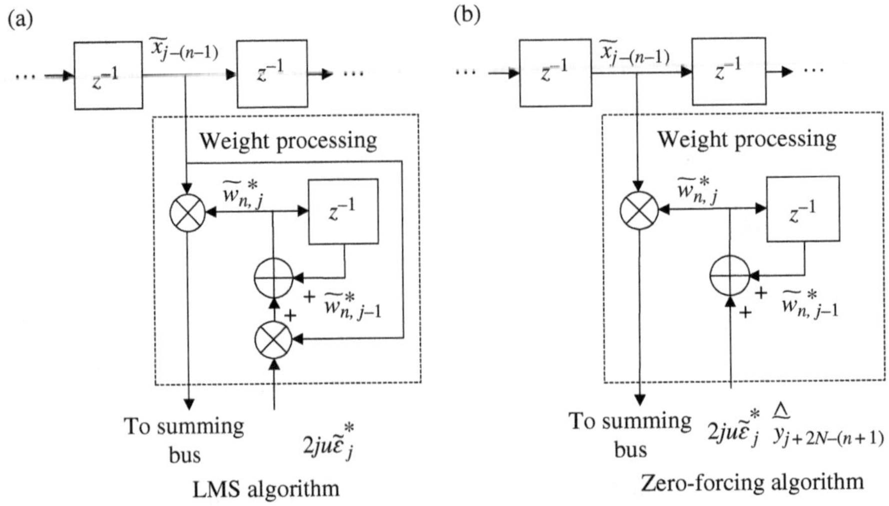

FIGURE 12.3 Adaptive filter weight update algorithms.

domain is discussed by Dentino, McCool, and Widrow [12] and by Bershad and Feintuch [13].

12.3.1 Least Mean-Square Algorithm

The weight updating in the preceding section, using the Wiener FIR-adaptive filter, is problematic, in that, the algorithm is dependent on the correlation matrix $\boldsymbol{R}$. To circumvent the sensitivity of the weight update algorithm on the correlation matrix, Widrow and Hoff [14] developed the LMS algorithm using real data. The LMS algorithm was subsequently modified by Widrow, McCool, and Ball [15] using the complex data. The LMS algorithm is an efficient method of updating the FIR-adaptive filter tap weights through the entire process of adaptation and estimation. The steepest-decent algorithm is used, in that, the filter tap weights are updated using a single-pole recursive filter that reduces the slope, or gradient vector $\nabla_{\tilde{w}}$, as given by the first equality in (12.30). However, in this case, the N-dimensional instantaneous square-error is defined as

$$J'(\tilde{w}) \triangleq \tilde{\varepsilon}_j(\tilde{w})\tilde{\varepsilon}^*_j(\tilde{w}) \quad \text{: instantaneous-squared-error} \quad (12.37)$$

and the gradient vector is evaluated as

$$\nabla_{\tilde{w}}{}' \triangleq \frac{\partial J'(\tilde{w})}{\partial \tilde{w}} = \tilde{\varepsilon}_j(\tilde{w})\frac{\partial \tilde{\varepsilon}^*_j(\tilde{w})}{\partial \tilde{w}} + \tilde{\varepsilon}^*_j(\tilde{w})\frac{\partial \tilde{\varepsilon}_j(\tilde{w})}{\partial \tilde{w}}$$
$$= -2\tilde{\varepsilon}_j(\tilde{w})\tilde{x}^* + 0 \quad (12.38)$$

The third equality in (12.38) is established using the rules for complex matrix differentiation discussed in Section 1.12.4. From (12.38), the filter tap weights are updated as

$$\tilde{\boldsymbol{w}}(j+1) = \tilde{\boldsymbol{w}}(j) - u\nabla_{\tilde{w}} = \tilde{\boldsymbol{w}}(j) + 2u\tilde{\varepsilon}_j(\tilde{\boldsymbol{w}})\tilde{\boldsymbol{x}}^* \quad (12.39)$$

The LMS tap weight update algorithm is identical for each tap weight and is shown in terms of the n-th complex weight $\tilde{w}_{n,j}$ in Figure 12.3. Although (12.39) is a convenient form in expressing the tap weight updating, an alternate expression is obtained by substituting (12.16) for $\tilde{\varepsilon}_j(\tilde{\boldsymbol{w}})$ and rearranging the terms as in (12.34) to obtain the algorithm for the n-th tap weight update expressed as

$$\tilde{w}_{n,j+1} = \left(1 - 2u\left|\tilde{x}_{j-(n-1)}\right|^2\right)\tilde{w}_{n,j} + 2u\tilde{y}_j\tilde{x}^*_{j-(n-1)} \quad \text{: implicit } \tilde{\varepsilon}_j \quad (12.40)$$

The tap weight update algorithm processing for the LMS feedforward equalizer and canceler, depicted respectively in Figures 12.6 and 12.8, is shown in Figure 12.3a. Figure 12.3b shows the weight update processing for the peak-distortion zero-forcing equalizer (ZFE) depicted in Figure 12.5.

In remainder of this section, the stability and convergence properties of the FIR-adaptive filter using the LMS algorithm are summarized based on the work by Haykin [5]. As before, the analysis focuses on the eigenvalues of the correlation matrix $\boldsymbol{R}$ and, based on the method of steepest-descent, the adaptive filter is stable and converges to the steady-state condition if the step-size satisfies the condition

$$0 < u < \frac{2}{\sum_{n=1}^{N}\lambda_n} \quad \text{: LMS condition for average MSE convergence} \quad (12.41)$$

The sum of the eigenvalues represents the mean-square value, or total power, of the filter taps. Since the sampled input represents a stationary process, the eigenvalues represent the zero lag correlation values $\sigma_x^2 = r_{ii}(0)$ of the

Hermitian matrix $\boldsymbol{R}$ and the left-hand-side (*lhs*) of (12.41) follows from the positive real values $\sigma_x^2 \geq 0$.

The *misadjustment* parameter [16] M is defined as the ratio of the average excess MSE $\bar{J}_{excess}$ to the MMSE J_{min}, expressed in (12.29), that is,

$$M \triangleq \frac{\bar{J}_{excess}}{J_{min}} = \frac{u\sum_{n=1}^{N}\lambda_n}{2-u\sum_{n=1}^{N}\lambda_n} \quad : \text{LMS misadjustment} \tag{12.42}$$

The parameter M corresponds to the steady-state condition evaluated as expressed in the second equality in (12.42) [16].

The average excess MSE is defined as

$$\bar{J}_{excess} \triangleq E[J(\infty)-J_{min}] = E\left[\boldsymbol{\varepsilon}^H(j)\boldsymbol{R\varepsilon}(j)|_{\lim j\to\infty}\right] \tag{12.43}$$

where $J(\infty)$ is steady-state mean-square convergence error of the LMS algorithm which exceeds J_{min}. In terms of the convergence samples j and the tap weight vector error $\varepsilon(j) = \boldsymbol{w}(j) - \boldsymbol{w}_o$, the LMS mean-square convergence error is $J(j) = J_{min} - \varepsilon^{\boldsymbol{H}}(j)\boldsymbol{R}\varepsilon(j)$ which leads to the second equality in (12.43).

Haykin's salient performance observations for the LMS algorithm are [17]:

- *The adaptive filter converges slowly for small values of u with a correspondingly low misadjustment. Conversely, for larger values of u the filter converges rapidly with a correspondingly large misadjustment.*
- *The convergence of the average MSE, as given by* (12.41), *also guarantees the convergence of the mean tap weight values given by* (12.36) *since* $\lambda_{max} \leq \sum_{n=1}^{N}\lambda_n$.
- *For increasing eigenvalue spreads, the convergence time of the excess MSE becomes slower and the average tap weight convergence is limited by the smallest eigenvalues.*

The simplicity of the LMS algorithm stems from the fact that it is based on the instantaneous squared error and does not require the expectation associated with the evaluation of the correlation functions or inversion of the correlation matrix. Considerable attention has been given to the theoretical evaluation and measurement of the symbol-error probability, with channel-induced ISI, using the LMS algorithm [18–22].

The adaptive filters are characterized as having a *primary* and *reference* input that determine the principal application of the adaptive processing. The primary input is associated with the received signal. For example, *equalization* uses the received signal as the primary input into the adaptive filter with the reference input derived from a locally generated replica of known data or data estimates derived from the received data. The reference data are often derived from a preamble or training sequence associated with the received data. With noise and interference *cancellation*, the reference input is a correlated replica of the interfering signal that is applied to the input to the adaptive filter. For sufficiently high interference-to-signal ratios (ISRs), the reference input may also be derived directly from the received signal. System or waveform *identification* is accomplished by applying the reference signal to the adaptive filter, possibly using multiple references that are suspected be correlated with the primary received signal.

12.4 INTERSYMBOL INTERFERENCE AND MULTIPATH EQUALIZATION

In this section, several commonly used equalizers are discussed with case studies of their performance presented in Sections 12.7–12.9.

12.4.1 Zero-Forcing Equalizer

The ZFE is among the first equalizers to be analyzed and implemented [23]. The equalizer operates on the output of the channel filter that introduces ISI distortion on the transmitted data-modulated symbols denoted by the baseband signal samples $\tilde{s}_j = \tilde{s}(jT_s)$. The channel filter is shown in Figure 12.4 with the output expressed as the convolutional sum

$$\tilde{x}_i = \sum_{\ell=0}^{L-1} \tilde{h}_\ell \tilde{s}_{i-\ell} \tag{12.44}$$

The peak distortion, at the output of the channel, is defined in terms of the channel impulse response as

$$D_o \triangleq \frac{1}{|\tilde{h}_{\ell'}|}\sum_{\substack{\ell=0 \\ \ell\neq\ell'}}^{L-1}|\tilde{h}_\ell| \tag{12.45}$$

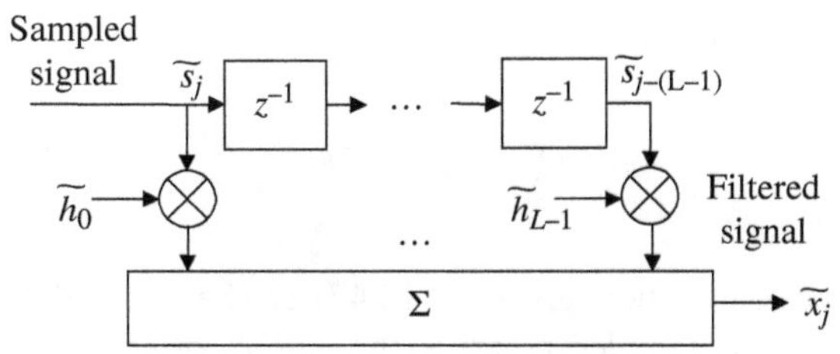

FIGURE 12.4 Channel filter.

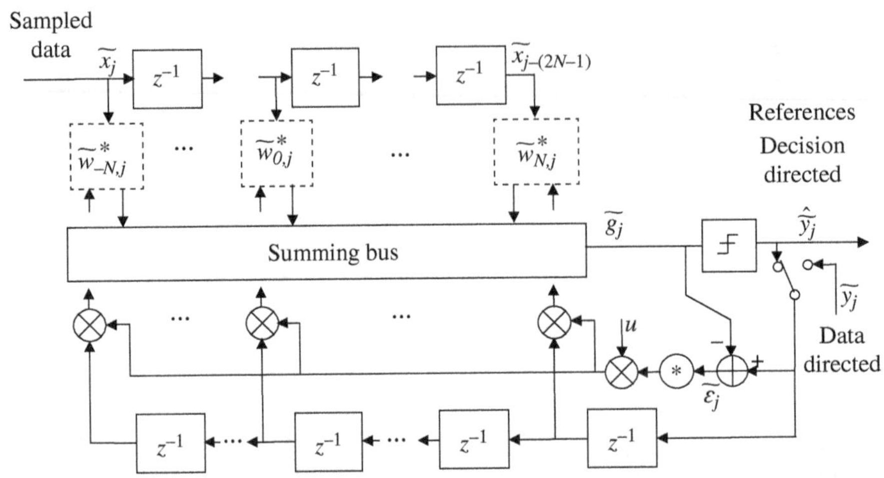

FIGURE 12.5 Zero-forcing equalizer.

where $\tilde{h}_{\ell'}$ is the maximum filter sample with the filter normalized such that $|\tilde{h}_{\ell'}| = 1$. The peak distortion is a measure of the noise-free dispersion about the data-modulated symbol *rest-points* resulting from the channel ISI. The dispersion between adjacent rest-points characterizes the *eye-opening* and when $D_o < 1$ the *eye* is said to be open [24].

The ZFE, shown in Figure 12.5, follows the channel filter and attempts to minimize the loss from the channel ISI. The equalizer output is evaluated as the convolutional sum

$$\tilde{g}_j = \sum_{n=-N}^{N} \tilde{w}_{n,j}\tilde{x}_{j-n} \tag{12.46}$$

The peak distortion at the output of the equalizer is defined in terms of the impulse response of the channel cascaded with the equalizer and is expressed as

$$D = \sum_{\substack{n=-N \\ n\neq 0}}^{N+L-1} |\tilde{g}'_n| \tag{12.47}$$

As indicated in Figures 12.4 and 12.5 the channel has L taps and the equalizer has $2N+1$ taps, so the extent or range of the non-zero sampled impulse response is $-N$ to $N+L-1$ for a total of $2N+L$ samples. The cascaded impulse response in (12.47) is sampled at the time intervals jT_s and computed as

$$\tilde{g}'_n = \sum_j \tilde{w}_{n,j}\tilde{h}_{n-j} \tag{12.48}$$

The peak distortion, given by (12.47), can be made arbitrarily small at the sampled points as the number of equalizer taps approaches infinity, and, although Lucky [25] has shown that D is a convex upward function of the equalizer coefficients, there is no assurance that the tap weights will converge for a finite number of taps. However, a solution does exist for a finite number of taps if the equalizer input peak distortion given by (12.45) satisfies the condition $D_o < 1$, but the distortion or decision error cannot be made arbitrarily small. The decision error at the output of the equalizer FIR filter is computed as

$$\tilde{\varepsilon}_j = \hat{\tilde{y}}_j - \tilde{g}_j \tag{12.49}$$

Using the data estimate $\hat{\tilde{y}}_j$ results in a decision-directed equalizer algorithm and using the known data $\tilde{y}_j$, as with a preamble or training data sequence, results in a data-directed equalizer algorithm. The relationship between the forward indexing of $\tilde{x}_{j-\alpha}$ and the backward indexing of $\tilde{y}_{j-\beta}$ is expressed as $\beta = -2N+\alpha+2 \; : \alpha = 0,\ldots,2N-1$. The weight update processing indicated in Figure 12.5 is based on the algorithm shown in Figure 12.3b.

12.4.2 Linear Feedforward Equalizer

The linear feedforward equalizer (LFFE) using the LMS tap weight update algorithm is shown in Figure 12.6. The LMS tap weight processing is indicated in Figure 12.3a. The complex tap weights are all initialized to zero and, as the sampled input data make its way through the FIR filter, the complex output of the summing bus is applied to the decision device to obtain the data decision estimate $\hat{\tilde{y}}_j$ that is used to form the decision error $\tilde{\varepsilon}_j$. The data decision estimate is based on a complex limiter that computes the real and imaginary parts of $\hat{\tilde{y}}_j$ based on the closest received waveform modulation rest states to the complex summation $\tilde{g}_j$. For example, with binary phase shift keying (BPSK), if $\mathrm{Re}\{\tilde{g}_j\} \geq 0$ then $\hat{\tilde{y}}_j = 1 + j0$

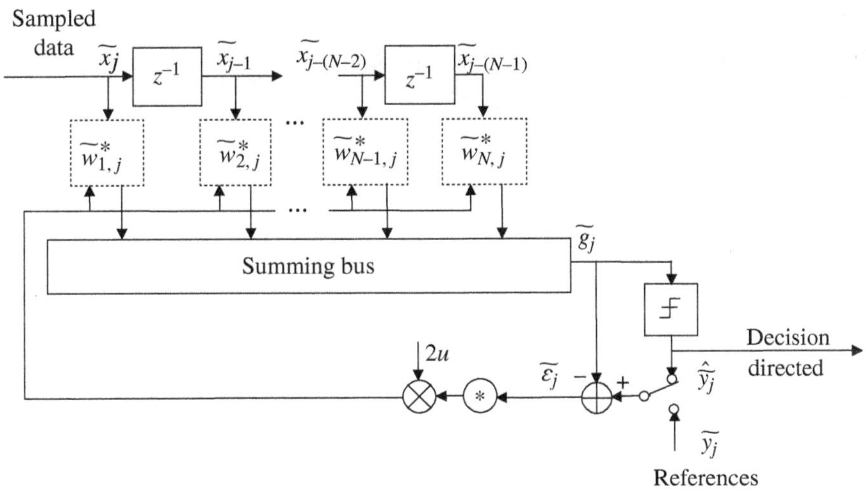

FIGURE 12.6 LMS linear feedforward equalizer.

otherwise $\hat{\tilde{y}}_j = -1 + j0$. Under the input signal condition of wide-sense stationarity,* the mean value of the decision error will approach zero as the filter weights are adjusted during the adaptation processing. The decision estimate $\hat{\tilde{y}}_j$ is subject to signal-to-noise ratio-dependent decision errors; however, when a preamble is used, the known preamble data $\tilde{y}_j$ is used to reduce the learning time.

The performance of the LFFE is examined in the case study in Section 12.7. In the study, the recovery of the bit-error performance loss resulting from ISI is examined for narrowband filtering of a BPSK-modulated waveform.

12.4.3 Nonlinear Decision Feedback Equalizer

The decision feedback equalizer (DFE) [26, 27], shown in Figure 12.7, augments the LFFE symbol equalizer by forming the weighted linear combination of the previous symbols and reducing their contributions to the ISI. The DFE processing is equivalent to that involving linear prediction as embodied in forward linear prediction (FLP), backward linear prediction (BLP), and forward–backward linear prediction (FBLP) algorithms described by Haykin [28]. Belfiore and Park [29] and Proakis [30] also provide detailed analyses and compare the implementation and performance of the ZFE, LFFE, the DFE implementations. The DFE is used to correct multipath interference and signal distortion resulting from frequency selective fading [31] that is prevalent in wideband direct-sequence spread-spectrum (DSSS) applications. Lee and Hill [32] discuss the application of the DFE in terms of maximum-likelihood sequence estimation (MLSE).

Proakis shows that, for moderate to severe multipath profiles, there is about a 2 dB loss in E_b/N_o performance between using known data decisions, as would occur during the reception of a learning or training data sequence, and using the locally detected data estimates. When using the LFFE, a severe multipath profile results in bit-error probabilities that asymptotically approach $P_{be} \cong 10^{-2}$ with increasing signal-to-noise ratio; however, with the DFE equalizer the bit-error probabilities continue to decrease. These results are demonstrated using symbol-spaced equalizer (SSE) taps with an optimally sampled matched filter that is matched to the waveform and channel. Typically the channel is unknown so using the fractionally-spaced equalizer (FSE) taps, as described in the following section, is necessary. The performance of the DFE is sensitive to the detected data because each decision error is shifted through the entire feedback path of the FIR filter so the selection of the step-size gain is a particularly important consideration.

12.4.4 Fractionally-Spaced Equalizers

In the preceding examples, the equalizer sample delay may correspond to the symbol interval of T seconds or a sample interval corresponding to $T_s = T/N_s$ seconds where N_s is the number of samples-per-symbol. The SSE must operate on the demodulator matched filter samples at $nT + T_\varepsilon \;:n = 0, 1, \ldots, s$ and only corrects for distortion aliased about the frequency $1/2T$; it is also sensitive to the demodulator sampling delay (T_ε) relative to the received symbols. On the other hand, the FSE with sample delay T_s will compensate for the sample delay error T_ε and the channel distortion within the sampled bandwidth of $1/2T_s$. In addition to the channel distortion compensation, the FSE provides the symbol matched filtering required for the optimum detection of the received signal when sampled at the optimum sampling time [33–36].

*The simulation results to follow involve wide-sense stationary Gaussian noise which is also strict-sense stationary.

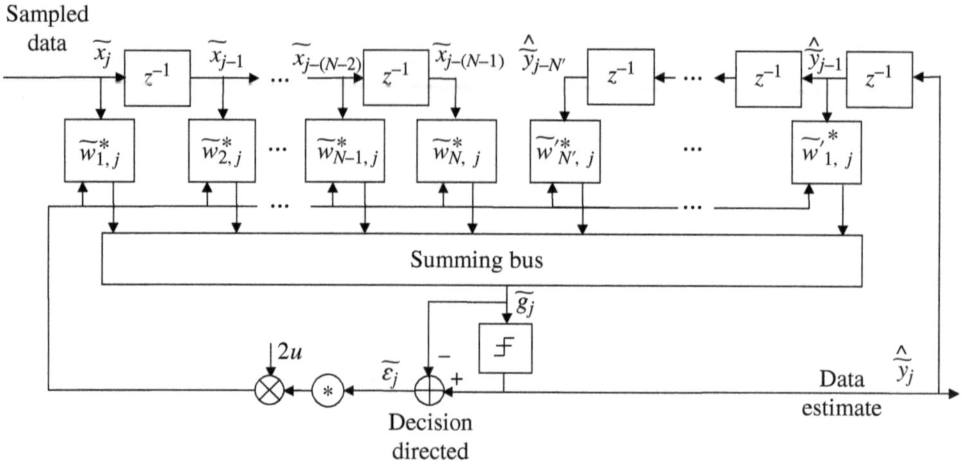

FIGURE 12.7 LMS nonlinear decision feedback equalizer.

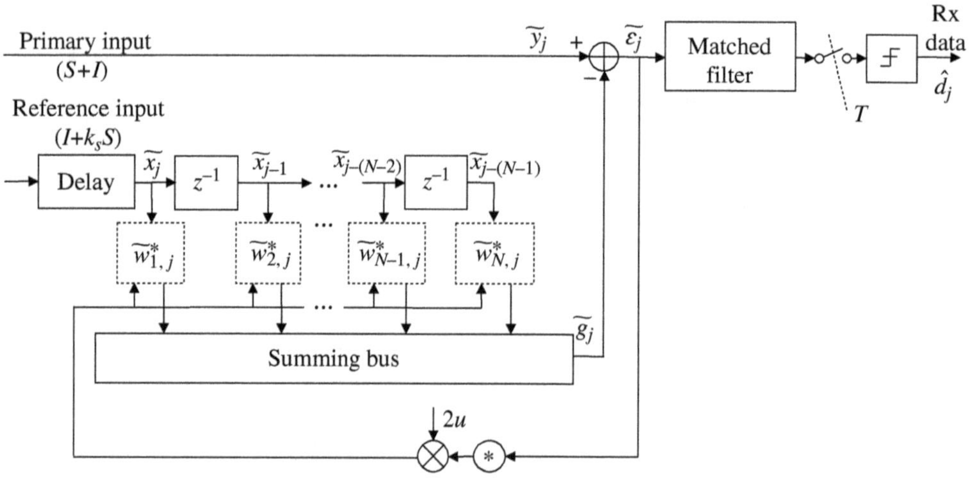

FIGURE 12.8 LMS-adaptive canceler.

12.4.5 Blind or Self-Recovering Equalizers

In many applications, it is desirable for the equalizer to adapt to the distorted received signal without the benefit of a preamble or special training sequence. Equalizers that function in this manner are referred to as blind or self-recovering equalizers [37–40]. This is a major undertaking, in that, knowledge of symbol timing and carrier phase must be resolved simultaneously with the equalizer weight adaptation. Godard [41] discusses an approach for equalizer adaptation independent of the carrier phase and the modulated waveform symbol constellation. Since this approach does not require demodulator data estimates for equalization, the carrier acquisition and tracking must be performed on the equalizer output.

12.5 INTERFERENCE AND NOISE CANCELLATION

The adaptive filter, used as an interference signal canceler, has many applications as discussed by Widrow et al. [42] in their comprehensive paper on the subject. The adaptive canceler requires two inputs, usually identified as the *primary* signal and the *reference* signal. The *primary* input to the system is the desired signal corrupted by an additive interfering signal and the *reference* input is a correlated version of the interfering signal, ideally derived from the source of the interference. The *reference* input may be generated internally as in the case of decision-directed applications or obtained locally in the case of nearby interfering signal and noise sources. The applications are nearly endless; however, include the broad subject of interference cancellation involving: antenna sidelobes, adaptive antenna arrays, electric motors, power lines, speech, and electrocardiography.

Figure 12.8 shows the general configuration of the LMS-adaptive canceler with the reference input passing through the FIR filter and subtracted from the primary signal forming the estimation error that is minimized by the recursive weight update processing. The canceler performance is evaluated in the case study of Section 12.8 where the primary input is a BPSK-modulated signal and a continuous wave (CW) interfering signal. In this case, the canceler reduces the interfering signal and the performance measure is the bit-error probability based on the optimally sampled output of the matched filter.

12.6 RECURSIVE LEAST SQUARE (RLS) EQUALIZER

The adaptive RLS equalizer [43] algorithm processes data block-by-block to update the filter weights and results in faster convergence than the previously discussed equalizers. The implementation of the RLS-adaptive equalizer is shown in Figure 12.9 and the following description of the processing follows the work of Haykin [44]. This configuration uses a known training sequence $\tilde{y}_j \ : j = 0, 1, \ldots$ which is stored in a separate array containing the M most current values. The values of $\tilde{y}_j$ can be generated locally at the receiver without the need for storage. During reception of the message data, the detected data estimates $\hat{\tilde{y}}_j$ are used resulting in a decision-directed equalizer.

Since the RLS algorithm processes a block of data, it is convenient to characterize the filter tap values and weights in term of the vectors $\tilde{\boldsymbol{x}}_j = (\tilde{x}_{0,j}, \ldots, \tilde{x}_{(N-1),j})^T$ and $\tilde{\boldsymbol{w}}_j = (\tilde{w}_{1,j}, \ldots, \tilde{w}_{N,j})^T$ respectively. The number of FIR filter taps (N) is selected to span the duration of the expected ISI of the received signal. The value of the M determines the relative location of the reference signal and is typically about $N/2$; however, if the duration of the pre- and post-symbol ISI is not equal, then M must be selected proportionately to encompass all of the ISI samples. The computations also require computing a gain vector $\tilde{\boldsymbol{k}}_j = (\tilde{k}_{1,j} \cdots \tilde{k}_{N,j})^T$ and the $N \times N$ matrix $\boldsymbol{C} = \Phi^{-1}$; where Φ is the $N \times N$ correlation matrix defined as

$$\boldsymbol{\Phi} j \triangleq \sum_{i=1}^{j} \lambda^{j-i} \tilde{x}_i \tilde{x}_i^H \qquad (12.50)$$

This definition of the correlation matrix is different from the correlation matrix $\boldsymbol{R}$ defined by (12.22), in that, it includes the exponential weighting factor $\lambda^{j-i} \ : i = 1, 2, \ldots, j$ where $\lambda < 1$ is a positive constant close to unity. The application of the scalar parameter λ forms a recursive-adaptive infinite impulse response (IIR) filter that increasingly diminishes the influence of the older data blocks with decreasing values of λ. From a filtering point-of-view, λ controls the bandwidth of the recursive tap adjustments, influences the convergence time of the weight processing, and the responsiveness to the channel dynamics. Conventional inversion of the correlation matrix is processing intense; however, to obtain the recursive solution for the tap weight updating, the matrix $\boldsymbol{C}$ is computed based on the *matrix inversion lemma* discussed by Haykin. These quantities are initialized as indicated in Table 12.1. The vector $\tilde{\boldsymbol{w}}_{j-1}^*$ and matrix $\boldsymbol{C}_{j-1}$ provide auxiliary data storage for computing the recursively updated values and, upon completion, the contents of the updated arrays are transferred to these arrays in preparation for the next iteration.

After initialization, the adaptive processing computations are described as follows. Upon reception of each received signal sample, the FIR filter output is computed as

$$\tilde{g}_j = \tilde{\boldsymbol{w}}_j^H \tilde{\boldsymbol{s}}_j = \sum_{n=1}^{N} w_{n,j}^* \tilde{x}_{j-(n-1)} \qquad (12.51)$$

The complex output of (12.51) is used for data detection and updating the adaptive filter weights for the next iteration. In

TABLE 12.1 Parameter Initializations

Parameter	Initial Value	Comments
λ	0.99	Recursive scalar coefficient (typically: 0.999–0.95)
$\tilde{\boldsymbol{w}}_{j-1}^*$	(0,0)	Complex weight vector
$\tilde{c}_{\ell,\ell'}$	(0,0): $\ell \neq \ell'$ $(\delta^{-1}$,0): $\ell = \ell'$	Elements of complex matrix $\boldsymbol{C}_{j-1}$ (small positive constant $\delta = 0.005$)

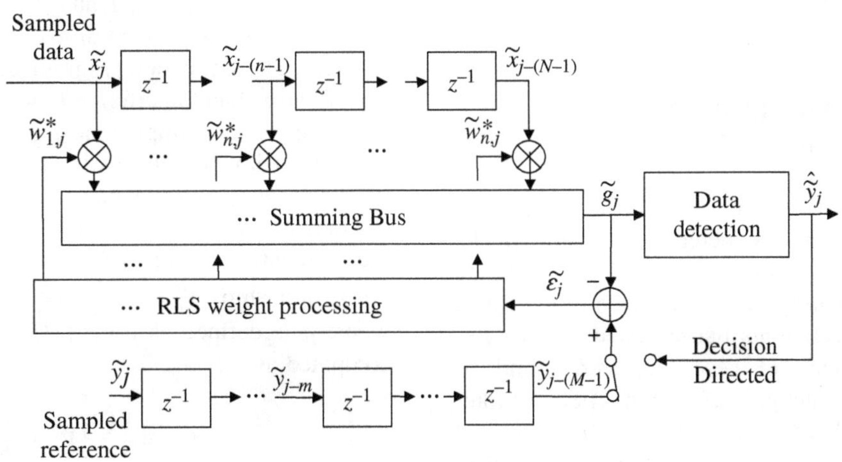

FIGURE 12.9 RLS-adaptive equalizer.

computing the new weight values, the estimation error is determined as

$$\tilde{\varepsilon}_j = \tilde{r}_{j-(M-1)} - \tilde{g}_j \tag{12.52}$$

The complex error is input to the weight processing function where the following operations are performed. Starting with the inverse correlation matrix, the gain vector is computed as

$$\tilde{k}_j = \frac{\tilde{v}_j}{(\lambda + u)} \tag{12.53}$$

where the vector $\tilde{v}_j$ and scalar u are computed, respectively, as $\tilde{v}_j = C_{j-1}\tilde{x}_j$ and $u = \tilde{x}_j^H \tilde{v}_j$. Using (12.52) and (12.53), the FIR filter taps are updated as

$$\tilde{w}_j^* = \tilde{w}_{j-1}^* + \tilde{k}_j \tilde{\varepsilon}_j^* \tag{12.54}$$

The final step involves updating the matrix inverse for the next iteration and the computation for this step is expressed as

$$\begin{aligned} C_j &= \frac{1}{\lambda}\left(I - \tilde{k}_j \tilde{x}_j^H\right) C_{j-1} \\ &= \frac{1}{\lambda}\left(C_{j-1} - \tilde{k}_j \tilde{v}_j^H\right) \end{aligned} \tag{12.55}$$

In preparation for the next iteration $\tilde{w}_{j-1}^* = \tilde{w}_j^*$ and $C_{j-1} = C_j$. Following these computations, the processing is repeated with the next received block of samples until receipt of the communication message is completed. The message must be prefixed with a training data sequence and possibly periodically repeated training sequences throughout the entire message. The inclusion and frequency of the periodic training sequences depends on the channel and system dynamics. As mentioned above, the training sequence reference bit must be replaced with the M-th tap value of the stored data upon declaring acquisition. With periodic training sequences, the equalizer must switch back-and-forth between the training sequence and the message data. A case study demonstrating the performance of the RLS equalizer is given in Section 12.9.

12.7 CASE STUDY: LMS LINEAR FEEDFORWARD EQUALIZATION

In this case study, the LMS algorithm is used with the LFFE to reduce the ISI loss of a coherently detected BPSK-modulated waveform. The performance with ISI is examined for three 0.4 dB ripple Chebyshev filters. Two of the filters have 6-poles with respective normalized bandwidths $BT =$ 1.5, 2.0; the phase response of these filters corresponds to the intrinsic Chebyshev filter phase function. The third filter has 4-poles, a normalized bandwidth of $BT = 3.0$, and includes quadratic phase equalization with 6.5° at the band edges. The filtered received signal spectrums are shown in Figure 12.10 for the three filters.

The bit-error performance results, shown in Figure 12.11, are based on Monte Carlo (MC) simulations using 1 M bits at each signal-to-noise ratio <8 dB, otherwise 10 M bits are used. The *dotted* curve is the theoretical or antipodal performance of BPSK and the *circled* data points verify the fidelity of the simulations program and correspond to the AWGN channel without the equalizer. The filters are symmetrical baseband filters and the received signal frequency error is zero. Ideal demodulator phase and bit-time tracking is used with an integrate-and-dump (I&D) matched filter so the performance degradation is solely based on the channel and equalizer. The *solid* and *dashed* cures in each figure correspond to the channel filter without and with the equalizer. The equalizer uses a nine tap bit-spaced FIR filter with a tap weight gain step-size $u = 0.002$. The learning time, before demodulation bit-error counting, is 400 bits. The simulation results demonstrate the performance improvement for the worst case filter shown in Figure 12.10a; however, for the less severe channel filters the equalizer has little effect. Reducing the equalizer step-size during the data detection will improve the equalizer bit-error performance while maintaining an acceptable learning time, although the channel dynamics will influence the misadjustment parameter restricting the lower limit of the step size.

12.8 CASE STUDY: NARROWBAND INTERFERENCE CANCELLATION

In this case study, the cancellation of narrowband interference is examined using CW interfering signals characterized as

$$s_i(t) = V_i \cos(\omega_i t + \phi_i) \tag{12.56}$$

where $V_i = k_v V_s$, $\omega_i = 2\pi f_i T$ and V_s is the peak carrier voltage of the received waveform and T is the modulation symbol duration. The simulation program provides an arbitrary number of narrowband interfering tones; however, the following results apply for a single tone centered on the desired signal spectrum with discrete phases relative to the phase of the received signal. The total power (P_{tot}) of the interfering tones, the modulated signal, and the receiver noise are adjusted to the automatic gain controlled power of P_{agc} according to the relationships $V_i' = \sqrt{2P_i\rho}$, $V_s' = \sqrt{2P_s\rho}$, and $\sigma_n = \sqrt{\sigma_n^2\rho}$ where ρ is defined as $\rho = P_{agc}/P_{tot}$ and the total power is computed as

$$P_{tot} = \frac{1}{2}\sum_{i=1}^{N_{cw}} V_i^2 + \frac{V_s^2}{2} + \sigma_n^2 \tag{12.57}$$

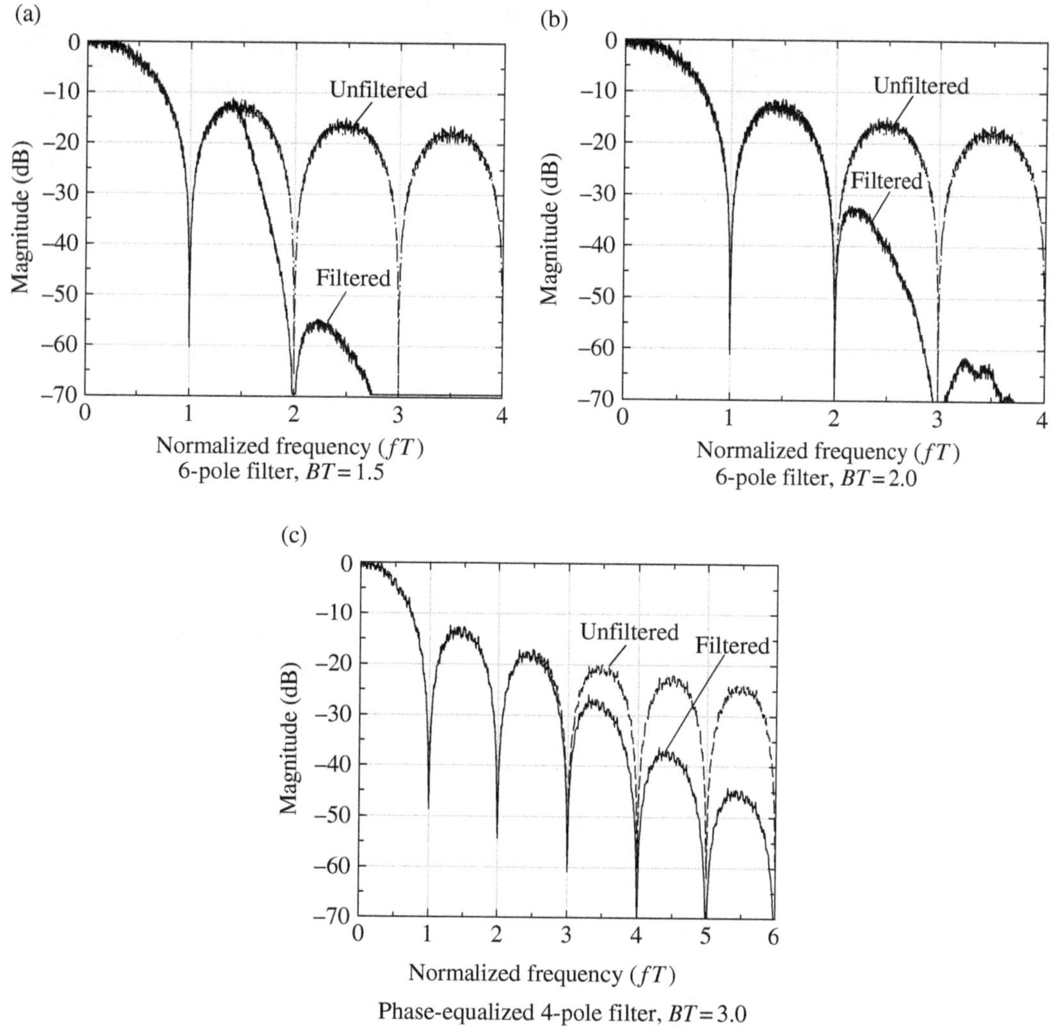

FIGURE 12.10 Filtered BPSK-modulated spectrums.

The noise power is established from the specified signal-to-noise ratio $\gamma_b = E_b/N_o$ according to the relationship

$$\sigma_n^2 = \frac{P_s}{\gamma_s} \tag{12.58}$$

where γ_s is the signal-to-noise ratio in the bandwidth of the transmitted symbol so $\gamma_s = kr_c\gamma_b$ with k bits/symbol and a forward error correction (FEC) code rate of r_c. The performance simulations use uncoded BPSK so $k = r_c = 1$. Using these normalized conditions results in a common received power level that provides for establishing a common adaptive filter gain, or step size u, independent of the number of interferers or the ISR; it also simplifies the specification of the analog-to-digital converter (ADC) range and the maximum amplitude level when the effect of finite amplitude quantization is being examined.

The model for evaluating the performance with the signal and interference is depicted in Figure 12.12. Under ideal conditions, the interfering signal is input as the canceler reference; however, the model provides for scenarios where the reference also contains the signal. To accommodate these situations, the parameter $k_s' \leq 1$ defines the signal level that is added to the canceler reference input. In the following simulated performance results, the parameter $k_s = 1/k_s'$ is specified as the signal attenuation or loss and is alternately used to denote the signal attenuation in decibels.

Based on Figure 12.12, the interference into the canceler can be characterized, in terms of the ISR, as

$$\begin{aligned} I_{in} &= I + \frac{S}{k_s} \\ &= I\left(1 + \frac{1}{k_s(I/S)}\right) \end{aligned} \tag{12.59}$$

Equation (12.59) is plotted in Figure 12.13 as a function of I_{in}/I for various values of the parameter k_s. The departure from the condition $I_{in} = I$ results from the influence of the

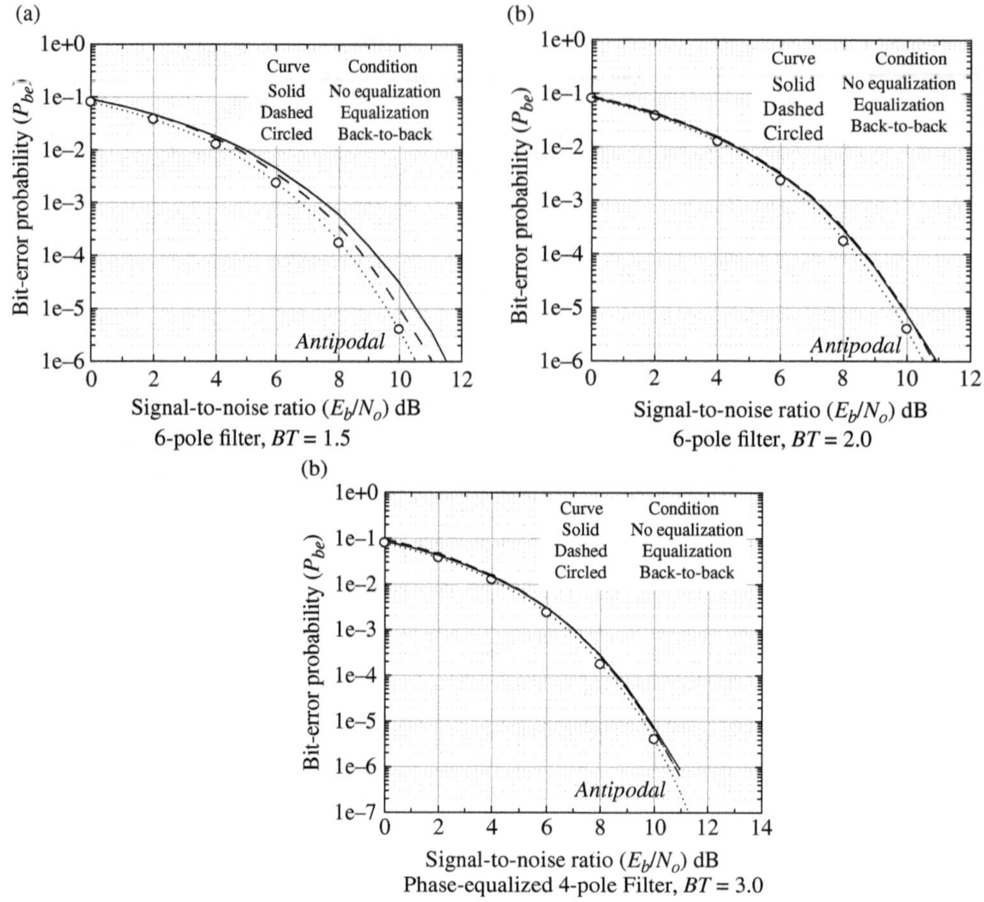

FIGURE 12.11 BPSK modulation with LMS linear equalization (9-tap FIR filter, $\Delta = 0.002$, training = 400 bits).

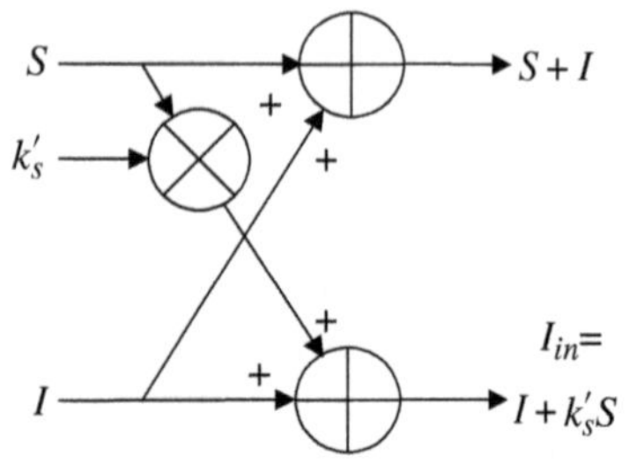

FIGURE 12.12 Model for signal and interference.

additive signal power to the interference with decreasing I/S. From Figure 12.12 it is seen that as k_s increases in value from 1 (0 dB), the interference power $P_{Iin} \rightarrow P_I$ which is a more authentic representation of the interference.

As defined in (12.57), the total receiver power, including the desired signal, interference, and channel noise, is adjusted to the AGC power of 0.5 W and the resulting signal carrier power is computed using $\gamma_b = E_b/N_o$ as

$$P_s = E_b R_b \tag{12.60}$$

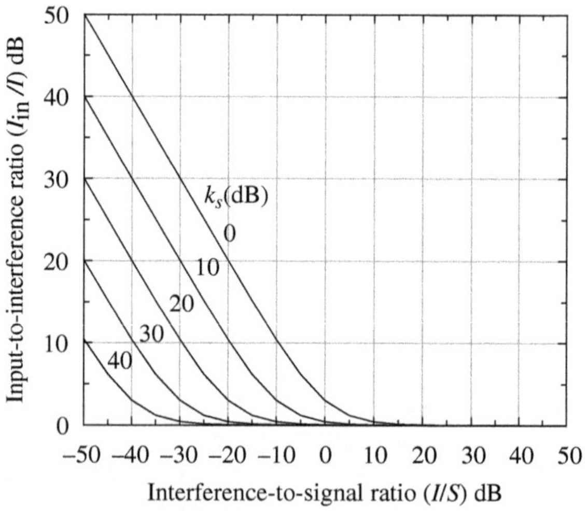

FIGURE 12.13 Dependence of input interference on the interference-to-signal ratio.

where R_b is the user bit rate. In the performance simulation program $R_b = 1$ kbps so the modulated carrier power is 30 dB above the energy-per-bit. For a single interfering *CW* tone, the power (P_{cw}) is given by

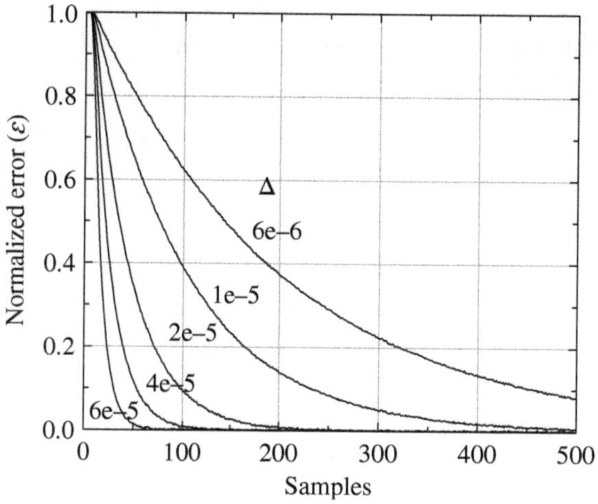

FIGURE 12.14 Theoretical LMS canceler convergence responses ($N_s = 1000$).

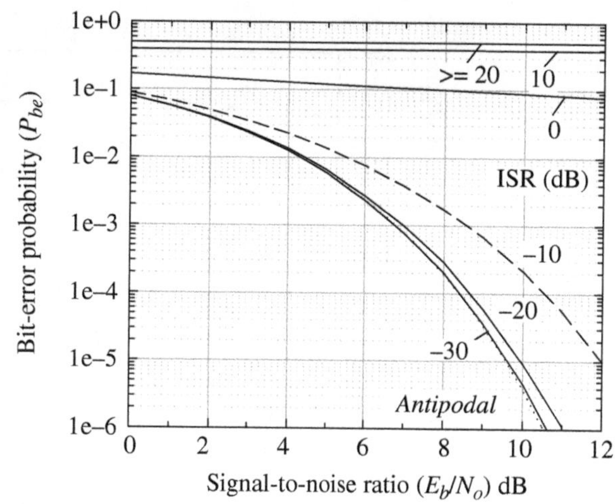

FIGURE 12.15 Performance of BPSK with CW interference without canceler.

$$P_{cw} = ISR(P_s) \tag{12.61}$$

As shown in Figure 12.14, the canceler gain adjustment, or step-size $\Delta = 2u$, is the key parameter that determines the canceler convergence characteristics given a specified FIR filter length. As seen from Figure 12.14, the canceler convergence time can be decreased by increasing Δ; however, the bit-error performance of the demodulator is improved by decreasing Δ following an acceptable convergence time. Consequently, a lower adjustment step-size is typically used when convergence is detected based on the statistical properties of the canceler error. In the following bit-error performance simulations, the bit-error counting is delayed until the canceler has reached, or nearly reached, the steady-state condition.

The canceler convergence is derived in Section 12.8.1, and Figure 12.14 is based on the generic normalized expression

$$\frac{\varepsilon_j}{\varepsilon_o} = (1 - N_s\Delta)^j \tag{12.62}$$

where N_s is the number of canceler taps, Δ is the canceler weight incremental adjustment, and j is the incremental time index. The bit-error performance of the demodulator with the CW interfering tone is shown in Figure 12.15 without the canceler. The ISR is indicated in decibels and the results clearly demonstrate the impact on the performance when the canceler is not used. To achieve a performance loss less than a few tenths of a decibel requires ISR ≤ −30 dB, which is virtually no interference.

With the exception of the performance in Figure 12.15 for ISR ≥ 0 dB, the MC simulations involve 1 M symbols or trials for signal-to-noise ratios <8 dB otherwise 10 M symbols for each signal-to-noise ratio are used. The range of signal-to-noise ratios is 0–14 dB in 1 dB steps. Furthermore, the phase of the interfering tone is indexed over 360° using ten deterministic phases given by $\phi_i = 36(i-1)^\circ$ for $i = 1, \ldots, 10$. Therefore, the bit-error performance involving 10 M MC trails results by averaging the bit errors over ten 10 M MC trials with each trial corresponding to a different signal phase. These simulation conditions apply to the simulated bit-error results in the remainder of this case study.

Figure 12.16 shows the bit-error performance under the steady-state conditions following the acquisition or learning time using a canceler step-size of $\Delta = 5 \times 10^{-5}$. The signal attenuation parameter is $k_s = \infty$ dB, that is, the reference signal into the canceler is the CW interference tone without any additive signal. Upon examining the expanded plot in Figure 12.16b, it is seen that the performance loss is negligible for ISR ≤ 0 dB and increases to 0.4 dB at $P_{be} = 10^{-6}$ when the ISR is increased to 30 dB; the corresponding loss is 0.23 dB at $P_{be} = 10^{-5}$. Consequently, for a step-size of $\Delta = 5 \times 10^{-5}$ the losses were reduced to that shown in Figure 12.16 for ISR ≤ 30 dB. These losses can be further reduced by decreasing the step-size.*

In Figure 12.17, a series of bit-error performance curves show the impact of the additive signal level to the canceler reference input as depicted in Figure 12.12. For a given ISR, the bit-error performance approaches that in Figure 12.16 as k_s approaches infinity. The objective is to characterize the minimum value of k_s (k_{smin}) to achieve a loss in $E_b/N_o \leq 0.2$ dB relative to that shown in Figure 12.16. Examination of the results in Figure 12.17, supplemented with the additional performance for ISR = 30 and 50 dB, indicates that $k_s \geq 20$ dB results in an E_b/N_o performance that is

*Lowering the step-size has other performance implications that are not addressed in this case study; for example, a decreased responsiveness to interfering signal dynamics and the increased sensitivity to demodulator quantization.

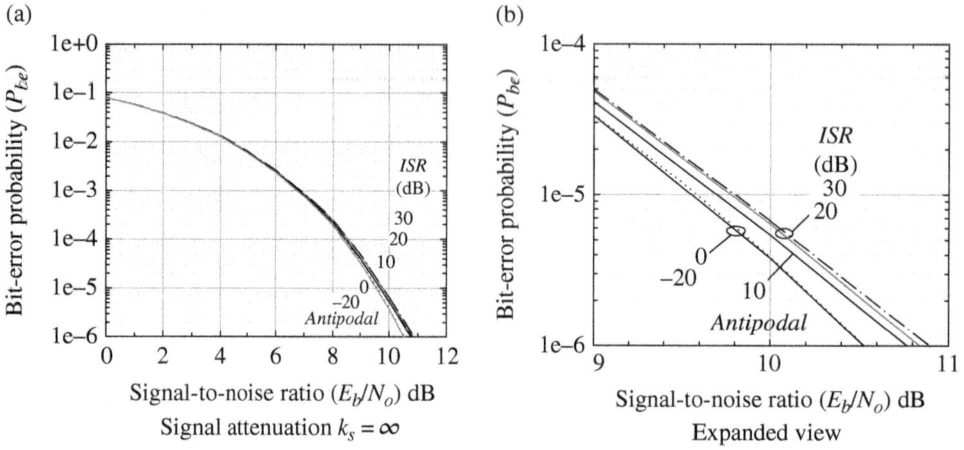

FIGURE 12.16 LMS performance of CW interference cancellation with constant signal attenuation.

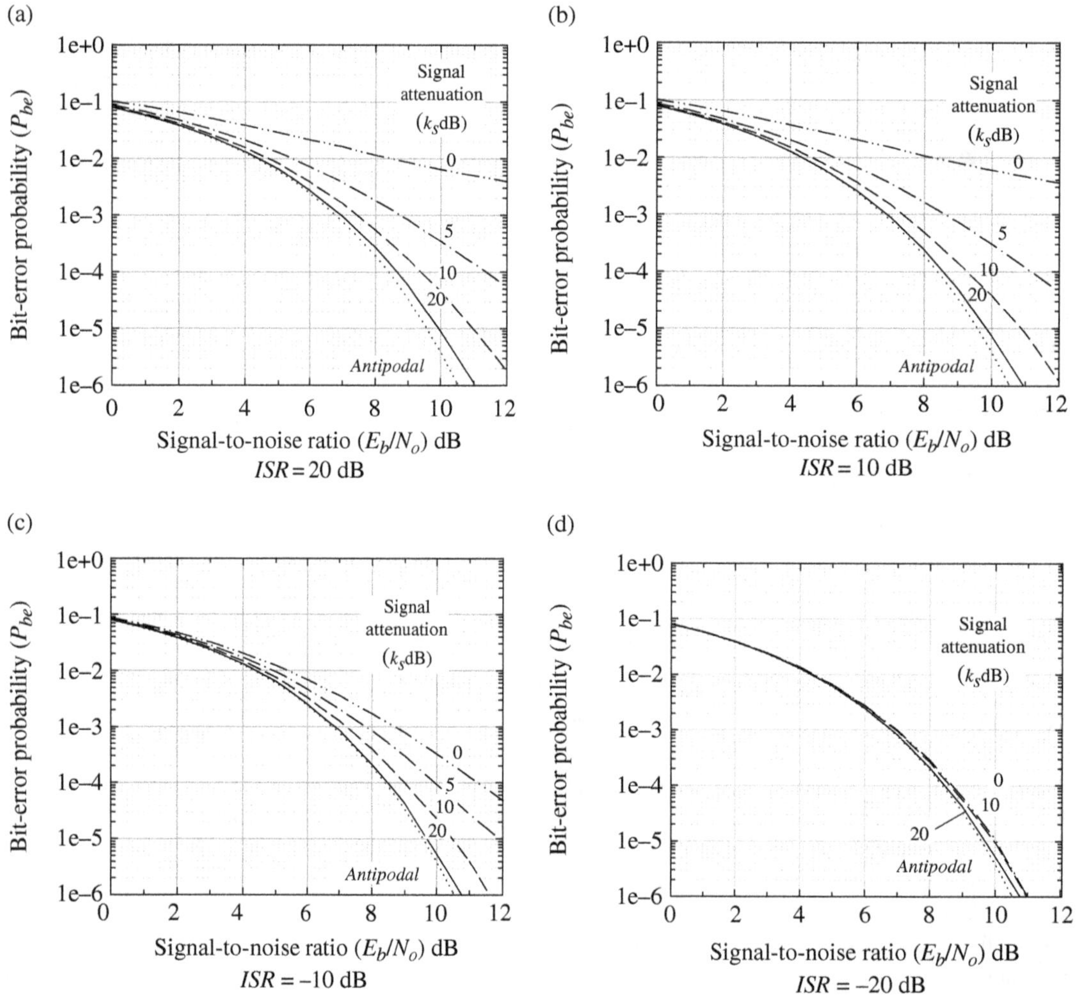

FIGURE 12.17 LMS performance of CW interference cancellation with constant ISR.

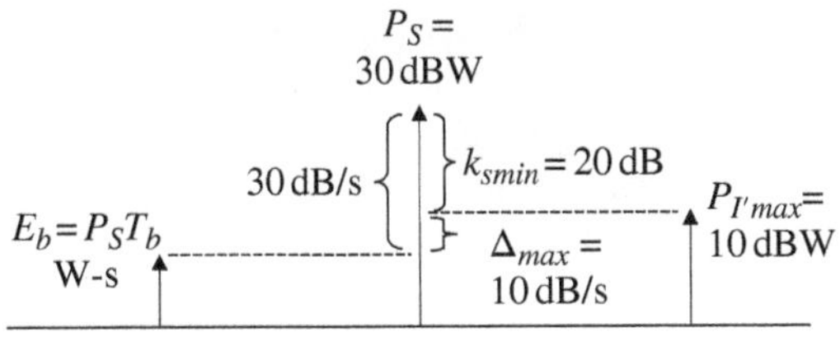

FIGURE 12.18 Interpretation of k_s in terms of received signal energy.

consistent with the above objective. These performance results are a direct consequence of the user bit-rate of $R_b =$ 1 kbps corresponding to the signal carrier power (P_s) being 30 dB-Hz above the energy-per-bit as expressed in (12.60).

The interpretation of these results is shown in Figure 12.18 with the signal carrier power $S = P_s = 30$ dBW. Referring to Figure 12.12 and using $k_s' = 1/k_s$ the canceler reference input is $I_{in} = I + I'$ where $I' = S/k_s$ ($P_{I'} = P_s/k_s$) is the interference input power resulting from the signal. Therefore, the signal-related interference power is k_s(dB) below the signal power and, from Figure 12.17, $k_s \geq 20$ dB so $k_{smin} = 20$ dB corresponding to $P_{I'max} = 10$ dB. Consequently, $P_{I'max}$ cannot exceed the maximum margin of $\Delta_{max} = 10$ dB/s relative to E_b, otherwise, the presence of the signal in the canceler reference degrades the bit-error probability. Noting that R_b and Δ have units of 1/s, these results can be generalized for BPSK modulation by the requirement $k_{smin} = R_b/\Delta_{max} = R_b/10$.

12.8.1 Theoretical Canceler Convergence Evaluation

This case study is concluded by examining the theoretical tap weight convergence based on the characteristic equation of the canceler. The analysis is based on the minimum shift keying (MSK) symbol matched filer expressed as

$$h(t) = \begin{cases} \cos(\omega_m t) & : |t| \leq \dfrac{T}{2} \\ 0 & : o.w. \end{cases} \tag{12.63}$$

where $\omega_m = 2\pi f_m$ with $f_m = 1/2T$ where T is symbol duration. The analysis considers the input as a CW interfering signal given by

$$\tilde{s}(t) = V_p \cos(\omega_\varepsilon t + \phi) + jV_p \sin(\omega_\varepsilon t + \phi) \tag{12.64}$$

The optimally sampled MSK filter outputs, $\tilde{s}_o(jT_s)$, are input to the canceler as $\tilde{x}_j = \tilde{s}_o(jT_s)$ and the canceler output is expressed as

$$\tilde{y}_j = \tilde{x}_j - \sum_{i=1}^{N} \tilde{w}_{ij}\tilde{x}_{j-i-M} \tag{12.65}$$

where N is the number to FIR filter tap weights $\tilde{w}_{ij}$ and $M = \lfloor N/2 \rfloor$ is the delay of the interfering tone into the canceler. The optimum weight vector $\mathbf{w}_o$ corresponds to the weights that minimize the MSE $E[\tilde{y}_i^+ \tilde{y}_i]$. The weight error is defined as the vector

$$v_j \triangleq w_j - w_o \tag{12.66}$$

Applying the method of steepest-decent, discussed in Section 12.3, the transient response of the weights is formulated in terms of the weight error vector as

$$v_j = (\boldsymbol{I} - \Delta\boldsymbol{R})^j v_o \tag{12.67}$$

where v_o is the initial weight error vector, $\Delta = 2u$ is the canceler gain step-size, $\boldsymbol{I}$ is an $N \times N$ identity matrix, and $\boldsymbol{R}$ is the $N \times N$ correlation matrix, given by

$$\boldsymbol{R} = E\left[x_j^* x_j^T\right] \tag{12.68}$$

The matrix $\boldsymbol{R}$ is Hermitian, so there exists N linearly independent characteristic vectors,* and, upon using the diagonal transformation $\boldsymbol{\Lambda} = \boldsymbol{Q}^{-1}\boldsymbol{R}\boldsymbol{Q}$ where $\boldsymbol{Q}$ is a matrix composed of the N independent characteristic vectors and Λ is a diagonal matrix of characteristic values. Using these results, (12.67) is expressed as

$$v_j = (\boldsymbol{I} - \Delta\boldsymbol{\Lambda})^j v_o \tag{12.69}$$

The solution to (12.69) is based on the trace tr[Λ], defined as the summation of the diagonal elements of Λ, that is,

$$\mathrm{tr}[\boldsymbol{\Lambda}] \triangleq \sum_{i=1}^{N} \lambda_i \tag{12.70}$$

so the normalized solution to (12.69) is

$$\frac{v_j}{v_o} = (\boldsymbol{I} - \Delta \mathrm{tr}[\Lambda])^j \tag{12.71}$$

At this point the analysis is simplified by assuming *real* signals corresponding to ω_ε and φ equal to zero, in which case, $s(t) = V_p$, independent of time, and the optimally sampled matched filter output is given by

$$x_j = \frac{4}{\pi} V_p \tag{12.72}$$

In this example, the corresponding characteristic values are identical and evaluated as

***Characteristic vectors* and *values* are also referred to as *eigenvectors* and *eigenvalues*.

$$\lambda_i = x_j^2 = \frac{4}{\pi} V_p^2 \quad : \forall i \tag{12.73}$$

and (12.71) becomes

$$\frac{v_j}{v_o} = \left(1 - \left(\frac{16}{\pi^2}\right) \Delta N V_p^2\right)^j \tag{12.74}$$

Equation (12.74) is plotted in Figure 12.19 with $V_p = 0.125$ V. The simulated performance, based on (12.72), is indicated by the *circled* data points.

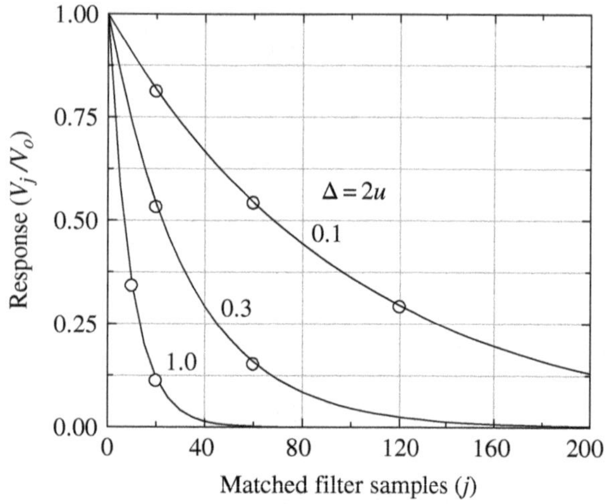

FIGURE 12.19 Theoretical convergence performance of 4-tap CW LMS canceler with simulated data points.

12.9 CASE STUDY: RECURSIVE LEAST SQUARES PROCESSING

The RLS processing is examined using MC computer simulations of the bit-error performance with MSK modulation using 10 M bits for each signal-to-noise ratio. The application involves the acquisition of the received signal when the signal phase and/or amplitude are unknown over the respective ranges $|\phi| \le 180°$ and $|A| \le A_m = \sqrt{2}(=3\text{ dB})$. The signal phase increment is $\Delta\phi = 18°$ with $\phi_i = -180 + \Delta\phi(i-1)$: $i = 1, \ldots, 11$ and the amplitude increment is $\Delta A = 2|A_m|/9$ with $A_i = -A_m + \Delta A(i-1) : i = 1, \ldots, 10$. For example, when both phase and amplitude are unknown, the simulation indexes over 100 K bits for each parameter result in 11 × 10 × 100 K = 11 M bits for each signal-to-noise ratio. At each signal-to-noise ratio, the acquisition processing is re-initiated for each parameter and the detected bit-error counting starts following completion of the training sequence. The training data sequence is generated using the same random process as used for generating the message data, so no attention is given to finding a *good* training sequence. The functions of the simulation processing are shown in Figure 12.20.

Upon completion of the training data sequence, the *acquisition switch* is changed to use either the fixed tap weights or the decision-directed RLS processing during receipt of the message data.

The channel is characterized by a cosine-weighted impulse response expressed as

$$h(nT_s) = 0.5\left[1 + \cos\left(\frac{2\pi\left(n - N_p'\right)}{K_s}\right)\right]$$
$$: n = 1, \ldots, N_p, N_p' = \left\lceil \frac{N_p}{2} \right\rceil \tag{12.75}$$

where T_s is the sampling interval, N_p = odd integer is the number channel impulses, and K_s determines the impulse spread. The characteristics of the four channels considered, ranging in severity from the *worst-case* to *best-case*, are summarized in Table 12.2.

TABLE 12.2 Channel Dispersion Characteristics

	Channel Impulses			
	$N_P = 5$		$N_P = 3$	
n	Ch-1 $k_s = 6$	Ch-2 $k_s = 3.5$	Ch-3 $k_s = 3.5$	Ch-4 $k_s = 2.9$
1	0.25	0.050	0.389	0.219
2	0.75	0.389	1.000	1.000
3	1.00	1.000	0.389	0.219
4	0.75	0.389		
5	0.25	0.050		

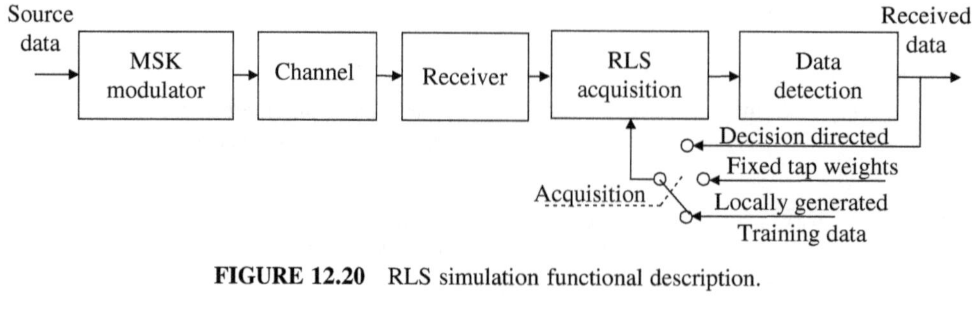

FIGURE 12.20 RLS simulation functional description.

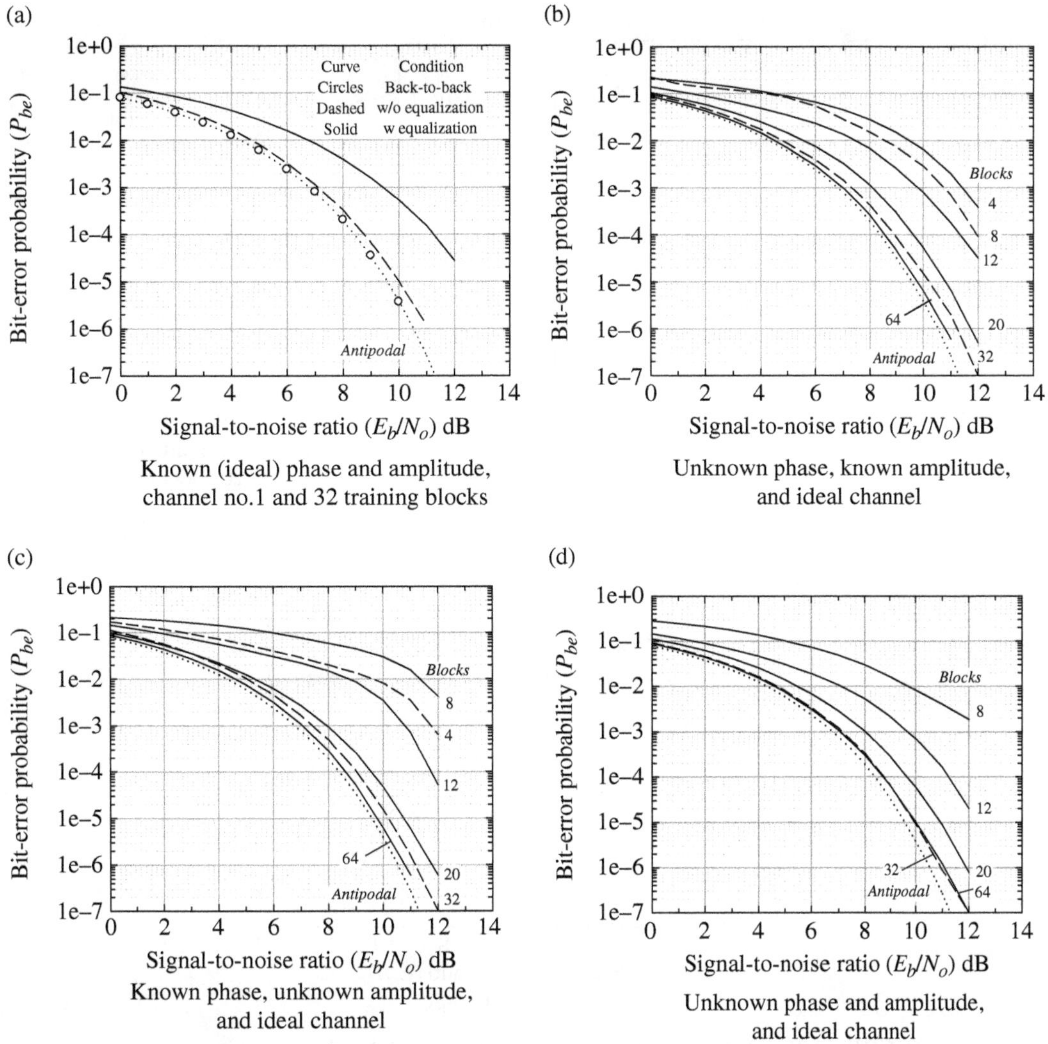

FIGURE 12.21 RLS performance with known and unknown parameters.

The details of RLS processing are described in Section 12.6 and the data detection includes the MSK matched filter detection and ideal tracking following the training sequence in use.

12.9.1 Performance with Fixed Weights Following Training

The bit-error performance results in this section characterize the acquisition and subsequent bit-error evaluation with combinations of known and unknown received signal phase and amplitude as described above. For these results the RLS algorithm uses $\delta = 0.1$ for initialization of the matrix $\boldsymbol{C}$ and $\lambda = 0.999$; this value of λ results a long memory of past events resulting in a narrow recursive bandwidth. Figure 12.21 shows the performance using 4 samples/bit and a block length of 2-bits or 8 samples. The training data ranges in length from 4 to 64 blocks with various combinations of known and unknown parameters. Figure 12.21a corresponds to the ideal conditions with zero signal phase and a fixed carrier amplitude using channel No.1. These known received signal phase and amplitude conditions are ideal so the evaluations in Figure 12.21a characterize the performance of the RLS processing as a channel equalizer. The *circled* data points represent the ideal simulation conditions corresponding to antipodal detection performance and the *dashed* curve represents the performance without equalization. The *solid* curve demonstrates the performance improvement when the RLS equalizer is used. The results in Figure 12.21b–d represent the performance improvements with increasing training blocks under the indicated received signal parameter conditions. In these cases, the ideal channel is used, that is without filtering, so the RLS processing estimates the unknown received signal phase and/or amplitude parameters. A training sequence of 64 blocks results in near optimum performance.

Figure 12.22a–c shows the performance results with unknown received signal phase and amplitude when using channels No. 1 and 3 for the respective training sequence

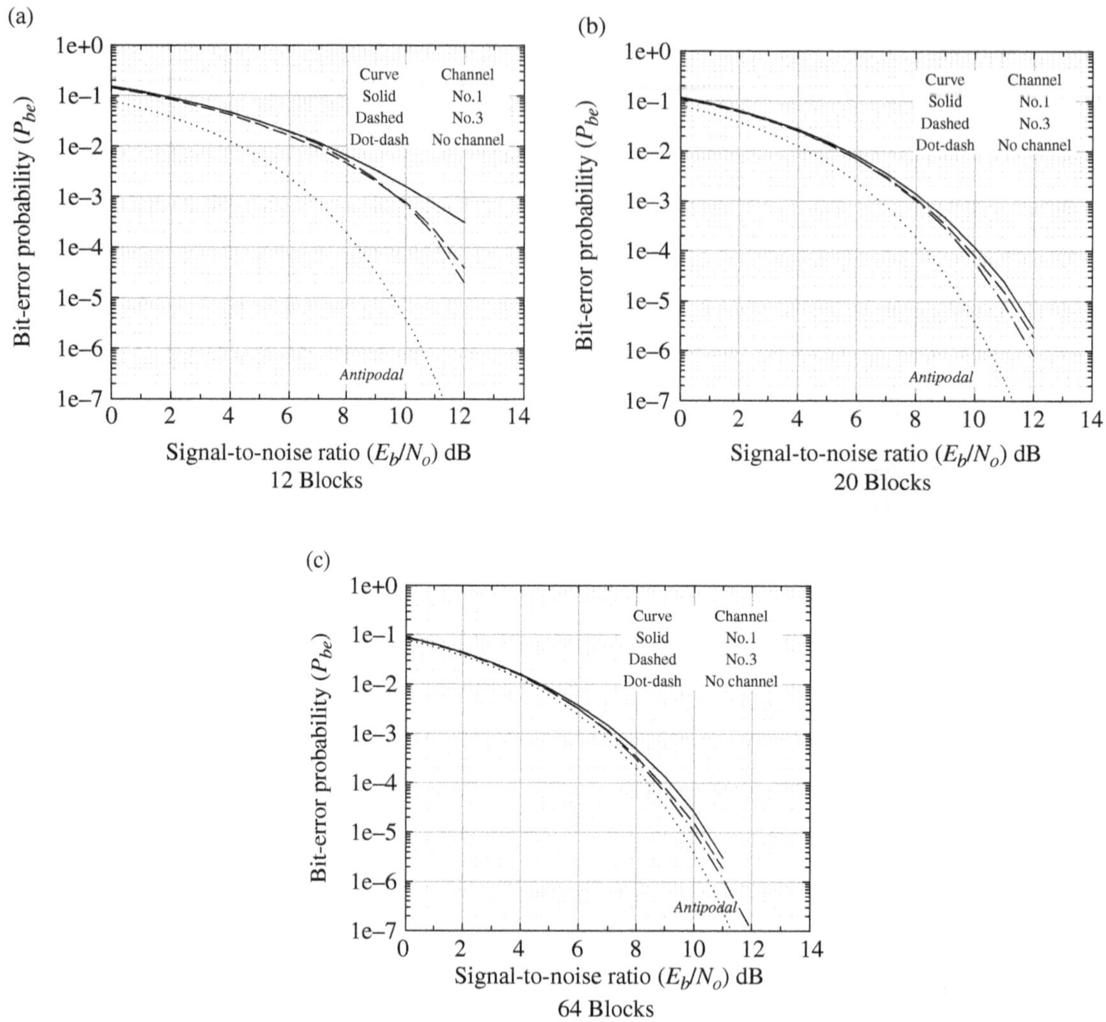

FIGURE 12.22 RLS performance with channels 1 and 2 and unknown parameters.

lengths of 12, 20, and 64 blocks. These results demonstrate the ability of the RLS-adaptive processing to perform equalization, parameter estimation and correction, and acquire a noisy received signal with sufficient fidelity to provide near optimal bit-error performance. However, as mentioned above, these results are based on $\lambda = 0.999$ which severely restricts the dynamic behavior of the input signal.

ACRONYMS

BLP	Backward linear prediction
BPSK	Binary phase shift keying
CW	Continuous wave
DFE	Decision feedback equalizer
DSSS	Direct-sequence spread-spectrum
FBLP	Forward–backward linear prediction
FIR	Finite impulse response (filter)
FLP	Forward linear prediction
FSE	Fractionally-spaced equalizer
I&D	Integrate-and-dump (filter)
IIR	Infinite impulse response (filter)
ISI	Intersymbol interference
ISR	Interference-to-signal ratio
LFFE	Linear feedforward equalizer
LMS	Least mean-square (algorithm)
LMSE	Linear mean-square estimation (algorithm, equalizer)
MC	Monte Carlo (simulation)
MLSE	Maximum-likelihood sequence estimation
MMSE	Minimum mean-square error (algorithm)
MSE	Mean-square error (algorithm)
MSK	Minimum shift keying
RLS	Recursive least-squares (equalizer)
SSE	Symbol-spaced equalizer
TDL	Tapped delay line
ZFE	Zero-forcing equalizer

PROBLEMS

1. Evaluate the MMSE for the optimum filter $h_o(t)$ developed in Section 12.2 using the relationship $\varepsilon_{min} = E[(y-\hat{y})y]$ expressed in (12.6). Express the result in terms of the respective spectrums $S(\omega)$, $S_{\tilde{x}\tilde{x}}(\omega)$, and $S_{\tilde{s}\tilde{x}}(\omega)$. Note that the Fourier transform of $R_{\tilde{s}\tilde{x}}(-\tau)$ is $S_{\tilde{s}\tilde{x}}(-\omega)$.

2. Consider that the optimally sampled matched filter output of an isolated symbol through a bandlimited channel is characterized as

$$g(jT) = d_n c_j \quad : |j| \le 1$$

where c_j is the matched filter sample with $c_0 \gg c_j : j \neq 0$ and $d_n = \pm 1$ is the isolated n-th source data symbol of duration T seconds, such that, $E[d_n d_m] = \delta_{nm}$ where δ_{nm} is the Kronecker delta function.

For the estimator shown in the following figure, determine the open-loop estimates $E_k[\bullet] : k = 1, 2, 3$ considering contiguous source data symbols corresponding to the matched filter samples

$$g(nT) = \sum_{j=-1}^{1} d_{n-j-1} c_j \quad : \forall n$$

Redraw the figure to include an ideal integrator following the appropriate estimators and a transversal filter with adjustable taps that will adaptively eliminate the ISI terms $c_j : j \neq 0$.

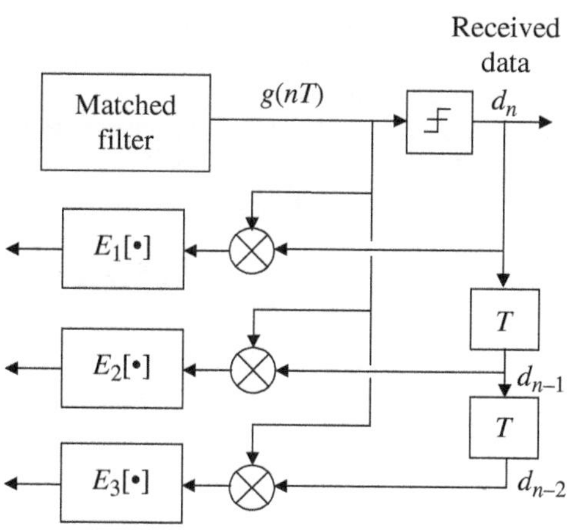

REFERENCES

1. R.W. Lucky, "Techniques for Adaptive Equalization of Digital Communication Systems," Bell Labs Technical Journal, Vol. 45, No. 2, pp. 255–286, February 1966.
2. D.W. Tufts, "Nyquist's Problem—The Joint Optimization of Transmitter and Receiver in Pulse Amplitude Modulation," Proceedings of the IEEE, Vol. 53, pp. 248–259, March 1965.
3. A. Papoulis, *Probability, Random Variables, and Stochastic Processes*, Chapter 11, Linear Mean-Square Estimation, pp. 385–429, Chapter 7, Functions of Two Random Variables, pp. 187–232, McGraw-Hill Book Company, New York, 1965.
4. Y.W. Lee, *Statistical Theory of Communication*, pp. 56–69 and pp. 367–393, John Wiley & Sons, Inc., New York, 1964.
5. S. Haykin, *Adaptive Filter Theory*, pp. 3–121, Prentice-Hall, Englewood Cliffs, NJ, 1986.
6. A. Papoulis, *Probability, Random Variables, and Stochastic Processes*, McGraw-Hill, New York, 1965, Section 11.5.
7. R.D. Gitlin, J.E. Mazo, M.G. Taylor, "On the Design of Gradient Algorithms for Digitally Implemented Adaptive Filters," IEEE Transactions on Circuit Theory, Vol. CT-20, No. 2, pp. 125–136, March 1973.
8. B. Widrow, "Adaptive filters," R.E. Kalman, N. DeClaris, Editors, *Aspects of Network and System Theory*, Holt, Rinehart and Winston, New York, 1970.
9. G. Ungerboeck, "Theory on the Speed of Convergence in Adaptive Equalizers for Digital Communication," IBM Journal of Research and Development, Vol. 16, pp. 546–555, November 1972.
10. R.D. Gitlin, S.B. Weinstein, "On the Required Tap-Weight Precision for Digitally Implemented, Adaptive, Mean-Squared Equalizers," Bell Labs Technical Journal, Vol. 58, No. 2, pp. 301–321, February 1979.
11. S. Haykin, *Adaptive Filter Theory*, pp. 197–218, Prentice-Hall, Englewood Cliffs, NJ, 1986.
12. M. Dentino, J. McCool, B. Widrow, "Adaptive Filtering in the Frequency Domain," Proceedings of the IEEE, Vol. 66, No. 12, pp. 1978–1979, December 1978.
13. N.J. Bershad, P.L. Feintuch, "Analysis of the Frequency Domain Adaptive Filter," Proceedings of the IEEE, Vol. 67, No. 12, pp. 1658–1659, December 1979.
14. B. Widrow, M.E. Hoff, "Adaptive Switching Circuits," IRE WESCON Conference Record, Part 4, pp. 96–104, 1960.
15. B. Widrow, J. McCool, M. Ball, "The Complex LMS Algorithm," Proceedings of the IEEE, Vol. 63, No. 4, pp. 719–720, April 1975.
16. S. Haykin, *Adaptive Filter Theory*, pp. 236–237, Prentice-Hall, Englewood Cliffs, NJ, 1986.
17. S. Haykin, *Adaptive Filter Theory*, pp. 238–240, Prentice-Hall, Englewood Cliffs, NJ, 1986.
18. B.R. Saltzberg, "Intersymbol Interference Error Bounds with Application to Ideal Bandlimited Signaling," IEEE Transactions on Information Theory, Vol. IT-14, pp. 563–568, July 1968.
19. E.Y. Ho, Y.S. Yeh, "A New Approach for Evaluating the Error Probability in the Presence of Intersymbol Interference and Additive Gaussian Noise," Bell Labs Technical Journal, Vol. 49, No. 9, pp. 2249–2265, November 1970.
20. F.E. Glave, "An Upper Bound on the Probability of Error Due to Intersymbol Interference for Correlated Digital Signals," IEEE Transactions on Information Theory, Vol. IT-18, pp. 356–362, May 1972.
21. O. Shimbo, M. Celebiler, "The Probability of Error Due to Intersymbol Interference and Gaussian Noise in Digital Communication Systems," IEEE Transactions on Communication Technology," Vol. COM-19, pp. 113–119, April 1971.
22. R. Lugannani, "Intersymbol Interference and Probability of Error in Digital Systems," IEEE Transactions on Information Theory, Vol. IT-15, pp. 682–688, November 1969.
23. R.W. Lucky, J. Salz, E.J. Weldon, Jr., *Principles of Data Communication*, McGraw-Hill, New York, 1968.

24. R.W. Lucky, J. Salz, E.J. Weldon, Jr., *Principles of Data Communication*, pp. 65–72, McGraw-Hill, New York, 1968.
25. R.W. Lucky, "Automatic Equalization for Digital Communication Systems," Bell Labs Technical Journal, Vol. 44, pp. 548–588, April 1965.
26. M.E. Austin, "Decision-Feedback Equalization for Digital Communication over Dispersive Channels," Technical Report No. 437, M.I.T. Lincoln Laboratory, Cambridge, MA, August 1967.
27. J. Salz, "Optimum Mean Square Decision Feedback Equalization," Bell Labs Technical Journal, Vol. 52, pp. 1341–1373, October 1973.
28. S. Haykin, *Adaptive Filter Theory*, pp. 234–376, Prentice-Hall, Englewood Cliffs, NJ, 1986.
29. C.A. Belfiore, J.H. Park, Jr., "Decision Feedback Equalization," Proceedings of the IEEE, Vol. 67, No. 8, pp. 1143–1156, August 1979.
30. J.G. Proakis, *Digital Communications*, McGraw-Hill, New York, 1989.
31. R.L. Bogusch, F.W. Guigliano, D.L. Knepp, "Frequency-Selective Scintillation Effects and Decision Feedback Equalization in High Data-Rate Satellite Links," Proceedings of the IEEE, Vol. 71, No. 6, pp. 754–767, June 1983.
32. W.U. Lee, F.S. Hill, Jr., "A Maximum-Likelihood Sequence Estimator with Decision–Feedback Equalization," IEEE Transactions on Communications, Vol. COM-25, No. 9, pp. 971–979, September 1977.
33. J.G. Proakis, *Digital Communications*, pp. 585–587, McGraw-Hill, New York, 1989.
34. R.D. Gitlin, S.B. Weinstein, "Fractionally-Spaced Equalization: An Improved Digital Transversal Equalizer," Bell Labs Technical Journal, Vol. 60, No. 2, pp. 275–296, February 1981.
35. S.U.H. Qureshi, G.D. Forney, Jr., "Performance and Properties of a *T*/2 Equalizer," National Telecommunications Conference Record, pp. 11.1.1–11.2.14, Los Angeles, CA, December 1977.
36. G. Ungerboeck, "Fractional Tap-Spacing Equalizer and Consequences for Clock Recovery in Data Modems," IEEE Transactions on Communications, Vol. COM-24, No. 8, pp. 856–864, August 1976.
37. D.D. Falconer, "Jointly Adaptive Equalization and Carrier Recovery in Two-Dimensional Communication Systems," Bell Labs Technical Journal, Vol. 55, No. 3, pp. 317–334, March 1976.
38. A. Benveniste, M. Goursat, "Blind Equalizers," IEEE Transactions on Communications, Vol. COM-32, No. 8, pp. 871–883, August 1984.
39. G. Picchi, G. Prati, "Blind Equalization and Carrier Recovery Using a 'Stop-and-Go' Decision-Directed Algorithm," IEEE Transactions on Communications, Vol. COM-35, No. 9, pp. 877–887, September 1987.
40. G.J. Foschini, "Equalizing Without Altering or Detecting Data," Bell Labs Technical Journal, Vol. 64, No. 8, pp. 1885–1911, October 1985.
41. D.N. Godard, "Self-Recovering Equalization and Carrier Tracking in Two-Dimensional Data Communications Systems," IEEE Transactions on Communications, Vol. COM-28, No. 11, pp. 1867–1875, November 1980.
42. B. Widrow, J.R. Glover, Jr., J.M. McCool, J. Kaunitz, C.S. Williams, R.H. Hearn, J.R. Zeidler, E. Dong, Jr., R.C. Goodlin, "Adaptive Noise Canceling: Principles and Applications," Proceedings of the IEEE, Vol. 63, No. 12, pp. 1692–1716, December 1975.
43. T. Vaidis, C.L. Weber, "Block Adaptive Techniques for Channel Identification and Data Demodulation Over Band-Limited Channels," IEEE Transactions on Communications, Vol. 46, No. 2, pp. 232–243, February 1998.
44. S. Haykin, *Adaptive Filter Theory*, Chapter 8, Prentice-Hall, Englewood Cliffs, NJ, 1986.

13

SPREAD-SPECTRUM COMMUNICATIONS

13.1 INTRODUCTION

The original motivation for spread-spectrum (SS) communications [1–8] was to mitigate the effect of electronic counter measures (ECMs), generally referred to as jamming signals intended to disrupt the flow of information between cooperating terminals. Solutions to the ECM problem were so widespread that the applications of SS to communication systems became known as anti-jam (AJ) communications [9]. The more general benefit is that SS techniques also provides for more robust communications under a variety of conditions including unintentional interference resulting from the increased demand on the available frequency spectrum. The key to successful SS communication systems is the design of the transmitted waveform and the demodulator matched filter. In effect, the relatively narrow information bandwidth is spread over a much wider bandwidth and transmitted. To provide an AJ capability, the transmitted waveform is spread in bandwidth using a unique code or signaling technique known only by the intended receivers. The receiver detection processing correlates or matched-filters the desired waveform while decorrelating the jammer or interfering signals at the detector output to provide reliable detection of the information. Regardless of whether the jamming signal is designed to concentrate the power in a narrow portion of the SS bandwidth or over the entire bandwidth, the jammer transmitted energy must be inordinately high to retain the ECM effectiveness. The notion of SS also has its roots in the design of radar signals referred to as pulse compression (PC) waveforms. The principles of PC waveform design are identical to those of SS waveform designs; however, in radar the emphasis is placed on the transmission of wide-bandwidth low average power signals of duration T_{pri} that, when matched filtered, retain the same detection capability and range resolution as radars that transmit short duration high peak-power pulses with a pulse repetition interval (PRI) of T_{pri} seconds.

Another application of SS communications involves covert communication between cooperating terminals referred to as low probability of intercept (LPI) communications. In these cases, the requirement of the adversary's ECM equipment is not to overtly jam but to detect and exploit the communications. In other words, instead of directing a radiating source at the communication receiver, the LPI ECM threat directs a sensitive receiver toward the communications transmitter. The definition of intercept includes signal detection and the ability to obtain information from the signal that can be used to disrupt or exploit the communications network.

Therefore, LPI communication requirements impose more stringent system requirements on the waveform design than does AJ. For example, the transmitted power is intentionally kept as low as possible requiring power control and possibly additional spectrum spreading. LPI includes the additional system specifications involving: low probability of detection (LPD) and low probability of exploitation (LPE). LPD techniques are used to prevent the adversary from establishing the existence of communication traffic, that is, detection of the signal, that may be used to: obtain knowledge about the carrier frequency and waveform that would allow exploitation or

Digital Communications with Emphasis on Data Modems: Theory, Analysis, Design, Simulation, Testing, and Applications,
First Edition. Richard W. Middlestead.

Companion website: www.wiley.com/go/digitalcommunications

better use of jamming capabilities. LPE techniques are directed toward denying the adversary the ability to determine cryptographic security, deception, decoys, and any operational procedures which can be used to deny the enemy knowledge of the system, its signal design, its use, and its users. For example, the communication terminal locations, the information flow that may lead to command and control identification, and strategic and tactical information that may lead to spoofing by injecting false information.

LPI communication systems may also require cryptographic message or information security (INFOSEC) involving: communication security (COMSEC) and transmission security (TRANSEC). COMSEC is provided by encrypting the baseband data using an authorized key generator that prevents an adversary from learning the message content and TRANSEC uses an approved code for scrambling the transmitted signal to ensure, for example, that the frequency hopping or message interleaving is not repetitive and detectable. COMSEC and TRANSEC increase the waveform acquisition time and complexity.

The application of SS techniques is a natural fit for code division multiple access (CDMA) communications [10], where a number of users share the same bandwidth in a cooperative manner. The various users cooperate to the extent that each uses a different spreading code selected from a set of orthogonal or uncorrelated codes. For the most part, these designs focus on the use of pseudo-noise (PN) sequences that exhibit orthogonally and low correlation sidelobes such as Gold codes [11] and Kasami codes [12]. As with AJ waveforms, the PN codes spread the information bits over a wide instantaneous bandwidth. To communicate with a particular user the receiver demodulates the waveform corresponding to the selected PN code while the other user waveforms occupying the same bandwidth appear as additive low-power noise sources. As the number of users increases the SS bandwidth must also increase to maintain the specified demodulator performance. CDMA took on a life of its own with the advance of commercial digital communications and cellular phones in the 1980s.

Normally, the application of SS techniques focuses on waveforms with very large time-bandwidth products T_bW where T_b and W represent the respective user's bit-duration and instantaneous bandwidth. Typical T_bW products range from 100 : 1 to 1,000,000 : 1 (20–60 dB). However, less demanding applications, requiring time-bandwidth products on the order of 10 : 1 (10 dB), provide low power-density transmissions to reduce spatial and adjacent channel interference. The application of SS using a PN code will reduce the power density by an amount equal to the time-bandwidth product.

There are three basic forms of SS waveform designs: direct sequence (DS), frequency hopping (FH), and time hopping (TH). There are also a variety of hybrid combinations [13] of these forms that offer additional performance advantages at the expense of processing complexity. The hybrid forms include: DS-FH, DS-TH, FH-TH, and DS-LFM. The linear frequency modulation (LFM) is analogous to the *chirp* radar PC waveform. The FH is typically generated using a digital frequency synthesizer (DFS) that can provide wider bandwidth spreading than with the direct-sequence spread-spectrum (DSSS) waveform. The various forms of SS waveforms are all characterized by an underlying PN random process that is known only by the intended communication receivers—this is a fundamental requirement for the waveform acquisition and subsequent data demodulation.

As discussed above, the time-bandwidth product of the SS waveform is a measure of the AJ or LPI system performance. The time-bandwidth product is also referred to as the processing gain of a SS waveform defined as the ratio of the bandwidth W of the SS waveform to the communicators information data rate R_b prior to application of the bandwidth spreading code, that is,

$$G_p \triangleq \frac{W}{R_b} \tag{13.1}$$

The bandwidth W is the bandwidth used to communicate an information bit in T_b seconds. The selection of the bandwidth W_j by the jammer is based on the jamming game and energy allocation strategies of the jammer and communicator [14]. These strategies are dependent on the communicators' selected waveform modulation. For example, the optimum communications bandwidth for a SS PN chip-rate* of R_c, corresponding to N_c chips-per-user bit T_b, is to select $W = R_c$. So, with $R_b = 1/T_b$, the processing gain is $G_p = R_c/R_b = T_bW = N_c$. Therefore, for this DSSS example, the processing gain can be expressed in terms of the output-to-input signal-to-noise ratios of a SS receiver, that is, $G_p = \gamma_o/\gamma_i$. To demonstrate this, consider the signal to noise-plus-jammer ratio (SNJR) at the SS receiver input given by

$$\gamma_i = \frac{P_s}{N + P_J} \tag{13.2}$$

where P_s and P_J are the respective received signal and jammer signal powers and N is the receiver noise power measured in the bandwidth $W = R_c$. The noise is considered to be white Gaussian noise with power density N_o watts/hertz. For simplicity, it is also assumed that the jammer signal is white Gaussian noise with power spectral density N_{oJ} watts/hertz. Substituting these results into (13.2) gives

$$\gamma_i = \frac{P_s}{(N_o + N_{oj})W} \tag{13.3}$$

*The PN code interval $T_c = 1/R_c$ is referred to as a *chip* in contrast to the *bit* interval of T_b.

Multiplying the numerator and denominator of (13.3) by the user bit-rate R_b, or equivalently by $T_b = 1/R_b$, results in

$$\gamma_i = \frac{P_s T_b}{(N_o + N_{oj}) T_b W} = \frac{\gamma_o}{Gp} \tag{13.4}$$

where γ_o is the SNJR defined by the bandwidth of the data rate at the output of the SS processor. So, for DSSS the processing gain is simply a measure of the improvement in SNJR, that is, $G_p = \gamma_o / \gamma_i$. Simon et al. [15] caution about evaluating the processing gain in this way with frequency-hopping spread-spectrum (FHSS) waveforms in favor of the definition (13.1). However, they also favor using the bit-error degradation as the AJ performance measure for a specified waveform modulation and jamming scenario rather than the processing gain. As mentioned above, the selection of the optimum jammer bandwidth is dependent upon the communication waveform modulation and this is examined for the selected waveforms in subsequent sections.

In the following section, the modulation and demodulation of DSFH, and time-hopping spread-spectrum (THSS) waveforms are discussed. The acquisition of SS waveforms is a time-consuming and processing intense undertaking and is also discussed. In Section 13.3, the scenarios of an intentional jammer and interceptor are described in terms of their respective ranges to the communication receiver and transmitter. This involves all of the elements of the communication range equation discussed in Chapter 15. Section 13.4 examines the characteristics of various communication interceptors. In Sections 13.5 through 13.9, the performance of SS waveforms using a variety of waveform symbol modulations and jammer strategies are discussed. The chapter concludes with the case study of terrestrial jammer link characterized by the *link-standoff ratio* (LSR).

13.2 SPREAD-SPECTRUM WAVEFORMS AND SPECTRUMS

A major, processing intense and time-consuming, issue with SS receivers and demodulators involves the waveform acquisition and synchronization. The three critical pieces of information required to facilitate these processes are: knowledge of the underlying SS waveform modulation, knowledge of the PN random process of the waveform, and knowledge of the epoch time stamp (ETS) relative to the local receiver/demodulator that identifies the starting conditions of the random processes. In addition, knowledge of the direction, range, and frequency of the transmitter will reduce the search time for authorized users.

Since the underlying SS waveform and PN random process are known to the authorized system users, the local epoch time is only known to be within the accuracy of the local clock and the range accuracy to the SS transmitter. The accuracy of the local ETS must be within a fraction of the correlation interval, T_{cor}, that is, $\delta T_{epoch} \ll T_{cor}$. Therefore, the range accuracy must satisfy the condition $\delta R_{range} \leq c\delta T_{epoch}$, where c is the propagation velocity through the media. Likewise, the local clock must maintain an accuracy much less than the reciprocal of the bandwidth of the power spectral density (PSD) of the SS waveform. For example, for DS waveforms, the clock error must satisfy $\delta T_{epoch} \ll T_{cor}$ and, for TH waveforms, the requirement is $\delta T_{clock} \ll T_h$, where T_c and T_h are, respectively, the chip and hop durations. If these requirements are satisfied, the locally stored PN code data are synchronized to the time-of-arrival (TOA) of the received SS waveform to provide an acceptable correlation loss, expressed as

$$L_{cor}(\text{dB}) \cong 10\log_{10}\left(1 - \frac{\delta T_{epoch}}{NT_c} - \frac{\delta T_{clock}}{T_c}\right) \tag{13.5}$$

The use of cesium atomic clocks and global position system (GPS) ranging provide the best capabilities for achieving the accuracies; however, if the estimates result in excessive correlator loss, the local-time base must be advanced by the uncertainty in the TOA with sequential acquisition processing performed in increments of $\delta T_{epoch} + \delta T_{clock}$ until acquisition is declared. In the following sections, the three most basic forms of SS waveforms (DS, FH, and TH) are discussed with example modulator and demodulator implementations.

13.2.1 Direct-Sequence Spread-Spectrum

The DSSS waveform is the simplest to modulate; however, the waveform acquisition is time-consuming and the tracking is processing intensive. The modulator simply multiplies (or exclusive ORs) the user bits by a sequence of randomly generated binary chips of duration $T_c = T_b/N$, so the chip rate is $R_c = NR_b$. The chips are then used to modulate a carrier using a variety of modulation techniques. If binary phase shift keying (BPSK) modulation is used then the transmitted spectrum has the form $sinc(f'T_c)$ where T_c is the transmitted symbol duration and $f' = f - f_c$ is the frequency relative to the carrier f_c. Although, in this case, the instantaneous bandwidth is infinite, the instantaneous noise bandwidth is generally taken to be $B_n = R_c$ so the number of chips/bit is $N = R_c/R_b = B_n/R_b = G_p$. If forward error correction (FEC) coding with code-rate* $r_c < 1$ is used, the code-bit rate is $R_{cb} = R_b/r_c$ corresponding to a decrease in the code-bit interval such that $T_{cb} < T_b$. If the FEC code-rate is a rational number, the processing gain can be selected by using Nr_c chips during each code-bit. With $N = T_b/T_c >> 1$ the parameters

*The FEC code-rate is denoted as r_c and the chip-rate as R_c.

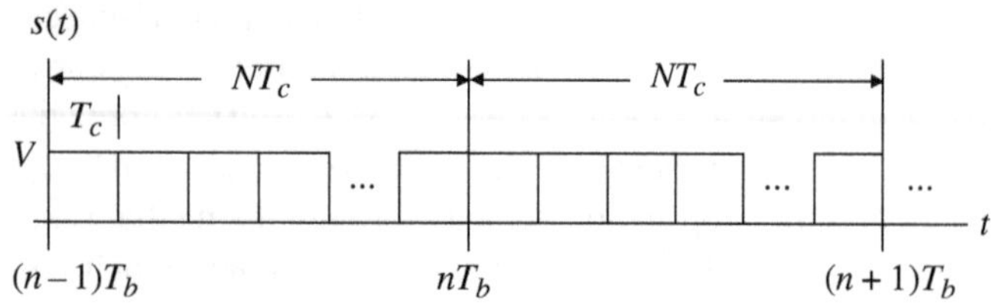

FIGURE 13.1 DSSS baseband waveform characteristics.

TABLE 13.1 Comparison of DSSS Advantages and Disadvantages

Advantage	Disadvantage
Very large bandwidths	Bandwidth limited by power amplifies
sinc(fT_c) frequency spectrum	Difficult to obtain and maintain synchronization
Low peak power density	Processing intensive PN correlators
Accurate range measurement	Susceptible to channel dispersion distortion
Inherently provides LPI	Long acquisition times
Constant amplitude modulations: MPSK, OQPSK, MSK, and MFSK	
Robust in multipath environment	
Minimum effect on narrowband co-channel receivers	
Suitable for multiple access applications (asynchronous)	

N, T_c, T_b, T_{cb}, and r_c, must be selected to result in an integer number of chips in T_{cb} and T_b without altering the FEC code decoding performance in an additive white Gaussian noise (AWGN) channel. This definition of the processing gain applies to the case where the jammer bandwidth is narrow with respect to the chip rate R_c. The DSSS waveform results in the instantaneous signal spectrum occupying the entire spectrum bandwidth providing robust AJ performance. In addition, because the power density is very low, there is an inherent LPI capability. Figure 13.1 shows a typical DSSS waveform with $N = T_b/T_c$ chips-per-bit and Table 13.1 lists some of the advantages and disadvantages of the waveform.

Figure 13.2 shows the implementation of the DSSS symbol modulators for BPSK, quadrature phase shift keying (QPSK), offset QPSK (OQPSK), and minimum shift keying (MSK) waveforms. The sample-and-hold (S/H) functions correspond to respective hold durations of T_c and $2T_c$ for BPSK/QPSK and OQPSK/MSK waveform modulations. These quadrature phase-modulated waveforms have the same E_c/N_o-dependent chip-error performance as BPSK with AWGN. Although bandwidth conservation is not a prime concern in SS applications, these modulations are attractive because the modulation features are less detectable making it more difficult for the jammer. The source data rate is R_b bits-per-second with $d_n = \pm 1$. Depending on the modulation, the source data are mapped to $d_n = dI_n$ and $d_{n-1} = dQ_n$ as indicated in Figure 13.2. The PN generator provides the random chips, at the rate R_c chip-per-second, that are used for the bandwidth spreading. By replacing the C_i chip multipliers in Figure 13.2b with independent inphase and quadrature (I/Q) channel chip multipliers with CI_i and CQ_i, the bandwidth spreading can be thought of as using two independent PN generators on each quadrature rail. Although not depicted, applying the same date d_n on each with quadrature rail with independent PN chip generation is advantageous with BPSK waveform modulation and continuous wave (CW) tone jamming.

The chip shaping function $p(t)$ is a unit-gain *rect*(t/T) function except for MSK, in which case, $p(t)$ is the unit gain cos ($\omega_m t$) function with $\omega_m = \pi R_c/2$. Following the mixing to the carrier frequency ω_c, the power amplifier (PA) produces the transmitted signal with constant voltage level of $\sqrt{2P}$ where P is the carrier power.

To simplify the description of the processing requirements, the following example uses the BPSK DSSS waveform with symbol function $p(t)$ equal to the unit gain *rect* (t/T) function. The data transitions corresponding to d_n occur at $t = nT$ and the PN chip transitions occur at $t = iT_c = iT/N$ corresponding to N chips-per-symbol; the DS chips, denoted as C_i, are referenced to the ETS. The source data (d_n) are also synchronized to the ETS modulo N chips so that an authorized terminal with knowledge of the exact local time relative to the chip ETS can identify the first chip associated the N-chip spreading of the data d_n.

The BPSK transmitted signal is expressed in terms of the lowpass analytic function $\tilde{s}_T(t)$, as

$$s_T(t) = \text{Re}\{\tilde{s}_T(t)e^{j\omega_c t}\} \tag{13.6}$$

where

$$\tilde{s}_T(t) = d_n C_i rect_{T_c}(t,i) rect_{T_b}(t,n) \tag{13.7}$$

To emphasize the fundamental processing requirements, the analysis does not consider channel noise or various sources of signal distortion that only tend to obscure the fundamentals of the signal detection; however, the received signal does contain a channel-induced phase ϕ. Also, to be consistent with the simplicity, the *rect*(x) functions are not carried forward in the following analysis. Following the signal acquisition processing, the channel phase is estimated and removed through the baseband phaselock loop (PLL) acquisition and tracking resulting in the phase error $\phi_\varepsilon = \phi - \hat{\phi}$ where the phase estimate is $\hat{\phi}$. Therefore, in the following

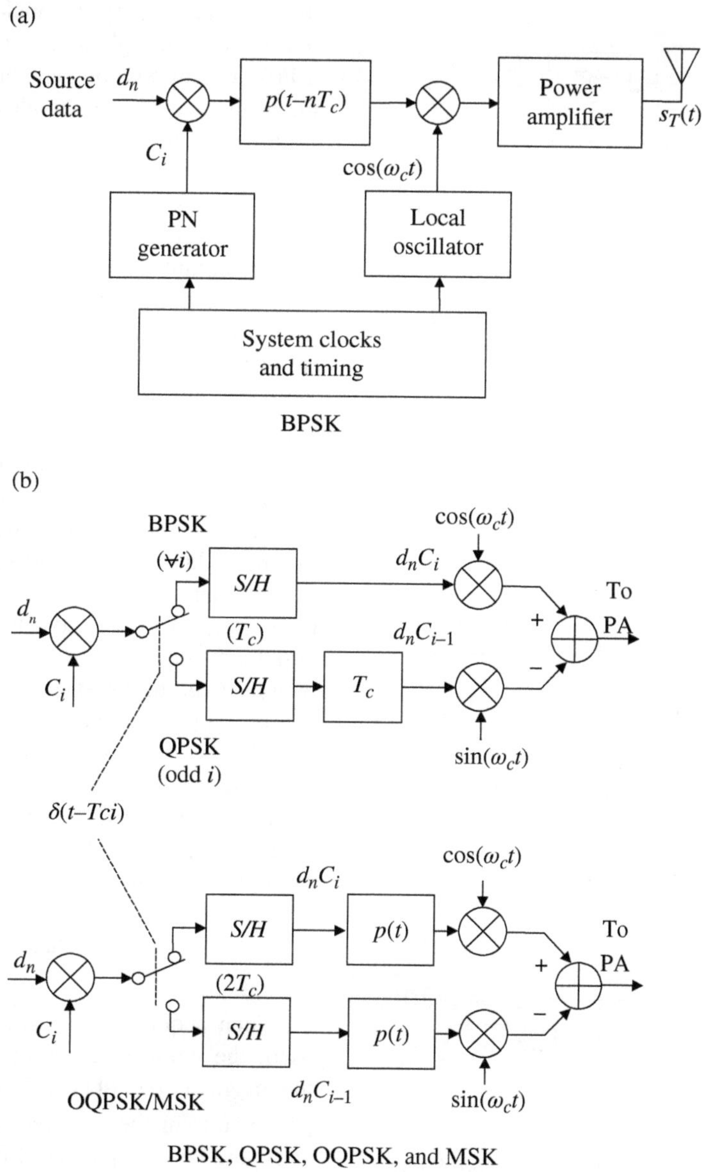

FIGURE 13.2 DSSS modulator implementations.

description, the phase error is equal to ϕ during the acquisition processing and results in a zero-mean phase function after phase-lock is achieved. Consequently, noncoherent processing must be used during acquisition. With these caveats, the received baseband BPSK-modulated sampled signal is described in terms of the sampling index i, with $t = i\Delta t$, as

$$\begin{aligned}\tilde{s}_r(i) &= \tilde{s}_T(i)e^{j\phi_\varepsilon} \\ &= d_nC_i\cos(\phi_\varepsilon) + jd_nC_i\sin(\phi_\varepsilon)\end{aligned} \tag{13.8}$$

The correlator-stored reference PN chips are denoted as

$$s_{ref}(i) = C_i \quad : 1 \le i \le N \tag{13.9}$$

Using (13.8) and (13.9), the PN chip correlator output for a correlation lag $\tau = \ell\Delta t$, with $t - \tau = (i-\ell)\Delta t$, is evaluated as

$$\begin{aligned}R_s(\ell) &= \sum_{i=1}^{N}\tilde{s}_T(i)s_{ref}(i-\ell) \\ &= d_n\sum_{i=1}^{N}C_iC_{i-\ell}\cos(\phi_\varepsilon) + jd_n\sum_{i=1}^{N}C_iC_{i-\ell}\sin(\phi_\varepsilon)\end{aligned} \tag{13.10}$$

Recognizing that the summation over i for $\sum C_iC_{i-\ell} = \delta \ll N$ and $\sum C_iC_i = N$, (13.10) is evaluated as

$$R_s(\ell) = \begin{cases}\delta d_n\cos(\phi_\varepsilon) + j\delta d_n\sin(\phi_\varepsilon) & : \ell \ne 0 \\ Nd_n\cos(\phi_\varepsilon) + jNd_n\sin(\phi_\varepsilon) & : \ell = 0\end{cases} \tag{13.11}$$

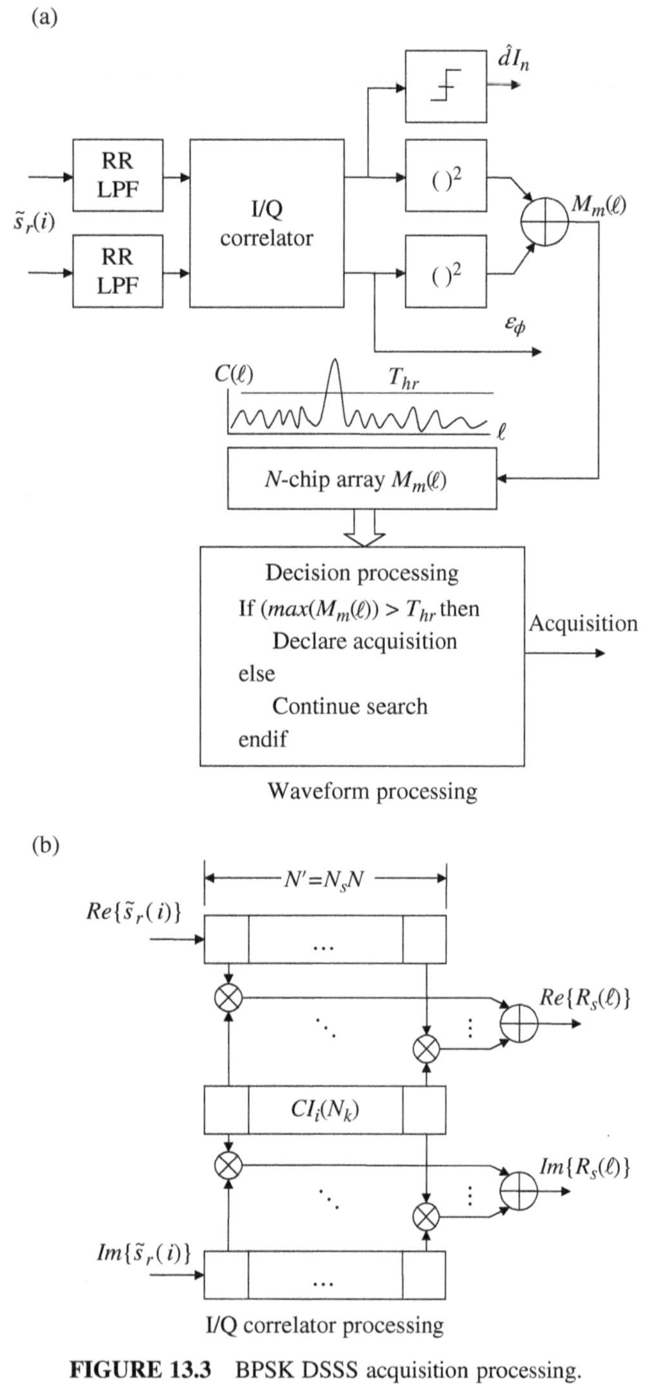

FIGURE 13.3 BPSK DSSS acquisition processing.

The zero-lag correlation $R_s(0)$ corresponds to the desired response. The imaginary part of (13.11) is used as a discriminator for phase tracking and the real part is used as the received data estimate. The phase-lock condition results in zero phase error corresponding to

$$R_s(0) = Nd_n \quad \text{: zero-lag phase-locked condition} \tag{13.12}$$

The demodulator processing represented by this BPSK example is shown in Figure 13.3. The baseband received signal is sampled by the analog-to-digital converter (ADC) and passed through a rate reduction (RR) lowpass filter (LPF) yielding two or four samples-per-chip. During acquisition, the use of four samples-per-chip ($N_s = 4$) reduces the scalloping loss of the PN correlator; however, after acquisition is declared, one-half chip early–late (E/L) gate correlation samples are used to generate the discriminator response for delaylock loop (DLL) chip tracking [16, 17].

The complex signal correlator details are shown in Figure 13.3b with the I/Q samples of the baseband signal shifted through the indicated $N_s N$ sample registers. The N-chip PN sequence $CI_i(N_k)$, stored in the registers, corresponds to the number of chip intervals, N_{chips}, between the ETS and the local time corresponding to $N_k = \text{ceiling}(N_{chips}/N) + 1$. The reference chips are stored as $CI_i = \{\pm 1\}$ so the multiplication simply involves changing the sign of the real and imaginary signal samples. Returning to the description in Figure 13.3a, because the received signal phase* is unknown, noncoherent detection is used by forming the squared magnitude, $M_m(\ell)$, of the correlator samples and shifting them into the N-chip† storage register. If the initial frequency error is unresolved to the extent that the correlation loss over the entire N-chips is undetectable then multiple frequency hypotheses correlations [18] must be performed.‡ The decision threshold, T_{hr}, is computed using the two-parameter, censored constant false-alarm rate (CFAR) and acquisition is declared at the chip ℓ_{max} corresponding to the largest $M_m(\ell)$ that exceeds the threshold; this is expressed as§

$$\ell_m = \max_{\ell}{}^{-1}(M_m(\ell) > T_{hr}) \tag{13.13}$$

When acquisition is declared and coherent detection is used, the phase discriminator error ε_ϕ is established from the imaginary part of (13.11) and the bipolar hard-limited real part results in the received data estimates $\hat{d}_n \in \{\pm 1\}$. The early and late samples used for chip tracking are

$$M_e = M_m(\ell_m + \Delta_s) \tag{13.14}$$

and

$$M_\ell = M_m(\ell_m - \Delta_s) \tag{13.15}$$

where $\Delta_s \le N_s$; $\Delta_s = N_s/2$ is referred to as a *one-delta* loop.

*In general, the phase function is a time-varying function involving the dynamics of the encounter.

†Since the $M_m(\ell)$ are generated sequentially in time, it is only necessary to store enough samples to form a reliable two-parameter censored CFAR threshold.

‡Refer to the discussed in Section 11.3.2.

§This notation picks ℓ_m corresponding to the location of the largest $M_m(\ell)$ exceeding the threshold.

13.2.1.1 DSSS Acquisition In the preceding description, the correlator squared-magnitudes are denoted as $M_m(\ell)$ that suggests more than one correlator response may be involved. Each N-chip correlator response is referred to a *dwell* and the acquisition algorithm may involve *single-dwell* ($m = 1$) or *multiple-dwells* ($m = 1,\ldots,M$) detections.

The *single-dwell* acquisition algorithm [19, 20] simply relies on one correlation interval to declare a signal acquisition and typically requires a suitably high signal-to-noise ratio and a nonfading channel. In this case, if a type II error or false detection occurs the acquisition will fail over the entire remaining search window before beginning a new search. If, however, a correct detection is declared the $C_i(N_{k+1})$ PN code is stored in the correlator array and the process is repeated leading to a verification with two consecutive detections* or a false detection. The detailed performance analysis of a serial search acquisition algorithm is based on a N_u-state Markov process for which a state-transition diagram is created with transition probabilities associated with each transition [21]. Using this approach, the mean and variance of the acquisition time for a *single-dwell* acquisition algorithm, based on an initial time uncertainty of $T_u = N_uT_c$ seconds with estimates made at $T_c/2$ s intervals† resulting in $2N_u$ estimations or cells, are given by [22]

$$\bar{T}_{acq} = \frac{(2-P_d)(1+KP_{fa})}{P_d} N_u T_{dwell} \tag{13.16}$$

and

$$\sigma^2_{T_{acq}} = (2N_uT_{dwell})^2(1+KP_{fa})^2\left(\frac{1}{12}+\frac{1}{P_d^2}-\frac{1}{P_d}\right) \tag{13.17}$$

where P_d and P_{fa} are the correct acquisition detection and false-alarm probabilities, $T_{dwell} = NT_c$ is the dwell time, and K is the number of dwell intervals consumed as a result of a false alarm. In consideration of the difference in the parameters P_d, P_{fa}, and T_{dwell}, (13.16) and (13.17) also apply for coherent and noncoherent detection.

The false-alarm probability is usually given as a system specification and the corresponding detection probability is evaluated, assuming Gaussian distributed random variables, as [23]

$$P_d = Q\left(\frac{\beta-\sqrt{N_B\gamma}}{\sqrt{1+2\gamma}}\right) \tag{13.18}$$

where $Q(x)$ is the Gaussian probability integral, $N_B \triangleq T_{dwell}/T_s = BT_{dwell}$, $T_s = 1/f_s$ where $f_s = 2B$, B is the noise bandwidth of the predetection LPF, and the signal-to-noise ratio is $\gamma = (A^2/2)/(N_oB)$. The parameter β is related to the false-alarm probability and is expressed as

$$\beta = Q^{-1}(P_{fa}) \tag{13.19}$$

The detection probability in (13.18) considers that the signal is contained in one chip interval or cell; however, with a sampling resolution of one-half chip, the signal could straddle two cells. Therefore, Simon et al. [6] outline a procedure for modifying the calculation of P_d to account for this practical condition; they also outline modifications to account for modulation distortion and the *smearing* effect of frequency and frequency-rate errors.

With *multiple-dwell* processing the N-chip accumulator, shown in Figure 13.3, makes detection decisions each $M_m(\ell) : m = 1,\ldots,M$ or on shorter correlation intervals $M_m(j\ell') : j = 1,\ldots,J; \ell' < \ell, J\ell' = N_sN$. A variety of strategies have been examined for implementing *multiple-dwell* detection systems [24] with a net reduction in the acquisition time relative to *single-dwell* detection systems. For example, using a majority logic or an m-out-of-M decision criterion and variable sub-correlation dwell intervals ℓ' that reject incorrect PN alignments sooner than with single-dwells of duration N_sN [25]; this can be viewed as a sliding-window correlator. In addition to strategies involving the implementation of PN correlation, search strategies that take advantage of nonuniformly distribution ETS over the search window reduces the mean acquisition time. These and other *multiple-dwell* acquisition search strategies are described and analyzed by Simon et al. [26] and Peterson et al. [27].

The acquisition of the maximum chip correlator output, as described above, can be viewed as a coarse chip alignment of the locally stored reference with the chip-modulated received signal. This coarse alignment is within $\pm T_c/2$ of the true alignment and the error is eliminated using the DLL and the E/L outputs as mentioned above. The chip timing error is computed as

$$\varepsilon_t = T_c\left(\frac{M_e - M_\ell}{M_eM_\ell}\right) \quad \text{: DLL chip timing error} \tag{13.20}$$

Furthermore, the demodulator signal phase is acquired and tracked using a PLL with the aid of the initial phase estimate computed, using (13.11), as

$$\hat{\phi}_o = \mathrm{atan}^{-1}\left(\frac{\mathrm{Im}(R_s(\ell_m))}{\mathrm{Re}(R_s(\ell_m))}\right) \quad \text{: initial phase estimate} \tag{13.21}$$

The E/L chip timing error expressed in (13.20) is used as the input to a closed-loop tracking filter configuration; however, the implementation is relatively sensitive to the initial

*A detection may be declared based on m-out-of-M detections.

†The time estimation accuracy of $\leq T_c/2$ is typically required to initialize the DDL for code tracking.

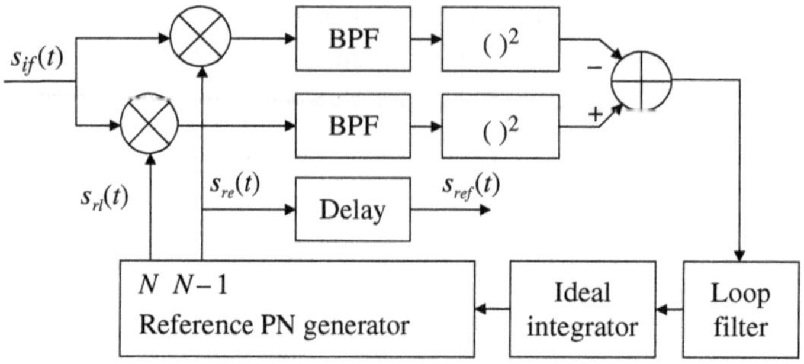

FIGURE 13.4 DSSS DLL chip tracking.

frequency estimate and, if an all-digital implementation is used, the chip sampling rate must be relatively high unless a Farrow interpolation filter is used. The commonly used, and extensively analyzed [28–30], implementation of PN chip tracking in DSSS systems is the DLL shown in Figure 13.4.* The DLL forms an E/L gate reference from the last two chips of the PN reference sequence; these correspond to the zero-lag or peak of the correlation output similar to M_e and M_ℓ formed using (13.14) and (13.15). However, the DLL also tracks the chip timing error $T_\varepsilon = T_d - \hat{T}_d$ between the received signal and the local reference and, ideally, zeros the mean tracking error. Alternate implementations of the DLL are the tau-dither loop [31] (TDL) and the double-dither loop [32] (DDL). The TDL implementation involves alternating the early and late references between successive chips thereby decreasing the processing load and the DDL implementation resolves the E/L channel gain unbalance sensitivity of the DLL and TDL implementations. The DDL averages the E/L gate sample unbalance by alternately associating the I/Q and Q/I samples with the E/L gate loop filter input.

Figure 13.4 depicts the commonly analyzed DLL configuration involving the input signal derived from the final IF stage and expressed as

$$s_{if}(t) = Vm(t-T_d)C(t-T_d)\cos(\omega_o t + \theta(t)) \qquad (13.22)$$

where V is the peak carrier voltage, T_d is the received chip timing delay relative to the demodulator reference PN sequence, $m(t-T_d) = d_n rect(T-T_d, n)$ is the data modulation function, $C(t-T_d) = C_i rect_{T_c}(T-T_d, i)$ is the chip modulation function, ω_o is the IF angular frequency, and $\theta(t)$ is the received signal phase function that includes a constant phase error. The early and late PN references are expressed as

$$s_{re}(t) = C\left(t - \hat{T}_d + \Delta T_c\right) \qquad (13.23)$$

and

$$s_{rl}(t) = C\left(t - \hat{T}_d - \Delta T_c\right) \qquad (13.24)$$

Typically, the E/L gate offsets correspond to† $\Delta = 1/2$ with the early and late gate time difference of one-chip. The delay in Figure 13.4 is ΔT_c and the delay-corrected reference $s_{ref}(t) = s_{re}(t - \Delta T_c)$ is used to recover the coherently detected data $\hat{d}I_n$ depicted in Figure 13.3. The bandpass filters (BPFs) are centered at ω_o with bandwidths much narrower than the chip bandwidth and selected, in consideration of the data and the input signal-to-noise ratio, to minimize the mean-square tracking jitter. The discriminator function formed by the difference in the E/L gate filters, shown in Figure 13.4, results in a linear discriminator error, D_ε, as a function $T_\varepsilon = T_d - \hat{T}_d$, expressed as

$$D_\varepsilon(T_\varepsilon) = 2T_\varepsilon \quad : |T_\varepsilon| \le \frac{T_c}{2} \qquad (13.25)$$

The equality limitation in (13.25) results in a quasi-optimum solution, in that, it results in the maximum linear range with the maximum peak response; smaller values result in a narrower linear range with lower peak values. Larger values tend to become increasingly nonlinear when $|T_\varepsilon| = T_c$ resulting in zero discriminator gain at $T_\varepsilon = 0$.

The bandwidth of the DLL baseband loop filter is also mush less than the chip rate and removes the $2\omega_o$ frequency term and PN self-noise generated by the squaring operations. The acquisition performance of noncoherent DDL and TDL PN code tracking loops is examined by Simon [28] under linear and nonlinear conditions. The loop bandwidth and squaring loss are examined and the performance is characterized in terms of the mean-square tracking jitter as a function of the signal-to-noise ratio measured in the symbol bandwidth.

*Spilker [16] was the first to analyze the DLL and Gill [29] used Spilker's results to compare the performance of several DLL applications. Nielsen [30] contributed to improvements in Spilker's implementation.

†As used here, $0 < \Delta \le 1$ and represents a fraction of T_c; however, many authors refer to Δ as the chip interval, that is, $\Delta = T_c$.

13.2.2 Frequency-Hopping Spread-Spectrum

Spreading the bandwidth of the source data with bandwidth R_b by randomly hopping the carrier frequency over a bandwidth $W_{ss} \gg R_b$ is another way to achieve coding gain in the presence of a jammer. The FH is often applied to $M = 2^k$-ary or multitone frequency shift keying (MFSK)-modulated waveforms with k bits-per-symbol period, that is, $T_s = kT_b$, with the k bits taken from the $M = 2^k$ binary set of symbols. In the following descriptions, the FH interval T_h is defined as the duration of one transmitted frequency tone. There are two regimes of FH, identified as: slow frequency hopping (SFH) and fast frequency hopping (FFH); defined for the MFSK underlying modulation as

$$T_h \geq T_s \quad : \text{slow frequency hopping} \tag{13.26}$$

and

$$T_h < T_s \quad : \text{fast frequency hopping} \tag{13.27}$$

Therefore, SFH applies when a message of MFSK symbols is hopped over the entire SS bandwidth W_{ss}. For example, the SFH strategy may hop a sequence of differentially coherent multiphase shift keying (MPSK) (DCMPSK) modulated symbols. However, FFH occurs when the messages symbols are individually hopped.

The MFSK and DCMPSK waveforms are easily decoded using robust noncoherent detection techniques. For example, with MFSK the demodulator provides a bank of orthogonally spaced symbol matched filters, perhaps generated using the fast Fourier transform (FFT), and makes a maximum-likelihood decision in estimating the correct symbol. For the DCMPSK waveform, the modulator attaches a leading phase reference symbol to a sequence of N information symbols and the demodulator uses the reference symbol to sequentially differentially detect the information symbols. In this case, the SFH interval is $T_h = (N+1)T_s$. Coherent FH acquisition is achieved with slow hopping using a preamble or midamble with known data for estimating signal phase during each hop interval and estimating the frequency using the known data from hop-to-hop.

Figure 13.5 illustrates the time and frequency characteristics of FHSS and Table 13.2 lists some of the advantages and disadvantages of the waveform.

Figure 13.6 shows the functional implementation of the FHSS modulator and transmitter and the corresponding receiver and demodulator. These implementations apply for any waveform modulator and demodulator and the following description is intended to be generic. To this extent the modulator maps the input source data to the unit-power modulation function $M_i(t)$. The modulator output is then mixed to the carrier frequency (f_c) of the local oscillator. The modulator timing is established by the *system clocks and timing* function that also synchronizes the PN code generator and the DFS to the i-th FH interval T_h. The resulting hopping frequency function $2\cos(\omega_{m_i} t)$ is weighted by the symbol function $p(t - iT_h) = rect_{T_h}(t, i)$, mixed with the modulated carrier frequency, and passed through a highpass filter (HPF) yielding the transmitted signal

$$s_T(t) = VM_i(t)\cos((\omega_c + \omega_{m_i})t) \quad : (i-1)T_h \leq t \leq iT_h \tag{13.28}$$

The received signal is characterized by a channel delay and phase function as

$$s_r(t) = VM_i(t - T_d)\cos((\omega_c + \omega_{m_i})(t - T_d) + \theta(t)) \\ : (i-1)T_h \leq t - T_d \leq iT_h \tag{13.29}$$

where $\theta(t)$ contains the channel frequency-related errors and a constant phase error θ_o. The received signal encounters the

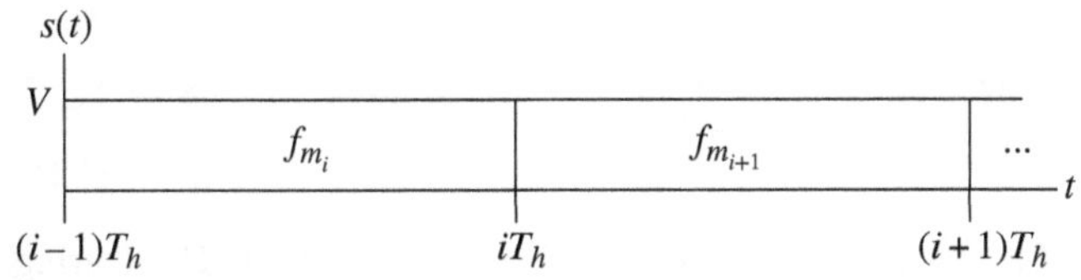

FIGURE 13.5 FH spread-spectrum waveform characteristics ($0 \leq f_{m_i} \leq W_{ss}$, $f_{m_i} = f - f_c$).

TABLE 13.2 Comparison of FHSS Advantages and Disadvantages

Advantage	Disadvantage
Very large bandwidths (limited by synthesizers)	Discrete frequency occupancy
Discontinuous frequencies	Synthesizer settling time overhead
Occupies continuous-time transmissions	High instantaneous power density
More tolerance of channel distortions than DSSS	Susceptible to detection and CW jammers
Constant amplitude MFSK waveform modulations	Requires network timing for multiple access
Robust noncoherent detection	Cost of digital synthesizers
Moderately robust in multipath environment when FEC is used	
Moderate effect on narrow co-channel receivers	
Suitable for multiple access applications	
Relatively easy to obtain and maintain synchronization	

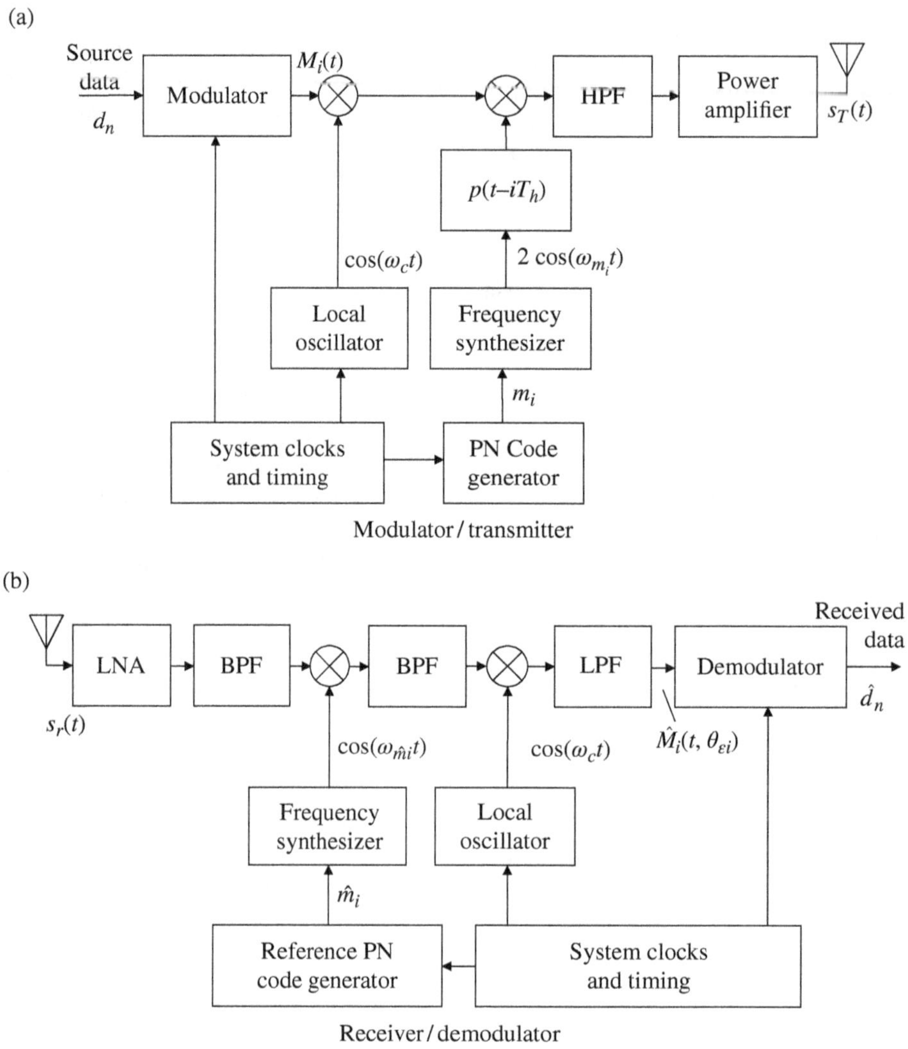

FIGURE 13.6 Functional implementation of FHSS system.

complementary functions in the receiver using the estimates $\hat{m}_i$, $\hat{T}_d$ and $\hat{\theta}(t)$. By virtue of the acquisition and synchronization processing, the timing delay and symbol frequency errors are considered to be negligible. Although, the phase error expressed as

$$\theta_{\varepsilon i} = \omega_{m_i}\left(T_d - \hat{T}_d\right) + \theta_o \tag{13.30}$$

is not negligible so noncoherent detection is used. The output of the demodulator LPF is the reconstructed baseband modulation function $\hat{M}_i(t, \theta_{\varepsilon i})$ from which the demodulated data estimate is $\hat{d}_n$.

13.2.2.1 Slow Frequency Hopping Typically, the frequency spacing between the MFSK symbol tones is chosen to correspond to the minimum orthogonal tone spacing, that is, $\Delta f_{min} = 1/T_s$, so that the spectrum corresponding to one M-ary symbol is defined as $W_s = M\Delta f_{min} = 2^k/T_s$ Hz. During the i-th FH interval, the frequency is selected from the M-ary set

$$f_{m_i} = m_i \Delta f_{min} \quad : m_i \in M \tag{13.31}$$

where $f_{m_i} \in W_s$. Considering that there are 2^L FH intervals, the resulting SS bandwidth is $W_{ss} = 2^L W_s$ and the processing gain is

$$G_p = \frac{W_{ss}}{W_s} = 2^L \quad : \text{slow frequency hopping} \tag{13.32}$$

These relationships are depicted in Figure 13.7 for k bits-per-symbol and 2^L frequency hops over the SS bandwidth W_{ss} Hz.

An alternative slow FH strategy is to randomly hop each of the M-ary tones over the entire spread bandwidth of W_{ss} in

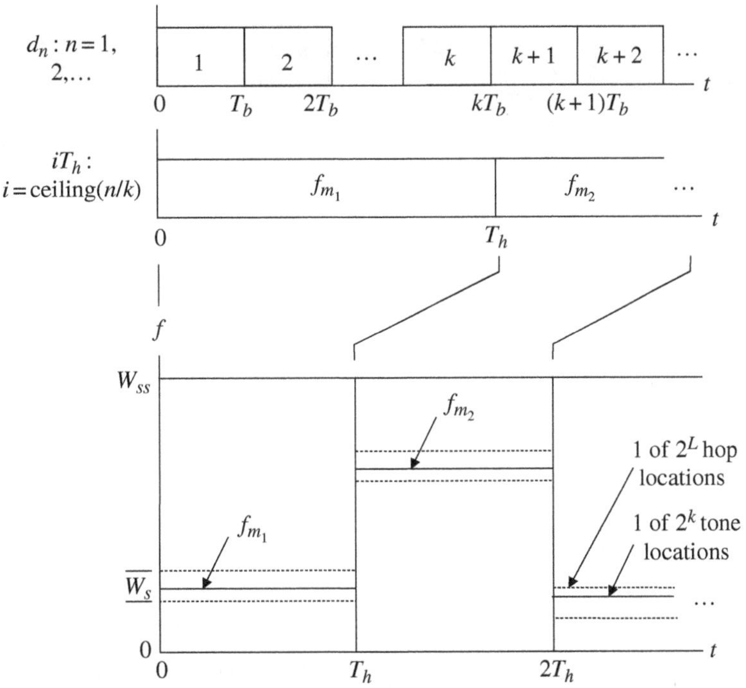

FIGURE 13.7 Data and frequency mapping for MFSK FHSS.

increments of $\Delta f_{min} = 1/T_s$. This results in the instantaneous symbol energy being confined to the narrow bandwidth Δf_{min}. Although this strategy does not alter the SS processing gain, it does provide a performance advantage with partial-band jamming and fading channels when used with FEC and frequency diversity combining (FDC).

13.2.2.2 Fast Frequency Hopping To generate a FFH SS waveform, consider the MFSK example in the previous section corresponding to one k-bit code word in each symbol interval $T_s = kT_b$ and the associated M-ary frequency tone f_{m1} shown in Figure 13.7. Instead of hopping the tones f_{mi}, FFH divides the symbol into Q frequency hopped increments of duration $T_h = T_s/Q$, that are independently hopped over the bandwidth W_{ss} using $m_{i,q} : q = 1, \ldots, Q$ generated by the PN code generator, where i denotes the symbol interval iT_s. In this case, the processing gain is unchanged from the SFH waveform; however, there is a performance improvement with jamming because it is unlikely that all, or even a majority, of the de-hopped symbols will be impacted by the jammer. This concept is similar to repetition coding of DSSS waveforms that results in significant performance improvements against the worst-case pulsed jammers as discussed in Section 13.5.4.

13.2.2.3 Frequency Hopping Acquisition The acquisition of the FHSS waveform is similar to that of DSSS acquisition [33]. For example, the acquisition processing involves two steps: FH acquisition and FH tracking. FH acquisition is declared when the locally generated hopping sequence is synchronized to the received signal sequence to within $T_h/2$ and FH tracking establishes the synchronization to within a fraction of $T_h/2$. The following description of the acquisition processing applies to the serial search single-dwell algorithm discussed in Section 13.2.1.1.

Consider the MFSK FFHSS waveform outlined in Section 13.2.2.2, with the local-time base advanced by the local-time error and range uncertainty relative to the PN spreading code generator ETS. The synchronization processing identifies the FH code estimate $\hat{m}_{i,q}$ in the symbol interval iT_s. Referring to Figure 13.8, the bank of M-ary demodulator integrate-and-dump (I&D) matched filters, separated in frequency by the tone spacing Δf, are used to detect the presence of the DFS down-converted, or de-hopped, received tones. However, prior to the arrival of the transmitted hop code $m_{i,q}$ at the receiver, the down-converter output, in the absence of a jammer, corresponds to the channel and receiver noise. The I&D matched filter sampling corresponds to the hop interval $T_h < T_s$ and the summation refers to summing the Q frequency hops-per-symbol. When the signal is present, the selection of the maximum input corresponds to the symbol energy* $P_s(T_s - T_\varepsilon)$ where $T_e = T_d$ is the unknown signal delay error. During the symbol tracking

*Frequency synthesizer settling time and energy from adjacent de-hopped signals is not included.

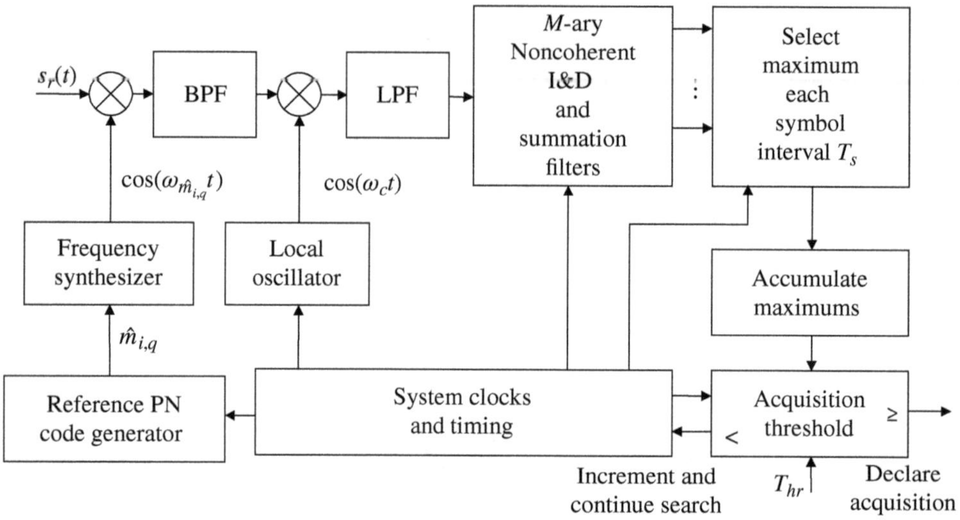

FIGURE 13.8 FHSS waveform acquisition processing.

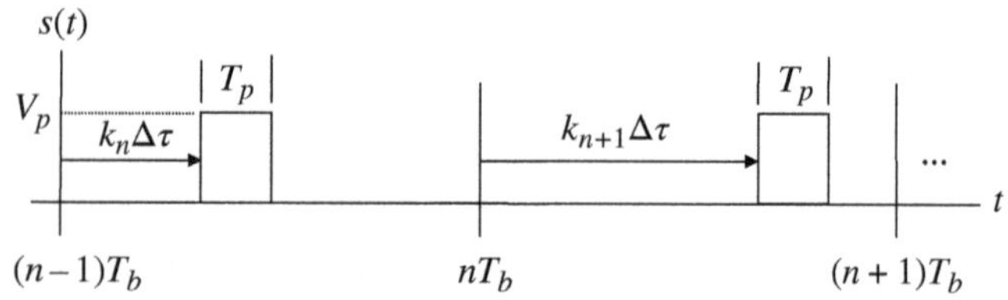

FIGURE 13.9 THSS baseband waveform characteristics.

the delay error is $T_\varepsilon = T_d - \hat{T}_d$ where the demodulator delay estimate is $\hat{T}_d$. Although this description focuses on FFH, the acquisition processing also applies to SFH by simply letting $Q = 1$. The function *accumulate maximums* provides for post-detection integration of the individual hops prior to the symbol acquisition decision based on a threshold T_{hr}.* If acquisition is declared, the FH tracking mode is entered; otherwise, the local epoch time is advanced in preparation for the next sequential acquisition attempt.

13.2.3 Time-Hopping Spread-Spectrum

Figure 13.9 illustrates the time characteristics of the time-hopping spread-spectrum (THSS) waveform and Table 13.3 lists some of the advantages and disadvantages. Relative to the DSSS waveform, the TH peak carrier voltage (V_p) is increased by $G_p = T_b/T_p$ so that the transmitted pulse energy in each bit interval is identical to that of the DSSS waveform. Although there are fundamentally N pulse time-slots associated with each information bit, the random TH increment k_n is based on a resolution of $\Delta\tau$ seconds where

*The detection may also be based on a q-out-of-Q majority count.

TABLE 13.3 Comparison of THSS Advantages and Disadvantages

Advantage	Disadvantage
Occupies large instantaneous bandwidth	Discrete-time transmissions
Moderate effect on narrow co-channel receivers	Requires network timing for multiple access
Suitable for multiple access applications	Difficult to obtain and maintain synchronization
Good near/far performance	Requires high peak power
	Susceptible to detection

k_n is uniformly distributed over the range $0 \le k_n \le N(N_r - 1)$ and $\Delta t = T_b/(N_r N)$, when $N_r = 1$ the time-hopped pulse is placed exactly in one of the N time-slots. Figure 13.10 depicts a functional implementation of the THSS modulator.

13.2.4 Spectral Characteristics of DS, FH, and TH Spread-Spectrum Waveforms

The spectral characteristics of each of the SS waveforms discussed above are shown in Figure 13.11; the spectrum of the unspread waveform is shown in Figure 13.11a for comparison. In these figures, the baseband frequency $f' = f - f_c$ represents the radio frequency (RF) shift from the carrier frequency. Each waveform uses an underlying *rect(t/T)* modulation function and the indicated processing gains. With a common source data rate R_b and $T_p = T_c$, the processing gain is $G_p = R_b T_c = R_b T_p$ and the corresponding *sinc*$(f'T_c)$ and *sinc* $(f'T_p)$ functions for the DSSS and THSS waveforms are identical. Referring to Figure 13.11a, the *sinc*$(f'T_b)$ spectrum of the unspread waveform has a first spectral null that

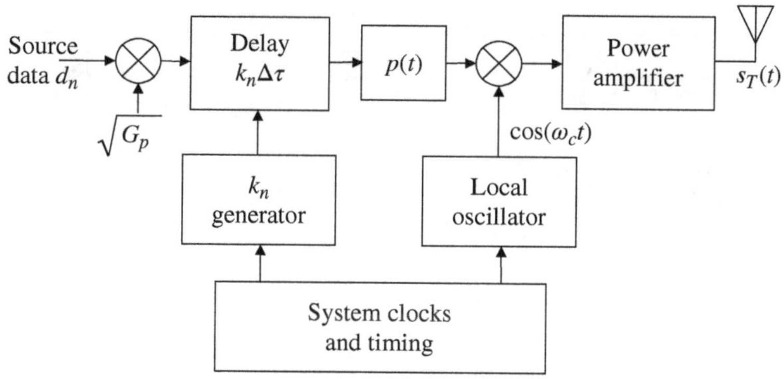

FIGURE 13.10 Functional implementation of THSS modulator.

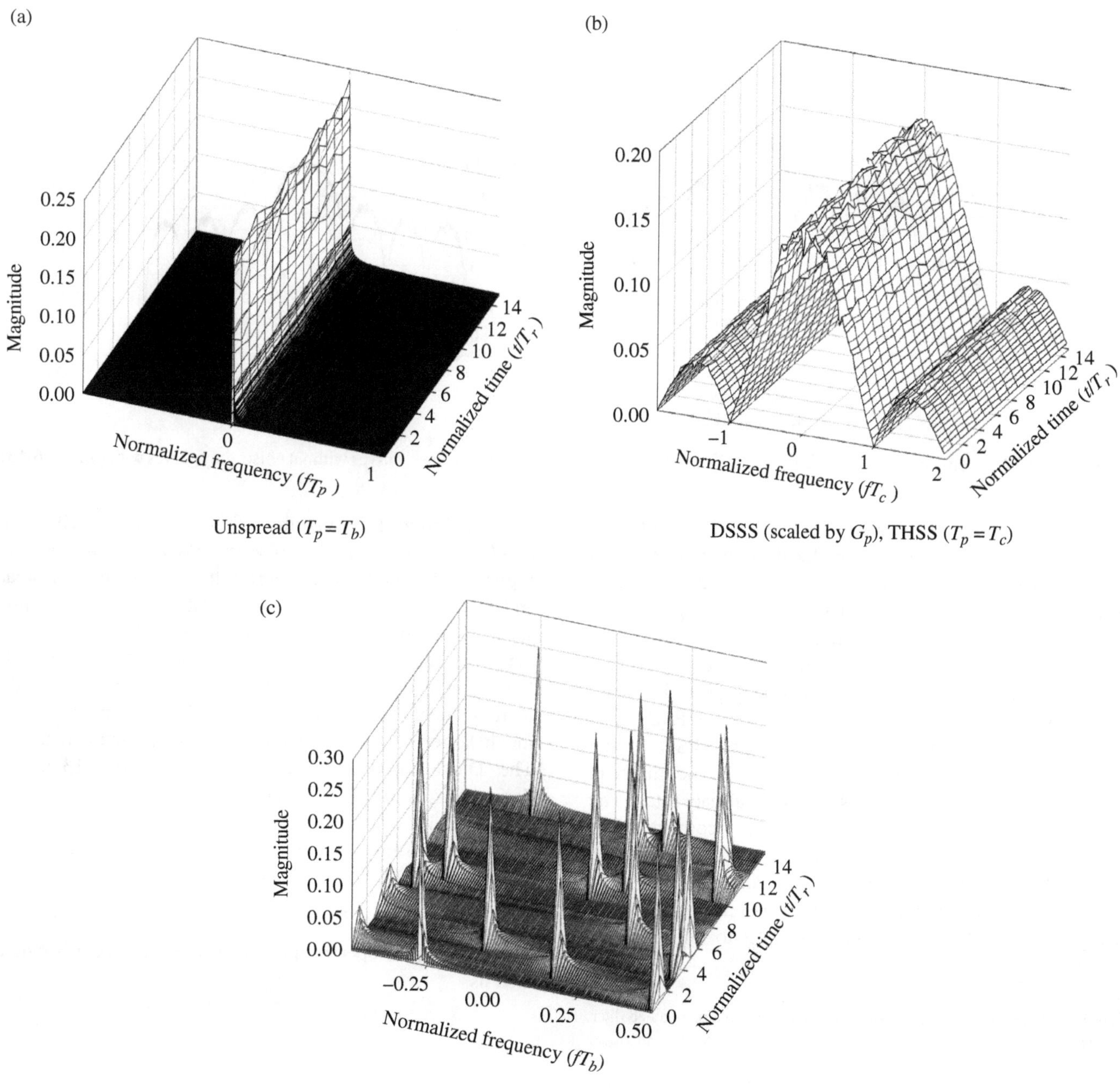

FIGURE 13.11 Illustration of SS spectral characteristics.

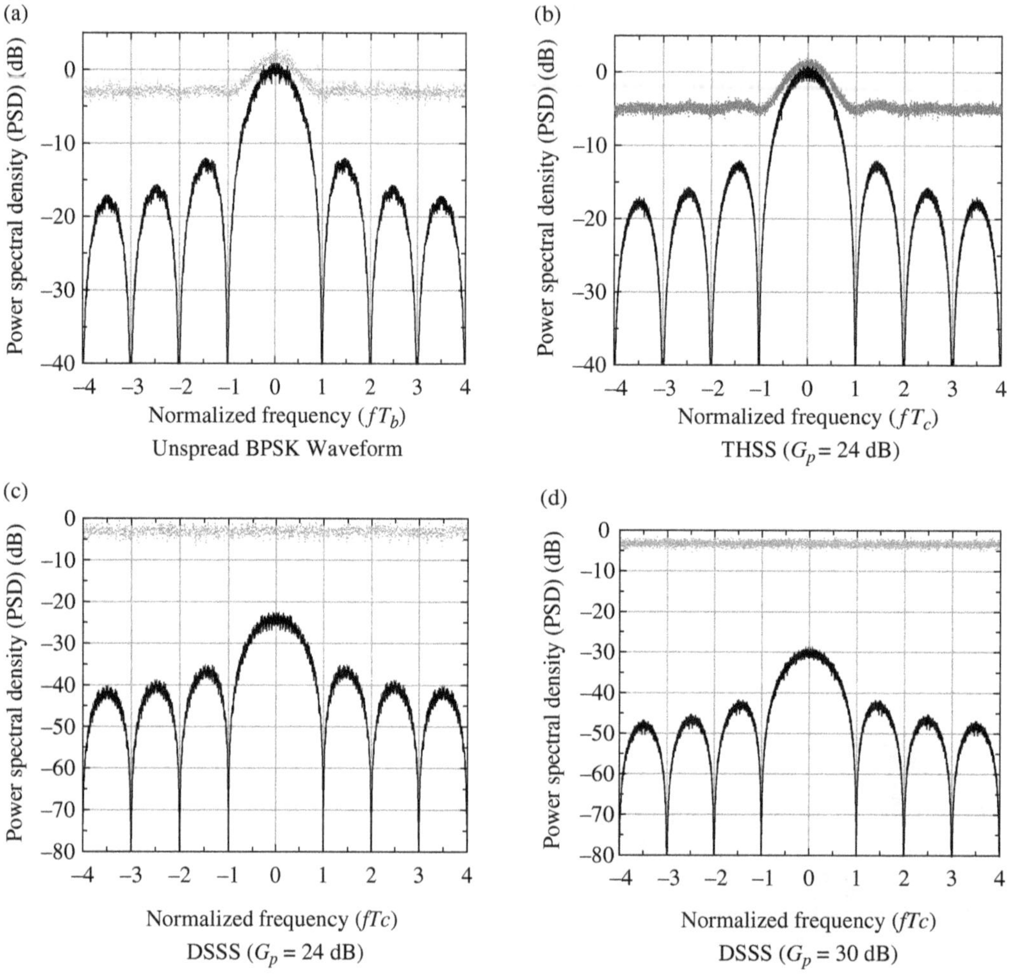

FIGURE 13.12 Comparison of unspread with DSSS and THSS waveform spectrums (black, curves without noise; gray curves with $E_b/N_o = 6$ dB).

corresponds to $f'T_b = 1/G_p = 0.0078$. On the other hand, with the FHSS waveform, the normalized frequency f'/B_h is hopped over the bandwidth $-B_h/2 \le f' \le B_h/2$ with $G_p = R_bT_c = R_b/B_h$ so that $B_h = 1/T_c$. The time axes in Figure 13.11 are normalized by the simulation record duration of T_r seconds where $T_r = N_{sm}T_{fft}$, T_{fft} is the duration of the FFT window, and N_{sm} is the number of periodograms, described in Sections 1.2 and 1.2.8, used in estimating the PSD.

Figure 13.12 shows the two-dimensional PSD of the DSSS and FHSS waveforms in comparison to the unspread BPSK waveform modulation using identical data rates. The SS processing gains are $G_p = 24$ and 30 dB corresponding to $N = 256$ and 1024 chips-per-bit and $N_r = 8$. The DSSS chip and the THSS pulse durations are identical. The signal-plus-noise plots, shown as the *gray* curves, is relative to the bit energy with $E_b/N_o = 6$ dB* corresponding to an optimally detected bit-error probability of about $P_{be} = 2 \times 10^{-3}$; the *black* curves represent the noise-free PSD of the waveforms. Figure 13.12c and d demonstrate that, unlike the unspread and THSS waveform, there is no visual evidence of their presents when looking and the noise level. A major distinction between Figure 13.12a and b is the difference in bandwidth. For example, the bandwidth of the THSS system is NR_b so the increase in the noise-floor of about 5 dB is spread over a much larger bandwidth making it more difficult to detect. The Dicke radiometer, discussed in Section 13.4.2, is designed specifically for the purpose of detecting slight increases in the noise-floor over large bandwidths.

The spectrum of the DSSS waveform, with $p(t) = rect((t - T_c/2)/T_c)$, is evaluated as

$$S_{DS}(f) = VT_c e^{-j2\pi f T_c} sinc(fT_c) \tag{13.33}$$

Similarly, the spectrum of the THSS waveform is evaluated as

$$\begin{aligned} S_{TH}(f) &= G_p VT_c e^{-j2\pi f(\tau_{kn} + T_c)} sinc(fT_c) \\ &= G_p e^{-j2\pi f \tau_{kn}} S_{DS}(f) \end{aligned} \tag{13.34}$$

*The noise is also adjusted by the bandwidth into demodulator matched filter, corresponding to a sampling rate of $N_s = 8$ samples/bit, resulting in $E_{samp}/N_o = -3$ dB.

so that

$$|S_{TH}(f)| = G_p|S_{DS}(f)| \tag{13.35}$$

The normalized squared magnitude of the spectrum $S(f)$ is defined as the periodogram that, when averaged, forms the estimate the PSD of the waveform.

Hybrid SS waveforms are used to complement the underling SS waveform by improving the AJ and/or LPD characteristics. For example, the relatively low LPD characteristic of FHSS and THSS waveforms can be improved using the respective hybrid forms FHSS/DSSS and THSS/DSSS. Referring to Figure 13.12b, a modest amount, 10 : 1 or so, of PN chip spreading applied to each TH pulse will lower the detestability.

13.3 JAMMER AND INTERCEPTOR ENCOUNTERS

13.3.1 Anti-Jam Spread-Spectrum Communications

The effectiveness of AJ communication systems is maximized by maximizing the receiver antenna gain in the direction of the communication transmitter while minimizing the receiver antenna gain in the direction of the jammer. The application of adaptive antenna null steering for this purpose may provide a 20–60 dB advantage in the presence of relatively narrowband jammers. Increasing the transmitter power and the antenna gain in the direction of the receiver is always beneficial for AJ systems. An encounter with a jammer is depicted in Figure 13.13 and the performance is evaluated in terms of the communications range equation applied to the jammer and communication paths.

The expression for the received signal power at the input to the receiver at a range R from a transmitting source is

$$P_r = \frac{P_t G_t G_r \lambda^2}{(4\pi)^2 L_s L_p R^2} \tag{13.36}$$

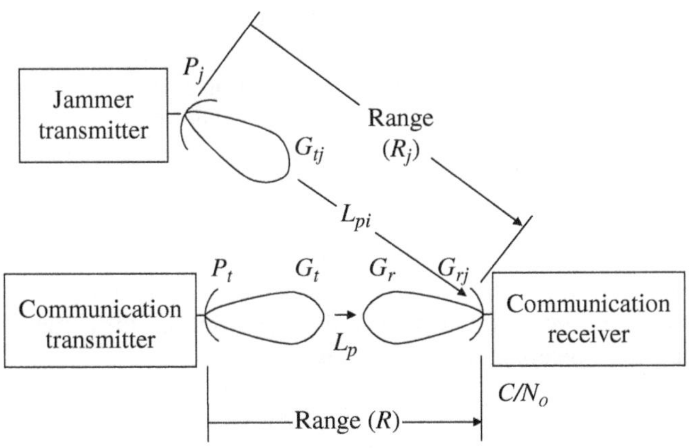

FIGURE 13.13 Communications jammer scenario.

where P_r represents the received power from the communication transmitter. The transmitter power into the transmit antenna with gain G_t is denoted as P_t so the transmitter system losses are accounted for in P_t. G_r denotes the receiver antenna gain in the direction of the communication transmitter. The loss L_s represents the receiver system losses including those associated with antenna pointing, tracking, and polarization and the loss L_p represents the propagation losses associated with the atmosphere, rain, and scintillation. In the following analyses the losses listed in Table 13.4 are used with equal propagation path and antenna losses for the communicator and jammer, such that, $L_p = L_{pj} = L_{atm} + L_{rain} + L_{scint} =$ 2.5 dB and $L_{ant} = L_{antj} = L_{pointing} + L_{pol} = 1$ dB. The communication system receiver and demodulator losses do not have jammer counterparts and are defined as $L_{rd} = L_{rec} + L_{comb} + L_{trk} = 2.3$ dB; in (13.36) $L_s = L_{ant} + L_{rd}$. The numerical values are considered to be generic losses; however, in practice they are dependent upon a variety of operational design conditions unique to each system. The receiver combining losses can be significant if noncoherent combining is required. The demodulator frequency, phase, and symbol time tracking losses are dependent on the waveform and the dynamics of the channel. The rain loss is typically characterized in terms of decibels/meter so the total loss depends on the propagation path through the region of rain; rain attenuation is strongly dependent on the operating frequency. With these thoughts in mind, the parameters listed in Table 13.4 might be better thought of as place-holders for consideration in an actual system design.

The ratio of the desired signal power to noise-plus-jammer power is evaluated as

$$\frac{P_r}{N + P_{rj}} = \frac{P_r}{kT_oB_nF_{ns} + P_{rj}} \tag{13.37}$$

where kT_oB_n is the noise power and F_{ns} is the receiver system noise figure. Multiplying both sides (13.37) by the receiver noise bandwidth B_n results in the expression for the receiver signal to noise-density ratio C/N_o', where $C = P_r$ denotes the desired received signal power and $N_o' = N_o + P_{rj}/B_n$ denotes the Gaussian noise-plus-jammer power density with

TABLE 13.4 Losses Associated with Range Equation

Parameter	Value (dB)	Description
$L_{pointing}$	0.5	Antenna pointing
L_{pol}	0.5	Antenna polarization
L_{rec}	1.5	Receiver RF losses
L_{comb}	0.5	Integration loss
L_{trk}	0.3	Demodulator tracking losses
L_{atm}	0.5	Atmospheric absorption
L_{rain}	1.0	Rain loss
L_{scint}	1.0	Scintillation

dimension of watts/hertz, so C/N_o' has units of Hz. For successful communications it is required that the received $C/N_o \geq (E_b/N_o)_{reqd} R_b$, where R_b is the communication data rate and $(E_b/N_o)_{reqd}$ is the minimum energy-per-bit to noise-density ratio required to satisfy the specified bit-error probability performance of the demodulator. Developing C/N_o' in terms of the *rhs* of (13.37) results in the expressions

$$\begin{aligned}\frac{C}{N_o'} &= \frac{B_n P_r}{kT_o B_n F_{ns} + P_{rj}} \\ &= \frac{B_n}{1 + kT_o B_n F_{ns}/P_{rj}} \left(\frac{P_r}{P_{rj}}\right) \\ &\geq \left(\frac{E_b}{N_o}\right)_{reqd} R_b\end{aligned} \tag{13.38}$$

Referring to Chapter 15 and substituting the expressions for the received power of the communication and jammer links operating at the same carrier frequency $(\lambda_j = \lambda)$ and using the previously defined losses, then, upon rearranging terms, the condition for a successful communication is expressed as

$$\left(\frac{1}{1 + \dfrac{kT_o B_n F_{ns}}{P_{rj}}}\right) \left(\frac{\text{EIRP}}{\text{EIRP}_j}\right) \left(\frac{G_r}{G_{rj}}\right) \left(\frac{L_{pj}}{L_p}\right) \left(\frac{L_{antj}}{L_{ant}}\right) \left(\frac{1}{L_{rd}}\right) \left(\frac{R_j}{R}\right)^2 \geq \left(\frac{E_b}{N_o}\right)_{reqd} \left(\frac{R_b}{B_n}\right) \tag{13.39}$$

Equation (13.39) delineates the influence of the communicator, jammer, and link geometry on the system performance. Upon rearranging terms and using the processing gain $G_p = B_n/R_b$ and assuming that $P_{rj} >> kT_o B_n F_{ns}$, the requirements for successful communications is expressed as

$$\left(\frac{\text{EIRP}}{\text{EIRP}_j}\right) \left(\frac{G_r}{G_{rj}}\right) \left(\frac{L_{pj}}{L_p}\right) \left(\frac{L_{antj}}{L_{ant}}\right) \left(\frac{1}{L_{rd}}\right) \left(\frac{R_j}{R}\right)^2 Gp = \left(\frac{E_b}{N_o}\right)_{rec} \geq \left(\frac{E_b}{N_o}\right)_{reqd} \quad \text{: successful communication} \tag{13.40}$$

Defining the link margin (LM) as

$$\text{LM} \triangleq \frac{(C/N_o)_{rec}}{(C/N_o)_{reqd}} = \frac{(E_b/N_o)_{rec}}{(E_b/N_o)_{reqd}} \tag{13.41}$$

the LM is expressed in term of the link parameters as

$$\text{LM} = \left(\frac{\text{EIRP}}{\text{EIRP}_j}\right) \left(\frac{G_r}{G_{rj}}\right) \left(\frac{L_{pj}}{L_p}\right) \left(\frac{L_{antj}}{L_{ant}}\right) \left(\frac{1}{L_{rd}}\right) \left(\frac{R_j}{R}\right)^2 \left(\frac{Gp}{(E_b/N_o)_{reqd}}\right) \geq 1 \quad \text{: successful communication} \tag{13.42}$$

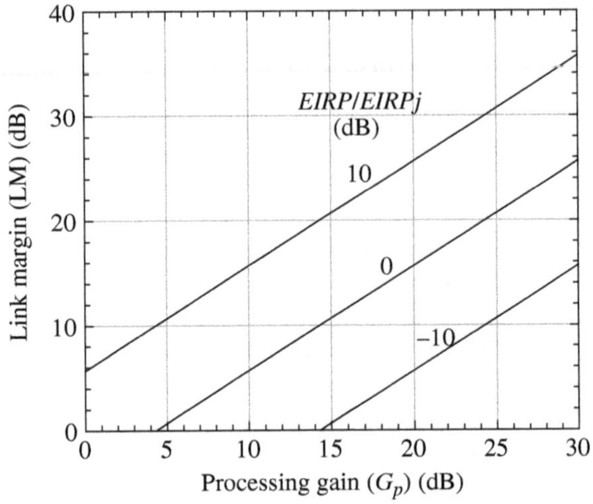

FIGURE 13.14 Communication and jammer characteristics for $E_b/N_o = 6$ dB $(L_{pj}/L_p = 0$ dB, $L_{ant}/L_{antj} = 0$ dB, $L_{rd} = 2.3$ dB, $G_r/G_{rj} = 10$ dB, $R_j/R = 2:1)$.

The LM is plotted in Figure 13.14 as a function of the processing gain (G_p) for several values of the effective isotropic radiated power (EIRP) and a required E_b/N_o of 6 dB. The condition LM ≥ 1 (0 dB) results in successful link communications. From these results, it is seen that the EIRP increases in direct proportion to the processing gain. In Figure 13.14, the path loss ratio is 0 dB and the communication link has a receiver gain advantage of 10 dB and range disadvantage of 6 dB. The LM performance for other parameter values is readily determined with the understanding that the range ratio is squared. For example, if the jammer range advantage increases by a factor of two, the communication LM will decrease by 6 dB.

13.3.2 Low Probability of Intercept Spread-Spectrum Communications

The effectiveness of LPI communication systems is maximized by maximizing the transmitter antenna gain in the direction of the communication receiver while minimizing the transmitter antenna gain in the direction of the interceptor. An encounter with an interceptor is depicted in Figure 13.15 and the performance is evaluated in terms of the communications range equation applied to the interceptor and communication paths.

The expression for the desired received signal power from a distant transmitter at range R is given by (13.36) and the received signal-to-Gaussian-noise power ratio at the communication receiver is expressed as

$$\frac{P_r}{N} = \frac{P_t G_t G_r \lambda^2}{(4\pi)^2 kT_o B_n F_{ns} L_p L_{ant} L_{rd} R^2} \tag{13.43}$$

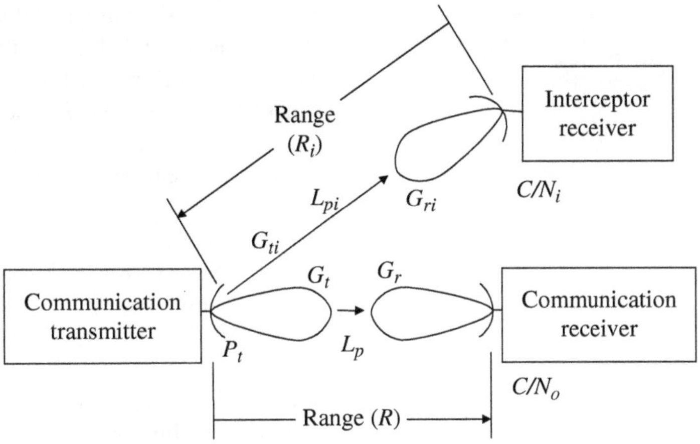

FIGURE 13.15 Communications intercept scenario.

where F_{ns} is the receiver noise figure and B_n is the communication SS bandwidth. In this case, the communicator and interceptor propagation and antenna losses and the communicator's receiver and demodulator loss are the same as with the jamming scenario in Figure 13.14; however, the interceptors' receiver and demodulator loss is $L_{rdi} = L_{reci} + L_{combi} = 2.0$ dB. The parameters in the following examples are based on a DSSS, BPSK-modulated SS waveform with a chip rate of R_c chips-per-second.

The received carrier to noise-density ratio is $(C/N_o)_{rec} = B_n P_r / N$ and the requirement to preserve the communication performance is $(C/N_o)_{rec} \geq (E_b/N_o)_{reqd} R_b$. The LM is defined as $LM = (C/N_o)_{rec}/(C/N_o)_{reqd}$, so the condition for successful communications with $\text{LM} \geq 1$ is $(C/N_o)_{rec} = LM(C/N_o)_{reqd}$. Using this result and solving (13.43) for the required transmitter power gives

$$P_t = \frac{(4\pi)^2 k T_o F_{ns} L_p L_{ant} L_{rd} R^2}{G_t G_r \lambda^2} \text{LM} \left(\frac{E_b}{N_o}\right)_{reqd} R_b \quad (13.44)$$

In terms of the interceptor, the received signal-to-noise ratio is given by

$$\left(\frac{P_r}{N}\right)_i = \frac{P_t G_{ti} G_{ri} \lambda_i^2}{(4\pi)^2 k T_o B_n F_{nsi} L_{pi} L_{anti L_{rdi}} R_i^2} \quad (13.45)$$

Substituting the transmitter power expressed in (13.44) required to maintain the communication link into (13.45), with $\lambda_i = \lambda$, results in

$$\left(\frac{P_r}{N}\right)_i = \left(\frac{G_{ti} G_{ri}}{G_t G_r}\right)\left(\frac{F_{ns} L_p L_{ant} L_{rd}}{F_{nsi} L_{pi} L_{anti} L_{rdi}}\right)\left(\frac{R}{R_i}\right)^2 \text{LM}\left(\frac{R_b}{B_n}\right)\left(\frac{E_b}{N_o}\right)_{reqd} \quad (13.46)$$

When attempting to detect a communication transmission, the interceptor has a time advantage similar to that of the communicator's SS frequency advantage. For example, an interceptor employing a *total-energy radiometer* detector integrates the signal-plus-noise power in an RF bandwidth $W = B_n = R_c$ over the time interval T_I and declares the presence of a signal when the integrated signal-to-noise ratio exceeds a threshold γ_T. For the case of noncoherently integrating a BPSK-modulated SS signal in additive noise over the interval T_I-seconds, the communications signal will be detected if the interceptor output signal-to-noise ratio exceeds the threshold, that is, $\gamma_{io} \geq \gamma_T$. For the total-energy radiometer, using the optimum RF bandwidth of $W = R_c$ and the optimum signal power factor* $\alpha = 0.773$, the interceptor output signal-to-noise ratio is given by

$$\gamma_{io} = 0.597 \left(\frac{E_b}{N_o}\right)_i^2 \left(\frac{G_{pi}}{G_p}\right) < \gamma_T \quad (13.47)$$

: undetected communications

where $G_{pi} = R_b T_I$ is the interceptor processing gain; the communication is undetected when (13.47) is satisfied. This inequality can be rearranged to express the requirement for the E_b/N_o at the interceptor input to remain undetected with the result

$$\left(\frac{E_b}{N_o}\right)_i < 1.29 \sqrt{\frac{\gamma_T G_p}{G_{pi}}} \quad \text{: undetected communications} \quad (13.48)$$

Recalling that $(P_r/N)_i$ is the received signal-to-noise ratio measured in the bandwidth B_n, then $(E_b/N_o)_i$ is evaluated as

$$\left(\frac{E_b}{N_o}\right)_i = \left(\frac{P_r}{N}\right)_i \frac{B_n}{R_b} = \left(\frac{G_{ti} G_{ri}}{G_t G_r}\right)\left(\frac{F_{ns} L_p L_{ant} L_{rd}}{F_{nsi} L_{pi} L_{anti} L_{rdi}}\right)\left(\frac{R}{R_i}\right)^2 \text{LM}\left(\frac{E_b}{N_o}\right)_{reqd} \quad (13.49)$$

*The interceptor power factor is discussed in Section 13.5.2.

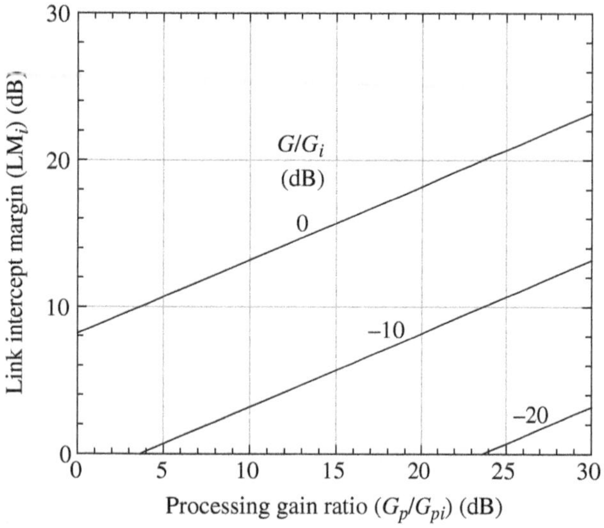

FIGURE 13.16 Communication link intercept margin with total-energy radiometer interceptor for $E_b/N_o|_{reqd} = 6$ dB, $LM = 0$ dB, $\gamma_T = 30$ (14.8 dB) ($L_p/L_{pi} = L_{ant}/L_{anti} = F_{ns}/F_{nsi} = 0$ dB, $L_{rd}/L_{rdi} = 0.3$ dB, $R_i/R = 2:1$).

Therefore, using the total-energy radiometer with RF bandwidth $B = R_c = B_n$, the condition for undetected communications corresponds to

$$\left(\frac{G_{ti}G_{ri}}{G_tG_r}\right)\left(\frac{F_{ns}L_pL_{ant}L_{rd}}{F_{nsi}L_{pi}L_{anti}L_{rdi}}\right)\left(\frac{R}{R_i}\right)^2 \text{LM}\left(\frac{E_b}{N_o}\right)_{reqd} < 1.29\sqrt{\frac{\gamma_T G_p}{G_{pi}}} \tag{13.50}$$

For convenience in plotting, (13.50) is normalized by the term on the *right-hand-side* (*rhs*) with the communication intercept LM$_i$ defined as*

$$\text{LM}_i \triangleq 1.29\left(\frac{G_tG_r}{G_{ti}G_{ri}}\right)\left(\frac{F_{nsi}L_{pi}L_{anti}L_{rdi}}{F_{ns}L_pL_{ant}L_{rd}}\right)\left(\frac{R_i}{R}\right)^2\left(\frac{\sqrt{\gamma_T}}{\text{LM}}\right)\left(\frac{E_b}{N_o}\right)^{-1}_{reqd}\sqrt{\frac{G_p}{G_{pi}}}$$

$$> 1 \quad \text{: undetected communications} \tag{13.51}$$

Figure 13.16 is a plot of the communication margin, expressed by (13.51), as a function of processing gain ratio G_p/G_{pi} for various values of $G/G_i = G_tG_r/(G_{ti}G_{ri})$ with the remaining parameters as indicated. When plotted in dB, positive values of the ordinate correspond to undetected communication. The interceptor threshold $\gamma_T = 18.58$ and corresponds to $P_d = 0.995$, $P_{fa} = 10^{-3}$, and interceptor output signal-to-noise ratio of $\gamma_o = 17.8$ dB. These values are determined using the chi-square distribution nomogram developed by Urkowitz [34] as described in Appendix C. The processing gain ratio is subject to the competing interests of increasing and decreasing the communication intercept LM$_i$. The performance is significantly influenced by the antenna gain ratio G/G_i where the interceptor gain G_{ti} is a function of the interceptor location and the beam width of the communication transmit antenna. To minimize the radiometers integration time, the interceptor receive antenna gain must overcome the G_{ti} loss and the free-space range loss corresponding to LSR $= R_i/R$; typically the communication system is designed to operate with LSR ≥ 2.

The total-energy radiometer discussed above is optimum in the sense that the optimum radiometer RF bandwidth for the DSSS waveform is $W = R_c$. Normally, however, the communication chip rate is unknown so the input RF bandwidth must be somewhat wider giving suboptimal interceptor performance. Also, when FHSS or frequency division multiplex (FDM) waveforms are to be detected, the RF bandwidth of the interceptor must be considerably greater than the instantaneous bandwidth occupied by the communication waveform. For example, with FHSS systems the interceptor bandwidth must be on the order of the hopping bandwidth, which is wider than the instantaneous signal bandwidth by a factor equal to the SS processing gain. Therefore, a more realistic interceptor encounter requires $W > W_{ss}$ where W_{ss} is the SS or FDM occupied bandwidth. In Section 13.4.1, the interceptor excess bandwidth is defined as $\rho = W/W_{ss}$ and the signal power factor is defined as $\alpha = P_{si}/P_{ri}$ and corresponds to output-to-input power ratio of the radiometer RF BPF. Referring to (13.59), the optimum intercept processing gain is effectively altered by the changes in the factor α^2/ρ. For example, the optimum DSSS performance results in Figure 13.16 correspond to $\rho = W/R_c = 1$ and $\alpha = 0.773$. If the interceptor power factor is reduced by 20%, the radiometer output signal-to-noise ratio is reduced and the communication LM to interception is reduced by 0.97 dB. The interceptor's probability of detection and false alarm are related to the input signal-to-noise ratio (SNR) and the threshold γ_T. If the interceptor makes a decision regarding the presence of a signal at the end of each integration interval T_I, the average time between false alarms is given by $T_{fa} = T_I/P_{fa}$.

13.4 COMMUNICATION INTERCEPTORS

Communication interceptors [35, 36] have a difficult job in attempting to determine, not only the presence of a communication signal but also to extract as much information as possible from the signal [37], to accomplish the goals of the interceptor. The goals of the interceptor may be well intentioned, as in performing rescue missions; however, when the communicator takes great pains to avoid being intercepted, the goals of the interceptor are more adversarial and may involve directing and optimizing jamming signals or intercepting messages to defeat the plans of the

*The normalization is inverted to result in a positive dB margin to detection by the interceptor; this is the same as normalizing by the *lhs* of (13.50).

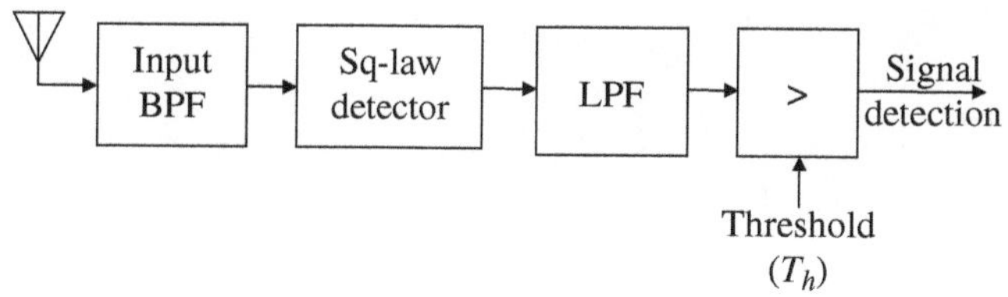

FIGURE 13.17 Total-energy radiometer functional diagram.

communicator. The subject of communication interceptor design is as varied as is the subject of communication waveform designs and is made more difficult if little is known about the communication waveform or its underlying characteristics. Interceptors generally look for energy exceeding an assumed background level or signal temporal and spectral features characteristic of the waveform modulation. Interceptors can be classified into several broad categories based on the detection processing [38]: radiometers, binary sliding-window [39], distribution free and nonparametric [40–42], and filter-bank comparative detection. A summary of commonly encountered types of interceptors and their optimum parameters is given by Ames et al. [13].

13.4.1 Total-Energy Radiometer

The total-energy radiometer [34] (TER) is the simplest interceptor to implement because it integrates the output of a square-law detector and compares the output signal-to-noise ratio based on a predetermined detection threshold.* With very low input carrier-to-noise density ratios, it is difficult to maintain a constant gain intercept receiver over the long integration times required by TER. The functional diagram of the total-energy radiometer is shown in Figure 13.17. The input BPF establishes the signal-to-noise ratio into the square-law detector with a sufficiently wide bandwidth to capture the signal energy. There is, however, an optimum input bandwidth that maximizes the signal detection in the presence of AWGN. In Figure 13.17, the LPF is an integrator with integration time of T_I seconds that determines the interceptor processing gain. In theory the interceptor can make T_I as long as necessary to overcome the communicator's SS processing gain and thus detect the signal.

The following performance analysis of the total-energy radiometer is based on a DSSS, BPSK-modulated waveform with a chip rate of $R_c = 1/T_c$, given by

$$s_r(t) = \sum_i s(t)\, rect\left(\frac{t - iT_c}{T_c}\right) \tag{13.52}$$

*Urkowitz [34] has developed a convenient nomogram for determining the detection threshold corresponding to specified detection and false-alarm probabilities.

The underlying modulated signal, with carrier power $A^2/2$ and data modulation $d_i = \{\pm 1\}$, is

$$s(t) = Ad_i \cos(\omega_c t) \tag{13.53}$$

The spectrum of this modulated waveform has the form $sinc(fT_c)$ and, considering a perfectly rectangular BPF with a RF bandwidth $W = 2B$, the signal power out of the BPF, centered on the carrier f_c, is given by $P_{so} = \alpha P_{ri} = \alpha A^2/2$ where α is evaluated as

$$\begin{aligned} \alpha &= 2\int_0^B |S(f)|^2 df \\ &= \frac{2}{R_c}\int_0^B \left|\frac{\sin(\pi f/R_c)}{\pi f/R_c}\right|^2 df \end{aligned} \tag{13.54}$$

Equation (13.54) is evaluated as

$$\alpha = \frac{2}{\pi}\left[S_i\left(\frac{\pi B}{R_c}\right) - \frac{\sin^2(\pi B/R_c)}{\pi B/R_c}\right] \tag{13.55}$$

where $S_i(x)$ is the sine integral [43]. Using these results, the signal power out of the square-law detector is

$$P_{so} = P_{si}^2 = \alpha^2\left(\frac{A^2}{2}\right)^2 \tag{13.56}$$

For white Gaussian input noise with one-sided power spectral density N_o, the noise power out of the square-law detector and integrator, ignoring the DC term, is evaluated by Davenport and Root [44] as

$$P_{no} = W\left(N_o^2 B_I\right) \tag{13.57}$$

where $B_I = 1/T_I$ and the output signal-to-noise ratio of the total-energy radiometer is expressed as

$$\begin{aligned} \gamma_{io} &= \frac{P_{so}}{P_{no}} = \alpha^2\left(\frac{A^2/2}{N_o}\right)^2 \frac{T_I}{W} \\ &= \alpha^2 \gamma_{ci}^2 \frac{T_I}{W} \end{aligned} \tag{13.58}$$

where γ_{ci} is equal to or exceeds the input carrier-to-noise density ratio. Multiplying the numerator and denominator of (13.58) by the square of the communication data rate and defining the communication and intercept processing gains as $G_p = R_c/R_b$ and $G_{pi} = T_I R_b$ respectively with the interceptor excess bandwidth defined as $\rho = W/R_c$, results in the expression

$$\gamma_{io} = \alpha^2\left(\frac{E_b}{N_o}\right)^2\left(\frac{G_{pi}}{\rho G_p}\right) \ge \gamma_T \quad : \text{communications detected} \tag{13.59}$$

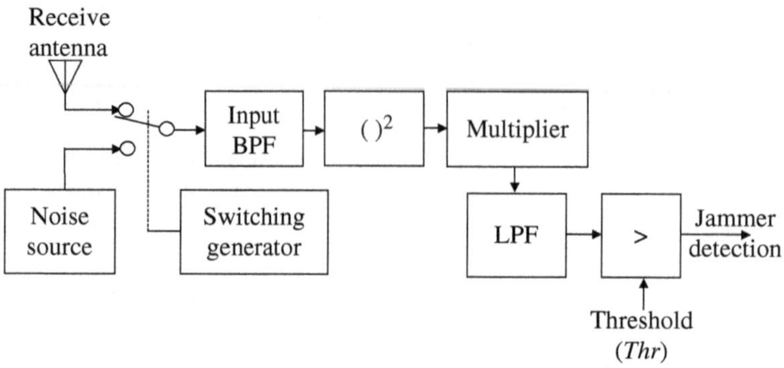

FIGURE 13.18 Dicke radiometer functional diagram.

When γ_{io} exceeds the predetermined threshold γ_T, the interceptor declares that the communication signal has been detected. The optimum interceptor RF bandwidth occurs when $W = R_c$ ($r = 1$) which results in $\alpha = 0.773$. As the excess bandwidth increases ($\rho > 1$), α approaches unity resulting in the maximum signal power; however, there is a disproportionate increase in the noise resulting in degraded detection. With the excess bandwidths $\rho < 1$, the opposite result occurs, still resulting in a net degradation in detection performance.

13.4.2 Dicke Radiometer

The Dicke [45] radiometer overcomes the sensitivity of the total-energy radiometer to the receiver gain variations during the long integration intervals. This is accomplished by switching the receiver input between the antenna terminal and a calibrated noise source. The switching rate is chosen to be sufficiently high so as not to interfere with the integration sensitivity. The integrated product of the two switched sources produces a DC voltage when the signal is present. The DC voltage is integrated and compared to the detection threshold. The original Dicke radiometer focused on square wave switching; however, sinusoidal and saw tooth switching [46] are also used. The functional implementation of the Dicke radiometer is shown in Figure 13.18.

Table 13.5 summarizes the parameters required for optimum interceptor detection using the total-energy and Dicke [47] radiometers. In this summary the interceptor performance is based on the communication carrier-to-noise power ratio ($\gamma_{io} = P_{so}/N_{no}$) at the output to the interceptor radiometer with the threshold (T_h) established from the specified detection and false-alarm probabilities.

TABLE 13.5 Summary of Optimum Interceptor Radiometer Parameters ($\gamma_c = P_c/N_o$ Hz; P_c Is Carrier Power)

Radiometer Type	Output SNR (γ_{io})	Comments
Total-energy	$\alpha^2\gamma_{ci}{}^2T_I/W$	DSSS: $W = B_n = R_c$; $\alpha = 0.773$
Dicke	$0.15\alpha^2\gamma_{ci}{}^2T_I/W$	DSSS: $W = R_c$
	$0.25\alpha^2\gamma_{ci}{}^2T_I/W$	FHSS and FDM

13.5 BIT-ERROR PERFORMANCE OF DSSS WAVEFORMS WITH JAMMING

In the following analyses, the communication performance with jamming is based on the SS waveforms occupying bandwidths much greater than the information bandwidth, that is, $W_{ss} \gg W$ where W is the symbol bandwidth; for binary antipodal signaling W is taken to be equal to the information rate R_b. Furthermore, the analyses focus on the desired and jamming signals that are transmitted through linear, nondispersive channels. The performance examples are based on single-tone, multitone, and noise jammers; however, Kullstam [48] discusses the performance with arbitrary interfering signals. In Section 13.8, these assumptions and conditions are also applied to the performance evaluation of FHSS waveforms.

The SNJR ratio is defined as

$$\gamma_{s+j} = \frac{E_b}{N_o + N_{oJ}} \tag{13.60}$$

where $E_b = P_sT_b$ and $N_o = NT_b$ and $N_{oJ} = N_JT_c$ are the AWGN densities of the receiver thermal and jammer noises respectively; N and N_J are the respective noise powers. Dividing the numerator and dominator of (13.60) by E_b results in the expression

$$\gamma_{s+j} = \frac{1}{N_o/E_b + N_{oJ}/E_b} \tag{13.61}$$

The first term in the denominator of (13.61) is simply the inverse signal-to-thermal noise ratio, defined as $\gamma_b = E_b/N_o$ and the second term is defined as the effective jammer-to-signal power ratio, expressed as

$$\rho_J \triangleq \frac{P_J}{P_sG_p} \tag{13.62}$$

where $G_p = T_b/T_c$ is the SS processing gain. Upon defining the signal-to-jammer power ratio as

$$\gamma_{sj} \triangleq \frac{E_b}{N_{oJ}} = \frac{1}{\rho_J} = \frac{P_s G_p}{P_J} \tag{13.63}$$

and substituting (13.63) in (13.61) and rearranging terms, the SNJR simplifies to

$$\gamma_{s+j} = \frac{\gamma_b \gamma_{sj}}{\gamma_b + \gamma_{sj}} \tag{13.64}$$

This expression is the *parallel combination* (see Problem 4) of the individual signal-to-noise ratios and, as one dominates the other, the resulting signal-plus-jammer ratio approaches the smaller of the two. This phenomenon results in the *flaring* of the bit-error probability with increasing γ_b when jammer power is added to the receiver thermal noise.

Equations (13.63) and (13.64) are used in the following sections to characterize the bit-error performance and the resulting E_b/N_o loss with jamming. The performance analysis is based on the signal and jammer powers at the input to the communication receiver and does not consider the transmitter power necessary to overcome range, propagation and other link losses as discussed in the preceding sections.

13.5.1 DSSS with BPSK and QPSK Modulation and CW Jamming

The CW jammer signal is the easiest jamming waveform to generate and the most effective CW jammer strategy is to concentrate the jammer power at the carrier frequency of the DSSS waveform; this is also the case for FHSS waveforms; however, locating the carrier frequency is more improbable. Consider the received signal characterized as a DSSS BPSK chip-modulated waveform at the carrier angular frequency ω_c radians-per-second and expressed in terms of the analytic signal as

$$\tilde{s}_r(t) = \sqrt{2P_s} d_k \sum_{i=0}^{N-1} c_i e^{j\phi_s} \quad \text{: signal only (no jamming)} \tag{13.65}$$

where P_s is the signal carrier power, $c_i = \{\pm 1\} : i = 0, \ldots, N-1$ represents the PN chips of duration T_c characterized as $rect((t - T_c(1+2i)/2)/T_c)$, $d_k = \{\pm 1\}$ is the source data of duration $T_b = NT_c$, synchronized with the PN chips, and ϕ_s is the received signal phase resulting from the channel. The analytic CW jamming signal is expressed as

$$\tilde{s}_J(t) = \sqrt{P_J}\left(e^{j\phi_J} + je^{j\phi_J}\right) \quad \text{: jamming signal} \tag{13.66}$$

The jammer power is distributed equally between the quadrature jammer signals. The demodulator correlates the received signal-plus-jammer using the locally generated reference signal, expressed as

$$\tilde{s}_{ref}(t) = \sqrt{\frac{2}{T_b}} \sum_{i=0}^{N-1} c_i rect\left(\frac{(t - T_c(1+2i)/2)}{T_c}\right) e^{j\hat{\phi}_s} \tag{13.67}$$

: local reference signal

where the locally generated PN sequence c_i is time synchronized with the received PN chips and $\hat{\phi}_s$ is the demodulator estimate of the received signal phase. Using these functions, the k-th *real* correlator output, corresponding to the bit interval kT_b, is evaluated as

$$\begin{aligned} y_{ck} &= \frac{1}{2}\left(\int_0^{T_b} \tilde{s}_r(t)\tilde{s}^*_{ref}(t)dt + \int_0^{T_b} \tilde{s}_J(t)\tilde{s}^*_{ref}(t)dt\right) \\ &= \frac{1}{2}\sqrt{\frac{4P_s}{T_b}} d_k \sum_{i=0}^{N-1} c_i^2 \int_{iTc}^{(i+1)Tc} dt\cos\left(\phi_s - \hat{\phi}_s\right) \\ &+ \frac{1}{2}\sqrt{\frac{2P_J}{T_b}} \sum_{i=0}^{N-1} c_i \int_{iTc}^{(i+1)Tc} dt\sqrt{2}\cos(\theta) \end{aligned} \tag{13.68}$$

With ideal correlator phase tracking, the correction phase corresponds to $\phi_s - \hat{\phi}_s = 0$ and letting $\theta = \phi_J - \hat{\phi}_s + \varphi$ with $\varphi = \pi/4$, and evaluating the correlator voltage outputs as $\sum_i c_i^2 = N$ and $\sum_i c_i = \sqrt{N}$,* (13.68) simplifies to

$$y_k = \sqrt{\frac{P_s}{T_b}} d_k N T_c + \sqrt{\frac{P_J}{T_b}} \sqrt{N} T_c \cos(\theta) \tag{13.69}$$

where the phase θ is unknown to the jammer; however, it can be changed in an attempt to achieve the worst-case jamming scenario. Using (13.69), the resulting signal-to-jammer power ratio is computed as

$$\gamma'_{sJ} = \begin{cases} \dfrac{E_b}{P_J T_c} = \dfrac{E_b}{N_J} & : \theta = 0;\ \text{worst-case jammer} \\ \dfrac{2E_b}{P_J T_c} = \dfrac{2E_b}{N_J} & : \text{independent of } \theta, \text{best-case jammer} \end{cases} \tag{13.70}$$

Simon et al. [49] use the notation in the second equality, which results in antipodal bit-error signaling performance when plotted in terms of E_b/N_J. Equation (13.70) can also be expressed as

$$\gamma'_{sJ} = \begin{cases} \dfrac{P_s G_p}{P_J} = \gamma_{sJ} & : \theta = 0 \\ 2\dfrac{P_s G_p}{P_J} = 2\gamma_{sJ} & : \text{independent of } \theta \end{cases} \tag{13.71}$$

*These results are obtained using the Gaussian *pdf* approximation resulting from the central limit theorem when summing over $N \gg 1$.

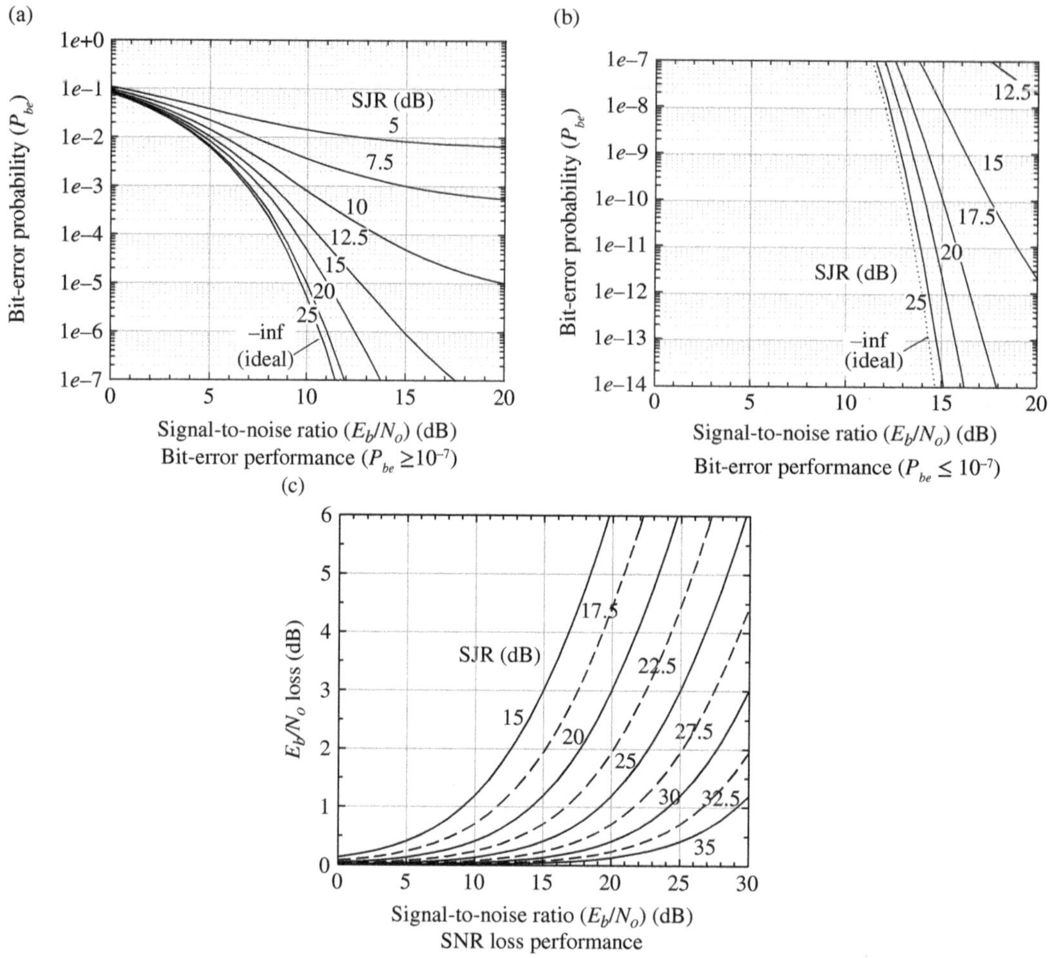

FIGURE 13.19 BPSK performance with wideband noise jammer.

Equations (13.70) and (13.71) conform to (13.63) in the preceding section. In general, the bit-error probability for the CW jammer is evaluated as

$$P_{be} = \frac{1}{2} erfc\left(\sqrt{\frac{\gamma_{sJ}}{2\cos^2(\theta)}}\right) \quad : \text{BPSK with CW jammer} \tag{13.72}$$

Kullstam [50] analyzes the use of DSSS QPSK chip modulation with identical source data on each rail as a means to minimize the impact of a CW jammer. For example, in the analysis leading to (13.72), the jammer can adjust the phase of the jamming signal to degrade the antipodal signaling performance; however, with DSSS chips and identical source data on each QPSK rail the performance corresponds to the best case independent of the jammer phase.

13.5.2 DSSS with BPSK Modulation and Pulsed Noise Jamming

With broadband noise jamming, the jammer power is spread over the entire SS bandwidth of W_{ss} Hz used by the communication waveform and the bit-error performance is evaluated as

$$P_{be} = \frac{1}{2} erfc\left(\sqrt{\frac{\gamma_b \gamma_{sj}}{\gamma_b + \gamma_{sj}}}\right) \quad : \text{broadband noise jammer} \tag{13.73}$$

Referring to (13.63), when the signal-to-jammer ratio, defined in decibels as $JSR = 10\log_{10}(\gamma_{sj})$, increases (13.73) approaches the familiar expression for BPSK bit-error probability in the AWGN channel, shown as the *dotted* curve in Figure 13.19a and b. These figures also demonstrate the *flaring* in the bit-error performance as the signal-to-jammer ratio decreases. In this regime, as γ_b increases the performance asymptotically approaches the bit-error performance based solely on γ_{sj}. Figure 13.19c shows the loss in $\gamma_b = E_b/N_o$ for various values of the signal-to-jammer ratio (SJR); the loss is computed as

$$\text{Loss}(\gamma_b) = -10\log_{10}\left(\frac{\gamma_{s+j}}{\gamma_b}\right) = -10\log_{10}\left(\frac{\gamma_{sj}}{\gamma_b + \gamma_{sj}}\right) \quad : \text{dB} \tag{13.74}$$

The loss in γ_b expressed in (13.74) is identical to the flaring loss in Figure 13.19a and b relative to the ideal performance at a specified P_{be} corresponding to SJR and E_b/N_o.

For the narrowband noise jammer, the jammer noise power is concentrated over the bandwidth $R_b < W'_{ss} < W_{ss}$. Furthermore, to achieve the greatest amount of peak power the noise jammer must operate as a pulsed jammer with a duty cycle of T_J/T_{pri} corresponding to an *on* interval with probability $u = T_J/T_{pri}$ and the *off* interval with probability $1-u \ :0 \le u \le 1$. The time T_{pri} is the PRI and is equal to the inverse of the pulse repetition frequency (PRF), that is, $f_{prf} = 1/T_{pri}$. The time T_J is the duration of the jammer-transmitted noise pulse. In this case, P_J is related to the average power that the noise jammer transmitter can sustain without overheating; therefore, a much larger peak power is achievable as long as the average power is not exceeded. The relationship between the peak and average power is obtained by the proportionality

$$P_{Jpeak} = P_J\left(\frac{T_{pri}}{T_J}\right) = \frac{P_J}{u} \tag{13.75}$$

For example, with $u = 0.0001$ the jammer gains an instantaneous power advantage of 40 dB although only for 0.01% of the time.

Applying these relationships to the pulsed noise jamming of the DSSS BPSK-modulated waveform, the bit-error probability is given by

$$P_{be} = \frac{u}{2}erfc\left(\sqrt{\frac{\gamma_b\gamma_{sj}u}{\gamma_b+\gamma_{sj}u}}\right) + \frac{1-u}{2}erfc\left(\sqrt{\gamma_b}\right) \tag{13.76}$$

: pulsed noise jammer

In this expression, when $u = 1$ the performance reverts to that of broadband jammer noise expressed by (13.73) and when $u = 0$ the performance corresponds to the unjammed BPSK waveform operating through an AWGN channel. The increase in the jammer peak power, due to the *prf*, is accounted for in the term $\gamma_{sj}\, u$ as described above. For example, referring to (13.63), $\gamma_{sj}u = P_sG_p/(P_J/u)$ and with $u \le 1$ the jammer peak power is increased thus decreasing γ_{sj}. Figure 13.20 depicts the bit-error performance as a function of E_b/N_o under various pulsed jammer conditions of u for specified values of SJR ranging from −300 dB, corresponding to jammer-dominated performance, and increasing to 40 dB with increasing communicator-dominated performance. For example, with SJR = 10 dB the impact of all conditions is slightly reduced; however, the constant $P_{be} = 0.5$ corresponding to $u = 1.0$ is significantly reduced with a *flaring* limit of $4e^{-6}$. With SJR = 22.5 dB the trend continues with the impact of $u = 1.0$ nearly negligible and the impact of $u = 0.04$ and 0.1 significantly reduced. With SJR = 50 dB the impact of all of the pulsed jammer conditions is significantly reduced with a *flaring* limit of $4e^{-10}$ with $u = 0.0001$. These improvements with increasing SJR are also depicted in Figure 13.19a and b in the context of a broadband noise jammer as described by (13.73).

13.5.3 Optimum Pulsed Noise Jammer Strategy against DSSS Uncoded BPSK Modulation without Side Information

Side information is used by the communicator to determine the presence of a jammer to improve the link performance. Side information is obtained from channel sensors not directly related to the communication signal; perhaps obtained from a sensor link or a network controller. The benefit of side information is demonstrated in (13.76) with the inclusion of the ideal unjammed bit-error performance and when side information is not available the performance is simply dependent on the signal-to-jammer ratio as expressed by (13.77).

To determine the optimum pulsed noise jammer strategy, the jammer focuses on the optimum choice of *prf* and the corresponding optimum value u_o that maximizes the communicator's bit-error probability. The expression for the bit-error probability with pulsed noise jamming without the benefit of side information is* expressed as

$$P_{be}(\gamma_{sj},u) = \frac{u}{2}erfc\left(\sqrt{\gamma_{sj}u}\right) \tag{13.77}$$

Equation (13.77) is plotted in Figure 13.21 as a function of SJR = $10\log_{10}(\gamma_{sj})$ with parametric dependence on u. With $u = 1$ the performance corresponds to that of the BPSK bit-error performance operating in a background of broadband noise; however, as u is decreased the noise becomes impulsive with increasingly high peak powers levels. In this event, for $u \le 0.0005$, the communicator must increase the signal power, P_s, or the processing gain, G_p, by about 40–10.5 dB = 29.5 dB to maintain $P_{be} = 10^{-6}$. Therefore, the pulsed noise jammer is very affective in defeating a DSSS-uncoded BPSK-modulated waveform.

Imagining the tangent to the performance curves in Figure 13.21, the optimum jammer strategy is to determine the value of u_o that matches the slope and is tangent to the extremes of the P_{be} performance curves with pulsed noise jamming. Consequently, the solution is to differentiate $P_{be}(\gamma_{sj},u)$ with respect to u and determine the conditions on u and γ_{sj} that satisfy these conditions. The details in determining the function $F(\gamma_{sj},u) = dP_{be}(\gamma_{sj},u)/du$ are relegated to Problem 6; however, the result is expressed as

*Note that (13.77) is the limiting form when $\gamma_{sj} \ll \gamma_b$.

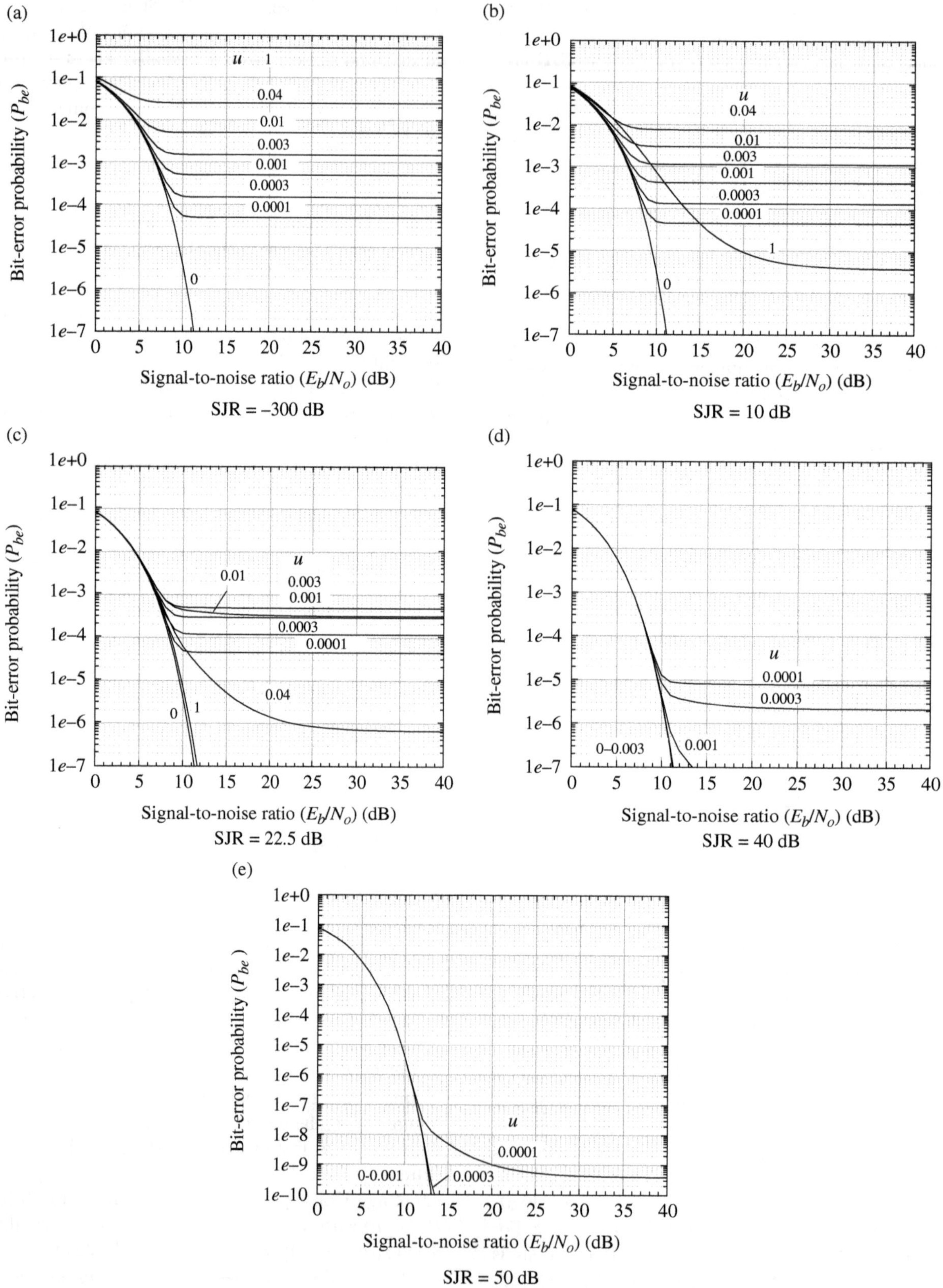

FIGURE 13.20 BPSK performance with pulsed noise jammer.

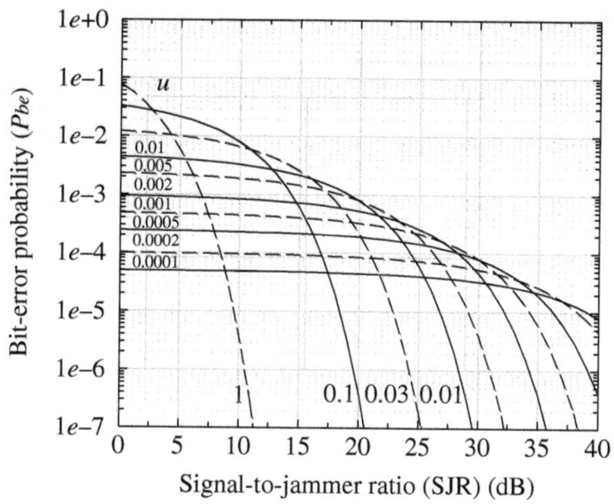

FIGURE 13.21 DSSS BPSK performance with pulsed noise jammer.

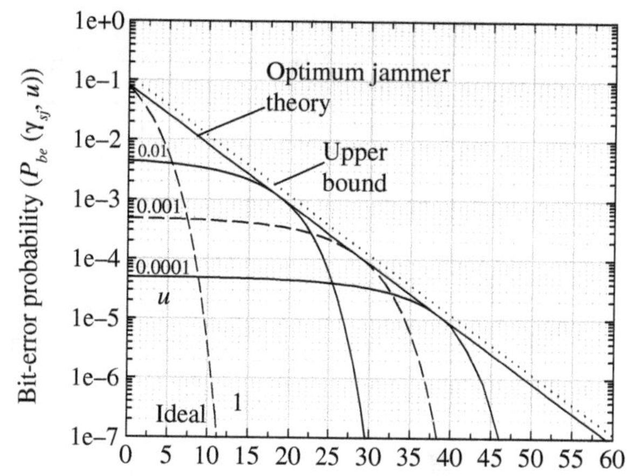

FIGURE 13.23 Optimum jammer strategy with pulsed noise jammer.

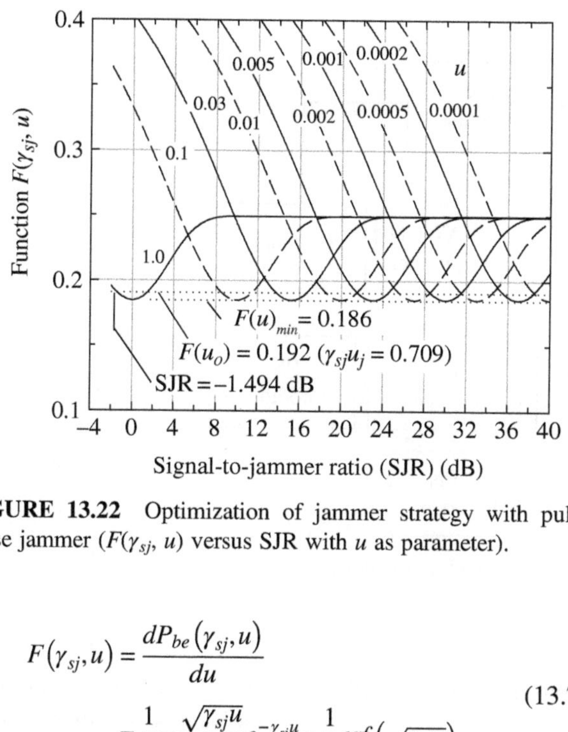

FIGURE 13.22 Optimization of jammer strategy with pulsed noise jammer ($F(\gamma_{sj}, u)$ versus SJR with u as parameter).

$$F(\gamma_{sj}, u) = \frac{dP_{be}(\gamma_{sj}, u)}{du} = \frac{1}{2} - \frac{\sqrt{\gamma_{sj}u}}{2\sqrt{\pi}} e^{-\gamma_{sj}u} - \frac{1}{4} erf\left(\sqrt{\gamma_{sj}u}\right) \quad (13.78)$$

Equation (13.78) is plotted in Figure 13.22 as a function of SJR for various values of the parameter u.

The optimum jammer criteria described above correspond to the slope $F(\gamma_{sj},u)$ evaluated at $u_o = 1.0$ and $\gamma_{sj} = 0.709$ (SJR = −1.494 dB*) resulting in the product $\alpha_o = u_o\gamma_{sj} = 0.709$; this dependence on $\alpha = u\gamma_{sj}$ is not surprising since (13.78) can be expressed as the function $F(\alpha)$. As seen from Figure 13.22, the optimum value of $\alpha_o = 0.709$ applies for $u \le 1$ so that

$$u_o = \frac{0.709}{\gamma_{sj}} \quad (13.79)$$

Upon substituting (13.79) into (13.77) the theoretical bit-error probability with optimum pulsed noise jamming corresponding to $\gamma_{sj}|_{u<1} > 1$ and $\gamma_{sj} > 0.709$, is expressed as

$$P_{be} = \frac{0.709}{2\gamma_{sj}} erfc\left(\sqrt{0.709}\right) = \frac{0.083}{\gamma_{sj}} \quad (13.80)$$

: pulse noise jammer theory

An upper bound on P_{be}, corresponding to $u = 1$ in (13.77), is expressed as [51]

$$P_{be} < \frac{1}{\sqrt{4\pi\gamma_{sj}}} e^{-\gamma_{sj}} \quad : \text{upper bound} (u = 1) \quad (13.81)$$

and, in terms of the optimum pulsed noise jammer, (13.81) becomes

$$P_{be} < \frac{\sqrt{0.709}}{\gamma_{sj}\sqrt{4\pi}} e^{-0.709} = \frac{0.117}{\gamma_{sj}} \quad : \text{pulse noise jammer bound} \quad (13.82)$$

Equations (13.80) and (13.82) are plotted in Figure 13.23 which demonstrates the optimum jammer objective of maximizing the communicator's bit-error performance. Otherwise, with $u = 1$ or $\gamma_{sj} \le 0.709$, the bit-error performance is given by

$$P_{be}(\gamma_{sj}, u = 1) = \frac{1}{2} erfc\left(\sqrt{\gamma_{sj}}\right) \quad : \text{unjammed performance} \quad (13.83)$$

*It may be of interest that, with $u_o = 1.0$, the value of $u\gamma_{sj} = 0.709000$ is accurate, with rounding, to six decimal places corresponding to SJR = −1.49354 dB and $F(u\gamma_{sj}) = 0.191536$.

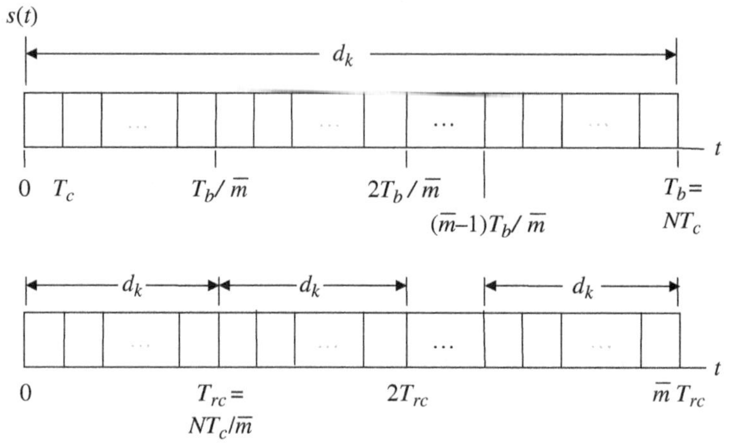

FIGURE 13.24 Repetition coded waveform mapping.

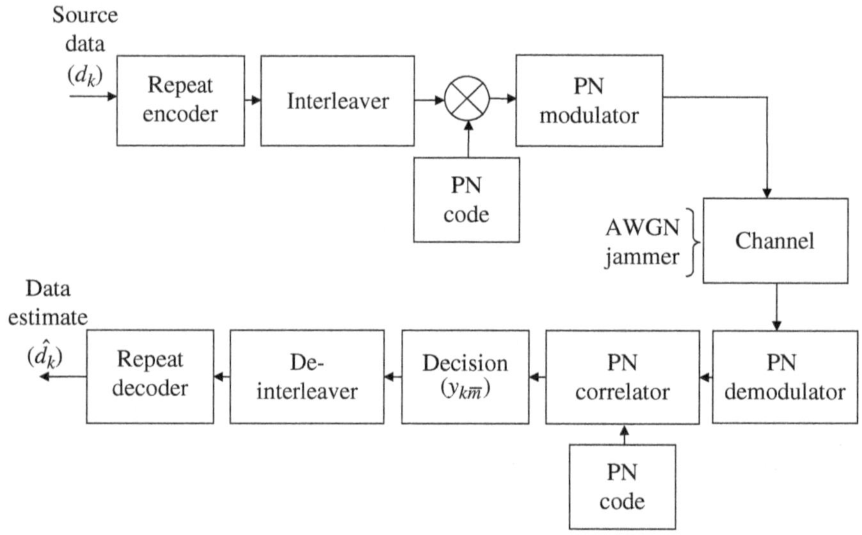

FIGURE 13.25 DSSS with repetition coding and interleaving.

13.5.4 DSSS with BPSK Modulation and Repetition Coding

In this section, the performance of DSSS BPSK-modulated chips is examined using repetition coding as depicted in Figure 13.24. The PN-coded chips are divided into $\bar{m}$ code symbols each having $N/\bar{m} \geq 1$ PN chips with the same source bit data d_k. The repetition-coded interval is $T_{rc} = NT_c/\bar{m}$. The functional processing is shown in Figure 13.25. The interleaver plays the fundamental role of scrambling the sourced coded symbols so the correlated channel disturbances, like fading and jamming, result in uncorrelated, or more specifically independent, disturbances among the deinterleaved code symbols. That is, the $\bar{m}$ decision variables $y_{k\bar{m}}$ are all independent random variables. Interleavers are recommended in all coding applications involving channel burst-error events.* Interleavers are discussed in Section 8.9 and are designed to ensure that channel errors at the deinterleaver output are independent among the coded symbols. In the following analysis, ideal interleavers and deinterleavers are assumed to exit that provide for the theoretical statistical independence between the $\bar{m}$ deinterleaved repetition codes regardless of the jammer correlation interval.

The following analysis of the repetition coding is based on the work of Simon et al. [52], from which, the following expressions are characterized for the bit-error probabilities

*Reed–Solomon and other burst-error correcting codes are specialized for specified burst-error patterns.

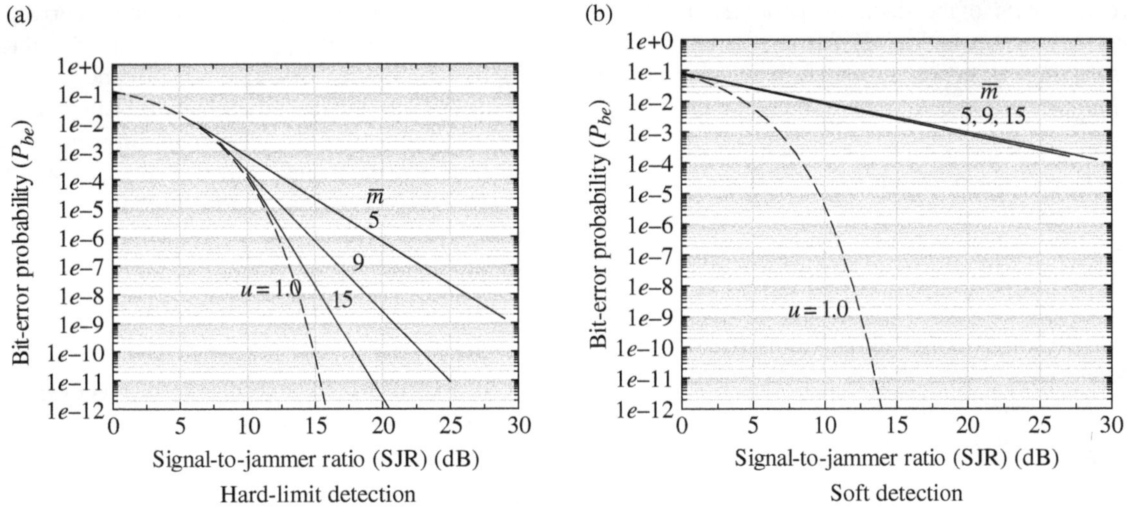

FIGURE 13.26 Worst-case repetition code performance with unknown jammer states.

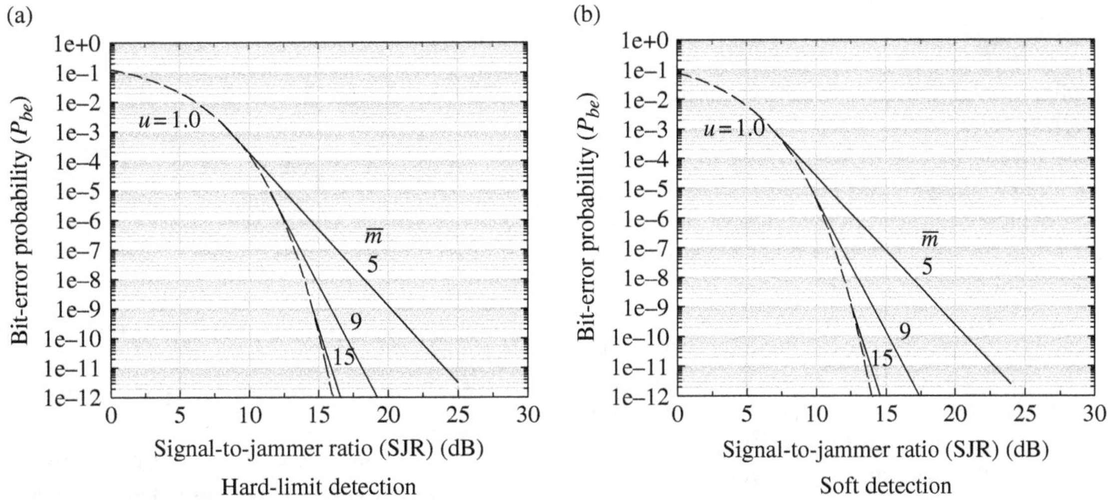

FIGURE 13.27 Worst-case repetition code performance with known jammer states.

under the indicated condition. These bit-error results are plotted in Figures 13.26a, b and 13.27a, b in terms of the worst-case performance using the pulsed noise jammer described in the preceding section.

$$P_{be} = \sum_{n=(\bar{m}+1)/2}^{\bar{m}} \binom{\bar{m}}{n} \varepsilon^n (1-\varepsilon)^{\bar{m}-n}$$

$$\varepsilon = uQ\left(\sqrt{\frac{2E_b}{\bar{m}N_J}u}\right) \tag{13.84}$$

: hard-decisions, unknown jammer state

$$P_{be} = \sum_{n=0}^{\bar{m}} \binom{\bar{m}}{n} u^n (1-u)^{\bar{m}-n} Q\left(\sqrt{\frac{2\bar{m}E_b}{nN_J}u}\right) \tag{13.85}$$

: soft-decisions, unknown jammer state

$$P_{be} = u^{\bar{m}} \sum_{n=(\bar{m}+1)/2}^{\bar{m}} \binom{\bar{m}}{n} \varepsilon^n (1-\varepsilon)^{\bar{m}-n}$$

$$\varepsilon = Q\left(\sqrt{\frac{2E_b}{\bar{m}N_J}u}\right) \quad \text{: hard-decisions, known jammer state} \tag{13.86}$$

and

$$P_{be} = u^{\bar{m}} Q\left(\sqrt{\frac{2E_b}{N_J}u}\right) \quad \text{: soft-decisions, known jammer state} \tag{13.87}$$

The *hard-decision detection* chooses $\hat{d}_{k\bar{m}}$ based on the sign of $y_{k\bar{m}}$ in accordance with the modulator mapping. An error in the data estimate $\hat{d}_k$ is based on having one-half or more of the n-of-$\bar{m}$ decisions being in error so the probability of error involves the *pdf* of the binomial distribution [53] with probability $p = \varepsilon$ as expressed in (13.84).

The *soft-decision detection* sums the $\bar{m}$ repetition code correlator outputs yielding $y_k = \sum_{\bar{m}} y_{k\bar{m}}$ and chooses $\hat{d}_k$ based on the sign of y_k in accordance with the modulator mapping. However, the probability that n-of-$\bar{m}$ repetitions are jammed is characterized by the *pdf* of the bionomical distribution, with the probability $p = u$ of being jammed, resulting in (13.85).

Knowledge of the jammer is applied to improve the communication performance. For example, through the use of channel monitoring, pulsed jammer transmissions can be determined and applied to the demodulator processing to excise their impact on the signal detection. In this event, a bit-error occurs with probability $u^{\bar{m}}$. With hard-decision detection, the n-of-$\bar{m}$ decision rule applies and the jammer is characterized by the bionomical *pdf* weighted by $u^{\bar{m}}$, leading to bit-error probability given by (13.86). With soft-decision detection, since all repetitions with jamming are ideally eliminated, a bit-error occurs when all $y_{k\bar{m}}$ repetitions are in error resulting in (13.87).

The assumptions involving ideal interleavers and knowledge of the jammer conditions result in an overly optimistic estimate of the performance. Furthermore, if practical statistical models of a selected interleaver and channel monitoring algorithm are included in the above bit-error analytical expressions, a computer simulation is in order to confirm or establish the actual bit-error performance.

Figure 13.26a and b show the worst-case bit-error probability using the pulsed noise jammer with the respective hard- and soft-decision detection algorithms for $\bar{m} = 5$, 9, and 15 repetition coding when the jammer state is unknown. The results using hard-decisions show significant improvement with increasing $\bar{m}$; however, the soft-decision case shows no improvement with repetition coding. The poor performance of the soft-decision detection results from summing all of the $y_{k\bar{m}}$ in establishing the overall decision static y_k and, with increasing jammer peak power, that is, decreasing u, the jammer increasingly dominates each data decision $\hat{d}_k$. Whereas, with hard-decision detection the decision is based on n-of-$\bar{m}$ hard-limited repetition code data decisions $\hat{d}_{k\bar{m}}$, with each of the repetition code decisions equally weighted. Figure 13.27a and b show the comparable performance when jammer state information is available. The hard-decision detection with known jammer conditions shows significant improvement over the unknown jammer case; however, the soft-decision detection performance exceeds that of the hard-decision detection with known jammer states.

13.5.4.1 Comment on the Worst-Case Pulsed Noise Jammer Performance

The worst-case bit-error probability performance, shown in Figure 13.23, is based on the theoretical optimum jammer strategy corresponding to the slope of $P_{be}(\gamma_{sj(\mathrm{dB})}, u)|_{u=1}$ as demonstrated in Figure 13.22. The worst-case performance in Figures 13.26 and 13.27 is based on a graphical assessment of the slope of the $P_{be}(\gamma_{sj(\mathrm{dB})}, u)|_{u=1}$ curve involving m repetitions of hard or soft decisions with unknown or known jammer states. The graphical procedure for determining the worst-case performance with jamming is discussed in Appendix 13A.

13.6 PERFORMANCE OF MFSK WITH PARTIAL-BAND NOISE JAMMING

The error performance of MFSK with partial-band noise jamming is evaluated using the expression (7.21) with the inclusion of the jammer pulsed noise duty factor $u = T_p/T_{pri}$. The result is expressed as

$$P_e = \left(\frac{2^{k-1}}{2^k-1}\right) P_{se}$$

$$= \left(\frac{2^{k-1}}{2^k-1}\right)\left[u\sum_{m=1}^{M-1}\binom{M-1}{m}\frac{(-1)^{m+1}}{m+1}e^{-u\gamma_{sj}m/(m+1)}\right] \tag{13.88}$$

where $k = \log_2(M)$ is the number of source bits-per-symbol. The signal-to-noise ratio in the bandwidth of the source bits is $\gamma_{bj} = E_b/N_{oJ} = \gamma_{sj}/k$, where γ_{sj} is the signal-to-noise ratio measured in the symbol bandwidth. By pulsing the noise power, the jammer creates a high peak power of duration T_p resulting in a partial-band noise jammer. The bit-error performance, characterized by (13.88), is plotted in Figure 13.28 for $M = 8$, 4, and 2.

The worst-case error performance, inflicted by the jammer, corresponds to the common tangent to all of the performance curves for u in the range of $0 \le u \le 1$. The jammer strategy is to use the optimum jamming parameter (u_o) that maximizes the communicator's bit-error performance. This worst-case condition for the communicator is determined by differentiating (13.88) with respect to u, setting the result equal to zero, and solving for u_o. For $k > 1$, the resulting solution involves a transcendental equation that can be approximated analytically using the Newton's method. An alternate solution involves determining the function $F(u) = dP_{se}(u)/du$, and noting that $F(u) = F(u\gamma_{sj})$. Upon letting $x = u\gamma_{sj}$, the function $F(x) = 0$ corresponds to the optimum jammer condition $x_o = u_o\gamma_{sj}$. The maximum bit-error probability, resulting from the jammers use of the optimum pulsed jammer condition u_o, is evaluated by substituting $u = u_o = x_o/\gamma_{sj}$ in (13.88). For a given $M = 2^k$, x_o is a constant

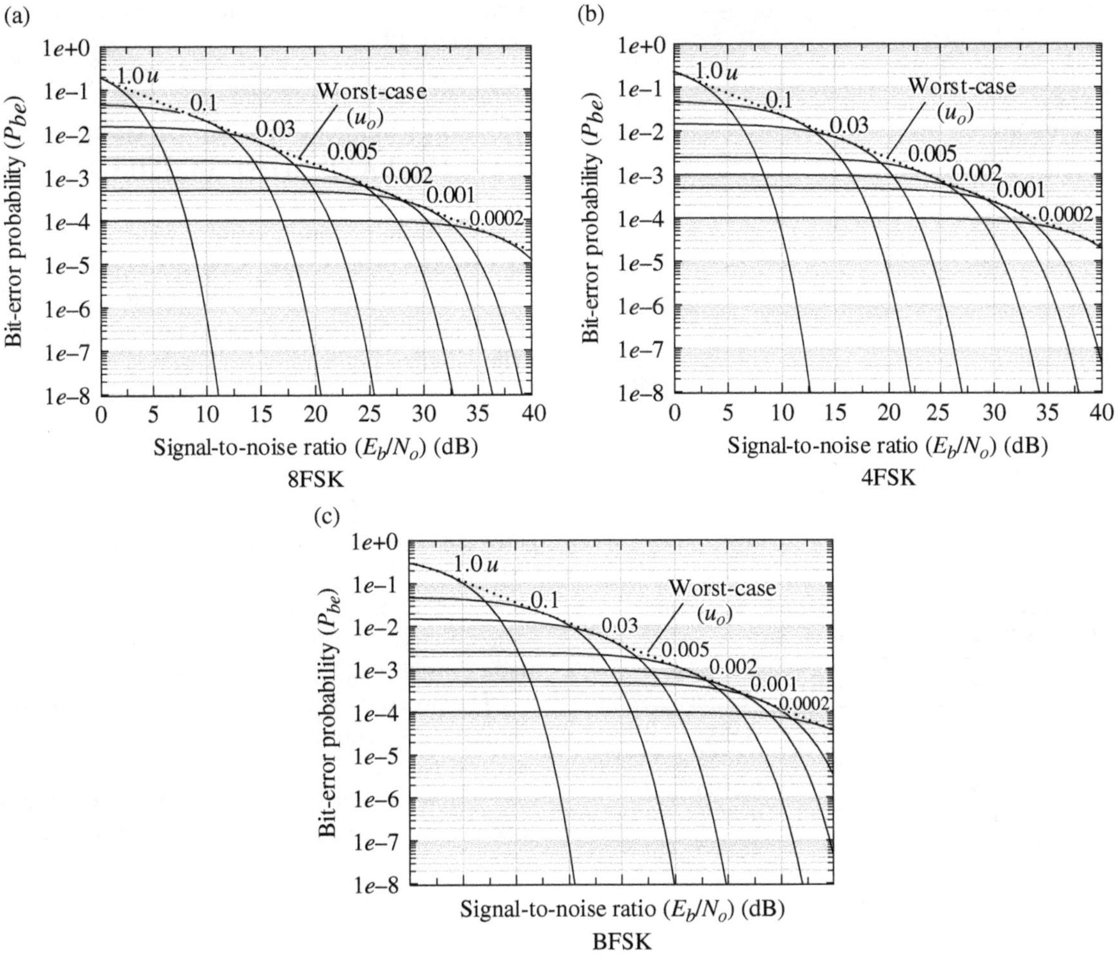

FIGURE 13.28 Noncoherent MFSK performance with partial-band noise jamming.

less than one,* so for jamming to occur the signal-to-jammer ratio must satisfy the condition $\gamma_{sj} > x_o$; otherwise, the ideal bit-error probability, given by (13.88) with $u = 1$, is achieved.

The function $F(x)$ is evaluated in terms of the derivative of the error probability, $dP_e(u)/du$, and is expressed as

$$F(x) = \frac{2^{k-1}}{2^k - 1}\left[\sum_{m=1}^{M-1}\binom{M-1}{m}\frac{(-1)^{m+1}}{m+1}e^{-xm/(m+1)}\left(1 - \frac{xm}{m+1}\right)\right] \tag{13.89}$$

Equation (13.89) is plotted in Figure 13.29 using increments of $\Delta x = 0.05$ with x_o determined using linear interpolation of the neighboring the points around $F(x_o) = 0$; the results are summarized in Table 13.6 with $x'_o = x_o/k = \gamma_{bj}$ and $\gamma_{bj\min}(\mathrm{dB}) = 10\log_{10}(x'_o)$. The maximum bit-error probability of the communicator is determined by substituting $u = u_o$ in (13.88) resulting in

$$P_e = \frac{2^{k-1}}{2^k - 1}\left[\frac{x_o}{\gamma_{sj}}\sum_{m=1}^{M-1}\binom{M-1}{m}\frac{(-1)^{m+1}}{m+1}e^{-x_o m/(m+1)}\right]$$

: worst-case

(13.90)

The worst-case bit-error performance, described by (13.90), is plotted as the *dotted lines* in Figure 13.28 as a function of E_b/N_o in dB with the parameter $k = 1$, 2, and 3.

Table 13.6 is in excellent agreement with Simon et al. [54], and the values of u_o correspond to the points at which the tangent of the unjammed P_{be} performance curve ($u = 1$) is equal to the slope of the linearly decreasing worst-case jammer performance. The parameter $x_o = u\gamma_{sj}$ corresponds to $u = 1$ and the signal-to-noise ratio γ_{sj} is measured in the symbol bandwidth, so $x'_o = \gamma_{bjmin} = x_o/k$ is the signal-to-noise ratio measured in the bandwidth equal to the bit-rate. The bit-error probability $P_{be}(\gamma_{bj\,min})$ corresponds to the unjammed performance at $\gamma_{bj\,min}$ and u_o is computed as

$$u_o = x'_o P_{be}(\gamma_{bjmin}) = \gamma_{bjmin} P_{be}(\gamma_{bjmin}) \tag{13.91}$$

*Actually, $u_o \le 1$; however, when $u_o = 1$ the communicator is not jammed.

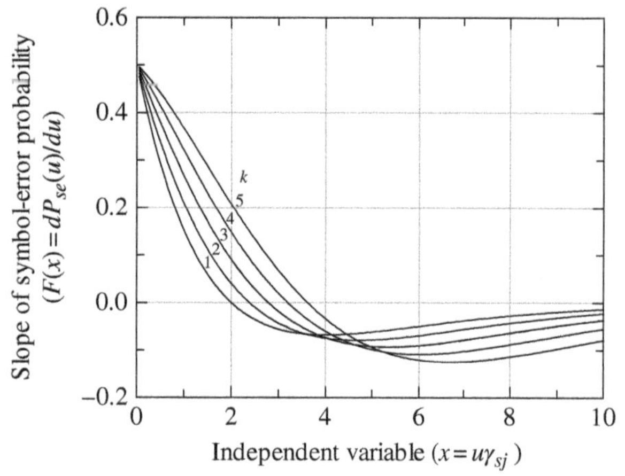

FIGURE 13.29 Determination of the jammer parameter $x_o=\gamma_{sj\,min}$ for MFSK waveform modulation with $u=1$.

TABLE 13.6 Summary of Optimum Jammer Parameter against MFSK ($M=2^k$)-Modulated Waveform

$k=\log_2(M)$	x_o	x_o'	$\gamma_{bj\,min}$ (dB)	$P_{be}(x_o')$	u_o
1	2.000	2.000	3.010	0.1839	0.3679
2	2.383	1.191	0.761	0.1955	0.2328
3	2.783	0.927	−0.326	0.2106	0.1953
4	3.193	0.798	−0.979	0.2271	0.1813
5	3.613	0.723	−1.409	0.2433	0.1758

Note: x_o' corresponds to $\gamma_{bj\,min}$ with $u=1$.

In Figure 13.28 the point of intersection between the unjammed and worst-case jammed performance occurs at $\gamma_{bj}=\gamma_{bj\,min}$ and for $\gamma_{bj}<\gamma_{bj\,min}$ the performance corresponds to the unjammed bit-error performance expressed in* (13.88) with $u=1$; otherwise, the performance corresponds to the worst-case condition based on the linearly decreasing bit-error probability with increasing γ_{bj}. These results are summarized in (13.92) as

$$P_{be}=\begin{cases}\left(\dfrac{2^{k-1}}{2^k-1}\right)\left[\displaystyle\sum_{m=1}^{M-1}\binom{M-1}{m}\frac{(-1)^{m+1}}{m+1}e^{-k\gamma_{bj}m/(m+1)}\right] & : u=1, \gamma_{bj}<\gamma_{bjmin}\\ \dfrac{u_o}{\gamma_{bj}} & : u<1,\ \gamma_{bj}\ge\gamma_{bjmin}\end{cases} \tag{13.92}$$

The preceding discussion concerning the evaluation of $\gamma_{bj\,min}$, $P_{be}(\gamma_{bj\,min})$, and u_o is based on theory from which P_{be} is evaluated using (13.92) and plotted as in Figure 13.28 for various value of k. However, the performance parameters for the worst-case jamming can also be generated graphically with remarkable accuracy as discussed in Appendix 13A.

13.7 PERFORMANCE OF DCMPSK WITH PARTIAL-BAND NOISE JAMMING

The bit-error performance of differentially coherent BPSK (DCBPSK) with partial-band noise jamming is evaluated using the bit-error probability expression (4.42) with the inclusion of the jammer pulsed noise duty factor $u=T_p/T_{pri}$. The bit-error probability is expressed as

$$P_{be}=\frac{u}{2}e^{-u\gamma_{bj}} \tag{13.93}$$

where $\gamma_{bj}=P_s/N_J$ is the signal-to-jammer power ratio measured in the bandwidth equal to the bit-rate, that is, $N_J=N_{oJ}R_b$. By pulsing the noise power, the jammer creates a high peak power of duration T_p resulting in a partial-band noise jammer. The bit-error performance, characterized by (13.93), is plotted in Figure 13.30 for various value of u.

The worst-case bit-error performance, inflicted by the jammer, corresponds to the common tangent to all of the curves for the range of $0\le u\le 1$. Upon differentiating (13.93) with respect to u, setting the result equal to zero, and solving for u_o, the optimum jammer strategy is to use

$$u_o=\frac{1}{\gamma_{bj}}\quad : \text{optimum jammer strategy} \tag{13.94}$$

Equation (13.94) corresponds to $\gamma_{bj}|_{u<1}>1$ and the worst-case communicator bit-error (BE) performance is evaluated by substituting (13.94) into (13.93) resulting in

$$P_{be}=\frac{e^{-1}}{2\gamma_{bj}}=\frac{0.184}{\gamma_{bj}}\quad : \gamma_{bj}>1,\ \text{worst-case BE performance} \tag{13.95}$$

Otherwise, $u=1$ and the communicator's BE performance is given by the optimum or theoretical expression

$$P_{be}=\frac{1}{2}e^{-\gamma_{bj}}\quad : \gamma_{bj}\le 1,\ \text{best-case BE performance} \tag{13.96}$$

Equation (13.95) is plotted as the *dotted curve* in Figure 13.30.

The symbol-error probability for DCMPSK is expressed by Simon [55] as

$$P_{se}=\frac{\sin(\pi/M)}{\pi}\int_0^{\pi/2}\frac{e^{-k\gamma_b(1-\cos(\pi/M)\cos(\alpha))}}{1-\cos(\pi/M)\cos(\alpha)}d\alpha\quad : \text{DCMPSK} \tag{13.97}$$

*The bit-error performance is plotted as a function of $\gamma_{bj}=\gamma_{sj}/k$.

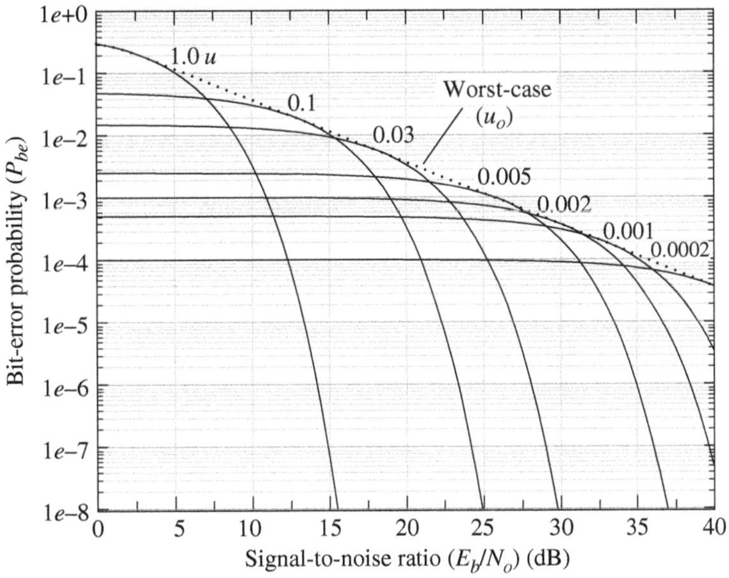

FIGURE 13.30 DCBPSK performance with partial-band noise jamming.

where $k=\log_2(M)$ and $\gamma_b=E_b/N_o$. The performance with partial-band noise jamming is expressed using (13.97)* by replacing γ_b with $\gamma_{bj}=E_b/N_{oJ}$ and then multiplying the integral and γ_{bj} by the parameter u. Upon making these modifications and converting from symbol-errors to bit-errors using random bit-to-symbol mapping, the bit-error performance is expressed as

$$P_{be}=\left(\frac{2^{k-1}}{2^k-1}\right)\gamma_{bj}^{-1}\left\{u\gamma_{bj}\left(\frac{\sin(\pi/M)}{\pi}\right)\int_0^{\pi/2}\frac{e^{-ku\gamma_{bj}(1-\cos(\pi/M)\cos(\alpha))}}{1-\cos(\pi/M)\cos(\alpha)}d\alpha\right\} \tag{13.98}$$

Equation (13.98) is placed in this form† to facilitate the computation of the optimum jammer strategy u_o that results in P_{be}(max). For example, the P_{se} expressed in (13.97) is evaluated from Table 5.5 in Lindsey and Simon [56] using various signal-to-noise ratios $\gamma_b=u\gamma_{bj}$; the resulting P_{se} is multiplied by $u\gamma_{bj}$ until P_{se}(max) is determined corresponding to $(u\gamma_{bj})_{max}$. Therefore, using $(u\gamma_{bj})_{max}$, the optimum jammer strategy is evaluated as

$$u_o=\gamma_{bj}^{-1}\left(u\gamma_{bj}\right)_{max} \quad \text{: DCMPSK optimum jammer strategy} \tag{13.99}$$

Based on (13.99), the worst-case communicator's bit-error (BE) performance is evaluated for $\gamma_{bj}|_{u<1}>\left(u\gamma_{bj}\right)_{max}$, as

$$P_{be}=\left(\frac{2^{k-1}}{2^k-1}\right)\gamma_{bj}^{-1}P_e(\max) \tag{13.100}$$

: DCMPSK, worst-case BE performance

The error probability P_e (max) is determined from the bracketed term {–} in (13.98) with $u\gamma_{bj}=\left(u\gamma_{bj}\right)_{max}$ and is expressed as

$$P_e(\max)=\left(u\gamma_{bj}\right)_{max}\left(\frac{\sin(\pi/M)}{\pi}\right)\int_0^{\pi/2}\frac{e^{-k\left(u\gamma_{bj}\right)_{max}(1-\cos(\pi/M)\cos(\alpha))}}{1-\cos(\pi/M)\cos(\alpha)}d\alpha \tag{13.101}$$

Otherwise, for $\gamma_{bj}|_{u=1}\le\left(u\gamma_{bj}\right)_{max}$, the theoretical unjammed bit-error performance expressed as

$$P_{be}=\left(\frac{2^{k-1}}{2^k-1}\right)P_{se} \quad \text{: DCMPSK, best-case BE performance} \tag{13.102}$$

is used where P_{se} is the unjammed symbol-error probability expressed in (13.97).

13.8 FHSS WAVEFORMS WITH MULTITONE JAMMING

The impact of a single-tone jammer on DSSS with BPSK and QPSK waveform modulation is discussed in Section 13.5.1 and found to be most effective when the jammer tunes the CW jammer signal to the communicator's carrier frequency.

*This expression does not have a closed-form solution and must be evaluated using numerical integration.

†This method is proposed by Simon et al. [6], vol. II, pp. 326–329.

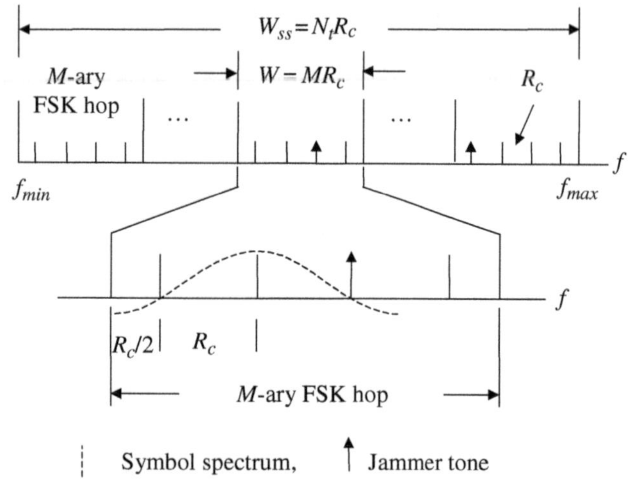

FIGURE 13.31 Band multitone jamming strategy against FH MFSK waveform (for e.g., $M=4$ with $n=1$ jammer tone/hop).

However, with FHSS systems this becomes increasingly difficult as the FH rate is increased. Consequently, the jammer is forced to use multiple jamming tones with the potential of concentrating more jamming signal energy in the communicator's detection filter bandwidth than with partial-band noise jamming.

The following analysis of multitone jamming of a FHSS system with MFSK waveform modulation and noncoherent detection is based on the work of Simon et al. [6]. Figure 13.31 depicts the parameters required to quantify the performance analysis.

The total jamming power is denoted as P_{JT} and the jammer places Q equal power CW tones given by

$$P_j = \frac{P_{JT}}{Q} \quad \text{: equal power CW jamming tones} \tag{13.103}$$

spread over the SS bandwidth of W_{ss} Hz. Each communication M-ary frequency hop (or slot) occupies an instantaneous bandwidth of $W = MR_c$ Hz, where R_c is the M-ary symbol rate corresponding to orthogonally spaced MFSK symbol tones. In this characterization* of the FH structure, the number of FH user carrier frequency locations available in the SS bandwidth is $N_t = M(W_{ss}/W) = W_{ss}/R_c$ and the fraction of the available FH-user frequency locations that are jammed is defined as

$$\rho \triangleq \frac{\text{jammed tones}}{\text{total tones}} = \frac{Q}{N_t} \tag{13.104}$$

*Brief summary of notation: N_t = user carrier frequency locations, N_t/M = user M-ary hops (or frequency hops) available, Q = CW jamming tone locations, and Q/n is the jammer hops available with $n \le M$ jammer tones/hop.

where Q is the number of jamming tone locations used to maximize the communicator's bit-error probability. Typically $Q \gg M$ and, defining the number of user M-ary hops as N_t/M with n jammer tones per FH hop, the Q/n ($Q/n \in$ integer) jammed hops available are randomly selected from among the available M-ary user hops. The preferred [57] definition of the SS processing gain with FHSS is $G_p = W_{ss}/R_b$, where R_b is the information bit-rate. This definition is independent of the bits-per-symbol and FEC code-rate; however, the definition of W_{ss} may vary with the FH plan. Furthermore, the fundamental measure of the SS waveform performance is the bit-error probability dependence on the signal-to-noise ratio (E_b/N_{oJ}) measured in the bandwidth equal to R_b Hz where the jammer noise density is defined as

$$N_{oJ} \triangleq \frac{P_{JT}}{W_{ss}} \tag{13.105}$$

A reasonable jammer strategy is to distribute the jammer tones over the bandwidth W_{ss} with n tones ($1 \le n \le M$) in each of the integer Q/n jammed hops available. For the Q/n hops that are jammed, a jammer tone in any one of the $M-1$ user frequency locations not containing the transmitted symbol energy may cause a symbol error if $P_J + N_J > P_s + N'_J$, where P_s is the M-ary carrier signal power and N_J and N'_J are the respective noise powers at the sampled output of the demodulator matched filters. To simplify the following analysis, P_s and P_J are considered to be much greater than the noise powers so the noise is neglected. Another simplifying assumption, which results in the worst-case communications performance, is that each of the jamming tone locations in the set $\{Q\}$ must coincide exactly in frequency with one of the N_t user carrier frequency locations. However, it is not necessary that the jammer tone locations be synchronized with the user FH transitions as long as the synchronization is commensurate with the previous requirement that the jammer tones always coincide with the N_t user carrier frequencies; in other words, the jammer may assume the FH synchronization to be $Th'_{sync} = Th_{sync} + \ell/R_c$, where ℓ is an integer and R_c is the M-ary symbol or chip rate.

The strategy discussed above, with the simplifying assumptions, is referred to as *band multitone jamming* as depicted in Figure 13.31 that uses the example for $n=1$ jammer tone in each of the Q/n jammed hops. The jamming tone in each hop is randomly selected in the range $1 \le \ell \le M$. The left-most M-ary hop in Figure 13.31 is representative of the $(N_t - Q)/M$ hops *not* containing a jammer tone and the *dotted curve*, in the expanded view, represents the spectrum of a FH–MFSK data symbol. The figure is also specialized for $M=4$ user frequencies/hop. In general, band multitone jamming uses $M = 2^k$ tones/hop with k bits per M-ary symbol and exactly n jammer tones in each jammed hop.

For the general case of placing n ($1 \le n \le M$) jamming tones in each of the Q/n jammed hops, total number of jammed hops equal to N_t/M and the fraction of jammed hops is defined as

$$u \triangleq \frac{\text{jammed hops}}{\text{total jammed hops}} = \left(\frac{Q/n}{N_t/M}\right) = \frac{MQ}{nN_t} = \frac{M}{n}\rho \quad (13.106)$$

Upon defining the parameter α as

$$\alpha \triangleq \frac{P_s}{P_J} \quad (13.107)$$

the jammer strategy optimizes parameter α by maximizing the error probability $P_e(\alpha)$ resulting in the *noise-free** worst-case performance with $\alpha = \alpha_0$ as characterized in (13.108).

$$P_e(\alpha) = \begin{cases} 0 & : \text{any}\,\alpha;\ \text{under condition 1} \\ 0 & : \alpha > 1;\ \text{under condition 2} \\ 1 & : \alpha \le 1;\ \text{under condition 3} \\ \frac{1}{\pi}\cos^{-1}(\sqrt{\alpha}/2) & : \alpha < 4;\ \text{under condition 4} \end{cases} \quad (13.108)$$

The conditions for the error probability associated with (13.108) are:

1. The data symbol or any of the other $M-1$ symbol tone locations are not jammed.
2. The data symbol is not jammed and any of the other $M-1$ symbol tone locations are jammed with $P_J < P_s$.
3. The data symbol is not jammed and any of the other $M-1$ symbol tone locations are jammed with† $P_J \ge P_s$.
4. The data symbol and any of the other $M-1$ symbol tone locations are simultaneously jammed, an error with the indicated probability $P_e(\alpha)$ results if $\cos(\phi) < -\sqrt{\alpha}/2$.

The third condition applies when the data symbol is not jammed and any of the other symbol locations are jammed with α in the range $0 \le \alpha \le 1$. The special case involving the phase difference ϕ between the desired symbol frequency and jammer tone frequency applies only when the data symbol and any of the other $M-1$ tone locations are simultaneously jammed. The phase difference is uniformly distributed over the range $0 \le \phi \le 2\pi$. The impact of the jammer phase on the detection filter is credited to Trumpis [58]; previously the range was considered to be $0 < \alpha \le 1$.

Equation (13.106) can be thought of as the probability of a symbol being jammed under condition 3 so the symbol-error probability is expressed as

$$P_{se} = u\left(\frac{M-1}{M}\right) \quad (13.109)$$

The bit-error probability, with random bit-to-symbol mapping, is expressed as

$$P_{be} = \frac{M}{2(M-1)}P_{se} = \frac{u}{2} \quad (13.110)$$

The parameter u is evaluated by substituting $Q = P_{JT}/P_J$ from (13.103) with $N_t = W_{ss}/R_c$ resulting in

$$u = \left(\frac{M}{n}\right)\frac{P_{JT}R_c}{P_J W_{ss}} = \left(\frac{M}{n}\right)\frac{N_{oJ}(R_b/k)}{(P_s/\alpha)} = \frac{M\alpha}{nk\gamma_{bj}} \quad (13.111)$$

The second equality in (13.111) results from using (13.105) and (13.107) and recognizing that the bit-rate for the M-ary symbol is $R_b = R_c/k$. The final result is based on the signal-to-noise ratio, measured in the bandwidth R_b expressed as $\gamma_{bj} = E_b/N_{oJ} = P_s T_b/N_{oJ}$. From (13.110), the resulting bit-error probability under condition 3 is expressed as

$$P_{be} = \frac{\alpha M}{2nk\gamma_{bj}} \quad : \text{band multitone jammer strategy}, \quad 1 \le n \le M,\ 0 \le \alpha \le 1 \quad (13.112)$$

13.8.1 Single Jammer Tone in Each Jammed M-ary Hop

With a single jammer tone ($n = 1$) assigned to each of the Q FH jammed hops, the bit-error probability is expressed as

$$P_{be} = \left\{ \frac{\alpha M}{2k\gamma_{bj}} \right. \quad : n = 1,\ \text{band multitone jammer strategy} \quad (13.113)$$

In this case, finding the optimum jammer condition α_0 that maximizes $P_r(J|Q)$ under the constraints‡ $\alpha \le 1$ and $P_r(J|Q) \le 1$ is fairly straightforward because of the linear relationship with α in (13.112). The result is the solution to $P_r(J|Q) = \alpha_0 M/(k\gamma_{bj}) \le 1$ with $\gamma_{bj} = M/k$ and, since $\alpha_0 \le 1$, this requires that the multitone jammer select α_0 according to the following signal-to-jammer ratio conditions,

$$\alpha_o = \begin{cases} \frac{k\gamma_{bj}}{M} & : \gamma_{bj} \le \frac{M}{k} \\ 1 & : \gamma_{bj} > \frac{M}{k} \end{cases} \quad (13.114)$$

*The noise-free condition simplifies the analysis by assuming that $P_J \gg P_{noise}$.

†The case $P_J = P_s$ is counted as an error.

‡The condition $\alpha = 1$ is considered to result in an error.

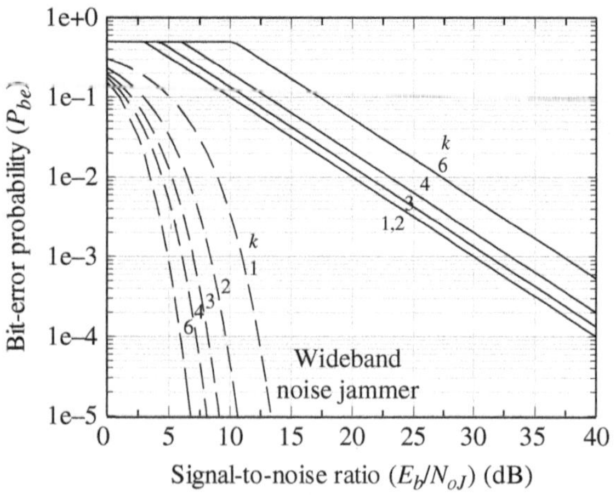

FIGURE 13.32 Worst-case performance of FH MFSK with band multitone jamming ($n = 1$).

Substituting (13.114) into (13.113) results in the worst-case communicator bit-error performance for $n = 1$ expressed as

$$P_{be} = \begin{cases} \dfrac{1}{2} & : \gamma_{bj} \le \dfrac{M}{k} \\ \dfrac{2^{k-1}}{k\gamma_{bj}} & : \gamma_{bj} > \dfrac{M}{k} \end{cases} \tag{13.115}$$

Equation (13.115) is plotted in Figure 13.32 as a function of γ_{bj} for several values of $k = \log_2(M)$. The *dashed curves* represent the corresponding unjammed performance of the MFSK waveform when N_{oJ} is considered to be AWGN. The worst-case results for $k = 1$ and 2 are identical because the threshold M/k is the same for both cases; otherwise, the loss increases with increasing k. As plotted in the figure, the bit-error performance decreases linearly with increasing signal-to-jammer ratio (γ_{bj}) above the *break-points*; defined as the intersection of the linearly decreasing and constant P_{be} performance curves.

The worst-case band multitone jamming bit-error performance given in (13.115) is significantly larger than that of partial-band noise jamming expressed in (13.92). Using these expressions, the signal-to-jammer ratio advantage of the band multitone jammer for $n = 1$, over that of a partial-band noise jammer is evaluated as

$$\Delta\text{dB} = 10\log_{10}\left(\frac{2^{k-1}}{ku_o}\right) \quad : \text{advantage of band multitone jammer} \, (n = 1) \tag{13.116}$$

where u_o is the optimum jammer constant used to maximize the communicators bit-error performance. Using the relationship between k and u_o in Table 13.6, the jammer advantage of band multitone jamming is tabulated in Table 13.7.

TABLE 13.7 FH MFSK Advantage of Band Multitone Jammer over Partial-Band Noise Jammer ($n = 1$)

k	u_o	ΔdB
1	0.3679	4.34
2	0.2328	6.33
3	0.1953	8.34
4	0.1813	10.43
5	0.1758	12.60

13.8.2 Multiple Jammer Tones in Each Jammed MFSK Hop

The worst-case performance analysis for $1 < n \le M$, where one of the jammer tones coincides with an MFSK data tone is considerably more involved, in that, the phase difference, ϕ, between the jammer and MFSK data tones, must be considered. The analysis in this case involves the probability of an error event identified by the third and fourth conditions in (13.108). The complexity in the analysis involves the error probabilities $P_e(\alpha) = 1$ in the range $0 < \alpha \le 1$ and $P_e(\alpha) = \arccos((\sqrt{\alpha}/2)/\pi)$ conditioned on $\cos(\phi) < -\sqrt{\alpha}/2$ in the range $1 < \alpha \le 4$. As in the case for $n = 1$ this analysis follows that of Simon et al. [59]. In this case, the symbol-error probability is expressed as

$$P_{se} = P_e\left(\frac{J}{Q}\right)\left[\left(1 - \frac{n}{M}\right)u'(1-\alpha) + \left(\frac{n}{M\pi}\right)\cos^{-1}\left(\frac{\sqrt{\alpha}}{2}\right)\right] \tag{13.117}$$

where the unit step function $u'(1-\alpha) = 1$ for $\alpha < 1$ and 0 otherwise. The bit-error probability is determined by substituting $P_e(J|Q)$ from (13.112) with uniform bit-to-symbol mapping and then maximizing the P_{be} with respect to α. The result is expressed as

$$P_{be} = \frac{Mc}{2\gamma_{bj}(M-1)}\max_{\alpha=\alpha_0}\left[\alpha\left(1 - \frac{n}{M}\right)u'(1-\alpha) + \left(\frac{n}{M\pi}\right)\alpha\cos^{-1}\left(\frac{\sqrt{\alpha}}{2}\right)\right] \tag{13.118}$$

where $c = M/nk$ and the optimum jammer strategy is obtained by maximizing P_{be} over the range $0 < \alpha \le \min(4, \gamma_{bj}/c)$ corresponding to $\alpha = \alpha_0$. Both of the additive terms in the *square brackets* of (13.118) increase monotonically for $\alpha < 1$; however, for $\alpha \ge 1$ only the *right-hand* additive term applies, reaching a maximum value at $\alpha = 2.52$ corresponding to

$$\left(\frac{\alpha}{\pi}\right)\cos^{-1}\left(\frac{\sqrt{\alpha}}{2}\right)\Big|_{\alpha=2.52} = 0.525 \quad : \text{maximum for } 0 < \alpha \le 4 \tag{13.119}$$

Simon et al. [60] have developed the conditions under which (13.118) is evaluated given the dependence of the bit-error performance on the regions over which the signal-to-jammer ratio (γ_{bj}) applies, the relative number of jammer tones-per-symbol ($\rho = n/M$), and the applicable conditions 3 and 4 in (13.108). Based on these dependencies and the conditions $1 \le n \le M$ and $c = M/nk$, the bit-error probability is evaluated as follows [60]:

if $(\gamma_{bj} \le c)$ then

$$\alpha_o = \frac{\gamma_{bj}}{c} \tag{13.120}$$

$$P_{be} = \frac{M}{2(M-1)}\left[1 - \frac{n}{M} + \frac{n}{M\pi}\cos^{-1}\left(\frac{\sqrt{\alpha_o}}{2}\right)\right] \tag{13.121}$$

$u = 1;\ \rho = n/M$! Parameter values

else

! The following simplifies the maximization of (13.118),
! with the discontinuity at $\alpha = 1$, by maximizing P_{be}
! separately over the ranges $0 < \alpha < 1$ and $1 \le \alpha \le$
! γ_{bj}/u and then choosing the greater of the two values:

$$P_{be} = \frac{Mc}{2(M-1)\gamma_{bj}}\max\left(1 - \frac{2n}{3M},\ \frac{\beta n}{M}\right) \tag{13.122}$$

if $(\gamma_{bj} \le 2.52c)$ then

$$\beta = \frac{\gamma_{bj}}{\pi c}\cos^{-1}\left(\frac{1}{2}\sqrt{\frac{\gamma_{bj}}{c}}\right)$$

else

$\beta = 0.525$! Maximum value at $\alpha = 2.52$

endif

$\gamma = \dfrac{1}{\beta + 2/3}$! Temporary constant

if $(n \le \gamma M)$ then

$$\alpha_o = 1 \tag{13.123}$$

$$P_{be} = \frac{c(M - 2n/3)}{2(M-1)\gamma_{bj}} \tag{13.124}$$

$u = \dfrac{c}{\gamma_{bj}};\ \rho = \dfrac{1}{(k\gamma_{bj})}$! Parameter values

else

$$\alpha_o = \min\left(2.52,\ \frac{\gamma_{bj}}{c}\right) \tag{13.125}$$

$$P_{be} = \frac{M\beta}{2k(M-1)\gamma_{bj}} \tag{13.126}$$

$u = \min\left(1,\ \dfrac{2.52c}{\gamma_{bj}}\right)$

! Parameter values

$\rho = \min\left(\dfrac{n}{M},\ \dfrac{2.52}{(k\gamma_{bj})}\right)$

endif

endif

The conditions and algorithms described and outlined above for the band multitone jammer are plotted in Figures 13.33, 13.34, and 13.35 as a function of the energy-per-bit to the jammer noise-density ratio for $k = 2$, 3, and 4 with the indicated ranges of the parameter n. Starting in the vicinity of $E_b/N_{oJ} = 5$ dB, there is an inverse linear logarithmic relationship with P_{be} as E_b/N_{oJ} increases. For a given k the jammer effectiveness decreases with increasing n; however, there is slight jammer advantage as k increases.

The jammer-to-signal power ratio (P_J/P_s) is also shown as a function of E_b/N_{oJ} corresponding to each figure. For negative values of $10\log_{10}(E_b/N_{oJ})$, there is an inverse logarithmic relationship in $10\log_{10}(P_J/P_s)$ with increasing signal-to-noise ratio up to about 0 dB. For positive values of the signal-to-noise ratio, the power ratio saturates at 0 or −4.01 dB depending upon the value of M and n. For example, when $n = 1$ the power ratio saturates at 0 dB and when $n = M$ the power ratio saturates* at −4.01 dB. This latter condition corresponds to the jammer power operating 4.01 dB below signal power; however, the condition $n = 1$ corresponds to the worst-case jammer. For $1 < n \le M - 1$, the P_J/P_s ratio corresponds to −4.01 dB for n in the vicinity of M otherwise the ratio saturates at 0 dB. To quantify these relationships, the processing for the worst-case band multitone jammer performance is depicted in Figure 13.36 for 32FSK for various values of n. For this case, it is seen from Table 13.8 that for values of $n \ge 27$ are observed to saturate at P_J/P_s ratios corresponding to −4.01 dB over the indicated range of signal-to-noise ratios. Similar evaluations were performed for the other values of M and n and the results are summarized in Table 13.8. Based on this analysis, the optimum jammer strategy for band multitone jamming would most likely be to choose $n = 1$.

Simon et al. [61] discuss and analyze the performance of *independent multitone jamming* that randomly distributes the Q jamming tones over the range N_t FH-tone locations without regard to the communicator's M-ary symbol tone locations. This jammer strategy also involves n over the range $0 < n \le M$ and the upper limit of $\alpha \le 4$ as noted by Trumpis. The independent multitone strategy performance is similar to the band multitone strategy, in that, the performance is upper-bounded by the piece-wise linear performance of the multitone jammer on either side of the *breakpoint*.

The authors also analyze and discuss various jamming counter measures including time-diversity repetition coding and convolutional, Reed–Solomon, and concatenated coding. Coding and interleaving are essential to maintain reliable communications with jamming and the theoretical results should be used as benchmarks for computer simulation of SS systems to verify the validity of the underlying

*This condition results from $\alpha_o = 2.52$ in (13.125) with $P_J/P_s = 1/\alpha_o$.

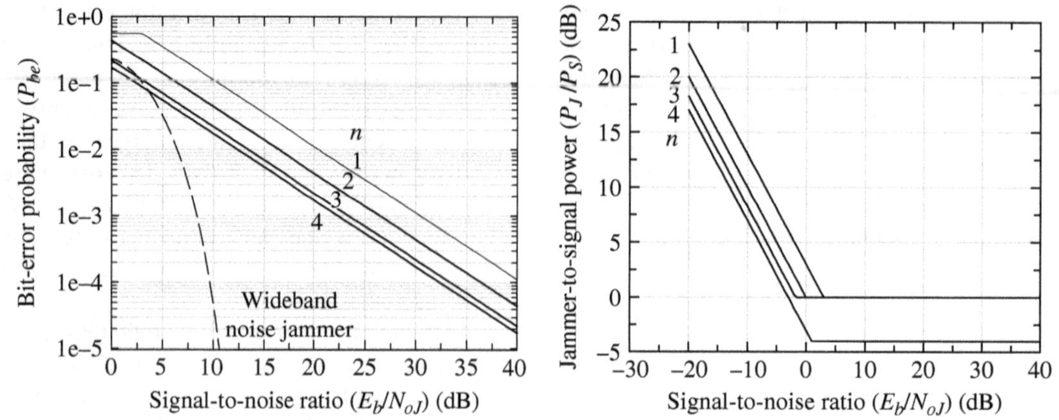

FIGURE 13.33 Worst-case performance of FH MFSK with band multitone jamming ($k = 2$, $1 \le n \le 4$).

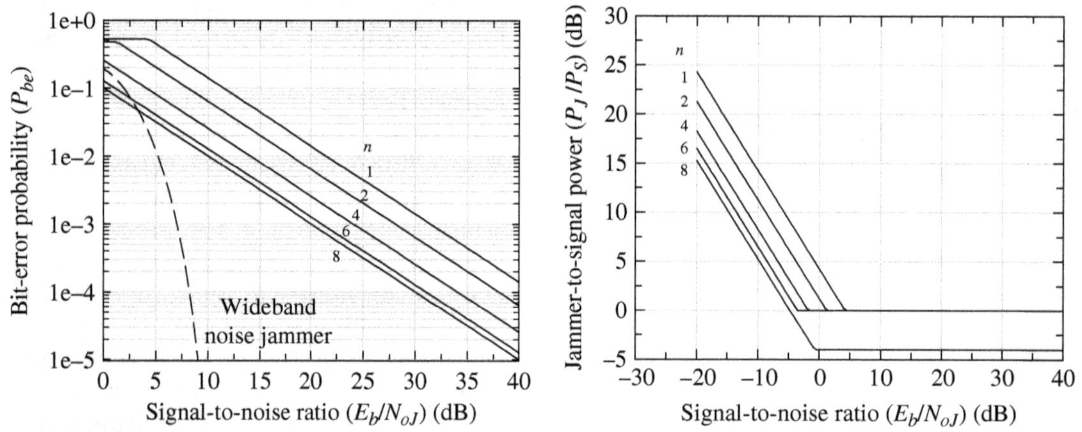

FIGURE 13.34 Worst-case performance of FH MFSK with band multitone jamming ($k = 3$, $1 \le n \le 8$).

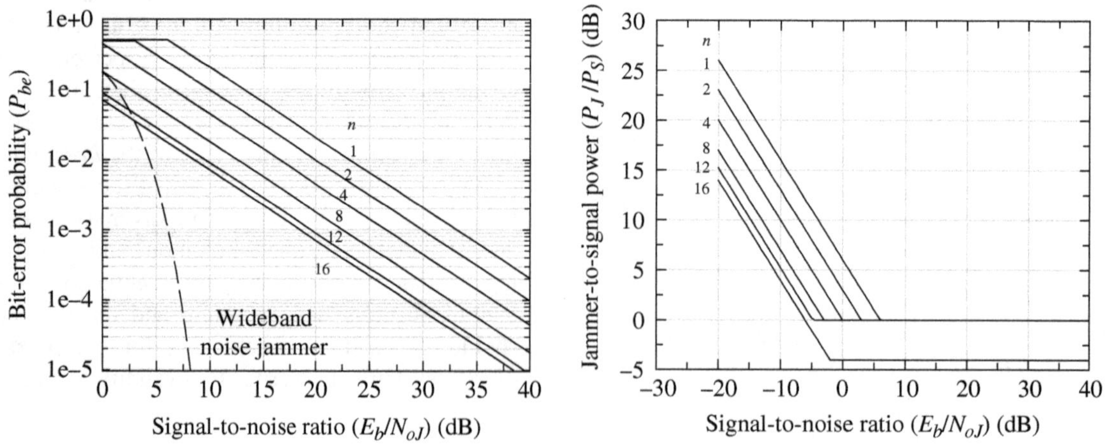

FIGURE 13.35 Worst-case performance of FH MFSK with band multitone jamming ($k = 4$, $1 \le n \le 16$).

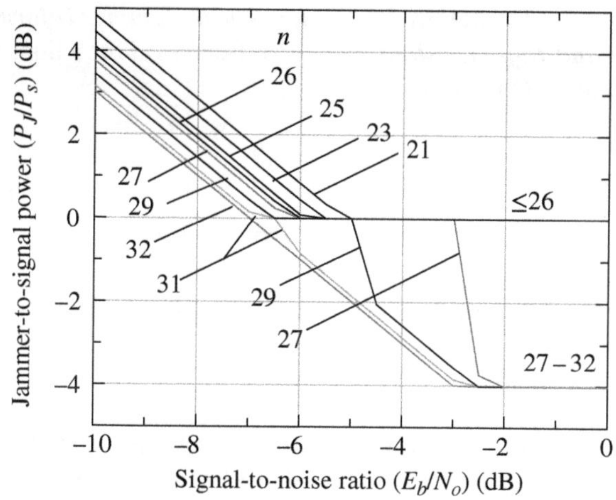

FIGURE 13.36 Limiting behavior of the P_J/P_s ratio for MFSK with $M = 32$ and various $n \le M$.

TABLE 13.8 Relationship between n and M and the Limiting Behavior of the P_J/P_s Ratio

M	2	4	8	16	32	64
n^a	=2	=4	≥7	≥14	≥27	≥54

[a]Value of n for P_J/P_s (dB) = −4.01 as γ_{bj} increases.

assumption of independence of the uncoded symbols or code symbol when coding is used.

13.9 APPROXIMATE PERFORMANCE WITH JAMMER THREATS

The simplified evaluation, using upper bounds on the performance of various modulations and coding configurations in a memoryless channel, is introduced in Section 3.4. The work of Simon et al. [62] provides a comprehensive description of the upper bound performance measures with tables summarizing the performance under a variety of communicator and jammer conditions. In these cases, the analysis is based on memoryless channels that are considered to exist by providing sufficient interleaving; in a practical design, this assumption must be substantiated through simulations. Nevertheless, the upper bound approximations are useful in providing estimates that are typically within a decibel of the theoretical performance. The summary tables [62] include the performance bounds for DSSS BPSK with CW jamming and pulsed noise jamming as discussed in Section 13.9.1 and for FH MFSK with partial-band noise jamming as discussed in Section 13.9.2. Each of the tables provided by the authors includes formulas for computing the upper performance bounds with and without jammer state information and with hard and soft demodulator detections. Knowledge of the jammer state is typically available from other sensor systems and is embodied in the analysis by including the unjammed or ideal performance. In the following sections, a brief introduction to the procedure of evaluating the upper bound performance is provided.

13.9.1 DS-BPSK with Pulsed Noise Jamming

In this case, the bit-error probability for DS-BPSK with pulsed noise jamming is upper-bounded [63] by*

$$P_{be} \le \frac{1}{2}D \quad : \text{DS} - \text{BPSK upper bound} \qquad (13.127)$$

with the computational code-rate evaluated as

$$R_o = 1 - \log_2(1 + D) \quad : \text{bits/symbol} \qquad (13.128)$$

The following two sections describe the parameters for evaluating the performance of DS-BPSK with pulse jamming under the indicated conditions with $M = 2$. The expressions for the bounded performance of MPSK are examined for $M = 2^k$ with $k > 1$ (see Problem 9). The jammer condition corresponding to $u = 1$ represents a continuous noise jammer with the signal-to-jammer noise ratio expressed as $\gamma_{sj} = kE_b/N_{oJ} = k(E_b/N_o)(N_o/N_{oJ})$ where $N_o = k'T_{sys} << N_{oJ}$ is the receiver system noise power-density.† In this situation, $\gamma_{sj} << \gamma_b$ and P_{be}, when plotted in as a function of γ_b, is degraded by $\log_{10}(N_{oJ}/N_o)$ dB.

13.9.1.1 DS-BPSK Performance with Jammer Information and Hard-Decisions

With the communicator's knowledge of the jammer state information and hard-decision detection, the parameter D is evaluated as

$$D = u\sqrt{4\varepsilon(1-\varepsilon)} \quad : \text{jammer information, hard-decisions} \qquad (13.129)$$

and, with $\gamma_{bj} = E_b/N_{oJ}$, ε corresponds to the jammed demodulator matched filter output error probability expressed as

$$\varepsilon = Q\left(\sqrt{2u\gamma_{bj}}\right) \quad : \text{jammer information, hard-decisions} \qquad (13.130)$$

Using (13.130), (13.129), and (13.127), the bit-error probability is upper-bounded by

*The factor of 1/2 is included to compensate for the excessive over-bounding of the Chernoff and union bounds when compared to maximum likelihood demodulator detection.

†The parameter k' is Boltzmann's constant with units of watts/(hertz–degree Kelvin) and T_{sys} is the receiver system noise temperature in degrees Kelvin.

$$P_{be} \le \frac{1}{2}u\left\{4Q\left(\sqrt{2u\gamma_{bj}}\right)\left(1-Q\left(\sqrt{2u\gamma_{bj}}\right)\right)\right\}^{1/2} \quad (13.131)$$
$$:\text{jammer information, hard-decisions}$$

13.9.1.2 DS-BPSK Performance with Jammer Information and Soft-Decisions

In this case, the parameter of interest is the upper bound on the demodulator matched filter output bit-error probability and the channel and demodulator detection metric is expressed as

$$D = ue^{-u\gamma_{bj}} \quad :\text{jammer information, soft-decisions} \quad (13.132)$$

and the bit-error probability is upper-bounded by

$$P_{be} \le \frac{1}{2}ue^{-u\gamma_{bj}} \quad :\text{jammer information, soft-decisions} \quad (13.133)$$

The performance bounds in (13.132) and (13.133) can be compared with the respective theoretical results expressed in (13.86) and (13.87) with $\bar{m}=1$ (see Problem 11).

13.9.2 FHSS MFSK with Partial-Noise Jamming

The bit-error probability of MFSK, with partial-band noise jamming and repetition diversity $\bar{m}$, is upper-bounded by

$$P_{be} \le \frac{MD^{\bar{m}}}{4} \quad :M\text{-ary FSK, upper bound} \quad (13.134)$$

and the computational code rate is

$$R_o = \log_2\left(\frac{M}{1+(M-1)D^{\bar{m}}}\right) \quad :\text{bits/symbol} \quad (13.135)$$

The following two sections describe the parameters for evaluating the performance of FHSS using MFSK with partial-band jamming under the indicated condition. In these cases, the signal-to-jammer noise ratio includes the bits-per-symbol k and the number of symbol repetitions $\bar{m}$ and is expressed as $\gamma_{sj} = k(E_b/N_{oj})/\bar{m}$. With $M=2$ and $\bar{m}=1$ these results correspond to FHSS BFSK waveform modulation and can be compared to the upper bound performance of DSSS BPSK discussed in the previous section.

13.9.2.1 FH MFSK Performance with Jammer Information and Hard-Decisions

With the communicator's knowledge of the jammer state information and hard-decision detection, the parameter D is evaluated as

$$D = u\sqrt{\frac{4\varepsilon(1-\varepsilon)}{M-1}} + u\left(\frac{M-2}{M-1}\right)\varepsilon \quad (13.136)$$
$$:\text{jammer information, hard-decisions}$$

where $\varepsilon = P_{se}/u$ and P_{se} is evaluated as the symbol-error probability in (13.88).

13.9.2.2 FH MFSK Performance with Jammer Information and Soft-Decisions

With soft-decision detection, the parameter D is evaluated as

$$D = \min_{0\le\lambda\le1}\left\{\frac{u}{1-\lambda^2}e^{-u\gamma_{sj}\lambda/(1+\lambda)}\right\} \quad (13.137)$$
$$:\text{jammer information, soft-decisions}$$

where λ is the optimization parameter used in determining the Chernoff bound [51, 63, 64].

13.10 CASE STUDY: TERRESTRIAL JAMMER ENCOUNTER AND LINK-STANDOFF RATIO

In this section, the jamming encounter shown in Figure 13.13 is examined for a specified LSR. The encounter corresponds to a terrestrial link with the range loss modeled using the Longley-Rice irregular terrain model (ITM) described in Section 19.7. The LSR is defined as

$$\text{LSR} \triangleq \frac{R_j}{R} \ge \frac{R_j(thr)}{R} \quad :\text{link standoff ratio} \quad (13.138)$$

where $R_j(thr)$ is the minimum acceptable range between the jammer and the communication receiver. For ranges $R_j > R_j(thr)$ the jammer threat is considered to be manageable and for ranges $R_j \le R_j(thr)$ the jammer is considered to be an electronic or physical threat requiring counter measures. The LSR threshold is defined in terms of the received jammer-to-signal power ratio expressed as

$$\gamma_{thr} \triangleq \frac{P_{rj}(\text{thr})}{P_r} \quad :\text{jammer detection threshold} \quad (13.139)$$

where $P_{rj}(thr)$ corresponds to the received jammer power at the range $R_j(thr)$. The details of the receiver and demodulator jammer detection processing are dependent on the communication and jammer waveforms. These topics are beyond the purpose of this case study which focuses on the signal power levels, gains, and ranges encountered in the terrestrial environment. Furthermore, the analysis considers noise-free received signals so the impact of the jammer on the demodulator bit-error probability is not evaluated. These subjects are, however, discussed in various chapters, for example, Section 13.3.1 for the formulation of the following link equations with noise, Sections 13.5, 13.6, 13.7, 13.8, and 13.9 for the evaluation of bit-error performance with jamming, and Sections 1.9 and 11.5 for the estimation of signals, signal parameters, and noise.

Figure 13.37 shows the functional implementation of the receiver and demodulator considered in the following analysis of the terrestrial jamming encounter. The relevant observations are that the jammer range is established by the LSR specification and the jammer received power P_{rj} is detected

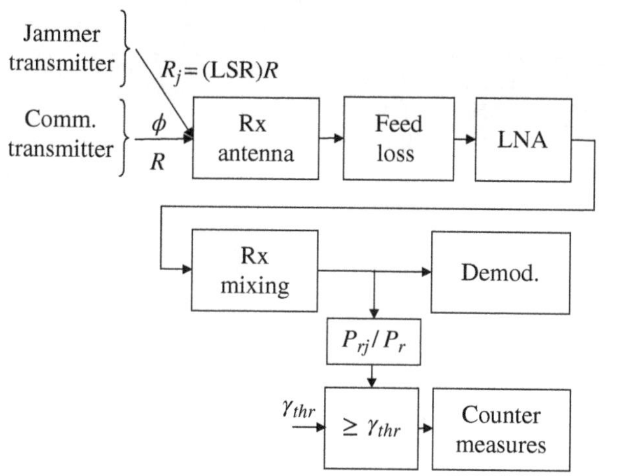

FIGURE 13.37 Communication receiver/demodulator functional implementation.

using the threshold γ_{thr}. The impact of these parameters on the link equation results in the received jammer power $P_{rj} \geq P_{rj}(thr)$ required for detection by the communication receiver. The jammer detection capability can be simplified and improved by intermittently silencing the communication transmitter at known intervals during which time, if the threshold based on receiver noise is exceeded, the appropriate counter measure procedures can be executed.

Referring to Figures 13.13 and 13.37, the link evaluation is based on the communication range equations developed in Chapter 15. With inclusion of the terrestrial loss (L_{ter}), the communication received power is expressed as

$$P_r = \frac{P_t G_t G_r}{L_{atm} L_{ter} L_{fs} L_{rs}} \tag{13.140}$$

In (13.140), the total propagation loss is defined as $L_p = L_{atm} L_{ter} L_{fs}$. The transmitter power required from the high-power amplifier (HPA) is $P_{PA} = L_{ts} P_t$, where L_{ts} is the transmitter system loss including the antenna and feed losses. The expression for the jammer power at the communication receiver is

$$P_{rj} = \frac{P_{tj} G_{tj} G_{rj}}{L_{atmj} L_{terj} L_{fsj} L_{rsj}} \tag{13.141}$$

Expressing (13.141) in terms of the required jammer transmitter power P_{tj} results in

$$P_{tj} = \frac{P_{rj} L_{atmj} L_{terj} L_{fsj} L_{rsj}}{G_{tj} G_{rj}} \tag{13.142}$$

The losses L_{rs} and L_{rsj} represent the receive system loss including the antenna and the antenna feed to the receiver low-noise amplifier (LNA); excluding the negligible effects of small carrier frequencies differences, these losses are considered to be the same for each input signal. At this point in the analysis, the free-space loss is expressed in terms of the link range and carrier frequencies as

$$L_{fs} = \left(\frac{4\pi R f_c}{c}\right)^2 \tag{13.143}$$

where R is the link range, f_c the carrier frequency, and c is the free-space velocity of light. The ratio of the jammer-to-communication free-space loss is

$$\frac{L_{fsj}}{L_{fs}} = \left(\frac{R_j}{R}\right)^2 \left(\frac{f_{cj}}{f_c}\right)^2 \tag{13.144}$$

TABLE 13.9 Case study of System Parameters

System Parameter	Description
Communication Tx power (P_t)	1.76 dBW
Jammer Tx power (P_{tj})	Computed
Jammer detection threshold (γ_{thr})	3 dB
Atmospheric loss ratio (L_{atmj}/L_{atm})	0 dB
Range between communication Tx and Rx (R)	(0.5–5 km)
Range between jammer and comm. Rx (R_j)	R(LSR)
Terrain loss ratio (L_{terj}/L_{ter})	Computed[a]
Carrier frequencies ($f_{cj} = f_c$)	0.1 and 1.0 GHz
Carrier frequency ratio (f_{cj}/f_c)	0 dB
Receiver system loss ratio (L_{rsj}/L_{rs})	0 dB
Communication Tx antenna gain (G_t)	10 dB
Jammer Tx gain antenna (G_{tj})	20 dB
Transmitter gain ratio (G_t/G_{tj})	−10 dB
Receiver antenna gain ratio (G_r/G_{rj}) given link angle ϕ	3 dB
Link-standoff ratio (LSR)	2.5 : 1 and 3 : 1

[a] L_{terj} is interpolated given $R_j = R$(LSR) using the function $L_{ter}(r)$: $0.5 \leq r \leq 5$ km.

TABLE 13.10 Case study of Terrain Parameters for Point-to-Point ITM Model

Terrain Parameters	Description
Communication and jammer Tx antenna physical heights (h_{tie}, h_{je})	10 m
Communication R_X antenna physical height (h_{re})	1 m
Effective antenna heights	$h_e = h_p$
Antenna polarization	1-Vertical
Terrain elevation change (Tx to Rx)	0—Flat terrain
Average elevation above sea level	0
Mean surface refractivity at sea level (N_s)	301
Climate code 5: continental temperate	
Effective curvature of Earth	4/3
Surface dielectric constant (ε_s)[a]	15 F/m
Surface conductivity (σ_s)[a]	0.005 mho/m
Reliability	90%
Confidence	50%

[a] Average ground conditions.

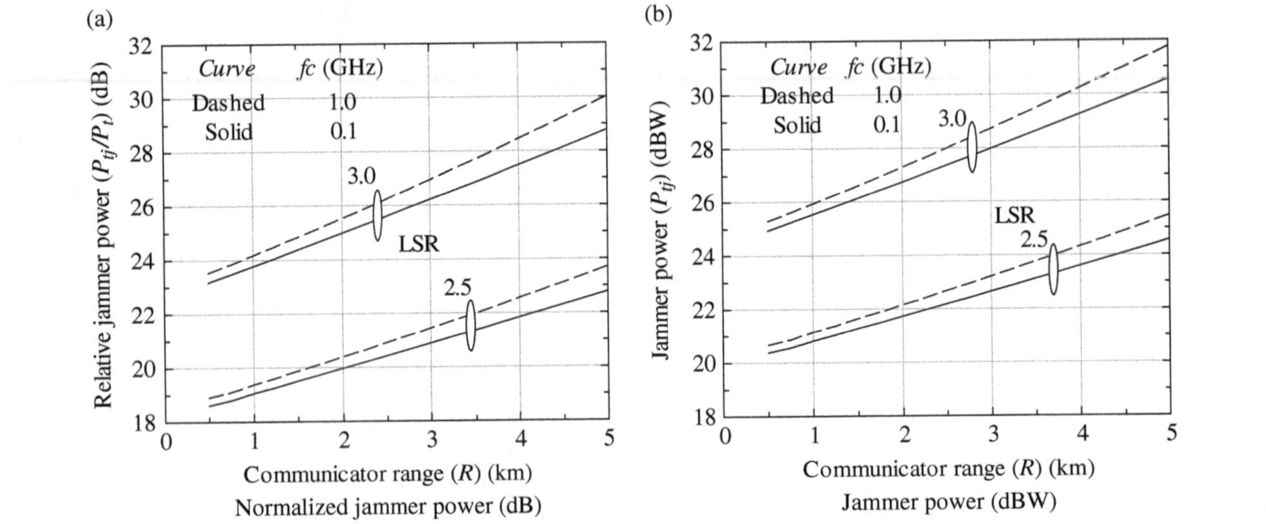

FIGURE 13.38 Jammer power to achieve $P_{rj}(thr)$ corresponding to the LSR specification $\left(h_{te}=h_{tje}=10m,\ h_{re}=1\,m, G_p=0\text{ dB}\right)$.

The jammer received power is expressed in terms of the jammer detection threshold and the communication received power using $P_{rj}=\gamma_{thr}P_r$ and P_r expressed by (13.140). Substituting these results with (13.144) and (13.138) into (13.142), the jammer transmit power is expressed as

$$P_{tj}=\gamma_{thr}\left(\frac{L_{atmj}}{L_{atm}}\right)\left(\frac{L_{terj}}{L_{ter}}\right)\left(\frac{f_{cj}}{f_c}\right)^2\left(\frac{L_{rsj}}{L_{rs}}\right)\left(\frac{G_t}{G_{tj}}\right)\left(\frac{G_r}{G_{rj}}\right)(\text{LSR})^2P_t \tag{13.145}$$

The LSR parameter appears explicitly in (13.145) representing the free-space range loss in (13.144), it also appears implicitly through the terrestrial loss ratio.

The parameters expressed in (13.145) are summarized Table 13.9 and the ITM model link parameters are summarized in Table 13.10. These results are used in a link simulation program to evaluate the jammer transmitter power for detection of $P_{rj}(thr)$ at the communication receiver. If the jammer increases the transmit power above $P_{rj}(thr)$, the probability of detection increases with respect to the specification. The results are shown in Figure 13.38a and b as the respective dependence of P_{tj}/P_t (dB) and P_{tj} (dBW) on the communication link range R.

The preceding analysis corresponds to an unspread communication waveform and when a SS waveform is used the jammer must respond by increasing P_{tj} in the same proportion as G_p using the appropriate jammer strategy. The inclusion of the spread-spectrum processing gain into the link equations is discussed in Sections 13.3.1 and 13.3.2 for AJ and LPI SS applications. The bit-error performance is dependent on the output of the demodulator's SS correlator (matched filter) in the presence of the jamming waveform and the communicator's AJ waveform, FEC coding, and interleaving processing that must be examined as discussed in the preceding sections.

ACRONYMS

ADC	Analog-to-digital converter
AJ	Anti-jam
AWGN	Additive white Gaussian noise (channel)
BE	Bit-error
BPF	Bandpass filter
BPSK	Binary phase shift keying
CDMA	Code division multiple-access
CFAR	Constant false-alarm rate (detection algorithm)
COMSEC	Communication security
CW	Continuous wave
DCBPSK	Differentially coherent BPSK
DCMPSK	Differentially coherent MPSK
DDL	Double-dither loop (PN sequence tracking)
DFS	Digital frequency synthesizer
DLL	Delaylock loop
DS	Direct sequence (PN code)
DS-LFM	Direct sequence with LFM chips
DS-FH	Direct sequence FH
DSSS	Direct-sequence spread-spectrum
DS-TH	Direct sequence TH
E/L	Early–late (sampling)
ECM	Electronic counter measure

EIRP	Effective isotropic radiated power
ETS	Epoch time stamp
FDC	Frequency diversity combining
FDM	Frequency division multiplex
FEC	Forward error correction
FFH	Fast frequency hopping
FFHSS	Fast frequency hopping SS
FFT	Fast Fourier transform
FH	Frequency hopping
FHSS	Frequency-hopping spread-spectrum
FH–TH	Frequency hopping TH
FSK	Frequency shift keying
GPS	Global position system
HPA	High-power amplifier
HPF	Highpass filter
I&D	Integrate-and-dump (filter)
I/Q	Inphase and quadrature (channels or rails)
INFOSEC	Information security
ITM	Irregular terrain model
LFM	Linear frequency modulation
LM	Link margin
LNA	Low-noise amplifier
LPD	Low probability of detection
LPE	Low probability of exploitation
LPF	Lowpass filter
LPI	Low probability of intercept
LSR	Link-standoff ratio
MFSK	Multitone frequency shift keying (*M*-ary FSK)
MPSK	Multiphase shift keying
MSK	Minimum shift keying
OQPSK	Offset QPSK
PC	Pulse compression
PLL	Phaselock loop
PN	Pseudo-noise
PRF	Pulse repetition frequency
PRI	Pulse repetition interval
PSD	Power spectral density
Q/I	Quadrature and inphase (channels or rails)
QPSK	Quadrature phase shift keying
RR	Rate reduction
SFH	Slow frequency hopping
S/H	Sample-and-hold
SJR	Signal-to-jammer ratio
SNJR	Signal to noise-plus-jammer ratio
SNR	Signal-to-noise ratio
SS	Spread-spectrum
TDL	Tau-dither loop (PN sequence tracking)
TER	Total-energy radiometer
TH	Time hopping
THSS	Time-hopping spread-spectrum
TOA	Time-of-arrival
TRANSEC	Transmission security

Note: CHIRP is not an acronym: it is the name applied to a linear frequency modulated waveform.

APPENDIX 13A
13A.1 GRAPHICAL DETERMINATION OF OPTIMUM JAMMER

The optimum jammer strategy for band multitone jamming against a noncoherently detected MFSK-modulated waveform is to choose an optimum jamming parameter, u_o, such that the communicator's bit-error probability becomes

$$P_{be} = \frac{u_o}{\gamma_{bj}} \quad : \text{for} \gamma_{bj} \geq \gamma_{bj\,min} \tag{13A.1}$$

The slope of the $P_{be}(u)$ versus γ_{bj} curves is selected as the tangent to each of the theoretical performance curves for $u =$ 1.0–0.0002 shown in Figure 13.28b for 4FSK ($k = 2$) and duplicated in Figure 13A.1. The curves for $u = 1.0$ and 0.0002 are emphasized because the extreme cases are the most important in constructing the desired response corresponding to the dotted curve in (13A.1).

To begin the process of establishing the desired optimum jammer strategy, select reasonable starting points corresponding to $\gamma'_{bj} = 40$ and 0 dB with the corresponding P'_{be} values that would appear to yield the desired linear slope tangent to each of the $P_{be}(u)$ responses. The points corresponding to (40, $\log(2e^{-5})$) and (0, $\log(2e^{-1})$) are reasonable first guesses. The two equations that must be solved correspond to

$$\begin{aligned} -4.699 &= a(40) + b \\ -0.699 &= a(0) + b \end{aligned} \tag{13A.2}$$

from which $b = -0.699$ and $a = (-4.699 + 0.699)/40 = -4.0$. The function

$$P'_{be} = -4\gamma'_{bj} - 0.0699 \tag{13A.3}$$

is plotted, as in Figure 13A.1 and examined for satisfying the condition of being tangent to each of the selected extreme $P_{be}(u)$ curves; if the plotted function is tangent to the two extreme cases it will also be tangent to the intermediate curves. Usually, two adjustments are necessary to obtain the desired function. First, the function can be made parallel to the desired tangent function typically by a small adjustment in the parameter a, with the second adjustment applied to b to result in the desired tangent response.

The response in Figure 13.28b was established using this procedure and it remains to determine $\gamma_{bj\,min}$ and u_o by expanding the plot of P_{be} ($u = 1$) in Figure 13A.1 as shown in Figure 13A.2. Visual determination of the exact point of tangency corresponding to $\gamma_{bj\,min}$ is difficult to obtain, so, in Figure 13A.2, the two extremes points at which the two curves appear to diverge from one another are identified as

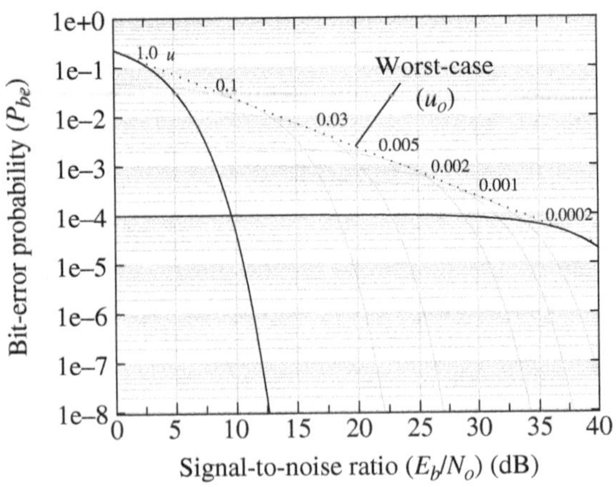

FIGURE 13A.1 Establishing the worst-case inverse linear performance ($k = 2$).

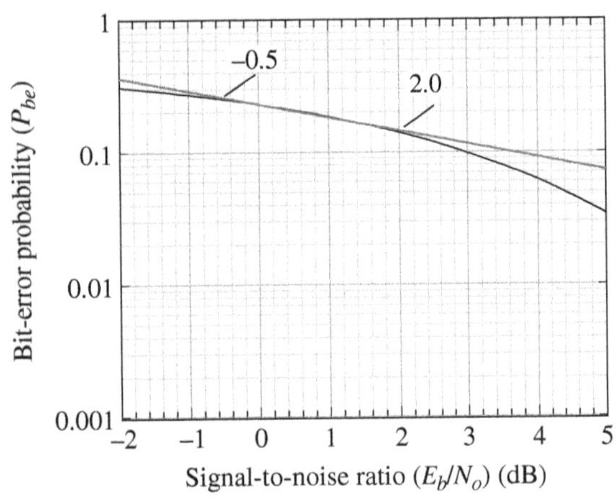

FIGURE 13A.2 Expanded view of Figure 13A.1 for $u = 1$.

$\gamma_{bj1} = -0.5$ and $\gamma_{bj2} = 2.0$. Using these results, $\gamma_{bj\,min}$ is computed as the average value

$$\gamma_{bjmin}(\text{dB}) = \frac{\gamma_{bj1} + \gamma_{bj2}}{2} = 0.75 \text{ dB} \tag{13A.4}$$

with $\gamma_{bjmin} = 1.188$. The corresponding bit-error probability* is $P_{be}(\gamma_{bj\,min}) = 0.1955$ and the optimum jammer parameter u_o is computed as

$$u_o = \gamma_{bjmin} P_{be}\left(\gamma_{bjmin}\right) = 0.2323 \tag{13A.5}$$

A similar analysis of Figure 13.28a, for $k = 3$, results in $\gamma_{bj1} = -1.5$, $\gamma_{bj2} = 0.7$, and yields $\gamma_{bjmin} = 0.912$, $P_{be}\left(\gamma_{bjmin}\right) = 0.2106$, and $u_o = 0.1921$. These values, corresponding to $k = 2$ and 3, are in reasonable agreement with those in listed in Table 13.6.

*As in Table 13.36, the value of $P_{be}(\gamma_{bj\,min})$ is linearly interpolated between the theoretically computed bit-error probabilities around $\gamma_{bj\,min}$; the theoretically computed bit-error probabilities are based on uniform sampling of γ_{bj} in increments of $\Delta\gamma_{bj} = 0.05$ dB.

PROBLEMS

1. Referring to Figure 13.2, the expression for the received analytic or baseband QPSK-modulated signal is

$$\tilde{s}_r(i) = (d_n C_i \cos(\phi) - d_n C_{i-1} \sin(\phi)) + j(d_n C_i \sin(\phi) + d_n C_{i-1} \cos(\phi))$$

where the constant phase angle ϕ is introduced by the channel and the samples i correspond to the demodulator sampling intervals $t = i\Delta t$. Evaluate the demodulator correlator response

$$R(\ell) = \tilde{s}_r(i)\tilde{s}^*_{ref}(i-\ell)$$

for $\ell = 0$ where $\tilde{s}_{ref}(i-\ell) = C_{i-\ell} + jC_{i-1-\ell}$ is the N-chip demodulator reference sequence. Note that the I and Q correlator channels are each summed over the interval $0 \le i \le N/2$. To complete this problem draw the functional implementation of the correlator for the quadrature-modulated waveform, similar to that in Figure 13.3b for BPSK modulation.

2. Given the MFSK SFH example in Section 13.2.2.1, suppose that K contiguous FSK symbols correspond to one hop interval of T_h seconds and that the SS bandwidth corresponds to 2^L frequency hops in increments corresponding to the baseband noise bandwidth (B_n) of the unhopped modulated waveform. Under these conditions, determine the following in terms of the applicable parameters: the source data rate R_b, $k = \log_2(M)$, K, and L. (i) The baseband bandwidth W_s, (ii) the SS bandwidth W_{ss}, (iii) the processing gain, (iv) the FH interval (T_h), and (v) the hopping rate (R_h).

3. Referring to Figure 13.9, consider a processing gain of $G_p = 128$ so that the pulse duration is $T_p = T_b/G_p$ and consider $N_{samp} = 4$ samples-per-pulse and an FFT window of $T_{fft} = 8T_p$ or eight bits. The modulated bits and pulses have the respective forms $rect(t/T_b)$ and $rect(t/T_p)$. Under these conditions, determine the numerical value of the seven (7) parameters associated with the spectrum in the following figure.

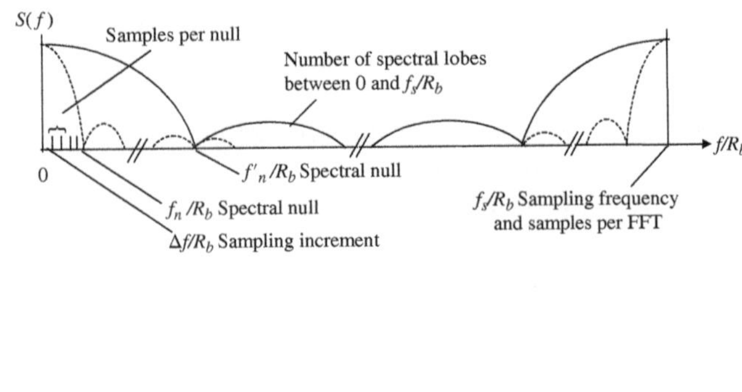

4. Consider a modulated signal with power P_s and modulation bit-rate R_b is operating in an AGWN channel with two additive noise sources with power spectral densities N_o and N'_o. Determine the composite signal-to-noise ratio γ in terms of the signal-to-noise ratios $\gamma_b = E_b/N_o$ and $\gamma'_b = E_b/N'_o$. Characterize γ for the cases: $\gamma_b \gg \gamma'_b$ and $\gamma_b \ll \gamma'_b$.

5. Referring to Problem 4, derive the expression for the loss in the signal-to-noise ratio $k = \gamma/\gamma_b$ under the conditions $\gamma > \gamma_b$ and $P_{be}(\gamma) = P_{be}(\gamma_b)$. Since $k \le 1$, the signal-to-noise ratio loss in dB is Loss (dB) $= -10\log_{10}(k)$. Using the expression for k, determine the Loss (dB) when $\gamma'_b = \gamma_b$.

6. Verify that $F(u,\gamma_{sj})$ is expressed as in (13.78). Hint: $erfc(z) = 1 - erf(z)$ and $derf(z)/dz = (2/\sqrt{\pi})H_o(z)e^{-z^2}$ where the Hermite function $H_o(z) = 1\ \forall\ z$.

7. Using the first equality in (13.68) and the definitions of the received and jammer signals, and the demodulator references given respectively in (13.65), (13.66), and (13.67), show the details in arriving at the second equality in (13.68). Note that the received signal corresponds to BPSK-modulated DSSS chips with chip weighting $p(t) = rect((t - T_c(1 + 2i)/2)/T_c)$ and the correlation integral has zero lag indicating ideal chip synchronization. Also, defining $M_J = 1/2\sqrt{2P_J/T_b}\sum_{i=0}^{N-1} c_i \int_{iTc}^{(i+1)Tc} dt$ as the magnitude of the jammer signal at the output of the correlator, the expression for the equivalent inphase jammer signal is $N_{Jc} = M_J(\cos(\phi_J - \hat{\phi}_s) - \sin(\phi_J - \hat{\phi}_s))$, so shown that $\cos(\phi_J - \hat{\phi}_s) - \sin(\phi_J - \hat{\phi}_s) = \sqrt{2}\cos(\theta)$ with $\theta = \phi_J - \hat{\phi}_s + \pi/4$.

8. Derive the expression for the correlator quadrature output of the signal-plus-jammer using (13.65), (13.66), and (13.67). In other words, evaluate the correlator quadrature output y_{sk} corresponding to y_{ck} expressed in (13.68).

9. For noncoherently detected BFSK with partial-band noise jamming, discussed in Section 13.6, show the detailed steps in verifying the parameters in the first row of Table 13.6 corresponding to $k = 1$. Hint: The unjammed bit-error performance of noncoherently detected BFSK is $P_{be} = 0.5e^{-\gamma bj/2}$.

10. Using the performance bound $P_{be} \le MD/4$, evaluate and compare the P_{be} versus γ_{bj} bounded performance of DS-BPSK modulation described in Section 13.9.1 with $k = 1$ ($M = 2$). Repeat the above using the bounded performance of FH-BFSK described in Section 13.9.2. Make these comparisons using the hard and soft decisions. Compare the DS-BPSK results with those in (13.86) and (13.87) with $\bar{m} = 1$.

11. Evaluate and compare the bounded P_{be} versus γ_{bj} performance of FH-BFSK using $P_{be} \le MD^{\bar{m}}/4$ as described in Section 13.9.2. Make the comparisons using repetition coding diversities of $\bar{m} = 1$, 2, and 4. Repeat the above using FH–MFSK for $M = 4$ and 8.

12. Under the conditions in Problems 9 and 11, evaluate and plot the computational cutoff rate R_o as a function of γ_{sj} in dB.

REFERENCES

1. R.C. Dixon, *Spread Spectrum Systems*, John Wiley & Sons, Inc., New York, 1976.
2. L.A. Gerhardt, R.C. Dixon, Editors, "Special Issue on Spread Spectrum Communications," IEEE Transactions on Communications, Vol. COM-25, No. 8, pp. 745–869, August 1977.
3. J.K. Holmes, *Coherent Spread Spectrum Systems*, John Wiley & Sons, Inc., New York, 1982.
4. C.E. Cook, F.W. Ellersick, L.B. Milstein, D.L. Schilling, Guest Editors, "Special Issue on Spread Spectrum Communications," IEEE Transactions on Communications, Part I of two parts, Vol. COM-30, No. 5, pp. 817–1072, May 1982.
5. C.E. Cook, F.W. Ellersick, L.B. Milstein, D.L. Schilling, Editors, "*Spread Spectrum Communications*," IEEE Press, New York, May 1983.
6. M.K. Simon, J.K, Omura, R.A. Scholtz, B.K. Levitt, *Spread Spectrum Communications*, Computer Science Press, Inc., Rockville, MD, 1985. (This is a three volume series: Introduction to Spread-Spectrum Communications; Classical Spread-Spectrum Communications; Synchronization of Spread-Spectrum Systems.).
7. D.L. Nicholson, *Spread Spectrum Signal Design: LPE and AJ Systems*, Computer Science Press, Inc., Rockville, MD, 1988.
8. R.L. Peterson, R.E. Ziemer, D.E. Borth, *Introduction to Spread Spectrum Communications*, Prentice-Hall, Inc., Upper Saddle River, NJ, 1995.
9. National Security Industrial Association Command, Control and Commutations Committee, Anti-Jam Policy Study Committee, Anti-Jam Policy Study Final Report, Vol. V, New Anti-Jam Systems, Washington, DC, May 1978.
10. A.J. Viterbi, *CDMA Principles of Spread Spectrum Communication*, Addison-Wesley Publishing Company, Reading, MA, 1995, Seventh Printing 1998.
11. R. Gold, *Properties of Linear Encoding Sequences*, Robert Gold Associates, Los Angeles, CA, July 1975.
12. E.H. Dinan, B. Jabbari, "Spreading Codes for Direct Sequence CDMA and Wideband CDMA Cellular Networks," IEEE Communications Magazine, Vol. 36, No. 9, pp. 48–54, September 1998.
13. J.W. Ames, R.C. Cumming, E.G. Ottoboni, A.B. Conner, "ADM Vulnerability Analysis," Stanford Research Institute, Technical Report, SRI Project 4061, Menlo Park, CA, September 1975.
14. M.K. Simon, J.K., Omura, R.A. Scholtz, B.K. Levitt, *Spread Spectrum Communications*, Vol. I, Ch. 1, "A Spread-Spectrum

Overview," Computer Science Press, Inc., Rockville, MD, 1985.
15. M.K. Simon, J.K., Omura, R.A. Scholtz, B.K. Levitt, *Spread Spectrum Communications*, Vol. II, pp. 66–69, Computer Science Press, Inc., Rockville, MD, 1985.
16. J.J. Spilker, Jr., "Delay-Lock Tracking of Binary Signal," IRE Transactions on Space Electronics and Telemetry, Vol. SET-9, No. 2, pp. 1–8, March 1963.
17. J.J. Spilker, Jr., *Digital Communications by Satellite*, Prentice-Hall, Inc., Englewood Cliffs, NJ, 1977.
18. W.K, Alem, C.L. Weber, "Acquisition Techniques for PN Sequences," NTC 1977 Conference Record, Vol. II, pp. 35:2-1–35:2-4, Los Angeles, CA, December 5–7, 1977.
19. J.K. Holmes, C.C. Chen, "Acquisition Time Performance of PN Spread Spectrum Systems," IEEE Transactions on Communications, Vol. COM-35, No. 8, pp. 778–783, August 1977.
20. D.M. DiCarlo, C.L. Weber, "Statistical Performance of Single Dwell Serial Synchronization Systems," IEEE Transactions on Communications, Vol. COM-28, No. 8, pp. 1382–1388, August 1980.
21. W. Feller, *An Introduction to Probability Theory and Its Applications*, John Wiley & Sons, Inc., New York, 1950.
22. M.K. Simon, J.K., Omura, R.A. Scholtz, B.K. Levitt, *Spread Spectrum Communications*, Vol. III, Sections 1.2.1 and 1.2.3, Computer Science Press, Inc., Rockville, MD, 1985.
23. M.K. Simon, J.K., Omura, R.A. Scholtz, B.K. Levitt, *Spread Spectrum Communications*, Vol. III, Sections 1.2.4 and 1.2.5, Computer Science Press, Inc., Rockville, MD, 1985.
24. P.M. Hopkins, "A Unified Analysis of Pseudo-Noise Synchronization by Envelope Correlation," IEEE Transactions on Communications, Vol. COM-25, No. 8, pp. 770–778, August 1977.
25. D.M. DiCarlo, C.L. Weber, "Multiple Dwell Serial Search: Performance and Application to Direct Sequence Code Acquisition," IEEE Transactions on Communications, Vol. COM-31, No. 5, pp. 650–659, May 1983.
26. M.K. Simon, J.K., Omura, R.A. Scholtz, B.K. Levitt, *Spread Spectrum Communications*, Vol. III, Ch. 1, "Pseudonoise Acquisition of Spread-Spectrum Systems," Computer Science Press, Inc., Rockville, MD, 1985.
27. R.L. Peterson, R.E. Ziemer, D.E. Borth, *Introduction to Spread Spectrum Communications*, Chapter 5, "Initial Synchronization of the Receiver Spreading Code," Prentice-Hall, Inc., Upper Saddle River, NJ, 1995.
28. M.K. Simon, "Noncoherent Pseudonoise Code Tracking Performance of Spread Spectrum Receivers," IEEE Transactions on Communications, Vol. COM-25, No. 3, pp. 327–345, March 1977.
29. W.J. Gill, "A comparison of Binary Delay-Lock Implementations," IEEE Transactions on Aerospace and Electronic Systems, Vol. AES-2, No. 4, pp. 415–424, July 1966.
30. P.T. Nielsen, "On the Acquisition Behavior of Binary Delay-Lock Loops," IEEE Transactions on Aerospace and Electronic Systems, Vol. AES-11, No. 3, pp. 415–418, May 1975.
31. H.P. Hartmann, "Analysis of a dithering loop for PN Code Tracking," IEEE Transactions on Aerospace and Electronic Systems, Vol. AES-10, No. 1, pp. 2–9, January 1974.
32. P.M. Hopkins, "Double Dither Loop for Pseudonoise Code Tracking," IEEE Transactions on Aerospace and Electronic Systems, Vol. AES-13, No. 6, pp. 644–650, November 1977.
33. M.K. Simon, J.K., Omura, R.A. Scholtz, B.K. Levitt, *Spread Spectrum Communications*, Vol. III, Chapter 3, "Time and Frequency Synchronization of Frequency-Hopped Receivers," Computer Science Press, Inc., Rockville, MD, 1985.
34. H. Urkowitz, "Energy Detection of Unknown Deterministic Signals," Proceedings of the IEEE, Vol. 55, No. 4, pp. 523–531, April 1967.
35. A.B. Glenn, "Low Probability of Intercept," IEEE Communications Magazine, Vol. 21, No. 4, pp. 26–33, July 1983.
36. M.K. Simon, J.K. Omura, R.A. Scholtz, B.K. Levitt, *Spread Spectrum Communications*, Vol. III, Part 5, Chapter 4, "Low Probability of Intercept Communications," pp. 283–345, Computer Science Press, Inc., Rockville, MD, 1985.
37. N.F. Krasner, "Maximum Likelihood Parameter Estimation for LPI Signals," MILCOM'82 Conference Proceedings, pp. 2.3-1–2.3-4, Boston, MA, October 17–20, 1982.
38. R.A. Dillard, "*Detection of Covert Signals*," Naval Electronics Laboratory Center, San Diego, CA, June 27, 1974.
39. G.M. Dillard, "A Moving-Window Detector for Binary Integration," IEEE Transactions on Information Theory, Vol. IT-13, No. 1, pp. 2–6, January 1967.
40. J.V. Bradley, *Distribution-Free Statistical Tests*, Prentice-Hall, Inc., Englewood Cliffs, NJ, 1968.
41. J.W. Carlyle, *Communication Theory*, Chapter 8, A.V. Balakrishnan, Editor, "Nonparametric Methods in Detection Theory," McGraw–Hill Book Company, New York, 1968.
42. J.B. Thomas, "Nonparametric Detection," Proceedings of the IEEE, Vol. 58, No. 5, pp. 623–631, May 1970.
43. M. Abramowitz, I.A. Stegun, *Handbook of Mathematical Functions with Formulas, Graphs, and Mathematical Tables*, National Bureau of Standards, Applied Mathematics Series 55, pp. 231, U.S. Government Printing Office, Washington, DC, November 1964.
44. D.B. Davenport, Jr., W.L. Root, *An Introduction to the Theory of Random Signals and Noise*, McGraw–Hill Book Co., New York, 1958.
45. R.H. Dicke, "The Measurement of Thermal Radiation at Microwave Frequencies," Review of Scientific Instruments, Vol. 17, pp. 268–275, July 1946.
46. S.J. Goldstein, "A Comparison of Two Radiometer Circuits," Proceedings of the IRE, Vol. 43, No. 2, Part 1, pp. 1663–1666, November 1955.
47. J.W. Ames, R.C. Cumming, E.G. Ottoboni, A.B Conner, "ADM Vulnerability Analysis," Stanford Research Institute, Technical Report, SRI Project 4061, pp. 126–127, Menlo Park, CA, September 1975.
48. P.A. Kullstam, "Spread Spectrum Performance Analysis in Arbitrary Interference," IEEE Transactions on Communications, Vol. COM-25, No. 8, pp. 848–853, August 1977.
49. M.K. Simon, J.K. Omura, R.A. Scholtz, B.K. Levitt, *Spread Spectrum Communications*, Vol. II, pp. 12–13, Computer Science Press, Inc., Rockville, MD, 1985.
50. P.A. Kullstam, "A Theoretical Investigation of Spread Spectrum Modulation Concepts and Performance," Ph.D. Thesis, Catholic University of America, May 1977.
51. J.M. Wozencraft, I.M. Jacobs, *Principles of Communication Engineering*, John Wiley & Sons, Inc., New York, 1967.

52. M.K. Simon, J.K. Omura, R.A. Scholtz, B.K. Levitt, *Spread Spectrum Communications*, Vol. I, pp. 151–168, Computer Science Press, Inc., Rockville, MD, 1985.
53. A. Papoulis, *Probability, Random Variables, and Stochastic Processes*, p. 102, McGraw–Hill Book Company, New York, 1965.
54. M.K. Simon, J.K., Omura, R.A. Scholtz, B.K. Levitt, *Spread Spectrum Communications*, Vol. II, p. 77, Table 2.1, Computer Science Press, Inc., Rockville, MD, 1985.
55. M.K. Simon, "The performance of M-ary DPSK/FH in the Presence of Partial-Band Multitone Jamming," IEEE Transactions on Communications, Special Issue on Spread Spectrum Communications, Vol. COM-30, No. 5, pp. 953–958, May 1982.
56. W.C. Lindsey, M.K. Simon, *Telecommunication Systems Engineering*, Chapter 5, "Phase-Coherent Detection with Perfect Reference Signals," Dover Publication, New York, 1973.
57. M.K. Simon, J.K., Omura, R.A. Scholtz, B.K. Levitt, *Spread Spectrum Communications*, Vol. II, pp. 66–67, Computer Science Press, Inc., Rockville, MD, 1985.
58. B.D. Trumpis, "*On the Optimum Detection of Fast Frequency Hopped MFSK Signals in Worst-Case Jamming*," TRW Internal Memorandum, TRW Space Technology Sector, Redondo Beach, CA, June 1981.
59. M.K. Simon, J.K., Omura, R.A. Scholtz, B.K. Levitt, *Spread Spectrum Communications*, Vol. II, pp. 85–91, Computer Science Press, Inc., Rockville, MD, 1985.
60. M.K. Simon, J.K., Omura, R.A. Scholtz, B.K. Levitt, *Spread Spectrum Communications*, Vol. II, pp. 86–87, Computer Science Press, Inc., Rockville, MD, 1985.
61. M.K. Simon, J.K., Omura, R.A. Scholtz, B.K. Levitt, *Spread Spectrum Communications*, Vol. II, pp. 78–79, Computer Science Press, Inc., Rockville, MD, 1985.
62. M.K. Simon, J.K., Omura, R.A. Scholtz, B.K. Levitt, *Spread Spectrum Communications*, Vol. I, Chapter 4, "Differentially Coherent Modulation Techniques," Tables 4.1, 4.2, and 4.3, pp. 245–247, Computer Science Press, Inc., Rockville, MD, 1985.
63. M.K. Simon, J.K., Omura, R.A. Scholtz, B.K. Levitt, *Spread Spectrum Communications*, Appendices 4A, p. 248 and 4B, p. 250, Computer Science Press, Inc., Rockville, MD, 1985.
64. A.J. Viterbi, J.K. Omura, *Principles of Digital Communication and Coding*, McGraw–Hill Book Company, New York, 1979.

14

MODEM TESTING, MODELING, AND SIMULATION

14.1 INTRODUCTION

Perhaps the most useful and certainly the most ubiquitous tool for evaluating any communication system is a good uniform number generator; a close runner-up is a good Gaussian number generator. Once this tool is available, it is used in virtually every simulation program to perform a variety of tasks. These tasks include: random source-data generation, pseudo-noise (PN) sequence generation, generation of various types of fading channels, and additive white Gaussian noise (AWGN) generation. Various methods of generating uniformly distributed random numbers are discussed in the following paragraph. With this background, the generation of other forms of random variables is examined and subsequent examples of system simulations focus on their applications in accurately characterizing system performance.

In terms of network terminology, the communication link is used here to include all of the functions of the physical layer between a data source and a data sink as depicted in Figure 14.1. In this context the channel may include, for example, a satellite repeater or *bent-pipe* that does not process the received uplink signal for information content. Also, whenever there is a linear relationship between the transmitter and channel filtering, it is often convenient to treat the cascaded response as one filter. When this is done, however, care must be taken in the manner that various noise sources are added throughout the system. For example, if the uplink includes adjacent channel interference, then the transmitter and satellite filtering must be included as separate filters. In general, it is necessary to model the system implementation as closely as possible to the real encounter.

Figure 14.1 is intended to aid in the description of the functions involved in simulating the performance of a communication system. The various *seeds* are typically used to generate uniformly distributed random numbers from which Gaussian, Rayleigh, and other random processes are generated. Using the same seeds for repeated simulations guarantees that the random conditions are duplicated so that performance variation can be attributed to changes in various system parameters. The various functions shown in Figure 14.1 provide inputs to the performance measure function that computes and formats the desired outputs for assessing the overall system performance. Example outputs are: received signal-to-noise ratio, number of block or bit errors, block and bit-error probability,* phase, frequency, and symbol tracking loop conditions, matched filter sampled outputs, and estimated parameter values.

Two significant advantages of system simulations are: the ability to evaluate the system performance using parametric condition to determine how gracefully the system degrades; the ability to alter system parameters like: demodulator bandwidths, noise generator seeds, and interference and channel conditions, and then perform otherwise identically repeated simulations to attribute the performance variations to the system parameter or changes in the random conditions.

*The bit-error probability as output from the simulation is more accurately referred to at the bit-error frequency, or bit-error rate, defined as the received bit-errors divided by the total number of received bits.

Digital Communications with Emphasis on Data Modems: Theory, Analysis, Design, Simulation, Testing, and Applications,
First Edition. Richard W. Middlestead.

Companion website: www.wiley.com/go/digitalcommunications

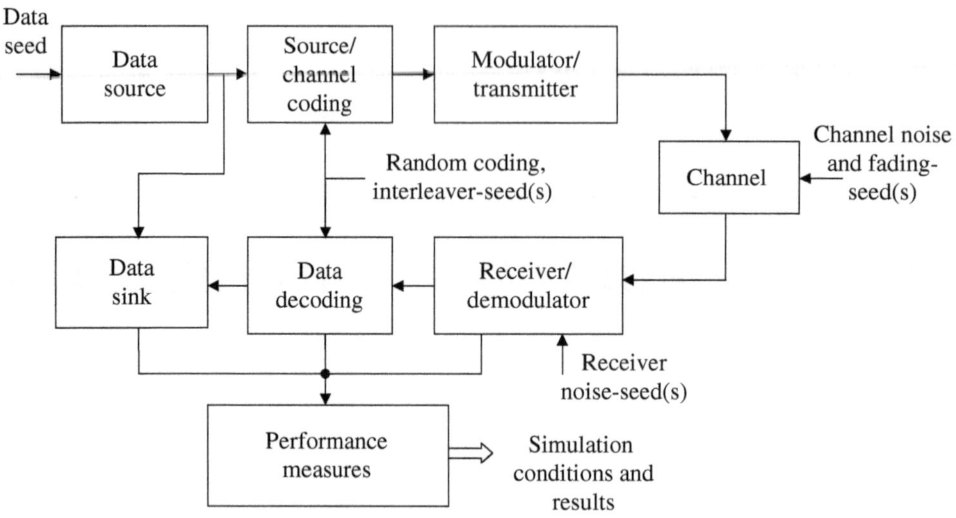

FIGURE 14.1 Functional diagram of a typical communication systems simulation.

This latter advantage is also a major simulation software debugging tool, in that, by repeating the same noise conditions in the various simulation *loops*, if the performance measures do not agree exactly then there is a problem with the simulation code. Typically this involves arrays or parameters not being re-initialized properly between successive simulations.

14.2 STATISTICAL SAMPLING

The first consideration in any Monte Carlo simulation [1] is to determine the number of Monte Carlo trials that are required to achieve a specified accuracy in the performance measure(s). Typically, the performance measure in a digital communication system is the received bit-error probability; however, symbol, character, and message-error probabilities are also used. Each Monte Carlo trial or event is independent of all others leading to the notion of Bernoulli trials [2]. Under these conditions, if the underlying probability of an event is given by p, then the probability of an event occurring k times in n independent trials is given by the binomial distribution

$$P(k) = \binom{n}{k} p^k q^{n-k} \tag{14.1}$$

where $q = p - 1$. The coefficient in (14.1) represents the number of possible combinations of n things taken k at a time. This is referred to as the binomial coefficient and is evaluated as

$$\binom{n}{k} = \frac{n!}{k!(n-k)!} \tag{14.2}$$

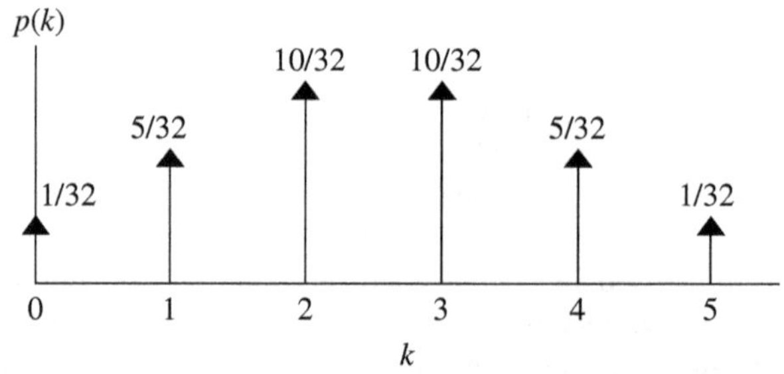

FIGURE 14.2 Discrete binomial probability distribution ($n = 5$, $p = 1/2$).

where $x!$ is the factorial of x. The binomial distribution characterizes the discrete random variable k as depicted, for example, in Figure 14.2 for $n = 5$ and $p = 1/2$.

14.2.1 Fixed-Sample Testing Using the Gaussian Distribution

Referring to Papoulis [2] it is shown that for large values of n, such that $npq \gg 1$, that (14.1) is approximated over the region $|k - np| \cong \sqrt{npq}$ by*

$$p(k) \sim \frac{1}{\sqrt{2\pi npq}} e^{-(k-np)^2/(2npq)} \tag{14.3}$$

This is the DeMoivre–Laplace theorem [3] and the right-hand-side is recognized as the Gaussian distribution with mean value $\bar{k} = np$ and variance $\sigma_k^2 = npq$. Upon defining

*In the remainder of this analysis the equality condition is used.

the variable $x = k/n$, the mean of the distribution $p(x)$ is p and the standard deviation is $\sigma_x = \sqrt{pq/n}$ so that

$$p(x) = \frac{1}{\sqrt{2\pi}\sigma_x} e^{-(x-p)^2/(2\sigma_x^2)} \tag{14.4}$$

Returning to the problem of determining the number of Monte Carlo trials to perform in a system simulation, consider the mean value p of the *pdf* $p(x)$ to be the system bit-error probability of interest. Then, after n trials or bits, the standard deviation in the estimation of p will be

$$\sigma = \sqrt{\frac{p(1-p)}{n}} \tag{14.5}$$

Because the underlying distribution is Gaussian, this result states that after n Monte Carlo trials the estimate of p will be within $p \pm \sigma$ with a confidence of 68.26%. Stated another way, if repeated Monte Carlo simulations are performed with independent underlying Gaussian random variables, then 68.26% of the measurements will be within $\pm\sigma$ of the expected value p. Higher confidence levels are readily determined by evaluating, for example, the $\pm 2\sigma$ or $\pm 3\sigma$ limits that correspond to confidence levels of 95.45 and 99.93% respectively.

It is often required to determine the number of trials, such that, the resulting measurement error is within a specified confidence level. Suppose, for example, that p is to be estimated within a two-sigma ($\pm k\sigma$: $k = 2$) confidence level of 95.45% and with an accuracy of $\leq$10% ($\eta = 0.1$). In this case, let $2\sigma = 1.1p$ and evaluate (14.5) for n with $\sigma = 1.1p/2$. In the following section, the accuracy and confidence of a Monte Carlo simulation are characterized in terms of the required number of trials, n, and the number of observed errors, e, in the simulation.

14.2.1.1 Number of Trials Based on Accuracy and Confidence of Test When the system bit-error probability is unknown it is reasonable to ask, "How many bits (trials) must be simulated before the performance can be adequately assessed?" The answer to this question is found in the context of the sampling criterion discussed in the previous section. For example, for a discrete randomly sampled process with all events having equal probabilities of occurrence p, the probability P_o of observing k events in n trials is approximated by [4]

$$\begin{aligned} P_o = P_r(0 \leq i \leq k) &= \sum_{i=0}^{k} \binom{n}{i} p^i (1-p)^{n-i} \\ &= \frac{1}{2} + \frac{1}{2} erf'\left(\frac{k-np}{\sqrt{np(1-p)}}\right) \end{aligned} \tag{14.6}$$

where $erf'(x)$ is the error function defined in Section 3.5 as

$$erf'(x) \triangleq \frac{2}{\sqrt{2\pi}} \int_0^x e^{-y^2/2} dy \tag{14.7}$$

The mean value of the density function is np and the parameter k is defined as $k = np_1$ where $p_1 \geq p$ is an upper bound on p corresponding to a specified measurement accuracy of $\eta = p_1/p - 1$. Using these results with $p_1 = p(1+\eta)$, the expression for P_o in (14.6) becomes

$$P_o = \frac{1}{2} + \frac{1}{2} erf'\left(\frac{n\eta p}{\sqrt{np(1-p)}}\right) \tag{14.8}$$

The parameter p_1 upper bounds the true error probability p. Using (14.8) the sample size, n, of the test can be evaluated in terms of the specified accuracy (η) and confidence level, P_o. The unknown probability p is approximated by the frequency interpretation $p \cong e/n$, where e is the number of error events after n trials. In the following analysis the argument of the error function in (14.8) is defined as

$$x = \frac{n\eta p}{\sqrt{np(1-p)}} \tag{14.9}$$

The value of x defined in (14.9) is evaluated for a specified P_o by applying Newton's method [5] to determine the solution to $f(x) = 0$, where

$$f(x) = P_o - \left(\frac{1}{2} + \frac{1}{2} erf'(x)\right) \tag{14.10}$$

The iterative solution is given by

$$x_{i+1} = x_i - \frac{f(x_i)}{f'(x_i)} \tag{14.11}$$

where the iteration continues until an acceptable error in (14.10) is achieved, for example, when $|f(x)| < 0.0001$. The derivative of $f(x)$ is evaluated as

$$f^{(1)}(x) = \frac{d}{dx} erf'(x) = \frac{2}{\sqrt{\pi}} H_0(x) e^{-x^2} \tag{14.12}$$

where H_n is the Hermite polynomial [5] and $H_0(x) = 1$ for all x. Having approximated the value of x, the number of events required to satisfy the conditions of the test is determined as

$$n = \left(\frac{x}{\eta}\right)^2 \frac{(1-p)}{p} \cong \left(\frac{x}{\eta}\right)^2 \frac{1}{p} \tag{14.13}$$

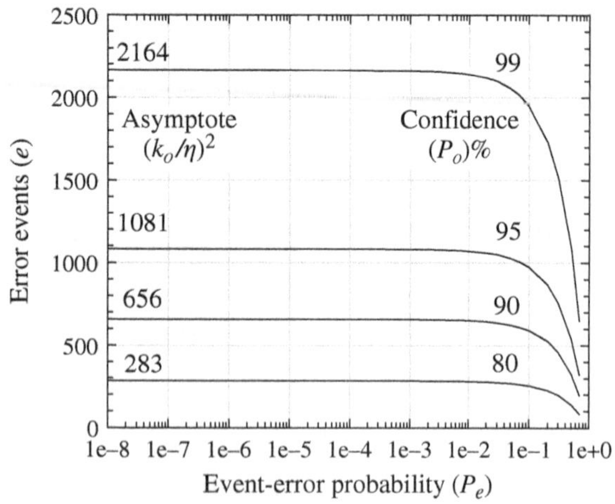

FIGURE 14.3 Required error events to achieve an error probability with 5% accuracy.

where the approximation applies when $p \ll 1$ corresponding to low error probabilities. Using $e = np$, the number of allowable errors during the test is

$$e = \left(\frac{x}{\eta}\right)^2 (1-p) \cong \left(\frac{x}{\eta}\right)^2 \tag{14.14}$$

Therefore, upon terminating the test after n events, the test fails if the number of errors exceeds e otherwise the test passes.*

As an example of the application of these results consider simulating the performance of a modem detection algorithm at a signal-to-noise corresponding to a bit-error probability of $p = 1 \times 10^{-3}$ with a confidence of 99% ($P_o = 0.99$) and an accuracy of 5% ($\eta = 0.05$) corresponding to an upper bound on the bit-error probability of $p_1 = 1.05p$. Using Newton's method, the value of x is 2.326282 and the number of bits required by the simulation is $n = 2{,}162{,}471$ and the maximum number of allowable bit errors is 2,163. Figure 14.3 shows the number of error events as a function of the error probability $P_e = p$ for various confidence levels with a test accuracy of 5%, that is, the test determines that the modem is operating with an error probability p with an accuracy of $1.05p$. This analysis leading to the number of error events, shown in Figure 14.3, results in a measurement accuracy of 5% regardless of the error probability; so, for very low error probabilities, the accuracy is commensurately low but the number of trials becomes very large. For example, when trying to measure an error event with a probability of $p = 10^{-9}$ with 5% accuracy requires a sample size of more than 10^{12} trials. These examples are based on extremely accurate test requirements and more practical specifications may require confidence levels on the order of 80–90% with accuracies on the order of 10, 30, or even 100%.

In all of these tests, it is assumed that the signal-to-noise ratio is established with sufficient precision so that the modem is operating at the desired bit-error probability; however, the accuracy of the test is influenced by the ability to accurately measure the signal-to-noise ratio. When examining modems with forward error correction (FEC) coding, the decoded bit-error performance is very sensitive to the signal-to-noise ratio. In these cases, it is often more practical to first test the uncoded error performance. In addition, FEC decoding results in bursts of errors that influence the test accuracy; the preceding and following analysis assumes independent random error events. In Section 14.2.2, the method of *sequential testing* is discussed that terminates a test that is bound to fail with considerably fewer trials.

The conditions in the preceding example ensure that the simulated event error probability is *less than* $1.05p$; however, it may be desirable to have the measurement accuracy *within* $\pm 0.05p$. This additional precision will require simulating more events resulting in longer simulation times. To assess the impact of this requirement the probability

$$\begin{aligned} P_o' &= P_r(0 \le i \le k_1) - P_r(0 \le i \le k_2) = \\ &\cong \frac{1}{2} + \frac{1}{2}erf'\left(\frac{k_1 - np}{\sqrt{np(1-p)}}\right) - \left\{\frac{1}{2} + \frac{1}{2}erf'\left(\frac{k_2 - np}{\sqrt{np(1-p)}}\right)\right\} \\ &= \frac{1}{2}erf'\left(\frac{k_1 - np}{\sqrt{np(1-p)}}\right) - \frac{1}{2}erf'\left(\frac{k_2 - np}{\sqrt{np(1-p)}}\right) \end{aligned} \tag{14.15}$$

is evaluated with $k_1 \ge k_2$. Substituting $k_1 = np_1 = n(1+\eta)p$ and $k_2 = np_2 = n(1-\eta)p$ results in

$$P_o' = erf'(x) \quad : |\text{measurement accuracy}| \le \eta \tag{14.16}$$

where x is given by (14.9). In this case the previous analysis procedures can be used by computing an equivalent value of P_o given by

$$P_o = \frac{1}{2}(1 + P_o') \quad : \text{measurement accuracy} \le \eta \tag{14.17}$$

where P_o is the confidence level of the $+x\sigma$ case corresponding to (14.8). For example, with $P_o' = 0.99$ the equivalent confidence level is $P_o = 0.995$ and, using $p = 1 \times 10^{-3}$ and $\eta = 0.05$, as in the previous example, the value of x is computed as 2.573709 so the number of bits required by the simulation is 2,646,942 and the number of bit errors is 2647.†

*In practice, and in the examples, the actual number of events and event errors is determined as $N_{test} = \lceil n \rceil$ and $E_{test} = \lceil e \rceil$.

†For modem design engineering performance simulations, these Monte Carlo trials are simply guidelines; however, for modem contractual acceptance testing, they attract considerably more attention.

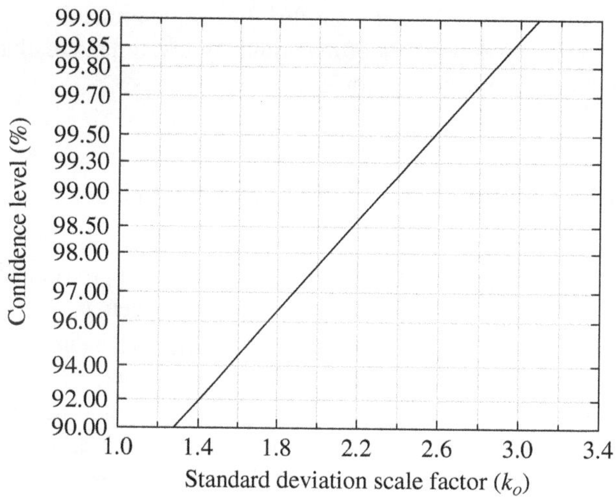

FIGURE 14.4 Confidence level dependence on scale factor k_o.

Returning to (14.5) and the following examples, the sampling requirements can be easily characterized by using tabulated confidence levels or probabilities $P_o(x)$ given the scaled standard deviation $x = k_o\sigma$. Typically [6] $P_o(x)$ is tabulated for a unit standard deviation with $x = k_o$. The confidence level is plotted in Figure 14.4 as a function of k_o and applies to the cases $+k_o\sigma$ and $\pm k_o\sigma$ as discussed above. The disadvantage in this approach is that these values must be input from tables or interpolated from curves, whereas, in the previous analysis they are computed based on a specified confidence level.

Therefore, using a desired confidence level, k_o is determined from Figure 14.4 and, following the above analysis, the standard deviation is computed as

$$\sigma = \frac{\eta p}{k_o} \tag{14.18}$$

Equating (14.18) and (14.5) and solving for the number of trails yields

$$n = \left(\frac{k_o}{\eta}\right)^2 \frac{(1-p)}{p} \tag{14.19}$$

and, from (14.14), the number of errors is

$$e = \left(\frac{k_o}{\eta}\right)^2 (1-p) \tag{14.20}$$

where the parameter k_o is determined using standard probability tables or Newton's method as discussed following Equation (14.9).

Frequently *error bars* of length $k_o\sigma$ or $\pm k_o\sigma$ about the probability p are included to indicate the accuracy of the simulation; an example using error bars is shown in Section 14.6. During the initial stages of the modem design and simulation, rough estimates of the system performance are needed. In this case, a test accuracy of 100% ($\eta = 1.0$) with $k_o = 2$ or 3 is recommended to reduce the number of simulation trials to k_o^2/p. As the modem algorithms are debugged and optimized, the accuracy and confidence level of the tests can be tightened for the final performance evaluations. As a final comment, the simulation or test time is a function of the event-rate (R) events-per-second and is given by $T_{test} = n/R$ seconds. It is very time-consuming to establish the number of trials to satisfy the performance requirements at a specified probability, say $p = 10^{-5}$, and then run the simulation or test for a range of probabilities, say for $p = 10^{-1}$ to 10^{-5}, using the same number of trials. This results in unnecessarily high accuracies at the lower error probabilities and can be avoided by terminating the simulation at each probability (p) when the number of event errors, as expressed in (14.14) and (14.20), is exceeded.

14.2.1.2 Measurement Accuracy with Signal-to-Noise Ratio Degradation A useful criterion, which results in shorter simulation run-times, is to require that the corresponding signal-to-noise ratio be within some specified accuracy $\Delta\gamma_b$(dB). In effect, this condition provides for a relative measurement accuracy that increases as the error probability decreases. For AWGN the bit-error probability for uncoded antipodal signaling is related to the signal-to-noise ratio $\gamma_b = E_b/N_o$ as

$$P_{be} = \frac{1}{2}\left(1 - erf\left(\sqrt{\gamma_b}\right)\right) \tag{14.21}$$

The variation in the bit-error probability is expressed as

$$\Delta P_{be} = \frac{\delta P_{be}}{\delta \gamma_b}\Delta\gamma_b \tag{14.22}$$

Letting $x = \sqrt{\gamma_b}$, the derivative in (14.22) is evaluated as

$$\frac{\delta P_{be}}{\delta \gamma_b} = \frac{\delta x}{\delta \gamma_b}\frac{\delta P_{be}(x)}{\delta x} = -\frac{e^{-\gamma_b}}{2\sqrt{\pi\gamma_b}} \tag{14.23}$$

with the result

$$\Delta P_{be} = -\frac{e^{-\gamma_b}}{2\sqrt{\pi\gamma_b}}\Delta\gamma_b \tag{14.24}$$

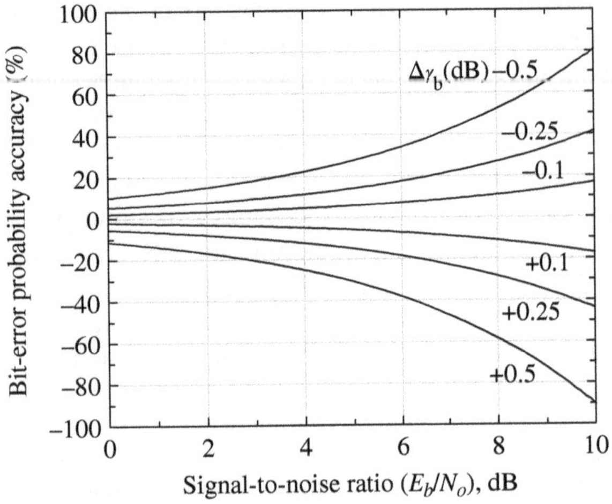

FIGURE 14.5 Measurement accuracy versus signal-to-noise ratio for various signal-to-noise accuracies.

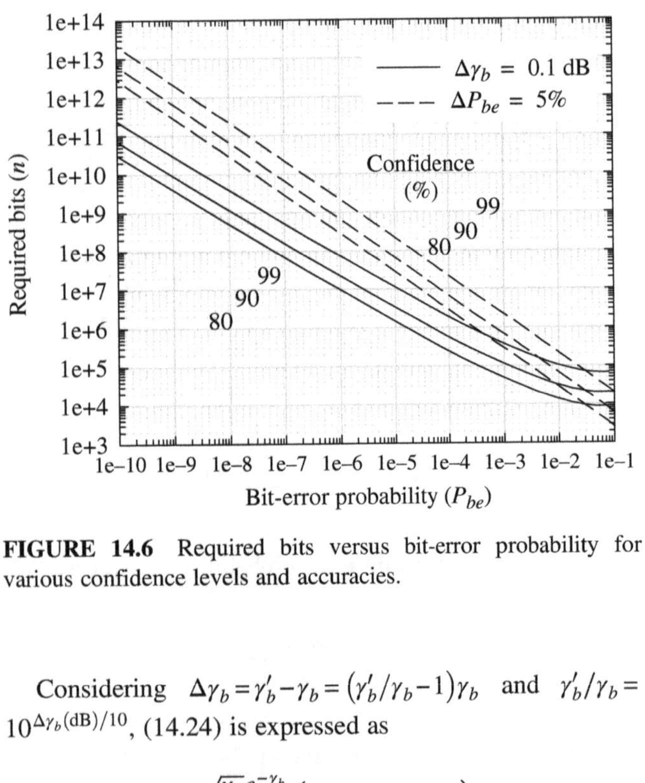

FIGURE 14.6 Required bits versus bit-error probability for various confidence levels and accuracies.

Considering $\Delta\gamma_b = \gamma'_b - \gamma_b = (\gamma'_b/\gamma_b - 1)\gamma_b$ and $\gamma'_b/\gamma_b = 10^{\Delta\gamma_b(\mathrm{dB})/10}$, (14.24) is expressed as

$$\Delta P_{be} = -\frac{\sqrt{\gamma_b}e^{-\gamma_b}}{2\sqrt{\pi}}\left(10^{\Delta\gamma_b(\mathrm{dB})/10} - 1\right) \tag{14.25}$$

Figure 14.5 is plotted $\Delta P_{be}/P_{be}$ in percent as a function of γ_b for various accuracies in the signal-to-noise ratio $\Delta\gamma_b$(dB). The required number of trials is shown as the *solid* curves in Figure 14.6 as a function of the bit-error probability for the indicated confidence levels and decibel accuracy in the measurement of E_b/N_o. For comparison, the number of trials associated with achieving a fixed accuracy in P_{be} is shown as the *dashed* curves in Figure 14.6. For low error probabilities, a saving of several orders of magnitude in test time is realized by specifying a fixed error in E_b/N_o.

14.2.2 Fixed-Sample Testing Using the Poisson Distribution

The preceding description using the Gaussian approximation to the binomial distribution applies only when $p_1 = k/n$ is in the vicinity of $\sigma = \sqrt{p(1-p)/n}$ around the mean value p. These conditions require that $n \gg 1, p \gg 1$, with the product $np \gg 1$. In the preceding sections, it was assumed that this vicinity extended over two or three standard deviations. Furthermore, the Gaussian distribution extends over the range $p_1 = (-\infty,\infty)$ that does not apply to the positive probability values and the finite sample size being considered. To overcome the positive and finite limitations of the sample variable k, the binomial distribution is approximated by the Poisson distribution expressed as [7]

$$\binom{n}{k} p^k (1-p)^{n-k} \cong \frac{(np)^k}{k!} e^{-np} \tag{14.26}$$

This result applies for k on the order of np and does not require that $np \gg 1$. A formal proof of (14.26) is provided by Feller [3]. The criterion for establishing the sample size and the number of event errors is based on hypothesis testing [8–10], where the hypothesis H_0 corresponds to the event probability p_0 and the hypothesis H_1 corresponds to the event probability p_1. In terms of the Poisson distribution, these hypotheses correspond to the discrete density functions*

$$H_0 : p_0(k) = \frac{(np_0)^k}{k!} e^{-np_0} \tag{14.27}$$

and

$$H_1 : p_1(k) = \frac{(np_1)^k}{k!} e^{-np_1} \tag{14.28}$$

14.2.2.1 *Single-Threshold Fixed-Sample Testing*

In this section, the Poisson distribution is applied to a fixed-sample test that determines the acceptance and rejection probabilities of the system under test. This fixed-sample test is similar to the Gaussian test, in that, the test is performed until all n samples are completed before a pass–fail decision is made.

*Strictly speaking the discrete probability density function is $p(k) = \sum_i p'_i \delta(k-i)$, where p'_i is the probability of the ith event.

The likelihood ratio (LR) for the Poisson distribution is defined as

$$\lambda_P = \frac{p_1(k)}{p_0(k)} \tag{14.29}$$

and the test threshold is established based on the log-likelihood ratio (LLR) evaluated as

$$\begin{aligned}\ln(\lambda_P) &= \ln\left(\frac{p_1(k)}{p_0(k)}\right) \\ &= k\ln\left(\frac{p_1}{p_0}\right) - n(p_1 - p_0)\end{aligned} \tag{14.30}$$

The optimum threshold is determined by solving (14.30) for k with $\lambda_P = 1$, yielding

$$Th_k = \frac{n(p_1 - p_0)}{\ln(p_1/p_0)} \tag{14.31}$$

The probabilities under the two hypotheses H_0 and H_1 are depicted in Figure 14.7; where it is assumed that a modem is being tested for a bit-error probability of p_0. As previously defined $p_1 = (1 + \eta)p_0$, where η is the accuracy of the test, that is, for a test length of n-bits, the modem fails and is declared *bad* if the number of bits-errors (k) exceeds the threshold Th_k. The various regions indicated in Figure 14.7 give rise to the acceptance and rejection probabilities associated with the test and two regions are identified for establishing the criterion for the test, they are:

$$\begin{aligned}P_{ca} &= 1 - \alpha = P_r(\text{correctly accepting a good modem}) \\ &= 1 - \sum_{Th_k+1}^{n} p_0(k) \quad : H_0\end{aligned} \tag{14.32}$$

and

$$\begin{aligned}P_{fa} &= \beta = P_r(\text{falsely accepting a bad modem}) \\ &= \sum_{k=0}^{Th_k} p_1(k) \quad : H_0\end{aligned} \tag{14.33}$$

The probabilities P_{ca} and P_{fa} are respectively referred to as the *confidence* and *significance* levels of the test and are typically expressed in percent. In statistical parlance, the probability α is a type I error, or error of the *first kind*, that rejects hypothesis H_0 when H_0 is true and the probability β is a type II error, or error of the *second kind*, that accepts hypothesis H_0 when H_1 is true.

For production testing the specification of P_{ca} and P_{fa}, or confidence and significance, depend upon the modem application and are typically agreed upon through a written contract. However, the respective values of 0.95 and 0.05 (95 and 5%) seem reasonable in noncritical applications or in situations where the receiver signal-to-noise has a build-in margin. In applications where the modem is costly it may be re-worked and/or re-tested. Prior to performing production testing, the modem must undergo extensive engineering testing to demonstrate that the fundamental algorithms are working correctly. For example, experience has shown that a modem can exhibit very infrequent bursts of errors that obliterate the error probability over hours of testing that would not be detected during a short test of n bits. These situations can often be traced to algorithm designs and related signal processing issues like: overflow and truncation.

The single-threshold Poisson test described above is evaluated and the performance is characterized in Figures 14.8 and 14.9 as a function of the product np_0 for various accuracies of the test expressed in percent. These results apply to any value of p_0 as long as the product np_0 is held constant.

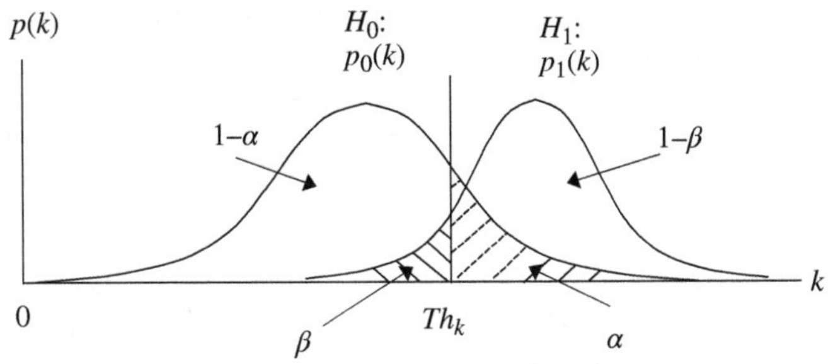

1−β = Probability of correctly rejecting a bad modem.
β = Probability of falsely accepting a bad modem (P_{fa}).
1−α = Probability of correctly accepting a good modem (P_{ca}).
α = Probability of falsely rejectin g a good modem.

FIGURE 14.7 Sequential test description.

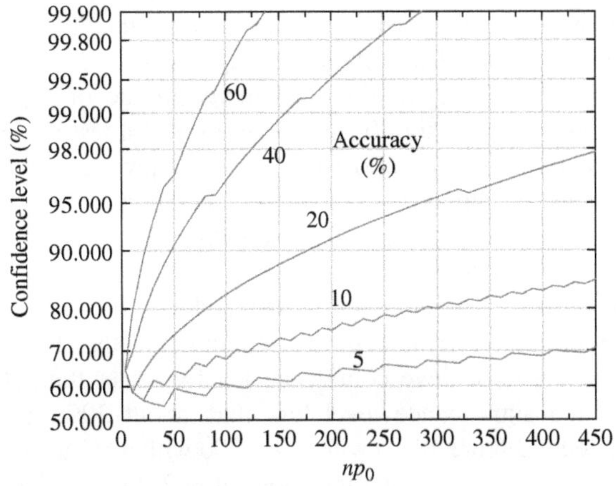

FIGURE 14.8 Single-threshold correct-acceptance probability.

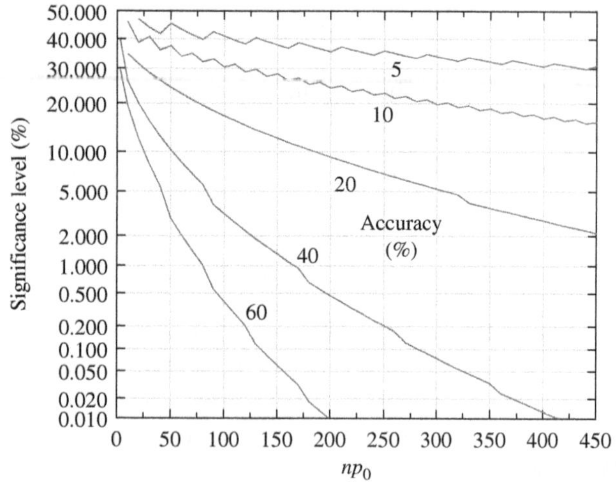

FIGURE 14.9 Single-threshold false-acceptance probability.

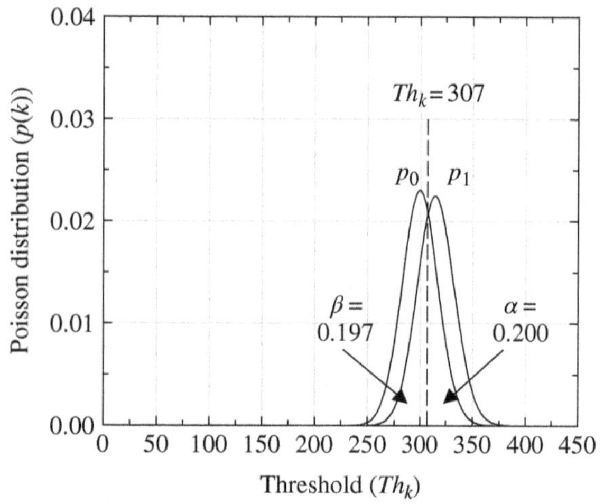

FIGURE 14.10 Poisson distributions with 5% accuracy and $np_0 = 300$.

The irregular or scalloped appearance of these curves results from the integer quantization of the threshold np_0. To achieve a confidence level of 95% and a significance level of 5% requires an accuracy of 40% with an np_0 product of 80; however, to achieve a significance level of 1%, a test accuracy 60% is required with about the same np_0 product. In these examples, the number of bits required in the test of a modem bit-error probability is $n = 80/p_0$, so with $p_0 = 1 \times 10^{-3}$ the test requires $n = 80$ K bits.

For test accuracies of about 10% and less, the underlying distributions for p_0 and p_1 are so close together that it is impossible to achieve confidence levels greater than about 90% and significance level less than about 20%; this is depicted in Figure 14.10 with $np_0 = 300$ with a 5% accuracy, that is, $np_1 = 315$, and an threshold of 307. In this case, the maximum confidence level is about 80% and the minimum significance is 19.7% and these represent limiting values as np_0 increases above 300.

14.2.3 Sequential Sample Testing Using the Binomial Distribution

Sequential-sample testing [11, 12] makes a pass–fail decision at each sample $n' : 1 \le n' \le n$ throughout the test. If at any sample, the number of errors crosses either of two thresholds the system fails the test, otherwise, the test continues until all n samples are completed; after the completion of the n samples, the system is accepted as *good*. The sequential test has the capability of rejecting *bad* systems early in the testing cycle resulting in considerably shorter test times. Because pass–fail decisions are made at each sample beginning at $n = 1$ the preceding approximations to the binomial distribution do not apply and the exact expression, given by (14.1), must be used. The two thresholds, Th_0 and Th_1, are depicted in Figure 14.11 and computed using the bounded LR λ_B for the binomial distributions; the conditions are:

$$\frac{\beta}{1-\alpha} \le \lambda_B \le \frac{1-\beta}{\alpha} \tag{14.34}$$

In this case, the LR, under the hypotheses H_0 and H_1, is expressed as

$$\lambda_B = \frac{P_1(k)}{P_0(k)} = \frac{p_1^k(1-p_1)^{n'-k}}{p_0^k(1-p_0)^{n'-k}} = \left(\frac{p_1 q_0}{p_0 q_1}\right)^k \left(\frac{q_1}{q_0}\right)^{n'} \tag{14.35}$$

Upon substituting (14.35) into (14.34) and taking the natural logarithm results in the bounded expression

$$\ln\left(\frac{\beta}{1-\alpha}\right) \le k \ln\left(\frac{p_1 q_0}{p_0 q_1}\right) + n' \ln\left(\frac{q_1}{q_0}\right) \le \ln\left(\frac{1-\beta}{\alpha}\right) \tag{14.36}$$

Using the upper bound and solving for k leads to the linear equation for the upper threshold as a function of n' expressed as

$$Th_1 = An' + B \tag{14.37}$$

where

$$A = -\frac{\ln(q_1/q_0)}{\ln(p_1 q_0/p_0 q_1)} \quad \text{and} \quad B = \frac{\ln((1-\beta)/\alpha)}{\ln(p_1 q_0/p_0 q_1)} \tag{14.38}$$

Similarly, using the lower bound results in the linear equation for the lower threshold

$$Th_0 = An' + C \tag{14.39}$$

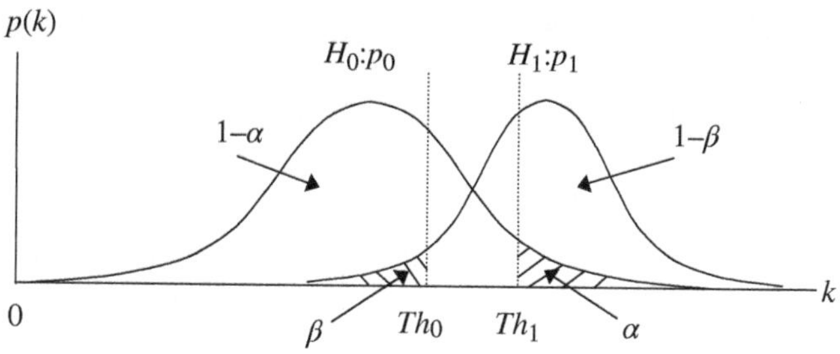

1–β = Probability of correctly rejecting a bad modem.
β = Probability of falsely accepting a bad modem (P_{fa}).
1–α = Probability of correctly accepting a good modem (P_{ca}).
α = Probability of falsely rejecting a good modem.

FIGURE 14.11 Sequential test description.

where

$$C = \frac{\ln(\beta/(1-\alpha))}{\ln(p_1 q_0/p_0 q_1)} \tag{14.40}$$

These decision boundaries are depicted in Figure 14.12 where the abscissa and ordinate intercepts are evaluated as

$$n'_o = \frac{\ln(\beta/(1-\alpha))}{\ln(q_1/q_0)} \tag{14.41}$$

and

$$k_o = \frac{\ln((1-\beta)/\alpha)}{\ln(p_1 q_0/p_0 q_1)} \tag{14.42}$$

The average number of samples is [11]

$$\bar{n}' = \frac{(1-\alpha)\ln(\beta/(1-\alpha)) + \alpha\ln((1-\beta)/\alpha)}{p_0\ln(p_1/p_0) + q_0\ln(q_1/q_0)} \tag{14.43}$$

It has been proven [11] that one or the other threshold will be crossed with probability 1 as $n' \to \infty$. However, it is common practice to limit the sampling and terminate the test when $n' = n$ where n, somewhat arbitrarily, is set equal to $\bar{n}'$ plus twice the size of the fixed-sample test, as determined from Figure 14.6, with the same accuracy using either ΔP_{be} or $\Delta\gamma_b$. The horizontal *solid* line in Figure 14.12 corresponds to the average number of errors at the termination point n under the two hypotheses, that is,

$$k_n = \frac{np_0 + np_1}{2} \tag{14.44}$$

If the test crossed the horizontal *solid* line, the modem fails; however, if the test terminates at the *solid* vertical line at n, the modem passes.

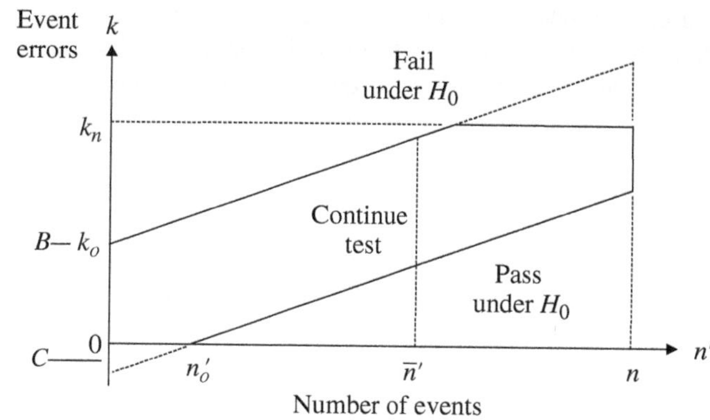

FIGURE 14.12 Pass–fail decision boundaries for sequential test.

TABLE 14.1 Sequential Sample Test Requirements and Parameters for Modem Testing

Test Requirements	
$p_0 = 1 \times 10^{-3}$	H_0: specified bit-error probability
Confidence 95%	H_0
Significance 5%	H_0
$\eta = 5\%$	Test accuracy
$p_1 = 1.05 \times 10^{-3}$	H_1: $p_1 = p_0\,(1 + \eta/100)^a$
Test Parameters	
$A = 1.0252 \times 10^{-3}$	Slope of decision boundaries
$B = 60.287$	Upper boundary constant
$C = -60.287$	Lower boundary constant
$k_o = 60.287$	Upper boundary ordinate intercept
$n'_o = 5.881 \times 10^4$	Lower boundary abscissa intercept
$\bar{n}' = 2.157 \times 10^6$	Average number of samples
$n = 4.557 \times 10^6$	Test termination samples

[a] If p_1 is specified then compute the test accuracy.

Table 14.1 summarizes the test requirements for an example modem production acceptance test and the corresponding sequential test parameters. This test requires a bit-error probability of $P_{be} = 10^{-3}$ at some specified signal-to-noise ratio. The test parameters are used to establish the pass–fail boundaries as shown in Figure 14.12 and the number of bit errors is recorded and plotted as a function of the number of received bits. In this case, referring to Figure 14.6, the fixed sample test under these conditions requires about 1.2×10^6 bits so the sequential test is terminated after $n = \bar{n}' + 2\ (1.2 \times 10^6) = 4.557 \times 10^6$. If the recorded data crosses the lower boundary, including the vertical boundary at n, the modem passes with the prescribed confidence and significance levels otherwise the modem fails the test.

14.3 COMPUTER GENERATION OF RANDOM VARIABLES

The generation of various types of random variables is paramount in the performance simulation of modems and the foundation of nearly all random variable generators is the uniform number generator. The most ubiquitous of the random variable generators is the Gaussian random number generator that is used to model receiver and modem input noise. In this context, the Gaussian noise generator is used to establish the receiver or modem signal-to-noise ratio against which the premier performance measures involving the modem error probability are evaluated. During the modem development it is a common practice to establish the modem performance in a back-to-back mode, that is, without a channel, where the modulator intermediate frequency (IF) output (or transmitter radio frequency (RF) output) is coupled directly to the demodulator IF input (or receiver RF input). In this configuration the modem performance is compared to theoretical performance curves under relatively ideal conditions like: oscillator phase noise, amplifier nonlinearities, intersymbol interference (ISI), and demodulator quantization noise will result in somewhat less than idea performance. With the exception of the quantization noise these transceiver and modem noise sources are modeled using specialized random number generators, for example, the ISI noise is influenced by the randomization of the source data that is generated using a binary uniform random data generator. To examine the modem performance through a communication channel, the channel is frequently modeled using an additive or multiplicative random number generator. For example, additive co-channel or adjacent channel interference is modeled using randomly modulated signals and the implementation of a fading channel simulator is modeled as a Rayleigh or Ricean random process.

In the following sections, computer implementation of various random number generators is discussed. The random variable U_i is used to denote uniformly distributed random variables; X_i is used to denote Gaussian distributed random variables, including correlated Gaussian variables; Y_i is used to denote other random variables and Z_i is used to denote the summation of independent *iid* random variables. Additional insights regarding the generation and testing of random numbers can be found in the literature [13–15].

14.3.1 Uniform Random Number Generation

As mentioned above, the uniform random number generator forms the foundation for the generation of nearly all other random variables. The uniform number generator generates a sequence of random numbers that are ideally* independent and uniformly distributed over the period of the generator; the period is determined by the computer's internal processing size and the specific algorithm being used to generate the uniform numbers. Ideally the uniform numbers are

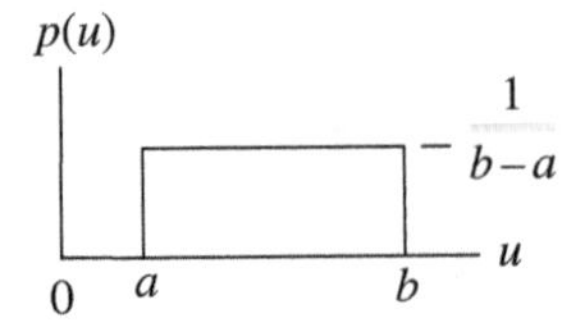

FIGURE 14.13 Uniformly distributed probability density function.

characterized by the uniform *pdf* shown in Figure 14.13 where the range of the random numbers is $a \le u_i \le b$ with a mean value of $(a + b)/2$.

Random numbers that represent a continuous *pdf* $p(u)$ with cumulative distribution $0 \le P(U) \le 1$ are generated using uniformly distributed random numbers $0 \le U_i \le 1, i = 0, 1, \ldots$ and the inverse function $u = P^{-1}(U_i)$. In these cases, the parameters a and b are: $a = 0$ and $b = 1$. To generate random bipolar binary source data, $d = \{-1,1\}$ the same procedure is used, however, the cumulative distribution $P(U)$ is a linear function of U and, using the parameters $a = -1/2$ and $b = 1/2$, the source data is generated as $d_i = \text{sign}(1, U_i)$, where $-1/2 \le U_i \le 1/2$, $i = 0, 1, \ldots$ are uniformly distributed random numbers. This description discusses the characteristics and application of uniformly distributed random numbers; however, methods for generating and testing them† [16] is the subject of the remainder of this section.

Computer language software often includes a function for generating uniformly distributed random numbers that presumably has been tested for various properties of randomness. The congruence method [13] or algorithm is commonly used for generating a sequence of random numbers $0 \le X_i \le C$ where C is the magnitude of the maximum number as determined by the number of computer bits used to represent a number. Uniformly distributed random numbers $0 \le U_i \le 1$ are determined as $U_i = X_i / C$.

The congruence method [13] is expressed as‡

$$X_{i+1} = AX_i + B \mod(C) \quad : i = 0, 1, \ldots \tag{14.45}$$

where B is chosen to be relatively prime to C and the constants A with B chosen to provide the desired random properties and the longest possible sequence of random numbers. The initial value, X_0, is the generator *seed* and is selected to be relatively prime to C. Theoretically each seed produces an independent random sequence and a unique seed is assigned to the various simulation functions. For example, unique uniform number generator seeds are assigned to generate: additive Gaussian noise, phase noise, desired and interfering channel source data, and channel fading. Although the congruence method described by (14.45) provides excellent

*Computer-generated random numbers are often referred to as pseudo-random numbers because they are subject to the idiosyncrasies of the selected algorithm.

†Reference 16 includes an extensive list of references.

‡The $x_{i+1} = y_i \mod(c)$ function denotes $x_{i+1} = y_i - \text{int}(y/c)$, where $y_i = ax_i + b$ and $\text{int}(z)$ denotes the integer part of z.

random properties and adequate sequence lengths, it is relatively slow compared to the implementation described by Knuth [17].

Finite-length pseudo-random sequences generated using maximal-length sequences exhibit excellent random properties and nearly ideal correlation responses; these are often referred to as PN sequences or M-sequences. PN sequences are introduced and discussed in Chapter 8.

Uniform random number generators and M-sequences can be tested against several relatively simple criteria to verify that their performance characteristics satisfy the independence and random properties. The first test involves the autocorrelation response of the random sequence and the resulting power spectral density (PSD). For a sequence of length n, the autocorrelation response is a single sample with amplitude n and sidelobe levels on the order of $1/n$. The PSD is determined by taking the Fourier transform of the correlation function that ideally corresponds to white noise characterized by a constant PSD frequency response. Correlation and PSD responses expose non-random behaviors involving initial seed repetitions and the existence of repeated patterns that do not follow the definition of Bernoulli trials and the underlying binomial distribution. This suggests a second test involving the binomial distribution expressed by (14.1) with $p = q = 1/2$, that is, each binary event occurs with equal probability. In this case, the binomial distribution becomes

$$p(k) = \binom{n}{k}\frac{1}{2^n} \tag{14.46}$$

Considering a binary sequence of length n, consisting of *mark* and *space* data bits* with respective probabilities $p = q = 1/2$, the binomial coefficients represent the number of times that k *mark*-bits will occur in the data sequence and (14.46) is the corresponding probability of occurrence. For example, using a binary sequence of length $n = 3$ bits the binomial coefficients for $k = 0, 1, 2, 3$ are $\{1, 3, 3, 1\}$ respectively. The binary data patterns are listed in Table 14.2 with the corresponding patterns containing k mark bits. The number of k mark-bit patterns is consistent with the binomial coefficients and the corresponding probabilities of occurrence are $\{1/8, 3/8, 3/8, 1/8\}$. The Bernoulli trial test is a relatively simple test to perform.

Other uniform number generator tests involve: examining the repetition of seeds during the sequence generation; sequence independence with different starting seeds; the length of the sequence; the periodic generation of subsequences. For example, when different starting seeds are applied to the generation of co-channel and adjacent channel interference, it is important to ensure that the random properties of these channels are independent from each other and the desired channel. Also, when simulating the performance of modems using turbo-coded FEC, run lengths of millions or perhaps billions of bits may be required so the length of the random sequence generators becomes important. Additional tests, involving the generation of uniform random numbers, must be performed to examine the generators' ability to duplicate specific random processes within the prescribed confidence levels; examples using the exponential and Gaussian random processes are given in the following sections.

*Unipolar *mark* data is identified as a binary 1 and *space* data as a binary 0.

TABLE 14.2 Binary Data Patterns in a Sequence of Length $n = 3$

Data	k
000	0
001	1
010	1
011	2
100	1
101	2
110	2
111	1

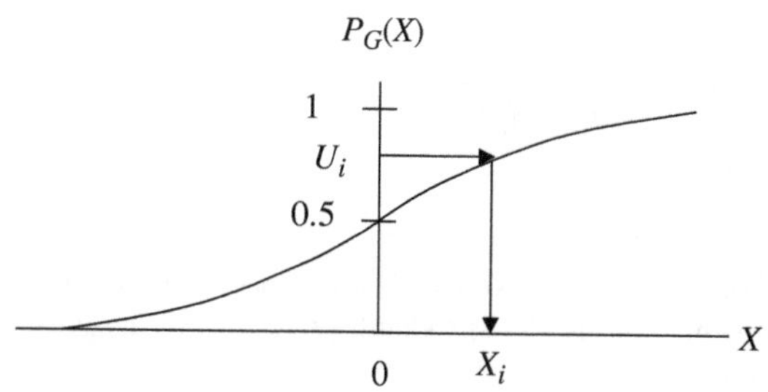

FIGURE 14.14 Generation of Gaussian random numbers using inverse cumulative probability.

14.3.2 Gaussian Random Number Generation

The generation of Gaussian random numbers is a close runner-up to uniform number generators in terms of its utility in modem performance simulations. Gaussian number generators use the inverse of the Gaussian cumulative distribution $P_G(X)$: $0 \le P_G(X) \le 1$ so, the uniformly distributed random number U_i: $0 \le U_i \le 1$, forms the basis for the generation of Gaussian random numbers using $X_i = P_G^{-1}(U_i)$. This is depicted in Figure 14.14 for a zero-mean, unit standard deviation Gaussian distributed random number. Gaussian random numbers with mean (m_x) and standard deviation (σ_x) are simply obtained as

$$X_i' = \sigma_x X_i + m_x \tag{14.47}$$

Initialize:

$c_0=2.515517$, $c_1=0.802853$, $c_2=0.010328$

$d_1=1.432788$, $d_2=0.189269$, $d_3=0.001308$

$0<U\le 1.0$
if $(U>0.5)$ then
$U'=1-U$
else
$U'=U$
endif

$$t=\sqrt{\ln(1/U')^2}$$

$$X=t-\frac{(c_0+c_1t+c_2t^2)}{1+d_1t+d_2t^2+d_3t^3}+e(U')$$

if $(U<0.5)$ $X=-X$

FIGURE 14.15 Approximate Gaussian random number generation. Abramowitz [5, p. 933]. Courtesy of U.S. Department of Commerce.

Although the inverse mapping $P_G^{-1}(U_i)$ of the Gaussian *cdf* describes the process for the generation of Gaussian random numbers, fast and efficient methods for their generation on computers are described by Muller [18]. The approach used in most of the simulation results presented in this book uses the approximation to the inverse function $X=Q^{-1}(U)$: $0<U\le 1.0$ given by Hastings [19, 20], with an approximation accuracy of $|e(U)|<4.5\times10^{-4}$. Hastings' approximation for Gaussian number generation is given in Figure 14.15.

Gaussian random numbers can also be generated by summing independent random variables and recognizing that the central limit theorem [21, 22] applies. One approach is to sum uniformly distributed numbers U_i with the result $Z=\sum_i U_i$ where the distribution of Z approaches a Gaussian distributed random number as the number of summations approaches infinity. Another application of the central limit theorem is the summation of sinusoidal functions with uniformly distributed phase, that is, with $-1\le U_i\le 1$ and $\phi_i=\pi U_i$ then $Z=\sum_i\cos(\phi_i)$ and Z approaches a Gaussian distribution as the summation index increases.

The utility of the Gaussian number generator in simulating the modem bit-error probability performance is determined by examining the number of simulation trials that fall within the confidence and accuracy limits given the required number of Monte Carlo samples-per-trial. The number of samples-per-trial is determined using (14.13) or the equivalent expression (14.19), repeated here as

$$n=\left(\frac{k_o}{\eta}\right)^2\left(\frac{1-p}{p}\right) \tag{14.48}$$

where η is the specified confidence level of the Monte Carlo simulation and, using Figure 14.4, k_o is determined as the corresponding standard deviation scale factor. This test is examined in the case study in Section 14.6 where it is demonstrated that the Gaussian random number generator, when used as a Gaussian noise source, performs within the expected statistical bounds of the bit-error probability $p=P_{be}$.

14.3.2.1 *Correlated Gaussian Random Number Generation*

Correlated Gaussian random numbers can be generated in several ways using X_i. For example, referring to Papoulis [23], the conditional distribution of a correlated Gaussian random variable X_{i+1} with correlation coefficient ρ, mean value ρX_i, and variance σ^2, for $i=0, 1, \ldots$ and $X_0=0$, is expressed as

$$p(X_{i+1}|X_i)=\frac{1}{\sqrt{2\pi(1-\rho)\sigma^2}}e^{-(X_{i+1}-\rho X_i)^2/(2(1-\rho)\sigma^2)} \tag{14.49}$$

The various expectations of X_i are given by

$$E[X_{i+1}X_i]=\rho\sigma^2 \tag{14.50}$$

$$E\left[X_{i+1}^2X_i^2\right]=\sigma^4(1-2\rho^2) \tag{14.51}$$

and

$$E[|X_{i+1}X_i|]=\frac{2\sigma^2}{\pi}(\cos(\alpha)+\sin(\alpha)) \tag{14.52}$$

where $\alpha=\sin^{-1}(\rho)$: $-\pi/2<\alpha\le\pi/2$. Upon generating a sequence of correlated Gaussian random samples, the correlation coefficient is computed from the normalized response of the autocorrelation function.

The correlated Gaussian random variables are also generated using the autoregressive (AR) model [24] expressed as

$$Z_i=\sum_{j=1}^{N}w_jZ_{i-j}+X_i \tag{14.53}$$

This is an N-th order AR process where the current value, Z_i, of the processes is a linear combination of the past N values plus an error term X_i characterized as independent Gaussian noise samples. The weights w_j are determined as the solution to N simultaneous equations.

The method of generating correlated random variables using a defined PSD is discussed in Chapter 20. A classical method used to generate correlated random samples that characterize the Doppler spread of the channel is Jakes' PSD function given by [25]

$$S(f)\begin{cases}=\dfrac{1}{\pi f_d\sqrt{1-(f/f_d)^2}} & :|f|\le f_d\\ =0 & :o.w.\end{cases} \tag{14.54}$$

where f is the frequency around f_d. An application using the phase PSD to generate inphase and quadrature samples for a Rayleigh fading channel is given in Section 20.7.

14.3.3 Ricean and Rayleigh Random Number Generation

Ricean random numbers Y_i are generated using two zero mean independent *iid* Gaussian random variables X_{Ii} and X_{Qi} with standard deviation σ_r,

$$Y_i = \sqrt{(V_s + X_{Ii})^2 + X_{Qi}^2} \quad \text{: Ricean random variable} \tag{14.55}$$

where V_s represents the received peak signal level or specular value of the received signal plus noise. Defining the specular to random signal-to-noise ratio as

$$\gamma_{sr} = \frac{V_s^2}{2\sigma_r^2} \tag{14.56}$$

and normalizing Y_i by σ_r results in the normalized Ricean random variables

$$\bar{Y}_i = \sqrt{\left(\sqrt{2\gamma_{sr}} + \bar{X}_{Ii}\right)^2 + \bar{X}_{Qi}^2} \tag{14.57}$$

where $\bar{X}_{Ii}$ and $\bar{X}_{Qi}$ are zero-mean unit-variance Gaussian random variables. Rayleigh random numbers result when the specular to random power ratio approaches zero, that is, when $\gamma_{sr} \to 0$ and, in the limit (14.55), becomes

$$Y_i = \sqrt{X_{Ii}^2 + X_{Qi}^2} \quad \text{: Rayleigh random variable} \tag{14.58}$$

Examples of Ricean and Rayleigh *pdfs* generated using (14.55) and (14.58) respectively are given in Section 18.2. Correlated Ricean and Rayleigh random numbers are obtained by using (14.49) in the generation of $\bar{X}_{Ii}$ and $\bar{X}_{Qi}$. An example of the correlation characteristics for the Ricean random numbers generated with specular-to-random component ratio of $\gamma_{sr} = 12$ dB is shown in Figure 14.16. These results are obtained using a sequence of 5000 Ricean generated samples using (14.55) with the underlying correlated Gaussian random variables generated using (14.49).

When these results are used in a system performance simulation, it is important to know the corresponding correlation time associated with the specified correlation coefficient ρ. The correlation interval is defined as correlation lag n_o where the correlation response equals the normalized correlation threshold of e^{-1}. For example, for the case $\gamma_{sr} = 12$ dB, the correlation lag n_o is determined from Figure 14.16 as a function of correlation parameters ρ. The correlation lag

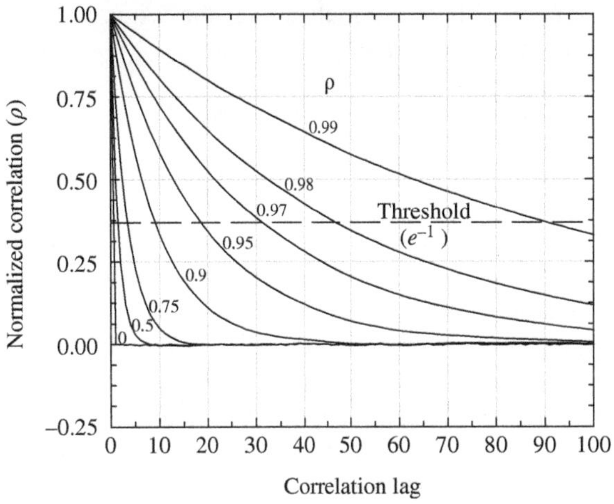

FIGURE 14.16 Correlation response for correlated Ricean random numbers ($\gamma_{sr} = 12$ dB).

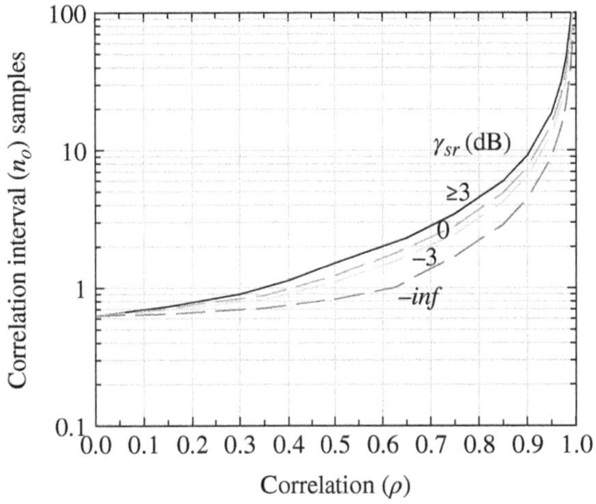

FIGURE 14.17 Correlation interval for Ricean random numbers generated using Equation 14.57.

characteristics are summarized in Figure 14.17 for the indicated values of γ_{sr}. The case $\gamma_{sr} = -\infty$ corresponds to correlated Rayleigh random variables and as $\gamma_{sr} \to +\infty$ and the random variable Y_i approaches the constant signal amplitude V_s. The decorrelation time is denoted as τ_o and is related to the sampling interval, $\Delta\tau$, associated with the random variable Y_i and is expressed as

$$\tau_o = n_o \Delta\tau \tag{14.59}$$

Fast fading environments are associated with signal fluctuations that are rapid relative to the transmitted symbol duration. Fast fading results in significant symbol distortion in the demodulator matched filter. Considering that a minimum of four samples-per-symbol is required to satisfy the Nyquist requirement with or without fading, the required sampling frequency is expressed as

$$f_s = \max\left(\frac{N_s}{T}, \frac{n_o}{\tau_o}\right) \qquad (14.60)$$

The application of these results to communication system operating in a fading environment starts with the specification of the parameters T, τ_o, ρ, and γ_{sr}. Using ρ and γ_{sr}, the value of n_o is determined from Figure 14.17. The de-correlation time is typically specified in terms of slow fading ($\tau_o \geq NT$) or fast fading ($\tau_o < NT$), where the selection of N is somewhat subjective; choosing $N = 1$ results in a matched filter distortion loss, so values of three or four are preferable. Normally the specified range of τ_o is several orders of magnitude. The selection of N_s depends on the symbol spectral sidelobes and the specified aliasing distortion, however, $N_s = 4$ is frequently used. Based on these observations, the sampling frequency is selected using (14.60).

14.3.4 Poisson Random Numbers

The generation of Poisson random numbers is based on the *pdf* expressed as

$$p_P(k) = \frac{\alpha^k}{k!}e^{-\alpha} \qquad (14.61)$$

Expressing the parameter α as $\alpha = \lambda\tau$, where $\tau \geq 0$ is the time between the Poisson distributed events, then (14.61) becomes

$$p_P(k,\tau) = \frac{(\lambda\tau)^k}{k!}e^{-\lambda\tau} \qquad (14.62)$$

As seen in the following section, the time interval between the Poisson distributed events is represented by the continuous exponentially distributed random variable τ that is used to generate the Poisson random variables.

14.3.5 Exponential and Poisson Random Number Generation

The generation of exponentially distributed random numbers uses the inverse probability mapping $Y_i = P_E^{-1}(U_i)$ where the cumulative distribution $P_E(Y): 0 \leq P_E(Y) \leq 1$ and U_i: $0 \leq X_i \leq 1$ is a uniformly distributed random number. Using the exponential *pdf*, given in Table 1.8 and expressed as,

$$p_Y(y) = \lambda e^{-\lambda y} \qquad (14.63)$$

the exponentially distributed random variables Y_i are computed as

$$Y_i = \frac{-\ln(1-U_i)}{\lambda} \qquad (14.64)$$

: exponentially distributed random variable

When simulating the performance of queuing systems [26] the random variable y represents the random time intervals $\Delta\tau$ between events in the queuing system and, in this application (14.64), is used to generate the time between the occurrences of events in the Poisson process. The exponential distribution is memoryless, in that, the past occurrences Y_n: $n = 0,\ldots, i-1$ do not influence the prediction of Y_i.* Because Y_i represents the interval between the occurrences of Poisson distributed events, the Poisson distributed random events or numbers are generated as

$$Z_k = \sum_{i=0}^{k} Y_i \quad : \text{Poisson random variable} \qquad (14.65)$$

Using k as the summation limit is significant, in that, this is the same variable used in (14.62) to characterize the Poisson distribution. Papoulis [27] shows that there are exactly k events in the interval τ where, upon letting $Y_i = \Delta\tau_i$, $\tau = Z_k$. Plots of the theoretical *pdf* for Poisson distributed random variables are shown in Figure 14.18 for $k = 0$–4; it is left as an exercise (see Problem 14) to plot the corresponding *pdfs* for Poisson pseudo-random variables generated using (14.64) and (14.65).

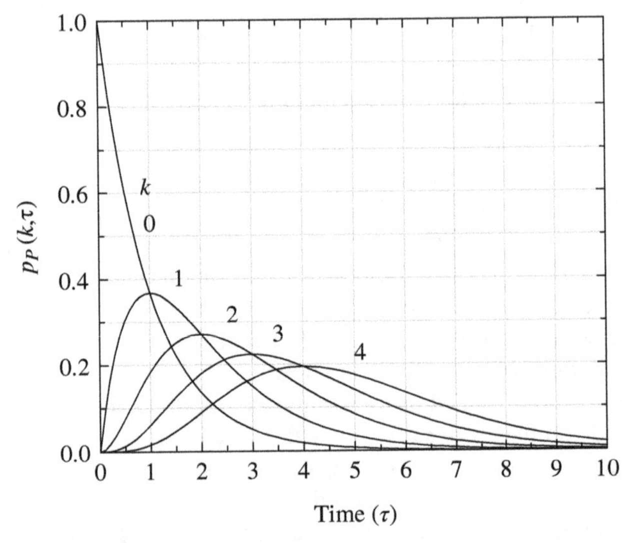

FIGURE 14.18 Theoretical *pdf* for the Poisson distributed random variables ($\lambda = 1.0$, $k \leq 4$).

*Kleinrock [26], pp. 66–67, points out that the exponential distribution is the only continuous distribution having this property.

14.3.6 Lognormal Distribution

The lognormal distribution characterizes an impulsive noise communication channel in terms of the received noise amplitude x, expressed by

$$p_X(x) = \frac{1}{\sqrt{2\pi}x\sigma} e^{-(\ln(x)-m)^2/(2\sigma^2)} \tag{14.66}$$

where x is a random variable for which the logarithm $y = \ln(x)$ is a normally distributed random variable, with mean $m = E[\ln(x)] = E[y]$ and variance $\sigma^2 = E\left[(\ln(x)-m)^2\right]$. Using the random variable y, (14.66) is expressed as the Gaussian or normal distribution $p_Y(y)$ given by

$$p_Y(y) = \frac{1}{\sqrt{2\pi}\sigma} e^{-(y-m)^2/(2\sigma^2)} \tag{14.67}$$

The random variable x is exponentially related to y as* $x = e^y$ and, unlike the Gaussian distribution, the lognormal distribution is skewed so that high values of x occur with relatively high probabilities as seen from the plots of (14.66) in Figure 14.19.

Because of the impulsivity of the random variable x, the lognormal distribution is used to generate impulse noise that is characterized or measured in terms of the ratio of the *rms* envelope to average envelope defined as

$$V_d \triangleq \frac{\sqrt{E[x^2]}}{E[x]} \tag{14.68}$$

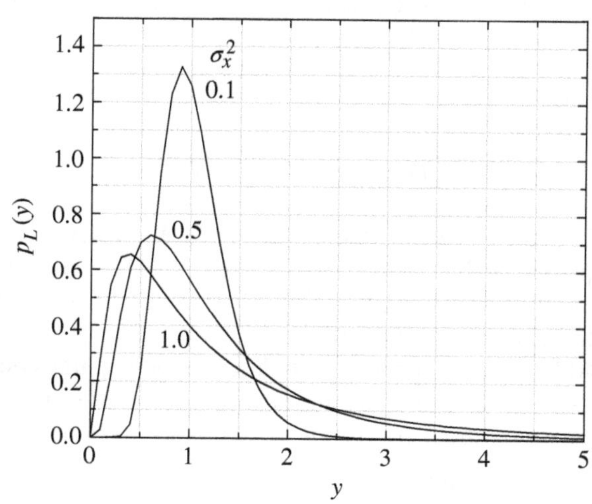

FIGURE 14.19 Theoretical *pdf* for the lognormal distributed random variables ($m_x = 0$; $\sigma_x^2 \leq 1, 0.5, 0.1$).

*In general, the exponential relationship $x = e^y$ characterizes a logarithmic distribution; however, when y is normally distributed it is referred to as a lognormal distribution.

The sampled lognormal noise appears at the input to the receiver and undergoes dispersion in the various receiver and modem filters. The filtering broadens the impulse response in time and simultaneously decreases impulsiveness resulting in a smaller value of V_d. For example, the limiting case of impulsive noise occurs when the noise energy, from worldwide storm activity, is randomly added and filtered resulting in Gaussian noise with a V_d value of 1.049 dB. The lognormal *pdf* is encountered in Section 19.10 as it relates to lightening-induced impulsive noise in the VLF and LF communication channel.

14.4 BASEBAND WAVEFORM DESCRIPTION

With few exceptions,† the simulation of communication systems involves a baseband characterization of the signal and channel. This characterization is embodied in the complex envelope, also referred to as the analytic signal representation of signals, channels, and various transmitter and receiver filters. The analytic signal representation is simply a baseband description of the signal that is obtained by a linear translation of a carrier-modulated signal to baseband as found in heterodyne or homodyne receivers. Typically, baseband simulations are valid in situations involving low instantaneous bandwidths, B, relative to the carrier frequency (f_c) that is, when $B/f_c \ll 1$. As in the case of establishing the Nyquist bandwidth of a signal, the bandwidth B must be defined with care. For example, with a rectangular modulated waveform the signal bandwidth is typically defined in terms of the equivalent noise bandwidth of the matched filter, that is, $1/T$ where T is the symbol duration. However, the resulting $sinc(fT)$ spectrum has considerable energy for frequencies $f > 1/T$ so, for an acceptable ISI distortion loss, the bandwidth B must be defined somewhat greater than $1/T$.

In general, a modulated waveform can be expressed as

$$\begin{aligned} s(t) &= s_c(t)\cos(\omega_c t + \theta(t)) - s_s(t)\sin(\omega_c t + \theta(t)) \\ &= [s_c(t)\cos(\theta(t)) - s_s(t)\sin(\theta(t))]\cos(\omega_c t) \\ &\quad - [s_c(t)\sin(\theta(t)) + s_s(t)\cos(\theta(t))]\sin(\omega_c t) \end{aligned} \tag{14.69}$$

where ω_c is the carrier radian frequency, $\theta(t)$ is an arbitrary signal phase function, and $s_c(t)$ and $s_s(t)$ are inphase and quadrature functions describing the waveform modulation. The time dependence of the arbitrary signal phase function allows inclusion of frequency errors and frequency-rate as might be encountered under dynamic channel conditions. The second expression for $s(t)$ in (14.69) simply isolates the carrier frequency term from the modulation and phase

†A notable exception is the evaluation of harmonic distortion resulting from carrier-modulated signals and nonlinear power amplifiers.

function terms. In practice, only *real* signals are encountered; however, $s(t)$ can be characterized as the real-part of a complex signal representation given by

$$\begin{aligned} s(t) &= \mathrm{Re}\left\{ (s_c(t) + js_s(t))e^{j(\omega_c t + \theta(t))} \right\} \\ &= \mathrm{Re}\{\tilde{s}(t)e^{j\omega_c t}\} \end{aligned} \tag{14.70}$$

where $\tilde{s}(t)$ is defined as the complex envelope or analytic signal characterized as

$$\begin{aligned} \tilde{s}(t) &= (s_c(t) + js_s(t))e^{j\theta(t)} \\ &= [s_c(t)\cos(\theta(t)) - s_s(t)\sin(\theta(t))] \\ &\quad + j[s_c(t)\sin(\theta(t)) + s_s(t)\cos(\theta(t))] \end{aligned} \tag{14.71}$$

The relationship of the real and imaginary terms of $\tilde{s}(t)$ to the expression of the real signal $s(t)$ given in (14.69) is evident, that is, the real part of $\tilde{s}(t)$ is associated with the *cosine* of the carrier frequency and the imaginary part is associated with the *sine* of the carrier frequency.

It is instructive to relate these expressions to the baseband output when using a heterodyne receiver as shown in Figure 14.20. The zonal filters are simply ideal lowpass filters with a linear phase response and are used to eliminate the $2\omega_c$ terms resulting from the mixing operation. Typically, the lowpass bandwidth of the zonal filters is considered to be ω_c so that negligible signal distortion is introduced when $B \ll f_c$.

Considering the input signal $s(t)$ described by (14.69), the baseband outputs shown in Figure 14.20 are evaluated as

$$\begin{aligned} s'_c(t) &= s(t)\cos(\omega_c t)|_{filtered} \\ &= (s_c(t)\cos(\omega_c t + \theta(t)) - s_s(t)\sin(\omega_c t + \theta(t)))\cos(\omega_c t)|_{filtered} \\ &= \frac{1}{2}s_c(t)\cos(\theta(t)) - \frac{1}{2}s_s(t)\sin(\theta(t)) \\ &= \frac{1}{2}\mathrm{Re}\{\tilde{s}(t)\} \end{aligned} \tag{14.72}$$

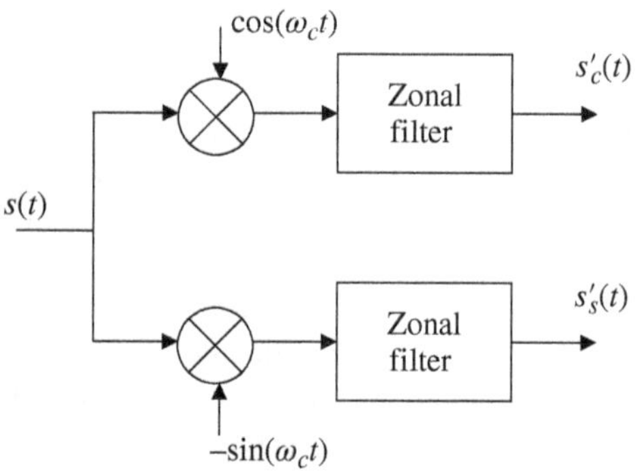

FIGURE 14.20 Simplistic view of heterodyne or homodyne receiver.

Similarly, the quadrature output is given by

$$\begin{aligned} s'_s(t) &= -s(t)\sin(\omega_c t)|_{filtered} \\ &= -(s_c(t)\cos(\omega_c t + \theta(t)) - s_s(t)\sin(\omega_c t + \theta(t)))\sin(\omega_c t)|_{filtered} \\ &= \frac{1}{2}s_c(t)\sin(\theta(t)) + \frac{1}{2}s_s(t)\cos(\theta(t)) \\ &= \frac{1}{2}\mathrm{Im}\{\tilde{s}(t)\} \end{aligned} \tag{14.73}$$

Therefore, there is simply a factor of 1/2 relating these outputs to their counterparts of the analytic signal. These results have assumed that the zonal filter has filtered all vestiges of the signal spectrum centered at $2\omega_c$ hence the restriction on the signal bandwidth relative to f_c stated previously as $B/f_c \ll 1$.

With discrete-time sampled data the sampling frequency is defined as the Nyquist frequency $f_s = 2f_N$, where f_N is the Nyquist bandwidth selected as $f_N \geq B$. In the following sections, computer simulation techniques are discussed to examine the impact of the sampling frequency selection on the discrete-time digital signal processing (DSP) of the received signal. The signal sampling requirements for software simulations are identical to those for the DSP hardware implementations. For example, choosing the lowest possible sampling frequency, commensurate with acceptable distortion, results in the shortest simulation times and the most efficient use of the DSP hardware capabilities.

The quadrature signals $s_c(t)$ and $s_s(t)$ are defined as

$$s_c(t) \triangleq \frac{V}{\sqrt{2}}u_c(t) \quad \text{and} \quad s_s(t) \triangleq \frac{V}{\sqrt{2}}u_s(t) \tag{14.74}$$

and

$$u_c(t) \triangleq d_c p_c(t) \quad \text{and} \quad u_s(t) \triangleq d_s p_s(t) \tag{14.75}$$

In (14.74) V is the peak voltage of the carrier frequency and in (14.75) d_c, $d_s = \{1,-1\}$. For quadrature phase shift keying (QPSK) modulation $p_c(t) = p_s(t) = p(t)$ with $p(t) = rect((t-T/2)/T)$.

Although it is convenient to describe the real transmitted signal in terms of the quadrature carrier terms $\cos(\omega_c t)$ and $\sin(\omega_c t)$, it is instructive to consider the polar form of (14.69) expressed, with $p_c(t) = p_s(t) = 1$, as

$$s(t) = M(t)\cos(\omega_c t + \psi(t)) \tag{14.76}$$

with magnitude and phase functions expressed as

$$M(t) = \sqrt{s_c^2(t) + s_s^2(t)} = V \tag{14.77}$$

and

$$\psi(t) = \tan^{-1}\left(\frac{s_c(t)\sin(\theta(t)) + s_s(t)\cos(\theta(t))}{s_c(t)\cos(\theta(t)) - s_s(t)\sin(\theta(t))}\right)$$
$$= \theta(t) + \tan^{-1}\left(\frac{d_s}{d_c}\right) \quad (14.78)$$

where (14.77) results from noting that $p^2(t) = 1 \forall t$ and (14.78) results from an *arctan* trigonometric identity. This result indicates that the waveform has a constant envelope corresponding to a peak carrier level of V volts. The data-dependent phase term *rests* at $n\pi/2$ radians during each symbol interval; however, the results also apply to offset quadrature phase shift keying (OQPSK) with a phase shift of $\pm\pi/2$ every bit interval. The modulated signal power is given by

$$P_s = \frac{1}{T}\int_{-\infty}^{\infty} s(t)^2 dt = \frac{1}{2T}\int_{-\infty}^{\infty} |\tilde{s}(t)|^2 dt \quad (14.79)$$

and, for the QPSK waveform with $M(t) = V$, the carrier power is $P_s = V^2/2$.

14.5 SAMPLED WAVEFORM CHARACTERIZATION

System performance simulations involve sampled descriptions of the waveforms. In this section, the waveform descriptions are re-written in terms of the Nyquist sampling requirements and various parameters are normalized with respect to the sampling frequency (f_s) and symbol duration (T). With the parameter normalization, the simulation results can be applied to virtually any symbol rate provided the channel conditions are not restricted to channel time constraints, that is, the channels are linear and time invariant. Under these conditions, the baseband simulation performance results with AWGN and linear filtering normalized to the signal bandwidth can be applied to any carrier frequency. However, other types of channels involving, for example, ocean waves, atmospheric noise, and fading are dependent on time constraints and carrier frequencies related to natural phenomena. In these cases, the performance simulations can often be performed at baseband; however, the results are specialized to the channel and the carrier frequency.

The first step in the sampled waveform characterization is to define the time variable (t) in terms of the sampling interval ΔT as $t = i\Delta T$: $i = 1, 2, \ldots$. The sampling frequency is related to the sampling interval by $f_s = 1/\Delta T$ and is conveniently normalized by the symbol rate $R_s = 1/T$ with the result $f_s/R_s = f_sT = T/\Delta T = 1/N_s$ where N_s is the number of samples-per-symbol. The time-varying phase function $\theta(t)$ of the received signal is described by the frequency error (f_d), the frequency-rate $(\dot{f}_d)$, various higher order frequency terms, and the constant phase term ϕ_o. These terms result from the dynamics of the encounter and the propagation through the channel.*

14.5.1 BPSK Waveform Simulation with AWGN

Considering a binary phase shift keying (BPSK)-modulated waveform and the AWGN channel, the received signal is expressed as

$$s(t) = Vd_{c\ell}rect\left(\frac{t - T/2 - \ell T}{T}\right)\cos\left((\omega_c + \omega_d)t + \left(\frac{\dot{\omega}_d}{2}\right)t^2 + \varphi_o\right) \quad (14.80)$$

where $\ell = 0, 1, \ldots$ represents the received data sequence and $d_{c\ell} = \{1,-1\}$ where the mapping of the binary data $b = \{0,1\}$ is $d_c = 1 - 2b$. Equation (14.80) is equated to (14.69) with $s_s(t) = 0$ and the equivalent analytic baseband signal is obtained using (14.71) with the result

$$\tilde{s}(t) = s_c(t)[\cos(\theta(t)) + j\sin(\theta(t))] \quad (14.81)$$

Substituting $t = i\Delta T$ and using the simplified notation $d_{c\ell}$ for the data sequence $d_{c\ell}rect((t - T/2 - \ell T)/T) : \ell = 0, 1, \ldots$, the normalized discrete-time sampled analytic baseband waveform becomes

$$\tilde{s}_i = \tilde{s}(i\Delta T)$$
$$= Vd_{c\ell}[\cos(\theta_i) + j\sin(\theta_i)] \quad (14.82)$$

where

$$\theta_i = \theta(i\Delta T)$$
$$= 2\pi f_d T\left(\frac{i}{N_s}\right) + \pi\dot{f}_d T^2\left(\frac{i}{N_s}\right)^2 + \phi_o \quad (14.83)$$

In this notation there are exactly N_s samples-per-symbol so that ℓ and i are related by $\ell = [i/N_s]$. This is not a restrictive condition because rate conversion techniques are generally applied to achieve a specified number samples-per-symbol at the input to the receiver matched filter. The simulation must also include the effect of the additive channel noise and this is accomplished by computing the standard deviation of the noise based on the specified signal-to-noise ratio γ. For example, the signal-to-noise ratio measured in the bandwidth

*Frequency errors, phase shifts, and filter distortion also result from frequency translations and filtering within the modulator/transmitter and receiver/demodulator; however, these are typically combined with similar affects arising from the channel.

$R_b = 1/T_b$ corresponding to the bit-rate is defined as γ_b (dB) = E_b/N_o (dB) and the linear signal-to-noise ratio is computed as

$$\gamma_b = 10^{(\gamma_b(\text{dB})/10)} \tag{14.84}$$

Since the simulation processes the signal and noise using the sampling frequency $f_s = N_s R_b$, the signal-to-noise ratio γ_b must be decreased by N_s so that*

$$\gamma_s = \frac{\gamma_b}{N_s} \tag{14.85}$$

The standard deviation of the noise in the sampling bandwidth is then computed as

$$\sigma_n = \sqrt{\frac{P_s}{\gamma_s}} = \frac{V}{\sqrt{2\gamma_s}} \tag{14.86}$$

and the simulation includes the AWGN to obtain the received analytic baseband signal plus noise expressed as

$$\tilde{y}_i = \tilde{s}_i + \tilde{n}_i \tag{14.87}$$

where $\tilde{n}_i = n_{ci} + jn_{si}$ and $n_{ci}, n_{si} \equiv N(0, \sigma_n)$.

An important distinction involves the manner in which the signal and noise are included in the simulation program in consideration of the peak carrier frequency voltage V and the standard deviation of the noise as calculated in (14.86). For example, if the baseband signal is established as in (14.82) and the noise standard deviation as in (14.86), then the quadrature noise samples must be generated using independent Gaussian samples denoted as $N(0,\sigma_n)$. This analytic signal and noise generation results in the correct signal-to-noise ratio; however, the signal and noise powers are a factor of two times that of the true signal and noise powers. This discrepancy will be problematic for automatic gain control (AGC) and in the estimation of the signal and noise powers as discussed in Section 11.5. On the other hand, if the signal is generated as the quadrature baseband components of the real signal expressed by (14.74), the quadrature noise samples must be generated using the independent Gaussian samples denoted as $N(0, \sigma_n/\sqrt{2})$ (see Problem 17). This approach results in the correct signal-to-noise ratio and the correct signal and noise power levels. This issue is directly related to the factor of two relating the real signals in (14.72) and (14.73) and the analytic signals described in (14.71).

*This example applies to uncoded BPSK; however, for MPSK modulation with $k = \log_2(M)$ and rate r_c FEC coding the signal-to-noise ratio in the sampling bandwidth is determined as $\gamma_s = k r_c \gamma_b / N_s$.

14.6 CASE STUDY: BPSK MONTE CARLO SIMULATION

In this case study, the bit-error performance of a BPSK-modulated waveform is evaluated using a Monte Carlo simulation. The emphasis in this case study is on the accuracy of the simulation in predicting the known theoretical performance given by

$$P_{be} = Q\left(\sqrt{\frac{2E_b}{N_o}}\right) \quad : \text{BPSK theoretical performance} \tag{14.88}$$

The error in the evaluation of P_{be} is measured as the percentage error relative to the theoretical value given a fixed number of Monte Carlo trials for each signal-to-noise ratio. Furthermore, because the Q-function describes the theoretical performance under ideal conditions of frequency, phase and bit timing, these results are based on the conditions: $f_d = \dot{f}_d = 0$ and $\phi_o = 0$. In addition, the simulation models the optimally sampled matched filter output so the exact symbol timing is known. With these ideal conditions, including the AWGN channel, it is convenient to use only one sample-per-symbol, that is, by letting $N_s = 1$ the simulation will run considerably faster and the performance will be identical to that for $N_s > 1$. Based on these conditions, the bit-error performance of the BPSK received signal plus noise is evaluated using $n = 1000$ bits or Monte Carlo trials for each signal-to-noise ratio. To examine the spread in the bit-error estimates the simulation is repeated 20 times using different noise generator seeds and the results are plotted in Figure 14.21. From the discussion in Section 14.2.1, the standard deviation in the estimation of P_{be} is given by

$$\sigma = \sqrt{\frac{P_{be}(1 - P_{be})}{n}} \tag{14.89}$$

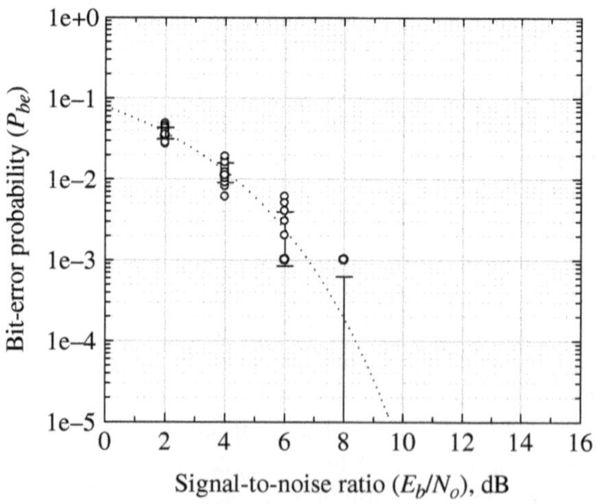

FIGURE 14.21 Fixed-length Monte Carlo simulation of BPSK performance in AWGN channel (n = 1 K bits for each signal-to-noise ratio with 20 independent repetitions).

TABLE 14.3 Monte Carlo Simulation Results for Confidence Testing ($n = 1000$ Samples/MC Simulation, 20 Independent Simulations)

E_b/N_o (dB)	Expected P_{be}	1-Sigma Error	Estimates Within Error Bars	Confidence of Test (%)	Theoretical Confidence (%)
2	3.75(−2)	6.01(−3)	13	65	68.26
4	1.25(−2)	3.51(−3)	14	70	68.26
6	2.3883(−3)	1.55(−3)	15	75	68.26
8	1.9091(−4)	4.37(−4)	18	90	84.13[a]

[a]Based on one-sided error bar.

and the error-bars, shown for signal-to-noise ratios of $E_b/N_o =$ 2, 4, 6, and 8 dB, reflect this 1-sigma range. In this case, σ is the accuracy of the test and the error-bars are computed as

$$P = P_{be} \pm \sigma \tag{14.90}$$

With the exception of the 8 dB case, this corresponds to a test confidence of 68.26%, that is, 68.26% of the simulations at the corresponding signal-to-noise ratio will fall within error-bars corresponding to (14.90). The 8 dB case is unique, in that, the lower error-bar is negative so any estimate less than the upper error-bar will pass the test; this is referred to as a one-sided test for which the confidence (P_o) is evaluated using (14.8) yielding a confidence of 84.13% as indicated in Table 14.3. Conversion between the confidence of the two-sided test, involving both upper and lower error-bars, and the one-sided test is given by (14.17).

Table 14.3 summarizes the confidence testing results from Figure 14.21 and compares the results with the theoretical values discussed in Section 14.2.1. For a given E_b/N_o, the theoretical accuracy in the measurement of P_{be} is determined by the confidence level of the test; however, the accuracy can be improved by averaging the number of error in repeated trails at each E_b/N_o. Normally, it is desired to achieve the accuracy in the simulations at the lowest error probability of interest, for example, by requiring an accuracy of 5% ($\eta = 0.05$) at $p = P_{be} = 10^{-5}$ with a 90% confidence level. Under these conditions the accuracy of the test is 0.05 and, from Figure 14.4, $k_o = 1.3$ and using (14.19) the number of Monte Carlo for each signal-to-noise ratio is computed as $n = 67.6$ M bits.

Typically, a simulation is executed to evaluate the performance over a range of signal-to-noise ratios. However, it is time-consuming and often unnecessary to evaluate the performance at the lower signal-to-noise ratios with the same number of Monte Carlo trials required to achieve a given accuracy at the higher signal-to-noise ratios. For example, when using a 67.6 *M*-bit Monte Carlo simulation the average number of errors at $P_{be} = 10^{-5}$ is 676 and the average number of errors at $P_{be} = 10^{-2}$ is 676 K with a commensurate increase in the estimation accuracy and confidence. However, the same measurement accuracy and confidence corresponding to the 10^{-5} simulation is achieved using a Monte Carlo simulation with only 67.6 K bits at $P_{be} = 10^{-2}$. Therefore, to preserve the same

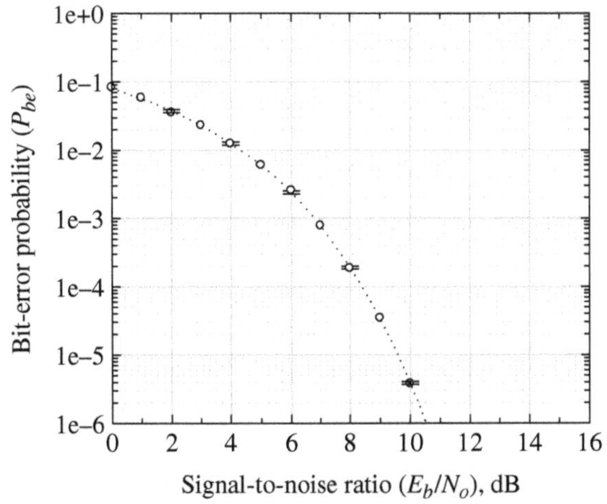

FIGURE 14.22 Fixed-error Monte Carlo simulation of BPSK performance in AWGN channel ($n = 67.6$ M bits for each signal-to-noise ratio).

accuracy and confidence for all signal-to-noise ratios the simulation can be terminated when a fixed number of errors have occurred for each signal-to-noise ratio. This fixed-error criterion is determined using Figure 14.3 or from (14.20) with $p = P_{be}$ and is approximated as

$$e = \left(\frac{k_o}{\eta}\right)^2 (1 - P_{be}) \cong \left(\frac{k_o}{\eta}\right)^2 \tag{14.91}$$

Figure 14.22 shows the simulated bit-error performance using a fixed number of bit errors $e = 676$ corresponding to the above example. Two-sided error-bars are shown for signal-to-noise ratios of 2, 4, 6, 8, and 10 dB and are evaluated using as $P = P_{be} \pm k_o\sigma = P_{be}\ (1 \pm \eta)$ and, while the results of only one Monte Carlo simulation is shown, it is claimed that 90% of the simulated bit-error results will fall within the error-bars. The 5% accuracy corresponds to $\eta = 0.05$ and, from the performance results in Figure 14.22, the simulation results are seen to accurately represent the theoretical performance.

The normalized cumulative central processing unit (CPU) time is summarized in Figure 14.23 for the fixed-error and fixed-length criteria using the above example parameters with $n = 67.6$ M bits for each signal-to-noise ratio. The normalization is based on the average of the fixed-length simulation time

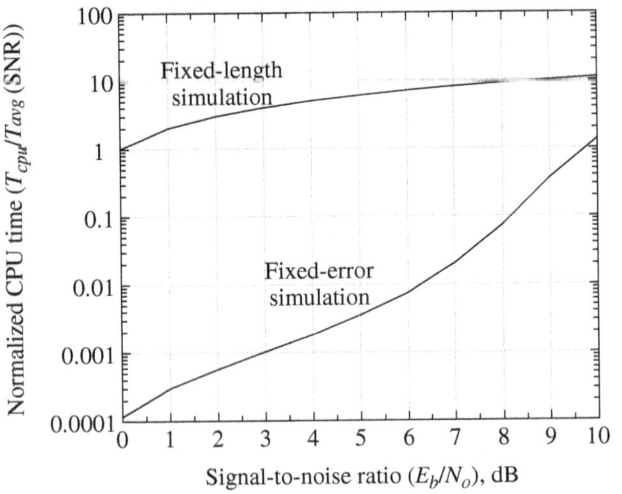

FIGURE 14.23 Monte Carlo simulation cumulative CPU time (fixed-length = 67.6 M bits, fixed-error = 676 bits).

for each signal-to-noise, denoted T_{avg}(SNR). The saving in the simulation time for $E_b/N_o = 0$ through 10 dB in 1 dB steps when using the fixed-error criterion is about 4000:1.*

14.7 SYSTEM PERFORMANCE EVALUATION USING QUADRATURE INTEGRATION

The Monte Carlo approach to communication system performance simulation discussed in the previous sections can be applied to evaluate algorithms involving quantization, coding, tracking, and the modeling of real-world channels. However, Monte Carlo simulations are generally time-consuming especially if accurate results are required at very low error probabilities. Evaluation of the demodulator detection error probability necessarily involves performing intricate integrations defined over a range of the integration variable. An alternative approach to the performance evaluation involves using numerical analysis techniques referred to as quadrature integration. Although limited to specific applications, these numerical techniques provide rapid and accurate assessments of the system performance and are useful in characterizing the performance sensitivities to various anomalies in the early state of the waveform selection and system design process. The numerical analysis technique of interest involves the approximate evaluation of integrals using quadrature integration [28].

*Monte Carlo simulation of error-probabilities terminated after a fixed number of errors typically do not result in visually appealing plots and for this reason nearly all of the Monte Carlo error-probability plots in this book use a fixed number of trials corresponding to the desired accuracy of the lowest error probability of interest. Unavoidable outliers are adjusted by increasing the number of trials or by manual adjustment equivalent to using a French curve applied to the previous data points taking into consideration possible flaring of the curve resulting from interference or channel nonlinearities.

The term *quadrature integration* results from the underlying quadrature polynomials used in the approximate evaluation of integrals of the form

$$\int_a^b w(x)f(x)dx \cong \sum_{i=1}^{N} A_i f(x_i) \qquad (14.92)$$

where $w(x)$ is a weighting function satisfying the integral equation

$$\int_a^b w(x)\Phi_i(x)\Phi_j(x)dx = \begin{cases} 0 & : i \neq j \\ k_i & : i = j \end{cases} \qquad (14.93)$$

In (14.93) $\Phi_n(x) : \forall n \leq i$ represents a set of orthogonal polynomials over the interval (a,b) with respect to the weighting function $w(x)$. The values x_i correspond to the zeros of the polynomial $\Phi_i(x)$ and the coefficients A_i are weights determined as

$$A_i = \int_a^b \frac{w(x)\lambda(x)}{(x-x_i)\lambda'(x_i)}dx \qquad (14.94)$$

where

$$\lambda(x) = \prod_{n=1}^{i}(x - x_n) \qquad (14.95)$$

The polynomials associated with various weighting functions are identified in Table 14.4.

These relationships are developed by Krylov [28] including the characterization of the error in the approximation of (14.92) and tables of abscissa x_i and coefficient values A_i for the various quadrature integration approximations. The Legendre polynomial results in Gauss-quadrature integration with $w(x) = 1$. In general, the integration limits are finite with range (a,b). However, by changing the integration variable to correspond to the range (−1, 1) the coefficients A_i are symmetrical in $\pm x_i$ as provided in most of tabulations. The values of x_i and A_i for the Chebyshev polynomials of the first kind are easily calculated as

$$x_i = \cos\left(\frac{(2i-1)\pi}{2N}\right) \text{ and } A_i = \frac{\pi}{N} \qquad (14.96)$$
$$\text{: Chebyshev polynomial of first kind}$$

where the coefficients A_i are constant, independent of i. The topic of orthogonal polynomials is also discussed in Chapter 22 of Abramowitz and Stegun [5] with extensive

TABLE 14.4 Summary of Some Orthogonal Polynomials for Quadrature Integration (Page Numbers Refer to Reference 28)

Polynomial	$\Phi_i(x)$	$w(x)$	k_i	(a,b)
Legendre (Gaussian), p. 26	$P_i(x)=\dfrac{1}{2^i i!}\dfrac{d^i(x^2-1)^i}{dx^i}$	1	$\dfrac{2}{2i+1}$	(−1,1)
Chebyshev (first kind), p. 27	$T_i(x)=\cos(i\theta)$: $\theta=\cos^{-1}(x)$	$\dfrac{1}{\sqrt{1-x^2}}$	$\dfrac{\pi}{2}$: $i\neq 0$; π : $i=0$	(−1,1)
Chebyshev (second kind), p. 29	$U_i(x)=\dfrac{\sin((i+1)\theta)}{\sqrt{1-x^2}}$: $\theta=\cos^{-1}(x)$	$\sqrt{1-x^2}$	$\pi/2$	(−1,1)
Chebyshev–Hermite, p. 33	$H_i(x)=(-1)^i e^{x^2}\dfrac{d^i}{dx^i}e^{-x^2}$	e^{-x^2}	$2^i\sqrt{\pi}\,i!$	(−∞,−∞)
Chebyshev–Laguerre[a], p. 34	$L_i^{(\alpha)}(x)=(-1)^i x^{-\alpha}e^x\dfrac{d^i}{dx^i}(x^{\alpha+i}e^{-x})$	$x^{\alpha}e^{-x}$	$i!\Gamma(i+\alpha+1)$	(0,−∞)

[a]For $x \geq 0$.

TABLE 14.5 Abscissas and Weights for Hermite Integration[a] (Symmetrical Weights[b] for $N = 20$)[c]

n	Abscissa ($\pm x_i$)	Weights (w_i)
1	0.24534 07083 009	4.62243 66960 06 (−1)
2	0.73747 37285 454	2.86675 50536 28 (−1)
3	1.23407 62153 953	1.09017 20602 00 (−1)
4	1.73853 77121 166	2.48105 20887 46 (−2)
5	2.25497 40020 893	3.24377 33422 38 (−3)
6	2.78880 60584 281	2.28338 63601 63 (−4)
7	3.34785 45673 832	7.80255 64785 32 (−6)
8	3.94476 40401 156	1.08606 93707 69 (−7)
9	4.60368 24495 507	4.39934 09922 73 (−10)
10	5.38748 08900 112	2.22939 36455 34 (−13)

[a]Also referred to as Chebyshev–Hermite polynomial.
[b]Attributed to Salzer, Zucker, and Capuano [30].
[c]Abramowitz [29]. Courtesy of U.S. Department of Commerce.

tables of x_i and A_i provided in Chapter 25. The tabulated data are usually given with 15–20 decimal places so the numerical evaluation should be used with a commensurate degree of precision. The abscissas and weight values for the Hermite integration used in the following case studies are listed in Table 14.5.

14.8 CASE STUDY: BPSK BIT-ERROR EVALUATION WITH PLL TRACKING

Based on the signal-to-noise ratio, the bit-error probability of the BPSK-modulated waveform, conditioned on a phase error, is given by

$$P(\text{error}|\phi) = \frac{1}{2}erfc_*\left[\sqrt{2\gamma_b}\cos(\phi)\right] \tag{14.97}$$

where $\gamma_b = E_b/N_o$ is the signal-to-noise ratio in the bandwidth equal to the bit rate R_b. The bit-error probability is then evaluated by removing the conditioning based on the *pdf* of the random phase resulting in

$$P_{be} = \int_{\varphi_\ell} P(\text{error}|\varphi)p(\varphi)d\varphi \tag{14.98}$$

where the range of integration is $|\varphi_\ell| \leq \pi$. The random phase function at the output of a second-order phaselock loop is characterized by the Tikhonov [31] *pdf* expressed as

$$p(\varphi) = \frac{e^{\sigma_\varphi^{-2}\cos(\varphi)}}{2\pi I_o\left(\sigma_\varphi^{-2}\right)} \quad : |\varphi| \leq \pi \ \text{ for BPSK} \tag{14.99}$$

where σ_φ^2 is the phase variance and $I_o(x)$ is the zero-order modified Bessel function of the first kind. Viterbi [32] also derives (14.99) and discusses its applicability to the second-order phaselock loop. The phase variance σ_φ^2 is related to the signal-to-noise ratio γ_L, measured in the phaselock loop bandwidth and the squaring loss S_L by the expression

$$\sigma_\varphi^2 = \frac{1}{\gamma_L S_L} \tag{14.100}$$

The signal-to-noise ratio in the loop bandwidth is computed as

$$\gamma_L = \gamma_b\left(\frac{B_b}{B_L}\right) \tag{14.101}$$

where B_L is the phaselock loop bandwidth and B_b is the bandwidth corresponding to the bit-rate R_b of the BPSK-modulated waveform; the ratio B_L/B_b is generally expressed as the time-bandwidth product (B_LT) of the loop where $T = 1/R_b$ for BPSK. The squaring loss of the phaselock loop is associated with the removal of the biphase data modulation. The phase distribution, expressed by (14.99), is based on a continuous wave (CW) received signal with AWGN and a second-order phaselock loop. However, inclusion of the squaring loss in evaluating the phase variance, as in (14.100), accounts for the loss associated with the binary data modulation. The squaring loss for BPSK modulation is discussed in Chapter 10 and, for matched filter detection, is expressed as

$$S_L = \frac{1}{1+2/(3\gamma_b)} \quad : \text{BPSK} \qquad (14.102)$$

14.8.1 BPSK Bit-Error Evaluation Using Tikhonov Phase Distribution

Substitution of (14.97) and (14.99) into (14.98) with $\bar{\gamma} = \gamma_L S_L$ results in the bit-error expression

$$P_{be} = \frac{1}{2}\int_{-\pi}^{\pi} erfc_*\left(\sqrt{2\gamma_b}\cos(\varphi)\right)\frac{e^{\cos(\varphi)/\bar{\gamma}}}{2\pi I_o(1/\bar{\gamma})}d\varphi \qquad (14.103)$$

This result can be put into the quadrature integration form, using the Chebyshev polynomial of the first-kind, through the variable substitution $x = \varphi/\pi$ with $d\varphi = \pi\, dx$. The resulting expression is

$$P_{be} = \frac{1}{4I_o(1/\bar{\gamma})}\int_{-1}^{1}\frac{f(x)}{\sqrt{1-x^2}}dx \cong \sum_{i=-N'}^{N'} A_i f(x_i) \qquad (14.104)$$

where

$$f(x) = erfc_*\left(\sqrt{\gamma_b}\cos(\pi x)\right)e^{\cos(\pi x)/\bar{\gamma}} \qquad (14.105)$$

The constants A_i and x_i are evaluated using (14.96) with x_i evaluated over the range $-N' \le i \le N'$, where $N' = (N-1)/2$ for N = odd integer and $N' = N/2$ for N = even integer.

14.8.2 BPSK Bit-Error Evaluation Using Gaussian Phase Approximation

The Tikhonov phase *pdf* approaches a Gaussian distribution as the input signal-to-noise increases resulting in

$$p(\varphi) \cong \frac{e^{-\varphi^2/(2\sigma_\varphi^2)}}{\sqrt{2\pi\sigma_\varphi^2}} \quad : \sigma_\varphi^2 \to 0 \ \text{ or } \ \gamma_L \to \infty \qquad (14.106)$$

Using (14.106) and (14.97) with (14.98), the quadrature integration using the Chebyshev–Hermite polynomial results in the following expression for the bit-error probability

$$P_{be} \cong \frac{1}{2\sqrt{\pi}}\int_{-\infty}^{\infty} f(x)e^{-x^2}dx \cong \frac{1}{2\sqrt{\pi}}\sum_{i=1}^{N} A_i f(x_i) \qquad (14.107)$$

where

$$f(x) = \frac{1}{2\sqrt{\pi}} erfc_*\left(\sqrt{2\gamma_b}\cos\left(\sqrt{2}\sigma_\varphi x\right)\right) \qquad (14.108)$$

The first approximation results because the range of $p(\varphi)$ is $(-\pi,\pi)$ and the integration limits are $(-\infty,\infty)$; this assumes that $p(\varphi)$ is essentially zero outside of the range $(-\pi,\pi)$ which is valid as $\sigma_\varphi^2 \to 0$.

Using (14.108), Equation (14.107) is evaluated for $N = 20$* abscissa and weight values for Hermite integration are listed in Table 14.5 and the results are plotted in Figure 14.24 as the *solid* lines for $B_LT = 0.5$ and 0.1. The *dotted* line represents the theoretical performance of antipodal signaling and the *circled* data points are evaluated using (14.107) with $\sigma_\varphi = 0$, independent of the signal-to-noise ratio. The *circled* data points are a special case used to examine the performance of antipodal signaling using quadrature integration and it is seen from the figure that the results are identical.

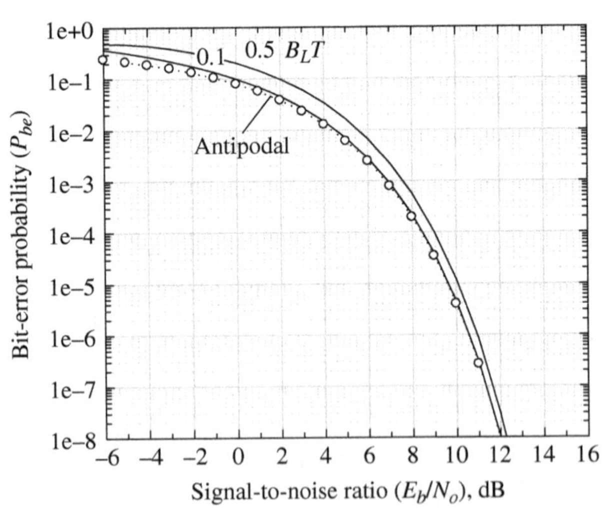

FIGURE 14.24 Performance of BPSK using Chebyshev–Hermite quadrature integration (Gaussian phase approximation).

*The tabulated coefficient values are symmetrical about the N zeros x_i of the orthogonal Hermite polynomials.

14.9 CASE STUDY: QPSK BIT-ERROR EVALUATION WITH PLL TRACKING

Based on the signal-to-noise ratio out of each baseband channel, the symbol-error probability conditioned on the phase error for QPSK modulation is given by

$$P(\text{error}|\varphi) = \frac{1}{2}erfc_*\left[\sqrt{2\gamma_b}(\cos(\varphi)-\sin(\varphi))\right] + \frac{1}{2}erfc_*\left[\sqrt{2\gamma_b}(\cos(\varphi)+\sin(\varphi))\right] \quad (14.109)$$

The symbol-error probability is evaluated by removing the conditioning as expressed by (14.98), substituting P_{se} for the QPSK waveform in place of P_{be}, and using either the Tikhonov phase *pdf* for QPSK expressed as

$$p(\varphi) = \frac{2e^{\cos(4\varphi)/\left(4\sigma_\varphi^2\right)}}{\pi I_o\left(1/\left(4\sigma_\varphi^2\right)\right)} \quad : |\varphi| \le \pi/4 \text{ for QPSK} \quad (14.110)$$

or the high signal-to-noise ratio Gaussian approximation of (14.106).

14.9.1 QPSK Bit-Error Evaluation Using Tikhonov Phase Distribution

In this case quadrature integration using the QPSK Tikhonov's phase *pdf* is performed using Legendre polynomials with weight $w(x) = 1.0$ and integration limits (−1,1). Upon making the substitution of variable $x = 4\phi/\pi$ this approach results in Gaussian integration, using the Legendre polynomial, expressed as

$$P_{se} = \frac{1}{4}\int_{-1}^{1} f_a(x)f_c(x)dx + \int_{-1}^{1} f_b(x)f_c(x)dx$$
$$\cong \frac{1}{4}\sum_{i=-N'}^{N'}[f_a(x_i)+f_b(x_i)]f_c(x_i) \quad : \text{Legendre polynomial} \quad (14.111)$$

where

$$f_a(x) = erfc_*\left[\sqrt{2\gamma_b}\left(\cos\left(\frac{\pi x}{4}\right) - \sin\left(\frac{\pi x}{4}\right)\right)\right] \quad (14.112)$$

$$f_b(x) = erfc_*\left[\sqrt{2\gamma_b}\left(\cos\left(\frac{\pi x}{4}\right) + \sin\left(\frac{\pi x}{4}\right)\right)\right] \quad (14.113)$$

and

$$f_c(x) = \frac{e^{\cos(\pi x)/\left(4\sigma_\varphi^2\right)}}{I_o\left(1/\left(4\sigma_\varphi^2\right)\right)} \quad (14.114)$$

In evaluating these integrals, or summations, the phase variance for the QPSK phaselock loop is given by

$$\sigma_\varphi^2 = \frac{1}{\gamma_L S_{L4}} \quad (14.115)$$

where γ_L is the signal-to-noise ratio in the loop bandwidth, B_L, and S_{L4} is the squaring loss resulting from the processing required to remove the quadrature phase modulation for the carrier recovery. For QPSK, the signal-to-noise ratio in the loop bandwidth is expressed as

$$\gamma_L = \frac{P}{N_oB_L} = \frac{E_s}{N_oB_LT} \quad (14.116)$$

where E_s is the symbol energy and T is the symbol duration. The squaring loss for QPSK is discussed in Chapter 10.

The abscissa and coefficient values are tabulated [33] for $N = 2$ through 96 corresponding to symmetrical summations on x_i from $-N/2 \le i \le N/2$ for N = even integer and $-(N-1)/2 \le i \le (N-1)/2$ for N = odd integer.

14.9.2 QPSK Bit-Error Evaluation Using Gaussian Phase Approximation

With this approximation the integral for the QPSK symbol error is approximated using the Chebyshev–Hermite polynomial integration as in Section 14.8.2. For this evaluation, let $x = \phi/\sqrt{2}\sigma_\phi$ and evaluate the integral as

$$P_{se} \cong \frac{1}{2\sqrt{\pi}}\int_{-\infty}^{\infty} f_\alpha(x)e^{-x^2}d\xi + \frac{1}{2\sqrt{\pi}}\int_{-\infty}^{\infty} f_\beta(x)e^{-x^2}dx$$
$$\cong \frac{1}{2\sqrt{\pi}}\sum_{i=1}^{N}A_i[f_\alpha(x_i)+f_\beta(x_i)] \quad (14.117)$$
$$: \text{Chebyshev–Hermite polynomial}$$

where

$$f_\alpha(x) = erfc_*\left[\sqrt{2\gamma_b}\left(\cos\left(\sqrt{2}\sigma_\varphi x\right) - \sin\left(\sqrt{2}\sigma_\varphi x\right)\right)\right] \quad (14.118)$$

$$f_\beta(x) = erfc_*\left[\sqrt{2\gamma_b}\left(\cos\left(\sqrt{2}\sigma_\varphi x\right) + \sin\left(\sqrt{2}\sigma_\varphi x\right)\right)\right] \quad (14.119)$$

The standard deviation of the phase is evaluated from (14.115) using the signal-to-noise ratio in the phaselock loop bandwidth γ_L and the squaring loss S_{L4} as discussed in the preceding section. The abscissa and coefficient values for the Chebyshev–Hermite polynomial integration are tabulated [32] for N = 2–20.

ACRONYMS

AR	Autoregressive (model)
AWGN	Additive white Gaussian noise
BPSK	Binary phase shift keying
CPU	Central processing unit (computer)
CW	Continuous wave
DSP	Digital signal processing (processor)
FEC	Forward error correction
IF	Intermediate frequency
ISI	Intersymbol interference
LR	Likelihood ratio
LLR	Log-likelihood ratio
OQPSK	Offset quadrature phase shift keying
PN	Pseudo-noise
PSD	Power spectral density
QPSK	Quadrature phase shift keying
RF	Radio frequency

PROBLEMS

1. Show that Equation (14.15) leads to (14.16) under the conditions $k_1 = n(1+\eta)p$ and $k_2 = n(1-\eta)p$.

2. Derive equation (14.17).

3. For a correct-acceptance criteria of 95%, the single-threshold *Poisson test* results in an accuracy of 40% with $np_o = 80$. Determine the np_o product for the single-threshold *Gaussian test* under the same conditions, that is, a test confidence of 95% and accuracy of 40%. Assuming a bit-error probability test with a bit-rate of 75 bits-per-second and $p_o = 10^{-3}$, what is the test time in minutes for each test?

4. Derive the optimum single-threshold for deciding between hypothesis H_0 and H_1 using the binomial distribution.

5. Using the binomial coefficient ${}_nC_k$, derive the recursive relationship for ${}_nC_{k+1}$.

6. Using the binomial density function $p(k)$, derive the recursive relationship for $p(k+1)$.

7. Design a sequential sampling test involving the flipping of a coin that is known to be either balanced with a probability of heads of $p_0 = 0.5$ under hypothesis H_0 or unbalance with a probability of heads of $p_1 = 0.7$ under hypothesis H_1. The coin is to be tested with a confidence level of 95% and a significance level of 5% under H_0. To complete this problem fill out Table 14.1 with the test requirements and parameters for this test.

8. For the test designed in Problem 7, sketch the test boundaries as in Figure 14.12 and flip a coin a number of times, indexing k by one if a tail appears and plotting the results on the sketch.

9. Express the mean value of the uniform distribution shown in Figure 14.13 in terms of the probability $0 \le p \le 1$ and $q = 1 - p$.

10. Plot or sketch the binomial distribution for a sequence of length 4 using the Bernoulli random number generator with $p = 0.2$, 0.5, and 0.8.

11. Write a computer code to implement the congruence method for generating uniformly distributed random numbers given by (14.45) using $A = 9$, $B = 25853$, and $C = 32768$. Replace the coin flipping example in Problem *10* with your random number generator code with binary 1 = heads and −1 = tails and repeat the sequential test procedure as outlined in Problem 10 under the following conditions:

    a. Using a balanced binary generator.

    b. Using an unbalanced binary generator with P_r (tails) = 0.6.

    Note: For the unbalanced binary generator, referring to Figure 14.13, shift the mean value of the uniform distribution between a and b to −0.1.

12. Applying the convolution integral to the uniform probability density function of Figure 14.1 with $a = -1$ and $b = 1$, form the *pdf* of the sum of three uniformly distributed probability random variables. Plot the resulting distribution and compare it with the Gaussian distribution with variance $\sigma^2 = 3\sigma_u^2$ where σ_u^2 is the variance of the uniform distribution.

13. Using the uniform random number generator code in Problem 11 with $p = 1/2$, form the summation

$$Y_j = \sum_{i=1}^{3} X_i$$

for $j = 1{,}1000$ and plot a histogram of the occurrences of Y_j ranging in amplitude from −3 to 3 in steps of 0.1. Compare the resulting histogram plot with the distributions in Problem 12.

14. Using the uniform random number generator code in Problem 11 generate simulated Poisson random variables for $k = 3$ and $0 \le \tau \le 10$ and plot the resulting *pdf*. To plot the *pdf* use a histogram with 101 bins (including the $\tau = 0$ bin), a bins size corresponding to $\tau/10$ and a sample size of 10,000 trials. Compare the simulated *pdf* with the corresponding theoretical *pdf* shown in Figure 14.18.

15. Referring to the (14.74) and (14.75) express the magnitude and phase functions in (14.77) and (14.78)

respectively in terms of the MSK-modulated waveform with $p_c(t) = \cos(\omega_m t)$ and $p_s(t) = \sin(\omega_m t)$. Noting that $f_m T = 1/2$, $d_{ci} = rect((t - iT)/T)$, and $d_{si} = rect((t - T/2 - iT)/T)$ evaluate and plot the phase function for $d_{ci}, d_{si} = \{1, -1, 1, -1\}$, that is, the d_{ci} and d_{si} correspond, respectively, to constant mark and space data.

16. Evaluate the carrier power P_s given by (14.79) for QPSK modulation waveform using the *real* signal $s(t)$ and the *analytic* signal $\tilde{s}(t)$. Repeat this evaluation for the MSK-modulated waveform. In each case, state the relationship between the *true* or actual signal power based on the *real* signal and that corresponding to the *analytic* signal.

17. Referring to the discussion following (14.87) consider a simulation program that implements a transmitted signal with independent inphase and quadrature data modulation functions $m_c(iT_s)$ and $m_s(iT_s)$ with N_s samples-per-symbol and a sampling interval of T_s seconds. The transmitted signal is implemented as

$$s_t(iT_s) = Vm_c(iT_s) + jVm_s(iT_s)$$

where V is the peak voltage of an equivalent modulated carrier frequency. The signal is passed through an AWGN channel and the resulting received signal is

$$s_r(iT_s) = Vm_c(iT_s) + jVm_s(iT_s) + n_c(iT_s) + jn_s(iT_s)$$

where $n_c(iT_s)$ and $n_s(iT_s)$ are *iid* Gaussian random variables characterized by $N(0,\sigma_n)$ where the noise standard deviation is computed as $\sigma_{ns} = V/\sqrt{2\gamma_s}$ where $\gamma_s = \gamma_b/N_s = 10^{\gamma_b(\mathrm{dB})/10}/N_s$. The corresponding carrier modulated power is $P_s = V^2/2$ and the noise power and signal-to-noise ratio in the bandwidth equal to bit rate R_b are $\sigma_n^2 = P_s/\gamma_b$ and γ_b respectively. Considering that $m_c(t) = d_c\ rect_T(t - T/2 - mT)$ and $m_s(t) = d_s\ rect_T(t - T/2 - mT)$ with $d_c, d_s = \{\pm 1\}$, determine the signal and noise powers and the signal-to-noise ratio at the output of the demodulator matched filter.

Replace V with $V/\sqrt{2}$ and the Gaussian noise $N(0,\sigma_n)$ with $N(0, \sigma_n/\sqrt{2})$ and repeat this problem.

18. Given the Gaussian random variable x described by $N(0,\sigma)$, *set-up* the problem to evaluate the mean value of $g(x)$ using quadrature integration where

$$g(x) = \frac{ae^x}{1 + ae^x}$$

Identify the underlying orthogonal polynomial.
Hint: The mean value of $g(x)$ is evaluated as

$$E[g(x)] = \frac{1}{\sqrt{2\pi\sigma^2}} \int_{-\infty}^{\infty} g(x) e^{-x^2/2\sigma^2} dx.$$

19. Using the Tikhonov phase *pdf* for BPSK and (14.105), evaluate (14.104) and plot the bit-error probability versus γ_b using the Chebyshev polynomial of the first-kind using coefficients A_i and zeros x_i given by (14.96). Note: The results should compare favorably with Figure 14.24.

20. Show that the Tikhonov phase *pdf* given by (14.99) is approximated by the zero mean Gaussian *pdf* of (14.106) as the phase variance σ_ϕ^2 approaches zero.

21. Derive (14.108) for the evaluation of P_{be} with BPSK modulation given by (14.107) using Gaussian approximation to the Tikhonov phase *pdf*.

22. Evaluate the bit-error probability for QPSK modulation using the Gaussian approximation to the Tikhonov phase *pdf*.

REFERENCES

1. K.S. Shanmugam, P. Balaban, "A Modified Monte Carlo Simulation Technique for the Evaluation of Error Rate in Digital Communication Systems," IEEE Transactions on Communications, Vol. COM-28, No. 11, pp. 1916–1924, November 1980.
2. A. Papoulis, *Probability, Random Variables, and Stochastic Processes*, p. 57, McGraw–Hill, New York, 1965.
3. W. Feller, *An Introduction to Probability Theory and Its Applications*, John Wiley & Sons, Inc., New York, 1957.
4. A. Papoulis, *Probability, Random Variables, and Stochastic Processes*, p. 69, McGraw–Hill, New York, 1965.
5. M. Abramowitz, I.A. Stegun, Editors, *Handbook of Mathematical Functions with Formulas, Graphs, and Mathematical Tables*, Applied Mathematics Series No. 55, National Bureau of Standards, Washington, DC, June 1964.
6. A. Papoulis, *Probability, Random Variables, and Stochastic Processes*, pp. 966–973, McGraw–Hill, New York, 1965.
7. A. Papoulis, *Probability, Random Variables, and Stochastic Processes*, pp. 71–73, McGraw–Hill, New York, 1965.
8. W.B. Davenport, Jr., W.L. Root, *An Introduction to the Theory of Random Signals and Noise*, Chapter 14, McGraw–Hill, New York, 1958.
9. H.D. Brunk, *An Introduction to Mathematical Statistics*, Blaisdell Publishing Company, Waltham, MA, 1965.
10. A. Wald, *Statistical Decision Functions*, John Wiley & Sons, Inc., New York, 1950.
11. H.D. Brunk, *An Introduction to Mathematical Statistics*, Chapter 16, "Other Sampling Methods," Blaisdell Publishing Company, Waltham, MA, 1965.
12. A. Wald, *Sequential Analysis*, John Wiley & Sons, Inc., New York, 1947.
13. M. Abramowitz, I.A. Stegun, Editors, *Handbook of Mathematical Function with Formulas, Graphs, and Mathematical Tables*, pp. 949–953, Applied Mathematics Series No. 55, National Bureau of Standards, Washington, DC, June 1964.
14. H.A. Meyer, *Symposium on Monte Carlo Methods*, John Wiley & Sons, Inc., New York, 1956.

15. A.O. Arthur, Random Digit Generation, Computing News, September 1956.
16. R.P. Chambers, "Random-Number Generation on Digital Computers," IEEE Spectrum, Vol. 4, pp. 48–56, February 1967.
17. D.E. Knuth, *The Art of Computer Programming, Volume 2: Semi-Numerical Algorithms*, pp. 170–173, Addison-Wesley, Reading, MA, 1981.
18. M.E. Muller, "An Inverse Method for the Generation of Random Normal Deviates on Large-Scale Computers," Mathematical Tables and Other Aids to Computation, Vol. 63, pp. 167–174, 1958.
19. C. Hastings, Jr., *Approximations for Digital Computers*, Princeton University Press, Princeton, NJ, 1955
20. M. Abramowitz, I.A. Stegun, Editors, *Handbook of Mathematical Function with Formulas, Graphs, and Mathematical Tables*, p. 933, Applied Mathematics Series No. 55, National Bureau of Standards, Washington, DC, June 1964. (Also includes many other approximations by C. Hastings, Jr.).
21. A. Papoulis, *Probability, Random Variables, and Stochastic Processes*, pp. 266–268, McGraw–Hill, New York, 1965.
22. E. Parzen, *Modern Probability Theory and Its Applications*, John Wiley & Sons, Inc., New York, 1960.
23. A. Papoulis, *Probability, Random Variables, and Stochastic Processes*, pp. 220–221, McGraw–Hill, New York, 1965.
24. S. Haykin, *Adaptive Filter Theory*, pp. 67–85, Prentice-Hall, Englewood, NJ, 1986.
25. W.C. Jakes, *Microwave Mobile Communications*, John Wiley & Sons, Inc., New York, 1975.
26. L. Kleinrock, *Queueing Systems, Volume 1: Theory*, John Wiley & Sons, Inc., New York, 1975.
27. A. Papoulis, *Probability, Random Variables, and Stochastic Processes*, pp. 556–560, McGraw–Hill, New York, 1965.
28. V.K. Krylov, *Approximate Calculation of Integrals*, Translated by A.H. Stroud, The Macmillan Company, New York, 1962.
29. M. Abramowitz, I.A. Stegun, Editors, *Handbook of Mathematical Function with Formulas, Graphs, and Mathematical Tables*, p. 924, Applied Mathematics Series No. 55, National Bureau of Standards, Washington, DC, June 1964.
30. M. Abramowitz, I.A. Stegun, Editors, *Handbook of Mathematical Functions with Formulas, Graphs, and Mathematical Tables*, Chapter 25, "Table of the Zeros and Weight Factors of the First Twenty Hermite Polynomials," p. 924, Applied Mathematics Series No. 55, National Bureau of Standards, Washington, DC, June 1964.
31. W.C. Lindsey, M.K. Simon, *Telecommunication Systems Engineering*, Dover Publications, Inc., New York, 1973.
32. A.J. Viterbi, *Principles of Coherent Communication*, pp. 86–96, McGraw–Hill, New York, 1966.
33. M. Abramowitz, I.A. Stegun, Editors, *Handbook of Mathematical Function with Formulas, Graphs, and Mathematical Tables*, Applied Mathematics Series No. 55, National Bureau of Standards, Washington, DC, pp. 916–919, June 1964.

15

COMMUNICATION RANGE EQUATION AND LINK ANALYSIS

15.1 INTRODUCTION

This chapter deals with numerous issues related to the design of the data modulator, transmitter, antennas, receiver, and demodulator subsystems as they relate establishing the system performance over a communication link. A communication link, as defined here, involves issues related to the transmission of data from one point to another and, in addition to the various systems, includes the transmission channel. The focus is on the received signal and noise powers and the resulting signal-to-noise ratio required to meet the overall system performance specification. The channel is generally viewed as the transmission medium between a transmitter and receiver location and the emphasis in this chapter is on wireless channels involving transmissions through the atmosphere and ionosphere. The link might include, for example, a satellite relay with unique uplink and downlink channels. The satellite relay may be thought of as a "bent pipe" with frequency translation and power amplification or a full processing satellite in which the data is demodulated and then remodulated for transmission on an independent downlink. The emphasis in this chapter is on the application of the communication range equation in analyzing the system performance over wireless channels. Chapter 18 introduces the subject of fading channels. Chapters 19 and 20 focus is on the impact of atmospheric and ionospheric propagation on the communication performance and discusses mitigation techniques to maintain reliable communications.

The communication range equation is reviewed in the remainder of this introductory section highlighting the major parameters regarding the transmitter, antenna, and receiver designs. In this characterization the impact of the channel is identified by the free-space signal loss L_{fs} and the atmospheric loss, L_{atm}. Sections 15.2–15.10 review a variety of system issues related to establishing the link budget, including antenna gains and patterns, and losses involving rain, polarization, multipath, component interfaces, and miscellaneous system losses. Section 15.11 discusses various aspects of system nonlinearities including solid-state power amplifiers (SSPAs) and traveling wave tube amplifiers (TWTAs) and their impact on the system performance. Sections 15.12 and 15.13 outline the modeling of the nonlinearities for computer simulations and Section 15.14 provides a case study of the impact of a SSPA on the bit-error performance of a communication link. This chapter concludes in Section 15.15 with an example link budget highlighting the various parameters involved in the link evaluation and their impact on meeting the system performance specification.

The following analysis makes extensive use of the various subsystem temperatures in degrees Kelvin (°K) and Table 15.1 provides conversion formula for degrees Celsius (Centigrade) and Fahrenheit. The standard noise temperature $T_o = 290°K$ is defined as the temperature at the input of a device or subsystem that corresponds to a standard noise power based on the bandwidth and noise figure as characterized, for example, by a receiver noise figure F_n and bandwidth B expressed in (15.11).

Digital Communications with Emphasis on Data Modems: Theory, Analysis, Design, Simulation, Testing, and Applications, First Edition. Richard W. Middlestead.

Companion website: www.wiley.com/go/digitalcommunications

TABLE 15.1 Commonly Used Temperature Conversion Formulas

deg-K = deg-C + 273.15	Celsius → Kelvin
deg-K = deg-F + 459.67	Fahrenheit → Kelvin
deg-C = (deg-F − 32)(5/9)	Fahrenheit → Celsius

Consider a transmitter that delivers a power of P_{PA} watts from the high-power amplifier (HPA) to an antenna having a peak gain of G_t in the direction of a receiver. The power density of the transmitted signal at a range R meters is given by

$$PD = \frac{P_{PA}G_t}{4\pi R^2} \quad : \mathrm{W/m^2} \tag{15.1}$$

where the factor of 4π radians accounts for the solid angle of the spherical radiation pattern of an ideal dipole antenna in free space. A receiver located at a range R from the transmitter will intercept a fraction of the incident energy. The received power intercepted by the antenna is simply the power density times the effective area, A_e, of the receiver antenna. The antenna effective area is defined as the physical area A_a of the antenna aperture times the aperture efficiency η_a, so the power intercepted by the receive antenna is

$$\begin{aligned} P_{ra} &= PD\eta_a A_a \\ &= \frac{P_{PA}G_t A_e}{4\pi R^2} \end{aligned} \tag{15.2}$$

The antenna gain is related to the physical and effective areas of the antenna by*

$$G = \frac{4\pi\eta_a A_a}{\lambda^2} = \frac{4\pi A_e}{\lambda^2} \tag{15.3}$$

Upon substituting $\eta_a A_a = A_e = G_r\lambda^2/4\pi$ into (15.3), the received power becomes

$$P_{ra} = \frac{P_{PA}G_t G_r \lambda^2}{(4\pi)^2 R^2} = \frac{P_{PA}G_t G_r}{L_{fs}} \tag{15.4}$$

The second expression in (15.4) is a convenient form and results from using the free-space loss expressed as

$$L_{fs} = \left(\frac{4\pi R}{\lambda}\right)^2 \tag{15.5}$$

*The antenna efficiency is defined as $\eta_{ant} = \eta_{rad}\eta_a$ where η_{rad} is the antenna radiation efficiency. In this context, the antenna gain is also expressed as $G = \eta_{rad}G_D$ where G_D is defined as the directive antenna gain; this characterization is discussed subsequently.

Although (15.4) is a concise expression for the received power, additional insight into the system design is obtained by including three additional sources of power loss. These losses account for transmit and receive system radio frequency (RF) losses (L_{ts}, L_{rs}) and the losses associated with atmospheric absorption (L_{atm}). Generally, the transmitter power is specified in terms of the HPA output power P_{PA}; however, there are transmitter and receiver antenna losses associated with waveguide feeds, rotary joints, radomes, and so on, that must also be considered. Eventually all of the losses will be accounted for; however, currently the losses of interest are as follows: L_{ts}, L_{atm}, L_{fs}, and L_{rs} and the received power at the receiver low-noise amplifier (LNA) input is expressed as

$$P_r = \frac{P_{ra}}{L_{ts}L_{atm}L_{rs}} = \frac{P_{PA}G_t G_r}{L_{ts}L_{atm}L_{fs}L_{rs}} = \frac{P_t G_t G_r}{L_{atm}L_{fs}L_{rs}} \tag{15.6}$$

where P_t is the power into the transmit antenna. The atmospheric absorption loss is a function of range and is normally specified in terms of the attenuation-per-kilometer (α dB/km), for which, the atmospheric loss is given by

$$L_{atm} = \left(10^{\alpha/10}\right)^{R/1000} = 10^{\alpha R_{km}/10} \tag{15.7}$$

where R_{km} is the range extent through the atmosphere in kilometers. For terrestrial links there is also a frequency-dependent terrain loss (L_{ter}) associated with terrain fluctuations, foliage, surface dielectric constant, and conductivity; for ELF through LF frequencies this loss is often denoted as a ground loss L_{gnd}. Terrestrial links are discussed in Chapter 19. A number of contributing sources to the system losses are identified in the following sections and, when included with the RF losses, provide further fidelity to the communication range equation in predicting the overall system performance. There are also several contributing sources to the atmospheric loss and, when taken collectively, these losses comprise the system loss budget.

Equation (15.6) is often used to determine the received signal power at various points throughout the receiver to evaluate the dynamic range requirements and possible saturation, especially at the input to the analog-to-digital converter (ADC). The ADC has an instantaneous dynamic range of about 6 dB/bit and is often designed to saturate or clip with input levels on the order of milli-watts. Therefore, with the aid of (15.6) the specification of the receiver dynamic range and AGC requirements can be identified. There are a number of variations in the communication range equation that are useful in predicting the system performance under different circumstances and these are introduced in the following sections.

15.1.1 EIRP and Power Aperture

Commonly used performance measures for the transmitter system are the effective isotropic radiated power (EIRP) and power aperture. The EIRP is defined as the product of the power into an isotropic antenna and the antenna gain in a given direction relative to the isotropic antenna gain. Usually the EIRP is specified in the direction of the maximum antenna gain. The power aperture is defined as the product $P_{PA}A_{ta}$ of the transmitter power and the transmit antenna area. The maximum gain along the antenna boresight axis corresponding to $G_t = G_t(\max) = G_t(0,0)$ and the EIRP* is defined as

$$\mathrm{EIRP} \triangleq P_t G_t(\max) = \left(\frac{P_{PA}}{L_{ts}}\right) G_t(\max) \tag{15.8}$$

To express the communication range equation in terms of the EIRP, the transmitter loss (L_{ts}) is removed from the link RF losses resulting in the communication range equation

$$P_r = \frac{\mathrm{EIRP}\,G_r}{L_{atm}L_{fs}L_{rs}} \tag{15.9}$$

The communication range equation is expressed in terms of the power-aperture product, $P_{PA}A_{ta}$, by substituting for G_t using (15.3), with the result

$$P_r = \frac{4\pi\eta_{ta}(P_{PA}A_{ta})G_r}{\lambda^2 L_{ts}L_{atm}L_{fs}L_{rs}} \tag{15.10}$$

In this expression $A_{ta} = A_a$ is the physical area of the transmit antenna. The power-aperture product is used as a performance measure of the transmitter when the physical size of the antenna is an important consideration, for example, when the transmit power and physical size of the antenna tend to dominate the cost of the transmit system.

15.1.2 Signal-to-Noise Ratio

Expressing the range equation in terms of the signal-to-noise ratio, γ, is necessary when relating the system performance to the signal detection and false-alarm probabilities or the acquisition and bit-error probabilities. The noise floor at the input to the receiver is characterized in terms of the thermal noise power in the receiver noise bandwidth B_n and is given by

$$N_t = kT_oB_nF_n \tag{15.11}$$

where $k = 1.38 \times 10^{-23}$ W/s-°K is Boltzmann's constant, T_o = 290°K is the standard temperature,† B_n is the noise bandwidth in Hertz, and F_n is receiver noise figure. The receiver noise figure accounts for the additional noise contributed by the receiver relative to the intrinsic kTB_n thermal noise. The noise bandwidth, as used here, is related to the RF bandwidth at the input to the LNA with prefiltering frequency response $H(f)$ and is defined as

$$B_n = \frac{\int_{-\infty}^{\infty} |H(f)|^2 df}{|H(f_o)|^2} \tag{15.12}$$

where f_o is the frequency corresponding to the maximum of $H(f)$.‡ This definition of the noise bandwidth corresponds to an ideal unit amplitude rectangular noise filter having the same area as $|H(f)|^2$. Using these results and (15.6), the communication range equation is used to express the received signal-to-noise ratio as

$$\gamma = \frac{P_r}{N_t} = \frac{P_tG_tG_r}{kT_oB_nF_{ns}L_{atm}L_{fs}L_{rs}} \tag{15.13}$$

where F_{ns} is the system noise figure that is somewhat higher than that of the receiver because of receiver antenna related noise sources as discussed in Section 15.2.3.

15.1.3 Maximum Range

The maximum range of a communication link is defined as the range beyond which the signal-to-noise ratio falls below the value required to maintain the specified system performance; typically this is viewed as the signal-to-noise ratio corresponding to the specified bit-error probability; however, a more stressful requirement might be the signal-to-noise ratio required to support the waveform acquisition probability. In any event, using (15.5) to express the free-space loss in terms of the range, (15.13) is then used to evaluate the maximum range to achieve the specified performance criteria in an additive white noise environment with the result

$$R_{max} = \frac{\lambda}{4\pi}\left\{\frac{P_tG_tG_r}{kT_oB_nF_{ns}L_{atm}L_{rs}\gamma_{min}}\right\}^{1/2} \tag{15.14}$$

*The effective radiated power (ERP) is defined as the radiated power relative to an antenna source other than an ideal isotropic radiator. For example, a half-wavelength dipole reference antenna has a gain that is 2.15 dB greater than an ideal isotropic radiating antenna. ERP may also be interpreted as the radiated power in an arbitrary direction so that ERP $= P_tG(\theta,\phi)$ where θ and ϕ are the antenna azimuth and elevation angles relative to boresight.

†The IEEE has defined the standard temperature as a reference for measuring thermal noise. Degrees Kelvin and Celsius are related as T°K $= T$°C + 273; therefore, 17°C (62.6°F) corresponds to 270°K.

‡The noise bandwidth is generally dominated by the received IF filtering and, with additive white noise, the noise power is reduced in direct proportion to the reduction of the noise bandwidth.

where γ_{min} is the minimum received signal-to-noise ratio corresponding to the system performance specifications.

15.2 RECEIVER AND SYSTEM NOISE FIGURES AND TEMPERATURES

The *noise figure* of a receiver [1] is defined as the ratio of the input and output signal-to-noise ratio as

$$F_n \triangleq \frac{\gamma_{in}}{\gamma_{out}} \tag{15.15}$$

In this section the noise figure of a receiver comprising a cascade of amplifiers, mixers, and attenuators is examined in terms of the noise characteristics of the individual components or devices. For all of the major components in the receiver, the key parameter used to characterize the noise is the effective noise temperature. For devices that contribute to the receiver gain, the effective noise temperature is determined from the device noise figure as defined in (15.15); however, for lossy devices it is determined from the physical temperature of the devise and the loss. The analysis that follows is primarily based on the work of Charton [2].

15.2.1 Receiver Noise Figure

The receiver noise figure is developed in terms of the effective noise temperatures of the individual components of the receiver starting with those that contribute to the system gain. In these cases the input and output signal-to-noise ratios are defined by

$$\gamma_{in} \triangleq \frac{P_s}{kT_oB_n} \tag{15.16}$$

and

$$\gamma_{out} \triangleq \frac{G_iP_s}{G_ikT_oB_n + \Delta P_{ni}} \tag{15.17}$$

where P_S is the signal power, $T_o = 290°\text{K}$ is the standard temperature, B_n is the noise bandwidth, G_i is the gain of the device, and k is Boltzmann's constant. The term $G_ikT_oB_n$ is the output noise power resulting from the input noise power kT_oB_n and ΔP_{ni} is the output noise power contributed by the device.* Using these results and (15.15), the noise figure of the device is expressed as

$$F_{ni} = \frac{\Delta P_{ni}}{G_ikT_oB_n} + 1 \tag{15.18}$$

Upon defining the effective noise temperature as $\bar{T}_i$ °K the device contribution to the output noise is defined as $\Delta P_{ni} = G_ik\bar{T}_iB_n$ so, in terms of the effective noise temperature, the device noise figure becomes

$$F_{ni} = \frac{\bar{T}_i}{T_o} + 1 \tag{15.19}$$

Charton points out that this characterization of the noise figure applies to a heterodyned receiver where the signal image is blocked or filtered and when the image is not blocked (15.19) is expressed as $F_{ni} = \bar{T}_i/T_o + 2$; the factor of two results when the image power equals the desired signal power.

Because the noise figure of a device is usually specified by the manufacturer, the effective noise temperature is computed as

$$\bar{T}_i = (F_{ni} - 1)T_o \tag{15.20}$$

The interpretation of the effective noise temperature is depicted in Figure 15.1 in terms of noisy and noiseless devices both with gain G_i. In Figure 15.1a the output temperature is computed as $T_{out} = (F_{ni} - 1)T_oG_i$, whereas, in the noiseless representation of Figure 15.1b, the output temperature is computed as $T_{out} = \bar{T}_iG_i$. Therefore, using the equivalent noise temperature, the output temperature is obtained simply by multiplying the input temperature by the device gain. It will be seen that this also applies for lossy devices when the gains are replaced by the reciprocal of the losses, that is, for a lossy device with gain $G_i \le 1$ the gain is replaced by $L_i = 1/G_i \ge 1$ so the output temperature becomes $T_{out} = \bar{T}_iL_i$.

Using these results, a cascade of successive amplifiers, such that $i = 1, 2, 3, \ldots$, with different gains and noise figures, results in an overall receiver noise figure given by

$$F_n = F_{n1} + \frac{F_{n2} - 1}{G_1} + \frac{F_{n3} - 1}{G_1G_2} + \cdots \tag{15.21}$$

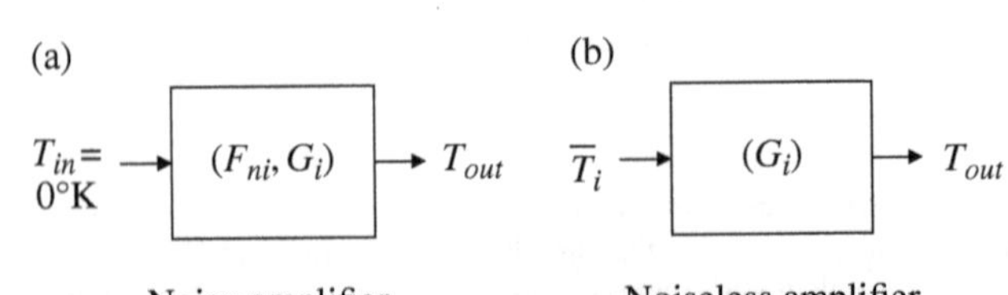

FIGURE 15.1 Comparison of noisy amplifier with equivalent noiseless amplifier.

*In the context of the receiver noise figure, the *device* corresponding to the subscript $i = 1$ is the input high-gain LNA that generally establishes the receiver noise figure. This is evident from (15.21) when $G_1 \gg F_{n2} - 1$.

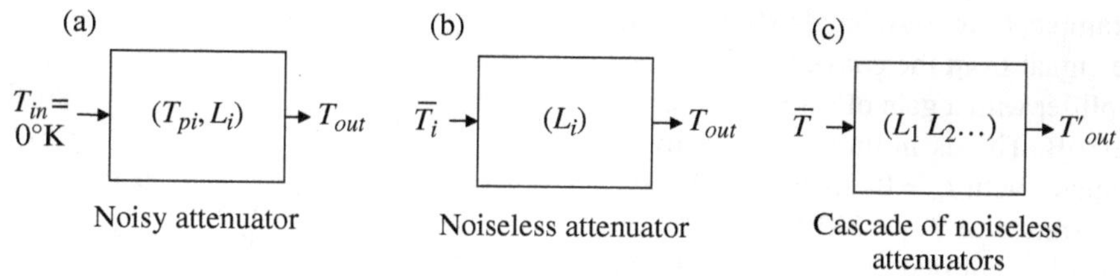

FIGURE 15.2 Comparison of noisy attenuator with equivalent noiseless attenuators.

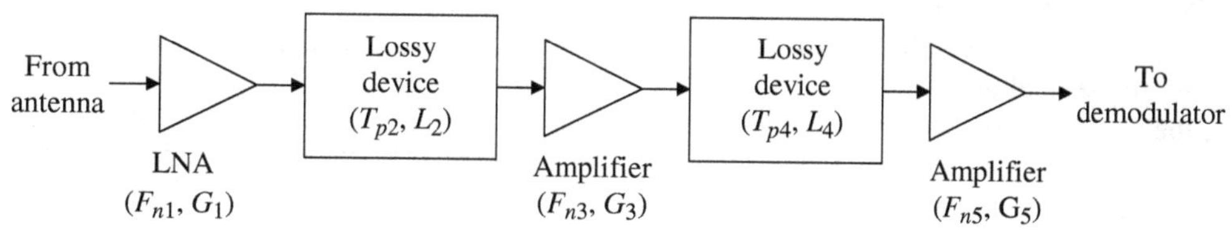

FIGURE 15.3 Receiver configuration.

or equivalently in terms of the effective noise temperatures

$$F_n = \left[\frac{\bar{T}_1}{T_o} + \frac{1}{G_1}\left(\frac{\bar{T}_2}{T_o}\right) + \frac{1}{G_1G_2}\left(\frac{\bar{T}_3}{T_o}\right) + \cdots\right] + 1 \\ = \frac{T_{rec}}{T_o} + 1 \qquad (15.22)$$

where T_{rec} is the effective receiver noise temperature given by

$$T_{rec} = \bar{T}_1 + \frac{\bar{T}_2}{G_1} + \frac{\bar{T}_3}{G_1G_2} + \cdots \qquad (15.23)$$

From these results it is apparent that the input device should be a high-gain, LNA, in which case it essentially determines the receiver noise figure. For this reason, LNAs with high gain are typically used at the receiver input.

The noise figure of a lossy device, such as an attenuator, is evaluated in a similar way; however, there are two major differences to consider: the gain of the lossy device is less than unity so the loss is given by $L_i = 1/G_i \geq 1$, the second consideration is that the *physical temperature* (T_{pi}) of the lossy device is typically used to characterize the *effective noise temperature* performance instead of the noise figure.* The equivalent noisy and noiseless attenuators are shown in Figure 15.2 in terms of the physical and effective temperatures. For the noisy attenuator, the output temperature is given by $T_{out} = T_{pi}(L_i - 1)/L_i$ and, for the equivalent noiseless attenuator, the output temperature is $T_{out} = \bar{T}_i/L_i$ so, upon equating these results, the equivalent noise temperature for a lossy device is

$$\bar{T}_i = (L_i - 1)T_{pi} \qquad (15.24)$$

Applying these results to a cascade of N attenuators $i = 1, 2, \ldots, N$, as shown in Figure 15.2c, each with loss L_i and physical temperature T_{pi}, the effective noise temperature of the cascaded attenuators is

$$\bar{T} = (L_1 - 1)T_{p1} + L_1(L_2 - 1)T_{p2} + \cdots + L_1L_2\ldots L_{N-1}(L_N - 1)T_{pN} \\ = \bar{T}_1 + L_1\bar{T}_2 + \cdots + L_1L_2\ldots L_{N-1}\bar{T}_N \qquad (15.25)$$

and the output temperature is computed as

$$T'_{out} = \frac{\bar{T}}{(L_1L_2\ldots L_N)} \qquad (15.26)$$

Usually the cascade of receiver devices alternates between amplifiers and passive filters or attenuators as shown in Figure 15.3. In this context, evaluation of the overall system noise temperature and noise figure involves the appropriate application of the results in this section.

15.2.1.1 Example: Evaluation of Receiver Noise Figure and Temperature As an example application of the results in Section 15.1.2, consider a receiver as shown in Figure 15.3 having a 20 dB gain LNA with a 2 dB noise figure. The LNA is located as close to the receive antenna output as possible, typically it is attached directly to the antenna waveguide, so that the loss prior to the LNA is associated directly with the antenna. The output of the LNA is coupled to the receivers first RF mixer, usually by a short length of coaxial cable. In the following example, the coaxial cable is assumed to have a physical temperature of $T_{p2} = 290°K$ and a loss of L_2 dB.

*The noise figure of a lossy device with loss L and temperature T_p is defined as $F_n = (L-1)T_p/290 + 1$ or, in terms of the effective temperature of the device, $F_n = \bar{T}_p/290 + 1$.

In examining this example, L_2 is considered to be the AGC control parameter. The signal from the coaxial cable is then input to the second amplifier with a gain of $G_3 = 25$ dB and a noise figure of $F_{n3} = 10$ dB. This is in turn followed by an AGC gain control attenuator with $L_4 = 0$–70 dB of gain control and a physical temperature of $T_{p4} = 290°$K. Finally, the output of the gain control is amplified by a $G_5 = 40$ dB gain amplifier with a noise figure of $F_{n5} = 10$ dB. The receiver noise figure for this configuration is shown in Figure 15.4 for LNA gains of 20 and 40 dB as a function of the AGC attenuator setting (L_4) for various conditions of loss (L_2) between the LNA and the second amplifier.

In this example, the system noise figure for an LNA gain of 20 dB and low values of AGC attenuation is degraded from the 2 dB LNA noise figure by 0.25–0.95 dB for feed losses (L_2) between 0 and 6 dB. These low AGC attenuation levels are associated with very weak signals relative to the thermal noise floor (kT_oB_n) at the receiver input. As the input signal becomes stronger the AGC lowers the overall gain of the receiver by increasing the AGC attenuation to keep the output level constant. However, in spite of the increasing receiver noise figure, the resulting impact on the system performance is not significant because of the correspondingly high output signal-to-noise ratio associated with the higher input signal levels.

For example, consider the case with $L_2 = 3$ dB, a minimum detectable signal with the AGC attenuation $L_4 = 0$ dB, and a system requirement that the minimum output signal-to-noise is $\gamma_o = 6$ dB. Under these conditions the input signal-to-noise ratio is $\gamma_i = \gamma_o F_n$ or $6 + 2.5 = 8.5$ dB. When the input signal level increases by 20 dB the ideally adjusted AGC attenuation becomes 20 dB and the receiver noise figure is about 2.6 dB. Under these conditions the corresponding output signal-to-noise ratio is $\gamma_o = \gamma_i / F_n$ or $8.5 + 20 - 2.6 = 25.9$ dB. As the input signal continues to increase another 20 dB the AGC attenuation becomes 40 dB and the receiver noise figure increases to 9 dB. In this case, the output signal-to-noise ratio is 39.5 dB. Therefore, in spite of the dramatic increase in the receiver noise figure from 2.5 to 9 dB, the system requirement that $\gamma_o \geq 6$ dB is more than satisfied.

The curves in Figure 15.4 for the 40 dB LNA gain demonstrate how the sensitivity of the system noise figure is reduced by increasing the LNA gain. Furthermore, with the higher gain LNA the system noise figure is hardly degraded for AGC attenuations up to 20 dB even with a feed loss of 6 dB following the LNA. The higher LNA gain also provided for a wider dynamic range which may be desirable if the feed loss L_2 becomes excessive. From these examples it is clear that the loss between the LNA and the second amplifier plays a significant role in determining the receiver noise figure; however, the focus should be on impact of the receiver noise figure with the minimum detectable input signal, that is, when the AGC attenuation is 0 dB. For example, when the loss L_2 decreases to 0 dB the effective noise temperature

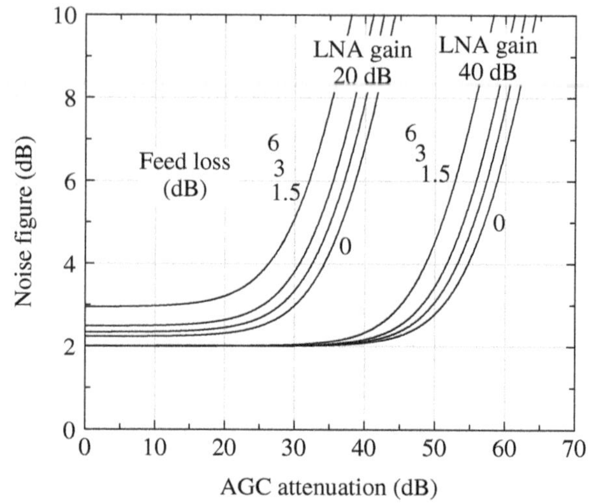

FIGURE 15.4 Example evaluation of receiver noise figure.

becomes $\bar{T}_2 = 0°$K even though the device remains at the physical temperature of T_{p2} °K. Therefore, the lossless feed does not influence the receiver noise performance. In this case, the receiver noise figure is influenced by the noise figure of the second amplifier, that is, F_{n3}, and, as the LNA gain is increased, this influence is diminished providing improved system performance with the minimum input signal power.

15.2.2 Antenna Temperature

With the evaluation of the receiver noise figure and temperatures *in hand*, the focus is turned to the antenna temperature T_a which is the remaining unknown parameter required to evaluate the system noise temperature and noise figure. The antenna temperature is characterized at the output of the antenna subsystem as indicated in Figure 15.5. The antenna is depicted as having a radome and a radiating dish with a feed horn as shown in Figure 15.5a. The feed horn is located at the focal point of the dish and, in the receive mode, is used to collect the incident electromagnetic wave energy. A short length of waveguide associated with the feed horn provides a flange for connecting the feed horn to the receiver subsystem. The radome protects the reflecting or radiating surface of the antenna from the elements of weather. Although the radome is optional, in Figure 15.5a the radome, reflecting dish, and antenna feed are considered to be an integral part of the antenna subsystem with their associated losses and physical temperatures. The overall antenna efficiency is made up of the aperture efficiency and the radiation efficiency. The radiation efficiency includes the spillover loss of a reflecting antenna, the reflector and feed ohmic losses and, if used, the radome loss [3]. These losses are included in the antenna gain and sidelobe measurements and are only used here to characterize the noise introduced by the antenna

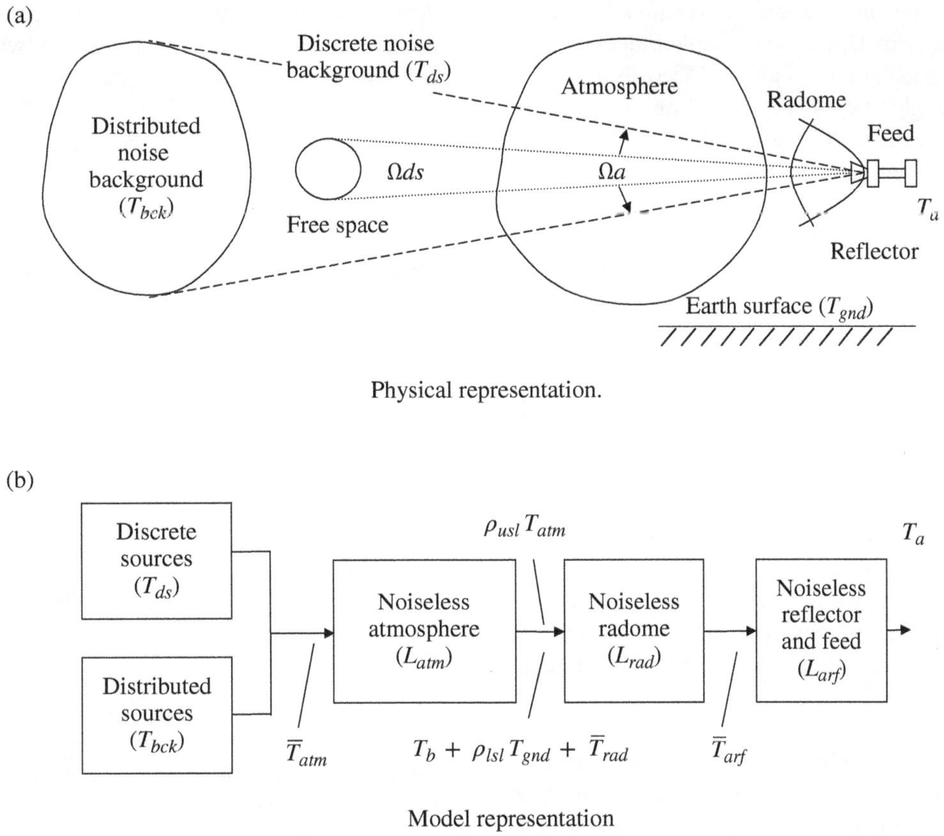

FIGURE 15.5 Contributors to antenna temperature.

subsystem. Combining the antenna reflector loss with the antenna feed loss as the product of their individual losses, that is, $L_{arf} = L_{ar}L_{af}$ is valid only if the reflector and feed are at the same temperature (see Problem 2). As indicated in Figure 15.5a, the external noise sources that are within the field of view of the antenna, including the antenna sidelobes, also contribute to antenna temperature. These noise sources include *discrete sources* (*ds*) such as stars and our Sun and Moon and *background* (*bck*) or distributed noise sources such as stellar gas, our galaxy, the Earth's surface and thermal noise resulting from atmospheric absorption.

To begin the evaluation of the antenna noise temperature, the noise temperature of all external noise sources is characterized at the input to the antenna as $T_{ain} = T_b + \rho_{lsl}T_{gnd} + \rho_{usl}T_{atm} + \bar{T}_{rad}$. These noise sources are indicated in Figure 15.5b and are identified as follows: T_b is the noise temperature of distant noise sources entering through the main antenna beam, ρ_{lsl} is the fraction of the ground noise temperature seen by the lower sidelobes (*lsl*) of the antenna, ρ_{usl} is the fraction of the atmospheric noise temperature T_{atm} seen by the upper sidelobes (*usl*) of the antenna, and $\bar{T}_{rad}$ is the effective noise temperature of the antenna radome. The notations T_{atm} and T_{gnd} represent the respective physical temperatures of the atmosphere and ground. With the exception* of $\bar{T}_{rad}$, all of the temperatures at the antenna radome are dependent on the grazing (or elevation) angle and the antenna beamwidth. For example, as the antenna viewing angle is changed the temperature T_b will change as different noise sources come into view. Also, as the grazing angle decreases the factor ρ_{lsl} increases with a commensurate increase in the ground noise at the antenna radome; the factor ρ_{usl} will also fluctuate impacting the atmospheric noise at the radome. The factors ρ_{lsl} and ρ_{usl} are evaluated by integrating the appropriate noise temperatures over the sidelobes of the antenna beam in consideration of the attenuation with range. At the elevation angle θ_m, corresponding to the maximum ground noise temperature seen by the lower sidelobes, the ground temperature contribution factor to the antenna temperature is approximated as

$$\rho_{lsl} \cong \frac{1}{G_r(\max)} \int_{\phi_m-\pi}^{\phi_m+\pi} G_r(\theta_m, \varphi)\,d\varphi \tag{15.27}$$

*The radome or reflector physical temperature will be impacted by the viewing angle relative to the Sun.

where $G_r(\theta_m,\varphi)$ is the antenna gain corresponding to θ_m relative to the peak gain $G_r(\max)$. Example values are [2]: $\rho_{lsl} = \rho_{usl} = 0.1$ for grazing anglcs abovc 15° with $\rho_{lsl} = 0.49$ and $\rho_{usl} = 0.1$ for angles below 15°. The link analysis program discussed in Section 15.15 uses (15.27) with $G_r(\theta, \phi)$ evaluated using the circular aperture antenna described in Section 15.3.2.

Referring once again to Figure 15.5b, the antenna temperature is characterized using the relationships developed in the preceding sections and is evaluated as

$$T_a = \frac{T_b + \rho_{lsl}T_{gnd} + \rho_{usl}T_{atm} + \bar{T}_{rad}}{L_{rad}L_{arf}} + \frac{\bar{T}_{arf}}{L_{arf}} = \frac{T_{ain}}{L_{rad}L_{arf}} + \frac{\bar{T}_{arf}}{L_{arf}} \quad (15.28)$$

where T_{ain} is the temperature of the external noise sources. Evaluation of T_b is discussed in the following two sections and the culmination of the general case is expressed in (15.36) in Section 15.2.2.2. Section 15.2.3.1 provides an example involving the computation of the antenna temperature.

15.2.2.1 Contribution of a Single Sky-Noise Source to the Antenna Temperature The analysis of the antenna temperature is simplified by considering one external noise source. In the context of Figure 15.5, T_b' is defined as the temperature of either a distributed or a single discrete noise source, that is, $T_b' = T_{bck}$ or T_{ds}. In evaluating the antenna noise temperature due to a distant radiating source, the radiation intensity or brightness (b) of the source is characterized by the Rayleigh–Jeans approximation [4] of the radiated flux density per steradian, expressed as*

$$b \cong \frac{2k\,T_b'}{\lambda^2} \quad : \mathrm{W/(Hz\text{-}m^2\text{-}steradian)}, \quad \text{Rayleigh-Jeans approximation} \quad (15.29)$$

where k is Boltzmann's constant. The approximation in (15.29) is valid if the ratio f/T_b' not to large which is the case for most applications. For example, the error is about 1% for $f = 150$ GHz and $T_b' = 300$°K; if the error is unacceptable then Planck's law must be used, for which the brightness is expressed as [5]

$$b = \frac{2hf^3B}{c^2\left(e^{hf/k\,T_b'} - 1\right)} \quad : \mathrm{W/(Hz\text{-}m^2\text{-}steradian)}, \text{ Planck's law} \quad (15.30)$$

*The electromagnetic radiation intensity of a hot blackbody at a temperature T_b is based on Planck's radiation law. The radiation intensity is also referred to as the brightness of the blackbody.

where $h = 6.63e^{-34}$ W-s^2 is Planck's constant, $c = 3e^8$ m/s is the free-space speed of light, the bandwidth $B \ll f$, and $k =$ Boltzmann's constant, previously defined along with the remaining parameters. The factor of two in (15.29) and (15.30) results from the Omni polarized electrometric waves emitted from the radiating source; for a linearly polarized receiver antenna these results must be divided by two.

Considering a lossless path, corresponding to $L_{atm} = 1$ in Figure 15.5, with an effective aperture of A_e m^2 and one polarization, the total power density at the receiver antenna is evaluated as

$$kT_b = \frac{b\Omega_s A_e}{2} \quad : \mathrm{W/Hz} \quad (15.31)$$

Upon substituting (15.29) into (15.31), the temperature of the source at the antenna input becomes

$$T_b = \frac{\Omega_s A_e}{\lambda^2} T_b' = \frac{\Omega_s}{\Omega_a} T_b' \quad : \Omega_s < \Omega_a, \text{ single polorization, lossless} \quad (15.32)$$

where Ω_s is the solid angle of the sky-noise source and $\Omega_a = \lambda^2/A_e$ is the solid angle of the received antenna beam. Equation (15.32) applies for an antenna using single polarization; with dual polarization this result must be doubled. Also, as indicated, it applies when the solid angle of the source is less than that of the antenna, that is, $\Omega_s < \Omega_a$; when $\Omega_s \geq \Omega_a$ the result is

$$T_b = T_b' \quad : \Omega_s \geq \Omega_a, \text{ single polorization, lossless} \quad (15.33)$$

If the medium between the radiation source and the antenna exhibits a loss, for example, when $L_{atm} > 1$ with an effective temperature $\bar{T}_{atm}$, then, using the previous results, the contribution of this external noise source at the antenna input is

$$T_b = \begin{cases} \left(\dfrac{\bar{T}_{atm}}{L_{atm}}\right) + \dfrac{1}{L_{atm}}\left(\dfrac{\Omega_s}{\Omega_a}\right)T_b' & : \Omega_s < \Omega_a \\ \left(\dfrac{\bar{T}_{atm}}{L_{atm}}\right) + \dfrac{1}{L_{atm}}T_b' & : \Omega_s \geq \Omega_a \end{cases} \quad : \text{single polarization, lossey} \quad (15.34)$$

15.2.2.2 Contribution of Multiple Sky-Noise Sources to the Antenna Temperature When all of the sky noise sources shown in Figure 15.5 are considered, the brightness temperature is evaluated by summing the background temperature with the temperatures of the discrete sources in consideration of overlapping regions as viewed by the antenna beam. To simplify this evaluation it is assumed that, except for the Sun, all of the discrete sources are completely within the antenna beam and do not overlap each other or the Sun.

Evaluating spaced-based cross-links where the Sun may occupy a significant portion the beam area is a special case. Encounters involving the Moon simply involve replacing the Sun's noise temperature with that of the Moon. With these considerations the brightness temperature of the noise sources with no atmospheric affects is computed as

$$T_b = T_{sun}\left(\frac{\Omega_{sun}}{\Omega_a}\right) + \sum_i T_{si}\frac{\Omega_{si}}{\Omega_a} + T_{bck}\left(1 - \frac{\Omega_{sun}}{\Omega_a} - \sum_i \frac{\Omega_{si}}{\Omega_a}\right) \quad : \Omega_{si}, \Omega_{sun} < \Omega_a \tag{15.35}$$

and, by analogy with (15.34), when atmospheric affects are involved this result becomes

$$T_b = \left(\frac{\bar{T}_{atm}}{L_{atm}}\right) + \frac{1}{L_{atm}}\left\{T_{sun}\left(\frac{\Omega_{sun}}{\Omega_a}\right) + \sum_i T_{si}\frac{\Omega_{si}}{\Omega_a} + T_{bck}\left(1 - \frac{\Omega_{sun}}{\Omega_a} - \sum_i \frac{\Omega_{si}}{\Omega_a}\right)\right\} \quad : \Omega_{si}, \Omega_{sun} < \Omega_a \tag{15.36}$$

The noise power of the Sun varies approximately as the inverse square of the frequency. The piecewise linear dependence of a quiet Sun's noise temperature on frequency is described by the approximate relationship

$$T_{sun}(°\mathrm{K}) \cong \begin{cases} 1 \times 10^6 & : 0.04 \le f_g < 0.4 \\ (-0.21525 f_g + 1.0861) \times 10^6 & : 0.4 \le f_g < 5 \\ 1 \times 10^4 & : 5 \le f_g \end{cases} \tag{15.37}$$

where f_g is the frequency in GHz. Equation (15.37) is based on the results of Kuiper [6] and Blake [5]. During sunspot activity, the noise temperature of the Sun may be 10^2–10^4 times higher than indicated by (15.37) lasting for several seconds followed by temperatures about 10 times higher lasting several hours.

The noise power from cosmic sources, such as our galaxy and interstellar gases, varies approximately as the inverse square of the frequency and the frequency dependence of the noise temperature varies approximately as [7] $1/f^{2.5}$. For example, denoting the galactic noise temperature at 100 MHz as T_{100}, the galactic temperature at another frequency is expressed as [8]

$$T_{gal} = T_{100}\left(\frac{100}{f_{\mathrm{MHz}}}\right)^{2.5} \tag{15.38}$$

Approximate maximum and minimum values of T_{100} are given by Brown and Hazard [9] as 18,650 and 500°K with a geometric mean value of 3,050°K. The dependence of cosmic noise sources on frequency is discussed in more depth by Hogg and Mumford [10], Ko [11], Smerd [12], and Strum [13]. The inclusion of other discrete sources depends upon the knowledge of their noise characteristics and location relative to the receiver antenna beam pointing coordinates.

The antenna noise temperature is also a function of the loss through the atmosphere as characterized by the atmospheric noise temperature which is dependent on the elevation or grazing angle of the antenna, the temperature, $T(r)$, and the absorption coefficient, $\alpha(r)$, at a range r from the antenna. The effective temperature of the atmosphere is evaluated using the relationship [14]

$$\bar{T}_{atm} = \int_0^\infty \alpha(r)T(r)\exp\left(-\int_0^r \alpha(\ell)d\ell\right)dr \tag{15.39}$$

A simplified evaluation of (15.39) results by using the average temperature and Gardner [15] has reported that this average temperature is approximately 84% of the Earth's surface temperature (T_{gnd}) so that $\bar{T}_{atm} \cong 0.84 T_{gnd}$.

15.2.3 System Noise Figure

The preceding analysis of the receiver and antenna subsystems, in terms of their noise temperatures and the receiver noise figure, was undertaken so that the system noise temperature and noise figure can be established to evaluate the performance of the entire receiver system. If the antenna is connected to the receiver input through the antenna to receiver feed subsystem, then the antenna temperature is transferred to the receiver input using the concepts developed in Section 15.2.1. In this case, the system noise figure at the receiver input is defined in terms of the corresponding system temperature so the sensitivity of the overall receiver performance is easily determined from knowledge of the antenna noise temperature and the antenna feed temperature and loss.

Consider the antenna and receiver system shown in Figure 15.6, where the antenna is connected to the receiver input amplifier or LNA using a coaxial cable or waveguide with a loss L_{ar} and physical temperature T_{par} °K. Using the antenna noise temperature (T_a) and the receiver noise figure (F_n), the receiver system temperature T_{rs} and the corresponding system noise figure F_{ns} at the input to the receiver are related as indicated in Figure 15.6. The relationship between T_{rs} and F_{ns} is similar to that expressed by (15.19), that is,

$$F_{ns} = \frac{T_{rs}}{T_o} + 1 \tag{15.40}$$

Therefore, upon evaluating the system noise temperature at the receiver input it is a simple matter to compute the system noise figure.

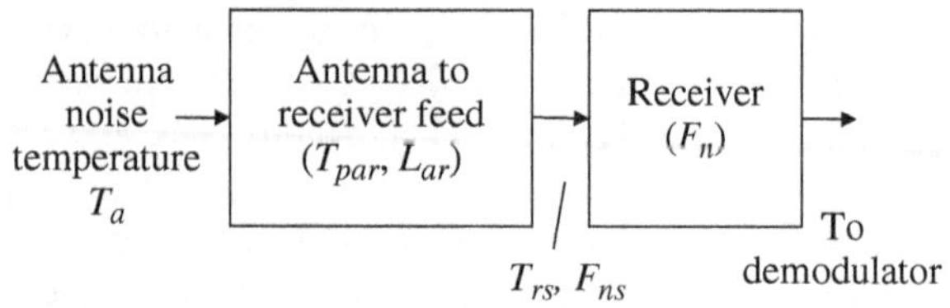

FIGURE 15.6 Antenna and receiver interface configuration.

The evaluation of T_{rs} follows the procedures discussed in the preceding section with the result

$$T_{rs} = \frac{T_a + \bar{T}_{ar}}{L_{ar}} + T_{rec} \tag{15.41}$$

where the effective noise temperature of the antenna-to-receiver feed is evaluated as

$$\bar{T}_{ar} = (L_{ar} - 1)T_{par} \tag{15.42}$$

The receiver noise temperature is evaluated using (15.23) or, more directly, in terms of the receiver noise figure F_n using (15.22) as

$$T_{rec} = (F_n - 1)T_o \tag{15.43}$$

Using (15.43) and (15.42) to compute T_{rs}, the system noise figure is evaluated using (15.40) and is plotted in Figure 15.7 as a function of the feed loss L_{ar} for several antenna temperatures. In this plot a receiver noise figure of 2 dB (1.58:1) is used and the physical temperature of the feed network is assumed to be $T_{parf} = 290°\text{K}$.

The asymptotic value of the system noise figure as L_{ar} approaches infinity is evaluated as $F_{ns} = F_n + T_{par}/T_o = 4.12$ dB so the system noise figure is degraded by the physical temperature of the feed; this is of academic interest because the received signal power also experiences this loss and approaches zero as the feed loss increases. A more informative plot is the loss in the receiver noise figure as a function of the antenna noise temperature; this is shown in Figure 15.8 for various losses in the antenna-to-receiver feed with $F_n = 2$ dB and $T_{par} = 290°\text{K}$.

The receiver noise figure loss is characterized by substituting (15.42) into (15.41) and recognizing that the increase in the receiver temperature relative to T_{rec} is

$$\Delta T_{rec} = \frac{T_a + (L_{ar} - 1)T_{par}}{L_{ar}} \tag{15.44}$$

The resulting loss in the receiver noise figure is defined as

$$LF_n \triangleq \frac{F_{ns}}{F_n} = \frac{\Delta T_{rec}}{T_o F_n} + 1 \tag{15.45}$$

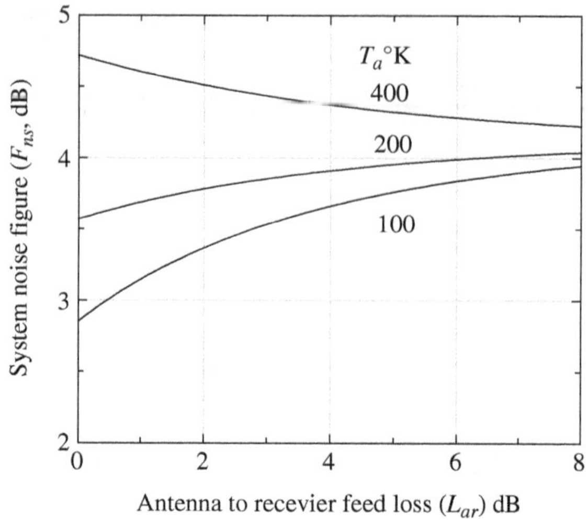

FIGURE 15.7 Example evaluation of system noise figure ($F_n = 2$ dB, $T_{par} = 290°\text{K}$).

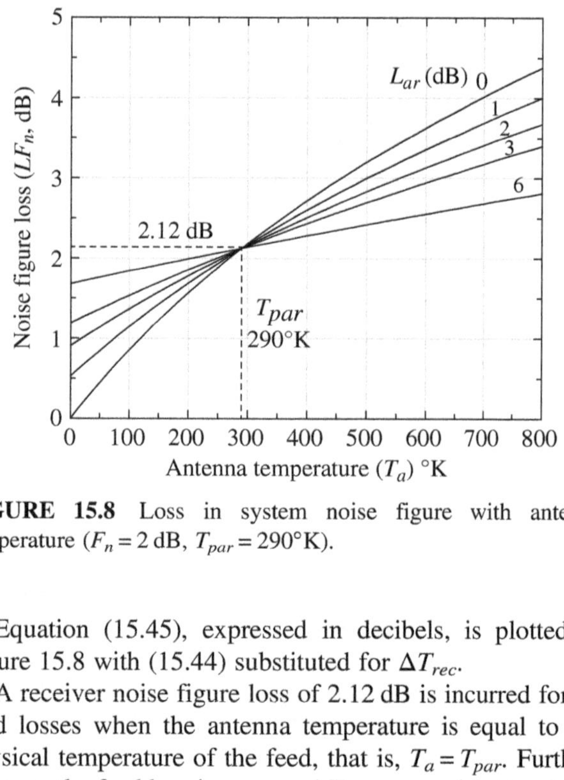

FIGURE 15.8 Loss in system noise figure with antenna temperature ($F_n = 2$ dB, $T_{par} = 290°\text{K}$).

Equation (15.45), expressed in decibels, is plotted in Figure 15.8 with (15.44) substituted for ΔT_{rec}.

A receiver noise figure loss of 2.12 dB is incurred for all feed losses when the antenna temperature is equal to the physical temperature of the feed, that is, $T_a = T_{par}$. Furthermore, as the feed loss increases, ΔT_{rec} approaches T_{par} independent of the antenna temperature so the noise figure loss corresponds to a constant loss given by

$$LF_n\big|_{L_{ar}=\infty} = \frac{T_{par}}{T_o F_n} + 1 \tag{15.46}$$

This result verifies the system noise figure in Figure 15.7, in that, as $L_{ar} \to \infty\, F_{ns} = F_n + 2.12 = 4.12\text{dB}$. Figures 15.7 and 15.8 are based on the unique condition that $T_{par} = T_o$.

On occasions it is useful to refer the system noise temperature, as computed by (15.41), to the antenna input using the relationship

$$T_{as} = L_{ar} T_{rs} \tag{15.47}$$

This section concludes by characterizing the gain temperature ratio (*G/T*) of the receive antenna under several conditions. The *G/T* ratio is a measure of the quality of the receive system much like the EIRP is a quality measure of the transmit system. The antenna *G/T* gain temperature figure of merit is typically expressed in units of dB/°K.

One application of the antenna system noise temperature referred to the antenna is to use the specification $G/T = G_r/T_{as}$. This definition characterizes the antenna performance under operational conditions including various sky and ground noise sources and is used in the design and evaluation of the antenna subsystem. Another definition is $G/T = G_r/T_a$ when $T_{ain} = 0$. This definition provides a figure of merit for the antenna subsystem with no external noise sources as might be required for a performance acceptance test of the antenna.

15.2.3.1 Example: Evaluation of Antenna Temperature and Receiver G/T and C/N$_o$ Ratios Consider an Earth station antenna with a gain of $G_a = 55$ dB and a spot beam with beamwidth $\theta_B = 1°$ (0.0174 rad) viewing a satellite at a low elevation angle. The antenna does not use a radome; however, the reflector and feed losses are assumed to be $L_{arf} = 0.5$ dB with a physical temperature of $T_{parf} = 290°$K. The Sun, at a temperature of $T_{sun} = 500{,}000°$K, is totally within the satellite beamwidth and is viewed against cosmic background noise at a temperature of $T_{bck} = 40°$K. The ground temperature in the vicinity of the antenna is $T_{gnd} = 290°$K and Gardner's approximation to the atmospheric temperature is used. The atmospheric loss is 2 dB, that is, $L_{atm} = 1.585$, and the lower and upper antenna sidelobe contribution factors to the ground and atmospheric noises are $\rho_{lsl} = \rho_{usl} = 0.1$. Furthermore, the antenna is connected to the receiver LNA through a feed with a physical temperature of $T_{ar} = 290°$K and a loss of 1 dB, that is, $L_{ar} = 1.26$; this loss results from a coaxial cable with an impedance mismatch. Applying the feed loss, gain, and AGC conditions used in the example in Section 15.2.1.1, with $L_2 = 3$ dB, LNA gain = 20 dB, and an AGC attenuation of 10 dB, the receiver noise figure is 2.5 dB.

Based on these conditions the antenna temperature (T_a), the antenna system noise temperature (T_{as}), and the receiver *G/T* ratio are evaluated. Some necessary constants are Sun's radius is $r_s = 6.96 \times 10^5$ km and the Sun's distance from Earth is $R_s = 1.495 \times 10^8$ km.

The first task is to determine the contribution of the sky noise temperature T_b at the antenna input. The solid angles of the Sun and antenna are

$$\Omega_{Sun} = \frac{\text{Area(Sun)}}{\text{Range}} = \frac{\pi\, r_s^2}{R_s^2} = 6.8 \times 10^{-5} \tag{15.48}$$

$$\Omega_a = \frac{\text{Area(beam)}}{\text{Range}} = \frac{\pi\, D_B^2/4}{R_s^2} = \frac{\pi}{4}\theta_B^2 = 2.4 \times 10^{-4} \tag{15.49}$$

where $D_B = R_s \theta_B$ is the diameter of the beam at a range of R_s; a spot beam has a circular pattern normal to the pointing direction. Because $\Omega_{sun}/\Omega_a = 0.28$ and $L_{atm} = 1.585$, the sky and atmospheric noise temperatures at the antenna input are computed using (15.36). In this computation, the effective atmospheric noise temperature, based on the ground noise, is approximated as

$$\bar{T}_{atm} = .84 T_{gnd} = 244°\text{K} \tag{15.50}$$

These results are used to compute the background noise temperature at the antenna input as

$$T_b = \frac{\bar{T}_{atm}}{L_{atm}} + \frac{T_{Sun}}{L_{atm}}\left(\frac{\Omega_{Sun}}{\Omega_a}\right) + \frac{T_{bck}}{L_{atm}}\left(1 - \frac{\Omega_{Sun}}{\Omega_a}\right) \\ = 88{,}556°\text{K} \tag{15.51}$$

The ground and atmospheric noise temperature entering through the antenna sidelobes are $\rho_{lsl} T_{gnd} = 27°$K and $\rho_{usl} T_{atm} = 66°$K where $T_{atm} = 660°$K is the computed physical temperature of the atmosphere. From these results and (15.28) the total temperature at the antenna input is $T_{ain} =$ 88,649°K.

Since there is no radome, only the effective antenna reflector and feed temperatures are involved in referring the antenna input temperature to the antenna output. Given the combined reflector and feed loss (L_{arf}) and physical temperature (T_{parf}), the effective antenna reflector and feed temperature is computed as

$$\bar{T}_{arf} = (L_{arf} - 1) T_{parf} = 35°\text{K} \tag{15.52}$$

Using (15.52) with $T_{ain} = 88{,}469°$K and (15.28) with $L_{rad} = 1.0$ the antenna temperature is computed as

$$T_a = \frac{T_{ain} + \bar{T}_{arf}}{L_{arf}} = 79{,}040°\text{K} \tag{15.53}$$

To determine the antenna system noise temperature the receiver noise temperature is required and, using (15.22) or (15.43) and the specified receiver noise figure, the result is

$$T_{rec} = (F_n - 1) T_o = 226°\text{K} \tag{15.54}$$

Using (15.53), (15.54), and (15.41), with a feed loss of 1 dB, temperature $T_{par} = 290°$K, $L_{ar} = 1.26$ and, from

(15.42), $\bar{T}_{ar} = 75.4°K$, the receiver system temperature is evaluated as

$$T_{rs} = \frac{T_a + \bar{T}_{ar}}{L_{ar}} + T_{rec} = 63{,}016°\text{K} \tag{15.55}$$

The Sun has significantly degraded the system sensitivity as seen from the increase in the system noise figure $F_{ns} =$ 23.4 dB. The system temperature referenced to the antenna input is computed using (15.47) with the result $T_{as} =$ 79,400°K and, using this result, the antenna-receiver system G/T_{as} ratio is found to be

$$G/T|_{\text{dB}} = G_a(\text{dB}) - 10\log_{10}(T_{as}) = 6\text{dB}/°\text{K} \tag{15.56}$$

This example demonstrates the procedures in the evaluation of the antenna G/T ratio as seen at the receiver input; it also demonstrates the horrific impact on the system performance when the antenna is pointed directly at the Sun. Problem 3 examines the impact of the Sun's temperature when the antenna beam is partially illuminated by the Sun.

The receiver carrier power-to-noise density ratio, C/N_o, is an important receiver parameter, in that, the demodulator E_b/N_o performance is determined as $E_b/N_o|_{\text{dB}} = C/N_o|_{\text{dB}} - 10\log_{10}(R_b)$ where R_b is the user data rate. From the viewpoint of the link budget, E_b/N_o includes the theoretical value, corresponding to a specified P_{be}, and includes the demodulator losses. Based on (15.13) the C/N_o ratio for a point-to-point link is expressed in decibels, using previously defined parameters, as

$$C/N_o|_{\text{dB}} = EIRP_{\text{dB}} + \left(\frac{G_r}{T_{rs}}\right)_{\text{dB}} - 10\log_{10}(L_{atm}L_{fs}L_{rs}) - 10\log_{10}(\kappa) \tag{15.57}$$

where $\kappa = k(T_o/T_{as})R_bF_{ns}$.

15.2.4 Remarks on the System Noise Figure

Often the minimum received power at the input to the LNA is specified along with a corresponding system performance requirement, for example, the bit-error probability at a given E_b/N_o(req'd) signal-to-noise ratio. This defines the required receiver sensitivity and allows for the design and development of the receiver and demodulator to proceed without the antenna subsystem. Defining the minimum received power as

$$P_r(\min) = \frac{P_{PA}G_tG_r}{L_{ts}L_{atm}L_{fs}L_{rs}} \tag{15.58}$$

the minimum signal-to-noise ratio is evaluated using (15.13) and is expressed as

$$\gamma_{min} = \frac{P_r(\min)}{kT_oB_nF_{ns}} \tag{15.59}$$

In this computation F_{ns} is the system noise figure so a specification of the system antenna temperature T_a must be provided. The temperature of the feed or cable required to connect the antenna to the LNA must also be considered in the computation of F_{ns}. With these caveats and the bandwidth expressed in terms of the bit rate $B_n = R_b$, and $N_o = kT_oF_{ns}$, (15.59) specifies the minimum (or theoretical) signal-to-noise ratio E_b/N_o(min) that is achievable based on the receiver sensitivity. Therefore, using these results, the receiver and demodulator design margin is

$$\frac{E_b}{N_o}(\text{margin}) = \frac{E_b}{N_o}(\text{req'd}) - \frac{E_b}{N_o}(\min) \tag{15.60}$$

If the specified E_b/N_o corresponds to the theoretical value of P_{be} based on the selected waveform modulation and coding, then this margin must include all receiver and demodulator losses and hopefully some to spare. Examples of various receiver and demodulator losses are given in Sections 15.4–15.10 and the magnitude of the demodulator losses is identified in the simulations discussed in Section 15.14.

15.3 ANTENNA GAIN AND PATTERNS

This discussion of antennas is intended to provide insight into the antenna patterns used in the evaluation of the communication link performance and, as such, draws upon the volumes of literature [3, 16–18] on the subject of antenna theory and performance. Because antennas that produce spot beam radiation patterns are in such widespread use in terrestrial and satellite communications, the circular aperture antenna pattern is of particular interest. The *antenna efficiency* (η_{ant}) includes the following efficiencies: the *aperture efficiency* (η_a), the *spillover efficiency* (η_s) of the reflector, and the *antenna reflector surface and feed efficiency* (η_{arf}). The aperture efficiency is defined as the ratio of the effective aperture to the physical aperture, such that, $A_e = \eta_aA_a$, and is dependent on weighting function $w(x)$ that is applied to the aperture illumination. The weighting function can significantly reduce the sidelobes at the expense of producing a wider beamwidth and reduced aperture efficiency. Uniform aperture weighting results in the highest aperture efficiency and lowest half-power beamwidth; however, the sidelobes are only 13 dB below the main lobe. The radiation efficiency is defined as the product of the antenna spillover efficiency and the antenna reflector surface and feed efficiency, that is, $\eta_r = \eta_s\eta_{arf}$. The reflector surface and feed efficiencies are related to the antenna ohmic loss, denoted as L_{arf}. The spillover efficiency is associated with the loss when the antenna feed pattern is not completely concentrated on the reflecting surface. The spillover loss accounts for a large part of the radiated energy behind the reflector. Blocking of the aperture by the antenna feed and support structures lower the gain and raise the sidelobes of the antenna pattern. The offset-feed antenna configuration eliminates aperture blocking.

In the previous sections, the antenna gain is related to the physical and effective antenna area by the relationships

$$G = \frac{4\pi\eta_a A_a}{\lambda^2} = \frac{4\pi A_e}{\lambda^2} \tag{15.61}$$

The antenna gain is defined in terms of the directive antenna gain by the relationship

$$G = \eta_r G_D \tag{15.62}$$

where $G_D = 4\pi/\Omega_a$ and Ω_a is defined as the solid angle of the antenna beam, expressed as

$$\Omega_a = \frac{\iint P(\theta,\phi)d\theta d\phi}{P(0,0)} \tag{15.63}$$

The angles θ and ϕ represent the orthogonal azimuth and elevation 3 dB beamwidth angles of the antenna pattern and the 3 dB beam area is approximated as $\Omega_a \cong \theta_B\phi_B$. Using this relationship with G_D as given earlier, the antenna gain is expressed as

$$G = \eta_r\left(\frac{4\pi}{\theta_B\phi_B}\right) \tag{15.64}$$

15.3.1 Rectangular Aperture Antenna Pattern

The antenna pattern for a linear one-dimensional aperture of length L_a is characterized in terms of the far-field electric intensity as

$$E(\phi) = \int_{-La/2}^{La/2} W(\ell)e^{j2\pi(\ell/\lambda)\sin(\phi)}d\ell \tag{15.65}$$

where, in general, $W(\ell)$ is the complex aperture weighting function and λ is the wavelength of the radiating frequency. The far field occurs when the range $r \gg \lambda$ and is formally characterized by the Fraunhofer region where the received electric field is represented by a plane wave; typically, the far field is considered to be $r \geq 2L_a^2/\lambda$. Under this condition, the phase error at the 3 dB points of a uniformly illuminated aperture antenna corresponds to $\lambda/16$ wavelength. Upon defining the spatial frequency as $\xi = \sin(\phi)/\lambda$, (15.65) is recognized as the inverse Fourier transform so the weighting function can be determined by taking the Fourier transform of a specified electric field intensity with respect to the differential $d\xi = d\sin(\phi)/\lambda$.

Evaluation of the electric field pattern for a uniformly illuminated aperture with constant phase, such that, $W(\ell) = 1/L_a$, results in the $\sin(x)/x$ electric field intensity pattern given by

$$E(\phi) = \frac{\sin(\pi(L_a/\lambda)\sin(\phi))}{\pi(L_a/\lambda)\sin(\phi)} \quad : \text{uniform aperture; } w(\ell) = 1 \ : \forall\ell \tag{15.66}$$

The magnitude-squared $|E(\phi)|^2$ of the electric field intensity results in the pattern of the radiated power that is used to characterize the antenna gain. The *solid* curve in Figure 15.9 is a plot of (15.66) in terms of $|E(\phi)|^2$ in decibels and is the familiar $sinc^2(x)$ response function.* The antenna gain is normalized for unit amplitude at $\phi = 0$ so the peak gain is 0 dB. The abscissa is plotted as the normalized angle $\theta_n = L_\lambda \sin(\phi)$ where $L_\lambda = L_a/\lambda$; however, the antenna beam angle is ϕ, evaluated as

$$\phi = \sin^{-1}\left(\frac{\theta_n}{L_\lambda}\right) \cong \frac{\theta_n}{L_\lambda} \tag{15.67}$$

where the approximation applies when $L_\lambda \gg 1$, that is, for large apertures.

The cosine weighted aperture function is expressed as

$$W(\ell) = \frac{\sqrt{2}}{L_a}\cos\left(\frac{\pi\ell}{L_a}\right) \quad : |\ell| \leq \frac{L_a}{2} \tag{15.68}$$

where the factor $\sqrt{2}$ results in the same aperture power as in the uniformly weighted case. Using (15.68) in the evaluation of (15.65) and defining $L_\lambda = \lambda/L_a$ results in the electric field intensity pattern

$$E(\phi) = \left(\frac{2\sqrt{2}}{\pi}\right)\frac{\cos(\pi L_\lambda \sin(\phi))}{1-(2L_\lambda \sin(\phi))^2} \quad : \text{cosine aperture} \tag{15.69}$$

Equation (15.69) is normalized by $2\sqrt{2}/\pi$ and plotted as the *dashed* curve in Figure 15.9. The response is identical to the previous applications of the cosine weighting function with a first spatial sidelobe level of −23 dB and null at $L_\lambda\sin(\phi) = 1.5$.

The final antenna aperture to be considered is the triangular aperture with the weighting function given by

$$W(\ell) = \frac{\sqrt{3}}{L_a}\left(-\frac{2}{L_a}|\ell| + 1\right) \quad : |\ell| \leq \frac{L_a}{2} \tag{15.70}$$

where, in this case, the $\sqrt{3}$ results in the same aperture energy as the uniformly weighted aperture. Upon evaluation of (15.65) for the triangular aperture results in the electric field intensity given by

*This radiation pattern, when looking or pointing normal to the array, is similar to that of a large one-dimensional array of N radiation elements with spacing $d = \lambda/2$; in this case $L_a = Nd$.

$$E(\phi) = \left(\frac{\sqrt{3}}{2}\right)\frac{\sin^2((\pi/2)L_\lambda \sin(\phi))}{((\pi/2)L_\lambda \sin(\phi))^2} \quad \text{: triangular aperture} \tag{15.71}$$

This result is plotted in Figure 15.9 as the *dotted* curve.

For the uniformly illuminated aperture antenna, the first spatial sidelobes are 13 dB below the main lobe; the distance between the first nulls, on either side of the main lobe, is λ/L_a radians; and the 3 dB beamwidth is $0.89\lambda/L_a$ radians (51°). These characteristics are summarized in Table 15.2 along with those of the cosine and triangular weighted apertures.

15.3.2 Circular Aperture Antenna Pattern

The circular aperture is evaluated by converting the aperture weighting function to polar coordinates with the result that the electric field intensity pattern is expressed as

$$E(\phi) = \int_0^{2\pi}\int_0^{D/2} W(r,\theta)e^{j2\pi(r/\lambda)\sin(\phi)\cos(\theta)}rdrd\theta \tag{15.72}$$

Normally the aperture weighting is independent of the angle θ, that is, the illumination amplitude is constant around the aperture for $0 \le r \le D/2$, and (15.72) simplifies to

$$\begin{aligned} E(\phi) &= 2\pi\int_0^{D/2} W(r)e^{j2\pi(r/\lambda)\sin(\phi)}rdr \\ &= 2\pi\int_0^{D/2} W(r)J_o\left(2\pi\left(\frac{r}{\lambda}\right)\sin(\phi)\right)rdr \end{aligned} \tag{15.73}$$

where $J_o(x)$ is the zero-order Bessel function of the first kind. Evaluation of this result for a uniformly illuminated aperture of diameter D results in the expression [18]

$$E(\phi) = \frac{2J_1(\pi D_\lambda \sin(\phi))}{\pi D_\lambda \sin(\phi)} \tag{15.74}$$

where $J_1(x)$ is the first-order Bessel function of the first kind and $D_\lambda = D/\lambda$. This result is normalized by $\pi D^2/4$ to yield unit gain at $E(\phi = 0)$. The normalized radiated pattern for the uniformly weighted circular antenna is shown as the *solid* curve in Figure 15.10. This pattern is used in the link evaluation

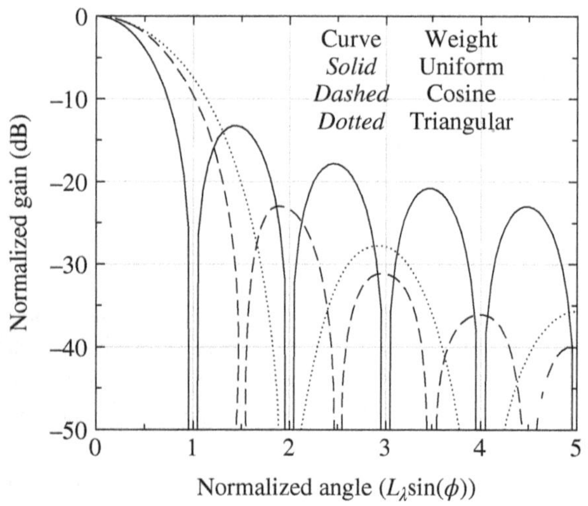

FIGURE 15.9 Normalized radiated power pattern for weighted linear (one-dimensional) aperture antenna.

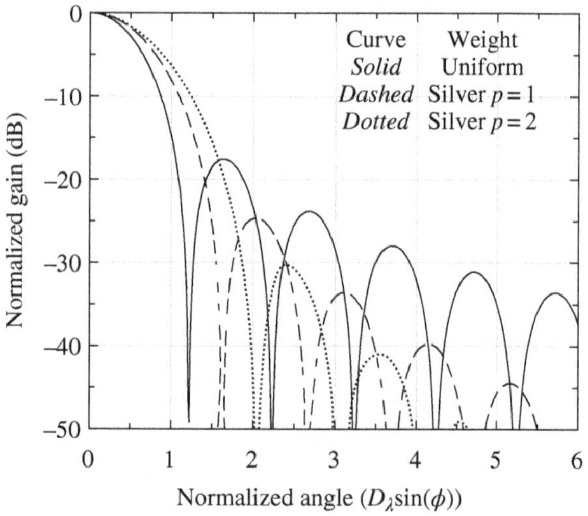

FIGURE 15.10 Normalized radiated power pattern for several weighted circular aperture antennas.

TABLE 15.2 Antenna Pattern Characteristics for Several Aperture Weighting Functions

Aperture	Aperture Weight $W(\ell)$: $\lvert\ell\rvert \le L_a/2$	Angle Between First Nulls (rad)	3 dB Beamwidth θ_B (rad)	First Sidelobe Level (dB)	Gain Loss (dB)
Uniform	1.0	$2.0/L_\lambda$	$0.89/L_\lambda$	−13.26	0.000
Cosine	$\frac{\sqrt{2}}{L_a}\cos\left(\frac{\pi\ell}{L_a}\right)$	$3.0/L_\lambda$	$1.19/L_\lambda$	−23.0	0.912
Triangular	$\frac{\sqrt{3}}{L_a}\left(-\frac{2}{L_a}\lvert\ell\rvert+1\right)$	$4.0/L_\lambda$	$1.27/L_\lambda$	−26.52	1.25

TABLE 15.3 Approximate Antenna Pattern Characteristics for Several Circular Aperture Weighting Functions

Aperture	Aperture Weight $w(r): 0 \le r \le D/2$	Angle Between First Nulls (rad)	3 dB Beamwidth θ_B (rad)	First Sidelobe Level (dB)	Gain (dB)
Uniform	1.0	$2.42/D_\lambda$	$1.02/D_\lambda$	−17.6	0.00
Silver, $p=1$	$1-4(r/D)^2$	$3.28/D_\lambda$	$1.27/D_\lambda$	−24.6	−1.25
Silver, $p=2$	$\left(1-4(r/D)^2\right)^2$	$4.02/D_\lambda$	$1.48/D_\lambda$	−30.6	−2.52

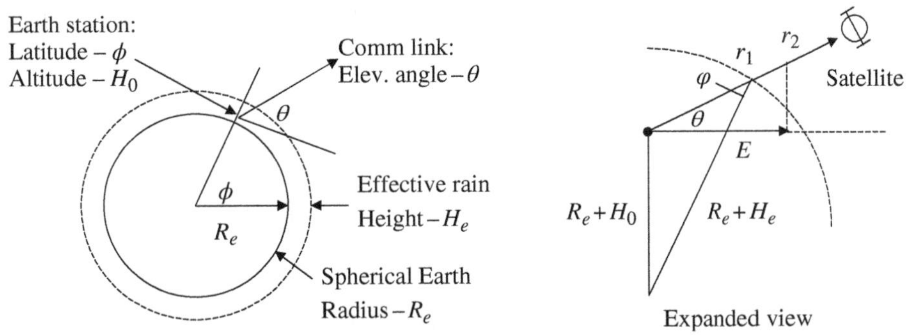

FIGURE 15.11 Communication link geometry used to evaluate rain attenuation.

program to determine the ground noise dependence on the antenna elevation angle as discussed in Section 15.2.2. The 3-dB beamwidth of this antenna is approximately $1.02/D_\lambda$ radians ($58.4/D_\lambda$ degrees) and the first sidelobe is 17.6 dB below the peak gain. The aperture efficiency relative to the uniformly weighted linear aperture is 0.865 (−0.63 dB).

Silver [17] has developed expressions for the circular aperture electric field intensities using an aperture weighting function described by

$$W(r)=\left(1-4\left(\frac{r}{D}\right)^2\right)^p \tag{15.75}$$

where $p \ge 0$ with the result

$$E(\phi)=\frac{2^p p!\,J_{p+1}(\pi D_\lambda \sin(\phi))}{(\pi D_\lambda \sin(\phi))^{p+1}} \tag{15.76}$$

This result is normalized by $\pi D^2/4$ so the value at $E(\phi=0)$ is relative to the uniformly weighted aperture. Table 15.3 summarizes the approximate antenna characteristics for the circular aperture weighting functions considered.

15.4 RAIN LOSS

Signal propagation through rain can result in a significant loss depending on the rain rate, usually measured in millimeter per hour; the system carrier frequency; and the propagation path through the rain region. The propagation path is influenced by the antenna elevation angle. This subject has received considerable attention over the years [19, 20] and has been reduced to several models that characterize the rain attenuation as a function of environmental and operational parameters. The Crane model [21] is the first comprehensive characterization of rain attenuation based on empirical worldwide rain statistics. Arnold and Kao [22] provide a summary of four models*: the Crane model, two CCIR models [23, 24], and the simple attenuation model (SAM) [25]. The link evaluation program discussed in Section 15.15 uses the SAM rain attenuation model† with the communication link geometry shown in Figure 15.11. This geometry is used to compute the path length $L(R)$ through the rain region, with rain rate R millimeter per hour, where $L(R)$ is given by

$$L(R)=\min(r_1,r_2) \tag{15.77}$$

The slant ranges r_1 and r_2 are shown in Figure 15.11; all distances are in kilometer. This characterization assumes that the slant range between the communication platforms is greater than $max(r_1, r_2)$ as encountered in satellite links; however, for terrestrial links, that may involve short ranges, $L(R)$ must be computed as $L(R)=\min(r_1,r_2,r_o)$ where r_o is the

*The system designer must select the appropriate model and verify the current specifications for the rain attenuation.

†The SAM model results in the same or slightly better predictions [25] than the CCIR models.

link range. The range r_1 is the propagation range from the ground station through the rain region to the effective height of the rain and is evaluated using the *law of sines* as

$$r_1 = (R_e + H_e)\left(\frac{\cos(\theta+\varphi)}{\cos(\theta)}\right) \quad (15.78)$$

where θ is the ground antenna elevation angle, H_e is the effective rain height, related to the 0°C isothermal height (H) relative to the Earth's surface, as

$$H_e = \begin{cases} H & : R \le 10\,\text{mm/h} \\ H + \log(R/10) & : R > 10\,\text{mm/h} \end{cases} \quad (15.79)$$

and

$$\varphi = \arcsin\left(\left(\frac{R_e + H_o}{R_e + H_e}\right)\cos(\theta)\right) \quad (15.80)$$

The isothermal height is a function of the latitude of the Earth station and is computed by Crane [26] as

$$H = \begin{cases} 4.8 & : \phi \le 30° \\ 7.8 - 0.1\phi & : \phi > 30° \end{cases} \quad (15.81)$$

The range r_2 is evaluated as $r_2 = E/\cos(\theta)$ where E is the empirically derived horizontal extent of the rain region given by [27, 28]

$$E = 41.4 - 23.5\log_{10}(p) \quad (15.82)$$

In (15.82), E is in kilometers and p is the rain rate in millimeter per hour. Using the path length L, in kilometers, the attenuation is computed in decibels as

$$A(\text{dB}) = \begin{cases} aR^b L(R) & : R \le 10\,\text{mm/h} \\ aR^b \dfrac{1 - exp[-b\lambda\ln(R/10)L(R)\cos(\theta)]}{b\lambda\ln(R/10)\cos(\theta)} & : R > 10\,\text{mm/h} \end{cases} \quad (15.83)$$

where the constant $\lambda = 1/14$ results in the best fit to the data; R is the rain rate, in millimeter per hour, along the communication path $L(R)$ in the selected rain region; and the constants a and b are frequency- and temperature-dependent constants, approximated as [20]

$$a = 4.21 \times 10^{-5} f^{2.42} \quad : 2.9 \le f < 54\,\text{GHz} \quad (15.84)$$

and

$$b = \begin{cases} 1.41 f^{-0.0779} & : 8.5 \le f < 25\,\text{GHz} \\ 2.63 f^{-0.272} & : 25 \le f \le 164\,\text{GHz} \end{cases} \quad (15.85)$$

The rain rate used by this model is associated with a world rain rate map and is identified by the geographic region of interest using a letter/number designation [29]. Table 15.4 shows the association of the letter designations of the geographic regions. For each geographic region, the selected rain rate is associated with a percent of time (PCT) that the attenuation will exceed the computed value. Table 15.4 also shows a similar correspondence between the PCT designations. For example, in the rain rate region A, a PCT of 0.001% will result in a computed rain attenuation that will be exceeded by only $0.00001T$, where T is in units of time representing 1 year [20]. That is, in this example, the attenuation will exceed the computed value no more than 5.24 min in 1 year. Figure 15.12 identifies the rain rate regions in the continental United States, Alaska, and Hawaii.

TABLE 15.4 Correspondence Between Rain-Rate Regions and PCT[a]

	Climate Region Rain Rates (mm/h)										
PCT	A	B_1	B_2	C	D_1	D_2	D_3	E	F	G	H
0.001	28.5	45.0	70.0	78.0	90.0	108.0	126.0	165.0	66.0	185.0	253.0
0.002	21.0	34.0	54.0	62.0	72.0	89.0	106.0	144.0	51.0	157.0	220.5
0.005	13.5	22.0	35.0	41.0	50.0	64.5	80.5	118.0	34.0	120.5	178.0
0.01	10.0	15.5	23.5	28.0	35.5	49.0	63.0	98.0	23.0	94.0	147.0
0.02	7.0	11.0	16.0	18.0	24.0	35.0	48.0	78.0	15.0	72.0	119.0
0.05	4.0	6.4	9.5	11.0	14.5	22.0	32.0	52.0	8.3	47.0	86.5
0.1	2.5	4.2	6.1	7.2	9.8	14.5	22.0	35.0	5.2	32.0	64.0
0.2	1.5	2.8	4.0	4.8	6.4	9.5	14.5	21.0	3.1	21.8	43.5
0.5	0.7	1.5	2.3	2.7	3.6	5.2	7.8	10.6	1.4	12.2	22.5
1.0	0.4	1.0	1.5	1.8	2.2	3.0	4.7	6.0	0.7	8.0	12.0
2.0	0.1	0.5	0.8	1.1	1.2	1.5	1.9	2.9	0.2	5.0	5.2
5.0	0.0	0.2	0.3	0.5	0.0	0.0	0.0	0.5	0.0	1.8	1.2

[a]Ippolito [20]. Courtesy of the National Aeronautics and Space Administration (NASA).
Notes: Region B is the average of B_1 and B_2 rounded up to one decimal place and region $D = D_2$.
Polar (A, dry; B, moderate), temperate (C, maritime; D, continental), subtropical (E, wet; F, arid), tropical (G, moderate; H, wet).

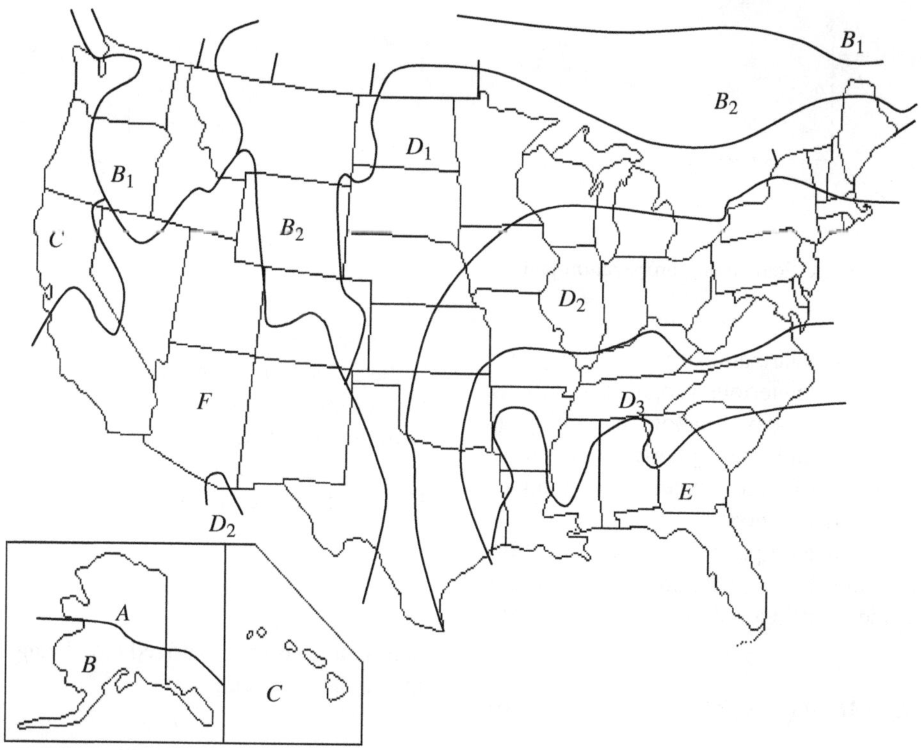

FIGURE 15.12 U.S.A. rain-rate climate regions. Ippolito [20]. Courtesy of National Aeronautics and Space Administration (NASA).

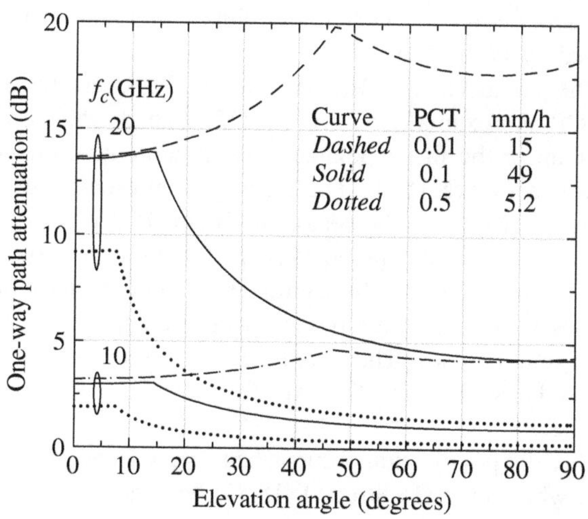

FIGURE 15.13 Examples of rain attenuation using the SAM model ($\phi = 45°N$, $H_0 = 0$, Region D_2).

Figure 15.13 shows some examples, based on the SAM model, of the rain attenuation as a function of the elevation angle for an Earth–satellite communication link operating at 10 and 20 GHz. For these plots, the geographic region is D_2 and the PCTs, 0.01, 0.1, and 0.5, correspond to respective rain rates of 49, 14.5, and 5.2 mm/h.

15.5 ELECTRIC FIELD WAVE POLARIZATION

Orthogonal antenna polarization provides for communication diversity without the need for bandwidth expansion. In general, the polarization of a wave in space is characterized as being elliptical in the plane normal to the direction propagation; linear polarization (LP) and circular polarization (CP) are special cases of elliptical polarization (EP) having widespread applications. Orthogonal linear polarized (LP) antennas transmit *vertical polarized* (VP) and *horizontal polarized* (HP) electric fields and CP antennas transmit either *right-hand circular* (RHC) or *left-hand circular* (LHC) electric fields depending upon parameter selections. The right- and left-hand rules apply to the direction of the circular rotation of the electric field, that is, when the *right-* or *left-hand thumb* points in the direction of the propagation, then the *curl of the fingers*, as viewed from the receiver, indicates either counterclockwise or clockwise rotation, respectively. Under ideal conditions, when the receiver antenna and electric field are polarized in the same manner the maximum signal power is received and the orthogonally polarized electric-field results in zero cross-polarization interference. Unfortunately, there are a number of practical issues that work against this ideal behavior, for example, in the case of LP it is difficult to keep the transmitter and receiver antennas exactly aligned

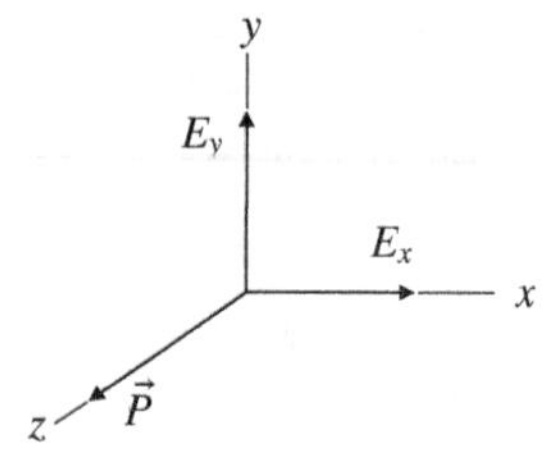

FIGURE 15.14 Wave electric fields along propagation path.

and the channel introduces polarization rotation along the propagation path. The most deleterious propagation affects are multipath interference, Faraday rotation, especially at HF and VHF frequencies, and atmospheric polarization resulting from various forms of water content, for example, rain, hail and, to a lesser extent, snow.

The electric fields of a wave propagating in the z-direction with Poynting vector* $\vec{P}$ are shown in Figure 15.14 with the time- and range-dependent orthogonal electric fields given by [30]

$$E_x = M_x \sin(\omega t - \beta z) \tag{15.86}$$

and

$$E_y = M_y \sin(\omega t - \beta z + \delta) \tag{15.87}$$

where M_x and M_y are the magnitudes of the electric fields, ω is angular carrier frequency of the wave, $\beta = 2\pi/\lambda$ is the phase-per-carrier frequency wavelength, and δ is the phase of the y-electric field relative to the x-electric field. The argument $\omega t - \beta z$ is derived from the delayed carrier $\omega(t - t_o)$ where $t_o = z/c$ is the range delay and c = velocity of light. The instantaneous electric field intensity, expressed as functions of the parameters t and z, is given by

$$E(t,z) = E_x(t,z) + jE_y(t,z) \tag{15.88}$$

Kraus [30] shows that (15.86) and (15.87), evaluated at $z = 0$, can be formulated into the expression of an ellipse given by

$$\left(\frac{1}{M_x^2 \sin^2(\delta)}\right) E_x^2 - \left(\frac{2\cos(\delta)}{M_x M_y \sin^2(\delta)}\right) E_x E_y + \left(\frac{1}{M_y^2 \sin^2(\delta)}\right) E_y^2 = 1 \tag{15.89}$$

The orientation of the ellipse relative to the x, y axes is shown in Figure 15.15.

*The Poynting vector represents the power density along the propagation path.

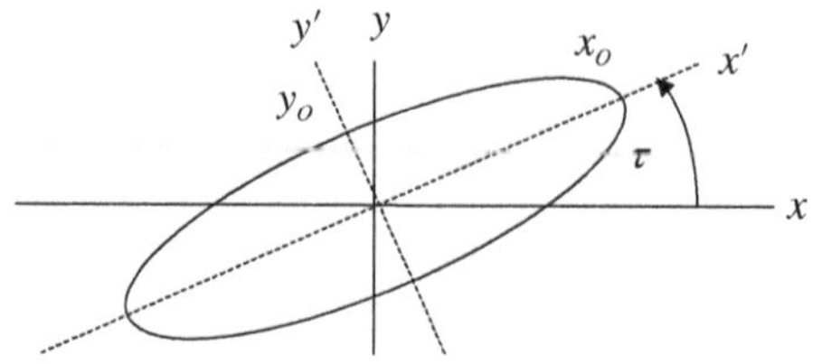

FIGURE 15.15 Polarization ellipse.

An important parameter in characterizing the polarization is the *axial ratio* (AR), defined as the ratio of the semimajor to semiminor axes of the ellipse. The polarization ellipse is completely defined by the *tilt angle* τ, the sense of the rotation of the electric field, and the AR. For the ellipse shown in Figure 15.15 the AR is

$$\mathrm{AR} = \frac{x_o}{y_o} \quad : \text{axial ratio} \tag{15.90}$$

and Kraus has characterized the tilt angle in terms of the parameters M_x, M_y, and δ as

$$\tau = \frac{1}{2}\tan^{-1}\left(\frac{2M_x M_y \cos(\delta)}{M_x^2 - M_y^2}\right) \quad : \text{tilt angle} \tag{15.91}$$

This expression for the tilt angle is relative to the x-axis and applies to either the major or minor axis of the ellipse; the last condition occurs if the ellipse in Figure 15.15 is rotated by $\pi/2$ radians.

The parameters M_x, M_y, and δ are used to characterize LP and CP. For example, referring to (15.86) and (15.87), when $\delta = 0$ and π the linear polarized waves are characterized as shown in Figure 15.16. The tilt angle represents the slope of the polarization. The vectors in Figure 15.16 are actually sinusoidal time and range varying functions as given by (15.86) and (15.87) with magnitudes M_x and M_y; these magnitudes vary with range based on the free space and other propagation losses. The *double tipped* vectors suggest the sinusoidal variations. The AR for VP and HP waves is infinite.

Circular polarization occurs when $\delta = \pm\pi/2$ and $M_y = M_x = M$ in which case (15.86) and (15.87) become

$$E_x = M \sin(\omega t - \beta z) \quad : \text{circular polarization} \tag{15.92}$$

and

$$E_y = \pm M \cos(\omega t - \beta z) \quad : \text{circular polarization} \tag{15.93}$$

and (15.89) reduces to the equation of a circle, that is,

$$E_x^2 + E_y^2 = M^2 \tag{15.94}$$

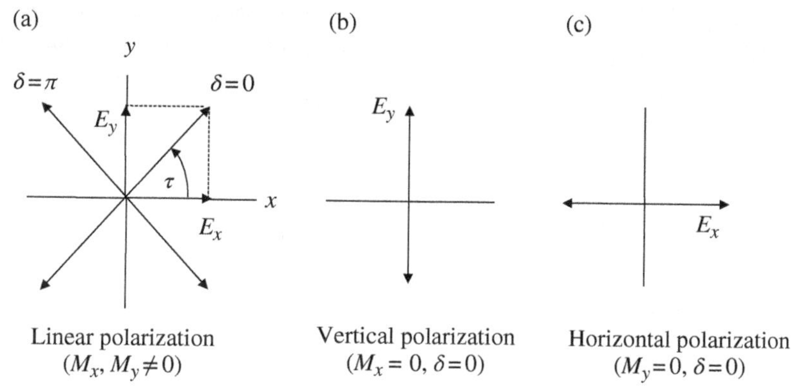

FIGURE 15.16 Linear polarizations viewed toward the Poynting vector.

The plus and minus sign in (15.93) represents the sense of the CP rotation when viewed toward the Poynting vector. For example, the phase of the composite sinusoidal *E*-field wave rotates clockwise as time advances when $\delta = \pi/2$ corresponding to the plus sign and counterclockwise when $\delta = -\pi/2$ corresponding to the negative sign. These definitions correspond to those of conventional positive and negative frequencies with respective positive and negative phase advances with increasing time. Applying the *right- and left-hand thumb* rules, described earlier, $\delta = \pi/2$ corresponds to *left-hand* CP (LHCP) and $\delta = -\pi/2$ corresponds to *right-hand* CP (RHCP). The AR for CP waves is unity.

15.5.1 Antenna Polarization Loss and Isolation

In this section, the analysis of the antenna power loss and the isolation of an orthogonally polarized interfering signal are examined based on the work of Ippolito [31]. In this analysis, the polarization state of the received wave at the input to the receiver antenna, that is, after propagation over the range distance z_{max} between the transmitter and receiver, is specified together with the receive antenna polarization state. The interaction of the received wave and the antenna is based on their polarization states* and the ARs as characterized by the polarization mismatch factor $0 \le m_p \le 1$. For arbitrary EP or LP states of the received wave and the receiver antenna, the polarization mismatch factor is given by

$$m_p(w,a) = \frac{1}{2} + \frac{4r_w r_a + (r_w^2 - 1)(r_a^2 - 1)\cos(2(\tau_a - \tau_w))}{2(r_w^2 + 1)(r_a^2 + 1)}$$

: mismatch factor

(15.95)

where r_x is the AR and $x = (w,a)$ refers to the received wave or antenna, respectively. The sign of the AR is negative for right-hand polarization and positive for left-hand polarization. For LP, the sign is always positive. The rotation of the polarization axis is $\pi/2$ radians for VP and zero radians for HP and right- and left-hand polarizations. The AR is expressed in decibels as

$$r_x(\text{dB}) = 20\log 10(|r_x|) \tag{15.96}$$

The power at the output of the receive antenna is evaluated as

$$P_r = PDA_e m_p(w,a) \tag{15.97}$$

where *PD* is the received power density and A_e is the effective area of the receive antenna normal to the Poynting vector. The antenna polarization power loss, in decibels relative to an ideally matched antenna, is expressed as

$$\text{Loss}(\text{dB}) = 10\log_{10}\left(\frac{1}{m_p(w,a)}\right) \tag{15.98}$$

The *polarization mismatch factor* for an elliptically polarized received wave with RHCP copolarization state and a vertically polarized receive antenna is shown in Figure 15.17 for wave ARs $r_w = -1$, -1.5, and -5 (0, 3.52, 40 dB). This example is not representative of an ideal situation, in that, a vertically polarized received wave† is required to match the vertically polarized receive antenna. Referring to Figure 15.17, it is seen that a RHCP received wave $r_w = -1$ and $|m_p(w,a)| = 0.5$ results in a power loss of 3 dB and, as r_w becomes larger, the minimum loss approaches 0 dB and corresponds to the vertical *y*-axis thus matching the vertically

*The polarization state of the receive antenna is the same as that of the transmit antenna but the time is reversed.

†With an ideal channel and with stable antenna orientations, the received wave polarization is identical to the transmit antenna polarization.

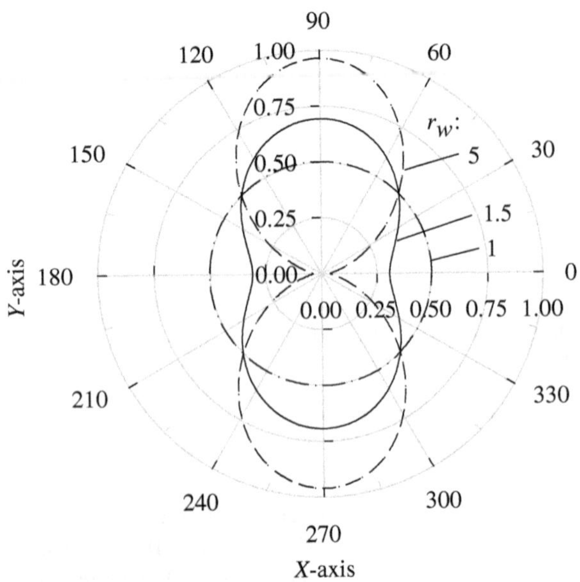

FIGURE 15.17 Polarization mismatch factor for elliptical wave with RHCP copolarization state and vertical polarized antenna.

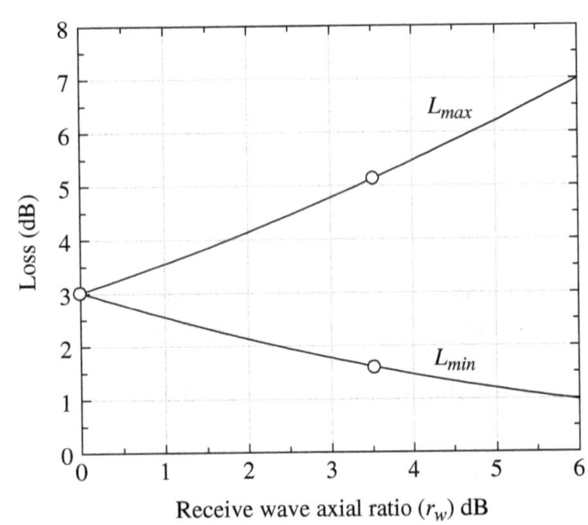

FIGURE 15.18 Polarization losses corresponding to Figure 15.17.

polarized receive antenna. An ideal circularly polarized antenna has an AR of $r_a = 1$ (0 dB) and a well-designed antenna will have an AR of $r_a = 1.19$ (1.5 dB) and an ideal LP antenna has an AR of $r_a = \infty$ (∞ dB) with practical values ranging from $r_a = 17.8$ to 31.6 (25–30 dB).

The *solid* lines in Figure 15.18 show the maximum and minimum polarization losses as a function of the wave AR; the losses corresponding to the polarization mismatch factors in Figure 15.17 shown as the *circled* data points. These losses are orthogonal to each other with the minimum loss corresponding to the VP or the copolarized state and the maximum loss corresponding to the cross-polarized state.

The antenna isolation is defined as the ratio of the power at the antenna copolarized output a_c and the power at the cross-polarized output a_x. Given the total received signal power density $PD_{\bar{w}}$ the isolation is expressed as

$$I_{isol} = \frac{A_e PD_{\bar{w}} m_p(\bar{w}, a_c)}{A_e PD_{\bar{w}} m_p(\bar{w}, a_x)} \tag{15.99}$$

Because any wave can be resolved into two orthogonal polarized states, denoted as w and w_o, (15.99) is rewritten as

$$\begin{aligned} I_{isol} &= \frac{A_e\left(PD_w m_p(w, a_c) + PD_{w_o} m_p(w_o, a_c)\right)}{A_e\left(PD_w m_p(w, a_x) + PD_{w_o} m_p(w_o, a_x)\right)} \\ &= \frac{xpd\; m_p(w, a_c) + m_p(w_o, a_c)}{xpd\; m_p(w, a_x) + m_p(w_o, a_x)} \end{aligned} \tag{15.100}$$

where the second equality is obtained in a straightforward way with the constant *xpd* defined as the cross-polarization discrimination, expressed as

$$\begin{aligned} xpd &\triangleq \frac{PD_w}{PD_{w_o}} = \frac{PD_{\bar{w}} m_p(\bar{w}, w)}{PD_{\bar{w}} m_p(\bar{w}, w_o)} \\ &= \frac{m_p(\bar{w}, w)}{m_p(\bar{w}, w_o)} \end{aligned} \tag{15.101}$$

Referring to (15.100) an ideal antenna occurs when all of the antenna co- and cross-polarized outputs are equal to the respective co- and cross-polarized inputs, such that, $m_p(w, a_c) = m_p(w_o, a_x) = 1$; this also leads to the conditions $m_p(w, a_x) = m_p(w_o, a_c) = 0$ with the result that the isolation for an ideal antenna is

$$I_{isol} = xpd \quad \text{: ideal antenna} \tag{15.102}$$

The antenna mismatch factor, cross-polarization discrimination, and isolation are examined in the following case study for an elliptical received wave with copolarization state LHCP and a receive antenna with a LHCP polarization state.

15.5.2 Case Study: Polarization Characteristics for a LHCP Antenna

This case study examines the antenna mismatch factor, the corresponding polarization loss, the cross-polarization discrimination, and the isolation for a LHCP antenna with an elliptical received wave with a copolarization state that is matched to the receive antenna. The antenna is not ideal so the isolation asymptotically approaches the cross-polarization discrimination function *xfd* for low values of *xfd*.

The antenna mismatch factor is evaluated using (15.95) for a specified r_a and parametric values of r_w with $\tau_a = 0$

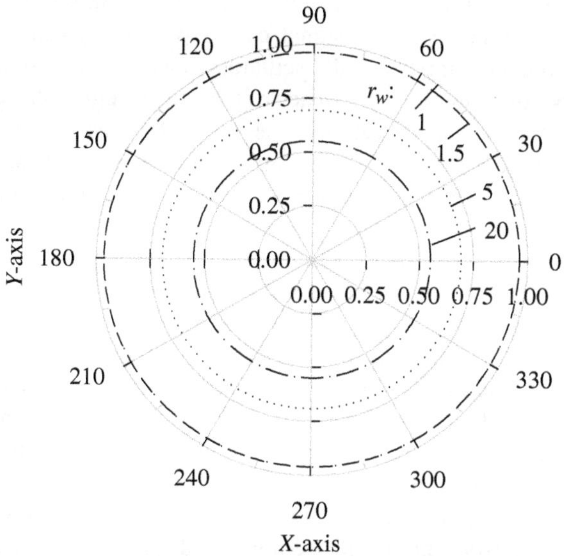

FIGURE 15.19 Antenna mismatch factor for LHCP antenna ($r_a = 1$) and elliptical received wave with LHCP copolarization.

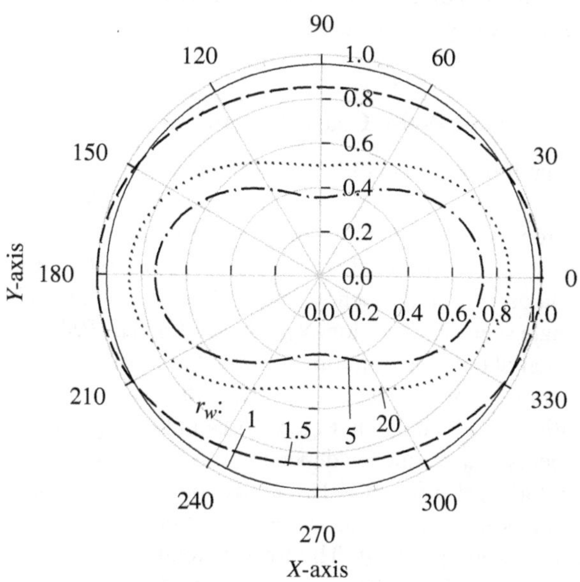

FIGURE 15.20 Antenna mismatch factor for elliptical antenna ($r_a = 1.5$) and elliptical received wave with LHCP copolarization.

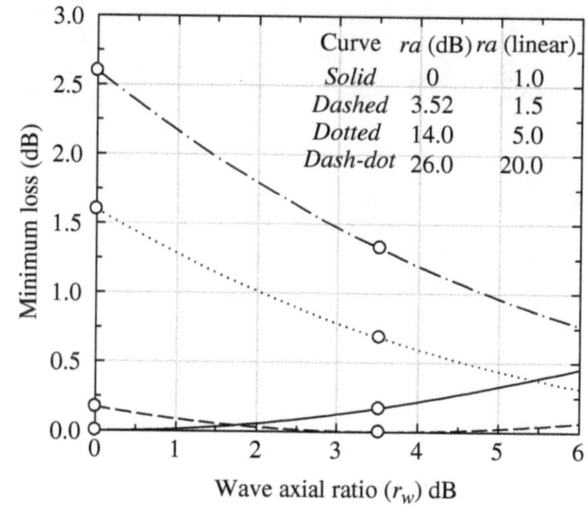

FIGURE 15.21 Minimum antenna mismatch factor losses.

radians and variable angular displacements $\tau_w = \theta^*: 0 \le \theta \le 2\pi$ radians. The results are plotted in Figures 15.19 and 15.20 corresponding to $r_a = 1$ and 1.5, respectively. With $r_a = 1$ the antenna mismatch factor appears as *circles* with radius ≤ 1. The case having unit radius corresponds the received wave being matched to the antenna with $r_w = r_a = 1$; the other cases result in a mismatched antenna and lower values of m_p corresponding to higher polarization losses. The results in Figure 15.20 represent a nonideal antenna with $r_a = 1.5$ and, although the mismatch factor for $r_w = 1$ is a circle it does not have unit magnitude and results in a polarization loss. The case with $r_w = r_a = 1.5$ is matched in the horizontal direction with increasing loss in the vertical direction. As the parameter r_w continues to increase the antenna mismatch factor decreases along both axes resulting in increasing polarization loss. The salient point is that the antenna mismatch factor results in zero loss when the received wave AR is equal to the antenna AR. The losses are shown in Figure 15.21 as a function of r_w, expressed in decibels, for $r_a = 1$ and 1.5. The *circled* data points represent the losses for the corresponding r_a and r_w conditions in the range of the abscissa.

Evaluation of the isolation of an ideal LHCP antenna involves evaluating the cross-polarization discrimination *xpd* as expressed in (15.101) as a function of the wave axial r_w. The numerator of (15.101) is the copolarized antenna mismatch factor corresponding to wave AR $r_a = 1$ and the denominator corresponds to the cross-polarized or orthogonal antenna mismatch factor with $r_a = -1$. In these evaluations, the antenna mismatch factors are independent of τ_a and τ_w and the resulting ideal antenna isolation is plotted in Figure 15.22 with the $xpd(\text{dB}) = 10\log_{10}(xpd)$ and the abscissa plotted in terms of the common logarithm as $20\log_{10}(r_w)$. From this result it is seen that the isolation increases as the wave AR approaches 0 dB or $r_w = 1$. The value of *xpd*, plotted in Figure 15.22, is used to evaluate the isolation of a nonideal antenna using (15.100). In evaluating (15.100), the choice of the antenna copolarized and the cross-polarized ARs r_c and r_x are based on the antenna design; the nonideal antenna isolation plotted in Figure 15.23 corresponds to $r_c = 1.122$ (1 dB) and various values of r_x in decibels. The selected values of the nonideal

*The parameter θ is used with the x,y axes for plotting $m_p(w,a)$ in polar form. When r_a or $r_w = 1$, $m_p(w,a)$ is a constant independent of the τ_a and τ_w and θ is used to generate a circular polar plot.

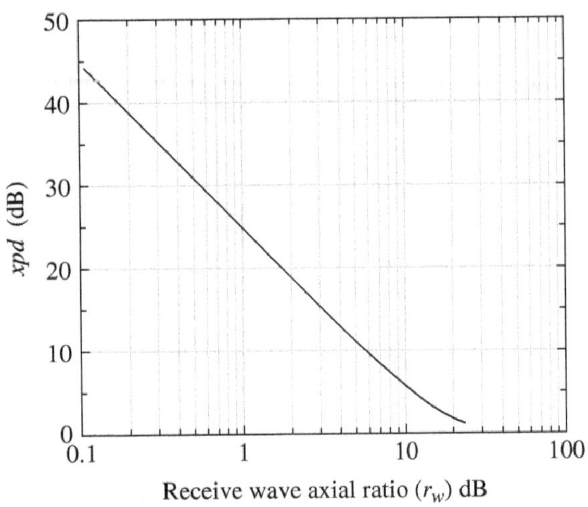

FIGURE 15.22 LHCP cross-polarization discrimination.

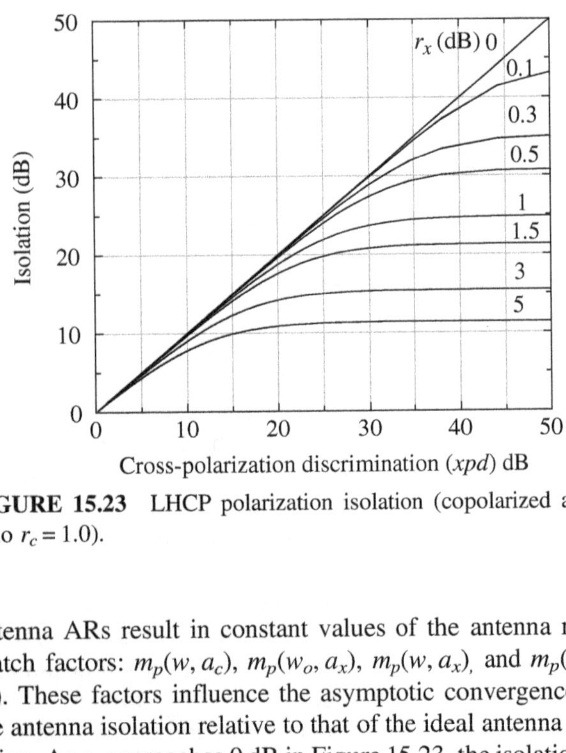

FIGURE 15.23 LHCP polarization isolation (copolarized axial ratio $r_c = 1.0$).

antenna ARs result in constant values of the antenna mismatch factors: $m_p(w, a_c)$, $m_p(w_o, a_x)$, $m_p(w, a_x)$, and $m_p(w_o, a_c)$. These factors influence the asymptotic convergence of the antenna isolation relative to that of the ideal antenna isolation. As r_x approaches 0 dB in Figure 15.23, the isolation is ultimately limited by the selection of r_c. The value of $r_c = 1.122$ results in a copolarization loss of 0.014 dB and a cross-polarization loss of 48.8 dB.

15.6 PHASE-NOISE LOSS

Phase noise results from intrinsic noise in the transceiver and modem signal sources typically from oscillators and frequency synthesizers used for frequency translation. The phase noise is characterized in terms of the phase-noise power spectral density (PSD) in dBc/Hz where the decibel level is relative to the signal carrier power. Typically the specification applies to all oscillators in a cascade of subsystems; for example, in a cascade of two subsystems with equal phase-noise specifications the overall phase noise is increased by 3 dBc/Hz. For higher order modulations, like multiphase shift keying (MPSK) and quadrature amplitude modulation (QAM), the oscillator phase noise is a dominant source of performance loss over the AWGN channel and results in an irreducible error probability with increasing E_b/N_o.

There is a wealth of information [32–38] on the theory, design, and applications of oscillators with an emphasis on understanding and minimizing the impact of phase noise on communication systems. The phase-noise variance is the principal performance measure and is defined as the integral of the phase-noise PSD. In the following sections, the characteristics of the phase-noise PSD are reviewed and, based on an acceptable phase-noise variance specification, the subsystem phase-noise density is established. In Section 15.6.2 the phase-noise variance is determined from a specification of the spectral density and the resulting performance loss is given for the MPSK-modulated waveform.

15.6.1 Phase-Noise Characterization

Oscillator phase noise [37, 39, 40] is characterized in terms of a log–log plot of the relative phase-noise[*] PSD ($S(f)$) as a function of the frequency deviation f from the oscillator frequency f_o. The power density $N_{o\phi}$ is defined relative to the oscillator carrier frequency power C and plotted with the ordinate specified as $10\log(N_{o\phi}/C)$ dBc/Hz. The PSD is symmetrical about f_o and, normally, only the positive frequency portion is shown. A generic plot of a phase-noise PSD specification $S_s(f)$ is shown in Figure 15.24 indicating distinct line segments with frequency dependence k_m/f^m corresponding to amplitude roll-offs of $10\,m$ in decibels per decade. The parameter k_m is a scaling factor that determines the dBc/Hz level for each segment. The lowest frequency contained in $S_s(f)$ is typically 1–10 Hz because accurate measurements below 1 Hz are difficult to obtain and the phase noise in these ranges is generally removed in the demodulator by the phaselock loop (PLL) filter. The upper frequency of interest in the phase-noise PSD is determined by the intermediate frequency (IF) filter bandwidth and ultimately by the symbol matched filter bandwidth.

Barnes [33] characterizes a low frequency segment for $m = 4$; however, this is seldom considered in applications involving a demodulator PLL.[†] The phase noise corresponding

[*]In the following discussions, a distinction is made between the phase-noise PSD $S(f)$ and the spectral density specification denoted as $S_s(f)$.

[†]Gardner [38] indicates that the $m = 4$ noise term appears to be associated with precision cesium clocks.

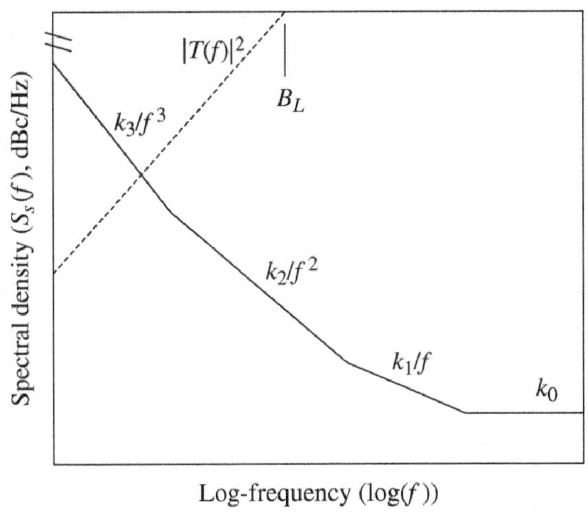

FIGURE 15.24 Generic phase-noise spectral density.

to $m = 3$ is referred to as *flicker* noise and is related to abrupt phase changes associated with the oscillators inherent feedback circuit as it attempts to control the phase of the oscillators carrier frequency. The phase noise associated with $m = 2$ results from white and flicker noise sources within the oscillator. The phase noise associated with $m = 1$ is high frequency flicker noise and influences the $m = 3$ and $m = 2$ phase-noise segments. The phase noise associated with $m = 0$ results from the receiver white noise density N_o and is the underlying noise term associated with the demodulators E_b/N_o ratio.

Typically, a coherent communication demodulator applies the received carrier-modulated waveform to a PLL for phase and frequency tracking [41] that also tracks and removes the lower frequency phase-noise terms that are within the loop bandwidth B_L. The PLL closed-loop frequency response is characterized by the low-pass function $H(f)$ and the output phase noise results from the equivalent high-pass response given by*

$$T(f) = 1 - H(f) \tag{15.103}$$

In the following analysis, the closed-loop response for a second-order loop, with damping factor $\varsigma = 1/\sqrt{2}$, is used resulting in

$$T(f) = \frac{f^2}{(f^2 - f_n^2) - j\sqrt{2} f_n f} \tag{15.104}$$

Referring to (10.38) the natural resonant frequency of the closed-loop response is related to the loop bandwidth B_L as

$$\begin{aligned} f_n &= \frac{\omega_n}{2\pi} = \frac{B_L}{\pi(\varsigma + 1/(4\varsigma))} \quad &: \text{Hz} \\ &= \frac{0.94281 B_L}{\pi} \quad &: \varsigma = \frac{1}{\sqrt{2}} \end{aligned} \tag{15.105}$$

*$T(s)$ is defined as the phase-error transfer function in (10.7).

The response $|T(f)|^2$ is shown as the *dotted* curve in Figure 15.24 with a band-reject frequency of $B_L \ll R_s$ Hz. Therefore, for low symbol rate modems the phase noise possesses a more severe performance issue. Typical values of the loop bandwidth are $B_L = R_s/10$ and $R_s/100$ for binary phase shift keying (BPSK) and quadrature phase shift keying (QPSK) modulations, respectively.

The variance of the phase noise is computed as

$$\begin{aligned} \sigma_{N\phi}^2 &= \int_0^\infty S(f)df \\ &= \sigma_3^2 + \sigma_2^2 + \sigma_1^2 + \sigma_0^2 \end{aligned} \tag{15.106}$$

where the individual noise variances are computed as

$$\sigma_m^2 = k_m \int_0^\infty \frac{|T(f)|^2}{f_m} df \tag{15.107}$$

The integral in (15.107) is evaluated for $m = 3$ and 2 using the integral formula given by Gradshteyn and Ryzhik [42] and using (15.105), with f_n evaluated in terms of the symbol rate and the time-bandwidth product $\rho_L = B_L T = B_L/R_s$, the noise variances are evaluated as

$$\sigma_3^2 = \frac{0.28125\pi^2 k_3}{\rho_L^2 R_s^2} \tag{15.108}$$

$$\sigma_2^2 = \frac{0.375\pi^2 k_2}{\rho_L R_s} \tag{15.109}$$

The integrals involving $m = 1$ and 0 with an infinite integration limit result in infinite variances so the integration must be performed over a finite bandwidth $B \geq R_s/2$ Hz where the equality corresponds to the Nyquist bandwidth. In this evaluation the integral in (15.107) is defined as†

$$I_m\left(\frac{B}{f_n}\right) = \int_0^B \frac{|T(f)|^2}{f_m} df \tag{15.110}$$

with the integration limit $B = R_s$ corresponding to the bandwidth of an integrate-and-dump (I&D) symbol detection filter. Therefore, for $m = 1$ and 0, (15.110) is evaluated using Mathcad‡ symbolic processing and the results are expressed as

†The designation $I_m(B/f_n)$ suggests that the integration in (15.110) is over the finite normalized bandwidth B/f_n.

‡Mathcad and Mathsoft are registered trademarks TM of Mathsoft Engineering & Education, Inc., January 2006.

$$I_1\left(\frac{B}{f_n}\right)=\frac{1}{4}\ln\left(\left(\frac{B}{f_n}\right)^4+1\right) \tag{15.111}$$

and

$$I_0\left(\frac{B}{f_n}\right)=\left(\frac{B}{f_n}\right)-\frac{\sqrt{2}}{8}\ln\left(\frac{(B/f_n)+\sqrt{2}(B/f_n)+1}{(B/f_n)-\sqrt{2}(B/f_n)+1}\right)-\frac{\sqrt{2}}{4}\tan^{-1}\left(\frac{\sqrt{2}(B/f_n)}{1+(B/f_n)^2}\right) \tag{15.112}$$

Using (15.105), the dependence on the normalized bandwidth B/R_s for $\varsigma=1/\sqrt{2}$ is obtained by substituting

$$\frac{B}{f_n}=\frac{1.06066\pi}{\rho_L}\left(\frac{B}{R_s}\right)\quad :\varsigma=\frac{1}{\sqrt{2}} \tag{15.113}$$

into Equations (15.111) and (15.112). The results are plotted in Figure 15.25 as a function of B/R_s for $\varsigma=1/\sqrt{2}$ and $\rho_L=0.1, 0.01$. The numerical solutions to (15.110) for $B/R_s=0.5$, 1, 4, and 8 are listed in Table 15.5.

Using these results, the phase-noise variances for $m=1$ and 0 are expressed as

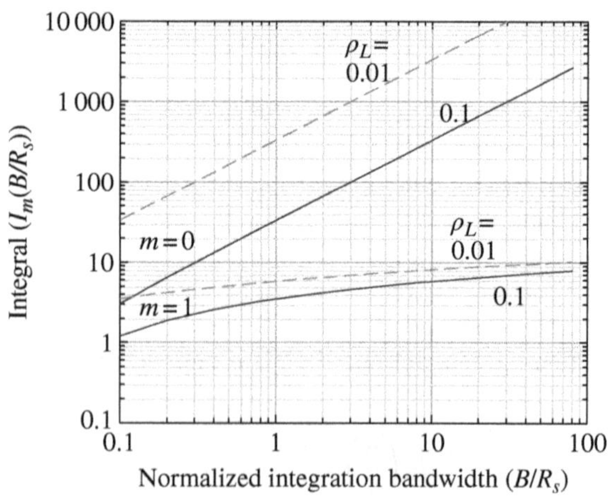

FIGURE 15.25 Integration results for $m=1$ and 0.

TABLE 15.5 Selected Integration Values $I_m(B/f_n)$ for $m=1$ and 0

	$\rho_L=0.1$		$\rho_L=0.01$	
B/R_s	$m=1$	$m=0$	$m=1$	$m=0$
0.5	2.8125	16.5915	5.1151	166.508
1	3.5056	33.2728	5.8082	333.025
4	4.8919	133.204	7.1945	1332.11
8	5.5851	266.418	7.8877	2664.22

$$\sigma_1^2=k_1I_1\left(\frac{B}{R_s}\right) \tag{15.114}$$

and

$$\sigma_0^2=k_0I_0\left(\frac{B}{R_s}\right) \tag{15.115}$$

The unknown parameters in these relationships are the scale factors k_m; however, with knowledge of the breakpoints and the additive constant phase-noise PSD, the piecewise linear PSD shown in Figure 15.24 can be constructed. In the following section, these results are applied to a system specification of the phase noise and the results are used to evaluate the impact on the bit-error probability of the communication link. The resulting performance loss is then used in establishing an overall link budget which is essentially the purpose of this chapter.

15.6.2 Phase-Noise Evaluation Using System Specifications

The phase noise is frequently specified as part of an overall system requirements specification as in Figure 15.26.* The specification discussed in this section is intended to represent the demodulator heterodyning to baseband and is 10 dB lower than that specified for the receive terminal which has multiple heterodyning stages. Because the received and demodulator subsystems are typically developed by different contractors, separate subsystem phase-noise specifications are required to provide for independent subsystem testing. The demodulator phase noise is intended not to impact the more sensitive phase noise of the receiver oscillators and frequency conversion stages.

These specifications apply to the $1/f^3$ and $1/f$ frequency dependencies so the composite linear spectral density is expressed as

$$S(f)=\frac{k_3}{f^3}+\frac{k_1}{f}+k_0 \tag{15.116}$$

The constants k_m, required for the evaluation of (15.116), are determined from the specifications as follows. Considering the specification shown in Figure 15.26 and repeated in Figure 15.27 as the *solid* lines; each linear segment

$$\ell=\frac{k_m}{f^m} \tag{15.117}$$

is plotted in terms of the common logarithm

*The phase-noise specifications characterize the single-sideband phase-noise PSD relative to the carrier power and have units of dBc/Hz.

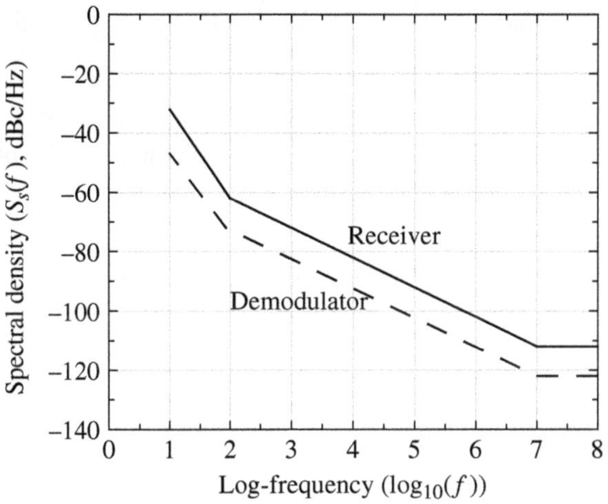

FIGURE 15.26 Phase-noise spectral density specifications. Defense Information Systems Agency (DISA) [43]. Courtesy of U.S.A. Department of Defense (DOD).

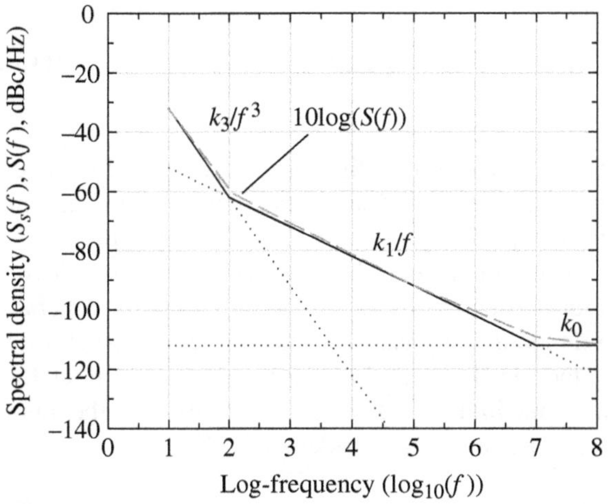

FIGURE 15.27 Phase-noise spectral density characteristics for receiver subsystem.

$$\begin{aligned} L_m &= 10\log(\ell) = 10\log(k_m) - 10m\log(f) \\ &= K_m - 10mF \end{aligned} \tag{15.118}$$

where $K_m = 10\log(k_m)$ and $F = \log(f)$ are used as the abscissa in the spectral density plots. Therefore, using these results with $m = 3$, k_3 is evaluated by solving for K_3 in (15.118) under the condition $F = 1$. The corresponding values of L_3 and K_3 are −32 dBc/Hz and −2 dBc-Hz^2; in general K_m has units of dBc-Hz^{m-1}. In a similar manner K_1 is found to be −42 dBc and $K_0 = -112$ dBc/Hz corresponding to the constant white noise spectral density. The corresponding values of k_m are as follows: $k_3 = 6.3095734(-1)$, $k_1 = 6.3095734$

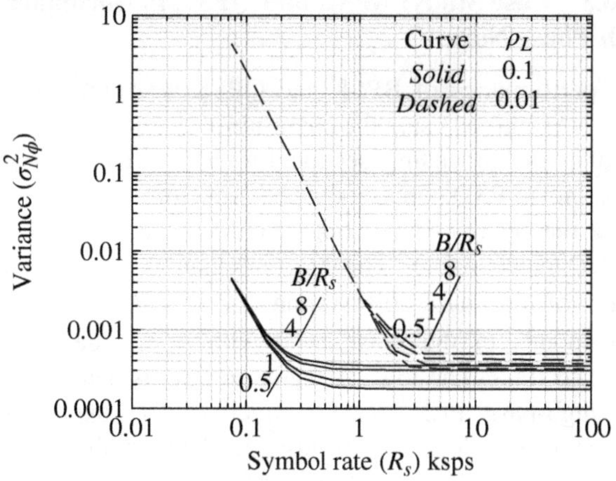

FIGURE 15.28 Total phase-noise variance for receiver subsystem.

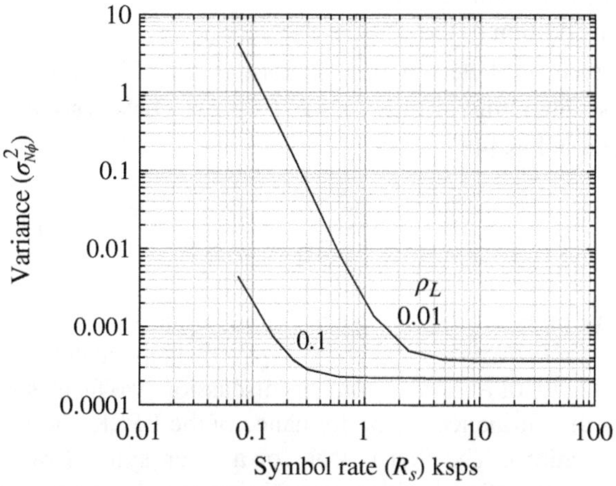

FIGURE 15.29 Total phase-noise variance at I&D MF output for receiver subsystem ($B/R_s = 1$).

(-5), and $k_0 = 6.3095734(-12)$. The composite phase-noise PSD, shown in Figure 15.27 as the *dashed* curve, is a plot of (15.116) using the computed values of k_m. The *dotted* curves represent the computed line segments based on K_m plotted over the range of abscissa values $1 \le F \le 8$ with portions of these line segments corresponding to those of the specification.

Referring to (15.116) and evaluating the phase-noise variances σ_m^2 in (15.108), (15.114), and (15.115) with the values of k_m computed earlier for the receiver specification, the total untracked phase-noise variance $\sigma^2_{N\phi}$ is evaluated using (15.106) with $\sigma_2^2 = 0$. The resulting untracked phase noise is plotted in Figure 15.28 with $\rho_L = 0.1$ and 0.01 for the indicated normalized bandwidth ratios B/R_s. Figure 15.29 shows the untracked phase-noise variance when $B = R_s$ corresponding to the bandwidth of the I&D symbol detection filter.

15.6.3 Case Study: BPSK and QPSK Performance with Phase Noise

The performance of BPSK and QPSK waveform modulations is examined using the phase-noise characteristics developed in the preceding section for the receiver subsystem phase-noise PSD. The modulated waveforms are generated using the *rect(t/T)* weighting functions and detected using an I&D matched filter. The upper integration limit in the evaluation of the phase-noise variance $\sigma^2_{N\phi}$ is $B = R_s$, where R_s corresponds to the noise bandwidth of the demodulator matched filter. The lower integration limit is determined by the high-pass PLL transfer function *T(f)* expressed in (15.104). The low-pass band-reject bandwidth of *T(f)* is dependent on the loop time-bandwidth product $\rho_L = B_L T$ where B_L is the closed-loop noise bandwidth. For BPSK and QPSK modulation, typical values* of ρ_L are 0.1 and 0.01, respectively; these values are used in the following performance evaluations.

In addition to the phase-noise variance $\sigma^2_{N\phi}$ resulting from the receiver heterodyning operations, the PLL introduces phase jitter that is characterized by the phase variance σ^2_ϕ expressed as [44]

$$\sigma^2_\phi = \frac{1}{2\gamma_L} = \frac{\rho_L}{2\gamma_b} \quad (15.119)$$

where γ_L is the signal-to-noise ratio in the PLL bandwidth and $\gamma_b = E_b/N_o$ is the signal-to-noise ratio in the matched filter bandwidth of R_s hertz. From this discussion two factors are in play that influence the performance of the BPSK and QPSK modulations. The first is that, for a given symbol rate, the lower integration limit in determining the phase-noise variance is decreased by ρ_L. For a given bit rate, this is exacerbated with QPSK modulation since $R_s = R_b/2$. These factors become less significant as the data rate is increased and the phase noise is eventually influenced solely by the oscillator white noise density corresponding to k_0. The second factor influencing the performance is that the PLL phase noise is decreased by ρ_L with the advantage going to QPSK. In consideration of these issues, the total phase-noise power measured in the matched filter bandwidth is given by

$$\sigma^2_\varphi = \sigma^2_{N\phi} + \sigma^2_\phi \quad (15.120)$$

Because the oscillator phase is influenced by several phase-noise sources, as indicated by terms giving rise to the k_m/f^n spectral density response, the phase-noise random variable φ is generally characterized by the Gaussian distribution

$$p(\varphi) \cong \frac{1}{\sqrt{2\pi\,\sigma^2_\varphi}} e^{-\varphi^2/(2\sigma^2_\varphi)} \quad : |\varphi| \le \pi \quad (15.121)$$

This approximation and the limits require that $\sigma_\varphi \ll 1$. The bit-error probability performance conditioned on the phase error φ for BPSK and QPSK is expressed, respectively, as

$$P_{be}(\gamma_b;\varphi) = \frac{1}{2} erfc\left(\sqrt{\gamma_b}\cos(\varphi)\right) \quad : \text{BPSK} \quad (15.122)$$

and

$$P_{be}(\gamma_b;\varphi) = \frac{1}{4} erfc\left(\sqrt{\gamma_b}(\cos(\varphi) - \sin(\varphi))\right) + \frac{1}{4} erfc\left(\sqrt{\gamma_b}(\cos(\varphi) + \sin(\varphi))\right) \quad : \text{QPSK} \quad (15.123)$$

Using (15.121) with either (15.122) or (15.123) the resulting bit-error probability is expressed as

$$P_{be}(\gamma_b) \cong \int_{-\pi}^{\pi} P_{be}(\gamma_b,\varphi)p(\varphi)d\varphi \quad (15.124)$$

The bit-error probability expressed in (15.124) is evaluated using a 96-term Gauss-quadrature integration and the results are depicted in Figure 15.30. The robustness of BPSK modulation is evidenced by a loss of less than 0.1 dB even for data rates as low as 150 bps. For the 1024 kbps data rate, the QPSK modulation has a maximum loss of about 0.1 dB; however, the sensitivity to the phase noise is evident at 150 bps with a maximum loss of about 0.3 dB for the range of signal-to-noise ratios considered.

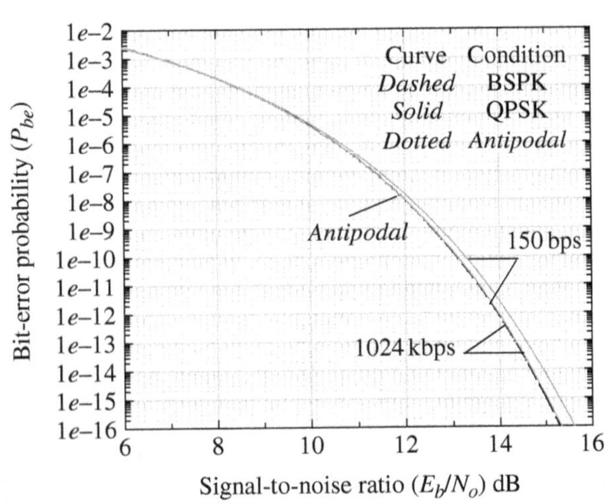

FIGURE 15.30 Performance of BPSK and QPSK modulations using the receiver subsystem phase-noise specification.

*The selection of ρ_L is largely dependent on the dynamics of the communications platforms and the channel.

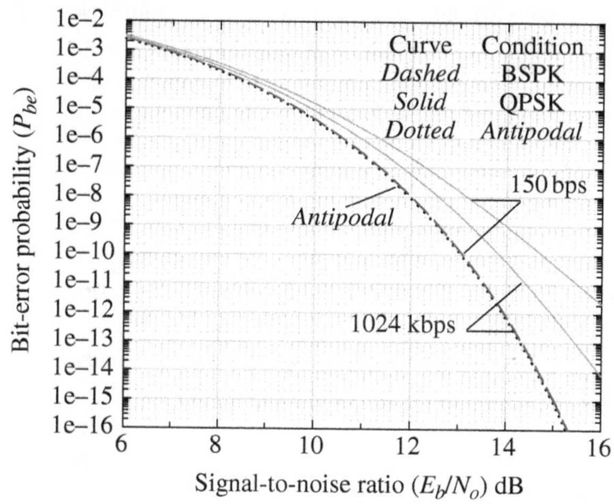

FIGURE 15.31 Performance of BPSK and QPSK modulations using a cascade of three subsystems corresponding to the receiver phase-noise specification.

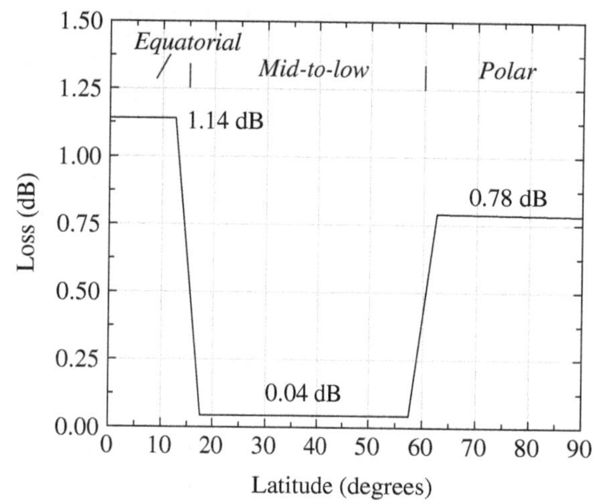

FIGURE 15.32 Scintillation loss in natural environment (90% confidence at 500 MHz).

TABLE 15.6 Absorption Loss Factors L_f (dB) in Natural Environment

	Latitude		
Confidence (%)	Equatorial	Mid to Low	Polar
50[a]	5.75e4	9.50e3	5.100e4
90	2.85e5	1.08e4	1.957e5
95	3.50e5	1.11e4	2.375e5
99	4.72e5	1.18e4	3.143e5

[a]Mean or average electron densities.

Figure 15.31 depicts the performance sensitivity when the phase noise results from the cascade of three subsystems: a transmitter, satellite repeater, and receiver, each with a phase-noise specification corresponding to receiver subsystem. In effect, the phase noise is increased by 4.77 dB from that used in Figure 15.30. In this case, the BPSK performance for the 1024 kbps data rate is essentially unchanged; however, the QPSK performance is degraded by about 0.2 dB at $P_{be} = 10^{-5}$ for 1024 kbps (512 ksps) and by about 0.5 dB for 150 bps (75 sps).

This procedure can be applied to higher order MPSK modulations ($M > 4$) with anticipated increases in the performance loss. In this regard, Baker [45] has analyzed the performance of 256-ary and 1024-ary QAM with phase noise in terms of the standard deviation σ_φ of the phase and, for $\sigma_\varphi = 0.3°$, the losses at $P_{be} = 10^{-8}$ are about 0.6 and 5.5 dB, respectively.

15.7 SCINTILLATION LOSS

The losses due to signal scintillation are examined in detail in Chapter 20 and the results in a natural environment are summarized in Table 15.6 in terms of the loss factor L_f. The loss factor allows for determining the loss at any carrier frequency $f > 2$ MHz* according to

$$L_a(f_{\text{MHz}}) = \frac{L_f}{f_{\text{MHz}}^2} \quad : \text{dB} \tag{15.125}$$

where f_{MHz} is carrier frequency expressed in MHz. The dependence of the loss on latitude results from the different electron density profiles in the three regions corresponding to equatorial (±15°), mid to low (15°–60°), and polar (>60°) latitudes. The loss factors correspond to the worst-case mean electron densities and variations under turbulent conditions in the natural environment. For example, the loss corresponding to a 90% confidence level at 500 MHz in the equatorial region is determined as $L_a(500) = 2.85\text{e}5/(500)^2 = 1.14$ dB. This example is shown in Figure 15.32 where the transitions between the latitude regions are plotted as 5° centered on the transitions at 15° and 60°. Figure 15.33 shows the scintillation losses at 250 MHz for the indicated confidence levels.

15.8 MULTIPATH LOSS

The maximum and average single-reflection multipath loss is defined in terms of the *multipath factor* and the results are plotted in Figure 15.34 for a low Earth orbit (LEO) 200 km altitude satellite receiver and a ground station with antenna height of 30 m. The ground station and satellite use uniformly weighted antennas; the satellite antenna gain is 10 dB and the transmitter antenna gain is varied as indicated in

*This condition is based upon a maximum electron collision frequency of $v = 1.25\text{e}6$ rad/s and the requirement that $\omega \gg v$.

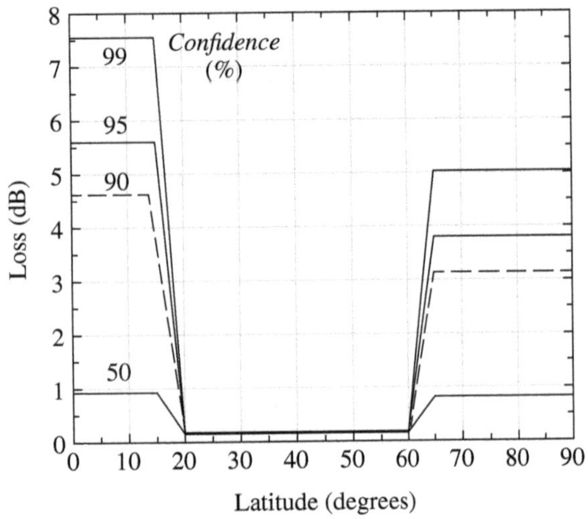

FIGURE 15.33 Scintillation loss in natural environment (250 MHz).

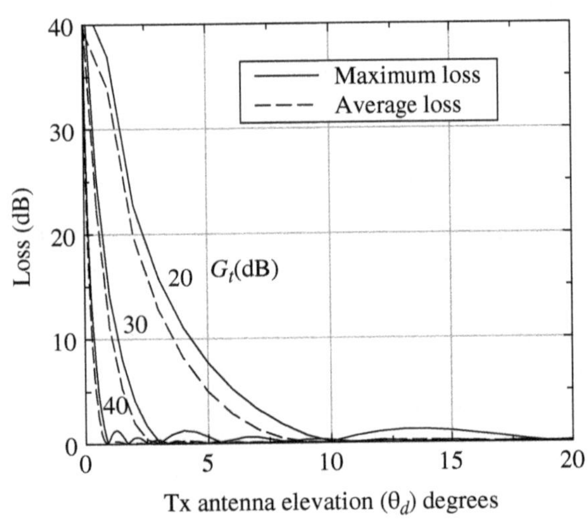

FIGURE 15.34 Single-reflection multipath loss.

Figure 15.34. The description of satellite link encounter with example multipath losses is given in Section 19.4. The impact of the antenna gains is evident, in that, multipath losses greater than 1 dB occur with transmit antenna elevation angles less than about 14°, 4°, and 1.5° for respective antenna gains of 20, 30, and 40 dB. These gains correspond to 3-dB transmitter antenna beam widths of 16.2°, 5.2°, and 1.64°. It is assumed that the ground station and satellite beams are ideally tracking each other along the line of sight (LOS) or direct signal path. A word of caution is in order regarding the inclusion of the multipath losses shown in Figure 15.34 directly into the link budget. The reason is that waveform designs and signal processing techniques can effectively mitigate the multipath losses. These techniques include forward error control (FEC) coding and interleaving, optimal combining techniques with data repetition, data equalization, and adaptive antenna null steering.

15.9 INTERFACE MISMATCH LOSS

The losses associated with the termination of connecting cables and devices within a transmitter or receiver are evaluated in terms of the voltage standing wave ratio (VSWR) or the voltage reflection coefficient (ρ_r) as measured at the interface between the components. These losses must be added to the measured or specified device losses that are normally determined under ideal termination conditions.

The VSWR is defined as [46] the ratio of the maximum-to-minimum voltage of the *standing wave* expressed as

$$\mathrm{VSWR} \triangleq \frac{V_{max}}{V_{min}} \tag{15.126}$$

The *standing wave ratio* (SWR) is simply the VSWR expressed in decibels, that is,

$$\mathrm{SWR} = 20\log(\mathrm{VSWR}) \tag{15.127}$$

The voltage reflection coefficient is defined as [46] the ratio of the reflected-to-incident voltages across the termination or load impedance Z_L and is expressed as

$$\rho_r \triangleq \frac{V_r}{V_i} = \frac{Z_L - Z_o}{Z_L + Z_o} \tag{15.128}$$

where Z_o is the characteristic impedance of the cable or the source impedance of the device. In general, ρ_r is a complex quantity, such that, $0 \le |\rho_r| \le 1$.

The relationships between VSWR and ρ_r are:

$$\mathrm{VSWR} = \frac{1 + |\rho_r|}{1 - |\rho_r|} \tag{15.129}$$

and

$$|\rho_r| = \frac{\mathrm{VSWR} - 1}{\mathrm{VSWR} + 1} \tag{15.130}$$

The return loss is defined as

$$L_r \triangleq \frac{1}{|\rho_r|^2} \tag{15.131}$$

Defining the incident power as $P_i = |V_i|^2$ and the reflected power as $P_r = |V_r|^2$ the power delivered to the load is $P_L = P_i - P_r$ and, using these relationships together with (15.128) and (15.131), the loss at the termination is computed as

$$L_t = \frac{P_i}{P_L} = \frac{1}{1-|\rho_r|^2} = \frac{L_r}{L_r - 1} \qquad (15.132)$$

The loss L_t is typically included in the device losses in the computation of the receiver noise figure discussed in Section 15.2.1.

15.10 MISCELLANEOUS SYSTEM LOSSES

If the component loss of a device at a frequency f_1 is known to be L_1 then the loss at another frequency can often be approximated simply by frequency scaling. For example, the loss at frequency f_2 is approximately

$$L_2 \cong \left(\frac{f_2}{f_1}\right)^2 L_1 \qquad (15.133)$$

15.10.1 Antenna Shaping Loss

Antenna shaping loss is based on aperture weighting function and is included in antenna gain.

15.10.2 Antenna Scallop Loss

Antenna scallop loss is associated with antenna beam loss during spatial antenna acquisition. Site loss is the antenna gain loss at a site location removed from the center of the beam. For Earth coverage fixed satellite beams, the site loss is compensated through beam shaping to ensure equal returns from all locations.

15.10.3 Frequency Scallop Loss

Frequency scallop losses are associated with frequency acquisition using discrete frequency scanning or a Fourier transform.

15.10.4 Signal Processing Loss

Signal processing losses are associated with discrete amplitude and time sampled quantization losses. Discrete amplitude sampling losses are minimized by using the maximum number of bits to describe the signal amplitude at various points along the processing path and the discrete-time sampling losses are minimized through filter design and sampling frequency selection. Other sources of signal processing loss are associated with the limitation of various algorithms that approximate theoretical models; for example, the computation of decision thresholds based on a limited sample size.

15.11 NONLINEAR POWER AMPLIFIER ANALYSIS AND SIMULATION

The following discussions and analysis refer to an amplitude-modulated (AM) and/or phase-modulated (PM) carrier signal. When a continuous wave (CW) carrier is applied to a nonlinear power amplifier (PA) the output will contain AM and possibly AM and PM distortion which is the subject of this section [47]. The intermodulation noise produced by a TWTA or SSPA nonlinearity is related to the input AM carrier signal level and the AM-to-AM (AM-AM) and AM-to-PM (AM-PM) characteristics of the amplifier. The operating point, or input drive level, of a TWTA is relative to the saturation level. For a SSPA, the operating point is relative to the 1 dB gain compression level. The input backoff (IBO) corresponds to the input power backoff from the saturation or the 1 dB gain compression powers of the respective devices required to achieve the specified system performance. As IBO is increased the devices approach the linear operating region resulting in decreasing levels of intermodulation noise and improved system performance. However, lowering the input power level also results in less efficient use of the overall power capabilities of the devices. Constant envelope-modulated waveforms, like BPSK and the various forms of QPSK, are more tolerant to lower IBO levels than waveforms with inherently large peak-to-rms levels, like QAM, frequency division multiplex (FDM), and orthogonal frequency division multiplexing (OFDM).

The intermodulation noise is characterized in terms of the carrier-to-intermodulation noise (C/I) ratio or as the carrier-to-intermodulation noise density (C/I_o) ratio. The intermodulation noise density is related to the intermodulation noise by the users channel bandwidth B Hz as $I_o = I/B$ W/Hz. The C/I performance parameter is discussed in the following sections in the context of the device AM-AM and AM-PM characteristics. In addition to the single-channel intermodulation noise, the nonlinear PA output will include the signals of other users in frequency division multiple access (FDMA) applications. This situation occurs, for example, when the PA is operating as a multichannel satellite relay downlink transmitter, in which case, the receiver and intermodulation noise from all of the channels influences the amplifier output level. The desired signal will include some degree of signal suppression depending upon the severity of the nonlinearity.

As a consequence of the nonlinear device transfer function and the various noise terms, the output backoff (OBO) conditions are defined in terms of the level of the desired signal saturation point or the 1 dB gain compression point as described earlier. The following example focuses on the saturation point of the TWTA, for which, the OBO is characterized as

$$\text{OBO} = \frac{P_{sat}}{G_{tot} P_{tot}} \qquad (15.134)$$

The parameter P_{tot} in (15.134) is the total power into the amplifier and G_{tot} is the corresponding gain. The gain G_{tot} is the input-to-output signal gain of the TWTA as determined by the input signal operating point defined by the IBO level. In the following sections, several methods of evaluating the intermodulation noise are examined and compared. These comparisons are characterized by *C/I* and OBO and their dependence on the input operating levels as defined by IBO.

15.11.1 Characterizing the TWTA Transfer Function

The characteristics of three TWTAs are examined with varying degrees of nonlinear signal distortion as identified by Saleh [48, 49]. The characteristics of interests are the AM-PM transfer function in which the AM carrier input signal results in a nonlinear PM output signal; and the AM-AM transfer function in which the AM carrier input signal results in a nonlinear AM output signal. These nonlinear transfer functions result in varying degrees of harmonic distortion at the output of the TWTA that cause co-channel and adjacent channel interference (ACI). For clarity the three TWTAs, characterized by Saleh, are denoted as: TWTA No. 0 that is the least severe with zero AM-PM phase distortion; TWTA No. 1 that exhibits a moderate AM-PM phase distortion; and TWTA No. 2 that has severe AM-PM phase distortion. All three result in nearly identical nonlinear AM-AM distortion.

The comparison of the three traveling wave tube amplifiers is based on Saleh's two-parameter in-phase and quadrature (I/Q) (or real and imaginary) curve-fit functions expressed as

$$P(r) = \frac{\alpha_p r}{1+\beta_p r^2}$$
$$Q(r) = \frac{\alpha_q r^3}{\left(1+\beta_q r^2\right)^2} \qquad (15.135)$$

where α_p, β_p, α_q, and β_q are the curve fitting coefficients given in Table 15.7 for the three representative traveling wave tube (TWT) characteristics. The parameter r is the composite input signal level and determines the operating point on the TWT transfer function. In terms of the amplitude and phase responses, the corresponding I/Q components are expressed as $P(r) = A(r)\cos(\Phi(r))$ and $Q(r) = A(r)\sin(\Phi(r))$.

Plots of the curve fit results for the quadrature functions expressed in (15.135) are converted to the amplitude and phase functions and shown in Figure 15.35 using the parameter values listed in Table 15.7.

TABLE 15.7 Two-Parameter Curve-Fit Values from Saleh[a]

	$P(r)$		$Q(r)$	
TWT	α_p	β_p	α_q	β_q
0	2.0	1.0	0.0	0.0
1	1.90947	1.07469	4.35023	2.33525
2	2.11075	2.22764	7.33959	2.11475

[a]Saleh [49]. Reproduced by permission of the IEEE.

15.11.2 Evaluation of *C/I* and *OBO*

The evaluation of *C/I* and OBO discussed in this section is applied to a FDMA satellite communication system and follows the work of Saleh [49] using the I/Q responses for TWTs 0, 1, and 2 discussed earlier. The input signal is considered to be the sum of n independent phase-modulated signals expressed as

$$x(t) = \sum_{i=1}^{n} V_i e^{j(\omega_i t + \varphi_i(t))} \qquad (15.136)$$

where ω_i represents carriers in different FDMA bandwidths (B) occupying a total bandwidth of $W = nB$ Hz. The output of the TWTA when operating in a FDMA network [50–52] is composed of distinct intermodulation tones with angular frequency $k_1\omega_1 + k_2\omega_2 + \cdots + k_n\omega_n$; the restriction $k_1 + k_2 + \cdots + k_n = 1$ ensures that all of the intermodulation products (IMPs) fall in the first spectral zone of each frequency band. The order of the IMP is given by $m = |k_1| + |k_2| + \cdots + |k_n| = 1, 3, 5, \ldots$, that is, only odd order IMPs are present. The value $m = 1$ represents the desired carrier term at the output of the nonlinear amplifier and all of the other terms, $m = 3, 5, \ldots$ are representative of output distortion tones in the users bandwidth. In the following analysis, the input carriers are assumed to have equal amplitudes with $V_i = V : \forall i$ and $n \gg m$.

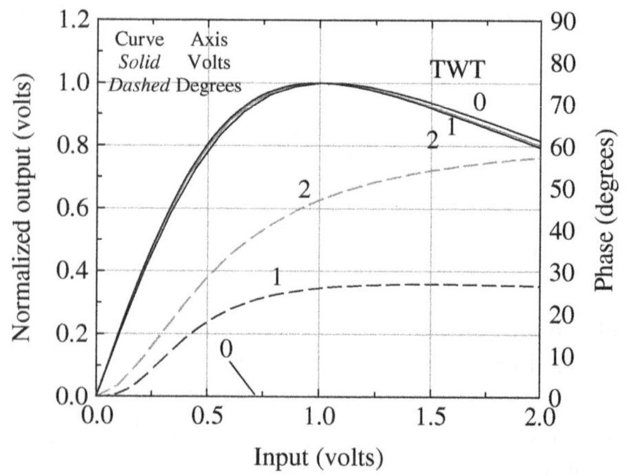

FIGURE 15.35 Normalized amplitude and phase characteristics. Saleh [49]. Reproduced by permission of the IEEE.

Based on this brief introduction, the C/I ratio at the output of the TWTA is given by

$$C/I(\nu)=\frac{P_{1,n}^2+Q_{1,n}^2}{\sum_{m=3,5,\ldots}N_{m,n}(\nu)\left[P_{m,n}^2+Q_{m,n}^2\right]} \quad (15.137)$$

where $\nu=0$ corresponds to the C/I ratio at the center of the users band and $\nu=\pm 1/2$ is the ratio at either edge of the users band. The parameter $N_{m,n}(\nu)$ is the number of dominant m-th order IMPs falling at location ν and is given by [53]

$$N_{m,n}(\nu)=\frac{n^{m-1}}{((m+1)/2)!((m-1)/2)!(m-1)!}\sum_{k=0}^{[m/2-\nu]}(-1)^k\binom{m}{k}\left(\frac{m}{2}-k-\nu\right)^{m-1} \quad :|\nu|\le\frac{m}{2} \quad (15.138)^*$$

Where $N_{1,n}(\nu)=1$ for $\nu=(1,\pm 1/2)$ and $N_{m,n}(\nu)=0$ for $|\nu|>m/2$. The terms $P_{m,n}$ and $Q_{m,n}$ represent the I/Q components of the m-th order IMP, for which, Saleh developed the expressions

$$P_{m,n}=\frac{\alpha_p}{\sqrt{\beta_p}n^{m/2}}\left\{\left[\sum_{k=0}^{(m-1)/2}A_{k,m}X^{2k+1}\right]-\left[\sum_{k=0}^{(m-1)/2}B_{k,m}X^{2k+3}\right]e^{X^2}E_1\left(X^2\right)\right\} \quad (15.139)$$

$$Q_{m,n}=\frac{\alpha_q}{\sqrt{\beta_q^3}\,n^{m/2}}\left\{\left[\sum_{k=0}^{(m+1)/2}C_{k,m}Y^{2k+1}\right]-\left[\sum_{k=0}^{(m+1)/2}D_{k,m}Y^{2k+3}\right]e^{Y^2}E_1\left(Y^2\right)\right\} \quad (15.140)$$

where

$$X=\left(U\sqrt{\beta_p}\right)^{-1},\quad Y=\left(U\sqrt{\beta_q}\right)^{-1},\quad U=\sqrt{n}V \quad (15.141)$$

and

$$A_{k,m}=(-1)^k\left(\frac{m+1}{2}\right)!\sum_{l=k}^{(m-1)/2}(-1)^l\binom{\frac{m+1}{2}}{l}\frac{(l-k)!}{(l+1)!} \quad (15.142)^\dagger$$

$$B_{k,m}=\frac{[(m+1)/2]!}{(k+1)!}\binom{\frac{m-1}{2}}{k} \quad (15.143)^\dagger$$

$$C_{k,m}=(k+1)A_{k,m}+B_{k-1,m} \quad (15.144)^\dagger$$

$$D_{k,m}=(k+2)B_{k,m}+B_{k-1,m} \quad (15.145)^\dagger$$

with $A_{k,m}=0$ for $k=(m+1)/2$ and $B_{k,m}=0$ for $k=-1$ or $(m+1)/2$. The parameter U is equal to the square root of the total average input power and the function $E_1(z)$ is the exponential integral [54]

$$E_1(z)=\int_z^\infty\frac{exp(-u)}{u}du \quad :|\arg z|<\pi \quad (15.146)$$

Saleh [49] provides an appendix that outlines a method for the numerical computation of (15.139) and (15.140) that avoids computational roundoff errors for large values of m. The solution involves rewriting the respective summations in the brackets { } of (15.139) and (15.140) as follows:

$$\left\{\left[\sum_{k=0}^{(m-3)/2}(A_{k,m}-B_{k,m})X^{2k+1}\right]+\left[\sum_{k=0}^{(m-1)/2}B_{k,m}X^{2k-1}\right]F(X^2)\right\} \quad (15.147)$$

and

$$\left\{\left[\sum_{k=0}^{(m-1)/2}(C_{k,m}-D_{k,m})Y^{2k+1}\right]+\left[\sum_{k=0}^{(m+1)/2}D_{k,m}Y^{2k-1}\right]F(Y^2)\right\} \quad (15.148)$$

with the respective limit on $m\cong 82/X+4.4$ and $96/Y-2.6$. The $F(z)$ is expressed as

$$F(z)=\int_0^\infty\frac{u\,exp(-u)}{(1+u/z)}du \quad (15.149)$$

and can be evaluated numerically using Laguerre–Gauss quadrature integration.

The OBO is defined as

$$OBO=\frac{\max\left(P^2(r)+Q^2(r)\right)}{n\left(P_{1,n}^2(r)+Q_{1,n}^2(r)\right)} \quad (15.150)$$

where $P(r)$ and $Q(r)$ are the nonlinear output values corresponding to the operating point r of the input composite signal. These values are computed as given in Section 15.11.1 using the appropriate α_p, β_p, α_q, and β_q values from Table 15.7.

*Saleh [49] provides solutions to (15.138) for $m=1, 3, 5, 7$ and for $\nu=1, \pm 1/2$.

†Saleh [49] provides solutions to these equations for m = 1, 3, 5, 7 and $k=0$–4.

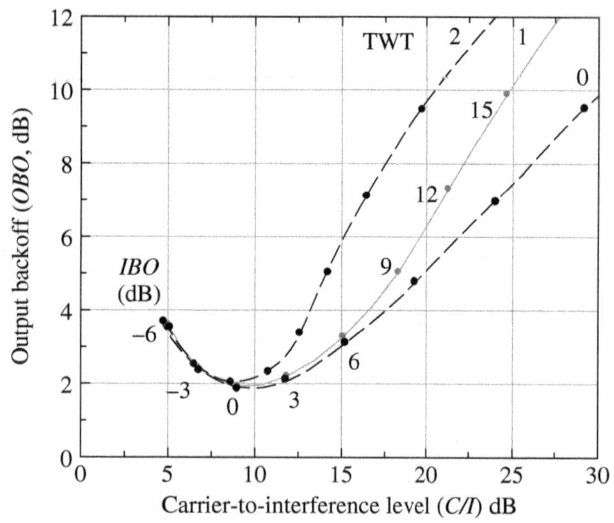

FIGURE 15.36 *C/I* and OBO performance corresponding to center of the band ($\nu = 0$). Saleh [49]. Reproduced by permission of the IEEE.

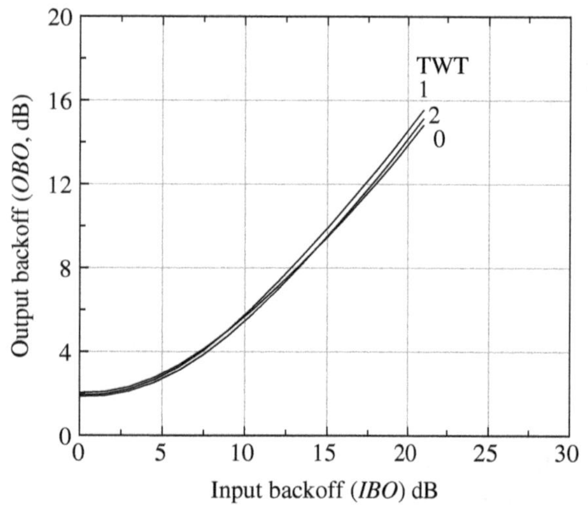

FIGURE 15.37 OBO vs. IBO performance corresponding to the center of the band ($\nu = 0$).

These relationships are evaluated for the TWTs 0, 1, and 2 and the results are shown in Figure 15.36 which relate the *OBO* dependence on *C/I* for the indicated IBO operating points. As the IBO increases the TWTs operate more in the linear region and away from saturation resulting in a higher *C/I* ratio. Figure 15.37 characterizes the OBO level as a function of the IBO for the three TWTs under consideration.

15.12 COMPUTER MODELING OF TWTA AND SSPA NONLINEARITIES

The nonlinear amplifier is generally characterized in terms of the devices AM-AM and AM-PM transfer functions. These transfer functions are typically available from the manufacturer as the measured output power and phase vs. the input power or the measured output voltage and phase vs. the input voltage. The manufacturers may also provide information on the bandwidth and temperature sensitivities.

The following examples involve the modeling of two SSPAs having slightly different transfer characteristics. These models serve to highlight some significant performance differences concerning the specified output power and the OBO required to achieve a specified performance measure. The performance measure may be the spectral regrowth or the degradation in the bit-error probability given as adjacent channel. This latter situation can also be characterized in terms of the ACI specification. Although these examples focus on the modeling of SSPAs, the techniques apply equally well to TWTAs with the primary exception being the definition of the maximum output power. The maximum output power of a SSPA is defined as the power corresponding to the 1-dB gain compression point, referred to as P_1, whereas, for the TWTA the maximum output power is the saturation power, referred to as P_{sat}. SSPA and TWTA amplifiers are referred to as soft limiters and are contrasted with the hard limiters discussed in Section 10.2. Soft limiters are characterized as having an output that is differentiable with respect to the input and approaches a constant value or limit as the input increases. The soft limiter response is modeled by various forms of sigmoid functions of which the error and arctan functions are commonly used examples.

15.12.1 SSPA with Soft Saturation Response

In this example, the SSPA transfer function is characterized as having AM-AM that is characterized in terms of the peak output voltage (V_{po}) as a function of the peak input voltage (V_{pi}). Because the AM-PM response is zero the SSPA is considered to be ideal. The *dashed* curve in Figure 15.38 shows the measured voltage transfer function and the *solid* curve is based on a seventh-order polynomial that is curve fit to the V_{po} vs. V_{pi} response. The modified response results in a constant gain over the linear range of the SSPA and a monotonically decreasing gain as the input signal level increases through the nonlinear range. The actual gain of the device is not important insofar as the performance simulation is concerned so the gain of the soft saturation response is purposely low to avoid dealing with large signal levels in the simulation program. The curve-fit transfer function is expressed as

$$V_{po} \cong 0.898407V_{pi} - 155.24V_{pi}^3 + 17552.5V_{pi}^5 - 751241V_{pi}^7 \quad (15.151)$$

with the corresponding gain evaluated as $20\log_{10}(V_{po}/V_{pi})$. Using (15.151), the third-order IMPs are characterized by the cubic term as

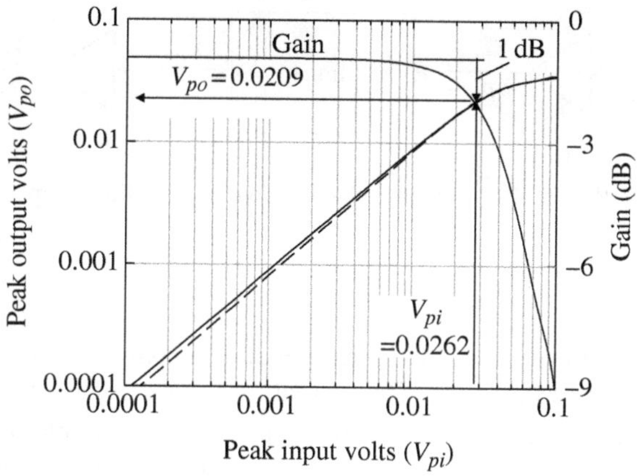

FIGURE 15.38 Characteristics for soft saturation response SSPA.

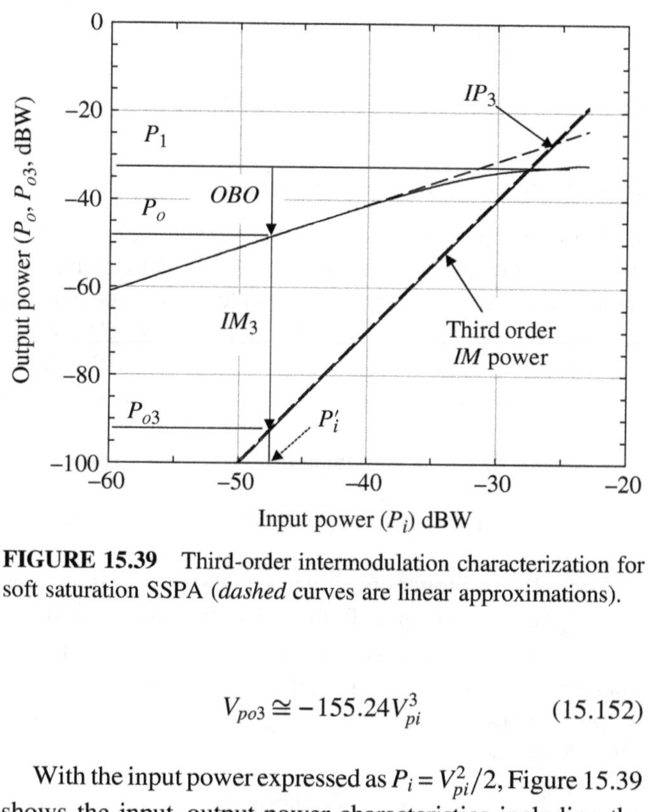

FIGURE 15.39 Third-order intermodulation characterization for soft saturation SSPA (*dashed* curves are linear approximations).

$$V_{po3} \cong -155.24V_{pi}^3 \tag{15.152}$$

With the input power expressed as $P_i = V_{pi}^2/2$, Figure 15.39 shows the input–output power characteristics including the third-order intermodulation power. The term $P_1 = V_{po1}^2/2$ is the output power corresponding to the 1-dB gain compression identified in Figure 15.38, $P_0 = P_1 10^{-(\text{OBOdB}/10)}$ is the primary output power corresponding to the carrier power P_i' of the input signal under test, IP_3 is the third-order intercept output power, and IM_3 is the ratio of the test signal carrier power to the third-order intercept power resulting from the nonlinearity. Using the linear approximations to these power transfer functions, the third-order intermodulation power ratio is defined as ratio P_o/P_{o3} and is expressed in decibels as

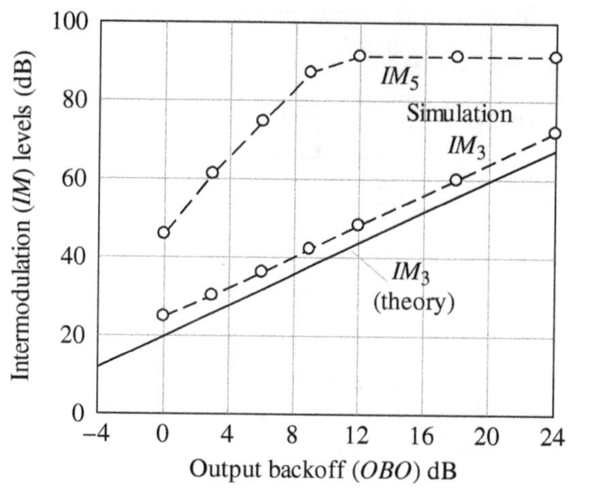

FIGURE 15.40 Intermodulation characteristics using soft saturation SSPA response (*circled* data is for SRRC-QPSK with $\alpha = 0.4$).

$$\begin{aligned} IM_3(\text{dB}) &\triangleq P_o(\text{dB}) - P_{o3}(\text{dB}) \\ &= P_1(\text{dB}) - \text{OBO}(\text{dB}) - P_{o3}(\text{dB}) \end{aligned} \tag{15.153}$$

Equation (15.153) is plotted in Figure 15.40 as the *dotted* theoretical curve for various OBO levels based on the SSPA characteristics described in Figures 15.38 and 15.39. The simulated intermodulation distortion discussed in Sections 15.13 and 15.14 is evaluated using a slightly different SSPA model and the theoretical and simulated intermodulation distortion are also shown in Figure 15.40 as the *solid* curve and the *circled* data points, respectively. Therefore, Figure 15.39 and Equation (15.153) serve as a general description for the evaluation of intermodulation distortion.

15.12.2 TWTA with Gain Compensated Response

The TWTA often use gain and phases compensation circuits to extend the linear range of the device. The resulting gain response rises to a slight peak before decreasing monotonically. The goal is to extend the 1 dB gain compression point and thereby extend the linear range. The gain compensated curve-fit response for a TWTA is expressed as

$$V_{po} \cong 1.04172V_{pi} + 116.862V_{pi}^3 - 53103.4V_{pi}^5 \tag{15.154}$$

and the phase transfer function is approximated as

$$\theta_o \cong 33.5914V_{pi} + 7009.96V_{pi}^3 - 340831V_{pi}^5 \tag{15.155}$$

It is left as an exercise (see Problem 13) to plot these responses and determine the peak input and output voltages that correspond to the 1 dB gain compression point.

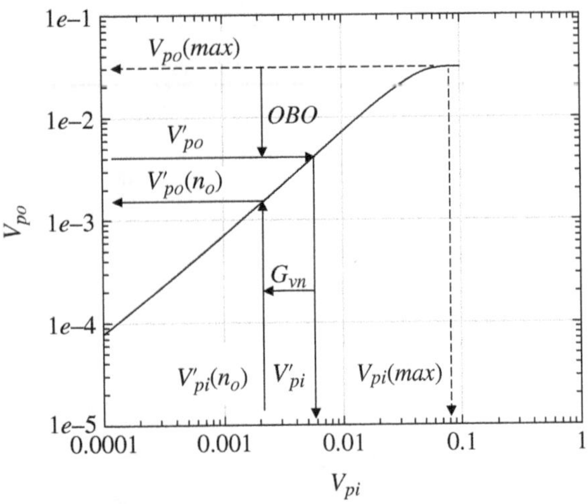

FIGURE 15.41 Soft limiter voltage gain characteristic.

15.13 ESTABLISHING SIGNAL LEVELS FOR SIMULATION MODELING

Consider the soft limiter modeled by curve fitting a fifth-order polynomial to the measured AM-AM gain characteristic of a nonlinear device. For this example, the polynomial coefficients are as follows: $C_m = (0.0, 0.687834, 6.070865, -190.532776, 2729.8845, -8057.7788)$ and the peak output voltage (V_{po}) is expressed in terms of the peak input voltage (V_{pi}) as

$$V_{po} = \sum_{m=0}^{5} C_m V_{pi}^m \tag{15.156}$$

This voltage gain characteristic is shown in Figure 15.41. The magnitude of the voltage levels and the limiter gain are not important for the purpose simulation. However, it is important to map the voltage level of the composite input signal to the specified operating range of the nonlinear model and, in turn, to map the limiter output level back to the levels used in the simulation program. The importance of the input mapping is to ensure that the signal is operating at the input level corresponding to a specified IBO. Similarly the importance of the output mapping is to ensure that the limiter output is mapped back to the expected simulation level. This procedure allows for the addition of postlimiter noise based on the known signal power level. Throughout this section the focus is on the modeling of a HPA for use in a performance simulation program with an emphasis placed on the HPA nonlinearity. In this regard, the HPA gain is only used to establish the received signal-to-noise ratio specified in the simulation program.

Before establishing the mapping, the operating point through the limiter must be identified. Referring to Figure 15.41, the maximum output is denoted $V_{po}(\max) =$

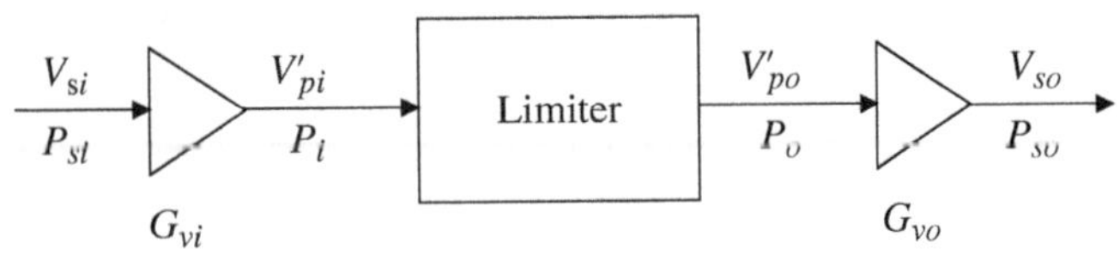

FIGURE 15.42 Mapping of simulation levels to limiter operating points.

0.03055 mV and the corresponding maximum input is $V_{pi}(\max) = 0.08$ mV. The output operating point (V'_{po}) is defined in terms of the output power backoff relative to $V_{po}(\max)$ and the corresponding input operating point is V'_{pi}. The OBO in decibels is: $\text{OBO(dB)} = 20\log(V_{po}(\max)/V'_{po})$ and V'_{pi} is determined as the inverse of the polynomial expression in (15.156). The voltages V'_{pi} and V'_{po} apply to the single-carrier operation discussed in the following section and the voltages $V'_{pi}(n_o)$ and $V'_{po}(n_o)$ apply to the multiple-carrier operation discussed in Section 15.13.2.

15.13.1 Single-Carrier Simulations

With a single carrier per channel (SCPC) simulation the limiter may occur in the transmitter HPA of a single-channel-modulated waveform or with multiple data channels that are time division multiplexed (TDM) onto a single carrier. The SCPC waveform also applies to a satellite HPA; however, the satellite often combines many carriers to form a FDM downlink.

Consider the SCPC-modulated waveform is expressed as

$$s(t) = V_{si} m(t) \cos(\omega_c t + \theta(t)) \tag{15.157}$$

The HPA or limiter operating points are characterized by (V'_{pi}, V'_{po}) and the mapping from (and to) the simulation signal levels is determined as shown in Figure 15.42. The limiter operating points are determined using the maximum peak output voltage of the limiter $V_{po}(\max)$ and a specified OBO level such that

$$V'_{po} = V_{po}(\max) 10^{-(\text{OBO(dB)}/20)} \tag{15.158}$$

The corresponding input V'_{pi} is determined using the inverse of the polynomial expression of the limiter transfer function as indicated in Figure 15.41.

With the modulation function $m(t)$ normalized to result in a unit power level, the power of the simulated waveform is simply the power in the carrier signal with peak voltage V_{si}, that is,

$$P_{si} = \frac{V_{si}^2}{2} \tag{15.159}$$

The mapping to the limiter input is determined by the voltage gain, expressed as,

$$G_{vi} = \sqrt{\frac{P_i}{P_{si}}} = \frac{V'_{pi}}{V_{si}} \tag{15.160}$$

Similarly, the output of the limiter is mapped *back* to the simulation signal level using the voltage gain

$$G_{vo} = \sqrt{\frac{P_{so}}{P_o}} = \frac{V_{so}}{V'_{po}} \tag{15.161}$$

It is convenient in the simulation to let the input and output signal levels be equal so that $P_{so} = P_{si}$. That is, when the limiter is not used, the HPA appears as an ideal unit gain amplifier. Neglecting the various propagation and channel distortion losses, the received signal-to-noise ratio (γ_b) is established at the limiter output and the received noise power is computed as

$$\sigma_n^2 = \frac{V_{so}^2}{2\gamma_b} \tag{15.162}$$

15.13.2 Multiple Carrier Simulations

Multiple carrier simulations will occur in the transmitter HPA when several modulated carriers are combined at the IF to form a FDM waveform; this is also typical of multichannel satellite repeater downlinks. In these cases, the simulation is generally focusing on the performance of a particular carrier or signal and the limiter operating point must be established for the signal of interest.

Consider the composite signal described as

$$s(t) = \sum_n s(t;n) \tag{15.163}$$

where

$$s(t;n) = V_{si}(n)m(t;n)\cos(\omega_c(n)t + \theta(t;n)) \tag{15.164}$$

Considering $s(t;n_o)$ to be the desired signal with the modulation term $m(t;n_o)$ normalized to result in unit power, the desired signal power is then the carrier power

$$P_{si}(n_o) = \frac{V_{si}^2(n_o)}{2} \tag{15.165}$$

The limiter operating point $\left(V'_{pi}, V'_{po}\right)$ for the composite signal is established as described earlier with the input power given by

$$P_{si} = \frac{V_{si}^2}{2} = \frac{1}{2}\sum_n V_{si}^2(n) \tag{15.166}$$

However, the operating point for the desired signal is $\left(V'_{pi}(n_o), V'_{po}(n_o)\right)$ so the input signal mapping to the limiter is determined by the voltage gain

$$G_{vi} = \sqrt{\frac{P_i(n_o)}{P_{si}}} = \frac{V'_{pi}(n_o)}{\sqrt{\sum_n V_{si}^2(n)}} = \frac{V'_{pi}G_{vn}}{\sqrt{\sum_n V_{si}^2(n)}} \tag{15.167}$$

where the voltage gain is

$$G_{vn} = \frac{V'_{pi}(n_o)}{V'_{pi}} \tag{15.168}$$

Equation (15.168) maps the limiter composite input signal to the desired signal as indicated in Figure 15.41. The limiter output is mapped back to the simulation signal levels using the voltage gain

$$G_{vo} = \sqrt{\frac{P_{so}}{P_o(n_o)}} = \frac{V_{so}}{V'_{po}(n_o)} \tag{15.169}$$

where the desired output signal voltage $V'_{po}(n_o)$ is determined directly from the limiter transfer function using the input voltage $V'_{pi}(n_o)$. In this case, the signal-to-noise ratio (γ_b) for the desired signal is established at the limiter output using the noise power

$$\sigma_n^2 = \frac{V_{so}^2}{2\gamma_b} \tag{15.170}$$

where

$$V_{so} = G_{vo}V'_{po}(n_o) \tag{15.171}$$

The previous steps, required to setup and initialize the limiter processing, are summarized as follows:

- Given $V_{po}(\max)$ determine the composite signal operating point V'_{po} at the limiter output using the specified OBO
- Determine the output voltage gain G_{vo} that maps V'_{po} to the simulation voltage V_{so}
- For single-carrier operation, compute the noise power σ_n^2 for the specified signal-to-noise ratio γ_b
- Determine the corresponding composite signal operating point V'_{pi} at the limiter input
- Determine the input voltage gain G_{vi} that maps the simulation voltage V_{si} to the limiter operating point V'_{pi}

For multiple carriers, the following additional steps are required:

- Determine the voltage gain G_{vn} between the signal of interest $V_{si}(n_o)$ and the composite signal V_{si}
- Using G_{vn} and V'_{pi} determine the limiter input operating point $V'_{pi}(n_o)$ for the desired signal
- Determine the limiter output voltage $V'_{po}(n_o)$ corresponding to the input $V'_{pi}(n_o)$

The final step involves computing the noise power using either (15.162) or (15.170) for a desired signal-to-noise ratio γ_b.

15.14 CASE STUDY: PERFORMANCE SIMULATION OF SRRC-QPSK WITH SSPA NONLINEARITY

In this case study, three major impacts of the HPA on the system performance are examined and related to the OBO of an SSPA. The three system performance measures are intermodulation (IM) noise, spectrum degradation, and ACI. The two-tone third and fifth-order intermodulation distortion terms are examined for various OBO conditions using computer simulation of the SSPA discussed in Section 15.12. The goal is to achieve the theoretical intermodulation IM performance shown as the *solid* curve in Figure 15.40. The *circled* data points in Figure 15.40 are based on the simulated IM performance and are used in this section. Following the description of the IM noise impact on the system performance, the spectral degradation of the spectral root-raised-cosine offset quadrature phase shift keying (SRRC-QPSK) modulated waveform is examined for various amounts of OBO. In the concluding section of this case study, the bit-error performance is examined for the output power backoff with and without adjacent channels. The results suggest that, to avoid significantly degrading the performance due to the ACI with Nyquist channel spacing, the OBO should be greater than 6 dB.

The mapping of the simulated signal levels into, and out of, the SSPA device is based on the descriptions given in Section 15.13. The waveforms used in examining the IM performance are based on SRRC-QPSK modulation with an excess bandwidth of $(1+\alpha)$ where α is the SRRC excess bandwidth parameter; $\alpha = 0.4$ is used in these evaluations. The simulation program generates the desired modulated waveform at baseband with SRRC-QPSK modulated upper and lower adjacent channels. The data rates, amplitude levels, carrier frequencies, and excess bandwidth for each of the channels are specified with the sampling frequency computed to satisfy the Nyquist criterion for the composite FDMA waveform. The spectrum plots in Figure 15.43 are examples of the three channels operating with symmetrical and asymmetrical adjacent channels. The asymmetrical adjacent channel spectrums, shown in Figure 15.43b, operate with symbol rates (R'_s) four times the desired channel symbol rate of R_s. In this case, the adjacent channel levels are decreased by $10\log_{10}(R'_s/R_s) = 6$ dB; although the level can be arbitrarily specified in the simulation.

The fundamentals of the signal processing are as follows: The desired (or center) channel waveform is centered $f_c = 0$ Hz and is sampled using 128 samples per symbol, so the Nyquist sampling frequency is $f_s = 128R_s$. The adjacent channel carrier frequencies are located at

$$|f'_c| = \frac{(1+\alpha)(R_s + R'_s)}{2} \tag{15.172}$$

and the maximum occupied bandwidth is

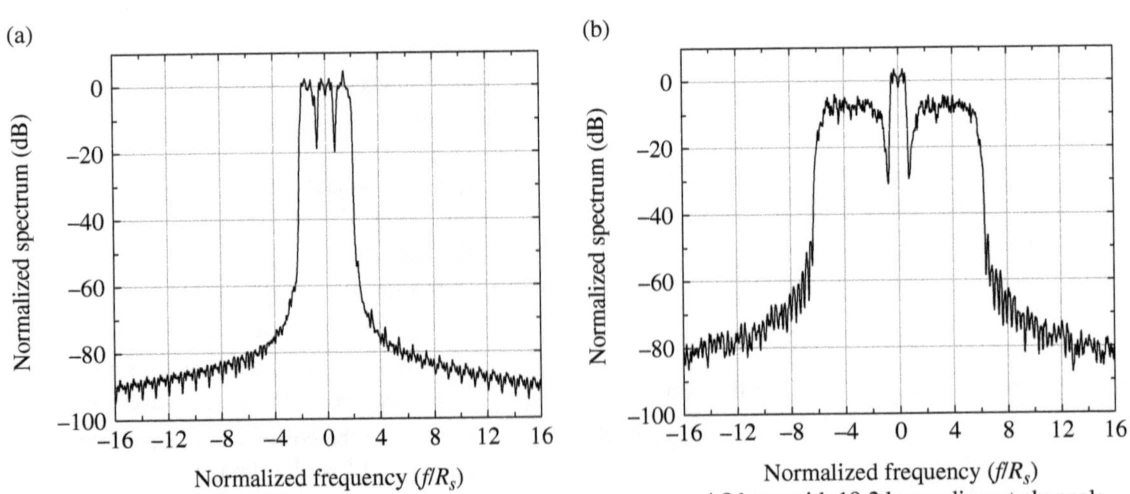

FIGURE 15.43 Spectrum of SRRC-QPSK waveforms (linear PA).

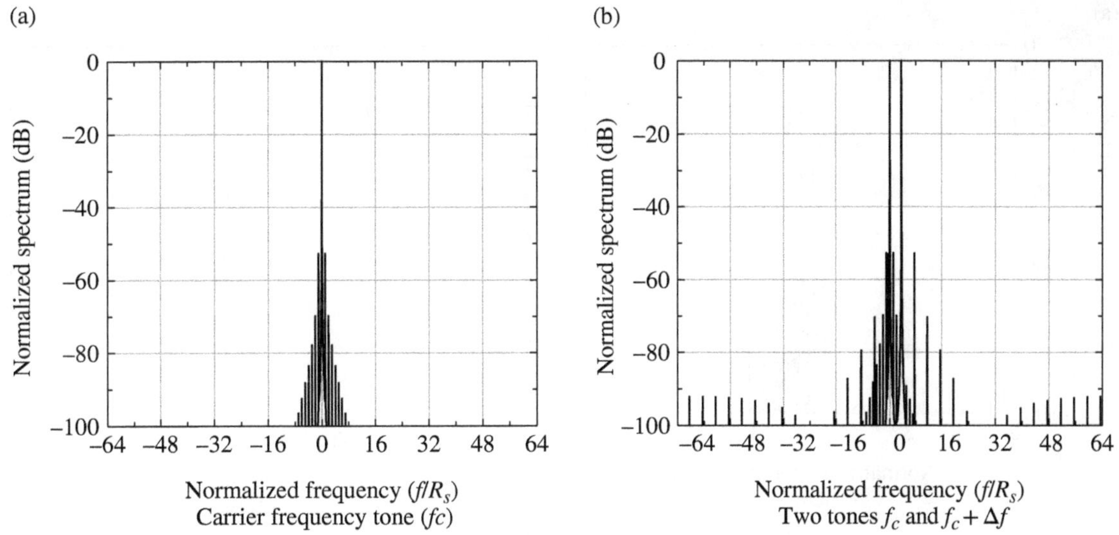

FIGURE 15.44 Performance with ideal linear PA (no SSPA).

$$f_{max} = (1+\alpha)\left(\frac{R_s}{2} + R'_{max}\right) + \Delta f_{max} \qquad (15.173)$$

where R'_{max} is the maximum of the upper and lower adjacent channel symbol rates and Δf_{max} is used, if necessary, to provide additional frequency separation to improve the system performance. Based on these conditions the Nyquist sampling criterion requires that $f_s \geq 2f_{max}$. The SRRC waveform is generated using a uniformly weighed span of 13 symbols with 6 leading and trailing symbols. On the other hand, the receiver matched filter is based on a total span of seven symbols; this results in a theoretical detection loss of less than 0.01 dB.

15.14.1 Simulation of Third- and Fifth-Order Intermodulation Distortions Terms

The two-tone intermodulation noise terms, resulting from the SSPA nonlinearity as characterized by (15.151), are evaluated as a function of OBO using the SSPA computer modeling described in Section 15.13. The third- and fifth-order intermodulation noise terms are evaluated by applying two CW signals to the SSPA and examining the spectrum of the SSPA output. The simulation uses a received *rms* carrier signal level of 1 V that is mapped through the nonlinearity and back to the simulation level of 1 V *rms*.

The simulation uses a rectangular windowed SRRC frequency response with the QPSK-modulated waveform operating at a baseband data rate of 4.8 K symbols per second with the adjacent channel at 19.2 Ksps. The adjacent channel is located at a positive carrier frequency in accordance with (15.172) using $\alpha = 0.4$. The received CW tones for the intermodulation evaluation are obtained by using mark-hold data on the I/Q rails. The intermodulation tones are clearly discernable in the inherently *harmonic-filled* spectrum.

The spectrums are generated using an equivalent of $N_s =$ 128 samples per symbol and averaging 20 spectrums with a cosine windowed fast Fourier transform (FFT) size of $N_{fft} =$ 16,384 samples; the corresponding frequency resolution is 0.0375 Hz. The results depicted in Figure 15.44 are obtained using a linear PA and show the spectral content using the mark-hold SRRC-QPSK waveform to generate the two tones. Figure 15.45 shows the resulting two-tone intermodulation performance for various *OBO* conditions; these results are plotted as the *circled* data points in Figure 15.40.

15.14.2 Spectrum Degradation with SSPA OBO

The spectrum of the SRRC-QPSK-modulated waveform is shown in Figure 15.46 for various OBO levels; the *linear channel* spectrum corresponds to an essentially infinite OBO condition. The spectrum results apply to an arbitrary symbol rate as suggested by the normalized abscissa. The OBO is defined relative to the 1 dB gain compression point of the SSPA and the 0 dB backoff case results in the peak carrier voltage being compressed by 1 dB with the average carrier voltage being 3 dB below the peak level. Therefore, a −3 dB OBO results in the average carrier voltage being compressed by 1 dB; this causes considerable compressing or clipping of the peak voltage. The increase in the modulation one-sided spectral bandwidth beyond the specified limit of $1.4R_s/2$ ($\alpha = 0.4$) is influenced by the SRRC modulator weighted window duration that spans 12 symbols as discussed in Section 4.4.4.1. The impact of the OBO level on the performance with the adjacent channels is examined in the following section; however, it appears as though the OBO should be greater than 6 dB to preserve the excellent

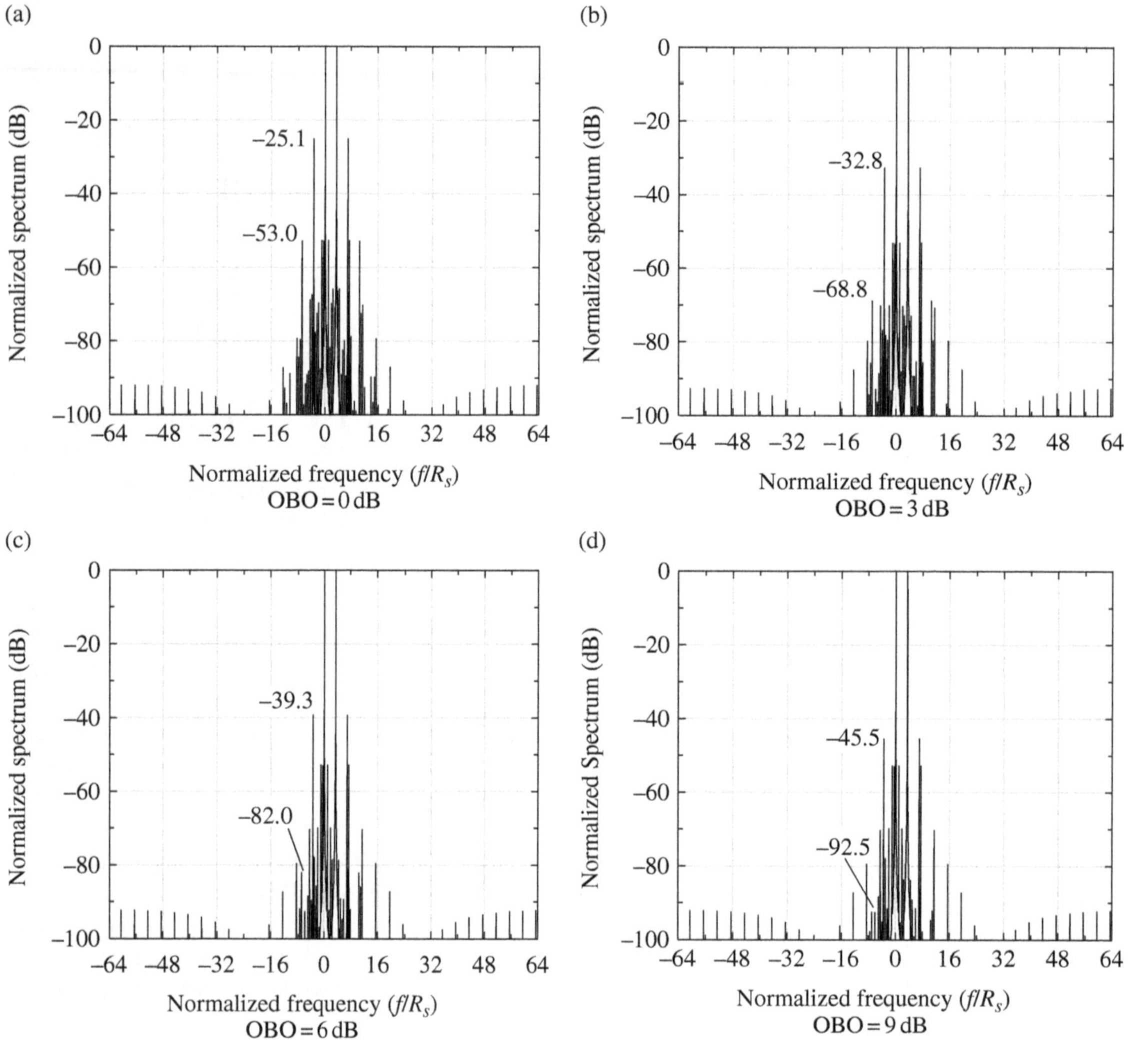

FIGURE 15.45 SSPA IM performance with various OBO levels.

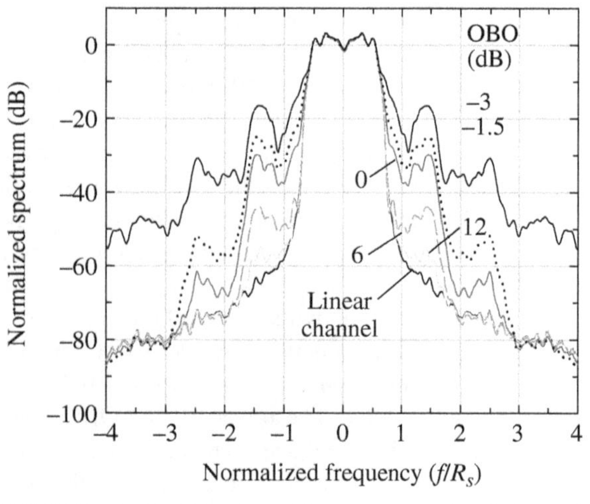

FIGURE 15.46 SRRC-QPSK spectrum with SSPA OBO.

Nyquist bandwidth properties of the SRRC-modulated waveform.

15.14.3 Bit-Error Performance with OBO and Adjacent Channels

In the final section of this case study, the bit-error performance of a SRRC-BPSK-modulated waveform is examined with and without adjacent channels. In this case the desired channel is centered at baseband with a symbol rate of $R_s = 32$ ksps and the adjacent channels are centered above and below the desired channel in accordance with (15.172) using $\alpha = 0.4$. Two adjacent channel symbol rate conditions are examined, one with symmetrical adjacent channels with $R_s' = R_s = 32$ksps and the other with asymmetrical adjacent channels with $R_s' = 4R_s = 128$ksps. The Monte Carlo simulations are based on using 100K symbols for each signal-to-noise ratio ≤6 dB, otherwise, 1 M symbols for each signal-to-noise ratio is used.

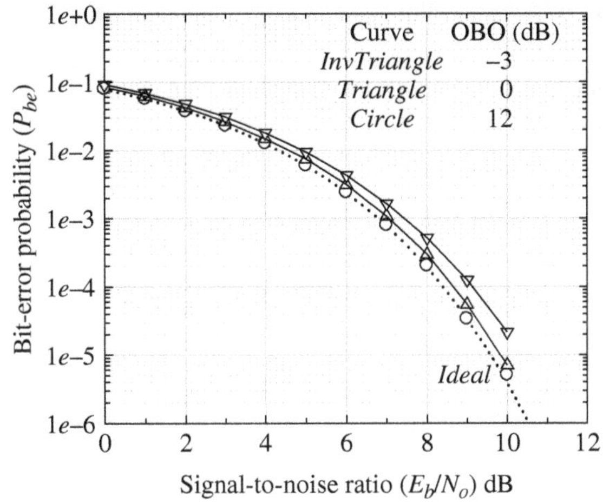

FIGURE 15.47 SRRC-BPSK single-channel performance with SSPA OBO.

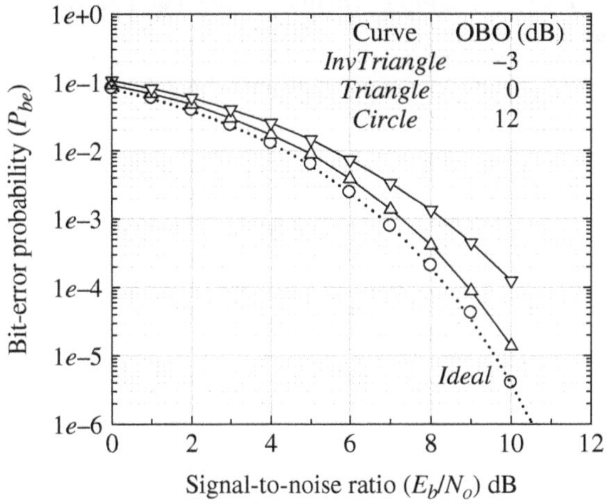

FIGURE 15.48 SRRC-BPSK symmetrical adjacent channel performance with SSPA OBO.

On occasions, when the Monte Carlo results appear inconsistent with the expected performance curve, 10 M symbols are used to establish the data point.

The performance in Figure 15.47 corresponds to the desired channel bit-error performance without the adjacent channels. The *dotted* performance curve represents ideal antipodal signaling and is shown as a reference point when using a linear PA; the performance with 12 dB OBO is nearly the same as the ideal curve. The performance with OBO = 0 dB is only degraded by 0.2 dB at $P_{be} = 10^{-5}$. The severe limiting with OBO = −3 dB is very much like that of an ideal hard limiter, in that, the SRRC AM is essentially removed, resulting in a performance degradation of about 0.9 dB.

In the performance evaluation with adjacent channels, each channel is generated using a separate transmitter with identical SSPAs and the channels are combined to form a FDMA channel; the SSPAs operate with identical OBO conditions. With this understanding, Figure 15.48 shows the performance with symmetrical adjacent channels with the 12 dB OBO condition having a negligible effect on the desired channel performance. The degradations with 0 and −3 dB OBO are about 0.5 and 1.8 dB, respectively.

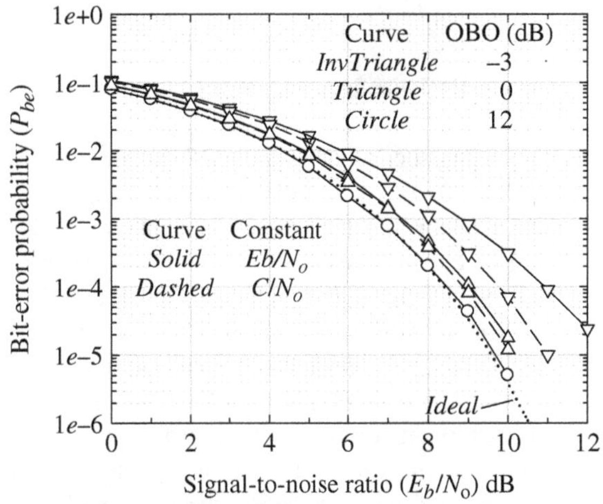

FIGURE 15.49 SRRC-BPSK asymmetrical adjacent channel performance with SSPA OBO.

The performance with asymmetrical adjacent channels with $R'_s = 4R_s$ is evaluated under the conditions of constant E_b/N_o and C/N_o. If the higher data rate channels are modulated on a carrier with the same power as the desired channel, then the channel spectrum will be as shown in Figure 15.43 with the adjacent channel spectral levels $10\log_{10}(R'_s/R_s) = 6\text{dB}$ lower than the desired channel spectral level. This corresponds to the constant C/N_o case that results in 6 dB less E_b/N_o than is available for the desired channel. This results in degradation of the bit-error performance with low levels of OBO, although, the impact on the desired channel performance is less severe. On the other hand, if the adjacent channel carriers are increased by $10\log_{10}(R'_s/R_s) = 6\text{dB}$, then the spectrums in Figure 15.43 all have the same level and correspond to the constant E_b/N_o case. The increase in the carrier power results in a greater loss in the desired channel performance with low levels of OBO. The performance results for these two cases are shown in Figure 15.49 as the *dashed* and *solid* curves corresponding to the constant E_b/N_o and C/N_o cases, respectively. The performance difference in the degradations is about 2 and 1 dB for OBO = −3 and 0 dB, respectively, and is negligible for an OBO of 12 dB. Furthermore, the absolute performance loss decreases with increasing OBO and is negligible for OBO = 12 dB. The best performance for all channels is obtained when operating in the linear range of the SSPA under the constant E_b/N_o condition. The performance

TABLE 15.8 Example Ground Terminal and Satellite Parameters

Parameter	Value	Units	Comments
Satellite			
Longitude ascending node	180	Degrees	Input (east is positive)
Argument of perigee	0	Degrees	Input (north is positive)
Satellite orbit angle	—	Degrees	Computed
Inclination	7	Degrees	Input
Eccentricity	0.003	—	Input
Ground Terminal Location			
Latitude	33	Degrees	Input (north is positive)
Longitude	−120	Degrees	Input (east is positive)
Elevation	0	Meters	Input

degradation can also be reduced, for a given OBO, by increasing the channel separation by Δf_{max} as indicated in (15.173). Therefore, there is a trade-off between power and bandwidth.

15.15 LINK BUDGET ANALYSIS

In this concluding section of Chapter 15 the previous results are brought into focus by completing a link budget to determine if the modem and transceiver designs are capable of establishing or closing the link in view of the system parameters and the communication channel characteristics. The examples in this section involve the link budget for a communication link between a ground terminal and a satellite and Table 15.8 lists the ground station's location, the satellite's orbit, and location in the orbit at an arbitrary initial time. The ground station and the satellite geometry change throughout the satellite orbit period in consideration of the rotating Earth and orbit dynamics as discussed in Chapter 16. The examples are specialized for an uplink signal that is received by the satellite.

The various parameters that makeup the link budget focus on the communication range equation as discussed in the introduction to this chapter. Since an ideal demodulator is used Table 15.9 focuses on the transceiver and channel characteristics; however, to satisfy the requirements of the entire communication system, the modem characteristics shown in Tables 15.10 and 15.11 must be included as part of the link budget. Identifying the modem characteristics separately from those of the transceiver and channel is convenient because the interface between the receiver and demodulator is often at an IF, as discussed in Chapter 2, and these subsystems are generally designed, built, and tested independently before the final system integration. In this regard, Tables 15.10 and 15.11 form the basis of an independent modem specification. The principal modem parameters

TABLE 15.9 Satellite System Link Budget (Ideal Demodulator)

Parameter	Value	Units	Comments
Data rate R_b	19.2	kbps	Input
Uplink frequency	2.0	GHz	Input
Link margin	2.0	dB	Input
HPA power	4.6	dBW	Input
Antenna feed loss	0.0	dB	Input
Antenna gain	10.0	dB	Input
Reflector loss	0.0	dB	Input
Radome loss	0.0	dB	Input
Wet antenna loss	0.0	dB	Input
Polarization	—	—	Input (LVP: AR = ∞)
EIRP	14.6	dBW	Computed
	Channel		
Path loss	0.214	dB	Computed (less F_s loss)
Atmosphere	0.014	dB	Input
Rain	0.2	dB	Input
Antenna tracking	0.0	dB	Input
Multipath	0.0	dB	Input[a]
Scintillation	0.0	dB	Input[a]
Polarization rotation	0.0	Degrees	Input
Sky temperature	20.0	°K	Input[b]
	Receiver		
Antenna gain	30.0	dBi	Input
Radome loss	0.0	dB	Input
Temperature	290	°K	Input[b]
Reflector loss	0.2	dB	Input
Temperature	290	°K	Input[b]
Polarization	—	RHC	Input
Axial ratio	0.0	dB	Input
Polarization loss	3.0	dB	Input RHC: AR = 0 dB
Antenna system temperature	506.8	°K	Computed
Receiver *G/T*	3.75	dB/°K	Input
Feed loss	0.0	dB	Input
Temperature	290	°K	Input[b]
Rx noise figure	—	dB	Input[c]
Receiver system temperature	506.8	°K	Computed
Receiver C/N_o	50.8	dB-Hz	Computed
Rx phase noise loss	0.2	dB	Input
Rx ISI/IF filter loss	0.1	dB	Input
Channel bandwidth *W*	4	MHz	Input
Receiver/user bandwidth	30.0	kHz	Input
Received E_b/N_o (req'd)	8.0	dB	Computed
Specified E_b/N_o	6.0	dB	Input
Margin	2.0	dB	Computed

[a]Based on waveform and demodulator algorithm analysis and simulation.
[b]These parameters are used to compute the system noise temperatures when antenna gain and *G/T* is not provided as input.
[c]The receiver noise figure is not needed when antenna gain and *G/T* is provided, otherwise, it is used as the starting point for computing the system noise temperatures.

TABLE 15.10 Modem Detection Budget

Parameter	Value	Units	Comments
Modulation	1	Bits/symbol	BPSK
Data rate R_b	19.2	kbps	Specified
Theoretical E_b/N_o[a,b]	2	dB	Input
Bandwidth B_{cb}[b]	57.6	kHz	Baseband
Link margin	1	dB	Input
Implementation loss	1.4	dB	Computed
Demod phase noise	0.2	dB	Input
PLL tracking	0.4	dB	Input
Symbol tracking	0.2	dB	Input
ACI/CCI	0.2	dB	Input
MF detection	0.1	dB	Input
Quantization	0.1	dB	Input
Signal processing	0.2	dB	Input
Modem spec. E_b/N_o	4.4	dB	Computed
Modem spec. $C/N_o\|_{reqd}$	47.2	dB-Hz	Computed

[a]At $P_{be} = 10^{-5}$.
[b]Rate 1/3 turbo code.

TABLE 15.11 Modem Acquisition Budget

Parameter	Value	Units	Comments
Data rate R_b	19.2	kbps	Specified
Theoretical E_{cb}/N_o[a]	−2.8	dB	Input
Bandwidth B_{cb}[b]	57.6	kHz	Baseband
Link margin	0	dB	Input
Implementation loss[c]	2.6	dB	Computed
Channel fading	0.4	dB	Input
ACI/CCI	0.2	dB	Input
Interference	0.2	dB	Input
Antenna scalloping			
Frequency scalloping	0.25	dB	Input
Time scalloping	0.25	dB	Input
NonCoh integration	1.0	dB	Input
Quantization	0.1	dB	Input
Signal processing	0.2	dB	Input
Acquisition spec. P_{acq}	0.95	—	Probability
Modem spec. E_{cb}/N_o	−0.2	dB	Computed
Modem spec. C/N_o	47.4	dB-Hz	Computed

[a]This is 0.8 dB lower than detection budget.
[b]Rate 1/3 turbo code.
[c]Noncoherent integration losses are overcome with increased acquisition time.

needed to establish the link budget are the required signal-to-noise ratio (E_b/N_o) and the user data rate (R_b). E_b/N_o is the required signal-to-noise ratio, measured in a bandwidth equal to R_b and is selected to achieve a specified bit-error probability based on the waveform modulation. The required or specified carrier-to-noise density ratio is computed as

$$\left.\frac{C}{N_o}\right|_{reqd} = \left(\frac{E_b}{N_o}\right)_{reqd} R_b = \left(\frac{E_b}{N_o}\right)_{theory} R_b\ LM \qquad (15.174)$$

Where $LM \geq 1$ is the selected link margin and is a significant specification. The link is established, or closed, when the received C/N_o satisfies the condition

$$\left.\frac{C}{N_o}\right|_{rec} \geq \left.\frac{C}{N_o}\right|_{reqd} \qquad (15.175)$$

The channel bandwidth W must support the transmitted symbol rate which may be significantly wider than the data rate with FEC coding and spread-spectrum processing. During acquisition the channel bandwidth may be selected to support the frequency uncertainty range of the received signal.

The modem link budget typically focuses on the message detection requirements as shown in Table 15.10. However, an equally important consideration is the modem acquisition link budget shown in Table 15.11. In both of these modem link budgets, the selection of the link margins is established; however, the theoretical E_b/N_o specification for acquisition may be selected one or two decibels lower than for data detection; an alternate method is to choose the acquisition link margin somewhat less than that selected for the modem detection budget.* The acquisition processing must detect the presence of the received signal and estimate the necessary parameters with sufficient accuracy to provide for the message detection. The link acquisition budget requirements include scallop losses associated with antenna scanning, symbol time and frequency uncertainties, and other modem-related losses as shown in the acquisition loss budget. During the initial search for signal detection, the carrier frequency and phase are unknown and noncoherent integration may be required to establish an adequate signal-to-noise ratio to meet the specified correct acquisition probability.

The channel scintillation and fading losses are influenced by the waveform design and the demodulator algorithms used to mitigate these losses. Therefore, although these losses are included as channel loss, they are established by analysis and simulation of the degradation in the bit-error performance for the selected waveform. For example, the channel may result in a scintillation or multipath fading loss of 10–20 dB; however, properly designed and analyzed waveforms, using various combinations of FEC coding, data interleaving, and frequency and spatial diversity, will very likely result in a loss of 1–2 dB. Similarly, the receiver intersymbol interference (ISI) loss is evaluated based on the waveform modulation and is typically determined by evaluation of the sensitivity

*The intent is that the acquisition processing should be more robust than the data detection processing; however, this may occur naturally due to the uniqueness of an acquisition preamble. As long as the specified correct and missed acquisition probabilities are satisfied the link will behave as designed. This is the major responsibility during the design and simulation of the demodulator.

of the modulated waveform to the filter bandwidth, amplitude ripple, and phase linearity. The performance loss of ACI and co-channel interference (CCI) is usually determined by analysis and verified by simulation of the improvements attainable through the use of adaptive canceling algorithms.

ACRONYMS

ACI	Adjacent channel interference
ADC	Analog-to-digital converter
AM	Amplitude modulation
AM-AM	Amplitude modulation to amplitude modulation (conversion)
AM-PM	Amplitude modulation to phase modulation (conversion)
AR	Axial ratio
BPSK	Binary phase shift keying
C/I	Carrier-to-intermodulation noise (ratio)
C/I_o	Carrier-to-intermodulation noise density (ratio)
CCI	Co-channel interference
CCIR	Consultative Committee on International Radio
CP	Circular polarized (polarization)
CW	Continuous wave
EIRP	Effective isotropic radiated power
EP	Elliptical polarization
ERP	Effective radiated power
FDM	Frequency division multiplex
FDMA	Frequency division multiple access
FEC	Forward error correction
FFT	Fast Fourier transform
G/T	Gain-temperature ratio (receive antenna)
HP	Horizontal polarized (polarization)
HPA	High-power amplifier
I&D	Integrate-and-dump
I/Q	In-phase and quadrature (rails)
IBO	Input backoff
IF	Intermediate frequency
IM	Intermodulation
IMP	Intermodulation product
ISI	Intersymbol interference
LEO	Low Earth orbit (satellite)
LHC	Left-hand circular
LHCP	Left-hand circular polarized (or polarization)
LNA	Low-noise amplifier
LOS	Line of sight
LP	Linear polarization
MPSK	Multiphase shift keying
OBO	Output backoff
OFDM	Orthogonal frequency division multiplex
OQPSK	Offset quadrature phase shift keying
PA	Power amplifier
PCT	Percent-of-time
PLL	Phaselock loop
PM	Phase modulation
PSD	Power spectral density
QAM	Quadrature amplitude modulation
QPSK	Quadrature phase shift keying
RF	Radio frequency
RHC	Right-hand circular
RHCP	Right-hand circular polarized (polarization)
SAM	Simple attenuation model
SCPC	Single carrier per channel
SRRC	Spectral root-raised-cosine
SSPA	Solid-state power amplifier
SWR	Standing wave ratio
TDM	Time division multiplex
TWT	Traveling wave tube
TWTA	Traveling wave tube amplifier
VP	Vertical polarized
VSWR	Voltage standing wave ratio

PROBLEMS

1. Given a receiver antenna with a physical temperature of $T_a = 15°C$, $T_{ain} = 0$, and a gain of $G = 50$ dB. The antenna is connected to a receiver (with a noise figure $F_n = 6$ dB) through a lossy 1 dB cable at a physical temperature of 15°C. Determine the antenna *G/T* ratio in dB/°K. Repeat this problem using $T_{ain} = 1000°K$ and, using (15.13), determine the increase in the received signal-to-noise ratio required to maintain system performance.

2. Derive the expression of the output temperature T_{out} for a cascade of two noisy filters shown in the following figure. Use the indicated losses and physical temperatures of the noisy devices; all of the temperatures are in °K. With $T_{in} = 0$, the effective input temperature of the two cascaded devices is defined as $\bar{T} = L_1 L_2 T_{out}$. Evaluate the effective input temperature of the cascaded devices and express the result in terms of the effective temperatures $\bar{T}_1$ and $\bar{T}_2$ of the two equivalent noiseless devices.

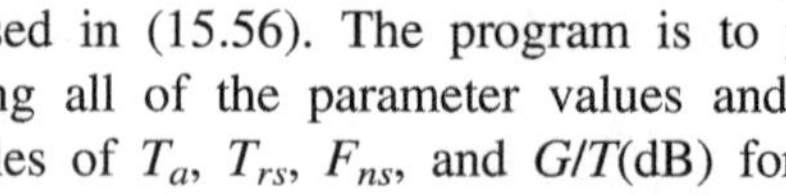

Cascade of two noisy filters.

3. Write a general simulation program for the example in Section 15.2.3.1 to evaluate the antenna *G/T* as expressed in (15.56). The program is to provide for inputting all of the parameter values and outputting data files of T_a, T_{rs}, F_{ns}, and *G/T*(dB) for $0 \le T_{Sun} \le$ 100,000°K in steps of 500°K; create the plots parametrically with G_a and θ_B. The decreasing Sun temperatures

reflect the percentage of the Sun's area in the antenna beamwidth for a given ratio Ω_{Sun}/Ω_a.

4. Referring to the antenna gain (G) as expressed in Equation 15.3, derive expressions for: (A) the diameter D of a circular (dish) antenna as a function of the gain G; (B) the 3-dB beamwidth θ_B for a uniformly weighted circular aperture as a function of G; (C) plot D in meters and θ_B in degrees as a function of the gain G in dB for carrier frequencies of 100 and 500 MHz.

5. Using the general expression for the elliptically polarized wave given by (15.89), derive the expression for E_x as a function of E_y and the parameters M_x and M_y for the LP, that is, for $\delta = 0$ and π.

6. For ideally polarized received waves and antennas with the indicated polarizations, determine the polarization mismatch factor and the polarization loss, expressed in decibels, for the following cases:

a. LHCP received wave and RHCP receiver antenna.
b. RHCP received wave and RHCP receiver antenna.
c. LHCP received wave and LP receiver antenna.
d. LP received wave and LP receiver antenna.
e. Elliptically polarized received wave and LP receiver antenna.

7. Develop a computer program and evaluate the cross-polarization discrimination, *xpd*, and the isolation I_{isol} for a nonideal VP antenna using a linearly polarized received wave with a VP copolarization state. This involves evaluating (15.100) using (15.101) and (15.95). Start by generating the ideal *xpd* vs. τ'_w curve using ARs of $r = 10^4$ for the received wave and ideal VP antenna. Then evaluate the nonideal antenna isolation under the four conditions: $r_{ax} = r_{ac} = 300$, 100, 30, and 10 for an ideal received wave using $r_{\bar{w}} = r_w = 10^4$; these ARs are linear values, not in dB. Plot the I_{isol} vs. *xpd* in decibels with each axis ranging from 0 to 50 dB.

Hints: For VP polarization the antenna and the VP copolarized wave the tilt angles are $\tau_w = \tau_{ac} = 90°$ so $\tau_{wo} = \tau_{ax} = 0°$. Also, in generating the ideal *xpd* curve use tilt angles $\tau'_w = 0(0.125)180°$, that is, 0° to 180° in increments of 0.125°; this will result in a good quality plot.

8. Develop a computer program and evaluate the cross-polarization discrimination, *xpd*, for an ideal VP antenna using an ideal linearly polarized received wave with a VP copolarization state under the conditions described in Problem 7. (This part is complete if Problem 7 was examined.) Now develop the expression for the theoretical *xpd* for the same VP wave and antenna and compare with the earlier computer-generated *xpd* by plotting both curves on the same graph.

9. Derive the expression for the noise figure of a cascade of N linear amplifiers with individual noise figures F_{ni}, gains G_i, and effective temperatures $\bar{T}_i$, $i = 1, \ldots, N$.

10. Derive the expression for the effective noise temperature of a cascade of N lossless attenuators each having physical temperature T_{pi} and loss L_i, $i = 1, \ldots, N$.

11. An antenna with effective temperature $\bar{T}_a$ is covered by a radome with a loss L_{rad} and effective temperature $\bar{T}_{rad}$. The antenna output is connected to an antenna feed with loss L_{rfd} and effective temperature $\bar{T}_{rfd}$. Express the effective temperature of a single equivalent device, with loss $L = L_{rad} L_{rfd}$, in terms of the temperatures: $\bar{T}_{rad}$, $\bar{T}_a$, and $\bar{T}_{rfd}$. Hint: Compute $\bar{T}_c$ in the following figure.

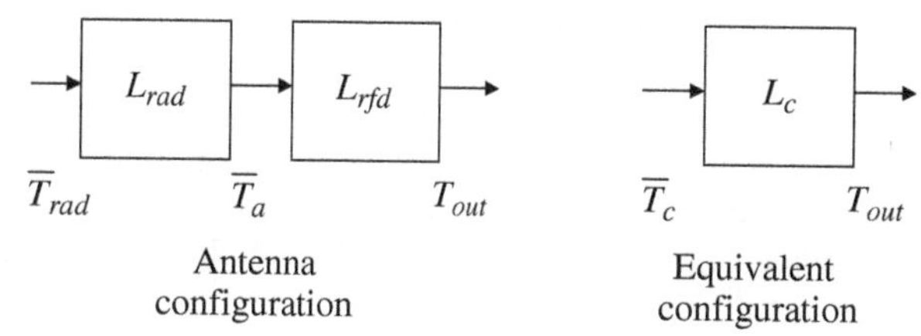

12. Consider that the antenna feed, described in Problem 11, is connected to the input of an LNA with a receiver system noise temperature T_{rs}. If the antenna gain is 60 dB derive the expression for the antenna system temperature and the antenna G/T ratio and indicate the units for each result.

13. For the gain compensated TWTA described in Section 15.12.2, plot the principal signal linear response and the third- and fifth-order intermodulation responses. Then identify the intercept points IP_3 and IP_5 and determine the intermodulation distortion terms IM_3 and IM_5 for two values of the OBO. Express all results in decibels.

REFERENCES

1. R.S. Berkowitz, Editor, *Modern Radar: Analysis, Evaluation, and System Design*, John Wiley & Sons, Inc., New York, 1965.
2. R.S. Berkowitz, Editor, *Modern Radar: Analysis, Evaluation, and System Design*, PART V, Radio Frequency Considerations, Chapter 3, S. Charton, "Radar System Sensitivity," pp. 415–430, John Wiley & Sons, Inc., New York, 1965.
3. M.I. Skolnik, *Introduction to Radar Systems*, Chapter 7, "Antennas," McGraw-Hill, New York, 1962.
4. R.S. Berkowitz, Editor, *Modern Radar: Analysis, Evaluation, and System Design*, PART V, Radio Frequency Considerations, Chapter 3, S. Charton, "Radar System Sensitivity," pp. 415–420, John Wiley & Sons, Inc., New York, 1965.

5. L.V. Blake, *Radar Range-Performance Analysis*, Munro Publishing Co., Silver Spring, MD, 1991.
6. G.P. Kuiper, Editor, *The Solar System*, Vol. 1, Chapter 7, "The Sun," p. 485, University of Chicago Press, Chicago, 1953.
7. J.D. Kraus, H.C. Ko, "Celestial Radio Radiation," Ohio State University Report 1 of Radio Frequency (RF) Project 673, U.S. Air Force Contract 19(604)-1591, Columbus, OH, May 1957.
8. L.V. Blake, *Radar Range-Performance Analysis*, pp. 160–162, Munro Publishing Co., Silver Spring, MD, 1991.
9. R.H. Brown, C. Hazard, "A Model of the Radio Frequency Radiation from the Galaxy," Philosophical Magazine (Philos. Mag.), Vol. 44, pp. 939–963, September 1953.
10. D.C. Hogg, W.W. Mumford, "The Effective Noise Temperature of the Sky," Microwave Journal, Vol. 3, No. 3, pp. 80–84, March 1960.
11. H.C. Ko, "The Distribution of Cosmic Radio Background Radiation," Proceeding of the IRE, Vol. 46, pp. 208–215, January 1958.
12. S.F. Smerd, "Radio-Frequency Radiation from the Quiet Sun," Australian Journal of Scientific Research, Vol. A3, p. 34, 1950.
13. P.D. Strum, "Considerations in High-Sensitivity Microwave Radiometry," Proceedings of the IRE, Vol. 46, Issue 1, pp. 43–53, January 1958.
14. R.S. Berkowitz, Editor, *Modern Radar: Analysis, Evaluation, and System Design*, Part V, Chapter 3, S. Charton, "Radar System Sensitivity," pp. 412–414, John Wiley & Sons, Inc., New York, 1965.
15. R. Gardner, "Final Engineering Report on Antenna Noise Temperature Study," AIL Report 3304-11, Contract DA-49-170-SC-1547, Airborne Instruments Laboratory, Mineola, NY, November 1957.
16. J.A. Stratton, *Electromagnetic Theory*, McGraw-Hill, New York, 1941.
17. S. Silver, Editor, *Microwave Antenna Theory and Design*, MIT Radiation Laboratory Series, Vol. 12, McGraw-Hill, New York, 1949.
18. J.D. Kraus, *Antennas*, McGraw-Hill, New York, 1949.
19. L.J. Ippolito, *Radiowave Propagation in Satellite Communications*, Van Nostrand Reinhold Co., Inc., New York, 1986.
20. L.J. Ippolito, "Radio-Wave Propagation for Space Communications Systems," NASA Technical Paper 1770, p. 29, Figure 12 and Table 9, NASA Office of Space and Terrestrial Applications, Scientific and Technical Information Branch, NASA National Aeronautics and Space Administration, Washington, DC, 1981.
21. R.K. Crane, "Prediction of Attenuation by Rain," IEEE Transactions on Communications, Vol. COM-28, No. 9, pp. 1717–1733, September 1980.
22. G.A. Arnold, T.W. Kao, "Models Describe Rain Attenuation in EHF Systems," Microwave and RF, Part 1, pp. 74–83, April 1996.
23. International Telecommunications Union-Radio (ITU-R), "Calculation of Long-Term Attenuation Statistics for Point Rainfall Rate," Report 564-4: Recommendations and Reports of the CCIR, Vol. V, Formerly: Consultative Committee on International Radio (CCIR), Geneva, 1986.
24. A.W. Dissanayake, "Proposed Amendment to Consultative Committee on International Radio (CCIR), Recommendation P618: Propagation Data and Prediction Methods Required for the Design of Earth-Space Telecommunication Systems, Prediction of Rain Attenuation," Document USSG5C/1T, International Telecommunications Union, Geneva, Switzerland, 1992.
25. W.L. Stutzman, K.M. Yon, "A Simple Rain Attenuation Model for Earth-Space Radio Links Operating at 10–35 GHz," Radio Science, Vol. 21, No. 1, pp. 65–72, January–February 1986.
26. R. K. Crane, "A Global Model for Rain Attenuation Prediction," EASCON'78 Record, IEEE Aerospace and Electronic Systems Society, Arlington, VA, pp. 391–395, September 25–27, 1978.
27. W.J. James, "The Effect of the Weather in Eastern England on the Performance of X-Band Ground Radars," Technical Note No. 655, Royal Radar Establishment, Malvern, England.
28. W. Holzer, "Atmospheric Attenuation in Satellite Communications," Microwave Journal, Vol. 8, pp. 119–125, March 1965.
29. L.J. Ippolito, *Propagation Effects Handbook for Satellite System Design*, Fourth Edition, NASA Reference Publication 1082(04), Section 3.4.2, Westinghouse Electric Corporation, Baltimore, MD, 1989.
30. J.D. Kraus, *Antennas*, Chapter 15, "Antenna Measurements," pp. 444–492, McGraw-Hill, New York, 1950.
31. L.J. Ippolito, *Propagation Effects Handbook for Satellite Systems Design*, Fourth Edition, NASA Reference Publication 1082(04), Chapter IV, "Polarization on Earth-Space Path," pp. 4-1–4-54, Westinghouse Electric Corporation, Baltimore, MD, 1989.
32. D.B. Leeson, "A Simple Model of Feedback Oscillators," Proceedings of the IEEE, Vol. 54, pp. 329–330, February 1966.
33. J.A. Barnes, "Models for the Interpretation of Frequency Stability Measurements," National Bureau of Standards, NBS Technical Note 683, U.S. Department of Commerce, Washington, DC, 1976.
34. G. Sauvage, "Phase Noise in Oscillators: A Mathematical Analysis of Leeson's Model," IEEE Transactions on Instrumentation and Measurement, Vol. IM-24, pp. 408–410, December 1977.
35. A. Hajimiri, T.H. Lee, *The Design of Low Noise Oscillators*, Kluwer Academic, Norwell, MA, 1999.
36. T.H. Lee, A. Hajimiri, "Oscillator Phase Noise: A Tutorial," IEEE Journal of Solid-State Circuits, Vol. 35, pp. 326–336, March 2000.
37. W.P. Robins, *Phase Noise in Signal Sources (Theory and Applications)*, Peter Peregrinus, Ltd., London, UK, 1996.
38. F.M. Gardner, *Phaselock Techniques*, Third Edition, Chapter 7, "Effects of Phase Noise," pp. 143–182, Chapter 9, "Oscillators," pp. 209–236, Wiley-Interscience, John Wiley and Sons, Inc., Hoboken, NJ, 2005.
39. A.R. Chi, Guest Editor, "Special Issue on Frequency Stability," Proceeding of the IEEE, Vol. 54, No. 2, pp. 101–338, February 1966.
40. J. Rutman, "Characterization of Phase and Frequency Instabilities in Precision Frequency Sources: Fifteen Years of Progress," Proceeding of the IEEE, Vol. 66, No. 9, September 1978.

41. J. Rutman, G. Sauvage, "Measurement of Frequency Stability in Time and Frequency Domains via Filtering of Phase Noise," IEEE Transactions on Instrumentation and Measurement, Vol. IM-23, pp. 515–518, December 1974.
42. I.S. Gradshteyn, I.M. Ryzhik, *Table of Integrals, Series and Products*," p. 292, Sect. 3.241, no. 2, Academic Press, New York, 1980.
43. U.S.A. Department of Defense (DOD), Interface Standard, MIL-STD-188-164B, Interoperability of SHF Satellite Communications Terminals, p. 19, Figure 1, Terminal Phase Noise, Defense Information Systems Agency (DISA), Fort Meade, MD, March 23, 2012.
44. F.M. Gardner, *Phaselock Techniques*, John Wiley & Sons, Inc., New York, 1967.
45. D. Baker, "The Effects of Phase Noise on High-Order QAM Systems," Communication Systems Design, Vol. 5, No. 10, pp. 35–42, October 1999.
46. H.J. Reich, J.G. Skalnik, P.F. Ordung, H.L. Krauss, *Microwave Principles*, D. Van Nostrand Company, Inc., Princeton, NJ, 1959.
47. A.R. Kaye, D.A. George, M.J. Eric, "Analysis and Compensation of Band-Pass Nonlinearities for Communications," IEEE Transactions on Communications, Vol. COM-20, pp. 965–972, October 1972.
48. A.A.M. Saleh, "Frequency-Independent and Frequency-Dependent Nonlinear Models of TWT Amplifiers," IEEE Transactions on Communications, Vol. COM-29, Issue 11, pp. 1715–1720, November 1981.
49. A.A.M. Saleh, "Intermodulation Analysis of FDMA Satellite Systems Employing Compensated and Uncompensated TWT's," IEEE Transactions on Communications, Vol. COM-30, Issue 5, pp. 1233–1242, May 1982.
50. O. Shimbo, "Effects of Intermodulation, AM-PM Conversion and Additive Noise in Multi-Carrier TWT Systems," Proceeding of the IEEE, Vol. 59, pp. 230–238, February 1971.
51. J.C. Fuenzalida, O. Shimbo, W.L. Cook, "Time-Domain Analysis of Intermodulation Effects Caused by Nonlinear Amplifiers," COMSAT Technical Review, Vol. 3, pp. 89–143, Spring 1973.
52. E. Imboldi, G.R. Stette, "AM-to-PM Conversion and Intermodulation in Nonlinear Devices," Proceedings of the IEEE, Vol. 61, pp. 796–797, June 1973.
53. F.E. Bond, H.F. Meyer, "Intermodulation Effects in Limiter Amplifier Repeaters," IEEE Transactions on Communications, COM-22, pp. 127–135, April 1970.
54. M. Abramowitz, I.A. Stegun, *Handbook of Mathematical Functions With Formulas, Graphs, and Mathematical Tables*, Second Printing, p. 228, eq. 5.1.1, November 1964, National Bureau of Standards Applied Mathematics Series, 55, U.S. Department of Commerce, Washington, DC, June 1964.

16

SATELLITE ORBITS

16.1 INTRODUCTION

The analysis and characterization of satellite orbits considered in this chapter are based entirely on Newton's laws of motion, his law of universal gravitation, and Kepler's laws of planetary motion. Briefly stated, Newton's laws of motion are as follows: a body will remain in its present state until influenced by an outside force; the rate of change of angular momentum of a body is proportional to and in the direction of an applied force; and every action has an equal and opposite reaction. Newton's law of universal gravitation states that the force of attraction between two bodies is proportional to the product of their masses and inversely proportional to the square of the distance between them. Similarly, Kepler's laws are summarized as follows: the planetary orbit is an ellipse with the Sun at one focus; for all of the planets, the line between the planet and the Sun sweeps through equal areas in equal time intervals; and the square of the orbital period is proportional to the cube of the mean distance to the Sun.

This chapter applies these laws to the Earth's elliptical satellite orbits with the Earth at one focus; the circular orbit being a special case. The emphasis is on characterizing communication links between the Earth and various satellites and cross-links between satellites. In addition to elliptical satellite orbits, the results also apply to ballistic missile trajectories. Parabolic and hyperbolic orbits are involved more with interplanetary travel and transfer between orbits as may be required, for example, to establish an elliptical or circular Earth orbit. Although these are important transitional orbit maneuvers, requiring communication links to be established and maintained, the system requirements and their impact on the communication links are significantly different from the established Earth satellite orbits and ballistic missile trajectories; therefore, parabolic and hyperbolic orbits are not considered. Furthermore, the gravitational attraction of the Moon and Sun and neighboring planets, the influence of the Earth's magnetic field and interior composition, solar radiation, charged and uncharged particles, and atmospheric drag are all ignored; however, these subjects are addressed by Roy [1].

In this chapter, the range, range-rate (velocity), and acceleration along the line of sight (LOS) between a satellite and ground station are evaluated and, based on the results, the range, Doppler, and Doppler rate are characterized. The following analysis focuses on Earth satellites and therefore applies to elliptical orbits defined in terms of the six fundamental orbital parameters ($\Omega, i, \omega, T, \varepsilon$, and τ) as shown in Figure 16.1. Alternative sets of orbital parameters can be chosen with a straightforward rearrangement of the equations; for example, on occasions it is more convenient to replace the orbit period (T) with the semimajor axis length (a) of the ellipse. The development of the relationships discussed in this chapter draws upon the works of Roy [1] and Bate, Mueller, and White [2]. Furthermore, the analysis assumes many ideal conditions embodied in the basic assumptions: two-body conditions, point masses, and free space. The first assumption ignores the influences of neighboring bodies like the Sun and Moon and other satellites. The second assumption ignores the influence on the orbit trajectories of the Earth's interior and nonspherical shape. The third assumption

Digital Communications with Emphasis on Data Modems: Theory, Analysis, Design, Simulation, Testing, and Applications, First Edition. Richard W. Middlestead.

Companion website: www.wiley.com/go/digitalcommunications

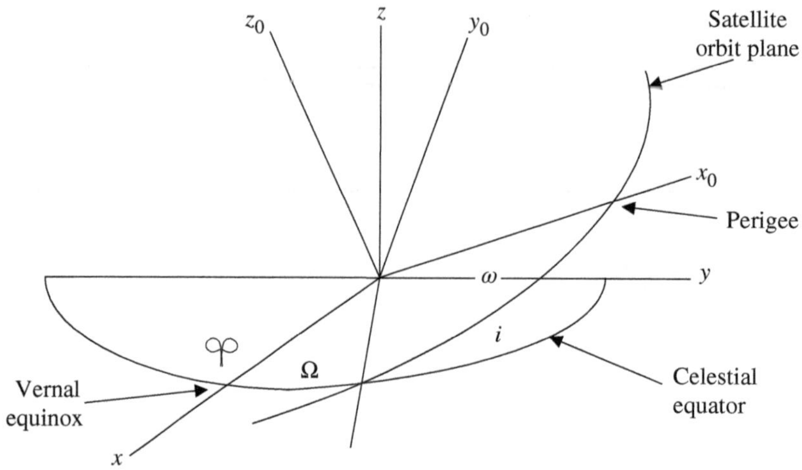

x, y, z :	Reference plane coordinates	Ω :	Longitude of ascending node
(x_0, y_0, z_0) :	Orbit plane coordinates	ω :	Argument of perigee
		i :	Orbit plane inclination

FIGURE 16.1 Satellite orbit plane, reference coordinates, and parameter definitions.

ignores the influence of atmospheric drag on the satellite trajectory. The elliptical shape and rotation of the Earth is, however, considered in determining the range and range-rate between satellites and the ground stations. Although the following analysis focuses on establishing the orbit for a single satellite, many applications involve several orbital planes with one or more satellites in each plane. In these cases, it is assumed that the mutual gravitational forces between the satellites is negligible and communication cross-links are used for relaying information from various Earth locations for which there is no direct LOS link. Roy [3] discusses the influence of the many-body problem when the mutual gravitational forces are not negligible, for example, in the case of an artificial satellite transitioning from an Earth-to-Moon orbit.

The analysis that follows focuses on Earth satellites and the xy-axes of the reference coordinates lie in the plane of the celestial equator, that is, the projection of the Earth's equator to the *great circle* on the celestial sphere, and the positive z-axis points northward through the North Pole. The positive x-axis passes through the vernal equinox also referred to as the First Point of Aries as shown in Figure 16.1. The satellite orbit plane is an ellipse with the focus located at the center of mass of the Earth. Table 16.1 provides a convenient reference for various constants and parameters used in this chapter.

The distinction between the sidereal and solar day is based on the celestial point of reference that is used to establish the time interval between two successive reference-point crossings of the observer's meridian. The sidereal day uses a distant celestial star and the solar day uses the Sun. The period of a satellite in a circular orbit in the plane of the Earth's equator is approximated as [4]

$$\begin{aligned} T &\cong 2\pi\sqrt{\frac{a^3}{u}}\left[1-\frac{3}{4}\left(\frac{R_{eq}}{a}\right)^2 J_2\right] &&: \text{circular equatorial orbit} \\ &\cong 2\pi\sqrt{\frac{a^3}{u}} &&: \text{seconds} \end{aligned} \tag{16.1}$$

where $a = R_{eq} + A$ and $J_2 = 1082.86e{-}6 \pm 10^{-7}$ is a constant that accounts for the second-order effect of the Earth's asymmetry between the Northern and Southern Hemispheres. The term inside the square bracket of (16.1) is approximately unity so the second approximation is typically used even for satellite orbit altitudes as low as $A = 180$ km. Using the equatorial Earth radius and satellite altitudes listed in Table 16.1, the satellite orbit period, based on a sidereal day, is about 4.023 min shorter than that of the 24 h solar day.

For an elliptical orbit with an arbitrary inclination, the orbit period is approximated as [4]

$$\begin{aligned} T &\cong 2\pi\sqrt{\frac{r^3}{u}}\left[1-\left(\frac{R_{eq}}{r}\right)^2\left(\frac{3}{2}J_2+\frac{\omega_e^2 R_{eq}^3}{2GM}\right)\right]^{1/2} &&: \text{elliptical orbit} \\ &\cong 2\pi\sqrt{\frac{r^3}{u}} &&: \text{seconds} \end{aligned} \tag{16.2}$$

where, referring ahead to Figure 16.2 and (16.8), the magnitude of the satellite radius vector at perigee,* corresponding

*When referring to Earth-satellite (Sun–Moon) orbits, perigee (perihelion) and apogee (aphelion) refer to the orbits location nearest and farthest, respectively, from the Earth (Sun).

TABLE 16.1 Earth and Satellite Orbit Constants

Constant	Value	Description
R_e	6,378.28 km	Earth's radius (average)
R_{eq}	6,378.145 km	Earth's radius (equator)
R_{ep}	6,356.74 km	Earth's radius (polar)
ε_e	0.08183	Earth's eccentricity
ε	Selectable	Orbit eccentricity
G	$6.66560515 \times 10^{-8}$ dyne-cm^2/g	Constant of gravitation
M_e	5.98×10^{27} g	Mass of Earth
M_s	Negligible	Mass of satellite
u	398,603.188 km^3/s^2	Gravitational constant $GM = G(M_e + M_s) \cong GM_e$
ω_e	π/1,800 rad/s	Angular rotation of Earth
$A_{geosidereal}$	35,784.334 km	Geostationary satellite altitude (sidereal day)
$R_{geosidereal}$	42,162.479 km	Radius of geostationary satellite ($R_{eq} + A_{geosidereal}$)
$A_{geosolar}$	35,863.047 km	Geostationary satellite altitude (solar day)
$R_{geosolar}$	42,241.192 km	Radius of geostationary satellite ($R_{eq} + A_{geosolar}$)

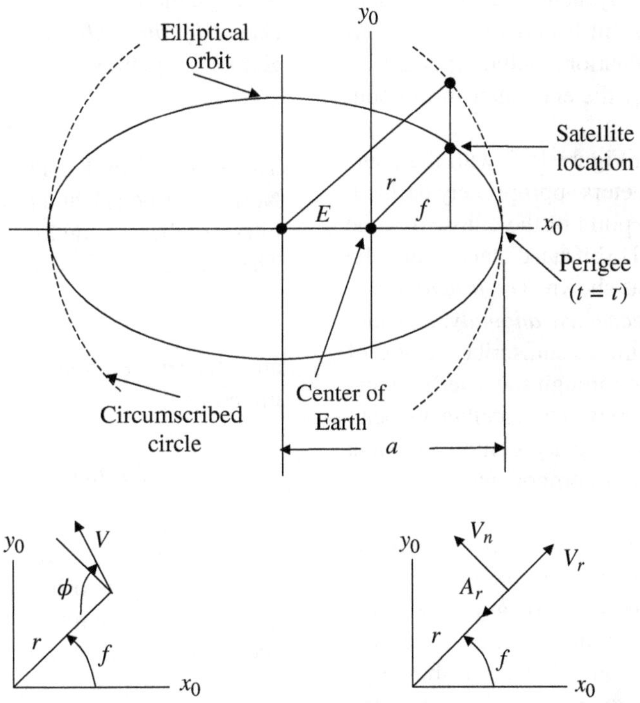

$V = V_r\boldsymbol{u}_r + V_n\boldsymbol{u}_n$: Velocity vector
V_r : Radial component of velocity
V_n : Normal component of velocity

r : Magnitude of satellite location vector
f : Satellite location angle (true anomaly)
E : Eccentric anomaly
a : Semi major axis of elliptical orbit

FIGURE 16.2 Satellite orbit, reference coordinates, and parameter definitions.

to $t=\tau$, is $r=a(1-\varepsilon)$ for which the second approximation in (16.2) is expressed as

$$T \cong 2\pi\sqrt{\frac{a^3(1-\varepsilon)^3}{u}} \quad : \text{elliptical orbit} \tag{16.3}$$

These considerations of the satellite orbit time lead to another important distinction in the application of the satellite orbits discussed in this chapter. The satellite orbit time is related to the time starting at $t=0$ corresponding to the geocentric longitude of the ascending node* as indicated in Figure 16.1. This satisfies the objective of evaluating satellite communication systems and link performance. However, it does not include the impact of the planets, the Moon, Sun, stars, and other galactic noise sources as they would appear in the antennas' field of view (FOV) based on data from the American Ephemeris and other almanacs. These effects, however, can be artificially inserted in the FOV and the system performance evaluated parametrically. The Satellite Orbit Analysis Program (SOAP) [5] includes celestial backgrounds in the FOV and other useful features for communication systems' performance evaluation, including data files and three-dimensional displays of the encounter throughout the orbit period.

The satellite orbit is shown in Figure 16.2 with the coordinate system and essential parameters appropriately defined. The Earth is located at the focus point of the ellipse and the satellite is located on the elliptical orbit at a distance r and the angle f, called the *true anomaly*, as shown. The *true anomaly* is determined with the aid of the *eccentric anomaly*, E, established using the intersection of the circumscribed circle of diameter $2a$ with the vertical line through the satellite location and perpendicular to the x_0-axis. The satellite velocity vector, the normal or tangential velocity component, and the radial velocity and acceleration components are shown in the two detailed depictions.

In the following evaluations, the orbit eccentricity is denoted as ε and the parameter u is defined as $u=G(M_e+M_s)$ where G is the constant of gravitation and M_e and M_s are the masses of the Earth and satellite, respectively. When $\varepsilon<1$, the orbit is an ellipse as shown in Figure 16.2, $\varepsilon=1$ corresponds to a parabolic orbit and $\varepsilon>1$ corresponds to a hyperbolic orbit. This chapter focuses exclusively on Earth satellites in elliptical orbits; however, the performance of ballistic missile communication and mono- and bi-static radar [6] systems has been examined using elliptical and parabolic trajectories.

*In Figure 16.1, the ascending node is labeled the *vernal equinox*, also referred to as the First Point of Aries. This is the point in the Earth's orbit nearest to the Sun. When referring to planetary orbits, this point is called "perihelion."

16.2 SATELLITE ORBITS

Based on the geometry in Figure 16.2, the position, velocity, and acceleration of the satellites are evaluated using the expression for the semimajor axis of the ellipse expressed as

$$a=\left[\left(\frac{T}{2\pi}\right)^2 u\right]^{1/3} \quad : \text{km} \tag{16.4}$$

The mean angular velocity

$$n=\frac{2\pi}{T}=\sqrt{\frac{u}{a^3}} \quad : \text{rad/s} \tag{16.5}$$

is used to compute the *mean anomaly* as

$$\bar{M}=n(t-\tau)=2\pi\left(\frac{t-\tau}{T}\right) \quad : \text{rad} \tag{16.6}$$

The mean anomaly is the orbital angle of the satellite from perigee. For example, if K satellites are spaced at equal angles in the same orbit, then $\bar{M}_i=2\pi i/K$: $i=0, 1, \ldots, K-1$. The *eccentric anomaly*, E, is determined as the solution to Kepler's equation expressed as the transcendental equation

$$E-\varepsilon\sin(E)=\bar{M}=n(t-\tau) \quad : \text{rad} \tag{16.7}$$

The solution to Kepler's equation is accomplished using Newton's rule [7], and the result is used to compute the magnitude of the satellite location and velocity vectors as follows. The magnitude of the location vector is

$$r=a[1-\varepsilon\cos(E)] \quad : \text{km} \tag{16.8}$$

and the *true anomaly* or satellite location angle is determined as

$$f=2\tan^{-1}\left[\sqrt{\frac{1+\varepsilon}{1-\varepsilon}}\tan\left(\frac{E}{2}\right)\right] \quad : \text{rad} \tag{16.9}$$

These relationships are all that is necessary for locating the satellite along the orbit path at any instant in time. Although not explicitly required in the following analysis, the angle ϕ, shown in Figure 16.2, is computed as

$$\phi=\frac{\pi}{2}+\cos^{-1}\left(\sqrt{\frac{a^2(1-\varepsilon^2)}{r(2a-r)}}\right)=\sin^{-1}\left(\sqrt{\frac{a^2(1-\varepsilon^2)}{r(2a-r)}}\right) \tag{16.10}$$

The velocity vector, $\mathbf{V}=V_r\mathbf{u_r}+V_n\mathbf{u_n}$, is tangent to the orbit path and is characterized in terms of the velocity components radial and normal to the satellite location vector. These velocities are given by

$$V_r=\dot{r}=\frac{na\varepsilon}{\sqrt{1-\varepsilon^2}}\sin(f) \quad : \text{km/s} \tag{16.11}$$

and

$$V_n = r\dot{f} = nr\left(\frac{a}{r}\right)^2\sqrt{1-\varepsilon^2} \quad :\text{km/s} \tag{16.12}$$

Equation (16.12) uses the expression for the angular velocity of the satellite location vector given by

$$\dot{f} = n\left(\frac{a}{r}\right)^2\sqrt{1-\varepsilon^2} \quad :\text{rad/s} \tag{16.13}$$

The magnitude of the velocity vector $V = |\mathbf{V}|$ is

$$V = \sqrt{u\left(\frac{2}{r}-\frac{1}{a}\right)} \quad :\text{km/s} \tag{16.14}$$

Similarly, the components of the acceleration vector $\mathbf{A} = A_r\mathbf{u_r} + A_n\mathbf{u_n}$ are evaluated as

$$\begin{aligned} A_r &= \ddot{r} - r\dot{f}^2 = -\frac{u}{r^2} \\ &= -\frac{G(M_e+M_s)}{r^2} \cong -\frac{GM_e}{r^2} \quad :\text{km/s}^2 \end{aligned} \tag{16.15}$$

and

$$A_n = \frac{1}{r}\frac{d}{dt}(r^2\dot{f}) = 0 \tag{16.16}$$

Equation (16.16) is based on Kepler's second law requiring that

$$r^2\dot{f} = h = \sqrt{ua(1-\varepsilon^2)} = \text{constant} \tag{16.17}$$

These relationships must be expressed in terms of the orbit coordinates (x_0, y_0, z_0). In this regard, the satellite location is expressed as the satellite range vector $\mathbf{R}$ as

$$\boldsymbol{R}_{os} = r\cos(f)\boldsymbol{u}_{ox} + r\sin(f)\boldsymbol{u}_{oy} \tag{16.18}$$

and the velocity and acceleration vectors are expressed as

$$\boldsymbol{V}_{os} = (V_r\cos(f) - V_n\sin(f))\boldsymbol{u}_{ox} + (V_r\sin(f) + V_n\cos(f))\boldsymbol{u}_{oy} \tag{16.19}$$

$$\boldsymbol{A}_{os} = A_r\cos(f)\boldsymbol{u}_{ox} + A_r\sin(f)\boldsymbol{u}_{oy} \tag{16.20}$$

Finally, the transformation of these vectors in the orbit coordinates to the reference coordinates of Figure 16.1 is accomplished using Euler's transformation, $\boldsymbol{T}_u$, as

$$\boldsymbol{R}_s = \boldsymbol{T}_u\boldsymbol{R}_{os};\ \boldsymbol{V}_s = \boldsymbol{T}_u\boldsymbol{V}_{os};\ \boldsymbol{A}_s = \boldsymbol{T}_u\boldsymbol{A}_{os} \tag{16.21}$$

where the Euler transform is expressed in terms of the Euler angles (Ω, $\imath$, ω) as

$$\boldsymbol{T}_u = \begin{pmatrix} \cos(\Omega)\cos(\imath)+\sin(\Omega)\sin(\omega)\sin(\imath) & \sin(\Omega)\sin(\omega)\cos(\imath)-\cos(\Omega)\sin(\imath) & \sin(\Omega)\cos(\omega) \\ \cos(\omega)\sin(\imath) & \cos(\omega)\cos(\imath) & -\sin(\omega) \\ \cos(\Omega)\sin(\omega)\sin(\imath)-\sin(\Omega)\cos(\imath) & \cos(\Omega)\sin(\omega)\cos(\imath)+\sin(\Omega)\sin(\imath) & \cos(\Omega)\cos(\omega) \end{pmatrix} \tag{16.22}$$

In the following analysis, the computations involving the range, velocity, and acceleration vectors of the satellites and Earth stations are performed in the Earth-fixed coordinates so the satellite vector data must be transformed to this system. Therefore, the following additional transformations will place the satellite data in the desired Earth-fixed coordinates.

$$\boldsymbol{R}_{es} = \boldsymbol{T}\boldsymbol{R}_s;\ \boldsymbol{V}_{es} = \boldsymbol{T}\boldsymbol{V}_s;\ \boldsymbol{A}_{es} = \boldsymbol{T}\boldsymbol{A}_s \tag{16.23}$$

where the transformation is given by

$$\boldsymbol{T} = \begin{pmatrix} \cos(\omega_e t) & -\sin(\omega_e t) & 0 \\ \sin(\omega_e t) & \cos(\omega_e t) & 0 \\ 0 & 0 & 1 \end{pmatrix} \tag{16.24}$$

16.3 EARTH STATIONS

The Earth stations are defined in terms of their latitude and longitude with respect to the Earth-centered Earth-fixed coordinate system (x_e, y_e, z_e) as shown in Figure 16.3. Only fixed Earth sites are considered in the following discussions; however, the inclusion of mobile sites (ships, airplanes, etc.) can be included with appropriate modifications. The xy-plane contains the Earth's equator and the positive x-axis is aligned with the Greenwich meridian with the z-axis passing through the North Pole. The Earth-fixed coordinates rotate with respect to the fixed reference coordinates (x, y, z) at an angular rate of $\omega_e = 2\pi/3600$ rad/s or one revolution in a 24 h period.

The location of the Earth station is shown in Figure 16.3 at point P. A more detailed depiction is shown in Figure 16.4 that explicitly shows the local altitude a_p of the Earth station

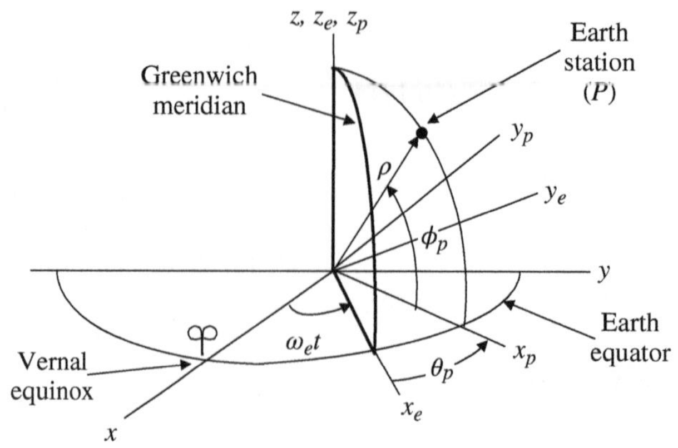

FIGURE 16.3 Earth station locations, reference coordinates, and parameter definitions.

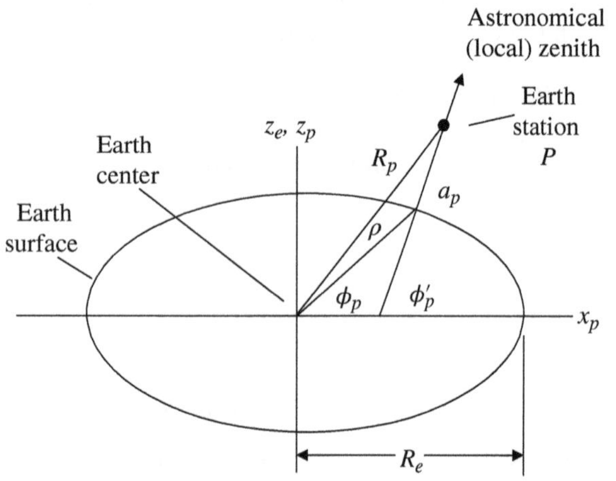

FIGURE 16.4 Elliptical Earth model to Earth coordinates.

and the resulting distance R_p from the Earth's center. Figure 16.4 also shows the elliptical shape of the Earth in the x_p– z_p plane containing the Earth station. To express the results in terms of the Earth-fixed coordinates, the (x_s, y_s, z_s) coordinates are simply rotated through the angle θ_p as indicated in Figure 16.3.

The vector $\boldsymbol{R_p}$ is the sum of the position vectors $\boldsymbol{\rho}$ and $\boldsymbol{a_p}$. The vector $\boldsymbol{\rho}$ is determined from the distance ρ, latitude ϕ_p, and longitude θ_p where positive latitudes are north from the equator and positive longitudes are measured east from the Greenwich meridian.

The Earth is elliptical, with eccentricity $\varepsilon_e = 0.08183$, so the astronomical zenith angle is a function of the latitude of the Earth station and is computed as

$$\phi'_p = \tan^{-1}\left[\frac{\tan(\phi_p)}{1-\varepsilon_e^2}\right] \tag{16.25}$$

The radial distance ρ is also related to the latitude of the Earth station and is expressed in terms of the astronomical zenith angle as

$$\rho = R_e\left[\frac{1-(2\,\varepsilon_e^2-\varepsilon_e^4)\sin^2\left(\phi'_p\right)}{1-\varepsilon_e^2\sin^2\left(\phi'_p\right)}\right]^{1/2} \tag{16.26}$$

Using these results, the vector $\boldsymbol{\rho}$ in Earth-fixed coordinates is given by

$$\boldsymbol{\rho} = \rho\cos(\phi_p)\cos(\theta_p)\boldsymbol{u}_{ex} + \rho\cos(\phi_p)\sin(\theta_p)\boldsymbol{u}_{ey} + \rho\sin(\phi_p)\boldsymbol{u}_{ez} \tag{16.27}$$

and similarly the altitude vector $\boldsymbol{a_p}$ is given by

$$\begin{aligned}\boldsymbol{a_p} &= a_p\cos\left(\phi'_p\right)\cos(\theta_p)\boldsymbol{u}_{ex}\\ &\quad + a_p\cos\left(\phi'_p\right)\sin(\theta_p)\boldsymbol{u}_{ey} + a_p\sin\left(\phi'_p\right)\boldsymbol{u}_{ez}\end{aligned} \tag{16.28}$$

From these results the distance from the Earth's center to the Earth station at altitude $\boldsymbol{a_p}$ is expressed in Earth-fixed coordinates by the range and velocity vectors

$$\begin{aligned}\boldsymbol{R_{ep}} &= \left(\rho\cos(\phi_p) + a_p\cos\left(\phi'_p\right)\right)\cos(\theta_p)\boldsymbol{u}_{ex}\\ &\quad + \left(\rho\cos(\phi_p) + a_p\cos\left(\phi'_p\right)\right)\sin(\theta_p)\boldsymbol{u}_{ey} + \left(\rho\sin(\phi_p)\right.\\ &\quad \left. + a_p\sin\left(\phi'_p\right)\right)\boldsymbol{u}_{ez}\end{aligned} \tag{16.29}$$

$$\begin{aligned}\boldsymbol{V_{ep}} &= -\left(\rho\cos(\phi_p) + a_p\cos\left(\phi'_p\right)\right)\omega_e\sin\left(\theta_p\right)\boldsymbol{u}_{ex}\\ &\quad + \left(\rho\cos(\phi_p) + a_p\cos\left(\phi'_p\right)\right)\omega_e\cos\left(\theta_p\right)\boldsymbol{u}_{ey} + 0\boldsymbol{u}_{ez}\end{aligned} \tag{16.30}$$

where the z-axis component of the velocity is zero because the Earth rotates about the z-axis and the Earth stations are considered to be fixed sites. Also, because the Earth is rotating at a constant angular velocity, the acceleration of the Earth station is zero, that is, $\boldsymbol{A}_{ep} = 0$, for the fixed sites being considered.

16.4 PATH LOSS, DOPPLER, AND DOPPLER-RATE

To determine the free-space path loss, Doppler, and Doppler rate encountered by a communications receiver, it is first necessary to compute the ground station-to-satellite range and the corresponding range-rate (velocity) and acceleration along the range vector. These parameters are evaluated in the Earth-fixed coordinates using the satellite and Earth station range, velocity, and acceleration vectors established in Sections 16.2 and 16.3.

The Earth station-to-satellite range is simply the magnitude of the range vector and is computed as

$$R_{ps} = \left|\boldsymbol{R}_{es} - \boldsymbol{R}_{ep}\right| \qquad (16.31)$$

The cosine of the angle between these vectors is expressed as

$$\cos(\theta_{ps}) = \frac{\boldsymbol{R}_{es} \bullet \boldsymbol{R}_{ep}}{|\boldsymbol{R}_{es}||\boldsymbol{R}_{ep}|} \qquad (16.32)$$

The velocity and acceleration along the range vector are computed as follows:

$$V_{ps} = \left|\boldsymbol{V}_{es} - \boldsymbol{V}_{ep}\right|\cos(\theta_{rv}) \qquad (16.33)$$

and

$$A_{ps} = \left|\boldsymbol{A}_{es} - \boldsymbol{A}_{ep}\right|\cos(\theta_{ra}) \qquad (16.34)$$

Here, θ_{rv} and θ_{ra} are the angles between the Earth station-to-satellite range vector and the corresponding velocity and acceleration vectors, respectively. These angles are evaluated using the relationships

$$\cos(\theta_{rv}) = \frac{\boldsymbol{R}_{es} \bullet \boldsymbol{V}_{ep}}{|\boldsymbol{R}_{es}||\boldsymbol{V}_{ep}|} \qquad (16.35)$$

and

$$\cos(\theta_{ra}) = \frac{\boldsymbol{R}_{es} \bullet \boldsymbol{A}_{ep}}{|\boldsymbol{R}_{es}||\boldsymbol{A}_{ep}|} \qquad (16.36)$$

From these results, the free-space loss, Doppler, and Doppler rate along the propagation path are readily determined to be

$$L_{fs} = \left(\frac{4\pi R_{ps} f_c}{c}\right)^2 \qquad (16.37)$$

and

$$f_D = \left(\frac{V_{ps}}{c}\right) f_c \qquad (16.38)$$

and

$$\dot{f}_D = \left(\frac{A_{ps}}{c}\right) f_c \qquad (16.39)$$

where $c = 3 \times 10^8$ m/s is the speed of light in free space and f_c is the carrier frequency of the communication link.

16.5 SATELLITE VIEWING

For arbitrary satellite orbits, the satellite is not always in view of the Earth station, and it is useful to determine when this condition exists. Also, when the satellite is in view, it is helpful to determine the viewing angle of the satellite from the position of the Earth station. These topics [8] are addressed in this section.

To determine whether the satellite is in view of the ground station the computations are simplified by assuming a spherical Earth to determine the observer's horizons as shown in Figure 16.5. Since the Earth is very nearly spherical, this assumption has little loss in significance unless it is necessary to determine very accurately the viewing horizons. Referring to Figure 16.5, the viewing angle of the local observer is $\pm(90 + \alpha)$ degrees where the angle α is given by

$$\alpha = 90 - \left(\frac{180}{\pi}\right)\sin^{-1}\left(\frac{R_e}{R_e + a_p}\right) \quad : \text{degrees} \qquad (16.40)$$

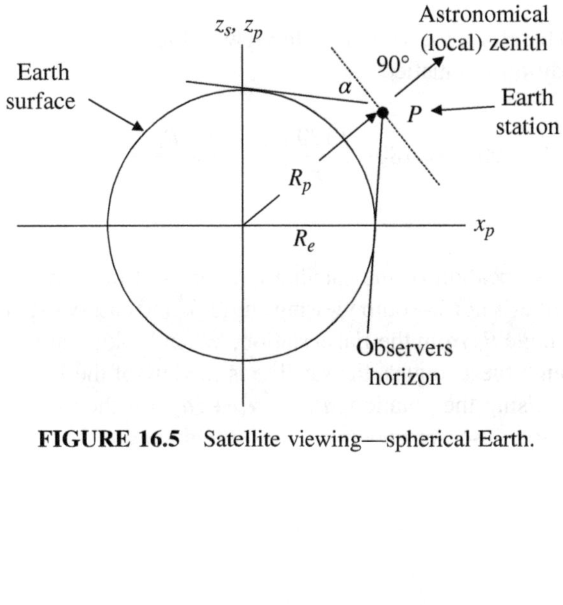

FIGURE 16.5 Satellite viewing—spherical Earth.

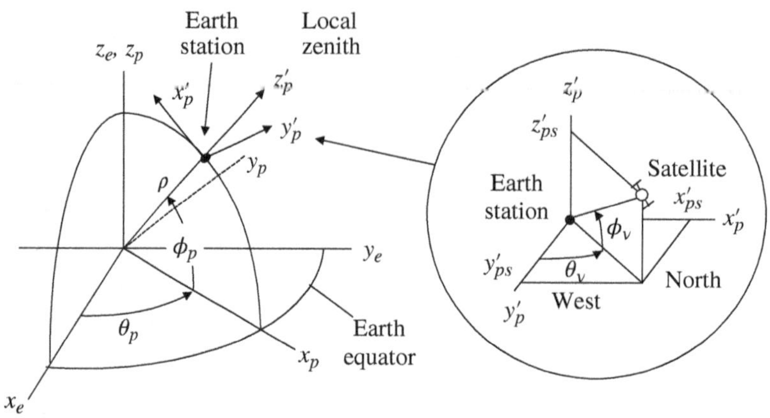

FIGURE 16.6 Earth station coordinates.

The Earth station-to-satellite range vector in the Earth coordinates is $\boldsymbol{R_{ps}} = \boldsymbol{R_{es}} - \boldsymbol{R_{ep}}$. However, the vector $\boldsymbol{R_{ps}}$ is relative to the Earth-fixed coordinates and must be transformed to the vector $\boldsymbol{R'_{ps}}$ with respect to the Earth station coordinates defined in Figure 16.6. The vector $\boldsymbol{R'_{ps}}$ is computed as

$$\boldsymbol{R'_{ps}} = \begin{pmatrix} -\sin(\phi_p) & 0 & \cos(\phi_p) \\ 0 & 1 & 0 \\ \cos(\phi_p) & 0 & \sin(\phi_p) \end{pmatrix} \begin{pmatrix} \cos(\theta_p) & \sin(\theta_p) & 0 \\ -\sin(\theta_p) & \cos(\theta_p) & 0 \\ 0 & 0 & 1 \end{pmatrix} \boldsymbol{R_{ps}} \tag{16.41}$$

where X'_{ps} and Y'_{ps} are aligned with the local south and east directions, respectively, and Z'_{ps} is aligned with the local zenith.

The vector $\boldsymbol{R'_{ps}}$ is the satellite position vector as viewed from the Earth station and $\boldsymbol{R_{ep}}$ represents the observers local zenith, so the angle between $\boldsymbol{R'_{ps}}$ and $\boldsymbol{R_{ep}}$ is the local observers viewing angle of the satellite given by

$$\Theta_v = \cos^{-1}\left(\frac{\boldsymbol{R'_{ps}} \bullet \boldsymbol{R_{ep}}}{|\boldsymbol{R'_{ps}}||\boldsymbol{R_{ep}}|}\right) \tag{16.42}$$

Therefore, the satellite is in view as long as the following condition is satisfied

$$|\Theta_v| \le 90 + \alpha = 180 - \left(\frac{180}{\pi}\right)\sin^{-1}\left(\frac{R_e}{R_e + a_p}\right) \quad : \text{degrees} \tag{16.43}$$

The location of the satellite is expressed in terms of the observer's north–south viewing angle ϕ_v and east–west viewing angle θ_v from the Earth station. When looking at the sky through these angles, the satellite is in view of the Earth station. Using the notation $x\boldsymbol{u}_{x'} + y\boldsymbol{u}_{y'} + z\boldsymbol{u}_{x'}$ for the range vectors, the observer viewing angles are given by

$$\phi_v = \tan^{-1}\left(\frac{X'_{ps}}{Z'_{ps}}\right) \quad : \text{north-south} \tag{16.44}$$

$$\theta_v = \tan^{-1}\left(\frac{Y'_{ps}}{Z'_{ps}}\right) \quad : \text{east-west} \tag{16.45}$$

where positive viewing angles represent north and east, respectively, from the Earth station location.

16.6 SATELLITE ORBIT SELECTION

The selection of the satellite orbit depends upon the intended mission of the satellite, and these generally fall into a few special categories or type of orbits. The period of an elliptical orbit follows directly from Kepler's third law and, referring to (16.1) and (16.3), is approximately expressed as

$$T \cong 2\pi\sqrt{\frac{a^3(1-\varepsilon)^3}{u}} \tag{16.46}$$

This result and a specification of the eccentricity ε are all that is necessary to characterize the orbit in the obit plane. The various orbit types and the parameter values required to establish them are discussed in Sections 16.6.1 through 16.6.4.

16.6.1 Geosynchronous and Geostationary Orbits

Geostationary orbits (GEOs) are characterized as circular orbits with $\varepsilon = 0$ and a 24 h period equal to that of the Earth's rotation. In practice, $\varepsilon \cong 0$ and newly established and well-maintained orbits will result in $\varepsilon \le 0.0008$; with deteriorated orbits $\varepsilon \le 0.03$. GEO orbits with zero inclination result in a satellite orbit plane that coincides with the Earth's equatorial plane and the satellite appears stationary when viewed from the Earth. The satellite orbit in this special case plays an important role in satellite communications. In theory the inclination ($\imath$) of a GEO satellite orbit is zero degrees, however, in practice $\imath \cong 0$. Typically, a good GEO will result in $\imath \leq 1.5°$ and for a deteriorated orbit $\imath \leq 3°$. Geosynchronous orbits (GSOs) with inclinations much greater than three degrees are used for high latitude and polar coverage. With GSO orbits, reliable coverage at higher latitudes, that is, obit planes with inclinations greater than about 30° is possible, however, a single Earth station cannot continuously view the satellite and must rely on satellite cross-linking for coverage. In general, on successive satellite orbits the longitude of the ascending node will drift due to the Earth's rotation.

In the preceding analysis, a GEO can be specified by simply letting $\varepsilon = 0$, which defines the circular orbit. In this case, because a circle has only one focal point, the mean orbit radius must also be specified as $a = R_{geosolar} = R_e + A_{geosolar}$. A more direct and accurate approach is to specify a 24 h synchronous orbit period corresponding to $T = 86{,}400$ s and use (16.46) to solve for the orbit radius $a = 42{,}241.192$ km; this agrees with the value $a = R_{geosolar}$ in Table 16.1.

16.6.2 Medium Earth Orbits

Medium Earth orbits (MEOs) range in altitude between low Earth orbits (LEOs) and GEO altitudes, typically from 2,000 km to below 35,784.334 km.

16.6.3 Low Earth Orbits

The LEO is a low-altitude orbit, typically circular, often used to provide coverage for communications, global positioning and navigation, space exploration, environmental monitoring, and intelligence gathering. For example, the National Astronautics and Space Administration (NASA) Space Shuttle is used for space exploration and other scientific studies, and the Global Positioning System (GPS) is used for precision location and navigation using multiple LEOs with different inclinations to provide worldwide coverage. Due to atmospheric drag, LEOs are limited to a minimum altitude of about 180 km and, for safety reasons, involving space debris and Van Allen belt radiation, long-term manned space flights are limited to a maximum altitude of about 560 km; however, LEO altitudes may extend to 2000 km.

To specify the circular LEO, simply choose $\varepsilon = 0$ and determine the desired value of the mean distance as $a = R_e + A_s = 6378.28 + A_s$ where $180\text{ km} \leq A_s \leq 560\text{ km}$; this upper limit is imposed for long-term manned space flight. From the value of a, the orbit period is computed using (16.46) and the other characteristics of the orbit plane are determined as described in the Introduction and Section 6.2.

16.6.4 Highly Elliptical Orbits

The highly elliptical orbit (HEO) is used for environmental monitoring and intelligence gathering. It has the advantage over LEOs, in that, the satellite hovers over an area of interest for a long period of time around the point of apogee gathering data from various sensors. Upon approaching the point of perigee, the information is downlinked to an Earth station and the satellite moves on to repeat the process. Although the viewing time is relatively long near apogee with long ranges to Earth, the transition through perigee is very fast involving relatively short ranges. So the satellite gathers information over a long period of time and downlinks the information rapidly over a short interval of time.

16.6.4.1 Molniya Orbits The Molniya orbit is a HEO with an inclination of 63.4° and a period of 12 h. It was originally developed to provide high elevation angle visibility of a satellite over the Soviet Union for a period of 8 h every second revolution. A constellation of three Molniya satellites provides constant coverage. With the orbit apogee strategically located over Russia, the high elevation viewing angles avoid multipath interference caused by signal reflections from the Earth's surface. Furthermore, the 63.4° inclination of the orbit plane, theoretically, results in zero drift of the argument of perigee with time due to the oblate Earth.

16.7 SATELLITE ORBIT POSITION ESTIMATION FROM PARAMETER MEASUREMENTS

The location of a satellite in its orbital is determined using the six orbital parameters as discussed in Sections 16.1 and 16.2. However, in reality, the knowledge of these parameters is only approximated from observations of the satellite position over long periods of time. The computed positions of a satellite are typically available in table form, referred to as ephemeris data that is published periodically with the latest satellite information.* The focus in this section is on the range, velocity, and time errors that impact the satellite's ability to acquire and track a received signal.

Consider that a large ground terminal is tracking and measuring a satellite's position in its FOV. Referring to Figure 16.2, the measurements, when referred to Earth-centered coordinates, are used to estimate satellite range $\hat{r}$, velocity magnitude $\hat{V}$, velocity angle $\hat{\phi}$, and the orbit time $\hat{t}$ relative to perigee. These parameters form the parameter set† required to determine the satellite's position in the orbit plane. The gravitational constant u is known. The raw measurements involve estimating the radial velocity $\hat{V}_r = \widehat{\dot{r}}$ and the angular rate $\widehat{\dot{f}}$ of the true anomaly, from which the normal

*For example, International Satellite Directory is published annually by Design Publishers and contains details of satellite orbits and operational parameters.

†Roy [1] (Chapter 4) discusses the application of the parameters set (a, ε, τ) to determine the satellite position in the orbit plane.

velocity estimate is computed as $\hat{V}_n = \widehat{\hat{r}f}$. Using these measurements, the velocity magnitude and angle are computed as

$$\hat{V} = \sqrt{\hat{r}^2 + \left(\widehat{\hat{r}f}\right)^2} \quad (16.47)$$

and

$$\hat{\phi} = \sin^{-1}\left(\frac{\hat{\dot{r}}}{\widehat{\hat{r}f}}\right) \quad (16.48)$$

With this information, the satellite orbit parameters are computed as follows:

- Compute the semimajor axis $\hat{a}$ using (16.14) as

$$\hat{a} = \frac{1}{\left(2/\hat{r} - \hat{V}^2/u\right)} \quad (16.49)$$

- Compute the mean angular velocity $\hat{n}$ using (16.5) as

$$\hat{n} = \sqrt{\frac{u}{\hat{a}^3}} \quad (16.50)$$

- Compute the eccentricity $\hat{\varepsilon}$ using (16.10) as

$$\hat{\varepsilon} = \sqrt{1 - \frac{\hat{r}(2\hat{a} - \hat{r})}{\hat{a}^2}\sin^2(\hat{\phi})} \quad (16.51)$$

- Compute the eccentric anomaly $\hat{E}$ using (16.8) as

$$\hat{E} = \cos^{-1}\left(\frac{1 - \hat{r}/\hat{a}}{\hat{\varepsilon}}\right) \quad (16.52)$$

- Compute the true anomaly $\hat{f}$ using (16.9)
- Compute the mean anomaly $\widehat{\bar{M}}$ using the first equality in (16.7)
- Compute the $\hat{\tau}$ using $\hat{n}$, $\widehat{\bar{M}}$, and (16.6) as

$$\hat{\tau} = \hat{t} - \widehat{\bar{M}}/\hat{n} \quad (16.53)$$

These computed orbital parameters, including the time estimate, are then used to establish the estimated satellite position in the orbital plane. The range and timing errors are based on the true satellite position and orbit time. The estimated orbit inclination is not included in these measurements, so the impact of the inclination on the range and timing errors can be evaluated parametrically. Roy [9] discusses the theory of orbit determination based on the methods of Laplace and Gauss, and Ash [10] discusses the orbit fitting computer program Planetary Ephemeris Program (PEP) developed at Lincoln Laboratory and the Massachusetts Institute of Technology. These methods use three-dimensional spatial observations over time that characterizes the satellite orbit plane in geocentric coordinates with the satellite location identified in the inclination of the orbit plane.

16.8 CASE STUDY: EXAMPLE SATELLITE ENCOUNTERS

In this section, the Doppler, Doppler rate, and ground station satellite viewing angles are examined for communication links involving two satellite orbits; the parameters for the two satellite orbits are summarized in Table 16.2. The details regarding the computation of these parameters are discussed in Sections 16.4 through 16.6 and Figure 16.7 summarizes the geometry of the encounters. The satellite orbit angle is determined as $\varpi = (\omega + f)$ modulo-2π; however, because the GEO is essentially circular the argument of perigee is used to identify the satellite orbit time.

The salient satellite orbit parameters are the range, the LOS velocity, and acceleration, respectively, that are used to determine the signal free-space loss, Doppler frequency, and Doppler rate. The satellite orbit simulation code, developed from the results in this chapter, is integrated into the overall link budget and performance simulation code discussed in Section 15.15. Figure 16.8a and b show the uplink and downlink Doppler and Doppler rate for the case 1 encounter listed in Table 16.2. Figure 16.8c shows the ground terminal satellite viewing requirements in terms of the local azimuth and elevation angles. These local viewing angles determine the antenna pointing requirements necessary to

TABLE 16.2 Summary of Ground and Satellite Parameters

	Ground Station				Satellite[a]		Orbit[b]		
Case	Lat. (deg)	Lon. (deg)	Elev. (Ft)	Uplink Freq (GHz)	Lon. (deg)	Downlink Freq (GHz)	Period (h)	Incl. (deg)	Ecc.
1	0	−120	0	2.3	−120	16	24	7	0.003
2	33	−120	0	2.3	170	16	24	7	0.003

Positive longitudes and latitudes are °*E* and °*N*.
[a]Longitude of ascending node.
[b]Satellite orbit angle = 0° at time zero.

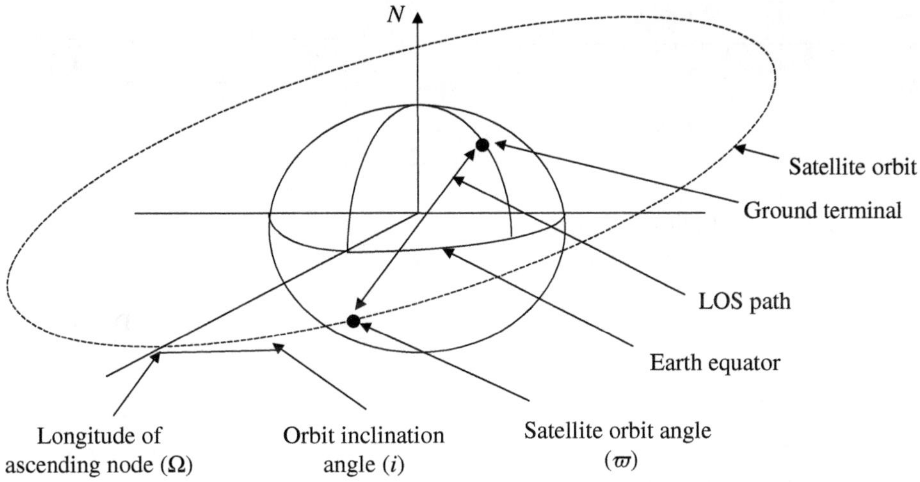

FIGURE 16.7 Satellite and ground station geometry.

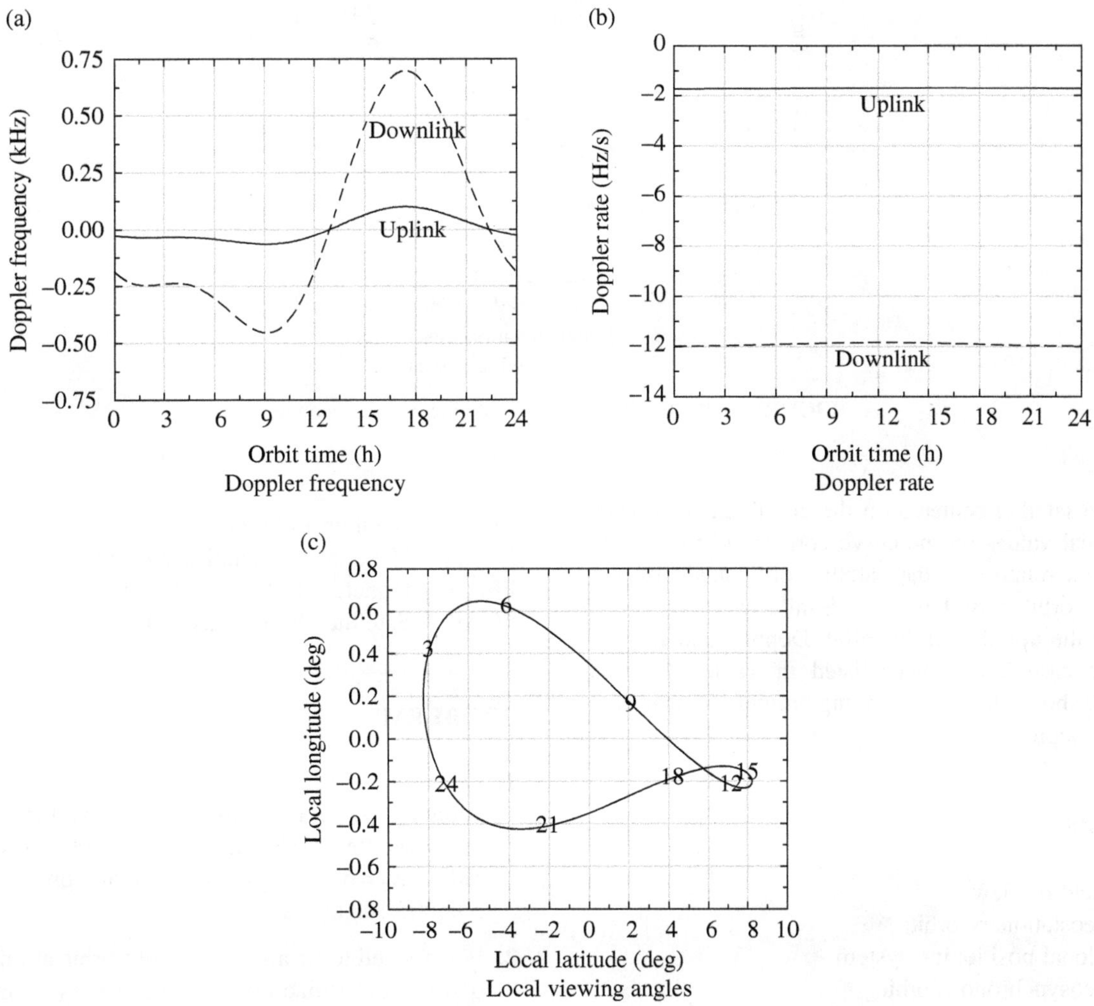

FIGURE 16.8 Earth station and satellite case 1 encounter.

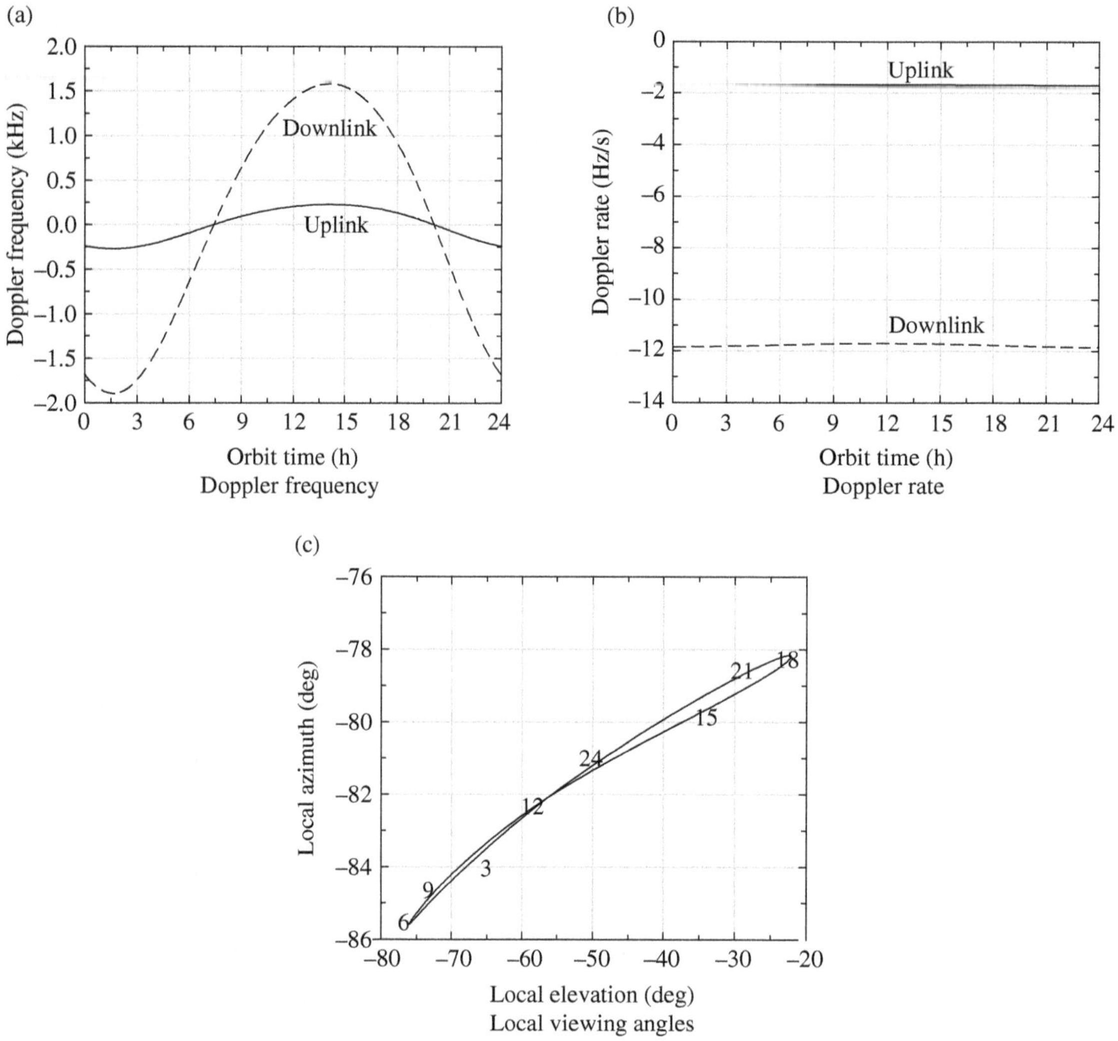

FIGURE 16.9 Earth station and satellite case 2 encounter.

maintain the satellite centered on the uplink antenna beam. The numerical values on the curve correspond to the orbit time in hours relative to the satellite orbit angle of zero degrees; the orbit period is 24 h. Similarly, Figure 16.9a and b show the uplink and downlink Doppler and Doppler rate for the case 2 encounter listed in Table 16.2, and Figure 16.9c shows the corresponding ground terminal satellite viewing angles.

ACRONYMS

FOV	Field of view
GEO	Geostationary orbit
GPS	Global positioning system
GSO	Geosynchronous orbit
HEO	Highly elliptical orbit
LEO	Low Earth orbit
MEO	Medium Earth orbit
NASA	National Aeronautics and Space Administration
PEP	Planetary Ephemeris Program
SOAP	Satellite Orbit Analysis Program

PROBLEMS

1. Determine the mean eccentric anomaly (E) and the true anomaly (f) of a satellite, located at the longitude of the ascending node, in a circular orbit. Use the Earth radius $R_e = 6{,}356.74$ km and gravitational constant $u = 398{,}603.188\ \text{km}^3/\text{s}^2$.

2. For a satellite in a circular LEO orbit at an altitude of 650 km, determine the orbit parameters: orbit period (T) in hours, the magnitude of the location vector (r), and the radial and normal velocities (V_r and V_n) and

accelerations (A_r and A_n). Use the Earth radius and gravitational constant given in Problem 1.

3. For a satellite in an elliptical orbit with a perigee altitude of 350 km and eccentricity $\varepsilon = 0.2$, determine following parameters: the orbit period (T) and the velocities (V_r and V_n) at perigee and apogee. Use the Earth radius and gravitational constant given in Problem 1.

Hint: Use Newton's method for solving Kepler's equation for the eccentric anomaly.

4. Show that $\pi/2 - \cos^{-1}(\phi)$ is equal to $\sin^{-1}(\phi)$.

REFERENCES

1. A.E. Roy, *The Foundations of Astrodynamics*, The Macmillan Company, New York, 1965.
2. R.R. Bate, D.D. Mueller, J.E. White, *Fundamentals of Astrodynamics*, Dover Publications, Inc., New York, 1971.
3. A.E. Roy, *The Foundations of Astrodynamics*, Chapter 5, "The Many-Body Problem," and Chapter 6, "General and Special Perturbations," The Macmillan Company, New York, 1965.
4. A.E. Roy, *The Foundations of Astrodynamics*, Chapter 7, "Artificial Satellites," The Macmillan Company, New York, 1965.
5. D.Y. Stodden, G.D. Galasso, "Satellite Orbit Analysis Program (SOAP)," The Aerospace Corporation, El Segundo, California, Rev 4, April 1996.
6. M.I. Skolnik, *Introduction to Radar Systems*, McGraw-Hill Book Company, Inc., New York, 1962.
7. M. Abramowitz, I.A. Stegun, *Handbook of Mathematical Functions with Formulas, Graphs, and Mathematical Tables*, National Bureau of Standards, Applied Mathematics Series 55, pp. 18, U.S. Government Printing Office, Washington, DC, 1964.
8. J.A. Eisele, S.A. Nichols, *Orbital Mechanics of General-Coverage Satellites*, NRL Report 7975, Space Systems Division, Naval Research Laboratory, Washington, DC, 1976.
9. A.E. Roy, *The Foundations of Astrodynamics*, Chapter 11, "Orbit Determination and Interplanetary Navigation," The Macmillan Company, New York, 1965.
10. M.E. Ash, Determination of Earth Satellites Orbits, Technical Note 1972–5, Lincoln Laboratory, Massachusetts Institute of Technology, Lexington, MA, 1972.

17

COMMUNICATIONS THROUGH BANDLIMITED TIME-INVARIANT LINEAR CHANNELS

17.1 INTRODUCTION

In this chapter, the analysis and simulation of bandlimited time-invariant linear communication channels is discussed. The channel is viewed in broad terms and may include the transmitter and receiver filtering as well as the transmission medium. The linear time-invariant channel is frequently encountered in practice and is relatively easy to analyze and simulate. Since these channels are often related to classical filter theory, the analysis and application of linear filters draw on the abundance of time-tested work related to filter theory and design.

In communication systems, the channel distortion manifests itself in two distinct ways: the transmitted waveform is distorted in such a way that the received symbol energy within the symbol interval is reduced leading to a loss in signal-to-noise ratio (SNR); the channel distortion results in dispersion of the symbol energy into adjacent symbols resulting in noise-like terms referred to as intersymbol interference (ISI). The channel distortion results from the amplitude and phase responses of the channel that becomes increasingly severe as the waveform bandwidth approaches the channel bandwidth. The ISI results in the closing of the demodulator *eye-opening* at the sampled matched filter output that reduces the probability of a correct decision. It also impacts the symbol timing and carrier phase tracking functions of the demodulator that often take advantage of distinct spectral properties of the modulated waveform.

Because the ISI under consideration results from a linear time-invariant channel, it is deterministic with respect to a known symbol; however, it appears to be random because of the randomly modulated stream of contiguous symbols. ISI cancellation is embodied in the subject of equalization that is based upon this deterministic nature of the ISI. Although the statistical distribution characterizing the ISI is not generally known by the demodulator, estimates, based on the demodulated data,* are formed to cancel the ISI, thereby, improving the system's performance [1].

In this chapter, the channels are characterized in the frequency domain in terms of amplitude and phase responses that can be arbitrarily specified. Typical channel characterizations include classical filters like the Butterworth and Chebyshev designs and dial-up wireline channels [2, 3] with or without various degrees of conditioning, for example, *C*1 and *C*2 designated conditioning. The phase response can be specified independently of the amplitude response and transmissions through phase-equalized channels demonstrate a marked improvement in performance. Although the analysis permits applying arbitrarily specified modulation functions, examples involving rectangular modulated waveforms applied to channels with quadratic and cubic phase responses are examined to demonstrate the effects of channel

*Message preambles or midambles and periodic insertion of known data sequences are often used to aid in the equalizer learning.

Digital Communications with Emphasis on Data Modems: Theory, Analysis, Design, Simulation, Testing, and Applications, First Edition. Richard W. Middlestead.

Companion website: www.wiley.com/go/digitalcommunications

distortion. The analysis is based on the work of Sunde [4] and Urkowitz [5]; in that, the inphase and quadrature (I/Q) responses of the low-pass equivalents of bandpass filters are developed. With digital signal processing, the lowpass filter is a computationally efficient way to perform bandpass filtering; however, in most applications, the lowpass or baseband filtered output is used directly for subsequent signal processing and matched filter detection. The analysis also highlights the use of complex notations, in which the I/Q components are the real and imaginary parts of complex functions referred to as *pre-envelopes* [6].

In Section 17.2, some basic filter concepts are established to generate *real* carrier and baseband modulated outputs based on an arbitrary input signal. In Section 17.3, an input carrier modulated signal is filtered using a baseband filter with appropriate frequency heterodyning. Sections 17.4 and 17.5 characterize a *rect(t/T)* input signal and develops the output response for an arbitrarily specified filter. The filter is expressed in terms of the I/Q amplitude and phase impulse responses. The parameters of the signal and filter response functions are then conveniently normalized for computer evaluation. In Sections 17.6 and 17.7, the responses of an ideal *rect(u/B)/B* filter and a single-pole bandpass filter are examined using the computer model. The chapter concludes with computer simulated examples relating the characteristics of the ISI to various channel phase and amplitude characteristics.

17.2 INPHASE AND QUADRATURE CHANNEL RESPONSE

In this section, generalized expressions for the bandpass channel impulse response are developed in terms of the I/Q responses of the equivalent lowpass filter. Expressing a bandlimited channel impulse response in this form provides considerable insight into the channel characteristics. Consider the time-invariant linear channel having a transfer function or frequency response given by $H(\omega)$. Because only *real* impulse responses are of interest, the transfer function exhibits an even real part and odd imaginary part in frequency.* The channel transfer function, expressed in terms of real and imaginary parts, is

$$H(\omega)=H_r(\omega)+jH_i(\omega) \tag{17.1}$$

and the impulse response is defined as the inverse Fourier transform, expressed as

*The property corresponds to $H(-\omega)=H^*(\omega)$ and is referred to as *conjugate symmetry*.

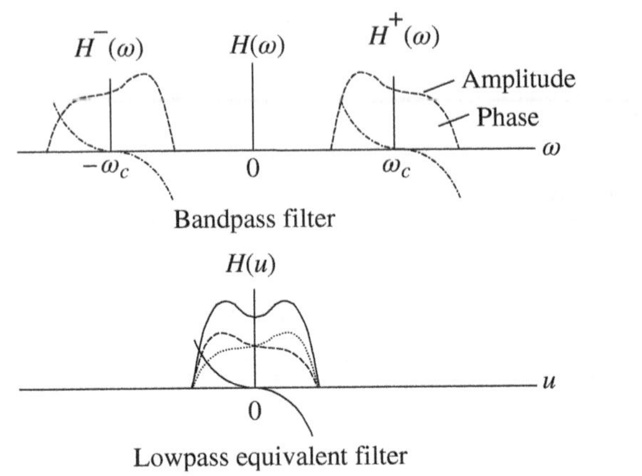

FIGURE 17.1 Channel filter characteristics.

$$h(t)=\frac{1}{2\pi}\int_{-\infty}^{\infty}H(\omega)e^{j\omega t}d\omega=\frac{1}{2\pi}\int_{-\infty}^{\infty}[H_r(\omega)+jH_i(\omega)]e^{j\omega t}d\omega \tag{17.2}$$

Because only *real* impulse responses are of interest, the functions $H_r(\omega)$ and $H_i(\omega)$ must be, respectively, even and odd functions of frequency. Expressing $H(\omega)$ in terms of positive and negative frequency functions, (17.1) is rewritten as

$$\begin{aligned}H(\omega)&=H_c(\omega-\omega_c)+jH_s(\omega-\omega_c) \quad :\omega>0\\ &=H_c(\omega+\omega_c)-jH_s(\omega+\omega_c) \quad :\omega<0\end{aligned} \tag{17.3}$$

where ω_c is any convenient reference frequency for $H(\omega)$; usually ω_c is chosen as the mid-band or the arithmetic mean of the upper and lower channel cutoff frequencies. Defining the lowpass frequency as $u=\omega-\omega_c$, the positive and negative frequency functions in (17.3) have equivalent lowpass functions centered at $\pm\omega_c$ as depicted in Figure 17.1.

Rewriting (17.2) in terms of the lowpass frequency $u=\omega-\omega_c$ with $du=d\omega$ results in

$$\begin{aligned}h(t)&=\mathrm{Re}\left\{\left[\frac{1}{2\pi}\int_{-\infty}^{\infty}H(u)e^{jut}du\right]e^{j\omega_c t}\right\}\\ &=\mathrm{Re}\left\{\left[\frac{1}{2\pi}\int_{-\infty}^{\infty}[H_r(u)+jH_i(u)]e^{jut}du\right]e^{j\omega_c t}\right\}\end{aligned} \tag{17.4}$$

Equating the equivalent lowpass expressions for the real and imaginary parts $H_r(u)$ and $H_i(u)$ of (17.3) results in $H_r(u)=H_c(u)+H_c(-u)$ and $H_i(u)=H_s(u)-H_s(-u)$. Upon using the even or odd symmetry property then either $H_r(u)=2H_c(u)$ or $H_i(u)=2H_s(u)$, so that (17.4) becomes

$$h(t) = \mathrm{Re}\left\{\left[\frac{1}{\pi}\int_{-\infty}^{\infty}[H_c(u)+jH_s(u)]e^{jut}du\right]e^{j\omega_c t}\right\} \tag{17.5}$$
$$= h_c(t)\cos(\omega_c t) - h_s(t)\sin(\omega_c t)$$

where

$$h_c(t) = \frac{1}{\pi}\int_{-\infty}^{\infty} H_c(u)\cos(ut)du - \frac{1}{\pi}\int_{-\infty}^{\infty} H_s(u)\sin(ut)du \tag{17.6}$$

and

$$h_s(t) = \frac{1}{\pi}\int_{-\infty}^{\infty} H_c(u)\sin(ut)du + \frac{1}{\pi}\int_{-\infty}^{\infty} H_s(u)\cos(ut)du \tag{17.7}$$

The functions $h_c(t)$ and $h_s(t)$ are, respectively, the lowpass I/Q impulse responses of the bandpass channel filter $H(\omega)$. In general, $H_c(u)$ and $H_s(u)$ will not possess either even or odd symmetry; however, if even or odd symmetry does exist, then either $h_c(t)$ or $h_s(t)$ will be zero.

The magnitude and phase functions of the baseband filter output are computed, respectively, as

$$A(u) = 2\sqrt{H_c(u)^2 + H_s(u)^2} \tag{17.8}$$

and

$$\phi(u) = \tan^{-1}\left(\frac{H_s(u)}{H_c(u)}\right) \tag{17.9}$$

Using (17.8) and (17.9), the lowpass frequency response is expressed as

$$H(u) = A(u)e^{j(\phi(u))} \tag{17.10}$$

Substituting (17.10) into (17.5), the quadrature responses in (17.6) and (17.7) are evaluated as

$$h_c(t) = \frac{1}{2\pi}\int_{-\infty}^{\infty} A(u)\cos(ut+\phi(u))du - \frac{1}{2\pi}\int_{-\infty}^{\infty} A(u)\sin(ut+\phi(u))du \tag{17.11}$$

and

$$h_s(t) = \frac{1}{2\pi}\int_{-\infty}^{\infty} A(u)\sin(ut+\varphi(u))du + \frac{1}{2\pi}\int_{-\infty}^{\infty} A(u)\cos(ut+\varphi(u))du \tag{17.12}$$

The lowpass equivalent filter, characterized in this section, is used in Section 17.3 to describe the lowpass filtering of an arbitrary carrier modulated input signal.

17.3 INPHASE AND QUADRATURE CHANNEL RESPONSE TO ARBITRARY SIGNAL

In Section 17.2, the impulse response of a bandpass channel, centered at an angular frequency of ω_c, is expressed in terms of I/Q lowpass channel functions. In this section, the lowpass channel response is examined for an arbitrary *real* input signal in terms of the equivalent lowpass filter functions. Figure 17.2 shows the functional implementation with the channel's lowpass impulse response functions, $h_c(t)$ and $h_s(t)$, developed in Section 17.2.

The arbitrary carrier modulated input signal $s(t)$ is expressed as

$$s(t) = s_c(t)\cos(\omega_s t) - s_s(t)\sin(\omega_s t) \tag{17.13}$$

where ω_s is the signal angular carrier frequency and $s_c(t)$ and $s_s(t)$ are the quadrature components that characterize the baseband modulation. The input carrier frequency is intentionally offset from the channel frequency so that the impact of the frequency error $\Delta\omega = \omega_s - \omega_c$ can be examined in the subsequent example applications.

A completely analogous but considerably easier implementation to analyze and less prone to analysis errors is based on the analytic or complex signal representations shown in Figure 17.3. The input carrier frequency of ω_s is offset from the channel filter's center frequency by $\Delta\omega$ and must be

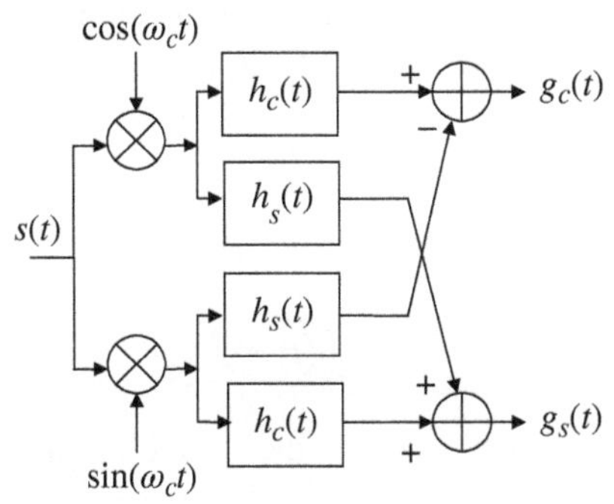

FIGURE 17.2 Lowpass filtering of carrier modulated signal.

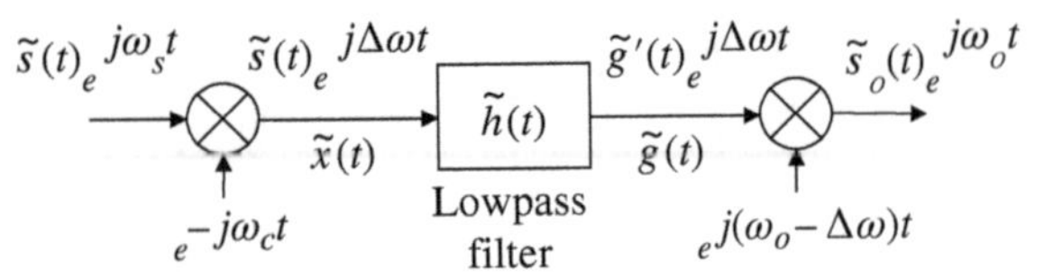

FIGURE 17.3 Analytic lowpass filtering of carrier modulated signal.

considered in evaluating the output of the lowpass filter, defined by the convolution integral

$$\tilde{g}(t) \triangleq \int_{-\infty}^{\infty} \tilde{s}(t-\lambda)e^{j\Delta\omega(t-\lambda)}\tilde{h}(\lambda)d\lambda = e^{j\Delta\omega t}\int_{-\infty}^{\infty} \tilde{s}(t-\lambda)\tilde{h}(\lambda)e^{-j\Delta\omega\lambda}d\lambda \qquad (17.14)$$

The integral in the second equality of (17.14) is defined as $\tilde{g}'(t)$, such that,

$$\tilde{g}'(t) \triangleq \int_{-\infty}^{\infty} \tilde{s}(t-\lambda)\tilde{h}(\lambda)e^{-j\Delta\omega\lambda}d\lambda \qquad (17.15)$$

Using these relationships, the composite lowpass filter is described as shown in Figure 17.3.

The final up-conversion to the output frequency ω_o may represent an intermediate frequency (IF) of a linear heterodyning receiver. In this case, all of the intervening filtering of the linear receiver must be included in the lowpass filter function $\tilde{h}(t)$. If the lowpass channel filter includes the entire receiver and demodulator filtering prior to the baseband conversion in the demodulator, then no up-conversion is necessary. However, the coherent demodulator Costas phaselock loop must remove the frequency error $\Delta\omega$. On the other hand, if the receiver expects the received signal frequency to be ω_s, using the up-conversion frequency ω_s, instead of $\omega_o - \Delta\omega$, results in the receiver input frequency $\omega_s + \Delta\omega$, in which case, the frequency error must be removed by the coherent demodulator phaselock loop.

In the following evaluation of the filter output, the analytic functions $\tilde{x}(t)$ and $\tilde{g}(t)$ are defined as

$$\tilde{x}(t) \triangleq \tilde{s}(t)e^{-j\Delta\omega t} \qquad (17.16)$$

and

$$\tilde{g}(t) \triangleq \tilde{x}(t) * \tilde{h}(t) = \tilde{g}'(t)e^{j\Delta\omega t} \qquad (17.17)$$

where the asterisk denotes the convolution of $\tilde{x}(t)$ and $\tilde{h}(t)$ that follows directly from (17.14) with $\tilde{g}'(t)$ defined in (17.15).

Therefore, expressing the analytic functions $\tilde{x}(t), \tilde{h}(t)$, and $\tilde{g}(t)$ in terms of their respective quadrature components, having the form $z(t) = z_c(t) + jz_s(t)$, the analytic lowpass filter output function, $\tilde{g}'(t)$, is evaluated using (17.14) and (17.15), and the result is expressed as

$$\begin{aligned}\tilde{g}'(t) = &\int_{-\infty}^{\infty} [s_c(t-\lambda)h_c(\lambda) - s_s(t-\lambda)h_s(\lambda)]\cos(\Delta\omega\lambda)d\lambda \\ &+ \int_{-\infty}^{\infty} [s_c(t-\lambda)h_s(\lambda) + s_s(t-\lambda)h_c(\lambda)]\sin(\Delta\omega\lambda)d\lambda \\ &+ j\int_{-\infty}^{\infty} [s_c(t-\lambda)h_s(\lambda) + s_s(t-\lambda)h_c(\lambda)]\cos(\Delta\omega\lambda)d\lambda \\ &- j\int_{-\infty}^{\infty} [s_c(t-\lambda)h_c(\lambda) - s_s(t-\lambda)h_s(\lambda)]\sin(\Delta\omega\lambda)d\lambda\end{aligned} \qquad (17.18)$$

The second and fourth terms in (17.18), that is, the terms involving $\sin(\Delta\omega\lambda)$, correspond, respectively, to the quadrature components of the third and first terms. For coherent demodulation, these terms are eliminated by the demodulator phaselock loop tracking prior to data detection. The quadrature components, $g_c'(t)$ and $g_s'(t)$, correspond, respectively, to the real and imaginary parts of $\tilde{g}'(t)$ and, upon substitution into (17.18), the complex lowpass filter output is expressed as

$$\begin{aligned}\tilde{g}(t) &= (g_c'(t) + jg_s'(t))e^{j\Delta\omega t} \\ &= (g_c'(t)\cos(\Delta\omega t) - g_s'(t)\sin(\Delta\omega t)) + j(g_c'(t)\sin(\Delta\omega t) \\ &\quad + g_s'(t)\cos(\Delta\omega t))\end{aligned} \qquad (17.19)$$

The *real* part of (17.19) simplifies to

$$g(t) = g_c'(t)\cos(\Delta\omega t) - g_s'(t)\sin(\Delta\omega t) \qquad (17.20)$$

and, when the transmitted signal carrier frequency is equal to ω_c with $\Delta\omega = 0$, (17.20) simplifies to

$$g(t) = \int_{-\infty}^{\infty} [s_c(t-\lambda)h_c(\lambda) - s_s(t-\lambda)h_s(\lambda)]d\lambda \quad : \Delta\omega = 0 \qquad (17.21)$$

However, when the output is mixed to the carrier frequency ω_o, as shown in Figure 17.3, the received signal is characterized as the *real* signal given by

$$g(t) = g_c'(t)\cos(\omega_o t) - g_s'(t)\sin(\omega_o t) \quad \text{: filter output at } \omega_o \tag{17.22}$$

where

$$g_c'(t) = \int_{-\infty}^{\infty} [s_c(t-\lambda)h_c(\lambda) - s_s(t-\lambda)h_s(\lambda)]d\lambda \tag{17.23}$$

and

$$g_s'(t) = \int_{-\infty}^{\infty} [s_c(t-\lambda)h_s(\lambda) + s_s(t-\lambda)h_c(\lambda)]d\lambda \tag{17.24}$$

17.3.1 Frequency Domain Characterization of Lowpass Filter Output

In this section, the lowpass channel filter response is described in terms of the frequency domain representation of the lowpass signal and the channel filter. The simulation processing discussed in Section 17.5 and the simulation results discussed in Sections 17.6 and 17.7 use the frequency response functions. In this description, the signal phase ϕ and the frequency error $\Delta\omega$ are introduced to characterize the *rect (t/T)* modulated signal phase at the input to the channel filter. This signal phase and frequency error are used in Section 17.4 in describing the signal.

For the linear filtering operations being considered, the spectrum of the lowpass filter output ($\tilde{G}(u)$) is the product of the signal and channel filter spectrums, expressed as

$$\tilde{G}(u) = \tilde{X}(u)\tilde{H}(u) \tag{17.25}$$

where the signal spectrum is based on the Fourier transform of $\tilde{x}(t)$ expressed as

$$\begin{aligned}\tilde{X}(u) &= \int_{-\infty}^{\infty} \tilde{s}(t)e^{j(\Delta\omega t+\phi)}e^{-ut}dt \\ &= e^{j\phi}\int_{-\infty}^{\infty}(s_c(t)+js_s(t))e^{-j(u-\Delta\omega)t}dt \\ &= e^{j\phi}(S_c(u-\Delta\omega)+jS_s(u-\Delta\omega))\end{aligned} \tag{17.26}$$

Substituting (17.26) into (17.25) with the channel filter spectrum equal to $H(u) = 2(H_c(u) + jH_s(u))$, as developed in Section 17.2, results in the channel filter spectrum

$$\begin{aligned}\tilde{G}(u) = 2e^{j\phi}[&S_c(u-\Delta\omega)H_c(u) - S_s(u-\Delta\omega)H_s(u) \\ &+ j(S_c(u-\Delta\omega)H_s(u) + S_s(u-\Delta\omega)H_c(u))]\end{aligned} \tag{17.27}$$

The channel output is evaluated as the inverse Fourier transform of (17.27) and is expressed as

$$\tilde{g}(t) = g_c(t) + jg_s(t) \tag{17.28}$$

where

$$\begin{aligned}g_c(t) = e^{j(\Delta\omega+\phi)}\Bigg\{&\frac{1}{\pi}\int_{-\infty}^{\infty}(S_c(u)H_c(u+\Delta\omega) \\ &-S_s(u)H_s(u+\Delta\omega))e^{jut}du\Bigg\}\end{aligned} \tag{17.29}$$

and

$$\begin{aligned}g_s(t) = e^{j(\Delta\omega+\phi)}\Bigg\{&\frac{1}{\pi}\int_{-\infty}^{\infty}(S_c(u)H_s(u+\Delta\omega) \\ &+S_s(u)H_c(u+\Delta\omega))e^{jut}du\Bigg\}\end{aligned} \tag{17.30}$$

17.4 PULSE MODULATED CARRIER SIGNAL CHARACTERISTICS

In this section, a binary phase shift keying (BPSK) modulated carrier signal is characterized by the *rect(t/T)* modulation function with amplitude A volts, binary data $d_c = \pm 1$, and angular carrier frequency ω_s. An isolated symbol is expressed as

$$\begin{aligned}s(t) &= Ad_c rect\left(\frac{t}{T}\right)\cos(\omega_s + \phi) \\ &= s_c(t)\cos(\omega_s t) - s_s(t)\sin(\omega_s t)\end{aligned} \tag{17.31}$$

Because BPSK modulation is being considered, the quadrature data $d_s = 0$, so the quadrature signal term $s_s(t)$ in (17.31) corresponds to the quadrature component of the inphase modulated symbol resulting from the signal phase error ϕ.

When the signal $s(t)$ is mixed to baseband, using the channel carrier frequency, ω_c, the resulting baseband symbol is expressed as

$$\begin{aligned}\tilde{x}(t) &= Ad_c rect\left(\frac{t}{T}\right)e^{j(\Delta\omega t+\phi)} \\ &= x_c(t) + jx_c(t)\end{aligned} \tag{17.32}$$

The spectrum of the baseband signal $\tilde{x}(t)$ is evaluated using the Fourier transform and is expressed as

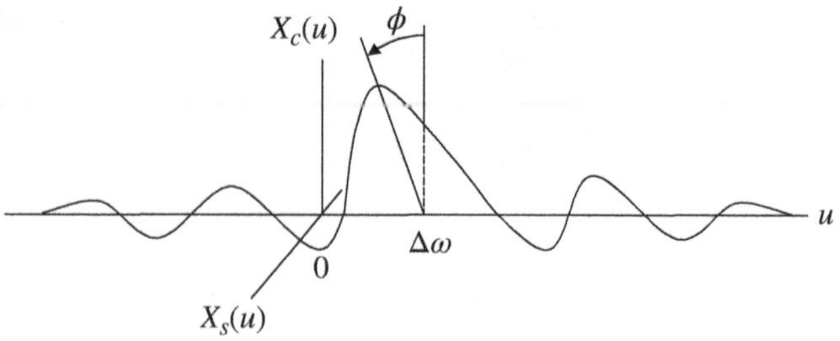

FIGURE 17.4 BPSK symbol baseband spectral characteristics ($\phi > 0$).

$$X(u) = Ad_c e^{j\phi} \int_{-T/2}^{T/2} \tilde{x}(t) e^{-j(u-\Delta\omega)t} dt \qquad (17.33)$$

$$= ATd_c e^{j\phi} \frac{\sin((u-\Delta\omega)T/2)}{(u-\Delta\omega)T/2}$$

and is shown in Figure 17.4.

Using this ideal, infinite bandwidth, form of the transmitted symbol does not detract from the generality of a performance simulation because the bandwidth limiting characteristics of a practical transmit filter can often be included as part of the underlying channel filter. However, without a transmit filter to contain the transmitted signal spectrum, the time–frequency product f_sT would have to meet the condition $f_sT \gg 1$ in order to contain the signal spectrum fold-over about the zero-frequency axis.*

In Section 17.5, these relationships, describing the signal, are combined with those describing the channel to characterize the channel output response under a variety of conditions.

17.5 CHANNEL RESPONSE TO A PULSED MODULATED WAVEFORM

The channel response to the pulse modulated carrier signal described in Section 17.4 is characterized in this section. If the receiver filtering functions are included in the channel description along with the transmitter filtering functions, as mentioned in Section 17.1, then the response can be viewed as the input to the demodulator detection filter. If the demodulator matched filter response is also included in the channel description, then the response to the modulated waveform can be optimally sampled for subsequent data decision processing. The lowpass channel output is evaluated using (17.28), repeated here as

$$\tilde{g}(t) = g_c(t) + jg_s(t) \qquad (17.34)$$

For the BPSK modulated signal described in Section 17.4, the quadrature components of (17.34) are evaluated using (17.29) and (17.30) with $S_s(u) = 0$. The resulting expressions are

$$g_c(t) = e^{j(\Delta\omega+\phi)} \left\{ \frac{1}{\pi} \int_{-\infty}^{\infty} S_c(u) H_c(u+\Delta\omega) e^{jut} du \right\} \qquad (17.35)$$

: BPSK modulation

and

$$g_s(t) = e^{j(\Delta\omega+\phi)} \left\{ \frac{1}{\pi} \int_{-\infty}^{\infty} S_c(u) H_s(u+\Delta\omega) e^{jut} du \right\} \qquad (17.36)$$

: BPSK modulation

These relationships are evaluated in Sections 17.6 and 17.7 to characterize the channel output for the pulse modulated symbol spectrum as described in Section 17.4. The amplitude and phase descriptions of the channel, as described by (17.8) and (17.9), respectively, are useful; in that, the magnitude and phase responses can be independently changed to examine the response under various channel conditions. When applied to the signal spectrum, the phase response can be altered to examine the effect of frequency errors and signal delays.

17.5.1 Normalized Channel Impulse Response

There are a number of advantages to normalize the channel response expressions for computer simulations. For example, the normalized expressions often apply to any system regardless of the carrier frequency or bandwidth.† The use of very large or small numbers is avoided making input easier and less prone to errors, and the parameters are combined resulting in fewer parametric results required to characterize the system performance. For these reasons, the preceding relationships involving the channel response are evaluated in normalized form for subsequent computer programming and performance evaluation.

In this section, the channel impulse responses expressed by (17.5), (17.6), and (17.7) are normalized for computer

*The spectral fold-over can be viewed as the negative frequency signal spectrum appearing in the positive frequency range. The spectral control provided by the transmitter filter does not preserve the *sinc(fT)* spectrum or the constant signal amplitude characteristics. With any time-limited modulation symbol, this is an issue; however, it is particularly pronounced with the *sinc(fT)* function because of the $1/f^2$ power roll-off (20 dB/decade) of the spectral sidelobes.

†This statement applies to time-invariant channels with AWGN; there are also many conditions in which this advantage does not apply, most notably with time-varying channels and multiplicative noise. In these cases, the symbol modulation and carrier frequency must be normalized to the dynamics of the channel.

simulation. The signal sampling rate must be selected to satisfy the Nyquist sampling criteria, so it is necessary to integrate the frequency over an adequate number of channel bandwidths relative to the center frequency. Although the baseband channel simulation is the primary interest, the simulation also characterizes the carrier modulated input and output signals which require satisfying the Nyquist bandpass sampling criterion. Defining the lowpass or one-sided baseband channel bandwidth as B, the normalized frequency variable is defined as $X = u/2\pi B = (f - f_c)/B$; the normalized dependent variable is defined as $Y = tB$. With these normalizations the differential of u is $du = 2\pi B dX$. The normalized parameters F_k and ρ are defined as $F_k = f_c/B$ and $\rho = BT$. Substituting these parameters into the impulse response expression given by (17.5) results in

$$h(Y) = h_c(Y)\cos(2\pi\rho F_k Y) - h_s(Y)\sin(2\pi\rho F_k Y) \quad (17.37)$$

where

$$h_c(Y) = 2B\int_{-\infty}^{\infty} A(X)\cos(2\pi\rho XY + \phi'(X) + \phi_c)dX \quad (17.38)$$

and

$$h_s(Y) = 2B\int_{-\infty}^{\infty} A(X)\sin(2\pi\rho XY + \phi'(X) + \phi_c)dX \quad (17.39)$$

17.5.2 Normalized Symbol Pulse Response

The parameter normalization is the same as for the channel filter defined in Section 17.5.1; however, in this case, there are two additional normalizations to consider. The normalized signal delay is defined as $Y_d = \tau/T$ and the normalized signal frequency shift is defined as $X_d = (f_s - f_c)/B$. The isolated symbol pulse is considered to have a positive amplitude with $d_c = 1$. The parameter normalizations are summarized in Table 17.1.

TABLE 17.1 Normalized Parameter Definitions

Parameter	Definition	Normalized Description
Y	t/T	Time
Y_d	τ/T	Symbol delay
X	$(f-f_c)/B$	Frequency, $u/2\pi B$
X_d	$(f_s-f_c)/B$	Signal frequency shift
dX	$\Delta f/B$	Differential frequency, $\Delta u/2\pi B$
F_k	f_c/B	Channel frequency
F'_k	f_s/B	Signal frequency
$\phi'(X)$	—	Frequency-dependent phase
ρ	BT	Channel-to-signal bandwidth

For the pulsed carrier modulated signal, the channel responses in (17.34), (17.35), and (17.36) are expressed in normalized form as

$$g(Y) = g_c(Y)\cos(2\pi F'_k(Y - Y_d) - \phi) - g_s(Y)\cos(2\pi F'_k(Y - Y_d) - \phi) \quad (17.40)$$

where

$$g_c(Y) = \int_{-\infty}^{\infty} S_c(X)A(X)\cos(2\pi(X + X_d)(Y - Y_d) + \phi'(X) + \phi)dX \quad (17.41)$$

and

$$g_s(Y) = \int_{-\infty}^{\infty} S_c(X)A(X)\sin(2\pi(X + X_d)(Y - Y_d) + \phi'(X) + \phi)dX \quad (17.42)$$

with

$$S_c(X) = A\frac{\sin(\pi(X + X_d))}{\pi(X + X_d)} \quad (17.43)$$

Defining the lowpass frequency variable $u' = \omega - \omega_c$ with respect to the signal frequency, the translation between the channel and signal frequencies is simply $u = u' - (\omega_s - \omega_c)$ or, in normalized form, $X = X' - X_d$.

17.6 EXAMPLE PERFORMANCE SIMULATIONS

In this section, the channel output is examined under several conditions using computer simulations and the normalized parameters defined in Table 17.1. Figure 17.5 shows the various parameters relative to the signal and channel implementation. The signal delay τ results in a propagation phase of $\varphi = -2\pi f_s \tau$ radians that is included in the signal phase ϕ. In the frequency domain, the delay is evident as a linear phase function with frequency. In these evaluations, the simulation program performs a numerical integration over a finite number of channel bandwidths covering the frequency range kB. Since the signal bandwidth is less than the channel bandwidth, the frequency sampling resolution of $\delta f = 1/k_1 T$ is used. The total number of frequency samples is $N_f = \rho k k_1$ where $\rho = BT$ is the number of signal bandwidths contained in the bandwidth of the filter. The sampling resolution and the total number of samples are chosen simply to provide for a high fidelity plot of the respective signal and filter

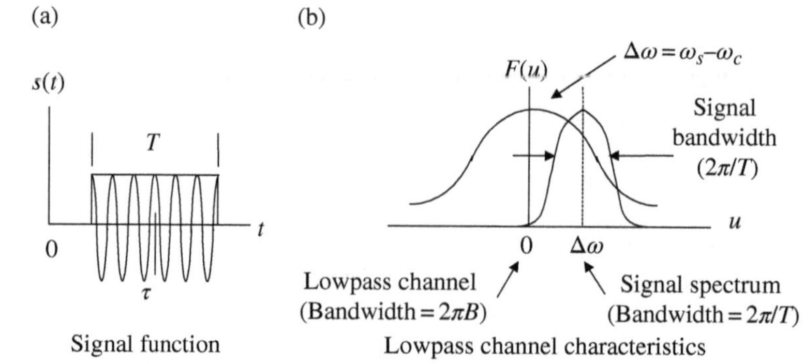

FIGURE 17.5 Signal and channel parameter descriptions.

spectrums and generally exceed the minimum sample rate required by the Nyquist criterion. If the carrier frequency is under sampled, because the time sample increment Δt is too large for the specified carrier frequency, the simulation outputs only the baseband responses $g_c(Y)$ and $g_s(Y)$.

17.7 EXAMPLE OF CHANNEL AMPLITUDE AND PHASE RESPONSES

In this section, several channel amplitude and phase functions are considered and example responses for the pulse modulated carrier input are examined. First, an ideal bandpass channel having constant amplitude and linear phase is considered, and the Gibbs phenomenon [7] is demonstrated in the computer simulation results. The Gaussian channel is then examined followed by a single-pole bandpass filter representation of the channel. The single-pole filter is representative of the antenna response for very-low-frequency (VLF) through high-frequency (HF) systems, and the resulting ISI is characterized for various filter (antenna) bandwidths. The final simulation examines the amplitude and phase response of a dial-up voice grade telephone circuit with and without delay equalization. In this case, several interesting characteristics of the ISI are examined. An example list of simulation conditions is provided in Table 17.2.

17.7.1 Ideal Bandpass Channel

The ideal bandpass channel is examined to demonstrate the nature of channel distortion under ideal conditions. The ideal bandpass channel is characterized as having a constant amplitude over the entire bandwidth with zero response otherwise. The phase function of the ideal bandpass channel is linear and results in a constant signal delay. In terms of the lowpass frequency variable $u = \omega - \omega_c$ the amplitude and phase responses are expressed as

$$A(u) = rect\left(\frac{u}{B}\right) \tag{17.44}$$

and

TABLE 17.2 Example Conditions for Channel Response

Parameter	Value	Description
f_c	2.4	kHz
B	1.2	kHz
Y_d	1.5	Signal delay
f_s	2.4	kHz
R_s	0.15	ksps
F_d	0.075	Constant delay
$\phi'(X)$	—	Frequency-dependent phase[a]
ϕ_c	0.0	Carrier phase (degrees)
F_{d1}	0.0	Linear delay coefficient[a]
F_{d2}	±5.0	Quadratic delay coefficient[a]

[a] See Sections 17.8.2 and 17.8.3.

$$\phi(u) = -u\tau + \phi_c \tag{17.45}$$

where ϕ_c is the channel phase shift at the channel frequency and the frequency dependent phase function, previously defined as $\phi'(u)$, is given by $\phi'(u) = -u\tau$. The normalized expressions used in the simulation are

$$A(X) = rect\left(\frac{X}{\rho}\right) \tag{17.46}$$

and

$$\phi'(X) = -2\pi Y_d X \tag{17.47}$$

where $Y_d = \tau/T$ is the normalized channel delay.

The response to a pulse modulated carrier is shown in Figure 17.6 for the conditions listed in Table 17.2. Figure 17.6a shows the channel and signal spectral characteristics, and Figure 17.6b and c shows the respective carrier and baseband responses. The oscillations in the responses about the pulse edges result from the Gibbs phenomenon and are caused by the finite number of terms in the Fourier expansion of the output resulting from the finite channel bandwidth. The normalized signal delay of 1.5 will normally center the response about $Y = 1.5$. However, the normalized channel

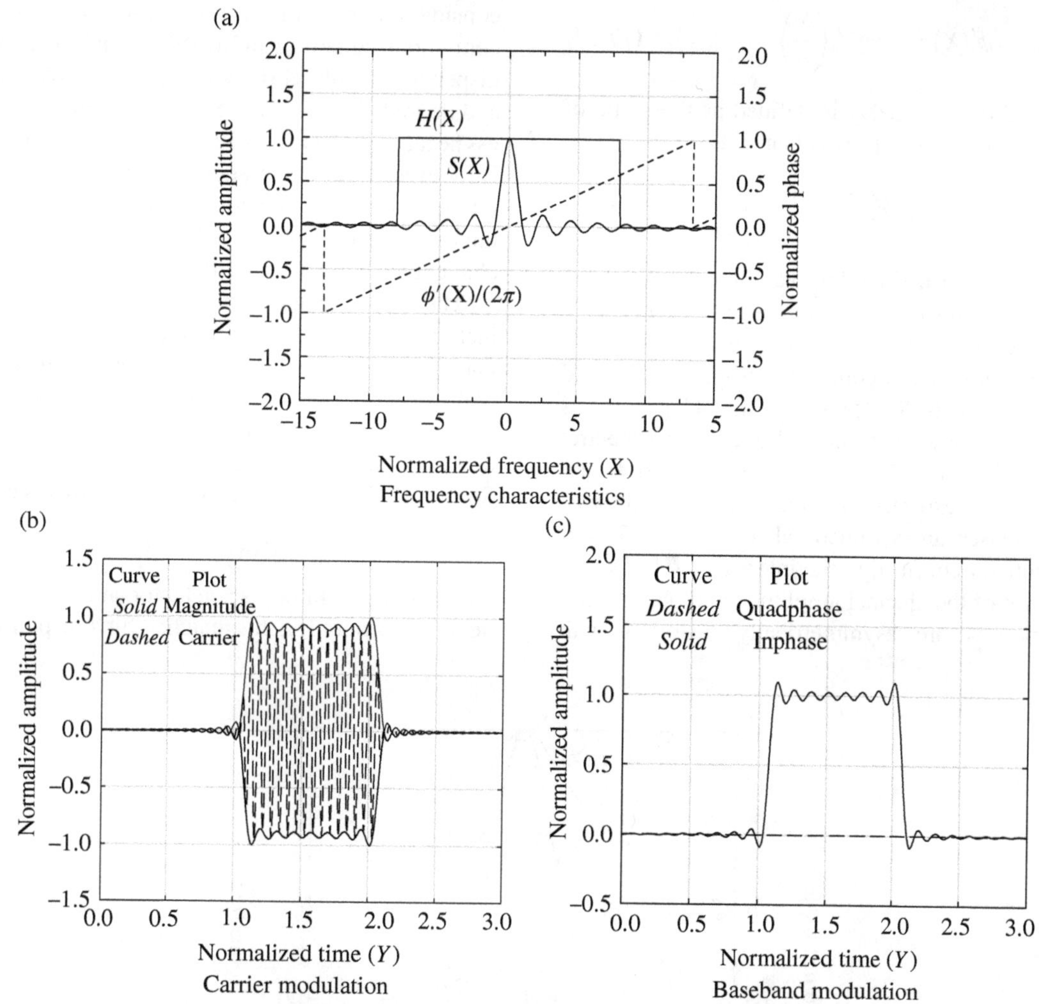

FIGURE 17.6 Response to ideal bandlimited channel.

delay of 0.075 associated with the slope of the phase functions results in the center of the response corresponding to $Y = 1.575$. The quadrature response in Figure 17.6c is zero because the channel is symmetrical with respect to the signal spectrum.

17.7.2 Single-Pole Channel Filter

Amplitude functions characterized as single-pole filters are often useful in characterizing channels modeled as VLF antennas. The response for a single-pole filter with a zero at the origin of the s-plane is given by

$$H(\omega) = \left. \frac{Ks}{(s - s_o)(s - s_o^*)} \right|_{s = j\omega} \tag{17.48}$$

where the poles are located at $s_o = -\alpha \pm j\omega_c$. By letting $K = 2\alpha/\sqrt{1 + (\alpha/\omega_c)^2}$, the positive frequency spectrum, expressed in terms of the lowpass frequency $u = \omega - \omega_c$, is evaluated as

$$H^{+}(u) = \frac{\alpha}{\sqrt{\alpha^2 + u^2}} e^{j(\phi(u) + \phi_c)} \tag{17.49}$$

where the frequency-dependent phase is $\phi'(u) = -\tan^{-1}(u/\alpha)$ and the channel constant phase shift is $\phi_c = \tan^{-1}(\alpha/\omega_c)$. The 3 dB bandwidth is readily evaluated as $B = \alpha/2\pi$ and the resulting normalized form of the amplitude and phase responses are given by

$$A(X) = \frac{\rho}{\sqrt{X^2 + \rho^2}} \tag{17.50}$$

and

$$\phi_c = \tan^{-1}\left(\frac{\rho}{Y_d}\right) \tag{17.51}$$

$$\phi'(X) = -\tan^{-1}\left(\frac{X}{\rho}\right) \tag{17.52}$$

The normalized channel delay is defined as the slope of the phase function at $u=0$ and is evaluated as

$$F_d = -\frac{d\phi'(X)}{dX} = \frac{1}{\rho} \tag{17.53}$$

Various responses to the single-pole channel model are shown in Figure 17.7 for the conditions listed in Table 17.2 with the following exceptions: the channel phase ϕ_c is computed using (17.51), $\phi'(X)$ is computed using (17.52), and the channel bandwidth is $B = 150$ Hz so the BT product is unity. Figure 17.7a shows the channel and signal spectral characteristics, and Figure 17.7b and c show the carrier and baseband responses, respectively. Under ideal conditions, the channel responses are symmetrical about $Y = 1.5$ with the energy essentially confined to the interval $|Y - 1.5| < 0.5$. However, because of the channel amplitude and phase functions, the responses are asymmetrically distorted with considerable signal energy exceeding $Y = 2$. In the case of a continuous stream of information symbols this distortion or dispersion results in predominately postsymbol interference and degraded performance in the information recovery process because of the ISI. In Section 17.8.2 the channel characteristics resulting in predominately pre- and postsymbol interference are examined. As in the case of the ideal channel, the quadrature response in Figure 17.7c is zero because the channel is symmetrical with respect to the signal spectrum.

The demodulator matched filter response can also be included in the channel characterization. In an additive white Gaussian noise (AWGN) environment, the frequency response of the matched filter is the amplitude scaled complex conjugate of the transmitted signal spectrum with an associated arbitrary delay. Considering the transmitted signal described in Section 17.4, the demodulator matched filer is characterized as

$$F(u) = KS_r^*(u)e^{-juT_o} \tag{17.54}$$

where K is an arbitrary scale factor and T_o is an arbitrary filter delay. Incorporating this matched filter response as part of the

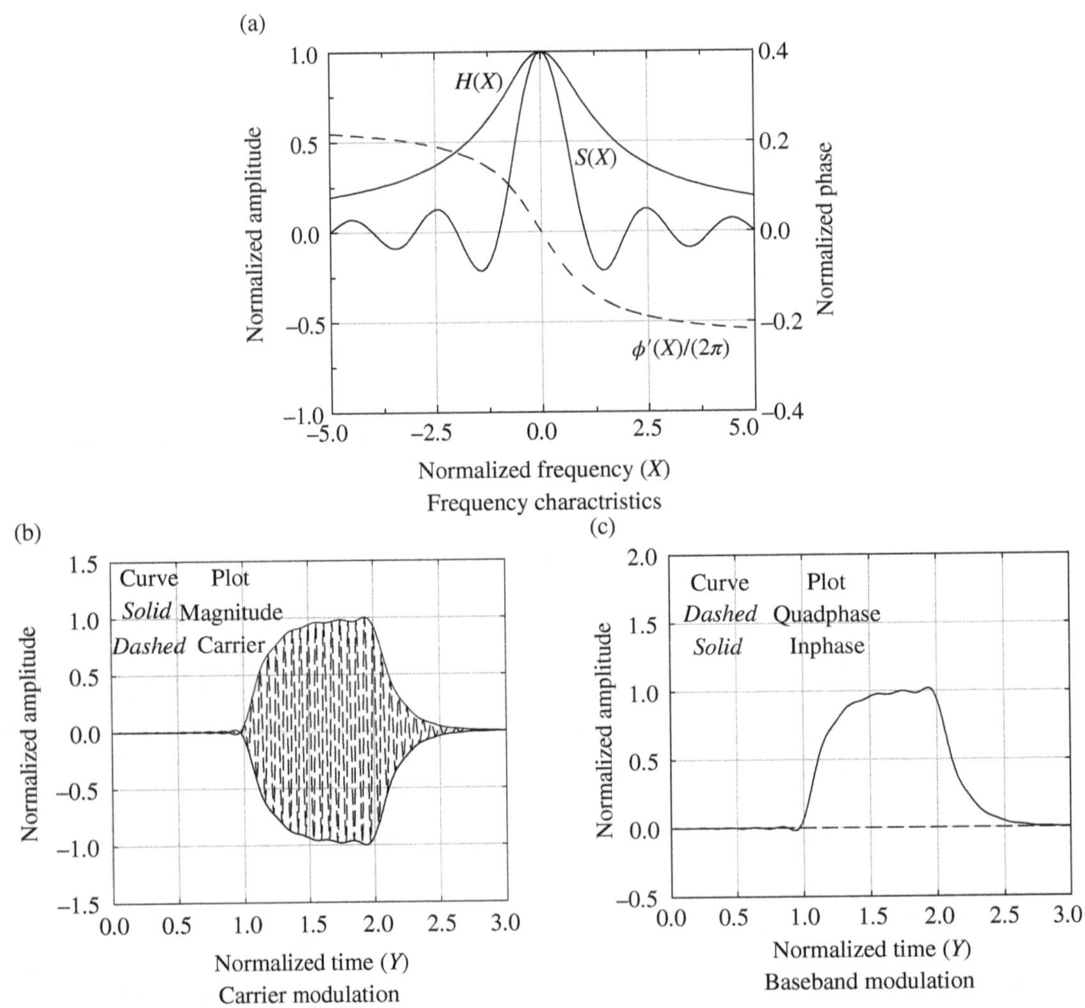

FIGURE 17.7 Response to single-pole channel model ($BT = 1$).

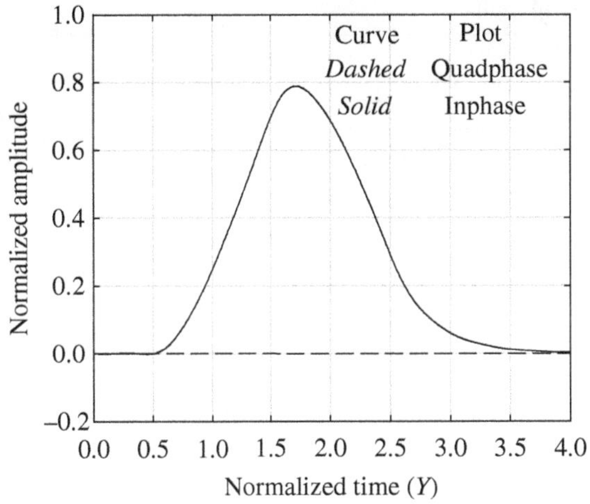

FIGURE 17.8 Response to single-pole channel and matched filter ($BT = 0.5$).

overall channel response, the demodulator baseband output for $BT = 0.5$ is shown in Figure 17.8. Aside from these changes, the channel conditions used in Figure 17.7 are identical. The optimally sampled matched filter output in Figure 17.8 corresponds to $Y = 1.715$. In a sequence of contiguous received symbols, the demodulator symbol synchronization processing must sample the matched filter at $1.715 \pm n$, for $n = 0, 1, \ldots$. The sampled output for $BT = 0.5$ is 0.788, which represents a loss in the SNR of 2.07 dB relative to the sampled matched filter output with an ideal channel. In addition to a loss of symbol energy, the channel distortion also results in a loss because of the ISI that appears as an interference noise source in random data. For the isolated symbol shown in Figure 17.8, the ISI level for values of $n \neq 0$ are listed in Table 17.3 for several values of the BT parameter. These results indicate that the pre- and postsymbol ISI, corresponding to $n = -1$ and 1, respectively, are appreciable but diminish as the channel bandwidth is increased. As $BT \rightarrow \infty$, the ideal matched filter response to an isolated symbol is a triangular function with a maximum value at $n = 0$ and zero response for $n \neq 0$.

17.8 EXAMPLE CHANNEL AMPLITUDE, PHASE, AND DELAY FUNCTIONS

17.8.1 Dial-Up Telephone Channel

The dial-up telephone channel* of interest is the former Bell Telephone Company's 3002 channel [8, 9] which is characterized by prescribed boundaries for the amplitude and delay responses. The 3002 channel is also provided with various amounts of amplitude and delay conditioning denoted as $C1$, $C2$, and $C4$. The line conditioning allows the channel, originally designed for voice grade communications, to be used for data rates up to 9600 bps. However, equalization is generally required for good performance above 2400 bps. An arbitrarily accessed dial-up channel will have a one-sigma attenuation response and a two-sigma delay response roughly equivalent to the respective 3002 line specifications. The response specifications are usually normalized to a 1700 Hz carrier and the total dial-up connection, including the local loops, can be modeled as a bandpass filter with a 5 dB bandwidth ranging between 300 and 2800 Hz. Usually, the modulated data is applied to carriers ranging between 1600 and 1800 Hz.

The amplitude and delay responses are of special interest for two reasons: they always exist to some extent on all connections, and therefore must be dealt with in one way or another. Because the amplitude and delay functions are time invariant for a given connection, the degradation caused by ISI can be minimized using fixed or adaptive equalizers. Medium-speed modems operating between 1200 and 2400 bps can use fixed or statistical equalizers, also called compromise equalizers, to compensate for the mean channel distortion. For these lower data rates, fixed equalizers operate quite well over the dial-up channel. On the other hand, high-speed modems operating greater than 2400 bps must look to more sophisticated adaptive equalization techniques to preserve the performance over the dial-up channel.

In this section, the response of a typical 3002 channel to a pulse modulated carrier is examined and the resulting ISI is quantified. The amplitude and delay boundaries for the basic 3002 dial-up line are shown in Figures 17.9 and 17.10, respectively. The $C1$ and $C2$ line conditioning specifications are also indicated in these figures.

The one-sigma amplitude and delay for the dial-up channel corresponds approximately to the 3002 channel so that about 84% for the dial-up connections are within the 3002 channel boundaries. In the following computer simulation examples, a fifth degree polynomial is curve-fit to the $C1$-conditioned amplitude boundaries and the resulting response is considered to be representative of a randomly selected $C1$-conditioned channel. The phase response used in the simulations is based on a quadratic delay characteristic as discussed in Section 17.8.2.

17.8.2 Quadratic Delay (Cubic Phase) Function

The expression for the cubic phase function centered about the center frequency of the channel is given by

$$\phi(u) = \phi_c - T_c u - \frac{\bar{\rho}}{2} u^2 - \frac{\bar{\rho}'}{3} u^3 \tag{17.55}$$

*This analysis was performed in the 1960s and is included for its technical insights and historical interest. In the mid-to-late 1970s, with the advancement of commercial cable, satellite, and wireless communications, conventional wireline communications came to an end.

TABLE 17.3 Sampled Matched Filter Output Results for the Single-Pole Channel

	n						
BT	−1	1	2	3	4	0	Loss (dB)[a]
0.25	0.093	0.217	0.044	0.010	0.003	0.632	3.98
0.5	0.058	0.147	0.007	—	—	0.788	2.07
1.0	0.032	0.077	—	—	—	0.891	1.00
1.5	0.021	0.052	—	—	—	0.927	0.65

[a]Loss in the symbol interval *T* excluding ISI.

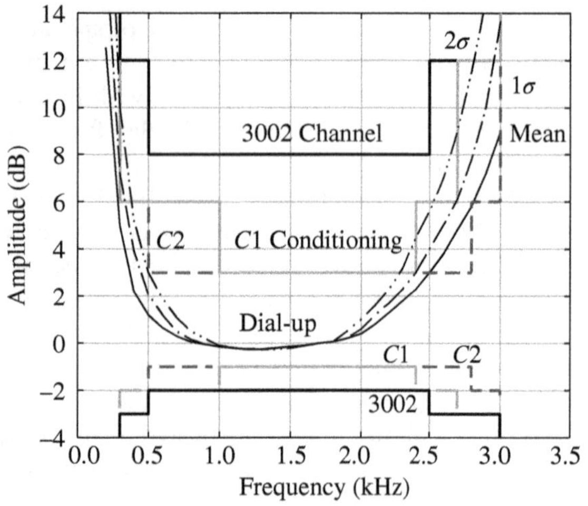

FIGURE 17.9 Dial-up line amplitude characteristics for 3002 connection with *C*1 and *C*2 conditioning.

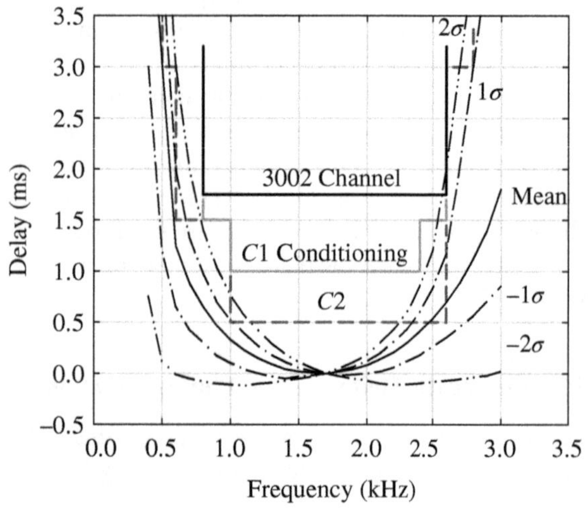

FIGURE 17.10 Dial-up line delay characteristics for 3002 connection with *C*1 and *C*2 conditioning.

and the corresponding quadratic delay function is given by

$$T(u) = -\frac{d\phi(u)}{du} = T_c + \bar{\rho}u + \bar{\rho}'u^2 \\ = T_c + T_1(u) + T_2(u) \tag{17.56}$$

The constants $\bar{\rho}$ and $\bar{\rho}'$ are defined in terms of a specified delay at the channel band edge, *B*, so that

$$T_1(B) = (2\pi B)\bar{\rho};\ \ T_2(B) = (2\pi B)^2\bar{\rho}' \tag{17.57}$$

and

$$\bar{\rho} = \frac{T_1(B)}{2\pi B} = \frac{F_{d1}}{2\pi BT}T^2 \quad : s^2/\text{rad} \tag{17.58}$$

$$\bar{\rho}' = \frac{T_2(B)}{(2\pi B)^2} = \frac{F_{d2}}{(2\pi BT)^2}T^3 \quad : s^3/\text{rad}^2 \tag{17.59}$$

where F_{d1} and F_{d2} are the normalized delays at the band edge for the quadratic and cubic phase terms, respectively. Substituting these results into the channel phase and delay functions and recalling that $uT = 2\pi X$, results in the following normalized expressions

$$\phi(X) = \phi_c - 2\pi\left(F_d X + \frac{F_{d1}}{2BT}X^2 + \frac{F_{d2}}{3(BT)^2}X^3\right) \tag{17.60}$$

$$\frac{T(X)}{T} = F_d + \frac{F_{d1}}{BT}X + \frac{F_{d2}}{(BT)^2}X^2 \tag{17.61}$$

Computer simulations are used to demonstrate the distortion associated with the *C*1 line characteristic, using the pulsed modulated carrier input signal. The conditions are identified in Table 17.2 with the following exceptions: $f_c = f_s = 1.7$ kHz, $F_d = 0$, and $R_s = 1.0$ ksps. The signal distortion through an all-pass filter with a linear phase response is examined in Problem 4. The various responses to the pulse modulated carrier are shown in Figure 17.11. Figure 17.11a shows the curve-fit amplitude and the quadratic channel delay responses relative to the signal spectrum. Figure 17.11b shows the carrier modulated response and the

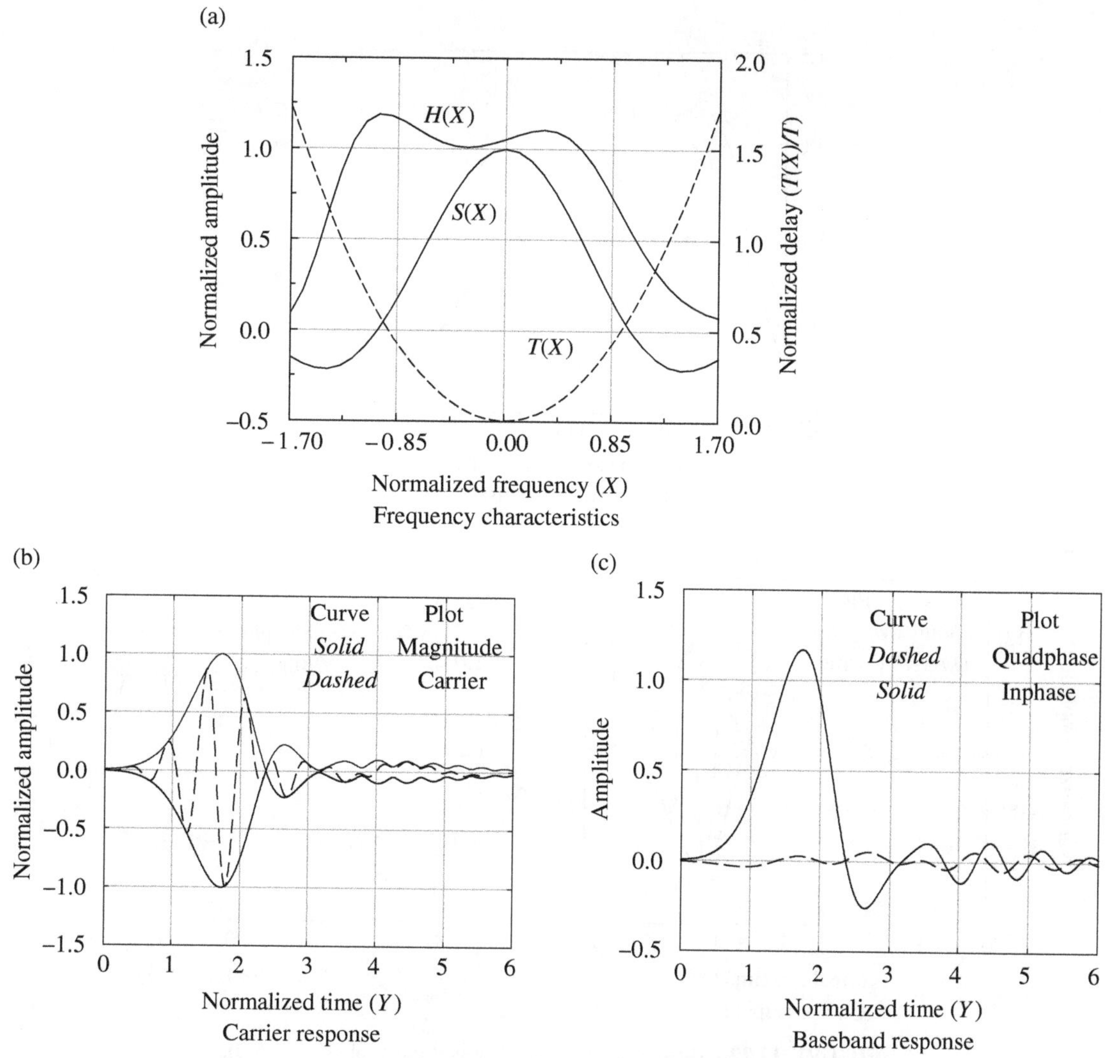

FIGURE 17.11 Response to $C1$ conditioned channel ($F_{d2} = 5.0$).

corresponding magnitude function, and Figure 17.11c shows the I/Q responses. The concave upward, or *smiling*, characteristic of the quadratic delay distortion results in predominately post ISI terms. Figure 17.12 shows the same sequence of plots for a concave downward, or *frowning*, quadratic delay distortion. It is seen that the frowning quadratic delay distortion simply *flips* the response in time with the distortion appearing as mostly presymbol interference. Channel equalizers must be capable of equalizing both presymbol and postsymbol interference.

17.8.3 Practical Interpretation of the Phase Function

The phase functions used for the dial-up channels are representative of the physical channels over certain regions of the frequency response. Suppose, for example, the actual phase function of the channel is similar to that shown in Figure 17.13. This satisfies the condition that the phase function has odd symmetry in frequency.

Suppose now that this phase function is expressed as

$$\phi(\omega) = -b_o\omega + b_1\sin(\omega T_o) \tag{17.62}$$

Expanding (17.62) in a Taylor series about ω_c gives

$$\phi(\omega) = \phi(\omega_c) + \dot{\phi}(\omega_c)(\omega - \omega_c) + \frac{\ddot{\phi}(\omega_c)}{2!}(\omega - \omega_c)^2 + \cdots \tag{17.63}$$

where

$$\phi(\omega_c) = -b_o\omega_c + b_1\sin(\omega_c T_o) \tag{17.64}$$

and

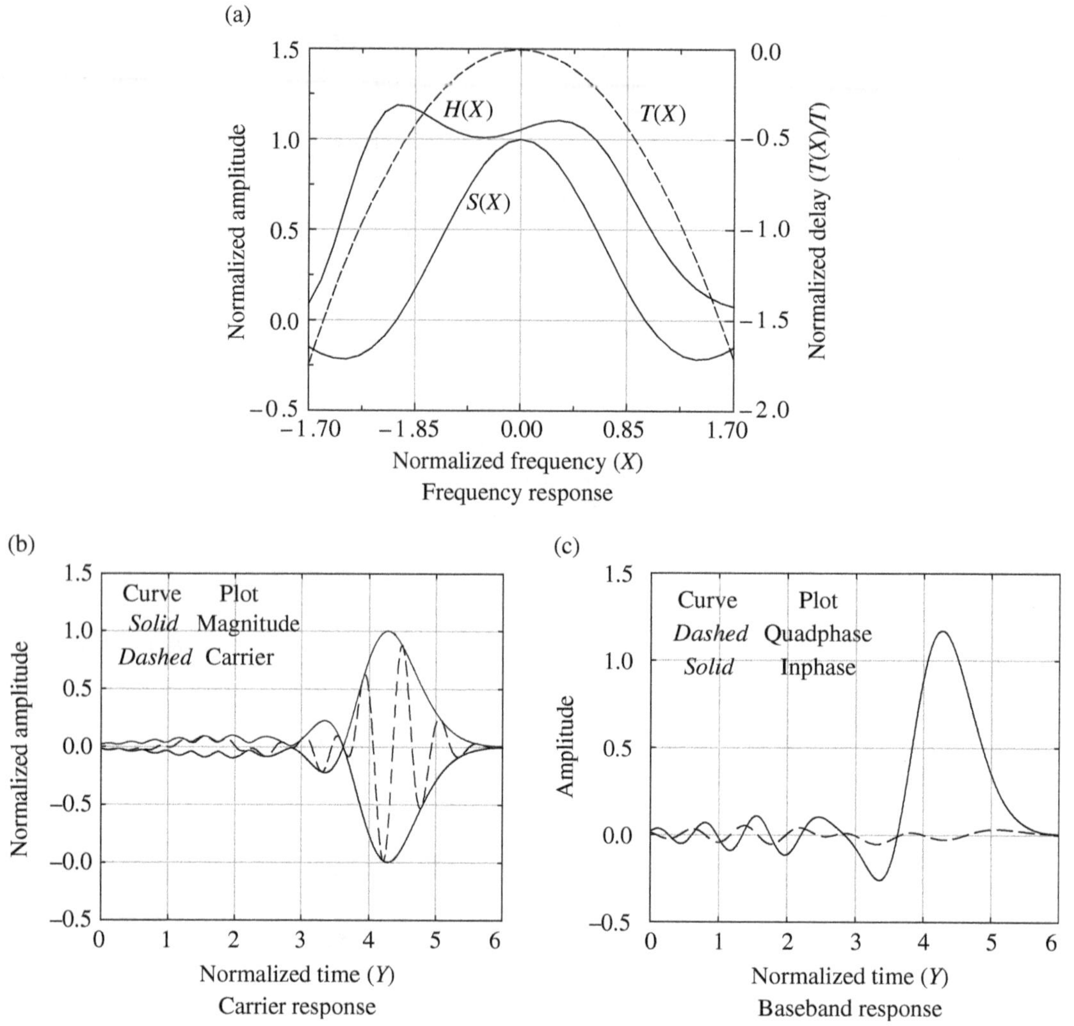

FIGURE 17.12 Response to $C1$ conditioned channel ($F_{d2} = -5.0$).

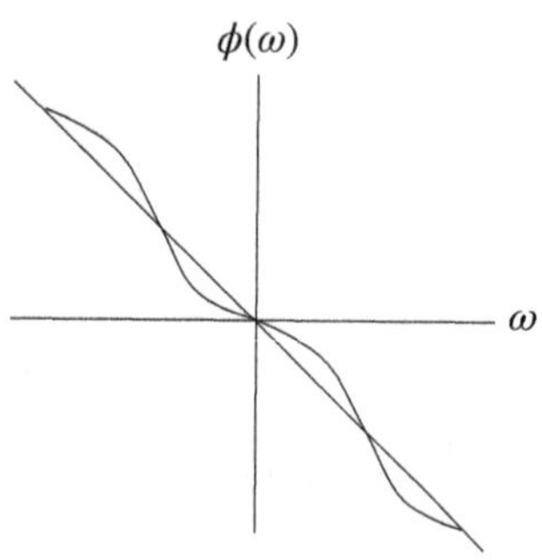

FIGURE 17.13 Example channel phase response.

$$\dot{\phi}(\omega_c) = -b_o + b_1 T_o \cos(\omega_c T_o) \qquad (17.65)$$

and

$$\ddot{\phi}(\omega_c) = -b_1 T_o^2 \sin(\omega_c T_o) \qquad (17.66)$$

and

$$\dddot{\phi}(\omega_c) = -b_1 T_o^3 \cos(\omega_c T_o) \qquad (17.67)$$

When these results are expressed in terms of the lowpass argument $u = \omega - \omega_c$ the phase function becomes

$$\phi(u) = \phi'(u) + \phi_c \qquad (17.68)$$

with

$$\phi_c = \phi(\omega_c) \qquad (17.69)$$

and

$$\phi'(u) = \dot{\phi}(\omega_c)u + \frac{\ddot{\phi}(\omega_c)}{2!}u^2 + \cdots \qquad (17.70)$$

Upon substituting (17.64) through (17.67) into (17.70), the phase function is expressed as

$$\phi'(u) = (b_o - b_1 T_o \cos(\omega_c T_o))u + \left(\frac{b_1 T_o^2}{2!}\sin(\omega_c T_o)\right)u^2 + \left(\frac{b_1 T_o^3}{3!}\cos(\omega_c T_o)\right)u^3 + \cdots \tag{17.71}$$

and the corresponding delay function is

$$T(u) = (b_o - b_1 T_o \cos(\omega_c T_o)) + (b_1 T_o^2 \sin(\omega_c T_o))u + \left(\frac{b_1 T_o^3}{2!}\cos(\omega_c T_o)\right)u^2 + \cdots \tag{17.72}$$

As an example of the cubic phase function, let $\omega_c = 2\pi/T_o$ and neglect powers of u greater than three so that (17.71) becomes

$$\phi'(u) \cong (b_o - b_1 T_o)u + \left(\frac{b_1 T_o^3}{3!}\right)u^3 \tag{17.73}$$

Upon equating coefficients and considering $F_{d1} = 0$, this result is identical to the cubic phase function used in Section 17.8.2. In a similar manner, the quadratic phase function is obtained when $\omega_c = \pi/T_o$. Therefore, the phase characteristics outlined in Section 17.8.2 can be approximated by considering selected regions of a realizable channel phase function. In general, more complex phase functions can be realized using a Fourier series representation involving more terms.

ACRONYMS

AWGN	Additive white Gaussian noise
BPSK	Binary phase shift keying
HF	High frequency
I/Q	Inphase and quadrature phase
ISI	Intersymbol interference
VLF	Very low frequency

PROBLEMS

1. Referring to (17.4), show that $H_r(\omega)$ and $H_i(\omega)$ must be, respectively, even and odd functions of ω for $h(t)$ to be a *real* function.

2. Referring to (17.6) and (17.7) show the even and odd conditions of $H_c(u)$ and $H_s(u)$ with respect to their arguments that result in either $h_c(t) = 0$ or $h_s(t) = 0$.

3. Show that the integration in (17.5) results in the I/Q lowpass filter functions $h_c(t)$ and $h_s(t)$ as expressed in (17.6) and (17.7). Then, using these results, show that the second equality in (17.5) applies.

4. For a pulse signal input expressed as $s(t) = Arect(t/T)$, compute the output response of a unit-gain all-pass filter with a linear phase response $\phi(\omega) = \omega\tau$. Does the out signal experience any distortion?

5. Part 1: Derive the expression for the spectrum of the signal in (17.31) at the output of a unit-gain all-pass channel with the phase function expressed as $\phi(\omega) = \omega\tau$.

Part 2: Sketch the three-dimensional spectrum (see Figure 17.4) of the filtered signal obtained in Part 1. Show the horizontal ω axis and the orthogonal axes $\mathrm{Re}\{S(\omega)\}$ and $\mathrm{Im}\{S(\omega)\}$. Label all pertinent parameters including A, T, ω_s, τ, and ϕ.

Part 3: Evaluate the expression for the equivalent lowpass signal $s_c(t)$.

6. Given the complex frequency translation of the channel input shown in Figure 17.3, express the frequency translated baseband signal $\tilde{x}(t)$ in terms of the quadrature components $s_c(t)$ and $s_s(t)$ for the BPSK modulated input signal $s(t)$ expressed in (17.31). Using the expression for $\tilde{x}(t)$ in terms of $s_c(t)$ and $s_s(t)$ that you have just developed, show that the first equally in (17.32) applies.

7. Repeat Problem 5 for the QPSK modulated signal using the isolated symbol with $d_s = 1$.

REFERENCES

1. R.W. Lucky, J. Salz, E.J. Weldon, *Principles of Data Communication*, McGraw Hill Book Company, New York, 1968.
2. A.A. Alexander, R.M. Gryb, D.W. Nast, "Capabilities of the Telephone Network for Data Transmission," Bell System Technical Journal, Vol. 39, p. 431, 1960.
3. Bell System Technical Journals on Conditioned Lines, American Telephone and Telegraph Company, Bedminster, NJ, June 1969–1971.
4. E.D. Sunde, "Theoretical Fundamentals of Pulse Transmission-I," Bell System Technical Journal, Vol. XXXIII, No. 3, pp. 721–788, May 1954.
5. H. Urkowitz, "Bandpass Filtering with Lowpass Filters," Journal of the Franklin Institute, Vol. 276, No. 1, July 1963.
6. J. Dugundji, "Envelopes and Pre-Envelopes of Real Waveforms," IRE Transactions on Information Theory, Vol. IT-4, pp. 53–57, March 1958.
7. H.H. Skilling, *Electrical Engineering Circuits*, John Wiley & Sons, Inc., New York, 1958.
8. "1969–1970 Switched Telecommunications Networks Connector Survey (Reprints of Bell System Technical Journal Articles)," Bell System Technical Reference, PUB-41007, April 1971.
9. "Data Communications Using the Switched Telecommunications Network," Bell System Technical Reference, PUB-41005, May 1971.

18

COMMUNICATIONS IN FADING ENVIRONMENTS

18.1 INTRODUCTION

The fading channel is characterized as a multiplicative disturbance in which the signal amplitude fluctuates with time depending upon the temporal characteristics of the channel. The term "channel" is used somewhat broadly in the sense that the amplitude fluctuations often result from many signal paths that appear as a signal with random phase variations at the receiver terminal. These multipath signals combine in such a way that the phases constructively and destructively combine resulting in random variations of the received signal amplitude. The amplitude of the received signal is often dominated by a direct or line of sight (LOS) path and the received signal amplitude fluctuations are related to the direct-path signal amplitude. The signals arriving at the receiver over direct and indirect paths are, respectively, referred to as the specular and random components of the received signal. The specular component corresponds to the received signal over the path without fading.

An important aspect of the fading environment is the temporal characteristic of the amplitude fluctuations. For example, if all of the signal paths through the channel were fixed or unchanging, the received signal amplitude would be a constant random variable independent of time. However, real channels are seldom stationary but fluctuate with time due to natural disturbances, such as atmospheric and ionospheric turbulence, Doppler velocities, Faraday rotation, and changes in ground reflection coefficients. An interesting phenomenon in the satellite channel is that the fluctuations or scintillations are a result of the diurnal changes in the electron density of the ionosphere, which are more severe in the equatorial and polar regions than at midlatitudes. The temporal fading is also sensitive to the motion of the transmitter and receiver terminals. Slow fading results when the channel correlation time is much greater than the symbol rate or the fundamental time constants of the receiver and demodulator. On the other hand, rapid fading results when the channel correlation times are less than the symbol durations or the system time constants.

Another important aspect of the fading environment is the frequency bandwidth over which the amplitude fading characteristics remain constant or correlated. For example, when the entire radio frequency (RF) bandwidth occupied by the communication signal fades identically, the signal level changes; however, the signal is undistorted; in this case, the fading is referred to as frequency nonselective fading, or simply nonselective or flat fading. When the channel coherence bandwidth is less than the occupied bandwidth, then frequency selective fading occurs resulting in independent fading over portions of the bandwidth resulting in severe signal distortion. Mobile radio communications [1, 2] in urban and rural areas with indoor and outdoor cell phone communications encounter small-scale fading effects leading to rapidly changing Rayleigh fading with large Doppler shifts and signal dispersion. These environments also encounter man-made interference [3] that compounds the system complexity in providing reliable communications.

Based on these introductory comments, the communication systems designer can generally improve the performance

Digital Communications with Emphasis on Data Modems: Theory, Analysis, Design, Simulation, Testing, and Applications,
First Edition. Richard W. Middlestead.

Companion website: www.wiley.com/go/digitalcommunications

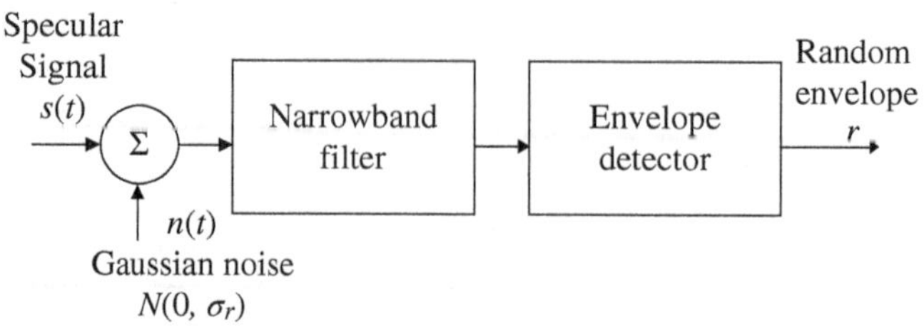

FIGURE 18.1 Ricean model: envelope of signal and narrowband Gaussian noise.

by applying robust waveforms, coding, interleaving, diversity combining,* and automatic repeat request (ARQ) protocols. Diversity combining is a technique of making the most of a bad situation, in that, the received waveforms to be combined must be separated by more than the fading correlation intervals either in time or frequency, resulting in increased data throughput and/or RF bandwidth. Several forms of diversity combining to improve the system performance are temporal, spatial, frequency, angle-of-arrival, and polarization diversity. These topics are also examined in Chapters 19 and 20.

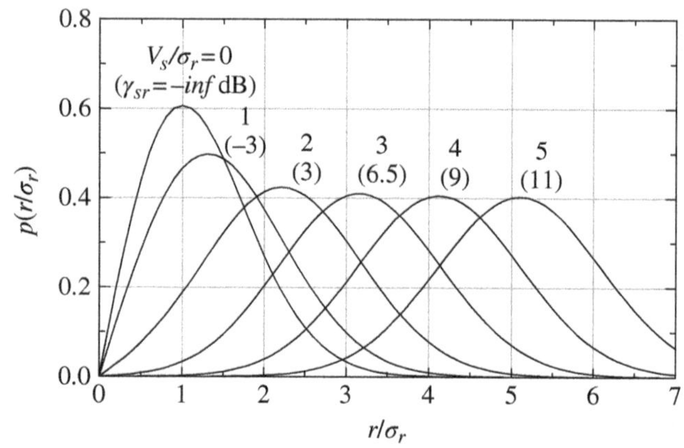

FIGURE 18.2 Normalized Ricean probability density function.

18.2 RICEAN FADING CHANNELS

The Ricean fading model characterizes the statistical fluctuations in the received signal amplitude in terms of the specular and random components. The analysis is based on examining the envelope of a sinusoidal signal and narrowband Gaussian noise as shown in Figure 18.1 where the signal is expressed as

$$s(t) = V_s \cos(\omega_c t + \varphi_s) \tag{18.1}$$

with peak voltage level V_s, carrier frequency $f_c = \omega_c/2\pi$ Hz, and arbitrary signal phase ϕ_s. The peak voltage of the input sinusoidal signal represents the specular component of the received signal power given by $P_s = V_s^2/2$ watts. The expression for the statistics of the signal envelope random variable r is the Ricean probability density function (*pdf*) that plays a dominate role in evaluating the system performance in a fading environment. The specular component in the Ricean model is the signal voltage, V_s, and the random component is characterized in terms of the standard deviation, σ_r, of the narrow band Gaussian noise. When the specular component is zero, the amplitude fading is characterized by the Rayleigh *pfd*.

The derivation of the Ricean distribution is developed in Section 1.5.5.1, and the result is expressed as

$$p_R(r) = \frac{r}{\sigma_r^2} e^{-(r^2+V_s^2)/(2\sigma_r^2)} I_o\left(\frac{rV_s}{\sigma_r^2}\right) \tag{18.2}$$

where $I_o(x)$ is the modified Bessel function of order zero [4]. It is useful to express the Ricean distribution in normalized form by letting $x = r/\sigma_r$ and $x_s = V_s/\sigma_r$ with the result

$$\begin{aligned} p_R(x) &= p_R(r)\frac{dr}{dx} \\ &= xe^{-(x^2+x_s^2)/2} I_o(xx_s) \end{aligned} \tag{18.3}$$

This normalized expression for the Ricean *pdf* is plotted in Figure 18.2 with $x_s = V_s/\sigma_r$ as a parameter. The specular-to-random noise power and the additive white noise-to-random noise power ratios are defined as

$$\gamma_{sr} \triangleq \frac{P_s}{\sigma_r^2} \quad \text{and} \quad \alpha \triangleq \frac{\sigma_n^2}{\sigma_r^2} \tag{18.4}$$

18.2.1 Rayleigh Fading Channels

When the specular component of the Ricean *pdf* is zero, the resulting distribution for the envelope r is the Rayleigh *pdf* expressed as

*Diversity *combining* involves the selection of one of several independently received signals or the selection of an optimally combined output.

$$p_R(r) = \frac{r}{\sigma_r^2} e^{-r^2/(2\sigma_r^2)} \tag{18.5}$$

18.2.2 Gaussian Limit

At the other end of the *spectrum* from the Rayleigh distribution, that is, as V_s approaches infinity, such that $V_s \gg \sigma_r$; the modified Bessel function is approximated as

$$I_o\left(\frac{rV_s}{\sigma_r^2}\right) \cong \frac{e^{-rV_s/\sigma_r^2}}{\sqrt{2\pi rV/\sigma_r^2}} \tag{18.6}$$

Substitution of (18.6) into the Ricean *pdf* described by (18.2) and completing the square in the exponent results in the expression

$$\begin{aligned} p_R(r) &\cong \frac{1}{\sqrt{2\pi\sigma_r^2}} \sqrt{\frac{r}{V_s}} e^{-(r-V_s)^2/(2\sigma_r^2)} \\ &\cong \frac{1}{\sqrt{2\pi\sigma_r^2}} e^{-(r-V_s)^2/(2\sigma_r^2)} \end{aligned} \tag{18.7}$$

Where the last approximation in (18.7) applies for values of r within several standard deviations of V_s, that is, for $r = V_s + k\sigma_r$ or $r/V_s = 1 + k\sigma_r/V_s$. Therefore, based on the condition $V_s \gg \sigma_r$, it follows that $r/V_s \cong 1$ for reasonably small values of k. Consequently, as the specular component approaches infinity, the Ricean *pdf* approaches the Gaussian *pfd* with mean value V_s and standard deviation σ_r.

18.3 RICEAN CUMULATIVE DISTRIBUTION

It is informative to examine the cumulative distribution function (*cdf*) of the Ricean *pdf* so that the probability of the received signal exceeding a specified level can be determined. The *cdf* is defined as

$$P_r(r \le r_T) = \int_0^{r_T} p_R(r)dr \tag{18.8}$$

However, it is convenient to examine the probability that $r > r_T$ corresponding to $P_r(r > r_T) = 1 - P_r(r \le r_T)$. Using the normalized parameters introduced in Section 18.2, the normalized threshold is $x_T = r_T/m_r$ and (18.8) becomes

$$P_r(x > x_T) = e^{-x_s^2/2} \int_{x_T}^{\infty} xe^{-x^2/2} I_o(xx_s)dx \tag{18.9}$$

Equation (18.9) is numerically integrated and the results are shown in Figure 18.3 for several values of the specular signal-to-noise ratio (SNR). As the specular component becomes large, the median value, *Med*[x] corresponding to $P_r(x > x_T) = 0.5$, approaches the specular SNR. For example, in Figure 18.3, when $\gamma_{sr} = 11$ dB, the abscissa, corresponding to $P_r(x > x_T) = 0.5$, is approximately 11 dB. This occurs because, for large values of γ_{sr}, the Ricean distribution approaches the Gaussian distribution in which case the median and mean values are equal.

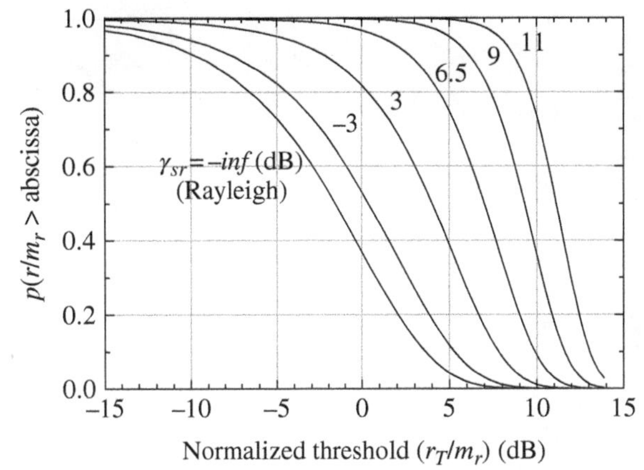

FIGURE 18.3 Normalized Ricean cumulative distribution function.

The mean value of the Ricean distribution, used as the normalizing value in Figure 18.3, is evaluated as

$$E[r] = \frac{e^{-V_s^2/(2\sigma_r^2)}}{\sigma_r^2} \int_0^{\infty} r^2 e^{-r^2/(2\sigma_r^2)} I_o\left(\frac{rV_s}{\sigma_r^2}\right) dr \tag{18.10}$$

or, in normalized form

$$E[x] = \sigma_r e^{-x_s^2/2} \int_0^{\infty} x^2 e^{-x^2/2} I_o(xx_s)dx \tag{18.11}$$

The normalized mean value, $E[x]/\sigma_r$, is plotted in Figure 18.4. The straight-line plot assumes the relationship $E[r] = V_s$ and this appears to be a good approximation for specular SNRs exceeding about 3 or 4 dB. For a zero specular component, that is, for $V_s = 0$, the normalized mean value is $\sqrt{\pi/2}$, which is simply the mean of the Rayleigh distribution.

18.4 APPLICATION OF RICEAN CHANNEL MODEL

18.4.1 Slow Fading

To apply the Ricean distribution to the slow fading channel, the received signal power is viewed as the random variable $P_r = r^2/2$ and, considering zero-mean additive white

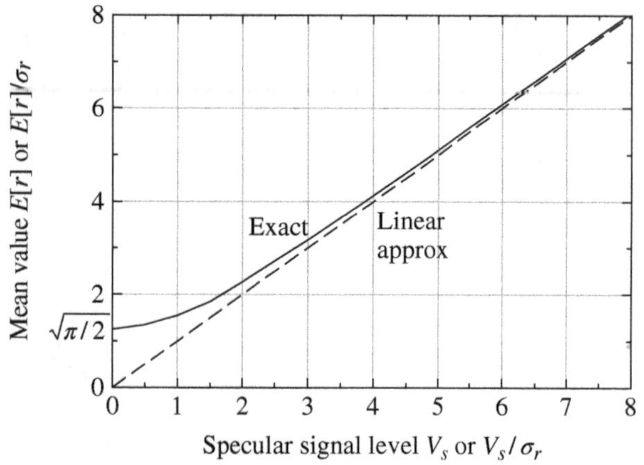

FIGURE 18.4 Mean value of Ricean *pdf*.

Gaussian receiver noise (AWGN) with power σ_n^2, the average receiver SNR is $\bar{\gamma} = \gamma_r + \gamma_s$. In addition, the random and specular SNRs are defined as $\gamma_r = P_r/\sigma_n^2$ and $\gamma_s = V_s^2/2\sigma_n^2 = P_s/\sigma_n^2$, respectively. Substituting these SNRs into the Ricean *pdf* results in the distribution of the received SNR in a Ricean fading environment, given by

$$p_R(\gamma) = p_R(r)\left(\frac{dr}{d\gamma}\right) = e^{-(\bar{\gamma}_r + \bar{\gamma}_s)} I_o\left(2\sqrt{\bar{\gamma}_r \bar{\gamma}_s}\right) \quad (18.12)$$

The performance of a communication system is usually characterized in terms of the bit-error probability, that is, expressed as a function of, or conditioned on, the receiver SNR as

$$P_{be}(\gamma) = P_r(\text{bit-error}|\gamma) \quad (18.13)$$

In a nonfading channel, the SNR is not a random number but is constant so that $P_{be} = P_{be}(\gamma)$; however, in a fading environment, the SNR is a random variable and the conditioning on γ must be removed using the *pdf* of the fading SNR. In general, the result is

$$P_{be} = \int P_r(\text{bit-error}|\gamma) p(\gamma) d\gamma \quad (18.14)$$

where the integration is over the applicable range of γ.

18.4.2 Fast Fading

In the fast fading environment, the temporal correlation of the fading is much faster than the minimum system time-constant, which is typically the bit or symbol duration of the transmitted waveform.* In this situation, the random component of the fading channel is considered to be additive noise, so the received SNR is defined as the ratio of the specular component to the random component plus receiver noise, that is,

$$\gamma = \frac{P_s}{P_r + \sigma_n^2} = \frac{\gamma_s}{\gamma_r + 1} \quad (18.15)$$

Expressed in terms of the average signal-to-noise ratio $\bar{\gamma} = \gamma_s + \gamma_r$ with $\beta \triangleq \gamma_s/\gamma_r$, (18.15) becomes

$$\gamma = \frac{\beta\bar{\gamma}}{\bar{\gamma} + \beta + 1} \quad (18.16)$$

Using (18.14) and (18.16) the bit-error probability $P_{be} = P_r(\gamma)$ is unique to the underlying modulation, for example, for CBPSK modulation the bit-error probability is expressed as

$$P_{be} = P_{be}(\gamma) = P_{be}\left(\frac{\beta\bar{\gamma}}{\bar{\gamma} + \beta + 1}\right) : \textit{CBPSK; Fast Ricean fading} \quad (18.17)$$

18.5 PERFORMANCE OF SEVERAL BINARY MODULATION WAVEFORMS WITH RICEAN FADING

In this section, the bit-error performance of coherent binary phase shift keying (BPSK), (CBPSK), differentially coherent BPSK (DCBPSK), and noncoherent (NC) binary frequency shift keying (BFSK) (NCBFSK) are examined in slow Ricean fading (SRF) and fast Ricean fading (FRF) environments. Noncoherent or NC refers to noncoherent detection processing in the demodulator.†

18.5.1 CBPSK with Slow Nonselective Ricean Fading

The bit-error probability of CBPSK, conditioned on the SNR is

$$P_s(\text{bit-error}|\gamma) = \frac{1}{2} erfc\left(\sqrt{\gamma}\right) \quad (18.18)$$

where $\gamma = P_s/\sigma_n^2 = E_b/N_o$ is measured in the data rate bandwidth of $1/T_b$. $E_b = P_s T_b$ is the energy-per-bit and N_o is the receiver noise power density. Using the concepts discussed in Section 18.4, the demodulated bit-error probability is evaluated as

*With FEC coding, the symbol duration is based on the code-bit and, with tracking, the system time-constant is limited by the tracking bandwidth.

†R.L. Bogusch, Ionospheric Scintillation Effects on Satellite Communications: Interim Technical Note, MRC-N-139, pp. 60–96, Mission Research Corporation, Santa Barbara, CA, June 1974.

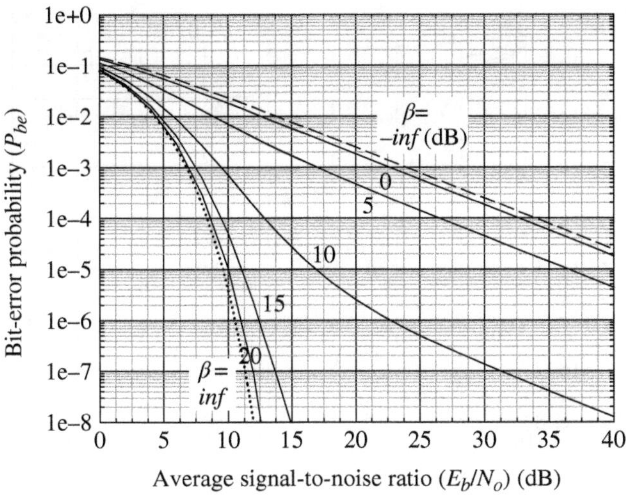

FIGURE 18.5 CBPSK with slow nonselective Ricean fading.

$$P_{be} = \frac{1}{2\gamma_r} \int_0^\infty erfc\left(\sqrt{\gamma}\right) e^{-(\gamma+\gamma_s)/\gamma_r} I_o\left(\frac{2\sqrt{\gamma\gamma_s}}{\gamma_r}\right) d\gamma \qquad (18.19)$$

It is convenient to evaluate (18.19) in terms of the average SNR $\bar{\gamma} = \gamma_s + \gamma_r$ with the ratio of the specular-to-random signal components defined as $\beta = \gamma_s/\gamma_r$. Substituting these results with $x = \gamma/\gamma_r$ into (18.19), the bit-error probability is expressed as

$$P_{be} = \frac{1}{2} \int_0^\infty x erfc\left(\sqrt{x\bar{\gamma}/2(\beta+1)}\right) e^{-(x^2+2\beta)/2} I_o\left(\sqrt{2\beta x}\right) dx \qquad (18.20)$$

Equation (18.20) is plotted in Figure 18.5 for various values of the parameter β. The case $\beta = \infty$ corresponds to the Gaussian channel with no fading and $\beta = 0$ results in the slow fading Rayleigh limit (SFRL) channel performance. The closed-form expression for the bit-error probability for $\beta = 0$ is evaluated using the integral (Reference 4, p. 303, Integral No. 7.4.19), resulting in the expression

$$P_{be} = \frac{1}{2}\left[1 - \sqrt{\frac{\bar{\gamma}}{\bar{\gamma}+1}}\right] \quad : \beta = 0, \text{CBPSK slow fading Rayleigh limit} \qquad (18.21)$$

18.5.2 Coherent BPSK with Fast Nonselective Ricean Fading

The performance of CBPSK in the fast fading Ricean channel is evaluated as

$$\begin{aligned} P_{be} &= P_r(\text{bit-error}|\gamma) \\ &= \frac{1}{2} erfc\left(\sqrt{\frac{\beta\bar{\gamma}}{\bar{\gamma}+\beta+1}}\right) \end{aligned} \qquad (18.22)$$

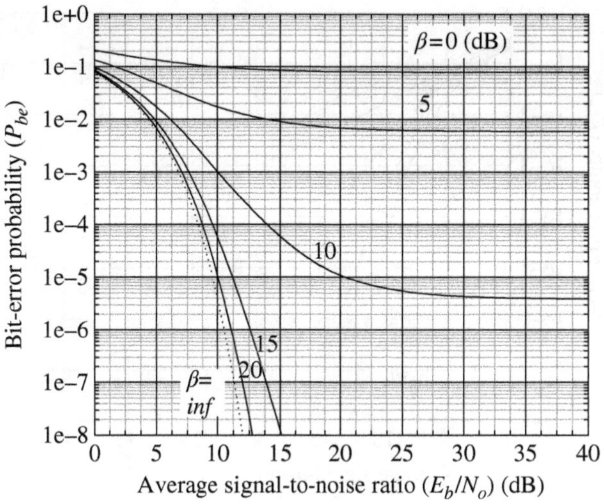

FIGURE 18.6 CBPSK with fast nonselective Ricean fading.

Equation (18.22) is plotted in Figure 18.6. When $\beta = \infty$, the result is simply the performance of ideally demodulated CBPSK; however, when $\beta = 0$, the bit-error probability is 0.5 corresponding to the fast fading Rayleigh limit (FFRL) with no information being conveyed by the signal.

$$P_{be} = \frac{1}{2} \quad : \beta = 0, \text{CBPSK fast fading Rayleigh limit} \qquad (18.23)$$

18.5.3 Differentially Coherent BPSK with Slow Nonselective Ricean Fading

The bit-error probability performance of DCBPSK in an AWGN channel is given by

$$P_{be} = \frac{1}{2} e^{-\gamma} \qquad (18.24)$$

and, in keeping with the previous discussions, the performance in the slow fading Ricean channel is evaluated as

$$\begin{aligned} P_{be} &= \frac{1}{2\gamma_r} \int_0^\infty e^{-\gamma} e^{-(\gamma+\gamma_s)/\gamma_r} I_o\left(\frac{2\sqrt{\gamma\gamma_s}}{\gamma_r}\right) d\gamma \\ &= \frac{e^{-\gamma_s/\gamma_r}}{2\gamma_r} \int_0^\infty e^{-\gamma(\gamma_r+1)/\gamma_r} I_o\left(\frac{2\sqrt{\gamma\gamma_s}}{\gamma_r}\right) d\gamma \end{aligned} \qquad (18.25)$$

Substituting the variable $x = \sqrt{\gamma(\gamma_r+1)/\gamma_r}$ results in the expression

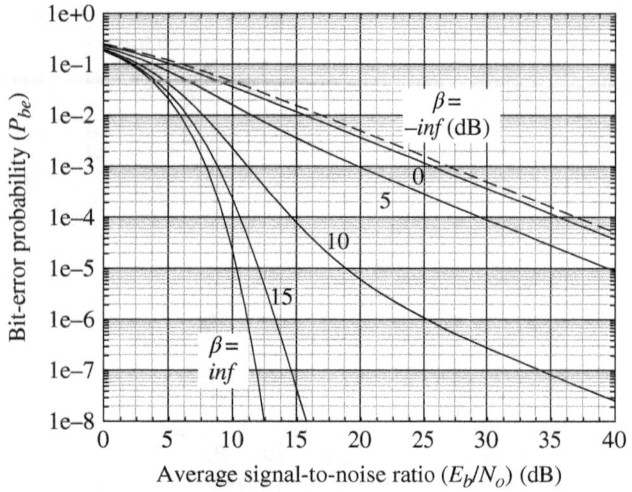

FIGURE 18.7 DCBPSK with slow nonselective Ricean fading.

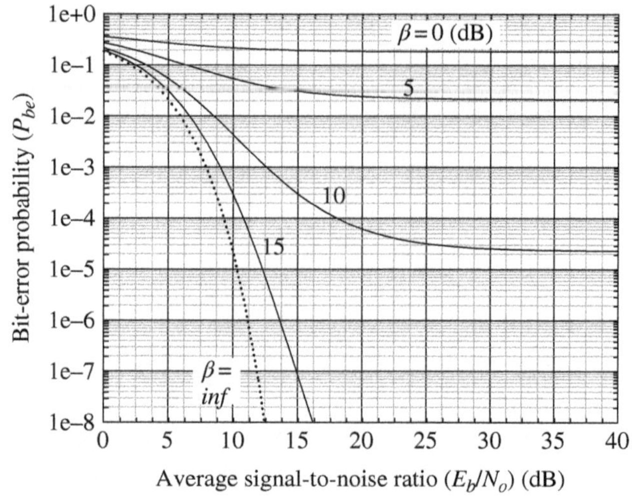

FIGURE 18.8 DCBPSK with fast nonselective Ricean fading.

$$P_{be} = \frac{e^{-\gamma_s/\gamma_r}}{(\gamma_r + 1)} \int_0^{\infty} xe^{-x^2} I_o(bx)dx \tag{18.26}$$

where $b = 2\sqrt{\frac{\gamma_s}{\gamma_r(\gamma_r+1)}}$. Performing this integration [5] yields

$$P_{be} = \frac{1}{2(\gamma_r + 1)} e^{-\gamma_s/(\gamma_r+1)} \tag{18.27}$$

and substituting for the previously defined parameters $\bar{\gamma}$ and β results in

$$P_{be} = \frac{\beta + 1}{2(\bar{\gamma} + \beta + 1)} e^{-\beta\bar{\gamma}/(\bar{\gamma}+\beta+1)} \tag{18.28}$$

Equation (18.28) is the bit-error performance of DCBPSK in a slow fading Ricean channel and is plotted in Figure 18.7. The SFRL is evaluated with $\beta = 0$ and is

$$P_{be} = \frac{1}{2(\bar{\gamma} + 1)} \quad : \beta = 0, \text{ DCBPSK slow fading Rayleigh limit} \tag{18.29}$$

18.5.4 Differentially Coherent BPSK with Fast Nonselective Ricean Fading

The performance of DCBPSK in the fast fading environment simply attributes the received noise as being the sum of the thermal noise and the random component of the channel noise resulting in the bit-error probability expression

$$P_{be} = \frac{1}{2} e^{-\gamma_s/(\gamma_r+1)} \tag{18.30}$$

or, in terms of the parameters $\bar{\gamma}$ and β,

$$P_{be} = \frac{1}{2} e^{-\beta\bar{\gamma}/(\bar{\gamma}+\beta+1)} \tag{18.31}$$

Equation (18.31) is the bit-error performance of DCBPSK in a fast fading environment and is plotted in Figure 18.8. The FFRL is evaluated with $\beta = 0$ and is

$$P_{be} = \frac{1}{2} \quad : \beta = 0, \text{ DCBPSK fast fading Rayleigh limit} \tag{18.32}$$

18.5.5 Noncoherent BFSK with Slow Nonselective Ricean Fading

The bit-error performance of NCBFSK in an AWGN channel is given by

$$P_{be} = \frac{1}{2} e^{-\gamma/2} \tag{18.33}$$

and, in keeping with the previous discussions, the performance in the slow fading Ricean channel is evaluated as

$$\begin{aligned} P_{be} &= \frac{1}{2\gamma_r} \int_0^{\infty} e^{-\gamma/2}\, e^{-(\gamma+\gamma_s)/\gamma_r} I_o\left(\frac{2\sqrt{\gamma\gamma_s}}{\gamma_r}\right) d\gamma \\ &= \frac{e^{-\gamma_s/\gamma_r}}{2\gamma_r} \int_0^{\infty} e^{-[\gamma(\gamma_r+2)]/2\gamma_r} I_o\left(\frac{2\sqrt{\gamma\gamma_s}}{\gamma_r}\right) d\gamma \end{aligned} \tag{18.34}$$

Substituting the variable $x = \gamma(\gamma_r + 2)/2$ results in the expression

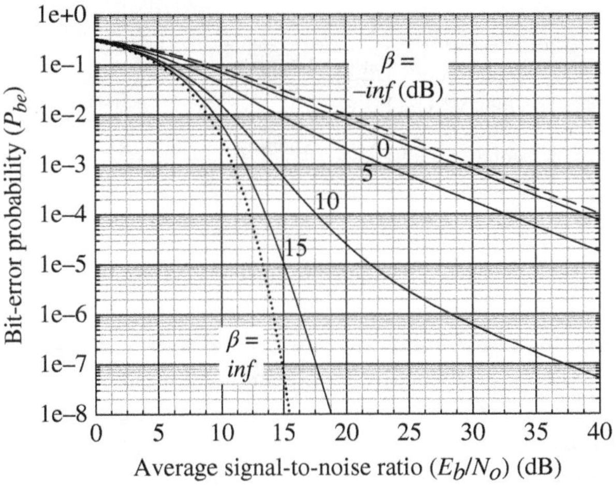

FIGURE 18.9 NCBFSK with slow nonselective Ricean fading.

$$P_{be} = \frac{e^{-\gamma_s/\gamma_r}}{\gamma_r+2}\int_0^\infty \frac{1}{\gamma_r}e^{-x/\gamma_r}I_o\left(\frac{2}{\gamma_r}\sqrt{\frac{2\gamma_s x}{\gamma_r+2}}\right)dx \quad (18.35)$$

The integrand in (18.35) can be expressed in terms of the *pdf* $p_R(x)$, in the form of (18.12), by substituting $b=2\gamma_s/(\gamma_r+2)$ in the argument of the Bessel function and adding and subtracting b to the exponent. The result is

$$\begin{aligned} P_{be} &= \frac{e^{-(\gamma_s-b)/\gamma_r}}{\gamma_r+2}\int_0^\infty \frac{1}{\gamma_r}e^{-(x+b)/\gamma_r}I_o\left(\frac{2\sqrt{xb}}{\gamma_r}\right)dx \\ &= \frac{e^{-(\gamma_s-b)/\gamma_r}}{\gamma_r+2}\int_0^\infty p_R(x)dx \end{aligned} \quad (18.36)$$

The integral in (18.36) is equal to unity, so the expression for the bit-error probability becomes simply

$$P_{be} = \frac{e^{-\gamma_s/(\gamma_r+2)}}{\gamma_r+2} \quad (18.37)$$

and, upon substituting for the previously defined parameters $\bar{\gamma}$ and β (18.37), results in

$$P_{be} = \frac{\beta+1}{2(\bar{\gamma}/2+\beta+1)}e^{-\frac{\beta\bar{\gamma}/2}{\bar{\gamma}/2+\beta+1}} \quad (18.38)$$

Equation (18.38) represents the performance of NCBFSK in a slow fading Ricean channel and is plotted in Figure 18.9. The SFRL limit is evaluated with $\beta=0$ and is

$$P_{be} = \frac{1}{\bar{\gamma}+2} \quad :\beta=0,\ \text{NCBFSK slow fading Rayleigh limit} \quad (18.39)$$

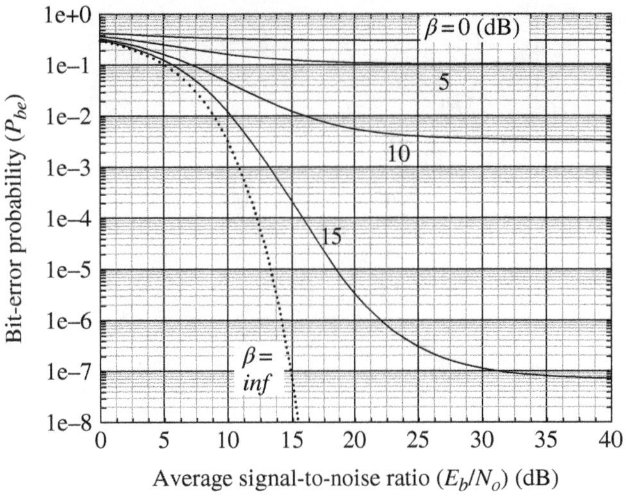

FIGURE 18.10 NCBFSK with fast nonselective Ricean fading.

18.5.6 Noncoherent BFSK with Fast Nonselective Ricean Fading

The bit-error performance of NCBFSK in an AWGN channel is given by

$$P_{be} = \frac{1}{2}e^{-\gamma/2} \quad (18.40)$$

Because the result is very similar to the performance for DCBPSK, the performance for NCBFSK in the fast fading Ricean channel is readily evaluated as

$$P_{be} = \frac{1}{2}e^{-\gamma_r/2(\gamma_r+1)} \quad (18.41)$$

or, in terms of the parameters $\bar{\gamma}$ and β,

$$P_{be} = \frac{1}{2}e^{-\beta\bar{\gamma}/2(\bar{\gamma}+\beta+1)} \quad (18.42)$$

Equation (18.42) is the bit-error performance of NCBFSK in a fast fading environment and is plotted in Figure 18.10. The FFRL is evaluated with $\beta=0$ and is

$$P_{be} = \frac{1}{2} \quad :\beta=0,\ \text{NCFSK fast fading Rayleigh limit} \quad (18.43)$$

18.6 GENERATION OF RICEAN RANDOM VARIABLES

Ricean distributed random variables are easily generated by using the expression for the magnitude of a narrowband signal plus AWGN given by

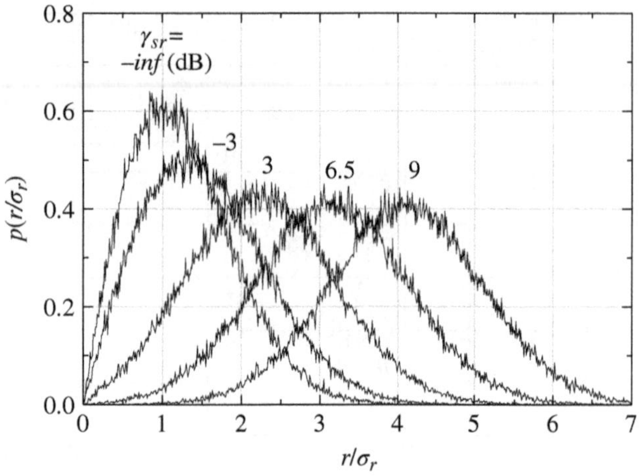

FIGURE 18.11 Normalized Ricean probability density functions (based on 100K computer-generated samples).

$$r=\sqrt{(V_s+n_I\cos(\phi))^2+(n_Q\sin(\phi))^2} \quad (18.44)$$

where n_I and n_Q are *iid* Gaussian random variables $N(0, \sigma_r)$ and ϕ is a uniformly distributed random phase over $\pm\pi$. The signal or specular value V_s is determined from the signal-to-noise specification, γ_{sr}, as

$$V_s=\sigma_r\sqrt{2\gamma_{sr}} \quad (18.45)$$

Figure 18.11 shows the *pdf* for the normalized random variable r/σ_r formed using 100K Ricean random variables r corresponding to SNRs $\gamma_{sr}=-\infty, -3, 3, 6.5$, and 9 dB. Considering that the responses represent a single record, or ensemble of a stochastic process, the plots are in reasonable agreement with the corresponding theoretical results shown in Figure 18.2.

Correlated Ricean random variables can also be generated based on the conditional *pdf** expressed as

$$p(n_i|n_{i-1})=\frac{1}{\sqrt{2\pi(1-\rho^2)\,\sigma_r^2}}e^{-n_i^2/2(1-\rho^2)\sigma_r^2} \quad (18.46)$$

Where n_i and n_{i-1} are the underlying zero-mean *iid* Gaussian random variables and ρ is the correlation coefficient. Using (18.46), the magnitude of the correlated random variables, corresponding to (18.44), is computed as

$$n_i=\sqrt{(V_s+n_{Ii}\cos(\phi_i))^2+(n_{Qi}\sin(\phi_i))^2}-\rho n_{i-1}$$
$$:i=1,2,\ldots \quad (18.47)$$

*See section "LINK The Bivariate Distribution—An Example of Conditional Distributions."

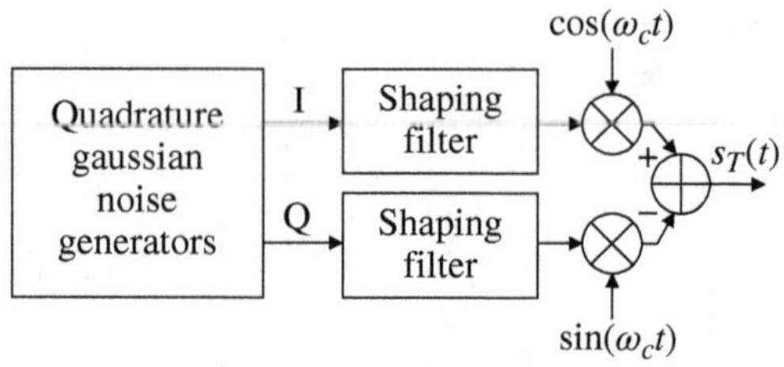

FIGURE 18.12 Multipath Rayleigh fading simulator.

where the initial conditions correspond to $n_{I0}=n_{Q0}=0$. The quadrature variables n_{Ii} and n_{Qi} can be thought of as the discrete-time *wss* stochastic process $n(t_i,\zeta)$ with ζ representing the quadrature outcomes. The received signal plus noise complex random process is characterized as

$$n_i=V_s+n_{Ii}\cos(\phi_i)+jn_{Qi}\sin(\phi_i) \quad (18.48)$$

where ϕ_i is a uniformly distributed random variable over $\pm\pi$ with *pdf* $p(\phi)=1/2\pi$. The complex random variables n_i are discrete-time Ricean distributed random variables with $t=i\Delta t$ corresponding to the sampled *wss* random process $n(t)$ with autocorrelation response given by

$$R_n(t)=\int_0^\infty n(t')n^*(t'-t)dt' \quad (18.49)$$

The range of the integration in (18.49) can be limited to T_I corresponding to a negligible correlation response. The normalized correlation response is expressed as

$$\rho_n(t)=\frac{R_n(t)}{\sigma_n^2} \quad (18.50)$$

The sampling interval Δt is established by the channel bandwidth B_{ch} necessary to support the modulated waveform, such that, $\Delta t\leq 1/2B_{ch}$.

18.6.1 Multipath Rayleigh Fading Simulator

A multipath Rayleigh fading simulator for mobile radio applications is described by the functional diagram in Figure 18.12 [6]. The quadrature noise generators provide zero-mean *iid* Gaussian random samples† denoted as $N(0,\sigma_n)$. The sampled output $s_T(t)$, in Figure 18.12, represents a real carrier-modulated waveform; however, the following analysis is based on the complex envelope given by

$$\tilde{s}_T(t)=\int h(t')\tilde{n}(t-t')dt' \quad (18.51)$$

†The author's implemented an analog simulator using a Zener noise diode.

In this context, the baseband signal is identical to (18.48) with $V_s = 0$ resulting in Rayleigh fading. The movement of the mobile radio results in a Doppler frequency shift with the envelope of the received signal fluctuating about the local mean of the Rayleigh distribution and a uniformly distributed random carrier frequency phase. The characterization of the frequency response of the shaping filters is the principal factor in the design of mobile radio channels.

The shaping filter spectrum is dependent on the directive gain of the mobile radio antenna pattern and can be designed to accommodate any directivity characteristic [7]. However, for the commonly used omnidirectional antenna, the theoretical spectral density of the complex envelope of the received signal is expressed as [8]

$$S(f) = \begin{cases} \dfrac{E^2}{2\pi f_d}\sqrt{1-(f/f_d)^2} & : f \le f_d \\ 0 & : o.w. \end{cases} \quad (18.52)$$

where E is the *rms* value of the signal envelope and f_d is the Doppler frequency given by

$$f_d = \frac{V}{\lambda} = \frac{V}{c} f_c \quad (18.53)$$

where V is the mobile vehicle velocity, $\lambda = c/f_c$ is the wavelength of the carrier frequency f_c, and c is the speed of light. The theoretical Doppler frequency is evaluated by defining the n-th moment of the spectrum, b_n, as

$$b_n \triangleq (2\pi)^n \int_0^\infty S(f) f^n df \quad (18.54)$$

Upon substituting (18.52) into (18.54), the Doppler frequency is evaluated as

$$f_d = \frac{1}{2\pi}\sqrt{\frac{2b_2}{b_0}} \quad (18.55)$$

The theoretical filter spectrum has an infinite response at f_d and zero response at higher frequencies and the practical or simulated filter spectrum, $\hat{S}(f)$, is designed to increase to a normalized peak of 6 dB at f_d with a roll-off of 18 dB/octave above f_d. The moments of the simulated spectrum are denoted as $\hat{b}_n$ and the simulated Doppler, $\hat{f}_d$, is computed using (18.55) with $\hat{b}_2$ and $\hat{b}_0$. The simulated vehicle speed is $\hat{V} = \lambda \hat{f}_d$.

Using this approach the simulated fade rate is related to the theoretical fade rate by equating second moments of the respective spectrums. The authors report that the measured Rayleigh envelopes are within 3 dB of the theoretical envelopes down to nearly 40 dB below the *rms* level. The fading level crossing rate and average fade duration were also compared to the respective theoretical values, expressed as [9]

$$N_\Gamma = \sqrt{2\pi} f_d \Gamma e^{-\Gamma^2} \quad : \text{theoretical positive crossings-per-second} \quad (18.56)$$

and

$$\bar{\tau}_\Gamma = \frac{e^{\Gamma^2} - 1}{\sqrt{2\pi} f_d \Gamma} \quad : \text{theoretical fade duration (ms)} \quad (18.57)$$

where Γ is the fade level relative to the *rms* level of the Rayleigh *pdf*. The authors report excellent agreement with theory in both of these performance measures when normalized by f_d and $\hat{f}_d$.

18.7 RELATIONSHIPS BETWEEN FADING CHANNEL PARAMETERS

For multipath channels with N discrete paths, the multipath delays are denoted as τ_n and the channel baseband impulse is expressed as the summation [10, 11]

$$h(t;\tau) = \sum_{n=0}^{N-1} a_n(t,\tau) e^{j\theta_n(t,\tau)} \delta(\tau - \tau_n(t)) \quad (18.58)$$

where $a_n(t, \tau)$ is the multipath time-dependent amplitude function and the corresponding phase function is

$$\theta_n(t;\tau) = 2\pi f_c \tau_n(t) + \phi_n(t,\tau) \quad (18.59)$$

where f_c is the carrier frequency. Considering the analytic baseband transmitted signal $\tilde{s}_T(t)$, the received baseband signal is described as the convolution

$$\tilde{r}(t;\tau) = \int_{-\infty}^{\infty} h(t',\tau) \tilde{s}_T(t-t') dt' \quad (18.60)$$

In the expressions $h(t; \tau)$ and $\tilde{r}(t;\tau)$, the parameter t represents the time variation of the channel resulting from the relative motion between the transmitter, receiver, multipath points of reflection, and the multipath delay $\tau = \tau_n(t)$ at time t corresponding to each reflection path. For the time-invariant channel, or a channel that is relatively constant over a period of time, the channel impulse response in (18.58) simplifies to

$$h(\tau) = \sum_{n=0}^{N-1} a_n(\tau) e^{j\theta_n(\tau)} \delta(\tau - \tau_n) \quad (18.61)$$

with

$$\theta_n(\tau) = 2\pi f_c \tau_n + \phi_n(\tau) \quad (18.62)$$

These relationships and their impact on wireless communications in urban and rural environments play an important role in the understanding and application of cellular communication technologies. Because of the complexity in analyzing the link performance, a common practice is to perform *in situ* measurements using test signals to evaluate the signal losses in the various environments, including indoor applications. Chapter 19 reviews the conditions and results for various *in situ* link models.

Although (18.60) is dependent on the discrete multipath though the implicit delta function $\delta(\tau-\tau_n(t))$, the expressions $\tilde{r}(t;\tau)$ and $h(t;\tau)$ are considered to be continuous in τ by applying the impulse $\delta(t-\tau)$ to the channel response [12]; this simplifies the characterization of the underlying multipath parameters. In the following descriptions, the channel is considered to be *wss* with autocorrelation response $X(t;\tau_1)$ with the corresponding two-dimensional Fourier transform* $H(f_d; f_c')$. These functions are expressed, respectively, as

$$X(t;\tau_1) = \int_{-\infty}^{\infty} h(t',\tau_1)h^*(t'-t,\tau_2)dt'\delta(\tau_1-\tau_2) = X(t;\tau_1,\tau_2)\delta(\tau_1-\tau_2) \quad (18.63)$$

and

$$H(f_d; f_c') = \int_{-\infty}^{\infty}\int_{-\infty}^{\infty} X(t,\tau_1)e^{-j2\pi(f_d t - f_c'\tau_1)}dtd\tau_1 \quad (18.64)$$

where the delta function $\delta(\tau_1-\tau_2)$ is used to denote that the multipath reflections are uncorrelated.

18.7.1 Channel Coherence Time and Doppler Spread

The *channel coherence time*, T_c, is characterized by the correlation response $X(t; 0)$ and is defined, by the normalized form, as $X(T_c; 0)/X(0; 0) = k$ where $0<k<1$. The value of k is selected based on an acceptable decorrelation point in the response, for example, $k = e^{-1} = 0.368$. However, T_c can also be determined from the *Doppler spread* as described in the remainder of this section.†

*The notations f_d and f_c' represent independent variables corresponding, respectively, to the Doppler and the channel coherence frequencies.

†The notation used for the coherence time and Doppler spread varies in the literature and T_c and B_d are adopted here. Propagation through the severely disturbed ionosphere is discussed in Chapter 20 where the decorrelation time is denoted as τ_o.

The Fourier transform, $H(f_d; 0)$, of the correlation response $X(t;0)$ characterizes the Doppler frequency corresponding to the time varying channel. This is consistent with the notion that the time dependence of the channel is the result of the changing path lengths due to the relative motion between the transmitter, receiver, and reflecting surfaces. The underlying principle is that the Doppler frequency is related to the velocity along the propagation path as

$$f_d = \left(\frac{v}{c}\right) f_c \quad (18.65)$$

where f_c is the carrier frequency, c is the free-space speed of light, and $v = \Delta d/t$ is the velocity, normal to the plane of the receiver antenna. The velocity results from a change in the propagation path length Δd that, in turn, results from variations in the multipath reflections. Based on the Fourier transform relationship, T_c and B_d are inversely related as

$$T_c \propto \frac{1}{B_d} \quad (18.66)$$

where B_d is the bandwidth beyond which the Doppler spread in $H(f_d; 0)$ is negligible. In practice it is found that (18.66) results in considerable signal decorrelation and intersymbol interference (ISI) and selecting T_c, such that $X(T_c; 0) = 0.5$ results in the approximation [13]

$$T_c \cong \frac{0.18}{B_d} \quad : B_d = f_d(\max) \quad (18.67)$$

where $f_d(max)$ is the maximum Doppler expressed in (18.65). However, Rappaport [14] points out that the coherence time in (18.67) is too restrictive and suggests a rule of thumb given by

$$T_c = \frac{0.423}{B_d} \quad : B_d = f_d(\max), \text{ rule of thumb} \quad (18.68)$$

The coherence time is an important system design parameter, in that, it represents the time over which contiguous received modulated symbols are correlated and is a principal channel parameter because data symbols separated by more than T_c seconds are affected independently by the channel. Therefore, T_c influences the selection of the forward error correction (FEC) coding, the span of data interleavers, and the type of noncoherent combining.

As discussed in this section, fast and slow fading are characterized by the Doppler spread in terms of the dynamics of the channel. However, in Sections 18.4 and 18.5, fast and slow fading are characterized in terms of the coherence time relative to the symbol rate, for example, if the coherence time is less than (or comparable to) the symbol duration, then fast fading occurs. Because of the inverse relationship between T_c and B_d, these parameters impact the system performance in the same way, that is, high B_d and low T_c result in fast fading and vice versa.* In either event, to improve the system performance, some combination of FEC coding, interleaving, combining, or other mitigation technique as outlined in Tables 20.13 and 20.14 is most likely required. The use of higher symbol rates to avoid the fast fading regime has its limits as discussed in Section 18.7.2.

18.7.2 Channel Coherence Bandwidth and Time Dispersion

The *channel multipath spread* is characterized by the correlation response $X(0;\tau)$ and is defined in the normalized form using $\tau = T_m$ and expressed as $X(0; T_m)/X(0; 0) = k$ where $0 < k < 1$. The parameter k is selected as a suitable decorrelation point in the response, for example, $k = e^{-1} = 0.368$.

The Fourier transform, $H(0; f)$, of the multipath spread correlation response $X(0;\tau)$ characterizes the *channel coherence bandwidth*, B_c.† This is consistent with the notion that the time dispersion of a signal can be characterized as delayed signal replicas‡ corresponding to the channel multipath delays [15]. Based on the Fourier transform relationship, T_m and B_c are inversely related as

$$B_c \propto \frac{1}{T_m} \tag{18.69}$$

The bandwidth B_c is the bandwidth beyond which the Doppler spread in $H(0;f)$ is negligible. For example, $H(0;B_c) = 0.368$; however, Rappaport [16] quantifies the delay dispersion for wideband wireless systems in terms of the mean and second moment of the excess delay and the *rms* delay spread of the *power delay profile*, defined, respectively, as

$$\bar{\tau} = \frac{\sum_n a_n^2 \tau_n}{\sum_n a_n^2}, \quad \overline{\tau^2} = \frac{\sum_n a_n^2 \tau_n^2}{\sum_n a_n^2} \tag{18.70}$$

and

$$\sigma_\tau = \sqrt{\overline{\tau^2} - \bar{\tau}^2} \tag{18.71}$$

where n corresponds to the number of signal paths arriving at the receiver, a_n is the power of the n-th multipath received signal, and the excess delays are measured relative to the first received signal. The maximum excess delay corresponds to the delay of the last multipath signal that exceeds a defined threshold relative to the maximum of the multipath levels; an example threshold is −10 dB. The power delay profile can be generated by examining the return of a narrow pulse, wide bandwidth test signal.

Defining the coherence bandwidth as the frequency for which $H(0;B_c) = 0.9$, then (18.69) is approximately related to the *rms* delay spread as [17]

$$B_c \cong \frac{1}{40\sigma_\tau} \quad : H(0;B_c) = 0.9 \tag{18.72}$$

and with $H(0;B_c) = 0.5$ the approximate relationship is

$$B_c \cong \frac{1}{4\sigma_\tau} \quad : H(0;B_c) = 0.5 \tag{18.73}$$

The results vary widely and the impact of the time-varying multipath on a particular waveform demodulator must be examined by simulation [13, 18, 19] using accurate multipath channel models [20, 21].

The coherence bandwidth is an important system design parameter, in that, modulated waveforms with bandwidths greater than the coherence bandwidth will experience severe delay dispersion distortion requiring processing intense equalizers. It is also used to limit the chip rate of spread-spectrum pseudo-noise (PN) codes in code division multiple access (CDMA) systems if equalization is to be avoided. The coherence bandwidth is the frequency domain equivalence of the coherence time. For example, the frequency-hopping spread-spectrum (FHSS) and frequency diversity (FD) systems must use carrier frequencies separated by at least the coherence bandwidth to provide a combining gain and a net improvement in performance.

18.8 DIVERSITY TECHNIQUES FOR FADING CHANNELS

As mentioned in the introduction, diversity combining is an important aspect in providing robust communications in fading environments. The various forms of diversity, for example, temporal, frequency, spatial, angle-of-arrival, and polarization, are typically combined in the communication receiver or demodulator using time and frequency domain processing based on estimates of relative power levels and SNRs. Temporal diversity includes a variety of techniques including channel coding, involving FEC coding, data interleaving, message repetition, and various methods of ARQ [22] involving error detection and correction (EDAC) coding. In many applications, the diversity combining takes place between multiple or parallel receivers and demodulators that are dedicated to an independent diversity channel or communication link. In these applications, smart transmit

*The system performance in fast and low fading is also impacted by the more restrictive bandwidths of the tracking loop, and for this reason noncoherent and differentially coherent detection is preferred in fast fading channels with commensurably low symbol rates.

†For propagation through the severely ionized ionosphere, discussed in Chapter 20, the decorrelation frequency is denoted as f_o.

‡See, for example, the paired-echo analysis in Sections 1.3.2 and 1.3.3.

and receive antennas [23] play an essential role in the diversity combining. Therefore, because of the common time and frequency domain processing, diversity techniques are generally applicable to a variety of fading channel conditions. In the slow nonselective Rayleigh and Ricean fading regimes, diversity combining is the most effective method of improving the system performance, providing significantly improved bit-error performance and a reduction in message throughput under severe time and frequency fading conditions.

In Sections 18.8.1 and 18.8.2, diversity combining is described in terms of the number, N, of independently received signals that are available. The basic concept involves choosing an appropriate system performance parameter and then determining the best diversity selection algorithm based on combining signals or selecting one of N signals. The system performance parameters and the diversity selection options are apparent in the following discussions of the commonly used forms of diversity combining [12, 24, 25]. The diversity performance is based on the underlying Rayleigh channel fading, described in terms of the SNR using (18.12) with $\gamma_s = 0$ and the average SNR $\bar{\gamma} = \gamma_r$. The *pdf* of the instantaneous SNR (γ) is expressed as

$$p(\gamma) = \frac{1}{\bar{\gamma}} e^{-\gamma/\bar{\gamma}} \tag{18.74}$$

The diversity selection algorithm must form the estimates of γ in the time equal to the reciprocal of the fade rate and, because of the relative short estimation times, the estimate is typically evaluated in terms of the signal-plus-noise to noise ratio or simply the signal-plus-noise. The automatic gain control (AGC) of each of the N receivers must establish identical receiver estimates for the diversity combining.

18.8.1 Frequency Nonselective (Flat) Fading

In this section, diversity selection techniques are discussed that are applicable to either frequency nonselective fading or frequency selective fading. In the former case, the combining interval must exceed the decorrelation time, whereas, with frequency selective fading the combining bandwidth must exceed the channel decorrelation frequency. The commonly used diversity techniques are *selection diversity* (SD) or *switching diversity*, *maximal ratio combining diversity*, and *equal gain diversity*. The diversity performance is based on the underlying SFRL, that is, the fading is slow relative to the transmitted symbol duration, and the reduction in the required E_b/N_o ratio to achieve a specified bit-error probability is examined for several waveform modulations.

18.8.1.1 Selection Diversity SD is the simplest type of diversity, in that, the demodulator uses the received signal with the largest estimate of the SNR. When the largest SNR is selected from among the N independent Rayleigh fading received signals, the underlying Rayleigh random variable (r) is transformed to the random variable r'. With the SNR of r' defined as $\gamma = r'^2/2\sigma_n^2$, the probability density function of the transformation is expressed as [26]

$$p(\gamma) = N\bar{\gamma}\left(1 - e^{-\gamma/\bar{\gamma}}\right)^{N-1} e^{-\gamma/\bar{\gamma}} \quad : \gamma \ge 0, \text{ selection diversity} \tag{18.75}$$

where $\bar{\gamma}$ is the average SNR of the Rayleigh *pdf* in (18.74). The derivation of (18.75) is rooted in the concept of order statistics discussed in Section 1.5.8.

SD Performance in a SFRL Channel The performance of SD in a SFRL channel is examined in terms of the bit-error probability of the DCBPSK and NCBFSK waveform modulations. The performance is based on (18.14) with P_r (bit-error|γ) equal to the underlying bit-error probability of the modulation, given, respectively, by (18.24) and (18.33), and $p(\gamma)$ is the channel SD *pdf* expressed by (18.75). Upon performing the integration in (18.14) over the range 0–∞, the resulting bit-error probability for each of the modulations is evaluated (see Problem 4) as

$$P_{be} = \begin{cases} \left(\dfrac{N}{2}\right)\dfrac{\Gamma(\bar{\gamma}+1)\Gamma(N)}{\Gamma(\bar{\gamma}+N+1)} & : \text{DCBPSK SFRL} \\[2ex] \left(\dfrac{N}{2}\right)\dfrac{\Gamma(\bar{\gamma}/2+1)\Gamma(N)}{\Gamma(\bar{\gamma}/2+N+1)} & : \text{NCBFSK SFRL} \end{cases} \tag{18.76}$$

where $\bar{\gamma} = \gamma_b = E_b/N_o$ and $\Gamma(x)$ is the gamma function. The bit-error performance for these modulations with SD is shown in Figure 18.13 for various values of N. For $N = 1$, (18.76) is simply the expression of the SFRL channel corresponding to $\beta = 0$ in Figures 18.7 and 18.9. The bit-error results corresponding to the *dashed* curve represents the ideal AWGN performance of the respective underlying modulations.

18.8.1.2 Maximal Ratio Combining Diversity The maximal ratio combining diversity (MRCD) algorithm generates a composite envelope based on the summation of the N gain-weighed and phase-adjusted individual received signals and is expressed as

$$r_N = \sum_{n=1}^{N} G_n r_n \tag{18.77}$$

In (18.77), the phase of each received signal is adjusted to zero so the summation represents a coherent integration.

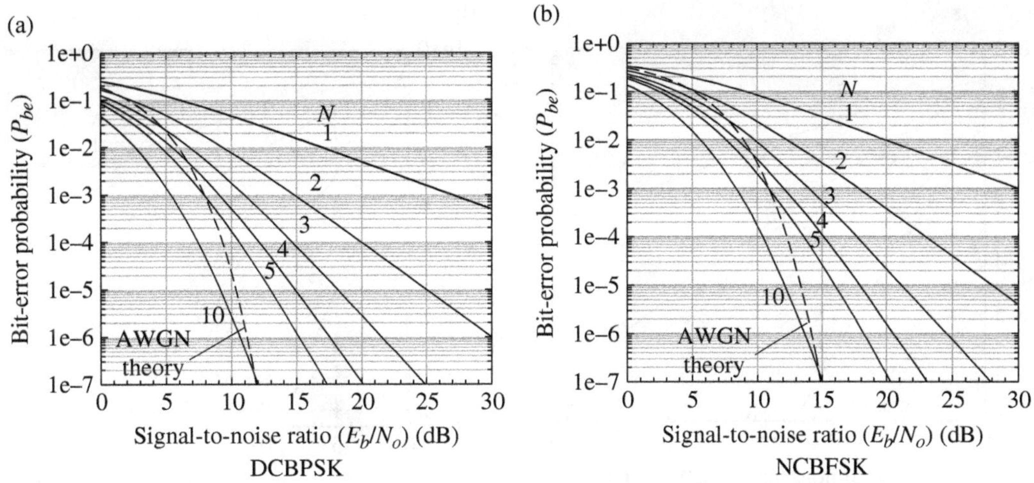

FIGURE 18.13 Performance of DCBPSK and NCBFSK with SD and SFRL.

If each receiver has the same average noise power σ_n^2, the SNR in each receiver is $\gamma_n = r_n^2/2\sigma_n^2$. Consequently, the composite noise and signal powers are $\sigma_T^2 = \sum_n \sigma_n^2$ and $P_T = Nr_n^2/2$ corresponding to the SNR $\gamma_T = P_T/\sigma_T^2$. Upon defining the gain as $G_n = r_n/N$, Couch [27] shows that the maximum SNR occurs when

$$\gamma = \sum_{n=1}^{N} \gamma_n \quad : \text{maximum signal-to-noise ratio} \tag{18.78}$$

and that the corresponding *pdf* of γ is expressed as

$$p(\gamma) = \frac{\gamma^{N-1} e^{-\gamma/\bar{\gamma}}}{\bar{\gamma}^N (N-1)!} \quad : \gamma \geq 0, \text{ maximal ratio combining} \tag{18.79}$$

where $\bar{\gamma}$ is the average SNR of the Rayleigh *pdf* expressed by (18.74). The individual channel SNRs (γ_n) are estimated using independent quadrature Gaussian samples characterized as $N(m, \sigma_n)$. Therefore, based on the summation of N independent SNR estimates, expressed by (18.78), the distribution in (18.79) describes the chi-square *pdf* with $2N$ degrees of freedom as discussed in Section 1.5.7.1.

Maximal Ratio Combining Diversity Performance in SFRL Channel The performance of MRCD in a SFRL channel is examined in terms of the bit-error probability of the DCBPSK and NCBFSK waveform modulations. The performance evaluation follows that described in Section 18.8.1.1 for SD; however, in this case, the maximal ratio combining *pdf* expressed by (18.79) is used. Therefore, with these substitutions, (18.14) is integrated over the range 0–∞ and the resulting bit-error probability for each of the modulations with MRCD is evaluated (see Problem 5) as

$$P_{be} = \begin{cases} \dfrac{1}{2(\bar{\gamma}+1)^N} & : \text{DCBPSK, SFRL, MRCD} \\[2ex] \dfrac{2^{N-1}}{(\bar{\gamma}+2)^N} & : \text{NCBFSK, SFRL, MRCD} \end{cases} \tag{18.80}$$

where $\bar{\gamma} = \gamma_b = E_b/N_o$. The bit-error performance for these modulations with MRCD is shown in Figure 18.14 for various values of N. For $N = 1$, (18.80) is simply the expression of the SFRL channel corresponding to $\beta = 0$ in Figures 18.7 and 18.9. The bit-error results corresponding to the *dashed* curve represents the ideal AWGN performance of the respective underlying modulation.

18.8.1.3 Equal Gain Diversity Equal gain diversity (EGD) is similar to MRCD with $G_n = 1\ \forall n$, that is, equal gains for each receiver, so it is easier to implement. However, the phase of each received signal is adjusted so that (18.77) represents a coherent integration. There is no known closed-form solution for the resulting *pdf* $p(\gamma)$; however, through testing and numerical evaluations, the performance is better than SD and slightly worse than that of MRCD [24].

18.8.2 Frequency Selective Fading

As mentioned previously, when the bandwidth of the modulated symbol is close to or exceeds the coherence bandwidth of the channel the received signal encounters a delay spread resulting in ISI distortion. In the absence of mitigation techniques, an irreducible degradation or *flaring* of the bit-error performance occurs as the E_b/N_o is increased. To mitigate

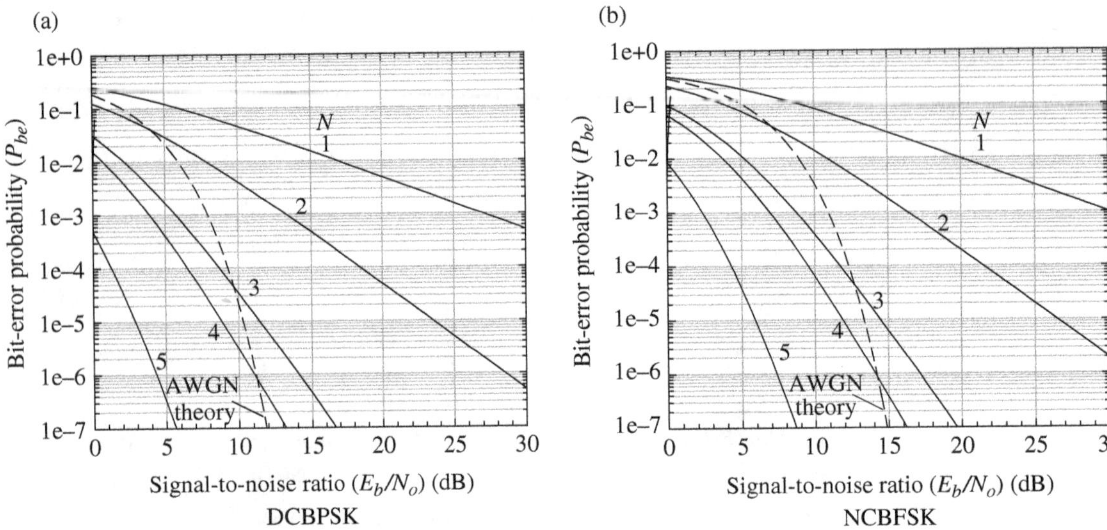

FIGURE 18.14 Performance of DCBPSK and NCBFSK with MRCD and SFRL.

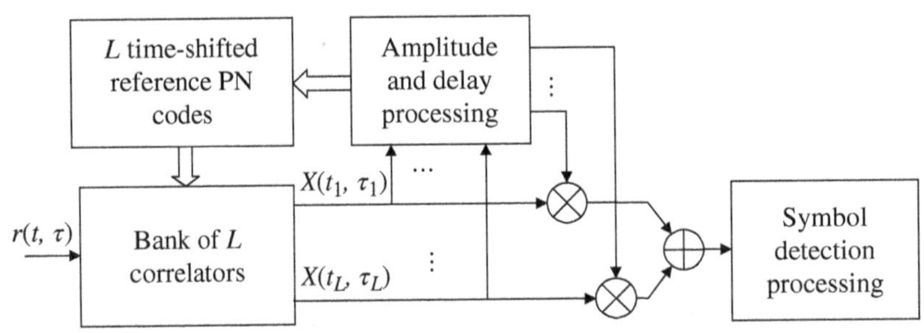

FIGURE 18.15 RAKE correlator processing of signals in frequency selective fading channels.

the performance degradation, the receiver or demodulator must use processing intense symbol equalization as discussed in Chapter 12. In urban wireless communication environments, the distortion appears as interfering signals from discrete multipath reflections. The reflections are characterized by the power delay profile with multipath delays τ_n: $n = 1,\ldots,$ N ranging over the multipath delay spread of T_m measured in μs. The impulse response of the discrete multipath channel is modeled as a tapped delay line as characterized by (18.61). For example, an average delay spread of 2 μs results in a coherence bandwidth of $B_c \cong 500$ kHz and, operating under this condition, the American Digital Cellular (ADC) system uses a distortion-free symbol rate of 48.6 ksps, approximately $B_c/10$, whereas, the Global System for Mobile Communications (GSM) signaling rate is 271 ksps and requires a 26-bit equalizer training sequence [28].

Another important aspect of signaling in a frequency selective channel is the use of direct-sequence spread-spectrum (DSSS) and CDMA modulated waveforms. These modulations divide the bit interval into N' PN code chips of a duration of $T_{ch} = T_b/N'$ and use a correlator or matched filter sampled at the peak or zero lag of the correlator output to detect the received bit. The sampled correlator output corresponds to a processing gain of $G_p = 10\log_{10}(N)$ dB relative to the correlation sidelobes so correlation lags in excess $\pm T_{ch}$ are essentially uncorrelated for sufficiently large values of N. Under these conditions, a DSSS or CDMA waveform will resolve multipath delayed reflections as long as $T_{ch} < \Delta\tau_n = \min(\tau_{n-1},\tau_n)$. The RAKE* correlator [29–31] or matched filter is the optimum demodulator detection filter for wideband PN spread signals. The processing involves $L \le N$ parallel correlators to resolve the largest L uncorrelated multipath delay returns that span the multipath spread interval T_m. The functional implementation is shown in Figure 18.15. Each RAKE correlator is implemented to locate and track an independent correlated multipath component of the received signal. This is accomplished by assigning a different and adjustable time-shifted reference sequence from that of the known PN code. The parallel correlators output the optimally sampled multipath responses corresponding to the peak of the respective RAKE correlator responses.

*The RAKE receiver is based on a correlator used for the detection and combining of multipath returns in spread-spectrum systems. It is not an acronym and refers to the similarities of a tapped delay line correlator to the common garden rake (see Reference 30, Vol. II, Chapter 1, p. 46 and Reference 31).

Defining the PN code chip bandwidth as $W = 1/T_{ch} \gg B_c$, the minimum number of correlator taps required to span the multipath spread is $N_{taps} = \text{ceiling}(T_m W)$ or $\cong \text{ceiling}(W/B_c)$.

The multipath components and the L correlator outputs represent weighted signal powers or SNR estimates and are combined to form an MRCD receiver as described in Section 18.8.1.2. The *pdf* of the combined multipath components is the chi-squared distribution described by (18.79) with $2L$ degrees of freedom.

ACRONYMS

ADC	American digital cellular
AGC	Automatic gain control
ARQ	Automatic repeat request
BFSK	Binary frequency shift keying
BPSK	Binary phase shift keying
CBPSK	Coherent binary phase shift keying
CDMA	Code division multiple access
DCBPSK	Differentially coherent binary phase shift keying
DSSS	Direct-sequence spread-spectrum
EDAC	Error detection and correction
EGD	Equal gain diversity
FD	Frequency diversity
FEC	Forward error correction (coding)
FFRL	Fast fading Rayleigh limit
FHSS	Frequency-hopping spread-spectrum
FRF	Fast Ricean fading
GSM	Global System for Mobile Communications
ISI	Intersymbol symbol interference
LOS	Line of sight
MRCD	Maximum ratio combining diversity
NC	Noncoherent (detection)
NCBFSK	Noncoherent binary frequency shift keying
PN	Pseudo-noise
RF	Radio frequency
SD	Selection diversity
SFRL	Slow fading Rayleigh limit
SRF	Slow Ricean fading

Note: RAKE is not an acronym; it is the name applied to a wide bandwidth multipath correlator.

PROBLEMS

1. Consider the analytic description of the baseband signal $\tilde{s}(t)$ with amplitude V_s, phase ψ, and additive narrowband zero-mean Gaussian noise $\tilde{n}(t)$ with uniformly distributed phase ϕ over $\pm\pi$. Show that the magnitude of the signal plus noise can be expressed by (18.44).

2. Write a program to compute correlated Ricean samples as described by (18.46) and (18.47) and compute and plot the normalized correlation responses $R_r(t)$ and $\rho_r(t)$ expressed, respectively, in (18.49) and (18.50) and determine the value of T_c corresponding to $\rho_r(T_c) = 0.368$. Use an underlying correlation of $\rho = 0.8$ and unit standard deviation $\sigma_r = 1$ in (18.46). Also, use a channel bandwidth of $B_{ch} = 100$ kHz and compute V_s using (18.45) for SNRs of $\gamma_{sr} = 0$ ($-\infty$ dB) corresponding to the Rayleigh fading channel. *Note*: Once the program is completed, it is a simple matter to evaluate the performance under a variety of conditions.

3. Evaluate the Fourier transform of the correlation response $R_r(t)$ in Problem 2 and determine the Doppler spread based on the spectral response $H_r(B_d) = 0.368$ and 0.1. What are the corresponding channel coherence times? What are the corresponding maximum symbol rates that the channel will support without distortion? In each case, what percentage of the symbol rate is the Doppler spread?

4. Evaluate the solution for DCBPSK waveform modulation with SD and the SFRL channel as described by (18.76). Hints: Perform a transformation of variables using $x = \gamma/\bar{\gamma}$, then use the solution to the following integral and the definition of the beta function given by [32]

$$\int_0^\infty e^{-ux}\left(1-e^{-x/\beta}\right)^{\nu-1} dx = \beta B(\beta u, \nu) \quad : u, \beta, \nu > 0$$

where the beta function is defined as

$$B(a,b) \triangleq \int_0^1 t^{a-1}(1-t)^{b-1} dt = \frac{\Gamma(a)\Gamma(b)}{\Gamma(a+b)}$$

The solution to the beta function in terms of gamma functions is found in Korn and Korn [33].

5. Evaluate the solution for DCBPSK waveform modulation with MRCD and the SFRL channel as described by (18.80). Hint: Use the following integral solutions given by Korn and Korn [34].

$$\int_0^\infty x^b e^{-ax^2} dx = \begin{cases} \dfrac{1*3\cdots(b-1)\sqrt{\pi}}{2^{(b/2)+1}a^{(b+1)/2}} & : a>0,\ b=0,2,4,\ldots \\ \dfrac{((b-1)/2)!}{2a^{(b+1)/2}} & : a>0,\ b=1,3,5,\ldots \end{cases}$$

REFERENCES

1. E. Ström, H. Hartenstein, P. Santi, W. Wiesbeck, Guest Editors, "Vehicular Communications," Special Issue: Proceedings of the IEEE, Vol. 99, No. 7, pp. 1158–1326, July 2011.
2. A.D. Kucar, J. Uddenfeldt, Editors, "The Mobile Radio Centennial," Special Issue: Proceedings of the IEEE, Vol. 86, No. 7, pp. 1303–1527, July 1998.

3. D. Middleton, "Man-Made Noise in Urban Environments and Transportation Systems: Models and Measurements," IEEE Transactions on Communications, Vol. COM-21, No. 11, pp. 1232–1241, November 1973.
4. M. Abramowitz, and I.A. Stegun, *Handbook of Mathematical Functions with Formulas, Graphs, and Mathematical Tables*, National Bureau of Standards, Applied Mathematics Series 55, United States Department of Commerce, Washington, DC, June 1964.
5. I.S. Gradshteyn, I.M. Ryzhik, *Table of Integrals, Series, and Products*, Corrected and Enlarged Edition, by A. Jeffrey, Academic Press, New York, 1980.
6. G.A. Arredondo, W.H. Chriss, E.H. Walker, "A Multipath Fading Simulator for Mobile Radio," IEEE Transactions on Communications, Vol. COM-21, No. 11, pp. 1325–1328, November 1973.
7. M.J. Gans, "A Power-Spectral Theory of Propagation in the Mobile-Radio Environment," IEEE Transactions on Vehicular Technology, Vol. VT-21, pp. 27–38, February 1972.
8. R.H. Clarke, "A Statistical Theory of Mobile-Radio Reception," Bell System Technical Journal, Vol. 47, pp. 957–1000, July/August 1968.
9. W.C. Jakes, Jr., "A Comparison of Specific Space Diversity Techniques for Reduction of Fast Fading in UHF Mobile Radio Systems," IEEE Transactions on Vehicular Technology, Vol. VT-20, pp. 81–92, November 1971.
10. T.S. Rappaport, *Wireless Communications, Principles and Practice*, Second Edition, Prentice-Hall PTR, Upper Saddle River, NJ, 2002.
11. G.L. Turin, F.D. Clapp, T.L. Johnston, S.B. Fine, D. Lavry, "A Statistical Model of Urban Multipath Propagation," IEEE Transactions on Vehicular Technology, Vol. VT-21, Issue 1, pp. 1–9, February 1972.
12. J.G. Proakis, *Digital Communications*, Second Edition, McGraw-Hill Book Company, New York, 1989.
13. R. Steele, Editor, *Mobile Radio Communications*, IEEE Press, New York, 1994.
14. T.S. Rappaport, *Wireless Communications, Principles and Practice*, Second Edition, pp. 203–205, Prentice-Hall PTR, Upper Saddle River, NJ, 2002.
15. W.C. Jakes, Jr., "An Approximate Method to Estimate an Upper Bound of the Effect of Multipath Delay Distortion on Digital Transmission," IEEE Transactions on Communications, Vol. COM-27, No. 1, pp. 76–81, January 1979.
16. T.S. Rappaport, *Wireless Communications, Principles and Practice*, Second Edition, pp. 199–203, Prentice-Hall PTR, Upper Saddle River, NJ, 2002.
17. W.C.Y. Lee, "Smaller Cells for Greater Performance," IEEE Communications Magazine, Vol. 9, No. 11, pp. 19–23, November 1991.
18. J. Chuang, "The Effect of Time Delay Spread on Portable Communications Channels with Digital Modulation," IEEE Journal on Selected Areas in Communications, Vol. SAC-5, No. 5, pp. 879–889, June 1987.
19. V. Fung, T.S. Rappaport, B. Thoma, "Bit Error Simulation for $\pi/4$ DQPSK Mobile Radio Communication Using Two-Ray and Measurement-Based Impulse Response Models," IEEE Journal on Selected Areas in Communications, Vol. SAC-11, No. 3, pp. 393–405, April 1993.
20. T.S. Rappaport, S.Y. Seidel, K. Takamizsawa, "Statistical Channel Impulse Response Models for Factory and Open Plan Building Radio Communication System Design," IEEE Transactions on Communications, Vol. COM-39, No. 5, pp. 794–806, May 1991.
21. B.D. Woerner, J.H. Reed, T.S. Rappaport, "Simulation Issues for Future Wireless Modems," IEEE Communications Magazine, Vol. 32, No. 7, pp. 42–53, July 1994.
22. S. Lin, D.J. Costello, Jr., M.J. Miller, "Automatic-Repeat-Request Error-Control Schemes," IEEE Communications Magazine, Vol. 22, No. 12, pp. 5–15, December 1984.
23. J. Kennedy, M.C. Sullivan, "Direction Finding and "Smart Antenna" Using Software Radio Architectures," IEEE Communications Magazine, Vol. 33, No. 5, pp. 62–68, May 1995.
24. M. Schwartz, W.R. Bennett, S. Stein, *Communication Systems and Techniques*, McGraw-Hill Book Co., New York, 1966.
25. T.S. Rappaport, *Wireless Communications, Principles and Practice*, Second Edition, pp. 380–394, Prentice-Hall PTR, Upper Saddle River, NJ, 2002.
26. W.C. Jakes, *New Techniques for Mobile Radio*, pp. 326–330, Bell Laboratories Record, New York, December 1970.
27. L.W. Couch, *Digial and Analog Communication Systems*, Fourth Edition, Macmillan, New York, 1993.
28. R.C.V. Macario, *Cellular Radio: Principles and Design*, p. 207, McGraw-Hill, Inc., New York, 1993.
29. R. Price, P.E. Green, "A Communication Technique for Multipath Channels," Proceedings of the IRE, Vol. 46, pp. 555–570, March 1958.
30. M.K. Simon, J.K. Omura, R.A. Scholtz, B.K. Levitt, *Spread Spectrum Communications*, Vol. II, Chapter 1, "Coherent Direct-Sequence Systems," p. 46, Computer Science Press, Inc., Rockville, MD, 1985.
31. R. Price, "The Detection of Signals Perturbed by Scatter Noise," IRE Transactions on Information Theory, PGIT-4, pp. 163–170, September 1954.
32. I.S. Gradshteyn, I.M. Ryzhik, *Table of Integrals, Series, and Products*, Corrected and Enlarged Edition, by A. Jeffrey, p. 305, No. 3.312-1, and p. 948, No. 8.380-1, Academic Press, New York, 1980.
33. G.A. Korn, T.M. Korn, *Mathematical Handbook for Scientists and Engineers: Definitions, Theorems, and Formulas for Reference and Review*, Second Edition, Section E-3, p. 973, No. 3, McGraw-Hill Book Company, New York, 1968.
34. G.A. Korn, T.M. Korn, *Mathematical Handbook for Scientists and Engineers: Definitions, Theorems, and Formulas for Reference and Review*, Second Edition, Section E-3, p. 977, No. 38, McGraw-Hill Book Company, New York, 1968.

19

ATMOSPHERIC PROPAGATION

19.1 INTRODUCTION

Communication links that propagate through the atmosphere encounter a number of effects that distort the received signal making detection and parameter estimation difficult and prone to errors. The signal distortion results because the atmosphere is an inhomogeneous medium with spatial and temporal variations that result in random behavior leading to: absorption or signal attenuation, inter- and intra-symbol distortion due to symbol dispersion, variations in range-delay, Doppler spreading, polarization rotation, and signal amplitude and phase fluctuations resulting from multipath interference. The principal regions of the atmosphere that impact electromagnetic wave propagation are the troposphere, and ionosphere, and the principal parameter that characterizes the performance in each region is the index of refraction.

The lower region of the atmosphere is the troposphere extending to about 30 km in altitude and the upper region is the ionosphere extending to several thousand kilometers. The troposphere is essentially an ion-free region consisting of about 99% oxygen with nitrogen and water vapor at the lower altitudes. In this region electromagnetic wave propagation characteristics are determined by the refractive index which is a function of pressure, temperature, and water vapor; other natural phenomena like dust, rain, and clouds must also be considered. Propagation in the ionosphere is influenced primarily by free-electrons and characterized by the electron content, the Earth's magnetic field, and diurnal variations influenced by the Sun. Variations in the atmosphere are influenced by location, season, weather, and the time-of-day (TOD). Typically, the mean and stand deviation of the influential parameters are given for a specified season and location.

The impact of these natural occurrences in the atmosphere on the communication signal is dependent upon the carrier frequency and bandwidth of the communication link as well as the antenna beamwidth, height, and the elevation angle of the encounter. The various phenomena resulting from the encounter with the atmospheric channel are characterized in terms of the operating modes as shown in Figure 19.1. The effective Earth radius r_e is chosen to be $(4/3)R_e$, where $R_e = 6371$ km is the average Earth radius. This selection of r_e provides for line of sight (LOS) radio frequency coverage to locations that would otherwise be over-the-horizon. This LOS condition applies for altitudes less than about 4 km and is discussed in more detail in the following section. The factor of 4/3 is computationally convenient because it is based on a linearly decreasing index of refraction with height, whereas, an exponentially decreasing function with height is in more agreement with *in situ* measurements and is discussed in Section 19.5.

In Section 19.3, the impact of multipath reflections from a smooth Earth is examined followed by a case study in Section 19.4. Signal refraction in the troposphere and ionosphere is discussed in Section 19.5 and diffraction from irregular surfaces is discussed in Section 19.6 with examples of irregular terrain and urban propagation losses given in Sections 19.7 and 19.8.

Digital Communications with Emphasis on Data Modems: Theory, Analysis, Design, Simulation, Testing, and Applications, First Edition. Richard W. Middlestead.

Companion website: www.wiley.com/go/digitalcommunications

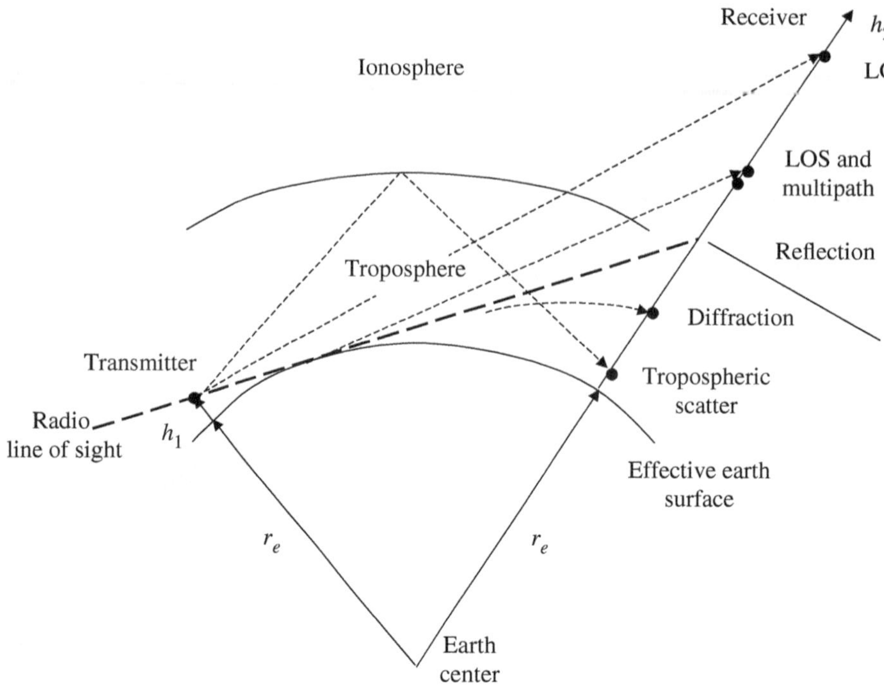

FIGURE 19.1 Various propagation modes depending on atmospheric conditions, frequency, and geometry.

19.2 COMMUNICATION LINK GEOMETRY FOR CURVED EARTH

The geometry for determining the propagation paths for LOS and multipath rays has been modeled under various simplifying assumptions involving a flat-Earth approximation and a spherical Earth with receivers at sufficiently long ranges such that the direct and reflected waves are assumed to be parallel. Although there is no known exact solution to the general case involving unrestricted ranges, transmitter, and receiver heights, Fishback [1] provides a solution involving a spherical Earth model, with effective Earth radius r_e to account for refraction and unrestricted transmitter and receiver heights. Fishback's analysis assumes the transmitter and receiver heights, h_1 and h_2, are given along with the length of the ground range, G, between the two points. Fishback's solution is somewhat restricted by assuming small grazing angles. Instead of following Fishback's solution, the relationships developed by Blake [2] are discussed because they are less prone to numerical and computational inaccuracies due to rounding and because Blake's solution simply requires the specification of h_1, h_2 and the LOS range, R_d, between the transmit and receive points and there is no restriction on the grazing angle. The geometry of the encounter is shown in Figure 19.2.

The following analysis outlines the procedures for the solution of the reflected wave at the receiver terminal in terms of R_1, R_2, and ψ. With these parameters, the amplitude and delay of the interfering signal can be determined relative to the LOS signal. To begin, the angle θ_d is computed as

$$\theta_d = \arcsin^{-1}\left(\frac{2r_e(h_2-h_1)+h_2^2-h_1^2-R_d^2}{2(r_e+h_1)R_d}\right) \cong \arcsin^{-1}\left(\frac{h_2-h_1}{R_d}-\frac{R_d}{2r_e}\right) \quad (19.1)$$

As shown in Figure 19.2 the angle θ_d corresponds to a negative angle, that is, when the transmitter is looking below the local horizon. The approximation in (19.1) applies when $h_1, h_2 \ll r_e$. The ground path, as derived by Blake, is determined using the direct path length R_d and is computed as

$$G = 2r_e \arcsin\sqrt{\frac{R_d^2-(h_2-h_1)^2}{4(r_e+h_1)(r_e+h_2)}} \cong \sqrt{\frac{R_d^2-(h_2-h_1)^2}{1+(h_1+h_2)/r_e}} \quad (19.2)$$

The ground range G is defined as $G = G_1 + G_2$ and Fishback has shown that G_1 is the solution to a cubic equation involving G, h_1, h_2, and r_e. When $h_1, h_2 \ll r_e$, the solution is

$$G_1 = \frac{G}{2} - p\sin\left(\frac{\xi}{3}\right) \quad (19.3)$$

where

$$p = \frac{2}{\sqrt{3}}\sqrt{r_e(h_1+h_2)+\left(\frac{G}{2}\right)^2} \quad (19.4)$$

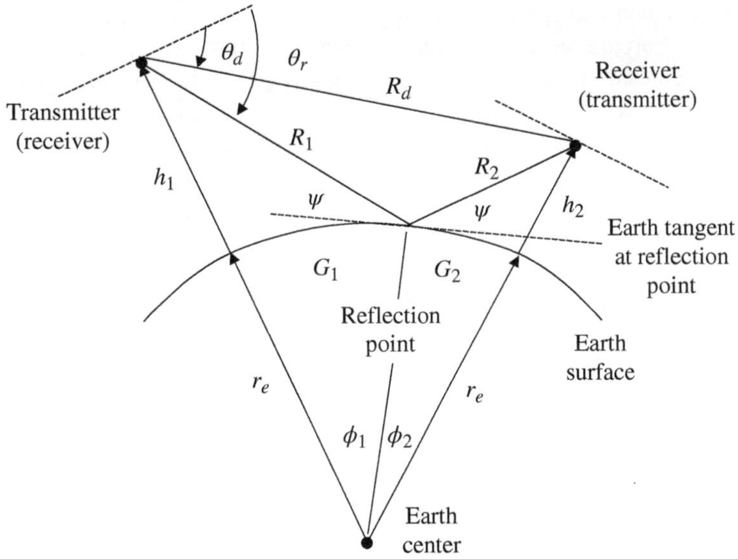

FIGURE 19.2 Spherical-earth encounter geometry.

and

$$\xi = \arcsin\left(\frac{2r_e G(h_2 - h_1)}{p^3}\right) \tag{19.5}$$

The angles ϕ_1 and ϕ_2 are then determined as

$$\phi_1 = \frac{G_1}{r_e} \tag{19.6}$$

and

$$\phi_2 = \frac{G_2}{r_e} = \frac{G - G_1}{r_e} \tag{19.7}$$

from which the ranges R_1 and R_2 are determined as

$$R_1 = \sqrt{h_1^2 + 4r_e(r_e + h_1)\sin^2\left(\frac{\phi_1}{2}\right)} \tag{19.8}$$

and

$$R_2 = \sqrt{h_2^2 + 4r_e(r_e + h_2)\sin^2\left(\frac{\phi_2}{2}\right)} \tag{19.9}$$

Using the range R_1, the depression angle of the reflected wave is determined as

$$\begin{aligned}\theta_r &= \arcsin\left(\frac{2r_e h_1 + h_1^2 + R_1^2}{2(r_e + h_1)R_1}\right)\\ &\cong \arcsin\left(\frac{h_2 - h_1}{R_d} - \frac{R_d}{2r_e}\right)\end{aligned} \tag{19.10}$$

and the grazing angle is determined as

$$\begin{aligned}\psi &= \arcsin\left(\frac{2r_e h_1 + h_1^2 - R_1^2}{2r_e R_1}\right) = \arcsin\left(\frac{2r_e h_2 + h_2^2 - R_2^2}{2r_e R_2}\right)\\ &\cong \arcsin\left(\frac{h_1}{R_1} - \frac{R_1}{2r_e}\right) = \arcsin\left(\frac{h_2}{R_2} - \frac{R_2}{2r_e}\right)\end{aligned} \tag{19.11}$$

Referring to Figure 19.2, the grazing angle is also expressed as $\psi = \theta_r - \phi_1$. The approximations in (19.10) and (19.11) apply when h_1, $h_2 \ll r_e$. Using these relationships, the difference between the direct path R_d and the reflected path $R_1 + R_2$ is determined as

$$\begin{aligned}\delta R &= R_1 + R_2 - R_d\\ &= \frac{4R_1 R_2 \sin^2\psi}{R_1 + R_2 + R_d}\end{aligned} \tag{19.12}$$

The second of these solutions [3] is preferred because it is not subject to numerical errors encountered when performing the subtraction in the first solution. Referring to Figure 19.1, the LOS criterion corresponds to a zero degree grazing angle or the equivalent requirement that $R_d = R_1 + R_2$. Therefore, using (19.11), the LOS criterion, involving h_1 and h_2, is evaluated as

$$\begin{aligned}R &= \sqrt{2r_e h_1 + h_1^2} + \sqrt{2r_e h_2 + h_2^2} \quad : \text{LOS criterion}\\ &\cong \sqrt{2r_e h_1} + \sqrt{2r_e h_2}\end{aligned} \tag{19.13}$$

Therefore, when $R_d > R$ the LOS path will exhibit ground reflections, otherwise, the over-the-horizon reception occurs from diffraction or refraction as depicted in Figure 19.1.

Although the analysis of the geometry in this section is correct for $R_d \leq R_1 + R_2$, the application to multipath interference based on reflections strictly applies for LOS paths a few degrees above the radio horizon such that $\delta R \gg \lambda/4$. For lower elevation angles, θ_d, an intermediate region is entered between the reflection and diffraction regions. When the Earth is modeled with a smooth surface, the region below the radio LOS, shown as the *heavy dashed* lines in Figure 19.1, is the diffraction region; however, the intermediate region just above and below the radio LOS requires special consideration as described by Fishback. Fishback's solution is outlined by Blake [4] for the first diffraction mode, that is, when $\delta R \leq \lambda/4$ and the solution of Burroughs and Atwood [5] is outlined by Barton [6]. In the following section, the impact of reflections from a smooth Earth is examined followed by a case study in Section 19.4 involving a low Earth orbit satellite.

19.3 REFLECTION

Electromagnetic wave propagation is characterized by Huygens' principle [7] which states that points along the wavefront depicted in Figure 19.3 behave like secondary sources of wavelets that spread out in all directions with velocities that are functions of the propagation medium. This principle provides for the analysis of wave reflection when the wavefront is incident to a reflecting surface. It also provided for the analysis of refractive wave bending and propagation beyond LOS conditions as discussed in Section 19.5.

Consider the incident wave shown in Figure 19.4 having uniform phase along the wavefront represented by the straight line intersecting the reflecting surface with an angle-of-incidence ϕ_1. The incident wave is traveling in the medium above the reflecting surface at a velocity $v = 1/\sqrt{\mu\varepsilon}$, where u and c are the permeability and permittivity of the medium. The angle ϕ_2 of the reflected wave is determined using Huygens' principle recognizing that the velocity of the reflected wave is also equal to v. For example, when a wavelet of the incident wave travels the distanced $d = vt$ and intersects the surface, the reflected wave t seconds earlier has traveled the same distance describing the reflected wavefront at the angle of reflection ϕ_2. Upon application of elementary geometry, it is found that the angle of reflection is equal to the angle of incidence, that is, $\phi_2 = \phi_1$. In many practical applications, the reflecting surface is not an ideal reflector and wave refraction occurs resulting in the incident wave propagating through the surface into the second medium with different electromagnetic properties; wave refraction is discussed in Section 19.5. In general, both wave reflection and refraction occur at the surface between two different medium.

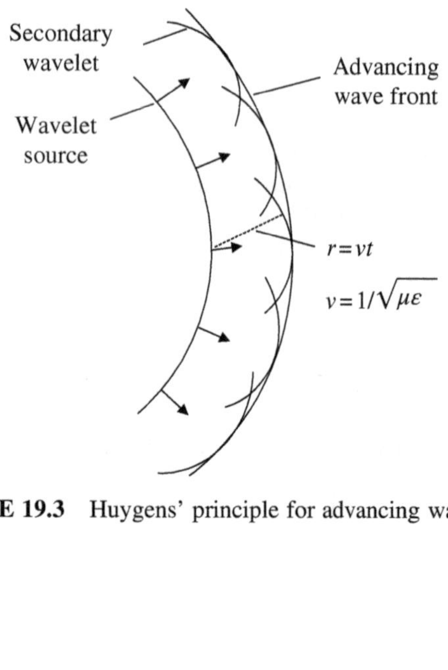

FIGURE 19.3 Huygens' principle for advancing wavefront.

19.3.1 Reflection from Earth's Surface

Signal reflection occurs when an incident electrometric wave intersects a large surface with dimensions much greater than the carrier wavelength. In long-distance communication links, signal reflections from the Earth's surface will combine with a direct path signal resulting in received signal variations referred to as multipath fading. On a smaller scale, signals in a mobile communication system will reflect off of building and other structures often without an identifiable direct signal path. In these situations, the received signal will fluctuate rapidly with the dynamics of the mobile unit. Signal reflections are characterized in terms of the complex reflection coefficient

$$\Gamma = \rho e^{j\Phi} \tag{19.14}$$

where $\rho = |E_r/E_i|$ is the magnitude of the reflected-to-incident electric field ratio and Φ is the corresponding angle of E_r/E_i. The ranges of ρ and Φ are: $0 \leq \rho \leq 1$ and $|\Phi| \leq \pi$ and when

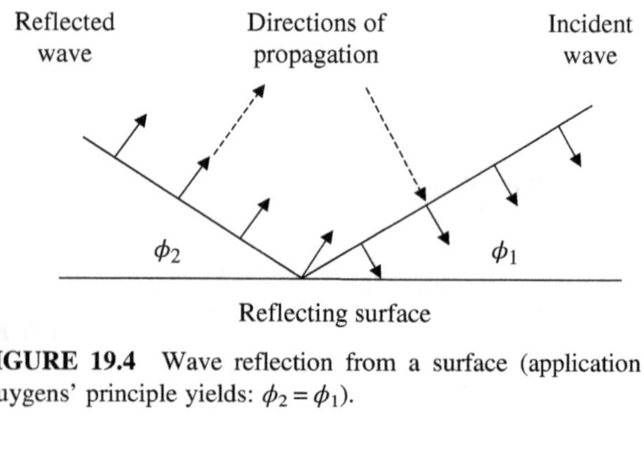

FIGURE 19.4 Wave reflection from a surface (application of Huygens' principle yields: $\phi_2 = \phi_1$).

$\Phi > 0$ then the reflected electric field lags the incident electric field. Ramo and Whinnery [8] provide a detailed discussion of the reflection of electromagnetic wave forming the basis of the following results.

The reflection coefficient also depends on the grazing angle and polarization of the incident electric field and for horizontally and vertically polarized waves the results are expressed as [9]

$$\Gamma_h = \rho_h e^{-j\Phi_h} = \frac{\sin(\psi) - \sqrt{\varepsilon_c - \cos^2(\psi)}}{\sin(\psi) + \sqrt{\varepsilon_c - \cos^2(\psi)}} \tag{19.15}$$

and

$$\Gamma_v = \rho_v e^{-j\Phi_v} = \frac{\varepsilon_c \sin(\psi) - \sqrt{\varepsilon_c - \cos^2(\psi)}}{\varepsilon_c \sin(\psi) + \sqrt{\varepsilon_c - \cos^2(\psi)}} \tag{19.16}$$

where ε_c is the complex dielectric constant given by

$$\varepsilon_c = \varepsilon_r - j\varepsilon_i \tag{19.17}$$

where ε_r is the ordinary dielectric constant and ε_i is given by

$$\varepsilon_i = 59.9584916\lambda\sigma \cong 60\lambda\sigma \tag{19.18}$$

where λ is the carrier wavelength and σ is surface conductivity.* When λ is in meters then σ has units of mhos per meter; values of ε_r and σ for various types of surfaces are given in Table 19.3.

The reflection coefficient for the vertically polarized wave is examined by substituting (19.17) into (19.16) and the resulting amplitude and phase functions are plotted in Figures 19.5 and 19.6 for the indicated conditions. The grazing angle corresponding to the minimum magnitude of the reflection coefficient in Figure 19.5 is referred to as the pseudo-Brewster angle. With a perfect dielectric, the minimum magnitude of the reflection coefficient is $|\Gamma_v| = 0$, that is, no reflection occurs, and the corresponding grazing angle is referred to as the polarizing angle or, in geometrical optics, the Brewster angle. The Brewster angle occurs only with a vertically polarized wave. With horizontal polarization, the reflection coefficient decreases slowly with grazing angle and the angle Φ_h is very close to π radians over the entire range of grazing angles.

Circularly polarized waves are described in terms of horizontally and vertically polarized waves that are in quadrature with one another. Right-hand circular polarization occurs when the horizontal polarized wave lags that of the vertically polarized wave, that is, when $\phi_h - \phi_v > 0$ and left-hand circular polarization occurs when $\phi_v - \phi_h > 0$. For right-hand or left-hand circularly polarized incident waves, the reflection coefficient with the same-sense as the incident wave is expressed as [2, 11]

$$\Gamma_{cs} = \frac{1}{2}(\Gamma_v + \Gamma_h) \tag{19.19}$$

and the reflection coefficient with the opposite-sense as the incident wave is

$$\Gamma_{co} = \frac{1}{2}(\Gamma_v - \Gamma_h) \tag{19.20}$$

Therefore, in general, both senses of the incident circular polarized wave appear at the receiver from the reflecting surface.

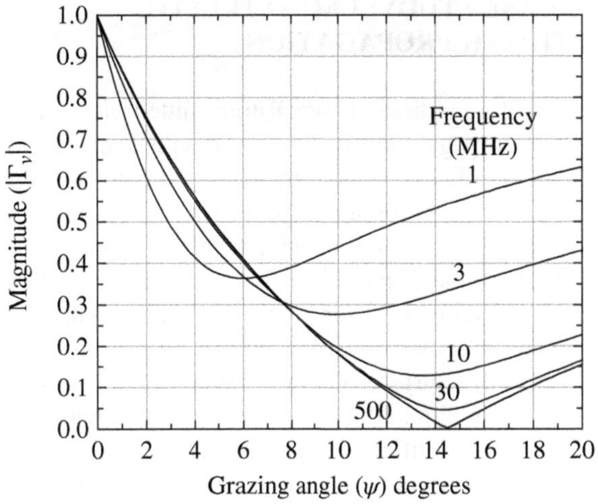

FIGURE 19.5 Magnitude of reflection coefficient for vertical polarization (conductivity = 0.005 mho/m, dielectric constant = 15).

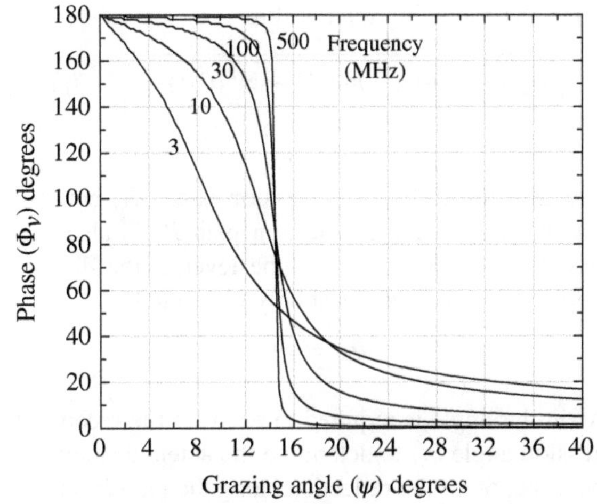

FIGURE 19.6 Phase of reflection coefficient for vertical polarization (conductivity = 0.005 mho/m, dielectric constant = 15).

*Equation (19.18) is in *mks* units and the constant k is evaluated as $1/(2\pi\varepsilon_o c)$, where $\varepsilon_o = 8.854187817 \times 10^{-12}$ F/m is the free-space electric constant and $c = 2.99792458 \times 10^8$ m/s is the velocity of light in free space; the product $\varepsilon_i = k\lambda\sigma$ is unit less. In optics the units of k are 1/Siemens.

19.4 CASE STUDY: LEO SATELLITE MULTIPATH PROPAGATION

Under ideal conditions, a signal transmitted along a noise-free direct propagation path is received with a constant signal amplitude and phase. However, with multipath propagation there are one or more signal propagation paths that result from reflections that interfere with the direct path signal causing amplitude and phase fluctuations. This scenario is depicted in Figure 19.7 for a communication link between a ground terminal and a satellite with a single-point reflection from the Earth's surface. The description of the various parameters and additional geometrical details regarding this encounter are identified in Figure 19.2.

The description of the spherical Earth geometry given in Section 19.2 uses the LOS range R_d as the independent variable; however, in this case study, the independent variable is the transmitter elevation angle θ_d. Using the parameters h_1 and h_2, the direct path range is computed as a function of θ_d using the following relationship

$$R_d = \sqrt{(r_e+h_2)^2-(r_e+h_1)^2\left(1-\sin^2(\theta_d)\right)}-(r_e+h_1)\sin(\theta_d) \qquad (19.21)$$

The transmitter beam depression angle θ_{bt} between the elevation angle and the reflection path R_1 is also required to determine the antenna sidelobe level in the direction of the reflection; this angle is simply computed as

$$\theta_{bt} = \theta_r + \theta_d \qquad (19.22)$$

A similar angle is required to evaluate the receiver beam depression angle θ_{br} to determine the antenna sidelobe level in the direction of the reflection along the path R_2 and, referring to Figure 19.2, this angle is computed as

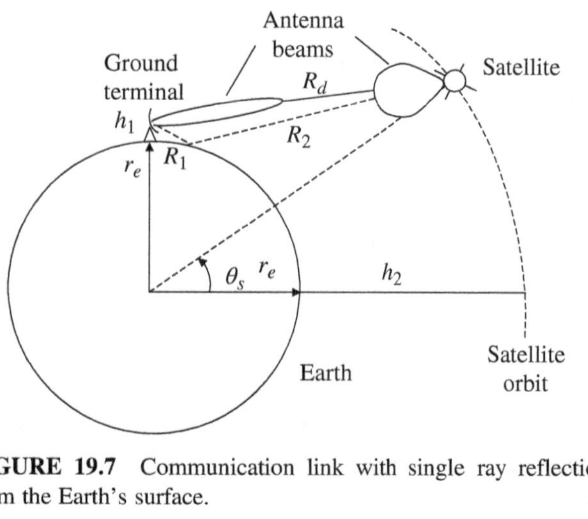

FIGURE 19.7 Communication link with single ray reflection from the Earth's surface.

$$\theta_{br} = \cos^{-1}\left(\frac{R_d^2+R_2^2-R_1^2}{2R_dR_1}\right) \qquad (19.23)$$

In the following analysis, the ground and satellite terminal antennas are modeled as uniformly weighted circular aperture antennas as described in Section 15.3.2 with specified gains G_t and G_r respectively. The antenna gains are related to the antenna area A_a or the effective area $A_e = n_aA_a$, where n_a is the antenna efficiency, and the gain is computed using (15.61) with the area for the circular aperture antenna given by

$$A_a = \frac{\pi}{4}D^2 \quad : \text{circular aperture} \qquad (19.24)$$

where D is the antenna diameter. From Table 15.3, the antenna beamwidth is given by

$$\theta_B = \frac{1.02\lambda}{D} \qquad (19.25)$$

where λ is the wavelength of the carrier frequency.

Neglecting channel and receiver noise, the signals $s_d(t)$ and $s_r(t)$ are the received signals at the satellite from the direct and reflected paths respectively. These signals are combined at the receiver to form the received signal $s(t)$ expressed as

$$\begin{aligned} s(t) &= s_d(t)+s_r(t) \\ &= \sqrt{P_t}f_de^{j\omega_c(t-T_d)}+\sqrt{P_t}f_r\Gamma e^{j\omega_c(t-T_r)} \end{aligned} \qquad (19.26)$$

where P_t is the transmitted signal power at the antenna input, ω_c is the carrier angular frequency, T_d, T_r and f_d, f_r are the delays and propagation factors along the respective paths and Γ is the complex reflection coefficient given by (19.14). Referring to (15.6), the propagation factors are evaluated as

$$f_d = \sqrt{\frac{G_t(\theta_t)G_r(\theta_r)}{L_{atmd}L_{rs}L_{fsd}}}e^{-j\beta_t} \qquad (19.27)$$

and

$$f_r = \sqrt{\frac{G_t(\theta_{bt})G_r(\theta_{br})}{L_{atmr}L_{rs}L_{fsr}}}e^{-j\beta_r} \qquad (19.28)$$

The antenna beam angles θ_{bt} and θ_{br} are in the direction of the transmitted and reflected waves along the respective paths R_1 and R_2 and the angles β_t and β_r are the corresponding antenna-dependent signal phase along these paths; typically β_t and β_r are either 0 or π radians depending on the sign of the antenna sidelobes. As depicted in Figure 19.7, ideal beam tracking corresponds to θ_t, $\theta_r = 0$. Upon substituting (19.27),

(19.28), and (19.14) into (19.26) and factoring out the direct path received signal, the satellite received signal is expressed as

$$s(t) = \sqrt{P_r} f_d e^{j\omega_c(t-T_d)} \left[1 + \rho \frac{|f_r|}{|f_d|} e^{-j(\omega_c(T_r - T_d) - \Phi - (\beta_r + \beta_t))} \right] \tag{19.29}$$

The magnitude of the additive term is evaluated as

$$\rho \frac{|f_r|}{|f_d|} = \rho \sqrt{G_{tn}(\theta_t) G_{rn}(\theta_r) \frac{L_{atmt} L_{fst}}{L_{atmr} L_{fsr}}} \tag{19.30}$$

In (19.30), the subscript "n" on the antenna gains indicate normalized gains with unit peak gain. The free-space loss ratio is given by

$$\frac{L_{fst}}{L_{fsr}} = \left(\frac{R_d}{R_1 + R_2} \right)^2 \tag{19.31}$$

Also, the carrier frequency phase term is expressed as

$$\omega_c(T_r - T_d) = 2\pi \left(\frac{\delta R}{\lambda} \right) \tag{19.32}$$

This phase term results in large, frequency-dependent, amplitude fluctuations as the range difference between the direct and reflected wave change. The magnitude of the term in the square brackets in (19.29) is defined as the multipath factor F_m given by

$$F_m \triangleq \left| 1 + \rho \frac{|f_r|}{|f_d|} e^{-j(2\pi\delta R/\lambda - \Phi + (\beta_r - \beta_t))} \right| \tag{19.33}$$

To demonstrate the severity of the multipath on the desired or direct path signal $s_d(t)$, the multipath factor is plotted in Figure 19.8 for a satellite in a LEO circular orbit, coplanar with the Earth terminal and using ideal omnidirectional antennas, that is, the gains are assumed to be identical in all directions. In this and the remaining plots, vertical antenna polarization is used, the atmospheric loss along each path is assumed to be identical, that is, $L_{atmr} = L_{atmt}$ and the parameter values listed in Table 19.1 are used.

Three things are noteworthy in Figure 19.8: the effect of the multipath loss is about the same over all elevation angles θ_d; the maximum level of the received signal is 6 dB*; the lower limit of received signal is between −42 and −43 dB. The upper and lower limits, corresponding to the *solid black* curves, result from the constructive and destructive interference of the two received signals based on the phase term in (19.33). When F_m is plotted as a function of the antenna elevation angle, as in Figure 19.8, the phase term varies more slowly at the higher elevation angles where δR changes slowly†; this is evident in Figure 19.7 when $\theta_d > 80°$.

Figure 19.9 shows the impact of the multipath interference using a transmitter antenna gain of 20 dB; this gain corresponds to a 16.4° antenna beamwidth. The variation in the upper and lower limits of the multipath factor is significantly reduced for elevation angles greater than about 10°. This multipath improvement results from the increasing attenuation of the multipath signal by the antenna sidelobes as the elevation angle increases.

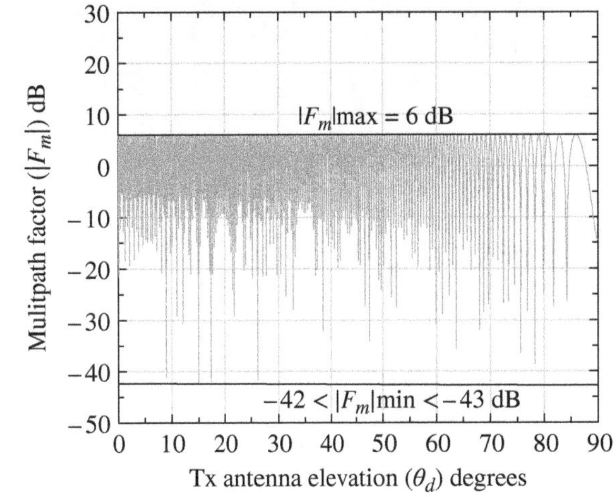

FIGURE 19.8 Communication link with single-ray reflection (omnidirectional antennas).

TABLE 19.1 Parameter Values for Multipath Factor (F_m) Evaluation

Parameter	Value	Comment
R_e	6371	Earth radius (km)
r_e	8494.66	4/3 Earth radius (km)
h_1	30	Transmitter antenna height (m)
h_2	200	LEO circular orbit[a] altitude (km)
G_t	Variable	Transmit antenna gain
G_r	10	Receiver antenna gain (dB)
η_{at}	80	Transmit antenna efficiency (%)
η_{ar}	80	Receiver antenna efficiency (%)
f_c	500	Carrier frequency (MHz)
σ	0.005	Conductivity (mho/m)
ε_r	15	Dielectric constant (F/m)

[a]Co-located with the plane of the ground terminal.

*The maximum level will be less than 6 dB because of $R_1 + R_2 > R_d$ and $\rho < 1$.

†This received signal phase variations with θ_d are influenced to a lesser extent by the phase of the complex reflection coefficient.

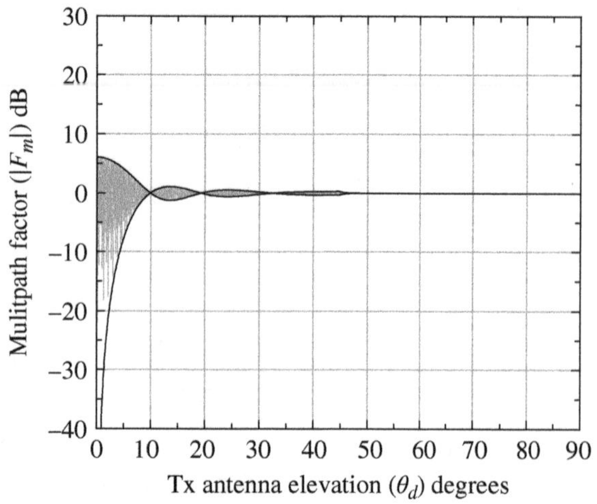

FIGURE 19.9 Communication link with single-ray reflection ($G_t = 20$ dB, $\theta_{Bt} = 16.2°$).

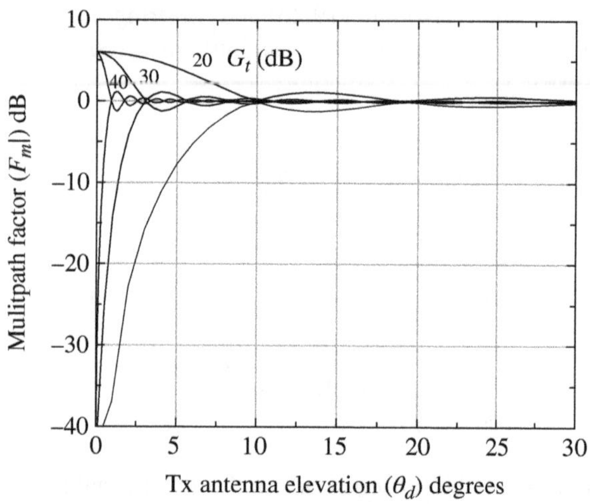

FIGURE 19.10 Communication link with single-ray reflection from the Earth's surface ($G_t = 20$, 30, and 40 dB; $\theta_{Bt} = 16.2$, 5.2, and 1.64°).

By eliminating the dependence of F_m on the phase function in (19.33), the upper and lower limits on the multipath factor are evaluated as

$$F_m(\max) = 1 + \frac{|f_r|}{|f_d|}\rho \quad \text{and} \quad F_m(\min) = 1 - \frac{|f_r|}{|f_d|}\rho \tag{19.34}$$

The squared inverse of the lower limit is defined as the multipath loss and is expressed as

$$L_m \triangleq F_m^{-2}(\min) \quad : \text{multipath loss} \tag{19.35}$$

The average multipath loss within the transmitter antenna beamwidth corresponding to a specified θ_d is denoted as $\bar{L}_m$. Equations (19.34) and (19.35) are used in Chapter 15 to establish the system link loss budget. The multipath loss can be effectively mitigated through the system and waveform designs and demodulator signal processing.

Figure 19.10 compares the maximum and minimum multipath factors for transmitter antenna gains of 20, 30, and 40 dB; the 20 dB gain case corresponds to Figure 19.9. The improvements are significant and are accompanied by a lower link power requirement; however, the cost and complexity associated with antenna beam pointing and tracking must be taken into account.

19.5 REFRACTION

Signal refraction occurs when an incident electrometric wave intersects irregularities in the transmission media. Refraction results in electromagnetic wave bending and allows for communications into the shadow region beyond the LOS. For example, wave refraction occurs in regions behind hills, mountains, and other obstructions that are impenetrable by a direct path link. Long-distance communications experience refraction due to varying dielectric constant in the atmosphere. As in the case of wave reflection, wave bending due to refraction is characterized by Huygens' principle that is introduced in Section 19.3.

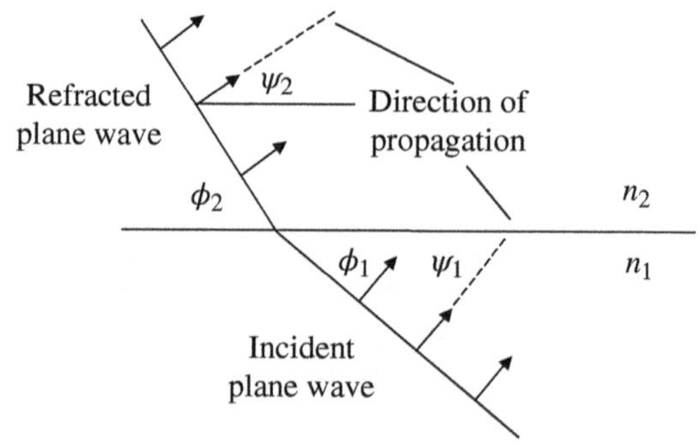

FIGURE 19.11 Wave bending in the atmosphere (shown for $n_1 > n_2$).

Application of Huygens' principle together with Snell's law is used to describe the bending of a plane wave as it propagates through the atmosphere with different refractive indices. This phenomenon is depicted in Figure 19.11 for an abrupt change in the index of refraction. Snell's law states that when a plane wave propagating in a medium with refractive index n_1 intersects a second medium at an angle of incidence ϕ_1 with refractive index n_2 then the *sine* of the angle of refraction, ϕ_2, is related to the *sine* of the angle of incidence ϕ_1 by the relationship

$$n_2 \sin(\phi_2) = n_1 \sin(\phi_1) \quad : \text{Snell's law} \tag{19.36}$$

The angles ψ_i, $i = 1, 2$, are the grazing angles of the plain waves and are related to the incident angles as $\psi_i = \pi/2 - \phi_i$ radians.

Figure 19.11 is suggestive of uplink propagation with the angle of refraction shown to be greater than the angle of incidence. This corresponds to decreasing electron densities with altitude, that is, $n_2 < n_1$, and a bending of the wave that tend to follow the curvature of the Earth giving rise to range delay and angle errors. When the angle of refraction satisfies the condition $\phi_2 > \pi/2$ radians, the incident wave is not propagated forward but reflected back through the medium containing the incident wave. Using the angle $\phi_2 = \pi/2$ radians, the critical angle of incidence for reflection is determined as

$$\phi_c = \sin^{-1}\left(\frac{n_2}{n_1}\right) \quad \text{: critical angle of incidence} \qquad (19.37)$$

19.5.1 Tropospheric Refraction

Because the troposphere does not contain free-electrons, signal propagation in the troposphere is influenced principally by water vapor content, air temperature, and pressure. Based on the standard atmosphere described by Kerr [1], the index of refraction in the troposphere is given by

$$n = 1 + \frac{a}{T}\left(p + \frac{be}{T}\right)10^{-6} \qquad (19.38)$$

where a and b are constants equal to 79°K/mb and 4800°K respectively; the respective values of a and b as determined by Smith and Weintraub [12] are 77.6°K/mb and 4810°K and by Campen and Cole [13] are 74.4°K/mb and 4973°K. The parameter T is the air temperature in °K, p is the air pressure in millibars (mb) and e is the partial pressure of water vapor in millibars.

Based on the analysis of Campen and Cole as outlined by Millman [14] the parameters describing the standard atmosphere are expressed as a function of altitude or height h in kilometers for $h \le 10$ km as

$$T = 288 - 6.5h \qquad (19.39)$$

$$p = p_o(1 - 0.02257h)^{5.2561} \qquad (19.40)$$

and

$$e = e_o(1 - 0.064h)^6 \qquad (19.41)$$

where $p_o = 1013$ mb is the standard temperature at sea level and e_o is the partial pressure of water vapor at the Earth's surface in mb. These expressions are essentially independent of frequency. The normalized form, $(n - 1)10^6$, of the refractive index is valid to within 0.5% for frequencies below about 30 MHz.

Upon curve fitting these results using a polynomial in h with units of km, Campen and Cole express the refractive index as a function of h through the troposphere for a completely wet environment with 100% relative humidity as

$$n = 1 + \left(338 - 50.9h + 4.39h^2 - 0.245h^3 + 0.0071h^4 - 0.00006h^5\right)10^{-6} \quad \text{: wet, } h \le 10 \text{ km} \qquad (19.42)$$

and for a completely dry environment with 0% relative humidity at refractive index is expressed as

$$n = 1 + \left(262 - 25.1h + 0.92h^2 - 0.016h^3 + 0.0001h^4\right)10^{-6} \quad \text{: dry, } h \le 10 \text{ km} \qquad (19.43)$$

Equations (19.42) and (19.43) apply for heights up to 10 km and decay exponentially from 10 km to the upper limit of the troposphere at about 30.5 km according to the relationships

$$n = 1 + 338e^{-h/82.02}10^{-6} \quad \text{: wet, } h > 10 \text{ km} \qquad (19.44)$$

and

$$n = 1 + 262e^{-h/82.02}10^{-6} \quad \text{: dry, } h > 10 \text{ km} \qquad (19.45)$$

These results are plotted in Figure 19.12 in terms of the parameter $N = (n - 1)1e^6$ referred to refractivity.

Millman uses these results to evaluate the refractive bending along the propagation path using direct integration and ray tracing. In his analysis the troposphere is divided into spherical 30.48 m layers of constant refractive index up to 3.048 km and 3.048 m layers up to 30.48 km; the refractive index above 30.48 km is unity. The refractive bending of the propagation path gives rise to antenna pointing errors, range delay errors, and Doppler frequency errors. An encounter between a ground terminal and an aircraft is depicted in Figure 19.13. The refraction angle and range errors are shown in Figures 19.14 and 19.15 respectively. These errors increase with increasing height and decreasing antenna grazing or apparent elevation angle. For example, with receiver heights above 1853.2 km and zero degree grazing angle the maximum refraction angular errors are 0.799° and 0.049° for the wet and dry conditions respectively. For the wet and dry conditions, the corresponding maximum range errors are about 381 and 290 ft (116 and 88.4 m).

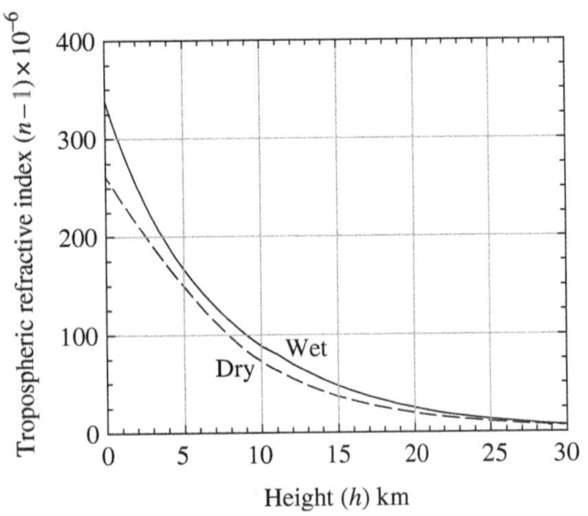

FIGURE 19.12 Refractive index in troposphere using the standard atmosphere.

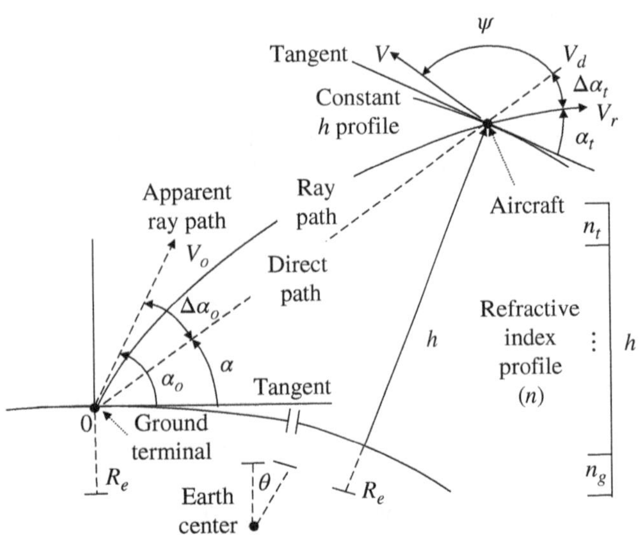

FIGURE 19.13 Tropospheric wave bending encounter. Millman [15]. Reproduced by permission of John Wiley & Sons, Inc.

The Doppler frequency is dependent on the angle ψ between the receiver terminal velocity vector and the direct propagation path to the receiver and is expressed as*

$$f_d = -\frac{Vf_c}{c}\cos(\psi) \tag{19.46}$$

where V is the magnitude of the receiver terminal velocity and f_c is the carrier frequency, and the free-space speed of light $c = 3e^8$ m/s. Referring to Figure 19.13, the velocity error is the difference between apparent path velocity (V_o) and the ray path velocity (V_r) and is evaluated as

*In radar applications, the Doppler frequency is a factor of two higher because of the two-way range measurement.

$$\Delta V = V_o - V_r = V\cos(\psi)\cos(\Delta\alpha_o) - V\cos(\psi + \Delta\alpha_t) \tag{19.47}$$

where $V_o = V_d\cos(\Delta\alpha_o)$, $V_d = V\cos(\psi)$, and $V_r = V\cos(\psi + \Delta\alpha_t)$. After separating terms and applying small argument approximations, Millman arrives at the following approximation to (19.47):

$$\Delta V \cong V\left[\sin(\psi)\Delta\alpha_t + \cos(\psi)\left(\frac{\Delta\alpha_t^2}{2} - \frac{\Delta\alpha_o^2}{2}\right)\right] \tag{19.48}$$

The angle $\Delta\alpha_t$ is determined based on the refractive indices n_g and n_t computed using the appropriate equations (19.42) through (19.45). The indices are used in the spherical earth model of the atmosphere that provides for evaluating $\Delta\alpha_t$ using the geometry of the encounter and Snell's law expressed as the ratio n_g/n_t. Millman ([14], pp. 326–328, 346) characterizes the evaluation of $\Delta\alpha_t$ as

$$\Delta\alpha_t = \cos^{-1}\left(\frac{R_e}{R_e+h}\cos(\alpha_o - \Delta\alpha_o)\right) - \cos^{-1}\left(\frac{n_g R_e}{n_t(R_e+h)}\cos(\alpha_o)\right) \tag{19.49}$$

The angle α_o is measured at the ground terminal and is expressed in terms of the an angle θ as

$$\Delta\alpha_o = \alpha_o - \alpha = \alpha_o - \cos^{-1}\left(\frac{R_e+h}{R}\sin(\theta)\right) \tag{19.50}$$

The angle α_o is the apparent ground antenna elevation angle observed at the ground terminal; however, the angle θ is unknown so Millman's profiles in Figure 19.14 must be used to determine refraction angle error $\Delta\alpha_o$ at the ground terminal.

Based on the velocity error in (19.48), the one-way Doppler frequency error is expressed as

$$\Delta f_d = \frac{f_c\Delta V}{c} \cong \frac{f_c V}{c}\left[\sin(\psi)\Delta\alpha_t + \cos(\psi)\left(\frac{\Delta\alpha_t^2}{2} - \frac{\Delta\alpha_o^2}{2}\right)\right] \tag{19.51}$$

As an example application in the determination of the Doppler frequency and the corresponding Doppler frequency error using (19.46) and (19.48), consider the following encounter conditions: Earth radius is $R_e = 6378.28$ km, $h = 12$ km, $\alpha_o = 5°$, $\psi = \pm 90°$, $f_c = 100$ MHz, $V = 0.3$ km/s, and $c = 3e^8$ m/s is the free-space speed of light. The results are also to be calculated with 100 and 0% humidity corresponding wet and dry conditions. To begin the solutions, the Doppler velocity can be computed using the specified parameters and (19.46). The next step is to use Figure 19.14 and verify that the values of $\Delta\alpha_o$ are: 0.15° and 0.104° respectively for 100 and 0% relative humidity. With these parameters and the specified data, compute the values of n_g and n_t using (19.42) through (19.45) under the appropriate

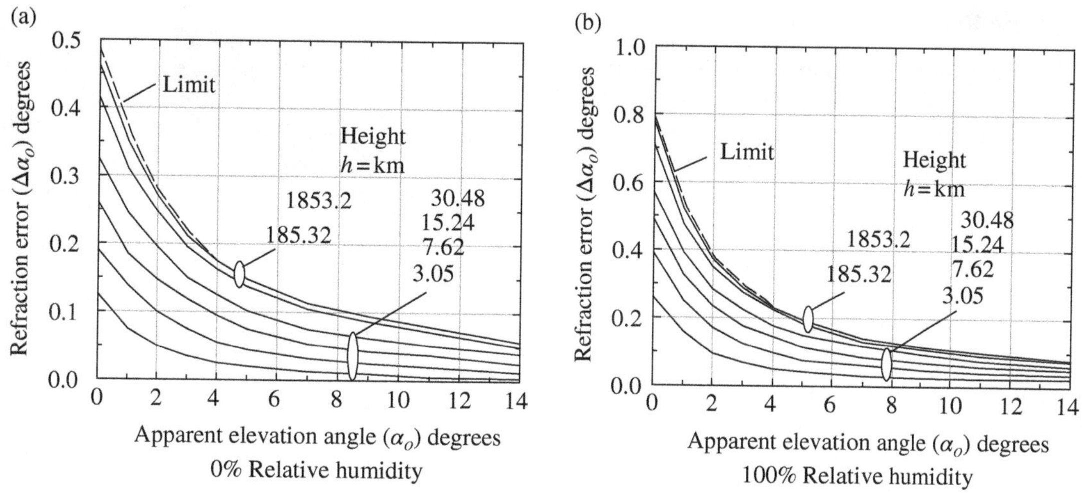

FIGURE 19.14 Tropospheric refraction angle measurement error ($\Delta\alpha$) for standard atmosphere. Millman [16]. Reproduced by permission of John Wiley & Sons, Inc.

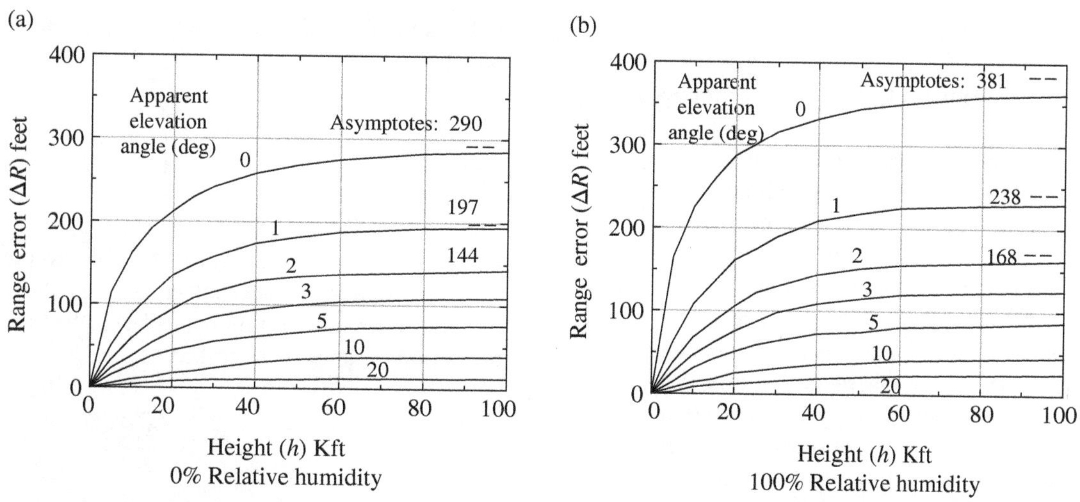

FIGURE 19.15 Tropospheric refraction one-way range measurement error (ΔR) for standard atmosphere. Millman [15]. Reproduced by permission of John Wiley & Sons, Inc.

humidity and height conditions. The next step involves evaluating $\Delta\alpha_t$ using (19.49) from which the velocity error is determined using (19.48). The final evaluation involves the determination of the one-way Doppler frequency error (Δf_d) using (19.51). The maximum Doppler error occurs when $\psi = 90°$ and, based on this example, the wet and dry conditions correspond to

$$\Delta f_d(\max) = \begin{cases} 0.170 \quad : \text{Hz} \ \ \alpha_o = 5°, \ h = 12 \text{ km (wet)} \\ 0.122 \quad : \text{Hz} \ \ \alpha_o = 5°, \ h = 12 \text{ km (dry)} \end{cases} \tag{19.52}$$

Because the tropospheric parameters are independent of the carrier frequency, (19.52) can also be formulated as

$$\Delta f_d(\max) = \begin{cases} 0.00170 f_c(\text{MHz}) : \text{Hz} \ \ \alpha_o = 5°, \ h = 12 \text{ km (wet)} \\ 0.00122 f_c(\text{MHz}) : \text{Hz} \ \ \alpha_o = 5°, \ h = 12 \text{ km (dry)} \end{cases} \tag{19.53}$$

This example uses a relatively slow airplane and with higher speeds, lower elevation angles, and higher operating frequencies increased Doppler frequency errors will be encountered. The two-way Doppler frequency error is of interest in radar applications that simply requires doubling the one-way results in this example. Additional reading on the subject of refraction is also provided by Skolnik [17] and Barton [18].

19.5.2 Ionospheric Refraction

Electromagnetic propagation through the ionosphere is influenced by two key parameters: the electron density, n_e, and the *rms* fluctuation in the electron concentration defined in terms of the parameter σ_e. Therefore, before the performance of a communication system can be evaluated, it is necessary to quantify these parameters throughout the ionosphere. The characterization of the ionosphere and its impact on communication link performance is discussed in Chapter 20. However, by way of introduction, the electron density profiles in the natural environment are characterized as a function of height using Chapman's analytic model. In a nuclear-disturbed environment, the electron densities are determined from experimental measurements and computer simulations. Ray bending in the ionosphere is not explicitly addressed in Chapter 20; however, an analysis is provided by Millman [14] and follows that given for troposphere in the preceding section.

19.6 DIFFRACTION

Diffraction of an electromagnetic signal is the process of constructive and destructive interference of signals or wavelets radiated, for example, from an antenna or an interfering object. By controlling the phase and amplitude of the radiation, as in the aperture design of an antenna, the wave interference results in a prescribed antenna pattern. The antenna pattern is the result of the combination of numerous wavelets and is described as a plane wave at a distance $d \geq D^2/\lambda$, were D is the size of the aperture and λ is the wavelength of the carrier frequency. However, diffraction resulting from the scattering of electromagnetic energy by interfering objects is not controlled and results in a distortion and bending of the electric field. Mountains, hills, and buildings are examples of obstructions that result in shadows and *blind spots* in communication links. Fortunately, however, diffractive ray bending provides some measure of communication visibility in the shadow regions as analyzed in the following section where the obstruction is characterized as a knife-edge.

19.6.1 Knife-Edge Diffraction

When an incident wave is obstructed by a knife-edge conductor, shown in Figure 19.16, secondary wavelets result in signal diffraction in the shadow region behind the knife-edge. The transmitted ray that just intersects the top of the knife-edge and all lower angle rays are blocked from the receiver. However, the higher angle rays result in secondary wavelets, some of which are directed behind the knife-edge providing coverage in the shadow of the obstruction. The direct path of length $d_1 + d_2$ is obstructed; however, for path lengths greater than the refraction path $D_1 + D_2$ an electric field is produced in the shadow region of the obstruction. The height h, shown as the *heavy solid* line in Figure 19.16, is the effective height of the knife-edge between the direct and minimum refraction path; the total height of the obstruction is defined as $h_o = h + h'$.

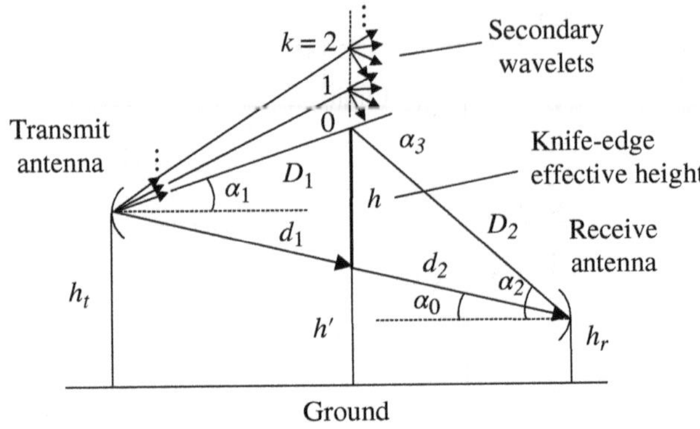

FIGURE 19.16 Knife-edge diffraction.

The electric field strength at a point in the shadow region is the sum of all the secondary wavelets and is computed as [19]

$$\frac{E(v)}{E_o} = \frac{(1+j)}{2} \int_v^\infty e^{-j\pi t^2/2} dt \tag{19.54}$$

where E_o is the electric field that would be produced at the receiver over a free-space path equal to the direct path length. The lower integration limit in (19.54) is the Fresnel–Kirchoff diffraction parameter expressed as

$$v = 2\sqrt{\frac{\Delta_k}{\lambda}} \tag{19.55}$$

where λ is the carrier frequency wavelength and Δ_k is the differences between the direct path and the refraction path and is evaluated as

$$\begin{aligned} \Delta_k &= D_1 + D_2 - (d_1 + d_2) \\ &\cong \frac{h^2(d_1 + d_2)}{2(d_1 d_2)} \end{aligned} \tag{19.56}$$

The approximation in (19.56) is based on the conditions: $\lambda \ll h \ll d_1$ and d_2.

Evaluation of the integral in (19.54) is performed in Section 20.6.2 involving the Fresnel integrals [20] $C(z)$ and $S(z)$ with the result

$$\frac{E(v)}{E_o} = \frac{1}{2}\left\{1 - C\left(\frac{\pi v}{2}\right) - S\left(\frac{\pi v}{2}\right)\right\} - j\frac{1}{2}\left\{C\left(\frac{\pi v}{2}\right) - S\left(\frac{\pi v}{2}\right)\right\} \tag{19.57}$$

The magnitude of (19.57), expressed in decibels, is plotted in Figure 19.17 as a function of the diffraction parameter v. When $v \leq 0$ there is a direct path of length $d_1 + d_2$ between the transmitter and receiver and the electric field variations can be thought of as resulting from knife-edge reflections. However, when $v = 0$ the peak of the knife-edge just intersects the direct path and the received electric field strength is reduced by 6 dB. For $v > 0$, the signal reception is due to diffraction into the shadow region.

The function plotted in Figure 19.17 is called the Fresnel diffraction pattern and the peaks and troughs of the ripples result from successive constructive and destructive interference from the received secondary wavelets shown in Figure 19.16 and identified for $k > 0$. These peaks and troughs occur because of path differences given by $\Delta_k = k\lambda/2$ for k = even and odd respectively. The plane perpendicular to Figure 19.16, that is, through the page and containing the knife-edge, contains unique secondary wavelets that intersect the plane forming annular rings with radius h_k that identify Fresnel zones. The Longley-Rice propagation model, discussed in Section 19.7, makes extensive use of reflection and knife-edge diffraction in the evaluation of signal loss over various terrain conditions.

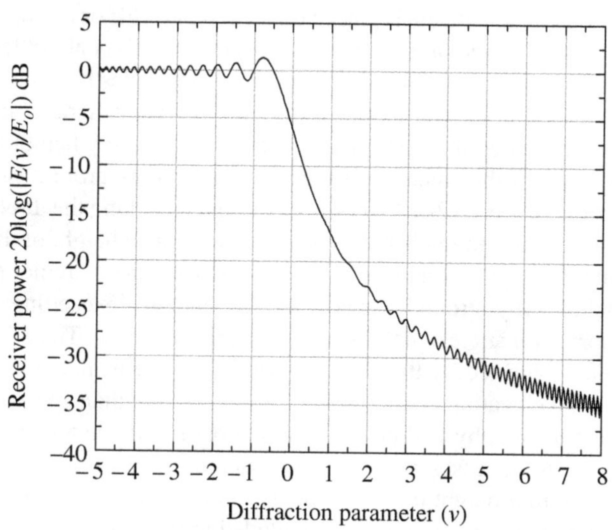

FIGURE 19.17 Knife-edge reflected and refracted electric field strength.

19.7 LONGLEY-RICE PROPAGATION LOSS MODEL

The propagation loss resulting from transmissions through the troposphere and over rough terrain has been modeled by Longley and Rice [21] and the irregular terrain model (ITM) simulation program is used to evaluate the losses under a variety of user specified conditions. Two distinct modes of the program are available: the point-to-point mode and the area prediction mode. The point-to-point mode allows for user-specified two-dimensional terrain variations between the transmitter and receiver sites; whereas, the area prediction mode allows for three-dimensional topographical maps to be used with the transmitter and receiver sites identified on the selected map. These modes are selected by the user along with other application-specific inputs. The program first conditions the input parameters for the user-specified environment, then computes the propagation losses between the transmitter and receiver locations. The computer code also provides for user-specified reliability and confidence specifications that are applied to the resulting loss computations; these statistical measures are based on the assumed normal distribution of the losses resulting from the central limit theorem involving various random processes used in the computations. The point-to-point mode corresponds to a fixed communication link and the reliability represents the time availability. The area prediction mode allows for four types of communication serves: single-message, individual, mobile, and broadcast. In these cases, the reliabilities correspond to time and location availability.

A typical point-to-point mode scenario is shown in Figure 19.18 where the terrain is specified by an array of height-range parameters (h_i, r_i) between the transmitter and receiver sites.

The terrain profile is entered using equal range increments $\Delta r = r_{i+1} - r_i$: $i = 1, \ldots, n_p$ with (h_o, r_o) corresponding to the transmit site and (h_{np}, r_{np}) corresponding to the receive site; the maximum range is then $r(\max) = n_p \Delta r$. In addition to the height-range parameters the user-specified parameters listed in Table 19.2 are also required. Table 19.3 provides numerical values for the dielectric constants and conductivities at the Earth's surface under various conditions.

The *physical height* of the antenna is simply the length of the antenna structure and the center-of-radiation is the

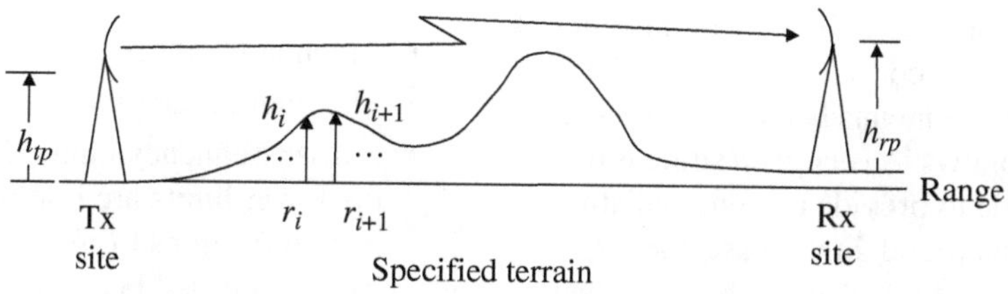

FIGURE 19.18 ITM simulation program point-to-point mode scenario.

TABLE 19.2 User Input Parameters for the Point-to-Point Mode[a]

Parameter	Description
Tx and Rx antenna heights	h_{tp} and h_{rp}
Effective antenna heights	h_{te} and h_{re}
Terrain elevation change (Tx to Rx)	h_i (m) [10] 0-Flat terrain or water 30-Plains 90-Hills (average) 200-Mountains 500-Rugged mountains
Range between Tx and Rx	r(max)
Carrier frequency	f_c (MHz)
Average elevation above sea level[b]	*zsys* (m)
Mean surface refractivity at sea level[c]	*Eno*
Antenna polarization	0-Horizontal, 1-vertical
Surface dielectric constant	ε_s (F/m) (see Table 19.3)
Surface conductivity	σ_s (mho/m) (see Table 19.3)
Reliability	0.1–99.9%
Confidence	0.1–99.9%
Effective curvature of Earth	k^d
Climate Codes	**Surface Refractivity (N_s) [10]**
1-Equatorial	360
2-Continental subtropic	320
3-Maritime subtropic	370
4-Desert	280
5-Contental temperate	301 average[d]
6-Maritime temperate (overland)	320
7-Maritime temperate (oversea)	350

[a]For a complete list of parameters with various precautions, including those for the area prediction mode, refer to user manual. More details and ordering information for the Longley-Rice ITM program are available online at: elbert.its.bldrdoc.gov and softwright.com.

[b]*zsys* is only used to adjust *eno* to determine the surface refractivity of the atmosphere (*ens*).

[c]When *zsys* = 0 then *ens* is determined directly as *ens* = *eno*.

[d]The surface refractivity impacts the amount of ray bending and the effective Earth curvature. The average value $N_s = 301$ results in the effective Earth radius $r_e = 4/3R_e$ corresponding to $k = 4/3$ (see Section 19.1).

TABLE 19.3 Electrical Constants for Earth Surface [10]

Condition	Surface or Ordinary Dielectric Constant (ε_s,ε_n F/m)	Conductivity (σ_s,σ mho/m)
Polar ice	1	0.0001
Poor ground (dry) and sea ice	4	0.001
Average ground	15	0.005
Good ground (wet)	25	0.020
Fresh water	81	0.010
Sea water	81	5.000

radiation height above the base of the antenna. The *effective antenna height* is defined as the height of the antenna center-of-radiation above the effective radiating plane that is determined by the intermediate foreground in the radial direction of the antenna. For the most part, the loss computations are based on the effective antenna height, so establishing the effective height is an important consideration. Guidelines [10] for establishing the effective antenna height are based on the antenna siting criteria as follows. A *very good* siting is one in which every attempt is made to provide a strong radiating signal as, for example, on high ground. In this case, the effective height should not exceed the physical height by more than 10 m. A *good* siting is one in which an attempt is made to locate the antenna on elevated sites, but not on hilltops or locations that would result in the strongest radiating signal. In this case, the effective height should not exceed the physical height by more than 5 m. The greatest benefit with *very good* and *good* sitings is that low physical antenna heights of less than about 10 m can be used. A *random* siting is one in which the siting is selected on factors other than signal strength, with the stipulation that there is no deliberate concealment, and there is an equal chance of having *good* or *poor* performance. In this case, this effective height should be equal to the physical height. When the antennas are concealed, a special loss factor based on site measurements must be included.

When the antenna is placed on the top of a cliff or mesa, or high on the side of a mountain, the question arises about the correct physical height of the antenna. Three rules have been suggested in answering this question [10]: The federal communication commission (FCC) uses a rule that any height at the base of the antenna greater than the average ground elevation measured 2–10 miles from the antenna along the direction of propagation is to be included as physical height of the antenna. The second rule is that ground elevations are not to include any ground viewed by the antenna exceeding a depression angle of 45° from the antenna boresight. The third rule, suggested by the authors [10], is that the evaluation of ground elevations should start at a point greater than 15 times the antenna physical height from the base of the antenna. In these special cases, the implication of the last two rules is that the ground height under the physical antenna exceeding the average ground level is to be included in the physical height of the antenna.

Hufford, Longley, and Kissick [10] suggest the following limitations on the application of the ITM model.

- Antenna height limits: 0.5 m to 3 km.
- Maximum range limits: 1–2000 km.
- Carrier frequency limits: 20–100 MHz $\leq f_c \leq$ 2–20 GHz; the lower limits are imposed because of possible dominate reflections from the ionosphere and the upper limits are imposed because of losses resulting from water vapor absorption.

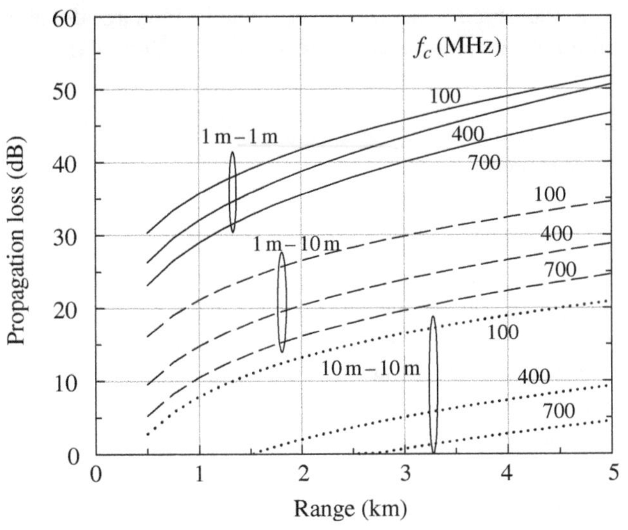

FIGURE 19.19 Longley-rice propagation loss (h_{tp} m – h_{rp} m with $h_e = h_p$).

TABLE 19.4 Example ITM Program Parameters Used with Figure 19.19 (Point-to-Point Mode)

Parameter	Description
Tx and Rx antenna physical heights	1 and 1 m
	1 and 10 m
	10 and 10 m
Effective antenna effective heights	$h_e = h_p$
Terrain elevation change (Tx to Rx)	0—Flat terrain
Range between Tx and Rx	0.5–5 km
Carrier frequency	100, 400, 700 MHz
Average elevation above sea level	0
Mean surface refractivity at sea level (N_s)	301
Antenna polarization	1-Vertical
Surface dielectric constant (ε_s)	15 F/m
Surface conductivity (σ_s)	0.005 mho/m
Reliability	90%
Confidence	50%
Effective curvature of Earth	4/3
Climate codes	Surface refractivity (N_s)
5-Contental temperate	301

- The model can be used for evaluating the loss from a ground site to an aircraft for aircraft altitudes ≤1 km (3290 ft).

Application of the model is not appropriate for examining the propagation losses for:

- LOS microwave links that are usually sited on mountain tops where there is little interaction with the terrain.
- Propagation paths involving the Ionosphere.
- Propagation paths with severe ducting.
- Urban and heavily forested environments; models for these environments are discussed in Section 19.8.

Figure 19.19 shows the propagation loss using a flat terrain, that is, with $h_i = 13$ m: $0 \le i \le 156$ and for three scenarios involving the transmitter and receiver antenna heights. In these cases, the effective antenna height is equal to the physical height. The simulation parameters are summarized in Table 19.4. The propagation losses in Figure 19.19 do not include the free space loss; the propagation loss is often plotted as a percentage of the free space loss (see, e.g., Problem *9*). The free space loss is included in the ITM model.

19.8 URBAN, SUBURBAN, AND RURAL ENVIRONMENT PROPAGATION LOSS MODELS

In this section, the propagation losses in urban, suburban, and rural areas are discussed focusing on applications involving mobile communications. There is a wealth of information in the literature on this subject and, although only a few propagation models are discussed in the following sections, the models and results of other researchers can be found in additional references at the end of this chapter. Rappaport [22] provides an in-depth discussion on the theoretical aspects of mobile radio propagation including various multipath fading models. In the following descriptions, the free space loss is computed as

$$L_{fs}(\text{dB}) = 20\log_{10}\left(\frac{4\pi d}{\lambda}\right) \qquad (19.58)$$

where d is the distance between the transmitter and receiver antennas and λ is the carrier frequency wavelength measured in the same units as d. Typically, the models are based on continuous wave (CW) data collected using omnidirectional antennas.

Examples of the losses for each of the following models and loss comparisons under similar condition are included in the problems at the end of this chapter.

19.8.1 Okumura Model for Urban Environments

The Okumura model [23] is based on curves that are fit to measured signal losses taken in an urban environment over frequency (f_c) and distance (d) ranges: $150\text{ MHz} \le f_c \le 1920\text{ MHz}$, $1\text{ km} \le d \le 100\text{ km}$ respectively and apply for transmitter effective antenna heights: $30\text{ m} \le h_{te} \le 1000\text{ m}$ and effective receiver antenna heights: $h_{re} \le 30$ m. The resulting median signal loss is expressed as

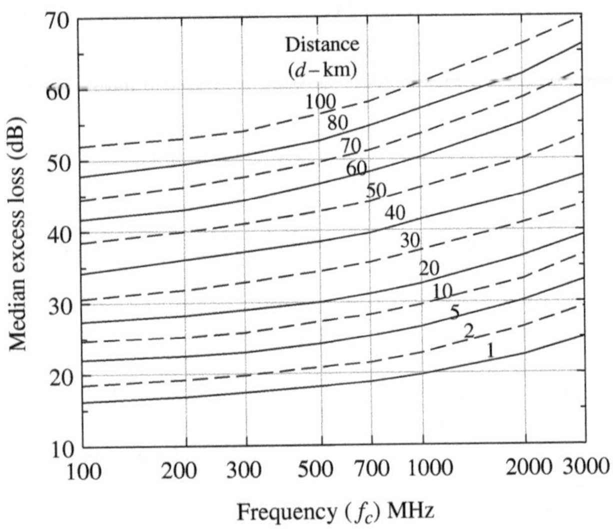

FIGURE 19.20 Median excess propagation loss. Okumura et al. [23]. Reproduced by permission of IEEE.

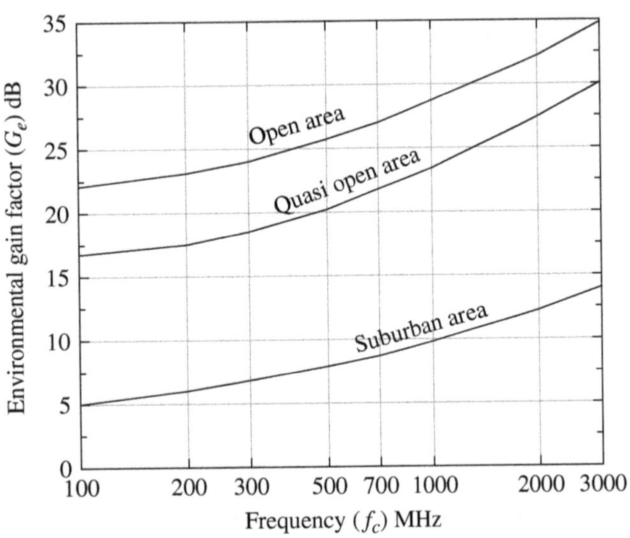

FIGURE 19.21 Environmental gain factor G_e. Okumura et al. [23]. Reproduced by permission of IEEE.

$$L(\text{dB}) = \begin{cases} L_{fs} + A_{mu}(f_c, d) - 20\log_{10}\left(\dfrac{h_{te}}{200}\right) - 10\log_{10}\left(\dfrac{h_{re}}{3}\right) \\ -G_e : h_{re} \leq 3\ \text{m} \\ L_{fs} + A_{mu}(f_c, d) - 20\log_{10}\left(\dfrac{h_{te}}{200}\right) - 20\log_{10}\left(\dfrac{h_{re}}{3}\right) \\ -G_e : 3\ \text{m} < h_{re} \leq 10\ \text{m} \end{cases} \tag{19.59}$$

where L_{fs} is the free-space signal loss, $A_{mu}(f_c, d)$ is the median excess propagation loss determined from Figure 19.20, and G_e is a frequency-dependent environmental gain factor determined from Figure 19.21. To attach a level of confidence to these measured values, the estimated standard deviation about the median losses varies between 10 and 14 dB.

19.8.2 Hata Model for Urban and Suburban Environments

The Hata model [24] is derived from measured path loss data and applies over the frequency range: 150 MHz $\leq f_c \leq$ 1500 MHz and for transmitter effective antenna heights: 30 m $\leq h_{te} \leq$ 200 m and effective receiver antenna heights: 1 m $\leq h_{re} \leq$ 30 m. There is no specified limit on the distance d; however, for reasonable agreement with the Okumura model requires that $d > 1$ km. The median loss for the urban environment is expresses as

$$\begin{aligned} L_u(\text{dB}) = 60.55 &+ 26.16\log_{10}(f_c) \\ &+ (44.9 - 6.55\log_{10}(h_{te}))\log_{10}(d) \\ &- 13.82\log_{10}(h_{te}) - a(h_{re}) \end{aligned} \tag{19.60}$$

The correction factor $a(h_{re})$ is a function of the cell area and for small to medium cities is given by

$$a(h_{re}) = (1.1\log_{10}(f_c) - 0.7)h_{re} - (1.56\log_{10}(f_c) - 0.8) \quad \text{: small-medium cities} \tag{19.61}$$

and for large cities

$$a(h_{re}) = \begin{cases} 8.29(\log_{10}(1.54h_{re}))^2 - 1.1 & : f_c \leq 300\ \text{MHz}, \\ 3.2(\log_{10}(11.75h_{re}))^2 - 4.97 & : f_c > 300\ \text{MHz} \end{cases} \quad \text{: large cities} \tag{19.62}$$

For suburban and open rural areas, the correction factors are

$$L(\text{dB}) = L_u(\text{dB}) - 2\left(1.1\log_{10}\left(\frac{f_c}{28}\right)\right)^2 - 5.4 \quad \text{: suburban} \tag{19.63}$$

and

$$L(\text{dB}) = L_u(\text{dB}) - 4.78(\log_{10}(f_c))^2 + 18.33\log_{10}(f_c) - 40.94 \quad \text{: open rural} \tag{19.64}$$

19.8.3 Erceg Model for Suburban and Rural Environments

The Erceg model [25] characterizes the loss for wireless mobile communication in suburban and rural areas. The results are based on curve-fit plots of loss vs. distance derived from 1.9 GHz experimental data collected in 95 macro cells across the United States. In this case, the data was collected using omnidirectional azimuth antennas with transmit and receive gains of 8.14 dB and 2.5 dB respectively. The receiver antenna height was fixed at $h_r = 2$ m.

This model is directed toward applications, like personal communication services (PCS) that involve smaller cells, lower transmit antenna heights and higher frequencies. The path loss applies for transmitter effective antenna heights: $10\text{ m} \le h_{te} \le 80\text{ m}$ and distances in the range: $0.1\text{ m} \le d \le 8$ km. The three environments include various terrain hill conditions and tree densities. The Erceg model loss is expressed as

$$L(\text{dB}) = 20\log_{10}\left(\frac{4\pi d_o}{\lambda}\right) + 10\gamma\log_{10}\left(\frac{d}{d_o}\right) + s \quad : d \ge d_o \tag{19.65}$$

where $d_o = 100$ m is the minimum *close-in* distance. The exponent γ is a Gaussian random variable expressed as*

$$\gamma = (a - bh_t + c/h_t) + x\sigma_\gamma \quad : 10\text{ m} \le h_t \le 80\text{ m} \tag{19.66}$$

where, the term in brackets, and σ_γ are the mean and standard deviation of γ respectively. The parameter x is a normalized Gaussian random variable characterized as $N(0,1)$.† The s term in (19.65) is a zero-mean Gaussian random shadow fading term characterized as $N(0,\sigma)$ and expressed as

$$s = y\sigma = y(\mu_\sigma + z\sigma_\sigma) \tag{19.67}$$

where $y = N(0,1)$, $z = N(0,1)$, and σ is a Gaussian random variable characterized by $N(\mu_\sigma,\sigma_\sigma)$. The variables x, y, and z are independent random variables. The constants a, b, c, σ_γ, μ_σ, and σ_σ are listed in Table 19.5 for each terrain category.

19.9 LAND MOBILE SATELLITE PROPAGATION LOSS MODELS

Land mobile communications through a satellite must consider the loss from the terrain surrounding the mobile location. For example, in suburban and rural areas, when traveling along roads or walking in forested regions, foliage attenuation from trees and vegetation may result in significant signal losses. Link margins of 20–25 dB are recommended at ultra-high frequency (UHF) for satellite viewing at elevation angles on the order of 20° or less.

*In this model the antenna heights are not characterized as effective heights so the physical height is assumed.

†The generation of Gaussian or Normal random variables characterized by $N(m,\sigma)$, with mean m and standard deviation σ, is discussed in Chapter 14.

TABLE 19.5 Numerical Values for Model Parameters[a]

	Terrain Category[b]		
Parameter	A	B	C
a	4.6	4.0	3.6
b (m^{-1})	0.0075	0.0065	0.005
c (m)	12.6	17.1	20.0
σ_γ	0.57	0.75	0.59
μ_σ	10.6	9.6	8.2
σ_σ	2.3	3.0	1.6

[a]Erceg et al. [25]. Reproduced by permission of IEEE.
[b]A, hilly with moderate to heavy tree density; B, hilly with light tree density or flat with moderate to heavy tree density; C, flat with light tree density.

Several link loss models for the suburban and rural areas are examined in the following sections. The underlying models are based on the modified exponential decay (MED) model introduced by Weissberger [26] with variations based on the recommendations of the Consultative Committee on International Radio‡ (CCIR) [27]. Barts and Stutzman [28] have also proposed modification to the MED model. The CCIR link margin model [29] for urban, suburban, and rural areas is also given in Section 19.9.4. Moraitis, Milas, and Constantinou [30] compare these and other land mobile satellite channel models.

In the following loss models the distance D_n is the propagation path length through the foliage, measured in meters, with the restriction $D_n \le 400$ m. Furthermore, the frequency f is the carrier frequency in megahertz in the range 200–95,000 MHz. The loss for each of the models is evaluated at 400, 1,200, and 16,000 MHz corresponding to the UHF, L, and K_u bands. The link margin evaluation in Section 19.9.4 is characterized in term of the elevation angle θ from the mobile site to the satellite. In this case, the frequency is denoted in gigahertz and the elevation angle in degrees. The models are based on *in situ* measured data with regression curve fitting applied to evaluate various parameter coefficients.

19.9.1 Modified Exponential Decay Model Link Loss

The MED model applies to suburban and rural environments consisting primarily of trees and vegetation. The total loss through the foliage is evaluated as

$$L_n = a_n D_n \quad : \text{dB} \tag{19.68}$$

‡The CCIR is currently referred to as the International Telecommunication Union-Radio (ITU-R).

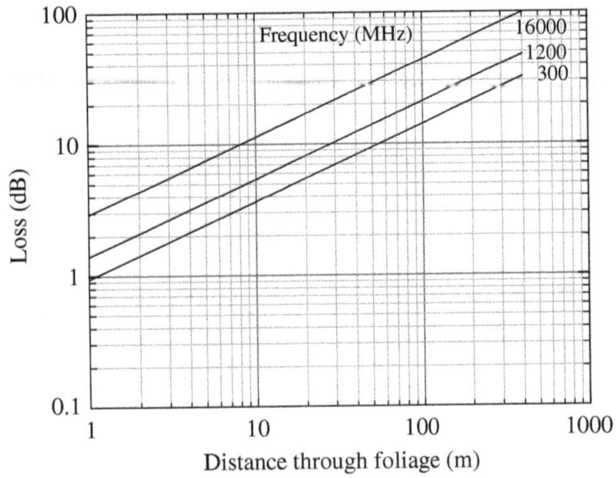

FIGURE 19.22 MED model link loss.

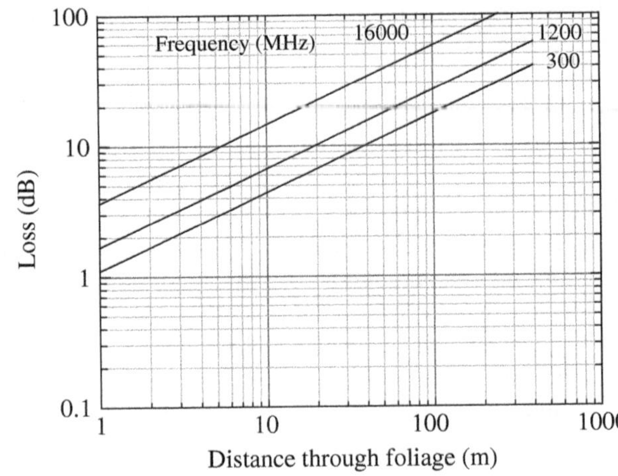

FIGURE 19.23 CCIR model link loss.

where a_n is the specific attenuation expressed as

$$a_n = 0.187 f^{0.284} D_n^{-0.412} \quad : \text{dB/m} \qquad (19.69)$$

The losses are shown in Figure 19.22 for the indicated frequencies.

19.9.2 CCIR Link Loss Model

The CCIR model is a modification of the MED model with the total loss given by

$$L_n = a_n D_n \quad : \text{dB} \qquad (19.70)$$

and the specific attenuation expressed as

$$a_n = 0.187 f^{0.3} D_n^{-0.4} \quad : \text{dB/m} \qquad (19.71)$$

The loss using the CCIR model is shown in Figure 19.23 and is somewhat higher than that predicted by the MED model.

19.9.3 Barts and Stutzman Link Loss Model

The Barts and Stutzman model is also a modification of the MED model for distances ≤14 m; otherwise, the loss predictions are identical to the MED model. The total loss through the foliage is evaluated as

$$L_n = a_n D_n \quad : \text{dB} \qquad (19.72)$$

and the specific attenuations for the different ranges are expressed as

$$a_n = 0.063 f^{0.284} \quad : \text{dB/m for } D_n \leq 14 \text{ m} \qquad (19.73)$$

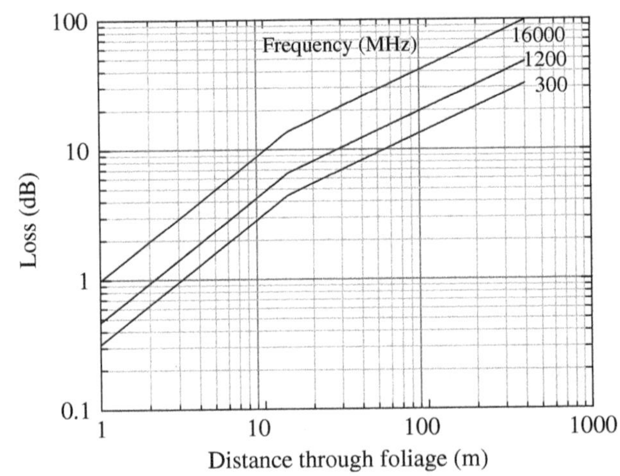

FIGURE 19.24 Barts and Stutzman model link loss.

and

$$a_n = 0.187 f^{0.284} D_n^{-0.412} \quad : \text{dB/m for } 14 < D_n \leq 400 \text{ m} \qquad (19.74)$$

The loss for the Barts and Stutzman model is shown in Figure 19.24.

19.9.4 CCIR Link Margin Model

The CCIR link margin model [29] data were collected in urban, semi-urban, suburban, and rural areas at 860 MHz and 1.55 GHz for elevation angles [31] ranging from 19° to 43°. In the following relationships, the angles are entered as degrees and the frequency as gigahertz. The link margin applies for a percentage availability of $P_a = 90\%$, that is, the received signal power exceeds the detection threshold 90% of the time; however, other percentages of availability can be evaluated by subtracting the loss $L = 0.1(90 - P_a)$

dB from the derived link margins.* The range of P_a is 50–90%. Furthermore, a factor K is included that relates the percentage of the surrounding locations or area (*area* %) for which the received power is expected to exceed the detection threshold; the values of K for *area*% are tabulated in Table 19.6.

The link margin, in dB, for the urban, suburban, and rural models are expressed as

$$M = 17.8 + 1.93f - 0.052\theta + K(7.6 + 0.053f + 0.040\theta) - L \quad \text{: urban} \qquad (19.75)$$

and

$$M = 12.5 + 0.17f - 0.17\theta + K(6.4 - 1.19f + 0.05\theta) - L \quad \text{: suburban/rural} \qquad (19.76)$$

Equations (19.75) and (19.76) are plotted in Figure 19.25 as a function of θ for various carrier frequencies in gigahertz with the indicated conditions of: link availability, percent of area, and $L = 0$ dB. With $P_a = 90\%$, $area\% = 50\%$, and $K = 0$, the link is available 90% of the time and in 50% of the surrounding area the received power is expected to exceed the detection threshold. If the availability is decreased to 50% then $L = -4$ dB and the link margin can be decreased by 4 dB. On the other hand, if the detection threshold is to be exceeded in over 99% of the surrounding area then $K = 2.35$ and the link margin must be increased by the additive term involving K in (19.75) and (19.76). Note that the range of the elevation angle in Figure 19.25 exceeds the stated range of the model by about 5° on each end of the abscissa. The frequency translations method of Goldhirsh and Vogel [32] is recommended to evaluate the link margin requirements at other frequencies.

TABLE 19.6 Values of the Factor *K* Given *area*%

area%	*K*
50	0
90	1.3
95	1.65
99	2.35

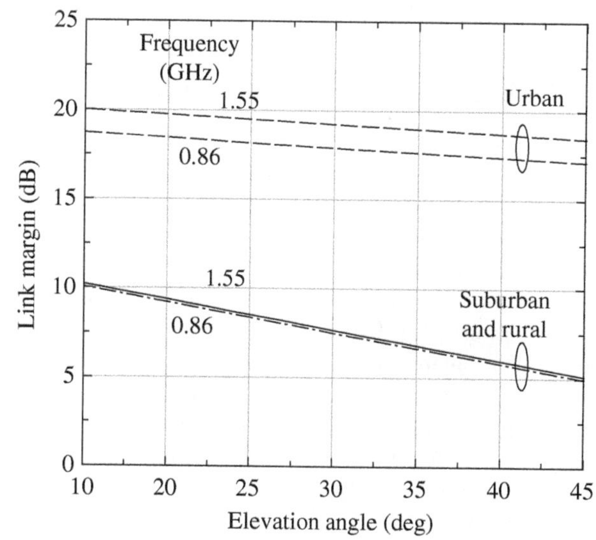

FIGURE 19.25 CCIR link margins for urban, suburban and rural areas ($P_a = 90\%$, $area\% = 50$).

19.10 IMPULSIVE NOISE CHANNEL

19.10.1 Introduction

Impulsive noise occurs from thunder storm activity around the world and is present in most regions as the energy from lightning flashes or strikes propagates through the natural wave guide between the earth surface and the ionosphere. The severity of the storm activity varies with geographic regions and seasons; however, the worldwide average rate of lightning strikes is on the order of hundreds-per-second. In regions near active storm centers the noise spikes are most pronounced, characterized as short high-energy pulses that disrupt communications. As the impulsive energy propagates farther from the storm center the wide bandwidth pulses undergo attenuation and dispersion and combine with similarly filtered impulses from storm centers in other regions of the globe. The global effects of lightning strikes resulting from storm activity are most evident at the lower frequencies, typically in the low frequency (LF) region and below. These effects become less troublesome at frequencies in the high frequency (HF); however, the HF region has unique issues [33] to contend with including time-varying multipath, ducting, and Faraday rotation.

The impulsivity measure V_d is introduced in Chapter 14 as the parameter that characterizes the severity of the impulsive noise and is defined as the ratio of the *rms* noise envelope to the average noise envelope. The V_d measure is expressed in decibels with the minimum value of 1.049 dB corresponding to minimum storm activity; larger values indicate increased storm activity. As the storm energy from around the globe is combined, the noise addition is subject to the central limit theorem and the impulsive noise approaches white Gaussian noise with the corresponding $V_d = 1.049$ dB. This condition is observed during periods of relatively calm worldwide storm activity and will change suddenly as a result of a distance

*The loss L is included in the link margin equation.

storm. The impulsivity is also characterized by the amplitude probability distribution (APD), defined as the probability that the noise envelope exceeds the abscissa [34]. The worldwide characterization of impulsive noise due to storm activity is published by the International Telecommunication Union (ITU) through the CCIR Report 322 [35] and associated reports [36]. These reports characterize the APD based on impulsive noise measurements corresponding to the V_d measure. The impulsive noise from lightning strikes is characterized as a nonstationary random process and the APD data are based on the ensemble average of recorded time sequences.

19.10.2 Lognormal Impulse Noise Model

To the casual observer, a lightning strike appears as a single flash of light; however, in many events each flash is actually composed of multiple strokes separated typically by 50–100 ms. The multiple strokes following the initial lightning strike are referred to as return strokes. The number and interval between the return strokes is modeled statistically based on observations [37]. Uman and Krider [38] have summarized the phenomenon of lightning strikes and, based on the studies of Mackerras [39], conclude that the number of return strokes is typically distributed between 2 and 8 resulting in a mean value of 5 return strokes for each lightning flash. Figure 19.26 shows Mackerras' results in terms of the probability that the number of return stokes exceeds the abscissa; the *dashed* curve represents the piece-wise linear approximation to the data expressed in (19.77) and is used in the computer simulations. Beach and George [40] and Uman [41] report on the time between return strokes based on the data collected by Schonland in South Africa. Figure 19.27 characterizes Schonland's data in terms of the probability that the time interval between return strokes exceeds the abscissa; the *dashed curve* corresponds to a piece-wide linear approximation in (19.78) used for computer simulations. Beach and George observed that the time between stokes corresponding to Schonland's data can be approximated using the Gamma *pdf* with $\alpha = 2$ and mean value $\alpha/\beta = 55$. This approximation is also applied to the data of Kitagawa, Brook, and Workman [42] for cloud-to-ground lightning using a mean value of $\alpha/\beta = 35$. Although these Gamma function *pdf* approximations are good fits to the data over regions about the mean values, over regions several standard deviations removed from the mean they are not as accurate, so the piece-wise linear approximations are used in the computer simulations.

The piece-wise linear approximations for the simulated probabilities are

$$P_r(\text{strokes } j > n) \cong \begin{cases} -0.1354n + 1.1354 & : 1 \le n \le 7.5 \\ -0.0184n + 0.2579 & : 7.5 < n \le 14 \\ 0 & : o.w. \end{cases} \tag{19.77}$$

and

$$P_r(t_j > t_n) \cong \begin{cases} -0.01292t_n + 1.1486 & : 11.5 \le t_n \le 68 \\ -0.004524t_n + 0.5776 & : 68 < t_n \le 110 \\ -0.001379t_n + 0.2317 & : 110 < t_n \le 147 \\ -0.000429t_n + 0.0916 & : 147 < t_n \le 200 \\ -0.000105t_n + 0.0267 & : 200 < t_n \le 254 \\ 0 & : o.w. \end{cases} \tag{19.78}$$

where j and n are integers and the time is in milliseconds.

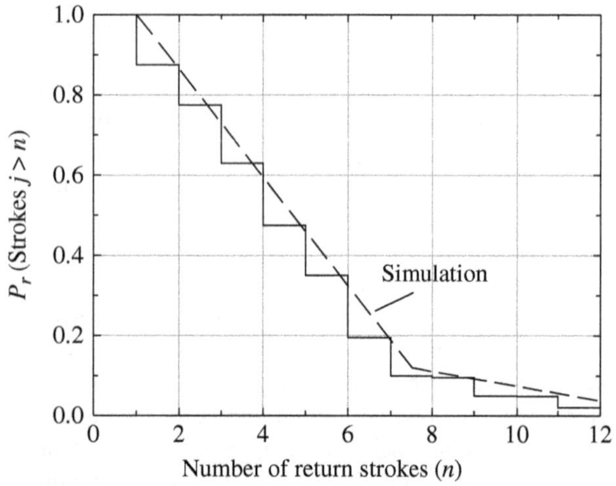

FIGURE 19.26 Number of return strokes. Mackerras [39]. Courtesy of the American Geophysical Union (AGU).

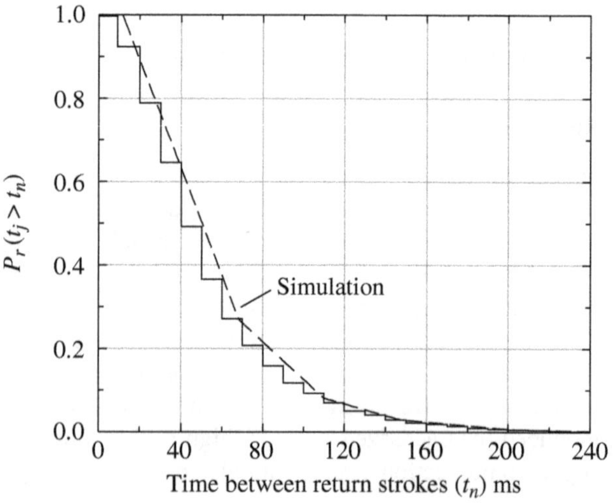

FIGURE 19.27 Time interval between return strokes.

The impulse noise for lightning strikes is modeled as shot noise [43] and expressed as the summation

$$n_p(t) = \sum_i N_p(t)\delta(t - t_i) \tag{19.79}$$

where $N_p(t)$ represents the main lightning stroke with the associated return strokes, $N'_p(t)$, and is expressed as

$$N_p(t) = A_p(t)(\cos(\varphi(t)) + j\sin(\varphi(t))) + \sum_j N'_p(t)\delta(t - t_j) \tag{19.80}$$

with

$$N'_p(t) = A'_p(t)(\cos(\varphi'(t)) + j\sin(\varphi'(t))) \tag{19.81}$$

The functions $N_p(t)$ and $N'_p(t)$ represent complex impulse noise processes with lognormal distributed amplitudes $A_p(t)$, $A'_p(t)$ and uniformly distributed phases, $\varphi(t)$, $\varphi'(t)$, respectively.

From the discussions in Section 14.3.6, the lognormal amplitude is $A_p = e^x$, where x is a normally distributed random variable with mean m_0, variance σ_0^2, and phase φ uniformly distributed between $-\pi$ and π. A zero mean Gaussian background noise term, $n_g(t)$, with variance σ_1^2 is added to the impulsive noise; the background noise results from quiescent worldwide thunder storm activity. Therefore, the total atmospheric noise at the input to the receiver antenna is described as

$$n(t) = n_g(t) + n_p(t) \tag{19.82}$$

Substituting (19.80) and (19.81) into (19.79) the total atmospheric noise at the receiver antenna input, characterized by (19.82), becomes*

$$n(t) = n_g(t) + \sum_i \{A_{pi}(\cos(\varphi_i) + \boldsymbol{j}\sin(\varphi_i))\delta(t - t_i) + \sum_{j=1}^{n} A'_{pij}\left(\cos\left(\varphi'_{ij}\right) + \boldsymbol{j}\sin\left(\varphi'_{ij}\right)\right)\delta(t - t_i - t_j)\} \tag{19.83}$$

The most commonly observed lightning strike or flashes occur as intra-cloud, cloud-to-cloud, and cloud-to-ground.† To the casual observers the cloud-to-ground lightning is the most spectacular. In cloud to ground flashes the first or main strike is preceded by a *stepped leader* that is followed by *dart leader* that propagates from the cloud to ground. The dart leader is immediately followed by a *return stroke* that propagates from ground to cloud and results in the visible lightning flash. Depending on the remaining charge and the electric field intensity, additional dart leaders followed by return strokes may occur resulting in a *multiple-stroke* flash. In the communication performance simulation program, the parameter λ is input to establish the mean lightning flash-rate. The time intervals $\Delta t_i = t_i - t_{i-1}$ between multiple return strokes are randomly distributed according to (19.78). As discussed in Section 19.10.3, the number of strokes and flashes over a recorded ensemble of atmospheric noise is adjusted to match the APD corresponding to the selected V_d (dB).

Two points are noteworthy regarding the noise description in (19.83). The implied bandwidth of the noise impulses is infinite and in areas of intense storm activity the parameter λ may be sufficiently high so that return strokes from several main strokes overlap. The bandwidth issue is handled by passing the impulse noise through the receiver intermediate frequency (IF) filter with one-sided bandwidth denoted by B. The resulting received atmospheric noise at the output of the IF filter is then evaluated as

$$\bar{n}(t) = n(t) * h(t) \tag{19.84}$$

where $h(t)$ is the impulse response of the IF filter where the asterisk (*) denotes convolution. Regarding the second point, parameter λ is selected to conform to measured data to result is the prescribed APD.

19.10.3 Fitting the Noise Model to Measured Data

In a simulation model the required sampling frequency f_s is chosen to conform to the receiver Nyquist sampling criterion. For example, when simulating the performance of a communication system using analytic signal representations, the sampling frequency is selected such that $f_s \geq 2B$, where B is the bandwidth of the received signal. In this regard, the sampling frequency is selected to result in an acceptably low loss resulting from the detected symbol energy and aliasing distortion. The received samples are processed in the demodulator for waveform acquisition and subsequently symbol and carrier tracking and data detection. In this context $t = kT_s$: $T_s = 1/f_s$ and (19.84) is the narrowband analytic representation of the received atmospheric noise. For evaluating the communication performance, the modulation symbol rate and the sampling frequency are related to the anti-aliasing filter bandwidth B as shown in Figure 19.28.

In the following evaluation of the impulse noise model, a minimum shift keying (MSK)-modulated waveform is used with a symbol rate of $R_s = 25$ sps, $B = 800$ Hz, $f_s = 2B$, and $N_s = 32$ samples-per-symbol. Because λ has units of impulses per second, the impulse rate is restricted to $\lambda \leq N_s R_s$ so the parameter N_s can be adjusted as necessary to fit the noise

*To avoid confusion the **bold** $\boldsymbol{j}$ is used to denote the complex root of unity.
†Uman's book [41] *Lightning* is suggested for additional reading.

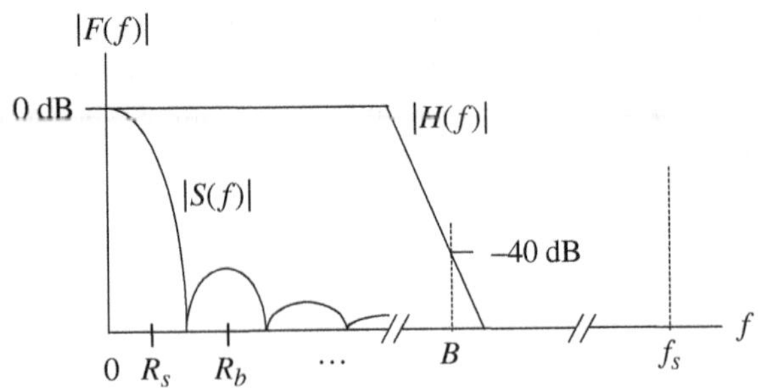

FIGURE 19.28 Relationship between f_s, B, and R_s for MSK modulation.

model to the measured data. These relationships are scaled to accommodate specific waveform modulations and data rates as discussed in the case study in Section 19.10.5.

Lightning strikes are characterized by the impulsivity measure V_d defined in (14.67) as

$$V_d \triangleq \frac{\sqrt{E[\bar{n}_k{}^2]}}{E[\bar{n}_k]} \tag{19.85}$$

where $\bar{n}_k$ represents the narrowband time-sampled receiver analytic noise. The system performance is typically characterized for a specified value of V_d; however, the APD for each V_d must conform to the corresponding measured APD in the ITU publication CCIR 322 [35]. To this end, the measured APD results are shown in Figure 19.29 for several values of V_d; these results are adapted from Gamble [44] and plotted using Rayleigh coordinates that result in a linear APD curve with slope −1/2 for Gaussian noise. The CCIR APD results are measured at the output of a receive antenna modeled as a single-pole filter with a noise bandwidth of $B_n = 243$ Hz. Therefore, when specifying V_d in (19.85) the parameters σ_1^2, σ_0^2, m_0, and λ of the filtered noise process characterized by (19.83) and (19.84) must be chosen to conform to the corresponding APD curve. The parameter λ is implicit in (19.83) through the random distribution of the time between the lognormal impulses. The parameters m_0 and σ_0^2 are implicit in the normally distributed random variable x denoted as $x = N(m_0, \sigma_0)$.

An analytic closed-form solution for V_d in terms of the model parameters is intractable because of the denominator term in (19.85) involving the expectation of the magnitude of the received noise. Therefore, V_d is evaluated numerically in terms of σ_1^2, σ_0^2, m_0, and λ with the background noise power based on a specified receiver signal-to-noise ratio* $\gamma_b = E_b/N_o$ as

$$\sigma_1^2 = \frac{V_r^2/2}{\gamma_b} \tag{19.86}$$

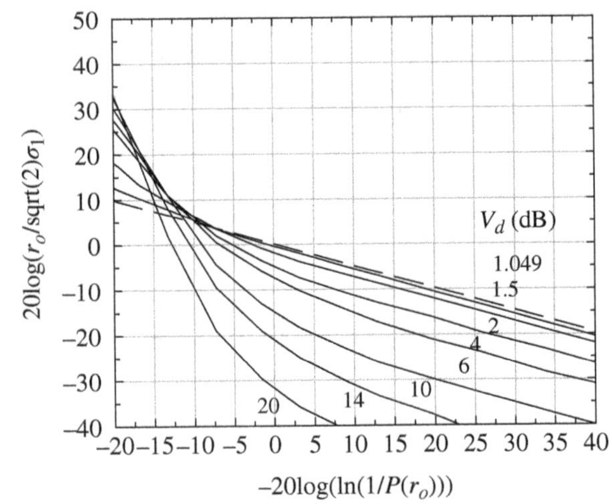

FIGURE 19.29 APD curves corresponding to CCIR-322 measured V_d values (243 Hz receive filter noise bandwidth).

where V_r is the peak voltage of the carrier-modulated received signal. Upon specifying σ_1^2, λ, and m_0, (19.85) is evaluated by indexing $\sigma_0^2 = n\Delta\sigma_0^2$: $n = 1, \ldots, N$ until the computed value of V_d just exceeds the specified value. Each evaluation involves a time series of one million samples to compute the expectations required in (19.85).† The starting value of σ_0^2, denoted as $\sigma_0^2(\text{start})$, is chosen to minimize the search time and $\Delta\sigma_0^2 = 0.2$ was found to provide sufficient coarse resolution to match the desired APD curve. However, to improve the estimation accuracy, a fine resolution interval of $\Delta\sigma_0^2/10$ is used with the indexing restarted at the previous value and continuing with $\sigma_0^2 = \sigma_0^2(\text{previous}) + n'\Delta\sigma_0^2/10$: $n' = 0, \ldots, 10$ until the computed value of V_d again exceeds the specified value. Upon completion, the value σ_0^2 is linearly interpolated between the ending and previous values and the simulation is run once again to verify the evaluation of the desired APD curve using the four parameter values.

The final simulation run uses the interpolated value of σ_0^2 and the numerically computed error in V_d in decibels is typically less than 0.1%. The final parameter sets for several V_d values are summarized in Table 19.7. The results of the V_d and APD evaluations are shown as the data points in Figure 19.30 for V_d values of 1.049, 2, 6, and 14 dB; the *solid* curves are taken from Figure 19.29 and represent the corresponding measured APD curves.

Figures 19.31, 19.32, and 19.33 show typical recordings of the sampled magnitude, $|\bar{n}_k|$, of the lognormal impulse noise for $V_d = 1.049$, 2.0, and 14.0 dB, respectively. These recordings are representative of the filtered samples at the

*The receiver E_b/N_o includes the receiver kTB noise and the power associated with the impulsive channel noise.

†The nonstationary characteristic of the channel noise is not an issue because the model is being fit to measure data that is based on the ensemble averages.

TABLE 19.7 Lognormal Parameter Sets Corresponding to Selected V_d Values

V_d (dB)	λ	m_0	γ_b	σ_1^2	σ_0^2	$\bar{\sigma}_{ch}^2$	$\bar{\sigma}_{mf}^2$ [a]
1.049	—	—	Any	Any	—	—	—
1.5	1150	0	36.5	1.120(−4)	0.2074	0.0163	0.0035
2.0	300	0	28.0	7.920(−4)	0.7625	0.1000	0.0212
6.0	3000	0	15.0	0.01581	2.218	2.0500	0.4400
14.0	1500	3	−30.0	500.00	4.726	5.483(3)	1.354(4)

[a] Applies to MSK matched filter.

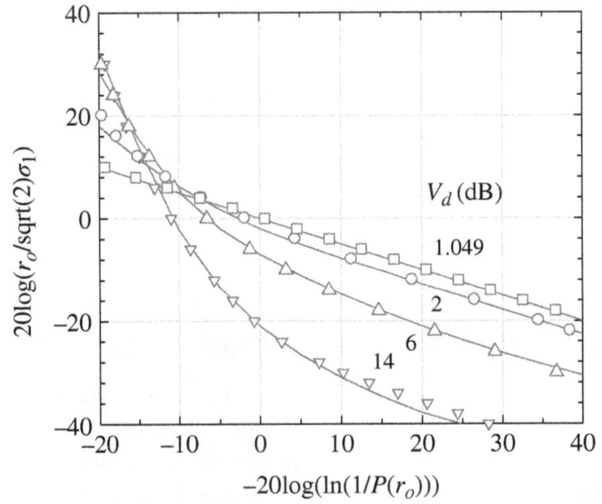

FIGURE 19.30 Simulated APD characteristics for several values of V_d using lognormal model (parameter sets from Table 19.7).

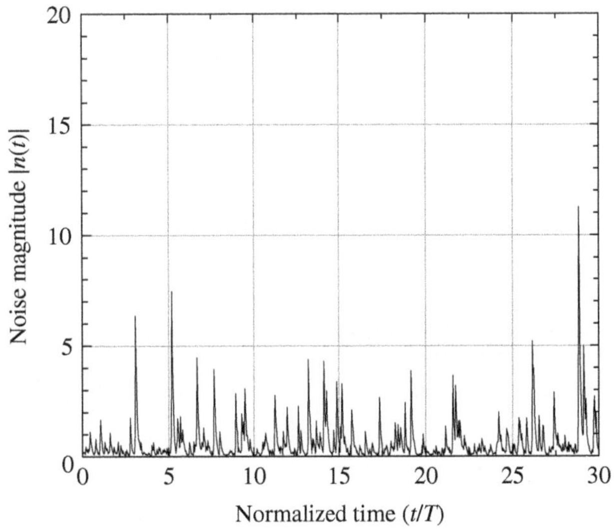

FIGURE 19.32 Channel impulse noise record: $V_d = 6$ dB.

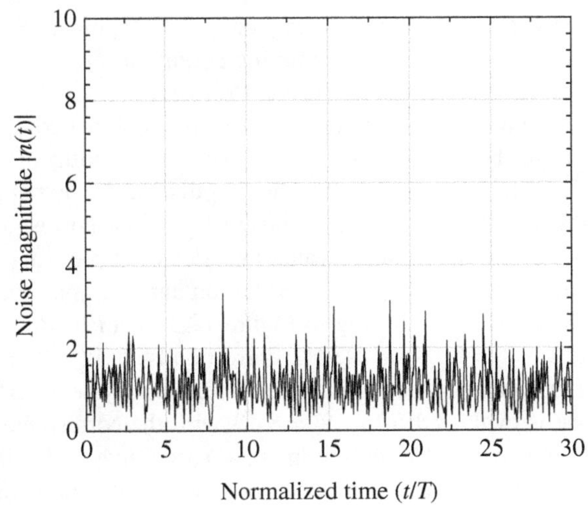

FIGURE 19.31 Channel impulse noise record: $V_d = 1.049$ dB.

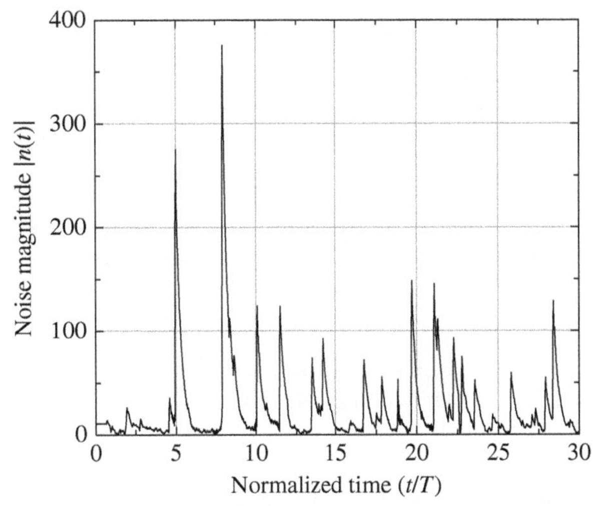

FIGURE 19.33 Channel impulse noise record: $V_d = 14$ dB.

output of the 243 Hz noise bandwidth receive antenna used to collect the CCIR-322 noise data.

The value of V_d will change as the filter bandwidth is changed but the CCIR-322 results apply only to the 243 Hz noise bandwidth filter [45]. Spaulding, Roubique, and Crichlow [46] have evaluated the conversion of V_d with bandwidth and their results are shown in Figures 19.34 and 19.35 for decreasing and increasing bandwidths respectively. $B_i = B_n = 243$ Hz is the noise bandwidth in which V_{di} is measured and B_o is the noise bandwidth of the desired filter.

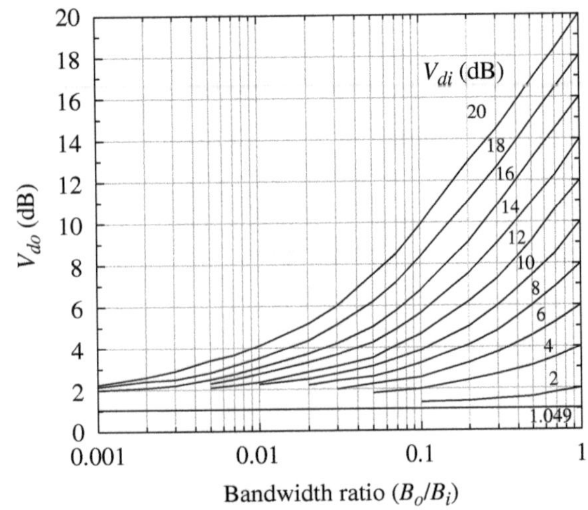

FIGURE 19.34 V_d dependence on bandwidth: $B_o/B_i \leq 1$. Spaulding et al. [46]. Courtesy Journal of Research of the National Bureau of Standards.

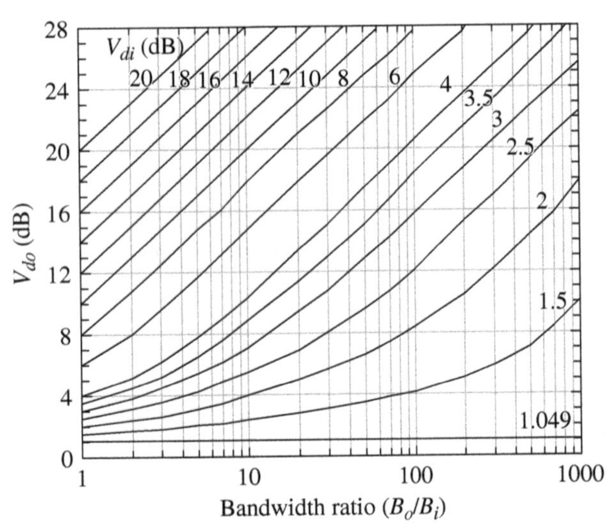

FIGURE 19.35 V_d dependence on bandwidth: $B_o/B_i \geq 1$. Spaulding et al. [46]. Courtesy Journal of Research of the National Bureau of Standards.

The case study in Section 19.10.5 uses $B_i = B_n = 243$ Hz with $B_o = B_n$ so no bandwidth conversion is necessary.*

19.10.4 Impulsive Noise Mitigation Techniques

Modem performance improvements can be achieved through the use of various impulsive noise mitigation techniques like: clipping [47], limiting [48], excision, and hole-punching [49]. These techniques are applied in the receiver or demodulator signal path prior to the matched filter detection; however, at the outset of the modem design, the proper selection of the waveform modulation and forward error correction (FEC) coding will result in significant performance advantages. When FEC is applied to the waveform, the use of interleaving is also an effective mitigation technique for impulsive noise. Gamble [44] provides a review of these subjects and summaries the performance results of various authors for constant amplitude waveform modulations: PSK, MSK, and continuous phase frequency shift keying (CPFSK). In Section 19.10.5, the performance improvement using clipping is examined using the MSK-modulated waveform. Time domain clipping is most effective when applied at a high IF frequency before significant pulse dispersion occurs; this can be achieved for constant envelope waveforms with bandpass limiting followed by narrowband filtering.

*The ratio of the noise bandwidth to 3 dB bandwidth for a single-pole filter is 1.57.

19.10.5 Case Study: Minimum Shift Keying Performance with Lognormal Impulse Noise

This case study involves simulating the performance of MSK operating as a bit-rate of $R_b = 50$ bps in lognormal noise representative of lightning strikes from worldwide storm centers. In the preceding sections, the parameter V_d is used to characterize the receiver noise described by (19.84). The application of (19.84) involves generating noise samples $\bar{n}_k = \bar{n}(kT_s)$, where T_s is the sampling interval, and combining the noise samples with the sampled signal (s_i) as shown in Figure 19.36. The sampling frequency $f_s = 1/T_s$ must be chosen to satisfy the Nyquist sampling condition.

The receiver filter in Figure 19.36 represents the cascade of receiver and demodulator IF filters with a composite noise bandwidth of B Hz. The bandwidth of the single-pole filter following the impulse noise generator establishes the impulsivity measure V'_d and the corresponding APD for the system evaluation. For example, suppose that the receiver being evaluated has an antenna noise bandwidth of 486 Hz, referring to Figure 19.35 with $B_o/B_i = 2$, a value of $V_d = 6$ dB in 243 Hz corresponds to $V'_d = 8$ dB in the receiver noise bandwidth. Therefore, the system performance is evaluated using $V'_d = 8$ dB with the corresponding APD response. However, in this case study, $B_i = B_n = 243$ Hz with $B_o = B_n$, so the measured V_d values can be used without applying bandwidth conversion. This avoids the uncertainty introduced in the bandwidth conversion processing [46].

The signal must be passed through the equivalent receiver filters and the signal and intersymbol interference (ISI) distortion losses must be considered. Although the simulated signal is characterized as an analytic signal, the modulated signal at the filter output has an equivalent carrier-modulated peak level of V_r volts corresponding to the signal power

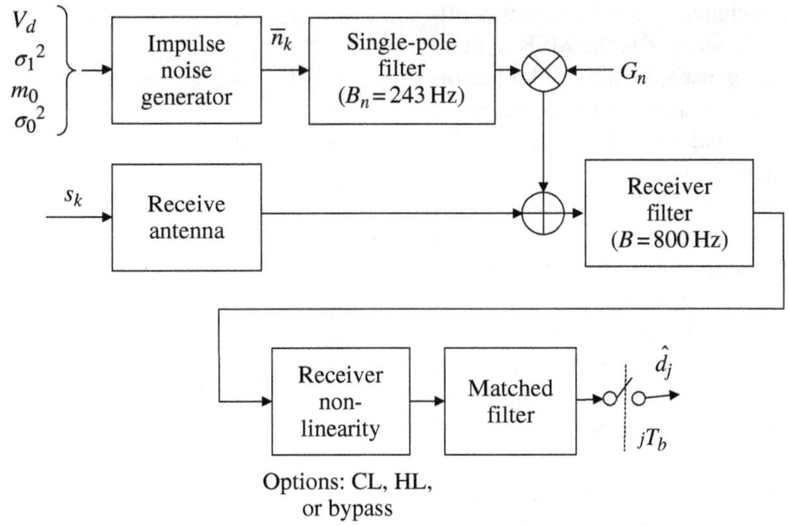

FIGURE 19.36 Application of impulse noise for modem performance evaluation.

$P_s = V_r^2/2$. Similarly, the power of the zero mean sampled noise is computed as

$$\sigma_{ch}^2 = \frac{1}{N}\sum_{k=1}^{N} |\bar{n}_k|^2 \tag{19.87}$$

This power is measured in the bandwidth of the sampling frequency and the signal-to-noise ratio in the sampling bandwidth is $N_s = f_s/R_s$ times higher than E_b/N_o. In an environment involving white noise the channel power can be scaled by the bandwidth ratio and adjusted to correspond to a specified E_b/N_o ratio. However, because the impulse noise does not have a constant power spectral density (PSD) the signal will necessarily involve the channel noise power (σ_{mf}^2) at the output of the symbol matched filter. For MSK modulation, the matched filter noise bandwidth is

$$B_{mf} = \left(\frac{\pi^2}{8}\right) R_S \quad \text{: MSK matched filter bandwidth} \tag{19.88}$$

The noise powers $\bar{\sigma}_{ch}^2$ and $\bar{\sigma}_{mf}^2$ are listed in Table 19.7 for the indicated values of V_d. The performance simulation evaluates the bit-error probability as a function of the signal-to-noise ratio, $\gamma_b = E_b/N_o$ measured in the bandwidth to the data rate R_b, that is,

$$\gamma_b = \frac{P_s}{\sigma_b^2} \tag{19.89}$$

Upon expressing (19.89) in terms of the noise power σ_{mf}^2 measured in the bandwidth B_{mf} and solving for σ_{mf}^2 results in

$$\sigma_{mf}^2 = \frac{P_s}{\gamma_b}\left(\frac{\pi^2}{16}\right) \tag{19.90}$$

Using (19.90) the voltage gain required to bring $\bar{\sigma}_{mf}^2$ up to the level σ_{mf}^2 for a specified $\gamma_b = E_b/N_o$ is

$$G_n = \sqrt{\frac{\sigma_{mf}^2}{\bar{\sigma}_{mf}^2}} = \frac{\pi}{4}\sqrt{\frac{P_s}{\gamma_b\,\bar{\sigma}_{mf}^2}} \tag{19.91}$$

This gain is applied to the simulated impulse noise as shown in Figure 19.36. As stated above, in the following simulations the input filter is the single-pole filter with a noise bandwidth of B_i = 243 Hz used to characterize the CCIR-322 data collection hardware.

Following the impulse noise gain adjustment, the received signal, receiver thermal noise, and impulsive channel noise are added and passed through the receive filter, the optional nonlinearity, and the MSK matched filter. The demodulator data estimates are determined from the optimally sampled matched filter and the errors are counted to determine the bit-error probability. The hard limiting (HL) is ideal, in that, it normalizes the complex baseband signal plus noise (s_{rk}) as

$$s'_{rk} = \frac{s_{rk}}{|s_{rk}|} V_r \tag{19.92}$$

When clipping is used, the clipping threshold is defined relative to the channel noise power $\bar{\sigma}_{ch}^2$ and, because the received signal power is held constant, the clipping level (CL) relative to the received signal voltage V_r is also constant. The matched filter is a complex inphase and quadrature (I/Q) filter with quadrature cosine weighed symbol integrators offset by the bit duration $T_b = T/2$, where T is the MSK symbol duration. Ideal symbol timing is used, however, the performance using ideal frequency and phase tracking is compared to the performance using a second-order Costas phaselock loop with various time-bandwidth products.

The following figures characterize the performance of the MSK-modulated waveform for various V_d values as described above. The modulation symbol rate of $R_s = 25$ sps has negligible distortion when passed through the single-pole equivalent receiver antenna filter. The receiver bandwidth of $B = 800$ Hz is sufficiently wide to provide for effective clipping of the impulse noise peaks and the associated reduction of the V_d through clipping, HL, or bandwidth reduction is reflected in the performance results.

Gamble [44] has evaluated the theoretical performance of coherently detected binary phase shift keying (BPSK) with a linear receiver for various values of V_d. Gamble also provides comparisons of the theoretical performance with published experimental results for PSK [50] with $V_d = 3.75$ dB measured in 33.3 Hz and for MSK [51] with $V_d = 5$ dB measured in 40 Hz. These comparisons indicate the Gambles theoretical results provide a lower-bound to the experimental results by 2–3 dB; however, the differences may be accounted for by measurement accuracies, tracking losses and other hardware anomalies. The simulation results in the following section focus on MSK modulation and compare reasonable well with Gambles results.

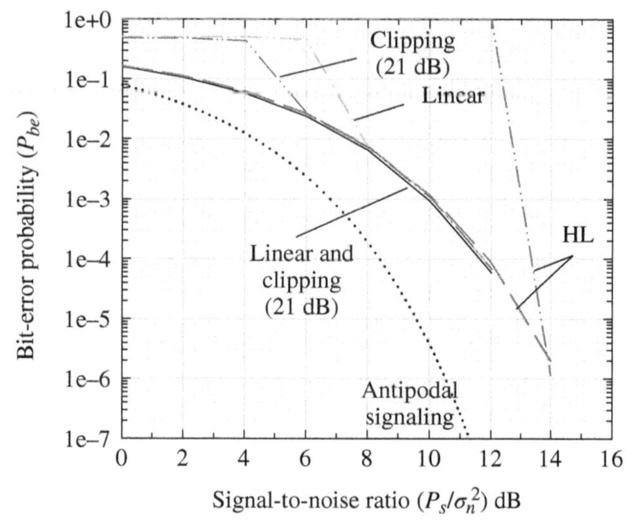

FIGURE 19.37 Simulated performance of MSK in atmospheric noise ($V_d = 1.5$ dB, clipping, hard limiting and ideal tracking).

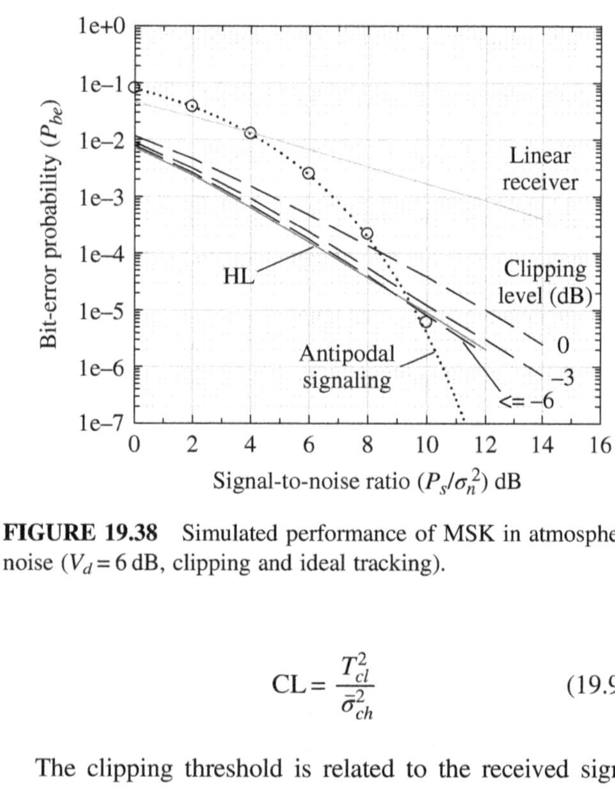

FIGURE 19.38 Simulated performance of MSK in atmospheric noise ($V_d = 6$ dB, clipping and ideal tracking).

19.10.5.1 *Simulated MSK Performance with* $V_d = 1.5$ dB

The simulated performance of MSK in atmospheric noise with $V_d = 1.5$ dB and ideal symbol and phase tracking is shown in Figure 19.37. The performance using a linear receiver is compared to a HL receiver and a receiver with various CLs ranging from 0 to 21 dB. The Monte Carlo simulations use 1 M bits (500K symbols) for each signal-to-noise ratio and, because the channel noise does not have a constant power spectral density, the abscissa is defined more generally as P_s/σ_n^2.* The *dotted* curve is the theoretical performance of MSK in additive white Gaussian noise (AWGN) and the *circled* data points (see Figure 19.38) result from the corresponding simulated performance. These ideal tracking performance results are provided as a baseline for comparison. The modem performance with impulsive noise is degraded from the AWGN performance by about 3.5 dB at $P_{be} = 10^{-5}$ with the linear receiver and an additional 0.2 dB with the HL receiver. The performance with CLs ranging from 0 to 21 dB fall between HL and linear performance.

The CL is defined in terms of the clipping threshold T_{cl} and the channel noise power $\bar{\sigma}_{ch}^2$ as

$$CL = \frac{T_{cl}^2}{\bar{\sigma}_{ch}^2} \tag{19.93}$$

The clipping threshold is related to the received signal power as

$$\frac{T_{cl}^2}{P_s} = CL\left(\frac{\bar{\sigma}_{ch}^2}{P_s}\right) \tag{19.94}$$

The signal power in the simulation is constant with $P_s = 0.5$ and, using the computed noise power from Table 19.7, $\bar{\sigma}_{ch}^2 = 0.0163$, the clipping threshold relative to P_s in decibels is

*In terms of the atmospheric impulsive noise power $\sigma_n^2 = \sigma_{mf}^2$.

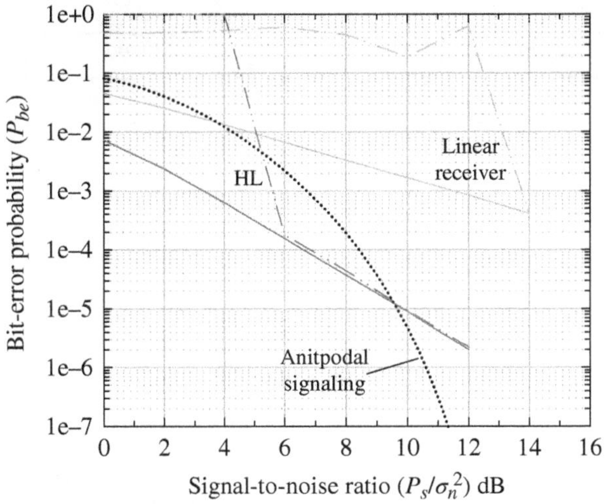

FIGURE 19.39 Simulated performance of MSK in atmospheric noise ($V_d = 6$ dB, hard limiting and PLL tracking).

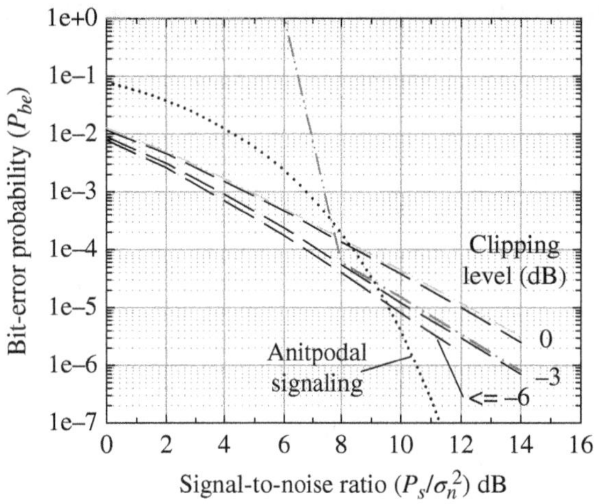

FIGURE 19.40 Simulated performance of MSK in atmospheric noise ($V_d = 6$ dB, clipping and PLL tracking).

$$\left.\frac{T_{cl}^2}{P_s}\right|_{\text{dB}} = \text{CL(dB)} - 14.87 \quad : V_d = 1.5 \text{ dB} \tag{19.95}$$

Figure 19.37 also shows the performance with phaselock loop tracking using the time-bandwidth product B_LT_b and lock detector $B_{LD}T_b$ products of 0.025 and an initial phase error of 10° with zero frequency error. The critical phaselock loop signal-to-noise ratio of the linear receiver is about 8 dB and for the HL receiver it is about 14 dB. Clipping levels ranging from 0 to 21 dB were examined and the phaselock loop never achieved phase-lock for CLs ≤ 18 dB. With CL = 21 dB phase-lock was achieve with a critical signal-to-noise ratio of about 6 dB. Considering the CL relative to the signal power, as given by (19.95), the 21 dB CL corresponds to a clipping threshold relative to the signal power of 6.13 dB, whereas, CLs ≤ 18 dB correspond to clipping thresholds ≤3.13 dB relative to the signal power. These results indicate that an optimum clipping threshold exits with phase tracking. In all cases, once phase-lock is achieved the performance loss is less than about 0.25 dB with tracking.

19.10.5.2 Simulated MSK Performance with V_d = 6 dB

The simulated MSK results for $V_d = 6$ dB are shown in Figures 19.38, 19.39, and 19.40. In these figures, the *dotted* curve is the theoretical performance of MSK in AWGN and the *circled* data point corresponded to the simulated MSK performance. The Monte Carlo simulations use 1 M bits (500K symbols) for each signal-to-noise ratio. The results in Figure 19.38 compare the performance of a linear receiver with one using HL and various CLs; all of these results use ideal symbol time and phase tracking. As predicted by Gamble the performance of the linear receiver is severely degraded; however, HL and CLs ≤ −6 dB provide significant improvements.

Referring to (19.94) and Table 19.7, $\bar{\sigma}_{ch}^2 = 2.05$ and the clipping threshold relative to the signal power, $P_s = 0.5$, is expressed in decibels as

$$\left.\frac{T_{cl}^2}{P_s}\right|_{\text{dB}} = \text{CL(dB)} + 6.13 \quad : V_d = 6 \text{ dB} \tag{19.96}$$

Therefore, a CL of −6 dB corresponds to a clipping threshold of relative to the signal power level of 0.13 dB. The CLs of −3 and 0 dB, shown in Figure 19.38, also exhibit considerable improvement relative to the linear receiver. These CLs correspond to the respective clipping thresholds of 3.13 and 6.13 dB relative to the signal power.

The simulation results in Figure 19.39 show the performance of the linear receiver and the HL receiver with phase tracking indicated by the *dash-dot-dot* curves. The phaselock loop (PLL) B_LT_b and lock detector $B_{LD}T_b$ products are 0.025 and the initial phase error is 10° with zero frequency error. The critical signal-to-noise ratio for phase-lock is about 14 dB for the linear system and 6 dB for the HL receiver.

Figure 19.40 shows the phase tracking performance with clipping. With CL = 0 dB phase-lock is achieved for signal-to-noise ratios less than 0 dB and with CL = −3 dB the critical signal-to-noise ratio is about 8 dB. For signal-to-noise ratios below about 6 dB the PLL never achieved phase-lock. With CLs ≤ −6 dB phase-lock was never achieved for any signal-to-noise ratio; this in contrasted with HL that did achieve phase-lock as mentioned above. The loop bandwidth is directly related to the signal level into the PLL and, because the receiver gain following the limiting and clipping is not adjusted, the low signal level associated with low CLs may be attributed to the failure to achieve phase-lock. In any event, the performance at lower signal-to-noise ratios is an

TABLE 19.8 Sea-State Number and Wind Wave Parameters

	Sea-State Number							
Parameter	1	2	3	4	5	6	7	8
Wind velocity (knots)	5	12	16	18	24	28	34	42
Average wave height (ft)	0.18	1.4	2.9	3.8	7.9	11	19	31
Average wave length (ft)	6.7	40	71	90	160	212	322	492
Average wave period (s)	1.4	3.4	4.6	5.1	6.8	7.9	9.7	12

important consideration with the use of FEC coding. In all cases, once phase-lock is achieved the performance loss is less than about 0.25 dB with tracking. The phaselock performance can be improved by reducing the time-bandwidth products at the expense of less tolerance to the dynamics of the encounter.

19.11 OCEAN WIND WAVE CHANNEL

19.11.1 Introduction

When communicating with submarines in deep ocean environments, the distortion introduced by surface waves driven by wind is a major factor that degrades the system performance. The impact of wind waves on the signal amplitude and phase is severe at very low frequency (VLF) and LF frequencies and in this section the signal distortion is analyzed and quantified in terms of the sea-state conditions. The performance degradation is examined in a case study in Section 19.11.5 and signal processing techniques to mitigate the impact of the wave distortion are discussed.

In the real-world environment, the received signal is corrupted by additive atmospheric noise as discussed in the preceding sections; however, the impulsivity of the atmospheric noise is decreased because of the narrowband filtering through the sea water. As the water depth increases, the V_d measure of the atmospheric noise approaches 1.049 dB corresponding to Gaussian noise. In addition to the filtered atmospheric noise, the noise at the receiver consists of additive thermal noise and man-made interference noise from equipment that falls within the bandwidth of the receiver IF stages. In these cases, the performance loss is mitigated through the use of robust symbol modulations, FEC coding, interleaving, and adaptive interference cancellation techniques [52].

A theoretical characterization of wind waves is given by Kinsman [53] where, in Chapter 8, the Pierson-Neumann theory is discussed leading to the Neumann wave-height energy spectral density (EDS) characterization. An alternate expression of the wave-height spectral density is based on the Pierson-Moskowitz theory and the two spectrums are compared. These energy spectrums characterize the wave height for a fully developed sea after a period of sustained winds and that is not influenced by the ocean floor or surrounding shore lines. In other words, the sustained winds are over deep oceans with a large surface area or fetch. In addition to the introduction given by Kinsman, the U.S. Navy [54] provides a description of the growth and decay of waves. The wind wave characteristics are identified in terms of a sea-state number [55, 56] as given in Table 19.8. In the following sections, the Neumann wave-height energy EDS is used to compute random wave-height temporal records; these records are seamlessly linked in a computer simulation program and used to distort the amplitude and phase of the communication waveform for the purpose of evaluating the communication link performance under various sea-state conditions.

19.11.2 Neumann Wave-Height Energy Spectrum

The EDS of wave height in a fully developed sea of unlimited depth, fetch, and time is characterized by Neumann [57] as

$$S(\varpi) = \begin{cases} \dfrac{4.791}{\varpi^6} e^{-2\left(\frac{g}{u\varpi}\right)^2} & : \varpi \geq 0 \text{ Neumann} \\ 0 & : o.w. \end{cases} \tag{19.97}$$

where g is the gravitational acceleration at sea level, found from experimental data to be 9.8106 m/s^2, u is the wind velocity in meter per second, and ϖ is the angular frequency of the energy spectrum in radian per second. The constant 4.791 is determined from experimental data and has dimension of m^2/s^5. $S(\varpi)$ has dimensions of square meter-second and is plotted in Figure 19.41 in terms of the frequency $f = \varpi/2\pi$. The severity of the sea condition is designated by the sea-state number and the corresponding wind velocity as indicated in Table 19.8. Figure 19.41 clearly demonstrates the impact of the wind velocity with an increase of five decades in the peak energy spectrum between sea-states 1 and 8. The alternate wind wave spectral density is based on the Pierson-Moskowitz wave theory [58] and expressed by (19.98) and shown in Figure 19.42.* The Neumann characterization is

*The U.S. Navy HIDRAS simulator uses the Pierson-Moskowitz spectral density.

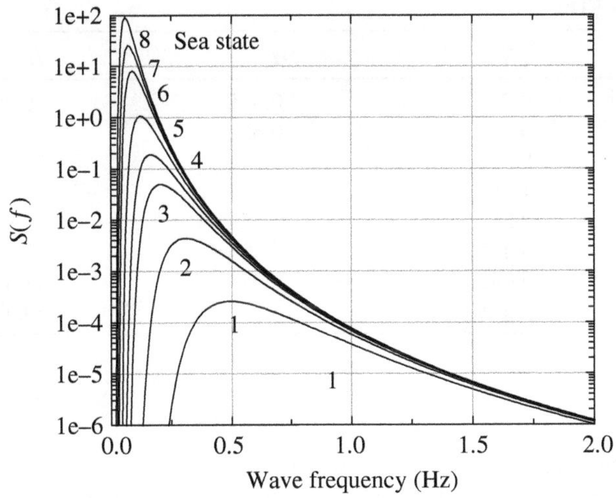

FIGURE 19.41 Neumann wave-height energy spectrums.

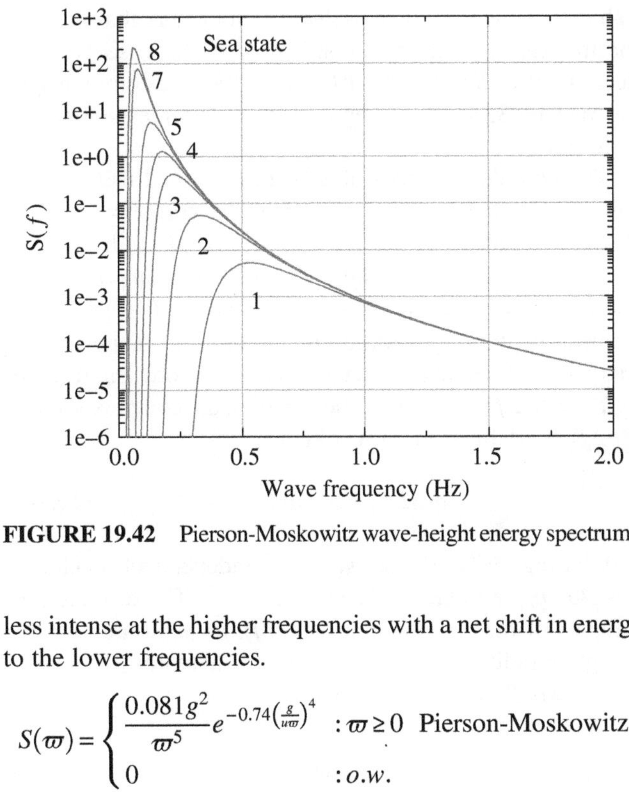

FIGURE 19.42 Pierson-Moskowitz wave-height energy spectrums.

less intense at the higher frequencies with a net shift in energy to the lower frequencies.

$$S(\varpi) = \begin{cases} \dfrac{0.081g^2}{\varpi^5} e^{-0.74\left(\frac{g}{u\varpi}\right)^4} & : \varpi \geq 0 \text{ Pierson-Moskowitz} \\ 0 & : o.w. \end{cases} \tag{19.98}$$

The integral of the wave-height EDS is the wave energy and, for the Neumann spectral density, is evaluated as

$$E = \int_0^\infty S(\varpi)d\varpi = 4.791 \int_0^\infty \frac{e^{-2\left(\frac{g}{u\varpi}\right)^2}}{\varpi^5} d\varpi \tag{19.99}$$
$$= 0.3176\sqrt{\pi}\left(\frac{u}{g}\right)^5$$

Commonly used statistical measures of the wave height are [59]: the average height $\bar{H}$, the average of the 1/3 highest wave $\bar{H}_{1/3}$, and the average of the 1/10 highest wave $\bar{H}_{1/10}$. Expressed in terms of the wave energy, these average wave-height measures are given by

$$\bar{H} = 1.77\sqrt{E} \tag{19.100}$$

$$\bar{H}_{1/3} = 2.83\sqrt{E} \tag{19.101}$$

and

$$\bar{H}_{1/10} = 3.60\sqrt{E} \tag{19.102}$$

The constant 4.791 in (19.99) applies for wave heights between $\bar{H}_{1/3}$ and $\bar{H}_{1/10}$. The wave energy and corresponding wave heights are listed in Table 19.9 for the sea-state conditions.

19.11.3 Wave-Height Generation Using Neumann's Energy Density Spectrum

In this section, the received signal phase and amplitude characteristics are examined at a point beneath the ocean's surface under various sea-state conditions using Neumann's wave-height EDS. The approach is similar to that described in Chapter 20 for the generation of Rayleigh fading signals and involves the uniform sampling of the wave-height EDS and then using the inverse Fourier transform (IFT) to generate discrete-time sampled wave data. The resulting record of finite wave data samples is extended to an arbitrary length by joining independently generated records in a seamless manner.*

The sampling conditions are depicted in Figure 19.43 using an N_{fft} – point radix-2 fast Fourier transform (FFT). The wave EDS is bandlimited to B Hz and sampled at the sampling frequency $f_s = N_{fft}\Delta f$ Hz with N samples over the EDS response, that is, $B = N\Delta f$. The temporal response of the sampled EDS is shown as the sampled wave record in Figure 19.43b. $T_B = 1/B$ is defined as the time-constant of the wave temporal response and is related to the correlation time of the wave; there are N time-constants for each record.

The parameter N is chosen such that there are about four samples within the 3-dB bandwidth (B_3) of the EDS so the frequency increment is $\Delta f \cong B_3/4$. The parameters B_3 and B are determined from Figure 19.41 and the value of Δf is chosen to satisfy the condition

*The HIDRAS channel simulator generates random time-varying stochastic signal representations by summing 26 sinusoidal signals with amplitude and phase functions dependent upon the wind wave height fluctuations. To avoid periodicities in the received signal, the frequency of each sinusoidal signal is randomly adjusted at 1 s intervals corresponding to a total repetition time of 26 s.

TABLE 19.9 Wave Energy and Average Wave Heights for Neumann EDS

Sea-State	u_k (knots)[a]	E (m^2)	$\bar{H}$[b]	$\bar{H}_{1/3}$[b]	$\bar{H}_{1/10}$[b]
1	5	0.0007	0.047	0.075	0.095
2	12	0.056	0.419	0.668	0.849
3	16	0.235	0.858	1.373	1.746
4	18	0.423	1.151	1.840	2.340
5	24	1.783	2.363	3.778	4.806
6	28	3.854	3.475	5.550	7.070
7	34	10.173	5.645	9.025	11.480
8	42	29.262	9.575	15.307	19.472

[a]In (19.99) wind velocity u is meter per second so $u = 0.5148\ u_k$.
[b]Meters.

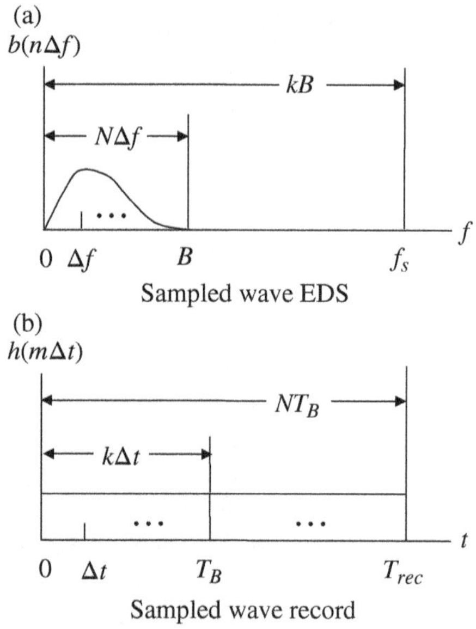

FIGURE 19.43 EDS sampling characteristics.

$$N = \left[\frac{B}{\Delta f}\right]_{int} = \left[\left(\frac{B}{B_3}\right)\left(\frac{B_3}{\Delta f}\right)\right]_{int} = 2^m \quad : m = 1, 2, \ldots \tag{19.103}$$

where B is chosen so that integrated alias distortion is more than 25 dB below the EDS in the bandwidth B. Using these results, the record duration T_{rec} is computed as $T_{rec} = 1/\Delta f$.

Referring to Figure 19.54a, the sampling frequency is selected to satisfy the Nyquist criteria $f_s = kB$ with $k \geq 2$. Using this result and (19.103), the size of the FFT is determined as

$$N_{fft} = \left[\frac{f_s}{\Delta f}\right]_{int} = \left[k\left(\frac{B}{\Delta f}\right)\right]_{int} = [kN]_{int} = 2^{m+n} \tag{19.104}$$

where $k = 2^n$: $n \geq 1$. The minimum value of k results in the smallest FFT size of 2^{m+1} samples and selecting larger values of k increases the size of the FFT by decreasing size sample interval Δt, however, Δf and T_{rec} remain unchanged. Therefore, for a given number of samples-per-symbol, increasing k results allows for more user symbols-per-record with fewer records required for the Monte Carlo performance simulations.* These relationships are dependent on the modulated symbol duration $T = 1/R_s$ that is typically less than T_B. Table 19.10 identifies the selected parameters that are used for the various sea-state conditions in the case study of Section 19.11.5 and Table 19.13 identifies the common symbol rate of 32 symbols-per-second used for each sea-state condition.

The complex samples of $S(f)$ are generated using identically distributed zero-mean independent complex Gaussian random variables $b(n\Delta f)$: $n = 0, \ldots, N-1$ with variance $\sigma_s^2(n\Delta f) = S(n\Delta f)$. Although $S(f)$ extents over all positive frequencies, the maximum frequency $(N-1)\Delta f$ is selected to result in negligible distortion due to aliasing. The sampling frequency increment Δf determines the record length given by $T_{rec} = 1/\Delta f$. To simplify the notation, the complex samples $b(n\Delta f)$ are defined in terms of b_n as

$$b(n\Delta f) \triangleq b_n = b_{nI} + jb_{nQ} \tag{19.105}$$

and, letting $i = \{I,Q\}$, the *pdf* of the independent quadrature samples b_{ni} are characterized by $N(0,\sigma_s)$. The discrete-time samples of the wave-height data $h(t)$, $t = m\Delta t$, are obtained using the radix-2 inverse fast Fourier transform (IFFT) with $M = 1/(\Delta t\Delta f)$ and are evaluated as

$$h(m\Delta t) \triangleq h_m = \Delta f \sum_{n=0}^{N-1} b_n e^{j2\pi nm/M} \quad : m = 1, \ldots, M \tag{19.106}$$

Because the wave function is real-valued, the negative frequency response, that is, the response from $f_s/2$ to $f_s = M\Delta f$, is the complex conjugate of the positive frequency response; this property simplifies the evaluation of (19.106).

*This is a natural application for using a discrete Fourier transform (DFT) with unrelated frequency and time domain samples; however, the computational efficiency of the FFT is preferred.

TABLE 19.10 Wave Energy Density Spectrum Sampling

Sea-State	B_3 (Hz)	B (Hz)[a]	Δf (Hz)	N	T_{rec} (s)	N_{fft}
1–3	0.150	2.0	1/32	64	32	4096
4–5	0.100	2.0	1/32	64	32	4096
6–8	0.043	0.5	1/64	32	64	8192

[a]Chosen for aliased distortion greater than 25 dB below the integrated EDS over B Hz.

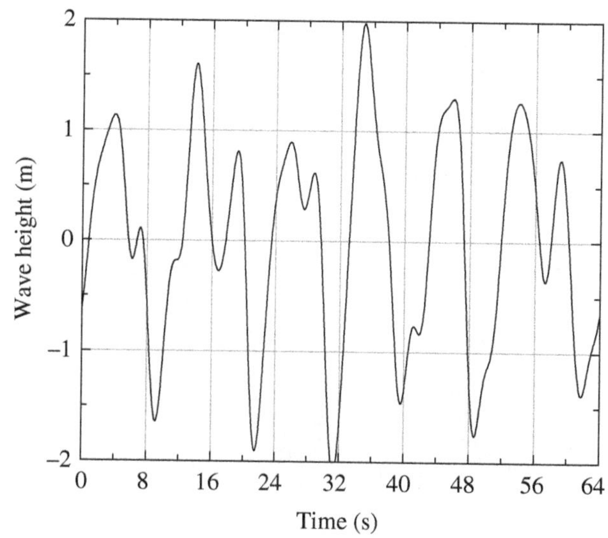

FIGURE 19.44 Sea-state 6 wave-height record ($N=32$, $M=8192$).

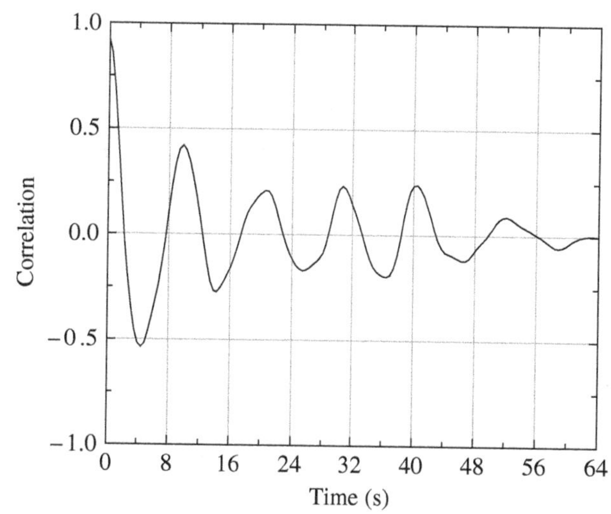

FIGURE 19.45 Sea-state 6 wave correlation response ($N=32$, $M=8192$).

Figure 19.44 shows a typical wave-height record. The M-sample record describing the wave heights h_m is concatenated with other independently generated records to create arbitrarily long temporal wave-height conditions for Monte Carlo performance simulations. The record concatenation uses a third-degree polynomial in t, such that, the amplitudes and slopes are equal at the record boundary points. The boundary points are separated by the wave-height correlation interval τ_c. The normalized correlation response is evaluated as

$$\rho(\tau)=\frac{R(\tau)}{R(0)} \tag{19.107}$$

and the correlation time is defined as the time at which

$$\rho(\tau_c)=e^{-1}=0.368 \tag{19.108}$$

With $\tau=\ell\Delta t$ and $S(n\Delta f)=S_n$ the discrete-time correlation function is evaluated as

$$R(\ell\Delta t)=R_\ell=\Delta f\sum_{n=0}^{N-1}S_n e^{-j2\pi n\ell/N} \quad :\ell=0,\ldots,N-1 \tag{19.109}$$

Using (19.107) through (19.109) to determine the correlation time, the number of interpolation samples required to concatenate the wave-height records is $N_c=\lceil\tau_c/\Delta t\rceil$.

Figure 19.45 shows a typical correlation response for sea-state 6.

To demonstrate the fidelity of the wave-height generation, the Fourier transform of the sea-state 6 wave correlation response in Figure 19.45 is compared to the theoretical Neumann EDS. The results are shown in Figure 19.46 where the *triangular* data points represent the re-constructed EDS for an average of 1000 ensembles or records. Figure 19.47 shows the reconstructed EDS with different sampling conditions for an average of 20 ensembles. Although a single ensemble will differ distinctly from the theoretical EDS, this is the nature of statistical sampling ensemble records as required for communications system performance evaluation.

19.11.4 Signal Generation using Temporal Wave-Height Records

Having characterized the wave height for different sea-state conditions, it is now necessary to solve for the sampled received signal with amplitude and phase temporal fluctuations [60] influenced by the changing height of the wind-driven waves. The solution is found in the characterization of an electric field traversing the surface of a conducting medium expressed as [61]

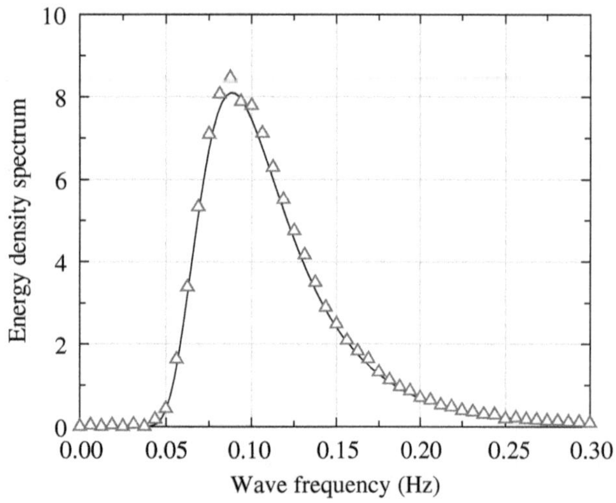

FIGURE 19.46 Re-constructed Neumann wave-height energy spectrum (SS-6, $N=250$, $M=512$, 1000 ensemble average).

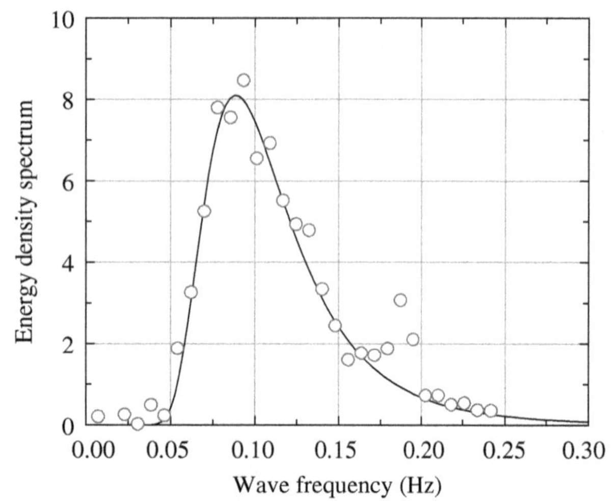

FIGURE 19.47 Re-constructed Neumann wave-height energy spectrums (SS-6, $N=32$, $M=8192$, 20 ensemble average).

$$E(t,\omega_c)=E_o(t,\omega_c)e^{-\sqrt{j\mu\sigma\omega_c}h(t)} \tag{19.110}$$

where μ is the permeability and σ is the conductivity of the medium and $\omega_c=2\pi f_c$, where f_c is the carrier frequency of the incident electric field $E_o(t,\omega_c)$ in hertz. Typical values of μ and σ for sea water are: $4\pi 10^{-7}$ H/m and 4 mho/m respectively; values of sea water conductivity range between 2 and 8 mho/m. Equation (19.110) is used to evaluate the fluctuations in the received signal with changes in the water depth due to the wave-height variations $h(t)$, described by (19.106). By noting that $\sqrt{j}=e^{j\pi/4}$, the exponent in (19.110) is expressed as

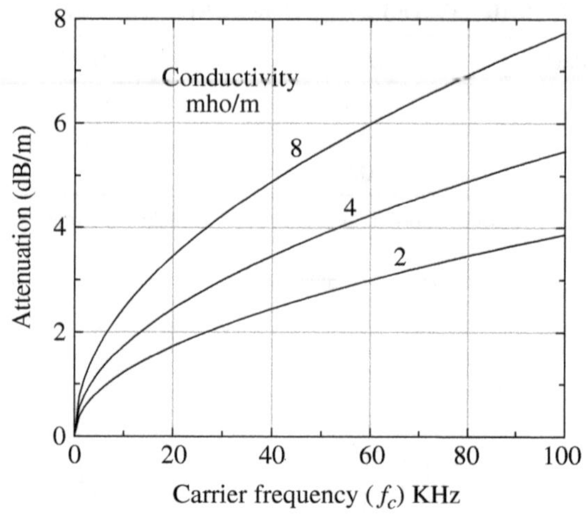

FIGURE 19.48 Static signal attenuation through sea water.

$$\sqrt{j\mu\sigma\omega_c}h(t)=\alpha h(t)+j\beta h(t) \tag{19.111}$$

where*

$$\begin{aligned}\alpha=\beta&=\sqrt{\frac{\mu\sigma\omega_c}{2}}\quad :\mathrm{m}^{-1}\\ &=3.974\times 10^{-3}\sqrt{f_c}\end{aligned} \tag{19.112}$$

The last equality in (19.112) is based on the above typical parameter values with f_c expressed in hertz. Using (19.110) through (19.112), the temporal expression of a CW signal is

$$g(t)=A(t)e^{-\theta(t)} \tag{19.113}$$

where $A(t)$ and $\theta(t)$ are the wind wave-dependent amplitude and phase functions given by

$$A(t)=e^{-\alpha h(t)} \tag{19.114}$$

and

$$\theta(t)=-\beta h(t) \tag{19.115}$$

The attenuation $A(t)$ results from the variation in the signal level due to the wave-height temporal variations, however, (19.114) also expresses the static loss as the water depth (d) increases beneath the surface of the sea. This static attenuation or loss, expressed in decibel per meter, is given by

$$\frac{L(d)}{d}=\sqrt{\frac{\mu\sigma\omega_c}{2}}\quad :\mathrm{dB/m} \tag{19.116}$$

Equation (19.115) is plotted in Figure 19.48 as a function of the carrier frequency for various sea water conductivities.

*The depth of penetration or *skin depth* is defined as $\delta=(\mu\sigma\omega_c/2)^{-1/2}$ m.

TABLE 19.11 Theoretical Signal Phase Characteristics with Sea-State[a]

	Sea-State Number							
Parameter	1	2	3	4	5	6	7	8
$\sigma_{\theta^{(0)}}$ (degrees)	1.04	9.31	19.11	25.65	52.66	77.42	125.8	213.4
$\sigma_{\theta^{(1)}}$ (deg/s)	4.59	17.07	26.28	31.36	48.28	60.84	81.42	111.8
$\sigma_{\theta^{(2)}}$ (deg/s^2)	35.00	54.22	62.61	66.40	76.68	82.83	91.27	101.4

[a]Neumann spectrum corresponding to $f_c = 30$ kHz.

The theoretical characteristics of the signal phase function are evaluated in terms of the derivatives of $\theta(t)$ using (19.115) with the result

$$\frac{\delta^n \theta(t)}{\delta t^n} = \theta^{(n)}(t) \tag{19.117}$$

The phase derivatives represent stochastic processes with correlation functions $R_\theta^{(n)}(\tau)$ and corresponding phase energy EDSs are expressed as

$$S_\theta^{(n)}(\omega) = \sqrt{\frac{\mu\sigma\omega_c}{2}} S^{(n)}(\omega) \tag{19.118}$$

where

$$S^{(n)}(\omega) = \omega^{2n} S(\omega) \tag{19.119}$$

The phase variance is defined as

$$\sigma^2_{\theta^{(n)}} \triangleq \int_0^\infty S^{(n)}(\omega) d\omega = \left(\frac{\mu\sigma\omega_c}{2}\right) E^{(n)} \tag{19.120}$$

where $E^{(n)}$ is the energy of the derivatives of the wave height. Using (19.119) $E^{(n)}$ is evaluated as

$$E^{(n)} = \int_0^\infty \omega^{2n} S(\omega) d\omega \tag{19.121}$$

Performing the integration in (19.121) using the Neumann EDS and substituting the result into (19.120) gives the standard deviation of the phase derivatives expressed as

$$\sigma_{\theta^{(n)}} = 1.0944 \sqrt{\frac{\Gamma(5/2-n)\mu\sigma\omega_c}{2^{5/2-n}} \left(\frac{u}{g}\right)^{5/2-n}} \tag{19.122}$$

The theoretical standard deviations expressed by (19.122) are listed in Table 19.11 for $n = 0$, 1, and 2 and Table 19.12 lists some corresponding results based on Monte Carlo simulation of the phase function as expressed by (19.115). The simulation results are based on eight wave-height samples within the wave time-constant T_B depicted in Figure 19.43 with samples taken over 5 N time-constants. Using a first-order polynomial to link the records, the simulated results for $\sigma_\theta^{(2)}$ resulted in discontinuities at the record boundaries. These discontinuities degrade the performance of PLL tracking with coherent demodulation; however, they can be eliminated by using a higher order polynomial for linking the records.*

TABLE 19.12 Simulated Signal Phase Characteristics with Sea-State[a]

	Sea-State Number		
Parameter	2	4	6
$\sigma_{\theta^{(0)}}$ (degrees)	11.26	20.54	37.98
$\sigma_{\theta^{(1)}}$ (deg/s)	16.87	14.10	31.85

[a]Neumann spectrum corresponding to $f_c = 30$ kHz.

The application of the wave amplitude and phase fluctuations, expressed in (19.114) and (19.115), respectively, to the received signal in a communication link is expressed in terms of the analytic signal representation as

$$\begin{aligned} \tilde{s}_r(t) &= \tilde{s}_t(t) g(t) \\ &= \sqrt{2P_s} A(t) m(t) e^{-j(\phi_m(t)+\theta(t))} \end{aligned} \tag{19.123}$$

where the transmitted analytic signal is expressed as

$$\tilde{s}_t(t) = \sqrt{2P_s} m(t) e^{j\phi_m(t)} \tag{19.124}$$

with

$$s_t(t) = \mathrm{Re}\{\tilde{s}_t(t) e^{j\omega_c t}\} \tag{19.125}$$

In these expressions, $m(t) = \sqrt{m_c^2(t) + m_s^2(t)}$ and $\phi_m(t) = \tan^{-1}(m_s(t)/m_c(t))$ are the modulation-dependent amplitude and phase functions and P_s is the transmitted signal power.

*The performance simulations, discussed in Section 19.11.5, use a third-order polynomial to concatenate multiple records.

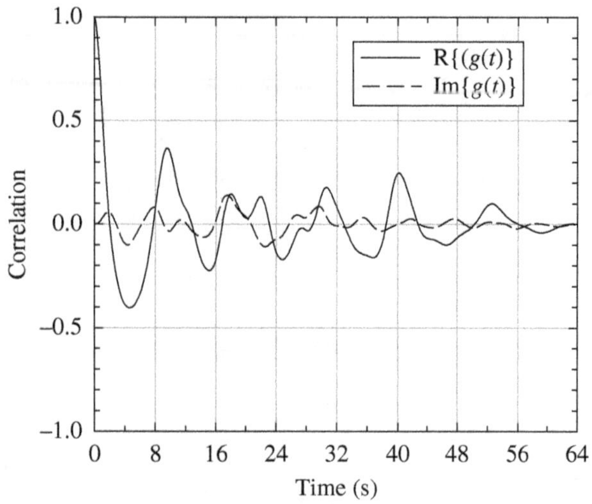

FIGURE 19.49 Sea-state 6 signal correlation characteristics (N = 32, M = 8192, f_c = 30 kHz).

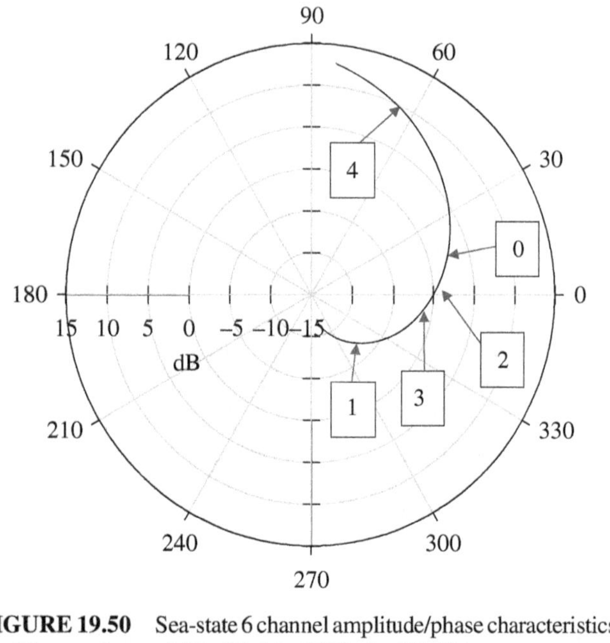

FIGURE 19.50 Sea-state 6 channel amplitude/phase characteristics for 64 second record (N = 32, M = 8192, f_c = 30 kHz).

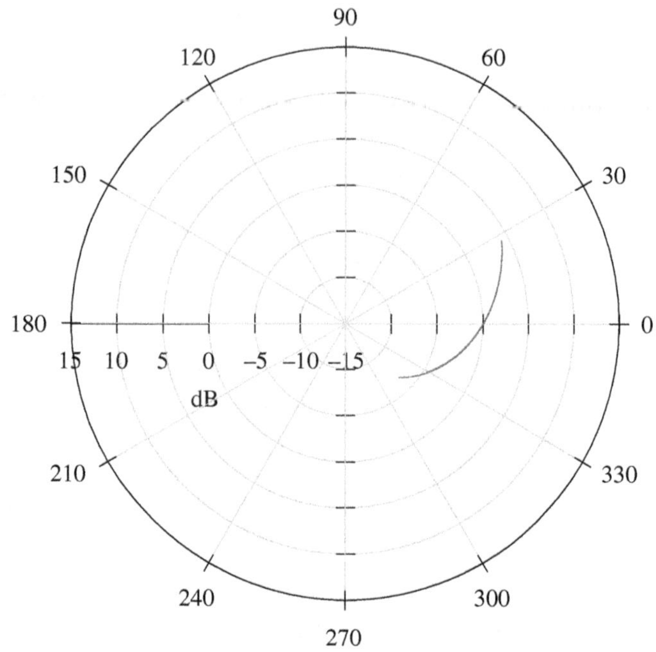

FIGURE 19.51 Sea-state 4 channel amplitude/phase characteristics for 64 second record (N = 32, M = 4096, f_c = 30 kHz).

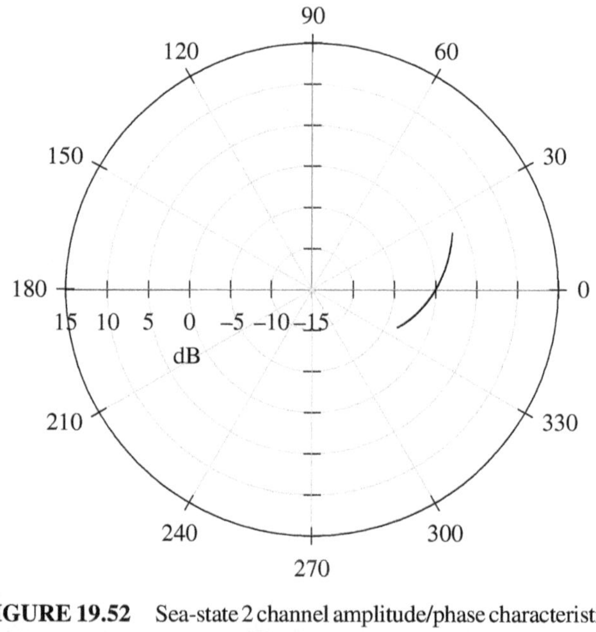

FIGURE 19.52 Sea-state 2 channel amplitude/phase characteristics for 64 second record (N = 32, M = 4096, f_c = 30 kHz).

The correlation response of a typical record of $g(t)$ for sea-state 6 is shown in Figure 19.49. Polar plots of the amplitude and phase fluctuations, in decibels and degrees, are shown in Figures 19.50, 19.51, and 19.52 for sea-states 6, 4, and 2 respectively. Along the polar trajectory there are numerous phase reversals that are not apparent, for example, the occurrence of the first four phase reversals is indicated by the *labeled arrows* in Figure 19.50; the label 0 represents the record starting position. These phase reversals occur in the first quarter of the record and the entire record has 12-phase reversals.

19.11.5 Case Study: Differentially Coherent BPSK Performance with Sea-State

This case study examines the performance of differentially coherent BPSK (DCBPSK) waveform modulation with sea-state distortion resulting from wind waves. The transmitted message is preceded by a single phase reference symbol that is used to detect the information bits. The ideal performance of

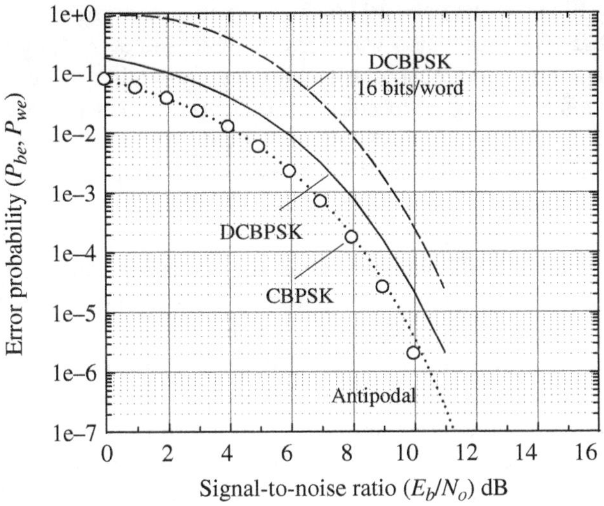

FIGURE 19.53 Performance of DCBPSK in AWGN channel with comparison to coherent BPSK detection.

DCBPSK in an AWGN channel is shown in Figure 19.53 and compared to the performance of coherent BPSK (CBPSK). The *circled* data points correspond to Monte Carlo simulated performance using 1 M bits for each signal-to-noise ratio. The following performance simulations with sea-state examine the probability of receiving a 16-bit word correctly so the word-error probability is also plotted in Figure 19.53 under the ideal conditions involving AWGN. FEC coding is not used. These performance plots using the ideal AWGN channel are used for subsequent comparisons with the performance under various sea-state conditions. The sea-state simulations correspond to a carrier frequency of $f_c = 30$ kHz.

The criteria for the sampling of the temporal wave-height records are outlined in Section 19.11.3 and the results are summarized in Table 19.10. Implicit in these requirements is the sampling of the communication waveform as determined by the selection of the FFT size. The most important consideration is that the selected sampling frequency results in an acceptable aliasing distortion loss. This condition requires that $f_s \geq f_N = 2(\rho R_s + |f_e|)$ where f_N is the Nyquist frequency, R_s is the transmitted symbol rate, f_e is the maximum received carrier frequency error, and ρ is the number of spectral sidelobes permitted in the principal bandwidth $|f| < f_N/2$. The aliased spectrum is depicted in Figure 19.54 as the *dashed* curves for the signal spectrum $S_s(fT) = sinc(fT)$ with $\rho = N_s = 4$ and $f_e = 0$.* Although the FFT size is identified in Table 19.10 for the generation of the wave-height samples, the selection criteria involve the symbol rate and the number of samples-per-symbol as characterized by

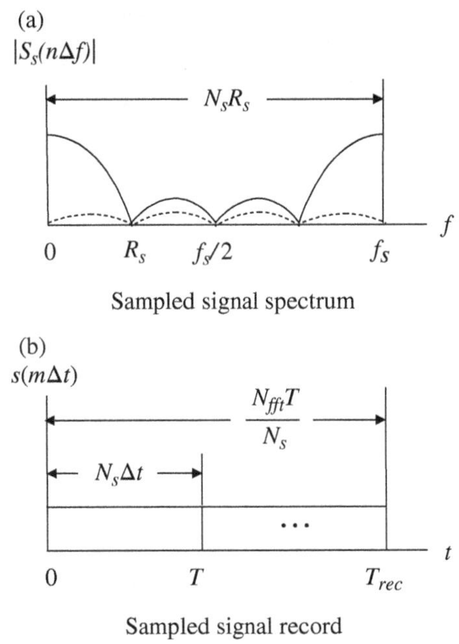

FIGURE 19.54 Modulated waveform sampling (shown for $N_s = 4$).

TABLE 19.13 Sampling Requirements and Modulation Symbol Rates

Sea-State	T_{rec} (s)	N_s	R_s (sym/s)	N_{fft}
1–3	32	4	32	4096
4–5	32	4	32	4096
6–8	64	4	32	8192

$$N_{fft} = N_s R_s T_{rec} \quad (19.126)$$

where the record length $T_{rec} = 1/\Delta f$ is determined in Section 19.11.3. Table 19.13 summarizes the selected FFT sizes for the various sea-state conditions and the corresponding symbol rates with $N_s = 4$ samples-per-symbol. With these sampling conditions, the following bit-error performance comparisons with sea-state use identical symbol rates. Using 4 samples-per-symbol, as in this case study, the symbol rate can be increased by increasing the FFT size; however, this requires generating a new set of data records. On the other hand, the symbol rate can be decreased by simply increasing the samples-per-symbol without having to generate new sea-state data records.† The use of an integer number of samples-per-symbol is simply for convenience and avoids the necessity of rate conversion in the demodulator processing.

*These conditions are used in this case study and allowances for larger values of ρ and $|f_e| > 0$ are accommodated by appropriate increases in the sampling frequency.

†Unlike the memoryless AWGN channel, the symbol interval (T) is dependent on the channel time constant (T_B) requiring that the performance simulation to be re-run for each change in the ratio T/T_B.

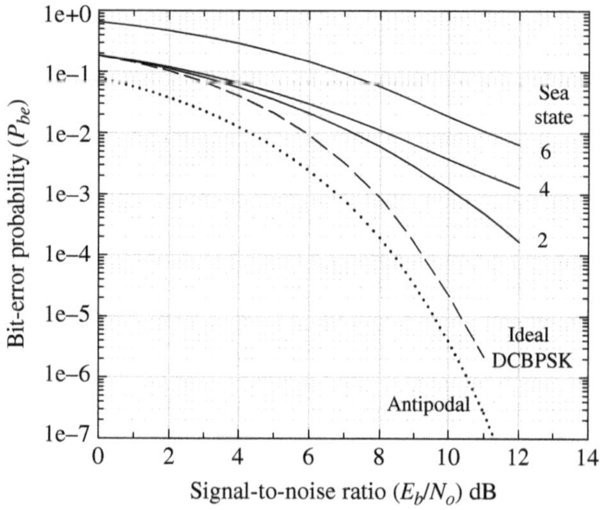

FIGURE 19.55 Performance of DCBPSK with 16-bit words with sea-state conditions 2, 4, and 6 ($f_c = 30$ kHz).

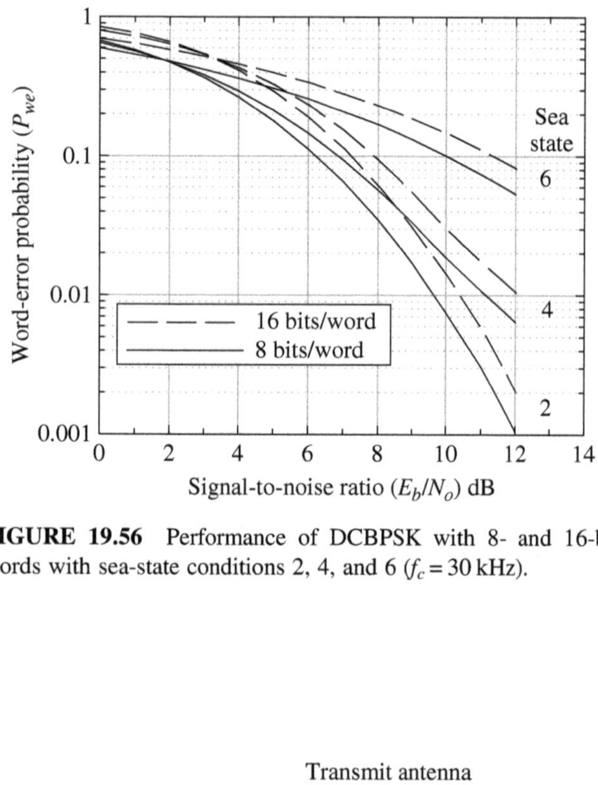

FIGURE 19.56 Performance of DCBPSK with 8- and 16-bit words with sea-state conditions 2, 4, and 6 ($f_c = 30$ kHz).

Figures 19.55 and 19.56 show the error performance for a DCBPSK-modulated waveform with sea-states 2, 4, and 6. Figure 19.55 corresponds to bit-error probability using 1 M bits for each signal-to-noise ratio. The performance of antipodal and ideal DCBPSK operating in the AWGN channel are included for reference. When the waveform amplitude distortion, given by (19.114), is removed in the simulation, it is found that the performance is essentially unchanged indicating that the differentially coherent detection is predominantly impacted by the sea-state phase distortion. The simulation conditions result in $R_sT_B = 16$, 32, and 64 symbols for each sea-state time-constant so increasing the symbol rate by about 8 : 1 or (32 : 1) should not result in a significant performance loss due to the phase distortion. The use of FEC coding and bit interleaving, with an interleaver span comparable to T_B, will mitigate the performance loss resulting from wave fluctuations. Figure 19.56 shows the performance of DCBPSK-coded message words with 8 and 16 bits-per-word.

19.12 LASER COMMUNICATIONS USING PHOTOMULTIPLIER DETECTOR

19.12.1 Introduction

A laser detector using a photomultiplier tube (PMT) [62] is shown in Figure 19.57. The collimated incident light waves (or photons) from the transmitter source are focused on the optical frequency filter and the desired filtered output illuminates the photosensitive surface of the photomultiplier tube releasing elections. The electrons are attracted to the photomultiplier anode by an electric field, created using the photomultiplier prime power, resulting in a current (i_s) through the load resistor R_L. The current fluctuates with the source modulation as does the voltage across R_L which is used to demodulate the source information. The gain of the photomultiplier is related to the number of free electrons at the anode relative to the energy of the incident light wave and gains on the order of 10^3 to 10^4 are not uncommon. The spatial filter eliminates light waves, entering the optical detector from directions other than from the communication transmitter [63, 64].

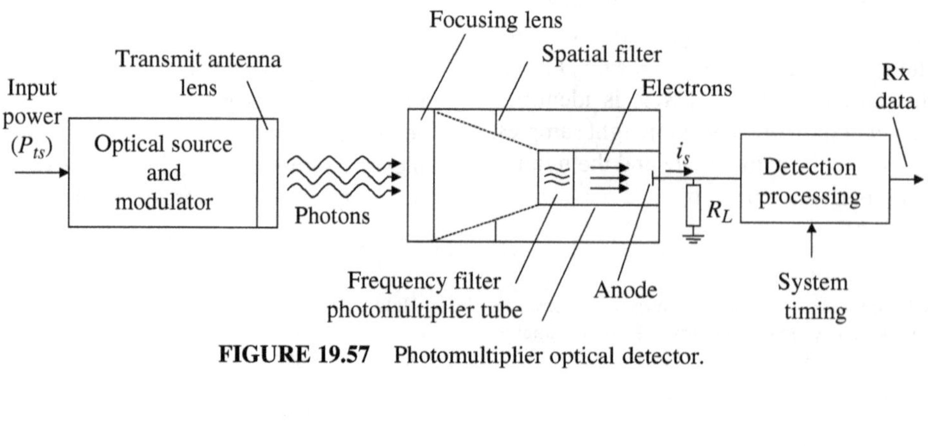

FIGURE 19.57 Photomultiplier optical detector.

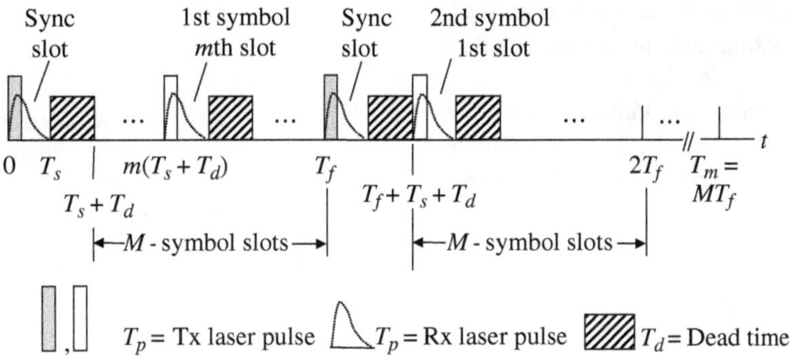

FIGURE 19.58 *M*-ary PPM waveform symbol frame.

The equivalent received signal power at the input to the PMT is denoted as P_s (Watts) and the current created by the PMT is determined as

$$i_s = P_s GS \tag{19.127}$$

where G is the gain of the PMT and S is the responsivity of the PMT defined as

$$S \triangleq \frac{q\eta}{h\nu} \quad : \text{A/W} \tag{19.128}$$

where q is the electron charge = 1.602×10^{-19} C, η is the efficiency of the PMT, h is Planck's constant = 6.62×10^{-34} J-s,* and ν is the frequency in hertz.

The received signal power at the output of the PMT is the power delivered to the load resister and is computed as $P_L = i_s^2 R_L$. Substituting (19.127) for i_s results in the received power

$$P_L = (P_s GS)^2 R_L \quad : \text{W} \tag{19.129}$$

19.12.1.1 Photomultiplier Tube Noise Sources The noise developed across the load resister results from several noise sources as described in this section.

DC or non-fluctuating noise is defined in terms of the noise current $i_{dc} = P_{dc} GS$; therefore, the DC power into the load resistor is

$$P_{Ldc} = (P_{dc} GS)^2 R_L \quad : \text{W} \tag{19.130}$$

that behaves like the received signal power.

Thermal noise is determined by the system temperature T_s and is unaffected by the PMT gain; the two-sided noise power density of the thermal noise is

$$N_{ot} = 2kT_s \quad : \text{W/Hz} \tag{19.131}$$

Signal-induced noise is shot-noise resulting from the random release of electrons from the PMT optical surface. The shot-noise is excited by the incident signal photons and results in the two-sided noise power density expressed as

$$N_{os} = qP_s G^2 SR_L \quad : \text{W/Hz} \tag{19.132}$$

Background-induced noise results from background light energy seen by the optical receiver. It consists of the Sun, Moon, bioluminescence, and other sources that are within the optical field of view (FOV); the two-sided background noise power density is

$$N_{ob} = qP_b G^2 SR_L \quad : \text{W/Hz} \tag{19.133}$$

Dark current noise results from shot-noise caused by the random release of photoelectrons from the PMT optical surface due to the surface temperature. Although dark current noise encounters the gain of the PMT and it is usually negligible compared to the background induced noise and is often neglected. The dark noise power density is denoted as N_{od}.

19.12.1.2 Demodulator Signal-to-Noise Ratio The following description is based on the *M*-ary pulse position modulation (PPM) waveform format as shown in Figure 19.58. The symbol frame consists of a synchronization slot, containing the *shaded* transmitted laser pulse, followed by a *dead-time* to provide for laser recharging, shown as the cross-hatched interval, and *M* symbol slots that are selected based on the unique ($\log_2 M$)-bit data symbol. The transmitted laser pulses encounter dispersion through the channel and are depicted as the *dotted* received pulses. Selecting the slot duration of T_s seconds involves a tradeoff between several system design considerations. For example, in consideration of the pulse dispersion, increasing the slot duration increases the average received pulse energy capability at the detriment of the frame efficiency and message throughput. The dead-time (T_d) determines the pulse repetition frequency (*prf*) of

*Note: 1 C = A-s and 1 J = W-s.

the laser and is a major contributor to the frame overhead.* The PPM parameters are examined in more detail in Section 19.12.3.

The demodulator post-detection bandwidth is the reciprocal of the slot duration, that is, $B_s = 1/T_s$. The demodulator signal-to-noise ratio is defined in terms of the pulse energy, $E_P = P_L T_s$, as

$$\gamma_{slot} = \frac{P_L T_s}{N_o} \tag{19.134}$$

The noise density is the sum of the noise sources discussed above, excluding the DC noise that is removed by the post-detection filter, so the noise density is expressed as

$$N_o = N_{os} + N_{ob} + N_{ot} + N_{od} \tag{19.135}$$

Upon substituting (19.129) for P_L, neglecting the dark and thermal noise sources, and substituting (19.135) with (19.132) and (19.133) into (19.134) results in

$$\gamma_{slot} = \frac{P_s^2 S^2 T_s}{q(P_s S + P_b S)} \quad : \text{W-s/(W/Hz)} \tag{19.136}$$

The responsivity (S) has units of amperes-per-watt, so the units of the product PS are amperes or coulombs-per-second. Therefore, since q is the electron charge with units of coulombs-per-electron, the quantity PS/q has units of electrons-per-second. In the following, the notation $\lambda = PS/q$ is used and (19.136) simplifies to

$$\gamma_{slot} = \frac{\lambda_s^2 T_s}{\lambda_s + \lambda_b} = \frac{N_s^2}{N_s + N_b} \tag{19.137}$$

where the second equality results by multiplying the numerator and denominator by T_s and recognizing that $N = \lambda T_s$ is the number of photoelectrons in T_s seconds. Equation (19.137) is expressed in terms of the signal-to-noise ratio in the bandwidth equal to the bit-rate, $R_b = 1/T_b$, using $\gamma_{slot} = \rho\gamma_b$, where

$$\rho = \frac{T_s}{T_b \log_2(M)} \tag{19.138}$$

Upon solving (19.137) for N_s and expressing the result in terms of γ_b, the signal photoelectron count is evaluated as

$$N_s = \frac{\rho\gamma_b}{2}\left(1 + \sqrt{1 + \frac{4N_b}{\rho\gamma_b}}\right) \tag{19.139}$$

*To simplify the demodulator time synchronization, the dead time is often defined as an integer number of slot time intervals.

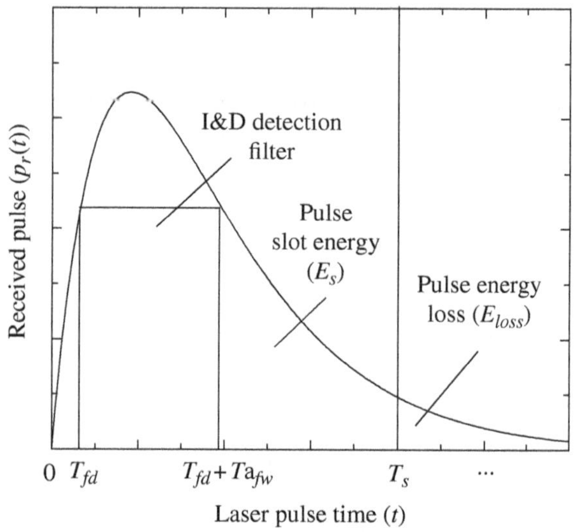

FIGURE 19.59 Received pulse shape through clouds.

These parameters are exactly those required to characterize the bit-error performance of the demodulator in terms of the Poisson distribution as discussed in the following case studies [65, 66]. The photoelectron counting is also modeled using the Nakagami m-distribution [67].

19.12.2 Pulse Dispersion in the Atmosphere

Based on experimental results [68–70], the generally accepted shape of a pulsed laser after emerging from clouds is shown in Figure 19.59 and characterized as

$$p_r(t) = te^{-\alpha t} \tag{19.140}$$

In the following analysis, the PPM slot duration, measured in microsecond, is selected to satisfy two conflicting requirements: increase T_s to capture more received signal energy (E_s) for detection or decrease T_s to provide the shortest PPM frame. The second of these options is dominated by the peak-to-average power ratio and the resulting duty cycle [71] of the laser transmitter. The laser pulse dispersion† in the atmosphere results from the propagation path length (L_{cp}) through clouds and is characterized principally by the geometric thickness (L_{cg}) and the mean scattering angle (θ_{cs}) of the cloud. The delay (T_{fd}) and width (T_{fw}) of the integrate-and-dump (I&D) detection filter must be selected to maximize the signal-to-noise ratio at the optimum sample time of the detection filter. Communication links from the atmosphere to underwater links encounter additional dispersion that is dependent on the geometry of the encounter and

†Dispersion occurs in any medium in which the velocity of a wave varies with wavelength.

the water type [72], for example [73], Jerlov II and III water[*] and bioluminescence is a major source of background noise.

Stotts [74] has developed a closed-form expression for laser pulse dispersion through clouds, for which, the pulse dispersion is expressed as

$$T_{cd}=\begin{cases}\dfrac{L_{cp}}{c}\left(\dfrac{0.3\left(1+2.25\omega_o\tau_c\theta_{cs}^2\right)^{1.5}-1}{\omega_o\tau_c\theta_{cs}^2}-1\right) & :\tau_c>10,\text{pulse dispersion}\\ 0 & :o.w.\end{cases} \quad (19.141)$$

where $c = 3e8$ m/s is the free-space speed of light, $\omega_o > 0.999$ is the cloud albedo[†] (in the following analysis $\omega_o = 1.0$), τ_c is the optical thickness of the cloud,[‡] and L_{cp} and θ_{cs} are in *meters* and *radians* respectively. The relationship between the geometric cloud thickness (L_{cg}) and the optical cloud thicknesses is defined as

$$L_{cg} \triangleq \frac{\tau_c}{\sigma_c} \quad (19.142)$$

where σ_c is defined as the average extinction (or attenuation) coefficient of the cloud with dimensions of meters^{-1}. The extinction coefficient is a measure of absorption of light in the cloud. In the following analysis $\sigma_c = 0.04\ \text{m}^{-1}$ which results in L_{cg} having dimensions of meters. The optical thickness of various types of clouds [75, 76] is listed in Table 19.14.

Figure 19.60 is a plot of (19.141) showing the dependence of the pulse dispersion on the propagation path L_{cp} for the indicated mean scattering angles and average extinction of the cloud.

The pulse dispersion is measured as the pulse duration between the half-power points of the pulse relative to the peak power P_{pk} that occurs at $t = T_{pk} = 1/\alpha$. Defining the half-power points of the pulse as occurring at T_1 and T_2, the pulse dispersion, expressed in (19.141), is $T_{cd} = T_2 - T_1$. Referring to Problem **23**, T_1 and T_2 are related to the parameter α as $T_1 = 0.232/\alpha$ and $T_2 = 2.677/\alpha$. Using these relationships, the pulse is completely characterized in terms of the dispersion through the cloud by the parameter α expressed as

$$\alpha=\frac{2.445}{T_{cd}} \quad :\text{pulse parameter dependence on cloud dispersion} \quad (19.143)$$

[*]Among other characteristics, the Jerlov water types include the optical backscatter and attenuation.

[†]Albedo is the ratio of the reflected to the incident radiant power falling on the surface of the cloud.

[‡]The optical thickness is equal to ln(T), where T is the transmittance of the cloud defined as the ratio of the incident to transmitted radiant power through the cloud. The notations τ, τ_c, or τ_{opt} are often used to denote the cloud optical thickness.

TABLE 19.14 Optical Thickness of Clouds[a]

	Optical Thickness	
Cloud Type	Range	Average
Cirrus, Altocumulus, Cumulus	0–3.6	1.3
Cirrostratus, Altostratus, Stratocumulus	3.6–23	9.4
Deep Convection, Nimbostratus, Stratus	23–379	60

[a]International Satellite Cloud Climatology Project (ISCCP) 1983–2010 [76]. Courtesy of National Aeronautics and Space Administration (NASA).

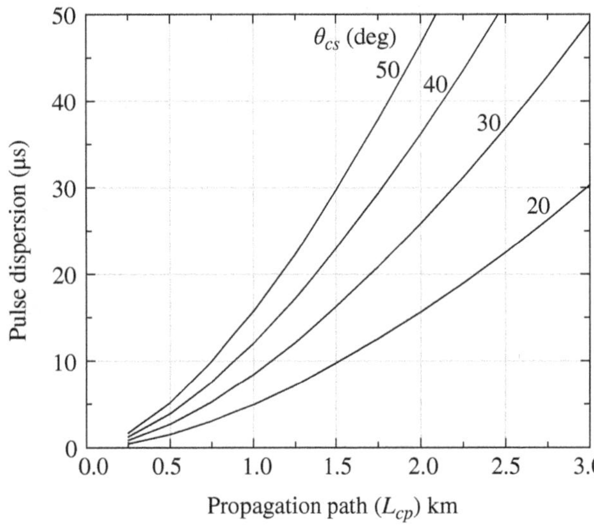

FIGURE 19.60 Laser pulse dispersion in clouds ($\sigma_c = 0.04\ \text{m}^{-1}$).

Figure 19.61 shows the received pulse responses for the indicated values of the parameter L_{cp}, in *kilometer*, corresponding to the mean scattering angles used in Figure 19.60. In these plots, the received pulses are normalized by the total pulse energy $E_p = 1/\alpha^2$ resulting in unit energy pulses for each condition. Therefore, the integral

$$E(T_{cd}) = \int_{T_1}^{T_2} te^{-\alpha t}dt \quad (19.144)$$

and $E(T_{cd})/E_p$ represents the normalized received pulse energy in the interval T_{cd}. Conversely, $1 - E(T_{cd})/E_p$ represents the loss of energy in decibels.

Evaluating (19.144) over the range $t = 0$ to T_s, the received pulse energy in the PPM slot is determined as E_s and is plotted in Figure 19.62 as a percent of the total received pulse energy. Figures 19.61 and 19.62 or Table 19.15 show that there is a large variation in T_s to cover the conditions; however, with $\theta_{cs} = 40°$ a slot duration of 23.3 μs will capture 95% of the pulse energy with $L_{cp} = 1$ km, whereas, a slot duration of 100 μs is required with $\theta_{cs} = 40$ and $L_{cp} = 2$ km

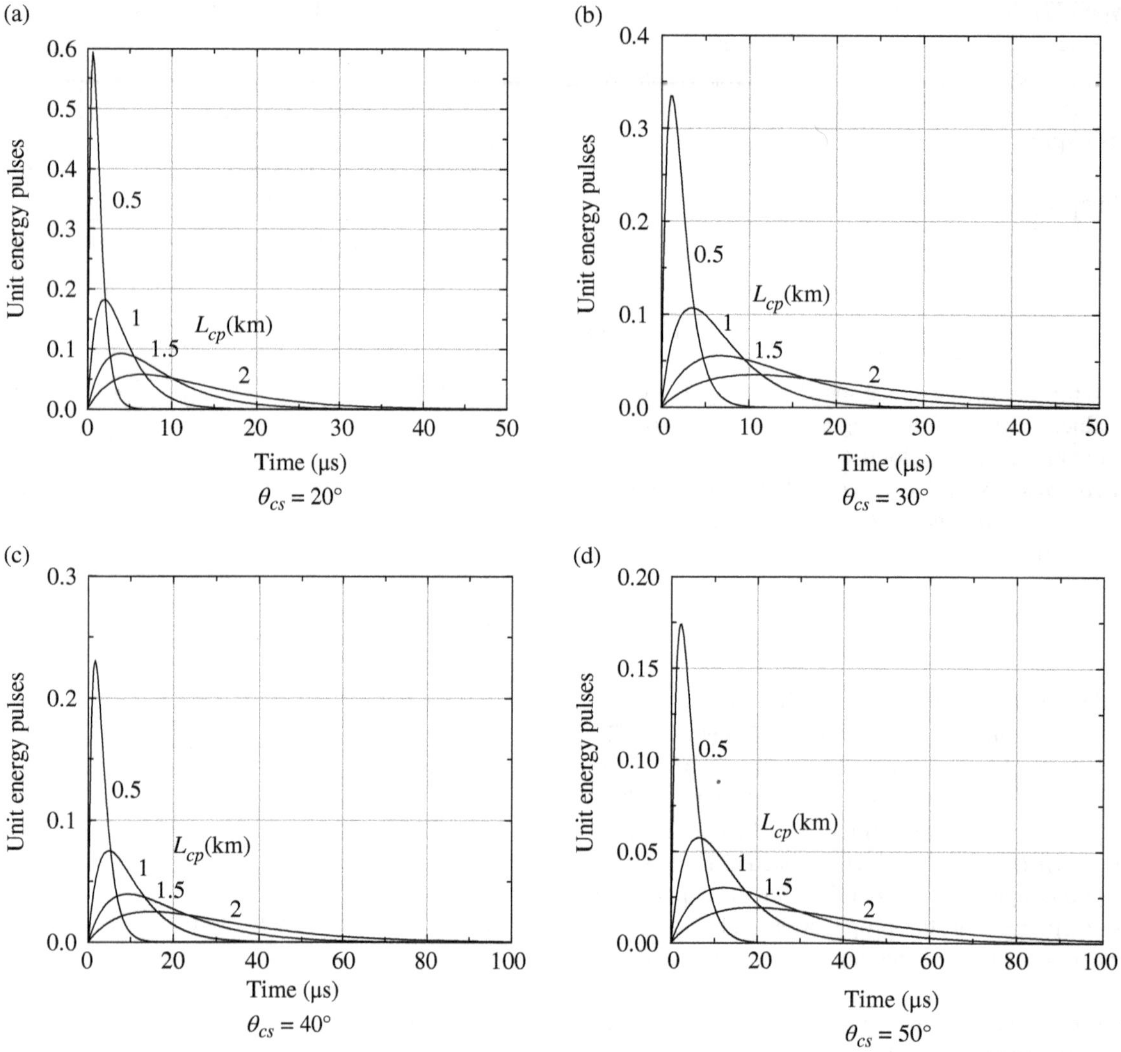

FIGURE 19.61 Laser pulse shape in clouds.

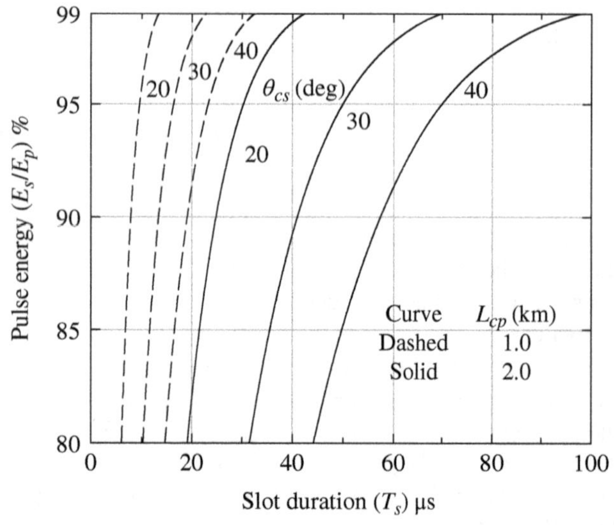

FIGURE 19.62 Percent of total pulse energy in slot duration.

TABLE 19.15 Slot Width to Capture Percent of Total Pulse Energy

		T_s (μs)					
		$L_{cp} = 1$ km			$L_{cp} = 2$ km		
E_s/E_p (%)	Loss (dB)	20[a]	30	40	20	30	40
80	0.97	6.1	10.3	14.7	19.1	31.6	44.2
85	0.71	6.8	11.6	16.6	22.1	35.5	49.8
90	0.45	7.9	13.4	19.1	24.9	41.0	57.4
95	0.22	9.6	16.3	23.3	50.0	50.0	70.0

[a]This row is θ_{cs} in degrees.

capture the same energy. If the transmitter can estimate the cloud conditions then T_s can be minimized and the receiver can form an estimate of the received slot duration during the waveform acquisition processing.

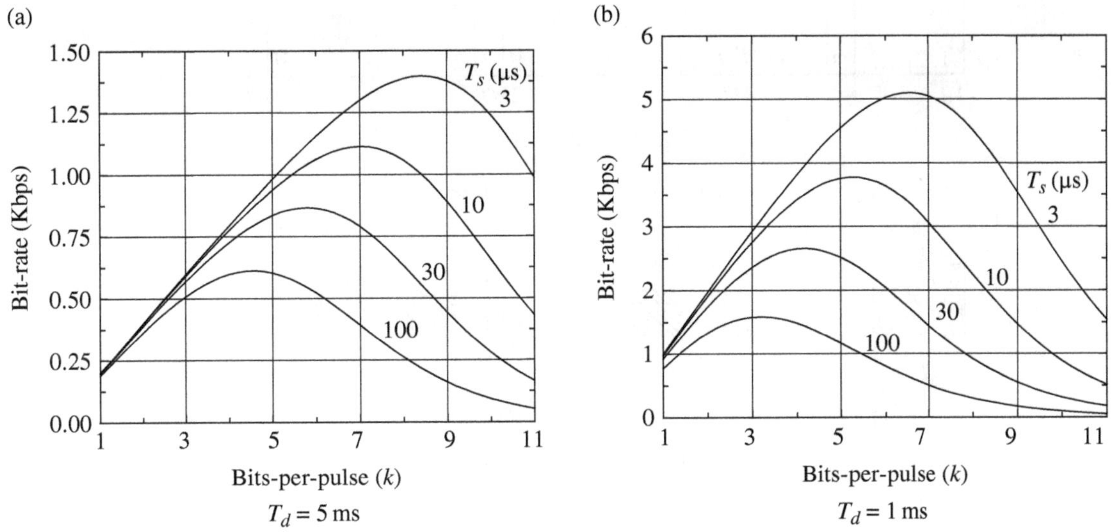

FIGURE 19.63 PPM bit-rate.

19.12.3 Pulse Position Modulation Waveform

The transmission of M-ary coded data [77, 78] using a pulsed laser power source is often transmitted using PPM as described in Figure 19.58. The slot synchronization pulses are shown to occur prior to each M-ary symbol location; however, for covert synchronization, they may occur at predetermined random locations known by the demodulator. Using periodic synchronization pulses simplifies the synchronize processing through the use of a narrowband bandpass filter centered at the synchronization pulse repetition rate; however, the use of FEC coding improves the bit-error performance at the expense of increased complexity [79, 80]. Example FEC configurations are: a rate r_c convolutional code with a prescribed constraint length, a Reed–Solomon code, for example a concatenated code with a constraint length K convolutional inner code and an M-ary Reed–Solomon outer code with $M = 2^k$ bits per Reed–Solomon symbol. Data interleavers should also be used with FEC coding; especially between concatenated codes.

The source bit-rate (R_b) of M-ary PPM, without FEC coding, is expressed in terms of the bits-per-pulse (k) as

$$R_b = \frac{k}{(1+2^k)T_s + T_d} \tag{19.145}$$

where $k = \log_2(M)$. Equation (19.145) is plotted in Figure 19.63a and b as a function of k for $T_d = 5$ and 1 ms respectively with the slot duration (T_s) as the parametric parameter. The laser dead-time and the slot duration are seen to have a significant impact on the maximum data rate that also dictates the selection of k. The PPM waveform design, including the selection of the slot duration, bits-per-slot, and maximum user bit-rate, are established using Figure 19.63. For example, using the 5 ms dead-time shown in Figure 19.63a, and selecting $k = 7$ and $T_s = 10$ μs results in a 128-ary PPM waveform with a maximum uncoded user bit-rate of 1.113 kbps.

Defining the slot duration as in Figure 19.59, the total received pulse energy in the slot and the percentage of pulse energy is determined from Figure 19.62 or Table 19.15. The optimum I&D detection filter delay and duration, shown in Figure 19.59, must be determined in the context of the statistical characterization of the PMT noise and other noise sources as discussed in Sections 19.12.1.1 and 19.12.1.2.

19.12.4 Differential Pulse Interval Modulation Waveform

With differential pulse interval modulation (DPIM) [81], the pulse intervals are synchronized to the transmission of an initial synchronization pulse followed by M-ary coded pulses based on the differential change in the time slots relative to the preceding pulse. DPIM modulation is depicted in Figure 19.64 using the M-ary coded waveform with symbol sequence (m, 5, 2, …). In this example, the DPIM-coded waveform transmits the first information pulse in slot m following the reference pulse and the associated dead-time, the second pulse is transmitted in slot five following the laser recovery of the first pulse, and the third pulse is transmitted in slot two following the laser recovery time. Therefore, the three information pulses are transmitted in a time equal to $(m + 8)T_s + 3T_d$ μs; whereas, the PPM-modulated waveform, shown in Figure 19.58, transmits the first three information pulses in $(3M + 1)T_s + 4T_d$ μs. The average time to transmit a frame of M-ary coded random data using the DPIM-modulated waveform is

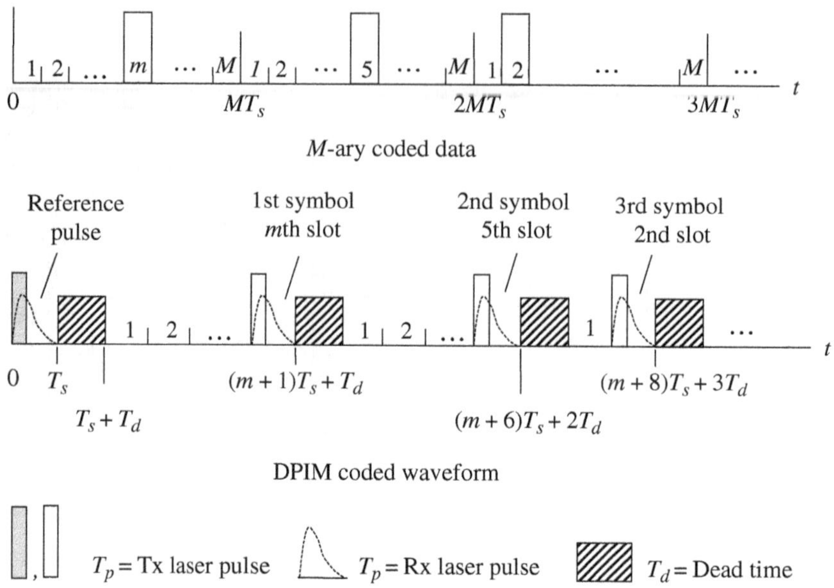

FIGURE 19.64 *M*-ary DPIM waveform symbol modulation.

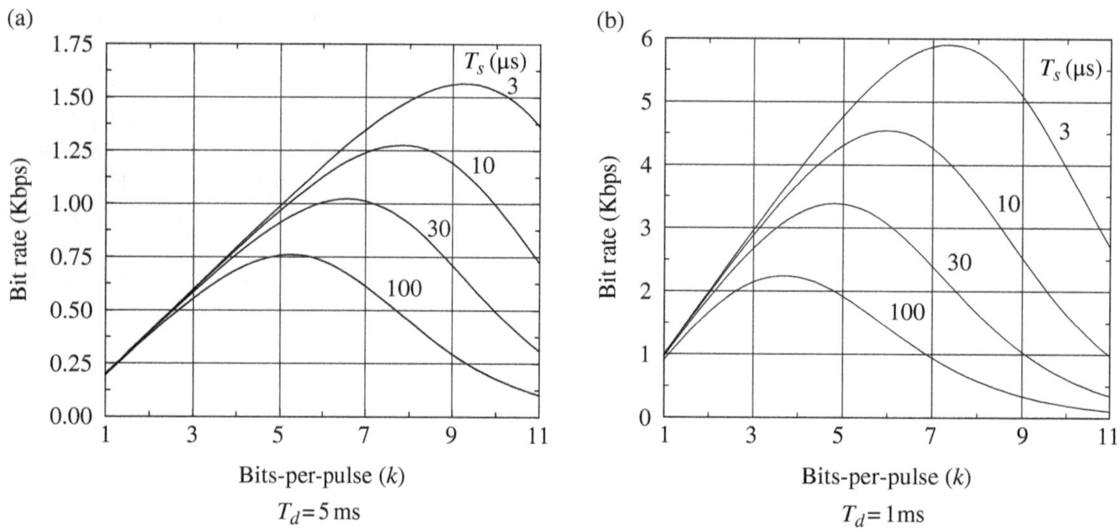

FIGURE 19.65 DPIM average bit-rate.

$$\bar{T}_f = \left(\frac{M}{2}\right) T_s + T_d \quad : \mu\text{s} \tag{19.146}$$

compared to $\bar{T}_f = MT_s + T_d$ μs for the PPM-modulated waveform.

Figure 19.65 shows the bit-rate of DPIM as a function of the bits-per-slot under similar conditions for the PPM modulation shown in Figure 19.63. The average bit-rate for DPIM is expressed as

$$\bar{R}_b = \frac{k}{2^{k-1} T_s + T_d} \quad : \text{DPIM modulation} \tag{19.147}$$

19.12.5 Case Study: PMT Demodulator Performance Using OOK Modulation

In this case study, the laser transmitter is either on-off keying (OOK) resulting in the binary hypothesis in which a binary *zero* (b_0) is represented by no signal energy being transmitted and a binary *one* (b_1) is represented by transmitting a signal energy of E_s. Under the maximum-likelihood (ML) decision rule, the a priori probabilities are represented by $P(b_0) = P(b_1) = 1/2$ and the decision rule is expressed in terms of the transition probabilities using the likelihood ratio as

$$\Lambda(p) = \frac{p(y|b_1)}{p(y|b_0)} \begin{cases} >1 & : \text{choose } b_1 \\ o.w. & : \text{choose } b_0 \end{cases} \tag{19.148}$$

The transition probabilities are characterized in terms of the Poisson distributions [82]

$$p(y|b_i) = \frac{(\lambda_i T_s)^k}{k!} e^{-(\lambda_i T_s)} = \frac{N_i^k}{k!} e^{-N_i} \tag{19.149}$$

where λ_i is the photoelectron rate and T_s is the interval in which the photoelectrons are counted, so N_i is the number of photoelectrons in T_s seconds. Under the conditions b_1 and b_0, the respective notations for the photoelectron counts are $N_1^k = N_s^k$ and $N_1^k = (N_s + N_b)^k$. T_s is the reciprocal of the receiver post-detection bandwidth and typically $T_s \cong T_b$ to allow for dispersion of the bit energy through the optical channel. Substituting (19.149) into (19.148) and rearranging the terms of the log-likelihood ratio (LLR), the log-likelihood ratio test (LLRT) is expressed, in terms of the threshold k_T, as

$$k_T \begin{cases} > \dfrac{N_s}{\ln((N_s + N_b)/N_b)} & : \text{choose } b_1 \\ o.w. & : \text{choose } b_0 \end{cases} \tag{19.150}$$

This threshold is the optimum threshold for making the decision that minimizes the bit-error probability, evaluated as

$$P_{be} = \frac{e^{-N_b}}{2} \sum_{k=k_T}^{\infty} \frac{N_b^k}{k!} + \frac{e^{-(N_s+N_b)}}{2} \sum_{k=0}^{k_T} \frac{(N_s + N_b)^k}{k!} \tag{19.151}$$

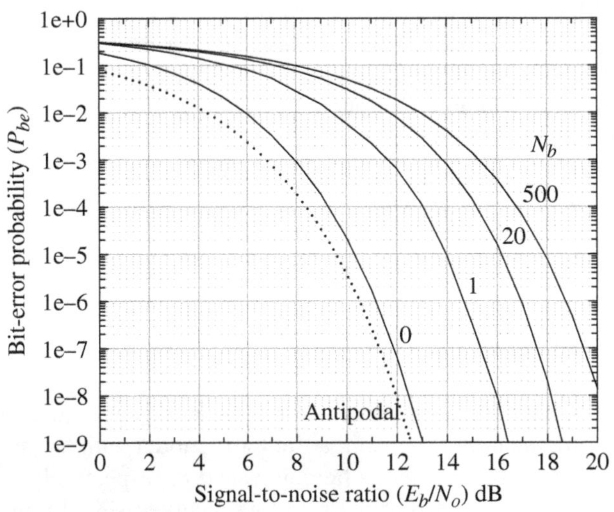

FIGURE 19.66 Bit-error performance of OOK optical detector ($T_s = T_b$).

P_{be} is evaluated in terms of γ_b by specifying the background photoelectron count N_b, and using (19.139) to determine N_s. The resulting bit-error performance is shown in Figure 19.66 under ideal channel conditions with $T_s = T_b$ and the indicated values of background noise. The bit-error performance for $N_b = 0$ results in $k_T = 0$ so P_{be} simplifies to $P_{be} = 0.5e^{-N_s}$. The slight ripple in the $N_b = 1$ curve results from the integer summations involving the Poisson distributions; for threshold levels exceeding 14 the Gaussian approximation to the Poisson distributions is use to evaluate P_{be}. The noncoherent OOK performance with AWGN is discussed in Chapter 6.

19.12.6 Case Study: PMT Demodulator Performance of *M*-ary PPM Waveform

In this case study, the bit-error performance is examined for the M-ary PPM waveform [77] as characterized in Figure 19.58 with $k = \log_2 M$ bits transmitted in one of M time slots during each symbol frame of duration $T_f = MT_s + T_d$ seconds. The laser pulse contained in the interval T_{sync} is used for frame synchronization and tracking. In this case, the ML decision rule uses equal a priori probabilities for each symbol with $P_m = 1/M$: $\forall\, m \in M$. The rule that minimizes the symbol-error probability selects the time slots having the largest post-detection filter output when the post-detection filter is matched to the received pulse. An upper bound on the symbol error is given by Bucher [83] as

$$P_{be} \le (M-1)e^{-N_s E_c(N_s/N_b)} \tag{19.152}$$

where the exponent term $E_c(N_s/N_b)$ is equal to

$$E_c\left(\frac{N_s}{N_b}\right) = \frac{\left(\sqrt{1 + N_s/N_b} - 1\right)^2}{N_s/N_b} \tag{19.153}$$

Solving (19.137) for N_s/N_b and substituting the result into (19.153) results in

$$E_c\left(\frac{N_s}{N_b}\right) = \left(\frac{N_s - \gamma_{slot}}{\gamma_{slot}}\right)\left(\sqrt{\frac{N_s}{N_s - \gamma_{slot}}} - 1\right)^2 \tag{19.154}$$

As in the preceding binary example, the following performance is characterized in terms of the background noise count N_b; however, in this case, the independent variable will be the signal-to-noise ratio as measured in the slot bandwidth. Referring to (19.139), the parameter N_s is expressed in terms of N_b with the signal-to-noise ratio (γ_{sym}) measured in the symbol bandwidth equal to $1/T$. Therefore, the signal photoelectron count, measured in the slot bandwidth, is computed as

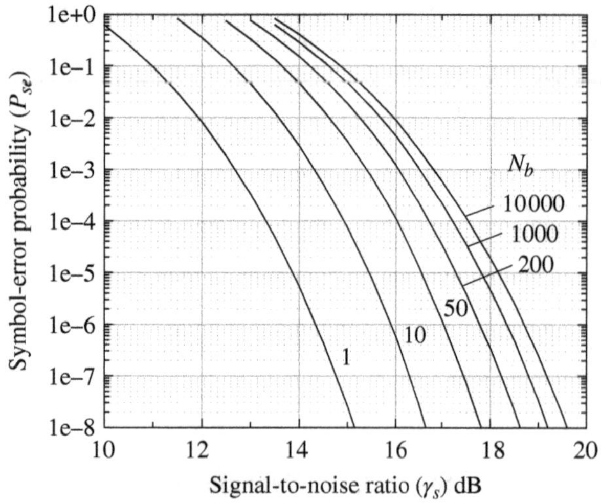

FIGURE 19.67 Symbol-error performance of 256-ary PPM optical detector ($T_s = T$).

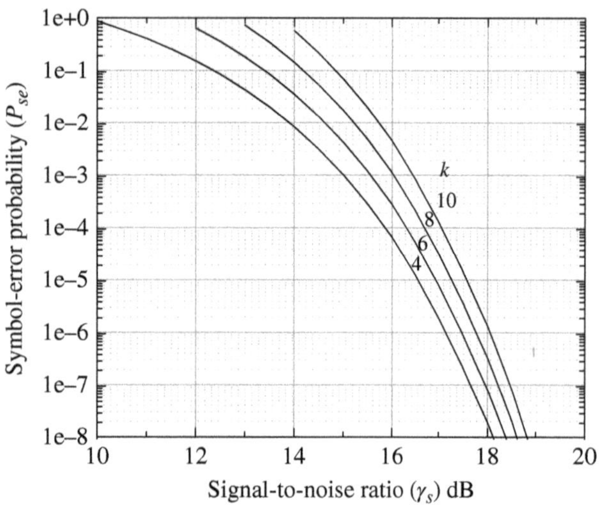

FIGURE 19.68 Symbol-error performance of M-ary PPM optical detector ($N_b = 200$).

$$N_s = \frac{\rho' \gamma_{sym}}{2}\left(1 + \sqrt{1 + \frac{4N_b}{\rho' \gamma_{sym}}}\right) \qquad (19.155)$$

where $\rho' = T_s/T$ converts the signal-to-noise ratio from the symbol bandwidth to the signal-to-noise ratio in slot bandwidth.

The M-ary PPM performance is shown in Figure 19.67 as a function of γ_s for various values the background noise count N_b. The symbol-error results are based on the union-bound [84] which is a *loose* upper bound on the performance; however, for a fixed M, the union-bound becomes increasingly tighter as γ_s increases. Figure 19.68 shows the union-bound symbol-error performance for the indicated values of $k = \log_2(M)$ using a background photoelectron count of $N_b = 200$.

ACRONYMS

AGU	American Geophysical Union
APD	Amplitude probability distribution
AWGN	Additive white Gaussian noise
BPSK	Binary phase shift keying
CBPSK	Coherent BPSK
CCIR	Consultative Committee on International Radio
CL	Clipping level
CPFSK	Continuous phase frequency shift keying
CW	Continuous wave
DCBPSK	Differentially coherent BPSK
DFT	Discrete Fourier transform
DPIM	Differential pulse interval modulation
EDS	Energy density spectrum
FEC	Forward error correction (coding)
FFT	Fast Fourier transform
FOV	Field of view
HF	High frequency
HL	Hard limiting
I/Q	Inphase and quadrature (channels or rails)
IF	Intermediate frequency
IFFT	Inverse fast Fourier transform
IFT	Inverse Fourier transform
ISCCP	International Satellite Cloud Climatology Project
ISI	Intersymbol interference
ITU	International Telecommunication Union
ITU-R	International Telecommunication Union-Radio
LF	Low frequency
LLR	Log-likelihood ratio
LLRT	Log-likelihood ratio test
LOS	Line of sight
MED	Modified exponential delay (loss model)
ML	Maximum-likelihood (decision rule)
MSK	Minimum shift keying
NASA	National Aeronautics and Space Administration
PCS	Personal communication services
PMT	Photomultiplier tube
PLL	Phaselock loop
PPM	Pulse position modulation
TOD	Time-of-day
UHF	Ultra-high frequency
VLF	Very low frequency

PROBLEMS

1. Establish the geometry for a ground-to-ground communication link using a flat Earth model with transmit and receive antenna heights h_t and h_r respectively. (a) Evaluate and plot the resulting received signal power density in terms of microwatts per square meter as a function of range for: $h_t = h_r = 3$ m, transmitter power

$P_t = 10$ W, $G_t = 3$ dB and $f_c = 1$ GHz. (b) Compare the results using the spherical Earth model in Figure 19.2 using the effective Earth radius $r_e = (4/3)R_e$; assume that the Earth reflected multipath ray is negligible.

2. Referring to Figures 19.1 and 19.2, given a transmitter height of h_1, determine the minimum elevation angle θ_d(min) for a radio line-of-light path.

3. Derive the expression for the average multipath loss $\bar{L}_m$ defined in Section 19.4 in terms of the parameter $x = \rho|f_r|/|f_d|$. Express the result as a function of x.

4. Using the geometry shown in Figures 19.2 and 19.7 for a circular LEO satellite orbit with altitude $h_2 = 200$ km and a ground antenna height of $h_1 = 30$ m, derive an expression for the antenna elevation angle θ_d as a function of the satellite orbit angle θ_s.
   a. In Figure 19.2 consider the satellite to be at the height h_2. At what angle, θ_d(min), will the satellite be in view of a ground antenna with $h_1 = 0$?
   b. Using (16.3), with $\varepsilon = 0$ for a circular orbit, compute the satellite orbit time rate in second per degree. What is the viewing time of the satellite by the ground antenna over the interval $\theta_s(\text{min}) \le \theta_s \le 90°$?
   c. Plot the satellite viewing time as a function of the satellite orbit angle θ_s.

5. By applying Huygens' principle to Figure 19.4 use geometry to prove that the angle of reflection ϕ_r is equal to the angle of incidence ϕ_i.

6. Evaluate the expressions (19.46), (19.48), and (19.52) for 0 and 100% relative humidity using an apparent elevation angle of $\alpha_o = 15°$, $\psi = 120°$ with $h_g = 0$ and $h_t = 50$ K ft. Using h_g and h_t compute the corresponding refractive indices n_g and n_t using (19.42) through (19.45) and then compute the refraction angle error $\Delta\alpha_o$ corresponding to the apparent elevation angle α_o using Figures 19.14 and 19.15. Using these newly computed parameters determine the refractive angle error $\Delta\alpha_t$ using (19.49); $\Delta\alpha_t$ must be in radians. The final steps involve computing the Doppler frequency and frequency Doppler frequency error using (19.46) and (19.48). The Doppler frequency error is also to be expressed in the normalized form of (19.52) using the normalizing parameters $f_{co} = 100$ MHz, $V_o = 6.096$ km/s. Note that these Doppler frequency expressions are what the aircraft terminal would encounter when attempting to receive a communication message; however, in a radar application the ground terminal radar would, because of reciprocity, experience twice these values over the two-way path. Additional parameters necessary to complete this problem are: $R_e = 6378$ km and $c = 3 \times 10^8$ m/s.

7. Using the differential path length $\Delta_k = k\lambda/2$ and the approximation given in (19.56): (a) derive the expression for the Fresnel zone radius h_k and (b) show that the maximum Fresnel zone radius occurs when $d_1 = d_2$.

8. For a knife-edge obstruction with $h_1 = h_2 = 30$ m and a carrier frequency of 500 MHz, determine: (a) the diffraction loss for $h = 12.76$, 4.26, 0, −2.13, and −4.26 m, (b) in each case determine the Fresnel zone, k, of the obstruction, (c) comment on the significance of the Fresnel zones for $h = 4.26$ and −4.26 being identical given that the losses diffraction losses are significantly different.

9. Using an effective transmit and receive antenna height of 1 and 10 m and a carrier frequency $f_c = 100$ MHz, plot the normalized propagation loss in percent ($100\,L_p/L_{fs}$) as a function of the range $r = (0.5, 1, 2, 3, 4, 5)$ where L_p is the propagation loss computed using the Longley-Rice Model and L_{fs} is the corresponding free space loss. For each range r use the following values of L_p corresponding to 100 MHz from Figure 19.19:

$$L_p(h_e = 1\text{ m}) = (30.4, 37.7, 41.7, 45.7, 49.0, 51.7)$$
$$L_p(h_e = 10\text{ m}) = (16.1, 21.2, 26.5, 29.8, 32.4, 34{,}5)$$

10. Choose realistic heights for the transmit and receive antennas for a wireless mobile link and, using a base-station or transmitter power of $P_t = 10$ W, and antenna gain $G_t = 3$ dB, compute, plot, and compare the received power density in microwatts per square meter as a function of distance d for the Okumura and Hata models described in Sections 19.8.1 and 19.8.2 using the suburban environments; for the Hata model use (19.63) for the suburban environment. Repeat the comparison using the open area environment; for the Hata model use the open rural environment loss expressed by (19.64).

11. Choose a realistic transmit antenna height (h_t) for a PCS link and use the ETAL model with a transmitter power of $P_t = 10$ W, antenna gain of $G_t = 8.14$ dB to evaluate and plot as a function of range the received signal power density in terms of microwatts per square meter for all three terrain categories listed in Table 19.5. Note: This model applies for a carrier frequency of $f_c = 1.9$ GHz and a receiver antenna height of $h_r = 2$ m.

12. Show that the Rayleigh distribution $p(r)$ results in a straight line with slope of −1/2 when plotted as $\log(r_o/2\sigma_o)$ as a function of $-\log(-\ln(P(r > r_o)))$.

13. Compute V_d for Gaussian noise.

14. Using the equation for G_n, determine the clipping threshold voltage, V_{th}, relative to the received signal voltage V_r given that the clipping threshold relative to the channel

noise standard deviation, σ_{ch}, is $k = V_{th}/\sigma_{ch}$, that is, determine $k' = V_{th}/V_r$.

15. Determine the noise bandwidth of the MSK matched filter.

16. Evaluate the wave energy of the Pierson-Moskowitz energy density function for sea-states 1 through 8 and compare with that of the Neumann energy density function in Table 19.9.

17. Beginning with equation (19.106) in the text, derive Equation (19.109).

18. Show that $\sqrt{u\sigma\omega_c}$ has dimensions of meters^{-1}.

19. Starting with the function $\theta^{(n)}(t)$, defined in (19.117), express the autocorrelation $R_{h^{(n)}}(\tau)$ in terms of the autocorrelation function $R_{\theta^{(n)}}(\tau)$.

20. Using the results of Problem 19, derive expression (19.119) in the text for the EDS $S^{(n)}(\omega)$.

21. Evaluate $E^{(n)}$ by performing the integration in (19.121) and using the result, verify the expression for the standard deviation $\sigma_{\theta^{(n)}}$ given by (19.122) in the text.

22. Derive the coefficients for a second-degree record linking polynomial $y(x_i)$ spanning m samples, such that $i = 0, \ldots, m$ with boundary conditions: $y(0) = y_o$, $\dot{y}(0) = \dot{y}_o$ where y_o and $\dot{y}_o$ are the magnitude and slope of the last samples of the previous record and $y(x_m) = y_m$, $\dot{y}(x_m) = \dot{y}_m$ where y_m and $\dot{y}_m$ are the magnitude and slope of the first samples of the next record.

23. Part a: Determine the time (T_{pk}) and the corresponding peak value $p_r(T_{pk})$ of the received laser pulse expressed by (19.140). Part b: Determine the times T_1 and T_2, in terms of the parameter α, corresponding to the half-power points $p_r(T_1)$ and $p_r(T_2)$ respectively.

Hint: Evaluate the resulting transcendental equation $f(x)$: $x = \alpha t$, for the two solutions using Newton's method using initial guesses to the solution of $x_1(\text{guess}) < 1$ and $x_2(\text{guess}) > 1$.

24. Show that the Poisson distribution approaches the Gaussian distribution as $M \to \infty$.

REFERENCES

1. D.E. Kerr, Editor, *Propagation of Short Radio Waves*, M.I.T. Radiation Laboratory Series, Vol. 13, McGraw Hill Book Company, New York, 1951.
2. L.E. Blake, *Radar Range-Performance Analysis*, Munro Publishing Co., Silver Springs, MD, 1991.
3. P.L. Rice, A.G. Longley, K.A. Norton, A.P. Barsis, "Transmission Loss Predictions for Tropospheric Communication Circuits," National Bureau of Standards Technical Note No. 101 Vol. 2, U.S. Department of Commerce, National Bureau of Standards, Washington, DC, May 7, 1965. Currently: National Institute of Standards and Technology (NIST).
4. L.E. Blake, *Radar Range-Performance Analysis*, pp. 271–274, Munro Publishing Co., Silver Springs, MD, 1991.
5. C.R. Burroughs, S.S. Atwood, *Radio Wave Propagation*, Academic Press, New York, 1949.
6. D.K. Barton, *Modern Radar System Analysis*, pp. 297–280 and Appendix A, Artech House, Norwood, MA, 1988.
7. J.D. Kraus, *Antennas*, McGraw-Hill Book Company, Inc., New York, 1950.
8. S. Ramo, J.R. Whinnery, *Fields and Waves in Modern Radio*, Chapter 7, John Wiley & Sons, Inc., New York, 1959.
9. L.E. Blake, *Radar Range-Performance Analysis*, p. 259, Munro Publishing Co., Silver Springs, MD, 1991.
10. G.A. Hufford, A.G. Longley, W.A. Kissick, "A Guide to the Use of the ITS Irregular Terrain Model in the Area Prediction Mode," U.S. Department of Commerce, Malcolm Baldrige, Secretary, April 1982.
11. E. Shotland, R.A. Rollin, "Complex Reflection Coefficient Over a Smooth Sea in the Micro and Millimeter Wave Bands for Linear and Circular Polarizations," Report FS-76-060, John Hopkins University Applied Physics Laboratory, Laurel, MD, May 1976.
12. E.K. Smith, Jr., S. Weintraub, "The Constants in the Equation for Atmospheric Refractive Index at Radio Frequencies," Proceedings of the Institute Radio Engineers (IRE), Vol. 41, No. 8, pp. 1035–1037, August 1953.
13. C.F. Campen, A.E. Cole, "Tropospheric Variations of the Refractive Index at Microwave Frequencies," Air Force Surveys in Geophysics, No. 79, Geophysical Research Directorate, October 1955.
14. R.S. Berkowitz, Editor, *Modern Radar, Analysis, Evaluation and System Design, Part V, Radio Frequency Considerations*, Chapter 1, G.H. Millman, "Atmospheric Effects on Radio Wave Propagation," John Wiley & Sons, Inc., New York, 1966.
15. R.S. Berkowitz, Editor, *Modern Radar, Analysis, Evaluation and System Design, Part V, Radio Frequency Considerations*, Chapter 1, G.H. Millman, "Atmospheric Effects on Radio Wave Propagation," p. 341, Figure 1-12, p. 342, Figure 1-13, p. 344, Figure 1-16, John Wiley & Sons, Inc., New York, 1966.
16. R.S. Berkowitz, Editor, *Modern Radar, Analysis, Evaluation and System Design, Part V, Radio Frequency Considerations*, Chapter 1, G.H. Millman, "Atmospheric Effects on Radio Wave Propagation," p. 331, Figure 1-4, p. 332, Figure 1-5, John Wiley & Sons, Inc., New York, 1966.
17. M.I. Skolnik, *Introduction to Radar Systems*, McGraw-Hill Book Company, Inc., New York, 1962.
18. D.K. Barton, *Modern Radar Systems Analysis*, Artech House, Inc., Norwood, MA, 1988.
19. W.C.Y. Lee, *Mobile Communications Engineering*, McGraw-Hill Publications, New York, 1985.
20. M. Abramowitz, I.A. Stegun, *Handbook of Mathematical Function with Formulas, Graphs, and Mathematical Tables*, National Bureau of Standards, Applied Mathematics Series

No. 55, U.S. Government Printing Office, Washington, DC, June 1964.
21. A.G. Longley, P.L. Rice, "Prediction of Tropospheric Radio Transmission Loss Over Irregular Terrain—A Computer Method," Institute for Telecommunications Sciences, Environmental Science Services Administration, Boulder, CO, Technical Report ERL 79-ITS 67, July 1968.
22. T.S. Rappaport, *Wireless Communications: Principles and Practice*, Prentice Hall PTR, Upper Saddle River, NJ, 2002.
23. T. Okumura, E. Ohmori, T. Kawano, K. Fukuda, "Field Strength and Its Variability in VHF and UHF Land Mobile Service," Review Electrical Communication Laboratory, Vol. 16, No. 9–10, pp. 85–873, September–October 1968. Also appears in: *Land-Mobile Communications Engineering*, Edited by D. Bodson, G.F. McClure, S.R. McConoughey, IEEE Press, Selected Reprint Series, pp. 19–67, 1984.
24. M. Hata, "Empirical Formula for Propagation Loss in Land Mobile Radio Services," IEEE Transactions on Vehicular Technology, Vol. 29, No. 3, pp. 317–325, August 1980.
25. V. Erceg, L.J. Greenstein, S.Y. Tjandra, S.R. Parkoff, A. Gupta, B. Kulic, A.A. Julius, R. Bianchi, "An Empirically Based Path Loss Model for Wireless Channels in Suburban Environments," IEEE Journal on Selected Areas in Communications, Vol. 17, No. 7, pp. 1205–1211, July 1999.
26. M.A. Weissberger, "An Initial Critical Summary of Models for Predicting the Attenuation of Radio Waves by Foliage," ECAC-TR-81-101, Electromagnetic Compatibility Centre, Annapolis, MD, August 1981.
27. Consultative Committee on International Radio (CCIR), "Influence of Terrain Irregularities and Vegetation on Tropospheric Propagation," Report 236-6, Vol. XIII-3, Geneva, Switzerland, 1986. Currently: International Telecommunication Union-Radio (ITU-R).
28. R.M. Barts, W.L. Stutzman, "Modeling and Simulation of Mobile Satellite Propagation," IEEE Transactions on Antennas and Propagation, Vol. 40, No. 4, pp. 375–382, April 1992.
29. Consultative Committee on International Radio (CCIR), "Factors Affecting the Choice of Antennas for Mobile Stations of the Land Mobile Satellite Service," Report 925-1, Vol. V, Geneva, Switzerland, 1986. Currently: International Telecommunication Union-Radio (ITU-R).
30. N. Moraitis, V. Milas, P. Constantinou, "On the Empirical Model Comparison for the Land Mobile Satellite Channel," Institute of Communications and Computer Systems, National Technical University of Athens, Athens, Greece, 2007.
31. G.C. Hess, "Land-Mobile Satellite Excess Path Loss Measurements," IEEE Transactions on Vehicular Technology, Vol. 29, No. 2, pp. 290–297, May 1980.
32. J. Goldhirsh, W.J. Vogel, "An Overview of Results Derived from Mobile Satellite Propagation Experiments," International Mobile Satellite Conference, Ottawa, Canada, 1990.
33. C. Clarke, "A Study of Atmospheric Radio Noise Received in a Narrow Bandwidth at 11 Mc/s," The Proceedings of the Institution of Electrical Engineers, May 1960.
34. W.Q. Crichlow, C.J. Roubique, A.D. Spaulding, W.M. Beery, "Determination of the Amplitude Probability Distribution of Atmospheric Radio Noise for Statistical Moments," Journal of Research of the National Bureau of Standards, Vol. 64D, No. 1, pp. 49–58, January–February 1960.
35. Consultative Committee on International Radio (CCIR), "World Distribution and Characteristics of Atmospheric Radio Noise," CCIR, Report 322, International Telecommunication Union (ITU), Geneva, Switzerland, 1964.
36. Consultative Committee on International Radio (CCIR), "Characteristics and Applications of Atmospheric Radio Noise Data," CCIR, Report 322-3, International Telecommunication Union (ITU), Geneva, Switzerland, 1968.
37. F. Horner, "The Design and Use of Instruments for Counting Local Lightning Flashes," The Proceedings of the Institution of Electrical Engineers, Vol. 107, Part B, No. 34, pp. 321–330, July 1960.
38. M.A. Uman, E.P. Krider, "A Review of Natural Lightning: Experimental Data and Modeling," IEEE Transactions on Electromagnetic Compatibility, Vol. EMC-24, No. 2, pp. 79–112, May 1982.
39. D. Mackerras, "A Comparison of Discharge Processes in Cloud and Ground Lightning Flashes," Journal of Geophysical Research, Vol. 73, No. 4, pp. 1175–1183, February 15, 1968.
40. C.D. Beach, D.C. George, "Error Performance of VLF and LF Receiving Systems with Nonlinear Atmospheric Noise Reduction," Westinghouse Electric Corporation, Geo-research Laboratory, Boulder, CO, September 1970.
41. M. A. Uman, *Lightning*, p. 94, Figure 3.20, Westinghouse Research Laboratories/McGraw-Hill Book Company, Pittsburgh, PA/New York, 1969.
42. N. Kitagawa, M. Brook, E.J. Workman, "Continuing Currents in Cloud-to-Ground Lightning Discharges," Journal of Geophysical Research, Vol. 67, No. 2, pp. 637–647, February 1962.
43. A. Papoulis, *Probability, Random Variables, and Stochastic Processes*, pp. 357–360, McGraw-Hill, New York, 1965.
44. J.T. Gamble, "An Analysis of Linear and Non-linear Coherent Detection in Atmospheric Noise at Very Low Frequency," Office of Information (OI), Rome Air Development Center (RADC), Available from National Technical Information Service (NTIS), Springfield, VA, November 1974.
45. H.R. Frederick, "Bandwidth Conversion of ELF Noise Parameters," IEEE Transactions on Communications, Vol. COM-32, No. 2, pp. 209–211, February 1984.
46. A.D. Spaulding, C.J. Roubique, W.Q. Crichlow, "Conversion of Amplitude-Probability Distribution Function for Atmospheric Radio Noise from One Bandwidth to Another," Journal of Research of the National Bureau of Standards—D. Radio Propagation, Vol. 66D, No. 6, pp. 713–620, National Bureau of Standards, Boulder, CO, November–December 1962.
47. J.K. Omura, P.D. Shaft, "Modem Performance in VLF Atmospheric Noise," IEEE Transactions on Communications, Vol. COM-19, No. 5, pp. 659–668, October 1971.
48. H.M. Hall, "A New Model for Impulsive Phenomena: Application to Atmospheric Noise Communication Channels," U.S. Air Force Contract AF 49 (638)-1517, Stanford Electronics Laboratories, Stanford, CA, August 1966.
49. C.O. Mallinckrodt, "Peak Limiter and Hole Puncher Noise Suppression Capabilities for Minuteman Receivers," AD425672, TRW Space Technology Laboratories, Redando Beach, CA, December 1963.

50. R.F. Linfield, R.W. Plush, "Measured Performance of VLF Coherently Detected PSK with Impulsive Noise," Contract No. NOBSR 85360, Interim Report 34-R-3, pp. 5–6–5–7, DECO Electronics, Inc., Boulder, CO, December 1963.
51. J.N. Entzminger, J.T. Gamble, S.J. Kraszewski, A.D. Paoni, "A Comparison of Frequency Shift Modulation Systems at Very Low Frequencies," RADC Technical Report 65-227, Rome Air Development Center, New York, November 1965.
52. B. Widrow, J.R. Glover, Jr., J.M. McCool, J. Kaunitz, C.S. Williams, R.H. Hearn, J.R. Zeidler, E. Dong, Jr., R.C. Goodlin, "Adaptive Noise Canceling: Principles and Application," IEEE Proceeding, Vol. 63, No. 12, pp. 1692–1716, December 1975.
53. B. Kinsman, *Wind Waves—Their Generation and Propagation on the Ocean Surface*, Dover Publications, Inc., New York, 1984.
54. U.S. Navy Hydrographic Office, "Techniques for Forecasting Wind Waves and Swell," Publication H.O. No. 604, John C. Stennis Space Center, Hancock County, MS, 1951.
55. J.P. Brett, "Computer Generated Random Sea Surfaces for Use with the Program CABUOY," Aero Electronic Technology Department, Naval Air Development Center, September 30, 1977.
56. H.T. Wang, "Investigation of Tension Drift and Mean Sea Level Drift Calculated by Program CABUOY," Naval Research Laboratory, Washington, DC, October 1978.
57. B. Kinsman, *Wind Waves—Their Generation and Propagation on the Ocean Surface*, Chapter 8, "The Pierson-Neumann Theory," Dover Publications, Inc., New York, 1984.
58. H.O. Berteaux, *Buoy Engineering*, John Wiley & Sons, Inc., New York, 1976.
59. M.S. Longuet-Higgins, "On the Statistical Distribution of the Height of Sea Waves," Journal of Marine Research, Vol. 11, No. 3, pp. 245–266, 1952.
60. R.K. Moore, "Radio Communication in the Sea," IEEE Spectrum, pp. 42–51, November 1967.
61. S. Ramo, J.R. Whinnery, *Fields and Waves in Modern Radio*, pp. 236–239, John Wiley & Sons, Inc., New York, 1953.
62. W.K. Pratt, *Laser Communication Systems*, John Wiley & Sons, Inc., New York, 1969.
63. R.M. Gagliardi, S. Karp, *Optical Communications*, John Wiley & Sons, Inc., New York, 1976.
64. M. Ross, *Laser Receivers*, John Wiley & Sons, Inc., New York, 1966.
65. I. Bar-David, "Communication under the Poisson Regime," IEEE Transactions on Information Theory, Vol. IT-15, No. 1, pp. 31–37, January 1969.
66. S.R. Laxpati, G. Lachs, "Closed-Form Solutions for the Photocount Statistics of Superposed Coherent and Chaotic Radiation," Journal of Applied Physics, Vol. 43, No. 11, pp. 4773–4776, November 1972.
67. M. Yoshikawa, N. Morinaga, T. Namekawa, "Photoelectron Counting Statistics and Error Performance of *m*-Distributed Optical Signal," IEEE Transactions on Aerospace and Electronic Systems, Vol. AES-14, No. 4, pp. 623–629, July 1978.
68. A. Ishimaru, "Theory of Optical Propagation in the Atmosphere," Optical Engineering, Vol. 20, No. 1, pp. 63–70, January/February 1981.
69. J.H. Shapiro, C. Warde, "Optical Communication through Low-Visibility Weather," Optical Engineering, Vol. 20, No. 1, pp. 76–83, January/February 1981.
70. E.A. Bucher, "Computer Simulation of Light Pulse Propagation for Communication through Thick Clouds," Applied Optics, Vol. 10, No. 4, pp. 2391–2400, October 1972.
71. G. Lachs, J.A. Quarato, "The Effect of Pulse Shape and Duty Cycle on Laser Binary Communication Systems," IEEE Transactions on Aerospace and Electronic Systems, Vol. AES-11, No. 2, pp. 266–270, March 1975.
72. L.B. Stotts, G.W. Beaghler, G.C. Mooradian, "Conceptual Design of the FY83 Downlink Communications Experiment," Naval Ocean Systems Center, San Diego, CA, 1983.
73. K.L. Mahoney, N.D. Allen, "Glider Observation of Optical Backscatter in Different Jerlov Water Types: Implications to U.S. Naval Operations," Naval Oceanographic Office, Stennis Space Center, MS, 2009.
74. L.B. Stotts, "Closed Form Expression for Optical Pulse Broadening in Multiple Scattering Media," Applied Optics, Vol. 17, No. 4, pp. 504–505, February 1978.
75. J.C. Barnard, C.N. Long, "A Simple Empirical Equation to Calculate Cloud Thickness from Shortwave Broadband Measurements," Thirteenth ARM Science Team Meeting Proceedings, Broomfield, CO, March 31–April 4, 2003.
76. International Satellite Cloud Climatology Project (ISCCP), "A Consortium of International Participants to Collect Data and Study the Global Distribution of Clouds, Definitions of Cloud Types," Global Processing Center, NASA Goddard Institute for Space Studies, New York, 1982–2010.
77. S. Karp, R.M. Gagliardi, "The Design of a Pulse-Position-Modulated Optical Communication System," IEEE Transactions on Communication Technology, Vol. COM-17, No. 6, pp. 670–676, December 1969.
78. R.M. Gagliardi, S. Karp, "M-ary Poisson Detection and Optical Communications," IEEE Transactions on Communication Technology, Vol. COM-17, No. 2, pp. 208–216, April 1969.
79. V.W.S. Chan, "Coding for the Turbulent Atmospheric Optical Channel," IEEE Transactions on Communications, Vol. COM-13, No. 1, pp. 269–274, January 1982.
80. J.H. Shapiro, "Imaging and Optical Communication Through Atmospheric Turbulence," J.W. Strohbehn, Editor, *Laser Beam Propagation Through the Atmosphere*, Springer Verlag, Berlin, 1978.
81. M. Ross, "Pulse Interval Modulation Laser Communication," Supplement to IEEE Transactions on Aerospace and Electronics Systems, Vol. AES-3, p. 324, 1967.
82. B. Reiffen, H. Sherman, "An Optimum Demodulator for Poisson Processes: Photon Source Detectors," Proceedings of the IEEE, Vol. 51, No. 10, pp. 1316–1320, October 1963.
83. E.A. Bucher, "Error Performance Bounds for Two Receivers for Optical Communication and Detection," Applied Optics, Vol. 11, No. 4, pp. 884–889, April 1972.
84. J.M. Wozencraft, I.M. Jacobs, *Principles of Communication Engineering*, John Wiley & Sons, Inc., New York, 1967.

ADDITIONAL WIRELESS MOBILE COMMUNICATION REFERENCES

K. Allesbrook, J.D. Parsons, "Mobile Radio Propagation in British Cities at Frequencies in the VHF and UHF Bands," IEEE transaction on Vehicular Technology, Vol. 26, No. 4, pp. 313–323, November 1977.

J.J. Egli, "Radio Propagation Above 40 Mc Over Irregular Terrain," Proceedings of the Institute of Radio Engineers (IRE), Vol. 45, No. 10, pp. 1383–1391, October 1957.

V. Erceg, S. Ghassemzadeh, M. Taylor, L. Dong, D.L. Schilling, "Urban/Suburban Out-of-Sight Propagation Modeling," IEEE Communications Magazine, Vol. 30, No. 6, pp. 56–61, June 1992.

B. Goldberg, *Communications Channels: Characterization and Behavior*, IEEE Press, New York, 1976.

W.C. Jakes, Editor, *Microwave Mobile Communications*, IEEE Press, New York, 1994.

20

IONOSPHERIC PROPAGATION

20.1 INTRODUCTION

The amplitude variations of a signal propagating through the ionosphere [1] result from the destructive and constructive interaction of the signal phase resulting from the numerous signal paths through the nonhomogeneous medium; this phenomenon is referred to as scintillation. In addition to scintillation, signal propagation through the ionosphere is subjected to anomalies characterized by time delay variations, angular errors caused by refractive bending, frequency shifts, dispersion, polarization rotation, and absorption that must be accounted for in the communication link budget. Refractive bending affects azimuth and elevation measurement accuracies while time delay and frequency variations result in range and velocity estimation errors. Dispersion gives rise to symbol broadening and intersymbol interference (ISI) that degrade the symbol-error performance while polarization rotation and absorption can significantly degrade the available link margin. These errors also impact antenna, symbol, and carrier tracking loops contributing to degraded communication performance.

Signal propagation through the ionosphere is characterized by the refractive index. The refractive index and its influence on the various aspects of a received signal is the subject of this chapter. The significant parameters that influence the signal propagation through the ionosphere are the electron density $n'_e(\ell)$ with units of electrons per cubic-centimeter and the total electron content (TEC or N_T) along the propagation path of length ℓ. The electron density is generally specified in terms of electrons/cubic-centimeter; however, the following analysis uses the *mks* system of units with the TEC specified in terms of electrons/square-meter. The TEC is computed as

$$N_T = \int_{Lp} n_e(\ell)d\ell \quad : \text{electrons/m}^2 \qquad (20.1)$$

where $n_e(\ell) = n'_e(\ell) \times 10^6$ electrons/m^3.

In Sections 20.2 and 20.3 the electron densities in the ionosphere are characterized for the natural and nuclear-disturbed environments and the influence of the electron densities on signal propagation is discussed in Section 20.4 in terms of the refractive index. With this background material the electron density is used to characterize signal scintillation in Section 20.5 in terms of the signal decorrelation time τ_o, frequency-selective bandwidth f_o, dispersion, and absorption. Although the focus in this section is on the nuclear environment, the results are also applicable to electron densities occurring naturally. In Section 20.6 the Rayleigh fading channel is described and the results are used to outline the development of a computer simulation program for simulating the performance of a communication link with various waveform modulations, forward error correction (FEC) coding, combining, and interleaving techniques. The chapter concludes with a case study of a scintillation scenario using a differentially coherent modulation with interleaving and combining.

Digital Communications with Emphasis on Data Modems: Theory, Analysis, Design, Simulation, Testing, and Applications, First Edition. Richard W. Middlestead.

Companion website: www.wiley.com/go/digitalcommunications

20.2 ELECTRON DENSITIES: NATURAL ENVIRONMENT

The electron density content of the natural ionosphere has been modeled by Chapman [2] as*

$$n'_e(h) = n'_e(h_m)e^{(1-Z-\sec(\theta)e^{-Z})/2} \qquad (20.2)$$

where h is the vertical height above the Earth's surface, $n'_e(h_m)$ is the maximum electron density where h_m is the height of maximum electron density, θ is the solar zenith angle, and Z is a normalized height parameter expressed as

$$Z = \frac{h-h_m}{H} \qquad (20.3)$$

where H is a scale height given by

$$H = 0.10197\frac{kT}{m_a g} \quad : \mathrm{m} \qquad (20.4)$$

Using the *mks* system of units, $k = 1.372 \times 10^{-23}$ J/°K is Boltzmann's constant, T is temperature in degrees Kelvin, m_a is the mean mass of an air molecule, and $g = 9.7538$ m/s^2 is the gravitational acceleration at the Earth's surface. The product $m_a g = 4.8 \times 10^{-26}$ kg is the mean weight of an air molecule and the factor 0.10197 converts joules to kilogram-meters. Using these parameters the scale height H given by (20.4) is in meters at the Earth's surface and, upon conversion to kilometers, is evaluated as $H = 0.029146T$ km or 8.452 km when $T = 290$°K. However, the parameters m_a, g, and T are functions of height and Davies [3] provides an approximation to (20.4) at a height h given by

$$H \cong 0.848\left(1+\frac{h}{R_e}\right)^2 \frac{T}{M} \quad : \mathrm{km} \qquad (20.5)$$

where h is in kilometers, $R_e \cong 6370$ km is the Earth's radius, and M is the molecular weight in grams/mole. The dependence of the parameters T and M on h is tabulated in Table 20.1 for the 1959 Air Research and Development Command (ARDC) model atmosphere [4]. The 1959 ARDC model atmosphere is a revision of the 1956 model atmosphere that includes new rocket and satellite data; the data up to 53 km is the same in each model. Additional details and various assumptions are also provided by Davies [3].

The following evaluations using the Chapman model are based on Millman [5] where H, h_m, and $n'_e(h_m)$ are characterized as daytime and nighttime parameters as shown in Table 20.2. Using these results, the electron density profiles for daytime and nighttime conditions are shown in Figure 20.1.

*This model uses the *cgs* system of units.

TABLE 20.1 Dependence of the Parameters T and M on Height h[a]

h (km)	T (K)	M (g/mol)
0	288	28.966
10	223	28.966
20	217	28.966
30	231	28.966
40	261	28.966
50	283	28.966
60	254	28.966
70	210	28.966
80	166	28.97
90	166	28.97
100	200	28.90
120	477	28.71
140	850	28.45
160	1207	28.04
180	1371	27.36
200	1404	26.32
300	1423	21.95
400	1480	19.56
500	1576	18.28
600	1691	17.52
700	1812	17.03

[a]Davies [3]. Courtesy of the U.S. Department of Commerce.

TABLE 20.2 Chapman Model Electron Density Parameters[a]

h_m (km)	H (km)	$n'_e(h_m)$ (electrons/cm^3)
Daytime		
100	10	1.5e4
200	40	3.0e5
300	50	12.5e5
Nighttime		
120	10	0.8e4
250	45	4.0e5

[a]Millman [6]. Reproduced by permission of John Wiley & Sons, Inc.

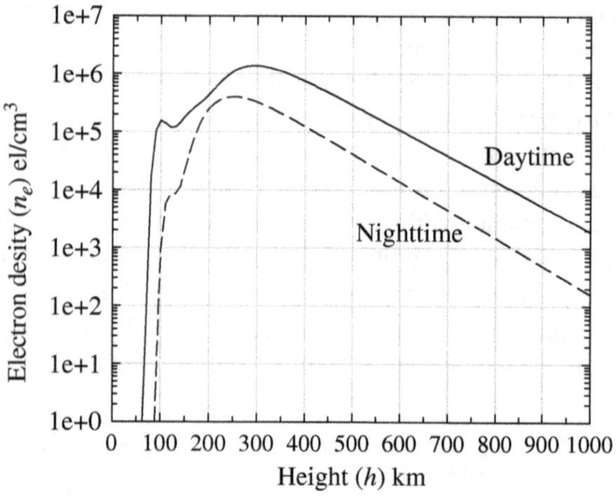

FIGURE 20.1 Chapman electron density profiles.

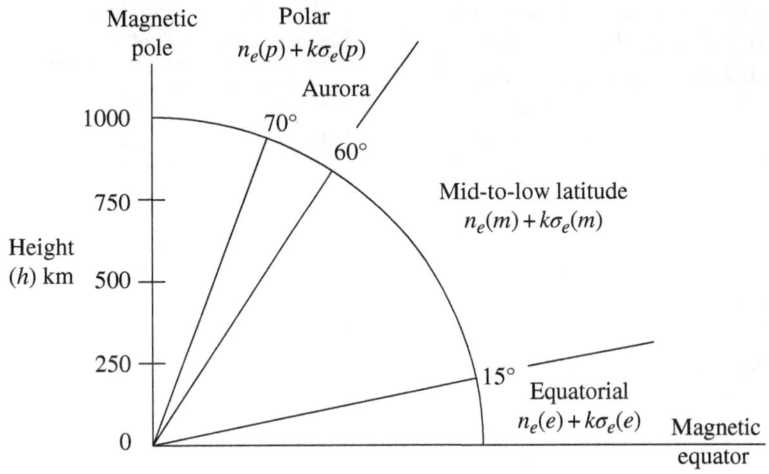

FIGURE 20.2 Characterization of regional electron density profiles.

The electron density profiles described earlier are typical densities that apply to daytime and nighttime conditions; however, during periods of sunrise and sunset the concentration of electrons generally increases due to the Sun–Earth solar activity. This increase is most notable in the polar region between latitudes of 60° and 70° and in the equatorial region between latitudes of ±15°.

Bogusch [7] and McClure and Hanson [8] have analyzed the results of various experiments to characterize the mean and variation of the electron densities in the ionosphere. The wide ranges in the parameters portray diurnal and seasonal variations as well as positional or longitude and latitude variations. In arriving at the inferred electron density profiles, Bogusch presents data based on agreement with observed scintillation results around the world. The approach in arriving at the inferred data is to adjust the mathematical model parameters, principally the electron density n_e, to match the statistical characteristics of the amplitude and phase of the model to those observed from test signals. For example, based on the tactical satellite (TACSAT) tests data taken in the equatorial zone at 250 MHz, the model infers that the electron density fluctuation is 10^4 electrons/cm^3. This is reported [9] to be a severe scintillation condition lasting 1.5 h/day during which normal communications were disrupted. An earlier test [10], using the same satellite, resulted in an inferred fluctuation of 2.5×10^4 electrons/cm^3. Wittwer [11] reports on electron density fluctuations ranging between 30 and 90% of the mean value as being typical in the Equatorial and Polar Regions. Also, inferred data taken from the INTELSAT network at 6 GHz and reported by Taur [12] indicated electron densities ranging between 4×10^4 and 10^5 electrons/cm^3 in the equatorial region. Because of the higher carrier frequency, measurable scintillation was observed only about one percent of the time. Figure 20.2 is a regional depiction of the electron densities and the corresponding standard deviations with numerical values provided for moderate and turbulent conditions listed Table 20.3. The results for the electron density fluctuations are based principally on the analysis of Bogusch in establishing inferred quantitative agreements between data obtained from radar observation throughout the world and computer models. Application of the mean and standard deviation of the electron densities listed in Table 20.3 to the Chapman model profiles provides a measure of confidence in the system performance parameter being examined. However, caution must be used when considering a particular satellite-to-Earth link that intersects a wide range of latitudes and the assumed underlying Gaussian statistics based on the mean and standard deviation.

TABLE 20.3 Regional Variations in Electron Density Concentrations

Region	n'_e (electrons/cm^3)	σ_e (electrons/cm^3)	Condition
Polar (p)	6.2e5	2.5e4	Moderate
	1.0e6	3.0e5	Turbulent
Mid-to-low latitude (m)	8.0e5	1.0e4	Moderate
Equatorial (e)	5.0e5	2.5e4	Moderate
	1.0e6	3.0e5	Turbulent

The *in situ* data taken with the orbiting geophysical observatory (OGO) OGO-6 satellite [7] provides electron density data taken in the upper portion of the F-region and the magnitude of the inferred fluctuations resulting from the model fall within the range of the *in situ* data.*

*This discussion concentrates on the accuracy of the mathematical model based on agreement between the inferred and *in situ* electron density fluctuation data. Bogusch presents data supporting the agreement based on the statistical characteristics of the received signal amplitude and phase. The Ricean channel model is in excellent agreement from both points of view.

The retarding potential analyzer can measure changes in ion concentration as small as 0.03% and, therefore, the large ranges presented in Figure 20.2 represent realistic variations over time and position. A quiet atmosphere exhibits electron density fluctuation less than about 0.2% and a moderate atmosphere will range as high as 2 or 3%, while in a turbulent atmosphere the fluctuations range up to about 30%. It is suggested that within a given evening the entire range may be encountered under turbulent conditions. The electron density characteristics in the F2-region, more specifically in the altitude range from 300 to 600 km, are the most variable and tend to dominate the scintillation characteristics of the Earth/satellite communication channel. For these reasons, most of the emphasis is placed on characterizing the electron density in the upper F-region. The electron density profiles for each of the regions, shown in Figure 20.2, are summarized in the following sections. It should be kept in mind that the results in Figure 20.2 are based on normal or average conditions and geomagnetic storms, ionospheric disturbances such as solar flares, and nuclear detonations will result in much larger variations in the ionospheric structure and considerably higher electron density concentrations.

20.2.1 Equatorial Region

The equatorial region ranges roughly between ±15° latitude and is characterized by increases in electron content during local sunrise and sunset. Although scintillation is generally encountered in this region, severe scintillation occurs between 6 p.m. and 1 a.m. local time with the most severe conditions occurring at the equinoxes. The extremely high longitudinal gradients that exist during these periods result in variations with short correlation times that correspond to sudden changes in conditions. Considerably more variations in the electron density occur in this region than in the mid-to-low latitude region. Extrapolation of scintillation data for evaluating communication links in the equatorial region is restricted because of the limited land masses where data can be collected. The equatorial electron density profile, based on Wittwer's model ionosphere [13], is given in normalized form in Table 20.4.

20.2.2 Mid-to-Low Latitude Region

The mid-to-low latitude region is generally *quiet* and allows for reliable communications. Detailed studies [14] have shown the hour-to-hour variations in the electron density are highly correlated which results in a relatively time-invariant channel. The mid-latitude region has a relatively high mean electron density; however, the low *k*-sigma variations result in reasonably predicable scintillation in this region. Taylor [15] presents data showing the seasonal variation of the noon-time TEC (N_T), at mid-latitude on quiet days. The results indicate that N_T reaches a maximum near periods of sunspot activity and this maximum is somewhat worse during the winter months (8×10^{17} electrons/cm^2) when compared to the summer months (5×10^{17} electrons/cm^2).

TABLE 20.4 Normalized Equatorial and Polar Ionospheric Electron Density Profiles

Equatorial		Polar	
h (km)	n'_e/n'_e(max)	h (km)	n'_e/n'_e(max)
135	0.03	130	0.04
250	0.24	160	0.16
278	0.67	285	0.35
290	1.00	300	1.00
345	1.00	325	1.00
425	0.67	380	0.65
470	0.45	413	0.35
520	0.24	475	0.17
580	0.11	600	0.08
600	0.09	—	—

20.2.3 Polar Region

The lower edged of the polar region is characterized by the aurora region where the most severe polar scintillation occurs. The aurora region drifts south from that shown in Figure 20.2 by about 10° between 9 a.m. and 9 p.m. local time with greater drifts occurring during geomagnetic storms. Because of the concentration of the Earth's magnetic field lines, the polar region results in the most severe signal polarization rotation.

The equatorial and polar electron density profiles, based on Wittwer's model ionosphere [13], are given in normalized form in Table 20.4 where $n'_e(\max) = 4.5\text{e}5$ electrons/cm^3 in the equatorial region and 3.1e5 electrons/cm^3 in the polar region.

The variations of the electron densities result in an inhomogeneous medium that gives rise to the signal scintillation and anomalies involving time delay, angular errors, frequency shifts, dispersion, polarization, and absorption.

Signal loss in the ionosphere results from the collision of free electrons with ions and neural particles resulting in a loss of energy or absorption of the signal as it propagates through the ionosphere. The parameter of interest in evaluating the loss is the electron collision frequency (ν) with units of radians per second. The collision frequency is also a function of the height as expressed by [5]

$$\nu(h) = \nu_m e^{(h_{\nu m} - h)/H_\nu} \tag{20.6}$$

where $h_{\nu m}$ is of the height of the maximum collision frequency ν_m in each homogeneous region and H_ν is a corresponding scale factor applied to each region.

TABLE 20.5 Chapman Model Electron Collision Frequency Parameters

h_{vm} (km)	H_v (km)	v_m (rad/s)
100	10	3.0e5
134	45	1.0e4

Millman [16]. Reproduced with permission of John Wiley & Sons, Inc.

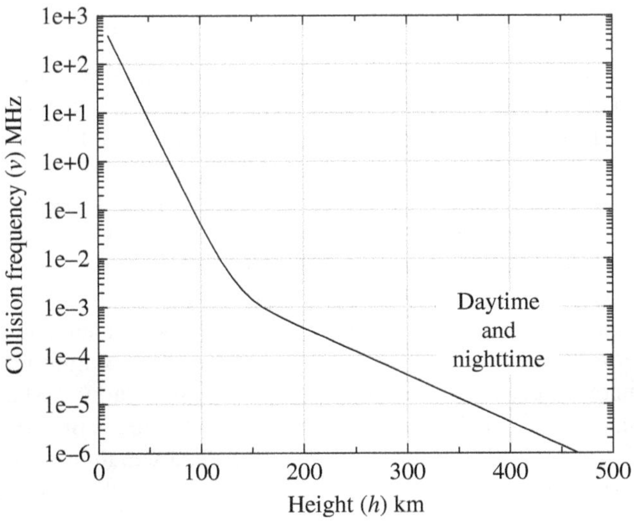

FIGURE 20.3 Electron collision frequency profile.

These parameters are quantified by Millman in Table 20.5 and the resulting electron collision frequency is shown in Figure 20.3. In Section 20.5.2 the collision frequency, expressed by (20.6), is included in the integrand of (20.39). So the absorption loss is determined by integrating over the communication link path through the ionosphere. The evaluation of (20.39) also includes the electron density characteristics given in Table 20.3 and the carrier frequency, so the absorption loss is quantified as a function of the collision frequency, TEC, and the operating frequency.

20.3 ELECTRON DENSITIES: NUCLEAR-DISTURBED ENVIRONMENT

The phenomenon of high concentrations of elections is not limited to the ionosphere, in that, free electrons resulting from a nuclear detonation are forced far into space above the ionosphere forming an ionized plume that follows the Earth's magnetic field lines. These extra-ionospheric plumes result in severe disruptions to otherwise benign satellite links including cross-links [17]. From the initial forces within a nuclear detonation the electron plume forms rapidly, within several minutes, resulting in a time-varying inhomogeneous medium. As the impact of the initial detonation diminishes

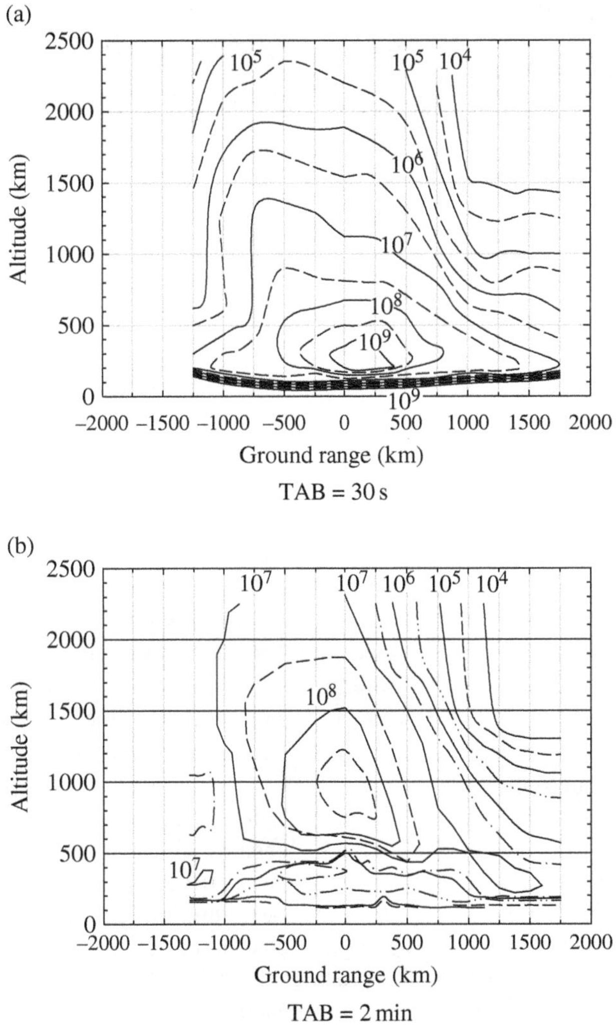

FIGURE 20.4 Example electron density contours n_e' following high-altitude high-yield nuclear detonation.Middlestead et al. [17]. Reproduced by permission of the IEEE.

the electrons recombine slowly, over many hours, within the ionosphere. In addition to the time variations resulting from the initial blast and subsequent electron recombination, the plume will move or drift due to normal atmospheric winds resulting in additional random time fluctuations. These natural effects and the dynamics of the communication platforms result in signal scintillation with varying correlation time and coherence bandwidth that require uniquely designed waveform for reliable communications. Signal scintillation resulting from a high-altitude nuclear detonation and the parameters that impact signal reception are discussed in Section 20.5 and waveform designs techniques that provide reliable communications are discussed in Section 20.8.

Example electron density profiles resulting from high-altitude nuclear detonations are shown in Figure 20.4 for time

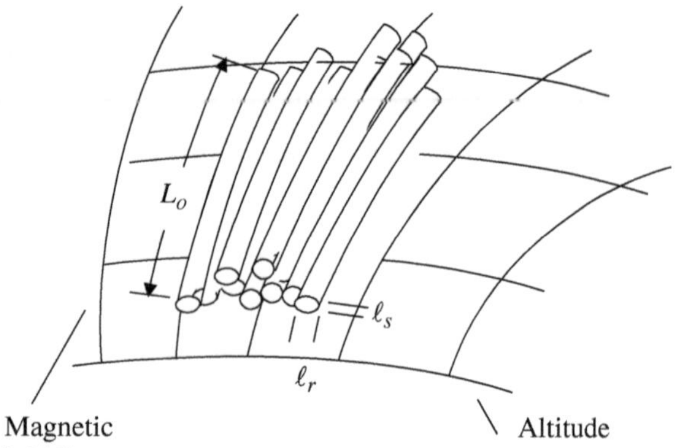

FIGURE 20.5 Irregularities formed by electron clusters around geomagnetic field lines.

after blast (TAB) corresponding to 30 s and 2 min. The plots are based on data from published contours based on weapon characteristics, location of the detonation, and environmental conditions. In these figures the high-altitude detonation occurs at ground range zero and the electron densities represent electrons/cubic-centimeter with the *solid* lines corresponding to $n_e' = 1 \times 10^n$ electrons/cm^3 and the *dashed* lines corresponding to $n_e' = 3 \times 10^n$ electrons/cm^3 for $n = 4$ through 9.

The profiles in Figure 20.4 represent a two-dimensional macro view of the electron densities; however, a three-dimensional electron profile is necessary to determine the TEC along the path of a communication link with an arbitrary antenna pointing angle. Because the electron concentrations form along the geomagnetic field lines, the electron profiles in geomagnetic coordinates provide the necessary source data for determining the profile along an arbitrary path in geographic coordinates; the coordinate transformations are described in Section 20.5 and APPENDIX 20A. The total electrons along the communication path, as expressed by (20.1), provide an important measure of the static propagation disturbances that change relatively slowly with the mean electron density. However, the tubular striations formed by electron clusters around the geomagnetic field lines result in irregularities as depicted in Figure 20.5 that result in small-scale size spatial variations in the electron concentrations. The dimension L_o is the outer scale size of the striation along the axis parallel to the magnetic field lines ranging between 1 and 10 km. The orthogonal dimensions l_s and l_r represent the inner scale sizes of the striations normal to the magnetic field lines with typical values ≤1 km. These small-scale size electron density variations give rise to dynamic disturbances or scintillation resulting from constructive and destructive signal phase combining as the signal propagates through the inhomogeneous medium. The static and dynamic propagation disturbances are listed in Table 20.6 and characterized in Sections 20.5 and 20.5.2 in terms for the mean and variation of the electron densities.

TABLE 20.6 Static and Dynamic Signal Propagation Disturbances

Static	Dynamic
Absorption	Amplitude scintillation
Noise	Phase scintillation
Dispersion	Angular scattering
Phase shift	Time delay jitter
Time delay	

20.4 THE REFRACTIVE INDEX AND SIGNAL PROPAGATION

A fundamental consideration in analyzing the propagation of a radio wave through the ionosphere is the characterization of the index of refraction under the system operating conditions, principally the carrier frequency and instantaneous bandwidth. The following analysis focuses on relatively high frequency communication links corresponding to carrier frequencies greater than about 1 GHz. The applications involve communications between ground and airborne terminals and satellites, including satellite cross-links [17]. Appleton's formulation of the refractive index [3, 18–22] is expressed as

$$n^2 = 1 - \frac{(\omega_p/\omega)^2}{1 - i(\nu/\omega) - \frac{1}{2}(eB_T/m'\omega)^2\left(1-(\omega_p/\omega)^2 - i(\nu/\omega)\right)^{-1} \pm \sqrt{\frac{1}{4}(eB_T/m'\omega)^4\left(1-(\omega_p/\omega)^2 - i(\nu/\omega)\right)^{-2} + (eB_L/m'\omega)^2}} \tag{20.7}$$

where ω_p is the plasma frequency given by

$$\omega_p = \sqrt{\frac{n_e e^2}{m'\varepsilon_o}} \quad : \text{rad/s} \tag{20.8}$$

Figure 20.6 is a plot of the plasma frequency; $f_p = \omega_p/2\pi$, dependence on the electron density; and Table 20.7 tabulates and describes the parameters.

The refractive index is also characterized in terms of the complex quantity

$$n = \mu - j\chi \tag{20.9}$$

Application of the index of refraction to a communication link yields the influence of the ionospheric propagation on the received signal. Consider, for example, a transmitted communication waveform expressed as $e_T(t)$ and after propagating a distance ℓ through a striated region of the ionosphere of path length L, the signal experiences an absorption and phase shift and is expressed as

$$\begin{aligned}\tilde{e}(t) &= \tilde{e}_T(t)e^{j\omega(t-n\ell/c)} \\ &= \tilde{e}_T(t)e^{-\omega\chi\ell/c}e^{j(\omega t-\omega\mu\ell/c)}\end{aligned} \tag{20.10}$$

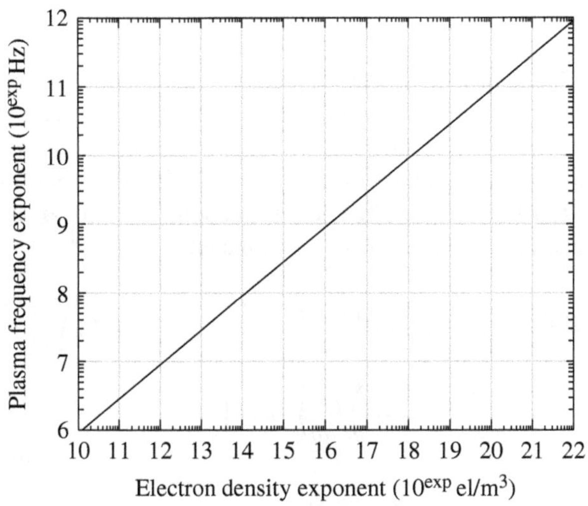

FIGURE 20.6 Dependence of plasma frequency on electron density.

The absorption coefficient κ is defined as

$$\kappa = \frac{\omega}{c}\chi \quad : \text{Np/m} \tag{20.11}$$

or, expressed in decibels, the absorption coefficient is 8.686κ dB/m. The signal phase shift introduced by the striated region gives rise to signal dispersion that is characterized by the channel phase constant denoted as $\beta(\ell)$ and expressed as

$$\beta(\ell) = \mu\ell\left(\frac{\omega}{c}\right) \tag{20.12}$$

The dependence of the phase constant on the communication path length ℓ is explicitly shown for a constant electron density. However, as will be seen in Section 20.6, the fluctuations over the path through the striated region are characterized by the range-dependent refractive index $u(\ell)$.

A general evaluation of the real and imaginary parts of Appleton's expression is difficult; however, several simplifying assumptions provide insight into the channel behavior as well as practical characterizations of the received signal. The first of these assumptions is that the electron collision frequency is negligible, that is, $\nu/\omega \ll 1$, and the second is that magnetic field effects are negligible, that is, $\boldsymbol{B} = 0$.

20.4.1 Magnetic Field and No Electron Collisions

In this case it is assumed that the electron collision frequency is much less than the carrier angular frequency so the imaginary term $j\nu/\omega$ in Appleton's expression is neglected leading to the result

TABLE 20.7 Ionospheric Channel Parameters and Constants

Parameter	Value	Units[a]	Description
$N(\ell)$	Computed	Electrons/m^3	Electron density
ℓ	Computed	Meters	Distance along propagation path
ω, ω_c	System parameters	Radians/second	Angular frequency[b]
e	1.602×10^{-19}	Coulombs/electron	Electron charge
ε_o	8.854×10^{-12}	Coulomb^{2}second^2/(K$_g$-m^3)	Free-space permittivity (dielectric constant)
m'	9.109×10^{-31}	K$_g$/electron	Electron mass
n_e	Channel parameter	Electrons/m^3	Electron density
n'_e	Channel parameter	Electrons/cm^3	Electron density
c	2.997925×10^8	Meters/second	Free-space velocity
ν	6.06×10^6 at 50 km 1.75×10^3 at 100 km	Radians/second	Electron collision frequency[c]
B_T	Computed	Gauss	Transverse magnetic induction
B_L	Computed	Gauss	Longitudinal magnetic induction
$\boldsymbol{B}$	0.5	Webers/m^2	Magnitude of magnetic induction[c]

[a] *mks* system of units.
[b] In the following f denotes the frequency in hertz and f_c denotes a selected carrier frequency.
[c] Numerical values based on the 1959 ARDC model atmosphere.

$$\left.\begin{aligned}\mu^2 &= 1 - \frac{2\left(1-(\omega_p/\omega)^2\right)(\omega_p/\omega)^2}{2\left(1-(\omega_p/\omega)^2\right) - (eB_T/m'\omega)^2 \pm \sqrt{(eB_T/m'\omega)^4 + 4\left(1-(\omega_p/\omega)^2\right)^2 (eB_L/m'\omega)^2}} \\ \chi &= 0\end{aligned}\right\} : \nu/\omega \ll 1 \qquad (20.13)$$

Although this is a realistic result for satellite links operating above about 1 GHz, it is difficult to evaluate, in part, because the ± term in the denominator leads to ordinary and extraordinary waves, respectively, which are, to varying degrees, dependent on the strength and orientation of the magnetic field and the direction of propagation. A convenient expression results if the influence of the magnetic field is ignored as indicated in the following two sections.

20.4.2 No Magnetic Field and No Electron Collisions

If the magnetic field is neglected if $B_T = B_L = 0$ and the index of refraction reduces to the simplest form given by

$$n = \mu = \sqrt{1 - \left(\frac{\omega_p}{\omega}\right)^2} \quad : \nu/\omega \ll 1, \;\; B_T = B_L = 0 \qquad (20.14)$$

20.4.3 No Magnetic Field with Electron Collisions

When the electron collisions are ignored the imaginary part of the index of refraction is zero so the absorption cannot be characterized in terms of the physical characteristics of the channel. However, to provide some insight into the absorption characteristics, it is convenient to ignore the effects of the magnetic field while permitting electron collisions. These conditions lead to the result

$$\begin{aligned} n^2 &= \frac{1-(\omega_p/\omega)^2 + (\nu/\omega)^2}{1+(\nu/\omega)^2} - j\frac{(\omega_p/\omega)^2(\nu/\omega)}{1+(\nu/\omega)^2} \quad : B_T = B_L = 0 \\ &= (\mu^2 - \chi^2) - j2\mu\chi \end{aligned} \qquad (20.15)$$

and equating the real and imaginary parts of these expressions yields

$$\mu^2 = \frac{1-(\omega_p/\omega)^2 + (\nu/\omega)^2}{1+(\nu/\omega)^2} + \chi^2 \qquad (20.16)$$

and

$$\chi = \frac{(\omega_p/\omega)^2(\nu/\omega)}{2\mu\left(1+(\nu/\omega)^2\right)} \qquad (20.17)$$

Solving for μ, using (20.16) and (20.17) under the condition $(\nu/\omega)^2 \ll 1$ results in the approximate expression

$$\mu \cong \sqrt{1 - \left(\left(\frac{\omega_p}{\omega}\right)^2 + \cdots\right) + \left(1 + \frac{1}{4}\left(\frac{\omega_p}{\omega}\right)^4 + \cdots\right)\left(\frac{\nu}{\omega}\right)^2 + \cdots} \qquad (20.18)$$

The ellipsis in (20.18) represents neglected terms involving powers of (ω_p/ω) and (ν/ω) greater than four and two, respectively. The approximation applies when $\nu \ll \omega$ which is a reasonable approximation when $f > 100$ MHz. Substituting (20.18) into (20.17), under the condition $(\nu/\omega)^2 \ll 1$, the signal absorption term is approximated as

$$\begin{aligned}\chi \equiv \frac{1}{2}\left(\frac{\omega_p}{\omega}\right)^2\left(\frac{\nu}{\omega}\right)\Bigg[&\left(1 + \frac{1}{2}\left(\frac{\omega_p}{\omega}\right)^2 + \cdots\right) - \left(\frac{1}{2} + \frac{3}{4}\left(\frac{\omega_p}{\omega}\right)^2 \right. \\ &\left. + \frac{17}{16}\left(\frac{\omega_p}{\omega}\right)^4 + \cdots\right)\left(\frac{\nu}{\omega}\right)^2 + \cdots\Bigg]\end{aligned} \qquad (20.19)$$

20.5 SIGNAL PROPAGATION IN SEVERE SCINTILLATION ENVIRONMENT

The principal parameters associated with scintillation in an ionized or striated channel are listed in Table 20.8.

The most commonly used measure of signal fading is the S_4 scintillation index defined as*

$$S_4 \triangleq \sqrt{\frac{\langle V^4\rangle - \langle V^2\rangle^2}{\langle V^2\rangle^2}} \qquad (20.20)$$

where V is the instantaneous *rms* signal voltage and the averaging time is much greater than the signal fade duration. The S_4 scintillation index saturates at unity corresponding to severe scintillation with Rayleigh signal amplitude fading; but S_4 index may exceed unity prior to saturation under some conditions. As discussed in Chapter 1, the Rayleigh amplitude *pdf* is characterized by independent quadrature Gaussian signals, $N(0,\sigma_n)$, with a uniform phase *pdf* over 2π radians. Rayleigh scintillation may persist for many hours affecting

*The S_4 scintillation index is also described (Reference 23, pp. 230–238) in terms of the parameters of the striated region.

TABLE 20.8 Principle Scintillation Dependent Parameters

Parameter	Name	Description
S_4	Scintillation parameter	Typically: $0 \geq S_4 \leq 1$
$\overline{\chi^2}$	Rytov parameter[a]	$\overline{\chi^2} \geq 0$
σ_e^2	Electron variance	Over propagation path
σ_ϕ^2	Signal phase variance	Over propagation path
σ_θ^2	Energy angle-of-arrival variance	Results in antenna loss
ℓ_o	Decorrelation length	Spatial correlation parameter
τ_o	Signal decorrelation time	Temporal correlation parameter
f_o	Signal decorrelation bandwidth	Frequency-selective bandwidth

[a]This notation should not be confused with the imaginary part of the refractive index.

TABLE 20.9 Ground Range Extent Affected by Severe Scintillation

Frequency Band	Longitude Ground Range (km)
EHF	480
X	1600
S	2400
L	2570
UHF	3200

High altitude detonation, TAB = 30 min.

communications over large geographical regions with longitude ground ranges listed in Table 20.9; the EHF band is sensitive to parameter uncertainties and may be less than indicated.

Because of the severity and extent of the scintillation, a robust communication system design must be capable to operating under severe scintillation conditions and, for this reason, the analysis, design, and system performance evaluations in Section 20.7 and following are based on a Rayleigh fading received signal. For $S_4 > 0.4$ the Nakagami *pdf** is a good approximation to the amplitude fading statistics which is, theoretically, equal to the Rayleigh *pdf* when $S_4 = 1.0$; however, $S_4 > 0.4$ corresponds to severe scintillation and it is recommended that system designs be based on Rayleigh fading statistics when $S_4 > 0.4$. As the scintillation index decreases the scintillation diminishes with $S_4 \leq 0.4$ corresponding to weak scintillation and, when $S_4 = 0$, the received signal does not exhibit scintillation; however, the signal propagation is influenced by phase distortion-related effects as discussed in Section 20.6.

*The Nakagami *pdf* in Table 1.8 corresponds to $p_Y(y)$ with the Nakagami parameters $m = S_4^{-2}$, $y = x^2$, and $\Omega = 1$.

The Rytov parameter is defined in terms of the parameters of the striated region and is approximated as [24]

$$\overline{\chi^2} \cong 1.138 \times 10^{-32} \frac{L_z L_p^2}{L_x L_y} \left(\frac{\sigma_e^2}{f_g^3} \right) \tag{20.21}$$

where the scale sizes (L_z, L_x, L_y) form an orthogonal coordinate system that is dependent on the geomagnetic field lines,† L_p is the propagation path length through the striated region, and f_g is the carrier frequency in gigahertz. The parameter $\sigma_e^2 = Var[N_T]$ is the variance of the TEC (N_T) expressed in (20.1). In the absence of an accurate estimate of the phase standard deviation, it is reasonable to use $\sigma_e = N_T$. The condition $\overline{\chi^2} > 0.1$ results in severe scintillation so Rayleigh fading statistics are to be applied under this condition.

In view of the uncertainty of parameters available in the open literature, the system design must be based on parametric performance evaluations. The uncertainties include the time-dependent electron density profiles; the computation of the scintillation index S_4 or $\overline{\chi^2}$; the parameters σ_e^2, σ_θ^2, ℓ_o, τ_o, f_o, and the signal losses L_{scat}, and L_a (described later). However, the recommendation that antiscintillation (AS) systems be designed to operate in the Rayleigh fading regime allowing the system design and performance evaluation to proceed based on specifications identifying the range of several parameters; most notability the electron density fluctuations, the channel decoration time (τ_o), and the frequency selective bandwidth (f_o). Waveform and system mitigation techniques are described in Section 20.8.

In the following descriptions, the communication link is modeled as a one-way path between a transmitting and receiving terminal with an ionized medium characterized as a plume of elections forming striations along the Earth's magnetic field lines. The terminals are typically thought to be Earth or airborne terminals communicating with a satellite or satellites communicating over cross-links. In general, the geometry is depicted as shown in Figure 20.7. In the evaluation of the communication link characteristics, the uplink and downlink asymmetry associated with $R_{gs} \neq R_{ss}$ is important. The following description considers a nearly vertical communication link between a ground terminal and a geosynchronous satellite with altitude or range $R = 35{,}784$ km. The communication link passes through a striated medium resulting from a nuclear detonation at an altitude (range) of $R_s = 400$ km having an extent along the line of sight (LOS) path of $R_{se} = 1000$ km. R_{se} is considered to be symmetrical about R_s. Furthermore, consider that the lower altitude of the striated region is $R_{gs} = 150$ km. If the communication link corresponds to an uplink, the parameters in Figure 20.7

†Appendix 20.A describes the transformation between geomagnetic and geographic coordinates.

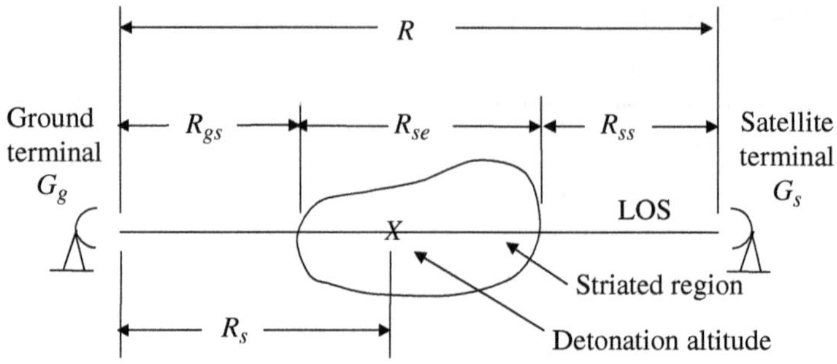

FIGURE 20.7 Communication link encounter through striated region.

become: $G_t = G_g$, $R_t = R_{gs} = 150$ km, $G_r = G_s$, and $R_r = R_{ss} = R - R_{gs} - R_{se} = 34{,}634$ km. Conversely, the communication downlink is evaluated by reversing the roles of the transmitter and receiver gains and ranges so that: $G_t = G_s$, $R_t = R_{ss} = 34{,}634$ km, $G_r = G_g$, and $R_r = R_{gs} = 150$ km. These examples will be used to emphasize the impact of the direction of transmission on various design parameters.

For strong localized scattering* the square of the spatial decorrelation distance, ℓ_o, is expressed as [25]

$$\ell_o^2 = \frac{2\,L_o'^2 (R_t + R_r)^2}{\ln\left(L_o'/\ell_i'\right) R_t^2\, \sigma_\phi^2} \tag{20.22}$$

where the parameters $R_t = R_{gs}$ and $R_r = R_{ss}$ are depicted in Figure 20.7 and σ_ϕ^2 is the variance of the carrier frequency phase over the range R_{se} through the striated region. In (20.22) the parameters L_o' and ℓ_i' represent the outer and inner scale sizes, respectively, normal to the propagation path through the striated region. These scales sizes are obtained through coordinate transformations from the geomagnetic field lines as described in APPENDIX 20A; the scale sizes in geomagnetic coordinates are depicted in Figure 20.5. Typically L_o' ranges from 1 to 10 km and ℓ_i' is about 1/15-th of L_o'. In addition to the coordinate transformations resulting in L_o' and ℓ_i', (20.22) implicitly includes the transformation from the magnetic field coordinates to the LOS vector containing propagation path L_p that is used in the computation of σ_ϕ^2. The implicit transformation factor $K(\Phi)$ is based on the penetration angle Φ between the geographic LOS path and geomagnetic field at the altitude of the strong localized scattering and is expressed as

$$K(\Phi) = \sqrt{\cos^2(\Phi) + (15\sin(\Phi))^2} \tag{20.23}$$

The relationship to the signal phase variance is $\sigma_\phi^2 \propto 1/K(\Phi)$. The expression (20.22) for ℓ_o strictly applies for unit-gain omnidirectional transmit and receive antennas, in which case, the variance of the signal energy angle-of-arrival is expressed as

$$\sigma_\theta^2 \triangleq \frac{\lambda^2}{2\pi^2 \ell_o^2} \tag{20.24}$$

where λ is the wavelength of the signal carrier frequency. The angle σ_θ is a measure of the angular deviation of the received multipath signal rays from the LOS path.

It is evident from (20.22) that ℓ_o is dependent upon the direction of transmission. For example, referring to Figure 20.7 and the conditions of the previous example, suppose that $R_t = R_{ss} = 34{,}634$ km with $R_t + R_r = 34{,}784$ km corresponding to a downlink transmission. In the case of an uplink transmission $R_t + R_s$ remains the same however $R_t = R_{gs} = 150$ km so the value of ℓ_o increases by a factor of about 232 : 1.

The signal decorrelation time is given by

$$\tau_o = \frac{\ell_o}{V} \tag{20.25}$$

where $V = V_{pr} + V_T$. The velocity V_{pr} is the velocity of the plasma or striated region and V_T is the terminal velocity contributions, where both are in the plane of the receive antenna. The plasma velocity is computed as

$$V_{pr} = \left(\frac{R_t + R_r}{R_t}\right) V_M \tag{20.26}$$

where V_M is the magnitude of the plasma velocity normal to the LOS propagation path and aligned with the plane of the receiver. Considering the plasma velocity vector in the magnetic field plane (MFP) to be $\vec{V}_{pm} = \left(V_{pm}u_x + V_{pm}u_y + V_{pm}u_z\right)$, the plasma velocity normal to the LOS is computed as

$$\begin{pmatrix} V_{px} \\ V_{py} \\ V_{pz} \end{pmatrix} = \begin{pmatrix} \text{Transformation} \\ \text{to propagation} \\ \text{coordinates} \end{pmatrix} \begin{pmatrix} \cos(\phi)\cos(\lambda) & 0 & 0 \\ 0 & \cos(\phi)\sin(\lambda) & 0 \\ 0 & 0 & \sin(\phi) \end{pmatrix} \begin{pmatrix} V_{pmx} \\ V_{pmy} \\ V_{pmz} \end{pmatrix} \tag{20.27}$$

*The striated region is relatively localized in altitude and is isotropic in the plane perpendicular to the LOS path.

Typically the magnitude of the plasma velocity in the MFP is $V_M = 1\text{–}20$ km/s. The first transformation in (20.27) is evaluated in APPENDIX 20A and results from the transformation of the geomagnetic polar coordinates (Φ,Λ) to the geographic polar coordinates (ϕ, λ) of a point in space P_x. The second transformation is required to rotate geographic coordinates of point P_x into alignment with the propagation path.

Using the expression (20.22) for ℓ_o and neglecting the terminal velocity, the decorrelation time is evaluated as

$$\tau_o = \frac{\sqrt{2}\,L'_o}{\ln\left(L'_o/\ell'_i\right)V_M\sigma_\phi} \quad \text{: neglecting terminal velocity } V_T \tag{20.28}$$

From (20.28) it is evident that τ_o is independent of the direction of transmission. The frequency-selective bandwidth is also dependent on ℓ_o through the following relationship [26]

$$f_o = \frac{\pi\,\ell_o^2 R_t f_c^2}{(cR_r(R_r+R_t))\sqrt{1+C_1^2}} \cong \frac{0.97\pi\,\ell_o^2 R_t f_c^2}{cR_r(R_r+R_t)} \tag{20.29}$$

The last expression results with the constant $C_1 = 0.25$; this is a practical constant bound for the relatively small effect of the time delay jitter and results in the scintillation being a function only of τ_o and f_o [27]. Upon substituting (20.22) for ℓ_o^2 into (20.29) the frequency-selective bandwidth is expressed as

$$f_o = \frac{2\pi\,L'^2_o f_c^2(R_t+R_r)}{c\ln\left(L'_o/\ell'_i\right)\sqrt{1+C_1^2 R_r R_t}} \cong \frac{1.94\pi\,L'^2_o f_c^2(R_t+R_r)}{c\ln\left(L'_o/\ell'_i\right)R_r R_t} \tag{20.30}$$

From (20.29) and (20.30) it is seen that f_o is dependent on the direction of transmission.

Estimates of reasonable worst-case ranges [23] of the signal decorrelation time and decorrelation bandwidth are shown, respectively, in Figures 20.8 and 20.9 as a function of the carrier frequency. The shaded areas correspond to the most severe or Rayleigh scintillation that transitions through Ricean scintillation to the channel conditions prior to the nuclear detonation. Depending on the geometry of the encounter, a blackout regime may be encountered prior to the Rayleigh regime. The blackout regime is generally defined as the time following the detonation when the received signal level is greater than 3 dB below the mean level of the Rayleigh fading signal. The signal decorrelation time and the frequency decorrelation bandwidth are defined as the point that the respective normalized correlations fall to e^{-1} of the peak correlation; these correlation responses are also referred to as the channel correlation responses.

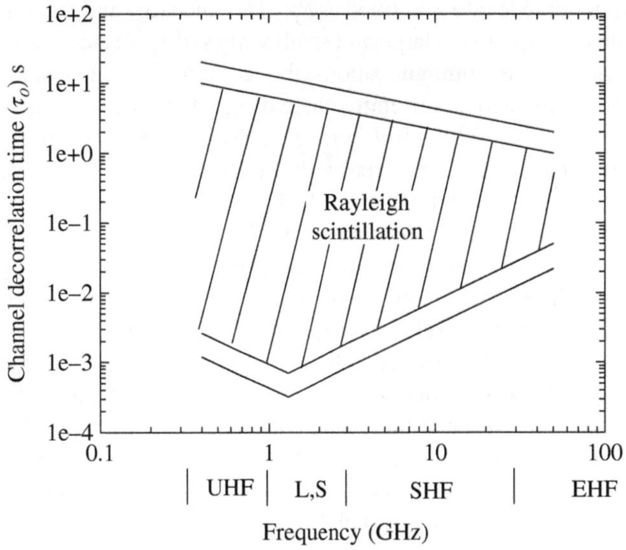

FIGURE 20.8 Reasonable worst-case channel decorrelation times.McDaniel [23]. Reproduced with permission of the Defense Threat Reduction Agency (DTRA).

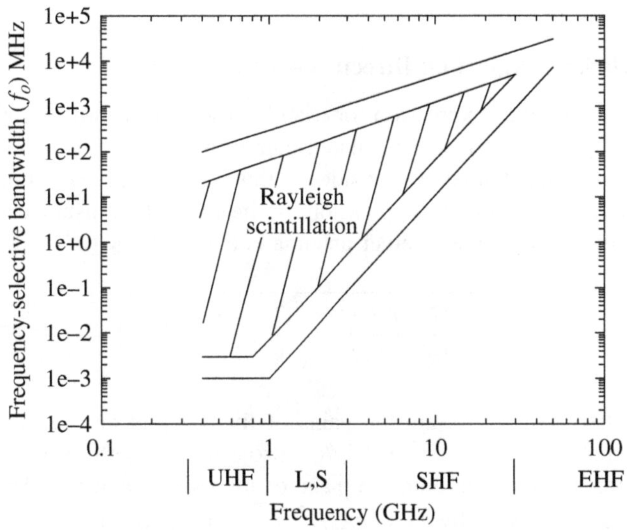

FIGURE 20.9 Reasonable worst-case channel decorrelation bandwidth (frequency-selective bandwidth). McDaniel [23]. Reproduced with permission of the Defense Threat Reduction Agency (DTRA).

The channel decorrelation time τ_o and the decorrelation bandwidth (or frequency-selective bandwidth) f_o are the two most influential channel parameters in the selection of the waveform and the system designs. For example, the channel fade rate is defined as $R_f = 1/\tau_o$ and slow and fast fading corresponds to large and small values of τ_o, respectively.* For reliable communications the communication systems must operate over the entire range of τ_o at the specified carrier frequency as shown in Figure 20.8. The range of decorrelation times places an increasingly heavy burden on the waveform selection and design of FEC coding, interleaving, and combining as the information rate increases; in some cases it may be prudent to use message repetition and combining.

If the instantaneous bandwidth of a transmitted symbol exceeds the decorrelation frequency f_o the signal will experience *frequency-selective* fading in which regions of the signal spectrum become uncorrelated resulting in severe signal distortion. However, if the signal bandwidth is sufficiently less than f_o, the entire spectrum is affected in the same way resulting in *frequency-nonselective* fading. With frequency-nonselective fading, signal FEC coding, interleaving, and combining are effective mitigations techniques.† Although it is always prudent to verify the performance using computer simulations, they are particularly important when the channel fading lies between frequency selective and nonselective regimes.

20.5.1 Impact on Directive Antenna Gain

The previous expression for ℓ_o and consequently those for τ_o and f_o are based on an ideal unit-gain isotropic radiator. When practical antennas are considered, that is, antennas exhibiting a directive gain, the expression for the correlation distance at the output of the receiver antenna is evaluated as [28]

$$\bar{\ell}_o = \ell_o \sqrt{1 + \frac{G_r}{(2\pi)^2}\left(\frac{\lambda}{\ell_o}\right)^2 + \frac{G_t}{(2\pi)^2}\left(\frac{R_r}{R_t}\right)^2\left(\frac{\lambda}{\ell_o}\right)^2} \tag{20.31}$$

where G_t and G_r are the gains of the transmit and receive antennas and the factor $(R_r/R_t)^2$ projects the aperture of the transmit antenna onto the plane of the receive antenna. The gain is given by $G = 4\pi A_e/\lambda^2$ and, for a parabolic dish antenna with radius r and efficiency η_a, the effective antenna aperture is given by $A_e = \pi\eta_a r^2$. The designation $\bar{\ell}_o$ is used to denote the correlation distance at the output of the receive antenna. The energy angle-of-arrival is defined as the angle, relative to the receiver antenna LOS axis, of the received signal emerging from the striated region. The standard deviation of the energy angle-of-arrival is given by

$$\sigma_\theta = \frac{1}{\sqrt{2}\pi}\left(\frac{\lambda}{\ell_o}\right) \tag{20.32}$$

Using this result, $\bar{\ell}_o$ is expressed in terms of the variance of the energy angle-of-arrival as

$$\bar{\ell}_o = \ell_o\sqrt{1 + \frac{G_r\sigma_\theta^2}{2} + \left(\frac{R_r}{R_t}\right)^2 \frac{G_t\sigma_\theta^2}{2}} = \ell_o\sqrt{1 + \frac{G_r\sigma_\theta^2}{2} + \frac{G_t\sigma'^2_\theta}{2}} \tag{20.33}$$

where σ'^2_θ is defined as the energy angle-of-departure from the transmit antenna and is given by

$$\sigma'^2_\theta = \left(\frac{R_r}{R_t}\right)^2 \sigma_\theta^2 \tag{20.34}$$

The signal decorrelation time and bandwidth and the antenna loss are impacted by the antenna directional gain in a similar manner. Upon substituting the channel decorrelation length $\bar{\ell}_o$ into expressions (20.22) and (20.25) for τ_o and f_o, respectively, the corresponding expressions for $\bar{\tau}_o$ and $\bar{f}_o$ at the receiver antenna output terminals are evaluated as

$$\begin{aligned}\bar{\tau}_o &= \frac{\bar{\ell}_o}{V} \\ &= \frac{\ell_o}{V}\sqrt{1 + \frac{G_r\sigma_\theta^2}{2} + \frac{G_t\sigma'^2_\theta}{2}}\end{aligned} \tag{20.35}$$

and

$$\begin{aligned}\bar{f}_o &\cong \frac{0.97\pi\,\ell_o^2 R_t f_c^2}{cR_r(R_r + R_t)}\left(1 + \frac{G_r\sigma_\theta^2}{2} + \frac{G_t\sigma'^2_\theta}{2}\right) \\ &= f_o\left(1 + \frac{G_r\sigma_\theta^2}{2} + \frac{G_t\sigma'^2_\theta}{2}\right)\end{aligned} \tag{20.36}$$

Although τ_o is not dependent on the direction of transmission, because of the different antenna gains and the asymmetry of the striated region along the transmission path, $\bar{\tau}_o$ is dependent on the direction of the transmission.

The antenna loss at the receiver, resulting from the Gaussian distributed ray scattering through the medium, is expressed as [26]

$$\begin{aligned}L_{scat} &= 1 + \frac{G_r\sigma_\theta^2}{2} + \left(\frac{R_r}{R_t}\right)^2\frac{G_t\sigma_\theta^2}{2} \\ &= 1 + \frac{G_r\sigma_\theta^2}{2} + \frac{G_t\sigma'^2_\theta}{2}\end{aligned} \tag{20.37}$$

Equation (20.37) is defined as the antenna scattering loss and must be combined with the absorption loss along the propagation path.

*Chapter 18 characterizes the bit-error probability of various waveform modulations with fast and slow fading channels in the Rayleigh and Ricean regimes.

†Mitigation techniques for these disturbances and others are discussed in Section 20.8.

20.5.2 Ionospheric Absorption

The signal absorption loss through the ionosphere is evaluated using (20.10) as

$$L_a = e^{-\omega_c \chi \ell / c} \tag{20.38}$$

or, upon substituting for χ using (20.19), which applies for $\nu \ll \omega$, and expressing the absorption loss in terms of decibels, the signal absorption loss is evaluated as

$$\begin{aligned} L_a(f_c)_{\text{dB}} &= 8.686 \frac{1}{2\,\omega_c^2 c} \int_L \omega_p^2(\ell)\nu(\ell)d\ell \\ &= 1.0858 \frac{e^2}{\pi^2 f_c^2 m c \varepsilon_o} \int_L N(\ell)\nu(\ell)d\ell \\ &= \frac{1.1677 \times 10^{-6}}{f_c^2} \int_L N(\ell)\nu(\ell)d\ell \end{aligned} \tag{20.39}$$

where* the second expression uses (20.8) to substitute for $\omega_p^2(\ell)$ and the last expression substitutes the constant values from Table 20.7. Upon expressing the electron collision frequency profile in (20.6) in terms of the height h above the Earth's surface, (20.39) becomes

$$L_a(f_c)_{\text{dB}} = \frac{1.1677 \times 10^{-6}}{f_c^2} \int_H N(h)\nu(h)f(h)dh \tag{20.40}$$

where $f(h)$ is a unit-less function of the height dependence on the antenna elevation angle θ_e as expressed by [5]

$$f(h) = \frac{R_e + h}{\sqrt{(R_e + h)^2 - (R_e \cos(\theta_e))^2}} \tag{20.41}$$

The total signal loss due to the medium and the antenna scatter power losses is

$$L_T(\text{dB}) = L_{scat}(\text{dB}) + L_a(f_c)_{\text{dB}} \tag{20.42}$$

The absorption loss in the natural environment, based on the Chapman electron density profiles, is given in Table 20.2. The electron collision profiles are listed in Table 20.5 and plotted in Figure 20.10 for carrier frequencies of 100 and 500 MHz under daytime and nighttime conditions. The loss is negligible at nighttime for frequencies greater than 100 MHz and at daytime for frequencies greater than 500 MHz.

The mean electron densities (n_e) given in Table 20.3 are used to evaluate the losses and the corresponding variations

*The carrier frequency f is explicitly denoted as f_c in this section.

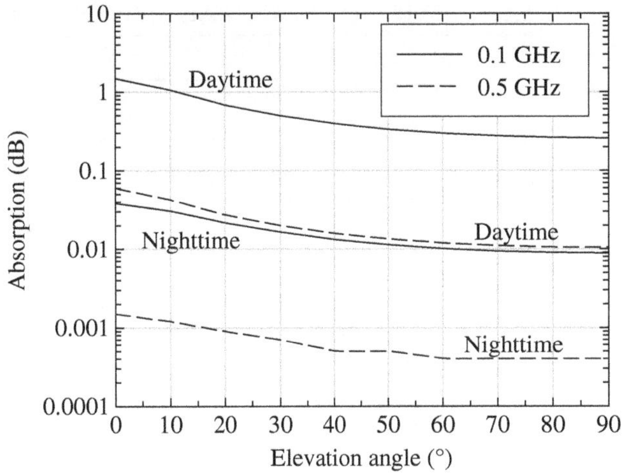

FIGURE 20.10 Absorption loss in natural environment (Chapman model with solar zenith angle = 0°).

(σ_e) are used to compute the associated confidence levels based on the Gaussian distributed loss variations denoted as $N(n_e, \sigma_e)$. The losses at a carrier frequency of 100 MHz and corresponding confidence levels are tabulated in Table 20.10 for the equatorial, mid-to-low latitude, and Polar Regions. A major source of uncertainty in computing the absorption loss is determining the value of the collision frequency. The greatest impact of the collision frequency on the absorption loss is in the lower ionospheric regions from about 50 to 120 m and, as the altitude increases, the collision frequency has a diminishing effect on absorption.

Referring to (20.40), the absorption loss scales inversely proportional to the square of the frequency; therefore, defining the losses in Table 20.10 as $L_a(100\text{ MHz})_{\text{dB}}$, the loss at an arbitrary frequency, expressed in megahertz, is determined as

$$L_a(f_{\text{MHz}})_{\text{dB}} = \frac{10^4 L_a(100\text{ MHz})_{\text{dB}}}{f_{\text{MHz}}^2} \tag{20.43}$$

20.5.3 Receiver Noise

The receiver noise is impacted by the increase in the receiver antenna noise temperature as a result of the fire ball from the detonation. The antenna noise is dependent beam width and the propagation LOS relative to the location of the detonation and is expressed as [29]

$$T_{\Delta A} = T_{FB}\left(1 - 10^{-L'_a(f_c)_{\text{dB}}/10}\right) \quad :^\circ\text{K} \tag{20.44}$$

where T_{FB} is the fireball temperature on the order of 1000°K and $L'_a(f_c)_{\text{dB}}$ is the propagation loss over the path between the fireball and the receiver.

TABLE 20.10 Absorption Losses (dB) in Natural Environment at 100 MHz

	Latitude				
	Equatorial		Mid-to-Low	Polar	
Confidence (%)	Turbulent	Moderate	Turbulent	Turbulent	Moderate
50[a]	5.75	2.88	0.95	5.10	2.55
90	28.52	4.77	1.08	19.57	3.76
95	35.10	5.32	1.11	23.75	4.11
99	47.19	6.33	1.18	31.43	4.77

[a]For mean or average electron density.

20.6 PROPAGATION DISTURBANCES FOLLOWING SEVERE ABSORPTION

The initial impact of a nuclear detonation on a communication link is a severe signal attenuation that may exceed several minutes in duration depending upon the operating frequency and the link path relative to the fireball of the detonation. This is referred to as the signal blackout regime and the only effective mitigation techniques are spatial diversity that uses another link path that is not impacted by the fireball. However, increasing the carrier frequency is only advantageous because of the lower absorption loss and susceptibility to scintillation following the severe fireball temperatures. As the blackout regime subsides, the signal level begins to recover and enters the scintillation regime. During the scintillation regime, the signal absorption has essentially diminished so that communications can resume if the underlying communication waveform is properly designed to mitigate the signal scintillation. In the scintillation regime, the received signal level fluctuations, or fading, results from carrier frequency phase constructive and destructive interference that cannot be overcome by increasing the power. This is an especially important concept in the design of frequency and time diversity waveform mitigation techniques. Therefore, because it is impractical to increase the signal power to overcome the increase in the system noise temperature resulting from the fireball or to overcome the signal phase cancellation effects, it is recommended that a 3 dB link margin be provided to aid in the link recovery during the transition to the scintillation regime with the principal mitigation techniques embodied in the network protocol and waveform design as discussed in Section 20.8.

Therefore, in this section, severe signal absorption is assumed to have subsided and the electron density of the ionosphere is considered to have a slowly varying average value with a diminishing, electron density variation about the mean value. In this regime the signal scintillation is referred to as resulting from *phase-only* affects; however, the signal amplitude continues to fluctuate about the mean value of the Rayleigh distribution with a uniformly distributed phase.

TABLE 20.11 Multipath Related Parameters

Parameter	Name	Description
T_d	Free-space delay	
t_d	Delay through striated region	Additional delay to free-space delay
T_{d1}	Quadratic delay distortion	Dispersion delay causing signal distortion
T_{d2}	Cubic delay distortion	Dispersion delay causing signal distortion
f_d	Doppler	Doppler spread
$\bar{L}_a$(dB)	Absorption loss (dB)	Path loss through striation region
T_a	Antenna temperature	Increase due to elevated temperature of plasma
θ_f	Faraday rotation	Linear polarization phase change

By considering the time dependence of the TEC either resulting from changes in the communication path or the electron density fluctuations along the path, the impact of time-varying ISI on the communication system performance is evaluated. In this context, the channel impulse response is examined and the resulting ISI is characterized in terms of the modulated waveform symbol rate. Based on these considerations, the analysis in this section involves traditional multipath phenomenon using the parameters identified in Table 20.11.

For this analysis the channel frequency response is characterized as

$$H(\omega) = e^{j\Phi(\omega)} \tag{20.45}$$

where ω is the instantaneous angular frequency. Considering the length of the communication path through the striated region of the ionosphere to be L meters, the channel phase function is evaluated as

$$\Phi(\omega) = \int_L \beta(\ell)d\ell \tag{20.46}$$

where $\beta(\ell) = \omega\mu(\ell)/c$ is the channel phase constant. For the previous simplifying assumptions, the real part of the refractive index is expressed by (20.14) as

$$\mu(\ell) = \sqrt{1 - \omega_p^2(\ell)/\omega^2} \tag{20.47}$$

The change in the channel phase relative to that of free-space propagation, that is, for $N(\ell) = 0$, results in the phase function

$$\begin{aligned} \Phi(\omega) &= \Phi_o + (\Phi(\omega) - \Phi_o) \\ &= \Phi_o + \phi(\omega) \end{aligned} \tag{20.48}$$

where $\Phi_o = -\omega L_p / c$ and L_p is the undisturbed propagation path length. The phase function through the disturbed region with path length L is expressed as

$$\phi(\omega) = -\frac{\omega}{c} \int_L \left(\sqrt{1 - \left(\frac{\omega_p(\ell)}{\omega}\right)^2} - 1 \right) d\ell \tag{20.49}$$

Expanding the radical in the integrand of (20.49) in terms of a power series with $\omega > \omega_p$ results in the approximation

$$\phi(\omega) \cong \frac{1}{2c\omega} \int_L \omega_p^2(\ell) d\ell \tag{20.50}$$

This approximation ignores the higher order terms: $-(\omega_p/\omega)^4/8 - (\omega_p/\omega)^6/16 - \cdots$. Expanding the function $f(\omega) = 1/\omega$ in (20.50) using a Taylor series about the carrier frequency ω_c results in

$$\phi(\omega) \cong \frac{1}{2c\omega_c} \left(1 - \frac{\omega - \omega_c}{\omega_c} + \frac{2(\omega - \omega_c)^2}{\omega_c^2} - \frac{6(\omega - \omega_c)^3}{\omega_c^3 + \cdots} \right) \int_L \omega_p^2(\ell) d\ell \tag{20.51}$$

Referring to (20.8), the integral in (20.51) is evaluated as

$$\int_L \omega_p^2(\ell) d\ell = \int_L \frac{n'(\ell) e^2}{m\varepsilon_o} d\ell \cong \bar{\omega}_p^2 L \tag{20.52}$$

where the dimension of $\bar{\omega}_p^2$ is expressed in radians^2/second^2. The last approximation assumes that the electron density is the average (or a weighted average) over the path length L. This is a reasonable assumption over short time intervals since there are no electron collisions and the electron plume is expanding from the force of the detonation and later contracting through electron recombining in the troposphere and lower ionosphere. Both of these events occur over relatively long periods of time compared to the typical communications message duration and *snapshots* of the electron profiles can be predicted. It is recommended that AS systems are to be designed for the worst-case scenario which favors the weighted average being biased toward the worse-case electron density. However, laying these details aside, in the following analysis the performance of the communication system is evaluated parametrically in terms of the plasma frequency $\bar{\omega}_p^2$.

Substituting (20.52) into (20.51) results in the approximate channel phase expression

$$\phi(\omega) \cong \frac{\bar{\omega}_p^2 L}{2c\omega_c} \left(1 - \frac{\omega - \omega_c}{\omega_c} + \frac{2(\omega - \omega_c)^2}{\omega_c^2} - \frac{6(\omega - \omega_c)^3}{\omega_c^3} + \cdots \right) \tag{20.53}$$

Referring again to (20.8), characteristic frequency of the plasma is evaluated as

$$f_p = \frac{1}{2\pi} \sqrt{\frac{\bar{n}_e e^2}{m' \varepsilon_o}} = 8.9779 \sqrt{\bar{n}_e} \quad : \text{Hz} \tag{20.54}$$

Using these results, the phase function in (20.48) is expressed as

$$\Phi(\omega) = \Phi_o + \frac{\bar{\omega}_p^2 L}{2c\omega_c} \left[1 - \frac{\omega - \omega_c}{\omega_c} + \frac{2(\omega - \omega_c)^2}{\omega_c^2} - \frac{6(\omega - \omega_c)^3}{\omega_c^3} + \cdots \right] + \text{higher order distortion terms} \tag{20.55}$$

It is convenient to characterize the phase function about the carrier frequency by defining* $u = \omega - \omega_c$ and, upon substitution into (20.55), the low-pass phase function is expressed as

$$\begin{aligned} \Phi(u) = &-\left(\frac{\omega_c L_p}{c} - \frac{\bar{\omega}_p^2 L}{2c\omega_c} \right) - \left(\frac{\omega_c L_p}{c} + \frac{\bar{\omega}_p^2 L}{2c\omega_c} \right) \left(\frac{u}{\omega_c} \right) \\ &+ \frac{\bar{\omega}_p^2 L}{c\omega_c} \left(\frac{u}{\omega_c} \right)^2 - \frac{3\,\bar{\omega}_p^2 L}{c\omega_c} \left(\frac{u}{\omega_c} \right)^3 \\ &+ \text{higher order distortion terms} \end{aligned} \tag{20.56}$$

where Φ_o is evaluated using the relationship $\Phi_o = -\omega L_p / c = -(\omega_c L_p / c + u L_p / c)$. The linear phase term in u simply represents a constant delay and the higher order terms contribute to the signal distortion.

20.6.1 Signal Delay and Dispersion

Considering the channel phase function $\Phi(u)$, the resulting signal delay function is given by

$$\begin{aligned} T_d(u) &= -\frac{\partial \Phi(u)}{\partial u} \\ &= \left(\frac{L_p}{c} + \frac{\bar{\omega}_p^2 L}{2c\omega_c^2} \right) - \frac{2\bar{\omega}_p^2 L}{c\omega_c^3} u + \frac{9\bar{\omega}_p^2 L}{c\omega_c^4} u^2 - \cdots \end{aligned} \tag{20.57}$$

The constant delay, resulting from the LOS path, is given by $T_o = T_d + t_d$ where $T_d = L_p/c$ is the delay from the

*The use of the parameter u with dimensions of radians/second should not be confused with the real part of the refractive index as expressed in (20.18).

undisturbed channel and, with $T_L = L/c$, the delay over the path L through the disturbed channel is evaluated using

$$t_d = \frac{\bar{\omega}_p^2 T_L}{2\,\omega_c^2} = 40.302\frac{\bar{n}_e T_L}{f_c^2} \quad : \text{plasma dependent delay in seconds} \tag{20.58}$$

These delays do not result in signal distortion; however, the quadratic and higher order frequency-dependent delay terms result in signal distortion. The linear and quadratic delay terms are evaluated as

$$T_{d1} = -4t_d\left(\frac{u}{\omega_c}\right) \quad : \text{linear delay} \tag{20.59}$$

and

$$T_{d2} = 18t_d\left(\frac{u}{\omega_c}\right)^2 \quad : \text{quadratic delay} \tag{20.60}$$

Expressing the frequency deviation from the carrier as $\Delta f = u/2\pi$ Hz and normalizing these delays by the symbol duration, the normalized delay distortion terms become

$$\frac{T_{d1}}{T} = -4\frac{t_d}{T}\left(\frac{\Delta f}{f_c}\right) \tag{20.61}$$

and

$$\frac{T_{d2}}{T} = 18\frac{t_d}{T}\left(\frac{\Delta f}{f_c}\right)^2 \tag{20.62}$$

20.6.2 Example of Signal Delay Distortion

In this example, the delay terms through the linear distortion term are considered, that is, the higher order distortion terms are neglected, so the channel frequency response is characterized as

$$H(u) = e^{j\left(\Phi_o + \phi_o + \frac{\rho}{2}\omega_c + (-T_o + \rho\omega_c)u + \frac{\rho}{2}u^2\right)} \tag{20.63}$$

where $u = \omega - \omega_c$ is the angular frequency about the carrier frequency. Referring to (20.48) the phase ϕ_o is evaluated as

$$\phi_o = -\left(\frac{\omega_c L_p}{c} - \frac{\bar{\omega}_p^2 L}{2c\omega_c}\right) = -\left(\omega_c T_p - \frac{\bar{\omega}_p^2 T_L}{2\omega_c}\right) \quad : \text{rad} \tag{20.64}$$

and from (20.57) T_o is evaluated as

$$T_o = \frac{L_p}{c} + \frac{\bar{\omega}_p^2 L}{2c\omega_c^2} = T_p + \frac{\bar{\omega}_p^2 T_L}{2\omega_c^2} \quad : \text{s} \tag{20.65}$$

$$\rho = \frac{\bar{\omega}_p^2 L}{c\,\omega_c^3} = \frac{\bar{\omega}_p^2 T_L}{\omega_c^3} \quad : \text{s}^2/\text{rad} \tag{20.66}$$

The channel impulse response is evaluated using the inverse Fourier transform as

$$h(t) = \frac{e^{j\left(\Phi_o + \phi_o + \frac{\rho}{2}\omega_c\right)}}{2\pi}\int_{-\pi W}^{\pi W} e^{j\left((-T_o + \rho\omega_c)u + \rho u^2/2\right)} e^{jut}\,du \tag{20.67}$$

where W is the radio frequency (RF) bandwidth centered on the carrier frequency f_c. In this analysis the magnitude of the channel impulse response is evaluated in terms of Fresnel integrals as [30]

$$|h(t)| = \frac{1}{2}\sqrt{\frac{2f_c}{T_L}}\left(\frac{\omega_c}{\bar{\omega}_p}\right)\sqrt{[C(z)]^2 + [S(z)]^2} \tag{20.68}$$

The upper and lower integration limits z_2 and z_1 of the Fresnel integrals are expressed as

$$\begin{pmatrix} z_2 \\ z_1 \end{pmatrix} = \sqrt{\frac{2f_c}{T_L}}\left(\frac{\omega_c}{\bar{\omega}_p}\right)(t - T_o + \rho\omega_c) \pm \sqrt{\frac{T_L}{2f_c}}\left(\frac{\bar{\omega}_p}{\omega_c}\right)W \tag{20.69}$$

These results are normalized by letting $y = tW$, $y_o = T_o W - \rho\omega_c W$, and $x = u/W$. Using these results the magnitude of the channel impulse response is evaluated in terms of the normalized parameters as

$$|h(y)| = \frac{W}{2}\sqrt{\frac{2f_c/W}{T_L W}}\left(\frac{\omega_c}{\bar{\omega}_p}\right)\sqrt{[C(\bar{z}_2) + C(\bar{z}_1)]^2 + [S(\bar{z}_2) + S(\bar{z}_1)]^2} \tag{20.70}$$

and the normalized arguments of the Fresnel integrals are expressed as

$$\begin{pmatrix} \bar{z}_2 \\ \bar{z}_1 \end{pmatrix} = \sqrt{\frac{2f_c/W}{T_L W}}\left(\frac{\omega_c}{\bar{\omega}_p}\right)(y - y_o) \pm \sqrt{\frac{T_L W}{2f_c/W}}\left(\frac{\bar{\omega}_p}{\omega_c}\right) \tag{20.71}$$

The channel impulse response simulation results, shown in Figure 20.11, correspond to $y_o = 0$, a path length of $L = 1$ km through the ionized medium, a carrier frequency of $f_c = 10$ GHz, and a normalized bandwidth parameter of $W/f_c = 2\text{e}{-7}$, which corresponds to a channel symbol rate of $R_s = 2$k symbols/s. The nonideal impulse response depicts the pulse dispersion caused by the quadratic phase distortion that is a direct result of the TEC (electrons/m^2) through the disturbed region. The results apply to noncoherent (NC) symbol detection and the range $1\text{e}11 \le \text{TEC} \le 1\text{e}12$ corresponds to those found in the natural environment. By way of comparison, for carrier frequencies of 60 and 300 GHz the respective TEC ranges are $1\text{e}13 \le \text{TEC} \le 1\text{e}14$ and $6\text{e}15 \le \text{TEC} \le 1\text{e}16$.

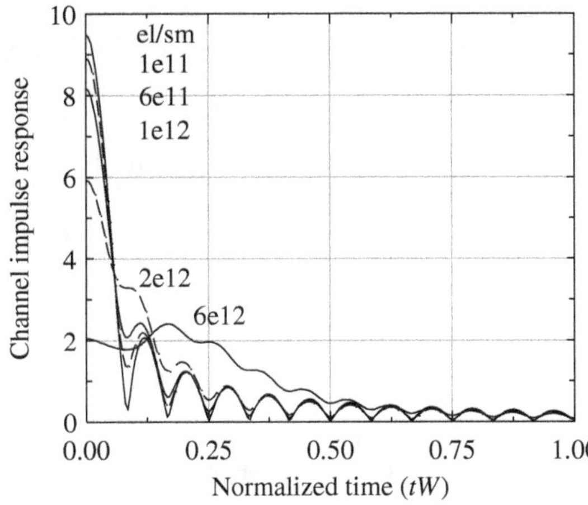

FIGURE 20.11 Ionospheric channel impulse response characteristics ($L = 1$ km, $f_c = 10$ GHz, $R_s = 2$k symbols/s).

The ISI must be evaluated further by examining the correlation response of the detection filter; however, the impact on the symbol-error performance must be examined using additional analysis or Monte Carlo simulations. These evaluations should include mitigation techniques including adaptive ISI cancellation and FEC. In this regard, this analysis has limited utility and may be considered as a first step in characterizing the impact of the TEC on the communication link performance. Although considerably more involved, this analysis can also be extended to initially examine the impact of the TEC on coherent detection and phaselock loop tracking.

20.7 RAYLEIGH SCINTILLATION CHANNEL MODEL

The amplitude variations or scintillation of a received signal propagating through a heavily ionized region of the ionosphere is the result of the constructive and destructive interaction of the signal phase resulting from numerous signal paths through the media. The phenomena of scintillation are described in Sections 20.2 and 20.3 that includes example electron density profiles for natural and nuclear-disturbed environments. In addition to the electron density concentrations, the dynamics of the channel and communication system will further influence the scintillation characteristics of the received signal.

Because the predominant influence of the media is upon the signal phase, the channel is characterized in terms of a phase power spectral density (PPSD) function. The analysis described in this section to characterize the received signal scintillation was proposed by Wittwer [31] and provides a relatively straightforward way to generate receiver amplitude and phase perturbations in a Rayleigh environment corresponding to severe scintillation. An alternate approach to that presented in this section is discussed by Knepp [32]. Using the PPSD also allows for generating receiver amplitude and phase fluctuations having correlation properties directly related to the physical parameters of the environment. The PPSD of interest, obtained from extensive research involving the modeling of observed phenomena [33], is expressed as

$$\Phi(\Omega) = \frac{\sigma_\phi^2 L_o}{2\left[1 + (\Omega L_o)^2\right]^{3/2}} \tag{20.72}$$

The variable $\Omega = 2\pi/\lambda = 2\pi\xi$ is the spatial angular frequency and λ represents the spatial wavelength. The parameter ξ is the spatial frequency having units of cycles/meter and L_o is referred to as the outer scale size and represents the length of the electron homogeneity in the structured ionosphere. This scale size ranges from 1 to 10 km and, as will be seen, is related to the spatial correlation length ℓ_c in the plain of the receiver.

The parameter σ_ϕ^2 is the signal phase variance and is related to the electron density fluctuation [34] σ_{Ne}^2. However, because the present analysis is concerned with Rayleigh amplitude statistics, the intensity fluctuation is, in a sense, saturated and σ_ϕ^2 simply becomes a scale factor. Expressed as a function of the spatial frequency ξ, the PPSD is given by

$$\Phi(\xi) = \frac{\sigma_\phi^2 L_o}{2\left[1 + (2\pi\xi L_o)^2\right]^{3/2}} \tag{20.73}$$

The electric field fluctuation in the plain of the receiver* is obtained by taking the inverse Fourier transform of the zero-mean complex Gaussian random variable

$$b(\xi) = b_I(\xi) + jb_Q(\xi) \tag{20.74}$$

where the quadrature components $i = \{I,Q\}$ are distributed as

$$p_{b_i(\xi)}(b_i) = \frac{1}{\sqrt{2\pi}\sigma_\phi(\xi)} e^{-b_i^2/2\sigma_\phi^2(\xi)} \tag{20.75}$$

with variance $\sigma_\phi^2(\xi) \propto \Phi(\xi)$. The components $b_i(\xi)$ are statistically independent in i and ξ.

Based on these characterizations, the received spatial electric field strength is given by

$$\varepsilon(\ell) = \frac{1}{2\pi} \int_{-\infty}^{\infty} b(\xi) e^{j2\pi\xi\ell} d\xi \tag{20.76}$$

*The *plane of the receiver* is defined as the plain normal to the receiver antenna pointing vector.

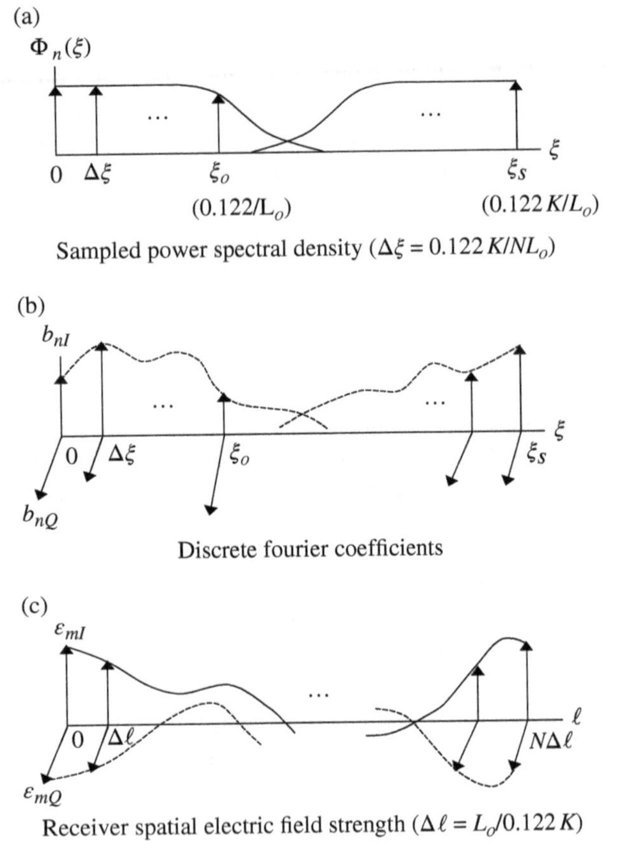

FIGURE 20.12 Representations of spectral and spatial functions.

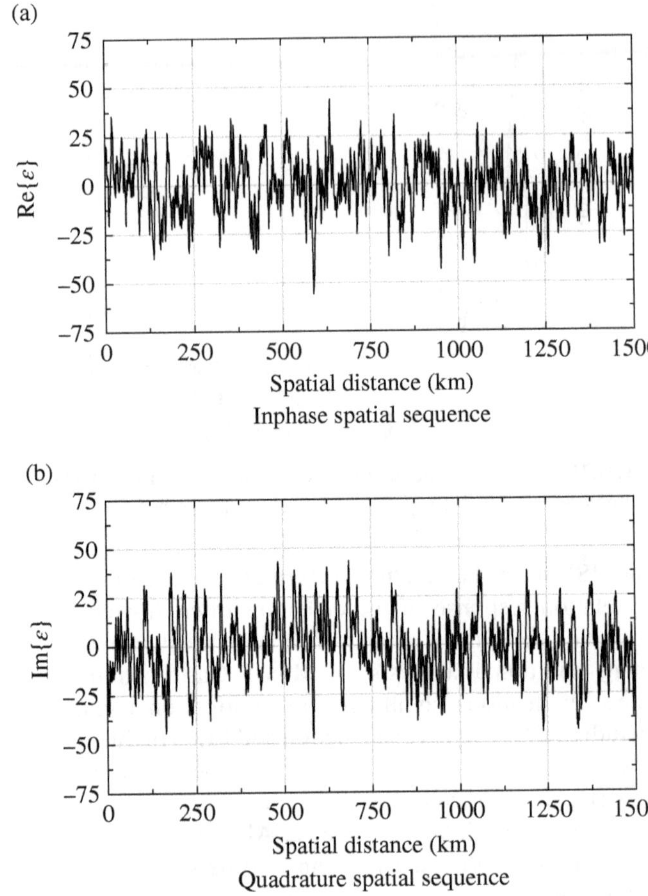

FIGURE 20.13 Typical computer-generated receiver spatial sequence.

To generate receiver sample functions or sequences for use in subsequent system simulations, the inverse FFT is used and the discrete form of (20.76) is expressed as

$$\varepsilon_m(\ell) = \frac{\Delta\xi}{2\pi}\sum_{n=0}^{N-1} b_n(\xi)e^{j2\pi nm/N} \tag{20.77}$$

where the n and m indices are defined as $\ell = m\Delta\ell : m = 1,\ldots,N$ and $\xi = n\Delta\xi : n = 1,\ldots,N$. Furthermore, using a radix-2 FFT of length N samples such that N is a power of two, results in $\Delta\xi\Delta\ell = 1/N$. The quadrature components of the complex function $\varepsilon_m = \varepsilon_{Im} + j\varepsilon_{Qm}$ are also *iid* zero-mean Gaussian random variables.

The 3-dB spatial frequency of the PPSD, normalized to $\Phi_n(0) = \sigma_\phi^2/2$, is evaluated as $\xi_o = 0.122/L_o$ and the spatial sampling frequency is $\xi_s = K\xi_o$, where K is selected to satisfy the Nyquist criterion. Figure 20.12 shows the sampling characteristics of the three functions of interest with the abscissa expressed in terms of the outer scale size L_o. Figure 20.13 shows a typical computed spatial sequence (or record) for $L_o = 3$ km, $\sigma_\phi = 1°$, with $K = 64$ and $N = 4096$.

20.7.1 Spatial Correlation of Receiver Electric Field Strength

The correlation distance ℓ_o of the receiver electric field is a significant parameter and is used to determine the required separation between receiver terminals for spatial combining diversity. It is also used to determine the scintillation decorrelation time τ_o used to determine FEC code length, interleaver length, and repeat message intervals for temporal combining. The correlation distance is defined in terms of the normalized autocorrelation coefficient $\rho_c(\ell) = R(\ell)/R(0)$ such that

$$\rho_c(\ell)|_{\ell_o} = e^{-1} = 0.368 \tag{20.78}$$

The correlation function $R(\ell)$ is evaluated as

$$R(\ell) = E\left[\varepsilon_{k+\ell}\varepsilon_k^*\right] \tag{20.79}$$

Evaluation of (20.79) using the expression for ε_k proceeds as follows:

$$R(\ell) = \frac{\Delta\xi^2}{(2\pi)^2}\sum_{n=0}^{N-1}\sum_{m=0}^{N-1} E\left[b_n b_m^*\right]e^{jn(k+\ell)/N}e^{-jmK/N} \tag{20.80}$$

Because b_n and b_m^* are orthogonal, that is, are independent normal random variables, $E[b_n b_m^*] = C\delta_{nm}$, where δ_{nm} is the Kronecker delta function. Using this result the correlation function becomes

$$R(\ell) = \frac{\Delta\xi}{2\pi}\sum_{n=0}^{N-1} E[b_n b_n^*] e^{jn\ell/N} \tag{20.81}$$

Using the expression for the complex *iid* zero-mean samples b_n results in

$$E[b_n b_n^*] = \frac{\sigma_\phi^2 L_o}{\left[1+(2\pi n\Delta\xi L_o)^2\right]^{3/2}} \tag{20.82}$$

and the desired expression for the correlation function becomes

$$R(\ell) = \frac{\Delta\xi\,\sigma_\phi^2 L_o}{2\pi}\sum_{n=0}^{N-1}\frac{e^{jn\ell/N}}{\left[1+(2\pi n\Delta\xi L_o)^2\right]^{3/2}} \tag{20.83}$$

Equation (20.83) is evaluated numerically to determine the correlation distance ℓ_o or by changing the integrand to a continuous function, normalizing by $R(0)$ and evaluating the resulting integral yields the normalized solution

$$\rho(x) = \frac{x}{L_o}K_1\left(\frac{x}{L_o}\right) \tag{20.84}$$

where $K_1(-)$ is the modified Bessel function of order one. Evaluation of this result at the decorrelation value of e^{-1} results in the correlation distance

$$\ell_o = 1.65 L_o \tag{20.85}$$

To check the fidelity of the simulation code in generating the sampled received electric field sequences, the correlation responses of the sequences shown in Figure 20.13 are evaluated and the results are shown in Figure 20.14 with the abscissa plotted in terms of the spatial distance $\ell = mL_o/0.122K$, $m = 1,\ldots, N-1$. From these results and using the decorrelation value of 0.368 defined earlier, the correlation distance for both the in-phase and quadrature channels are nearly the same and equal to $\ell_o = 4.61$ km. The theoretical value, derived earlier is $\ell_o = 1.65(3\text{ km}) = 4.95$ km so the simulated results are within 7.4% of the theoretical value. As the number of independently generated received sequences increases the average correlation response converges to the theoretical response expressed by (20.83).

As a further verification of the simulation code, the power spectral density of the received field strength is computed and shown in Figure 20.15. The *circled* data points represent the

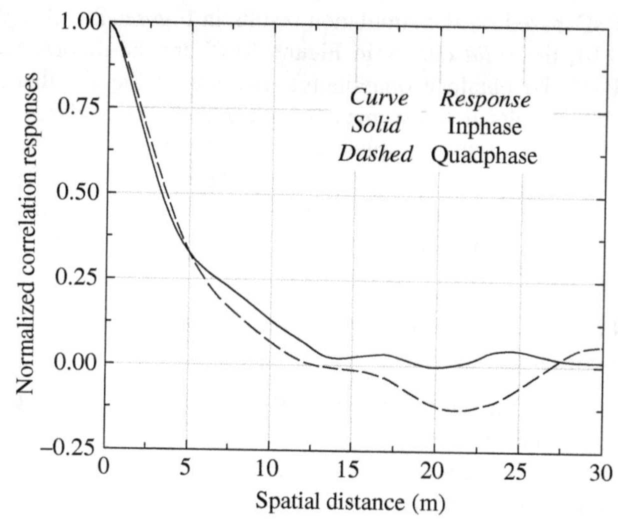

FIGURE 20.14 Correlation characteristics of simulated receiver sequence.

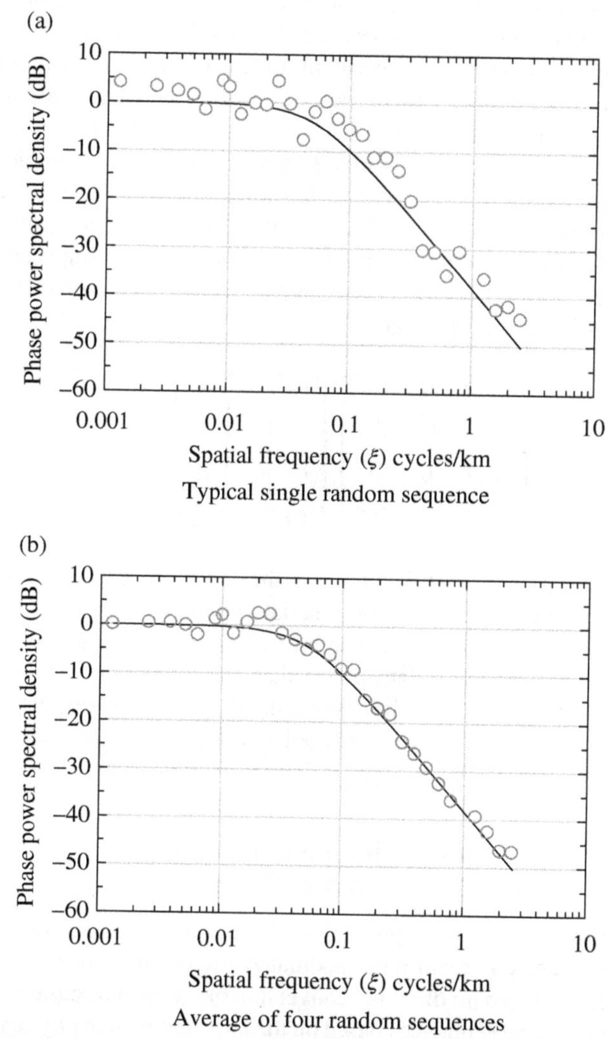

FIGURE 20.15 Reconstructed PPSD based on the simulation-generated sequences of the receiver electric field strength.

PPSD based on the simulation results in Figures 20.13 and 20.14; the *solid* curves in Figure 20.15 are the theoretical PPSD. To obtain a quantitative measure of the simulated PPSD results a linear mean-square regression analysis is computed for the samples in Figure 20.15b in the range $0.2 \le \xi \le 2.17$. This analysis indicates that slope of the PPSD curve and 3 dB intercept point are within 3.6 and 4.5% of the respective theoretical values.

20.7.2 Concatenation of Computer-Generated Scintillation Records

In this and the following section, the independently generated orthogonal receiver electric field strength sequences described in the preceding section are stored as *data records* and used as required in a system simulation. Typically a system simulation will require the concatenation of many data records over the time spanned by the communication data or message. Each data record is generated using different noise seeds so the in-phase and quadrature scintillation samples in each record and between the records are independent. To concatenate data records in a seamless manner without severe amplitude and phase discontinuities, the records to be joined are separated in distance by one correlation interval (ℓ_o) and interpolation is used to generate additional samples to *fill-in* the *gap* of length ℓ_o. The interpolation is based on a third-degree polynomial in ℓ with coefficients computed based on the boundary conditions yielding equal amplitude and slopes of the two records being joined. This is equivalent to solving the equation

$$\begin{pmatrix} 1 & 0 & 0 & 0 \\ 1 & \ell_c & \ell_c^2 & \ell_c^3 \\ 0 & 1 & 0 & 0 \\ 0 & 1 & 2\ell_c & 3\ell_c^2 \end{pmatrix} \begin{pmatrix} c_o \\ c_1 \\ c_2 \\ c_3 \end{pmatrix} = \begin{pmatrix} \varepsilon_o \\ \varepsilon_1 \\ \varepsilon'_o \\ \varepsilon'_1 \end{pmatrix} \tag{20.86}$$

where ε_o is the last data point in the current record, assumed to be at $\ell = 0$ and ε_1 is the first data point in the new record, assumed to be at $\ell = \ell_o$. The primed values are the slopes of the corresponding data points that are computed using the next to the last and the second data point in the respective records. The number of data points used to span the correlation interval ℓ_o is $N_c = \lfloor \ell_o/\Delta\ell \rfloor = \lfloor 0.2013\ K \rfloor$.

20.7.3 Spatial-to-Temporal Conversion of Computer-Generated Data Records

In this section, the application of the spatial data records to the received signal data-modulated waveform is described. The first step involves the conversion of the spatial sequences to temporal sequences based on the knowledge of the dynamics of the electron plume and the communication platform. If the encounter were completely stationary, that is, the communication platforms were *fixed* in position and the Earth's rotation and electron plume were *frozen* in place, then the received signal amplitude would be based on a random selection from the Rayleigh amplitude distribution. If, however, the selected amplitude resulted in a deep amplitude null, exceeding the available link margin then communications would be impossible. In this situation if, for example, the receiver were to move by about one spatial correlation interval ℓ_o, then it would be likely that the resulting signal amplitude would be high enough of to establish communications. The phenomena of this idealized example is exactly what is taking place; however, the Earth and electron plume is continually moving and the communication platforms themselves exhibit velocity components that contribute to the Rayleigh received signal amplitude fluctuations. During the time immediately following the detonation, the electron plume will exhibit extremely high velocity components due to the forces of the initial blast. Following the initial detonation, the energy dissipates resulting in less dynamic interactions and lower velocities. During this period of electron recombining in the ionosphere and troposphere much longer lasting scintillation effects continue due to the influence of wind and other natural dynamic forces that influence the direction and velocity of the electron plume.

The velocity of interest is the velocity vector in the plane of the receiver that results from all of the relative motions of the encounter. For example, consider that the resulting velocity in the plane of the receiver is $V = \ell/t$, where ℓ is the distance that a point on the received spatial sequence moves in t seconds. As mentioned previously, the time for this fictitious point to move one correlation interval of length ℓ_o is defined as the temporal decorrelation time τ_o, so that, the velocity is described as

$$V = \frac{\ell}{t} = \frac{\ell_o}{\tau_o} \tag{20.87}$$

Using the right-hand equality in this expression and substituting for the correlation distance in terms of L_o results in the normalized expression

$$\frac{t}{\tau_o} = \frac{\ell}{\ell_o} = \frac{\ell}{1.65L_o} \tag{20.88}$$

The maximum time spanned by a computer-generated sequence is determined from the maximum length ℓ_{max} and, referring to Figure 20.12, $\ell_{max} = 1/\Delta\xi = NL_o/0.122K$ so that

$$\frac{t_{max}}{\tau_o} = \frac{\ell_{max}}{1.65L_o} = \frac{N}{0.2013K} \tag{20.89}$$

Using the example parameters: $N = 4096$, $K = 64$ and $L_o = 3$ km, the number of decorrelation time constants spanned by a computer-generated record is $t_{max}/\tau_o = 317.9$. This is an important consideration, in that, the accuracy and resulting confidence of a Monte Carlo performance simulation must take into account the relatively long times involved in characterizing the impact of the channel fluctuations. The corresponding sampling interval of the computer-generated records is

$$\Delta t_s = \frac{t_{max}}{N} = \frac{\tau_o}{0.2013K} = \frac{\ell_o}{0.2013KV} = \frac{8.1967L_o}{KV} \tag{20.90}$$

where the *link* between τ_o and ℓ_o is through the velocity, that is, $V = \ell_o/\tau_o$. In terms of the computer-generated records, referring to Figure 20.12c, the sample interval is

$$\Delta\ell = \Delta t_s V = \frac{L_o}{0.122K} \tag{20.91}$$

Consider that a modulated symbol of duration T seconds ($R_s = 1/T$ symbols/s) is received and sampled using N_s samples per symbol. Under these conditions, the required receiver sampling interval is

$$\Delta T_s = \frac{T}{N_s} \tag{20.92}$$

Except for the unique case when $\Delta T_s = \Delta t_s$, it will be necessary to interpolate between the scintillation record samples and the required receiver sampling instant. To aid in the description of the sampling, Figure 20.16 shows the temporal samples (g_k) that have been transformed from the spatial samples (ε_m) as described earlier. For convenience g_k is considered to be a real signal and it is implicit that sampling and interpolation is applied to both the real and imaginary parts of the complex scintillation process. The interpolation of the symbol sample is based on linear interpolation. For example, considering $k\Delta t_s \le i\Delta T_s \le (k+1)\Delta t_s$ the symbol sample g_i corresponding to $i\Delta T_s$ is computed using the nearest record samples $k\Delta t_s$ and $(k+1)\Delta t_s$ as

$$g_i = g_k + \frac{(g_{k+1} - g_k)}{\Delta t_s}(i\Delta T_s - k\Delta t_s) = g_k + (g_{k+1} - g_k)\left(i\frac{\Delta T_s}{\Delta t_s} - k\right) \tag{20.93}$$

The slow and fast scintillation conditions are defined in terms of the decorrelation time relative to the symbol duration. For the slow fading condition shown in Figure 20.16a, there are typically many symbol samples relative to the record samples and the parameters of the interpolation g_k and g_{k+1} are not updated very often. For very slow

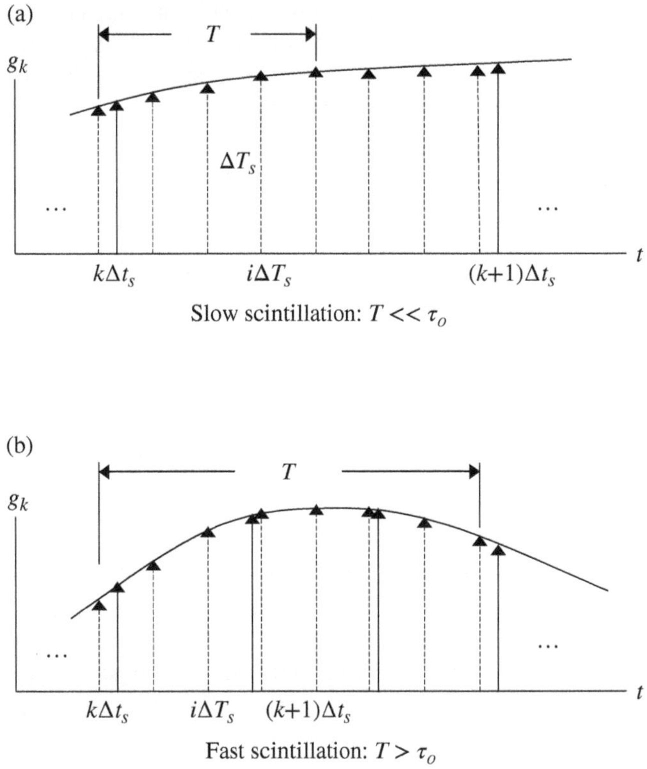

FIGURE 20.16 Temporal scintillation record sampling (*solid* lines) and interpolated samples (*dashed* lines).

scintillation, the record samples or symbol amplitudes are virtually constant over many symbols leading to a considerable saving in simulation time. In the case of fast scintillation, the record sampling must ensure that the symbol sampling does not result in aliasing. This can be accomplished by selecting the symbol sampling frequency f_s such that $f_s = \max(1/\Delta t_s, 1/\Delta T_s)$ where $\Delta T_s = T/\bar{N}_s$ and $\bar{N}_s$ is the required samples per symbol to satisfy the Nyquist sampling frequency in the additive white Gaussian noise (AWGN) channel. Using this and the previous results the required sampling frequency with scintillation can be expressed as

$$f_s = \max\left(\frac{0.2013K}{\tau_o}, \frac{\bar{N}_s}{T}\right) = \frac{1}{T}\max\left(\frac{0.2013K}{\tau_o/T}, \bar{N}_s\right) \tag{20.94}$$

Although not necessary, it is often convenient to choose the number of samples per symbol to be an integer in which case the condition $f_s = N_s/T$ applies and

$$N_s = \max\left(\left\lceil\frac{0.2013K}{\tau_o/T}\right\rceil, \bar{N}_s\right) \tag{20.95}$$

where $\lceil\,\rceil$ signifies the ceiling or the smallest integer greater than the argument and, in this case, is the required samples per symbol to satisfy the Nyquist condition with channel scintillation. For a given scenario the ratio $\Delta T_s/\Delta t_s$ in the interpolation equation is a constant and, in view of the preceding comments, is generally expressed as

$$\frac{\Delta T_s}{\Delta t_s} = \frac{1}{f_s \Delta t_s} = \begin{cases} \dfrac{0.2013K}{N_s}\left(\dfrac{T}{\tau_o}\right) & : \text{slow and moderate scintillation} \\ 1 & : \text{fast scintillation} \end{cases} \tag{20.96}$$

In the last equality of (20.96), the symbol sampling frequency is equal to the scintillation record sampling frequency and no interpolation is necessary. In this case, the channel symbol sampling frequency must be increased so that $\Delta T_s = \Delta t_s$ and rate conversion or downsampling in the receiver processing may be applied to result in a convenient number of samples per symbol for demodulation and tracking. Of course, the rate conversion must satisfy the Nyquist sampling criterion in consideration of the received signal and channel bandwidths.

20.7.4 Additional Simulation Considerations

Upon inclusion of the signal scintillation data records described earlier into a simulation program, it is necessary to establish and verify various operational conditions to ensure reliable and accurate results. Several parameters requiring calibration and verification are discussed in this section.

20.7.4.1 Establishing the Receiver Signal-to-Noise Ratio The Rayleigh fading channel introduces absorption and other losses that result in a net loss in the received signal power relative to the AWGN channel. The simulation code, however, does not need to account for these detailed losses as long as the received signal-to-noise ratio is properly established. In fact, because of various normalizations, the Rayleigh fading records result in a net channel gain that must be determined to establish the received signal-to-noise ratio. The power gain through the Rayleigh fading channel is simply the second moment of the Rayleigh *pdf* and is given by

$$E\left[|\varepsilon_m|^2\right] = 2\sigma_\phi^2 \tag{20.97}$$

where σ_ϕ^2 is the variance of the underlying Gaussian noise process used to generate the Rayleigh samples. For a finite-length record, the channel power gain will vary from record to record and Table 20.12 lists the statistical parameters associated with five records each with 2048 complex samples ε_m. The received signal-to-noise ratio in an AWGN channel is computed as $\gamma_n = P/\sigma_n^2$ where P is received signal power and σ_n^2 is the AWGN power. Therefore, the received signal-to-noise ratio from the scintillating channel is determined as

TABLE 20.12 Computed Parameters for Five Typical Rayleigh Scintillation Records

Record	$E[\lvert\varepsilon_m\rvert^2]$	$E[\lvert\varepsilon_m\rvert]$	$\sigma[\lvert\varepsilon_m\rvert]$
1	463.47	19.116	9.902
2	502.793	19.892	10.349
3	555.349	20.850	10.983
4	489.232	19.448	10.536
5	512.193	20.234	10.138
Composite results			
All	504.607	19.908	10.383

$$\gamma_c = \frac{\rho P}{\sigma_n^2} \tag{20.98}$$

where $\rho = E[|\varepsilon_m|^2] = 504.607$ is the average of five independently generated scintillation data records* as listed in Table 20.12.

20.7.4.2 Generating Uncorrelated Samples When performing a simulation it may be necessary to evaluate the received symbols under independent or uncorrelated fading conditions. This requirement applies, for example, with frequency hopping where groups of transmitted symbols are associated with each hop interval and the frequency hopping is greater than the channel correlation bandwidth f_o. In these situations the scintillation record is *run-up* or advanced by one correlation interval between each frequency hop thus simulating an uncorrelated scintillation condition from hop to hop. To achieve this result the scintillation record must be advanced by N_{so} samples where

$$N_{so} = \left\lceil \frac{\ell_o}{\Delta \ell} \right\rceil = \lceil 0.2013K \rceil \tag{20.99}$$

For the example records discussed previously $K = 64$ so $N_{so} = 13$.

20.7.4.3 Verifying Record Generation As in all simulations, it is necessary to evaluate the simulation code before committing to lengthy system performance runs. A convenient check of the channel sampling relative to the modulated symbol is to choose the number of channel samples to be

*Averaging additional data records resulted in a difference of less than 0.1 dB in the E_b/N_o performance specification; however, the averaging of different record sizes and scintillation parameters should also be examined.

equal to the Nyquist samples that are required to avoid aliasing in the AWGN channel. That is, by choosing

$$\frac{0.2013K}{\tau_o/T} = \bar{N}_s \qquad (20.100)$$

Or alternately by choosing

$$\frac{\tau_o}{T} = \frac{0.2013K}{\bar{N}_s} \qquad (20.101)$$

the symbol samples will correspond to the scintillation record samples which are easily verified.

20.8 SCINTILLATION MITIGATION TECHNIQUES

The techniques used to mitigate communications signal scintillation involve the system operating conditions, medium access control (MAC) protocols, and waveform design. The applicable mitigation techniques depend on the system mission requirements, for example, multiple links and satellites, selection of satellite orbits, ground support systems, communication networks and protocols, long transmissions involving video and mapping data, and short emergency action messages (EAMs). The waveform mitigation techniques include time diversity, FEC coding, data interleaving, and a form of robust symbol modulation like noncoherent or differentially coherent. Table 20.13 summarizes the system mitigation techniques to provide reliable communications through a scintillating channel.

TABLE 20.13 System Scintillation Mitigation Techniques

Mitigation Technique	Comments
Increase link margin	Increase EIPR
	Decrease the system noise figure
	Decrease the data rate
	Increase FEC coding and interleaving
	Increase antenna gain (limited by ℓ_o)
Frequency selection	Increasing frequency is an effective way to mitigate scintillation
Spatial diversity	Multiple satellites
	Satellite cross-links
	Multiple antennas
Network protocols	Full or partial satellite processing
	Satellite up and downlink switching
	Network timing
	Data priorities
	Bit count integrity (BCI)
	Automatic repeat request (ARQ)
Satellite orbit selection	GSO, LEO, and HEO nonstationary satellites
	Satellite orbit control provides spatial diversity and survivability

Increasing the EIRP has a limited effect on the link margin because the signal variations result from phase cancellations. Typically a 3 dB link margin is provided to assist the link recovery as the absorption recovers from a sudden loss of signal due to a blackout event or deep fade condition. Increasing the EIRP by increasing the antenna gain is limited by acquisition and beam tracking issues and angular scattering through the channel. From a system point of view spatial diversity is effective, however, is problematic with geostationary orbits (GEOs) in terms of networking and protocol requirements for systems operating at lower frequencies. For example, if an ultra-high frequency (UHF) GEO satellite link to a ground site is affected by severe scintillation it will most likely remain so for several hours. Relay links through neighboring satellites or ground stations separated in distance greater than $2\ell_o$ is effective but requires robust and flexible protocols. The geosynchronous orbits (GSOs), low Earth orbits (LEOs), and highly elliptical orbits (HEOs) inherently provide spatial diversity by virtue of the satellite motion relative to fixed earth coordinates. An issue with these orbits is reacquiring the satellite emerging from a severely stressed region and providing protocols that handle lost data such as ARQ and BCI. BCI requires that demodulator symbol timing and carrier tracking states be preserved during a loss-of-signal condition, keeping an accurate count of lost bits when signal detection is resumed. Links that can be switched to other satellites not affected by the stressed environment are effective in preserving communications but require robust and flexible protocols. Increasing the carrier frequency is arguably the single most effective system mitigation technique.

Table 20.14 summarizes waveform and demodulator mitigation techniques that provide reliable communications through a scintillating channel.

Robust waveform modulations include noncoherent M-ary frequency shift keying (NC M-ary FSK), alternately referred to as NC multifrequency shift keying (NCMFSK) and differentially coherent multiphase shift keying (DCMPSK); however, to minimize the matched filter detection loss, the symbol duration must be less than the minimum channel decorrelation time, that is, $T < \tau_o(\text{min})$; for differently coherent modulation $T < \tau_o(\text{min})/2$. Coherent modulation requires that the carrier tracking phaselock loop bandwidth be much greater than the reciprocal of minimum channel decorrelation time, that is, $B_L \geq 10/\tau_o(\text{min})$. As discussed in Chapter 10, the symbol rate for phaselock loop tracking must satisfy the requirement $R_s > B_L/k$ where $k \ll 1$ is dependent on the waveform modulation. Suppose, for example, that a message consists of N_s symbols; in this case, the entire message can be coherently tracked during the channel decorrelation time if the symbol rate satisfies the condition $R_s > 10\,N_s/(k\tau_o(\text{min}))$. The probability of a correct

TABLE 20.14 Waveform and Demodulator Scintillation Mitigation Techniques

Mitigation Technique	Comments
Modulation selection	Noncoherent Differentially coherent Differentially encoded coherent
Time diversity	Time diversity $\geq\tau_o$, bit interleaving, message repetition, message piecing
Adaptive processing	Decision feedback equalization (DFE)
FEC coding	Block coding (BCH or RS coding) Convolution coding Turbo, turbo-like, and LDPC coding Data randomizers and interleavers
Diversity combining	Selection diversity (SD) Maximum ratio combining diversity (MRCD) Equal gain diversity (EGD)
Frequency diversity	Requires frequencies $\geq f_o$
Spatial diversity	Multiple antennas with beam steering multiple-in multiple-out (MIMO) antennas

message reception and delivery time depends on the communication protocol and the statistical characteristics of τ_o. During deep signal fades the signal-to-noise ratio will be degraded resulting in degraded detection and tracking; however, the noncoherent and differentially coherent modulations will inherently recover from these deep fades, whereas, coherent modulation may require reacquisition processing to resume detection. With the aid of signal-to-noise ratio estimation algorithms the tracking loops can be locked during deep fades and unlocked as the signal level recovers. Differentially encoded coherent modulation has the advantage of resolving phaselock ambiguities that may occur during phase acquisition and tracking. The most sensitive demodulator issues in a scintillation environment are related to the antenna, symbol, and carrier tracking loops.

Bit interleaving provides a measure of time diversity and is necessary with random error correcting FEC coding; however, interleaving over the maximum channel decorrelation time τ_o(max) results in long delays and large data storage requirements. Burst-error correction convolutional codes and Reed–Solomon block codes provide robust error correction performance with scintillation. Frequency diversity is effective as long as the frequency separation exceeds the frequency-selective bandwidth, that is, $\Delta f > f_o$(max). Because of the large frequency shifts required to satisfy this condition, noncoherent frequency combining is often impractical. The use of spread-spectrum modulated waveforms with RAKE demodulator processing is an effective mitigation in multipath environments. Spatial diversity is discussed earlier, in the context of system operations; the use of multiple receiver antennas or antenna steering is also applicable to the receiver and demodulator

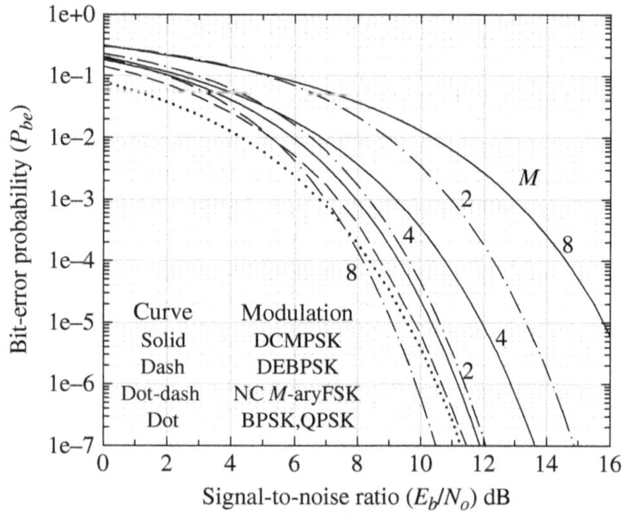

FIGURE 20.17 AWGN performance of several robust waveform modulations.

processing. For example, the application of MIMO antenna structures [35–38] with multiple-receiver signal combining is an effective implementation of spatial diversity.

20.9 CASE STUDY: BPSK AND DCBPSK PERFORMANCE IN RAYLEIGH FADING CHANNEL

In this case study, the performance of several robust waveform modulations is examined using Monte Carlo simulations and computer-generated Rayleigh channel scintillation records as described in Section 20.7. Based on the Rayleigh model, the channel fading conditions correspond to slow and fast Rayleigh fading that are defined in terms of the modulated symbol rate. For example, when the channel decorrelation time is much greater than the symbol duration then slow fading occurs and when it is much less than the symbol duration then fast fading is encountered. A major advantage of performing Monte Carlo simulations is that the system performance can be evaluated in the intermediate range involving $T/k_1 \leq \tau_o \leq k_2T$ where k_1 and k_2 depend on the modulation with typical values of $k_1 = 10$ and k_2 ranging from ten to several hundred; in this region analytical models are, at best, questionable. Robust waveform modulations, without FEC and interleaving, that offer the least complexity and performance degradation in a severely fading channel* are differently encoded binary phase shift keying (DEBPSK), DCMPSK, and NCMFSK. As a point of reference, the bit-error performance of these waveform modulations, operating in the AWGN channel without FEC coding and combining, is depicted in Figure 20.17; in this context, the most useful

*A severely fading channel is a channel that experiences the range of τ_o or f_o shown in Figures 20.8 and 20.9.

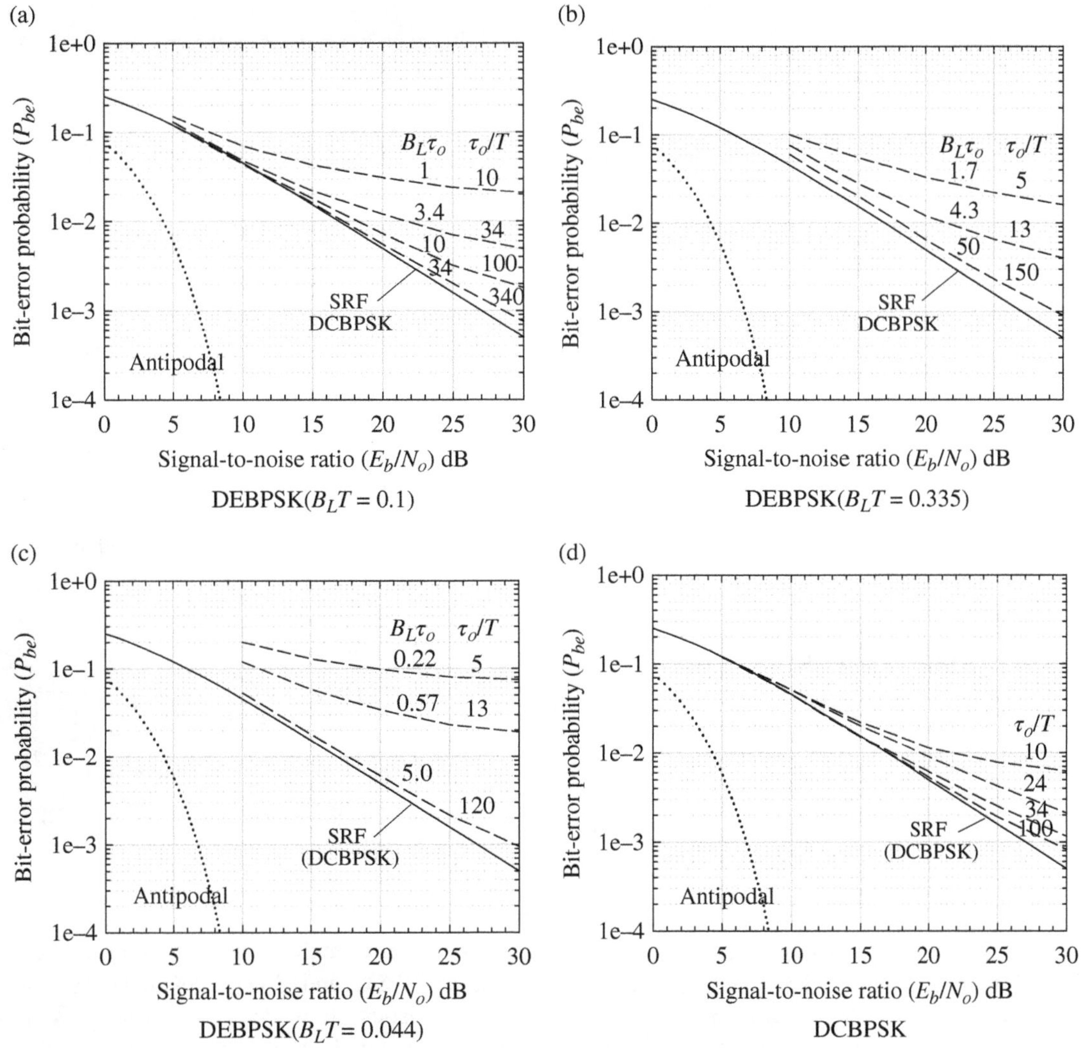

FIGURE 20.18 Performance of DEBPSK and DCBPSK in a slow Rayleigh fading channel ($\beta = 0$).

implementation of DCMPSK is DCBPSK. The *dotted* performance curves represent coherently detected binary phase shift keying (BPSK) and quadrature phase shift keying (QPSK) modulations and are shown only as a point of reference; these modulations do not perform well in severe channel scintillation due the phase tracking requirement. In Section 20.9.1, the performance of these modulated waveforms is evaluated with slow fading channels using Monte Carlo simulation under various conditions of the normalized channel decorrelation time τ_o/T and, in the case of DEBPSK, the phaselock loop time bandwidth products $B_L\tau_o$ and B_LT.

20.9.1 Performance of Robust Modulations in Rayleigh Fading Channel

The simulated bit-error performance of DEBPSK and differently coherent binary phase shift keying (DCBPSK) in a severe Rayleigh channel with slow fading is shown in Figure 20.18. In this example, the channel corresponds to frequency nonselective or flat fading. The performance depicted by the *solid* curve is the theoretical performance of DCBPSK with slow fading Rayleigh limit (SFRL) from Figure 18.7. The *dashed* curves show the Monte Carlo simulated performance under the indicated normalized channel decorrelation times. FEC coding, interleaving, or combining is not used in these examples and the received signal frequency error is zero. The Monte Carlo simulations involve a minimum of 100 channel decorrelation intervals for each E_b/N_o. The DEBPSK waveform requires phase tacking in the demodulator. However, the π radian ambiguity, that may result after the tracking loop recovers from the channel fading, does not result in catastrophic errors but an error multiplication results due to the differential bit-to-bit phase decoding.

Figure 20.18a–c shows the impact on the performance of DEBPSK for the indicated conditions of normalized loop bandwidth (B_LT) of the second-order, hard-limiting, Costas

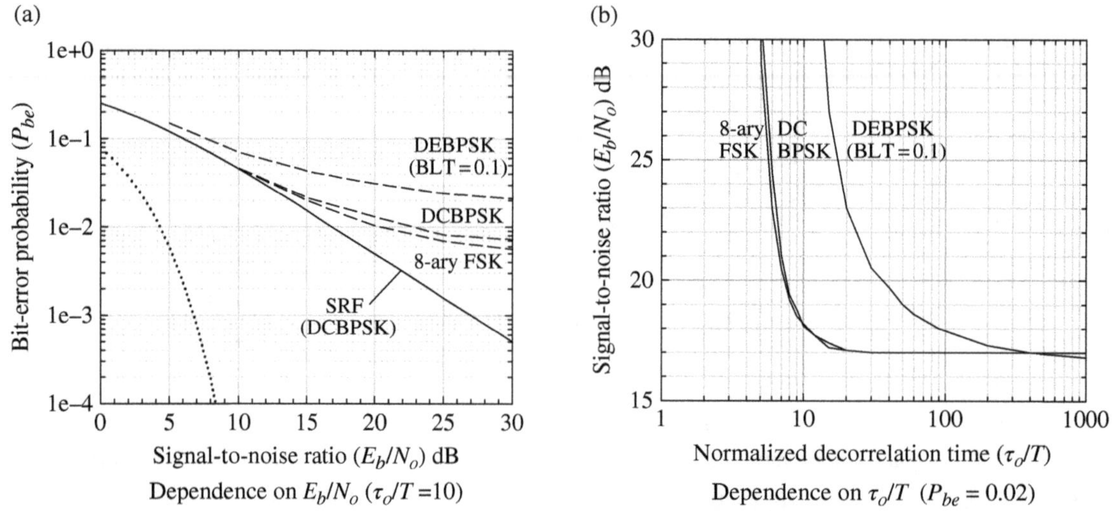

FIGURE 20.19 Performance Comparison of DEBPSK, DCBPSK, and NC 8-ary FSK in a slow Rayleigh fading channel ($\beta = 0$).

phaselock loop. The DCBPSK demodulator performance is shown in Figure 20.18d. In this case, a first-order frequency tracking loop is used that recovers rapidly from a severe channel fade. In general, Figure 20.18 shows that the bit-error performance approaches an irreducible error rate with increasing E_b/N_o and, with decreasing E_b/N_o, the bit-error performance approaches the theoretical SFRL bit-error probability asymptotically from above.* Furthermore, as the normalized channel decorrelation time (τ_o/T) increases the bit-error probability becomes asymptotically closer to the theoretical SFRL performance with a correspondingly lower irreducible bit-error probability with increasing E_b/N_o. The impact of the SFRL channel on the phaselock loop tracking is evident; however, it is complicated by the influence of the random data and Rayleigh channel fluctuations on the hard-limiting phaselock loop and the manner in which the differentially encoded data recovers from deep fades. Figure 20.18b and c demonstrates performance symmetry in terms of τ_o/T about the selection of $B_LT = 0.1$ shown in Figure 20.18a. However, in terms of the product $B_L\tau_o$, using $B_LT = 0.044$ is preferred because of the tolerance to lower channel decorrelation times.

Figure 20.19 compares the performance of these waveforms, including the 8-ary noncoherently detected FSK waveform, under the indicated conditions. Part a is similar to the performance in Figure 20.18 and part b shows the dependence of E_b/N_o on the normalized decorrelation time τ_o/T corresponding to a bit-error probability of $P_{be} = 0.02$. The rapid increase in the required E_b/N_o ratio as τ_o/T decreases results from the fast fading Rayleigh limit (FFRL) relative to the symbol duration. These results indicate that DCBPSK and 8-ary FSK have similar performance characteristics; however, the 8-ary FSK provides three bits per symbol at the expense of a wider transmission bandwidth.

*The curves are labeled SRF; however, since $\beta = 0$ the channel corresponds to the SFRL condition.

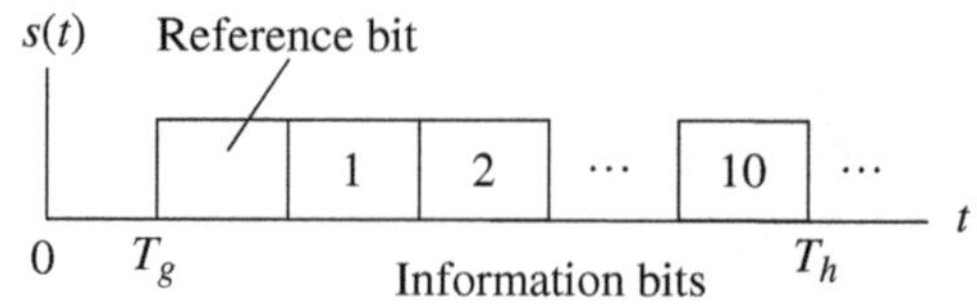

FIGURE 20.20 Frequency-hopped DCBPSK modulated information bits.

20.9.2 Performance of Frequency Hopped DCBPSK with Combining in Rayleigh Fading Channel

In this section the performance of DCBPSK is examined in a frequency-hopping application with a hopping capability that exceeds the decorrelation bandwidth f_o. In this case, 10 differentially coherent information bits are appended to a reference bit and a guard interval $T_g = 0.1T_b$ is used to provide for frequency synthesizer settling. The hopping format, of duration $T_h = 11.1T_b$, is shown in Figure 20.20.

To realize a combining gain with nonselective frequency fading, the information bits are repeated with the synthesizer frequency exceeding the previous hop frequency by the decorrelation bandwidth f_o.†

†In a baseband simulation, this simply requires that the simulated fading record be advanced by one decorrelation interval to encounter an independent hoping interval.

The resulting demodulator performance is shown, as the *dashed* curves, for the SFRL channel in Figure 20.21 using SD with $N_c = 2$ and 3. This corresponds to the nonselective frequency combining of two and three hops with repetitions of the information bits. These results are based on Monte Carlo simulations with a minimum of 1000 independently fading hops at each signal-to-noise ratio using the simulation and hop combining techniques discussed in Sections 20.7.2 through 20.7.4. The theoretical aspects of diversity combing techniques are discussed in Section 18.8 and the simulated performance results are in good agreement with the theoretical results shown in Figure 18.13. The *circled* data points correspond to the SFRL and AWGN channels and demonstrate the accuracy of the DCBPSK waveform simulation under these respective channel conditions.

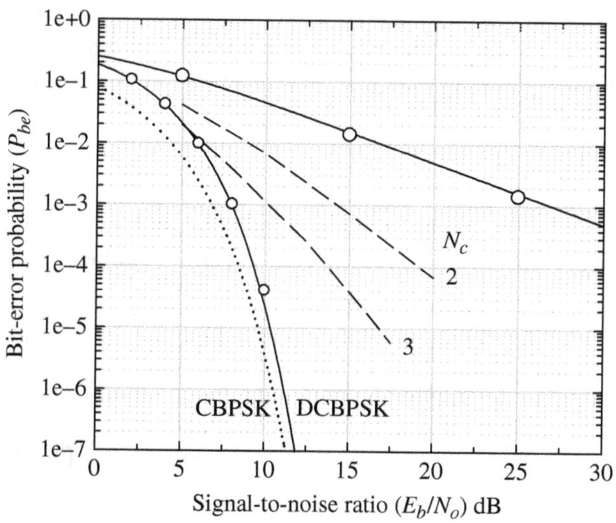

FIGURE 20.21 Performance of frequency-hopped DCBPSK waveform. (10 bits/hop, $\tau_o/T_h = 10$, $\tau_o/T_b = 111$, $\beta = 0$).

20.9.3 Performance of Convolutional Coded DCBPSK with Interleaving in Rayleigh Fading Channel

This example uses FEC coding, data interleaving, and DCBPSK waveform modulation as shown in Figure 20.22. The details of the FEC coding, interleaving, and modulation are given in Table 20.15. The output of the convolutional coder is used to fill the block interleaver column by column and, when the interleaver is filled, the row-by-row output is sent to the differentially coherent modulator following the initial reference bit. On average, the errors are separated by $T_{sep} = T_{row} + 1 = 32$ bits. In terms of the channel decorrelation time $\tau_o/T_{sep} = 34/32 = 1.0625$ so the interleaver spans about one decorrelation interval. The bit-error performance in the SFRL channel for this coding configuration is shown as the *dashed* curve in Figure 20.23. The *solid* curves with the *circled* data points represent, respectively, the theoretical and simulated performance in the SFRL and AWGN channels without coding. The *solid* curve, labeled *coded*, is the simulated performance of the convolution coded waveform in the AWGN channel. The *solid* curves are included as a reference or performance goal. This coding and modulation configuration performs somewhat better, at the expense of bandwidth and complexity, than the $N_c = 3$ SD combining performance shown in Figure 20.21.

20.9.4 Performance of Concatenated Convolutional and Reed–Solomon Block Codes with DCBPSK and DCQPSK in Rayleigh Fading Channel

The code concatenation in this example uses the rate 1/2 convolutional inner code, as described in the preceding section, with a rate 1/2, 6-symbol-error correcting (24,12) 64-ary shortened Reed–Solomon outer code [39]. The intent is to examine the capability of the 6-symbol (6-bits/symbol) burst-error correction capability of the Reed–Solomon block code without using an interleaver between the codes. The

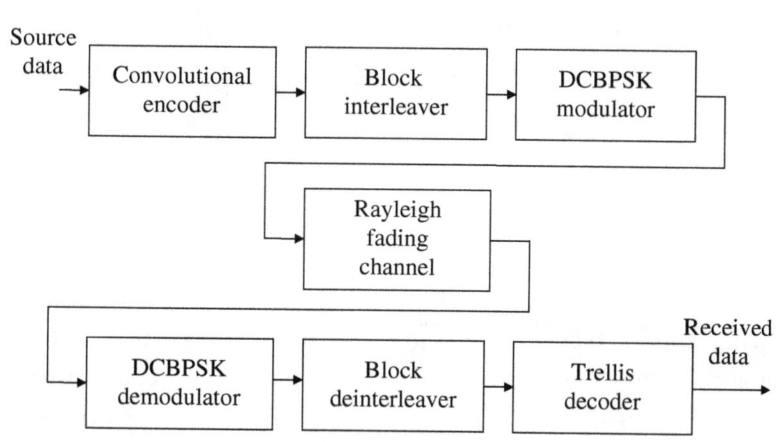

FIGURE 20.22 Implementation of FEC coded and interleaved DCBPSK waveform.

performance of this configuration is characterized as contours of E_b/N_o versus τ_o for constant block-error probabilities of $P_e(Blk) = 1/3$, 1/30, and 1/300 as shown in Figure 20.24 for DCBPSK and differentially coherent QPSK (DCQPSK) modulations. These plots are obtained by plotting families of $P_e(Blk)$ versus E_b/N_o, similar to the P_{be} versus E_b/N_o in Figure 20.23, for a number of channel decorrelation times and then plotting E_b/N_o versus τ_o that intersect the selected constant $P_e(Blk)$ condition. These plots reveal some interesting characteristics concerning the performance in the fading channel. For example, as τ_o decreases the signal-to-noise ratio approaches the FFRL and as τ_o increases the signal-to-noise ratio approaches the SFRL of the coding configuration. The region exhibiting a relatively constant signal-to-noise ratio is the range over which the Reed–Solomon code is effective in correcting the errors from the convolutional inner code.

The influence of τ_o is dependent on the symbol rate, so the performance curves in Figure 20.24 will shift left or right corresponding, respectively, to higher or lower symbol rates. The symbol rate is also a function of the FEC code rate, in that, $R_s = r_c R_b$. However, for a given symbol rate, the region of constant signal-to-noise performance can be expanded to the right, accommodating larger decorrelation times, by increasing the span (N) of the block interleaver for a given K, thus increasing the parity bits, of the (N,K) Reed–Solomon code.

TABLE 20.15 FEC Coding, Data Interleaving, and Waveform Parameters

Function	Characteristics
FEC coding	Rate $r_c = 1/2$, constraint length $K = 7$ outer convolutional code
	Generators: $g_1 = 1111001$, $g_2 = 1011011$
	Viterbi 64 state trellis decoder with: 3-bit soft decisions and 32-bit depth
Data interleaving	(M,N) block interleaver: $M = 31$, $N = 56$
DCBPSK	$\tau_o/T_b = 34$

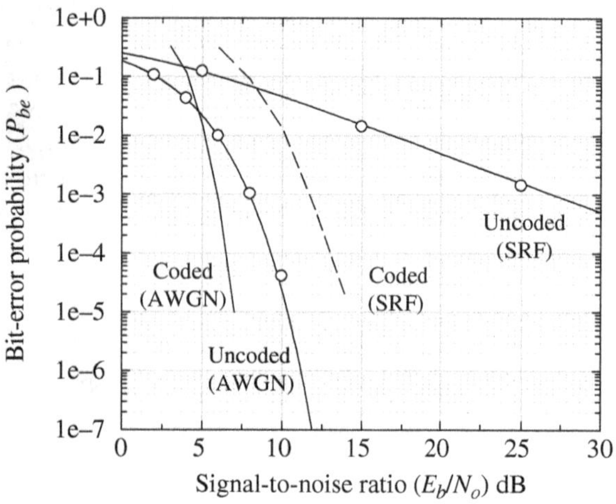

FIGURE 20.23 Performance of DCBPSK with FEC coding and interleaving ($\tau_o/T_b = 34$, $\tau_o/T_{sep} = 1.06$, $\beta = 0$).

20.9.5 Performance of Uncoded Noncoherent M-ary FSK Modulation with Rayleigh Fading Channel

An upper bound on the bit-error performance of uncoded noncoherently detected M-ary FSK with diversity ℓ is given by [40, 41]

$$P_{be} \le 2^{k-1} p^{\ell} \sum_{i=0}^{\ell-1} \binom{\ell - 1 + i}{\ell} (1-p)^i \tag{20.102}$$

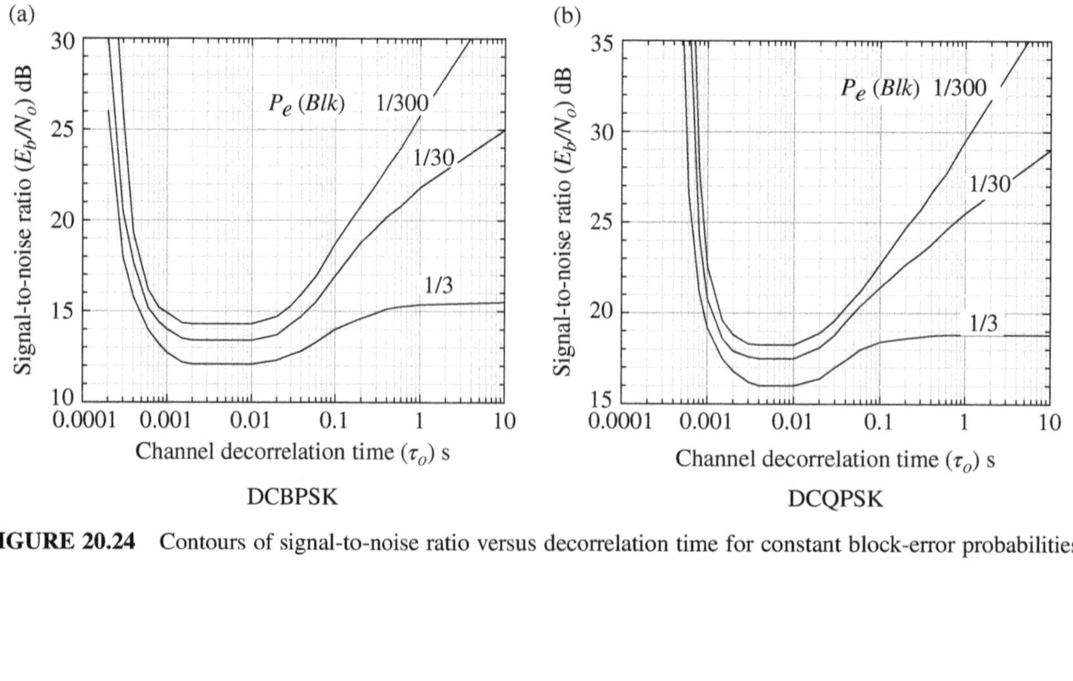

FIGURE 20.24 Contours of signal-to-noise ratio versus decorrelation time for constant block-error probabilities.

where $k = \log_2(M)$ is the number of bits per symbol and the probability p is given by

$$p = \left(2 + \frac{k}{\ell}\left\langle\frac{E_b}{N_o}\right\rangle\right)^{-1} \tag{20.103}$$

Evaluation of (20.102) is shown in Figure 20.25 for various values of ℓ. The average received signal-to-noise ratio from the fading channel is $<E_b/N_o>$. For bit-error probabilities above about 10^{-5} there is a performance loss between $\ell = 16$ and 20 and further increases in diversity only provide performance gains at bit-error probabilities less than about 10^{-5}. Upon comparing the uncoded 8-ary FSK diversity performance with that of the dual-3 convolutional coded 8-ary FSK performance in Figure 8.50, the dual-3 code performance for $\ell = 1$ and 2 is about the same as the respective uncoded performance for $\ell = 6$ and 12. Although the performance of the dual-3 code is significantly better with diversities of 3 and 4, the decision influencing the choice between the two approaches may involve the computational complexity.

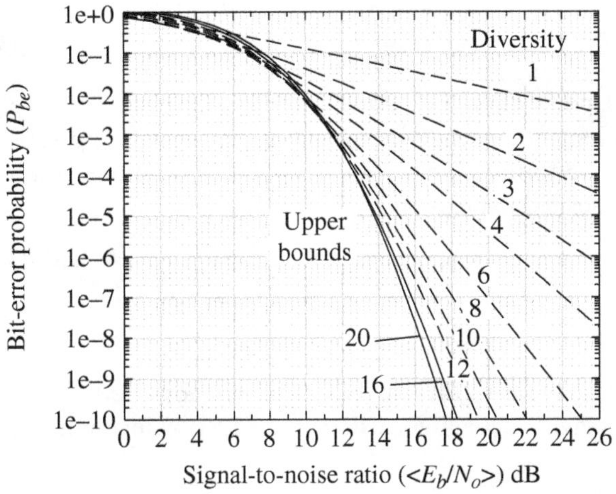

FIGURE 20.25 Uncoded noncoherent 8-ary FSK performance in slow Rayleigh fading.

APPENDIX 20A GEOGRAPHIC AND GEOMAGNETIC COORDINATE TRANSFORMATIONS

This appendix outlines the geometry required to convert a point described in the geographic coordinate system to the geomagnetic coordinate system and vice versa. The points are described in terms of the respective latitude and longitude polar angles (ϕ, λ) and (Φ, Λ). Positive latitudes correspond to the northern hemisphere and positive longitudes correspond to degree east of the Greenwich Meridian; negative longitudes are measured to the west. The geographic system axes are defined as (x, y, z) where the positive z-axis in the direction of the North Pole and the positive x-axis intersect the Greenwich Meridian at the geographic equator and the positive y-axis corresponds to 90° east longitude. A spherical Earth with radius R is assumed and the polar angles of the geomagnetic system are defined in terms of the geographic system as (ϕ_m, λ_m) where $\phi_m = 78.6°$ and $\lambda_m = -69.8°$. The geomagnetic system axes are defined as (x_m, y_m, z_m). These relationships are depicted in Figure 20A.1 where the spherical triangle [42] (P, P_m, P_x) is used to evaluate the transformation between the polar coordinates as described in the following sections.

The location of the point P_x with geomagnetic polar coordinates (Φ, Λ) is transformed into geographic coordinates using

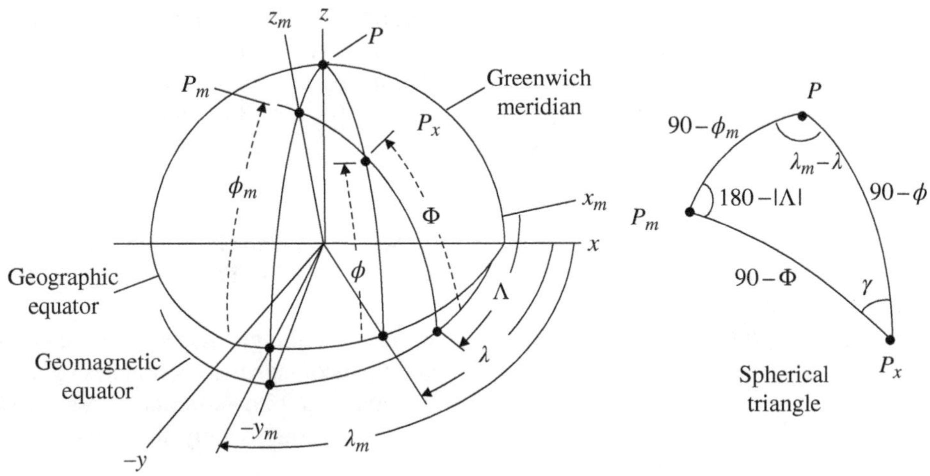

FIGURE 20A.1 Relationships between geographic and geomagnetic coordinates.

$$\begin{pmatrix} x \\ y \\ z \end{pmatrix} = \begin{pmatrix} \cos(\phi)\cos(\lambda) & 0 & 0 \\ 0 & \cos(\phi)\sin(\lambda) & 0 \\ 0 & 0 & \sin(\phi) \end{pmatrix} \begin{pmatrix} x_m \\ y_m \\ z_m \end{pmatrix} \tag{20A.1}$$

where the geographic polar coordinates (ϕ, λ) are computed as described in Section 20A.2 and the geomagnetic location of P_x is determined as

$$\begin{pmatrix} x_m \\ y_m \\ z_m \end{pmatrix} = R \begin{pmatrix} \cos(\Phi)\cos(\Lambda) & 0 & 0 \\ 0 & \cos(\Phi)\sin(\Lambda) & 0 \\ 0 & 0 & \sin(\Phi) \end{pmatrix} \tag{20A.2}$$

The transformation from geographic-to-geomagnetic coordinates is accomplished by the appropriate interchange of the axes and polar angles.

20A.1 GEOGRAPHIC-TO-GEOMAGNETIC COORDINATE TRANSFORMATION

In this case the geographic coordinates (ϕ, λ) are known and the geomagnetic coordinates (Φ, Λ) are to be computed. Referring to the spherical triangle in Figure 20A.1 and applying the *law of cosines** results in the expression

$$\cos(90-\Phi) = \cos(90-\phi_m)\cos(90-\phi) + \sin(90-\phi_m)\sin(90-\phi)\cos(\lambda-\lambda_m) \tag{20A.3}$$

that simplifies to

$$\sin(\Phi) = \sin(\phi_m)\sin(\phi) + \cos(\phi_m)\cos(\phi)\cos(\lambda-\lambda_m) \tag{20A.4}$$

Using (20A.4) the geomagnetic latitude is determined as

$$\Phi = \arcsin(\sin(\phi_m)\sin(\phi) + \cos(\phi_m)\cos(\phi)\cos(\lambda-\lambda_m)) \tag{20A.5}$$

The geomagnetic longitude is determined using the *law of sines* for spherical geometry that results in the expression

$$\frac{\sin(180-|\Lambda|)}{\sin(90-\phi)} = \frac{\sin(\lambda-\lambda_m)}{\sin(90-\Phi)} \tag{20A.6}$$

*There are two law-of-cosine formulas, one involving the solution of a side in terms of the adjacent sides and opposite angle and the other involving the solution of an angle in terms of the adjacent angles and opposite side; in this analysis the former solution is used.

When simplified and rearranged (20A.6) results in the expression for $\sin(|\Lambda|)$

$$\sin(|\Lambda|) = \frac{\cos(\phi)\sin(\lambda-\lambda_m)}{\cos(\Phi)} \tag{20A.7}$$

Equation (20A.7) can be used to solve for Λ by using the trigonometric identity

$$\cos(\Phi) = \sqrt{1-\sin^2(\Phi)} \tag{20A.8}$$

Substituting (20A.4) to eliminate the dependence on Φ results in an involved expression and using (20A.7) results in a quadrant or sign ambiguity that needs to be resolved. A more satisfying approach is to use the *law of cosines* involving the longitude angle Λ expressed as

$$\cos(90-\phi) = \cos(90-\phi_m)\cos(90-\Phi) + \sin(90-\phi_m)\sin(90-\Phi)\cos(180-|\Lambda|) \tag{20A.9}$$

Upon simplifying (20A.9), solving for $\cos(|\Lambda|)$, then eliminating the dependence on $\sin(\Phi)$ by substituting (20A.4), and performing some additional simplifications results in the desired expression

$$\cos(|\Lambda|) = \frac{\sin(\phi_m)\cos(\phi)\sin(\lambda-\lambda_m) - \cos(\phi_m)\sin(\phi)}{\cos(\Phi)} \tag{20A.10}$$

Equations (20A.7) and (20A.10) are now used to compute the magnetic longitude using the atan2(y, x) function with the result

$$\Lambda = \mathrm{atan2}(\cos(\phi)\sin(\lambda-\lambda_m), \sin(\phi_m) \cos(\phi)\sin(\lambda-\lambda_m) - \cos(\phi_m)\sin(\phi)) \tag{20A.11}$$

The dependence on Φ in (20A.11) is eliminated by the formation of the ratio $\sin(|\Lambda|)/\cos(|\Lambda|)$ inherent in the atan2( ) function. The advantage in using the atan2( ) function is that the angles are computed in the range 0 to $\pm 180°$† so that angles $0 \le \Lambda \le 180°$ correspond to east longitudes or $0 \ge \Lambda \ge -180°$ correspond to west longitudes.

20A.2 GEOMAGNETIC-TO-GEOGRAPHIC COORDINATE TRANSFORMATION

In this case the geomagnetic coordinates (Φ, Λ) are known and the geomagnetic coordinates (ϕ, λ) are to be computed. The analysis follows similar lines of reasoning as in the preceding section so only the principal equations are given.

†The atan2( ) function in FORTRAN returns the angle in radians that must be converted to degrees to be consistent with units used in this appendix.

Using the *law of cosines* the geographic latitude is evaluated as

$$\phi = \arcsin(\sin(\phi_m)\sin(\Phi) - \cos(\Phi)\cos(\phi_m)\cos(|\Lambda|)) \tag{20A.12}$$

Using the *law of sines* $\sin(\lambda - \lambda_m)$ is evaluated as

$$\sin(\lambda - \lambda_m) = \frac{\cos(\Phi)\sin(|\Lambda|)}{\cos(\phi)} \tag{20A.13}$$

Applying the law of cosines $\cos(\lambda - \lambda_m)$ is evaluated, after substitutions and simplifications similar to those described in arriving at (20A.10), as

$$\cos(\lambda - \lambda_m) = \frac{\sin(\Phi)\cos(\phi_m) + \sin(\phi_m)\cos(\Phi)\cos(|\Lambda|)}{\cos(\phi)} \tag{20A.14}$$

Using (20A.13) and (20A.14) and the atan2($\sin(\lambda-\lambda_m)$, $\cos(\lambda-\lambda_m)$) function, the geographic longitude is computed as

$$\lambda = \lambda_m + \operatorname{atan2}(\cos(\Phi)\sin(|\Lambda|),\ \sin(\Phi)\cos(\phi_m) + \sin(\phi_m)\cos(\Phi)\cos(|\Lambda|)) \tag{20A.15}$$

ACRONYMS

a.m.	Before noon
ARDC	Air Research and Development Command
ARQ	Automatic repeat request
AS	Antiscintillation
AWGN	Additive white Gaussian noise
BCI	Bit count integrity
BPSK	Binary phase shift keying
DCBPSK	Differentially coherent BPSK
DCMPSK	Differentially coherent multiphase shift keying
DCQPSK	Differentially coherent QPSK
DEBPSK	Differentially encoded BPSK
EAM	Emergency action message
EGD	Equal gain diversity
FEC	Forward error correction (coding)
FFRL	Fast fading Rayleigh limit
GEO	Geostationary orbit
GSO	Geosynchronous orbit
HEO	Highly elliptical orbit
INTELSAT[a]	International telecommunications satellite
ISI	Intersymbol interference
LEO	Low Earth orbit
LOS	Line of sight
MAC	Medium access control (protocol)
NC *M*-ary FSK	Noncoherent *M*-ary frequency shift keying or NCMFSK (detection)
MFP	Magnetic field plane
MIMO	Multiple-in multiple-out (antenna)
MRCD	Maximum ratio combining diversity
NC	Noncoherent (detection)
NCMFSK	Noncoherent multifrequency shift keying or NC *M*-ary FSK (detection)
OGO	Orbiting geophysical observatory
p.m.	After noon
PPSD	Phase power spectral density
QPSK	Quadrature phase shift keying
RF	Radio frequency
SD	Selection diversity
SFRL	Slow fading Rayleigh limit
TAB	Time after blast
TACSAT	Tactical satellite (UHF)
TEC or N_T	Total electron content (electrons/m^2)
UHF	Ultra-high frequency

Note: RAKE is not an acronym; it is the name applied to a wide bandwidth multipath correlator.

[a]INTELSAT© Corporation is a communications satellite services provider.

PROBLEMS

1. Consider the inverse Fourier transform of the filter function $H(u)$ expressed in the indefinite form

$$h(t) = \frac{1}{2\pi}\int H(u)e^{jut}du = \frac{1}{2\pi}\int e^{j((\rho/2)u^2 - T_o u + \varphi)}e^{jut}du$$

$$= \frac{1}{2\pi}\int e^{j((\rho/2)u^2 + (t - T_o)u + \varphi)}du$$

$$= \frac{1}{2\pi}\int e^{j(au^2 + 2bu + c)}du$$

The general solution, expressed in terms of the parameters a, b, and c and the Fresnel integrals, is [30]

$$h(t) = \sqrt{\frac{\pi}{2a}}\left\{ \begin{array}{l} \cos\left(\dfrac{b^2 - ac}{a}\right)C\left[\sqrt{\dfrac{2}{a\pi}}(au+b)\right] + \sin\left(\dfrac{b^2 - ac}{a}\right)S\left[\sqrt{\dfrac{2}{a\pi}}(au+b)\right] \\ +j\cos\left(\dfrac{b^2 - ac}{a}\right)S\left[\sqrt{\dfrac{2}{a\pi}}(au+b)\right] - j\sin\left(\dfrac{b^2 - ac}{a}\right)C\left[\sqrt{\dfrac{2}{a\pi}}(au+b)\right] \end{array} \right\}$$

Considering the definite integral with lower and upper integration limits $-\pi W$ and πW and noting that $a = \rho/2$, $b = (t - T_o)/2$, and $c = \varphi$:

a. Derive the expression for the magnitude of $h(t)$ and express the result in terms of the Fresnel integrals and the parameters ρ, t, T_o, φ, and W.

b. Derive the expression for the phase of $h(t)$ and express the result in terms of the Fresnel integrals and the parameters ρ, t, T_o, φ, and W.

2. Using the spherical triangle in Figure 20A.1 write the expression for the *law of cosines* for determining the geographic latitude ϕ and write the simplifying equations leading to (20A.12).

3. Using the spherical triangle in Figure 20A.1 write the expression for the *law of sines* for determining $\sin(\lambda - \lambda_m)$ and write the simplifying equations leading to (20A.13).

4. Using the spherical triangle in Figure 20A.1 write the expression for the *law of cosines* for determining $\cos(\lambda - \lambda_m)$ and write the simplifying equations and manipulations leading to (20A.14).

Hint: Use the trigonometric identity $\cos(\phi) = \sqrt{1 - \sin^2(\phi)}$.

REFERENCES

1. M.G. Morgan, Guest Editor, "The Nature of the Ionosphere – An IGY Objective," Special Issue on the International Geophysical Year Objectives, Proceedings of the IRE, Vol. 47, No. 2, pp. 129–351, February 1959.
2. S. Chapman, "The Absorption and Disassociative Ionizing Effect of Monochromatic Radiation in an Atmosphere of a Rotating Earth," Proceeding of the Physical Society, Vol. 43, pp. 26–45, 1931. Proceeding of the Physical Society, London, England, 1931.
3. K. Davies, *Ionospheric Radio Propagation*, National Bureau of Standards Monograph, No. 80, United States Department of Commerce, National Bureau of Standards, Washington, DC, April 1, 1965.
4. R.A. Minzner, K.S.W. Champion, H.L. Pond, *The ARDC Model Atmosphere*, AFCRC-TR-59-267, Air Force Surveys in Geophysics, No. 115, Photochemistry Laboratory, Geophysics Research Directorate, Air Force Cambridge Research Center, Air Research and Development Command, Air Force Surveys in Geophysics, Bedford, MA, August 1959.
5. R.S. Berkowitz, Editor, *Modern Radar, Analysis, Evaluation and System Design*, Part V, Chapter 1, G.H. Millman, "Atmospheric Effects on Radio Wave Propagation," John Wiley & Sons, Inc., New York, 1966.
6. R.S. Berkowitz, Editor, *Modern Radar, Analysis, Evaluation and System Design*, Part V, Chapter 1, G.H. Millman, "Atmospheric Effects on Radio Wave Propagation," John Wiley & Sons, Inc., New York, 1966, p. 321, Tables 1-2 and 1-3.
7. R.L. Bogusch, "Ionospheric Scintillation Effects on Satellite Communications," Interim Technical Note MRC-N-139, Mission Research Corporation, Santa Barbara, CA, June 1974.
8. J.P. McClure, W.E. Hanson, "A Catalog of Ionospheric F-Region Irregularity Behavior Based on Ogo 6 Retarding Potential Analyzer Data," Journal of Geophysical Research, Vol. 78, No. 31, pp. 7431–7440, November 1973.
9. A.L. Johnson, "The Effects of Ionospheric Scintillation on Satellite Communications," Air and Space University Review, Vol. 24, No. 6, pp. 26–31, November–December 1973, U.S. Air Force, Maxwell AFB, AL.
10. M.R. Paulson, R.U.F. Hopkins, "Effects of Equatorial Scintillation Fading on SATCOM Signals," NELC/TR 1875, Naval Electronics Laboratory Center, San Diego, CA, May 8, 1973.
11. L.A. Wittwer, "Satellite Communications in a Scintillated Environment," Air Force Weapons Laboratory, Report AFWL-TR-75-240, January 1976. Also available from NTIS AD/A-021-091.
12. R.R. Taur, "Ionospheric Scintillation at 4 and 6 GHz," COMSAT Technical Review, Vol. 3, No. 1, pp. 145–163, Spring 1973.
13. L.A. Wittwer, "Satellite Communications in a Scintillated Environment," Air Force Weapons Laboratory, Report AFWL-TR-75-240, p. 39, January 1976.
14. C.M. Rush, "An Ionospheric Observation Network for Use in Short-Term Propagation Predictions," Telecommunications Journal, Vol. 43, pp. 544–548, August 1976.
15. G.N. Taylor, "The Seasonal Anomaly in Electron Content," J. Frihagen, Editor, Proceedings of NATO Advanced Studies Institute, pp. 543–548, Ohrid, Bulgaria, 1965.
16. R.S. Berkowitz, Editor, *Modern Radar, Analysis, Evaluation and System Design*, Part V, Chapter 1, G.H. Millman, "Atmospheric Effects on Radio Wave Propagation," John Wiley & Sons, Inc., New York, 1966, p. 322, Table 1-4.
17. R.W. Middlestead, R.E. LeLevier, M.D. Smith, "Satellite Crosslink Communications Vulnerability in a Nuclear Environment," IEEE Journal on Selected Area in Communications, Vol. SAC-5, No. 2, pp. 138–145, February 1987.
18. J.A. Stratton, *Electromagnetic Theory*, p. 327, McGraw Hill Book Company, Inc., New York, 1941.
19. J.A. Ratcliffe, *The Magneto-Ionic Theory and Its Application to the Ionosphere*, Cambridge University Press, New York, 1959.
20. K.G. Budden, *Radio Waves in the Ionosphere*, Cambridge University Press, New York, 1959.
21. K.G. Budden, *The Wave-Guide Mode Theory of Wave Propagation*, Logos Press/Prentice-Hall, London/Englewood Cliffs, NJ, 1961.
22. J.M. Kelso, *Radio Ray Propagation in the Ionosphere*, McGraw-Hill Co. Inc., New York, 1964.
23. D.R. McDaniel, "Proceedings of the DNA/LASL High-Altitude Nuclear Weapons Effects Summer Study (U), Vol. II—Working Group Reports (U)," DNA 4736P-3, SRI International, Menlo Park, CA, November 1, 1978.
24. L.A. Wittwer, "Radio Wave Propagation in Structured Ionization for Satellite Applications," Report DNA 5304D, Defense Nuclear Agency, Washington, DC, December 1979.
25. RDA Staff, "Nuclear Effects Handbook for Satellite Communications (U)," Research and Development Associates (RDA), pp. 2–23, Marina Del Rey, CA, May 1986.
26. RDA Staff, "Nuclear Effects Handbook for Satellite Communications (U)," Research and Development Associates (RDA), pp. 2–26, Marina Del Rey, CA, May 1986.

27. RDA Staff, "Nuclear Effects Handbook for Satellite Communications (U)," Research and Development Associates, pp. 2–21, Marina Del Rey, CA, May 1986.
28. RDA Staff, "Nuclear Effects Handbook for Satellite Communications (U)," Research and Development Associates, pp. 2–24, 25, Marina Del Rey, CA, May 1986.
29. W.S. Knapp, K. Schwartz, "Aids for the Study of Electromagnetic Blackout(U)," Report, DNA 3499H, General Electric—TEMPO, Santa Barbara, CA, February 1975.
30. M. Abramowitz, I.A. Stegun, *Handbook of Mathematical Functions with Formulas, Graphs, and Mathematical Tables*, Applied Mathematics Series, No. 55, p. 304, U.S. Department of Commerce, Washington, DC, June 1964.
31. L.A. Wittwer, "The Propagation of Satellite Signals Through a Turbulent Media," Air Force Weapons Laboratory, Report AFWL-TR-77-183, Kirkland Air Force Base, NM, January 1978.
32. D.L. Knepp, "Multiple Phase-Screen Calculation of the Temporal Behavior of Stochastic Waves," Proceeding of the IEEE, Vol. 71, No. 6, pp. 722–737, June 1983.
33. D.L. Knepp, "Multiple Phase-Screen Propagation Analysis for Defense Satellite Communications Systems," Report DNA-4424T, Defense Nuclear Agency, Washington, DC, September 1977.
34. R.L. Bogusch, F.W. Guigliano, "Nuclear Survivability Criteria for Satellite Links(U)," Final Report, AFWL-TR-82-18, Mission Research Corporation, Santa Barbara, CA, October 1982.
35. M. Torlak, T.M. Duman, "MIMO Communication Theory, Algorithms, and Prototyping," Proceedings of the 20th Signal Processing and Communications Applications Conference (SIU), Mugla, April 18–20, 2012.
36. S. Choi, Y.C. Ko, E.J. Powers, "Optimization of Switched MIMO Systems Over Rayleigh Fading Channels," IEEE Transactions on Vehicular Technology, Vol. 56, No. 1, pp. 103–114, January 2007.
37. M. Chiani, M.Z. Win, S. Hyundong, "MIMO Networks: The Effects of Interference," IEEE Transactions on Information Theory, Vol. 56, No. 1, pp. 336–349, January 2010.
38. L. Hanzo, M. El-Hajjar, O. Alamri, "Near Capacity Wireless Transceivers and Cooperative Communications in the MIMO Era: Evolution of Standards, Waveform Design, and Future Perspectives," Proceeding of the IEEE, Vol. 99, No. 8, pp. 1343–1385, August 2011.
39. R.W. Middlestead, "Performance of DC-PSK Modulation with Concatenated Coding in a Nuclear Environment," Air Force Weapons Laboratory, AFWL-TR-81-191, Rockwell International, Newport Beach, CA, February 1982.
40. A.J. Viterbi, I.M. Jacobs, "Advances in Coding and Modulation for Noncoherent Channels Affected by Fading, Partial Band, and Multiple-Access Interference," A.J. Viterbi, Editor, *Advances in Communication Systems*, Vol. 4, pp. 279–308, Academic Press, New York, 1975.
41. J.P. Odenwalder, "Error Control Coding Handbook," Prepared Under USAF Contract No. F44620-76-C-0056, Final Report, pp. 24–28, Linkabit Corporation, San Diego, CA, July 15, 1976.
42. O.W. Eshbach, *Handbook of Engineering Fundamentals*, Wiley Engineering Handbook Series, John Wiley & Sons, Inc., New York, 1952.

APPENDIX A
CLASSICAL FILTERS AND APPLICATIONS

A.1 INTRODUCTION

In this chapter, the application of analog filter designs in the evaluation of systems involving discrete-time sampled data simulations is described. This approach is straightforward, in that, the use of the impulse-invariant z-transform is used in the system simulation where $z = e^{sT_s}$ and T_s is the sampling interval. The parameter $s = \sigma + j\omega$ is the Laplace transform variable. The filter designs are sufficiently band-limited and the system simulation sampling frequency is sufficiently high so that aliasing is negligible. This is the case in system simulations where the sampling frequency is generally several times the symbol rate to provide for frequency error and receiver timing resolution during acquisition and tracking.

The analog filter designs used in this chapter are classical and appear in a number of references [1–8], so the derivation of the filter design details are not examined. However, the results presented are intended to document the implementation details as embodied in filter and system simulation programs. The filters examined in this appendix are the Butterworth filter with maximally flat amplitude (Section A.2.1), the Bessel filter with maximally flat delay (Section A.2.3), the Chebyshev filter with equal passband ripple (Section A.2.2), and the Elliptic filter with equal passband and stopband ripple (Section A.2.4).

A.2 CLASSICAL FILTER POLE LOCATIONS

This section reviews and documents the s-plane pole and zero locations for the filters indentified in Section A.1. In Section A.3, the frequency domain characteristics are used in a simulation program to examine the amplitude and phase responses. The resulting impulse responses are then generated for inclusion in various communication system simulation programs. Performance characteristics regarding the filtered symbol and intersymbol interference distortion attributed to the phase and amplitude responses are examined. The filter characteristics, including those described in Appendix B, are used throughout this book in simulation performance evaluations involving the generation and demodulation of a variety of communication waveforms through various channels.

A.2.1 Butterworth Filter Pole Locations

The Butterworth filters are characterized as having equally spaced poles on a unit circle in the s-plane as shown in Figure A.1.

For an N-pole filter, the pole locations are computed for $n = 0, 1, 2, \ldots, 2N - 1$ as

$$s_n = \begin{cases} e^{j(2n+1)\pi/2N} & : N = \text{even} \\ e^{jn\pi/N} & : N = \text{odd} \end{cases} \tag{A.1}$$

Digital Communications with Emphasis on Data Modems: Theory, Analysis, Design, Simulation, Testing, and Applications, First Edition. Richard W. Middlestead.

Companion website: www.wiley.com/go/digitalcommunications

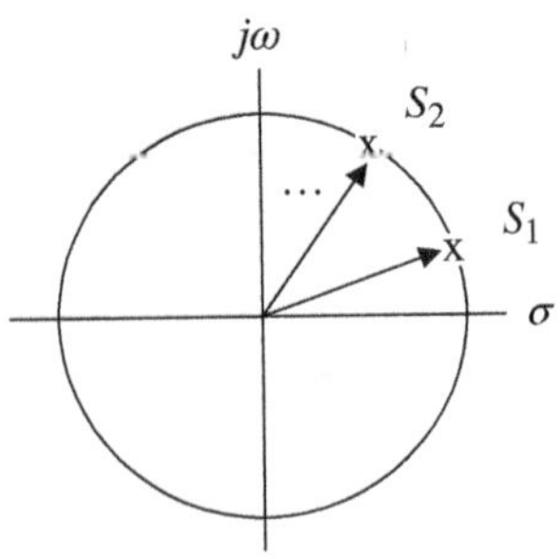

FIGURE A.1 Butterworth s-plane poles (shown for $N=$ even).

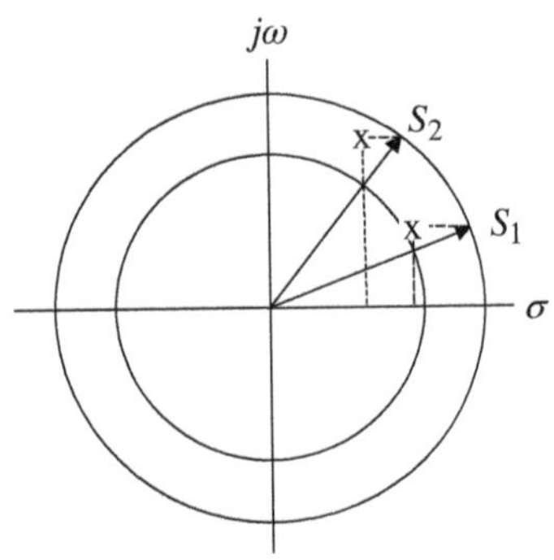

FIGURE A.2 Chebyshev s-plane poles (shown for $N=$ even).

The radius of the unit circle determines the normalized cutoff frequency Ω_c, so a unit radius results in a cutoff frequency of 1 rad/s.

A.2.2 Chebyshev Filter Pole Locations

The Chebyshev filter is characterized as having equal amplitude ripple in the passband with a monotonically increasing attenuation in the stopband. The ripple is characterized by the parameter ε and is related to the ripple specification R_{dB} as $\varepsilon=\sqrt{10^{R_{\text{dB}}/10}-1}$. The Chebyshev poles are related to the Butterworth pole locations as characterized in Figure A.2 and are described in the following text.

The Chebyshev poles are located on an ellipse [9] with semimajor and semiminor axes given by

$$a=\sinh\left(\frac{1}{n}\sinh^{-1}\left(\frac{1}{\varepsilon}\right)\right) \tag{A.2}$$

and

$$b=\cosh\left(\frac{1}{n}\sinh^{-1}\left(\frac{1}{\varepsilon}\right)\right) \tag{A.3}$$

Equations (A.2 and A.3) represent the radius of two concentric circles as shown in Figure A.2. The angles of the radial lines correspond to the angles of the Butterworth pole locations. The intersections of the equally spaced radial lines with the two concentric circles (denoted by the *dotted* lines) are used to determine the Chebyshev pole locations. The real part of a Chebyshev pole is equal to the real part of the radial vector to the inner circle, and the imaginary part of a Chebyshev pole is equal to the imaginary part of the radial vector to the outer circle.

For an N-pole Chebyshev filter, the pole locations are computed for $n=0, 1, 2, \ldots, 2N-1$ as

$$\begin{aligned} s_n &= \sinh\left(\frac{1}{n}\sinh^{-1}\left(\frac{1}{\varepsilon}\right)\right)\text{Re}\{s'_n\} \\ &\quad + j\cosh\left(\frac{1}{n}\cosh^{-1}\left(\frac{1}{\varepsilon}\right)\right)\text{Im}\{s'_n\} \end{aligned} \tag{A.4}$$

where s'_n are the pole locations for the Butterworth filter. For this characterization of the Chebyshev filter the attenuation at $\Omega'_c=1$ is equal to the specified ripple. To adjust the cutoff frequency to correspond to the 3 dB attenuation point, the frequency must be scaled as

$$\Omega_c=\Omega'_c\cosh\left(\frac{1}{n}\sinh^{-1}\left(\frac{1}{\varepsilon}\right)\right) \tag{A.5}$$

with the result that $\Omega_c=1$ corresponds to the 3 dB attenuation point. The normalized bandwidth can be altered by scaling Ω_c as discussed for the Butterworth filter.

For these filters, the left-half-plane (*lhp*) poles give rise to the realizable filter frequency transfer function expressed as

$$F(\omega)=\frac{1}{\prod_{n=1}^{N}(j\omega-s_{n+(N-1)/2})} = \begin{cases} \dfrac{1}{\prod_{n=1}^{N/2}(j\omega-s_{n+(N-1)/2})(j\omega-s^*_{n+(N-1)/2})} & : N=\text{even} \\ \dfrac{1}{(j\omega-s_N)\prod_{n=1}^{N/2}(j\omega-s_{n+(N-1)/2})(j\omega-s^*_{n+(N-1)/2})} & : N=\text{odd} \end{cases} \tag{A.6}$$

When the right-half-plane (*rhp*) poles are multiplied by *minus one*, they are then identical to the desired *lhp* poles, in which case, the frequency transfer function is expressed as

$$F(\omega)=\begin{cases} \dfrac{1}{\prod_{n=1}^{N/2}(j\omega+s_{n-1})(j\omega+s^*_{n-1})} & : N=\text{even} \\ \dfrac{1}{(j\omega+s_o)\prod_{n=1}^{N/2}(j\omega+s_n)(j\omega+s^*_n)} & : N=\text{odd} \end{cases} \tag{A.7}$$

In terms of the real and imaginary components of the complex pole locations $s_n = \sigma_n + j\omega_n$, the frequency transfer function becomes

$$F(\omega) = \begin{cases} \dfrac{1}{\prod_{n=1}^{N/2}[(\sigma_n^2 + \omega_n^2 - \omega^2) + j2\sigma_n\omega]} & : N = \text{even} \\ \dfrac{1}{(\sigma_o + j\omega)\prod_{n=1}^{N/2}[(\sigma_n^2 + \omega_n^2 - \omega^2) + j2\sigma_n\omega]} & : N = \text{odd} \end{cases} \tag{A.8}$$

Using (A.8), the expressions for the magnitude, phase, and delay functions are evaluated as

$$|F(\omega)| = \begin{cases} \dfrac{1}{\prod_{n=1}^{N/2}\sqrt{(\sigma_n^2 + \omega_n^2 - \omega^2)^2 + (2\sigma_n\omega)^2}} & : N = \text{even} \\ \dfrac{1}{\sqrt{\sigma_o^2 + \omega^2}\prod_{n=1}^{N/2}\sqrt{(\sigma_n^2 + \omega_n^2 - \omega^2)^2 + (2\sigma_n\omega)^2}} & : N = \text{odd} \end{cases} \tag{A.9}$$

$$\Phi(\omega) = \begin{cases} -\sum_{n=1}^{N/2}\tan^{-1}\left(\dfrac{2\sigma_n\omega}{\sigma_n^2 + \omega_n^2 - \omega^2}\right) & : N = \text{even} \\ -\tan^{-1}\left(\dfrac{\omega}{\sigma_o}\right) - \sum_{n=1}^{N/2}\tan^{-1}\left(\dfrac{2\sigma_n\omega}{\sigma_n^2 + \omega_n^2 - \omega^2}\right) & : N = \text{odd} \end{cases} \tag{A.10}$$

and

$$T(\omega) = -\frac{d\Phi(\omega)}{d\omega} = \begin{cases} \sum_{n=1}^{N/2}\dfrac{2\sigma_n(\sigma_n^2 + \omega_n^2 + \omega^2)}{(\sigma_n^2 + \omega_n^2 - \omega^2)^2 + (2\sigma_n\omega)^2} & : N = \text{even} \\ \dfrac{\sigma_o}{\sigma_o^2 + \omega^2} + \sum_{n=1}^{N/2}\dfrac{2\sigma_n(\sigma_n^2 + \omega_n^2 + \omega^2)}{(\sigma_n^2 + \omega_n^2 - \omega^2)^2 + (2\sigma_n\omega)^2} & : N = \text{odd} \end{cases} \tag{A.11}$$

An important consideration in a communication simulation program is the requirement to adjust the receiver timing to match the received bit or symbol timing, that is, the receiver timing must be synchronized to the received waveform to properly detect the data. In this regard, the filter delay, resulting from a cascade of transmitter and receiver filters and the communication channel, will result in unnecessary simulation processing to search for the optimum matched filter sampling and subsequent bit-error performance evaluation. For example, if the filters and channel result in a delay, relative to the demodulator symbol timing of $(N + \delta)T$, where $|\delta| < 1$ and T is the symbol duration, the demodulator simulation must resolve the timing error prior to the performance evaluation. This may involve using rate conversion to resolve the timing error if δT is not integrally related to the sampling interval T_s. This additional simulation complexity is not necessary if the filter phase delay* is removed or compensated in the filter design. A major advantage in providing phase delay compensation is in implementing a delay to capture the lead-in transit response resulting in a causal response; if the delay compensation is chosen as an integer number of symbols, then the simulated demodulator optimum matched filter timing is easy and accurately determined.

Using (A.11), the zero-frequency delay is readily determined as

$$T(0) = \begin{cases} \sum_{n=1}^{N/2}\dfrac{2\sigma_n}{(\sigma_n^2 + \omega_n^2)} & : N = \text{even} \\ \dfrac{1}{\sigma_o} + \sum_{n=1}^{N/2}\dfrac{2\sigma_n}{(\sigma_n^2 + \omega_n^2)} & : N = \text{odd} \end{cases} \tag{A.12}$$

The constant delay term, expressed by (A.12), is characterized by a linear-phase function with frequency, so that the delay corrected phase response of the filter is

$$\Phi'(\omega) = \Phi(\omega) + T(0)\omega \tag{A.13}$$

and the delay compensated filter delay response becomes

$$T'(\omega) = T(\omega) - T(0) \tag{A.14}$$

These filter characteristics are examined in Sections A.3 and A.4 with an emphasis on the application of the filtering functions in communication systems performance simulation programs.

A.2.3 Bessel Filter Pole Locations

The Bessel filter is a variation of the Butterworth filter, in that, it provides maximally flat delay response. For example, the delay response of the Bessel filter is essentially constant over the entire 3 dB bandwidth and increases beyond the 3 dB bandwidth in proportion to the number of poles. A major contributor to signal distortion is the nonlinear phase response, which is directly related to the deviation from a constant

*The filter phase delay is expressed in the frequency domain as a linear phase function with frequency. The linear phase function results in a delayed and undistorted replica of the filter input.

TABLE A.1 Left-Half-Plane Bessel Pole Locations

Poles (*N*)	Pole Locations		Normalizing Factor
1	−1.0 + j0.0		1.0
2	−1.101598 ± j0.636008		1.361658
3	−1.322656 + j0.0	−1.047393 ± j0.999249	1.755699
4	−1.370078 ± j0.410253	−0.995216 ± j1.257115	2.113902
5	−1.471989 + j0.0	−1.353001 ± j0.703417	2.477423
	−0.938344 ± j1.441427		
6	−1.571492 ± j0.320897	−1.381860 ± j0.971473	2.703392
	−0.930658 ± j1.661865		
7	−1.684893 + j0.0	−1.612541 ± j0.589428	2.950802
	−1.379333 ± j1.191938	−0.910151 ± j1.837024	
8	−1.757420 ± j0.272869	−1.636950 ± j0.822801	3.179596
	−1.373850 ± j1.388366	−0.892876 ± j1.998339	
9	−1.856491 + j0.0	−1.807064 ± j0.512354	3.391893
	−1.652299 ± j1.031329	−1.367508 ± j1.567641	
	−0.878348 ± j2.149674		
10	−1.927545 ± j0.241614	−1.842125 ± j0.727230	3.59119
	−1.661746 ± j1.221053	−1.360640 ± j1.733439	
	−0.865724 ± j2.292516		

delay. For the Butterworth and Chebyshev filters, the delay response fluctuates substantially within the 3 dB bandwidth resulting in signal distortion even for relatively narrowband signals. It is not uncommon to provide additional all-pass filters to compensate or equalize the delay responses of these filters. The advantage of providing a filter with an inherent maximally flat delay response is of interest in many practical applications; however, the Bessel filters are not without their drawbacks, in that, the attenuation beyond the 3 dB bandwidth is not very sharp which results in wide transition bands compared to other filter types. Also, the passband amplitude roll-off is more severe. For example, an amplitude loss of 0.5 dB occurs at about 40% of the 3 dB bandwidth. Because of these trade-offs, the signal distortion must be examined and compared with other filter types for a particular application.

The ideal delay transfer function for the Bessel filter is given by [10]

$$T(s) = e^{-s} = \frac{1}{\cosh(s) + \sinh(s)} \tag{A.15}$$

It is desirable to approximate (A.15) by a Hurwitz polynomial* with a specified number of poles. Simply expanding e^{-s} in a Maclaurin series does not result in a Hurwitz polynomial for $N > 4$, so another approach must be used. The method outline by Balabanian [11] recognizes that the denominator in the second expression for $T(s)$ involves even and odd functions in s, so the ratio cosh(s)/sinh(s) is expanded about the origin using a Taylor series. The result is then expanded in an infinite continued fraction that is then truncated to N terms resulting in the approximation

$$\frac{\cosh(s)}{\sinh(s)} \cong \frac{m(s)}{n(s)} = \frac{1}{3} + \cfrac{1}{\cfrac{3}{s} + \cfrac{1}{\cfrac{5}{s} + \ddots + \cfrac{1}{(2N-1)/s}}} \tag{A.16}$$

With this expansion, the sum $m(s) + n(s)$ is a Hurwitz polynomial, so the ideal Bessel transfer function is approximated by the realizable function

$$T(s) \cong \frac{1}{m(s) + n(s)} \tag{A.17}$$

Balabanian recombines the expansion in (A.16) and shows that†

$$T(s) \cong \frac{b_o}{b_o + b_1 s + \cdots + b_N s^N} \tag{A.18}$$

where

$$b_n = \frac{(2N-n)!}{2^{N-n}(N-n)!n!} \tag{A.19}$$

Equations (A.18 and A.19) are used to determine the Bessel poles in the *lhp* as shown in Table A.1 for $N \le 10$. Normalizing factors are used to scale the poles so that the resulting 3 dB bandwidth of the filter occurs at unit

*A Hurwitz polynomial has all of its zeros in the *lhp* including the $j\omega$ axis.

†The resulting denominator polynomial $P_D(s)$ is related to the Bessel polynomial $P_B(s)$ as $P_D(s) = s^N P_B(1/s)$, hence the filter name.

frequency, that is, at $\Omega_c = 1$. The pole locations listed in Table A.1 are included in a software program for evaluation of the amplitude, phase, and delay responses of the Bessel filter. For more than 10 poles, the software program uses the approximation of Williams [12] that involves equally spacing the imaginary parts of the poles such that

$$\mathrm{Im}\{s_n\} = \begin{cases} \dfrac{2n}{N} & : N \in \text{even} \\ \dfrac{2(n-1)}{N} & : N \in \text{odd} \end{cases} \tag{A.20}$$

and, upon scaling to provide a unit magnitude, the real part is evaluated as,

$$\mathrm{Re}\{s_n\} = \sqrt{1 - \mathrm{Im}\{s_n\}^2} \tag{A.21}$$

These relationships are also included in a filter simulation program that applies additional scaling to result in the 3 dB amplitude response at $\Omega_c = 1$.

A.2.4 Elliptic Filter Pole Locations

The bandpass response of an elliptic filter [13, 14] is similar to that of the Chebyshev with the amplitude ripple parameter (ε) defined as

$$\varepsilon = \sqrt{10^{R_{dB}/10} - 1} \tag{A.22}$$

where R_{dB} is the bandpass ripple specification in decibels. However, the elliptic filter, unlike the Chebyshev filter, places zero in the stopband that results in a minimum stopband attenuation of A_{dB} and rapid attenuations above the normalized cutoff frequency $\Omega_c = 1$; this results in very low transition bands.* The cutoff frequency corresponds to the R_{dB} attenuation filter bandwidth. The amplitude response, in decibels, is expressed in terms of ε and the nth order elliptic function $Z_n(\Omega)$ as

$$|F(\Omega)|_{dB} = 10\log_{10}\left(1 + \varepsilon^2 Z_n^2(\Omega)\right) \tag{A.23}$$

The elliptic function characterizes the location of the n poles and zeros as

$$Z_n(\Omega) = \frac{\Omega\left(a_2^2 - \Omega^2\right)\left(a_4^2 - \Omega^2\right)\cdots\left(a_m^2 - \Omega^2\right)}{\left(1 - a_2^2\Omega^2\right)\left(1 - a_4^2\Omega^2\right)\cdots\left(1 - a_m^2\Omega^2\right)} \quad : n \in \text{odd},\ m = \frac{(n-1)}{2} \tag{A.24}$$

*No other filter of the same order can have a narrower transition bandwidth than the elliptical filter.

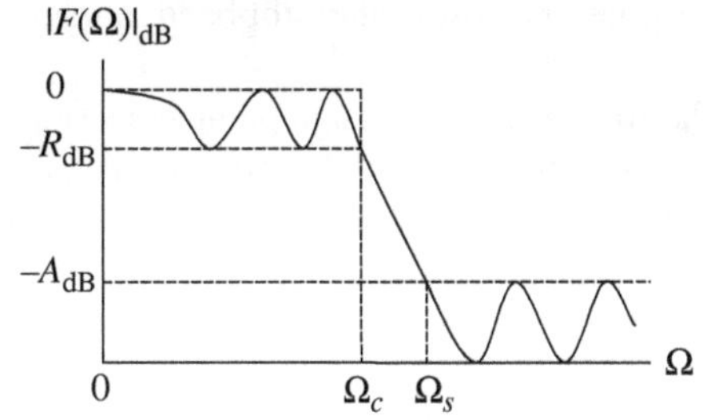

FIGURE A.3 Magnitude of elliptic filter frequency response.

and

$$Z_n(\Omega) = \frac{\left(a_2^2 - \Omega\right)\left(a_4^2 - \Omega^2\right)\cdots\left(a_m^2 - \Omega^2\right)}{\left(1 - a_2^2\Omega^2\right)\left(1 - a_4^2\Omega^2\right)\cdots\left(1 - a_m^2\Omega^2\right)} \quad : n \in \text{even},\ m = \frac{n}{2} \tag{A.25}$$

The poles and zeros corresponding to A.24 and A.25 are difficult to determine; however, Zverev [15] provides extensive tables of the pole and zero locations for filters of complexity (or order) n ranging from three to seven. Given the specified filter parameters R_{dB}, A_{dB}, and Ω_s, the elliptical filter is completely characterized for the normalized cutoff frequency $\Omega_c = 1$. Based on these parameters, the elliptical filter frequency response is shown in Figure A.3.

By convention, the elliptical filter is specified by the parameter set $C(n, \rho, \theta)$ where C denotes the Cauer filter, after Wilhelm Cauer,† and ρ is the reflection coefficient given by

$$\rho = \sqrt{\frac{\varepsilon^2}{1 + \varepsilon^2}} \tag{A.26}$$

and

$$R_{dB} = -10\log_{10}\left(1 - \rho^2\right) \tag{A.27}$$

The angle θ determines the location of the stopband frequency given by

$$\Omega_s = \frac{1}{\sin(\theta)} \quad : \theta < 90^\circ\,(\text{odd order}) \tag{A.28}$$

The relationship between Ω_s and θ for even-order filters is tabulated in Zverev's tables. The passband ripple and stopband attenuation can be independently specified, and as θ approaches 90° the stopband frequency approaches the cutoff frequency resulting in a narrower transition bandwidth; however, for a given n, the stopband attenuation $|A_{dB}|$ decreases as $\Omega_s \rightarrow \Omega_c$. On the other hand, for a specified θ and n, as the

†The elliptical filter is also referred to as the Cauer or Cauer–Chebyshev filter.

passband ripple increases, the stopband attenuation $|A_{dB}|$ increases.

From the specified set of design parameters (R_{dB}, A_{dB}, Ω_s), the order of the filter can be approximated by examining Zverev's tabulated results. The plots in Figures A.4 and A.5 are based on Zverev's data and the approximate filter order is chosen based on the best match to the three design parameters. For example, usually the inband ripple and stopband attenuation are specified, and the filter order is selected based on the maximum allowable transition band. Therefore,

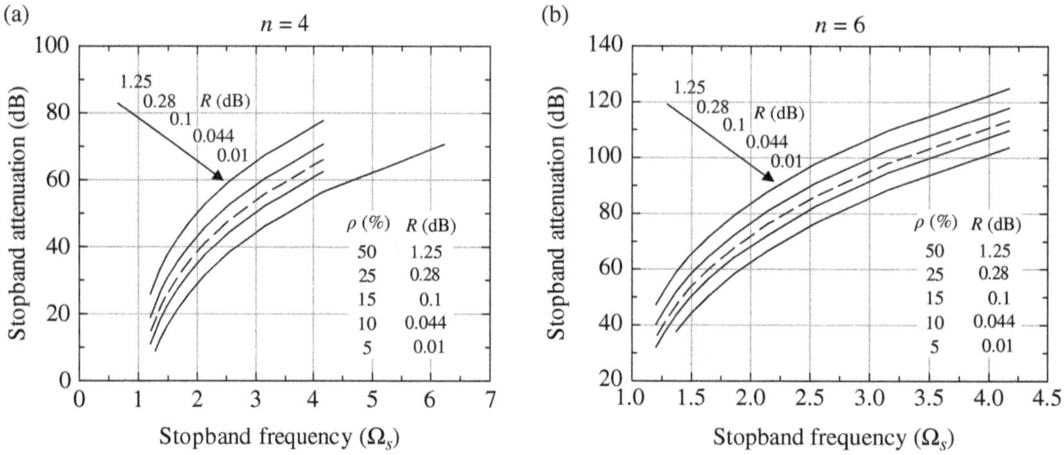

FIGURE A.4 Parameters for even-order elliptic filters.

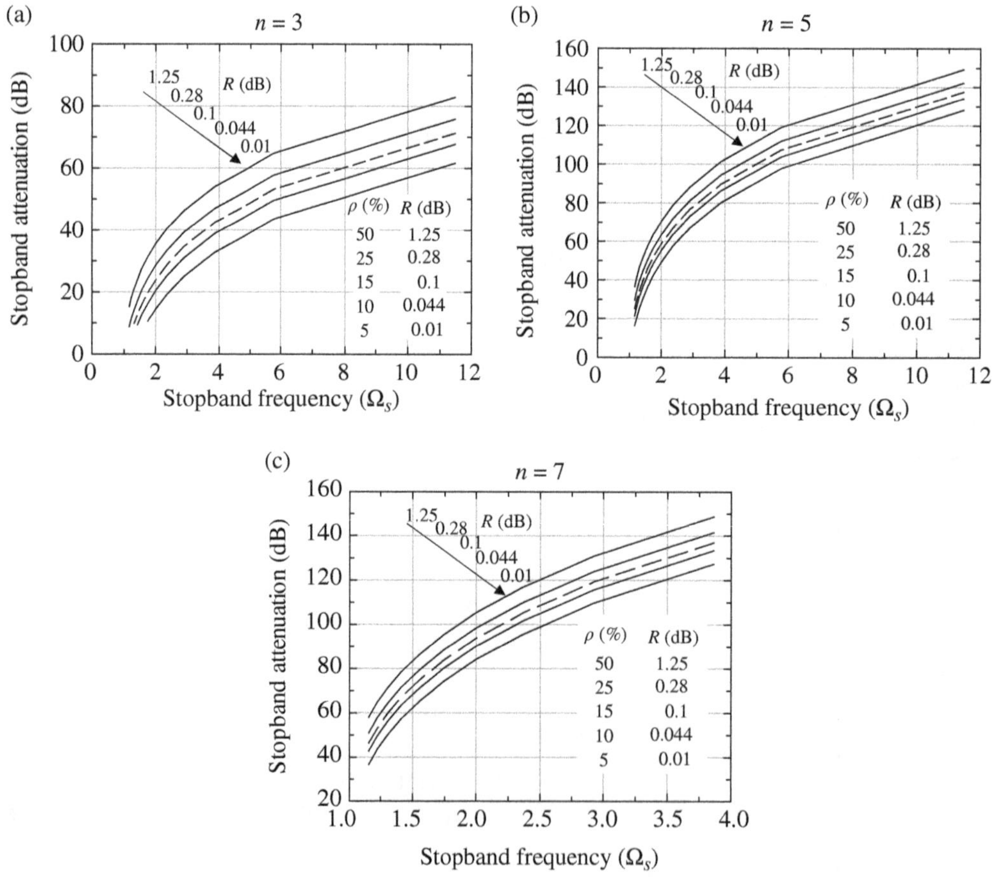

FIGURE A.5 Parameters for odd-order elliptic filters.

TABLE A.2 Selected Third-Order Elliptical Filter Poles and Zeros[a]

θ (degrees)	Ω_s	A_{min}	Poles		Zeros
$\rho = 5\%$					
2	28.6537	85.50	−1.56503	−0.78023 ± j1.60733	±j33.0839
5	11.4737	61.61	−1.57370	−0.77256 ± j1.60793	±j13.2424
8	7.1853	49.33	−1.59006	−0.75840 ± j1.60887	±j8.2868
11	5.2408	41.00	−1.61458	−0.73784 ± j1.60986	±j6.0377
16	3.6280	31.14	−1.67561	−0.68984 ± j1.61037	±j4.1688
23	2.5593	21.53	−1.81308	−0.59641 ± j1.60383	±j2.9256
$\rho = 10\%$					
3	19.1073	80.98	−1.17362	−0.58425 ± j1.33427	±j22.0595
7	8.2055	58.88	−1.18218	−0.57713 ± j1.33524	±j9.4661
10	5.7588	49.55	−1.19326	−0.56807 ± j1.33636	±j6.6370
14	4.1336	40.72	−1.21468	−0.55108 ± j1.33817	±j4.7552
20	2.9238	31.29	−1.26269	−0.51532 ± j1.34065	±j3.3505
30	2.0000	20.40	−1.39666	−0.43027 ± j1.33901	±j2.2701
$\rho = 15\%$					
3	19.1073	84.56	−0.97343	−0.48502 ± j1.20800	±j22.0595
8	7.1853	58.96	−0.98143	−0.47863 ± j1.20922	±j8.2868
11	5.2408	50.63	−0.98986	−0.47201 ± j1.21042	±j6.0377
16	3.6280	40.77	−1.01038	−0.45636 ± j1.21296	±j4.1688
23	2.5593	31.13	−1.05443	−0.42488 ± j1.21687	±j2.9256
34	1.7883	20.53	−1.17044	−0.35393 ± j1.21947	±j2.0199
$\rho = 25\%$					
4	14.3356	81.67	−0.74475	−0.37060 ± j1.07941	±j16.5483
9	6.3925	60.50	−0.75052	−0.36626 ± j1.08071	±j7.3700
13	4.4454	50.86	−0.75848	−0.36038 ± j1.08243	±j5.1166
19	3.0716	40.84	−0.77640	−0.34752 ± j1.08593	±j3.5224
28	2.1301	30.44	−0.81884	−0.31909 ± j1.09248	±j2.4231
41	1.5243	19.88	−0.92480	−0.25887 ± j1.10072	±j1.7044
$\rho = 50\%$					
5	11.4737	82.84	−0.45426	−0.22589 ± j0.95127	±j13.2424
12	4.8097	59.95	−0.45928	−0.22246 ± j0.95340	±j5.5386
17	3.4203	50.78	−0.46555	−0.21826 ± j0.95597	±j3.9277
25	2.3662	40.50	−0.48069	−0.20841 ± j0.96178	±j2.6999
36	1.7013	30.51	−0.51387	−0.18827 ± j0.97268	±j1.9165
51	1.2868	20.41	−0.59255	−0.14753 ± j0.99074	±j1.4137

[a]Zverev [15]. Reproduced by permission of John Wiley & Sons, Inc.

considering the design parameters (0.1, 60.0, <1.7), the filter order that satisfies these conditions with minimum complexity is $n = 6$.

The elliptic filter pole and zero locations for third- and sixth-order elliptic filter designs, based on the work of Zverev, are given in Tables A.2 and A.3. The filters are categorized in terms of the order (n) of the polynomial, which also corresponds to the number of poles in the transfer function, and the reflection coefficient (ρ) in percent. For each of these two filter categories, the modular angle (θ) in degrees, the normalized stopband frequency (Ω_s), the minimum stopband attenuation ($A_{\min}$) in dB, and the corresponding pole and zero locations are specified. In all of the designs, the normalized cutoff frequency corresponds to $\Omega_c = 1$. The dependence on the reflection coefficient of the inband ripple (ε) and ripple attenuation (R) in decibels are determined from (A.26) and (A.27) and are shown in Figure A.6.

For each filter order, Zverev provides the data for reflection coefficients $\rho = 1$ through 5, and 8, 10, 15, 20, 25, and 50%; however, Tables A.2 and A.3 contain the data corresponding to the conditions $\rho = 5$, 10, 15, 25, and 50%. Although Zverev limits the filter order to $n \leq 7$, filters for orders 6, 7, 8, 9, 10, and 11 for $\rho = 20\%$ are characterized by Saal and Ulbrich [16]; their results are also provided by Williams and Taylor [17]. Golden and Kaiser [18] also present the pole and zero locations for an 11-order elliptic lowpass filter for $\rho = 33\%$ corresponding to the parameter set (0.5, 76.5, 1.066). The frequency responses of the elliptic filters with order $n = 3$ and 11 are plotted and compared in Section A.3.

TABLE A.3 Selected Sixth-Order Elliptical Filter Poles and Zeros[a]

θ (degrees)	Ω_s	A_{min}	Poles	Zeros
$\rho = 5\%$				
16	3.751039	100.33	−0.1608733 ± j1.1494074	±j5.452491
			−0.4551016 ± j0.8582356	±j3.888329
			−0.6439170 ± j0.3205477	
23	2.642462	81.05	−0.1519894 ± j1.1441778	±j3.800865
			−0.4466624 ± j0.8724571	±j2.735370
			−0.6571745 ± j0.3332466	
28	2.196422	70.43	−0.1438020 ± j1.1392246	±j3.128134
			−0.4383048 ± j0.8857904	±j2.270699
			−0.6699592 ± j0.3457954	
33	1.890370	61.43	−0.1341590 ± j1.1332159	±j2.660243
			−0.4276726 ± j0.9017789	±j1.951268
			−0.6857864 ± j0.3617553	
40	1.597615	50.63	−0.1183982 ± j1.1229522	±j2.203049
			−0.4081924 ± j0.9285830	±j1.644814
			−0.7137060 ± j0.3911003	
47	1.399891	41.24	−0.1003316 ± j1.1104248	±j1.883335
			−0.3819741 ± j0.9603185	±j1.436822
			−0.7495140 ± j0.4310823	
10%				
16	3.751039	106.39	−0.1287903 ± j1.0862083	±j5.452491
			−0.3626602 ± j0.8073507	±j3.888329
			−0.5106815 ± j0.3001183	
26	2.353536	80.50	−0.1194234 ± j1.0821294	±j3.366027
			−0.3540182 ± j0.8246522	±j2.434463
			−0.5252927 ± j0.3148009	
31	2.000308	70.93	−0.1131204 ± j1.0792925	±j2.829163
			−0.3477100 ± j0.8365374	±j2.066092
			−0.5356186 ± j0.3254371	
37	1.708362	61.10	−0.1041827 ± j1.0751337	±j2.377602
			−0.3380039 ± j0.8537396	±j1.760893
			−0.5510221 ± j0.3417186	
45	1.449216	49.87	−0.0900608 ± j1.0682079	±j1.964398
			−0.3205571 ± j0.8817939	±j1.488829
			−0.5774697 ± j0.3708879	
52	1.296076	41.17	−0.0758386 ± j1.0607352	±j1.708356
			−0.2997691 ± j0.9111777	±j1.326965
			−0.6073282 ± j0.4057781	
$\rho = 15\%$				
19	3.174064	100.87	−0.1085781 ± j1.0549572	±j4.595218
			−0.3086973 ± j0.7866868	±j3.288476
			−0.4389051 ± j0.2934291	
28	2.196422	80.07	−0.1011126 ± j1.0525534	±3.128134
			−0.3018417 ± j0.8021407	±j2.270699
			−0.4509322 ± j0.3062188	
33	1.890370	71.06	−0.0957513 ± j1.0507639	±j2.660243
			−0.2965234 ± j0.8134652	±j1.951268
			−0.4599693 ± j0.3160388	
39	1.632449	61.70	−0.0881974 ± j1.0481473	±j2.258218
			−0.2884052 ± j0.8297591	±j1.681350
			−0.4733341 ± j0.3309038	
47	1.399891	50.87	−0.0762933 ± j1.0437773	±j1.883335.
			−0.2738962 ± j0.8562900	±j1.436872
			−0.4961281 ± j0.3572451	
55	1.244853	41.23	−0.0624534 ± j1.0382660	±j1.618888
			−0.2537453 ± j0.8885543	±j1.272479
			−0.5259504 ± j0.3937182	

TABLE A.3 (*continued*)

θ (degrees)	Ω_s	A_{min}	Poles	Zeros
20%				
20	3.020785	100.72	−0.0953830 ± j1.0360245	±j4.366743
			−0.2718256 ± j0.7725704	±j3.129050
			−0.3873832 ± j0.2881628	
29	2.126320	80.73	−0.0888491 ± j1.0345596	±j3.021559
			−0.2658847 ± j0.7874644	±j2.197588
			−0.3980997 ± j0.3002073	
35	1.793769	70.37	−0.0831672 ± j1.0332226	±j2.510774
			−0.2602723 ± j0.8007063	±j1.850277
			−0.4078732 ± j0.3114315	
41	1.564662	61.42	−0.0764372 ± j1.0315581	±j2.150602
			−0.2530228 ± j0.8167609	±j1.610227
			−0.4200613 ± j0.3257616	
50	1.334577	49.73	−0.0644331 ± j1.0283499	±j1.774078
			−0.2381839 ± j0.8464780	±j1.367782
			−0.4437381 ± j0.3547149	
57	1.214564	41.52	−0.0536101 ± j1.0251582	±j1.564571
			−0.2221885 ± j0.8745806	±j1.240135
			−0.4677485 ± j0.3856848	
$\rho = 25\%$				
21	2.882384	100.16	−0.0851298 ± j1.0228980	±j4.160091
			−0.2433104 ± j0.7630799	±j2.985065
			−0.3477418 ± j0.2847363	
30	2.061105	80.92	−0.0792423 ± j1.0220953	±j2.922133
			−0.2379847 ± j0.7777435	±j2.129549
			−0.3575389 ± j0.2963942	
36	1.749781	70.84	−0.0741577 ± j1.0213515	±j2.442321
			−0.2329898 ± j0.7906943	±j1.804254
			−0.3664041 ± j0.3071573	
42	1.533460	62.08	−0.0681492 ± j1.0204083	±j2.100682
			−0.2265582 ± j0.8063633	±j1.577454
			−0.3774193 ± j0.3208259	
51	1.314859	50.56	−0.0574350 ± j1.0185369	±j1.740561
			−0.2134172 ± j0.8353762	±j1.346891
			−0.3987646 ± j0.3482958	
59	1.187032	41.31	−0.0462836 ± j1.0162932	±j1.514011
			−0.1968618 ± j0.8672164	±j1.210630
			−0.4239216 ± j0.3824605	
50%				
16	3.751039	121.57	−0.0553928 ± j0.9898055	±j5.452491
			−0.1549694 ± j0.7309754	±j3.888329
			−0.2167725 ± j0.2699331	
24	2.537873	100.00	−0.0530381 ± j0.9904476	±j3.643786
			−0.1530115 ± j0.7395913	±j2.626475
			−0.2207029 ± j0.2762365	
34	1.840548	81.01	−0.0487356 ± j0.9915926	±j2.583292
			−0.1491489 ± j0.7556703	±j1.899195
			−0.2281772 ± j0.2883904	
41	1.564662	70.45	−0.0448057 ± j0.9926048	±j2.150602
			−0.1452634 ± j0.7707651	±j1.610227
			−0.2353704 ± j0.3003021	
47	1.399891	62.48	−0.0408237 ± j0.9935952	±j1.883335
			−0.1409304 ± j0.7864940	±j1.436822
			−0.2430637 ± j0.3132867	
57	1.214564	50.55	−0.0329166 ± j0.9954440	±j1.564571
			−0.1308980 ± j0.8191971	±j1.240135
			−0.2598029 ± j0.3424698	

[a]Zverev [15]. Reproduced by permission of John Wiley & Sons, Inc.

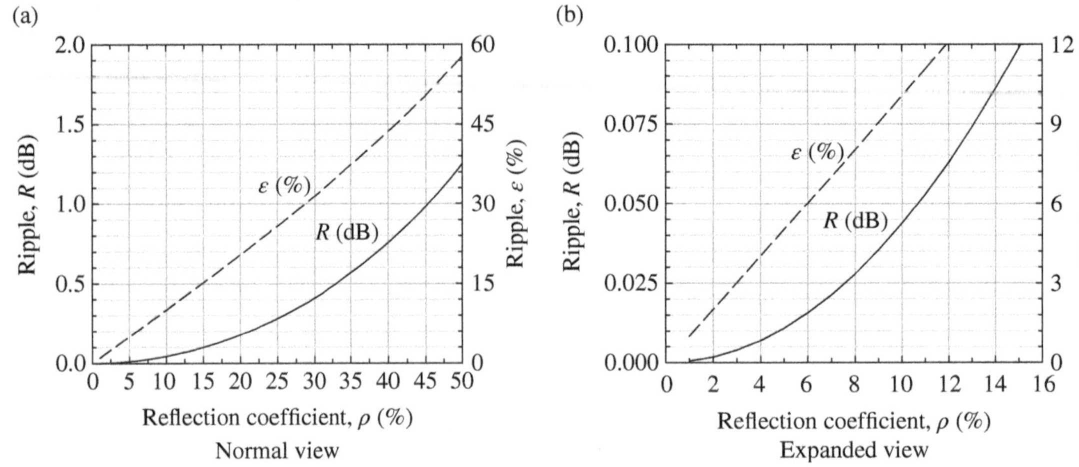

FIGURE A.6 Dependence of ε and R on reflection coefficient.

A.3 COMPUTER SIMULATION RESULTS AND APPLICATIONS

The filter pole-zero locations discussed in Section A.2 are programmed in a filter simulation, and evaluation program and several examples of the filter characteristics are discussed in this section. The filters are represented as discrete-time and discrete-frequency sampled digital filters. Consider the sampled frequency response $F(i\Delta\omega)$, $i = 1, 2, \ldots, M$ where the simulation sampling frequency is $f_s = M\Delta\omega/2\pi$. Furthermore, consider the normalized frequency function $F(i\Delta\omega/\omega_c) = F(i\Delta\Omega)$ so that $i\Delta\Omega = \Omega_c = 1$ corresponds to the 3 dB attenuation point of the lowpass filter. Under this condition, there are $M_c = \Omega_c/\Delta\Omega$ samples over the filter bandwidth. Suppose now that the normalized sampling frequency is* $\Omega_s' = 4\Omega_c$, that is, the simulation sampling frequency is four times the filter cutoff frequency or twice the Nyquist frequency yielding a total of $M = 4M_c$ frequency domain samples. In the computer simulations, the expressions for the amplitude and delay functions are normalized by substituting the normalized angular frequency Ω for ω and using $M_c = 100$ corresponding to $\Delta\Omega = 0.01$. The resulting responses are shown in Figures A.7 and A.8 for the Butterworth and Chebyshev filters, respectively. In these figures, $\Omega_c = 1$ and the normalized sampling frequency corresponds to $\Omega_s' = 4$.

Considering the four-pole filters in these examples, at the folding point, corresponding to the Nyquist normalized band $\Omega_N = 2$, the aliasing is about 25 and 35 dB below the peak value for the Butterworth and Chebyshev filters, respectively. The distortion caused by the aliasing is an important consideration in the selection of the number of filter poles, and the

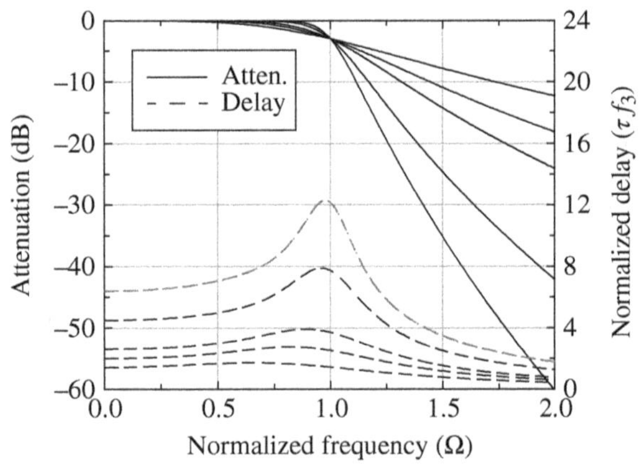

FIGURE A.7 Butterworth lowpass responses.†

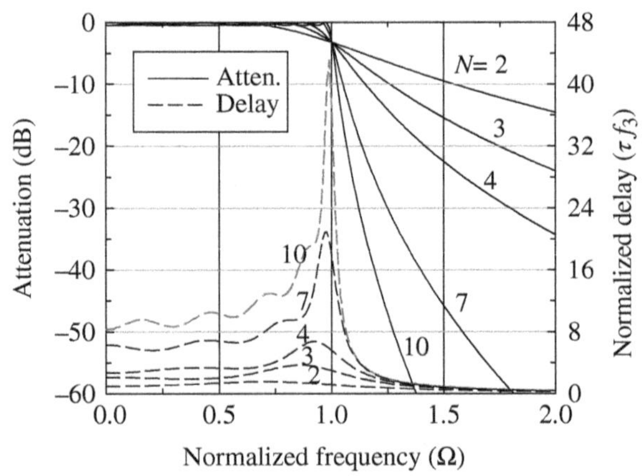

FIGURE A.8 0.5 dB Ripple Chebyshev lowpass responses.

*The sampling frequency is denoted as Ω_s' and is distinguished from the stopband frequency Ω_s.

†The delay response in the simulation program is normalized as $\tau f_3 = \tau w_3/2\pi$, so a factor 2π is removed in Figures A.7 – A.9.

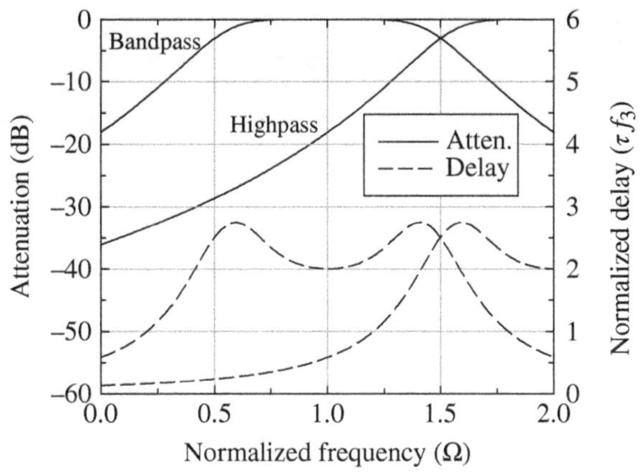

FIGURE A.9 Simulated three-pole Butterworth bandpass and highpass responses.

TABLE A.4 An Eleventh-Order Elliptical Filter Poles and Zeros (Ripple = 0.5 dB, A_{min} = 76.504 dB, and Transition Ratio = 0.937917)[a]

θ (degrees)	Ω_s	A_{min}	Poles	Zeros
$\rho = 33\%$				
69.7	1.066	76.504	−0.0069130 ± j1.0010752	±j1.0695414
			−0.0257616 ± j0.9756431	±j1.1009005
			−0.0615122 ± j0.9063786	±j1.1946271
			−0.1269215 ± j0.7504391	±j1.4652816
			−0.2142976 ± j0.4483675	±j2.5031313
			−0.2611853 + j0.	

[a]Golden and Kaiser [18]. Reproduced by permission of Alcatel-Lucent USA, Inc.

sampling frequency and attenuations greater than or equal to 25 dB at the folding point are generally considered to be acceptable.

The sampled responses must be mirrored about the Nyquist band, so the sampled analog responses must be modified to reflect the required mirroring as follows. For $M \in$ even, the $M/2$ samples obtained directly from the analog expressions are mirrored as

$$F((M/2-i+2)\Delta\Omega) = F(i\Delta\Omega) \quad : i = 2, 3, \ldots, \frac{M}{2} \tag{A.29}$$

and for $M \in$ odd, the index i extends to $i = 2, 3, \ldots, M/2 + 1$. The mirroring obtained in this way is ideal, in that, the filter responses are implicitly predistorted to compensate for the aliasing. If the aliasing is included in the mirroring or folding process, the analog expressions must be evaluated for all M samples and the samples corresponding to $i > M/2$ (or $M/2 + 1$ for $M \in$ odd) must be included in Equation (A.29). In this case, the magnitude of the response at Ω_N will be increased by 3 dB, and the 3 dB cutoff frequency of the filter will be altered slightly depending upon the amount of aliasing occurring at Ω_c.

It is evident from the preceding discussion that the normalized 3 dB bandwidth of the filter can be increased or decreased for a fixed sampling frequency simply by choosing larger or smaller values of M_c. Similarly, a bandpass or highpass filter can be realized by simply shifting the normalized frequency scale. For example, a bandpass filter centered at $\Omega_N/2$ is obtained by letting

$$F'(i\Delta\Omega) = F((i-M_N/2)\Delta\Omega) \tag{A.30}$$

and a highpass filter is obtained as

$$F'(i\Delta\Omega) = F((i-M_N)\Delta\Omega) \tag{A.31}$$

Examples of these filters are shown in Figure A.9. Normally, the gain of the bandpass filter will be 3 dB lower than the equivalent sampled lowpass or highpass filters; however, in these results the gains are all normalized to unity.

The following examples of elliptic filter responses correspond to the third-order filter of Table A.2 using the set of design parameters (0.1, 50.63, 5.2408) with $\rho = 15\%$ and the 11-order filter of Table A.4 using the set of design parameters (0.5, 76.504, 1.066) with $\rho = 33\%$ (Figures A.10 and A.11). Upon comparison with the theoretically expected frequency response in Figure A.3, these responses are in remarkable agreement with the specified design parameters. These two examples show the wide range of frequency responses achievable with elliptic filters; of special interest is the very narrow transition bands compared to the other filters.

A.4 CASE STUDY: FILTER APPLICATION

In this case study, the response of a five-pole, 0.05 dB ripple Chebyshev filter* is characterized in terms of the in phase and quadrature responses. The filter response is examined using the unit-impulse and rectangular pulse inputs. The evaluation program provides for zeroing the intrinsic phase response of the filter and inserting an equalized quadratic phase function with a prescribed band-edge phase error. The quadratic phase function is expressed as

$$\phi(\omega) = \phi_s\left(\frac{\omega}{\omega_m}\right)^3 \tag{A.32}$$

*This filter model is used to characterize the amplitude response of a satellite channel based on the average 3-dB bandpass bandwidth of several measured channel filters. The filters are phase equalized with a phase of 6° at a specified frequency deviation from the carrier.

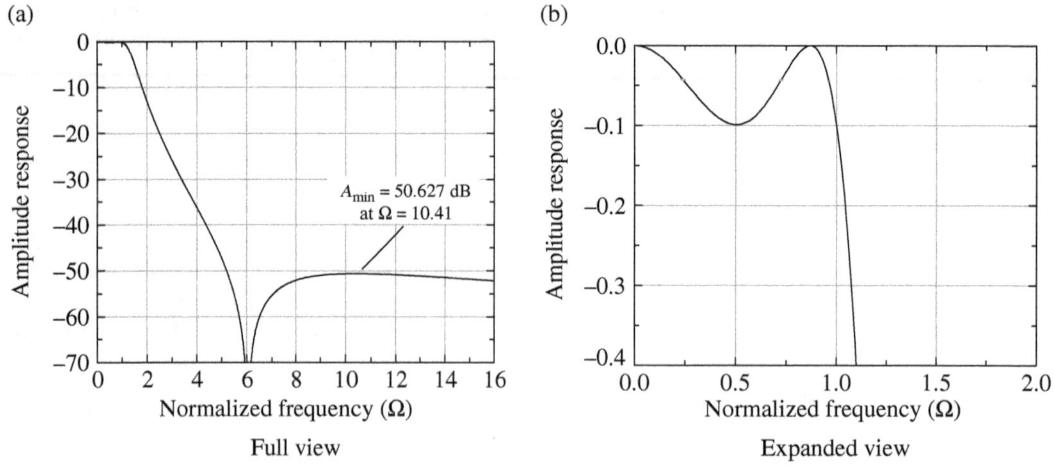

FIGURE A.10 Third-order elliptic filter frequency response.

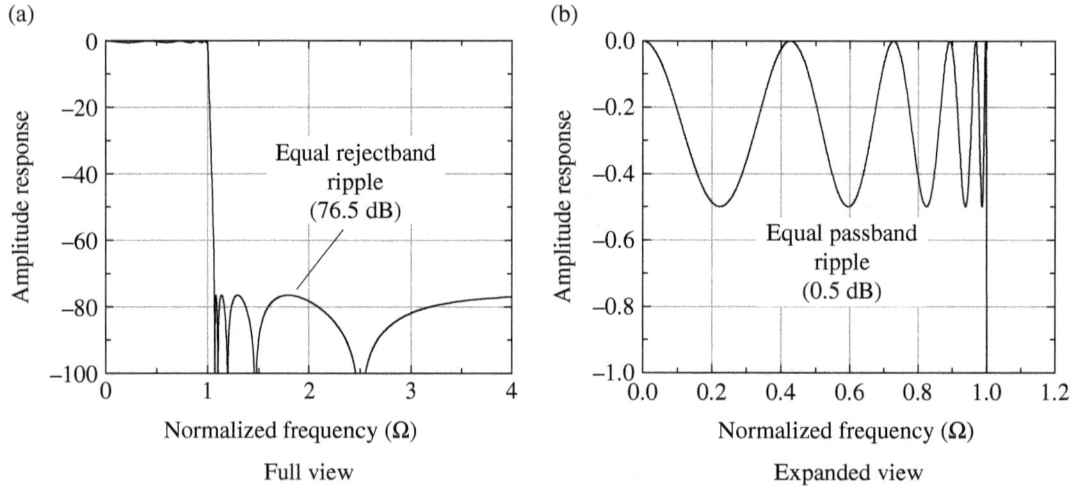

FIGURE A.11 Eleven-pole elliptic filter frequency response.

where ϕ_s is in radians and, in this example, corresponds to a 6° phase error at the frequency $\omega_m = 2\pi f_m$ where $f_m = 10$ kHz is the frequency corresponding to the specified phase error. Using this result, the phase error at the 3 dB band-edge frequency $f_3 = 15$ kHz is $\phi_{be} = 20.25\pi/180$ rad and (A.32) is expressed as

$$\phi(\omega) = \phi_{be}\left(\frac{\omega}{\omega_3}\right)^3 \quad \text{(A.33)}$$

The group delay of the equalized filter is defined as*

$$\tau_g(\omega) = -\frac{d\phi(\omega)}{d\omega} \quad \text{(A.34)}$$

*The quadratic phase error is negative for positive frequencies, so this definition results in a positive delay error.

and the resulting normalized group delay, expressed in Hz, is

$$\tau_g(f)f_3 = \frac{3\phi_{be}}{2\pi}\left(\frac{f}{f_3}\right)^2 \quad \text{(A.35)}$$

The filter impulse responses are shown in Figures A.12, A.13, and A.14 for B/R_s ratios of 2.0, 1.5, and 1.0, respectively. The evaluation of the response uses 16 samples per symbol and, to capture the entire impulse response, that is, for a causal response, the responses are delayed by two symbols or 32 samples except for the $B/R_s = 1$ case that uses a delay of three symbols or 48 samples. The pulse responses represent the filter response to a unit amplitude symbol pulse of duration $T = 1/R_s$ seconds. The filters are symmetrical in amplitude and asymmetrical in phase about zero frequency so the quadrature response is zero. The pulse responses shown in these figures correspond to an isolated $rect((t - T_d)/T)$ input symbol with normalized delays of $T_d/T = 2.5$, 2.5, and 3.5. The filters are considered to be demodulator matched filters.

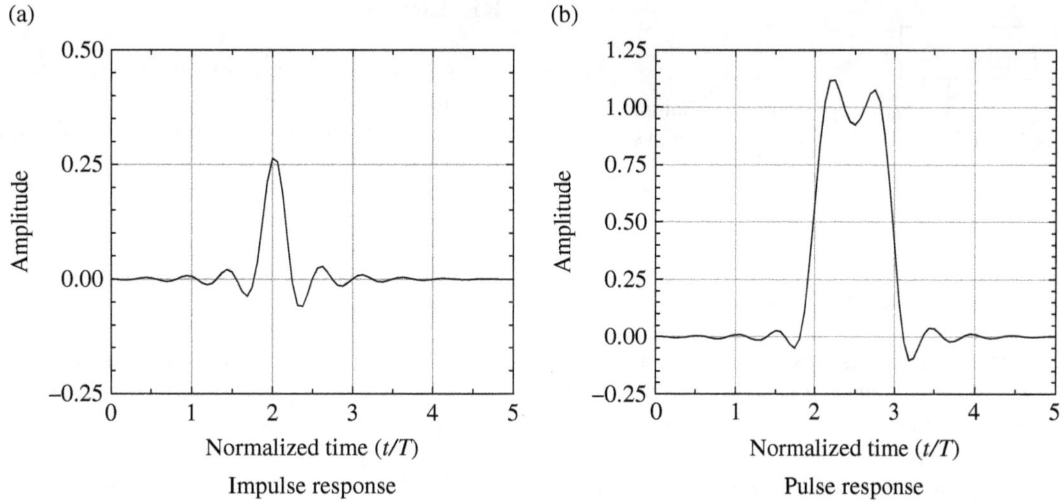

FIGURE A.12 Responses for $B/R_s = 2.0$, filter loss = 0.434 dB.

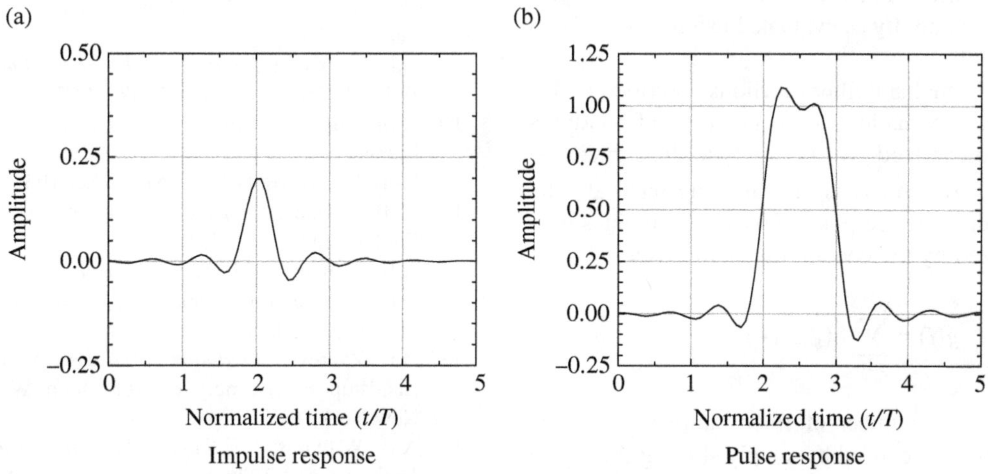

FIGURE A.13 Responses for $B/R_s = 1.5$, filter loss = 0.586 dB.

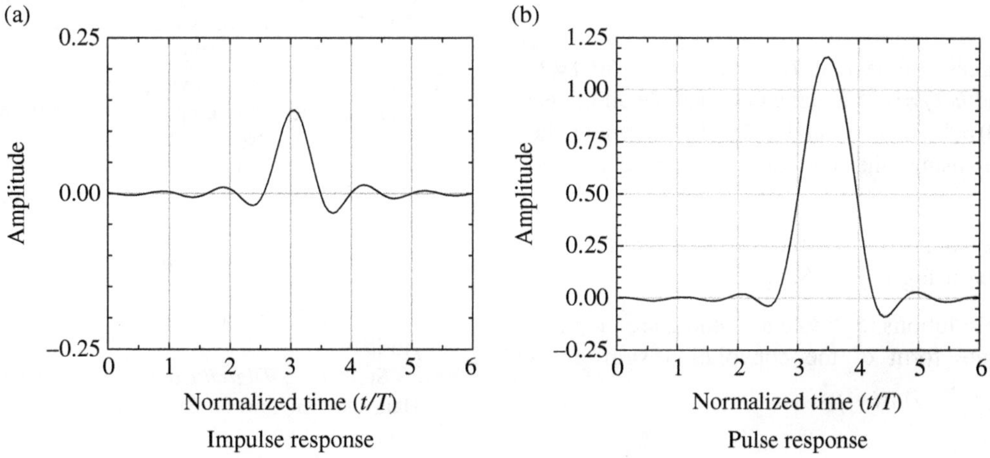

FIGURE A.14 Responses for $B/R_s = 1.0$, filter loss = 0.89 dB.

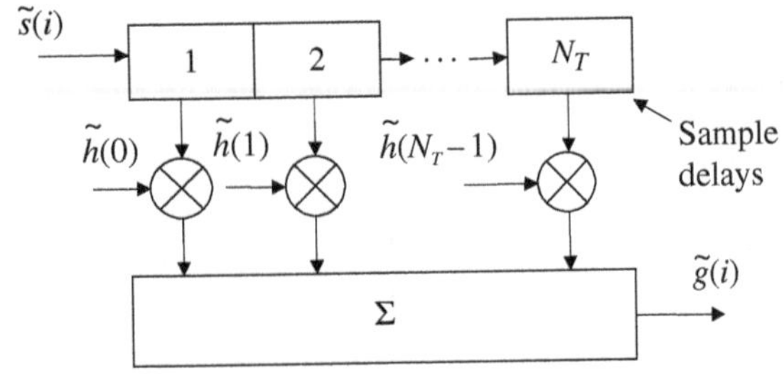

FIGURE A.15 Filter implementation using tapped delay line.

When the matched filter impulse responses are convolved with an isolated symbol of duration T seconds, the losses at the optimum sampling times for the three filters are 0.434, 0.58, and 0.89 dB, respectively. These losses represent the loss in symbol energy over the symbol duration with ideal symbol timing and do not reflect the effects of intersymbol interference (ISI) with a contiguous sequence of randomly modulated symbols. The total loss including the ISI loss is typically characterized in terms of the E_b/N_o loss at a prescribed bit-error probability as evaluated using Monte Carlo simulations [19, 20].

In general, the signal and filter functions described in this chapter are represented as lowpass equivalents of bandpass functions as characterized by the analytic functions $\tilde{s}(i)$ and $\tilde{h}(i)$ where $t = iT_s$ and T_s is the sampling interval. The computation of the filter response $\tilde{g}(i)$ to the input signal $\tilde{s}(i)$ is characterized by the convolution sum given by

$$\tilde{g}(i) = \sum_{j=0}^{N_T-1} \tilde{h}(j)\tilde{s}(i-j) \qquad \text{(A.36)}$$

where N_T is the total number of filter samples. The time reverse of the signal is realized by simply shifting the signal samples as they are generated through a weighted tapped delay line filter as shown in Figure A.15.

PROBLEMS

1. Using the three sample real filter responses $h(0) = 1$, $h(1) = 0.5$, and $h(2) = 0.25$ and the three sample unit pulse responses $s(0) = 1$, $s(1) = 1$, and $s(2) = 1$, determine the response $g(i)$ using the convolution sum expressed by (A.36).

2. Show that the solution to Problem 1 is identical to the implementation using Figure A.15.

3. Show that the solutions to Problems 1 and 2 are identical to the alternate form of the convolution sum given by $\tilde{g}(i) = \sum_{j=0}^{N_T-1} \tilde{s}(j)\tilde{h}(i-j)$.

REFERENCES

1. A.I. Zverev, *Handbook of Filter Synthesis*, John Wiley & Sons, Inc., New York, 1967.
2. N. Balabanian, *Network Synthesis*, Prentice-Hall, Inc., Englewood Cliffs, NJ, 1958.
3. A.B. Williams, F.J. Taylor, *Electronic Filter Design Handbook*, Second Edition, McGraw-Hill Book Company, Inc., New York, 1988.
4. C.S. Lindquist, *Active Network Design*, Steward and Sons, Santa Cruz, CA, 1977.
5. A.V. Oppenheim, R.W. Schafer, *Digital Signal Processing*, Prentice-Hall, Inc., Englewood Cliffs, NJ, 1975.
6. K.W. Henderson, W.H. Kautz, "Transient Responses of Conventional Filters," IRE Transactions on Circuit Theory, Vol. 5, Issue 4, pp. 333–347, December 1958.
7. Y. Peless, T. Murakami, "Analysis and Synthesis of Transitional Butterworth-Thomson Filters and Bandpass Amplifiers," RCA Review, Vol. XVIII, No. 1, pp. 60–94, March 1957.
8. M.J. Di Toro, "Phase and Amplitude Distortion in Linear Networks," Proceedings of the IRE, Vol. 36, No. 1, pp. 24–36, January 1948.
9. A.I. Zverev, *Handbook of Filter Synthesis*, p. 408, John Wiley & Sons, Inc., New York, 1967.
10. L. Storch, "Synthesis of Constant Time Delay Ladder Networks Using Bessel Polynomials," Proceeding of the IRE, Vol. 42, Issue 11, pp. 1666–1676, November 1954.
11. N. Balabanian, *Network Synthesis*, p. 393, Prentice-Hall, Inc., Englewood Cliffs, NJ, 1958.
12. A.B. Williams, F.J. Taylor, *Electronic Filter Design Handbook*, Second Edition, pp. 2–54, McGraw-Hill Book Company, Inc., New York, 1988.
13. A.I. Zverev, *Handbook of Filter Synthesis*, Design of Bandstop Filters, pp. 163–168, John Wiley & Sons, Inc., New York, 1967.
14. A.B. Williams, F.J. Taylor, *Electronic Filter Design Handbook*, Second Edition, pp. 2-70–2-77, McGraw-Hill Book Company, Inc., New York, 1988.
15. A.I. Zverev, *Handbook of Filter Synthesis*, pp. 169–289, John Wiley & Sons, Inc., New York, 1967.
16. R. Saal, E. Ulbrich, "On the Design of Filters by Synthesis," IRE Transactions on Circuit Theory, Vol. 5, Issue 4, pp. 284–327, December 1958.
17. A.B. Williams, F.J. Taylor, *Electronic Filter Design Handbook*, Second Edition, pp. 11-84–11-110, McGraw-Hill Book Company, Inc., New York, 1988.
18. R.M. Golden, J.F. Kaiser, "Design of Wideband Sampled-Data Filters," Bell System Technical Journal, Vol. 43, Issue 4, pp. 1533–1546, July 1964.
19. J.J. Jones, "Filter Distortion and Intersymbol Interference Effects on PSK Signals," IEEE Transactions on Communications and Technology, Vol. 19, Issue 2, pp. 120–132, April 1971.
20. J.J Spilker, Jr., *Digital Communications by Satellite*, Prentice-Hall, Inc., Englewood Cliffs, NJ, 1977.

APPENDIX B
DIGITAL FILTER DESIGN AND APPLICATIONS

B.1 INTRODUCTION

In this appendix, several methods for the design of digital filters are reviewed. Typical applications involve rate conversion filtering, transmitter waveform filtering for spectral control, demodulator filtering for coherent and noncoherent data detection, image rejection filtering following receiver heterodyning, and a variety of bandpass and lowpass filter applications involved in waveform detection, acquisition, and verification. The two fundamental types of digital filters considered are characterized as infinite impulse response (IIR) and finite impulse response (FIR) filters. The IIR filters involve feedback in which the output is the summation of the input with weighed samples of previous outputs; whereas, the FIR is strictly a feedforward implementation in which the output is the summation of a finite number of weighed input samples. The FIR implementation is also referred to as a transversal filter (TF), tapped delay line (TDL) filter, and sequential filter (SF) in which the input signal samples are sequentially stored in the memory of a TDL.

These filters exhibit a unit sample response $h(n)$ that is linear and shift-invariant [1] and, for an input $x(n)$, the response $y(n)$ is characterized by the convolution sum

$$y(n) = \sum_{k=-\infty}^{\infty} x(k)h(n-k) \tag{B.1}$$

The condition for a linear shift-invariant filter to be causal requires that $h(n) = 0$ for $n < 0$ and the condition for stability requires that

$$\sum_{n=-\infty}^{\infty} |h(n)|^2 < \infty \quad \text{: condition for filter stability} \tag{B.2}$$

It is often desirable that the filter has unit gain which requires

$$\sum_{n=-\infty}^{\infty} |h(n)|^2 = 1 \quad \text{: condition for unit gain response} \tag{B.3}$$

In these expressions, the filter response may be complex valued. In many applications, the linear shift-invariant property is characterized by the Nth-order linear constant-coefficient difference equation given by [1]

$$\sum_{k=0}^{N} a_k y(n-k) = \sum_{r=0}^{M} b_r x(n-r) \tag{B.4}$$

Equation (B.4) has a unique solution under the condition of causality with $x(n) = y(n) = 0$: $n < n_o$, in which case, the linear constant-coefficient difference equation corresponds to a linear shift-invariant filter and (B.4) provides a unique relationship between $x(n)$ and $y(n)$. Under the causality condition, solving (B.4) for $y(n)$ results in the expression

$$y(n) = -\frac{1}{a_0}\sum_{k=1}^{N} a_k y(n-k) + \frac{1}{a_0}\sum_{r=0}^{M} b_r x(n-r) \quad \text{: IIR filter} \tag{B.5}$$

Digital Communications with Emphasis on Data Modems: Theory, Analysis, Design, Simulation, Testing, and Applications, First Edition. Richard W. Middlestead.

Companion website: www.wiley.com/go/digitalcommunications

The first term of (B.5) describes an IIR filter and, when $N=0$, the response is that of a FIR filter of length M described as

$$y(n) = \frac{1}{a_0}\sum_{r=0}^{M} b_r x(n-r) \quad : \text{FIR filter} \qquad \text{(B.6)}$$

Typically, the filter response samples $h(n)$ are derived from the description of the desired filter frequency response using the inverse discrete Fourier transform. The Fourier transform, applied to discrete-time linear shift-invariant sampled functions, is referred to as the z-transform* with the complex frequency variable $z = re^{j\omega T_s}$ with $|z| = r \le 1$. The sampling duration T_s is often normalized to the unit interval of 1 s. Describing the sampled sequence as the desired impulse response $h(n)$ of the filter, the z-transform is defined as

$$H(z) \triangleq \sum_{n=-\infty}^{\infty} h(n) z^{-n} \qquad \text{(B.7)}$$

With $z = re^{j\omega T_s}$ and $r = 1$,† (B.7) is expressed as

$$H(e^{j\omega}) \triangleq \sum_{n=-\infty}^{\infty} h(n) e^{-j\omega n T_s} \qquad \text{(B.8)}$$

The frequency response $H(e^{j\omega})$ is a periodic function in ω and can be described in terms of a Fourier series‡ with the coefficients $h(n)$ evaluated as

$$h(n) = \frac{1}{2\pi}\int_{-\pi}^{\pi} H(e^{j\omega}) e^{j\omega n T_s} d\omega \qquad \text{(B.9)}$$

The discrete Fourier transform relationship between $H(e^{j\omega})$ and $h(n)$ is based on the linear shift-invariant property of $h(n)$. These relationships form the basis of the following digital filter design techniques. A more in-depth discussion of these relationships is given by Oppenheim and Schafer.

*The z-transform is contrasted to the Laplace transform, which applies to continuous-time functions and is defined in terms of the complex frequency variable $s = \alpha + j\omega$.

†The summation in (B.8) determines the region of convergence in the z-plane. The sampled data filters discussed in this appendix are based on poles within the unit circle of the z-plane or sampled sequences that are known to converge. On the important topic of convergence regions for other sequences, refer to Oppenheim and Schafer [1] (Chapter 2, "The Z-Transform").

‡The Fourier series representation of a periodic function in ω is discussed in Chapter 1.

B.2 DIGITAL FILTER DESIGN USING s-PLANE POLE–ZERO LOCATIONS

As discussed in Appendix A, analog filters are characterized in terms of their s-plane poles and zeros, and these filters can be implemented as sampled digital filters by expressing the equivalent z-plane pole and zero locations [2]. Stable s-plane filter designs require that all poles are in the *left half* of the s-plane corresponding to $s_p = -\sigma_p \pm j\omega_p$. The condition $\sigma_p = 0$ represents a constant amplitude continuous wave (CW) frequency at $\pm j\omega_p$; zeros may be located in either half of the s-plane and on the $j\omega$ axis. The mapping to the z-plane results in the poles within the unit circle described by $z = e^{j\omega T_s}$ with $T_s = 1/f_s$ where f_s is the sampling frequency $f_s = \omega_s/2\pi$ and $\omega_s T_s = 2\pi$ represents a complete revolution about the unit circle. The Nyquist frequency band $f_N = f_s/2$ corresponds to $z = e^{j\pi} = -1$; consequently, the entire filter frequency response is contained within the angular range 0 to $\pm\pi$.

There are several methods of mapping the s-plane poles and zero to the z-plane. The bilinear transform [3, 4] mapping from the s- to z-plane is expressed as

$$s = \frac{2}{T_s}\left(\frac{1-z^{-1}}{1+z^{-1}}\right) \qquad \text{(B.10)}$$

and $H(z)$ is obtained by substituting (B.10) into $H(s)$. The bilinear transform maps the entire *left half* of the s-plane, that is, the entire region of s-plane stability, into the unit circle of the z-plane resulting in a stable digital filter that is completely contained within the principal Nyquist frequency band.

A much simpler transformation, and the one used in most of the narrowband filter designs used in the preceding chapters, is the impulse-invariant transformation (IIT) expressed as $z_k = e^{s_k T_s}$ where $s_k = -\sigma_k \pm j\omega_k$ represents an s-plane pole or zero. A shortcoming of the IIT method is that the filter is not uniquely contained within the Nyquist frequency band $f_N = f_s/2$, which results in spectral aliasing about f_N. This aliasing can be reduced to an acceptable amount by selecting a filter with a monotonically increasing stopband loss with frequency using a low transition bandwidth with a high stopband attenuation and/or by selecting a high sampling frequency, such that, $f_s \gg B$ where B is the lowpass filter bandwidth.

As discussed in Section B.2, for a realizable and stable filter design [5], the complex s-plane poles and zeros appear as complex conjugates. Figure B.1 depicts the poles in the *left half* of the s-plane, with the poles and zeros expressed as

$$s_{x_k}, \bar{s}_{x_k} = -\varpi_{x_k}\cos(\varphi_{x_k}) \pm j\varpi_{x_k}\sin(\varphi_{x_k}) \quad : x = \{p, z\} \qquad \text{(B.11)}$$

where $x = p$ and z denote, respectively, poles and zeros, and ϖ_{xk} is the magnitude pole or zero vector as shown.

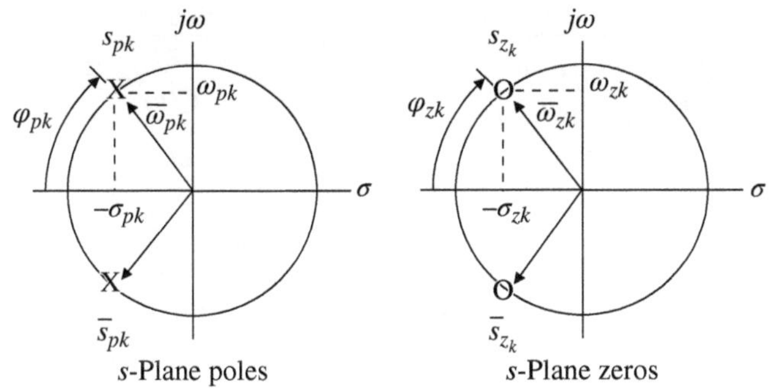

FIGURE B.1 s-Plane poles and zeros.

In this notation, the complex conjugate locations, $\bar{s}_{x_k} = s^*_{x_k}$, correspond to the negative sign of the imaginary part in (B.11) and the angular locations are relative to the negative σ-axis as shown in the figure. Using the IIT, the corresponding z-plane pole locations are evaluated as

$$z_{p_k}, \bar{z}_{p_k} = e^{-\varpi_{p_k} T_s \cos(\varphi_{p_k})} e^{\pm j\varpi_{p_k} T_s \sin(\varphi_{p_k})} \tag{B.12}$$

and the z-plane zero locations are evaluated as

$$z_{z_k}, \bar{z}_{z_k} = e^{-\varpi_{z_k} T_s \cos(\varphi_{z_k})} e^{\pm j\varpi_{z_k} T_s \sin(\varphi_{z_k})} \tag{B.13}$$

Using the notation N_{2p} and N_{2z} for the number of pole and zero pairs, N_{1p} and N_{1z} for the number of single poles and zeros, and N_{0z} to represent the number of zeros at the origin of the s-plane, the s-plane transfer function $F(s)$ is expressed as

$$F(s) = \frac{\prod_k^{N_{2z}} (s-s_{z_k})(s-\bar{s}_{z_k}) \prod_{k+1}^{N_{1z}} (s-s_{z1_k}) \prod_{k+1}^{N_{0z}} s}{\prod_k^{N_{2p}} (s-s_{p_k})(s-\bar{s}_{p_k}) \prod_{k+1}^{N_{1p}} (s-s_{p1_k})} \tag{B.14}$$

Using the IITs (B.12) and (B.13), the corresponding z-plane transfer function is expressed as

$$F(z) = \frac{\prod_k^{N_{2z}} (z-z_{z_k})(z-\bar{z}_{z_k}) \prod_{k+1}^{N_{1z}} (z-z_{z1_k}) \prod_{k+1}^{N_{0z}} (z-1)}{\prod_k^{N_{2p}} (z-z_{p_k})(z-\bar{z}_{p_k}) \prod_{k+1}^{N_{1p}} (z-z_{p1_k})} \tag{B.15}$$

A typical z-plane complex pair of roots is evaluated as

$$\begin{aligned}(z-z_{x_k})(z-\bar{z}_{x_k}) &= z^2 - (z_{x_k} + \bar{z}_{x_k})z + z_{x_k}\bar{z}_{x_k} \\ &= z^2 - 2\cos\left(\varpi_{x_k} T_s \sin\left(\varphi_{x_k}\right)\right) e^{-\varpi_{x_k} T_s \cos(\varphi_{x_k})} z \\ &\quad + e^{-2\varpi_{x_k} T_s \cos(\varphi_{x_k})}\end{aligned} \tag{B.16}$$

and a pair of real roots is evaluated using (B.16) with $\varphi_{xk} = 0$ so that

$$\begin{aligned}(z-z_{x_{1k}})^2 &= \left(z-e^{-\varpi_{x_{1k}} T_s}\right)^2 \\ &= z^2 - 2e^{-\varpi_{x_{1k}} T_s} z + e^{-2\varpi_{x_{1k}} T_s}\end{aligned} \tag{B.17}$$

A single real root is evaluated as $z - e^{-\varpi_{x_{1k}} T_s}$. When each of these roots or root pairs are substituted into (B.15), the transfer function becomes

$$F(z) = \frac{\prod_k^{N_{2z}} (z^2 - a_k z + b_k) \prod_{k+1}^{N_{1z}} (z - a'_k) \prod_{k+1}^{N_{0z}} (z-1)}{\prod_k^{N_{2p}} (z^2 - c_k z + d_k) \prod_{k+1}^{N_{1p}} (z - c'_k)} \tag{B.18}$$

where the coefficients a_k and b_k are associated with those in (B.16) and (B.17) corresponding to the zeros ($x = z$) and the coefficients c_k and d_k are associated with those in (B.16) and (B.17) corresponding to the poles ($x = p$). Defining N as the order of the denominator polynomial in (B.18) and multiplying the numerator and denominator by 2^{-N}, the filter transfer function is expressed in terms of the unit delay operator z^{-1}. This characterization of the transfer function provides for a causal implementation of the sampled linear shift-invariant filter corresponding to (B.5).

As an example, consider the single-pole bandpass filter with a zero at the origin of the s-plane described, respectively, as $s_p, \bar{s}_p = -\sigma \pm j\omega_c$ and $s_z = 0$ with the transfer function expressed as

$$\begin{aligned}H(s) &= \frac{s - s_z}{(s-s_p)(s-\bar{s}_p)} \\ &= \frac{s}{(s+\sigma_p - j\omega_c)(s+\sigma_p + j\omega_c)}\end{aligned} \tag{B.19}$$

The z-plane pole and zero roots are evaluated as discussed in Section B.2 and the resulting transfer function corresponding to (B.19) is

$$\begin{aligned}F(z) &= \frac{z - z_z}{(z-z_p)(z-\bar{z}_p)} \\ &= \frac{z-1}{z^2 - cz + d}\end{aligned} \tag{B.20}$$

where c and d are evaluated as

$$\begin{aligned}c &= 2\cos\left(\varpi_p T_s \sin(\varphi)\right) e^{-\varpi_p T_s \cos(\varphi_p)} \\ &= 2\cos(\omega_c T_s) e^{-\sigma_p T_s}\end{aligned} \tag{B.21}$$

$$\begin{aligned}d &= e^{-2\varpi_p T_s \cos(\varphi_p)} \\ &= e^{-2\sigma_p T_s}\end{aligned} \tag{B.22}$$

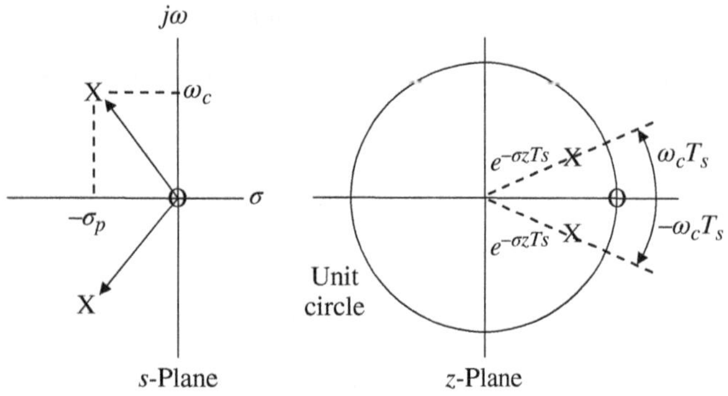

FIGURE B.2 s- and z-plane poles and zeros for example single-pole BPF.

and the angle φ_p is

$$\varphi_p = \tan^{-1}\left(\frac{\varpi_p}{\sigma_p}\right) \tag{B.23}$$

Using (B.19) and (B.20), the respective pole-zero plots in the s-plane and z-plane are shown in Figure B.2.

From the mapping of the poles in Figure B.2, it can be seen that the poles will cluster in the vicinity $z = 1$ as the sampling frequency $f_s = 1/T_s$ is increased and the bandwidth is decreased [6]; the filter bandwidth is directly proportional to σ_p.

Multiplying the numerator and denominator of (B.20) by z^{-2} results in the filter transfer function

$$H(z) = \frac{E_o(z)}{E_i(z)} = \frac{z^{-1}(1-z^{-1})}{1-cz^{-1}+dz^{-2}} \tag{B.24}$$

Equation (B.24) can be used to determine the filter response as a function of the normalized frequency $fT_s = \omega T_s/2\pi$ by substituting $z^{-1} = e^{-j2\pi fTs}$ with the result

$$H(fT_s) = \frac{e^{-j2\pi fT_s}\left(1-e^{-j2\pi fT_s}\right)}{1-ce^{-j2\pi fT_s}+de^{-j4\pi fT_s}} \tag{B.25}$$

Filters described in the form of rational functions, as in (B.24), result in sampled inputs and outputs related by the convolution sum or the difference equation, as expressed, respectively, in (B.1) and (B.4). Therefore, the difference equation corresponding to (B.24) is expressed as

$$\sum_{k=0}^{2} a_k e_o(n-k) = \sum_{r=0}^{2} b_r e_i(n-r) \tag{B.26}$$

where $(a_0, a_1, a_2) = (0, 1, -1)$ and $(b_0, b_1, b_2) = (1, -c, d)$. Using (B.26) to solve for the output $e_o(n)$, as in (B.5), results in the expression

$$e_o(n) = e_i(n-1) - e_i(n-2) + ce_o(n-1) - de_o(n-2) \tag{B.27}$$

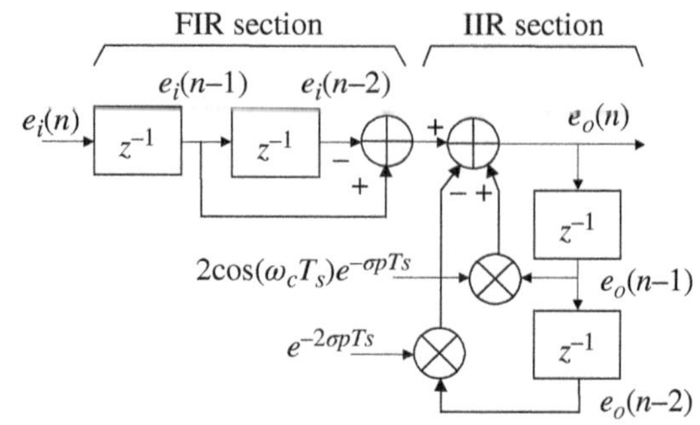

FIGURE B.3 Implementation of example single-pole BPF.

Equation (B.27) is implemented as shown in Figure B.3. The first input delay simply delays the entire filter output; however, it provides an exact match in the delay, following the transient response, between an input CW tone at the center frequency f_c and the corresponding output. The subtraction of the adjacent delay outputs in the FIR section results in a zero output for a constant input; this is a result of the z-plane zero at $z = 1$, which *blocks* the direct current (DC) input. The implementation of the IIR section is typical for complex conjugate poles and can be replicated for multiple complex poles using the appropriate pole location coefficients. The implementations of the FIR and IIR sections in Figure B.3 are referred to as direct-form implementations. Other implementations and the impact of coefficient quantization on the filter response and stability are discussed in the literature [7–13].

B.2.1 Cascade of N Synchronously Tuned Single-Pole Filters

This analysis involves the cascade of N synchronously tuned single-pole lowpass filters with the s-plane transfer function for the cascaded filters given by

$$H(s) = \frac{G(s)}{E(s)} = \left(\frac{a}{s+a}\right)^N \tag{B.28}$$

The s-plane and z-plane poles and zero are shown in Figure B.4.

This filter has unit gain and the 3-dB angular frequency, ω_3, is determined from the solution to the magnitude-squared response of (B.28) with $s = j\omega$, that is, the solution to

$$|H(\omega_3)|^2 = \frac{1}{\left[(\omega_3/a)^2+1\right]^N} = \frac{1}{2} \tag{B.29}$$

Solving (B.29) for the parameter a results in

$$a = \frac{\omega_3}{\sqrt{2^{(1/N)}-1}} \tag{B.30}$$

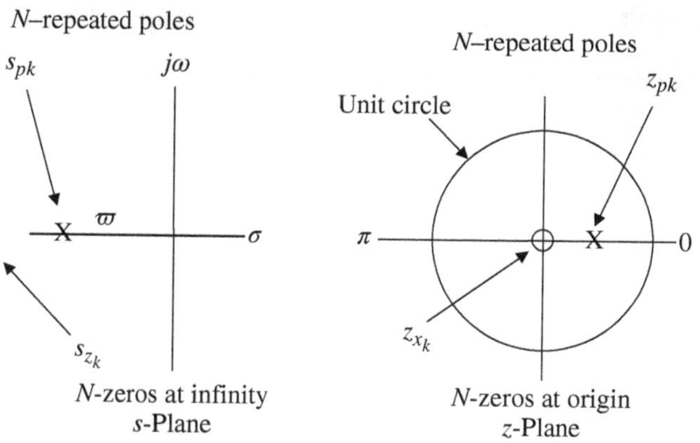

FIGURE B.4 s-Plane and z-plane poles and zeros.

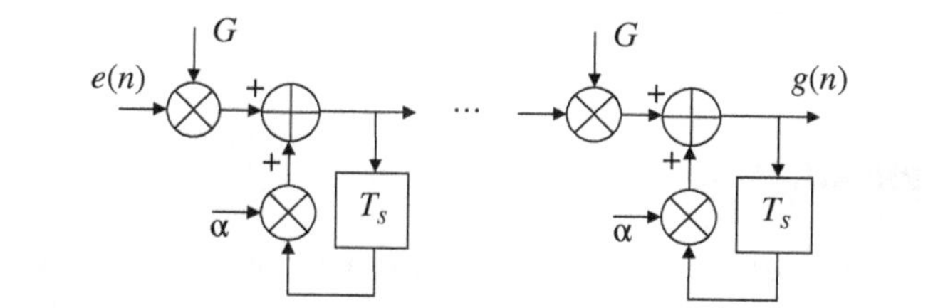

FIGURE B.5 Cascade of N synchronously tuned single-pole filters.

The parameter a is the 3-dB frequency for a single filter. Considering a sampling frequency of $f_s = 1/T_s$ and using the impulse-invariant z-transform $z = e^{sT_s}$, the z-plane filter coefficient for a single-pole filter with an s-plane pole at $s = -a$ is

$$\alpha = e^{-aT_s} = \exp\left(\frac{-\omega_3 T_s}{\sqrt{2^{(1/N)}-1}}\right) \quad \text{(B.31)}$$

For a unit gain response, the gain of each single-pole filter is $G = 1 - \alpha$. The cascade of N identical synchronously tuned filters is shown in Figure B.5. The filter attenuation increases at the rate of 6 N dB/octave.

A cascade of two synchronously tuned filters is used in Chapter 10 as the lock-detection filter for the phaselock loop. For a unit-step input, $e(t) = u(t)$, the filter output response is expressed as

$$g(t) = u(t)(1 - e^{-at}) \quad \text{(B.32)}$$

and the time to reach an output level $g(t_o) = V$ volts ($V < 1$) is

$$t_o = \frac{-\sqrt{2^{(1/N)}-1}}{2\pi f_3}\ln(1 - V) \quad \text{(B.33)}$$

B.3 DIGITAL FILTER DESIGN USING WINDOWS

A variety of window functions are discussed in Section 1.11, and their application in the design of digital filters [14] is discussed in this section. Several examples are given that demonstrate the design procedures and provide insight into the resulting filter responses. The filters are implemented using FIR filters, that is, finite-length transversal filters that exhibit a linear-phase response. The focus of this section is on lowpass filter designs.

The design procedures involve selecting a window function that satisfies the attenuation requirements in the stopband and then determining the required number of filter taps based on the specification of the sampling frequency, the cutoff frequency, and the transition band frequency. Using these basic parameters, the filter design is completed by evaluating the number of filter taps and the tap values are described in the following.

The design procedure begins by establishing a target frequency response for the desired digital filter and, for this purpose, the ideal linear-phase filter with amplitude response given by

$$H(f) = \frac{1}{2f_B} rect\left(\frac{f}{2f_B}\right) \quad \text{: ideal target freqeuecy response} \quad \text{(B.34)}$$

is selected. In this evaluation, f_B is the one-sided baseband or lowpass bandwidth. The corresponding impulse response of the target frequency function is described as the $sinc(2f_B t)$ function. Considering discrete-time sampling at the sample frequency $f_s = 1/\Delta T$ with $t = n\Delta T$, the sampled impulse response corresponding to the target frequency response is

$$h(n) = sinc\left(\frac{nf_B}{f_s}\right) \quad \text{(B.35)}$$

Equation (B.35) extends over the infinite range of n and is unrealizable; however, by applying a finite duration window to $h(n)$, a practical implementation can be achieved that results in an acceptable approximation to the ideal target frequency response. Therefore, using a finite sampled window $w(n)$, the target frequency response (B.34) is approximated by

$$\bar{H}(f) = \Delta T \sum_{n=-N}^{N} w(n)h(n)e^{-j2\pi n\Delta Tf} \quad \text{(B.36)}$$

In terms of discrete frequency, domain sampling with $\Delta f = f_s/M$, where M is the number of discrete Fourier transform (DFT) samples and (B.36) is expressed as

$$\bar{H}(m) = \Delta T \sum_{n=-N}^{N} w(n)h(n)e^{-j2\pi nm/M} \quad \text{(B.37)}$$

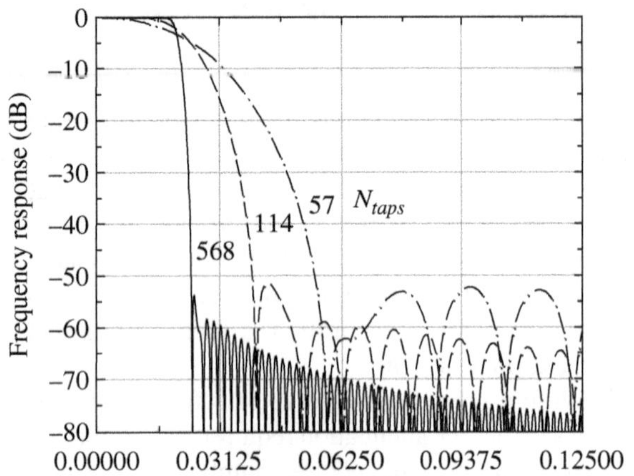

FIGURE B.6 Simulated hamming window FIR filter responses ($f_c/f_s = 0.02$).

Based on these results the FIR filter tap weights are given by

$$\bar{h}(n) = w(n)h(n) \quad : |n| \le N \tag{B.38}$$

Several methods for the selection of N are described in Section 2.8.2; however, in the following examples Harris' method is used. Given the sampling frequency, f_s, the transition bandwidth, f_T, and the stopband attenuation, A_{dB}, the approximate number of taps is expressed as

$$N_{taps} \cong \frac{f_s}{f_T}\left(\frac{A_{dB}}{22}\right) \tag{B.39}$$

For an odd number of taps, $N = (N_{taps} - 1)/2$.

B.3.1 Example of Digital Filter Designs Using Hamming Window

Three FIR filter responses are evaluated using the Hamming window and Harris' approximation to the number of taps expressed by (B.39). The ideal $sinc(nf_b/f_s)$ target frequency impulse response is used where f_B is the baseband bandwidth and f_s is the sampling frequency. The three filters use a common normalized cutoff frequency of $f_c/f_s = 0.02$ and a minimum stopband attenuation of 50 dB. Two comments are worth noting: first, the baseband bandwidth is the cutoff frequency of the target filter, that is, $f_c = f_B$, which is also used as the 3 dB cutoff frequency of the FIR filter; second, the inherent minimum attenuation of the Hamming window sidelobes is about 45 dB; however, when convolved with the ideal target filter the resulting stopband attenuation is about 53 dB so the 50 dB specification is a reasonable requirement.

TABLE B.1 Summary of Simulation Conditions and Results ($f_c/f_s = 0.02$, $A_{dB} = 50$)

Conditions		Simulated Results		
(f_T/f_s)	(N_{taps})	f_c/f_s	f_{50}/f_s	f_T/f_s
0.004	568	0.0201	0.0237	0.0036
0.02	114	0.0215	0.0395	0.018
0.04	57	0.0195	0.0555	0.036

The FIR filter responses are evaluated for three conditions of the normalized transition bandwidth $f_T/f_s = 0.004$, 0.02, and 0.04, and the results are shown in Figure B.6 with the resulting conditions listed in Table B.1. The DFT is used to evaluate the spectrums and, for the $N_{taps} = 568$ filter, the normalized frequency resolution is 1.25e−4.

PROBLEMS

1. The four-pole Butterworth lowpass filter with bandwidth $f_B = \varpi/2\pi$ has two complex s-plane pole pairs corresponding to the *lhp* angular locations $\varphi_{p1} = \pm\pi/8$ and $\varphi_{p2} = \pm 3\pi/8$ as defined in Figure B.1. The s-plane pole locations are:

$$s_{p1}, \bar{s}_{p1} = -\varpi\cos(\varphi_{p1}) \pm j\varpi\sin(\varphi_{p1})$$
$$= -0.383\varpi \pm j0.925\varpi$$

and

$$s_{p2}, \bar{s}_{p2} = -\varpi\cos(\varphi_{p2}) \pm j\varpi\sin(\varphi_{p2})$$
$$= -0.925\varpi \pm j0.383\varpi$$

Determine the corresponding impulse-invariant z-plane poles under the conditions $f_B = 1$ with $f_s = 10$ and 100. Repeat the computation of the z-plane locations using the bilinear transformation given in (B.10). Sketch or plot the z-plane pole locations for each transformation. Note: The units of the frequencies may be Hz, kHz, and so on, as long as f_B and f_s have the units.

2. Express the z-plane transfer function $H(z)$ using the z-plane pole locations found in Problem 1; and sketch the implementation of each filter, corresponding to the case $f_s = 10$; and identify the coefficients as in Figure B.3.

3. Using the filter coefficients c and d expressed in (B.21) and (B.22) with $\sigma_p = 1.0$ and $\omega_c = 2\pi f_c$ with $f_c = 30$ and $T_s = 1/f_s$ with $f_s = 200$, evaluate and plot the filter frequency response expressed in (B.25) in terms of the normalized frequency, fT_s.

4. For the cascade of N synchronously tuned single-pole baseband filters, plot or sketch the function $t_o f_3$, expressed using (B.33), as a function of $N = 1$ through 5 with the parameter values $V = 0.99$, 0.98, 0.95, and 0.9.

REFERENCES

1. A.V. Oppenheim, R.W. Schafer, *Digital Signal Processing*, Prentice-Hall, Inc., Englewood Cliffs, NJ, 1975.
2. C.M. Rader, B. Gold, "Digital Filter Design Techniques in the Frequency Domain," Proceeding of the IEEE, Vol. 55, No. 2, pp. 149–171, February 1967.
3. A.V. Oppenheim, R.W. Schafer, *Digital Signal Processing*, pp. 206–226, Prentice-Hall, Inc., Englewood Cliffs, NJ, 1975.
4. C.R. Wylie, Jr., *Advanced Engineering Mathematics*, Second Edition, pp. 630–641, McGraw-Hill Book Company, Inc., New York, 1960.
5. J.F. Kaiser, *System Analysis by Digital Computer*, Chapter 7, F.F. Kuo, J.F. Kaiser, Editors, "Digital Filters," John Wiley & Sons, Inc., New York, 1966.
6. A.V. Oppenheim, R.W. Schafer, *Digital Signal Processing*, pp. 203–206, Prentice-Hall, Inc., Englewood Cliffs, NJ, 1975.
7. Y. Chow, E. Cassignol, *Linear Signal-Flow Graphs and Applications*, John Wiley & Sons, Inc., New York, 1960.
8. A.V. Oppenheim, R.W. Schafer, *Digital Signal Processing*, Chapter 4, "Discrete Hilbert Transforms," Prentice-Hall, Inc., Englewood Cliffs, NJ, 1975.
9. D.S.K. Chan, L. Rabiner, "Analysis of Quantization Errors in the Direct Form for Finite Impulse Response Digital Filters," IEEE Transactions on Audio and Electroacoustics, Vol. AU-21, pp. 354–366, August 1973.
10. R.M. Golden, J.F. Kaiser, "Design of Wideband Sampled-Data Filters," The Bell System Technical Journal, Vol. 43, No. 4, pp. 1533–1546, July 1964.
11. M.R. Campbell, R.E. Crochiere, L.R. Rabiner, "An Algorithmic Procedure for Designing Hybrid FIR/IIR Digital Filters," The Bell System Technical Journal, Vol. 55, No. 1, pp. 89–108, January 1976.
12. L.R. Rabiner, "The Design of Finite Impulse Response Digital Filters Using Linear Programming Techniques," The Bell System Technical Journal, Vol. 51, No. 6, pp. 1177–108, July–August 1972.
13. O. Herrmann, L.R. Rabiner, D.S.K. Chan, "Practical Design Rules for Optimum Finite Impulse Response Low-Pass Digital Filters," The Bell System Technical Journal, Vol. 52, No. 6, pp. 769–799, July–August 1973.
14. N.J. Fliege, *Multirate Digital Signal Processing: Multirate Systems, Filter Banks, Wavelets*, John Wiley & Sons, Inc., New York, 1994.

APPENDIX C

DETECTION OF SIGNALS IN NOISE

C.1 INTRODUCTION

This appendix evaluates the detection and false-alarm characteristics for the reception of single and multiple pulses using coherent and noncoherent processing techniques. The performance of the square-law envelope detector is contrasted with that of the linear detector, and it is shown that the detection characteristics of the two techniques are nearly the same for sufficiently high signal-to-noise ratios (SNRs). The practical result of this observation is that the relatively straightforward analysis involving square-law detection can be used to establish the performance of linear detectors under a wide variety of applications involving combining several received pulses. The classical work of Marcum and Swerling is used to characterize the detection and false-alarm performance with multiple-pulse combining in slow and fast fading Rayleigh environments. Although this appendix derives most of the results from radio detection and ranging (radar) signal detection applications, the analysis and statistical characteristics are directly applicable to communication systems analysis involving signal presence detection, waveform acquisition, diversity combining, and spread-spectrum signal detection.

In Section C.2, the statistical performance is examined for coherent pulse integration and detection. In radar applications [1–3], coherent combining is referred to as predetection integration and generally applies to the combining of multiple returns at an intermediate frequency (IF) where the signal phase is preserved. The coherent combining must be performed over an interval of time in which the desired signal returns are correlated, and for a scanning radar, this corresponds approximately to the time that a target appears in the antenna beam width during one scan of the antenna. In Section C.3, noncoherent detection is examined using the square-law and linear detectors. In Section C.3.1, postdetection square-law combining is examined for constant and fluctuating signals with multiple-pulse combining. The fluctuating signals are characterized by slow and fast Rayleigh fading. The slow fading corresponds to Swerling's case 1 in which the received pulse amplitudes are constant during each antenna scan but fade independently from scan to scan. The fast fading corresponds to Swerling's case 2 in which the received pulse amplitudes fluctuate from pulse to pulse. Swerling's cases 3 and 4 are characterized by chi-square fading with four degrees of freedom. Case 3 corresponds to a constant received signal with complete correlation, and case 4 corresponds to pulse-to-pulse fluctuations that are completely uncorrelated. In Section C.3.2, linear detection is examined for a constant signal and Rayleigh fading signal in Gaussian noise.

C.2 COHERENT DETECTION

Conceptually, coherent detection involves the matched filter detection of a single pulse. In this situation, the matched filter performs a coherent integration and a decision

Digital Communications with Emphasis on Data Modems: Theory, Analysis, Design, Simulation, Testing, and Applications,
First Edition. Richard W. Middlestead.

Companion website: www.wiley.com/go/digitalcommunications

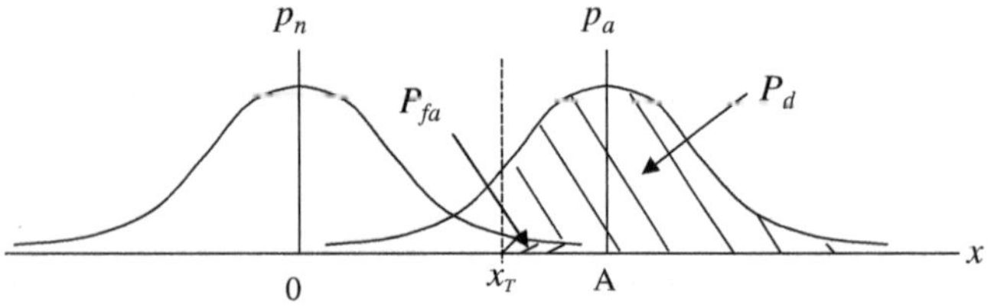

FIGURE C.1 Coherent detection of signal with additive noise.

regarding the presence or absence of the pulse is based upon the SNR of the optimally sampled matched filter output. In the radar case, the single-pulse SNR must be increased by coherently combining several pulses to improve the SNR for a reliable detection. However, because the range is unknown the coherent integration must be performed over all unambiguous range cells of interest as determined by the pulse repetition frequency (PRF). In effect, the received signal power will increase by N^2, while the noise power will increase by N with a net improvement in the SNR of $10\log_{10}(N)$ dB. Perhaps, the closest analogy in communications is the acquisition processing required to detect the presence of a frequency-hopped signal where the receiver synthesizer is fixed at a known future hop frequency and coherent integration is performed at successive time hypotheses looking for a signal detection. Although coherent integration can be performed over a received pulse interval, if the signal phase is unknown, an IF or baseband envelop (or video) detector is required to make a decision. However, with coherent detection the signal phase is known leading to an improved detection or coherent predetection pulse-to-pulse combining.

In the context of these introductory comments, the coherent detection performance involves two detection hypotheses: noise only and signal plus noise. With additive white Gaussian noise (AWGN), these choices are depicted in Figure C.1. The two underlying Gaussian distributions are characterized as

$$p_n(x) = \frac{1}{\sqrt{2\pi}\sigma_n} e^{-x^2/2\sigma_n^2} \quad : \text{noise only} \tag{C.1}$$

and

$$p_{sn}(x) = \frac{1}{\sqrt{2\pi}\sigma_n} e^{-(x-A)^2/2\sigma_n^2} \quad : \text{signal plus noise} \tag{C.2}$$

where A is the peak carrier signal voltage level and σ_n^2 is the noise power such that the SNR is defined as $\gamma = A^2/2\sigma_n^2$. Because the a priori probabilities P_n and P_{sn} of noise only and signal plus noise only are unknown, the optimum [4] detection criterion is based on the Neyman–Pearson criterion [5] that establishes a threshold x_T to realize a specified

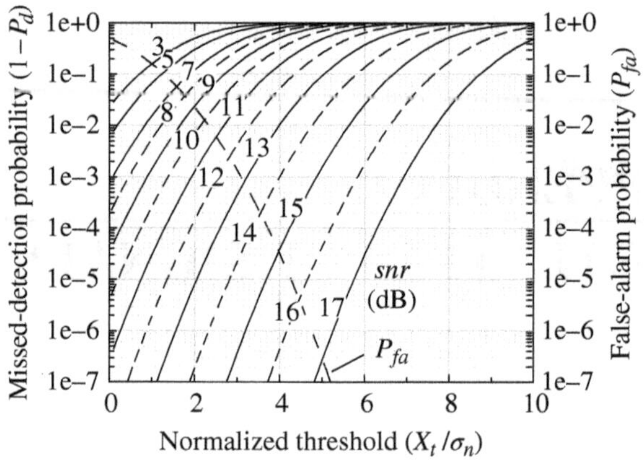

FIGURE C.2 Coherent detection performance (constant signal, single pulse). Solid—odd; dashed—even SNR.

false-alarm probability.* The corresponding detection probability is then based on the decision statistic x exceeding the threshold as indicated in Figure C.1.

C.2.1 False-Alarm Probability

Based on the Neyman–Pearson criterion or test, the false-alarm probability is computed as

$$\begin{aligned} P_{fa} &= \frac{1}{\sqrt{2\pi}\sigma_n} \int_{x_T}^{\infty} e^{-x^2/2\sigma_n^2} dx = \frac{1}{\sqrt{\pi}} \int_{x_T}^{\infty} e^{-y^2} dy \\ &= \frac{1}{2} erfc\left(\frac{x_T}{\sqrt{2}\sigma_n}\right) \end{aligned} \tag{C.3}$$

where the normalized variable $y = x/\sqrt{2}\sigma_n$ is used for notational convenience. The false-alarm probability is shown in Figure C.2 as the *dot–dashed* curve corresponding to the *right-side* ordinate with the abscissa representing the normalized threshold.

C.2.2 Detection Probability

In a similar manner, the detection probability is computed as

$$\begin{aligned} P_d &= \frac{1}{\sqrt{2\pi}\sigma_n} \int_{x_T}^{\infty} e^{-(x-A)^2/2\sigma_n^2} dx = \frac{1}{\sqrt{\pi}} \int_{\frac{x_T - A}{\sqrt{2}\sigma_n}}^{\infty} e^{-y^2} dy \\ &= \frac{1}{2} erfc\left(\frac{x_T - A}{\sqrt{2}\sigma_n}\right) \end{aligned} \tag{C.4}$$

*The Neyman–Pearson criterion or test is considered to be optimum regardless of the actual values of the unknown a priori probabilities.

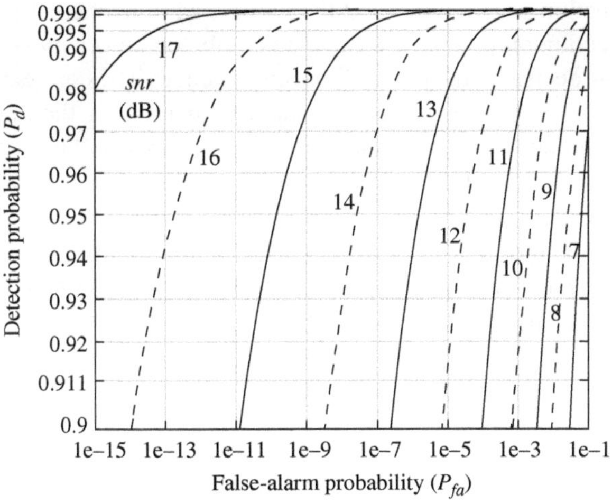

FIGURE C.3 ROC performance for coherent detection (constant signal, single pulse). Solid—odd; dashed—even SNR.

where the change in variables $y=(x-A)/\sqrt{2}\sigma_n$ is made for notational convenience. The detection probability is computed using (C.4) and is plotted in Figure C.2 in terms of the missed detection probability $P_m = 1 - P_d$ as the *left-side* ordinate for a family of even and odd SNR-dependent *dashed* and *solid* curves.

Using the fixed threshold, corresponding to the specified false-alarm probability, the minimum SNR, required to achieve a desired detection probability $P_d = 1 - P_m$, is established as the intersection of P_m and the fixed threshold. The receiver operating characteristic (ROC) plot, shown in Figure C.3, is a standard way of showing the dependence of the detection on the false-alarm performance. The thresholds required to achieve the detection and false-alarm results are implicit, and the corresponding SNR must be interpolated.

With coherent detection combining of N pulses, the SNR increases by a factor of N so the detection results can be viewed as the performance corresponding to the output SNR after combining. For example, if the results in Figure C.2 correspond to combining N pulses, then the SNR of a single pulse is $10\log_{10}(N)$ dB lower than the SNR obtained from Figure C.2 as described in Section C.2.

C.3 NONCOHERENT DETECTION

When a sinusoidal signal plus narrowband Gaussian noise is applied to a linear envelope detector, the probability density function or *pdf* of the detector output is characterized by the Ricean *pdf*, expressed as

$$p_R(r) = \frac{r}{\sigma_n^2} e^{-\left(r^2+A^2\right)/2\sigma_n^2} I_o\left(\frac{rA}{\sigma_n^2}\right) \quad : U(r) \qquad \text{(C.5)}$$

where A is the peak voltage of the carrier, σ_n^2 is the noise power in the bandwidth B at the input to the detector, and $I_o(x)$ is the modified Bessel function of order zero. The step function $U(r)$ is used to emphasize that the distribution applies for $r \geq 0$ and is zero otherwise. The SNR at the input to the detector is $\gamma_i = A^2/2\sigma_n^2$. This result is used in the following analyses to evaluate the probability of a false alarm and detection with various degrees of noncoherent combining.

There are two fundamentally different forms of noncoherent detection: linear detection and square-law detection. Linear detection is characterized by the direct application of (C.5), whereas the square-law detector applies the transformation $y = r^2$ or, in normalized form $y = r^2/2\sigma_n$. The linear detector is more difficult to analyze; however, for modest amounts of post detection combining, the false-alarm and detection probabilities are very nearly identical.

In the following analysis, the characteristic function of the *pdf* is defined as

$$C_R^1(p) = \int p_R(r) e^{pr} dr \qquad \text{(C.6)}$$

where $p = jv$* is used extensively. The superscript 1 indicates that a single random variable is involved in the detection process. When N random variables are summed, as in the case of noncoherent combining, the random variable w is formed as

$$w = \sum_{i=1}^{N} y_i \qquad \text{(C.7)}$$

where y_i is a normalized form of the *iid* random variables r_i. The resulting characteristic function for w is then expressed as

$$C_W^N(p) = \int p_W(w) e^{pw} dw = \left[C_R^1(p)\right]^N \qquad \text{(C.8)}$$

The following descriptions are based on the original work of Marcum in his classic RAND memorandum [7] and the associated mathematical appendix [8]. The analysis involving a fluctuating or fading signal is based on the original work of Swerling in his classic RAND memorandum [9]. Marcum and Swerling's work has been published in a special monograph issue of the Institute of Radio Engineers (IRE) transactions on information series [10]. In addition, many references are made to the work of Meyer and Mayer [11], the mathematical functions given by Abramowitz and Stegun

*Determination of the characteristic function $C(v)$ of $p(r)$ is similar to taking the Fourier transform if the integration variable is $v = -\omega$. Therefore, the characteristic functions are evaluated using the Fourier transform pairs given by Campbell and Foster [6] by substituting the variable $p = -j\omega$ or $-jv$ in their transforms.

[12], the Fourier transform pairs of Campbell and Foster [6], and the work of Urkowitz [13].

The *false-alarm number* is used as a convenience parameter in identifying the underlying false-alarm probability associated with each ROC plot. With multiple-pulse combining, the ROC plots presented by Meyer and Mayer are characterized as plots of detection probability (P_d in percent) versus the number (N) of pulses combined with the family of cures identified by the SNR parameter; in this case, each ROC plot is identified by a specific false-alarm number. For $N = 1$, the abscissa in the ROC plot is the false-alarm number that identifies the false-alarm probability; this is more consistent with the usage in Figure C.3. Meyer and Mayer [14] review several definitions of the false-alarm number including those of Marcum (n_m) and Fehlner (n_f) [15]. Meyer and Mayer exclusively use Fehlner's definition expressed as

$$n_f = \frac{\ln(2)}{P_{fa}} \tag{C.9}$$

This is related to Marcum's definition as $n_m = Nn_f$. Table C.1 relates the false-alarm probability and Fehlner's false-alarm number; the third column is included as a convenient parameter used in subsequent plots.

C.3.1 Signal Detection Using Square-Law Detector

This section focuses exclusively on nonlinear square-law envelope detection with constant and fading signals. Signal detection using linear envelope detection is discussed in Section C.3.2.

C.3.1.1 *Constant Signal in Gaussian Noise (Case 0)* In this case, the square-law detector is based on the condition $y = r^2$ or the normalized condition $y = r^2/2\sigma_n^2$ where r is the output of the linear detector having a Ricean *pdf* as described in Section C.3.1. The performance of the square-law detector is characterized in terms of the false-alarm and false-

TABLE C.1 False-Alarm Number and Probabilities.

n_f	P_{fa}	$\text{Log}_{10}(n_f)$
10	6.931e–2	1.0000
12	5.776e–2	1.0792
20	3.466e–2	1.3010
40	1.733e–2	1.6020
100	6.931e–3	2.0000
120	5.776e–3	2.0792
200	3.466e–3	2.3010
400	1.733e–3	2.6021
1K	6.931e–4	3.0000
10K	6.931e–5	4.0000
100K	6.931e–6	5.0000
1000K	6.931e–7	6.0000

detection probabilities and the number of noncoherent post-detection outputs ($N \geq 1$) that are combined before making a final decision. In this section, the characteristics are examined for a constant or nonfluctuating signal, that is, the carrier amplitude A is constant and, in Section C.3.1.2, the characteristics of a fluctuating signal are examined.

a. *Single-pulse false-alarm probability* With noise only, the Ricean distribution reduces to the Rayleigh distribution given by

$$p_R(r) = \frac{r}{\sigma_n^2} e^{-r^2/2\sigma_n^2} \quad : U(r) \tag{C.10}$$

Using the square-law detector decision statistic y as defined in Section C.3.1, the *pdf* of y is determined as

$$p_Y(y) = p_R(y)\left|\frac{dr}{dy}\right| = e^{-y} \quad : U(y) \tag{C.11}$$

The normalized detection threshold is $y_T = r_T/2\sigma_n^2$, and the false-alarm probability is related as

$$P_{fa} = \int_{y_T}^{\infty} p_Y(y)dy = e^{-y_T} \quad : \text{single pulse} \tag{C.12}$$

Normally the false-alarm probability is specified and then the detection probability is computed based on the corresponding threshold. For the single pulse, the threshold is easily determined from a given false-alarm probability as

$$y_T = \ln\left(\frac{1}{P_{fa}}\right) \qquad \text{C.13}$$

The dependence of the false-alarm probability is plotted in Figure C.4 as a function of the normalized threshold using (C.12).

The analysis of the performance with noncoherent integration of N pulses uses the characteristic function of $p_Y(y)$ evaluated as

$$C_Y^1(p) = \int_0^{\infty} e^{-y} e^{py} dy = \frac{1}{1+p} \tag{C.14}$$

b. *Single-pulse square-law detection probability (case 1)* Normalizing the Ricean *pdf* for the decision statistic y defined in Section C.3.1 and defining the input SNR as $\gamma = A^2/2\sigma_n^2$, where the signal power is $P_s = A^2/2$, the normalized *pdf* for the square-law detector output becomes

$$p_Y(y) = e^{-(y+\gamma)} I_o\left(2\sqrt{y\gamma}\right) \quad : U(y) \tag{C.15}$$

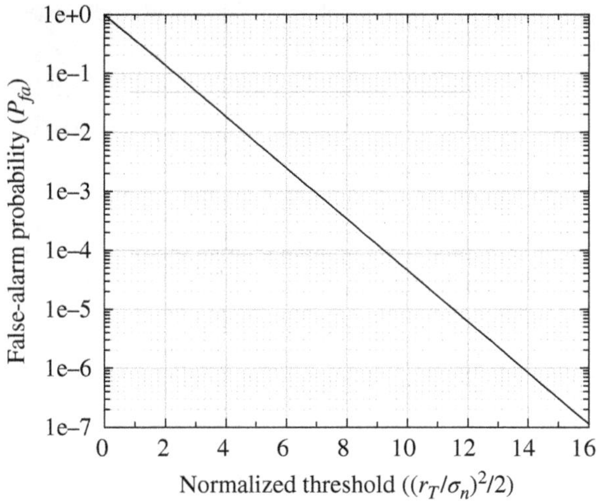

FIGURE C.4 False-alarm probability versus normalized threshold (constant signal, single pulse—case 0).

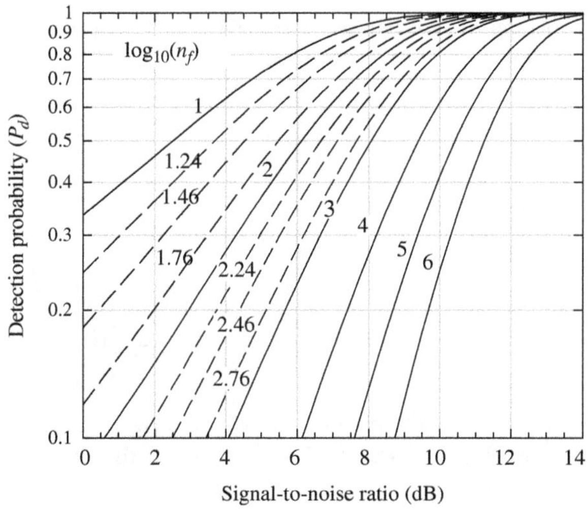

FIGURE C.5 Square-law detection performance (constant signal, single pulse—case 1).

Using this result, and the threshold y_T evaluated using (C.13), the detection probability is evaluated as

$$P_d = e^{-\gamma} \int_{y_T}^{\infty} e^{-y} I_o\left(2\sqrt{y\gamma}\right) dy \qquad \text{(C.16)}$$

: single pulse, constant signal

Equation (C.16) is evaluated for various false-alarm probabilities using numerical integration, and the results are shown in Figure C.5 as a function of the SNR.

The characteristic function corresponding to $p_Y(y)$ is determined using integral pair No. 655.1 from Campbell and Foster with the result

$$C_Y^1(p;\gamma) = e^{-\gamma} \frac{e^{\gamma/(p+1)}}{p+1} \quad \text{: constant signal} \qquad \text{(C.17)}$$

c. *Multiple-pulse false-alarm probability (case 0)* The random variable w is formed by summing N of the *noncoherently* detected pulses, and the resulting *pdf* is determined from the characteristic function

$$C_W^N(p) = \left[C_Y^1(p)\right]^N = \frac{1}{(p+1)^N} \qquad \text{(C.18)}$$

Using the integral pair No. 431 from Campbell and Foster, the desired *pdf* is found to be

$$p_w(w) = \frac{w^{N-1} e^{-w}}{(N-1)!} \quad : U(w) \qquad \text{(C.19)}$$

The probability of a false alarm is determined as follows:

$$P_{fa} = \int_{w_T}^{\infty} \frac{w^{N-1} e^{-w}}{(N-1)!} dw \qquad \text{(C.20)}$$

Upon making the substitution $\xi = w - w_T$, (C.20) becomes

$$P_{fa} = \frac{e^{-w_T}}{(N-1)!} \int_0^{\infty} (\xi + w_T)^{N-1} e^{-\xi} d\xi \qquad \text{(C.21)}$$

The first term in the integrand of (C.21) is expanded using binomial coefficients and is expressed as

$$(\xi + w_T)^{N-1} = \sum_{i=0}^{N-1} \binom{N-1}{i} \xi^{N-1-i} w_T^i \qquad \text{(C.22)}$$

and upon substitution into (C.21) results in

$$P_{fa} = \frac{e^{-w_T}}{(N-1)!} \sum_{i=0}^{N-1} \binom{N-1}{i} w_T^i \int_0^{\infty} \xi^{N-1-i} e^{-\xi} d\xi \qquad \text{(C.23)}$$

Recognizing that the integral in (C.23) is equal to the gamma function $\Gamma(N-i)$ and combining this result with the binomial coefficient, the false-alarm probability is expressed as

$$P_{fa} = e^{-w_T} \sum_{i=0}^{N-1} \frac{w_T^i}{i!} \quad \text{: multiple pulses} \qquad \text{(C.24)}$$

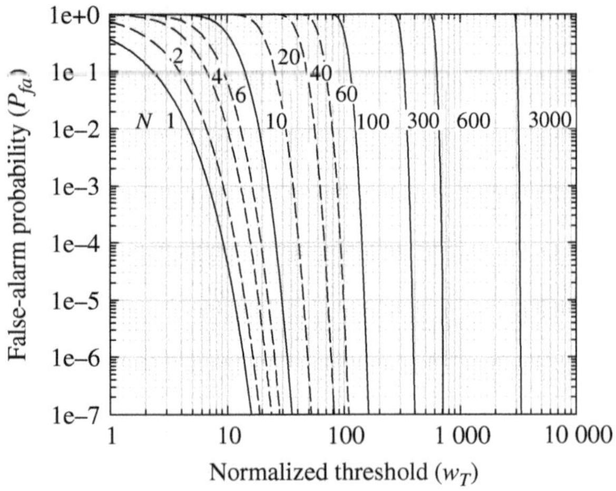

FIGURE C.6 False-alarm probability versus normalized threshold (constant signal, multiple pulses—case 0).

Marcum [16] expresses this result in terms of Pearson's incomplete gamma function as [17, 18]

$$P_{fa} = 1 - I\left(w_T/\sqrt{N}, N-1\right) \tag{C.25}$$

where Pearson's form of the incomplete gamma function is defined as

$$I(u,a) = \frac{1}{\Gamma(a+1)} \int_0^{u\sqrt{a+1}} e^{-t} t^a dt \tag{C.26}$$

The false-alarm probability is evaluated using (C.24) and plotted in Figure C.6 as a function of the normalized threshold w_T. In this evaluation, the summation in (C.24) is performed using logarithms to increase the computational precision (see Problem 1). For large values of N, a threshold value equal to N results in a false-alarm probability of about 50%, and as N increases the slope of the curves increases dramatically. The false alarm corresponding to $N=1$ is the same as that shown in Figure C.4 for a single received pulse.

d. *Multiple-pulse detection probability* For the multiple-pulse detection case, it is necessary to sum N *iid* random variables y_i to form the new random variable w. The *pdf* for w is most easily determined by taking the N-th power of the characteristic function (C.17) with the result

$$\begin{aligned} C_W^N(p;\gamma) &= \left[C_Y^1(p;\gamma)\right]^N \\ &= e^{-N\gamma} \frac{e^{N\gamma/(p+1)}}{(p+1)^N} \end{aligned} \tag{C.27}$$

Using pair No. 650.0 of Campbell and Foster, the inverse transform is found to be

$$P_W(w) = e^{-N\gamma} \left(\frac{w}{N\gamma}\right)^{(N-1)/2} e^{-w} I_{N-1}\left(2\sqrt{N\gamma w}\right) U(w) \tag{C.28}$$

The detection probability for the threshold corresponding to a specified false-alarm probability is

$$P_d = e^{-N\gamma} \int_{w_T}^{\infty} \left(\frac{w}{N\gamma}\right)^{(N-1)/2} e^{-w} I_{N-1}\left(2\sqrt{N\gamma w}\right) dw \tag{C.29}$$

Upon letting $\xi = w - w_T$ and rearranging terms, (C.29) becomes

$$P_d = \frac{e^{-(w_T+N\gamma)}}{(N\gamma)^{(N-1)/2}} \int_0^{\infty} (\xi + w_T)^{(N-1)/2} e^{-\xi} I_{N-1}\left(2\sqrt{N\gamma(\xi + w_T)}\right) d\xi \tag{C.30}$$

Marcum [19] expresses (C.30) in terms of the incomplete Toronto function* as

$$P_d = 1 - T_{\sqrt{w_T}}\left(2N-1, N-1, \sqrt{N\gamma}\right) \tag{C.31}$$

where the incomplete Toronto function is defined as [20, 21]

$$T_B(m,n,r) = 2r^{n-m+1} e^{-r^2} \int_0^B t^{m-n} e^{-t^2} I_N(2rt) dt \tag{C.32}$$

In terms of the parameters N, γ, and w_T, the detection probability, expressed in terms of the incomplete Toronto function, becomes

$$P_d = 1 - 2(N\gamma)^{-(N-1)/2} e^{-N\gamma} \int_0^{\sqrt{w_T}} t^N e^{-t^2} I_{N-1}\left(2\sqrt{N\gamma}t\right) dt \tag{C.33}$$

The following detection performance is evaluated by solving (C.30) using 32 abscissa values and weights for Gaussian–Laguerre quadrature numerical integration [22], which provides excellent agreement with the results of Swerling for values of N up to 200 before

*Marcum [19] provides graphs of detection probability (P_d) versus threshold (w_T) with parameter $N\gamma$ based on the incomplete Toronto function for selected values of N ranging from $N=1$ to 3000.

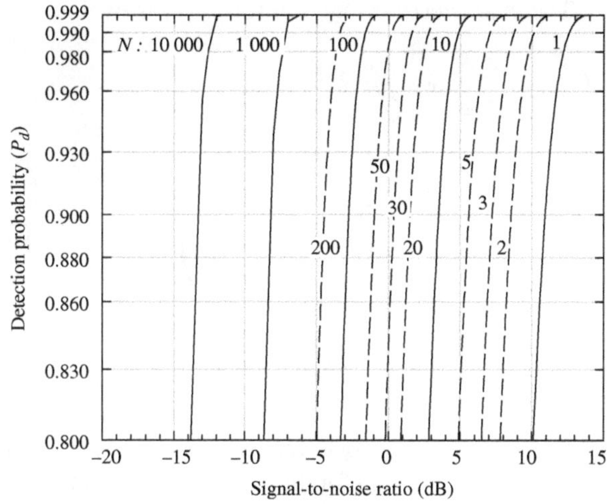

FIGURE C.7 Noncoherent detection (constant signal, multiple-pulse, $n_f = 1000$).

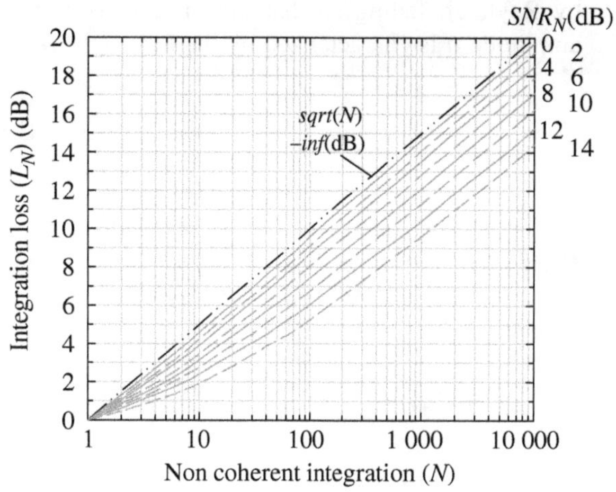

FIGURE C.8 Noncoherent integration loss (constant signal, multiple-pulse).

arithmetic precision* errors occur resulting in degraded performance.† The results are plotted in Figure C.7 for a false-alarm probability of $P_{fa} = 6.931\text{e}{-4}$, which corresponds to a false-alarm number of $n_f = 1000$; the detection results for $N > 200$ are based on the central limit theorem leading to the Gaussian distribution; the mean and variance are determined from the corresponding characteristic functions.

Figure C.7 represents the noncoherent detection probability for a false-alarm probability of 6.93e–4; the false-alarm probability determines the detection threshold w_T based on the solution of (C.20). Meyer and Mayer have published extensive curves of the detection probabilities for the Swerling's case 0 two-parameter chi-square signal and noise model. These curves cover 92 pages and apply for false-alarm probabilities ranging from 6.93e–2 to 6.92e–11, N ranging from 1 to 10,000 and SNRs ranging from –20 to +15 dB in 2 dB steps. The detection performance is plotted by Meyer and Mayer as a function of N for each SNR and, by way of comparison, the performance in Figure C.7 covers only a fraction of two of the 92 pages.

An important consideration regarding the noncoherent integration of multiple pulses is the correlation loss relative to coherent integration. The worst-case noncoherent integration loss is $10\log_{10}\sqrt{N}$ dB that occurs as the linear SNR approaches zero. The integration loss, in decibels, is defined as

$$L_N \triangleq 10\log_{10}(N) - (\text{SNR}_1 - \text{SNR}_N) \quad : \text{dB} \qquad (\text{C.34})$$

where SNR_1 and SNR_N are the SNRs in decibels corresponding, respectively, to the integration of a single pulse and N pulses at a given detection probability.‡ For example, referring to Figure C.7, for $N = 100$ and $P_d = 0.8$, the corresponding values of SNR_1 and SNR_{100} are –3.26 and 10.12 dB, respectively, so the noncoherent combining loss is $L_i(100) = 6.62$ dB. This procedure is followed using the Swerling's case 0 plots in Meyer and Mayer, and the results are shown in Figure C.8. The integration loss is independent of the false-alarm probability.

C.3.1.2 Rayleigh Fading Signal in Gaussian Noise (Case 1)

The Rayleigh fading described in this section corresponds to the chi-square *pdf* with two degrees of freedom.

a. *Single-pulse false-alarm probability* Because a fading signal in a constant noise background is being considered, the false-alarm probability is the same as expressed in Equation (C.12) for the constant signal case.

b. *Single-pulse detection probability (case 1 or 2)* In this section, the signal fluctuations are characterized

*The numerical precision used in the computations is approximately 33 decimal digits based on the *kind type parameter* 16 with range: 10^{-4931} to 10^{4932}.

†For $N > 250$, Urkowitz[13] suggests using the Gaussian approximation for the central and noncentral chi-square distributions denoted, respectively, as $N(N, \sqrt{2N})$ and $N(N+\gamma, \sqrt{2(N+2\gamma)})$. The Gaussian notation used here is $N(0, \sigma)$; however, Urkowitz uses the notation $N(0, \sigma^2)$ where σ is the noise standard deviation.

‡Skolnik (Reference 1, pp. 35–38) also outlines a procedure for evaluating the noncoherent integration loss.

by Rayleigh fading so that the instantaneous signal amplitude a is characterized by the Rayleigh* *pdf*

$$p_A(a) = \frac{a}{\sigma_a^2} e^{-a^2/2\sigma_a^2} \quad : U(a) \tag{C.35}$$

The mean signal power is evaluated as $\bar{P}_s = E[a^2] = 2\sigma_a^2$ so the mean SNR is $\bar{\gamma} = \bar{P}_s / 2\sigma_n^2$. In terms of these parameters, the Rayleigh distribution is expressed as

$$p_A(a) = \frac{a}{\sigma_n^2 \bar{\gamma}} e^{-a^2/2\sigma_n^2 \bar{\gamma}} \quad : U(a) \tag{C.36}$$

In the following analyses, it is assumed that the signal amplitude changes slowly so that a is constant during the detection interval T_I. In general, slow fading corresponds to the case where a is considered constant over the detection interval $T_I = NT$ where T is the pulse duration and N is the number of pulses combined, otherwise fast fading occurs. With this understanding, the single-pulse case being considered corresponds to slow fading. Furthermore, a is considered to be independent from one detection interval to another. This situation is reasonable for radar applications where individual pulses are separated by the pulse repetition interval (PRI) or the scan duration. However, for communication systems, where the detections are based on a contiguous sequence of independently modulated data symbols, the distinction between fast and slow fading and independence is based on the modulated symbol interval.

Defining the instantaneous SNR as $\gamma = a^2/2\sigma_n^2$, the received Rayleigh fading signal can be expressed in terms of the instantaneously received SNR as

$$p(\gamma) = \frac{1}{\bar{\gamma}} e^{-\gamma/\bar{\gamma}} \quad : U(\gamma) \tag{C.37}$$

For the constant signal case, (C.15) is used to expressed the *pdf* of the normalized variable y in terms of the SNR as

$$p_Y(y|\gamma) = e^{-(y+\gamma)} I_o\left(2\sqrt{y\gamma}\right) \quad : U(y) \tag{C.38}$$

where the conditional dependence on the SNR is explicit in (C.38). The average probability density function, $\bar{p}_Y(y)$, is defined as the marginal *pdf*

$$\bar{p}_Y(y) \triangleq E_\gamma[p(y|\gamma)] = \int_{-\infty}^{\infty} p(y|\gamma) p(\gamma) d\gamma \tag{C.39}$$

*Because a single pulse is being detected, this analysis also applies to Swerling's fast fading case 2.

The characteristic function for $\bar{p}_Y(y)$ is evaluated as

$$\begin{aligned} \bar{C}_Y^1(p,\bar{\gamma}) &= \int_{-\infty}^{\infty} \bar{p}(y) e^{py} dy \\ &= \int_{-\infty}^{\infty} \left(\int_{-\infty}^{\infty} p_Y(y|\gamma) p(\gamma) d\gamma \right) e^{py} dy \\ &= \int_{-\infty}^{\infty} C_Y^1(p,\gamma) p(\gamma) d\gamma \end{aligned} \tag{C.40}$$

Using the previously computed characteristic function expressed in (C.17) for the constant signal single-pulse case, the characteristic function for the fluctuating signal is evaluated as

$$\begin{aligned} \bar{C}_Y^1(p,\bar{\gamma}) &= \frac{1}{\bar{\gamma}(p+1)} \int_0^{\infty} e^{-\gamma/\bar{\gamma}} \left(e^{-\gamma} e^{\gamma/(p+1)} \right) d\gamma \\ &= \frac{1}{\bar{\gamma}(p+1)} \int_0^{\infty} e^{-\gamma\left(\frac{p(\bar{\gamma}+1)+1}{\bar{\gamma}(p+1)}\right)} d\gamma \\ &\qquad : \text{slow or fast Rayleigh fading} \\ &= \frac{1}{p(\bar{\gamma}+1)+1} \end{aligned} \tag{C.41}$$

Using transform pair No. 438 from Campbell and Foster, the single-pulse fluctuating signal *pdf* becomes

$$p_Y(y) = \frac{1}{\bar{\gamma}+1} e^{-y/(\bar{\gamma}+1)} \quad : U(y) \tag{C.42}$$

Upon specifying a false-alarm probability, the single-pulse threshold w_T is determined as the solution to (C.24) and letting $y_T = w_T$ the single-pulse detection probability is evaluated as

$$P_d = e^{-y_T/(\bar{\gamma}+1)} \quad : \text{single pulse, slow fading} \tag{C.43}$$

Equation (C.43) is plotted in Figure C.9 as a function of the average SNR for various thresholds corresponding to the indicated false-alarm probabilities.

C.3.1.3 Slow Rayleigh Fading Multiple-Pulse Detection Probability (Swerling's Case 1) The multiple-pulse detection discussed in this section corresponds to the noncoherent integration of N square-law detected pulses. The characteristic function of the multiple-pulse *pdf* is simply the Nth power of the characteristic function of the

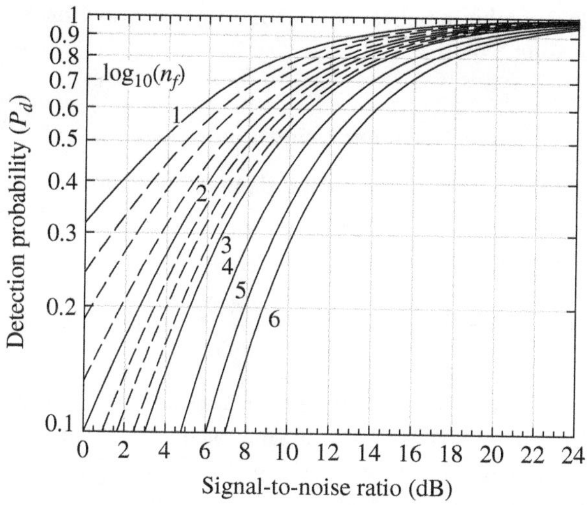

FIGURE C.9 Single-pulse noncoherent detection characteristics for slow Rayleigh fading signal. The curve labels are identical to those in Figure C.5.

single-pulse case, so, referring to (C.17) and (C.37), the characteristic function for this case is expressed as

$$\begin{aligned}\bar{C}_Y^N(p) &= \int_{-\infty}^{\infty} [C_Y^1(p,\gamma)]^N p(\gamma)d\gamma \\ &= \frac{1}{\bar{\gamma}(p+1)^N}\int_0^{\infty}\left[e^{-N\gamma}e^{N\gamma/(p+1)}\right]e^{-\gamma/\bar{\gamma}}d\gamma \\ &= \frac{1}{(1+N\bar{\gamma})}\left[\frac{1}{(p+1)^{N-1}(p+1/(1+N\bar{\gamma}))}\right]\end{aligned} \tag{C.44}$$

Using Campbell and Forster's transformation pair No. 581.7, the *pfd* is found to be

$$\begin{aligned}p_Y(y) &= \frac{1}{1+N\bar{\gamma}}\left\{\frac{1}{\Gamma(N-1)(1-1/(1+N\bar{\gamma}))^{N-1}}e^{-y/(1+N\bar{\gamma})}\right. \\ &\qquad \left.\gamma\left(N-1,\frac{N\bar{\gamma}y}{1+N\bar{\gamma}}\right)\right\} \quad :U(y) \\ &= \frac{1}{1+N\bar{\gamma}}\left(\frac{1+N\bar{\gamma}}{N\bar{\gamma}}\right)^{N-1}\left(\frac{e^{-y/(1+N\bar{\gamma})}}{\Gamma(N-1)}\right)\gamma\left(N-1,\frac{N\bar{\gamma}y}{1+N\bar{\gamma}}\right) \\ &\qquad :U(y)\end{aligned} \tag{C.45}$$

where $\gamma(\ )$ is the incomplete gamma function defined as

$$\gamma(\alpha,x) = \int_0^x e^{-t}t^{\alpha-1}dt \quad \alpha>0 \tag{C.46}$$

This function is evaluated using several relationships found in Abramowitz and Stegun [12, 17]. First, using (Equation 6.5.3, p. 260) [12], the incomplete gamma function is expressed in terms of the gamma function as

$$\gamma(\alpha,x) = \Gamma(\alpha)-\Gamma(\alpha,x) \tag{C.47}$$

For integer values of α, such that $\alpha=n+1$, the gamma function $\Gamma(\alpha,x)$, using (Equation 5.1.46, p. 230) [12], is expressed as

$$\Gamma(n+1,x) = \frac{\alpha_n(x)}{x^{-n-1}} \quad :n>0 \tag{C.48}$$

where $\alpha_n(x)$ is defined as the exponential integral (Equation 5.1.5, p. 228) [12]

$$\alpha_n(x) = \int_1^{\infty} t^n e^{-xt}dt: \quad n=0,1,2,\ldots; \ x>0 \tag{C.49}$$

Using (Equation 5.1.8, p. 228) [12], $\alpha_n(x)$ is evaluated in terms of the finite sum

$$\alpha_n(x) = n!x^{-n-1}e^{-x}\left(1+x+\frac{x^2}{2!}+\cdots+\frac{x^n}{n!}\right) \tag{C.50}$$

Combining these results yield the relatively straightforward solution to (C.47)

$$\gamma(n+1,x) = \Gamma(n-1)\left[1-e^{-x}\sum_{i=0}^{n}\frac{x^i}{i!}\right] \tag{C.51}$$

Equation (C.51) follows directly from (C.47), (C.48), and (C.50); the ratio $\gamma(n+1,x)/\Gamma(n-1)$ is recognized as Pearson's form of the incomplete gamma function. Substituting (C.51) into (C.45), with $n+1=N-1$, $b=N\bar{\gamma}/(1+N\bar{\gamma})$, and $x=by$, the expression for $p_Y(y)$ is evaluated as

$$p_Y(y) = \frac{1}{1+N\bar{\gamma}}\left(\frac{1+N\bar{\gamma}}{N\bar{\gamma}}\right)^{N-1}e^{-y/(1+N\bar{\gamma})}\left[1-e^{-by}\sum_{i=0}^{N-2}\frac{(by)^i}{i!}\right] \quad :U(y) \tag{C.52}$$

The detection probability is now evaluated as

$$\begin{aligned}P_{dN} &= \frac{1}{1+N\bar{\gamma}}\left(\frac{1+N\bar{\gamma}}{N\bar{\gamma}}\right)^{N-1}\int_{y_T}^{\infty}e^{-y/(1+N\bar{\gamma})}\left[1-e^{-by}\sum_{i=0}^{N-2}\frac{(by)^i}{i!}\right]dy \\ &= \frac{1}{1+N\bar{\gamma}}\left(\frac{1+N\bar{\gamma}}{N\bar{\gamma}}\right)^{N-1}\left\{\int_{y_T}^{\infty}e^{-y/(1+N\bar{\gamma})}dy-\int_{y_T}^{\infty}e^{-y/(1+N\bar{\gamma})}e^{-by}\sum_{i=0}^{N-2}\frac{(by)^i}{i!}dy\right\} \\ &= \frac{1}{1+N\bar{\gamma}}\left(\frac{1+N\bar{\gamma}}{N\bar{\gamma}}\right)^{N-1}\left\{(1+N\bar{\gamma})e^{-y_T/(1+N\bar{\gamma})}-\int_{y_T}^{\infty}e^{-y}\sum_{i=0}^{N-2}\frac{(by)^i}{i!}dy\right\}\end{aligned} \tag{C.53}$$

The threshold y_T is determined from the false-alarm probability as expressed in Equation (C.13) or (C.25).

The detection probability is shown in Figure C.10 for selected values of N, and the results indicate that a substantially higher SNR is required to achieve the same detection performance using the nonfading signal. For example, for $P_d = 0.9$ and $N = 100$, 10, and 1, the Rayleigh fading signal requires an 8 dB higher SNR to achieve the same detection probability. The performance in Figure C.10 strictly applies for the false-alarm number $n_f = 1000$.

Using (C.34) and the associated procedures for determining the noncoherent combining loss for constant signals, the noncoherent combining loss for Rayleigh fading signals is shown in Figure C.11. In this case, the improvement in the combining loss diminishes significantly for SNRs exceeding 8 dB.

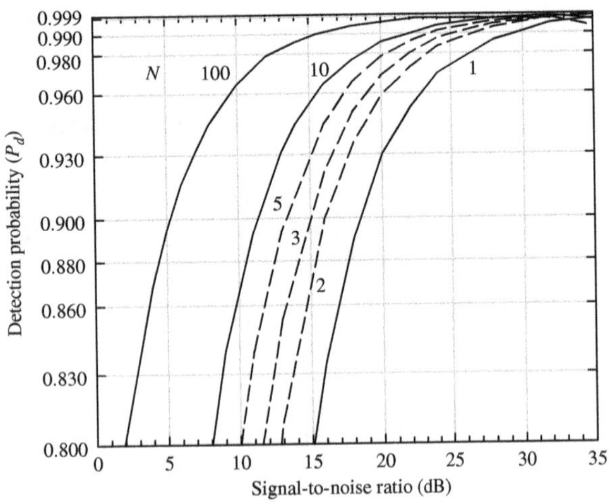

FIGURE C.10 Multiple-pulse noncoherent detection in slow Rayleigh fading (Swerling's case 1, $n_f = 1000$).

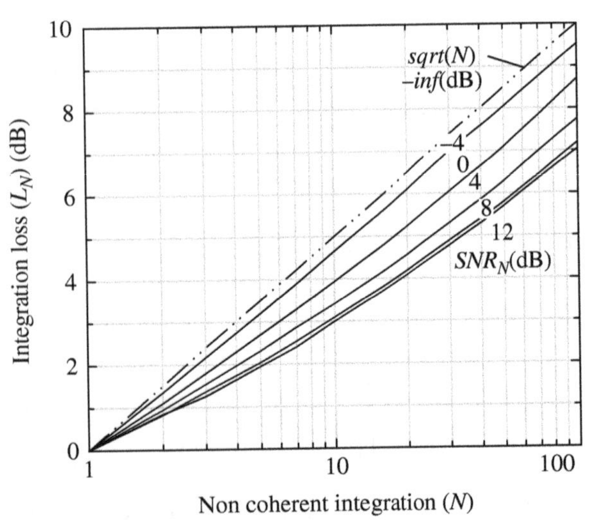

FIGURE C.11 Multiple-pulse noncoherent integration loss in slow Rayleigh fading (Swerling's case 1).

C.3.2 Signal Detection Using Linear Detector

In this case, the output of the demodulator linear detector is described by the Ricean *pdf* expressed in (C.5) and repeated here as

$$p_R(r) = \frac{r}{\sigma_n^2} e^{-(r^2+A^2)/2\sigma_n^2} I_o\left(\frac{rA}{\sigma_n^2}\right) \quad : U(r) \tag{C.54}$$

where r is the output of the linear detector. It is convenient to use the normalized transformation of (C.54) by letting $y = r/\sigma_n$ and defining the SNR as $\gamma = A^2/2\sigma_n^2$. The resulting normalized *pdf* is expressed as

$$p_Y(y) = ye^{-(y^2+2\gamma)/2} I_o\left(\sqrt{2\gamma}y\right) \quad : U(r) \tag{C.55}$$

The performance of the linear detector is characterized in terms of the false-alarm and false-detection probabilities and the number of noncoherent postdetection outputs ($N \geq 1$) that are combined before making a final decision. In Section C.3.2.1, the characteristics are examined for a constant or nonfluctuating signal, that is, the carrier amplitude A is constant. The intent of this section is to demonstrate that the detection performance of a signal with AWGN using a square-law detector approaches that of a linear detector as the SNR increases. Because the SNR increases with noncoherent combining of N pulses, it is only necessary to demonstrate that the performance difference between the square-law and linear detectors diminishes with increasing SNR with $N = 1$, that is, without noncoherent combining.

C.3.2.1 Constant Signal in Gaussian Noise The solutions to the false-alarm and detection probabilities discussed in Section C.4 involve Marcum's Q-function defined in Sections 5.3 and 5.5.3.

a. *Single-pulse false-alarm probability* With noise only, the Ricean distribution reduces to the Rayleigh distribution expressed in the normalized form of (C.55), with the normalized decision statistic $y = r/\sigma_n$ and $\gamma = 0$, the noise only *pdf* is expressed as

$$p_R(r) = ye^{-y^2/2} \quad : U(r) \tag{C.56}$$

The false-alarm probability is computed as

$$P_{fa} = \int_{y_T}^{\infty} ye^{-y^2/2} dy = Q(0, y_T) = e^{-y_T^2/2} \tag{C.57}$$

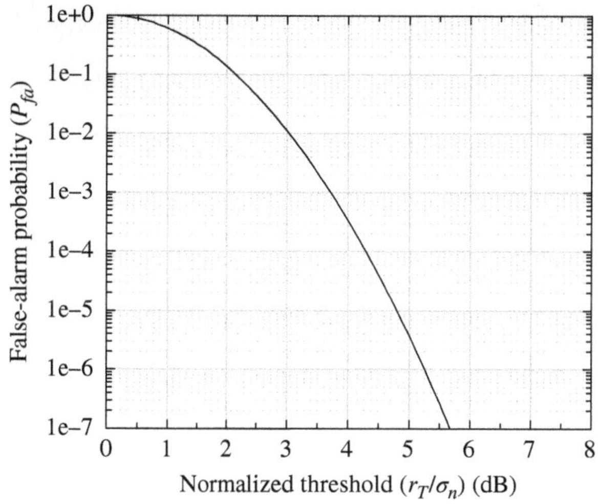

FIGURE C.12 Linear detector false-alarm performance.

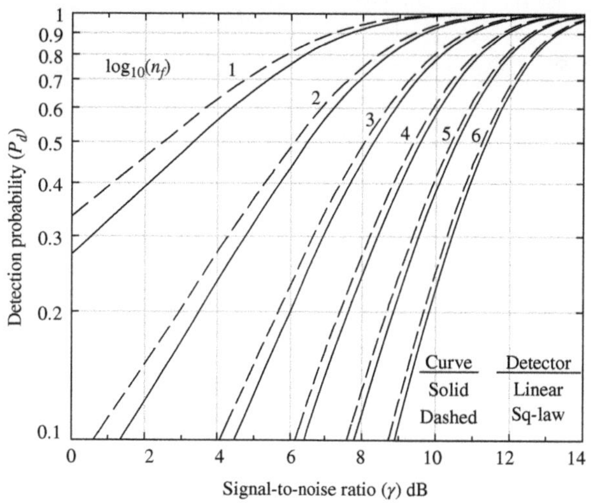

FIGURE C.14 Comparison of linear and square-law detection performance.

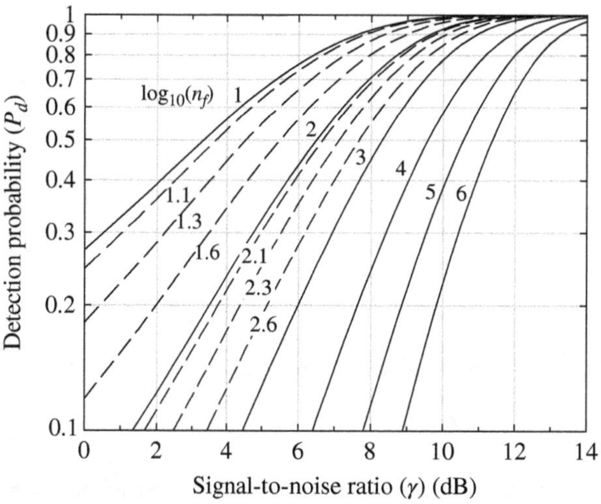

FIGURE C.13 Single-pulse linear detection performance.

where $y_T = r_T/\sigma_n$ is the normalized false-alarm threshold. The false-alarm probability is plotted as a function of $y_T = x_T/\sigma_n$ in Figure C.12.

b. *Single-pulse detection probability* In a similar manner, the probability of detection is evaluated using (C.55) with the normalized decision statistic $y = r/\sigma_n$ and the result is

$$P_d = \int_{y_T}^{\infty} y e^{-(y^2+2\gamma)/2} I_o\left(\sqrt{2\gamma y}\right) dy = Q\left(\sqrt{2\gamma}, y_T\right) \tag{C.58}$$

The Q-function in (C.58) is not nearly as easy to evaluate as the expression in (C.57) for the linear false-alarm probability or the square-law detection expression in (C.16). However, Johansen [23] has developed computer algorithms for the general solution to Marcum's Q-function that have been programmed, and the results are plotted as the *solid* and *dashed* curves in Figure C.13; these curves have the same labels as used for the square-law detection curves in Figures C.5 and C.9. Figure C.14 compares the detection performance of the linear (*solid* curves) and square-law (*dashed* curves) detectors. For a given SNR, linear detection performs better; however, there is very little performance difference with increasing SNR and false-alarm number.

C.4 EVALUATION OF THE DETECTION PROBABILITY AND FALSE-ALARM THRESHOLD FOR CHI-SQUARE DISTRIBUTED RANDOM VARIABLES

In this section, the detection probability is evaluated for the chi-square probability distribution* function given the degrees of freedom ($2WT$), the desired SNR (γ_o) at the detector output, and the false-alarm probability (P_{fa}). The procedure [24] is based on the nomogram, developed by Urkowitz [25], and is shown in Figure C.15.

The application of the nomogram is described using the following example parameters: $2WT = 10, \gamma_o = 60\,(17.8\text{dB})$, and $P_{fa} = 1e^{-3}$. There are two passes through the nomogram, the first pass connects the point corresponding to $2WT$ and

*The chi-square distribution function is introduced in Section 1.5.7.

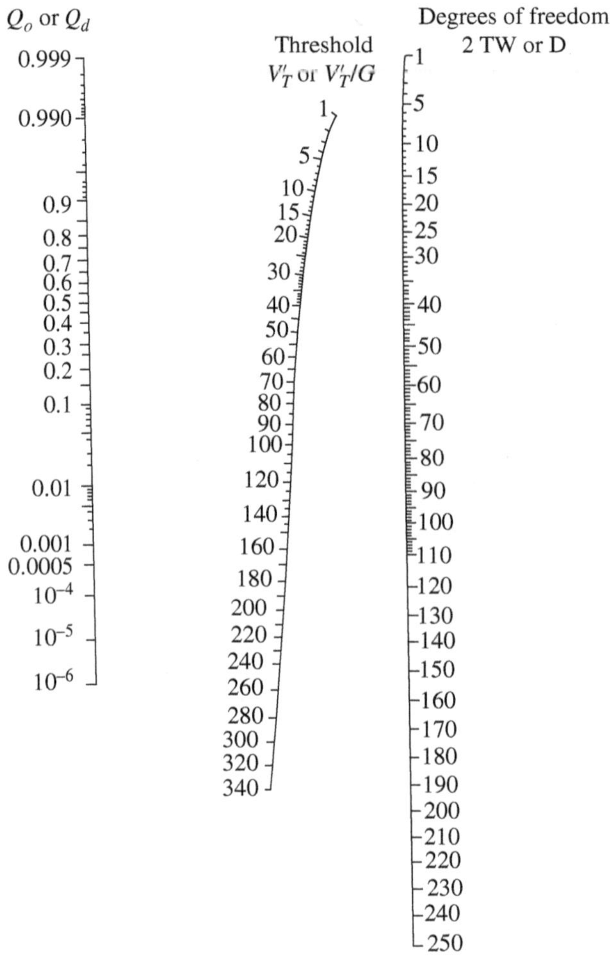

FIGURE C.15 Nomogram of the chi-square probability function.Urkowitz [26]. Reproduced by permission of the IEEE).

$Q_o = P_{fa}$ with a straight line passing through and identifying the threshold V'_T. The parameters D and G are computed as

$$D = \frac{(2WT + \gamma_o)^2}{(2WT + 2\gamma_o)} \quad : \text{degrees of freedom} \tag{C.59}$$

and

$$G = \frac{(2WT + 2\gamma_o)}{(2WT + \gamma_o)} \quad : \text{threshold adjustment gain} \tag{C.60}$$

For the second pass through the nomogram, a new threshold is computed as

$$V''_T = \frac{V'}{G} \tag{C.61}$$

and connecting the new degrees of freedom D and threshold V''_T with a straight line extending to Q_d that identifies the probability of detection $P_d = Q_d$. For the set of example parameters, the desire parameters are found to be $V'_T = 30$, $D = 37.7$, $G = 1.86$, $V''_T = 15.8$, and $P_d = 0.9989$. The threshold $V'_T = 30$ is used as the detection threshold in the subsequent demodulator processing.

For larger degrees of freedom, that is, values of $2WT > 250$, the chi-square distribution is approximated by the Gaussian distribution and P_{fa} and P_d are computed as

$$P_{fa} = \frac{1}{2} erfc\left(\frac{V'_T - 2WT}{2\sqrt{2}\sqrt{WT}}\right) \quad : \text{for distribution } N(2WT,\ 4WT) \tag{C.62}$$

and

$$P_d = 1 - \frac{1}{2} erfc\left(\frac{V'_T - (2WT + \gamma_o)}{2\sqrt{2}\sqrt{2WT + \gamma_o}}\right)$$

$$: \text{for distribution } N(2WT + \gamma_o, 4(WT + \gamma_o)) \tag{C.63}$$

PROBLEMS

1. The following are example applications in the use of the characteristic function. The characteristic functions are based on a received signal with zero-mean additive Gaussian noise with variance σ_n^2. In each case, evaluate the mean value $E(x)$, the second moment $E(x^2)$, and the variance σ_x^2 where $x = y$ without combining ($N = 1$) or $= w$ with N pulse integration.

a. Using the characteristic function $C_Y^1(p)$, expressed by (C.14)

b. Using the characteristic function $C_Y^1(p, \gamma)$, expressed by (C.17)

c. Using the characteristic function $C_W^N(p)$, expressed by (C.18)

d. Using the characteristic function $C_W^N(p, \gamma)$, expressed by (C.27)

e. Using the characteristic function $\bar{C}_Y^N(p, \bar{\gamma})$, expressed by (C.41), for a Rayleigh fading signal

2. Using the results in Problem 1D and E, express the detection probability P_d in terms of the error function *erf*( ) and plot P_d as a function of the SNR for $N = 100$ and compare the results with those in Figures C.7 and C.10, respectively.

REFERENCES

1. M.I. Skolnik, *Introduction of Radar Systems*, McGraw-Hill Book Company, Inc., New York, 1962.
2. R.S. Berkowitz, Editor, *Modern Radar: Analysis, Evaluation, and System Design*, John Wiley & Sons, Inc., New York, 1965.
3. D.K. Barton, *Modern Radar System Analysis*, Artech House, Inc., Norwood, MA, 1988.
4. D. Middleton, "Statistical Criteria for the Detection of Pulsed Carriers in Noise Parts I and II," Journal of Applied Physics, Vol. 24, pp 371–391, April 1953.
5. M.I. Skolnik, *Introduction of Radar Systems*, pp. 423–427, McGraw-Hill Book Company, Inc., New York, 1962.
6. G.A. Campbell, R.M. Foster, *Fourier Integrals for Practical Applications*, Fourth Printing November 1957, D. Van Nostrand Company, Inc., Princeton, NJ/New York, 1948.
7. J.I. Marcum, *A Statistical Theory of Target Detection by Pulsed Radar*, RAND Research Memorandum RM-754, RAND Corporation, Santa Monica, CA, December 1, 1947.
8. J.I. Marcum, *Mathematical Appendix*, RAND Research Memorandum RM-753, RAND Corporation, Santa Monica, CA, July 1, 1948.
9. P. Swerling, *Probability of Detection for Fluctuating Targets*, RAND Memorandum RM-1217, RAND Corporation, Santa Monica, CA, March 17, 1947.
10. J.I. Marcum, P. Swerling, "Studies of Target Detection by Pulsed Radar," IRE Transactions on Information Theory, Special Monograph Series, Vol. IT-6, No. 2, April 1960.
11. D.P. Meyer, H.A. Mayer, *Radar Target Detection: Handbook of Theory and Practice*, Academic Press, New York, 1973.
12. M. Abramowitz, I.A. Stegun, *Handbook of Mathematical Functions with Formulas, Graphs, and Mathematical Tables*, National Bureau of Standards, Applied Mathematical Series 55, U.S. Government Printing Office, Washington, DC, June 1964.
13. H. Urkowitz, "Energy Detection of Unknown Deterministic Signals," Proceedings of the IEEE, Vol. 55, No. 4, pp. 523–531, April 1967.
14. D.P. Meyer, H.A. Mayer, *Radar Target Detection: Handbook of Theory and Practice*, pp. 38–41, 68–71, 83, 107, 108, Academic Press, New York, 1973.
15. L.F. Fehlner, "Marcum's and Swerling's Data on Target Detection by a Pulsed Radar," Appendix A, R.G. Roll and G.T. Trotter, *Method for Digital Machine Computation of Detection Probability*, Applied Physics Laboratory, pp. 23–34, The John Hopkins University, Silver Springs, MD, July 1962.
16. J.I. Marcum, *Mathematical Appendix*, p. 165, Equation 41, RAND Research Memorandum RM-753, RAND Corporation, Santa Monica, CA, July 1, 1948.
17. M. Abramowitz, I.A. Stegun, *Handbook of Mathematical Functions with Formulas, Graphs, and Mathematical Tables*, National Bureau of Standards, Applied Mathematical Series 55, p. 262, Equation 6.5.6, U.S. Government Printing Office, Washington, DC, June 1964.
18. D.P. Meyer, H.A. Mayer, *Radar Target Detection: Handbook of Theory and Practice*, Appendix, pp. 478–487, A-2, "The Incomplete Gamma Function", pp. 478–479, Academic Press, New York, 1973.
19. J.I. Marcum, *Mathematical Appendix*, RAND Research Memorandum RM-753, p. 182, Equation 100c, RAND Corporation, Santa Monica, CA, July 1, 1948.
20. J.I. Marcum, *Mathematical Appendix*, RAND Research Memorandum RM-753, p. 182, Equation 100b, RAND Corporation, Santa Monica, CA, July 1, 1948 (see Reference 11, p. 509 for the relationship to the confluent hypergeometric function *M(a,b,z)*).
21. R.S. Berkowitz, "Probability Density and Distribution Functions," W.W. Weinstock, Editor, *Modern Radar: Analysis, Evaluation, and System Design*, Part II, Chapter 3, p. 188, John Wiley & Sons, Inc., New York, 1965.
22. V.I. Krylov, *Approximate Calculation of Integrals*, Translated by A.H. Stroud, pp. 130, 132 and Appendix C, "Gaussian-Laguerre Quadrature Formulas", pp. 347–352, The Macmillan Company, New York, 1960.
23. D.E. Johansen, "New Techniques for Machine Computation of the *Q*-Function, Truncated Normal Deviates and Matrix Eigenvalues," Scientific Report No. 2, United States Air Force Contract AF19(604)-7237, Applied Research Laboratory, Sylvania Electronic Systems, Waltham, MA, July 1961.
24. J.A. Greenwood, H.O. Hartley, *Guide to Tables on Mathematical Statistics*, Princeton University Press, Princeton, NJ, 1962.
25. H. Urkowitz, "Energy Detection of Unknown Deterministic Signals," Proceedings of the IEEE, Vol. 55, No. 4, pp. 526–528, April 1967.
26. H. Urkowitz, "Energy Detection of Unknown Deterministic Signals," Proceedings of the IEEE, Vol. 55, No. 4, p. 528, Figure 7, April 1967.

INDEX

Note: Underlined page numbers emphasize the index topic.

Digital Communications with Emphasis on Data Modems: Theory, Analysis, Design, Simulation, Testing, and Applications, First Edition. Richard W. Middlestead.

Companion website: www.wiley.com/go/digitalcommunications

www.ingramcontent.com/pod-product-compliance
Lightning Source LLC
Chambersburg PA
CBHW081214180926
51613CB00003B/239